De Vries · Kolb

Wörterbuch der Chemie
und der
chemischen Verfahrenstechnik

Dictionary of Chemistry
and Chemical Engineering

L. De Vries · H. Kolb

Dictionary of Chemistry and Chemical Engineering

2nd revised and enlarged edition

Volume I
German/English

verlag chemie

Weinheim
Deerfield Beach, Florida
Basel

L. De Vries · H. Kolb

Wörterbuch der Chemie und der chemischen Verfahrenstechnik

2., überarbeitete und erweiterte Auflage

Band 1
Deutsch/Englisch

Weinheim
Deerfield Beach, Florida
Basel

1. Auflage 1970
2., überarbeitete Auflage 1978
   1. Nachdruck, 1983, der 2., überarbeiteten Auflage 1978

CIP-Kurztitelaufnahme der Deutschen Bibliothek

DeVries, Louis:
Wörterbuch der Chemie und der chemischen
Verfahrenstechnik / L. DeVries; H. Kolb. –
Weinheim: Deerfield Beach, Florida; Basel:
Verlag Chemie
   Parallelsacht.: Dictionary of chemistry and
   chemical engineering. – Früher mit d. Verl.-
   Orten: Weinheim, New York
   Vol. 2 u. d. T.: DeVries, Louis: Dictionary
   of chemistry and chemical engineering
NE: Kolb, Helga; HST
Bd. 1 Deutsch, englisch. – 1. Nachdr. d. 2., überarb.
u. erw. Aufl. 1978. – 1983.
   ISBN 3-527-25635-0

© Verlag Chemie, GmbH, D-6940 Weinheim, 1978, 1983
Alle Rechte, insbesondere die der Übersetzung in fremde Sprachen, vorbehalten. Kein Teil dieses Buches darf ohne schriftliche Genehmigung des Verlages in irgendeiner Form – durch Photokopie, Mikroverfilmung oder irgendein anderes Verfahren – reproduziert oder in eine von Maschinen, insbesondere von Datenverarbeitungsmaschinen, verwendbare Sprache übertragen oder übersetzt werden. Die Wiedergabe von Warenbezeichnungen, Handelsnamen oder sonstigen Kennzeichen in diesem Buch berechtigt nicht zu der Annahme, daß diese von jedermann frei benutzt werden dürfen. Vielmehr kann es sich auch dann um eingetragene Warenzeichen oder sonstige gesetzlich geschützte Kennzeichen handeln, wenn sie nicht eigens als solche markiert sind.
All rights reserved (including those of translation into foreign languages). No part of this book may be reproduced in any form – by photoprint, microfilm, or any other means – nor transmitted or translated into a machine language without written permission from the publishers. Registered names, trademarks, etc. used in this book, even without specific indication thereof, are not to be considered unprotected by law.
Satz: W. Girardet, Graphischer Betrieb, D-5600 Wuppertal
Druck: betz-druck gmbh, D-6100 Darmstadt
Bindung: Josef Spinner, D-7583 Ottersweier
Printed in the Federal Republic of Germany

Vorwort zur zweiten, überarbeiteten Auflage

Gegenüber der ersten Auflage des „Wörterbuches der Chemie und der chemischen Verfahrenstechnik", die zusammen mit dem inzwischen verstorbenen Herausgeber Prof. Louis De Vries erarbeitet wurde, ist der Wortschatz der vorliegenden Auflage durch Aufnahme zahlreicher Termini auf dem Gebiet der reinen und angewandten Chemie erweitert worden. Außerdem wurden wichtige Begriffe aus Fachgebieten, mit denen der Chemiker heute häufig zu tun hat, wie der Elektronik, Meßgeräteausrüstung, Mathematik, Datenverarbeitung, Medizin und anderen, aufgenommen. Die neuen Termini wurden aufgrund sorgfältiger Lektüre eines umfangreichen Schrifttums, einschließlich Lehrbüchern, Journalen, Patentschriften etc. gesammelt.

Bei der Überarbeitung sind Druckfehler und Ungenauigkeiten der ersten Auflage korrigiert worden; außerdem wurde versucht, Entbehrliches wegzulassen. In diesem Zusammenhang sei jedoch vermerkt, daß es durchaus berechtigt ist, einige der „obsoleten Ausdrücke" beizubehalten, da man auf diese in der älteren Literatur gelegentlich noch stößt. Dies trifft ebenfalls für chemische Begriffe zu, die in den beiden Sprachen ähnlich oder gleich sind, da nicht jeder Benutzer des Wörterbuches ein Fachmann auf chemischem Gebiet ist.

Der Anhang zum Wörterbuch wurde durch die Aufnahme eines Periodensystems nach neuestem Stand und einer Umrechnungstabelle für Einheiten des SI-Systems erweitert.

Benutzer des Wörterbuches haben in Zuschriften und Rezensionen bestätigt, daß das vorliegende Wörterbuch ein zuverlässiges Hilfsmittel bei deutsch/englischen Übersetzungen chemischer Texte darstellt. Dadurch ermutigt, wurde mit großer Mühe versucht, das Wörterbuch inhaltlich weiter zu bereichern, um es zu einem Standard-Referenzwerk für Chemie und chemische Verfahrenstechnik zu machen. Für Hinweise und Vorschläge, die zur weiteren Verbesserung zukünftiger Auflagen beitragen können, sind wir dankbar.

Dank gebührt all denen, die direkt oder indirekt zur Erarbeitung des Werkes beigetragen haben. Besonderen Dank verdienen Herr Arthur Stimson und seine Kollegen für ihre sehr wertvolle Mitarbeit.

Helga Kolb

Foreword to the Second, Revised Edition

The principles adopted in the first edition (compiled in collaboration with the late Professor Louis De Vries) have been adhered to and developed. The vocabulary has been expanded by inclusion of numerous expressions in the fields of pure and applied chemistry. Also included in this new edition of the dictionary are a number of useful terms from such diverse allied scientific disciplines as electronics, instrumentation, mathematics, data processing, medicine, etc. All the new terms and expressions have been carefully selected and compiled on the basis of a direct search of a wide range of the chemical literature, including textbooks, journals, patents, etc.

In the course of revising the first edition of this work special attention has been paid to the correction of printing errors, and an effort has been made to dispense with superfluous material. Naturally, the retention of some "obsolete words" is justifiable, since the older literature has to be consulted at times. Moreover, some obvious chemical terms, such as those which are similar or the same in the two languages have been retained, since some of the users of the dictionary may not be expert in chemistry.

The Appendix has also been amended, and now includes an up-to-date Periodic Table of the Elements and a Conversion Table for the newly adopted SI units.

The first edition of this dictionary has already been widely acclaimed as a reliable aid in the translation of German/English chemical literature. Encouraged by this response every effort has been made to enhance the value of the contents and to make the dictionary a standard work of reference in chemistry and chemical engineering. We would be very grateful for comments and suggestions which might contribute to further improvement of future editions.

We express our sincere thanks to all those who have directly or indirectly contributed to the compilation of this work. Especial thanks are also due to Mr. Arthur Stimson and his colleagues for their invaluable assistance.

Helga Kolb

Vorwort zur ersten Auflage

Das vorliegende deutsch-englische Fachwörterbuch ist mit seinem umfangreichen chemischen und technischen Wortschatz für Studenten der Naturwissenschaften, für Wissenschaftler, insbesondere für Chemiker, für Ingenieure und Fachübersetzer gleichermaßen geeignet. Es behandelt das Vokabular der organischen, anorganischen, physikalischen und pharmazeutischen Chemie sowie der Elektrochemie. Außerdem wurde sehr sorgfältig der Wortschatz der gesamten Verfahrenstechnik bearbeitet, unter besonderer Berücksichtigung der Metall-, Erdöl-, Gummi- und Textilindustrie. Hinzu kommen die wichtigsten Ausdrücke aus der Färberei.

Für den Chemiker ist eine strikte Begrenzung seines Fachgebietes heute kaum noch möglich. Aus diesem Grunde wurde besonderer Wert darauf gelegt, in diesem Wörterbuch neben dem chemischen auch das gängige Vokabular der Physik, Biologie, Medizin, Mathematik, Mineralogie und Kristallographie wiederzugeben.

Bei der endgültigen Bearbeitung dieses Wörterbuches wurden auch modernste, zum Teil erst in den letzten Jahren entstandene Ausdrücke einer Reihe von Fachgebieten erfaßt; diese sind der neueren Literatur der Biochemie, Physiologie, Molekularbiologie, Biophysik, Atomphysik, Spektroskopie, Stereochemie, Reaktortechnik und Elektronik entnommen.

Im Rahmen dieses Wörterbuches wurde eine Anzahl von chemischen Trivialnamen aufgenommen, und zwar – soweit es sich um gebräuchliche Begriffe handelt – auch dann, wenn der Ausdruck im Englischen die gleiche oder eine sehr ähnliche Form wie im Deutschen hat.

Werden für ein deutsches Stichwort mehrere englische Synonyme aufgeführt, so sind diese durch Kommata voneinander getrennt, während ein Semikolon immer dann gesetzt wurde, wenn ein prägnanter Bedeutungsunterschied besteht. Bei zahlreichen Fachausdrücken ist das Spezialgebiet, dem der Ausdruck angehört, angegeben.

Infolge der ständigen Wandlung chemischer Terminologie wurde der Versuch gemacht, veraltete deutsche Ausdrücke als solche zu kennzeichnen (obs) und in der englischen Übersetzung zusätzlich den derzeit gültigen Ausdruck aufzuführen. Allgemein wurde in diesem Wörterbuch der amerikanischen Schreibweise der Vorrang gegeben; jedoch wird auf Abweichungen in der Schreibweise (s. unter dem Stammwort) sowie auf terminologische Unterschiede im amerikanischen und britischen Englisch hingewiesen. Ein Ausdruck in runden Klammern dient der Erläuterung. Eckige Klammern schließen beliebige Weglassungen ein, wie z. B. tyrosin[e]. Markenschutz wurde berücksichtigt, jedoch erfolgt dies ohne Gewähr.

Begriffe, die sich aus mehreren Wörtern zusammensetzen, z. B. "absoluter Alkohol", sind unter dem Substantiv wiedergegeben. Allgemein sind Substantive durch maskulin (m), feminin (f) oder neutrum (n) gekennzeichnet. Unregelmäßige Plurale werden in Klammern angegeben.

Die neuere Schreibweise von Oxid (mit „i") wurde im Wörterbuch einheitlich durchgeführt, ebenso die Schreibung von „Eszet".

Im Anhang des Buches findet der Benutzer eine alphabetische Aufstellung der chemischen Elemente, eine Übersicht über deutsche Maße und Gewichte sowie eine Temperaturumrechnungstabelle und Umrechnungsregeln.

Die Verfasser haben bei ihren Arbeiten eine Vielzahl von Quellen benutzt, darunter zahlreiche deutsch-englische und englisch-deutsche Fachlexika sowie eine umfangreiche Fachliteratur. Aus diesem Grunde wurde von der Aufstellung eines detaillierten Quellenverzeichnisses abgesehen. Die Verfasser wollen jedoch an dieser Stelle allen Herausgebern von deutsch- und englisch-sprachigen Fachwörterbüchern, die ihnen Anregungen und Belehrungen vermittelten, ihren Dank aussprechen.

Louis De Vries   Helga Kolb

Foreword to the First Edition

The present German/English technical dictionary with its comprehensive vocabulary of chemical and technical terms is equally suitable for students of science, practising scientists, particularly chemists, and for engineers and technical translators. The fields of organic, inorganic, physical, pharmaceutical and electro-chemistry have been dealt with thoroughly and many new terms have been listed. In addition great care has been taken in the compilation to ensure a detailed coverage of terms used in process technology, particularly in the metal, mineral oil, rubber, and textile industries; the most important words in use in the dye industry are also included.

For the chemist today, a rigid delineation of his scientific discipline is sarcely possible so great importance has been attached to the inclusion of words of common usage in other related branches of science such as physics, biology, medicine, mathematics, mineralogy, and crystallography.

In the final compilation of the vocabulary a large number of up-to-date newly-coined words have been carefully selected from recent literature on topics such as biochemistry, physiology, molecular biology, biophysics, atomic physics, spectroscopy, stereochemistry, reactor technology, and electronics thus enabling the user to cope with the more specialized problems of translation.

Numerous trivial names of chemical substances still in common use have been included, even in those cases where the English and German forms are similar or the same. When several English synonyms can be listed for a German entry they are separated by commas, whereas semicolons are employed for separation of alternative meanings. Further, when several meanings are given for a German word the particular subject field to which they belong is clearly indicated.

In view of the continual change in chemical terminology an attempt has been made to indicate those German words which are now considered obsolete (obs) and to give the presently valid English translation. American spelling has been used throughout; however, the variations in spelling (see under root word) as well as differences in American-English and British-English terminology are also included. Explanations are given in round brackets; parts of words or words which may be used or omitted as best suits the context are set off in square brackets, e.g. tyrosin[e]. Trade names have been taken into consideration; however, no guarantee can be given in this respect.

Expressions comprising several words e.g. "absoluter Alkohol" are also listed under the noun. All nouns are followed by an indication of gender: masculin (m), feminine (f), and neuter (n).

Irregular plurals are given in brackets, the modern spelling of Oxid (with i) has been used throughout, as has also that of „eszet", i.e. ß.

An alphabetical list of the chemical elements, a tabulation of German weights and measures, as well as a table and rules for temperature conversion and calculation, are given in the Appendix.

In their compilation of this dictionary the Authors have searched through a wide range of technical literature and consulted numerous German/English and English/German dictionaries. A detailed list of the sources from which information has been obtained is not given; however, the Authors wish to express their sincere thanks to all the editors of German and English dictionaries who have assisted in and contributed towards the materialization of this work.

Louis De Vries   Helga Kolb

Inhaltsverzeichnis
Table of Contents

Deutsch-Englisches Wörterverzeichnis
German-English Dictionary 1

Anhang
Appendix 839

Verzeichnis der in diesem Wörterbuch verwandten Abkürzungen
Abbreviations used in this Dictionary 841

Chemische Elemente
Chemical Elements 843

Deutsche Maße und Gewichte
German Measures and Weights 845

Temperaturumrechnungstabelle
Conversion Table of Temperatures 847

Temperaturumrechnungsregeln
Rules for Converting Temperatures 847

Umrechnungstabelle nicht mehr zugelassener Einheiten in Einheiten des SI-Systems
Conversion Table for Obsolete Units into Newly Adopted SI-Units 849

Periodensystem der Elemente
Periodic Table of the Elements 851

A

Aa-Lava *f* (Geol) aa lava
abänderlich alterable, modifiable, variable
abändern to alter, to change, to modify, to vary
Abänderung *f* alteration, change, modification, variation, (Biol) variant
Abänderungspatent *n* (Jur) reissue patent
Abänderungsvorschlag *m* (Jur) amendment
abätmen (Metall) to remove [metal] by etching
abätzen to corrode, to eat away by corrosion, (with caustics) to burn off, (Med) to cauterize
Abätzen *n* corroding off
Abaka *m* abaca, Manila hemp
Abakus *m* (Math) abacus
abarbeiten to rough work, to work off, (sich abarbeiten) to overwork, to tire out
Abart *f* degenerate species, modification, variant, variation, variety
abarten to degenerate, to deviate, to vary
Abasin *n* abasin
Abbau *m* (Anal) analysis, (Bergb) exploitation, working, (Chem) decomposition, degradation, depolymerization, disintegration, (Physiol) catabolism, ~ **einer Maschine** dismounting of a machine, ~ **von Eiweiß** proteolysis, **enzymatischer** ~ enzymatic degradation, **hydrolytischer** ~ hydrolysis, **thermischer** ~ thermal degradation or decomposition
abbauen to break down, (Anal) to analyse, (Bergb) to exploit, to work, (Chem) to decompose, to degrade, to disintegrate
Abbaugrad *m* (b. Zerkleinerung) reduction ratio
Abbaumethode *f* (Bergb) mining method
Abbaumittel *n* decomposing agent, disintegrant
Abbauprodukt *n* decomposition product, degradation product, (Biol) catabolic product
Abbauschacht *m* (Bergb) working pit
Abbaustufe *f* (Chem) stage of decomposition
Abbautemperatur *f* breaking-down temperature, cracking temperature, decomposition temperature
abbauverhindernd antidisintegrating, stabilizing
Abbau-Vortriebsmaschine *f* (Min) cutter-loader
abbauwürdig (Bergb) suitable for exploiting or working, workable
abbeizen (Gerb) to dress, (Med) to cauterize, (Met) to dip, to pickle
Abbeizen *n* (Met) pickling
Abbeizmittel *n* pickling agent; paint remover, paint stripper, varnish remover
Abbeizung *f* cleansing, removal of paint, (Med) cauterization, (Met) dipping, pickling, scouring
Abbeizzusatz *m* stripping auxiliary
Abbe-Kondensor *m* (Opt) Abbe condenser
Abbesche Sinusbedingung *f* (Opt) Abbe sine condition
Abbesche Theorie *f* **für das Auflösungsvermögen eines Mikroskops** (Opt) Abbe theory of the resolution of a microscope
Abbesche Zahl *f* (Opt) Abbe number
abbiegen to bend off, to branch off, to deflect, to deviate
Abbiegung *f* bending off, deflection
Abbild *n* copy, image, likeness, representation
abbilden to copy, to illustrate, to reproduce
Abbildung *f* diagram, figure, illustration, picture, representation, (Projektion) projection, **flächentreue** ~ equal-area projection, **punktförmige** ~ point image
Abbildungsfehler *m* (Opt) image fault, optical aberration
Abbildungsgleichung *f* image equation
Abbildungsvermögen *n* (eines Mikroskops) resolving power (of a microscope)
Abbindefähigkeit *f* setting quality
abbinden to cure, to harden; (abschnüren) to tie up, (losbinden) to detach, to loosen, to unbind; (Zement, Klebstoffe) to set
Abbindeverzögerer *m* setting retarder
Abbindewärme *f* (Zement) heat of setting
Abbindezeit *f* curing period, (Zement) setting time (of cement)
Abbindung *f* (Leim) setting
abblättern to chip off, to flake, to peel off, (Bot) defoliate, (Med) to scale off, to shell off
Abblättern *n* flaking, scaling, (Met) spalling (wear particularly in roller bearings when flakes of metal are removed at the surface)
Abblätterung *f* exfoliation
Abblätterungsmittel *n* exfoliative
Abblasedruck *m* blow-off pressure
Abblasehahn *m* blow-off cock, vent
Abblasekolonne *f* desorption column or tower
Abblaseleitung *f* blow-down piping
abblasen (Dampf) to blow off (steam), ~ **von Luft** to blow off air
Abblasen *n* blowing off, discharging
Abblaserohr *n* blow-off pipe
Abblasetank *m* blow-down tank
Abblaseventil *n* blow-off valve, blow-off cock, discharge valve, exhaust valve, vent
abblenden (Opt) to dim, to screen off, to turn down
Abblenden *n* (Düsen in Extrusion) choking down
Abblendlicht *n* dimmed light
Abblendschalter *m* dimming switch
Abblendvorrichtung *f* dimmer, dimming device
Abbrand *m* calcination, cinder, furnace loss, melting loss, residues, roasted ore, scaling loss, waste (of metals), (Atom) burn-up, (Elektr) fusing

Abbrandförderanlage – Abdichtungsscheibe

Abbrandförderanlage *f* cinder handling plant
Abbrandtransport *m* cinder handling
abbrechen to break off, to discontinue, to interrupt
abbremsen to brake, to decelerate, to retard, to slow down, (Atom) moderate
Abbremsprozeß *m* (Atom) slowing-down process
Abbremsung *f* braking, slowing-down, (Atom) moderation, (Verzögerung) retardation
Abbremsungslänge *f* (Atom) relaxation distance, relaxation length
abbrennbar combustible
Abbrennbarkeit *f* combustibility, deflagrability
Abbrennbürste *f* contact breaking brush, sparking contact brush
Abbrenneinrichtung *f* pickling plant
abbrennen to blaze off, to burn off (or away), (Chem) to deflagrate, (Farbe) to burn off paint, (Gips) to calcine, (Met) to refine, (Metall) to dip, to pickle, to scour, (Stahl) to harden, to temper
Abbrennen *n* calcining; blazing off, deflagrating
Abbrennglocke *f* deflagrating jar
Abbrennkessel *m* dipping pan, pickling vat
Abbrennkontaktstück *n* sparking contact piece
Abbrennlampe *f* blow lamp
Abbrennlöffel *m* deflagrating spoon
Abbrennschweißung *f* arc welding, constant-temperature pressure welding
Abbrennung *f* calcination; deflagration
abbröckeln to chip off, to crumble off, to scale off
Abbruch *m* fragments, debris; demolition, pulling down, (Met) scrap, (Unterbrechung) breaking off, discontinuance
Abbrucharbeit *f* demolition work
Abbruchlinie *f* (Geol) line of cleavage
Abbruchreaktion *f* terminating reaction
abbruchreif due for demolition, (Masch) due to be scrapped
abbrüchig brittle, crumbly, friable
abbrühen to boil, to parboil, to scald, to seethe
Abbrühkessel *m* scalding kettle
ABC-Waffen *pl* (atomare, biologische und chemische Waffen) ABC weapons
abdämpfen to deaden, (Akust) to attenuate, to muffle, (Feuer) to quench, (Licht) to subdue, (Schwingung) to damp[en]
Abdämpfung *f* (Akust) attenuation, muffling, (Licht) subduing, (Schwingung) damp[en]ing
Abdampf *m* exhaust steam, dead steam, waste steam
Abdampfapparat *m* evaporating apparatus, evaporator
Abdampfdruckregler *m* exhaust-steam pressure regulator
Abdampfeintritt *m* exhaust steam inlet
abdampfen to evaporate, to boil down, to steam off

Abdampfen *n* evaporation, volatilization, ~ der Trennflüssigkeit (Gaschromatographie) bleeding the column
Abdampfentöler *m* exhaust-steam oil separator
Abdampfgefäß *n* evaporating vessel
Abdampfheizung *f* waste steam heating
Abdampfkasserolle *f* evaporating pan
Abdampfkessel *m* evaporating boiler
Abdampfkolben *m* evaporating flask
Abdampfkondenswasserabscheider *m* exhaust steam separator
Abdampfleitung *f* exhaust pipe, exhaust steam main
Abdampfmaschine *f* evaporator
Abdampfofen *m* (Keram) slip kiln
Abdampfpfanne *f* evaporating pan
Abdampfrohr *n* exhaust steam pipe, steam trap, vent pipe
Abdampfrohrnetz *n* exhaust steam piping
Abdampfrückstand *m* evaporation residue
Abdampfsammelstück *n* exhaust steam collector
Abdampfschale *f* evaporating dish, evaporating pan
Abdampftemperatur *f* evaporation temperature
Abdampftrichter *m* evaporating funnel
Abdampfturbine *f* exhaust steam turbine
Abdampfung *f* evaporation, volatilization
Abdampfungsrückstand *m* evaporation residue
Abdampfverflüssiger *m* exhaust steam collector
Abdampfverwertung *f* waste steam utilization
Abdampfvorrichtung *f* evaporating equipment, evaporator
abdarren (Früchte) to desiccate, (Hopfen) to kiln-dry, (Met) to cure, to liquate
Abdarrtemperatur *f* (Brau) finishing temperature
Abdeckblende *f* (Opt) shutter mask
abdecken to cover; to uncover
Abdecklack *m* masking coating, peel-off coating
Abdeckleiste *f* cover strip
Abdeckmittel *n* masking material
Abdeckpapier *n* masking paper
Abdeckplatte *f* cover[ing] plate
Abdeckscheibe *f* cover[ing] disc
Abdeckvorrichtung *f* covering device
abdekantieren to decant, to draw off
Abderhaldensche Reaktion *f* Abderhalden reaction
abdestillieren to distil[l] [off]
abdichten to seal, to make tight, to pack, to tighten, (kalfatern) caulk, (wasserdicht machen) to waterproof
Abdichten *n* sealing, waterproofing
Abdichtung *f* sealing, caulking, packing; seal, (Dichtungsring) gasket, washer
Abdichtungseigenschaften *pl* barrier properties
Abdichtungsmaterial *n* packing or sealing material
Abdichtungsring *m* packing washer
Abdichtungsscheibe *f* packing washer

abdicken to boil down, to concentrate, to inspissate, to thicken
Abdiffundierung *f* outward diffusion
abdörren to desiccate, to dry up, (Met) to roast, to sweat
Abdörren *n* desiccating, drying, (Met) roasting
Abdörrofen *m* (Kupfer) liquation hearth, (Met) refining furnace
Abdörrprozeß *m* drying-up process
Abdomen *n* (Anat) abdomen
Abdrängeffekt *m* (Stereochem) effect of crowding
abdrehen (abschalten) to switch off, to turn off or down, (abschrauben) to unscrew
Abdrehmaschine *f* finishing machine, lathe
Abdrehspäne *pl* turnings
Abdrehstahl *m* turning tool
abdrosseln to throttle, to choke, to shut off, (teilweise) to throttle down
Abdrosseln *n* choking, throttling
Abdrosselung *f* choking, throttling
Abdruck *m* copy, impress[ion], mark, stamp, **erster** ~ (Buchdr) first proof
abdrucken to copy, to print
Abdruckmasse *f* molding material
Abdruckverfahren *n* (Buchdr) method of copying
abdrücken to make a mold, to take an impression, (Gewehr) to fire a gun, to pull the trigger
Abdrücken *n* **von oben** (Abstreifen vom Stempel) top ejection
Abdrückschraube *f* lifting screw, set screw, push screw
abdunkeln to darken, to dim, to dull, (Färb) to sadden
Abdunkelung *f* darkening, blackout, dimming
abdunsten to evaporate, to vaporize
Abdunsten *n* evaporating, vaporizing
Abdunstung *f* evaporation
Abdunstungsbad *n* evaporation bath
Abelit *n* (Sprengstoff) abelite
Abelmoschusfaser *f* abelmosk fiber
Abelmoschuskorn *n* amber seed, musk seed
Abelsche Gleichung *f* (Mech) Abel equation
Abelscher Satz *m* Abel's theorem
Abelsches Feld *n* Abelian field
Abeltest *m* Abel test
Abendfarbe *f* color in artificial light, evening shade
Aberration *f* (Opt) aberration, **sphärische** ~ spherical aberration
Abfälleverwertung *f* recycling, waste utilization
abfärben to bleed, to come off, to destain, to lose color, to rub off, to stain off
abfärbig discolorable, liable to lose color
Abfall *m* garbage, litter, refuse, rubbish, waste, (Abnehmen) decrease, (Met, Holz) by-product[s], cuttings, filings, scraps, shavings, turnings, waste product[s], (Min) tailings, (Stein, Holz etc.) chippings, chips, (Verlust) loss, ~ **der Schallintensität** attenuation of sound, ~ **des Stromes** (Elektr) decay of current
Abfallanlage *f* waste utilization plant
Abfallauge *f* spent lye, waste lye
Abfallbehälter *m* garbage can, refuse tank
Abfallbeseitigung *f* waste disposal
Abfallbrennstoff *m* waste fuel, refuse fuel
Abfalleisen *n* scrap iron, waste iron
abfallen to be wasted; to drop, to fall [off], (abnehmen) to decrease, to diminish, (Straße etc.) to decline, to slope
Abfallende *n* waste end
Abfallenergie *f* waste energy
Abfallerzeugnis *n* by-product, waste product
Abfallfett *n* waste fat, waste grease
Abfallgrenze *f* critical limit
Abfallgummi *m* scrap rubber, waste rubber
Abfallgut *n* waste, (Min) material recovered from waste, waste washings
Abfallholz *n* offal timber, waste wood
Abfallkohle *f* waste coal
Abfallmoment *n* break-down moment, breakdown torque
Abfallöl *n* waste oil
Abfallpapier *n* waste paper
Abfallprodukt *n* waste product, by-product, residuary product
Abfallquote *f* waste quota
Abfallrohr *n* discharge pipe, drain pipe, suction pipe
Abfallsäure *f* waste acid, acid sludge, recovered acid, spent acid
Abfallsalz *n* salt refuse, waste salt
Abfallsammelanlage *f* refuse collecting plant
Abfallstoff *m* waste material, waste product
Abfallstoffaufbereitung *f* (Pap) preparation of waste for pulping
Abfallstutzen *m* drain
Abfallvernichtung *f* refuse disposal or destruction
Abfallvernichtungsanlage *f* refuse destruction plant, incinerator, refuse destructor, refuse disposal plant
Abfallverwertung *f* recycling, scrap recovery, utilization of waste
Abfallverwertungsanlage *f* waste utilization plant
Abfallwärme *f* waste heat
Abfallware *f* reject[s], waste material
Abfallwasser *n* waste water
Abfallwolle *f* droppings, waste wool
abfangen to catch, to seize, to trap, (Atom) to capture, (Met) to tap
Abfanggraben *m* (Wasser) pick-up carrier
Abfangplatte *f* (Metall) cupola supporting frame
abfasern to chamfer, to fray, to lose fibers
abfedern to spring, to provide with springs

abfeilen – Abgasüberhitzer

abfeilen to file off
Abfeimen *n* (Glas) skinning
abfertigen to dispatch; to complete, to finish, to process
Abfertigung *f* dispatch; completion
Abfett *n* waste fat, fat skimmings
abfetten to degrease, to skim off the fat
abfeuern (Schuß) to fire
abfiedeln to skim off, to take off the litharge
abfiltern to filter off, to separate by filtering, to strain
Abfiltern *n* filtering
abfiltrieren to filter off
Abfiltrierung *f* filtration, filtering off
abflachen to flatten, to bevel, to even up, to level off, to smooth down, (abstumpfen) to truncate
Abflachung *f* flattening
Abflachungsdrossel *f* attenuation coil, smoothing choke
abflauen to abate, to become weaker, to decelerate, to diminish, to fade, (Min) to rinse, to scour, to wash
Abflauen *n* abatement, (Min) rinsing, washing
abfleckecht fast to staining
abflecken to stain, to make stains; to remove spots
abfließen to flow off, to discharge, to drain, to run off
abfluchten to align, to mark out, to sight
Abflug *m* (Luftf) departure, take-off
Abflugbahn *f* (Luftf) runway
Abfluggeschwindigkeit *f* (Luftf) take-off speed
Abfluß *m* discharge, drain, effluent, outlet tube, waste water, (Gully) gutter, sink
Abflußgeschwindigkeit *f* discharge velocity
Abflußkanal *m* drain, waste channel
Abflußkühler *m* efflux condenser
Abflußleckleitung *f* drain
Abflußleitung *f* discharge pipe, drain [pipe]
abflußlos having no outlet
Abflußmenge *f* discharge quantity
Abflußrinne *f* gutter, [outlet] trough
Abflußrohr *n* discharge pipe, drain pipe, (Gully) gutter, sink
Abflußschacht *m* gully
Abflußstopfen *m* drain plug
Abflußventil *n* discharge valve, escape valve, outlet valve
Abflußwasser *n* discharge water, sewage, waste water
Abfolge *f* sequence, succession
abformen to form, to mold, to shape
Abformmasse *f* molding compound
abfräsen to mill off
abfressen to corrode, to eat away, to remove by caustics
abführen to remove, (Dampf) to exhaust, (Gase) to draw off, (Wärme) to conduct, to dissipate, to eliminate

Abführmittel *n* (Pharm) cathartic, laxative, lenitive, purgative
Abführung *f* outlet; discharge, evacuation, removal
Abfüllanlage *f* (f. Fässer) racking plant, (f. Flaschen) bottling plant, (Flaschen) bottling line, (f. Pulver etc.) filling plant
Abfüllapparat *m* filling device, racking apparatus
Abfüllbetrieb *m* packaging plant
Abfüllbütte *f* racking square
Abfülldatum *n* date of bottling (filling or racking)
abfüllen (Brau) to draw off, to fill, to pour off, (in Flaschen) to bottle
Abfüllheber syphon
Abfüllmaschine *f* (f. Flaschen) bottling machine, (f. Pulver etc.) filling machine
Abfüllpipette *f* delivery pipet[te]
Abfüllpumpe *f* (Brauw) racking pump
Abfüllschlauch *m* siphon tube
Abfüll- und Dosiermaschine *f* filling and dosing machine
Abfüll- und Verschließmaschinen *pl* filling and sealing machinery
Abfüll- und Zählmaschinen *pl* filling and counting machinery
Abfüllung *f* bottling, **aseptische** ~ aseptic filling
Abfüllvorrichtung *f* filling device, dispenser
Abfüllwaage *f* fill-weigher, weighing and filling machine
abfüttern to line
Abfuhr *f* disposal, elimination, removal
Abfurchung *f* division, segmentation
Abgabe *f* delivery, loss, output, waste, yield, (Chem) generation, (Energie) release, (Phys) emission, ~ **von Elektronen** donation of electrons
Abgaberohr *n* (Abgaberöhre) delivery tube, escape tube, outlet tube
Abgabespannung *f* discharge voltage
Abgabetemperatur *f* delivery temperature
abgängig waste
abgären to ferment
Abgang *m* loss, rejects, tailings, (Gas) escape
Abgangsdampf *m* dead steam, exhaust steam
Abgangspapier *n* waste paper
Abgangsrohr *n* waste pipe
Abgangssäure *f* spent acid, waste acid
Abgas *n* discharge gas, exhaust gas, flue gas, waste gas
Abgasabführungsrohr *n* exhaust tail pipe
Abgasanalyse *f* waste-gas analysis
Abgasanlage *f* exhaust gas plant
Abgasausgleichleitung *f* exhaust balance line
Abgasprüfgerät *n* exhaust gas analyzer, flue gas analyzer
Abgasstrahlschub *m* exhaust gas jet thrust
Abgasüberhitzer *m* waste gas superheater

Abgasverwertung *f* waste gas utilization
Abgasvorwärmer *m* economizer, waste gas feed heater
abgebaut degraded
abgeben (Chem) to generate, to liberate, (Energie) to release, (Phys) to deliver, to dissipate, to emit, to give off, to yield
abgebimst pumiced
abgeblaßt (verblaßt) faded
abgeblendet siehe abblenden
abgebogen s. abbiegen
abgebrannt calcined, refined, (Chem) blazed off, deflagrated
abgebrochen siehe abbrechen
abgedichtet pressure-tight, sealed, water-tight
abgedroschen commonplace, stale, trite
abgefettet degreased, freed from fat, skimmed off
abgefiltert filtered, strained
abgeflacht bevelled, flattened, shallow
abgeflaut s. abflauen
abgefressen corroded, decayed
abgeführt drained off, drawn off, removed
abgefüllt filled, packed
abgegärt fermented
abgegeben given off, developed, generated
abgeglichen balanced, equalized
abgeglüht (Met) annealed, heated thoroughly, made red hot, tempered
abgegriffen worn
abgehärtet (Met) hardened, tempered
abgehen to come off; to pass off
abgehoben lifted off, separated, siphoned off
abgeklärt clarified, cleared from dregs, filtered
abgekocht s. abkochen
abgekühlt cooled [down], refrigerated
abgeläutert clarified, filtered, purified, refined, (Leder) lime-washed
abgelagert (Holz) seasoned, (Niederschlag) deposited, (reif) mature[d], mellow, stored
abgelassen s. ablassen
abgelaugt steeped in lye, lixiviated; rinsed off, washed out
abgeleitet s. ableiten
abgelenkt (Opt) deflected
abgelöscht quenched, (Kalk) slaked
abgelöst detached, eliminated
abgeloht excorticated
abgelotet sounded
abgemessen measured; precise
abgenutscht suction-filtered, filtered by suction, sucked off
abgenutzt s. abnutzen
abgeplattet flattened, oblate
abgepreßt pressed out, squeezed out
abgequickt purified, refined, separated, washed
abgerahmt skimmed off
abgeraucht (Chem) evaporated to dryness; vaporized

abgerben to take off by tanning
abgerundet rounded [off]
abgesättigt s. absättigen
abgesäuert acidified, acidulated
abgesaugt s. absaugen
abgeschält (Früchte etc.) pared, peeled, (ohne Rinde oder Hülse) decorticated
abgeschäumt skimmed off
Abgeschäumte[s] *n* scum, skimmings
abgeschaltet cut off, disconnected
abgeschieden isolated, separated, (niedergeschlagen) precipitated
abgeschirmt shielded, guarded, screened
abgeschlackt cleared from dross, freed of slag, slagged out
abgeschlämmt decanted, elutriated, washed out
abgeschleimt slimed, clarified
abgeschlossen closed
abgeschmirgelt s. abschmirgeln
abgeschmolzen s. abschmelzen
abgeschrägt bevelled, chamfered, tapered
abgeschreckt quenched
abgeschwächt s. abschwächen
abgeschwelt calcined, roasted
abgesehen apart [from]
abgeseiht filtered, strained, (dekantieren) decanted
abgesetzt precipitated, settled, (verkauft) sold
abgesondert extracted, isolated, segregated, separated, (Biol) abstricted
abgespalten split off, eliminated
abgespannt released
abgesperrt isolated; shut off, turned off
abgespült s. abspülen
abgestanden decayed, flat, matured, stale, (Gips) dead, (Stahl) perished
abgestellt turned off
abgestochen (Met) tapped
abgestorben dead, perished
abgestrichen scraped off, scummed, skimmed
abgestuft graded, graduated
abgetrieben driven off; assayed (by cupelling), refined, (Chem) retorted
abgewogen s. abwiegen
abgezogen (Dest) decanted, distilled
abgießen to decant, to pour off
Abgießen *n* decantation, decanting
abglänzen to scour
abglätten to polish, to smooth
Abglanz *m* reflection
Abgleich *m* balance, balancing, equilibrium
abgleichen to adjust, to align, to balance, to equalize, to equilibrate, to level, to smooth, (Gewichte) to counterpoise
Abgleichfehler *m* (Elektr) adjusting error, balance error, tuning error
Abgleichfrequenz *f* balancing frequency
Abgleichkondensator *m* trimming condenser
Abgleichprüfer *m* balancing tester

Abgleichsbereich *m* balance range
Abgleichschreibfolge *f* synchro-balance printing
Abgleichverfahren *n* balance method
Abgleichverstärker *m* continuous balance amplifier
abgleiten to slip off
Abgleitung *f* plastic shear
abglühen (Met) to anneal, to heat thoroughly, to make red hot, to temper
Abglühen *n* (Met) annealing, heating, tempering
abgraten to deburr, to trim
Abgratmaschine *f* burring machine, trimming machine
Abgratschrott *m* trimmings
Abgratwerkzeug *n* trimming tool
abgrenzbar definable, limitable
abgrenzen to define, to delimit, to delineate, to demarcate, to mark off
Abgrenzung *f* border, boundary, delimitation, demarcation, limit
Abgriff *m* reading, recording
Abguß *m* decanting, pouring off, (Gieß) casting; mold, replica
Abgußkasten *m* trap
Abgußmasse *f* casting compound, molding compound
Abgußtiefe *f* depth of casting
abhängen to depend [on or upon], (aushängen) to disconnect, to take off
abhängig dependent, (abschüssig) inclined
Abhängigkeit *f* dependence, reliance
abhärten to harden, to toughen
Abhärten *n* (Met) tempering
abhäuten to skin, (Met) to scum, to skim
abhalten to detain, to hold back, to keep away [from], to keep off, to protect against
Abhaltung *f* hindrance, impediment
abhandeln to discuss, to treat
abhanden missing
Abhandlung *f* article, dissertation, essay, paper, treatise
Abhang *m* slope, declivity
abharzen to remove the resin [from]
Abhaspelmaschine *f* reeling machine
abhaspeln to reel off, to wind off
abhauen to cut off, to hew down
abhebbar removable
Abhebemaschine *f* lifting type machine
abheben to lift off, to take off, (Flüssigkeiten etc.) to separate, (sich abheben) to contrast [with]
abhebern to siphon off
Abhebern *n* siphonage
Abhebestift *m* guide pin
Abhebevorrichtung *f* lifting device
abheften to file
abhelfen to correct; to remedy, to satisfy
abhetzen to overwork, to rush
Abhilfe *f* remedy

Abhitze *f* lost heat, waste heat
Abhitzekanal *m* waste heat flue
Abhitzekessel *m* waste heat boiler
Abhitzeverwertung *f* utilization of waste heat
abhobeln to plane [off]
Abhobeln *n* planing
abhören to examine, to monitor
abholen to call for, to fetch
abholzen to clear of trees, to cut off, to deforest, to untimber
Abhub *m* scum, skimmings, waste, (Met) dross, scoria
abhülsen to shell
Abichit *m* (Min) abichite
Abiesöl *n* abies oil
Abietat *n* (Salz oder Ester der Abietinsäure) abietate
Abietin *n* abietin
Abietinsäure *f* abietic acid
Abietinsäureäthylester *m* ethyl abietate
Abietinsäureanhydrid *n* abietic anhydride
Abietit *n* abietite
Abiogenese *f* (Biol) abiogenesis, spontaneous generation
abiogenetisch (Biol) abiogenetic, abiogenous
Abiose *f* (Biol) abiosis
abiotisch (Biol) abiotic
Abirrung *f* deviation, aberration
abisolieren to insulate
abkalken to unlime, to delime, to remove lime [from]; to treat with lime
abkanten to level, to shear off, to trim, (abrunden) to round off, (abschrägen) to bevel, to chamfer, (Blech) to fold
Abkanten *n* (Blech) folding
Abkantklappe *f* creasing support
Abkantmaschine *f* folding machine, folding press, (Met) bending machine
Abkantpresse *f* trimming press
Abkantschiene *f* heated bar
Abkantstab *m* creasing bar
Abkant- und Falzmaschine *f* bending and folding machine, bevelling machine
Abkantung *f* bent, chamfer, chamfered edge, (Met) fold
Abkantwinkel *m* creasing angle
abkappen to cut off, to decapitate
abkapseln to seal off, to wall off, (sich abkapseln) to become encapsulated or encysted
Abkapselung *f* incapsulation, (Med, Biol) encystation
abkeltern to press
abketten (Text) to cast off, to fasten the stitches
abklären to clarify, to clear, to decant, to elutriate, to filter
Abklären *n* clarifying, clearing, decanting, filtering

Abklärflasche *f* aspirator, decanting bottle, decanting flask
Abklärgefäß *n* decanter, decanting vessel
Abklärung *f* clarification, decantation, filtration
Abklärungsflasche *f* decanting flask
Abklärungsgefäß *n* decanting vessel
Abklärungsmaschine *f* decanting apparatus
Abklärungsmethode *f* method of clarification
Abklärungsmittel *n* clarifying agent
abklatschecht fast to staining
abklatschen to marking off (of prints), to knock off, to strike off, (Buchdr) to copy, to print [off], to reproduce, (Techn) to squeeze
abklemmen to clamp, to pinch [off], (Elektr) to disconnect
Abklingcharakteristik *f* (Opt) decay characteristic, persistence characteristic
abklingen (Akust) to die away, to fade, (Katalysator) to deactivate
Abklingen *n* **der Schwingung** dying-out of the oscillation
Abklingkonstante *f* damping constant, decay constant
Abklingkurve *f* (Radioaktivität) decay curve
Abklingzeit *f* (Radioaktivität) decay period, decay time
abklopfen to beat; to descale, to scale off
abknallen to explode
abknicken to bend sharply
Abkochbad *n* boiling bath
abkochbar decoctible
abkochecht fast to boiling
abkochen to decoct, to extract, (Milch) to scald
Abkochen *n* boiling down
Abkochmittel *n* decoction medium
Abkochung *f* decoction
Abkömmling *m* descendant, (Chem) derivative
Abkohlung *f* decarbonization, decarburization, decoking
Abkommen *n* agreement, arrangement, contract, settlement
abkratzen to scrape off, to scratch off
abkreidefest non-chalking
Abkreidefestigkeit *f* chalking resistance
Abkröpfung *f* bending at angles
Abkühlapparat *m* cooler, refrigerator
abkühlen to refrigerate, to chill, to cool [down], to quench, (Glas) to anneal
Abkühlen *n* refrigerating, chilling, cooling down, quenching
Abkühler *m* cooler
Abkühlfaß *n* cooling vat, cooling vessel
Abkühlgeschwindigkeit *f* cooling speed
Abkühlkessel *m* cooling vessel, refrigerator
Abkühlmittel *n* cooling agent, refrigerant
Abkühlofen *m* (Glas) annealing furnace
Abkühlung *f* chilling, cooling [down], quenching, refrigeration

Abkühlungsfläche *f* cooling surface, refrigeration surface
Abkühlungskurve *f* (Therm) cooling curve
Abkühlungsmittel *n* cooling agent, refrigerant
Abkühlungsoberfläche *f* cooling surface
Abkühlungsspannung *f* cooling strain, cooling stress
Abkühlungsversuch *m* cooling test
Abkühlungszeit *f* cooling time, cool down time, cooling period
Abkühlverlust *m* loss from cooling, condensation loss
abkürzen to abbreviate, to shorten
Abkürzsäge *f* cross-cut saw
Abkürzung *f* abbreviation
Abkürzungsverfahren *n* rapid (or accelerated) method
Abkunft *f* descent, origin
abkuppeln to disconnect, to uncouple
Abladehahn *m* dump valve
abladen to unload, to discharge, to dump
Abladeplatz *m* dumping ground
Ablader *m* discharger, unloader
Abladung *f* unloading, discharge
Abläuterfaß *n* washing tub
abläutern to clarify, to clear, to filter, to lime-wash, to purify, to saw-dust, to strain, (Erze) to wash, (Zucker, Glas) to refine
Abläutern *n* filtering, purifying, refining
Abläuterung *f* clarification, filtration, lime-washing, purification, saw-dusting, (Erze) washing, (Zucker, Glas) refining process
Ablage *f* depot, storage place, (Ablagetisch) stand, (Akten) file
ablagern to age, to mature; to store, (Holz) to season, (sich niederschlagen) to deposit, to sediment
Ablagerung *f* deposit, sediment, (Geol) sedimentary rock, (Med) concretion, (Niederschlag) sedimentation, (Vorrat) storage, ~ **von Oxidkrusten** deposit of oxide films
Ablagerungsbereich *m* depositonal environment
Ablagerungskunde *f* sedimentology
Ablagerungsmenge *f* amount of deposit, quantity of deposit
Ablangmaschine *f* **für Laufstreifen** tread skiver
Ablaß *m* outlet, discharge, drain, escape
Ablaßdruck *m* blow-off pressure
ablassen to anneal (steel), to decant, to discharge, to draw off, to dump, to let off [steam], to siphon off
Ablassen *n* discharging, drawing off, ~ **der Schlacke** (Met) tapping of the slag, ~ **des Natriummetalls** drawing off the sodium metal
Ablaßhahn *m* blow off cock, delivery cock, discharge cock, drainage tap, drain cock, [drain] tap

Ablaßleitung – ablöschen

Ablaßleitung *f* blow off pipe, delivery pipe
Ablaßöffnung *f* outlet, vent
Ablaßrohr *n* outlet tube
Ablaßschieber *m* drain sluice valve
Ablaßschraube *f* bleeder screw, drain plug, drain screw
Ablaßstopfen *m* plug
Ablaßtemperatur *f* drawing temperature
Ablaßtrichter *m* emptying funnel
Ablaßventil *n* blow off valve, drain valve, dump valve, escape valve, exhaust valve, vent
Ablationskühlung *f* (Schmelzkühlung) ablation cooling
Ablauf *m* (Ausfluß) discharge, efflux, outlet, (Beendigung) expiration, lapse; (Dest) bottom product, (Verlauf) course
Ablaufapparat *m* launching cradle
Ablaufbrett *n* drain board
ablaufen to drain off, to flow out, to run off, (fortlaufen) to go on, (Mechanismus) to run down, (von einer Spule) to unwind, (zeitlich) to elapse, to expire, to run out, to terminate, ~ **lassen** to drain
Ablaufen *n* running out
ablaufend discharging, draining, running out
Ablaufende *n* discharge end
Ablaufflüssigkeit *f* discharge liquid
Ablaufgestell *n* let-off stand
Ablaufkolben *m* drain flask
Ablauföffnung *f* outlet
Ablauföl *n* expressed oil, run oil, used oil
Ablaufrinne *f* drainage channel, gully, gutter
Ablaufrohr *n* discharge pipe, drain pipe, drip pipe
Ablaufschacht *m* drain shaft
Ablaufstutzen *m* (Säule) downcomer
Ablauftemperatur *f* outlet temperature
Ablauftrichter *m* draining funnel, condensing funnel, discharge hopper, outlet hopper
Ablaufwasser *n* discharged water
Ablauge *f* spent liquor, waste liquor, waste lye
ablaugen to lixiviate, to pickle, to steep in lye
Ablaugen *n* lixiviating, pickling
Ablaugmittel *n* cauterizing medium
ablegen to deposit, to lay off
ablehnen to refuse, to reject, to turn aside
Ablehnung *f* refusal, rejection
ablehren to dress, to gauge, to trim (grinding wheels)
ableitbar derivable
Ableitelektrode *f* (Elektr) bleeder electrode
ableiten to branch off, to divert, to draw off, to trace back, (Folgerung) conclude, deduce, (Math) derive [from], (Wärme) abduct
Ableiter *m* conductor
Ableiterohr *n* delivery tube
Ableitstrom *m* leakage current
Ableitung *f* branch, by-pass, discharge piping, drain, drainage channel, exhaust, outlet, (Elektr) branch circuit, (Math) derivation, ~ **einer Funktion** (Math) derivative of a function, **höhere** ~ higher derivative, **logarithmische** ~ logarithmic derivative, **partiale** ~ partial derivative
Ableitungselektrode *f* reference electrode
Ableitungsgesetz *n* derivation law
Ableitungskanal *m* drain
Ableitungsmittel *n* draining medium
Ableitungsrinne *f* drain pipe
Ableitungsröhre *f* delivery pipe, conduit pipe
Ableitungsrohr *n* delivery pipe, discharge pipe
Ableitungsstrom *m* (Elektr) leakage current
Ableitungsverlust *m* loss by leakage
Ableitwiderstand *m* leakage resistance, (Telev) bleeder resistance
Ablenkastigmatismus *m* deflection astigmatism
Ablenkbarkeit *f* deflectability
Ablenkelektrode *f* deflecting electrode, deflector
ablenken to deflect, to deviate, to divert, to turn away
Ablenker *m* deflector
Ablenkfehler *m* deflection aberration
Ablenkfeld *n* deflector field
Ablenkfrequenz *f* (Telev) sweep frequency
Ablenkregistrierung *f* deflection registration
Ablenkspule *f* orbit shift coil
Ablenksystem *n* deflecting system
Ablenkung *f* deflection, deviation, diffraction, (Magnetnadel) declination
Ablenkungsband *n* (Elektr) sweep band
Ablenkungskraft *f* deflecting force
Ablenkungsmagnet *m* deflecting magnet
Ablenkungsmagnetfeld *n* deflecting magnetic field
Ablenkungsmundstück *n* deflecting nozzle
Ablenkungsvervielfacher *m* deflection multiplier
Ablenkungswinkel *m* angle of deflection
ablesbar readable
Ablesbarkeit *f* readability
Ablesefehler *m* error in reading, reading error
Ablesegenauigkeit *f* accuracy in reading
Ablesegerät *n* direct-reading instrument
Ableselinse *f* reading lens
Ableselupe *f* reading glass
Ablesemikroskop *n* reading microscope
ablesen to take readings, to read off, (Früchte etc.) to gather, to pick
Ablesestrich *m* graduation mark
Ablesevorrichtung *f* reading device
Ablesung *f* reading
Ablesungsfehler *m* error in reading
ablichten to make a duplicate; to shade
Ablichtung *f* photostatic copy, projection
ablösbar removable
ablöschbar quenchable, temperable, (Kalk) slakable
ablöschen to quench, to temper, (Kalk) to slake

Ablöschflüssigkeit *f* tempering liquid, hardening liquid
Ablösearbeit *f* (eines Elektrons) energy required to detach an electron
ablösen to break loose, to break off, to come off, to detach, to loosen, to release, to separate, (abschälen) to peel off, to scale off
Ablösung *f* detachment, elimination, (b. Strömungen) separation of flow, ~ **von Elektronen** detachment of electrons
Ablösungsfläche *f* cleavage surface
Ablösungsrichtung *f* (Krist) cleavage plane, plane of cleavage
ablöten to unsolder
ablohen to bark
Abluft *f* exhaust air, exhaust gas, foul air, outgoing air, used air, waste air
Abluftkamin *m* exhaust gas stack
Abluftkanal *m* air outlet conduit
Abluftlutte *f* exhaust air duct
Abluftrohr *n* air outlet conduit, air outlet pipe
Abluftzeit *f* flash-off period (time between spraying and stoving)
ablutieren to unlute
Ablutieren *n* unluting
abmachen (losmachen) to detach, to remove, to take off, (vereinbaren) to arrange
Abmachung *f* agreement, arrangement, contract, settlement, stipulation
Abmagerung *f* emaciation
abmaischen to finish mashing
Abmaischtemperatur *f* (Brauw) final mashing temperature
Abmaß *n* off-size, (Maßabweichung) tolerance
abmeßbar measurable
abmessen to measure, to gauge
Abmessung *f* measurement, dimension, size
abmontieren to disassemble, to dismantle, to dismount, to remove, to take apart
Abmontierung *f* dismantling, dismounting
Abmusterung *f* color matching, sampling
Abmusterungsversuch *m* color matching experiment
abnagen to gnaw off
Abnahme *f* taking away, taking off, (Empfang) acceptance, (Prüfung) inspection, (Schrumpfung) shrinkage, (Verminderung) decline, decrease, diminution, drop, loss, subtraction, withdrawal
Abnahmebeamter *m* inspector
Abnahmebedingungen *pl* test conditions
Abnahmeboden *m* (Kolonne) outlet plate, outlet tray
Abnahmeelektrode *f* collector electrode, output electrode
Abnahmeprüfung *f* buyer's inspection
Abnahmeversuch *m* acceptance test
Abnahmevorschriften *f* specifications
Abnahmewalze *f* take-up roll

abnehmbar detachable, dismountable, removable
abnehmen (empfangen) to accept, (entfernen) to remove, to take off, (geringer werden) to decrease, to diminish, to reduce, to subtract, (Met etc.) to skim
Abnehmen *n* diminishing, reducing
Abnehmerbürste *f* collecting brush
Abneigung *f* (Abgeneigtheit) disinclination, dislike, (Med) antipathy, aversion
abnorm abnormal, irregular
Abnormität *f* abnormality, anomaly, irregularity
abnutschen to drain off, to filter by suction, to suck off
Abnutschen *n* filtering by suction
Abnutzbarkeit *f* wear, wearability, wearing capacity, wearing quality
abnutzen (abnützen) to wear out
Abnutzung *f* (Abnützung) utilization, wear, wear und tear, ~ **der Werkzeuge** wear on the tools, ~ **durch Abrieb** abrasive wear, ~ **durch Reibung** abrasion, wear by friction, ~ **eines Getriebes** runout of a gear
Abnutzungsbeständigkeit *f* resistance to wear, wearability
Abnutzungserscheinung *f* sign of wear
Abnutzungsfläche *f* wearing surface
Abnutzungshärte *f* passive hardness, wear hardness
Abnutzungsprobe *f* wear test
Abnutzungsprüfung *f* wearing test
Aböl *n* waste oil
abölen to oil; to remove oil
Abonnement *n* subscription
Abonnent *m* subscriber
abonnieren to subscribe
Aborin *n* aborine
Abpackbetrieb *m* contract packager, contract packaging plant
Abpackbeutel *m* packaging bag
abpacken to pack, to package
Abpackmaschine *f* packaging machine, packing machine
Abpackung *f* packaging, packing
Abperleffekt *m* water repellent effect
abplatten to flatten
Abplattung *f* ellipticity, flattening, oblateness
Abplattungsringe *pl* rings for flattening
abprägen to make an impress[ion]
Abprall *m* bounce, rebound, (Akust) reverberation
abprallen to bounce, to rebound, to recoil, (Akust) to reverberate
Abprallwinkel *m* angle of deflection
abpressen to press out, to squeeze out
Abpressen *n* pressing out, squeezing out
Abpreßmaschine *f* (Buchdr) nipping press
abpuffern to buffer
abpumpen to pump out, to drain by pumping
Abputz *m* plaster

abputzen to wipe off, to clean, to cleanse; to plane; (Hauswand) to plaster; to rough-cast
Abquetscheffekt *m* squeezing effect, degree of hydroextraction
abquetschen to crush, to squeeze [off], to squeeze out
Abquetschfläche *f* flash edge, flash groove, mating surface, shear edge, (eines Werkzeuges) land (of a mold)
Abquetschform *f* flash [overflow] mold
Abquetschgrat *m* flash, land
Abquetschnute *f* flash groove
Abquetschrand *m* flash ridge, flash ring, land area, shear edge
Abquetschwalze *f* doctor roll[er], squeeze roll
Abquetschwerkzeug *n* flash mold
Abquickbeutel *m* filter bag
abquicken to purify, to refine, to separate
Abquickung *f* purification, separation, washing
abräuchern to fumigate, to smoke
abräumen to clear, to remove
abrahmen to skim off
abraspeln to rasp off, to scrape off
Abrastol *n* abrastol, asaprol
abrauchen to evaporate, to fume off, to vaporize
Abrauchen *n* evaporating
Abrauchraum *m* evaporating chamber
Abrauchschale *f* evaporating dish
Abraum *m* refuse, rubbish, waste
Abraumsalze *n* abraum salts, saline deposits
Abraumstoff *m* residue, waste
Abrazit *m* (Min) abrazite, gismondite
abrechnen (abziehen) to deduct, to subtract, (ausgleichen) to balance, to settle
Abregung *f* (Atom) deexcitation
Abregungsphoton *n* deexcitation photon
Abreibefestigkeit *f* resistance to abrasion
abreiben to rub off, to scour, to scrape off, (abschleifen) to abrade, to sand, to sandpaper, (durch Abnutzung) to wear down
Abreibewiderstand *m* abrasion resistance
Abreibfestigkeit *f* abrasion resistance
Abreibung *f* abrasion, attrition, rubbing off
Abreibungsmittel *n* abrasive
Abreicherung *f* (Atom) depletion
abreißen to detach, to tear off
Abreißkapsel *f* tear-off cap
Abreißkegel *m* (verjüngter Angußkanal) restricted gate
Abreißlasche *f* tear tab
Abreißverpackung *f* tear-off package
Abreißverschluß *m* tear-off closure, pull-off closure, strip-seal
Abreißzünder *m* friction igniter
Abreißzündung *f* make and break ignition
Abrichtelauge *f* weak caustic liquor
abrichten to fit, to train
Abrichter *m* (Techn) dressing tool

Abrichthammer *m* dressing hammer, planishing hammer
Abrichthobel *m* smoothing plane, smooth planing machine, surface planing fixture
Abrichtlineal *n* straightedge, ruler
Abrieb *m* (durch Abnutzung) abrasion, abrasion loss, attrition, dust, grindings, grit, wear, (Kohle) fines, ~ **der Lauffläche** (Reifen) abrasion of tire tread
abriebbeständig wear resistant
abriebfest abrasion-proof
Abriebfestigkeit *f* abrasion resistance, abrasion wear, nonabrasiveness, nonabrasive quality, scuff resistance
Abriebkorrosion *f* fretting
Abriebwiderstand *m* abrasion resistance
Abriebzahl *f* index of abrasion
abriegeln to bar, to barricade, to bolt
abrieseln to trickle down
Abrin *n* abrin, jequiritin
abrinden to bark, to debark, to pare, to peel off, to strip off the bark, (abkorken) to decorticate
abrinnen to run off, to trickle away
Abrinnen *n* running off
Abriß *m* abstract, design, draft, outline, plan, sketch, summary
abrösten to roast [thoroughly]
Abrohr *n* discharge pipe, outlet pipe
Abrollapparat *m* dispenser, dispensing machine
Abrolldeckel *m* key-opening end, wind-open end
abrollen to uncoil, to unroll, to unwind, (wegrollen) to roll off
Abrollen *n* (in Kugelmühle) cascading
Abrollwalze *f* take-off roll
abrosten to corrode, to rust off
abrunden to make round, (Techn) to true, (Zahl) to round off
Abrundung *f* curvature, (Zahl) rounding off, ~ **von Übergängen** fillet
Abrundungsfehler *m* truncation error, (Math) rounding-off error
Abrundungshalbmesser *m* fillet radius
Abrundwerkzeug *n* rounding tool
abrußecht fast to rubbing
abrußen to rub off (colored goods)
abrutschen to slide off, to slip [off]
Abrutschkegel *m* charging cone
Absackanlage *f* bagging plant, sack-filling plant
absacken to sink down, (Bauten etc.) to sag, (in Säcke füllen) to sack
Absackmaschine *f* bagging machine, sacking machine
Absackung *f* bagging
Absackwaage *f* bagging scales, bagging and weighing machine
Absättigung *f* saturation, (Säure) neutralization
Absätzen, in ~ gradually, in stages, intermittently

absäuern to acidify, to acidulate, to make acidic
absanden to grind with blast, to sandblast
Absatz *m* (Niederschlag) deposit, precipitate, sediment, (Verkauf) sale, turnover
Absatzbassin *n* settling bassin, settling tank
Absatzbecken *n* settling basin, settling tank
Absatzfläche *f* settling area
Absatzforschung *f* marketing research
Absatzgefäß *n* precipitating vessel, settling tank
Absatzgestein *n* sedimentary rock
absatzweise fractional, gradually, intermittently, stagewise, **~ bewegt** actuated intermittently
Absaug[e]anlage *f* suction apparatus, dust extraction plant, exhauster, suction plant
Absaug[e]exsikkator *m* vacuum desiccator
Absaug[e]flasche *f* filtering flask
Absaug[e]haube *f* suction hood
Absaug[e]kolben *m* suction flask, draining flask
Absaug[e]leitung *f* suction piping
absaugen to suck off, to exhaust by suction, to filter by vacuum, to filter with suction, (Flüssigkeit) to drain off, to hydroextract, (Gas) to aspirate
Absaugen *n* exhausting by suction, sucking-off
Absaugeöffnung *f* [suction] outlet
Absauger *m* exhauster, hydroextractor, suction box
Absaug[e]rohr *n* outlet tube, suction pipe
Absaug[e]schlot *m* discharge pipe
Absaug[e]stutzen *m* connection for exhaust
Absaug[e]trockenofen *m* vacuum drying oven
Absaug-Filterzentrifuge *f* air-knife or suction discharge centrifuge
Absaugpumpe *f* exhaust pump, vacuum pump
Absaugung *f* removal by suction, exhaustion
Absaugungsanlage *f* exhaust plant
Absaugventilator *m* exhaust fan
Absaugvorrichtung *f* exhausting device
Abschab[e]eisen *n* scraper, doctor, scraping iron
Abschabemesser *n* scraper blade, doctor blade
abschaben to abrade, to grind off, to scrape off
Abschabsel *n pl* scrapings, shavings
abschälen to peel off, to exfoliate, to pare, to shell, (entrinden) to debark
Abschälen *n* peeling off, chipping off, flaking, scaling off, (des Lackes) [lacquer] peeling, (Entrinden) decorticating
Abschälverhältnis *n* cut (in separation processes)
abschätzen to appraise, to estimate, to value
Abschätzen *n* appraising, rating
abschäumen to scum, to skim off
Abschäumen *n* scumming, skimming off
Abschäumer *m* skimmer, skimming device
Abschäumlöffel *m* skimming ladle, skimmer
Abschäummittel *n* skimming agent
Abschäumsieb *n* skimming sieve
Abschäumung *f* scumming, skimming off
abschaffen to abolish
abschalten to cut off, to disconnect, to shut off, to switch off, to turn off

Abschalter *m* (Elektr) circuit breaker
Abschaltmagnet *m* disconnecting magnet
Abschaltrelais *n* cut-off relay, overflow relay
Abschaltung *f* disconnection, switching off
Abschaltventil *n* by-pass valve
abschattieren to shade off
Abschaum *m* dross, refuse, scum, skimmings
abscheidbar separable, (als Niederschlag) precipitable
Abscheidegefäß *n* precipitating vessel, separating vessel, separator
Abscheidekalorimeter *n* separating calorimeter
Abscheidekammer *f* condensation chamber
Abscheideleistung *f* efficiency of separation
abscheiden to separate, (Chem) to form a precipitate, to precipitate, (Galv) to deposit, (Met) to refine, (Physiol) to excrete, to secrete
Abscheiden *n* separating, (als Niederschlag) precipitating
Abscheider *m* separator, trap, (Atom) stripper, (Dispersion) settler
Abscheidung *f* separation, (Chem) precipitation
Abscheidungsmittel *n* precipitant, separating medium
Abscheidungspotential *n* deposition potential
Abscheidungsprodukt *n* deposit; (Chem) precipitate, separated product, (Physiol) secretion
Abscheidungsverfahren *n* method of separation, separating process
Abscheidungsvorrichtung *f* separator
abschelfe[r]n to scale off
abscheren to shear [off], to cut off
Abscheren *n* shearing
Abscherfestigkeit *f* shear[ing] strength
Abscherfläche *f* shear area
Abscherung *f* cutting [off], shear, shearing
Abscherungsbeanspruchung *f* shearing stress
Abscherversuch *m* shearing test
abscheuern (durch Abnutzung) to rub off, to wear off, (reinigen) to cleanse by scrubbing, to scour off
abschichten to separate into layers, to layer
abschiefern to scale off, to flake, to peel
abschießen to fire off, (entfärben) to lose color, (Rakete) to launch
abschilfern to chip, to flake, to peel, to scale [off]
Abschirmdublett *n* screening doublet
Abschirmeffekt *m* screening effect
abschirmen to shield, to block off, to cover, to protect, to screen
Abschirmkammer *f* shielding chamber
Abschirmkonstante *f* screening constant, shielding constant
Abschirmkorrektur *f* screening correction
Abschirmnetz *n* screen
Abschirmschicht *f* barrier sheet
Abschirmung *f* screen[ing], shielding
Abschirmungseffekt *m* screening effect

Abschirmungskonstante *f* screening constant
Abschirmungszahl *f* screening number
Abschirmwirkung *f* screening effect
abschlacken to remove slag [from], to clear from dross, to rake out slag, to slag out, to tap off the slag
Abschlacken *n* removing slag, slacking
Abschlacköffnung *f* cinder notch, cinder tap, slag notch
Abschlackrinne *f* slagging spout
Abschlackung *f* removal of slag
abschlämmen to remove mud, to decant, to elutriate, to wash
Abschlämmen *n* decanting, elutriation, washing
Abschlämmung *f* clearance of mud, decantation, elutriation, (Zucker) clarification
Abschläg *m* fragments, chips
abschlagen to strike off, to chip off, to knock off; (ablehnen) to refuse
Abschlagen *n* uncoupling
Abschlagevorrichtung *f* uncoupling device
Abschlagformkasten *m* snap flask
Abschlagschiene *f* knock-over bar
abschleifen to abrade, to grind off, to rub down, to sand, to sandpaper, to sharpen, (polieren) to polish
Abschleiffestigkeit *f* abrasion resistance
Abschleifung *f* abrasion, attrition, grinding off, sharpening
Abschleifversuch *m* abrasion test
abschleimen to clarify, to rid of slime
Abschleimung *f* clarification, purification
abschleppen to tow away, to carry away, to draw off
Abschleppgerät *n* towing equipment
Abschleudermaschine *f* centrifuge, hydroextractor
abschleudern to hydroextract, to throw off, (zentrifugieren) to centrifuge
abschlichten to plane, to polish, to smooth
abschließen to complete; to close, to prevent access to, to seal, to shut off
abschließend conclusive, final
Abschließung *f* closing off; isolation
Abschluß *m* conclusion; closing device, seal, shut-off, shutting-off device, **luftdichter** ~ air-tight seal
Abschlußbericht *m* final report
Abschlußblech *n* protecting screen
Abschlußdichtung *f* seal
Abschlußdüse *f* shut-off nozzle
Abschlußglas *n* glass plug
Abschlußhahn *m* stop cock
Abschlußmeldung *f* final report
Abschlußscheibe *f* closing disk
Abschlußschieber *m* cover slide, slide valve
Abschlußventil *n* closing valve, paddle valve
Abschlußvorrichtung *f* closing device, seal
Abschlußzapfen *m* delivery plug

abschmecken to taste, to test by tasting, (würzen) to season
Abschmelzdauer *f* (Sicherung) time of fusing
Abschmelzdraht *m* fuse wire, safety fuse
abschmelzen to seal, to separate by smelting, (Met) to melt off, to fuse, to smelt, (Sicherung) to blow out
Abschmelzkonstante *f* fusing coefficient, fusing constant
Abschmelzschweißung *f* fusion welding, flash welding
Abschmelzsicherung *f* electric fuse, safety cutout, safety fuse
Abschmelzstreifen *m* fuse link, fuse strip
Abschmelzstromstärke *f* fusing current
Abschmelztisch *m* melting table
Abschmelzung *f* melting, fusion, separation by smelting
Abschmelzverfahren *n* flash welding method, fusion welding
Abschmelzzeit *f* time of fusion
abschmieren to grease, to lubricate
abschmirgeln to grind with emery, to polish with emery, to sandpaper
Abschneideapparat *m* cutter, cutting apparatus
Abschneidefrequenz *f* cut-off frequency
abschneiden to clip, to cut off
Abschneideradius *m* cut-off radius
Abschnellfeder *f* starting spring
Abschnitt *m* part, portion; section, segment, (Geom) intercept, (Rohling) blank
abschnüren to tie off
abschöpfen to ladle out, to scoop off, to scum, to skim off
abschrägen to bevel off, to chamfer, to incline, to slant, to slope
Abschrägmaschine *f* chamfering machine
Abschrägrost *m* sloping grate
Abschrägung *f* slant, slope, (Techn) bevel, bevelling, chamfer[ing]
abschrauben to screw off, to unscrew
Abschreckalterung *f* quench aging
Abschreckbad *n* quenching bath
Abschreckbehälter *m* quench tank
Abschreckbiegeprobe *f* bending test in tempered state
Abschreckbottich *m* quenching tank
abschrecken to quench, to harden, to plunge in water, to chill, to cool suddenly, (Stahl) to temper
Abschrecken *n* chilling, quenching
Abschreckform *f* chill mold
Abschreckhärtung *f* quench hardening
Abschreckmethode *f* quenching method
Abschreckmittel *n* quenching agent, cooling agent
Abschrecktemperatur *f* quenching temperature
Abschreckung *f* chilling, hardening, quenching
Abschreckzeit *f* chilling time, quenching time

abschroten to chop off, to clip, to grind roughly, (Holz) to crosscut
abschruppen to scour, (Holz) to plane off
Abschürfung *f* scratch
abschüssig precipitous, steep
abschütten to pour off
Abschützvorrichtung *f* throttle arrangement
abschuppen to scale off, to flake, to laminate, to peel off
abschwächen to weaken, to diminish, to soften, (Akust) to tune down, (Elektr) to attenuate, (Phot) to reduce
Abschwächer *m* (Phot) reducing medium, clearing agent, reducer
Abschwächung *f* weakening, decline, diminution, reduction, (Elektr) attenuation
Abschwächungsfaktor *m* attenuation factor
Abschwächungskoeffizient *m* (Atom) attenuation coefficient, attenuation factor
Abschwächungsmittel *n* diluent, reducer, thinner
abschwärzen to blacken; to lose black color
abschwarten to bark, to decorticate
abschwefeln to desulfurize, to free of sulfur; to impregnate with sulfur, (Erze) to calcine
Abschwefeln *n* desulfurizing, desulfurating
Abschwefelung *f* desulfuration, desulfurization
abschwelen to calcine, to carbonize under vacuum, to roast
abschwemmen to elutriate, to rinse off, to wash away
abschwenkbar capable of being swivelled
abschwenken to swivel
abschwitzen to sweat off
absedimentieren to sediment out
absehbar observable, perceivable
abseifen to clean with soap, to soap
Abseifung *f* soaping, cleaning with soap; cleansing from soap
abseigern to separate by fusion, to liquate; to measure the perpendicular height, to plumb
Abseihebeutel *m* filtering bag
abseihen to strain, to decant, to filter, to percolate
Abseihen *n* straining, decanting, filtering
Abseihküpe *f* settling vat
Abseihung *f* elutriation, clarification; decantation, filtration
absengen to singe off
absenken to lower, to sink, to subside
Absenken *n* lowering, subsidence
Absenkung *f* lowering, settlement
Absenkungsfaktor *m* (Atom) disadvantage factor
Absenkvorrichtung *f* (Techn) height regulating device
absetzbar (entfernbar) removable, (verkäuflich) marketable
Absetzbassin *n* settling tank
Absetzbehälter *m* draining tank, settler, settling tank, settling vessel

Absetzbottich *m* settling vat
absetzen to deposit, to settle, (Kontrast geben) to contrast, (niederschlagen) to precipitate, (reduzieren) to reduce, (verkaufen) to sell
Absetzen *n* depositing, sedimentation, (Kontrast geben) contrasting, (Niederschlagen) settling
Absetzenlassen *n* sedimentation, settling
Absetzgefäß *n* settling tank, settling vessel
Absetzung *f* settling, deposition, precipitation
Absetzungsteich *m* settling trough
Absetzverhinderungsmittel *n* antisettling agent, sedimentation inhibitor
Absetzverhütungsmittel *n* antisettling agent
Absetzzeit *f* settling time
Absetzzentrifuge *f* sedimentation centrifuge
Absicht *f* purpose
absickern to trickle [down]
absieben to sieve, to screen, to sift [off]
absieden to boil, to extract by boiling, (Chem) to decoct
Absieden *n* boiling, decocting
absinken to sink [away]
Absinth *m* absinth[e], absinthium
absinthartig absinthial, absinthine
Absinthelixier *n* elixir of absinth
Absinthessenz *f* essence of absinth
Absinthsäure *f* absinthic acid
Absinthwermut *m* absinth bitter
absiphonieren to siphon off, draw off by siphoning
Absitzbehälter *m* settling tank, settling vessel
Absitzbütte *f* settling tub
absitzen to deposit, to settle, ~ **lassen** to allow to settle
Absitzgrube *f* settling pit
Absitzrohr *n* deposition tube, settling cylinder
Absitzverfahren *n* gravity settling process
Absitzvolumen *n* settling volume
absolut absolute, ~ **sicher** foolproof, ~ **trocken** absolutely dry, anhydrous
Absolutbewegung *f* absolute motion
Absolutdämpfung *f* absolute damping
Absolutdruck *m* absolute pressure
Absolute *f* (Math) absolute
Absoluteinheit *f* absolute unit
Absolut-Filter *n* air filter
Absolutgeometrie *f* absolute geometry, hyperbolic geometry
Absolutgeschwindigkeit *f* absolute velocity
Absolutgröße *f* absolute magnitude, absolute quantity
Absolutierung *f* dehydration
Absolutraum *m* absolute space
Absolutrechnung *f* absolute calculus
Absolutschwarz *n* absolute black
Absoluttrockengewicht *n* absolute dry weight
Absolutwert *m* absolute value

absondern to separate, to abstract, to detach, to isolate, to segregate, (ausscheiden) to discharge, to excrete, to secrete
Absondern *n* isolating, separating
Absonderung *f* (Abtrennung) separation, (Sekretion) secretion
Absorbens *n* (pl. Absorbentia) absorbent [agent]
Absorber *m* (Kältemaschine) absorber
Absorberelement *n* (Reaktor) absorber device
Absorberregelung *f* (Kältemaschine) absorber control
absorbierbar absorbable
Absorbierbarkeit *f* absorbability
absorbieren to absorb, to suck up, (Gase) to occlude
Absorbieren *n* absorbing
absorbierend absorbent, absorbing
Absorbierungsvermögen *n* absorption capacity
Absorptiometer *n* absorptiometer
Absorptiometrie *f* absorptiometry
absorptiometrisch absorptiometric
Absorption *f* absorption, **dielektrische** ~ (Elektr) dielectric absorption
Absorptionsachse *f* absorption axis
Absorptionsanlage *f* absorption equipment, absorption installation, absorption plant
Absorptionsapparat *m* absorption apparatus
Absorptionsbande *f* (pl Absorptionsbande) absorption band
Absorptionsbehälter *m* absorption cell
Absorptionsbereich *m* absorption region
absorptionsfähig absorbent, absorptive, capable of absorbing
Absorptionsfähigkeit *f* absorptive capacity, degree of absorption
Absorptionsfaktor *m* absorption factor
Absorptionsfarbe *f* absorption color
Absorptionsfilter *n* absorption filter
Absorptionsflasche *f* absorption bottle, absorption flask
Absorptionsflüssigkeit *f* absorption liquid, washing liquid
Absorptionsgefäß *n* absorption vessel
Absorptionsgeschwindigkeit *f* absorption velocity
Absorptionsgesetz *n* law of absorption
Absorptionsgewebe *n* absorbent tissue, absorbing tissue
Absorptionsgleichgewicht *n* absorption equilibrium
Absorptionsgrenze *f* absorption limit
Absorptionshygrometer *n* absorption hygrometer
Absorptionsindex *m* (Opt) absorption index
Absorptionskältemaschine *f* absorption refrigeration machine
Absorptionskante *f* (Atom) absorption edge, absorption limit, (Spektr) absorption discontinuity, absorption edge, absorption limit
Absorptionskeil *m* absorption wedge

Absorptionskoeffizient *m* absorption coefficient
Absorptionskohle *f* absorptive charcoal
Absorptionskolonne *f* absorption column, washing column
Absorptionskraft *f* absorptive power
Absorptionsküvette *f* absorption cell
Absorptionsleistung *f* absorptive capacity
Absorptionslinie *f* absorption line
Absorptionsmaschine *f* absorption machine
Absorptionsmeßgerät *n* absorptiometer
Absorptionsmessung *f* absorption measurement
Absorptionsmittel *n* absorbent, absorber, absorbing medium
Absorptionspipette *f* absorption pipet[te]
Absorptionsquerschnitt *m* absorption cross section
Absorptionsraum *m* absorption chamber
Absorptionsröhre *f* absorption tube
Absorptionsschlange *f* absorption coil
Absorptionsspektralanalyse *f* absorption-spectrum analysis
Absorptionsspektrum *n* absorption spectrum
Absorptionssprung *m* absorption discontinuity
Absorptionsstreifen *m* absorption band
Absorptionsstrom *m* (Elektr) absorption current
Absorptionsturm *m* absorption column, absorption tower
Absorptionsverbindung *f* (Chem) absorption compound
Absorptionsverfahren *n* method of absorption, process of absorption
Absorptionsverlust *m* absorption loss
Absorptionsvermögen *n* absorptive capacity, absorption factor, absorptive power, absorptivity
Absorptionswärme *f* heat of absorption
Absorptiv *n* absorbate
abspachteln to scrape [off]
abspänen to chip
abspaltbar cleavable, detachable, separable
Abspaltbarkeit *f* (des Elektrons) splitting off (of the electron)
abspalten to split off, to cleave, to crack, to eliminate, to remove, to separate
Abspalten *n* splitting off, cracking, eliminating, separating
abspaltend cleaving, cracking, splitting
Abspaltung *f* splitting off, cleavage, elimination, removal, separation
Abspannen *n* **eines Werkzeuges** stripping (of a mold)
absperrbar capable of being locked
Absperrdüse *f* shut-off nozzle
absperren to bar, to block, to cut off, to isolate, to shut off
absperrend blocking
Absperrflüssigkeit *f* sealing liquid, confining liquid
Absperrglied *n* shut-off device, sluice, valve

Absperrhahn *m* shut-off cock, shut-off nozzle, stop cock, stop valve
Absperrmittel *n* insulating agent, sealer
Absperrorgan *n* shut-off device
Absperrschieber *m* gate valve, slide valve, sluice valve
Absperrschraube *f* plug screw, stopper screw
Absperrung *f* blockade
Absperrventil *n* check valve, cut-off valve, cut-out valve, gate valve, shut-off valve, stop valve
Absperrvorrichtung *f* closing device
abspiegeln to mirror, to reflect
abspielen (sich abspielen) to happen, to occur, to take place
absplittern to clip, to flake, to peel, to scale
Absplittern *n* splintering off
Absplitterung *f* (von Neutronen) spallation (of neutrons)
absprechen to discuss
abspreizen to spread [apart], (absteifen) to brace, to strut, (Bergb) to prop, to stay
absprengen to blast off, to blow off, to burst off, (Glas) to break off, to crack
Absprenger *m* (Techn) cutter
Absprengung *f* chipping
abspringen to chip off; to leap off, to rebound, to snap off, to spring off
Abspritzdruck *m* discharge pressure
abspritzen to cleanse, to spray, to wash off
Abspritzrohr *n* spray pipe, squirting pipe
Abspritzverfahren *n* cleansing process; spray painting process
Absprung *m* leaping off, (Phys) reflection
Abspülbad *n* rinsing bath
abspülen to cleanse, to rinse, to wash
Abspülen *n* cleansing, rinsing
abspulen to reel off, to uncoil, to unwind
abstammen to be derived from, to descend
Abstammung *f* derivation, origin
Abstammungslehre *f* genetics, theory of the origin of species
Abstand *m* distance, clearance, gap, interval, space, spacing, ~ **nehmen von** to abandon, to object to, to refrain from, **interatomarer** ~ interatomic distance
Abstandbuchse *f* distance piece
Abstandhalter *m* spacer
Abstandsfaktor *m* (Stereochem) strain factor
abstandsgleich equidistant
Abstandshalterung *f* spacer
Abstandsklappe *f* bending support
Abstandsplatte *f* spacer
Abstandsstück *n* spacing piece, distance piece
abstatten to render, to make, to pay
abstauben to dust [off]
Abstauben *n* dusting [off]
Abstechautomat *m* (Techn) automatic cutting-off machine

abstechen (Kontrast geben) to contrast, (Metall) to draw off, to tap
abstechend contrasting
Abstecher *m* etcher, tapper
Abstechmesser *n* paint scraper
abstecken to mark out, to stake off, to trace
abstehen to become stagnant; to stand apart or aside; (Metall) to leave molten steel quiescent, ~ **lassen** to let stand, to allow to settle, (Metall) to dead melt
abstehend projecting
absteifen to reinforce, to stiffen, to support
absteigen to descend, to dismount to descend
Abstellbrett *n* storage shelf
abstellen to stop; to annul; to cut off, to disconnect; to set down; to store; to switch off, to turn off, (z. B. Maschine) to shut off
Abstellen *n* **des Gebläses** shutting off the blast, stopping the blast
Abstellhahn *m* stop cock
Abstellhebel *m* stop lever
Abstellplatte *f* side rest
Abstellschalter *m* stop switch
Abstelltisch *m* stand
Abstellventil *n* cut-off valve
Abstellvorrichtung *f* cut-off device, stop mechanism
absterben to die off, to expire, to fade, to wither
Abstich *m* drawing off, tapping (blast furnace), tapping operation, ~ **des Metalls** tapping the metal
Abstichbrust *f* (Metall) tapping front
Abstichgaserzeuger *m* slagging gas producer
Abstichloch *n* (Metall) tap[ping] hole, discharge aperture, metal notch
Abstichöffnung *f* (Metall) tap hole
Abstichpause *f* tapping interval
Abstichprobe *f* tapping sample
Abstichrinne *f* (Metall) tapping spout, runner
Abstichschaufel *f* tapping shovel
Abstichschlacke *f* tapping slag
Abstichsohle *f* pouring level
Abstichtätigkeit *f* (Metall) tapping operation
Abstieg *m* descent
Abstimmanzeigerohr *n* tuning indication tube
Abstimmbarkeit *f* (Phot) gradation
abstimmen (Akust) to tune; (Konto) to balance, (Techn, zeitlich) to synchronize, (z. B. Arbeiten) to coordinate
Abstimmknopf *m* (Radio) tuning knob
Abstimmkondensator *m* tuning capacitor
Abstimmkreis *m* tuned circuit, tuning circuit
Abstimmschärfe *f* clearness or selectivity or sharpness of tuning
Abstimmskala *f* tuning dial, tuning scale
Abstimmspule *f* synchronizing coil, tuning coil
Abstimmung *f* tuning, modulation, resonance, selection, synchronization
Abstimmvorrichtung *f* tuning apparatus

Abstoppen *n* **der Strömung** stoppage of the flow
Abstoß *m* push-off
abstoßen to knock off, to reject, to repel, to repulse, (abnützen) to fray, to wear off, (abwerfen) to cast off, ~ **gleichnamige Pole stoßen einander ab** like poles repel each other
Abstoßenergie *f* (Phys) repulsion energy
Abstoßfett *n* fat scrapings
Abstoßkraft *f* repellent force
Abstoßung *f* casting off, repulsion, **elektrostatische** ~ electrostatic repulsion, **gegenseitige** ~ mutual repulsion
Abstoßungseffekt *m* effect of repulsion
Abstoßungsenergie *f* repulsive energy
Abstoßungskraft *f* repelling power, repulsion, repulsive force
Abstoßungspotential *n* repulsive potential
abstrahieren to abstract
Abstrahl *m* (Phys) reflected ray
abstrahlen to radiate; to emit, (mit Sandstrahlgebläse) to sandblast, (Opt) to reflect
Abstrahler *m* radiator; sandblast
Abstrahlung *f* radiation, reflection
abstrakt abstract
Abstraktion *f* abstraction
Abstrebekraft *f* (Zentrifugalkraft) centrifugal force
abstrecken (Met) to iron
Abstrecken *n* (Metall) ironing
Abstreckziehen *n* ironing
Abstreicheisen *n* scraper
abstreichen to scrape off, to scum, to skim, to wipe off
Abstreichholz *n* straightedge, (Metall) strike
Abstreichlöffel *m* skimmer, skimming ladle
Abstreichmesser *n* doctor bar, scraper, scraping-off knife
Abstreifbacke *f* guide cheek, stock guide
abstreifen to strip off, to doctor off, (Buchdr) to squeegee
Abstreifen *n* (vom Stempel) stripping
Abstreifer *m* [ingot] stripper, scraper, skimmer, stripping fork, wiper, (f. Schlacke) scummer, skimmer
Abstreiferform *f* stripper plate mold
Abstreifergut *n* skimmings
Abstreiferplatte *f* stripper plate
Abstreifmeißel *m* guard
Abstreifmesser *n* doctor blade, scraping knife
Abstreifplatte *f* stripper plate, stripping plate
Abstreifring *m* wiping ring, wiper
Abstreiftisch *m* doctor table
Abstreifwalze *f* doctor roll
Abstrich *m* dross, litharge, scum, skimming, smear (of microscopic preparations), (Bakt) streak
Abstrichblei *n* (Metall) crude litharge, lead scum, lead skim

Abstrichlineal *f* straightedge
abströmen to flow off, to pass off
abstürzen to fall down, to hurl down
abstützen (Bergb) to stay, (Techn) to prop, to support, (versteifen) to reinforce, (verstreben) to brace, to strut
Abstützung *f* (Techn) bracing, support
abstufen to grade, to graduate, (Farbe) to shade
Abstufung *f* degree, gradation, shading
abstumpfen (Chem) to buffer, (Math) truncate, (Säure) to neutralize, (schwächen) to deaden, to dull, to weaken, (Spitze) to blunt
Abstumpfung *f* (Math) truncation, (Säure) neutralization, (Schneide) bluntness, dullness
Absturz *m* fall, crash, dive, plunge
Absturzhalde *f* dumping ground
Absturzplatz *m* dumping place
absuchen to scan, to search, to test
Absuchen *n* search
Absud *m* (Brau) decoction, suds, (Chem) extract, mordant
Absüßbottich *m* (Zuck) edulcorating tub
absüßen to sweeten, to edulcorate, (Pharm) to purify, to wash out
Absüßkessel *m* (Zuck) edulcorating tank
Absüßschale *f* (Zuck) edulcorating basin
Absüßspindel *f* saccharometer, sweet water spindle
Absüßung *f* (Zuck) sweetening [off]
Absüßwanne *f* (Zuck) edulcorating basin
Abszisse *f* (Geom) abscissa
Abszissenachse *f* (Geom) axis of abscissa
abtakeln to clear, to dismantle, to strip
Abtastband *n* (Telev) scanning belt
Abtasteinrichtung *f* scanning device
Abtastelektrode *f* exploring electrode
abtasten to feel, to make contact with, to palpate, (Elektr) to scan, (Techn) to cal[l]iper
Abtasten *n* scanning
Abtaster *m* feeler, (Phys) scanner
Abtastfeld *n* scanning field
Abtastfläche *f* scan area
Abtastfrequenz *f* scanning frequency
Abtastgerät *n* scanner, scanning device
Abtastgeschwindigkeit *f* pick-up velocity, scanning speed
Abtastlinearität *f* (Telev) scanning linearity
Abtastmikroskop *n* scanning microscope
Abtastöffnung *f* (Telev) scanning aperture
Abtastscheibe *f* (Telev) scanning disk
Abtastspannung *f* (Telev) scanning voltage
Abtaststrahl *m* scanning beam
Abtasttheorem *n* (nach Shannon) Shannon's sampling theory
Abtastung *f* scanning
Abtastvorrichtung *f* scanner
Abtastzeile *f* scanning line
Abtauautomatik *f* (Kühlschrank) automatic defroster

abtauen to defrost, to melt, to thaw off
abteilen to divide, to graduate; to separate
Abteilung f compartment, department, division, section
Abteilungsleiter m departmental manager, superintendent
Abteilventil n block valve
Abteufarbeit f sinking work
abteufen to bore, to deepen, to drill, to sink
Abteufgerüst n sinking trestle
Abteufkabel n pit cable
abtönen (Farbe) to tone, (gleichschalten) to synchronize, (schattieren) to shade, to tint
Abtönen n coloring, (Färb) tinting
Abtönfarbe f (Färb) tint, tinting color
Abtönpaste f colorant in paste form
Abtönungsbad n toning bath
Abtönvermögen n tinting power
abtöten to kill, to destroy
Abtöten n killing, destruction, extinction
abtränken to impregnate, to saturate, to soak
Abträufeln n dripping
Abtrag m excavation, cutting; excavated material
abtragen to carry away, (Erdmassen) to level, (Geol) to erode
Abtragung f clearing away, (Geol) denudation, erosion, (Techn) removal
Abtreibapparat m evaporating apparatus, decanter, still, stripper
Abtreibearbeit f (Metall) refining by cupellation
Abtreibeherd m (Metall) refining hearth
abtreiben (Chem) to decant, to distill off, to expel, (läutern) to cupel, to cupellate, to separate, to strip, (Metall) to refine
Abtreibofen m (Metall) cupeling furnace, refining furnace
Abtreiber m decanter, still
Abtreibesäule f stripping column
Abtreibescherbe f small cupel furnace, test furnace
Abtreibgas n expelled gas
Abtreibkapelle f (Metall) refining cupel
Abtreibkolonne f stripping column
Abtreibschmelze f (Metall) refining melt
Abtreibung f (Metall) cupellation, refining, stripping
abtrennbar separable
abtrennen to cut off, to detach; to dissociate; to isolate, to separate; to sever
Abtrenner m separator
Abtrennschalter m splitting key
Abtrennung f separation; cleaving, cutting; dissociation
Abtrennungsarbeit f work of separation
abtreten to cede, to transfer
Abtretung f assignment, cession, surrender, transfer
Abtrieb m downward pressure

Abtriebsäule f distilling column, separating column, stripping column
Abtriebsteil n stripping zone
abtriefen to drip, to trickle down, to trickle off in drops
Abtrift f deviation, drift, leeway
Abtriftmesser m drift gauge, drift meter
abtrocknen to desiccate, to dry, to wipe off
Abtrocknen n desiccating, drying
abtröpfeln to drip off, to drain off
Abtröpfeln n dripping
Abtropfblech n draining board, drip pan
abtropfen to drip off, ~ **lassen** to drain off
Abtropfer m drainer, drip board
Abtropfgefäß n drainer
Abtropfgestell n dish drainer, draining stand
Abtropfkasten m (Met) drain table, (Pap) drain chest
Abtropfpfanne f drain pan, list pot
Abtropfschale f draining dish, drip pan
Abtropfständer m draining stand
Abtrübungsfarbe f darkening dye, dulling dye
abtun to discard, to dismiss, to dispose, to put aside, to take off
abtupfen to dab, to wipe off
abtuschieren to lap
Abutsäure f abutic acid
abwägen to weigh out, to balance, (bedenken) to consider, to ponder, to weigh (words)
Abwägung f weighing
Abwärme f waste heat
Abwärmekraftmaschine f waste heat engine
Abwärmeofen m waste heat kiln
Abwärmeverlust m waste heat loss
Abwärmeverwertung f waste heat utilization
Abwärmeverwertungsanlage f waste heat plant, waste heat recovery plant
abwärts downward, ~ **bewegen** to descend, ~ **gleiten** to descend, to drop, ~ **transformieren** to reduce, to step down
Abwärtsbewegung f downward motion
Abwärtstransformator m (Elektr) step down transformer, voltage reducing transformer
Abwärtswandler m step down transformer
Abwässer pl, **radioaktive** ~ radioactive effluents
abwässern to drain, to free from water
Abwässerung f drainage
abwalzen to roll off, to compact with roller, to make smooth by rolling, to roll down
Abwalzen n hobbing, smoothing off
Abwalzfederbock m slide shacklespring carrier
Abwalzfräser m generating milling cutter
abwandelbar modifiable, variable
abwandeln to alter, to change, to modify
abwandern to migrate, to deviate, to move
Abwandlung f modification, variation
abwarten to await, to expect, to wait for
abwaschbar resistant to scrubbing, washable
Abwaschbarkeit f washability

abwaschen – Abziehmittel

abwaschen to wash [off], to cleanse, to rinse
Abwaschen *n* washing, ~ **von Anstrichen durch Regen** abrasion by rain
Abwaschmittel *n* cleansing agent, detergent
Abwaschtisch *m* washing table
Abwaschung *f* washing
Abwaschwanne *f* keeler
Abwasser *n* waste water, effluent, sewage
Abwasseranlage *f* sewage disposal plant
Abwasserbehandlung *f* effluent water treatment
Abwasserbeseitigung *f* waste water disposal
Abwasserklärung *f* sewage clarification, waste water clarification
Abwasserkontrolle *f* supervision of sewage
Abwasserleitung *f* drain [pipe], overflow pipe
Abwasserorganismen *m pl* waste water organisms
Abwasserreinigung *f* purification of sewage disposal, waste water purification
Abwasserreinigungsanlage *f* waste water purifying plant
Abwasserrohr *n* drain pipe, waste pipe
Abwassertechnik *f* effluent treatment
Abwasseruntersuchung *f* effluent water testing
Abwasserverwertung *f* salvage of sewage, utilization of waste water
Abwasserzufluß *m* sewage flow
abwechseln to vary, to alternate, to change, to fluctuate
abwechselnd alternating, intermittent, periodical, variable, varying
Abwechslung *f* variety, alternation, change
Abwehrferment *n* (Biochem) protective ferment or enzyme
Abwehrmechanismus *m* defence mechanism
Abwehrmittel *n* (Med) prophylactic, preventive
Abwehrpatent *n* defensive patent
Abwehrreaktion *f* (Biol) defence reaction
Abwehrstoff *m* antibody
abweichen to vary, to deflect, to deviate, to diverge; (einweichen) to macerate, to soak, to soften, to steep
abweichend anomalous, deviating, divergent, varying
Abweichung *f* deviation, difference, discrepancy, divergence, tolerance, variance, (Chem) anomaly, (Opt) aberration, (Phys) deflection, **magnetische** ~ magnetic deviation, **zulässige** ~ permissible aberration, permissible tolerance
Abweichungskompaß *m* declination compass
abweisen to refuse, to reject, to repel, to repulse
abweisend repellent
abweißen to lose whiteness; to whiten
abwenden to avert, to turn off
abwerfbar droppable
abwerfen to discard, to drop, to dump, to give off, to release, to shed, to throw off, to yield
Abwerfofen *m* refining furnace

Abwerfpfanne *f* list pot, refining pan
abwerten to devaluate, to devalue
Abwertung *f* devaluation
abwesend absent, missing
Abwetter *n* (Bergb) exhaust air
Abwickelbock *m* let-off stand, unwind stand, **gebremster** ~ friction let-off stand
abwickeln to perform; to reel off, to uncoil, to unroll, to unwind
Abwickelwalze *f* take-off roll
Abwiegemaschine *f* weighing machine, (Dosiermaschine) dosing machine
abwiegen to weigh [out], to dose
abwinden to unwind
Abwischbausch *m* wiping pad
abwischen to wipe off, to clean, to wash
Abwischen *n* wiping
Abwischlösemittel *n* wiping solvent
abwracken to break up, to dismantle, to scrap, to wreck
abwringen to wring out
Abwurf *m* discharge, release, throw off, (Bauw) rough cast
Abwurfrolle *f* tripper
Abwurftrichter *m* discharge hopper
Abwurftrommel *f* head drum, head pulley, tripper
Abwurfvorrichtung *f* tripper
abzählbar countable
abzählen to count [out or off]
abzapfen (Flüssigkeit) to draw off, to tap
Abzapfhahn *m* drain valve, draw-off cock
abzehren to waste away, to consume, to corrode, to emaciate
Abzehrung *f* wasting away, consumption, corrosion
Abzeichen *n* mark, sign
abzeichnen to mark off; to copy, to sketch
Abziehapparat *m* racking apparatus
Abziehbad *n* stripping bath, degumming bath
Abziehbild *n* transfer picture
Abziehblase *f* alembic, retort, still
Abzieheffekt *m* stripping effect
abziehen (abgießen) to siphon, (absaugen) to draw off, (destillieren) to distill, (entrinden) to strip, (entweichen) to escape, (Gerb) to pare, to scrape, (Leder abziehen) to rub leather pumice-stone, (Math) to deduct, to subtract, (Phot) to print, (Rauch) to escape, (wegziehen) to pull off
Abziehen *n* rectification, distillation, (Text) discoloring
Abziehformkasten *m* box with movable part
Abziehhülse *f* withdrawal sleeve
Abziehkeller *m* racking cellar
Abziehkolben *m* alembic, retort
Abziehmaschine *f* shaving machine
Abziehmittel *n* stripping agent

Abziehpapier *n* stripping paper, (Buchdr) proof[ing] paper
Abziehschraube *f* puller screw
Abziehstein *m* grindstone, hone, oil stone, whetstone
Abziehvorrichtung *f* honing appliance, take-off conveyor, take-up mechanism
Abziehzünder *m* friction igniter
Abzug *m* (Buchdr, Phot) copy, print, (Extrusion) nip rolls, (für Abgase) drain, dross hood, exhaust, flue, fume cupboard, fume hood, outlet, vent, (Metall) scum, slag
Abzugband *n* discharge belt
Abzughahn *m* (einer Spritzpistole) trigger
Abzugkanal *m* main flue
Abzugrohr *n* discharge pipe
Abzugsblei *n* lead obtained from dross
Abzugsdampf *m* exhaust steam
Abzugsgas *n* flue gas, burnt gas, chimney gas, exit gas, waste gas
Abzugsgrube *f* sink hole, trap
Abzugshaube *f* fume hood
Abzugshebel *m* trigger
Abzugskanal *m* discharge pipe, drain, escape channel, sewer
Abzugsklappe *f* flow-off valve, outlet valve
Abzugskupfer *n* copper obtained from dross
Abzugsöffnung *f* outlet, vent
Abzugspapier *n* duplicating paper
Abzugsrohr *n* drain pipe, delivery tube, outlet, vent pipe, waste pipe
Abzugsschacht *m* flue
Abzugsschlacke *f* (Metall) scoria
Abzugsschlot *m* flue
Abzugsschrank *m* fume cupboard, hood
Abzugstemperatur *f* exit temperature, temperature of escaping gases
Abzugsventil *n* outlet valve
Abzugsvorrichtung *f* ventilator, discharge apparatus, escape device, pulling device, take-off device
Abzugswärme *f* exit heat, heat loss, waste heat
Abzugswalze *f* draw-off roll, nip roll, pull roll, take-off roll
Abzugwalzer *m* haul-off device
abzundern to scale off
Abzweigdose *f* branch box, junction box
Abzweigdraht *m* shunt wire
abzweigen to branch off, to bifurcate, to tap, (Elektr) to shunt off
abzweigend branching
Abzweigkasten *m* branch box, (Elektr) distributing box
Abzweigklemme *f* (Elektr) branch terminal
Abzweigleitung *f* branch [line]
Abzweigmuffe *f* branch sleeve, branch socket
Abzweigrohr *n* branch pipe
Abzweigschalter *m* tapping switch
Abzweigspule *f* tapped coil

Abzweigstecker *m* distribution plug, tap plug
Abzweigstelle *f* branch, branching-off point
Abzweigstrom *m* derived current, shunted current
Abzweigstromkreis *m* derived circuit
Abzweigstück *n* branch piece
Abzweigstutzen *m* branch piece
Abzweigung *f* branch, branching; derivation; tapping
Abzweigungsverhältnis *n* branching ratio
Abzweigwiderstand *m* shunt resistance, leak resistance
abzwicken to nip off, to pinch off
Acacatechin *n* acacatechin
Acacetin *n* acacetine
Acaciin *n* acaciine
Acadialith *m* (Min) acadialite
Acanthit *m* (Min) acanthite
Acanthoidin *n* acanthoidine
Acanthoin *n* acanthoine
Acaricid *n* acaricide, mite killer
Acaroidharz *n* acaroid resin, accroides gum, black-boy gum, xanthorrhea resin
Accracopalinsäure *f* accracopalinic acid
Accracopalsäure *f* accracopalic acid
ACD-Lösung *f* (Med) ACD solution, acid citrate dextrose solution
Aceanthren *n* aceanthrene
Aceanthrenchinon *n* aceanthrenequinone
Aceconitsäure *f* aceconitic acid
Acenaphthalid *n* acenaphthalide
Acenaphthalsäure *f* acenaphthalic acid
Acenaphthazin *n* acenaphthazine
Acenaphthen *n* acenaphthene, ethylenenaphthene
Acenaphthenchinon *n* acenaphthenedione, acenaphthenequinone
Acenaphthenol *n* acenaphthenol
Acenaphthenon *n* acenaphthenone
Acenaphthenpikrat *n* acenaphthene picrate
Acenaphthinden *n* acenaphthindene
Acenaphthochinolin *n* acenaphthoquinoline
Acenaphthoylpropionsäure *f* acenaphthoylpropionic acid
Acenaphthylen *n* acenaphthylene
Acenaphthylessigsäure *f* acenaphthylacetic acid
Acenocumarol *n* acenocoumarol
Aceperimidin *n* aceperimidine
Aceperinaphthan *n* see under Acephenalan
Acephenalan *n* acephenalan, aceperinaphthane
Acepleiadan *n* acepleiadane
Acepleiadien *n* acepleiadiene
Acepromazine *n* acepromazine
Acer[i]tannin *n* aceritannin
Acerit[ol] *n* aceritol
Acetacetat *n* acetoacetate
Acetal *n* acetal
Acetaldazin *n* acetaldazine

Acetaldehyd *n* acetaldehyde, acetic aldehyde, aldehyde, ethanal, ethyl aldehyde
Acetaldehydammoniak *n* acetaldehyde ammonia
Acetaldehyddiäthylacetal *n* 1,1-diethoxyethane, acetaldehyde diethyl acetal
Acetaldehydharz *n* acetaldehyde resin
Acetaldol *n* acetaldol, aldol
Acetaldoxim *n* acetaldoxime, aldoxime
Acetalharz *n* acetal resin
Acetamid *n* acetamide, acetic acid amine, ethanamide
Acetamidin *n* acetamidine
Acetamido- acetamido-, acetamino-, acetylamino-
Acetaminomalonsäure *f* acetaminomalonic acid
Acetaminosalol *n* acetaminosalol
Acetanhydrid *n* acetic anhydride, acetyl oxide
Acetanilid *n* acetanilide, antifebrin, N-phenylacetamide
Acetanisidin *n* acetanisidine, methacetin, p-methoxyacetanilide
Acetarsol *n* acetarsol
Acetat *n* acetate, salt or ester of acetic acid
Acetatfaser *f* acetate fiber
Acetatfolie *f* acetate film, acetate sheeting
Acetatkunstseide *f* acetate rayon
Acetatregenerierungsanlage *f* acetate regenerating plant
Acetatreyon *n* acetate rayon
Acetatseide *f* acetate rayon
Acetatthiokinase *f* (Biochem) acetate thiokinase
Acetatzellwolle *f* spun acetate rayon
Acetazolamid *n* acetazolamide
Acetenyl- acetenyl-, ethinyl-
Acetessigäther *m* acetoacetic ether
Acetessigester *m* acetoacetic ester, ethyl acetoacetate
Acetessigesterenol *n* acetoacetic ester enol
Acetessigsäure *f* acetoacetic acid, acetyl acetic acid
Acetessigsäureäthylester *m* (Acetessigester) acetoacetic ester, ethyl acetoacetate
Acethydrazid *n* acethydrazide
Acethydroxamsäure *f* acethydroxamic acid
Acethydroximsäure *f* acethydroximic acid
Acetidin *n* acetidin[e], ethyl acetate
Acetiminoester *m* acetimidic ester
Acetin *n* acetin, glyceryl monoacetate, monacetin
Acetinblau *n* acetin blue
Acetmethylanilid *n* acetmethylanilide
Acetnaphthylamid *n* acetnaphthylamide
Acetobromglucose *f* acetobromoglucose, acetylated glucosyl bromide, acetylglucosyl bromide
Acetogenine *pl* (Biochem) acetogenins
Acetoglyceral *n* acetoglyceral, glycerol ethylidene ether
Acetoguanamin *n* acetoguanamine

Acetoin *n* acetoin, 3-hydroxy-2-butanone, acetylmethylcarbinol
Acetol *n* acetol, 1-hydroxy-2-propanone, acetone alcohol, acetonyl alcohol, acetyl carbinol, hydroxyacetone
Acetolyse *f* acetolysis
Acetometer *n* acetometer
acetometrisch acetometric
Acetomycin *n* acetomycin
Aceton *n* acetone, 2-propanone, dimethylketon, ketopropane
Acetonämie *f* (Med) aceton[a]emia
Acetonalkohol *m* acetone alcohol, 1-hydroxy-2-propanone
Acetonamin *n* acetonamine
Acetonaphthon *n* acetonaphthone, naphthylmethyl ketone
Acetonbisulfit *n* acetone bisulfite
Acetonbromoform *n* acetone-bromoform
Acetoncarbonsäure *f* acetone carboxylic acid
Acetonchloroform *n* 1,1,1,-trichloro-2-methyl-2-propanol, acetone chloroform, chlorobutanol
Acetoncyanhydrin *n* acetone cyanhydrin, 2-metyl-lactonitrile
Acetondicarbonsäure *f* acetone dicarboxylic acid, β-ketoglutaric acid
Acetondiessigsäure *f* acetone diacetic acid, hydrochelidonic acid
Acetonentwässerung *f* acetone dehydration
Acetonextraktion *f* acetone extraction
Acetonharz *n* acetone resin
Acetonin *n* acetonine
Acetonitril *n* acetonitrile, methyl cyanide
Acetonketal *n* acetoneketal
Acetonkörper *m* (Med) acetone body
acetonlöslich acetone-soluble
Acetonmannit *m* acetone mannite, mannitol isopropylidene ether
Acetonnachlauf *m* second runnings from acetone
Acetonnatriumbisulfit *n* acetone sodium bisulfite
Acetonsäure *f* acetone acid, acetonic acid
Acetonsulfit *n* acetone sulfite
Acetontrockenpulver *n* acetone powder
Acetonyl- acetonyl-
Acetonylaceton *n* 1,2-diacetylethane, acetonyl acetone
Acetonylharnstoff *m* acetonylurea
Acetoperoxid *n* acetic peroxide, diacetyl peroxide
Acetopersäure *f* peracetic acid, peroxyacetic acid
Acetophenon *n* acetophenone, acetyl benzene, phenyl methyl ketone
Acetopiperon *n* acetopiperone
Acetopseudocumol *n* acetopseudocumene
Acetoresorcin *n* acetoresorcinol
Acetothienon *n* acetothienone
Acetoxim *n* acetoxime, acetone oxime

Acetoxy- acetoxy-
Acetoxyanthracen *n* acetoxyanthracene
Acetoxybenzpyren *n* acetoxybenzpyrene
Acetoxybutyraldehyd *m* acetoxybutyraldehyde
Acetoxymethylcholanthren *n* acetoxymethylcholanthrene
Acetoxynaphthochinon *n* acetoxynaphthoquinone
Acetozon *n* acetozone, acetylbenzoyl peroxide, benzozone
Acetpersäure *f* peracetic acid, peroxyacetic acid
Acetphenetid *n* acetophenetide
Acetsäure *f* acetic acid
Acetum *n* acetum, vinegar
Acetursäure *f* aceturic acid, acetylglycine
Acetyl- acetyl-
Acetylaceton *n* acetylacetone
Acetylacetonäthylendiimin *n* acetylacetone ethylenediimine
Acetylacetonat *n* acetylacetonate
Acetylamin- acetylamino-
Acetylbestimmung *f* acetyl determination
Acetylbromid *n* acetyl bromide, ethanoyl bromide
Acetylcelluloid *n* acetyl celluloid
Acetylcellulose *f* cellulose acetate
Acetylchlorid *n* acetyl chloride, ethanoyl chloride
Acetylcholin *n* acetyl choline
Acetylcholinchlorid *n* acetyl choline chloride
Acetyl-CoA *n* (Biochem) acetyl CoA, acetyl coenzyme A
Acetylcoenzym *n* acetyl coenzyme
Acetyldigitoxin *n* acetyldigitoxin
Acetyldiisopropylamin *n* acetyl diisopropylamine
Acetylen *n* acetylene, ethine, ethyne
Acetylenanstalt *f* acetylene plant
Acetylenbindung *f* acetylene linkage, triple bond
Acetylenbrenner *m* acetylene burner
Acetylenbunsenbrenner *m* acetylene Bunsen burner
Acetylendicarbonsäure *f* acetylene dicarboxylic acid
Acetylendichlorid *n* acetylene dichloride
Acetylendiurein *n* acetylene diurein, glycoluril
Acetylenentwickler *m* acetylene generator
Acetylenerzeuger *m* acetylene generator
Acetylengas *n* acetylene gas
Acetylengebläse *n* oxyacetylene blowpipe, oxyacetylene torch
Acetylengebläselampe *f* acetylene blowpipe lamp
Acetylenharnstoff *m* acetylene urea
Acetylenkohlenwasserstoff *m* acetylene hydrocarbon
Acetylenkupfer *n* copper acetylide
Acetylenlampe *f* acetylene lamp
Acetylenmagnesiumbromid *n* acetylene magnesium bromide

Acetylenoxid *n* acetylene oxide
Acetylenreihe *f* acetylene series
Acetylenreinigungsmasse *f* acetylene purifier, acetylene purifying mass
Acetylenruß *m* acetylene black
Acetylensauerstoffbrenner *m* oxyacetylene blowpipe
Acetylensauerstoffschweißung *f* oxyacetylene welding
Acetylen-Schnittbrenner *m* oxyacetylene cutting device
Acetylenschweißanlage *f* acetylene welding plant
Acetylenschweißbrenner *m* oxyacetylene welding torch
Acetylenschweißung *f* acetylene welding
Acetylentetrabromid *n* acetylene tetrabromide, Muthmann's liquid, tetrabromoethane
Acetylentetrachlorid *n* acetylene tetrachloride, tetrachloroethane
Acetylenyl- acetylenyl-, ethynyl-
Acetylessigsäure *f* acetoacetic acid
Acetylesterase *f* (Biochem) acetylesterase
Acetylharnstoff *m* acetylurea
Acetylhydrocellulose *f* hydrocellulose acetate
Acetylid *n* acetylide
Acetyliden *n* acetylidene
Acetylidenoxid *n* acetylidene oxide
acetylieren to acetylate
Acetylieren *n* acetylating
Acetylierung *f* acetylation
Acetylierungsgemisch *n* acetylating mixture
Acetylierungsmittel *n* acetylating agent, acetylating medium
Acetylierungsverfahren *n* acetylating method, acetylation process
Acetyljodid *n* acetyliodide
Acetylmethadol *n* acetylmethadol
Acetylmethionin *n* acetylmethionine
Acetylnitrocellulose *f* acetylated cellulose nitrate, acetyl cellulose nitrate
Acetylphosphat *n* acetylphosphate
Acetylpropionsäure *f* acetopropionic acid
Acetylsalicylsäure *f* acetylsalicylic acid, aspirin
Acetylschwefelsäure *f* acetylsulfuric acid
Acetyltannin *n* acetyltannin, tannigen
Acetyltransacylase *f* (Biochem) acetyl transacylase
Acetylzahl *f* acetyl number, acetyl value
Achat *m* (Min) agate, **bunter** ~ mochastone, **schwarzer** ~ silicious schist
achatähnlich agate-like, agatine
achatartig agate-like, agatine
Achatbohrer *m* agate drill
Achatfeuerstein *m* agate flint
achatförmig agatiform
Achatglättmaschine *f* stone burnisher
Achatglas *n* (Alabaster-, Opalglas) alabaster glass
achathaltig agatiferous

Achatmörser *m* agate mortar
Achatonyx *m* (Min) agate onyx
Achatpapier *n* (Techn) agate paper
Achatporzellan *n* agateware
Achatrolle *f* agate cylinder
Achatschale *f* agate dish, agate mortar
Achatschellack *m* agate shellac, shellac substitute
Achatschleifer *m* agate grinder
Achatschneide *f* agate edge, agate knife edge
Achatseidenpapier *n* agate tissue paper
Acherit *m* (Min) acherite
Achillea *f* (Bot) milfoil
Achilleasäure *f* achilleaic acid, aconitic acid
Achillein *n* achilleine
Achillenöl *n* achillea oil
Achmatit *m* (Min) achmatite, achmite
Achmit *m* (Min) acmite
Achrematit *m* achrematite
Achrodextrin *n* achrodextrin
Achroit *m* (Min) achroite
Achromasie *f* achromatism, ~ **für die Linien D und G** (Opt) D-G achromatism
Achromat *m* (Opt) achromat, achromatic lens
achromatisch achromatic
achromatisieren to achromatize
Achromatisierung *f* achromatization
Achromatismus *m* achromatism, **sphärischer** ~ spherical achromatism
Achromatopsie *f* (Med) achromatopsia
Achroodextrin *n* achroodextrin
Achsabstand *m* axle base, wheel base, (Math) distance between axes
Achsaufhängung *f* axle suspension
Achsbelastung *f* axle load
Achsbüchse *f* axle bearing, axle box
Achsbüchsenführung *f* axle box guide
Achsdruck *m* axial pressure, axle load
Achse *f* arbor, axle, shaft, spindle, (Geom) axis, (Math) coordinate, (Projektionsachse) reference line, **freie** ~ free axis, **große** ~ (Ellipse) major axis, **horizontale** ~ horizontal axis, **kleine** ~ (Ellipse) minor axis, **kristallographische** ~ crystallographic axis, **magnetische** ~ magnetic axis, **neutrale** ~ neutral zone, **optische** ~ optic axis, **parallaktische** ~ equatorial or parallactic axis, **sich um seine eigene** ~ **drehen** to rotate about one's own axis, **vertikale** ~ vertical axis
Achsel *f* shoulder
Achselstück *n* shoulder strap
Achsenabschnitt *m* [axis] intercept, **Gesetz der rationalen** ~ **e** law of rational intercepts
Achsenbild *n* (Interferenzfigur) interference figure
Achsendrehbank *f* axle lathe
Achsendrehung *f* circular motion, rotation of axes
Achsendruck *m* axle pressure

Achsenebene *f* (Krist) axial plane
Achsengehäuse *n* axle casing, axle sleeve
achsengleich coaxial
Achsenkreuz *n* intersection of axes, system of coordinates
Achsenlager *n* axle bearing, journal bearing
Achsennagel *m* linch pin
Achsenneigung *f* inclination of an axis
Achsenöl *n* axle oil
Achsenregler *m* shaft governor
Achsenreibung *f* axle friction
Achsenrichtung *f* axial direction
Achsenring *m* axle ring
Achsenrohr *n* axial tube
Achsenschmiere *f* axle grease
Achsenschnitt *m* section through the axis
Achsensymmetrie *f* (Krist) axial symmetry
achsensymmetrisch axially symmetrical
Achsensystem *n* system of coordinates
Achsenverhältnis *n* axial ratio, **kristallographisches** ~ crystallographic axial ratio
Achsenwinkel *m* axial angle
Achsfeder *f* axle spring
Achsfederung *f* axle suspension
Achsfett *n* axle grease
Achsgleitplatte *f* axle box liner
achsig axial
Achslager *n* axle bearing, journal bearing, journal box
Achsmitte *f* **des Bohrlochs** center of bore hole
Achswelle *f* arbor shaft
Achszapfen *m* axle end pivot, axle journal
Achtaragit *m* (Min) achtaragite
achtbasisch octabasic
achtbindig octavalent
Achteck *n* octagon
achteckig octagonal, eightangled
Achtecksäule *f* octagonal column
Achtel *n* eighth (part)
achten to esteem, to regard, to respect
Achter *m* figure of eight, (Atom) quadruple twin
Achterschale *f* (Atom) octet, octet shell, ring of eight electrons
Achterspant *n* after frame
achtfach eightfold
achtfältig eightfold
achtflächig (Krist) octahedral
Achtflächner *m* (Krist) octahedron
achtgeben to pay attention
achtkantig octagonal
Achtkantschlüssel *m* octagonal spanner
Achtkantstab *m* octagon bar
achtlos careless, negligent, unmindful
Achtring *m* (Chem) eight-membered ring
achtsam careful
Achtsamkeit *f* care
achtseitig eight-sided

Achtundvierzigflächner m (Krist) hexakisoctahedron
Achtung! attention!, danger!, look out!, take care!, warning!, watch out!
Achtung f respect
achtungsvoll respectful
achtwertig octavalent
Achtwertigkeit f octavalence
Achtzehnerschale f 18-shell, ring of eighteen electrons
achtzehnflächig octodecimal
Aci- s. a. unter Azi-
Acidalbumin n acid albumen
Acidalbuminat n acid albuminate
Acidifikation f acidification
acidifizieren to acidify
Acidifizierung f acidification
Acidimeter n acidimeter
Acidimetrie f acidimetry
acidimetrisch acidimetric
Acidität f acid capacity, acidity
Aciditätsbestimmung f determination of acidity
Acidobutyrometer n acidobutyrometer
Acidobutyrometrie f acidobutyrometry
Acidol n acidol, betaine hydrochloride
Acidolchromfarbstoff m acidol chrome color
Acidoligand m acido ligand, anionic ligand
Acidomycin n acidomycin, actithiazic acid
Acidose f (Med) acidosis, acidemia
Acidum acetylsalicylicum n acetylsalicylic acid
Acidum arsenicosum n arsenous acid
Acidum benzoicum n benzoic acid
Acidum boricum n boric acid
Acidum carbolicum n carbolic acid
Acidum citricum n citric acid
Acidum formicum n formic acid
Acidum hydrochloricum n hydrochloric acid
Acidum nitricum n nitric acid
Acidum phosphoricum n phosphoric acid
Acidum sulphuricum n sulfuric acid
aci-Form f (Stereochem) aci-form
Acinitrazol n acinitrazole, aminitrozole
Acker m acre, farm land, field
Ackerbau m agriculture, cultivation of the land, farming
Ackerbauchemie f agricultural chemistry
Ackerbaukunde f agronomy
Ackerboden m arable soil, surface soil
Ackererde f arable soil
Ackerfrucht f fruit of the soil
Ackerschädling m field pest, injurious insect
Acocantherin n acocantherin
Acofriose f acofriose
Acoin n acoine
Acolycitin n acolycitine
Acolytin n acolytine
Aconellin n aconelline
Aconin n aconine
Aconinextrakt m extract of aconine

Aconit n aconite
Aconitase f (Biochem) aconitase
Aconitat n aconitate, salt or ester of aconitic acid
Aconitin n aconitine, acetylbenzoyl aconine
Aconitsäure f aconitic acid, propene-1,2,3-tricarboxylic acid
Aconittinktur f aconitic tincture
Aconsäure f aconic acid
Acoranon n acoranone
Acoron n acorone
Acovenose f acovenose
Acrammerin n acrammerin
Acrichin n acrichine
Acridan n acridan
Acridin n acridine
Acridinfarbstoff m acridine dye
Acridingelb n acridine yellow
Acridinorange n acridine orange
Acridinsäure f acridic acid
Acridizin n acridizine
Acridol n acridol
Acridon n acridone
Acriflavin n acriflavine
Acrifolin n acrifoline
Acrilan n acrilan
Acrit m acritol
Acrolein n acrolein, acrylaldehyde, propenal
Acronycidinsäure f acronycidinic acid
Acronycinsäure f acronycinic acid
Acrose f acrose
Acrosom n acrosome
Acrylaldehyd m acrolein, acrylaldehyde, propenal
Acrylat n acrylate
Acrylester m acrylate, acrylic ester
Acrylglas n acryl glass
Acrylharz n acrylic resin, acrylate resin
Acrylhydroxamsäure f acrylhydroxamic acid
Acrylkautschuk m acrylic rubber
Acrylkunststoffe m pl acrylic plastics
Acrylmethylester m methyl acrylate
Acrylmischpolymere n pl acrylic copolymers
Acrylnitril n acrylonitrile, vinyl cyanide
Acrylsäure f acrylic acid, propenoic acid
Acrylsäureäthylester m ethyl acrylate
ACS n (antiretikulär-zytotoxisches Serum) anticytotoxic serum
Actamer n actamer
ACTH n (Adrenocorticotropin) adrenotropic or adrenocorticotropic hormone
Actin n actin
Actinamin n actinamine
Actinide pl actinides
Actinidin n actinidine
Actinium n actinium (Symb. Ac)
Actinocin n actinocin
Actinocinin n actinocinin
Actinomycin n actinomycin

Actinon *n* (radioaktive Ausströmung von Actinium) actinon
Actinospectinsäure *f* actinospectinoic acid
Actiphenol *n* actiphenol
Actithiazsäure *f* actithiazic acid, acidomycin
Actol *n* actol, silver lactate
Actomyosin *n* actomyosin
Actomyosinkomplex *m* actomyosin complex
acyclisch acyclic
Acylgruppe *f* acyl-group
acylieren to acylate
Acyloin-Kondensation *f* acyloin condensation
Acylphosphat *n* acyl phosphate
Adalin *n* adaline
Adamantan *n* adamantane
Adamantoblast *m* (Schmelzzelle) adamantoblast, ameloblast, enamel builder, ganoblast
Adamin *m* (Min) adamine
Adamsit *m* (Kampfstoff) adamite adamsite, diphenylamine-arsine chloride
Adams-Katalysator *m* Adams catalyst
Adantverfahren *n* (Zuck) Adant cube process
Adapter *m* (Phot) adapter
adaptieren to adapt
Adaptierung *f* adaption
Adaptometer *n* (Opt) adaptometer
Adatom *n* adatom, adsorption atom
addieren to add up, to sum up, to total
addierend additive
Addiermaschine *f* adding machine
Addierwerk *n* adding device
Addisonsche Krankheit *f* (Med) Addison's disease
Addition *f* adding up, addition, **elektrophile** ~ electrophilic addition, **ionische** ~ ionic addition, **nucleophile** ~ nucleophilic addition, **stereospezifische** ~ stereospecific addition
additionsfähig capable of addition
Additionsfähigkeit *f* additive power
Additionsfarbe *f* addition color
Additionsfehler *m* mistake in adding
Additionskonstante *f* addition constant
Additionspolymerisat *n* addition polymer
Additionspolymerisation *f* addition polymerization
Additionsprodukt *n* addition product
Additionsreaktion *f* additive reaction
Additionssatz *m* addition theorem
Additionstheorem *n* addition theorem
Additionsverbindung *f* addition compound
Additionszeichen *n* plus sign, addition sign
additiv additive
Additiv *n* additive
Additivität *f* additivity
Additivitätsbeziehung *f* additivity relation
Addukt *n* adduct, inclusion complex
Adduktor *m* adductor
Adelit *m* (Min) adelite
Adelpholit *m* (Min) adelpholite

Adenase *f* (Biochem) adenase
Adenin *n* 6-aminopurine, adenine
Adeninhexosid *n* adenine hexoside
Adeninribosid *n* adenine riboside, adenosine
Adenitis *f* (Med) adenitis
Adenocarpin *n* adenocarpine
Adenose *f* adenose
Adenosin *n* adenine riboside, adenosin[e]
Adenosindiphosphat *n* adenosine diphosphate
Adenosinmonophosphat *n* adenosine monophosphate, adenylic acid, **zyklisches** ~ cyclic adenosine monophosphate, cyclic adenylic acid, cyclic AMP
Adenosintriphosphat *n* adenosine triphosphate
Adenosintriphosphatase *f* adenosine triphosphatase
Adenosylhomocystein *n* adenosyl homocysteine
Adenosylmethionin *n* adenosyl methionine
Adenovirus *m* adenovirus
Adenylatcyclase *f* (Biochem) adenyl cyclase
Adenylsäure *f* adenylic acid, adenosine monophosphate
Adenylsäurepyrophosphat *n* s. unter Adenosintriphosphat
Adeps *m* adeps, animal fat, animal grease
Ader *f* strand, wire, (Bergb) lode, seam, (Holz) grain, streak, (Med) vein
Aderhaut *f* (des Auges) choroid membrane (of the eye)
Aderholz *n* grain wood
aderig streaked, veined
Adermin *n* (Pyridoxin, Vitamin B$_6$) adermin, pyridoxin
Aderung *f* (Holz) veining, (Marmor) marbling, marmoration
Aderverkalkung *f* (Med) arteriosclerosis, atherosclerosis
adhärieren to adhere
Adhäsion *f* adherence, adhesion
Adhäsionsfähigkeit *f* adhesiveness
Adhäsionsfett *n* adhesive grease or fat
Adhäsionsgrenze *f* limit of adhesion
Adhäsionskraft *f* adhesive power, force of adhesion
Adhäsionsmasse *f* adhesive [substance]
Adhäsionsspannung *f* adhesion tension
Adhäsionsvermögen *n* adhesive capacity, adhesiveness, adhesive power
adhäsiv adhesive
Adhumulinon *n* adhumulinone
Adiabate *f* adiabatic curve
Adiabatenexponent *m* adiabatic exponent
Adiabatenhypothese *f* adiabatic hypothesis
Adiabatensatz *m* adiabatic theorem
adiabatisch adiabatic
adiabetisch (Med) adiabetic
adiaktinisch adiactinic
adiatherm adiathermal
Adion *n* adion

Adiowanöl *n* ajava oil, ajowan oil
Adiowansamen *m* ajowan
Adiphenin *n* adiphenine
Adipinaldehyd *m* adipic aldehyde
Adipinat *n* adipate, salt or ester of adipic acid
Adipinsäure *f* 1,4-butanedicarboxylic acid, adipic acid, hexanedioic acid
Adipinsäureamid *n* adipamide
Adipinsäuredinitril *n* adipic dinitrile
Adipinsäureester *m* adipate, adipic ester
Adipiodon *n* adipiodone
Adipocellulose *f* adipo-cellulose
Adipocerit *n* adipocerite
adipös adipose, obese
Adipoin *n* adipoin, β-hydroxy cyclohexanone
Adipomalsäure *f* adipomalic acid, α-hydroxy-α-methylglutaric acid
Adipoweinsäure *f* adipotartaric acid, dihydroxyadipic acid
adjungiert adjoint
Adjunktion *f* adjunction
Adjustage *f* adjustment
adjustieren to adjust, to regulate, to set
Adjustieren *n* adjusting
Adjustiertisch *m* adjusting table
Adjustierung *f* adjustment
Adjustierwaage *f* (Techn) adjusting balance
Adjuvans *n* adjuvant
Adlerholz *n* eagle wood
Adlerstein *m* eagle stone, aetites
Adlumidin *n* adlumidine
Adlupulon *n* adlupulone
Admissionsdruck *m* admission pressure
Admissionsspannung *f* admission tension
Admittanz *f* (Rad) admittance
Adnamin *n* adnamine
Adonidin *n* adonidin
Adonin *n* adonin
Adonit *m* adonite, adonitol
Adonitoxigenin *n* adonitoxigenin
adoptieren to adopt
adoucieren to anneal, to decarbonize; to edulcorate, to sweeten, (Farben) to soften
Adouciergefäß *n* annealing pot; edulcorating vessel
Adoucierofen *m* annealing furnace, tempering furnace
ADP siehe Adenosindiphosphat
Adrenalektomie *f* (Med) adrenalectomy, excision of an adrenal gland
Adrenalin *n* adrenalin[e], epinephrine, suprarenin
Adrenalon *n* adrenalone
Adrenochinon *n* adrenoquinone
Adrenochrom *n* adrenochrome
Adrenocorticotropin *n* adrenotropic or adrenocorticotropic hormone, ACTH
Adrenodoxin *n* adrenodoxin
Adrenosteron *n* adrenosterone

Adressant *m* consignor, sender, writer
Adressat *m* consignee, person addressed
Adronolacetat *n* adronol acetate
Adsorbat *n* adsorbate, adsorbed substance
Adsorbens *n* adsorbent, adsorbing substance
Adsorber *m* adsorber
adsorbierbar adsorbable
Adsorbierbarkeit *f* adsorbability
adsorbieren to adsorb
Adsorbieren *n* adsorbing
Adsorption *f* adsorption, adherence, **aktivierte** ~ activated adsorption, **apolare** ~ (Phys) apolar adsorption, **bevorzugte** ~ preferential adsorption, **isotherme** ~ adsorption isotherm, **negative** ~ negative adsorption, **selektive** ~ selective or preferential adsorption, **van der Waalsche** ~ van der Waals adsorption
Adsorptionsanalyse *f* (Chromatographie) adsorption analysis, chromatographic analysis
Adsorptionseigenschaft *f* adsorptive property
Adsorptionserscheinung *f* adsorption phenomenon
Adsorptionsexponent *m* adsorption exponent
Adsorptionsfähigkeit *f* adsorptive capacity, adsorptiveness
Adsorptionsgleichgewicht *n* adsorption equilibrium, **das** ~ **ändern** to displace the adsorption equilibrium
Adsorptionsgrad *m* degree of adsorption
Adsorptionshäutchen *n* adsorbed film
Adsorptionshaut *f* adsorbed film
Adsorptionshülle *f* adsorption shell
Adsorptionsindikator *m* adsorption indicator
Adsorptionsisotherme *f* adsorption isotherm, **unstetige** ~ discontinuous adsorption isotherm
Adsorptionskatalyse *f* adsorption catalysis
Adsorptionskohle *f* activated charcoal
Adsorptionskraft *f* adsorptive power
Adsorptionsmittel *n* adsorbent, ~ **für Gase** adsorbent for gases
Adsorptionspotential *n* adsorption potential
Adsorptionspumpe *f* adsorption pump
Adsorptionssäule *f* adsorption column
Adsorptionsschicht *f* adsorbed layer; adsorbent layer, adsorption layer
Adsorptionsverbindung *f* (Chem) adsorption compound
Adsorptionsverdrängung *f* adsorption displacement
Adsorptionsvermögen *n* adsorbing capacity, adsorptive power
Adsorptionsvorgang *m* adsorption process
Adsorptionswärme *f* heat of adsorption
Adsorptionszentrum *n* adsorption center
Adsorptiv *n* adsorbate
Adstringens *n* (pl Adstringentien) astringent
Adstringenz *f* (Med) astringency
adstringierend astringent

A-Düse f A die
Adular m (Min) adularia
Adynerin n adynerin
Aegelin n egeline
Ägirin m (Min) aegirine, aegirite, acmite
Ägirit m (Min) aegirite, acmite, aegirine
Ägyptisch Blau n Egyptian blue
ähneln to resemble
ähnlich like, resembling, similar, (vergleichbar) analogous, comparable
Ähnlichkeit f likeness, resemblance, similarity, similitude
Ähnlichkeitsgeometrie f geometry of general similarity
Ähnlichkeitsgesetz n (Phys) law of similarity
Ähnlichkeitskennzahlmethode f dimensional analysis
Ähnlichkeitsmechanik f similarity principle
Ähnlichkeitsprinzip n (Phys) principle of similitude
Ähnlichkeitsregel f (Phys) similarity rule
Ähnlichkeitssatz m (Phys) law of similarity, similarity theorem
Ähnlichkeitstheorie f principle of similarity
Ähnlichkeitstransformation f (Math) similar transformation of matrices
Ähnlichkeitsverhältnis n ratio of similitude
Ähre f spica, (Getreide) ear, head
Ährensilber n (Min, ährenförmiges Graukupfererz) spicular silver
ändern to alter, to change, (umkehren) to reverse
Änderung f alteration, change, modification
Änderungsgeschwindigkeit f rate of change
Äpfelsäure f malic acid, hydroxybutanedioic acid
Äpfelsäurediäthylester m diethyl malate
Äpfelsäuresalz n malate, salt of malic acid
Äquator m equator
äquatorial equatorial
äquianharmonisch equianharmonic
äquidistant equidistant
Äquilibrierung f equilibration
äquimolekular equimolecular
Äquipotential n (Phys) equipotential
Äquipotentialfläche f (Phys) equipotential surface
Äquipotentiallinie f equipotential line, contour line
Äquipotentialverbindung f equalizing connection
äquipotentiell equipotential
äquivalent equivalent
Äquivalent n equivalent
Äquivalentgewicht n equivalent weight
Äquivalentkonzentration f equivalent concentration
Äquivalentleitfähigkeit f equivalent conductance, equivalent conductivity
Äquivalentleitvermögen n equivalent conductivity
Äquivalentreaktionsgeschwindigkeit f equivalent rate of reaction
Äquivalentverhältnis n equivalent proportion
Äquivalenz f equivalence
Äquivalenzeinheit f equivalence unit
Äquivalenzgesetz n equivalence law
Äquivalenzgewicht n equivalent weight
Äquivalenzladung f equivalent charge
Äquivalenzprinzip n equivalence principle
Äquivalenzpunkt m end point, equivalence point
aerob aerobic
Aerobe f (Biol) aerobe, aerobian
Aerobiologie f aerobiology
Aerobiont m (Biol) aerobic organism
aerobiontisch aerobic
Aerobiose f aerobiosis
Aerodynamik f aerodynamics
Aerodynamiker m aerodynamicist
aerodynamisch aerodynamic
Aeroelastizität f aero-elasticity
Aerofallmühle f aerofall mill
Aerogel n (Phys) aerogel
aerogen (Phys) aerogenic, aerogenous
Aerogengas n aerogene gas
Aerografie f aerography
Aerolith m (Min) aerolite
Aeromechanik f aeromechanics
Aerometer n aerometer, hydrometer
Aeronautik f aeronautics
aeronautisch aeronautic
aerophysikalisch aerophysical
Aerophyt m (Biol) epiphyte
Aerosol n aerosol
Aerosoldose f aerosol can
Aerosoltreibgas n aerosol propellant gas
Aerosolverpackung f aerosol package, pressure pack, pressurized pack
Aerostat m aerostat
Aerostatik f aerostatics
aerostatisch aerostatic
Ärzteschaft f medical profession
ärztlich medical
Äscher m (Gerb) tanner's pit, lime pit; slaked lime
Äscherbrühe f lime liquor
Äscherfaß n liming tub
Äschergrube f lime pit, tanner's pit
Äscherkalk m lime
äschern to reduce to ashes, (Leder) to lime
Äschern n chalk-liming, liming
Äscherofen m (Keram) calcining oven
Äschersatz m limeflux
Äscherung f chalk liming, liming
Äscherverfahren n lime process
Aeschynit m (Min) echinite, eschynite
Äsculetin n esculetin
Aesculin n aesculin, esculin
Äsculussäure f esculetinic acid
Aethacridin n ethacridine

Äthal *n* ethal, cetyl alcohol
Äthan *n* ethane
Äthanal *n* acetaldehyde, acetic aldehyde, ethanal
Äthanalkohol *m* ethanol
Äthandiol *n* glycol
Äthanol *n* ethanol, ethyl alcohol
Äthanolamin *n* ethanolamine, colamine
Äthansäure *f* acetic acid, ethanoic acid
Äthebenin *n* ethebenine
Äthebenol *n* athebenol
Äthen *n* ethene, ethylene
Äthenyl- ethenyl-, vinyl-
Äther *m* ether
ätherähnlich ethereal, ether-like
Ätheranlage *f* ether plant
Ätherart *f* kind of ether
ätherartig ethereal
Ätherat *n* etherate
Ätherauszug *m* ethereal extract
Ätherbildung *f* ether formation, etherification
Ätherdampf *m* ether vapor
ätherförmig etheriform
Ätherharz *n* ether resin
Ätherin *n* (obs) etherin, ethylin
ätherisch volatile, essential, ethereal
ätherisierbar etherifiable, etherizable
ätherisieren (Med) to anesthetize, to etherize
Äthermaschine *f* ether machine
Äthermitbewegung *f* ether drift
Äthernarkose *f* (Med) ether narcosis
Ätherospermin *n* atherospermine
Ätherprober *m* ether tester
Ätherschwefelsäure *f* acid alkyl sulfate, alkylsulfuric acid, hydrogen alkyl sulfate
Äthervergoldung *f* ether gilding
Ätherwaschverfahren *n* ether washing process
Ätherweingeist *m* solution of ether in alcohol, spirit of ether
Ätherweinsäure *f* (obs) ethyltartaric acid
Ätherzerstäuber *m* ether spray
Äthiden *n* ethidene, ethylidene
Äthin *n* acetylene, ethyne
Äthinylierung *f* ethynylation
Äthionsäure *f* ethionic acid, ethylene sulfonic acid
Äthoxalyl- ethoxalyl
Äthoxy- ethoxy-, ethoxyl
Äthoxyäthan *n* diethyl ether, ethyl ether, ethyl oxide
Äthoxychrysoidinhydrochlorid *n* ethoxychrysoidine hydrochloride
Äthoxykoffein *n* ethoxycaffeine
Äthoxyl- ethoxy-, ethoxyl-
Äthoxylierung *f* ethoxylation
Äthoxylin *n* ethoxylene
Äthoxylinharz *n* epoxy resin, ether resin
Äthoxysalicylaldehyd *m* ethoxysalicylic aldehyde

Aethusanol *n* ethusanol
Aethusin *n* ethusin
Äthyl- ethyl-
Äthyl *n*, **carbaminsaures** ~ ethyl carbamate, urethane, **chloressigsaures** ~ ethyl chloroacetate, ethyl chloracetate, **essigsaures** ~ ethyl acetate, acetic ester
Äthylacetanilid *n* ethyl acetanilide
Äthylacetat *n* ethyl acetate
Äthylacetoacetat *n* ethyl acetoacetate, acetoacetic ester
Äthyläther *m* ethyl ether, diethyl ether, ethoxyethane, ethyl oxide
Äthyläthoxysilan *n* ethyl ethoxysilane
Äthylal *n* acetaldehyde, ethylal
Äthylaldehyd *m* acetaldehyde, ethyl aldehyde
Äthylalkohol *m* ethanol, ethyl alcohol
äthylalkoholisch ethanolic, ethylalcoholic
Äthylamin *n* ethylamine
Äthylaminobenzoat *n* ethyl aminobenzoate, anesthesin, benzocaine
Äthylaminobenzoesäure *f* ethyl aminobenzoic acid
Äthylanilin *n* ethylaniline
Äthylarsin *n* ethyl arsine
Äthylat *n* ethylate
Äthylbenzoat *n* ethyl benzoate
Äthylbenzol *n* ethyl benzene
Äthylbenzolsulfonat *n* ethyl benzene sulfonate
Äthylbenzylanilin *n* ethyl benzyl aniline
Äthylbenzylketon *n* ethyl benzyl ketone
Äthylbromid *n* ethyl bromide, bromoethane
Äthylbutylmalonat *n* ethyl butyl malonate
Äthylbutyrat *n* ethyl butyrate
Äthylcarbamat *n* ethyl carbamate
Äthylcarbinol *n* ethyl carbinol, normal propyl alcohol, propanol
Äthylcarbylamin *n* ethylcarbylamine
Äthylcellosolve *f* ethyl cellosolve
Äthylcellulose *f* ethyl cellulose
Äthylchloracetat *n* ethyl chloroacetate
Äthylchlorformiat *n* ethyl chloroformate
Äthylchlorid *n* ethyl chloride, chloroethane
Äthylchlorsilan *n* ethylchlorosilane
Äthylchlorzinnsäure *f* ethyl chlorostannic acid
Äthylcinnamat *n* ethyl cinnamate
Äthylcyanid *n* ethyl cyanide, propane nitrile, propionitrile
Äthyldimethylsulfoniumsalz *n* ethyldimethylsulfonium salt
Äthyldisulfid *n* diethyl disulfide, ethyl disulfide
Äthylen *n* ethylene, ethene
Äthylenalkohol *m* 1,2-dihydroxyethane, 1,2-ethanediol, [ethylene] glycol
Äthylenbenzol *n* ethylene benzene
Äthylenbindung *f* ethylene linkage
Äthylenbromid *n* 1,2-dibromoethane, ethylene bromide
Äthylencarbonat *n* ethylene carbonate

Äthylenchlorhydrin − Ätiophyllin

Äthylenchlorhydrin *n* ethylene chlorohydrin, 2-chloroethanol
Äthylenchlorid *n* 1,2-dichloroethane, ethylene [di]chloride
Äthylencyanhydrin *n* ethylene cyanohydrin
Äthylencyanid *n* 1,2-dicyanoethane, ethylene [di]cyanide, succinonitrile
Äthylendiamin *n* 1,2-diaminoethane, 1,2-ethanediamine, ethylenediamine
Äthylendiamintetraessigsäure *f* (EDTA) ethylene diamine tetraacetic acid
Äthylendibromid *n* 1,2-dibromoethane, ethylene [di]bromide
Äthylengewinnungsanlage *f* ethylene recovery plant
Äthylenglykol *n* 1,2-dihydroxyethane, 1,2-ethanediol, ethylene glycol
Äthylenhydrid *n* ethane, ethylene hydride
Äthylenimin *n* ethylene imine
Äthylenjodid *n* 1,2-diiodoethane, ethylene [di]iodide
Äthylenketal *n* ethylene ketal
Äthylenkunststoffe *pl* ethylene plastics
Äthylenoxid *n* 1,2-epoxyethane, ethylene oxide, dimethylene oxide
Äthylenreihe *f* ethylene series
Äthylenrest *m* ethylene radical
Äthylenthioketal *n* ethylene thioketal
Äthylfluorid *n* ethyl fluoride
Äthylformiat *n* ethyl formate
Äthylglucosid *n* ethyl glucoside
Äthylglykolacetat *n* ethyl glycol acetate
Äthylhexanol *n* ethyl hexanol
Äthyliden *n* ethylidene
Äthylidenbromid *n* 1,1-dibromoethane, ethylidene [di]bromide
Äthylidenchlorid *n* 1,1-dichloroethane, ethylidene [di]chloride
Äthylidenharnstoff *m* ethylidene urea
Äthylidenjodid *n* 1,1-diiodoethane, ethylidene [di]iodide
Äthylidenmilchsäure *f* ethylidene lactic acid, 2-hydroxypropanoic acid, [fermentation] lactic acid
äthylieren to ethylate
Äthylieren *n* ethylation
Äthylierung *f* ethylation
Äthylisobutyrat *n* ethyl isobutyrate
Äthylisosuccinat *n* ethyl isosuccinate
Äthyljodid *n* ethyl iodide, iodoethane
Äthylkohlensäurechininester *m* quinine ethylcarbonate
Äthylmalonat *n* ethyl malonate, malonic acid ethyl ester
Äthylmalonsäure *f* ethylmalonic acid
Äthylmercaptan *n* ethyl mercaptan, ethanethiol, ethyl hydrosulfide
Äthylmethylacetsäure *f* ethyl methylacetic acid
Äthylmorphin *n* ethyl morphine
Äthylmorphinhydrochlorid *n* ethylmorphine hydrochloride, dionine
Äthylnitrat *n* ethyl nitrate
Äthylnitrit *n* ethyl nitrite
Äthylnitrobenzoat *n* ethyl nitrobenzoate
Äthylorthoformiat *n* ethyl orthoformate
Äthyloxid *n* [di]ethyl ether, diethyl oxide, ethoxy ethane
Äthylpentan *n* ethyl pentane
Äthylperoxid *n* diethyl peroxide, ethyl peroxide
Äthylphenylacetat *n* ethyl phenylacetate
Äthylphenyläther *m* ethyl phenyl ether, ethoxybenzene, phenetole
Äthylphenylcarbonat *n* ethyl phenyl carbonate
Äthylphenylketon *n* ethyl phenyl ketone
Äthylphthalat *n* [di]ethyl phthalate
Äthylpropionat *n* ethyl propionate
Äthylpropyläther *m* ethyl propyl ether
Äthylpropylcarbinol *n* ethyl propyl carbinol
Äthylpropylketon *n* ethyl propyl ketone
Äthylpyruvat *n* ethyl pyruvate
Äthylracemat *n* ethyl racemate
Äthylresorcin *n* ethyl resorcinol
Äthylrhodanid *n* ethyl thiocyanate
Äthylrot *n* ethyl red
Äthylsalicylat *n* ethyl salicylate
Äthylschwefelsäure *f* ethyl sulfuric acid, acid ethyl sulfate, ethyl hydrogen sulfate
äthylschwefelsauer ethyl sulfate
Äthylsenföl *n* ethyl isothiocyanate, ethyl mustard oil
Äthylsuccinat *n* ethyl succinate, diethyl succinate
Äthylsulfat *n* [di]ethyl sulfate
Äthylsulfhydrat *n* ethanethiol, ethyl hydrogen sulfide, ethyl mercaptan
Äthylsulfid *n* [di]ethyl sulfide
Äthylsulfocarbonsäure *f* ethyl sulfocarbonic acid, ethyl thiocarbonic acid
Äthylsulfocyanat *n* ethyl thiocyanate
Äthylsulfonsäure *f* ethyl sulfonic acid, ethanesulfonic acid
Äthyltartrat *n* [di]ethyl tartrate
Äthyltoluolsulfonamid *n* ethyl toluene sulfonamide
Äthyltoluolsulfonat *n* ethyl toluene sulfonate
Äthylurethan *n* ethylcarbamate, [ethyl]urethane
Äthylverbindung *f* ethyl compound
Äthylvinylketon *n* ethyl vinyl ketone
Äthylviolett *n* ethyl violet
Äthylwasserstoff *m* ethane, ethyl hydride
Äthylweinsäure *f* ethyltartaric acid
Äthylxanthogenat *n* ethyl xanthate, ethyl xanthogenate
Äthylzinnsäure *f* ethyl stannic acid
Ätioergosterin *n* etioergosterol
Ätiologie *f* (Med) etiology
Ätiomesoporphyrin *n* etiomesoporphyrin
Ätiophyllin *n* etiophyllin

Ätioporphyrin *n* etioporphyrin
Ätit *m* (Min) etite, eagle stone
Ätzalkali *n* caustic alkali
Ätzalkalilösung *f* caustic alkaline solution
ätzalkalisch caustic alkaline
Ätzammoniak *n* ammonia water, ammonium hydroxide, caustic ammonia
Ätzartikel *m* etching article
Ätzbad *n* etching bath
ätzbar (Chem) corrodible, (Metall) capable of being etched, (Text, Farben) dischargeable
Ätzbarkeit *f* (Chem) corrodibility, (Text, Farben) dischargeability
Ätzbaryt *m* barium hydroxide, caustic barytes, hydrated barium oxide
Ätzbeizdruck *m* discharge printing
Ätzbeize *f* (Färb) discharge mordant
ätzbeständig resisting caustic
Ätzbild *n* etch pattern
Ätzbrett *n* etching board
Ätzdruck *m* etching, discharge printing
Ätze *f* caustic water, aqua fortis
ätzen to bite, (anätzen) to cauterize, (auf Zink) to zinkograph, (Chem) to corrode, (einätzen) to etch, (Text, Farben) to discharge
Ätzen *n* (Chem) corroding, (Einätzen, Buchdr) etching
ätzend caustic, corroding
Ätzer *m* etcher
Ätzfarbe *f* discharge color
Ätzfigur *f* etching figure; corrosion figure
Ätzflüssigkeit *f* caustic liquid, etching acid
Ätzgift *n* caustic poison
Ätzgrübchen *n* etched pit
Ätzgrund *m* etching ground, etching surface; etching varnish
Ätzgrundierung *f* self-etching primer
Ätzhilfsmittel *n* discharging auxiliary
Ätzhügel *m* etching relief
Ätzkali *n* caustic potash, potassium hydroxide
Ätzkalianlage *f* caustic potash plant
Ätzkalk *m* caustic lime, **gebrannter** ~ calcium hydroxide, calcium oxide, quicklime
Ätzkalklösung *f* lime water
Ätzkasten *m* etching board
Ätzkraft *f* causticity, corrosiveness
Ätzkunst *f* art of etching
Ätzlack *m* discharge lake
Ätzlauge *f* caustic liquor, caustic lye, caustic solution; potassium hydroxide solution, (Buchdr) etching lye
Ätzlösung *f* caustic solution (Galvanik) conditioner
Ätzmagnesia *f* caustic magnesia, magnesium hydroxide
Ätzmittel *n* caustic, chemical discharge, corrosive, discharging agent
Ätznadel *f* etching needle
Ätznatron *n* caustic soda, sodium hydroxide

Ätznatronanlage *f* caustic soda plant
Ätznatronelektrolyse *f* electrolysis of caustic soda
Ätznatronlösung *f* caustic soda lye, caustic soda solution, sodium hydroxide solution
Ätznatronschmelze *f* caustic soda melt, fused caustic soda
Ätzpaste *f* corrosive paste, etching paste
ätzpolieren to etch-polish, to polish with acid
Ätzpolieren *n* etching polishing
Ätzpolitur *f* etch-polish
Ätzprobe *f* etching test, corroding proof, etching sample
Ätzpulver *n* caustic powder, etching powder
Ätzsalz *n* caustic salt, corrosive salt
Ätzschliff *m* ground section for etching
Ätzsilber *n* silver nitrate, lunar caustic
Ätzstein *m* caustic stone; caustic potash; lunar caustic
Ätzstift *m* caustic stick
Ätzstoff *m* caustic, corrosive
Ätzsublimat *n* corrosive sublimate, mercuric chloride, mercury dichloride, mercury(II) chloride
Ätztinte *f* caustic ink, etching ink
Ätzung *f* etching, cauterization, corrosion, **galvanische** ~ electro-engraving, electro-etching
Ätzverfahren *n* caustic process, etching process, (Text) discharging method
Ätzwasser *n* caustic water, aqua fortis, nitric acid
Ätzweiß *n* white discharge
Ätzwirkung *f* discharge action
Ätzzeichnung *f* etching
äußern to express, to manifest, to utter
Äußerung *f* expression, manifestation
Affe *m* monkey, ape
Affekt *m* emotion
Affenbrotbaum *m* baobab tree
Affenbrotöl *n* baobab oil
affin affine
Affinade *f* (Zuck) affinated sugar, washed raw sugar
Affination *f* refining, refinement
affinierbar affinable
affinieren to refine
Affinieren *n* refining
Affinierung *f* refining
Affinierungsverfahren *n* refining process
Affinin *n* affinin
Affinität *f* affinity, **chemische** ~ chemical affinity, **Elektro-** ~ electro-affinity
Affinitätseinheit *f* unit of affinity
Affinitätskonstante *f* affinity constant
Affinitätskraft *f* affinity force
Affinitätslehre *f* doctrine of affinity
Affinitätsrest *m* affinity residue
Affinitätsrichtung *f* direction of affinity

affizieren (Med, krankhaft verändern) to affect
Afghangelb n Afghan yellow
AFL (Antifibrinolysin) antifibrinolysin
Afridol n afridol
Afrikagrün n African green, hydrated chromium sesquioxide
Afromosin n afromosin
AFT (Antifibrinolysintest, Med) antifibrinolysin test
Afterbildung f false formation, malformation
Afterhorn m (Bot) sycamore
Afterkegel m (Math) conoid
Afterkohle f slack [coal]
Afterkristall m pseudomorphous crystal
Afterkugel f (Math) spheroid
Aftermehl n coarse flour, pollard
Afterschörl m (Min) axinite
Aftersilber n silver containing dross
Aftertopas m (Min) Bohemian brown topas
Afzelin n afzelin
Agalit m agalite
Agallol n (Pflanzenschutzmittel) agallol
Agalmatolith m (Min) agalmatolite, lard stone, steatite
Agaphit m (Min) agaphite, blue turquoise
Agar-Agar n agar-agar, Bengal gelatin, Ceylon or Chinese isinglass or gelatin, Japanese gelatin, Macassargum
Agaricin n agaricin
Agaricinsäure f agaric acid, agaricic acid, laricic acid
Agaritin n agaritine
Agarlösung f solution of agar
Agarobiose f agarobiose
Agarol n agarol
Agarplatte f agar plate
Agarythrin n agarythrine
Agathalen n agathalene
Agathalin n agathalene
Agathan n agathane
Agatholsäure f agatholic acid
Agathsäure f agathic acid
Agave f (Bot) agave fibre agave fiber, aloe fiber
Agavenbranntwein m pulque
Agavenfaser f agave fiber, sisal
Agavenhanf m agave hemp, aloe hemp
Agavensaft m agave sap
Agavose f agavose
Agens n agent, principle, **chemisches** ~ chemical agent, **oxydierendes** ~ oxidant, oxidizing agent, **reduzierendes** ~ reducing agent, reductant
Agent m agent, salesman
Agentur f agency
Ageratochromen n ageratochromene
Ageusie f (Med, Geschmacksverlust) ageusia, loss of the sense of taste
Aggerlit m (HN, Edelstahlformguß) aggerlit (HN), high grade steel castings

Agglomerat n agglomerate, sinter cake
Agglomeratbildung f agglomeration
Agglomeration f agglomeration
agglomerieren to agglomerate
Agglomerieren n agglomerating
Agglutination f agglutination
agglutinieren to agglutinate
Agglutinin n agglutinin
Agglutinogen n agglutinogen
Agglutinoid n agglutinoid
Aggregat n (Elektr) charging set, (Techn) set of machines, unit, (Techn, Phys, Biol) aggregate
Aggregatform f (Phys) physical form
Aggregation f aggregation
Aggregatradreifen m wheel and tire assembly
Aggregatzustand m state of aggregation, condition of matter, physical condition, physical state, **fester, flüssiger, gasförmiger** ~ solid, liquid, gaseous state of aggregation
Aggregatzustandsänderung f change of matter, change of state
aggregieren to aggregate
aggressiv aggressive; corrosive; offensive
Aggressivität f aggressiveness
Aglucon n aglucone
Aglykon n aglycone
Agmatin n agmatine
Agnolith m agnolite
Agnosterin n agnosterol
Agnosterol n agnosterol
Agoniadin n agoniadin
Agricolit m (Min) agricolite
Agrikultur f agriculture
Agrikulturchemie f agricultural chemistry
Agrikulturphysik f agricultural physics
Agrimonol n agrimonol
Agrochemikalien pl agricultural chemicals
Agroclavin n agroclavine
Agronom m agronomist
Agropyren n agropyrene
Aguilarit m (Min) aguilarite
Agurin n agurin, aguirin
Aguttan n aguttane
AHG (antihämatophiles Globulin) antih[a]emophilic globulin
Ahistan n ahistan
Ahle f awl, punch
Ahming f depth gauge
ahnden to punish
Ahornholz n maple wood
Ahornholzöl n maple wood oil
Ahornhonig m maple honey
Ahornlack m maple varnish
Ahornmelasse f maple molasses
Ahornsäure f aceric acid
Ahornsaft m maple juice, maple sap
Ahornsirup m maple syrup
Ahornzellstoff m maple pulp
Ahornzucker m maple sugar

Aikinit *m* (Min) aikinite
Airoform *n* airoform, airogen, airol, bismuth oxyiodide subgallate
Airogen *n* airoform, airol, bismuth oxyiodide subgallate
Airol *n* airol, bismuth oxyiodide subgallate
Airstat *m* (Warmluftthermostat) airstat
Ajacin *n* ajacine
Ajarmin *n* ajarmine
Ajmalicin *n* ajmalicine
Ajmalicinsäure *f* ajmalicinic acid
Ajmalidin *n* ajmalidine
Ajmyrin *n* ajmyrine
Ajourware *f* (Text) ajour fabric
Ajowan *n* ajowan
Ajowanöl *n* ajowan oil
Ajugose *f* ajugose
Akademie *f* academy, college
Akajou *m* (Nierenbaum) acajou, cashew, white mahogany
Akajoubalsam *m* cashew [nut] oil
Akajougummi *m* acajou gum
Akajouharz *n* cashew resin
Akajounuß *f* cashew nut
Akajouöl *n* cashew nut oil
Akanthit *m* (Min) acanthite
Akaricid *n* (Akarizid) acaricide, mite killer
akarizid acaricidal
Akaroidharz *n* acaroid resin
Akaustan *n* (HN, Papierhilfsmittel) akaustan
Akazie *f* (Bot) acacia
Akaziengummi *m* acacia gum, gum Arabic
Akazienöl *n* acacia oil
Akaziensaft *m* gum Arabic
Akazienschote *f* (Bot) acacia pod
Akazin *n* gum Arabic
Akazingummi *m* gum Arabic
Akermanit *m* (Min) akermanite
Akklimatisation *f* acclimatization
akklimatisieren to acclimatize
Akklimatisierung *f* acclimatization
Akkommodationsgebiet *n* accommodation range
Akkommodationskoeffizient *m* (Mech) accommodation coefficient
Akkommodationsvermögen *n* power of accommodation
Akkommodationszentrum *n* accommodation center
akkommodativ accommodative
akkommodieren to accommodate, (Opt) to focus, (Techn) to adjust
Akkord *m* job work, piecework, (Akust) chord
Akkordarbeit *f* contract work, piece work
Akkordlohn *m* piece-rate wages
Akku *m* siehe Akkumulator
Akkuflasche *f* accumulator jar
Akkukasten *m* accumulator box
Akkumulation *f* accumulation

Akkumulator *m* accumulator, secondary battery, storage battery, ~ **mit Laugenfüllung** (Nickeleisenbatterie) nickel-iron battery, ~ **mit Säurefüllung** (Bleibatterie) lead-acid accumulator, **einen ~ laden** to charge an accumulator, **hydraulischer ~** hydraulic accumulator
Akkumulatorbleiplatte *f* accumulator lead plate
Akkumulatorenbatterie *f* storage battery
Akkumulatorenelement *n* storage battery cell
Akkumulatorentladung *f* battery discharge
Akkumulatorkasten *m* accumulator box, battery case
Akkumulatorladung *f* battery charge; battery charging
Akkumulatorplatte *f* accumulator plate
Akkumulatorprüfer *m* accumulator tester
Akkumulatorsäure *f* accumulator acid, battery acid
Akkumulatorschrott *m* battery scrap
Akkumulatorspannung *f* cell voltage, storage battery voltage
Akkumulatorstreichmassen *pl* accumulator pastes
Akkumulatorunterstation *f* accumulator substation
Akkumulatorzelle *f* storage battery cell
akkumulieren to accumulate, to pile up
aklinisch (Geogr) aclinic
Akmegelb *n* acme yellow, chrysoidin
Akmit *m* (Min) acmite, aegirite
A-Kohle *f* siehe Aktivkohle
Akonit *m* (Bot, Pharm) aconite, monk's hood, wolf's bane
Akonitin *n* (Pharm) aconitine
Akonitinchlorhydrat *n* aconitine hydrochloride
Akonitinsäure *f* aconitinic acid
Akonitknolle *f* (Bot) aconite root
Akonitsäure *f* aconitic acid, achilleic acid, adonic acid, citridic acid
Akonitwurzel *f* (Bot) aconite root
Akratothermen *pl* acratothermal springs
Akridin *n* acridine
Akridinfarbstoff *m* acridine color
Akridingelb *n* acridine yellow
Akridinsäure *f* acridic acid, 2,3-quinoline dicarboxylic acid
Akridon *n* acridone
Akriflavin *n* acriflavine, trypaflavine
Akrit *n* (HN, Hartlegierung) akrit
Akrolein *n* acroleine, acrylaldehyde, propenal
Akromegalie *f* (Med) acromegaly
Akryl- s. auch Acryl-
Akrylamid *n* acryl amide
Akrylamid-Gel-Elektrophorese *f* acryl amide gel electrophoresis
Akrylglas *n* (Plexiglas, HN) acrylic glass, plexiglass
Akte *f* file, document, record

Aktedon *n* (Benzedrin) benzedrine
Aktendeckel *m* folder, file, file cover
Aktennotiz *f* memo, memorandum
Aktennummer *f* file number, reference number
Aktenpapier *n* record paper
Aktenschrank *m* filing cabinet
Aktenvermerk *m* memorandum
Aktenzeichen *n* file number
Akticit *n* (Gummiindustriehilfsmittel) akticit
Aktiniden *pl* actinides
Aktinidenreihe *f* actinide series
aktinisch actinic
Aktinismus *m* actinism
Aktinit *m* actinite
Aktinität *f* actinity
Aktinium *n* (Symb. Ac) actinium, ~ **enthaltend** actiniferous
Aktiniumemanation *f* (Atom) actinium emanation, actinon
Aktiniumreihe *f* actinium series
Aktinium-Zerfallsreihe *f* (Atom) actinium radioactive series
Aktinochemie *f* actinochemistry
aktinoelektrisch actinoelectric
Aktinoelektrizität *f* actinoelectricity
Aktinogramm *n* actinogram
Aktinograph *m* actinograph
Aktinolith *m* (Min) actinolite
Aktinolithschiefer *m* (Min) actinoslate
Aktinologie *f* actinology
Aktinometer *n* actinometer
Aktinometrie *f* actinometry
aktinometrisch actinometric
Aktinon *n* actinon
Aktinoskopie *f* actinoscopy
Aktinouran *n* actinouranium
Aktion *f* action
Aktionskonstante *f* action constant
Aktionspotential *n* action potential
Aktionsradius *m* range of action
Aktionsspektrum *n* (Festkörperphysik) action spectrum
Aktionsstrom *m* action current
Aktionsturbine *f* (Mech) impulse turbine, impulse wheel, Pelton wheel
Aktionsverlauf *m* course of action
aktiv active
Aktivator *m* activator
aktivieren to activate
Aktivieren *n* activating
Aktivierung *f* activation, ~ **durch Deuteronen** deuteron-induced activation
Aktivierungsanalyse *f* (Atom) activation analysis
Aktivierungsdetektor *m* activation detector
Aktivierungsenergie *f* activation energy
Aktivierungsenthalpie *f* enthalpy of activation
Aktivierungsentropie *f* entropy of activation
Aktivierungsfolie *f* foil detector
Aktivierungsmittel *n* activating agent

Aktivierungsspannung *f* activation stress
Aktivierungswärme *f* heat of activation
Aktivierungszahl *f* activation number
Aktivin *n* activin
Aktivität *f* activity, **amylolytische** ~ amylolytic activity, **chemische** ~ chemical activity, **enzymatische** ~ enzymatic activity, **optische** ~ optical activity, rotary power, **tryptische** ~ tryptic activity
Aktivitätsfaktor *m* activity factor
Aktivitätsgefälle *n* activity gradient
Aktivitätskoeffizient *m* activity coefficient
Aktivitätsmessung *f* measurement of activity
Aktivitätsverteilung *f* (Atom) distribution of activity
Aktivkieselsäure *f* silica gel
Aktivkohle *f* activated charcoal, active carbon
aktuell actual, effective, important, topical
Akundarsäure *f* akundaric acid
Akustik *f* (Phys) acoustics
akustisch acoustic[al], aural
akut acute
Akzeleration *f* acceleration
Akzelerator *m* accelerator
Akzelerin *n* accelerin, accelerator globulin, factor VI
akzentuieren to accentuate
akzeptieren to accept
Akzeptor *m* acceptor
Akzeptorniveau *n* acceptor level
akzessorisch accessorial
Akzidenzfarbe *f* jobbing ink
Alabandin *m* (Min) alabandine, native manganese sulfide
Alabandit *m* (Min) alabandite
Alabaster *m* (Min) alabaster
Alabastergips *m* gypseous alabaster, gypsum cement, plaster of Paris
Alabasterglas *n* alabaster glass
Alabasterverfahren *n* alabastrine process
Alabasterweiß *n* alabaster white
Alabastrit *m* (Min) alabastrite
Alait *m* (Min) alaite
Alakreatin *n* alacreatine, lactyl guanidine
Alakreatinin *n* alacreatinine
Alalit *m* (Min) alalite
Alamosit *m* (Min) alamosite
Alangin *n* alangine
Alanin *n* alanine, aminopropionic acid
Alaninamid *n* alanine amide
Alaninchelat *n* alanine chelate
Alaninol *n* alaninol
Alanintransaminase *f* (Biochem) alanine transaminase
Alant *m* (Bot) elecampane
Alantanhydrid *n* alantic anhydride
Alantcampher *m* elecampane camphor, alantin, helenin
Alantin *n* alantin, inulin

Alantöl *n* elecampane oil
Alantolakton *n* alantolactone
Alantolsäure *f* alantolic acid
Alantsäure *f* alantic acid
Alantstärkemehl *n* alantin, alant starch, inulin
Alantwurzel *f* (Bot) elecampane root
Alanyl- alanyl-
Alanylalanin *n* alanylalanine
Alanylglycin *n* alanylglycine
Alarmschalter *m* alarm switch
Alaskait *m* (Min) alaskaite
Alaun *m* alum, **basischer** ~ basic alum,
  entwässerter ~ dehydrated alum,
  gebrannter ~ burnt alum, calcined alum,
  exsiccated alum
alaunartig aluminous
Alaunbad *n* alum bath
Alaunbeize *f* alum bath, aluminous mordant
Alaunbildung *f* formation of alum
Alaunblumen *pl* flowers of alum
Alaunbrühe *f* alum bath, alum mordant, alum pickle, alum steep
alaunen to steep in alum, to alum, to aluminate
Alaunerde *f* alum earth, alumina, aluminum oxide
Alaunerz *n* (Min) alum stone, alunite
Alaunerzeugung *f* production of alum
Alaunfaß *n* alum vat
Alaunfels *m* (Min) alunite
Alaunfestigkeit *f* alum resistance
alaungar (Gerb) dressed with alum, alum-tanned, steeped, tawed
Alaungehalt *m* alum content
Alaungerber *m* tawer
Alaungerberei *f* tawery, tawing
Alaungips *m* artificial marble
Alaungrube *f* alum mine
alaunhaltig aluminiferous, aluminous
Alaunhütte *f* (Alaunsiederei) alum works
alaunig aluminiferous, aluminous, alumish
alaunisieren to alum
Alaunit *m* (Min) alum stone, alunite
Alaunkies *m* (Min) aluminous pyrites
Alaunkristall *m* alum crystal
Alaunkuchen *m* (Papier) alum cake
Alaunlauge *f* alum liquor
Alaunleim *m* (Papier) alum glue
Alaunmehl *n* alum flour, alum powder, powdered alum, precipitated alum
alaunsauer aluminous
Alaunschiefer *m* (Min) alum schist, alum slate, alunite
Alaunschiefererz *n* (Min) alum shale, alum slate
Alaunseife *f* aluminous soap
Alaunsieden *n* alum boiling, alum making, preparation of alum
Alaunsieder *m* alum boiler
Alaunsiederei *f* alum works

Alaunspat *m* (Min) aluminous limestone, alum stone, alunite
Alaunstein *m* (Min) aluminous limestone, alum stone, alunite
Alaunstift *m* (Med, Pharm) styptic pencil
Alaunton *m* alum clay, alum earth
Alaunwasser *n* aluminous water, alum water
Alaunwurzel *f* (Bot) alum root
Alaunzucker *m* alum sugar
Albafix *n* (Füllstoff) albafix
Albamin *n* albamine, alabamine
Alban *n* albane
Albaspidin *n* albaspidin
Alberenstein *m* alberene stone
Albertit *m* (Min) albertite
Albertol *n* (Lackindustrie) albertol
Albertotypie *f* (Lichtdruckverfahren) alber[t]type, collotype
Albidur *n* albidur, iron aluminum alloy
Albigen *n* (Textilhilfsmittel) albigen
Albigensäure *f* albigenic acid
Albinismus *m* albinism
Albino-Ratte *f* albino rat
Albit *m* (Min) albite
albitartig albitic
albithaltig albitiferous
Albizziin *n* albizziine
Albocarbon *n* albocarbon, naphthalene
Albolith *m* (Min) albolite
Albomycin *n* albomycin
Albumen *n* albumen
Albumin *n* albumin
albuminartig albuminoid
Albuminat *n* albuminate
Albuminfaden *m* albumin filament
Albumingehalt *m* albumin content
albuminhaltig containing albumin, albuminiferous, albuminous
Albuminin *n* albuminine
albuminisieren to albuminize
Albuminisieren *n* albuminizing
Albuminisierung *f* albuminization
Albuminkupfer *n* copper albuminate
Albuminleim *m* albumin glue
Albuminoid *n* albuminoid
Albuminometer *n* albuminometer
Albuminon *n* albuminone
Albuminose *f* albuminose
Albuminpapier *n* (Phot) albuminized paper, albumin paper
Albuminprozeß *m* (Phot) albumin process
Albuminurie *f* (Med) albuminuria
Albuminverfahren *n* albumin process
Albumose *f* albumose
Alchimie *f* alchemy
Alchimist *m* alchemist
alchimistisch alchemistic
Aldazin *n* aldazine
Aldehyd *m* aldehyde

Aldehydammoniak *n* aldehyde ammonia
Aldehydgerbung *f* (Gerb) aldehyde tannage
Aldehydgrün *n* aldehyde green
aldehydhaltig aldehydic, containing aldehyde
Aldehydharz *n* aldehyde resin
Aldehydin *n* aldehydine, 2-ethyl-5-methylpyridine
aldehydisch aldehydic
Aldehydkondensation *f* aldehyde condensation
Aldehydoxydase *f* (Biochem) aldehyde oxidase
Aldehydsäure *f* aldehyde acid, aldehydic acid
Aldehydverbindung *f* aldehyde compound
Aldesulfon *n* aldesulfone
Aldim *n* aldime
Aldimin *n* aldimine
Aldiminchelat *n* aldimine chelate
Aldobiuronsäure *f* aldobiuronic acid
Aldohexose *f* aldohexose
Aldoketen *n* aldoketene
Aldol *n* aldol; acetaldol, oxybutyric aldehyde
Aldolalphanaphthylamin *n* aldol alpha-naphthylamine
Aldolase *f* (Biochem) aldolase
Aldolkondensation *f* aldol condensation
Aldomedon *n* aldomedone
Aldonsäure *f* aldonic acid
Aldopentose *f* aldopentose
Aldose *f* aldose
Aldosteron *n* aldosterone
Aldotripiperidein *n* aldotripiperideine
Aldoxim *n* aldoxime
Aldrin *n* (Insektenmittel) aldrin
Alectoronsäure *f* alectoronic acid
Alepit *n* alepite
Alepopinsäure *f* alepopinic acid
Aleppokammwolle *f* (Text) Aleppo combings
Alethein *n* aletheine
Alethin *n* alethine
Aleudrin *n* aleudrine
Aleuritinsäure *f* aleuritic acid
Aleurometer *n* aleurometer
Aleuron *n* (Biol) aleurone
Aleuronat *n* aleuronate
aleuronhaltig aleuronic
Alexandrit *m* (Min) alexandrite
Alexin *n* alexin, cytase
Alfa *f* (Bot) alfa [grass], esparto [grass]
Alfagras *n* alfa [grass], alfalfa
Alfalfasaponin *n* alfalfasaponin
Alfalfol *n* alfalfol
Alfapapier *n* (Buchdr) esparto paper
Alfenid *n* alfenide
Algamagrün *n* algama green
Algarobilla *f* (Gerb) algaroba, algarobilla
Algarotpulver *n* algaroth powder, antimony oxychloride, basic antimony chloride
Alge *f* (Bot) alga (pl. algae), seaweed
Algebra *f* (Math) algebra
algebraisch algebraic

Algenbekämpfungsmittel *n* algicide
Algenbildung *f* formation of algae
Algenfaser *f* seaweed fiber
Algenniederschlag *m* deposit of algae
Algenschleim *m* mucus of algae
Algerit *m* algerite
Algin *n* algin, alginic acid
Alginat *n* alginate
Alginatfaser *f* algin fiber
Alginsäure *f* alginic acid
Algizid *n* algicide
Algodonit *m* (Min) algodonite
Algolblau *n* algol blue
Algolfarbe *f* algol-color
Algolin *n* algoline
alicyclisch alicyclic
Alikantesoda *f* alicant soda
Alimemazin *n* alimemazine
aliphatisch aliphatic
aliquant aliquant
aliquot aliquot, proportional
Alisonit *m* (Min) alisonite
Alit *n* (Min) alite
alitieren to alitize; to aluminize
Alitieren *n* alitizing, aluminum diffusion coating
Alival *n* alival
Alizarin *n* 1,2-dihydroxy-anthraquinone, alizarin
Alizarinaltrot *n* Turkey red
Alizarinblau *n* alizarin blue, anthracene blue
Alizarinbraun *n* alizarin brown, alizarin bordeaux, anthracene brown
Alizarinfarbe *f* alizarin dye
Alizarinfarblack *m* alizarin lake
Alizaringelb *n* alizarin yellow
Alizarinkrapplack *m* alizarin madder lake
Alizarinlack *m* alizarin lake
Alizarinmonosulfonsäure *f* alizarin monosulfonic acid
Alizarinneurot *n* alizarin new red
Alizarinreinblau *n* alizarin sky blue
Alizarinrot *n* alizarin red
Alizarinsäure *f* (obs) alizarinic acid, phthalic acid
Alizarinschwarz *n* alizarin black
Alizarinsulfonsäure *f* alizarinsulfonic acid
Alizurol *n* (Farbstoff, HN) alizurol
alizyklisch alicyclic
Alkaleszenz *f* alkalescence
Alkali *n* alkali
Alkalialbuminat *n* alkali albuminate
Alkaliamalgam *n* alkali amalgam
alkaliarm poor in alkali
Alkaliatom *n* alkali atom
alkalibeständig alkaliproof, alkali-resistant, resistant to alkali
alkalibildend alkaligenous
Alkalibindemittel *n* alkali-binding agent
Alkaliblau *n* alkali blue

Alkalicellulose *f* alkali cellulose
Alkalichlorid *n* alkali chloride
Alkalichloridelektrolyseur *m* alkali chloride electrolyser
Alkalicyanid *n* alkali cyanide
alkaliecht fast to alkali
Alkaliechtfarbe *f* alkali fast color
Alkaliechtheit *f* alkali fastness
Alkaliechtrot *n* alkali fast red
alkaliempfindlich sensitive to alkali
Alkalien *pl* alkalis
alkalifest alkali-proof, alkali-resistant, alkali-resisting
alkalifrei alkali-free, non-alkali
Alkaligehalt *m* alkali content, alkaline strenght, alkalinity
Alkaligestein *n* alkali rock
Alkalihalogenid *n* alkali halide
Alkalihalogenidkontinua *pl* continuous spectra of alkali halides
Alkalihalogenidkristall *m* alkali halide crystal
Alkalihalogenidschicht *f* alkali halide film
alkalihaltig alkaline, containing alkali
Alkaliherstellung *f* alkali production
Alkalihumat *n* alkali humate
Alkalihydrat *n* alkali hydroxide
Alkalihydroxid *n* alkali hydroxide
Alkaliindustrie *f* alkali industry
Alkalikarbonat *n* alkali carbonate
Alkalilauge *f* alkali lye
alkalilöslich soluble in alkali
Alkalilösung *f* alkali solution
Alkalimenge *f* amount of alkali
Alkalimesser *m* alkalimeter
Alkalimessung *f* alkalimetry
Alkalimetall *n* alkali metal
Alkalimetallchelat *n* alkali metal chelate
Alkalimetallion *n* alkali metal ion
Alkalimeter *n* alkalimeter
Alkalimetrie *f* alkalimetry
alkalimetrisch alkalimetric
Alkalinität *f* alkalinity
Alkaliphenolat *n* alkali phenate
Alkaliphosphat *n* alkali phosphate
Alkaliphotozelle *f* alkali photocell
Alkaliquellung *f* swelling caused by alkali
alkaliraffiniert alkali-refined
Alkalireserve *f* alkali reserve
Alkalirückstand *m* alkali residue
Alkalisator *m* alkalizer
alkalisch alkaline, ~ **machen** to render alkaline, to alkalize
Alkalischmelze *f* alkali fusion
Alkalisieranlage *f* alkali treatment plant
alkalisierbar alkalizable
alkalisieren to alkalinize, to alkalize, to render alkaline, to treat with alkali
Alkalisieren *n* alkalizing
Alkalisierung *f* alkalization

Alkalisilikat *n* alkali silicate
Alkalistannat *n* alkali stannate
Alkalität *f* alkalinity
Alkalitherapie *f* alkali-therapy
Alkalitoleranzversuch *m* alkali tolerance test
Alkaliverlust *m* loss in alkali
Alkalizelle *f* alkali cell
Alkalizellstoff *m* alkali cellulose
Alkalizellulose *f* alkali cellulose
Alkaloid *n* alkaloid
alkaloidartig alkaloidal, alkaloid-like
Alkaloidlösung *f* alkaloidal solution
Alkaloidsalz *n* alkaloid salt
Alkalose *f* (Med) alkalosis
Alkamin *n* alkamine, amino alcohol
Alkan *n* alkane
Alkanisierung *f* alkanization
Alkannaextrakt *m* (Färb) alkanna extract
Alkannarot *n* alkanna red, anchusin
Alkannawurzel *f* alkanet
Alkannin *n* alkannin, anchusin
Alkansulfonsäure *f* alkane sulfonic acid
Alkaptonurie *f* alkaptonuria
Alkargen *n* alkargen, cacodylic acid
Alkazidlauge *f* alkazid solution
Alkazidverfahren *n* alkazid process
Alken *n* alkene
Alkin *n* alkine
Alkogel *n* alcogel
Alkohol *m* alcohol, **mehrwertiger** ~ polyhydric alcohol, **primärer** ~ primary alcohol, **sekundärer** ~ secondary alcohol, **tertiärer** ~ tertiary alcohol, **wasserfreier** ~ absolute alcohol
Alkoholäther *m* alcoholic ether
Alkoholanlage *f* alcohol plant
alkoholartig alcoholic, alcohol-like
Alkoholat *n* alcoholate
Alkoholauszug *m* alcoholic extract
Alkoholbestimmung *f* alcoholometry
Alkoholbildung *f* alcoholization
Alkoholdampf *m* alcohol vapor
Alkoholderivat *n* alcohol derivative
alkoholfest alcohol-proof
Alkoholfraktionierung *f* alcohol fractionation
alkoholfrei alcohol free
Alkoholgärung *f* alcoholic fermentation
Alkoholgehalt *m* alcohol content
Alkoholgruppe *f* alcohol group
alkoholhaltig alcoholic, containing alcohol
alkoholisch alcoholic
alkoholisch-wässerig aqueous alcoholic
alkoholisierbar alcoholizable
alkoholisieren to alcoholate, to alcoholize
Alkoholisieren *n* alcoholizing
Alkoholisierung *f* alcoholization
Alkoholismus *m* alcoholism
alkohollöslich soluble in alcohol
Alkoholmesser *m* alcohol meter, alcoholometer

Alkoholometer n alcoholometer
Alkoholometerskala f alcoholometric scale
Alkoholometrie f alcoholometry
alkoholometrisch alcoholometric
Alkoholprobe f alcohol test
alkoholreich containing much alcohol, rich in alcohol
Alkoholsäure f alcohol acid
Alkoholspiegel m alcohol content
Alkoholsulfonat n alcohol sulfonate
alkoholunlöslich insoluble in alcohol
Alkoholverbindung f alcohol compound
Alkoholvergällung f denaturing of alcohol
Alkoholyse f alcoholysis
Alkoholzusatz m addition of alcohol
Alkosol n alcosol
Alkoxysilan n alkoxysilane
Alkydal n (Lackrohstoff, HN) alkydal
Alkydharz n alkyd resin
Alkydlack m alkyd resin varnish
Alkyl n alkyl
Alkyläther m alkylic ether
Alkylalkoxysilan n alkylalkoxysilane
Alkylamin n alkylamine
alkylaromatisch alkylaromatic
Alkylarylsilicon n alkyl aryl silicone
Alkylbromid n alkyl bromide
Alkylchlorid n alkyl chloride
Alkylderivat n alkyl derivative
Alkylen n alkylene
Alkylenpolysulfid n alkylene polysulfide
Alkylester m alkyl ester
Alkylfluorid n alkyl fluoride
Alkylgruppe f alkyl group
Alkylhalogenid n alkyl halide
Alkylhaloid n alkyl halide
Alkylharzfarbe f alkyl enamel
Alkyliden n alkylidene
alkylieren to alkylate
Alkylieren n alkylating
alkyliert alkylated
Alkylierung f alkylation
Alkylierungsmittel n alkylating agent
Alkyljodid n alkyl iodide
Alkylmagnesiumhalogenid n alkylmagnesium halide
Alkylnitrat n alkyl nitrate
Alkylnitrit n alkyl nitrite
Alkyloxy- alkoxy-
Alkylphenolharz n alkylphenolic resin
Alkylradikal n alkyl radical, alkyl residue
Alkylrest m alkyl residue, alkyl group
Alkylschwefelsäure f alkylsulfuric acid
Alkylsilan n alkylsilane
Alkylsilicandiol n alkylsilicanediol
Alkylsilicon n alkylsilicone
Alkylsiliconharz n alkylsilicone resin
Alkylsulfhydrat n alkyl hydrogen sulfide, alkyl sulfhydrate

Alkylsulfid n alkyl sulfide, dialkyl sulfide
Alkylsulfosäure f alkylsulfonic acid
Alkyltriäthoxysilan n alkyltriethoxysilane
Alkyltrichlorsilan n alkyltrichlorosilane
Alkyltrihalogensilan n alkyltrihalosilane
Alkylverbindung f alkyl compound
Allacit m allacite
Allaktit m allactit
Allanit m (Geol) allanite, orthite
Allansäure f allanic acid
Allantoin n allantoin
Allantoinsäure f allantoic acid
allantoisch allantoic
Allantoxaidin n allantoxaidine
Allantoxansäure f allantoxanic acid
Allantursäure f allanturic acid
Alleinbeschleuniger m sole accelerator
Alleingerbstoff m selftannin
Alleingerbung f selftannage
Alleinhandel m monopoly
Alleinhersteller m sole producer
alleinstehend isolated
Alleinvertreter m sole distributor
Allel n (Biol) allel[e]
allelomorph allelomorphic
Allelomorphismus m allelomorphism
Allelopathikum n allelopathic
allelotrop allelotropic
Allelotropie f allelotropy
Allemontit m (Min) allemontite
Allen n allene, propadiene
Allen-Enantiomerie f (Stereochem) allene enantiomerism
Allenolsäure f allenolic acid
Allergie f allergy
Allerleigewürz n allspice, mixed spice, pimento
Alleskleber m all-purpose adhesive
Alles-oder-Nichts-Gesetz n all-or-none law, (Med) Bowditch's law
Alles-oder-Nichts-Mechanismus m all-or-none mechanism
Allethrin n allethrin
Allethrolon n allethrolone
allfarbig (Phot) panchromatic
Allgebrauchsthermometer n general purpose thermometer
allgemein general, (im allgemeinen) in general
Allgemeingültigkeit f generality, general validity
Allgemeinkosten pl general expenses
Allgemeinreaktion f general reaction
Allgemeinwirkung f general action, general effect
Allgemeinwissen n general knowledge
Allglasausführung f allglass construction
Allheilmittel n cure-all
Allicin n allicin
Alligation f (Math) alligation
Alligationsrechnung f (Math) calculation of alligation

Alligationsregel f (Math) rule of alligation
Allihnsches Rohr n Allihn filter tube
alliieren to ally
Alliin n alliin
Allit m allitol, allodulcitol
Allithiamin n allithiamine
Allitursäure f allituric acid
Alliuminosid n alliuminoside
allmählich gradual[ly]
Allnetzgerät n (Elektr) all-mains receiver
Alloaromadendren n alloaromadendrene
Allobarbital n allobarbital
Allocain n allocaine
Allochlorophyll n allochlorophyll
Allocholansäure f allocholanic acid
Allocholesterin n allocholesterol
Allocholsäure f allocholic acid
Allochroit m (Min) allochroite
allochromatisch allochromatic
Allocinchonin n allocinchonine
Allocolchicein n allocolchiceine
Allocolchicin n allocolchicine
Allocupreid n allocupreide
Allocyanin n allocyanine, neocyanine
Allodalbergin n allodalbergin
Allodulcit m allodulcitol, allitol
Allodunnion n allodunnione
Alloechitamin n alloechitamine
Alloevodion n alloevodione
Allogeigersäure f allogeigeric acid
Allogibbersäure f allogibberic acid
Allogonit m allogonite
Alloheptulose f alloheptulose
Alloibogain n alloibogaine
Alloimperatorin n alloimperatorin
Alloinosit m alloinositol
Alloisoleucin n alloisoleucine
Alloisomerie f alloisomerism, stereo-isomerism
Alloit m alloite
Allokaffein n allocaffeine
Allokainsäure f allokainic acid
Allokryptopin n allocryptopine
Allolactose f allolactose
Allomaltol n allomaltol
Allomatridin n allomatridine
Allomerie f allomerism
allomerisch allomeric
Allomerismus m allomerism
Allomethadion n allomethadione
allomorph (Min) allomorphic, allomorphous
Allomorphismus m (Min) allomorphism
Allomorphit m (Min) allomorphite
Allomuscarin n allomuscarine
Allonge f (Buchdr) adapter, fly leaf
Allonsäure f allonic acid
Alloocimen n alloocimene
Allopalladium n (Min) allopalladium
Alloperiplocymarin n alloperiplocymarin
Allophan m (Min) allophane

Allophanat n allophanate
Allophansäure f allophanic acid, carbamyl carbamic acid
allophansauer allophanate
Allophit m (Min) allophite
Allopimaran n allopimarane
Allopregnan n allopregnane
Allosamin n allosamine
Alloschleimsäure f allomucic acid
Allose f allose
Allosedamin n allosedamine
Allosedridin n allosedridine
allosterisch (Biochem) allosteric
Alloteloidin n alloteloidine
Allothreonin n allothreonine
Allothreooxazolin n allothreooxazoline
allotriomorph (Krist) allotriomorphic
allotrop allotropic
Allotropie f allotropism, allotropy
allotropisch allotropic
Allotropismus m allotropism
Alloxan n alloxan
alloxan-diabetisch (Med) alloxan-diabetic
Alloxansäure f alloxanic acid
alloxansauer alloxanate
Alloxanthin n alloxanthin
Alloxantin n alloxantin
Alloxanylharnstoff m alloxanylurea
Alloxazin n alloxazin[e]
Alloxurbase f alloxuric base, purine base
Alloxurkörper m alloxuric body, purine base
Alloyohimbol n alloyohimbol
Allozimtsäure f allocinnamic acid
Allreiniger m all-purpose cleaner
allseitig from all sides, polydimensional, universal
Alluaudit m (Min) alluaudite
Allulose f allulose
Alluransäure f alluranic acid
alluvial (angeschwemmt) alluvial
Alluvialgold n alluvial gold, placer gold
Alluvialschutt m (Geol) wash
Alluvium n (Geol) alluvial soil, alluvium
Allyl- allyl-
Allylaceton n allylacetone, 5-hexene-2-one
Allylaldehyd m allyl aldehyde, acrolein
Allylalkohol m allyl alcohol
Allylamin n allylamine, 2-propenylamine
Allylbenzol n allylbenzene
Allylbromid n allyl bromide
Allylchlorid n allyl chloride
Allylcyanid n allyl cyanide, 3-butanenitrile
Allylderivat n allyl derivative
Allyldichlorsilan n allyldichlorosilane
Allyldisulfid n allyl disulfide
Allylen n allylene, propyne
Allylessigsäure f allylacetic acid
Allylharz n allyl resin
Allylharzkunststoffe pl allyl plastics

Allyljodid *n* allyl iodide, 3-iodopropene
Allylphenylmethyläther *m* allyl phenyl methyl ether, anethol[e], anise camphor
Allylpyridin *n* allyl pyridine
Allylrhodanid *n* allyl thiocyanate
Allylsenföl *n* allyl mustard oil, allyl isothiocyanate
Allylsulfid *n* allyl sulfide, diallyl sulfide
Allylsulfocarbamid *n* allyl thiocarbamide, allylthiourea
Allylsulfocyanid *n* allylthiocyanate
Allylthioharnstoff *m* allyl thiourea
Allyltriäthoxysilan *n* allyltriethoxysilane
Allyltrichlorsilan *n* allyltrichlorosilane
Allylumlagerung *f* allylic rearrangement
Allylverbindung *f* allyl compound
Allzweck- general purpose
Allzweckglaselektrode *f* general purpose electrode
Allzwecksonde *f* general purpose probe
Almagrerit *m* (Min) almagrerite
Almandin *m* (Min) almandine
Almandinspat *m* (Min) eudialite
Almandinspinell *m* (Min) almandine spinel
Almaska *n* almasca
Almedin *n* almedine
Almeriit *m* almeriite
Alnusen *n* alnusene
Aloe *f* (Bot) aloe
Aloealkaloid *n* aloin
aloeartig aloid
Aloeauszug *m* aloe extract, aloetic gum, aloin
Aloebitter *n* aloetic gum, aloin
Aloebitterstoff *m* aloetic gum, aloin
Aloeemodin *n* aloe emodin
Aloeextrakt *m* aloe extract, extract of aloes
Aloefaser *f* aloe fiber
aloehaltig aloetic
Aloehanf *m* aloe hemp, aloe fiber
Aloeharz *n* aloetic resin
Aloeholz *n* aloe-wood
Aloelatwerge *f* (Bot) aloetic electuary
Aloemittel *n* (Med) aloetic preparation
Aloepräparat *n* (Med) aloetic preparation
Aloerot *n* aloe red
Aloesäure *f* aloetic acid
Aloesaft *m* aloe juice
Aloesol *n* aloe sol
Aloetinsäure *f* aloetic acid
aloetisch aloetic
Aloin *n* aloin
Aloinose *f* aloinose
Aloinsäure *f* aloic acid, tetranitro-anthraquinone
Aloxit *n* (HN) aloxite, aluminum oxide
Alpaka *n* (Leg, Text) alpaca
Alpakawolle *f* (Text) alpaca hair
Alpenbeifußöl *n* artemisia oil
Alpenmehl *n* lycopodium powder

alpha-aktiv (Atom) alpha-active, alpha-radioactive
Alpha-Aktivität *f* (Atom) alpha-activity, alpha-radioactivity
Alphabeschuß *m* (Atom) alpha-particle bombardment
Alphabetschloß *n* combination lock
Alphacellulose *f* alpha cellulose
Alphachloracrylat *n* alpha chloroacrylate
Alphaeisen *n* alpha iron
Alphageschoß *n* (Atom) alpha projectile
Alpha-Helix *f* alpha-helix
Alpha-Instabilität *f* (Atom) alpha instability
Alphalinie *f* alpha line
Alphaprodin *n* alphaprodine
alpha-radioaktiv (Atom) alpha-active, alpha-radioactive
Alphastellung *f* alpha position
Alphastrahlen *pl* alpha rays
alphastrahlend (Atom) alpha-emitting
Alphastrahlenmesser *m* alpha meter
Alphastrahlenquelle *f* (Atom) alpha-ray source
Alphastrahlenspektrometer *n* (Atom) alpha-ray spectrometer
Alphastrahler *m* alpha ray emitter
Alphateilchen *n* alpha particle
Alphateilchenbeschuß *m* alpha bombardment
Alphateilchenmasse *f* alpha particle mass
Alphateilchenquelle *f* alpha particle source
Alphateilchenzähler *m* alpha particle counter
Alphatron *n* (Atom) alphatron [gauge]
Alphaüberwachungsgerät *n* alpha survey meter
Alphazerfall *m* (Atom) alpha decay, alpha[-particle] disintegration
Alphazerfallsenergie *f* alpha disintegration energy
Alphazustand *m* alpha state
Alphenicum *n* alphenic, alphenix
Alphol *n* alphol, α-naphthyl salicylate
Alphyl *n* alphyl, alkyl-phenyl
Alpinol *n* alpinol
Alraun *m* (Bot) mandrake
Alraunwurzel *f* (Bot) mandragora
Alsinit *n* (gesintertes Aluminiumoxid, HN) alsinit (sintered alumina)
Alsol *n* alsol, aluminum acetotartrate
Alstonidin *n* alstonidin
Alstonin *n* alstonine, chlorogenine
Alstonit *m* (Min) alstonite, bromlite
Altait *m* (Min) altaite, native lead telluride
Alteisen *n* scrap iron
Alterans *n* alterant
altern to grow old, to age; (Techn) to ageharden, to quench-age, (Wein etc.) to mature, to ripen
Altern *n* ag[e]ing, fatiguing, growing old
Alternariol *n* alternariol
Alternarsäure *f* alternaric acid
Alternationsgesetz *n* (Spektr) law of alternation
Alternativ-Verbot *n* exclusion rule

Alternidin *n* alternidine
alternieren to alternate
Alternieren *n* alternating
alternierend alternating, alternate
Altersbestimmung *f* age determination, dating
Altertumskunde *f* archaeology
Alterung *f* **durch Verwittern** weather-ageing, ~ **eines Zählrohrs** hysteresis of a counter
Alterungsanlage *f* aging plant
Alterungsbelag *m* age coating
Alterungsbeständigkeit *f* resistance to aging
Alterungsbomben *pl* aging bombs
Alterungsdauer *f* period of aging
Alterungsempfindlichkeit *f* susceptibility to aging
Alterungsgrenze *f* limit of brittleness, maximum brittleness
Alterungsprüfung *f* aging test
Alterungsschutzmittel *n* age protector, age resister, anti-ager, antioxidant
Alterungsverfahren *n* aging process
Alterungsversuch *m* aging test
Alterungsvorgang *m* aging process
Altgold *n* old gold
Altgummi *m* scrap rubber, used rubber, waste rubber
Altgummimehl *n* rubber crumb
Altheeblätter *pl* marsh mallow leaves
Altheesirup *m* althaea syrup, marsh mallow syrup
Altheewurzel *f* (Bot) althaea, marsh mallow root
Althein *n* (obs) altheine, asparagine
Altimeter *n* altimeter, height indicator
Altkupfer *n* scrap copper
Altmalz *n* stored malt
Altmaterial *n* junk, old material, used material
Altmaterialverwertung *f* recycling, utilization of scrap
Altmetall *n* old metal, scrap metal
Altmetallegierung *f* secondary alloy
Altmetallverhüttung *f* smelting of scrap metal
Altöl waste oil
Altosid *n* altoside
Altpapier *n* waste paper
Altrit *m* altritol
Altroheptit *m* altroheptite
Altroheptulose *f* altroheptulose
Altronsäure *f* altronic acid
Altrose *f* altrose
Altruronsäure *f* altruronic acid
Altsand *m* old sand
Altsandaufbereitung *f* (Gießerei) sand cleaning
Aluchi *n* aluchi
Aluchibalsam *m* aluchi balsam
Aluchiharz *n* aluchi resin
Aludel *n* aludel
Aludelofen *m* aludel furnace
Aludrin *n* aludrine
Alulegierung *f* aluminum alloy

Alumen *n* alum[en]
alumetieren to plate with aluminum
Alumetieren *n* aluminum plating
Alumian *n* alumian, basic aluminum sulfate
Aluminat *n* aluminate
Aluminatlauge *f* aluminate liquor
Aluminatlösung *f* aluminate solution
Aluminbad *n* alumina bath
Aluminieren *n* aluminizing
Aluminit *m* (Min) aluminite
Aluminium *n* aluminum (Symb. Al), aluminium (Br. E.)
Aluminiumacetat *n* aluminum acetate
Aluminiumacetotartrat *n* aluminum acetotartrate
Aluminiumalkoholat *n* aluminum alcoholate
Aluminiumalkyl *n* aluminum alkyl
Aluminiumbeize *f* aluminum mordant
Aluminiumblech *n* aluminum sheet, sheet aluminum
Aluminiumblock *m* aluminum block
Aluminiumboranat *n* aluminum borohydride
Aluminiumborat *n* aluminum borate
Aluminiumborid *n* aluminum boride
Aluminiumbromid *n* aluminum bromide
Aluminiumbronze *f* aluminum bronze
Aluminiumcarbid *n* aluminum carbide
Aluminiumchlorid *n* aluminum chloride
Aluminiumdraht *m* aluminum wire
Aluminiumdruckguß *m* aluminum pressure casting
Aluminiumeisen *n* ferroaluminum
Aluminiumerz *n* aluminum ore
Aluminiumfarbe *f* aluminum paint
Aluminiumfeilspan *m* aluminum filing
Aluminiumfluorid *n* aluminum fluoride
Aluminiumfluorsilikat *n* aluminum silicofluoride, aluminum fluosilicate
Aluminiumfluorwasserstoffsäure *f* fluo-aluminic acid
Aluminiumfolie *f* aluminum foil
Aluminiumformiat *n* aluminum formate
Aluminiumgehalt *m* aluminum content
Aluminiumgestell *n* aluminum frame
Aluminiumgießerei *f* aluminum foundry
Aluminiumgrieß *m* aluminum shot
Aluminiumguß *m* aluminum casting, cast aluminum
Aluminiumhydrat *n* aluminum hydroxide
Aluminiumhydrid *n* aluminum hydride
Aluminiumhydroxid *n* aluminum hydroxide
Aluminiumjodat *n* aluminum iodate
Aluminiumjodid *n* aluminum iodide
Aluminiumkabel *n* aluminum cable
Aluminiumkaliumsulfat *n* aluminum potassium sulfate, potash alum, potassium alum
Aluminiumkarbid *n* aluminum carbide
Aluminiumkontakt *m* aluminum contact
Aluminiumlackfarbe *f* aluminum enamel
Aluminiumlaktat *n* aluminum lactate

Aluminiumlegierung – Amboßbahn

Aluminiumlegierung *f* aluminum alloy
Aluminiummennige *f* burnt island red
Aluminiummessing *n* aluminum brass
Aluminiumnaphtholsulfonat *n* aluminum-2-naphtholsulfonate, alumnol
Aluminiumnatriumchlorid *n* aluminum sodium chloride
Aluminiumnatriumfluorid *n* sodium aluminum fluoride, sodium fluoaluminate, (Min) cryolite
Aluminiumnitrat *n* aluminum nitrate
Aluminiumnitrid *n* aluminum nitride
Aluminiumofen *m* aluminum furnace
Aluminiumoxid *n* (Tonerde) aluminum oxide, alumina
Aluminiumoxid-Einkristallfasern *pl* saphire whiskers
Aluminiumoxidhydrat *n* aluminum hydroxide
Aluminiumprofil *n* aluminum section
Aluminiumpulver *n* aluminum powder
Aluminiumrhodanid *n* aluminum thiocyanate
Aluminiumsalz *n* aluminum salt
Aluminiumschweißung *f* aluminum welding
Aluminiumsilicat *n* aluminum silicate
Aluminiumspritzguß *m* die-cast aluminum
Aluminiumstange *f* aluminum rod
Aluminiumstearat *n* aluminum stearate
Aluminiumsulfat *n* aluminum sulfate
Aluminiumsulfathydrat *n* hydrated aluminum sulfate
Aluminiumsulfocyanat *n* aluminum sulfocyanate, aluminum thiocyanate
Aluminiumüberzug *m* aluminum coating
Aluminiumwalzwerk *n* aluminum rolling mill
Aluminiumzellengleichrichter *m* aluminum rectifier
Aluminon *n* aluminon
Aluminothermie *f* aluminothermics, aluminothermy
aluminothermisch aluminothermic
Alumnol *n* alumnol, aluminum-2-naphtholsulfonate
Alumocalcit *m* (Min) alumocalcite
Alumogel *n* alumina gel
Alumosulfat *n* alumosulfate
Alundum *n* (HN) alundum
Alunit *m* (Min) alunite, alum stone
Alunitisierung *f* alunitization
Alunogen *m* (Min) alunogen
Alutherm-Verfahren *n* thermit welding of aluminum
alveolar (Med) alveolar
Alveolit *m* alveolite
Alvit *m* (Min) alvite
Alvolen *n* albolene, medicinal paraffin oil, white petrolatum
Alypin *n* alypine
Amadori-Umlagerung *f* Amadori rearrangement
Amalgam *n* amalgam, mercury alloy

Amalgamation *f* amalgamation [process]
Amalgambad *n* amalgamating bath
Amalgambildung *f* formation of amalgams
Amalgamelektrode *f* amalgam electrode
Amalgamfüllung *f* amalgam filling
amalgamierbar amalgamable
amalgamieren to amalgam[ate], to amalgamize
Amalgamieren *n* amalgamating
Amalgamierfaß *n* amalgamating barrel
Amalgamiermaschine *f* amalgamator
amalgamiert amalgamated
Amalgamierung *f* amalgamation
Amalgamierungsflüssigkeit *f* amalgamating liquid
Amalgamkonzentration *f* amalgam concentration
Amalgamsilber *n* silver amalgam
Amalgamverfahren *n* amalgamation process
Amalgamzersetzung *f* amalgam decomposition
Amalinsäure *f* amalic acid, amalinic acid, tetramethyl alloxantine
Amandin *n* (Globulin) amandin
Amanitin *n* amanitine
Amanozin *n* amanozine
Amarant *m n* amaranth
Amarantbordeaux *n* amaranth bordeaux
Amarantholz *n* purplewood
Amarantit *m* amarantite
Amarantrot *n* amaranth red
Amarin *n* amarine
Amaron *n* amaron, benzoin imide, tetraphenyl-p-pyrimidine, tetraphenyl pyrazine
Amarsäure *f* amaric acid
Amaryl *m* (Min) artificial green sapphire
Amaryllidin *n* amaryllidine
Amausit *m* emausite
Amazonenstein *m* (Min) amazon stone, amazonite
Amazonit *m* (Min) amazonite, amazon stone
Ambazon *n* ambazone
Amber *m* (Ambra) ambergris, (Bernstein) amber
Amberfett *n* ambergris fat, ambrein
Amberfettsäure *f* ambreic acid
Amberglaselektrode *f* amber glass electrode
Amberglimmer *m* (Min) amber mica, magnesium mica, phlogopite
Ambergsieb *n* (Histol) perforated cup
Amberharz *n* ambrein
Amberit *n* amberite
Amberkraut *n* (Bot) cat thyme
Amberlitkationenaustauscherharz *n* (HN) amberlite cation exchange resin
Amberöl *n* amber oil
Amblygonit *m* (Min) amblygonite
Amblystegit *m* amblystegite, enstatite
Ambosid *n* amboside
Amboß *m* anvil
Amboßbahn *f* anvil plate, anvil face

Ambra *f* ambergris
Ambrain *n* ambrein(e)
Ambran *n* ambrane
Ambraöl *n* ambergris oil, sperm oil
Ambrasalz *n* ambergris salt
Ambrein *n* ambrein
Ambreinolid *n* ambreinolide
Ambrettekörner *pl* (Bot) amber [or musk] seed
Ambrettemoschus *m* ambrette musk
Ambretteöl *n* ambrette oil
Ambrettolid *n* ambrettolide, hexdecen-6-olide
Ambrettolsäure *f* ambrettolic acid
Ambrinol *n* ambrinol
Ambrit *m* (Min) ambrite
Ambroin *n* ambroin
Ambrosia *f* (Bot) ambrosia
Ambrosiaöl *n* ragweed oil
Ambrosin *n* ambrosine
Ambucain *n* ambucaine
ambulant ambulant
Ameise *f* ant
Ameisenaldehyd *m* formaldehyde, formic aldehyde, methanal
Ameisenpersäure *f* performic acid, peroxyformic acid
Ameisensäure *f* formic acid
Ameisensäureäthylester *m* ethyl formate
Ameisensäureanhydrid *n* formic anhydride
Ameisensäureanlage *f* formic acid plant
Ameisensäurenitril *n* formonitrile, hydrocyanic acid, hydrogen cyanide
Ameisensäuretinktur *f* formic tincture
ameisensauer formate
Americium *n* americium (Symb. Am)
Amesit *m* (Min) amesite
Amethyst *m* (Min) amethyst
amethystartig amethystine, amethyst-like
Amethystfarbe *f* amethyst color
amethystfarben amethyst-colored
amethystfarbig amethyst-colored
Amiant *m* (Min) amianthus, silky asbestos
amiantartig amianthine
amiantförmig amianthine
Amicardin *n* amicardine
Amicetamin *n* amicetamine
Amicetaminol *n* amicetaminol
Amicetin *n* amicetin
Amicetose *f* amicetose
Amichin *n* amichin
Amid *n* amide
Amidase *f* (Biochem) amidase
Amidbindung *f* amide bond, amide linkage
Amidgruppe *f* amido group, amidogen
amidieren to amidate, to convert into an amide
Amidieren *n* amidating
amidiert amidated
Amidierung *f* amidation
Amidin *n* amidine
Amidinomycin *n* amidinomycin

Amidogen *n* amidogen
Amidokohlensäure *f* amidocarbonic acid, carbamic acid
Amidol *n* amidol, 3,4-diaminophenol hydrochloride
Amidonbad *n* amidon bath
Amidopyrin *n* amidopyrine
Amidoschwefelsäure *f* amidosulfuric acid, amidosulfonic acid, sulfamic acid
Amidosulfonsäure *f* amidosulfonic acid, amidosulfuric acid, sulfamic acid
Amidsäure *f* amide acid
Amidstickstoff *m* amide nitrogen
Amidulin *n* amidulin
Amikron *n* amicron
amikroskopisch amicroscopical
Amin *n* amine, **primäres** ~ primary amine, amino-base, **quartäres** ~ quaternary amine, tetra-alkyl ammonium base, **sekundäres** ~ secondary amine, imino-base, **tertiäres** ~ tertiary amine, nitrile-base
aminartig amine-like
Aminase *f* (Biochem) aminase
Aminbase *f* amine base
aminieren to aminate, to convert into an amine
Aminieren *n* aminating
Aminierung *f* amination
Aminitrozol *n* aminitrozole
Aminoacetanilid *n* amino acetanilide
Aminoacetophenon *n* aminoacetophenone
Aminoacridin *n* aminoacridine
Aminoäthanol *n* aminoethanol, ethanolamine, β-hydroxyethylamine
Aminoäthylbenzoesäure *f* aminoethylbenzoic acid
Aminoäthylnitrat *n* aminoethyl nitrate
Aminoalkohol *m* amino alcohol
Aminoalkylierung *f* aminoalkylation
Aminoazobenzol *n* aminoazobenzene
Aminoazobenzolbase *f* aminoazobenzene base
Aminoazobenzolchlorhydrat *n* aminoazobenzene hydrochloride
Aminoazobenzoldisulfonsäure *f* amino azobenzene disulfonic acid
Aminoazotoluol *n* aminoazotoluene
Aminobarbitursäure *f* aminobarbituric acid
Aminobenzoesäure *f* aminobenzoic acid
Aminobenzol *n* aminobenzene, aniline
Aminobernsteinsäure *f* aminosuccinic acid, aspartic acid
Aminobuttersäure *f* aminobutyric acid
Aminobuttersäurechelat *n* aminobutyric acid chelate
Aminocaprolactam *n* aminocaprolactam
Aminocolchicid *n* aminocolchicide
Aminoessigsäure *f* aminoacetic acid, glycine, glycocoll
Aminoform *n* aminoform, hexamethylenetetramine, urotropine

Aminoformyl n aminoformyl, carbamyl
Aminoglutarsäure f aminoglutaric acid, glutamic acid
Aminogruppe f amino group
Aminoketon n amino ketone
Aminokohlensäure f amidocarbonic acid, carbamic acid
Aminomethylphosphonsäure f aminomethylphosphonic acid
Aminometradin n aminometradine
Aminonaphthalin n aminonaphthalene, naphthylamine
Aminonaphthochinon n aminonaphthoquinone
Aminonaphthol n aminonaphthol
Aminonaphtholdisulfonsäure f aminonaphtholdisulfonic acid
Aminonaphtholsulfonsäure f aminonaphtholsulfonic acid
Aminooxin n aminooxine
Aminopeptidase f (Biochem) aminopeptidase
Aminophenol n aminophenol
Aminophyllin n aminophylline
Aminoplaste pl aminoplastics
Aminoplastharz n aminoplastic resin
aminoplastisch aminoplastic
Aminoplastkunststoffe pl amino plastics
Aminopolycarbonsäure f aminopolycarboxylic acid
Aminopromazin n aminopromazine
Aminopropanol n aminopropanol
Aminopropionsäure f aminopropionic acid, alanine
Aminopterin n aminopterin
Aminopurin n aminopurine, adenine
Aminopyrazolin n aminopyrazoline
Aminopyridin n aminopyridine
Aminosäure f amino acid, **C-terminale** ~ carboxy terminal [amino acid], C-terminal [amino acid], **ketogene** ~ ketogenic amino acid, **lebenswichtige** ~ (**essentielle**) essential amino acid, **N-terminale** ~ amino terminal [amino acid], N-terminal [amino acid], **physiologische** ~ physiological amino acid, **proteinogene** ~ proteinogenic amino acid, **radioaktiv markierte** ~ radiolabelled amino acid
Aminosäureamid n amino acid amide
Aminosäureaufnahme f amino acid uptake
Aminosäureaustausch m (Mol. Biol) amino acid replacement
Aminosäuredecarboxylase f (Biochem) amino acid decarboxylase
Aminosäureoxydase f (Biochem) amino acid oxidase
Aminosäurerest m amino acid residue
Aminosäuresequenz f amino acid sequence
Aminosäurezusammensetzung f amino acid composition
Aminosalicylsäure f aminosalicylic acid

Aminosulfonsäure f amidosulfonic acid, amidosulfuric acid, sulfamic acid
Aminothiazol n aminothiazole
Aminotoluol n aminotoluene, toluidine
Aminoverbindung f amino compound
Aminozucker m amino sugar
Aminschwarz n amine black
Amiphenazol n amiphenazole
Amitose f (Biol) amitosis, direct cell division
Amizol n amizol
Ammelid n ammelide
Ammelin n ammeline
Ammidin n ammidin
Ammin n (Ammoniakat) ammine
Ammiol n ammiol
Ammiolith m (Min) ammiolite
Ammoidin n ammoidin
Ammolin n ammoline
Ammon n (Ammonium) ammonia, ammonium
Ammoncarbonisierung f reaction with ammonium carbonate
Ammoneisenalaun m ammonium iron alum
Ammoniak n ammonia, **doppelkohlensaures** ~ bicarbonate of ammonia, **gebundenes** ~ fixed ammonia
Ammoniakabscheider m ammonia separator
Ammoniakalaun m ammonium aluminum sulfate, ammonia alum, ammonium alum
ammoniakalisch ammoniacal
Ammoniakanlage f ammonia plant
ammoniakarm poor in ammonia
ammoniakartig ammoniacal
Ammoniakat n ammine, ammoniate
Ammoniakausscheidung f ammonia excretion
Ammoniakbehälter m ammonia tank
Ammoniakbestimmung f determination of ammonia
ammoniakbindend combining with ammonia
Ammoniakdämpfe pl ammonia vapors
Ammoniakdestillation f ammonia distillation
Ammoniakdestillationsapparat m ammonia distillation apparatus
Ammoniakdüngung f ammonia manuring
Ammoniakentwicklung f formation of ammonia
Ammoniakflasche f ammonia carboy
Ammoniakgummi m gum ammoniac
ammoniakhaltig containing ammonia, ammoniacal
Ammoniakharz n gum ammoniac
Ammoniakkältemaschine f ammonia refrigerating machine
Ammoniakkühlschlange f ammonia refrigerating coil
Ammoniaklaugung f ammonia leaching
Ammoniakleitung f ammonia line; conduction of ammonia
Ammoniakmesser m ammonia meter
Ammoniakphosphat n ammonium phosphate
Ammoniakprüfer m ammonia tester

Ammoniakregelventil *n* ammonia regulating valve
ammoniakreich rich in ammonia
Ammoniakrest *m* ammonia residue
Ammoniakrohr *n* ammonia pipe, ammonia tube
Ammoniaksalpeter *m* ammonium nitrate
Ammoniaksalz *n* ammonium salt
Ammoniakscheidung *f* ammonia separation
Ammoniakseife *f* ammonia soap
Ammoniaksoda *f* ammonia soda, Solvay soda
Ammoniaksodaprozeß *m* ammonia soda process, Solvay process
Ammoniakstickstoff *m* ammonia nitrogen
Ammoniaksynthese *f* ammonia synthesis
Ammoniaküberhitzer *m* ammonia superheater
Ammoniakverbindung *f* ammonia compound, ammonium compound
Ammoniakverdampfungsrohr *n* ammonia expansion pipe
Ammoniakverdichter *m* ammonia compressor
Ammoniakverdichtung *f* ammonia compression
Ammoniakverdichtungsanlage *f* ammonia compression system
Ammoniakverfahren *n* ammonia process
Ammoniakverflüssiger *m* ammonia condenser
Ammoniakverteiler *m* ammonia distributor
Ammoniakwäsche *f* ammonia scrubbing
Ammoniakwascher *m* ammonia washer, ammonia scrubber
Ammoniakwasser *n* ammonium hydroxide, ammoniacal liquor, ammonia water, aqueous ammonia
Ammoniakweinstein *m* ammonium potassium tartrate
Ammoniakzerfall *m* decomposition of ammonia
Ammoniakzusatz *m* addition of ammonia
Ammonin *n* ammonine
Ammoniometer *n* ammoniometer
Ammonit *n* (Sprengstoff) ammonite, ammonium nitrate explosive
Ammonium *n* ammonium, **thiocyansaures** ~ ammonium thiocyanate
Ammoniumacetat *n* ammonium acetate
Ammoniumalaun *m* ammonium alum, ammonium aluminum sulfate
Ammoniumarsenat *n* ammonium arsenate
Ammoniumbase *f* ammonium base
Ammoniumbenzoat *n* ammonium benzoate
Ammoniumbicarbonat *n* ammonium bicarbonate
Ammoniumbichromat *n* ammonium bichromate, ammonium dichromate
Ammoniumbifluorid *n* acid ammonium fluoride, ammonium bifluoride, ammonium hydrogen fluoride, matt salt
Ammoniumbioxalat *n* ammonium bioxalate, acid ammonium oxalate, ammonium binoxalate

Ammoniumbiphosphat *n* ammonium biphosphate, diacid ammonium phosphate, monoammonium phosphate
Ammoniumbisulfit *n* ammonium bisulfite, ammonium hydrogen sulfite
Ammoniumborat *n* ammonium borate
Ammoniumbromcamphersulfonat *n* ammonium bromocamphor sulfonate
Ammoniumbromid *n* ammonium bromide
Ammoniumcamphorat *n* ammonium camphorate
Ammoniumcarbamat *n* ammonium carbamate
Ammoniumcarbonat *n* ammonium carbonate
Ammoniumchlorid *n* ammonium chloride, sal ammoniac, salmiac
Ammoniumchlorostannat *n* ammonium chlorostannate
Ammoniumchlorplatinat *n* ammonium chloroplatinate
Ammoniumchromat *n* ammonium chromate
Ammoniumcitrat *n* ammonium citrate
Ammoniumdichromat *n* ammonium dichromate
Ammoniumdihydrogenphosphat *n* ammonium dihydrogen phosphate
Ammoniumdisulfid *n* ammonium disulfide
Ammoniumeisenalaun *m* ammonium iron alum, ammonium iron sulfate
Ammoniumferricyanid *n* ammonium cyanoferrate (III), ammonium ferricyanide
Ammoniumferrisulfat *n* ammonium ferric sulfate, ferriammonium sulfate
Ammoniumferrosulfat *n* ammonium ferrosulfate, ferroammonium sulfate, Mohr's salt
Ammoniumhydrosulfid *n* ammonium hydrosulfide, ammonium sulfhydrate
Ammoniumhydroxid *n* ammonium hydroxide
Ammoniumhypophosphit *n* ammonium hypophosphite
Ammoniumhyposulfit *n* ammonium hyposulfite
Ammoniumiridiumchlorid *n* ammonium iridium chloride, ammonium iridichloride
Ammoniumjodid *n* ammonium iodide
Ammoniummagnesiumarsenat *n* ammonium magnesium arsenate
Ammoniummagnesiumphosphat *n* ammonium magnesium phosphate
Ammoniummetaborat *n* ammonium meta-borate
Ammoniummolybdat *n* ammonium molybdate
Ammoniummonovanadat *n* ammonium monovanadate
Ammoniumnatriumphosphat *n* ammonium sodium phosphate
Ammoniumnitrat *n* ammonium nitrate
Ammoniumnitrit *n* ammonium nitrite
Ammoniumpalladiumchlorid *n* ammonium palladium chloride, ammonium chloropalladate (IV), ammonium palladichloride
Ammoniumpermanganat *n* ammonium permanganate

Ammoniumperoxydisulfat — Amphifluorochinon

Ammoniumperoxydisulfat *n* ammonium persulfate
Ammoniumpersulfat *n* ammonium persulfate
Ammoniumphosphat *n* ammonium phosphate
Ammoniumphosphormolybdat *n* ammonium phosphomolybdate
Ammoniumphosphorwolframat *n* ammonium phosphotungstate, ammonium phosphowolframate
Ammoniumplatinchlorid *n* ammonium chloroplatinate(IV), ammonium platinic chloride
Ammoniumpolysulfid *n* ammonium polysulfide
Ammoniumradikal *n* ammonium radical
Ammoniumrest *m* ammonium residue
Ammoniumrhodanid *n* ammonium thiocyanate, ammonium rhodanate
Ammoniumrhodanür *n* (obs) ammonium thiocyanate
Ammoniumsalicylat *n* ammonium salicylate
Ammoniumsalpeter *m* ammonium nitrate
Ammoniumsalz *n* ammonium salt
Ammoniumsilicofluorid *n* ammonium fluosilicate, ammonium silicofluoride
Ammoniumsulfat *n* ammonium sulfate
Ammoniumsulfid *n* ammonium sulfide
Ammoniumsulfit *n* ammonium sulfite
Ammoniumsulfocyanat *n* ammonium sulfocyanate, ammonium thiocyanate
Ammoniumsulfocyanid *n* ammonium thiocyanate
Ammoniumsulfoichthyolat *n* ammonium sulfoichthyolate, ammonium ichthyolsulfonate, ichthyol
Ammoniumthalliumalaun *m* ammonium thallium alum, ammonium thallium sulfate
Ammoniumthioarsenat *n* ammonium thioarsenate
Ammoniumthioarsenit *n* ammonium thioarsenite
Ammoniumthiosulfat *n* ammonium thiosulfate
Ammoniumuranat *n* ammonium uranate
Ammoniumvanadat *n* ammonium vanadate
Ammoniumwismutcitrat *n* ammonium bismuth citrate
Ammoniumwolframat *n* ammonium tungstate, ammonium wolframate
Ammoniumzinnchlorid *n* tin ammonium chloride, ammonium chlorostannate(IV)., ammonium stannic chloride, pinksalt
Ammonjodid *n* ammonium iodide
Ammonkarbonat *n* ammonium carbonate
Ammonmagnesiumarsenat *n* ammonium magnesium arsenate
Ammonnatriumphosphat *n* ammonium sodium phosphate
Ammonolyse *f* ammonolysis
ammonotelisch (Biol) ammonotelic
Ammonpurpurat *n* murexide
Ammonquecksilberchlorid *n* ammonium chloromercurate(II), ammonium mercuric chloride
Ammoresinol *n* ammoresinol
Amnesie *f* (Med) amnesia
Amnionflüssigkeit *f* amniotic fluid
Amnionsäure *f* amniotic acid
Amnionwasser *n* amniotic fluid
Amobarbital *n* amobarbital
Amodiachin *n* amodiaquine
Amöbe *f* (Biol) amoeba, ameba
Amöbizid *n* amebicide
Amolanon *n* amolanone
Amopyrochin *n* amopyroquine
amorph amorphous
Amorphismus *m* amorphism
Amosamin *n* amosamine
Amosaminol *n* amosaminol
Amoxecain *n* amoxecaine
AMP *n* **Adenosinmonophosphat** adenosine monophosphate, AMP
Ampelin *n* ampeline
Ampelit *m* (Min) ampelite, channel coal
Ampelopsidin *n* ampelopsidine
Ampelopsin *n* ampelopsine
Ampeloptin *n* ampeloptin
Ampère *n* (Elektr) ampere
Ampèremesser *m* amperemeter, ammeter, amperometer
Ampèrestunde *f* ampere-hour
Ampèrestundenleistung *f* ampere-hour capacity
Ampèrestundenzähler *m* ampere-hour counter
Ampèrewindung *f* ampere turn, ampere winding
Ampèrewindungsfaktor *m* ampere winding coefficient
Ampèrewindungszahl *f* ampere turns
Ampèrezahl *f* amperage
Ampèrometrie *f* amperometry
amperometrisch amperometric
Amphetamin *n* amphetamine
Amphetaminphosphat *n* amphetamine phosphate
Amphetaminum phosphoricum *n* amphetamine phosphate
Amphetaminum sulfuricum *n* amphetamine sulfate
Amphibien *fpl* amphibians, amphibious animals
Amphibienstein *m* (Min) amphibolite
Amphibol *m* (Min) amphibole, amphibolite, horn blende
amphibolartig amphibolic
amphibolhaltig amphiboliferous
Amphibolit *m* (Min) amphibolite
Amphibolreihe *f* amphibolic series
amphichroitisch amphichroic
amphichromatisch amphichromatic
Amphidromiepunkt *m* amphidromic point
Amphidsalz *n* amphide salt
Amphifluorochinon *n* amphifluoroquinone

Amphigenit *m* amphigenite
Amphilogit *m* amphilogite
Amphiole *f* small ampoule
Amphioxus *m* amphioxus, (Zool) lancelet
amphipatisch amphipathic
amphiphil amphiphilic
amphiprotisch amphiprotic
Ampholyt *m* ampholyte
amphoter amphoteric
Amphotericin *n* amphotericin
Amphotropin *n* amphotropine
Amplitude *f* (Polarwinkel) azimuth, polar angle, (Schwingungsweite) amplitude
Amplitudenänderung *f* variation of amplitude
Amplitudenaufschaukelung *f* amplitude build-up
Amplitudenbegrenzer *m* amplitude limiter
Amplitudenentzerrung *f* correction of amplitudes
Amplitudenmodulation *f* amplitude modulation
Amplitudenregelung *f* amplitude regulation, gain control
Amplitudenschwankung *f* amplitude fluctuation
Amplitudenschwund *m* amplitude fading
Amplitudenverzerrung *f* amplitude distortion
Amplitudenweite *f* amplitude
Ampulle *f* ampul[e], ampoule, vial
ampullenförmig ampulliform
Ampullensäge *f* ampul saw
Amsonin *n* amsonine
amtlich official
Amurensin *n* amurensin
Amygdalat *n* amygdalate, salt or ester of amygdalic acid
Amygdalin *n* amygdalin
Amygdalinsäure *f* amygdalic acid
amygdalinsauer amygdalate
amygdaloid amygdaloid, almond-shaped
Amygdalose *f* amygdalose
Amyl- amyl-, pentyl-
Amylacetat *n* amyl acetate
Amylacetatlampe *f* amyl acetate lamp
Amyläther *m* amyl ether, amyl oxide, diamyl ether, pentyl oxide
Amyläthylketon *n* amyl ethyl ketone
Amylaldehyd *m* amylaldehyde, valeraldehyde
Amylalkohol *m* amyl alcohol, **tertiärer** ~ tertiary amyl alcohol
Amylamin *n* 1-amino-pentane, amylamine, pentylamin
Amylaminchlorhydrat *n* amylamine hydrochloride
Amylan *n* amylan
Amylanilin *n* amylaniline
Amylase *f* (Biochem) amylase
Amylat *n* amylate
Amylbenzoat *n* amyl benzoate
Amylbenzyläther *m* amyl benzyl ether
Amylbromid *n* amyl bromide
Amylbutyrat *n* amyl butyrate
Amylchlorid *n* amyl chloride

Amylcinnamat *n* amyl cinnamate
Amylen *n* amylene
Amylenchloral *n* amylene chloral, dormiol
Amylenhydrat *n* amylene hydrate, 1-dimethyl-propyl-alcohol
Amylenhydrid *n* amylene hydride, pentane
Amylester *m* amyl ester
Amyljodid *n* amyl iodide
Amylnitrit *n* amyl nitrite, pentyl nitrite
Amylobiose *f* amylobiose
Amylobiosedextrin *n* amylodextrin
Amylobioseform *n* amyloform
Amyloerythrin *n* amyloerythrine
Amylogen *n* amylogen
amyloid (stärkeähnlich) amyloid, (stärkehaltig) amylaceous
Amyloid *n* amyloid, amylaceous substance
Amyloidentartung *f* amyloid degeneration
Amyloidfaden *m* amyloid filament
Amyloin *n* amyloin, maltodextrin
amyloklastisch amyloclastic
Amylolyse *f* amylolysis
amylolytisch amylolytic, amyloclastic
Amylometer *n* amylometer
Amylopektin *n* amylopectin
Amylopsin *n* (Pankreasamylase) pancreatic amylase
Amylose *f* amylose
Amylotriose *f* amylotriose
Amyloverfahren *n* amylo process
Amyloxalat *n* amyl oxalate
Amyloxid *n* (obs) amyl ether, amyl oxide
Amyloxidhydrat *n* (obs) amyl alcohol
Amylphenylacetat *n* amylphenyl acetate
Amylrhodanid *n* amyl thiocyanate, amyl rhodanate
Amylrhodanür *n* (obs) amyl thiocyanate
Amylsalicylat *n* amyl salicylate
Amylsilicon *n* amyl silicone
Amylsulfid *n* amyl sulfide, diamyl sulfide
Amylsulfocyanat *n* amyl thiocyanate
Amylum *n* amylum, corn starch, starch
Amylvalerianat *n* amyl valerate
Amylverbindung *f* amyl compound
Amyranon *n* amyranone
Amyrenol *n* amyrenol
Amyrin *n* amyrin
Amyrol *n* amyrol
Anabasin *n* anabasine, neonicotine
Anabiose *f* (Biol) anabiosis
anabiotisch anabiotic
anabolisch anabolic
Anabolismus *m* (Biol) anabolism
Anacardienöl *n* anacardium oil
Anacardinsäure *f* anacardic acid
anacardinsauer anacardate
Anacardium *n* anacardium, caje nut, cashew nut
Anacardsäure *f* anacardic acid
Anacidität *f* anacidity

Anacyclin *n* anacyclin
anaerob anaerobic
Anaerobien *pl* anaerobic bacteria, anaerobionts
Anaerobier *m* anaerobe, anaerobiont, anaerobium
Anaerobiont *m* anaerobe, anaerobiont, anaerobium
Anaerobiose *f* anaerobiosis, anaerobism, anoxybiosis
Anaerophyt *m* anaerophyte
Anästhesie *f* (Med) anesthesia
Anaesthesin *n* anaesthesin[e], benzocaine
Anästhetikum *n* (Med) anesthetic
Anästhol *n* anesthol
anätzen to begin to corrode, to cauterize, to corrode, to etch
Anätzen *n*, **elektrolytisches** ~ anodic etching
Anätzung *f* cauterization
Anaferin *n* anaferine
Anagyrin *n* anagyrine
Anagyrinhydrobromid *n* anagyrine hydrobromide
anaklastisch anaclastic
Analcim *m* (Min) analcime, analcite
Analcit *m* (Min) analcime, analcite
Analeptikum *n* analeptic
Analgen *n* analgen[e], 5-benzamido-8-ethoxyquinoline
Analgesie *f* (Med) analgesia
Analgesin *n* analgesine, antipyrine
Analgetikum *n* (Med) analgetic
analog analogous
Analogie *f* analogy
Analogon *n* analog, analogon
Analog-Rechner *m* (Math) analog computer
Analysator *m* analyzer
Analyse *f* analysis (pl analyses), ~ **auf nassem Wege** analysis by wet process, fluid analysis, ~ **auf trockenem Wege** analysis by dry process, ~ **Bio-** bioanalysis, **anorganische** ~ inorganic analysis, **biochemische** ~ biochemical analysis, **derivatographische** ~ derivative gravimetric analysis, **Elektro-** ~ electro-analysis, **Elementar-** ~ elementary analysis, ultimate analysis, **Gas-** ~ gas-analysis, **gravimetrische** ~ gravimetric analysis, **jodometrische** ~ iodimetric analysis, **klinische** ~ clinical analysis, **kolorimetrische** ~ colorimetric analysis, **konduktometrische** ~ conductometric analysis, **Mikro-** ~ micro-analysis, **qualitative** ~ qualitative analysis, **quantitative** ~ quantitative analysis, **refraktometrische** ~ analysis by refraction, **Spektral-** ~ spectrum-analysis, **spektroskopische** ~ spectroscopic analysis, **technische** ~ technical analysis, **thermische** ~ thermal analysis, **Tüpfel-** ~ spot test, **volumetrische** ~ volumetric analysis

Analysenbefund *m* result of analysis, analytical finding, analytical result
Analysenbericht *m* analytical report, report of analysis
Analyseneinwaage *f* weighed portion for analysis
Analysenfehler *m* analytical error
Analysenfehlergrenze *f* limit of analytical error, limit of error of analysis
analysenfertig ready for analysis
Analysenfilter *n* analytical filter
Analysenformel *f* analysis formula, empirical formula
Analysengang *m* course of analysis, process of analysis
Analysengewicht *n* analytical weight
Analysenkosten *pl* cost of analysis
Analysenlampe *f* quartz lamp
Analysenmethode *f* analytical method, method of analysis
Analysenprobe *f* sample for analysis
Analysenquarzlampe *f* analytical quartz lamp
analysenrein analytically pure
Analysenschnellwaage *f* analytical rapid-weighing balance
Analysensubstanz *f* substance to be analyzed
Analysentrichter *m* analytical funnel
Analysenunterschied *m* difference in analysis
Analysenvorschrift *f* analytical instruction
Analysenwaage *f* analytical balance
analysierbar analyzable
analysieren to analyze
Analysieren *n* analyzing
Analysierung *f* analyzation
Analytik *f* analytics
Analytiker *m* analyst
analytisch analytical
Analzim *m* (Min) analcime
Anamesit *m* anamesite
Anamirtin *n* anamirtin
anamorph (Phys) anamorphous
Anamorphismus *m* anamorphism
Anamorphose *f* anamorphosis
Ananasäther *m* pineapple essence, ethyl butyrate
Ananasessenz *f* pineapple essence, ethyl butyrate
Ananasfaser *f* pineapple fiber
Anaphase *f* (Biol) anaphase
Anaphorese *f* anaphoresis
Anaphylaxie *f* (Med) anaphylaxis
anaplerotisch anaplerotic
anastigmatisch (Opt) anastigmatic
Anatabin *n* anatabine
Anatas *m* (Min) anatase, octahedrite
Anatexis *f* (Geol) anatexis
Anatomie *f* anatomy
anatomisch anatomical[ly]
Anatoxin *n* anatoxin
Anatto *n* (Bot, Färb) anatto
Anauxit *m* (Min) anauxite
anbacken to bake lightly; to stick on

Anbauversuch *m* (Agr) field trial
anbeizen to mordant
anbelangen to concern, to relate to
Anbetracht, in ~ in consideration [of], in view [of]
anbieten to offer
anbinden to bind, to tie
Anbindezettel *m* tag
anbläuen (Färb) to blue, to color blue
anblaken to blacken, to smoke, to soot
Anblaken *n* blackening with soot
anblasen to blow in, to fan, (Schachtofen) to put on the blast
Anblasen *n* blowing in, igniting, putting in blast, starting
Anblick *m* view
anbluten (Farben) to bleed (colors)
Anbluten *n* staining
anbohren to bore, to drill a hole [into], to perforate; (Faß) to tap, (zum Zentrieren) to center
Anbohrer *m* center bit
Anbohrschelle *f* service saddle, side-outlet [pipe] clamp
Anbrachen *n* repair of damaged furs
anbräunen to brown
Anbräunen *n* browning
anbrennen to begin to burn, to burn, to calcine, to light, to set on fire
Anbrennen *n* scorching
anbringen to install, to place, to put
Anbringung *f* placing
Anbruch *m* incipient crack, (Bruchfläche) fracture
Anbruchslager *n* weigh-out store
anbrüchig decayed, rotten
Anbrüchigkeit *f* decay, moldiness, putrescence, putridity, rot
anbrühen to scald
Anbrühen *n* scalding
anchoinsauer anchoate, azelate
Anchovisöl *n* anchovy oil
Anchusasäure *f* anchusic acid
Anchusin *n* anchusin
Andalusit *m* (Min) andalusite
Andaöl *n* anda oil
andauern to continue, to last, to persist
andauernd continuous, lasting, steady
andersartig different
andersgestaltet heteromorphic
Anderson-Meßbrücke *f* Anderson bridge
anderthalb one and a half, (Chem) sesqui
anderthalbarsensauer sesquiarsenate
anderthalbbasisch sesquibasic
anderthalbchlorsauer sesquichlorate
anderthalbchromsauer sesquichromate
anderthalbfach one and a half times
Andesin *m* (Min) andesine
Andesit *m* (Min) andesite

andeuten to indicate
Andeutung *f* indication
Andirin *n* andirin
Andisin *n* andisine
andörren to dry, to wither
Andorit *m* (Min) andorite
Andradit *m* (Min) andradite
Andrang *m* rush
andrehen to screw on, (Elektr) to switch on, (Schraube) to tighten, (Techn) to start, (Wasser, Gas) to turn on
Andrehkurbel *f* starting crank
Andreolith *m* (Min) andreolite, chiastolite
Andrewsit *m* (Min) andrewsite
Androgen *n* androgen
Andrographolid *n* andrographolide
Andromedotoxin *n* andromedotoxin
Androsin *n* androsine
Androstadien *n* androstadiene
Androstan *n* androstane
Androstanolon *n* androstanolone, androsterone
Androstendiol *n* androstenediol
Androsteron *n* androsterone, androkin, androstanolone, androtin
Andruck *m* (Buchdr) proof print
Andruckfeder *f* leaf spring, pressure roller
Andruckmaschine *f* (Buchdr) proof press
Andruckplatte *f* (Phot) pressure plate
Andruckregler *m* pressure regulator
Andruckrolle *f* pressure roll
andrücken to press on
Andrückwalze *f* pressure roller
andunkeln to darken
aneignen to appropriate, to take possession of s. th., (Kenntnisse) to acquire
aneinanderfügen to join [together]
aneinandergrenzen to be adjacent, to be in contact with, to come in contact with
aneinandergrenzend adjacent
aneinanderhaften to stick together
Aneinanderlagerung *f* juxtaposition
Aneinanderreiben *n* rubbing against each other
aneinanderschließen to connect
anelektrisch anelectric
Anelektrolyt *m* non-electrolyte
anellieren to anellate
Anellierung *f* anellation
Anemometer *n* wind gauge
Anemoncampher *m* anemone camphor
Anemonin *n* anemonin
Anemoninsäure *f* anemoninic acid
Anemonolsäure *f* anemonolic acid
Anemonsäure *f* anemonic acid
Anemosit *m* (Min) anemosite
anerkennen to acknowledge, to admit, to appreciate, to recognize
anerkennenswert worthy of recognition
Aneroidbarometer *n* (Meteor) aneroid barometer
Anesin *n* anesine

Anethin *n* anethine
Anethol *n* anethol, anise camphor, p-allylphenyl methyl ether
Aneurin *n* aneurin[e], thiamin, vitamin B₁
Aneurindiphosphat *n* aneurine diphosphate
Aneurindisulfid *n* aneurine disulfide
Aneurinhydrochlorid *n* aneurine hydrochloride
Aneurinmononitrat *n* aneurine mononitrate
Aneurinthiol *n* aneurinethiol
anfachen to blow, to fan, to incite, to kindle
anfächeln to fan
anfällig prone, susceptible
Anfälligkeit *f* proneness, susceptibility
anfänglich incipient, initial, original
anfärbbar dyeable, colorable
Anfärbbarkeit *f* dye affinity, behavior towards dyes
anfärben to color, to dye slightly, to stain superficially, to tinge
Anfärbevermögen *n* colorability, coloring power, dye affinity
Anfahrbereich *m* approach (e. g. to a large building)
anfahren to start; to approach; to attack; to bring up
Anfahren *n* **des Ofens** starting up the furnace
Anfahrenergie *f* starting energy
Anfahrhebel *m* starting lever
Anfahrkraft *f* starting energy, tractive force
Anfahrmischung *f* mixture to start up the furnace
Anfahrstrom *m* starting current
Anfahrstufe *f* (einer Rakete) preliminary stage of propulsion (of a rocket)
Anfahrvorgang *m* start up procedure
Anfahrwirbel *m* initial vortex
Anfahrzeit *f* time required to start the furnace
Anfall *m* yield, amount formed, (Med) attack
anfallen to accumulate; to attack; to yield
Anfallstelle *f* gathering point
Anfangsanreicherung *f* initial enrichment
Anfangsausschlag *m* initial amplitude
Anfangsbahn *f* initial orbit
Anfangsdaten *pl* initial data
Anfangsdosis *f* (Med) initial dose
Anfangsdruck *m* initial pressure
Anfangsdurchschlag *m* initial breakdown
Anfangsenergie *f* initial energy
Anfangsergebnis *n* initial result, initial yield
Anfangserzeugnis *n* first product, initial product
Anfangsgeschwindigkeit *f* initial velocity
Anfangsglied *n* first member, initial member, (Math) leading term
Anfangsintensität *f* original intensity
Anfangskaliber *n* first pass
Anfangskapazität *f* initial capacity
Anfangskonzentration *f* initial concentration
Anfangslage *f* initial position
Anfangslösung *f* initial solution, original solution

Anfangsniveau *n* initial level
Anfangsphase *f* initial phase
Anfangsprodukt *n* initial product, first product, (Dest) first runnings
Anfangspunkt *m* initial point, origin, starting point, zero
Anfangsquerschnitt *m* first section, earliest section
Anfangsskala *f* basic range
Anfangsspannung *f* (Elektr) initial voltage, sparking potential
Anfangsstadium *n* initial phase
Anfangsstück *n* initial part
Anfangstemperatur *f* initial temperature
Anfangsverstärkung *f* (Elektr) initial amplification
Anfangsvolumen *n* original volume
Anfangswärmegrad *m* initial temperature
Anfangswertaufgabe *f* initial value problem
Anfangswertprobleme *pl* (Math) initial-value problems
Anfangszustand *m* initial condition, initial state
anfassen to seize, to touch
anfaulen to begin to rot, to decay
Anfaulen *n* putrefaction, putrescense
anfechten to attack, to challenge
anfertigen to manufacture, (Chem) to prepare
Anfertigung *f* manufacture, production
Anfettapparat *m* greasing apparatus
anfetten to grease, to lubricate, to oil
anfeuchten to damp[en], to humidify, to moisten, to soak
Anfeuchter *m* moistener, damper
Anfeuchtgrube *f* (Papier) sizing trough, sizing vat
Anfeuchtmaschine *f* (Pap) damping rolls
Anfeuchtung *f* damping, humectation, moistening, wetting
Anfeuchtwalze *f* damping roller
anfeuern to inflame, to light, to prime, (Kessel) to heat, (Ofen) to fire
Anfeuern *n* firing
Anfeuerungstopf *m* priming pan
anflanschen to couple, to flange on
Anflanschrohr *n* flange-type pipe
Anflug *m* film, thin layer, (Min) efflorescence
anfordern to claim, to demand; to require
Anforderung *f* demand, requirement
Anfrage *f* inquiry, invitation for tender
Anfragespezifikation *f* enquiry specification
Anfraß *m* corrosion, erosion
anfressen to attack, to eat away, to erode, to start corroding, (Gestein) to score, (Met) to corrode
Anfressen *n* corroding, decaying
Anfressung *f* corrosion, corrosive attack
anfrieren to freeze [on]
Anfrieren *n* freezing, liquid solidification

anfrischen (Metall) to reduce, to refresh, to refine
Anfrischen *n* (Metall) refreshing
Anfrischer *m* workman reducing litharge
Anfrischfeuerung *f* (Metall) furnace for refining iron
Anfrischgefäß *n* refining vessel
Anfrischherd *m* refining furnace
Anfrischofen *m* refining furnace
Anfrischschlacke *f* (Metall) refining cinders
Anfrischtrog *m* refining trough
Anfrischung *f* freshening, reduction, refinement
anfügen to add, to annex, to attach, to join
anführen to lead; to cite, to mention, to quote
Anführungszeichen *n* quotation mark
anfüllen to charge, to fill up
Anfüllen *n* filling, priming
Anfüllventil *n* priming valve
Angabe *f* indication, information; (Darstellung) representation, (Weisung) direction[s]; (Zitierung) citation, quotation
Angaben *pl* data, particulars
Angärung *f* (Chem) preliminary fermentation
angeätzt slightly corroded
angeben to declare; to indicate; to quote, to state
angeblakt blackened with smoke
angeboren innate
Angebot *n* offer, quotation
Angebotsvergleich *m* bid evaluation
angebracht applicable, fitting, suitable
angebrütet (Ei) embryonated
angebunden bound
angefacht induced
angehörig belonging [to]
angekörnt punch-marked
Angel *f* hinge, axis
Angelactinsäure *f* angelactic acid
Angelardit *m* (Min) angelardite
angelassen annealed, tempered
angelatinieren to plasticize on the surface
angelaufen (Met) tarnished, blued, coated, oxidized
Angelegenheit *f* affair, concern
angelernt semi-skilled
Angelicin *n* angelicin
Angelicon *n* angelicone
Angelika *f* (Bot) angelica, lungwort
Angelikaaldehyd *m* angelic aldehyde
Angelikaextrakt *m* extract of angelica
Angelikalakton *n* angelica lactone
Angelikasäure *f* angelic acid, angelicic acid, α-methylisocrotonic acid
Angelikaspiritus *m* angelica spirit
Angelin *n* angelin, rhatanin
angelötet soldered
Angelpunkt *m* pivot, turning point
angemessen adequate, conformable [to], proportionate [to], suitable
angenähert approximate

angenietet riveted
angenommen supposing, assuming; hypothetical
angepaßt appropriate; adapted, matched, suited
angequollen swollen
Angerbung *f* superficial tanning
angeregt excited
angereichert concentrated, enriched
angerostet rusted
angesäuert acidified, acidulated
angeschimmelt [slightly] moldy
angeschlossen connected
angeschmaucht blackened with smoke
angeschwollen swollen, bulged
angesichts in view of
Angestellter *m* employee
angestrengt intense; forced
angewärmt preheated, warmed, (lauwarm) tepid
angewandt applied, employed
angewiesen sein auf to be dependent on
angießen to cast on, to melt on
Angiom *n* (Med) angioma
Angiotensin *n* angiotensin
Angiotonin *n* angiotonin
angleichen to adapt, to adjust, to compensate, (Biol) to imitate
Angleichen *n* adjustment
Angleicher *m* rectifier
Angleichfeder *f* adapter spring
Angleichung *f* adaptation, adjustment, assimilation
Anglesit *m* (Min) anglesite
angliedern to join, to attach, (Organisation) to affiliate, to incorporate
Angliederung *f* attachment, (Organisation) affiliation, incorporation
anglühen to bring to red heat, to glow, to incandesce
Angolensin *n* angolensin
Angophorol *n* angophorol
Angorawolle *f* (Text) angora wool
Angosturaalkaloid *n* angostura alkaloid
Angosturabitter *m* angostura bitters
Angosturarinde *f* angostura bark
Angosturin *n* angosturine
angreifbar attackable, susceptible, (Met) corrodible
Angreifbarkeit *f* (Met) corrodibility
angreifen to affect, to attack; to corrode, to eat into
angrenzen to border on
angrenzend adjacent, contiguous
Angriff *m*, **chemischer** ~ chemical attack, corrosion, **ebenmäßiger** ~ uniform attack, **in** ~ **nehmen** to take up
Angriffsfläche *f* working surface
Angriffslinie *f* line of attack
Angriffsmittel *n* attacking agent, corrosive

Angriffspunkt *m* point of attack, point of application, (Abnutzung) point of wear, ~ **der Kraft** force impact point
Angriffsschenkel *m* actuating arm
Angström-Einheit *f* (Maß), Angstrom unit
Anguß *m* sprue, gate, gating, lug, (Gieß) feedhead; **den ~ entfernen** to degate, **direkter ~** direct gate, **entfernen des ~ es** to degate, **ringförmiger ~** ring gate, **verjüngter ~** restricted gate
Angußabreißer *m* sprue puller
Angußabschneideventil *n* gate cutoff valve
Angußabstanzer *m* gate cutter
Angußauswerferstift *m* sprue lock pin
Angußbuchse *f* sprue bushing; feed bush, injection molding nozzle, sprue bush
Angußdruckstift *m* sprue lock pin
Angußdrückstift *m* sprue ejector
Angußfarbe *f* (Keram) colored clay
Angußkanal *m* sprue
Angußkegel *m* sprue [slug]
Angußkreuz *f* four flow paths at 90°
Angußsteg *m* inlet
Angußstelle *f* gate mark
Angußstück *n* sprue slug
Angußtrichter *m* sprue
Angußverteiler *m* fan gate, (Einspritzkanal) runner
Angustifolin *n* angustifoline
Angustion *n* angustione
Angustose *f* angustose
Anhänger *m* (Auto) trailer
anhäufen to accumulate, to aggregate
anhäufend accumulative, cumulative
Anhäufung *f* accumulation, agglomeration, aggregation
anhaften to adhere, to be attached, to cling, to stick
anhaftend adhering, adhesive, clinging, sticking
Anhaftung *f* adherence, adhesion, attachment
Anhalamin *n* anhalamine
Anhalin *n* anhaline
Anhalmin *n* anhalmine
Anhalonidin *n* anhalonidine
Anhalonin *n* anhalonine
Anhalt *m* pause, stop; support
anhalten to stop
anhaltend continuous, persistent
Anhaltevorrichtung *f* stopping device
Anhaltpotential *n* stopping potential
Anhaltspunkt *m* criterion
Anhaltswert *m* approximate value
Anhang *m* annex, appendix, attachment, supplement
Anharmonizität *f* (Mech) anharmonicity
Anharmonizitätskonstante *f* anharmonicity constant, constant of anharmonicity
anheben to lift

Anhebung *f* **eines Elektrons** excitation of an electron
anheften to attach, to fasten [to]
anheimstellen to submit
anheizen to start heating
Anheizungsdauer *f* warming-up period
Anheizvorrichtung *f* heating arrangement
Anhydrämie *f* (Med) anhydremia
Anhydrid *n* anhydride
Anhydridbildung *f* anhydride formation
anhydrisch anhydrous
anhydrisieren to dehydrate, to become anhydrous, to eliminate water [from], to render anhydrous
Anhydrisierungsmittel *n* dehydrating agent
Anhydrit *m* anhydrous calcium sulfate, overburnt gypsum, (Min) anhydrite
Anhydro- anhydro-
Anhydroalkannin *n* anhydro alkannin
Anhydrobase *f* anhydro base
Anhydrocarminsäure *f* anhydrocarminic acid
Anhydrocymarigenin *n* anhydro cymarigenin
Anhydrodigitoxigenin *n* anhydro digitoxigenin
Anhydrodigitsäure *f* anhydrodigitic acid
Anhydroformaldehydanilin *n* anhydro formaldehyde aniline
Anhydroformaldehydparatoluidin *n* anhydroformaldehyde paratoluidine
Anhydrogitalin *n* anhydrogitalin
Anhydrolycorin *n* anhydrolycorine
Anhydrosäure *f* anhydroacid
Anibin *n* anibine
Anil *n* anil
Anileridin *n* anileridine
Anilid *n* anilide
Anilin *n* aniline, aminobenzene, phenylamine, **salzsaures ~** aniline hydrochloride, **schwefelsaures ~** aniline sulfate
Anilinblau *n* aniline blue, gentian blue, spirit blue
Anilinbraun *n* aniline brown, triaminoazobenzene
Anilinchlorat *n* aniline chlorate
Anilinchlorhydrat *n* aniline hydrochloride
Anilinchlorzink *n* aniline zinc chloride
Anilindruck *m* aniline printing
Anilindruckfarbe *f* aniline printing ink
Anilinfabrik *f* aniline works
Anilinfarbe *f* aniline dye; aniline color
Anilinfarbenindustrie *f* aniline dye industry
Anilinfarbenrückstand *m* residue of aniline dyes
Anilinfarbstoff *m* aniline dye
Anilingelb *n* aniline yellow, p-aminoazobenzene
Anilinharz *n* aniline resin
Anilinhydrochlorid *n* aniline hydrochloride
Anilinnitrat *n* aniline nitrate
Anilino- anilino-
Anilinöl *n* aniline oil
Anilinoxalat *n* aniline oxalate

Anilinrosa *n* aniline pink
Anilinrot *n* aniline red, fuchsin, magenta
Anilinsalz *n* aniline salt
Anilinschwarz *n* aniline black, nigrosine
Anilinsulfat *n* aniline sulfate
Anilinsulfonsäure *f* anilinesulfonic acid
Anilintinte *f* aniline ink
Anilinvergiftung *f* aniline poisoning, anilinism
Anilinviolett *n* aniline violett, regina purple
Anilipyrin *n* anilipyrine
Anilismus *m* (Med) aniline poisoning, anilinism
animalisch animal
Animalisieren *n* (Wolle) animalizing
Anime *n* animé, animi gum, animi resin
Animegummi *m* animi gum
Animeharz *n* animé resin, animi resin
Animikit *m* (Min) animikite
Aninsulin *n* aninsulin
Anion *n* anion
anionaktiv anion-active, anionic
Anionenaustauscher *m* anion exchanger
Anionenfehlstelle *f* anion vacancy
Anionenleitfähigkeit *f* anion conductivity
Anionfarbstoff *m* anionic dye
anionisch anionic
Anionkomplex *m* anion complex
Anionotropie *f* anionotropy
Anionsäure *f* anionic acid
Anis *m* (Bot) anise, aniseed
Anisal *n* anisal, anisylidene
Anisaldehyd *m* anisaldehyde
Anisalhydantoin *n* anisalhydantoin
Anisalkohol *m* anis alcohol, anisyl alcohol, p-methoxy-benzalcohol
Anisapfel *m* spice apple
anisartig like anise, anisate
Anisbranntwein *m* anisette
Aniscampher *m* anise camphor, anethol
Anisett *m* anisette
Anisgeist *m* anisette, spirit of anise
Anisholz *n* aniseed wood
Anishydramid *n* anishydramide
Anisidid *n* anisidide
Anisidin *n* anisidin[e]
Anisidino- anisidino-
Anisil *n* anisil, bianisoyl
Anisilsäure *f* anisilic acid
Anislikör *m* anisette
Anisöl *n* aniseed oil, anise oil
anisoelastisch anisoelastic
Anisoin *n* anisoin, dimethoxybenzoin
Anisol *n* anisol[e], methoxybenzene, methylphenyl ether
Anisolessigester *m* anisoylacetic ester
anisomer (Phys) anisomeric
anisometrisch anisometric
Anisomycin *n* anisomycin
anisotrop anisotropic, non isotropic, (Phys) anisotrope

Anisotropie *f* anisotropy
Anissamen *m* (Bot) aniseed
anissauer anisate
Aniswasser *n* anisette
Anisyl- anisyl-
Anisylcyclon *n* anisylcyclone
Anisyliden *n* anisal, anisylidene
ankalken to limewash, to whitewash
Anker *m* anchor [bolt], (Magnet, Dynamo) armature
Ankerbüchse *f* (Elektr) armature spider
Ankerisolierscheibe *f* armature end-plate
Ankerit *m* (Min) ankerite
Ankerkern *m* armature core
Ankerkörper *m* armature body
Ankernut *f* armature slot
Ankerrührer *m* anchor bolt, anchor screw
Ankerspannung *f* (Elektr) armature voltage
Ankerspannungsabfall *m* armature drop
Ankerstahl *m* anchor steel
Ankerstrom *m* armature current
Ankerwicklung *f* armature winding
anketteln to stitch on
ankitten to cement, to fasten with cement, to fasten with mastic
anklammern to fasten [with]
ankleben to stick on, to attach, to glue on, to paste on
Ankleben *n* adherence, adhesion
ankleistern to paste on
Anklemmrührer *m* clamp-on [high speed] stirrer, clip-on stirrer, portable mixer
anknüpfen to connect, to fasten, to join
Anknüpfungspunkt *m* point of contact, starting point
ankochen to boil partly, to par boil
Ankörnbohrer *m* countersinking bit
ankörnen to center-punch
Ankörner *m* center punch
Ankörnung *f* punch mark, center mark
ankohlen to char partially
Ankohlen *n* partial charring
Ankohlung *f* partial charring
ankreiden to chalk
ankündigen to announce
ankuppeln to couple, to attach, to connect, to engage
Ankylit *m* (Min) ancylite
anläßlich occasionally
Anlage *f* plant, equipment, establishment, installation, unit, (Investition) investment, (Veranlagung) talent, tendency, ~ **geregelte** controlled system
Anlagekosten *pl* cost of construction
Anlagengrenzen *pl* battery limits
anlagern (anhäufen) to accumulate, to add, to store up, (Chem) to bind, to combine, to form a complex
Anlagern *n* **von Kohlenstoff** addition of carbon

Anlagerung – annektieren

Anlagerung *f* accumulation, (Chem) addition, (Geol) agglomeration
Anlagerungserzeugnis *n* addition product, additive compound
anlagerungsfähig capable of addition
Anlagerungskomplex *m* addition complex
Anlagerungsreaktion *f* addition reaction
Anlagerungsverbindung *f* addition compound
Anlagesicherung *f* major loss prevention
Anlaß *m* cause, motive, occasion, occurrence, reason, stimulus, (Techn) temper, **ohne jeden ~** without any motive
Anlaßätzung *f* temper etching
Anlaßbeständigkeit *f* retention of hardness
Anlaßdauer *f* (Met) duration of tempering
Anlaßdüse *f* starter nozzle
anlassen (Metall) to anneal, to temper, (Motor) to start
Anlassen *n* (Met) annealing, tempering, (Motor) starting
Anlasser *m* starter, starting switch
Anlaßfarbe *f* annealing color, temper[ing] color
Anlaßgenerator *m* starting dynamo
Anlaßhärte *f* tempering hardness
Anlaßhärtung *f* temper hardening
Anlaßhebel *m* starting lever
Anlaßkraftstoff *m* primer
Anlaßmaschine *f* starting machine
Anlaßmittel *n* tempering medium
Anlaßmotor *m* starter
Anlaßofen *m* annealing oven, tempering furnace
Anlaßrohrleitung *f* primer tubing
Anlaßsalz *n* tempering salt
Anlaßschalter *m* starting switch
Anlaßtemperatur *f* temperature of tempering
Anlaßtransformator *m* (Elektr) starting transformer
Anlaßventil *n* primer valve, starting valve
Anlaßwiderstand *m* starter resistance
Anlaßwirkung *f* annealing effect
Anlauf *m* swelling, (Met) tarnish, (Motor)) starting, warming up
Anlaufbeständigkeit *f* resistance to heat-tinting, resistance to tarnish, tarnish proofness
anlaufen (Met) to bloom, to oxidize, to tarnish
Anlaufen *n* blooming, hazing, tarnishing, **~ bei erhöhter Temperatur** heat-tinting
Anlauffarbe *f* (Met) annealing color, heat indicating paint, heat sensitive paint, temperature indicating paint, temper color, tempering color, thermo-recording paint
Anlaufgebiet *n* (Atom) counter range
Anlauflack *m* annealing lacquer
Anlaufrechnung *f* preliminary calculation
Anlaufschutzbad *n* anti-tarnishing bath
Anlauftemperatur *f* annealing temperature, tempering temperature
Anlaufvorgang *m* **des plastischen Fließens** incipient plastic flow

Anlaufwert *m* (Regeltechn) reaction value
Anlaufzeit *f* control lag; response time, starting period, starting time, **~ der Wasserstoffentwicklung** (Metall) pickle lag
Anlegeapparat *m* (Buchdr) feeding apparatus
Anlegegoniometer *n* protractor
Anlegemaßstab *m* contact rule
anlegen (Spannung) to apply
anlehnen to lean against
anleimen to glue [on]
Anleimmaschine *f* gluing machine
anleiten to guide, to instruct, to introduce
Anleitung *f* instruction, introduction
anleuchten to shine light on, to spot with a beam of light
anliefern to deliver, to supply
Anliegen *n* (Gesuch) request
anliegend adjacent
anlösen to solubilize, to solvate
anlöten to solder on
Anlöten *n* soldering on
Anmach[e]bottich *m* mixing vat
anmachen (befestigen) to attach, to fasten, (Gastr) to prepare, (Licht) to switch on, (mischen) to mix
Anmachwasser *n* mixing water
anmalen to paint
anmelden to announce, (Patent) to apply for a patent
Anmeldung *f* (Patent) patent application
anmengen to blend, to mix
anmischen to mix
Anmischen *n* mixing
Annabergit *m* (Min) annabergite, nickel bloom, nickel ocher
annähern to approach, to approximate
Annäherung *f* approach, approximation
Annäherungsaufstellung *f* approach disposition
Annäherungsformel *f* (Math) approximate formula
Annäherungsgeschwindigkeit *f* approach rate
Annäherungsgrad *m* degree of approximation
Annäherungsrechnung *f* approximate calculation
Annäherungsverfahren *n* method of approximation, trial and error method
annäherungsweise approximately
Annäherungswert *m* approximate value, approximation
annässen to damp, to moisten
Annässen *n* damping, moistening
Annaglas *n* uranium glass
Annahme *f* assumption, hypothesis
Annahmegrenze *f* acceptable quality level
Annalen *pl* annals
Annalin *n* (Pap) annaline
Annatto *n* annatto
annehmen to assume, to suppose, (empfangen) to accept
annektieren to annex

Annelid *m* (Zool) annelid
Annerödit *m* (Min) annerodite, anneroedite
Annidalin *n* annidalin, diiodothymol
annieten to rivet
Annihilation *f* (Atom) annihilation
Annihilationsstrahlung *f* annihilation radiation
Annit *m* annite
Annivit *m* (Min) annivite
Annofolin *n* annofoline
Annotinin *n* annotinine
annullieren to annul, to cancel
Anobial *n* anobial
Anode *f* anode, positive electrode, **sperrende** ~ anode with insulating surface
Anodenbatterie *f* anode battery, high-voltage battery, plate battery
Anodenbelastung *f* anode loading
Anodendichte *f* anode density
Anodendunkelraum *m* anode dark space
Anodeneffekt *m* anode effect
Anodenentladung *f* anode discharge
Anodengleichrichter *m* (Elektr) anode rectifier
Anodengleichrichtung *f* anode rectification
Anodenhalter *m* anode holder
Anodenkapazität *f* plate capacitance
Anodenklemme *f* anode clamp, anode terminal
Anodenkorb *m* anode cage
Anodenkreis *m* anode or plate circuit
Anodenleitung *f* plate conduction
Anodenplatte *f* anode plate
Anodenrahmen *m* anode frame
Anodenraum *m* anode region
Anodenschatten *m* heel effect (X-ray tubes), (Röntgenröhren) heel effect
Anodenschlamm *m* (Kupfergewinnung) anode mud, anode slime, anode sludge
Anodenschutznetz *n* anode screen
Anodenschwamm *m* (Bleigewinnung) anode sponge
Anodenspannung *f* anode potential
Anodenspeisung *f* plate supply
Anodenspitzenspannung *f* plate peak voltage
Anodenstange *f* anode bar
Anodenstecker *m* anode plug
Anodenstrahl *m* anode ray
Anodenstrahlung *f* positive rays
Anodenstrom *m* anode or plate current
Anodenüberzug *m,* **sperrender** ~ insulating anodic coating
Anodenverlustleistung *f* plate dissipation
Anodenwiderstand *m* plate resistance, plate resistor
anodisch anodic, ~ **behandeln** to anodize
Anodyn *n* anodyne
Anodynin *n* anodynin, antipyrine
anölen to oil, to coat with oil
Anogen *n* anogen, mercurous iodobenzene-p-sulfonate
Anol *n* anol, p-propenylphenol

Anolobin *n* anolobine
Anolyt *m* anolyte
anomal anomalous
Anomalie *f* anomaly
anomer anomeric
Anomer *n* anomer
Anomeriezentrum *n* **Stereochem** anomeric center
Anomit *m* (Min) anomite
Anonain *n* anonaine
Anonol *n* anonol
Anophorit *m* (Min) anophorite
anordnen to regulate; to set up
Anordnung *f* regulation, arrangement, instruction, order, ~ **im Raum** configuration, **ebene** ~ planar position, **gekreuzte** ~ (Opt) crossed position
Anordnungsschema *n* layout
Anorganik *f* inorganics
Anorganiker *m* inorganic chemist
anorganisch inorganic
anormal abnormal, anomalous
Anorthit *m* (Min) anorthite, christianite
Anorthoklas *m* (Min) anorthoclase
Anosmie *f* (Med) absence of the sense of smell, anosmia
anoxydieren to oxidize partially
anpassen to adapt, to adjust, to match, to suit
Anpaßform *f* forming die
Anpaßstück *n* adapter
Anpassung *f* adaptation, adjustment, matching
Anpassungsfähigkeit *f* adaptability
Anpassungsstab *m* (Atom) shim rod
Anpassungtransformator *m* matching transformer
Anprall *m* impact, collision, impingement
anprallen to impinge, to strike
anpressen to press against
anquicken (obs) to amalgamate
Anquicken *n* (obs) amalgamating
Anquickfaß *n* amalgamating barrel
Anquicksilber *n* silver amalgam
Anquickung *f* (obs) amalgamation
anräuchern to fumigate
Anräucherung *f* fumigation
anraten to advise, to recommend
anrauchen to smoke
anrauhen to roughen
anrechnen to charge, to count
Anrecht *n* claim, right
Anregekristalle *m pl* seed crystals, shock seed
anregen to stimulate, to activate, to excite; to mention, to suggest
anregend exciting, stimulating
Anreger *m* (Zement) accelerator
Anregung *f* stimulation, (Chem, Elektr) excitation, **thermische** ~ (Phys) thermal excitation
Anregungsbedingung *f* excitation condition
Anregungsenergie *f* excitation energy

Anregungsintensität – Anschlagbutzen

Anregungsintensität *f* excitation intensity
Anregungsmittel *n* (Pharm) stimulant
Anregungsniveau *n* excitation level
Anregungsspannung *f* excitation potential, excitation voltage
Anregungsstoß *m* excitation collision
Anregungsstrom *m* excitation current
Anregungsstufe *f* excitation stage
Anregungsvorgang *m* excitation process
Anregungszustand *m* excited state
Anreibemaschine *f* grinding machine
anreiben to rub on, (Farbe) to grind
Anreiben *n* grinding, milling
Anreibeversilberung *f* silvering by rubbing on
anreichern to concentrate, to enrich, to strengthen
Anreicherung *f* accumulation, concentration, enhancement, (Met) enrichment, ~ **der Erze** *n pl* ore-dressing
Anreicherungsanlage *f* (Atom) enrichment plant
Anreicherungsfaktor *m* (Atom) enrichment factor
Anreicherungsprozeß *m* accumulative process, enrichment
anreihen to add
Anreißvorrichtung *f* marking device
Anreiz *m* incentive, stimulant
anreizen to stimulate
anrichten to bring about, to perform, to cause; (Farben) to mix, (Gastr) to dress, to prepare
Anrichten *n* preparing, mixing
Anrichter *m* assayer
Anriß *m* crack, flaw, hairline
Anrißsucher *m* crack detector
anritzen to scratch, to slit slightly
anrösten to roast slightly
Anroller *m* (Reifenwickelei) face stitcher
Anrollvorrichtung *f* (Reifenwickelei) face stitcher, ply down stitcher
anrosten to begin to rust, to corrode
Anrostung *f* corrosion, rusting, ~ **des Rohres** superficial rusting of the pipe
anrühren to mix, to stir
anrußen to coat with soot, to smoke, to soot
ansäuern to acidify, to acidulate, (Gastr) to sour
Ansäuern *n* acidification, acidifying, acidulation, souring
Ansäuerung *f* **des Bades** acidification of the bath
ansalzen to salt lightly
ansammeln to accumulate, to collect, to gather
Ansammlung *f* accumulation
Ansatz *m* (Belag) clinker coating, clinker crust (cement), deposit, incrustation, sediment, (Chem) composition, preparation, (Math) formulation, (Med) appendage, (Techn) batch, charge, ingredients, (Verlängerung) extension, (Zapfen) spigot
Ansatzbad *n* initial bath, starting bath
Ansatzbottich *m* preparing vessel

Ansatzlösung *f* starting solution
Ansatzpunkt *m* point of attachment, point of insertion; starting point
Ansatzrohr *n* connecting tube, attached tube, extension pipe, insert tube, nozzle
Ansatzschraube *f* setscrew
Ansatzstück *n* attached piece, connecting piece, extension piece
Ansatztisch *m* extension table
Ansaugdruck *m* suction pressure
Ansaugehub *m* suction stroke
ansaugen to absorb, to aspirate, to suck in, to suck up
Ansaugen *n* suction, aspiration, sucking, ~ **von Luft** drawing in air
Ansaugevolumen *n* aspirated volume
Ansauggebläse *n* suction fan
Ansaugheber *m* siphon
Ansaugleistung *f* suction capacity
Ansaugluft *f* induction air
Ansaugöffnung *f* suction orifice
Ansaugpumpe *f* priming pump, suction pump
Ansaugrohr *n* suction pipe, intake pipe, suction tube
Ansaugventil *n* intake valve, suction valve
Ansaugvolumen *n* aspirated volume
Ansaugvorrichtung *f* suction apparatus
Ansa-Verbindung *f* (Stereochem) ansa compound
anschärfen to sharpen, to prime, (Bad) to strengthen
anschaffen to provide, to purchase
Anschaffungskosten *pl* first cost, initial cost, prime cost
anschalten (andrehen) to switch on, (anschließen) to connect
anschaulich descriptive, intuitive
Anschauung *f* opinion, view, (Vorstellung) conception, idea
Anschauungsmaterial *n* illustrative material
Anschauungsweise *f* standpoint
Anschein *m* appearance
anscheinend apparent
anschichten to pile in layers, to stratify
anschieben to push, to shove
anschießen to shoot crystals into, to crystallize
Anschießfaß *n* crystallizer
Anschießgefäß *n* crystallizer, crystallizing vessel
anschimmeln to begin to mold, to grow moldy
Anschimmeln *n* growing moldy
anschlämmen to become muddy, to deposit mud; to make into a paste, to paste, to smear; to suspend; to wash, (Chem) to elutriate
Anschlämmen *n* depositing of mud
Anschlämmung *f* (Chem) elutriation
Anschlag *m* stroke, beat, impact; (Techn) projection, stop, stop collar
Anschlagbolzen *m* stop dog, stop pin
Anschlagbutzen *m* stop button

anschlagen to strike [at]
Anschlagpunkt *m* stop
Anschlagraste *f* quantity stop
Anschlagrolle *f* tappet roller
Anschlagschiene *f* stock rail, striker bar
Anschlagschraube *f* stop screw
Anschlagspindel *f* stop spindle
Anschlagstift *m* detent pin, stop pin
Anschlagwert *m* estimated value
anschließen to join, to connect; to fasten
anschließend subsequent
Anschluß *m* (Elektr) contact, (Klemme) terminal, (Leitung) connection, (Masch) coupling, (Techn) joint, junction
Anschlußbuchse *f* socket [for plug]
Anschlußdose *f* connecting box
Anschlußflansch *m* joining flange
Anschlußkabel *n* connecting cable
Anschlußkasten *m* outlet box
Anschlußklemme *f* terminal, terminal clamp
Anschlußleitung *f* connecting cable
Anschlußschraube *f* connecting screw
Anschlußstecker *m* connecting plug
Anschlußstück *n* connection
Anschlußstutzen *m* connecting piece, connection, connection branch
Anschlußverschraubung *f* connection screw
Anschlußwert *m* connected value, contact value
anschmauchen to smoke, to soot
Anschmelzherd *m* smelting furnace
anschmieden to forge on
anschmieren to smear
Anschnallen *n* strapping
Anschnitt *m* (Fläche) cutting face, (Metall) feed orifice, gate, (Techn) starting cut
anschrauben to screw on, to bolt on, to fasten with screws
Anschrift *f* address
anschüren to stir up
anschütten to pour on
anschwängern to impregnate, to saturate
Anschwängerung *f* impregnation, saturation
Anschwänzapparat *m* (Brau) sprinkler
anschwänzen (Brau) to sprinkle
anschwärzen to blacken, to smoke
anschwefeln to fumigate with sulfur, to treat with sulfur
anschweißen to weld [on]
Anschweißen *n* welding
Anschweißung *f* welding
anschwellen to swell, to increase
Anschwellprobe *f* swelling test, bulging test
Anschwemmfilter *m* precoat filter, settling filter, suspension filter
Anschwemmschicht *f* precoat
Anschwödemaschine *f* sulfide paint spraying machine
anschwöden (Gerb) to cleanse with lime water, to lime

Ansehen *n* reputation
ansehnlich considerable
Anserin *n* anserine, N-β-alanyl-1-methyl-histidine
ansetzbar attachable
ansetzen to establish; (anbauen) to attach, (Chem) to prepare, (Math) to arrange, (Rost etc.) to become covered with, (Schmutz) to deposit, (Termin) to set
Ansetzen *n* **eines Bades** setting a bath, ~ **von Kristallen** accretion (of crystals)
Ansetzung *f* (Rost etc.) accretion, deposition
Ansicht *f* opinion, view, ~ **von der Seite** side view, ~ **von oben** top view, ~ **von unten** bottom view, worm's eye view
ansieden to blanch, to boil
Ansieden *n* (Metall) hot galvanizing
Ansiedeprobe *f* (Chem) scorification test
Ansiedescherben *m* scorifier
Ansiedesilberbad *n* hot silvering bath
Ansiedeverfahren *n* hot galvanizing method
ansintern to sinter, to form sinter, to frit
anspalten to begin to cleave
anspannen to strain, to stress, to subject to tension, to tighten
Ansperrschieber *m* goggle valve
anspitzen to point, to sharpen
Ansporn *m* incentive
Ansprechempfindlichkeit *f* sensitivity
Ansprechvermögen *n* response, (Atom) counter efficiency
Ansprechverzögerungszeit *f* transient response
Ansprechwert *m* (Regeltechn) input resolution, resolution sensitivity (Am. E.) or sensitiveness (Br. E.)
Ansprechzeit *f* (Atom) sensitive time
Anspruch *m* claim, demand, **in** ~ **nehmen** to engage
anspruchslos modest, unassuming, unpretentious
anstählen to steel, to caseharden, to point with steel, to steel-face
Anstählen *n* steeling
anstandslos unhesitating, without hesitation
anstechen to pierce, to tap [off]
Anstechventil *n* priming valve
Ansteckblech *n* float plate
Ansteckdose *f* socket
anstecken (befestigen) to fasten, to stick on, (infizieren) to infect
ansteckend (Med) contagious, infectious
Ansteckungsgefahr *f* (Med) risk of infection
Ansteckungsherd *m* center of infection
Ansteckungskeim *m* germ
Ansteckungsquelle *f* source of infection
ansteigen to ascend, to increase, to mount, to rise
Anstellbottich *m* (Brau) pitching vessel
anstellen to start, (beschäftigen) to employ, to engage
Anstellhefe *f* store ferment, pitching yeast

Anstellmechanismus — Anthranilsäure

Anstellmechanismus *m* screw-down mechanism
Anstellschraube *f* positioning bolt
Anstelltemperatur *f* initial temperature, (Brau) pitching temperature
Anstellverstärker *m* positioner booster
Anstellwinkel *m* setting angle
Anstich *m* (Brau) broaching, tap
Anstichmethode *f* puncture method
Anstieg *m* elevation; increase, rise
Anstiegsleitfähigkeit *f* slope conductivity
anstiften to contrive
Anstoß *m* (Ärgernis) offence, (Anprall) collision; (Anregung) impulse, cause, impetus, initiative, (Elektr) trigger pulse, (Stoß) push
anstoßen to impinge [against], to push, to strike against
anstoßend (angrenzend) adjacent, contiguous
anstrahlen to illuminate, to irradiate
Anstrebekraft *f* centripetal force
anstreben to strive toward
anstreichen to paint, to coat, to color, to varnish
Anstreichgerät *n* painting tool
anstrengen to exert, to strain
Anstrich *m* paint coat, coat of paint; color, paint, painting, tint, **erster** ~ base coat, priming, **letzter** ~ finishing coat, top coat, **wasserfester** ~ waterproof painting, **wetterfester** ~ weatherproof painting
Anstrichaufbau *m* paint system
Anstrichfarbe *f* coating composition, coating compound, paint, **schalldämpfende** ~ acoustic paint, antinoise paint, sound absorbing paint, sound deadening paint
Anstrichfilm *m* paint film
Anstrichgerät *n* painting utensils, painter's tools
Anstrichmangel *m* paint film defect
Anstrichmittel *n* coating compound, coating material, painting material, varnish
Anstrichprüfgeräte *n pl* paint testing equipment
Anstrichsystem *n* painting system
anstücken to connect, to join
Ansud *m* boiling
ansüßen to edulcorate, to sweeten slightly
Antabus *n* antabuse
Antagonismus *m* (Med, Biol) antagonism
Antagonist *m* antagonist
antagonistisch antagonistic
Antarktis *f* Antarctic
antarktisch antarctic
anteeren to tar
Anteigemittel *n* pasting auxiliary
anteigen to paste
Anteil *m* component, constituent; fraction (rehtification); interest; part, portion, share
anteilig proportionate
Antenne *f* aerial, antenna
Antennenkabel *n* aerial cable
Anthanthron *n* anthanthrone
Anthelminthikum *n* (Pharm, Wurmmittel) anthelmintic, vermifuge
Anthemen *n* anthemene, octadecylene
Anthemol *n* anthemol, chamomile camphor
Anthemolcampher *m* anthemol, chamomile camphor
Antheraxanthin *n* antheraxanthin
Anthesterin *n* anthesterine
Anthochroit *m* (Min) anthochroite
Anthocyan *n* anthocyan
Anthocyanidin *n* anthocyanidin
Anthocyanin *n* anthocyanin
Anthophyllit *m* (Min) anthophyllite
Anthophyllitamphibol *m* anthophyllite amphibol
Anthosiderit *m* (Min) anthosiderite
Anthoxanthin *n* anthoxanthin
Anthracen *n* anthracene, anthracin, para-naphthalene
Anthracenblau *n* anthracene blue, alizarin blue
Anthracenfarbstoff *m* anthracene dye
Anthracenöl *n* anthracene oil
Anthracenpech *n* anthracene pitch
Anthrachinoketen *n* anthraquinoketene
Anthrachinolin *n* anthraquinoline
Anthrachinon *n* anthraquinone
Anthrachinonabkömmling *m* anthraquinone derivative
Anthrachinonacridin *n* anthraquinone acridine, naphthacridinedione
Anthrachinondisulfonsäure *f* anthraquinone disulfonic acid
Anthrachinonfarbstoff *m* anthraquinone dye
Anthrachinonfluorescein *n* anthraquinone fluoresceine
Anthrachinonsulfonsäure *f* anthraquinone sulfonic acid
Anthrachryson *n* anthrachrysone
Anthracyl- anthracyl-
Anthradichinon *n* anthradiquinone, anthracene tetrone
Anthraflavin *n* (Färb) anthraflavine
Anthraflavinsäure *f* anthraflavic acid, 2,6-dihydroxyanthraquinone
Anthraflavon *n* anthraflavone
Anthrafuchson *n* anthrafuchsone
Anthragallol *n* anthragallol
Anthrahydrochinon *n* anthrahydroquinone
Anthrakonit *m* (Min) anthraconite
Anthrakose *f* (Med) anthracosis, miner's lung
Anthraldehyd *m* anthraldehyde
Anthralur *n* (HN, Aktivkohle) anthralur (activated carbon)
Anthramin *n* anthramine, aminoanthracene, anthrylamine
Anthranil *n* anthranil, anthranilic acid lactam
Anthranilaldehyd *m* anthranil aldehyde, o-aminobenzaldehyde
Anthranilsäure *f* anthranilic acid, o-aminobenzoic acid

Anthranol *n* anthranol, 9-hydroxyanthracene
Anthranon *n* anthranone, anthrone, dihydroketoanthracene
Anthranyl- anthranyl-, anthryl-
Anthraphenon *n* anthraphenone
Anthrapurpurin *n* 1,2,7-trihydroxyanthraquinone, anthrapurpurin
Anthrapyridin *n* anthrapyridine
Anthrarobin *n* anthrarobin, 2,3,9-trihydroxyanthracene
Anthrarufin *n* 1,5-dihydroxyanthraquinone, anthrarufin
Anthrathiazin *n* anthrathiazine
Anthrazen *n* anthracene, para-naphthalene
Anthrazenöl *n* anthracene oil
Anthrazin *n* anthrazine
Anthrazit *m* (Min) anthracite
anthrazitartig anthracitic, anthracitoid
Anthrazitbildung *f* anthracitization
anthrazithaltig anthracitic
Anthrazithochofen *m* anthracite blast furnace
anthrazitisch anthracitic
Anthrazitkohle *f* anthracite
Anthrazitroheisen *n* anthracite pig iron
Anthrazyl- anthracyl-
Anthricin *n* anthricin, silicocolin
Anthrodianthren *n* anthrodianthrene
Anthroesäure *f* anthroic acid, anthracene carboxylic acid
Anthrol *n* anthrol, hydroxyanthracene
Anthron *n* anthrone, anthranone
Anthropobiologie *f* anthropobiology
anthropogen anthropogenic
Anthropogenese *f* anthropogenesis, anthropogeny, evolution of man
Anthropologe *m* anthropologist
Anthropologie *f* anthropology
Anthroxan *n* anthroxan, anthranil, o-aminobenzoic acid lactam
Anthroxanaldehyd *m* anthroxan aldehyde
Anthroxansäure *f* anthroxanic acid
Anthryl- anthryl-
Antiabsetzmittel *n* anti-settling agent, suspension agent
Antiadiposita *pl* (Pharm) antiadipogenics
Antiagglutinin *n* antiagglutinin
Antiaggressin *n* antiaggressin
Antialbumose *f* antialbumose
Antialdoxim *n* antialdoxime
Antialkoholikum *n* (Pharm) alcohol deterrent
Antiallergikum *n* (Pharm) antiallergic agent
Antiar *n* antiar
Antiarigenin *n* antiarigenin
Antiarin *n* antiarin
Antiarol *n* antiarol
Antiarose *f* antiarose
Antiautomorphismus *m* antiautomorphism
antibakteriell antibacterial, bactericidal

Antibase *f* anti-base
Antibindungszustand *m* antibonding state
Antibiose *f* (Med, Biol) antibiosis
Antibiotikum *n* (pl Antibiotika) antibiotic
antibiotisch antibiotic
Antibronze *f* patina
Antichlor *n* antichlor
Antideuteron *n* anti-deuteron
Antidiabetikum *n* (Pharm) antidiabetic
Antidiazotat *n* antidiazotate
Antidiuretikum *n* (Pharm) antidiuretic
Antidot *n* (Pharm) antidote
Antidröhnlack *m* acoustic paint, anti-noise paint, sound-absorbing paint, sound-deadening composition
Antidröhnmasse *f* antidrumming compound
Antiemetikum *n* (Pharm) antiemetic
Antiemulgierungsmittel *n* anti-skimming agent
Antienzym *n* antienzyme
Antiepileptikum *n* (Pharm) antiepileptic
Antifäulnisfarbe *f* anti-fouling composition
Antifebrin *n* antifebrin, acetanilide
Antiferroelektrika *pl* antiferroelectrics
antiferroelektrisch antiferroelectric
Antiferroelektrizität *f* antiferro electricity
Antiferromagnetismus *m* antiferro magnetism
Antifibrinolysin *n* AFL, antifibrinolysin
Antifibrinolysintest *m* AFT, antifibrinolysin test
Antiflußspatgitter *n* (Krist) antifluorite lattice
Antifriktionslager *n* antifriction bearing
Antifriktionsmetall *n* antifriction metal
Antifungin *n* antifungin, magnesium borate
antigen antigenic
Antigen *n* antigen, (bei Hautkrankheiten) dermatogen
Antigen-Antikörper-Reaktion *f* antigen-antibody reaction
Antigenität *f* antigenicity
Antigenizität *f* antigenicity
Antigenkörper *m* antigen
Antigorit *m* (Min) antigorite
Antigraufaktor *m* anti-gray [hair] factor, pantothenic acid
antihämorrhagisch antihemorrhagic
Antihautmittel *n* (Lack) antiskinning agent
Antihelix *f* antihelix, anthelix
Antihistamin *n* (Pharm) antihistamine
Antihistaminpräparat *n* (Pharm) antihistamine preparation
Antihormon *n* antihormone
Antiisotypie *f* antiisotypism
Antikamin *n* antikamine
Antikatalysator *m* anticatalyzer
antikatalytisch anticatalytic
Antikathode *f* anticathode, target
Antikathodenleuchten *n* anticathode luminescence
Antiklopfbrennstoff *m* antiknock fuel
Antiklopfeigenschaft *f* anti-knock property

Antiklopfmittel – Antimonpersulfid 58

Antiklopfmittel *n* (Mot) anti-knock agent, anti-knock compound
Antikoagulierungsmittel *n* anti-coagulant, anti-coagulin
Antikodon *m* anticodon
Antikörper *m* antibody
Antikörperbildung *f* antibody formation
Antikörperreaktion *f* antigen-antibody reaction
Antikoinzidenzanalysator *m* anticoincidence analyzer
Antikoinzidenzanordnung *f* anticoincidence arrangement
Antikoinzidenzstufe *f* anticoincidence stage
Antikonzeptionsmittel *n* (Pharm) contraceptive
Antikorrosionsmittel *n* anti-corrosive
Antikrackmittel *n* anticracking agent
Antiloch *n* antihole
antilog (antimer, enantiomer) antilog, antimer, enantiomer
antilogarithmisch (Math) antilogarithmic
Antilogarithmus *m* antilogarithm
Antiluetin *n* antiluetin, potassium ammonium antimonic bitartrate
Antilysin *n* antilysin
Antilysinwirkung *f* antilysis
antimagnetisch antimagnetic, non-magnetic
Antimaterie *f* (Phys) antimatter
antimer (Opt) antimer
Antimer *n* antimere
Antimetabolit *m* antimetabolite
anti-metrisch antimetric
anti-mikrobiell antimicrobial
Antimon *n* antimony, (obs) stibium (Symb. Sb)
Antimonalblei *n* antimonial lead
Antimonamalgam *n* antimony amalgam
Antimonammoniumfluorid *n* antimony ammonium fluoride, ammonium fluoantimonate
Antimonarsen *n* (Min) allemontite
Antimonarsenfahlerz *n* (Min) tetrahedrite
antimonartig antimonial
Antimonat *n* antimonate
Antimonbad *n* antimony bath
Antimonblei *n* antimonial lead
Antimonbleiblende *f* (Min) boulangerite
Antimonbleikupferblende *f* bournonite
Antimonbleispat *m* (Min) bindheimite
Antimonblende *f* (Min) antimony blende, kermesite, red antimony
Antimonblüte *f* (Min) antimony bloom, valentinite
Antimonbutter *f* (Antimon(III)-chlorid) antimony(III) chloride, butter of antimony
Antimonchlorid *n* antimony chloride
Antimonchlorür *n* (Antimon(III)-chlorid) antimonous chloride, antimony(III) chloride, antimony trichloride, butter of antimony
Antimondämpfe *pl* antimony vapors
Antimonelektrode *f* antimony electrode

Antimonerz *n* (Min) antimony ore
Antimonfahlerz *n* (Min) antimonial gray copper, tetrahedrite
Antimongelb *n* antimony yellow
Antimonglanz *m* (Min) antimony glance, antimonite
Antimonglas *n* glass of antimony
Antimongoldschwefel *m* gold-colored antimony sulfide, golden antimony sulfide
Antimonhalogen *n* antimony halide
antimonhaltig containing antimony, antimonial, antimoniferous
Antimoniat *n* antimonate
antimonig antimonial, antimonous
Antimonige Säure *f* antimonious acid
Antimonigsäureanhydrid *n* antimonous anhydride, antimony(III) oxide, antimony trioxide
antimonigsauer antimonite
Antimonin *n* antimonine, antimony lactate
Antimonit *m* (Min) antimonite
Antimonjodür *n* (Antimon(III)-jodid) antimony triiodide
Antimonkaliumsalz *n* antimony potassium salt
Antimonkaliumtartrat *n* potassium antimonyl tartrate, tartar emetic
Antimonkermes *m* (Min) kermes mineral
Antimonkupferglanz *m* (Min) antimonial copper glance, bournonite, chalcostibite, wolchite
Antimonlaktat *n* antimonine, antimony lactate
Antimonlegierung *f* antimony alloy
Antimonmetall *n* antimony metal, metallic antimony
Antimonnickel *n* (Min) antimonial nickel, breithauptite
Antimonnickelglanz *m* (Min) nickel stibine, ullmannite
Antimonnickelkies *m* (Min) antimonial nickel, nickel stibine, ullmannite
Antimonnitrat *n* antimony nitrate
Antimonocker *m* antimony ocher, cervantite, stibiconite
Antimonoxalat *n* antimony oxalate
Antimonoxid *n* antimony oxide
Antimonoxychlorid *n* antimony oxychloride, (Min) algaroth
Antimonoxychlorür *n* antimony oxychloride
Antimonpentachlorid *n* antimony pentachloride, antimonic chloride, antimony(V) chloride
Antimonpentajodid *n* antimony pentaiodide
Antimonpentasulfid *n* antimony pentasulfide, antimonic sulfide, antimony(V) sulfide
Antimonpentoxid *n* antimony pentoxide, antimonic oxide, antimony(V) oxide
Antimonperchlorid *n* antimony pentachloride, antimony(V) chloride
Antimonpersulfid *n* antimony(V) sulfide

Antimonpräparat n preparation containing antimony
Antimonregulus m antimony regulus
Antimonsäure f antimonic acid
Antimonsäureanhydrid n antimonic anhydride, antimony pentoxide
Antimonsafran m antimonial saffron, antimonous oxysulfide, crocus of antimony
Antimonsalz n antimony salt
antimonsauer antimonate
Antimonsesquioxid n antimony sesquioxide
Antimonsilber n (Min) antimonial silver, dyscrasite
Antimonsilberblende f (Min) pyrargyrite
Antimonsilberglanz m (Min) black antimonial silver glance, stephanite
Antimonspiegel m antimony mirror
Antimonsulfid n antimony sulfide
Antimonsulfür n (Antimon(III)-sulfid) antimonous sulfide, antimonious sulfide, antimony(III) sulfide, antimony trisulfide
Antimonsuperchlorid n antimony pentachloride
Antimonsupersulfid n antimony pentasulfide
Antimontannat n antimony tannate
Antimontetroxid n antimony tetroxide
Antimonthiosulfatchelat n antimonous thiosulfate chelate
Antimontrichlorid n antimonous chloride
Antimontrijodid n antimony triiodide, antimonous iodide, antimony(III) iodide
Antimontrioxid n antimonous oxide, antimony trioxide
Antimontrisulfid n antimony trisulfide
Antimonwasserstoff m antimony hydride, stibine
Antimonweiß n antimony white
Antimonyl- antimonyl-
Antimonylchlorid n antimonyl chloride
Antimonylkaliumtartrat n antimonyl potassium tartrate
Antimonyloxalat n antimonous oxalate, antimonyl oxalate
Antimonylrest m antimonyl radical
Antimonzinklegierung f antimonial zinc alloy
Antimonzinnober m (Min) antimonial cinnabar, kermesite
Antimutagen n antimutagen
Antimycin n antimycin
Antimykotikum n (Pharm) antimycotic, fungicide
antinarkotisch antinarcotic
Antinervin n antinervine
Antineuralgiebalsam m (Pharm) antineuralgic balsam
Antineuralgikum n antineuralgic
Antineutrino n antineutrino
Antineutron n (Atom) antineutron
Antinonnin n antinonnin, sodium-o-dinitro cresolate

Antinosin n antinosin, sodium tetraiodo phenolphthalein
Antiodorin n antiodorin
Antiosmose f antiosmosis
Antioxid n anoxide, rust preventative
Antioxydans n (pl Antioxydantien) antioxidant, age resistor
Antioxydationsmittel n antioxidant, antioxidizing agent
Antioxygen n antioxygen, age resistor
antiparallel antiparallel
Antiparasitikum n antiparasitic, parasiticide
Antipellagrafaktor m antipellagra factor
Antiperthit m antiperthite
Antiphein n antiphein
antiphlogistisch (obs) antiphlogistic
Antipode m (Chem) antipode, optically opposite form
Antipoden pl, **optische** ~ optical antipodes, antilogs, antimers
Antipodenpunkt m antipodal point
Antipodentrennung f optical resolution, resolution of optical forms
antipodisch antipodal
Antiprisma n antiprism
Antiproton n (Atom) antiproton
Antipyonin n antipyonin
Antipyretikum n (Pharm) antipyretic
antipyretisch antipyretic
Antipyrin n (HN) antipyrine
Antischaumemulsion f antifoaming emulsion
Antischaummittel n antifoaming agent, antifoam, antifrothing agent, defoamer, foam suppressor
Antischimmelmittel n anti-mildew agent
Antischrumpfbehandlung f anti-shrink process
Antisepsin n antisepsin, acetbromanilide
Antisepsis f (Med) antisepsis
Antiseptik f (Med) antisepsis
Antiseptikum n (Pharm) antiseptic
Antiseptin n antiseptin
antiseptisch antiseptic
Antiseptol n antiseptol
Antiserum n antiserum
Anti-Sigma-Minus-Hyperon n (Elementarteilchen, 2300fache Masse e. Elektrons) anti-sigma-minus-hyperon
antiskorbutisch antiscorbutic
Antispasmin n antispasmin, narcein sodium salicylate
Antispasmodikum n (Pharm) antispasmodic, spasmolysant
Antispastikum n antispastic
Antistatikum n anti-static agent
Antistatin n (HN, Textilhilfsmittel) antistatin
antistatisch antistatic
Antistatischmittel n antistatic agent
Antistellung f anticonfiguration
Antistokessche Linie f antistokes line

Antistreptolysin *n* antistreptolysin
Antisymmetrie *f* antisymmetry
antisymmetrisch antisymmetric
Antiteilchen *n* (Atom) antiparticle
Antitetanusserum *n* antitetanic serum
Antithermin *n* antithermin
Antithixotropie *f* anti-thixotropy
Antithrombin *n* antithrombin
Antithrombinfaktor *m* antithrombinic factor
Antithromboplastin *n* antithromboplastin
Antithrombosin *n* antithrombosin
Antitoxin *n* antitoxin
Antitoxineinheit *f* antitoxic unit, AU
Antiweinsäure *f* antitartaric acid, mesotartaric acid
Antiwuchsstoff *m* anti-growth substance
Antizentrum *n* anticenter
Antizyklon *m* (Meteor) anticyclone
Antodin *n* antodine
antönen to stain slightly
Antogorit *m* (Min) antogorite
Antophyllit *m* (Min) anthophylite
anträufeln to drip upon
antreffen to encounter, to find, to meet with
Antreibeholz *n* wood for refining furnaces
antreiben to accelerate, to actuate, to operate, to start
Antrieb *m* drive, driving gear; impulse, **regelbarer** ~ adjustable speed drive
Antriebhaspel *f* power take reel
Antriebkolben *m* driving plunger
Antriebmaschine *f* driving engine, driving machine
Antriebsachse *f* driving axle
Antriebsart *f* mode of driving
Antriebsdruck *m* driving pressure
Antriebskraft *f* driving force, driving power, motive power
Antriebsleistung *f* power input
Antriebsmaschine *f* driving engine, prime mover
Antriebsmittelpunkt *m* center of thrust
Antriebsmotor *m* driving motor
Antriebsreibungskupplung *f* friction starting clutch
Antriebsritzel *n* pinion
Antriebsscheibe *f* drive pulley
Antriebsschnecke *f* worm
Antriebsstrahl *m* propulsive jet
Antriebssystem *n* drive system
Antriebsteil *n* driving gear, thrust section
Antriebstrommel *f* head drum, drive pulley, head pulley
Antriebstrumm *n* drive reach
Antriebswelle *f* driving shaft, main shaft
Antriebswerk *n* driving mechanism
Antrimolith *m* (Min) antrimolite
antrocknen to begin to dry, to dry on
antröpfeln to drip upon
antropfen to drip upon

Antwerpenerblau *n* Antwerp blue
anvisieren to aim at, to sight at
Anvulkanisation *f* prevulcanization, precure, precuring, scorching
anvulkanisieren to precure, to prevulcanize
anwachsen to grow, to increase
Anwachsen *n* growth, increase, rise
anwärmen to warm up, to begin to heat
Anwärmen *n* warming up
Anwärmer *m* preheater, economizer, feed heater
anweichen to soak, to soften, to steep
anweisen to instruct
anweißen (tünchen) to whiten, to whitewash
Anweisung *f* instruction, direction, (Platz etc.) assignment
anwendbar applicable, adaptable, practical
Anwendbarkeit *f* applicability, practicability
anwenden to apply, to employ, to use
Anwendung *f* application, employment, use, utilization, **betriebliche** ~ technique
Anwendungsbeispiel *n* example of application, example of operation, sample application
Anwendungsbereich *m* field of application, range of application, range of uses
Anwendungsgebiet *n* field of application, range of uses
Anwendungsgrenze *f* limit of applicability
Anwendungsmöglichkeit *f* applicability
Anwendungsweise *f* mode of application
anwerfen to start, to throw on
anwesend present
Anwesenheit *f* presence
Anwitterung *f* oxidation, incrustation
Anwuchs *m* (junge Pflanzen) growth
Anwurf *m* (Bauw) rough cast, plaster[ing]
Anzahl *f* number, quantity
Anzahlfunktionen *pl* permutations and combinations
anzapfen (Brau) to tap
Anzapfstelle *f* tapping point; bleeding point
Anzapfstrom *m* bleeder current, leakage current
Anzapfturbine *f* tapping turbine
Anzapfung *f* tapping, bleeding
Anzapfventil *n* bleeder valve
Anzeichen *n* evidence, indication, sign
anzeichnen to indicate, to mark, to note
Anzeige *f* indication, (Gerät) reading, (Werbung) advertisement;, **alleinstehende** ~ (Werbung) solus ad, **Geschäfts-** ~ (Werbung) commercial ad, **gruppierte** ~ (Werbung) group ad, **Klein-** ~ (Werbung) classified ad, **kleine Text-** ~ (Werbung) short paragraph
Anzeigebereich *m* indicating range
Anzeigeempfindlichkeit *f* indication sensitivity
Anzeigefehler *m* instrument error
Anzeigegerät *n* indicating instrument, indicator gauge
Anzeigeglimmlampe *f* indicator glow lamp
Anzeigeinstrument *n* indicating instrument

anzeigen to indicate, to announce, to inform, to show
anzeigepflichtig notifiable, reportable
Anzeiger *m* indicator, gauge, pointer, registering or recording instrument
Anzeigeträgheit *f* indicator lag, inertia
Anzeigeverfahren *n* recording process
Anzeigevorrichtung *f* indicating device
Anziasäure *f* anzic acid
Anziehbolzen *m* clamping bolt
anziehen to attract, to absorb, to draw [in], to set (cement), (festigen) to tighten
Anziehung *f* attraction, adhesion, attractive force, **elektrostatische** ~ electrostatic attraction
Anziehungseffekt *m* effect of attraction
Anziehungsgesetz *n* law of attraction
Anziehungskraft *f* attractive force, adhesive power, attractive capacity
Anziehungspotential *n* attraction potential
Anziehungszahl *f* attraction constant
Anziehungszentrum *n* center of attraction
anzinnen to coat with tin
anzuckern to sugar
anzünden to light, to ignite, to kindle, to set on fire
Anzünden *n* lighting, igniting
Anzünder *m* igniter, lighter, lighting device
Anzündung *f* ignition, lighting
Anzugskraft *f* initial power, tractive force
Anzugsvermögen *n* pickup, (Auto) getaway power
Anzugsverzögerung *f* delay in starting
Anzugszeit *f* lap time
Apatelit *m* (Min) apatelite
Apatit *m* (Min) apatite, phosphorite
APC-Viren *pl* (Adenoviren) APC viruses
aperiodisch aperiodic
Aperiodizität *f* aperiodicity
Apertur *f* (Opt) aperture, opening
Aperturbegrenzung *f* (Opt) aperture limitation
Aperturblende *f* diaphragm
Apertur-Vorteil *m* (Spektr) „Jacquinot" advantage, „throughput" advantage
Apex *m* (Astr) apex
Apfeläther *m* malic ether
Apfelblattsauger *m* apple sucker
Apfelblütenstecher *m* (Zool) apple blossom weevil
Apfelbranntwein *m* apple brandy, apple jack
Apfeleisenextrakt *m* ferrated extract of apples
Apfellaus *f* (Zool) apple aphid
Apfelmost *m* sweet cider
Apfelöl *n* apple oil, amyl valerate
Apfelpresse *f* cider press
Apfelsägewespe *f* (Zool) apple sawfly
apfelsauer malate
Apfelsinensaft *m* orange juice
Apfelsinenschalenöl *n* orange peel oil, essence of orange
Apfelwein *m* cider
Apfelwickler *m* (Zool) codling moth
Aphanin *n* aphanin
Aphanit *m* (Min) aphanite, diorite
aphanithaltig aphanitic
aphanitisch aphanitic
Aphrit *m* (Min) aphrite
Aphrodisiakum *n* aphrodisiac
Aphrodit *m* aphrodite
Aphrosiderit *m* aphrosiderite
Aphthalose *f* aphthalose
Aphthonit *m* (Min) aphthonite
Aphtit *n* (Leg) aphtite
Aphyllidin *n* aphyllidine
Aphyllin *n* aphylline
Aphyllinalkohol *m* aphylline alcohol
Apigenidin *n* apigenidin
Apigenin *n* apigenin
Apiin *n* apiin
Apinol *n* apinol
Apiol *n* apiol, apiole
Apiolaldehyd *m* apiol aldehyde
Apiolsäure *f* apiolic acid
Apionol *n* apionol
Apiose *f* apiose
Aplanat *m* (Phot) aplanatic system
aplanatisch (Phot) aplanatic
Aplit *m* (Min) aplite
Aplom *m* (Min) aplome
Aplotaxen *n* aplotaxene
Apoaromadendron *n* apoaromadendrone
Apoaspidospermin *n* apoaspidospermine
Apoatropin *n* apoatropine, apatropine, atropamine
Apobornylen *n* apobornylene
Apocamphan *n* apocamphane
Apocamphenilon *n* apocamphenilone
Apocampher *m* apocamphor
Apocampholensäure *f* apocampholenic acid
Apocampholsäure *f* apocampholic acid
Apocardol *n* apocardol
Apocarnitin *n* apocarnitine
Apocarotinal *n* apocarotenal
Apochinin *n* apoquinine
Apocholsäure *f* apocholic acid
Apochromat *m* (Opt) apochromat lens
apochromatisch (Opt) apochromatic
Apocinchen *n* apocinchen
Apocinchonidin *n* apocinchonidine
Apocinchonin *n* apocinchonine
Apocitronellol *n* apocitronellol
Apocodein *n* apocodeine
Apoconessin *n* apoconessine
Apocyanin *n* apocyanine
Apocyclen *n* apocyclene
Apocynin *n* apocynin
Apoenzym *n* apoenzyme

Apoephedrin n apoephedrine
Apoerythralin n apoerythraline
Apogalanthamin n apogalanthamine
Apogelsemin n apogelsemine
Apoglucinsäure f apoglucic acid
Apogossypol n apogossypol
Apoharmin n apoharmine
Apoharmyrin n apoharmyrine
Apoholarrhenin n apoholarrhenine
Apohydrochinin n apohydroquinine
Apohyoscin n apohyoscine
Apoisoborneol n apoisoborneol
Apokaffein n apocaffeine
Apokodein n apocodeine
Apokrensäure f apocrenic acid
Apolysin n apolysin
Apomethylbrucin n apomethylbrucine
Apomorphimethin n apomorphimethine
Apomorphin n apomorphine
Apomorphinhydrochlorid n apomorphine hydrochloride
Aponal n aponal
Aponarcein n aponarceine
Apophedrin n apophedrin
Apophyllensäure f apophyllenic acid
Apophyllit m (Min) apophyllite
Apophyllitgruppe f group of apophyllites
Apopinen n apopinene
Apoplexie f (Med) apoplexy, apoplectic fit, stroke
Apopseudojonon n apopseudoionone
Aporeidin n aporeidine
Aporotaminsäure f aporotaminic acid
Aporphin n aporphine
Aposafranin n aposafranine
Aposafranon n aposafranone
Aposepedin n aposepedine
Aposorbinsäure f aposorbic acid
Apospermostrychnin n apospermostrychnine
Apoterramycin n apoterramycin
Apotheke f pharmacy, chemist's shop (Br E), drugstore (Am E)
Apotheker m pharmacist, dispensing chemist (Br E), druggist (Am E), pharmaceutical chemist
Apothekergewicht n apothecaries' weight
Apothekerkunst f apothecary's art, pharmaceutical science, pharmaceutics
Apothekerordnung f dispensatory
Apothekerwaage f druggist's scale
Apothekerwaren f pl drugs, pharmaceuticals
Apotheobromin n apotheobromine
Apothesin n apothesine
Apothiopyronin n apothiopyronine
Apotricyclol n apotricyclol
Apoyohimbinsäure f apoyohimbic acid
Apoyohimboasäure f apoyohimbic acid
Apparat m apparatus, device

Apparatebau m apparatus construction, construction of apparatus, manufacture of apparatus
Apparatebrett n instrument board, instrument panel
Apparateraum m instruments room
Apparateteile pl apparatus parts
Apparatewesen n engineering, equipment
apparativ pertaining to apparatus
Apparatklemme f wander plug
Apparatur f apparatus, device, equipment
Appetitzügler m anoretic, anorexigenic
Applikate f applicate, Z-coordinate
Applikatenachse f Z-axis
Applikation f (Med) application, injection, medication
applizieren to administer, to apply
appretieren to finish, to dress, to size
Appretieren n finishing, dressing
Appretur f finish, dressing, finishing, size
Appreturbad n finishing bath
appreturecht unaffected by finishing
Appreturleim m dressing size
Appreturmaschine f dressing machine, finishing machine
Appreturmasse f sizing material
Appreturmittel n finishing agent, chemical dressing medium, dressing agent
Appreturöl n dressing oil
Appreturverfahren n finishing process
Appreturzusatzmittel n chemical dressing agent
approbieren (Med) to qualify
a-p-Projektion f (Atom) AP projection
Approximation f im Mittel mean approximation
approximativ approximate
Aprikosenäther m apricot kernel oil, volatile oil of apricot kernels
Aprikosenessenz f apricot essence
Aprikosenkernöl n apricot kernel oil
aptieren to adapt, to adjust
Apyonin n apyonin, auramine
apyrisch apyrous, fireproof, incombustible
Apyrit m (Min) apyrite, rubellite
Apyron n apyron, lithium acetyl salicylate
Aqua destillata (Lat) distilled water, ~ **fortis** (Lat) aqua fortis, nitric acid, ~ **regia** (Lat) aqua regia, mixture of hydrochloric and nitric acid, nitro-hydrochloric acid
Aquamarin m (Min) aquamarine
Aquamycin n aquamycin
Aquarellfarbe f water color
Aquastat m (Tauchthermostat) aquastat
Aquocobalamin n aquocobalamin, hydroxocobalamin, vitamine $B_{12}a$
Aquoverbindung f aquo compound
Ar n (Maß) are (approx 119 square yards)
Arabin n arabin
Arabinal n arabinal
Arabingummi n gum arabic, gum acacia

Arabinose *f* arabinose
Arabinoson *n* arabinosone
Arabinsäure *f* arabic acid
Arabinzucker *m* arabinose
Arabischbalsam *m* balm of gilead, mecca balsam
Arabit *m* arabitol, arabite
Araboflavin *n* araboflavine
Arabonsäure *f* arabonic acid
Arabonylglycin *n* arabonylglycine
Arabulose *f* arabulose
Araburonsäure *f* araburonic acid
Arachidonsäure *f* arachidonic acid
Arachin *n* arachin
Arachinalkohol *m* arachic alcohol, eicosyl alcohol
Arachinsäure *f* arachic acid, arachidic acid
Arachisöl *n* arachis oil, peanut oil
Arachylalkohol *m* arachyl alcohol
Aräometer *n* areometer, hydrometer, (Dichtemesser) densimeter, (für Silbernitrat, Phot) argentometer
Aräometrie *f* areometry, hydrometry
aräometrisch areometric, hydrometric
Aräosaccharimeter *n* areosaccharimeter
Araeoxen *m* araeoxen
Aragonit *m* (Min) aragonite
Araroba *n* araroba, goa powder
Ararobinol *n* ararobinol
Arasan *n* arasan
Arbeit *f* job, work
arbeiten to operate, to work
Arbeiter *m* worker, laborer, operator, **angelernter** ~ semi-skilled worker, **gelernter** ~ skilled worker, **qualifizierter** ~ qualified worker, **ungelernter** ~ unskilled worker
Arbeitgeber *m* employer
Arbeitnehmer *m* employee
Arbeitsablauf *m* sequence of operations
Arbeitsäquivalent *n* mechanical equivalent
arbeitsam busy, industrious
Arbeitsamkeit *f* industriousness
Arbeitsanweisung *f* working instruction
Arbeitsanzug *m* overall, lab coat, work dress
Arbeitsaufwand *m* expenditure of work
Arbeitsausschuß *m* executive committee, working committee
Arbeitsband *n* continuous assembly belt
Arbeitsbedingung *f* operating condition, working condition
Arbeitsbelastung *f* working load, working pressure
Arbeitsbereich *m* range of operation, (Dest) turn-down ratio
Arbeitsbühne *f* working platform
Arbeitsdampf *m* working steam
Arbeitsdiagramm *n* working diagram
Arbeitsdruck *m* working pressure
Arbeitseinheit *f* unit of work

Arbeitseinsparung *f* labor-saving
Arbeitseinstellung *f* cessation of work, strike
Arbeitselektrode *f* operational electrode, working electrode
arbeitsfähig capable of working
Arbeitsfläche *f* working surface
Arbeitsflüssigkeit *f* working fluid, working liquid
Arbeitsgang *m* operation, procedure, working process, **in einem einzigen ~ auftragen** to apply in one pass
Arbeitsgeschwindigkeit *f* working speed, working velocity
Arbeitsgröße *f* amount of work, quantity of work
Arbeitsherd *m* hearth, reaction hearth, working hearth
Arbeitshub *m* working stroke
Arbeitshypothese *f* working hypothesis
Arbeitsinhalt *m* energy storage, potential energy, (Kolonne) hold-up
Arbeitsintensität *f* working intensity
Arbeitskapazität *f* work capacity
Arbeitskörper *m* working substance
Arbeitskolben *m* working piston
Arbeitskraft *f* working power
Arbeitskreis *m* (Elektr) load circuit, (Gruppe) working group
Arbeitskurve *f* (Atom) operating line
Arbeitsleistung *f* performance of work, (Techn) efficiency, working capacity
Arbeitslinie *f* operating line
Arbeitsmethode *f* working method
Arbeitsordnung *f* working regulations
Arbeitsraum *m* work room, (Chem) laboratory, (Werkstatt) workshop
Arbeitsschema *n* flow sheet
Arbeitsschutz *m* protective appliance, safety provisions
Arbeitsspannung *f* working voltage
arbeitssparend labor-saving
Arbeitsstrom *m* working current
Arbeitsstunde *f* man-hour
Arbeitsteilung *f* division of labor
Arbeitstisch *m* worktable, (Chem) laboratory table, (Techn) workbench
Arbeitstür *f* puddling door, (Schmelzofen) charging door
Arbeitsüberlastung *f* overwork, (Techn) overload
Arbeitsunterbrechung *f* interruption of work
Arbeitsverfahren *n* working method, working process, manufacturing process
Arbeitsverlust *m* loss of work
Arbeitsvermögen *n* working capacity, (Phys) [kinetic] energy
Arbeitsverrichtung *f* performance of work
Arbeitsvorbereitung *f* preparation[s] for work
Arbeitsvorgang *m* operation

Arbeitsvorschrift f procedure
Arbeitsweise f function; method of working, mode of operation; process
Arbeitswert m **der Wärme** mechanical equivalent of heat
Arbeitswiderstand m (Elektr) load resistance, working resistance
Arbeitswissenschaft f work studies
Arbeitszacken m fore plate
Arbeitszylinder m working cylinder
Arborescin n arborescin
Arboricin n arboricine
Arborin n arborine
Arbusterin n arbusterol
Arbutin n arbutin
Arcain n arcaine
Arcanit m (Min) arcanite, glaserite
Arcatomschweißung f atomic hydrogen arc welding
Arc-Funktion f (Math) inverse trigonometrical function
Archebiose f archebiosis, archegenesis
Archegenese f archegenesis, archebiosis
Archimedesschnecke f helix of Archimedes
Archimedisches Prinzip n (Phys) Archimedes' principle
Archiv n archives, records
Arctigenin n arctigenin
Arctiin n arctiin
Arctuvin n arctuvine, hydroquinone
ARC-Verfahren n auto-refrigerated cascade
Ardennit m (Min) ardennite
Ardometer n ardometer
Areafunktion f area function, (Math) inverse hyperbolic function
Areal n area
Arecaidin n arecaidine
Arecain n arecaine
Arecanuß f (Bot) areca nut, betel nut
Arecin n arecin
Arecolidin n arecolidine
Arecolin n arecoline
Arecolinhydrobromid n arecoline hydrobromide
Arecolon n arecolone
Arekanuß f (Bot) areca nut, betel nut
Arekolinhydrobromid n arecoline hydrobromide
Arekolinhydrochlorid n arecoline hydrochloride
Arendalit m (Min) arendalite
Arenobufagin n arenobufagin
Areolatin n areolatin
Areoxen n areoxene
Aretan n (HN, Naßbeizmittel) aretan
Arfvedsonit m (Min) arfvedsonite
Argal m argal, crude tartar
Argand-Diagramm n argand diagram
Argandlampe f argand lamp
Argentamin n argentamine
Argentan n argentan, german silver
Argentanband n argentan strap

Argentindruck m metal printing
Argentit m (Min) argentite, argyrite, silver glance
Argentobismuthit m (Min) argentobismutite
Argentometer n argentometer
Argentometrie f argentometry
argentometrisch argentometric
Argentopyrit m (Min) argentopyrite
Argentum n (Lat) argentum, silver
Argillit n (Leg) argillite
Arginase f (Biochem) arginase
Arginin n arginine
Argininbernsteinsäurekrankheit f argininosuccinic acidemia
Argininhydrochlorid n arginine hydrochloride
Arginobernsteinsäure f argino succinic acid
Arginylarginin n arginylarginine
Argochrom n argochrome
Argoflavin n argoflavine
Argon n (Symb. Ar) argon (Symb. A)
Argonalgleichrichter m argonal rectifier
Argonarc-Verfahren n argonarc welding process
Argon-Clathrat n argon-clathrate
Argongehalt m argon content
Argonhochstrombogen m high current argon arc
Argonin n argonin, silver caseinate
Argonschweißung f argon-arc welding
Argument n argument
Argyrescin n argyrescine
Argyrit m (Min) argyrite, argentite, silver glance
Argyrodit m (Min) argyrodite
Argyrofelt n (HN, Haarbeize) argyrofelt
argyrometrisch argyrometric
Argyropyrit m (Min) argyropyrite
Aribin n aribine, harman
Aricin n aricine, cusconine
Arin n arine
Aristochin n aristochin, aristoquin
Aristochinin n aristochinine
Aristol n aristol, thymol iodide
Aristolochiasäure f aristolochic acid
Aristolochin n aristolochine
Arit m arite
Arithmetik f arithmetics
arithmetisch arithmetic[ally]
Arizonit m (Min) arizonite
Arjunolsäure f arjunolic acid
Arkanit m (Min) arcanite
Arkansit m (Min) arkansite, brookite
Arkatomschweißen n atomic hydrogen welding
Arkose f (Geol) arcose
Arksutit m (Min) arcsutite
Arktis f arctic
arktisch arctic
Arktisit m arctisite
Arkusfunktion f (Math) function of arcus
Arkuskosinus m arc cosine, inverse cosine
Arkuskotangens m arc cotangent, inverse cotangent

Arkussekans *m* arc secant, inverse secant
Arkussinus *m* arc sine, inverse sine
Arkustangens *m* (Math) arc tangent, inverse tangent
arm poor [in], (Erz) low-grade, (Gas) lean, ~ **an Sauerstoff** oxygen-deficient, lean in oxygen
Arm *m* (eines Hebels) arm (of a lever), (Träger) arm, support bearer
Armalak *m* armalak
Armatur *f* fitting, mounting, (Magnet) armature
Armaturen *f pl* fittings, plumbing fixtures, ~ **für Betriebe** industrial fittings, ~ **für Labors** laboratory fittings, ~ **mit Auskleidungen** lined fittings
Armaturenbrett *n* instrument panel
Armblei *n* lead free from silver, refined lead
Armco-Eisen *n* (HN) Armco-iron, ingot iron
Armepavin *n* armepavine
Armerz *n* low-grade ore
Armgas *n* lean gas
Armhängelager *n* bracket drop hanger
armieren reinforce (concrete), (Kabel) to sheath
Armierung *f* armoring, reinforcement
Armierungsspachtel *f* reinforced surfacer
Armkran *m* hand crane
Armkreuz *n* spider
Armoxide *pl* (Blei) skimmings
Armprisonstift *m* arm dowel pin
Armreichschalter *m* mixture control
Armstein *m* (Min) poor silver ore
armtreiben to concentrate
Arnicin *n* arnicine
Arnika *f* (Bot) arnica
Arnikablüten *f pl* (Bot) arnica flowers
Arnikaextrakt *m* extract of arnica
Arnikakraut *n* (Bot) arnica, arnica leaves
Arnikaöl *n* arnica oil
Arnikatinktur *f* arnica tincture
Arnikawurzel *f* (Bot) arnica root, arnica rhizome
Arnimit *m* (Min) arnimite
Arnoldsche Probe *f* (Chem) Arnold's test
Arnotta *n* arnotta
Arobier *m* (Biol) aerobe, aerobian
Arochlor *n* arochlor
Aroma *n* aroma, fragrance, perfume, scent
Aromadendral *n* aromadendral
Aromadendren *n* aromadendrene
Aromadendrin *n* aromadendrine
Aromaentwicklung *f* development of flavor
Aromakonzentrat *n* flavoring essence
Aromat *m* (Chem) aromatic substance
aromatisch aromatic
aromatisieren to aromatize, to perfume, to scent
Aromatisierung *f* aromatization, perfuming, scenting
Aromolin *n* ar[o]moline
Aronsknolle *f* (Bot) arum root
Aronsstärke *f* arum starch
Aronswurzel *f* (Bot) arum root
Aroyl- aroyl-
Aroylbenzoesäure *f* aroylbenzoic acid
Arquerit *m* (Min) arquerite
Arrak *m* arrack
arretieren to arrest, to lock, to stop
Arretierschraube *f* locking screw
Arretierstift *m* locking pin
Arretierung *f* arrest, lock[ing], stopping device
Arretierungsknopf *m* locking button
Arretiervorrichtung *f* arresting device, locking device, stopping device
Arrhenal *n* arrhenal
Arrhenius-Gesetz *n* Arrhenius law
Arrhenius-Gleichung *f* Arrhenius equation
Arsacetin *n* arsacetin, acetylatoxyl, sodium p-acetylamino-phenylarsonate
Arsafluorinsäure *f* arsafluorinic acid
Arsalyt *n* arsalyte
Arsanilsäure *f* arsanilic acid, atoxylic acid, p-aminobenzene-arsonic acid
arsanilsauer arsanilate
Arsanthracen *n* arsanthracene
Arsanthren *n* arsanthrene
Arsanthrensäure *f* arsanthrenic acid
Arsanthridin *n* arsanthridine
Arsedin *n* arsedine
Arsen *n* arsenic (Symb. As), **metallisches** ~ arsenic metal, metallic arsenic
Arsenabstrich *m* arsenic skimmings
Arsenantimon *n* antimony arsenide, (Min) allemontite
Arsenantimonnickelglanz *m* (Min) corynite, korynite
Arsenat *n* arsenate, salt or ester of arsenic acid
Arsenbad *n* arsenic bath
Arsenbestimmung *f* arsenic determination
Arsenbestimmungsrohr *n* tube for arsenic determination
Arsenbleierz *n* (Min) mimetisite, mimetite
Arsenblende *f* arsenic blende, **rote** ~ red arsenic, realgar, red arsenic sulfide, ruby sulfur
Arsenblüte *f* (Min) arsenic bloom, arsenolite
Arsenbromid *n* (Arsen(III)-bromid) arsenic(III) bromide, arsenic tribromide
Arsencalcit *m* calcium arsenate
Arsenchlorid *n* (Arsen(III)-chlorid) arsenic(III) chloride, arsenic trichloride
Arsendampf *m* arsenic vapor
Arsendimethyl *n* arsenic dimethyl
Arsendisulfid *n* arsenic disulfide, (Min) arsenic ruby, realgar
Arseneisen *n* (Min) arsenical iron, iron arsenide, loellingite
Arsenerz *n* (Min) arsenic ore
Arsenfahlerz *n* (Min) tennantite
Arsenfleck *m* arsenic stain
arsenfrei arsenic-free, free from arsenic
Arsengehalt *m* arsenic content

Arsengitter *n* (Krist) arsenic lattice
Arsenhalogenid *n* arsenic halide
arsenhaltig arsenical, arsenic-containing
Arsenid *n* arsenide
arsenig arsenous, arsenious
Arsenige Säure *f* arsenous acid
Arsenigsäureanhydrid *n* arsenic(III) oxide, arsenic trioxide, arsenous acid anhydride, arsenous oxide, white arsenic
Arsenigsäuresalz *n* arsenite, salt of arsenous acid
arsenigsauer arsenite
Arsen(III)-oxid *n* arsenic trioxide, arsenous oxide
Arsenik *n* (Arsen(III)-oxid) arsenic(III) oxide, arsenic trioxide, arsenous oxide, white arsenic s. auch **Arsen**
Arsenikäscher *m* arsenic lime
Arsenikalien *pl* arsenicals
arsenikalisch arsenical
Arsenikantimon *n* (Min) allemontite, antimony arsenide
Arsenikbleispat *m* (Min) mimetite
Arsenikblüte *f* arsenic bloom
Arsenikblume *f* (Min) arsenolite
Arsenikbromid *n* arsenic(III) bromide, arsenic tribromide, arsenous bromide
Arsenikbutter *f* arsenic butter, arsenous chloride
Arsenikeisen *n* arsenical iron, iron arsenide, loellingite
Arsenikerz *n* (Min) arsenic ore
Arsenikfahlerz *n* (Min) tennantite
Arsenikgegengift *n* arsenic antidote
arsenikhaltig containing arsenic, arsenical, arseniferous
Arsenikhütte *f* arsenic works
Arsenikjodür *n* (Arsen(II)-jodid) arsenic diiodide, arsenic(II) iodide
Arsenikkalk *m* arsenolite, calcium arsenate
Arsenikkies *m* (Min) arsenical pyrites, arsenopyrite
Arsenikkobalt *n* (Min) native iron cobalt arsenide, safflorite, skutterudite
Arsenikkobaltkies *m* (Min) native iron cobalt arsenide, safflorite, skutterudite
Arsenikkobaltoxid *n* cobalt arsenite, cobaltic arsenite
Arsenikkupfer *n* copper arsenide
Arseniknickel *n* (Min) niccolite, nickel arsenide
Arseniknickelkies *m* (Min) chloanthite, leucopyrite
Arsenikpräparat *n* arsenical preparation
Arsenikrubin *m* (Min) realgar
Arsenik s. a. **Arsen**
Arsenikschwarz *n* arsenic black
Arseniksilber *n* (Min) arsenical silver, silver arsenide
Arseniksilberblende *f* (Min) proustite
Arseniksinter *m* (Min) scorodite
Arsenikverbindung *f* arsenic compound
Arseniopleit *m* (Min) arseniopleite

Arseniosiderit *m* (Min) arseniosiderite
Arsenit *n* arsenite, salt of arsenous acid
arseniziert arsenic cured
Arsenjodid *n* (Arsen(III)-jodid) arsenic(III) iodide, arsenic triiodide, arsenous iodide
Arsenjodür *n* (Arsen(II)-jodid) arsenic diiodide, arsenic(II) iodide
Arsenkies *m* (Min) arsenical pyrites, arsenopyrites, mispickel
Arsenkobalt *n* (Min) native iron cobalt arsenide, safflorite, skutterudite
Arsenkupfer *n* copper arsenide, (Min) domeykite
Arsenlegierung *f* arsenic alloy
Arsenmetall *n* arsenic metal, metallic arsenic
Arsenmetallglanz *m* (Min) gersdorffite
Arsennickel *n* nickel arsenide
Arsennickelglanz *m* (Min) gersdorffite
Arsennickelkies *m* (Min) natural nickel arsenide, niccolite
Arsenobenzoesäure *f* arsenobenzoic acid
Arsenobenzol *n* arsenobenzene
Arsenoferrit *m* (Min) arsenoferrite
Arsenohippursäure *f* arsenohippuric acid
Arsenolamprit *m* (Min) arsenolamprite
Arsenolith *m* (Min) arsenolite
Arsenomethylalkohol *m* arsenomethanol
Arsenophenol *n* arsenophenol, dihydroxyarsenobenzene
Arsenopyrit *m* (Min) arsenopyrite, mispickel
Arsenostibinobenzol *n* arsenostibinobenzene
Arsenostovain *n* arsenostovaine
Arsenpentachlorid *n* arsenic pentachloride
Arsenpentafluorid *n* arsenic pentafluoride, arsenic(V) fluoride
Arsenpentasulfid *n* arsenic pentasulfide
Arsenpentoxid *n* arsenic pentoxide, arsenic acid anhydride, arsenic(V) oxide
Arsenpräparat *n* arsenical preparation
Arsenprobe *f* test for arsenic
Arsenrohr *n* arsenic tube
Arsenrot *n* red arsenic
Arsenrotgültigerz *n* (Min) light red silver ore, proustite
Arsenrubin *m* (Min) realgar
Arsensäure *f* arsenic acid
Arsensäureanhydrid *n* arsenic [acid] anhydride, arsenic pentoxide, arsenic(V) oxide
Arsensalz *n* arsenic salt
arsensauer arsenate
Arsensilber *n* (Min) huntilite, silver arsenide
Arsensilberblende *f* (Min) arsenical silver glance, arsenic silver blende, proustite
Arsentribromid *n* arsenic tribromide, arsenous bromide
Arsentrichlorid *n* arsenic trichloride, arsenic butter, arsenic(III) chloride, arsenous chloride

Arsentrioxid *n* arsenic(III) oxide, arsenic trioxide, arsenous acid anhydride, arsenous oxide, white arsenic
Arsentrisulfid *n* arsenic(III) sulfide, arsenic trisulfide, arsenous sulfide
Arsentrisulfidhydrat *n* hydrated arsenic trisulfide
Arsenür *n* (obs) arsenide
Arsen(V)-oxid *n* arsenic(V) oxide, arsenic acid anhydride, arsenic pentoxide
Arsenwasserstoff *m* arsine, arsenic hydride, arsenous hydride, hydrogen arsenide
Arsindolin *n* arsindoline
Arsphenamin *n* arsphenamine, 3,3-diamino-4,4-dihydroxylarsenobenzene, salvarsan
Arsthinol *n* arsthinol
Arsulin *n* arsuline
Arsulolidin *n* arsulolidine
Arsulolin *n* arsuloline
Art *f* manner; nature, way, (Biol) species, (Gattung) category, class, kind, sort, type
Artabotrin *n* artabotrine
Artabsin *n* artabsin
Artamin *n* artamine
Artarin *n* artarine
Artebufogenin *n* artebufogenin
Artefakt *n* artefact, artifact
arteigen characteristic
Artemazulen *n* artemazulene
Artemetin *n* artemetin, artemisetin
Artemisetin *n* artemisetin, artemetin
Artemisiaketon *n* artemisia ketone
Artemisin *n* artemisin
Artemissäure *f* artemisic acid
Arterenol *n* arterenol
Arterenon *n* arterenone
Arterhaltung *f* (Biol) preservation of the species, propagation of the species
Arterhaltungstrieb *m* preservation of the species
Arterie *f* (Med) artery
arteriell arterial
Arterienblut *n* arterial blood
Arteriengewebe *n* (Med) arterial tissue
Arterienverkalkung *f* (Med) arteriosclerosis, atherosclerosis
Arterienwand *f* (Med) arterial wall
Arteriole *f* (Med) arteriole
Arteriosklerose *f* (Med) arteriosclerosis, atherosclerosis
artfremd alien, foreign, heterozoic
Artgewicht *n* specific weight
artgleich homogeneous, homozoic, identic, of the same species
Arthranitin *n* arthranitin, cyclamin
Arthritis *f* (Med) arthritis
Arthritismittel *n* (Pharm) antarthritic
Artkreuzung *f* (Biol) cross-breeding
Artocarpetin *n* artocarpetin

Artosin *n* artosine
artspezifisch species-specific, specific to the species
Artspezifität *f* species specificity
Artunterschied *m* species difference
Artwärme *f* specific heat
Arubren *n* (HN, Gummiindustrie) arubren
Arum *n* arum
Aryl- aryl-
Arylamin *n* arylamine
Arylbuttersäure *f* arylbutyric acid
Arylhalogenid *n* aryl halide
arylieren to arylate
Arylierung *f* arylation
Arylmethylbenzoesäure *f* arylmethylbenzoic acid
Arylpropionsäure *f* arylpropionic acid
Arylsilicon *n* aryl silicone
Arylsulfonat *n* aryl sulfonate
Arznei *f* drug, medicament, medicine
Arzneibuch *n* dispensatory, pharmacop[o]eia
Arzneifläschchen *n* vial
Arzneiflasche *f* medicine bottle
Arzneiformel *f* medical formula
Arzneigewicht *n* officinal weight
Arzneikapsel *f* (Pharm) medicine capsule
Arzneikugel *f* bolus
Arzneikunde *f* pharmacy
Arzneimittel *n* drug, pharmaceutical
Arzneimittelforschung *f* pharmacological research
Arzneimittelgesetzgebung *f* pharmaceutical legislation
Arzneimittelgewöhnung *f* drug tolerance
Arzneimittelindustrie *f* pharmaceutical industry
Arzneimittelkunde *f* pharmacology, pharmaceutics
Arzneimittellehre *f* pharmacology, pharmaceutics
Arzneimittelmißbrauch *m* drug abuse
Arzneimittelresistenz *f* drug resistance
Arzneimittelstandardisierung *f* standardization of drugs
Arzneimittelsucht *f* drug addiction, pharmacomania, pharmacophilia
Arzneimittelsynthese *f* drug synthesis
Arzneimittelträger *m* excipient, menstruum
Arzneimittelvergiftung *f* drug intoxication, drug poisoning
Arzneipflanze *f* medical plant, medicinal plant
Arzneiwissenschaft *f* pharmacology
Arzneizubereitung *f* dispensing
Arzrunit *m* (Min) arzrunite
Arzt *m* doctor, medical man, physician
A-Säure *f* A acid
Asa Foetida *f* (Pharm) asafetida, asafoetida
Asafötidaöl *n* asafetida oil, asafoetida oil
Asant *m* (Pharm) asafetida
Asantöl *n* asafetida oil

Asaprol *n* asaprol, abrastol, calcium-β-naphthol-α-sulfonate
Asarin *n* asarin, asarone
Asarinin *n* asarinin
Asaron *n* asarone, asarin, propenyl-2,4,5-trimethoxybenzene
Asaronsäure *f* asaronic acid, 2,4,5-trimethoxybenzoic acid
Asarumcampher *m* asarum camphor, asarone
Asarumöl *n* asarum oil
Asaryl- asaryl-
Asarylsäure *f* (Asaronsäure) asarylic acid
Asbest *m* asbestos; amianthus, mountain cork, stone flax
Asbestader *f* vein of asbestos
Asbestanzug *m* asbestos suit
asbestartig asbestoid, asbestiform, asbestous
Asbestaufbereitung *f* dressing of asbestos
Asbestaufschlämmung *f* asbestos milk, asbestos suspension
Asbestbekleidung *f* asbestos covering
Asbestdichtung *f* asbestos joint
Asbestdrahtnetz *n* asbestos-coated wire gauze, asbestos wire gauze, asbestos wire net
Asbesteinlage *f* asbestos layer
Asbestfaser *f* asbestos fiber
Asbestfaserbündel *n* asbestos fiber bundle
Asbestfaserplatte *f* asbestos fiber sheet
Asbestfilter *n* asbestos filter
Asbestflocken *f pl* asbestos wool
Asbestfüllstoff *m* asbestos filler
Asbestgewebe *n* asbestos fabric, asbestos cloth
Asbestgrube *f* asbestos quarry
Asbesthandschuh *m* asbestos glove
Asbestin *n* asbestin
Asbestisolierplatte *f* asbestos insulating plate
Asbestlager *n* asbestos bed
Asbestnetz *n* asbestos fabric, asbestos cloth
Asbestofen *m* asbestos stove
Asbestolith *m* short-fibered asbestos
Asbestpackung *f* asbestos packing, asbestos jointing
Asbestpapier *n* asbestos paper
Asbestpappe *f* asbestos board
Asbestplatte *f* asbestos sheet
Asbestring *m* asbestos [jointp] ring
Asbestscheibe *f* asbestos washer
Asbestschicht *f* asbestos layer, (Geol) asbestos stratum
Asbestschichtstoff *m* laminate glass fabric
Asbestschiefer *m* asbestos slate
Asbestschirm *m* asbestos screen
Asbestschnur *f* asbestos cord
Asbestschwamm *m* asbestos sponge
Asbestwaren *pl* asbestos goods
Asbestwicklung *f* asbestos wrapping
Asbestzement *m* asbestos cement
Asbolan *m* (Min) asbolane, asbolite, earthy cobalt

Asbolin *n* asbolin
Ascaridinsäure *f* ascaridic acid, ascaridolic acid
Ascaridol *n* ascaridol
Ascarosid *n* ascaroside
Ascarylit *m* ascarylitol
Ascarylose *f* ascarylose
Aschan-Dichlorid *n* Aschan's dichloride
Ascharit *m* (Min) ascharite
Aschblau *n* zaffer
Aschblei *n* blacklead, native bismuth
Asche *f* cinders, (Chem) ash, ashes, (Techn) sullage
aschefrei ashless, free from ashes
Aschegehalt *m* ash content, percentage of ash
Aschegehaltsbestimmung *f* ash determination
aschen to ash
Aschen *n* ashing
Aschenabfuhr *f* removal of ashes
aschenarm poor in ash content
Aschenauswerfer *m* ash ejector
Aschenbad *n* ash bath
Aschenbestandteil *m* ash constituent
Aschenbestimmung *f* ash determination
Aschenermittlung *f* determination of ash
aschenfleckig sullage-spotted
Aschengrube *f* ash pit
Aschenhalde *f* ash heap, slag heap
Aschenkasten *m* (Ofen) ash pan
Aschenlauge *f* lye from ashes
Aschenofen *m* (Glasherstellung) ash furnace
Aschenraum *m* ash pit
aschenreich rich in ash content
Aschensack *m* ash pocket
Aschenverflüssigung *f* fusing of the ashes (clinkering)
Aschenzacken *m* back plate
aschereich rich in ash
aschfarben ash-colored, ashy
aschig ashy, cineraceous
Aschkern *m* hearth ashes containing silver
Ascinin *n* (HN, Lackhilfsstoff) ascinin
Asclepiol *n* asclepiol
Ascorbigen *n* ascorbigen
Ascorbinsäure *f* ascorbic acid
Ascorbinsäuredehydrogenase *f* (Biochem) ascorbic acid dehydrogenase
Ascorbinsäureoxydase *f* (Biochem) ascorbic acid oxidase
Ascorbylpalmitat *n* ascorbylpalmitate
Ascosterin *n* ascosterol
Asduana *f* asduana
Asebogenin *n* asebogenin
Asebotin *n* asebotin
Asellinsäure *f* asellic acid
Asepsis *f* (Med) asepsis
Aseptik *f* (Med) aseptic technique, asepticism
Aseptin *n* (Med) aseptine
aseptisch aseptic
Aseptol *n* aseptol, ortho-phenolsulfonic acid

Asiaticosid *n* asiaticoside
Asiatsäure *f* asiatic acid
Asiderit *m* (Min) asiderite
Asimina *f* asimina
Asmanit *m* (Min) asmanite, tridynite
Asordin *n* (HN, Lösungsmittel) asordin
Asparacemsäure *f* asparacemic acid, dl-aspartic acid
Asparagin *n* asparagine, amino-succinamic acid, asparamide, aspartamic acid
Asparaginase *f* (Biochem) asparaginase
Asparaginsäure *f* aspartic acid, aminosuccinic acid, asparaginic acid
asparaginsauer aspartate
Asparamid *n* asparagine, asparamide
Aspartal *n* aspartal
Aspartase *f* (Biochem) aspartase
Aspartat *n* aspartate
Aspasiolith *m* (Min) aspasiolite
Aspekt *m* aspect
Aspergillsäure *f* aspergillic acid
Aspergillus *m* (Pilz) aspergillus
Asperolith *m* asperolite
Asperthecin *n* asperthecin
asphärisch aspherical
Asphalin *n* asphalin
Asphalit *m* asphalite
Asphalt *m* asphalt, asphaltum, bitumen, mineral pitch
Asphaltanlage *f* bitumen plant
asphaltartig asphaltic, bituminous
Asphaltbeton *m* asphalt concrete
Asphaltbrei *m* asphalt pulp
Asphaltdraht *m* asphalted wire, compounded wire
Asphaltfeinbeton *m* fine asphaltic concrete
Asphaltfilz *m* asphalt-impregnated felt
Asphaltfirnis *m* asphalt varnish
Asphaltgestein *n* crude asphalt
asphalthaltig asphalt bearing, asphaltic, bituminous
asphaltieren to asphalt, to coat with asphalt
Asphaltieren *n* asphalting
asphaltisch asphaltic
Asphaltisolierfilz *m* insulating asphalt felt
Asphaltkitt *m* asphalt cement, asphalt mastic
Asphaltlack *m* asphalt varnish, bituminous varnish
Asphaltmastix *m* asphaltic cement, asphalt mastic, bituminous mastic
Asphaltnegativ *n* asphaltotype
Asphaltpapier *n* asphalt-laminated kraft paper
Asphaltpappe *f* asphalted paper
Asphaltpech *n* bituminous pitch
Asphaltprüfgerät *n* asphalt testing apparatus
Asphaltstein *m* native asphalt
Asphaltverfahren *n* bitumen process
Asphaltzement *m* asphaltic cement
Asphodill *m* (Bot) asphodel

Asphyxie *f* (Med) asphyxia
Aspidin *n* aspidin
Aspidinol *n* aspidinol
Aspidocarpin *n* aspidocarpine
Aspidolith *m* aspidolite
Aspidosamin *n* aspidosamine
Aspidosin *n* aspidosine
Aspidospermin *n* aspidospermine
Aspirator *m* aspirator
aspirieren to aspire
Aspirin *n* (HN) aspirin, acetylsalicylic acid
Asplit *n* (HN, Kunstharzkitte) asplit (synthetic resin cements)
Assamar *n* assamar
Assimilation *f* (Biol) assimilation
Assimilationsprozeß *m* assimilation [process], assimilative process
assimilatorisch assimilatory
assimilierbar assimilable, capable of being assimilated
Assimilierbarkeit *f* assimilability
assimilieren to assimilate
Assistent *m* assistent
assortieren to assort, to sort
assouplieren to boil partially, to render pliable
Assoziation *f* combination, (Chem) association
Assoziationsgrad *m* degree of association
assoziieren to associate
Astacein *n* astacein
Astacin *n* astacin
Astat *n* astatin[e]
Astatin *n* astatin[e]
astatisch (Magnet) astatic, (unstetig) not stable
Astatisierung *f* astatization
Asteran *n* asterane
Asterin *n* asterine
Asterismus *m* (Phys) asterism
Asterrsäure *f* asterric acid
Asterubin *n* asterubin
Asthenosphäre *f* asthenosphere
astigmatisch astigmatic
Astigmatismus *m* astigmatism
Astilbin *n* astilbin
Astragalin *n* astragalin
Astragelb *n* astra yellow
Astrakanit *m* (Min) astrakanite, astrochanite
Astralit *m* astralite
Astraphloxin *n* astraphloxine
Astrol *n* astrol
Astronom *m* astronomer
Astronomie *f* astronomy
Astrophyllit *m* (Min) astrophyllite
Astrophysik *f* astrophysics
Astropyrin *n* astropyrine
Astrozyt *m* astrocyte, spider cell
Astrozytom *n* astrocytoma
Asulgan *n* (HN, Pelzindustrie) asulgan (fur industry)
Asurol *n* asurol

Asymmetrie *f* asymmetry, **atomare** ~ (Stereochem) atomic asymmetry, ~ **infolge Rotationsbehinderung** (Stereochem) asymmetry due to restricted rotation, **konformative** ~ (Stereochem) conformational asymmetry, **molekulare** ~ (Stereochem) molecular asymmetry
asymmetrisch asymmetric
Asymptote *f* (Math) asymptote
Asymptoten-Abbildung *f* asymptotic image
Asymptoten-Brennweite *f* asymptotic focal length
Asymptotik *f* asymptotic behavior
asymptotisch asymptotic
asynchron asynchronous, non-synchronous
Asynchronmotor *m* asynchronous motor, induced motor, induction motor
Atakamit *m* (Min) atacamite, copper oxychloride, remolinite
ataktisch atactic
Atebrin *n* (HN, Pharm) atebrin, atabrine
Atelestit *m* atelestite
Atelit *m* (Min) atelite
Atemfilter *m* respiration filter
Atemfrequenz *f* respiratory frequency
Atemgift *n* respiratory poison
Atemluft *f* breathable air
Atemmaske *f* oxygen mask
Atemnot *f* shortness of breath
Atemschlauch *m* [folding] respirator hose
Atemschutzgerät *n* gas mask, (Techn) respirator
Atemweg *m* respiratory tract
Atemzug *m* breath
Atephen *n* (Kunstharz) atephen
Athamantin *n* athamantin
Atheriastit *m* atheriastite
atherman athermanous
athermisch athermal
Atidin *n* atidine
Atisin *n* atisine
AT-Kristallschnitt *m* (Krist) AT-cut crystal
Atlanton *n* atlantone
Atlasblau *n* atlas blue
Atlaserz *n* (Min) false emerald, fibrous malachite
Atlasgips *m* fibrous gypsum
atlasglänzend satiny
Atlasit *m* (Min) atlasite
Atlaspapier *n* glazed paper, satin paper
Atlasspat *m* (Min) calcite, satin spar
Atlasstein *m* satin spar
Atlasvitriol *n* (Min) white vitriol
Atmen *n* respiration, ~ **eines Werkzeuges** swelling of a mold
Atmolyse *f* atmolysis
Atmosphäre *f* atmosphere
Atmosphärendruck *m* atmospheric pressure
Atmosphärenrückstand *m* residual atmosphere
Atmosphärilien *pl* atmospheric substances, atmospheric influences
atmosphärisch atmospheric
Atmung *f* breathing, respiration
Atmungsferment *n* (Biochem) respiratory enzyme
Atmungsgeschwindigkeit *f* respiration rate
Atmungshemmung *f* inhibition of respiration
Atmungskette *f* (Biochem) respiratory chain
Atmungszentrum *n* respiratory center
Atom *n*, **angeregtes** ~ excited atom, **geladenes** ~ charged atom, **gespaltenes** ~ disintegrated atom, **hochionisiertes** ~ stripped atom
Atomabsorptionsanalyse *f* atomic absorption analysis
Atom-Absorptionsspektrometrie *f* atomic absorption spectrometry
Atomabstand *m* interatomic distance
Atomaffinität *f* atomic affinity, atomic valence
Atomantrieb *m* atomic propulsion
Atomaufspaltung *f* atom splitting
Atombau *m* atomic structure
Atombeschießung *f* atomic bombardment
Atombeschuß *m* atomic bombardment
atombetrieben atomic-powered
Atombewegung *f* atomic motion, movement of atoms
Atombindung *f* atomic bond, atomic linkage
Atombindungskraft *f* atomic combining power
Atombombe *f* atomic bomb
Atombrennstoff *m* atomic fuel
Atomchemie *f* atomic chemistry
Atomdichte *f* atomic density
Atomdispersion *f* atomic dispersion
Atomdurchmesser *m* atomic diameter
Atomeigenstrahlung *f* characteristic atom radiation
Atomenergie *f* nuclear energy
Atomenergieerzeugung *f* atomic energy generation
Atomerregung *f* atom excitation
Atomexplosion *f* atomic explosion, atomic burst
Atomflugmotor *m* atomic aeroengine
Atomformfaktor *m* atomic scattering factor
Atomforscher *m* atomic scientist
Atomforschung *f* atomic research
Atomfrequenz *f* atomic frequency
Atomgebiet *n* field of atomic energy
Atomgeschoß *n* atomic projectile
Atomgeschoßteilchen *n* atomic bombardment particle
Atomgewicht *n* atomic weight
Atomgewichtsbestimmung *f* determination of atomic weight
Atomgewichtstabelle *f* table of atomic weights
Atomgewichtstafel *f* atomic chart, table of atomic weights

Atomgitter *n* atomic lattice, atomic space lattice
Atomgramm *n* gram atom
Atomhalbmesser *m* atomic radius
atomhaltig atomic, atomiferous
Atomhaufen *m* cluster of atoms
Atomhülle *f* atomic envelope, atomic shell
Atomhüllenniveau *n* atomic level
Atomigkeit *f* atomicity
atomisch atomic
atomisieren to atomize
Atomisierung *f* atomization
Atomisierungswärme *f* heat of atomization
Atomismus *m* atomicity, atomic structure
Atomistik *f* atomicity, atomistics
atomistisch atomistic, atomic
Atomizität *f* atomicity
Atomkalotte *f* stereomodel, Stuart model
Atomkanone *f* atom gun
Atomkern *m* atomic nucleus, atom nucleus, (ohne Elektronen) stripped atom, nuclear atom
Atomkerndichte *f* nuclear density
Atomkerne *pl* **aufbauen** to build up the nuclei of atoms
Atomkernform *f* nuclear shape
Atomkerngröße *f* nuclear size
Atomkernspaltung *f* nuclear fission
Atomkernsprengung *f* atomic nucleus explosion
Atomkraft *f* atomic power, nuclear power
Atomkraftanlage *f* atomic power plant, nuclear power station
Atomkraftbändiger *m* harnesser of atomic energy
Atomkraftmaschine *f* atomic energy machine
Atomkraftstation *f* atomic power station
Atomkraftwerk *n* atomic power plant, nuclear power station
Atomkrieg *m* nuclear war
Atomladung *f* atomic charge
Atomlage *f* atomic position
Atomleitfähigkeit *f* atomic conductance
Atommagnetismus *m* atomic magnetism
Atommasse *f* atomic mass
Atommasseeinheit *f* atomic mass unit
Atommassenkonstante *f* atomic mass constant
Atommedizin *f* atomic medicine
Atommeiler *m* nuclear pile
Atommodell *n* atomic model, space filling atom model
Atommüll *m* radioactive waste
Atommüllbeseitigung *f* removal of atomic waste
Atomnummer *f* (Chem) atomic number
Atomorbital *n* atomic orbital
Atomphysik *f* nuclear physics, atomics
Atomphysiker *m* nuclear physicist
Atomradius *m* atomic radius
Atomreaktor *m* atomic pile, nuclear reactor
Atomrefraktion *f* atomic refraction
Atomregen *m* radioactive fall-out
Atomrest *m* atomic residue
Atomring *m* ring of atoms
Atomrotation *f* atomic rotation
Atomrumpf *m* atomic residue
Atomschale *f* atomic shell
Atomschlacke *f* radioactive waste
Atomschmelzwärme *f* atomic heat of fusion
Atomschwingung *f* atomic oscillation
Atomspaltung *f* atom fission
Atomspektrum *n* atomic spectrum
Atomsprengkopf *m* atomic warhead
Atomsprengstoff *m* atomic explosive
Atomstaub *m* atomic dust, radioactive fall-out
Atomstrahl *m* atomic ray
Atomstrahlung *f* atomic radiation
Atomstrahlungsschäden *pl* injury caused by atomic radiation
Atomstruktur *f* atomic structure
Atomsynthese *f* atomic synthesis
Atomtechnik *f* nuclear technology
Atomtheorie *f* atom theory
Atom-U-Boot *n* atomic-powered submarine
Atomuhr *f* atomic clock, atomic timing device
Atomumwandlung *f* atom conversion, atomic transformation
Atomvalenz *f* atomic valence
Atomverband *m* union of atoms, atomic union
Atomverhältnis *n* atomic ratio
Atomverkettung *f* atomic linkage, linking of atoms
Atomverschiebung *f* atomic displacement
Atomverschmelzung *f* atomic fusion
atomverseucht radioactively contaminated
Atomversuch *m* atomic test, nuclear test
Atomversuchsgelände *n* nuclear testing ground
Atomvolumen *n* atomic volume
Atomwärme *f* atomic heat
Atomwaffe *f* atomic weapon, nuclear weapon
Atomwertigkeit *f* atomic valence
Atomwiderstand *m* atomic resistivity
Atomwissenschaft *f* atomic science, atomistics
Atomzahl *f* atom number, atomic index
Atomzeichen *n* atomic symbol, atomic sign
Atomzeitalter *n* atomic age, nuclear age
Atomzerfall *m* atomic disintegration
Atomzerlegung *f* atomic decomposition
Atomzertrümmerung *f* atom smashing, atom splitting
Atomzertrümmerungsmaschine *f* atom smasher, atom smashing machine
Atomzertrümmerungsversuch *m* atom smashing experiment
Atomziffer *f* atomic index
Atomzustand *m* atomic state
Atophan *n* atophan
Atopit *m* (Min) atopite
Atoxicocain *n* atoxicocaine
atoxisch non-toxic, atoxic

Atoxyl *n* (HN) atoxyl, sodium arsanilate, sodium p-aminobenzenearsonate
Atoxylsäure *f* arsanilic acid, atoxylic acid, p-aminobenzenearsonic acid
ATP siehe Adenosintriphosphat
Atractylen *n* atractylene
Atractylol *n* atractylol
Atramentstein *m* ferrous sulfate, inkstone, melanterite
Atramentverfahren *n* atrament process
Atranol *n* atranol
Atranorin *n* atranorin
Atranorinsäure *f* atranorinic acid
Atranorsäure *f* atranoric acid
Atrolactinsäure *f* atrolactic acid, α-phenyllactic acid
Atromentin *n* atromentin
Atromentinsäure *f* atromentic acid
Atropamin *n* atropamine, apoatropine
Atropasäure *f* atropic acid, α-phenylacrylic acid
Atrophie *f* (Med) atrophy
Atropin *n* atropine, daturine, dl-hyoscyamine
Atropinbrommethylat *n* atropine methylbromide
Atropinhydrobromid *n* atropine hydrobromide
Atropinmethylbromid *n* atropine methylbromide
Atropinmethylnitrat *n* atropine methyl nitrate
Atropinsalicylat *n* atropine salicylate
Atropinsulfat *n* atropine sulfate
Atropinvergiftung *f* (Med) atropine poisoning, atropism
Atropisomerie *f* (Stereochem) atropisomerism
Atropurol *n* atropurol
Atroscin *n* atroscine
Atrovenetin *n* atrovenetin
Attest *n* certificate
Attrappe *f* dummy, mock-up
Attritor *m* (Sandmühle) sand or attrition mill
atü (Atmosphärenüberdruck) gauge pressure in atmospheres, excess pressure, gauge pressure
Aucubin *n* aucubin
Audiogramm *n* audiogram
Audionempfänger *m* audion receiver
Auerbachit *m* (Min) auerbachite
Auerbrenner *m* Welsbach burner
Auerlicht *n* Welsbach light
Auerlith *m* (Min) auerlite
Auermetall *n* Auer metal, Welsbach metal
Auerstrumpf *m* Welsbach mantle
aufätzen to etch, to cauterize, to corrode upon
aufarbeiten to process; to use up; to work [up]
Aufarbeitung *f* working up, reprocessing
aufbäumen (Metall) to show metallic luster
Aufbau *m* (Bauw) erection, (Chem) structure, synthesis, (eines Moleküls) molecular structure, ~ **der Materie** structure of matter
aufbauen to synthetize, to build up; to erect
aufbauend constructive; synthetic
Aufbaufaktor *m* build-up factor
Aufbaugußring *m* adjusting tubing

Aufbauprinzip *n* construction principle
aufbauschen to puff up; to swell
Aufbauschung *f* swelling, (Übertreibung) exaggeration
Aufbaustoff *m* compounding ingredient, material used for synthesis, (Waschmittel) builder
Aufbaustoffwechsel *m* (Biol) anabolism, constructive metabolism
Aufbautrommel *f* (Reifen) building-up former
Aufbauzeit *f* time of formation
aufbeizen to corrode, to etch upon, (Möbel) to restain
aufbereitbar (Erze) washable
Aufbereitbarkeit *f* washability (of ores)
aufbereiten (Chem) to regenerate, (Kohle) to prepare, (Metall) to concentrate, (Metall, Häute) to dress, (Wasser) to refine
Aufbereitung *f* treatment, (Chem) regeneration, (Metall, Häute) dressing, ~ **der Kohle** dressing of the coal, ~ **des Formsandes** working up of the molding sand, ~ **durch Maschinen** mechanical dressing
Aufbereitungsanlage *f* (f Kohle) preparation plant, (Metall) dressing plant, (Trinkwasser) processing plant
Aufbereitungsart *f* manner of preparation, dressing or processing
Aufbereitungsgut *n* (Met) product of dressing
Aufbereitungskunde *f* (Metall) science of dressing
Aufbereitungsprodukt *n* concentrate
Aufbereitungs- und Absetzbehälter *m* treatment and settling vessel
Aufbereitungsverfahren *n* (Met) ore dressing
Aufbereitungsverlust *m* loss due to dressing, loss due to screening, loss from preparation
aufbessern to improve
aufbewahren to keep, to preserve, (lagern) to store
Aufbewahrung *f* conservation, preservation, (Lagerung) storage
Aufbewahrungsbehälter *m* storage vessel
Aufbewahrungsdauer *f* duration of storage, time of storage
Aufbewahrungsflasche *f* storage bottle
aufbiegen to bend up, to turn up
aufblähen to bulge out, to swell up
Aufblähung *f* blowing, inflation, swelling
Aufblättern *n* (v. Schichtstoffen) cleavage
aufblasen to blow up, to inflate
Aufblasen *n* inflation
Aufblaskonverter *m* top-blown converter
Aufblitzen *n* flash[ing]
aufbocken to jack up
aufbohren to bore
aufbrauchen to consume, to use up
aufbrausen to effervesce, to ferment
Aufbrausen *n* effervescence, ebullition

aufbrausend effervescent, fermenting
aufbrechen to break open, to break up, to burst
Aufbrechen *n* breaking up
aufbrennen (Email) to bake
Aufbringen *n* **des Abstichloches** opening the tap hole
aufbrodeln to boil up, to bubble up
Aufbruchhammer *m* heavy breaker
aufbrühen to bring to boil
Aufchromen *n* chrome-plating
aufdampfen to rise as steam; to vaporize on, to vapor-metallize
Aufdampfen *n* coating, vapor deposition, ~ **im Vakuum** vacuum coating
Aufdampfkontakt *m* evaporated contact
Aufdampfschicht *f* evaporated film
aufdecken to disclose, to uncover
Aufdockbleiche *f* roll bleaching
aufdörren to desiccate, to dry, (Malz) to kiln-dry
Aufdornprobe *f* drifting test
Aufdornversuch *m* expanding or drift test
Aufdornwerkzeug *n* reamer
aufdrängen to force upon
aufdrehen to turn on; to screw open, to unscrew, to untwist
aufdringlich penetrating, obtrusive
Aufdruck *m* imprint, printing, stamp
aufdrucken to imprint, to impress, to stamp
Aufdrückdeckel *m* (Dose) press-on cap
aufdüsen to spray
aufdunsten to vaporize, to evaporate
Aufeinanderfolge *f* sequence, series, succession
aufeinanderfolgend consecutive, successive
Aufeinanderprall *m* collision
aufeinanderschichten to stack
aufeinanderwirken to influence each other, to interact
Aufenthaltswahrscheinlichkeit *f* sojourn probability
Aufenthaltszeit *f* residence time
Auffächern *n* fanning
Auffänger *m* collecting vessel, receiver; target, ~ **aus Eis** ice target
auffärben to dye again, to redye
auffallend striking, conspicuous, distinctive, remarkable
Auffallwinkel *m* angle of incidence
auffalten (auseinanderfalten) to unfold, (Geol) to fold upward
Auffangbehälter *m* collecting vessel, receiver, sampler
Auffangblech *n* baffle sheet
Auffangelektrode *f* collector electrode
auffangen to collect, to catch, to gather
Auffangflasche *f* receiver bottle
Auffanggefäß *n* collecting vessel, catch pan, pan
Auffangglas *n* drip pan, object glass
Auffangglocke *f* drip cup

Auffangkolben *m* collecting flask, receiving flask
Auffangquerschnitt *m* absorption cross section
Auffangröhre *f* receiving tube
Auffangrohr *n* collecting cylinder
Auffangschale *f* drip pan, collecting basin
Auffangschirm *m* target
Auffangtasse catchpot
Auffangtrichter *m* collecting funnel
auffasern to separate into fibers
auffassen to comprehend, to conceive, to grasp, to interpret, to understand
Auffassung *f* apprehension, comprehension, conception
auffeuchten to dampen, to moisten, to wet
auffinden to detect, to discover
aufflackern to deflagrate, to flare up
Aufflackern *n* deflagration
aufflammen to flame up, to flare up
auffrischen to freshen up, to regenerate, to renew, to restore, to revive
Auffrischen *n* refreshing, regeneration, ~ **des Bades** regeneration of the bath liquor
Auffrischung *f* regeneration
aufführen (Bauw) to erect, to raise
auffüllen replenish, to fill up, (einen Akkumulator) to charge
Auffüllung *f* filling [up], replenishment, topping up
Aufgabeband *n* feeding belt
Aufgabeblech *n* feeding plate
Aufgabeboden *m* feed plate
Aufgabestelle *f* loading point
Aufgabetrichter *m* feed funnel, feed hopper
Aufgabevorrichtung *f* charging device, charging mechanism, feeder, feeding device
Aufgabewert *m* (Regeltechn) desired value
aufgären to effervesce, to ferment
Aufgeben *n* (Metall) charging
Aufgeber *m* feeder, forwarder
Aufgebesohle *f* charging platform
Aufgebevorrichtung *f* charging equipment or device
aufgedampft evaporated
aufgedruckt impressed
aufgedunsen inflated, puffed up, swollen
aufgefalzt seamed-on
aufgehen to swell
Aufgehen *n* **des Kalkes** ebullition of lime, ~ **des Werkzeuges** swelling of a mold
aufgerollt coiled
aufgeschmolzen fused on, melted on
aufgeschoben postponed
aufgespalten split, cloven
aufgespeichert stored
aufgewickelt coiled
aufgichten (Metall) to charge
Aufgichtvorrichtung *f* charging device
aufgießen to infuse, to pour on

Aufgießer *m* feeder
Aufgießlöffel *m* feeding ladle
Aufgipsen *n* cementing
aufgischen to foam up, to bubble over, to ferment
Aufglasur *f* overglaze
Aufglasurfarbe *f* overglaze color
aufgliedern to classify
aufglimmen to flicker up
aufglühen to flame up, to blaze, to flare up
aufgraben to dig up
Aufguß *m* infusion
Aufgußapparat *m* (Brau) sparger
Aufgußgefäß *n* infusion vessel, digester
Aufgußprobe *f* pour test
Aufhängebügel *m* (Techn) suspension hook
Aufhängedraht *m* suspension wire
aufhängen to suspend
Aufhängenasen *f pl* suspension lugs
Aufhängepunkt *m* (Waagschale) suspension point
Aufhängevorrichtung *f* suspension device
Aufhängung *f* (Mech) suspension
aufhäufen to heap up, to pile up, (anhäufen) to accumulate
Aufhaltekraft *f* power of resistance
aufhalten to detain, to stop
aufhaspeln to wind up
aufheben to pick up, (ausgleichen) to compensate, to counterbalance, to neutralize, (hochheben) to raise, (konservieren) to preserve, (Math) to cancel
Aufhebung *f* **der Polarität** depolarisation
aufheizen to heat
Aufheizpastenkathode *f* self-heating oxide cathode
aufhellen brighten [up], to lighten, (Flüssigkeiten) clarify, to clear up
Aufheller *m* brightening agent, **optischer** ~ optical bleach, optical bleaching agent, optical brightener
Aufhellung *f* brightening, (Anfärbung) tint tone (Lack) reduction
Aufhellungsmittel *n* (Mikroskopie) brightening agent
Aufhellungswert *m* tinting value
Aufhellvermögen *n* lightening power, reducing power
aufhöhen (Farbe) to raise
aufholen to catch up
Aufkalandrieren *n* calender-coating
aufkeilen to key on
aufkippen to tilt up, to tip up
aufkitten to cement on
aufklären to clarify, to clear up; to explain; to illuminate
aufklappbar capable of opening
aufklappen to swing open
Aufklebeetikett *n* adhesive label

aufkleben to paste on
Aufkleber *m* label, sticker
Aufklebezettel *m* label
aufkleistern to paste on
aufknacken to crack [open]
aufkochen to boil
Aufkochen *n* boiling, ~ **des Bades** boiling of the bath
Aufkocher *m* boiler, boiling vessel
Aufkochgefäß *n* boiler, boiling vessel
aufkohlen (Metall) to carburize
Aufkohlen *n* carbon penetration, carburization, cementation
Aufkohlung *f* carbonization
Aufkohlungsmittel *n* carburizer
Aufladeeinrichtung *f* supercharging equipment
aufladen to load, (Batterie) to charge, (Motor) to boost, to supercharge
Aufladen *n* **des Bleiakkumulators** charging of the lead accumulator
Auflader *m* (Elektr) charger, (Luftf) booster
Aufladestrom *m* charging current
Aufladewiderstand *m* (Kondensator) charging resistance
Aufladezeit *f* time required for charging
Aufladung *f* charge; loading, (Elektr) charging property, **elektrostatische** ~ electrostatic charge
Auflage *f* (Belag) coating, cover, (Buch) edition, (Galv) deposit, (Mech) arbor, bracket, rest, support
Auflagebedingung *f* (Mech) condition of support
Auflagebock *m* supporting block
Auflageentfernung *f* distance between supports
Auflagefläche *f* bearing surface, contact surface, seating surface
Auflageflansch *m* supporting flange
Auflagegewicht *n* coating weight
Auflagekette *f* suspending chain
Auflageplatte *f* bed plate
Auflagepunkt *m* point of contact, point of support
Auflageschuh *m* (Seilbahn) saddle
Auflagetisch *m* supporting table
Auflauf *m* (Herbicid) emergence
auflaufen to rise, to swell
Auflaufwalze *f* (Extrusion) chill roll
auflegieren to alloy up
Auflegieren *n* **der Elektroden** electrode pick-up
aufleimen to glue on
Auflicht *n* incident light, reflected light, vertical illumination, (Elektronenmikroskop) direct light
Auflichtbeleuchtung *f*, **direkte** ~ (Opt) direct illumination
auflockern to loosen [up], to disintegrate
Auflockerung *f* loosening; breaking up; disintegration; spongification
Auflockerungsmittel *n* dispersing agent

auflodern to flame up, to flare up
auflösbar (Chem) soluble, dissolvable; (Math) solvable, (Opt) resolvable
Auflösbarkeit f (Chem) solubility; (Opt) resolvability
Auflösebehälter m dissolver
Auflöseholländer m breaker beater
auflösen (Chem) to dissolve, (Math) to resolve, (schmelzen) to melt, (zersetzen) to decompose
auflösend dissolving
Auflöser m solvent, dissolver
Auflösezeit f resolving time
Auflösung f (Chem) solution, dissociation, dissolution, **spektrale** ~ (Opt) spectral dispersion
auflösungsfähig soluble
Auflösungsfähigkeit f solubility, solvent power
Auflösungsflüssigkeit f dissolving liquid
Auflösungsgefäß n dissolving vessel
Auflösungsgeschwindigkeit f (Chem) velocity of dissolution
Auflösungsgrenze f (Opt) limit of resolution
Auflösungskraft f dissolving power; (Opt) resolving power
Auflösungsmittel n solvent, dissolvent
Auflösungspfanne f clarifier
Auflösungsprozeß m process of disintegration
Auflösungsvermögen n (Chem) solvent power, dissolving capacity, (Opt) resolving power
Auflösungswärme f heat of dissolution
auflöten to solder on; (hartlöten) to braze
Auflötflansch m soldered flange, brazed flange
Aufmaischen n (Zuck) second mashing
aufmontieren to mount, to assemble, to erect, to set up
Aufnahme f absorption, (Aufzeichnung) recording, (Empfang) reception, (im Körper) intake, uptake, (Nahrung) ingestion, (Phot) picture, shot, (Stoffwechsel) assimilation, ~ **Röntgen-**~ radiogram, ~ **von Elektronen** acceptance of electrons
Aufnahmebehälter m receptacle
Aufnahmebereich m absorption range
Aufnahmebereitschaft f susceptibility
aufnahmefähig absorbable, capable of absorbing
Aufnahmefähigkeit f absorbability, absorptivity; receptivity, (Chem) absorption power or capacity
Aufnahmegeschwindigkeit f absorption rate
Aufnahmekammer f receiver
Aufnahmekolben m absorption flask
Aufnahmeprüfung f entrance examination
Aufnahmevermögen n (Chem) absorption capacity; (geistig) receptivity; (Magnetismus) susceptibility
Aufnahmeverstärker m recording amplifier
Aufnahmewalze f pick-up roll
aufnehmen to hold, to receive, (beginnen) to begin, to start, (Chem) to absorb, to adsorb

aufnehmend absorbing, absorptive
Aufnehmer m absorber, receiver
Aufnehmerdiagramm n receiver diagram
aufoxydieren to oxidize
aufpassen to watch, (Techn) to fit on
aufpfropfen to graft, to splice
aufplatzen to burst [open], to explode
aufprägen to impress [on]; to imprint
Aufprall m collision, impact
aufprallen to bound, to rebound, to strike on
Aufprallen n bounce
Aufprallfläche f (Elektronen) target area
aufpumpen to inflate, to pump up
aufquellen to swell up
Aufquellung f swelling
aufräumen to clear away, to tidy
Aufräumung f clearing up
Aufräumungsarbeiten pl salvage work
aufrahmen (Latex) to cream [up]
Aufrahmung f creaming [up]
aufrauhen to roughen, to buff, to rough up
Aufrauhen n roughening, buffing
Aufrauhmaschine f buffing machine
aufrecht straight
aufrechterhalten to maintain; to keep alive
Aufrechterhaltung f maintenance, support
aufrechtgießen to top cast
aufregen to excite, to arouse, to enrage
Aufreißband n rip band
aufreißen to tear open, to crack open, to split
Aufreißstreifen m tab, tear strip
aufrichten to set up, to erect, to raise
Aufrichtung f (Anlagenbau) erection, ~ **einer Doppelbindung** opening (of a double bond)
Aufriß m front view, design, sketch, (Math) vertical projection
aufrösten to roast slightly
aufrollen to reel [up], to wind [up]
aufrühren to agitate, to stir
Aufrühren n stirring
aufrütteln to shake up, to stir
aufrunden (Math) to round up
aufsammeln to collect, to gather
Aufsatz m attachment, fixture, head, head-piece, (Destillation) still head
Aufsatzrahmen m filling frame
Aufsatzschlüssel m socket wrench
Aufsatzzeiger m sight indicator
aufsaugbar absorbable
Aufsaugbarkeit f absorbability
Aufsaugefähigkeit f absorptivity
Aufsaugeflüssigkeit f absorbing liquid
aufsaugen to absorb, to suck up
Aufsaugen n absorption
aufsaugend absorbent, absorbing, absorptive
Aufsauger m absorber
Aufsaugevermögen n absorption capacity, absorption power
Aufsaugmittel n absorbent

Aufsaugung f absorption
Aufsaugungsfähigkeit f absorptive ability, degree of absorption
Aufsaugungsverfahren n method of absorption, process of absorption
Aufsaugungsvermögen n absorptive power
Aufsaugungswärme f heat of absorption
Aufsaugverfahren n absorption process
Aufsaugvermögen n absorptivity
Aufschärfung f re-sharpening
aufschäumen to foam up, to effervesce, to ferment, to froth
Aufschaukelvorgang m building-up process
aufschichten to pile up, to arrange in layers, to layer, to stack, (Geol) to stratify
Aufschichtung f piling up, (Geol) stratification
aufschlämmen to suspend, to make into a paste
Aufschlämmen n slurrying
Aufschlämmung f suspension, slime, sludge, slurry
Aufschlämmverfahren n flotation process
Aufschlag m percussion, impact; (Zuschlag) increase
aufschlagen to strike open; to unfold
Aufschlaggeschwindigkeit f striking velocity, velocity of impact
Aufschlagzünder m impact detonator, percussion fuse
aufschleifen to grind on
aufschließbar (Chem) hydrolyzable
Aufschließbarkeit f (Pap) pulpability
aufschließen to open; to break up, to triturate, (Chem) to decompose, to disintegrate, to hydrolyze, to solubilize
Aufschließen n (Anal) decomposition, fusion
Aufschließmaschine f dissolving machine; crusher
Aufschließung f breaking up, decomposition, hydrolyzation, (Atom) digestion
aufschlitzen to rip [open], (z. B. Reifen) to slash
Aufschluß m breaking up, (Chem) hydrolysis, decomposition; disintegration, solubilization, (Nahrungsmittel) digestion; (Papier) pulping, **alkalischer** ~ alkaline decomposition, **nasser** ~ wet extraction, **saurer** ~ acid decomposition, **Sulfit-** ~ sulfite digestion, **trockener** ~ dry decomposition
Aufschlußbohrtätigkeit f exploration drilling
Aufschlußbohrung f exploration well
Aufschlußmittel n decomposing agent, disintegrating agent, hydrolytic agent, means of attack, solubilizer
aufschlußreich informative
Aufschlußverfahren n decomposition process, disintegration process, hydrolyzation process
aufschmelzen to dissolve by heat; to fuse on, to melt on; (Chem) to be melted down
Aufschmelzüberzug m hot melt-coating
aufschneiden to cut open

aufschöpfen to scoop up
aufschrauben to screw on; to screw open, to unscrew
aufschreiben to record, to write down
Aufschreiber m recorder
Aufschrift f label, inscription
aufschrumpfen to shrink on
Aufschrumpfen n shrink coating
Aufschub m adjournment, delay, putting off, suspension
aufschüren to poke, to stir up, to stoke
Aufschüttdichte f apparent density
aufschütteln to shake
aufschütten to charge, to feed, to pour on
Aufschüttrichter m feeding hopper, hopper
Aufschüttung f charging, firing, pouring on, stoking; (Geol) debris, deposit
Aufschüttungsmasse f filling material
Aufschweißbund m slip-on collar
aufschweißen to fuse, to weld on, to weld together
Aufschweißflansch m slip-on flange
Aufschweiß-Hartlegierung f cemented carbide
aufschwellen to swell up
aufschwemmen to suspend; to bloat
Aufschwung m improvement, recovery
Aufseher m foreman, supervisor
aufsetzbar attachable
aufsetzen to set up, to affix, to mount
Aufsetzgewicht n fractional weight
Aufsetzvorrichtung f catch device, safety catch
Aufsicht f supervision, control, inspection, survey, **in der** ~ viewed in reflected
Aufsichtschwärzungsmesser m reflection densitometer
aufsieden to boil up, to bring to boil
Aufsieden n ebullition
Aufsilizieren n siliconizing
aufsintern to sinter
Aufsitzfläche f [valve] seat
aufspachteln to trowel
aufspalten to burst, to break down, to cleave, to split, to split [up], (abblättern) to delaminate, (Chem) to decompose, (Leder, Fell) to skive
Aufspaltreaktion f fission reaction
Aufspaltung f cleavage, cracking, decomposition, fission, splitting, (Atom) fission, (Chem) dissociation
Aufspaltungstemperatur f (NMR-Spektroskopie) coalescence temperature
Aufspannapparat m fixing apparatus
Aufspannblock m vice
aufspannen to strain, to stress, to stretch, to tighten, (Techn) to fix
Aufspannfutter n chuck
Aufspannmaschine f expanding machine
Aufspannplatte f adapter plate, backing plate, clamping plate, mounting plate, top plate, upper plate

Aufspannvorrichtung f clamping device; work-holding fixture
aufspeichern to stockpile, to store up, (Phys) to accumulate
Aufspeicherung f accumulation; storage
aufsperren to unlock, to open; to spread apart
aufsplittern to split up
Aufsplitterung f splitting
Aufsprechfeld n recording field
Aufsprechkopf m recording head
Aufsprechkopfspalt m recording head gap
aufsprengen to blow up, to force open
aufspritzen to spray on
Aufspritzen n spraying, (Metall) metallization
aufsprudeln to bubble up, to effervesce
aufsprühen (Flüssigkeit) to spray
aufspüren to trace
aufspulen to coil, to reel, to wind up
aufstäuben to dust; to spray
Aufstampfboden m stamping board
aufstampfen to stamp
aufstapeln to heap, to pile [up], to stack, (Vorräte) to stock, to store up
aufstechen to pierce open, to puncture
aufsteckbar attachable
aufstecken to attach, to put on, to set up
Aufsteckfahrantrieb m (Kran) slip-on travelling gear
Aufsteckfassung f adapter rim
Aufsteckgetriebe n power take-off
Aufsteckglas n slip-on lens
Aufsteckkurbel f cranked handle
Aufsteckring m slip-on ring
Aufsteckschlüssel m socket spanner (Br. E.), socket wrench (Am. E.)
Aufstecktülle f adapter spout
aufsteigen to ascend, to mount, to rise, (Gasblasen) to bubble up
aufsteigend ascending, rising
Aufsteigschichtverdampfer m climbing film evaporator
aufstellen to set up, to assemble, to erect, to place in position
Aufstellung f arrangement, graph; schedule, set up, (Maschine etc.) erection, installation
Aufstellungsskizze f layout sketch
Aufstieg m ascent, rise
aufstreichen to brush on, to coat, to spread on
Aufstreichen n spread coating
aufstreuen to sprinkle, to strew
Aufstrich m spreading on, brushing on; coat, coating
Aufströmung f fluidization
Aufströmungswirksamkeit f fluidization efficiency
Aufstrom m updraft, upward stream
Aufstromklassierer m elutriator
aufstutzen to prop up, to support
auftanken to refuel

auftauen to defrost, to thaw
Auftaumittel n thawing agent
aufteilen to distribute, to allot
aufträufeln to drop on, to pour on drop by drop, to trickle down in drops
Auftrag m coat, coating; commission, order
auftragen to spread, to coat, to lay on; (Kurven etc.) to draw, to plot
Auftragen n application, ~ **eines Striches** (Pap) application of a coat
Auftragkalander m calender coater
Auftragmaschine f coater, coating machine
Auftragsbearbeitung f order processing
Auftragschweißung f deposit welding
Auftragsdicke f (z. B. Lack) coating thickness
Auftragsforschung f contract research
Auftragsgewicht n coat weight
Auftragspinsel m inking brush
Auftragsschweißen n built-up welding
Auftragung f application
Auftragwalze f applicator roll, casting roll, doctor roll, pick-up roll, spreader roll
auftreffen to strike against
Auftreffen n (Geschoß) impact
Auftreffenergie f energy of impact, striking energy
Auftreffwinkel m (Mech) angle of impact, (Opt) angle of incidence
auftreiben to ream, (aufblähen) to distend, to enlarge, to inflate
Auftreiber m reamer
Auftreibung f blowing up; distention, inflation
auftrennen to rip open; to separate, to sever
Auftreten n occurrence
Auftrieb m buoyancy, ascending force, ascending power, upward thrust
Auftriebkoeffizient m buoyancy coefficient, lift coefficient
Auftriebmittel n buoying agent, swelling agent
Auftriebsachse f lift axis
Auftriebsformel f equation of lift
Auftriebskraft f lift force
Auftriebsmittelpunkt m center of buoyancy
Auftriebswaage f baroscope
aufvulkanisieren to vulcanize on
Aufwachsverfahren n (Met) filament growth method
aufwärmen to heat, to warm up
Aufwärmer m (Atom) reboiler
aufwärts upward
Aufwärtsbewegung f upward motion
Aufwärtstransformator m step-up transformer, voltage raising transformer
aufwallen to boil up, to bubble, to effervesce, to well up
Aufwallen n ebullition, boiling
Aufwallung f ebullition, effervescence
Aufwalzen n (Folien) roller application
Aufwalzflansch m expanded flange

Aufwalzvorrichtung f roll coater
Aufwand m expenditure
aufweichen to soak, to soften; (Metall) to temper
aufweichend emollient, softening
Aufweichmittel n emollient, softener
aufweisen to exhibit, to show
aufweiten to bulge, to expand, to stretch
Aufweiten n bulging
Aufweitversuch m (an Rohren) bulging test
aufwenden to devote; to employ; to spend
aufwerfen to throw up
Aufwerfhammer m helve hammer, lift hammer, tilt hammer
Aufwickelbock m wind-up stand
Aufwickelmaschine f winding machine, rewinding machine
aufwickeln to wind [up], to coil up, to reel up, to roll up, (Verpacktes) to unwrap
Aufwickelung f take-up
Aufwickelwalze f wind-up roll, material roll, take-up roll
aufwiegen to balance, to counterbalance
Aufwind m updraft
aufwirbeln (Wirbelschichtverfahren) to fluidize
Aufwuchs m growing up
aufzählen to enumerate, to itemize, to number
Aufzehrgrad m degree of absorption
Aufzehrung f consumption, absorption, utilization, ~ **der Elektrode** destruction of the electrode, ~ **des Kohlenbodens** consumption of the hearth electrode
aufzeichnen to plot, to trace, (aufschreiben) to mark, to record, to register, (skizzieren) to draw
aufzeichnend recording
Aufzeichnung f design, diagram; record
aufziehen to draw up, to raise up
Aufziehen n (Flüssigkeit) penetration
Aufziehfenster n sash window
Aufziehgeschwindigkeit f (Färb) absorption rate
Aufziehkurve f affinity curve
Aufziehvermögen n absorptive power, affinity, uptake
Aufziehwerk n winding-up mechanism
aufzischen to fizz
Aufzug m (Buchdr) blanket, (Fahrstuhl) elevator, lift, (Hebevorrichtung) hoist, (Kran) crane
Aufzugsführungsschiene f lift guide rail
Aufzugsseil n elevator cable, hoisting rope, lift rope
aufzwängen to force on, to press on
Aufzweigung f (Atom) branching, multiple decay
aufzwingen to force on, to press on
Augbolzen m eyebolt
Augenachat m (Min) cat's eye, eye agate
Augenblick m instant, moment
augenblicklich instantaneous
Augenglas n eyepiece

Augenheilkunde f ophthalmology
Augenlinse f eye lens, eyepiece
Augenmarmor m (Min) spotted marble
Augenmuschel f (Atom) eyeshield
Augennichts n nihil album, sublimated zinc oxide, white tutty
Augenreizstoff m lacrimator
Augenstein m (Min) eye stone, white copperas, white vitriol, zinc sulfate
Augenzeuge m eye-witness
Augereffekt m Auger effect
Auger-Elektron n (Atom) Auger electron
Auger-Spektroskopie f Auger electron spectroscopy
Augerübergang m Auger transition
Augit m (Min) augite, malacolite
augitartig augitic
augithaltig augitic
Augitit m (Min) augitite
Augitmasse f pyroxenic mass
Augitporphyr m augite porphyry
Augitreihe f augite series
Aulamin n aulamine
Auramin n auramine
Auraminbase f auramine base
Auraminhydrochlorid n auramine hydrochloride
Auranetin n auranetin
Aurantia n aurantia
Aurantiacin n aurantiacin
Aurantin n aurantine
Aurantiogliocladin n aurantiogliocladin
Aurat n aurate
Aurein n aureine
Aureolin n aureolin, primulin
Aureomycin n aureomycin
Aureomycinsäure f aureomycinic acid
Aureonamid n aureone amide
Aureothinsäure f aureothinic acid
Aureothricin n aureothricin
Aureusidin n aureusidin
Aureusin n aureusin
Aurichalcit m (Min) aurichalcite, green calamine
Aurichlorid n (Gold(III)-chlorid) auric chloride, gold(III) chloride, gold trichloride
Aurichlorwasserstoffsäure f (Gold(III)-chlorwasserstoffsäure) aurichlorohydric acid, chloroauric acid
Auricyanid n (Gold(III)-cyanid) auric cyanide
Auricyanwasserstoffsäure f cyanoauric acid
Aurihydroxid n (Gold(III)-hydroxid) auric hydroxide, gold(III) hydroxide
Aurin n aurin
Aurioxid n (Gold(III)-oxid) auric oxide, gold(III) oxide
Auripigment n (Min) auripigment, arsenous sulfide, orpiment
Aurirhodanwasserstoffsäure f aurithiocyanic acid, thiocyanatoauric(III) acid

Aurisulfid *n* (Gold(III)-sulfid) auric sulfide, gold trisulfide
Auriverbindung *f* (Gold(III)-verbindung) auric compound
Aurocantan *n* aurocantane
Aurochin *n* aurochin, quinine p-aminobenzoate
Aurochlorid *n* (Gold(I)-chlorid) aurous chloride, gold monochloride
Aurochlorwasserstoffsäure *f* chloroauric(I) acid, chloroaurous acid
Aurocyanid *n* (Gold(I)-cyanid) aurous cyanide, gold(I) cyanide
Aurocyanwasserstoffsäure *f* aurocyanic acid, cyanoauric(I) acid
Aurofelt *n* (Haarbeize) aurofelt
Aurokaliumcyanid *n* potassium aurocyanide, potassium cyanoaurate(I)
Auron *n* aurone
Auronalfarbe *f* auronal dye
Aurooxid *n* gold(I) oxide, aurous oxide
Aurophenin *n* aurophenine
Aurora-Linie *f* (Spektr) auroral line
Aurorhodanwasserstoffsäure *f* aurothiocyanic acid, thiocyanatoauric(I) acid
Aurotin *n* aurotine
Auroverbindung *f* (Gold(I)-Verbindung) aurous compound, gold(I) compound
Auroxanthin *n* auroxanthin
Aurum *n* (Lat) gold
ausäthern to extract with ether, to etherize, to shake out with ether
Ausäthern *n* extracting with ether
ausätzen to cauterize; to destroy by caustics, to discharge
Ausätzung *f* cauterization
ausarbeiten to complete, to finish, to perfect, to work out
ausarten to degenerate
ausatmen to exhale, to expire
Ausatmung *f* expiration
ausbalancieren to equilibrate, to balance, to compensate, to counterbalance, to counterpoise
Ausbalancierung *f* balancing, counterbalancing, equilibration
Ausbau *m* development, completion, extension
ausbauchen to emboss, to hollow out, to swell
Ausbauchung *f* bulge, camber, widening
ausbauen to complete, to improve; (Teile) to disassemble, to dismount
ausbedingen to reserve, to stipulate
ausbeizen to remove with corrosive
ausbessern to repair
ausbesserungsbedürftig in want of repairs
Ausbesserungsmasse *f* lining material for repairs
Ausbesserungswerkstatt *f* repair shop
ausbeulen to round out, to swell out
Ausbeute *f* conversion (polymerization); efficiency, gain, output, profit, (Chem) yield, **photoelektrische ~** photoelectric yield, photoelectric emissivity
Ausbeuteerhöhung *f* increase in yield
Ausbeutegleichung *f* (Atom) gain equation
Ausbeutekurve *f* yield curve
Ausbeutematrix *f* efficiency matrix
Ausbeutemessung *f* yield measurement
Ausbeutetensor *m* efficiency tensor
Ausbeutung *f* exploitation, utilization, (Bergbau) winning, working
ausbiegen to bend out, to deflect, to turn out
Ausbiegung *f* deflection
ausbilden to develop, to improve
Ausbildung *f* formation; development; education, **~ der Asymmetrie** formation of asymmetry
Ausblasedampf *m* exhaust steam
Ausblasehahn *m* blow-off cock, drain cock
Ausblaseleitung *f* escape pipe, blow-off main, blow-off pipe
ausblasen (Dampf) to blow off, to exhaust, (Kerze) to blow out
Ausblasen *n* blowing out
Ausblaseventil *n* blow-off valve
ausbleiben to vanish, to disappear
Ausbleiben *n* absence, disappearance
ausbleichen to discolor, to bleach by decoloring, to fade, to lose color
Ausbleichen *n* bleaching, **~ durch Abgase** *f pl* gas fume fading
Ausbleichverfahren *n* bleaching-out process
ausbleien to line with lead
ausblenden (Elektr) to shield, (Opt) to collimate
Ausblick *m* outlook, prospect
ausblühen (Chem) to effloresce
Ausblühen *n* bloom, blooming, efflorescing
Ausblühung *f* bloom, efflorescence, (Schweiß) blistering
Ausbluten *n* bleeding (of colors)
Ausblutung *f* bleeding (of colors)
ausbohren to bore out, to drill
ausbrechen to break out
Ausbrechen *n* break-away
ausbreiten to spread out, to display, to extend, to flatten; to permeate
Ausbreiteprobe *f* flattening test, flow test, hammering test
Ausbreitung *f* diffusion; flattening out; propagation
Ausbreitungsfeld *n* (Comp) propagate field
Ausbreitungsgeschwindigkeit *f* velocity of propagation
Ausbreitungsparameter *m* propagation parameter
Ausbreitungsproblem *n* propagation problem
Ausbreitungsrichtung *f* direction of propagation
Ausbreitungswiderstand *m* diffusion resistance, resistance to spreading

ausbrennen – Ausfällung

ausbrennen to anneal, to bake, to burn; to eliminate; to purify
Ausbrennen *n* fritting
Ausbrennverhalten *n* burn-out feature
ausbringen to obtain, to produce, to yield
Ausbringen *n* **des Metalls** removing the metal
Ausbringung *f* output, production, yield
Ausbröckeln *n* (Grübchenbildung) pitting
Ausbruch *m* outburst
ausbrühen to scald, to seethe
ausdampfen to smoke out, to evaporate, to steam [out], to vaporize
Ausdampfnukleon *n* evaporation nucleon
Ausdampfung *f* evaporation, vaporization
Ausdampfwasser *n* steaming out liquor
Ausdauer *f* endurance, perseverance
ausdauernd enduring
ausdehnbar ductile, expansible, expansive, (Gase) dilatable, (Techn) extensible
Ausdehnbarkeit *f* (Länge) extensibility, (Phys) dilatability
ausdehnen (erweitern) to expand, (Gas) dilate, (strecken) to stretch, (verlängern) to extend
Ausdehnung *f* elongation, expansion, extension; extent, scope, thermal expansion,
  adiabatische ~ (Therm) adiabatic expansion,
  elastische ~ elastic extension
Ausdehnungsarbeit *f* work done on expansion
Ausdehnungsarmatur *f* expansion fitting
Ausdehnungsbogen *m* expansion joint
Ausdehnungsdrang *m* tendency to expand
Ausdehnungsfähigkeit *f* dilatability, expansibility
Ausdehnungsfeld *n* (Comp) propagate field
Ausdehnungsgefäß *n* expansion vessel
Ausdehnungskoeffizient *m* coefficient of expansion, expansion coefficient, **linearer** ~ coefficient of linear expansion
Ausdehnungskraft *f* force of expansion
Ausdehnungskupplung *f* expansion joint, flexible coupling
Ausdehnungsmesser *m* dilatometer, extensometer
Ausdehnungsraum *m* expansion space; expansion chamber
Ausdehnungsringstück *n* expansion ring
Ausdehnungsstück *n* expansion piece
Ausdehnungsverflüssiger *m* expansion liquefier
Ausdehnungsvermögen *n* expansibility, extensibility, (Met) ductility
Ausdehnungszahl *f* expansion coefficient
Ausdehnungsziffer *f* coefficient of expansion
ausdörren to desiccate, to dry up, to parch, to scorch
ausdrehen to switch off, to turn off
Ausdrehstahl *m* boring tool
Ausdruck *m* (Math) expression, **algebraischer** ~ (Math) algebraic expression,
  asymptotischer ~ asymptotic expression
Ausdruckbolzen *m* ejector connecting bar

ausdrucken to finish printing, to print
Ausdrückbalken *m* knock-out bar
Ausdrückbolzen *m* ejector rod
Ausdrückbolzenfeder *f* (Rückzugfeder) return spring
ausdrücken to eject; to express; (z. B. Frucht) to squeeze out
Ausdrücken *n* ejection, ~ **von unten** (aus dem Gesenk) bottom ejection, **Hilfsvorrichtung zum** ~ extractor
Ausdrücker *m* ejector
Ausdrücker-Führung *f* ejector frame guide
Ausdrücker-Verbindungsstange *f* ejection tie-bar
Ausdrückgehänge *n* pullrod
Ausdrückkolben *m* ejection ram
Ausdrückmaschine *f* (Kokerei) mechanical pusher
Ausdrückplatte *f* ejection plate, ejector plate
Ausdrückrahmen *m* ejector frame, knock-out frame
Ausdrückstift *m* ejector pin, knock-out pin, (mit Federkraft betätigt) spring ejector
Ausdrückstiftplatte *f* knock-out pin plate
Ausdrückvorrichtung *f* knock-out
ausdünsten to evaporate, to exhale, (Biol) to transpire
Ausdünstung *f* evaporation, exhalation; steam, vapor, (Schweiß) perspiration
ausduften to evaporate, to exhale; to fill with odor
Ausduftung *f* exhalation
Ausdunstungsapparat *m* evaporating apparatus
ausegalisieren to equalize
Auseinanderbrechen *n* break-down
auseinanderbringen to separate
auseinanderfallen to disintegrate, to fall apart, to fall into pieces
auseinanderfalten to unfold
auseinanderhalten to distinguish
Auseinanderlaufen *n* divergence
auseinandernehmen to take to pieces, to disassemble, to dismantle, to dismount
Auseinandernehmen *n* disassemblage
auseinandersetzen to explain
auseinanderstrebend divergent
ausentwickeln to develop fully
Ausentwicklung *f* full development
Ausfällapparat *m* precipitator
ausfällbar precipitable
Ausfällbarkeit *f* precipitability
ausfällen to precipitate
Ausfällen *n* precipitating, precipitation
Ausfällmittel *n* precipitating agent
Ausfällung *f* precipitate, deposit; deposition, precipitation, ~ **von Verunreinigungen** precipitation of impurities

ausfärben to finish dyeing
Ausfärbvorrichtung *f* color extractor
Ausfahren *n* **der Elektrodenbündel** *n pl* withdrawal of the electrode bundles
Ausfall *m* breakdown, fall-out, (Haar) falling out, thinning, (Mangel) deficiency, deficit, lack, loss, ~ **von Zeichen** (Comp) dropout
Ausfalleisen *n* off-grade iron
ausfallen to be lacking; to deposit, to result, to turn out, (Chem) to precipitate
Ausfallen *n* (Pigmente) fall-out
Ausfallerscheinung *f* deficiency symptom, (Med) disturbed function, (Muskel) atrophy
Ausfallöffnung *f* discharge opening
Ausfallwinkel *m* angle of reflection
ausfasern to feaze, to ravel out, to unravel
Ausfasern *n* fringing
ausfertigen to make out
ausfetten to extract fat or grease
ausfiltern to filter [out]
ausfixieren to fix completely
ausfließen to flow out, to deliver, to discharge, to escape, to run out
ausfließend effluent, outflowing
Ausflockbarkeit *f* flocculation tendency
ausflocken to flocculate, to separate in flakes
Ausflockung *f* flocculation, coagulation, separation in flakes
Ausflockungspunkt *m* flocculation point
Ausfluß *m* discharge, efflux, escape, outflow, ~ **unter Wasser** submerged discharge
Ausflußapparat *m* flow apparatus
Ausflußgeschwindigkeit *f* discharge velocity, outflow velocity
Ausflußhahn *m* discharge cock, outlet cock
Ausflußkanal *m* output channel
Ausflußkoeffizient *m* coefficient of discharge, efflux coefficient
Ausflußloch *n* discharge opening, outlet
Ausflußmenge *f* discharge quantity
Ausflußöffnung *f* discharge opening, nozzle
Ausflußrohr *n* discharge pipe, escape pipe, outlet pipe
Ausflußstrahl *m* discharge jet
Ausflußströmung *f* discharge flow
Ausflußventil *n* discharge valve
Ausflußzahl *f* discharge coefficient
Ausflußzeit *f* time of discharge
ausforschen to investigate, to search out
ausfräsen to mill out, to countersink, to edge, to ream, to shape, (Masch) to bead
Ausfräsen *n* routing
ausfragen to question, to examine
ausfressen to corrode, to erode
Ausfressung *f* act of corrosion
ausfrieren to freeze out, to concentrate by freezing, to congeal
Ausfrieren *n* freezing, freezing out, ~ **der Erze** *n pl* freezing out the ore

Ausfrierfalle *f* freezing trap, cooling trap
Ausfriertemperatur *f* freezing-out temperature
Ausfrierverfahren *n* freezing method, freezing process
Ausfriervorrichtung *f* freezing assembly
ausführbar feasible, practicable, (exportierbar) exportable
Ausführbarkeit *f* feasibility
ausführen to carry out, to perform, (exportieren) to export; (Met) to assay
Ausführung *f* carrying out; design; performance
Ausführungsform *f* form of construction
Ausführungsrohr *n* discharge tube, outlet pipe
ausfüllen to fill, to stuff
Ausfüllmasse *f* filling, packing, stuffing
Ausfüllstoff *m* filling, packing material
ausfüttern to line, to coat
Ausfütterung *f* lining
Ausfuhr *f* export, exportation
ausgären to ferment, to throw off [by fermentation]
Ausgärzeit *f* fermentation period, (Stahl) quiescent period
Ausgangsatom *n* parent atom
Ausgangsbasis *f* starting point
Ausgangsdruck *m* initial pressure
Ausgangserzeugnis *n* initial product, original product, primary product, starting product
Ausgangsgeschwindigkeit *f* initial speed
Ausgangsgleichung *f* initial equation, starting equation
Ausgangsisotop *n* parent isotope
Ausgangslage *f* original or initial position
Ausgangslösung *f* initial solution
Ausgangsmaterial *n* original material, base material, charge material, feed material, primary material, raw material, starting material
Ausgangsposition *f* initial position
Ausgangspotential *n* (Elektr) initial potential
Ausgangsprodukt *n* base product, initial material, initial product, original constituent, original product, starting product
Ausgangspunkt *m* base, origin, starting point
Ausgangsrichtung *f* initial direction
Ausgangsrohr *n* outlet pipe
Ausgangsspalt *m* (Opt) exit slit
Ausgangsspannung *f* output voltage
Ausgangsstellung *f* initial position
Ausgangsstoff *m* base material, basic material, feed, initial material, original material, primary material, starting material
Ausgangssubstanz *f* original material, raw material
Ausgangstemperatur *f* initial temperature, original temperature
Ausgangsverbindung *f* starting compound
Ausgangswerkstoff *m* raw material
Ausgangswert *m* output value

Ausgangszustand *m* initial state, parent state
ausgaren (Metalle) to refine
ausgeartet degenerate
ausgebaucht bulged, concave
ausgedehnt extensive, large
ausgeglichen balanced
Ausgeglichenheit *f* compensated state
ausgeglüht annealed
ausgekehlt chamfered, fluted, grooved
ausgekerbt dented, indented, notched, serrated
ausgekleidet lined
ausgekocht boiled out; concentrated by boiling
ausgekohlt carbonized
ausgelaugt lixiviated, extracted, leached out
ausgeleuchtet lighted
ausgemauert brick-walled
ausgeprägt characteristic, distinct, marked, particular, peculiar
ausgerben to finish tanning
ausgerüstet provided with, to be well equipped [for]
ausgeschieden precipitated, deposited
ausgeseigert segregated, separated out
Ausgesiebte[s] *n* screenings
ausgesprochen decided, outspoken, pronounced
Ausgestaltung *f* development; equipment
ausgestanzt punched out
ausgewachsen fully grown
ausgewalzt cogged (ingot)
ausgewogen (im Gleichgewicht befindlich) balanced
ausgezeichnet distinguished
ausgezogen extracted; drawn out, stretched, (Linie) continuous (line)
ausgiebig abundant, productive
Ausgiebigkeit *f* abundance, extensiveness, productiveness
ausgießen to pour [out], to decant, to empty
ausgipsen to fill with plaster
ausgischen to cease foaming
Ausgleich *m* compensation; balancing; equalization, ~ **der Ladungen** *f pl* equalization of charges, **Regelstrecke mit** ~ self-regulating process (Am. E.), system with inherent regulation (Br. E.)
Ausgleichaggregat *n* compensation set
Ausgleichaufhängung *f* compensating suspension
Ausgleichbehälter *m* equalizing tank
Ausgleichdraht *m* equalizing wire
Ausgleichdruck *m* equalizing pressure
Ausgleichdüse *f* compensating jet
Ausgleichdynamo *m* balance dynamo
ausgleichen to adjust; to equilibrate, to balance, to compensate, to equalize; to flatten, to level up; to settle
Ausgleichentwickler *m* (Phot) compensation developer
Ausgleicher *m* compensator, equalizer

Ausgleichfeder *f* compensating spring
Ausgleichfilter *m* balancing filter, equalizing filter
Ausgleichgetriebe *n* differential driving gear
Ausgleichgewicht *n* counterbalance weight, counterweight
Ausgleichgrube *f* soaking pit
Ausgleichleitung *f* balancing network, equalizer, equalizing mains
Ausgleichmagnet *m* compensating magnet
Ausgleichrohr *n* compensating pipe
Ausgleichsanzeiger *m* balance indicator, balance recorder
Ausgleichsbehälter *m* expansion tank, expansion vessel, fluid reservoir
Ausgleichsdruck *m* equalizing pressure
Ausgleichsfehler *m* corrective error
Ausgleichsgefäß *n* equalizing reservoir
Ausgleichsgewicht *n* counterweight
Ausgleichskurve *f* compensating curve
Ausgleichslehrsatz *m* compensation theorem
Ausgleichsleitung *f* compensation line
Ausgleichsmasse *f* gap-filling material
Ausgleichsmessung *f* compensation measurement
Ausgleichspannung *f* transient voltage, compensating voltage; equalizing pressure
Ausgleichsrad *n* compensating gear
Ausgleichsrechnung *f* calculation of most probable values, method of least squares
Ausgleichsspule *f* rectifying coil
Ausgleichsstrom *m* balancing current, compensating current, equalizing current
Ausgleichsverdampfer *m* equalizing evaporator
Ausgleichsvorgang *m* balance mechanism
Ausgleichsvorrichtung *f* compensator
Ausgleichswaage *f* adjusting scales
Ausgleichswärme *f* compensating heat
Ausgleichswelle *f* differential shaft
Ausgleichswert *m* compensation value
Ausgleichszulage *f* compensating allowance
Ausgleichszustand *m* equilibrium state
Ausgleichtemperatur *f* equalizing temperature
Ausgleichtransformator *m* balancing transformer
Ausgleichung *f* equilibration, equalization; equilibrium
Ausgleichungsrechnung *f* (Spektr) regression analysis
Ausgleichungsströmung *f* compensating current, equalizing current
Ausgleichvermögen *n* (Text) levelling
Ausgleiten *n* skid, slip
ausglühen to anneal, to calcine, to glow, to heat red hot, to roast
Ausglühen *n* annealing
Ausglüher *m* annealer
Ausglühflammofen *m* annealing furnace
Ausglühofen *m* heating furnace

Ausglühtopf *m* annealing pot
Ausglühung *f* annealing; reheating
Ausgrabung *f* excavation
Ausguß *m* sink; discharge connection, lip, outlet, spout
Ausgußbecken *n* sink basin
Ausgußkasten *m* discharge pipe
Ausgußleitung *f* discharge piping
Ausgußloch *n* drain hole
Ausgußmasse *f* casting composition
Ausgußmörser *m* lipped mortar
Ausgußpfanne *f* ingot mold
Ausgußrinne *f* pouring spout
Ausgußrohr *n* delivery pipe, drain pipe
Ausgußschnauze *f* pouring spout or lip, pouring nozzle
Ausgußstopfen *m* sink plug
Ausgußstutzen *m* discharge connection
Ausgußventil *n* discharge valve
Ausgußwasser *n* waste water
Aushängebühne *f* finger or derrickman's platform
aushärtbar age-hardenable
aushärten to harden, to quench-age, to temper (steel), (Kunststoff) to cure
Aushärten *n* curing, age hardening, cure
Aushärtung *f* cure hardening, (Metall) precipitation hardening
Aushärtungsgrad *m* (Kunststoff) degree of cure
Aushärtungskatalysator *m* curing agent
ausharzen to exude resin
aushauen to cut [out], to chip out
ausheben to draw out
Ausheben *n* lifting; raising; withdrawal, ~ **der Abstiche** *m pl* lifting out the tappings, ~ **des Blocks** lifting out the block
Aushebeplatte *f* ejector pad
Ausheber *m* ejector, trowel
Ausheberbolzen *m* ejection ram, ejector rod
aushebern to siphon out
Ausheberplatte *f* ejector plate
Ausheberrahmen *m* ejector frame
Ausheberstift *m* ejector pin, knock-out pin
Ausheilen *n* recuperation (from mechanical damage)
ausheizen to anneal
Aushilfsmittel *n* auxiliary
aushilfsweise temporarily
aushobeln to plane
Aushöhlung *f* scoop
Aushub *m* (Baugruben) spoil
auskalten to chill, to cool thoroughly
auskehlen to chamfer, to flute, to groove
Auskehlung *f* channel, groove, recess
auskellen to ladle out
Auskerbung *f* groove, notch
auskippen to dump out, to pour out
auskitten to cement, to fill with cement
ausklauben to pick [out]

Auskleidefolie *f* film for lining
auskleiden to line, to coat, to face
Auskleidung *f* covering, hearth lining, jacket, ~ **des Schachtes** shaft lining, **feuerfeste** ~ refractory lining, **säurefeste** ~ acidproof lining, acid seal paint
ausklinkbar disengageable
ausklinken to disengage, to release, to unlatch, to unlock
Ausklopfmaschine *f* beating machine
auskochen to boil out, to extract, to separate by boiling
Auskochen *n* decocting
Auskocher *m* boiler, bucking kier, extractor
Auskochung *f* decoction
Auskohlbad *n* carbonizing bath
auskohlen to carbonize
Auskohlen *n* carbonizing
Auskohlung *f* carbonization
Auskohlungsanlage *f* carbonizing plant
Auskohlungsofen *m* carbonizing stove
Auskohlungsverfahren *n* carbonizing process
Auskohlvorrichtung *f* carbonizing apparatus
Auskopierprozeß *m* (Phot) print-out process
Auskragung *f* lip
auskratzen to scrape out, to scratch out
Auskristallisation *f* crystallization, formation of crystals
auskristallisieren to crystallize [out], to form crystals
Auskristallisierung *f* crystallization, formation of crystals
auskühlen to cool completely, to cool thoroughly
Auskunft *f* information, reference
auskuppeln to disconnect, to uncouple
Auskuppeln *n* disconnecting
ausladen to unload, to discharge
Ausladen *n* unloading, discharging
Ausladung *f* unloading, discharge, (Kran) action radius, reach
Ausläufer *m* tail
auslagern (Bier) to settle, (Duralum) to age, to harden
Auslagersystem *n* goods-out system
Auslandspatent *n* foreign patent
Auslaß *m* outlet, exhaust valve
auslassen to discharge, to exhaust; to omit, to skip
Auslaßenergie *f* exhaust energy
Auslaßhahn *m* discharge cock
Auslaßöffnung *f* port
Auslaßrohr *n* delivery pipe
Auslaßstutzen *m* nozzle box
Auslaßventil *n* escape valve, exhaust valve, outlet valve
Auslauf *m* leak; discharge, drain, outlet, (Maschine) slowing down, ~ **des rohen Laufstreifens** tread skirt
Auslaufbecher *m* viscosity cup

auslaufen to stop running; (ausrinnen) to flow out, to leak, to run out; (s. verjüngen) to taper, (zeitlich) to expire
Auslaufflasche f overflow flask
Auslaufhahn m drain cock
Auslaufofen m flashing furnace
Auslaufpipette f delivery pipet[te]
Auslaufprobe f pouring test
Auslaufrille f spew channel
Auslaufrohr n discharge pipe, outlet tube
Auslaufspitze f discharge tip
Auslaufventil n tap
Auslaufviskosimeter n efflux viscometer, efflux viscosimeter
Auslaufzeit f outflow time
Auslaugeapparat m extraction apparatus
Auslaugebehälter m lixiviating tank
Auslaugeflüssigkeit f lixiviating bath
Auslaugehülse f extraction thimble
Auslaugekasten m lixiviating vat
auslaugen to leach, to extract, to lixiviate, to steep in lye
Auslaugen n leaching, bucking, lixiviation, (Zucker) maceration
Auslauger m extraction apparatus, extractor, leacher
Auslaugeturm m diffusor
Auslaughülse f extraction thimble
Auslaugtrichter m extraction funnel
Auslaugung f leaching, lixiviation, extraction, wet extraction
ausleeren to drain, to empty, to evacuate
Auslegearm m cantilever arm
Ausleger m jib, **Gitter-** ~ lattice jib, **Kasten-** ~ box jib, **Katzen-** ~ trolley jib, **Laufkatzenknick-** ~ articulated trolley jib, **Nadel-** ~ luffing jib, **Seiten-** ~ side jib, **Spitzen-** ~ flying jib
Auslegerdrehkran m jib rotary crane
Auslegerkran m jib crane
Auslegeschrift f patent application published for opposition
Auslegung f explanation, interpretation
auslenken to deflect, to incline
Auslenkhärte f deflection hardness
Auslenkung f deflection from true path, distance from center, inclination
Auslese f selection
Ausleser m selector; separator, sorter
ausleuchten to illuminate
auslochen to punch, to mortise
auslöschen to extinguish, (austilgen) to obliterate, to wipe out, (Feuer) to quench
Auslöschung f extinction, quenching
Auslöseeinrichtung f release mechanism
Auslöseelektron n triggering electron
Auslösehebel m release lever, coupling lever
Auslöseimpuls m triggering impulse
Auslöseknopf m release button

Auslösemagnet m trigger magnet
Auslösemechanismus m releasing mechanism
Auslösemuffe f release sleeve
auslösen to trigger, to induce, to release
Auslöser m trigger
Auslöserelais n trip relay
Auslöseschalter m trip switch
Auslösestrom m trigger current, release current
Auslöseverzug m trigger lag
Auslösevorrichtung f trigger [mechanism], disengaging gear, release [mechanism]
Auslösung f release, liberation
ausloten to level
auslüften to ventilate, to air
ausmahlen to powder, to grind out, to pulverize
Ausmaß n amount, dimension, extent, measure
Ausmauern n lining
Ausmauerung f brick lining
ausmergeln to emaciate, to exhaust
ausmerzen to eliminate, to eradicate; to pick out, to sort out
ausmessen to measure, to gauge, to survey
Ausmessung f measuring, surveying
ausmitteln to form the average
ausmittig eccentric, off center
Ausmündung f mouth, orifice, outlet
ausmultiplizieren to multiply
Ausnahme f exception
ausnahmslos without exception
ausnahmsweise exceptionally
ausnutzbar utilizable, exhaustible
ausnutzen (ausnützen) to exhaust, to make the most of, to make use, to use up, to utilize
Ausnutzung f exploitation; consumption, use, utilization
Ausnutzungsfaktor m utilization factor
Ausnutzungskoeffizient m utilization coefficient
ausölen to oil
auspacken to unpack, to unwrap
auspeilen to sound
auspichen to pitch, to tar
ausplatten to flatten out
auspressen to press out, to squeeze out, (Werkstoffe) to extrude
Auspreßkneter m plodder
Auspreßmaschine f squeezer, squeezing factor
ausprobieren to test, to try
Auspuff m exhaust
Auspuffdampf m exhaust steam, escape steam
auspuffen to exhaust
Auspuffflammendämpfer m exhaust flame damper
Auspuffgas n exhaust gas, burnt gas
Auspuffhaube f exhaust head
Auspuffhub m exhaust stroke
Auspuffilter m exhaust filter
Auspuffleitung f exhaust pipe, exhaust main
Auspuffmaschine f noncondensing engine
Auspufföffnung f exhaust port

Auspuffrohr *n* exhaust pipe, exhaust manifold
Auspuffsammler *m* exhaust collector
Auspuffseite *f* exhaust side
Auspuffstutzen *m* exhaust port
Auspuffventil *n* exhaust valve
auspumpen to pump out, to evacuate
Auspumpen *n* evacuation
ausquadrieren (Math) to square out
ausquetschen to squeeze out, to pump out, to wring out
ausradieren to obliterate, to rub out, to scrape out, (Schrift etc.) to erase
ausräuchern to fumigate, to smoke out
Ausräucherung *f* fumigation
Ausräumen *n* clearing
ausrechnen to calculate, to compute
Ausrechnung *f* calculation, computation
ausrecken (Häute) to set out, (Metall) to elongate, to extend, to stretch
ausreichend adequate, satisfactory, sufficient
ausreifen *n* curing
ausreißen to pull out, to tear out
Ausreißer *m* (Bauw) ripper, (Statist) outlier
Ausreißfestigkeit *f* tear resistance
ausrichten to adjust, to align, to coordinate, to straighten, to true [up], (planieren) to level
Ausrichtmaschine *f* centering molding machine
Ausrichtrahmen *m* centering frame
Ausrichtung *f* alignment, **schlechte ~** misalignment
Ausrichtwalzen *f pl* straightening rolls
ausrotten to exterminate, to extirpate
Ausrottung *f* extermination, extirpation
ausrückbar disconnectable
ausrücken to disconnect, to disengage, to throw out [of gear], to uncouple
Ausrücker *m* disengaging gear, shifter
Ausrückergabel *f* disengaging fork
Ausrückerhebel *m* disengaging lever, throw-out lever
Ausrückhebel *m* clutch lever, disengaging lever
Ausrückkupplung *f* disengaging clutch
Ausrückmuffe *f* disengaging clutch, sliding sleeve
Ausrückstellung *f* disengaged position
Ausrückvorrichtung *f* shifting device, disengaging gear
Ausrückwelle *f* disengaging shaft
ausrüsten to equip, to supply
Ausrüster *m* (v Kranen) rigger
Ausrüstung *f* equipment, outfit, (Text) textile finish
ausrunden to round out
Ausrundung *f* fillet, rounding-off, smoothing out
aussäuern to precipitate by addition of acid
Aussage *f* statement
Aussagefähigkeit *f* reliability
aussalzbar capable of being salted out
Aussalzeffekt *m* salting out effect

aussalzen to salt out, to separate by adding salt, (Seife) to grain
Aussalzen *n* (Seifenherstellg) salting out
Aussalzung *f* salting out, separation by addition of salt
aussaugen to suck out, to drain, to exhaust
Aussaugpumpe *f* suction pump
ausschärfen to deaden, to dull, to neutralize, (Techn) to bevel, to scarf
Ausschärffräsmaschine *f* scarf milling machine
ausschäumen to cease frothing
ausschalten to switch off; to disconnect, to disengage; to eliminate, to exclude, to isolate
Ausschalter *m* circuit breaker, cutout switch, disconnecting switch
Ausschalthebel *m* disconnecting lever
Ausschaltkontakt *m* disconnecting contact
Ausschaltstellung *f* off-position
ausscharten to notch
ausscheiden to eliminate; to exclude, to separate, (absondern) to secrete; (Chem) to deposit, to precipitate; (Feuchtigkeit) to exude, (Med) to discharge, to excrete, (Techn) to sort out
Ausscheiden *n* precipitation, separation
Ausscheider *m* (Med) carrier
Ausscheidung *f* (Drüse) excretion; secretion, (Ablagerung) deposit, (Chem) precipitate, precipitation, (Entfernung) elimination; (Freisetzung) liberation, (Med) excreted substance, (Techn) separation, (v. Feuchtigkeit) exudation
Ausscheidungsglasur *f* precipitation glaze
Ausscheidungshärtung *f* age hardening, precipitation hardening, (Leichtmetall) artificial aging
Ausscheidungskühlung *f* separation cooling
Ausscheidungsmittel *n* separating agent, (Chem) precipitant
Ausscheidungsprodukt *n* precipitate, by-product, precipitated product, separated product, separation product
Ausscheuerung *f* hollow wear
ausschiebbar telescopic
ausschieben to push out, to displace
ausschießen to eject, to cast out, to discharge
ausschirren to ungear
ausschlacken to clear of clinker, to clinker, to slack
ausschlämmen to clean from slime; to elutriate, to wash out
Ausschlämmen *n* elutriation
Ausschlag *m* (Nadel eines Instrumentes) deflection deflection [of the needle], (Pendel) amplitude, swing, (Waage) turn
Ausschlageisen *n* piercer, pounding tool, punch
ausschlagen punch out, (ablehnen) to refuse, (Zeiger) to deflect
ausschlaggebend decisive, determinative, determining

Ausschlag[s]weite *f* amplitude, maximum value of swing
Ausschlagwinkel *m* angle of deflection, angle of deviation
ausschleifen to grind out, to hollow-grind, to polish fine, to whet
ausschleudern to centrifuge; to eject; to expel by centrifugal force, to hydroextract, (Atom) emanate
Ausschleußöffnung *f* feed opening
ausschließen to exclude
ausschließlich except; exceptional
Ausschließung *f* disqualification, exclusion
Ausschließungsprinzip *n* (Atom) [Pauli] exclusion principle
Ausschließungsregel *f* exclusion rule
Ausschlußbereich *m* exclusion sphere
Ausschlußprinzip *n* exclusion principle
ausschmelzen to melt out, to fuse [off], to purify by smelting
Ausschmelzgießerei *f* investment foundry
Ausschmelzung *f* melting, smelting
Ausschmelzverfahren *n* melting out process
ausschmieden to forge [out], to hammer
ausschmieren to smear
ausschmoren to extract by stewing
ausschneiden to cut out, to excise
Ausschneidewerk *n* batch off mill, feed mill
Ausschnitt *m* cutout, trimming (blockmaking), (Detail) detail, (Kerbe) notch, (Kreis) section, sector, (Zuschnitt) blank
Ausschöpfen *n* **des Metalls** ladling out the metal
Ausschöpfkelle *f* ladle, scoop
Ausschöpfung *f* drainage
ausschrauben to screw out, to unscrew
Ausschruppen *n* roughing out
Ausschub *m* exhaust; oscillation
Ausschubwinkel *m* angle of oscillation
ausschütteln to extract, to shake out
Ausschütteln *n* extraction, shaking out
ausschütten to pour out, to dump out, to empty, to pour out
Ausschuß *m* committee; culls, reject, scrap, waste, (Gieß) spoiled casting
Ausschußblech *n* sheet iron scraps
Ausschußpapier *n* waste paper
ausschwefeln to sulfur, to bloom, to fumigate with sulfur
Ausschwefeln *n* sulfur fumigation, blooming
Ausschwefelung *f* sulfur bloom; (Chem) sulfurization
ausschweißen to weld out, to clean iron
ausschwemmen to flood, to flush, to rinse
Ausschwemmen *n* flushing
ausschwenkbar swing-out, swivel[l]ing
ausschwenken to shake [out], to swing out, to swivel
Ausschwimmen *n* (Pigmente) floating, flooding

Ausschwimmverhütungsmittel *n* antiflotation agent
ausschwingen (im Schwung nachlassen) to cease swinging, to decay, to die down, to fade out, (voll schwingen) to swing out
Ausschwingmaschine *f* centrifuge
Ausschwingung *f* swinging out
Ausschwingungsversuch *m* free vibration test
ausschwitzen to bleed (color); to exude, to sweat
Ausschwitzen *n* chalking, exudation, sweating, ~ **des Gleitmittels** lubricant exudation
Ausschwitzung *f* exudation
aussedimentieren to deposit, to deposit as sediment, to sediment
Aussehen *n* appearance
ausseigern to liquate, to separate by liquation
Ausseigern *n* liquation
außen exterior, out, outer
Außenabmessung *f* external dimension, outside dimension
Außenanode *f* outer anode
Außenanstrich *m* exterior coating
Außenanstrichfarbe *f* exterior paint, outdoor paint, paint for outside use
Außenatom *n* peripheric atom
Außenbackenbremse *f* external shoe brake, external contracting brake
Außenbahn *f* (Elektron) outer orbit
Außenbahnkomplex *m* outerorbital complex
Außenbehälter *m* outer container
Außenberieselung *f* outside sprinkling
Außenbeschaffenheit *f* finish appearance
Außenbewitterung *f* outdoor exposure
aussenden to send out, to emit, to radiate
Aussender *m* emitter
Außendruck *m* atmospheric pressure, outside pressure
Aussendung *f* emission
Außendurchmesser *m* outside diameter
Außenelektrode *f* outward electrode
Außenelektron *n* outer electron
Außenelektronenschale *f* outer electron shell
Außenfarbe *f* exterior paint, outdoor paint, paint for outside use
Außenfeuerung *f* external furnace
Außenfräser *m* hollow mill
Außengewinde *n* external thread, male thread, outside thread
Außengummi *m* hose cover
Außenhaut *f* skin
Außenkorrosion *f* external corrosion
Außenkreis *m* external circuit
Außenlack *m* exterior varnish
Außenläufermotor *m* motor of external rotor type
Außenleiter *m* (Elektr) outer mains
Außenleitung *f* external line, outer circuit
Außenluft *f* atmosphere, external air
Außenlufttemperatur *f* outside-air temperature

Außenlunker *m* surface blowhole, surface defect
Außenmantel *m* outer casing, shell (heat exchanger)
Außenmaß *n* outer dimension, outside dimension
Außenmetallisierung *f* outer metallization
außenmittig eccentric, off center
Außenpol *m* external pole
Außenpolanker *m* outerpole armature
Außenpolgenerator *m* external pole generator
Außenrohr *n* outer tube
Außenrundschleifmaschine *f* external grinder
Außenschale *f* (Atom) outermost shell, valence shell
Außenschmarotzer *m* (Biol) ectoparasite
außenseitig external, outdoor
Außenstehender *m* outsider
Außentaster *m* outside calipers
Außenteiler *m* divider
Außentemperatur *f* outdoor temperature, outside temperature, surface temperature
Außentiegel *m* outer crucible
Außenwand *f* outer wall
Außenwegschaltung *f* two position by-pass
Außenwelt *f* outside world
Außenwiderstand *m* (Elektr) external resistance
Außenwinkel *m* (eines Dreiecks) exterior angle (of a triangle)
Außenwirtschaft *f* foreign trade
Außenzarge *f* shell of basket
außerachsial extra axial
außerachsig eccentric
Außerbetriebsetzung *f* shut-down, shutting down
außermittig eccentric
Außermittigkeit *f* eccentricity
außerordentlich extraordinary
außerstande unable
aussetzen to expose; to subject to, to submit; (Pause) to interrupt; to suspend
Aussetzen *n* (Pause) intermittence
aussetzend discontinuous, intermittent
Aussetzer *m* (Comp) dropout
Aussicht *f* prospect, outlook, view
aussichtslos hopeless, useless
aussichtsreich promising, prospective
aussieben to screen out, to filter out, to sift out
aussieden to boil out, to blanch
Aussoltechnik *f* leaching (caverns in rocks)
aussondern to separate; to eliminate; to excrete, to secrete
Aussonderung *f* segregation
aussortieren to sort out
ausspannen to stretch, to extend; to relax, to slacken
aussparen to recess, to hollow out, to isolate, to remove material from
Aussparen *n* **des Niederschlags** keeping back the deposit

Aussparung *f* recess, cavity, clearance, groove, notch, opening, relief, slot
aussprengen to blast; to sprinkle
ausspritzen to eject, (Flüssigkeit) to spurt, to squirt
aussprühen to throw out
ausspülen to wash out, to flush, to rinse, to scour, to wash away
Ausspülung *f* elutriation, wash out
ausstäuben to dust, to winnow
ausstaffieren to equip, to furnish
ausstanzen to punch [out], to blank, to die-cut, to stamp out
Ausstanzen *n* punching with a die, cutting with a die, ~ **von Hand** freehand blanking
Ausstanzer *m* clicker press
ausstatten to equip, to provide
Ausstattung *f* equipment, outfit
Austausch *m* substitute; transfer
ausstechen to cut out
aussteifen to brace, to stiffen, to strengthen
Aussteifungsring *m* stiffening ring
ausstellen to display, to exhibit, to lay out
Ausstellung *f* exhibition
Ausstellungsraum *m* exhibition room, show room
Ausstemmung *f* chiseling out
aussterben to become extinct, to die out
aussteuern (Elektron) to control, to modulate
Aussteuerung *f* modulation
Ausstoß *m* (Phys) discharge, ejection, (Techn) output, production, throughput
ausstoßen to eject, to expel, to throw out, to thrust out
Ausstoßer *m* knockout
Ausstoßleistung *f* output
Ausstoßprodukt *n* waste product
Ausstoßrohr *n* expulsion tube
Ausstoßsystem *n* cleansing system
Ausstoßung *f* (Teilchen) ejection
Ausstoßvolumen *n* volumetric rate of discharge
Ausstoßvorrichtung *f* ejector device
ausstrahlen to emit, to radiate
ausstrahlend emissive
Ausstrahlung *f* emission, irradiation, radiation
Ausstrahlungsfläche *f* radiating surface
Ausstrahlungsintensität *f* intensity of radiation
Ausstrahlungsvermögen *n* emissive power, radiation capacity
Ausstrahlverlust *m* radiation loss
ausstrecken to stretch, to extend, to spread
Ausstreichmesser *n* leveling blade, scraper
ausstreuen to disseminate
ausströmen to discharge, to emanate, to flow [out], (Dampf) to escape, to exhaust
Ausströmen *n* discharge
ausströmend effluent
Ausströmgeschwindigkeit *f* outflow velocity, rate of discharge

Ausströmkanal *m* exhaust port
Ausströmraum *m* volute chamber
Ausströmung *f* discharge, efflux, outflow
Ausströmungsmesser *m* effusiometer
Ausströmungsöffnung *f* escape orifice
Ausströmungsregulator *m* throttle valve
Ausströmungsrohr *n* exhaust pipe, delivery pipe, discharge pipe
Ausstülpung *f* bulge, protrusion
aussüßen to edulcorate, to wash
Aussüßglas *n* wash bottle
Aussüßpumpe *f* leaching pump
Aussüßrohr *n* washing tube
Aussüßvorrichtung *f* washing arrangement, washing bottle
austarieren to calibrate, to tare
Austastung *f* blanking
Austausch *m* exchange, interchange; (Ersatz) replacement, substitution
Austauschalgorithmus *m* exchange algorithm, gauss algorithm
Austauschazidität *f* exchange acidity
austauschbar exchangeable, interchangeable
Austauschbarkeit *f* exchangeability
Austauschbeschränkung *f* exchange limitation
Austauschboden *m* (Rektifikation) plate, tray
Austauschchromatographie *f* exchange chromatography
Austauscheinheit *f* (Rektifikation) transfer unit
Austauscheinrichtungen *pl* mass transfer equipment
austauschen to exchange, to interchange, (ersetzen) to replace
Austauschenergie *f* interchange energy
Austauscher *m* exchanger
Austauscherharz *n* ion exchange resin
Austauscherwirkungsgrad *m* efficiency of exchanger
Austauschfläche *f* (Dest) interfacial area
Austauschgerade *f* (Rektifikation) operating line
Austauschgerbstoff *m* replacement tannin, integral tannin
Austauschgröße *f* exchange coefficient
Austauschhalbwertzeit *f* exchange half-life
Austauschkorrektur *f* exchange correction
Austauschmaschine *f* stand-by machine
Austauschpotential *n* exchange potential
Austauschprodukt *n* substitute product
Austauschreaktion *f* exchange reaction
Austauschröhre *f* spare tube
Austauschstoff *m* substitute
Austauschstromdichte *f* current density
Austauschteil *n* interchangeable part
Austauschterm *m* exchange term
Austauschwerkstoff *m* substitute material
Austauschwerkzeug *n* substitute tool
Austauschwirkung *f* exchange action, (Wärmeübergang) efficiency
austeilen to distribute

Austenit *m* gamma iron, (Min) austenite
austenitisch (Met) austenitic
Austenitstahl *m* austenitic steel
Australen *n* australene, pinene
Australgerbung *f* Australian tannage
Australit *m* (Min) australite
Australol *n* australol
austreiben to drive off [or out], to expel, to regenerate
Austreiben *n* desorption; driving off, expelling; regeneration; stripping
Austreiber *m* generator (gases from liquids)
Austreibung *f* expulsion; desorption, regeneration; stripping
Austreibungswärme *f* heat of expulsion
austreten to emerge, to flow out, to project, to protrude, (Dampf) to escape, to exhaust
Austrieb *m* lateral waste; spew, (Kunststoff) flash
Austriebnut *f* groove spew (e. g. of a mo[u]ld)
Austriebrille *f* spew groove
Austritt *m* (Austreten v. Flüssigkeiten) efflux, (Austrittstelle) exit, issue, outlet, (Dampf) escape
Austrittgeschwindigkeit *f* exit velocity
Austrittleitung *f* outlet piping, outlet tubing
Austrittsarbeit *f* (Atom) electron affinity work function, (Festkörperphysik) work function
Austrittsblende *f* (Opt) exit slit
Austrittseite *f* exit side, catcher's side
Austrittsenergie *f* total exhaust energy
Austrittsöffnung *f* orifice, outlet
Austrittspotential *n* (Elektron) work function
Austrittsschlitz *m* outlet port
Austrittstelle *f* der Luft air leakage
Austrittstemperatur *f* outlet temperature
Austrittsventil *n* outlet valve
Austrittswert *m* output value
Austrittverlust *m* discharge loss
austrocknen to dry, to desiccate, to drain, to exsiccate, (Holz) to season
Austrocknen *n* drying, desiccation, draining, (Holz) seasoning
Austrocknung *f* drying, desiccation, exsiccating
Austrocknungspunkt *m* drying up point
auströpfeln to cease dripping; to drip out
austuschen to ink
ausüben to practise, to carry on, to exert
Ausvulkanisation *f* full cure, tight cure
auswachsen to germinate, to grow, to sprout
auswägen to weigh [out]
auswählen to choose, to select
auswärmen to anneal, to heat, to roast, to warm
Auswärmeofen *m* annealing furnace
auswässern to water, to soak, to steep
Auswahl *f* choice, selection
Auswahlprinzip *n* selection principle
Auswahlregel *f* selection rule
Auswahlverfahren *n* selective procedure

auswalzen to roll [out]
auswaschbar removable by washing; washable
Auswaschbottich *m* rinsing vat, wash bowl
auswaschen to wash out, to edulcorate, to scour, to scrub
Auswaschleitung *f* washout connection
Auswaschung *f* washing out; (Geol) erosion
Auswaschzahl *f* (Füllkörperkolonne) height equivalent to transfer unit (HTU)
auswechselbar exchangeable, changeable, detachable, interchangeable, removable
Auswechselbarkeit *f* exchangeability
auswechseln to exchange, to interchange, (erneuern) to renew, (ersetzen) to replace
Auswechselplatte *f* duplicate plate
Auswechslung *f* exchange, exchanging, interchange, interchanging, renewing
Ausweg *m* outlet
ausweichen to deviate; to escape, to slip away; to yield
Ausweichen *n* deviation; escape, evasion
ausweichend elusive, evasive
Ausweichlösung *f* alternate solution
Ausweichpunkt *m* yield point
Ausweichreaktion *f* escape reaction
Ausweichung *f* deflection; deviation; evasion
ausweitbar stretchable, dilatable, expanding, extensible
ausweiten to extend, to distend, to widen
auswendig exterior, outer
auswerfen to eject, to throw out
Auswerfen *n* ejection
Auswerfer *m* ejector, ejector tool; knock-out
Auswerferbolzen *m* ejector rod
Auswerferkolben *m* ejector ram
Auswerfernocke *f* knock-out cam
Auswerferplatte *f* knock-out pin plate
Auswerferring *m* knock-out ring
Auswerferstift *m* ejector pin, knock-out pin
Auswerfertisch *m* reject tray
Auswerfung *f* rejection
auswertbar evaluable, (Patent) utilizable
auswerten to analyze, to evaluate, to interpret
Auswerteverfahren *n* evaluation method
Auswertung *f* evaluation, interpretation
auswichten to balance, to center
auswinden to wring out
Auswirkung *f* effect
auswischen to wipe out, to eradicate, to rub out
auswittern to weather, (Chem) to effloresce, (Holz) to season
Auswitterung *f* weathering, (Chem) efflorescence, (Holz) seasoning
auswringen to wring out, to squeeze out
Auswuchs *m* outgrowth, (Biol, Med) excrescence
Auswuchtdorn *m* balancing mandrel
auswuchten to balance, to compensate, to counterbalance
Auswuchtpaste *f* balance dough

Auswuchtung *f* balancing
Auswuchtungsverhältnis *n* mechanical balance
Auswuchtvorrichtung *f* balancing device, balancing fixture
Auswurf *m* discharge, ejection, refuse, trash
Auswurfgestein *n* eruptive rocks, volcanic rocks
Auswurfleitung *f* discharge piping
Auswurfrohr *n* discharge pipe
Auswurfvorrichtung *f* ejecting device
auszacken to indent, to jag, to notch, to serrate
Auszackung *f* serration
auszählen to count
auszeichnen to distinguish; to label, to mark out
Auszeichnung *f* distinction; label
auszentrifugieren siehe zentrifugieren
ausziehbar extensible, (Metall) ductile, (zusammenlegbar) collapsible, telescopic
Ausziehbarkeit *f* (Met) ductility
ausziehen to bleed (color), (herausziehen) to exhaust, to extract, (langziehen) to extend, to stretch
Ausziehen *n* extending; extracting
Ausziehröhre *f* telescopic tube
Ausziehrohr *n* telescopic tube
Ausziehschnecke *f* drawing-out worm
Ausziehtisch *m* extension table
Ausziehtusche *f* drawing ink, indian ink
Ausziehung *f* extraction; (Mech) extrusion
Ausziehverfahren *n* (Färberei) exhaust process
Ausziehwalze *f* sheeting out roller
Auszug *m* essence, extract, quintessence, (Pharm) tincture
Auszuggehalt *m* content of extract
Auszugmehl *n* superfine flour
Auszwirn *m* cable twist, cord
Autan *n* autan
Autoabgas *n* automobil exhaust gas
Autoagglutination *f* auto-agglutination
autobarotrop autobarotropic
Autobarotropie *f* autobarotropy
Autochrom *n* autochrome
Autochromdruck *m* autochrom(e) printing
Autochromverfahren *n* (Phot) autochrome process
Autodigestion *f* autodigestion, self-digestion
Autodyn *n* autodyne
Autoelektronenemission *f* auto electronic emission
Autofining-Verfahren *n* autofining process
autogen autogenous
Autogengas *n* dissolved acetylene
Autogengerät *n* gas welding apparatus
Autogenmühle *f* autogenous tumbling mill
Autogenschweißverfahren *n* autogenous welding process
Autographie *f* autography
autographisch autographic
Autohämolyse *f* autoh[a]emolysis
Autohämolysin *n* autoh[a]emolysin

autohämolytisch autoh[a]emolytic
Autointoxikation *f* autointoxication, self-intoxication
Autoionisation *f* autoionization
Autokatalysator *m* autocatalyst
Autokatalyse *f* autocatalysis
autokatalytisch autocatalytic
Autoklav *m* autoclave, cooker, digester, pressure tank, sterilizer, ~ **für Leitfähigkeitsmessungen** *f pl* high pressure conductance cell
Autoklaveneinzelheizer *m* autoclave single heater
Autoklavenverfahren *n* autoclave method
Autoklavpresse *f* steam dome press
Autokollimator *m* (Opt) autocollimator
Autokorrelationsfunktion *f* autocorrelation function
Autokran *m* truck crane
Autolack *m* automobile finish, motor car enamel, motor car lacquer
Autolackiererei *f* car paint shop
Autoluftschlauch *m* automobile inner tube
Autolyse *f* autolysis
Automat *m* automat, automatic machine, (Verkaufsautomat) slot machine, vending machine
Automatenstahl *m* free-cutting steel
Automation *f* automation
automatisch automatic, self-acting, ~ **betätigt** automatically controlled, ~ **registrierend** self-recording
automatisieren to automate
Automatisierung *f* automation, automatic control
Automolith *m* (Min) automolite
automorph automorphic
Automorphie *f* automorphism
autonom autonomic
autooxydationsfähig autoxidizable
Autooxydationsfähigkeit *f* autoxidizability
Autopolitur *f* car polish
Autopolymerisation *f* auto-polymerization, self-polymerization
Autor *m* author
Autoracemisierung *f* autoracemization
Autoradiogramm *n* (autoradiogram)
Autoradiograph *m* radioautograph
Autoradiographie *f* radioautography
Autorenanweisungen *pl* instructions to authors
Autorenregister *n* author index
Autospachtel *f* sanding sealer
Autostereoregulierung *f* autostereo-regulation
Autotrophie *f* (Biol) autotrophy
Autotypie *f* autotype, halftone engraving, process engraving
autoxydabel autoxidizable
Autoxydation *f* autoxidation, self-oxidation
Autoxydator *m* autoxidizer
Autunit *m* (Min) autunite, calcouranite

Auxin *n* auxin
auxochrom auxochromic
Auxochrom *n* (Opt) auxochrome
Ava *f* ava, kava, kavakava, kawa
Avalit *m* avalite
Avenin *n* avenin, legumin
Aventurin *m* (Min) aventurine
Aventurinfeldspat *m* (Min) aventurine feldspar
Aventuringlas *n* (Techn) aventurine glass
Aventurinquarz *m* (Min) aventurine quartz
Aventurinstein *m* (Min) aventurine feldspar
aventurisieren to aventurize
Avertin *n* (HN) avertin
Avicin *n* avicine
Avicularin *n* avicularin
Avidin *n* avidin
Avidität *f* (obs) avidity
Avitaminose *f* vitamin deficiency disease, (Med) avitaminosis, hypovitaminosis
Avivage *f* reviving, brightening, finish, finishing, livening, (Seide) scrooping
avivierecht not affected by reviving, (Seide) fast to scrooping
Avivierechtheit *f* (Text) brightening fastness
avivieren to revive, to brighten, to improve, to restore, (Seide) to scroop
Aviviermittel *n* brightening agent
Avogadrosche Konstante *f* Avogadro's constant
Avogadrosche Zahl *f* Avogadro's constant
Avogadrosches Gesetz *n* Avogadro's law
Awadscharidin *n* avadharidine
Awaruit *m* (Min) awaruite
Axerophthen *n* axerophthene
Axerophthol *n* axerophthol, retinol, vitamine A$_1$
axial axial, (in der Wannenform) boat axial
Axialbeanspruchung *f* axial stress
Axialdissymmetrie *f* (Stereochem) axial dissymmetry
Axialdruck *m* axial pressure
Axialkraft *f* axial force
Axialpumpe *f* axial-flow pump
Axialschlag *m* wheel flap, ~ **eines Rades** lateral runout
Axialschub *m* axial thrust
Axialspannung *f* axial stress
Axialstromturbine *f* axial flow turbine
axialsymmetrisch axial symmetric
Axialturbine *f* axial flow turbine
Axialverdichter *m* axial-flow compressor
Axin *n* axin
Axinit *m* (Min) axinite, thumite
Axinsäure *f* axinic acid
Axiometer *n* axiometer, telltale gauge
Ayanin *n* ayanin
Ayapanin *n* ayapanin
Ayapin *n* ayapin
Azacyclonol *n* azacyclonol
Azafrin *n* azafrin
Azaleatin *n* azaleatin

Azalein *n* azaleine, fuchsin
Azamethonium *n* azamethonium
Azarsin *n* azarsine
Azaserin *n* azaserine
Azaurolsäure *f* azaurolic acid
Azelain *n* azelain
Azelainhalbaldehyd *m* azelaic semialdehyde
Azelainsäure *f* azelaic acid
Azelaoin *n* azelaoin
Azelaon *n* azelaone
Azelat *n* azelate
Azenaphthen *n* acenaphthene
Azenaphthylen *n* acenaphthylene
azentrisch acentric
azeotrop azeotropic
Azeotropie *f* azeotropy
Azeotroppunkt *m* azeotropic point
Azepin *n* azepine
Azetal *n* siehe Acetal
Azetat *n* siehe Acetat
Azetidin *n* azetidine
Azetobakter *m* azetobacter
Azeton *n* siehe Aceton
Azetonkörper *m* acetone body
Azetyl- siehe Acetyl-
Azetylen siehe Acetylen
Azi- s. auch Aci-
Aziäthan *n* aziethane, diazoethane
Azibenzil *n* azibenzil
Azid *n* azide
Azidimetrie *f* acidimetry
Azidinechtgelb *n* azidine fast yellow
Azidinfarbe *f* azidine color
Azidität *f* acidity (s. auch Acidität)
Azido- azido-, triazo-
Azidoessigsäure *f* azidoacetic acid
azidolytisch acidolytic
Azimethan *n* azimethane, diazomethane
Azimino- azimino-
Aziminobenzol *n* aziminobenzene, benzotriazole
Azimut *n* (Opt) azimuth [angle]
Azimutalschwingung *f* azimuthal mode
Azimutkompaß *m* azimuth compass
Azin *n* azine
Azinfarbstoff *m* azine color, azine dye, azine dye-stuff
Azingrün *n* azine green
Azipyrazol *n* azipyrazole
Azipyridil *n* azipyridil
Aziridin *n* aziridine
Azlacton *n* azlactone
Azlactonsynthese *f* azlactone synthesis
Azlonfaser *f* azlon fiber
Azoameisensäure *f* azoformic acid, azodicarbonic acid
Azoanisol *n* azoanisole
Azobenzol *n* azobenzene, azobenzol
Azoblau *n* azoblue
Azodicarbonamid *n* azodicarbonamide, azoformamide

Azodicarbonsäure *f* azodicarboxylic acid
Azodimethylanilin *n* azodimethylaniline
Azodruck *m* azoic print
Azodulcin *n* azodulcin
Azoechtfarbe *f* fast azo color, fast azo dye
Azo-Entwicklungsbad *n* azoic developing bath
Azofarbe *f* azo color, azo dye, azo dyestuff
Azofarbstoff *m* azo dye
Azoflavin *n* azoflavin, Indian yellow
Azoformamid *n* azoformamide, azodicarbonamide
Azogelb *n* azoflavin, Indian yellow
Azogruppe *f* azogroup
Azoimid *n* azoimide, diazoimide, hydrazoic acid
Azokarmin *n* azo carmine
Azokörper *m* azo body, azo compound, azo derivative
Azol *n* azole
Azolitmin *n* azolitmin
Azolitminpapier *n* azolitmin paper
Azometer *n* azometer
Azomethan *n* azomethane
Azomethinchelat *n* azomethine chelate
Azomycin *n* azomycin
Azonaphthalin *n* azonaphthalene
Azophenetol *n* azophenetole
Azophenol *n* azophenol
Azophosphin *n* azophosphin
Azophthalsäure *f* azophthalic acid
Azopiperonal *n* azopiperonal
Azopiperonylsäure *f* azo piperonylic acid
Azorit *m* (Min) azorite
Azosäureblau *n* azo acid blue
Azosäuregelb *n* azo acid yellow, azoflavin, Indian yellow
Azosäuregerbstoff *m* acid azo color, acid azo dye
Azosalicylsäure *f* azosalicylic acid
Azotid *n* azotide
Azotin *n* azotine
Azotobakterium *n* azotobacter
Azotoluol *n* azotoluene
Azotometer *n* azotometer
Azoverbindung *f* azo compound, azo derivative
Azoxyanisol *n* azoxyanisole
Azoxybenzoesäure *f* azoxybenzoic acid
Azoxybenzol *n* azoxybenzene
Azoxynaphthalin *n* azoxynaphthalene
Azoxyphenetol *n* azoxyphenetole
Azoxyphenol *n* azoxyphenol
Azoxyverbindung *f* azoxy compound, azoxy derivative
Azoxyzimtsäure *f* azoxycinnamic acid
Azozimtsäure *f* azocinnamic acid
azozyklisch azocyclic
Aztechin *n* aztequine
Azulen *n* azulene
Azulin *n* azulin
Azulminsäure *f* azulmic acid, azulmin

Azurblau *n* azure, skyblue
Azurin *n* azurine
Azurit *m* (Min) azurite, chessylite, lapis lazuli, lazulite, mountain blue
Azurolblau *n* azurol blue
Azurstein *m* (Min) azurite, chessylite, lapis lazuli, lazulite
A-Zustand *m* A-state
azyklisch acyclic

B

Bababudanit *m* (Min) bababudanite
Babassuöl *n* babassu oil
Babbitmetall *n* babbitt metal, bearing metal
Babelquarz *m* (Min) babel-quartz
Babingtonit *m* (Min) babingtonite
Bablahschote *f* bablah gall
Babylonquarz *m* babel-quartz
Bachbildung *f* (Füllkörperkolonne) channeling
Bacille *f* (s. a. Bazille) bacillus
Bacillus *m* bacillus, ~ **botulinus** Bacillus botulinus, Clostridium botulinum, ~ **coli** Escherichia coli, ~ **tetani** Clostridium tetani
Bacitracin *n* bacitracin
backbrassen (Mar) to brace back
Backe *f* cavity plug, cheek, jaw, (Bremse) brake shoe
backen to bake, to cement (steel), to coke, to fire (pottery), to fry, (brennen) to burn
Backen *n* caking, clinkering, coking, ~ **der Kohle** clinkering of coal
Backenbrecher *m* jaw crusher
Backenbremse *f* shoe brake
Backenform *f* split-cavity mold, split-follower mold
Backenknochen *m* cheekbone
Backen-Kreiselbrecher *m* roll jaw crusher
Backenwerkzeug *n* bar mo[u]ld
Backenzahn *m* molar tooth
backfähig capable of baking or caking
Backfähigkeit *f* bakability, baking characteristics, caking property, cementing property
Backfett *n* cocking fat
Backhefe *f* baker's yeast
Backkohle *f* bituminous coal, caking coal
Backofen *m* [baking] oven
Backofenpyrometer *n* bakery pyrometer
Backpulver *n* baking powder
Backsteinofen *m* brick kiln
Backsteintrog *m* brick trough
Backströmit *m* (Min) backstromite
Bactericid *n* (s. a. Bakterizid) bactericide
Bacterie *f* bacterium (siehe auch Bakterie)
Bad *n* bath, (Färb) dip, steep, ~ **bei steter (unveränderlicher) Temperatur** constant-temperature bath, **das ~ entkupfern** to decopper the bath, **ein ~ ansetzen** to prepare a bath, **galvanisches ~** galvanic bath, electroplating bath
Badbeschickung *f* preparation of the bath
Badbewegung *f* agitation of the bath
Badbottich *m* vat
Baddeleyit *m* (Min) baddeleyite
Badesalz *n* bath salt
Badfeder *f* flat spring

Badflüssigkeit *f* bath liquid, bath solution
Badfutter *n* trough lining
Badian *m* star anise
Badianöl *n* star anise oil
Badische Säure *f* 2-naphthylamine-8-sulfonic acid, baden acid, badische acid
Badkasten *m* (Galv) electrolytic cell
Badprobe *f* bath sample
Badreaktion *f* bath reaction
Badschmelze *f* molten electrolyte
Badspannung *f* (Galv) bath potential, cell voltage
Badstromregler *m* (Galv) bath current regulator
Badunterabteilung *f* bath compartment
Badverarmung *f* impoverishment of the bath
Badwiderstand *m* (Galv) bath resistance
Badzusatz *m* bath addition agent
Baeckeol *n* baeckeol
bähen to foment, to sweat charcoal
bälgen to skin
Bändsel *n* (Mar) lashing, seizing
Baeomycessäure *f* beomycesic acid
Bär *m* (Metall) skull
Bärenfett *n* bear's grease
Bärenklauöl *n* acanthus oil, bear's breech oil
Bärentraube *f* (Gerb) common bearberry
Bärenwurzel *f* saxifrage, spicknel
Bärgewicht *n* ram weight
Bärlapp *m* clubmoss, lycopodium
Bärlappmehl *n* clubmoss seed, lycopodium, vegetable sulfur
Bärlappsamen *m* lycopodium powder, lycopodium seed, vegetable sulfur
Bärlappsporen *pl* lycopodium powder, lycopodium seed
Bärlauch *m* (Bot) ramson
Bärlauchöl *n* broad-leaved garlic oil
Bärwurzöl *n* bear's wort oil, saxifrage oil, spicknel oil
Bäumen *n* (Web) warping
Bäumlerit *m* (Min) baumlerite
Baeyersche Probe *f* Baeyer test
Baeyersche Spannungstheorie *f* (Chem) Baeyer strain theory
Baffle *n* baffle [plate]
Bagasse *f* bagasse, cane trash, crushed sugar cane
Bagassefeuerung *f* bagasse furnace
Bagassezellstoff *m* bagasse cellulose
Bagatelle *f* bagatelle, trifle
Bagger *m* digger, dredger, excavator
Baggerbetrieb *m* dredging
Baggerseil *n* dredge rope
Baggertorf *m* dredged peat
Bagrationit *m* (Min) bagrationite
Bahiagummi *n* bahia rubber, mangabeira rubber

Bahn *f* path, road, trajectory, ~ **des Lichtzentrums** photocentric orbit
Bahnachse *f* axis parallel to the path
Bahnanalyse *f* analysis of path
Bahnbewegung *f* (Atom) orbital motion
bahnbrechend epoch-making
Bahnbrecher *m* pioneer
Bahndrehimpuls *m* (Atom) orbital angular momentum, orbital moment, path spin
Bahndurchmesser *m* (Atom) orbital diameter
Bahnebene *f* orbital plane
Bahnelektron *n* orbital electron
Bahnellipse *f* (Astr) orbital ellipse
bahnen to clear a way, to make a path, to smooth a way
Bahnenführung *f* (Web) web guide
Bahnenmaterial *n* sheeting
bahnentartet (Atom) orbitally degenerate
Bahnentrockner *m* tenter drier, web drier
Bahngeschwindigkeit *f* (Atom) orbital velocity
Bahnhammer *m* face hammer
Bahnimpuls *m* linear momentum, (Atom) orbital momentum
Bahnkorrektur *f* orbit correction
Bahnkrümmung *f* track curvature
Bahnkurve *f* trajectory
Bahnmagnetismus *m* (eines Elektrons) orbital moment (of an electron)
Bahnmoment *n* orbital moment
Bahnmomentanteile *pl* orbital moment components
Bahnradius *m* orbital radius, (Atom) orbital radius
Bahnspur *f* track formation
Bahnspurenphotometrie *f* track photometry
Bahnübergang *m* (eines Elektrons) orbital transition (of an electron)
Bahnumdrehung *f* orbital revolution
Bahnumkehr *f* path reversal
Baicalein *n* baicalein
Baicalin *n* baicalin
Baikalit *m* (Min) baikalite
Baikiain *n* baikiaine
Bajonettanschluß *m* bayonet fitting
Bajonetthülse *f* bayonet socket
Bajonettrohr *n* bayonet tube
Bajonettverschluß *m* bayonet lock or joint
Bakelit *n* (HN) bakelite
Bakelitlack *m* bakelite varnish
Bakeln *n* rotary ironing
Bakerit *m* (Min) bakerite
Bakteriämie *f* (Med) bacteriemia
Bakterie *f* bacterium, **aerobe** ~ aerobic bacterium, **anaerobe** ~ anaerobic bacterium
bakteriell bacterial
Bakterien *pl* bacteria, **denitrifizierende** ~ denitrifying bacteria, **gramnegative** ~ gram-negative bacteria, **grampositive** ~ gram-positive bacteria, **thermophile** ~ thermophil[ic] bacteria
bakterienartig bacteroid[al], bacterial, bacterioid
Bakterienaufschwemmung *f* bacterial suspension
Bakterienbeize *f* bacterial bate
Bakterienfärbung *f* staining of bacteria
bakterienfeindlich antibacterial, bactericidal, germicidal
bakterienfest bacteria-resistant, mildew-proof
bakterienförmig bacteriform
bakterienfrei free of bacteria
Bakteriengift *n* bacterial toxin, bactericide, bacteriotoxin
Bakterienkrieg *m* bacterial warfare
Bakterienkunde *f* bacteriology
Bakterienlehre *f* bacteriology
Bakterienmikroskop *n* microbioscope
Bakteriennachweis *m* demonstration of bacteria
Bakteriennährlösung *f* solution for bacterial cultures
Bakterienresistenz *f* bacteria resistance
Bakterienstamm *m* bacterial strain
bakterientötend bactericidal
Bakterientoxin *n* bacterial toxin
Bakterienträger *m* carrier of bacteria
bakterienvernichtend bactericidal
Bakterienwachstum *n* bacterial growth
bakterienwachstumshemmend antibacterial, antimicrobic, bacteriostatic
Bakterienwachstumshemmung *f* bacteriostasis
Bakterienzählung *f* bacterial count
Bakterienzelle *f* bacterial cell
Bakteriochlorin *n* bacteriochlorine
bakteriogen bacteriogenic, bacteriogenous
Bakteriohämolysin *n* bacterioh(a)emolysin
Bakteriologe *m* bacteriologist
Bakteriologie *f* bacteriology
bakteriologisch bacteriological
Bakteriolyse *f* bacteriolysis
Bakteriolysin *n* bacteriolysin
bakteriolytisch bacteriolytic
Bakteriophäophytin *n* bacteriopheophytin
Bakteriophage *m* (Biol) bacteriophage
Bakteriose *f* (Bot) bacteriosis
Bakterioskopie *f* bacterioscopy
bakterioskopisch bacterioscopic
Bakteriostase *f* bacteriostasis
Bakteriostatikum *n* bacteriostatic
bakteriostatisch bacteriostatic
bakteriotherapeutisch bacteriotherapeutic
Bakterium *n* (pl Bakterien) bacterium (pl. bacteria), germ
Bakteriurie *f* (Med) bacteriuria
bakterizid bactericide, bactericidal
Bakterizid *n* bactericide
bakteroid bacteroid, bacteroidal
Bakuin *n* bakuin
Bakuol *n* bakuol
BAL siehe British-Anti-Lewisit

Balance *f* balance
Balancier *m* balancing arrangement, beam
balancieren to balance
Balancierspant *n* balance frame
Balanophorin *n* balanophorin
Balasrubin *m* (Min) balas ruby
Balata *f* balata
Balatagummi *n* balata gum
Balatariemen *m* balata belt
Baldrian *m* valerian
baldrianartig valerianaceous
Baldrianextrakt *m* extract of valerian, valerian extract
Baldrianisoamylester *m* isoamyl valerate
Baldrianöl *n* valerian oil
Baldriansäure *f* isovaleric acid
Baldriansäureäthylester *m* ethyl isovalerate
baldriansauer isovalerate
Baldriantinktur *f* valerian tincture
Baldrianwurzel *f* (Bot) valerian rhizome, valerian root
Balfourodin *n* balfourodine
Balg *m* skin, (Hülsenfrucht) husk, (Orgel) bellows
Balgauszug *m* bellows extension
Balglinse *f* bellow-ring packing
Balken *m* beam, girder, timber, **ausladender** ~ (oder freitragender ~) cantilever [beam], semi-beam
Balkenarretierung *f* beam arresting device
Balkenbelastung *f* beam loading
Balkenbiegung *f* deflection of a beam
Balkeneisen *n* kamacite
Balkenende *n* beam end
Balkenknie *n* beam end
Balkenmikrowaage *f* beam microbalance
Balkenwaage *f* beam scale
Ball *m* ball, globe, sphere
Ballast *m* ballast
Ballastbehälter *m* ballast tank
Ballastmaterial *n* inert material
Ballastwiderstand *m* loading resistance
ballen to bale, to conglomerate, to form balls
Ballen *m* bale, pack
ballenförmig bale-shaped
Ballenlänge *f* length of body
Ballenpresse *f* baling press
Ballenschneidemaschine *f* bale cutting machine
ballförmig ball-shaped, globular, spherical
ballig bulged
Ballistik *f* ballistics
ballistisch ballistic
Ballistit *n* ballistite
Ballon *m* (Chem) carboy, balloon, balloon flask, demijohn
Ballonabfüller *m* carboy filling apparatus
Ballonentleerer *m* carboy emptier
Ballonfänger *m* balloon control device
Ballonfilter *n* carboy filter

Ballongestell *n* carboy holder
Ballonheber *m* carboy pump
Ballonkipper *m* carboy inclinator, carboy tipper
Ballonkorb *m* carboy hamper
Ballonreifen *m* balloon tire, doughnut tire (Am. E)
Ballonschutzkorb *m* carboy protective basket
Ballonstoff *m* balloon fabric
Ballung *f* agglomerate, balling
Ballungsfähigkeit *f* balling property
Ballungszentrum *n* conurbation, megalopolis
Balmerserie *f* (Spektr) Balmer series
Balsaholz *n* balsa wood
Balsam *m* balm, balsam, **kanadischer** ~ Canada balm
Balsamapfel *m* balm apple, balsam apple
balsamerzeugend balsamiferous
Balsamharz *n* balsamic resin
Balsamholz *n* balm wood
balsamieren to embalm
balsamisch balmy, balsamic
Balsamtanne *f* balsam fir
Baltimorit *m* (Min) baltimorite
Balyrohr *n* Baly tube
Balysche Zelle *f* (Spektr) Baly cell, Baly tube
Bambus *m* bamboo
Bambusgras *n* bamboo grass
Bambusrohr *n* bamboo cane
Bambuszucker *m* bamboo sugar
banal common, commonplace
Bananenstecker *m* (Elektr) banana plug
Bananenwachs *n* pisang wax
Bancazinn *n* (Leg) banca tin
Bancora-Verfahren *n* (Wolle) Bancora process
Band *m* volume
Band *n* band, belt, ribbon, strip, tape; (Fließband) assembly line; ~ **der zweiten Harmonischen** (Spektr) second harmonic band, first overtone band, ~ **ohne Ende** endless band, endless belt
Bandabsetzer *m* discharge belt
Bandabstand *m* (Festkörperphysik) band gap
Bandachat *m* (Min) banded agate, ribbon agate
Bandagewalzwerk *n* strip mill, tire mill
Bandanguß *m* tab gate
Bandarbeit *f* moving belt production
Bandbreite *f* (Spektr) band width
Bandbreiteneinstellung *f* (Spektr) band width control
Bandbreitenregelung *f* band-width control, frequency-range control
Bandbreitenschalter *m* band selector switch
Bandbremse *f* band brake, strap brake
Banddrehrohrofen *m* grate-kiln system
Bande *f* (Spektr) band
Bandeisen *n* band iron, hoop iron, strap iron, (Verpackung) metal strapping
Bandenanalyse *f* band analysis
Bandenkopf *m* (Spektr) band head

Bandenreihe *f* (Spektr) progression
Bandenspektrum *n* band spectrum
Bandextrakteur *m* belt extractor
Bandfilter *n* belt filter
Bandförderer *m* belt carrier, belt conveyor
Bandförderung *f* belt conveyance
Bandheizkörper *m* (Heizband) heater band, Heizung mit ~ strip heating
Bandheizung *f* strip heating
Bandjaspis *m* (Min) ribbon jasper
Bandklammer *f* clasp
Bandkupplung *f* belt coupling
Bandlack *m* ribbon lacquer
Bandlackierverfahren *n* coil coating
Bandmagnet *m* band magnet, strip magnet
Bandmaß *n* measuring tape, tape measure
Bandmaterial *n* strip material
Bandmischer *m* ribbon mixer
Bandolin *n* bandoline
Bandrührer *m* spiral agitator
Bandsäge *f* band saw
Bandscheibe *f* (Med) intervertebral disk
Bandscheider *m* belt separator
Bandschleifmaschine *f* belt grinding machine
Bandschnitt *m* ribbon cutting
Bandseil *n* flat cable, flat rope
Bandseparator *m* belt separator
Bandsiegeln *n* tape sealing
Bandsintermaschine *f* up-draught sintering machine
Bandspan *m* continuous chip
Bandspule *f* ribbon coil
Bandstahl *m* band steel, hoop steel, steel strip, strip steel
Bandstahlschnitt *m* steel rule die
Bandströmung *f* laminar flow, streamline flow
Bandtransporteur *m* belt conveyor
Bandtrockner *m* belt drier
bandumwickelt taped
Bandwaage *f* weight belt feeder
Bandwickelmaschine *f* strip winding machine
Bandwurm *m* (Med) tapeworm
Bandzellenfilter *n* travelling pan filter
Bangreagenz *n* Bang reagent
Bangsche Lösung *f* Bang's solution
Banisterin *n* banisterine
Bank *f* bench; (Geol) bed, (Mar) bank
Bankazinn *n* (Leg) banca tin
Bankschraubstock *m* bench vice
bannen to avert, to banish
Baobaböl *n* baobab oil
Baphiin *n* baphiin
Baphiniton *n* baphinitone
Baptifolin *n* baptifoline
Baptisia tinctoria *f* baptisia tinctoria, wild indigo
Baptisin *n* baptisin
Bar *n* bar (international unit of pressure . ,7 bear), international unit of pressure

Barbaloin *n* barbaloin
Barbatolsäure *f* barbatolic acid
Barbierit *m* (Min) barbierite
Barbital *n* barbital, diethylbarbituric acid
Barbitursäure *f* barbituric acid, malonyl urea
Barcenit *m* (Min) barcenite
Barech *m* kelp
Baregin *n* baregin
Barilla *n* barilla soda
Barium *n* (Symb Ba) barium, **schwefelsaures** ~ barium sulfate
Bariumacetat *n* barium acetate
Bariumaluminat *n* barium aluminate
Bariumborat *n* barium borate
Bariumboratglas *n* barium borate glass
Bariumbrei *m* barium meal
Bariumbromat *n* barium bromate
Bariumcarbid *n* barium carbide, barium acetylide
Bariumcarbonat *n* barium carbonate
Bariumchromat *n* barium chromate
Bariumcyanid *n* barium cyanide
Bariumdioxid *n* barium dioxide, barium binoxide, barium peroxide, barium superoxide
Bariumfeldspat *m* (Min) barium feldspar, hyalophane
Bariumfluorsilikat *n* barium fluosilicate, barium silicofluoride
Bariumflußspat *m* (Min) barytic fluorspar, barytofluorite
Bariumhydrid *n* barium hydride
Bariumhydroxid *n* barium hydroxide, caustic baryta
Bariumhydrür *n* (obs) barium hydride
Bariumhyperoxid *n* barium dioxide, barium peroxide
Bariumhypophosphat *n* barium hypophosphate
Bariumhypophosphit *n* barium hypophosphite
Bariumhyposulfat *n* barium dithionate, barium hyposulfate
Bariumlack *m* barium lake
Bariumlegierung *f* barium alloy
Bariummanganat *n* barium manganate, (Min) manganese green
Bariummonosulfid *n* barium monosulfide, barium sulfide
Bariumnitrat *n* barium nitrate
Bariumnitrit *n* barium nitrite
Bariumoxid *n* barium oxide, barium monoxide, barium protoxide, baryta
Bariumoxidhydrat *n* barium hydroxide, barium oxide hydrate
Bariumpermanganat *n* barium permanganate
Bariumperoxid *n* barium peroxide, barium binoxide, barium dioxide, barium superoxide
Bariumperoxidhydrat *n* barium peroxide hydrate

Bariumplatincyanür *n* barium cyanoplatinate(II), barium cyanoplatinite, barium platinocyanide
Bariumpyrophosphat *n* barium pyrophosphate
Bariumselenat *n* barium selenate
Bariumsilicat *n* barium silicate
Bariumsilicofluorid *n* barium fluosilicate, barium silicofluoride
Bariumsulfat *n* barium sulfate; barite white, blanc fixe, permanent white, (Min) barytes, heavy spar
Bariumsulfid *n* barium sulfide
Bariumsuperoxid *n* barium superoxide, barium binoxide, barium dioxide, barium peroxide
Bariumsuperoxidhydrat *n* barium peroxide hydrate
Bariumthiosulfat *n* barium thiosulfate
Bariumuranit *n* barium uranite
Bariumwolframat *n* barium tungstate
Bariumzement *m* barium plaster
Barkometer *n* barkometer
Barn *n* (Flächeneinheit des Kernquerschnitts) barn
Barnhardtit *m* (Min) barnhardtite
Barograph *m* barograph, recording barometer
baroklin baroclinic
Baroklinievektor *m* baroclinicity vector
Barometer *n* barometer, **luftleeres** ~ aneroid barometer
Barometerablesung *f* barometer reading
Barometerblume *f* barometer flower
Barometerdose *f* barometer box
Barometerdruck *m* barometric pressure
Barometerhöhe *f* barometric height
Barometersäule *f* barometric column
Barometerschwankung *f* barometrical variation
Barometerstand *m* barometric height, barometric pressure
Barometrie *f* (Meteor) barometry
barometrisch barometric
Baroselenit *m* (Min) baroselenite
Baroskop *n* baroscope, barometer
baroskopisch baroscopic, barometric
Barostat *m* barostat
barotrop barotropic
Barotropie *f* barotropy
Barrandit *m* (Min) barrandite
Barras *m* (Harz) barras
Barre *f* bar
Barrelen *n* barrelene
Barren *m* ingot, bar, pig
Barreneinguß *m* ingot mold
Barrensilber *n* silver ingot
Barriere *f* barrier
Barringtogensäure *f* barringtogenic acid
Barthit *m* (Min) barthite
Bartholomit *m* (Min) bartholomite
Barutin *n* barutin
Barylit[h] *m* (Min) barylite

Baryon *n* baryon
Barysilit *m* (Min) barysilite
Barystrontianit *m* (Min) barystrontianite
Baryt *m* barium oxide, barium monoxide, barium protoxide, baryta
barytartig barytic
Baryterde *f* barium monoxide, barium oxide, barium protoxide
Barytfeldspat *m* (Min) baryta feldspar
barytführend barytic, barytiferous
Barytgelb *n* barium yellow, baryta yellow
Barytharmotom *m* (Min) barytharmotome
Barythydrat *n* barium hydroxide, caustic baryta, hydrate of baryta
Barytlauge *f* barium hydroxide solution, baryta solution, baryta water
Barytlösung *f* barium hydroxide solution, baryta solution, baryta water
Barytocalcit *m* (Min) barytocalcite
Barytpapier *n* baryta paper
Barytphyllit *m* (Min) barytophyllite
Barytsalpeter *m* barium nitrate, nitrobarite
Barytspat *m* (Min) barite, barytes, heavy spar, native barium sulfate
Barytstein *m* (Min) barite, barytes, heavy spar, native barium sulfate
Baryturanit *m* (Min) uranocircite, uranyl-barium phosphate
Barytwasser *n* barium hydroxide solution, baryta solution, baryta water
Barytweiß *n* barite white, blanc fixe, permanent white
baryzentrisch (Phys) barycentric
Baryzentrum *n* (Phys) center of gravity
Barzylinder *m* hammer cylinder
Basalstoffwechsel *m* (Med) basal metabolic rate
Basalt *m* (Min) basalt
basaltähnlich basaltic, basaltoid
basaltartig basalt-like
Basalteisenerz *n* basaltic iron ore
Basaltemperatur *f* (Med) basal temperature
Basaltfelsen *m* basaltic rock
basaltförmig basaltiform
Basaltgestein *n* basaltic rock
Basaltglas *n* (Min) basalt glass
basalthaltig basaltic, containing basalt
basaltieren to convert slag into a material resembling basalt
Basaltin *m* (Min) basaltine
basaltisch basaltic
Basaltjaspis *m* (Min) basaltic jasper
Basaltporphyr *m* (Min) porphyric basalt
Basaltsand *m* basalt sand
Basaltschiefer *m* schistous basalt
Basaltsteingut *n* basalt ware, black ware, wedgwood ware
Basaltwolle *f* basalt wool
Basalzelle *f* (Biol) basal cell
Basanit *m* (Geol) basanite

Base — Bauchspeicheldrüse

Base *f* (Chem) base
Basedowsche Krankheit *f* (Med) Graves's (or Basedow's) disease
Basekatalyse *f* basic catalysis
Baseler Grün *n* Basle green, cupric acetoarsenite, Paris green, Schweinfurt green
Basenanalogon *n* base analog
Basenanhydrid *n* base anhydride
Basenaustausch *m* base exchange, exchange of bases
Basenaustauscher *m* base exchanger, base exchanging compound
basenbildend base-forming, basifying, basigenous
Basenbildner *m* base former, basifier
Basenkatalyse *f* base catalysis
Basenpaar *n* (Mol. Biol) base pair
Basenpaarung *f* (Mol. Biol) base pairing
Basenüberschuß *m* excess base
Basenwert *m* base value
Basicität *f* siehe Basizität
Basidie *f* (Bot) basidium
Basidienpilze *m pl* (Bot) basidiomycetes
Basidiomyceten *pl* (Bot) basidiomycetes
Basidiospore *f* basidiospore
Basidium *n* (Bot) basidium
basieren to base
Basilienöl *n* basil oil
Basilikum *n* basil
Basilikumkraut *n* basil
Basilikumöl *n* basil oil
Basilikumsalbe *f* basilicon, resin cerate
Basis *f* (Math) base, basis
basisch basic
basischchromsauer basic chromate
Basischmachen *n* basification
basischsalpetersauer basic nitrate, subnitrate
basischschwefelsauer basic sulfate
Basiseinstellung *f* base adjustment, setting of the base
Basisentfernungsmesser *m* base range finder
Basisfläche *f* basal area, basal surface, (Krist) basal plane
basisflächenzentriert (Krist) end-centered
Basiskiste *f* (112 Tafeln Blech 20x14″) basebox
Basislinie *f* base line
Basisvektor *m* unit cell vector
Basiszelle *f* unit cell
basiszentriert (Krist) base centered
Basizität *f* basic capacity, basicity
Basizitätstheorie *f* theory of valences
Basizitätszahl *f* basicity number, basicity value
Baslerblau *n* Basle blue
basophil (Färb) basophilic, basophilous
Basozyt *m* (basophiler Leukozyt) basocyte, basophil leucocyte
Bassanit *m* (Min) bassanite
Bassetit *m* (Min) bassetite
Bassia *f* bassia

Bassiafett *n* bassia oil
Bassiaöl *n* bassia oil
Bassiasäure *f* bassic acid
Bassin *n* basin, reservoir
Bassoragummi *n* bassora gum
Bassorin *n* bassorin
Bassorinsäure *f* bassoric acid
Bast *m* bast, gum of silk, inside bark, raffia
Bastard *m* bastard, hybrid cross
Bastardfeile *f* bastard file
Bastern *m* raw sugar
Basternzucker *m* bastard sugar, raw sugar
Bastfaser *f* bast fiber
Bastit *m* (Min) bastite
Bastnäsit *m* (Min) bastnasite, hamartite
Bastpapier *n* bast paper, manila paper
Bastschwarz *n* bast black
Bastzelle *f* bast cell
Batate *f* sweet potato
Batat[in]säure *f* batatic acid
Batchelorit *m* (Min) batchelorite
Bathmetall *n* Bath metal
bathochrom bathochromic
Bathochrom *n* (Opt) bathochrome
Batist *m* (Text) batiste
Batrachit *m* (Min) batrachite
batschen to batch jute, to ret flax
Batschen *n* batching
Batterie *f* battery, pile, **Blei-** ~ lead storage battery, **galvanische** ~ galvanic battery, voltaic battery
Batterieanschluß *m* (Elektr) battery connection
Batterieanschlußklemme *f* battery terminal
Batterieaufladung *f* battery charging
batteriebetrieben battery-operated
Batterieelement *n* battery cell
Batterieentladung *f* battery discharge
Batteriegalvanometer *n* battery gauge
Batteriegehäuse *n* battery box
Batterieglas *n* battery jar
Batteriekapazität *f* battery capacity
Batterieklemme *f* battery terminal
Batterieladegerät *n* battery charger
Batterieladesatz *m* battery charger
Batteriemassen *f pl* accumulator pastes
Batteriespannung *f* battery voltage
Batteriewiderstand *m* battery resistance
Batteriezelle *f* battery cell
Batteriezündung *f* battery ignition
Batylalkohol *m* batyl alcohol
Bau *m* structure, constitution, construction; cultivation; formation
Bauarbeiten *pl* (Bauw) construction work
Bauart *f* construction, design, structure, style, type
Bauchfell *n* (Med) peritoneum
bauchig bulbous, bulged
Bauchspeichel *m* pancreatic juice
Bauchspeicheldrüse *f* (Med) pancreas

Baudisserit *m* (Min) baudisserite
Baueisen *n* constructional iron, structural iron
Bauelement *n* constructional unit
bauen to build, to construct; to cultivate; to make
Bauerde *f* tillable soil
Bauerenol *n* bauerenol
baufällig out of repair, defective, dilapidated
baufest firm, solid
Baufestigkeit *f* structural strength
Bauform *f* design
Baugewerbe *n* building trade
Baugrube *f* trench
Bauguß *m* structural casting
Bauholz *n* lumber, timber
Bauindustrie *f* building industry
Baukastensystem *n* unit construction system
Baukeramik *f* structural ceramics
Baulänge *f* overall length
baulich constructional, structural
Baulit *m* (Min) baulite
Baum *m* tree, arbor, (Mech) beam
Baumachat *m* (Min) dendritic agate, arborescent agate, dendrachate, moss agate
baumähnlich arborescent, dendritic, treelike
Baumannsche Schwefelprobe *f* Baumann sulfur test
baumartig arborescent
Baumaschinen *pl* building machinery
Baumaterial *n* building material
Baumégrad *m* Baumé degree
Bauméspindel *f* Baumé spindle
Bauméstandard *m* Baumé standard (American standard scale for density)
baumförmig arborescent, arboriform, dendritic
Baumharz *n* tree resin, wood resin
Baumhauerit *m* (Min) baumhauerite
Baumrinde *f* bark, cortex
Baumwachs *n* grafting wax
Baumwollabfall *m* cotton waste
baumwollartig cottonlike
Baumwollaufbereitung *f* cotton dressing
Baumwollazofarbe *f* cotton azodye
Baumwollbeize *f* cotton mordant
Baumwolle *f* cotton, calico
baumwollen cotton
Baumwollentkörner *m* ginner
Baumwollfaden *m* cotton thread
Baumwollfaser *f* cotton fiber
Baumwollfaserpapier *n* cotton fiber paper
Baumwollflanell *m* (Text) cotton beaver
Baumwollflocken *f pl* cotton flocks
Baumwollgarn *n* cotton yarn
Baumwollgewebe *n* cotton cloth
Baumwollkapsel *f* (Bot) cotton boll
Baumwollkapselkäfer *m* (Zool) boll weevil
Baumwollkernöl *n* cotton seed oil
Baumwollköper *m* (Text) cotton twill, jean
Baumwollöl *n* cotton oil, cotton seed oil

Baumwollpflanze *f* cotton-plant
Baumwollpfropfen *m* cotton plug, cotton stopper
Baumwollputzwolle *f* cotton waste
Baumwollriemen *m* cotton belt
Baumwollsaatöl *n* cotton seed oil
Baumwollsamen *m* cottonseed
Baumwollsamenöl *n* cotton seed oil
Baumwollscharlach *m* cotton scarlet
Baumwollwatte *f* cotton padding, cotton wadding
Baupappe *f* building board
Bauplatte *f* building board, panel
Bausch *m* pad, compress, plug, wad
Bauschelastizität *f* bulk resilience
bauschen to bulge, to protrude, to swell out
Baustahl *m* structural steel, building block, legierter ~ structural alloy steel
Baustein *m* particle
Bausteine *pl* (Comp) hardware
Bausteinsystem *n* modular system
Baustoff *m* building material
Baustoffprüfung *f* building materials testing
Bauteilchen *n* particle
Bautenfarbe *f* concrete paint
Bautenschutz *m* constructional proofing
Bautenschutzmittel *n* building preservative agent, building proofing material
Bauvorschriften *pl* building regulations
Bauwerk *n* building, edifice, structure
Bauwinde *f* winch
bauwürdig profitable, workable
Bauxit *m* (Min) bauxite, native alumina hydrate, native aluminum hydroxide
Bauxitlaugerei *f* bauxite lixiviation
Bauxitmühle *f* bauxite mill
Bauxitofen *m* bauxite kiln
Bauxitziegel *m* bauxite brick
Bayerprozeß *m* Bayer process
Bayersche Säure *f* (Bayer-Säure, HN) bayer's acid, 2-naphthol-8-sulfonic acid
Bayin *n* bayin
Bayöl *n* bay oil, myrcia oil, oil of bay, volatile laurel
Bayrisch Blau *n* Bavarian blue
Bayrum *m* (Kosm) bay rum
Baysalz *n* bay salt, coarse grained salt, sea salt
Bazille *f* bacillus
Bazillenausscheider *m* carrier
Bazillenfärbung *f* staining of bacilli
bazillenfrei free from bacilli
Bazillenherd *m* bacilli focus
Bazillenkultur *f* [bacilli-]culture
Bazillenlehre *f* bacteriology
Bazillenreinkultur *f* pure culture
Bazillenspore *f* bacillary spore, spore of a bacillus
Bazillenstamm *m* strain of bacilli, bacillary strain

bazillentötend bacillicidal, bacteri[o]cidal, germicidal
Bazillenträger *m* bacillus carrier, [germ] carrier
Bazillenzüchtung *f* bacilli culture
Bazillus *m* bacillus, (pl. Bazillen) bacillus (pl. bacilli)
Bazzit *m* (Min) bazzite
Bdellium *n* bdellium
beabsichtigen to intend, to contemplate, to plan
beachtenswert noteworthy, remarkable
beanspruchen to require, to claim; to load, to strain, to stress
Beanspruchung *f* requirement, claim; duty, load, strain, stress, tension, wear and tear, ~ **auf Abscherung** shearing strain, ~ **auf Biegung** bending strain, ~ **auf Biegung durch den Winddruck** bending stress by wind pressure, ~ **auf Druck** compression strain, ~ **auf Schlagfestigkeit** blow stress, shock stress, ~ **auf Verdrehen** torsional stress, torsional loading, ~ **auf Zug** tensile strain, tensile stress, **auf gleichmäßige ~ hin konstruiert** designed for stress uniformity, **elastische ~** elastic strain, **hohe ~** severe service condition, severe strain, **innere ~** internal strain, stress, **starke ~** high working rate, **stoßweise ~** intermittent stress, pulsating stress, **zulässige ~** permissible load, safe working stress
Beanspruchungsart *f* kind of stress
Beanspruchungsbereich *m* range of loading
Beanspruchungs-Charakteristik *f* load spectrum under service conditions
Beanspruchungsgrenze *f* limiting range of stress, limit of stress
beanstanden to object [to], to complain, to protest against, to refuse, to reject
beantragen to apply for, to demand
beantworten to reply [to]
bearbeitbar workable, machinable, treatable
Bearbeitbarkeit *f* workability, **spanabhebende ~** free-cutting machinability
bearbeiten to manipulate, to handle, to machine, to process, to treat, to work
Bearbeitung *f* manipulation, handling, machining, processing, treatment, working
Bearbeitungsmethode *f* manufacturing method, (Techn) machining method
Bearbeitungsstufe *f* machining step
Bearbeitungsverfahren *n* fabricating technique, working method
Bearbeitungsvorrichtung *f* tooling fixture
Beaufschlagung *f* impact, pressure, stress
beaufsichtigen to supervise, to control, to inspect, to superintend, to survey
Beaufsichtigung *f* supervision
beauftragen to authorize, to commission, to entrust
Beaumontit *m* (Min) beaumontite

Beaverit *m* (Min) beaverite
Bebeerin *n* bebeerine, bibirine
Bebeerurinde *f* bebeeru bark
beben to tremble, to quake, to vibrate
Bebirin *n* bebeerine, bibirine
Bebrütung *f* incubation
Becher *m* beaker, cup, goblet, (Techn) bucket
Becheraufzug *m* bucket conveyor
Becherbandförderer *m* bucket elevator
Becherfließzahl *f* cup flow figure
becherförmig cupshaped
Becherglas *n* (Chem) beaker
Becherrad *n* bucket wheel
Becherwerk *n* bucket conveyor, bucket elevator, elevator
Becherwerksextrakteur *m* bucket-conveyor extractor
Bechilit *m* (Min) bechilite
Beckelith *m* (Min) beckelite
Becken *n* basin, pool, reservoir, tank, vessel
Beckmannsche Umlagerung *f* Beckmann molecular transformation, Beckmann rearrangement
Beckmannthermometer *n* Beckmann thermometer
Beclamid *n* beclamide
Becquerel-Effekt *m* Becquerel effect, photovoltaic effect
Becquerelit *m* (Min) becquerelite
Becquerelstrahlen *m pl* Becquerel rays
bedampfen to treat with vapor, to vapor-blast
Bedampfung *f* vapor-blast process
Bedampfungsanlage *f* vaporization plant
Bedarf *m* demand, consumption, need, requirement, **den ~ decken** to satisfy the demand, **geringster ~** minimum requirement
Bedarfsartikel *m* commodity, requisite, **photographischer ~** photo supply
Bedarfsgegenstand *m* necessary article, requirement, requisite
bedecken to cover
Bedeckung *f* cover, covering
Bedeckungsgrad *m* degree of coverage
Bedeckungsmaschine *f* coating machine
Bedenken *n* consideration, hesitation, objection
bedenklich questionable, critical, doubtful, hazardous, risky, serious
bedeuten to mean, to denote, to indicate, to signify
bedeutend significant, considerable, important
bedeutsam significant, important, imposing
Bedeutung *f* importance, meaning, signification
bedeutungslos meaningless, inconsiderable, insignificant
Bedeutungslosigkeit *f* insignificance
bedienbar manipulable, operable
bedienen to serve, to attend; to handle, to manipulate, to operate

Bedienung *f* operation, attendance, charging, feeding, manipulation, service
Bedienungsanleitung *f* operating instructions, working instructions
Bedienungsanweisung *f* instruction manual
Bedienungsbühne *f* operating platform
Bedienungsgleichung *f* equation of condition
Bedienungsgriff *m* control knob
Bedienungshahn *m* working valve
Bedienungshebel *m* control lever, operating lever
Bedienungsknopf *m* control knob
Bedienungsmann *m* operator, attendant
Bedienungspult *n* control panel
Bedienungsschalter *m* control switch
Bedienungsstand *m* operator's stand
Bedienungsvorschrift *f* direction for use, operating instruction
bedingen to bring about; to cause, to necessitate, to require
bedingt conditional; limited, ~ **richtig** (Math) conditionally correct
Bedingung *f* term, condition, requirement, stipulation, (Voraussetzung) [pre]requisite
bedingungslos unconditional
bedrahten to wire
Bedruckbarkeit *f* **von** adhesion of printing inks to
bedrucken to print [upon]
Bedrucken *n* printing, overprinting
Bedruckung *f* printing; impression
bedürfen to need, to be in need [of], to require
Bedürfnis *n* necessity, requirement, want
Beegerit *m* (Min) beegerite
beehren to favor, to honor
beeinflußbar influenceable, susceptible
beeinflussen to influence, to affect
Beeinflussung *f* influence
Beeinflussungstest *m* well interference test
beeinträchtigen to impair, to affect, to encroach upon, to injure
beenden to conclude, to finish
Beerenfarbstoff *m* berry pigment
Beerengrün *n* sap green
befähigen to enable, to qualify
Befall *m* attack, infestation
befallen to affect, to attack, to befall
befangen embarrassed, prejudiced
befassen to attend [to], to concern oneself [with], to deal with, to handle
Befehl *m* command, order
befestigen to fasten, to attach, to fix, to secure, (anstecken) to clip, (Kitt) to cement, to glue, (Klammer) to clamp, (Klammern) to cramp
Befestigung *f* attachment, fastening
Befestigungsbolzen *m* mounting stud
Befestigungsbügel *m* sling
Befestigungsleiste *f* cover strip
Befestigungsmittel *n* fixing agent
Befestigungsmutter *f* retaining nut

Befestigungsplatte *f* fixing plate, mounting plate
Befestigungsschelle *f* fastening clip, loop type clamp
Befestigungsschraube *f* clamping bolt, fixing bolt, set screw
Befestigungsstelle *f* point of attachment
Befestigungsvorrichtung *f* fixing device
befeuchten to moisten, to humidify, to wet
Befeuchtung *f* moistening, dampening
Befeuchtungsanlage *f* moistening plant
Befeuchtungskammer *f* moistening chamber
Befeuchtungskasten *m* moistening chamber
Befeuchtungsluft *f* dampening air
Befeuchtungsmittel *n* moistening agent, wetting agent
Befeuchtungstrommel *f* wetting drum
befeuern to fire, to heat
Befeuerung *f* firing, heating
Befilmen *n* film coating
Befinden *n* condition, state
befindlich existing, present, situated
beflecken to stain, to soil, to spot
befleißigen to endeavor
Beflocken *n* (ganze Oberfl.) flock-coating, (Muster) flock-printing
befördern to transport, to convey, to forward, to promote; to ship; to transfer
Beförderung, *f* transportation, forwarding; promotion, **bodenfreie** ~ overhead handling promotion
Beförderungsmittel *n* means of transportation
Beförderungsvorrichtung *f* transporting installation
befolgen to comply with
befreien to liberate, to clear, to disengage, to free from, to relieve, to rescue, to rid, **von Kohlenstoff** ~ to decarbonize
befriedigen to satisfy
befriedigend satisfactory
Befriedigung *f* satisfaction
befruchten to fecundate; to fertilize
Befruchtung *f* fertilization
befühlen to touch, to feel
befürworten to recommend, to advocate, to support
befugen to authorize, to entitle, to license
Befugnis *f* authorization, competence, power, privilege, warrant
Befund *m* findings, data, report, result
Begabung *f* talent
begasen to fumigate, to gas
Begasen *n* (v. Flüssigkeiten) mixing gases and liquids
Begebenheit *f* occurrence
begeistern to inspire, to enthusiasm
begichten (Metall) to burden, to charge
Begichtungsanlage *f* charging plant
Begichtungsbühne *f* charging floor
Begichtungsvorrichtung *f* charging device

begierig desirous, eager
begießen to irrigate, to moisten, to wet
Beginn *m* beginning, outset, start
beginnend incipient
beglasen to glaze
beglaubigen to attest, to certify, to confirm, to testify
Beglaubigungsschreiben *n* credentials
begleichen to settle, to balance, (Konto) to square
Begleichung *f* payment; settlement
begleiten to accompany, to attend, to go with
Begleiterscheinung *f* accompanying phenomenon, attendant phenomenon, secondary phenomenon
Begleitfarbe *f* illuminating color
Begleitlinie *f* satellite line
Begleitmetall *n* accompanying metal, foreign metal, metal impurity
Begleitstoff *m* impurity, accompanying body, accompanying substance
begnügen to be content, to be satisfied
begreifen to comprehend, to conceive, to understand
begrenzen to border, to limit, to restrict, to set a limit, (Aufgabe) to circumscribe
Begrenzer *m* limiter
begrenzt limited, restricted
Begrenzung *f* boundary, limit, limitation, restriction
Begrenzungsfläche *f* boundary surface, (Techn) surface of contact
Begrenzungskegel *m* shadow cone
Begrenzungslinie *f* boundary line
Begrenzungsoberfläche *f* boundary surface
Begriff *m* definition, concept[ion], notion
begrifflich conceptual
Begriffsbestimmung *f* definition
Begriffsfestsetzung *f* definition
begründen to confirm; to constitute, to establish; to found, to set up, (Handlungsweise) to account for; to substantiate
Begründung *f* establishment, justification; proof, reason
begünstigen to favor, to promote
Begußmasse *f* engobe (ceramics), slip
Begutachtung *f* approval, expert evidence
Behälter *m* container, basin, receiver, receptacle, reservoir, tank, vessel, (Erz) bin, **halbstarrer ~** semirigid container
Behälterblech *n* tank sheet iron
Behälterwagen *m* tank wagon
behaftet affected, afflicted
behalten to keep, to retain
behandeln to handle, to manipulate, to operate, to process, to treat, to work
Behandlung *f* manipulation, operation, processing, treatment, working, **~ mit Ozon** ozonization, **thermische ~** heat-treatment

Behandlungsmittel *n* agent, reagent
Behandlungsverfahren *n* method of treatment, procedure
Behandlungsvorschrift *f* working instructions
Behandlungsweise *f* method of treatment, procedure
beharren to persist, to persevere, (Phys) to remain inert
Beharrung *f* persistence
Beharrungskraft *f* inertia
Beharrungsprinzip *n* inertia principle
Beharrungstemperatur *f* equilibrium temperature
Beharrungsvermögen *n* inertia
Beharrungswirkung *f* inertia effect
Beharrungszustand *m* state of inertia, equilibrium state, permanence, settled state, (Regeltechn) steady-state condition
Beharzungsmaschine *f* resin smearing machine
behaupten to assert, to state
Behauptung *f* assertion, declaration, statement
beheben to eliminate, to overcome, to remove
beheizen to heat
Beheizung *f* heating
Beheizungsart *f* heating method
Behelf *m* emergency device, expedient, makeshift, (Ersatz) substitute
Behelfskonstruktion *f* makeshift construction
Behelfslösung *f* expedient
behelfsmäßig improvised, makeshift, temporary
Behelfsmittel *n* expedient
Behennuß *f* behen nut, ben nut
Behenöl *n* behen oil, ben oil
Behenolsäure *f* behenolic acid
Behenon *n* behenone
Behensäure *f* behenic acid
beherrschen to command, to control, to manage, to master, to rule
Beherrschung *f* control
behilflich helpful, useful
behindern to hamper, to hinder, to impede, to inhibit
Behinderung *f* hindering, hindrance, impediment, restraint, (Chem) inhibition
Behinderungsfaktor *m* hindrance factor
Behörde *f* authority
behördlich official
behüten to guard
behutsam careful, cautious
Beiblatt *n* supplement[ary sheet]
beibringen (lehren) to teach, (Zeugen, etc) to produce
beiderseitig bilateral; **~ aufgetragen** coated on both sides
beifügen to add, to annex, to attach, to enclose, to include
Beifuß *m* (Pharm) artemisia, mugwort, wormwood
Beifußöl *n* mugwort oil
Beifuttermittel *n* auxiliary feeding stuff

Beigabe *f* addition, supplement
beige beige
beigeben to add [to]
Beigeschmack *m* flavor, aftertaste, off-flavor, tang
Beilage *f* addition, supplement, (Anhang) appendix, (Anlage) enclosure
Beilby-Schicht *f* Beilby layer
beilegen to add, to attach, to enclose, (Streit) to settle
Beilegering *m* spacing collar
Beilegscheibe *f* shim, washer
beiliegend enclosed
Beilstein *m* (Min) axstone, jade, nephrite
Beilsteinprobe *f* Beilstein test
Beiluft *f* admixed air, secondary air
beimengen to add, to admix, to mix with
Beimengen *n* admixing
Beimengung *f* addition, additive, admixture, impurity
beimessen to attribute
beimischen to add, to admix, to mix with
Beimischung *f* addition, admixture, impurity
Bein *n* (Gestell) branch, leg, (Knochen) bone
Beinasche *f* bone ash, impure calcium phosphate
Beinmehl *n* bone meal
Beinschwarz *n* animal charcoal, bone black, bone charcoal
Beintürkis *m* (Min) bone turquoise, odontolite
beipflichten to agree, to approve, to assent
Beirat *m* adviser, advisory board
Beisatz *m* addition, admixture, (Metall) alloy
Beisein *n* presence
Beispiel *n* example
beispiellos unprecedented
beispielsweise for example
beißend biting, pungent, sharp
Beißzange *f* pliers, nippers, (f. Draht) wire cutters
Beistand *m* support, assistance, assistant; supporter
beistehen to assist, to stand by
beisteuern to contribute, to assist
beistimmen to agree, to assent, to concur
Beitrag *m* contribution, share, (Zahlung) subscription fee
beitragen to contribute, to assist
beitreten to join, to participate, to take part
Beiwerk *n* accessory, attachment
Beiwert *m* factor, coefficient, mathematical constant, parameter
Beizabwasser *n* waste pickling water
Beizanlage *f* pickling plant
Beizbad *n* corroding bath, blanching bath, disinfecting bath, drenching bath, steeping bath, (Filz) carroting bath, (Gerb) bating bath, puering bath, (Met) pickling bath
Beizbaderwärmer *m* pickling bath heater
Beizbehandlung *f* (Met) pickling

beizbrüchig brittle, short
Beizbrüchigkeit *f* corrosion brittleness, acid brittleness
Beizbrühe *f* caustic liquor, drench, (Gerb) bate
Beize *f* mordant; caustic, corrosive, (Galv) pickle; pickling bath, (Gerb) mastering, ooze, **die ~ annehmen** to seize the mordant
beizen to corrode, to mordant, (Gerb) to swell, to tan, (Holz) to stain, to tinge, (Met) to pickle, (Tabak) to sauce tobacco
Beizen *n* (Färb) mordanting, (Filz) carroting, (Gerb) bating, drenching; puering, (Holz) staining, (Met) etching, pickling, (Saatgut) disinfecting, (Techn) corroding
beizend caustic, corroding, corrosive, pickling, (Färb) mordant
Beizendruck *m* mordant printing
Beizenfärben *n* mordant dyeing
beizenfärbend dyeing on a mordant
Beizenfarbstoff *m* mordant dye, mordant dyestuff
Beizenfetter *m* pickling inhibitor
Beizengelb *n* mordant yellow
Beizenklotzartikel *m* mordant padded style
Beizerei *f* pickling house
Beizfarbe *f* mordant color, stain
Beizflüssigkeit *f* caustic liquor, corrosive liquid, stain, steeping liquor, (Gerb) bate, drench, (Met) pickle, (Saatgut) disinfecting liquid
Beizgeräte *n pl* pickling appliances
Beizgrundierung *f* wash primer
Beizhilfsmittel *n* mordant auxiliary
Beizkasten *m* pickling vat
Beizkorb *m* pickling basket
Beizkraft *f* causticity, corrosive power, disinfecting power, mordant action
Beizkufe *f* (Gerb) puering vat, tan vat
Beizlauge *f* pickling liquor
Beizmittel *n* caustic, corrosive, mordant, (Holz) stain, (Met) pickling agent, (Saatgut) disinfectant, seed dressing
Beizofen *m* scaling furnace
Beizprozeß *m* (Gerb) bating process
Beizrad *n* pickling wheel
Beizsprödigkeit *f* acid brittleness, corroding brittleness, hydrogen embrittlement
Beizstein *m* (Gerb) bate stone
Beizstoff *m* mordant, mordanting substance
Beiztank *m* pickling tank
Beizung *f* mordanting, (Filz) carroting, (Gerb) bating, drenching, puering, (Holz) staining, (Met) pickling, (Saatgut) disinfecting, (Tabak) saucing, (Techn) corroding
Beizwasser *n* caustic liquor, corrosive liquid
Beizwertbestimmung *f* determination of the strength of bating
Beizwirkung *f* mordant action
Beizzeit *f* conditioning time, etching time
Beizzusatz *m* pickling compound

bekämpfen – Belegungsfunktion

bekämpfen to combat, to counteract, to oppose
Bekämpfungsmaßnahme *f* combat measure, control measure
Bekämpfungsmethode *f* method of control
Bekämpfungsmittel *n* means of attack, means of combat
Bekanntgabe *f* public announcement, publicity
bekanntgeben to announce, to publish
Bekanntmachung *f* announcement, notice, publication
bekappen (Holz) to lop, to top
bekleben to label, to paste
bekleckern to blot, to blur
bekleiden to coat, to clothe, to cover, (Amt) to occupy
Bekleidung *f* coating, casing, covering, (Innen-) lining
Bekleidungsindustrie *f* clothing industry
Bekohlung *f* coaling, supplying with coal
Bekohlungsanlage *f* coal plant
bekräftigen to emphasize, to assert, to confirm
Bekräftigung *f* emphasis, confirmation
bekritteln to criticize
bekrusten to crust, to cover with a crust, to incrust
Bekrustung *f* incrustation
Beladefähigkeit *f* (Adsorption) adsorption capacity
beladen to load, to burden, to charge
Beladeplan *m* loading diagram
Beladestation *f* loading station
Beladungsreaktion *f* priming reaction
belästigen to annoy, to bother, to molest, to trouble
Belag *m* coating, covering, lining, plating, **geerdeter** ~ earthed coating, **leitender** ~ conducting coat
Belagkorrosion *f* deposit attack
Belang *m* importance, consequence
belangen to accuse, to bring action against, to concern
belanglos inconsiderable, irrelevant, negligible, unimportant
belassen to leave
belastbar chargeable
Belastbarkeit *f* load carrying capacity, [loading] capacity, (Dest) capacity, (Mech) permissible working stress, (Med) maximum stress
belasten to load, to burden, to charge, (Med) to stress
Belastung *f* loading, application of load, charging, load; load factor; strain, (Elektr) current density, (Med) stress; ~ **durch Eigengewicht** dead load, bearing stress, **bleibende** ~ basic load, **drehsymmetrische** ~ axially symmetrical load, **elastische** ~ elastic strain, **induktive** ~ inductance load, inductive load, **kapazitive** ~ condenser load, **künstliche** ~ (Elektr) dummy load, **ruhende** ~ dead load, **spezifische** ~ unit stress, **ständige** ~ permanent load, **stoßweise** ~ intermittent shock load, pulsating load, **wechselnde** ~ live load, **zulässige** ~ safe load, **zunehmende** ~ increasing load
Belastungsänderung *f* (Elektr) load variation
Belastungsausgleich *m* load compensation, load equalization
Belastungsdauer *f* duration of load application, load duration, loading time
Belastungsdiagramm *n* load diagram
Belastungsdichte *f*, **mittlere** ~ (Elektr) average density of charge
Belastungsfähigkeit *f* load capacity, carrying capacity
Belastungsfaktor *m* load factor
Belastungsfläche *f* load area
Belastungsfolgen *pl* loading sequence, load spectrum
Belastungsgrad *m* load factor
Belastungsgrenze *f* load limit, **obere** ~ loading point
Belastungskoeffizient *m* load factor
Belastungskurve *f* load curve
Belastungsphase *f* load phase
Belastungsprobe *f* load test, static test, (Med) tolerance test, ~ **bis zum Bruch** breaking load test, load test up to breaking
Belastungsring *m* ring weight, weight ring
Belastungsschwankung *f* load fluctuation, variation of load
Belastungsspannung *f* (Elektr) load voltage
Belastungsspitze *f* peak load
Belastungsventil *n* loaded valve
Belastungsverhältnis *n* ratio of load, turndown ratio
Belastungsvermögen *n* loading capacity
Belastungsversuch *m* static test
Belastungswechsel *m* load fluctuation, load variation
beleben to activate, to quicken, to revive; to stimulate
Belebtschlamm *m* activated sludge (sewage disposal), biological sludge
Belebung *f* vivification
Belebungsanlage *f* activated sludge plant
Belebungsmittel *n* activator, (Pharm) restorative, stimulant
Belebungsverfahren *n* activated sludge process, bio-aeration
Beleg *m* document, evidence, proof, verification
Beleganalyse *f* check analysis, documentary analysis
belegen to prove, to coat; to cover, (Vorlesung) to enrol[l], to sign up [for]
Belegleserfarben *pl* optical reader printing inks
Belegschaft *f* personnel, crew, employees, staff
Belegung *f* coating, covering, seizure
Belegungsfunktion *f* surface distribution

belehren to advise, to instruct
Belehrung f advice, instruction
Belemnit m (Min) belemnite, thunderstone
beleuchten to illuminate, to elucidate, to throw light [upon]
Beleuchtung f illumination, lighting, (Elektr) lighting equipment
Beleuchtungsanlage f lighting installation
Beleuchtungsarmaturen f pl light fittings
Beleuchtungseinrichtung f illuminating equipment, illuminating plant, lighting installation
Beleuchtungsfenster n exposure window
Beleuchtungskörper m light fitting
Beleuchtungskohle f carbon of an arc lamp
Beleuchtungslinse f illuminating lens
Beleuchtungsmast m lamp pole
Beleuchtungsmesser m photometer
Beleuchtungsmittel n illuminant
Beleuchtungsschirm m reflector
Beleuchtungsstärke f degree of brightness, light intensity
Beleuchtungstafel f lighting panel
belichten (Phot) to expose
Belichtung f illumination, (Phot) exposure [to light]
Belichtungsautomat m automatic exposure timer
Belichtungsdauer f time of exposure
Belichtungsdifferenz f exposure difference
Belichtungskarte f exposure chart
Belichtungslampe f exciter
Belichtungsmesser m (Phot) lightmeter, exposure meter
Belichtungsschieber m exposure lid, exposure shutter
Belichtungsstärke f intensity of exposure to light, intensity of irradiation
Belichtungsstufe f stage of exposure
Belichtungstabelle f exposure table
Belichtungswert m exposure value
Belichtungszeit f (Phot) time of exposure
Belit m (Min) belite
Belladin n belladine
Belladonna f (Bot) belladonna, deadly nightshade
Belladonnaalkaloid n belladonna alkaloid
Belladonnablätter n pl (Bot) belladonna leaves
Belladonnin n belladonnine
Bellamarin n bellamarine
Belmontit m (Min) belmontit
Belonit m (Min) belonite
belüften to ventilate, to air, (klimatisieren) to air-condition
Belüftung f ventilation, aeration; (Klimatisierung) air-conditioning;, unterschiedliche ~ differential aeration
Belüftungsanlage f ventilation plant, aeration plant; ventilation system

Belüftungsapparat m aeration apparatus, aeration device
Belüftungselement n (Korr) differential aeration cell
Belüftungshaube f ventilation hood
Belüftungsrohr n aeration tube
Belüftungsstutzen m air admission tube
Belüftungsventil n air release valve
Bemegrid n bemegride
Bementit m (Min) bementite
bemerken to remark, to note, to notice, to observe
bemerkenswert noteworthy, noticeable, remarkable
Bemerkung f remark, note, observation
bemessen to measure, to adjust by measuring, (dosieren) to dose, (Leistung) to rate, (Zeit) to calculate
Bemidon n bemidone
bemühen to endeavor, to strive
bemustern to ornament, to dye to pattern, to furnish with a pattern; to sample
benachbart adjacent, neighboring, proximate, vicinal
benachrichtigen to inform, to notify
benachteiligen to disadvantage, to disfavor, to handicap, to harm; to injure
Benachteiligung f disadvantage, detriment, injury
Bence-Jonesscher Eiweißkörper m (Bence-Jones-Protein) Bence-Jones protein
Benedicts-Lösung f Benedict's solution
Benediktinerlikör m benedictine
benennen to name, to call, to designate
Benennung f designation, name, naming, specification, (Math) denomination, term, (Nomenklatur) nomenclature
benetzbar moistenable, wettable
Benetzbarkeit f wettability
benetzen to moisten, to dampen, to sprinkle, to wet
Benetzen n moistening, sprinkling, wetting
Benetzung f wetting
Benetzungseigenschaften f pl wetting properties
Benetzungsfähigkeit f moistening power
Benetzungskraft f wetting power
Benetzungsmittel n wetting agent
Benetzungspulver n wetting powder
Benetzungswärme f moistening heat, heat of humidification, heat of wetting
Benetzvorrichtung f sprinkling apparatus
Bengalblau n bengal blue
Bengalisches Feuer n Bengal light or fire
Bengalrosa n bengal rose, rose bengal
Benihiol n benihiol
Benincopalinsäure f benincopalinic acid
Benincopalsäure f benincopal acid
Benitoit m (Min) benitoite
Benitzucker m barley sugar

benötigen to need
Bentonit *m* (Emulgiermittel) bentonite
Bentonitbinder *m* bentonite binder
benutzbar usable, (anwendbar) applicable
Benutzbarkeit *f* usability
benutzen to utilize, to employ, to use
Benutzer *m* user
Benutzung *f* use
Benutzungsvorschriften *pl* instructions for use
Benzacridin *n* benzacridine
Benzal- benzal, benzylidene
Benzalaceton *n* benzalacetone, benzylidene acetone
Benzalacetophenon *n* benzalacetophenone
Benzalanilin *n* benzalaniline, benzylidene aniline
Benzalazin *n* benzalazine, benzaldazine, benzylidene azine, dibenzalhydrazine
Benzalcampher *m* benzalcamphor
Benzalchlorid *n* benzal chloride, benzylidene chloride
Benzaldazin *n* benzalazine, benzaldazine, benzylidene azine, dibenzalhydrazine
Benzaldehyd *m* benzaldehyde
Benzaldehydcyanhydrin *n* benzaldehyde cyanohydrin, benzalcyanohydrin, mandelonitrile
Benzaldoxim *n* benzaldehyde oxime, benzaldoxime
Benzalgrün *n* benzal green, malachite green
Benzalizarin *n* benzalizarin
Benzalphenylhydrazon *n* benzalphenylhydrazone
Benzamaron *n* benzamarone
Benzamid *n* benzamide
Benzamidin *n* benzamidine
Benzaminblau *n* benzamine blue
Benzaminsäure *f* benzaminic acid
Benzanalgen *n* benzanalgen, analgen
Benzanilid *n* benzanilide
Benzanisid *n* benzaniside
Benzanthracen *n* benzanthracene
Benzanthrachinon *n* benzanthraquinone
Benzanthren *n* benzanthrene
Benzanthron *n* benzanthrone
Benzatropin *n* benzatropine
Benzaurin *n* benzaurine
Benzazimid *n* benzazimide
Benzedrin *n* benzedrine
Benzenyl- benzenyl
Benzerythren *n* benzerythrene
Benzhydrazid *n* benzhydrazide
Benzhydrol *n* benzhydrol, diphenylcarbinol
Benzhydroxamsäure *f* benzohydroxamic acid
Benzhydroxyanthron *n* benzhydroxyanthrone
Benzhydryl- benzhydryl
Benzhydryläther *m* benzhydryl ether, dibenzhydryl ether
Benzhydrylamin *n* benzhydrylamine

Benzidin *n* benzidine, 4,4-diamino-biphenyl, 4,4-diamino xenene
Benzidinbraun *n* benzidine brown
Benzidindisulfonsäure *f* benzidine disulfonic acid
Benzidinhydrochlorid *n* benzidine hydrochloride
Benzidinsulfonsäure *f* benzidine sulfonic acid
Benzidinumlagerung *f* benzidine rearrangement
Benzil *n* benzil, bibenzoyl, diphenyl diketone, diphenylglyoxal
Benzildianil *n* benzildianil
Benzildioxim *n* benzildioxime
Benzilid *n* benzilide
Benziliminoxim *n* benzilimine oxime
Benzilmethyliminoxim *n* benzilmethylimine oxime
Benzilsäure *f* benzilic acid
Benzilsäureumlagerung *f* benzilic acid rearrangement
Benzimidazol *n* benzimidazole
Benzimido- benzimido
Benzin *n* gas[oline] (Am. E.), petrol (Br. E.), (Leichtbenzin) benzin[e]
Benzinabscheider *m* gasoline trap
Benzinanlage *f* gasoline plant, petrol plant
Benzinanzeiger *m* (Auto) fuel gauge indicator
Benzinbrenner *m* benzine burner
Benzindampf *m* gasoline vapor, benzine vapor, petrol vapor
Benzindruckmesser *m* gasoline pressure indicator
Benzinersatz *m* gasoline substitute, benzine substitute, petrol substitute
Benzinfilter *n* gasoline filter
Benzinkanister *m* gasoline can
Benzinkohlenwasserstoffe *pl* petroleum hydrocarbons
Benzinleck *n* fuel leak
Benzinleitung *f* fuel pipe line, gasoline line
Benzinlötlampe *f* benzine blowtorch
Benzinluftmischung *f* gasoline air mixture
Benzinmotor *m* petrol engine (Br. E.), (Am. E.) gasoline engine
Benzin-Öl-Gemisch *n* gasoline-oil mixture
Benzinpumpe *f* gasoline pump (Am. E.), petrol pump (Br. E.)
Benzinraffinerieanlage *f* gasoline refining plant
Benzinreiniger *m* gasoline filter
Benzinschwimmer *m* gasoline float
Benzinseife *f* benzine soap
Benzinstand *m* gasoline level
Benzinstandmesser *m* gasoline meter (Am. E.), petrol gauge (Br. E.)
Benzintank *m* gasoline tank, ~ zum Abwerfen jettison tank
Benzinverbrauch *m* gasoline consumption
Benzinvergaser *m* petrol carburettor
Benzinzuführung *f* fuel feed
Benzo- benzo

Benzoat *n* benzoate
Benzoazurin *n* benzoazurine
Benzocain *n* benzocaine, ethyl-aminobenzoate
Benzochinhydron *n* benzoquinhydrone, quinhydrone
Benzochinol *n* benzoquinol, quinol
Benzochinolin *n* benzoquinoline
Benzochinon *n* benzoquinone, quinone
Benzodioxan *n* benzodioxan
Benzodioxol *n* benzodioxole
Benzoe *f* benzoin gum, benzoin resin
Benzoeäther *m* benzoic ether
Benzoebaum *m* benzoin tree
Benzoechtfarbe *f* benzo fast color
Benzoegummi *n* benzoin gum, benzoin resin
benzoehaltig benzoic
Benzoeharz *n* benzoin gum, benzoin resin
Benzoelorbeeröl *n* spice wood oil
Benzoesäure *f* benzoic acid
Benzoesäureäther *m* benzoic ether, ethyl benzoate
Benzoesäureäthylester *m* ethyl benzoate, benzoic ether
Benzoesäureanhydrid *n* benzoic anhydride
Benzoesäurebenzylester *m* benzyl benzoate
Benzoesäurechlorid *n* benzoyl chloride, benzenecarbonyl chloride, chloride of benzoic acid
Benzoesäureester *m* benzoate, benzoic ester, ester of benzoic acid
Benzoesäurelösung *f* solution of benzoic acid
Benzoesäuremethylester *m* methyl benzoate
Benzoesäuresalz *n* benzoate, salt of benzoic acid
Benzoesäureseife *f* soap of benzoic acid
Benzoesäuresulfimid *n* benzosulfimide, saccharin
benzoesauer benzoate
Benzoeschmalz *n* benzoated lard
Benzoetinktur *f* tincture of benzoin
Benzoflavin *n* benzoflavine
Benzofuran *n* benzofuran
Benz[o]furazan *n* benzofurazan
Benz[o]furoxan *n* benzofuroxan
Benzoguanamin *n* benzoguanamine
Benzoin *n* benzoin
Benzoinkondensation *f* benzoin condensation
Benzoinoxim *n* benzoinoxime
Benzokupferfarbstoff *m* benzo copper dye
Benzol *n* benzene, benzol, phenyl hydride, ~ abtreiben to debenzolize
Benzolabkömmling *m* benzene derivative
Benzolanlage *f* benzene plant
Benzolarsonsäure *f* benzenearsonic acid
Benzolazonaphthol *n* benzeneazonaphthol
Benzolazophenol *n* benzeneazophenol
Benzolbindung *f* benzene linkage
Benzolboronsäure *f* benzene boronic acid, phenylboric acid
Benzolderivat *n* benzene derivative

Benzoldiazoniumchlorid *n* benzenediazonium chloride
Benzoldiazoniumhydroxid *n* benzenediazonium hydroxide
Benzoldiazosäure *f* benzenediazoic acid
Benzoldiazotat *n* benzene diazotate
Benzoldisulfonsäure *f* benzene disulfonic acid
Benzolhexabromid *n* benzene hexabromide
Benzolhexachlorid *n* benzene hexachloride
Benzolkern *m* benzene nucleus
Benzolkohlenwasserstoff *m* aromatic hydrocarbon, benzene hydrocarbon
Benzollack *m* benzol varnish
Benzolpentacarbonsäure *f* benzene pentacarboxylic acid
Benzolreihe *f* benzene series
Benzolrest *m* benzene residue
Benzolring *m* benzene ring
Benzolrückgewinnung *f* benzol recuperation
Benzolsulfinsäure *f* benzenesulfinic acid
Benzolsulfochlorid *n* benzenesulfonyl chloride
Benzolsulfonamid *n* benzene sulfonamide
Benzolsulfonsäure *f* benzenesulfonic acid
benzolsulfonsauer benzenesulfonic
Benzolsulfonylchlorid *n* benzenesulfonyl chloride
Benzoltetracarbonsäure *f* benzene tetracarboxylic acid
Benzoltinktur *f* benzol tincture
Benzoltricarbonsäure *f* benzene tricarboxylic acid
Benzolvergiftung *f* benzene poisoning
Benzolvorprodukt *n* crude benzol
Benzolwäsche *f* benzol recovery by washing
Benzolwäscher *m* benzol washer, benzol scrubber
Benzolwaschöl *n* benzol wash oil
Benzonaphthol *n* benzonaphthol
Benzonatat *n* benzonatate
Benzonitril *n* benzonitrile
Benzooxin *n* benzooxine
Benzopersäure *f* perbenzoic acid, peroxybenzoic acid
Benzophenol *n* benzophenol
Benzophenon *n* benzophenone, diphenylketone
Benzophenonsulfon *n* benzophenone sulfone
Benzopurpurin *n* benzopurpurin
Benzopyran *n* benzopyran
Benz[o]pyren *n* benzopyrene
Benzopyron *n* benzopyrone
Benzoreingelb *n* benzo pure yellow
Benzoresorcin *n* benzoresorcinol
Benzosalin *n* benzosalin
Benzoschwarzblau *n* benzo black blue
Benzosol *n* benzosol, guaiacol benzoate
Benzotetronsäure *f* benzotetronic acid
Benzothialen *n* benzothialene
Benzothiazin *n* benzothiazine
Benz[o]thiazol *n* benzothiazole

Benzothiophen n benzothiophene, thionaphthene
Benz[o]triazol n benzotriazole
Benzotrichlorid n benzotrichloride, benzenyl chloride, phenylchloroform
Benzoxazin n benzoxazine
Benzoxdiazin n benzoxdiazine
Benzoxdiazol n benzoxdiazole
Benzoxthiol n benzoxthiole
Benzoyl- benzoyl-
Benzoylaceton n benzoylacetone
Benzoylacetonamin n benzoylacetone amine
Benzoylaminomalonester m benzoylaminomalonic ester
Benzoylanisylcarbinol n benzoylanisylcarbinol
Benzoylbenzoesäure f benzoylbenzoic acid
Benzoylbrenztraubensäure f benzoylpyruvic acid
Benzoylcampher m benzoylcamphor
Benzoylchlorid n benzoyl chloride, chloride of benzoic acid
Benzoylecgonin n benzoylecgonine
Benzoylen- benzoylene
Benzoylenharnstoff m benzoylene urea, diketotetrahydroquinazoline
Benzoylglykokoll n benzoylglycine, benzoylglycocoll, hippuric acid
benzoylieren to benzoylate, to introduce benzoyl [into]
Benzoylierung f benzoylation
Benzoylleucin n benzoylleucine
Benzoylperoxid n benzoyl peroxide, dibenzoyl peroxide
Benzoylpseudotropein n benzoylpseudotropeine, tropacocaine
Benzoyl-Radikal n benzoyl radical
Benzoylsalicin n benzoylsalicin, populin
Benzoylsuperoxid n benzoyl peroxide, dibenzoyl peroxide
Benzoyltrifluoraceton n benzoyltrifluoroacetone
Benzoylwasserstoff m benzoyl hydride, benzaldehyde
Benzpinakol n benzopinacol
Benzpinakon n benzopinacol
Benzpyren n benzpyrene
Benzselenazol n benzoselenazole
Benzselendiazol n benzoselenodiazole
Benzthiazin n benzothiazine
Benzthiazol n benzothiazole
Benzthiazon n benzothiazone
Benzthiophenid n benzthiophenide
Benztriazin n benzotriazine
Benztriazol n benzotriazole
Benzyl- benzyl-
Benzylacetat n benzyl acetate
Benzyläther m benzyl ether, dibenzyl ether
Benzyläthyläther m benzyl ethyl ether
Benzyläthylanilin n benzyl ethylaniline
Benzylalkohol m benzyl alcohol, phenylcarbinol
Benzylamin n benzylamine

Benzylanilin n benzylaniline
Benzylbenzoat n benzyl benzoate
Benzylbenzol n benzylbenzene
Benzylbromid n benzyl bromide
Benzylbutyladipinat n benzyl butyl adipate
Benzylbutylphthalat n benzyl butyl phthalate
Benzylcarbinol n benzylcarbinol, phenylethyl alcohol
Benzylcellulose f benzyl cellulose
Benzylchlorid n benzyl chloride
Benzylcinnamat n benzyl cinnamate
Benzylcyanid n benzyl cyanide, phenylacetonitrile
Benzylen- benzylene
Benzylester m benzyl ester
Benzyleugenol n benzyleugenol
Benzyliden- benzylidene, benzal
Benzylidenmalonsäure f benzylidenemalonic acid
Benzylidenmalonsäureester m benzylidenemalonic ester
Benzylierung f benzylation
Benzylisoeugenol n benzylisoeugenol
Benzylmalonsäure f benzylmalonic acid
Benzylmalonsäureester m benzylmalonic ester
Benzylmercaptan n benzyl mercaptan
Benzylmethylanilin n benzylmethylaniline
Benzylmethylglyoxim n benzylmethylglyoxime
Benzyloctyladipat n benzyl octyl adipate
Benzylorange n benzyl orange
Benzyloxyacetophenon n benzyloxyacetophenone
Benzylphenol n benzylphenol, hydroxydiphenylmethane
Benzylresorcin n benzylresorcinol
Benzylsenföl n benzyl isothiocyanate
Benzylsiliciumtrichlorid n benzylsilicochloroform, benzyl silicon trichloride, benzyltrichlorosilane
Benzylsilicon n benzyl silicone
Benzylsuccinat n benzyl succinate, succinic acid dibenzyl ester
Benzylsulfid n benzyl sulfide, dibenzyl sulfide
Benzyltrichlorsilan n benzyl trichlorosilane
Benzyn n benzyne
beobachten to observe
Beobachter m observer, **punktförmiger ~** point observer
Beobachtung f observation, detection, examination, study, **spektroskopische ~** spectroscopic observation
Beobachtungsdaten pl observational data
Beobachtungsergebnis n observation data
Beobachtungsfehler m error of observation, personal error
Beobachtungsfeld n field of observation
Beobachtungsmethode f observational method
Beobachtungsmikroskop n viewing microscope
Beobachtungsmittel n tool for observation

Beobachtungsrohr *n* observation tube
Beobachtungsstelle *f* observation point
Beobachtungstatsache *f* observation data
Beobachtungswert *m* observed value
bepacken to burden, to charge, to load, to pack
bepflastern to pave
bepichen to pitch
Beplankung *f* planking, sheeting with plywood
Beplattung *f* plating
Berapp *m* roughcast, rough plaster
berappen to roughcast
beraten to advise, to consult
beratschlagen to consult
Beratung *f* consultation, advice, conference
Beratungsstelle *f* information center
Beraunit *m* (Min) beraunite
berauschen to intoxicate
Berauschungsmittel *n* intoxicant
Berbamin *n* berbamine
Berban *n* berban
Berberal *n* berberal
Berberiden *n* berberidene
Berberidinsäure *f* berberidic acid
Berberilsäure *f* berberilic acid
Berberin *n* berberine
Berberinal *n* berberinal
Berberinbase *f* berberine base
Berberinhydrat *n* berberine hydrate
Berberinhydrochlorid *n* berberine hydrochloride
Berberiniumhydroxid *n* berberinium hydroxide
Berberinol *n* berberinol
Berberinsulfat *n* berberine sulfate
Berberisgelb *n* berberine yellow
Berberisrinde *f* (Bot) barberry bark
Berberitze *f* (Bot) barberry, berberis
Berberonsäure *f* berberonic acid, 2,4,5-pyridinetricarboxylic acid
Berberrubin *n* berberrubine
Berberrubinol *n* berberrubinol
Berberrubinon *n* berberrubinone
Berberrubinsäure *f* berberrubic acid
Berbin *n* berbine
berechenbar calculable, computable
Berechenbarkeit *f* computability, calculability
berechnen to compute, to calculate, to charge
Berechnung *f* calculation, computation
Berechnungsformel *f* formula of computation
Berechnungsgrundlage *f* basis for calculation
Berechnungsnorm *f* computational specifications
Berechnungsweise *f* method of calculation, mode of calculation
berechtigen to authorize, to entitle
berechtigt entitled
Berechtigung *f* authorization, justification
Beregnung *f* overhead irrigation, spraying, sprinkling
Bereich *m* range, domain, region, zone, **kritischer** ~ critical range
bereinigen to settle

Bereinigung *f* settlement
bereiten to prepare
Bereitschaft *f* readiness, preparedness
bereitstellen to make available, to provide
Bereitstellung *f* provision
Bereitung *f* preparation, manipulation
bereitwillig ready, willing
Bereitwilligkeit *f* readiness, willingness
Berengelit *m* (Min) berengelite
Beresit *m* (Min) beresite
Beresowit *m* (Min) beresovite
Bergalaun *m* (Min) alum stone, alunite, rock alum
Bergamottessenz *f* bergamot essence, bergamot oil
Bergamottin *n* bergamottin
Bergamottöl *n* oil of bergamot
Bergapten *n* bergapten
Bergaptenchinon *n* bergaptenquinone
Bergaptin *n* bergaptin, bergamottin
Bergarbeit *f* mining
Bergarbeiter *m* miner
Bergasche *f* (Min) azurite, chessylite, hydrated basic copper carbonate, mountain blue
Bergbalsam *m* mountain balm, bear's weed, eriodictyon
Bergbau *m* mining
Bergbausprengmittel *n* mining explosive
Bergblau *n* mountain blue, azurite, chessylite, hydrated basic copper carbonate
Bergbraun *n* brown iron ore, limonite, umber
Bergbutter *f* impure iron alum
bergen to save
Bergenin *n* bergenine
Bergerz *n* crude ore
Bergfarbe *f* hydrated ferric oxide, ocher
Bergfett *n* mineral tallow, earth wax, hatchettin, mineral wax, mountain tallow, ozokerite
Bergflachs *m* (Min) mountain flax, amianthus, asbestos, **holziger** ~ ligneous amianthus, **holziger** ~ ligneous mountain flax, wood rock
Bergfluß *m* calcium fluoride, fluorite, fluorspar
Berggelb *n* mountain yellow, yellow ocher
Bergglas *n* rock crystal, transparent quartz
Bergglimmer *m* (Min) margarite
Berggrün *n* (Min) mountain green, basic copper carbonate, malachite
Berggur *f* diatomaceous earth, diatomite, infusorial earth, kieselguhr, mountain flour, mountain meal
Berghaar *n* amianthus, asbestos
Bergharz *n* bitumen, mineral pitch
bergharzig bituminous
Bergholz *n* amianthus, ligniform asbestos, rockwood, xylotile
Bergiusverfahren *n* Bergius process
Bergkalk *m* mountain limestone, rock lime
Bergkiesel *m* rock flint, chert, felsite

Bergkork *m* mountain cork, amianthus, asbestos
Bergkreide *f* mountain chalk, powdered dolomite, rock lime
Bergkristall *m* rock crystal, transparent quartz
Bergkupfer *n* native copper
Bergleder *n* mountain leather, amianthus, asbestos
Bergmann *m* miner, pitman
Bergmannit *m* (Min) bergmannite
Bergmehl *n* diatomaceous earth, diatomite, infusorial earth, kieselguhr, mountain flour, mountain meal
Bergmelisse *f* calamint
Bergmilch *f* mountain milk, agaric mineral
Bergnaphtha *f n* mineral naphtha, petroleum, rock oil
Bergöl *n* rock oil, mineral naphtha, petroleum
Bergpech *n* asphalt, earth pitch, mineral pitch
Bergrot *n* native cinnabar
Bergsalz *n* rock salt, common salt, halite
Bergschwefel *m* native sulfur
Bergseife *f* mountain soap, rock soap
Bergtalg *m* earth wax, hatchettin, mineral tallow, mineral wax, mountain tallow, ozokerite
Bergteeöl *n* (Bot) checkerberry oil, gaultheria oil, teaberry oil, wintergreen oil
Bergteer *m* maltha, soft tarlike bitumen, viscous bitumen
Bergtorf *m* mountain peat
Bergung *f* salvage
Bergungsanlage *f* salvage plant
Bergvitriol *m* (Min) native vitriol
Bergwachs *n* (Min) earth wax, hatchettin, mineral tallow, mineral wax, mountain tallow, ozokerite
Bergwerk *n* mine, pit
Bergwolle *f* amianthus, asbestos, mineral wool
Bergzinn *n* cassiterite, native tin dioxide
Bergzinnerz *n* (Min) cassiterite, native tin dioxide, tin ore, tinstone
Bergzinnober *m* (Min) native cinnabar
Bergzunder *m* asbestos, mountain tinder
Bergzundererz *n* jamesonite
Bericht *m* report, information, notice, record
berichtigen to correct, to adjust, to rectify, to set right
Berichtigung *f* correction, adjustment, amendment, rectification
Berichtigungsbeiwert *m* corrective factor
Berichtigungsschraube *f* adjusting screw
Berichtigungstafel *f* correction table
Berichtigungsverfahren *n* correction method, theory of approximations
berieseln to irrigate, to spray
Berieselung *f* irrigation, spraying, sprinkling, watering
Berieselungsanlage *f* irrigation plant
Berieselungsapparat *m* sprayer, sprinkler
Berieselungsfläche *f* irrigation surface
Berieselungskühler *m* irrigation cooler, spray cooler, (Benzolgewinnung) wash tower, ~ **mit Rundrohrflächen** *f pl* round-tube spray cooler
Berieselungsturm *m* scrubbing tower, spray tower, washer, washing column
Berieselungsverflüssiger *m* spray condenser
Berieselungsvorrichtung *f* spraying device
Berilsäure *f* berilic acid
Berkelium *n* berkelium (Symb. Bk)
Berlambin *n* berlambine
Berliner Blau *n* Berlin blue, ferric ferrocyanide, Prussian blue
Berliner Grün *n* Prussian green
Berliner Rot *n* Berlin red, red ferric oxide
Berliner Weiß *n* Prussian white
Berlinit *m* (Min) berlinite
Berlsattel *m* (Dest) Berl saddle
Bernoullische Gleichung *f* Bernoulli's equation
Bernstein *m* amber, succinite, **schwarzer** ~ jet
Bernsteinalaun *m* aluminous amber
Bernsteinaldehyd *m* succinaldehyde
Bernsteinaldehydsäure *f* succinaldehydic acid
Bernsteinamidsäure *f* succinamic acid
bernsteinartig amberlike, succinous
Bernsteinerde *f* amber-containing earth, blue earth
bernsteinfarben amber[-colored]
Bernsteinfett *n* ambrain
Bernsteinfirnis *m* amber varnish
bernsteinhaltig containing amber
Bernsteinharz *n* amber resin
Bernsteinlack *m* amber varnish
Bernsteinnitrilsäure *f* succinonitrilic acid
Bernsteinöl *n* amber oil
Bernsteinsäure *f* succinic acid
Bernsteinsäureanhydrid *n* succinic anhydride
Bernsteinsäurechlorid *n* succinyl chloride
Bernsteinsäurediäthylester *m* diethyl succinate
Bernsteinsäurenitril *n* succinonitrile
Bernsteinsalz *n* salt of succinic acid, succinate
bernsteinsauer succinate
Berosten *n* rusting, getting rusty
Berstdruck *m* bursting strength, bursting pressure
Berstdruckfestigkeit *f* bursting strength
Berstdruckprüfer *m* bursting tester
bersten to burst, to explode
Bersten *n* bursting, explosion
Berstfestigkeit *f* bursting strength
Berstscheibe *f* bursting disc, rupture disc, safety diaphragm
Berstwiderstand *m* bursting strength
Berthierit *m* (Min) berthierite
Bertrandit *m* (Min) bertrandite
berücksichtigen to consider, to bear in mind, to regard
Berücksichtigung *f* consideration, regard
berührend tangent

Berührung f contact
Berührungsdichtung f contacting seal
Berührungsebene f (Math) tangential plane
Berührungselektrizität f contact electricity
Berührungsfläche f contact surface, interface
Berührungsheizung f contact heating
Berührungsinsektizid n contact insecticide
Berührungskathode f contact cathod
Berührungskorrosion f contact corrosion, crevice corrosion at a contact with non-metallic material
Berührungslinie f (Math) tangent
Berührungspunkt m point of contact, (Math) tangent point
Berührungsspannung f contact voltage
Berührungsstelle f place of contact, point of contact
Berührungstransformation f contact transformation
Berührungszündung f (einer Mine) contact firing
Berührungszwilling m contact twin
Beruf m job, occupation, profession, vocation
Berufsausbildung f professional training
Berufskrankheit f professional disease
berufsmäßig professional
beruhigen to calm, to reassure, to soothe, to stabilize, (Metall) to deoxidize, to kill, to quiet
beruhigt (Stahl) killed
Beruhigungsmittel n (Metall) killing agent, (Pharm) sedative, tranquilizer
berußen to [cover with] soot, to smoke
Berußen n covering with soot
berußt sooty
Beryll m (Min) beryl, **blättriger** ~ cyanite, disthene
beryllartig like beryl, berylline
Beryllat n beryllate
Beryllerde f beryllia, beryllium oxide, glucina
Beryllid n beryllide
Beryllium n beryllium (Symb. Be), glucinum
Berylliumbromid n beryllium bromide, glucinum bromide
Berylliumcarbid n beryllium carbide
Berylliumchlorid n beryllium chloride
Berylliumderivat n beryllium derivative
Berylliumfluorid n beryllium fluoride, glucinum fluoride
Berylliummethylsalicylat n beryllium methylsalicylate
Berylliumnitrat n beryllium nitrate, glucinum nitrate
Berylliumnitroacetatchelat n beryllium nitroacetate chelate
Berylliumorthosilikat n beryllium orthosilicate
Berylliumoxid n beryllium oxide, beryllia, glucina, glucinum oxide (obs)
Berylliumreaktor m beryllium reactor
Berylliumreflektor m beryllium reflector

Berylliumsalz n beryllium salt, glucinum salt (obs)
Berylliumsulfat n beryllium sulfate, glucinum sulfate (obs)
Beryllon n beryllon
Beryllonit m (Min) beryllonite
Berzelianit m (Min) berzelianite
Berzeliit m (Min) berzeliite
Berzelius'sche Weingeistlampe f Berzelius spirit lamp
besäumen to trim
Besäummaschine f trimming cutter
Besamung f (Biol) insemination, **künstliche** ~ artificial insemination
Besatz m (Kleid etc) trimming
beschädigen to damage, to harm, to hurt, to impair, to mar, to ruin, to spoil
beschädigt damaged, defective
Beschädigung f damage, harm, injury
beschäftigen to engage
Beschäftigung f occupation, business, employment
beschäumen to cover with foam
beschaffen to provide, to procure, to supply
Beschaffenheit f nature, quality, state, **gallertartige** ~ gelatinousness, **glasartige** ~ glassiness, vitreousness
beschallen to sonicate, to irradiate acoustically
Beschallung f sonication, acoustic irradiation
Beschaufelung f blading
Bescheid m answer, decision, information
bescheinigen to certify, to attest
beschichten to apply a coat [to], to coat
Beschichten n coating
Beschichtung f coat, coating;, **wasserabweisende** ~ water-repellant coating
Beschichtungsanlage f coating unit
Beschichtungsmaschine f coater, coating machine
Beschichtungsmasse f coating compound
Beschichtungsmaterial n coating, coating compound
Beschichtungsmittel n zum Verhindern des Gleitens anti-slip surfacing compound
Beschichtungsstoff m coating material
beschicken to charge, to feed, to load
Beschicken n charging, feeding, loading
Beschickkran m charging crane
Beschickmaschine f charging device, charging machine
Beschickschwengel m balanced peel
Beschickung f charge, burden, feed, feeding; loading, (Text) load
Beschickungsanlage f charging equipment, feeder, feeding installation, feeding plant, mechanical charger
Beschickungsansatz m batch
Beschickungsbühne f charging platform, charging floor

Beschickungsdauer *f* charging period
Beschickungsfolge *f* sequence of charging
Beschickungsgut *n* charge, melting stock
Beschickungshöhe *f* charge level
Beschickungsklappe *f* charge door, feed hopper door
Beschickungskran *m* charging crane
Beschickungslöffel *m* charging spoon, charging peel
Beschickungsmaschine *f* charging machine
Beschickungsmaterial *n* charge, melting stock
Beschickungsmulde *f* charging box
Beschickungsoberfläche *f* stock line
Beschickungsperiode *f* charging period
Beschickungsprobe *f* charging sample
Beschickungssäule *f* stock column
Beschickungstrichter *m* feed hopper
Beschickungstür *f* charge door, charging door, feed hopper door
Beschickungsverzögerung *f* charging delay
Beschickungsvorrichtung *f* charging device, feeding plant
Beschickungswagen *m* charge wagon
Beschickungsweise *f* method of charging
Beschickungszeit *f* charging period
Beschickungszone *f* feed zone, feed section
Beschickwagen *m* charge wagon
beschießen (Atom) to bombard
Beschießung *f* bombardment
Beschirmung *f* protection
Beschläge *m pl* mountings
Beschläge *pl* fittings, (Chem) incrustations, sublimates
Beschlag *m* armature, fitting, mounting, (Auskleidung) lining, (Chem) deposit, (Überzug) coating; ~ **auf Metall** tarnish, **kalkiger** ~ chalking
Beschlagbändsel *n* (Mar) furling line
beschlagen, sich ~ to become coated, to grow moldy, (Met) to tarnish; (mit einer Kruste) to effloresce
beschlagnahmen to confiscate, to seize
Beschlagteile *pl* fittings
Beschlagverhinderungsmittel *n* (Glas) antifogging agent
beschleimen to coat with slime
beschleunigen to accelerate, to hasten, to quicken, to speed up
Beschleuniger *m* activator, (Atom) accelerator, (Chem) catalyst; **chemischer** ~ catalyst
Beschleunigung *f* acceleration, speeding up, **katalytische** ~ catalytic acceleration, **lineare** ~ linear acceleration, **mittlere** ~ (Mech) average acceleration, **negative** ~ deceleration, retardation
Beschleunigungsanlage *f* (Atom) accelerator
Beschleunigungselektrode *f* accelerator electrode
Beschleunigungsgesetz *n* law of acceleration
Beschleunigungsgitter *n* accelerator grid

Beschleunigungskammer *f* (Atom) accelerating chamber
Beschleunigungskraft *f* accelerator force
Beschleunigungsmesser *m* accelerometer
Beschleunigungsmoment *n* moment of acceleration
Beschleunigungsnerv *m* (Med) accelerator nerve
Beschleunigungsregistrierapparat *m* accelerograph
Beschleunigungsregler *m* speed regulator
Beschleunigungsröhre *f* (Atom) accelerating tube
Beschleunigungstrocknung *f* forced drying
Beschleunigungszeit *f* period of acceleration
beschließen to determine, to decide, to resolve, (abschließen) to conclude, to finish
beschmieren to smear, to grease
beschmutzen to soil, to stain
beschneiden to cut, to curtail, to shear, to shorten, to trim
Beschneiden *n* trimming
Beschneidmaschine *f* (Buchdr, Metall) trimming cutter
Beschnitt *m* (Buchdr) trim (original size minus size after trimming)
Beschränkung *f* restriction, confinement, limitation, reduction
Beschreibung *f* description, outline, sketch, specification
beschriften to label, to inscribe
Beschriftung *f* label, inscription
beschützen to protect, to defend
Beschuß *m* (Atom) bombardment
Beschußteilchen *n* (Atom) bombarding particle
beschwefeln to fumigate
Beschwefeln *n* fumigating
beschweren to charge, to load, to weight, **sich** ~ (reklamieren) to complain
Beschweren *n* charging, loading, weighting
Beschwerung *f* charge, loading; (Text, Pap) weighting
Beschwerungsmittel *n* pigment, weighting agent, weighting material
Beschwerungsverfahren *n* (Text) weighting method
beseitigen to eliminate, to remove
Beseitigung *f* elimination, disposal, removal, ~ **radioaktiver Abfälle** *m pl* radioactive waste disposal
Besenborsten *pl* sweeper bristles
Besetzungsgrad *m* **der Banden** *f pl* degree of filling of bands
Besetzungsverbot *n* (Chem) exclusion principle
Besetzungsvorschrift *f* occupation rule
Besetzungszahl *f* occupation number, ~ **der Elektronen** *pl* **in einem Niveau** extent to which level is populated by electrons
besichtigen to examine, to inspect, to view
Besohlen *n* (Reifen) recapping, top capping, top treading

Besonderheit *f* particularity, peculiarity
besonders in particular, particular; peculiar, special
besorgen to provide
Bespannstoff *m* covering material
besprengen to moisten, to spray, to sprinkle
bespritzen to spatter, to spray, to sprinkle
Bespritzung *f* spraying, sprinkling
besprühen to spray
Bessel-Funktionen *pl* Bessel functions
Bessel-Interpolationsformel *f* Bessel's interpolation formula
Bessemerbirne *f* bessemer converter
Bessemerei *f* bessemer foundry, bessemer plant
Bessemereisen *n* bessemer steel, acid steel
Bessemerflußeisen *n* bessemer ingot iron
bessemern (Metall) to bessemerize, to bessemer iron
Bessemern *n* bessemerization
Bessemerprozeß *m* (Metall) bessemer process
Bessemerroheisen *n* bessemer pig iron, hematite pig iron
Bessemerschlacke *f* bessemer slag
Bessemerstahl *m* bessemer steel, acid steel
Bessemerstahlwerk *n* bessemer steel works
Bessemerverfahren *n* bessemerization, acid method, bessemer process
Bessisterin *n* bessisterol
beständig (dauernd) lasting, permanent, (fest) stationary, (gleichbleibend) constant, (stabil) proof, stable, (widerstandsfähig) resistant
Beständigkeit *f* durability, resistance, stability, ~ **der Frequenz** frequency stability
bestärken to confirm, to fortify, to strengthen
bestätigen to authenticate, to confirm; to verify
Bestätigungsurteil *n* confirmatory decision
bestäuben to dust
Bestäubung *f* dusting
Bestand *m* (Dauerhaftigkeit) durability, stability, (Existenz) existence, (Fortbestehen) continuance, duration, permanence, (Rest) remainder, rest, (Vorrat) reserve
Bestandteil *m* (Chem) component, constituent, ingredient, **flüchtiger** ~ volatile constituent, volatile matter, **unbrennbarer** ~ incombustible constituent
Bestandteile *pl*, **sich in seine** ~ **auflösen** to disintegrate, **strukturelle** ~ structural members
bestausgerüstet best-equipped
Besteck *n* cutlery
bestehen to exist, ~ **aus** to be composed of, to consist of
bestellen to order, to appoint; (Agr) to farm, to till
Bestellmaß *n* specified dimension
Bestellschein *m* requisition, order form
Bestellung *f* order
bestgeeignet most suitable

bestimmbar determinable
bestimmen to decide, (anordnen) to order, (Chem) to analyse, to determine, (definieren) to define, (identifizieren) to identify
Bestimmung *f* decision, definition, determination; (Vorschrift) specification, **kalorimetrische** ~ calorimetric determination, **polarographische** ~ polarography
Bestimmungen *f pl* regulations, rules
Bestimmungsapparat *m* testing apparatus, determining apparatus
Bestimmungsgleichung *f* (Math) defining equation, determinental equation
Bestimmungsgröße *f* defining quantity, (Techn) specification factor
Bestimmungsmethode *f* method of determination
Bestimmungsort *m* destination
Bestimmungsrohr *n* determination tube
Bestleistung *f* best performance, record
Bestoßmaschine *f* wood trimmer
Bestoßzeug *n* trimming tool
bestrahlen to irradiate, to expose to rays
Bestrahlung *f* irradiation, exposure to rays, (Beleuchtung) illumination, (Med) radiation treatment, ~ **des Brutstoffes** blanket material exposure
Bestrahlungsdauer *f* (Atom) duration of irradiation, (Med) duration of exposure
Bestrahlungserholung *f* radiation annealing
Bestrahlungsgeräte *n pl* irradiation equipment
Bestrahlungskanal *m* exposure hole, irradiation hole
Bestrahlungslampe *f* irradiation lamp
Bestrahlungsmutation *f* radiation mutation
Bestrahlungsprodukt *n* radiation product
Bestrahlungsschaden *m* irradiation damage
Bestrahlungsstärke *f* (Opt) irradiance, radiant flux density
Bestrahlungssterilisierung *f* radiation sterilization
Bestrahlungszeit *f* time of exposure
bestreichen to coat, **mit Asphalt** ~ to asphalt, **mit Gummilösung** ~ to gum
Bestreichen *n* coating, painting
bestreiten to dispute, (Kosten) to defray
bestreuen to powder, to sprinkle, to strew
Bestreuen *n* powdering, sprinkling, strewing
bestückt (Werkzeug) tipped
Bestwert *m* optimum
besudeln to contaminate, to soil
beta-aktiv (Atom) beta-active, beta-radioactive
Betadickenmesser *m* (Atom) beta gauge
Betadickenmeßgerät *n* beta thickness gauge
Betaeisen *n* beta iron
betaempfindlich (Atom) beta-sensitive
Betaempfindlichkeit *f* (Atom) beta sensitivity

betätigen (maschinell) to operate, to actuate, to control, (von Hand) to manipulate, to set going
Betätigung f operation, actuation, control, manipulation
Betätigungshebel m operating lever
Betätigungsknopf m start and stop button
Betätigungsschalter m control switch
Betätigungsvorrichtung f operating gear
Betäubungsmittel n (Med) narcotic
Betafit m (Min) betafite
Betaglobulin n beta globulin
Betain n betain[e], (Trimethylglykokoll) trimethyl-glycocoll
Betainchlorhydrat n betaine hydrochloride
Betainchlorid n betaine hydrochloride
Betainformel f betain formula
Betainhydrochlorid n betaine hydrochloride
Beta-Instabilität f (Atom) beta instability
betalgen to tallow
Betaprodin n betaprodine
beta-radioaktiv (Atom) beta-radioactive, beta-active
Betasäure f beta acid
Betaspektrometer n beta ray spectrometer
Betaspektroskopie f (Atom) beta-ray spectroscopy
Betaspektrum n (Atom) beta-particle spectrum, beta-ray spectrum
betastrahlend beta-emitting
Betastrahlenmeßgerät n beta ray gauge
Betastrahlenquelle f beta-ray source
Betastrahlentherapie f beta-ray therapy
Betastrahler m beta emitter
Betastrahlspektrometer n (Atom) beta-ray spectrometer
Betastrahlung f beta emission
Betastung f touching, (Med) palpation
Betateilchen n beta particle
Betatron n (Atom) betatron
Betatronstrahlung f betatron radiation
Beta-Übergang m (Atom) beta transition
Beta-Umwandlung f (Atom) beta transformation, beta disintegration
Beta-Untergrund m beta background
Beta-Uran n beta uranium
Beta-Zählrohr n (Atom) beta counter
Beta-Zählung f beta counting
Beta-Zerfall m beta decay, beta disintegration, beta transformation
Beta-Zerfallsenergie f (Atom) beta disintegration energy
Betazol n betazole
beteeren to tar
Beteiligung f investment, partnership; (Anteilnahme) participation
Betelpfeffer m (Bot) betel pepper
Bethogenin n bethogenin
Beting f (Mar) bitts, carrick bitt

Betingknie n (Mar) carrick bitt
Betol n betol, 2-naphthyl salicylate
Beton m concrete, **gestampfter** ~ rammed concrete, **haufwerkporiger** ~ no-fines concrete
Betonabschirmung f (Reaktor) concrete shield
Betonbau m concrete construction
Betondichtungsmittel n concrete sealing agent
Betoneisen n reinforcing iron
betonen to accentuate, to emphasize
Betonfarbe f concrete paint
Betonfertigteil n precast concrete unit
Betonhartstoff m concrete hardening material
Betonicin n betonicine
Betonienkraut n (Bot) betony
betonieren to concrete, to cover with concrete
Betonierung f concreting
Betonmischmaschine f concrete mixer
Betonplatte f cement slab
Betonschicht f layer of concrete
Betonsockel m concrete foundation
Betonstahl m concrete steel
Betonstein m concrete building brick, concrete stone
Betonunterlage f concrete base, concrete subfloor
Betonverflüssiger m concrete thinner
Betonverschalung f concrete shuttering
Betorcinol n betorcinol
Betracht m, **in** ~ **kommen** to be considered, to be of interest
beträchtlich considerable
Betrag m amount, sum
betreffen to concern
betreffend concerning, respective
betreiben to carry on, to conduct
Betrieb m plant, establishment, factory, operation; performance, working, **in** ~ **sein** to be at work, to be in operation, **in** ~ **setzen** to set into action, to set to work
betrieblich operational, operative, working
Betriebsablauf m operational procedure
Betriebsabteilung f production department, operation section
Betriebsangaben f pl shop data
Betriebsanlage f installation, plant, works
Betriebsanleitung f operating instructions
Betriebsanweisung f maintenance instructions, operating instructions
Betriebsarbeit f industrial labor
Betriebsarzt m factory doctor
Betriebsbeanspruchung f operating load, service condition, service stress
betriebsbedingt operational
Betriebsbedingungen f pl conditions of operation, performance
Betriebschemiker m industrial chemist, works chemist
Betriebsdämpfung f effective attenuation, operative attenuation

Betriebsdampf *m* process steam
Betriebsdaten *pl* operating data
Betriebsdauer *f* time of operation
Betriebsdirektor *m* managing director, works manager
Betriebsdrehzahl *f* normal speed
Betriebsdruck *m* operating pressure, service pressure, working pressure, (Hydr) line pressure
Betriebsdruckleitung *f* main [pressure] line
Betriebsdynamik *f* dynamic range
Betriebseinrichtung *f* operating installation
Betriebseinstellung *f* shut-down of work
Betriebserfahrung *f* practical experience
Betriebserfordernis *f* service requirement
Betriebsergebnis *n* operating result
betriebsfähig operable, workable
Betriebsfähigkeit *f* working condition
Betriebsfeld *n* field of operation
betriebsfertig ready for use
Betriebsforschung *f* industrial research, production research
betriebsführend managing
Betriebsführer *m* manager
Betriebsführung *f* management
Betriebsgas *n* feed gas
Betriebshygiene *f* factory hygiene
Betriebsingenieur *m* operating engineer, industrial engineer, manufacturing engineer, plant engineer
Betriebsinhalt *m* (Reaktionsführung) hold-up
Betriebskennzahlen *f pl* operating figures, process data
Betriebskessel *m* plant boiler, working boiler
Betriebskontrolle *f* supervision of instrumentation, control of operations; process control; production control
Betriebskontrollingenieur *m* instrumentation engineer
Betriebskosten *pl* operating cost, operating expenses, working expenses
Betriebskraft *f* motive power
Betriebsleistung *f* production, operational efficiency
Betriebsleiter *m* superintendent, plant manager
Betriebsmaßstab *m* plant scale
Betriebsorganisation *f* works management
Betriebsperiode *f* period of service
Betriebsraum *m* work shop
Betriebsschema *n* flow sheet
Betriebs-Schwingversuch *m* fatigue test under service condition (at mixed amplitudes, at operational stresses or at working stress levels)
betriebssicher reliable in operation, safe to operate
Betriebssicherheit *f* working safety, safety in operation
Betriebsspannung *f* working voltage, operating voltage, operational tension
Betriebsstillegung *f* shut-down
Betriebsstörung *f* breakdown, operating trouble, shut-down
Betriebsstoff *m* fuel, power fuel
Betriebsstoffverbrauch *m* fuel consumption
Betriebsstrom *m* operating current
Betriebstechniker *m* industrial engineer, operator
Betriebstemperatur *f* working temperature
Betriebsüberschüsse *m pl* surplus
Betriebsüberwachung *f* production control, production supervision
Betriebsunkosten *pl* working expenses
Betriebsunterbrechung *f* delay in production, interruption of operation, shut-down
Betriebsverhältnisse *n pl* working conditions, plant conditions
Betriebsverhalten *n* performance
Betriebsvorschriften *f pl* operating instructions, specifications
Betriebswasser *n* water for industrial use
Betriebsweise *f* method of operation
Betriebswissenschaft *f* business administration
Betriebszeit *f* working hours, shift, working period
Betriebszustand *m* working condition, working order, **stationärer** ~ steady-state operation
Betriebszuverlässigkeit *f* reliability in service
Betriebszweig *m* branch of manufacture
Betruxinsäure *f* betruxic acid
Bettfilter *n* deep bed filter
Bettibase *f* Betti's base
Betulafolientriol *n* betulafolientriol
Betulen *n* betulene
Betuligenol *n* betuligenol
Betulin *n* betulinol, birch camphor
Betulinsäure *f* betulinic acid
Betulol *n* betulol
Betulosid *n* betuloside
betupfen to spot, to dab
Beuche *f* buck, bucking lye
beuchen to steep or boil in lye, to bowk, to buck, to kier boil
Beuchen *n* bowking, bucking, kier boiling
Beuchfaß *n* bucking tub
Beuchkessel *m* bucking kier
Beuchlauge *f* bucking lye
Beuchwasser *n* buck, steep
Beudantit *m* (Min) beudantite
Beuge *f* bend, bow
beugen to bend, to bow, (Opt) to diffract
beugsam flexible, pliable, pliant
Beugsamkeit *f* flexibility, pliability
Beugung *f* flection, flexion (Br. E.), (Opt) diffraction, refraction
Beugungsbild *n* (Opt) diffraction pattern
Beugungsdiagramm *n* (Opt) diffraction pattern
Beugungserscheinung *f* phenomenon of diffraction

Beugungsfarben *f pl* prismatic colors
Beugungsfigur *f* diffraction pattern
Beugungsgitter *n* diffraction grating
Beugungsmesser *m* diffractometer
Beugungsspektroskop *n* diffraction spectroscope
Beugungsspektrum *n* (Spektr) diffraction spectrum, normal spectrum
Beugungsstreifen *m* (Opt) diffraction fringe
Beugungstheorie *f* diffraction theory
Beugungswinkel *m* angle of refraction
Beugungszahl *f* (Opt) order of interference
Beugungszentrum *n* center of diffraction
Beugungszone *f* diffraction zone
Beule *f* dent, hammer mark
Beulspannung *f* denting stress
Beulsteifigkeit *f* blister resistance
beurkunden to attest, to authenticate, to legalize
Beurkundung *f* certification, verification by documentary evidence, (notarielle) notarization
beurteilen to estimate, to rate, to value
Beurteilung *f* evaluation, rating
Beutel *m* bag, sac, (Med) pouch
Beutelfilter *n* (Zuck) bag filter
beuteln to bolt, to sift
Beuteln *n* bolting, sifting
Beutelschließen *n* bag sealing
Beutelschweißen *n* bag sealing
Beutelsieb *n* (Müllerei) bolter, bolting sieve
Beuteltuch *n* bolting cloth
Beutelung *f* sieving
Beutelverschließgerät *n* bag sealer
Beutelzeug *n* bolting apparatus, bolting cloth
Bevatron *n* (Atom) bevatron
Bevölkerungsgenetik *f* population genetics
bevollmächtigen to authorize, to empower
Bevollmächtigter *m* authorized agent, trustee
Bevollmächtigung *f* authorization
bevorzugt preferential
Bevorzugung *f* preference
Bewachung *f* supervision, custody
bewähren, sich ~ to prove
bewährt approved, successful, tested, tried
bewältigen to master, to handle, to overcome
bewässern to irrigate, to sprinkle, to water
Bewässerung *f* irrigation
bewahren to keep, to preserve, to protect
bewegen to move, to actuate
beweglich mobile, flexible, moveable, (tragbar) portable
Beweglichkeit *f* mobility, movability, ~ **der Stromträger** *m pl* current carrier mobility
Beweglichkeitsgleichung *f* mobility equation
Beweglichkeitsgrenze *f* (Krist) limitation of mobility
Beweglichkeitstensor *m* mobility tensor
Bewegung *f* movement, migration, motion, **auf- und niedergehende** ~ up and down motion, **erzwungene** ~ (Mech) constrained motion, **fortschreitende** ~ continuous motion, **gedämpfte harmonische** ~ damped harmonic motion, **gegenläufige** ~ double motion, **hin- und hergehende** ~ alternate back and forward motion, reciprocating motion, to and fro motion, **rotierende** ~ rotation, revolution, rotary motion, **ruckweise** ~ (Med) shock motion, **thermische** ~ thermal agitation, thermic motion
Bewegungsachse *f* axis of rotation
Bewegungsantrieb *m* motion drive, motive power
Bewegungsbilanz *f* statement of application of funds
Bewegungsenergie *f* kinetic energy
Bewegungsfreiheit *f* freedom of movement
Bewegungsgeschwindigkeit *f* rate of motion
Bewegungsgesetz *n* law of motion
Bewegungsgleichung *f* motion equation
Bewegungskraft *f* motive power
Bewegungslehre *f* mechanics, kinematics
Bewegungslosigkeit *f* immobility, immovability
Bewegungsrichtung *f* direction of movement
Bewegungssitz *m* free fit, medium fit
Bewegungsumkehr *f* reversal of motion
Bewegungsvorrichtung *f* operating mechanism, working gear
Bewegungszustand *m* state of motion
bewehren to reinforce, to strengthen
Bewehrung *f* stiffener, stiffening
Beweis *m* evidence, proof
Beweisartikel *m* argument
Beweisaufnahme *f* evidence
beweisen to prove, to demonstrate, to verify
Beweisführung *f* argumentation, demonstration, method of proof
beweiskräftig convincing, conclusive
Beweismaterial *n* evidence
Beweisstück *n* proof, concrete evidence
Bewerber *m* applicant, candidate
Bewerbung *f* application
bewerkstelligen to accomplish, to carry out, to effect, to perform
bewerten to estimate, to evaluate, to rate
Bewerten *n* estimating, evaluating, rating
Bewertung *f* estimation, evaluation, qualification, rating, valuation
Bewertungsausgleich *m* valuation adjustment
bewettern (Bergb) to ventilate
Bewetterung *f* weathering, air conditioning, ventilation
bewickeln to envelop, to wind, to wrap around
Bewicklung *f* winding
Bewilligung *f* permission, concession, license, permit
bewirken to cause, to effect
bewirtschaften to manage
Bewirtschafter *m* manager, operator
Bewitterung *f* weathering
Bewitterungsleitung *f* airing tube

Bewitterungsprüfstand *m* exposure rack
Bewitterungsprüfung *f* natural weathering test
Bewitterungsstelle *f* exposure site
Bewitterungsversuche *pl* exposure trials
Beyrichit *m* (Min) beyrichite
bezeichnen to mark, (angeben) to indicate, (bedeuten) to signify, to stand for, (benennen) to designate
bezeichnenderweise characteristically
Bezeichnung *f* marking, description, designation, indication, notation, specification, symbol, term
Bezeichnungsschild *n* label
Bezeichnungssystem *n* notation system
Bezeichnungsweise *f* method of notation
Beziehung *f* relation, correlation, respect; (Bezugnahme) reference
beziehungsweise respectively
beziffern to number, to figure
Bezifferung *f* numbering, numeration
Bezoar *m* bezoar
Bezoarpulver *n* bezoar powder
Bezoarstein *m* bezoar stone
Bezoarwurzel *f* contrayerva
bezogen auf related to, relative to
Bezug *m* cover, covering, reference; relation
Bezugnahme *f* reference
Bezugsbedingungen *pl* terms of purchase
Bezugseinheit *f* unit
Bezugselektrode *f* reference electrode
Bezugselement *n* reference element
Bezugsfrequenz *f* (Elektr) reference frequency
Bezugsgröße *f* reference magnitude
Bezugskraftstoff *m* reference fuel
Bezugskreis *m* reference circle
Bezugslänge *f* sampling length
Bezugslinie *f* (Meßtechnik) reference line
Bezugslösung *f* reference solution
Bezugsnormal *n* reference standard
Bezugspunkt *m* reference point
Bezugsquelle *f* source of supply
Bezugsquellenverzeichnis *n* buyer's guide
Bezugssterne *m pl* reference stars
Bezugssystem *n* reference system
Bezugstemperatur *f* reference temperature
Bezugswert *m* reference value, relative value
bezuschussen to subsidize
Biacen *n* biacene
Biacenon *n* biacenone
Bialamicol *n* bialamicol
Biallyl *n* biallyl
Bianthron *n* bianthrone
biaxial biaxial
Bibergeil *n* castoreum
Bibergeilcampher *m* castorin
Bibernelltinktur *f* tincture of pimpinella
Bibernellwurzel *f* (Bot) pimpernel root
Bibirinchloraurat *n* biberine chloroaurate, biberine gold chloride

Bibirinchlorhydrat *n* biberine hydrochloride
Bibirinchloromercurat *n* biberine chloromercurate, biberine mercury chloride
Bibirinchloroplatinat *n* biberine chloroplatinate, biberine platinum chloride
Bibirinsäure *f* bebeeric acid
Bibliolit *m* (Min) bibliolite, book stone
Bibrocathol *n* bibrocathol
Bicarbonat *n* bicarbonate, acid carbonate, hydrogen carbonate
Bichlorbenzol *n* dichlorobenzene
Bichromat *n* bichromate, dichromate
Bichromatanlage *f* bichromate plant
Bichromatgefäß *n* bichromate cell
Bichromatschwefelsäure *f* mixture of dichromate and sulfuric acid
Bichromattitration *f* bichromate titration
Bicucin *n* bicucine
Bicucullin *n* bicuculline
bicyclisch bicyclic
Bicyclobutan *n* bicyclobutane
Bicycloheptanol *n* bicycloheptanol
Bicycloheptanon *n* bicycloheptanone
Bicyclohexan *n* bicyclohexane
Bicyclohexanon *n* bicyclohexanone
Bicyclononatrien *n* bicyclononatriene
Bicyclooctadien *n* bicyclooctadiene
Bidestillat *n* bidistillate
Bidestillator *m* bidistiller
Bidesyl *n* bidesyl
Bieberit *m* (Min) bieberite
Biebricher Scharlach *m* Biebrich scarlet
biegbar flexible, pliable, pliant, (Met) ductile
Biegbarkeit *f* flexibility, pliability
Biegebeanspruchung *f* bending stress, flexural strain, flexural stress
Biegebelastung *f* bending strain, transverse loading
Biegedauerfestigkeit *f* **im Schwellbereich** fatigue strength under repeated (pulsating, fluctuating) bending stresses
Biegedorn *m* bending mandrel
Biegedruck *m* bending pressure, bending strain
Biegeeigenschwingung *f* natural frequency in bending
Biegeermüdung *f* flex fatigue
Biegefähigkeit *f* bending property, flexibility
Biegefaktor *m* elasticity of flexure
biegefest resistant to bending
Biegefestigkeit *f* resistance to bending, bending strength, cross breaking strength, fatigue strength, flexural strength, transverse strength
Biegeformung *f* bending or flexural strain
Biegegesenk *n* forming die
Biegegröße *f* bending coefficient
Biegehalbmesser *m* bending radius
Biegekraft *f* bending force
Biegemaschine *f* bending machine, flexing machine

Biegemoment *n* bending moment
biegen to bend, to bow, to curve, (Phys) to diffract
Biegepresse *f* bending press
Biegeprobe *f* bending test
Biegeprüfer *m* bend test apparatus
Biegeprüfung *f* bending test, (Gußeisen) transverse test, ~ **in der Kälte** cold-bend test
Biegerisse *pl* flex cracking
Biegeschlagversuch *m* bending impact test
Biegeschwellfestigkeit *f* pulsating fatigue strength under bending stress
Biegespannung *f* bending stress, flexural strain
Biegespannungsmaschine *f* bending stress tester
biegesteif resistant to bending
Biegesteifigkeit *f* bending resistance
Biegestelle *f* bending point
Biegestempel *m* (Biegeprüfung) loading nose
Biegeversuch *m* [transverse] bending test, crossbreaking test, deflection test
Biegevorrichtung *f* bending device, bending fixture, bending-test shackle
Biegewalze *f* bending roll
Biegewechselfestigkeit *f* fatigue strength under reversed bending stresses, Wöhler fatigue strength
Biegezahl *f* bending value
Biegfestigkeit *f* bending strength, flexural strength, rigidity
biegsam flexible, pliable, pliant
Biegsamkeit *f* flexibility, pliability, (Met) ductility
Biegung *f* bend, bending; deflection
Biegungsbeanspruchung *f* bending strain, flexural stress
Biegungs-Dauerschwingfestigkeit *f* flexural fatigue strength
Biegungsebene *f* plane of flexure
Biegungseinräumung *f* bend allowance
Biegungselastizität *f* elasticity of flexure, flexural elasticity
Biegungsermüdung *f* bending fatigue
biegungsfähig flexible, pliable, (Met) ductile
biegungsfest moment resisting
Biegungsfestigkeit *f* bending strength, strength of flexure
Biegungshalbmesser *m* bending radius
Biegungsmesser *m* deflection indicator, deflectometer
Biegungsmoment *n* bending moment, moment of flexure
Biegungsschwingung *f* oscillation due to deflection, flexure mode
Biegungsspannung *f* bending stress
Biegungswelle *f* flexural wave
Biegungswiderstand *m* resistance to bending
Bienengift *n* bee venom
Bienenharz *n* bee glue, propolis
Bienenwabe *f* honey comb
Bienenwabenstruktur *f* honey-comb structure
Bienenwachs *n* beeswax
Bienenzucht *f* beekeeping
Bier *n* **vom Faß** draft beer, **abgelagertes** ~ vatted beer, **dunkles** ~ dark beer, **helles** ~ light beer, **obergäriges** ~ top fermentation beer, **untergäriges** ~ bottom fermentation beer, low fermentation beer
Bierbrauen *n* beer brewing
Bierbrauerei *f* brewery
Bieressig *m* beer vinegar, malt vinegar
Bierfaß *n* beer barrel
Bierhefe *f* (Brau) brewer's yeast, barm
Bierschaum *m* beer froth
Biersteinentferner *m* beer scale destroying agent
Biertrübung *f* beer haze
Bierwürzekühler *m* beer wort cooler
Bietamiverin *n* bietamiverine
bieten to offer
Bifidus-Faktor *m* bifidus factor
bifilar bifilar, double-wound, ~ **gewickelt** double wound, wound in duplicate
Bifilargalvanometer *n* bifilar galvanometer
Biflorin *n* biflorine
bifokal (Opt) bifocal
Biformin *n* biformin
bifunktionell bifunctional
Biglandulinsäure *f* biglandulinic acid
Biguanid *n* biguanide
Biindon *n* biindone
Bikarbonat *n* bicarbonate s a Bicarbonat
Bikathodenröhre *f* bicathode tube
Bikhaconitin *n* bikhaconitine
bikonkav biconcave, concavo-concave
bikonvex biconvex, convexo-convex
Bilanz *f* balance
Bilanzlinie *f* (McCabe-Thiele-Diagramm) operating line
bilateral bilateral
Bild *n* picture, figure, image; (Phot) photo[graph]; **latentes** ~ latent image, **latentes photographisches** ~ (Phot) latent photographic image, **optisches** ~ optical image, **reelles** ~ (Opt) real image, **umgekehrtes** ~ (Opt) inverted image, **virtuelles** ~ (Opt) virtual image
Bildabtaster *m* (Telev) scanning device
Bildauflösung *f* resolution
Bildauflösungsvermögen *n* resolving power
Bildaufrichtung *f* image erection
Bildaufzeichnung *f* image recording
Bildebene *f* focal plane, plane of projection
Bildeinstellung *f* image focus[s]ing
bilden to form, to fashion, to model, to shape
Bilden *n* formation
Bilderachat *m* (Min) figured agate
Bilderlack *m* picture lacquer
Bildfehler *m* image aberration, distortion

Bildfeld *n* (Opt) image field, view, (Telev) picture screen
Bildfeldwölbung *f* (Opt) curvature of image field
Bildformat *n* size of print
Bildfunktion *f* (Math) transform
bildlich figurative
Bildmarmor *m* figured marble, variegated marble
Bildmesser *m* view meter
Bildmittelpunkt *m* center of image
Bildnachleuchten *n* afterglow
Bildröhre *f* (Telev) cathode ray tube, picture tube, television tube
bildsam flexible, plastic, (Met) ductile
Bildsamkeit *f* flexibility, plasticity, (Met) ductility
Bildschärfe *f* (Phot) image sharpness
Bildschirm *m* projection screen, picture screen
Bildstein *m* (Min) agalmatolite, figure stone, pagodite
Bildstörung *f* (Telev) picture disturbance
Bildsucher *m* view finder
Bildung *f* development, formation, generation, (Ausbildung) education, ~ **von Kristallisationskeimen** *m pl* nucleation
Bildungsdauer *f* formative time
Bildungsenergie *f* energy of formation, energy of generation
Bildungsenthalpie *f* formation enthalpy
Bildungsgeschwindigkeit *f* speed of formation
Bildungsgleichung *f* equation of formation, structural equation
Bildungsgrad *m* degree of formation
Bildungskonstante *f* formation constant
Bildungsvorgang *m* process of formation
Bildungswärme *f* heat of formation
Bildungsweise *f* mode of formation
Bildverzerrung *f* image distortion
Bildweite *f* (Phot) distance from image to lens
Bildwerfer *m* epidiascope
Bildwiedergabe *f* projection, reproduction
Bildzeichen *n* (Telev) picture signal
Bildzerleger *m* image dissector
Bilgenpumpe *f* bilge pump
Bilgenrohr *n* bilge pipe
Bilgenwasser *n* bilge, bilge water
Bilharzia *f* (Zool, Med) bilharzia, schistosome
Bilharziose *f* (Med) bilharziasis, bilharziosis, schistosomiasis
Biliansäure *f* bilianic acid
Bilicyanin *n* bilicyanin, cholecyanin
Biliflavin *n* biliflavin
Bilifuscin *n* bilifuscin
Biligenese *f* biligenesis
Bilihumin *n* bilihumin
Bilin *n* bilin
Bilineurin *n* bilineurine
Bilinigrin *n* bilinigrin
Biliobansäure *f* biliobanic acid
Biliprasin *n* biliprasin
Bilipurpurin *n* bilipurpurin[e], choleh[a]ematin[e]
Bilirubin *n* bilirubin
Bilirubingehalt *m* bilirubin content, **erhöhter** ~ **im Blut** hyperbilirubin[a]emia, **erhöhter** ~ **im Urin** hyperbilirubinuria
Bilirubinmangel *m* hypobilirubin[a]emia
Bilirubinsäure *f* bilirubinic acid
Bilirubinspiegel *m* bilirubin level
Bilisoidansäure *f* bilisoidanic acid
Biliverdin *n* biliverdin
Biliverdinsäure *f* biliverdic acid
Bilixanthin *n* bilixanthine, choletelin
Billiarde *f* one thousand billions (Br. Engl.), one thousand trillions (Am. E.); quadrillion (Am. E.)
billigen to approve
Billigung *f* approval, approbation
Billion *f* billion (Br. E.), trillion (Am. E.)
Billitonit *m* (Min) billitonite
Bilobol *n* bilobol
Biloidansäure *f* biloidanic acid
Bilsenkrautblätter *n pl* (Bot) henbane leaves
Bilsenkrautextrakt *m* extract of henbane
Bilsenkrautöl *n* henbane oil
Bilsenkrautsamen *m* (Bot) henbane seed
Bimalonester *m* bimalonic ester
Bimetall *n* bimetal, duplex metal
Bimetallaktinometer *n* bimetal actinometer
Bimetalldraht *m* bimetallic wire
bimetallisch bimetallic
Bimetallschalter *m* bimetal switch, thermoswitch
Bimetallstreifen *m* bimetallic strip
Bimetallthermometer *n* bimetal thermometer
bimolekular bimolecular
Bimsbeton *m* pumice concrete
bimsen to pumice, to rub with pumice stone
Bimskies *m* pumice gravel
Bimssand *m* pumice sand
Bimsstein *m* pumice, pumice stone
bimssteinähnlich pumiceous, pumice stone-like, pumiciform
Bimssteinmehl *n* pumice powder
Bimssteinpapier *n* pumice stone paper
Bimssteinseife *f* pumice soap
Bimssteintuch *n* pumice cloth
binär binary
Binärcode *m* (Comp) binary code
Binärkies *m* (Min) marcasite
Binde *f* band, bandage; ligature
Bindedraht *m* binding wire
Bindefähigkeit *f* binding property
Bindegewebe *n* connective tissue, **elastisches** ~ elastic tissue
Bindeglied *n* connecting link, connecting member
Bindehaut *f* (Biol) conjunctiva
Bindekraft *f* binding power, combining power
Bindematerial *n* binding material

Bindemittel – Biostatik

Bindemittel *n* binder, binding agent, binding material, bonding agent, cement, cementing agent
Bindemitteldispersion *f* dispersion binder
Bindemittelemulsion *f* emulsion binder
binden to bind, to combine, to fix, to link, to tie, to unite, (Mörtel) to set
Binden *n* binding, (Säure) neutralizing
Bindenaht *f* weld line
Binder *m* binder
Binderfarbe *f* emulsion paint
Bindeton *m* white plastic clay, ball clay, bonding clay
Bindevermögen *n* binding capacity, binding power, binding property, cementing power, combining power
Bindezeit *f* setting time
Bindfaden *m* string, cord
Bindheimit *m* (Min) bindheimite
Bindschedlers Grün *n* Bindschedler's green
Bindung *f* link, linkage, tie, (Chem) binding, bond, (Fixierung) fixation; (Phys) absorption, fusion, **äquatoriale** ~ (Stereochem) equatorial bond, **Atom-** ~ atomic bond, **axiale** ~ (Stereochem) axial bond, **heteropolare** ~ heteropolar bond, **homöopolare** ~ homopolar bond, **konjugierte** ~ conjugated bond, **koordinative** ~ coordinate bond, **Molekül-** ~ molecular bond, **nicht räumlich gerichtete** ~ spatially undirected bond, **semipolare** ~ semipolar bond
Bindungsabstand *m* bond length
Bindungsart *f* bond type
Bindungsbruch *m* bond breaking
Bindungselektron *n* linkage electron
Bindungsenergie *f* binding energy, bond energy
Bindungsenthalpie *f* enthalpy of bonding
bindungsfähig bondable
Bindungsfehler *m* bonding fault
Bindungsform *f* form of bond
Bindungsgrad *m* bond order
Bindungskraft *f* binding force, bonding strength, linkage force
Bindungslänge *f* bond length
Bindungsmoment *n* bond moment
Bindungsrefraktion *f* bond refraction
Bindungsstärke *f* bonding strength
Bindungsstruktur *f* bond structure
Bindungstyp *m* (Chem) bond type
Bindungswärme *f* heat of combination
Bindungsweise *f* mode of combination, mode of linkage, mode of union
Bindungswinkel *m* bond angle
Binghamsches Medium *n* Bingham body
Binitrobenzol *n* dinitrobenzene
Binitrochlorbenzol *n* dinitrochlorobenzene
Binitrotoluol *n* binitrotoluene, dinitrotoluene
Binnendruck *m* internal pressure
Binnenklima *n* (Meteor) continental climate

Binnit *m* (Min) binnite, tennantite
Binodalfläche *f* binodal surface
Binodalkurve *f* binodal curve
Binode *f* binode, double diode
binokular binocular
binomial binomial
Binomialformel *f* (Math) binomial formula
Binomialgleichung *f* (Math) binomial equation
Binomialkoeffizient *m* (Math) binominal coefficient
Binomialreihe *f* (Math) binomial series
Binomialsatz *m* (Math) binomial theorem
Binomialverteilung *f* Bernoulli distribution, binomial distribution
binomisch binomial
Bio-Analyse *f* bio-assay
Biochanin *n* biochanin
Biochemie *f* biochemistry
Biochemikalien *pl* biochemicals
Biochemiker *m* biochemist
biochemisch biochemical
Biocytin *n* biocytin
Biodynamik *f* biodynamics
biodynamisch biodynamic
bioelektrisch bio-electric
Bioenergetik *f* bioenergetics
Biofilter *n* biofilter
Biogas *n* fermentation gas
Biogenese *f* biosynthesis, (Biol) biogenesis
biogenetisch biogenetic[al]
Biokatalysator *m* biocatalyst
Bioklimatik *f* bioclimatology
Bioklimatologie *f* bioclimatology
Biologe *m* biologist
Biologie *f* biology
biologisch biological
Biolumineszenz *f* bioluminescence
Biolyse *f* biolysis
Biomechanik *f* biomechanics
Biometrie *f* biometry
biometrisch biometric
Biomikroskopie *f* biomicroscopy
Biomorphose *f* biomorphosis
Bionik *f* bionics
Biopathologie *f* biopathology
Biophen *n* biophene
Biophor *m* biophore, bioplast
Biophysik *f* biophysics
biophysikalisch biophysical
Biopolymer *n* biopolymer
Biopsie *f* (Med) biopsy
Biopterin *n* biopterin
bioptisch bioptic
Biorhythmus *m* biorhythm
Biose *f* biose, disaccharide
Bioskopie *f* bioscopy
bioskopisch bioscopic[al]
Biosphäre *f* biosphere
Biostatik *f* biostatics

Biostimulation *f* biostimulation
Biosynthese *f* (Biochem) biosynthesis
Biotechnik *f* bioengineering
Biotin *n* biotin
Biotinol *n* biotinol
biotisch biotic[al]
Biotit *m* (Min) biotite
Biotitreihe *f* biotite series
Biotropismus *m* biotropism
Biotypus *m* biotype
Bioxid *n* bioxide, dioxide
Bioxyanthrachinon *n* dihydroxyanthraquinone
Bioxybernsteinsäure *f* dihydroxysuccinic acid, tartaric acid
Biozid *n* biocide
Biozyklus *m* biocycle
Biphenyl *n* biphenyl, xenene
Bipindalosid *n* bipindaloside
Bipindogenin *n* bipindogenin
Bipiperidin *n* bipiperidine
bipolar bipolar
Bipolarität *f* bipolarity
Bipotentialgleichung *f* bipotential equation
Biprismaversuch *m* biprism method
Bipyramide *f* (Krist) bipyramid
Bipyridyl *n* bipyridyl
Biradikal *n* biradical
Birkencampher *m* birch camphor, betulin
Birkenkohle *f* birch charcoal
Birkenöl *n* birch oil, betula oil
Birkenrindenöl *n* betula oil, birch oil
Birkensaft *m* birch juice
Birkenteer *m* birch tar
Birkenwasser *n* birch water
Birne *f* (Elektr) bulb, (Metall) converter, **die ~ aufrichten** (Metall) to raise the converter, to tilt the converter
Birnenauswurf *m* (Metall) converter waste, discharge from the converter
Birnenbodenform *f* converter bottom mold
Birnenessig *m* pear vinegar
Birnenhalle *f* converter shed, (Metall) converter shop
Birnenkühler *m* (Chem) pear-shaped condenser
Birnenöl *n* pear oil, isoamyl acetate
Birnenring *m* trunnion ring
Birnenständer *m* converter stand, converter support
Birnenverfahren *n* (Metall) converter process
Birnenwein *m* perry
Birnenzapfen *m* trunnion
birnförmig pear-shaped, pyriform
Birotation *f* birotation, mutarotation
Bisabolen *n* bisabolene
Bisamkorn *n* abelmosk seed, musk seed
Bisanthen *n* bisanthene
Bisanthranil *n* bisanthranil
Bisazobenzol *n* bisazobenzene
Bisbeeit *m* (Min) bisbeeite

Bischofit *m* (Min) bischofite
Bisdiazoessigsäure *f* bisdiazoacetic acid
Bismarckbraun *n* Bismarck brown, triaminoazobenzene
Bismit *m* (Min) bismite, bismuth ocher
Bismutin *m* (Min) bismuth glance, bismuthine, bismuthinite
Bismutit *m* (Min) bismutite, basic bismuth carbonate
Bismutoferrit *m* (Min) bismutoferrite
Bispidin *n* bispidine
Bister *m n* bister
Bisterbraun *n* bister
bisubstituiert disubstituated
Bisulfat *n* bisulfate, acid sulfate, hydrogen sulfate
Bisulfit *n* bisulfite, acid sulfite, hydrogen sulfite
Bisyringyl *n* bisyringyl
Bitartrat *n* bitartrate, acid tartrate, hydrogen tartrate
Bittererde *f* (Min) bitter earth, magnesia, magnesium oxide
Bitteresche *f* bitter ash, quassia
Bitterfenchelöl *n* common fennel oil
Bitterharz *n* bitter resin, arnicin
Bitterholz *n* bitter wood, quassia wood
Bitterkalk *m* (Min) dolomite, magnesia lime stone, pearl spar
Bitterling *m* bitter mineral water
Bittermandelessenz *f* essence of bitter almond
Bittermandelgeist *m* spirit of bitter almond
Bittermandelöl *n* oil of bitter almond, **künstliches ~** artificial bitter almond oil, benzaldehyde
Bittermandelseife *f* bitter almond soap
Bitterrinde *f* bitter bark, amargoso bark, goatbush bark
Bittersäure *f* (obs) picric acid, trinitrophenic acid
Bittersalz *n* bitter salt, epsom salt, magnesium sulfate
Bittersalzwasser *n* bitter salt water
Bitterspat *m* (Min) bitter spar, dolomite, magnesite, magnesite spar
Bitterstein *m* (Min) picrolite
Bitterstoff *m* bittering, bitter principle
Bittersüß *n* bittersweet, woody nightshade
Bittersüßextrakt *m* extract of bittersweet
Bitterwasser *n* bitter [mineral] water
Bitterwein *m* absinthiated wine
Bitterwurz *f* (Pharm) gentian
Bitumen *n* asphalt, bitumen
Bitumendecke *f* bituminous surface, (Straßenbau) bitumen pavement
Bitumenfarbe *f* bituminous paint
Bitumenisolierpappe *f* bituminous roofing paper
bitumenkaschiert bitumen-laminated
Bitumenlack *m* bituminous varnish

Bitumenpapier — Blasensäure

Bitumenpapier *n* asphalt paper, bituminous paper
Bituminisierung *f* bituminization
bituminös bituminous
Bitumol *n* bituminol, ammonium ichthyolsulfonate
Bityit *m* (Min) bityite
Biurat *n* acid urate, biurate, hydrogen urate
Biuret *n* biuret, carbamylurea
Biuretreaktion *f* biuret reaction
bivalent bivalent, divalent
Bivektor *m* bivector
Bixbyit *m* (Min) bixbyite, ferrous manganite
Bixin *n* (Färb) bixin
Bjelkit *m* (Min) bjelkite
Blachenstoff *m* awning, canvas, duckcloth
Bladhianin *n* bladhianin
blähen to expand, to inflate, to swell
Blähglimmer *m* exfoliated mica
Blähmittel *n* blowing agent, foaming agent
Bläschen *n* small bubble, blister, pimple, pustule; vesicle
bläschenartig blister-like
Bläschenbildung *f* (Med) vesiculation
bläschenförmig vesicular
bläschenziehend blistering, vesicating, vesicatory
Blättchen *n* small leaf, flake, lamella, lamina
blättchenartig lamellar, lamelliform, laminated
Blättchenpulver *n* flake powder, leaf powder
Blätteraugit *m* (Min) foliated augite
Blätterblende *f* blende of zinc
Blätterbruch *m* lamellar cleavage
Blätterbürste *f* foil brush, laminated brush
Blättererz *n* (Min) nagyagite, foliated tellurium
blätterförmig laminated
Blättergelb *n* (obs) xanthophyll
Blättergips *m* (Min) selenite
blätterig foliated, lamellar, lamelliform, laminated, leafy, scaly
Blätterkernspule *f* laminated iron core coil
Blätterkies *m* (Min) lamellar pyrites, marcasite
Blätterkohle *f* foliated coal, slate coal
Blätterkondensator *m* plate condenser
Blättermagnet *m* laminated magnet, compound magnet
blättern to exfoliate, to flake, to laminate, to peel, to scale
Blätterschiefer *m* (Min) bibliolite, book stone
Blätterserpentin *m* (Min) antigorite
Blätterspat *m* (Min) foliaceous spar
Blätterstein *m* variolite
Blättertellur *n* (Min) foliated tellurium, nagyagite
Blätterton *m* foliated clay, slaty clay
Blättertorf *m* lamellated peat
Blätterzeolith *m* (Min) foliated zeolite, heulandite
blättrig foliated, lamellar, lamelliform, laminated, leafy, scaly

Bläue *f* blueness
Bläuebefall *m* (Holz) blue staining of wood
Bläuen *n* blueing
bläulich bluish
blaken to smoke
Blanc fixe *n* blanc fixe, barite, barite white, barium sulfate, pearl white, permanent white
blanchieren to blanch, (Leder) to whiten
Blanchimeter *n* blanchimeter
blank clear, bright, clean, ~ **beizen** to dip metals, to pickle, ~ **reiben** to polish, to scour
Blankätzen *n* electrolytic polishing
Blankdrahtelektrode *f* bare wire electrode
blankgeglüht bright annealed
blankgezogen bright drawn, cold drawn
Blankglühen *n* bright annealing
Blankglühofen *m* non-oxidizing annealing furnace
Blankholz *n* campeachy wood, logwood
blankkochen (Zuck) to boil down without graining sugar
Blankleder *n* harness leather
Blankmachen *n* **des Metalls** pickling the metal
Blankprofil *n* bright section
Blasapparat *m* blast apparatus
Blasbalg *m* bellows
Blasdruck *m* blowing pressure
Blasdüse *f* blast nozzle, discharge nozzle
Blase *f* (Anat) bladder, (Chem) alembic, still, (Embryologie) vesicle, (Folienblase) bubble, (Gieß) blowhole, flaw, hollow, (Glas) nodule, (Luft, Gas) bubble, (Pathol) blister, bulla, vesicle, (Schweiß) blister, blowhole, ~ **in Metall** blasthole
Blasebalg *m* bellows
blasen to blow, to inject steam, to smelt in a blast furnace
Blasen *n* blowing, blow forming
Blasen *pl* **bilden** to blister
Blasenaufstieg *m* bubble rising
Blasenaufstiegsgeschwindigkeit *f* bubble rising velocity
Blasenbildung *f* bubbling, (Kunststoff) bubble formation, (Oberflächenfehler) blister formation, blistering, (Pathol) pimpling, ~ **in Feststoff** occlusion of gases, (Gieß) formation of blowholes
Blasendestillation *f* batch distillation
blasenfrei non-porous, dense
Blasenkammer *f* bubble chamber, (Farben) blister box
Blasenkupfer *n* blister copper
Blasenmodell *n* (v. Fließbetten) bubble model
Blasenpflaster *n* blister plaster
Blasenraum *m* blowhole
Blasenregelung *f* bubble control
Blasenrückstand *m* residue
Blasensäule *f* bubble-cap columns
Blasensäure *f* (obs) uric acid

Blasensieden *n* nucleate boiling
Blasenstahl *m* blister steel, cement steel, converted steel
Blasenströmung *f*, **laminare** ~ bubble flow, **turbulente** ~ froth flow
Blasentang *m* bladder wrack, fucus vesiculosus
Blasenverdampfung *f* nucleate boiling, (Öl) bubble vaporization
Blasenwachstum *n* bubble growth
blasenweise bubble by bubble
Blasenzähler *m* bubble counter, bubble gauge
blasenziehend blistering, vesicant, vesicatory
Blaseprobe *f* bubble test
Blaserohr *n* blow pipe, blast pipe, blast tube
Blasetisch *m* blow pipe table, blastlamp table
Blasfolie *f* blown film
Blasform *f* blow mold, tuyere
Blasformung *f* blow forming
Blasgeschwindigkeit *f* rate of blowing
blasig blistery, blistered, bubbly, honeycombed, spongy, vesicular, ~ **werden** to get blisters
Blasigkeit *f* blistered condition, blistered state
Blaskopf *m* blow[ing] head, **seitwärts gespeister** ~ side-fed die
Blasofen *m* blast furnace, (Glasfabrikation) blowing furnace
Blasprobe *f* (Zuck) bubble test
Blasrohr *n* blow pipe, blast pipe, blast tube
blaß pale
blaßgelb pale yellow
Blasspritzen *n* **von Filmen** *m pl* blow extrusion
Blastmycin *n* blastomycin
Blastmycinsäure *f* blastomycic acid
Blastogenese *f* (Biol) blastogenesis, germination
Blastolyse *f* blastolysis
Blastomere *f* (Biol) blastomere, morula cell
Blasverfahren *n* blowing process, blow molding, blow mo[u]lding
Blasverformung *f* blow forming
Blasverlust *m* blow-off loss
Blasvorgang *m* blowing process
Blaswirkung *f* blowing effect, action of blast
Blatt *n* (Bot) leaf, lamina, (Papier) sheet, (Anat) layer, membrane, sheet, (Messer) blade
Blattaluminium *n* leaf aluminum, beaten aluminum
Blattbildungsapparat *m* sheet forming apparatus
Blattblau *n* (obs) phyllocyanin
Blattblei *n* sheet lead
Blattdüngung *f* leaf fertilization
Blatteisen *n* sheet iron
Blattelektrometer *n* leaf electrometer
Blattfeder *f* flat spring, laminated spring, leaf spring, plate spring
Blattfilter *n* leaf filter
Blattgelb *n* (obs) xanthophyll
Blattgold *n* gold leaf, leaf gold
Blattgoldgrundöl *n* gold size
Blattgrün *n* chlorophyll

Blattgummi *m* sheet rubber
Blattkohle *f* foliated or paper coal, slate coal
Blattkupfer *n* copper foil, sheet copper
Blattlack *m* shellac
Blattmetall *n* sheet metal, leaf metal, metal foil
Blattpigment *n* leaf coloring matter
Blattrührer *m* blade mixer
Blattsäge *f* band saw
Blattschichtung *f* stratification
Blattsilber *n* leaf silver, beaten silver, silver foil, silver leaf
Blattstahl *m* sheet steel
Blattvergoldung *f* gilding with gold leaf, gold blocking, plating with gold leaf
Blattwinkel *m* blade angle
Blattzapfen *m* slipper block, slipper shoe
Blattzinn *n* tin foil
Blaualgen *pl* blue-green algae
Blauanlaufen *n* blooming, blueing
Blauasche *f* saunders blue
Blaubleierz *n* (Min) blue lead ore, pyromorphite
Blaubruch *m* blue brittleness, blue shortness
Blaubruchversuch *m* blue fracture test
blaubrüchig blue-short, brittle at blue heat
Blaubrüchigkeit *f* (Metall) blue brittleness, tempering brittleness
Blaucarmin *n* indigo carmine
Blaudruck *m* blueprint, blueprinting, blueprint process
Blaueisenerde *f* blue iron earth
Blaueisenerz *n* (Min) blue iron ore, hydrated ferrous phosphate, vivianite
blauempfindlich sensitive to blue
Blauerz *n* (Min) blue ore, decomposed siderite, decomposed spathic iron ore, vivianite
Blaufärbung *f* blue color, blue coloration, blue coloring
Blaufäule *f* (Holz) blue disease, blue stain
Blaufarbe *f* cobalt blue, smalt
Blaufarbenglas *n* smalt
Blaufilter *n* blue filter
Blaugas *n* blue gas, oil gas
Blauglühen *n* (Metall) blue annealing
blauglühend blue hot
blaugrau bluish gray
blaugrün bluish green
Blauholz *n* campeachy wood, logwood
Blauholzbaum *m* campeachy tree, logwood
Blauholzextrakt *m* logwood extract, hematine
Blauholzschwarz *n* logwood black
Blauholzspäne *pl* logwood chips
Blauholztinktur *f* logwood liquor
Blauküpe *f* (Färb) blue vat
Blaumühle *f* smalt mill
Blauofen *m* blast furnace, shaft furnace, wind furnace
Blaupapier *n* blue carbon paper
Blaupause *f* blueprint, blueprint tracing
Blauprobe *f* blue test

blaurot — Bleiauskleidung

blaurot bluish red, purple, violet
Blaurot *n* purple, violet
Blausäure *f* hydrocyanic acid, prussic acid
Blausäuregas *n* hydrogen cyanide
blausäurehaltig containing hydrocyanic acid
Blausäureverbindung *f* compound of hydrocyanic acid
Blausäurevergiftung *f* hydrocyanic acid poisoning, poisoning with hydrocyanic acid
Blausalz *n* (obs) potassium ferrocyanide, yellow potassium prussiate, yellow prussiate of potash
Blausand *m* coarsest smalt
blausauer cyanide, (obs) prussiate
Blauschörl *m* (Min) blue aluminum silicate, cyanite
blauschwarz bluish black
Blauspat *m* (Min) azure spar, blue spar, lazulite
Blauspiegel *m* blue light mirror
Blaustein *m* (Min) bluestone, blue vitriol, copper vitriol, native hydrated copper sulfate
Blaustich *m* bluish tint, bluish cast, bluish tinge
Blausucht *f* (Med) cyanosis
blauviolett blue violet
Blauvitriol *n* blue vitriol, copper vitriol, hydrated copper sulfate
Blauwasser *n* ammoniacal copper oxide solution, ammoniacal solution of blue vitriol, cuprammonium hydroxide solution
Blech *n* sheet metal, plate, sheet, **differenzverzinntes** ~ dual coated plate, **doppelt reduziertes** ~ double-reduced plate, **elektrolytisch verzinntes** ~ electrolytic tinplate, **feuerverzinntes** ~ hot-dipped tinplate, **gelochtes** ~ perforated plate, punched plate, **getriebenes** ~ worked plate, **gewalztes** ~ rolled sheet metal, **PVC-beschichtetes** ~ skin plate, **verzinktes** ~ galvanized sheet metal, **Weiß-** ~ tin plate
Blechabfälle *m pl* clippings, sheet metal scrap
Blechbearbeitung *f* sheet metal working
Blechbehälter *m* sheet metal container
Blechbiegemaschine *f* plate bending machine
Blechbund *m* batch of sheet metal
Blechdose *f* can
Blechdruck *m* sheet metal printing, tinplate printing
Blecheisen *n* sheet iron
Blechemail *n* sheet iron porcelain enamel
blechern of metal plate, of tin
Blechfalzmaschine *f* seaming machine
Blechflasche *f* tin bottle
Blechgefäß *n* tin vessel, steel shell
Blechglühofen *m* furnace for heating plates, plate heating furnace
Blechgurtförderer *m* steel belt apron conveyer
Blechhalter *m* blank holder
Blechkanal *m* sheet steel duct
Blechkante *f* sheet-metal edge
Blechkasten *m* tin box, tin case
Blechkern *m* laminated core
Blechkernspule *f* (Elektr) laminated core
Blechkonstruktion *f* sheet construction, plate construction
Blechlack *m* tinplate lacquer, sheetmetal varnish, tinplate enamel, tinplate varnish
Blechlehre *f* plate gauge, sheet gauge
Blechmantel *m* sheet metal casing, iron covering, iron shell, sheet iron shell
Blechröhre *f* sheet iron pipe
Blechrohr *n* sheet metal tube
Blechschachtel *f* tin box
Blechscheibe *f* sheet plate
Blechschere *f* plate shears, sheet shearing machine
Blechschneidemaschine *f* plate cutting machine
Blechschweißung *f* plate welding
Blechstanzen *pl* sheet trimmings
Blechstraße *f* plate mill
Blechstrecke *f* plate mill
Blechstreifen *m* sheet metal strip
Blechtafel *f* sheet iron, iron plate
Blechtrichter *m* tin funnel
Blechverarbeitung *f* sheet metal working
Blechverzinnungsofen *m* plate tinning furnace
Blechwärmofen *m* sheet furnace
Blechwalze *f* plate roll
Blechwalzen *n* sheet rolling, plate rolling
Blechwalzwerk *n* (Feinblech) sheet [rolling] mill, (Grobblech) plate [rolling] mill
Blei *n* lead (Symb. Pb), **gediegenes** ~ native lead, **gewalztes** ~ rolled lead, sheet lead, **metallisches** ~ native lead, **schwefelsaures** ~ lead sulfate, lead(II) sulfate, lead vitriol, plumbous sulfate, **silberhaltiges** ~ argentiferous lead
Bleiabfälle *pl* lead waste
Bleiabgang *m* lead dross, lead scoria
Bleiabschirmung *f* lead shielding
Bleiacetat *n* lead acetate
Bleiader *f* lead lode, lead vein
Bleiakkumulator *m* lead accumulator, lead storage battery
Bleiamalgam *n* amalgam of lead
Bleianode *f* lead anode
Bleiantimonat *n* lead antimonate
Bleiantimonerz *n* (Min) lead antimony glance, zin[c]kenite
Bleiantimonglanz *m* (Min) lead antimony glance, zin[c]kenite
Bleiantimonlegierung *f* lead antimony alloy
Bleiarbeit *f* plumbing, lead smelting
Bleiarmaturen *pl* lead fittings
Bleiarsenat *n* lead arsenate
bleiartig leadlike, leady, plumbeous
Bleiasche *f* lead ash, lead dross, lead monoxide, lead oxide, litharge, massicot
Bleiauskleidung *f* lead lining

Bleibad *n* lead bath
Bleibarren *m* lump of lead
Bleibaum *m* lead tree, arbor saturni
Bleibelastung *f* (Toxikologie) lead body burden
bleibend permanent
Bleibenzin *n* leaded gasoline, tetraethyl lead fuel
Bleiblech *n* sheet lead
Bleiblock *m* lead brick
Bleiblüte *f* (Min) mimetite
Bleiborat *n* lead borate
Bleiboratglas *n* lead borate glass
Bleibrikett *n* lead brick
Bleibrocken *m* lump of lead
Bleibügel *m* lead spring
Bleichanlage *f* bleachery, bleaching plant
Bleichanstalt *f* bleaching works, bleachery
Bleichapparat *m* bleaching apparatus
Bleichbad *n* bleaching bath, bleaching liquor
Bleichbeize *f* bleaching mordant
Bleichchlor *n* active chlorine
Bleiche *f* bleachery, bleaching, **chemische** ~ chemical bleaching; **natürliche** ~ grass bleaching
Bleichechtheit *f* resistance to bleach
Bleicheffekt *m* bleaching effect
bleichen to bleach, to blanch, to whiten, (verblassen) to fade
Bleichen *n* bleaching, decolorizing, ~ **des Zellstoffes** bleaching of cellulose
Bleicher *m* bleacher
Bleicherde *f* bleaching earth, bleaching clay, fuller's earth
Bleicherdebehandlung *f* clay treatment
Bleicherei *f* bleaching plant, bleachery
Bleichfähigkeit *f* bleachability
Bleichfaß *n* bleaching vat
Bleichfleck *m* bleach spot, bleaching stain
Bleichflotte *f* bleaching liquor
Bleichflüssigkeit *f* bleaching liquor
Bleichgefäß *n* bleaching vessel
Bleichgut *n* material to be bleached
Bleichhilfsmittel *n* bleaching assistant, bleaching auxiliary
Bleichholländer *m* (Pap) poacher, poaching engine
Bleichkalk *m* bleaching powder, calcium hypochlorite, chloride of lime, chlorinated lime
Bleichkasten *m* bleaching vat
Bleichkraft *f* bleaching power
Bleichlauge *f* bleaching liquor, bleaching lye, bleaching solution
Bleichmaschine *f* bleaching machine
Bleichmittel *n* bleaching agent, decolorant
Bleichprodukt *n* bleaching product
Bleichprozeß *m* bleaching process
Bleichpulver *n* bleaching powder
Bleichrahmen *m* bleaching frame
Bleichromat *n* lead chromate

Bleichsäure *f* bleaching acid, hypochlorous acid
Bleichsalz *n* bleaching salt
Bleichseife *f* bleaching soap
Bleichsoda *f* bleaching soda, sodium hypochlorite
Bleichtrommel *f* tumbler
Bleichturm *m* bleaching tower
Bleichung *f* bleaching, fading
Bleichverfahren *n* bleaching process
Bleichvermögen *n* bleaching capacity, bleaching power
Bleichwasser *n* chlorine water, eau de javelle, solution of sodium hypochlorite
Bleichwirkung *f* bleaching action, bleaching effect
Bleicyanamid *n* lead cyanamide
Bleidampf *m* lead fume, lead vapor
Bleidichlorid *n* lead dichloride
Bleidichtung *f* lead gasket, lead joint, lead packing
Bleidioxid *n* lead dioxide, lead peroxide
Bleidraht *m* lead wire
Bleidruse *f* lead druse
Bleieinfassung *f* leading
Bleieinlage *f* inside lead lining
Bleieisenstein *m* iron lead
Bleielektrolyse *f* electrolytic refining of lead
bleiempfindlich lead responsive
Bleierde *f* earthy cerussite, earthy lead carbonate, earthy white lead ore
bleiern lead, leaden
Bleierz *n* lead ore
Bleiessig *m* lead vinegar, basic lead acetate solution, Goulard's extract
Bleiextrakt *m* Goulard's extract, lead vinegar
Bleifahlerz *n* (Min) bournonite
Bleifarbe *f* lead paint, lead color, lead pigment
bleifarben lead-colored
bleifarbig lead-colored
Bleifassung *f* (Techn) lead mount
Bleiflasche *f* lead bottle
Bleifluorid *n* lead fluoride
Bleifluß *m* crystallized lead, (Keram) colorless enamel
Bleifolie *f* lead foil, beaten lead
Bleiform *f* lead mold
bleiführend lead bearing, plumbiferous
Bleigang *m* (Bergb) lead lode, lead vein
Bleigans *f* pig-lead
Bleigehalt *m* lead content
Bleigelb *n* lead chromate, yellow lead
Bleigewicht *n* lead weight
Bleiglätte *f* lead monoxide, lead ocher, lead protoxide, litharge, ~ **frischen** to revive the litharge
Bleiglätteanlage *f* litharge plant
Bleiglanz *m* (Min) lead glance, galena, galenite, native lead sulfide, **falscher** ~ pseudogalena,

Bleiglanz – Bleistein

sphalerite, **gemeiner** ~ galena, native lead sulfide
bleiglanzhaltig galenic
Bleiglas *n* lead crystal, lead glass
Bleiglasur *f* lead glaze
Bleiglaszähler *m* lead glass counter
Bleiglimmer *m* (Min) micaceous cerussite
bleigrau lead-colored, lead-gray
Bleigrau *n* lead color
Bleigrieß *m* lead gravel, lead grit
Bleigrundierung *f* lead primer
Bleihärtung *f* lead hardening
bleihaltig plumbiferous, containing lead, plumbic
Bleihaube *f* lead hood
Bleihornerz *n* (Min) native lead chloride carbonate, phosgenite
Bleihütte *f* lead refining plant, lead smeltery
Bleihydroxid *n* lead hydroxide
Bleihyperoxid *n* lead dioxide, lead peroxide
bleiig plumbous, plumbic
Bleiion *n* lead ion
Bleijodid *n* lead iodide
Bleikammer *f* lead chamber
Bleikammerkristalle *m pl* lead chamber crystals
Bleikammerverfahren *n* lead chamber process
Bleikarbonat *n* lead carbonate
Bleikönig *m* lead regulus
Bleikolik *f* (Med) lead colic
Bleikrätze *f* lead dross, lead waste
Bleilasur *f* linarite
Bleilegierung *f* lead alloy
Bleilöterei *f* lead soldering, plumbing
Bleilötung *f* lead soldering
Bleilötverfahren *n* lead-soldering process
Bleilot *n* lead solder, plumb [line], plummet
Bleimanganat *n* lead manganate
Bleimantel *m* lead jacket, lead sheath[ing] or covering
Bleimehl *n* lead meal, lead dust, lead powder
Bleimennige *f* minium, plumbous-plumbic oxide, red lead, red lead oxide
Bleimennigeanstrich *m* red lead paint
Bleimennigefarbe *f* red lead paint
Bleimetaphosphat *n* lead metaphosphate
Bleimetasilikat *n* lead metasilicate
Bleimonoxid *n* lead monoxide
Bleimuffe *f* lead sleeve
Bleimulde *f* lead pig
Bleimulm *m* black lead ore, black-colored cerussite
Bleinaphthenat *n* lead naphthenate
Bleiniederschlag *m* precipitated lead
Bleiniere *f* (Min) bindheimite, native hydrated lead antimonate
Bleinitrat *n* lead nitrate
Bleiocker *m* (Min) lead ocher, litharge, massicot, native lead monoxide
Bleiofen *m* lead furnace, lead tempering bath

Bleiorthosilikat *n* lead orthosilicate
Bleioxalat *n* lead oxalate
Bleioxid *n* lead oxide, lead monoxide, lead protoxide, litharge
Bleioxychlorid *n* lead oxychloride
Bleioxydul *n* lead suboxide
Bleioxyduloxid *n* minium, plumbous-plumbic oxide, red lead, red lead oxide
Bleipapier *n* (Anal) lead paper
Bleiperoxid *n* lead peroxide, brown lead oxide, lead dioxide
Bleipersulfat *n* lead persulfate
Bleipflaster *n* lead plaster
Bleipflastersalbe *f* diachylon ointment, lead plaster ointment
Bleiplombe *f* lead seal
Bleiraffination *f* lead refining
Bleiraffinerie *f* lead refinery
Bleirauch *m* lead fume, lead smoke
Bleiregulus *m* lead regulus
Bleirhodanid *n* lead thiocyanate
Bleiröhre *f* lead pipe, lead tube
Bleiröstprozeß *m* lead roasting process
Bleirohr *n* lead pipe, lead tube
Bleirot *n* minium, red lead, red lead oxide
Bleisafran *m* orange lead
Bleisalbe *f* lead ointment, cerate of lead subacetate, unguentum plumbi
Bleisalpeter *m* lead nitrate
Bleisalz *n* lead salt
Bleisammler *m* lead accumulator, lead storage battery
bleisauer plumbate
Bleischale *f* lead dish, lead basin
Bleischaum *m* lead scum, lead dross
Bleischirm *m* lead screen
Bleischlacke *f* lead slag
Bleischlamm *m* (Akku) lead deposit
Bleischlich *m* lead slime
Bleischmelzherd *m* lead smelting hearth
Bleischrot *m* lead shot
Bleischwärze *f* black lead, graphite, plumbago
Bleischwamm *m* lead sponge, litharge, spongy lead
Bleischweif *m* (Min) compact galena
Bleiselenid *n* lead selenide
Bleisicherung *f* lead fuse, fusible plug
Bleisikkativ *n* lead siccative, lead drier
Bleisilikat *n* lead silicate
Bleisilikofluorid *n* lead silicofluoride
Bleispat *m* (Min) lead spar, cerussite, native lead carbonate, **gelber** ~ yellow lead spar, native lead molybdate, wulfenite, yellow lead ore
Bleispiegel *m* (Min) specular galena
Bleistaub *m* lead dust
Bleistaubakkumulator *m* lead dust accumulator, lead dust storage battery
Bleistearat *n* lead stearate
Bleistein *m* lead dross

Bleisubacetat *n* lead subacetate, monobasic lead acetate
Bleisulfat *n* lead sulfate, lead vitriol, (Min) anglesite
Bleisulfatrückstand *m* lead sulfate residue
Bleisulfid *n* lead sulfide, (Min) galena, galenite, lead glance
Bleisuperoxid *n* lead peroxide, brown lead oxide, lead dioxide
Bleisuperoxidplatte *f* lead peroxide plate, brown lead oxide plate, lead dioxide plate
Bleitannat *n* lead tannate
Bleitellurid *n* lead telluride
Bleitetraacetat *n* lead tetraacetate
Bleitetraäthyl *n* tetraethyl lead
Bleitetrachlorid *n* lead tetrachloride
Bleithiosulfat *n* lead thiosulfate
Bleititanat *n* lead titanate
Bleitube *f* lead tube
Bleiumhüllung *f* lead covering
Bleiverbindung *f* lead compound
Bleivergiftung *f* (Med) lead poisoning
Bleiverhüttung *f* lead smelting
Bleivitriol *n* lead vitriol
Bleiwasser *n* lead water, aqueous solution of basic lead, Goulard's lotion, Goulard water
Bleiwasserstoff *m* lead hydride
Bleiweiß *n* white lead, basic lead carbonate
Bleiweißersatz *m* white lead substitute
Bleiweißfarbe *f* white lead paint
Bleiweißkitt *m* white lead putty
Bleiweißsalbe *f* (Pharm) lead carbonate ointment
Bleiwismutlegierung *f* lead bismuth alloy
Bleiwolframat *n* lead tungstate
Bleizinkakkumulator *m* lead-zinc accumulator, lead-zinc storage battery
Bleizinnlegierung *f* lead-tin alloy
Bleizinnober *m* minium, red lead, red lead oxide
Bleizucker *m* lead acetate, sugar of lead
Bleizuckersalbe *f* lead acetate ointment
Blende *f* (Phot) circular slide valve, diaphragm, orifice
blenden to blind, to dazzle
Blendenebene *f* (Opt) diaphragm plane
Blendeneinstellung *f* (Phot) aperture setting, diaphragm setting
Blendenhebel *m* (Opt) shutter lever
Blendenöffnung *f* lens opening, screen opening
Blendenpaar *n* pair of diaphragms
Blendenrad *n* rotating vane
Blendenskala *f* diaphragm scale
Blendenstellung *f* (Opt) diaphragm position, diaphragm setting
Blenderöstofen *m* blende roasting furnace
Blendleiste *f* cover strip
Blendschutzanstrich *m* shading paint
Blendung *f* dazzle, dazzlement, glare
Bliabergit *m* (Min) bliabergite

Blickpunkt *m* (Opt) visual point
Blicksilber *n* crude silver, lightened silver
blind (Metall) dull, mat, tarnished, (Spiegel) clouded
Blinddarm *m* appendix
Blinddarmentzündung *f* (Med) appendicitis
Blindenergie *f* reactive energy
Blindflansch *m* blank flange, blind flange
Blindgänger *m* blind shell, dud
Blindhärten *n* blank hardening
Blindkaliber *n* blind pass, false pass
Blindkohle *f* underburned charcoal
Blindleistung *f* (Elektr) reactive power
Blindleitwert *m* susceptance
Blindleitwertsmatrix *f* susceptance matrix
Blindloch *n* dummy hole
Blindmaterial *n* filling material
Blindprobe *f* (Analyse) blank test
Blindscheibe *f* blank flange
Blindspannung *f* reactive voltage
Blindstrom *m* wattless current, blind current, idle current, reactive current
Blindtitration *f* blank titration
Blindversuch *m* blank test, blank trial, **einen ~ durchführen** to run a blind (or blank test)
Blindwerden *n* dulling, loss of gloss, tarnishing
Blindwert *m* numerical result of blank test
Blindwiderstand *m* reactance
Blindwiderstandsmatrix *f* reactance matrix
blinken to glitter, to sparkle
Blinkfeuer *n* signal fire
Blitz *m* lightning flash
Blitzableiter *m* lightning conductor, lightning arrester, lightning rod
Blitzbeleuchtung *f* flash illumination
Blitzgerät *n* strobe light
Blitzleuchte *f* flash gun
Blitzlicht *n* (Phot) flash light
Blitzlichtaufnahme *f* (Phot) flash light photo
Blitzlichtmischung *f* flashlight mixture
Blitzlicht-Photolyse *f* flash photolysis
Blitzpulver *n* lycopodium, lycopode
Blitzröhre *f* fulgurite, lightning tube, thundertube
Blitzsinter *m* fulgurite
Blitzstrahl *m* lightning flash
Bloch-Wand *f* Bloch wall
Block *m* block, pulley block, (Barren) ingot, (Met) pig, (Rohmetall) slug, **den ~ einspannen** to clamp the ingot, to grip the ingot
Blockabstreifer *m* (Metall) ingot stripper
Blockanlage *f* block system
Blockausdrücker *m* ingot pusher
Blockbetrieb *m* block process
Blockbildung *f* block formation
Blockblei *n* pig lead
Blockdreherei *f* ingot turning
Blockdrücker *m* ingot pusher, ingot pushing device

Blockeinfrierung *f* solidification of the block
Blockeinsetzkran *m* ingot charging crane
Blockeinspannvorrichtung *f* ingot holding device
Blockeis *n* ice blocks
Blockeisen *n* ingot iron
blocken to block, (Metall) to bloom (Am. E.), to cog (Br. E.)
Blocken *n* blocking
Blockende *n* bloom end
Blockförderkran *m* ingot crane
Blockgefüge *n* macrostructure of an ingot
Blockgerüst *n* blooming mill stand
Blockguß *m* ingot casting
Blockhahn *m* stop cock
Blockhebetisch *m* ingot tilter
blockieren to block, to jam
blockierend blocking
Blockierung *f* blocking, stoppage
Blockkaliber *n* blooming pass
Blockkette *f* block chain
Blockkokille *f* ingot mold
Blockkondensator *m* block condenser
Blockkopfbeheizung *f* ingot head heating device
Blockkupfer *n* ingot copper
Blocklager *n* ingot storage
Blocklagerung *f* storage in blocks
Blocklehm *m* boulder clay
Blockleitersystem *n* block conductor pattern
Blockmethode *f* block method
Blockmühle *f* block mill
Blocknickel *n* pig nickel
Blockofen *m* crucible furnace
Blockpolymer *n* block polymer
Blockpolymerisat *n* block polymer
Blockpolymerisation *f* block polymerization, bulk polymerization, mass polymerization
Blockpresse *f* block press
Blockpreßverfahren *n* fluid compression process
Blockpunkt *m* blocking point, blocking temperature
Blockreifen *m* solid tire
Blockrelaxation *f* block relaxation
Blockscheibe *f* block sheave, pulley block
Blockschneidmaschine *f* block slicing machine
Blockstahl *m* (Metall) ingot steel
Blockstraße *f* blooming mill, cogging mill, roughing mill
Blockteilmaschine *f* block slicing machine, ingot slicing machine
Blockverwerfung *f* block faulting
Blockwalze *f* cogging roll, bloomer, ingot rolling mill
Blockwalzkaliber *n* blooming pass, bloom pass, cogging pass
Blockwalzwerk *n* blooming mill, cogging mill, roughing mill
Blockzange *f* pincher
Blockziehkran *m* crane for the reheating furnace, soaking pit crane
Blockzinn *n* block tin
Blödit *m* (Min) blodite, astrakanite, bloedite
Blomstrandin *m* (Min) blomstrandite
bloßlegen to expose, to lay bare, to reveal, to strip
blühen to bloom
Blühen *n* blooming
Blüte *f* (Bot) flower, blossom
Blütenfarbstoff *m* flower pigment
Blütenöl *n* blossom oil, essence of flowers
Blütenstaub *m* (Bot) pollen
Blume *f* flower, (Wein) aroma, bloom
Blumenbachit *m* (Min) alabandite
Blumengelb *n* (obs) anthoxanthin
Blumenseite *f* (Gerb) outside of the skin
Blut *n*, **arterielles** ~ arterial blood, **defibriniertes** ~ defibrinated blood, **geronnenes** ~ clot [blood], clotted blood, coagulated blood, **konserviertes** ~ stored blood, **peripheres** ~ peripheral blood, **sauerstoffarmes** ~ venous blood, **sauerstoffreiches** ~ arterial blood
Blutachat *m* (Min) blood-colored agate, hemagate
Blutader *f* (Med) blood vessel, vein
Blutagglutination *f* h[a]emoagglutination
Blutalbumin *n* blood albumin
Blutalkohol *m* blood alcohol, percentage of alcohol in the blood
Blutalkoholbestimmung *f* blood alcohol determination, determination of the alcohol content of the blood
Blutalkoholgehalt *m* blood alcohol concentration
Blutanalyse *f* blood analysis, h[a]emanalysis
blutarm anemic, bloodless
Blutarmut *f* anemia
Blutbahn *f* blood circulation, blood stream
Blutbank *f* blood bank
Blutbankkühlschrank *m* blood bank refrigerator
Blutbestandteil *m* blood constituent, haematic substance
Blutbild *n* blood picture, blood count, h[a]emogram, **rotes** ~ red blood count, R.B.C., **weißes** ~ white blood count, W.B.C.
blutbildend blood-forming, h[a]ematogenic, h[a]ematogenous
Blutbildner *m* h[a]ematogen[e]
Blutbildung *f* formation of blood cells, h[a]emo[cyto]genesis
Blutbildungsmittel *n* hematogenic agent
Blutbilirubin *n* h[a]emobilirubin
Blutcalciumspiegel *m* blood-calcium level
Blutdruck *m* (Med) blood pressure
blutdrucksenkend hypotensive
Bluteiweiß *n* serum protein
bluten to bleed
Bluten *n* (Anstrich) bleeding
Bluterguß *m* hemorrhage

Blutersatz *m* artificial serum preparation, blood substitute
Blutfarbstoff *m* blood pigment
Blutfaserstoff *m* fibrin
Blutfibrin *n* blood fibrin
Blutfleck *m* blood stain
Blutgefäß *n* blood vessel
blutgerinnend h[a]emocoagulative
Blutgerinnsel *n* (Med) blood clot, thrombus
Blutgerinnung *f* blood clotting, blood coagulation
Blutgift *n* blood toxin
Blutgruppe *f* (Biol) blood group
Blutgruppenabstimmung *f* blood matching
Blutgruppenbestimmung *f* blood-grouping, blood-typing
Blutgruppendissonanz *f* incompatibility of the blood
Blutgruppeneinteilung *f* blood-group classification
Blutgruppenfeststellung *f* blood-grouping, blood-typing
Blutgruppenunverträglichkeit *f* blood group incompatibility
Blutgruppenverträglichkeit *f* blood group compatibility
Blutholz *n* campeachy wood, logwood
Blutjaspis *m* (Min) bloodstone
Blutklumpen *m* blood clot
Blutkörperchen *n* blood cell, blood corpuscule, **rotes** ~ erythrocyte, h[a]ematid, red blood corpuscule, **weißes** ~ leucocyte, white blood corpuscule
Blutkörperchenzählung *f* blood count
Blutkohle *f* blood charcoal
Blutkonservierung *f* banking of blood, blood conservation
Blutkrankheit *f* blood disease
Blutkreislauf *m* blood circulation
Blutkristall *m* blood crystal
Blutkuchen *m* blood clot, coagulum
Blutlaugensalz *n*, **gelbes** ~ ferrous potassium cyanide, potassium ferrocyanide, yellow prussiate of potash, **rotes** ~ ferric potassium cyanide, potassium ferricyanide, red prussiate of potash
Blutmangel *m* an[a]emia, blood deficiency
Blutmehl *n* dried blood, blood meal, blood powder
Blutmelasse *f* blood molasses
Blutpfropf *m* blood clot, thrombus
Blutplättchen *n* blood platelet, thrombocyte
Blutprobe *f* blood sample, blood specimen, blood test
Blutpuffer *m* blood buffer
blutreinigend depurative
Blutsauerstoff *m* circulating oxygen [in the blood], oxygen contained in the blood
Blutsenkung *f* blood sedimentation
Blutserum *n* blood serum
Blutserumflasche *f* blood serum bottle
Blutspender *m* blood donor
Blutspiegelwert *m* blood level
Blutstein *m* (Min) bloodstone, hematite
blutstillend hemostatic, blood-stanching, styptic
Blutsturz *m* (Med) haemorrhage
Blutsystem *n* circulation system of the blood, vascular system
Bluttransfusion *f* (Med) blood transfusion
Blutübertragung *f* (Med) blood transfusion
Blutung *f* bleeding, haemorrhage
Blutuntersuchung *f* (Med) blood test
Blutverarbeitungsanlage *f* blood processing plant
Blutwurzel *f* tormentille root
Blutzellenbildung *f* h[a]emocytopoiesis
Blutzersetzung *f* decomposition of the blood
Blutzucker *m* blood sugar
Blutzuckeranstieg *m* hyperglyc[a]emia, increase of the blood sugar
Blutzuckerbestimmung *f* blood-sugar determination
Blutzuckergehalt *m* blood-sugar level, sugar content of the blood
Blutzuckerspiegel *m* blood sugar level
Blutzufuhr *f* blood supply
Blutzusammensetzung *f* blood composition
Bobbierit *m* (Min) bobbierite
Bobine *f* bobbin, reel
Bock *m* frame, truss
Bockasche *f* coal ashes
Bockformmaschine *f* upright molding machine
Bockkran *m* gantry crane
Bockseife *f* mountain soap
Bockshornsamen *m* fenugreek
Bockstalg *m* buck-tallow
Boden *m* (Erde) soil, (Fußboden) floor, earth, ground, (Rektifikation) plate, tray, ~ **Zahl der einzubauenden Böden** number of actual trays, **angeschwemmter** ~ (Geol) alluvial deposits, **durchlässiger** ~ pervious ground, **eisenhaltiger** ~ ferriferous earth, **kalkiger** ~ chalky soil, limy soil
Bodenabstand *m* (Dest) [tray] spacing
Bodenäquivalent *n* (Dest) height equivalent of theoretical plate, height of transfer unit
Bodenanalyse *f* soil analysis
Bodenanstrich *m* floor painting, antifouling composition
Bodenart *f* type of soil
Bodenbakterie *f* ground bacterium
Bodenbeanspruchung *f* soil requirement
Bodenbelag *m* floor covering, flooring, **rutschfester** ~ non-skid flooring
Bodenberührungsfläche *f* bottom contact area
Bodenbeschaffenheit *f* nature of the soil
Bodenbewachsung *f* vegetation

Bodenelektrode – Bohren

Bodenelektrode *f* bottom electrode, lower electrode
Bodenelemente *pl* (Rektifikation) tray elements
Bodenfeuchtigkeit *f* ground humidity
Bodenhefe *f* dregs, grounds, lees
Bodenheizung *f* bottom heating
Bodenhöhe *f* floor level
Bodenhydraulik *f* tray hydraulics
Bodenit *m* (Min) bodenite
Bodenkörper *m* (Chem) precipitate, sediment, solid phase
Bodenkolloid *n* soil colloid
Bodenkolonne *f* plate column, plate tower, (Dest) tray column
Bodenkorrosion *f* soil corrosion
Bodenkühlung *f* cooling of furnace floor
Bodenkupfer *n* copper bottom
Bodenlängendifferential *n* (Math) differential equation for length of arc
Bodenlehre *f* soil science
Bodenmehl *n* farine, fecula, starch
Bodennährstoff *m* soil nutrient
Bodennutzung *f* soil utilization
Bodenplatte *f* base plate, bottom plate
Bodenreaktion *f* soil reaction
Bodensatz *m* precipitate, bottoms, bottom sediments, deposit, dregs, sediment, (nach der Destillation von Rohöl) crude bottom, residue, (Rückstand) residue, residuum, settlement, settling, ~ **ablagern** to precipitate, to deposit
Bodensatzbildung *f* formation of deposit, sediment formation
Bodenschätze *pl* (Min) mineral resource or wealth, wealth under ground
Bodenscheibe *f* bottom disc
Bodenstampfmaschine *f* plug ramming machine
Bodenstein *m* bottom stone, base block, hearth block, (Mühle) bed stone
Bodentechnologie *f* soil technology
Bodentemperatur *f* ground temperature
Bodenuntersuchung *f* soil examination, soil investigation
Bodenventil *n* bottom blow valve
Bodenvermörtelung *f* soil [cement] stabilization
Bodenverstärkung *f* (Rektifikation) enrichment
Bodenverstärkungsverhältnis *n* tray [or plate] efficiency
Bodenwasser *n* ground water
Bodenwelle *f* ground wave
Bodenwirksamkeit *f* (Rektifikation) tray efficiency
Bodenwirkungsgrad *m* (Rektifikation) tray efficiency
Bodenzahl *f* (Rektifikation) number of plates, **äquivalente theoretische** ~ equivalent theoretical number of trays
Bördel *n* flange
Bördelarbeit *f* flanged sheetwork
Bördelautomat *m* automatic flanging machine
Bördelblech *n* flanged plate
Bördelmaschine *f* beading machine, flanging machine
bördeln to flange, to border, to edge, to rim, (abkanten, falzen) to seam, (sicken) to bead
Bördeln *n* beading, flanging, seaming
Bördelpresse *f* flanging press
Bördelrand *m* rim, vial mouth
Bördelring *m* beading ring
Bördelrolle *f* beading roll
Bördelversuch *m* bead test
Bördelwalze *f* flanging roll
bösartig ill-natured, (Med) malignant
Bösartigkeit *f* (Med) malignancy
böschen to slant, to slope
Böschung *f* slope, ramp
Böschungsfläche *f* surface of constant slope
Böttcher *m* cooper
Böttgersche Probe *f* Boettger's test
Bogen *m* arch, bend, curve; (Elektr) arc, (Pap) sheet, ~ **mit Rohransatz** bend with vent
Bogenbildung *f* arcing
Bogeneinheit *f* radian
Bogenentladung *f* arc discharge
Bogenflamme *f* arc flame
Bogenkalander *m* (Papier) sheet calender
Bogenkaschiermaschine *f* sheet lining machine
Bogenlampe *f* arc lamp
Bogenlicht *n* arc light
Bogenlichtkohlen *f pl* arc-lamp carbons
Bogenschweißen *n* arc welding
Bogenschweißung *f* arc welding
Bogensieb *n* sieve bend
Bogenspektrum *n* (Spektr) arc spectrum
Bogenstück *n* bend, curved piece
Bogenwärme *f* temperature of the arc
Bogenzahnkupplung *f* dihedral gear coupling
Bogenzirkel *m* wing callipers
Bogheadkohle *f* boghead coal
Bohle *f* plank
Bohlenzaun *m* piling
Bohnenerz *n* (Min) bean ore, pea ore
Bohnengallerte *f* bean curd
Bohnenkraut *n* (Bot) savory
Bohnenkrautöl *n* savory oil
Bohnenmehl *n* bean flour, bean meal
bohnern to wax
Bohnerwachs *n* floorwax, polishing wax
Bohnerz *n* pea ore
Bohranlage *f* drilling rig
Bohrautomat *m* automatic boring machine
Bohrbank *f* drilling bench, boring bench, boring mill
Bohrdeckel *m* hole cover
Bohreisen *n* bit, drill
bohren to bore, to drill
Bohren *n* boring, drilling, (Öl) well drilling

Bohrer *m* drill, auger, (Gesteinsbohrer) rock drill, (Gewindebohrer) tap, (Holzbohrer) wood bit, (Nagelbohrer) gimlet
Bohrerspitze *f* drill bit
Bohrfutter *n* boring socket, drill chuck
Bohrgestänge *n* boring rod, drill poles
Bohrgestell *n* bore frame, boring frame
Bohrhaken *m* tool hook
Bohrkern *m* drill core
Bohrknarre *f* ratchet drill, rack brace
Bohrkopf *m* drilling head
Bohrkran *m* drilling rig
Bohrkurbel *f* crank brace
Bohrloch *n* bore hole, drilled hole
Bohrlochachse *f* center of bore
Bohrlochmessungen *pl* logging
Bohrlochpumpe *f* bore hole pump
Bohrlochschieber *m* borehole gate valve
Bohrlochuntersuchung *f* borehole logging
Bohrmaschine *f* boring machine, drilling machine, drill press
Bohröl *n* cutting oil
Bohrprobe *f* drill test
Bohrröhre *f* boring tube
Bohrrohr *n* casing, casing pipe
Bohrsche Theorie *f* Bohr theory
Bohrscher Kreis *m* (Atom) Bohr orbit
Bohrscher Wasserstoffradius *m* Bohr's hydrogen radius
Bohrsches Atommodell *n* Bohr's atom model
Bohrsches Magneton *n* (Atom) Bohr's magneton
Bohrschlamm *m* drilling mud, drilling slime
Bohrspäne *pl* borings
Bohrspitze *f* auger bit
Bohrstahl *m* boring tool
Bohrstock *m* boring stock, borer
Bohrturm *m* boring tower, derrick
Bohrung *f* borehole, drill hole, drilling, duct, (Zylinder) bore, **kalibermäßige** ~ dowel bushel
Bohrungsgruppe *f* cluster of wells
Bohrungstoleranz *f* boring tolerance
Bohrvorrichtung *f* boring apparatus
Bohrwelle *f* boring spindle, cutter bar
Boivinose *f* boivinose
Boldin *n* boldine
Boleit *m* boleite
Boletol *n* boletol
Boletsäure *f* boletic acid
Bologneser Flasche *f* Bologna flask
Bologneser Spat *m* Bologna spar
Bolometer *n* bolometer
Boltonit *m* (Min) boltonite
Boltzmannkonstante *f* Boltzmann's constant
Boltzmannsche Funktion *f* Boltzmann's function
Bolus *m* (Min) bolus
bolusartig bolar, bolary
Boluserde *f* bole, lemnian earth, sealed earth
Bolzen *m* bolt, peg, pin, stud

Bolzenkeil *m* bolt key
Bolzenmutter *f* nut for a bolt
Bolzenscheibe *f* washer for a bolt
Bolzenschraube *f* expansion bolt
Bolzensetzer *m* explosive powered tool
Bolzensetzwerkzeug *n* explosive powered tool
Bombage *f* (Dose) [hydrogen] swelling, swelling
Bombardement *n* bombardment
bombardieren (beschießen) to bombard
Bombaycachou *n* bombay cachou
Bombe *f* bomb, (Chem) steel cylinder, **kalorimetrische** ~ bomb calorimeter
Bombenaufschluß *m* bomb fusion process
Bombenkalorimeter *n* bomb calorimeter
Bombenkörper *m* shell of bomb
Bombenrohr *n* bomb tube, Carius tube
Bombierdampf *m* shaping steam
bombieren to swell, to blow, (Dose) to bulge
Bombieren *n* (Techn) cambering, roll bending
Bombierung *f* camber, ~ **der Kalanderwalze** crown of calender roll
Bombykol *n* bombycol
Boninsäure *f* boninic acid
Boothit *m* (Min) boothite
Bootslack *m* boat varnish
Bor *n* boron (Symb. B), **graphitartiges** ~ graphitic boron
Boracit *m* (Min) boracite
Boräquivalent *n* boron equivalent
Boräthan *n* borethane, boroethane
Boräthyl *n* borethyl, boron ethyl, boron triethyl, triethylborine, triethylboron
Boral *n* boral, aluminum borotartrate
Borameisensäure *f* boroformic acid
Boran *n* borane, boron hydride
borartig boric
Borat *n* borate
Borax *m* borax, sodium tetraborate, **gebrannter** ~ boiled borax
Boraxanlage *f* borax plant
Boraxglas *n* borax glass
Boraxhonig *m* borax honey
Boraxkalk *m* calcium borate
Boraxperle *f* (Anal) borax bead
Boraxsäure *f* boric acid, boracic acid
boraxsauer borate
Boraxsee *m* borax lake
Boraxseife *f* borax soap
Boraxspat *m* (Min) boracite
Boraxweinstein *m* potassium borotartrate
Borazaren *n* borazarene
Borazit *m* (Min) boracite
Borbutan *n* borobutane, tetraborane
Borcarbid *n* boron carbide
Borchelat *n* boron chelate
Borchlorid *n* boron chloride, boron trichloride
Bordeauxrot Bordeaux red, claret
Bordflansch *m* flange
bordfrei free on board, f.o.b.

Boretsch *m* (Bot) borage
Borfluorid *n* boron trifluoride
Borfluorwasserstoffsäure *f* borofluoric acid, borofluohydric acid, hydroborofluoric acid, hydrofluoboric acid
Borflußsäure *f* borofluoric acid, hydroborofluoric acid, borofluohydric acid, fluoboric acid, hydrofluoboric acid
borflußsauer fluoborate
borgefüllt boron-filled
Borgehalt *m* boron content
Borhydrid *n* boron hydride, hydroboron
Borickit *m* (Min) borickite
Borieren *n* boriding
Borin *n* borine
Borinsäure *f* borinic acid
Bor-Ionisations-Kammer *f* boron ionization chamber
Borjodid *n* boron iodide, boron triiodide
Borkalk *m* calcium borate
Borkarbid *n* boron carbide
Borkarbidstab *m* boron-carbide rod
Borkenflechte *f* [crustose] lichen
Borkenkrepp *m* bark-like crepe
Borkristalle *m pl* boron crystals, crystallized boron
Bormethyl *n* bormethyl, boron methyl, boron trimethyl, trimethylborine, trimethylboron
Born *m* brine, salt pit
Bornan *n* bornane
Borneocampher *m* borneo-camphor, d-borneol
Borneol *n* borneol, borneo-camphor, bornyl alcohol
Bornesit *m* bornesitol
Born-Haber-Kreisprozeß *m* Born-Haber cycle
Bornit *m* (Min) bornite, erubescite, peacock ore
Bornitrid *n* boron nitride
Bornyl- bornyl-
Bornylamin *n* bornylamine
Bornylchlorid *n* bornyl chloride
Bornylen *n* bornylene
Bornylon *n* bornylone
Bornylvalerianat *n* bornyl valerate
Bornyval *n* bornyval, gynoval
Borocalcit *m* (Min) borocalcite
Borocarbid *n* boron carbide
Borocitrat *n* borocitrate
Boroglycerid *n* boroglyceride, glyceryl borate
Borol *n* borol
Boronatrocalcit *m* (Min) boronatrocalcite
Boronsäure *f* boronic acid
Borotannat *n* borotannate
Borotartrat *n* borotartrate
Borovertin *n* borovertine
Borowolframsäure *f* borotungstic acid, tungstoboric acid
Borsäure *f* boric acid, boracic acid
Borsäureanhydrid *n* anhydrous boric acid, boric anhydride, boron oxide
Borsäurechelat *n* boric acid chelate
borsäuregetränkt soaked in boric acid
Borsäurelösung *f* boric acid solution
Borsäuremagnesia *f* magnesium borate
Borsäuremanganoxydul *n* manganous borate
Borsäuremethylester *m* boric acid methyl ester
Borsäureseife *f* boric acid soap
Borsäureweinstein *m* potassium borotartrate
Borsalbe *f* boric acid ointment
Borsalicylsäure *f* borosalicylic acid
Borsalicylsäurechelat *n* borosalicylic acid chelate
borsauer borate
Borschirm *m* boron shield, boron target
Borsilicid *n* boron silicide
Borstahl *m* (Atom) boron steel
Borste *f* bristle
Borstenscheibe *f* buffing brush
Borstickstoff *m* boron nitride
borstig bristly
Bort *m* (Industriediamant) bort [stone]
Borte *f* braid, border, edging
Bortribromid *n* boron tribromide, boron bromide
Bortrichlorid *n* boron trichloride, boron chloride
Bortrifluorid *n* boron trifluoride, boron fluoride
Bortrijodid *n* boron triiodide, boron iodide
Bortrioxid *n* boron trioxide, anhydrous boric acid, boric anhydride
Borverbindung *f* boron compound
Borwasser *n* boric acid solution
Borwasserstoff *m* borane, boron hydride
Borwatte *f* boric cotton [wool]
Borwolframsäure *f* borotungstic acid, tungstoboric acid
Boryl- boryl-
Borylweinsäure *f* boryltartaric acid
Boson *n* boson
Bosse *f* (Holz) white guarea
Bossierwachs *n* molding wax
Boswellinsäure *f* boswellic acid
Boswellonsäure *f* boswellonic acid
Botanik *f* botany
Botaniker *m* botanist
botanisch botanical
Botogenin *n* botogenin
Botryolith *m* (Min) botryolite
Bottich *m* vat, container, tank, tub
Bottichgärung *f* tub fermentation
Bottichkühler *m* attemperator
Bottichthermometer *n* [mash] tub thermometer
Bottone's Härteskala *f* Bottone's scale of hardness
Botulismus *m* botulism
Boudouard-Gleichgewicht *n* Boudouard equilibrium
Bougie *f* bougie
Bouillon *f* broth [culture]

Boulangerit *m* (Min) boulangerite
Bournonit *m* (Min) bournonite, cogwheel ore
Bouteillenstein *m* (Min) boldawite
Bovist *m* (Bot) bovista, puffball
Bovogenin *n* bovogenin
Bovokryptosid *n* bovocryptoside
Bovosid *n* bovoside
Bovosidol *n* bovosidol
Bovovaccin *n* bovovaccine
Bovovakzine *f* cow vaccine
Bowdenlitze *f* bowden stranded wire
Bowdenseil *n* bowden rope
Bowenit *m* (Min) bowenite
Bowmanit *m* (Min) bowmanite
Boxer-Kolbenverdichter *m* double-piston compressor
Boyle-Mariottesches Gesetz *n* Boyle-Mariotte's law
Brachiose *f* brachiose
Brachistochrone *f* (Math) brachistochrone
Brachyachse *f* (Krist) brachyaxis
Brachyprisma *n* brachyprism
Brackebuschit *m* brackebuschite
Brackettlinien *pl* (Spektr) Brackett series
Brackett-Serie *f* (Spektr) Brackett series
Brackgut *n* damaged goods, refuse, rubbish
Brackholz *n* decayed wood
brackig brackish
Brackwasser *n* brackish water
Bradykinin *n* bradykinin
bräunen (Färb) to brown, to dye brown, (Gieß) to burnish
Bräunen *n* browning
bräunlich brownish
Bräunung *f* brown coloring, browning
Braggebene *f* (Krist) Bragg plane
Braggmethode *f* (Krist) Bragg method
Braggscher Winkel *m* (Opt) Bragg angle
Braggsches Gesetz *n* Bragg's law
Bragit *m* (Min) bragite
Bramme *f* flat bloom, iron slab, slab bloom, slab ingot
Brammenkokille *f* chill mold
Brand *m* fire, (Keram) baking, burning, (Meltau) blight, (Mutterkorn) ergot, (Verbrennung) burning, combustion, ~ **eines Kalkofens** charge of a limekiln
Brandbalsam *m* ointment for burns
Brandbombe *f* fire bomb, incendiary bomb
Brandeisen *n* burnt iron
Branderz *n* inflammable ore
brandfest fireproof
brandgelb reddish yellow colored
Brandgold *n* refined gold
Brandharz *n* empyreumatic resin
Brandisit *m* (Min) brandisite
Brandkitt *m* fireproof cement, fireproof lute
Brandmauer *f* fire wall
Brandmesser *m* pyrometer

Brandmittel *n* (Pharm) remedy for burns
Brandöl *n* empyreumatic oil
Brandprobe *f* fire test, fire assay
Brandpuder *m* powder for burns
Brandsalbe *f* (Pharm) burns ointment
Brandsatz *m* incendiary composition
Brandschaden *m* fire damage
Brandschiefer *m* bituminous shale, coal slate
Brandschutz *m* fire protection
Brandschutzanstrichfarbe *f* fire-retarding paint, fireproof paint, fireresisting paint
Brandsilber *n* refined silver
Brandstein *m* brick
Brandtit *m* (Min) brandtite
Brandversicherung *f* fire insurance
Brandwunde *f* burn
Brannerit *m* (Min) brannerite
Brannthefe *f* spent yeast
Branntwein *m* brandy, spirit, ~ **brennen** to distill brandy
Branntweinblase *f* alembic, still
Branntweinbrenner *m* distiller
Branntweinessig *m* brandy vinegar
Branntweingeist *m* rectified spirit
Branntweinhefe *f* alcohol ferment
Branntweinprüfer *m* alcoholometer
Branntweinvergiftung *f* alcoholic poisoning
Branntweinwaage *f* alcoholometer
Brasan *n* brazan
Brasanchinon *n* brazanquinone
Brasildiamant *m* Brazil brilliant
Brasilein *n* brazilein
Brasilholz *n* Brazil wood, fernambuco wood
Brasilin *n* (Färb) brazilin
Brasilinsäure *f* brazilinic acid
Brasilkopalinsäure *f* brazilcopalinic acid
Brasilkopalsäure *f* brazilcopalic acid
Brasilsäure *f* brasilic acid
Brasse *f* (Mar) brace
Brassenblock *m* brace block
Brassicasterin *n* brassicasterol
Brassidinsäure *f* 12-docosenoic acid, brassidic acid, brassic acid, isoerucic acid
Brassidon *n* brassidone
Brassinsäure *f* 12-docosenoic acid, brassic acid, brassidic acid, isoerucic acid
Brassylsäure *f* brassylic acid
braten to roast, to broil, to grill, (Techn) to calcine
Braubottich *m* (Brau) brewing vat
brauchbar useful
Brauchbarkeit *f* usefulness, adaptability, practicability, serviceability, utility, workability
Brauchwasser *n* process water
brauen to brew
Brauerei brewery
Brauereimaschine *f* brewing machine
Brauerpech *n* (Brau) brewer's pitch

Braumalz *n* brewing malt
Braumethode *f* brewing method
Braun *n* brown
Braunalgen *pl* (Bot) brown algae
Braunbleierz *n* (Min) pyromorphite
Brauneisen *n* (Min) brown iron ore
Brauneisenerz *n* (Min) brown iron ore, limonite
Brauneisenocker *m* (Min) brown iron ocher
Brauneisenstein *m* (Min) brown iron ore, brown hematite, limonite
Brauneisensteinknollen *f pl* (Min) brown iron ore nodules
Braunerz *n* (Min) brown [iron] ore, limonite
Braunfärbung *f* brown coloring
Braunfäule *f* (Bot) brown rot
braungelb brownish yellow, yellowish brown
Braunglas *n* amber glass, mica, potash mica
Braunholz *n* logwood
Braunit *m* (Min) braunite
Braunkalk *m* brown spar
Braunkohle *f* brown coal, lignite, soft coal, **ölreiche** ~ bituminous lignite
Braunkohlenanlage *f* brown coal plant
Braunkohlenaufbereitungsanlage *f* lignite dressing plant
Braunkohlenbenzin *n* boghead naphtha
Braunkohlenbergwerk *n* lignite mine
Braunkohlenbitumen *n* brown coal bitumen
Braunkohlenbrikett *n* brown coal briquet[te], lignite briquet[te]
Braunkohlenfeuerung *f* lignite firing, lignite fired furnace, lignite furnace
Braunkohlenförderung *f* brown-coal mining
braunkohlenhaltig lignitic, lignitiferous
Braunkohlenklein *n* lignite breeze
Braunkohlenkoks *m* brown coal coke, lignite coke
Braunkohlenlager *n* lignite bed
Braunkohlenlösche *f* lignite dust
Braunkohlenpech *n* lignite pitch
Braunkohlenpulver *n* Cassel brown, Cologne earth
Braunkohlensandstein *m* brown coal grit
Braunkohlenschiefer *m* (Min) lignite shale
Braunkohlenschwelerei *f* lignite carbonization plant, lignite coking plant
Braunkohlenschwelgas *n* lignite distillation gas
Braunkohlenschwelkoks *m* lignite low temperature coke
Braunkohlentagebau *m* lignite open cut
Braunkohlenteer *m* brown coal tar, lignite tar, tar from lignite
Braunkohlenteerpech *n* lignite pitch
Braunkohlentrocknung *f* lignite drying
Braunkohlenwachs *n* brown coal wax
Braunlack *m* brown lake
Braunmanganerz *n* (Min) manganite
Braunocker *m* brown ocher
braunrot brownish red

Braunrotglühhitze *f* dark-red heat
Braunsche Röhre *f* Braun tube, cathode ray tube
Braunschliff *m* brown wood pulp, (Papier) brown mechanical pulp, steamed mechanical wood pulp
braunschwarz brown black, dark brown
Braunspäne *m pl* logwood chips
Braunspat *m* (Min) brown pearlspar, brown spar
Braunstein *m* (Chem) manganese dioxide, (Min) native black manganese oxide, pyrolusite, **kobalthaltiger** ~ asbolane, asbolite, **roter** ~ (Min) rhodochrosite
Braunsteinblende *f* (Min) alabandine
Braunsteinelement *n* Leclanché cell
Braunsteinkies *m* (Min) native manganese sulfide
Braunsteinkiesel *m* manganese silicate, rhodonite
Braunsteinrahm *m* bog manganese
Braunsteinschaum *m* bog manganese
Brauntran *m* blubber, thick cod oil
Brause *f* effervescence, shower, spray
Brausemischung *f* effervescent mixture
brausen to effervesce, to fizz
Brausen *n* effervescing, ~ **der Flamme** roaring of the flame
Brausepulver *n* effervescent powder
Brausesalz *n* effervescent salt
Brauseton *m* bituminous clay
Brausewein *m* sparkling wine
Bravaisit *m* (Min) bravaisite
Brazilit *m* (Min) brazilite
Breccie *f* (Geol) breccia
Breccienachat *m* (Min) brecciated agate
breccienartig brecciated
Brecciengefüge *n* breccia structure
Breccienmarmor *m* brecciated marble
Brechachse *f* breaking shaft
Brechanlage *f* crushing plant
Brechbacke *f* jaw breaker
brechbar brittle
Brechbarkeit *f* breakability, brittleness
Brecheisen *n* crowbar, wrecking bar
brechen to break, to crush, to fracture, to pulverize, to rupture, (Opt) to refract
Brecher *m* breaker, cracker, crusher, grinding mill, **Backen-** ~ jaw crusher, **Backen-Kreisel-** ~ roll jaw crusher, **Einschwingen-** ~ single-toggle crusher, **Hammer-** ~ hammer crusher, **Kegel-** ~ cone crusher, **Kniehebel-** ~ toggle crusher, **Mantel-** ~ rolling ring crusher, **Messer-** ~ knife crusher, **Prall-** ~ impact crusher, **Rotex-** ~ Rotex [eccentric roll] crusher, **Rund-** ~ gyratory crusher, **Schlag-** ~ impact crusher, **Symons-** ~ Symons standard cone crusher, **Walzen-** ~ double-roll crusher
Brecheranlage *f* crushing plant
Brecherräder *n pl* crusher wheels

Brecherströmung *f* slug flow
Brechertyp *m* breaker type
Brechgraupen *pl* crystallized tin oxide
Brechholländer *m* breaker beater
Brechkapsel *f* breaker
Brechkoks *m* crushed coke
Brechkraft *f* (Opt) power of refraction
Brechmaul *n* breaker mouth
Brechmittel *n* (Med) emetic
Brechnuß *f* nux vomica, vomic nut
Brechnußextrakt *m* extract of nux vomica
Brechnußtinktur *f* tincture of nux vomica
Brechplatte *f* bursting plate, rupture disc, safety diaphragm
Brechprobe *f* breaking test
Brechpunkt *m* point of break
Brechreiz *m* nausea
Brechsand *m* crushed sand
Brechscheibe *f* bursting disc, rupture disc, safety diaphragm
Brechstange *f* crowbar, pinchbar, pointed rabble, puddling bar
Brechstoff *m* (Pharm) emetic
Brechtopf *m* breaker
Brechung *f* breaking, (Opt) diffraction, refraction, ~ **der Flußlinien** *f pl* (Magn) flux refraction, ~ **der Gasströmung** interruption of the gas flow, **diffuse** ~ (Opt) diffuse refraction, **ideale** ~ (Opt) standard refraction, **molare** ~ molar refraction, **ordentliche** ~ ordinary refraction, regular refraction
Brechungsabweichung *f* refractive aberration
Brechungsachse *f* axis of refraction
Brechungsebene *f* (Opt) plane of refraction
Brechungsexponent *m* refractive index
Brechungsgesetz *n* law of refraction
Brechungsindex *m* refractive index, index of refraction, **modifizierter** ~ (Opt) modified index of refraction, **spezifischer differentieller** ~ (Opt) specific dispersivity
Brechungsindexbestimmung *f* refractometry
Brechungsindexrealteil *m* real part of the refractive index
Brechungsindexsprung *m* leap of refractive index
Brechungskoeffizient *m* (Opt) refractive index
Brechungsmesser *m* (Opt) refractometer
Brechungsschicht *f* refracting layer
Brechungsstrahl *m* refracted ray
Brechungsvermögen *n* refractive power, refraction, refractivity, **molares** ~ molar refraction power, molar refractivity
Brechungswinkel *m* angle of refraction
Brechungszahl *f* refractive index
Brechwalzwerk *n* crushing mill
Brechwein *m* antimonial wine, emetic wine, wine of antimony
Brechweinstein *m* antimony potassium tartrate, potassium antimonyltartrate, tartar emetic
Brechweinsteinersatz *m* tartar emetic substitute

Brechwerk *n* crusher, grinder
Brechwurz *f* (Bot) ipecac
Bredtsche Regel *f* (Stereochem) Bredt rule
Breechtsches Doppelsalz *n* Breecht's double salt, potassium bimagnesium sulfate
Brei *m* (Aufschlämmung) slurry, (Pap) mash, paste, pulp, **zu** ~ **zerquetschen** to mash
breiartig pasty, semifluid
breiig pasty
Brein *n* brein
Breislakit *m* (Min) breislakite
breit broad, wide, (Buchdr) extended, ~ **fett** (Buchdr) extended extrabold, ~ **mager** (Buchdr) extended standard
Breitbandbeizanlage *f* wide strip pickling plant
Breitbandverstärker *m* wide band amplifier
Breitbeil *n* broad axe
Breitbleiche *f* open width bleach
Breitbleichmaschine *f* open width bleaching machine
Breite *f* width, breadth, **lichte** ~ inside width, width in the clear
Breiteisen *n* broad tool, cross chisel, flat billet
Breitengrad *m* degree of latitude
Breitenmetazentrum *n* transverse metacenter
Breitfelge *f* wide base rim
breitflanschig wide-flanged
Breitflanschträger *m* wide-flanged beam
Breithalter *m* (Text) temple textile
Breithauptit *m* (Min) breithauptite
Breitreckvorrichtung *f* (Film) "across" stretching device
Breitschlitzdüse *f* sheeting die, slot die, **einfache** ~ single manifold die, **Mehrschicht-** ~ multi-manifold die
Breitwaschmaschine *f* open width washing machine
Brekzie *f* (Geol) breccia
Brekzienachat *m* (Min) broken agate
Bremerblau *n* Bremen blue, blue copper carbonate, verditer blue
Bremergrün *n* Bremen green, Prussian green
Bremsäquivalent *n* stopping equivalent
Bremsantrieb *m* brake control
Bremsarbeit *f* braking effect
Bremsausrüstung *f* brake equipment
Bremsbacke *f* brake shoe, brake clutch
Bremsbahn *f* braking distance, brake distance
Bremsband *n* brake band
Bremsbandimprägnieranlage *f* brake band impregnation plant
Bremsbandspannung *f* brake band tension
Bremsbandstütze *f* brake band support
Bremsbehälter *m* brake cylinder
Bremsbelag *m* brake lining
Bremsbelastung *f* brake load
Bremsdichte *f* (Atom) slowing-down density
Bremsdruck *m* brake pressure
Bremsdynamometer *n* absorption dynamometer

Bremse – Brennofen

Bremse f brake, (Atom) moderator
Bremselektron n stopping electron
Bremsenergie f energy of braking
Bremsfeld n reflecting field (vacuum tube), retarding field (electron-oscillation tube)
Bremsfeldelektrode f retarding-field or reflecting electrode
Bremsfläche f slowing-down area, braking area
Bremsflüssigkeit f brake fluid, recoil liquid
Bremsformel f stopping formula
Bremsfußhebel m foot brake pedal
Bremsfutter n brake lining
Bremsgelenk n braking articulation
Bremsgewicht n counterbalance
Bremsklotz m brake block, brake shoe
Bremsklotzsohle f brake block holder
Bremskraft f brake horsepower, brake pressure
Bremslänge f slowing-down length
Bremsleistung f brake horsepower
Bremsleuchte f brake light
Bremsmittel n (Flotation) depressant
Bremsmoment n braking moment
Bremspotential n stopping potential
Bremsrad n brake wheel
Bremsregulierlager n brake adjusting bearing
Bremsreibung f braking friction
Bremsscheibe f brake disc, brake pulley
Bremsschuh m brake shoe
Bremsstange f brake rod
Bremsstrahlung f (Atom) bremsstrahlung, retardation radiation
Bremssubstanz f (im Reaktor) moderating material, moderator
Bremsung f braking, deceleration, retardation, ~ **von Neutronen** moderation of neutrons
Bremsvermögen n retarding power, stopping power
Bremswelle f brake shaft
Bremswirkung f brake effect
Bremszahl f stopping number
Bremszeit f (Reaktor) slowing-down time
Bremszylinder m brake cylinder
Brennachse f (Opt) focal axis
Brennätzverfahren n pyrography
Brennapparat m distilling apparatus, still
Brennarbeit f burning
brennbar combustible, inflammable
Brennbarkeit f inflammability, combustibility, deflagrability, flammability
Brennbarkeitsprobe f burning test
Brennbereich m firing range
Brennblase f alembic, still
Brennblock m crabbing jack
Brenndauer f duration of combustion
Brenndüse f nozzle
Brenne f dip, pickle
Brennebene f (Opt) focal plane
Brennelement n fuel element, (Atom) fuel rod

brennen to burn, (Alkohol) to distill, (Keram) to bake, (z. B. Kalk) to calcine
Brennen n burning, calcining, refining; (Alkohol) distillation, ~ **der Leimfuge** burning of glue line
brennend burning, ardent, (Geruch) pungent
Brenner m burner, (Schweiß) blowpipe, torch, **Bunsen-** ~ Bunsen burner, **Flach-** ~ fishtail burner, **Hochfrequenzplasma-** ~ high-frequency plasma torch, **Injektor-** ~ premix burner, **Kaskaden-** ~ plasma cascade torch, **Lichtbogenplasma-** ~ plasma arch torch, **Mekker-** ~ Mekker burner, **Misch-** ~ circular burner, **Parallelstrom-Gas-** ~ parallel-flow gas burner, **Plasma-** ~ plasma torch, **Staub-** ~ pulverized coal burner, **Strahl-** ~ multiple-intertube burner, **Teclu-** ~ Teclu burner, **Venturi-** ~ venturi burner, **Vielfach-** ~ multiple tube burner, **Vormisch-** ~ premix burner, **Wirbel-** ~ vortex burner
Brennerausrüstung f burner equipment
Brennerdüse f burner nozzle
Brennerei f **für Branntwein** distillery, ~ **für Kalk** lime works, ~ **für Kohle** charcoal burning plant
Brennereiabwässer pl distillery waste water
Brennereianlage f distilling plant
Brennerkopf m burner tip
Brennesselkraut n (Bot) common nettle
Brennfleckausdehnung f focal spot characteristic
Brenngas n combustible gas, fuel gas
Brenngemisch n combustible mixture
Brenngerste f distilling barley
Brenngeschwindigkeit f burning rate, rate of combustion
Brennglas n burning glass, magnifying glass
Brenngut n material to be burnt
Brennhaus n baking plant
Brennhelm m still head
Brennherd m hearth, refining furnace
Brennholz n fire-wood, burning wood
Brennhütte f refinery
Brennkammer f combustion chamber
Brennkapsel f (Keram) sagger
Brennkegel m fusible cone, melting cone, pyrometric cone
Brennkolben m alembic, distilling flask, still
Brennkraft f intensity of combustion
Brennlinie f (Opt) focal line
Brennluft f air for combustion, inflammable air
Brennmaterial n combustible material, fuel
Brennmaterialausnutzung f fuel utilization
Brennmaterialverbrauch m fuel consumption
Brennöl n fuel oil, burning oil, lamp oil
Brennölrückstand m fuel oil residue
Brennofen m (Erz) calcining furnace, calcining kiln, roasting furnace, (Keram) kiln, (Techn) burning oven

Brennperiode *f* combustion period
Brennpfanne *f* roasting pan
Brennprobe *f* calcining test
Brennprozeß *m* calcining process
Brennpunkt *m* (Opt) focus, focal point, **astigmatischer** ~ (Opt) astigmatic focus, **in den** ~ **bringen** to bring into focus, **nicht im** ~ out of focus
Brennpunktabstand *m* (Opt) focal distance, focal length
Brennpunktebene *f* (Opt) focal plane
Brennpunktprüfer *m* ignition point tester
Brennrohrmundstück *n* burner nozzle
Brennsäure *f* pickling acid
Brennschicht *f* fuel bed
Brennschichthöhe *f* thickness of fuel bed
Brennschneiden *n* arc cutting, flame cutting, gas cutting
Brennschneider *m* oxyacetylene cutter
Brennschwindung *f* shrinkage in firing
Brennsilber *n* amalgam for silvering
Brennspiegel *m* concave mirror
Brennspiritus *m* raw alcohol for burning
Brennstahl *m* blister steel, cemented steel, converted steel
Brennstahlbereitung *f* cementation
Brennstahlofen *m* converting furnace
Brennstaub *m* pulverized fuel
Brennstoff *m* fuel, combustible, **flüssiger** ~ liquid fuel, **gasförmiger** ~ gaseous fuel, **hochwertiger** ~ high grade fuel, **klopffester** ~ antiknock or antidetonating fuel, **leicht entzündbarer** ~ easily inflammable fuel, **leicht flüchtiger** ~ light [volatile] fuel, **minderwertiger** ~ low-grade fuel, **schwer flüchtiger** ~ heavy fuel, non-volatile fuel
Brennstoffaufarbeitung *f* fuel regeneration
Brennstoffaufwand *m* consumption of fuel
Brennstoffausnutzung *f* fuel utilization
Brennstoffbehälter *m* fuel tank
Brennstoffchemie *f* fuel chemistry
Brennstoffdurchfluß *m* fuel flow
Brennstoffeinspritzpumpe *f* fuel-injection pump
Brennstofferneuerung *f* refueling
Brennstofffilter *n* fuel strainer
Brennstoffförderpumpe *f* fuel feed pump
Brennstoffgicht *f* charge of fuel
Brennstoffkanal *m* fuel channel
Brennstoffkreislauf *m* (Atom) fuel cycle
Brennstoffladeröhre *f* (Atom) fuel charge tube
Brennstoffluftgemisch *n* fuel-air mixture
Brennstoffpatrone *f* (Atom) fuel cartridge
Brennstoffpumpe *f* fuel pump
Brennstoffschicht *f* fuel bed
Brennstoffstab *m* fuel rod
Brennstofftechnik *f* fuel technology
Brennstofftrichter *m* fuel hopper
Brennstoffventil *n* fuel valve
Brennstoffverbrauch *m* fuel consumption
Brennstoffvorwärmer *m* fuel heater
Brennstoffwirtschaft *f* fuel economy
Brennstoffzelle *f* fuel cell
Brennstoffzuführung *f* fuel feed
Brennstoffzufuhr *f* fuel supply, transport of fuel
Brennstunde *f* (Elektr) burning hour, lighting hour
Brenntemperatur *f* temperature of combustion
Brennversuch *m* burning test
Brennwärme *f* heat of combustion
Brennweite *f* focal distance, focal length, ~ **des Okulars** focal length of the eyepiece, **dingseitige** ~ (Opt) front focal length
Brennweiteneinstellung *f* setting of the focal length
Brennweitenmesser *m* focimeter
Brennwert *m* calorific value, fuel value
Brennwertbestimmung *f* calorimetry; determination of calorific value
Brennwiderstand *m* combustion drag
Brennzeit *f* burning period, duration of combustion, period of roasting
Brennziegel *m* firebrick, fuel briquet[te]
Brennzone *f* zone of combustion, zone of incandescence
Brenzapfelsäure *f* maleic acid
Brenzcain *n* brenzcain
Brenzcatechin *n* 1,2-dihydroxybenzene, catechol, pyrocatechin, pyrocatechol
Brenzcatechindimethyläther *m* 1,2-dimethoxybenzene, pyrocatechol dimethyl ether, veratrole
Brenzcatechinphthalein *n* pyrocatechol phthalein
Brenzcitronensäure *f* pyrocitric acid
Brenze *pl* combustibles
Brenzgallussäure *f* 1,2,3-benzenetriol, 1,2,3-trihydroxybenzene, pyrogallic acid, pyrogallol
Brenzharz *n* empyreumatic resin
Brenzkatechin *n* pyrocatechine, pyrocatechol
Brenzkatechingerbstoff *m* catechol tannin
brenzlig burnt, empyreumatic, tarry
Brenzöl *n* empyreumatic oil
Brenzreaktion *f* pyrolysis
Brenzsäure *f* pyroacid
Brenzschleimsäure *f* pyromucic acid
brenzschleimsauer pyromucate
Brenzterebinsäure *f* pyroterebic acid, 4-methyl-2-pentenoic acid
Brenztraubenaldehyd *m* pyruvic aldehyde, pyroracemic aldehyde, pyruvaldehyde
Brenztraubenalkohol *m* pyroracemic alcohol
Brenztraubensäure *f* pyruvic acid, pyroracemic acid
Brenztraubensäurealdehyd *m* pyroracemic aldehyde, pyruvaldehyde
Brenzweinsäure *f* pyrotartaric acid, methylbutanedioic acid, methylsuccinic acid, pyrovinic acid

brenzweinsteinsauer pyrotartrate
Brett *n* board
Brettfallhammer *m* board drop hammer
Breunnerit *m* (Min) breun[n]erite
Brevifolin *n* brevifolin
Brevium *n* brevium
Brewsterit *m* (Min) brewsterite
Brewstersche Streifen *pl* (Opt) Brewster fringes
Briefumschlagform *f* (Stereochem) envelope form
Briggscher Logarithmus *m* Brigg's logarithm, common logarithm, decimal logarithm
Brigls Anhydrid *n* Brigl's anhydride
Brikett *n* briquet[te]
Brikettbindemittel *n* briquette cement
brikettieren to briquet[te]
Brikettierfähigkeit *f* briquetting property
Brikettierung *f* briquetting
Brikettierungsanlage *f* briquetting plant
Brikettierverfahren *n* briquetting process
Brillant *m* brilliant
Brillantalbuminpapier *n* double albuminized paper
Brillantalizarinblau *n* brilliant alizarin blue
Brillantazurin *n* brilliant azurine
Brillantcarmoisin *n* brilliant carmoisine
Brillantcrocein *n* brilliant crocein
Brillantfarbstoff *m* brilliant dye
Brillantgelb *n* brilliant yellow
Brillantine *f* brilliantine
Brillantkarmin *n* brilliant carmine
Brillantkongo *n* brilliant congo
Brillantkresylblau *n* brilliant cresyl blue
Brillantlack *m* high-gloss varnish
Brillantpapier *n* glazed paper
Brillantrosa *n* brilliant pink
Brillantsäureblau *n* brilliant acid blue
Brillantsäuregrün *n* brilliant acid green
Brillantschwarz *n* brilliant black
Brille *f* glasses, spectacles
Brillenofen *m* furnace with two pits, spectacle furnace
Brillouin-Zonen *pl* (Krist) Brillouin zones
Brinell-Einheit *f* Brinell unit
Brinellhärte *f* Brinell hardness
Brinell-Kugeldruckprobe *f* Brinell ball-hardness test
Brinell-Presse *f* (mit Hebelwaage) dead-weight Brinell machine, lever-type Brinell machine
Brinellsche Härtezahl *f* (Brinell-Zahl) Brinell hardness number
Brinellzahl *f* Brinell hardness number
brisant highly explosive
Brisanz *f* brisance, explosive power, shattering power
Brisanzmunition *f* high explosive ammunition
Brisanzstoff *m* high explosive
Britholith *m* (Min) britholith
Britisch Gummi *n* British gum, dextrin

Brochantit *m* (Min) brochantite
Brocken *m* clot, crumb, lump, (Met) slug
Brockengestein *n* breccia
Brockenmessing *n* brass in lumps
Brockenstahl *m* steel in thin bars, superfine steel
brodeln to boil, to bubble
Brodeln *n* boiling, bubbling, ebullition
Brodem *m* steam, exhalation, vapor
bröckelig brittle, crumbly, friable
Bröckeligkeit *f* brittleness, friableness
bröckeln to crumble
Bröckelspan *m* broken chips
Bröckelstärke *f* lump starch
bröcklig brittle, crumbly, friable
Bröggerit *m* (Min) bröggerite
Brokat *m* (Text) brocade
Brokatell *m* brocatelle
Brokatfarbe *f* brocade color, brocade dye, bronze color
Brokatleder *n* brocade leather
Brom *n* bromine (Symb. Br)
Bromacetol *n* bromacetol, 2,2-dibromopropane
Bromaceton *n* bromoacetone
Bromacetylen *n* bromoacetylene, bromoethyne
Bromadditionsprodukt *n* additive compound of bromine
Bromäthyl *n* ethyl bromide, bromoethane
Bromäthylen *n* bromoethylene, bromoethene
Bromäthylformin *n* bromalin, bromethylformin, hexamethylenetetramine bromethylene
Bromal *n* bromal, 2,2,2-tribromoethanal, tribromoacetaldehyde
Bromalhydrat *n* bromal hydrate, 2,2,2-tribromo-1,1-ethanediol
Bromalin *n* bromalin, bromethylformin, hexamethylenetetramine bromethylene
Bromaluminium *n* aluminum bromide
Bromamid *n* bromamide, 2,4,6-tribromoaniline
Bromammonium *n* ammonium bromide
Bromanil *n* bromanil
Bromanilin *n* bromaniline
Bromanilsäure *f* bromanilic acid
Bromantimon *n* antimony bromide
Bromargyrit *m* (Min) bromargyrite
Bromarsen *n* arsenic bromide
Bromat *n* bromate, salt or ester of bromic acid
Bromatätze *f* bromate discharge
Bromatographie *f* bromatography
Bromatologie *f* bromatology
Bromatom *n* bromine atom
bromatometrisch bromatometric
Brombarium *n* barium bromide
Brombenzoesäure *f* bromobenzoic acid
Brombenzol *n* bromobenzene, phenyl bromide
Brombernsteinsäure *f* bromosuccinic acid
Bromcadmium *n* cadmium bromide
Bromcalcium *n* calcium bromide
Bromcampher *m* bromocamphor, monobromated camphor

Bromcarmin *n* bromcarmine
Bromcholinbromid *n* bromocholine bromide
Bromcyan *n* bromocyanogen, cyanogen bromide
Bromdampf *m* bromine vapor
Bromderivat *n* bromine derivative
Bromdifluormethan *n* bromodifluoromethane
Bromein *n* bromeine
Bromeisen *n* iron bromide
Bromelin *n* bromelin
Bromessigsäure *f* bromoacetic acid, bromacetic acid
Brometon *n* brometone, 2,2,2-tribromo-1-methyl-1-propanol
Bromflasche *f* bromine bottle
Bromfluorid *n* bromine fluoride
Bromgehalt *m* bromine content
Bromgold *n* gold bromide
Bromgoldkalium *n* potassium bromoaurate
Bromguanid *n* bromoguanide
bromhaltig containing bromine
Bromhydrin *n* bromohydrin
Bromhydrochinon *n* bromohydroquinol
Bromid *n* bromide
Bromidpapier *n* bromide paper
bromieren to bromate, to brominate
Bromieren *n* bromation, bromination
Bromierung *f* bromation, bromination
Bromierungsreaktion *f* bromination reaction
Bromipin *n* bromipin, brominated sesame oil
Bromipinkapsel *f* bromipin capsule
Bromismus *m* (Med) bromine poisoning, brominism, bromism
Bromisocapronsäure *f* bromoisocaproic acid
Bromisocaproylchlorid *n* bromoisocaproyl chloride
Bromisoval *n* bromisoval
Bromit *m* (Min) bromite, bromyrite
Bromjod *n* iodine bromide
Bromjodbenzol *n* bromiodobenzene
Bromjodnaphthalin *n* bromoiodonaphthalene
Bromkadmium *n* cadmium bromide
Bromkalium *n* potassium bromide
Bromkohlenstoff *m* carbon tetrabromide
Bromkresolgrün *n* bromocresol green
Bromkresolpurpur *n* bromocresol purple
Bromkupfer *n* copper bromide
Bromlit *m* (Min) bromlite
Bromlithium *n* lithium bromide
Bromlösung *f* bromine solution
Brommagnesium *n* magnesium bromide
Bromnaphthalin *n* bromonaphthalene
Bromnaphthochinon *n* bromonaphthoquinone
Bromnatrium *n* sodium bromide
Bromnickel *n* nickel bromide
Bromnitrobenzol *n* bromonitrobenzene
Bromochinal *n* bromoquinal
Bromocodid *n* bromocodide
Bromocoll *n* bromocoll
Bromocollösung *f* bromocoll solution
Bromocollsalbe *f* bromocoll ointment
Bromocollseife *f* bromocoll soap
Bromoform *n* bromoform, tribromomethane
Bromol *n* bromol, 2,4,6-tribromophenol
Bromometrie *f* bromometry
bromometrisch bromometric
Bromomorphid *n* bromomorphide
Bromoniumion *n* bromonium ion
Bromopikrin *n* bromopicrin, nitrobromoform, tribromonitromethane
Bromoplatinat *n* bromoplatinate
Bromopyrin *n* bromopyrine
Bromphenanthrolin *n* bromophenanthroline
Bromphenetol *n* bromophenetol, bromoethoxybenzene
Bromphenol *n* bromophenol
Bromphenolblau *n* bromophenol blue
Bromphenylendiamin *n* bromophenylenediamine
Bromphenylhydrazin *n* bromophenylhydrazine
Bromphosgen *n* bromophosgene, carbonyl bromide
Bromphosphor *m* phosphorus bromide
Brompikrin *n* bromopicrin, nitrobromoform, tribromonitromethane
Brompräparat *n* bromine preparation
Bromradium *n* radium bromide
Bromsäure *f* bromic acid
Bromsalz *n* bromide, salt of hydrobromic acid
bromsauer bromate
Bromschwefel *m* sulfur bromide
Bromsilber *n* silver bromide
Bromsilbergelatine *f* (Phot) gelatin containing silver bromide, silver bromide gelatin
Bromsilbergelatinepapier *n* silver bromide gelatin paper
Bromsilberkollodium *n* collodion containing silver bromide, silver bromide collodion
Bromsilberpapier *n* (Phot) silver bromide paper
Bromsilberverfahren *n* silver bromide process
Bromsilicium *n* silicon bromide
Bromspat *m* (Min) bromyrite
Bromstickstoff *m* nitrogen bromide
Bromstyrol *n* bromostyrene
Bromtannin *n* bromotannin
Bromthymolblau *n* bromothymol blue, bromthymol blue
Bromthymolblau-Grundnährboden *m* bromthymol blue broth base
Bromtoluol *n* bromotoluene
Bromüberträger *m* bromine carrier
Bromür *n* (obs) bromide
Bromural *n* bromural, 2-bromoisovaleryl urea
Bromverbindung *f* bromine compound
Bromvergiftung *f* bromine poisoning, bromism
Bromwaschflasche *f* bromine washing bottle
Bromwasser *n* bromine water
Bromwasserstoff *m* hydrogen bromide, hydrobromic acid

Bromwasserstoffsäure *f* hydrobromic acid, hydrogen bromide
Bromwasserstoffverbindung *f* bromide, compound of hydrogen bromide
Bromwismut *n* bismuth bromide
Bromxylenolblau *n* bromoxylenol blue
Bromyrit *m* (Min) bromyrite
Bromzahl *f* bromine number, bromine value
Bromzink *n* zinc bromide
Bromzinn *n* tin bromide
Broncholytikum *n* (Pharm) broncholytic preparation
Brongniardit *m* (Min) brongniardite
Bronze *f* bronze, **gelbe** ~ yellow bronze, tombac powder, **rote** ~ red bronze, copper powder, **weiße** ~ white bronze, silver powder
bronzeartig bronze-like, bronzy
Bronzebad *n* bronze cyanide bath
Bronzebezug *m* bronze coating, bronze plating, bronze sleeve
Bronze-Diabetes *m* (Med) bronze diabetes
Bronzedraht *m* bronze wire
Bronzedruck *m* (Buchdr) bronze printing, metallic printing
Bronzefarbe *f* bronze color
bronzefarbig bronze-colored
Bronzefutter *n* bronze packing
Bronzegießer *m* bronze founder
Bronzegießerei *f* bronze foundry
Bronzeglanz *m* bronze luster, bronzing
Bronzekrankheit *f* (Med) Addison's disease
Bronzelack *m* bronze lacquer, bronze varnish, bronzing lacquer, bronzing liquid
Bronzemetall *n* bronze metal
Bronzepapier *n* bronze paper
Bronzepulver *n* bronze powder
Bronzespießglanz *m* (Min) native antimony bromide
Bronzetinktur *f* bronze liquor, bronzing liquid
Bronzevergoldung *f* gilding by amalgamation of bronze
bronzieren to bronze
Bronzieren *n* bronzing
Bronzierflüssigkeit *f* bronzing liquid
Bronzierpulver *n* bronze powder, bronzing powder
Bronziersalz *n* bronzing salt, antimony trichloride
Bronzierung *f* bronzing
Bronzin *n* bronzine
Bronzit *m* (Min) bronzite
Brookit *m* (Min) brookite
Broparöstrol *n* broparoestrol
Brotgärung *f* panary fermentation
Brotraffinade *f* (Brotzucker) loaf-sugar
Brotteig *m* [bread] dough
Brotzucker *m* loafsugar
Brownsche Molekularbewegung *f* (Phys) Brownian movement

Brownsches Teilchen *n* Brownian particle
Bruch *m* fracture, break, bursting, rupture, (Math) fraction, **echter** ~ (Math) common fraction, simple fraction, **feinfaseriger** ~ finely fibrous fracture, **körniger** ~ granulated fracture, **muscheliger** ~ conchoidal fracture, **unechter** ~ (Math) improper fraction
Brucharbeit *f* energy of fracture
Bruchart *f* method of fracture
bruchartig fracture-like
Bruchausbreitung *f* propagation of fracture
Bruchaussehen *n* appearance of fracture
Bruchbeanspruchung *f* breaking stress
Bruchbeginn *m* initiation of fracture
Bruchbelastung *f* breaking load, maximum load, ultimate stress
Bruchbildung *f* fracture formation
Bruchblei *n* broken lead, scrap lead
Bruchdehnbarkeit *f* elongation at rupture
Bruchdehnung *f* extension at break, breaking tension, elongation at break, elongation at rupture, stretch
Brucheisen *n* broken iron, scrap iron
Bruchentstehung *f* fracture origin
bruchfest crack-proof, breakage-proof, break-resistant, resistant to fracture, (zäh) tenacious, tough
Bruchfestigkeit *f* break resistance, breaking strength, bursting strength, tenacity, tensile strength, toughness, transverse strength
Bruchfestigkeitsgrenze *f* breaking-down point
Bruchfestigkeitsprüfer *m* breaking tester
Bruchfläche *f* fractured surface, surface of fracture, **faserige** ~ fibrous fracture, **irreguläre** ~ irregular or uneven fracture, **muschelige** ~ conchoidal fracture
Bruchflächenaussehen *n* appearance of fracture
Bruchgefahr *f* risk of fracture
Bruchgefüge *n* character of fracture, fracture pattern
Bruchgeschwindigkeit *f* fracture velocity, tear speed
Bruchgewicht *n* fractional weight
Bruchglas *n* broken glass, cullet
Bruchgold *n* broken gold
Bruchgrenze *f* breaking limit or strain, point of maximum load, rupture limit, strength limit, ultimate stress limit
Bruchkoks *m* broken coke
Bruchkraut *n* burstwort
Bruchkupfer *n* broken copper, scrap copper
Bruchlast *f* breaking load, load at break
Bruchlastspielzahl *f* number of load cycles to failure
Bruchmechanik *f* fracture mechanics
Bruchmessing *n* brass scrap, broken brass
Bruchmetall *n* scrap metal, metal scrap
Bruchmodul *m* modulus of rupture

Bruchmoment *n* breaking moment, moment of rupture
Bruchneigung *f* tendency to fracture
Bruchprobe *f* breaking test
Bruchrechnung *f* (Math) fractional mathematics
bruchsicher break-resistant, unbreakable
Bruchsicherheitsfaktor *m* rupture safety factor
Bruchsilber *n* broken silver, scrap silver
Bruchspannung *f* breaking stress, ultimate stress
Bruchstein *m* broken stone, quarry stone
Bruchstelle *f* location of fracture, spot of fracture
Bruchstrich *m* (Math) fractional line
Bruchstück *n* broken piece, fragment, scrap, shred, splinter
Bruchstücke *n pl* fission products
Bruchteil *m* fraction, portion
Bruchvorgang *m* theory of fractures
Bruchwiderstand *m* breaking strength
Brucidin *n* brucidine
Brucin *n* brucine
Brucinchinon *n* brucinequinone
Brucinchlorhydrat *n* brucine hydrochloride
Brucinlösung *f* brucine solution
Brucinnitrat *n* brucine nitrate
Brucinolon *n* brucinolone
Brucinolonsäure *f* brucinolonic acid
Brucinolsäure *f* brucinolic acid
Brucinonsäure *f* brucinonic acid
Brucinsäure *f* brucic acid, brucinic acid
Brucinsulfat *n* brucine sulfate
Brucit *m* (Min) brucite
brüchig brittle, cracky, fragile, friable
Brüchigkeit *f* brittleness, fragility, friability, shortness
Brüchigwerden *n* embrittlement
Brückenatom *n* (Stereochem) bridge atom, bridging atom
Brückenbindung *f* cross linking
Brückenkopf *m* bridge head
Brückenkopfatom *n* (Stereochem) bridgehead atom
Brückenkopf-Carbanion *n* (Stereochem) bridgehead carbanion
Brückenkopf-Radikal *n* (Stereochem) bridgehead radical
Brückenkran *m* bridge crane, overhead travelling crane
Brückenrelais *n* bridge relay
Brückenringstruktur *f* (Stereochem) bridge-ring structure
Brückenringsystem *n* (Stereochem) bridge-ring system
Brückensauerstoff *m* bridging oxygen, connecting oxygen
Brückenschaltung *f* (Elektr) bridge circuit, bridge method
Brückenstecker *m* bridging plug
Brückensystem *n* (Stereochem) bridged system

Brüden *m* vapor
Brüdenhaube *f* vapor hood
Brüdenkondensator *m* evaporator condenser
Brüdenverwertung *f* waste steam utilization
Brüdenwasser *n* condenser water
Brühe *f* liquor, lye, solution
brühen to scald, to seethe
Brühen *n* scalding
Brüherz *n* (Min) yellow copper ore
Brühfaß *n* boiling tub
brühheiß scalding hot
Brühmesser *m* (Lohmesser, Gerb) barkometer
Brühwunde *f* scald
Brünierbeize *f* bronzing pickle
Brüniereisen *n* (Techn) burnisher, burnishing iron
brünieren (Metall) to bronze, to brown, to burnish
Brünieren *n* browning, burnishing
Brünierer *m* burnisher
Brünierstein *m* burnishing stone, bloodstone
Brüstung *f* screen
Brütbarkeit *f* fertility
Brüten *n* (Kerntechnik) breeding
Brüter *m* breeder
Brugnatellit *m* (Min) brugnatellite
brummfrei (Elektr) ripple free
Brummspannung *f* (Elektr) ripple voltage
Brunnen *m* well
Brunnensalz *n* brine salt, well salt
Brunnenwasser *n* spring water, well water
Brunnerit *m* (Min) magnesite
Brunswigit *m* brunsvigite
Brushit *m* (Min) brushite
Brustalant *m* (Bot) elecampane
Brustdrüse *f* mammary gland
Brustöffnung *f* (Metall) hearth hole, hearth opening
Brusttee *m* (Pharm) pectoral tea
Brustwurzel *f* (Bot) angelica root
Brustzacken *m* fore plate
Brutapparat *m* breeding apparatus, incubator
Brutgewinn *m* (Atom) breeding gain
Brutkasten *m* incubator
Brutmantel *m* (Reaktor) blanket
Brutreaktor *m* (Atom) breeder, breeding reactor
Brutschrank *m* incubator
Brutstoff *m* breeder material, fertile material
Brutthermometer *n* incubator thermometer
brutto gross, overall
Bruttoformel *f* gross molecular formula, empirical [molecular] formula
Bruttogewicht *n* gross weight
Bruttopreis *m* gross price
Bruttoreaktion *f* gross reaction, over-all reaction
Bruttotonnengehalt *m* gross tonnage
Brutverhältnis *n* breeding ratio
Brutvorgang *m* breeding
Brutzyklus *m* (Biol) breeding cycle

Bruyèreholz *n* brier wood
Bruzin *n* brucine
Bryogenin *n* bryogenine
Bryonin *n* bryonin
Bryonol *n* bryonol
Bryoresin *n* bryoresin
BT-Kristallschnitt *m* (Krist) BT-cut crystal
Buccobitter *n* diosmine
Buccoblätter *pl* bucco leaves, buchu
Buccocampher *m* buccocamphor, buchu camphor
Buchdruck *m* letterpress printing
Buchdruckerfarbe *f* printer's ink, printing ink
Buchdruckerfirnis *m* printer's varnish
Buchdruckerschwärze *f* printer's ink
Buchdruckfarben *pl* letterpress printing inks
Buche *f* (Bot) beech tree
Buchecker *f* (Bot) beechnut
Bucheckernöl *n* beechnut oil
Buchenasche *f* beechwood ashes
Buchengallapfel *m* beech gall
Buchenholzpech *n* beech pitch
Buchenholzteerkreosot *n* beech tar creosote
Buchenholzteeröl *n* beech tar oil
Buchholzit *m* (Min) bucholzite, fibrolite
Buchöl *n* beechnut oil
Buchse *f* (Elektr) socket
Buchsring *m* axle ring, bushing; packing ring
Buchstabensäuren *pl* letter acids
Buchweizen *m* (Bot) buckwheat
Buckelschweißung *f* projection welding
Bucklandit *m* (Min) bucklandite
Buclizin *n* buclizine
Büchner-Trichter *m* Büchner funnel, (Chem) Buchner funnel
Büchse *f* box, case, (Konservendose) can (Am. E.), tin (Br. E.)
Büchsenfilter *m n* (an Gasmasken) canister filter
Büchsenfleisch *n* canned meat, tinned meat
Büchsengemüse *n* canned vegetables, tinned vegetables
Büchsenmetall *n* bearing metal, bush metal
Büchsenmilch *f* canned milk, evaporated milk, tinned milk
Büchsenöffner *m* can opener
Büchsenstein *m* (Min) iron pyrites
Bügelgriff-Tasche *f* bag with mo[u]lded handle
Bügeln *n* ironing
Bügelverschluß *m* lever closure
Bügelzurichtung *f* (Leder) plated finish
Bündel *n* pack, (Licht) beam
Bündelschraube *f* clamp screw
Bündeltetrode *f* beam tetrode
Bündelung *f* beaming, bunching, concentrating, grouping, (Opt) focus[s]ing
Bündelungselektrode *f* (Radar) focusing electrode
Bündelungslinse *f* focusing lens
Bürette *f* buret, burette

Bürettenausflußspitze *f* burette tip
Bürettengestell *n* burette stand
Bürettenhahn *m* burette cock, burette valve
Bürettenhalter *m* burette holder
Bürettenkappe *f* burette head
Bürettenquetschhahn *m* burette clip
Bürettenständer *m* buret[te] stand
bürgen to give security [for], to vouch [for]
Bürgschaft *f* guarantee
Bürste *f* brush
Bürstenabzug *m* brush proof, galley [proof]
Bürstenkontakt *m* (Elektr) brush contact
Bürstenstreichmaschine *f* brush spreader
Bürstenstreichverfahren *n* brush spreading
Büschelentladung *f* brush discharge
Bütte *f* vat, tub
Büttenfärbung *f* pulp coloring, vat coloring
büttengefärbt vat colored
Büttenleimung *f* pulp sizing, vat sizing
Büttenpapier *n* handmade paper, vat paper
Büttenpresse *f* vat press
Büttenruß *m* chimney soot, lamp black
Bufagin *n* bufagin
Bufalin *n* bufalin
Bufogenin *n* bufogenin
Bufotalin *n* bufotalin
Bufotalsäure *f* bufotalic acid
Bufotenidin *n* bufotenidine
Bufotenin *n* bufotenine
Bufothionin *n* bufothionine
Bufotoxin *n* bufotoxine
Bulbocapnin *n* bulbocapnine
Bullatenon *n* bullatenone
Buna *n* (HN) buna, synthetic rubber
Bunagummi *n* buna rubber
Bundesforschungsanstalt *f* Federal Research Institute
Bundesgesundheitsamt *n* Federal Health Office
Bunker *m* bunker, bin, shelter
Bunkeraufsatzfilter *n* bin filter (in pneumatic conveyor systems)
Bunkerbandsystem *n* bunker belt system
Bunkerkohle *f* bunker coal
Bunkerwaage *f* bin scale
Bunsenbrenner *m* Bunsen burner
Bunsenelement *n* (Elektr) Bunsen cell
Bunsenflamme *f* Bunsen flame
Bunsenin *n* bunsenine, bunsenite
Bunsenit *m* (Min) bunsenite
Bunsenkohle *f* retort carbon, carbon for the Bunsen cell
Bunsenschirm *m* Bunsen screen
Bunsenstativ *n* retort stand
Bunsenventil *n* Bunsen valve
Buntachat *m* (Min) variegated agate
Buntätze *f* colored discharge
Buntbleiche *f* bleaching of colored goods
Buntbleierz *n* (Min) green lead ore, variegated pyromorphite

Buntdruck *m* (Buchdr) color printing, (Erzeugnis) color print
Buntebürette *f* Bunte gas burette
Buntesalze *pl* Bunte salts
buntfärben to dye, to color, to stain
buntgewebt colored woven, fancy woven
Buntkupfererz *n* (Min) purple copper ore, bornite, variegated copper ore
Buntkupferkies *m* (Min) variegated copper ore
Buntmetall *n* nonferrous metal
Buntpapier *n* colored paper, (Glanzpapier) glossy paper
Buntsandstein *m* (Min) variegated sandstone
buntsandsteinartig poecilitic
buntschillernd iridescent
Buntton *m* motley clay
Buphanidrin *n* buphanidrine
Buphanisin *n* buphanisine
Buphanitin *n* buphanitine
Burawoy-Keton *n* Burawoy ketone
burnettisieren to burnettize, to impregnate wood with zinc chloride
Burseracin *n* burseracine
Busch *m* bush
Buschbohne *f* dwarf runner bean
Bushy-Stunt-Virus *m* (Biol) bushy stunt virus
Bustamit *m* (Min) bustamite
Busulfan *n* busulfan
Busycon *n* busycon
Butacain *n* butacaine
Butadien *n* butadiene, bivinyl, divinyl
Butadienchinon *n* butadiene quinone
Butadienkautschuk *m* butadiene caoutchouc
Butadiensulfon *n* butadiene sulfone
Butalbital *n* butalbital
Butan *n* butane
Butanol *n* butanol, butyl alcohol
Butanolacetonvergärung *f* butanol acetone fermentation
Butanolgewinnungsanlage *f* butanol production plant
Butantetracarbonsäure *f* tetracarboxy butane
Buten *n* butene, butylene
Butenandt-Keton *n* Butenandt ketone
Butenandt-Säure *f* Butenandt acid
Butenolid *n* butenolide
Butenyl- butenyl
Butesin *n* butesin, n-butyl-p-aminobenzoate
Buthalital *n* buthalital
Butin *n* butine, butyne
Butindiol *n* butine diol, butyne diol
Butolan *n* butolane
Butrin *n* butrin
Butropin *n* butropine
Butter *f*, **ungesalzene** ~ sweet butter
butterähnlich buttery, butyraceous
butterartig buttery, butyraceous
Butterbrotpapier *n* wax paper
Butterei *f* creamery

Butterfett *n* butter fat
Buttergelb *n* butter yellow, annatto
butterhaltig containing butter, butyraceous
Buttermesser *m* butyrometer
Buttermilch *f* butter milk
Buttermilcherz *n* (Min) earthy hornsilver
Butteröl *n* (obs) butyric oil
Butterpersäure *f* perbutyric acid, peroxybutyric acid
Buttersäure *f* butyric acid
Buttersäureanhydrid *n* butyric anhydride
Buttersäuregärung *f* butyric fermentation
Buttersäureradikal *n* butyryl, radical of butyric acid
buttersauer butyrate
Buttersilber *n* (Min) earthy hornsilver
Butterwasserwaage *f* balance for testing butter
Butyl- butyl
Butylacetat *n* butyl acetate
Butylacetylaceton *n* butylacetylacetone
Butylaldehyd *m* butanal, butyl aldehyde, butyral, butyraldehyde
Butylaldol *n* butyraldol
Butylalkohol *m* butanol, butyl alcohol
Butylamin *n* butylamine
Butylbenzoat *n* butyl benzoate
Butylbutyrat *n* butyl butyrate
Butylcellosolve *f* butyl cellosolve
Butylchloralhydrat *n* butylchloral hydrate, 2,2,3-trichlorobutane-1,1-diol, trichlorobutylidene glycol
Butylcyanid *n* butyl cyanide
Butylen *n* butylene, butene
Butylenglykol *n* butylene glycol
Butylensulfid *n* butylene sulfide, tetrahydrothiophene
Butylgruppe *f* butyl group
Butylhalogenid *n* butylhalide
Butylhypnal *n* butylhypnal
Butyljodid *n* butyl iodide
Butylkautschuk *m* butyl rubber
Butyllactat *n* butyl lactate
Butylmalonat *n* butyl malonate, dibutyl malonate
Butylmalonester *m* butylmalonic ester
Butylmalonsäure *f* butylmalonic acid
Butylphenyl *n* butylphenyl
Butylphthalat *n* butyl phthalate, dibutyl phthalate
Butylresorcin *n* butyl resorcinol
Butylsalicylat *n* butyl salicylate
Butylsilicon *n* butyl silicone
Butyltartrat *n* butyl tartrate
Butyltrichlorsilan *n* butyl trichlorosilane
Butyn *n* butine, butyne
Butyraldehyd *m* butyraldehyde, butanal, butyral
Butyramid *n* butyramide, butanamide
Butyramidin *n* butyramidine
Butyrat *n* butyrate

Butyrin *n* butyrin, glycerol tributyrate, tributyrin
Butyrobetain *n* butyrobetaine
Butyroin *n* butyroin, 5-hydroxy-4-octanone
Butyrolacton *n* butyrolactone, 4-hydroxybutanoic acid lactone
Butyrometer *n* butyrometer
Butyron *n* butyrone, 4-heptanone, dipropyl ketone
Butyronitril *n* butyronitrile, propylcyanide
Butyropersäure *f* perbutyric acid, peroxybutyric acid
Butyrophenon *n* butyrophenone, phenyl-n-propyl ketone
Butyrospermadien *n* butyrospermadiene
Butyryl- butyryl
Butyrylresorcin *n* butyrolresorcinol
Buxin *n* buxine
Byakangelicin *n* byakangelicin
Byakangelicol *n* byakangelicol
Byssolith *m* byssolite, asbestoid
Bytownit *m* (Min) bytownite

C

Cabrerit *m* (Min) cabrerite
Cabreuvaöl *n* cabreuva oil
Cacaobohne *f* cocoa bean
Cacaobutter *f* cocoa butter
Cacaofett *n* cocoa butter
Caceresphosphat *n* caceres phosphate
Cachou *n* cachou, catechu, cutch
Cachoubraun *n* catechu brown
cachoutieren to dye with catechu
Cacogenin *n* cacogenin
Cadalin *n* cadalene, 3,8-dimethyl-5-isopropylnaphthalene
Cadaverin *n* cadaverine
Cadeöl *n* cade oil, juniper tar oil
Cadetsche Flüssigkeit *f* Cadet's fuming liquid
Cadinan *n* cadinane
Cadinen *n* cadinene
Cadinöl *n* cade oil
Cadinol *n* cadinol
Cadmat *n* cadmate
Cadmieren *n* cadmium plating
Cadmin *n* cadmine
Cadmium *n* cadmium (Symb. Cd)
Cadmiumacetat *n* cadmium acetate
Cadmiumäthylendiaminchelat *n* cadmium ethylenediamine chelate
Cadmiumamalgam *n* cadmium amalgam
Cadmiumarsenat *n* cadmium arsenate
Cadmiumarsenid *n* cadmium arsenide
Cadmiumblende *f* (Min) greenockite
Cadmiumbromid *n* cadmium bromide
Cadmiumcyanwasserstoff *m* cyanocadmic acid
Cadmiumelektrode *f* cadmium electrode
Cadmiumfarbe *f* cadmium color
Cadmiumfluorid *n* cadmium fluoride
Cadmiumgehalt *m* cadmium content
Cadmiumgelb *n* cadmium yellow, cadmium sulfide
Cadmiumgoldchlorid *n* cadmium aurichloride, cadmium chloroaurate(III)
cadmiumhaltig cadmiferous, containing cadmium
Cadmiumhydroxid *n* cadmium hydroxide
Cadmiumjodid *n* cadmium iodide
Cadmiumkontrollstab *m* (Atom) cadmium control rod
Cadmiumlegierung *f* cadmium alloy
Cadmiumlösung *f* cadmium solution
Cadmiummetall *n* cadmium metal, metallic cadmium
Cadmiummetasilikat *n* cadmium metasilicate
Cadmium-Normalelement *n* Standard Weston cadmium cell
Cadmiumoxidhydrat *n* cadmium hydroxide
Cadmiumsalicylat *n* cadmium salicylate
Cadmiumsalz *n* cadmium salt

Cadmiumselenid *n* cadmium selenide
Cadmiumsuboxid *n* cadmium suboxide
Cadmiumsulfat *n* cadmium sulfate
Cadmiumsulfid *n* cadmium sulfide
Cadmiumtellurid *n* cadmium telluride
Cadmiumverbindung *f* cadmium compound
Cadmiumwolframat *n* cadmium tungstate
Cadmiumzement *m* cadmium cement
Cadogel *n* cadogel
Caerulin *n* soluble indigo blue
Caeruloplasmin *n* ceruloplasmin
Caesium *n* cesium, caesium (Symb. Cs), cesium (Symb. Cs), (siehe Cäsium)
Cäsiumalaun *m* cesium alum, cesium aluminum sulfate
Cäsiumazid *n* cesium azide
Cäsiumcarbonat *n* cesium carbonate
Cäsiumchlorid *n* cesium chloride
Cäsiumchloridgitter *n* cesium chloride lattice
Cäsiumchloridstruktur *f* cesium chloride structure
Cäsiumhydrid *n* cesium hydride
Cäsiumjodat *n* cesium iodate
Cäsiumjodid *n* cesium iodide
Cäsiumlichtbogen *m* cesium arc
Cäsiumnitrat *n* cesium nitrate
Cäsiumphotozelle *f* cesium cell
Cäsiumsalicylaldehyd *m* cesium salicylaldehyde
Cäsiumsilicofluorid *n* cesium fluosilicate, cesium silicofluoride
Cäsiumsulfat *n* cesium sulfate
Cäsiumsulfid *n* cesium sulfide
Cäsiumsuperoxid *n* cesium peroxide
Cäsiumverbindung *f* cesium compound
Cafestadien *n* cafestadiene
Caffein *n* caffeine
Caffeinsalz *n* caffeine salt
Cailcedrarinde *f* cailcedra bark
Caincasäure *f* cahincic acid, caincic acid, caincin
Caincawurzel *f* (Bot) cahinca root, cainca root
Caincin *n* caincin, cahincic acid, caincic acid
Cairngorm *m* (Min) cairngorm [stone]
Caissonkrankheit *f* (Med) caisson disease
Cajeputgeist *m* spirit of cajeput
Cajeputin *n* cajeputene
Cajeputöl *n* cajeput oil, cajuput oil, cajuputol, eucalyptol
Calabarbohne *f* calabar bean
Calabarbohnenextrakt *m* extract of calabar bean
Calabarin *n* calabarine
Calabarol *n* calabarol
Calafatit *m* calafatite
CA-Lagerung *f* controlled atmosphere storage
Calainsäure *f* calaic acid
Calambakholz *n* calambac

Calamen n calamene
Calamendiol n calamenediol
Calamenen n calamenene
Calameon n calameone, calmus camphor
Calamin m calamine
Calamit m calamite
Calaverit m (Min) calaverite
Calcibram n calcibram
Calciferol n calciferol, vitamin D$_2$
Calcifikation f calcification
calcifizieren to calcify
Calciglycin n calciglycine
Calcihyd n calcihyde
Calcimeter n calcimeter
Calcimurit m (Min) calcimurite
Calcination f calcination
calcinierbar calcinable
calcinieren (s. a. kalzinieren) to calcine
Calcinieren n calcination, calcining
Calcinierherd m calcining hearth
Calciniertopf m calcining crucible
Calciniertrommel f calcinating furnace
Calcinierung f calcination
Calcioferrit m (Min) calcioferrite
Calciostrontianit m (Min) calciostrontianite
Calcit m (Min) calcite (s. a. Kalzit)
Calcium n calcium (Symb. Ca), **kohlensaures** ~ calcium carbonate
Calciumacetat n calcium acetate
Calciumacetosalicylicum n calcium acetosalicylicum, calcium acetylsalicylate
Calciumacetylsalicylat n calcium acetylsalicylate
Calciumaluminat n calcium aluminate
Calciumaminoäthylphosphonsäure f calcium aminoethylphosphonic acid
Calciumaminoäthylsulfonsäure f calcium aminoethylsulfonic acid
Calciumarsenid n calcium arsenide
Calciumarsenit n calcium arsenite
Calciumascorbat n calcium ascorbate
Calciumbenzoat n calcium benzoate
Calciumbernsteinsäurechelat n calcium succinate chelate
Calciumbichromat n calcium bichromate, calcium dichromate
Calciumbisulfit n acid calcium sulfite, calcium bisulfite, calcium hydrogen sulfite
Calciumborat n calcium borate
Calciumbromid n calcium bromide
Calciumcarbid n calcium carbide, calcium acetylide
Calciumcarbonat n calcium carbonate, carbonate of lime, (Min) limestone
Calciumchelat n calcium chelate
Calciumchelatbildungsvermögen n calcium chelating power
Calciumchlorat n calcium chlorate
Calciumchlorid n calcium chloride
Calciumchromat n calcium chromate
Calciumcitrat n calcium citrate, citrate of lime
Calciumcyanamid n calcium cyanamide, lime nitrogen, nitrolim[e]
Calciumcyanid n calcium cyanide
Calciumcyclamat n calcium cyclamate
Calciumferricyanid n calcium ferricyanide
Calciumferrocyanid n calcium ferrocyanide
Calciumfluorid n calcium fluoride, calcium difluoride, fluoride of lime
Calciumformiat n calcium formate
Calciumgehalt m calcium content
Calciumglycerophosphat n calcium glycerophosphate
Calciumhippurat n calcium hippurate
Calciumhydrid n calcium hydride
Calciumhydrogencarbonat n acid calcium carbonate, calcium bicarbonate, calcium hydrogen carbonate
Calciumhydrosulfid n calcium hydrogen sulfide, calcium hydrosulfide
Calciumhydrosulfit n calcium hydrosulfite
Calciumhydroxid n calcium hydroxide, hydrated lime, slaked lime
Calciumhypobromit n calcium hypobromite
Calciumhypochlorit n calcium hypochlorite
Calciumhypophosphit n calcium hypophosphite
Calciumjodid n calcium iodide
Calciumlactat n calcium lactate
Calciumlegierung f calcium alloy
Calciumlignosulfat n calcium lignosulfate
Calciummalonat n calcium malonate
Calciummalonsäurechelat n calcium malonate chelate
Calciummetall n calcium metal, metallic calcium
Calciummetaphosphat n calcium metaphosphate
Calciummetasilikat n calcium metasilicate
Calciummethylamindiessigsäurechelat n calcium methylamine diacetate chelate
Calciummonosulfür n (obs) calcium monosulfide
Calciumnaphtholsulfonat n calcium naphtholsulfonate
Calciumnitrat n calcium nitrate
Calciumnitrid n calcium nitride
Calciumnitrit n calcium nitrite
Calciumnitroessigsäure f calcium nitroacetate
Calciumorthoplumbat n calcium orthoplumbate
Calciumorthosilikat n calcium orthosilicate
Calciumoxalat n calcium oxalate
Calciumoxid n calcium oxide, burnt lime
Calciumpantothenat n calcium pantothenate
Calciumperborat n calcium perborate
Calciumpermanganat n calcium permanganate
Calciumperoxid n calcium peroxide
Calciumphenolsulfonat n calcium phenolsulfonate
Calciumphosphat n calcium phosphate
Calciumphosphid n calcium phosphide
Calciumphospholactat n calcium phospholactate

Calciumplumbat *n* calcium plumbate, lime plumbate
Calciumpräparat *n* calcium preparation
Calciumpyrophosphat *n* calcium pyrophosphate
Calciumrhodanid *n* calcium rhodanide
Calciumsaccharat *n* calcium saccharate
Calciumsalicylat *n* calcium salicylate
Calciumsalz *n* calcium salt
Calciumselenat *n* calcium selenate
Calciumsilicid *n* calcium silicide
Calciumsilicofluorid *n* calcium fluosilicate, calcium silicofluoride
Calciumsilicotitanit *n* calcium silicotitanite, sphene, titanite
Calciumsilikat *n* calcium silicate, (Min) wollastonite
Calciumspiegel *m* calcium level
Calciumstearat *n* calcium stearate
Calciumsuccinat *n* calcium succinate
Calciumsulfat *n* calcium sulfate, (Dihydrat) gypsum, (Sesquihydrat) plaster of Paris, (wasserfrei) anhydrite
Calciumsulfhydrat *n* calcium hydrogen sulfide, calcium hydrosulfide, calcium sulfhydrate
Calciumsulfid *n* calcium sulfide
Calciumsulfit *n* calcium sulfite
Calciumsuperoxid *n* calcium peroxide
Calciumteilchen *n* calcium particle
Calciumthiosulfat *n* calcium thiosulfate
Calciumtitriplex *n* calcium titriplex
Calciumverbindung *f* calcium compound
Calciumwolframat *n* calcium tungstate
Calcoferrit *m* (Min) calcoferrite
Caldariomycin *n* caldariomycin
Caledonischbraun *n* Caledonian brown
Caledonit *m* (Min) caledonite
Calendulin *n* calendulin
Caliche *m* caliche, chile saltpeter
Californit *m* (Min) californite
Californium *n* californium (Symb. Cf)
Calinlegierung *f* calin alloy
Calisayarinde *f* calisaya bark, chinona
Callistephin *n* callistephin[e]
Callitrol *n* callitrol
Callopisminsäure *f* callopisminic acid
Calmonal *n* calmonal
Calomel *n* calomel, mercurous chloride, mercury(I) chloride
Calomelelektrode *f* calomel electrode
Calomelol *n* colloidal calomel
Calophyllolid *n* calophyllolide
Calophyllsäure *f* calophyllic acid
Calorescenz *f* calorescence
Calorie *f* calorie, calory (s. a. Kalorie)
Calorienwert *m* calorific value
Calorimeter *n* calorimeter
Calorimetergefäß *n* calorimeter vessel
Calorimeterwasser *n* calorimeter water
Calorimetrie *f* calorimetry

calorimetrieren to measure with a calorimeter
Calorimetrierung *f* determination of the calorific value
calorimetrisch calorimetric
calorisch caloric, thermal, thermic
Calotropagenin *n* calotropagenin
Calsil *n* calsil
Calsolenöl *n* calsolene oil
Calspirin *n* calspirine
Calutron *n* calutron
Calvacin *n* calvacin
Calycanin *n* calycanine
Calycanthin *n* calycanthine
Calycanthosid *n* calycanthoside
Calycin *n* calycine
Calycopterin *n* calycopterin
Calythron *n* calythrone
Cambium *n* cambium
Cambopinensäure *f* cambopinenic acid
Cambopinonsäure *f* cambopinonic acid
Cambriebinde *f* cambric bandage
Camellin *n* camellin
Campbellverfahren *n* Campbell method
Campecheholz *n* campeachy wood, logwood
Campestadienol *n* campestadienol
Campestanol *n* campestanol
Campesterin *n* campesterol
Camphalin *n* camphaline
Camphan *n* camphane
Camphanol *n* camphanol
Camphansäure *f* camphanic acid
Camphen *n* camphene, camphine
Camphenchlorhydrin *n* camphene chlorohydrin
Camphenglykol *n* camphene glycol
Camphenhydrid *n* camphene hydride
Camphenilan *n* camphenilane
Camphenilanaldehyd *m* camphenilaic aldehyde
Camphenilansäure *f* camphenilaic acid, camphenilanic acid
Camphenilen- camphenilene
Camphenilol *n* camphenilol
Camphenilolsäure *f* camphenilolic acid
Camphenilon *n* camphenilone
Camphenilonsäure *f* camphenilonic acid
Camphenilsäure *f* camphenilic acid
Camphenlauronolsäure *f* camphenelauronolic acid
Camphenol *n* camphenol
Camphenolsäure *f* camphenolic acid
Camphenon *n* camphenone
Camphenonsäure *f* camphenonic acid
Camphensäure *f* camphenic acid
Campher *m* camphor (s. a. Kampfer), **behandeln mit** ~ to camphorate
Campheraldehydsäure *f* camphoraldehydic acid
Campheramidsäure *f* camphoramic acid
campherartig camphoraceous, camphor-like
Campherazin *n* camphorazine
Campherblumen *pl* flowers of camphor

Campherchinon *n* camphor quinone, 2,3-diketocamphane, 3-ketocamphor
Campherdestillationsanlage *f* camphor distillation plant
Campherdibromid *n* camphor dibromide, dibromated camphor, dibromocamphor
Campherdichlorid *n* camphor dichloride, dichlorated camphor, dichlorocamphor
Campherersatz *m* camphor substitute
Campherersatzmittel *n* camphor substitute
Camphergeist *m* camphorated spirit
campherhaltig camphorated, camphorous, containing camphor
Campherharz *n* enosmite
Campherholz *n* camphor wood
Campherholzöl *n* camphor wood oil
campherieren to camphorate
Campherliniment *n* camphorated oil, camphor liniment
Camphermilch *f* camphorated emulsion
Camphernitrilsäure *f* camphoric acid nitrile, camphornitrilo acid
Campheröl *n* camphorated oil, camphor liniment
Campherol *n* campherol
Campheroxalsäure *f* camphorated oxalic acid
Campherphoron *n* camphor phorone
Campherpinakon *n* camphor pinacone
Camphersäure *f* camphoric acid
Camphersäureanhydrid *n* camphoric anhydride
Camphersäuremethylester *m* methyl camphorate
camphersauer camphorate
Campherseife *f* camphor soap
Campherspiritus *m* camphorated spirit, spirit of camphor
Camphersulfonsäure *f* camphorsulfonic acid
Campherwasser *n* camphor water
Campherwein *m* camphorated wine
Campherweinsäure *f* camphovinic acid
campherweinsauer camphovinic
Camphidin *n* camphidine
Camphidon *n* camphidone
Camphocamphersäure *f* camphocamphoric acid
Camphocarbonsäure *f* camphocarboxylic acid
Camphoceensäure *f* camphoceenic acid
Camphogen *n* camphogen, cymene
Camphoglykuronsäure *f* camphoglycuronic acid
Camphoid *n* camphoid
Camphoindol *n* camphoindole
Camphoketen *n* camphoketene
Campholacton *n* campholactone
Campholalkohol *m* camphol alcohol
Campholamin *n* campholamine
Campholansäure *f* campholic acid
Campholen *n* campholene
Campholenaldehyd *m* campholene aldehyde
Campholensäure *f* campholenic acid
Campholid *n* campholide
Campholonsäure *f* campholonic acid
Campholsäure *f* campholic acid
Campholytsäure *f* campholytic acid
Camphon *n* camphone, campholene aldehyde
Camphonansäure *f* camphonanic acid
Camphonen *n* camphonene
Camphonensäure *f* camphonenic acid
Camphonitril *n* camphonitrile
Camphonolsäure *f* camphonolic acid
Camphononsäure *f* camphononic acid
Camphorat *n* camphorate
Camphoren *n* camphorene
Camphoron *n* camphorone
Camphoronsäure *f* camphoronic acid
Camphotamid *n* camphotamide
Camphren *n* camphrene
Camphrensäure *f* camphrenic acid
Camphyl- camphyl
Camphylamin *n* camphylamine
Camphylglykol *n* camphylglycol
Camphylsäure *f* camphylic acid
Campnospermonol *n* campnospermonol
Camptonit *m* camptonite
Camwood *n* camwood
Canadabalsam *m* canada balsam
Canadin *n* canadine
Canadiniumhydroxid *n* canadinium hydroxide
Canadinol *n* canadinol
Canadinolsäure *f* canadinolic acid
Canadinsäure *f* canadic acid
Canadol *n* canadol, impure hexane
Canadolsäure *f* canadolic acid
Canalin *n* canaline
Canangaöl *n* cananga oil
Canarienfarbe *f* canarin, canary color
Canarin *n* canarin
Canavalin *n* canavaline
Canavanin *n* canavanine
cancerös cancerous
cancerogen cancerogenic, carcinogenic
Cancerogenese *f* carcinogenesis
Cancrinit *m* (Min) cancrinite
Candelillapflanze *f* (Bot) candelilla plant
Candelillawachs *n* candelilla wax
Candelnußöl *n* candlenut oil
Candicin *n* candicine
Candiolin *n* candioline
Candogenin *n* candogenin
Canellarinde *f* (Bot) canella bark
Canelle *f* (Bot) canella, white cinnamon
Canellenöl *n* canella oil
Canescin *n* canescine
Canescinsäure *f* canescic acid
Canfieldit *m* (Min) canfieldite
Cannabidiol *n* cannabidiol
Cannabin *n* cannabin
Cannabinol *n* cannabinol
Cannabinon *n* cannabinone
Cannabiscetin *n* cannabiscetin
Cannastärke *f* canna starch

Cannelkohle *f* cannel coal
Cannizzarosche Reaktion *f* Cannizzaro reaction
Cannogenin *n* cannogenin
Cantharen *n* cantharene, dihydroxylene
Cantharenol *n* cantharenol
Cantharidenäther *m* cantharic ether
Cantharidencampher *m* cantharidin
Cantharidentinktur *f* cantharides tincture
Cantharidenvergiftung *f* cantharides poisoning
Cantharidin *n* cantharidin[e]
Cantharinsäure *f* cantharidic acid
Cantharolsäure *f* cantharolic acid
Cantharsäure *f* cantharic acid
Canthaxanthin *n* canthaxanthin
Canthin *n* canthine
Cantonit *m* (Min) cantonite
Capauridin *n* capauridine
Capaurin *n* capaurine
Caperatsäure *f* caperatic acid
capillar s. kapillar
Capillen *n* capillene
Capnoidin *n* capnoidine
Caporcianit *m* caporcianite
Cappelinit *m* (Min) cappelenite
Caprarsäure *f* capraric acid
Capreomycin *n* (Pharm) capreomycin
Capriblau *n* Capri blue
Caprin *n* caprin, glycerol caprate, glyceryl tricaprate
Caprinat *n* caprate
Caprinon *n* caprinone
Caprinsäure *f* capric acid
Caprinsäureäthylester *m* ethyl caprate
Caprinsäureisoamylester *m* isoamyl caprate
Caprinsäuremethylester *m* methyl caprate
caprinsauer caprate
Caproamid *n* caproamide
Caproin *n* caproin, glycerol caproate, glyceryl tricaproate
Caprolactam *n* caprolactam
Caprolacton *n* caprolactone
Capron *n* caprone, 6-hen-decanone
Capronaldehyd *m* caproaldehyde, hexanal
Capronamid *n* capronamide
Capronfett *n* caproin, glycerol caproate
Capronitril *n* amyl cyanide, capronitrile
Capronoin *n* capronoine
Capronophenon *n* capronophenone
Capronsäure *f* caproic acid, capronic acid, hexanoic acid
Capronsäureäthylester *m* ethyl caproate
Capronsäuremethylester *m* methyl caproate
capronsauer caproate
Caproyl- caproyl
Capryl- capryl, hexyl
Caprylaldehyd *m* caprylaldehyde
Caprylalkohol *m* capryl alcohol
Caprylamin *n* caprylamine
Caprylat *n* caprylate

Caprylen *n* caprylene, octene
Capryliden *n* caprylidene, 1-octyne
Caprylon *n* caprylone
Capryloyl- capryloyl
Caprylsäure *f* caprylic acid, n-octanoic acid
Caprylsäureäthylester *m* ethyl caprylate
Capsacutin *n* capsacutine
Capsaicin *n* capsaicin
Capsanthin *n* capsanthin
Capsanthon *n* capsanthone
Capsicin *n* capsicine
Capsicol *n* capsicol
Capside *pl* capsids
Capsomer *n* capsomere
Capsorubin *n* capsorubin
Capsorubon *n* capsorubone
Capsularin *n* capsularin
Captan *n* captan
Captol *n* captol
Caput mortuum *n* caput mortuum, red ferric oxide
Caracurin *n* caracurine
Carajurin *n* carajurin
Caramel *m* caramel (s. a. Karamel)
Caramelan *n* caramelan
Caramelen *n* caramelene
Caramelgeruch *m* caramel smell
Caramelin *n* caramelin
Caramelisierung *f* caramelization
Caramin *n* caramine
Caramiphen *n* caramiphen
Caran *n* carane
Caranin *n* caranine
Carannabalsam *m* caran[n]a balsam
Carannagummi *n* caran[n]a resin
Carannaharz *n* caran[n]a resin
Carapafett *n* carap[a] oil
Carapaöl *n* carap[a] oil
Carbachol[in] *n* carbachol
Carbäthoxy- carbethoxy
Carbäthoxycyclopentanon *n* carbethoxycyclopentanone
Carbäthoxyl- carbethoxyl
Carballylsäure *f* carballylic acid
Carbamat *n* carbamate
Carbamid *n* carbamide
Carbamidsäure *f* carbamic acid, amidocarbonic acid, aminoformic acid
Carbaminat *n* carbamate, carbaminate
Carbaminsäure *f* carbamic acid
Carbaminsäureäthylester *m* ethyl carbamate, urethane
carbaminsauer carbamate, carbaminate
Carbamyl- carbamyl
Carbanil *n* carbanil, phenyl isocyanate
Carbanilid *n* carbanilide, diphenylurea
Carbanilsäure *f* carbanilic acid, phenylcarbamic acid

Carbanion *n* carbanion, ~ **an einem Brückenkopf** (Stereochem) carbanion at bridgehead, ~ **durch Sprengung einer C-C-Bindung** (Stereochem) carbanion by cleavage of a C-C-bond, **intermediäres** ~ intermediary carbanion
Carbarson *n* carbarsone
Carbazid *n* carbazide
Carbazin *n* carbazine
Carbazinsäure *f* carbazinic acid
Carbazochrom *n* carbazochrome
Carbazol *n* carbazole
Carbazolblau *n* carbazole blue
Carbazolgelb *n* carbazole yellow
Carbazolkalium *n* potassium carbazolate
Carbazylsäure *f* carbazylic acid
Carben *n* carbene
Carbeniat-Ion *n* carbeniate ion
Carbeniat-Struktur *f* carbeniate structure
Carbenium-Ion *n* carbenium ion
Carbenium-Kation *n* carbenium cation
Carbenium-Salz *n* carbenium salt
Carbenium-Struktur *f* carbenium structure
Carbid *n* carbide, carbonide
Carbidgas *n* acetylene, ethyne
Carbidkohle *f* carbide carbon
Carbidnest *n* carbide pocket
Carbidofen *m* carbide furnace
Carbidschmelze *f* carbide fusion
Carbidverfahren *n* carbide process
Carbimazol *n* carbimazole
Carbimid *n* carbimide, isocyanate
Carbindigo *n* carbindigo
Carbinol *n* carbinol
Carbinoxamin *n* carbinoxamine
Carbitol *n* carbitol
Carboanhydrase *f* (Biochem) carboanhydrase, carbonic anhydrase
Carboazotin *n* carboazotine
Carbobenzoxychlorid *n* carbobenzoxy chloride
Carbobenzoxyderivat *n* carbobenzoxy derivative
Carbobenzoxyglutaminsäure *f* carbobenzoxy glutamic acid
Carbobenzoxysynthese *f* carbobenzoxy synthesis
Carbocyanin *n* carbocyanine
carbocyclisch carbocyclic
Carbodiimid *n* carbodiimide
Carbodiphenylimid *n* carbodiphenylimide, diphenylcarbodiimide
Carbodynamit *n* carbodynamite
Carboferrit *m* (Min) carboferrite
Carbohydrase *f* (Biochem) carbohydrase
Carbohydrazid *n* carbohydrazide
Carbol *n* carbolic acid, phenol
Carbolfuchsin *n* carbol fuchsin, Ziehl's stain
Carbolglycerinseife *f* carbolic glycerol soap
Carbolignum *n* carbolignum, wood charcoal
Carbolin *n* carboline
Carbolineum *n* carbolineum
Carbolineumanstrich *m* coat of carbolineum
Carbolkalk *m* calcium phenolate, calcium phenate, carbolated lime, phenate of lime
Carbolöl *n* carbolated oil, carbolic acid oil, phenol oil
Carbolpulver *n* carbolic powder
Carbolsäure *f* carbolic acid, phenic acid, phenol, **tränken mit** ~ to carbolate, to carbolize
Carbolsäurebad *n* carbolic acid bath
Carbolsäureflasche *f* carbolic acid carboy
Carbolsäurelösung *f* solution of carbolic acid
Carbolsäureseife *f* carbolic acid soap
Carbolsalbe *f* (Pharm) phenol ointment
carbolsauer carbolate, phenate, phenolate
Carbolschwefelsäure *f* sulfocarbolic acid, sulfophenic acid
carbolschwefelsauer sulfocarbolate, sulfophenate
Carbolseife *f* carbolic soap
Carbolwasser *n* aqueous solution of phenol
Carbolxylol *n* carbolxylene
Carbomethoxy- carbomethoxy
Carbomethoxyl- carbomethoxyl
Carbomycin *n* carbomycin
Carbonado *m* (Min) carbonado, anthracite diamond, bort
Carbonat *n* carbonate, salt or ester of carbonic acid
Carbonathärte *f* carbonate hardness, hardness due to carbonate(s)
carbonathaltig containing carbonate
Carbonation *f* carbonation, carbonization
Carbonat-Ion *n* carbonate ion
Carbonatochelat *n* carbonato chelate
Carbonatotetramin *n* carbonato tetramine
Carbonatwasser *n* carbonate water
Carboneum *n* carboneum
Carbonid *n* carbide, carbonide
carbonieren to carbonate, to carbonize
Carbonisation *f* carbonation, carbonization
Carbonisationswirkung *f* carbonizing action
Carbonisieranlage *f* carbonizing plant
Carbonisieranstalt *f* carbonizing works
Carbonisierbad *n* carbonizing bath
carbonisieren to carbonate, to carbonize
Carbonisieren *n* carbonation, carbonization, conversion into carbon; (Wolle) carbonizing
Carbonisierflüssigkeit *f* carbonizing liquor
Carbonisiertemperatur *f* carbonizing temperature
Carbonisierung *f* carbonization, (mit CO_2 versetzen) carbonation
Carbonit *m* carbonite
Carbonium-Ion *n* carbonium ion
Carbonsäure *f* carboxylic acid
Carbonschwarz *n* carbon black
Carbonyl- carbonyl
Carbonylammoniak *n* carbonyl ammonia
Carbonylchlorid *n* carbonyl chloride, phosgene
Carbonylgruppe *f* carbonyl group

Carbonylierung f carbonylation
Carbonylsauerstoff m carbonyl oxygen
Carbonylsulfid n carbonyl sulfide
Carbonylverbindung f carbonyl compound
Carbopyrrid n carbopyrride
Carboran n carborane
Carborundum n carborundum (s. a. Karborundum)
Carbosilan n carbosilane
Carbosiliciumoxid n siloxicon
Carbostyril n carbostyril, 2-hydroxyquinoline
Carbothialdin n carbothialdine
Carbothiocyanin n carbothiocyanin
Carboxoniumsalz n carboxonium salt
Carboxybenzoylessigsäure f carboxylbenzoylacetic acid
Carboxyl- carboxy, carboxyl
Carboxylase f (Biochem) carboxylase
Carboxylgruppe f carboxyl group
Carboxylierung f carboxylation
Carboxyllignin n carboxyllignin
Carboxymethylcellulose f carboxymethyl cellulose
Carboxypeptidase f (Biochem) carboxypeptidase
Carboxyphenylacetonitril n carboxyphenylacetonitrile
carbozyklisch carbocyclic
Carbromal n carbromal
Carbür n (obs) carbide
Carburation f carburization, carburation, carbureting
Carburationsapparat m carburetor
Carburator m carbureter, carburetor, carburetter (Br. E.)
carburieren to carburize, to carburate, to carburet
Carburieren n carburating, carburizing
Carburierung f carburization, carburation, carbureting
Carburometer n carburometer
Carbylamin n carbylamine, isocyanide
Carbylsulfat n carbyl sulfate
carcinogen carcinogenic, cancerogenic
Carcinogen n carcinogen
Carcinom n (Med) carcinoma
Cardanolid n cardanolide
Cardiaka pl (Pharm) cardiac stimulants
Cardiazol n cardiazol
Cardiolipin n cardiolipin
Cardiotonin n cardiotonine
Cardol n cardol
Caren n carene
Carisson n carissone
Carliermaschine f rounding machine
Carlinaoxid n carlina oxide
Carlinsäure f carlic acid
Carlossäure f carlosic acid
Carlsäure f carlic acid
Carmin m, **blauer** ~ indigo carmine, solid blue

Carmin n carmine, crimson
Carminativum n (Pharm) carminative
Carminazarin n carmine azarine
Carminazarinchinon n carmine azarine quinone
carminblau indigo-carmine colored
Carminblau n indigo carmine, solid blue
Carminfarbe f carmine
Carminlack m carmine lake
Carminnaphtha n carmine naphtha
Carminochinon n carminoquinone
Carminon n carminone
Carminpapier n carmine paper
carminrot carmine, carmine red, ~ **färben** to dye in carmine
Carminsäure f carminic acid
carminsauer carminate
Carminspat m (Min) carmine spar, carminite
Carmoisin n carmoisine, crimson
Carnallit m (Min) carnallite
Carnat m (Min) kaolinite
Carnaubaersatz m carnauba substitute
Carnaubasäure f carnaubic acid
Carnaubawachs n carnauba wax
Carnaubon n carnaubon
Carnaubylalkohol m carnaubyl alcohol
Carnegieit m (Min) carnegieite
Carnegin n carnegine
Carneol m (Min) carnelian, carneol
Carnin n carnine
Carnitin n carnitine
Carnitin-Fettsäure-Acyltransferase f (Biochem) carnitine fatty acyl transferase
Carnomuscarin n carnomuscarine
Carnosidase f carnosidase
Carnosin n carnosine
Carnosinase f carnosinase
Carnotit m (Min) carnotite
Carnotscher Kreisprozeß m (Therm) Carnot cycle
Carnotscher Wirkungsgrad m Carnot efficiency
Carolathin m carolathine
Carolinsäure f carolinic acid
Carolsäure f carolic acid
Caron n carone
Caronsäure f caronic acid
Carosche Säure f Caro's acid, peroxysulfuric acid, persulfuric acid
Carotin n carotene, carotin, provitamin A
Carotingelb n carotin yellow
Carotinoid n carotenoid, carotinoid
Carotol n carotol
Carpain n carpaine
Carpamol n carpamol
Carpamsäure f carpamic acid
Carpilin n carpiline
Carpilinsäure f carpilic acid
Carpogenin n carpogenin
Carpyrinsäure f carpyrinic acid
Carquéjol n carquejol

Carrabiose – Catecholamin

Carrabiose *f* carrabiose
Carrachenmoos *n* carragheen moss, Irish moss
Carraramarmor *m* Carrara marble
Carthamin *n* carthamin, carthamic acid, safflor red
Carthaminextrakt *m* extract of carthamin
Carthaminsäure *f* carthamic acid, carthamin
Carthamintinktur *f* carthamin tincture
Carthamon *n* carthamone
Carube *f* caro-bean, locust bean
Carubinkaffee *m* carubin coffee
Carvacrol *n* carvacrol, 2-methyl-5-isopropylphenol
Carvacrolchinon *n* carvacrol quinone
Carvacrolphthalein *n* carvacrol phthalein
Carvacrolsulfophthalein *n* carvacrol sulfophthalein
Carvacryl- carvacryl
Carvacrylamin *n* carvacrylamine, cymidine
Carvacrylarabinosid *n* carvacrylarabinoside
Carvacrylxylosid *n* carvacrylxyloside
Carvelon *n* carvelone
Carven *n* carvene, d-limonene
Carvenen *n* carvenene
Carvenol *n* carvenol
Carvenolid *n* carvenolide
Carvenolsäure *f* carvenolic acid
Carvenon *n* carvenone
Carveol *n* carveol
Carvestren *n* carvestrene
Carvin *n* carvine
Carviolin *n* carviolin
Carvol *n* carvol, carvone
Carvomenthan *n* carvomenthane
Carvomenthol *n* carvomenthol
Carvomenthon *n* carvomenthone
Carvon *n* carvol, carvone
Carvonborneol *n* carvone borneol
Carvoncampher *m* carvone camphor
Carvopinon *n* carvopinone
Carvotanaceton *n* carvotanacetone
Carvyl- carvyl
Carvylamin *n* carvylamine
Caryin *n* caryin
Caryinit *m* (Min) caryinite
Caryolan *n* caryolane
Caryolysin *n* caryolysine
Caryophyllan *n* caryophyllane
Caryophyllen *n* caryophyllene
Caryophyllin *n* caryophyllin, oleanolic acid
Caryophyllinsäure *f* caryophyllinic acid
Caryophyllol *n* caryophyllol
Caryoterpin *n* caryoterpine
Carzenid *n* carzenide
Cascararinde *f* (Bot) cascara bark
Cascarasagrada *f* cascara sagrada
Cascarilla *f* (Bot) cascarilla, sweetwood bark
Cascarillaextrakt *m* cascarilla extract
Cascarillaöl *n* cascarilla oil

Cascarillarinde *f* (Bot) cascarilla bark
Cascarillin *n* cascarilline
Cascarol *n* cascarol
Casealutin *n* casealutine
Caseanin *n* caseanine
Caseïn *n* caseïn
Caseïnammoniak *n* ammonium caseînate
Caseïnat *n* caseînate
Caseindeckfarbe *f* casein coating color
Caseïneisen *n* iron caseînate
Caseinkunststoff *m* caseîn plastic
Caseïnnatrium *n* sodium caseînate
Caseïnogen *n* caseînogen
Caseïnsalbe *f* caseîn ointment
Caseïnseide *f* caseîn rayon
Casimiroedin *n* casimiroedine
Casimiroin *n* casimiroin
Casimiroinol *n* casimiroinol
Casimiroitin *n* casimiroitine
Casimirosäure *f* casimiroic acid
Casing-Verfahren *n* (Kunststoffe) Casing process
Cassaidinsäure *f* cassaidic acid
Cassain *n* cassaine
Cassainsäure *f* cassaic acid
Cassamin *n* cassamine
Cassaminsäure *f* cassamic acid
Cassansäure *f* cassanic acid
Cassawastärke *f* tapioca
Casselergelb *n* cassel yellow
Cassia *f* (Bot) cassia
Cassiaöl *n* cassia oil
Cassiarinde *f* (Bot) cassia bark
Cassienblütenöl *n* cassia oil
Cassienmark *n* cassia pulp
Cassinit *m* cassinite
Cassinsäure *f* cassic acid
Cassiopeium *n* cassiopeium, lutecium
Cassiterit *m* (Min) cassiterite (s. a. Kassiterit)
Cassuispurpur *m* cassius gold purple
Castanit *m* (Min) castanite
Castelagenin *n* castelagenine
Castelamarin *n* castelamarin
Castelin *n* casteline
Castelnaudit *m* castelnaudite
Castillianerseife *f* castile soap
Castillit *m* castillite
Castnerprozeß *m* Castner's process
Castorat *n* castorate
Castoreum *n* castoreum
Castorin *n* castorin
Castorinsäure *f* castoric acid
castorinsauer castorate
Castornuß *f* castor bean
Castoröl *n* castor oil
Catechin *n* catechin
Catechinsäure *f* catechin, catechuic acid
Catechol *n* catechin, pyrocatechol
Catecholamin *n* catecholamine

Catechu *n* catechu, cashoo, cutch
Catechudiamin *n* catechudiamine
Catechugerbsäure *f* catechutannic acid, catechu tannin
Catechusäure *f* catechin, catechol, catechuic acid
Catechutinktur *f* catechu tincture
Catechuulmin *n* catechu-ulmine
Catenarin *n* catenarin
Cathartinsäure *f* cathartic acid
Cathartomannit *m* cathartomannitol
Cathin *n* cathine
Catin *n* catine
Cativinsäure *f* cativic acid
Cat-Ox-Verfahren *n* Cat-Ox-process
Cauchy-Multiplikation *f* Cauchy multiplication (of convergent series)
Caulophyllin *n* caulophylline
Caulophyllosapogenin *n* caulophyllosapogenin
Caulophyllosaponin *n* caulophyllosaponin
Caulosapogenin *n* caulosapogenin
Caulosaponin *n* caulosaponin
Causticum *n* caustic, causticum
Cayennepfeffer *m* capsicum, cayenne pepper
Cayennezimt *m* cayenne cinnamon
Cearakautschuk *m* ceara rubber, pernambuco rubber
Cederncampher *m* cedar-camphor
Cedernharz *n* cedrium
Cedernholzöl *n* cedar oil
Cedernöl *n* cedar oil
Cedren *n* cedrene
Cedrengurjunen *n* cedrene guriunene
Cedrenol *n* cedrenol
Cedrenolsäure *f* cedrenolic acid
Cedrensäure *f* cedrenic acid
Cedrin *n* cedrin
Cedrinsäure *f* cedric acid
Cedriret *n* cedriret
Cedrium *n* cedrium
Cedrol *n* cedrene camphor, cedrol
Cedrolsäure *f* cedrolic acid
Cedron *n* cedron
Celaxanthin *n* celaxanthin
Cellase *f* (Biochem) cellase
Cellobial *n* cellobial
Cellobiit *m* cellobiitol
Cellobionsäure *f* cellobionic acid
Cellobiose *f* cellobiose, cellose
Cellobiuronsäure *f* cellobiuronic acid
Celloidin *n* celloidin, collodion wool
Cellon *n* cellon, tetrachlorethane
Cellonlack *m* cellon varnish, cellulose acetate lacquer
Cellophan *n* (HN) cellophane
Cellose *f* cellobiose, cellose
Cellosolve *f* cellosolve
Cellotetrose *f* cellotetrose
Cellotriose *f* cellotriose
Cellotropin *n* cellotropin

Cellulase *f* (Biochem) cellulase
Cellulith *m* cellulith
Celluloid *n* celluloid, xylonite, zylonite
Celluloidin *n* celluloidine
Celluloidlack *m* celluloid varnish
Celluloidpapier *n* celluloid paper
Celluloidschale *f* celluloid dish
Cellulose *f* cellulose
Celluloseabkömmling *m* cellulose derivative
Celluloseacetat *n* acetylcellulose, cellulose acetate
Celluloseacetatflocken *f pl* cellulose acetate flakes
Celluloseacetatfolie *f* cellulose acetate sheeting
Celluloseacetatlack *m* cellulose acetate laquer
Celluloseacetobutyrat *n* cellulose acetate butyrate
Cellulosebestandteil *m* cellulose constituent
Cellulosederivat *n* cellulose derivative
Celluloseester *m* cellulose ester
Cellulosefolie *f* cellophane
Cellulosehydrat *n* cellulose hydrate
Cellulose-Ionenaustauscher *m* cellulose ion exchanger
Cellulosenitrat *n* cellulose nitrate
Cellulosepulver *n* cellulose powder
Cellulosetetracetat *n* cellulose tetracetate
Cellulosethiocarbonat *n* cellulose thiocarbonate
Cellulosetriacetat *n* cellulose triacetate
Cellulosexanthogenat *n* cellulose xanthate, cellulose xanthogenate
Celosiaöl *n* celosia oil
Celsian *m* (Min) celsian
Celsiusgrad *m* Celsius degree, degree centigrade
Celsiusskala *f* centigrade (scale)
Celsiusthermometer *n* centigrade thermometer
Celtium *n* (obs) celtium, hafnium
Celtrobiose *f* celtrobiose
Cembren *n* cembrene
cementieren to cement, to caseharden (s. a. zementieren)
Cementit *m* (Min) cementite
Centaur *n* centaur
Centaureidin *n* centaureidin
Centaurein *n* centaureine
Centaurin *n* centaurin
Centipoise *n* (Maß) centipoise
Centistoke *n* (Maß) centistoke
Centralit *n* centralit
Centriole *f* (Biol) centriole, central body, centrosome
Ceolat *n* ceolate
Ceolatpulver *n* ceolate powder
Cephaelin *n* cephaeline
Cephalanthin *n* cephalanthin
Cephalin *n* cephalin, kephalin
Cephalinsäure *f* cephalic acid
Cephalosporin *n* cephalosporin
Cer *n* cerium (Symb. Ce)

Cerain – Chalkanthit

Cerain *n* ceraine
Cerainsäure *f* ceraic acid
Ceramid *n* ceramide
Cerammoniumnitrat *n* ceric ammonium nitrate
Cerammoniumsulfat *n* ceric ammonium sulfate
Ceran *n* cerane, isohexacosane
Cerargyrit *m* (Min) cerargyrite, hornsilver
Cerasin *n* cerasin, cerasein
Cerasit *m* (Min) cerasite
Cerat *n* cerate, cerotate, salt or ester of cerotic acid
Ceratin *n* ceratin, keratin
Ceratit *m* ceratite
Ceratpapier *n* cerate paper
Cerberin *n* cerberin
Cercarbid *n* cerium acetylide, cerium carbide
Cerchlorid *n* ceric chloride, cerium chloride
Cerdioxid *n* ceric oxide, cerium dioxide, cerium(IV) oxide
Cereawachs *n* carnauba wax
Cerebrin *n* cerebrin
Cerebrinsäure *f* cerebric acid
Cerebron *n* cerebron
Cerebronsäure *f* cerebronic acid
Cerebrose *f* cerebrose
Cerebrosid *n* cerebroside, galactoside
Cerebrosterin *n* cerebrosterol
Cereit *m* (Min) cereite
Cerenox *n* cerenox
Cererz *n* (Min) cerium ore, cerite
Ceresin *n* ceresin, earth wax, mineral wax, ozocerite, ozokerite
Ceresit *n* ceresit
Cerfluorid *n* cerium fluoride
Cer(III)-salz *n* cerium(III) salt, cerous salt
Cer(III)-verbindung *f* cerium(III) compound, cerous compound
cerimetrisch cerimetric
Cerin *n* cerin
Cerioxid *n* ceric oxide, cerium dioxide, cerium(IV) oxide
Cerisulfat *n* ceric sulfate, cerium(IV) sulfate
Cerit *m* (Min) cerite
Ceriverbindung *f* (Cer(IV)-Verbindung) ceric compound, cerium(IV) compound
Cer(IV)-salz *n* ceric salt, cerium(IV) salt
Cer(IV)-verbindung *f* ceric compound, cerium(IV) compound
Cermetall *n* cerium metal, metallic cerium
Cernitrat *n* ceric nitrate
Cernuin *n* cernuin
Cernuosid *n* cernuoside
Cerolein *n* cerolein
Ceropten *n* ceroptene
Cerosin *n* cerosin, cerosinyl cerosate
Ceroten *n* cerotene
Cerotin *n* cerotin, ceryl cerotate
Cerotinsäure *f* cerotic acid, cerotin, cerotinic acid, heptacosanoic acid

Ceroverbindung *f* (Cer(III)-Verbindung) cerium(III) compound, cerous compound
Ceroxalat *n* cerium oxalate
Ceroxid *n* ceria, ceric oxide, cerium dioxide, cerium(IV) oxide
Cersilicid *n* cerium silicide
Cersulfat *n* cerium sulfate
Cersulfid *n* cerium sulfide
Cerulin *n* cerulin
Cerussit *m* (Min) cerussite
Cervantit *m* (Min) cervantite
Ceryl- ceryl
Cerylalkohol *m* ceryl alcohol, 1-hexacosanol, cerotin, cerotol
Cesol *n* cesol
Cetan *n* cetane
Cetanzahl *f* cetane number
Ceten *n* cetene, cetylene, hexadecylene
Cetin *n* cetin, cetylcetylate
Cetobemidon *n* cetobemidone
Cetoleinsäure *f* cetoleic acid
Cetrarin *n* cetrarin, cetrarinic acid
Cetrarsäure *f* cetraric acid
Cetyl- cetyl, hexadecyl
Cetylalkohol *m* cetyl alcohol, 1-hexadecanol, ethal, hexadecyl alcohol
Cetylen *n* cetylene, cetene, hexadecylene
Cetyljodid *n* 1-iodohexadecane, cetyl iodide
Cetylsäure *f* cetylic acid, palmitic acid
Cevadin *n* cevadine, veratrine
Cevagenin *n* cevagenine
Cevan *n* cevane
Cevanthridin *n* cevanthridine
Cevin *n* cevine
Cevinilsäure *f* cevinilic acid
Ceylanit *m* (Min) ceylanite
Ceylonmoos *n* (Bot) Ceylon moss
Ceylonöl *n* Ceylon oil
Ceylonrinde *f* (Bot) Ceylon cinnamon
Ceylonrubin *m* (Min) Ceylon ruby
Ceylonzimt *m* Ceylon cinnamon
Chabasit *m* chabazite, phacolite
Chaconin *n* chaconine
Chacotriose *f* chacotriose
Chakranin *n* chakranine
Chaksin *n* chaksine
Chaksinsäure *f* chaksinic acid
Chalcanthit *m* (Min) chalcanthite
Chalcedon *m* (Min) chalcedony
chalcedonartig chalcedonic
Chalcedonchalcit *m* (Min) chalcedonic chalcite
chalcedonhaltig chalcedoniferous
Chalcit *m* (Min) chalcite
Chalcographie *f* chalcography
Chalcopyrit *m* (Min) chalcopyrite
Chalcose *f* chalcose
Chalilith *m* chalilite
Chalinasterin *n* chalinasterol
Chalkanthit *m* (Min) chalcanthite

chalkographisch chalcographic
Chalkolamprit *m* (Min) chalcolamprite
Chalkolith *m* (Min) chalcolite, copper uranite
Chalkom *n* chalcome
Chalkomenit *m* (Min) chalcomenite
Chalkon *n* chalcone
Chalkophanit *m* (Min) chalcophanite
Chalkophyllit *m* (Min) chalcophyllite
Chalkopyrit *m* (Min) chalcopyrite
Chalkosiderit *m* (Min) chalcosiderite
Chalkosin *m* chalcocite, chalcosine
Chalkosphäre *f* chalcosphere
Chalkostibit *m* (Min) chalcostibite
Chalkotrichit *m* (Min) chalcotrichite, plush copper
Chalkpyrit *m* (Min) chalcopyrite, yellow copper ore
Chalmersit *m* (Min) chalmersite
Chalybit *m* (Min) chalybite
Chamäleon *n* chameleon, (Min) chameleon mineral
Chamäleonlösung *f* potassium permanganate solution
Chamazulen *n* chamazulene
Chamenol *n* chamenol
Chaminsäure *f* chaminic acid
Chamois *n* (Leder) chamois, shammy leather
Chamomillaalkohol *m* chamomilla alcohol
Chamomillaester *m* chamomilla ester
Chamomillol *n* chamomillol
Chamosit *m* (Min) chamo[i]site
Chamotte *f* chamotte, fireclay
Chamottestein *m* firebrick
Chamottetiegel *m* fireclay crucible
Chamotteton *m* fireclay
Chamotteziegel *m* firebrick
Champacol *n* champacol
Champagner *m* champagne
Champagnerwurz *f* (Bot) ipecacuanha root, white veratrum
Chamsäure *f* chamic acid
Chanarcillit *m* chanarcillite
Chanoclavin *n* chanoclavine
charakterisieren to characterize
Charakterisierungsfaktor *m* (Therm) characterization factor
Charakteristik *f* (Math) characteristic
Charakteristikenverfahren *n* characteristics method
Charakteristikum *n* characteristic feature
Chardonnetseide *f* Chardonnet silk, (Text) viscose rayon
Charge *f* batch, charge, load
Chargenautoklav *m* batch retort
Chargenbetrieb *m* batch production
Chargengröße *f* charge quantity
Chargenmischer *m* batch mixer
Chargenprozeß *m* batch process
Chargenzeit *f* charging period

Charge-Transfer-Bande *f* charge-transfer band
Charge-Transfer-Komplex *m* charge-transfer complex
Chargierbühne *f* charging floor, charging platform
chargieren to charge, to load
Chargierfolge *f* sequence of charging
Chargiergeschwindigkeit *f* charging rate
Chargierkran *m* charging crane, loading crane
Chargierkübel *m* charging bucket
Chargierleistung *f* charging rate
Chargierlöffel *m* charging spoon, charging peel
Chargiermaschine *f* charging machine, charging device
Chargiermulde *f* charging box
Chargiermuldentransportwagen *m* charging box carrier
Chargierschwengel *m* balanced peel
Chargiertür *f* feeding door
Chargiervorrichtung *f* charging device, charging machine
Chargierzeit *f* charging period
Charltonweiß *n* Charlton white, Griffith white, lithopone, Orr white
Charpie *f* lint
Chassis *n* chassis, trough
Chassisrahmen *m* chassis frame
Chaudronelement *n* Chaudron's thermopile
Chaulmoograbutter *f* chaulmoogra butter
Chaulmoograöl *n* chaulmoogra oil, gynocardia oil
Chaulmoograsäure *f* chaulmoogric acid
Chaulmosulfon *n* chaulmosulfone
Chavibetol *n* chavibetol
Chavicin *n* chavicine
Chavicinsäure *f* chavicic acid, chavicinic acid
Chavicol *n* chavicol
Chavosot *n* chavosot
Chebulsäure *f* chebulic acid
Cheddit *n* cheddite
Cheilanthifolin *n* cheilanthifoline
Cheiranthin *n* cheirantin
Cheiranthussäure *f* cheiranthic acid
Cheirolin *n* cheiroline
Chelalbin *n* chelalbine
Chelat *n* chelate, ~ **zweiwertiger Metalle** *n pl* bivalent metal chelate
Chelatbildner *m* chelating agent
Chelatbildung *f* chelation
Chelatdonatorgruppe *f* chelate donor group
Chelatkomplex *m* chelate complex
Chelatometrie *f* chelatometry
Chelatring *m* chelate ring
chelatstabilisierend chelate stabilizing
Chelatstabilität *f* chelate stability
Chelatstabilitätskonstante *f* chelate stability constant
Chelerythrin *n* chelerythrine
Cheleutit *m* (Min) cheleutite

Chelidamsäure *f* chelidamic acid
Chelidonin *n* chelidonine, swallowstone
Chelidonsäure *f* chelidonic acid
Chellolglucosid *n* chellolglucosid
Chelonit *m* (Min) chelonite
Chemie *f* chemistry, **analytische** ~ analytical chemistry, **angewandte** ~ applied chemistry, **anorganische** ~ inorganic chemistry, **Astro-** ~ astro-chemistry, **Bio-** ~ biochemistry, **experimentelle** ~ experimental chemistry, **Gerichts-** ~ judicial chemistry, **Histo-** ~ histo-chemistry, **Mikro-** ~ micro-chemistry, **Nahrungsmittel-** ~ food-chemistry, **organische** ~ organic chemistry, **pharmazeutische** ~ pharmaceutical chemistry, **Photo-** ~ photo-chemistry, **physikalische** ~ physical chemistry, **physiologische** ~ physiological chemistry, **präparative** ~ preparative chemistry, **Stereo-** ~ stereo-chemistry, **theoretische** ~ theoretical chemistry, **toxikologische** ~ toxicological chemistry
Chemiefaden *m* chemical fiber, man-made filament, synthetic fiber
Chemiefaser *f* artificial fiber, chemical fiber, man-made fiber, synthetic fiber
Chemiefasererzeugung *f* manufacture of man-made fibers, manufacture of synthetic fibers
Chemieingenieur *m* chemical engineer
Chemie-Ingenieur-Technik *f* chemical engineering, chemical and process engineering
Chemielaborant *m* laboratory assistent
Chemiephysik *f* chemical physics, chemico-physics
Chemiespinnfaser *f* man-made fiber, staple fiber
Chemigraphie *f* chemigraphy, photoengraving
Chemikalie *f* chemical
chemikalienbeständig resistant to chemicals
Chemiker *m* chemist
Chemilithotrophie *f* chemolithotrophy
Chemilumineszenz *f* chemiluminescence
Chemilumineszenzkontinuum *n* chemiluminescence continuum
Chemischblau *n* chemic blue
chemisch gebunden chemically combined or fixed, ~ **rein** chemically pure, ~ **widerstandsfähig** chemically resistant
Chemischgelb *n* cassel yellow
Chemischgrün *n* chemic green
Chemisch-Reinigen *n* dry cleaning
Chemischrot *n* venetian red
Chemismus *m* chemical mechanism, chemism
Chemisorption *f* chemisorption, chemosorption
Chemitypie *f* chemitypy
Chemokoagulation *f* chemocoagulation
Chemopathologie *f* (Med) pathological chemistry
Chemoprophylaxe *f* (Med) chemoprophylaxis
Chemoreflex *m* chemoreflex
chemoresistent chemoresistant
Chemoresistenz *f* chemoresistance
Chemorezeption *f* chemoreception
Chemorezeptor *m* chemoreceptor
chemosensibel chemosensitive
Chemosensibilität *f* chemosensitivity
Chemosynthese *f* chemosynthesis
Chemotaxis *f* chem[i]otaxis
Chemotaxonomie *f* chemotaxonomy
Chemotechniker *m* chemical technician, laboratory technician
Chemotherapie *f* chemotherapy
Chemotropismus *m* chemotropism
Chemurgie *f* chemurgy
Chenevixit *m* (Min) chenevixite
Chenocholensäure *f* chenocholenic acid
Chenocholsäure *f* chenocholic acid
Chenodesoxycholsäure *f* chenodeoxycholic acid
Chessylit[h] *m* (Min) chessylite, azurite, hydrated basic copper carbonate, mountain blue
Chesterlith *m* chesterlite
Chevillieren *n* chevilling, glossing, polishing, softening process
Chevilliermaschine *f* glossing machine
Chiastolith *m* (Min) chiastolite
Chicagosäure *f* chicago acid
Chichipegenin *n* chichipegenin
Chiclafluavil *n* chiclafluavil
Chiclalban *n* chiclalbane
Chicorsäure *f* chicoric acid
Chigadmaren *n* chigadmarene
Childrenit *m* (Min) childrenite
Chileit *m* (Min) chileite
Chilekupfer *n* **in Barren** *m pl* chile copper bars
Chilenit *m* (Min) chilenite
Chilesalpeter *m* chile saltpeter, chile niter, sodium nitrate
Chillagit *m* (Min) chillagite
Chimaphyllin *n* chimaphilin
Chimylalkohol *m* chimyl alcohol
Chinaalkaloid *n* cinchona alkaloid
Chinabase *f* cinchona base, quinine base
Chinabaum *m* cinchona tree
Chinablau *n* Chinese blue
Chinacetophenon *n* quinacetophenone
Chinaclay *m* china clay
Chinacridin *n* quinacridine
Chinacridon *n* quinacridone
Chinaessenz *f* essence of cinchona
Chinäthylin *n* quinethyline
Chinaextrakt *m* aqueous extract of cinchona
Chinagerbsäure *f* quinotannic acid
Chinagras *n* cambric grass, ramie
Chinaldin *n* quinaldine
Chinaldinäthyljodid *n* quinaldinium ethyl iodide, N-ethylquinaldinium iodide
Chinaldinat *n* quinaldinate

Chinaldinblau n quinaldine blue
Chinaldinsäure f quinaldic acid, quinaldinic acid
Chinalizarin n quinalizarin
Chinamicin n quinamicine
Chinamin n quinamine
Chinan n quinane, desoxyquinine
Chinanisol n quinanisol
Chinanthridin n quinanthridine
Chinarinde f cinchona bark, Peruvian bark
Chinarindenbaum m cinchona bark tree
Chinarindenextrakt m extract of Peruvian bark
Chinarindenrot n cinchona red
Chinarot n cinchona red
Chinasäure f quinic acid, chinic acid, kinic acid
chinasauer quinate, chinate, kinate
Chinasilber n China silver
Chinatinktur f tincture of cinchona
Chinatoxin n cinchona toxin
Chinawachs n Chinese wax
Chinawein m quinine wine
Chinawurzel f China root
Chinazolin n quinazoline
Chinazolon n quinazolone
Chindolin n quindoline
Chinen n quinene
Chineonal n chineonal, quinine diethylbarbiturate
Chinesischgrün n Chinese green
Chinesischrot n Chinese red
Chinetum n quinetum
Chinhydron n quinhydrone
Chinhydronelektrode f quinhydrone electrode
Chinicin n quinicine
Chinid n chinide
Chinidamin n quinidamine
Chinidan n quinidane
Chinidin n quinidine
Chinidinsulfat n quinidine sulfate
chinieren to cloud, to print chiné, to weave chiné
chiniert (Text) chiné
Chinierung f printing chiné, weaving chiné
Chinin n quinine
Chininacetat n quinine acetate
Chininäthylcarbonat n quinine ethyl carbonate
Chininalkaloid n quinine alkaloid
Chininanisat n quinine anisate
Chininantimonat n quinine antimonate
Chininarsenat n quinine arsenate
Chininaspirin n quinine aspirin
Chininbisulfat n acid quinine sulfate, quinine bisulfate, quinine hydrogen sulfate
Chininbromhydrat n quinine hydrobromide
Chininchlorhydrat n quinine hydrochloride
Chininchlorid n quinine hydrochloride
Chininchromat n quinine chromate
Chinincitrat n quinine citrate
Chininden n quinindene
Chinindol n quinindole
Chinindolin n quinindoline
Chininferricyanid n quinine ferricyanide
Chininhydrobromid n quinine hydrobromide
Chininhydrochlorid n quinine hydrochloride
Chininhydrojodid n quinine hydroiodide
Chininlactat n quinine lactate
Chininnitrat n quinine nitrate
Chininon n quininone
Chininphosphat n quinine phosphate
Chininsäure f quinic acid, chinic acid, kinic acid
Chininsalicylat n quinine salicylate
Chininsuccinat n quinine succinate
Chininsulfat n quinine sulfate
Chininvalerianat n quinine valerate
Chiniofon n chiniofon
Chinisatin n quinisatin
Chinisatinsäure f quinisatinic acid
Chinisocain n quinisocaine
Chinit n 1,4-cyclohexanediol, quinite, quinitol
Chinizarin n quinizarin
Chinizaringrün n quinizarin green
Chinodimethan n quinodimethane
chinoid quinoid
Chinoidin n quinoidine
Chinol n quinol, hydroquinone
Chinolin n quinolin[e]
Chinolinaldehyd m quinoline aldehyde
Chinolinalkaloid n quinoline alkaloid
Chinolinbase f quinoline base
Chinolinblau n quinoline blue, cyanine
Chinolincarbonsäure f quinoline carboxylic acid
Chinolinchinon n quinoline quinone
Chinolinderivat n quinoline derivative
Chinolinfarbstoff m quinoline dye
Chinolingelb n quinoline yellow
Chinolinhydrochlorid n quinoline hydrochloride
Chinoliniumhydroxid n quinolinium hydroxide
Chinolinol n quinolinol
Chinolinrot n quinoline red
Chinolinsäure f quinolinic acid, pyridine-2,3-dicarboxylic acid
Chinolinsalicylat n quinolin salicylate
Chinolinstrychenon n quinoline strychenone
Chinolinsulfat n quinoline sulfate
Chinolizidin n quinolizidine
Chinolizidon n quinolizidone
Chinolizin n quinolizine
Chinolizon n quinolizone
Chinolmethyläther m quinol methyl ether
Chinologe m quinologist
Chinologie f quinology
Chinolon n quinolone
Chinolyl- quinolyl, quinoyl-
Chinolylpyridin n quinoylpyridine
Chinomethan n quinomethane
Chinon n benzoquinone, quinone
Chinondiimin n quinone diimine
Chinondioxim n quinone dioxime
Chinonimin n quinonimine

Chinonmonoxim *n* quinone monoxime
Chinophthalon *n* quinophthalone
Chinopyrin *n* quinopyrine
Chinosol *n* quinosol, chinosol, potassium hydroxyquinoline sulfate
Chinoticin *n* quinoticine
Chinotidin *n* quinotidine
Chinotin *n* quinotine
Chinotinon *n* quinotinone
Chinotoxin *n* quinotoxine, chinotoxine
Chinotropin *n* quinotropine, urotropine quinate
Chinova *n* quinova, kinova
Chinovabitter *n* quinova bitter, quinovin
Chinovasäure *f* quinovic acid
Chinoven *n* quinovene
Chinovin *n* quinovin, chinovin
Chinovose *f* quinovose
Chinoxalin *n* quinoxaline, phenpiazine, quinazine
Chinoxanthen *n* quinoxanthene
Chinoyl *n* quinoyl, quinolyl, quinone group
Chinrhodin *n* quinrhodin
Chintz *m* chintz
Chintzkalander *m* chintzing calender
Chinuclidin *n* quinuclidine
Chiolith *m* (Min) chiolite
Chiralität *f* chirality
Chiralkol *n* chiralkol
Chirurgie *f* surgery
chirurgisch surgical
Chitarsäure *f* chitaric acid
Chitenin *n* chitenine
Chitin *n* chitin
Chitinase *f* chitinase
Chitobiose *f* chitobiose
Chitonsäure *f* chitonic acid
Chitopyrrol *n* chitopyrrol
Chitosamin *n* chitosamine, glucosamine
Chitosan *n* chitosan
Chitosazon *n* chitosazone
Chitose *f* chitose
Chiviatit *m* (Min) chiviatite
Chladnit *m* (Min) chladnite
Chloanthit *m* (Min) chloanthite
Chlor *n* chlorine, **aktives** ~ active chlorine
Chlorableitung *f* chlorine outlet
Chloracetamid *n* chloroacetamide, chloroethanamide
Chloracetat *n* chloroacetate
Chloracetobrenzcatechin *n* chloroacetylpyrocatechol, chloroacetyl-1,2-benzenediol
Chloracetol *n* chloroacetol, chloropropanolone
Chloraceton *n* chloroacetone
Chloracetyl- chloroacetyl, chloroacetyl
Chloracetylaceton *n* chloroacetylacetone
Chloracetylchlorid *n* chloroacetyl chloride, chloroacetyl chloride

Chloracetylglycylglycin *n* chloroacetylglycylglycine
chlorähnlich chlorinous, like chlorine
Chloräpfelsäure *f* chloromalic acid
Chloräthylen *n* chloroethylene, chlorethene, vinyl chloride
Chloräthylenchlorid *n* 1,1,2-trichlorethane, chloroethylene chloride, as-trichloroethane
Chloräthyliden *n* chloroethylidene, ethylidene chloride
Chloral *n* chloral, trichloroacetaldehyde, trichloroethanal
Chloralalkoholat *n* chloral alcoholate
Chloralaun *m* chloralum
Chloralcoffein *n* chloral caffeine
Chloralformamid *n* chloralformamide, chloralamide
Chloralhydrat *n* chloral hydrate
Chloralid *n* chloralide
Chloralimid *n* chloralimide
Chloralkali *n* alkali chloride
Chloralkalielektrolyse *f* chlorine-alkali electrolysis
Chlorallyl- chlorallyl
Chloralose *f* chloralose
Chloraluminit *m* (Min) chloraluminite
Chloraluminium *n* aluminum chloride
Chloralurethan *n* chloral urethane, uralin
Chloralvergiftung *f* chloral poisoning
Chlorambucil *n* chlorambucil
Chloramid *n* chloral formamide, chloramide
Chloramin *n* chloramine
Chloraminbraun *n* chloramine brown
Chloraminfarbstoff *m* chloramine dye
Chloramingelb *n* chloramine yellow
Chloraminotoluol *n* chloroaminotoluene
Chlorammonium *n* ammonium chloride, sal ammoniac
Chlorammoniumlösung *f* ammonium chloride solution, solution of sal ammoniac
Chloramphenicol *n* chloramphenicol
Chloramyl *n* amyl chloride
Chloranil *n* chloranil, tetrachloro-p-benzoquinone
Chloranilin *n* chloroaniline, chloroaminobenzene, chlorophenylamine
Chloranilinhydrochlorid *n* chloroaniline hydrochloride
Chloranilsäure *f* chloranilic acid, 2,5-dichloro-3,6-dihydroxy quinone
Chloranlage *f* chlorine plant
Chloranol *n* chloranol, tetrachloroquinol
Chloranthrachinon *n* chloroanthraquinone
Chlorantimon *n* antimony chloride, antimony trichloride
Chlorapatit *m* (Min) chlorapatite
Chlorargyrit *m* (Min) chlorargyrite, argentum cornu
Chlorarsen *n* arsenic chloride

Chlorarsenik *n* arsenic chloride, arsenic trichloride, arsenous chloride
chlorartig chlorinous, like chlorine
Chlorastrolith *m* (Min) chlorastrolite
Chlorat *n* chlorate, salt or ester of chloric acid
Chloratätze *f* chlorate discharge
Chloratanilinschwarzbase *f* chlorate aniline black base
Chloratranorin *n* chloratranorin
Chloratsprengstoff *m* chlorate explosive
Chlorazanil *n* chlorazanil
Chlorazodin *n* chlorazodin
Chlorazol *n* chlorazol
Chlorbarium *n* barium chloride
Chlorbenzanthron *n* chlorobenzanthrone
Chlorbenzoesäure *f* chlorobenzoic acid
Chlorbenzol *n* chlorobenzene, phenyl chloride
Chlorbenzoyl *n* benzoyl chloride, chlorobenzoyl
Chlorbenzyl *n* benzyl chloride, chlorobenzyl
Chlorbernsteinsäure *f* chlorosuccinic acid
Chlorblei *n* lead chloride
Chlorbleiche *f* chlorine bleaching
Chlorbleispat *m* (Min) phosgenite
Chlorbrom *n* bromine chloride
Chlorbromessigsäure *f* chlorobromacetic acid
Chlorbromsilber *n* silver chloride bromide, (Min) embolite
Chlorbromsilberpapier *n* silver chloride bromide paper
Chlorcadmium *n* cadmium chloride
Chlorcalcium *n* calcium chloride
chlorcalciumhaltig containing calcium chloride
Chlorcalciumlösung *f* calcium chloride solution
Chlorcalciumröhre *f* calcium chloride tube
Chlorcalciumzylinder *m* calcium chloride cylinder
Chlorchinaldol *n* chlorquinaldol
Chlorchrom *n* chromium chloride
Chlorchromsäure *f* chlorochromic acid, chromic oxychloride, chromyl chloride
chlorchromsauer chlorochromate
Chlorcodid *n* chlorocodide
Chlorcrotonsäure *f* chlorocrotonic acid
Chlorcyan *n* chlorine cyanide, cyanogen chloride
Chlorcyanid *n* chlorine cyanide, cyanogen chloride
Chlorcyclizin *n* chlorocyclizine
Chlordan *n* chlordane
Chlordimorin *n* chlordimorine
Chlordioxid *n* chlorine dioxide, chlorine peroxide
Chlordioxidanlage *f* chlorine dioxide plant
chlorecht fast to chlorine
Chloreisen *n* iron chloride
Chloreisentinktur *f* iron chloride tincture
Chlorella *f* chlorella
chloren to treat with chlorine, to chlorinate, (Garne) to chemick

Chloren *n* chlorination, chloridizing, chlorinating
Chlorentwickler *m* chlorine generator
Chlorentwicklung *f* generation of chlorine
Chlorentwicklungsapparat *m* chlorine generator
Chlorentwicklungsflasche *f* chlorine generating flask
Chlorentwicklungskolben *m* chlorine generating flask
Chloressigsäure *f* chloroacetic acid, carboxy methyl chloride, chloracetic acid
Chloressigsäureäthylester *m* chloroacetic ethyl ester, ethyl chloroacetate
chloressigsauer chloroacetate
Chloreton *n* 1,1,1-trichloro-2-methyl propanol, chloretone, acetone chloroform, anesin, aneson, chlorobutanol
Chloretyl *n* chloretyl
chlorfrei free from chlorine
Chlorgas *n* chlorine gas, gaseous chlorine
Chlorgaselement *n* chlorine gas cell
Chlorgehalt *m* chlorine content, chlorinity
Chlorgeruch *m* chlorine odor
Chlorgold *n* gold chloride
Chlorgoldkalium *n* potassium chloroaurate(III), gold potassium chloride, potassium auric chloride
Chlorgoldnatrium *n* sodium chloroaurate(III), gold sodium chloride, sodium auric chloride
Chlorhämin *n* chlorhemine
chlorhaltig containing chlorine
Chlorheptoxid *n* chlorine heptoxide, perchloric anhydride
Chlorhexidin *n* chlorhexidine
Chlorhydrat *n* hydrochloride, chlorhydrate, chlorine hydrate
Chlorhydrin *n* chlorohydrin, chlorhydrin
Chlorhydrochinon *n* chlorohydroquinol, chlorohydroquinone
Chlorid *n* chloride
Chloridakkumulator *m* chloride accumulator
Chloridion *n* chloride ion
chlorieren to chlorinate, to chloridize
Chlorieren *n* chlorination, chloridizing, chlorinating
Chlorierer *m* chlorinator
chloriert chlorinated, chloridized
Chlorierturm *m* chlorinating tower
Chlorierung *f* chlorination, chloridization
chlorierungsfähig chloridizable
Chlorierungsmittel *n* chlorinating agent
chlorig chlorous
Chlorige Säure *f* chlorous acid
chlorigsauer chlorite
Chlorin *n* chlorine
Chlorionenmesser *m* chlorine ion determination apparatus
Chloriridium *n* iridium chloride
Chlorisatin *n* chlorisatin

Chlorisatinsäure *f* chlorisatic acid
Chlorisocyanursäure *f* chloroisocyanuric acid
Chlorisondamin *n* chlorisondamine
Chlorit *m* (Min) chlorite
Chlorit *n* chlorite, salt or ester of chlorous acid
chloritartig chloritous
Chloritbleiche *f* chlorite bleaching
chloritführend chloritous
chlorithaltig chloritous, containing chlorite
Chloritoid *m* (Min) chloritoid
Chloritschiefer *m* (Min) chlorite slate
Chloritspat *m* (Min) masonite, spathic chlorite
Chlorjod *n* iodine chloride
Chlorjodverbindung *f* chlorine iodine compound
Chlorkalium *n* potassium chloride
Chlorkalk *m* bleaching powder, chloride of lime
Chlorkalkanlage *f* chloride of lime plant
Chlorkalkbleiche *f* chloride of lime bleaching, bleaching with bleaching powder, chemicking
Chlorkalklösung *f* bleaching powder solution, solution of chlorinated lime
Chlorkammer *f* chlorine compartment
Chlorkautschuk *m* chlorinated rubber
Chlorknallgas *n* chlorine detonating gas
Chlorknallgasexplosion *f* chlorine-hydrogen explosion
Chlorkobalt *n* cobalt chloride
Chlorkohlenoxid *n* carbonyl chloride
Chlorkohlenoxidäther *m* ethyl chlorocarbonate, ethyl chloroformate
Chlorkohlensäure *f* chlorocarbonic acid, chloroformic acid
Chlorkohlensäureäthylester *m* ethyl chlorocarbonate, ethyl chloroformate
Chlorkohlensäureester *m* chlorocarbonate, chlorocarbonic ester, chloroformate, chloroformic ester
chlorkohlensauer chlorocarbonate, chloroformate
Chlorkohlenstoff *m* carbon tetrachloride, carbon chloride, phenoxin, tetrachloromethane
Chlorkohlenwasserstoff *m* chlorinated hydrocarbon
Chlorkresol *n* chlorocresol
Chlorkupfer *n* copper chloride
Chlorlauge *f* sodium hypochlorite solution
Chlorlithium *n* lithium chloride
Chlormagnesium *n* magnesium chloride
Chlormangan *n* manganese chloride
Chlormercuriphenol *n* chloromercuriphenol
Chlormerodrin *n* chlormerodrine
Chlormesser *m* chlorometer, chloridometer
Chlormessung *f* chlorimetry
Chlormethin *n* chlormethine
Chlormethyl- chlormethyl
Chlormethylierung *f* chloromethylation
Chlormethylmenthyläther *m* chloromethyl menthyl ether
Chlormethylnaphthalin *n* chloromethyl naphthalene
Chlormezanon *n* chlormezanone
Chlormonoxid *n* chlorine monoxide, hypochlorous anhydride
Chlornaphazin *n* chlornaphazine
Chlornaphthalin *n* chloronaphthalene, naphthyl chloride
Chlornatrium *n* common salt, sodium chloride
chlornatriumhaltig containing sodium chloride, containing common salt
Chlornatriumlösung *f* sodium chloride solution, common salt solution
Chlornatronlösung *f* sodium chloride solution, common salt solution
Chlornickel *n* nickel chloride
Chlornickelammoniak *n* nickel ammonium chloride
Chlornitrobenzol *n* chloronitrobenzene
Chlornitrobenzolsulfonsäure *f* chloronitrobenzenesulfonic acid
Chlornitroparaffin *n* chloronitroparaffin
Chloroäthylchlorid *n* chloroethyl chloride, dichlorethane
Chlorobenzal *n* chlorobenzal, benzal chloride, benzylidene chloride, chlorobenzylidene
Chlorobenzil *n* dichlorobenzil
Chlorobrombenzol *n* chlorobromobenzene
Chlorobutanol *n* chlorobutanol, chloreton
Chlorocalcit *m* (Min) chlorocalcite
Chlorochin *n* chloroquine
Chlorocodid *n* chlorocodide
Chlorocruorin *n* chlorocruorin
Chloroform *n* chloroform, trichloromethane
chloroformartig chloroformic
chloroformieren to chloroform
Chloroformieren *n* chloroforming
Chloroformierung *f* chloroformation
Chlorogallol *n* chlorogallol
Chlorogenin *n* chlorogenine, alstonine
Chlorogensäure *f* chlorogenic acid
Chlorolyse *f* chlorinating pyrolysis
Chloromelanit *m* (Min) chloromelanite
Chlorometer *n* chlorometer
Chlorometrie *f* chlorometry
chlorometrisch chlorometric
Chloromorphid *n* chloromorphide
Chloromycetin *n* chloromycetin
Chloropal *m* (Min) chloropal
Chlorophait *m* chlorophaite
Chlorophan *m* (Min) chlorophane
Chlorophorin *n* chlorophorin
Chlorophyll *n* chlorophyll
Chlorophyllan *n* chlorophyllane
Chlorophyllfarbstoff *m* chlorophyll coloring matter
chlorophyllführend chlorophyll-bearing
chlorophyllhaltig chlorophyllaceous

Chlorophyllid *n* chlorophyllide, chlorophyllin ester
Chlorophyllin *n* chlorophyllin
Chlorophyllinpaste *f* chlorophyllin paste
Chlorophyllit *m* (Min) chlorophyllite
chlorophyllreich rich in chlorophyll
Chloroplast *m* chloroplast
Chloroplatinat *n* chloroplatinate
Chloroporphyrin *n* chloroporphyrin
Chloropren *n* chloroprene, chlorobutadien
Chloropyramin *n* chloropyramine
Chloropyridin *n* chloropyridine
Chlorose *f* (Med) chlorosis
Chlorosmiumkalium *n* osmium potassium chloride, potassium chloroosmate(IV)
Chlorospinell *m* (Min) chlorospinel, magnesia iron spinel
Chlorothiazid *n* chlorothiazide
Chlorotil *m* (Min) chlorotile, hydrated copper arsenate
Chlorotrianisen *n* chlorotrianisene
Chloroxaläthylin *n* chloroxalethyline, chloroxalic acid
Chloroxalsäure *f* chloroxalic acid, chloroxalethyline
Chloroxid *n* chlorine oxide
Chloroxin *n* chlorooxine
Chloroxydul *n* chlorine monoxide, hypochlorous anhydride
Chlorozon *n* chlorozone
Chlorparaffin *n* chlorinated paraffin, chloroparaffin
Chlorphenamin *n* chlorphenamine
Chlorphenantrol *n* chlorophenanthrol
Chlorphenesin *n* chlorphenesin
Chlorphenoctium *n* chlorphenoctium
Chlorphenol *n* chlorophenol
Chlorphenylendiamin *n* chlorophenylenediamine
Chlorphosphor *m* phosphorus chloride
Chlorphosphorstickstoff *m* phosphonitryl chloride, phosphorus chloronitride
Chlorphthalsäure *f* chlorophthalic acid
Chlorpikrin *n* chloropicrin, aquinite, nitrochloroform, trichloronitromethane
Chlorplatin *n* platinum chloride
Chlorplatinsäure *f* chloroplatinic acid
Chlorpolypropylen *n* chloropolypropylene
Chlorprocain *n* chloroprocaine
Chlorpromazin *n* chlorpromazine
Chlorpropamid *n* chlorpropamide
chlorpropionsauer chloropropionate
Chlorpyren *n* chloropyrene
Chlorpyridin *n* chloropyridine
Chlorpyrilen *n* chloropyrilene
Chlorquecksilber *n* mercury chloride
Chlorräucherung *f* chlorine fumigation, fumigation with chlorine
Chlorricin[in]säure *f* chlororicinic acid
Chlorsäure *f* chloric acid

Chlorsäureanhydrid *n* chloric anhydride, chlorine(V) oxide
Chlorsalicylaldehyd *m* chlorosalicylaldehyde
Chlorsalol *n* chlorosalol, chlorophenyl salicylate
chlorsauer chlorate
Chlorschwefel *m* sulfur chloride
Chlorseife *f* chlorine soap
Chlorsilan *n* chlorosilane
Chlorsilber *n* silver chloride
Chlorsilberbad *n* silver chloride bath
Chlorsilberelement *n* silver chloride battery
Chlorsilbergelatine *f* silver chloride gelatine
Chlorsilbergelatinepapier *n* silver chloride gelatine paper
Chlorsilberkollodium *n* silver chloride collodion
Chlorsilberkruste *f* incrustation of silver chloride
Chlorsilicium *n* silicon tetrachloride
Chlorspießglanz *m* (Min) antimonic chloride
Chlorstickstoff *m* nitrogen chloride
Chlorstrom *m* current of chlorine
Chlorstrontium *n* strontium chloride
Chlorsuccinat *n* chlorsuccinate
Chlorsulfamin *n* chlorsulfamin, chloramine-T
Chlorsulfonierung *f* (Sulfochlorierung) chlorosulfonation
Chlorsulfonsäure *f* chlorosulfonic acid
Chlorsuperoxidbleiche *f* chlorine-peroxide bleach
Chlortetracyclin *n* chlorotetracycline, aureomycine
Chlorthion *n* chlorthion
Chlorthymol *n* chlorothymol
Chlortitan *n* titanium chloride
Chlortoluol *n* chlorotoluene
Chlorüberträger *m* chlorine carrier
Chlorür *n* (obs) chloride, protochloride
Chlorung *f* chlorination
Chlorungsanlage *f* chlorinating plant
Chlorurethan *n* chlorourethane
Chlorvanadium *n* vanadium chloride
Chlorventil *n* chlorine valve
Chlorverbindung *f* chlorine compound
Chlorverflüssigung *f* chlorine liquefaction
Chlorverflüssigungsanlage *f* chlorine liquefying plant
Chlorwasser *n* aqueous solution of chlorine, chlorine water
Chlorwasserpumpe *f* chlorine water pump
Chlorwasserstoff *m* hydrochloric acid, hydrogen chloride
Chlorwasserstoffanlage *f* hydrochloric acid plant
Chlorwasserstoffentwicklung *f* generation of hydrogen chloride
Chlorwasserstoffsäure *f* hydrochloric acid, (muriatic acid)
chlorwasserstoffsauer chloride, hydrochloride
Chlorwismut *n* bismuth chloride
Chloryl- chloryl

Chlorylen *n* chlorylene, trichloroethylene
Chlorzelle *f* chlorine cell
Chlorzink *n* zinc chloride
Chlorzinkanlage *f* zinc chloride plant
Chlorzinkcelluloselösung *f* zinc chloride cellulose solution
Chlorzinklauge *f* zinc chloride solution
Chlorzinn *n* tin chloride
Chlorzinnbad *n* tin chloride bath
Choladien *n* choladiene
Choladiensäure *f* choladienic acid
Cholalsäure *f* cholalic acid, cholic acid
cholalsauer cholate, cholalate
Cholan *n* cholane
Cholanamid *n* cholanamide
Cholansäure *f* cholanic acid
Cholanthren *n* cholanthrene
Cholatrien *n* cholatriene
Cholatriensäure *f* cholatrienic acid
Cholein *n* choleine
Choleinsäure *f* choleic acid, cholenic acid
Cholera *f* (Med) cholera
Choleragift *n* cholera toxin
Choleraschutzimpfung *f* cholera inoculation
Cholestan *n* cholestane
Cholestandiol *n* cholestanediol
Cholestanol *n* cholestanol
Cholestanon *n* cholestanone
Cholestanonol *n* cholestanonol
Cholesten *n* cholestene
Cholestenon *n* cholestenone
Cholesterat *n* cholesterate
Cholesterin *n* cholesterol, cholesterin
cholesterinähnlich cholesterol-like
Cholesterinester *m* cholesterol ester
cholesterinsauer cholesterate, cholesterinate
Cholesterinspiegel *m* cholesterol level
Cholesterinstoffwechsel *m* cholesterol metabolism
Cholesterinwachs *n* cholesterol wax
Cholesterol *n* cholesterol, cholesterin
Cholesterylen *n* cholesterylene
Cholestrophan *n* cholestrophan, dimethyl parabanic acid
Cholin *n* choline, sinkaline
Cholinäther *m* choline ether
Cholinbase *f* choline base
Cholinbitartrat *n* choline bitartrate
Cholinchlorid *n* choline chloride
Cholindehydrase *f* (Biochem) choline dehydrogenase
Cholindihydrogencitrat *n* choline dihydrogen citrate
Cholinesterase *f* (Biochem) choline esterase
Cholinsalz *n* choline salt
Choloidansäure *f* choloidanic acid
Cholsäure *f* cholic acid, cholalic acid
cholsauer cholate
Chondocurarin *n* chondocurarine
Chondocurin *n* chondocurine
Chondodendrin *n* chondodendrine
Chondridin *n* chondridine
Chondrillasterin *n* chondrillasterol
Chondrin *n* chondrin
Chondrit *m* (Min) chondrite
Chondroarsenit *m* (Min) chondroarsenite
Chondrodin *n* chondrodine
Chondrodit *m* (Min) chondrodite, humite
Chondroitin *n* chondroitin
Chondroitinsulfat *n* chondroitin sulfate
Chondroninsäure *f* chondroninic acid
Chondronsäure *f* chondronic acid
Chondrosamin *n* chondrosamine
Chondrosaminsäure *f* chondrosamic acid
Chondrosin *n* chondrosin
Chondrosinsäure *f* chondrosinic acid
Chonemorphin *n* chonemorphine
Chonikrit *m* chonicrite
Chovisminsäure *f* chovismic acid
Christianit *m* (Min) anorthite, christianite
Christpalmöl *n* castor oil
Christwurzel *f* (Bot) hellebore root
Chroicolyt *n* chroicolyte
Chrom *n* chromium, chrome (Symb. Cr)
Chromacetat *n* chromium acetate
Chromalaun *m* chrome alum, chromium potassium sulfate
Chromalaunlauge *f* chrome alum solution
Chroman *n* chroman
Chromanol *n* chromanol
Chromanon *n* chromanone
Chromat *n* chromate, **saures** ~ bichromate, dichromate
Chromatdruck *m* (Buchdr) chromatic printing
Chromatin *n* (Biol, Med) chromatin
chromatisch chromatic
Chromatizität *f* (Opt) chromaticity, **komplementäre** ~ (Opt) complementary chromaticity
Chromatogramm *n* chromatogram
Chromatographie *f* chromatography, **absteigende** ~ descending chromatography, **aufsteigende** ~ ascending chromatography
chromatographieren to chromatograph
chromatographisch chromatographic
Chromatophor *n* (Bot) chromatophore
Chromatoskop *n* chromatoscope
Chromatschwarz *n* chromate black
Chromazurol *n* chromazurol
Chrombeize *f* chrome mordant
chrombeizen to chrome-mordant
Chromblau *n* chrome blue
Chromblei *n* (Min) crocoite, native lead chromate
Chrombraun *n* chrome brown
Chromcarbid *n* chromium carbide
Chromchlorat *n* chromium chlorate

Chromchlorid *n* chromium chloride, (Chrom(III)-chlorid) chromic chloride, chromium(III) chloride
Chromchlorür *n* (Chrom(II)-chlorid) chromium(II) chloride, chromous chloride
Chromdichlorid *n* chromium dichloride, chromium(II) chloride, chromous chloride
Chromdifluorid *n* chromium difluoride, chromium(II) fluoride, chromous fluoride
chromecht fast to chrome
Chromechtschwarz *n* fast chrome black
Chromeisen *n* (Min) chrome iron ore, chromite
Chromeisenerz *n* (Min) chrome iron ore, chromite
Chromeisenstein *m* (Min) chrome iron ore, chromite
Chromelement *n* bichromate cell, chromic acid cell, dichromate cell
Chromen *n* chromene
Chromerz *n* (Min) chromium ore
Chromfarbe *f* chrome color
Chromfluorid *n* (Chrom(III)-fluorid) chromic fluoride, chromium(III) fluoride
chromgar chrome-tanned
Chromgehalt *m* chromium content, percentage of chrome
Chromgelatine *f* bichromated gelatin, chromatized gelatin, dichromated gelatin
Chromgelb *n* chrome yellow, lead chromate
Chromgerbeextrakt *m* chrome tanning extract
Chromgerberei *f* chrome tannery
Chromgerbung *f* chrome tannage, chrome tanning
Chromglimmer *m* (Min) chrome mica, fuchsite
Chromgranat *m* (Min) calcium-chromium garnet, ouvarovite, uvarovite, uwarowite
Chromgrün *n* chrome green
chromhaltig chromiferous, containing chromium
Chromhydroxid *n* (Chrom(III)-hydroxid) chromic hydroxide, chromium(III) hydroxide
Chromhydroxydul *n* (Chrom(II)-hydroxid) chromium(II) hydroxide, chromous hydroxide
Chromiacetat *n* (Chrom(III)-acetat) chromic acetate, chromium(III) acetate
Chromiak *n* chrome ammine
Chromichlorid *n* chromic chloride, chromium(III) chloride
Chromicyanwasserstoffsäure *f* chromicyanic acid, cyanochromic(III) acid
chromieren (Text) to chrome
Chromierfarbstoff *m* chrome dyestuff
Chromihydroxid *n* (Chrom(III)-hydroxid) chromic hydroxide, chromium(III) hydroxide
Chrom(III)-Salz *n* chromic salt, chromium(III) salt
Chrom(III)-Verbindung *f* chromic compound, chromium(III) compound
Chrom(II)-Salz *n* chromium(II) salt, chromous salt
Chrom(II)-Verbindung *f* chromium(II) compound, chromous compound
Chrominanz *f* (Opt) chrominance
Chromirhodanwasserstoffsäure *f* chromithiocyanic acid, thiocyanatochromic(III) acid
Chromisalz *n* (Chrom(III)-Salz) chromic salt, chromium(III) salt
Chromisulfat *n* chromic sulfate, chromium(III) sulfate
Chromisulfocyansäure *f* chromithiocyanic acid, thiocyanatochromic(III) acid
Chromit *m* (Min) chrome iron ore, chromite
Chromitstein *m* chromite brick
Chromiverbindung *f* (Chrom(III)-Verbindung) chromic compound, chromium(III) compound
Chromkarbid *n* chromium carbide
Chromlackleder *n* chrome-tanned patent leather
Chromlauge *f* chrome lye
Chromleder *n* chrome leather, chrome-tanned leather
Chromlederfarbe *f* chrome leather color
Chromlederschwarz *n* chrome leather black
Chrommetall *n* chromium metal, metallic chromium
Chrommetaphosphat *n* chromium metaphosphate
Chromnatron *n*, **gelbes** ~ sodium chromate, **rotes** ~ sodium bichromate, sodium dichromate
Chromnickelstahl *m* chromium-nickel steel
Chromnitrat *n* chromium nitrate
Chromoacetat *n* (Chrom(II)-acetat) chromium(II) acetate, chromous acetate
Chromoaristopapier *n* baryta paper
Chromochlorid *n* chromium(II) chloride, chromous chloride
Chromocker *m* (Min) chrome ocher, native chromium oxide
Chromocyclit *m* (Min) chromocyclite
Chromodiessigsäure *f* chromodiacetic acid
Chromoessigsäure *f* chromoacetic acid
Chromoform *n* chromoform
Chromogen *n* chromogen
Chromograph *m* chromograph
Chromohydroxid *n* (Chrom(II)-hydroxid) chromium(II) hydroxide, chromous hydroxide
Chromoion *n* chromium(II) ion, chromous ion
Chromoisomer *n* chromoisomer
Chromoisomerie *f* chromoisomerism
Chromolithograph *m* chromolithographer
Chromolithographie *f* chromolithography
Chromolyse *f* chromolysis
Chromomer *n* chromomere
Chromomonoessigsäure *f* chromomonoacetic acid
Chromon *n* 1,4-benzopyrone, chromone
Chromonol *n* chromonol

Chromopapier *n* (Buchdr) chromopaper
chromophor chromophoric, chromophorous
Chromophor *m* (Opt) chromophore, color carrier
Chromophorelektronen *pl* chromophoric electrons
Chromophotographie *f* chromophotography
Chromophototypie *f* chromophototypy
Chromophotoxylographie *f* chromophotoxylography
Chromophotozinkographie *f* chromophotozincography
Chromoplast *m* (Biol) chromoplast
Chromoproteid *n* chromoprotein
Chromopyrometer *n* chromopyrometer
Chromorange *n* chrome orange
Chromosantonin *n* chromosantonin, photosantonin
Chromosom *n* (Gen) chromosome
chromosomal chromosomal
Chromosomenabnormität *f* chromosomal abnormality
Chromosomenanordnung *f* arrangement of chromosomes
Chromosomenaustausch *m* exchange of chromosomes
Chromosomenbild *n* chromosome map
Chromosomenpaar *n* pair of chromosomes
Chromosomenpaarung *f* pairing of chromosomes
Chromosomensatz *m* set of chromosomes, genome
Chromosomenschleife *f* chromosome loop
Chromosomenverdopplung *f* reduplication of chromosomes
Chromosomenvereinigung *f* fusion of chromosomes
Chromosomenverschmelzung *f* fusion of chromosomes
Chromosomenzahl *f* chromosome count, number of chromosomes
Chromosomenzusammenballung *f* clumping of chromosomes
Chromosomenzusammensetzung *f* chromosomal make-up
Chromosphäre *f* chromosphere
chromosphärisch chromospheric
chromotrop chromotropic
Chromotropsäure *f* chromotropic acid
Chromotypie *f* chromotypy
Chromotypographie *f* chromotypography
Chromoverbindung *f* (Chrom(II)-Verbindung) chromium(II) compound, (Chromo(II)-Verbindung) chromous compound
Chromoxid *n* chromium oxide, (Chrom(III)-oxid) chromic oxide, chromium(III) oxide
Chromoxidgrün *n* chrome green, chrome oxide green, chromic oxide

Chromoxidhydrat *n* chromium hydroxide, chromium(III) hydroxide
Chromoxidhydratgrün *n* hydrated chrome oxide green
Chromoxidnatron *n* sodium chromite
Chromoxidsalz *n* chromic salt, chromium(III) salt
Chromoxychlorid *n* chromium oxychloride, chromyl chloride
Chromoxydul *n* (Chrom(II)-oxid) chromium(II) oxide, chromous oxide
Chromoxydulverbindung *f* chromium(II) compound, chromous compound
Chromoxyfluorid *n* chromium oxyfluoride, chromyl fluoride
Chromoxylograph *m* chromoxylograph
Chromoxylographie *f* chromoxylography
Chromperoxid *n* chromium peroxide
Chromphosphid *n* chromium phosphide
Chromphotolithographie *f* chromophotolithography
Chrompicotit *m* chromopicotite
Chromresinat *n* chrome resinate, chromium resinate
Chromrot *n* chrome red, basic lead chromate
Chromsäure *f* chromic acid
Chromsäureanhydrid *n* chromic anhydride, chromium trioxide, chromium(VI) oxide
Chromsäureelement *n* chromic acid cell, bichromate cell, dichromate cell
Chromsäureflüssigkeit *f* chromic acid solution
Chromsäuresalz *n* chromate, salt of chromic acid
Chromsäureverbindung *f* chromate, compound of chromic acid
Chromsalpetersäure *f* chromonitric acid
Chromsalz *n* chromium salt
chromsauer chromate
Chromschwarz *n* chrome black
Chromschwefelsäure *f* chromosulfuric acid, potassium dichromate / sulphuric acid mixture
Chromsesquioxid *n* (Chrom(III)-oxid) chromic oxide, chromium(III) oxide, chromium sesquioxide
Chromsilicid *n* chromium silicide
Chromsilicofluorid *n* chromium fluosilicate, chromium silicofluoride
Chromstahl *m* chrome steel, chromium steel
Chromsubchlorid *n* basic chromium chloride, chromium oxychloride, chromium subchloride, chromyl chloride
Chromsubsulfat *n* basic chromium sulfate, chrome subsulfate, chromium oxysulfate, chromyl sulfate
Chromsulfat *n* chromium sulfate
Chromsulfid *n* chromium sulfide
Chromsulfit *n* chromium sulfite

Chromsulfür *n* (Chrom(II)-sulfid) chromium(II) sulfide, chromous sulfide
Chromtrichlorid *n* chromic chloride, chromium(III) chloride, chromium trichloride
Chromtrifluorid *n* chromic fluoride, chromium(III) fluoride, chromium trifluoride
Chromtrioxid *n* chromic anhydride, chromium trioxyde
Chromtrockengerbung *f* dry chrome-tanning
Chromvanadiumstahl *m* chrome-vanadium steel
Chromverbindung *f* chromium compound
Chromviolett *n* chrome violet
Chromwolframat *n* chromium tungstate
Chromwolframstahl *m* chrome-tungsten steel
Chromyl- chromyl
Chromylchlorid *n* chromyl chloride, chromium oxychloride
Chromzinnober *m* chromate of mercury protoxide, mercurous chromate, mercury(I) chromate
chronisch chronic
Chronometer *n* chronometer
Chronopotentiometrie *f* chronopotentiometry
Chronoskop *n* chronoscope, time-meter
Chronotron *n* chronotron
Chrysalidenöl *n* chrysalis oil
Chrysamin *n* chrysamine, flavophenine
Chrysamminsäure *f* chrysammic acid
Chrysanilin *n* chrysaniline
Chrysanilsäure *f* chrysanilic acid
Chrysanissäure *f* chrysanisic acid, 3,5-dinitro-4-aminobenzoic acid
Chrysanthemalkohol *m* crysanthemumyl alcohol
Chrysanthemaxanthin *n* crysanthemaxanthin
Chrysanthemin *n* chrysanthemine
Chrysanthem[um]säure *f* chrysanthem[um]ic acid
Chrysanthenon *n* chrysanthenone
Chrysarin *n* chrysarine
Chrysarobin *n* chrysarobin
Chrysarobinseife *f* chrysarobin soap
Chrysaron *n* chrysarone
Chrysatropasäure *f* chrysatropic acid
Chrysazin *n* chrysazin
Chrysean *n* chrysean
Chrysen *n* chrysene, benzophenanthrene
Chrysenchinon *n* chrysene quinone
Chrysensäure *f* chrysenic acid
Chryseolin *n* chryseoline
Chrysidan *n* chrysidan
Chrysidin *n* chrysidine
Chrysin *n* chrysin
Chrysoberyll *m* (Min) chrysoberyl
Chrysocal *n* chrysocale
Chrysocetrarsäure *f* chrysocetraric acid
Chrysochinon *n* chrysoquinone, cyresen quinone
Chrysoeriol *n* chrysoeriol
Chrysofluoren *n* chrysofluorene
Chrysogen *n* chrysogen
Chrysographie *f* chrysography

Chrysoid *n* chrysoid
Chrysoidin *n* chrysoidine
Chrysoidinorange *n* crysoidine orange
Chrysoin *n* chrysoine
Chrysoinbraun *n* chrysoine brown
Chrysoketon *n* chrysoketone
Chrysokoll *m* (Min) chrysocolla
Chrysolin *n* chrysoline
Chrysolith *m* (Min) chrysolite, olivine
Chryson *n* chrysone
Chrysonetin *n* chrysonetine
Chrysophanol *n* chrysophanol, chrysophanic acid
Chrysophansäure *f* chrysophanic acid, chrysophanol
Chrysophenin *n* chrysophenine
Chrysopheninsäure *f* chrysophenic acid
Chrysophyll *n* chrysophyll
Chrysophyscin *n* chrysophyscin
Chrysopikrin *n* chrysopicrin, vulpic acid
Chrysopontin *n* chrysopontine
Chrysopras *m* (Min) chrysoprase
Chrysopterin *n* chrysopterin
Chrysorhaminin *n* chrysorhaminine
Chrysotil *m* (Min) chrysotile
Chrysotilasbest *m* (Min) amianthus, chrysotile
Chrysotoxin *n* chrysotoxin
Churchit *n* (Min) churchite
Chydenanthin *n* chydenanthine
Chylomikronen *pl* chylomicrons
Chylus *m* chyle
Chymosin *n* chymosin
Chymotrypsin *n* (Biochem) chymotrypsin
Chymotrypsinogen *n* (Biochem) chymotrypsinogen
Chymus *m* (Med) chyme
Chyralin *n* chyraline
Cichorie *f* (Bot) chicory
Cichorigenin *n* cichorigenin
Cichoriin *n* cichoriin
Cicutin *n* cicutine
Cicutol *n* cicutol
Cicutoxin *n* cicutoxine
Cidarit *m* (Min) warstone
Cider *m* cider
Ciderbranntwein *m* cider brandy
Cideressig *m* cider vinegar
Cidertrester *pl* cider marc
Ciderwein *m* cider
Cignolin *n* cignolin, anthrarobin, dioxyanthranol
Ciliansäure *f* cilianic acid
Ciloidansäure *f* ciloidaic acid
Cimicifugin *n* cimicifugin, macrotin
Cimicin *n* cimicine
Cimicinsäure *f* cimicic acid
Cimolit *m* (Min) cimolite
Cinchain *n* cinchaine
Cinchamidin *n* cinchamidine

Cinchen *n* cinchene
Cinchocain *n* cinchocaine
Cincholidin *n* cincholidine
Cincholoipon *n* cincholoipone
Cincholoiponsäure *f* cincholoiponic acid
Cincholsäure *f* cincholic acid
Cinchomeronsäure *f* cinchomeronic acid
Cinchona *f* (Bot) cinchona
Cinchonabase *f* cinchona base
Cinchonamin *n* cinchonamine
Cinchonhydrin *n* cinchonhydrin
Cinchonicin *n* cinchonicine, cinchotoxine
Cinchonidin *n* cinchonidine, chinidine
Cinchonidinbisulfat *n* cinchonidine bisulfate, cinchonidine hydrogen sulfate
Cinchonidinbromhydrat *n* cinchonidine hydrobromide
Cinchonidinchlorhydrat *n* cinchonidine hydrochloride
Cinchonidinhydrochlorid *n* cinchonidine hydrochloride
Cinchonidinsulfat *n* cinchonidine sulfate
Cinchonifin *n* cinchonifine
Cinchonigin *n* cinchonigine
Cinchonilin *n* cinchoniline
Cinchonin *n* cinchonine
Cinchoninal *n* cinchoninal
Cinchoninbisulfat *n* cinchonine bisulfate, cinchonine hydrogen sulfate
Cinchoninchlorhydrat *n* cinchonine hydrochloride
Cinchoninhydrochlorid *n* cinchonine hydrochloride
Cinchoninnitrat *n* cinchonine nitrate
Cinchoninon *n* cinchoninone
Cinchoninsulfat *n* cinchonine sulfate
Cinchoninvergiftung *f* cinchonism
Cinchoniretin *n* cinchoniretine
Cinchonirin *n* cinchonirine
Cinchonismus *m* quinism
Cinchonsäure *f* cinchoic acid, cinchonic acid
Cinchophen *n* cinchophen, atophan
Cinchotenin *n* cinchotenine
Cinchoticin *n* cinchoticine
Cinchotidin *n* cinchotidine
Cinchotin *n* cinchotine
Cinchotinon *n* cinchotinone
Cinchotintoxin *n* cinchotine toxine
Cinchotoxin *n* cinchotoxine
Cinchotoxol *n* cinchotoxol
Cinen *n* cinene, terpene
Cinensäure *f* cinenic acid
Cineol *n* cineol[e], eucalyptol[e]
Cineolsäure *f* cineolic acid
Cinerin *n* cinerin
Cinerolon *n* cinerolone
Cineron *n* cinerone
Cinnabarin *n* cinnabarine
Cinnabarit *m* (Min) cinnabarite

Cinnachinon *n* cinnaquinone
Cinnamal- cinnamal
Cinnamaldehyd *m* cinnamaldehyde, 3-phenylpropenal, cinnamic aldehyde
Cinnamat *n* cinnamate, salt or ester of cinnamic acid
Cinnamein *n* cinnamein, benzyl cinnamate
Cinnamen *n* cinnamene, phenethylene, phenylethylene, styrene, styrol, styrolene, vinylbenzene
Cinnamenyl- cinnamenyl, styryl
Cinnamoin *n* cinnamoin
Cinnamoyl- cinnamoyl
Cinnamyl- cinnamyl
Cinnamylaldehyd *m* cinnamic aldehyde, 3-phenylpropenal, benzalacetaldehyde, cinnamaldehyde
Cinnamylalkohol *m* cinnamyl alcohol, 3-phenyl-2-propene-1-ol, styrolene alcohol, styryl alcohol
Cinnamylchlorid *n* cinnamyl chloride
Cinnamylcocain *n* cinnamyl cocaine
Cinnamylecgonin *n* cinnamyl ecgonine
Cinnamyliden- cinnamylidene
Cinnamylidenaceton *n* cinnamylidene acetone
Cinnamylsäure *f* cinnamic acid, benzal acetic acid, styrylformic acid
Cinnarizin *n* cinnarizine
Cinnolin *n* cinnoline
Cinnolinsäure *f* cinnolic acid, cinogenic acid
Cinobufagin *n* cinobufagin
Circulardichroismus *m* (Spektr) circular dichroism
Circumanthracen *n* circumanthracene
Cirrholit *m* (Min) cirrolite
cis-Effekt *m* (Stereochem) cis effect
Cismollan *n* cismollan
cis-Regel *f* cis rule
Cissoide *f* (Math) cissoid
cis-Stellung *f* cis position
Cis-Trans-Isomerie *f* cis-trans isomerism
Cis-Trans-Isomerisierung *f* (Stereochem) cis-trans isomerization
Cistron *n* cistron
Citarin *n* citarin, sodium anhydro methylene citrate
Citraconanil *n* citracone anil
Citraconsäure *f* citraconic acid, methylbutenedioic acid, methylmaleic acid
Citraconsäureanhydrid *n* citraconic anhydride, methylbutenedioic anhydride, methylmaleic anhydride
Citraconsäureester *m* citraconic ester
Citraconyl- citraconyl
Citral *n* citral, geranial
Citramalsäure *f* citramalic acid
Citranilid *n* citranilide
Citrat *n* citrate, salt or ester of citric acid
Citratcyclus *m* citrate cycle

Citraurin *n* citraurin
Citrazinsäure *f* citrazinic acid
Citren *n* citrene
Citreorosein *n* citreorosein
Citridinsäure *f* citridic acid, aconitic acid
Citrifoliol *n* citrifoliol
Citrin *m* (Min) citrine
Citrinin *n* citrinin
Citromycetin *n* citromycetin
Citronat *n* candied lemon peel
Citronellal *n* citronellal
Citronellasäure *f* citronellic acid
Citronellöl *n* citronella oil
Citronellol *n* citronellol
Citronellsäure *f* citronellic acid
Citronenöl *n* lemon oil
Citronensäure *f* citric acid
Citronensäurecyclus *m* citric [acid] cycle, Krebs cycle, tricarboxylic acid cycle
citronensauer citrate
Citronetin *n* citronetin
Citronin *n* citronine, dinitrodiphenylamine
Citronyl *n* citronella oil, citronyl
Citrophen *n* citrophen
Citropten *n* citroptene
Citrostadienol *n* citrostadienol
Citrostanol *n* citrostanol
Citrovorumfaktor *m* citrovorum factor, coenzyme F
Citroxanthin *n* citroxanthin
Citrullin *n* citrulline
Citrullol *n* citrullol
Citryl- citryl
Cladestin *n* cladestine
Cladestinsäure *f* cladestic acid
Cladinose *f* cladinose
Cladonin *n* cladonine
Claisenaufsatz *m* Claisen stillhead
Claisenkolben *m* Claisen flask
Claisensche Umlagerung *f* (Chem) Claisen rearrangement
Claisen-Schmidt-Kondensation *f* Claisen-Schmidt condensation
Clandestinin *n* clandestinine
Clarit *m* clarite
Clarkelement *n* (Elektr) Clark cell
Claron *n* clarone
Clathrat *n* clathrate
Claudetit *m* (Min) claudetite
Clausius-Clapeyronsche Gleichung *f* (Therm) Clausius-Clapeyron equation
Clauskatalysator *m* Claus catalyst
Clausthalit *m* (Min) clausthalite
Clavatin *n* clavatin
Clavatol *n* clavatol
Clavicepsin *n* clavicepsin
Claviformin *n* claviformin
Clavolonin *n* clavolonine
Clay *m* clay

Cleavelandit *m* (Min) cleavelandite
Clemizol *n* clemizole
Cleveit *m* (Min) cleveite
Clevesäure *f* Cleve's acid, 1-naphthylamine-6-sulfonic acid
Clidinium *n* clidinium
Cliftonit *m* (Min) cliftonite
Clingmannit *m* (Min) clingmannite
Clintonit *m* (Min) clintonite
Clionasterin *n* clionasterol
Clivonin *n* clivonine
Cloisonnéemail *n* cloisonne enamel
Clovan *n* clovane
Clovandiol *n* clovanediol
Cloven *n* clovene
Clovensäure *f* clovenic acid
Clupanodensäure *f* clupanodenic acid
Clupein *n* clupein
Clusiussches Trennrohr *n* Clusius column
Cluster-Ion *n* cluster ion
Cluthalit *m* (Min) cluthalite
Cluytianol *n* cluytianol
Cluytiasterin *n* cluytiasterine
Cluytiasterol *n* cluytiasterol
Cluytinsäure *f* cluytic acid
Cluytylalkohol *m* cluytyl alcohol
Cnicin *n* cnicin
Cnidiumlacton *n* cnidium lactone
Cnidiumsäure *f* cnidionic acid
Coagulum *n* coagulum
Cobalamin *n* cobalamine, vitamin B$_{12}$
Cobalt siehe Kobalt
Cobaltum *n* cobalt, cobaltum (s. a. Kobalt)
Cobamid *n* cobamide
Cobamid-Coenzym *n* cobamide coenzyme
Cobamsäure *f* cobamic acid
Cobinsäure *f* cobinic acid
Coca *f* coca
Cocablätter *pl* coca leaves
Cocain *n* cocaine, **behandeln mit** ~ to cocainize
Cocainchlorhydrat *n* cocaine hydrochloride
Cocainersatz *m* cocaine substitute
Cocainhydrochlorid *n* cocaine hydrochloride
Cocainvergiftung *f* cocaine poisoning, cocainism
Cocamin *n* cocamine
Cocarboxylase *f* (Biochem) cocarboxylase, cozymase II, TPP
Cocasäure *f* cocaic acid
Coccelsäure *f* coccelic acid
Coccerylalkohol *m* cocceryl alcohol
Coccin *n* coccine
Coccinin *n* coccinine
Coccinit *m* (Min) coccinite
Coccinon *n* coccinone
Coccinsäure *f* coccic acid
Cocculin *n* cocculin
Cochalsäure *f* cochalic acid
Cochenille *f* cochineal
Cochenillenbad *n* scarlet dye

Cochenillenfarbstoff — Collodin

Cochenillenfarbstoff *m* cochineal
Cochenillenscharlach *m* cochineal scarlet
Cochenillentinktur *f* cochineal tincture
Cochenillerot *n* cochineal red
Cochenillesäure *f* cochenillic acid
Cochenillin *n* cochenilline
Cochinchinawachs *n* cochinchina wax
Cochinöl *n* cochin oil
Cocin *n* cocine
Cocinin *n* cocinine
cocinsauer cocinic
Coclanolin *n* coclanoline
Coclaurin *n* coclaurine
Cocolith *m* (Min) cocolite
Cocoonase *f* cocoonase
Cocosbutter *f* coconut butter
Cocosit *n* cocosite, cocositol
Cocositol *n* cocositol
Cocosnußmilch *f* coconut milk
Cocosnußöl *n* coconut oil
Cocosnußölseife *f* coconut soap
Cocosöl *n* coconut oil
Cocostalg *m* coconut tallow
Cocostalgsäure *f* cocinic acid
Codecarboxylase *f* (Biochem) codecarboxylase
Codehydrase I *f* (Biochem) codehydrase I, coenzyme I, Co I, cozymase I, DPN, NAD
Codehydrase II *f* (Biochem) codehydrase II, Co II, NADP, TPN
Codein *n* codeine, methylmorphine
Codeinchlorhydrat *n* codeine hydrochloride
Codeinhydrobromid *n* codeine hydrobromide
Codeinhydrochlorid *n* codeine hydrochloride
Codeinon *n* codeinone
Codeinphosphat *n* codeine phosphate
Codeonal *n* codeonal
Codimer *n* codimer
Codimerisation *f* codimerisation
Codiran *n* codiran
Codöl *n* cod liver oil, cod oil
Codon *m* codon
Coelestin *m* (Min) celestite
Cölestin *n* celestine
Cölestinblau *n* celestine blue
Coenzym A *n* CoA, coenzyme A
Coenzym F *n* (Folinsäure) coenzyme F
Coenzym I *n* (Biochem) codehydrase I, coenzyme I, Co I, cozymase I, DPN, NAD
Coenzym Q *n* coenzyme Q
Coenzym R *n* coenzyme R, beta-biotine, vitamine H
Cöramidonin *n* ceramidonine
Cöroxonon *n* ceroxonone
Coerthian *n* cerothian
Coerthien *n* cerothiene
Coerthion *n* cerothione
Cörulein *n* cerulein, coerulein
Cöruleolactin *n* ceruleolactine
Coeruleum *n* (Färb) ceruleum

Cörulignol *n* cerulignol
Cörulignon *n* cerulignone
Cörulin *n* ceruline
Cofaktor *m* cofactor
Coferment *n* coferment
Coffamin *n* coffamine
Coffearin *n* coffearine
Coffein *n* caffeine, coffeine
Coffeinaldehyd *m* caffeine carboxaldehyde
Coffeinbromhydrat *n* caffeine hydrobromide
Coffeinnatriumbenzoat *n* caffeine sodium benzoate
Coffeinnatriumsalicylat *n* caffeine sodium salicylate
Coffeinoxalat *n* caffeine oxalate
Coffeinvalerianat *n* caffeine valerate
Coffeinvergiftung *f* caffeine poisoning, caffeinism
Cohenillerot *n* crimson lake
Cohenit *m* (Min) cohenite
Cohumulon *n* cohumulone
Coixol *n* coixol
Cola *f* cola nut, kola seed
Colafluidextrakt *m* liquid extract of cola
Colamin *n* colamine
Colanuß *f* cola nut, kola seed
Colatin *n* colatannin, colatin
Colatür *f* filtrate
Colchamin *n* colchamine
Colchicamid *n* colchicamide
Colchicein *n* colchiceine
Colchicid *n* colchicide
Colchicin *n* colchicine
Colchicinsäure *f* colchicinic acid
Colchicosid *n* colchicoside
Colchicumextrakt *m* colchicum extract
Colchicumpräparat *n* colchicum preparation
Colchicumtinktur *f* tincture of colchicum seeds
Colchicumwein *m* colchicum wine
Colchid *n* colchide
Colchinol *n* colchinol
Colchinsäure *f* colchic acid
Colcothar *n* colcothar, English red
Colemanit *m* (Min) colemanite
Coleol *n* coleol
Colitit *m* colititol
Colitose *f* colitose
Collagen *n* collagen
Collargol *n* collargol
Collargolsalbe *f* collargol ointment
Collatolon *n* collatolone
Collatolsäure *f* collatolic acid
Collaurin *n* collaurin, colloidal gold
Collemplastrum *n* collemplastrum, sticking plaster
Collidin *n* collidine
Collinol *n* collinol
Collocynthein *n* colocynthein
Collodin *n* collodine, gluten

Collodium *n* collodion, collodium
Collodiumlösung *f* collodion solution, flexible collodion
Collodiumpapier *n* collodium paper
Collodiumwolle *f* collodion cotton, collodion wool, gun cotton, nitrocellulose
Colloid siehe Kolloid
Colloturin *n* colloturine
Colloxylin *n* collodion, colloxylin
Colombosäure *f* calumbic acid, columbic acid
Colombowurzel *f* calumba
Colophen *n* colophene
Colophonium *n* colophonium, rosin
Coloradoit *m* (Min) coloradoite
Colorimeter *n* colorimeter
Colorimetrie *f* colorimetric analysis, colorimetry
colorimetrisch colorimetric
Colorin *n* colorin
Colostrum *n* colostrum
Colubridin *n* colubridine
Colubrin *n* colubrine
Columbamin *n* columbamine
Columbaminbase *f* columbamine base
Columbaminhydrat *n* columbamine hydrate
Columbaminhydroxid *n* columbamine hydroxide
Columbeisen *n* (Min) columbite, tantalite
Columbiagelb *n* columbia yellow
Columbiakopalinsäure *f* columbia copalinic acid
Columbiakopalolsäure *f* columbia copalolic acid
Columbiakopalsäure *f* columbia copalic acid
Columbiaschwarz *n* columbia black
Columbin *n* columbin
Columbit *m* (Min) columbite, niobite
Columbium *n* columbium (Symb. Cb), niobium
Columbobitter *n* columbin
Colupulon *n* colupulone
Colzaöl *n* colza oil
Combinal *n* combinal
Combitron *n* combitron
Combretumblätter *n pl* combrette leaves
Compoundkern *m* (Atom) compound nucleus
Compoundkernmodell *n* (Atom) compound nuclear model
Compoundmaschine *f* compound engine, (Elektr) compound generator
Compoundöl *n* compounded oil
Compoundpapier *n* compound impregnated paper
Compoundstahl *m* compound steel, casehardened soft steel
Comptoneffekt *m* Compton effect
Comptonelektron *n* Compton electron
Comptoniaöl *n* sweet fern oil
Comptonit *m* (Min) comptonite
Compton-Simonscher Versuch *m* (Atom) Compton-Simon experiment
Compton-Strahlung *f* Compton radiation
Comptonstreuung *f* Compton scattering
Comptonverschiebung *f* Compton shift

Compton-Wellenlänge *f* Compton wave-length
Computer *m* computer, ~ **für die direkte digitale Verfahrensregelung** on-line digital process control computer
con- siehe auch kon-
Conamin *n* conamine
Conanin *n* conanine
Conarrhimin *n* conarrhimine
Conchairamidin *n* conchairamidine
Conchinamin *n* conquinamine
Conchinin *n* conchinine, quinidine
Conchit *n* conchite
Condurangin *n* condurangin
Condurangofluidextrakt *m* condurango liquid extract
Condurangorinde *f* condurango bark
Condurit *m* condurite
Conephrin *n* conephrin
Conessidin *n* conessidine
Conessin *n* conessine
Conessirinde *f* conessi bark
Conglomeron *n* conglomerone
Congoblau *n* Congo blue
Congocidin *n* congocidine
Congogelb *n* Congo yellow
Conhydrin *n* conhydrine, conydrine
Conhydrinon *n* conhydrinone
Conicein *n* coniceine
Conicin *n* conicine, coniine
Conicinbalsam *m* conicine balsam
Conidendrin *n* conidendrin
Conidin *n* conidine
Coniferin *n* coniferin
Coniferosid *n* coniferoside
Coniferyl- coniferyl
Coniferylaldehyd *m* coniferaldehyde, ferulaldehyde
Coniferylalkohol *m* coniferol, coniferyl alcohol
Coniin *n* coniine, conine
Coniinextrakt *m* coniine extract
Coniinhydrobromid *n* coniine hydrobromide
Coniinsalbe *f* coniine ointment
Conima *n* conima
Conimaharz *n* conima
Conin *n* coniine, conine
Coniumsäure *f* conic acid, conicic acid, coniic acid
Conkurchin *n* conkurchine
Connellit *m* (Min) connellite
Connesin *n* connesine
Conopharyngin *n* conopharyngine
Conopharynginsäure *f* conopharynginic acid
Conteben *n* conteben, thioacetazone
Contergan *n* (Pharm) contergan, (Pharm, HN) thalidomid
Conus *m* cone
Convallamarin *n* convallamarin
Convallamarogenin *n* convallamarogenin
Convallaretin *n* convallaretin

Convallarin *n* convallarin
Convallatoxin *n* convallatoxin
Convallosid *n* convalloside
Conveyoranlage *f* conveyor plant, transporter plant
Convicin *n* convicin
Convolamin *n* convolamine
Convolvidin *n* convolvidine
Convolvin *n* convolvine
Convolvulin *n* convolvulin, rhodeorhetin
Convolvulinsäure *f* convolvulic acid
Conwayschale *f* Conway micro diffusion dish
Conydrin *n* conhydrine, conydrine
Conylen *n* 1,4-octadiene, conylene
Conyrin *n* conyrine
Cookeit *m* (Min) cookeite
Coolgardit *m* coolgardite
Coorongit *m* (Min) coorongit
Copaen *n* copaene
Copaiva *f* copaiva, copaiba
Copaivabalsam *m* copaiva balsam
Copaivabalsamöl *n* copaiva oil
Copaivaöl *n* copaiva oil
Copaivasäure *f* copahuvic acid, copaivic acid
Copal *m* copal
Copaldryer *m* copal dryer
Copalharz *n* copal
Copalin *n m* copalin, kopalin
Copellidin *n* copellidine
Copiapit *m* (Min) copiapite
Copolykondensation *f* copolycondensation
Copolymer *n* copolymer
Copolymerisation *f* copolymerization
Coptisin *n* coptisine
Copyrin *n* copyrine
Coquimbit *m* (Min) coquimbite
Coracit *m* (Min) coracite
Corallin *n* corallin, aurine
Coralydin *n* coralydine
Coralydiniumhydroxid *n* coralydinium hydroxide
Coralyn *n* coralyn
Coralynbase *f* coralyne base
Coralynhydrat *n* coralyne hydrate
Coralyniumhydroxid *n* coralynium hydroxide
Coramin *n* coramine
Coranen *n* coranene
Corazol *n* corazole, cardiazole, pentetrazole
Corbadrin *n* corbadrine
Corbisterin *n* corbisterol
Corchor[gen]in *n* corchorigenin
Corchorolsäure *f* corchorolic acid
Corchsularose *f* chorchsularose
Cordgewebe *n* (Text) cord fabric
Cordierit *m* (Min) cordierite, dichroite, iolite
Cordit *n* cordite
Cordnummer *f* cable size
Cordpreßmasse *f* fabric-filled molding compound

Cordrastin *n* cordrastine
Cordreifen *m* cord type tire
Cordycepin *n* cordycepin
Cordycepinsäure *f* cordycepic acid
Cordycepose *f* cordycepose
Cordylit *m* (Min) cordylite
Coregonin *n* coregonin
Coreopsin *n* coreopsin
Corepressor *m* (Mol. Biol) corepressor
Coreximin *n* coreximine
Coriamyrtin *n* coriamyrtin
Coriander *m* coriander seeds
Corianderöl *n* coriander oil
Coriandrol *n* coriandrol, linalool
Coridin *n* coridine
Corioliskraft *f* coriolis force
Corkit *m* (Min) corkite
Corlumin *n* corlumine
Cornicularin *n* cornicularine
Cornin *n* cornic acid, cornin
Cornubianit *m* cornubianite
Cornwallit *m* (Min) cornwallite
Coroglaucigenin *n* coroglaucigenin
Coronadit *m* (Min) coronadite
Coronaentladung *f* corona discharge
Coronaridin *n* coronaridine
Coronillin *n* coronillin
Coronium *n* (Leg) coronium
Coroniumlinie *f* coronium line
Coronopsäure *f* coronopic acid
Corotoxigenin *n* corotoxigenin
Corpaverin *n* corpaverine
Corphin *n* corphin
Corpus-luteum-Hormon *n* corpus luteum hormone, progesterone
Correllogenin *n* correllogenin
Corrin *n* corrin
Corrodkote-Verfahren *n* corrodkote process
Corrol *n* corrole
Corrosivum *n* corrosive
Corsit *m* (Min) corsite, napoleonite
Cortex *m* (Biol) cortex
Corticin *n* corticin
Corticinsäure *f* corticinic acid
Corticoid *n* corticoid
Corticosteron *n* corticosterone
Corticotropin *n* corticotropin, ACTH
Corticrocin *n* corticrocin
Cortisalin *n* cortisalin
Cortisol *n* cortisol
Cortison *n* cortisone
Cortol *n* cortol
Cortolon *n* cortolone
Corund *m* corundum (s. a. Korund)
Corybulbin *n* corybulbine
Corycavidin *n* corycavidine
Corycavin *n* corycavine
Corycavinmethin *n* corycavinemethine
Corydaldin *n* corydaldine

Corydalin *n* corydaline
Corydalisalkaloid *n* corydalis alkaloid
Corydin *n* corydine
Corydinsäure *f* corydinic acid
Corylin *n* coryline
Corylopsin *n* corylopsin
Corynantheal *n* corynantheal
Corynanthean *n* corynantheane
Corynanthein *n* corynantheine
Corynantheinsäure *f* corynantheic acid
Corynanthidin *n* corynanthidine
Corynanthin *n* corynanthine
Corynanthinsäure *f* corynanthic acid
Corynanthyrin *n* corynanthyrine
Corynein *n* coryneine
Corynomycolensäure *f* corynomycolenic acid
Corynomycolsäure *f* corynomycolic acid
Corynoxein *n* corynoxeine
Coryocavidin *n* coryocavidine
Coryocavin *n* coryocavine
Corypallin *n* corypalline
Corypalmin *n* corypalmine
Corytuberin *n* corytuberine
Corytuberinmethin *n* corytuberinemethine
Cosalit[h] *m* (Min) cosalite
Cosinus *m* (Math) cosine
Cosmen *n* cosmene
Cosmeticum *n* cosmetic
Cosmotron *n* (Atom) cosmotron
Cossait *m* cossaite
Cossasalz *n* cossa salt
Cossyrit *m* (Min) cossyrite
Costaclavin *n* costaclavine
Costen *n* costene
Costol *n* costol
Costunolid *n* costunolide
Costuslacton *n* costus lactone
Costusöl *n* costus oil
Costussäure *f* costic acid, costusic acid
Costylchlorid *n* costyl chloride
Cotarnin *n* cotarnine
Cotinin *n* cotinine
Cotoin *n* cotoin
Cotorinde *f* coto bark
Cotta *f* bloom, cake, lump
Cotton-Effekt *m* cotton effect, **mehrfacher** ~ multiple cotton effect
cottonisieren (Text) to cottonize
Cottonöl *n* cotton oil, cottonseed oil
Cottonwirkmaschine *f* (Text) flat knitting machine
Cottrell-Blockierung *f* (Krist) Cottrell locking
Cottrell-[Entstaubungs-]Verfahren *n* Cottrell [electrical precipitation] process
Cotunnit *m* (Min) cotunnite
Couette-Strömung *f* Couette flow
Coulomb *n* (Maß) coulomb
Coulomb-Energie *f* Coulomb energy
Coulomb-Kräfte *f pl* Coulomb forces
Coulombmeßgerät *n* coulombmeter, coulometer, voltameter
Coulombmeter *n* coulometer, voltameter
Coulombsche Abstoßung *f* Coulombian repulsion
Coulombsches Gesetz *n* Coulomb's law
Coulombsche Waage *f* (Elektr) torsion balance, Coulomb's balance
coulometrisch coulometric
Coumonöl *n* coumon oil
Courantfokussierung *f* strong focusing
Couseranit *m* (Min) couseranite
Covellin *m* (Min) covelline, covellite, indigo copper
Covellit *m* (Min) covelline, covellite
Cowper *m* Cowper stove, hot-air stove
Cowper-Coles-Methode *f* Cowper-Coles (coldgalvanizing method)
Cozymase I *f* cozymase I, codehydrase I, coenzyme I, Co I, DPN, NAD
Cozymase II *f* cozymase II, cocarboxylase, thiaminpyrophosphate, TPP
Crackanlage *f* cracking plant
Crackbenzin *n* cracked gasoline
Crackdestillation *f* cracking distillation, destructive distillation
cracken (Erdöl) to crack
Cracken *n* cracking
Crackprozeß *m* cracking process
Crag *m* crag
Cragformation *f* crag formation
Craigverteilung *f* countercurrent distribution
Crandallit *m* (Min) crandallite
Crataegin *n* crataegin
Creatin *n* creatine (s. a. Kreatin)
Creatinin *n* creatinine
Crebanin *n* crebanine
Credesches Silber *n* collargol, colloidal silver
Crednerit *m* (Min) crednerite
Creme *f* cream
cremefarben cream-colored
Cremor Tartari *m* acid potassium tartrate, cream of tartar, potassium bitartrate, potassium hydrogen tartrate
Creolin *n* creolin, analgin
Cresatin *n* cresatin
Cresidin *n* cresidine, aminohydroxytoluene
Cresineol *n* cresineol
Cresyl- cresyl
Cresylit *n* cresylite
Crichtonit *m* (Min) crichtonite
Crinamin *n* crinamine
Crinan *n* crinane
Crinidin *n* crinidine
Crinin *n* crinine, crinidine
Crispatsäure *f* crispatic acid
Cristobalit *m* (Min) cristobalite
Cristophit *m* (Min) cristophite
Crithmen *n* crithmene

Crithminsäure f crithminic acid
Criwellin n criwelline
Crocein n crocein
Croceinorange n crocein orange
Croceinsäure f croceic acid, 2-naphthol-8-sulfonic acid
Croceinscharlach m crocein scarlet
Crocetan n crocetane
Crocetin n crocetin
Crocetindialdehyd m crocetin dialdehyde
Crocin n crocin
Crocose f crocose
Crocosit m crocosite
Crookesit m crookesite
Crossit m crossite
Crotalotoxin n crotalotoxin
Crotamiton n crotamiton
Crotonaldehyd m crotonaldehyde, 2-butenal
Crotonalkohol m crotonyl alcohol, 3-buten-1-ol, crotyl alcohol
Crotonase f (Biochem) crotonase
Crotonat n crotonate, salt or ester of crotonic acid
Crotonbetain n croton betaine
Crotonharz n croton resin
Crotonin n crotonin
Crotonkörner n pl croton seeds
Crotonlacton n croton lactone
Crotonöl n croton oil
Crotonolsäure f crotonolic acid
Crotonsaat f croton seed
Crotonsäure f 1-carboxypropylene, crotonic acid
Crotonsäureester m crotonic ester
crotonsauer crotonate
Crotonyl- crotonyl
Crotonylen n crotonylene, dimethylacetylene
Crotophenon n crotonophenone
Crotylalkohol m crotyl alcohol, butenol
Crotylchlorid n crotyl chloride
Croweacin n croweacin
Croweacinsäure f croweacic acid
Crucilith m crucilite
Crurin n crurin, quinoline bismuth thiocyanate
Crusocreatinin n crusocreatinine
Crustaceenhormon n crustacea's hormone
Cryofluoran n cryofluorane
Cryptaustolin n cryptaustoline
Cryptograndosid n cryptograndoside
Cryptomeren n cryptomerene
Cryptomeriol n cryptomeriol
Cryptometer n cryptometer
Crypton n cryptone
Cryptopin n cryptopine
Cryptopinon n cryptopinone
Cryptowollin n cryptowolline
C-Säure f c-acid
Cuban n cubane
Cubanit m (Min) cubanite
Cubeben n cubebene
Cubebenextrakt m extract of cubeb
Cubebenöl n cubeb oil
Cubebenpfeffer m cubebs
Cubebin n cubebin
Cubebinäther m cubebin ether
Cubebinol n cubebinol
Cubebinolid n cubebinolide
Cubebinsäure f cubebic acid
Cubicit m (Min) cubicite
Cuboit m (Min) cubicite, cuboite
Cubosilicit m (Min) cubosilicite
Cucolin n cucoline
Cucurbitaceenöl n cucurbitaceac oil
Cucurbitacin n cucurbitacin
Cucurbitol n cucurbitol
Cuiteseide f boiled-off silk, cuite-silk, degummed silk
Cularimin n cularimine
Cularin n cularine
Culilawanöl n culilaban oil
Culsageeit m (Min) culsageeite
Cumalin n coumalin, cumalin
Cumalinsäure f coumalic acid, 2-oxo-1,2-pyran-5-carboxylic acid
Cumaraldehyd m coumaraldehyde
Cumaran n coumaran
Cumaranon n coumaranone
Cumarazon n coumarazone
Cumarilsäure f coumarilic acid
Cumarin n 1,2-benzopyrone, coumarin, cumarin
Cumarinchinon n coumarin quinone
Cumarinolin n coumarinoline
Cumarinsäure f coumaric acid, coumarinic acid, hydroxycinnamic acid
Cumaron n coumarone, benzofuran
Cumaronharz n coumarone resin
Cumaronpikrat n coumarone picrate
Cumarsäure f coumaric acid, coumarinic acid, hydroxycinnamic acid
Cumengeit m (Min) cumengeite
Cumenol n cumenol, cuminol
Cumidin n cumidine
Cumidinsäure f cumidic acid
Cumin n cumin
Cuminaldazin n cuminaldazine
Cuminaldehyd m cuminaldehyde, cumaldehyde, cumic aldehyde, p-isopropylbenzaldehyde
Cuminalkohol m cumic alcohol
Cuminil n cuminil, bicuminal, dicuminal
Cuminöl n cumin oil
Cuminoin n cuminoin, diisopropyl benzoin
Cuminol n cuminol, cumin alcohol, isopropylbenzyl alcohol
Cuminon n cuminone
Cuminsäure f cumic acid, cuminic acid
Cuminsamen m cumin seed
Cuminyl n cuminyl
Cummingtonit m (Min) cummingtonite
Cumochinol n cumoquinol

Cumochinon *n* cumoquinone
Cumöstan *n* coumestan
Cumöstrol *n* coumestrol
Cumol *n* cumene
Cumolen *n* cumolene
cumulusartig cumuliform
Cumyl- cumyl, cumenyl
Cumylsäure *f* cumylic acid, durylic acid
Cuorin *n* cuorin
Cuparen *n* cuparene
Cuparensäure *f* cuparenic acid
Cupferron *n* cupferron, copperone, nitrosophenyl hydroxylamine
Cupolstein *m* cupola brick
Cupral *n* cupral
Cupramminbase *f* cuprammine base, ammoniacal copper oxide, cuprammonia, cuprammonium hydroxide, tetrammine copper(II) hydroxide
Cuprat *n* cuprate
Cuprean *n* cupreane
Cupreidan *n* cupreidane
Cupreidin *n* cupreidine
Cuprein *n* cupreine
Cupren *n* cuprene, carbene
Cupretenin *n* cupretenine
Cupricarbonat *n* copper(II) carbonate, cupric carbonate
Cuprichlorid *n* copper(II) chloride, cupric chloride
Cuprichromat *n* copper(II) chromate, cupric chromate
Cupricyanwasserstoffsäure *f* cupricyanic acid, cyanocupric acid
Cuprifikation *f* cuprification
Cuprioxalat *n* (Kupfer(II)-oxalat) cupric oxalate
Cuprioxid *n* cupric oxide
Cuprisalz *n* (Kupfer(II)-Salz) cupric salt
Cuprisulfid *n* cupric sulfide
Cuprit *m* (Min) cuprite, ruberite, ruby copper
Cupriverbindung *f* (Kupfer(II)-Verbindung) copper(II) compound, cupric compound
Cupriweinsäure *f* cupritartaric acid
Cuproadamit *m* (Min) cuproadamite
Cuprochlorid *n* (Kupfer(I)-chlorid) copper(I) chloride, cuprous chloride
Cuprocuprisulfit *n* cuprocupric sulfite
Cuprocyanwasserstoffsäure *f* cuprocyanic acid, cyanocuprous acid
Cuprodekapierbad *n* cuprous pickling bath
Cuprodescloizit *m* (Min) cuprodescloizite
Cuprofulminat *n* cuprofulminate
Cuprohämol *n* cuprohemol
Cuproion *n* copper(I) ion, cuprous ion
Cupromagnesit *m* (Min) cupromagnesite
Cupromangan *n* cupromanganese
Cupromercurijodid *n* cuprous iodomercurate(II), cuprous mercuric iodide
Cupron *n* cupron, α-benzoin-oxime
Cupronelement *n* copper oxide cell, cupron cell
Cupronin *n* cupronine
Cuprooxid *n* (Kupfer(I)-oxid) copper(I) oxide, cuprous oxide
Cuproplumbit *m* cuproplumbite
Cupropyrit *m* (Min) cupropyrite
Cuprosalz *n* (Kupfer(I)-Salz) copper(I) salt, cuprous salt
Cuproscheelit *m* (Min) cuproscheelite
Cuprosulfit *n* copper(I) sulfite, cuprous sulfite
Cuprotungstit *m* (Min) cuprotungstite
Cuprotypie *f* cuprotypy
Cuproverbindung *f* (Kupfer(I)-Verbindung) copper(I) compound, cuprous compound
Cuprum *n* (Lat) copper
Curacao *m* curacao
Curacaoschale *f* curacao peel
Curacaoschalenöl *n* curacao peel oil
Curacose *f* curacose
Curare *n* curare, curara, curari, woorara
Curarin *n* curarine
Curarisation *f* curarization
Curaroid *n* curaroid
Curbin *n* curbine
Curcasöl *n* curcas oil
Curcuma *f* (Bot) curcuma, turmeric
Curcumagelb *n* curcumin
Curcumaöl *n* turmerol
Curcumapapier *n* curcuma paper, turmeric paper
Curcumaprobe *f* turmeric test
Curcumasäure *f* curcumic acid
Curcumawurzel *f* (Bot) turmeric
Curcumen *n* curcumene
Curcumin *n* curcumin
Curcumon *n* curcumone
Curie *n* (Phys) curie, Curie unit
Curiegehalt *m* Curie strength
Curiegraph *m* curiegraph
Curiekonstante *f* Curie constant
Curiepunkt *m* Curie point
Curiesches Gesetz *n* Curie's law
Curieskopie *f* curiescopy
Curie-Temperatur *f* curie temperature
Curin *n* curine
Curium *n* curium (Symb. Cm)
Curiumisotop *n* curium isotope
Curral *n* curral
Curry *m* curry
Curtiusscher Abbau *m* (Chem) Curtius degradation
Curtius-Umlagerung *f* Curtius rearrangement
Cuscamidin *n* cuscamidine
Cuscohygrin *n* cuscohygrine, hygrine
Cusconidin *n* cusconidine
Cuscorinde *f* cusco bark, red bark
Cuskhygrin *n* cuscohygrine
Cusparcin *n* cusparcine
Cusparein *n* cuspareine

Cuspariarinde f cusparia bark
Cusparidin n cusparidine
Cusparin n cusparine
Cuspidin m (Min) cuspidine
Custerit m (Min) custerite
Cusylol n cusylol
Cutinase f (Biochem) cutinase
Cutininsäure f cutinic acid
Cutinsäure f cutic acid
Cutose f cutose
Cyamelid n cyamelide
Cyan n cyanogen
Cyanacetamid n cyanacetamide, cyanoacetamide
Cyanäther m ethyl cyanide, hydrocyanic ether, propanenitrile, propionitrile
Cyanäthylierung f cyanoethylation
Cyanalkali n alkali cyanide
Cyanalkyl n alkyl cyanide, cyanoalkyl
Cyanamid n cyanamide
Cyanamidcalcium n calcium cyanamide
Cyanamidnatrium n sodium cyanamide
Cyanamidokohlensäure f cyanamidocarboxylic acid, cyanocarbamic acid
Cyanammonium n ammonium cyanide
Cyananil n cyananil
Cyananilid n cyananilide, cyanoanilide
Cyananilsäure f cyananilic acid
Cyananthren n cyananthrene, cyanoanthrene
Cyanat n cyanate, salt or ester of cyanic acid
Cyanazid n cyanogen azide
Cyanbad n cyanide bath
Cyanbarium n barium cyanide
Cyanbenzol n cyanobenzene
cyanblau cyanic
Cyanblau n cyan blue
Cyanbromid n cyanogen bromide
Cyanchlorid n cyanogen chloride, chlorine cyanide, cyanchloride
Cyandiskrepanz f cyanogen discrepancy
Cyandoppelsalz n double cyanide
Cyaneisen n iron cyanide
cyaneisenhaltig cyanoferric
Cyaneisenverbindung f iron cyanogen compound
Cyanessigester m cyanoacetic ester
Cyanessigsäure f cyanoacetic acid
Cyanessigsäurechlorid n cyanoacetyl chloride
cyanessigsauer cyanoacetate
Cyangas n cyanogen gas, cyanogas
Cyangold n gold cyanide
Cyangoldkalium n potassium auricyanide, potassium cyanoaurate(III)
Cyangruppe f cyano group
Cyanhärtung f cyanide hardening, cyaniding
cyanhaltig containing cyanogen, cyanous
Cyanhydrin n cyanohydrin, cyan alcohol
Cyanhydrinsynthese f cyanohydrin synthesis
Cyanid n cyanide
Cyanidan n cyanidan

Cyanidenolon n cyanidenolone
cyanidfrei free from cyanide
Cyanidhämatoporphyrin n cyanide hematoporphyrin
Cyanidin n cyanidin
Cyanidlauge f cyanide liquor
Cyanidlaugung f cyanide process, cyanidation
Cyanidmesoporphyrin n cyanide mesoporphyrin
Cyanidylin n cyanidylin
Cyanierung f cyanation
cyanig containing cyanogen, cyanous
Cyanin n cyanine, quinoline blue
Cyaninfarbstoff m cyanine, Leitch's blue, quinoline blue
cyanisieren to cyanize
Cyanisierung f cyanization
Cyanit m (Min) cyanite
Cyanjodid n cyanogen iodide, iodine cyanide
Cyankali n potassium cyanide
Cyankalium n potassium cyanide
Cyankaliumlauge f potassium cyanide liquor
Cyankobalt n cobalt cyanide
Cyankohlensäure f cyanocarbonic acid, cyanoformic acid
Cyankupfer n copper cyanide
Cyanmetall n metal cyanide
Cyanmethin n cyanmethine
Cyanmethyl n methyl cyanide, acetonitrile
Cyannatrium n sodium cyanide
Cyanocobalamin n cyanocobalamin
Cyanoform n cyanoform, tricyanomethane
Cyanogen n cyanogen, cyan cyanide
Cyanogensäure f cyanogenic acid
Cyanoguanidin n cyanoguanidine, dicyandiamide
Cyanol n cyanol, aniline
Cyanolgrün n cyanol green
Cyanomaclurin n cyanomaclurin
Cyanometer n cyanometer
Cyanophyllin n cyanophylline
Cyanoporphyrin n cyanoporphyrine
Cyanose f (Med) cyanopathy, cyanosis
Cyanosilan n cyanosilane
Cyanosin n cyanosine
cyanotisch cyanotic
Cyanotypie f blueprinting, cyanotype process, ferrotypy
Cyanotyppapier n cyanotype paper
Cyanphenanthren n cyanphenanthrene
Cyanphenyl- cyanphenyl
Cyanplatin n platinum cyanide
Cyanpräparat n cyanide preparation
Cyanpropionsäure f cyanopropionic acid
Cyanpyronin n cyanopyronine
Cyanquecksilber n mercury cyanide
Cyansäure f cyanic acid, cyanhydroxide
cyansauer cyanate
Cyanseife f cyanogen soap

Cyansenföl *n* cyanated mustard oil, cyanomustard oil
Cyansilber *n* silver cyanide
Cyansilberbad *n* silver cyanide bath
Cyanstickstoff *m* cyanonitride
Cyanstickstofftitan *n* titanium cyanonitride
Cyanthiamin *n* cyanothiamine
Cyantoluol *n* cyanotoluene
Cyantypverfahren *n* blueprinting, cyanotype process, ferrotypy
Cyanür *n* (obs) cyanide
Cyanur- cyanur
Cyanurat *n* cyanurate
Cyanurchlorid *n* cyanuric chloride, trichloro-s-triazine
Cyanursäure *f* cyanuric acid, s-triazinetriol, tricyanic acid
cyanursauer cyanurate
Cyanurtriamid *n* cyanurotriamide, 2,4,6-triamino-s-triazine, melamine
Cyanurtricarbonsäure *f* cyanurotricarboxylic acid, s-triazinetricarboxylic acid
Cyanurtrichlorid *n* cyanurotrichloride, cyanuric chloride, trichloro-s-triazine
Cyanurtricyanid *n* cyanurotricyanide, cyanuric cyanide, tricyano-s-triazine
Cyanurtrihydrazid *n* cyanurotrihydrazide, cyanuric hydrazide, trihydrazino-s-triazine
Cyanverbindung *f* cyanogen compound
Cyanwasserstoff *m* hydrocyanic acid, hydrogen cyanide, Prussic acid
Cyanwasserstoffsäure *f* hydrocyanic acid, hydrogen cyanide, Prussic acid
cyanwasserstoffsauer cyanide
Cyanzink *n* zinc cyanide
Cyarsal *n* cyarsal
Cybertron *n* cybertron
Cycasin *n* cycasin
Cyclamat *n* cyclamate
Cyclamen *n* cyclamen
Cyclamenaldehyd *m* cyclamenaldehyde
Cyclamin *n* cyclamin, arthranitin
Cyclaminsäure *f* cyclamic acid
Cyclamose *f* cyclamose
Cyclandelat *n* cyclandelate
Cyclanolin *n* cyclanoline
Cyclazin *n* cyclazin
Cycleanin *n* cycleanine
Cyclen *n* cyclene
Cyclisierung *f* cyclization, ring formation
Cyclizin *n* cyclizine
Cycloaddition *f* cycloaddition
cycloaliphatisch cycloaliphatic
Cycloalkan *n* cycloalkane
Cycloalkannin *n* cycloalkannine
Cycloalliin *n* cycloalliin
Cycloartan *n* cycloartane
Cycloartenol *n* cycloartenol
Cycloartenon *n* cycloartenone

Cyclobarbital *n* cyclobarbital
Cyclobutadien *n* cyclobutadiene, cyclobutylene
Cyclobutan *n* cyclobutane
Cyclobutanon *n* cyclobutanone
Cyclobuten *n* cyclobutene
Cyclobutyl- cyclobutyl
Cyclocampher *m* cyclocamphor
Cyclocamphidin *n* cyclocamphidine
Cyclocholestan *n* cyclocholestan
Cyclocitral *n* cyclocitral
Cyclocoloranol *n* cyclocoloranol
Cyclocolorenon *n* cyclocolorenone
Cyclodecan *n* cyclodecane
Cyclodecandion *n* cyclodecandione
Cyclodiastereomerie *f* (Stereochem) cyclodiastereomerism
Cyclodimerisation *f* cyclodimerisation
Cycloenantiomer *n* (Stereochem) cycloenantiomer
Cycloenantiomerie *f* (Stereochem) cycloenantiomerism
Cycloeucalan *n* cycloeucalane
Cyclofenchen *n* cyclofenchene
Cycloform *n* cycloform, isobutyl-p-aminobenzoate
Cyclogallipharaol *n* cyclogallipharaol
Cyclogeraniol *n* cyclogeraniol
Cyclogeraniolen *n* cyclogeraniolene
Cyclogeraniumsäure *f* cyclogeranic acid
Cycloheptadecanon *n* cycloheptadecanone
Cycloheptadien *n* cycloheptadiene
Cycloheptan *n* cycloheptane, suberane
Cycloheptanol *n* cycloheptanol, suberol
Cycloheptanon *n* cycloheptanone, suberone
Cycloheptatrien *n* cycloheptatriene
Cyclohepten *n* cycloheptene, suberene
Cycloheptin *n* cycloheptine
Cyclohexadien *n* cyclohexadiene
Cyclohexan *n* cyclohexane, hexamethylene
Cyclohexandiol *n* cyclohexanediol
Cyclohexandiondioxim *n* cyclohexanedionedioxime
Cyclohexanol *n* cyclohexanol
Cyclohexanon *n* cyclohexanone, pimelinketone
Cyclohexanonoxim *n* cyclohexanone oxime
Cyclohexatrien *n* cyclohexatriene, benzene
Cyclohexen *n* cyclohexene
Cyclohexenol *n* cyclohexenol
Cyclohexenon *n* cyclohexenone
Cyclohexenyl- cyclo hexenyl
Cycloheximid *n* cycloheximide, naramycine
Cyclohexin *n* cyclohexine
Cyclohexyl- cyclohexyl
Cyclohexylacetat *n* cyclohexyl acetate
Cyclohexyläthylamin *n* cyclohexylethylamine
Cyclohexylamin *n* cyclohexylamine
Cyclohexyliden *n* cyclohexylidene
Cyclohexylmethacrylat *n* cyclohexyl methacrylate

Cyclohexylsilicon – Cytidin

Cyclohexylsilicon *n* cyclohexyl silicone
Cyclohexyltrichlorsilan *n* cyclohexyl trichlorosilane
Cyclohomocitral *n* cyclohomocitral
Cyclohomogeraniol *n* cyclohomogeraniol
Cyclohomogeraniumsäure *f* cyclohomogeranic acid
Cycloide *f* (Math) cycloid
Cyclolavandulol *n* cyclolavandulol
Cyclomethycain *n* cyclomethycaine
Cyclon *n* cyclone, tetracyclone
Cyclonit *n* cyclonite, RDX
Cyclononan *n* cyclononane
Cyclonovobiocinsäure *f* cyclonovobiocic acid
Cyclooctadien *n* cyclo-octadiene
Cyclooctan *n* cyclooctane
Cyclooctanon *n* cyclo-octanone
Cyclooctatetraen *n* cyclo-octatetraene
Cycloocten *n* cyclo-octene
Cycloolefin *n* cyclo-olefine
Cyclooligomerisation *f* cyclooligomerisation
Cyclopaldsäure *f* cyclopaldic acid
Cycloparaffin *n* cycloparaffin
Cyclopentadecanon *n* cyclopentadecanone
Cyclopentadien *n* cyclopentadiene
Cyclopentamethylen *n* cyclopentamethylene, cyclopentane
Cyclopentamin *n* cyclopentamine
Cyclopentan *n* cyclopentane, cyclopentamethylene
Cyclopentancarbonsäure *f* cyclopentanecarboxylic acid
Cyclopentanol *n* cyclopentanol
Cyclopentanon *n* cyclopentanone
Cyclopenten *n* cyclopentene
Cyclopentenon *n* cyclopentenone
Cyclopentyl- cyclopentyl
Cyclophorase *f* cyclophorase
Cyclophosphamid *n* cyclophosphamide
Cyclopin *n* cyclopin
Cyclopit *m* cyclopite
Cyclopolsäure *f* cyclopolic acid
Cyclopolymerisation *f* cyclopolymerization
Cyclopregnol *n* cyclopregnol
Cyclopropan *n* cyclopropane
Cyclopropen *n* cyclopropene
Cyclopropenon *n* cyclopropenone
Cyclopropyl- cyclopropyl
Cyclopterin *n* cyclopterin
Cyclopyrethrosin *n* cyclopyrethrosin
Cycloschwefel *m* cyclic sulfur
Cycloserin *n* cycloserine
Cyclosilan *n* cyclosilane
Cyclosiloxan *n* cyclosiloxane
Cyclosolanidan *n* cyclosolanidane
Cyclostereoisomerie *f* (Stereochem) cyclostereoisomerism
Cyclosteroid *n* cyclosteroid
Cyclothreonin *n* cyclothreonine
Cyclotron *n* cyclotron
Cycloundecan *n* cycloundecane
Cyclus *m* cycle (s. a. Zyklus)
Cycrimin *n* cycrimine
Cyklonit *n* (Sprengstoff) cyclonite, RDX
Cymarigenin *n* cymarigenin
Cymarin *n* cymarin
Cymarinsäure *f* cymaric acid
Cymarol *n* cymarol
Cymarose *f* cymarose
Cymatolith *m* cymatolite
Cymidin *n* cymidine, aminocymene
Cymobrenzcatechin *n* cymopyrocatechol
Cymogen *n* cymogene
Cymol *n* cymene, 1-methyl-4-isopropylbenzene
Cymometer *n* cymometer
Cymophan *m* (Min) cymophane, chrysoberyl
Cymoscop *n* cymoscope
Cymylamin *n* cymylamine
Cynanchotoxin *n* cynanchotoxine
Cynapin *n* cynapine
Cynocannosid *n* cynocannoside
Cynoctonin *n* cynoctonine
Cynodontin *n* cynodontin
Cynoglossofin *n* cynoglossophine
Cyperblau *n* Cyprus blue
Cyperen *n* cyperene
Cyperon *n* cyperone
Cypressenöl *n* cypress oil
Cyprinin *n* cyprinine
Cyprischer Vitriol *m* (Min) copper vitriol
Cyrtolith *m* (Min) cyrtolite
Cyrtominetin *n* cyrtominetin
Cyrtopterinetin *n* cyrtopterinetin
Cystamin *n* cystamine
Cystathion *n* cystathione
Cystathionase *f* (Biochem) cystathionase
Cystathionin *n* cystathionine
Cystazol *n* cystazol
Cysteamin *n* cysteamine, mercaptamine
Cystein *n* cysteine
Cysteindesulfhydrase *f* (Biochem) cysteine desulfhydrase
Cysteinhydrochlorid *n* cysteine hydrochloride
Cystein-Reduktase *f* (Biochem) cysteine reductase
Cysteinsäure *f* cysteic acid
Cystin *n* cystine
Cystinal *n* cystinal
Cystinamin *n* cystamine, cystineamine
Cystinhydantoin *n* cystine hydantoin
Cystinol *n* cystinol
Cystinose *f* cystinosis
Cystinurie *f* (Med) cystinuria
Cystolith *m* cystolite
Cystopurin *n* cystopurine
Cystosin *n* cystosine
Cytase *f* (Biochem) cytase
Cytidin *n* cytidine

Cytidindiphosphat *n* cytidine diphosphate, CDP
Cytidinmonophosphat *n* cytidine monophosphate, CMP
Cytidintriphosphat *n* cytidine triphosphate, CTP
Cytidylsäure *f* cytidylic acid
Cytisin *n* cytisine
Cytisolidin *n* cytisolidine
Cytisolin *n* cytisoline
Cyto- (s. auch Zyto-)
Cytochemie *f* cytochemistry
Cytochrom *n* cytochrome
Cytochromoxydase *f* (Biochem) cytochrome oxidase
Cytochromreduktase *f* (Biochem) cytochrome reductase
Cytode *f* cytode
Cytodeuteroporphyrin *n* cytodeuteroporphyrin
Cytogenese *f* cytogenesis
Cytogenetik *f* cytogenetics
Cytoglobin *n* cytoglobin
Cytokinese *f* cytokinesis
Cytologie *f* cytology
Cytolyse *f* cytolysis
Cytolysin *n* cytolysin
Cytoplasma *n* (Biol) cytoplasm, cytoplast
Cytopyrrolsäure *f* cytopyrrolic acid
Cytosamin *n* cytosamine
Cytosin *n* cytosine
Cytosol *n* cytosol
Cytostatikum *n* antitumor agent
Cytosylsäure *f* cytosylic acid

D

Dachanguß *m* tab gate
Dachlack *m* roof paint
Dachpappe *f* roofing paper, asphalted cardboard, asphalt felt, roofing felt
Dachsfett *n* badger fat
Dachziegel *m* [roofing] tile
Dacit *m* (Min) dacite
Dacren *n* dacrene
Dacron *n* dacron
Dacryden *n* dacrydene
Dactylin *n* dactylin
Dämmblatt *n* cleaner, sleeker
Dämmbrett *n* sleeker, smoother
dämmen to dam, to block, to curb
Dämmholz *n* pegging rammer
Dämmplatte *f* sound insulation board, sound insulation sheet
dämpfbar absorbable
Dämpfbarkeit *f* absorbability
dämpfen (Geräusche) to absorb, to muffle, (Licht) to damp, to dim, (Schwingungen) to absorb, to attenuate, to suppress, (Stoß) to absorb, to cushion
Dämpfen *n* absorbing, damping down
Dämpfer *m* damper, damping device, steam cooker
Dämpferflügel *m* damper wing
Dämpferflügelarm *m* arm of damping vane
Dämpferkammer *f* damping chamber
Dämpfkasten *m* cottage steamer
Dämpfung *f* absorption, quenching, suppression, (Lärm) muffling, (Licht) dampening, (Schwingungen) attenuation, (Stoß) cushioning, **~ je Längeneinheit** attenuation per unit length, attenuation constant, **aperiodische ~** dead beat, **mechanische ~** mechanical resistance or impedance
Dämpfungsausgleich *m* attenuation equalization
Dämpfungsbereich *m* attenuation region
Dämpfungsentzerrung *f* attenuation equalization
Dämpfungsfähigkeit *f* damping capacity
Dämpfungskonstante *f* attenuation coefficient
Dämpfungskreis *m* damping circuit
Dämpfungsmagnet *m* damping magnet
Dämpfungsmesser *m* (Elektr) decremeter
Dämpfungsmittel *n* sound damper
Dämpfungsparameter *m* damping parameter
Dämpfungsregler *m* damping regulator
Dämpfungsstromkreis *m* damping circuit
Dämpfungsvermögen *n* damping power
Dämpfungsverzerrung *f* distortion of amplitude
Dämpfungsvorrichtung *f* damper, damping device
Dämpfungszylinder *m* dash pot
Dänisch Weiß *n* Danish white
Dagingolsäure *f* dagingolic acid
Dagingoresen *n* dagingoresene
Dahliaviolett *n* dahlia violet
Dahlie *f* (Bot) dahlia
Dahlin *n* dahlin
Dahllit *m* (Min) dahllite
Dahlsche Säure *f* Dahl's acid
Dahmenit *n* dahmenite
Daidzein *n* daidzein
Daidzin *n* daidzin
Dakin-Oxydation *f* Dakin oxidation
Dakin-Reaktion *f* Dakin reaction
Dakinsche Lösung *f* Dakin's solution
Dalbergin *n* dalbergin
Dalton *n* (Maß) dalton
Dalzin *n* dalzin
Damascenin *n* damascenine
Damasceninsäure *f* damasceninic acid
Damaststahl *m* damascened steel
Damaszener Stahl *m* damask steel
damaszieren to damascene, to damask
damasziert damasked
Damaszierung *f* damascening
Dambonit *m* dambonitol
Dambose *f* dambose
Damiana *f* damiana
Damianaöl *n* damiana oil
Dammar *n* dammar
Dammaran *n* dammarane
Dammaranol *n* dammaranol
Dammarfirnis *m* dammar varnish
Dammarharz *n* dammar, dammar resin
Dammaroresen *n* dammaroresene
Dammarsäure *f* dammaric acid
Dammaryl *n* dammaryl
Dammerde *f* black earth, humus, mold
Dammgans *f* dross conduit
Dammgrube *f* foundry pit, molding hole
Damnacanthal *n* damnacanthal
Damnacanthol *n* damnacanthol
Damourit *m* (Min) damourite
Dampf *m* steam, fume, smoke, vapor, **~ erzeugen** to generate steam, to produce steam, **den ~ ablassen** to let the steam off or out, **direkter ~** live steam, **gedrosselter ~** throttled steam, **gesättigter ~** saturated steam, **überhitzter ~** superheated steam
Dampfabblaserohr *n* blow-off pipe, steam escape pipe
Dampfabführung *f* steam discharge
Dampfableitung *f* steam discharge
Dampfableitungsrohr *n* blow-off pipe, waste-steam pipe
Dampfabpreßmaschine *f* steam press
Dampfabscheider *m* steam separator
Dampfabsperrschieber *m* steam cut-off valve, steam gate valve

Dampfabsperrung *f* steam cut-off
Dampfabsperrventil *n* steam stop valve
Dampfadsorption *f* vapor adsorption
Dampfanlage *f* steam plant
Dampfapparat *m* steam apparatus
Dampfarmaturen *pl* steam fittings
Dampfaufnahmerohr *n* steam pipe
Dampfausdehnung *f* steam expansion
Dampfausgang *m* exhaust passage, exhaust way, vapor outlet
Dampfauslaßrohr *n* exhaust steam pipe
Dampfausnützung *f* utilization of steam
Dampfauspuff *m* steam exhaust
Dampfausströmung *f* steam outlet
Dampfaustrittsöffnung *f* steam exhaust port
Dampfbad *n* steam bath
Dampfbarometer *n* manometer
Dampfbedarf *m* steam consumption, steam requirement
Dampfbehälter *m* steam chamber
Dampfbelastungsfaktor *m* vapor capacity factor
Dampfbetrieb *m* steam power, operation by steam
Dampfbildung *f* formation of steam, steam generation
Dampfbildungswärme *f* steam-generating heat
Dampfblase *f* steam bubble
Dampfblasen *n* distillation with steam
Dampfblasenbildung *f* (in Kraftstoffsystemen) vapor lock (in fuel systems)
Dampfbleicherei *f* steam bleaching
Dampfboden *m* jacketed bottom
Dampfbrennerei *f* steam distillery
Dampfdarre *f* steam kiln
Dampfdestillation *f* steam distillation
Dampfdestilliersäule *f* steam distillation column
dampfdicht steam proof, steam-tight, vapor-tight
Dampfdichte *f* steam density, vapor density
Dampfdichtigkeit *f* steam density, vapor density
Dampfdichtigkeitsmesser *m* condenser gauge, vacuum gauge
Dampfdruck *m* steam compression, steam pressure, vapor pressure
Dampfdruckdiagramm *n* vapor-pressure diagram
Dampfdruckerei *f* steam printing
Dampfdruckerniedrigung *f* lowering of vapor pressure
Dampfdruckmesser *m* manometer, pressure gauge, steam gauge
Dampfdruckpumpe *f* steam pressure pump
Dampfdruckschmiedepresse *f* steam power forging press
Dampfdruckschreiber *m* vapor pressure recorder
Dampfdruckthermometer *n* vapor-pressure thermometer
Dampfdüse *f* steam nozzle
Dampfdurchsatz *m* (Dest) vapor throughput
Dampfdurchtritt *m* steam flow, steam flow opening
Dampfdurchtrittsöffnung *f* (Dest) vapor riser
Dampfdynamo *m* steam-driven generator, steam dynamo
dampfecht fast to steam
Dampfeinlaßschieber *m* steam slide valve
Dampfeinlaßventil *n* steam admission valve, steam stop valve
Dampfeinströmung *f* steam inlet
Dampfeinströmungskanal *m* steam inlet port
Dampfeintritt *m* steam inlet
Dampfelektrisiermaschine *f* hydro-electrostatic machine
dampfen to steam, to fume, to smoke
Dampfen *n* steaming, evaporation, vaporization
Dampfenergie *f* energy of steam
Dampfentnahme *f* withdrawal of steam, drawing-off of steam, steam outlet
Dampfentöler *m* oil separator for steam
Dampfentölung *f* separation of oil from steam
Dampfentwässerer *m* steam desiccator, steam trap
Dampfentwässerungsapparat *m* steam desiccator, steam-drying device
Dampfentwickler *m* steam developer, steam generator, vaporizer
Dampfentwicklung *f* steam generation
Dampfentwicklungsapparat *m* steam generator
dampferzeugend steam generating
Dampferzeuger *m* steam generator
Dampferzeugung *f* steam generation
Dampferzeugungsapparat *m* steam generator
Dampfesse *f* chimney shaft, funnel, steam-pipe
Dampffarbe *f* steam color
Dampffaß *n* autoclave, digestor, steam sterilizer
Dampffeuchtigkeit *f* moisture of vapor, wetness of steam
Dampffilter *m* steam filter
dampfflüchtig volatile in steam
dampfförmig vaporous
Dampfgas *n* superheated steam
Dampfgebläse *n* steam-blower
dampfgeheizt steam heated
Dampfgeschwindigkeit *f* vapor velocity
Dampfgeschwindigkeitsmesser *m* steam governor
dampfgetrocknet steam-dried
Dampfgewicht *n* weight of steam
Dampfhahn *m* steam cock
Dampfhammer *m* steam hammer
Dampfheizkörper *m* steam radiator
Dampfheizrohr *n* steam heating pipe
Dampfheizschlange *f* steam coil, superheater coil
Dampfheizung *f* steam-heating
Dampfheizungsvorrichtung *f* steam-heating apparatus
Dampfholzschliff *m* steamed mechanical wood pulp
Dampfhülle *f* steam jacket, vaporous envelope

dampfig — Dampfzuführungsrohr

dampfig steamy, vapory
Dampfkamin *m* (Dest) vapor riser
Dampfkammerform *f* jacketed mold
Dampfkessel *m* steam boiler, steam generator
Dampfkesselfeuerung *f* boiler firing
Dampfkesselspeisepumpe *f* boiler feed pump
Dampfkesselspeisewasser *n* steam boiler feed water
Dampfkochapparat *m* steam cooking apparatus
Dampfkocherei *f* steam boiling
Dampfkochkessel *m* autoclave
Dampfkochtopf *m* steamer
Dampfkolben *m* steam piston
Dampfkompressionskältemaschine *f* compressed vapor refrigerator
Dampfkraft *f* steam power
Dampfkraftanlage *f* steam plant
Dampfkraftgebläse *n* steam blower
Dampfkraftwerk *n* steam power station
Dampfkran *m* steam crane
Dampfkühler *m* steam cooler
Dampfkugel *f* (obs) aeolipyle
Dampflagerungsverfahren *n* heating in the steam autoclave
Dampfleim *m* steam-glue
Dampfleitung *f* steam piping, vapor supply pipe
Dampfleitungsrohr *n* steam line, steam pipe
Dampfmantel *m* steam jacket
Dampfmantelheizung *f* steam jacket heating
Dampfmaschine *f* steam engine
Dampfmaschinenanlage *f* steam plant
Dampfmesser *m* pressure gauge, manometer, steam gauge, steam meter
Dampfmessung *f* steam measuring
Dampfniederschlagsrohr *n* steam condenser pipe
Dampfofen *m* steam-heated oven
Dampfpackung *f* steam packing
Dampfphasenkorrosion *f* vapor-phase corrosion
Dampfproduktion *f* production of steam
Dampfpumpe *f* steam diaphragm pump, steam pump
Dampfpurpur *m* steam purple
Dampfraum *m* steam space
Dampfreduzierventil *n* steam reducing valve
Dampfregulativ *n* steam regulator
Dampfreibung *f* steam friction
Dampfreservoir *n* steam chamber
Dampfröstofen *m* steam roasting furnace
Dampfrohr *n* steam pipe, steam tube
Dampfsack *m* steam pocket
Dampfsammler *m* steam collector, steam chamber
Dampfsandstrahlgebläse *n* steam jet sand blast
Dampfsauger *m* steam exhaustor
Dampfschaufelbetrieb *m* steam shovel excavation
Dampfschieber *m* steam slide valve
Dampfschlange *f* steam coil
Dampfschlauch *m* steam hose

Dampfschmierung *f* steam lubrication
Dampfschreiber *m* steam recorder
Dampfschwarz *n* steam black
Dampfsicherheitsventil *n* steam safety valve
Dampfspannung *f* vapor tension
Dampfspannungsröhre *f* vapor pressure tube
Dampfspeicher *m* steam accumulator
Dampfsperre *f* vapor barrier
Dampfstiefel *m* steam cylinder
Dampfstimulation *f* steam injection
Dampfstoßverfahren *n* steam injection process
Dampfstrahl *m* steam jet
Dampfstrahlapparat *m* steam injector
Dampfstrahldiffusionspumpe *f* vapor jet diffusion pump
Dampfstrahler *m* vapor diffusion pump
Dampfstrahlgebläse *n* steam jet blower, injector
Dampfstrahlkältemaschine *f* steam-jet refrigeration machine
Dampfstrahlpumpe *f* injector
Dampfstrahlsauger *m* steam jet aspirator
Dampfstrahlzerstäuber *m* steam jet atomizer, steam jet sprayer
Dampfstrom *m* steam flow, (Atom) overhead
Dampfstrommesser *m* steam flow meter
Dampftabelle *f* steam table
Dampftemperatur *f* temperature of steam
Dampftopf *m* steam pot, autoclave, digester, steam sterilizer
Dampftrichter *m* steam funnel
Dampftrockenapparat *m* steam dryer
Dampftrockenheit *f* dryness of steam
Dampftrockner *m* steam dryer
Dampfturbine *f* steam turbine
Dampfüberdruck *m* excess vapor pressure
Dampfüberführung *f* conveyance of steam
Dampfüberhitzer *m* steam superheater
Dampfüberhitzung *f* steam superheating
Dampfüberleitung *f* conveyance of steam
Dampfuhr *f* steam gauge, steam meter
Dampfumformer *m* steam converter
Dampfumwälzeinrichtung *f* steam circulation system
Dampfventil *n* steam valve, vapor nozzle
Dampfverminderung *f* reduction of steam pressure
Dampfverteiler *m* steam distributor
Dampfvolumen *n* volume of steam
Dampfwasser *n* condensate
Dampfwasserabscheider *m* steam separator
Dampfwassertopf *m* steam trap
Dampfweg *m* steam flow, steam pipe, steam port
Dampfwinde *f* steam winch
Dampfzähler *m* recording steam meter
Dampfzeiger *m* steam dial, steam gauge, steam meter
Dampfzucker *m* sugar refined with steam
Dampfzuführung *f* steam supply
Dampfzuführungsrohr *n* steam supply pipe

Dampfzulaßventil n steam regulating valve
Dampfzuleitung f steam supply
Dampfzuleitungsrohr n steam inlet pipe, steam supply pipe
Dampfzustand m state of steam
Dampfzylinder m column steam apparatus
Damsin n damsin
Danain n danain
Danait m (Min) danaite
Danalith m (Min) danalite
Danburit m (Min) danburite
Daniell-Element n Daniell cell
Dannemorit m (Min) dannemorite
Dansyl-Aminosäure f dansylamino acid
Dansylchlorid n dansyl chloride
Danthron n danthron
Danziger Blau n Danzig blue
Danziger Goldwasser n Danzig brandy, Danzig water
Daphnandrin n daphnandrine
Daphneöl n daphne oil
Daphnetin n daphnetin, 7,8-dihydroxycoumarin
Daphnin n daphnin
Daphnit m (Min) daphnite
Daphnolin n daphnoline
Darapskit m (Min) darapskite
Darm m, **künstlicher** ~ synthetic gut
Darmbakterien pl (Bakt) intestinal flora
Darranlage f kiln plant
Darrarbeit f kiln-drying, liquation (copper)
Darraum m drying chamber, kiln chamber
Darrboden m drying floor; kiln floor
Darrbrett n drying board
Darre f kiln-drying, drying, drying room, kiln
darren to dry, to kiln-dry; to liquate or smelt copper
Darren n drying, kiln-drying, liquation (copper); torrefaction
Darrfläche f drying surface
Darrgekrätz n dross of copper; washing slag
Darrhürde f hurdle to dry malt
Darrkammer f drying chamber, drying house, drying stove
Darrkupfer n liquated copper
Darrmalz n cured malt, kiln-dried malt
Darrofen m drying oven, liquation hearth
Darrschrank m drying cupboard, drying oven
Darrsohle f dross of copper
Darrstaub m maltdust
darstellen to make, (als Kurve) to graph, to plot [a curve], (Chem) to prepare, (graphisch) to represent, (Techn) to manufacture, to produce, **einen Wert in Abhängigkeit von einem anderen** ~ to plot a value against another
Darstellung f representation, graph, manufacture, preparation, production, **graphische** ~ chart; diagram, **logarithmische** ~ logarithmic plotting, **vektorielle** ~ vector representation

Darstellungsverfahren n process of manufacture, process of preparation, process of production, (Chem) synthesis
darübergelagert superimposed, overlying
darüberliegen to lie on top of s. th.
darunterliegend subjacent, underlying
Darwinische Theorie f (Biol) Darwinian theory
Darwinismus m darwinism
Darwinit m (Min) darwinite
Dasymeter n dasymeter
Daten pl data
Datenerfassung f data gathering
Datenerfassungsgerät n data input equipment
Datensichtgerät n reader
Datenspeicher m data logger
Datensteuerung f numerical control
Datentechnik f, **Anlagen der mittleren** ~ office computers
Datenübertragungsrate f data transfer rate
datenverarbeitend data-processing
Datenverarbeitung f data processing
Datenverarbeitungsanlage f data-processing plant
Datierung f dating
Datiscagelb n datiscin
Datiscetin n datiscetin
Datiscin n datiscin
Datolith m (Min) datolite, datestone
Dattel f date
Dattelwein m date wine
Dattelzucker m date sugar
Datumsangabe f date
Datumsaufdruck m date mark
Daturaöl n datura oil
Daturin n daturine, hyoscyamine
Daturinsäure f daturic acid, heptadecanoic acid, heptadecylic acid, margaric acid
Daubreeit m (Min) daubreeite
Daubreelith m (Min) daubreelite
Daucol n daucol
Daucosterin n daucosterol
Daueranriß m fatigue crack
Dauerbeanspruchung f continuous load, fatigue loading, permanent load, permanent stress, repetition of stress
Dauerbetrieb m continuous operation
Dauerbiegebeanspruchung f repeated flexural stress, repetition of bending stress
Dauerbiegefestigkeit f flexing life, flexural endurance properties, flexural strength, reversed fatigue strength
Dauerbiegeversuch m endurance bending test, fatigue bending test
Dauerbiegung f fatigue bending
Dauerbruch m fatigue [vibration] failure, lasting rupture
Dauerdruck m continuous pressure
Dauereigenschaften pl permanent properties
Dauereinsatz m continuous use

Dauerelektrode *f* continuous electrode, self-backing electrode
Dauerentladung *f* continuous discharge
Dauerexperiment *n* longterm run or experiment
Dauerfestigkeit *f* endurance limit of stress, creep strength, limiting range of stress, ~ **im Druck-Schwellbereich** fatigue strength under pulsating [oscillating, fluctuating] compressive stress, ~ **im Zug-Wechselbereich** fatigue strength under alternating tensile stress
Dauerfestigkeitsgrenze *f* endurance limit, fatigue limit
Dauerfestigkeitsprüfmaschine *f* endurance-testing machine
Dauerfestigkeitsschaubild *n* fatigue strength diagram
Dauerfestigkeits-Verhältnis *n* fatigue ratio
Dauerform *f* permanent mold
Dauergeschwindigkeit *f* (Kfz) maintainable speed
Dauergleichgewicht *n* (Atom) radioactive equilibrium, secular equilibrium
dauerhaft permanent, durable, enduring, lasting
Dauerhaftigkeit *f* durability, permanence, stability
Dauerhaltbarkeit *f* fatigue durability, fatigue strength (of large structures), service life
Dauerknickversuch *m* repeated flexural test
Dauerlagerversuch *m* long-time storage test
Dauerleistung *f* continuous output
Dauermagnet *m* permanent magnet
Dauermagnetstahl *m* permanent-magnet steel
Dauerpräparat *n* long-life preparation
Dauerprüfung *f* continuous test, endurance test, fatigue test
Dauerschablone *f* duplicating stencil
Dauerschlagarbeit *f* repeated impact energy
Dauerschlagfestigkeit *f* fatigue impact strength
Dauerschlagprobe *f* continuous impact test
Dauerschlagversuch *m* fatigue impact test
Dauerschlagwerk *n* repeated impact testing machine
Dauerschlagzugversuch *m* repeated-impact tension test
Dauerschwingbeanspruchung *f* alternating [cyclic] stress, fatigue loading
Dauerschwingbruch *m* fatigue [vibration] failure
Dauerschwingfestigkeitsversuch *m* metal fatigue testing
Dauerschwingungsfestigkeit *f* endurance limit of stress, fatigue strength
Dauerschwingversuch *m* fatigue test, Woehler test, ~ **in der Kälte** fatigue test at low temperature, ~ **in der Korrosion** fatigue test under corrosion, ~ **in der Wärme** fatigue test at elevated temperature, ~ **unter Reibkorrosion** fatigue test under fretting corrosion

Dauerstandfestigkeit *f* creep resistance, creep strength, endurance properties (dimensional)
Dauerstandfestigkeitsprüfung *f* dynamic creep test
Dauertauchversuch *m* (Korr) total immersion test
Dauerversuch *m* long-duration test, endurance test, fatigue test
Dauerversuchsmaschine *f* fatigue-testing machine
Dauerwärmefestigkeit *f* thermal endurance properties
Dauerwellenpräparat *n* permanent wave preparation
Dauerwirkung *f* lasting effect
Dauerzugapparat *m* permanent set apparatus
Dauerzugversuch *m* endurance-tension test, repeated-direct-stress test, repeated tension test
Daumen *m* (an Sperrhebel) tappet, (Nocken) cam
Daumenantrieb *m* cam gear
Daumennagelprobe *f* finger nail test
Daumennocke *f* cam
Daumenrad *n* cam wheel
Daumenscheibe *f* cam disc
Daumenschraube *f* thumb screw
Daumentrommel *f* cam drum, tappet drum
Daumenwelle *f* camshaft
Daunomycin *n* daunomycin
Dauphinit *m* (Min) dauphinite
Dauricin *n* dauricine
Davyn *m* (Min) davyn
Davysche Sicherheitslampe *f* (Bergb) Davy lamp, Davy's safety lamp
Dawsonit *m* (Min) dawsonite
dazwischenliegend intermediary, intermediate
dazwischenschalten to interpose
Deacon-Prozeß *m* Deacon process
Deacylierung *f* deacylation
Debilinsäure *f* debilic acid
de Broglie Wellenlänge *f* de Broglie wavelength
Debye-Hückel-Theorie *f* Debye-Hückel theory
Debye-Scherrer-Ring *m* Debye-Scherrer ring or circle
Decaleszenz *f* (Therm) decalescence
Decamethonium *n* decamethonium
Decamethylenglykol *n* decamethylene glycol
Decan *n* decane
Decandiol *n* decane diol
Decanol *n* decanol, decyl alcohol
decarbonisieren to decarbonize, to decarburize
Decarbonisieren *n* decarbonizing, decarburizing
Decarbonisierung *f* decarbonization, decarburization
Decarboxylase *f* (Biochem) decarboxylase
Decarboxylierung *f* decarboxylation
Decatylen *n* decatylene
Decendicarbonsäure *f* decenedicarboxylic acid

Decevinsäure f decevinic acid
Dechenit m (Min) dechenite
dechlorieren to dechlorinate
Dechlorieren n dechlorinating
Dechsel f adze
Decibel n (Phys, Maßeinheit f. Dämpfung) decibel, one-tenth of a bel (unit of attenuation)
Decin n decine, decinene, decyne
Deckanstrich m finish, finishing coat, top coat
Deckanstrichharz n surface coating resin
Deckappretur f (Leder) top finish
Deckband n lapping strip, wrap
Deckblatt n surface sheet, top lamination, (Zigarre) wrapper
Deckbogen m cover sheet, surface sheet
Decke f blanket, cover, covering, hood, (Biol) integument, (Plafond) ceiling, **abgehängte** ~ false, drop or counter ceiling
Deckel m lid, cap, cover, top, **aufgerollter** ~ (Dose) rolled-on cap
Deckeldichtung f cover joint
Deckelhalter m rim closure
Deckeln n (Tabletten) capping
Deckelschieber m cover valve
Deckelschraube f cover bolt
Deckelstopfen m flat stopper
Deckelverschraubung f cover joint, head bolting
decken to cover, **sich mit** ~ (Geom) to coincide with, **sich nicht** ~ to differ, to diverge
Deckendurchführung f ceiling duct
Deckeneinbauleuchten pl illuminated ceiling
Deckentransmission f overhead transmission
Deckenventilator m ceiling fan
deckfähig of good covering power, opaque
Deckfähigkeit f covering power, hiding power, opacity
Deckfarbe f body color, covering color, finishing paint, top coat
Deckfirnis m covering varnish, protecting varnish, varnish coating
Deckflansch m covering flange
Deckgeflecht n protective tissue
Deckglas n cover-glass, glass cover
Deckglastaster m cover glass gauge
Deckgrün n chrome green, opaque green
Deckgrund m undercoat
Deckkläre f covering liquor, wash liquor
Deckkraft f body, covering power, hiding power, opacity
Decklack m covering varnish, finishing lacquer, surface lacquer
Deckmittel n covering medium
Deckname m code name, code word, (Handelsname) trade name
Deckplatte f covering slab, top plate
Decksatinage f (Papier) board glazing
Deckscheibe f cover disk

Deckschicht f finishing coat, surface layer, top coat, upper layer, (Geol) upper stratum
Deckschichtenbildung f top layer formation
Deckschraube f (Techn) cover bolt
Deckstopfen m flanged stopper
Deckung f, **sich zur** ~ **bringen lassen** (Stereochem) to be superimposable
deckungsgleich (Math) congruent
Deckvermögen n covering power, effective coverage, hiding power, opacifying power
Deckweiß n opaque white, zinc white
Deckzelle f cover cell, surface cell
Deckziegel m cover tile
Decoct n decoction, siehe auch Dekokt
Decogenin n decogenin
Decolorimeter n decolorimeter
Decosid n decoside
Decussatin n decussatin
Decylaldehyd m decanal, decylaldehyde
Decylalkohol m decanol, decyl alcohol
Decylen n decene, decylene
Decylensäure f decylenic acid
Decylsäure f decylic acid, capric acid, decatoic acid, n-decanoic acid, n-decoic acid
Deeckeit m (Min) deeckeite
Defäkation f defecation, clarification, clearing
Defekt m defect, failure
Defektelektron n defect electron, hole [in electron valence band]
Defektelektronenbeweglichkeit f electron hole mobility
Defektelektronenhaftstelle f hole trap
Defektleiter m defect conductor
defibriniert defibrinated
definieren to define
Definition f definition, ~ **der Säuren und Basen nach Brönsted und Lowry** Bronsted-Lowry definition of acids and bases
Definitionsbereich m definition range
definitionsgemäß according to definition, as defined
Definitionsweise f notation
Deflagration f deflagration
deflagrieren to deflagrate
Defohärte f Defo hardness
defokussieren to defocus
Deformation f deformation, change of form, distortion, **ungleichmäßige** ~ (Med) non-uniform strain
Deformationsgeschwindigkeit f rate of deformation
Deformationsmaß n measure of deformation
Deformationsmechanismus m mechanism of deformation
Deformationsverteilung f strain distribution
Deformationswiderstand m resistance to deformation
deformierbar deformable
Deformierbarkeit f deformability

deformieren to deform, to distort, to strain
Deformierung *f* deformation, distortion
Defowert *m* deformation value
Degeneration *f* degeneration
Degenerationserscheinung *f* phenomenon of degeneration
degenerieren to degenerate
Degeroit *m* degeroite
Degras *n* degras, sod oil, stuff
degrassieren to deoil, to remove grease, to scour
Deguelidiol *n* deguelidiol
Deguelin *n* deguelin
degummieren (Rohseide) to degum
Dehalogenierung *f* dehalogenation
dehnbar elastic, dilatable, expansible, flexible, tensible, (Met) ductile
Dehnbarkeit *f* elasticity, dilatability, extensibility, flexibility, stretching property, (Met) ductility
Dehnbarkeitsmesser *m* dilatometer, extensometer
dehnen to stretch, to dilate, to draw out, to elongate, to expand, to extend
Dehnfähigkeit *f* extensibility, flexibility
Dehnfestigkeit *f* tensile strength
Dehngrenze *f* offset yield strength
Dehnkraft *f* expansibility, power of expansion
Dehnlinienverfahren *n* stress coating
Dehnung *f* elongation, expansion, extension, strain, (unter Wärmeeinwirkung) dilation
Dehnungsausgleicher *m* expansion compensator, expansion joint, expansion loop
Dehnungsbogen *m* expansion arch
Dehnungselastizität *f* elasticity of extension
Dehnungsgeschwindigkeit *f* rate of extension
Dehnungsgrenze *f* elastic limit, ultimate elongation, ultimate strength
Dehnungskoeffizient *m* coefficient of elasticity, coefficient of extension
Dehnungsmesser *m* extensometer
Dehnungsmeßstreifen *m* strain gauge
Dehnungsmodul *m* (Techn) modulus of elasticity
Dehnungsstopfbuchse *f* slip-type expansion joint
Dehnungszahl *f* modulus of elasticity
Dehnverlauf *m* (Prüftechnik) creep curve
Dehnvermögen *n* expansibility
Dehnzahl *f* strain coefficient (in tension)
Dehydracetsäure *f* dehydracetic acid
Dehyrase *f* (Dehydrogenase) dehydrase
Dehydratase *f* dehydratase
Dehydratation *f* dehydration
dehydratisieren to dehydrate
Dehydratisierung *f* dehydration
Dehydratisierungsmittel *n* dehydrating agent
dehydrieren (entwässern) to dehydrate, (Entzug v Wasserstoff) to dehydrogenate, to dehydrogenize

Dehydrierung *f* (Entwässerung) dehydration, (Entzug v Wasserstoff) dehydrogenation, dehydrogenization
Dehydrierungsanlage *f* dehydrogenation plant
Dehydroabietinsäure *f* dehydroabietic acid
Dehydroamarsäure *f* dehydroamaric acid
Dehydroascorbinsäure *f* dehydroascorbic acid
Dehydrobenzol *n* dehydrobenzene
Dehydroberberin *n* dehydroberberine
Dehydrobilinsäure *f* dehydrobilic acid
Dehydrobrucinolon *n* dehydrobrucinolone
Dehydrocamphenilansäure *f* dehydrocamphenilanic acid
Dehydrocamphensäure *f* dehydrocamphenic acid
Dehydrocamphenylsäure *f* dehydrocamphenylic acid
Dehydrocamphersäure *f* dehydrocamphoric acid
Dehydrochinacridon *n* dehydroquinacridone
Dehydrochinin *n* dehydroquinine
Dehydrocholeinsäure *f* dehydrocholeic acid
Dehydrocholesterin *n* dehydrocholesterol
Dehydrocholsäure *f* dehydrocholic acid
Dehydrocorticosteron *n* dehydrocorticosterone
Dehydrocortison *n* dehydrocortisone
Dehydrocorycavamin *n* dehydrocorycavamine
Dehydrocorydalinbase *f* dehydrocorydaline base
Dehydrocyclisierung *f* dehydrocyclization
Dehydrodesoxycholsäure *f* dehydrodesoxycholic acid
Dehydrodimerisation *f* dehydrodimerization
Dehydroemetin *n* dehydroemetine
Dehydroemodinanthranol *n* dehydroemodineanthranol
Dehydroepiandrosteron *n* dehydroepiandrosterone
Dehydrofenchocamphersäure *f* dehydrofenchocamphoric acid
Dehydrofenchosäure *f* dehydrofenchoic acid
Dehydrofluorindin *n* dehydrofluorindine
Dehydrogenase *f* (Biochem) dehydrogenase
Dehydroglaucin *n* dehydroglaucine
Dehydrohalogenierung *f* dehydrohalogenation
Dehydrohydantoinsäure *f* dehydrohydantoic acid
Dehydroisodypnopinakol *n* dehydroisodypnopinacol
Dehydroisolapachon *n* dehydroisolapachone
Dehydrokaupren *n* dehydrocauprene
Dehydrokortikosteron *n* dehydrocorticosterone
Dehydrokryptopiden *n* dehydrocryptopidene
Dehydrolaurolensäure *f* dehydrolaurolenic acid
Dehydrolithocholsäure *f* dehydrolithocholic acid
Dehydromovrasäure *f* dehydromovraic acid
Dehydronaphthol *n* dehydronaphthol
Dehydronerolidol *n* dehydronerolidol
Dehydrophellandren *n* dehydrophellandrene
Dehydroprogesteron *n* dehydroprogesterone
Dehydroschleimsäure *f* dehydromucic acid
Dehydrothiotoluidin *n* dehydrothiotoluidine
Dehydrothymol *n* dehydrothymol

Dehydroxykodein *n* dehydroxycodeine
Deionisation *f* deionization
Deionisiervermögen *n* deionizing property
Dekacyclen *n* decacyclene
Dekadenrheostat *m* decimal rheostat
Dekadenuntersetzer *m* decade scaler, decascaler
Dekadenwiderstand *m* (Elektr) decimal resistance
Dekadenzähler *m* decade counter
dekadisch decadic
Dekaeder *m* decahedron
dekaedrisch decahedral
Dekagramm *n* decagram
Dekahydrat *n* decahydrate
Dekalen *n* decalene
Dekaleszenz *f* decalescence
Dekalin *n* decalin, decahydronaphthalene, dekalin
Dekalol *n* decalol
Dekameter *n* decameter, ten meter
Dekanaphthen *n* decanaphthene
Dekanaphthensäure *f* decanaphthenic acid
Dekanteur *m* (Zucker) defecator
Dekantierapparat *m* decantation apparatus
dekantieren to decant
Dekantieren *n* decanting
Dekantiergefäß *n* decanter, decanting vessel
Dekantierkorbzentrifuge *f* bowl centrifugal decanter
Dekantierung *f* decantation
Dekantierungsgefäß *n* decanter, decanting vessel
Dekantierzentrifuge *f* centrifugal decanter, decanting centrifuge
Dekapierbad *n* (Met) pickling bath
dekapieren to deoxidize, to descale, to pickle
Dekapieren *n* dipping, pickling
Dekapierflüssigkeit *f* pickling solution
dekarbonisieren to decarbonize, to decarburize
Dekatieranstalt *f* decatizing establishment
Dekatierapparat *m* decatizing apparatus
dekatieren to decatize, to hot-press, to steam
Dekatieren *n* decatizing
Dekatiermaschine *f* decatizing machine
Dekatur *f* decatizing
dekaustizieren to decausticize
Deklination *f* declination
Deklinationsachse *f* (Astr) declination axis
Deklinationskarte *f* declination chart, declination map
Deklinationstiden *pl* declination tides
Deklinograph *m* declinograph
Dekokt *n* decoction
Dekoktpresse *f* decoction press
Dekontamination *f* (Radioakt) decontamination
Dekorationslack *m* decorative paint, varnish
Dekorationsplatte *f* decorative laminate, decorative sheet
Dekorationsstoff *m* drapery, furnishing fabric
Dekorfilm *m* (Pap) decoration film

Dekortikation *f* decortication
Dekrement *n* decrement
Dekrepitieren *n* decrapitating
Dekreszenz *f* decrease, decrescence
Dekrolin *n* decrolin
Dekupiersäge *f* nibbling saw, scroll saw
Delafossit *m* (Min) delafossite
Delatynit *m* delatynite
Delessit *m* (Min) delessite
Delle *f* dent
Delokalisationsenergie *f* delocalisation energy, resonance energy
delokalisieren to delocalize
Delokalisierung *f* delocalization
Delokansäure *f* delocanic acid
Delorenzit *m* (Min) delorenzite
Delphidan *n* delphidan
Delphinat *n* delphinate, salt or ester of delphinic acid
Delphinblau *n* delphin blue
Delphinidin *n* delphinidine
Delphinin *n* delphinin
Delphinit *m* delphinite
Delphinöl *n* dolphin oil
Delphinoidin *n* delphinoidine
Delphinsäure *f* delphinic acid
Delphinsalz *n* salt of delphinic acid
delphinsauer delphinate
Delphintran *m* dolphin oil
Delphisin *n* delphisine
Delphonin *n* delphonine
Delsin *n* delsine
Deltafunktion *f* (Math) delta function
Deltaisolator *m* (Elektr) delta high-tension insulator
Deltalin *n* deltaline
Deltametall *n* delta metal
Deltamin *n* deltamine
Deltapurpurin *n* delta purpurine
Deltaring *m* delta ring seal
Deltastrahlen *pl* (Phys) delta rays
Deltoeder *n* (Krist) deltohedron
Deltoiddodekaeder *n* (Krist) deltoiddodecahedron
Deltruxinsäure *f* deltruxic acid
Delvauxit *m* (Min) delvauxite
Demagnetisierungsfaktor *m* demagnetizing factor
Demagzug *m* electric pulley block
Demant *m* diamond
Demantblende *f* eulytine, eulytite
Demantoid *m* (Min) demantoid
Demantspat *m* (Min) adamantine spar, corundum
Demargarinierungsprozeß *m* demargarinating process
Demarkationslinie *f* demarcation line
Demecolcin *n* demecolcine
Demerol *n* demerol

demethylieren to demethylate
Demethylierung *f* demethylation
Demidowit *m* demidowite
Demissidin *n* demissidine
Demissin *n* demissine
demonstrieren to demonstrate
Demontage *f* disassembly, dismounting
demontierbar collapsible
demontieren to disassemble, to dismantle, to dismount, to take to pieces
Demulgator *m* defoaming agent, demulsifier
Denaturant *m* denaturant
denaturieren to denature, to denaturize
Denaturieren *n* denaturing, denaturizing
denaturiert denatured
Denaturierung *f* denaturation, denaturing, denaturization, denaturizing
Denaturierungsmittel *n* denaturant, denaturizing agent
Dendrachat *m* dendritic agate
Dendrit *m* (Min) dendrite
Dendritenachat *m* (Min) dendritic agate
Dendritenbildung *f* dendritic formation
Dendritenstruktur *f* dendritic structure
Dendroketose *f* dendroketose
Denier *m* (Text) denier
Denitrieranlage *f* denitrating plant
Denitrierbad *n* denitrating bath
denitrieren to denitrate
Denitrieren *n* denitrating
Denitriermittel *n* denitrating agent
Denitrierung *f* denitration
Denitrifikation *f* denitrification
denitrifizieren to denitrify
Denitrifizierung *f* denitrification
Denkansatz *m* food for thought
Denkmodell *n* model for further discussion
Densimeter *n* densimeter, hydrometer
densimetrisch densimetric
Densipimarsäure *f* densipimaric acid
Densitometer *n* densitometer
Densobinde *f* (Densoschutzbinde) denso band
Densograph *m* densograph
Dentalgips *m* dental plaster
Dentalith *m* dentalite
Dentalmaterial *n* dental material
Dentin *n* dentine
dentinbildend (Zahnmed) dentinogenic, dentinogenous
Dentinbildung *f* dentinification
Dephanthansäure *f* dephanthanic acid
Dephanthsäure *f* dephanthic acid
Dephlegmation *f* dephlegmation
Dephlegmator *m* dephlegmator
dephlegmieren to dephlegmate
Depilation *f* depilation
Depilatorium *n* (Kosm) depilatory
Deplacement *n* displacement
Deplacementsberechnung *f* computation of displacement
Deplacementsschwerpunkt *m* center of displacement
Deplacementsskala *f* scale of displacement
deplacieren to displace
Deplacierung *f* displacement
Depolarisation *f* depolarization
Depolarisationsfähigkeit *f* depolarizing ability
Depolarisationsgrad *m* depolarization factor
Depolarisator *m* depolarizer
depolarisieren to depolarize
Depolymerisation *f* depolymerization
depolymerisieren to depolymerize
Depolymerisierung *f* depolymerization
deponieren to deposit
Depoteffekt *m* (Pharm) sustained release action
Depoteisen *n* depot iron
Depoteiweiß *n* depot protein
Depotfett *n* depot fat
Depotinsulin *n* depot insulin
Depotmittel *n* retardant vehicle
Depotpräparat *n* (Pharm) depot drug
Depot-Tabletten *pl* drug retardant, tablets for sustained release action
Depression *f* depression
Depressionswert *m* der Spannung lowered value of voltage
Depressortest *m* depressor test
Depsan *n* depsane
Depsen *n* depsene
Depsid *n* depside
Depsidan *n* depsidan
Depsipeptid *n* depsipeptide
Dequalinium *n* dequalinium
derb compact, firm, solid
Derberz *n* (Min) massive ore
Derbgehalt *m* solid contents
Derbheit *f* coarseness, roughness, rudeness, solidity
Derbylith *m* (Min) derbylite
Derbyrot *n* derby red
Derepression *f* de-repression
Dericinöl *n* dericin oil
Derivat *n* derivate, (Chem) derivative
derivieren to derive
Dermatin *n* dermatin
Dermatol *n* dermatol, basic bismuth gallate, bismuth subgallate
Dermatologe *m* (Med) dermatologist, skin specialist
Dermatologie *f* (Med, Hautlehre) dermatology
Dermatomyzes *m* (Bakt) cutaneous fungus
Dermatophyt *m* (Hautpilz, Bakt) cutaneous fungus, dermatomyces
Dermatose *f* dermatosis
Dermocybin *n* dermocybine
Derrickkran *m* derrick (crane)
Derrid *n* derrid

Derrsäure f derric acid
Desaggregation f disaggregation, degradation, depolymerization, disintegration
desaktivieren to deactivate
Desaktivierung f deactivation
desamidieren to deamidate, to deamidize
Desaminierung f deamination
Desaminochondrosamin n desaminochondrosamine
Desarogenin n desarogenin
Desaromatisierung f dearomatization
Desaspidin n desaspidin
Desaspidinol n desaspidinol
Desaurin n desaurine
Descloizit m (Min) descloizite
Desemulgierung f breaking of an emulsion
Desensibilisator m desensitizer
desensibilisieren to desensitize, to make insensitive
Desensibilisierung f (Opt) desensitization
Deserpidinol n deserpidinol
Desinfektion f disinfection
Desinfektionsapparat m disinfecting apparatus
Desinfektionsbad n germicidal bath
Desinfektionskraft f disinfecting power
Desinfektionsmittel n disinfectant, sanitizer
Desinfektionspapier n disinfecting paper
Desinfektionsseife f (Pharm) disinfectant soap
Desinfektionswasser n disinfecting liquor
Desinfektor m disinfector
Desinfiziens n disinfectant
desinfizieren to disinfect, to sterilize
Desinfizierung f disinfection
Desintegrator m disintegrator
desintegrieren to disintegrate
Desmin m (Min) desmin
Desmoenzym n (Biochem) desmo-enzyme
Desmolase f (Biochem) desmolase
Desmosin n desmosine
Desmosterin n desmosterol
desmotrop desmotropic
Desmotropie f desmotropism, desmotropy
Desmotropoartemisin n desmotropoartemisine
Desmotroposantonin n desmotroposantonine
De[s]odorantien pl deodorants
Desodorieranlage f deodorizing plant
desodorieren to deodorize
Desodorisationsmittel n deodorant, deodorizer
Desodorisierung f deodorization
Desomorphin n desomorphine
desorbieren to desorb
Desorbieren n desorption
Desorption f desorption
Desosamin n desosamine
Desoxin n desoxine
Desoxindigo n desoxindigo
Desoxy- deoxy, desoxy
Desoxyalizarin n de[s]oxyalizarin, anthrarobin

Desoxyallokaffursäure f de[s]oxyallocaffuric acid
Desoxyandrographolid n de[s]oxyandrographolide
Desoxyanisoin n de[s]oxyanisoine
Desoxybiliansäure f de[s]oxybiliaic acid
Desoxybiliobansäure f de[s]oxybiliobanic acid
Desoxycantharidin n de[s]oxycantharidine
Desoxycantharidinsäure f de[s]oxycantharidic acid
Desoxycarminsäure f de[s]oxycarminic acid
Desoxycholsäure f de[s]oxycholic acid
Desoxycorticosteron n de[s]oxycorticosterone
Desoxydation f deoxidation, removal of oxygen
Desoxydationsmittel n deoxidant, deoxidating agent, deoxidizer, deoxidizing agent
desoxydieren to deoxidate, to deoxidize
Desoxydieren n deoxidating, deoxidizing
Desoxydierung f deoxidation, removal of oxygen
Desoxyflavopurpurin n de[s]oxyflavopurpurine
Desoxyfulminursäure f deoxyfulminuric acid
Desoxyhämatoporphyrin n de[s]oxyhematoporphyrin
Desoxyhämoglobin n deoxyhemoglobin
Desoxyinosintriphosphat n deoxyinosine triphosphate
Desoxykodomethin n de[s]oxycodomethine
Desoxymesityloxid n de[s]oxymesityl oxide
Desoxymyoglobin n deoxymyoglobin
Desoxy-nor-ephedrin n de[s]oxy-nor-ephedrine
Desoxyribonuclease f (DNase, Biochem) deoxyribonuclease
Desoxyribonucleinsäure f de[s]oxyribonucleic acid
Desoxyribose f deoxyribose
Desoxythebacodin n desoxythebacodine
Desoxyxanthin n deoxyxanthine
Desoxyzucker m deoxy sugar
Dessin n design, pattern
Destillat n distillate, distillation product
Destillatfettsäure f distilled fatty acid
Destillation f distillation, ~ **mit Wasserdampf** steam distillation, **azeotrope** ~ azeotropic distillation, **extraktive** ~ extractive distillation, **fraktionierte** ~ fractional distillation, **schonende** ~ non-destructive distillation, **trockene** ~ destructive distillation
Destillationsanlage f distillation plant, distilling plant
Destillationsapparatur f distillation apparatus, distilling apparatus
Destillationsaufsatz m [distillation] head
Destillationsbenzin n straight run gasoline
Destillationsblase f still
Destillationsgas n distilled gas
Destillationsgefäß n alembic, distilling vessel, still
Destillationskolben m distillating flask
Destillationsofen m distilling stove

Destillationsprodukt – Dezimalzahl

Destillationsprodukt *n* distillation product
Destillationsrohr *n* distilling pipe
Destillationsrückstand *m* distillation residue
Destillationsschnitt *m* distillation cut
Destillationsthermometer *n* distillation thermometer
Destillationstiegel *m* crucible for distillation
Destillationsvorlage *f* distillation receiver
Destillatmaische *f* distillery mash
Destillerie *f* distillery
Destillierapparat *m* distillation apparatus, alembic, distilling apparatus, still
Destillieraufsatz *m* [distillation] head
destillierbar distillable
Destillierbarkeit *f* distillability
Destillierblase *f* alembic, still
Destillierbrücke *f* bridge-shaped stillhead, vertical recovery bend
destillieren to distil[l], to rectify
Destillieren *n* distilling
Destillierfilter *m* distilling filter
Destilliergefäß *n* distilling flask
Destillierhaus *n* distillery
Destillierhelm *m* distilling head
Destillierholzkohle *f* distillery charcoal
Destillierkolben *m* distilling flask, still
Destillierkolonne *f* distilling column, distilling tower
Destillierspinne *f* bend with multiple connection
destilliert distilled
Destilliertopf *m* distillation pot, distillation vessel
Destillierung *f* distillation
Destilliervorlage *f* fraction receiver
Destilliervorstoß *m* receiver adapter
Destimulator *m* (Korr) destimulator
Destrictasäure *f* destrictic acid
Destrictinsäure *f* destrictinic acid
Desulfinase *f* (Biochem) desulfinase
Desulfonierung *f* desulfonation
Desulfurierung *f* (Entschwefelung) desulfurization
Desyl- desyl
Desylchlorid *n* desyl chloride
Detachur *f* stain removal
Detachurmittel *n* stain removing agent, spotting agent
Detektor *m* detector, scanner
Detergentien *pl* detergents
Determinante *f* (Math) determinant, **Entwicklung v. ~ n** (Math) expansion of determinants, **Unter- ~** (Math) minor determinant
Detonation *f* detonation
Detonationsdruck *m* blast pressure
Detonationsgeschwindigkeit *f* detonation velocity
Detonationswelle *f* detonation wave
Detonator *m* detonator
detonieren to detonate, to explode

Detritus *m* debris, detritus, rubbish
Deul *m* (Metall) [iron] ball
deuterieren to deuterate
Deuterieren *n* deuteration
Deuterium *n* deuterium, heavy hydrogen
Deuterium-Austausch *m* deuterium exchange
Deuteriumoxid *n* deuterium oxide
Deuterohämin *n* deuterohemin
Deuterolyse *f* deuterolysis
Deuteron *n* (Atom) deuteron, deuton
Deuteronenbildung *f* deuteron formation
Deuteronenstrahl *m* (Atom) deuteron beam
Deuteronphotoeffekt *m* deuteron photodisintegration
Deuteroporphyrin *n* deuteroporphyrin
Deuton *n* deuteron, deuton
Deutoplasma *n* deutoplasma
Deutsche Industrie-Norm *f* German industrial standards
Deutsches Arzneibuch *n* (DAB, Pharm) German Pharmacop[o]eia
Deutungsversuch *m* attempt to interpret
Devardasche Legierung *f* Devarda's alloy
Dewargefäß *n* Dewar [vessel]
Deweylit *m* (Min) deweylite
Dexamethason *n* dexamethasone
Dexamphetaminum *n* dexamphetamine
Dextran *n* dextran[e]
Dextrantriose *f* dextrantriose
Dextrin *n* dextrin, amylin, artificial gum, British gum, gommelin, starch gum
Dextrinase *f* (Biochem) dextrinase
Dextrinsirup *m* dextrin syrup
Dextrinstärke *f* dextrinized starch
Dextrinzucker *m* dextrin sugar
Dextroamphetaminsulfat *n* dextroamphetamine sulfate
dextrogyr dextrogyric, dextrorotatory
Dextromethorphan *n* dextromethorphan
Dextromoramid *n* dextromoramide
Dextronsäure *f* dextronic acid, d-gluconic acid
Dextropimaren *n* dextropimarene
Dextropimarin *n* dextropimarine
Dextropimarol *n* dextropimarol
Dextropimarsäure *f* dextropimaric acid
Dextropropoxyphen *n* dextropropoxyphene
Dextrorphan *n* dextrorphan
Dextrose *f* dextrose, d-glucose, grape sugar
Dezentralisation *f* decentralization
Dezidekameter *n* decidecameter
Dezigramm *n* decigram
dezimal decimal
Dezimalbruch *m* decimal fraction
Dezimalklassifikation *f* decimal classification
Dezimallogarithmus *m* common logarithm
Dezimalstelle *f* decimal place
Dezimalsystem *n* (Math) decimal system
Dezimalwaage *f* decimal balance
Dezimalzahl *f* decimal number

Dezimeter *m n* decimeter
Dezimilligramm *n* tenth of a milligram
Dezinormalelektrode *f* deci standard electrode
DFP siehe Diisopropylfluorphosphat
DHD-Anlage *f* dehydrogenation plant
Dhurrin *n* dhurrin
Dia *n* slide
Diabantit *m* (Min) diabantite
Diabas *m* (Min) diabase
Diabasmandelstein *m* (Min) amygdaloidal greenstone
Diabasschiefer *m* (Min) greenstone-slate
Diabetes *m* (Med) diabetes, **echter** ~ true diabetes, **neurogener** ~ neurogenous diabetes, **wechselnder** ~ temporary diabetes
Diabetesdiät *f* diabetic diet
Diabetesfaktor *m* diabetogenic factor
Diabeteskoma *n* diabetic coma
Diabetiker *m* diabetic
Diabetikerbrot *n* diabetic bread
Diabetikerdiät *f* diabetic diet
Diabetikermilch *f* diabetic milk
Diabetin *n* diabetin, fructose, fruit sugar, levulose
diabetogen (Med) diabetogenic
Diabetometer *n* diabetometer
Diabolin *n* diaboline
Diacetamid *n* diacetamide
Diacetat *n* diacetate
Diacetimid *n* diacetimide
Diacetin *n* diacetin, glycerol diacetate
Diacetonalkohol *m* diacetone alcohol
Diacetonamin *n* diacetonamine, β-aminoisopropylacetone
Diacetonid *n* diacetonide
Diacetonsorbose *f* diacetone sorbose
Diacetyl *n* diacetyl, 2,3-butanedione, biacetyl
Diacetyldiphenolisatin *n* diacetyldiphenolisatin
Diacetylen *n* diacetylene, biacetylene, butadiine
Diacetylmorphin *n* diacetylmorphine, heroine
Diacetylmutase *f* diacetylmutase
Diacetylphenylendiamin *n* diacetylphenylenediamine
Diacetyltannin *n* diacetyltannin, tannigen
Diachylonpflaster *n* diachylon, diachylum
Diachylonsalbe *f* (Pharm) diachylon ointment, lead plaster ointment
Diadelphit *m* (Min) diadelphite
diadoch diadoch
Diadochit *m* (Min) diadochite
Diät *f* diet
Diätetik *f* dietetics
diätetisch dietetic
Diätformel *f* diet formula
Diäthanolamin *n* diethanolamine
Diätherat *n* dietherate
Diäthyläther *m* diethyl ether
Diäthylamin *n* diethylamine

Diäthylaminchlorhydrat *n* diethylamine hydrochloride
Diäthylaminoäthanol *n* diethylaminoethanol
Diäthylanilin *n* diethylaniline
Diäthylbarbitursäure *f* diethylbarbituric acid
Diäthylbenzol *n* diethylbenzene
Diäthylbromacetamid *n* diethylbromacetamide
Diäthylcarbamazin *n* diethylcarbamazine
Diäthylcarbocyaninjodid *n* diethylcarbocyanine iodide
Diäthylcyanjodid *n* diethylcyanine iodide
Diäthyldiäthoxysilan *n* diethyldiethoxysilane
Diäthyldichlorsilan *n* diethyldichlorosilane
Diäthyldiphenylharnstoff *m* diethyldiphenylurea
Diäthyldithiocarbaminsäure *f* diethyl dithiocarbamic acid, DDTC
Diäthylendiamin *n* diethylenediamine, piperazine
Diäthylendisulfid *n* diethylenedisulfide, dithiane
Diäthylenglykol *n* diethylene glycol
Diäthylentriamin *n* diethylenetriamine
Diäthylessigsäure *f* diethyl acetic acid
Diäthylmalonat *n* diethyl malonate
Diäthylmalonsäure *f* diethyl malonic acid
Diäthylmesotartrat *n* diethyl mesotartrate
Diäthylmetanilsäure *f* diethyl metanilic acid
Diäthylnitrophenylthiophosphat *n* (E 605, HN) diethyl nitrophenyl thiophosphate
Diäthylphthalat *n* diethyl phthalate
Diäthylsuccinat *n* diethyl succinate
Diäthylsulfat *n* diethyl sulfate
Diäthylsulfid *n* diethyl sulfide
Diäthyltartrat *n* diethyl tartrate
Diäthyltoluidin *n* diethyl toluidine
Diätkunde *f* dietetics, sitology
Diätlehre *f* dietetics, sitology
Diagnostik *f* (Med) diagnostic[s]
diagnostisch diagnostic
diagnostizieren (Med) to diagnose
diagonal bias, diagonal
Diagonalband *n* diagonal tie
Diagonale *f* diagonal
Diagonalisierung *f* (Math) diagonalising, ~ **v. Matrizen** (Math) diagonalizing (of matrices)
Diagonalmatrize *f* (Math) diagonal matrix
Diagonalschiene *f* diagonal rider, diagonal tie plate
Diagonalschneider *m* diagonal cutter, bias cutting machine
Diagramm *n* diagram, graph, plot, **doppeltlogarithmisches** ~ log-log plot
Diagrammpapier *n* millimeter graph paper
Diagrammscheibe *f* circular chart
Diagrammstreifen *m* strip chart
Diagrammvorschub *m* chart speed
Diakenese *f* (Biol) diakinesis
Diaklasit *m* (Min) diaclasite
diaktinisch diactinic
Dialdehyd *m* dialdehyde

Dialin *n* dialin, dihydronaphthalene
Dialkyl *n* dialkyl
Dialkylamin *n* dialkylamine
Dialkylen *n* dialkylene
Diallag *m* (Min) diallage
Diallyden *n* diallydene
Diallyl *n* 1,5-hexadiene, diallyl, biallyl
Diallylbarbitursäure *f* diallylbarbituric acid
Diallyldichlorsilan *n* diallyldichlorosilane
Diallylen *n* diallylene, biallylene
Diallylmorphimethin *n* diallylmorphimethine
Dialogit *m* dialogite
Dialurat *n* dialurate
Dialursäure *f* dialuric acid, 5-hydroxybarbituric acid
Dialysator *m* dialyzer
Dialyse *f* dialysis
dialysierbar dialyzable
dialysieren to dialyze
Dialysieren *n* dialysing
Dialysierpapier *n* dialysing paper
Dialysierschlauch *m* dialysing tube
dialytisch dialytic, dialytical
diamagnetisch diamagnetic
Diamagnetismus *m* diamagnetism
Diamant *m* diamond
Diamantblau *n* diamond blue
Diamantbohrer *m* (Techn) diamond drill
Diamantfuchsin *n* diamond fuchsine
diamantführend diamantiferous, diamondiferous
Diamantgelb *n* diamond yellow
Diamantgewicht *n* carat
Diamantgitter *n* (Krist) diamond lattice
Diamantglanz *m* adamantine luster
diamanthaltig diamantiferous
Diamantmörser *m* diamond mortar
Diamantoeder *n* (Krist) hexoctahedron
Diamantpulver *n* diamond dust
Diamantschleifer *m* diamond cutter
Diamantschneider *m* (Techn) diamond cutter
Diamantschwarz *n* diamond black
Diamantspat *m* (Min) adamantine spar, corundum
Diamantstahl *m* chisel steel, extra hard steel, tool steel
Diamantstruktur *f* (Krist) diamond structure
Diamanttinte *f* diamond ink
Diamantwaage *f* diamond scale
Diamantzement *m* diamond cement
diametral diametrical
Diamid *n* diamide
Diamidobenzol *n* diaminobenzene, phenylenediamine
Diamidotoluol *n* diaminotoluene
Diaminblau *n* diamine blue
Diaminechtrot *n* diamine fast red
Diaminoacridin *n* diaminoacridine
Diaminobenzol *n* diaminobenzene, phenylenediamine
Diaminocyclopentan *n* diaminocyclopentane
Diaminodecan *n* diaminodecane
Diaminodiäthylamin *n* diaminodiethylamine
Diaminodiäthylsulfid *n* diaminodiethylsulfide
Diaminodiphenylharnstoff *m* diaminodiphenylurea
Diaminodiphenylmethan *n* diaminodiphenylmethane
Diaminodiphenylsulfon *n* diaminodiphenylsulfone
Diaminofluoren *n* diaminofluorene
Diaminogenblau *n* diaminogene blue
Diaminomethoxybenzol *n* diaminomethoxybenzene
Diaminophenol *n* diaminophenol
Diaminoresorcin *n* diaminoresorcinol
Diaminostilben *n* diaminostilbene
Diaminotoluol *n* diaminotoluene
Diaminreinblau *n* diamine pure blue
Diammoniumphosphat *n* diammonium phosphate
Diamol *n* diamol
Diamyläther *m* diamyl ether
Diamylamin *n* diamyl amine
Diamylketon *n* diamyl ketone
Diamylose *f* diamylose
Diamylphthalat *n* diamyl phthalate
Dianenbaum *m* (Bot) arbor dianae
Dianilblau *n* dianil blue
Dianisidin *n* dianisidine
Dianisidinblau *n* dianisidine blue
Dianol *n* dianol
Dianthin *n* dianthine
Dianthrachinon *n* dianthraquinone
Dianthrachinonyl *n* dianthraquinonyl
Dianthracyl *n* dianthracyl, bianthracyl
Dianthranilid *n* dianthranilide
Dianthranol *n* bianthranol, dihydrodianthrone
Dianthranyl *n* dianthranyl, bianthranyl, bianthryl, dianthryl
Dianthrenblau *n* dianthrene blue
Dianthrimid *n* dianthrimide
Dianthron *n* dianthrone
Dianthryl *n* bianthryl, dianthryl
diaphan diaphanous, transparent
Diaphanometer *n* diaphanometer
Diaphanoskop *n* diaphanoscope
Diaphenylsulfon *n* diaphenylsulfone
Diaphorase *f* (Biochem) diaphorase
Diaphorit *m* (Min) diaphorite
Diaphragma *n* diaphragm
Diaphragmapumpe *f* diaphragm pump
Diaphragmaverfahren *n* diaphragm process
Diaphragmenstrom *m* diaphragm current
Diaphragmenverfahren *n* diaphragm process
Diapositiv *n* (Phot) diapositive, slide
Diapurin *n* diapurine
Diarrhoe *f* (Med) diarrhoea
Diarsenobenzol *n* diarsenobenzene

Diaspirin *n* diaspirin, succinylsalicylic acid
Diaspor *m* (Min) diaspore
Diastase *f* (Biochem) diastase
diastatisch diastatic
diastereomer diastereo[iso]meric
Diastereomer *n* (Stereochem) diastereoisomer, diastereomer
Diastereomerie *f* (Stereochem) diastereomerism
Diastrophismus *m* diastrophism
Diaterebinsäure *f* diaterebinic acid
diatherm (Wärme durchlassend) diathermic (s. a. diatherman)
diatherman diathermic, capable of transmitting heat, diathermal, diathermanous, transcalent
diatomeenartig diatomaceous
Diatomeenerde *f* diatomaceous earth, infusorial earth, kieselguhr
Diatomit *m* (Min) diatomite
Diatomitschicht *f* diatomite layer
diatonisch diatonic
Diatophan *n* diatophane
Diatretin *n* diatretyne
Diazen *n* diazene, diazete
Diazepin *n* diazepine
Diazid *n* diazide
Diazin *n* diazin[e]
Diazinblau *n* diazine blue
Diaziridin *n* diaziridine
Diazoaceton *n* diazo acetone
Diazoäthan *n* diazoethane
Diazoameisensäure *f* diazoformic acid
Diazoaminobenzol *n* 1,3-diphenyltriazene, diazoaminobenzene
Diazoaminoverbindung *f* diazoamino compound
Diazobenzoesäure *f* diazobenzoic acid
Diazobenzol *n* diazobenzene
Diazobenzolchlorid *n* benzene diazonium chloride, diazobenzene chloride
Diazobenzosulfonsäure *f* diazobenzene sulfonic acid
Diazobernsteinsäure *f* diazosuccinic acid
Diazocampher *m* diazocamphor
Diazocystin *n* diazocystine
Diazodinitrophenol *n* diazodinitrophenol
Diazodruck *m* diazotype printing
Diazoechtfarbe *f* fast diazo color
Diazoessigester *m* diazoacetic ester
Diazoessigsäure *f* diazoacetic acid
Diazofarbstoff *m* diazo dye
Diazoimid *n* diazoimide, hydrazoic acid
Diazoisatin *n* diazoisatin
Diazokupplung *f* diazo coupling
Diazol *n* diazole
Diazolösung *f* diazo solution
Diazomalonsäure *f* diazomalonic acid
Diazometallverbindung *f* diazo metal compound, diazotized metal compound
Diazomethan *n* diazomethane, azimethane, azimethylene

Diazonaphtholsulfo[n]säure *f* diazonaphtholsulfonic acid
Diazoniumhydroxid *n* diazonium hydroxide
Diazoniumsalz *n* (Färb) diazonium salt
Diazoniumverbindung *f* diazonium compound
Diazooxalessigsäure *f* diazooxalacetic acid
Diazooxid *n* diazooxide
Diazopapier *n* diazotype paper
Diazoparaffin *n* diazoparaffin
Diazophenol *n* diazophenol
Diazoschwarz *n* diazo black
Diazotat *n* diazotate
Diazotetrazol *n* diazotetrazole
diazotierbar diazotizable
diazotieren to diazotize
Diazotieren *n* diazotizing
Diazotierung *f* diazotization
Diazotierungsfarbstoff *m* diazotizing dyestuff
Diazotypie *f* diazotype
Diazoverbindung *f* diazo compound
Dibemethin *n* dibemethine
Dibenzacridin *n* dibenzacridine, naphthacridine
Dibenzamid *n* dibenzamide, benzoylbenzamide
Dibenzanthracen *n* dibenzanthracene
Dibenzanthrachinon *n* dibenzanthraquinone
Dibenzanthron *n* dibenzanthrone, violanthrone
Dibenzanthronyl *n* dibenzanthronyl
Dibenzcarbazol *n* dibenzcarbazole
Dibenzcoronen *n* dibenzcoronene
Dibenzfluoren *n* dibenzfluorene
Dibenzopyron *n* dibenzopyrone, xanthone
Dibenzothiazyldisulfid *n* dibenzo thiazyl disulfide
Dibenzoyl- dibenzoyl
Dibenzoyl *n* benzil, dibenzoyl
Dibenzoyldinaphthyl *n* dibenzoyldinaphthyl
Dibenzoylheptan *n* dibenzoylheptane
dibenzoylieren to dibenzoylate
Dibenzoylperoxid *n* dibenzoyl peroxide
Dibenzoylperylen *n* dibenzoylperylene
Dibenzphenanthren *n* dibenzphenanthrene
Dibenzpyrenchinon *n* dibenzpyrenequinone
Dibenzyl- dibenzyl
Dibenzyl *n* (Diphenyläthan) 1,2-diphenylethane, dibenzyl, bibenzyl
Dibenzyläther *m* dibenzyl ether
Dibenzyldichlorsilan *n* dibenzyldichlorosilane
Dibenzylketon *n* dibenzyl ketone
Dibenzylsilandiol *n* dibenzylsilanediol
Diboran *n* diborane
Dibornyl *n* dibornyl, bibornyl
Dibortrioxid *n* diboron trioxide
Dibrasanchinon *n* dibrazanquinone
Dibromacetophenon *n* dibromoacetophenone
Dibrommäthan *n* dibromoethane
Dibromanthracen *n* dibromoanthracene, dibromanthracene
Dibromanthrachinon *n* dibrom[o]anthraquinone
Dibrombenzoesäure *f* dibromobenzoic acid

Dibrombenzol — Dichtegradient-Zentrifugation

Dibrombenzol *n* dibromobenzene
Dibrombernsteinsäure *f* dibromosuccinic acid
Dibrombuten *n* dibromo butene
Dibromchinonchlorimid *n* dibromoquinone chlorimide
Dibromdihydroxyanthracen *n* dibromo dihydroxyanthracene
Dibromdiphenyl *n* dibromodiphenyl
Dibromfumarsäure *f* dibromo fumaric acid
Dibromhydrin *n* dibromohydrin
Dibromid *n* dibromide
Dibromindigo *n* dibromindigo
Dibromketodihydronaphthalin *n* dibromo ketodihydronaphthalene
Dibrommaleinsäure *f* dibromo maleic acid
Dibrommenthon *n* dibromo menthone
Dibrommethan *n* dibromo methane
Dibromnaphthalin *n* dibromo naphthalene
Dibromnaphthol *n* dibromo naphthol
Dibromnaphthylamin *n* dibromo naphthylamine
Dibromoxychinolin *n* dibromo hydroxyquinoline
Dibrompropamidin *n* dibrompropamidine
Dibrompropan *n* dibromo propane
Dibromsalicylsäure *f* dibromo salicylic acid
Dibromtoluidin *n* dibromo toluidine
Dibromtyrosin *n* dibromo tyrosine
Dibutyladipinat *n* dibutyl adipate
Dibutyläther *m* dibutyl ether
Dibutylketon *n* dibutyl ketone
Dibutylphthalat *n* dibutyl phthalate
Dibutyrin *n* dibutyrin, glyceryl dibutyrate
Dicadisäure *f* dicadic acid
Dicalciumphosphat *n* dicalcium [ortho]phosphate, secondary calcium phosphate, calcium hydrogen orthophosphate, dibasic calcium phosphate
Dicamphen *n* dicamphene, bicamphene
Dicamphoketon *n* dicamphoketone
Dicamphoryl *n* dicamphoryl, bicamphoryl
Dicarbonsäure *f* dicarboxylic acid
Dicarboxylat *n* dicarboxylate
Dicentrin *n* dicentrin[e]
Dicetyl *n* bicetyl, dicetyl, dotriacontane
Dicetylsulfon *n* dicetyl sulfone
Dichinaldin *n* diquinaldine, biquinaldine
Dichinidin *n* diquinidine, diconchinine
Dichinolin *n* diquinoline
Dichinolyl *n* biquinolyl, diquinolyl
Dichloraceton *n* dichloro acetone
Dichloräthan *n* dichloro ethane
Dichloräther *m* dichloro ether
Dichloräthyläther *m* dichloro ethyl ether
Dichloräthylen *n* dichlor ethylene, dichloro ethylene
Dichloräthylsulfid *n* dichloro ethyl sulfide
Dichloranilin *n* dichloro aniline
Dichlorbenzalchlorid *n* dichloro benzal chloride
Dichlorbenzaldehyd *m* dichloro benzaldehyde
Dichlorbenzidin *n* dichloro benzidine
Dichlorbenzoesäure *f* dichloro benzoic acid
Dichlorbenzol *n* dichloro benzene
Dichlorbenzolsulfonsäure *f* dichloro benzene sulfonic acid
Dichlorbernsteinsäure *f* dichloro succinic acid
Dichlorcarben *n* dichlorcarbene
Dichlordiäthyläther *m* dichloro diethyl ether
Dichlordifluormethan *n* dichloro difluoromethane
Dichloressigsäure *f* dichloro acetic acid, dichloro ethanoic acid
Dichlorhydrin *n* dichlorohydrin
Dichlorid *n* dichloride
Dichlorisatin *n* dichlorisatin
Dichlormethan *n* dichloro methane, methylene chloride
Dichlorodiphenylessigsäure *f* dichloro diphenylacetic acid
Dichlorodiphenyltrichloroäthan *n* (Insektenvertilgungsmittel) dichlorodiphenyl trichloroethane (insecticide)
Dichlorophen *n* dichlorophene
Dichlorophenarsin *n* dichlorophenarsine
Dichlorpentan *n* dichloropentane
Dichlorphenoxyessigsäure *f* dichloro phenoxyacetic acid
Dichlorphenylmercaptan *n* dichloro phenylmercaptan, dichloro thiophenol
Dichlorphthalsäure *f* dichloro phthalic acid
Dichlorpropan *n* dichloropropane
Dichlorresorcin *n* dichloro resorcinol
Dichlorsilan *n* dichloro silane
Dichlortoluol *n* dichloro toluene
Dichlorvos *n* (Insektizid) dichlorvos
Dichroin *n* dichroine
Dichroismus *m* (Zweifarbigkeit) dichroism
Dichroit *m* (Min) dichroite, iolite
dichroitisch (Krist) dichroic
Dichromat *n* dichromate, bichromate, salt of dichromic acid
Dichromsäure *f* dichromic acid
dichromsauer dichromate
Dichroskop *n* dichroscope
dicht dense, compact, firm, massive, tight
dichtbrennen to vitrify
Dichte *f* firmness, tightness; (Phys) density, specific gravity, **kritische** ~ critical density
Dichteänderung *f* alteration of density
Dichteanisotropie *f* density anisotropy
Dichtebestimmung *f* density determination, determination of density
Dichtefeld *n* density field
Dichteflasche *f* specific gravity bottle
Dichtegradient *m* density gradient
Dichtegradient-Zentrifugation *f* density-gradient centrifugation

Dichtemesser *m* densimeter, picnometer, specific gravity bottle, (für Flüssigkeiten) areometer, hydrometer
dichten to render tight, to lute, to pack, to seal, (verstemmen) to caulk
Dichteverteilung *f* density distribution
Dichtewaage *f* density balance
Dichtewert *m* density value
Dichtezahl *f* density number
Dichthammer *m* caulking hammer
Dichtigkeit *f* denseness, density, tightness
Dichtigkeitseigenschaften *pl* impermeability
Dichtigkeitsprüfgerät *n* leak detector
Dichtmachen *n* rendering tight, compression, packing
Dichtmaschine *f* caulking machine
dichtpolen to toughen by poling
Dichtung *f* gasket, joint, seal, stuffing, washer, **Form-** ~ self-energized gasket, **Gewindewellen-** ~ threaded shaft or hydrodynamic seal, **Gleitring-** ~ slip ring seal, **Labyrinth-** ~ labyrinth seal, **Lippen-** ~ cylindrical gasket, **Manschetten-** ~ concentric gasket, **Membran-** ~ diaphragm packing, **Spalt-** ~ controlled gap seal
Dichtungsbahn *f* damp-proof sheeting
Dichtungsband *n* jointing ring
Dichtungsfett *n* joint grease
Dichtungsfläche *f* packing surface, sealing joint
Dichtungsfuge *f* packing joint, sealing joint
Dichtungskitt *m* lute, luting agent
Dichtungsklappe *f* packing valve
Dichtungslack *m* sealing lacquer
Dichtungsleiste *f* sealing ledge
Dichtungsmanschette *f* gasket
Dichtungsmasse *f* sealing compound
Dichtungsmaterial *n* caulking material, packing material, sealing compound
Dichtungsmittel *n* sealing compound, caulking material, joint grease, packing material, proofing
Dichtungsmuffe *f* sealing sleeve
Dichtungsnaht *f* caulking seam
Dichtungspappe *f* packing cardboard
Dichtungsplatte *f* packing sheet
Dichtungsprofil *n* profiled joint, sealing profile
Dichtungsring *m* washer, gasket, gasket ring, joint ring, packing ring, sealing ring
Dichtungsscheibe *f* washer, gasket sheet, packing ring
Dichtungsschraube *f* sealing screw
Dichtungsschweißung *f* caulk welding
Dichtungsstoff *m* packing material
Dichtungsstreifen *m* sealing tape
Dichtungstiefe *f* depth of packing
Dichtungswalze *f* packing drum
Dichtverschluß *m* throat stopper
Dichtwerg *n* oakum, packing hemp

Dicinnamyliden *n* bicinnamylidene, dicinnamylidene
dick thick, bulky, dense, (beleibt) corpulent, stout, (Blut) clotted, coagulated, (fett) fat, obese, (geschwollen) swollen, (Milch) curdled, ~ **werden** to thicken
Dick *n* (Kampfgas) dick, ethyl dichloroarsine
Dickauszug *m* inspissated extract
Dicke *f* thickness, consistency
Dickenabweichung *f* difference in thickness
Dickengradient *m* thickness gradient
Dickenlehre *f* thickness gauge
Dickenmesser *m* thickness gauge, calipers, micrometer, thickness tester
Dickfarbe *f* body color
dickflüssig syrupy, viscid, viscous
Dickflüssigkeit *f* high viscosity, syrupy consistency, viscidity
Dickmaische *f* thick mash
Dickmittel *n* thickener, thickening agent
Dicköl *n* heavy oil, bodied oil, heavy ends, linseed oil, oxidized oil, stand oil
Dickpfanne *f* concentration pan
Dicksaft *m* syrup, thick juice
Dickstoffverdampfer *m* viscous material evaporator
dickwandig heavy-section, thick-walled
Dickwerden *n* thickening
Dickzirkel *m* calipers, thickness gauge
Dicodein *n* dicodeine
Dicodid *n* dicodide, dihydrocodeinone
Dicrotalin *n* dicrotaline
Dictamnin *n* dictamnine
Dictamninsäure *f* dictamnic acid
Dictyosom *n* (Biol) dictyosome
Dicumarol *n* dicoumarol
Dicyan *n* cyanogen, dicyan, ethane dinitrile, oxalonitrile
Dicyandiamid *n* dicyandiamide, cyanoguanidine, param
Dicyandiamidin *n* dicyandiamidine, guanylurea
Dicyanimid *n* dicyanimide
Dicyanin *n* dicyanin[e]
Dicyclohexylbenzol *n* dicyclohexylbenzene
Dicyclohexylphthalat *n* dicyclohexyl phthalate
Dicyclopentadien *n* dicyclopentadiene
Dicycloverin *n* dicycloverine
Dicymol *n* dicymol, bicumenyl, bicymol, dicumenyl
Dicymylamin *n* dicymylamine
Didecylphthalat *n* didecyl phthalate
Didehydrocorycavamin *n* didehydrocorycavamine
Didehydrocorycavin *n* didehydrocorycavine
Didesyl *n* didesyl, bidesyl, dibenzoylbibenzyl
didodekaedrisch (Krist) didodecahedral
Didym *n* (obs) didymium
Didymit *m* (Min) didymite
Didymolit *m* (Min) didymolite

Dieckmann-Ester *m* Dieckmann ester
Dieder *m* dihedral
Dielektrikum *n* dielectric
dielektrisch dielectric
Dielektrizitätskonstante *f* dielectric constant, permittivity, specific inductive capacity
Diels-Alder-Reaktion *f* Diels-Alder-reaction
Diels-Kohlenwasserstoff *m* Diels hydrocarbon
Dielssäure *f* Diels' acid
Dien *n* diene, **konjugiertes** ~ conjugated diene
Dienöstrol *n* dienestrol
Dienol *n* dienol
Diensynthese *f* diene synthesis
Dieselaggregat *n* diesel generating set
Dieselantrieb *m* diesel drive
Dieselkraftfahrzeug *n* diesel automobile
Dieselkraftstoff *m* diesel fuel
Dieselmotor *m* diesel engine, compression ignition engine
Dieselöl *n* diesel fuel or oil
Diessigsäure *f* diacetic acid
Diethazin *n* diethazine
Dietrichit *m* (Min) dietrichite
Diferulasäure *f* diferulaic acid
Differential *n* (Math) differential, increment
Differentialbarometer *n* differential barometer
Differentialblutbild *n* differential blood picture, h(a)emogram
Differentialbrechung *f* differential refraction
Differential[dreh]kondensator *m* differential condenser
Differentialdruckmesser *m* differential manometer
Differentialeilpumpe *f* differential high-speed pump
Differentialflaschenzug *m* differential chain block
Differentialformel *f* (Math) differential formula
Differentialgetriebe *n* differential gear
Differentialgleichung *f* differential equation, ~ **erster Ordnung** differential equation of the first order, ~ **höherer Ordnung** higher order differential equation, ~ **zweiter Ordnung** second order differential equation, **elliptische** ~ elliptical differential equation, **gewöhnliche** ~ ordinary differential equation, **homogene** ~ homogeneous differential equation, **hyperbolische** ~ hyperbolic differential equation, **lineare** ~ linear differential equation, **nichtlineare** ~ non-linear differential equation, **Ordnung einer** ~ order of a differential equation, **parabolische** ~ parabolic differential equation
Differentialkolben *m* differential piston
Differentialkupplung *f* differential coupling
Differentialmanometer *n* differential manometer or gauge
Differentialquotient *m* (Math) differential quotient or coefficient

Differentialrechnung *f* differential calculus
Differentialrelais *n* differential relay
Differentialschaltung *f* differential system
Differentialschraube *f* differential screw
Differentialspule *f* differential coil
Differentialthermoanalyse *f* differential thermal analysis
Differentialthermometer *n* differential thermometer, aethrioscope
Differentialwinde *f* differential windlass
Differentialzentrifugation *f* differential centrifugation
Differentiation *f* (Math) differentiation
Differenz *f* difference
Differenzdruck *m* differential pressure
Differenzenschemaverfahren *n* difference scheme method
differenzierbar (Math) differentiable
Differenzierbarkeit *f* (Math) differentiability
differenzieren to differentiate
Differenzierschaltung *f* (Rdr) differentiating circuit
Differenzierspule *f* peaker strip
Differenzierung *f* diversification, (Math) differentiation
Differenzmanometer *n* differential gauge
Differenzmessung *f* (Elektr) differential measurement
Differenzstrom *m* differential current
Differenzthermostat *m* differential thermostat
Differenzzugmesser *m* differential draft gauge
Diffractasäure *f* diffractaic acid
Diffraktion *f* (Opt) diffraction
Diffraktionsgitter *n* diffraction grating
Diffraktionswinkel *m* diffraction angle
Diffraktometer *n* diffractometer
diffundieren to diffuse
diffus diffuse
Diffusat *n* (Chem) diffusate
Diffuseur *m* diffuser, diffusion apparatus
Diffusion *f* diffusion, ~ **bei gleichzeitiger Reaktion** diffusion accompanied by reaction, ~ **freier Moleküle** *pl* free molecule diffusion, Knudsen flow, ~ **in Festkörpern** *m pl* diffusion in solids
Diffusionsanalyse *f* (Phys) diffusion analysis
Diffusionsapparat *m* diffuser
Diffusionsbetriebswasser *n* working fluid for diffusion
Diffusionselektrophorese *f* diffusion electrophoresis
diffusionsfähig diffusible
Diffusionsfähigkeit *f* diffusibility, diffusivity
Diffusionsflammenreaktor *m* diffusion flame reactor
Diffusionsgeschwindigkeit *f* diffusion rate, velocity of diffusion
Diffusionsgleichung *f* diffusion equation

Diffusionsgrenzstrom *m* limiting diffusion current
Diffusionsintegralkern *m* diffusion kernel
Diffusionskoeffizient *m* diffusion coefficient
Diffusionskolonne *f* diffusion column
Diffusionsmessung *f* osmometry
Diffusionsnebelkammer *f* diffusion cloud chamber
Diffusionspotential *n* diffusion potential
Diffusionspumpe *f* diffusion pump
Diffusionsquerschnitt *m* diffusion cross section
Diffusionssprung *m* diffusional jog
Diffusionstensor *m* diffusion tensor
Diffusions-Thermoeffekt *m* (Dufour-Effekt) Dufour effect (inverse thermodiffusion)
Diffusionstransport *m* diffusion transport
Diffusionstrennverfahren *n* diffusion separation method
Diffusionsverfahren *n* diffusion process, osmosis
Diffusionsvermögen *n* diffusibility, diffusivity, **thermisches** ~ (Therm) thermal diffusivity
Diffusionswärme *f* diffusion heat
Diffusionswand *f* (Phys) diffusion barrier
Diffusionsweg *m* diffusion path
Diffusionszelle *f* diffusion cell
Diffusor *m* (Mot, Zuck) diffuser
Difluan *n* difluan
Difluoräthylen *n* difluoroethylene
Difluordiphenyl *n* difluorobiphenyl, difluorodiphenyl, difluoroxenene
Difluorenyl *n* difluorenyl
Difluorenyliden *n* bifluorenylidene, difluorenylidene
Diformamid *n* diformamide
Diformyl *n* biformyl, diformyl, glyoxal
Digalen *n* digalene
Digallussäure *f* digallic acid
Digenit *m* (Min) digenite
Digentisinsäure *f* digentisic acid
digerieren to digest
Digerieren *n* digesting
Digerierflasche *f* digestion bottle, digestion flask
Digerierofen *m* digestive apparatus
Digerierung *f* digestion
Digestion *f* digestion
Digestionskolben *m* digesting flask, digestion flask
Digestivsalz *n* digestive salt
Digestor *m* (Brau) digestor
Digestorium *n* (Abzug) fume cupboard, hood
Diginatigenin *n* diginatigenin
Diginatin *n* diginatin
Diginin *n* diginin
Diginose *f* diginose
Digipan *n* digipan
digital digital
Digitalanzeige *f* digital display
Digitalein *n* digitalein
Digitaligenin *n* digitaligenine

Digitalin *n* digitalin
Digitalisblätter *pl* (Bot) digitalis leaves
Digitalisglucosid *n* digitalis glucoside
Digitalisglykosid *n* digitalis glycoside
Digitalisvergiftung *f* digitalism
Digitalonsäure *f* digitalonic acid
Digitalose *f* digitalose
Digitalrechenmaschine *f* digital computer
Digitalrechner *m* digital computer
Digitalsalz *n* digitalic salt
Digitalschreiber *m* digital recorder
Digitan *n* digitane
Digitogenin *n* digitogenin
Digitogensäure *f* digitogenic acid
Digitolein *n* digitoleine
Digitolutein *n* digitolutein
Digitonid *n* digitonide
Digitonin *n* digitonin
Digitophyllin *n* digitophylline
Digitosäure *f* digitoic acid
Digitoxigenin *n* digitoxigenin
Digitoxin *n* digitoxin
Digitoxose *f* digitoxose
Digitsäure *f* digitic acid
Diglycerin *n* diglycerol, diglycerin
Diglycerinphosphorsäure *f* diglycerophosphoric acid
Diglycylcystin *n* diglycylcystine
Diglycylglycin *n* diglycylglycine
Diglykol *n* diglycol
Diglykolsäure *f* diglycollic acid, oxydiethanoic acid
Digoxigenin *n* digoxigenin
Diguanid *n* biguanide, guanylguanidine
Diguanylamin *n* diguanylamine
Diguanyldisulfid *n* diguanyldisulfide
Dihalogenid *n* dihalide
Diharnstoff *m* diurea, biurea
Diheptylamin *n* diheptylamine
diheteroatomig diheteroatomic
Dihexaeder *n* (Krist) dihexahedron
dihexaedrisch (Krist) dihexahedral
dihexagonal (Krist) dihexagonal
dihexagonalpyramidal (Krist) dihexagonalpyramidal
Dihexyl- dihexyl
Dihexyl *n* dihexyl, dodecane
Dihexyläther *m* dihexyl ether
Dihexyverin *n* dihexyverine
Dihydracrylsäure *f* dihydracrylic acid
Dihydralazin *n* dihydralazine
Dihydrat *n* dihydrate
Dihydrit *m* (Min) dihydrite
Dihydroanthracen *n* dihydroanthracene
Dihydroanthrachinonazin *n* dihydroanthraquinoneazine
Dihydrobenzol *n* dihydrobenzene
Dihydroberberin *n* dihydroberberine
Dihydrochinin *n* dihydroquinine

Dihydrodimerisation *f* dihydrodimerization
Dihydroergocristin *n* dihydroergocristine
Dihydroergotamin *n* dihydroergotamine
Dihydroergotamintartrat *n* dihydroergotamine tartrate
Dihydroergotoxin *n* dihydroergotoxine
Dihydrofolsäure *f* dihydrofolic acid
Dihydrofolsäurereduktase *f* (Biochem) dihydrofolic acid reductase
Dihydrofuran *n* dihydrofuran
Dihydroliponsäure *f* dihydrolipoic acid
Dihydronaphthalin *n* dihydronaphthalene
Dihydroorotase *f* (Biochem) dihydroorotase
Dihydroorotsäure *f* dihydroorotic acid
Dihydroorotsäurereduktase *f* (Biochem) dihydroorotate dehydrogenase
Dihydrophenanthren *n* dihydrophenanthrene
Dihydrophytylbromid *n* dihydrophytyl bromide
Dihydrostreptit *m* dihydrostreptitol
Dihydrostreptomycin *n* dihydrostreptomycin, DHS
Dihydrotachysterin *n* dihydro tachysterol
Dihydrotachysterol *n* dihydrotachysterol
Dihydrothiamin *n* dihydrothiamine
Dihydrotoxiferin *n* dihydrotoxiferine
Dihydroxyaceton *n* dihydroxyacetone
Dihydroxyacetonphosphat *n* dihydroxyacetone phosphate
Dihydroxyanthracen *n* dihydroxyanthracene
Dihydroxybenzoesäure *f* dihydroxy benzoic acid
Dihydroxybenzol *n* dihydroxybenzene
Dihydroxydiäthylstilben *n* dihydroxydiethylstilbene
Dihydroxydibenzanthracen *n* dihydroxydibenzanthracene
Dihydroxydinaphthyl *n* dihydroxydinaphthyl
Dihydroxydiphenylmethan *n* dihydroxydiphenylmethane
Dihydroxynaphthalin *n* dihydroxynaphthalene
Dihydroxyperylen *n* dihydroxyperylene
Dihydroxyphenylalanin *n* dihydroxyphenylalanine
Dihydroxyphenylessigsäure *f* dihydroxyphenylacetic acid
Dihydroxypyren *n* dihydroxypyrene
Dihydroxystearinsäure *f* dihydroxystearic acid
Dihydroxyviolanthron *n* dihydroxyviolanthrone
Dihypnal *n* dihypnal
Diimid *n* diimide
Diinden *n* diindene
Diindol *n* diindole
Diindolindigo *n* diindole indigo
Diindolyl *n* diindolyl
Diindon *n* diindone
Diindyl *n* diindyl, dindyl
Diisatogen *n* diisatogen
Diisoamyläther *m* diisoamyl ether
Diisobutylamin *n* diisobutylamine
Diisobutylketon *n* diisobutyl ketone
Diisochavibetol *n* diisochavibetol
Diisocyanat *n* diisocyanate
Diisodecyladipinat *n* diisodecyl adipate
Diisodecylphthalat *n* diisodecyl phthalate
Diisoeugenol *n* diisoeugenol
Diisooctyladipinat *n* diisooctyl adipate
Diisooctylazelat *n* diisooctyl azelate
Diisooctylphthalat *n* diisooctyl phthalate
Diisopropyläther *m* diisopropyl ether
Diisopropylamin *n* diisopropylamine
Diisopropylbenzol *n* diisopropyl benzene
Diisopropylcarbinol *n* diisopropyl carbinol
Diisopropylfluorphosphat *n* diisopropyl fluorophosphate, DFP
Diisopropylketon *n* diisopropyl ketone
Dijodacetylen *n* diiodoacetylene
Dijodcarbazol *n* diiodocarbazole
Dijodhydrin *n* diiodohydrin
Dijodid *n* diiodide, diiodid
Dijododithymol *n* diiododithymol
Dijodoform *n* diiodoform, tetraiodoethylene
Dijodsalicylsäure *f* diiodosalicylic acid
dijodsalicylsauer diiodosalicylate
Dijodtyrosin *n* diiodotyrosine
Dijodurotropin *n* diiodourotropine, diiodo-hexamethylenetetramine
Dikabutter *f* dika butter, dika fat
Dikaliumphosphat *n* dipotassium phosphate, dibasic potassium phosphate, monoacid potassium phosphate, potassium hydrogen phosphate
Dikarbonat *n* dicarbonate
Diketen *n* diketene
Diketoapocamphersäure *f* diketoapocamphoric acid
Diketocamphersäure *f* diketocamphoric acid
Diketocholansäure *f* diketocholanic acid
Diketon *n* diketone
Diketopiperazin *n* diketopiperazine
Diketoximchelat *n* diketoxime chelate
Dikodid *n* dicodid
Dikresol *n* dicresol, bicresol
Dilactamidsäure *f* dilactamic acid
Dilactylsäure *f* dilactic acid, lactolactic acid, lactyl-lactate
dilatant dilatant
Dilatanz *f* dilatancy
Dilatation *f* dilatation
Dilatometer *n* dilatometer
Dilatometrie *f* dilatometry
Dilaurin *n* dilaurin
Dilaurylamin *n* dilaurylamine
Dilimonen *n* dilimonene, bilimonene
Dilitursäure *f* dilituric acid, 5-nitrobarbituric acid
Dillnit *m* dillnite
Dillöl *n* dill seed oil, (Min) dill oil
Diluvium *n* (Geol) diluvium, drift
Dimazol *n* dimazole

Dimedon n dimedone, dimethyl cyclohexanedione
Dimenoxadol n dimenoxadole
Dimension f dimension
dimensional dimensional
dimensionieren to dimension
Dimensionierung f dimensioning, design, sizing
Dimensionsanalyse f (Phys) dimension analysis
dimensionslos dimensionless, nondimensional
Dimensionszahl f dimension number
Dimenthen n bimenthene, dimenthene
Dimenthyl n dimenthyl
Dimepheptanol n dimepheptanol
dimer dimeric
Dimer n dimer
Dimercaprol n dimercaprol
Dimercaptopropanol n dimercaptopropanol
Dimercaptothiodiazol n dimercaptothiodiazol
Dimercaptotoluol n dimercaptotoluene
Dimerisation f dimerization
dimerisieren to dimerize
Dimerisierung f dimerization
Dimesityl n bimesityl, dimesityl
Dimetanilsäure f bimetanilic acid, dimetanilic acid
Dimethacrylsäure f bimethacrylic acid, dimethacrylic acid
Dimethiodal n dimethiodal
Dimethoxyacetophenon n dimethoxy acetophenone
Dimethoxyanthrachinon n dimethoxy anthraquinone
Dimethoxybenzaldehyd m dimethoxy benzaldehyde
Dimethoxybenzidin n dimethoxy benzidine
Dimethoxybenzochinon n dimethoxy benzoquinone
Dimethoxybernsteinsäure f dimethoxy succinic acid
Dimethoxylepidin n dimethoxylepidine
Dimethylacetal n dimethyl acetal, ethylidene dimethyl ether
Dimethylacetophenon n dimethyl acetophenone
Dimethylacetylen n dimethyl acetylene
Dimethylacridin n dimethyl acridine
Dimethylacrylsäure f dimethyl acrylic acid
Dimethyläther m dimethyl ether, methoxymethane, methyl ether
Dimethyläthylen n dimethylethylene
Dimethyläthylendiamin n dimethylethylenediamine
Dimethyläthylmethan n dimethylethylmethane
Dimethyläthylpyrrol n dimethylethylpyrrole
Dimethylamin n dimethylamine
Dimethylaminchlorhydrat n dimethylamine hydrochloride
Dimethylaminhydrochlorid n dimethylamine hydrochloride
Dimethylaminoazobenzol n dimethylaminoazobenzene, butter yellow
Dimethylaminobenzaldehyd m dimethylaminobenzaldehyde
Dimethylaminobenzalrhodanin n dimethylaminobenzalrhodanine
Dimethylaminobenzol n dimethylaminobenzene
Dimethylaminobenzophenon n dimethylaminobenzophenone
Dimethylaminophenol n dimethylaminophenol
Dimethylanilin n dimethylaniline
Dimethylanthracen n dimethylanthracene
Dimethylanthrachinon n dimethylanthraquinone
Dimethylarsinsäure f alkargen, cacodylic acid, dimethylarsinic acid
Dimethylbenzacridin n dimethylbenzacridine
Dimethylbenzanthracen n dimethylbenzanthracene
Dimethylbenzimidazol n dimethylbenzimidazole
Dimethylbenzol n (Xylol) dimethylbenzene, xylene
Dimethylbenzthiophanthren n dimethylbenzthiophanthrene
Dimethylbutadien n dimethylbutadiene
Dimethylbutadienkautschuk m dimethylbutadiene caoutchouc
Dimethylbutan n dimethylbutane
Dimethylbutin n dimethylbutyne
Dimethylbutylchinolin n dimethylbutylquinoline
Dimethylbutylsulfoniumjodid n dimethylbutylsulfonium iodide
Dimethylcarbinol n dimethylcarbinol, isopropanol, isopropyl alcohol
Dimethylchinolin n dimethylquinoline
Dimethylchrysen n dimethylchrysene
Dimethylcyclohexandion n dimethylcyclohexanedione
Dimethylcyclopentan n dimethylcyclopentane
Dimethyldiäthoxysilan n dimethyldiethoxysilane
Dimethyldichlorsilan n dimethyldichlorosilane
Dimethyldihydroxyanthracen n dimethyldihydroxyanthracene
Dimethyldiphenyl n dimethyldiphenyl
Dimethyldiphenylharnstoff m dimethyldiphenylurea
Dimethyldipyrrylmethen n dimethyldipyrrylmethene
Dimethyldisilan n dimethyldisilane
Dimethylendiamin n dimethylenediamine, ethylenediamine
Dimethylformamid n dimethyl formamide
Dimethylgelb n dimethyl yellow
Dimethylglucose f dimethylglucose
Dimethylglyoxal n dimethylglyoxal
Dimethylglyoxim n dimethylglyoxime, diacetyl dioxime
Dimethylharnstoff m dimethylurea
Dimethylhydantoinformaldehyd m dimethyl hydantoin formaldehyde

Dimethylhydrochinon — Dinitrophenylaminosäure

Dimethylhydrochinon *n* dimethylhydroquinol
dimethyliert dimethylated
Dimethylisobutenylcyclopropan *n* dimethylisobutenylcyclopropane
Dimethylketon *n* dimethyl ketone, acetone
Dimethylmaleinsäureanhydrid *n* dimethylmaleic anhydride
Dimethylmalonsäure *f* dimethylmalonic acid
Dimethylmandelsäure *f* dimethylmandelic acid
Dimethylnaphthalin *n* dimethylnaphthalene
Dimethylnaphthalinpikrat *n* dimethylnaphthalene picrate
Dimethylnaphthidin *n* dimethylnaphthidine
Dimethylnaphthochinon *n* dimethylnaphthoquinone
Dimethyloxalessigsäure *f* dimethyloxaloacetic acid
Dimethyloxamid *n* dimethyloxamide
Dimethyloxazol *n* dimethyloxazole
Dimethylpentan *n* dimethylpentane
Dimethylphenol *n* dimethylphenol
Dimethylphenylendiamin *n* dimethylphenylendiamin
Dimethylphenylpyrazolon *n* dimethylphenylpyrazolone
Dimethylphosphin *n* dimethylphosphine
Dimethylphthalat *n* dimethyl phthalate
Dimethylpiperazin *n* dimethylpiperazine
Dimethylpyridin *n* dimethylpyridine
Dimethylpyron *n* dimethylpyrone
Dimethylpyrrol *n* dimethylpyrrole
Dimethylresorcin *n* dimethylresorcinol
Dimethylsalicylaldehyd *m* dimethylsalicylaldehyde
Dimethylsebacat *n* dimethyl sebacate
Dimethylsilan *n* dimethylsilane
Dimethylsilandiol *n* dimethylsilanediol
Dimethylsilicon *n* dimethylsilicone
Dimethylsiloxan *n* dimethylsiloxane
Dimethylsulfat *n* dimethyl sulfate
Dimethylsulfoxid *n* dimethyl sulfoxide
Dimethyltetrahydrobenzaldehyd *m* dimethyltetrahydrobenzaldehyde
Dimethylthallium *n* dimethylthallium
Dimethylthetin *n* dimethylthetin
Dimethylthiambuten *n* dimethylthiambutene
Dimethylthreonamid *n* dimethylthreonamide
Dimethyltoluidin *n* dimethyltoluidine
Dimethylxanthin *n* dimethylxanthine, theobromine
Dimilchsäure *f* dilactic acid
Dimit *m* dimite
dimorph (Krist) dimorphic, dimorphous
Dimorphecolsäure *f* dimorphecolic acid
Dimorphie *f* (Krist) dimorphism
Dimorphismus *m* (Krist) dimorphism
Dimoxylin *n* dimoxyline
Dimyristin *n* dimyristin

DIN (Deutsche Industrie-Normung) German Industrial Standards
Dinaphthacridin *n* dinaphthacridine
Dinaphthanthradichinon *n* dinaphthanthradiquinone
Dinaphthanthron *n* dinaphthanthrone
Dinaphthazin *n* dinaphthazine
Dinaphthaziniumhydroxid *n* dinaphthazinium hydroxide
Dinaphthochinon *n* binaphthoquinone, dinaphthoquinone
Dinaphthofluorindin *n* dinaphthofluorindine
Dinaphthol *n* dinaphthol
Dinaphthoxanthen *n* dinaphthoxanthene
Dinaphthyl *n* binaphthyl, dinaphthyl
Dinaphthyläther *m* dinaphthyl ether
Dinaphthylen *n* dinaphthylene
Dinaphthylin *n* dinaphthyline
Dinaphthylphenylendiamin *n* dinaphthylphenylenediamine
Dinasstein *m* dinas brick
Dinaston *m* dinas clay
Dinasziegel *m* dinas brick
Dinatriumarsenat *n* disodium arsenate, acid sodium arsenate, disodium hydrogen arsenate
Dinatriumcitrat *n* acid sodium citrate, disodium citrate
Dinatriumhydrophosphat *n* dibasic sodium phosphate, disodium orthophosphate, disodium phosphate, monoacid sodium phosphate
Dinatriumsalz *n* disodium salt
Dinatriumtetraborat *n* disodium tetraborate
Dinicotinsäure *f* dinicotinic acid, pyridinedicarboxylic acid
Dinit *m* dinite
dinitrieren to introduce two nitro groups
Dinitroaminophenol *n* dinitroaminophenol
Dinitroanilin *n* dinitroaniline
Dinitroanisol *n* dinitroanisol
Dinitroanthrachinon *n* dinitroanthraquinone
Dinitrobenzol *n* dinitrobenzene
Dinitrochlorbenzol *n* dinitrochlorobenzene
Dinitrodiphensäure *f* dinitrodiphenic acid
Dinitrodiphenyl *n* dinitrodiphenyl
Dinitrodiphenylamin *n* dinitrodiphenylamine
Dinitrofluorbenzol *n* dinitrofluorobenzene (DNFB)
Dinitroglycerin *n* dinitroglycerol, dinitroglycerin, glycerol dinitrate
Dinitroglyzerinsprengstoff *m* dinitroglycerol blasting charge, glycerol dinitrate explosive
Dinitromethylanilin *n* dinitromethylaniline
Dinitronaphthalin *n* dinitronaphthalene
Dinitronaphthol *n* dinitronaphthol
Dinitrophenol *n* dinitrophenol
Dinitrophenylacridin *n* dinitrophenylacridine
Dinitrophenylaminosäure *f* dinitrophenyl amino acid

Dinitrophenylhydrazin n dinitrophenylhydrazine
Dinitrophenylosazon n dinitrophenyl osazone
Dinitrotoluol n dinitrotoluene
Dioctylphthalat n dioctyl phthalate
Dioctylsebazat n dioctyl sebacate
Diode f diode, two-electrode tube
Diodengleichrichter m diode rectifier
Diodenstrecke f diode path
Diodon n diodone
Diönanthsäure f bienanthic acid, dienanthic acid
Diogenal n diogenal
Dioktaeder n (Krist) dioctahedron
dioktaedrisch (Krist) dioctahedral
Diol n diol, glycol
Diolefin n diolefine
Diolein n diolein, glycerol dioleate
Dioleopalmitin n dioleopalmitine, glycerol dioleate palmitate
Dioleostearin n dioleostearin, glycerol dioleate stearate
Dionin n dionine
Diopsid m (Min) diopside
Dioptas m (Min) dioptase, emerald copper
Dioptrie f (Phys) diopter
Dioptrik f dioptrics
Diorchit m (Min) diorchite
Diorit m (Min) diorite, greenstone
diorithaltig dioritic
Diorsellinsäure f diorsellinic acid, lecanoric acid
Diorthotolylguanidin n diorthotolyl guanidine
Dioscin n dioscin
Dioscorin n dioscorine
Dioscorinol n dioscorinol
Diosgenin n diosgenin
Diosmetin n diosmetine
Diosmin n diosmin
Diosphenol n diosphenol, buchu camphor
Diospyrol n diospyrol
Dioxaborol n dioxaborole
Dioxan n (Diäthylendioxid) dioxan[e]
Dioxaphospholan n dioxaphospholane
Dioxazin n dioxazine
Dioxazol n dioxazole
Dioxen n dioxene
Dioxid n (Dioxyd) dioxide
Dioxim n dioxime
Dioxin n dioxin
Dioxindol n dioxindole
Dioxol n dioxole
Dioxolan n dioxolane
Dioxopiperazin n dioxopiperazine
Dioxyaceton n dihydroxyacetone
Dioxyanthrachinon n (Alizarin) dihydroxyanthraquinone
Dioxybenzoesäure f dihydroxybenzoic acid
Dioxybenzol n dihydroxybenzene, dioxybenzene
Dioxybernsteinsäure f 2,3-dihydroxysuccinic acid, tartaric acid

Dioxychinolin n dihydroxyquinoline, dioxyquinoline
Dioxychinon n dihydroxyquinone, dioxyquinone
Dioxydiphenyl n dioxydiphenyl
Dioxydiphenylurethan n dioxydiphenyl urethane
Dipalmitin n dipalmitin, glycerol dipalmitate
Dipalmitostearin n dipalmitostearin, glycerol dipalmitate stearate
Dipenten n dipentene
Dipeptid n dipeptide
Dipeptidase f (Biochem) dipeptidase
Diperodon n diperodon
Diphanit m diphanite
Diphemanil n diphemanil
Diphenaldehyd m bibenzaldehyde, diphenaldehyde
Diphenan n diphenan
Diphenanthryl n biphenanthryl, diphenanthryl
Diphenazyl n diphenazyl, dibenzoylethane
Diphenetidin n biphenetidine, diphenetidine
Diphenhydramin n diphenhydramine
Diphenin n diphenine
Diphenol n diphenol
Diphenolsäure f diphenolic acid, DPA
Diphenonchinon n diphenoquinone, bibenzenone
Diphensäure f diphenic acid, bibenzoic acid
Diphenyl n diphenyl, biphenyl, xenene
Diphenyläther m diphenyl ether
Diphenyläthylamin n diphenylethylamine
Diphenyläthylen n diphenylethylene
Diphenylamin n diphenylamine, phenylaniline
Diphenylaminblau n diphenylamine blue
Diphenylarsenchlorür n diphenylarsenous chloride, diphenylchloroarsine
Diphenylbenzochinon n diphenyl benzoquinone
Diphenylborchlorid n diphenylboron chloride, diphenylchloroborine
Diphenylbutadien n diphenylbutadiene
Diphenylcarbamid n diphenylcarbamide
Diphenylcarbonat n diphenyl carbonate, phenyl carbonate
Diphenylchinaldinblau n diphenylquinaldine blue
Diphenylchinoxalin n diphenylquinoxaline
Diphenyldecapentaen n diphenyldecapentaene
Diphenyldecatetraen n diphenyl decatetraene
Diphenyldiazomethan n diphenyldiazomethane
Diphenyldichlorsilan n diphenyldichlorosilane
Diphenylen n diphenylene
Diphenylendioxid n diphenylene dioxide, dibenzodioxin
Diphenylenimid n diphenylene imide, carbazol
Diphenylenoxid n diphenylene oxide, dibenzofuran, dibenzofurfuran
Diphenylensulfid n diphenylene sulfide
Diphenylensulfon n diphenylene sulfone
Diphenylglyoxim n diphenylglyoxime

Diphenylguanidin n diphenylguanidine
Diphenylhexadien n diphenylhexadiene
Diphenylhexatrien n diphenylhexatriene
Diphenylhydrazin n diphenylhydrazine
Diphenylin n diphenyline
Diphenylisomerie f diphenyl isomerism
Diphenyljodoniumhydroxid n diphenyliodonium hydroxide
Diphenyljodoniumjodid n diphenyliodonium iodide
Diphenylketon n diphenyl ketone, benzophenone
Diphenylmethan n diphenylmethane, benzylbenzene
Diphenylmethanfarbstoff m diphenylmethane dyestuff
Diphenylnaphthylmethyl n diphenylnaphthylmethyl
Diphenylnitrosamin n nitrosodiphenylamine
Diphenyloctatetraen n diphenyloctatetraene
Diphenylphenylendiamin n diphenyl phenylene diamine, DPPD
Diphenylpolyen n diphenylpolyene
Diphenylsilandiol n diphenylsilanediol
Diphenylsulfon n diphenyl sulfone
Diphenylthiocarbazon n diphenylthiocarbazone
Diphenylthioharnstoff m diphenylthiourea, sulfocarbanilide
Diphenylyl n biphenylyl, diphenylyl, xenyl
Diphonie f, dynamische ~ (Akust) dynamic diphonia
Diphosgen n diphosgene, surpalite, trichloromethyl chloroformate
Diphosphoglycerinaldehyd m diphosphoglyceraldehyde
Diphosphoglycerinsäure f diphosphoglyceric acid
Diphosphoinositid n diphosphoinositide
Diphosphopyridinnucleotid n diphosphopyridine nucleotide, codehydrase I, coenzyme I, cozymase I, DPN, NAD, **reduziertes ~** reduced diphosphopyridine nucleotide, DPNH
Diphosphorsäure f diphosphoric acid, pyrophosphoric acid
Diphosphothiamin n diphosphothiamine
Diphthalsäure f diphthalic acid, biphthalic acid
Diphthalyl n biphthalyl, diphthalyl
Diphtherie f (Med) diphtheria
Diphtheriebazillus m diphtheria bacillus, Corynebacterium diphtheriae, Klebs-Loeffler's bacillus
Diphyscion n diphyscion
Dipicolinat n dipicolinate
Dipicolinsäure f dipicolinic acid, pyridinedicarboxylic acid
Dipikrinsäure f dipicric acid
Dipikrylamin n dipicrylamine, hexanitrodiphenylamine
Dipipanon n dipipanone
Dipipecolinsäure f dipipecolinic acid
Diploeder n diploid
diploid (Biol) diploid
Diplomchemiker m certified chemist
Diplosal n diplosal, disalicylic acid, salicylosalicylic acid
Diploschistessäure f diploschistesic acid
Diplospartyrin n diplospartyrine
Dipol m dipole, **permanenter ~** permanent dipole
Dipolachse f dipole axis
Dipolanordnung f dipole array
dipolar dipolar
Dipoldipolverbreiterung f dipole-dipole broadening
Dipoldipolwechselwirkung f dipole-dipole interaction
Dipolflächendichte f dipole surface density
Dipolion n dipolar ion
Dipolmolekül n dipole molecule
Dipolmoment n dipole moment
Dipolnäherung f dipole approximation
Dipol-Regel f (Stereochem) dipole rule
Dipolübergang m dipole transition
Dippelsöl n Dippel's oil
Dipren n diprene
Dipropargyl n dipropargyl, bipropargyl
Dipropionylmethan n dipropionyl methane
Dipropyläther m dipropyl ether
Dipropylamin n dipropylamine
Dipropylbarbitursäure f dipropylbarbituric acid
Dipropylentriamin n dipropylenetriamine
Dipropylketon n dipropyl ketone, 4-heptanone, butyrone
Dipropylmalonylharnstoff m dipropylmalonylurea, dipropylbarbituric acid
Dipsomanie f (Med) alcoholism, dipsomania
Dipterin n dipterine
Dipterocarpol n dipterocarpol
Dipterocarpon n dipterocarpone
Dipyr m (Min) dipyre
Dipyrazolanthronyl n dipyrazolanthronyl
Dipyridindiamminplatin n dipyridinediammine platinum
Dipyridinhämatoporphyrin n dipyridine hematoporphyrin
Dipyridinkoproporphyrin n dipyridine coproporphyrin
Dipyridinmesoporphyrin n dipyridine mesoporphyrin
Dipyridinprotoporphyrin n dipyridine protoporphyrin
Dipyridyl n dipyridyl, bipyridine
Dipyrrylmethen n dipyrrylmethene
Direktdruck m direct printing
Direktfarbstoff m direct dyestuff, substantive dye
Direktkopieren n direct printing

Direktrix f (Math) directrix
Direktschaltung f direct connection
Direktschwarz n direct black
Direktspinnverfahren n direct spinning
Direktstrahlung f direct radiation, head-on radiation
Direktwerbung f direct [mail] advertising
Diresorcin n biresorcinol, diresorcinol
Diresorcylsäure f diresorcylic acid
Dirhein n dirhein
Dirhodan n dirhodan
Diricinolein n diricinolein, glycerol diricinoleate
Diricinolsäure f diricinolic acid
Disaccharid n disaccharide
Disalicylid n disalicylide, salosalicylide
Disalicylsäure f disalicylic acid, diplosal, salicylosalicylic acid
Dischwefelsäure f disulfuric acid, pyrosulfuric acid
Dischwefeltrioxid n sulfur sesquioxide
Disgregationswärme f internal heat of evaporation
Disilan n disilane
Disiloxan n disiloxane
Diskatol n discatol
diskontinuierlich discontinuous[ly], intermittent
Diskontinuität f discontinuity, intermittence
Diskontinuitätsfläche f surface of discontinuity
Diskrasit m (Min) dyscrasite
Diskrepanz f discrepancy
Diskretisierungsfehler m discretisation error
Diskriminante f discriminant
Diskussion f discussion
diskutieren to discuss
Dislokation f (Metallographie) dislocation
Dismutation f dismutation
dispensieren to dispense, to excuse, to exempt
Dispergator m dispersing agent
Dispergens n dispersing agent, emulsifier
dispergieren to disperse
Dispergieren n dispersing
dispergierend dispersive
Dispergiergerät n dispersing tool
Dispergiermittel n dispersing agent
dispers disperse, dispersed
Dispersion f dispersion, ~ **des Schalles** acoustic dispersion, ~ **von festen Teilchen** n pl **in einer Flüssigkeit** (Phys) soliquid, sosoloid
Dispersionseffekt m dispersion effect
Dispersionsfarbe f emulsion paint
Dispersionsfarbstoff m dispersed dyestuff
Dispersionsgrad m degree of dispersion, dispersity
Dispersionskleber m adhesive dispersion
Dispersionskneter m dispersion kneader
Dispersionskolloid n dispersion colloid
Dispersionskraft f dispersion force
Dispersionsmessung f dispersion measurement
Dispersionsmethode f dispersion method

Dispersionsmittel n dispersant, dispersing agent, dispersion medium
Dispersionsvermögen n (Opt) dispersive power
Dispersionswirkung f dispersing action
Dispersoid n disperse system, dispersoid
Dispiroverbindung f dispiro compound
disponieren to dispose, to manage
Disposition f disposition, arrangement
disproportionieren to disproportionate
Disproportionierung f disproportionation
disrotatorisch disrotatory
disruptiv disruptive
Dissertation f thesis
Dissimilation f dissimilation
Dissipation f dissipation
Dissipationsenergie f (Mech) degradation energy
Dissipationswärme f heat of dissipation
Dissipator m dissipator
Dissolution f dissolution
Dissousgas n dissolved acetylene [gas]
Dissoziation f dissociation, ionization, **elektrolytische** ~ electrolytic or ionic dissociation
Dissoziationsenergie f dissociation energy
Dissoziationsformel f dissociation formula
Dissoziationsgleichgewicht n dissociation equilibrium, equilibrium of dissociation
Dissoziationsgleichung f dissociation equation
Dissoziationsgrad m dissociation degree
Dissoziationsgrenze f dissociation limit
Dissoziationsisotherme f dissociation isotherm
Dissoziationskonstante f coefficient of dissociation, dissociation constant
Dissoziationsrückgang m retrogression of dissociation
Dissoziationstemperatur f dissociation temperature
Dissoziationstheorie f dissociation theory
Dissoziationsvermögen n dissociation power
Dissoziationsvorgang m dissociation process
Dissoziationswärme f heat of dissociation
Dissoziationszustand m dissociation condition
dissoziativ dissociative
dissoziierbar dissociable
dissoziieren to dissociate, to ionize
Dissoziierung f dissociation
Dissymmetrie f dissymmetry, ~ **durch räumliche Überlappung** (Stereochem) dissymmetry due to molecular overcrowding, **axiale** ~ (Stereochem) axial dissymmetry
Distanzblock m spacer block
Distanzbuchse f spacer bushing
Distanzring m packing piece, spacing ring
Distanzrolle f spacer
Distanzscheibe f spacer
Distanzschraube f distance bolt
Distanzstück n spacer, stopper
Distearin n distearin, glycerol distearate

Distearopalmitin *n* distearopalmitin, glycerol distearate palmitate
Distemonanthin *n* distemonanthin
Disthen *m* (Min) disthene, **blauer** ~ cyanite, kyanite
Distichin *n* distichine
Distickstoffoxid *n* dinitrogen monoxide, laughing gas, nitrous oxide
Distickstoffpentoxid *n* nitric anhydride, nitrogen pentoxide
Distickstofftetroxid *n* dinitrogen tetroxide
Distickstofftrioxid *n* dinitrogen trioxide, nitrous anhydride
Distilben *n* distilbene
Distributivgesetz *n* distributive law
Distributivität *f* (Math) distributivity
Distylin *n* distylin
Distyrinsäure *f* distyrinic acid
Distyrol *n* bistyrene
Disubstituent *m* disubstituent
Disulfaminsäure *f* disulfamic acid
Disulfat *n* disulfate, bisulfate, pyrosulfate
Disulfid *n* disulfide
Disulfidbildung *f* disulfide formation
Disulfidbrücke *f* disulfide bridge
Disulfidspaltung *f* disulfide cleavage
Disulfiram *n* disulfiram
Disulfosäure *f* disulfonic acid
Ditain *n* ditaine, echitamine
Ditamin *n* ditamine
Ditarinde *f* dita bark
Diterpene *pl* diterpenes
ditetraedrisch (Krist) ditetrahedral
ditetragonalbipyramidal (Krist) ditetragonalbipyramidal
ditetragonalpyramidal (Krist) ditetragonalpyramidal
Ditetralyl *n* ditetralyl, bitetralyl
Dithian *n* dithiane, diethylene disulfide
Dithiazol *n* dithiazole
Dithietan *n* dithietane
Dithiin *n* dithiin
Dithioameisensäure *f* dithioformic acid
Dithiobenzoat *n* dithiobenzoate
Dithiobiuret *n* dithiobiuret, dithiocarbamylurea
Dithiocarbamat *n* dithiocarbamate
Dithiocarbamidsäure *f* dithiocarbamic acid
Dithiocarbazidsäure *f* dithiocarbazic acid
Dithiocumarin *n* dithiocoumarin
Dithiodiglykolsäure *f* dithiodiglycolic acid
Dithioerythrit *m* dithioerythritol
Dithiofluorescein *n* dithiofluorescein
Dithiokohlensäure *f* dithiocarbonic acid
Dithiol *n* dithiol, disulfole
Dithiolan *n* dithiolane
Dithion *n* dithion
Dithionaphthenindigo *n* dithionaphthene indigo
Dithionat *n* dithionate, salt or ester of dithionic acid
dithionig dithionous, hyposulfurous
Dithionigsäure *f* dithionous acid, hyposulfurous acid
Dithionit *n* dithionite, hyposulfite
Dithionsäure *f* dithionic acid, hyposulfuric acid
Dithiooxalatchelat *n* dithiooxalate chelate
Dithiophosphorsäure *f* dithiophosphoric acid
Dithiosalicylsäure *f* dithiosalicylic acid
Dithiothreitol *n* dithiothreitol
Dithiourethan *n* dithiourethane, ethyl dithiocarbamate
Dithizon *n* dithizone, diphenylthiocarbazone
Dithizonat *n* dithizonate
Dithizonchloroform *n* dithizone chloroform
Dithranol *n* dithranol
Dithymol *n* bithymol, dithymol
Ditolyl *n* bitolyl, ditolyl
Ditolyldichlorsilan *n* ditolyldichlorosilane
ditrigonal (Krist) ditrigonal
ditrigonalbipyramidal (Krist) ditrigonalbipyramidal
ditrigonalpyramidal (Krist) ditrigonalpyramidal
Dittmarit *m* (Min) dittmarite
Diuranat *n* diuranate
Diureid *n* diureide
Diuretikum *n* (Pharm) diuretic
Diuretin *n* diuretine, theobromin sodium salicylate
Divalonsäure *f* divalonic acid
Divaricatsäure *f* divaricatic acid
Divarin *n* divarinol
Divarsäure *f* divaric acid
Divergenz *f* **einer Reihe** (Math) divergence of a series
Divergenzverlust *m* (Akust) divergence loss
divergieren to diverge [from]
divergierend divergent, diverging
Diversin *n* diversine
Divicin *n* divicine
Dividend *m* (Math) dividend
dividieren to divide
Dividivi *pl* divi-divi, libi-dibi
Divinyl *n* divinyl, bivinyl, butadiene
Divinylacetylen *n* divinylacetylene
Divinyldichlorsilan *n* divinyldichlorosilane
Divinylkautschuk *m* butadiene rubber, divinyl caoutchouc
Division *f* (Math) division
Divisor *m* (Math) divisor
Divostrosid *n* divostroside
Diwasserstoff *m* dihydrogen
Diweinsäure *f* ditartaric acid
Diwolframcarbid *n* ditungsten carbide
Diwolframsäure *f* ditungstic acid
Dixanthogen *n* dixanthogen
Dixanthylen *n* dixanthylene
Dixylenol *n* bixylenol
Dixylyl *n* dixylyl, bixylyl
Dizimtalkohol *m* dicinnamic alcohol

Dizimtsäure *f* bicinnamic acid
Djenkolsäure *f* djenkolic acid
D-Linie *f* (Spektr) D-line
DNS (Desoxyribonukleinsäure) de[s]oxyribonucleic acid
Docht *m* wick
Dochtkohle *f* cored carbon
Dochtschmierer *m* wick lubricator, siphon lubricator
Dochtschmierung *f* wick-oiling
Dochtverdampfer *m* wick carburettor
Docke *f* (Text) hank
Docosan *n* docosane
Dodecaeder *n* (Krist) dodecahedron
dodecaedrisch (Krist) dodecahedral
Dodecahedran *n* dodecahedrane
Dodecahydrotriphenylen *n* dodecahydrotriphenylene
Dodecan *n* dodecane
Dodecansäure *f* dodecanic acid, dodecanoic acid, lauric acid
Dodecin *n* dodecine, dodecyne
Dodecyl- dodecyl
Dodecylaldehyd *m* dodecanal, lauraldehyde, lauric aldehyde
Dodecylalkohol *m* dodecyl alcohol, n-dodecanol
Dodecylen *n* dodecylene, l-dodecene
Dodecylensäure *f* dodecylenic acid
Dodecylsäure *f* dodecanoic acid, lauric acid
Dodekaeder *n* (Krist) dodecahedron
Dodekaedergleitung *f* dodecahedral slip
Döbners Violett *n* Doebner's violet
Döpperstahl *m* steel for dies and stamps
Dörnerschlacke *f* puddling slag, tap cinder
dörren to dry, to dehydrate, to desiccate, to exsiccate, (verwelken) to wither
Doggerstahl *m* dogger steel
Dohexacontan *n* dohexacontane
Dohlen *m* sewer, sink, small culvert
Doisynolsäure *f* doisynolic acid
Dokosansäure *f* docosanoic acid, behenic acid
Dokosylalkohol *m* docosyl alcohol, docosanol
Doktor *m* **der Medizin** (Dr. med.) doctor of medicine (M. D), ~ **der Naturwissenschaften** *f pl* (Dr. rer. nat.) doctor of philosophy (Ph. D), doctor of science (D. Sc), ~ **der Pharmazie** doctor of pharmacy (Phar. D), ~ **der Zahnheilkunde** (Dr. med. dent.) doctor of dental surgery (D. M. D)
Doktorand *m* candidate for a doctor's degree
Doktorarbeit *f* dissertation, [doctoral] thesis
Dokument *n* document, record
Dokumentation *f* documentation
dokumentieren to document
Dolantin *n* dolantin
Doldenblütler *pl* umbelliferae
Dolerit *m* (Min) dolerite
dolerithaltig doleritic
Dolerophanit *m* (Min) dolerophanite

Dolomit *m* (Min) dolomite, bitter spar, **eisenhaltiger** ~ siderocalcite, **totgebrannter** ~ deadburned dolomite
Dolomitbildung *f* (Geol) dolomitization
Dolomitbrennofen *m* dolomite calcining kiln
dolomithaltig dolomitic
Dolomitmasse *f* dolomite mass
Dolomitmehl *n* dolomite powder
Dolomitmergel *m* magnesium marlstone
Dolomitmischer *m* dolomite mixer
Dolomitmühle *f* dolomite mill
Dolomitsand *m* dolomite sand
Dolomitstein *m* dolomite brick
Dom *m* dome
Domänenspeicher *m*, **magnetischer** ~ magnetic domain memory
Domesticin *n* domesticine
Domeykit *m* (Min) domeykite
Domingit *m* (Min) domingite
Domit *m* domite
Donarit *n* (Kampfstoff) donarite
Donator *m* donor
Donatoratom *n* donor atom
Donatorenniveau *n* donor level
Donatorenwanderung *f* donor migration
Donatorenzym *n* (Biochem) donor enzyme
Donaxin *n* donaxine
Donnan-Effekt *m* Donnan effect
Donnan-Gleichgewicht *n* Donnan equilibrium
Donnerstein *m* belemnite
Dopa *n* dopa, 3,4-dihydroxyphenylalanine
Dopamin *n* dopamine
Dopol-Ofen *m* Dopol furnace
doppelachtflächig (Krist) dioctahedral
Doppelachtflächner *m* (Krist) dioctahedron
Doppelakkumulator *m* two-cell accumulator
Doppelarmkneter *m* double-arm kneader
doppelatomig biatomic, diatomic
Doppelbad *n* double bath
Doppelbandschreiber *m* double recorder
doppelbasisch dibasic, bibasic
Doppelbestimmung *f* repeated test
Doppelbezeichnung *f* synonym
Doppelbifilargravimeter *n* double bifilar gravimeter
Doppelbild *n* double image
Doppelbindung *f* double linkage, double union, (Chem) double bond, **konjugierte** ~ conjugated double bond, **semipolare** ~ semi-polar double bond
Doppelblendenmethode *f* (Opt) double-slit method
Doppelboden *m* double bottom, false bottom, jacket, (Dest) double tray
doppelbrechend (Opt) birefractive, birefringent, double-refracting
Doppelbrechung *f* (Opt) birefraction, birefringence, double refraction, **allogyrische** ~ (Opt) allogyric birefringence

Doppelbruch *m* compound fraction
Doppelchlorid *n* bichloride, dichloride, double chloride
doppelchromsauer bichromate, dichromate
Doppeldeutigkeit *f* ambiguity
Doppeldrahtzwirnmaschine *f* two-for-one twisting machine
Doppeldüsenvergaser *m* double jet carburet[t]or
Doppelelektrode *f* diode
Doppelender *m* double-ended boiler
doppelendig double-ended
Doppelerzscheider *m* double ore separator
Doppelfalz *m* (Dose) double seam
doppelfarbig dichroic, dichromatic
Doppelfarbigkeit *f* dichroism
Doppelflammenofen *m* double flame furnace
doppelgängig bifilar, bifurcate, branching, forked
Doppelgenerator *m* double producer
Doppelgewinde *n* double thread
Doppelgitter *n* bigrid, dual grid
Doppelgleitung *f* duplex slip
Doppelhelix *f* double helix
Doppelintegral *n* (Math) double integral
Doppeljodquecksilber *n* biniodide of mercury, mercuric iodide, mercury diiodide, mercury(II) iodide
Doppelkarbonat *n* double carbonate
Doppelkegelfilter *m* doublecone filter
Doppelkegelkreiselmischer *m* doublecone impeller mixer
Doppelkegelkreiselrührer *m* doublecone impeller
Doppelkegelkupplung *f* double-tapered muff coupling
Doppelkegelschleuder *f* double cone centrifuge
Doppelkegeltrommelmischer *m* doublecone mixer
Doppelkeil *m* gib and cotter
Doppelkessel *m* double boiler, battery boiler, double-ended boiler, jacketed pans
Doppelklemme *f* double clamp, double connector, double terminal
doppelkohlensauer acid carbonate, bicarbonate, hydrogen carbonate
Doppelkolbenpresse *f* double ram press
Doppelkolbenpumpe *f* double piston pump
Doppelkollektormaschine *f* double commutator engine
Doppelkommutatormaschine *f* double commutator engine
Doppelkonusmischer *m* double cone blender, double cone mixer
Doppelkraftmessung *f* double measurement of charge
Doppelkreislauf *m* dual cycle
Doppelkristall *m* (Krist) twin crystal
Doppelladeportal *n* twin loading portal
Doppelleerstelle *f* double void

Doppelleiter *m* twin conductor
Doppelleitung *f* duplicate main, duplicate pipe line, duplicate wire
Doppellinie *f* (Opt) doublet
Doppelmetallstange *f* bimetallic bar
Doppelmolekül *n* double molecule
Doppelmuffelofen *m* double muffle furnace
Doppelmuldenkneter *m* divided trough kneader
Doppelmutter *f* double nut
Doppelneutron *n* dineutron
Doppelobjektiv *n* double lens
Doppelofen *m* double furnace
Doppeloxid *n* dioxide, binoxide, double oxide
Doppelpapier *n* duplex paper
Doppelpendelklappe *f* double gate (in pneumatic conveyor systems)
Doppelpoliermaschine *f* double buffing machine
doppelpolig bipolar, two-polar
Doppelprisma *n* bi-prism
Doppelprofilringwaage *f* duplex edgewise pattern ring balance
Doppelpuddelofen *m* double puddle furnace, double puddling furnace
Doppelpyramide *f* (Krist) bipyramid
Doppelriemen *m* double-layer belt
Doppelröntgenblitzröhre *f* double X-ray flash tube
Doppelrohr *n* double pipe
Doppelrohrkondensator *m* double-pipe condenser
Doppelrohrwärmeaustauscher *m* double-pipe heat exchanger
Doppelrührwerk *n*, **gegenläufiges** ~ double motion agitator, double motion mixer
Doppelsalz *n* double salt
doppelschauflig double blade
Doppelschelle *f* twin clamp
Doppelschellenverbindung *f* double clamp connection
Doppelschicht *f* double layer, double striation, **elektrochemische** ~ electric double layer
Doppelschichtelektrode *f* double skeleton electrode
Doppelschleifmaschine *f* double-sided grinder
Doppelschleuderpumpe *f* twin centrifugal pump
Doppelschlußerregung *f* compound excitation
Doppelschlußgenerator *m* (Elektr) compound[-wound] dynamo
Doppelschmiege *f* combination bevel
Doppelschnecke *f* double screw, dual worm, twin screw, twin worm
Doppelschneckenmischer *m* double spiral mixer, twin-worm mixer
Doppelschrauben-Druckmischer *m* twin screw pressure mixer
Doppelschreiber *m* dual recorder
Doppelschwefeleisen *n* iron disulfide
Doppelschwefelsäure *f* disulfuric acid, pyrosulfuric acid

doppelschwefelsauer bisulfate, disulfate, pyrosulfate
doppelsechsflächig (Krist) dihexahedral
Doppelsechsflächner m (Krist) dihexahedron
doppelseitig bilateral
Doppelsessel-Konformation f (Stereochem) chair-chair conformation
Doppelsilicat n acid silicate, bisilicate, double silicate
Doppelsitzventil n double-seated valve, double seat valve
Doppelspalt m (Opt) double slit
Doppelspannung f double voltage
Doppelspat m (Min) iceland spar
Doppelsteckdose f (Elektr) twin socket
Doppelstecker m double plug
Doppelstein m (Min) calcareous spar
Doppelstempelform f double force mold
Doppelsterne m pl (Astr) binaries
Doppelsternkomponente f binary component
Doppelsternschaltung f (Elektr) duplex star connection
doppelstrangig double-stranded, twin-threaded
Doppelstrang-Kettenförderer m mit Tragrahmen double-strand platform conveyor
Doppelstreuversuch m double scattering experiment
Doppelstrich m double touch
Doppelstrom m double current
Doppelstromerzeuger m double current dynamo
Doppelstück n duplicate
Doppelsulfat n bisulfate, double sulfate
Doppelsulfit n acid sulfite, bisulfite, double sulfite
Doppelsuperphosphat n double superphosphate
doppelt double, duplex, twice, twofold
doppeltborsauer acid borate, biborate
Doppeltchlorzinn n tin(IV) chloride, stannic chloride
doppeltchromsauer bichromate, dichromate
Doppel-T-Eisen n double T-iron, H-iron
Doppelthermoskop n double thermoscope
doppeltkonisch biconical
Doppeltonfarbe f two-tone color
doppeltprismatisch diprismatic
Doppeltransportverfahren n double transfer [carbon] process
Doppeltrichterapparat m cup and cone charger
Doppeltropfenfänger m double bulb sloping splash head
Doppeltschwefeleisen n iron disulfide
doppeltschwefelsauer bisulfate
doppeltschwefligsauer bisulfite
Doppel-T-Träger m double T-girder, I-beam, I-girder
doppeltweinsauer acid tartrate, bitartrate
Doppelunterbrecher m commutator with two directions
Doppelventil n two-way valve
Doppelverband m double bandage
Doppelverhältnis n cross ratio
Doppelwägung f double weighing
Doppelwalzenpresse f paired compacting rolls
doppelwandig double-walled
Doppelwasserglas n potassium sodium silicate, silicate of potash and soda
Doppelweggleichrichter m full-wave rectifier
Doppelweghahn m two-way cock
Doppelweiche f double switch
Doppelwendelkühler m double spiral condenser
doppelwertig bivalent
doppelwirkend double acting
doppelwürfelig amphihexahedral
Doppelzentner m (Maß) one tenth of a metric ton
Doppelzersetzung f double decomposition
Doppelzünder m combination fuse
Doppelzündmagnet m dual magneto
Doppelzündung f double ignition
Doppelzweipunktregelung f two-zone control
Doppelzylinderpumpe f duplex pump
Dopplereffekt m Doppler's principle [or effect]
Dopplerit m (Min) dopplerite
Doppler-Verbreiterung f (Spektr) Doppler broadening
Doppler-Verschiebung f (Spektr) Doppler displacement, Doppler shift
Doremol n doremol
Doremon n doremone
Dormiol n dormiol, amylene chloral
Dormonal n dormonal
Dorn m thorn, arbor, core, mandrel
Dornbiegeprobe f mandrel test; rod bending test
Dornhalter m core spider, mandrel carrier
Dornhaltersteg m spider leg
Dornhalterung f mandrel carrier
Dornpresse f arbor press
Dornprüfung f plunger test
Dornwechsler m core changer
Dorschleberöl n codliver oil
Dose f can, box, tin [can], ~ **aus Stahl- und Aluminiumblech** bimetallic can, ~ **im Flachformat** squat can, ~ **mit Eindrückdeckel** lever lid tin, ~ **mit Einsatz** compartment can, **gesickte** ~ beaded can, **gezogene** ~ drawn can, **nahtlose** ~ seamless can, **undichte** ~ oozer (Am. E.), **unlackierte** ~ plain can
Dosenaneroid n aneroid barometer
Dosenbarometer n aneroid barometer
Dosenbier n canned beer
Dosenfüllmaschine f can filling machine
Dosengalvanometer n circular galvanometer
Dosenindustrie f can making industry
Dosenkonserve f canned food
Dosenlibelle f [round] spirit level
Dosenmilch f canned milk
Dosenöffner m can opener
Dosieranlage f dosing apparatus, dosing plant

Dosierbandwaage *f* continuous scales for dosing
Dosiereinrichtung *f* metering device
dosieren to dose, to determine
Dosieren *n* dosing, metering
Dosierkolbenpumpe *f* reciprocating proportioning pump
Dosiermaschine *f* dosing machine
Dosierpumpe *f* fractionating pump, metering pump, proportioning pump
Dosiertank *m* metering tank
Dosierung *f* dosage, dosing
Dosierungsfehler *m* wrong dosage
Dosierungskunde *f* dosimetry
Dosierungspumpe *f* dose pump
Dosierwalze *f* doctor roll
Dosimeter *n* weighing and measuring machine, dosimeter, quantimeter
Dosimetrie *f* dosimetry
Dosis *f* dose, **höchstzulässige ~** maximum permissible dose, **kleinste therapeutische ~** curative dose, **kleinste wirkungsvolle ~** minimum effective dose, **nichttödliche ~** sublethal dose, **tödliche ~** lethal amount, lethal dose
Dostenöl *n* origan oil, origanum oil
Dostkraut *n* (Bot) origan, origanum
Dotieren *n* (Elektr) doping
Dotriacontan *n* dotriacontane
Dotterhaut *f* vitelline membrane, yolk membrane
Doublé *n* double
Douglasit *m* (Min) douglasite
Dove-Prisma *n* (Opt) Dove prism, reversing prism
Dowexkationenaustauschharz *n* (HN) dowex cation exchange resin
Doxylamin *n* doxylamine
Dozent *m* lecturer, dozent
Drachenblut *n* dragon's blood
Drachenblutgummi *n* dragon's blood gum
Dracin *n* dracina
Dracorhodin *n* dracorhodin
Dracorubin *n* dracorubin
Dracosäure *f* dracoic acid
dränieren to drain
Dragée *n* dragee, sugar-coated tablet
dragieren to coat, to sugar-coat
Draht *m* wire, monofilament, thread, twist, **blanker ~** bare wire, **einen ~ auf Elastizität prüfen** elongation test, **einen ~ auf Zähigkeit prüfen** bending test, **geflochtener ~** braided wire, **isolierter ~** insulated wire, **verzinnter ~** tinned wire
Drahtarbeit *f* wire work
Drahtauslöser *m* wire release
Drahtbearbeitung *f* wire working
Drahtbewehrung *f* wire sheathing
Drahtbürste *f* wire brush
Drahtdreieck *n* wire triangle

Drahteinlage *f* wire core, wire reinforcement
Drahteisen *n* wire rod
Drahtemaille *f* wire enamel
drahten to wire
Drahtgaze *f* wire gauze
Drahtgeflecht *n* wire gauze, wire netting, wire work
drahtgeheftet wire-stitched
Drahtgewebe *n* meshing, wire cloth, wire gauze, wire mesh, wire netting
Drahtgitter *n* wire grating, wire lattice
Drahtglas *n* wire glass
Drahtkern *m* wire bead
Drahtkernummantelung *f* bead cover fabric
Drahtkernumspritzung *f* bead wire insulation
Drahtklammer *f* staple
Drahtkorb *m* wire basket
Drahtkorn *n* steel wire grain, wire shot
Drahtlack *m* wire enamel, wire lacquer
Drahtlehre *f* wire gauge
drahtlos wireless
Drahtnetz *n* network of wires, wire gauze, wire network
Drahtnetzdiaphragma *n* wire gauze diaphragm
Drahtnetzelektrode *f* wire gauze electrode
Drahtnetzgewebe *n* wire mesh
Drahtnetzkathode *f* wire gauze cathode
Drahtnummer *f* gauge of wire
Drahtprüfung *f* wire testing
Drahtring *m* coil
Drahtschere *f* wire shears
Drahtschleife *f* wire loop
Drahtschneider *m* wire cutter[s]
Drahtseil *n* wire cable, wire rope
Drahtseilbahn *f* cable railway, rope railway
Drahtsieb *n* wire mesh screen, wire sieve
Drahtsorte *f* quality of wire
Drahtspanner *m* wire stretcher
Drahtspirale *f* wire spiral, (Elektr) solenoid
Drahtstärke *f* diameter of wire
Drahtstift *m* wire nail
Drahtstrecke *f* looping mill, rod mill, wire mill
Draht-Trennrohr *n* hot-wire thermogravitational column
Drahtummantelung *f* wire covering
Drahtverzinnerei *f* wire tinning plant
Drahtwalzen *n* wire rolling
Drahtwickelmaschine *f* wire winding machine
Drahtwiderstand *n* resistance of a wire
Drahtwurm *m* wire worm
Drahtzange *f* wire cutters, cutting pliers
Drahtziehbank *f* wire drawing bench
drahtziehen to draw a wire
Drahtziehen *n* wire drawing
Drainrohr *n* drain pipe
Drainwasser *n* drainage water
Drall *m* **des Seiles** twist of the rope
Drallfreiheit *f* absence of twist

Drallkammerdüse f (Hohl- und Vollkegel) [hollow & full] cone spray nozzle
Drallklappen pl guide vanes
Drallmoment n (Atom) spin momentum
Drallrohr-Trockner m spiral tube pneumatic drier
Drallschichtfilter n granular bed dust separator
Drallvektor m torsion vector
Dralon n (HN, Text) dralon
Dravit m (Min) dravite
Drechselbank f turning bench
drechseln to turn
Dreelit m (Min) dreelite
Drehachse f rotating spindle, axis of rotation, axle, pin
Drehautomat m automatic lathe
Drehbandkolonne f laboratory column with rotating belt, spinning band column
Drehbank f lathe
Drehbankspitze f lathe center
drehbar rotary, turnable, twistable, ~ **aufgehängt** suspended in a rotating position
Drehbarkeit f rotation
Drehbeanspruchung f torsional strain, torsional stress
Drehbewegung f rotation, revolution, rotary motion, rotating motion, turning motion
Drehbohrer m rotary drill, fulcrum pin, pivot pin, twist drill
Drehbunker m rotary bin
Drehbunkeraufgabe f rotary bin feeder
Drehdorneisen n turning tool, chisel, gouge
Drehdornpresse f arbor press
Drehdrossel f variometer
Drehdurchführung f rotary transmission
Drehebene f revolving plane
Drehelektron n spinning electron
drehen to revolve, to rotate, to turn, to twist
Drehen n revolving, rotating, turning, twisting, ~ **der Birne** tipping the converter
Dreher m latheman, turner
Drehergewebe n breaker fabric
Drehfederwaage f torsion balance
Drehfeld n rotating field, rotation field, (Elektr) three-phase field
Drehfeldgeschwindigkeit f speed of the rotating field
Drehfeldmotor m three-phase motor
Drehfeldrichtung f direction of the rotating field
Drehfestigkeit f torsional strength, twisting strength
Drehfilter m rotary filter
Drehfläche f surface of revolution
Drehflammofen m revolving reverberatory furnace
Drehgelenk n rotating joint
Drehgeschwindigkeit f rotational speed, velocity of rotation
Drehgeschwindigkeitstensor m vorticity tensor

Drehgestell n bogie truck
Drehgestellfederung f torsion spring, torsion support, torsion suspension
Drehgriff m **für Gaszug** grip for hand throttle
Drehhalter m spider
Drehhaspel f winch, capstan, windlass
Drehimpuls m rotary impulse, angular momentum, rotary momentum, vorticity
Drehimpulsänderung f (Atom) spin variation
Drehimpulsoperator m angular momentum operator
Drehimpulsquantenzahl f secondary quantum number
Drehinversionsachse f rotary inversion axis
Drehkalzinierofen m rotating calcining oven
Drehklappe f butterfly valve, hinged valve
Drehknauf m handle
Drehknopf m control knob, switch knob
Drehko m (Drehkondensator) rotating plate condenser, variable condenser
Drehkolben m rotary piston
Drehkolbenpumpe f rotary [piston] pump
Drehkolbenverdichter m rotary compressor
Drehkondensator m rotary condenser
Drehkraft f torque, torsion, torsional load, turning force, twisting force
Drehkran m rotary crane, swing crane, turning crane
Drehkranz m rotary rim
Drehkreuz n spider
Drehkristall m revolving crystal
Drehkristallmethode f (Krist) rotating-crystal method
Drehkurbel f crank
Drehmagnet m rotary magnet
Drehmaschine f lathe
Drehmeißel m turning chisel
Drehmoment n moment of rotation, moment of torsion, torque, torsional moment, turning moment, twisting moment, ~ **eines Zahnrades** tangent line load
Drehmomentausgleich m torque compensation
Drehmomentverstärker m torque amplifier
Drehmühle f gyratory crusher
Drehmuffe f rotatable clamp
Drehofen m revolving furnace, rotary furnace, rotary kiln, (Metall) converter
Drehpendel n torsion pendulum
Drehpol m rotating field pole
Drehpolarisation f rotary polarization
Drehpotentiometer n rotary potentiometer
Drehpuddeln n revolving puddling
Drehpuddelofen m revolving puddling furnace
Drehpunkt m turning point, center of motion, fulcrum, pivot
Drehrichtung f direction of rotation
Drehrohrofen m revolving cylindrical furnace, rotary kiln, rotary tube furnace
Drehrollerbahn f roller conveyor turntable

Drehrost *m* revolving grate
Drehrostgenerator *m* rotary head generator, rotary grate gas producer
Drehschalter *m* rotary switch
Drehschaufelbagger *m* automatic grab
Drehscheibe *f* turning platform, pulley, rotating carrier, turntable
Drehscheibenmonochromator *m* rotating-disc monochromator
Drehschieber *m* cylindrical rotary valve, rotary slide valve
Drehschieberprinzip *n* rotary vane principle
Drehschieberpumpe *f* rotary vane pump
Drehschnelle *f* rotary velocity
Drehschwingung *f* torsional oscillation, oscillation around a center, torsional mode
Drehschwingungsrahmen *m* tilting stage
Drehschwingversuch *m* oscillation or alternating torsion test
Drehsinn *m* direction of rotation
Drehspäne *pl* shavings, turnings
Drehspannung *f* torsional tension, intensity of torsional stress, torque, torsional stress
Drehspiegel *m* swing mirror
Drehspiegelachse *f* axis of symmetry, rotary alternating axis
Drehspule *f* (Elektr) moving coil, revolving coil
Drehspulenstrommesser *m* moving-coil ammeter
Drehspulgalvanometer *n* moving coil galvanometer, pointer galvanometer
Drehspulinstrument *n* moving coil instrument
Drehspulspiegelgalvanometer *n* moving coil mirror galvanometer
Drehstabfeder *f* torsion bar
Drehstahl *m* chisel, cutting tool, lathe tool, steel tool, turning tool
drehsteif stiff against torsion
Drehströmungsentstauber *m* vortex classifier
Drehstrom *m* (Elektr) three-phase current
Drehstromanker *m* three phase armature
Drehstromanlage *f* three-phase [current] plant
Drehstromdynamo *m* three-phase generator, triphaser
Drehstromgenerator *m* three-phase alternator, three-phase [current] generator
Drehstrominduktionsmotor *m* three-phase asynchronous motor
Drehstromleistung *f* three-phase current output
Drehstromleitung *f* three-phase wire
Drehstrommotor *m* three-phase motor
Drehstromofen *m* alternating current furnace
Drehstromsystem *n* three-phase system
Drehsymmetrie *f* rotational symmetry
Drehteller *m* circular shelve
Drehtisch *m* rotary table, turntable
Drehtischformmaschine *f* molding machine with rotary table
Drehtischofen *m* rotary table furnace
Drehtransformator *m* adjustable transformer

Drehtrockner *m* rotary dryer, agitated dryer
Drehung *f* rotation, revolution, torsion, turning, twist, (Mech) gyration, ~ **des Achsenkreuzes** rotation of axes, ~ **des Cordfadens** cable twist, ~ **des Vorzwirns** ply twist, **entgegengesetzte** ~ reverse rotation, **optische** ~ optical rotation, **spezifische** ~ specific rotation
Drehungsachse *f* axis of rotation, turning axis, **feste** ~ (Mech) fixed axis of rotation
Drehungsdispersion *f* rotary dispersion
Drehungsebene *f* plane of rotation
Drehungsfaktor *m* twist factor
Drehungsfeder *f* torsion spring
Drehungsfläche *f* surface of rotation
Drehungskoma *n* rotation coma
Drehungskraft *f* rotatory force
Drehungsmoment *n* moment of rotation
Drehungssinn *m* direction of rotation
Drehungsträgheit *f* inertia due to rotation
Drehungsvermögen *n* rotatory power, **molekulares** ~ molecular rotation, **optisches** ~ optical rotation
Drehungswinkel *m* angle of rotation
Drehungszähler *m* twist counter
Drehvermögen *n* rotary power
Drehverschluß *m* swivel closure
Drehversuch *m* torsion test
Drehwaage *f* torsion balance
Drehwerkzeug *n* turning tool
Drehwinkel *m* angle of rotation, torsion angle
Drehwippkran *m* rotary luffing crane
Drehzahl *f* number of revolutions, [rotational] speed, velocity
Drehzahlenschreiber *m* recording tachometer
Drehzahlmesser *m* revolution counter
Drehzahlregelung *f* speed regulation
Drehzahlregler *m* speed regulating device
Drehzapfen *m* pivot
Drehzentrum *n*, **momentanes** ~ (Mech) instantaneous center of rotation
dreiachsig triaxial
dreiadrig triple core
dreiatomig triatomic
dreibasig tribasic
Dreibein *n* (Stativ) tripod
Dreibeinstativ *n* tripod [stand]
Dreibereichsverfahren *n* (Färb) tristimulus method
dreibindig (Chem) trivalent
Dreibrenner *m* triple burner
dreidimensional three-dimensional
Dreieck *n* triangle, (Elektr) delta, ~ **der Leitwerte** *n pl* diagram of conductance, **gleichschenkliges** ~ isosceles triangle, **gleichseitiges** ~ equilateral triangle, **rechtwinkliges** ~ rectangular triangle, **sphärisches** ~ spherical triangle, **ungleichseitiges** ~ (Math) scalenous triangle
Dreieckfeder *f* triangular spring

Dreieckschaltung *f* delta connection
Dreiecksdiagramm *n* triangular diagram
Dreieckselektrode *f* triangular electrode
Dreieckskoordinate *f* trilinear coordinate
Dreieckslehre *f* trigonometry
Dreiecksmatrize *f* (Math) triangular matrix
Dreieckspannungsmethode *f* delta voltage sweep method
Dreiecksquerschnitt *m* triangular cross section
Dreiecksspannung *f* delta voltage
Dreieckstrom *m* delta current
Dreiecksungleichung *f* triangular inequality
Dreielektrodenröhre *f* triode
Dreierstoß *m* three-body collision, three-fold collision, triple collision
Dreietagenofen *m* three-storied furnace
dreifach threefold, treble, triple
Dreifachaufspaltung *f* triple split
dreifachbasig tribasic
Dreifachbindung *f* triple linkage, (Chem) triple bond
Dreifachchlorantimon *n* antimony trichloride
Dreifachchlorjod *n* iodine trichloride
Dreifachkoinzidenz *f* triple coincidence
Dreifachkühler *m* triple condenser
Dreifachröhre *f* triple tube
Dreifachschreiber *m* triple recorder
Dreifadenlampe *f* (Elektr) three-filament lamp
Dreifarbendruck *m* three-color print[ing]
Dreifarbenfilter *n* trichromatic filter
Dreifarbenkolorimeter *n* trichromatic colorimeter
Dreifarbenmeßgerät *n* visual colorimeter
Dreifarbenphotographie *f* three-color photography
Dreifarbenraster *m* three-color screen
dreifarbig three-colored
Dreifarbigkeit *f* trichroism
Dreifingerregel *f* (Elektr) three-finger rule, right hand rule
dreiflächig three-faced, trihedral
Dreiflächner *m* (Krist) trihedron
dreiflammig three-flame, triple-flame
dreifüßig three-footed, tripedal
Dreifuß *m* tripod
Dreifußanlasser *m* tripod starter
Dreifußgestell *n* tripod
Dreifußstativ *n* tripod stand
dreigestaltig trimorphous
Dreigestaltigkeit *f* trimorphism
dreigliedrig three-membered, (Math) trinomial
Dreihalskolben *m* three-necked flask
Dreikant *m n* (Math) trihedral
Dreikantfeile *f* three-square file, triangular file
dreikantig triangular, three-cornered
Dreikantnut *f* triangular groove
Dreikantrührer *m* triangular mixer
Dreikantschaber *m* three-square scraper
dreikernig with three nuclei

Dreikreiselanordnung *f* three-gyro system
Dreileiterkabel *n* three-core cable
Dreileitersystem *n* three-wire system
Dreileitertransformator *m* transformer for a three-wire system, balancing transformer
Dreimalschmelzerei *f* triple melting-down process
Dreiphasendreileitersternschaltung *f* three-phase star connected system
Dreiphasendynamo *m* three-phase generator
Dreiphasenfeld *n* three-phase field
Dreiphaseninduktionsmotor *m* three-phase asynchronous motor
Dreiphasenkarbidofen *m* three-phase carbide furnace
Dreiphasenmotor *m* three-phase motor
Dreiphasenschaltung *f* three-phase circuit
Dreiphasensechsleitersystem *n* three-phase six wire system
Dreiphasenstrom *m* three-phase current, triphase current
Dreiphasenstromerzeuger *m* three-phase generator
Dreiphasensystem *n* three-phase system, triphase system
Dreiphasentransformator *m* three-phase transformer
Dreiphasenvierleitersystem *n* three-phase four-wire system
Dreiphasenwechselstrom *m* three-phase current
dreiphasig (Elektr) three-phase
Dreiphononenprozeß *m* three-phonon process
Dreiphononenwechselwirkung *f* three-phonon interaction
Dreipolanordnung *f* three-pole construction
Dreiprismensatz *m* set of three prisms
Dreipunktaufstellung *f* three-point mounting
Dreipunkt-Kontakt *m* (Enzym-Substrat, Stereochem) three-point contact
Dreiquantenvernichtung *f* three-photon annihilation
Dreiring *m* (Chem) three-membered ring
Dreisalzpigment *n* triple-salt pigment
Dreisatz *m* (Math) rule of three
dreischenkelig three-legged
Dreischenkelrohr *n* three-limb tube
dreischichtig three-ply
dreiseitig three-sided, triangular, trilateral
dreißigflächig (Krist) triacontahedral
Dreißigflächner *m* (Krist) triacontahedron
dreistellig (Math) three-digit, three-figure
Dreistiftstecker *m* three-pin plug
Dreistoffgemisch *n* ternary mixture, ternary system
dreistoffig ternary
Dreistofflegierung *f* ternary alloy, three-component alloy
Dreistoffsystem *n* ternary system, three-component system

Dreistrahlproblem *n* three-beam problem
Dreistrommotor *m* three-phase motor
Dreistufenmotor *m* three-speed motor
dreiteilen to trisect
dreiteilig three-part
Dreiviertelkonserve *f* three-quarter preserve
Dreiwalzenreibemaschine *f* triple roller grinding mill
Dreiwalzenringmühle *f* ring-roller mill
Dreiwalzenständer *m* triple roller mill stand
Dreiwalzenstuhl *m* three-roll mill, triple roller mill
Dreiwalzwerk *n* triple roller mill
Dreiwegehahn *m* three-way cock, three-way stop-cock
Dreiwegehahnküken *n* plug of three-way cock
Dreiwegeverteilerventil *n* three-way diverting valve
Dreiwegstück *n* three-way piece, three-limb tube, T-piece
Dreiwegventil *n* three-way valve
Dreiwegverbindung *f* three-way connection
Dreiwegverteilerventil *n* three-way diverting valve
dreiwertig (Chem) trivalent, (Math) three-valued
Dreiwertigkeit *f* (Chem) trivalence
dreiwinkelig three-cornered, triangular
Dreizack *m* trident
dreizählig ternary, threefold, triple, (Krist) trigonal
Dreizentrenbindung *f* three center bond
Dreizentren-Orbital *n* three-centered orbital
Dreizugkessel *m* three-gas-pass boiler
Dreizylindermotor *m* three-cylinder motor
Driftgeschwindigkeit *f* drift velocity
Driftkraft *f* drifting force
Drillbohrer *m* spiral drill
drillen (Faden) to drill
Drillich *m* (Text) drill, drilling, twill
Drilling *m* (Krist) trilling
Drillingmaschine *f* three-cylinder engine
Drillingswirkung *f* triple effect
Drillpendel *n* torsional pendulum
Drillschwingung *f* torsional wave
Drimanol *n* drimanol
Drimansäure *f* drimanic acid
Drimenol *n* drimenol
Drimon *n* drimone
Drimsäure *f* drimic acid
dringend urgent
dringlich urgent
Dringlichkeit *f* urgency
Drittelsilber *n* aluminum silver alloy, (Min) dyscrasite
Droge *f* (Pharm) drug
Drogenkunde *f* pharmacology
Drogenmißbrauch *m* abuse of drugs
Drogerie *f* drugstore
Drogist *m* druggist

Dromoran *n* dromoran
Droseron *n* droserone
Drosophilin *n* drosophilin
Drossel *f* choke, damper, (Mach) throttle, throttling valve
Drosseldiagramm *n* throttle diagram
Drosseleffekt *m* integral throttle expansion, **isenthalpischer** ~ isenthalpic expansion
Drosselflansch *m* throttle flange
Drosselkette *f* low pass filter
Drosselklappe *f* throttle valve, damper
drosseln to throttle, to baffle, to choke
Drosselregulator *m* throttle governor
Drosselrelais *n* high-impedance relay
Drosselscheibe *f* throttle diaphragm
Drosselschieber *m* throttle slide
Drosselspule *f* choke, choking coil, impedance coil
Drosselstelle *f* constriction
Drosselung *f* throttling, choking, constriction
Drosselventil *n* throttle valve, butterfly valve, restrictor valve
Drosselverfahren *n* throttling
Drosselvorrichtung *f* (f. Kältetechn) refrigerant liquid controls
Druck *m* force, thrust, (Buchdr) printing, (Phys) pressure, compression, ~ **der ruhenden Flüssigkeit** hydrostatic pressure, ~ **im Walzenspalt** nip pressure, ~ **in Millimeter Quecksilbersäule** pressure in terms of millimeters of mercury, ~ **nach unten** downward pressure, ~ **und Gegendruck** action and reaction, **atmosphärischer** ~ atmospheric pressure, pressure of the air, **axialer** ~ thrust, **den** ~ **wegnehmen** to decompress, **gleichbleibender** ~ constant pressure, **gleicher** ~ constant pressure, equal pressure, **hydrostatischer** ~ hydrostatic pressure, **innerer** ~ (Mech) internal pressure, **kritischer** ~ critical pressure, **osmotischer** ~ osmotic pressure, **verminderter** ~ reduced pressure
Druckabfall *m* pressure drop
druckabhängig being a function of pressure, pressure-dependent, pressure-responsive
Druckabnahme *f* reduction of pressure
Druckänderung *f* change of pressure, variation in pressure
Druckansammlung *f* stress concentration
Druckanstieg *m* increase in pressure, pressure rise
Druckanzeiger *m* manometer, pressure gauge, pressure indicator
Druckaufnehmer *m* pressure gauge
Druckausbreitung *f* pressure spread, propagation of pressure, transmission of pressure

Druckausgleich *m* balance of pressure, equalization of pressure, pressure equalization, pressure relief
druckausgleichend pressure-compensating
Druckautomat *m* automatic pressure apparatus
Druckbalken *m* ejector connecting bar
Druckbeanspruchung *f* compression, compression stress, compressive stress, pressure strain
Druckbehälter *m* pressure pack, pressure vessel
Druckbelastung *f* compression load
Druckbereich *m* pressure range
Druckbeständigkeit *f* (z. B. Rohr) resistence to pressure
Druckbestäubungspuder *m* anti-set-off powder
Druckbirne *f* acid egg, blow case
Druckblattfilter *n* pressure leaf filter
Druckbuchstabe *m* block letter
Druckdestillation *f* pressure distillation
druckdicht pressure-proof, tight
Druck-Dichte-Beziehung *f* pressure-density relation
Druckdichtung *f* high-pressure seal
Druckdiffusionsverfahren *n* pressure diffusion
Druckdose *f* pressure cylinder, pressure element or box
Druckdrehfilter *n* rotary pressure filter
Druckeinheit *f* pressure unit
Druckelastizität *f* elasticity of compression
druckelektrisch (piezoelektrisch) piezoelectric
Druckelektrizität *f* electricity produced by pressure, piezo-electricity
druckempfindlich sensitive to pressure
drucken (abziehen) to print
Druckentlastung *f* release of pressure
Druckentwicklung *f* generation of pressure, pressure increase
Drucker *m* printer
Druckerfarbe *f* printer's ink
Druckerhitzung *f* heating under pressure
Druckerhöher *m* intensifier
Druckerhöhung *f* increase of pressure
Druckerhöhungsanlage *f* pressure supercharger
Druckerniedrigung *f* decrease of pressure, depression
Druckerschwärze *f* printer's ink, Frankfort black, printing ink
Druckerweichungsprüfung *f* (Email) load test
Druckerzeuger *m* thrust generator
Druckerzeugung *f* generation of pressure
Druckextraktion *f* extraction under pressure
Druckfarbe *f* printing color, printing ink
Druckfarbentrocknung *f* printing ink drying
Druckfaß *n* autoclave
Druckfeder *f* pressure spring, compression spring
druckfest resistant to pressure
Druckfestigkeit *f* compression elasticity, compressive strength, pressure resistance
Druckfigur *f* (Krist) pressure figure
Druckfilter *m* pressure filter
Druckfiltration *f* pressure filtration
Druckfinne *f* loading nose (bending test)
Druckfirnis *m* lithographic varnish
Druckfläche *f* area of pressure
Druckflasche *f* pressure bottle, digester, pressure flask
Druckfließverhalten *n* compressive creep behavior
Druckflüssigkeit *f* pressure fluid, hydraulic medium
Druckfüllung *f* (Aerosoldose) pressure filling
Druckgärung *f* fermentation under pressure
Druckgas *n* compressed gas, pressure gas
Druckgasanlage *f* pressure gas plant
Druckgaserzeuger *m* pressure gas producer
Druckgasflasche *f* compressed gas cylinder
Druckgefälle *n* pressure drop, pressure gradient, (Meteor) isobaric slope
Druckgefäß *n* pressure pot, pressure vessel, vessel under pressure
Druckgeneratorgasanlage *f* pressure gas plant
Druckgleichgewicht *n* pressure equilibrium, **osmotisches** ~ osmotic equilibrium
Druckglocke *f* pressure bell
Druckguß *m* pressure die casting
Druckhärtung *f* (von Kunststoffen) pressure setting (of plastics)
Druckhalter *m* pressurizer
Druckheber *m* siphon
Druckhöhe *f* pressure altitude
Druckhöhengefälle *n* hydraulic gradient
Druckhydrierung *f* hydrogenation under pressure
Druckkammer *f* pressure chamber
Druckkanal *m* pressure duct, pressure passage, pressure port
Druckkattun *m* (Text) printed calico, chintz
Druckkessel *m* autoclave, pressure reservoir, pressure tank
Druckknopf *m* patent fastener, (Elektr) press button, push-button
Druckknopfschalter *m* push-button switch
Druckknopfsteuerung *f* push-button control [device]
Druckkocher *m* pressure digester
Druckkoeffizient *m* pressure coefficient
Druckkolben *m* clamping plunger, piston, pressure ram, (Chem) pressure flask
Druckkolbenmanometer *n* rotary piston manometer
Druckkraft *f* compressive force, crushing stress, pressure
Druckkühler *m* press cooler
Druckkühlung *f* forced-draft cooling
Druckkugellager *n* thrust ball bearing
Druckkupplung *f* compression coupling
Druckkurve *f* pressure curve

Drucklager – Druckvermittler

Drucklager *n* thrust bearing
Drucklast *f* pressure, working load
Drucklehre *f* pressure gauge
Druckleiste *f* (Druckscheibe) pressure pad
Druckleitung *f* pressure line, delivery pipe, discharge piping, high-pressure piping, pressure piping
Drucklinie *f* pressure curve
drucklos without pressure
Druckluft *f* (Preßluft) compressed air
Druckluftanlage *f* compressed air plant
Druckluftarmaturen *pl* compressed air fittings
Druckluftausrüstung *f* compressed air equipment
Druckluftbehälter *m* compressed air receiver
druckluftbetätigt pneumatic
Druckluftbohrer *m* pneumatic drill
Druckluftbremse *f* air-brake
Druckluftentölung *f* compressed air deoiling
Druckluftflasche *f* compressed air bottle
Drucklufthammer *m* pneumatic hammer
Drucklufttheber *m* air jet lift, air lift pump
Druckluftkühlung *f* forced draft cooling
Druckluftleuchte *f* compressed air lamp
Druckluftpegel *m* compressed air water gauge
Druckluft-Pistole *f* blow gun
Druckluftschlauch *m* compressed air tubing
Druckluftspritzapparat *m* compressed air spraying apparatus
Druckluftstutzen *m* compressed air pipe, compressed air valve
Druckmaschine *f* (Buchdr) printing machine
Druckmesser *m* (Manometer) manometer, pressure gauge
Druckmeßgerät *n* **für Vakuumtechnik** vacuum gauge
Druckmessung *f* pressure measurement
Druckminderungsventil *n* (Druckreduzierventil) [pressure] reducing valve, pressure reduction valve
Druckminderventil *n* pressure reducing valve
Druckmischer *m* pressurized mixer
Druckmittelpunkt *m* center of pressure
Drucknutsche *f* pressure filter, vacuum filter
Drucköl *n* oil under pressure, (Buchdr) printing oil
Druckölabfluß *m* oil pressure discharge
Druckölwäsche *f* pressure oil wash, scrubbing with oil under pressure
Druckpackung *f* aerosol packing, (Aerosoldose) pressurized pack
Druckpaste *f* (Buchdr) print paste
Druckpolymerisation *f* pressure polymerization
Druckprobe *f* pressure test, hydraulic test
Druckpumpe *f* pressure pump, force pump
Druckraum *m* pressure chamber
Druckraumdiagramm *n* pressure volume diagram
Druckreduzierventil *n* reducing pressure valve
Druckregler *m* pressure controller, pressure regulator
Druckregulator *m* pressure regulator
Druckring *m* thrust collar
Druckrohr *n* pressure pipe, delivery pipe, discharge pipe, gun
Druckrolle *f* pressure roll
Druckrückgewinn *m* pressure recovery
Druckscheibe *f* thrust washer, (Druckleiste) pressure pad
Druckschlauch *m* pressure tube, thick-walled rubber tubing
Druckschmierung *f* pressure lubrication, forced feed lubrication
Druckschraube *f* set screw
Druckschwankung *f* fluctuation of pressure
Druckschweißung *f* pressure welding
Druckschwingung *f* compressional vibration
Druckseite *f* pressure side
Drucksinterofen *m* press sintering furnace
Druckspaltung *f* compressive cleaving
Druckspannung *f* compression strain, compression stress, compressive stress, crushing load, tension due to pressure
Druckspritzverfahren *n* pressure spray process
Drucksteigerung *f* pressure increase, pressure rise
Druckstift *m* ejector pin
Druckstiftplatte *f* latch plate
Druckstrahlpumpe *f* injector
Druckstufe *f* pressure stage
Druckstufen *pl* **für Rohrleitungen** pressure code for pipelines
Druckstutzen *m* pressure connection, discharge branch, exhaust port, pressure joint
Drucktaste *f* push-button
Drucktastenbedienung *f* push-button control
Druckteig *m* (Buchdr) printing paste
Drucktopf *m* autoclave
Drucktrommelfilter *n* pressure drum filter
Drucktuch *n* offset blanket
Druckübersetzer *m* pressure transmitter
Druckunterschied *m* pressure difference
Druckunterschiedsmesser *m* pressure difference meter
Druckventil *n* pressure valve
Druckverbreiterung *f* (Spektr) pressure broadening
Druckverfahren *n* (Buchdr) printing method
Druckvergasungsanlage *f* pressure gasification plant
Druckverhältnis *n* pressure ratio
Druckverlust *m* pressure drop, **spezifischer ~** (Dest) specific pressure drop
Druckverlustziffer *f* (v. Rohren) skin friction coefficient
Druckverminderung *f* pressure decrease, pressure drop
Druckvermittler *m* pressure equalizer

Druckverschiebung *f* (Spektrallinien) pressure shift (spectral lines)
druckverschweißt pressure-welded
Druckversuch *m* compression test
Druckverteilung *f* distribution of pressure
Druckvolumen *n* pressure volume
Druckvolumendiagramm *n* pressure volume diagram
Druckwalze *f* pressure cylinder, pressure roll, pressure roller, (Buchdr) printing roller
Druckwandler *m* pressure converter
Druckwasser *n* hydraulic water, pressure water
Druckwasseranlage *f* mit Druckluftbelastung hydraulic plant with air bottle system, ~ mit **Druckpumpen** *pl* direct pumping hydraulic plant
Druckwasserbetrieb *m* hydraulic working
Druckwasser-Druckluftanlage *f* hydro-pneumatic plant
Druckwasserformmaschine *f* hydraulic molding machine
Druckwasserformpresse *f* hydraulic molding press
Druckwassernietung *f* hydraulic riveting
Druckwasserreaktor *m* pressurized water reactor
Druckwassersammler *m* hydraulic accumulator
Druckwasserschmiedepresse *f* hydraulic forging press
Druckwasserspeicher *m* hydraulic accumulator
Druckwassersteinpresse *f* hydraulic brick press
Druckwechsel *m* change in pressure
Druckwechselverfahren *n* (Dest) pressure swing process, (Füllkörperkolonne) pressure swing process
Druckwelle *f* pressure wave, strain wave, (Explosion) shock wave, (Techn) thrust shaft
Druckwellenreaktor *m* (Gastechnologie) shock wave reactor, shock tube
Druckwerk *n* press, (Buchdr) printer
Druckwindkessel *m* delivery air vessel, [delivery] air chamber
Druckzerstäuber *m* fingertip dispenser, pressure dispenser
Druckzone *f* compression section
Druckzug *m* pressure draft
Druckzunahme *f* pressure increase, pressure rise
Druckzwillingsbildung *f* compressive twin formation
Druckzylinder *m* pressure cylinder, (Buchdr) impression cylinder
Druckzylinderzwirnmaschine *f* cradle roll
Drudenfuß *m* (Bot) club moss
Drudesche Gleichung *f* (Opt) Drude equation
drücken (einen Druck ausüben) to press, to force, to jam, to push, to ram
Drücker *m* (Gewehr) trigger, (z. B. Tür) handle
Drüse *f* (Med) gland, **endokrine** ~ endocrine gland, **exokrine** ~ exocrine gland, open gland, **Lymph-** ~ lymphatic gland, lymph node, **milchgebende** ~ lactiferous gland, mammary gland, milk gland, **Schleim-** ~ mucous gland, **schweißabsondernde** ~ sudoriparous gland, sweat gland, **Speichel-** ~ salivary gland
Drüsenabsonderung *f* glandular secretion
Drüsenkrankheit *f* glandular disease
Drüsenlappen *m* (Med) gland lobule
Drüsensekret *n* (Med) glandular secretion
Drüsenzelle *f* glandular cell
Drüsenzyste *f* glandular cyst
Drummondsches Licht *n* drummond's light
Druse *f* (Min) cluster of crystals, druse, hollow druse, **linsenförmige** ~ lenticular vug
Drusen *pl* dregs, husks, lees
Drusenasche *f* calcined wine lees, potash from burnt lees of wine
drusenförmig drusy
Drusenkobalt *n* (Min) drusy cobalt
Drusenmarmor *m* (Min) shell marble
Drusenöl *n* oil made from dregs of wine
Druvatherm-Reaktor *m* ploughblade mixer
Dualin *n* dualin
Dualindynamit *n* dualin dynamite
Dualismus *m* dualism, ~ **von Welle und Teilchen** dualism of wave and particle
dualistisch dualistic
Dualität *f* duality
Dualitätsprinzip *n* duality principle
Dualsystem *n* binary system (of notation or numerals)
Dualzahl *f* binary number
Duante *f* (eines Zyklotrons) duant (of a cyclotron)
Dubamin *n* dubamine
Dubatol *n* dubatol
Dubinidin *n* dubinidine
Dublee *n* rolled gold
Dubleegold *n* rolled gold
Dublettabstand *m* doublet spacing
Dublettaufspaltung *f* doublet splitting
Dublette *f* doublet
Dublettlinie *f* doublet line
Dublettspektrum *n* doublet spectrum
Dublettstruktur *f* doublet structure
Dublierstein *m* concentrated metal
Duboisin *n* duboisine, l-hyoscyamine
Duckstein *m* (Geol) calcareous tufa, trass
Dudleyit *m* (Min) dudleyite
Dübel *m* dowel, dowel pin, plug
Dübelschweißung *f* slot welding
Düker *m* siphon
Dünensand *m* down sand
Düngeerde *f* compost, humus, mold, vegetable earth
Düngeharnstoff *m* fertilizer urea
Düngejauche *f* liquid manure
Düngekalk *m* lime fertilizer
Düngemittel *n* fertilizer

Düngemittelfabrik *f* fertilizer factory, manure works
düngen to fertilize
Düngen *n* fertilizing, manuring
Düngepulver *n* powdered manure
Dünger *m* fertilizer, fertilizing substance, manure, **künstlicher** ~ artificial fertilizer, **natürlicher** ~ natural manure
Düngeranlage *f* fertilizer plant
Düngesalz *n* fertilizer salt, dung-salt, manuring salt, saline manure
Düngung *f* fertilizing, composting, manuring
dünn thin, fine, subtle
Dünnbettverfahren *n* thin bed process
Dünnblechschweißung *f* light gauge sheet steel welding
Dünndarm *m* small intestine
Dünndarmentzündung *f* (Med) enteritis
Dünne *f* thinness, fluidity
Dünnfilmsublimator *m* thin-film sublimer
dünnflüssig fluid, (Öl) highly fluid, light, liquid, low-viscosity, thin-bodied, watery
Dünnflüssigkeit *f* fluidity, liquidity
dünngewalzt thinly-rolled
Dünnsaft *m* (Zuck) thin juice
Dünnschichtchromatographie *f* thin-layer chromatography, **zweidimensionale** ~ two-dimensional thin-layer chromatography
Dünnschichtelektrophorese *f* thin-layer electrophoresis
Dünnschicht-Ofenversuch *m* thin-film oven test
Dünnschichtreaktor *m* wetted wall column or thin-film reactor
Dünnschichtrektifikator *m* (Dest) thin-film recitifier
Dünnschichtsiebung *f* thin layer screening
Dünnschichttrockner *m* film dryer
Dünnschichtverdampfer *m* film evaporator, thin-film evaporator
Dünnschichtverdampfung *f* film evaporation
Dünnschichtwärmeaustauscher *m* thin-film heat exchanger
Dünnschichtzelle *f* thin-layer cell
Dünnschichtzentrifuge *f* conical screen centrifuge with differential conveyor
Dünnschnitt *m* thin section
Dünnschnittverfahren *n* (Mikroskopie) microtomy
Dünnstein *m* table diamond, thin matte
dünntafelig foliated, thin-sheeted
dünnwandig thin-walled
Düse *f* (Brenner) tip, (Gaschromat) flame tip, (Mundstück, Einspritzvorrichtung) nozzle, (Schachtofen) tuyère [hole], (Vergaser) jet, **axial gespeiste** ~ axial-fed die
Düsenabsperrschieber *m* tuyère gate
Düsenaggregat *n* jet engine
Düsenanschlußstück *n* die adapter system
Düsenantrieb *m* jet propulsion
Düsenaustritt *m* orifice, die orifice, orifice relief
Düsenauswerfer *m* nozzle ejector
Düsenblasverfahren *n* blast drawing
Düsenblock *m* nozzle block
Düsenboden *m* bottom of melting pan
Düsenflansch *m* front shoe
Düsenflugzeug *n* jet-propelled plane
Düsengehäuse *n* die body
Düsenkanal *m* die channel
Düsenkörper *m* die base, die body
Düsenkraftstoff *m* jet fuel
Düsenkühler *m* jet radiator
Düsenmesser *m* nozzle meter
Düsenmotor *m* jet engine
Düsenmund *m* die orifice
Düsenöffnung *f* die orifice
Düsenpaßstück *n* nozzle adapter
Düsenplatte *f* nozzle plate, diaphragm, front shoe
Düsenprofil *n* die ring
Düsenquerschnitt *m* nozzle section
Düsenregler *m* nozzle regulator
Düsenreihe *f* row of tuyères
Düsenring *m* die ring, jet ring
Düsenschreiber *m* recording steam meter
Düsenstock *m* nozzle pipe, blast connection
Düsen-Tellerzentrifuge *f* nozzle discharge centrifuge
Düsentreibstoff *m* jet fuel
Düsentrockner *m* jet drier, nozzle drier
Düsenunterteil *n* die bottom
Düsenventil *n* nozzle valve
Düsenverschluß *m* (Spritzguß) nozzle shut-off device
Düsenwäscher *m* jet scrubber
Düsenziehverfahren *n* continuous filament process, mechanical drawing
Dufrenit *m* (Min) dufrenite
Dufrenoysit *m* (Min) dufrenoysite
Duft *m* fragrance, odor, scent
Duftdrüse *f* (Zool) scent gland
Duftessig *m* aromatic vinegar
duftig fragrant, odorous
duftlos odorless, inodorous, scentless
Duftöl *n* aromatic oil, fragrant oil
duftreich perfumed
Duftstoff *m* odorous substance, aromatic principle, perfume
Dugongöl *n* dugong oil
Duisburger Natriumsulfat *n* crystalline sodium sulfate, Glauber salt
duktil ductile
Duktilität *f* ductility
Dulcamarin *n* dulcamarin[e]
Dulcin *n* dulcin
Dulcit *m* dulcite, dulcitol
Dulong-Petitsches Gesetz *n* Dulong and Petit's law
Dumasin *n* dumasin

Dumortierit *m* dumortierite
Dumperblock *m* pitch block
dumpf (abgedämpft) muffled, (hohlklingend) dull, hollow
Dundasit *m* (Min) dundasite
Dung *m* dung, manure
Dunit *m* dunite
dunkel black, dark, (Färb) deep
dunkeladaptiert dark-adapted
Dunkeladaption *f* (Opt) dark adaptation (Br. Engl.), dark adaption (A. Engl.)
Dunkelanpassung *f* (Opt) dark adaptation
Dunkelanpassungsfähigkeit *f* darkness adaptability
dunkelblau dark blue
dunkelbraun dark brown
Dunkelentladung *f* dark discharge
Dunkelfärbung *f* darkening, deepening in color
dunkelfarbig dark-colored
Dunkelfeldabbildung *f* (Opt) dark field illumination
Dunkelfeldverfahren *n* (Opt) dark contrast method
dunkelgelb dark yellow
Dunkelglühhitze *f* dark-red heat
dunkelgrün dark green
Dunkelkammer *f* (Phot) darkroom
Dunkelkammergerät *n* darkroom equipment
Dunkelleitfähigkeit *f* dark conductivity
Dunkelleitung *f* (Elektr) dark conduction
Dunkelleuchte *f* safelight
dunkeln to darken, to sadden
Dunkelöl *n* dark oil
Dunkelreaktion *f* dark reaction
dunkelrot dark red
Dunkelrotgiltigerz *n* (Min) dark red silver ore
Dunkelrotglut *f* low red heat, deep cherry glow
Dunkelscharlach *n* venice scarlet
Dunkelstrahler *m* dark body radiator
Dunkelstrom *m* (Photozelle) dark current
Dunkelstromimpuls *m* dark pulse
Dunkelvernicklung *f* dark nickeling
Dunkelwiderstand *m* (Photozelle) dark resistance
Dunnion *n* dunnione
Dunst *m* (Dampf) steam, vapor, (Nebel) mist, (Schleier) haze
Dunstabzug *m* hood
Dunstabzugsrohr *n* vent pipe
Dunstatmosphäre *f* haze atmosphere
Dunstbad *n* vapor bath
Dunstfang *m* hood
Dunstglocke *f* layer of smog
Dunsthaube *f* vapor hood
Dunsthülle *f* vaporous envelope, (Atmosphäre) atmosphere
dunstig steamy, vapory
Dunstigkeit *f* haziness, vaporousness
Dunstkreis *m* atmosphere
Dunstloch *n* vent

Dunstmesser *m* atmometer
Dunstsilber *n* grains of silver ashes
Dunstvulkanisation *f* vapor cure
Duodenum *n* (Med) duodenum
Duodezimalsystem *n* duodecimal system
Duotal *n* duotal, guaiacol carbonate
Duowalze *f* twin rollers
Duowalzwerk *n* two-roller mill
Duplexkarton *m* duplex board
Duplexpressen *n* duplex molding
Duplexschaltung *f* duplex connection
Duplexsystem *n* duplex system
Duplikat *n* duplicate
Duplikator *m* (Elektr) duplicator
Duplizitätstheorie *f* (des Auges) duplicity theory (of the eye)
Duponol *n* duponol
Duralium *n* (HN) duralium
Duraluminium *n* duralumin
Duramycin *n* duramycin
Duranametall *n* durana metal
Durangit *m* (Min) durangite
Duranglas *n* duran glass
Duratol *n* duratol
Duraxglas *n* durax glass
durchätzen to corrode, to eat through
durcharbeiten to work through, to work up
durchbiegen to bend through, to deflect, to sag
Durchbiegeversuch *m* deflection test
Durchbiegung *f* deflection, sagging, **bleibende** ~ permanent set, **elastische** ~ transverse elasticity
durchbiegungsfähig flexible, deflectable
Durchbiegungsfestigkeit *f* transverse bending strength
Durchbiegungsgeschwindigkeit *f* rate of deflection
Durchbiegungsspannung *f* strain, transverse stress
Durchbildung *f* development, formation, perfection
Durchblaseventil *n* (Techn) blow valve
durchbluten to supply with blood
Durchblutung *f* blood circulation
durchblutungsfördernd stimulating the circulation of blood
Durchblutungsstörung *f* disturbed circulation
durchbohren to bore through, to core, to pierce, (durchlöchern) to perforate, to punch
durchbrechen to break through [or in two], to collapse, to open
Durchbrechen *n* channel[l]ing
durchbrennen to burn through, (Lampen) to burn out, (Sicherung) to blow, to fuse
Durchbrennen *n* burning through
durchbrochen perforated, (Text) open-work
Durchbruch *m* breaking through, breakthrough, cutting through, opening up; (Bergb) opening,

(Geol) eruption, (Rohr) chase in walls (for pipes)
Durchbruchkurve *f* (Adsorption) break through curve
Durchbruchsfeldstärke *f* breakdown field
durchdringbar penetrable, permeable
Durchdringbarkeit *f* permeability, penetrability, penetrableness
durchdringen to penetrate, to ooze through, to percolate, to pervade
Durchdringen *n* penetration, impregnating
Durchdringung *f* penetration, impregnation, (Infiltration) infiltration
Durchdringungsfähigkeit *f* penetration capacity
Durchdringungskomplex *m* (Chem) penetration complex
Durchdringungsstelle *f* penetration point, merging point
Durchdringungsvermögen *n* penetrating power, (Röntgenstrahlen) penetrating power
Durchdringungszwillinge *m pl* (Krist) penetration twins
durchdrücken to press through
Durchdrückpackung *f* push-out package
Durchdrungenheit *f* state of impregnation
durcheinanderfliegen to fly around
durcheinandermischen to mix together
Durchfärbemittel *n* penetrative dyeing agent
Durchfärbung *f* penetration dyeing, (Text) penetration
Durchfahrtechnik *f* (Galv) method without rejigging, one-jig process, straight-through process
durchfallen to fall through, (Licht) to be transmitted
durchfaulen to putrefy, to rot through
durchfeuchten to soak thoroughly, to wet
Durchfeuchtung *f* penetration of dampness
durchfiltrieren to filter through, to percolate
Durchfiltrieren *n* filtering, percolation
durchfließen to flow through, to diffuse
Durchfluß *m* flow, flowing through, passage, (Geol) percolation, (Techn) discharge
Durchflußelektrode *f* flow-type electrode
Durchflußerhitzer *m* continuous flow heater
Durchflußextraktor *m* direct flow extractor
Durchflußgeschwindigkeit *f* flow velocity, rate of flow
Durchflußkette *f* flow chain
Durchflußmenge *f* flow
Durchflußmengenmeßgerät *n* flowmeter
Durchflußmesser *m* flowmeter
Durchflußmessung *f* flow measurement, ~ **aus Druckverlust an Einbauten** flow measurement by pressure drop, ~ **mit Schwimmkörpern** flow measurement with rotameters
Durchflußmischer *m* flow mixer, pipe-line mixer
Durchflußöffnung *f* discharge opening

Durchflußquerschnitt *m* cross section of passage, flow area, section of waterway
Durchflußrate *f* flow rate
Durchflußreaktor *m* flow reactor
Durchflußregler *m* flow controller, (Gaschromat) restrictor
Durchflußrichtung *f* direction of flow
Durchflußversuch *m* continuous flow test
Durchflußzeit *f* time of flow
Durchflußzelle *f* flow cell
durchführbar practicable, feasible
Durchführbarkeit *f* practicability, feasibility
durchführen to carry out, to carry on, to execute, to lead through
Durchführungshülse *f* grommet
Durchführungsrohr *n* wall tube, (Elektr) grommet
durchgängig penetrable, permeable, pervious
durchgären to ferment sufficiently
Durchgang *m* passage, (Elektr) flow, (Phys) transmission, (Walze) pass
Durchgangshahn *m* straight-way cock, two-way cock
Durchgangsloch *n* through hole
Durchgangsöffnung *f* passage opening, port-slot
Durchgangsquerschnitt *m* cross sectional area
Durchgangsschraube *f* through bolt
Durchgangsventil *n* straight through valve, through-way valve
Durchgangsversuch *m* transmission experiment
Durchgangswiderstand *m* flow resistance, volume resistance
Durchgangszeit *f* transit time
durchgehen (Reaktor) to run away
Durchgehen *n* (Reaktor) runaway, ~ **der Maschine** racing of the machine
durchgehend continuous[ly]
durchgerben to tan through
durchgeröstet roasted thoroughly
durchgeschmiert greased
Durchgeseihtes *n* colature, filtrate
durchgießen to filter, to pour through, to strain
durchglühen to heat to red heat, to anneal, to burn out
durchgreifend intensive, penetrative; radical
Durchgriff *m* penetration coefficient, transconductance, (Elektron) transgrid action
durchhämmern to hammer well
durchhängen to sag
durchhärten to temper thoroughly
durchheizen to heat thoroughly
durchkneten to knead thoroughly
durchkreuzen to cross, to intersect, to traverse
Durchkreuzungszwilling *m* (Krist) cruciform twin group, penetration twin
durchlässig penetrable, permeable, pervious, porous, transmissive
Durchlässigkeit *f* permeability, penetrability, perviousness, (Elektr) transmissibility,

(Porosität) porosity, ~ **für ultrarote Strahlen** diathermancy, **magnetische** ~ magnetic inductivity, permeability
Durchlässigkeitskoeffizient m permeability coefficient, penetration probability, transmission factor
Durchlässigkeitsmesser m (Elektr) permeameter, transmissiometer
Durchlässigkeitsversuch m permeability test
durchlassen to leak, (Licht) to let pass, to transmit, (Opt, Akust) to allow to pass through
Durchlaß m passage
Durchlaßfarbe f transmitted color
Durchlaßgitter n transmission grate
Durchlaßgrad m (Opt) transmission of light, transmittance
Durchlaßkennlinie f forward characteristic
Durchlaßrichtung f **eines Gleichrichters** forward direction
Durchlaßspannung f forward voltage
durchlaufen to pass [through], to run through
Durchlaufentgasungsanlage f continuous degassing column
Durchlauferhitzer m continuous flow heater
Durchlaufkessel m natural-circulation boiler
Durchlaufkolbenmesser m jacket piston meter
Durchlauflager n flow warehouse, transit warehouse
Durchlaufreaktor m straight-through reactor
Durchlaufregal n flow rack
Durchlaufstation f (Förderer) passing station or post
Durchlauftrockner m continuous dryer, tunnel dryer
Durchlauftropfentgasung f stream drop degassing
Durchlaufzeit f machining time
durchlaugen to steep thoroughly in lye
durchleiten to pass through, to conduct
Durchleitung f passage
durchleuchten to illuminate, (Med) to x-ray
Durchleuchtung f transillumination, (mit Röntgenstrahlen) x-raying
Durchleuchtungsapparat m X-ray apparatus
Durchlicht n transmitted light
durchlochen to perforate, to pierce, to punch
durchlöchern to perforate, to pierce, to punch, to puncture
durchlüften to aerate, to air, to ventilate
Durchlüftung f aeration, airing, ventilation
durchmessen to measure, to pass through
Durchmesser m diameter, **lichter** ~ inside diameter, internal diameter
durchmischen to blend, to mix thoroughly
Durchmischen n thorough mixing
Durchmischung f intermixing, ~ **der Lauge** intermixture of the liquor
durchnässen to soak, to steep, to wet
durchölen to oil thoroughly

durchpausen to trace
durchpelzen to curry well, to tan thoroughly
durchperlen to bubble through
Durchperlungselektrode f bubbling-type electrode
durchpressen to press through
durchprobieren to test, to try out
durchquetschen to squeeze through
durchräuchern to fumigate, to smoke
durchrechnen to calculate, to evaluate
durchregnen (Rektifizierboden) to weep
Durchregnen n (Rektifizierboden) weeping
durchreißen to break, to tear asunder, to tear up
durchrosten to rust through
Durchsatz m throughput, output, quantity passed, weight rate of flow
Durchsatzleistung f throughput
Durchsatzmenge f flow rate
Durchsatzofen m tunnel furnace
Durchsatzvolumen n volumetric rate of discharge
durchsaugen to suck through
Durchschaltzeit f (Elektr) rise time
durchscheinen to shine through
Durchscheinen n translucency
durchscheinend translucent, transparent
durchschicken to send [through]
durchschimmernd translucent
Durchschlag m (Durchschrift) copy, duplicate, (Elektr) breakdown, (Geschoß) penetration, piercing, (großes Sieb) colander, (Locheisen) piercer, punch, (Sicherung) blowout, **dielektrischer** ~ (Elektr) dielectric breakdown, **elektrischer** ~ electric breakdown
Durchschlagboden m perforated bottom
durchschlagen to filter, to sieve, to strain, (durchlöchern) to punch, to puncture, (Elektr) to break down, to spark, (Farbe) to show through, (Sicherung) to blow out
Durchschlagen n (Elektr) sparking; (Klebstoff) bleeding through (e.g. on an adhesive)
Durchschlagfeldstärke f dielectric strength
Durchschlagpapier n carbon paper
Durchschlagsbedingung f dielectric breakdown condition
Durchschlagsfestigkeit f breakdown resistance, breakdown strength, dielectric strength, disruptive strength, electrical insulation value
durchschlagsicher (Elektr) puncture-proof
Durchschlagskanal m breakdown filament
Durchschlagskraft f penetrative power, percussion force
Durchschlagspotential n breakdown potential
Durchschlagsprobe f (Elektr) breakdown test
Durchschlagspunkt m point of penetration, point of intersection
Durchschlagssicherung f (Elektr) puncture cutout

Durchschlagsspannung *f* breakdown voltage, disruptive voltage, flashover potential, puncture or rupturing voltage
Durchschlagsstrom *m* spark current
durchschmelzen to melt through, (Sicherung) to blow out
Durchschmelzen *n* melting, fusing, (Sicherung) blowing
durchschmieden to forge, to hammer
durchschmoren (Kabel) to char through, to scorch
durchschneiden to intersect, to cross, to cut across, (auseinanderschneiden) to cut through
Durchschneiden *n* cutting through, shearing
Durchschnitt *m* cut, section, (Math) intersection, (Mittelwert) average, (Profil) profile, (Querschnitt) cross section
Durchschnittgewicht *n* average weight
durchschnittlich average, mean
Durchschnittsausbringung *f* average output
Durchschnittsdruckrelais *n* averaging relay
Durchschnittsenergieverlust *m* average loss of energy
Durchschnittsgeschwindigkeit *f* average speed, mean velocity
Durchschnittsgewicht *n* average weight
Durchschnittsleistung *f* average performance
Durchschnittsprobe *f* average sample
Durchschnittsqualität *f* average quality
Durchschnittstemperatur *f* mean or average temperature
Durchschnittswert *m* average value, mean value
Durchschnittszahl *f* average [number], mean
Durchschütteln *n* shaking
Durchschuß *m* (Buchdr) interleaf, (Web) weft, woof
durchschwefeln to sulfur thoroughly
durchschweißen to weld thoroughly
durchschwelen to cause to smolder, to smolder
durchschwitzen to perspire, to ooze through
durchseihen to filter, to percolate, to strain
Durchseihen *n* straining, filtering, percolation
Durchseihung *f* colature, filtration
durchsetzen to infiltrate; to intermingle, to permeate, to streak through, to travel through
Durchsetzzeit *f* period of working charge
Durchsicht *f* inspection, looking through, view, **in der ~** [viewed] by transmitted light
durchsichtig transparent, clear, limpid, translucent
durchsichtig *n* **wie Glas** hyalescent
Durchsichtigkeit *f* transparence, translucency, transparency, **~ im Feuer** pyrophanousness, **~ im Wasser** hydrophanousness
durchsickern to trickle through, to percolate, to strain
Durchsickern *n* percolation
durchsieben to sieve, to screen, to sift
durchstechen to pierce, to cut, to stab

Durchsteckschraube *f* [through] bolt
durchstoßen to push through, to perforate, to pierce
Durchstoßfestigkeit *f* puncture resistance, total penetration energy [at failure]
Durchstoß-Test *m* punching test
Durchstoßversuch *m* impact penetration test
durchstrahlen to irradiate, to radiate, to shine through
Durchstrahlung *f* irradiation
Durchstrahlungsspektrograph *m* transmission spectrograph
Durchstrahlungsverfahren *n* transmission beam method
durchströmen to flow through, to pass through
Durchströmungsgeschwindigkeit *f* rate of flow
Durchströmungsmethode *f* flow method
durchtränken to impregnate, to infiltrate, to saturate
Durchtränken *n* impregnating, wetting out
Durchtränkung *f* impregnation, saturation
Durchtränkungssäure *f* impregnating acid
durchtreiben to drive through
Durchtritt *m* penetration, passage through
durchtrocknen to dry through
Durchtrocknungszeit *f* hard drying time
Durchtröpfeln *n* percolation
durchwachsen non-homogeneous
Durchwachsungszwilling *m* (Krist) penetration twin
durchwärmen to warm thoroughly
Durchwärmzeit *f* (Holz) additional pressing time
durchwässern to drench
Durchweichung *f* soaking
durchwerfen to throw through, to screen, to sift
Durchwirbelung *f* swirling motion, turbulence
durchwürzen to season thoroughly
Durchwürzen *n* seasoning
Durchwurf *m* riddle, screen, sieve
Durchziehboden *m* removable cylindrical plug
Durchziehformmaschine *f* stripping plate molding machine
Durchzug *m* draft, girder, passage
Durchzugformmaschine *f* stripping plate molding machine
Durchzugplatte *f* stripping plate
Durchzugring *m* centering frame; centering ring
Durenol *n* 1,2,4,5-tetramethyl-3-hydroxybenzene, durenol
Durexruß *m* durex carbon black
Durferrittiegelofen *m* durferrit pot furnace
Duridin *n* duridine, 3-amino-1,2,4,5-tetramethyl benzene, aminodurene
Duriron *n* duriron
Durochinon *n* duroquinone
Durohydrochinon *n* durohydroquinol
Durol *n* 1,2,4,5-tetramethylbenzene, durol, durene

Durometer *n* durometer
Duroplast *n* thermosetting plastic, duroplast, hardenable plastic
duroplastisch duroplastic, thermosetting
Duryl- duryl
Durylsäure *f* durylic acid, 2,4,5-trimethylbenzoic acid, cumylic acid
Dusche *f* shower, spray
Dutzend *n* dozen
Dyade *f* dyad
Dyakisdodekaeder *n* (Krist) dyakisdodecahedron
dyakisdodekaedrisch (Krist) dyakisdodecahedral
Dyclonin *n* dyclonine
Dymal *n* dymal, didymium salicylate
Dyn *n* (Maß) dyne
Dynamdonstein *m* dynamdon brick
Dynamik *f* dynamics, (Magnetband) dynamic range
dynamisch dynamic
Dynamit *n* dynamite
Dynamitexplosion *f* explosion of dynamite
Dynamitladung *f* dynamite charge
Dynamitstange *f* stick of dynamite
Dynamo *m* dynamo, generator
Dynamoanker *m* generator armature
Dynamoantrieb *m* method of driving the dynamo
Dynamobürste *f* dynamo brush
Dynamogröße *f* size of dynamo
Dynamolager *n* generator bearing
Dynamoleitung *f* dynamo mains
Dynamomaschine *f* dynamo
Dynamometer *n* dynamometer
Dynamoöl *n* dynamo oil
Dynamoraum *m* dynamo room, generator room
Dynamoregel *f* dynamo rule
Dynamostrom *m* generator current
Dynein *n* (Biochem) dyneine
Dynodenstruktur *f* dynode structure
Dypnon *n* dypnone
Dyrophanthin *n* dyrophanthine
Dysbakterie *f* (Med) abnormal bacterial flora, disturbed growth of bacteria
Dyscrasit *m* (Min) dyscrasite
Dysluit *m* (Min) dysluite
Dysodil *m* dysodil
Dysprosium *n* dysprosium (Symb. Dy)
Dysprosiumchlorid *n* dysprosium chloride
Dystektikum *n* dystectics
Dystomglanz *m*, **hemiprismatischer** ~ plagionite, **hexaedrischer** ~ stannine, **tetraedrischer** ~ fahlore, tetrahedrite

E

Eau de Cologne *n* eau de Cologne
Eau de Javelle *n* eau de Javelle, Javelle water, potassium hypochlorite solution
Eau de Labarraque *n* Labarraque solution, sodium hypochlorite solution
Ebbe *f* ebb [tide], falling tide, low tide, ~ **und Flut** [high and low] tides
eben (flach) even, plane, (horizontal) level
Ebene *f* (Geol) level country, plain, (Math) plane, ~ **der Windformen** *f pl* tuyere level, ~ **des polarisierten Lichtes** plane of polarized light, **schiefe** ~ inclined plane
Ebenen *pl,* **zwischen den** ~ interplanar
Ebenenabstand *m* interplanar spacing
Ebenenpaar *n* pair of planes
Ebenholz *n* ebon-wood, ebony, **echtes** ~ black ebony
Ebenholzöl *n* ebony wood oil
ebenieren (Holz) to ebonize
Ebenmaß *n* harmony, proportion, symmetry
ebenpolarisiert plane polarized
Ebereschenzucker *m* mountain ash sugar, sorbin
Ebereschöl *n* mountain ash oil
Eberwurzel *f* carline root
Eberwurzelöl *n* carlina oil
Eblanin *n* eblanin, paraxanthine
ebnen to even, to level, to plane
Ebnen *n* **der Blechtafel** straightening of the sheet
Ebonit *n* ebonite, vulcanite
Ebonitrohr *n* ebonite tube
Ebonitstange *f* ebonite rod
Ebulliometer *n* ebulliometer
Ebullioskop *n* ebullioscope
Ebullioskopie *f* ebullioscopy
ebullioskopisch ebullioscopic
Eburican *n* eburicane
Eburicosäure *f* eburicoic acid
Eburin *n* eburine
Ecgonidin *n* ecgonidine
Ecgonin *n* ecgonine, tropine carboxylic acid
Ecgoninchlorhydrat *n* ecgonine hydrochloride
Echappeöl *n* recovered oil
Echicerin *n* echicerin
Echinacein *n* echinacein
Echinatin *n* echinatine
Echinenon *n* echinenone
Echinit *m* echinite, fossil echinoid
Echinochrom *n* echinochrome
Echinomycin *n* echinomycin
Echinopsin *n* echinopsine
Echinopsöl *n* echinops oil, thistle seed oil
Echinulin *n* echinulin
Echitamin *n* echitamine, ditaine
Echiumin *n* echiumine
Echo *n* (Phys) echo
echt (Chem) genuine, pure, (Farbe) fast

Echtbase *f* fast base, fast color base
Echtbaumwollblau *n* fast cotton blue
Echtbeizenfarbe *f* fast mordant color
Echtblau *n* fast blue
Echtbraun *n* fast brown
Echtdampffarbe *f* fast steam color
Echtfärben *n* fast dyeing
Echtfarbe *f* fast color, ingrained color
Echtgelb *n* fast yellow, aniline yellow
Echtgrün *n* fast green
Echtheit *f* genuineness, purity, (Farbe) fastness, **farbmetrische** ~ (Opt) colorimetric purity
Echtheitsbeweis *m* proof of authenticity
Echtheitsgrad *m* degree of fastness
Echtheitsprüfung *f* verification, (Färb) fastness test
Echtheitszeugnis *n* certificate of authenticity
Echtneutralviolett *n* fast neutral violet
Echtorange *n* fast orange
Echtpergamentpapier *n* vegetable parchment
Echtponceau *n* fast ponceau
Echtsäurefuchsin *n* fast acid fuchsin
Echtscharlach *m* fast scarlet
Echtschwarz *n* fast black
Echujin *n* echujin
Eckblock *m* corner block
Ecke *f* corner, edge
Eckeisen *n* angle iron
Eckhahn *m* angle cock
eckig cornered, angled, angular
Ecknaht *f* (beim Schweißen) corner weld
Eckpfeiler *m* corner pillar
Ecksäule *f* corner column, prism
Eckstoß *m* (beim Schweißen) corner joint
Eckventil *n* [right] angle valve
Eckverband *m* edge joint, edge bond
Economiser *m* economizer, waste gas feed heater
EDA-Komplexe *m pl* (Abk. für Elektronen-Donator-Akzeptor-Komplexe)
edel precious, (Gas) inert, rare, (Met) noble
Edelerde *f* rare earth
Edelerz *n* rich ore
Edelgalmei *m* (Min) smithsonite
Edelgamander *m* germander
Edelgas *n* inert gas, noble gas, rare gas
Edelgasgleichrichter *m* rare gas rectifier
Edelgaskonfiguration *f* noble gas configuration
Edelgaskontinua *pl* continuous rare gas spectra
Edelgasschale *f* (Chem) inert gas shell
Edelgasstruktur *f* inert gas structure
Edelgestein *n* precious stones
Edelglanz *m* dull silvery luster
Edelguß *m* special cast iron
Edelholz *n* fine wood, precious wood
Edelkamille *f* (Bot) large c[h]amomile
Edelkastanie *f* sweet chestnut

Edelkorund *m* special fused alumina
Edelkunstharz *n* synthetic resin, cast resin
Edelmarderfett *n* pine marten fat
Edelmetall *n* noble metal, precious metal
Edelmetallkontakt *m* precious metal contact
Edelopal *m* (Min) precious opal, hydrophane
Edelpassung *f* close fit, wringing fit
Edelporzellan *n* hard porcelain
Edelputz *m* patent plaster
Edelrost *m* patina
Edelsalz *n* refined salt
Edelspat *m* (Min) adularia
Edelsplitt *m* fine broken stones
Edelstahl *m* high-grade steel, fine steel, high-quality steel, precious steel, refined steel, stainless steel, superior alloy steel, **korrosionsbeständiger** ~ high-quality stainless steel
Edelstein *m* jewel, precious stone
edelsteinartig gem-like
Edelsteingewicht *n* carat
Edeltannenöl *n* pine needle oil, pine oil
Edestin *n* edestin
Edetate *pl* (Kurzbez. f. Salze der Äthylendiamintetraessigsäure) edetates
Edetinsäure *f* (Äthylendiamintetraessigsäure) edetic acid
EDG (Elektrodermatogramm) electrodermogram
Edingtonit *m* (Min) edingtonite
Edinol *n* edinol
Edison-Akkumulator *m* (Elektr) Edison-battery
Edman-Abbau *m* Edman degradation
Edrophonium *n* edrophonium
EDTA (Abk. für Äthylendiamintetraessigsäure)
Edulein *n* edulein
EEG (Elektroenzephalogramm) electro-encephalogram, E E G or e e g
Efeu *m* (Bot) ivy
Efeubitter *n* hederine
Efeusäure *f* hederic acid
Effekt *m* effect, result, ~ **zweiter Ordnung** second-order effect, **äußerer lichtelektrischer** ~ external photoelectric effect, Hallwachs effect, photoemissive effect, **bathochromer** ~ bathochromic effect, **elastischer** ~ elastic effect, **induktiver** ~ inductive effect, **lichtelektrischer** ~ photoelectric effect, **photoelektrischer** ~ photoelectric effect, **thermoelektrischer** ~ thermoelectric effect, Seebeck effect
effektiv effective, efficient
Effektivgeschwindigkeit *f* root-mean-square velocity
Effektivhöhe *f* effective height
Effektivitätsfaktor *m* (heterogene Katalyse) effectiveness factor
Effektivladung *f* effective charge
Effektivstrom *m* (Elektr) effective current

Effektivwert *m* effective value, root-mean-square value, virtual value
Effektkohle *f* (Elektr) flame carbon
Effektlack *m* effect varnish
Effektoren *pl* effectors
Efferveszenz *f* effervescence
efferveszieren to effervesce, to ferment, to froth
Effloreszenz *f* efflorescence
effloreszieren to effloresce, to bloom
Effluviographie *f* effluviography
Effluvium *n* effluvium
Effusiometer *n* effusiometer
Effusion *f* effusion
Effusivgestein *n* effusive rock, volcanic rock
Egalfärbevermögen *n* level dyeing property
egalisieren to equalize, to even, to level
Egalisieren *n* levelling
Egalisierer *m* equalizer, level[l]ing agent
Egalisierfarbstoff *m* level[l]ing dye-stuff
Egalisierhilfsmittel *n* level dyeing auxiliary
Egalisiermaschine *f* blending machine
Egalisiermittel *n* level[l]ing agent
Egalisierungsvermögen *n* equalizing power, level[l]ing power
EGG (Elektrogastrogramm, Med) electrogastrogram
Egge *f* disc harrow
Egonol *n* egonol
EGR (elektrische Gasreinigungsanlage) electrical gas purification plant
Ehlit *m* (Min) ehlite
Ehrlichs Aldehydreagens *n* Ehrlich's reagent, dimethylaminobenzaldehyde
Eialbumin *n* egg albumin, ovalbumin
Eibischblätter *n pl* (Bot) marshmallow leaves
Eibischblüten *f pl* (Bot) marshmallow flowers
Eibischsirup *m* althea syrup
Eibischwurzel *f* (Bot) marshmallow root
Eichamt *n* bureau of standards, calibration office
Eiche *f* oak
Eichelkaffee *m* acorn coffee
Eichelstein *m* petrified acorn
Eichelzucker *m* quercite
eichen to calibrate, to gauge, to grade, to measure, to standardize
Eichen *n* calibrating, gauging, standardizing
Eichengerbsäure *f* quercitannic acid
eichengerbsauer quercitannate
Eichenholzextrakt *m* oak extract
Eichenholzmehl *n* oak dust
Eichenholzöl *n* oak tree oil
Eichenlohe *f* tanner's bark, tan
Eichenmehl *n* ground oak bark
Eichenrinde *f* oak bark
Eichenrindeabkochung *f* decoction of oak bark
Eichenrot *n* oak red
Eicher *m* calibrater, gauger, tester

eichfähig capable of being calibrated, adjustable, capable of adjustment, standardizable
Eichfaktor *m* calibration constant
Eichfehler *m* calibration error
Eichgruppe *f* gauge group
Eichinvarianz *f* gauge invariance
Eichkurve *f* calibrating plot
Eichmaß *n* standard measure, gauge
Eichmeister *m* calibrater, gauger, tester
Eichmetall *n* sterro metal
Eichnagel *m* gauge mark
Eichprotokoll *n* calibration record
Eichreiz *m* reference stimulus
Eichschein *m* certificate of calibration
Eichstrich *m* calibration mark
Eichsubstanz *f* gauge substance
Eichtransformation *f* gauge transformation
Eichtreue *f* permanent calibration
Eichung *f* standardization, calibration, gauging
Eichwert *m* calibration value
Eicosan *n* eicosane
Eicosantriensäure *f* eicosantrienonic acid
Eidotter *m* egg yolk
Eieralbumin *n* egg albumin, ovalbumin
Eierkonserve *f* egg preservative
Eierkonservierung *f* egg preservation
eiern to wabble
Eierstein *m* egg stone, oolite
eiersteinartig oolitic
eiersteinförmig oolitic
eiersteinhaltig oolitiferous
Eiform *f* egg shape, ovoid
Eigelb *n* egg yolk
eigen characteristic, individual, intrinsic, natural, own, proper
Eigenabsorption *f* individual absorption
Eigenabstoßung *f* **der Elektronen** repelling action of electron charge
Eigenarbeit *f* no-load work
Eigenarbeitsvermögen *n* specific energy
Eigenart *f* characteristic, individuality, nature, peculiarity
Eigenbelastung *f* dead load
Eigenbewegung *f* characteristic motion, individual motion
Eigendämpfung *f* (Elektr) self-modulation, (Mech) self-damping
Eigendrehbewegung *f* characteristic rotation, spin
Eigendrehimpuls *m* characteristic rotational momentum
Eigendrehung *f* individual rotation
Eigenenergie *f* characteristic energy, specific energy
eigenerregt self-excited
Eigenerregung *f* self-excitation, individual excitation
Eigenerwärmung *f* heat build-up
Eigenfarbe *f* characteristic color, natural color

eigenfarbig self-colored
Eigenfederung *f* automatic resilience, inherent spring-like action
Eigenfilterung *f* inherent filtration
Eigenfrequenz *f* characteristic frequency, individual frequency, natural frequency
Eigenfunktion *f* characteristic, eigenfunction
Eigengeschwindigkeit *f* characteristic speed, proper speed
Eigengewicht *n* dead weight, own weight, specific gravity
Eigengröße *f* characteristic variable
eigenhändig personal
Eigenhalbleiter *m* intrinsic semiconductor
Eigeninduktion *f* (Elektr) self-induction
Eigenkapazität *f* (Elektr) natural capacitance
Eigenkonvektion *f* (Therm) natural convection
Eigenkreiselwirkung *f* self-gyro-interaction
Eigenladung *f* space charge
Eigenlast *f* dead load
Eigenleitung *f* (Phys) intrinsic conduction
eigenmagnetisch selfmagnetic
Eigenmasse *f* (Phys) proper mass
Eigenmodulation *f* self modulation
Eigenparität *f* intrinsic parity
Eigenperiode *f* characteristic period
Eigenreaktanz *f* self-reactance
Eigenreibung *f* internal friction
Eigenschaft *f* property, feature, peculiarity, quality, **additive** ~ (Phys) additive property, **chemische** ~ chemical property, **halbmetallische** ~ metalloid property, **konzentrationsabhängige** ~ concentration-dependent property
Eigenschaften *pl*, **konstitutive** ~ constitutional properties, constitutive properties
Eigenschaftsveränderungen *f pl* change of properties
Eigenschwingung *f* free oscillation, individual vibration, natural vibration
Eigenspannung *f* internal stress
Eigenstabilisierung *f* self-stabilization
Eigenstrahlung *f* characteristic radiation, natural radiation
Eigensymmetrie *f* characteristic symmetry
Eigentemperatur *f* characteristic temperature
Eigentum *n* property
Eigenvektor *m* characteristic vector
Eigenverbrauch *m* internal consumption
Eigenvergrößerung *f* actual enlargement, actual magnification, real magnification
Eigenviskosität *f* intrinsic viscosity, internal viscosity
Eigenvolumen *n* specific volume
Eigenwärme *f* specific heat, total heat
Eigenwert *m* characteristic number, eigen-value, inherent value, (Math) eigenvalue
Eigenwertaufgabe *f* characteristic value problem
Eigenwertproblem *n* eigenvalue problem

Eigenwiderstand *m* internal resistance
Eigenzustand *m* characteristic state, proper state, (Chem) eigenstate
Eiglobulin *n* ovoglobulin
eignen (sich) to qualify, to be appropriate, to suit
Eignungsprüfung *f* qualifying examination
Eigröße *f* egg size
Eikonalgleichung *f* (Opt) eiconal equation
Eikonogen *n* eikonogen
Eikosan *n* eicosane, eikosane
Eikosansäure *f* eicosanic acid
Eikosen *n* eicosene
Eikosyl- eicosyl
Eikosylalkohol *m* eicosanol, arachidic alcohol, eicosyl alcohol
Eikosylen *n* eicosylene
Eikosylsäure *f* eicosylic acid
Eileiter *m* (Med) oviduct
Eimer *m* bucket
Eimerwerk *n* bucket elevator
einachsig (Krist) monaxial, uniaxial
Einachsigkeit *f* (Krist) uniaxiality
einadrig single core
einäschern to incinerate, to burn to ashes, to calcine
Einäschern *n* incinerating, calcining, ~ **des Filters** burning the filter, incineration of filter paper
Einäscherung *f* incineration, calcination
einätzen to etch
Einätzen *n* etching
Einätzung *f* etching
Einankerumformer *m* (Drehumformer) rotary converter
einarbeiten to incorporate
Einarbeiten *n*, **gegenseitiges** ~ coordination of new methods
einatmen to breathe in, to inhale
Einatmung *f* inhalation
einatomig monoatomic
Einatomigkeit *f* monoatomicity
Einbadgerbung *f* (Gerb) one-bath tanning
Einbadverfahren *n* single-bath process
einbalsamieren to embalm
Einbalsamieren *n* embalming
Einbalsamierung *f* embalmment
Einbalsamierungsflüssigkeit *f* embalming fluid
Einband *m* cover, binding
einbasig monobasic
einbasisch monobasic
Einbauaggregat *n* built-in unit
einbauen to incorporate, to install
Einbauflankenspiel *n* backlash
Einbauhöhe *f* mold opening, **lichte** ~ (Presse) daylight
Einbaumaß *n* installation dimension
Einbauring *m* flush ring
Einbaurührer *m* built-in stirrer
Einbaustelle *f* point of insertion

Einbauten *pl* installations, internal fittings
Einbegrenzungseffekt *m* pinch effect
einbehalten to keep back, to retain, to withhold
Einbehaltung *f* detention
einbeizen to etch
einbetten to embed, to imbed, (Elektr) to encapsulate, (Geol) to intercalate
Einbetten *n* embedding, potting
Einbettmasse *f* potting medium
Einbettung *f* embedding
Einbettungsmittel *n* embedding medium
Einbeulapparat *m* bulge testing apparatus
einbeulen to bend in, to deflect inwards, to hump inwards
Einbeulung *f* dent, indentation
einbiegen to bend inward[s]
Einbiegung *f* curvature, deflection
einbinden (Buchdr) to bind, (Med) to bandage
Einbindestelle *f* splice point
einbindig univalent
Einblasedruck *m* injection pressure
Einblasegefäß *n* injection air receiver
einblasen to blow in, to insufflate, (Dampf) to inject
Einblattelektrometer *n* single-leaf electrometer
einblenden (Elektron) to gate, (Farben) to blend
Einbrand *m* penetration by burning
Einbrandkerbe *f* penetration notch
Einbrennemaille *f* baking enamel, stoving enamel
einbrennen to anneal, to brand, to burn in, (Lacke etc.) to bake, (Leder) to hot-stuff, (Schweiß) to penetrate, (Text) to crab
Einbrennen *n* annealing, firing, hot stuffing (leather), (Text) crabbing, ~ **der Glasur** glaze baking, ~ **des Emails** firing the enamel
Einbrennfarbe *f* annealing color, baked varnish coat, baking finish, (Porzellan) color for porcelain painting
Einbrennfirnis *m* stove varnish
Einbrennlack *m* annealing lacquer, baking enamel, baking varnish, stoving enamel, stoving finish, stoving lacquer
Einbrennlackierung *f* baking varnishing, ceramic varnish
Einbrennmetallisierung *f* metallization by burning in
Einbrenntemperatur *f* baking temperature, stoving temperature
einbringen to yield
Einbringen *n* introduction
Einbruchstelle *f* breach
Einbuchtung *f* concavity, inward bulge, inward curvature
Eindampfanlage *f* concentration plant, evaporation plant
Eindampfapparat *m* evaporator, evaporating apparatus

eindampfen to boil down, to concentrate by evaporation, to thicken, ~ **zur Trockne** to evaporate to dryness
Eindampfen *n* concentrating, evaporating, steaming, thickening by evaporation, ~ **der Lauge** evaporating the sud
Eindampfer *m* evaporator
Eindampfkessel *m* evaporator, vaporizer
Eindampfrückstand *m* residue after evaporation
Eindampfschale *f* dryer, drying cup, evaporating pan, evaporator
Eindampfung *f* concentration, evaporation
eindeutig unambiguous, well defined, (Math) unique
Eindeutigkeit *f* unambiguity, uniqueness
Eindeutigkeitsprinzip *n* (von Pauli) Pauli's exclusion principle
Eindickanlage *f* concentrating plant
eindickbar condensable
eindicken to concentrate, to inspissate, to thicken, (von Ölen) to body
Eindicken *n* concentrating, inspissating, thickening, (Farbe) setting up, ~ **der Badschmelze** thickening of the melt
Eindicker *m* thickener, thickening machine
Eindickung *f* concentration, inspissation
Eindickungsmittel *n* thickening substance
Eindickzylinder *m* concentrator
eindimensional one-dimensional, unidimensional
eindörren to dry up, to exsiccate, to shrink
eindosen to can
Eindosen *n* canning
Eindrahtaufhängung *f* unifilar suspension
Eindrehen *n* **der Kaliber** roll turning
eindringen to enter [into], to infiltrate, to penetrate, to soak into
Eindringen *n* (des Leims) [glue] penetration
Eindringtiefe *f* penetration depth
Eindringungstiefe *f* depth of impression, depth of penetration
Eindruck *m* impression, indentation
Eindruckfarbe *f* grounding-in color
Eindruckfläche *f* area of impression
Eindruckhärte *f* indentation hardness, hardness by indentation
Eindruckkalotte *f* indentation cup
Eindruckschmierung *f* shot-lubrication system
Eindruckstempel *m* indenting tool
eindunsten to evaporate
Eindunsten *n* evaporating
Eindunstung *f* evaporation
Einebnung *f* levelling
Einelektronenbindung *f* singlet linkage
Einelektronenzustand *m* one-electron state
einengen to compress, to confine, to narrow down, (Chem) to concentrate
Einengen *n* (Chem) concentrating

Einengung *f* concentrating (solutions etc), narrowing, ~ **der Probe** reduction of bulk sample
Einer *m* digit, unit
Einerstelle *f* unit place
Einerstufe *f* digit place, units place
einfach (einzeln) single, (ursprünglich) elementary, primitive, ~ **geladen** (Chem) single-charged
einfachbasisch monobasic
Einfachbindung *f* (Chem) single bond
einfachbrechend single-refracting
Einfachbrechung *f* (Opt) single refraction
Einfachbromjod *n* iodine monobromide
Einfachchlorjod *n* iodine monochloride
Einfachchlorschwefel *m* sulfur monochloride
Einfachchlorzinn *n* stannous chloride, tin protochloride
einfachchromsauer chromate
Einfachfilter *n* one-layer filter
Einfachform *f* single-cavity mold, single-impression mold
Einfachgleitung *f* simple glide
Einfachionisation *f* single ionization
Einfachkessel *m* single boiler
einfachkohlensauer neutral carbonate, normal carbonate
einfachlogarithmisch semilog
Einfachsaugrohr *n* single throat venturi tube
Einfachschicht *f* monolayer
Einfachschwefelammonium *n* ammonium monosulfide, ammonium sulfide
Einfachschwefelcalcium *n* calcium monosulfide, calcium sulfide
Einfachschwefelkalium *n* potassium monosulfide, potassium sulfide
Einfachschwefelkupfer *n* copper(I) sulfide, cuprous sulfide
Einfachschwefelzinn *n* stannous sulfide, tin protosulfide
Einfachwechselventil *n* spring loaded shuttle valve
Einfachwerkzeug *n* single impression mo[u]ld
einfachwirkend single acting
Einfaden *m* monofilament
Einfadenaufhängung *f* (Elektr) unifilar suspension
Einfadenlampe *f* (Elektr) single-filament lamp
einfärben to dye, to ink
Ein-Faktor-Methode *f* (bei Optimierung) one-factor-at-a-time method
Einfall *m* idea, incidence (rays)
einfallen (Strahlen) to fall in
einfallend incident
Einfallicht *n* (Opt) incident light
Einfallsebene *f* (Opt) incidence plane
Einfallslot *n* axis of incidence, perpendicular
Einfallsrichtung *f* direction of arrival
Einfallstelle *f* sink mark, sink spot

Einfallswinkel *m* angle of incidence
Einfallvorrichtung *f* engaging clutch
Einfang *m* capture, trapping
einfangen to capture, to trap
Einfanggammastrahlen *m pl* capture gammarays
Einfangprozeß *m* (Atom) capturing process, radiative capture
Einfangquerschnitt *m* capture cross section
Einfangreaktion *f* capture
Einfangwahrscheinlichkeit *f* capture probability
einfarbig monochromatic
Einfarbpunktschreiber *m* single-color dotted line recorder
Einfassung *f* rim, fringe
Einfeiler *m* groover
Einfeilung *f* notch
einfetten to grease, to lubricate, to oil
Einfetten *n* greasing, lubricating, oiling
Einfettung *f* lubrication
einflammig single-flamed
Einflammrohrkessel *m* Cornish boiler, one-flue boiler
einflanschen to flange
Einflanschen *n* flanging
einflechten to interweave, to weave in
einfließen to flow in
einflößen to infuse, to instil[l]
Einflügelradzähler *m* turbine water meter, vane water meter
Einfluidumtheorie *f* one-fluid theory
Einfluß *m* (Einfließen) influx, (Einwirkung) effect, influence, **differenzierend wirkender** ~ (Regeltechn) derivative action, rate action
Einflußgröße *f* variable
Einflußrinne *f* loading trough, trench
Einflußrohr *n* in-pipe, suction pipe
Einflußschleuse *f* inlet sluice
einförmig monotonous, uniform
einformen to mold
einfrieren to freeze, to congeal
Einfrieren *n* freezing, congelation, liquid solidification, ~ **des Bades** solidification of the bath, ~ **des Spannungszustandes** stress freezing
Einfriertemperatur *f* freezing-in temperature
einfügen to fit in, to insert, to splice
Einfügungsgewinn *m* insertion gain
einführen to introduce
Einführung *f* introduction
Einführungsstelle *f* point of introduction, feed plate (distillation)
einfüllen to fill, to charge, to feed
Einfüllen *n* feed, ~ **der Erze** charging of ores
Einfüller *m* feeder, loader
Einfüllkasten *m* fill box
Einfüllöffnung *f* feed opening
Einfüllstutzen *m* inlet pipe, filling vent
Einfülltopf *m* replenishing cup

Einfülltrichter *m* charging funnel, charging hopper, feed hopper, funnel pass, hopper, loading hopper
Eingabe *f* (Gaschromat) injection port
Eingabewerk *n* input
Eingabezeichnung *f* drawing to support application for official approval
eingängig single-threaded
Eingang *m* (Elektr) input
Eingangsbahn *f* entry line
Eingangsdaten *n pl* input data
Eingangsenergie *f* (Elektr) input power
Eingangskreis *m* gate circuit
Eingangsleistung *f* input power
Eingangsleitwert *m* input admittance
Eingangsspalt *m* (Opt) entrance slit
Eingangsspannung *f* input voltage
Eingangsstrom *m* input current
Eingangswert *m* input value
Eingangswiderstand *m* input resistance
eingedickt concentrated
eingefangen trapped
eingefroren frozen
eingehen to enter, (schrumpfen) shrink, **eine Verbindung** ~ (Chem) to enter into combination
Eingehen *n* **des Gewebes** (Text) shrinkage of cloth
eingehend thoroughly
eingeißelig (Zool) monotrich, monotrichous
eingekalkt dressed with lime
eingekeilt wedged in
eingekerbt notched
eingelaugt lixiviated, steeped in lye
eingemacht pickled, salted
eingepökelt salted
eingerostet rusted
eingeschliffen ground-in
eingeschmolzen fused
Eingeschwindigkeitsmethode *f* one-velocity method
eingezuckert preserved in sugar
eingießen to pour in, to infuse
Eingießen *n* pouring in
eingipsen to fix in plaster, to incorporate in plaster
Eingipsung *f* plaster fixation
Eingitterröhre *f* single-grid tube (Br. E. valve)
eingraben to dig in, to trench
eingreifen to act, to catch, to interlock, to lock
eingreifend radical
eingrenzen to localize, to locate
Eingrenzung *f* localization
Eingriff *m* (Med) operation, ~ **der Zähne** *m pl* gearing of teeth
Eingriffsfläche *f* zone of contact
Eingriffsfunktion *f* influencing function
Eingriffsstrecke *f* (Getriebe) line of action (Am. E.), path of contact (Br. E.)

Eingriffstiefe – Einlagerungsstrukturen

Eingriffstiefe *f* depth of engagement of gears
Eingriffsverhältnisse *pl* (Getriebe) meshing
Eingriffswinkel *m* **der Zahnflanke** pressure angle of tooth profile
Einguß *m* filling, (Metall) feeder, sprue, ~ **mit Vormulde** pouring head with sunk basin
Eingußkanal *m* channel; gate, inlet; pouring head, sprue
Eingußkasten *m* feed box
Eingußloch *n* aperture for filling, (Bauw) grouting hole
Eingußmodell *n* gate pin, runner pin
Eingußöffnung *f* pouring-in hole
Eingußrohr *n* central gate
Eingußsieb *n* inlet strainer
Eingußtechnik *f* pouring practice
Eingußtrichter *m* downgate, feeding head, runner gate, sprue
Einhängefilter *m* hanging filter, suspension filter
Einhängehaken *m* suspension hook
Einhängekühler *m* cold finger, hanging condenser, immersion condenser
Einhängen *n* suspension
Einhaken *n* clasping, clenching, hitching
Einhalt *m* interruption, restraint
einhalten to interrupt, to restrain, (Bedingung, Diät) to keep
einhauen to cut into, to cut open
Einheit *f* (Maß) unit, (Zahl, Identität) unity;, ~ **der Stromstärke** unit of the strength of current, ~ **des Leitvermögens** unit of conductivity, **absolute** ~ absolute unit, **magnetische** ~ magnetic unit
einheitlich (gleichartig) homogeneous, uniform, unitary, (vereinheitlicht) unified
Einheitlichkeit *f* (Gleichartigkeit) homogeneity, uniformity
Einheitsbauart *f* (Arch) standard type construction
Einheitsbelastung *f* basic load
Einheitsbogen *m* unit arc, (Radian) radian
Einheitsdiagramm *n* unit diagram
Einheitsdruck *m* unit pressure
Einheitsflansch *m* unit flange
Einheitsfunktion *f* unit function
Einheitsgewinde *n* standard thread, unified thread
Einheitskristall *m* unit crystal
Einheitskugel *f* unit sphere
Einheitsladung *f* unit charge
Einheitsmaß *n* (Techn) standard measure
Einheitsmethode *f* standard method
Einheitspackung *f* standard packing
Einheitspol *m* unit magnetic pole
Einheitspreis *m* unit price
Einheitssystem *n* standard system, uniform system
Einheitstensor *m* unit or identity tensor

Einheitsvektor *m* **für Polarisation** (Elektr) polarization unit vector
Einheitsverfahren *n* unit operation
Einheitszelle *f* unit cell
einheizen to heat, to light a fire
einhellig unanimous
Einhiebverfahren *n* impact hardness test
einholen to haul in, to obtain
Einhüllbedingung *f* condition of enveloping
einhüllen to wrap, to cover; to embed, to envelop
Einhüllende *f* (einer Kurvenschar) envelope curve
Einhülsen *n* (Atom) canning
einhydratig monohydrate
Einigungskitt *m* cement, mastic, putty
einimpfen to inoculate
einkalken to lime, to soak in lime water, to treat with lime
einkapseln to encapsulate
Einkapselung *f* encapsulation
Einkauf *m* purchase
einkaufen to purchase
einkeilen to jam, to wedge in
Einkerbbiegeversuch *m* notched-bar bending test
einkerben to groove, to indent, to notch
Einkerbung *f* notch
einkernig mononuclear, uninuclear
Einkernreifen *m* single bead tire
einkitten (mit Glaserkitt) to putty, (mit Zement) fasten with cement
Einklang *m* (Harmonie) harmony, sympathy
einkleiden to coat
einklemmen to jam
einkochen to boil down, to evaporate, to thicken
Einkochen *n* boiling down, thickening by boiling
Einkochring *m* ring for preserving jars
Einkochung *f* concentration by boiling, evaporation by steaming
Einkomponentenkleber *m* one-component adhesive
Einkomponentensystem *n* one-component system
Einkristall *m* monocrystal, single crystal
Einkristallfäden *pl* whiskers
Einkristalloberfläche *f* single-crystal surface
Einkristallprobe *f* single-crystal specimen
Einkristallspitze *f* monocrystal point
einkugeln to treat in the ball mill
einkuppeln to couple, to engage, to throw into gear
Einlage *f* filling, insertion, investment, (Brief etc.) enclosure, (Med) implant, (Reifen) ply
Einlagenumschlag *m* (Reifen) tie-in
einlagern to deposit, to infiltrate, to intercalate, to store
Einlagersystem *n* goods-in system
Einlagerungsfremdatome *n pl* interstitial impurities
Einlagerungsstruktur *f* interstitial structure
Einlagerungsstrukturen *pl* interstitial structures

Einlagerungsverbindung f intercalation compound, interstitial compound
Einlageteil n insert
Einlagevlies n interlining felt
Einlaß m admission, inlet, insertion, **doppelseitiger** ~ double inlet
Einlaßemaille f wiping paint
Einlaßgrund m sealer, sealing primer
Einlaßöffnung f feed inlet, inlet, inlet port, port
Einlaßrohr n inlet pipe
Einlaßsteuerung f admission gear
Einlaßstutzen m inlet tube, fluid inlet
Einlaßsystem n (Massenspektr) inlet system
Einlaßventil n feed valve, inlet valve
Einlaßventilkammer f intake valve chamber
Einlauf m inlet, intake, (Text) shrinkage
Einlaufboden m feed plate, inlet plate
einlaufecht shrink-resistant; unshrinkable
einlaufen to arrive, to enter, to run in, to shrink
Einlauffestigkeit f (Text) resistance to shrinkage
Einlaufseiher m inlet strainer
Einlaufspur f run-in
Einlauftrompete f bellmouth intake
einlaugen to lixiviate, to steep in lye
Einlaugen n steeping in lye
Einlegearbeit f inlay work
einlegen to put in, to salt, to soak, to steep
einleiten to initiate; to introduce; to start, (Chem) to pass [into]
Einleiten n introducing, opening, (Med) inducing, (Techn) feeding
Einleitung f introduction
Einleitungsrohr n delivery tube, inlet pipe
einleuchten to be clear, to be evident
einlinsig (Opt) single-lens
Einlochdüse f single hole nozzle
einlöten to solder in
einmachen to preserve
Einmachessig m spiced vinegar
einmaischen to mash
einmalig unique, single
Einmalschmelzerei f single refining, once melting down process
einmarinieren to pickle
einmauern to brick
Einmauerung f embedding
einmitten to adjust, to center
Einmitten n (Opt) centering
einmolekular monomolecular, unimolecular
einmünden to flow into, to discharge
einnehmen to fill, to occupy, to take up, **einen anderen Platz** ~ to occupy another place
Einniveauformel f single-level formula
Ein- oder Zweikreisgoniometer n (Krist) one or two circle goniometer
einölen to grease, to lubricate, to oil
einordnen to arrange, to classify
einpacken to pack, (einwickeln) to wrap [up]
Einpackpapier n wrapping paper

einpassen to adjust, to fit
Einpassen n adjusting, fitting, (Buchdr) shimming
Einpaßverfahren n method of adjustment
Einpaßzugabe f margin of manufacture
Einpegelung f leveling
Einperldrossel f bubbling throttle
einpfeffern to season with pepper
Einphasenanker m single-phase armature
Einphasendynamo m single phase generator
Einphasengenerator m single-phase alternator
Einphaseninduktionsmotor m single-phase induction motor
Einphasenkollektormotor m single-phase commutator motor
Einphasenleitung f single-phase wiring
Einphasenmotor m single-phase shunt motor
Einphasennebenschlußmotor m single-phase shunt motor
Einphasenofen m single-phase furnace
Einphasenserienmotor m single-phase series motor
Einphasenstrom m single-phase current
Einphasenstromanlage f single-phase [current] plant
Einphasensystem n monophase system, one-phase system, single-phase system
Einphasenwechselstrom m single-phase current
einphasig monophase, single-phase
einpökeln to corn, to cure, to pickle in salt
Einpoldynamo m unipolar dynamo
einpolig unipolar
einprägen to impress, to imprint, to stamp
einpressen to press in, to force in, to squeeze in, (Keil) to drive in
Einpressen n injecting
Einpreßteil n insert, **durchgehendes** ~ through-type insert
einpudern to powder
einpumpen to pump in
Einquellbottich m soaking tub, steeping trough
einquellen to soak, to steep
Einquellwasser n steeping water
einquetschen to jam, to squeeze
einräuchern to fumigate
einräumen to yield, to allow, to clear up, to furnish, to stow away
einrahmen to frame
Einrastknopf m lock knob
einrechnen to take into account; to add [in], to comprise
einregeln to adjust
einregulieren to adjust, to regulate
Einreibemittel n (Pharm) embrocation, liniment
einreiben to rub in, to smear
Einreibung f embrocation, liniment, rubbing in, smearing
einreichen to present, to submit
einreihig single row, single series

einreißen – einschließen

einreißen to tear down, to demolish, (Papier) to tear
Einreißfestigkeit *f* tearing strength, tear resistance
einrichten to install, (anordnen) to arrange, (ausrüsten) to equip, (errichten) to establish, to set up
Einrichtung *f* arrangement, equipment, establishment, facility, installation, outfit, set up, (Möbel) furniture
Einriß *m* crack, fissure, flaw, rent
einritzen to scratch, to etch, to score
Einrollenmühle *f* single roller ring mill
Einrollstoff *m* runner, wrapper
einrosten to rust
Einrückhebel *m* control lever, engaging lever
Einrückkupplung *f* engaging coupling
Einrückvorrichtung *f* engaging gear
einrühren to mix and stir, to mix in
einrußen to cover with soot
einsacken to bag, to put into a sack, to sack
Einsacken *n* bagging
Einsackmaschine *f* sack filler
Einsackstelle *f* sink mark, sunk spot
Einsackwaage *f* bagging scales
einsäuern to acidify, to pickle, (Futter) to ensilage
Einsäuerungsbad *n* acid bath
einsäurig monoacid
einsalben to rub with ointment, to anoint
Einsalzeffekt *m* salting-in effect
Einsalzen *n* salting
Einsalzung *f* salting
Einsatz *m* holder, insert, insertion, inventory, support, unit mold, (Anteilnahme) participation; (Ofen) charge, **kalter** ~ cold charge; cold charging, **warmer** ~ hot charge, molten charge
Einsatzaufstreupulver *n* casehardening powder
einsatzbereit prepared, ready for action, ready for use
Einsatzenergie *f* (Phys) threshold energy
Einsatzfilter *m* extraction filter, filter for insertion
Einsatzgewicht *n* weight of charge
Einsatzhärtemittel *n* casehardening agent
einsatzhärten to caseharden, (aufkohlen) to carbonize
Einsatzhärten *n* casehardening, pack hardening, (Aufkohlen) carbonizing
Einsatzhärtepulver *n* casehardening powder
Einsatzhärtung *f* casehardening, surface converting, ~ **durch Aufkohlung** carburization
Einsatzkasten *m* annealing box, hardening trough
Einsatzkessel *pl* set of boilers
Einsatzkühler *m* insertion condenser
Einsatzmaterial *n* charge, feed-stock
Einsatzmittel *n* casehardening compound

Einsatzöffnung *f* charging opening
Einsatzofen *m* hardening furnace
Einsatzprodukt *n* charge stock, feed-stock
Einsatzpulver *n* carburizing agent or material, casehardening compound, casehardening powder, cementing powder
Einsatzpunkt *m* starting point
Einsatzring *m* adapter ring, fitting ring
Einsatzroheisen *n* (Metall) charge pig iron
Einsatzschaufel *f* charge shovel
Einsatzspannung *f* starting voltage
Einsatzstahl *m* casehardening steel
Einsatzstück *n* inserted piece, distance piece, insert, insert mold
Einsatzstutzen *m* connecting branch
Einsatztiegel *m* extraction crucible
Einsatztür *f* charging door, firing door
Einsatzunterbrechung *f* charging delay
Einsatzzylinder *m* cylinder liner
einsaugen to soak up, to absorb, to imbibe
Einsaugen *n* absorbing
Einsaugmittel *n* (Med) absorbent
Einsaugung *f* absorption, suction
einschaben to grind in, to machine to fit
Einschabung *f* machining
einschätzen to evaluate, to calculate, to estimate
Einschätzung *f* appreciation, assessment, rating
Einschaleisen *n* casing iron, framing iron; steel framework
einschalen to case, to encase
einschalig single pan
Einschaltdynamometer *n* transmission dynamometer
einschalten to insert, to intercalate, (Elektr) to put in circuit, to switch on, (Motor) to put in, (Rohmaterial) to feed
Einschalten *n* switching on
Einschalter *m* switch, circuit closer
Einschalthebel *m* starting lever
Einschaltung *f* insertion, making a connection, switching on, turning on
Einschalung *f* casing; enclosure, mold
einschauflig single-blade
einschichten to arrange in layers
einschichtig single-layered
Einschichtpapier *n* (Buchdr) self-contained paper
Einschichtpolarisator *m* layer polarizer
einschieben to insert, to put in, to shove in
Einschiebung *f* insertion
Einschiebungsreaktion *f* (Chem) insertion reaction
einschlägig pertinent, appropriate, competent
Einschlämmung *f* deposition, sedimentation, settlement
Einschlagpapier *n* wrapping paper
einschleifen to grind in, to engrave, to machine
Einschleifpaste *f* grinding paste
einschließen to encapsulate, to occlude, to trap

228

Einschließung *f* occlusion
Einschließungssatz *m* inclusion theorem, (v. Temple) Temple's estimation of eigenvalues
Einschluß *m* enclosure, inclusion, occlusion, seal
Einschlußrohr *n* sealed tube
Einschlußthermometer *n* enclosed thermometer
Einschlußverbindung *f* inclusion compound, clathrate
einschmauchen to fumigate
einschmelzen to melt down, to fuse, to seal, (Metall) to smelt
Einschmelzen *n* fusing-in, melting down, ~ **des Elektrolyten** melting down of the electrolyte
Einschmelzglas *n* fusible glass, combustion glass
Einschmelzkolben *m* melting flask
Einschmelzlegierung *f* sealing alloy
Einschmelzrohr *n* carius tube, sealing tube
Einschmelzschlacke *f* meltdown slag
Einschmelzstelle *f* **des Glühfadens** filament seal
Einschmelzung *f* seal
einschmieren to grease, to lubricate, to oil
Einschmieren *n* greasing, lubricating, oiling
Einschmierung *f* greasing, lubrication, oiling
Einschnappfeder *f* catch spring
Einschneckenmaschine *f* single-screw extruder
Einschneckenpresse *f* single-screw extruder
einschneiden to cut in[to], to incise
einschneidend considerable, incisive
Einschnitt *m* cut, incision, (Kerbe) kerf, notch
Einschnittarbeit *f* cut work, cutting
einschnüren to throttle, to bind up, to constrict
Einschnürung *f* constriction, **lineare** ~ linear pinch
Einschnürungseffekt *m* pinch-effect
einschränken to confine, to limit, to restrain
Einschränkungsmaßnahme *f* restrictive measure
einschrauben to screw in
Einschraubhülse *f* screw-in jacket
einschreiben to inscribe, to note, to register
einschrumpfen to shrink, to shrivel
Einschubrahmen *m* panel mounting, ~ **als Träger gedruckter Schaltungen** solid-state board
einschütten to pour in
Einschüttöffnung *f* (Metall) charging hole
einschwärzen to blacken, to ink
Einschwärzfarbe *f* ink
Einschwefeleisen *n* (Eisen(II)-sulfid) ferrous sulfide, iron protosulfide
einschwefeln to sulfur, to sulfurize
einschweißen to weld
Einschweißkrümmer *m* welding elbow
einschwenkbar retractable
einschwingen to oscillate
Einschwingenbrecher *m* single-toggle crusher
Einschwingverhalten *n* transient response
Einschwingvorgang *m* (Elektr) transient phenomenon
Einschwingzeit *f* (Elektr) transient time

einschwöden to daub with ashes and lime
einseifen to soap
einseitig one-sided, unilateral, ~ **abgeschrägt** single slope, ~ **aufgetragen** coated on one side
einsenken to dip, to recess
einsetzen to charge, (beginnen) to begin, to set in, (substituieren) to substitute, (Techn, einfügen) to insert, to put in
Einsetzen *n* beginning, onset, ~ **des Tiegels** placing the crucible, setting in the crucible
Einsetzgewichte *n pl* nest of weights
Einsetzkran *m* charging crane
Einsetzmaschine *f* charging device, charging machine
Einsetztür *f* charging door, firing door
einsickern to infiltrate, to soak in
Einsitzventil *n* single-seated valve
Einspannbacke *f* chuck jaw, clamp, clamping jaw
Einspanndruck *m* clamping force
einspannen to clamp, to chuck
Einspannen *n* clamping, ~ **der Massel** inserting the pig
Einspannklaue *f* jaw of clamp
Einspannklemme *f* jaw of clamp
Einspannlänge *f* distance between grips
Einspannschweißvorrichtung *f* clamp welding machine
Einspannvorrichtung *f* clamping device, gripping arrangement
Einsparung *f* cost reduction, saving
einspindlig single spindle
Einsprengen *n* sprinkling
Einsprengling *m* (Krist) phenocryst
Einsprengmaschine *f* spray damping machine
Einsprengung *f* dissemination
einspringen to catch (lock etc.), to leap in
Einspritzanlasser *m* engine primer
Einspritzdruck *m* injection pressure
Einspritzdüse *f* injection nozzle
einspritzen to inject, to squirt in
Einspritzen *n* injecting
Einspritzhahn *m* injection cock
Einspritzkanal *m* runner
Einspritzkondensator *m* jet condenser
Einspritzkühler *m* injection cooler, jet cooler
Einspritzkühlung *f* injection cooling
Einspritzmotor *m* injection-type engine
Einspritzstrahl *m* condensing jet
Einspritzung *f* injection
Einspritzverdampfer *m* continuous-coil evaporator
Einspritzverfahren *n* (Polyester) resin-injection mo[u]lding
Einspritzvorrichtung *f* (Motor) primer
Einstabglaselektrode *f* single-rod glass electrode
Einständerpresse *f* single-column press

einstäuben – einteilen

einstäuben to dust, to grind to dust or powder, to powder
Einstäubung *f* dusting
einstampfen to stamp, to compress, to ram down, to stamp in
Einstampfen *n* stamping
Einstampfmaschine *f* (Papier) pulper, pulping machine
Einstampfpapier *n* waste paper
Einstandspreis *m* cost price, prime cost
Einstaubverfahren *n* powder process
Einstauchmaschine *f* upsetting machine
einstechen to pierce, to puncture
einstecken to plug in, to stick in
Einsteckklebeverbindung *f* plastic-cemented sleeve joint, shrunk-on pipe joint
Einsteckthermometer *n* stabbing thermometer
Einsteigöffnung *f* manhole
Einsteigschacht *m* manhole
Einstein *n* (Maß) einstein
Einsteinium *n* einsteinium
Einsteinsche Gleichung *f* für spezifische Wärme (Therm) Einstein equation for heat capacity
Einsteinsche Masse-Energie-Äquivalentgleichung *f* Einstein equation of mass-energy equival
Einsteinsches photochemisches Äquivalentgesetz *n* Einstein's law of photo-chemical equi
Einsteinsche Verschiebung *f* Einstein shift
Einstellanschlag *m* adjustable stop
einstellbar adjustable, controllable
Einstellbereich *m* adjustment range
einstellen to set, (anpassen) to adapt, to adjust, (auftreten) to appear, to set in, (halten) to stop, (Opt) to focus, **einen Apparat auf eine bestimmte Stelle** ~ to adjust an apparatus to a certain position, **eine Tätigkeit** ~ to stop
Einstellen *n* alignment, setting up
Einsteller *m* regulator
Einstellhebel *m* adjusting lever
einstellig (Math) one-digit, one-figure
Einstellimpuls *m* timing pulse
Einstellknopf *m* adjusting knob, control knob
Einstellmuffe *f* adjusting sleeve
Einstellschraube *f* adjusting screw, focussing screw, levelling screw, setscrew
Einstellskala *f* adjustment scale
Einstelltuch *n* focussing cloth
Einstellung *f* adjustment, alignment, focussing, regulation, setting, timing
Einstellupe *f* (Opt) focussing magnifier
Einstellvorrichtung *f* setting device
Einstellzeit *f* adjustment time
Einstich *m* puncture
Einstichmeßkette *f* puncture gauging chain
Einstichpyrometer *n* (Extrusion) stock pyrometer
Einstichstelle *f* puncture
einstöpseln to plug in
Einstoffsystem *n* one-component system, single-phase system

einstoßen to push, to thrust
Einstrahlung *f* incoming radiation, [ir]radiation
Einstrangverfahren *n* single line process
Einstreichmittel *n* **für Reifenformen** tire mold lubricant
einstreuen to strew, (Zitat etc.) to intersperse
Einströmboden *m* feed plate, inlet plate
Einströmdruck *m* flow-in pressure
einströmen to flow in, to pour in, to run in, to stream in
Einströmenlassen *n* allowing to enter
Einströmkasten *m* inlet case
Einströmungsrohr *n* inlet pipe, live steam pipe, main steam pipe
Einströmventil *n* admission valve, inlet valve, intake valve
Einstromturbine *f* single-flow turbine, uniflow turbine
Einsturzgeschwindigkeit *f* re-entry velocity
Einstufenrakete *f* single-stage rocket
Einstufenrückführung *f* single-stage recycle
einstufig one-stage, single-stage
einsüßen to edulcorate, to sugar, to sweeten
einsumpfen to soak
Eintagsfliege *f* (Zool) dayfly, ephemera
eintauchen to dip, to immerse, to plunge
Eintauchen *n* dipping, immersion, plunging
Eintaucher *m* dipper, plunger, sinker
Eintauchkette *f* immersion chain
Eintauchlöten *n* dip soldering
Eintauchmeßzelle *f* immersion measuring cell
Eintauchnutsche *f* filter stick, immersion filter
Eintauchpyrometer *n* dipping pyrometer
Eintauchrefraktometer *n* dipping or immersion refractometer
Eintauchschweißen *n* dip welding
Eintauchthermometer *n* total-immersion thermometer
Eintauchthermostat *m* insertion thermostat
Eintauchtiefe *f* (Dest) depth of immersion
Eintauchtrog *m* dip tank
Eintauchverfahren *n* immersion method
Eintauchvergoldung *f* dip gilding
Eintauchzeit *f* immersion time
einteeren to tar
Einteeren *n* tarring
einteigen to make into a paste or dough
einteilbar dividable
Einteilchenmodell *n* single particle model, independent particle model, nuclear model
Einteilchenniveau *n* single particle level
Einteilchenschalenmodell *n* single particle shell model
Einteilchenübergang *m* single particle transition
Einteilchenwellenfunktion *f* single particle wave function
einteilen (einordnen) to classify, (kalibrieren) to calibrate, (unterteilen) to divide, to graduate, (verteilen) to distribute

Einteilung *f* classification, division, graduation, subdivision
Einteilungsbogen *m* divided sheet
Einträgerkran *m* single-girder crane
einträglich lucrative, profitable
einträufeln to drop in, to instil
Eintrag *m* entry, (Web) weft, woof
eintragen to enter, to add, to introduce, to register
Eintragen *n* entering, introducing
Eintragung *f* entry, record, registration
Eintragungsöffnung *f* charging door
eintreten to enter into, to occur, to set in, to take place
Eintritt *m* inlet, (Geschehen) occurrence; (Kolonne) feed plate, inlet plate, (Strahl) incidence
Eintrittsöffnung *f* inlet hole
Eintrittsphase *f* entrance phase
Eintrittsrohr *n* admission pipe, inlet pipe, supply pipe
Eintrittsschlitz *m* inlet port
Eintrittsseite *f* entry side, inlet
Eintrittstemperatur *f* inlet temperature
Eintrittsverlust *m* entrance loss
Eintrittswinkel *m* entering angle, incidence angle
eintrocknen to desiccate, to dry
Eintrocknen *n* desiccation, drying
Eintrocknung *f* desiccation, drying
Eintrocknungsprozeß *m* drying process
eintröpfeln to drop in, to instil
Eintröpfeln *n* instilling
Ein- und Ausschalter *m* on-off switch
einverleiben to embody, to incorporate
Einverleibung *f* embodiment, incorporation
Einverständnis *n* agreement
Einwaage *f* weighed sample, initial weight, test portion
Einwägelöffel *m* weighing-in spoon
einwägen to weigh [in]
einwärts gekrümmt dished (e. g. surface)
einwässern to soak, to steep, to water
Einwässern *n* steeping or soaking in water
Einwässerung *f* steeping in water
einwalzen to roll in
Einwalzenmaschine *f* single-roller mill
Einwalzentrockner *m* drum drier
einwandfrei faultless, perfect
Einwebung *f* crimp
Einweckapparat *m* sterilizer, sterilizing apparatus
Einweckglas *n* preserving jar
Einwegflasche *f* nondeposit bottle, nonreturnable bottle
Einweghahn *m* one-way stopcock
Einwegtasse *f* one-use cup, disposable hot drink cup
Einwegverpackung *f* disposable package

Einwegverpackungen *pl* non-returnable containers, disposable containers
einweichen to soak, to macerate, to steep
Einweichen *n* soaking, softening by steeping
Einweichgrube *f* (Text) retting pit
Einweichmittel *n* soaking agent
Einweichung *f* maceration
Einwerfen *n* **von Schlacke** charging of slag
einwertig (Chem) monovalent, univalent, (Math) single-valued
Einwertigkeit *f* (Chem) monovalence, univalence
Einwickelmaschine *f* wrapping machine
einwickeln to wrap [up]
Einwickelpapier *n* wrapping paper
Einwickler *m* wrapper
einwiegen to weigh in
einwintern to protect against frost
einwirken to influence, to act, to affect, to react, ~ **lassen** to allow to react, **aufeinander** ~ to interact
Einwirkung *f* action, effect, (Chem) reaction, (Einfluß) influence, (Wechselwirkung) interaction, **gegenseitige** ~ mutual influence, mutual interaction, **innere** ~ internal action
Einwirkungsdauer *f* duration of action
Einwirkungszeit *f* induction period, reaction time
Einwurf *m* objection, reply
einzahnen to indent, to dovetail, to tooth
einzapfen to mortise
einzehren to lose by evaporation
Einzehrung *f* loss by evaporation
Einzeichnung *f* plotting
Einzelanfertigung *f* single-piece work
Einzelantrieb *m* single drive, self-contained drive
Einzelarbeit *f* output per unit
Einzelaushubtisch *m* single lifting table
Einzelbestimmung *f* determination by separate operation
Einzelblech *n* single plate
Einzeldüse *f* single nozzle
Einzeleinsatz *m* intermittent use
Einzelelektrodenpotential *n* single electrode potential
Einzelelektron *n* (Atom) lone electron
Einzelfaden *m* monofilament
Einzelfall *m* individual case, particular case
Einzelform *f* unit mold
Einzelfrequenz *f* individual frequency
Einzelheit *f* detail
Einzelheizer *m* single heater
Einzelkolonnenanordnung *f* (Gaschromat) single column
Einzelkraft *f* point force, concentrated force
Einzellast *f* (Mech) concentrated load
Einzeller *m* (Biol) monocellular organism, (Zool) protozoon

einzellig – Eisenbedarf

einzellig single-celled, unicellular
Einzelmatrix *f* unit matrix
Einzelmündung *f* single opening
einzeln single, individual, isolated
Einzelöler *m* separate lubricator
Einzelpotential *n* single potential
Einzelprobe *f* spot check
Einzelregelung *f* single regulation
Einzelriemen *m* single belt
Einzelschieber *m* single valve
Einzelschmierung *f* separate oiling
Einzelschritt *m* individual stage
Einzelschrittverfahren *n* single-step method
Einzelspannung *f* single potential
Einzelstoß *m* single collision
Einzelteil *n* single part
Einzelteilverfahren *n* single dividing method
Einzeltransformator *m* (Elektr) single transformer
Einzelverband *m* single bandage, single bond
Einzelversetzung *f* single dislocation
Einzelvorgang *m* single process
Einzelwirkungsgrad *m* efficiency per unit, individual efficiency; point efficiency
einzementieren to grout [in]
einziehbar retractable, retractile
einziehen to draw in, to absorb, (Flüssigkeit) to soak in
Einziehung *f* absorption, infiltration
Einziehvorrichtung *f* retracting mechanism
Einzonenreaktor *m* single-zone reactor
einzuckern to preserve in sugar
Einzweckmaschine *f* (Techn) single-purpose machine
einzwingen to force into
einzylindrig mono-cylindrical, single-cylinder
Eiplasma *n* ovoplasm
Eireifungsteilung *f* (Biol) maturation division
Eis *n* ice, (Speiseeis) ice-cream
Eisachat *m* (Min) translucent agate, uncolored agate
eisähnlich glacial, ice-like
Eisalabaster *m* (Min) translucent alabaster
Eisalaun *m* (Min) rock alum
Eisansatz *m* deposit of ice, layer of ice
eisartig glacial, ice-like
Eisbärfett *n* polar bear fat
Eisbelag *m* coating of ice
Eisbereitung *f* ice-making
Eisbildung *f* ice formation, frosting
Eisblumenbildung *f* frosting
Eisblumenlack *m* crystal finish, crystal lacquer, frosting lacquer
Eiscreme *f* ice cream
Eisen *n*, **entkohltes** ~ decarburized iron, **faulbrüchiges** ~ burned iron, **galvanisiertes** ~ galvanized iron, **gediegenes** ~ pure iron, **gekohltes** ~ carburized iron, **gemeines** ~ white pig iron, **geschmeidiges** ~ ductile iron, soft iron, **geschmiedetes** ~ forged iron, wrought iron, **getriebenes** ~ wrought iron, **gewöhnliches** ~ white pig iron, **grobkörniges** ~ coarse-grained iron, **großluckiges** ~ very open-grained pig iron, **hartgeschlagenes** ~ cold- or cool-hammered iron, hammer-hardened iron, **heißbrüchiges** ~ hot brittle iron, red short iron, **hochsiliziertes** ~ high-silicon iron, **kadmiertes** ~ cadmiated iron, **keilförmiges** ~ wedge-shaped iron, **kleinluckiges** ~ close-grained pig iron, **rostfreies** ~ rustless iron, **rotbrüchiges** ~ hot brittle iron, red short iron, **saures** ~ acid iron, bessemer steel, **schmiedbares** ~ forging iron, wrought or malleable iron, wrought steel, **sehniges** ~ fibrous iron, **sprödes** ~ brittle iron, **technisches** ~ commercial iron, **verrostetes** ~ rusted iron, **verzinktes** ~ galvanized iron, **weiches** ~ soft iron
Eisenabbrand *m* iron loss, iron waste
Eisenabfall *m* iron filings, iron scrap, scrap iron
Eisenabflußrinne *f* iron runner
Eisenabgang *m* iron waste, iron scrap
Eisenabscheider *m* iron separator
Eisenabscheidung *f* iron separation
Eisenabstich *m* tapping of iron
Eisenacetat *n* iron acetate
Eisenacetatlösung *f* iron acetate solution
eisenähnlich ferruginous, iron like
Eisenalaun *m* iron alum, ferric alum, potassium ferric sulfate
Eisenalbuminat *n* iron albuminate
Eisenamiant *m* (Min) iron asbestos
Eisenammonalaun *m* ammoniacal iron alum, ammonium ferric sulfate, ammonium iron sulfate
Eisenammoncitrat *n* iron ammonium citrate, ammonium ferric citrate
Eisenammoniakalaun *m* ammonium iron sulfate, ferric ammonium sulfate, iron ammonium sulfat
Eisenantimon *n* ferruginous antimony, iron antimonide
Eisenantimonerz *n* (Min) berthierite
Eisenantimonglanz *m* berthierite
Eisenapatit *m* (Min) triplite
Eisenarsenik *n* iron arsenide
eisenartig ferruginous, iron-like
Eisenarznei *f* (Pharm) ferruginous remedy
Eisenasbest *m* (Min) iron asbestos
Eisenatom *n* iron atom
Eisenausscheider *m* iron separator
Eisenausscheidung *f* separation of iron
Eisenbad *n* iron bath
Eisenbakterien *pl* iron bacteria
Eisenballen *m* iron ball, puddle ball
Eisenband *n* hoop iron, iron hoop
Eisenbedarf *m* iron requirement

Eisenbeize f iron mordant
Eisenbergwerk n iron mine, iron pit
Eisenbeschwerung f iron weighting
Eisenbeton m reinforced concrete, armored concrete, concrete steel, ferroconcrete
Eisenbitterkalk m ferruginous magnesia lime-stone, (Min) ferromagnesian limestone
Eisenblätterkern m laminated iron core
Eisenblaudruck m cyanotype process, ferroprussiate process
Eisenblauerde f (Min) earthy vivianite
Eisenblauerz n (Min) vivianite
Eisenblausäure f ferrocyanic acid
Eisenblauspat m (Min) vivianite, **faseriger** ~ (Min) crocidolite
Eisenblech n sheet iron, iron lamina, iron plate, ~ **abbrennen** to dip sheet iron into melted tin, **getriebenes** ~ embossed or dished iron plate
Eisenblechanode f sheet iron anode
Eisenblechmantel m sheet iron shell
Eisenblechring m sheet iron belt, sheet iron ring
Eisenblechtafel f sheet steel plate
Eisenblock m block of iron, bloom, ingot
Eisenblüte f (Min) aragonite, needle spar
Eisenblumen pl ferric chloride, iron flowers
Eisenborid n iron boride
Eisenbraunkalk m (Min) ferruginous bitter spar, ferruginous dolomite
Eisenbromür n (Eisen(II)-bromid) ferrous bromide, iron(II) bromide
Eisenbronze f iron bronze, ferrobronze
Eisenbrucit m (Min) ferruginous brucite
Eisenbrühe f iron liquor, iron mordant
Eisencarbid n iron carbide
Eisencarbonat n iron carbonate
Eisencarbonyl n iron carbonyl
Eisenchinincitrat n iron quinine citrate, ferruginous quinine citrate
Eisenchlorid n iron chloride, (Eisen(III)-chlorid) ferric chloride, iron(III) chloride
Eisenchloridwatte f (Pharm) ferric chloride wool, styptic wool
Eisenchlorür n (Eisen(II)-chlorid) ferrous chloride, iron(II) chloride
Eisenchlorürchlorid n (Eisen(II)(III)-chlorid) iron(II)(III) chloride, ferrosoferric chloride
Eisenchrom n chromic iron, chromite, ferrochrome, ferrochromium
Eisenchromat n iron chromate
Eisenchrysolith m (Min) iron chrysolite, fayalite, hyalosiderite
Eisencitrat n (Eisen(III)-citrat) ferric citrate, iron(III) citrate
Eisencyanfarbe f ironcyanogen pigment
Eisencyanid n iron cyanide, (Eisen(III)-cyanid) ferric cyanide, iron(III) cyanide
Eisencyanür n (Eisen(II)-cyanid) ferrous cyanide, iron(II) cyanide
Eisencyanverbindung f ironcyanogen compound

Eisendialysat n dialyzed iron
Eisendibromid n ferrous bromide, iron(II) bromide
Eisendichlorid n ferrous chloride, iron(II) chloride
Eisendifluorid n ferrous fluoride, iron(II) fluoride
Eisendijodid n ferrous iodide, iron(II) iodide
Eisendisulfid n iron disulfide
Eisendraht m iron wire
Eisendrehspäne pl iron shavings
Eisendruse f (Min) iron druse, iron geode
Eiseneinlage f core iron, iron core
Eisenelektrode f iron electrode
Eisenelement n iron cell
Eisenerde f (Min) ferruginous earth, iron earth
Eisenerz n (Min) iron ore, iron stone, **derbes** ~ compact iron ore, **faseriges** ~ fibrous iron ore, **manganhaltiges** ~ manganiferous iron ore, **mulmiges** ~ friable iron ore, **nickelhaltiges** ~ nickeliferous iron ore, **poriges** ~ porous iron ore, **toniges** ~ argillaceous iron ore, **zinkhaltiges** ~ zinciferous iron ore
Eisenerzgrube f iron ore mine
Eisenerzlagerstätte f iron ore deposit
Eisenerzvorkommen n iron ore deposit
Eisenextrakt m iron extract
Eisenfarbe f iron color
eisenfarbig iron-colored
Eisenfaß n iron drum
Eisenfeile f iron filings
Eisenfeilicht n iron filings
Eisenfeilspäne pl iron filings
Eisenfeilstaub m iron dust
Eisenfeinschlacke f iron refinery slag, refining cinders
Eisenfirnis m varnish for iron
Eisenfleck m iron spot, iron stain, rust mark
Eisenfluorid n iron fluoride, (Eisen(III)-fluorid) ferric fluoride, iron(III) fluoride
eisenfrei iron-free
Eisenfrischerei f iron refinery
Eisenfrischflammofen m puddling furnace
Eisenfrischschlacke f finery cinders
eisenführend ferriferous, ferruginous, iron-bearing
Eisenfunke m iron spark
Eisengallustinte f iron gallate ink
Eisengalvanoplastik f iron galvanoplastic process
Eisengang m (Bergb) iron lode, iron ore vein
Eisengans f iron pig
Eisengehalt m iron content
Eisengelb n iron yellow
Eisengestell n iron frame, iron stand
Eisengewerbe n iron industry
Eisengewinnung f (Metall) iron production
Eisengießer m iron founder

Eisengießerei f iron foundry
Eisengilbe f yellow ocher
Eisengitter n iron grating
Eisenglanz m (Min) iron glance, specular iron, specular iron ore
Eisenglas n (Min) fayalite
Eisenglasur f iron glazing
Eisenglimmer m (Min) micaceous iron ore, micaceous iron oxide
Eisenglimmerschiefer m (Min) itabirite
Eisenglycerophosphat n iron glycerophosphate, ferric glycerophosphate
Eisengranat m (Min) iron garnet, almandite, melanite, **brauner** ~ allochroite, **roter** ~ almandine
Eisengrau n (Min) iron gray
Eisengraupen f pl granular bog iron ore
Eisengrund m iron liquor, iron mordant
Eisengummi n iron rubber
Eisenguß m iron casting
Eisengußware f hardware, cast iron-ware
Eisengymnit m deweylite
Eisenhäm n ferriheme
Eisenhämochromogen n ferrihemochromogen
Eisenhämoglobin n ferrihemoglobin
eisenhaltig containing iron, ferriferous, ferruginous
Eisenhammer m forge hammer
Eisenhammerschlag m iron hammer scale, iron scales
Eisenhandel m iron trade
Eisenhart m ferriferous gold sand
Eisenhaube f iron cap
Eisenhochofen m iron blast furnace
Eisenhochofenschlacke f blast furnace cinder
Eisenholz n (Bot) iron wood
Eisenhütte f ironworks, forge
Eisenhüttenkunde f iron metallurgy
Eisenhut m (Bot) aconite
Eisenhutextrakt m aconite extract
Eisenhutknolle f (Bot) aconite root
Eisenhutkraut n (Bot) aconite
Eisenhuttinktur f tincture of aconite
Eisenhydroxid n iron hydroxide, (Eisen(III)-hydroxid) ferric hydroxide, iron(III) hydroxide
Eisenhydroxydul n ferrous hydroxide, iron(II) hydroxide
Eisen(III)-oxid n ferric oxide, iron(III) oxide
Eisen(III)-Salz n ferric salt, iron(III) salt
Eisen(III)-Verbindung f ferric compound, iron(III) compound
Eisen(II)-Salz n ferrous salt, iron(II) salt
Eisen(II)-Verbindung f ferrous compound, iron(II) compound
Eiseninduktion f induction in iron
Eisenindustrie f iron industry, ferrous metallurgy
Eisenjaspis m (Min) ferruginous jasper
Eisenjodat n iron iodate
Eisenjodid n iron iodide, (Eisen(III)-jodid) ferric iodide, iron(III) iodide
Eisenjodür n (Eisen(II)-jodid) ferrous iodide, iron(II) iodide
Eisenjodürjodid n ferrosoferric iodide, iron(II)(III) iodide
Eisenjodürsirup m syrup of ferrous iodide
Eisenkalium n potassium ferrate
Eisenkaliumalaun m ferric potassium sulfate, iron potassium alum
Eisenkarbid n iron carbide
Eisenkarbonat n iron carbonate, carbonate of iron
Eisenkasten m iron casing
Eisenkeil m iron wedge
Eisenkern m (eines Magneten) iron core (of a magnet)
Eisenkernspule f iron core coil
Eisenkies m (Min) iron pyrites, native iron sulfide, **kupferreicher** ~ cupriferous iron pyrites
Eisenkiesel m iron flint, ferruginous quartz
Eisenkitt m iron cement, iron glue, iron putty, iron rust cement
Eisenklumpen m iron block, iron pig, sow
Eisenknebelit m (Min) knebelite
Eisenkobalterz n cobaltite, safflorite, spathiopyrite
Eisenkohlenoxid n iron carbonyl, iron carboxide
Eisenkohlenstoff m iron carbide
Eisenkonstantanelement n iron constantan element
Eisenkonstruktion f iron building, structural ironwork
Eisenkorn n iron grain
Eisenkraftfluß m iron magnetic flux
Eisenkranz m mantle bracket
Eisenkraut n (Bot) verbena
Eisenkrautöl n verbena oil
Eisenkristall m iron crystal
Eisenlack m iron lacquer, iron varnish
Eisenlactat n iron lactate
Eisenlager n iron storage
Eisenlebererz n (Min) hepatic iron ore
Eisenlebertran m ferrated cod liver oil
Eisenleder n iron-tanned leather
Eisenlegierung f iron alloy
Eisenlikör m iron liquor, iron acetate liquor
Eisenlunge f (Med) siderosis
Eisenmanganerz n manganiferous iron ore
Eisenmanganpeptonat n ferromanganese peptonate
Eisenmangansaccharat n ferromanganese saccharate
Eisenmangantitan n titaniferous ferromanganese
Eisenmanganwolframat n ferromanganese tungstate, iron manganese tungstate
Eisenmangelanämie f hypoferric an[a]emia

Eisenmann *m* (Min) scaly hematite
Eisenmantel *m* iron mantle, steel jacket
Eisenmehl *n* iron meal
Eisenmennige *f* iron minium, iron ocher, red ocher
Eisenmetasilicat *n* iron metasilicate
Eisenmittel *n* iron tonic
Eisenmodifikation *f* iron modification
Eisenmohr *m* (Min) black iron oxide, earthy magnetite
Eisenmolybdän *n* ferromolybdenum
Eisenmonosulfid *n* (Eisen(II)-sulfid) ferrous sulfide, iron(II) sulfide, iron monosulfide
Eisenmonoxid *n* ferrous oxide, iron(II) oxide, iron monoxide
Eisenmulm *m* (Min) earthy iron ore
Eisennatriumpyrophosphat *n* iron sodium pyrophosphate
Eisennatrolit *m* (Min) iron natrolite
Eisennickel *n* ferronickel
Eisennickelakkumulator *m* iron nickel accumulator
Eisennickelkies *m* (Min) pentlandite
Eisenniederschlag *m* iron precipitate
Eisenniere *f* kidney ore, eaglestone, hematite
Eisennitrat *n* iron nitrate
Eisennitrid *n* iron nitride
Eisennitrit *n* iron nitrite
Eisennuß *f* blood stone, red hematite
Eisenocker *m* iron ocher
Eisenofen *m* smelting furnace
Eisenolivin *m* (Min) fayalite, iron chrysolite
Eisenoolith *m* (Min) oolitic ironstone
Eisenopal *m* (Min) ferruginous opal
Eisenoxalat *n* iron oxalate, ferric oxalate
Eisenoxidammonsulfat *n* (obs) ammoniacal iron alum, ammonium ferric sulfate
eisenoxidarm poor in ferric oxide
Eisenoxidfarbe *f* iron oxide pigment
Eisenoxidgelb *n* ferrite yellow, iron yellow
eisenoxidhaltig containing ferric oxide
Eisenoxidhydrat *n* ferric hydroxide, iron(III) hydroxide
Eisenoxidkaliumsulfat *n* (obs) ferric potassium sulfate, iron potassium alum
Eisenoxidoxydul *n* ferrosoferric oxide, iron(II)(III) oxide
eisenoxidreich rich in iron oxide
Eisenoxidrot *n* iron oxide red
Eisenoxidschicht *f* iron oxide layer
Eisenoxidschwarz *n* iron oxide black
Eisenoxydul *n* (Eisen(II)-oxid) ferrous oxide, iron(II) oxide, iron protoxide
Eisenoxydulacetat *n* (obs) ferrous acetate, iron(II) acetate
Eisenoxydulammonsulfat *n* (obs) ammonium iron(II) sulfate, ferrous ammonium sulfate
Eisenoxydulcarbonat *n* (obs, Eisen(II)-carbonat) ferrous carbonate, iron(II) carbonate

Eisenoxydulhydrat *n* (obs) ferrous hydroxide, iron(II) hydroxide
Eisenoxyduloxid *n* (Eisen(II)(III)-oxid) ferriferrous oxide, ferrosoferric oxide, iron(II)(III) oxide, magnetic iron
Eisenoxydulsalz *n* (Eisen(II)-Salz) ferrous salt, iron(II) salt
Eisenoxydulsulfat *n* (obs) iron(II) sulfate, ferrous sulfate, green copperas, green vitriol
Eisenoxydulverbindung *f* (Eisen(II)-Verbindung) ferrous compound, iron(II) compound
Eisenoxyfluorid *n* basic iron fluoride; iron oxyfluoride
Eisenpastille *f* iron pastille
Eisenpecherz *n* (Min) limonite, pitticite, triplite
Eisenpentacarbonyl *n* iron pentacarbonyl
Eisenpeptonat *n* iron peptonate
Eisenpeptonlösung *f* solution of iron peptonate
Eisenperidot *m* (Min) fayalite
Eisenphosphat *n* iron phosphate
Eisenphosphid *n* iron phosphide, ferrophosphorus
Eisenphosphor *m* ferrophosphorus, iron phosphide
Eisenphyllit *m* (Min) vivianite
Eisenplatte *f* iron plate, sole plate
Eisenplattenbekleidung *f* steel plating
Eisenportlandzement *m* iron Portland cement, siderurgical cement, slag Portland cement
Eisenpräparat *n* iron preparation
Eisenprobe *f* iron sample, iron test
Eisenpulver *n* iron powder
Eisenpulverkernspule *f* iron powder coil
Eisenquarz *m* (Min) iron quartz, ferriferous quartz; iron flint
Eisenrahm *m* (Min) limonite, porous variety of hematite
eisenreich rich in iron
Eisenreihe *f* iron group
Eisenresinit *m* (Min) humboldtine
Eisenrhodanid *n* (Eisen(III)-rhodanid) ferric thiocyanate, iron(III) thiocyanate
Eisenrhodanür *n* (Eisen(II)-rhodanid) ferrous thiocyanate, iron(II) thiocyanate
Eisenrogenstein *m* (Min) oolitic iron stone
Eisenrohr *n* iron tube, hollow iron
Eisenrost *m* iron rust
Eisenrostfarbe *f* ferruginous color
Eisenrostwasser *n* iron liquor, iron mordant
Eisenrostzement *m* iron rust cement
Eisenrot *n* colcothar
Eisensaccharat *n* iron saccharate
Eisensäge *f* metal saw
Eisensäuerling *m* chalybeate water
Eisensäure *f* ferric acid
Eisensafran *m* crocus of iron
Eisensalmiak *m* iron ammonium chloride
Eisensalz *n* iron salt
Eisensand *m* ferruginous sand

Eisensau *f* (Metall) iron sow, bear, iron block, salamander, skull
eisensauer ferrate
eisenschaffend iron-producing
Eisenschaum *m* (Metall) kish, refined iron froth, (Min) porous form of hematite
Eisenschefferit *m* (Min) schefferite
Eisenscheider *m* iron separator, magnetic separator
Eisenschlacke *f* finery cinders, iron slag
Eisenschlamm *m* ferruginous mud
Eisenschlick *m* muddy iron ore
Eisenschmelzhütte *f* iron foundry
Eisenschmiede *f* forge
Eisenschörl *m* (Min) bog iron schorl
Eisenschrot *m* scrap iron
eisenschüssig (Min) ferriferous, ferrous, ferruginous
Eisenschutz *m* protection of iron
Eisenschutzfarbe *f* anti-corrosive paint, iron protecting paint
Eisenschutzlack *m* anti-corrosive paint or varnish
Eisenschwärze *f* black lead powder, currier's ink, earthy magnetite, ground graphite
Eisenschwamm *m* iron sponge [ore], porous iron, spongy iron
Eisenschwarz *n* (Min) graphite, iron black, lampblack
Eisenselenür *n* ferrous selenide, iron(II) selenide
Eisenseparator *m* iron separator
Eisensesquioxid *n* (Eisen(III)-oxid) ferric oxide, iron(III) oxide, iron sesquioxide
Eisensesquisulfid *n* (Eisen(III)-sulfid) ferric sulfide, iron(III) sulfide, iron sesquisulfide
Eisensilberglanz *m* (Min) sternbergite
Eisensilicat *n* iron silicate
Eisensilicid *n* iron silicide
Eisensinter *m* iron dross, iron scale, (Min) pitticite
Eisenspäne *m pl* iron filings, iron turnings
Eisenspat *m* (Min) siderite, spathic iron ore
Eisenspiegel *m* (Min) specular iron, hematite
Eisenspießglanzerz *n* (Min) berthierite
Eisenspinell *m* (Min) black spinel, ceylonite
Eisenstahlbad *n* steel bath
Eisenstein *m* iron ore, iron stone
Eisensteinmark *n* lithomarge containing iron
Eisensteinröstofen *m* kiln for roasting iron ore
Eisensublimat *n* ferric chloride, iron sublimate
Eisensulfat *n* iron sulfate
Eisensulfatanlage *f* iron sulfate plant
Eisensulfid *n* iron sulfide, (Eisen(III)-sulfid) ferric sulfide, iron(III) sulfide
Eisensulfür *n* (Eisen(II)-sulfid) ferrous sulfide, iron(II) sulfide
Eisensumpferz *n* (Min) bog iron ore, brown hematite, lake ore
Eisentabletten *pl* (Pharm) iron lozenges

Eisentartrat *n* iron tartrate
Eisentinktur *f* ferruginous tincture, tincture of iron
Eisentinte *f* iron ink
Eisentitan *n* ferrotitanium, (Min) ilmenite
Eisenton *m* clay ironstone, iron clay
Eisentongranat *m* (Min) almandite
Eisenträger *m* steel stanchion
Eisentrichlorid *n* ferric chloride, iron(III) chloride
Eisentrifluorid *n* ferric fluoride, iron(III) fluoride
Eisentüte *f* iron assaying crucible
Eisenuntersuchung *f* iron analysis, iron examination, iron testing
Eisenuntersuchungsapparat *m* iron testing apparatus
Eisenvalerianat *n* (Eisen(III)-valerianat) ferric valerate
Eisenvanadium *n* ferrovanadium
eisenverarbeitend metal-working
Eisenverbindung *f* iron compound
Eisenverhüttung *f* (Metall) iron smelting
eisenverkleidet iron-clad
Eisenverlust *m* iron loss
Eisenvitriol *n* iron vitriol, copperas, green vitriol, hydrous ferrous sulfate
Eisenvitriollösung *f* iron vitriol solution
Eisenwalzwerk *n* iron rolling mill
Eisenwaren *pl* hardware, ironware
Eisenwarenhandlung *f* hardware store
Eisenwasser *n* chalybeate water, iron water
Eisenweinstein *m* ferrotartrate
eisenweinsteinsauer ferrotartrate
Eisenwerk *n* iron structure
Eisenwolfram *n* ferrotungsten
Eisenzement *m* iron cement
Eisenzeug *n* iron ware
Eisenzinkblende *f* (Min) marmatite
Eisenzinkspat *m* (Min) ferruginous calamine, monkheimite
Eisenzinnerz *n* (Min) ferriferous cassiterite
Eisenzucker *m* ferric saccharate
eisern iron
Eiserzeugung *f* ice production
Eiserzeugungsapparat *m* ice-making machine, refrigerator
Eisessig *m* glacial acetic acid
eisgekühlt ice-cooled
Eisglas *n* frosted glass
eisig icy, freezing, glacial
Eiskalorimeter *n* ice calorimeter
eiskalt ice-cold
Eiskristall *m* ice crystal
Eiskühlung *f* cooling with ice
Eismaschine *f* ice making machine, refrigerator
Eispunkt *m* freezing point
Eisschrank *m* refrigerator

Eisspat *m* (Min) crystalline feldspar, glassy feldspar, sanidine
Eistrichter *m* ice funnel
Eiszapfen *m* icicle
Eiszone *f* ice zone
Eiszonenschmelzen *n* ice zone melting
Eiter *m* (Med) pus
Eiterzelle *f* pyocyte
Eiweiß *n* (das Weiße des Eies) egg white, (Protein) protein, **körperfremdes** ~ foreign protein
Eiweißabbau *m* protein degradation, proteolysis
Eiweißabbauprodukt *n* protein degradation product
eiweißarm poor in protein
Eiweißart *f* variety of protein
eiweißartig protein-like, albumiform, albuminoid, albuminous
Eiweißausscheidung *f* (im Urin) proteinuria
Eiweißbedarf *m* protein requirement
Eiweißdiät *f* protein diet
Eiweißfaser *f* (Text) protein fiber
Eiweißgehalt *m* protein content
Eiweißgerinnung *f* protein coagulation
eiweißhaltig containing protein
Eiweißharnen *n* (Med) albuminuria
Eiweißhaushalt *m* protein metabolism
Eiweißkörper *m* protein
Eiweißleim *m* albumin glue, gluten protein
Eiweißmangel *m* protein deficiency
Eiweiß-Mineralstoffwechsel *m* protein-mineral metabolism
Eiweißminimum *n* protein minimum
Eiweißmolekül *n* protein molecule
Eiweißpapier *n* albuminized paper, albumin paper
eiweißreich rich in protein
eiweißspaltend proteolytic
Eiweißspaltprodukt *n* product of proteolysis
Eiweißspaltung *f* protein degradation, proteolysis
Eiweißspiegel *m* protein level
Eiweißstickstoff *m* protein nitrogen
Eiweißstoffwechsel *m* (Biochem) protein metabolism
Eiweißtrübung *f* protein turbidity
Eiweißverbindung *f* protein
Eizelle *f* egg cell, ovocyte, ovum, **befruchtete** ~ fertilised ovum, zygocyte, zygote
Ejektor *m* ejector, jet pump
Ekaaluminium *n* eka-aluminum, gallium
Ekabor *n* eka-boron, scandium
Ekasilicium *n* eka-silicon, germanium
Ekbolin *n* ecboline
Ekdemit *m* (Min) ecdemite
ekelerregend disgusting, nauseating
EK-Filtration *f* (Entkeimungsfiltration) sterile filtration

EKG *n* (Elektrokardiogramm, Med) electrocardiogram, ECG
Ekgonidin *n* ecgonidine
Ekgonin *n* ecgonine
Ekgoninol *n* ecgoninol
Ekgoninsäure *f* ecgoninic acid
Ekkain *n* eccaine
ekliptisch (Konformation) eclipsed
Eklogit *m* (Min) eclogite
Eklogitschale *f* eclogite shell
Ekrasit *n* ecrasite
Ekrüseide *f* ecru silk, raw silk
Eksantalal *n* ecsantalal
Eksantalol *n* ecsantalol
Eksantalsäure *f* ecsantalic acid
Ektogan *n* ektogan
Ektoparasit *m* (Zool) ectoparasite
Ektophyt *m* (pflanzlicher Hautparasit) ectophyte
Ektoplasma *n* ectoplasm, exoplasm, plasma membrane
ektoplasmatisch ectoplasmatic, ectoplasmic
Ektosit *m* (Darmschmarotzer) ectoparasite, ectosite
Ektozoon *n* (tierischer Hautparasit) ectozoon
Ektylcarbamid *n* ectylcarbamide
Ekzem *n* (Med) eczema
Elaeagnin *n* elaeagnine
Eläolith *m* (Min) elaeolite, nephelite
Eläomargarinsäure *f* elaeo-margaric acid
Eläopten *n* elaeoptene
Eläostearin *n* glycerol eleostearate
Eläostearinsäure *f* eleostearic acid
Elaïdin *n* elaidin, glycerol elaidate
Elaïdinisierung *f* elaidinization
Elaïdinsäure *f* elaidic acid, trans-9-octadecenoic acid
elaïdinsauer elaidate
Elaïdinseife *f* elaidic soap
Elaidon *n* elaidone
Elaidylalkohol *m* elaidyl alcohol
Elaïn *n* (obs) elain, ethylene
Elainschwefelsäure *f* ethylsulfuric acid
Elaiomycin *n* elaiomycin
Elarson *n* elarsone
Elarsonsäure *f* elarsonic acid
Elastase *f* (Biochem) elastase
Elastifizierungsmittel *n* elasticator
Elastikator *m* elasticator
Elastikbereifung *f* cushion tire equipment
Elastin *n* elastin
elastisch elastic, flexible
Elastizität *f* elasticity, flexibility, **adiabatische** ~ (Therm) adiabatic elasticity, **unvollkommene** ~ imperfect elasticity, **vollkommene** ~ perfect elasticity
Elastizitätsbeiwert *m* coefficient of elasticity
Elastizitätsgesetz *n* law of elasticity, Hooke's law

Elastizitätsgrad *m* degree of elasticity, elastic ratio
Elastizitätsgrenze *f* elastic limit, limit of elasticity
Elastizitätsgrundgesetz *n* fundamental law of elasticity
Elastizitätskoeffizient *m* coefficient of elasticity
Elastizitätsmesser *m* elastometer
Elastizitätsmodul *m* modulus of elasticity, ~ **berechnet aus dem Biegeversuch** Young's modulus in flexure
Elastizitätstheorie *f* theory of elasticity
Elastkuppelung *f* flexible coupling
Elastodynamik *f* dynamic elasticity
elastomer elastomeric
Elastomere *n* elastomer
Elastomere *pl*, **synthetische** ~ synthetic elastomers
Elateridin *n* elateridine
Elateridochinon *n* elateridoquinone
Elaterin *n* elaterin
Elateringlykosid *n* elaterin glycoside
Elaterit *m* elaterite, elastic bitumen, elaterite, mineral caoutchouc
Elaterium *n* elaterium
Elateron *n* elaterone
Elatinolsäure *f* elatinolic acid
Elatinsäure *f* elatinic acid
Elatsäure *f* elatic acid
Elaylchlorid *n* (obs) 1,2-dichlorethane, ethylene [di]chloride
Elaylgas *n* (obs) elayl, ethylene
Elbon *n* elbon
Elbs-Reaktion *f* Elbs reaction
Elchfett *n* elk fat
Electuarium *n* (Pharm) electuary
Eledonin *n* eledonine
Elefantenhautbildung *f* (Gummi) crazing
Elektriker *m* electrician
elektrisch electric[al], ~ **beheizt** electrically heated, ~ **negativ** electronegative, ~ **pneumatisch** electric pneumatic, ~ **positiv** electropositive
elektrisierbar electrifiable
Elektrisierbarkeit *f* electric susceptibility, electrifiableness
elektrisieren to electrify
Elektrisieren *n* electrifying
Elektrisiermaschine *f* electrostatic generator
Elektrisierung *f* electrification
Elektrizität *f* electricity, **dynamische** ~ dynamic electricity, **galvanische** ~ galvanic electricity, voltaic electricity, **gleichnamige** ~ electricity of same sign, **hochgespannte** ~ hightened electricity, **ungleichnamige** ~ electricity of opposite sign
Elektrizitätsansammlung *f* storage of electrical energy

Elektrizitätsanzeiger *m* electroscope, electrification detector
Elektrizitätsauffüller *m* recharger
Elektrizitätsentlader *m* discharger
Elektrizitätsentladung *f* electrostatic discharge
elektrizitätserregend electrogeneous
Elektrizitätserregung *f* electrification, high-frequency oscillation
Elektrizitätserzeugung *f* generation of electricity
Elektrizitätsmenge *f* quantity of electricity
Elektrizitätsquelle *f* source of electricity
Elektrizitätsrechnung *f* electric bill
Elektrizitätsspeicher *m* storage battery
Elektrizitätsverlust *m* leakage of electricity
Elektrizitätsvermehrer *m* electric multiplier
Elektrizitätsverteilung *f* distribution of electricity
Elektrizitätswerk *n* power station
Elektrizitätszähler *m* electricity meter
Elektrizitätszerstreuung *f* dissipation of electricity
Elektroabscheider *m* electrostatic precipitator
Elektroabscheidung *f* electroprecipitation
Elektroaffinität *f* electroaffinity
Elektroanalyse *f* electroanalysis, electrolytic analysis
Elektroantrieb *m* electric drive
Elektrobiologie *f* electrobiology
Elektrochemie *f* electrochemistry
Elektrochemiker *m* electrochemist
elektrochemisch electrochemical
Elektrode *f* electrode, ~ **erster Art** electrode of the first type, primary electrode, ~ **zweiter Art** electrode of the second type, secondary electrode, **Bezugs- od. Vergleichs-** ~ reference electrode, **Doppelschicht-** ~ double skeleton electrode, **fluidisierte** ~ fluidized electrode, **Gegenstrom-** ~ countercurrent electrode, **Gittervibrator-** ~ vibrating grid electrode, **Janus-** ~ Janus electrode, **Kalomel-** ~ calomel electrode, **negative** ~ cathode, negative electrode, **polarisierbare** ~ polarizable electrode, **positive** ~ anode, positive electrode, **Quecksilber-** ~ mercury electrode, **Redox-** ~ redox electrode, **Riesel[film]-** ~ dropping electrode, **strombelastete** ~ working electrode, **ummantelte** ~ coated or sheathed electrode, covered electrode, **unpolarisierbare** ~ unpolarizable electrode, **Vibrator-** ~ vibrator electrode, **Wasserstoff-** ~ hydrogen electrode
Elektrodekantation *f* electrodecantation
Elektrodenabstand *m* electrode gap, spark gap
Elektrodenanordnung *f* arrangement of electrodes
Elektrodenbelastung *f* loading of the electrodes
Elektrodendurchführung *f* electrode passage
Elektrodenhalter *m* electrode holder

Elektrodenkasten *m* electrode box
Elektrodenkette *f* electrode chain
Elektrodenkinetik *f* electrode kinetics
Elektrodenkohle *f* electrode carbon
Elektrodenkontakt *m* electrode contact
Elektrodenmantel *m* electrode jacket
Elektrodenoberfläche *f* electrode surface
Elektrodenpotential *n* electrode potential
Elektrodenpresse *f* electrode press
Elektrodenreaktion *f* electrode reaction
Elektrodenregelung *f* adjustment of the electrodes
Elektrodenröhre *f* electrode tube
Elektrodenschutzglocke *f* electrode shield
Elektrodenschweißung *f* electrowelding
Elektrodenspannung *f* electrode potential
Elektrodenstromdichte *f* current density at the electrode
Elektrodensystem *n* system of electrodes
Elektrodenwalze *f* electrode roller
Elektrodenwechsel *m* changing the electrodes
Elektrodermatogramm *n* (Med) electrodermogram
Elektrodialyse *f* electrodialysis
elektrodisch electrodic
Elektrodispersion *f* electrodispersion
Elektrodynamik *f* electrodynamics
elektrodynamisch electrodynamic[al]
Elektroeisen *n* electrolytic iron
Elektroendosmose *f* electroendosmosis, electrokinesis
Elektroenergie *f* electrical energy
Elektroenzephalogramm *n* (Med) electroenzephalogram
Elektroerosion *f* electroerosion
Elektrofahrwerk *n* electric traversing gear
Elektrofahrzeug *n* electric vehicle
Elektrofaßpumpe *f* electric barrel pump
Elektrofilter *n* electric separator, electrostatic filter or precipitator
elektrogalvanisch electrogalvanic
Elektrogastrogramm *n* (Med) electrogastrogram, EGG
Elektrogerät *n* electric appliance
Elektrographit *m* electrographite, synthetic graphite
Elektrograviermaschine *f* electric engraving machine
Elektrograviervorrichtung *f* electrograving apparatus
Elektrogravimetrie *f* electrogravimetry
Elektrohochofen *m* electric shaft furnace, electric smelting furnace
elektroinduktiv electro-inductive
Elektroindustrie *f* electrical industry
Elektroingenieur *m* electrical engineer
Elektroisolierlack *m* electrical insulating lacquer
Elektroisoliermaterial *n* electrical insulating material

Elektrokardiogramm *n* (EKG, Med) electrocardiogram, ECG
Elektrokardiograph *m* electrocardiograph
Elektrokardiographie *f* electrocardiography
Elektrokatalyse *f* electrocatalysis
Elektrokinetik *f* electrokinetics
elektrokinetisch electrokinetic
Elektrokopierverfahren *n* electrocopying process
Elektrokorund *m* electrocorundum
Elektrolichtbogenofen *m* electric arc furnace
Elektrolumineszenz *f* electroluminescence
Elektrolyse *f* electrolysis, ~ **im Schmelzfluß** electrolysis of fused salts, ~ **ohne Strom** internal electrolysis
Elektrolysenbad *n* electrolytic bath or cell
Elektrolysendauer *f* duration of electrolysis
Elektrolysenschlamm *m* electrolytic slime
Elektrolyseprozeß *m* electrolytic refining process
Elektrolysierbecher *m* electrolysis beaker
elektrolysieren to electrolyze
Elektrolysiervorrichtung *f* electrolyzing apparatus
Elektrolyt *m* electrolyte, **dreiioniger** ~ triionic electrolyte, **einwertiger** ~ univalent electrolyte, **geschmolzener** ~ fused electrolyte, molten electrolyte, **kolloidaler** ~ (Elektr) colloidal electrolyte, **schwacher** ~ weak electrolyte, **starker** ~ strong electrolyte
Elektrolytbehälter *m* electrolyte tank
Elektrolytbeize *f* electrolytic pickling
Elektrolytblech *n* electrolytic tinplate
Elektrolytbleiblech *n* electrolytic lead sheet
Elektrolytbrücke *f* electrolyte bridge
Elektrolyteisen *n* (Elektrochem) electrolytic iron
Elektrolytgleichrichter *m* electrolytic rectifier
Elektrolytgold *n* electrolytic gold
Elektrolythaushalt *m* electrolyte metabolism
elektrolytisch electrolytic, ~ **aufgebracht** electrolytically deposited
Elektrolytkoagulation *f* electrolytic coagulation
Elektrolytkupfer *n* (Elektrochem) electrolytic copper
Elektrolytkupferblech *n* electrolytic sheet copper
Elektrolytlösung *f* solution of electrolytes
Elektrolytsilber *n* electrolytic silver
Elektrolytunterbrecher *m* electrolytic interrupter
Elektrolytvolumen *n* electrolyte volume
Elektrolytwiderstand *m* electrolyte resistance
Elektrolytzelle *f* electrolytic photocell, photoelectrolytic cell
Elektrolytzink *n* electrolytic zinc
Elektrolytzusatz *m* addition of electrolyte[s]
Elektromagnet *m* electromagnet
elektromagnetisch electromagnetic
Elektromagnetismus *m* electromagnetism
Elektromagnetofen *m* electromagnetic furnace
Elektromechanik *f* electromechanics
Elektromechaniker *m* electrician, electromechanic

elektromechanisch electromechanical
Elektromerie f electromerism
Elektrometallurgie f electrometallurgy
Elektrometer n electrometer
elektrometrisch electrometric
Elektromigrationsuntersuchung f electromigration studies
Elektromobil n electric vehicle, electromobile
Elektromotor m electromotor
elektromotorisch electromotive
Elektron n electron, (Metall) electron, **ausgestoßenes** ~ disintegration electron, **kernfernes** ~ conduction electron, orbital electron, outer level electron, peripheral electron, planetary electron, valence electron, **positives** ~ positive electron, positron, **schweres** ~ heavy electron, barytron, dynatron, mesotron, penetron
elektronegativ electronegative
Elektronegativität f electronegativity
Elektronen pl, **ungepaarte** ~ unpaired electrons
Elektronenabbildung f electron image
Elektronenabgabe f emission of electrons, **glühelektrische** ~ thermionic emission of electrons
Elektronenabgabefläche f electron emission area
Elektronenablösung f detachment of electrons
Elektronenabsorption f electron absorption
Elektronenaffinität f electron affinity
Elektronenakzeptor m electron acceptor
Elektronenanlagerungs-Massenspektrographie f electron-attachment mass spectrography
Elektronenanordnung f electron arrangement, electronic configuration
Elektronenanregung f excitation of electrons
Elektronenauffang m electron capture
Elektronenaufnahme f electron acceptance
elektronenaussendend electron-emitting
Elektronenausstrahlung f emission of electrons
Elektronenaustritt m releasing of electrons
Elektronenaustrittsarbeit f work function of electrons
Elektronenbahn f electron path, orbit
Elektronenbandenspektrum n (Spektr) electronic band spectrum
Elektronenbeschleuniger m electron accelerator, (Betatron) betatron, (Rheotron) rheotron
Elektronenbeschuß m electron bombardment
Elektronenbeugung f electron diffraction
Elektronenbeugungsgeräte pl electron diffraction equipment
Elektronenbildröhre f electron-image tube
Elektronenblitzgerät n electronic flash
Elektronenbremsung f deceleration of electrons, retardation of electrons
Elektronenbündel n cathode beam, electron beam
Elektronenbündelung f electron focus[s]ing

Elektronendefektleitung f electron defect conduction
Elektronendichte f electron density
Elektronendichteverteilung f distribution of electron density
Elektronendonator m electron donor
Elektronendrall m electron spin
Elektronendruck m electron pressure
Elektronenduett n electron pair
elektronendurchlässig electron-transmitting
Elektroneneigendrehimpuls m electron spin
Elektroneneinfang m electron capture
Elektroneneinfang-Detektor m (Gaschromat) electron capture detector
Elektronenemission f electron emission, **glühelektrische** ~ electron evaporation, temperature emission of electrons, thermionic emission, **lichtelektrische** ~ photoelectric emission, Hallwachs effect, photoemissive or external photoelectric effect
Elektronenenergie f electron energy
Elektronenentladung f electron discharge
Elektronenfänger m electron trap
Elektronenfluß m drift of electrons, flow of electrons
Elektronenformel f electronic formula
Elektronenfreigabe f releasing of electrons
Elektronengas n electron gas, **entartetes** ~ degenerate electron gas
Elektronengehirn n electronic brain
Elektronengenerator m electron generator
Elektronenhülle f electron shell
Elektroneninjektor m (Atom) electron injector
Elektronenionisierung f ionization by electrons
Elektronenisomerie f electronic isomerism
Elektronenkanone f electron gun
Elektronenkonfiguration f electronic configuration
Elektronenladedichte f electron charge density
Elektronenladung f electronic charge
Elektronenlaufzeit f electron transit time
Elektronenlawine f electron avalanche
Elektronenleerstelle f electron deficit position
Elektronenlehre f electronics
Elektronenleitung f electronic conductivity
Elektronenlinse f electronic lens
Elektronenlücke f electron gap, electron vacancy
Elektronenmangel m electron deficiency
Elektronenmangelbindung f (Chem) electron deficient binding
Elektronenmangelleitung f electron-defect conductivity
Elektronenmangelverbindung f electron deficient compound
Elektronenmikroskop n electron microscope
Elektronenmikroskopie f electron microscopy
elektronenmikroskopisch electron microscopic
Elektronenniveau n electron level
Elektronenoptik f electron optics

elektronenoptisch electron optical
Elektronenordnung *f* electron arrangement, electron system
Elektronenpaar *n* electron pair, **einsames** ~ lone-pair electrons
Elektronenpolarisierbarkeit *f* electronic polarizability
Elektronenquelle *f* electron source
Elektronenreibung *f* electron friction
Elektronenrelais *n* electron relay
Elektronenresonanz *f* electron resonance
Elektronenröhre *f* thermionic tube
Elektronenschar *f* electron swarm
Elektronenschleuder *f* betatron
Elektronenschwingung *f* electron oscillation
Elektronensextett *n* electron sextette
Elektronenspektroskopie *f* electron spectroscopy
Elektronenspin *m* electron spin, electron angular momentum
Elektronenspinquantenzahl *f* electron spin quantum number
Elektronenspinresonanz *f* electron spin resonance
Elektronensprung *m* electron jump
Elektronensprungspektrum *n* electron jump spectrum
Elektronenstauung *f* electron accumulation
Elektronensteuerung *f* electronic control
Elektronenstoß *m* electron collision, electron impact
Elektronenstrahl *m* cathode beam, electron beam or ray, electron stream
Elektronenstrahl-Mikroanalyse *f* electron probe microanalysis
Elektronenstrahloszillograph *m* cathode-ray oscillograph
Elektronenstrahlschweißverfahren *n* electron beam welding
Elektronenstrahlung *f* electron emission
Elektronenstreuung *f* electron scattering
Elektronensynchrotron *n* electron synchrotron
Elektronentechnik *f* electronics
Elektronenteilchendichte *f* electron particle density
Elektronenterm *m* electron term
Elektronentheorie *f* theory of electrons
Elektronentransport *m* electron transport
Elektronenüberführung *f* electron transfer
Elektronenübergang *m* electron transition, electron promotion
Elektronenübergangswahrscheinlichkeit *f* electron transition probability
elektronenüberschußleitend electron excess semiconducting
Elektronenüberschußleitung *f* electron excess conduction
Elektronenüberträger *m* electron carrier
Elektronenübertragung *f* electron transfer

elektronenunterschußleitend electron defect semiconducting
Elektronenverschiebung *f* electron displacement
Elektronenverstärkerröhre *f* electron multiplier valve
Elektronenverteilung *f* electron distribution
Elektronenvervielfältiger *m* electron multiplier
Elektronenvolt *n* electron volt
Elektronenwolke *f* cloud of electrons, electron cloud
Elektronenzerfall *m* electron decay
Elektronenzertrümmerung *f* electrodisintegration
Elektronenzustand *m* electron state, **entarteter** ~ degenerate electron state
elektroneutral electrically neutral
Elektroneutralitätsprinzip *n* electroneutrality principle
Elektroniederschlag *m* electrodeposit
Elektronik *f* electronics
Elektron-Ion-Rekombination *f* electron ion recombination
Elektronionwandrekombination *f* electron ion wall recombination
elektronisch electronic
Elektron-Loch-Paar *n* electron-hole pair
Elektron-Neutron-Wechselwirkung *f* electron neutron interaction
Elektron-Positron-Paar *n* electron-positron pair
Elektroofen *m* electric furnace
Elektroosmose *f* (Biochem) electric osmosis, electroosmosis
Elektropathie *f* electropathy, electrotherapeutics
Elektropherogramm *n* electropherogram
Elektropherographie *f* electropherography
elektrophil electrophilic
Elektrophon *n* electrophone
Elektrophordeckel *m* electrophorus disc
Elektrophorese *f* electrophoresis, **trägerfreie** ~ carrierless electrophoresis
Elektrophoreseapparatur *f* electrophoresis equipment
Elektrophorkuchen *m* electrophorus disc
Elektrophorteller *m* electrophorus bottom plate
Elektrophotographie *f* electrophotography
Elektrophysik *f* electrophysics
Elektrophysiologie *f* electrophysiology
elektrophysiologisch electrophysiological
elektroplattieren to electroplate
Elektroplattierung *f* electroplating
Elektroplattierverfahren *n* electroplating process
elektropneumatisch electropneumatic
elektropositiv electropositive
Elektroraffination *f* electro-refining
Elektroregler *m* electro-regulator
Elektroretinogramm *n* (Med) electroretinogram
Elektroschachtofen *m* electric shaft furnace
Elektro-Schlacke-Umschmelzverfahren *n* electroslag remelting

Elektroschnellförderer *m* electric highspeed conveying plant
Elektroschweißung *f* electric welding
Elektroskop *n* electroscope
Elektroskopie *f* electroscopy
elektroskopisch electroscopic
Elektrostahl *m* electric [furnace] steel
Elektrostahlerzeugung *f* electric steel manufacture
Elektrostahlformguß *m* electric steel casting
Elektrostahlofen *m* electric melting furnace
Elektrostapler *m* electric piler, electric stacker
Elektrostatik *f* electrostatics
elektrostatisch electrostatic
Elektrostriktion *f* (Elektr) electrostriction
Elektrotauchlack *m* electrodeposition paint
Elektrotauchlackierung *f* electrophoretic painting
Elektrotechnik *f* electrical engineering, electrotechnics
Elektrotechniker *m* electrical engineer
Elektrothermie *f* electrothermics
elektrothermisch electrothermal, electrothermic
Elektrothermostat *m* electrothermostat
Elektrotiegelofen *m* electric crucible furnace
Elektrotype *f* electrotype
Elektrovalenz *f* electrovalence, electrovalency
Elektrowärme *f* electrical heat
Elektrozähler *m* electricity meter
Elektrum *n* (Metall) electrum
Elemadienolsäure *f* elemadienolic acid
Eleman *n* elemane
Elemen *n* elemene
Element *n* (Chem) element, (Elektr) battery, (Galv) galvanic cell, galvanic pile, **elektronegatives** ~ electronegative element, **elektropositives** ~ electropositive element, **galvanisches** ~ galvanic cell, **gasförmiges** ~ gaseous element, **künstliches** ~ artificial element, **künstlich radioaktives** ~ artificially radioactive element, **linienreiches** ~ (Spektr) element with a large number of spectral lines, **radioaktives** ~ radioactive element, **thermoelektrisches** ~ thermocouple
elementar (grundlegend) elementary, (naturgewaltig) elemental
Elementaranalyse *f* elementary analysis
Elementarbaustein *m* elementary building block
Elementardipol *m* elementary dipole
Elementarereignis *n* (Statist) elementary event
Elementarfaden *m* monofilament
Elementarladung *f* elementary charge
Elementarmagnet *m* molecular magnet
Elementarphysik *f* elementary physics
Elementarprozeß *m* elementary process
Elementarquantum *n* elementary quantum
Elementarreaktion *f* elementary reaction, **photochemische** ~ elementary photochemical reaction

Elementarschwefel *m* elementary sulfur
Elementarteilchen *n* fundamental particle
Elementarvorgang *m* basic process, elementary process
Elementarzelle *f* unit cell, elementary cell, (Biol) embryonal cell
Elementumwandlung *f* transformation of elements
Elemi *n* elemi
Elemicin *n* elemicin
Elemiharz *n* gum elemi
Elemiöl *n* elemi oil
Elemisäure *f* elemic acid
Elemol *n* elemol
Elemolsäure *f* elemolic acid
Elemonsäure *f* elemonic acid
Eleolith *m* (Min) eleolite
Eleonorit *m* (Min) eleonorite
Eleutherin *n* eleutherin
Eleutherinol *n* eleutherinol
Eleutherolsäure *f* eleutherolic acid
Elevationswinkel *m* angle of elevation
Elevator *m* elevator, (Hebevorrichtung) hoist
Elevatoreimer *m* elevator bucket
Elfenbein *n* ivory, **gebranntes** ~ ivory black
elfenbeinartig eburnean
elfenbeinern ivory, eburnean
elfenbeinfarbig ivory-colored
Elfenbeingelb *n* ivory yellow
Elfenbeinnuß *f* ivory nut, vegetable ivory
Elfenbeinpapier *n* ivory paper
Elfenbeinporzellan *n* ivory porcelain
Elfenbeinschwarz *n* ivory black, bone black, ebony black
Elfenbeinsurrogat *n* ivoride
elfenbeinweiß ivory white
elfflächig hendecahedral
Elfflächner *m* hendecahedron
Elgadienol *n* elgadienol
Elgenol *n* elgenol
Eliasit *m* (Min) eliasite
Elimination *f* (Math) elimination
eliminieren to eliminate
Eliminierung *f* (Chem) elimination, **bimolekulare** ~ bimolecular elimination, **intramolekulare** ~ intramolecular elimination
Elixier *n* elixir
Ellagengerbsäure *f* ellagic acid
Ellagorubin *n* ellagorubin
Ellagsäure *f* ellagic acid
ellagsauer ellagate
Elle *f* (Maß) ell
Ellipse *f* ellipse
Ellipsenbahn *f* elliptical orbit or path
Ellipsenzirkel *m* elliptic compass, trammel
Ellipsoid *n* ellipsoid
Ellipsoideigenschaft *f* ellipsoidal property
Ellipticin *n* ellipticine
elliptisch elliptic

Elliptizität *f* ellipticity
Ellipton *n* elliptone
ELO *n* Abk. für epoxidiertes Leinöl
Elon *n* elon
Eloxalverfahren *n* anodizing, (Elektrochem) electrolytic oxidation process, eloxal process
Eloxieranlage *f* anodizing plant
eloxieren to anodize, to oxidize electrolytically
Eloxieren *n* anodic treatment
Eloxierung *f* anodic treatment, anodizing, electrolytic oxidation
Elozyöl *n* elozy oil
Elpidit *m* (Min) elpidite
Elsholtzion *n* elsholtzione
Eluat *n* (Chromatogr) eluate
eluieren to elute
Elumeter *n* elumeter
Elution *f* elution, extract
Elymoclavin *n* elymoclavine
Elytal-Verfahren *n* elytal process
Email *n* enamel
Emailfarbe *f* enamel color
Emaillack *m* enamel varnish
Emaille *f* enamel, siehe auch Email-
emaillieren to enamel
Emaillieren *n* enamelling
Emaillierofen *m* enamelling furnace
Emailliersoda *f* enamelling soda
emailliert enamelled
Emaillierung *f* enamelling, enamel work
Emailmischmaschine *f* enamel mixing machine
Emailmühle *f* enamel mill
Emailschicht *f* coat of enamel
Emailschmelzfarbe *f* enamel vitrifying color
Emanation *f* emanation
Emanium *n* (obs) emanium, actinium
Emanometer *n* emanometer
Emanon *n* (obs) emanon, radon
Emballage *f* bale, box, packing [material]
emballieren to bale, to box, to pack
Embarin *n* embarine
Embeliasäure *f* embel[l]ic acid
Embelin *n* embelin
Embolie *f* (Med) embolism
Embolit *m* (Min) embolite
Embonsäure *f* embonic acid
Embrithit *m* (Min) embrithite
Embryo *m* embryo
Embryologie *f* embryology
embryonal embryonic
Emeraldin *n* emeraldine
Emerylith *m* (Min) emerylite
Emetäthylin *n* emetethyline
Emetallylin *n* emetallyline
Emetamin *n* emetamine
Emetikum *n* (Pharm) emetic
Emetin *n* emetine
Emetinhydrochlorid *n* emetine hydrochloride
Emetiniumhydroxid *n* emetinium hydroxide

Emetinmethin *n* emetinemethine
emetisch emetic
Emetolin *n* emetoline
Emetpropylin *n* emetpropyline
Emission *f* emission
Emissionsbande *f* emission band
Emissionsfähigkeit *f* emission capability, emissive power, emissivity
Emissionsgeschwindigkeit *f* velocity of emission
Emissionskoeffizient *m* emission coefficient, emissivity
Emissionskontinuum *n* continuous emission spectrum
Emissionskristallspektrum *n* emission spectrum of crystals
Emissionsspektrum *n* emission spectrum
Emissionssteuerung *f* emission control
Emissionsstrom *m* emission current, cathode current, electronic current
Emissionsvermögen *n* emissive power, emissivity, emittance, **photoelektrisches** ~ photoelectric emissivity, photoelectric yield
emittieren to emit
EMK (elektromotorische Kraft) electromotive force, EMF
Emmonit *m* (Min) emmonite
Emodin *n* emodin
Emodinanthranol *n* emodinanthranol
Emodinol *n* emodinol
Emodinsäure *f* emodic acid
E-Modul *m* (Phys) modulus of elasticity
Emolliens *n* emollient
Empfänger *m* receiver, receiving set
Empfängerkammer *f* radiation receiver
Empfängerstromkreis *m* receiving circuit
Empfangsantenne *f* receiving aerial
Empfangsgerät *n* receiver
Empfehlung *f* reference
empfindlich sensitive, ~ **machen** to sensitize
Empfindlichkeit *f* sensitivity, ~ **der Waage** balance sensitivity, ~ **des photografischen Films** photographic film speed, **monochromatische** ~ monochromatic sensitivity, **spektrale** ~ color sensitivity or response, spectrophotoelectric sensitivity
Empfindlichkeitsbereich *m* range of sensitivity
Empfindlichkeitsgrad *m* degree of sensitivity
Empfindlichkeitsgrenze *f* sensitivity limit
Empfindlichkeitsmaß *n* sensitivity measure
Empfindlichkeitsmesser *m* (Opt) sensitometer
Empfindlichkeitsmessung *f* (Opt) sensitometry
Empfindlichkeitsregler *m* (Gaschromat) attenuator
Empfindlichkeitsstufe *f* degree of sensitivity
Empfindlichkeitszentrum *n* (Phot) sensitivity center
Empfindlichkeitsziffer *f* sensitivity factor
Empfindung *f* sensation, (Akust) perception
Empirie *f* empirical knowledge

empirisch empiric, empirical, ~ **bestimmen** to determine empirically
Emplektit *m* (Min) emplectite
Empressit *m* (Min) empressite
Empyroform *n* empyroform
Empyroformpaste *f* empyroform paste
Emscherbrunnen *m* Imhoff tank
Emulgator *m* emulsifier
Emulgens *n* emulgent
emulgieren to emulsify
Emulgieren *n* emulsification
Emulgierfähigkeit *f* emulsifiability
Emulgiergeräte *n pl* emulsification equipment
Emulgiermaschine *f* emulsifier, emulsifying machine, homogenizer
Emulgiermittel *n* emulsifying agent
Emulgierung *f* emulsification
Emulgierungsmittel *n* emulsifying agent
Emulgierzentrifuge *f* emulsifying centrifuge
Emulsin *n* emulsin, synaptase
Emulsion *f* emulsion, **deuterierte** ~ deuterium-loaded emulsion
Emulsionsbildung *f* emulsification
emulsionsfähig emulsifiable
Emulsionsfarbe *f* emulsion paint
Emulsionsguß *m* (Phot) emulsion coating
Emulsionskolloid *n* colloidal emulsion
Emulsionsmischpolymerisation *f* emulsion copolymerization
Emulsionsmörser *m* emulsion mortar
Emulsionspolymerisation *f* emulsion polymerization
Emulsionsrührer *m* emulsification stirrer
Emulsionssalbe *f* emulsion ointment
Emulsionsstabilität *f* emulsion stability
Emulsionsverfahren *n* emulsion process
Emulsionszerstörung *f* emulsion breaking
Emulsoid *n* colloidal emulsion
Enamine *pl* enamines
Enantat *n* enantate
enantiomer enantiomer
Enantiomerie *f* (Stereochem) enantiomerism
enantiotrop (Krist) enantiotropic
Enantiotropie *f* (Krist) enantiotropy, enantiotropism
Enargit *m* (Min) enargite
Endabnahme *f* final test
Endbearbeitung *f* finishing process
Endbildleuchtschirm *m* final image luminescent screen
Enddruck *m* ultimate pressure, discharge pressure
Ende *n* end, (Techn) tail
Endeiolith *m* (Min) endeiolite
Endergebnis *n* final result
endergon endergonic
Enderzeugnis *n* final product
Endfläche *f* (Krist) base of a crystal, end face, (Math) end plane

Endform *f* final shape
Endgas *n* tail gas
Endgeschwindigkeit *f* final velocity, terminal velocity
Endglied *n* terminal member
Endgruppe *f* end group, terminal group
Endgruppenbestimmung *f* end group analysis, terminal group analysis
Endhaken *m* suspension hook
Endiole *pl* enediols
Endkern *m* product nucleus
Endkühler *m* final condenser
Endlage *f* end position, final position
Endlauge *f* final liquor, foots, spent solution
Endleitfähigkeit *f* limiting conductivity
Endlichit *m* (Min) endlichite
Endlösung *f* final solution
endlos endless, continuous
Endlosdruck *m* continuous stationary form printing
Endlosgarn *n* filament yarn
Endodesoxyribonuklease *f* (Biochem) endo deoxyribonuclease
Endoenzym *n* (Biochem) endoenzyme
endo-exo-Isomerie *f* endo-exo-isomerism
endogen endogenic, endogenous
Endoiminotriazol *n* endoiminotriazole
endokrin endocrine
Endokrinologe *m* endocrinologist
Endokrinologie *f* endocrinology
endokrinologisch endocrinological
Endomitose *f* (Biol) endomitosis
endomorph endomorphic
Endomorphismus *m* endomorphism
Endonuklease *f* (Biochem) endonuclease
Endopeptidase *f* (Biochem) endopeptidase
Endoperoxid *n* endoperoxide
endoplasmatisch endoplasmic
Endosmose *f* endosmosis
endosmotisch endosmotic
Endosperm *n* (Bot) endosperm
Endotablette *f* endo tablet
Endothelwanderzelle *f* endothelial leucocyte, endothelial phagocyte, endotheliocyte
Endothelzelle *f* endothelial cell
Endothelzellenvermehrung *f* (Path) endotheliosis
endotherm endothermal, consuming heat, endothermic
Endothianin *n* endothianin
Endothiotriazol *n* endothiotriazole
Endotoxin *n* (Bakt) endotoxin
Endoxytriazol *n* endoxytriazole
Endpartialdruck *m* ultimate partial pressure
Endpentode *f* output pentode
Endphase *f* end phase
Endplatte *f* end plate
Endprodukt *n* end product, final product
Endprodukthemmung *f* (Biochem) feed-back inhibition

Endproduktrepression f (Biochem) end-product repression
Endpunkt m end point, final point
Endquerschnitt m final section
Endresultat n final result, net result
Endrin n endrin
Endschlacke f finishing slag
Endstabilität f ultimate strength
endständig terminal
Endstauchung f ultimate crushing stress
Endstellung f end position
Endstück n end piece, tail, (Spitze) tip
Endstufe f output stage
Endtemperatur f final temperature
Endtotaldruck m ultimate total pressure
Endübertrag m carry over
Endumschalter m limit switch
Endvakuum n final vacuum
endvergären to ferment completely
Endvergrößerung f final magnification
Energetik f energetics
Energie f energy, force, power, **chemische** ~ chemical energy, **empithermische** ~ epithermal energy, **freie** ~ free energy;, **innere** ~ internal energy, **kinetische** ~ kinetic energy, **potentielle** ~ potential energy, **verschluckte** ~ absorbed energy, **wahre** ~ intrinsic energy, **zurückgewonnene** ~ regenerated or restored energy
Energieabfall m energy drop
Energieabgabe f energy output
Energieabnahme f energy decrease
Energieabstand m energy spacing
energiearm energy-poor
Energieaufnahme f energy uptake
Energieaufspeicherung f accumulation of energy, energy storage
Energieaufwand m energy consumption, energy required
Energieausbeute f energy yield; power output
Energieausnutzung f utilization of energy
Energieausstrahlung f radiation of energy
Energieband n energy band, **erlaubtes** ~ allowed energy band, **verbotenes** ~ forbidden energy band, **vollbesetztes** ~ filled band
Energiebedarf m demand for energy
Energieberg m energy hill or barrier
Energiebetrag m amount of energy
Energiebilanz f energy balance, power balance
Energieblock m power supply unit
Energiedekrement n, **logarithmisches** ~ logarithmic energy decrement
Energiedichte f density of energy
Energie-Direktumwandlung f direct energy conversion
Energiedissipationsdichte f energy dissipation density
Energiedynamo m energy dynamo
Energieeinheit f energy unit
Energieemission f emission of energy
Energieentwicklung f development of energy
Energieerhaltung f energy conservation or preservation
Energieersparnis f saving in energy
Energiefluß m energy flow
Energieform f form of energy
Energiefreigabe f energy release
Energiefreisetzung f liberation of energy
Energiegewinn m energy gain
Energiegewinnungseinrichtung f power-harnessing equipment
Energiegleichung f energy equation
energiehaltig energy-containing
Energiehaushalt m energy balance
Energieimpulstensor m energy momentum tensor
Energieinhalt m energy content
Energiekreislauf m energy cycle
Energielehre f energetics
Energielieferung f liberation of energy
Energielinie f energy gradient
Energiemassenbeziehung f energy-mass relation
Energiemesser m ergometer, wattmeter
Energiemessung f measuring of energy
Energiemulde f energy depression
Energieniveau n energy level
Energieniveaudiagramm n (Mech) energy level diagram
Energieprinzip n energy principle
Energiequantelung f energy quantization
Energiequant[um] n energy quantum
Energiequelle f source of energy
energiereich energy rich, high-energy
Energiesatz m energy theorem
Energieschwelle f energy barrier, threshold of energy
Energiespeicherung f energy storage
Energiespektrum n energy spectrum
Energiestrom m energy flow
Energiestufe f energy level, energy term
Energieterm m (Atom) energy [state] term
Energieübertragung f energy transfer
Energieumformer m energy converter
Energieumwandler m energy converter
Energieumwandlung f energy conversion, transformation of energy
Energieverbrauch m power consumption
Energieverlust m loss of energy, loss of power
Energieverminderung f degradation of energy
Energieverschwendung f waste of energy
Energieversorgung f power supply, (Chem Ind) utilities
Energieverteilungskurve f curve of energy distribution
Energiewall m energy barrier
Energiezerstreuung f energy dissipation, power dissipation
Energiezufuhr f addition of energy

Energiezustand – Entfärbungskohle

Energiezustand *m* energy state
Energiezuwachs *m* gain of energy
energisch energetic, rigorous
Enesol *n* enesol, mercury salicylarsenite
eng (dicht) close, (schmal) narrow
Engelitin *n* engelitin
Engelwurzel *f* (Bot) angelica root
Engelwurzelöl *n* angelica oil
Enghalsflasche *f* narrow neck bottle, narrow-necked bottle, narrow-neck flask
Engländer *m* (Techn) monkey wrench, universal screw spanner
Englischblau *n* English blue, royal blue
Englischbraun *n* English brown
Englischgelb *n* lead oxychloride, patent yellow, Turner's yellow
Englischgrün *n* Paris green, patent green
Englischrot *n* English red, colcothar, red iron oxide
engmaschig close-meshed, narrow-meshed
Engobe *f* engobe
Engobefarbe *f* engobe color
engobieren to coat with engobe
engspaltig narrow spaced
Enhydros *m* (Min) enhydros
Enkaustik *f* encaustic
enkaustisch encaustic
Enkrinit *m* (Geol) encrinite, stonelily
Enkrinitenkalk *m* (Geol) encrinitic rock
Enlevagedruck *m* discharge printing
Enlevagegelb *n* discharge yellow
Enneakosan *n* enneacosane
Enol *n* enol
Enolase *f* (Biochem) enolase
Enolation *n* enolate ion
Enolform *f* enol form
Enolisierung *f* enolization
Enomorphon *n* enomorphone
Enoylhydratase *f* (Biochem) enoylhydratase
Enstatit *m* (Min) enstatite
entaktivieren to deactivate, to render inactive
Entaktivierung *f* deactivation
entalkoholisieren to dealcoholize, to extract alcohol [from], to free from alcohol
entalkylieren to dealkylate
entarretieren to release, to unlock
Entarsenieranlage *f* dearsenification plant
Entarsenierung *f* dearsenification, removal of arsenic
entarsenizieren to remove arsenic
entarten to degenerate, to deteriorate
Entartung *f* degeneration, degeneracy, (Verkümmerung) atrophy, ~ **des Grundzustandes** degeneracy of ground state, ~ **von Spinzuständen** degeneracy of spin states, **zufällige** ~ accidental degeneracy
Entaschung *f* (Treibstoff) ash removal, ash discharge
Entaschungsanlage *f* ash removal plant

Entasphaltierung *f* deasphaltization
entbasten to boil off, to degum, to scour
Entbasten *n* (Text) degumming, scouring
Entbastungsausbeute *f* (Text) degumming yield
Entbastungsbad *n* (Text) degumming bath
Entbastungsflotte *f* degumming bath
Entbastungsmittel *n* scouring agent
entbehrlich dispensable
Entbenzolierung *f* extraction of benzene
entbinden to disengage, to untie, (Chem) to release, to free, (Med) to deliver, (Wärme, Energie) to liberate
Entbinden *n* discharging, disengaging, releasing
Entbindung *f* discharge, release, (Med) delivery
entbittern to deprive of bitterness, to disembitter
entbittert disembittered
entbleien deled, to extract lead [from], to remove lead
entblößen to uncover, to bare, to deprive, to expose, to strip
entblößt deprived [of], naked
entbromen to debrominate
entbündeln to defocus
entbunden disengaged, evolved, liberated
entchloren to dechlorinate
Entchloren *n* **von Trinkwasser** dechlorination of potable water
Entchlorung *f* dechlorination
entdampfen to clear from vapor
Entdunstungsanlage *f* damp dispersion plant, devaporizing plant
Ente *f* (Metall) crucible
enteignen to dispossess, to expropriate
enteisen to defrost, to deice
Enteisung *f* deferrization, deironing, iron extraction, iron removal
Enteisungsanlage *f* defroster
Entemaillieren *n* de-enamelling, enamel-stripping
Entemaillierungsanlage *f* enamel stripping plant
entemulgieren to disemulsify
entemulsionieren to disemulsify
Enterokinase *f* (Biochem) enterokinase
Enterosan *n* enterosane
Enteroseptyl *n* enteroseptyl
Enterovirus *m* enterovirus
Enterozoon *n* (pl. Enterozoen) enterozoon (pl enterozoa), parasite of the intestine
Entfärbelösung *f* de-inking solution
entfärben to decolor, to bleach, to decolorize
Entfärben *n* decoloring
entfärbend bleaching, decolorizing
Entfärber *m* decolorant, decolorizing agent, de-inking solution
entfärbt discolored
Entfärbung *f* discoloration, bleaching, decolorizing
Entfärbungsfilter *n* decolorizing filter
Entfärbungskohle *f* decolorizing carbon

Entfärbungskraft f bleaching power
Entfärbungsmesser m decolorimeter
Entfärbungsmittel n decolorant, decolorizing agent
Entfärbungspulver n bleaching powder
Entfärbungsverfahren n decolorization process
Entfärbungsvermögen n decolorizing power
Entfärbungszahl f (von Öl) discoloration factor (of oil)
Entfall m discard, scrap, waste
entfallen (Gedächtnis) to escape
entfalten to display, to evolve, to unfold
Entfaltungspunkt m point of development
entfernen to remove, to eliminate
Entferner m remover
entfernt away, distant, remote
Entfernung f elimination, removal, (Abstand) distance, ~ **des Schwefels** desulfurization
Entfernungskreis m distance circle
Entfernungsmesser m distance meter, range finder, telemeter
entfesseln to release, to set free, to unbind
entfetten to degrease, to extract fat, to ungrease
Entfetten n degreasing, fat elimination, scouring
entfettend degreasing, scouring
entfettet degreased, scoured
Entfettung f extraction of grease, removal of fat
Entfettungsanlage f degreasing plant, fat removing plant, grease extracting plant
Entfettungsapparat m degreasing apparatus, grease extracting apparatus
Entfettungsbad n degreasing bath
Entfettungsmittel n degreasing agent
Entfettungsprozeß m degreasing process
Entfettungsvorgang m degreasing process
entflammbar flammable, ignitable, inflammable
Entflammbarkeit f flammability, ignitability, inflammability
entflammen to inflame, to flash, to kindle
Entflammung f inflammation
Entflammungsprobe f flash test
Entflammungspunkt m flash point, ignition temperature
Entflammungstemperatur f flash point
entflocken to deflocculate
entformen to release the mold
Entformung f mold release
Entformungsmittel n release agent, mold medium, mo[u]ld
Entformungsvorrichtung f mold release device
entfritten to decohere
Entfritter m decoherer
Entfrittung f decoherence
Entfroster m de-icer
entfuseln to remove fusel oil [from], to rectify
Entfuseln n removal of fusel oil
entgasen to decontaminate, to degas, to dry distil, to take off the gas

Entgasen n degassing, destructive distillation, distillation, outgassing
Entgaser m gas expeller
Entgasung f degasification, degassing, destructive distillation, outgassing, ~ **des Ofens** degassing of the furnace
Entgasungsanlage f degassing plant
Entgasungserzeugnis n distillation product
Entgasungsmittel n degasification agent
Entgasungsofen m degassing furnace
Entgasungswärme f heat of degasification
entgegen dem Uhrzeigersinn counter-clockwise
entgegengesetzt contrary, inverse, opposed, opposite, ~ **geladen** oppositely charged
entgegenlaufen to run counter to
entgegenrichten to oppose
entgegensehen to expect, to look forward to
entgegensetzen to contrast, to oppose
Entgegensetzung f opposition
entgegenwirken to counteract, to act counter to
entgegnen to reply, to respond
entgehen to avoid, to escape
entgeisten to dealcoholize, to reduce the alcoholic content
entgerben to de-tan
entgiften to detoxicate, to decontaminate, to disinfect
Entgiftung f decontamination, detoxication, detoxification, disinfection, extraction of poison
Entgiftungsanlage f decontamination plant, detoxication plant
Entgiftungsmittel n decontaminating agent, detoxicating agent, disinfectant
entglasen to devitrify
entgolden to remove gold
Entgoldung f gold stripping, removal of gold, ungilding
Entgraten n trimming
Entgratungsmaschine f deflashing machine
entgummieren to degum
enthaaren to depilate
Enthaaren n depilating
enthaarend depilatory
Enthaarung f depilation
Enthaarungsmittel n depilatory
enthärten to soften
Enthärtung f softening
enthalogenisieren to dehalogenate
Enthalpie f enthalpy, **Bildungs-** ~ enthalpy of formation, **Bindungs-** ~ enthalpy of bonding, **freie** ~ free enthalpy, **Mischungs-** ~ enthalpy of mixing, **Nullpunkts-** ~ enthalpy at absolute zero, **Reaktions-** ~ enthalpy of reaction, **Standard-** ~ standard enthalpy
Enthalpieänderung f enthalpy change
Enthalpie-Effekt m enthalpy effect
enthalten to contain

entharzen to deprive of resin, to deresinify, to remove resin
Entharzung f deresinification, extraction of resin, freeing from resin
enthülsen to shell
Enthülser m sheller
enthydratisieren to dehydrate
entionisieren to de-ionize
Entionisierung f de-ionization
Entionisiervermögen n de-ionizing property
entkalken to decalcify, to delime, to extract lime
Entkalken n deliming, unliming
entkalkt decalcified
Entkalkung f deliming, unliming
Entkalkungsmittel n deliming agent
Entkarbonisierung f decarbonation
entkarboxylieren to decarboxylate
Entkarboxylierung f decarboxylation
entkeimen to disinfect, to sterilize, (z. B. Getreide) to degerminate
Entkeimung f disinfection, sterilization, (Getreide) degermination, (Milch) pasteurization
Entkeimungsanlage f sterilization plant, sterilizing plant
Entkeimungsfilter n germ-proof filter
Entkeimungsmittel n disinfectant
entkieseln to desilicate, to desilicify
Entkieseln n desilicating
Entkieselung f desilicification
Entkieselungsverfahren n desilicification process
entkohlen to decarbonize, to decarburize
Entkohlen n decarbonizing
Entkohlung f decarbonization, decarburization
Entkompressionshebel m decompression lever
Entkoppelung f decoupling, disconnection
Entkoppler m (Biochem) uncoupler
entkrusten to descale
entkupfern to decopper, to free from copper
Entkupferung f decopperization
Entladebeanspruchung f (Elektr) strain of discharge
Entladedauer f (Elektr) duration of discharge
Entladegeschwindigkeit f (Elektr) rate of discharge
Entladekapazität f (Elektr) discharge capacity
entladen to unload, (Elektr) to discharge
Entladen n discharging, unloading
Entladeprobe f discharge test
Entladespannung f discharge voltage
Entladestärke f strength of discharge
Entladestrom m discharge current
Entladestromstärke f strength of discharge current
Entladevorgang m (Elektr) discharging process
Entladevorrichtung f discharging apparatus
Entladezeit f time of discharge
Entladung f unloading, (Elektr) discharge, ~ **ohne Funkenbildung** dark discharge, ~ **stille**
‰**Entladung** corona, **aperiodische** ~ aperiodic discharge, dead-beat discharge, impulsive discharge, non-oscillatory discharge, **gleichgerichtete** ~ unidirected discharge, **kathodische** ~ cathodic discharge, **oszillierende** ~ oscillation discharge, oscillatory discharge, **selbständige** ~ self-sustained discharge, spontaneous discharge, unassisted discharge, **stille** ~ silent discharge, **unselbständige** ~ assisted discharge, non-self-sustained discharge
Entladungsapparat m discharge device
Entladungsenergie f energy of discharge
Entladungsfunke m (Elektr) discharge spark
Entladungsgeschwindigkeit f velocity of discharge
Entladungsgesetz n emission law, three halves power law
Entladungskanal m breakdown channel
Entladungsnachglimmen n discharge afterglow
Entladungspotential n discharge potential
Entladungsraum m discharge space
Entladungsröhre f discharge tube
Entladungsrohr n discharge tube, outlet tube
Entladungsschlag m (Elektr) discharge stroke, shock of discharge
Entladungsspannung f discharge potential
Entladungsstoß m (Elektr) discharge pulse
Entladungsvorgang m, **kathodischer** ~ cathodic discharge process
Entlassen n dismissing, disengaging, releasing
entlasten to unload, to discharge, to release
Entlastung f unloading, balancing
Entlastungsschieber m balanced slide valve
Entlastungsventil n pressure reducing valve, blow-off valve, easing valve; relief valve, unloading valve
Entlaubungsmittel n defoliant
entleeren to empty, to clear, to discharge, (Behälter etc.) to drain
Entleerung f discharging, draining, emptying, ~ **des Kastens** tipping the box, ~ **des Ofens** emptying the furnace
Entleerungseffekt m (Atom) dispersal effect
Entleerungskammer f discharging chamber
Entleerungstür f discharge door
entleimen to degum, to scour, to wash
Entleimung f delamination
entlüften to de-aerate, to deair, to remove air, to vent, to ventilate
Entlüfter m de-aerator
Entlüftung f ventilation, air extraction; deaeration, (Chem) evacuation
Entlüftungsanlage f ventilation equipment
Entlüftungsdruck m venting pressure
Entlüftungshahn m air cock
Entlüftungsleitung f exhaust duct
Entlüftungspause f dwell
Entlüftungsrohr n vent pipe

Entlüftungsschraube f (Druckluftbremse) bleeder screw
Entlüftungsstutzen m (Pkw Kraftstoffbehälter) breather
Entlüftungsventil n air release valve, relief valve
Entlüftungsvorrichtung f de-aerating plant
Entlüftungszweck m venting purpose
Entlüfung f flash-off
Entlüfungseinsätze pl vents (in mo[u]ld)
Entlüfungsloch n vent (of a mo[u]ld)
entmagnetisieren to demagnetize, (Schiff) to degauss
Entmagnetisierung f demagnetization, (Schiff) degaussing
Entmagnetisierungsfaktor m demagnetizing factor
Entmanganung f demanganization
Entmaterialisierung f (Atom) dematerialization
entmethylieren to demethylate
entmineralisieren to demineralize
Entmineralisierung f demineralization
entmischen to separate into components, to demix, to disintegrate, (Metall) segregate, (Öl) to demulsify
Entmischen n separating into components, unmixing, (Öl) demulsifying
Entmischung f separation, disintegration, (Krist) coring, (Öl) demulsification, ~ **des Elektrolyten** segregation of the electrolyte
Entmischungsvorgang m separation process
Entmodulator m demodulator
Entnahme f withdrawal, (aus einem Lager) dispatch
Entnahmebuchse f (Elektr) output terminal
Entnahmegang m dispatch aisle
Entnahmerohr n outlet pipe
Entnahmesonde f sampling probe, tapping probe
Entnahmestelle f sampling point
Entnahmestutzen m distributor tube
entnebeln to clear of fumes, to devaporize, to eliminate the mist
Entnebeln n devaporizing, mist eliminating
Entnebelung f devaporation, mist elimination
Entnebelungsanlage f demisting plant, devaporizing plant, entrainment separator (destillation), fume dispersion installation, mist eliminator
entnehmen to withdraw, to take away
entnikotinisieren to denicotinize
entölen to de-oil, to remove oil, to separate oil
Entöler m oil separator
Entölung f deoiling
Entölungsanlage f deoiling plant
Entölungsapparat m oil separator
Entölungsgrad m oil recovery factor
Entomologe m entomologist
Entomologie f entomology
Entoparasit m endoparasite, entoparasite

Entoplasma n (innere Plasmaschicht) endoplasm, entoplasm
Entparaffinieranlage f dewaxing plant
Entparaffinierung f dewaxing
Entparaffinierungsvorrichtung f dewaxer
entpesten to disinfect
Entpestung f disinfection
Entphenolung f dephenolization
Entphenolungsanlage f dephenolation plant
entphosphoren to dephosphorate, to dephosphorize
Entphosphorung f dephosphorization
Entphosphorungsperiode f dephosphoring period, dephosphorizing period
entpolarisieren to depolarize
Entpolarisierung f depolarization
Entpolymerisation f depolymerization
Entpolymerisierung f depolymerization
entpülpen (Zucker) to depulp
entrahmen to skim off
Entrastung f disengaging
Entrochit m (Min) entrochite
Entrochitenkalk m (Min) entrochial limestone
Entropie f entropy, ~ **bei ungeordneter Verteilung** (Phys) entropy of disorder, **unveränderliche** ~ constant entropy, **virtuelle** ~ (Atom) virtual entropy, practical entropy, **kalorimetrische** ~ calorimetric entropy, **Mischungs-** ~ entropy of mixing, **Normal-** ~ normal entropy, **Nullpunkts-** ~ entropy at absolute zero, **spektroskopische** ~ spectroscopic entropy, **Standard-** ~ standard entropy
Entropieänderung f entropy change
Entropiediagramm n temperature entropy diagram
Entropiedichte f entropy density
Entropie-Effekt m entropy effect
Entropiegleichung f entropy equation
Entropieprinzip n entropy principle
Entropiesatz m (Therm) law of degradation of energy
entrosten to derust, to free from rust
Entrosten n de-rusting
Entrostung f rust removal
Entrostungsmittel n rust remover, rust-removing agent
entsäuern to free from acid, to deacidify
Entsäuern n deacidifying, elimination of acid
entsäuert de-acidified
Entsäuerung f deacidification, extraction of acid, ~ **von Abwässern** neutralization of waste water
Entsäuerungsapparat m acid extractor
entsaften to extract the juice [from]
entsalzen to desalt, (Quellwasser) to demineralize
Entsalzen n **von Rohöl** crude oil desalting

Entsalzung – Entwässerungsvorrichtung

Entsalzung *f* demineralization, desalination, desalting
Entsalzungsvorrichtung *f* demineralizer
Entschädigung *f* compensation
entschärfen to render safe, to unprime
entschäumen to remove the scum from, to skim
Entschäumer *m* antifoam, defoaming agent
Entschäumung *f* prevention of foaming
Entschäumungsmittel *n* defoamer, defoaming agent
Entschalungsöl *n* release agent
entscheiden to decide, to settle
entschimmeln to free from mildew
entschlacken to remove slag, to clinker, to deslag
Entschlacken *n* deslagging
Entschlackung *f* deslagging
Entschlackungsanlage *f* slag removal plant
entschlammen to clear from mud, to remove sludge
Entschlammungsanlage *f* sludge removal plant
Entschlammungsrohr *n* drain tube
entschleimen to deslime, to free from slime, to remove slime
Entschlichtung *f* (Text) desizing
Entschlickerung *f* drossing
entschwefeln to desulfurize, to free from sulfur
Entschwefelung *f* desulfurization, desulfurizing, sulfur removal
Entschwefelungsmittel *n* desulfurizing agent, sulfur remover
Entschwefelungsofen *m* desulfurizing furnace
Entschwefelungsprozeß *m* process of desulfurization
Entschwefelungsschlacke *f* desulfurizing slag
entschweißen to scour
Entschweißungsvorrichtung *f* steeping bowl
entseifen to free from soap, to rinse
Entseifen *n* cleansing from soap
entseift cleansed from soap
entseuchen to disinfect, to decontaminate
Entseuchung *f* disinfection, **radioaktive** ~ radioactive decontamination
Entseuchungsfaktor *m* (Atom) decontamination factor (or index)
entsilbern to desilver, to free from silver, to remove silver from
entsilizieren to desiliconize
Entsilizierung *f* desiliconization
entsintern to desinter
entspannen to release the tension, (Gas) to expand, to reduce the pressure (of a gas), (z. B. Seil) to slacken
Entspannen *n* removal of stress, unstressing
Entspannung *f* pressure release, (Gas) expansion
Entspannungsabkühlung *f* cooling due to expansion
Entspannungsbohrlöcher *n pl* stab holes
Entspannungsgas *n* let-down gas
Entspannungsgefäß *n* let down vessel

Entspannungsgeschwindigkeit *f* stress relaxation rate
Entspannungsglühen *n* stress-relieving anneal
Entspannungskriechgrenze *f* stress relaxation yield limit
Entspannungskühler *m* expansion cooler
Entspannungskühlung *f* cooling by expansion
Entspannungsmaschine *f* expander, ~ **für Tieftemperaturtechnik** cryogenic expander
Entspannungsventil *n* expansion valve, pressure reducing valve, reduction valve, relief valve
Entspannungsverdampfung *f* flash evaporation
Entspannungswiderstand *m* stress relaxation resistance
Entspannungszeit *f* stress relaxation time
entsprechen to conform to, **einer Gleichung** ~ to satisfy an equation
entsprechend corresponding [to], (homolog) homologous
entstählen to soften
entstärken to unstarch
Entstäuber *m* dust collector
entstauben to dust off, to extract dust from, to free from dust
Entstauben *n* dust removal, freeing from dust
Entstaubungsanlage *f* dust removal plant, dust extraction plant, dust separator
entstehen to arise, to originate, to result
entstehend nascent
Entstehung *f* formation, origin
Entstehungsart *f* mode of origin
Entstehungsbedingung *f* formation condition
Entstehungsbrand *m* early stages of a fire
Entstehungszustand *m* nascent state, status nascendi
entstellen to disfigure, to distort, to mar
Entstörung *f* interference elimination
entströmen to flow out, to issue, to stream
entteeren to deprive of tar, to detar
Entteerung *f* detarring
entwässerbar drainable
entwässern to drain, to concentrate, to dewater, to extract water from, (Chem) to dehydrate, (im Exsikkator) to desiccate
Entwässern *n* concentration, drainage, hydroextraction, water removal, (Chem) dehydration, (im Exsikkator) desiccation
entwässert anhydrous, dehydrated
Entwässerung *f* dehydration, dewatering, drainage, extraction of water, hydroextraction, (im Exsikkator) desiccation
Entwässerungsanlage *f* drainage, (Chem) dehydration plant
Entwässerungshahn *m* drain cock
Entwässerungspresse *f* (Pap) defibrator
Entwässerungsrohr *n* drain pipe
Entwässerungssystem *n* system of drainage
Entwässerungstopf *m* water separating vessel
Entwässerungsvorrichtung *f* water separator

250

Entwässerungszentrifuge *f* dewatering centrifuge
entweichen to escape, (Gas) to leak
Entweichen *n* escaping, (Gas) leakage
entweichend escaping
Entweichung *f* escape
Entweichungshahn *m* discharge cock
Entweichungsventil *n* delivery valve
entwerfen to design, to outline, to sketch
entwerten to reduce in value, to depreciate
entwickelbar developable
entwickeln to develop, (Chem) to generate, to liberate, (Film) to process, (Wärme) to produce
Entwickeln *n* (Phot) developing
entwickelt developed
Entwickler *m* generator, (Phot) developer
Entwicklung *f* development, evolution, (Chem) formation, generation, (Math) development, expansion;, ~ **im Durchlauf** (Phot) continuous development, **asymptotische** ~ asymptotic expansion, asymptotic series
Entwicklungsabteilung *f* department of development
Entwicklungsbad *n* (Phot) developing bath, developer
Entwicklungsdruck *m* generating pressure
entwicklungsfähig developable
Entwicklungsfarbstoff *m* developing dyestuff
Entwicklungsflüssigkeit *f* (Phot) developer
Entwicklungsgebiet *n* area of development
Entwicklungsgefäß *n* generating vessel
Entwicklungsgeschichte *f* (Biol) biogenesis
Entwicklungskeim *m* (Phot) development center
Entwicklungslinie *f* line of development
Entwicklungssatz *m* (Math) expansion theory
Entwicklungsstadium *n* stage of development
Entwicklungstempo *n* rate of development
Entwicklungsverfahren *n* (Phot) process of development
entwinden to extricate, to wrest from
entwöhnen to disaccustom, to wean
entwürzen to deprive of spicy flavor, to unseason
Entwurf *m* design, outline, plan, plot, (Skizze) sketch
Entwurfsbestandteile *m pl* design elements
Entwurfsmodell *n* plot layout model
Entwurfsskizze *f* lay-out sketch, rough sketch
Entzerrer *m* compensating network preamplifier
Entzerrungsanordnung *f* antidistortion device
Entzerrungsgerät *n* rectifier
entziehen to deprive [of], to extract, to take away, to withdraw
Entziehen *n* depriving, abstracting
Entziehung *f* deprivation; abstraction, drawing off, extraction, ~ **von Sauerstoff** removal of oxygen, deoxygenation, ~ **von Wasser** dehydration
Entziehungsrohr *n* outlet tube
entziffern to decipher, to decode

Entzinkung *f* dezincing, dezincking
entzinnen to detin, to remove the tin, to untin
Entzinnung *f* detinning, stripping of tin
Entzinnungsanlage *f* detinning plant
Entzinnungsbad *n* detinning bath
Entzinnungsgrad *m* degree of detinning
entzuckern to extract sugar, to dedulcify
Entzuckerung *f* desaccharification, extraction of sugar
entzündbar ignitable, inflammable
Entzündbarkeit *f* ignitability, [in]flammability
entzünden to ignite, to inflame, to kindle
Entzünden *n* igniting, kindling
entzündlich inflammable, (Med) inflammatory
Entzündung *f* ignition, (Med) inflammation
Entzündungsfunke *m* ignition spark
Entzündungsgefahr *f* danger of ignition
Entzündungsgemisch *n* ignition mixture
Entzündungspunkt *m* flash point, ignition point, ignition temperature
Entzündungstemperatur *f* ignition temperature
Entzündungszeit *f* ignition time
Entzug *m* **von Elektronen** removal of electrons
entzundern to descale
Entzundern *n* descaling
Entzunderung *f* descaling, scaling off
Enzian *m* (Bot) bitter wort, gentian
Enzianbitter *n* gentian bitter
Enzianbranntwein *m* gentian spirit
Enzianextrakt *m* extract of gentian
Enzianöl *n* gentian oil
Enziantinktur *f* tincture of gentian
Enzianwurzel *f* (Bot) gentian root
Enzym *n* (Biochem) enzyme, ferment, **adaptives** ~ adaptive enzyme, **allosterisches** ~ allosteric enzyme, **extrazelluläres** ~ extracellular enzyme, **hydrolytisches** ~ hydrolytic enzyme, **induzierbares** ~ inducible enzyme, **intrazelluläres** ~ intracellular enzyme, **konstitutives** ~ constitutive enzyme
Enzymaktivierung *f* enzyme activation
Enzymaktivität *f* enzyme activity
enzymatisch enzymatic
Enzymbestimmung *f* enzyme assay
Enzymeinheit *f* enzyme unit
Enzymenthaarung *f* (Gerb) enzymatic depilation
Enzymhemmung *f* enzyme inhibition
Enzyminduktion *f* enzyme induction
Enzym-Inhibitor-Komplex *m* enzyme-inhibitor complex
Enzymology *f* enzymology
Enzymreinigung *f* enzyme purification
Enzymrepression *f* enzyme repression
Enzymspezifität *f* enzyme specificity
Enzym-Substrat-Komplex *m* enzyme-substrate complex
Enzymumsatzzahl *f* enzyme turnover number
Eosin *n* eosin[e]
Eosinsäure *f* eosine acid

Eosphorit – Erdabplattung

Eosphorit *m* (Min) eosphorite
Eperusäure *f* eperuic acid
Ephedrin *n* ephedrine
Ephedrinhydrochlorid *n* ephedrine hydrochloride
Ephedrinsulfat *n* ephedrine sulfate
Epiallomuscarin *n* epiallomuscarine
Epiandrosteron *n* epiandrosterone
Epiarabinose *f* epiarabinose
Epiarabit *m* epiarabite
Epiarabonsäure *f* epiarabonic acid
Epiasarinin *n* epiasarinin
Epiblast *n* (Biol) epiblast
Epiborneol *n* epiborneol
Epiboulangerit *m* (Min) epiboulangerite
Epibromhydrin *n* epibromohydrin
Epicampher *m* epicamphor
Epicarisson *n* epicarissone
Epicatechin *n* epicatechol
Epichinamin *n* epiquinamine
Epichinidin *n* epiquinidine
Epichitosamin *n* epichitosamine
Epichitose *f* epichitose
Epichlorhydrin *n* epichlorohydrin
Epichlorit *m* epichlorite
Epicholestanol *n* epicholestanol
Epidemie *f* (Med) epidemic
Epidesmin *n* epidesmin
Epidiaskop *n* epidiascope
Epidichlorhydrin *n* epidichlorohydrin, 2,3-dichloropropene
Epididymit *m* (Min) epididymite
Epidosit *m* (Min) epidosite
Epidot *m* (Min) epidote, pistacite, **grüner** ~ arendalite
epidotähnlich epidotic
epidotführend epidotiferous
Epierythrose *f* epierythrose
Epifucose *f* epifucose
Epigalactose *f* epigalactose
Epigallocatechin *n* epigallocatechol
Epigenit *m* (Min) epigenite
Epigitoxigenin *n* epigitoxigenin
Epiglykose *f* epiglycose
Epiguanin *n* epiguanine, 2-amino-6-methyl-8-hydroxypurine
Epigulit *n* epigulite
Epigulose *f* epigulose
Epihämanthidin *n* epihaemanthidine
Epihydrinalkohol *m* epihydrin alcohol, glycidol
Epiinosit *m* epiinositol
Epikadmium *n* epicadmium
Epikoprostanol *n* epicoprostanol
Epikoprosterin *n* epicoprosterol
Epimer *n* epimer
Epimerisierung *f* epimerization
Epinephrin *n* epinephrine, adrenaline
Epinin *n* epinine
Epiöstriol *n* epiestriol

Epiquercit *m* epiquercitol
Epirockogenin *n* epirockogenin
Episamogenin *n* episamogenin
Episesamin *n* episesamin
Episkop *n* episcope, reflecting projector
Episterin *n* episterol
Epistilbit *m* (Min) epistilbite
Epistolit *m* (Min) epistolite
Epitaxie *f* (Krist) epitaxy
Epitestosteron *n* epitestosterone
Epithelschutzvitamin *n* (Axerophthol) axerophthol
epithermisch epithermal
Epitruxillsäure *f* epitruxillic acid
Epixylose *f* epixylose
Epizentrum *n* (eines Erdbebens) epicenter (of an earthquake)
Epizeorin *n* epizeorin
Epizuckersäure *f* episaccharic acid
epizyklisch epicyclic
Epizykloid *n* epicycloid
EPL Abk. für essentielle Phospholipide
Epoxyharz *n* epoxy resin
Epsomit *m* (Min) epsomite
Equilenan *n* equilenane
Equilenin *n* equilenin
Equisetrin *n* equisetrin
Equisetsäure *f* equisetic acid, aconitic acid
Erbanlage *f* genetic trait
Erbeinheit *f* (Biol) gene
Erbfaktor *m* gene, genetic factor, hereditary factor, inheritance
Erbfehler *m* hereditary defect
Erbforschung *f* genetics
Erbinerde *f* erbia, erbium oxide
Erbium *n* erbium (Symb. Er)
Erbiumgehalt *m* erbium content
Erbiumoxid *n* erbium oxide, erbia
Erbiumpräparat *n* erbium preparation
Erbiumsulfat *n* erbium sulfate
Erbiumverbindung *f* erbium compound
Erbkrankheit *f* hereditary disease
Erblehre *f* genetics
erblich hereditary, [in]heritable
Erbmasse *f* (Biol) genotype
Erbse *f* pea
Erbsenerz *n* (Min) pea ore, pisiform iron ore
erbsenförmig pealike, pea-shaped, pisiform
erbsengroß pea-sized
erbsengrün pea green
Erbsenstein *m* (Min) pea-stone, aragonite, granular limestone, pisolite
erbsensteinhaltig pisolitic
Erbsensteinmarmor *m* (Min) pisolitic marble
Erbtypus *m* (Biol) genotype
Erdableitung *f* earth connection (Br. E.), ground connection (Am. E.)
Erdabplattung *f* earth flattening, earth's oblateness

Erdalkali *n* alkaline earth
Erdalkaliatom *n* alkaline earth atom
Erdalkalichelat *n* alkaline earth chelate
Erdalkalihalogenid *n* alkaline earth halide
Erdalkalikontinua *n pl* continuous alkaline earth spectra
Erdalkalimetall *n* alkaline earth metal
Erdalkalioxid *n* alkaline earth oxide
Erdalkalisalz *n* alkaline earth salt
erdalkalisch earth alkaline
Erdbalsam *m* naphtha, rock oil
Erdbeben *n* earthquake
Erdbebenherd *m* earthquake focus
Erdbebenkunde *f* seismology
Erdbebenmesser *m* seismograph, seismometer
Erdbebenstrahl *m* seismic ray
Erdbebenwarte *f* seismological observatory
Erdbebenwelle *f* seismic wave
Erdbeeraldehyd *m* strawberry aldehyde
Erdbeschleunigung *f* gravitational acceleration, acceleration of gravity
Erdboden *m* soil
Erdbodenthermometer *n* soil thermometer
Erdbohrung *f* earth boring
Erdbrand *m* subterraneous fire
Erdchemie *f* geochemistry
Erddamm *m* earth bank
Erde *f* (Boden) ground, soil, (Chem, Elektr) earth, (Planet) earth
erdehaltig containing earth
erden (Elektr) to ground, to earth
Erden *pl,* **seltene** ~ rare earths
erdfahl earth-colored
Erdfarbe *f* (Min) earth color, mineral color
erdfarben earth-colored
erdfarbig earth-colored
Erdfeld *n* earth's field
Erdgammastrahlung *f* terrestrial gamma radiation
Erdgas *n* natural gas
Erdgasanlage *f* natural gas plant
Erdgasleitung *f* gas pipeline
Erdgasvorkommen *n* source of natural gas
Erdgelb *n* yellow ocher
Erdgeruch *m* earthy smell
Erdgeschichte *f* geology
erdgeschichtlich geological
Erdglas *n* (Min) selenite
Erdgrün *n* green copper ore, green verditer, mineral green
Erdharz *n* bitumen, elaterite, fossil resin, mineral caoutchouc, **gelbes** ~ amber, **verwandeln in** ~ to bituminize
erdharzartig asphaltic, bituminous
Erdharzfarbe *f* color of bitumen
erdharzhaltig asphaltic, bituminiferous, bituminous
erdig earthy

Erdin *n* erdin
Erdinnere *n* interior of the earth
Erdkabel *n* subterranean cable, underground cable
Erdkalk *m* (Min) marly limestone
Erdkobalt *m* (Min) earthy cobalt, asbolite, **grüner** ~ annabergite
Erdkohle *f* brown coal, earthy coal, lignitic earth
Erdkunde *f* geography, (Erdgeschichte) geology
Erdleiter *m* (Elektr) earthing conductor
Erdleitung *f* (Elektr) earth connection
Erdmagnetfeld *n* geomagnetic field, magnetic field of the earth
erdmagnetisch earth magnetic, geomagnetic
Erdmagnetismus *m* earth magnetism, geomagnetism, terrestrial magnetism
Erdmannit *m* (Min) earthmannite
Erdmehl *n* fossil meal, white stone marl
Erdmessung *f* geodesy
Erdmetall *n* earth metal
Erdnußöl *n* peanut oil, arachis oil
Erdnußsäure *f* arachic acid, arachidic acid, eicosanoic acid
Erdoberfläche *f* surface of the earth
Erdöl *n* crude oil, petroleum, rock oil
Erdölanlage *f* petroleum plant
Erdölchemie *f* oil chemistry, petroleum chemistry
Erdöldestillat *n* distillation product of naphtha, petroleum distillate
Erdölerzeuger *m* oil producer
Erdölerzeugnis *n* petroleum product
Erdölförderung *f* oil production
erdölhaltig containing petroleum, oil-bearing
Erdölindustrie *f* oil industry
Erdölkohlenwasserstoff *m* petroleum hydrocarbon
Erdöllager *n* mineral oil reservoir
Erdölleitung *f* oil piping
Erdölprodukt *n* oil product
Erdölraffinerie *f* oil refinery
Erdölrückstand *m* naphtha residue, residue of rock oil
Erdölverarbeitung *f* oil refining
Erdpech *n* asphalt, bitumen, mineral pitch, **verwandeln in** ~ to bituminize
erdpechartig bituminous
erdpechhaltig asphaltic, bituminous
Erdpotential *n* earth potential
Erdrinde *f* crust of the earth
Erdsalz *n* (Min) rocksalt
Erdschellack *m* acaroid resin, yellow balsam
Erdschierling *m* (Bot) hemlock
Erdschlacke *f* earthy slag
Erdschluß *m* (Elektr) earth connection, earth[ing], ground
Erdschwefel *m* club moss, lycopodium, vegetable sulfur
Erdseife *f* mountain soap

Erdstein *m* (Min) aetites, eagle stone
Erdtalk *m* earthy talc, mineral tallow, ozocerite
Erdteer *m* mineral tar, maltha
Erdtorf *m* peat
Erdumdrehung *f* rotation of the earth
Erdumlaufbahn *f* earth orbit
Erdung *f* (Elektr) earth, earth connection, earthing, ground [connection]
Erdungsdraht *m* earth wire, ground wire
Erdungsleitung *f* earth wire, ground wire
Erdungsschalter *m* earth switch
erdverlegt underground
Erdwachs *n* earth wax, mineral tallow, mineral wax, native paraffin, ozocerite
Erdwärme *f* heat of the earth
Erdwärmemesser *m* geothermometer
Ereignisfeld *n* (Statist) event field
Eremakausie *f* eremacausis, slow combustion
Eremophilon *n* eremophilone
Erepsin *n* erepsin
Erfahrung *f* experience
Erfahrungsaustausch *m* exchange of information, exchange of knowhow
Erfahrungsergebnis *n* empirical result
erfahrungsgemäß from experience
erfahrungsmäßig empirical
Erfahrungssatz *m* empirical law
Erfahrungstatsache *f* empirical fact
Erfahrungswert *m* empirical value
Erfahrungswissenschaft *f* empirical science
Erfassungsgrenze *f* identification limit
erfinden to invent
Erfinderpatent *n* patent
Erfinderschutz *m* protection of inventors
Erfindung *f* invention
erfindungsgemäß according to the invention
Erfordernis *n* requirement
erforschen to investigate, to discover, to explore
Erforscher *m* discoverer, investigator
Erforschung *f* exploration, investigation
erfrieren to freeze
Erfrieren *n* freezing
erfrischen to freshen, to refresh
ERG *n* (Elektroretinogramm, Med) electroretinogram
ergänzen to complete, to supplement, (auffüllen) to replenish, (ersetzen) to replace
Ergänzungsfarbe *f* complementary color
Ergänzungsfläche *f* complementary surface
Ergänzungsgleitung *f* glide relaxation
Ergänzungskugel *f* supplementary ball
Ergänzungsverordnung *f* supplementing ordinance
ergeben to result, to amount to, to issue in, to yield
Ergebnis *n* result, effect, product, (Ausbeute) yield
Ergiebigkeit *f* economy in use, yield; (Farbe) tinctorial strength

ergießen to discharge, to pour out
Ergin *n* ergine
erglühen to glow, to inflame
Ergmeter *n* ergmeter
Ergobasin *n* ergobasine
Ergocalciferol *n* ergocalciferol
Ergocristin *n* ergocristine
Ergodensatz *m* ergodic hypothesis
Ergokryptin *n* ergocryptine
Ergokryptinin *n* ergocryptinine
Ergolin *n* ergoline
Ergometrin *n* ergometrine, ergotocin
Ergometrinin *n* ergometrinine
Ergometrinmaleat *n* ergometrine maleate
Ergon *n* (Quant) ergon
Ergonovin *n* ergonovine
Ergosin *n* ergosine
Ergosomen *pl* (Biol) ergosomes
Ergostan *n* ergostane
Ergostanol *n* ergostanol
Ergosterin *n* ergosterol, ergosterin
Ergosteron *n* ergosterone
Ergostetrin *n* ergostetrine
Ergotachysterin *n* ergotachysterol
Ergot-Alkaloid *n* ergot alkaloid
Ergotamin *n* ergotamine
Ergotaminbitartrat *n* ergotamine bitartrate
Ergothionein *n* ergothioneine
Ergotin *n* ergotine, ergot extract
Ergotinin *n* ergotinine
Ergotismus *m* ergotism
Ergotoxin *n* ergotoxine
ergreifen to seize
Ergußgestein *n* (Geol) effusive rock
erhaben (konvex) convex, ~ **ausarbeiten** to emboss, ~ **polieren** to relief polish
erhältlich available
erhärten to harden, to set
Erhärten *n* hardening, setting
Erhärtungsbeginn *m* beginning of setting
Erhärtungsende *n* finish of setting
Erhärtungsfähigkeit *f* coagulating property
erhalten (bekommen) to get, to obtain, to receive, (bewahren) to conserve, to preserve
Erhaltung *f* conservation, preservation, ~ **der Energie** conservation of energy, ~ **des Impulses** conservation of momentum
Erhaltungssatz *m* conservation law
Erhaltungsstoffwechsel *m* conservation metabolism
Erhaltungstrieb *m* instinct of self-preservation
erhitzen to heat, (Metall) to anneal
Erhitzen *n* heating, making hot, reheating
Erhitzer *m* heater
erhitzt heated
Erhitzung *f* heating, ~ **durch Induktion** induction heating
Erhitzungsgeschwindigkeit *f* rate of heating
Erhitzungswiderstand *m* heat resistance

erhöhen to raise, to elevate, to heighten, (zunehmen) to increase
erhöht elevated, increased, raised, (Phys) enhanced
Erhöhung *f* raising, elevation, (Zunahme) increase
Erholung *f* recovery, (Verstärkung) intensification, ~ **der Mischung** (Gummi) rubber relaxation
Erholungseffekt *m* recovery effect
Erholungserscheinung *f* recovery process
Erholungsgeschwindigkeit *f* recovery rate
Erholungszeit *f* (eines Geiger-Zählrohrs) recovery time (of a Geiger counter)
Erianthin *n* erianthin
Erichsenprüfer *m* Erichsen cupping machine
Erichsentiefziehprobe *f* Erichsen cupping or ductility
Erichsenversuch *m* Erichsen test
Ericit *m* ericite
Erigeronöl *n* erigeron oil
Erinit *m* (Min) erinite
Eriochromblauschwarz *n* blue eriochrome black
Eriochromcyanin *n* eriochromcyanine
Eriochromschwarz *n* eriochrome black
Eriochromverdon *n* eriochrome verdone
Eriocitrin *n* eriocitrin
Eriodictyol *n* eriodictyol
Erionit *m* (Min) erionite
Erkältung *f* (Med) common cold
Erkältungskrankheit *f* (Med) common cold
erkalten to cool
Erkalten *n* cooling
erkaltend chilling, cooling, refrigerating
Erkenntnis *f* insight, knowledge, understanding
Erkennung *f* detection, recognition
Erkennungsmarke *f* identification tag
erklären to explain, to elucidate
Erklärung *f* explanation, elucidation
Erkrankung *f* disease, illness, sickness
Erläuterung *f* description, explanation
Erlanger Blau *n* Erlanger blue, Prussian blue
erlaubt permissible, permitted
erleichtern to facilitate, to alleviate, to ease, to lighten, to make easier
erleiden to suffer, to undergo
Erlenkohle *f* alder charcoal
Erlenmeyerkolben *m* conical flask, Erlenmeyer flask, **enghalsiger** ~ narrow-mouth Erlenmeyer flask, **weithalsiger** ~ wide-mouth Erlenmeyer flask
erlöschen to go out, to be extinguished
ermäßigen to reduce
ermitteln to determine, to ascertain
Ermitt[e]lung *f* search; discovery
Ermittlungsverfahren *n* method of determination, **empirisches** ~ trial-and-error method
Ermüdung *f* tiring, weariness, (Techn) fatigue

Ermüdungsbeständigkeit *f* fatigue strength
Ermüdungsbruch *m* fatigue fracture, fatigue failure
Ermüdungserscheinung *f* fatigue phenomenon, fatigue sign
Ermüdungsfestigkeit *f* fatigue strength
ermüdungsfrei fatigue-proof
Ermüdungsgrenze *f* endurance limit, fatigue limit
Ermüdungsriß *m* fatigue crack
Ermüdungsrisse *pl* (Reifen) flex cracking
Ermüdungsschutz *m* anti-flex cracking properties
Ermüdungsschutzmittel *n* anti-flex cracking agent
Ermüdungsstriemen *pl* fatigue striations
ermüdungstüchtig flex resisting
Ermüdungsverluste *m pl* damage rate in fatigue
Ermüdungsversuch *m* endurance test, fatigue test
ernähren to feed, to nourish
Ernährung *f* nourishment, nutrition
Ernährungsfehler *m* malnutrition, false diet
Ernährungsforschung *f* alimentary research
Ernährungskunde *f* science of nutrition, (Diätetik) dietetics
Ernährungsmangel *m* nutritional deficiency
Ernährungsstörung *f* malnutrition
erneuern to renew, to renovate, to restore, (Farben) to refresh
Erneuerung *f* renewal
erneut again, renewed
erniedrigen to decrease, to lower, to reduce
Erniedrigung *f* decrease, lowering, reduction
Ernte *f* harvest
Ernteverfrühen *n* early cropping
Eroberung *f* conquest
erodieren to erode
Erosion *f* erosion
Erosnin *n* erosnin
erproben to test, to try out
erprobt experienced
errechnet calculated
erregbar excitable
erregen to excite, to induce, to stimulate, (Elektr) to energize
erregend stimulating
Erreger *m* exciter
Erregerarbeit *f* work of excitation
Erregerenergie *f* field energy
Erregerfeld *n* (Elektr) exciting field
Erregerfrequenz *f* excitation frequency
Erregergleichrichter *m* (Elektr) exciting rectifier
Erregerkreis *m* exciting circuit, (Elektr) energizing circuit
Erregermasse *f* excitant
Erregerspannung *f* energizing voltage, exciter voltage
Erregerspule *f* (Elektr) exciter coil
Erregerstrom *m* energizing current, exciting current, induction current

Erregerstromkreis – Erweichungsmittel 256

Erregerstromkreis *m* exciting circuit
erregt excited
Erregung *f* excitation, agitation, commotion, (Physiol) stimulation
Erregungsart *f* method of excitation
Erregungsfehler *m* excitation defect
Erregungsleitung *f* (Nerv) stimulus conduction
Erregungsspannung *f* excitation voltage
erreichbar attainable
errichten to erect
Errichtung *f* erection
Errungenschaft *f* achievement
Ersatz *m* substitute, replacement
Ersatzbolzen *m* spare bolt
Ersatzheizplatte *f* spare heater plate
Ersatzkessel *m* spare boiler, stand-by boiler
Ersatzpräparat *n* (Pharm) substitute, surrogate
Ersatzrolle *f* spare roll[er]
Ersatzschaltbild *n* equivalent circuit
Ersatzschaltung *f* equivalent circuit
Ersatzstoff *m* substitute [material]
Ersatzstück *n* spare piece, reserve piece
Ersatzteil *n* replacement part, spare part
Ersbyit *m* (Min) ersbyite
Erscheinung *f* phenomenon, symptom
Erscheinungsform *f* form of matter
erschließen to discover, to disclose
erschöpfen to exhaust, to wear out
erschöpfend exhaustive
Erschöpfung *f* exhaustion
Erschöpfungshypothese *f* exhaustion hypothesis
Erschütterung *f* concussion, shock, vibration
Erschütterungsdämpfung *f* shock absorption
erschütterungsfrei concussion-free, shock-absorbent, vibrationless
erschweren to aggravate, to complicate, to make difficult
Erschwerung *f* aggravation, complication, difficulty
ersetzbar replaceable
ersetzen to replace, to substitute
Ersetzung *f* replacement, substitution
ersichtlich evident
erstarren to solidify, to become rigid, to freeze, to harden
Erstarren *n* solidification, freezing, setting
erstarrend coagulating, congealing
erstarrt solidified
Erstarrung *f* solidification, freezing, hardening, (z. B. Fett) congelation
Erstarrungsbad *n* coagulating bath
Erstarrungsbereich *m* solidification range
Erstarrungsbild *n* diagram of solidification
Erstarrungsdiagramm *n* diagram of solidification
Erstarrungsflüssigkeit *f* coagulating liquid
Erstarrungsgefüge *n* solidification structure
Erstarrungsgestein *n* igneous rock, (Geol) igneous rock

Erstarrungsintervall *m* crystallization interval
Erstarrungskammer *f* hardening room
Erstarrungskurve *f* freezing-point curve, solidification curve
Erstarrungspunkt *m* solidification point, setting point, ~ **am rotierenden Thermometer** continental solid point
Erstarrungsraum *m* hardening room
Erstarrungsstopfen *m* freeze plug
Erstarrungstemperatur *f* solidification or congealing temperature
Erstarrungswärme *f* [latent] heat of solidification
Erstarrungszeit *f* setting time
Erstausführung *f* prototype
Erstausrüstung *f* original equipment
Erstbelastung *f* initial load
ersticken to choke, to suffocate, (durch Gas) to asphyxiate
Ersticken *n* choking, suffocating
erstickend choking, suffocating
Erstickung *f* suffocation, asphyxiation, choking
Erstickungserscheinung *f* symptom of suffocation
erstrecken to extend, to stretch
Erstverdampfer *m* first evaporator
Erstwicklung *f* primary coil winding
Erträglichkeitsgrenze *f* limit of tolerance
Ertrag *m* yield, output
Ertragsaustausch *m* transfer of profits
Ertragskraft *f* viability
Erucasäure *f* erucic acid, cis-13-docosenoic acid
Erucin *n* erucine
Erucon *n* erucone
Erucylalkohol *m* erucyl alcohol
eruptiv (Geol) eruptive
Eruptivgänge *pl* (Geol) igneous dikes or veins
Eruptivgestein *n* (Geol) eruptive rocks, igneous rocks
Ervasin *n* ervasine
Erwähnung *f* reference
erwärmen to heat, to warm, (gelinde, mäßig) to warm gently
Erwärmen *n* heating, calefaction, warming
erwärmend warming, calefactory
Erwärmung *f* heating, calefaction, **dielektrische** ~ dielectric heating
Erwärmungsfehler *m* fault due to heating
Erwärmungsverlust *m* thermal or heat loss, Joulean effect
Erwärmungszeit *f* heating period
erwartungsgemäß as expected
Erwartungswert *m* expectation value
erweichen to soften, to mollify, to plastify
Erweichen *n* softening
erweichend softening
Erweicher *m* softener
Erweichung *f* softening, mollification, sintering
Erweichungsbereich *m* softening range
Erweichungsmittel *n* emollient

Erweichungspunkt *m* softening point; sintering point; sticking temperature
Erweichungstemperatur *f* fusing temperature, upsetting temperature
Erweichungszone *f* softening range
erweisen to prove, to demonstrate
erweitern to widen, to expand, to extend, (Pupille) to dilate, (Techn) to ream
Erweiterung *f* expansion, extension, (Pupille) dilation, ~ **des Düsenkanals** orifice relief
Erweiterungsbohrtätigkeit *f* development drilling
erwerben to acquire, to earn, to gain, to obtain
Erysodin *n* erysodine
Erysothiopin *n* erysothiopine
Erythralin *n* erythraline
Erythramin *n* erythramine
Erythren *n* erythrene, bivinyl
Erythrenglykol *n* erythrene glycol
Erythrenkautschuk *m* erythrene caoutchouc
Erythricin *n* erythricine
erythrisch erythric
Erythrische Säure *f* erythric acid
Erythrit *m* erythrite, erythritol, erythrol
Erythritol *n* erythrol, 3-butene-1,2-diol, erythrite, erythritol
erythro-Anordnung *f* erythro arrangement
Erythroaphin *n* erythroaphin
Erythrocruorin *n* erythrocruorin, hemoglobin
Erythrodextrin *n* erythrodextrin
Erythrogensäure *f* erythrogenic acid
Erythroglaucin *n* erythroglaucin
Erythroidin *n* erythroidine
erythro-Isomere *pl* erythro isomers
Erythronsäure *f* erythronic acid
Erythrophloein *n* erythrophloeine
Erythrose *f* erythrose, trihydroxybutyraldehyde
Erythrosephosphat *n* erythrose phosphate
Erythrosiderit *m* (Min) erythrosiderite
Erythrosin *n* erythrosin
Erythroson *n* erythrosone
Erythrozinkit *m* (Min) erythrozincite
Erythrozyt *m* erythrocyte, red blood cell, **farbloser** ~ achromacyte, achromatocyte, ghost corpuscle, shadow corpuscle
Erythrozytenresistenz *f* erythrocyte resistance
Erythrozytensenkung *f* erythrocyte sedimentation
Erythrozytenverminderung *f* diminution of the red cells, lowered red cell count, oligocyth[a]emia
Erythrozytenzählung *f* red blood count, R.B.C., red cell count
Erythrozytenzerfall *m* destruction of red blood cells, erythrocytoschisis
Erythrulose *f* erythrulose
Erz *n* ore, **abbauwürdiges** ~ pay ore, **geröstetes** ~ roasted ore, calcined ore, **gezogenes** ~ drawn ore, **hochhaltiges** ~ high-grade ore, **kieselreiches** ~ siliceous ore, **leicht reduzierbares** ~ easily reducible ore, **leicht schmelzbares** ~ easily fusible ore, **magnetisches** ~ magnetic ore, **minderhaltiges** ~ low-grade ore, **reichhaltiges** ~ high-grade ore, **selbstgehendes** ~ self-fluxing ore, **sulfidisches** ~ sulfide ore, **ungeröstetes** ~ raw ore, unroasted ore, **unvollkommen geröstetes** ~ incompletely roasted ore
Erzabfälle *m pl* tailings
Erzader *f* ore vein, lode seam, ore lode
Erzanreicherung *f* enrichment of ore
Erzart *f* type of ore
Erzaufbereitung *f* ore dressing, preparation of ore
Erzaufbereitungsanlage *f* ore-dressing plant
Erzbergbau *m* ore mining
Erzbergwerk *n* ore mine
Erzbrecher *m* ore breaker, ore crusher
Erzbrikettierung *f* ore briquetting
Erze *n pl* **brechen** to crush ores
erzeugen to produce, to make, (Chem) to generate, (Waren) to manufacture (goods)
Erzeuger *m* generator, producer
Erzeugerkolben *m* generating flask
Erzeugnis *n* product
Erzeugung *f* production, output; (Chem) generation
Erzeugungsmenge *f* output
Erzeugungsoperator *m* production operator
Erzeugungswärme *f* heat of generation
Erzfarbe *f* bronze color
Erzförderung *f* ore mining
Erzförderwagen *m* ore wagon
Erzformation *f* ore formation
Erzfrischverfahren *n* ore process, direct process
erzführend ore bearing
Erzgemenge *n* mixture of ores
Erzgemisch *n* mixture of ores
Erzgewinnung *f* ore mining
Erzgicht *f* charge of ore, ore charge
Erzglühfrischen *n* refining with iron ore
erzhaltig ore bearing, metalliferous
Erzladevorrichtung *f* ore loading device
Erzlagerplatz *m* ore storing place
Erzlaugegrube *f* ore leaching pit
Erzlaugerei *f* ore leaching plant
Erzmöllerung *f* ore burdening
Erzpocheisen *n* bucking iron
Erzpreßstein *m* ore briquette
Erzpuddeln *n* ore puddling
Erzquetsche *f* ore crusher, stone breaker
erzreich rich in ore
Erzröstofen *m* ore-drying kiln
Erzröstung *f* ore roasting
Erzsatz *m* charge of ore
Erzschaum *m* scoria, cinder, slag
Erzscheider *m* ore separator, ore screener
Erzscheidung *f* sorting of ores

Erzschlacke *f* ore slag
Erzschlämmen *n* washing of ores
Erzschlamm *m* ore slime, ore sludge
Erzschlich *m* concentrates of ore, dressed ore
Erzschmelzen *n* ore smelting, pig-and-ore process
Erzschmelzofen *m* ore-smelting furnace
Erzstahl *m* ore steel
Erzstaub *m* fines, ore dust
Erzstock *m* ore body, solid ore deposit
Erztagebau *m* open-pit ore mining
Erztasche *f* ore storing reservoir
Erztrübe *f* ore pulp, waste ore
Erzverteiler *m* ore distributer
Erzwäsche *f* ore washing
Erzwagen *m* ore wagon
Erzwalzwerk *n* ore crushing mill
Erzzerkleinerung *f* ore crushing
Erzzerkleinerungsanlage *f* ore-breaking plant, ore-crusher plant
Erzzerreiber *m* ore grinder
Erzziegel *m* ore briquette
Erzzusatz *m* addition of ore
Eschenholzöl *n* ash tree oil
Eschenrinde *f* (Bot) ash bark
Eschenwurzel *f* (Bot) dittany root, white dittany
Escherichia coli bacterium coli, coli-bacillus, E. coli, Escherich's bacillus
Escholerin *n* escholerine
ESE (elektrostatische Einheit) electrostatic unit (esu.)
Eseräthol *n* eseretholе
Eseridin *n* eseridine
Eserin *n* eserine, physostigmine
Eserinöl *n* eserine oil
Eserinsalicylat *n* eserine salicylate
Eserinsulfat *n* eserine sulfate
Eserolin *n* eseroline
Eskobeize *f* (HN) esco mordant
ESMA Abk. für Elektronenstrahl-Mikroanalyse
Esmarkit *m* (Min) esmarkite
ESO Abk. für epoxidiertes Sojaöl
ESR Abk. für Elektronenspinresonanz
eßbar edible
Esse *f* forge, chimney, hearth
Essenkanal *m* chimney flue
Essenkohlenkoks *m* forge coal coke
essentiell essential
Essenz *f* essence
Essig *m* vinegar, **verwandeln in** ~ to acetify
Essigäther *m* (obs) acetic ether, acetic ester, acetidin, ethyl acetate
Essiganlage *f* vinegar plant
essigartig vinegar like, acetic, acetous
Essigbakterien *pl* acetobacteria
Essigbereitung *f* vinegar making
Essigbildung *f* acetification
Essigdämpfe *m pl* vinegar vapors
Essigessenz *f* vinegar essence

Essigester *m* ethyl acetate
Essigesterlösung *f* solution in ethyl acetate
Essiggärung *f* acetic fermentation, acetification
essighaltig containing vinegar
Essighonig *m* oxymel
Essigkonserve *f* pickle
Essigmesser *m* acetometer
Essigpilz *m* vinegar mother
Essigsäure *f* acetic acid, ethanoic acid, ethyl acid
Essigsäureäther *m* (obs) acetic ether, acetic ester, ethyl acetate
Essigsäureäthylester *m* ethyl acetate, acetic ester, (obs) acetic ether
Essigsäureamylester *m* amyl acetate
Essigsäureanhydrid *n* acetic anhydride
Essigsäureanlage *f* acetic acid plant
Essigsäurebazillus *m* (Bakt) acetobacter
Essigsäurebereitung *f* acetification
Essigsäureeisenbeize *f* iron acetate mordant
Essigsäuregärung *f* acetic fermentation
Essigsäuremesser *m* acetometer
Essigsäuremessung *f* acetometry
Essigsalz *n* acetate
essigsauer acetate
Essigschaum *m* flower of vinegar
Essigsprit *m* triple vinegar, vinegar essence
Essigwasser *n* vinegar water
essigweinsauer acetotartrate
Eßkohle *f* forge coal
Ester *m* (Chem) ester
Esterase *f* (Biochem) esterase
Esterbildung *f* ester formation, esterification
Estergummi *n* ester gum
esterifizieren to esterify
Esterkondensation *f* ester condensation
Esterpyrolyse *f* ester pyrolysis
Esterrückgewinnungsanlage *f* ester recovery plant
Estersäure *f* ester acid
Esterverseifung *f* saponification of esters
Esterzahl *f* ester number
Estomycin *n* estomycin
Estragol *n* estragole
Estragon *m* (Bot) tarragon
Estragonessig *m* tarragon vinegar
Estragonöl *n* tarragon oil
Estrich *m* plaster
Estrichgips *m* estrich gypsum, flooring plaster, hydraulic gypsum
ESU-Verfahren *n* electroslag remelting
Etagenofen *m* multiple hearth roaster
Etagenpresse *f* multiple opening press, daylight press, multidaylight press, multiplaten press
Etagenröstofen *m* multiple-hearth roaster
Etagen-Schiebetischanlage *f* (Kunststoffe) automatic multiple-station sealing system (for plastics)
Etamiphyllin *n* etamiphyllin
Etamycinsäure *f* etamycin acid

Etardreaktion *f* Etard's reaction
Eternit *n* (HN, Asbestzement) asbestos cement, eternit
Ethaverin *n* ethaverine
Ethinamat *n* ethinamate
Ethypicon *n* ethypicone
Etikett *n* label
Etikettenbedruckmaschine *f* label printing machine
Etikettenlack *m* label lacquer
etikettieren to label
Etikettiermaschine *f* label[l]ing machine
Etoxeridin *n* etoxeridine
Ettringit *m* (Min) ettringite
Etui *n* box
Eucain *n* eucaine
Eucainhydrochlorid *n* eucaine hydrochloride
Eucalyptol *n* eucalyptole, cineole
Eucalyptusblätter *n pl* eucalyptus leaves
Eucalyptusöl *n* eucalyptus oil
Eucarv[e]ol *n* eucarv[e]ol
Eucarvon *n* eucarvone
Eucatropin *n* eucatropine
Euchinin *n* euchinine, quinine ethyl carbonate
Euchlorin *n* euchlorine
Euchroit *m* (Min) euchroite
Euchronsäure *f* euchroic acid
Euchrysin *n* euchrysine
Eucupin *n* eucupine
Eucupinsäure *f* eucupinic acid
Eudalin *n* eudalene
Eudesman *n* eudesmane
Eudesmen *n* eudesmene
Eudesmin *n* eudesmine
Eudesminsäure *f* eudesmic acid
Eudesmol *n* eudesmol
Eudialyt *m* (Min) eudialite
Eudidymit *m* (Min) eudidymite
Eudiometer *n* (Phys) absorption tube, eudiometer
eudiometrisch eudiometric
Eudnophit *m* (Min) eudnophite
Eudoxin *n* eudoxine, bismuth tetraiodophenolphthalein
Eugallol *n* eugallol, pyrogallol monoacetate
Eugenglanz *m* (Min) eugene glance, polybasite
Eugenin *n* eugenin
Eugenitin *n* eugenitin
Eugenol *n* 1-allyl-3-hydroxy-4-methoxybenzene, eugenol
Eugenon *n* eugenon
Eugensäure *f* eugenic acid
Eugentiogenin *n* eugentiogenin
Eugenylacetat *n* eugenyl acetate
Eugenylbenzoat *n* eugenyl benzoate
Eugenylbenzyläther *m* eugenyl benzyl ether
Euglobulin *n* euglobulin
Eugoform *n* eugoform
Eukain *n* eucaine

Eukairit *m* (Min) eucairite
Eukalyptol *n* eucalyptol, cajeputole, cineole
Eukalyptusöl *n* eucalyptus oil
Eukamptit *m* (Min) eucamptite
Eukanolbinder *m* (HN) eukanol binder
Eukasin *n* eucasin, ammonium caseinate
Euklas *m* (Min) euclase
Eukodal *n* eucodal, dihydrohydroxy codeinone
Eukolit *m* (Min) eucolite
Eukolloid *n* (Phys) eucolloid
Eukrasit *m* eucrasite
Eukryptit *m* (Min) eucryptite
Eulachonöl *n* eulachon oil, candle fish oil
Eulatin *n* eulatin
Eulaxan *n* eulaxane
Eulersche Gleichungen *f pl* (Mech) Euler-Lagrange equations, Lagrange equations
Eulysit *m* (Min) eulysite
Eulytin *m* (Min) eulytin
Eumenol *n* eumenol
Eumydrin *n* eumydrine, methylatropine nitrate
Eunatrol *n* eunatrol, sodium oleate
Eupatorin *n* eupatorin
Euphorbin *n* euphorbin
Euphorbium *n* euphorbium
Euphorin *n* euphorine, phenylurethane
Euphosterol *n* euphosterol
Euphthalmin *n* euphthalmine
Euphyllin *n* euphylline
Euphyllit *m* euphyllite
Euporphin *n* euporphine
Eupyrchroit *m* (Min) eupyrchroite
Eupyrin *n* eupyrine
Euralith *m* (Min) euralite
Euratom (Europäische Gemeinschaft für Atomenergie) European Atomic Energy Community
Euresol *n* (HN) euresol, resorcinol monoacetate
Eurhodin *n* eurhodine
Eurit *m* eurite
Europhen *n* europhen
Europium *n* europium (Symb. Eu)
Eusantonan *n* eusantonane
Eusapyl *n* eusapyl
Eustenin *n* eustenin, theobromine sodium iodide
Eusynchit *m* (Min) eusynchite
Eutannin *n* eutannin
Eutectan *n* eutectane
Eutektikum *n* eutectic, eutectic alloy, eutectic mixture
eutektisch eutectic
Eutektoid *n* eutectoid
Euthiochronsäure *f* euthiochronic acid
Euxanthan *n* euxanthane
Euxanthin *n* euxanthin
Euxanthinsäure *f* euxanthic acid, euxanthinic acid
Euxanthogen *n* euxanthogene

Euxanthon n 1,7-dihydroxyxanthone, euxanthone, porphyric acid
Euxenit m (Min) euxenite
evakuieren to evacuate
evakuiert evacuated, exhausted
Evakuierung f evacuation
Evansit m (Min) evansite
Evaporator m evaporator, vaporizer
evaporieren to evaporate
Evatromonosid n evatromonoside
Evektion f (Astr) evection
Evektionstiden fpl evectional tides
Everninsäure f everninic acid
Evernsäure f evernic acid
Evernursäure f evernuric acid
Evigtokit m (Min) evigtokite
Evipan n (HN) evipan
Evoden n evodene
Evodiamin n evodiamine
Evodin n evodine
Evodionol n evodionol
Evolatin n evolatine
Evolitrin n evolitrine
Evolution f evolution
Evolutionstheorie f (Biol) theory of evolution
Evolvente f (Geom) involute
Evolventenpumpe f involute pump, volute pump
Evomonosid n evomonoside
Evoxanthin n evoxanthine
Exalgin n exalgin, methylacetanilide
Exalton n exaltone
Exemplar n specimen
exemplarisch exemplary
exergon[isch] exergonic
Exhaustdampf m exhaust steam
Exhaustergebläse n exhauster
Exhaustor m exhauster, dust catcher, ventilator
Eximin n eximine
existenzfähig capable of existing
existieren to be, to exist, to subsist
Exkrement n excrement
Exkretion f excretion
exocyclisch exocyclic
Exoelektronen pl exoelectrons
Exoenzym n exo-enzyme
exogen exogenic
Exonuklease f (Biochem) exonuclease
Exopeptidase f (Biochem) exopeptidase
Exoplasma n ectoplasm, exoplasm
Exosmose f (Phys) exosmosis
exosmotisch exosmotic
exotherm (exothermisch) exothermal, exothermic, giving heat
Exotoxin n exotoxin
expandieren (Gas) to expand
Expandierraum m expansion chamber
Expansin n expansin
Expansion f dilatation, expansion

Expansionsadiabate f adiabatic curve of expansion
Expansionskammer f (Atom) expansion chamber
Expansionskraft f expansive force
Expansionskühlung f (Therm) dynamic cooling
Expansionsmaschine f expansion engine
Expansionsturbine f expansion turbine, turbo expander
Expansionsventil n expansion valve
Expansionsvermögen n expansive capacity, expansivity
Expedition f forwarding office
Expektorans n (Pharm) expectorant
Experiment n experiment, test, **langes** ~ longterm run
Experimentalchemie f experimental chemistry
Experimentalphysik f experimental physics
Experimentator m experimentalist, experimentor
experimentell experimental
experimentieren to experiment
Experimentierschnur f connecting lead, experiment lead
Experimentiertisch m experimenting table
Experimentiertransformator m demonstration transformer
explizit explicit
explodierbar explosive
Explodierbarkeit f explosiveness
explodieren to explode, to burst, to detonate
explodierend explosive, exploding
explosibel explosive, explodable
Explosion f explosion, detonation
explosionsartig explosive, ~ **verbrennen** to deflagrate
Explosionsdruck m explosion pressure
explosionsfähig explodable, explosive
Explosionsfähigkeit f explosive property
Explosionsgefahr f danger of explosion, explosion hazard
Explosionsgemisch n explosive mixture
Explosionsgrenze f explosion limit, **obere** ~ upper explosion limit, **untere** ~ lower explosion limit
Explosionskalorimeter n bomb calorimeter, combustion bomb
Explosionskraft f explosive force
Explosionskugel f explosion bulb
Explosionsmotor m combustion engine
Explosionsreaktion f explosion reaction
explosionssicher explosion proof
Explosionssicherheit f safety against explosion accidents
Explosionsspektrum n explosion spectrum
Explosionsstoß m explosive impact, explosion impulse
Explosionstrocknung f puff-drying
Explosionsursache f cause of explosion
Explosionsvorgang m explosive process
Explosionswelle f explosion wave

Explosionswirkung *f* explosive effect
explosiv explosive
Explosivkraft *f* explosive power
Explosivstoff *m* explosive
Explosivstoffchemie *f* chemistry of explosives
Exponent *m* (Math) exponent, **gebrochener** ~ (Math) fractional exponent
exponential exponential
Exponentialfunktion *f* exponential function
Exponentialgesetz *n* exponential law
Exponentialgleichung *f* (Math) exponential equation
Exponentialgröße *f* exponential
Exponentialintegral *n* (Math) exponential integral
Exponentialkurve *f* exponential curve
Exponentialreihe *f* exponential series
exponentiell exponential
exponieren to expose
exportieren to export
Expositionsfaktor *m* exposure factor
Exsikkator *m* desiccator
Exsikkatoraufsatz *m* desiccator cover
Exsikkatoreinsatz *m* desiccator plate
Exsikkatorfett *n* desiccator grease
Exsikkatorfüllung *f* desiccant
Exsudat *n* exudate
extemporieren to extemporize, to improvise
Extender *m* coupler
Extinkteur *m* extinguisher
Extinktion *f* extinction
Extinktionskoeffizient *m* extinction coefficient, total-reflection coefficient, **molarer** ~ molar extinction coefficient
Extinktionskurve *f* extinction curve, absorption curve
Extraausbeute *f* extra yield
extrahierbar extractable, extractible
Extrahierbarkeit *f* extractibility
extrahieren to extract, to eliminate, to separate
Extrahieren *n* extracting, **flüssig-flüssig** ~ liquid-liquid extraction
Extrakt *m* extract
Extraktbrühe *f* extracted liquor
Extrakteur *m*, **Band-** ~ belt extractor, **Becherwerks-** ~ bucket-conveyor extractor, **Karussell-** ~ rotary extractor, **Korbband-** ~ basket extractor, **Schnecken-** ~ screw-conveyor extractor, **Zellenrad-** ~ bucket-wheel extractor
Extraktgehalt *m* extract content
Extraktion *f* extraction, **kontinuierliche** ~ continuous extraction
Extraktionsanalyse *f* extraction analysis
Extraktionsanlage *f* extraction plant
Extraktionsapparat *m* extraction apparatus

Extraktionsbatterie *f* extraction battery
Extraktionsbenzin *n* extraction naphtha
Extraktionsdauer *f* time required for extraction
Extraktionsgefäß *n* extraction vessel
Extraktionshülse *f* extraction thimble, fitter cup
Extraktionskolonne *f* extraction column
Extraktionsmittel *n* extracting agent
Extraktionszentrifuge *f* extraction centrifuge
extraktiv extractive
Extraktivstoff *m* extractive matter, extractive principle
Extraktor *m* extractor
Extraktoraufsatz *m* extractor attachment
Extraktstoff *m* extractive substance
extramolekular extramolecular
Extraordinariat *n* associate professorship
Extraordinarius *m* associate professor
Extrapolation *f* extrapolation
Extrapolationsformel *f* extrapolation formula
Extrapolationskammer *f* extrapolation chamber
extrapolieren to extrapolate
Extraresonanzenergie *f* extra resonance energy
extraterrestrisch extraterrestrial
extrazellulär extracellular, exocellular
Extrem *n* extreme
Extremfall *m* extreme case
Extremität *f* extremity
Extremwerte *pl* (v. Funktionen) extreme values
Extruder *m* extruder
Extruderventile *pl* controllable valve arrangement
Extrudieren *n* extruding
Extrusionsdruck *m* head pressure
Extrusivgestein *n* extrusive rocks
Exzenter *m* eccentric, eccentric attachment, (Auto) cam
Exzenterantrieb *m* eccentric drive
Exzenterbolzen *m* eccentric bolt
Exzenterbügel *m* eccentric hoop, eccentric strap
Exzenterdrehbank *f* eccentric lathe
Exzentergabel *f* eccentric fork
Exzenterpresse *f* eccentric breaker, eccentric press
Exzenterregulator *m* eccentric governor
Exzenterring *m* eccentric ring
Exzenterscheibe *f* eccentric sheave
Exzenterschleifmaschine *f* eccentric grinder
Exzenterschneckenpumpe *f* eccentric single-rotor screw pump
Exzenterstange *f* eccentric rod
Exzenterstanze *f* eccentric punch
Exzenterwelle *f* camshaft
exzentrisch eccentric
Exzentrizität *f* eccentricity
Exzerpt *n* excerpt, extract
Exzeßgrößen *pl* excess factors (for non-ideality of mixtures)

F

Fabiatrin *n* fabiatrin
Fabrik *f* factory, mill, plant, works
Fabrikabwasser *n* industrial waste water
Fabrikanlage *f* factory plant
Fabrikant *m* manufacturer
Fabrikarbeiter *m* factory worker, factory hand, industrial worker
Fabrikat *n* brand, article, manufacture
Fabrikation *f* manufacture, production
Fabrikationsabfall *m* waste
Fabrikationsanlage *f* large-scale plant
Fabrikationsfehler *m* factory defect
Fabrikationskontrolle *f* production control
Fabrikationsleiter *m* production manager
Fabrikationsnummer *f* production lot
Fabrikationspartie *f* lot
Fabrikationsplan *m* manufacturing program, manufacturing scheme
Fabrikationsprogramm *n* manufacturing program, production scheme
Fabrikationsprozeß *m* manufacturing process
fabrikatorisch industrial
Fabrikatur *f* fabrication, manufacture, production
Fabrikbetrieb *m* factory operation, industrial plant
Fabrikkante *f* (Karton) body joint, manufacturer's joint
fabrikmäßig by factory methods, industrially; on a factory scale, ~ **herstellen** to manufacture on an industrial scale
Fabrikmarke *f* brand, trademark
fabrikneu brand-new
Fabriknormen *f pl* works standards
Fabrikofen *m* furnace
Fabrikpraxis *f* factory practice
Fabrikpreis *m* factory price, price ex works, [works] cost price
Fabrikware *f* manufactured article
Fabrikwesen *n* manufacturing industry
Fabrikzeichen *n* trade mark
fabrizieren to make, to manufacture, to produce
Facette *f* (eines Kristalls) facet (of a crystal)
Facettenauge *n* (Zool) compound eye
facettiert faceted
fach- expert..., -fold, specialist..., technical...
Fach *n* compartment, division, partition, (Fachgebiet) branch, special subject, (Schrank) row, shelf
Facharbeiter *m* technical worker, skilled worker, specialist
Fachausbildung *f* professional education, technical training
Fachausdruck *m* technical expression, technical term

Fachausdrücke *m pl* terminology, **chemische** ~ chemical terminology
Fachausschuß *m* committee of experts
Fachberater *m* technical adviser
Fachblatt *n* technical journal
Fachgebiet *n* field, special branch
fachgemäß workman-like
Fachgestell *n* rack
Fachkenntnis *f* expert knowledge, business knowledge, special knowledge
fachkundig competent, expert
fachlich professional, technical
Fachliteratur *f* scientific literature, technical literature
fachmännisch expert, professional, technical
fachmäßig professional
Fachmann *m* expert, specialist
Fachpersonal *n* experienced personnel
Fachrichtung *f* nature of activity
Fachschule *f* technical college
Fachsprache *f* technical terminology, technical terms
Fachstudium *n* technical studies
Fachwerk *n* framework, latticework, panel[l]ing
Fachwissenschaft *f* special branch of science
Fachzeitschrift *f* technical journal
Fackel *f* torch, flare
Fackelkohle *f* cannel coal
fackeln to blaze, to flare
Façon *f* s. Fasson
fade insipid, flat, stale
Faden *m* thread, (Bindfaden) string, (Faser) fiber (Am. E.), fibre (Br. E.), (Glühlampe) filament, (Gummifaden) elastic, **gezogener** ~ extruded strand
fadenähnlich filamentous
Fadenaufhängung *f* fiber suspension, thread suspension
Fadenbakterium *n* (Bakt) trichobacterium, threadlike bacterium, trichobacterium
Fadenbruch *m* (Text) yarn break
Fadendichte *f* (Reifen) cord count, number of ends per inch
Fadenelektrometer *n* thread electrometer
fadenförmig thread-like, fibrous, filamentary, filiform
Fadenführer *m* thread guide, pigtail guide
Fadengalvanometer *n* thread galvanometer, Einthoven's galvanometer, string galvanometer
Fadengitter *n* chain lattice
Fadenglas *n* reticulated glass, spun glass
Fadengold *n* gold thread
Fadenhygrometer *n* hair hygrometer
Fadenkorrektur *f* (Thermometer) stem correction

Fadenkorrosion *f* filiform corrosion
Fadenkreuz *n* (aus Draht) cross-lines, cross-wires, (aus Haar) cross-hairs
Fadenlänge *f* length of thread
Fadenlunker *m* pinhole
Fadenmikrometer *n* (Opt) micrometer eye-piece
Fadenmolekül *n* fiber molecule, linear macromolecule, linear molecular chain
Fadenpilz *m* hyphomycete
Fadenprobe *f* (Zuck) string test
fadenscheinig threadbare
Fadenschwefel *m* sulfur threads
Fadensilber *n* silver thread
Fadenstein *m* (Min) inolite
Fadenstoff *m* (Reifen) creel fabric, weftless fabric
Fadenstrahlrohr *n* fine beam tube
Fadenthermometer *n* faden thermometer
Fadentransistor *m* (Elektr) filamentary transistor
Fadenwächter *m* (Text) thread guide
Fadenwinkel *m* (Reifen) cord angle, crown angle
Fadenwurm *m* threadworm, filaria
Fadenzähler *m* (Text) thread counter
Fadenzahl *f* (Text) thread count
Fadenziehen *n* (Lack) cobwebbing
fadenziehend ropy, stringy
Fadeometer *n* fadeometer
Fading *n* (Radio) fading
Fadingregulation *f* (Radio) fading control
Fäces *pl* feces
fächeln to fan
Fächer *m* fan
Fächerbrenner *m* fantail burner
Fächerstein *m* (Min) prochlorite, ripidolite
Faecosterin *n* fecosterol
Fähigkeit *f* ability, capability, capacity, faculty, potentiality, power
Fährte *f* trace, track, trail
fäkal faecal, fecal
Fäkalien *pl* fecal substances, feces
Fäkalstoffe *m pl* f[a]eces, fecal substances
Fäkalwasser *n* fecal water, sewage water
Fäkulometer *n* feculometer
Fällanlage *f* refining plant
Fällapparat *m* precipitation apparatus
Fällbad *n* electrolytic bath, precipitating bath, refining bath
fällbar (Chem) precipitable
Fällbarkeit *f* (Chem) precipitability
Fällbottich *m* precipitating vat
fällen (ausfällen) to precipitate, (umfällen) to cut, to fell
Fällen *n* (Chem) precipitating
Fäller *m* precipitation apparatus, precipitator
Fällflüssigkeit *f* (Chem) precipitating liquid
fällig due, mature
Fälligkeitstermin *m* dead line, date of maturity
Fällkessel *m* precipitating vessel

Fällmethode *f* precipitation method
Fällmittel *n* (Chem) precipitant, precipitating agent
Fällprodukt *n* (Chem) precipitate
Fällsilber *n* deposited silver, precipitated silver
Fällung *f* (Chem) precipitation, (Niederschlag) precipitate, **elektrolytische** ~ precipitation by electrolysis, electrodeposition, **fraktionierte** ~ fractionated precipitation
Fällungsanalyse *f* volumetric precipitation analysis
Fällungsmittel *n* precipitating agent, coagulant, precipitant
Fällungspolymerisation *f* precipitation polymerization
Fällungsprodukt *n* precipitate, deposit
Fällungsreaktion *f* precipitation reaction
Fällungsregel *f* law of precipitation
Fällungsunterlage *f* substrate
Fällungswert *m* precipitation value
fälschen to counterfeit, to falsify, (z. B. Wein) to adulterate
fälschlich false[ly], incorrect
Fälschung *f* falsification, forging, (z. B. Wein) adulteration
Fälschungsmittel *n* adulterant
Fälschungsstoff *m* adulterant
Fängerglocke *f* valve guard
färbbar colorable, dyeable, stainable
Färbbarkeit *f* colorability, dyeability, stainability
Färbeanlage *f* tinting plant
Färbebad *n* dye bath, dye fluid
Färbebecher *m* dye beaker, test beaker for dyers
Färbefähigkeit *f* absorbing power for dyes, receptivity for dyes
Färbeflechte *f* (Bot) capeweed, orseille weed
Färbeflotte *f* (Färb) dye liquor
Färbeflüssigkeit *f* dye fluid, dye liquor
Färbegerberei *f* dye tanning
Färbegeschwindigkeit *f* rate of dyeing
Färbeginster *m* (Bot) dyer's broom, green weed
Färbehilfsmittel *n* dyeing auxiliary, coloring agent
Färbeindex *m* (Farbindex, Med) color index, blood quotient, C.I.
Färbejigger *m* (Färb) dye jig
Färbekraft *f* coloring power, dyeing capacity, tinctorial power, tinctorial strength
Färbekufe *f* (Färb) dyeing vat
Färbemalz *n* high kilned malt
Färbemaschine *f* dyeing machine
Färbemittel *n* coloring matter, pigment dye
färben to dye, to color, to tinge, (Kristall) to stain
Färben *n* dyeing, ~ **von Zinkographien** (pl) chromozincotypy
Färbenapf *m* staining block
färbend coloring, dyeing

Färber *m* dyer
Färberalkanna *f* (Bot) alkanet, dyer's bugloss
Färberbaum *m* (Bot) fustet, Venice sumach
Färberbaumholz *n* sumach
Färberbeere *f* (Bot) purging buckthorn
Färberblume *f* (Bot) yellow camomile
Färberbraun *n* rocella brown
Färberdistel *f* (Bot) bastard saffron, dyeing carthamus, dyer's safflower
Färberei *f* dye house, (Gewerbe) dyeing
Färbereiche *f* (Bot) black oak, dyer's oak, quercitron
Färbereihilfsmittel *n* (Färbehilfsmittel) dyeing auxiliary
Färbereimaschine *f* dyeing machinery
Färberflechte *f* (Bot) dyer's moss, orseille weed
Färberflotte *f* dye fluid, dyer's bath
Färbergallwespe *f* (Zool) ink gallfly
Färberginster *m* (Bot) dyer's broom, green broom
Färbergras *n* (Bot) wild woad
Färberkamille *f* (Bot) yellow camomile
Färberkraut *n* base broom, dyer's broom
Färberkreuzdorn *m* (Bot) dyer's buckthorn
Färberküpe *f* dye beck, dye tub
Färberlack *m* lac dye, lake dye
Färbermaulbeerbaum *m* dyer's mulberry
Färbermoos *n* dyer's lichen, dyer's moss
Färbernessel *f* (Bot) gunnera
Färberöl *n* dyer's oil
Färberpfrieme *f* (Bot) dyer's broom, green broom
Färberrinde *f* quercitron bark
Färberröte *f* dyer's madder, root of madder
Färberrot *n* alizarin, madder
Färbersaflor *m* (Bot) bastard saffron, dyer's safflower
Färberscharte *f* dyer's sawwort
Färbersüßblatt *n* (Bot) sweetleaf
Färbersumach *m* fustet, Venice sumach
Färberwaid *m* dyer's woad, wild indigo
Färberwaldmeister *m* (Bot) dyer's woodruff
Färberwurzel *f* (Bot) dyer's madder
Färbevermögen *n* coloring power
Färbewanne *f* dye bath
Färbung *f* coloration, color[ing]
Fässerheizmantel *m* barrel heating mantle
Fässerpumpe *f* barrel pump
Fässerrüttelmaschine *f* barrel shaking machine
Fässerstapler *m* barrel stacking machine
Fäulanlage *f* fermenting plant
Fäule *f* putrefaction, rot, rottenness, (Brand) blight
Fäulnis *f* decay, decomposition, putrefaction, putridness, rot, rottenness, sepsis
Fäulnisalkaloid *n* putrefactive alkaloid, ptomaine
Fäulnisbakterien *pl* putrefactive bacteria

Fäulnisbeständigkeit *f* rotproofness, imputrescibility
fäulnisbewirkend putrefactive
fäulnisecht rot-proof
Fäulniserreger *m* putrefactive agent, (Käseherstellung) bacterial starter
Fäulnisgärung *f* putrefactive fermentation, putrefaction
Fäulnisgeruch *m* putrefactive odor, putrid odor, rotten odor
fäulnishemmend aseptic, antiseptic
Fäulniskeim *m* putrefaction germ
Fäulnismittel *n* antiseptic, germicide, preservative
Fäulnisprodukt *n* decomposition product, product of putrefaction
Fäulnisprozeß *m* process of decomposition, putrefaction process
fäulnisverhindernd antifouling, antiputrefactive, antiseptic, rot-preventing
fäulnisverhütend antifouling, antirot, antiseptic, preservative
fäulniswidrig antifouling, antiseptic, rotproof
Fagaramid *n* fagaramide
Fagarasterol *n* fagarasterol
Fagarin *n* fagarine
Fagarol *n* fagarol
Fagarsäure *f* fagaric acid
Fageren-Zelle *f* Fageren flotation cell
Fagin *n* fagine
fahl fallow, dun
fahlblau pale blue
Fahlerz *n* (Min) fahl ore, fahlerz, grey copper ore, tetrahedrite
fahlfarben fallow colored
fahlgelb fallow, pale yellow
fahlgrau grayish, greyish
Fahlheit *f* fallowness, paleness
Fahlkupfererz *n* (Min) fahl copper ore
Fahlleder *n* russet leather, shaft leather
fahlrot fawn colored, pale red
Fahlstein *m* (Min) light gray slate, pale gray slate
Fahne *f* (Masch) lug
Fahnenabzieher *m* proof puller
Fahnenabzug *m* (Buchdr) galley [proof]
Fahrantrieb *m* travelling gear
fahrbar mobile, movable, transportable
Fahrenheit-Skala *f* Fahrenheit scale
Fahrenheit-Thermometer *n* Fahrenheit thermometer
Fahrenwald-Zelle *f* Fahrenwald flotation cell
Fahrgestell *n* chassis, undercarriage
fahrlässig careless, negligent
Fahrlässigkeit *f* carelessness, negligence
Fahrpark fleet of vehicles
Fahrrohre *pl* forwarding [or transmission] tubes
Fahrschalter *m* (am Motor) controller
Fahrstuhl *m* elevator (Am. E.), lift (Br. E.)

Fahrwerk *n* traversing gear, undercarriage
Fahrzeug *n* vehicle
Fahrzeugvolumen-Ausnutzung *f* exploitation of vehical capacity
Fairfieldit *m* (Min) fairfieldite
Fajans-Soddyscher Verschiebungssatz *m* displacement law of Soddy, Fajans and Russel
Faksimiledruck *m* (Buchdr) facsimile printing
Faksimileschreiber *m* facsimile recorder
Faktis *m* (HN) factice, rubber substitute, **weißer** ~ white [rubber] substitute
faktisch actual, real
Faktismasse *f* mass of factice
Faktor *m* factor, **sterischer** ~ steric factor
Faktorenanalyse *f* (Math) factor analysis
Faktorentabelle *f* (Math) table of factors
Faktorenzerlegung *f* factorization
Faktorgruppe *f* factor group
Faktorisierungsmethode *f* (Math) factorization method
Faktoröl *n* sulfurized oil
Fakultät *f* faculty, (Math) factorial
falb fallow, pale, pale yellow
Falcatin *n* falcatine
Fall *m* decline, drop, fall, (Beispiel) instance, (Umstand) case, **auf jeden** ~ in any case, **auf keinen** ~ on no account, **freier** ~ free fall
Fallbehälter *m* gravity tank
Fallbeschleunigung *f* (Phys) acceleration of fall, gravitational acceleration
Fallblock *m* halliard block
Fallbolzentest *m* dropping dart test
Fallbügelregler *m* chopper bar controller
Falle *f* trap
Falleisen *n* latch
Falleitblock *m* halliard leading block
fallen to drop, to fall, (absteigen) to descend, ~ **lassen** to drop
Fallfilm *m* falling film
Fallfilmkolonne *f* wetted-wall column
Fallfilmkratzverdampfer *m* falling film scraped surface evaporator
Fallfilmkühler *m* falling film cooler
Fallfilmverdampfer *m* [falling] film evaporator, falling film vaporizer
Fallgeschwindigkeit *f* (Phys) rate of fall, velocity of a falling body
Fallgesetz *n* (Phys) law of gravitation
Fallhärte *f* impact ball hardness
Fallhärteprüfer *m* impact ball hardness tester
Fallhärteprüfung *f* impact ball hardness test, scleroscope test
Fallhammer *m* drop hammer
Fallhammerform *f* drop hammer die
Fallhammergerät *n* drop hammer tester
Fallhöhe *f* height of fall, height or length of drop
Fallinie *f* fall line

Fallklappe *f* flap
Fallkugel *f* drop weight
Fallkugelmethode *f* dropping ball method
Fallkurve *f* trajectory
Fallmoment *m* (Phys) falling moment
Fallparabel *f* (Phys) trajectory parabola
Fallprobe *f* fall test
Fallrohr *n* downcomer, down main, downpipe, service pipe, vertical pipe, waste pipe
Fallrohrpumpe *f* gravity tube pump
falls in case, in the event that, provided that
Fallschacht *m* chute
Fallschirm *m* parachute
Fallstromschwelung *f* downdraft gasification
Fallstromverdampfer *m* falling film evaporator, falling film vaporizer
Falltür *f* discharge door
Fallversuch *m* impact test, shock test
Fallweg *m* downward path
Fallwerk *n* pig breaker, pile driver, stamp
Fallzeit *f* falling time
falsch false, incorrect, wrong, (künstlich) artificial, imitated
falschlastig incorrectly trimmed
Falschlicht *n* (Opt) false light, stray light
Falschluft *f* excess air, infiltrated air
Falschweisung *f* error in indication
Faltblattstruktur *f* pleated sheet structure
Falte *f* crease, fold, pleat, wrinkle
falten to fold
Faltenbalg *m* bellows
Faltenbalgsatz *m* spring and bellows assembly
Faltenband-Förderer *m* articulated or bucket-belt conveyor
Faltenbildung *f* wrinkle formation, wrinkling
Faltenfilter *m* (Chem) folded filter
faltenlos wrinkle-free
faltenreich creased, folded, wrinkled
Faltenschlauch *m* corrugated rubber tubing
Faltensiebzentrifuge *f* fluted screen centrifuge
Faltenwerfen *n* creasing, crimping, puckering
Falter *m* creaser, folder, (Zool) moth
faltig wrinkled, puckered
Faltkarte *f* folder
Faltkiste *f* **aus Wellpappe** corrugated box
Faltpunkt *m* plait point
Faltpunktkurve *f* plait point curve
Faltschachtel *f* (aus Karton) folding carton
Faltung *f* **einer Funktion** (Math) convolution of a function
Faltungsintegral *n* (Math) convolution integral
Faltungspunkt *m* plait point, point of folding
Faltversuch *m* bend test, folding test
Falunit *m* (Min) falunite
Falz *m* fold, (Dose) seam, (Zarge) groove, notch
Falzbaupappe corrugated board
Falzdeckeldose *f* open top can
falzen to fold, to crease, (Blechkanten) to bead, to seam

Falzfestigkeit – Farbenlage

Falzfestigkeit *f* folding endurance
Falzmaschine *f* folding machine
Falzverbindung *f* (Blech) seamed joint, (Bördelkante) bead
Falzwiderstand *m* folding endurance, folding resistance
Falzzange *f* pliers
Famatinit *m* (Min) famatinite
Familienpackung *f* economy size pack
Fangblech *n* baffle [plate], bank guard
Fangchinolin *n* fangchinoline
Fangdamm *m* cofferdam
Fangdraht *m* guard wire
fangen to capture, to catch, to trap
Fangraumniederschlagselektrode *f* pocket collecting electrode
Fangstoff *m* stuff from the save-all
Fangvorrichtung *f* safety catch
Farad *n* (Elektr) farad
Faraday *n* (Elektr) faraday
Faraday-Äquivalent *n* (elektrochem. Äquivalent) Faraday equivalent
Faraday-Gesetz *n* **der elektromagnetischen Induktion** (Elektr, Magn) Faraday law of electro-magnetic induction
Faradaykonstante *f* (Elektr) Faraday constant, F.
Faradayscher Käfig *m* Faraday cage
Faradiol *n* faradiol
Faratsihit *m* faratsihite
Farbabstufung *f* color gradation
Farbabweichung *f* chromatic aberration, color deviation
Farbabzug *m* (Phot) color print
Farbänderung *f* color change
Farbangleichung *f* color matching
Farbanpassung *f* (Opt) color matching
Farbauftragswalze *f* fountain roll
Farbausbeute *f* dyestuff yield
Farbbad *n* dye bath
Farbband *n* (Schreibmaschine) typewriter ribbon
Farbbase *f* color base
Farbbasenaufschluß *m* solubilized color base
Farbbeize *f* mordant
Farbbeständigkeit *f* color fastness, color stability
Farbbild *n* (Phot) color photo[graph]
Farbbluten *n* bleeding
Farbbrühe *f* dye liquor
Farbcharge *f* color batch
Farbdichte *f*, **spektrale** ~ (Opt) excitation purity
Farbdreieck *n* (Opt) chromaticity diagram
Farbdüse *f* dye nozzle
Farbe *f* color (Am. E.), colour (Br. E.), dye, paint, (Biol) pigment, ~ **erzeugend** chromogenic, color producing, **adjektive** ~ adjective color, **deckende** ~ body color, **feuerfeste** ~ fire-proof paint, **komplementäre** ~ complementary color, **plastische** ~ stipple paint, **substantive** ~ substantive color, **zusammengesetzte** ~ compound color
farbecht colorproof, of fast color, unfading
Farbechtheit *f* color fastness, color stability
Farbechtheitsprüfer *m* fadeometer
Farbechtheitsprüfung *f* color fastness test
Farbeffekt *m* color effect
Farbempfinden *n* color perception
farbempfindlich color-sensitive
Farbempfindlichkeit *f* color sensitivity
Farbenabbeizmittel *n* color remover, paint remover, paint removing agent
Farbenabstufung *f* gradation of colors
Farbenänderung *f* alteration of color
Farbenauftrag *m* laying-on of colors
Farbenbehandlung *f* color treatment
Farbenbestandteil *m* color constituent
Farbenbildung *f* formation of colors
Farbenbindemittel *n* color binder
farbenblind color-blind
Farbenblindheit *f* color blindness
Farbenbrechung *f* (Opt) color refraction
Farbenbrühe *f* coloring liquor
Farbenbuchdruck *m* chromotypography
Farbenchemie *f* color chemistry
Farbendiapositiv *n* (Phot) color slide
Farbendreieck *n* color triangle
Farbendruck *m* color printing, chromatic printing, lithochromatics
farbenempfindlich color sensitive, sensitive to colors
Farbenempfindlichkeit *f* color sensitivity, dye sensitivity
Farbenempfindung *f* color perception, color sensation
Farbenerde *f* colored earth
Farbenerscheinung *f* chromatic phenomenon
farbenerzeugend chromatogenic
Farbenerzeuger *m* chromogen
Farbenerzeugung *f* production of colors
Farbenfabrik *f* paint factory
Farbenfehler *m* color defect, chromatic defect
Farbenfilter *n* (Opt) color filter
Farbengang *m* color pit
Farbengebung *f* coloration, art of coloring
Farbenglanz *m* color brilliancy
Farbenglas *n* stained glass
Farbengrund *m* ground color
Farbenholz *n* dye wood
Farbenholzschnitt *m* chromoxylography
Farbenhomogenisiermaschine *f* color homogenizer, color homogenizing machine
Farbenindustrie *f* color and dye industry, dyestuffs industry
Farbenintensität *f* intensity of color
Farbenkörper *m* pigment, coloring matter
Farbenlack *m* color varnish, lake
Farbenlage *f* color coat

Farbenlehre *f* chromatics, chromatology, theory of colors
Farbenleiter *f* color scale
Farbenlithographie *f* chromolithography
Farbenmalz *n* roasted malt
Farbenmesser *m* colorimeter, chromatometer
Farbenmessung *f* colorimetry
Farbenmischung *f* mixture of colors
Farbenmischzylinder *m* paint mixer
Farbenmühle *f* color mill
Farbenofen *m* enamelling furnace, muffle
Farbenphotographie *f* color photography, (Aufnahme) color print
farbenprächtig brilliant
Farbenprobe *f* dye test
Farbenreiben *n* color grinding
Farbenreiber *m* color grinder
Farbenreibmaschine *f* color milling machine
Farbenreibstein *m* color grinding stone
farbenreich rich in color, colorful
Farbenreichtum *m* richness in color, variety of colors
Farbenschattierung *f* hue, shading
Farbenscheibe *f* color disc
Farbenschicht *f* color coat
Farbenschiller *m* iridescence, brilliancy of colors
farbenschillernd iridescent
Farbenschirm *m* (Opt) color screen
Farbenschmelzofen *m* enamelling furnace, muffle
Farbenschwund *m* fading
Farbensehen *n* color perception
Farbensieb *n* color sieve
Farbenskala *f* color chart, color scale
Farbenspektrum *n* chromatic spectrum
Farbenspiel *n* brilliancy of colors, iridescense, opalescence
Farbenstärke *f* color depth, color strength
Farbensteindruck *m* chromolithography
Farbenstufe *f* color gradation, color graduation, color shade, tinge
Farbenstufenmesser *m* tintometer
Farbentafel *f* color chart
Farbentheorie *f* theory of colors
Farbentrockner *m* color drier
Farbentüchtigkeit *f* true color perception
Farbenumschlag *m* (Chem) color change
Farben- und Lackfabrikation *f* paint and lake manufacture
Farbenveränderung *f* discoloration
Farbenvertauschung *f* allochromasy
Farbenwechsel *m* change of colors, allochroism
Farbenwurzel *f* (Bot) dyer's madder
Farbenzelle *f* chromatophore
Farbenzentrifuge *f* color centrifuge
Farbenzerlegung *f* analysis of color, color dispersion
Farberde *f* Armenian bole, colored clay, mineral pigment

Farbfehler *m* chromatic aberration, color defect, **longitudinaler ~** (Opt) longitudinal chromatic aberration
Farbfernsehen *n* color television
Farbfilm *m* color film
Farbfilter *n* (Opt) color filter, color screen
Farbflotte *f* dye bath
Farbfreilegungspapier *n* color layoff paper
Farbgebung *f* coloring
Farbgitter *n* color grating, color grid
Farbglas *n* colored glass, stained glass
Farbglaskomposition *f* stained glass compound
Farbglasur *f* colorizing glaze
Farbgut *n* goods for dyeing
Farbholz *n* dye wood
Farbholzextrakt *m* dye wood extract
farbig colored, stained, (Opt) chromatic
Farbindex *m* color index
Farbintensität *f* depth of color, intensity of color
Farbkochapparat *m* color boiler, color boiling apparatus
Farbkörper *m* pigment, coloring matter, coloring substance
Farbkraft *f* coloring power, tinctorial value, tinting strength
Farblack *m* color lake, lacquer, lacquer enamel, varnish
Farblösung *f* dye solution, coloring solution, color solution, (Bakterien) staining solution
farblos colorless, without color, (achromatisch) achromatic, (durchsichtig) transparent
Farblosigkeit *f* absence of color, achromatism, colorlessness
Farbmesser *m* colorimeter
Farbmetrik *f* colorimetry
Farbmischungskurve *f* (Opt) color mixture curve
Farbmittel *n* coloring agent
Farbmühle *f* color grinding mill
Farbnuance *f* shade
Farbpaste *f* color paste
Farbphotographie *f* color photography
Farbpigment *n* pigment
Farbprüfung *f* staining test
Farbraster *m* color screen
Farbrasterverfahren *n* (Opt) screen process
Farbreaktion *f* (Chem) color reaction
Farbreibmühle *f* color grinding mill
Farbreibstuhl *m* color grinding mill
Farbrezeptberechnung *f* computer color matching
Farbringsystem *n* color cycle system
Farbruß *m* color black
Farbschattierung *f* color shade
Farbschlierenverfahren *n* color schlieren method
Farbschwelle *f* (Opt) color threshold
Farbskala *f* color scale
Farbspritzanlage *f* paint spraying plant
Farbstärke *f* coloring strength, tinctorial strength, tinting power, tinting strength

Farbstein *m* dye stone
Farbstich *m* color cast
Farbstift *m* color pencil
Farbstiftmine *f* colored pencil lead
Farbstoff *m* dyestuff, coloring material, dye, pigment, **indigoider** ~ indigoid coloring matter, **substantiver** ~ direct dye, substantive dye
Farbstoffabrik *f* dyestuff factory
Farbstoffaffinität *f* dye affinity
Farbstoffaufnahme *f* (Text) dye absorption
farbstoffhaltig pigmentous
Farbstoffkarte *f* color chart
Farbstofflösemittel *n* dyestuff solvent
Farbstofflösung *f* dye solution
Farbstufe *f* color shade
Farbsubstanz *f* coloring matter, coloring substance
Farbtafel *f* chromaticity diagram, color chart
Farbtemperatur *f* (Opt) color temperature
Farbtemperaturskala *f* (Opt) temperature color scale
Farbtiefe *f* depth of color, depth of shade
Farbtönung *f* shade
Farbton *m* shade, hue, tint, tone
Farbtonänderung *f* change of shade
farbtonbeständig fast [to light]
Farbtonverschiebung *f* change in shade
Farbträger *m* chromophore, color carrier, substrate
Farbtrennung *f* color separation
Farbumkehrfilm *m* (Phot) color reversal film
Farbumschlag *m* (Chem) change in color
Farbunruhe *f* uneven dyeing
Farbunterschied *m* difference in color
Farbwertanteile *pl* chromaticity coordinates
Farbwiedergabe *f* color rendering, color reproduction
Farbzahl *f* color value, tinting value
Farbzentrum *n* color center
Farbzusammensetzung *f* color combination
Farbzusatz *m* addition of color
Farbzwischennegativ *n* (Phot) color internegative
Farinose *f* farinose
Farin[zucker] *m* brown sugar
Farnesal *n* farnesal
Farnesen *n* farnesene
Farnesensäure *f* farnesenic acid
Farnesiferol *n* farnesiferol
Farnesinal *n* farnesinal
Farnesol *n* farnesol
Farnesylbromid *n* farnesyl bromide
Farnesylsäure *f* farnesylic acid
Farnkraut *n* (Bot) fern
Farnkrautöl *n* fern oil
Farnochinon *n* farnoquinone, vitamin K$_2$
Farnwurzel *f* (Bot) fern root, male fern
Faroelith *m* (Min) faroelite
Farrerol *n* farrerol

Faschine *f* (Bauw) fascine, hurdle
Faschinenausfüllung *f* fascine filling, fagot filling
Fase *f* (Kantenabschrägung) bevel
Faser *f* fiber (Am. E.), fibre (Br. E.), filament, thread, (Holz) grain, **natürliche** ~ natural fiber, **synthetische** ~ synthetic fiber, man-made fiber
faserähnlich fibrous, filamentous
Faseralaun *m* (Min) fibrous alum, halotrichite
Faseraragonit *m* (Min) fibrous aragonite, flos ferri
faserartig fiber-like, fibrous, filamentous
Faserasbest *m* fibrous asbestos, (Min) mineral flax
Faserasche *f* fiber ash
Faseraufbau *m* fiber structure
Faseraufschlämmung *f* fiber slurry
Faserbaryt *m* (Min) fibrous heavy spar
Faserbildung *f* fiber formation, (Metall) fibrillation
Faserblende *f* (Min) zinc blende
Faserbraunkohle *f* fibrous lignite
Faserbrei *m* fiber slurry, pulp slurry
Faserbrücke *f* filament bridge
Fasercölestin *m* (Min) fibrous celestine
Faserende *n* fiber end, fiber tip
Fasererz *n* (Min) fibrous zeolite
Faserfänger *m* (Zuck) pulp catcher
Faserfärbung *f* color of the fiber
Faserfilz *m* mat
faserförmig fiber-like, fibrous, filamentous
Fasergehalt *m* fiber content
Fasergewebe *n* fibrous tissue
Fasergewinnung *f* production of fiber
Fasergips *m* fibrous gypsum
Faserhaut *f* fibrous membrane
faserig fibrous, filamentous, stringy
Faserisolation *f* fiber insulation
Faserkalk *m* (Min) fibrous limestone, agalite, aragonite, fibrous calcite
Faserkiesel *m* fibrolite, fibrous quartz, sillimanite
Faserkohle *f* fibrous coal
faserkristallin fibro-crystalline
Faserleder *n* leather-board
faserlos non-fibrous
Fasermalachit *m* (Min) fibrous malachite
Fasermaterial *n* fiber material
Fasermetallurgie *f* fiber metallurgy
Fasermuster *n* fiber pattern
fasern to fray, to fuzz, to tease
Faseroptik *f* fiber optic
Faserpappe *f* structural fiber, insulation board
Faserplatte *f* fiber board
Faserprotein *n* (Biochem) fibrous protein
Faserpulver *n* powdered fiber
Faserquarz *m* (Min) fibrous quartz
Faserquerschnitt *m* cross-section of a fiber

Faserrichtung *f* fiber direction, (Holz) grain
Fasersalz *n* fibrous salt
Faserspritzverfahren *n* spray-up method (polyesters)
Faserstaub *m* fiber dust
Faserstein *m* (Min) fibrolite, bucholzite, fibrous quartz, sillimanite
Faserstoff *m* fibrous material, fibrous substance, (Fibrin) fibrin
Faserstoffisolation *f* fibrous insulation
Faserstoffplatte *f* fiber board
Faserstruktur *f* fiber pattern, fiber structure, texture
Fasertorf *m* fibrous peat
Faserung *f* (Biol) fibrillation, (eines Gewebes) texture, (Fasern) teasing, (Holz) grain, (Metall) fibering
Faserverstärkung *f* fiber reinforcement
Faservlies *n* fiber fleece
Faserwerkstoff *m* fibrous material
Faserzelle *f* (Bot) fibrocellule
Faserzeolith *m* (Min) fibrous zeolite, natrolite
fasrig (faserig) fibrous, filamentous, stringy
Faß *n* barrel, cask, keg
Fassade *f* front wall, siding (Am. E.)
Fassadenfarbe *f* outside house paint
Fassadenfüllfarbe *f* architectural filler paint
Fassadenputz *m* house front plaster
Faßausbeulmaschine *f* barrel straightening machine
Faßbärme *f* (Brau) barm, yeast
Faßbier *n* draft beer, draught beer
fassen (begreifen) to comprehend, to conceive, (einfassen) to frame, to mount, to set, (enthalten) to contain, to hold, to include, (ergreifen) to seize, to take
Faßgärung *f* (Brauw) cask fermentation
Faßgelager *n* impure cask deposits
Faßgeschmack *m* taste of the cask
Faßheber *m* cask hoist
Faßhefe *f* (Brau) barm, yeast
Fasson *f* style, cut, fashion, shape, style
Fassonarbeit *f* (Techn) profiling [work], shaping [work]
Fassondrehbank *f* forming lathe
Fassoneisen *n* profile iron, section iron
fassonieren to form, to contour, to profile, to shape, (Text) to fashion
Fassonschlüssel *m* special spanner
Faßpech *n* cooper's pitch, pitch in casks
Faßreifen *m* hoop
Faßtalg *m* tallow in casks
Fassung *f* (Brille) frame, (Edelstein) mounting, (Elektr) holder, socket, (Lit) wording, (Opt) setting, ~ **optischer Instrumente** pl mounting of optical instruments
Fassungskraft *f* power of comprehension
Fassungslamelle *f* leaf of holder
fassungslos confused, disconcerted

Fassungsraum *m* capacity, volume
Fassungsrohr *n* (einer Quelle) well tube
Fassungsvermögen *n* power of comprehension, (Lastwagen) holding capacity, (Rauminhalt) volumetric capacity, (z. B. Lastwagen) carrying capacity
faßweise in barrels, in casks
Faßzargenrichtmaschine *f* barrel chime and bead machine
Faßzwickel *m* try cock
Fastage *f* barrels, casks
Faujasit *m* (Min) faujasite
faul indolent, lazy, putrid, rotten, ~ **schmecken** to taste rotten, ~ **werden** to turn putrid
faulbar putrescible
Faulbaumbitter *n* frangulin
Faulbaumrinde *f* black aldertree bark
Faulbruch *m* (Metall) brittleness, shortness
faulbrüchig brittle, short
Faulbütte *f* fermenting trough, rotting vat
faulen to putrefy, to become putrid, to cause to decompose, to decay, to rot, to spoil, (gären) to ferment, (Wasser) to stagnate
Faulen *n* decomposing, causing to putrefy, putrefying, rendering putrid, rotting
faulend putrescent, rotting, septic
Faulfleck *m* putrid spot
faulfleckig having putrid spots
Faulgas *n* digester gas, sewer gas
Faulgeruch *m* strong smell during fermentation
Faulgrube *f* fermenting pit, septic tank
Faulschlamm *m* (Geol) sapropel, (Techn) sludge
Faulverfahren *n* (Seide) fermentation method
Faulvorgang *m* fermentation process
Faulwerden *n* decay, putrefaction, putrescence, rotting
Fauna *f* fauna
Fauserit *m* (Min) fauserite
Faustformel *f* thumb rule
Faustgröße *f* fist size
Faustregel *f* rule of thumb
Fawcettiin *n* fawcettiine
Fayalit *m* (Min) fayalite, iron olivine
Fayence *f* faience, delf[t], fine pottery
Fazit *n* (Math) result, sum total
Febrifugin *n* febrifugin
Fechtsäure *f* Fecht acid
Feder *f* feather, (Keil) spline, (Techn) spring
Federachat *m* (Min) feather agate
federähnlich feather-like, feathery, (Techn) spring-like
Federalaun *m* (Min) feather alum, holotrichite
Federamiant *m* (Min) flexible amianthus
Federangel *f* spring hinge
Federanschluß *m* spring clamp
Federantrieb *m* spring action
federartig feather-like, feathery, springy, (Techn) spring-like
Federasbest *m* (Min) flexible asbestos

Federaufhängung *f* spring suspension
Federausgleicher *m* spring equilibrator
Federbarometer *n* aneroid barometer
Federbein *n* shock absorber leg
Federbelastung *f* spring load
Federbiegegrenze *f* bending limit of spring plate
Federblech *n* spring steel
Federbund *m* spring clip, spring shackle
Federdämpfer *m* concussion spring
Federdichtung *f* spring washer
Federdruck *m* spring pressure
Federelastizität *f* springiness
Federerz *n* (Min) feather ore, heteromorphite, jamesonite
Federgalvanometer *n* spring galvanometer
Federgehäuse *n* spring case
Federgips *m* fibrous gypsum, striate gypsum
Federgleitblock *m* spring guide block
Federhärte *f* spring hardness, elasticity
Federhammer *m* spring hammer
Federharz *n* elastic gum, caoutchouc, elaterite
Federharzbaum *m* (Bot) common caoutchouc tree, common rubber tree
Federheber *m* spring lever
Federklinke *f* spring latch
Federkonstante *f* spring constant
Federkraft *f* elastic force, elasticity, resilience, springiness
Federkraft-Kugelmühle *f* contrarotation ball-race-type pulverizing mill
Federlehre *f* pressure gauge
federleicht feather-weight, feather-light
Federmanometer *n* spring pressure gauge, ~ **für Vakuumtechnik** bellows vacuum gauge
federn to be elastic, to be flexible
federnd elastic, resilient
Federnute *f* groove, spline
Federpendel *n* spring pendulum
Federregler *m* spring-loaded governor
Federregulator *m* spring regulator
Federriegel *m* spring bolt
Federring *m* spring washer, lock washer
Federrollenlager *n* spring roller bearing
Federrollenmühle *f* ring-roller mill
Federschalter *m* clip spring switch
Federschraube *f* spring-loaded bolt
Federsicherung *f* safety spring
Federspannung *f* spring tension
Federspat *m* (Min) radiated spar
Federspitze *f* pen point
Federstahl *m* spring steel
Federteller *m* spring washer
Federung *f* elasticity, resilience, resiliency, spring suspension
Federungskörper *m* flexible connection
Federventil *n* spring valve
Federverschlußbolzen *m* spring couple pin
Federwaage *f* spring balance
Federweiß *n* fibrous gypsum, French chalk

Federwelle *f* flexible shaft
Federwirkung *f* spring action
Federwismut *n* (Min) needle-shaped sulfuretted bismuth
Federzange *f* forceps, pincers, tweezers
Federzirkel *m* spring dividers, spring bows
Federzugmesser *m* spring dynamometer
Fedorwit *m* fedorwite
Fege *f* riddle, screen, sieve
Fegesalpeter *m* saltpeter sweepings
Fegesand *m* scouring sand
Fegeschober *m* scum pan
Fegsel *n* sweepings
Fehlabgleichung *f* faulty alignment
Fehlaustrag *m* (Siebtechn) misplaced material
Fehlbestand *m* shortage, discrepancy
Fehleinstellung *f* misadjustment
fehlen to be absent, to be lacking, to lack
Fehlen *n* absence, deficiency
Fehler *m* error, mistake, (Elektr) trouble, (Holz) blemish, (Makel) defect, fault, flaw, (Opt) aberration, (Unzulänglichkeit) deficiency, (Versagen) failure, (Verzerrung) distortion, **mittlerer** ~ mean error, **systematischer** ~ systematic error, **wahrscheinlicher** ~ probable error, **zufälliger** ~ casual error
Fehlerabgleichsmethode *f* consummation of error
Fehlerabschätzung *f* estimation of error
Fehlerauffindung *f* fault detection
Fehlerausgleichung *f* adjustment of errors, compensation of errors
Fehlerbegrenzung *f* fault localization
Fehlerberichtigung *f* correction [of mistakes]
Fehlerbestimmung *f* determination of error
Fehlereingrenzung *f* location of mistakes
Fehlerellipsoid *n* ellipsoid of error
Fehlerfortpflanzung *f* propagation of error
Fehlerfortpflanzungsgesetz *n* law of propagation of error
fehlerfrei faultless, perfect
Fehlerfunktion *f* (Math) error function, (Opt) aberration function
Fehlergesetz *n* law of errors
Fehlergleichung *f* error equation
Fehlergrenze *f* limit of error, limit of inaccuracy, (Techn) tolerance
fehlerhaft defective, faulty, imperfect
Fehlerhaftigkeit *f* defectiveness, faultiness, incorrectness
Fehlerintegral *n* (Math) error integral
Fehlerlinie *f* curve of error
Fehlerlosigkeit *f* faultlessness, flawlessness
Fehlermatrix *f* (Math) error matrix
Fehlermöglichkeit *f* possibility of error
Fehlerorthogonalität *f* orthogonality of error
Fehlerortsmessung *f* localization test
Fehlerquadrate *pl*, **Methode der kleinsten** ~ method of least squares

Fehlerquadratmethode *f* (Math) least squares method
Fehlerquelle *f* source of error
Fehlerrechnung *f* error calculation, theory of errors
Fehlerstelle *f* defective portion
Fehlersuche *f* detection of troubles, trouble shooting
Fehlerverteilung *f* error distribution
Fehlerwahrscheinlichkeit *f* probability of error
Fehlgriff *m* mistake, blunder
Fehlguß *m* spoiled casting
Fehlingsche Lösung *f* Fehling's solution
Fehlingsches Reagenz *n* Fehling reagent
Fehlkorn *n* (Siebtechn) misplaced size
fehlleiten to misdirect, to mislead
Fehlordnung *f* defect, disarrangement, dislocation, disorder
Fehlordnungsstreuung *f* disorder scattering
Fehlschlag *m* failure
fehlschlagen to fail
Fehlschluß *m* wrong conclusion
Fehlsequenz *f* (Biochem) failure sequence
Fehlstelle *f* defect, imperfection, (Chem) vacancy, (Schweiß) crack, flaw, ~ **im Kristall** fault point, **durch das Werkzeug verursachte** ~ mold mark, **durch die Preßplatte verursachte** ~ platen mark
Fehlstellenerzeugung *f* defect generation
Fehlstellenkonzentration *f* **in Flüssigkeiten** hole density
Fehlstöße *m pl* spurious pulses
Fehlverteilung *f* (Füllkörperkolonne) maldistribution
Feige *f* (Bot) fig
Feigenkaffee *m* fig coffee
Feigenlack *m* fig gum lac
Feigenwein *m* fig wine
Feigenweinessig *m* fig vinegar
feilbar capable of being filed
Feile *f* file
feilen to file
Feilenhärteofen *m* file hardening furnace
Feilenhauer *m* file cutter
Feilenheft *n* file handle
Feilenhieb *m* file cut
Feilenzahn *m* file tooth
Feilicht *n* filings
Feilichtkette *f* chain of filings
Feilkloben *m* filing vise
Feilmaschine *f* filing machine
Feilspäne *m pl* borings, filings, turnings
Feilstaub *m* file dust, filings
Feilstrich *m* stroke of a file
fein (Korn, Faden) fine, ~ **zerteilt** finely disintegrated
Feinablesung *f* fine reading, vernier reading
Feinarbeit *f* (Metall) fine smelting, fine working, refining, (Techn) precision work

Feinaufspaltung *f* fine structure splitting
Feinbereichsbeugung *f* fine range diffraction, selected area diffraction
Feinblau *n* fine blue
Feinblech *n* sheet metal, thin gauge plate, thin plate, thin sheet
Feinblei *n* chemical lead
Feinbrecher *m* fine crusher
feinbrennen to refine
Feinbrennen *n* refining
Feinbrenner *m* refiner
Feinbruch *m* fine granulation
Feinbürette *f* (Anal) precision burette
Feinchemikalien *pl* fine chemicals
Feindrehbank *f* precision lathe
Feindruckmanometer *n* micromanometer
Feindruckmesser *m* differential manometer
Feine *f* fineness, delicacy
Feineinstellung *f* fine adjustment
Feineisen *n* refined iron, finer's metal
Feineisenstraße *f* small section rolling mill, small mill
Feinen *n* (Metall) refining
Feinerzgehalt *m* contents of fine ore
feinfädig fine threaded
feinfaserig fine-fibered, fine-fibrous, fine-grained
Feinfeuer *n* refining furnace, charcoal finery, refinery
Feinfilter *n* fine filter
Feinfolie *f* film
Feinfraktionierung *f* superfractionation
Feingefüge *n* fine structure, microstructure
Feingehalt *m* fineness, **gesetzlich festgelegter** ~ standard
Feingehaltsstempel *m* plate mark, (England) hall mark
feingemahlen finely ground, triturated
feingepulvert finely pulverized, finely crushed
Feingewinde *n* fine thread
Feingleitung *f* microslip
Feingold *n* fine gold, pure gold
Feingoldkathode *f* refined gold cathode
Feinguß *m* first quality casting
Feinheit *f* fineness, elegance, (Einzelheit) detail, (Gewinde) rate, (Metall) purity, (Text) gauges
Feinheitsanalyse *f* particle size analysis
Feinheitsgrad *m* degree of fineness
Feinherd *m* (Metall) refining hearth
Feinhubgeschwindigkeit *f* creep hoist speed
Feininstrument *n* precision instrument
Feinkies *m* fine ore, fines
feinkörnig (Phot) fine-grained
Feinkohle *f* fine coal, fines, slack, slack coal, small coal
Feinkorn *n* (Phot) fine grain
Feinkorneisen *n* close-grained iron, fine grained iron

Feinkornentwickler *m* (Phot) fine grain developer
Feinkornfilm *m* (Phot) fine grain film
Feinkorngrenzenwanderung *f* subgrain growth
feinkristallin fine-crystalline
Feinkupfer *n* fine copper, refined copper
feinlochig small-holed, finely porous
Feinlunter *m* rover
feinmahlen to grind fine, to pulverize, to triturate
Feinmahlen *n* fine grinding, fine crushing, powdering, pulverizing
Feinmahlmaschine *f* fine grinder
feinmaschig fine-meshed
Feinmechanik *f* fine mechanics, precision mechanics
feinmechanisch fine-mechanical
feinmehlig finely pulverized
Feinmesser *m* micrometer, vernier
Feinmeßgerät *n* precision instrument
Feinmeßlehre *f* micrometer cal[l]iper
Feinmeßtechnik *f* technique of precision measurement
Feinmessung *f* precision measurement
Feinmetall *n* fine metal
Feinmühle *f* (Pap) finishing machine
Feinofen *m* refining furnace, refining forge
Feinpassung *f* tight fit, wringing fit
Feinperiode *f* slag forming period
Feinplatin *n* refined platinum
Feinpore *f* ultrapore
feinporig finely porous
Feinprofilierung *f* (Reifen) safety slotting
Feinprozeß *m* refining process
Feinpuddeln *n* (Metall) refining puddling
feinpulverisiert finely powdered
feinpulvrig finely powdered, pulverulent
Feinregelung *f* fine control
Feinregulierung *f* fine adjustment, precise adjustment
Feinregulierventil *n* fine regulation valve
Feinschlacke *f* refinery slag, rich fining slag
feinschleifen to fine-grind
Feinschliff *m* fine-grinding, fine-ground finish
Feinschraublehre *f* micrometer cal[l]iper, micrometer screw gauge
Feinschraubzirkel *m* micrometer callipers
Feinschrot *m* fine groats
Feinseife *f* toilet soap, fancy soap
Feinsieb *n* fine sieve
Feinsilber *n* fine silver
Feinsoda *f* finely pulverized soda
Feinsprit *m* rectified alcohol
Feinstbearbeitung *f* microfinish, superfinishing
Feinsteinzeug *n* first quality stoneware
Feinstellschraube *f* fine-adjustment screw, micrometer cal[l]iper, micrometer screw gauge
Feinstellventil *n* precision adjusting valve
Feinstkohle *f* fines
Feinstmahlung *f* pulverization

Feinstmahlvorrichtung *f* micronizer
Feinstrahlmethode *f* microbeam technique
Feinstraße *f* (Metall) sheet rolling mill
Feinstrecke *f* (Metall) small mill, small section rolling mill
Feinstruktur *f* fine structure
Feinstrukturkonstante *f* (Spektr) fine-structure constant
feinstückig small pieces
Feinstzerstäuber *m* atomizer
Feintalg *m* refined tallow
Feinung *f* refining
Feinvakuum *n* medium high vacuum
feinverteilt finely dispersed, finely distributed, finely powdered
Feinwaage *f* precision balance
Feinwalze *f* finishing rolling mills
feinzellig finely cellular
Feinzerkleinerung *f* fine crushing, fine grinding
feinzerrieben finely ground
feinzerteilt finely divided
Feinzeug *n* (Pap) stuff
Feinzeugholländer *m* beating machine, finisher
Feinzink *n* refined zinc
Feinzinn *n* fine tin, grain tin, head tin
Feinzucker *m* refined sugar
Feists Säure *f* Feist's acid
Feld *n* field, ground, **elektromagnetisches** ~ electromagnetic field, **erdmagnetisches** ~ geomagnetic field, **gleichnamiges oder ungleichnamiges** ~ field of same or opposite polarity, **inhomogenes** ~ non-uniform field, **magnetisches** ~ magnetic field
Feldabhängigkeit *f* field dependence
Feldberechnung *f* computation of fields
Felddesorption *f* field desorption
Felddichte *f* intensity of field, strength of field
Feldelektron *n* field electron
Feldelektronenmikroskop *n* field emission electron microscope
Feldemission *f* field emission
Feldemissionsmikroskop *n* field emission microscope
Feldemissionsquelle *f* field emitter source
Felderregung *f* field excitation
Feldgleichung *f* field equation
Feldintensität *f* field intensity, resultant field strength
Feldionen-Massenspektrometer *n* field ion mass spectrometer
Feldionenmassenspektroskopie *f* field ion mass spectroscopy
Feldkamille *f* (Bot) common camomile
Feldmagnet *m* field magnet
Feldquant *n* (Atom) field quantum
Feldquantelung *f* field quantization
Feldschwankung *f* field fluctuation
Feldspannung *f* field voltage

Feldspat *m* (Min) feldspar, fieldspar, **edler** ~ flesh colored feldspar, **gemeiner** ~ [common] feldspar, microcline, orthoclase, potash feldspar, **heterotomer** ~ albite, pericline
feldspatähnlich feldspathic
Feldspatgestein *n* (Min) feldspathic rock
feldspathaltig containing feldspar, feldspathic
Feldspatporzellan *n* hard porcelain
Feldsprung *m* field change
Feldstärke *f* field intensity, field strength, intensity of field, magnetizing force, strength of field, ~ **im freien Raum** (Elektr) free-space field intensity, **elektrostatische** ~ electrostatic field strength, **magnetische** ~ intensity of magnetic field, magnetic field strength
Feldstein *m* boulder, landmark, (Min) adinole, compact feldspar, petrosilex
Feldsystem *n* field system, **mehrpoliges** ~ multipolar field system
Feldteilchen *n* (Atom) field particle
Feldthymiankraut *n* (Bot) mother of thyme
Feldthymianöl *n* wild thyme oil
Feldumkehr *f* (Elektr) field reversal
Feldverstärkung *f* (Elektr) strengthening of field
Feldversuch *m* field experiment
Feldverzerrung *f* field distortion
Feldzypresse *f* (Bot) field cypress
Felge *f* felloe, rim, **dreiteilige** ~ three-piece rim
Felgenband *n* (Reifen) flap
Felgenboden *m* rim base
Felgenhorn *n* flange of the rim, rim flange
Felgenhornring *m* removable rim flange
Felgenkörper *m* base rim
Felgenmaulweite *f* rim width between the flanges
Felgenprofil *n* rim contour
Felgenschulter *f* bead seat, rim ledge
Felicur *n* felicur
Felinin *n* felinine
Felith *m* (Min) felite
Fell *n* hide, skin, (gegerbt) fur, (ungegerbt) pelt
Fellabfälle *m pl* skin parings
fellgar duly dressed
Fellitin *n* fellitine
Fellspritzmaschine *f* extruder slabber, slab extruder
Fels *m* rock
Felsart *f* kind of rock
Felsenachat *m* (Min) rock agate
Felsenalaun *m* (Min) rock alum, alumina
Felsenglimmer *m* (Min) mica
Felsenguano *m* rock guano
felsenhart hard as rock, flinty
Felsenöl *n* petroleum, rock oil
felsig rocky
Felsit *m* (Min) felsite, petrosilex, **bestehend aus** ~ petrosilicious
Felsobanyit *m* (Min) felsobanyite
Feminell *n* feminell
Fenchan *n* fenchane

Fenchansäure *f* fenchaic acid
Fenchel *m* fennel
Fenchelapfel *m* (Bot) fennel apple, spice apple
Fenchelbranntwein *m* fennel brandy
Fenchelgeruch *m* fennel odor
Fenchelholz *n* sassafras wood
Fenchelöl *n* fennel oil
Fenchen *n* fenchene
Fenchenhydrat *n* fenchene hydrate
Fenchenonsäure *f* fenchenonic acid
Fenchensäure *f* fenchenic acid
Fenchenylansäure *f* fenchenylanic acid
Fenchlorphos *n* (Insektizid) fenchlorphos
Fenchobornylen *n* fenchobornylene
Fenchocamphersäure *f* fenchocamphoric acid
Fenchocamphorol *n* fenchocamphorol
Fenchol *n* fenchol, fenchyl alcohol
Fencholan *n* fencholane
Fencholauronolsäure *f* fencholauronolic acid
Fencholenamin *n* fencholene amine
Fencholensäure *f* fencholenic acid
Fencholsäure *f* fencholic acid
Fenchon *n* fenchone
Fenchosantenon *n* fenchosantenone
Fenchyl- *n* fenchyl
Fenchylalkohol *m* fenchol, fenchyl alcohol
Fenchylen *n* fenchylene
Fenethazin *n* fenethazine
Fensterdichtung *f* weather strip
Fensterglas *n* window glass
Fensterglimmer *m* (Min) muscovite, phengite
Fensterkitt *m* putty
Fensterleder *n* chamois [leather], wash leather
Ferberit *m* (Min) ferberite
Ferghanit *n* (Min) ferghanite
Fergusonit *m* (Min) fergusonite
Ferment *n* enzyme, ferment (s. auch Enzym), **abbauendes** ~ digestive ferment, **extrazelluläres** ~ exo-enzyme, extracellular enzyme
Fermentation *f* fermentation
Fermentationsmaschine *f* (Tabak) redrying machine
fermentativ fermentative
Fermentdiagnostikum *n* enzyme diagnostic, ferment diagnostic
fermentieren to ferment
Fermentieren *n* fermenting
Fermentierungsvorgang *m* fermentation process
Fermentwirkung *f* enzymatic action
Fermigrenze *f* Fermi limit
Fermikonstante *f* Fermi constant
Fermi-Kontakt-Term *m* Fermi contact term
Fermi-Niveau *n* (Phys) fermi level
Fermireaktor *m* (Phys) fermipile
Fermium *n* fermium (Symb. Fm)
Fermiumisotop *n* fermium isotope
Fermizeittafel *f* Fermi age model
Fermorit *m* (Min) fermorite

Fernablesung *f* remote instrument reading
Fernandinit *m* (Min) fernandinite
Fernantrieb *m* long distance operation
Fernanzeiger *m* remote indicator
fernbedient remote-controlled
Fernbedienung *f* distance-control, remote-control, remote handling
Fernbedienungsgerät *n* remote operation equipment, longhanded tool, remote handling equipment, remote manipulating equipment
Fernbedienungszange *f* extension tongs
Fernbeobachtung *f* remote viewing
Fernbestrahlung *f* long distance radiation
Ferndrehzahlmesser *m* distant reading tachometer
Ferneinstellung *f* remote control
Fernfokus *m* telefocus
Fernfokuskathode *f* telefocus cathode
Ferngasverdichter *m* booster for long-distance gas mains
Ferngeber *m* remote transmitter
Ferngespräch *n* long-distance call, trunk call
ferngesteuert remote controlled
Fernglas *n* telescope
Fernheizung *f* central heating system, district heating
Fernholz-Säure *f* Fernholz acid
Fernleitung *f* trunk line
Fernleitungskabel *n* trunk cable
Fernlenkung *f* remote control
Fernmeldetechnik *f* telecommunication[s] engineering, telephone and telegraph engineering
Fernmeßeinrichtung *f* remote control measuring installation
Fernordnung *f* long range order
Fernphotographie *f* telephotography
Fernrohr *n* telescope
Fernschaltapparat *m* remote switch
Fernschalter *m* distance switch
fernschreiben to telewrite
Fernschreiber *m* telewriter, teleprinter, telex
Fernschutzwirkung *f* (Korr) cathodic protection
Fernsehapparat *m* television set
Fernsehen *n* television
Fernsehgerät *n* television set
Fernsehröhre *f* television tube
Fernsteueranlage *f* remote control equipment
Fernsteuereinrichtung *f* (Regeltechn) remote control equipment
Fernsteuertafel *f* remote control panel
Fernsteuerung *f* remote control, **drahtlose ~** radio control, telecontrol, **mechanische ~** mechanical remote control
Ferntherapie *f* (Med) teletherapy
Fernthermometer *n* distance thermometer, remote control thermometer, telethermometer
Fernübertragung *f* remote transmission

Fernüberwachungssystem *n* remote area monitoring system
Fernwirkung *f* action at a distance, remote action, remote effect
Fernzählwerk *n* telecounter
Feronialacton *n* feronialactone
Ferralum *n* ferralum
Ferrarizement *m* Ferrari cement, iron-ore cement
Ferrat *n* ferrate
Ferratin *n* ferratin
Ferredoxin *n* ferredoxin
Ferreirin *n* ferreirin
Ferrescasan *n* ferrescasane
Ferri- (Eisen(III)-) ferric
Ferriacetat *n* (Eisen(III)-acetat) ferric acetate, iron(III) acetate
Ferriammoncitrat *n* (Eisen(III)-ammoncitrat) ammonium iron(III) citrate, ferriammonium citrate, ferric ammonium citrate
Ferriammonsulfat *n* (Eisen(III)-ammonsulfat) ferriammonium sulfate, ammoniacal iron alum, ammonioferric sulfate, ammonium ferric alum
Ferriarsenit *n* (Eisen(III)-arsenit) iron(III) arsenite, ferriarsenite, ferric arsenite
Ferribromid *n* ferric bromide, iron(III) bromide
Ferrichlorid *n* ferric chloride, iron(III) chloride
Ferrichlorwasserstoff *m* ferrichloric acid
Ferrichromat *n* ferric chromate, iron(III) chromate
Ferricitrat *n* ferric citrate, iron(III) citrate
Ferricyan *n* ferricyanogen
Ferricyaneisen *n* ferroferricyanide, ferrous cyanoferrate(III), ferrous ferricyanide
Ferricyanid *n* cyanoferrate(III), ferric cyanide, ferricyanide, iron(III) cyanide
Ferricyankalium *n* potassium ferricyanide, red potassium prussiate
Ferricyannatrium *n* sodium cyanoferrate(III), sodium ferricyanide
Ferrieisencyanür *n* ferric cyanoferrate(II), ferric ferrocyanide, ferriferrocyanide, Prussian blue
Ferriferrocyanid *n* ferric cyanoferrate(II), ferric ferrocyanide, ferriferrocyanide, Prussian blue
Ferriferrojodid *n* ferrosoferric iodide, iron(II)(III) iodide
Ferriferrooxid *n* ferrosoferric oxide, iron(II)(III) oxide
Ferrihydroxid *n* ferric hydroxide, iron(III) hydroxide
Ferriion *n* ferric ion
Ferrijodat *n* ferric iodate, iron(III) iodate
Ferrikaliumsulfat *n* potassium iron(III) sulfate, ferric potassium sulfate, potassium iron(III) alum
Ferrilactat *n* ferric lactate, iron(III) lactate
Ferrinitrat *n* ferric nitrate, iron(III) nitrate
Ferrioxalat *n* ferric oxalate, iron(III) oxalate

Ferrioxamin n ferrioxamine
Ferrioxid n ferric oxide, iron(III) oxide, iron sesquioxide
Ferriphosphat n ferric phosphate, iron(III) phosphate
Ferripyrin n ferripyrine
Ferrirhodanid n ferric thiocyanate, ferric rhodanide, iron(III) rhodanide, iron(III) thiocyanate
Ferrisalz n ferric salt, iron(III) salt
Ferrisulfat n ferric sulfate, iron(III) sulfate
Ferrisulfid n ferric sulfide, iron(III) sulfide
Ferrit m ferrite
Ferritannat n (Eisen(III)-tannat) ferric tannate, iron(III) tannate
Ferritartrat n (Eisen(III)-tartrat) ferric tartrate, iron(III) tartrate
Ferritin n ferritin
ferritisch (Stahl) ferritic
Ferritkugel f ferrite sphere
Ferritplatte f ferrite slab
Ferritstab m ferrite rod
Ferriverbindung f (Eisen(III)-Verbindung) ferric compound, iron(III) compound
Ferrivin n ferrivine
Ferro- (Eisen(II)-) ferrous
Ferroacetat n (Eisen(II)-acetat) ferrous acetate, iron(II) acetate
Ferroammonsulfat n ammonium ferrous sulfate, ammonium iron(II) sulfate, ferroammonium sulfate
Ferroarsenat n ferrous arsenate, iron(II) arsenate
Ferrobor n ferroboron, iron boride
Ferrobromid n ferrous bromide, iron(II) bromide
Ferrocarbonat n ferrous carbonate, iron(II) carbonate
Ferrocarbonyl n ferrocarbonyl, iron carbonyl
Ferrocen n (Dicyclopentadienyleisen) ferrocene
Ferrochlorid n ferrous chloride, iron(II) chloride
Ferrochrom n ferrochromium, ferrochrome
Ferrocyan n ferrocyanogen
Ferrocyaneisen n ferric cyanoferrate(II), ferric ferrocyanide, ferriferrocyanide, Prussian blue
Ferrocyanid n cyanoferrate(II), ferrocyanide, ferrous cyanide, iron(II) cyanide
Ferrocyankalium n potassium cyanoferrate(II), potassium ferrocyanide, yellow potassium prussiate
Ferrocyannatrium n sodium cyanoferrate(II), sodium ferrocyanide, yellow sodium prussiate
Ferrocyanwasserstoff m cyanoferrous acid, ferrocyanic acid
Ferroelektrika n pl ferroelectrics, ferroelectric substances
ferroelektrisch ferroelectric[ally]
Ferroelektrizität f ferroelectricity

Ferroferricyanid n ferroferricyanide, ferrous cyanoferrate(III)
Ferroglycerinphosphat n iron(II) glycerophosphate
Ferrohämol n ferrohemol
Ferrohydroxid n ferrous hydroxide, iron(II) hydroxide
Ferroinlösung f ferroin solution
Ferroion n (Eisen(II)-Ion) ferrous ion, iron(II) ion
Ferrojodid n ferrous iodide, iron(II) iodide
Ferrokaliumsulfat n ferrous potassium sulfate, potassium iron(II) sulfate
Ferrokaliumtartrat n ferrous potassium tartrate
Ferrokupfercyanid n copper ferrocyanide, copper(II) cyanoferrate(II), cupric cyanoferrate(II), cupric ferrocyanide
Ferrolactat n ferrous lactate, iron(II) lactate
Ferrolegierung f ferro-alloy
Ferromagnesium n ferromagnesium
Ferromagnetika n pl ferromagnetics, ferromagnetic substances
ferromagnetisch ferromagnetic
Ferromagnetismus m ferromagnetism
Ferromangan n ferromanganese
Ferromangansilizium n ferromanganese silicon, ferrosilico manganese, iron manganese silicide
Ferromangantitan n ferromanganese titanium
Ferromolybdän n ferromolybdenum
Ferronickel n ferronickel
Ferronitrat n ferrous nitrate, iron(II) nitrate
Ferrooxalat n ferrous oxalate, iron(II) oxalate
Ferrooxid n ferrous oxide, iron(II) oxide, iron monoxide
Ferrophosphat n ferrous phosphate, iron(II) phosphate
Ferrophosphor m ferrophosphorus
Ferropyrin n ferropyrine
Ferrosalz n ferrous salt, iron(II) salt
Ferrosilizium n ferrosilicon, silicon iron
Ferrosiliziumabstich m ferrosilicon tapping
Ferrosiliziumerzeugung f ferrosilicon production
Ferrosiliziumofen m ferrosilicon furnace
Ferrosulfat n ferrous sulfate, iron(II) sulfate
Ferrosulfid n ferrous sulfide, iron(II) sulfide
Ferrotitan n ferrotitanium
Ferrotitanit m ferrotitanite
Ferrotypie f (Phot) ferrotype process
Ferrovanadium n ferrovanadium
Ferroverbindung f (Eisen(II)-Verbindung) ferrous compound, iron(II) compound
Ferroverdin n ferroverdin
Ferrowolfram n ferrotungsten
Ferroxyl-Indikator m ferroxyl indicator
Ferrozirkonium n ferrozirconium
Ferruginol n ferruginol
Fertigbeton m precast concrete, ready-mixed concrete

fertigen to manufacture, to process
Fertigerzeugnis *n* final product, finished product
Fertigfabrikat *n* finished article
Fertigfrischen *n* (Met) final purification
Fertiggericht *n* ready-to-eat dish
Fertigguß *m* finished casting
Fertigkaliber *n* (Walzwerk) finishing groove
fertigmachen to finish
Fertigmachen *n* finishing
Fertigmischung *f* fully compounded stock
Fertigmörtel *m* ready-mixed mortar
Fertigmontage *f* final assembly
Fertigprodukt *n* finished product
Fertigröstung *f* final roasting
Fertigschleifen *n* finish grinding
fertigstellen to complete, to finish
Fertigstellraum *m* finishing-off room
Fertigstellung *f* completion, finishing
Fertigstraße *f* finishing rolls
Fertigstrecke *f* finishing rolls
Fertigteil *n* finished product, prefabricated unit
Fertigung *f* finishing, fabrication, manufacture, production
Fertigungsanlage *f* plant
Fertigungsfluß *m* process flow
Fertigungsplanung *f* production engineering
Fertigungsprogramm *n* production program
Fertigungsstraße *f* production line
Fertigungstechnik *f* manufacturing engineering
Fertigwalze *f* finishing roll
fertigwalzen to roll for finishing
Fertigwalzen *n* final rolling
Fertigwalzwerk *n* finishing rolls
Fertigware *f* fully fashioned goods
Fertigwaschtrommel *f* finishing drum
Ferulaaldehyd *m* ferulaldehyde
Ferulasäure *f* ferulic acid
Ferulen *n* ferulene
Fervenulin *n* fervenuline
Fesogenin *n* fesogenin
fest (dicht) tight, (endgültig) definite, (haltbar, stark) fast, firm, rigid, strong, (stabil) stable, (ständig) permanent, steady, (starr) compact, solid, (Zeitpunkt) fixed, ~ **angeordnet** fixed
festbacken to cake together, to form clinkers
Festbett *n* (Katalysator) solid bed, fixed bed packing
Festbettabsorptionsanlage *f* fixed-bed adsorption unit
Festbettelektrode *f*, **bipolar arbeitende** ~ bipolar particulate bed electrode
Festbraun *n* solid brown
festfressen to jam, (Lager) to seize
Festfressen *n* jamming, seizing (of a mold)
festfrieren to freeze hard
festgefressen (Lager) seized
festgefroren frozen
Festgehalt *m* solid content[s]
festgelagert stationary

Festhalteplatte *f* clevis plate
festigen to consolidate, to make firm, to stabilize
Festigkeit *f* firmness, compactness, consistency, rigidity, tightness, (Haltbarkeit) stability, sturdiness, (Mech) resistance, (Stärke) strength, (Starrheit) solidity, stiffness, ~ **des Werkstoffes** resistance of the material, **dielektrische** ~ dielectric strength, **dynamische** ~ dynamic strength, **elektrische** ~ electrical strength
Festigkeitsberechnung *f* stress calculation
Festigkeitseigenschaft *f* mechanical property, stress property
Festigkeitsgrad *m* degree of firmness
Festigkeitsgrenze *f* limit of resistance, limit of stability
Festigkeitsguß *m* high strength cast iron
Festigkeitskoeffizient *m* coefficient or modulus of resistance
Festigkeitsmodul *m* breaking modulus
Festigkeitsprüfer *m* strength tester
Festigkeitsprüfung *f* strength test
Festigkeitsverlust *m* loss of strength
Festigkeitswert *m* mechanical strength value, tensile value
Festigkeitszahl *f* consistency factor, modulus or coefficient of resistance, stability ratio, strength coefficient
Festkautschuk *m* solid rubber
festkeilen to wedge, to fasten by wedges, to jam
festkitten to cement
festklammern to clamp
festkleben to glue on
festklemmen to clamp firmly
Festkörper *m* solid
Festkörperchemie *f* solid-state chemistry
Festkörperdiffusion *f* solid diffusion
Festkörperextraktion *f* extraction of solids
Festkörpermodell *n* model of solid
Festkörperphysik *f* solid-state physics
Festkörperreaktion *f* solid-state reaction
Festkörperterminologie *f* solid-state terminology
Festkopfplattenspeicher *m* (Comp) fixed head storage unit
Festkraftstoff *m* solid fuel
festlegen to determine, to fix, to place
festmachen to fasten, to fix
Festmachen *n* fastening, fixation
Festmeter *n* theoretical cubic meter
festnageln to fasten by nails, to nail
Festphase *f* solid phase
Festphasentechnik *f* solid phase technique
Festpunkt *m* fixed point
Festrolle *f* fixed roll
Festscheibe *f* fast pulley, tight pulley
Festschmierstoffe *m pl* solid lubricants
Festsitz *m* tight fit
festspannen to clamp tightly
feststampfen to stamp

feststehend fixed, stationary, well-established
feststellbar ascertainable, determinable
feststellen to ascertain, to determine
Feststellhebel m lock[ing] lever
Feststellschraube f lock[ing] screw, set screw
Feststoff m solid matter
Feststoff-Flüssigkeitsrakete f solid-liquid rocket
Feststoffgehalt m solids content
Feststoffgeschwindigkeit f solid feed rate
Feststoffkatalysator m fixed catalyst
Feststoff-Probengabe f solid sampler
Feststoffrakete f solid propellant rocket
Festsubstanz f solid matter
Festuclavin n festuclavine
festwerden to solidify, (gerinnen) to coagulate, to curdle, (Tumor) to consolidate
Festwerden n solidification, coagulation, curdling, (Blut) clotting
Festwert m constant [value], standard value
Festwertregelung f (Regeltechn) regulation with fixed set
Festwiderstand m fixed resistance
fett fat, fatty, greasy, (Buchdr) extrabold
Fett n fat, grease, ~ **extrahieren** to extract fat, ~ **raffinieren** to purify grease, ~ **spalten** to cleave the fat, to split the fat, ~ **verseifen** to saponify fat, to hydrolyze fat, **festes** ~ solid fat, **öliges** ~ oily fat
Fettabbau m (Physiol) lipocatabolism
Fettabfälle m pl refuse fat
Fettablagerung f deposition of fat, (Med) adiposis
Fettabscheider m de-oiler, grease extractor, grease separator
fettähnlich fat-like, lipoid, resembling fat
Fettalkohol m fatty alcohol
Fettanlagerung f storage of fat
Fettappretur f (Leder) oil dressing
fettarm lean, poor in fat
fettartig fat-like, fatty, greasy, lipoid
Fettauffanggefäß n fat catching vessel, fat trap, grease catcher
Fettausscheider m de-oiler, grease extractor, grease separator
fettbeständig grease-proof
Fettbestimmung f determination of fat
Fettbildung f adipogenesis, fat formation
fettdicht fat-tight, grease-proof
fetten to grease, to oil
Fettentziehung f fat extraction, defatting, degreasing, fat removal
Fetterin n fetterine
Fettextrahierung f fat extraction
Fettextraktion f fat extraction
Fettextraktionsmittel n fat extracting agent
Fettextraktionsverfahren n method of fat extraction
Fettfarbe f fatty ink
Fettfarbstoff m fat dyestuff

fettfest grease-proof
Fettfleck[en] m grease spot, greasy stain
Fettfleckphotometer n grease-spot photometer
fettgar dressed with oil, oil-tanned
Fettgas n oilgas
Fettgasanstalt f oil gasworks
Fettgehalt m fat content
Fettgehaltsbestimmung f determination of the fat content
Fettgerbung f oil tanning
Fettgeruch m odor of grease
Fettgeschmack m taste of fat, taste of grease
Fettgewebe n adipose tissue
Fettgift n alantotoxicum
fettglänzend having a grease luster
Fettglanz m greasy luster
Fetthärtung f fat hardening, hardening of liquid fats, hydrogenation of fats
fetthaltig containing fat
Fettharz n oleoresin
Fetthaushalt m lipometabolism
Fetthefe f fatty yeast
Fetthydrierung f hydrogenation of fats
fettig fatty, greasy, like fat
Fettigkeit f greasiness, fatness, oiliness
Fettkalk m fat lime, rich lime
Fettkessel m fat pot, tallow pot
Fettkörper m fatty substance
Fettkohle f bituminous coal, cannel coal, fat coal, rich coal
Fettlappen m greasy rag
Fettlicker m (Gerbung) fat liquor
fettlösend fat-dissolving, lipolytic
fettlöslich fat-soluble, liposoluble, soluble in fat
Fettlösungsmittel n fat dissolver, fat solvent, grease solvent
Fettmangel m fat deficiency, hypolipsis
Fetton m Fuller's earth, smectite
Fettoxydation f oxidation of fat
Fettpech n fatty acid pitch
Fettpresse f grease press
Fettpuddeln n (Metall) slag puddling, pig boiling
Fettpuddelofen m slag puddling furnace
Fettquarz m (Min) greasy quartz
Fettreif m (Schokolade) bloom
Fettreihe f (Chem) aliphatic series
Fettreserve f depot fat
Fettsäure f fatty acid, **gesättigte** ~ saturated fatty acid, **mehrfach ungesättigte** ~ polyunsaturated fatty acid, **ungesättigte** ~ unsaturated fatty acid, **unveresterte** ~ nonesterified fatty acid
Fettsäuredestillationsanlage f fatty acid distillation plant
Fettsäureemulsion f emulsion of fatty acid[s]
Fettsäureester m ester of a fatty acid, fatty acid ester
Fettsäuresynthetase f (Biochem) fatty acid synthetase

Fettsäurethiokinase − Feuerkugel

Fettsäurethiokinase *f* (Biochem) fatty acid thiokinase
Fettsalbe *f* pomade
fettsauer of or combined with a fatty acid
Fettschicht *f* layer of grease
Fettschmiere *f* fat liquor, stuffing
Fettschmierpresse *f* grease gun
Fettschmierung *f* grease lubrication
Fettschweiß *m* fatty sweat, greasy yolk
Fettspaltanlage *f* fat splitting plant
fettspaltend fat-cleaving, fat-splitting, lipolytic
Fettspalter *m* fat cleaving agent, fat splitting agent, lipolytic agent
Fettspaltung *f* fat cleavage, fat splitting, hydrolysis of fat, lipolysis
Fettstein *m* (Min) nepheline, nephelite
Fettstift *m* wax pencil, pencil for marking glass, wax crayon
Fettstoff *m* fatty substance, fat, fatty matter
Fettstoffwechsel *m* (Biochem) fat metabolism, lipometabolism
Fettsubstanz *f* fatty substance
Fettsucht *f* (Med) obesity
Fettsynthese *f* fat synthesis
Fettung *f* creasing, lubrication
Fettverderb *m* fat deterioration
Fettverseifung *f* saponification of fat[s]
Fettwachs *n* adipocere
fettwachsartig adipoceriform, adipocerous
Fettwolle *f* grease wool
Fettzelle *f* adipose cell, fat cell
Fetzenkalander *m* (Gummi) rags calender
Fetzenmischung *f* (Gummi) rags compound
feucht damp, moist, wet
Feuchtegraderhöhung *f* humidification
Feuchteregler *m* humidity controller
Feuchtigkeit *f* dampness, humidity, humor, moisture, **gebundene** ~ bound moisture
Feuchtigkeitsanzeiger *m* hygroscope
Feuchtigkeitsaufnahme *f* absorption of humidity, moisture pick-up, moisture regain
Feuchtigkeitsausdehnung *f* moisture expansion
Feuchtigkeitsbedingung *f* humidity condition
feuchtigkeitsbeständig moisture-resistant
Feuchtigkeitsbeständigkeit *f* water resistance
Feuchtigkeitsdurchschlag *m* penetration of moisture
Feuchtigkeitsentzug *m* dehumidification, drying
Feuchtigkeitsfühler *m* humidity feeler
Feuchtigkeitsgehalt *m* amount of moisture, degree of wetness, moisture content, percentage of moisture
Feuchtigkeitsgrad *m* degree of humidity
Feuchtigkeitskorrosion *f* aqueous corrosion
Feuchtigkeitsmesser *m* hygrometer, moisture determination apparatus, psychrometer
Feuchtigkeitsmeßinstrument *n* hygrometer
Feuchtigkeitsmessung *f* hygrometry, psychrometry

Feuchtigkeitsniederschlag *m* deposit of moisture
Feuchtigkeitsprüfer *m* moisture tester
Feuchtigkeitsregelung *f* moisture control
Feuchtigkeitsregler *m* hygrostat
Feuchtigkeitsschreiber *m* hygrograph
Feuchtigkeitsschwankung *f* variation of humidity
Feuchtigkeitszustand *m* hygrometric condition
Feuchtkammer *f* humidity chamber, moistening chamber
Feuchtlabilität *f* hygroscopic instability
Feuchtmahlung *f* wet grinding
Feuchtraumleitung *f* damp-proof installation cable
Feuchttrockenthermometer *n* wet dry thermometer
Feuchtwalze *f* damping roller
Feuchtwerden *n* humidification
Feuer *n* fire
Feueralarm *m* fire alarm
Feueranzünder *m* kindler, lighter
feuerartig igneous
feuerbeständig fire-proof, fire-resisting, refractory
Feuerbeständigkeit *f* fireproofness, fire resistant quality, refractoriness, resistance against fire
Feuerblech *n* fire box plate
Feuerbrücke *f* fire bridge, fire stop
Feuerbüchse *f* fire box
feuerfangend inflammable, ignitable
feuerfarbig fire-colored
feuerfest fireproof, heat-proof, incombustible, refractory
Feuerfestigkeit *f* fireproofness, fire resistant quality, refractoriness
Feuerfestmaterialien *n pl* refractory materials
Feuerfläche *f* heating surface
Feuerflecken *m* fire stain
feuerflüssig liquid at high temperature[s]
Feuerfünkchen *n* sparklet
Feuerfunke *m* spark
Feuergarbe *f* fire sheaf
Feuergas *n* combustion gas, burnt gas, flue gas, furnace gas, waste gas
feuergefährlich combustible, dangerously inflammable
Feuergefährlichkeit *f* inflammability, liability to catch fire
Feuergefahr *f* danger of fire, risk of fire
Feuergewölbe *n* coking arch, fire arch
Feuerglut *f* blazing glow
Feuergrube *f* ash pit
feuerhemmend fire-retarding, fire-resisting, fire-retardant, flame-retardant
Feuerherd *m* hearth, (Brandherd) source of fire
Feuerholz *n* fire wood
Feuerkammer *f* fire box
Feuerkitt *m* fireproof cement, fireclay
Feuerkrücke *f* rake
Feuerkugel *f* fireball

Feuerlack *m* fireproof varnish
Feuerlöschapparat *m* fire extinguisher
Feuerlöschen *n* fire extinguishing
Feuerlöscher *m* fire extinguisher
Feuerlöschmittel *n* fire extinguishing substance
Feuerluft *f* air for combustion, furnace gas, inflammable gas, oxygen
Feuermeldeanlage *f* fire-alarm system
Feuermelder *m* fire alarm
feuern to fire
Feueropal *m* (Min) fire opal
feuerpolieren to fire-polish
feuerpoliert fire polished, flame-polished
Feuerpolitur *f* (Glas) fire polish
Feuerporzellan *n* refractory porcelain
Feuerprobe *f* fire test
Feuerraum *m* combustion chamber, fire chamber, (Dampfkessel) firebox, (Schmelzofen) fireplace, furnace
Feuerregen *m* cascade of fire
Feuerrohr *n* flue, firearm, fire tube
Feuerrohrkessel *m* fire tube boiler, tubular boiler
Feuerrost *m* grate
feuerrot fiery red
Feuerscheibe *f* rough emery grinding wheel
feuerschleifen to rough polish
Feuerschutz *m* fire protection
Feuerschutzanlage *f* firefighting installation
Feuerschutzanstrich *m* fireproof coating
Feuerschutzmittel *n* fire protecting agent
Feuerschwaden *m* fire damp
Feuerschwamm *m* (Techn) punk, tinder
Feuerschweißen *n* forge welding
feuersicher fireproof
Feuersprüher *m* squib
Feuerstein *m* flint, rock flint
feuersteinartig flinty
Feuerton *m* fire clay
Feuertrockner *m* fire drying apparatus
Feuerung *f* fire, firing, heating, ~ **mit Hochofengas** waste gas furnace
Feuerungsanlage *f* firing plant, furnace installation
Feuerungsart *f* type of firing
Feuerungsbedarf *m* consumption of fuel
Feuerungsgewölbe *n* furnace arch
Feuerungsgrad *m* degree of heat
Feuerungsmaterial *n* combustibles, fuel
Feuerungsraum *m* firebox, fuel chamber
Feuerungstechnik *f* technique of stoking
feuervergoldet fire-gilt
Feuervergoldung *f* fire gilding
Feuerverhütung *f* fire prevention
Feuerversilberung *f* fire silvering
feuerverzinkt hot-galvanized
Feuerverzinkung *f* hot galvanizing
Feuerverzinnen *n* (Metall) hot-tinning
Feuerverzinnung *f* (Metall) hot-tinning
Feuerwerk *n* fire work

Feuerwerkskörper *m pl* fireworks
Feuerwiderstandsfähigkeit *f* fire resisting property
Feuerzeug *n* lighter
Feuerzeugbenzin *n* lighter fuel
Feuerzug *m* flue, fire tube
Feulgentest *m* Feulgen method
feurig fiery
Feurigrot *n* fiery red, red hot
F-Faktor *m* (Dest) liquid load factor
Fiber *f* fiber
Fiberrad *n* fiber pinion
Fiberritzel *n* fiber pinion
Fiberscheibe *f* fiber disc
Fibrille *f* (Med, Biol) fibril
Fibrillierung *f* (Papier) fibrillation
Fibrin *n* fibrin
fibrinbildend fibrinogenous
Fibrinbildung *f* fibrinogenesis
Fibrinferment *n* fibrin ferment, thrombin
Fibringerinnsel *n* fibrin clot
Fibrinmangel *m* (Blut) hypinosis
Fibrinogen *n* fibrinogen
Fibrinogenese *f* fibrinogenesis
Fibrinogenmangel *m* hypofibrinogen[a]emia (aemia), lack or deficiency of fibrinogen
Fibrinolyse *f* fibrinolysis
Fibrinolysin *n* plasmin
fibrinoplastisch fibrinoplastic
Fibrinpfropf *m* fibrin plug
Fibrinüberschuß *m* hyperinosis
Fibroblast *m* (Histol) fibroblast, connective-tissue cell, fibrocyte, inoblast, inocyte
Fibroblastenkultur *f* fibroblast culture
fibrös fibrous
Fibroferrit *m* (Min) fibroferrite
Fibroin *n* fibroin
Fibrolith *m* (Min) fibrolite, sillimanite
Fibrolysin *n* fibrolysin
Fichtenharz *n* pine resin, rosin, spruce resin
Fichtenholzkohle *f* pine charcoal, spruce charcoal
Fichtenholzöl *n* fir wood oil
Fichtenholzzellstoff *m* pine wood cellulose
Fichtennadelextrakt *m* pine needle extract
Fichtennadelöl *n* pine needle oil
Fichtennadelpräparat *n* pine needle preparation
Fichtenöl *n* pine oil
Fichtenpech *n* pine resin, rosin, spruce resin
Fichtenrinde *f* pine bark
Fichtenruß *m* pine soot
Fichtensäure *f* pinic acid
Fichtenspanreaktion *f* (Chem) pine wood reaction
Fichtenteer *m* pine tar
Fichtenzucker *m* pinite
Ficin *n* ficin
Ficksches Gesetz *n* Fick's law

Ficocerylalkohol – Filterpressenrahmen

Ficocerylalkohol *m* ficoceryl alcohol
Ficusin *n* ficusin
Fieberarznei *f* (Med) antipyretic, febrifuge
Fieberklee *m* marsh trefoil, bog bean, buck bean
fiebermildernd (Med) antipyretic, febrifugal
Fiebermittel *n* (Med) antipyretic, febrifuge
Fieberrinde *f* ague bark, cinchona, Peruvian bark
Fieberthermometer *n* clinical thermometer
Fiedlerit *m* (Min) fiedlerite
fiktiv fictitious, imaginary
Filiale *f* branch
Filicin *n* filicin, filicic acid anhydride
Filicinsäure *f* filicinic acid
Filigran *n* filigree [work]
Filipin *n* filipin
Filixsäure *f* filixic acid
Fillowit *m* (Min) fillowite
Film *m* film, coat, layer, **gegossener** ~ cast film
Filmaron *n* filmaron
Filmaufziehgerät *n* film applicator
filmbildend film forming
Filmbildner *m* film former, film-forming agent
Filmdosimeter *n* dosifilm
Filmdosimetrie *f* film dosimetry
Filmdruck *m* film printing, screen printing
Filmgießen *n* film casting
Filmkolonne *f* film column
Filmkondensation *f* film condensation
Filmogen *n* filmogen, acetone collodion
Filmregenerierung *f* (Phot) film regeneration
Filmsieden *n* film boiling
Filmströmung *f* film flow
Filmträger *m* (Phot) film base
Filmunterlage *f* (Phot) film base
Filmverdampfer *m* film-type evaporator, agitated film evaporator
Filmverdampfung *f* film evaporation
Filmwärmeaustauscher *m* thin film heat exchanger
Filmwascher *m* falling-film scrubber
Filter *n* filter, screen, strainer, ~ **mit Doppelboden** gravity filter, **Band-** ~ belt filter, **Bandzellen-** ~ travelling pan filter, **Bett-** ~ deep bed filter, **Blatt-** ~ leaf filter, **Druckblatt-** ~ pressure leaf filter, **Druckdreh-** ~ rotary pressure filter, **Drucktrommel-** ~ pressure drum filter, **Falten-** ~ folded filter, **Gas-** ~ gas filter, **Glas-** ~ glass filter, **Innenzellen-** ~ top feed filter, **Kapillarband-** ~ travelling pan filter, **Karussellnutschen-** ~ tilting pan filter, **Kerzen-** ~ cartridge filter, **Kippwannen-** ~ tilting pan filter, **monochromatisches** ~ (Opt) monochromatic filter, **Nutschen-** ~ nutsch and pan filter, **Oben-Aufgabe-** ~ top feed filter, **Plan-** ~ table filter, **Platten-** ~ filter press, **Plattenpreß-** ~ hydraulic filter press, **Redler-** ~ drag chain, **Rohrdruck-** ~ tubular pressure filter, **Scheiben-** ~ disc filter, **Schichten-** ~ sheet filter, **Schüssel-** ~ horizontal rotary filter, **Sieb-** ~ strainer, **Siebtrommel-** ~ hydraulic filter press, **Tauch-** ~ vacuum leaf filter, **Teller-** ~ table filter, **Tiefen-** ~ deep bed filter, **Trommel-** ~ drum filter, **Trommelsaug-** ~ multi-compartment drum filter, **Trommelzellen-** ~ multi-compartment drum filter, **Ultra-** ~ ultra filter, **Vakuum-** ~ vacuum filter, **Vakuum-Trommel-** ~ vacuum drum filter, **Zentrifugalreinigungs-** ~ pressure leaf filter with centrifugal sluicing
Filteranlage *f* filtration plant
Filterbandzentrifuge *f* filter belt centrifuge
filterbar filterable
Filterbeutel *m* filter bag
Filterboden *m* filter bottom
Filterbottich *m* filtering tub
Filterdauer *f* time of filtering
Filterelement *n* filter cell
Filterentkeimung *f* filter sterilisation
Filterfläche *f* filter area
Filterflasche *f* filtering flask
Filterflocken *pl* filter flakes
Filtergaze *f* filter gauze
Filtergehäuse *n* filter casing
Filtergestell *n* filter stand, funnel holder
Filtergewebe *n* filter fabrics
Filterglas *n* absorption glass, filter glass
Filterhalter *m* filter holder, filter ring
Filterhilfsmittel *n* filtering auxiliary, filter aid
Filterkammer *f* filter chamber, straining chamber
Filterkasten *m* filter box
Filterkegel *m* filter cone
Filterkerze *f* candle filter, filter candle, filter cartridge
Filterkies *m* filter gravel
Filterkissen *n* filter pad
Filterkohle *f* filter charcoal
Filterkolben *m* Buchner funnel, filter flask
Filterkonus *m* filter cone
Filterkorb *m* filtering basket
Filterkuchen *m* filter cake, press cake
Filterkuchenabnahmevorrichtung *f* filter-press-cake removing device
Filtermasse *f* filter mass
Filtermaterial *n* filtering material, filter medium
Filtermittel *n* filter aid, filter media
filtern to filter, to filtrate, to percolate, to strain
Filtern *n* filtering, percolation
Filterpapier *n* filter paper
Filterpassierer *m* (Bakt) filter passer
Filterplatte *f* filter disc, filter plate
Filterpresse *f* filter press, **elektroosmotische** ~ electroosmotic filter press
Filterpressenplatte *f* filter press plate
Filterpressenrahmen *m* filter press frame

280

Filterpreßring *m* filter press ring
Filterpumpe *f* filter pump
Filterrahmen *m* filter frame
Filterraum *m* filtration chamber
Filterreinigungsmaschine *f* filter cleaning machinery
Filterröhre *f* filtering tube, filter tube
Filterrückstand *m* filtration residue, residue on the filter
Filtersack *m* filter bag
Filtersand *m* filter sand
Filterschale *f* filtration disc
Filterscheibe *f* filter plate
Filterschicht *f* filter bed, filtering layer
Filterschichtenwäscher *m* filter bed scrubber
Filterschieber *m* filter shifter, filter slide
Filterschlamm *m* filter mud
Filterschlauch *m* filter hose
Filtersieb *n* filtering sieve, strainer
Filterstäbchen *n* filter stick
Filterstativ *n* filter stand
Filterstechheber *m* filtering pipette
Filterstein *m* porous stone
Filterstoff *m* filtering material, filter cloth
Filtertasche *f* filter pocket, filter cell
Filtertiegel *m* filter crucible
Filterträger *m* filter holder
Filtertrichter *m* filter funnel
Filtertuch *n* filter cloth
Filtertuchdiaphragma *n* filter cloth diaphragm
Filterturm *m* filter tower
Filterung *f* filtration
Filterungsdauer *f* time of filtering
Filterveraschung *f* incineration of filter paper
Filterwäger *m* filter weigher
Filterwand *f* filter bed, filter plate
Filterwatte *f* filter wadding
Filtrat *n* filtrate
Filtration *f* filtration, **Druck-** ~ pressure filtration, **Klär-** ~ filter-medium filtration, **Kuchen-** ~ cake filtration, **Saug-** ~ suction filtration, **Sieb-** ~ straining filtration, **Tiefen-** ~ deep bed filtration, **Ultra-** ~ ultrafiltration, **Vakuum-** ~ vacuum filtration
Filtrationseinrichtung *f* filtering apparatus
Filtrationsenzym *n* enzymatic filter aid
filtrationsfähig filterable, capable of being filtered
Filtrationsgeschwindigkeit *f* rate of filtration
Filtrierapparat *m* filtering apparatus, filter
Filtrieraufsatz *m* filter attachment
filtrierbar filterable, capable of being filtered
Filtrierbarkeit *f* filtering property
Filtrierbecher *m* filter beaker
Filtrierbeutel *m* filtering bag, percolator
filtrieren to filter, to filtrate, to percolate, to strain
Filtrieren *n* filtering, straining
Filtriererde *f* filtering earth

Filtrierfläche *f* filtering area
Filtrierflasche *f* filtering flask
Filtriergeschwindigkeit *f* rate of filtering
Filtriergestell *n* filter stand
Filtrierkalkstein *m* filtering stone, dripstone
Filtriermaterial *n* filtering material
Filtrierpapier *n* filter paper
Filtrierplatte *f* filter plate, drainer plate
Filtrierpresse *f* filter press
Filtrierpumpe *f* filter pump
Filtrierring *m* filter ring
Filtrierrohr *n* filter tube
Filtriersack *m* filter bag, percolator, straining bag
Filtriersand *m* filter[ing] sand
Filtrierschicht *f* filter bed
Filtriersieb *n* filter sieve
Filtrierstativ *n* filter stand, funnel holder
Filtrierstein *m* filtering stone, dripstone
Filtrierstütze *f* filtering flask
filtriert filtered, strained
Filtriertiegel *m* filter crucible
Filtriertrichter *m* colander, strainer
Filtriertuch *n* straining cloth, filter[ing] cloth, percolator
Filtrierung *f* filtration, percolation, straining
Filtron *n* filtron
Filz *m* felt, (Bot) tomentum
Filzdichtung *f* felt washer
Filzführung *f* felt guide
filzig felt-like, felted, (Bot) tomentous
Filzisolierplatte *f* felt insulation plate
Filzpolierscheibe *f* felt polishing wheel, felt buffing wheel
Filzschlauch *m* felt jacket
Filzschreiber *m* felt tip marker, felt point marker
Filzüberzug *m* felt cover
Filzunterlage *f* felt blanket
Finger *m* finger, (Mech) grappling iron
Fingerabdruck *m* finger print
Fingerfräser *m* [end] milling cutter
Fingerhut *m* thimble, (Bot) fox glove
Fingerhutblätter *n pl* (Bot) digitalis leaves, foxglove leaves
Fingerhutextrakt *m* (Bot) digitalis extract
Fingerhuttinktur *f* digitalis tincture
Fingerkontakt *m* (Elektr) finger-type contact
Fingerplatte *f* finger plate
Fingerrührer *m* finger paddle agitator, finger paddle mixer
Fingerspitzengefühl *n* instinct
Fingerstein *m* (Min) belemnite, fingerstone
Finne *f* (Med) acne, pimple, (Techn) peen (of a hammer), (Zool) fin
Fiolaxglas *n* fiolaxglass
Fiorit *m* (Min) fiorite
Firnis *m* boiled linseed oil, colorless oil coating, varnish, **überziehen mit** ~ to varnish, to glaze, to gloss

firnisartig varnish-like
Firnisersatz *m* varnish substitute
Firnisfarbe *f* varnish color
Firnislack *m* lac varnish
Firnispapier *n* glazed paper
firnissen to varnish, to glaze, to gloss
Firpen *n* firpene
Firstenbohrloch *n* back borehole
Fischaugen *pl* (Metall) fish eyes
Fischaugenstein *m* (Min) apophyllite, fish eye stone
Fischbein *n* baleen
Fischdünger *m* fish manure
Fischerit *m* (Min) fischerite
Fischer-Tropsche Benzinsynthese *f* Fischer-Tropsch gasoline synthesis
Fischgift *n* fish poison
Fischkörner *n pl* Indian berries
Fischleim *m* fish glue, ichthyocolla, isinglass
Fischmehl *n* fish meal
Fischöl *n* fish oil
Fischölseparator *m* fish oil separator
Fischrechen *m* fish grid
Fischschuppe *f* fish scale
Fischsilbereffekt *m* pearl effect
Fischsilberpigmente *pl* nacreous particles, nacreous pigment, pearl essence pigment
Fischtran *m* fish oil, train oil
Fisetin *n* fisetin, fisetic acid
Fisetinidin *n* fisetinidin
Fisetinidol *n* fisetinidol
Fisetol *n* fisetol
Fisettholz *n* young fustic
Fisidan *n* fisidan
Fisidin *n* fisidin
Fisidylin *n* fisidylin
Fixateur *m* (Parfümerie) fixative
Fixativ *n* (Phot) fixer, fixative, fixing solution
Fixatorlösung *f* fixing solution
Fixbleiche *f* bleaching chloride of lime
Fixierbad *n* (Phot) fixing bath
fixierbar fixable
fixieren to fix, to harden, to set, (Seide) to mordant
Fixieren *n* fixation, fixing
Fixierflotte *f* fixing liquor
Fixierflüssigkeit *f* fixing liquid, fixing bath, fixing liquor
Fixierlösung *f* fixer, fixing solution
Fixiermittel *n* fixative, fixing agent
Fixiernatron *n* fixing salt, sodium thiosulfate
Fixiernatronlösung *f* sodium thiosulfate solution
Fixierprozeß *m* (Histol) fixation process
Fixiersalz *n* (Phot) fixing salt
Fixiersalzzerstörer *m* fixing salt destroyer
Fixierstift *m* locating pin
Fixiertemperatur *f* fixing temperature
Fixierton *m* fixing clay

Fixierung *f* (Phot) fixation, fixing, hardening, (Präparat) mounting
Fixierungsmittel *n* fixative, fixer, fixing agent
Fixierungspaar *n* standard couple
Fixpunkt *m* fixed point
Fixwert *m* fixed value
Fizeausches Rad *n* chopper disk
flach flat, even, level; plane, smooth
Flachabdampfschale *f* filter evaporating pan
Flachbagger *m* surface excavator
Flachbettdruck *m* flatbed printing
Flachbettfelge *f* flat base rim, straight side rim
Flachbeutel *m* flat bag
Flachbiegeversuch *m* flat bending test
Flachbrenner *m* fishtail burner, wing burner
Flachdraht *m* flat wire
flachdrehen (Mech) to surface
flachdrücken to flatten
Flacheisen *n* flat [bar] iron, flat steel
Flacheisenverankerung *f* retaining plate
flachen to flatten, to level, to plane
flachgängig square threaded
flachgeschweißt lap welded
Flachgewinde *n* square thread
flachgewunden planispiral
Flachglas *n* plate glass
Flachheit *f* flatness
Flachherdmischer *m* flat furnace mixer
Flachkaliber *n* box pass
Flachkeil *m* flat wedge, parallel key
Flachkopfschraube *f* countersunk screw, flathead screw
Flachkreuz *n* flat cross
Flachkupfer *n* flat copper
Flachland *n* (Geol) plain
Flachlasche *f* flat fish plate
Flachnaht *f* flush weld
Flachnutenfräsen *n* slab milling
flachpressen to flatten
Flachrost *m* flat grate, horizontal grate
Flachs *m* flax
Flachsack *m* pillow sack
Flachschaber *m* flat scraper
Flachschieber *m* flat slide valve
Flachschnitt *m* horizontal section
Flachschultertrommel *f* flat drum
Flachschweißung *f* lap welding
Flachseil *n* flat rope
flachsfarben flaxen, flaxy
flachsgelb flaxen
Flachssamenöl *n* flaxseed oil, linseed oil
Flachsstein *m* (Min) asbestos
Flachstahl *m* flat bar steel
Flachstampfer *m* flat rammer
Flachstrahldüse *f* flat spray nozzle
Flachswachs *n* flax wax
Flachtafel *f* flat sheet
Flachteilspule *f* flat coil
Flachtrommel *f* flat faced pulley

Flachunterseil *n* flat balance rope
Flachwulsteisen *n* flat bulb iron
Flachzange *f* flat pliers
Flackerkontakt *m* fluttering contact
flackern to flare, to flicker
Flacon *m* flacon, phial, small bottle
fladerig mottled, streaked, veined
Fladerschnitt *m* irregular vein section, knotty section
Fläche *f* area, space, (Ebene) level, (Krist) facet, (Oberfläche) surface, **~ gleichen Potentials** equipotential surface, **waagrechte ~** horizontal plane
Flächenausdehnung *f* surface expansion
Flächenausdehnungszahl *f* coefficient of superficial thermal expansion
Flächendichte *f* surface density
Flächeneinheit *f* unit of area
Flächeneinheitslast *f* loading per unit area
Flächenelement *n* surface element
Flächenfilter *n* screen cloth filter
Flächenfräser *m* (Techn) face-milling cutter
Flächengewicht *n* (Pap) basis weight, (Papier) basis weight
Flächengewichtswaage *f* quadrant balance
Flächengröße *f* area
flächenhaft plane
Flächeninhalt *m* area
Flächenintegral *n* (Math) surface integral
Flächenkathode *f* plate-shaped cathode
Flächenkontakt *m* broad area contact
Flächenkrümmung *f* curvature of a surface
Flächenmaß *n* square measure
Flächenmesser *m* planimeter
Flächenmessung *f* planimetry, surface measurement
Flächenmittelpunkt *m* center of area
Flächennormale *f* normal of surface
Flächenraum *m* area, surface
flächenreich polyhedral
Flächensatz *m* (Mech) law of areas
Flächenschleifmaschine *f* surface grinder
Flächenschnittwinkel *m* interfacial angle
Flächenschwerpunkt *m* center of gravity of a surface
Flächenstrom *m* (Elektr) surface current
Flächenstück *n* unit area, surface element
Flächensymmetrie *f* plane symmetry, symmetry of surface
Flächenteilchen *n* surface element
flächentreu area preserving; of equal area
Flächenverhältnis *n* area ratio
Flächenwinkel *m* angle of surface
Flächenzahl *f* number of faces, number of facets, square number
flächenzentriert (Krist) face-centered
Flächenzentrierung *f* centered face
Flagellat *m* (Zool) flagellate
Flagellenmembran *f* flagellar membrane
Flagellin *n* flagellin
Flakon *m n* phial, small bottle
flammbar flammable, inflammable
Flammbarkeit *f* inflammability
Flammbehandlungsverfahren *n* flame treatment process
Flamme *f* flame, light, **aufleuchtende ~** luminous flame, **nicht leuchtende ~** roaring flame, **nicht rußende ~** sootless flame, **oxydierende ~** oxidizing flame, **reduzierende ~** reducing flame, carbonizing flame
flammen to flame, to be in flames, to blaze
Flammenbehandlungsverfahren *n* flame treatment process
Flammenbogen *m* electric arc, flame arc, luminous arc
flammend flaming, blazing
Flammendurchschlag *m* flashback
Flammenemissionskontinuum *n* flame emission continuum
Flammenfärbung *f* flame coloration
Flammenfläche *f* flame surface
Flammenfront *f* flame front
Flammenhärtung *f* (Metall) flame hardening
Flammenionisationsdetektor *m* flame ionization detector
Flammenkegel *m* flame cone
Flammenloch *n* flame hole
flammenlos flameless
Flammenofen *m* flame furnace, reverberatory furnace
Flammenopal *m* (Min) fire opal
Flammenphotometer *n* flame photometer
Flammenrückschlag *m* backfiring, flashback
Flammenrückschlagsicherung *f* flame arrestor
Flammenschweißen *n* flame welding
flammensicher flameproof
Flammensonde *f* flame tester
Flammenspektrometrie *f* flame spectrometry
Flammenspektrum *n* flame emission spectrum
Flammenspritzen *n* flame spraying
Flammenstulpe *f* upturned flame duct
Flammentrosten *n* flame cleaning, flame descaling
Flammenüberwachung *f* flame control device
Flammenwerfer *m* flame thrower
Flammfestausrüstung *f* flame-resistant impregnation
Flammfeuer *n* blazing fire
Flammfeuerung *f* flame fire
flammgespritzt flame sprayed
Flammkohle *f* coal burning with flame, flaming coal, open burning coal, steam coal
Flammofen *m* reverberatory [puddling] furnace, **im ~ frischen** to puddle
Flammofenarbeit *f* reverberatory smelting
Flammofenschlacke *f* air furnace slag

Flammofenstahl *m* steel of a reverberatory furnace
Flammpolieren *n* flame polishing
Flammpunkt *m* point of ignition, flash point
Flammpunktbestimmung *f* flash test
Flammpunktprüfer *m* flash point apparatus, flash point tester, inflammation tester
Flammrohr *n* fire tube
Flammrohrkessel *m* [fire] flue boiler, radiant-type boiler
Flammruß *m* furnace black, lamp black
Flammschutzfarbe *f* flame retardant paint
Flammschutzmittel *n* fire-retardant paint, flame retardant
Flammspritzen *n* flame spraying
Flammspritzverfahren *n* flame spraying
Flammstrahlen *n* flame cleaning, flame descaling
Flammwidrigkeit *f* flame resistance
Flanellpolierscheibe *f* flannel polishing wheel
Flankendruck *m* (Keilriemen) face pressure
Flankenspiel *n* backlash (of gear), backlash of screw thread
Flankenstreuung *f* side leakage
Flankenwinkel *m* thread angle
Flansch *m* flange
Flanschbefestigung *f* flange coupling
Flanschdichtung *f* gasket
Flanschenaufwalzmaschine *f* flanging machine, flanging roll
Flanschenbolzen *m* flange bolt
Flanschendichtung *f* sealing flange, flanged packing, joint packing
Flanschendrehbank *f* flange lathe
Flanschenkupplung *f* flanged coupling
Flanschenmuffenstück *n* (E-Stück) flanged socket
Flanschenring *m* flanged ring
Flanschenrohr *n* flanged pipe
Flanschenrohrformstück *n* flanged fitting
Flanschenschraube *f* flange bolt, bolt of flange
Flanschenwalze *f* flanging roll
Flanschlager *n* flanged bearing, side bracket bearing
Flanschring *m* flange ring
Flanschverbindung *f* flanged connection, flanged coupling, flanged joint
Flanschverschluß *m* flange closure
Flanschwelle *f* flange shaft
Flanschwulsteisen *n* bulb rail
Flasche *f* bottle, (Chem) flask, (kleine Flasche) phial, **Leydener** ~ Leyden jar, **weithalsige** ~ wide-mouthed bottle, jar, **Woulfe'sche** ~ Woulfe's bottle
Flaschenabfüllraum *m* bottling room
Flaschenbürste *f* bottle brush
Flaschenfüllapparat *m* bottle filling apparatus
Flaschenfüller *m* bottling machine
Flaschenfüllmaschine *f* bottling machine

Flaschenfüllung *f* bottling
Flaschengärung *f* fermentation in the bottle
Flaschengas *n* bottled gas, cylinder gas
Flaschenglas *n* bottle glass
Flaschenhals *m* bottleneck
Flaschenkappe *f* bottle cap, bottle top
Flaschenkapsel *f* bottle cap
Flaschenkapsellack *m* bottle capsule lacquer
Flaschenkorb *m* hamper
Flaschenreinigungsmaschine *f* bottle cleaner
Flaschenschild *n* label
Flaschenventil *n* cylinder valve
Flaschenverschluß *m* bottle cap, bottle seal, stopper
Flaschenzug *m* block and tackle, pulley block, set of pulleys
Flaser *f* (Holz) streak
flattern to wobble, to wabble
Flatterruß *m* lamp black, soot black
flaumig fluffy
Flavacidin *n* flavacidin
Flavacol *n* flavacol
Flavan *n* flavan
Flavanilin *n* flavaniline
Flavanol *n* flavanol
Flavanomarein *n* flavanomarein
Flavanon *n* flavanone
Flavanookanin *n* flavanookanin
Flavanthren *n* flavanthrene
Flavanthrin *n* flavanthrine
Flavanthron *n* flavanthrone
Flavaspidsäure *f* flavaspidic acid
Flavazin *n* flavazine
Flavazol *n* flavazole
Flaveanwasserstoff *m* flaveanic acid
Flavellagsäure *f* flavellagic acid
Flaven *n* flavene
Flaviansäure *f* flavianic acid, 2,4-dinitro-1-naphthol-7-sulfonic acid
Flavicin *n* flavicin
Flavin *n* flavin[e]
Flavinadenindinucleotid *n* flavine adenine dinucleotide, FAD
Flavindehydrogenase *f* (Biochem) flavin dehydrogenase
Flavindulin *n* flavinduline
Flavinmononucleotid *n* flavin[e] mononucleotide
Flavinreduktion *f* flavin reduction
Flavinsynthese *f* flavin synthesis
Flaviolin *n* flaviolin
Flavipin *n* flavipin
Flavochrom *n* flavochrome
Flavocorylin *n* flavocoryline
Flavogallol *n* flavogallol
Flavoglaucin *n* flavoglaucin
Flavognost *n* flavognost
Flavokinase *f* (Biochem) flavokinase
Flavol *n* 1,3-anthradiol, flavol
Flavomycoin *n* flavomycoin

Flavon *n* flavone
Flavonol *n* flavonol
Flavopereirin *n* flavopereirine
Flavophen *n* flavophene
Flavoprotein *n* flavoprotein
Flavopurpurin *n* 1,2,6-trihydroxyanthraquinone, flavopurpurin
Flavotin *n* flavotine
Flavoxanthin *n* flavoxanthin
Flechte *f* (Bot) lichen, moss
flechten to braid, to plait
Flechten *n* plaiting
Flechtenbitter *n* cetrarin, cetrarinic acid, picrolichenin
Flechtenfarbstoff *m* (Bot) lichen coloring matter
Flechtenrot *n* lichen red, orcein
Flechtenseife *f* herpes soap
Flechtenstärkemehl *n* lichen starch, lichenin, moss starch
Flechtenstein *m* (Min) lichenite
Flechtmaschine *f* braiding machine
Fleck *m* speck, spot, stain
Flecken *m* spot, speck, stain
Fleckenentferner *m* stain remover
Fleckenentfernungsmittel *n* spot remover, stain remover
fleckenfrei stainless
Fleckenporphyr *m* (Min) tufaceous porphyry
Fleckenreiniger *m* spot remover, stain remover
Fleckenwasser *n* spot remover, stain remover
Fleckfieber *n* (Med) jail fever, typhus
fleckig spotted, stained, ~ **werden** to spot, to soil, to stain
Fleckigwerden *n* staining, soiling
Fleckschiefer *m* (Min) spilosite, fleckschiefer
Fleckstein *m* clay for stain removal, scouring stone
Fleckstift *m* scouring stick
Fleckverformung *f* spot distortion
Fleisch *n* meat, (Früchte) pulp, (lebendes) flesh
Fleischextrakt *m* meat extract
Fleischfarbe *f* flesh color
fleischfarben flesh-colored
fleischfressend carnivorous
Fleischkohle *f* animal charcoal, flesh charcoal
Fleischkonserven *f pl* canned meat, preserved meat, tinned meat
Fleischleim *m* sarcocol
Fleischmehl *n* meat meal
Fleischmehlmühle *f* meat meal mill
Fleischmilchsäure *f* (Biochem) sarcolactic acid
fleischrot flesh-colored
Fleischsaft *m* meat juice
Fleischstempelfarben *pl* meat marking inks
Fleischzerfaserungsmaschine *f* meat shredding machine
Fleischzucker *m* inosite, inositol, muscle sugar
Flexinin *n* flexinine
Flieder *m* (Bot) elder, elder flowers

Fliederöl *n* elder oil, lilac oil
Fliedersaft *m* elder juice, elder sirup
Fliedertee *m* elderberry tea
Fliederwein *m* elderberry wine
Fliegenfalle *f* fly trap
Fliegengift *n* fly poison
Fliegenholz *n* bitter wood, quassia wood
Fliegenpapier *n* fly paper
Fliegenpilz *m* (Bot) fly agaric, toadstool
Fliegenpulver *n* fly powder
Fliegenstein *m* (Min) fly stone, native arsenic
Fliegenstreifen *m* fly paper
Fliegerbenzin *n* aviation gasoline
Fliehbeschleunigung *f* (Phys) centrifugal acceleration
Fliehkraft *f* (Phys) centrifugal force
Fliehkraft-Kugelmühle *f* centrifugal ball mill
Fliehkraftstaubabscheider *m* cyclone dust collector
Fliehpendel *n* centrifugal pendulum
Fliese *f* tile, Dutch tile
Fließarbeit *f* continuous operation, continuous production
Fließarbeitstisch *m* conveying table
Fließbalken *m* restrictor bar
Fließband *n* assembly line, conveyor belt
Fließbandarbeit *f* assembly line work
Fließbandmontage *f* belt assembly
fließbar flowable
Fließbedingung *f* yield condition
Fließbetrieb *m* (Reaktionstechnik) continuous process
Fließbett *n* fluid bed, fluidized bed
Fließbettkatalysator *m* fluidized catalyst
Fließbettkatalyse *f* fluid bed catalysis
Fließbettmischer *m* fluidized bed mixer
Fließbettreformieranlage *f* fluid type reforming plant
Fließbetttrockner *m* fluidized bed dryer
Fließbettverfahren *n* fluid bed process
Fließbild *n* flow sheet
Fließdiagramm *n* flow chart, flow pattern
Fließdruck *m* flow pressure, hydraulic pressure
Fließeigenschaft *f* flow property, rheological property
Fließeinschränkung *f* yield restriction
fließen to flow, to melt, to run, (Tinte) to blot
Fließen *n* flow, flowing, running, **laminares** ~ telescopic flow, **plastisches** ~ plastic flow, **pseudostationäres** ~ pseudosteady plastic flow
fließend fluid
Fließerweichung *f* pseudoplastic flow, quasiviscous flow
Fließfähigkeit *f* flowability, fluidity, tendency to flow
Fließfestigkeit *f* flow resistance
Fließfigur *f* strain figure
Fließgelenk *n* yield hinge
Fließgeschwindigkeit *f* rate of flow

Fließglätte – Flosse

Fließglätte *f* wet litharge
Fließgold *n* stream gold, wash gold
Fließgrenze *f* flow limit, yield point
Fließharz *n* turpentine
Fließkatalysator *m* fluid catalyst
Fließkohle *f* coal oil
Fließkoksverfahren *n* fluid coking
Fließkontakt *m* fluidized catalyst
Fließkraft *f* yield force
Fließleder *n* poromeric material
Fließlehre *f* rheology
Fließlinie *f* flow line, flow mark
Fließmarkierung *f* flow mark
Fließmischer *m* continuous mixer, flow mixer
Fließmittel *n* (Gaschromat) mobile phase, (Gummi) plasticizer
Fließofen *m* pyrites kiln
Fließpapier *n* absorbent paper, blotting paper
Fließprüfer *m* flowmeter
Fließpunkt *m* flow point, yield point
Fließrinne *f* air-activated gravtiy conveyor
Fließrinnenförderung *f* air-activated gravity conveying
Fließschema *n* flow sheet
Fließschmelztemperatur *f* fusing point, softening point
Fließspannung *f* yield stress
Fließsprung *m* change in flow stress
Fließtemperatur *f* flow temperature
Fließverfestigung *f* resistance to flow
Fließverhältnis *n* yield stress ratio
Fließverhalten *n* flow behavior
Fließvermögen *n* flow [properties]
Fließvorgang *m* flow process
Fließware *f* bonded fiber fabrics
Fließwert *m* yield value
Fließwiderstand *m* flow stress, resistance to flow, yield resistance
Flimmer *m* glitter, spangle, tinsel
flimmern to glisten, to glitter, to shimmer, to sparkle
flimmernd glimmering, glistening, shining, sparkling
Flimmerphotometer *n* (Phys) flicker photometer
Flimmerschein *m* scintillating luster
Flindersiamin *n* flindersiamine
Flindersin *n* flindersine
Flindersinsäure *f* flindersinic acid
Flinkit *m* (Min) flinkite
Flint *m* flint, rock flint
Flintglas *n* flint glass
Flintglasisolator *m* flint glass insulator
Flintglasprisma *n* flint glass prism
Flintmehl *n* crushed flint, flint meal
Flintstein *m* flint
Flinz *m* (Min) siderite, sparry iron ore, sphaerosiderite
Flitter *m* gold leaf, silver leaf, spangle, tinsel
Flitterdraht *m* tinsel wire
Flittererz *n* (Min) glittering ore; ore in glittering laminas
Flitterglas *n* crushed glass, pounded glass
Flittergold *n* clinquant, Dutch metal, foliated brass, tombac
flittern to glisten, to glitter, to sparkle
Flittersand *m* micaceous sand
Flittersilber *n* tinsel
Flockdruck *m* (Text) flock printing
Flocke *f* (Schnee) flake, (Wolle) flock
flocken to flake, to flocculate, to form flakes
flockenartig flake-like, flaky, flocculent
Flockenasbest *m* (Min) flaked asbestos
Flockenbildung *f* flocculation, formation of flakes
Flockenerz *n* (Min) lead arsenate, native massicot
Flockengraphit *m* (Min) flake graphite
Flockenpapier *n* flock paper
Flockensalpeter *m* efflorescent saltpeter
Flockenstruktur *f* flocculent structure
flockig flaky, flocculent, flocky, fluffy
Flockung *f* flocculation
Flockungsklärbecken *n* reactor-clarifier
Flockungsmittel *n* flocculant, flocculating agent, sedimentation aid
Flockungspunkt *m* flocculation point
Flöz *n* (Bergb) seam, layer, lode, rock bed, stratum, vein
Flözasche *f* clay marl
Flözdolomit *m* (Min) secondary dolomite
Flözerz *n* ore in beds, ore in veins
Flözgranit *m* (Min) secondary granite
Flözgrünstein *m* (Min) dolerite
Flözkalk *m* bedded limestone
Flözporphyr *m* (Min) secondary porphyry
Flözsandstein *m* (Min) red sand stone
Flözvergasung *f* underground gasification
Flohkraut *n* (Bot) fleabane
Flokit *m* (Min) flokite
Flor *m* gauze, (Bot) bloom
Flora *f* (Bot) flora
Florantyron *n* florantyrone
Florencit *m* (Min) florencite
Florentiner Flasche *f* Florentine receiver
Florentiner Lack *m* Florentine lake, carmine lake
Florgewebe *n* pile fabric
Floridableicherde *f* Florida bleaching earth, floridin
Floridaphosphat *n* Florida phosphate
Floridosid *n* floridoside
Florigen *n* florigen
Floropryl *n* floropryl, DFP.
Florsafran *m* safflower
Floß *n* (Metall) pig iron, pig metal, **lückiges** ~ open grain pig, open white pig, porous pig iron
Flosse *f* fin

Floßenrohr n finned tube
Floßloch n mouth of a furnace
Flotation f flotation
Flotationsabgänge pl tailings
Flotationsanlage f flotation plant
flotationsfähig floatable
Flotationsmittel n flotation agent
Flotationsstoffänger m (Papier) flotation save-all
Flotationsverfahren n flotation method
flotieren to float
Flotte f (Färb) batch, dye bath, dye liquor
Flottenansatz m initial bath
Flottenaufnahme f liquor pick-up
Flottenkonzentration f liquor concentration
Flottenverhältnis n liquor ratio, volume of liquor
Flottenverlust m loss in liquor
Flottenzirkulation f liquor circulation
Flottstahl m ingot steel, run steel
Fluat n fluate, (Fluorosilicat) fluate, (Fluorsilikat) fluosilicate
fluatieren to weatherproof with fluosilicate
Fluatlösung f fluosilicate solution
Fluavil n fluavil
fluchten to align
Fluchtlinie f center line, sight line, straight line, (Opt) vanishing line
Fluchtliniennomogramm n alignment nomogram
Fluchtpunkt m (Opt) vanishing point
Fluchtungsfehler m misalignment
Fludrocortisonacetat n fludrocortisone acetate
flüchtig (Chem) volatile, ~ **entwerfen** to make a rough outline, to sketch, ~ **mit Wasserdampf** distillable with [water] steam
Flüchtigkeit f (Chem) volatility
Flüchtigkeitsprodukt n volatility product
Flüchtigmachung f volatilization
Flügel m wing, fan, (Rührwerk) blade, (Windmühle) vane
Flügelblatt n blade
Flügelgebläse n fan blower
Flügelhaut f wing covering
Flügelkopf m winged head
Flügelmutter f butterfly nut, finger nut, thumb nut, wing nut
Flügelradanemometer n vane anomometer
Flügelrädchen n paddle wheel
Flügelscheide f wing sheath
Flügelschraube f thumb screw, wing screw
Flügelspitze f wing tip
Flügelspitzenschwimmer m wing tip float
Flügelstellung f vane setting
Flügelventil n butterfly valve
Flügelwelle f fan shaft
Flügelzellenpumpe f sliding-vane pump
Fluellit m (Min) fluellite
flüssig fluid, liquid, ~ **machen** to liquefy, to melt
Flüssigbleiben n remaining in the liquid state
Flüssig-flüssig-Extraktion f liquid-liquid extraction

Flüssiggas n liquefied gas, liquid petrol gas (PG or LP gas)
Flüssiggasanlage f liquid gas plant
Flüssigkeit f fluid, **benetzende** ~ wetting liquid, **ideale** ~ ideal liquid, **lichtbrechende** ~ refractive fluid, **Newtonsche** ~ Newtonian liquid, **Nicht-Newtonsche** ~ non-Newtonian liquid, **optisch reine** ~ (Opt) optically void liquid, **reinviskose** ~ Newtonian fluid, **thixotrope** ~ thixotropic fluid, **unterkühlte** ~ supercooled liquid
Flüssigkeiten pl, **mischbare** ~ miscible liquids, **nicht mischbare** ~ immiscible liquids
Flüssigkeitsanzeiger m [liquid] level indicator
Flüssigkeitsaufgeber m distributor
Flüssigkeitsaufnahme f (Physiol) fluid intake, (Zelle) pinocytosis
Flüssigkeitsausscheidung f fluid output
Flüssigkeitsbahn f path of the liquid
Flüssigkeitsbedarf m (Physiol) fluid requirement
Flüssigkeitsbewegung f motion of a liquid
Flüssigkeits-Chromatographie f liquid chromatography
Flüssigkeitsdämpfer m liquid damper, viscous damper
Flüssigkeitsdämpfung f liquid damping
Flüssigkeits-Dampfverhältnis n liquid-vapor ratio
Flüssigkeitsdichtemesser m liquid densitimeter
Flüssigkeitsdichteschreiber m liquid density recorder
Flüssigkeitsdruck m fluid pressure, hydraulic pressure, hydrostatic pressure
Flüssigkeitseingabeort m (Gaschromat) injection port
Flüssigkeitseinschlüsse pl sealed liquids
Flüssigkeitsersatz m fluid replacement
Flüssigkeitsförderung f raising of liquids
Flüssigkeitsgehalt m fluid content, moisture content
Flüssigkeitsgemisch n liquid mixture
Flüssigkeitsgewicht n weight of a fluid
Flüssigkeitsgrad m degree of consistency, degree of viscosity
Flüssigkeitsgradient m (Bodenkolonne) hydraulic gradient, liquid gradient
Flüssigkeitsgradmesser m viscosimeter
Flüssigkeitshaushalt m (Physiol) fluid balance, intake and output of fluids
Flüssigkeitsheber m siphon
Flüssigkeitskühler m liquid cooler
Flüssigkeitskühlung f liquid cooling
Flüssigkeitsmaß n liquid measure
Flüssigkeitsmasse f bulk liquid
Flüssigkeitsmenge f quantity of liquid
Flüssigkeitsmesser m fluid meter, liquid meter
Flüssigkeitsmischapparat m liquid mixing apparatus
Flüssigkeitsmischer m flow mixer

Flüssigkeitsoberfläche *f* liquid surface
Flüssigkeitspumpe *f* pump for liquids
Flüssigkeitsreibung *f* friction of a liquid
Flüssigkeitsring-Pumpe *pl* liquid-seal pump, water-ring pump
Flüssigkeitssäule *f* liquid column
Flüssigkeitsschicht *f* liquid layer
Flüssigkeitssperrung *f* liquid seal
Flüssigkeitsspiegel *m* liquid level, surface of a liquid
Flüssigkeitsspiegelreflexion *f* liquid mirror reflection
Flüssigkeitsspindel *f* hydrometer
Flüssigkeitsstand *m* level of liquid, liquid level
Flüssigkeitsstandanzeiger *m* [liquid] level indicator
Flüssigkeitsstandmessung *f* liquid level measurement
Flüssigkeitsstrahl *m* jet of liquid
Flüssigkeits-Szintillationszähler *m* liquid scintillation counter
Flüssigkeitsthermometer *n* liquid-filled thermometer
Flüssigkeitsumlauf *m* liquid circulation
flüssigkeitsundurchlässig impermeable to liquids
Flüssigkeitsverdrängung *f* fluid displacement
Flüssigkeitsverteiler *m* (Dest) liquid distributor
Flüssigkeitsvorlage *f* liquid buffer, receiver
Flüssigkeitswaage *f* areometer, hydrometer
Flüssigkeitswärme *f* heat of a liquid
Flüssigkeitszähler *m* flow meter, fluid meter
Flüssigkeitszerstäuber *m* atomizer, liquid fuel nozzle, spray tank
Flüssigmachen *n* liquefaction
Flüssigmachung *f* liquefaction, melting
Flüssigmetallkühlung *f* liquid-metal cooling
Flüssigmetallreaktor *m* liquid metal fuelled reactor
Flüssigphase-Krackverfahren *n* liquid-phase cracking
Flüssigphasenoxidation *f* liquid-phase oxidation
Flüssigspaltstoffreaktor *m* fluidized fuel reactor
Flüssigwerden *n* liquefaction
Flüsterpumpe *f* whispering pump
Flugasche *f* flue ashes, flue dust, fly ash, light ashes
Flugascheablagerung *f* deposit of flue dust
flugaschenfrei free from volatile ashes
Flugbahn *f* line of flight, trajectory
Flugbenzin *n* aviation gasoline
Flugfeuer *n* fires caused by flying brands, flying brands
Flughafer *m* (Bot) wild oats
Flugkraftstoff *m* aviation fuel
Fluglinie *f* line of flight, trajectory
Flugplatzlager *n* (für flüss. Brennstoffe) airport storage tank

Flugstaub *m* air-borne dust, flue dust, fluidized dust, metallic dust, ~ **der Bleiöfen** *pl* lead fume, lead smoke
Flugstaubkammer *f* dust catcher
Flugstaubkondensator *m* smoke condenser
Flugwasser *n* spray
Flugzeitmessung *f* time-of-flight measurement
Flugzeitmethode *f* time-of-flight method
Flugzeitspektrometer *n* time-of-flight spectrometer
fluhen (betonieren) to cement, to concrete
Fluid *n* fluid, liquid
Fluidextrakt *m* fluid extract
Fluidifikation *f* fluidification
Fluidifikationsverfahren *n* fluidized process
fluidifizieren to fluidize
Fluidisation *f*, inhomogene ~ bubbling or aggregative fluidization
Fluidität *f* fluidity
Fluidoplaste *pl* liquid plastic masses
Fluidum *n* fluid, liquid
Fluktuation *f* fluctuation
fluktuieren to fluctuate
Fluocerit *m* (Min) fluocerite
Fluoflavin *n* fluoflavine
Fluolith *m* (Min) fluolite
Fluor *n* fluorine
Fluoraden *n* fluoradene
Fluoräthylen *n* fluoroethylene
Fluoraluminium *n* aluminum fluoride
Fluorammonium *n* ammonium fluoride
Fluoran *n* fluoran[e]
Fluoranil *n* fluoranil
Fluoranthen *n* fluoranthene
Fluoranthenchinon *n* fluoranthenequinone
Fluorantimon *n* antimony fluoride
Fluorapatit *m* (Min) fluorapatite
Fluorarsen *n* arsenic fluoride
Fluorazen *n* fluorazene
Fluorazol *n* fluorazol
Fluorbarium *n* barium fluoride
Fluorbenzol *n* fluorobenzene, fluobenzene
Fluorblei *n* lead fluoride
Fluorborsäure *f* fluoboric acid
fluorborsauer fluoborate
Fluorbrom *n* bromine fluoride
Fluorcalcium *n* calcium fluoride
Fluorchrom *n* chromium fluoride
Fluorcitrat *n* fluorocitrate
Fluoreisen *n* iron fluoride
Fluoren *n* fluorene
Fluorenacen *n* fluorenacene
Fluorenaphen *n* fluorenaphene
Fluorenon *n* fluorenone, diphenylene ketone
Fluorenoncarbonsäure *f* fluorenone carboxylic acid
Fluorenyl- fluorenyl
Fluorescein *n* fluorescein
Fluorescin *n* fluorescin

Fluorescyanin *n* fluorescyanine
Fluoressigsäure *f* fluoroacetic acid
Fluoreszenz *f* fluorescence
Fluoreszenzanalyse *f* fluorescence analysis
Fluoreszenzansprechvermögen *n* fluorescent response
Fluoreszenzaufnahme *f* fluorography
Fluoreszenzausbeute *f* fluorescence yield
Fluoreszenzfarbe *f* fluorescent paint, luminous paint
Fluoreszenzgeräte *n pl* fluorescence equipment
Fluoreszenzlampe *f* fluorescent lamp
Fluoreszenzleuchten *n* fluorescent glow
Fluoreszenzlöschung *f* quenching of fluorescence
Fluoreszenzmesser *m* fluorimeter, fluorometer
Fluoreszenzschirm *m* fluorescent screen
Fluoreszenzspektroskopie *f* fluorescence spectroscopy
Fluoreszenzspektrum *n* fluorescence spectrum
Fluoreszenzstrahlung *f* fluorescent radiation
fluoreszieren to fluoresce
Fluoreszieren *n* fluorescence
fluoreszierend fluorescent
Fluorgehalt *m* fluorine content
fluorhaltig containing fluorine
Fluorhydrocortison *n* fluorhydrocortisone
Fluorid *n* fluoride
Fluoridierung *f* fluoridation
Fluoridkomplex *m* fluoride complex
fluorieren to fluorinate
Fluorierung *f* fluorination
Fluorimeter *n* fluorimeter, fluorometer
Fluorimetrie *f* fluorimetry, fluorometry
fluorimetrisch fluorimetric, fluorometric
Fluorinden *n* fluorindene
Fluorindin *n* fluorindine
Fluorit *m* (Min) fluorite, fluorspar
Fluorjod *n* iodine fluoride
Fluorkalium *n* potassium fluoride
Fluorkalzium *n* calcium fluoride
Fluorkieselsäure *f* silicofluoric acid, fluosilicic acid
Fluorkohlenstoff *m* carbon tetrafluoride, fluorocarbon, tetrafluoromethane
Fluorlithium *n* lithium fluoride
Fluormagnesium *n* magnesium fluoride
Fluornatrium *n* sodium fluoride
Fluorocyclen *n* fluorocyclene
Fluoroform *n* fluoroform, fluoform, trifluoromethane
Fluoroid *n* fluoroid
Fluorometer *n* fluorimeter, fluorometer
Fluoron *n* fluorone
Fluorophor *n* fluorophore
Fluoropren *n* fluoroprene
Fluoroprenkautschuk *m* fluoroprene rubber
Fluoroskopie *f* fluoroscopy
Fluorphosphat *n* fluophosphate
Fluorphosphor *m* phosphorus fluoride

Fluorsalicylaldehyd *m* fluorosalicylaldehyde
Fluorsalz *n* fluoride, salt of hydrofluoric acid
Fluorschwefel *m* sulfur fluoride
Fluorselen *n* selenium fluoride
Fluorsilber *n* silver fluoride
Fluorsilikat *n* silicofluoride, fluosilicate, salt or ester of fluosilicic acid, salt or ester of silicofluoric acid
Fluorsilizium *n* silicon fluoride
Fluorsiliziumverbindung *f* silicofluoride, fluosilicate
Fluortantalsäure *f* fluotantalic acid
fluortantalsauer fluotantalate
Fluortellur *n* tellurium fluoride
Fluortitan *n* titanium fluoride
Fluortoluol *n* fluorotoluene, fluotoluene
Fluorubin *n* fluorubine
Fluorür *n* (obs) fluoride
Fluorverbindung *f* fluorine compound, fluoride
Fluorwasserstoff *m* hydrofluoric acid, hydrogen fluoride, phthoric acid (obs)
Fluorwasserstoffgas *n* gaseous hydrofluoric acid, hydrogen fluoride
Fluorwasserstoffsäure *f* hydrofluoric acid
fluorwasserstoffsauer fluoride, hydrofluoric
Fluoryliden *n* fluorylidene
Fluorzink *n* zinc fluoride
Fluorzinn *n* tin fluoride
Flurfördermittel *n* (Flurförderer) floorborne vehicles, surface floor-based handling equipment
Flurförderzeuge *pl* industrial trucks
Flusen *f pl* (Web) slubs
Flushingverfahren *n* (Web) flushing [process]
flusig downy, furry
Flusigkeit *f* furriness
Fluß *m* current, (Elektr, Metall) flux, (fließende Bewegung) flow, (Strom) river, stream, **kalter ~** (Prüftechnik) cold flow, creep
Flußabsenkung *f* flux depression
Flußäther *m* fluoric ether
Flußdichte *f* flux density
Flußeisen *n* low carbon steel, converter iron, fluid metal, ingot metal, soft or mild steel, structural iron
Flußeisenblech *n* mild steel plate
Flußeisenerzeugung *f* manufacture of steel by fusion
Flußeisenrohr *n* mild steel pipe
Flußerde *f* (Min) earthy fluorite, fluor earth
Flußharz *n* gum animé, gum copal
Flußkies *m* river gravel
Flußkieselsäure *f* silicofluoric acid, fluosilicic acid
flußkieselsauer fluosilicate, silicofluoride
Flußmetall *n* fluid metal
Flußmeter *n* fluxmeter
Flußmittel *n* flux, fluxing agent, fluxing medium
Flußpulver *n* flux powder

Flußsäure – Folgerung

Flußsäure *f* hydrofluoric acid
Flußsäureanlage *f* hydrofluoric acid plant
Flußsand *m* river sand
flußsauer fluoride
Flußschlamm *m* river silt
Flußspat *m* (Min) fluorite, fluor spar, **grüner** ~ pseudo emerald, **violetter** ~ pseudo amethyst
Flußspaterde *f* (Min) earthy fluorite
Flußspatgitter *n* (Krist) fluorite lattice
flußspathaltig fluoritic
Flußspatpulver *n* fluor spar powder
Flußspatsäure *f* (obs) hydrofluoric acid, hydrogen fluoride
Flußstahl *m* carbon steel, ingot steel, soft steel
Flußstahldraht *m* ingot steel wire, mild steel wire
Flußstein *m* (Min) compact fluor spar, fluorite, native calcium fluoride
Flußvektor *m* flow vector
Flußverfahren *n* flow coating
Flußverkettung *f* flux linkage
Flußverstärker *m* (Atom) doughnut
Flußverzerrung *f* flux distortion
Flut *f* flood
Flutbeginn *m* (Dest) [commencement of] flooding
Flutbelastung *f* flooding point
Fluten *n* (Dest) flooding
Flutgrenze *f* (Dest) flooding point
Flutlackieren *n* flow coating
Flutmesser *m* (Schiff) tide gauge
Flutpunkt *m* flood point, (Dest) flooding point
Flutwechsel *m* (Schiff) turn of the tide
Fluxen *n* magnaflux non-destructive testing
FMN Abk. für Flavinmononucleotid
föhnen to apply warm air
Förderanlage *f* conveyer, conveyer plant, conveyer system, conveying plant, conveyor, load carrying equipment
Förderapparat *m* conveyer
Förderband *n* conveyer belt, belt conveyer, production line
Förderbandwaage *f* belt weigher
förderbar transportable
Förderbrücke *f* hauling bridge
Förderdruck *m* delivery pressure
Förderer *m* transporter
Fördererz *n* (Min) run-of-mine ore
Fördergefäß *n* conveyer bucket
Fördergerät *n* conveyer
Fördergerüst *n* elevator frame, hoist frame, lift frame
Fördergeschwindigkeit *f* (b. Rühren) discharge rate
Fördergrus *m* (Bergb) rough dross
Fördergurt *m* conveyer belt, belt conveyer
Fördergut *n* conveying stock
Förderhaken *m* hoisting hook
Förderhaspel *f* (Bergb) winch
Förderhöhe *f* (Pumpe) delivery head

Förderkette *f* (Techn) coveyer chain
Förderkohle *f* mine coal, pit coal, rough coal, run-of-mine coal
Förderleistung *f* conveying capacity
Förderleitung *f* conveyer pipe line
förderlich advantgeous, beneficial, conducive, promotive
Fördermenge *f* quantity delivered
Fördermittel *n* conveying means
fördern to convey, to forward, (Bergb) to haul
Förderorgan *n* conveying agent
Förderpumpe *f* conveying pump, feed pump
Förderquantum *n* output
Förderriemen *m* conveying belt
Förderrinne *f* conveyer trough, conveying chute, conveying trough
Förderrohr *n* Archimedian screw, conveyer pipe
Förderrutsche *f* conveying chute
Förderschale *f* elevator bucket, elevator cage
Förderschnecke *f* screw conveyer, worm conveyor
Förderseil *n* hoisting cable
Förderstrecke *f* run
Fördertechnik *f* materials handling
Förderturm *m* elevator frame, hoist frame, lift frame
Förderung *f* conveyance, transfer, transport, (Bergb) haulage, (Biol) promotion, **pneumatische** ~ air lift, **unterirdische** ~ underground conveyance or hauling
Förderwagenrüttler *m* tub shaker
fötal (Med) fetal
Fötus *m* (Med) fetus
Fokometer *n* (Phys) focimeter, focometer
Fokus *m* focus
Fokuslänge *f* focal length
Fokusmesser *m* focimeter
fokussieren to focus
Fokussierung *f* focus[s]ing
Fokustiefe *f* (Phot) depth of focus
Fokusweite *f* focal length
Folge *f* sequence, series, succession, (Math) series
Folgeanalyse *f* sequence analysis
Folgebedingung *f* consequential condition
Folgeelement *n* (Atom) daughter atom
Folgeerscheinung *f* consequence, sequel
Folgekern *m* (Atom) daughter nucleus
Folgeprüfung *f* sequential test
Folgereaktion *f* (Chem) secondary reaction, consecutive reaction, consequent reaction
Folgeregelung *f* cascade control, (Regeltechn) cascade control, servo-control
Folgeregler *m* (Regeltechn) follower controller, servo-follower, submaster controller
Folgereihe *f* order of succession, sequence, series
folgerichtig consequent[ly], logical[ly]
Folgerit *m* (Min) folgerite
Folgerung *f* conclusion

Folgesatz *m* (Informatik) overflow record, (Math) corollary
Folgezeiger *m* direction indicator
Folgezeigersystem *n* follow-up system
Folgezeitschalter *m* sequence timer
Folie *f* foil, film, sheet, (Met) leaf metal, ~ **eines Spiegels** leaf tin, mirror foil, tin foil, **formbare** ~ postforming sheet, **gegossene** ~ cast film, **geprägte** ~ embossed sheet, **geschälte** ~ sliced film, **geschweißte** ~ welded film, **selbsttragende** ~ unsupported film
Folienabdeckung *f* diaphragm seal
Folienabdruckverfahren *n* (Mikroskopie) replica technique
Folienaktivierung *f* foil activation
Folienblase *f* bubble
Foliendicke *f* film thickness
Folienhaube *f* shrink wrap film
Folienkalander *m* sheeting calender
Folienkunstleder *n* artificial leather sheets
Folienlack *m* foil lacquer, strippable coating
Folienmaterial *n* sheeting
Folienpackung *f* flexible package
foliieren to cover with foil, to silver
Folinreagenz *n* folin reagent
Folinsäure *f* folinic acid, citrovorumfactor, CoF
Folio *n* (Papierformat) folio
Follikelhormon *n* (Biochem) follicular hormone
Follikelreifungshormon *n* follicle-stimulating hormone, F.S.H.
Follikulin *n* folliculin, [o]estrogenic hormone
Folsäure *f* folic acid
Folsäuremangel *m* folic acid deficiency
Forcherit *m* (Min) forcherite
Forcierungsversuch *m* forcing test, overload test
Forcit *n* forcite
Fordbecher *m* Ford cup, Ford viscosimeter
Forelleneisen *n* mottled white pig iron
forensisch forensic
Foresit *m* (Min) foresite
Form *f* (Gestalt) shape, (Gieß-Form) form, (Guß-Form) die, ~ **mit Abstreiferplatte** stripper plate mold, ~ **mit Doppelstempel** double force mold, **bleibende** ~ permanent mold, **cis-** ~ cis-form, **ekliptische** ~ (Stereochem) eclipsed form, **gestaffelte** ~ (Stereochem) staggered form, **gewölbte** ~ doming mold, **gußeiserne** ~ cast iron mold, chill mold, **massive** ~ solid form, **mehrteilige** ~ split cavity mold, split mold
Formänderung *f* change of shape, deformation, dimensional change, **bleibende** ~ permanent distortion, permanent set, plastic strain, **elastische** ~ elastic deformation, **gesamte** ~ range of deformation, **homogene** ~ homogeneous strain
Formänderungsarbeit *f* deformation energy, work of deformation

Formänderungsbereich *m* region of change of form, region of deformation
Formänderungsfähigkeit *f* deformability, (Met) ductility
Formänderungsfestigkeit *f* mean tensile strength
Formänderungsvermögen *n* capacity for deformation, plasticity
Formänderungswiderstand *m* deformation resistance
Formal *n* formal, dimethoxymethane
Formaldazin *n* formaldazine
Formaldehyd *m* formaldehyde, formic aldehyde, methanal, methylene oxide
Formaldehydbisulfit *n* formaldehyde bisulfite, formaldehyde hydrogen sulfite
Formaldehydhärtung *f* (Gerb) fixing with formaldehyde
Formaldehydharz *n* formaldehyde resin
Formaldehydhydrosulfit *n* formaldehyde hydrosulfite
Formaldehydlösung *f* formaldehyde solution
Formaldehydnatriumbisulfit *n* formaldehyde sodium bisulfite, formaldehyde sodium hydrogen sulfite
Formaldehydsulfoxylat *n* formaldehyde sulfoxylate
Formaldehydsulfoxylsäure *f* formaldehyde sulfoxylic acid
Formaldoxim *n* formaldoxime
Formalin *n* formalin, aqueous formaldehyde solution
Formalingerbung *f* (Gerb) formalin tanning
Formamid *n* formamide, formyl amine, methanamide
Formamidin *n* formamidine
Formanilid *n* formanilide, formylaniline, phenylformamide
Formartikel *m* molded article, molded good
Format *n* measure, size
Formatkreissäge *f* dimensional saw
Formazan *n* formazan
Formazyl *n* formazyl
formbar moldable, plastic, shapeable
Formbarkeit *f* moldability, plasticity, shapability
formbeständig dimensionally stable, resistant to deformation
Formbeständigkeit *f* dimensional stability, stability of shape
Formblatt *n* form
Formblech *n* profiled sheet iron
Formblock *m* mold block, former block
Formdraht *m* small section wire
Formeinheit *f* unit mold
Formeinsatz *m* mold insert
Formeisen *n* figured bar iron, section iron, structural steel
Formel *f* (Chem) formula, **empirische** ~ empirical formula
Formelausdruck *m* formula

Formelbestimmung *f* determination of formula
Formelbild *n* (Chem) chemical formula, structural formula
Formelgewicht *n* formula weight
Formelumsatz *m* formula conversion
Formelzeichen *n* symbol, formula sign
formen to form, to mold, to shape
Formenaustrieb *m* (Reifen) overflow
Formenbau *m* mold construction
Formenbuch *n* profile album
Formeneinsatz *m* insert mold, matrix
Formeneinstreichmittel *n* mold lubricant, mold dope
Formenkonstrukteur *m* mold designer
Formenöl *n* oil for molds
Formenschließdruck *m* mold closing pressure
Formenschluß *m* mold clamping
Formenschwindmaß *n* (Plast) mold[ing] shrinkage
Formentrennmittel *n* mold release
Formentrockner *m* mold drier
Former *m* former, molder
Formerde *f* molding clay
Formerei *f* molding hall, molding shop
Formergerätschaft *f* molder's tools
Formerspaten *m* molder spade
Formerstift *m* molding pin, stake
Formerton *m* molding clay
Formerwerkzeug *n* molder's tools
Formerz *n* (Min) rich silver ore
Formfaktor *m* shape factor (of particles)
Formfräsen *n* (Techn) profile milling
Formfräser *m* profile milling cutter
Formfräsmaschine *f* profile miller
Form-, Füll- und Schließsystem *n* form-seal equipment
Formgebung *f* shaping, fashioning, molding, profiling, styling, **spanabhebende** ~ shaping by machine tool with removal of chips, **spanlose** ~ noncutting shaping
Formgleichheit *f* conformity, (Krist) isomorphism
Formguß *m* die casting, mold casting
Formgußteil *n* die casting
Formheizung *f* mold curing
Formhydrazidin *n* formhydrazidine
Formhydroxamsäure *f* formhydroxamic acid
Formiat *n* formate, salt or ester of formic acid
Formicin *n* formicin, formaldehyde acetamide
Formieren *n* (v. Pigmenten) finishing conversion of pigments to powders, pastes, etc.
Formiersäure *f* forming acid
Formimidchlorid *n* formimidyl chloride
Formin *n* formin, glycerol formate
Forminitrazol *n* forminitrazole
Formkaliber *n* shaping pass
Formkasten *m* casting box, molding box, **hölzerner** ~ wooden box, wooden frame
Formkitt *m* molding clay

Formkörper *m* molded article
Formkohle *f* molded cylinders of activated charcoal
Formkohleplatte *f* carbon welding plate
Formlehm *m* molding loam
Formling *m* formed piece, molded article
formlos shapeless, (Krist) amorphous
Formlosigkeit *f* amorphousness, formlessness, shapelessness
Formmaschine *f* molding machine
Formmaschinenanlage *f* molding machinery
Formmasse *f* molding material, molding compound
Formmaterialien *n pl* molding materials
Formnase *f* (Gieß) tuyere
Formocholin *n* formocholine
Formöffnung *f* tuyere opening
Formoguanamin *n* formoguanamine
Formoguanin *n* formoguanine
Formol *n* formol
Formononetin *n* formononetin
Formopersäure *f* performic acid, peroxyformic acid
Formose *f* formose, i-fructose
Formosulfit *n* formosulfite
Formotanninstreupulver *n* formotannin strewing powder
Formoxim *n* formoxime, formaldehyde oxime, formaldoxime
Formplatte *f* molding board, plate with patterns, shaper plate
Formpresse *f* mold[ing] press
Formpressen *n* compression molding
Formpreßholz *n* molded plywood
Formpreßstoff *m* compression molding material
Formpreßverfahren *n* briquetting method
Formrahmen *m* mold frame, holder block, mold base
Formrüssel *m* tuyere nose
Formsand *m* foundry sand, molding sand, **fetter** ~ fat sand, loamy sand
Formsandanlage *f* casting sand plant
Formsandprüfgeräte *n pl* casting sand testing equipment
Formsattel *m* tuyere saddle
Formschablone *f* (Gieß) mold
Formschlacke *f* tuyere slag
Formschließkraft *f* locking pressure
Formschwingung *f* contour vibration
formstabil dimensionally stable, shape-retaining
Formstabilität *f* natural stability
Formstahl *m* section steel, structural steel
Formstanzen *n* die pressing, pressure forming
Formstaub *m* founder's dust
Formstoff *m* molding [material], mold composition
Formstück *n* molded piece, adapting piece, make-up piece, section iron, structural

member, ~ **aus Schichtstoff** laminated molding
Formteil *n* casting, formed piece; molded article; molded part, **nachgeformtes** ~ postformed molding, **nicht ausgeformtes** ~ short molding
Formton *m* molding clay
Formtrennmittel *n* mold parting agent, mold lubricant
formulieren to formulate
Formulierung *f* formulation
Formung *f* shaping, briquetting, **nachträgliche** ~ postforming
Formungsanlage *f* **mit Oberstempel** downstroke press
Formungsmethode *f* shaping method
Formveränderung *f* deformation, morphological change
Formveränderungsvermögen *n* deformability, plastic workability
Formwagen *m* mold carriage
Formwand *f* tuyere wall
Formwerkzeug *n* forming die
Formyl- formyl
Formylcampher *m* formylcamphor
Formylcellulose *f* formylcellulose, cellulose formate
Formylfluorid *n* formyl fluoride
formylieren to formylate
Formylierung *f* formylation
Formylsäure *f* methanoic acid, (obs) formic acid, formylic acid, methane acid
Formzacken *m* (Gieß) tuyere plate
Formzahl *f* form quotient
Forscher *m* researcher, research worker
Forschung *f* research
Forschungsabteilung *f* research department
Forschungsanstalt *f* research institute
Forschungsarbeit *f* research [work]
Forschungsauftrag *m* research contract
Forschungschemiker *m* research chemist
Forschungsergebnis *n* result of research
Forschungsinstitut *n* research institute
Forschungslaboratorium *n* research laboratory
Forschungsmethode *f* research method
Forschungsprogramm *n* research program
Forschungsprojekt *n* research project
Forschungsrat *m* Science Research Council (British equivalent)
Forschungsreaktor *m* research pile, research reactor
Forschungsrichtung *f* line of research
Forschungsstipendium *n* fellowship
Forschungszentrum *n* research center
Forschungszweck *m* research purpose
Forsterit *m* (Min) forsterite
Forstwirtschaft *f* forestry
Fortbestand *m* continuation

fortbestehen to continue to subsist, to endure, to last, to persist
Fortbewegungsfeld *n* propagation field
Fortbildung *f* further training
Fortdauer *f* continuance, duration
fortlaufend continuous[ly]
Fortoin *n* fortoin
fortpflanzen to transmit, (Biol) to regenerate, to reproduce, (Phys) to propagate
Fortpflanzung *f* (Biol) regeneration, reproduction, (Phys, Biol) propagation, ~ **elektrischer Wellen** propagation of electrical waves
Fortpflanzungsgeschwindigkeit *f* (Phys) propagation velocity
Fortpflanzungsgröße *f* propagation constant
Fortpflanzungsrichtung *f* direction of propagation
Fortschaltungsmechanik *f* trip free mechanism
Fortschritt *m* progress
fortschrittlich progressive
Fortschrittsgeschwindigkeit *f* velocity of underground flow
Fortsetzung *f* continuation
fortspülen to wash away
fossil fossilized
Fossil *n* fossil
fossilführend (fossilienhaltig) fossiliferous
Fossilienbildung *f* fossilization, fossilification
fotochemisch photochemical
Fotokathode *f* cathode of a phototube
Fotokopiergerät *n* photocopier
Fotoschicht *f* photosensitive layer, light-sensitive or photoelectric surface
Fototropie *f* phototropy
Fotowiderstand *m* photo resistance
Fotozelle *f* photoelectric cell or tube, light-sensitive tube
Fotozellenverstärker *m* photoelectric cell amplifier
Foulard *m* (Text) foulard, padding machine, padding mangle
foulardieren to pad
Fouqueit *m* (Min) fouqueite
Fourier-Analyse *f* (Math) Fourier analysis
Fouriersche Reihe *f* (Math) Fourier series
Fourmarierit *m* (Min) fourmarierite
Fowlerit *m* (Min) fowlerite
Fowlersche Lösung *f* (Pharm) Fowler's solution
Fracht *f* freight
Fräsarbeit *f* milling work
Fräsautomat *m* (Techn) automatic milling machine
fräsen to mill, to cut, (Holz) to shape
Fräsen *n* milling
Fräserschleiflehre *f* cutter clearance gauge
Fräslader *m* (Bergb) cutter-loader
Fräsmaschine *f* milling machine, (Holz) shaper
Fräsvorrichtung *f* milling fixture

Fräswerkzeug *n* milling cutter, mill grinder
Fragarin *n* fragarin
Fragment *n* fragment
fragmentarisch fragmentary
Fragmentierung *f* fragmentation
Fraktion *f* fraction, **hochsiedende** ~ high boiling fraction, **niedrigsiedende** ~ low[-boiling] fraction
Fraktionierapparat *m* fractionating apparatus
Fraktionieraufsatz *m* (Dest) fractionating attachment, distilling tube
Fraktionierbürste *f* fractionating brush
Fraktioniereinrichtung *f* fractionating device
fraktionieren (Chem) to fractionate
Fraktionieren *n* fractionating
Fraktionierkolben *m* fractionating flask
Fraktionierkolonne *f* (Dest) fractionating column, cracking tower, fractionating tower
Fraktionierrohr *n* fractionating tube
fraktioniert destillieren to fractionate
Fraktionierturm *m* fractionating tower, rectifying column
Fraktionierung *f* fractionation
Fraktionsaufsatz *m* fractionating attachment, distilling trap
Fraktionskolben *m* fractional distillation flask
Fraktionsröhre *f* fractional distillation tube
Fraktionssammler *m* fraction collector
Fraktionsschnitt *m* cut in distillation
Francium *n* francium (Symb. Fr)
Franckeit *m* (Min) franckeite
Francolith *m* (Min) francolite
Frangula-Emodin *n* frangula emodin
Frangulin *n* frangulin
Frangulinsäure *f* frangulic acid, frangulinic acid
Frankfurter Schwarz *n* Frankfort black, vine black
Franklandit *m* (Min) franklandite
Franklinit *m* (Min) franklinite
Franzbranntwein *m* (Pharm) cheap brandy for embrocations, surgical spirits
Franzgold *n* French leaf gold
Franzium *n* (Francium) francium (Symb. Fr)
Franzosenharz *n* guaiacum, gum guaiac
Franzosenholz *n* (Bot) boxwood, guaiacum wood, lignum vitae
Franztopas *m* (Min) smoky topaz
Fraueneis *n* (Min) mica, selenite, sparry gypsum
Frauenspat *m* (Min) selenite
Fraunhoferlinien *f pl* Fraunhofer lines
Fraxin *n* fraxin
frei an Bord free on board, F.O.B., ~ **Hütte** free at works, ~ **Waggon** free on trucks
Freiberger Aufschluß *m* (Chem) fusion with sulfur and alkali
Freibergit *m* (Min) freibergite
Freidampf *m* direct steam
Freidampfheizung *f* (Gummi) open steam cure
Freidampfvulkanisation *f* open steam curl

Freieslebenit *m* (Min) freieslebenite
Freifallmischer *m* free falling mixer
Freifallscheider *m* plate-type electrostatic separator
Freifallstromklassierer *m* free-settling hydraulic classifier
Freiformschmiedestück *n* hammer forging
freigeben to release, to set free
freihängend suspended
Freiheit *f*, **in** ~ **setzen** to disengage, (Chem) to liberate
Freiheitsgrad *m* degree of freedom, ~ **des freien Elektrons** electronic mode of the free electron
Freilagerversuch *m* natural exposure test, natural weathering test
freilaufend freewheeling
Freilegung *f* **des Lichtbogens** exposing the arc
Freileitung *f* open air piping, overhead line
freimachen to liberate, to set free
freischwebend airborne
Freisetzung *f* liberation, (Gas) evolution
Freispiegelleitung *f* gravity line
Freispiegeltunnel *m* grade tunnel
Freistrahl *m* free jet
Freistrahlturbine *f* (Mech) impulse turbine, impulse wheel, Pelton wheel
Freistrahlzentrifuge *f* impulse centrifuge
Freistrompumpe *f* non-clogging pump
freitragend cantilever, self-supporting
Freiwerden *n* liberation
freiwillig voluntary, (Zerfall) spontaneous
Freiwinkel *m* lip relief angle (drilling), relief [or clearance] angle
freizügig mobile
Fremdantrieb *m* separate drive
Fremdatom *n* foreign atom, impurity atom
Fremdatomzusatz *m* addition of impurities, impurity addition
Fremdbelüftung *f* external ventilation
Fremdbestandteil *m* foreign matter
Fremdeiweiß *n* foreign protein
Fremdelektron *n* stray electron
fremderregt separately excited
Fremderregung *f* foreign excitation, separate excitation
Fremdführung *f* separate commutation
Fremdgas *n* foreign gas
Fremdgasdiffusion *f* (Sieb, Lochblende, poröse Zwischenwand usw.) mass diffusion
Fremdgasschreiber *m* foreign gas recorder
fremdgesteuert separately controlled
Fremdkapital *n* total liabilities
Fremdkeim *m* foreign nucleus
Fremdkörper *m* foreign body, foreign matter, foreign substance
Fremdmetall *n* foreign metal
Fremdrost *m* rust from external sources
Fremdsalz *n* foreign salt
Fremdspannung *f* external voltage

Fremdstoff *m* by-product, foreign matter, impurity
Fremdstoffeinfluß *m* impurity effect
Fremdstrom *m* stray current
Fremdzustatz *m* impurity addition
Fremontit *m* (Min) fremontite
Frenzelit *m* (Min) frenzelite
Freon *n* freon
frequentieren to frequent
Frequenz *f* frequency, periodicity
frequenzabhängig frequency-dependent
Frequenzanalogwandler *m* analogue frequency changer
Frequenzangaben *f pl* frequency data
Frequenzband *n* frequency band
Frequenzbereich *m* frequency range
Frequenzeinheit *f* frequency unit
Frequenzgang *m* (b. Transformationen) frequency response [function], (Regeltechn) frequency response (Am. E.), harmonic response (Br. E.)
Frequenzgemisch *n* frequency spectrum
Frequenzgleichung *f* frequency equation
Frequenzmesser *m* frequency meter, ondometer, wavemeter
frequenzmoduliert frequency modulated
Frequenznormal *n* frequency standard
Frequenzregler *m* frequency controller
Frequenzschwankung *f* variation of frequency
Frequenzstabilisierung *f* frequency control
Frequenzstrahlung *f* frequency radiation
Frequenzumkehrung *f* frequency inversion
Frequenzuntersetzer *m* frequency divider
Frequenzverhalten *n* frequency dependence
Frequenzwiedergabe *f* frequency response [characteristic]
Frequenzwobbelverfahren *n* frequency sweep method
Fresnelsche Beugung *f* (Opt) Fresnel diffraction
Fresnelspiegel *m* Fresnel's mirror
fressen to decay, (Met) to corrode
Fressen *n* (Friktion) galling, (Futter) feed, fodder, forage, (Geol) scoring, (Techn) scuffing, seizure
Freundlichsche Adsorptionsisotherme *f* Freundlich adsorption isotherm
Freyalith *m* (Min) freyalite
Friedel-Craftssche Reaktion *f* Friedel-Crafts reaction
Friedelin *n* friedelin
Friedelit *m* (Min) friedelite
Friedonsäure *f* friedonic acid
Frieren *n* freezing, liquid solidification
Frieseit *m* (Min) frieseite
Friesreaktion *f* Fries reaction
Friessche Regel *f* Fries rule
Friessche Umlagerung *f* (Chem) Fries rearrangement
Frigen *n* freon, frigen

Friktion *f* friction
Friktionieren *n* friction coating
Friktionskalander *m* (Gummi) friction calender
Friktionskupplung *f* friction clutch
Friktionslagermetall *n* friction bearing
Friktionsmesser *m* friction meter
Friktionsmischung *f* (Gummi) friction compound
Friktionspresse *f* friction press
Friktionspulver *n* fulminating powder
Friktionsrad *n* friction gear
Friktionsscheibe *f* friction disc
Friktionswalze *f* friction roll
Frischarbeit *f* (Blei) reducing process, (Metall) fining process, puddling process, refining
Frischbirne *f* (Metall) converter
Frischblei *n* refined soft lead
Frischdampf *m* live steam
Frischdampfleitung *f* live steam line, live steam piping
Frischdampfrohr *n* live steam pipe
Frischeisen *n* (Metall) refined iron, bloom iron, finery iron
frischen (Blei) to reduce, (Metall) to refine, to decarburize, to oxidize, to purify
Frischen *n* (Metall) refining, decarburization, fining, oxidation
Frischer *m* refiner
Frischerei *f* (Metall) finery, refinery
Frischereiroheisen *n* (Metall) charcoal hearth cast iron, forge pig iron
Frischerz *n* raw ore
Frischfeuer *n* bloomery fire, finery fire, forge hearth
Frischfeuerbetrieb *m* finery process
Frischfeuereisen *n* charcoal hearth iron, fined iron, refined iron
Frischfeuerroheisen *n* charcoal hearth cast iron, pig iron for refining
Frischfeuerschlacke *f* refining forge slag
Frischgas *n* inlet gas, make-up gas, unburnt gas
Frischgestein *n* solid rocks
Frischgestübe *n* coaldust
Frischglätte *f* hard litharge
Frischhaltepackung *f* preserving package
Frischhaltetücher *pl* refreshing and cleaning tissues
Frischhalteverpackung *f* fresh food packaging
Frischhaltung *f* conservation, preservation
Frischherd *m* forge hearth, refining hearth
Frischherdverfahren *n* (Metall) refinery or purification process
Frischhütte *f* refining works, smelting finery
Frischluft *f* fresh air
Frischmethode *f* (Metall) fining process, refining
Frischofen *m* refining furnace
Frischprozeß *m* finery process
Frischsand *m* fresh sand

Frischschlacke *f* finery cinder, oxidizing slag, rich slag
Frischstahl *m* natural steel, purified steel
Frischung *f* (Metall) refining, fining, puddling, purification
Frischungsofen *m* refining furnace
Frischungsprozeß *m* (Metall) oxidizing or converting process, purifying method, refining or fining process
Frischverfahren *n* (Metall) fining process, oxidizing process, refining process
Frischwasser *n* fresh water
Frischzellen *pl* living cells
Frischzellentherapie *f* embryonal fresh cell therapy, Niehans' therapy
Fritfliege *f* frit fly
Fritte *f* frit, sintered glass plug
fritten to frit, to sinter, (Radio) to cohere
Frittenporzellan *n* frit porcelain, soft porcelain
Fritterelektrode *f* electrode of coherer
Fritterklemme *f* coherer terminal
Fritterröhre *f* coherer tube
Fritterwiderstand *m* coherer resistance
Frittofen *m* frit kiln, calcar arch, fire arch
Frittporzellan *n* frit porcelain, soft porcelain
Frittung *f* fritting, sintering, (Radio) coherence
Fritzscheit *m* (Min) fritzscheite
Frontalangriff *m* frontal attack
Frontgabelstapler *m* front-end forklift [truck]
Frontlader *m*, **knickgelenkter** ~ articulated front-end loader
Frosch *m* frog, (Metall) cam shoe
Froschklemme *f* draw tongs
Froschschenkel *m* (Zool) frog's leg
Froschstein *m* (Min) batrachite
frostbeständig frost-proof, frost resistant
Frostbeständigkeit *f* cold resisting property, resistance to frost
Frosteinwirkung *f* influence of frost
frostempfindlich frost sensitive, sensitive to frost
Frostgrad *m* degree below zero
Frostpunkt *m* freezing point
Frostsalbe *f* (Pharm) anti-frostbite, cold cream, salve for chilblains
Frostschutzmittel *n* antifreeze
Frostschutzscheibe *f* frost shield
frostsicher resistant to frost
Frostzeit *f* frost period
Frottee *n* (Text) terry cloth
Frucht *f* fruit, (Biol) embryo, foetus
Fruchtäther *m* fruit essence
fruchtartig fruit-like, fruity
fruchtbar fertile
Fruchtbarkeit *f* fertility, productiveness
Fruchtboden *m* (Bot) thalamus
Fruchtbranntwein *m* fruit brandy
Fruchteis *n* fruit ice
Fruchterde *f* humus
Fruchtessenz *f* fruit essence

Fruchtessig *m* fruit vinegar
Fruchtfleisch *n* fruit pulp
Fruchtfolge *f* rotation of crops
Fruchtgerbstoff *m* fruit tannin
Fruchtsäure *f* fruit acid
Fruchtsaft *m* fruit juice
Fruchtschale *f* peel [of fruit]
Fruchtzucker *m* fructose, fruit sugar
Fructokinase *f* (Biochem) fructokinase
Fructosamin *n* fructosamine
Fructosazin *n* fructosazine
Fructosazon *n* fructosazone
Fructose *f* fructose, fruit sugar
Fructosebisphosphat *n* fructose biphosphate
Fructosebisphosphatase *f* (Biochem) fructose biphosphatase
Fructosediphosphorsäure *f* fructose diphosphoric acid
Fructoson *n* fructosone
Früchteverarbeitung *f* fruit processing
Früherkennung *f* (Med) early diagnosis
Frühzündung *f* pre-ignition, advanced ignition, spark advance, (fehlerhafte) backfire
Frugardit *m* frugardite
Fuchs *m* (Techn) air duct, flue
Fuchsbrücke *f* flue bridge
Fuchsdecke *f* flue roof
Fuchsfett *n* fox fat
Fuchsgas *n* flue gas
Fuchsin *n* fuchsin, aniline red, magenta red, rosaniline hydrochloride, rosein
Fuchsinschweflige Säure *f* fuchsin sulfurous acid, Schiff's solution
Fuchsit *m* (Min) fuchsite
Fuchson *n* fuchsone
Fuchsonimin *n* fuchsonimine
Fuchsschwanz *m* crosscut saw, hand saw, pad saw
Fucit *m* fucitol
Fuconsäure *f* fuconic acid
Fucose *f* 2,3,4,5-tetrahydroxyhexanal, fucose
Fucoson *n* fucosone
Fucosterin *n* fucosterol
Fucoxanthin *n* fucoxanthin
Fuculose *f* fuculose
Fucusol *n* fucusol
Fühler *m* tracer
Fühlerlehre *f* feeler gauge
Fühlgerät *n* detector
Fühlglied *n* detector, sensing element, sensitive element
Fühlkopf *m* tracer head
Fühlstab *m* stick gauge
Führerstand *m* driver's stand
Führungsbahn *f* way
Führungsbohrung *f* dowel hole
Führungsbüchse *f* cylinder liner, dowel bush, guide pin bushing
Führungselement *n* guide

Führungsfeld *n* guiding field
Führungsgröße *f* (Regeltechn) command reference input
Führungslager *n* guide bearing
Führungsloch *n* dowel hole, feed hole, guide perforation
Führungsregler *m* master controller, (Regeltechn) master controller
Führungsrinne *f* trough line guide
Führungsrolle *f* contact roller, guide roller
Führungssäule *f* dowel, leader pin
Führungsschiene *f* rail
Führungsstange *f* guide bar, guide rod
Führungsstift *m* guide pin, dowel
Führungswalze *f* guide roller
Führungszapfen *m* guide pilot
Füllanlage *f* filling plant
Füllansatz *m* inlet fitting, inflation sleeve
Füllapparat *m* charging device, filling apparatus
füllbar capable of being filled
Füllblech *n* loading tray
Füllbrett *n* panel board
Fülldichte *f* packing density
Fülldruck *m* filling pressure
Fülle *f* abundance, fullness
Füllegrad *m* degree of fullness
Fülleisen *n* calking chisel
füllen to fill, to charge, to feed
Füllen *n* filling, feeding
Füller *m* filler, charger, loader, (Füllfederhalter) fountain pen
Füllfaktor *m* bulk factor, filling factor
Füllfarbe *f* brushable plaster for outdoors
Füllfederhalter *m* fountain pen
Füllflasche *f* filling cylinder
Füllform *f* positive mold, positive type mold, ~ mit **Abquetschfläche** landed plunger mold
Füllgas *n* filler gas
Füllgewicht *n* net weight, shot weight
Füllgut *n* charge
Füllkammer *f* loading chamber
Füllklappe *f* charging door
Füllkörper *pl* filling material, packing [material]
Füllkörperkolonne *f* packed column
Füllkörperrohr *n* packed tube
Füllkörpersäule *f* packed column, packed tower
Füllkörperturm *m* packed tower, packed column
Füllkoks *m* heating coke
Füllkonstante *f* bulk factor
Füllkraft *f* body, **gute** ~ full body
Füllmaschine *f* stuffing machine
Füllmasse *f* stuffing, filling, filling compound, sealing composition, (Zucker) massecuite
Füllmaterial *n* stuffing, filler, filling material, (Beschwerungsmittel) loading material, (Chem) packing [material]
Füllmischung *f* filling stock
Füllmittel *n* filler, filling; filling material
Füllochdose *f* vent hole can

Füllöffel *m* filling ladle
Füllöffnung *f* charging hole, charging opening
Füllplatte *f* filler plate
Füllraum *m* charge cavity, charge space, loading cavity, loading chamber, loading space, loading well, material well
Füllraum-Werkzeug *n* positive mold
Füllrohr *n* filling pipe, spout
Füllrumpf *m* loading hopper
Füllsäure *f* accumulator acid, electrolyte acid
Füllschlauch *m* inflation tube
Füllschraube *f* filler plug, filling plug, filling screw, priming plug
Füllschutzglocke *f* filling attachement for culture tubes
Füllsel *n* stuffing
Füllspant *n* filling timber
Füllstandsanzeiger *m* level indicator
Füllstoff *m* filler, filling material
Füllstoffaktivator *m* activator for fillers
füllstoffarm lightly loaded
Füllstoffnest *n* filler speck
füllstoffreich heavily loaded
Füllstreifen *m* filler strip
Füllstrich *m* filling level
Füllstück *n* filling piece, liner
Füllstutzen *m* filler cap, gas cylinder manifold
Fülltablett *n* charging tray, loading tray
Fülltrichter *m* feed chute, funnel, hopper
Füllung *f* stuffing, filling, packing, panel, ~ **mit Raschigringen** ring packing
Füllungsdichte *f* loading density
Füllungsgrad *m* degree of admission, degree of charge
Füllungsverhältnis *n* filling ratio
Füllvlies *n* high-bulk [or high-loft] wadding, nonwover quilting
Füllvorrichtung *f* charging hopper, loading tray
Füllzylinder *m* feed cylinder, charging bin, charging hopper
fünfatomig pentatomic
fünfbasisch pentabasic
Fünfeck *n* pentagon
fünfeckig pentagonal
fünffach fivefold, quintuple
Fünffachchlorantimon *n* (Antimon(V)-chlorid) antimonic chloride, antimony pentachloride, antimony(V) chloride
Fünffachchlorphosphor *m* (Phosphorpentachlorid, Phosphor(V)-chlorid) phosphorus pentachloride, phosphorus(V) chloride
Fünffachschwefelantimon *n* (Antimon(V)-sulfid) antimony pentasulfide
fünfflächig pentahedral
Fünfflächner *m* pentahedron
fünfgliedrig five-membered
fünfkantig pentagonal
Fünfphasensystem *n* five-phase system

Fünfpolendröhre *f* out-put pentode
Fünfpolröhre *f* (Elektr) five-electrode tube, pentode
Fünfring *m* five-membered ring
fünfseitig five-sided, pentahedral
Fünfsteckersockel *m* five-prong base
fünfstellig five-figure, five-place
Fünfstoffsystem *n* five component system
fünfteilig five-piece
fünfwertig pentavalent
Fünfwertigkeit *f* pentavalence, quinquevalence
Fünfzigstellösung *f* (Chem) solution of 1 in 50
füttern to feed, (auskleiden) to case, to line
Fütterungsstreifen *m* feed strip
Fugazität *f* fugacity
Fugazitätskoeffizient *m* fugacity coefficient
Fuge *f* joint, (Naht) seam, (Nut) groove
fugen to join[t]
fugenfüllend gap-filling
Fugengelenk *n* articulation
Fugenkitt *m* gap-filling adhesive
Fugenleim *m* splicer adhesive
fugenlos jointless
Fuggerit *m* fuggerite
Fugmörtel *m* jointing mortar
Fukose *f* fucose
Fulgensäure *f* fulgenic acid
Fulgid *n* fulgide, anhydride of fulgenic acid
Fulgurit *m* fulgurite
Fulmenit *n* fulmenite
Fulminat *n* fulminate
Fulminursäure *f* fulminuric acid, 2-cyano-2-nitroethanamide, cyanonitroacetamide, isocyanuric acid
Fulvalen *n* fulvalene
Fulvanol *n* fulvanol
Fulven *n* fulvene
Fulverin *n* fulverin
Fulvinsäure *f* fulvic acid
Fulwabutter *f* fulwa butter
Fumaramid *n* fumaramide
Fumarase *f* (Biochem) fumarase
Fumardialdehyd *m* fumaraldehyde, fumardialdehyde
Fumarin *n* fumarine, protopine
Fumarsäure *f* fumaric acid, trans-1,2-ethylene dicarboxylic acid, trans-butenedioic acid
fumarsauer fumarate
Fumarursäure *f* fumaruric acid
Fumarylchlorid *n* fumaryl chloride
Fumigatin *n* fumigatin
Fundament *n* foundation
fundamental elementary, fundamental
Fundamentalbaustein *m* fundamental building block
Fundamentaleinheit *f* fundamental unit
Fundamentalgesetz *n* (Math) fundamental theorem
Fundamentaltensor *m* fundamental tensor

Fundamentalversuch *m* pioneering experiment
Fundamentanker *m* anchor bolt, foundation bolt
Fundamentschraube *f* anchor bolt
fundieren to lay a foundation
Fundort *m* place of discovery
fungieren to function, to act, to behave, to work
Fungin *n* fungin, fungus cellulose, metacellulose
Fungistatikum *n* fungistatic substance
fungistatisch fungistatic
Fungisterin *n* fungisterin, fungisterol
Fungit *m* fungite
fungizid fungicidal
Fungizid *n* fungicide
fungös (pilzartig, schwammig) fungoid, fungous, spongy
Fungus *m* (pl Fungi) fungus (pl fungi), mycete
Funk *m* radio, wireless
Funkanlage *f* radio station
funkeln glitter, scintillate, to sparkle
Funkeln *n* sparkling, scintillation
funkelnd scintillating, sparkling
funken to radio, to transmit, to wireless, (Radio) to broadcast
Funken *m* spark, ignition spark
funkenähnlich spark-like
Funkenbildung *f* formation of sparks, sparking
Funkenentladung *f* spark discharge
Funkenentziehvorrichtung *f* spark drawing device
Funkenesse *f* chimney hood
Funkengeber *m* sparking plug, spark transmitter
Funkeninduktor *m* spark coil, induced coil, induction coil
Funkenkammer *f* spark chamber
Funkenkontinuum *n* spark discharge continuum
Funkenlänge *f* spark length
Funkenleiter *m* spark conductor
Funkenlöscher *m* spark arrestor, spark extinguisher, spark killer
Funkenlöschkreis *m* arc suppressor
Funkenlöschscheibe *f* spark arrestor, spark extinguisher
Funkenlöschspule *f* blow-out coil
Funkenlöschung *f* blow-out
Funkenlöschvorrichtung *f* spark quenching apparatus
funkenlos sparkless
Funkenlosigkeit *f* sparklessness
Funkenpotential *n* (Elektr) disruptive potential, sparking potential, spark potential
Funkenspannung *f* sparking voltage
Funkenspektrum *n* (Opt) spark spectrum
Funkensprühen *n* emitting of sparks, sparking
funkensprühend emitting sparks, scintillating
Funkenstrecke *f* spark gap, spark path
Funkenverdichter *m* spark condenser
funkenverhindernd spark suppressing
Funkenzieher *m* spark drawer
Funkenzündung *f* spark ignition

Funkinduktor *m* pinhole detector
Funkit *m* funkite
Funkmeßgerät *n* radar
Funkmeßtechnik *f* radar [direction finding]
Funkschwund *m* face-out
Funksprechgerät *n* radio telephone
Funktion *f* function, ~ **mehrerer Variabler** function of several variables, **algebraische** ~ algebraic function, **beschränkte** ~ finite function, **elementare** ~ elementary function, **elliptische** ~ elliptical function, **explizite** ~ (Math) explicit function, **ganze rationale** ~ integral rational function, **gebrochen rationale** ~ rational fractional function, **gerade** ~ even function, **geschlossen integrierbare** ~ continuous integrable function, **implizite** ~ (Math) implicit function, **integrierbare** ~ (Math) integrable function, **lineare** ~ linear function, **monoton fallende** ~ monotonic descending function, **monoton steigende** ~ monotonic ascending function, **Parameterdarstellung einer** ~ parametric representation of a function, **stetige** ~ continuous function, **transzendente** ~ transcendental function, **Umkehr-** ~ inverse function, **ungerade** ~ odd function, **unstetige** ~ (Math) discontinuous function, **willkürliche** ~ arbitrary function
Funktionalableitung *f* functional derivative
Funktionaldeterminante *f* functional determinant
Funktionaldifferentialgleichung *f* functional differential equation
Funktionalität *f* functionality
Funktionalmatrix *f* functional matrix
Funktionaltransformation *f* functional transformation
funktionell functional
Funktionenoptimierung *f* optimization of functions
Funktionensystem *n* set of functions, **orthogonales** ~ orthogonal set of functions, **orthonormiertes** ~ orthonormal set of functions, **vollständiges** ~ complete set of functions
Funktionentheorie *f* theory of functions
funktionieren to function, to work well
Funktionieren *n* fuctioning
Funktionsablauf *m* sequence of operations
Funktionsleiter *f* nomogram
Funktionsschreiber *m* function plotter
Funktionszentrum *n* functional center
Furaldehyd *m* furaldehyde, furfuraldehyde
Furan *n* furan, furfuran, tetrol
Furanalkohol *m* furfuralcohol, furfuryl alcohol, furyl carbinol
Furanidin *n* furanidine
Furankern *m* furan nucleus
Furanosering *m* furanose ring

Furanosid *n* furanoside
Furanplaste *pl* furan plastics
Furanring *m* furan ring
Furazan *n* furazan, azoxazole, oxdiazole
Furche *f* channel, groove
Furchenspatel *f* grooved spatula
Furchung *f* grooving, furrowing, segmentation
Furfuracrolein *n* furfuracrolein
Furfuracrylsäure *f* furfuracrylic acid
Furfural *n* furfural, 2-furaldehyde, 2-furancarbonal, furfuraldehyde, furfurol, furfuryl aldehyde, furol
Furfuraldazin *n* furfuraldazine
Furfuraldehyd *m* (s. auch Furfural) furfural, 2-furancarbonal
Furfuralkohol *m* furfuralcohol, furancarbinol, furfuryl alcohol
Furfuramid *n* furfuramide, furfuryl amide
Furfuran *n* furan, furfuran
Furfurol *n* furfurol[e], 2-furancarbonal, 2-furfuraldehyde, furfuryl aldehyde, furol
Furfurolgewinnungsanlage *f* furfurol production plant
Furfurolharz *n* furfurol resin
Furfuroyl- furfuroyl, furoyl
Furfuryl- furfuryl
Furfurylalkohol *m* furfuryl alcohol
Furfuryliden *n* furfurylidene
Furil *n* furil, difurylglyoxal
Furilsäure *f* furilic acid
Furin *n* furin[e]
Furmethid *n* furmethide
Furnier *n* veneer, **ein** ~ **aufbringen** veneering
furnieren to veneer
Furnierfugenverleimung *f* tapeless splicing of veneers
Furnierholz *n* veneer wood
Furnierleim *m* veneer glue, veneering adhesive
Furnierpresse *f* veneering press
Furocumarinsäure *f* furocoumarinic acid
Furodiazol *n* furodiazole, oxdiazole
Furoesäure *f* furoic acid
Furoin *n* furoin
Furol *n* furol, 2-furancarbonal, furfural
Furonsäure *f* furonic acid
Furostilben *n* furostilbene
Furoxan *n* furoxane
Furoyl- furfuroyl, furoyl
Furoylaceton *n* furoylacetone
Furoylbenzoylmethan *n* furoylbenzoylmethane
Furoylierung *f* furoylation
Furtrethonium *n* furtrethonium
Furyl- furyl
Furyliden *n* furylidene
Fusain *n* fusain, mineral charcoal, mother of coal
Fusanol *n* fusanol
Fusarin[in] *n* fusarinine
Fuscin *n* fuscin

Fuscinsäure *f* fuscinic acid
Fuscit *m* fuscite
Fusel *m* bad liquor, fusel oil, poor-quality spirits
fuselfrei free from fusel oil
Fuselgeruch *m* odor of fusel oil
Fuselgeschmack *m* taste of fusel oil
fuselhaltig containing fusel oil
fuselig containing fusel oil
Fuselöl *n* fusel oil
Fuselölbranntwein *m* spirit containing fusel oil
Fusion *f* fusion
Fusionierung *f* (Chem) fusion
Fusionsenergie *f* (Atom) fusion energy
Fusionspunkt *m* fusing point, melting point
Fusionsreaktor *m* (Atom) fusion reactor
Fusionsstoff *m* fusionable material
Fußboden *m* floor
Fußbodenanstrich *m* floor finish
Fußbodenbelag *m* floor covering
Fußbodenfliese *f* floor tile
Fußbodengrundfarbe *f* floor paint, floor stain
Fußbodenkitt *m* flooring cement
Fußbodenlack *m* floor finish
Fußbodenpflegemittel *n* floor polish
Fußbodenplatte *f* floor panel, floor tile
Fußgestell *n* base plate
Fußglätte *f* black impure litharge
Fußheizring *m* mold ring
Fußkasten *m* **mit Muffe** base box with flange
Fußklemme *f* terminal
Fußnote *f* footnote
Fußpendelpresse *f* kick press
Fußplatte *f* sole plate
Fußpunkt *m* **des Lotes** foot of a perpendicular
Fußpunktkurve *f* (Math) pedal locus
Fußring *m* base ring
Fußrolle *f* foot roller, guide roller
Fußstück *n* pedestal
Fußventil *n* foot valve
Fußwinkel *m* **des Kegelrads** addendum angle of bevel gear
Fustage *f* barrels, casks
Fustikholz *n* fustic wood
Fustin *n* fustin
Futter *n* lining, casing, chuck, coating, feed, fodder, forage
Futterautomat *m* (Techn) automatic chuck lathe
Futtermittel *n* animal feed
Futtermittelindustrie *f* food processing industry
Futterpflanze *f* forage plant
Futterrohr *n* casing, (Reaktor) liner
Futterrohrflansch *m* casing flange
Futterstoff *m* lining
Futterwalze *f* side roll
Futterzusatz *m* feed supplement

G

Gabanholz *n* (Bot) camwood, sandal wood
Gabardine *f* (Text) gabardine, gaberdine
Gabbro *m* (Geol) gabbro, diallage-rock, gabbronite
Gabbronit *m* (Geol) gabbronite
Gabel *f* fork, (Stimmgabel) tuning fork
Gabelbieger *m* shaft bender
Gabelblitz *m* forked lightning
Gabeleinguß *m* double branch gate
gabelförmig bifurcated, forked
Gabelführung *f* fork guide
Gabelgelenk *n* fork joint
Gabelhubstapler *m* fork lift truck
Gabelhubwagen *m* pallet truck
Gabelkneter *m* paddle mixer
Gabelpfanne *f* ladle for ring carrier
Gabelrohr *n* bifurcated pipe, forked tube
Gabelstapler *m* fork [lift] truck
Gabelstütze *f* forked prop
Gabiragummi *m* gumkino
Gabrielsche Phthalimidsynthese *f* Gabriel's phthalimide synthesis
Gadoleinsäure *f* gadoleic acid
Gadolinerde *f* gadolinia, gadolinium oxide
Gadolinit *m* (Min) gadolinite
Gadolinium *n* gadolinium (Symb. Gd)
Gadoliniumbromid *n* gadolinium bromide
Gadoliniumchlorid *n* gadolinium chloride
Gadoliniumnitrat *n* gadolinium nitrate
Gadoliniumoxid *n* gadolinium oxide
Gadoliniumsulfat *n* gadolinium sulfate
Gänsefett *n* goose fat
Gänsekotigerz *n* (Min) ganomatite, goosedung ore
Gäranlage *f* fermentation plant
gärbar fermentable
Gärbarkeit *f* fermentability
Gärbeisen *n* wrought iron
gärben (Metall) to refine
Gärben *n* refining
Gärbottich *m* fermenter, fermenting tub, fermenting vat
Gärbstahl *m* refined steel, wrought steel
Gärbütte *f* fermenter, fermenting tub, fermenting vat
Gärchemie *f* fermentation chemistry, zymurgy
Gärdauer *f* duration of fermentation, time of fermentation
Gäre *f* fermentation
gären to ferment
Gären *n* fermenting
Gärerzeugnis *n* fermentation product
Gärführung *f* method of fermentation
Gärgefäß *n* fermenting vessel
Gärkammer *f* fermenting room
Gärkasten *m* fermentation vat

Gärkeller *m* fermentation cellar, fermenting cellar
Gärkraft *f* fermenting power
Gärmesser *m* zymometer
Gärmittel *n* ferment, leaven, yeast
Gärprobe *f* fermentation sample, fermentation test
Gärprodukt *n* fermentation product
Gärraum *m* fermenting house
Gärsäure *f* fermenting acid
Gärsalz *n* fermentation salt
Gärsilo *m* silo, silage pits
Gärstoff *m* leaven, yeast
Gärtätigkeit *f* fermentative activity
Gärtank *m* fermenting tank
Gärtemperatur *f* fermenting temperature
Gärtnerbazillus *m* Gärtner's bacillus, Salmonella enteritidis
Gärung *f* fermentation, bubbling, effervescence, **alkoholische** ~ alcoholic fermentation, **faulende** ~ putrefactive fermentation, **in** ~ **übergehen** to enter into fermentation
Gärungsalkohol *m* fermentation ethyl alcohol
Gärungsamylalkohol *m* fermentation amyl alcohol, mixture of 2- and 3-methylbutanol
Gärungsbottich *m* fermenting vat
Gärungsbuttersäure *f* fermentation butyric acid
Gärungschemie *f* fermentation chemistry, zymochemistry, zymology, zymurgy
gärungserregend fermentative, causing fermentation, zymogenic
Gärungserreger *m* ferment
gärungsfähig fermentable, capable of fermenting
Gärungsfähigkeit *f* fermentability, fermenting capacity
gärungsfördernd zymogenic, zymogenous
Gärungsfuselöl *n* fusel oil
gärungshemmend antifermentative, antiseptic, antizymotic, stopping fermentation
Gärungskraft *f* fermentative power, fermenting power
Gärungsküpe *f* fermentation vat
Gärungslehre *f* zymology
Gärungsmesser *m* zymometer
Gärungsmilchsäure *f* fermentation lactic acid
Gärungsmittel *n* ferment, fermenting agent, leavening agent
Gärungspilz *m* (Bakt) fermentation fungus
Gärungsprodukt *n* fermentation product
Gärungsprozeß *m* fermentation [process]
Gärungsstoff *m* ferment, fermenting agent, leavening
Gärungstechnik *f* zymotechnology, zymotechnics
gärungstechnisch zymotechnical
gärungsunfähig unfermentable

Gärungsverfahren *n* method of fermentation
gärungsverhindernd antifermentative, antizymotic
Gärungsvorgang *m* fermentation process
gärungswidrig antifermentative, antizymotic
Gärungszeit *f* duration of fermentation
Gärverfahren *n* fermentation process
Gärvermögen *n* fermentability, fermentative power, fermenting capacity
Gärvorgang *m* fermentation process
Gärwärme *f* heat of fermentation
Gafrinin *n* gafrinin
Gahnit *m* (Min) gahnite, vitreous zinc aluminate, zinc spinel
Gaidinsäure *f* gaidic acid
Galactonsäure *f* galactonic acid
Galactosamin *n* galactosamine
Galaheptit *m* galaheptite
Galaheptonsäure *f* galaheptonic acid
Galaheptose *f* galaheptose
Galaktal *n* galactal
Galaktamin *n* galactamine
Galaktan *n* galactan, gelose
Galaktarsäure *f* galactaric acid
Galaktase *f* (Biochem) galactase
Galaktid *n* galactide
Galaktin *n* galactin
Galaktit *m* galactite
Galaktobiose *f* galactobiose
Galaktochloralose *f* galactochloralose
Galaktochloralsäure *f* galactochloralic acid
Galaktoflavin *n* galactoflavine
Galaktoheptulose *f* galactoheptulose
Galaktokinase *f* (Biochem) galactokinase
Galaktometasaccharin *n* galactometasaccharine
Galaktometasaccharinsäure *f* galactometasaccharic acid
Galaktometer *n* galactometer
Galaktosaminsäure *f* galactosaminic acid
Galaktosan *n* galactosan
Galaktose *f* galactose, cerebrose
Galaktosegalakturonsäure *f* galactose galacturonic acid
Galaktosid *n* galactoside, cerebroside
Galaktosidase *f* (Biochem) galactosidase
Galaktoson *n* galactosone
Galakturonsäure *f* galacturonic acid
Galambutter *f* galam-butter, sheabutter
Galamgummi *n* gum arabic
Galangin *n* galangin
Galanginidin *n* galanginidine
Galanginidiniumhydroxid *n* galanginidinium hydroxide
Galanthamin *n* galanthamine
Galanthaminon *n* galanthaminone
Galanthaminsäure *f* galanthaminic acid
Galanthidin *n* galanthidine
Galanthin *n* galanthine
Galbacin *n* galbacin

Galban *n* galban
Galbanharz *n* galban resin, galbanum, gum galbanum
Galbanumöl *n* galbanum oil
Galbulin *n* galbulin
Galcatin *n* galcatin
Galegin *n* galegine
Galenit *m* (Min) galena, galenite, lead glance
Galenobismutit *m* (Min) galenobismuthite
Galgant *m* (Bot) galingale
Galgantöl *n* galanga oil, galingale oil
Galgantwurzel *f* galanga root, galingale
Galgravin *n* galgravin
Galicid *n* galicide
Galilei-Zahl *f* Galilean number
Galiosin *n* galiosin
Galipidin *n* galipidine
Galipin *n* galipine
Galipoidin *n* galipoidine
Galipot *m* (Bot) Bordeaux turpentine, galipot
Gallacetein *n* gallaceteine
Gallacetophenon *n* gallacetophenone
Galläpfelabkochung *f* gallnut decoction
Galläpfelbeize *f* gallsteep
Galläpfelextrakt *m* gallnut extract
Galläpfelgerbsäure *f* gallotannic acid, gallotannin, tannic acid, tannin
Galläpfelsäure *f* gallotannic acid, gallotannin, tannic acid, tannin
galläpfelsauer gallotannate
Galläpfeltinktur *f* tincture of gallnuts
Gallal *n* gallal, aluminum gallate
Gallamid *n* gallamide
Gallaminblau *n* gallamine blue
Gallanilid *n* gallanilide, gallanol, gallinol
Gallanol *n* gallanilide, gallanol, gallinol
Gallapfel *m* (Bot) gallnut
Gallat *n* (Chem) gallate
Galle *f* (Med) bile, gall
Gallein *n* gallein, gallin, pyrogallolphthalein
gallenbitter bitter as gall
Gallenblase *f* gall bladder
Gallenexkretion *f* biliary excretion
Gallenfarbstoff *m* bile pigment, biliary pigment, bilirubin
Gallenfarbstoffausscheidung *f* excretion of bile pigments
Gallensäure *f* bile acid
Gallenstein *m* (Med) gall stone, biliary calculus
gallern to refine metal
Gallert *n* gel, gelatin-like substance, glue, jelly, **verwandeln in** ~ to convert into gelatin or jelly, to gelatinate, to gelatinize
gallertähnlich gelatinous, colloidal, jelly-like
gallertartig gelatinous, colloidal, jelly-like, resembling gelatin, ~ **werden** to become gelatinous, to jellify
Gallerte *f* gel, gelatin[e], glue, jelly
Gallertfilter *m* colloid filter

Gallertmasse f gelatinous substance, jelly-like mass
Gallertmoos n (Bot) carrageen, sea moss
Gallertsäure f pectic acid
Gallertsubstanz f gelatinous substance, jelly-like substance
Gallichlorid n (Gallium(III)-chlorid) gallic chloride, gallium(III) chloride, gallium trichloride
Gallicin n gallicin, methyl gallate
gallieren to gall, to treat with gallnut extract
gallig bitter, biliary, bilious, gall-like
Gallihydroxid n (Gallium(III)-hydroxid) gallic hydroxide, gallium(III) hydroxide
Gallioxid n gallic oxide, gallium(III) oxide, gallium sesquioxide
Gallipharsäure f gallipharic acid
Gallipinsäure f gallipinic acid
Gallipotharz n galipot, galipot resin
Gallisalz n (Gallium(III)-Salz) gallic salt, gallium(III) salt
Gallisin n gallisine
Gallium n gallium (Symb. Ga)
Galliumantimonid n gallium antimonide
Galliumarsenid n gallium arsenide
Galliumchlorid n gallium chloride, (Gallium(III)-chlorid) gallic chloride, gallium(III) chloride, gallium trichloride
Galliumchlorür n (Gallium(II)-chlorid) gallium dichloride, gallium(II) chloride, gallous chloride
Galliumgehalt m gallium content
Galliumion n gallium ion
Galliummonoxid n (Gallium(II)-oxid) gallium(II) oxide, gallium monoxide, gallous oxide
Galliumoxid n gallium oxide, (Gallium(III)-oxid) gallic oxide, gallium(III) oxide, gallium sesquioxide
Gallocatechin n gallocatechol
Gallochlorid n (Gallium(II)-chlorid) gallium dichloride, gallium(II) chloride, gallous chloride
Gallocyanin n gallocyanin
Galloflavin n galloflavin
Gallone f (Maß) gallon
Gallonitril n gallonitrile
Gallooxid n gallium(II) oxide, gallium monoxide, gallous oxide
Gallotanninsäure f gallotannic acid, gallotannin, tannic acid, tannin
Galloyl- galloyl
Galloylbenzophenon n galloylbenzophenone
Gallusgerbsäure f gallotannic acid, gallotannin, tannic acid, tannin
Gallussäure f gallic acid, 3,4,5-trihydroxybenzoic acid
Gallussäuregärung f gallic fermentation
Gallussäuremethylester m methyl gallate

gallussauer gallate
Gallustinte f gallnut ink
Galmei m (Min) galmei, calamine, **edler** ~ calamine, zinc spar, **gewöhnlicher** ~ silicious calamine
Galmeiblende f (Min) calamine blende
galmeihaltig containing calamine
Galuteolin n galuteolin
Galvani-Potential n single electrode potential
Galvanisation f (Elektrochem) electroplating
galvanisch galvanic, voltaic, ~ **niederschlagen** to electrodeposit, ~ **plattieren** to electroplate, to galvanize, ~ **versilbern** to electroplate with silver, to electrosilver
Galvaniseur m galvanizer
Galvanisieranlage f galvanizing plant, electroplating plant
Galvanisierapparat m galvanizer, galvanizing apparatus
galvanisieren to galvanize, to electroplate
Galvanisieren n electrodeposition, (Med) galvanizing
Galvanisierglocke f galvanizing kettle
Galvanisierprozeß m electrodeposition, galvanizing process
galvanisiert galvanized
Galvanisierung f electrodeposition, galvanization
Galvanisiervorgang m galvanizing [process], plating [process]
Galvanismus m galvanism, voltaism
Galvano- electro-
Galvano m galvanograph
Galvano n electrotype
Galvanobronze f electrobronze
Galvanochromie f galvanic coloring
Galvanographie f electrography, galvanography
galvanographisch galvanographic, electrotype
Galvanokaustik f galvanocautery
galvanomagnetisch galvanomagnetic
Galvanomagnetismus m electromagnetism, galvanomagnetism
Galvanometer n galvanometer
Galvanometrie f galvanometry
galvanometrisch galvanometric
Galvanoplastik f galvanoplastic process, electrometallurgy, electrotyping process, galvanoplastic art
Galvanoplastikbad n plating bath
galvanoplastisch galvanoplastic, metalloplastic
Galvanoplattierung f electroplating, electro-deposition
Galvanoskop n galvanoscope, detector
galvanostatisch galvanometric
Galvanostegie f galvanostegy, electrodeposition, electroplating, galvanoplastics
galvanostegisch electroplating
Galvanotechnik f electro-deposition, electroplating

Galvanotypie f electrotypy
Galyl n galyl
Gamabufogenin n gamabufogenin
Gamander m (Bot) germander
Gamatin n gamatin
Gambin n gambin, nitrosonaphthol
Gambir n gambir, yellow catechu
Gambircatechin n gambir catechol
Gambogebutter f gamboge butter
Gametoblast m (Biol) gametoblast, sporozoite
Gametophyt m (Bot) gametophyte
Gametozyt m gametocyte, gamont
Gametozytenträger m gametocyte carrier
gamma-aktiv (Atom) gamma-active, gamma-radioactive
Gamma-Aktivität f (Atom) gamma-activity, gamma-radioactivity
Gammabestrahlung f gamma irradiation
Gammaceran n gammacerane
Gammaeisen n gamma iron
gammaempfindlich gamma sensitive
Gammaempfindlichkeit f (Atom) gamma sensitivity
Gammaenergie f (Atom) gamma-ray energy
Gamma-Globulin n gamma-globulin
Gammagraphie f gammagraphy
Gammaquant n gamma ray quantum, gamma quantum, gamma ray photon
gamma-radioaktiv (Atom) gamma-active, gamma-radioactive
Gammaradiographie f gamma radiography, gammagraphy
Gamma-Resonanzspektroskopie f (Spektr) gamma-resonance spectroscopy
Gammasäure f gamma acid, 7-amino-1-naphthol-3-sulfonic acid
Gammaspektrometer n gamma-ray spectrometer
Gamma-Spektroskopie f (Atom) gamma-ray spectroscopy
Gammaspektrum n gamma-ray spectrum
Gammaspiegel m gamma background
Gammastellung f (Chem) gamma position
Gammastrahl m gamma ray
Gammastrahlaktivität f gamma-ray activity
Gammastrahlenspektrum n gamma-ray spectrum
Gammastrahlentherapie f gamma-ray therapy
Gammastrahlphoton n (Atom) gamma-ray photon
Gammastrahlstreuung f gamma-ray scattering
Gammastrahlung f gamma radiation
Gammateilchen n gamma particle
Gamma-Uran n (Atom) gamma uranium
Gammazähler m gamma-sensitive counter
Gammazählrohr n gamma counter tube
Gamma-Zeitkurve f (Phot) development curve
Gamma-Zerfall m (Atom) gamma decay, gamma disintegration

Gammexan n (Insektenvertilgungsmittel) gammexane, gamma isomer of benzene hexachloride
Gang m aisle, (Auto) gear, speed, (einer Maschine) motion, movement, running, working, (Lauf) course, (Min) lode, vein, (Schraube) helix, thread, turn (of a screw), in ~ setzen to set in motion, **toter** ~ backlash
Gangart f matrix, (Min) gangue, (Web) gang
Gangerz n (Min) vein ore
Ganggestein n (Min) gangue, gang
Ganghöhe f axial spacing of turns, lead, pitch, **abnehmende** ~ decreasing pitch, **zunehmende** ~ progressive pitch
Ganglienzelle f (Histol) gangliocyte, ganglion cell
Gangliosid n ganglioside
Gangschaltung f (Auto) gearshift[ing]
Gangsteigung f (Gewinde) lead
Gangtiefe f channel depth, depth of a screw thread, thread depth
Gangverschiebung f phase displacement
Ganister m (Min) ganister
Ganomalith m (Min) ganomalite
Ganomatit m (Min) ganomatite
Ganophyllit m ganophyllite
Gans f (Metall) pig iron, block pig
Gantrisin n gantrisin
ganz whole, entire, integer, integral
Ganzfabrikat n finished product
ganzflächig (Krist) holohedral
Ganzholz n unhewn timber
Ganzmetallgehäuse n all-metal container
Ganzreifenregenerat n whole tire reclaim
Ganzrost m grate in one section
Ganzstoff m pulp
ganzzahlig integer, integral
Ganzzahligkeit f integralness
Ganzzeug n (Papier) pulp
gar purified, (Häute) dressed, (Met) refined
Garanlage f distillery
Garantie f guarantee, warranty
Garanzin n garancine
Gararbeit f fining work, refining
Garbe f sheaf, (Met) fagot, pile
garbrennen to fire to maturity
Garbrühe f dressing liquor, finishing liquor
Gardenin n gardenin
Gare f (Gerb) tanning agent, (Metall) refined state
Gareisen n refined iron
garen to decarburize, to dress, to finish, to refine
Garen n decarburization, ~ **der Metalle** refining metals
Garerz n roasted ore
Garfaß n dressing tub, dressing vat
Garfeuer n roasting fire
garfrischen (Metall) to refine thoroughly
Garfrischen n (Metall) white iron fining

Garfrischentkohlungsperiode *f* decarburizing period
Gargang *m* refining process
Gargekrätz *n* refinery slag
Garheit *f* finished state, (Metall) refined state
Garherd *m* (Metall) refining hearth
Garkrätze *f* refinery slag
Garkupfer *n* refined copper
Garkupferschlacke *f* copper slag
Garleder *n* dressed leather
garmachen to dress, to finish, to refine
Garmachen *n* **des Kupfers** copper refining
Garnfärberei *f* yarn dyeworks
garngefärbt yarn dyed
garnieren to garnish, to trim
Garnierit *m* (Min) garnierite, noumeite
Garnierung *f* garnishing, trimming
Garnitur *f* fitting, mounting, trimming
Garnnummer *f* count of yarn, yarn count
Garofen *m* refining furnace
Garpfanne *f* large refining pan
Garprobe *f* (Metall) refining assay, refining test
Garrösten *n* finishing roasting
Garryin *n* garryine
Garschaum *m* kish, refining foam, scum
Garscheibe *f* (Metall) disk of refined copper
Garschlacke *f* refining slag, rich slag, tap cinder
Garschlackenboden *m* slag bed, slag bottom
garschmelzen (Met) to melt completety
Garsein *n* finished state
Garspan *m* coating of metal on trial rod
Garstück *n* final product, purified product
Gartenbau *m* gardening, horticulture
Garung *f* (Metall) carbonizing
Garungsdauer *f* (Metall) carbonizing period
Garungszeit *f* (Metall) carbonizing period, coking period
Gas *n* gas, ~ **absaugen** to draw off gas, ~ **aufdrehen** to turn on the gas, **absorbiertes** ~ absorbed gas, **brennbares** ~ combustible gas, **geringwertiges** ~ lean gas, poor gas, **hochwertiges** ~ rich gas, **ideales** ~ ideal gas, **indifferentes** ~ inert gas, **inertes** ~ inert gas, **ionisiertes** ~ ionized gas, **komprimiertes** ~ compressed gas, **permanentes** ~ (Phys) permanent gas, **schädliches** ~ noxious gas, toxic gas, **unverbrennbares** ~ incombustible gas, **verbranntes** ~ burnt gas
Gasableitungsrohr *n* gas outlet pipe, exhaust tube, gas exit tube, gas pipe
Gasableitungsstutzen *m* gas discharge branch
Gasabsaugungsanlage *f* gas suction plant
gasabsorbierend gas-absorbing
Gasabsorption *f* gas absorption
Gasabsorptionsmesser *m* gas absorptiometer
Gasabzug *m* gas flue, gas outlet
Gasabzugsrohr *n* gas off-take
Gasadsorption *f* gas adsorption
gasähnlich gaseous

Gasanalyse *f* gas analysis
Gasanalysenapparat *m* gas analyzing apparatus
gasanalytisch gas analytical
Gasanfall *m* gas yield
Gasangriff *m* gas attack
Gasansammlung *f* accumulation of gas
Gasanschluß *m* gas connection
Gasanstalt *f* gas works
Gasanteil *m*, **relativer** ~ (Dest) gas hold-up
Gasanzünder *m* gas lighter
Gasapparat *m* gas apparatus
Gasarmaturen *pl* gas fittings
Gasart *f* type of gas
gasartig gaseous, gasiform, gassy
Gasartigkeit *f* gaseousness
Gasatmosphäre *f* gaseous atmosphere
Gasaufnahme *f* absorption of gas, gas absorption
Gasausbeute *f* gas yield
Gasausbringen *n* extraction of gas, gas extraction
Gasausströmung *f* escape of gas, gas escape
Gasballastpumpe *f* gas ballast pump
Gasbatterie *f* (Elektr) gas battery
Gasbehälter *m* gas container, gasometer, gas reservoir, gas tank
gasbeheizt gas-fired, gas-heated
Gasbeleuchtung *f* gas lighting
Gasbereitung *f* gas production, gasification
Gasbereitungsapparat *m* gas generator
Gasbeschaffenheit *f* quality of the gas
Gasbeschießung *f* (Atom) gas bombardment
Gasbestandteil *m* gas constituent
Gasbeton *m* aerated concrete
gasbildend gas-forming, producing gas
Gasbildung *f* gas formation, gasification, gas production
Gasbläschen *n* gas bubble
Gasblase *f* gas bubble, (Gieß) blowhole, (Schweißen) gas pocket
Gasblasenströmungsmesser *m* bubble ga[u]ge
Gasblasenwäscher *m* bubbling column
Gasbleiche *f* gas bleaching, potching (pulp)
Gasbrenner *m* gas burner
Gasbürette *f* gas burette
Gaschromatogramm *n* gas chromatogram
Gaschromatographie *f* gas chromatography
Gasdetektor *m* gas detector
gasdicht gas-proof, gas-tight
Gasdichte *f* gas density
Gasdichteschreiber *m* gas density recorder
Gasdichtigkeit *f* impermeability to gas
Gasdiffusion *f* gas diffusion
Gasdiffusionsanlage *f* gas diffusion plant
Gasdrosselgestänge *n* gas throttle controls
Gasdruck *m* gas pressure
Gasdruckmesser *m* manometer, gas gauge, pressure gauge
Gasdruckregler *m* gas pressure regulator

Gasdruckregulator *m* gas pressure regulator
Gasdüse *f* gas jet
Gasdurchgang *m* gas passage
gasdurchlässig permeable to gas
Gasdurchlässigkeit *f* permeability to gas
Gasdurchlässigkeitsgerät *n* gas permeability tester
Gasdurchtrittshals *m* (Dest) vapor riser or downcomer
Gasdynamo *m* gas dynamo, gas-driven generator
Gase *pl*, **nitrose** ~ nitrous fumes, nitrous gases, **reale** ~ real gases
Gaseinleitungskerze *f* gas candle
Gaseinleitungsrohr *n* gas distribution tube
Gaseinrichtung *f* gas fittings
Gaseinschluß *m* gas occlusion, gas pocket
Gaseinsteller *m* gas regulator
Gaselektrode *f* gas electrode
Gaselement *n* (Elektr) gas battery, gas cell
gasen to gas
Gasentladung *f* (Elektr) gas discharge
Gasentladungsgefäß *n* gas discharge tube
Gasentladungslampe *f* gas-discharge lamp
Gasentladungsröhre *f* gas-discharge tube
Gasentweichung *f* escape of gas
Gasentwickler *m* gas generator
Gasentwicklung *f* formation of gas, gassing, generation of gas, production of gas
Gasentwicklungsapparat *m* gas generating apparatus, gas generator
Gasentwicklungsflasche *f* flask for developing gas
Gasentzündung *f* ignition of gas
gaserzeugend producing gas
Gaserzeuger *m* gas generator, gas producer
Gaserzeugung *f* gas generation, gasification, production of gas
Gaserzeugungsanlage *f* gas producing plant
Gaserzeugungsapparat *m* gas generator, gas producer
Gasexplosion *f* gas explosion
Gasfabrikation *f* production of gas
Gasfang *m* gas catcher, gas collector, gas exit pipe
Gas-Festkörperchromatographie *f* gas-solid chromatography
Gasfeuerung *f* gas-fired furnace, gas firing, gas heating
Gasflamme *f* gasflame, gaslight
Gasflammenelement *n* gasflame thermopile
Gasflammofen *m* gas-fired reverberatory furnace
Gasflasche *f* gas bottle, gas cylinder
Gas-Flüssig-Chromatographie *f* gas-liquid chromatography
Gas-Flüssig-Extraktion *f* gas-liquid-extraction
gasförmig gaseous, gasiform, gas-like
Gasförmigkeit *f* gaseousness
gasfrei free of gases

Gasfrischen *n* (Metall) gas puddling
Gasgebläse *n* gas blast apparatus, gas blowing engine
gasgefüllt gas-filled
Gasgehalt *m* gas content
gasgeheizt gas-heated
Gasgemenge *n* gas[eous] mixture, mixture of gases
Gasgemisch *n* gas mixture
Gasgemischregler *m* gas mixture controller
Gasgenerator *m* gas generator, gas producer
Gasgeneratoranlage *f* power gas generating plant
Gasgeruch *m* smell of gas
Gasgesetz *n* gas law
Gasgewicht *n* weight of gas
Gasgewinnung *f* gas production
Gasgleichung *f*, **adiabatische** ~ adiabatic gas equation
Gasglocke *f* gas bell, gas holder, gasometer
Gasglühlicht *n* incandescent gas light
Gasglühlichtkörper *m* gas mantle
Gasglühlichtstrumpf *m* gas mantle
Gashärteofen *m* gas-fired furnace for hardening
Gashahn *m* gas tap
Gashalter *m* gas bell, gas holder, gasometer
gashaltig gas-containing, gaseous
Gashauptleitung *f* gas main pipe
Gasheizung *f* gas heating
Gasherd *m* gas stove
Gashülle *f* gas envelope
gasig gaseous
Gaskälteprozeß *m* gas refrigeration cycle, **stilisierter** ~ isochoric gas refrigeration cycle
Gaskalk *m* gas lime
Gaskalorimeter *n* gas calorimeter
Gaskampfstoff *m* poison gas, war gas
Gaskette *f* gas battery, gas cell
gaskinetisch gaskinetic
Gaskocher *m* gas burner
Gaskohle *f* gas coal, gas retort carbon, high-volatile coal, smith's coal
Gaskoks *m* gas coke
Gaskondensator *m* gas condenser
Gaskonstante *f* gas constant
Gaskonzentrationsmesser *m* gas concentration apparatus, gas concentration tester
Gaskraftmaschine *f* gas power engine, gas motor
Gaskühler *m* gas cooler, gas condenser
Gaslampe *f* gas lamp, gas burner
Gasleitung *f* gas pipe line, gas piping, (Koksofen) gas flue
Gaslicht *n* gas light
Gasmaschine *f* gas engine, gas motor
Gasmaske *f* gas mask
Gasmenge *f* quantity of gas
Gasmesser *m* gas meter
Gasmeßrohr *n* eudiometer, gas-measuring tube
Gasmessung *f* gas measurement, gasometry

Gasmethanisieranlage *f* gas methanizing plant
Gasmischkammer *f* gas mixing chamber
Gasmischung *f* gas mixture, **nicht-ideale** ~ non-ideal gas mixture
Gasmolekül *n* gas molecule
Gasmotor *m* gas engine
Gasmuffelofen *m* gas-fired muffle furnace
Gasodorierung *f*
Gasöl *n* gas oil
Gasofen *m* gas heater, gas stove
Gasogen *n* gasogene, charcoal gas
Gasolin *n* gasoline
Gasometer *m* gasometer, gas tank
Gasometrie *f* gasometry
gasometrisch gasometric
Gasphase *f* gas phase, vapor phase
Gasphasechlorierung *f* vapor-phase chlorination
Gasphase-Hydrierung *f* vapor-phase hydrogenation
Gasphase-Krackverfahren *n* vapor-phase cracking
Gaspipette *f* gas pipette
Gaspolarisation *f* gas polarization
Gaspore *f* gas pore
Gasprobe *f* gas sample, gas test, gas testing
Gasprüfer *m* eudiometer, gas tester
Gasprüfvorrichtung *f* gas tester; gas testing apparatus
Gasraum *m* gas phase, gas space
Gasraumvolumen *n* volume of gas space
Gasreaktion *f* gaseous reaction
Gasregler *m* gas governor
Gasreiniger *m* gas cleaner, gas purifier, gas purifying apparatus
Gasreinigung *f* gas cleaning, gas conditioning, gas purification
Gasreinigungsanlage *f* gas cleaning plant
Gasreinigungsapparat *m* gas cleaner
Gasrest *m* gas residue
Gasretorte *f* gas retort
Gasrohr *n* gas pipe
Gasrückstand *m* gas residue, residual gas
Gasruß *m* gas black, carbon black, channel black
Gassammelröhre *f* gas collecting tube
Gassammler *m* gas collector, gasholder, gasometer
Gasschicht *f*, **molekulare** ~ gaseous film
Gasschmelzschweißen *n* gas welding
Gasschutz *m*, **unter** ~ **verpacken** gas packing
Gasschweißen *n* autogenous welding, gas welding
Gasschweißofen *m* gas welding furnace
Gasselbstzünder *m* automatic gas lighter
Gassenbildung *f* (Füllkörperkolonne) channeling
gassicher gasproof
Gasspannung *f* gas pressure
Gasspürgerät *n* gas tester

Gasspurenmesser *m* gas trace apparatus
Gasspurenschreiber *m* recording gas trace analyzer
Gasspurgerät *n* gas analyser
Gasstrahl *m* gas jet
Gasströmungsregelung *f* gas flow control
Gasstrom *m* current of gas, gas current
Gasstrumpf *m* gas mantle
Gastaldit *m* (Min) gastaldite
Gastechnik *f* gas engineering
gastechnisch relating to gas engineering
Gasteer *m* gas tar
Gasteilungsröhre *f* gas tee piece
Gastemperatur *f* gas temperature
Gastemperaturregler *m* gas temperature regulator
Gastheorie *f* theory of gases, **kinetische** ~ kinetic theory of gases
Gasthermometer *n* gas thermometer, air thermometer
Gastrenner *m* gas separator
Gastrennung *f* separation of gases
Gastrockner *m* gas drying apparatus, pneumatic dryer
Gasturbine *f* gas turbine
Gasüberdruck *m* positive gas pressure
Gasuhr *f* gas meter
Gasumlaufpumpe *f* gas circulating pump
gasundurchlässig gastight, impermeable to gas
Gas- und Wasserversorgung *f* gas and water supply
Gasuntersuchung *f* gas analysis
Gasventil *n* gas valve
Gasverbrauch *m* gas consumption
Gasverdichter *m* gas compressor
Gasverflüssigung *f* liquefaction of gases
Gasverflüssigungsanlage *f* gas-liquefying plant
Gasvergiftung *f* poisoning by gas
Gasverteiler *m* spurger
Gasverteilung *f* gas distribution
Gasvolumen *n* gas volume
Gasvolumeter *n* gas volumeter
gasvolumetrisch gasometric, gas-volumetric
Gaswaage *f* gas balance
Gaswäsche *f* gas scrubbing
Gaswäscher *m* [gas] washer, [gas] scrubber
Gaswaschflasche *f* gas-washing bottle
Gaswasser *n* gas liquor, gas water
Gasweg *m* gas duct, gas passage
Gaswerk *n* gas works
Gaszähler *m* gas meter
Gaszentrifuge *f* gas centrifuge
Gaszerlegung *f* gas separation
Gaszerlegungsanlage *f* gas separation plant
Gaszünder *m* gas lighter
Gaszündung *f* gas ignition
Gaszuführung *f* gas inlet
Gaszuführungskapillare *f* gas inlet capillary tube
Gaszuleiter *m* gas inlet tube

Gaszuleitungsröhre *f* gas inlet pipe
Gaszustand *m* gaseousness, gaseous state, gaseous condition
Gaszylinder *m* gas cylinder
Gatsch *m* crude paraffin, crude scale wax, slack wax
Gatter *n* lattice, grate, trellis
Gattergewebe *n* creel fabric, weftless fabric
Gattermannsche Aldehydsynthese *f* Gattermann aldehyde synthesis
Gattersäge *f* frame saw
gattieren to make up the charge, to calculate a charge, to classify, to sort
Gattieren *n* **der Rohmaterialien** mixing of the raw materials
Gattierung *f* (Metall) mixture of ores, sorting
Gattung *f* kind, sort, variety
Gattungsbegriff *m* generic term
Gattungsname *m* (Biol, Bot) generic name
Gaufrage *f* (Text, Papier) embossing
gaufrieren to emboss, to print, (kräuseln) to goffer
Gaufrieren *n* embossing
Gaufrierkalander *m* embossing calender
Gaufriermaschine *f* embossing machine
Gaultheriaöl *n* gaultheria oil, wintergreen oil
Gaultherin *n* gaultherin
Gaultherolin *n* gaultherolin, methyl salicylate
Gauß *n* (Maßeinheit des Magnetismus) gauss
Gauß-Okular *n* (Opt) Gauss eyepiece
Gaußsche Linsenformel *f* (Opt) Gaussian lens formula
Gaußsche Verteilung *f* Gaussian distribution
Gaußscher Bildpunkt *m* (Opt) Gaussian image point
Gaußsches Fehlergesetz *n* error law of Gauss
Gautsche *f* coucher, suction couch roll
gautschen to couch
Gautschmantel *m* coucher jacket
Gayerde *f* (Min) native saltpeter earth, saltpeter sweepings
Gay-Lussacsches Gesetz *n* Gay-Lussac law
Gay-Lussac-Turm *m* (Chem) Gay-Lussac tower
Gaylussit *m* (Min) gaylussite, native sodium calcium carbonate
Gaysalpeter *m* (Min) gay saltpeter
Gaze *f* gauze, cheese cloth
Gazeband *n* gauze ribbon
Gazefilter *n* gauze filter
GDCh Abk. für Gesellschaft Deutscher Chemiker
GDP Abk. für Guanosindiphosphat
gealtert aged
Geamin *n* geamine
Gearksutit *m* (Min) gearksutite
Gebälk *n* beams
Geber *m* controller
Gebiet *n* field, area, region, (Fachgebiet) field, (Math) domain, **abgetastetes** ~ scanned area,

achromatisches ~ achromatic locus,
epithermisches ~ epithermal region,
sichtbares ~ (Opt) visible region
Gebietskollokation *f* domain collocation
Gebilde *n* structure, formation, (Konfiguration) configuration
Gebirgsbewegung *f* tectonic movement
Gebläse *n* blast, blast apparatus, blower, fan, ventilator, (Blasbalg) bellows
Gebläsebrenner *m* blowpipe burner
Gebläselampe *f* blowlamp; blowpipe lamp, blowtorch
Gebläseluft *f* blast air, forced draft
Gebläsemaschine *f* blast engine, blower, blowing machine, fan engine
Gebläsemesser *m* blast meter, draft meter
Gebläsemühle *f* blowing mill
Gebläseofen *m* blast furnace
Gebläsepumpe *f* bellow[s] pump
Gebläseröhre *f* blast pipe, (Gieß) tuyere
Gebläseschachtofen *m* blast furnace
Gebläseschlägermühle *f* pressure pulverizer
Gebläsevorrichtung *f* blast apparatus
Gebläsewerk *n* blast apparatus, blower
Gebläsewind *m* blast, draft
Gebläsezylinder *m* blast cylinder
gebogen bent, curved, inflected
Gebräu *n* beverage, drink
Gebrauch *m* use
Gebrauchsanweisung *f* direction[s] for use, operating instructions, working instruction[s]
Gebrauchsartikel *m* article of daily use
Gebrauchsbedingungen *f pl* service conditions
Gebrauchseigenschaften *f pl* performance characteristics
gebrauchsfähig usable, serviceable
gebrauchsfertig ready for use, ready-made, working
Gebrauchsfestigkeit *f* service durability
Gebrauchsgegenstand *m* commodity
Gebrauchsgraphiker *m* designer
Gebrauchsgut *n* commodity
Gebrauchsmuster *n* utility model
Gebrauchsstellung *f* operative position
Gebrauchsstück *n* piece for use
Gebrauchsvorschrift *f* direction[s] for use
Gebrauchswasser *n* water for general use, service water, tapwater
Gebrauchswertprüfung *f* service evaluation test
gebrochen (Licht) refracted, (zerbrochen) broken
Gebrodel *n* boiling, bubbling, foaming
Gebühr *f* charge, fee
gebündelt bundled
gebunden bound, combined, fixed, linked
Geburtshelfer *m* obstetrician
Geburtshilfe *f* obstetrics
Gedächtnisspeicher *m* (Rechenmaschine) memory location, storage unit
Gedankenaustausch *m* exchange of ideas

Gedankenexperiment *n* imaginary experiment
Gedankengang *m* train of thought
gedeihen to grow, to proceed
gediegen genuine, solid, virgin, (Met) pure; (Min) native
Gediegeneisen *n* native iron
Gediegengold *n* native gold
Gedrit *m* (Min) gedrite
Geduld *f* endurance, patience
geeicht calibrated
geeignet [für] suitable [for]
geerdet earthed, grounded
gefährden to endanger, to imperil
Gefährdung *f* danger, hazard, **biologische** ~ biological hazard
Gefährdungskoeffizient *m* danger coefficient
Gefälldraht *m* measuring wire
Gefälle *n* (einer Straße) descent, grade, slope, (eines Flusses) drop (of a river), fall, (Math) gradient, slope
Gefällemesser *m* clinometer
Gefällestrecke *f* descent
Gefälleverlust *m* head loss
gefärbt colored, dyed
Gefäß *n* vessel, container, jar, receptacle
Gefäßabdichtung *f* reduction of capillary permeability
Gefäßbarometer *n* bulb barometer, cistern barometer
Gefäßdiffusion *f* jar diffusion
Gefäßerweiterung *f* (Med) vasodilatation
Gefäßerweiterungsmittel *n* (Pharm) vasodilator
Gefäßverengung *f* (Med) vasoconstriction
Gefäßverengungsmittel *n* (Pharm) vasoconstrictor
Gefäßverkalkung *f* (Med) vascular calcification
Gefäßverschluß *m* occlusion of a vessel
Gefahr *f* danger
Gefahrenbereich *m* danger zone
Gefahrensignal *n* danger signal
Gefahrpunkt *m* critical point, danger point
gefasert fibrous
gefettet greased, oiled
Geflacker *n* flickering
geflammt (Keram, Holz) mottled, spotted
gefleckt spotted, (Bot) maculate, (Keram) mottled
Geflimmer *n* glistening, glittering, shimmering
Gefrieranlage *f* freezing plant
Gefrierapparat *m* freezing apparatus, refrigerator
gefrierbar freezable, congealable
Gefrierbarkeit *f* congealability
Gefrierbrand *m* freezer burn
gefrieren to freeze, to congeal
Gefrieren *n* freezing, liquid solidification, **schnelles ~ durch Besprühen mit flüssigem Stickstoff** snap freezing
Gefrierer *m* freezer, congealer

Gefrierfach *n* (Kühlschrank) freezer, freezing compartment
Gefrierfleisch *n* frozen meat
gefriergetrocknet freeze-dried, lyophilized
Gefrierkammer *f* freezing chamber
Gefrierkern *m* freezing nucleus
Gefriermahlung *f* deep-freeze grinding
Gefriermischung *f* freezing mixture
Gefrierpackung *f* deep-freeze package
Gefrierprobe *f* frost test
Gefrierpunkt *m* freezing point, (Wasser) ice point
Gefrierpunktsbestimmungsapparat *m* freezing point apparatus
Gefrierpunktserniedrigung *f* lowering of the freezing point
Gefrierpunktsmesser *m* cryometer, cryoscope
Gefrierrohr *n* freezing pipe
Gefriersalz *n* freezing salt, ammonium nitrate
Gefrierschnitt *m* freezing section, frozen section
Gefrierschutzmittel *n* anti-freeze
Gefriertemperatur *f* freezing temperature
Gefrierthermometer *n* freezing-point thermometer
Gefriertrockenapparat *m* freeze drying apparatus, lyophilizer
gefriertrocknen to freeze-dry, to lyophilize
Gefriertrockner *m* vacuum freeze dryer
Gefriertrocknung *f* freeze drying, lyophilisation
Gefriertrocknungsanlage *f* freeze drying plant, lyophilizer
Gefriertunnel *m* tunnel freezer
Gefrierversuch *m* freezing test
Gefriervorrichtung *f* quick freezing instrument, congealer
gefrischt (Metall) refined
gefroren frozen, congealed
Gefüge *n* structure, grain, (Metall) texture, (Techn) joint, miter, **dichtes ~** close-grained structure, **gleichmäßiges ~** homogeneous texture, **strahliges ~** radiating structure
Gefügeänderung *f* structural change
Gefügeangaben *pl* structural data
Gefügeanordnung *f* structural arrangement
Gefügeart *f* type of structure
Gefügeaufbau *m* structural composition, structural constitution
Gefügebestandteil *m* portion of the texture
gefügelos structureless
Gefügeneubildung *f* structural transformation, phase change
Gefügeumwandlung *f* structural transformation, phase change
Gefügeuntersuchung *f* structural examination
Gefügeveränderung *f* structural change
gefüllt filled
gegabelt bifurcated, forked
gegärt fermented
Gegenbewegung *f* counter-movement

Gegendampf *m* countersteam
Gegendiagonale *f* counter diagonal, counterbrace
Gegendruck *m* counter pressure, backlash, back-pressure
Gegendruckfläche *f* area of counter pressure
Gegendruckkolben *m* dummy piston
Gegendruckmaschine *f* back-pressure engine
Gegendruckplatte *f* presser plate
Gegendruckventil *n* counter-pressure valve
Gegendruckwalze *f* backing roll
Gegenelektrode *f* counter-electrode
gegenelektromotorisch counter-electromotive
Gegenfärbung *f* contrast coloring
Gegenfarbe *f* complementary color
Gegenflansch *m* counter flange
Gegenflußsystem *n* return flow system
Gegengewicht *n* counter balance, counterpoise, counterweight
Gegengift *n* antidote, antitoxin, counterpoison, (Schlangenserum) antivenin, **chemisches ~** chemical antidote
Gegenguß *m* counter cast
Gegenhalter *m* (Niederhalter) hold-down
Gegenhormon *n* antihormone
Gegeninduktion *f* mutual inductance, mutual inductivity
Gegenion *n* counterion
Gegenkathode *f* anticathode
Gegenklopfmittel *n* (Brennstoff) anti-knock
Gegenkolben *m* counter piston
Gegenkraft *f* counteracting force, counter force, counter stress, opposing force, **elektromotorische ~** (Elektr) counter electromotive force
Gegenkrümmung *f* reverse curve
gegenläufig contra-rotating, countercurrent, counter-rotating, oppositely directed
Gegenlauf *m* anticlockwise motion, countercurrent, counter rotation, reverse motion
gegenlaufen to run anticlockwise, to run counter to
Gegenlauffräsen *n* up-milling
Gegenleistung *f* equivalence
Gegenlicht *n* incident light, (Phot) back-lighting
Gegenmaßnahme *f* countermeasure
Gegenmittel *n* (Pharm) antidote, preventive, remedy
Gegenmutter *f* check nut, lock nut
Gegenphase *f* antiphase
Gegenpol *m* antipol
Gegenreaktion *f* back reaction, reverse reaction
Gegensatz *m* antithesis, contradistinction, contrast
Gegenschaltung *f* counter connection, opposite connection
Gegenschräge *f* counterbrace
gegenseitig mutual; reciprocal

Gegensinn *m* opposite direction
gegensinnig opposing, counter rotating, irrational
Gegenstand *m* object, article, matter; (Thema) subject
Gegenstoß *m* counterthrust, repercussion
Gegenstrom *m* countercurrent, reversed current
Gegenstromapparat *m* countercurrent apparatus
Gegenstrombremsung *f* (Mech) counter-current braking (Am. counter-current braking), plugging (Am. E.)
Gegenstromdestillation *f* countercurrent distillation
Gegenstromelektrolyse *f* countercurrent electrolysis
Gegenstromextraktion *f* counter-current extraction, counterflow extraction
Gegenstromextraktor *m* countercurrent extractor
Gegenstromionophorese *f* countercurrent ionophoresis
Gegenstromkessel *m* countercurrent boiler
Gegenstromkolonne *f* countercurrent column
Gegenstromkondensator *m* counter-current condenser
Gegenstromkühler *m* countercurrent condenser, countercurrent cooler
Gegenstromprinzip *n* countercurrent principle, counterflow principle, counterstream principle
Gegenstromtellermischer *m* countercurrent pan mixer
Gegenstromverfahren *n* countercurrent process
Gegenstromverteilung *f* countercurrent distribution
Gegenstromwärmer *m* countercurrent feed heater
Gegenstück *n* counterpart, match
Gegenstützvorrichtung *f* counter stay
Gegentakt *m* push-pull
Gegentaktsender *m* push pull oscillator
Gegenteilchen *n* antiparticle
gegenüberstehen to stand opposite to, to be exposed to, to face
gegenüberstellen to compare, to confront
Gegenversuch *m* control experiment, control test
Gegenwart *f* presence
Gegenwartswert *m* present value
gegenwirken to counteract, to react
gegenwirkend antagonistic, counteractive
Gegenwirkung *f* counteraction, reaction
Gegenzug *m* back draft
gegliedert articulate
gegoren fermented
gegossen cast, poured
Gehänge *n* suspension
gehärtet hardened, (Met) casehardened, tempered
Gehäuse *n* casing, housing, shell

Gehäuseboden *m* bottom of shell
Gehäuseeisen *n* shell iron
Gehalt *m* (Anteil) content[s], percentage, proportion, (Fassungsvermögen) capacity, (Inhalt) content[s], (Konzentration) concentration, ~ **an Schiefer** content of slate, ~ **einer Flüssigkeit** content[s] of a solution, strength of a solution, titration standard
Gehalt *n* (Lohn) salary, wages
gehaltreich valuable
Gehaltsbestimmung *f* analysis, assay, determination of the content[s]
Geheimtinte *f* invisible ink
Gehirnsubstanz *f* brain substance, **graue** ~ grey brain, **weiße** ~ white brain
Gehirnzelle *f* brain cell
Gehirnzentrum *n* brain center
Gehlenit *m* (Min) gehlenite
Gehre *f* wedge, bevel, miter
Gehrung *f* miter [cut]
Gehrungsfuge *f* miter joint
Gehrungshobel *m* miter plane
Gehrungsschweißung *f* miter weld
Geierit *m* (Min) geierite
geifern to drivel, to slaver, to splutter
Geigenharz *n* colophony, rosin
Geigerin *n* geigerin
Geiger-Müller-Zähler *m* Geiger Mueller counter
Geiger-Müller-Zählrohr *n* Geiger Mueller counting tube
Geigersäure *f* geigeric acid
Geigerscher Spitzenzähler *m* Geiger counter with end window
Geigerschwelle *f* Geiger threshold
Geigerzähler *m* Geiger counter
Geijerin *n* geijerin
Geijerinsäure *f* geijerinic acid
Geikielith *m* (Min) geikielite, ilmenite
Gein *n* gein
Geinsäure *f* geinic acid
Geissoschizin *n* geissoschizine
Geissoschizol *n* geissoschizol
Geist *m* ingenuity, mind, (Alkohol) spirit, volatile liquid
geistig mental, spirituous, (alkoholisch) alcoholic
geistreich clever, ingenious
geistvoll ingenious
Geißbartkraut *n* (Bot) goatsbeard
Geißel *f* scourge, (Bakt) flagella
Geißeltierchen *n* (Zool) flagellatum (pl flagellata)
Geißelzelle *f* (Histol) ciliated cell, flagellate cell
gekerbt notched
geklärt clarified, purified
geknäuelt (Molekül) contorted
Geknister *n* decrepitation, crackling, crunching, rustling
gekörnt granular, grained, granulated, pearly

gekoppelt coupled
gekordelt milled
Gekrätz *n* dross, refuse, waste
Gekrätzprobentiegel *m* ashing crucible
gekröpft (Techn) cranked, goose-necked, offset
gekrümmt curved, (gebogen) bent
gekühlt cooled
Gel *n* gel, **elastisches** ~ elastic gel, plastic gel, **zusammenhängendes oder kohärentes** ~ coherent gel
geladen charged
Geländearbeit *f* field operation
geländegängig (Fahrzeug) cross-country
geläppt (Schnittflächen) lapped (cutting edges)
Gelatinase *f* (Biochem) gelatinase
Gelatine *f* gelatin[e], **japanische** ~ agar agar, Japanese gelatine
Gelatineagar *m* (Bakt) gelatin agar
gelatineartig gelatinous
Gelatinebad *n* gelatin bath
Gelatinecarbonit *m* gelatin carbonite
Gelatinedynamit *n* gelatin dynamite
Gelatineemulsion *f* gelatine emulsion
Gelatinefaden *m* gelatin filament
Gelatinefolie *f* gelatin foil, sheet gelatin
Gelatinekapsel *f* gelatin capsule
Gelatineleim *m* gelatin [glue]
Gelatinenährboden *m* gelatin culture medium
Gelatineprozeß *m* gelatin process
Gelatineschwefelbad *n* gelatinosulfurous bath
gelatinieren to gelatinate, to convert into a gel, to gelatinize
Gelatinierung *f* gelatinization
Gelatinier[ungs]mittel *n* gelatinizing agent
Gelatinier[ungs]vermögen *n* gelatinizing property
gelatinös gelatinous
gelb yellow, ~ **färben** to color yellow, to fade, to yellow
Gelbakazienöl *n* yellow acacia oil
Gelbantimonerz *n* (Min) cervantite
Gelbbeeren *f pl* (Bot) Persian berries
Gelbbeize *f* buff liquor
Gelbbeizen *n* yellowing
Gelbbleierz *n* (Min) yellow lead ore, native lead molybdate, wulfenite, yellow lead spar
gelbbraun buff-colored, tawny, yellowish brown
Gelbbrennanlage *f* brass dipping plant
Gelbbrenne *f* dipping liquid for brass, (Met) pickle
Gelbbrennen *n* (Met) dipping, pickling
Gelbbrenner *m* (Met) pickler
Gelbbrennsieb *n* dipping sieve (for metals)
Gelbbrenntopf *m* dipping pot (for metals)
Gelbeisenerz *n* (Min) yellow clay iron-stone, yellow iron ore
Gelbeisenkies *m* (Min) fool's gold, native iron sulfide, pyrite
Gelbeisenstein *m* (Min) limonite, xanthosiderite, yellow ironstone

Gelberde *f* (Min) yellow ocher
Gelber Körper *m* (Physiol) corpus luteum, yellow body
Gelberz *n* (Min) limonite
Gelbfärbung *f* yellow coloring, yellow dyeing, yellowing
gelbfarbig yellow[ish]
Gelbfieber *n* (Med) yellow fever
Gelbfilter *m* (Phot) yellow filter
gelbgar tanned
gelbgießen to cast in brass
Gelbgießer *m* brass founder
Gelbgießerei *f* brass foundry
Gelbglut *f* yellow heat, yellow incandescence
gelbgrau yellowish gray
gelbgrün yellowish green
Gelbguß *m* brass and bronze, yellow brass
Gelbharz *n* yellow resin
Gelbheit *f* yellowness
Gelbholz *n* (Bot) fustic wood
Gelbholzextrakt *m* fustic extract, yellow wood extract
Gelbildner *m* gelling agent
Gelbildung *f* gel formation, gelatinization, gelation, gelling, jellying
Gel-Bildungstemperatur *f* gelling point
Gelbkörperbildungshormon *n* interstitial cell-stimulating hormone, luteinizing hormone
Gelbkörperhormon *n* (Biochem) corpus luteum hormone
Gelbkupfer *n* crude copper, yellow copper
Gelbkupfererz *n* (Min) chalcopyrite, copper pyrites, yellow copper ore
gelblich yellowish
Gelbmetall *n* yellow metal
Gelbquarz *m* (Min) false topaz
gelbrot yellowish red
Gelbschleier *m* yellow stain, pyro stain
Gelbstich *m* yellow hue, yellow cast
Gelbsucht *f* jaundice
gelbweiß cream-colored, yellowish white
Gelbwurzel *f* (Bot) curcuma, Indian saffron, turmeric
Gelchromatographie *f* gel chromatography
Gelee *n* jelly
geleimt glued, sized
Gelelektrophorese *f* gel electrophoresis
Gelenk *n* link, (Masch) hinge, joint
Gelenkanschluß *m* hinge fitting
Gelenkbolzen *m* hinge pin, joint pin, link bolt
Gelenkechirurgie *f* arthroplasty
Gelenkkuppelung *f* flexible coupling
Gelenklager *n* (Kugelgelenk) ball-and-socket joint
Gelenkquarz *m* (Min) flexible sandstone, itacolumite
Gelenkrohr *n* articulated pipe, hinged pipe, joint pipe, pipe with swivel elbows

Gelenkschere *f* (od Nürnberger Schere) lazy tongs
Gelenkschraube *f* hinged bolt, swing bolt, swing screw
Gelenkverbindung *f* (Techn) articulated joint, universal joint
Gelenkwelle *f* universal joint shaft, (Techn) cardan shaft, universal shaft
Gelf *m* (Min) pyrites containing silver
Gelferz *n* (Min) copper pyrites, yellow copper ore
Gelfiltration *f* gel filtration
Gelfkupfer *n* (Min) copper pyrites, yellow copper ore
gelieren to jelly, to coagulate, to curdle, to gelatinize, to set, to turn into jelly
Gelierhilfe *f* jell[y]ing agent
Geliermittel *n* gelling agent
Gelierpunkt *m* gelling point
Gelierung *f* gelling
Gelierzeit *f* gel time, setting time
Gelkautschuk *m* gel rubber
gelöscht quenched, (Kalk) slaked
Gelöste *n* solute
Gelöste[s] *n* dissolved substance
Gelsemin *n* gelsemine
Gelseminin *n* gelseminine
Gelseminsäure *f* gelsemic acid, chrysatropic acid, gelseminic acid, scopoletin
Gelseminwurzel *f* (Bot) gelsemium root, yellow jasmine root
Gelsemium *n* gelsemium
Geltungsbereich *m* range of validity
gemäßigt moderate, temperate
gemein common, ordinary, vulgar
gemeinnützig for general use
gemeinsam common, joint
Gemeinschaftsarbeit *f* cooperative effort
Gemeinschaftsforschung *f* cooperative research
Gemeinschaftswerbung *f* collective advertising
Gemenge *n* heterogeneous mixture, (Geol) conglomerate
Gemengeanteil *m* constituent part or ingredient of a mixture
Gemengegestein *n* conglomerate
Gemengestoff *m* constituent of a mixture
Gemengeteil *m* ingredient
Gemisch *n* mixture, mix, **brennbares** ~ combustible mixture, inflammable mixture, **einheitliches** ~ homogeneous mixture, **eutektisches** ~ eutectic mixture
Gemischklappe *f* mixture shutter
Gemischkraftstoff *m* composite fuel
Gemischregler *m* mixture regulator
gemischt mixed, combined
Gemischt-Phase-Krackverfahren *n* mixed phase cracking
Gemischwirbelung *f* bubbling or aggregative fluidization

Gemischzusammensetzung *f* composition of mixture
gemittelt average
Gemmatin *n* gemmatin
gemustert with a design, (Text) printed
Gen *n* (Biol) gene
Genänderung *f* (Biol) mutation
Genaktivität *f* gene activity
genarbt (Leder) grained
genau accurate, correct, exact, precise
Genauigkeit *f* accuracy, correctness, exactness, precision, **erreichte** ~ attained accuracy, **fünfstellige** ~ five-figure accuracy, **verlangte** ~ required accuracy
Genauigkeitsarbeit *f* precision work
Genauigkeitsgrad *m* degree of accuracy
Genauigkeitsgrenze *f* limit of accuracy
Genaustausch *m* (Biol) crossing over
Genehmigung *f* agreement, approval
geneigt inclined, tilted
generalisieren to generalize
Generalnenner *m* (Math) common denominator
Generator *m* generator, (Chem) [gas] producer, (für Lichtmaschine) dynamo, ~ **für Kraftstrom** power generator, ~ **zur Gaserzeugung** gas generator, ~ **zur Verstärkung** booster generator
Generatorbetrieb *m* producer operation
Generatorgas *n* generator gas
Generatorgasanlage *f* producer gas plant
Generatorgasprozeß *m* producer gas process
Generatortemperatur *f* producer temperature
Generatorwelle *f* generator shaft
Genese *f* (Biol) genesis
Geneserethol *n* geneserethol
Geneserin *n* geneserine
Geneserolin *n* geneseroline
Geneserolinmethin *n* geneseroline methine
Genetik *f* genetics
Genetiker *m* geneticist
genetisch genetic
Genfer Nomenklatursystem *n* (Chem) Geneva system of nomenclature
genial ingenious
genießbar edible, palatable
Genießbarkeit *f* palatability
genietet riveted
Genistein *n* genistein, prunetol
Genistin *n* genistin
Genitalzyklus *m* genital cycle
Genkarte *f* gene map
Genkwanin *n* genkwanin
Genmutation *f* (Biol) gene mutation
genormt standardized, standard
genotypisch (Biol) genotypical
Genotypus *m* (Biol) genotype
Genregulation *f* gene regulation
Genrepression *f* gene repression
Genthit *m* (Min) genthite, nickel gymnite

Gentiacaulin *n* gentiacauline
Gentiamarin *n* gentiamarine
Gentianaviolett *n* gentian violet
Gentianin *n* gentianin, gentisin
Gentianinsäure *f* gentianinic acid
Gentianose *f* gentianose
Gentiobionsäure *f* gentiobionic acid
Gentiobiose *f* gentiobiose
Gentiogenin *n* gentiogenin
Gentiol *n* gentiol
Gentiopikrin *n* gentiopricrin
Gentiosid *n* gentioside
Gentisein *n* gentisein
Gentisin *n* gentisin, gentianin
Gentisinaldehyd *m* gentisaldehyde
Gentisinchinon *n* gentisinquinone
Gentisinsäure *f* gentianic acid, gentisic acid, gentisinic acid
genuin genuine
Genußmittel *n pl* semi-luxuries
Genveränderung *f* mutation, gene change
Genwirkung *f* gene action
Geocerinsäure *f* geoceric acid, geocerinic acid
Geochemie *f* geochemistry
geochemisch geochemical
Geochronologie *f* geochronology
Geodäsie *f* (Math) geodesy
geodätisch parallel geodesic parallel
Geode *f* (Krist) geode
Geoffroyarinde *f* geoffroya bark
Geoffroyin *n* geoffroyine
Geoidbestimmung *f* geoid determination
Geoidundulation *f* geoid warping
Geokronit *m* (Min) geocronite
Geologe *m* geologist
Geologie *f* geology
Geometrie *f* geometry, **analytische** ~ analytic geometry, **darstellende** ~ descriptive geometry
Geometriefaktor *m* geometrical efficiency
Geophysik *f* geophysics
geophysikalisch geophysical
Geosot *n* geosote, guaiacol isovalerate
geothermisch geothermic
geozentrisch geocentric
gepaart coupled, joined, paired
gepuffert buffered
gepunktet dotted
gequantelt quantized
gerade straight, ~ **richten** to align
Gerade *f* (Math) straight line
Geradeausempfänger *m* super-regenerative receiver
Geradeisen *n* planishing knife
Geradenbüschel *n* family of lines
Geradeverzahnung *f* spur tooth
geradfaserig straight-grained
Geradhängevorrichtung *f* pitching tool
geradkettig straight-chain

geradlinig linear, rectilinear, straight-lined
Geradsichtprisma n direct vision prism
geradzahlig even-numbered
Gerät n (Apparat) apparatus, instrument, (Ausrüstung) equipment, (Vorrichtung) device, (Werkzeug) implement, tool, utensil
Geräteglas n apparatus glass
Gerätelack m implement finish
Gerätlehre f tool gauge
Gerätschaft f instruments
Geräuschdämpfung f noise suppression
geräuschlos noiseless, silent
Geräuschpegel m noise level
geräuschsicher soundproof
Geranial n geranial, citral
Geraniin n geraniine
Geraniol n geraniol, geranyl alcohol
Geraniolen n geraniolene
Geraniumessenz f geranium essence
Geraniumöl n geranium oil
Geraniumsäure f geranic acid
Geranyl- geranyl
Geranylacetat n geranyl acetate
Geranylalkohol m geraniol, geranyl alcohol
Geranylamin n geranyl amine
Geranylmethyläther m geranyl methyl ether
gerauht (Leder) sueded
Gerbanlage f tannery, tanning plant
Gerbauszug m tanning extract
gerbbar tannable
Gerb[e]brühe f tanning liquor, tan liquor, tanner's liquor, tan ooze
Gerbemittel n tanning material, tan, tanning agent
gerben to tan, to curry, (Met) to polish, (Stahl) to pile and weld, to tilt
Gerben n tanning, currying
Gerber m tanner
Gerberbaumholz n (Bot) sumac
Gerberbeize f bate
Gerberei f tannery
Gerbereiabfälle n pl tanner's waste
Gerbereihilfsprodukte pl tanning auxiliaries
Gerberfett n drainoil scourings, fat skimming
Gerberkalk m slaked lime
Gerberlohe f tanner's bark, tan bark
Gerbertalg m tanner's tallow
Gerbextrakt m tanning extract
Gerbfaß n tanning drum
Gerbgrube f tanning pit
Gerbhilfsmittel n tanning auxiliary
Gerbkufe f tanning vat
Gerbleim m tannic acid glue
Gerblohe f tan bark, tan liquor
Gerbmittel n tanning agent, tan, tanning material
Gerbpflanze f (Bot) tanniferous plant
Gerbprozeß m tanning process
Gerbsäure f tannic acid, tannin

Gerbsäurebeschwerung f weighting with tannin
gerbsäurehaltig tanniferous
Gerbsäuremesser m barkometer, tannometer
Gerbstahl m burnisher, polishing steel, shear steel, tilted steel (s. auch Gärbstahl), wrought steel
Gerbstoff m tannic acid, tannin, tanning agent
gerbstoffartig tannic
Gerbstoffauszug m tanning extract, tannin
Gerbstoffbad n tannic acid bath
Gerbstoffbeize f tannin mordant
Gerbstoffbestimmung f determination of tanning matter
Gerbstoffextrakt m tanning extract, tannin
Gerbstofflösung f tannic acid solution, tannin solution
Gerbtrog m tan vat
Gerbung f tanning, tannage, (Lohgerbung) bark tanning, (Weißgerbung) alum tanning
Gerhardtit m (Min) gerhardtite, native basic copper nitrate
Geriatrie f (Med) geriatrics
gerichtet (Valenz, Kristallisation etc.) directional
Gerichtschemie f forensic chemistry, legal chemistry
gerichtsmedizinisch medicolegal
gerieft grooved, channeled
Geriesel n trickling
geriffelt channeled, corrugated, (genutet) grooved, (gerippt) ribbed, (gezahnt) serrated
gerillt grooved
geringfügig trivial
geringwertig inferior in value, low-grade, poor
gerinnbar coagulable, congealable
Gerinnbarkeit f coagulability
Gerinne n gutter, channel, drain, (Bergb) launder
gerinnen to coagulate, to congeal, to curdle, to gel, to jelly, (in Klumpen) to clot
Gerinnsel n coagulated mass, coagulum, curd
Gerinnstoff m coagulant, coagulator
Gerinnung f coagulation, congealing, curdling, gelling
Gerinnungsanalyse f (Med) coagulation analysis
gerinnungsbegünstigend inducing coagulation
Gerinnungsdefekt m defective coagulation
gerinnungsfähig coagulable, congealable
Gerinnungsfähigkeit f coagulability
Gerinnungsfaktor m clotting factor, coagulation factor
gerinnungshemmend anticoagulant
Gerinnungsmasse f coagulum
Gerinnungsmittel n coagulant, coagulating agent, coagulator, curdler
Gerinnungsverzögerung f clotting delay
Gerinnungsvorgang m clotting process
Gerippe n skeleton
gerippt ribbed
Germacran n germacrane

Germacrol *n* (früher Germacron) germacrane, germacrol
Germanicen *n* germanicene
Germanicol *n* germanicol
Germanicon *n* germanicone
Germanin *n* (HN) germanin
Germanit *m* (Min) germanite
Germanium *n* germanium (Symb. Ge)
Germaniumchlorid *n* (allg) germanium chloride, (Germanium(IV)-chlorid) germanic chloride, germanium(IV) chloride, germanium tetrachloride
Germaniumchloroform *n* germanium chloroform
Germaniumdioxid *n* germanic oxide, germanium dioxide, germanium(IV) oxide
Germaniumfluorwasserstoffsäure *f* fluogermanic acid
Germaniumoxydul *n* (Germanium(II)-oxid) germanium(II) oxide, germanium monoxide, germanous oxide
Germaniumsäure *f* germanic acid
Germaniumsulfid *n* germanium sulfide, germanic sulfide, germanium disulfide, germanium(IV) sulfide
Germaniumwasserstoff *m* germane, germanium hydride
Germerwurzel *f* (Bot) white hellebore root
Germicid *n* germicide
Germin *n* germine
Geröll *n* gravel, pebbles
geronnen coagulated, congealed
Geronsäure *f* geronic acid
gerottet retted, steeped
Gersdorffit *m* (Min) gersdorffite, nickel glance, plessite
Gerste *f* barley, ~ **keimen** to soak the barley
Gerstenmalz *n* barley malt
Gerstenmehl *n* barley flour
Gerstensamenöl *n* barley seed oil
Gerstenschrotbeize *f* barley dressing
Gerstenstoff *m* hordein
Gerstenwasser *n* barley water
Gerstenzucker *m* barley sugar
Geruch *m* smell, odor (Am. E.), odour (Br. E.), scent, **stechender** ~ pungent smell
geruchfrei odorless, free from smell or odor, inodorous, scentless
Geruchfreimachen *n* deodorization
geruchlos free from smell or odor, inodorous, odorless, scentless
Geruchlosigkeit *f* inodorousness
Geruchlosmachung *f* deodorization
geruchreich fragrant, odorous, scented
Geruchsbehandlung *f* (Parfümieren) perfuming
Geruchsbekämpfung *f* deodorization
geruchsbeseitigend deodorizing
Geruchssinn *m* sense of smell, smell
Geruchsverbesserung *f* deodorization
geruchverbreitend fragrant, odorous

Gerüst *n* scaffold
Gerüstsilikat *n* silicate with framework structure
Gerüstsubstanzen *f pl* builders
gesättigt saturated, impregnated
gesalzen salted
Gesamtalkalität *f* total alkalinity
Gesamtanalyse *f* total analysis
Gesamtanschluß *m* total connection
Gesamtansicht *f* full view
Gesamtarbeit *f* gross work, total work
Gesamtausbeute *f* total yield
Gesamtausgabe *f* complete edition
Gesamtbadwiderstand *m* total resistance of bath
Gesamtbreite *f* overall width
Gesamtdehnung *f* total expansion, total extension
Gesamtdosierung *f* cumulative dose
Gesamtdrehimpuls *m* total angular momentum
Gesamtdruck *m* total pressure
Gesamtdurchmesser *m* overall diameter
Gesamtenergie *f* total energy
Gesamterzeugung *f* total capacity, total production
Gesamtexplosion *f* total explosion
Gesamtfehler *m* total error
Gesamtgehalt *m* **an Feststoffen** total solids
Gesamthärte *f* total hardness
Gesamthäufigkeit *f* general average
Gesamtheit *f* totality, whole
Gesamtintensität *f* total intensity
Gesamtionenkonzentration *f* total ionic concentration
Gesamtkohlenstoffgehalt *m* total carbon content
Gesamtkraftlinienänderung *f* total flux variation
Gesamtkrümmung *f* total curvature
Gesamtladung *f* (Chem) net charge
Gesamtlänge *f* overall length
Gesamtleuchtkraft *f* overall luminosity
Gesamtlösliches *n* total soluble matter
Gesamtmenge *f* total quantity
Gesamtnutzeffekt *m* total efficiency
Gesamtquantenzahl *f* total quantum number
Gesamtquerschnitt *m* total cross section
Gesamtreaktion *f* overall reaction, total reaction
Gesamtreflexion *f* integral reflection
Gesamtrückstand *m* total residue
Gesamtsäure *f* total acid
Gesamtschrittverfahren *n* total step method
Gesamtschwefel *m* total sulfur
Gesamtspannung *f* (Elektr) total voltage
Gesamtstickstoff *m* total nitrogen
Gesamtstoffwechsel *m* total metabolism
Gesamtstrahlungsdosimeter *n* total radiation fluxmeter
Gesamtstrom *m* total current
Gesamtsystem *n* compound system
Gesamtumsatz *m* overall conversion
Gesamtverbrauch *m* total consumption
Gesamtverlust *m* total loss

Gesamtwärme – Gesetz

Gesamtwärme *f* total heat
Gesamtwärmegefälle *n* total heat drop
Gesamtwertigkeit *f* total valence
Gesamtwiderstand *m* total resistance, (Mech) total drag
Gesamtwirkung *f* overall effect
Gesamtwirkungsgrad *m* overall efficiency, total efficiency
Gesamtwirkungsquerschnitt *m* total cross section
Gesamtzahl *f* total number
Gesamtzerfall *m* total disintegration
Gesarol *n* gesarol, DDT.
Geschabsel *n* scrapings
Geschäftsanzeige *f* commercial ad
geschichtet laminated
geschickt ingenious, skilled
Geschiebelehm *m* boulder clay
Geschirr *n* utensils, ware, **irdenes** ~ earthenware
Geschirrspülkörbe *m pl* frames for dishes in dishwashing machines
Geschirrspülmittel *n* dish washing agent
Geschlechtschromosom *n* sex chromosome, allosome
Geschlechtsgen *n* sex-linked gene
Geschlechtshormon *n* sex hormone, **männliches** ~ male sex hormone, testosterone, **weibliches** ~ female sex hormone, ovarian hormone
geschlossen closed, sealed, **in sich** ~ self-contained
Geschmack *m* flavor (Am. E.), flavour (Br. E.), taste, **brennender** ~ burning taste
geschmackfrei flavorless, free from taste, tasteless, (fade) insipid
geschmacklich with respect to taste
geschmacklos flavorless, tasteless, (fade) insipid
Geschmacklosigkeit *f* tastelessness, unsavoriness
Geschmacksnerv *m* gustatory nerve
Geschmacksorgan *n* gustatory organ
Geschmacksprüfung *f* organoleptic testing
Geschmackssinn *m* gustation, sense of taste
Geschmackstoff *m* flavoring, ~ **für Nahrungsmittel** food flavoring
Geschmacksverbesserung *f* taste improvement
Geschmackszelle *f* gustatory cell, taste cell
geschmeidig flexible, (Met) ductile, malleable
Geschmeidigkeit *f* flexibility, pliability, suppleness, (Met) ductility, malleability
geschmiedet forged
geschmolzen fused, melted, molten
Geschoß *n* missile, projectile
Geschoßgeschwindigkeit *f* speed of projectiles
Geschoßteilchen *n* (Atom) bombarding particle, bombardment particle
Geschützbronze *f* gunmetal
Geschützmetall *n* gunmetal
geschützt protected
geschwefelt sulfuret[t]ed, sulfurized
geschweißt welded

Geschwindigkeit *f* speed, velocity, **durchschnittliche** ~ mean or average speed, **gleichförmige** ~ constant speed, uniform speed, **kritische** ~ critical speed, **mit hoher** ~ at high speed, **mittlere** ~ average velocity, **ungerichtete** ~ random velocity, **wahre** ~ true speed
Geschwindigkeitsabfall *m* loss of speed
Geschwindigkeitsabhängigkeit *f* velocity dependence
Geschwindigkeitsabnahme *f* decrease of velocity
Geschwindigkeitsauflösung *f* velocity resolution
geschwindigkeitsbestimmend rate-determining
Geschwindigkeitsdreieck *n* velocity triangle
Geschwindigkeitseinheit *f* unit of speed
Geschwindigkeitsgeber *m* speed transmitter
Geschwindigkeitsgefälle *n* velocity gradient
geschwindigkeitsgesteuert speed-controlled
Geschwindigkeitsgetriebe *n* speed gear
Geschwindigkeitsgradient *m* velocity gradient
Geschwindigkeitsgrenze *f* limit of speed
Geschwindigkeitskomponente *f* velocity component
Geschwindigkeitskonstante *f* velocity constant
Geschwindigkeitskurve *f* velocity curve
Geschwindigkeitsmesser *m* speedometer
Geschwindigkeitsraum *m* velocity space
Geschwindigkeitsregler *m* speed governor
Geschwindigkeitsselektor *m* velocity selector
Geschwindigkeitsstufe *f* velocity stage
Geschwindigkeitsüberschreitung *f* exceeding the speed limit
Geschwindigkeitsverhältnis *n* speed ratio
Geschwindigkeitsverteilung *f* velocity distribution
Geschwindigkeitswechsel *m* rate change
Geschwür *n* abscess, ulcer
Geschwulst *f* swelling, tumor
Gesenk *n* cavity, die, female die, hollow; press tool, punch
Gesenkeinsatz *m* insert mold, mold insert
Gesenkgußstück *n* die casting
Gesenkplatte *f* cavity retainer plate
Gesenkschmieden *n pl* die forging, swaging
Gesenkschmiedestück *n* swaged forging
Gesenkteile *pl* split cavity blocks, splits
Gesenktiefe *f* depth of recess
Gesetz *n* law, principle, rule, ~ **der kleinsten Stromwärme** law of least heating effect, ~ **der multiplen Proportionen** (Chem) law of multiple proportions, ~ **der rationalen Achsenabschnitte** law of rational intercepts, ~ **der Spannungsreihe** (Elektrochem) law of contact series, ~ **von der Erhaltung der Masse** law of conservation of matter, ~ **von Guldberg und Waage** Guldberg and Waage's law, law of mass action

gesetzlich legal, ~ **geschützt** protected by law, (durch ein Patent) patented, (z. B. Warenzeichen) registered
gesetzmäßig according to rule
Gesetzmäßigkeit f regularity
Gesichtspunkt m point of view
Gesichtswasser n (Kosm) face lotion
Gesichtswinkel m point of view
Gesims n (Arch) cornice
Gesimshobel m molding plane
Gesnerin n gesnerin
gespalten bifurcated, split
gespannt stretched, tense, tight, under tension
gesprenkelt mottled
Gestänge n structure, lever system, support
Gestängerohr n drill pipe
gestaffelt (Chem) staggered
Gestalt f (Form) form, shape, **von gleicher** ~ homomorphous
gestalten to design, to shape
Gestaltfestigkeit f design strength
gestaltlos amorphous, formless, shapeless
Gestaltlosigkeit f amorphism
Gestaltsveränderung f deformation, (Biol) metamorphosis
Gestaltung f design
Gestank m bad smell, malodor, stench
gestanzt punched out
Gestehungspreis m cost of production, costprice
Gestein n rock, rocky mineral, stones, **erzführendes** ~ ore-bearing rock, **faules** ~ brittle stones
Gesteinsart f species of rock
Gesteinsbohren n rock drilling
Gesteinsbohrer m rock drill
Gesteinsbohrmaschine f rock drilling machine
Gesteinskunde f mineralogy, petrology
Gesteinsmantel m lithosphere
Gesteinsmehl n stone powder, mineral filler, powdered stone
gesteinsphysikalisch petrophysical
Gesteinsstaub m mineral dust
Gestell n mounting, (Ablage) rack, (Maschine) frame, (Regal) shelves, (Ständer) stand, (Stütze) support
Gestellmantel m (Ofen) hearth jacket
Gestellteil n attachment
Gestellweite f (Ofen) diameter of hearth
gesteuert controlled
gestört disturbed, out of order, perturbed
gestreift striped, striated
gestrichelt broken, dashed
Gestübe n mixture of clay and culm
gestutzt truncated
Gesundheit f health, soundness
gesundheitlich hygienic[al], sanitary
gesundheitsgefährdend injurious to health
gesundheitshalber for reasons of health

Gesundheitspflege f sanitation, **öffentliche** ~ public health
Gesundheitspflegeartikel m sanitation article
gesundheitsschädlich injurious to health, insanitary, unhealthy
Gesundheitstechnik f sanitary engineering
Gesundheitswesen n sanitation and hygiene
Getäfel n wainscot, inlaying
Getöse n noise, din
Getränk n, **alkoholfreies** ~ soft drink, **geistiges** ~ alcoholic beverage
Getreide n cereals, grain
Getreidebrand m blight, smut
Getreidebranntwein m corn brandy, malt spirits
Getreidedarre f kiln for drying grains
Getreideheber m grain elevator
Getreidekeimöl n cereal seed oil
Getreidekümmel m spirits made from grain
Getreideprober m grain balance
Getreideschädling m cereal pest
Getreidesilo m grain silo
Getreidespeicher m granary
Getreidetrockner m grain drier
Getriebe n gear [unit], gearing, **dreistufiges** ~ three-speed gear
Getriebeeinheit f gear unit
Getriebegang m gear
Getriebegehäuse n gear case
Getriebekupplung f gear[ed] clutch
Getriebemotor m geared motor
Getriebeöl n gear lubricant, transmission oil
Getrieberad n gear wheel
Getriebeschema n diagram of drive
Getriebespiel n backlash
getrocknet dried, desiccated, exsiccated
getrübt turbid, cloudy, (Glas) frosted, opaque
Gewächs n plant
Gewächshaus n greenhouse
Gewähr f security
gewährleisten to assure, to guarantee
Gewährleistung f assurance, guarantee, warranty
Gewässerkunde f hydrology
Gewässerverschmutzung f water pollution
gewalzt rolled
gewaschen washed, scoured
Gewebe n (Biol) tissue, (Web) fabric, mesh, weave, **feines** ~ (Web) tight weave, **getränktes** ~ (Web) impregnated fabric, **intermuskuläres** ~ (Biol) intermuscular tissue, **schußloses** ~ weftless fabric, **weitgestelltes** ~ open fabric
Gewebeatmung f tissue respiration, (Biol) tissue respiration
Gewebeauflösung f (Biol) histodialysis
Gewebebildung f (Biol) histogenesis
Gewebebruch m cord rupture, (Reifen) carcass break
Gewebechemie f histochemistry
Gewebeeinlage f fabric insert

Gewebeeinstellung — Gewindesteigung

Gewebeeinstellung *f* (Web) number of ends per inch
Gewebeeiweiß *n* tissue protein
Gewebeextrakt *m* tissue extract
Gewebeflüssigkeit *f* (Med) tissue fluid
Gewebeführung *f* (Web) web guide
Gewebekultur *f* (Biol) tissue culture
Gewebelage *f* ply
Gewebelehre *f* (Med) histology
Gewebeneubildung *f* tissue regeneration
Gewebepackung *f* (Dest) braided metal packing
Gewebereparatur *f* sectional repair
Gewebeschaden *m* (Biol) tissue injury
Gewebeschneidmaschine *f* bias cutter
Gewebeschnitt *m* (Biol) tissue slice
Gewebeschnitzel *pl* (Web) fabric clippings
Gewebespätschaden *m* (Biol) latent tissue injury
Gewebestoffwechsel *m* metabolism of the tissues, tissue metabolism
Gewebetyp *m* (Web) fabric type
Gewebeunterbau *m* (Reifen) body of the tire, carcass
Gewebeverpflanzung *f* transplantation of tissue
Gewebezerstörung *f* histolysis, tissue disease
Gewebsextrakt *m* tissue extract
Gewebskultur *f* tissue culture
gewellt corrugated
Gewerbe *n* occupation, profession, trade
Gewerbehygiene *f* industrial hygiene
Gewerbesalz *n* common salt, industrial salt
gewerblich industrial
Gewicht *n* (Last) load, weight, (Schwere) gravity, (Wichtigkeit) importance, **spezifisches** ~ specific weight, density, specific gravity, **von gleichem** ~ equiponderous
gewichtig weighty, heavy, important
gewichtlos without weight, unimportant
Gewichtsabnahme *f* decrease in weight, loss in weight
Gewichtsänderung *f* change in weight, variation of weight
Gewichtsakku[mulator] *m* weight accumulator
Gewichtsalkoholometer *n* weight alcoholometer
Gewichtsanalyse *f* analysis by weight, gravimetric analysis
gewichtsanalytisch gravimetric
Gewichtsanzeige *f* weight indicator
Gewichtsausgleichung *f* balancing
Gewichtsbelastung *f* weighting, weight loading
Gewichtsbestimmung *f* determination of weight
Gewichtsdosierung *f* weight batching, gravimeter batching, weight feeding
Gewichtseinbuße *f* loss of weight
Gewichtseinheit *f* unit of weight
Gewichtserhöhung *f* weight increase
Gewichtsersparnis *f* weight saving
Gewichtskasten *m* weights case
Gewichtskonstanz *f* constant weight
Gewichtskontrolle *f* weight control

Gewichtskonzentration *f* mass concentration
Gewichtsmenge *f* weight
Gewichtsprozent *n* percent by weight, weight percentage
gewichtsprozentig percentage by weight
Gewichtssatz *m* set of weights
Gewichtsschwankung *f* weight variation
Gewichtsstück *n* weight
Gewichtssteil *m* part by weight
Gewichtsverhältnis *n* proportion by weight, weight ratio
Gewichtsverlust *m* loss in weight
Gewichtsverminderung *f* decrease in weight
Gewichtszunahme *f* increase in weight, weight gain
Gewinde *n* thread, ~ **fräsen** to mill threads, ~ **spritzen** to diecast threads, **doppelgängiges** ~ double thread, **dreigängiges** ~ triple thread, **einfaches** ~ (eingängig) single thread, **gefrästes** ~ milled thread, **geschnittenes** ~ cut thread, **gewalztes** ~ rolled thread, **linksgängiges** ~ left-handed thread, **metrisches** ~ metric thread, **zweigängiges** ~ double thread
Gewindeanschluß *m* screwed connection
Gewindeauslauf *m* end of the thread
Gewindebohrer *m* screw tap
Gewindebohrstahl *m* steel for screw taps
Gewindebuchse *f* screw socket, threaded bush
Gewindedrehbank *f* thread-cutting lathe
Gewindedurchmesser *m* thread diameter, major diameter, pitch diameter
Gewindeeisen *n* screw plate
Gewindeflanke *f* flank of thread
Gewindeflansch *m* threaded flange
Gewindeformdrehstahl *m* single-point threading tool
Gewindefräser *m* thread milling cutter
Gewindefräsmaschine *f* thread milling machine
Gewindefuß *m* thread base
Gewindegang *m* thread of a screw
Gewindeglas *n* glass tube with screw cap, screwed glass bottle
Gewindekluppe *f* screw stock, holder for screwing dies, screwing die stock
Gewindelänge *f* length of thread
Gewindelehre *f* thread gauge
Gewindeloch *n* threaded hole
Gewindematrix *f* male die
Gewindemuffe *f* screwed socket, threaded sleeve
Gewindenippel *m* screwed nipple
Gewindeprofil *n* form of thread
Gewindeschneidbacken *f pl* screw dies
Gewindeschneiden *n* thread cutting
Gewindeschneidmaschine *f* threading machine, screw cutting machine, tapping machine
Gewindestahl *m* screw tool
Gewindesteigung *f* pitch of thread, screw pitch

Gewindestift *m* screwed pin, core pin, headless screw
Gewindestopfen *m* screw plug
Gewindestrehler *m* chaser, chasing tool
Gewindestück *n* threaded coupling
Gewindetiefe *f* depth of thread, channel depth
Gewindewellendichtung *f* threaded shaft seal
gewinkelt angular, angled, at an angle
Gewinn *m* profit, (Vorteil) advantage
gewinnen (erlangen) to obtain, (erzeugen) to produce
Gewinnfaktor *m* gain factor
Gewinnung *f* obtaining, (durch Extraktion) extraction, (Rückgewinnung) recovery
Gewinnungsanlage *f* production plant
Gewinnungsweise *f* manner of preparation, manner of production
Gewirr *n* complication, confusion, entanglement
gewöhnen to acclimatize, to accustom, to familiarize, to habituate
gewöhnlich common, commonplace, customary, ordinary, usual
Gewöhnung *f* acclimatization, accustoming, habituation
Gewölbe *n* (Arch) arch, vault, (Kuppel) dome
Gewölbedeckel *m* arch lid
Gewölbeplatte *f* arch plate, crown plate
Gewölbeschlußstein *m* (Arch) keystone
gewölbt convex, doming (mold)
Gewürz *n* spice, aromatics, condiment
gewürzartig spicy, aromatic
Gewürzessig *m* aromatic vinegar
Gewürzextrakt *m* aromatic extract
Gewürznelkenöl *n* oil of cloves
Gewürzstoff *m* aromatic, spice
Gewürztinktur *f* aromatic tincture
Gewürzwein *m* aromatic wine, spiced wine
gewunden spiral, coiled, wound
Geyserit *m* (Min) geyserite
Gezähe *n* (Bergb) puddling tools
gezahnt cogged, serrated, toothed, (Bot) dentate, dented
Gezeiten *pl* tides
Gezeitenrechenmaschine *f* tide computing machine
GFC Abk. für Gas-Fest-Chromatographie
GFK Abk. für glasfaserverstärkte Kunststoffe
GG Abk. für Gamma-Globulin
Gheddasäure *f* gheddaic acid, ghetta acid
Gibberellensäure *f* gibberellenic acid
Gibberellsäure *f* gibberellic acid
Gibberen *n* gibberene
Gibberenon *n* gibberenone
Gibbersäure *f* gibberic acid
Gibbs-Helmholtz'sche Gleichung *f* Gibbs-Helmholtz equation
Gibbsit *m* (Min) gibbsite
Gibbssche Funktion *f* Gibbs function
Gibbssche Phasenregel *f* Gibbs' phase rule
Gibbssches Reagens *n* Gibbs reagent
Gicht *f* (Med) gout, (Metall) furnace throat, (Schmelzgut) smelting charge, **leere** ~ dead charge
Gichtanzeiger *m* stock line gauge
Gichtaufzug *m* charging apparatus
Gichtbeförderung *f* conveying of charge, transport of charge
Gichtbelag *m* charging gallery foot plates
Gichtbrücke *f* charge bridge
Gichtbühne *f* charging floor, charging platform (of a blast furnace)
Gichtebene *f* plane of throat
gichten (Metall) to charge the furnace
Gichten *n* (Metall) charging
Gichtflamme *f* throat flame, top flame
Gichtgas *n* [blast] furnace gas, top gas
Gichtgasabzugsrohr *n* downcomer [tube], gas offtake
Gichtgasentziehung *f* taking off the gases
Gichtgasleitung *f* [blast] furnace gas main
Gichtgasmaschine *f* blast furnace gas engine
Gichtgasreinigung *f* blast furnace gas cleaning
Gichtgasstaub *m* flue dust
Gichtgaswärme *f* waste gas heat
Gichtglocke *f* furnace-top bell
Gichtmittel *n* (Med) remedy for gout
Gichtöffnung *f* throat opening
Gichtpapier *n* (Med) gout paper
Gichtring *m* furnace top ring
Gichtschwamm *m* furnace cadmia, incrustation near furnace top, tutty, zinc deposit
Gichtstaub *m* blast furnace dust, dust from throat of furnace, flue dust
Gichtstaubabscheidung *f* flue-dust separation
Gichtstaubbrikettierungsanlage *f* flue-dust briquetting plant
Gichtstaubverminderung *f* reduction of furnace dust
Gichttemperatur *f* throat temperature
Gichtung *f* (Metall) charging the furnace
Gichtverschluß *m* throat stopper, furnace top distributor
Gichtzacken *m* baffle plate
gieren (Kfz) to yaw
Gieseckit *m* (Min) gieseckite
Gießanlage *f* casting implements
gießbar capable of being cast, castable, pourable
Gießbarkeit *f* (Metall) castability, pourability
Gießbett *n* casting bed, pig bed
Gießbuckel *m* casting cone
gießen to pour, (abgießen) to decant, (Gieß) to cast, to mold, **in Kokillen** ~ to chill
Gießen *n* casting, (allg) pouring, (Lack) flow coating
Gießer *m* founder, caster, molder
Gießerei *f* foundry
Gießereieisen *n* foundry pig iron
Gießereierzeugnis *n* foundry product

Gießereiflammofen *m* foundry reverberatory furnace
Gießereiform *f* mold (Am. E.), mould (Br. E.)
Gießereikoks *m* foundry coke
Gießereikupolofen *m* foundry cupola
Gießereiroheisen *n* cast iron, pig iron for castings
Gießereisand *m* molding sand
Gießereischwärze *f* molder's black, black wash
Gießereitechnik *f* foundry practice
Gießerz *n* bronze
gießfähig capable of being poured
Gießfähigkeit *f* (Metall) castability, pourability
Gießfolie *f* (Plast) cast film
Gießform *f* [casting] mold
Gießgeschwindigkeit *f* casting speed
Gießgrube *f* casting pit
Gießhalle *f* casting house
Gießharz *n* cast[ing] resin
Gießkanne *f* watering can
Gießkasten *m* casting mold
Gießkelle *f* casting ladle
Gießkopf *m* deadhead, flow head, jet, runner
Gießkran *m* casting crane
Gießlöffel *m* casting ladle
Gießmaschine *f* casting machine, curtain machine
Gießmetall *n* casting metal
Gießmodell *n* casting pattern
Gießmutter *f* matrix, mold
Gießofen *m* founding furnace
Gießpfanne *f* casting ladle, foundry ladle, pouring ladle
Gießrad *n* casting wheel
Gießrinne *f* gutter, sink
Gießsand *m* molding sand
Gießschweißen *n* molten metal welding
Gießtechnik *f* casting practice
Gießtemperatur *f* casting temperature
Gießtiegel *m* melting pot
Gießtrommel *f* casting wheel
Gießunterlage *f* casting base
Gießverfahren *n* casting process
Gießvorrichtung *f* pouring device
Gießwagen *m* ladle carriage, ladle truck
Gift *n* poison, toxin, (Tiergift) venom, **lähmendes** ~ paralysing poison, sedative poison, **schleichendes** ~ slow poison
giftabtreibend antidotal, antitoxic
giftartig poisonous
Giftbeschreibung *f* toxicography
Giftdrüse *f* poison gland
Gifterz *n* (Min) arsenic ore
gifterzeugend toxicogenic
Giftfang *m* tower to catch the arsenic
giftfrei non-poisonous, non-toxic
Giftgas *n* lethal gas, poison gas, toxic gas
gifthaltig poisonous, toxic
giftig toxic, poisonous, venomous

Giftimmunität *f* antitoxic immunity
Giftkies *m* (Min) arsenical pyrites, arsenopyrites
Giftkorn *n* (Bot) poison grain
Giftkunde *f* toxicology
Giftlattich *m* (Bot) wild lettuce
Giftmehl *n* flowers of arsenic, powdered arsenic trioxide, white arsenic powder
Giftmittel *n* (Pharm) antidote
Giftpaste *f* poison paste
Giftpille *f* (Pharm) antidotal pill
Giftrauch *m* sublimed arsenic trioxide
Giftstein *m* arsenical cadmia, arsenic trioxide, white arsenic
Giftstoff *m* poison, poisonous matter, toxic agent
Giftsumachextrakt *m* extract of poison oak
Giftturm *m* poison tower
Giftwasser *n* poisoned water
Giftweizen *m* (gegen Ratten) poisoned wheat
Giftwirkung *f* poisonous action, poisonous effect
Gigantolith *m* (Min) gigantolite
Gilbe *f* yellow [iron] ocher
gilben to turn yellow
Gilbertit *m* (Min) gilbertite
gilbig ocherous
Gillingit *m* (Min) gillingite
Gilsonit *m* (Min) gilsonite
Giltstein *m* (Min) steatite
Gin *m* gin
Gingerol *n* gingerol
Ginkgolsäure *f* ginkgolic acid
Ginnol *n* ginnol
Ginnon *n* ginnone
Ginsterblüten *f pl* (Bot) broom flowers
Giobertit *m* (Min) giobertite
Giorgiosit *m* (Min) giorgiosite
Gipfel *m* top, climax, peak, (Berg) summit
gipfeln to culminate, to reach a climax
Gipfelpunkt *m* culmination point, summit
Gips *m* gypsum, calcium sulfate dihydrate, **blättriger** ~ gypseous spar, **gebrannter** ~ plaster of Paris, **grober** ~ unsifted plaster, **in** ~ **legen** to plaster, to set in plaster, **körniger** ~ compact gypsum, **orthopädischer** ~ orthopaedic plaster, **schuppig-körniger** ~ foliated granular gypsum, **totgebrannter** ~ dead-burnt gypsum, overburnt gypsum
Gipsabdruck *m* plaster cast, gypsoplast
Gipsabguß *m* plaster cast
gipsartig gypseous
Gipsblume *f* gypseous spar, sparry gypsum
Gipsbrei *m* gypsum paste
Gipsbrennen *n* gypsum burning
Gipsbrennerei *f* gypsum calcination
Gipsbrennofen *m* plaster kiln
Gipsdruse *f* (Min) crystallized gypsum, gypsite
Gipselektrode *f* plaster electrode

gipsen to plaster
Gipserde f (Min) earthy gypsum, gypseous earth
Gipserhärtung f hardening of plaster
Gipsform f plaster mold
Gipsguß m plaster casting
gipshaltig gypsiferous, containing calcium sulfate, containing gypsum
Gipsindustrie f gypsum industry
Gipskalk m plaster lime
Gipskeil m gypsum wedge
Gipskitt m plaster cement
Gipskristall m crystallized gypsum
Gipslösung f calcium sulfate solution
Gipsmalerei f fresco painting
Gipsmarmor m imitation marble
Gipsmehl n powdered gypsum
Gipsmergel m gypseous marl
Gipsmodell n gypsum model, plaster model
Gipsmörtel m gypsum mortar, stucco
Gipsniederschlag m precipitate of calcium sulfate
Gipssand m gypseous sand
Gipssinter m stalactitical gypsum
Gipsspat m (Min) sparry gypsum, gypsum spar, selenite
Gipsstein m (Min) gypseous stone, hard calcium sulfate deposit, plaster stone
gipssteinartig gypseous
Girard-Reagens n Girard reagent
Girgensonin n girgensonine
Girlandentrockner m loop drier
gischen to foam, to froth
Gismondin m (Min) gismondine, gismondite
Gitalin n gitalin
Gitaloxigenin n gitaloxigenin
Gitigenin n gitigenine
Gitin n gitine
Gitogenin n gitogenine
Gitogensäure f gitogenic acid
Gitonin n gitonine
Gito[ro]sid n gito[ro]side
Gitoxigenin n gitoxigenin
Gitoxin n gitoxin
Gitoxosid n gitoxoside
Gitter n (Drahtnetz) iron fence, net, screen, wire netting, (Elektron, Techn) grid, (Krist) lattice, (Opt) grating, (Rost) grate, **flächenzentriertes** ~ (Krist) face-centered lattice, **konkaves** ~ (Opt) concave grating, **kubisches** ~ (Krist) cubic lattice, **kubisch-flächenzentriertes** ~ (Krist) face-centered cubic lattice, **kubisch-raumzentriertes** ~ (Krist) body-centered cubic lattice, **raumzentriertes** ~ (Krist) body-centered lattice, **reziprokes** ~ (Krist) reciprocal lattice, **weitmaschiges** ~ wide-meshed grid, coarse grid, open grid
Gitterableitwiderstand m grid leak
Gitterabmessungen pl (Krist) lattice dimensions

Gitterabsorptionskante f lattice absorption edge
Gitterabstand m lattice spacing
Gitterakkumulator m grid accumulator
Gitteranodenkapazität f grid anode capacity
Gitteranordnung f lattice arrangement, lattice array
Gitteraufbau m (Atom) atomic arrangement
Gitterausleger m lattice jib
Gitterbau m lattice structure
Gitterbaustein m (Phys) lattice point
Gitterbindung f lattice bond
Gitterblende f grating diaphragm
Gitterboden m (Rektifikation) grid plate
Gitterdynamik f (Krist) lattice dynamics
Gitterebenenabstand m (Krist) interlattice-plane distance
Gittereinheit f elementary cell, lattice unit, unit cell
Gitterelektrode f grid electrode
Gitterelektronen pl lattice electrons
Gitteremission f grid emission
Gitterenergie f (Phys) lattice energy, (Radio) grid energy
Gitterenergiebänder pl **besetzte** filled lattice energy bands
Gitterfehler m lattice imperfection or defect
Gitterfehlordnung f lattice defect
Gitterfehlstelle f lattice vacancy, lattice defect, lattice imperfection
Gitterfehlstellenerzeugung f lattice defect production
gitterförmig lattice-like
Gitterformel f grating formula
gittergesteuert grid-controlled
Gittergleichrichter m (Elektr) grid detector or rectifier
Gitterhohlraumatom n (Krist) interstitial atom
Gitterhohlraumverbindung f (Krist) interstitial compound
Gitterionisationskammer f grid ionization chamber
Gitterkapazität f (Elektr) input capacitance
Gitterkennlinie f grid characteristic
Gitterkomplex m lattice complex
Gitterkondensator m grid condenser
Gitterkonstante f lattice constant, (Krist) lattice constant or parameter, (Phys) grating constant, (Radio) grid constant
Gitterkräfte pl lattice forces
Gitterkreis m grid circuit
Gitterleerstelle f lattice vacancy
Gitterleitfähigkeit f lattice conductivity
Gitterlücke f lattice vacancy
Gittermessung f grating measurement
Gittermodulation f (Elektr) grid modulation
Gitterordnung f lattice arrangement
Gitterparameter m lattice parameter
Gitterplatz m lattice site
Gitterpunkt m grid point

Gitterraumladung *f* space grid charge
Gitterreaktor *m* lattice reactor
Gitterrelaxation *f* lattice relaxation
Gitterrührer *m* gate paddle mixer
Gitterschnittprüfung *f* square-cut adhesion test
Gitterschwingung *f* lattice vibration
Gitterschwingungsüberlagerung *f* lattice vibration superposition
Gitterspannung *f* (Elektr) grid voltage, grid potential
Gitterspat *m* (Min) grated spar
Gitterspektrometer *n* grating spectrometer
Gitterspektroskop *n* diffraction-grating spectroscope
Gitterspektrum *n* diffraction spectrum, lattice spectrum
Gitterstelle *f* lattice position
Gittersteuerung *f* (Elektr) grid control
Gitterstörung *f* lattice dislocation, lattice distortion; (Krist) lattice imperfection
Gitterstreuung *f* lattice scattering
Gitterventil *n* (Dest) grid-iron valve
Gitterverlagerung *f* lattice rearrangement
Gittervibratorelektrode *f* vibrating grid electrode
Gittervorspannung *f* bias voltage, grid bias
Gitterwärme *f* lattice heat
Gitterwelle *f* lattice wave
Gitterwerk *n* trellis [work], chequered portion, chequer work
Gitterwerte *m pl* grid data
Gitterwiderstand *m* biasing resistor, (Elektr) grid resistance
Gitterzellenvolumen *n* unit cell volume of a lattice
Glacégerbung *f* alum tanning
Glacéleder *n* kid glove leather, kid leather
Glacépapier *n* glazed paper
Gladiolsäure *f* gladiolic acid
Glänze *f* polishing material
glänzen to glisten, to glitter, to shine, to sparkle
glänzend glossy, lustrous, shiny
Glänzer *m* burnisher, finisher
gläsern glassy, hyaline, vitreous
Glättahle *f* broach
Glättanfrischen *n* reduction of litharge
Glätte *f* smoothness, evenness, (Bleiglätte) lead(II) oxide, lead monoxide, lead protoxide, litharge, massicot, plumbous oxide, (Politur) polish
Glätteisen *n* smoothing iron
glätten to smooth, to burnish, to flatten, to plane, to polish, (Papier) to calender, to glaze
Glätten *n* smoothing
Glätter *m* burnisher, polisher, smoother
Glättfrischen *n* reduction of litharge
Glättmaschine *f* buffer, calender, glazing machine
Glättung *f* finish, smoothing
Glättungstiefe *f* average roughness number

Glättvorrichtung *f* smoothing equipment
Glättwalze *f* smoothing roll
Glättwerk *n* calender stack
Glättwerkzeug *n* polishing tool
Glättzahn *m* polishing tool
Glagerit *m* (Min) glagerite
Glanz *m* brilliance, glaze, glitter, gloss, luster, sheen, **fettiger** ~ unctuous luster
Glanzarsenikkies *m* (Min) lollingite
Glanzblende *f* (Min) alabandine, alabandite
Glanzbraunstein *m* (Min) hausmannite, manganomanganite
Glanzbrenne *f* burnishing bath, final pickle, polishing pickle
Glanzdruckfarbe *f* glossy printing ink
Glanzeffekt *m* brilliance, gloss, luster
Glanzeisen *n* high silicon pig iron, pig iron rich in silicon
Glanzeisenerz *n* (Min) hematite, iron glance, specular iron ore
Glanzerhaltung *f* gloss retention
Glanzerz *n* (Min) argentite, silver glance, vitreous silver
Glanzerzschwärze *f* earthy silver glance
Glanzfarbe *f* brilliant color, gloss color ink
Glanzfirnis *m* glazing varnish
Glanzgold *n* bright gold, brilliant gold, burnished gold, gold foil, imitation leaf-gold
Glanzhammer *m* planishing hammer
Glanzkalander *m* glazing calender
Glanzkarton *m* glazed cardboard
Glanzkobalt *m* (Min) cobalt glance; cobaltite, smaltite
Glanzkohle *f* anthracite
Glanzlack *m* brilliant varnish, glazing varnish
Glanzlackfarbe *f* gloss paint
Glanzlederlack *m* patent leather enamel
Glanzleinwand *f* glazed linen, buckram
glanzlos lusterless, dead, dim, dull, mat
Glanzlosigkeit *f* dullness, lack of luster
Glanzmanganerz *n* (Min) gray manganese ore, hydrated manganic oxide, manganite
Glanzmarmor *m* shining marble
Glanzmasse *f* luster
Glanzmesser *m* gloss tester
Glanzmessing *n* polished brass
Glanzmetall *n* speculum metal
Glanzöl *n* gloss oil, brilliant oil, flooring oil
Glanzpapier *n* glazed paper
Glanzpappe *f* glazed cardboard
Glanzpigmente *n pl* luster pigments
Glanzplatin *n* bright platinum, brilliant platinum, burnished platinum, platinum glance
Glanzrot *n* colcothar, English red
Glanzruß *m* lustrous form of soot, lampblack, shining soot
Glanzscheibe *f* fine polishing disc
glanzschleifen to burnish, to polish

Glanzschleifen *n* polishing
Glanzschmelzen *n* (Wiederaufschmelzen z. B. der Verzinnung von Blech) flow-brightening
Glanzseite *f* bright side
Glanzsilber *n* (Min) argentite, bright silver, silver glance, vitreous silver
Glanzspat *m* (Min) glossy feldspar, sillimanite
Glanzstärke *f* gloss starch
glanzsteigernd gloss-improving
Glanzstoff *m* (HN) art silk, cuprammonium rayon, cuprated silk, glazed fabric
Glanzstoßen *n* glazing
Glanzvergoldung *f* bright gilding, glazed gilding
Glanzverlust *m* dulling, loss of gloss, tarnishing
Glanzversilberung *f* bright silverplating
Glanzverstärker *m* gloss improver
glanzvoll brilliant, lustrous, splendid
Glanzweiß *n* brilliant white, talc
Glanzwinkel *m* (Opt) glancing angle
Glanzwinkelreflexion *f* (Opt) specular reflectance
Glanzwirkung *f* luster effect
Glanzzahl *f* gloss number
Glanzzusatz *m* brightener, gloss additive
Glas *n* glass, **Blei-** ~ lead-glass, **böhmisches** ~ Bohemian glass, **buntes** ~ stained glass, **Chrom-** ~ chromium-glass, **Flaschen-** ~ bottle-glass, **gegossenes** ~ cast glass, **gehärtetes** ~ hardened glass, **geläutertes** ~ refined glass, **geschliffenes** ~ ground glass, **Jenaer** ~ Jena glass, **Kobalt-** ~ cobalt-glass, **komprimiertes** ~ compressed glass, **Milch-** ~ milk-glass, **Pyrex-** ~ (HN) pyrex-glass, **schwer schmelzbares** ~ hardly fusible glass, **spannungsloses** ~ strainfree glass
Glasabfall *m* broken glass, cullet, glass waste
Glasachat *m* (Min) obsidian, vitreous lava, volcanic glass
glasähnlich vitreous, glass-like, glassy, hyaloid
Glasätzen *n* hyalography
glasätzend hyalographic
Glasätzung *f* glass etching
Glasapparat *m* glass apparatus
glasartig glass-like, vitreous
Glasartigkeit *f* hyalescence, vitreousness
Glasballon *m* balloon flask, (Korbflasche) carboy, demijohn
Glasbearbeitung *f* glass working
Glasbereitung *f* glassmaking
Glasbildung *f* formation of glass; (Verglasung) vitrification
Glasbläser *m* glassblower
Glasbläserlampe *f* glassblower's lamp
Glasblase *f* bubble in glass
Glasblasen *n* glassblowing
Glasblaseröhre *f* blowpipe for glass
Glasbrennen *n* glass annealing
Glasbrocken *m* broken glass, cullet
Glasbüchse *f* glass box

Glaschemie *f* chemistry of glass
Glasdiamant *m* imitation diamond
Glasdose *f* glass box
glaselektrisch positive electric, vitroelectric
Glaselektrizität *f* vitreous electricity
Glaselektrode *f* glass electrode
Glasemballage *f* glass packing
Glaser *m* glazier
Glaserde *f* siliceous earth, vitrifiable earth
Glaserdiamant *m* glazier's diamond
Glaserit *m* (Min) glaserite, aphthitalite, arcanite
Glaserkitt *m* glazier's putty
Glasersatz *m* glass substitute
Glaserz *n* (Min) argentite, silver glance, vitreous silver
Glaserzschwärze *f* earthy silverglance
Glasfabrik *f* glass factory, glassworks
Glasfabrikation *f* glass manufacture
Glasfaden *m* glass fiber, glass thread
Glasfärben *n* glass staining
Glasfaser *f* glass fiber, (HN) fiberglas
Glasfaserkitt *m* polyester putty
Glasfaserkunststoffe *pl* glass fiber plastics
Glasfaserschichtstoff *m* (Plast) glassfiber laminate
glasfaserverstärkt glass-fiber reinforced
Glasfaserverstärkung *f* fiberglas reinforcement
Glasfilternutsche *f* glass filter funnel, sintered disc filter funnel
Glasfilterplatte *f* glass filter disc
Glasfiltertiegel *m* glass filter crucible
Glasflasche *f* glass bottle, glass flask
Glasfluß *m* glass flux
glasförmig glass-like, vitriform
Glasfritte *f* glass frit
Glasgalle *f* glass gall, sandiver
Glasgefäß *n* glass jar, glass vessel
Glasgerät *n* glass ware
Glasgewebe *n* glass cloth
Glasglanz *m* vitreous luster, frost, pounded glass
Glasglocke *f* bell jar, glass bell
Glasgranat *m* mock garnet
Glasgrün *n* bottlegreen
Glashärte *f* glass hardness
Glashärten *n* tempering of glass
Glashafen *m* glass crucible, glass pot
Glashafenton *m* glass crucible clay
Glashahn *m* glass cock
Glashalterung *f* glass stem
glashart hard as glass, glass-hard
Glashaus *n* greenhouse
Glasherstellung *f* glass manufacture
Glashütte *f* glass works
glasieren to glaze, to varnish, (Gastr) to ice, (Metall) to enamel, (Ziegel) to vitrify
Glasieren *n* glazing, enameling
glasiert glazed, enameled
Glasierung *f* enamel, glaze

glasig glass-like, glassy, hyaline, hyaloid, vitreous, ~ **werden** to become vitreous
Glasindustrie f glass industry
Glasinkrustation f glass incrustation, incrusted glass
Glasinstrument n glass instrument
Glasisolator m glass insulator
Glaskalk m glass gall, sandiver
Glaskapillare f glass capillary
Glaskappe f glass cap
Glaskasten m glass box, glass case
Glaskitt m glass cement, putty
glasklar glass clear (PVC bottles)
Glaskolben m glass flask
Glaskopf m (Min) hematite, **gelber** ~ limonite, **grüner** ~ green hematite, **schwarzer** ~ psilomelane
Glaskügelchen n glass bead
Glaskugel f glass ball, glass bulb
Glaskupfererz n (Min) vitreous copper ore
Glaslack m glass varnish
Glaslava f (Geol) volcanic glass, obsidian
Glasleinen n glass cloth, spun glass
Glaslinse f glass lens
Glasmalerei f glass painting
Glasmalz n brittle malt
Glasmehl n crushed glass, glass powder
Glasmembran f glass membrane
Glasmergel m vitrifiable marl
Glasmesser n glass cutter, glass knife
Glasofen m glass furnace
Glasopal m (Min) hyalite, volcanic glass
Glaspapier n glass paper
Glaspaste f glass paste, artificial gem
Glaspech n hard pitch, stone pitch
Glasperle f glass bead, glass marble
Glasperleneinschmelzung f glass bead seal[ing]
Glasplatte f glass plate
Glasporzellan n vitreous porcelain
Glaspulver n glass powder
Glasquarz m (Min) hyaline quartz
Glasrekristallisation f glass devitrification
Glasresonator m glass resonator
Glasretorte f glass retort
Glasröhre f glass tube
Glasröhrenhalter m glass tube holder
Glasrohr n glass tube, glass tubing
Glassalz n glass gall, sandiver
Glassand m vitreous sand
Glassatz m batch, frit
Glasschaum m glass gall, sandiver
Glasscheibe f glass pane, glass plate
Glasscherbe f broken glass, cullet
Glasschlacke f dross
Glasschleifen n glass grinding
Glasschleifer m glass grinder
Glasschliff m ground glass joint
Glasschmelz m enamel
Glasschmelze f glass melt
Glasschmelzhafen m glass pot
Glasschmelzofen m glass furnace, ash furnace
Glasschmutz m glass gall
Glasschneider m glass cutter
Glasschörl m (Min) axinite
Glasschraube f glass screw
Glasschrot m n glass beads
Glasseide f (Text) glass filaments, glass silk
Glasseideneinzelfaden m glass monofilament
Glasseidenfaden m glass filament
Glasseidengarn n glass yarn
Glasseidengewebe n glass fabric, woven glass fiber cloth
Glasseidenmatte f glass fiber mat, chopped mat, chopped strands mat, glass fiber mat
Glasseidenstrang m glass fiber roving
Glasselenit m (Min) vitrifiable selenite
Glasspannungsprüfer m glass stress tester
Glasspat m (Min) glass spar
Glasspitze f glass point
Glassplitter m fragment of broken glass, glass splinter, piece of broken glass
Glasspritze f glass syringe
Glasstab m glass rod
Glasstaub m glass dust, powdered glass
Glasstein m glass brick, (Min) axinite
Glasstift m glass pin
Glasstöpsel m glass stopper
Glasstopfen m glass stopper
Glastafel f glass plate
Glastiegel m glass pot
Glastinte f glass marking ink
Glastopf m glass crucible, glass pot, melting pot
Glastrichter m glass funnel
Glastrübung f clouding of the glass
Glasur f (Glanzüberzug) gloss, (Keram) glaze, glazing, (Metall) enamel, **überziehen mit** ~ to enamel, to glaze
Glasurasche f glaze ash
Glasurblau n zaffer
Glasurbrand m glaze baking
Glasurenfabrik f glaze factory
Glasurerz n lead glance, alquifou, potter's ore
Glasurfarbe f glaze color
Glasurfritte f glaze frit
Glasurmasse f glazing mass
Glasurmühle f glazing mill
Glasurofen m glazing kiln
Glasursand m glaze sand
Glasurschicht f layer of glaze
Glasurwelle f glaze wave
Glasvergoldung f glass gilding
Glasversilberung f glass silvering
Glasvlieslochbahnen pl perforated glass-fiber felts
Glaswanne f glass trough
Glaswaren f pl glassware
Glaswatte f glass [mineral] wool
Glaswolle f glass wool, spun glass

Glaszeolith *m* (Min) vitreous zeolite
Glasziehen *n* glass drawing
Glaszustand *m* vitreous state
Glaszylinder *m* glass cylinder, lamp glass
glatt smooth, polished, slippery
Glattbrand *m* glaze burn
Glattbrennen *n* hardening of the glaze
Glattbrennofen *m* glaze kiln; glazing kiln
Glattfeile *f* smoothing file
Glattpresse *f* calender, smoothing press
Glattrohrbündel-Wärmeaustauscher *m* shell and tube heat exchanger
Glattscherbe *f* potsherd
Glattschleifer *m* burnisher, polisher
Glattstrich *m* (Verputz) finishing coat
Glattwalze *f* glazing roller, plain roll, smoothing roller
Glauberit *m* (Min) glauberite, native calcium sodium sulfate
Glaubersalz *n* crystalline sodium sulfate, Glauber's salt, sodium sulfate decahydrate, **calciniertes** ~ calcined Glauber's salt, salt cake
Glaucentrin *n* glaucentrine
Glaucherz *n* poor ore, sterile ore
Glaucin *n* glaucine
Glauciumöl *n* glaucium oil
Glauciumsäure *f* glaucinic acid
Glauconsäure *f* glauconic acid
Glaukobilin *n* glaucobilin
Glaukochroit *m* (Min) glaucochroite
Glaukodot *m* (Min) glaucodot
Glaukolith *m* (Min) glaucolite
Glaukonit *m* (Min) glauconite, bravaisite
Glaukophan *m* (Min) glaucophane
Glaukophyllin *n* glaucophylline
Glaukoporphyrin *n* glaucoporphyrin
Glaukopyrit *m* (Min) glaucopyrite
Glaukosin *n* glaucosin
gleichachsig coaxial, equiaxial
gleicharmig equal-armed
gleichartig similar, [a]like, homogeneous, homomorphous, indistinguishable, uniform
Gleichartigkeit *f* similarity, homogeneity, homogeneousness, ~ **der Kristallform** hom[o]eomorphism
gleichatomig equiatomic, homoatomic
gleichbedeutend equivalent, synonymous
Gleichbelastung *f* uniformly distributed load
gleichbleiben to remain constant, to remain unchanged
gleichbleibend constant, steady, uniform
Gleichdruckrad *n* impulse wheel
Gleichdruckturbine *f* (Mech) impulse turbine, impulse wheel, Pelton wheel
gleichen to be equal [to], to match, to resemble
gleichfarbig of the same color, (Phys) homochromous, isochromatic
gleichflächig equiareal, isohedral

gleichförmig monotonous, similar, steady, uniform, (Min) homomorphic
Gleichförmigkeit *f* equiformity, regularity, uniformity, (Min) hom[o]eomorphism
Gleichförmigkeitsgrad *m* degree of uniformity
Gleichgang *m* even speed
gleichgekörnt even-grained
gleichgerichtet equidirectional, of like orientation, unidirected, (Elektr) rectified
gleichgestaltig isomorphous
Gleichgestaltigkeit *f* isomorphism
Gleichgewicht *n* equilibrium, balance, ~ **der Ruhelage** static balance, **das** ~ **herstellen** (Elektr) to balance circuits, **dynamisches** ~ dynamic equilibrium or balance, **einphasiges** ~ monophase equilibrium, **im** ~ **halten** to balance, to equilibrate, **indifferentes** ~ neutral equilibrium, **ins** ~ **bringen** to equilibrate, **kinetisches** ~ kinetic equilibrium, **labiles** ~ unstable equilibrium, **metastabiles** ~ metastable equilibrium, **nicht im** ~ unbalanced, **radioaktives** ~ radioactive equilibrium, **scheinbares** ~ apparent equilibrium, false equilibrium, **stabiles** ~ stable equilibrium, **stationäres** ~ stationary equilibrium, **unfreies** ~ nonvariant equilibrium, **vorgelagertes** ~ preceding equilibrium, **zweiphasiges** ~ biphase or diphase equilibrium
Gleichgewichtsapparat *m* equilibrium apparatus
Gleichgewichtsbedingung *f* condition of equilibrium
Gleichgewichtsdiagramm *n* equilibrium diagram
Gleichgewichtseinstellung *f* establishment of equilibrium
Gleichgewichtsgleichung *f* equilibrium equation
Gleichgewichtskonstante *f* equilibrium constant
Gleichgewichtskonzentration *f* equilibrium concentration
Gleichgewichtskurve *f* equilibrium curve
Gleichgewichtslage *f* equilibrium position, condition of equilibrium
Gleichgewichtslehre *f* statics, theory of equilibrium
Gleichgewichtspotential *n* equilibrium potential
Gleichgewichtspunkt *m* center of gravity
Gleichgewichtsreaktion *f* equilibrium reaction
Gleichgewichtsspannung *f* equilibrium voltage, equilibrium potential
Gleichgewichtsstörung *f* displacement of equilibrium, unbalancing, (Med) dizziness, vertigo
Gleichgewichtsumlagerung *f*, **asymmetrische** ~ asymmetric transformation of equilibrium
Gleichgewichtsverhältnis *n* equilibrium ratio
Gleichgewichtsverschiebung *f* displacement of equilibrium
Gleichgewichtszustand *m* state of equilibrium
gleichgültig indifferent

Gleichheit – Gleichzeitigkeit

Gleichheit *f* equality, identity, likeness, ~ **der Flächen** *f pl* equality of surfaces
Gleichheitsprüfer *m* regularity tester
gleichionisch isoionic
gleichkernig equinuclear
Gleichlauf *m* parallelism, synchronism, ~ **und Friktion** even and friction motion
gleichlaufend parallel, synchronized, synchronous
Gleichlauffräsen *n* down milling
Gleichlaufstöße *m pl* correcting currents
gleichmäßig (einheitlich) even, uniform, (glatt) smooth, (gleichartig) homogeneous, (gleichbleibend) constant, steady
Gleichmäßigkeit *f* steadiness, constancy, smoothness, uniformity
Gleichmaß *n* right proportion, symmetry
gleichmolekular equimolecular
gleichnamig like, similar
gleichphasig (Elektr) cophasal
Gleichpolgenerator *m* (Elektr) homopolar generator
gleichpolig homopolar
Gleichpolinduktion *f* homopolar induction
gleichrichten (ausrichten) to align, to straighten, (Elektr) to rectify
Gleichrichter *m* (Elektr) rectifier, **gasgefüllter** ~ gas-filled rectifier, **thermionischer** ~ thermionic valve
Gleichrichteranlage *f* rectifier plant
Gleichrichterelement *n* bridge rectifier
Gleichrichterkennlinie *f* rectifier characteristic
Gleichrichterkristall *m* rectifying crystal
Gleichrichterröhre *f* rectifier valve
Gleichrichtung *f* alignment, (Elektr) rectification, ~ **der Moleküle** *n pl* molecular alignment
gleichschalten to coordinate, to synchronize
gleichschenklig (Dreieck) having equal sides, isosceles
gleichsetzen to equalize, (Math) to equate
Gleichsetzung *f* equalization
Gleichspannung *f* (Elektr) direct current voltage
Gleichspannungsfeld *n* steady electric field
Gleichspannungskompensator *m* (Elektr) d. c. potentiometer
gleichstellen to compare, to coordinate, to equalize, to equate, to synchronize
Gleichstrom *m* direct current, continuous current, (Hydrodynamik) parallel flow, **pulsierender** ~ pulsating current, **stationärer** ~ continuous current
Gleichstromanker *m* direct current armature
Gleichstromanlage *f* direct current plant, continuous current plant
Gleichstrombelastbarkeit *f* capacity of direct current
Gleichstrombogenlampe *f* continuous current arc lamp, direct current arc lamp

Gleichstromdampfmaschine *f* uniflow steam engine
Gleichstromdrehstromumformer *m* continuous to threephase converter
Gleichstromdynamo *m* continuous current generator
Gleichstromelektromotor *m* direct current motor
Gleichstromenergie *f* direct current electric energy
Gleichstromerregung *f* continuous current excitation
Gleichstromgenerator *m* continuous current generator
Gleichstromgröße *f* direct current value, magnitude of direct current
Gleichstromkessel *m* concurrent boiler
Gleichstromklemme *f* direct current terminal
Gleichstromkreis *m* continuous current circuit, direct current system
Gleichstromleitung *f* continuous current line
Gleichstromlichtbogen *m* continuous current arc
Gleichstrommotor *m* direct current motor
Gleichstromprinzip *n* parallel flow principle
Gleichstromsystem *n* continuous current system, direct current system
Gleichstromumformer *m* continuous current converter, direct current converter
Gleichstromverstärker *m* direct current [linear] amplifier
Gleichstromwärmeaustauscher *m* parallel flow heat exchanger
Gleichteilung *f* equipartition
Gleichung *f* equation, ~ **dritten Grades** equation of the third order, ~ **erster Ordnung** first order or linear equation, **abgeleitete** ~ (Math) derived equation, **allgemeine** ~ general equation, **bestimmte** ~ (Math) determinate equation, **chemische** ~ chemical equation, **eine** ~ **ansetzen** to form an equation, **eine** ~ **nach x auflösen** to solve an equation with respect to x, **einer** ~ **entsprechen** to fit an equation, **kollineare** ~ (Opt) collinear equation, **lineare** ~ first order or linear equation, **quadratische** ~ (Math) quadratic equation
Gleichungsansatz *m* (Math) setup of an equation
Gleichungssystem *n* system of equations
Gleichungstyp *m* type of equation
Gleichverteilung *f* equipartition
Gleichverteilungssatz *m* law of equipartition
Gleichwerden *n* equalization
gleichwertig equivalent, equal, of equal value
Gleichwertigkeit *f* equivalence
gleichwinklig equiangular, having equal angles
Gleichwinkligkeit *f* equiangularity
gleichzeitig at the same time, simultaneous, synchronous, (zeitlich zusammenfallend) coincident
Gleichzeitigkeit *f* simultaneousness

gleichzellig isocellular
Gleis *n* track
Gleisanlage *f* rail installation
Gleishammer *m* drop hammer
gleißen to glisten, to glitter, to shine
Gleiswaage *f* wagon balance
Gleitarmaturen *pl* slip fittings
Gleitausrüstungsteile *pl* slip fittings
Gleitbacken *f pl* sliding block
Gleitbahn *f* chute, slide, (für Werkzeugmaschinen und Förderanlagen) rail
Gleitband *n* glide band, slip band
Gleitbandabstand *m* slip band spacing
Gleitbandbildung *f* glide band formation
Gleitbandverteilung *f* slip band distribution
Gleitbewegung *f* slip motion
Gleitblock *m* sliding block
Gleitebene *f* glide plane, slip plane
Gleitebenenblockierung *f* slip plane blocking
Gleitelement *n* slide
gleiten to glide, to slide, (ausgleiten) to slip
Gleiten *n* **des Riemens** slipping of the belt
Gleitfläche *f* chute, gliding plane, slipping surface, slip plane
Gleitfuge *f* slip joint
Gleitklotz *m* slide block
Gleitkontakt *m* (Elektr) sliding contact
Gleitlager *n* slide bearing, plain bearing
Gleitlinien *pl* slip lines
Gleitlinienbild *n* slip line pattern
Gleitlinienfeld *n* slip line field
Gleitlinienlänge *f* slip line length
Gleitliniennetz *n* net of slip lines
Gleitmagnet *m* sliding magnet
Gleitmittel *n* lubricant, parting compound, release agent
Gleitmodul *m* transverse modulus of elasticity
Gleitmuffe *f* sleeve
Gleitmutter *f* sliding nut
Gleitplatte *f* sliding plate
Gleitreibungskoeffizient *m* (Mech) coefficient of sliding friction
Gleitriemen *m* sliding belt
Gleitring *m* bearing ring, slip ring
Gleitrolle *f* roller
Gleitschiene *f* slide bar, guide, slide rail
Gleitschutz *m* anti-skid device
Gleitschutzprofil *n* non-skid design
Gleitsicherheit *f* slip resistance
Gleitspannung *f* sliding pressure
Gleitspiegelebene *f* (Krist) glide plane
Gleitspiegelung *f* glide reflection
Gleitspur *f* slide bar, slip line
Gleitstange *f* guide
Gleitung *f* shearing strain
Gleitungsausbreitung *f* propagation of glide
Gleitventil *n* slide valve
Gleitverbindung *f* slip joint
Gleitverschleiß *m* wear caused by sliding friction

Gleitwiderstand *m* sliding resistance
Gleitwinkel *m* (Mech) angle of slip
Gleitzone *f* slip zone
Gletschersalz *n* glacier salt, bitter salt, epsom salt, magnesium sulfate
Gliadin *n* gliadin
Glied *n* (Math) term
Gliederbahnförderer *m* apron conveyor, articulated-belt conveyor
Gliederfüßer *pl* arthropods
Gliederfuge *f* joint, articulation
Gliederkessel *m* sectional boiler
gliedern (in Abschnitte zerlegen) divide into segments, (klassifizieren) to classify, (organisieren) to organize
Gliederriemen *m* link belt
Gliederschnecke *f* sectional screw
Gliederung *f* articulation, jointing
Gliederwelle *f* articulated shaft
glimmen to glimmer, to glow, (unter der Asche) to smolder
Glimmen *n* glimmering, glowing, (Schwelen) smoldering
Glimmentladung *f* glow discharge, brush discharge, corona, silent discharge
Glimmentladungspotential *n* (Elektr) glow potential
Glimmentladungsstrom *m* luminous discharge current
Glimmer *m* (Min) glimmer, mica, **auf ~ aufgewickelt** wound on mica, **schwarzer ~** biotite, black mica, **weißer ~** muscovite, potash mica
Glimmeramphibolit *m* (Min) micaceous amphibolite
glimmerartig micaceous
Glimmerblättchen *n* micaceous lamina
Glimmerbrille *f* mica spectacles
Glimmerfolie *f* sheet mica
glimmerhaltig micaceous, containing mica
Glimmerkalk *m* micaceous lime
Glimmerklebemaschine *f* mica gluing machine
Glimmerkondensator *m* mica condenser
Glimmerkupfer *n* (Min) micaceous copper
glimmern to glimmer, to glitter like mica
glimmernd glimmering
Glimmerplatte *f* mica plate
Glimmerquarz *m* (Min) aventurine, venturine
glimmerreich rich in mica
Glimmersand *m* micaceous sand
Glimmerscheibe *f* mica sheet
Glimmerschiefer *m* mica schist, micaslate
Glimmerstein *m* (Min) micaceous rock
Glimmerton *m* micaceous clay
Glimmhaut *f* surface glow, **anodische ~** surface glow on the anode
Glimmlampe *f* glow lamp
Glimmlicht *n* glow light, **negatives ~** negative glow light, **positives ~** positive glow light

Glimmspannung — Glühasche 328

Glimmspannung *f* glow potential, glow point
Glimmstrom *m* glow current
Glimmverluste *pl* corona losses
Glimmzelle *f* photoemissive gas-filled cell, photoglow tube
Glinkit *m* (Min) glinkite
glitschig slippery
Glitsch-Ventil *n* (Dest) Glitsch valve
glitzern to glisten, to gleam, to glitter
Globin *n* globin
Globoidschneckentrieb *m* globoidal worm gear
Globol *n* globol
Globucid *n* globucid
Globulin *n* globulin, antihämatophiles ~ antih[a]emophilic globulin
Globulit *m* globulite
Globulol *n* globulol
Glocke *f* bell, (Dest) bubble cap, (Glasglocke) bell jar
Glockenboden *m* (Dest) bubble-cap plate, bubble-cap tray, bubble plate, bubble tray
Glockenbodenkolonne *f* bubble-cap column
Glockenbronze *f* bell metal
glockenförmig bell-shaped
Glockenform *f* bell shape
Glockengut *n* bell metal
Glockenmagnet *m* bell-shaped magnet
Glockenmetall *n* bell metal
Glockenmühle *f* bell crusher, cone mill, conical grinder, gyratory crusher, rotary crusher
Glockenspeise *f* bell metal, bronze for bells
Glockenspiel *n* chime
Glockentierchen *pl* vorticellae
Glockenventil *n* bell valve
Glockenverfahren *n* bell method
Glockenwinde *f* bell winch
Glockenzählrohr *n* bell counter
Glockerit *m* (Min) glockerite
Glomellifersäure *f* glomelliferic acid
Glonoin *n* glonoin, glyceryl nitrate, nitroglycerin, nitroglycerol, trinitrin
Gloversäure *f* Glover acid
Gloverturm *m* Glover tower
Glucagon *n* glucagon
Glucal *n* glucal
Glucamin *n* glucamine
Glucazidon *n* glucazidone
Glucinerde *f* (Min) beryllia, beryllium oxide, glucina, glucinum oxide
Glucinium *n* (obs) beryllium, glucinum
Glucinsäure *f* glucic acid, 3-hydroxypropenoic acid, acrolactic acid
glucinsauer glucate
Glucit *m* glucitol
Gluckertopf *m* bubbler
Glucoalyssin *n* glucoalyssin
Glucoarabin *n* glucoarabin
Glucoaubrietin *n* glucoaubrietin
Glucoberteroin *n* glucoberteroin

Glucocapparin *n* glucocapparin
Glucocheirolin *n* glucocheirolin
Glucocholsäure *f* glucocholic acid
Glucocochlearin *n* glucocochlearin
Glucoconringiin *n* glucoconringiin
Glucocorticosteroid *n* glucocorticosteroid
Glucocymarol *n* glucocymarol
Glucogallin *n* glucogallin
Glucogenkwanin *n* glucogenkwanin
Glucoheptonsäure *f* glucoheptonic acid
Glucoheptulose *f* glucoheptulose
Glucoiberin *n* glucoiberin
Glucokinase *f* (Biochem) glucokinase
Glucomannose *f* glucomannose
Glucometasaccharin *n* glucometasaccharin
Glucometer *n* glucometer
Glucomethylose *f* glucomethylose
Gluconamid *n* gluconamide
Gluconeogenese *f* (Biochem) gluconeogenesis
Gluconolaceton *n* gluconolacetone
Gluconsäure *f* gluconic acid, dextronic acid, glycogenic acid, pentahydroxyhexoic acid
Glucoproteid *n* glucoprotein
Glucosaccharin *n* glucosaccharin, saccharin
Glucosamin *n* glucosamine
Glucosaminsäure *f* glucosaminic acid
Glucosan *n* glucosan
Glucose *f* glucose, dextrose, grape sugar
Glucose-1-phosphat *n* glucose-1-phosphate
Glucoseaufnahme *f* glucose uptake
Glucosecyanhydrin *n* glucose cyanhydrin
Glucosederivat *n* glucose derivative
Glucoseoxidase *f* (Biochem) glucose oxidase
Glucoseoxim *n* glucose oxime
Glucosephenylhydrazon *n* glucose phenylhydrazone
Glucosephosphat *n* glucose phosphate
Glucosephosphorylierung *f* glucose phosphorylation
Glucosestoffwechsel *m* glucose metabolism
Glucosetransport *m* glucose transport
Glucoseüberträger *m* glucose carrier
Glucosid *n* glucoside
Glucosidase *f* (Biochem) glucosidase
Glucosometer *n* glucosometer
Glucoson *n* glucosone
Glucosulfamid *n* glucosulfamide
Glucosulfon *n* glucosulfone
Glucosurie *f* (Med) glucosuria
Glucovanillin *n* glucovanillin, vanilloside
Glucurolacton *n* glucurolactone, glucurone
Glucuron *n* glucurolactone, glucurone
Glucuronidase *f* (Biochem) glucuronidase
Glucuronsäure *f* glucuronic acid
Glühanlage *f* annealing installation
Glühanlaßschalter *m* glow plug switch (for diesel)
Glühapparat *m* heating apparatus
Glühasche *f* embers, glowing ashes, hot ashes

Glühaufschluß *m* decomposition by ignition
Glühbehandlung *f* annealing [treatment]
glühbeständig stable at red heat
Glühbetrieb *m* annealing practice
Glühbirne *f* incandescent bulb, light bulb
Glühbirnenfassung *f* lamp holder
Glühdraht *m* incandescent filament
Glühdrahtröhre *f* glowing filament tube
Glüheisen *n* glowing iron, red-hot iron
glühelektrisch thermionic
Glühelektrode *f* hot cathode
Glühelektron *n* glow electron, thermion
Glühelektronenemission *f* thermionic emission
Glühelektronenentladung *f* thermionic discharge
Glühelektronenquelle *f* thermionic electron source
glühelektronisch thermionic
Glühemissionseigenschaft *f* thermionic emission property
Glühemissionskonstante *f* thermionic constant
glühen to glow, (Metalle) to anneal, to temper, (Minerale) to calcine, to roast
Glühen *n* glowing, (Met) annealing, tempering, (Min) calcination, roasting, (Rotglut) red heat, (Weißglut) incandescence, white heat, **spannungsfreies** ~ strain-relief anneal treatment
glühend glowing, (rotglühend) red-hot, (weißglühend) incandescent
Glühfaden *m* incandescent filament
Glühfadentemperatur *f* filament temperature
Glühfarbe *f* flame color, glowing red color, temper color
Glühfeuer *n* annealing furnace, glowing fire, smoldering fire
glühfrischen (Metall) to malleabl[e]ize
Glühfrischverfahren *n* annealing process
Glühhaube *f* hot bulb
Glühhaut *f* (Metall) rolling mill scale
Glühherd *m* annealing hearth
Glühhitze *f* glowing heat
Glühkasten *m* annealing box
Glühkathode *f* incandescent cathode, incandescent valve
Glühkathodenentladung *f* hot cathode discharge
Glühkathodenröhre *f* thermionic tube
Glühkathodenverstärker *m* thermionic amplifier
Glühkerze *f* (Techn) heater plug
Glühkörper *m* incandescent body, incandescent filament, incandescent mantle
Glühkolben *m* retort
Glühkopf *m* hot bulb
Glühlampe *f* light bulb, incandescent lamp
Glühlampenfaden *m* incandescent filament
Glühlampenlack *m* electric bulb lacquer, varnish for incandescent lamps
Glühlampenphotometer *n* incandescent lamp photometer
Glühlampenwiderstand *m* glowlamp resistance

Glühlicht *n* incandescent light
Glühlichtbelastung *f* glowlamp load
Glühlichtbrenner *m* incandescent burner
Glühlichtstrumpf *m* incandescent mantle
Glühofen *m* annealing furnace, heating stove, [re]heating furnace, reverberatory furnace
Glühphosphat *n* high-temperature phosphate
Glühretorte *f* muffle furnace
Glühring *m* crucible triangle
Glühröhrchen *n* small combustion pipe
Glühröhrchenprobe *f* heating in a closed tube
Glührohrzündung *f* tube ignition
Glührückstand *m* residue from glowing, ignition residue, (Metall) hot scale
Glühsand *m* refractory sand
Glühschälchen *n* combustion cupel, ignition capsule, incinerating capsule
Glühschale *f* cupel, roasting dish
Glühschiffchen *n* combustion boat
Glühsonde *f* hot probe
Glühspan *m* hammer scale, iron scale, mill scale
Glühspanschicht *f* oxide scale
Glühstahl *m* malleable cast iron
Glühstoff *m* incandescent material
Glühstrumpf *m* gas mantle, incandescent mantle
Glühtemperatur *f* annealing temperature
Glühtiegel *m* crucible
Glühtopf *m* annealing box, cementing box
Glühung *f* annealing, calcination, glowing, (Glühfrischen) malleabl[e]izing, **erste** ~ first annealing, black annealing, **zweite** ~ second annealing, white annealing
Glühverfahren *n* annealing method, annealing process
Glühverlust *m* loss at red heat
Glühwachs *n* gilder's wax
Glühzone *f* zone of incandescence
Glühzündung *f* ignition by low tension fuses
Glukonsäure *f* gluconic acid
Glukose *f* (s. auch Glucose) glucose, grape sugar
Glut *f* glow, heat, incandescence, (Rotglut) red heat, (Weißglut) white heat
Glutaconaldehyd *m* glutaconaldehyde
Glutaconsäure *f* glutaconic acid, pentenedioic acid
Glutamal *n* glutamal
Glutamatdehydrogenase *f* (Biochem) glutamate dehydrogenase
Glutamat-Pyruvat-Transaminase *f* (Biochem) glutamate pyruvate transaminase
Glutamin *n* glutamine
Glutaminase *f* (Biochem) glutaminase
Glutaminsäure *f* glutamic acid, 2-amino glutaric acid, glutaminic acid
Glutamylglutaminsäure *f* glutamylglutamic acid
Glutaraldehyd *m* glutaraldehyde, glutaric aldehyde
Glutardialdehyd *m* glutardialdehyde
Glutaroin *n* glutaroin

Glutarsäure – Glykolaldehyd

Glutarsäure *f* glutaric acid, pentanedioic acid
Glutarsäurediäthylester *m* diethylglutarate
Glutasche *f* embers, hot ashes
Glutathion *n* glutathione
Glutazin *n* glutazine
Glutelin *n* glutelin
Gluten *n* gluten
Glutesse *f* blast forge
Glutethimid *n* glutethimide
Glutfestigkeit *f* glow resistance, spread of flame
Glutiminsäure *f* glutiminic acid
Glutinen *n* glutinene
Glutinol *n* glutinol
Glutinsäure *f* glutinic acid
Glutmesser *m* pyrometer
Glutokyrin *n* glutokyrine
Glutol *n* glutol
Glutose *f* glutose
glutrot glowing red
Glutsicherheit *f* flame resistance, resistance to smoldering
Glutwolke *f* hot avalanche
Glutzone *f* zone of incandescence
Glybuthiazol *n* glybuthiazol
Glycamid *n* glycamide
Glycarbylamid *n* glycarbylamide
Glycerid *n* glyceride
Glyceridisomerie *f* glyceride isomerism
Glycerin *n* 1,2,3-propanetriol, glycerol, glycerin
Glycerinaldehyd *m* glyceraldehyde, 2,3-dihydroxypropanal, glyceric aldehyde, glyceryl aldehyde
Glycerinaldehydphosphatdehydrogenase *f* (Biochem) glyceraldehyde phosphate dehydrogenase
Glycerinchlorhydrin *n* glycerol chlorohydrin
Glycerinester *m* glycerol ester
Glyceringelatine *f* glycerin gelatine
Glycerinkinase *f* (Biochem) glycerol kinase
Glycerinnitrat *n* glyceryl nitrate, agioneurosin, blasting oil, glonoin, nitroglycerin, nitroglycerol, nitroleum, trinitrin, trinitroglycerin
Glycerinöl *n* glycerin oil
Glycerinphosphorsäure *f* glycerophosphoric acid
glycerinphosphorsauer glycerophosphate
Glycerinsäure *f* glyceric acid, 2,3-dihydroxypropanoic acid
Glycerinsalbe *f* glycerin ointment
Glycerinschwefelsäure *f* glycerosulfuric acid
Glycerinseife *f* glycerin soap
Glycerintriacetat *n* glyceryl triacetate, triacetin
Glycerintrichlorhydrin *n* glycerol trichlorohydrin
Glycerintrinitrat *n* glycerol trinitrate
Glycerintriphosphorsäure *f* glycerotriphosphoric acid
Glycerophosphorsäure *f* glycerophosphoric acid
Glycerosazon *n* glycerosazone

Glycerose *f* glycerose
Glyceroson *n* glycerosone
Glyceryl- glyceryl
Glycid[ol] *n* glycidol, 2,3-epoxy-1-propanol, glycide
Glycidsäure *f* glycidic acid, epihydrinic acid, epoxypropionic acid
Glycidverbindung *f* glycide
Glycin *n* glycine, aminoacetic acid, glycocoll
Glycinaldehyd *m* glycinaldehyde, aminoacetaldehyde
Glycinamid *n* glycinamide
Glycinanhydrid *n* glycine anhydride, 2,5-piperazinedione
Glycinat *n* glycinate
Glycinerde *f* (Min) beryllia, beryllium oxide, glucina
Glycinnitril *n* glycine nitrile
Glycinoxidase *f* (Biochem) glycine oxidase
Glycinsäure *f* glycinic acid
Glycol *n* glycol
Glycolsäure *f* glycol[l]ic acid, hydroxyacetic acid, hydroxyethanoic acid
Glycosin *n* glycosine
Glycyl- glycyl
Glycylcholesterin *n* glycylcholesterol
Glycylhistamin *n* glycylhistamine
Glycyrrhetinsäure *f* glycyrrhetinic acid
Glycyrrhizin *n* glycyrrhizin
Glykämie *f* (Med) glyc[a]emia
Glykalhydrat *n* glycal hydrate
Glykamin *n* glucamine, glycamine
Glykane *pl* glycans
Glykocholeinsäure *f* glycocholeic acid
Glykocholsäure *f* glycocholic acid
glykocholsauer glycocholate
Glykocyamidin *n* glycocyamidine
Glykocyamin *n* glycocyamine
Glykogallussäure *f* glycogallic acid
Glykogen *n* glycogen, glucogen, hepatin, liver sugar
Glykogenese *f* (Biochem) glycogenesis
Glykogenkorn *n* glycogen granule
Glykogenolyse *f* glycogenolysis
Glykogenphosphorylase *f* (Biochem) glycogen phosphorylase
glykogenspaltend glycogenolytic
Glykogenspeicherkrankheit *f* (Med) glycogenosis, glycogen storage disease, von Gierke's disease
Glykogenstoffwechsel *m* glycogen metabolism
Glykogensynthetase *f* (Biochem) glycogen synthetase
Glykokoll *n* glycocoll, aminoacetic acid, aminoethanoic acid, glycine
Glykol *n* 1,2-ethanediol, ethylene glycol, dihydroxyethane, glycol
Glykolacetal *n* glycol acetal
Glykolaldehyd *m* glycolaldehyde

Glykolcephalin *n* glycol cephalin
Glykolchlorhydrin *n* glycol chlorohydrin
Glykolid *n* glycolic anhydride, glycolide
Glykolipid *n* glycolipid
Glykolithocholsäure *f* glycolithocholic acid
Glykolmonoacetat *n* glycol monoacetate
Glykolnitril *n* glycolonitrile
Glykolsäure *f* glycolic acid
Glykolsäurebutylester *m* butyl glycolate
Glykolschwefelsäure *f* glycolsulfuric acid, 2-hydroxyethylsulfuric acid
Glykoluril *n* glycoluril, acetylene diureine
Glykolursäure *f* glycoluric acid, carbamido acetic acid, carbamyl glycine, hydantoic acid
Glykolyl *n* glycolyl
Glykolyse *f* (Biochem) glycolysis
glykolytisch glycolytic, glycoclastic
Glykometer *n* glucometer
Glykonsäure *f* gluconic acid, glycogenic acid, glyconic acid, pentahydroxyhexoic acid
glykophor glycophore
Glykoproteid *n* glycoprotein
Glykoprotein *n* glycoprotein, glucoprotein
Glykosamin *n* glucosamine, glycosamine
Glykosazon *n* glucosazone, glycosazone
Glykose *f* glucose
Glykosid *n* glucoside
Glykosidase *f* (Biochem) glycosidase
Glykosidbindung *f* glycosidic linkage
Glykosin *n* glycosine, bisglyoxaline
Glykoson *n* glucosone, glycosone
Glykosphingosid *n* glycosphingoside
Glykosurie *f* (Med) glycosuria
Glykuronsäure *f* glucuronic acid, glycuronic acid
Glyoxal *n* glyoxal, biformyl, diformyl, ethanedial, oxalaldehyde
Glyoxalase *f* (Biochem) glyoxalase
Glyoxalatzyklus *m* glyoxalate cycle
Glyoxalin *n* 1,3-diazole, glyoxaline, imidazole
Glyoxalon *n* glyoxalone, imidazolone
Glyoxalsäure *f* glyoxalic acid, ethanal acid, glyoxylic acid, oxalaldehydic acid, oxoethanoic acid
Glyoxalsulfat *n* glyoxal sulfate
Glyoxim *n* glyoxime
Glyoxyl- glyoxyl
Glyoxylsäure *f* glyoxylic acid, ethanal acid, glyoxalic acid, oxalaldehydic acid, oxoethanoic acid
Glyphogen *n* glyphogene
Glyphographie *f* glyphography
glyphographisch glyphographic
Glyphyllin *n* glyphylline
Glyptal *n* glyptal
Glyzerinphosphatase *f* (Biochem) glycerophosphatase
Gmelinit *m* (Min) gmelinite

Gmelins Salz *n* Gmelin's salt, potassium cyanoferrate(III), potassium ferricyanide
GMP Abk. für Guanosin-5'-monophosphat
Gneis *m* (Min) gneiss
gneisig gneissic, gneissoid
gnomonisch (Krist) gnomonic
Gnoscopin *n* gnoscopine, dl-narcotine
Goapulver *n* goa powder, araroba
Goethit *m* (Min) goethite
Goitrin *n* goitrin
Goitrogen *n* goitrogen
gold- und silberhaltig auri-argentiferous
Gold *n* gold (Symb. Au), **gediegenes** ~ native gold, **kolloidales** ~ colloidal gold, **silberhaltiges** ~ electrum
Goldabfall *m* gold chips, gold parings
Goldader *f* (Bergb) golden vein
goldähnlich goldlike
Goldamalgam *n* gold amalgam
Goldanstrich *m* gilding
goldartig goldlike
Goldausziehung *f* gold extraction
Goldbad *n* [boiling] gold bath, gold toning bath
Goldbarren *m* gold ingot
Goldbergwerk *n* gold mine
Goldberyll *m* (Min) chrysoberyl
Goldblatt *n* gold foil, gold leaf
Goldblattelektrometer *n* gold leaf electroscope or electrometer
Goldblattelektroskop *n* gold leaf electroscope
goldbraun gold-brown, auburn
Goldbromid *n* gold bromide, (Gold(III)-bromid) auric bromide, gold(III) bromide, gold tribromide
Goldbromür *n* (Gold(I)-bromid) aurous bromide, gold monobromide
Goldbronze *f* gold bronze
Goldbronzepigment *n* gold bronze pigment
Goldcarbid *n* gold carbide
Goldchlorid *n* gold chloride, (Gold(III)-chlorid) auric chloride, aurichloride, gold(III) chloride, gold trichloride
Goldchlorür *n* (Gold(I)-chlorid) aurous chloride, gold(I) chloride, gold monochloride
Goldchlorwasserstoffsäure *f* chloroauric acid
Goldcyanid *n* gold cyanide, (Gold(III)-cyanid) auric cyanide, gold(III) cyanide, gold tricyanide
Goldcyanür *n* (Gold(I)-cyanid) aurous cyanide, gold(I) cyanide, gold monocyanide
Golddraht *m* gold wire
golden golden
Golderde *f* (Min) auriferous earth
Golderz *n* (Min) gold ore
Goldfällungsmittel *n* gold precipitant
Goldfarbe *f* auripigment, gold color, orpiment
goldfarben gold-colored
Goldfieldit *m* (Min) goldfieldite
Goldfirnis *m* gold varnish

Goldfolie *f* gold foil
goldführend auriferous
Goldgalvanoplastik *f* gold galvanoplastic process
Goldgehalt *m* gold content, number of carats
goldgelb gold-colored, aureate
Goldgeschirr *n* gold plates
Goldgespinst *n* spun gold
Goldgewicht *n* gold weight, troy weight
Goldgewinnung *f* gold extraction
goldglänzend shining like gold
Goldglätte *f* gold litharge, red litharge
Goldglanz *m* gold luster
Goldglimmer *m* (Min) yellow mica
Goldgranat *m* (Min) succinite, topazolite
Goldgraphit *m* mixture of gold and graphite
Goldgrieß *m* gold dust
Goldgrundfirnis *m* gold priming varnish, gold size
goldhaltig auriferous, containing gold
Goldhydroxid *n* gold hydroxide, (Gold(III)-hydroxid) auric hydroxide, gold(III) hydroxide
Gold(I)-cyanwasserstoffsäure *f* aurocyanic acid, cyanoauric(I) acid
goldig golden
Gold(III)-cyanwasserstoffsäure *f* auricyanic acid, cyanoauric(III) acid
Gold(III)-rhodanwasserstoffsäure *f* aurithiocyanic acid, thiocyanatoauric(III) acid
Gold(III)-Salz *n* auric salt, gold(III) salt
Gold(III)-Verbindung *f* auric compound, gold(III) compound
Gold(I)-rhodanwasserstoffsäure *f* aurothiocyanic acid
Gold(I)-Salz *n* aurous salt, gold(I) salt
Gold(I)-Verbindung *f* aurous compound, gold(I) compound
Goldjodid *n* gold iodide, (Gold(III)-jodid) auric iodide, gold(III) iodide, gold triodide
Goldjodür *n* (Gold(I)-jodid) aurous iodide, gold(I) iodide, gold monoiodide
Goldkaliumbromür *n* potassium aurobromide, potassium bromoaurate(I)
Goldkaliumcyanür *n* potassium aurocyanide, potassium cyanoaurate(I)
Goldkalk *m* gold calx
Goldkies *m* auriferous pyrites, gold quartz
Goldklumpen *m* nugget of gold
Goldkochkölbchen *n* assay flask for gold separation, gold boiling flask
Goldkorn *n* gold grain
Goldkrätze *f* goldsmith's wash, gold sweepings
Goldkristall *m* gold crystal
Goldkupfer *n* Mannheim gold, pinchbeck
Goldlack *m* gold varnish
Goldlackimitation *f* ormolu varnish

Goldlaugerei *f* gold extraction, gold leaching plant
Goldlegierung *f* gold alloy
Goldleim *m* gold size
Goldleiste *f* gilt cornice
Goldleistenlack *m* gilt cornice varnish
Goldlösung *f* gold solution
Goldlot *n* gold solder
Goldmetall *n* gold ore
Goldmünze *f* gold coin, gold medallion
Goldniederschlag *m* gold precipitate
Goldocker *m* finest yellow ocher
Goldoxid *n* gold oxide, (Gold(III)-oxid) gold(III) oxide, gold trioxide
Goldoxidnatron *n* (obs) sodium aurate
Goldoxidsalz *n* (obs) aurate, auric salt, gold(III) salt
Goldoxidverbindung *f* (obs) auric compound, gold(III) compound
Goldoxydul *n* (obs, Gold(I)-oxid) aurous oxide, gold(I) oxide, gold monoxide, gold protoxide
Goldoxydulverbindung *f* (obs) aurous compound, gold(I) compound
Goldpapier *n* gold paper, gilt paper
Goldphosphid *n* gold phosphide
Goldpigment *n* (Keram) porcelain gilding
goldplattieren to gold-plate
goldplattiert gold-plated
Goldplattierung *f* gold plating
Goldprobe *f* gold assay
Goldprobenglühapparat *m* gold assay heating apparatus
Goldprobenkochapparat *m* gold assay boiling apparatus
Goldprobentafel *f* gold assaying table
Goldpulver *n* gold powder, German gold
Goldpurpur *m* gold purple, purple of cassius
Goldquarz *m* (Min) auriferous quartz
Goldraffination *f* gold refining
goldreich auriferous
Goldsäure *f* auric acid, auric hydroxide, gold(III) hydroxide
Goldsand *m* auriferous sand
Goldschale *f* gold cup, gold dish
Goldscheideanstalt *f* gold refinery
Goldscheiden *n* gold refining, gold parting
Goldscheider *m* gold refiner
Goldscheideverfahren *n* process of gold refining
Goldscheidewasser *n* aqua regia
Goldscheidung *f* gold refining, gold parting
Goldschlag *m* gold foil, gold leaf
Goldschlich *m* gold slime, pounded gold ore
Goldschnitt *m* gilt edge, gold edge
Goldschrift *f* gold lettering, chrysography
Goldschwefel *m* antimony pentasulfide, golden antimony sulfide, golden sulfide of antimony, gold sulfide
Goldschwefelkies *m* (Min) auriferous iron pyrites

Goldselenid *n* auric selenide, gold(III) selenide, gold triselenide
Goldsiegellack *m* aventurin
Goldsilber *n* argentiferous gold, electrum
Goldsilbersulfid *n* goldsilver sulfide
Goldsilbertellurid *n* goldsilver telluride
Goldsinter *m* auriferous sinter
Goldstaub *m* gold dust
Goldstein *m* auriferous stone, goldstone
Goldstreichen *n* gold test
Goldsud *m* boiling gold bath, hot gold bath
Goldsulfid *n* gold sulfide, (Gold(III)-sulfid) auric sulfide, gold(III) sulfide, gold trisulfide
Goldsulfür *n* (Gold(I)-sulfid) aurous sulfide, gold(I) sulfide, gold monosulfide
Goldtalk *m* yellow talc
Goldtellurid *n* gold telluride
Goldthioschwefelsäure *f* aurothiosulfuric acid, thiosulfatoauric(I) acid
Goldtinktur *f* tincture of gold
Goldtopas *m* (Min) oriental topaz
Goldtropfen *m pl* ethereal ferric chloride tincture
Goldverbindung *f* gold compound
Goldwaage *f* gold balance
Goldweinstein *m* gold tartrate
Goldzahl *f* gold number
Goldzieher *m* goldwire drawer
Goldzusatz *m* addition of gold, gold doping
Golgi-Apparat *m* (Biol) Golgi apparatus
Gonan *n* gonan
Goniometer *n* goniometer
Goniometrie *f* goniometry
goniometrisch goniometric
Goochtiegel *m* Gooch crucible
Gorceixit *m* (Min)) gorceixite
Gorlisäure *f* gorlic acid
Goslarit *m* (Min) goslarite
Gosse *f* drain, gutter
Gossenstein *m* sink
Gossypetin *n* gossypetin
Gossypetonsäure *f* gossypetonic acid
Gossypin *n* gossypin, cotton cellulose
Gossypiton *n* gossypitone
Gossypitrin *n* gossypitrine
Gossypitron *n* gossypitrone
Gossypol *n* gossypol
Goudron *m* soft asphalt, tar
Goudronüberzug *m* tar enamel coating
Goyazit *m* (Min) goyazite
Grabeisen *n* graving tool
Graben *m* trench, canal, ditch, main runner, sewer
Grad *m* (Celsius) centigrade, (Math) power, (Temperatur, Winkel) degree, **10 ~ unter Null** ten degrees below zero, **bei null ~** at zero [degree]
Gradabteilung *f* [graduated] scale, graduation

Gradationskurve *f* (Phys) exposure-density relationship, Hurter and Drieffield curve
Gradbogen *m* graduated arc
Gradeinteilung *f* graduation, scale
Gradient *m* gradient, **hydraulischer ~** hydraulic gradient
Gradientenelution *f* gradient elution
Gradientenmethode *f* gradient method, **~ f. Funktionenoptimierung** gradient method for optimization of functions
Gradierapparat *m* (Chem) graduator
Gradiereisen *n* chisel, graving tool
gradieren to grade, to graduate [a scale]
Gradierhaus *n* graduation house
Gradierofen *m* graduation furnace
Gradierung *f* grading, graduation
Gradierwaage *f* brine gauge, water balance
Gradierwerk *n* graduation apparatus, graduation works
gradkettig straight-chain
Gradlaufapparat *m* stabilizer
Gradleiter *f* scale of degrees
Gradmesser *m* graduator
Gradmessung *f* measuring of degrees, (Geol) graduation
Gradteilung *f* graduation scale
graduiert graduated
gradzahlig even-numbered
Grädigkeit *f* **der Sole** density of the brine
Gräte *f* fish-bone
Graftonit *m* (Min) graftonite
Grainieren *n* (Pap) graining
Grainierkalander *m* embossing calender
Gram-Färbung *f* (Biol) gram stain[ing]
Gramicidin *n* gramicidin
Gramin *n* gramine
Gramm *n* gram (Am. E.), gramme (Br. E.)
Grammäquivalent *n* gram-equivalent
Grammatit *m* (Min) grammatite
grammatithaltig grammatitiferous
Grammatom *n* gram atom
Grammatomgewicht *n* gram-atomic weight
Grammcalorie *f* gram calorie, small calorie
Grammflasche *f* gram bottle, pycnometer, specific gravity bottle
Grammgewicht *n* gram weight
Grammion *n* gram ion
Grammit *m* grammite
Grammkalorie *f* gram calorie, small calorie
Grammol *n* gram molecule
Grammolekül *n* gram molecule
Grammolekülgewicht *n* gram-molecular weight
Grammolekülvolumen *n* gram-molecular volume
gram-negativ (Bakt) gram-negative
gram-positiv (Bakt) gram-positive
Gramsche Färbung *f* Gram's method, Gram's stain
Granalien *pl* granulated metal, shot

Granat *m* (Min) garnet, **gelber** ~ topazolite, **gemeiner** ~ common garnet, **grüner** ~ grossular
Granatal *n* granatal
Granatan *n* granatane
Granatanin *n* 1,5-imino cyclooctane, granatanine
granatartig garnet-like
Granatdodekaeder *n* (Krist) rhombic dodecahedron
Granatenin *n* granatenine
Granatfluß *m* artificial garnet
Granatgruppe *f* (Min) garnet group
granathaltig garnetiferous
Granatin *n* granatine
Granatit *m* (Min) granatite
Granatoeder *n* (Krist) rhombic dodecahedron
Granatolin *n* granatoline
Granatonin *n* granatonine
Granatstein *m* (Min) garnet rock
Grancoldverfahren *n* Grancold pellet bonding
Grand *m* coarse sand, fine gravel
Grandidierit *m* grandidierite
grandig containing gravel, gravelly
Grandmehl *n* coarse meal
Grandstein *m* (Min) granite
granieren (Pap) to grain, (Techn) to granulate
Granilit *m* (Min) granilite
Granit *m* (Min) granite
granitähnlich granitoid
granitartig granite-like, granitic, granitiform
Granitbildung *f* (Geol) granitification
granitfarbig granite-colored
Granitfels *m* granite rock, granitic rock
granitförmig granitiform, granite-like, granitic
Granitformation *f* (Geol) granite formation
Granitgestein *n* granitic rock
Granitisolator *m* granitic insulator
Granitit *m* (Min) granitite
Granitporphyr *m* (Min) porphyroid granite
Granitquarz *m* (Min) granitic quartz
Granitsand *m* (Min) granite sand
Grantianin *n* grantianine
Grantianinsäure *f* grantianic acid
Granulat *n* granulated or granular material, granules, pellets
Granulation *f* granulation
Granulationsanlage *f* granulating plant
Granulator *m* granulator
Granulatschleuder *f* pellet drier
Granulieranlage *f* granulating plant
Granulierapparat *m* granulating machine
granulieren to granulate
Granuliermaschine *f* pelletizer
Granuliermühle *f* granulating crusher
granuliert granular, granulated
Granulierung *f* granulation
Granulit *m* (Min) granulite, whitestone
granulitisch granulitic

Granulometrie *f* granulometry
Granulose *f* granulose
Granulozyt *m* granular leucocyte, granulocyte
graphisch graphic, graphical, ~ **darstellen** to represent graphically
Graphit *m* (Min) graphite, black lead, plumbago, **blättriger** ~ foliated graphite, **entflockter** ~ deflocculated graphite
Graphitablagerung *f* deposit of graphite
graphitähnlich graphite-like, graphitoidal
Graphit-Alkali-Verbindungen *pl* alkali metal graphites
Graphitanlasser *m* graphite starter
graphitarm poor in graphite
graphitartig graphite-like, graphitic, graphitoidal
Graphitausbildung *f* graphite formation, graphitization
Graphitausscheidung *f* separation of graphite
Graphitblock *m* graphite block
Graphitbogen *m* graphite arc
Graphitbremsmasse *f* graphite moderator
Graphitelektrode *f* graphite electrode
Graphiteutektikum *n* graphite eutectic
graphithaltig graphitic
graphitieren to blacklead, to coat with graphite, to graphitize
Graphitieren *n* coating with graphite
Graphitierung *f* (Metall) graphitization, graphitizing
graphitisch graphitic
Graphitit *m* (Min) graphitite
Graphitkohle *f* graphite carbon
Graphitlager *n* graphite bearing
Graphitmoderator *m* (Atom) graphite moderator
graphitmoderiert (Reaktor) graphite-moderated
Graphitmühle *f* graphite mill
Graphitöl *n* graphite oil
Graphitölschmierapparat *m* graphite oil lubricating apparatus
Graphitpulver *n* graphite powder
Graphitpyrometer *n* graphite pyrometer
Graphitreaktor *m* graphite[-moderated] reactor
graphitreich rich in graphite
Graphitrohr *n* graphite tube
Graphitsäure *f* graphitic acid
Graphitschicht *f* graphite layer
Graphitschmelztiegel *m* graphite crucible
Graphitschmiere *f* graphite lubricant
Graphitschmiermittel *n* graphite lubricant
Graphitschmierung *f* graphite lubrication
Graphitstaub *m* powdered graphite
Graphitstift *m* black lead, graphite pencil, lead pencil
Graphitthermometer *n* graphite pyrometer
Graphittiegel *m* graphite crucible
Graphitüberzug *m* graphitic film
Graphitwasser *n* graphite water
Graphitwiderstandsofen *m* graphite resistance furnace

Grasbakterium *n* bacillus of timothy grass
grasgrün grassgreen
Grasöl *n* grass oil, citronella oil
Grat *m* (Bergrücken) ridge, (Gewinde) featheredge, (Gieß) fin, (Kante) edge, (Schmiedegrat) flash, (Techn) burr
Gratbildung *f* formation of a shoulder
Grateinguß *m* circular gate runner
Gratiolin *n* gratiolin
Gratiolirhetin *n* gratiolirhetine
Gratiosolin *n* gratiosolin
Gratlinie *f* (Geom) edge of regression, (z.B. Preßgratlinie) flash line
Gratring *m* flash ring
grau gray, grey
graublau gray-blue
graubraun gray-brown, dun-colored
Graueisen *n* gray pig iron
Graueisenerz *n* (Min) marcasite, white iron pyrite
Grauerz *n* (Min) galena
Graufäule *f* gray rot
Graufahlerz *n* (Min) chalcocite, gray copper ore
graufarben gray-colored
graufarbig gray-colored
graugelb grayish yellow
Graugießerei *f* gray iron foundry
Graugolderz *n* (Min) nagyagite
Graugültigerz *n* (Min) gray copper ore, tetrahedrite
Grauguß *m* gray cast iron, pig iron casting
Graugußlöten *n* cast iron brazing
Graukalk *m* gray chalk, gray lime, pyrolignite of lime
Graukeil *m* (Opt) neutral wedge
Graukobalterz *n* (Min) gray cobalt ore, jaipurite
Graukupfererz *n* (Min) chalcocite, tennantite, tetrahedrite
Graulit *m* (Min) graulite
Graumanganerz *n* (Min) gray manganese ore, manganite
Graupe *f* grain
Graupel *f* soft hail
Graupenerz *n* (Min) granular ore
Graupenkobalt *n* (Min) cobalt pyrites, native cobalt diarsenide, smaltine, smaltite
Graupenschörl *m* (Min) aphrizite
Grauschimmel *m* grey mold
grauschwarz grayish black
Grausilber *n* (Min) gray silver, silver carbonate
Grauspießglanz *m* (Min) antimonite, gray antimony ore, stibnite
Grauspießglanzerz *n* (Min) antimonite, antimony blende, antimony glance
Graustein *m* (Min) graystone
Graustrahler *m* grey body emitter
grauviolett gray-violet
Grauwacke *f* (Geol) graywacke, transitional rock

Grauwackenkalkstein *m* (Geol) transition limestone
Grauwackenquarz *m* (Min) quartzite
Grauwackensandstein *m* (Geol) trap sandstone
grauweiß gray-white, gypseous
Graveolin *n* graveoline
Graveur *m* engraver
gravieren to engrave
Gravierfräsmaschine *f* engraving machine
Gravierung *f* embossing
Gravierungstiefe *f* non-skid depth
Gravimeter *n* (Phys) gravimeter
Gravimetrie *f* gravimetry
gravimetrisch gravimetric, gravimetrical
Gravitation *f* gravitation, gravity
Gravitationsanziehung *f* gravitational attraction
Gravitationsdrehwaage *f* gravitation torsion balance
Gravitationserregung *f* gravitational excitation
Gravitationsfeld *n* gravitational field
Gravitationsfluß *m* gravitational flux
Gravitationsgesetz *n* law of gravitation
Gravitationskonstante *f* gravitational constant
Gravitationskraft *f* gravitation
Gravitationspotential *n* gravitation potential
Gravitationstheorie *f* theory of gravitation
Gravitationsverschiebung *f* **nach Rot** (Spektr) gravitational red shift
gravitieren (Phys) to gravitate
Gravur *f* engraving
Greenockit *m* (Min) greenockite, native cadmium sulfide
Greenovit *m* (Min) greenovite
Greifbagger *m* automatic grab, clam-shell dredge
Greifer *m* grab, gripping device, **ferngesteuerter** ~ remote manipulator
Greiferbügel *m* bucket grab
Greiferkran *m* grabbing crane
Greifer-Wippkran *m* grabbing [luffing] crane
Greiffähigkeit *f* gripping power
Greifkran *m* crane with grab gear
Greifstein *m* (Min) griphite
Greifvorrichtung *f* catcher
Greifwerkzeug *n* gripping device or tool
Greifzange *f* gripping tongs
grell bright, dazzling, glaring
Grenadin *n* grenadine
Grenzalbedo *f* (Opt) albedo limiting value
Grenzamplitude *f* oscillation amplitude limit
Grenzbedingung *f* boundary condition, limiting condition
Grenzbelastung *f* limit load
Grenzbeweglichkeit *f* limiting mobility
Grenzbiegespannung *f* flexural strength at maximum deflection, ultimate flexural stress
Grenze *f* border, boundary, frontier, limit, (Beschränkung) limitation, ~ **des Auflösungsvermögens** (Opt) limit of resolution

grenzen to border, to limit
Grenzenergie *f* boundary energy
grenzenlos infinite, unbounded, unlimited
Grenzfall *m* borderline case, limiting case
Grenzfläche *f* boundary layer, boundary surface, contact surface, surface area, (Krist) interface, **spezifische** ~ specific area of interface
grenzflächenaktiv active at the interface
Grenzflächenaufladung *f* (Phys) charge separation of the interface
Grenzflächenbedingung *f* boundary condition
Grenzflächendiffusion *f* boundary diffusion, interfacial diffusion
Grenzflächenenergie *f* interfacial surface energy
Grenzflächenerscheinung *f* interfacial phenomenon
Grenzflächenpolarisation *f* interfacial polarization
Grenzflächenpotential *n* interface potential
Grenzflächenreibung *f* boundary layer friction
Grenzflächen-Reibungsarbeit *f* interfacial work
Grenzflächenspannung *f* interfacial tension
Grenzflächenwelle *f* boundary wave
Grenzflächenwinkel *m* interfacial angle
grenzflächenwirksam interface-active
Grenzform *f*, **mesomere** ~ contributing form
Grenzfrequenz *f* limiting frequency, cut-off frequency, dispersional frequency
Grenzgebiet *n* border region, limiting area
Grenzgeschwindigkeit *f* limiting velocity
Grenzkohlenwasserstoff *m* alkane, paraffin, saturated hydrocarbon
Grenzkonzentration *f* limiting concentration
Grenzkurve *f* limit curve
Grenzleitfähigkeit *f* limiting conductivity
Grenzlinie *f* boundary line, limiting line, line of demarcation
Grenzmaß *n* limiting size, limit size, maximum diameter
Grenzorbital *n* frontier orbital
Grenzpotential *n*, **lichtelektrisches** ~ photoelectric limiting potential
Grenzradius *m* boundary radius
Grenzschicht *f* boundary layer
Grenzschichtablösung *f* boundary break-away separation, layer separation
Grenzschichtbeeinflussung *f* boundary-layer influence
Grenzschichttheorie *f* boundary layer theory
Grenzschraube *f* setscrew, stop screw
Grenzschwingung *f* marginal vibration
Grenzspannung *f* limiting stress
Grenzstrahl *m* limiting ray
Grenzströmung *f* boundary flow, limiting flow
Grenzstrom *m* (Elektr) diffusion current
Grenzstromdichte *f* limiting current density
Grenzstruktur *f*, **mesomere** ~ resonating structure
Grenzstrukturen *pl* resonance structures

Grenztemperatur *f* limiting temperature
Grenzübergang *m* (Math) limiting process, passage to the limit
Grenzverbindung *f* terminal compound, terminal member
Grenzviskositätszahl *f* Staudinger index
Grenzwelle *f* boundary wave, threshold wave
Grenzwellenlänge *f* cut-off wavelength, wavelength limit
Grenzwert *m* (Math) limes, limit, (Techn) limiting value, end value, **lichtelektrischer** ~ photoelectric threshold
Grenzwerte *pl* range finding data
Grenzwertsatz *m* (Math) limit theorem
Grenzwiderstand *m* boundary resistance
Grenzwinkel *m* limiting angle, critical angle
Grenzzustand *m* limiting condition, limiting state
Grieben *pl* sieve residues
Griesheimer Rot *n* Griesheim red
Grieß *m* semolina, coarse powder, coarse sand, (Techn) granules, gravel, grits
grießeln to become granular, to crumble
Grießholz *n* nephritic wood
grießig grainy, granular, gravelly, gritty, sandy
Grießkohle *f* dust coal, pea coal, small coal, smalls
Grießwurzel *f* (Bot) pareira root
Grifa *n* grifa, apyron, hydropyrine, lithium acetyl salicylate, litmopyrine, tyllithin
Griff *m* handle
Griffel *m* style, stylus, (Bot) pistil, style
Griffelschiefer *m* pencil slate
Griffigkeit *f* grip
Griffithit *m* (Min) griffithite
Griffloch-Tasche *f* **mit Bodenfalt** bag with punched hand grip and gussetted bottom
Griffplatte *f* dial disk, holding plate
Griffstopfen *m* grip stopper
Grignardsche Reaktion *f* Grignard reaction
Grignard-Verbindung *f* Grignard compound, organo-magnesium compound
Grilon *n* grilon
Grindelienkraut *n* (Bot) grindelia
Grindelol *n* grindelol
Grindwurzel *f* (Bot) bitterdock root, lapathy root
Grippe *f* (Med) influenza
Grippevirus *m* influenza virus
Griqualandit *m* (Min) griqualandite
Grisan *n* grisan
Griseofulvin *n* griseofulvin
Griseofulvinsäure *f* griseofulvic acid
grob coarse, rough
Grobabstimmung *f* (Radio) coarse tuning
Grobätzung *f* coarse etching
Grobbetrieb *m* coarse adjustment
Grobblech *n* heavy plate, thick plate, thick sheet iron

grobbrechen to break up into pieces, to grind coarsely
Grobbrecher *m* primary crusher
Grobbruch *m* coarse granulation
grobdispers coarsely disperse, low disperse
grobdrahtig coarse-threaded
Grobe *n* (Siebüberlauf) shorts
Grobeinstellung *f* coarse adjustment
Grobeisen *n* large iron bars, blooms
Grobeisenstraße *f* big mill, breaking-down mill
grobfädig coarse-threaded
grobfaserig coarse-fibered
Grobfeile *f* coarse file
Grobflotation *f* roughing flotation
Grobgefüge *n* coarse or gross structure, macrostructure
grobgepulvert coarsely powdered, coarsely crushed, coarsely ground
Grobgewicht *n* gross weight
Grobkalk *m* coarse limestone
grobkörnig coarse-grained, coarsely granular, large-grained
Grobkorn *n* coarse grain
Grobkorneisen *n* coarse grained iron
grobkristallin coarse-crystalline
Groblehre *f* plate gauge
grobporig coarsely porous
Grobpumpleitung *f* roughing pump line
Grobregelung *f* coarse control
Grobreinigung *f* primary cleaning
Grobsand *m* gravel
grobschleifen to grind roughly
Grobsieb *n* wide-meshed sieve
Grobspiegeleisen *n* coarsely crystalline pig iron, coarse spiegeleisen
Grobstrecke *f* big mill, breaking-down mill train
Grobstrukturanalyse *f* macro-structure analysis
Grobstrukturuntersuchung *f* macroscopy
Grobteilung *f* rough scale
Grobvakuum *n* coarse vacuum
Grobwäsche *f* heavy laundering
grobzerkleinern to crush, to coarse crush, to coarse grind
Grobzerkleinerung *f* coarse grinding or crushing
Grobzerkleinerungsmühle *f* crushing mill
Grochauit *m* (Min) grochauite
Groddeckit *m* (Min) groddeckite
Grönhartin *n* groenhartine
Größe *f* (Ausdehnung) dimension, extent, magnitude, (Höhe) height, tallness, (Nummer) size, (Wert) value, **geregelte** ~ controlled variable, **gerichtete** ~ directed quantity, oriented quantity, **thermodynamische** ~ thermodynamic quantity, **unbekannte** ~ unknown quantity, **zu vernachlässigende** ~ negligible quantity
Größenänderung *f* dimensional change
Größenfaktor *m* size factor
Größenklasse *f* magnitude

Größenklassendiagramm *n* magnitude diagram
Größenmaß *n* dimension
Größenordnung *f* order of magnitude
Größenverhältnis *n* proportion of dimensions, proportion of sizes
Größenverteilung *f* size distribution
Größenzunahme *f* increase in size
größtenteils for the most part
Größtwert *m* maximal value
Groppit *m* (Min) groppite
Groroilith *m* (Min) groroilite
Gros *n* gross, twelve dozen
Großabbau *m* large scale workings
Großanlage *f* plant unit
Großapparat *m* large scale plant unit
Großbessemerei *f* converting in large converters
großblätterig large-leaved
Großfeldabschirmung *f* broadbeam shielding
Großfeldabsorption *f* (Atom) broad beam absorption
Großfeldmessung *f* broadbeam measurement
Großhandel *m* wholesale trade
Großindustrie *f* big industry
Großkern *m* macronucleus
Großmaschinenbau *m* heavy engineering
Großproduktion *f* large-scale production
Großraumbau *m* open planning
Großraumbüro *n* open-plan or open-style office
Großrohrleitung *f* large diameter pipeline
Großschaufelradbagger *m* high-capacity bucket-wheel excavator
Großstruktur *f* (Geol) geotectonic structure
Großtat *f* great achievement
großtechnisch on an industrial scale
Großteil *m* greater part, majority
Grossular *m* (Min) grossularite, gooseberry stone, green garnet
Großversuch *m* large-scale test, pilot plant
Großversuchsmaßstab *m* pilot plant scale
Großwinkelkorngrenze *f* large angle grain boundary
Grothit *m* (Min) grothite
Groveelement *n* grove cell
Grubenausbau *m* drift construction, supports
Grubenausbauprofil *n* section for shaft building
Grubenbau *m* mining
Grubenbewetterung *f* (Bergb) mine ventilation
Grubenbrand *m* (Bergb) pit fire
Grubendampf *m* choke damp
Grubenerz *n* rough ore from the mine
Grubenexplosion *f* (Bergb) mine explosion
Grubenfahrzeug *n* mining vehicle
Grubengas *n* mine gas, choke damp, fire damp, marsh gas, methane
Grubengasanzeiger *m* methanometer
Grubengasmeßgerät *n* methanometer
Grubengasprüfer *m* firedamp tester
Grubengerbung *f* pit tannage
Grubengut *n* minerals

Grubenhelm — Grundlack

Grubenhelm *m* pit helmet
Grubenkalk *m* pit-lime
Grubenkies *m* pit gravel
Grubenklein *n* fines, lump ore, smalls
Grubenkohle *f* pitcoal
Grubenlampe *f* miner's lamp
Grubensand *m* pit sand
Grubenschiene *f* mining rail
Grubenschwarz *n* mineral black
Grubenstaub *m* mine dust
Grubenstempel *m* pitprop
Grubenverkohlung *f* pit burning
Grubenwasser *n* pit water
Grubenwetter *n* (Bergb) mine damp
Grude *f* embers, hot ashes
Grudeherd *m* hearth heated with hot ashes
Grudekoks *m* coke breeze, granular coke from lignite, lignite coke, semicoke
Grübchenbildung *f* (Korr) pitting
Grün *n* green [color]
Grünalgen *pl* (Bot) green algae
grünblau greenish blue
Grünbleierz *n* (Min) green lead ore, [green] mimetite, green pyromorphite, mimetesite
grünbraun greenish brown
gründlich profound[ly], thorough[ly]
Gründüngen *n* green manuring, manuring with vegetables
Gründünger *m* green manure, vegetable manure
Grüneisenerde *f* (Min) dufrenite, green iron ore
Grüneisenerz *n* (Min) dufrenite
Grünerde *f* (Min) green earth, celadonite, glauconite
Grünfuttertrockner *m* green fodder drier
grüngelb greenish yellow
grüngrau greenish gray
Grünkalk *m* gas lime
grünlich greenish, virescent
Grünmalz *n* (Brau) green malt
Grünrost *m* cupric acetate, cupric carbonate, verdantique, verdigris
grünrostig aeruginous
Grünsalz *n* greensalt
Grünsand *m* glauconite, greensand
grünsandhaltig glauconitiferous
Grünschiefer *m* (Min) green slate
grünschwarz greenish black
Grünspan *m* copper rust, verdigris, **blauer** ~ blue verdigris
grünspanähnlich aeruginous, eruginous
Grünspanblumen *pl* crystallized verdigris, crystals of verdigris
Grünspat *m* (Min) diopside variety, malacolite
Grünstein *m* (Min) diabase, green porphyry, greenstone, **basaltischer** ~ aphanite
Grünsteinschiefer *m* (Min) diabase, schistous diorite
grünstichig greenish
Grünvergoldung *f* green gilding

Grütze *f* grits, groats
Grund *m* (Boden) bottom, (Erdboden) ground, soil, (Grundlage) base, basis, foundation, (Ursache) case, reason
Grundablaß *m* bottom discharge conduit
Grundanstrich *m* first coat, ground color, primer, priming coat, undercoat
Grundausrüstung *f* basic equipment
Grundbaurahmen *m* foundation frame
Grundbaustein *m* fundamental building block, fundamental structural unit
Grundbegriff *m* basic concept
Grundbestandteil *m* basic component, primary component, (Chem) element
Grundbrett *n* base board, mounting panel
Grundchemikalien *pl* basic chemicals
Grundeigenschaften *f pl* fundamental properties
Grundeinheit *f* fundamental unit
Grundeis *n* ground ice
Grundemail *n* base enamel
Grundenergiequelle *f* basic power source
Grundentwurf *m* basic design
Grunderfindung *f* basic invention
Grundfarbe *f* bottom color, priming color
Grundfirnis *m* priming varnish
Grundfläche *f* basal surface, floor space, [surface] area, (Math) base, basis
Grundform *f* basic form
Grundfrequenz *f* fundamental frequency
Grundgebiet *n* basic region
Grundgedanke *m* basic idea
Grundgefüge *n* structure
Grundgerät *n* basic instrument
Grundgesetz *n* fundamental law
Grundgestein *n* (Geol) bedrock, primary rock
Grundgestell *n* ground frame
Grundgewebe *n* (Histol) stroma, ~ **für Nadelvliesbodenbelag** carrier material for needled felt floor-covering, ~ **für Teppiche** carpet backing
Grundgitter *n* (Krist) fundamental lattice
Grundgleichung *f* basic equation
Grundieranstrich *m* first coat, priming coat
grundieren to prime, to give the first coat, to ground, to stain, (Buchdr) to size
Grundierfarbe *f* primer, priming paint
Grundierfirnis *m* priming varnish, filler
Grundierlack *m* primer
Grundiermittel *n* primer, wash primer
Grundiersalz *n* preparing salt, sodium stannate
Grundierschicht *f* priming coat
grundiert prime-coated
Grundierung *f* first coat, priming coat, bottom layer, impregnation, primer, undercoat, ~ **für Weich-PVC** key coat
Grundkapitel *n* capital stock
Grundkegel *m* pitch cone
Grundkurs *m* basic course
Grundlack *m* primer

338

Grundlackierung f base coating
Grundladung f base charge
Grundlage f base, foundation
Grundlagenforschung f basic research, fundamental research
Grundlager n main bearing
grundlegend fundamental, basical[ly], fundamental[ly]
Grundlegierung f base alloy
Grundlehre f fundamental doctrine
Grundlinie f basis, (Math) base line
Grundliniendrift f base drift
grundlos unfounded, baseless, causeless
Grundmauer f foundation [wall]
Grundmischung f base mix, master batch
Grundnahrungsmittel n staple food
Grundöl n base oil
Grundoperation f unit operation
Grundpartikel n basic particle
Grundplatte f base plate, base board, bed plate, bottom plate
Grundprinzip n basic pattern
Grundpunkt m basic fundamental point, (einer Linse, Opt) cardinal point (of a lens)
Grundregel f fundamental rule, principle, (Axiom) axiom
Grundriß m outline, ground plan, ground sketch, horizontal section, layout
grundsätzlich fundamental, ~ **verschieden** fundamentally different, diametrically opposed
Grundsatz m law, principle, (Axiom) axiom
Grundschaltung f basic or fundamental circuit
Grundschuß m ground weft
Grundschwingung f fundamental oscillation, fundamental vibration, (Spektr) fundamental band
Grundspannung f fundamental voltage, first harmonic voltage
Grundspektrum n fundamental spectrum, persistent spectrum
Grundstärke f **des Laufstreifens** undertread
Grundstein m (Arch) cornerstone
Grundstellung f normal position
Grundstoff m (Element) element, (Rohstoff) basic material, raw material
Grundstoffindustrie f basic materials industry, primary industry
Grundstoffwechsel m basal metabolism
Grundstrahlung f, **kosmische** ~ cosmic-ray background
Grundstrich m base coat, undercoat, ~ **mit porenfüllenden Eigenschaften** primer sealer
Grundstrom m fundamental current, background current, first harmonic current
Grundsubstanz f basic substance
Grundtranslationsvektor m unit cell vector
Grundumsatz m (Med) basal metabolic rate

Grundumsatzbestimmung f determination of the basal metabolic rate
Grundventil n foot valve
Grundverfestigung f basic workhardening
Grundversuch m fundamental experiment, fundamental test
Grundvorschub m basic chart speed
Grundwasser n ground water, underground water
Grundwasserspiegel m level of ground water
Grundwerkstoff m base material
Grundwert m basic or fundamental value
Grundzahl f base number, prime number, unit, (Kardinalzahl) cardinal number
Grundzustand m basic state, ground state
Grunerit m (Min) grunerite
Grunlingit m (Min) grunlingite
Gruppe f group, series, **auxochrome** ~ auxochrome group, **bathochrome** ~ bathochrome group, **chromophore** ~ chromophoric group, **eigenschaftsbestimmende** ~ (funktionelle) functional group, **hydrophile** ~ hydrophilic group, **hydrophobe** ~ hydrophobic group, **hypsochrome** ~ hypsochrome group, **polare** ~ hydrophilic group, polar group, **prosthetische** ~ prosthetic group, **unpolare** ~ non-polar group
Gruppenaktivierung f group activation
Gruppenalgebra f group algebra
Gruppenauswahl f (Statist) stratified sampling
gruppenfremd (Blut) incompatible
Gruppengeschwindigkeit f group velocity
gruppengleich (Blut) of the same group
Gruppenreagens n group reagent
Gruppentransferreaktion f group-transfer reaction
Gruppierung f, **dissymmetrische** ~ (Stereochem) dissymmetry grouping
Grus m breeze, fines, slack, slack coal, small coal
grushaltig containing coal dust
Gruskakao m cocoa husks
Gruskohle f slack, slack coal, small coal
Guacin n guacin
Guacoblätter n pl (Bot) guaco leaves
Guadalcazarit m (Min) guadalcazarite
Guaethol n guaethol, ajacol, ajakol, guaethol, thanatol
Guajac m guaiacum, gum guaiac
Guajacessenz f guaiacin
Guajacharz n guaiac resin
Guajacharzlösung f guaiac tincture
Guajacharztinktur f tincture of guaiac resin
Guajacholz n box-wood, guaiacum wood, pock wood
Guajacholzöl n guaiac wood oil
Guajacin n guaiacin
Guajacinsäure f guaiacic acid

Guajacöl *n* guaiac, guaiacum oil
Guajacol *n* 1-hydroxy-2-methoxybenzene, guaiacol, o-hydroxyanisole
Guajacolbenzoat *n* guaiacyl benzoate, benzosol
Guajacolcarbonat *n* guaiacol carbonate
Guajacolphosphat *n* guaiacyl phosphate
Guajacolphthalein *n* guaiacol phthalein
Guajacolsalicylat *n* guaiacyl salicylate
Guajaconsäure *f* guaiaconic acid
Guajacseife *f* guaiac soap
Guajactinktur *f* guaiac tincture
Guajadol *n* guaiadol
Guajak[harz]säure *f* guaiacic acid, guaiaretic acid
Guajan *n* guaiane
Guajaretsäure *f* guaiacic acid, guaiaretic acid
Guajazulen *n* guaiazulene
Guajol *n* guaiol, 2-methyl-2-butenal, tiglaldehyde, tiglic aldehyde
Guanamin *n* guanamine
Guanase *f* (Biochem) guanase
Guanazin *n* guanazine
Guanazol *n* guanazole
Guanazyl *n* guanazyl
Guanidin *n* guanidine, aminomethanamidine, carbamidine, uramine
Guanidinbase *f* guanidine base
Guanidincarbonat *n* guanidine carbonate
Guanidino- guanidino
Guanidinsulfat *n* guanidine sulfate
Guanidinsulfocyanat *n* guanidine thiocyanate
Guanidinthiocyanat *n* guanidine thiocyanate
Guanidylessigsäure *f* guanidylacetic acid
Guanin *n* guanine
Guaninnucleotid *n* guanine nucleotide
Guaninpentosid *n* guanine pentoside
Guanit *m* (Min) guanite
Guano *m* guano, bird manure
guanoführend guaniferous
Guanomineral *n* guanomineral
Guanosin *n* guanosin[e]
Guanosindiphosphat *n* guanosine diphosphate, GDP
Guanosinmonophosphat *n* guanosine monophosphate, GMP, **zyklisches** ~ cyclic guanosine monophosphate
Guanosinphosphorsäure *f* guanosinphosphoric acid, guanylic acid
Guanosintriphosphat *n* guanosine triphosphate, GTP
Guanosuperphosphat *n* guano superphosphate
Guanovulit *m* (Min) guanovulite
Guanyl- guanyl
Guanylglycin *n* guanylglycine
Guanylguanidin *n* guanylguanidine, biguanide
Guanylharnstoff *m* guanylurea, carbamylguanidine, dicyandiamidine
Guanylsäure *f* guanylic acid
Guanylthioharnstoff *m* guanylthiourea

Guarana *f* (Bot) Brazilian cocoa, guarana
Guaranin *n* guaranine
Guarinit *m* (Min) guarinite
Gülle *f* liquid manure
gültig valid
Gültigkeit *f* validity
Gültigkeitsbereich *m* region of validity
Gültigkeitsgrenze *f* limit of validity
Gürtel *m* belt, girdle, (Faß) hoop, (Geogr) zone
Güte *f* quality
Güteanforderung *f* requirement as to quality
Gütedaten *n pl* efficiency data
Güteeigenschaft *f* property influencing quality, quality
Gütefunktion *f* (in Statistik) power function (in testing hypothesis)
Gütegrad *m* degree of quality
Gütenorm *f* quality standard
Güteprüfung *f* quality test
Gütesteigerung *f* improvement of quality
Gütestufe *f* grade
Gütetest *m* quality test
Güteüberwachung *f* quality control
Gütevorschrift *f* buyer's specification, specification of quality
Gütewert *m* efficiency value
Gütezahl *f* quality factor, index value
Gütezeichen *n* quality seal, standardization mark
Güteziffer *f* quality index, (Dest) efficiency factor
Guignetgrün *n* Guignet's green, emerald green, viridian
Guineagrün *n* guinea green
Guineapfeffer *m* paradise grains
Guitermanit *m* (Min) guitermanite
Gulit *m* gulitol
Guloheptonsäure *f* guloheptonic acid
Gulonsäure *f* gulonic acid
Gulose *f* gulose
Guloson *n* gulosone
Guluronsäure *f* guluronic acid
Gummi *n* caoutchouc, (Bot, Klebstoff) gum, mucilage, (natürliches) rubber, india rubber, **geschwefeltes** ~ vulcanized india rubber, **regeneriertes** ~ regenerated rubber, **vulkanisiertes** ~ vulcanized rubber
Gummiabfall *m* rubber scrap
Gummiaderleitung *f* rubber-covered cable
Gummianimé *n* animé gum, animi resin
Gummiarabicum *n* gum arabic
gummiartig rubber-like, elastomeric, gum-like, gummy, rubbery
Gummiartikel *m pl*, **technische** ~ mechanical rubber goods
Gummiauflösung *f* rubber solution
Gummiball *m* rubber ball
Gummiband *n* elastic, elastic band, rubber band
Gummibandweberei *f* elastic web manufacture

Gummibereitungsmaschine *f* indiarubber machine
Gummibleispat *m* (Min) native aluminum lead phosphate, plumbogummite
Gummidichtung *f* rubber gasket, rubber packing, rubber seal
Gummidichtungsring *m* rubber gasket
Gummidruck *m* flexographic printing
Gummidruckfarben *pl* flexographic printing inks
Gummielasticum *n* gum elastic
gummieren to gum, to size, (mit Gummi imprägnieren) to rubberize
Gummierkalander *m* impregnating calender, rubberizing machine, sizing calender
Gummiersatz *m* rubber substitute
gummiert (klebend) gummed, (mit Gummi versehen) rubber-coated, rubber-lined
Gummierung *f* rubber-coating, rubber proofing, (klebende Fläche) gummed surface, (Klebstoff) gumming
Gummierz *n* (Min) gummite
Gummifaden *m* elastic thread, rubber thread
Gummifarben *pl* rubber colors
Gummifirnis *m* rubber varnish
Gummiformteil *n* molded rubber part
gummigefedert rubber-cushioned
Gummigift *n* rubber poison
Gummigutt *n* (Pharm) gamboge
gummihaltig containing rubber, containing gum, gummy
Gummihandschuh *m* rubber glove
Gummiharz *n* rubber resin
gummiharzig gumresinous
gummiisoliert rubber insulated
Gummiisolierung *f* rubber insulation
Gummikappe *f* rubber cap
Gummikissen *n* rubber cushion, forming pad, pressure pad, rubber pad
Gummikitt *m* india rubber cement
Gummikleber *m* rubber adhesive
Gummikneter *m* rubber kneader
Gummilack *m* rubber varnish, shellac
Gummiladanum *n* ladanum
Gummilinse *f* (Opt) variable focus lens, zoomar lens
Gummilösung *f* rubber solution, gum solution, mucilage
Gummimanschette *f* rubber gasket, rubber ring, rubber sleeve
Gummimembran *f* rubber diaphragm
Gummimembrankammer *f* rubber diaphragm chamber
Gummimetallbindung *f* rubber bonded to metal
Gummimetallverbindung *f* rubber-metal bond
Gummimilch *f* latex
Gummimuffe *f* rubber sleeve
Gummipackung *f* rubber jointing, rubber packing
Gummipapier *n* gumpaper

Gummipflaster *n* (Pharm) gummed diachylon
Gummipfropfen *m* rubber plug, rubber stopper
Gummiplatte *f* rubber plate
Gummipreßverfahren *n* pressure pad forming
Gummiprüfgerät *n* rubber testing equipment
Gummiquetschwalze *f* roller squeegee
Gummiriemen *m* rubber belt
Gummiring *m* india rubber ring
Gummirohr *n* rubber tube
Gummisackverfahren *n* rubber bag molding, vacuum bag molding
Gummisäure *f* arabic acid, arabitic acid, d-tetrahydroxyvaleric acid
Gummisauger *m* rubber suction ball
Gummischeibe *f* rubber washer
Gummischlauch *m* rubber hose, rubber tubing
Gummischleim *m* [acacia] mucilage
Gummischürze *f* rubber apron
Gummispat *m* (Min) native aluminum lead phosphate, plumbogummite
Gummistöpsel *m* rubber stopper
Gummistoff *m* gummite
Gummistopfen *m* rubber stopper
Gummit *n* gummite, gummy matter, rubber cloth
Gummitragant *m* (Med) gum tragacanth
Gummituch *n* rubber blanket
Gummiverbindung *f* rubber adapter
Gummiwachs *n* gumwax
Gummiwasser *n* gum water
Gummiwischer *m* rubber policeman
Gummizucker *m* arabinose
Gumnit *n* gumnite
Gundelrebe *f* (Bot) ground ivy
Gundermannöl *n* ground ivy oil
Gur *f* guhr, kieselguhr
Gurdynamit *n* guhr dynamite
Gurgelrohr *n* throttle pipe
Gurgelwasser *n* (Pharm) gargle
Gurhofian *m* gurhofian
Gurjunbalsam *m* gurjun [balsam]
Gurjunbalsamöl *n* gurjun balsam oil
Gurjunen *n* gurjunene
Gurjunenalkohol *m* gurjunene alcohol
Gurjunenketon *n* gurjunene ketone
Gurjuresen *n* gurjuresene
Gurtband *n* rubber belt, webbing, webs
Gurtförderband *n* belt conveyor
Gurtförderer *m* belt conveyor
Gurttaschenförderer *m* pocket belt conveyor
Gurttransporteur *m* belt conveyor
Gurttrommel *f* bobbin
Guß *m* casting, pouring, (Werkstoff) cast iron, ~ **in grünem Sand** greensand casting, **feuerbeständiger** ~ heatproof cast iron, **säurebeständiger** ~ acidproof cast iron, **schmiedbarer** ~ malleable cast iron, tempered casting, **schwammiger** ~ spongy cast iron, **steigender** ~ bottom casting, bottom pouring, **undichter** ~ porous cast iron

Gußaluminium *n* cast aluminum
Gußanode *f* cast anode
Gußarbeit *f* casting
Gußart *f* kind of casting
Gußasphalt *m* mastic asphalt
Gußbeton *m* cast concrete, concrete sufficiently wet to flow
Gußblase *f* blowhole, air bubble, flaw
Gußblei *n* cast lead
Gußblock *m* ingot
Gußbronze *f* cast bronze
Gußdraht *m* cast-steel wire
Gußeisen *n* cast iron, ~ **mit Stahlschrottzusatz** semi-steel, **gares** ~ gray metal, **heiß erblasenes** ~ hot-blast pig iron, **hochwertiges** ~ high-test or high-quality cast iron, **schmiedbares** ~ malleable pig iron, **weißes** ~ malleable hard cast iron, unannealed malleable iron
Gußeisengehäuse *n* cast iron shell
Gußeisengitter *n* cast iron grid
Gußeisenkranz *m* cast iron crown, cast iron flange, cast iron rim
Gußeisensäule *f* cast iron column
Gußeisenscheibe *f* cast iron disc
Gußeisenspäne *pl* cast iron turnings
Gußeisensplitter *m* cast iron chip, cast iron splinter
gußeisern cast iron
gußfähig castable
Gußfehler *m* casting defect, flaw in the casting
Gußfestigkeit *f* cast-iron strength
Gußflasche *f* molding flask
Gußform *f* frame, mold, **eine** ~ **aufstampfen** to ram up a mold
Gußgefüge *n* structure of cast
Gußgehäuse *n* cast [iron] housing
Gußhaut *f* skin of a casting
Gußkern *m* casting core
Gußkokille *f* cast iron mold
Gußlegierung *f* casting alloy
Gußloch *n* casting hole, tapping hole, **das** ~ **verstopfen** to fill up the casting hole
Gußmasse *f* pouring compound
Gußmessing *n* cast brass
Gußmetall *n* cast metal
Gußmodell *n* casting pattern
Gußmörtel *m* casting cement, casting concrete
Gußmutter *f* matrix
Gußnaht *f* casting burr
Gußnarbe *f* fault, flaw
Gußpfanne *f* ladle
Gußsatz *m* set of castings
Gußschaden *m* defective casting, fault, flaw
Gußschale *f* cast iron mold, chill mold
Gußschrott *m* casting scrap, iron shot
Gußspäne *pl* cast iron turnings
Gußspannung *f* casting strain or stress

Gußstahl *m* cast steel, crucible [cast] steel, **schweißbarer** ~ soft cast steel
Gußstahlblech *n* cast steel plate
Gußstahldrahtbürste *f* cast steel wire brush
Gußstahltiegel *m* cast steel crucible
Gußstein *m* drain, sink
Gußstruktur *f* cast-iron structure, macrostructure
Gußstück *n* cast piece, [metal] casting
Gußteil *n* casting
Gußtiegel *m* casting ladle, crucible, skillet
Gußtrichter *m* casting funnel
Gußwachs *n* casting wax
Gußwaren *f pl* castings
Gußwerk *n* cast, casting, cast metalwork
Gußzink *n* cast zinc
Gut *n*, **verhüttbares** ~ product ready for smelting
Gutachten *n* expert opinion, judgement, [legal] advice
Gutfeuchte *f*, **kritische** ~ critical moisture content
Guthaben *n* credit, deposit
Gutschein *m* coupon
Guttapercha *f n* guttapercha, **gereinigtes** ~ refined guttapercha
Guttaperchabereitungsmaschine *f* guttapercha machine
Guttaperchaflasche *f* guttapercha bottle
Guttaperchaform *f* guttapercha matrix
Guttaperchaharz *n* guttapercha resin
Guttaperchakitt *m* guttapercha mastic
Guttaperchalösung *f* guttapercha solution
Guttaperchamischung *f* guttapercha mixture
Guttaperchapapier *n* guttapercha paper, guttapercha sheet
Guttaperchapresse *f* guttapercha covering machine
Gutzeitsche Probe *f* (Anal) Gutzeit test
Guvacin *n* guvacine
Guvacolin *n* guvacoline
Gymnemasäure *f* gymnemic acid
Gymnit *m* (Min) gymnite
Gymnogrammen *n* gymnogrammene
Gynäkologie *f* (Med) gynaecology (Br. E.), gynecology (Am. E.)
Gynaminsäure *f* gynaminic acid
Gynocardiaöl *n* gynocardia oil
Gynocardin *n* gynocardin
Gynocardinsäure *f* gynocardinic acid
Gynocardsäure *f* gynocardic acid
Gynoval *n* gynoval, isobornyl isovalerate
Gyrator *m* (Elektr) gyrator
Gyrilon *n* gyrilone
Gyrofrequenz *f* gyro frequency
Gyrolith *m* (Min) gurolite, gyrolite
Gyrolon *n* gyrolone
gyromagnetisch gyromagnetic
Gyrometer *n* gyrometer, speed meter for liquids

Gyrophorin *n* gyrophorine
Gyrophorsäure *f* gyrophoric acid

Gyrostat *m* gyrostat

H

Haar *n* hair, (Borste) bristle, (Wolle) wool
Haaralaun *m* capillary alum
Haaramethyst *m* (Min) capillary amethyst
Haarbeize *f* depilatory
Haarerz *n* (Min) capillary ore
Haarfarbe *f* color of the hair
Haarfaser *f* (Biol) capillary filament
Haarfestiger *m* hair fixature, hair setting lotion
haarförmig capillary, (Bot) capilliform
Haargewebe *n* hair cloth, structure or texture of hair
Haargold *n* (Min) capillary gold
Haarhygrometer *n* hair hygrometer
haarig hairy, hirsute
Haarkalk *m* hair grout
Haarkanal *m* capillary duct
Haarkeratin *n* hair keratin
Haarkies *m* (Min) capillary pyrites, hair pyrites, millerite
Haarkreuz *n* cross hairs
Haarkupfer *n* capillary copper
Haarnadel *f* hair pin
Haarnadelgalvanometer *n* galvanometer with hairpin shaped-magnet
Haarnickel *n* capillary nickel
Haarparasit *m* (Zool) hair parasite
Haarpflege *f* hair-dressing
Haarpflegemittel *n* (Kosmetik) hair tonic or lotion
Haarriß *m* capillary fissure, craze, hair cracking, hairline crack, microflaw
Haarrißbildung *f* crazing
haarrissig crazed
Haarröhrchen *n* capillary tube
Haarsalbe *f* (Pharm) pomade, pomatum
Haarsalz *n* (Min) hair salt
haarscharf absolutely accurate, precise or correct
Haarschwefel *m* (Min) capillary sulfur, fibrous native sulfur
Haarsieb *n* horse hair sieve
Haarsilber *n* (Min) capillary silver, filamentous natural silver
Haarstein *m* (Min) crystallized hyaline quartz
Haarstrangwurzel *f* (Bot) hair cord root, peucedanum root
Haarstrich *m* hairline
Haarvitriol *n* (Min) capillary epsomite, native vitriol in filaments
Haarwaschmittel *n* shampoo
Haarwasser *n* (Kosm) hair tonic
Haarwurzel *f* root of the hair
Haarzange *f* tweezers
Haarzeolith *m* (Min) fibrous zeolite
HAB Abk. für Homöopathisches Arzneibuch
Haber-Bosch-Verfahren *n* Haber-Bosch process

Habilitation *f* habilitation (formal admission of an academical lecturer into the faculty to which he desires to attach himself)
Habilitationsschrift *f* probationary treatise embodying the results of original research that is submitted to the faculty in order to be recognized as a ‚Privatdozent' at the university
habilitieren to acquire the right of giving lectures at universities as a qualified academic teacher
Habitus *m* (Biol) habit, physical appearance
Habituswirkung *f* habit effect
Hackblock *m* chopping-block
Hackbolzen *m* rag bolt
Hackbrett *n* chopping-board
hacken to chop
Hacker *m* chopper, chipper
Hackfleisch *n* chopped meat, ground meat, minced meat
Hackmaschine *f* chipper, chopper, chopping machine, guillotine cutter, mincer, (Pap) chopping machine
Hackmesser *n* chipper knife, guillotine knife
Haderlumpen *pl* rags
Hadern *pl* (Papier) rags
Haderndrescher *m* (Pap) duster, willow
Hadernkocher *m* (Pap) rag boiler
Hadernschneider *m* (Pap) rag cutter, rag cutting machine
Hadernsortierer *m* (Pap) rag sorter
hadrig deficient for welding
Häcksel *n* chaff, chopped straw
Häkchen *n* little hook, crochet, hooklet
häkeln to crochet
Hälfte *f* half, moiety
Häm *n* heme
Hämafibrit *m* (Min) hemafibrite
Hämagglutination *f* h[a]emagglutination, clotting of the red cells
Hämalbumin *n* hemalbumin
Hämanthamin *n* hemanthamine
Hämanthidin *n* haemanthidine
Hämanthin *n* hemanthine
Hämataminsäure *f* hemataminic acid
Hämatein *n* hematein, haematein
Hämatin *n* hematin, haematin
Hämatinsäure *f* hematinic acid
Hämatit *m* (Min) hematite, bloodstone, haematite, hematite iron, red iron ore, red mine stone
Hämatiteisen *n* hematite iron
Hämatiterz *n* (Min) hematite ore
Hämatitroheisen *n* (Metall) hematite pig iron
Hämatitvorkommen *n* hematite deposit
Hämatogen *n* hematogen
Hämatoidin *n* hematoidin, bilirubin
Hämatokonit *m* hematoconite

Hämatolith *m* (Min) hematolite
Hämatommsäure *f* hematommic acid
Hämatopan *n* hematopan
Hämatoporphyrin *n* hematoporphyrin
Hämatostibiit *m* (Min) hematostibiite
Hämatoxylin *n* hematoxylin, hematine, hematoxylic acid
Hämatoxylinlösung *f* hematoxylin solution
Hämatozyt *m* h[a]ematocyte, h[a]emocyte
Hämatozytologie *f* h[a]emocytology
Hämatozytolyse *f* h[a]ematocytolysis, h[a]emocytolysis
Hämerythrin *n* hemerythrin
Hämiglobin *n* hemiglobin
Hämin *n* hemin, hematin chloride, Teichmann's crystals
hämmerbar (heiß) forgeable, (kalt) malleable
Hämmerbarkeit *f* forgeability, malleability
hämmern to hammer, to forge
Hämoagglutinogen *n* h[a]emagglutinogen
Hämobilirubin *n* h[a]emobilirubin
Hämochinin *n* hemoquinine
Hämochinsäure *f* hemoquinic acid
Hämochromogen *n* hemochromogen[e]
Hämocuprein *n* hemocuprein
Hämocyanin *n* hemocyanin, haemocyanin
Hämogenese *f* h[a]emogenesis, h[a]ematogenesis
Hämoglobin *n* hemoglobin, ~ **enthaltend** hemoglobiniferous
Hämoglobinbestimmung *f* h[a]emoglobin determination, h[a]emoglobinometry
Hämogramm *n* (Blutbild) h[a]emogram
Hämolyse *f* hemolysis
Hämoporphyrin *n* hemoporphyrin
Hämopyrrol *n* hemopyrrole
Hämopyrrolenphthalid *n* hemopyrrolenephthalide
Hämopyrrolidin *n* hemopyrrolidine
Hämopyrrolin *n* hemopyrroline
Hämorrhoidalmittel *n* (Pharm) hemorrhoid cure
Hämosiderin *n* h[a]emosiderin
Hämostyptikum *n* (Pharm) hemostyptic
Hämotoxin *n* (Blutgift) h[a]emotoxin
Hämovanadin *n* hemovanadium
Hämozoon *n* (Blutparasit) h[a]ematozoon, h[a]emozoon
Hämozytogenese *f* (Blutkörperchenbildung) h[a]emocytogenesis
Hämozytolyse *f* (Blutkörperchenauflösung) h[a]emocytolysis
Händler *m* dealer
Hängebahn *f* suspension track
Hängebahnlaufkatze *f* (Techn) overhead conveyer trolley
Hängebalken *m* suspension girder
Hängebock *m* double hanger frame, drop hanger frame
Hängeeisen *n* hanger
Hängegefäß *n* swing bucket

Hängegerüst *n* hanging stage, scaffold
Hängekorbzentrifuge *f* suspended centrifuge
Hängekran *m* overhead crane
Hängekran-Hängebahn-Anlage *f* overhead crane and track
Hängen *n* **der Gicht** hanging of the charge, scaffolding of the charge
hängend hanging, pendent, suspended
Hängering *m* suspension ring
Hängeschlitten *m* suspended knife sledge
Hängetrockner *m* festoon drier, loop drier
Hängewerk *n* truss
härtbar capable of being hardened or cemented, hardenable, temperable, ~ **durch Wärme** thermosetting
Härtbarkeit *f* hardening capacity
Härtbarkeitskurve *f* hardness cooling rate curve
Härte *f* hardness, rigidity, rigor, (Festigkeit) solidity, (Met) temper, ~ **des Stahles** temper of steel, ~ **des Wassers** hardness of water, **passive** ~ passive hardness, wear hardness, **permanente** ~ permanent hardness, **veränderliche** ~ temporary hardness
Härteanlage *f* hardening plant, tempering plant
Härtebad *n* casehardening bath, quenching bath, tempering bath
Härtebeständigkeit *f* retentivity of hardness
Härtebestimmung *f* determination of hardness
Härtebildner *m* hardening constituent
Härteeignungseigenschaft *f* degree of suitability for hardening
Härteeinheit *f* unit of hardness
Härteflammofen *m* tempering flame furnace
Härteflüssigkeit *f* hardening liquid, tempering liquid
Härtegefäß *n* hardening trough
Härtegrad *m* degree of hardness, (Metall) degree of temper, temper, temper of steel
Härtehitze *f* hardening heat
Härtekohle *f* hardening carbon
Härtemesser *m* durometer, hardness gauge, (Durchschlagskraft) penetrometer, (Ritzhärtemesser) sclerometer
Härtemessung *f* determination of hardness, measurement of hardness
Härtemittel *n* hardener, hardening agent, hardening mixture, tempering agent
härten to harden, to set, (Kunststoffe) to cure, to thermoset, (Metall) to quench, to temper, **im Einsatz** ~ to caseharden
Härten *n* hardening, solidifying, (Kunststoffe) curing, (Met) tempering, ~ **des Stahls** hardening of steel, tempering of steel
Härteöl *n* quenching oil, steel-hardening oil, tallow oil
Härteofen *m* hardening furnace, baking oven, furnace for hardening, (im Einsatz) casehardening furnace, (Kunststoffe) curing

Härteofen — Haftspannung

oven, (Metall) tempering furnace, tempering stove
Härtepräparat *n* hardening preparation
Härteprobe *f* hardness test, pressure test
Härteprüfapparat *m* hardness testing apparatus
Härteprüfer *m* hardness tester, durometer
Härteprüfgerät *n* hardness tester, durometer, sclerometer
Härteprüfmaschine *f* hardness testing machine
Härteprüfstab *m* hardness testing rod
Härteprüfung *f* hardness test[ing]
Härteprüfverfahren *n* hardness test procedure
Härtepulver *n* cementation powder, cementing powder, tempering powder
Härter *m* curing agent, hardener, (Kunststoffe) setting agent
Härtereihe *f* scale of hardness
Härteriß *m* fissure, flaw, hardening crack, heat-treatment crack
Härtesalz *n* hardening salt, carburizing salt, salt causing hardness
Härteschmiedefeuer *n* tempering forge
Härteskala *f* hardness scale, scale of hardness
Härtespannung *f* hardening strain
Härtestreuung *f* dispersion of hardness value
Härtetemperatur *f* setting temperature
Härtetiefe *f* depth of hardening
Härtetrog *m* hardening trough
Härteverzug *m* distortion due to hardening
Härtevorgang *m* hardening operation, process of hardening
Härtezahl *f* coefficient of hardness, hardness number
Härtezeit *f* cure time, curing time, setting time
Härtling *m* hard slag
Härtung *f* hardening, (Stahl) tempering, **unterbrochene** ~ interrupted or broken hardening
Härtungsautoklav *m* (Fette) hydrogenerator
Härtungsbiegeprobe *f* bending test in tempered state
Härtungsfähigkeit *f* hardening capacity
Härtungskohle *f* hardening carbon, temper carbon
Härtungsmittel *n* hardening agent
Härtungsspannung *f* hardening strain, internal hardening stress, quenching stress
Härtungsverfahren *n* hardening process, (für Fette) hydrogenation process (of fats)
Härtungsvermögen *n* hardening capacity
Härtungszeit *f* setting period
häufeln to heap, to pile, to stack, to staple
häufen, sich ~ to accumulate, to pile up
häufig frequent
Häufigkeit *f* abundance, frequency
Häufigkeitsanomalie *f* abundance anomaly
Häufigkeitsfaktor *m* (Statist) frequency factor
Häufigkeitsfunktion *f* (Math) frequency function
Häufigkeitskurve *f* (Statist) abundance curve, frequency curve, probability curve
Häufigkeitspolygon *n* (Statist) frequency polygon
Häufigkeitsverhältnis *n* (Statist) abundance ratio
Häufigkeitsverteilungskurve *f* frequency distribution curve
Häufigkeitswert *m* (Statist) frequency value
Häufung *f* accumulation, heaping, piling
Häufungspunkt *m* point of accumulation
Häufungsstelle *f* (Mech) accumulation point
Häutchen *n* film, pellicle
Häutungshormon *n* moulting hormone
Hafer *m* oats
Haferflocken *f pl* oat flakes
Haferkleber *m* gluten of oats
Hafermehl *n* oatmeal
Haferquetsche *f* oat crusher
Haferstärke *f* oat starch
Hafner *m* potter
Hafnererz *n* (Min) potter's ore, fine galena
Hafnium *n* hafnium, (obs) oceanum (Symb. Hf)
Haft *f* (Gewahrsam) custody, arrest
haftbar accountable, liable, responsible
Haftbarkeit *f* liability, responsibility, (Phys) adhesion
Haftbrücke *f* bonding coat
haften (garantieren) to guarantee; (kleben) to adhere, to cling to, to stick, (verantworten) to be liable, to be responsible
Haften *n* (Phys) adherence; adhesion
Haftenbleiben *n* adherence, adhering
haftfähig adhesive, sticky
Haftfähigkeit *f* adherence, adhesion, (Klebrigkeit) tackiness
Haftfestigkeit *f* bonding strength, adhesion, adhesive strength, peel strength (electroplated articles), tenacity, (Klebrigkeit) tackiness
Haftgrund[ier]mittel *n* etching primer, wash primer
Haftgrundmittel *n* wash primer
Haftintensität *f* intensity of adhesion
Haftklebeband *n* pressure sensitive tape
Haftkleber *m* contact adhesive, pressure-sensitive adhesive
Haftkraft *f* adhesion
Haftladung *f* sticking charge
Haftlinse *f* (Opt) contact lens
Haftmasse *f* adhesive
Haftmittel *n* adhesive
Haftnäpfchen *n* retention cup
Haftpflicht *f* liability, responsibility
Haftpotential *n* (Elektr) sticking potential
Haftreibung *f* adhesive friction, cohesive friction, static friction
Haftreibungskoeffizient *m* (Mech) coefficient of static friction
Haftsitz *m* tight fit, wringing fit
Haftspannung *f* adhesive stress, bond stress

Haftstärke *f* degree of adhesion, intensity of retention
Haftstelle *f* point of attachment, (Phys) trap
Haftstellen *f pl* **für Elektronen** *n pl* electron traps
Haftterm *m* adhesion term
Haftthermometer *n* sticking thermometer
Haftung *f* (Phys) adhesion, adsorption, (Verantwortung) liability, ~ **am Erdboden** low center of gravity, stability
Haftvermittler *m* adhesion promoter
Haftvermögen *n* adhesion, adhesive capacity, adhesiveness, adhesive power
Haftwasser *n* connate water
Haftwert *m* bond value
Hagebutte *f* (Bot) hip berry
Hagedorn *m* (Bot) hawthorn
Hagel *m* hail
Hagemannit *m* (Min) hagemannite
Hahn *m* faucet, tap, (Chem) cock, stopcock, (Rohrleitung) plug valve, ~ **mit zwei Wegen** two-way cock, **eingeschliffener** ~ ground-in stopcock, **selbsttätiger** ~ self-acting cock
Hahnbrücke *f* stopcock manifold
Hahnfassung *f* switch socket
Hahnfett *n* stopcock grease
Hahngriff *m* valve handle
Hahnmesser *m* cock meter
Hahnregulierung *f* stopcock control
Hahnschieber *m* plug valve
Hahnschlüssel *m* cock or tap wrench, plug of a cock
Hahnschraube *f* cock pin or nail
Hahnsteuerung *f* regulation by stopcock
Hahnventil *n* cock
Hahnverschluß *m* valve
Haidingerit *m* (Min) haidingerite
Haidinger-Ringe *pl* (Opt) Haidinger fringes
Hainit *m* (Min) hainite
haken to hook, to clamp, to clasp
Haken *m* hook, catch, clasp, ~ **eines Laufkranes** bar of a traveling crane
Hakeneisen *n* hook iron
hakenförmig hooked, hook-like
Hakengeschirr *n* hook fittings
Hakenkette *f* hook link chain, ladder chain
Hakennagel *m* spike
Hakenplatte *f* hook plate
Hakenschloß *n* hook lock
Hakenschlüssel *m* hooked wrench, hook spanner
Hakenschraube *f* bolt hook, clip bolt, screw hook
Hakenzange *f* hook pliers
Halazon *n* (HN) halazone, p-dichlorosulfamyl-benzoic acid
Halbacetal *n* hemiacetal
Halbachse *f* (Geom) semiaxis
Halbaddierwerk *n* half adder
halbaktiv semiactive, semi-reinforcing

Halbalaun *m* impure alum
Halbamplitude *f* semi-amplitude
Halbanthrazit *m* semianthracite
halbautomatisch semiautomatic, semimechanical
halbberuhigt (Metall) balanced, semi-killed
Halbbleiche *f* half bleach
Halbchlorschwefel *m* (obs) sulfur monochloride, sulfur subchloride
Halbchromosom *n* (Biol) chromatid
Halbdrehung *f* half-turn
halbdurchlässig semi-permeable, ~ **verspiegelt** (Opt) half silvered
Halbdurchmesser *m* radius
halbdurchscheinend semi-transparent
halbdurchsichtig semi-transparent
Halbdurchsichtigkeit *f* semitransparency
Halbe *n* half
Halbedelstein *m* semiprecious stone
halbeirund semioval, semiovoid
Halbelement *n* (Elektr) half-cell, half-element, single-electrode system
halbempirisch semiempirical
Halbfabrikat *n* intermediate product, semimanufactured article
Halbfertigerzeugnis *n* semimanufactured product
Halbfertigprodukt *n* semifinished product
Halbfertigwaren *f pl* semifinished goods
halbfest semisolid
halbfett (Buchdr) bold, (Kohle) semi-bituminous
halbflächig (Krist) hemihedral
Halbflächner *m* (Krist) hemihedron
Halbfließbandfertigung *f* semiproduction basis
halbflüssig semifluid, semiliquid
Halbflüssigkeit *f* semiliquidity
halbgar half-cooked, underdone
Halbgas *n* semigas
Halbgasfeuerung *f* semigas furnace
halbgeleimt (Pap) half-sized
halbgesäuert semi-acidified
halbgeschlossen semi-enclosed
halbgeschmolzen semi-fused
halbgesintert half-fused, semisintered
Halbglanz *m* semigloss
Halbgold *n* imitation gold, Mannheim gold, similor
Halbgranit *m* (Min) demigranite
Halbgut *n* tin lead alloy, tin with high lead content
Halbhadernpapier *n* 50% rag-content paper
Halbharz *n* crude resin, resinoid, subresin
Halbholländer *m* (Pap) washer, washing engine, worker
Halbholz *n* split log
Halbhydrat *n* hemihydrate
halbieren to bisect, to halve
Halbierende *f* (Math) s. Halbierungslinie
Halbierungslinie *f* (Math) bisector, bisectric line, bisectrix

Halbierungspunkt *m* point of bisection
Halbkettenantrieb *m* half-track drive
Halbkochen *n* parboiling, partial boiling
Halbkoks *m* semicoke, coalite
Halbkokung *f* partial carbonization of coal, semicoking
Halbkolloid *n* semicolloid
Halbkonserve *f* [semi-]preserve
halbkontinuierlich semibatch, semicontinuous
Halbkreis *m* semicircle
halbkreisförmig semicircular
halbkreisig semicircular
Halbkreuzriemen *m* quarterturn belt
Halbkristall *m* (Krist) semicrystal
halbkristallin semicrystalline
Halbkugel *f* hemisphere
halbkugelförmig hemispherical
Halbkugelgestalt *f* hemispherical shape
halbkugelig hemispherical, semispherical
Halbkupfererz *n* (Min) chalcopyrite
Halbkuppel *f* half dome, vaulted passage
Halblast *f* half load
Halbleinen *n* half-linen
halbleitend semiconducting
Halbleiter *m* (Elektr) semiconductor
Halbleitereigenschaft *f* semiconduction property
Halbleitergleichrichter *m* semi-conductor rectifier
Halbleiteroberfläche *f* semiconductor surface
Halbleiterphotozelle *f* barrier-plane photocell, rectifier photocell
Halbleiterschicht *f* semi-conductor layer
Halbleiterspeicher *m* semiconductor memory
Halbleiterzelle *f* semi-conductor cell
Halbliterkolben *m* (Chem) half-liter flask
halbmahlen to break up roughly, to half-crush, to half-grind
halbmatt semidull, semigloss
Halbmattlack *m* semigloss paint
halbmechanisch semimechanical, semiautomatic
Halbmesser *m* radius, **wirksamer dynamischer** ~ rolling radius, **wirksamer statischer** ~ static loaded radius
Halbmesserlehre *f* radius gauge
Halbmetall *n* semimetal
Halbmetallglanz *m* submetallic luster
halbmetallisch semimetallic
halbmikro- semimicro
Halbmikroanalyse *f* semimicro analysis
Halbmikroapparat *m* semimicro apparatus
halbnaß semiwet
Halbnitril *n* seminitrile
halbnormal seminormal
Halböffnungswinkel *m* (Opt) semiapertural angle
Halbopal *m* (Min) semiopal
Halbparabelfachwerkträger *m* semiparabolic superstructure

Halbperiode *f* halfperiod, alternation, half cycle, semicycle, semioscillation, semiperiod
halbpolar semipolar
Halbportalkran *m* semiportal crane
Halbporzellan *n* hard crockery, feldspathic stoneware, halfchina, semiporcelain
Halbprisma *n* (Krist) hemiprism
halbprismatisch (Krist) hemiprismatic
Halbpyritschmelzen *n* (Metall) semipyritic or partial pyritic process
halbracemisch semiracemic
Halbraum *m* half space
Halbreduktion *f* (Chem, Metall) partial reduction, semireduction
Halbringelektromagnet *m* semiannular electromagnet, semicircular electromagnet
halbrund half-round, semicircular
Halbrundeisen *n* (Techn) half-round iron
Halbrundfeile *f* half-round file
Halbrundkopf *m* buttonhead
Halbrundkopfschraube *f* button head screw
Halbrundschraube *f* roundheaded screw
halbsauer semiacid
Halbscharlachfarbe *f* bastard scarlet
Halbschatten *m* half shade, half tone, partial shadow
Halbschattenapparat *m* half-shade apparatus, half-shadow analyzer, half-shadow apparatus
Halbschattenplatte *f* (Opt) half-shade plate, quarter wave length plate
Halbschattenpolarimeter *n* half-shade polarimeter
Halbschattenpolarisationsapparat *m* half-shadow polarimeter
Halbschattenquarzplatte *f* half-shadow quartz plate
Halbschattensystem *n* half-shadow system
Halbschleife *f* half loop
halbschnittig (Zuck) semi-slicing
Halbschwefeleisen *n* iron hemisulfide
Halbschwefelkupfer *n* copper(I) sulfide, cuprous sulfide
halbselbsttätig semiautomatic, semimechanical
Halbsesselform *f* (Stereochem) half-chair form
Halbspannweite *f* semispan
Halbstahl *m* semisteel, malleable cast iron
halbstarr semirigid
Halbstoff *m* (Pap) half-stuff
Halbstoffholländer *m* (Papier) breaker
Halbstufenpotential *n* halfwave potential
halbtechnisch semi-industrial
Halbtonverfahren *n* halftone process
halbtrocken half dry, semidry
Halbtrockenverfahren *n* (Zement) mixed process
halbvulkanisiert semicured, semivulcanized
Halbwahlsystem *n* semiautomatic system, semimechanical system
Halbwalze *f* granulating roll

Halbwassergas *n* semiwater gas, dowson gas, mixed gas
Halbwattlampe *f* halfwatt lamp
Halbweggleichrichter *m* (Elektr) half-wave rectifier
Halbweggleichrichtung *f* half-wave rectification
halbwegs halfway
Halbwelle *f* half-wave, half cycle, half period
Halbwellenantenne *f* (Elektr) half-wave antenna
Halbwellengleichrichter *m* (Elektr) half-wave rectifier
Halbwertbreite *f* half-power [band] width
Halbwert[s]breite *f* half width, ~ **der Interferenz** (Opt) half-width of interference maxima
Halbwert[s]periode *f* half decay period, half value period
Halbwert[s]schicht *f* half-value layer
Halbwert[s]tiefe *f* half-value depth
Halbwertzeit *f* (Atom) half-life, half-life period, **biologische** ~ biological half-life, **effektive** ~ effective half-life
halbzahlig half-integer, half-integral
Halbzeit *f* half-life, half-life period, half-period
Halbzelle *f* (Elektr) half-cell
Halbzellstoff *m* half-stuff, semi-chemical pulp
Halbzellulose *f* hemicellulose
Halbzeug *n* semifinished goods, semifinished material, semifinished products, semimanufactured goods, semiproducts, (Pap) half-stuff
Halbzeugholländer *m* (Pap) half-stuff engine
Halbzinn *n* tin-lead alloy, tin with high lead content
Halbzirkel *m* semicircle
Halbzwölfflächner *m* (Krist) hemidodecahedron
halbzylindrisch semicylindrical
Halde *f* (Bergb) waste dump, waste heap
Haldenerz *n* waste heap ore
Haldenschlacke *f* dump slag, waste slag
Haldensturz *m* dumping ground
Haldenvermischung *f* mixing stockpiles
Halit *m* (Min) halite, common salt, rock salt
Hall *m* resonance, sound
Hallachrom *n* hallachrome
hallen to sound, to resound, to reverberate
Hallenbauten *pl* bays
Hallerit *m* (Min) hallerite
Halloysit *m* (Min) halloysite, halloylite
Halluzinogen *n* (Pharm) hallucinogen
Halmbruchpilz *m* eye spot fungus
Halochemie *f* halochemistry, chemistry of salts
halochemisch halochemical
Halochromie *f* (Färb) halochromism
Haloformreaktion *f* haloform reaction
Halogen *n* halogen
Halogenalkyl *n* alkyl halide, halogenalkane
Halogenaryl *n* aryl halide
Halogenatom *n* halogen atom
Halogenbenzoesäure *f* halobenzoic acid

Halogenchinon *n* haloquinone, halogenated quinone, halogen-substituted quinone
Halogenderivat *n* halogen derivative
Halogenid *n* halide
halogenidfrei halide-free
Halogenidkristall *m* halide crystal
halogenierbar halogenatable, capable of being halogenated
halogenieren to halogenate
Halogenierung *f* halogenation
Halogenindigo *m* halogenated indigo, halogen-substituted indigo, haloindigo
Halogenkohlenstoff *m* carbon halide
Halogenkohlenwasserstoff *m* halogenated hydrocarbon
Halogenlampe *f* halogen bulb
Halogenleckdetektor *m* halide leak detector
Halogenlöschung *f* halogen quenching
Halogenmetall *n* metal halide, metallic halide
Halogenpyridin *n* halopyridine
Halogenquecksilber *n* mercury halide
Halogensäure *f* halogen acid
Halogenschwefel *m* sulfur halide
Halogensilber *n* silver halide
Halogensubstitution *f* halogen substitution
Halogenüberträger *m* halogen carrier
Halogenür *n* (obs) halide
Halogenverbindung *f* halogen compound
Halogenwasserstoff *m* hydrogen halide
Halogenwasserstoffsäure *f* hydrohalic acid
Halogenwasserstoffspaltung *f* dehydrohalogenation
Halogenwasserstoffverbindung *f* hydrohalogen compound
Halogenzählrohr *n* halogen[-quenched] counter
Halographie *f* halography
Halohydrin *n* halohydrin
Haloid *n* halide, haloid, haloid salt
Haloidderivat *n* haloid derivative, halide
Haloidsalz *n* haloid salt, halide, haloid, salt of hydrohalic acid
Haloidsilber *n* haloid silver, silver halide
Haloidwasserstoff *m* hydrogen halide
Halometer *n* (Phys) halometer
halometrisch halometric
Halophyt *m* (Bot) halophyte
Halostachin *n* halostachine
Halothan *n* halothane
Halotrichit *m* (Min) halotrichite, alunogen
Haloxylin *n* haloxylin
Haloxylol *n* halogenated xylene, halogen-substituted xylene, haloxylene
Halozon *n* halozone
Halphensäure *f* halphen acid
Hals *m* neck, collar, throat, ~ **der Bessemerbirne** neck of a converter, ~ **einer Röhre** socket end of a pipe, ~ **einer Welle** neck or throat of a shaft
Halsansatz *m* projection (of a pipe etc.)

Halskerbe – handbedient

Halskerbe *f* neck groove
Halslager *n* collar bearing, journal bearing, neck journal
Halssenker *m* counterbore
Halsteilung *f* neck graduation
Halt *m* stop, halt, (Techn) hold, support, ~ geben to support
haltbar durable, fast, lasting, permanent, solid; stable, strong, ~ **machen** to sterilize, to conserve, to preserve, to stabilize, **eine Farbe** ~ **machen** to fix a color, **Nahrungsmittel** ~ **machen** to preserve food
Haltbarkeit *f* durability, endurance, pot life, service life, (Lagerfähigkeit) shelf life; (Stabilität) solidity, stability, ~ **von Farben** fastness
Haltbarkeitsgrenze *f* endurance limit
Haltbarmachen *n* sterilizing, conserving, preserving, (Techn) stabilizing
Haltbarmachung *f* conservation, preservation
Haltefeder *f* retaining spring
Haltefläche *f* supporting surface
Haltegerüst *n* support
Haltegriff *m* handle, grab; strap
Haltekabel *n* anchoring cable
Haltekerbe *f* hold-down groove, pick-up groove
Haltekontakt *m* (am Wendeschütz) auxiliary contactor
Haltekraft *f* (Phys) cohesion
Haltemagnet *m* holding magnet
Haltemutter *f* clamping or retaining nut
halten (unterstützen) to prop, to support, to sustain
Haltenute *f* hold-down groove, pick-up groove
Halteplatte *f* hold-down plate, latch plate, retainer
Haltepumpe *f* holding pump
Haltepunkt *m* stopping point, break, change point, halting point, stop-over, (Phys) critical point
Halter *m* holder, support
Halterahmen *m* clamping frame
Haltering *m* adapter ring, retainer ring
Halterungsvorrichtung *f* supporting facility
Halteschraube *f* holding screw, check screw, clamping bolt
Haltespule *f* holding-on coil, restraining coil
Haltestift *m* carrier pin, insert pin
Haltestrebe *f* tie rod
Haltevorrichtung *f* locking device, clamp, die adapter, (Fassung) collet
Haltewalze *f* supporting roll, nip roll
Haltezeit *f* halt, pause, residence time
Haltung *f* posture; attitude, (Agr) breeding, keeping, rearing
Halurgie *f* halurgy
Hamamelisblätterfluidextrakt *m* (Pharm) liquid extract of hamamelis
Hamamelisrinde *f* hamamelis bark

Hamamelissalbe *f* (Pharm) hamamelis ointment
Hamamelitannin *n* hamamelitannine
Hamamelonsäure *f* hamamelonic acid
Hamamelose *f* hamamelose
Hamartit *m* (Min) hamartite, bastnasite
hamatitartig hematitic
Hambergit *m* (Min) hambergite
Hamburger Blau *n* ferric ferrocyanide, Hamburg blue, Prussian blue
Hamburger Weiß *n* Hamburg white
Hamilton-Operator *m* Hamilton operator
Hamlinit *m* (Min) hamlinite
Hammelfleisch *n* mutton
Hammeltalg *m* mutton tallow
Hammer *m* hammer, (hölzerner) mallet
Hammeranlage *f* forge shop, hammer shop
Hammerauge *n* handle or shaft hole of a hammer
Hammerbär *m* hammer head, hammer tup, striker
Hammerbahn *f* hammer face
Hammerband *n* tilt brace
Hammerbrecher *m* hammer crusher
Hammereffekt *m* hammer effect
Hammereisen *n* forged or hammered iron, wrought iron
Hammerfinne *f* hammer peen, striking face of a hammer
hammergar tough, tough pitch, ~ **machen** to pole tough pitch
Hammergarmachen *n* poling tough pitch
Hammerhütte *f* forge shop, hammer shop
Hammerinduktor *m* hammer inductor
Hammerkopf *m* hammerhead
Hammerlack *m* hammer finish, hammer metal finish
Hammermühle *f* (Techn) hammer mill
Hammerschlag *m* [iron] scale, forge scale, stroke with hammer, (Chem, Techn) [iron] hammer scale, mill scale
Hammerschlageffekt *m* hammer effect
Hammerschlaglack *m* hammer finish, hammer metal finish
Hammerschmied *m* forge smith, hammer smith
Hammerschmiedestück *n* drop forging article
Hammersteuerung *f* hammer valve gear piston
Hammerstiel *m* hammer handle
Hammerunterbrecher *m* hammer interrupter
Hammerwerk *n* forge, hammer mill, ironworks, (Blechbearbeitung) pressworks
Hammonit *n* hammonite
Hampdenit *m* (Min) hampdenite
Hancockit *m* (Min) hancockite
Handanode *f* movable anode
Handapparat *m* hand apparatus, hand set
Handarbeit *f* manual labor, manual work
Handauslösung *f* manual release
handbedient hand-operated

Handbedienung *f* hand control, manual manipulation, manual operation
handbeschickt supplied by hand, handfed, handfired
handbetätigt worked by hand
Handbetrieb *m* hand operation, manual operation
handbetrieben manually controlled, manually operated
Handblasebalg *m* hand bellows
Handbohrer *m* (Techn) gimlet
Handbohrmaschine *f* hand drilling machine
Handbremse *f* hand brake
Handbuch *n* handbook, compendium, manual
Handbücherei *f* reference library
Handdruck *m* block printing
Handdurchschreibepapier *n* pencil carbon
Handeimer *m* hand bucket
Handel *m* commerce, trade
handeln to deal
Handelsantimon *n* commercial antimony
Handelsartikel *m* commercial article, commodity
Handelsbenzol *n* commercial benzene, commercial benzol
Handelsbezeichnung *f* trademark, trade designation, trade name
Handelsblei *n* commercial lead
Handelschemiker *m* commercial chemist
Handelseisen *n* commercial iron or steel, merchant bar iron
Handelsform *f* sales packing
Handelsgewicht *n* avoirdupois
Handelsguß *m* commercial cast iron or casting
Handelskautschuk *m* commercial rubber, raw rubber
Handelsmarke *f* trademark
Handelsname *m* trade name, trademark
Handelsphthalsäure *f* commercial phthalic acid
Handelsqualität *f* commercial quality, grade
Handelssorte *f* commercial grade
handelsüblich commercial, pertaining to commercial customs
Handelsware *f* commercial product
Handelszeichen *n* trademark, trade name
Handfeger *m* hand brush
handfest strong, stout, sturdy
Handfeuerlöscher *m* hand fire extinguisher
Handfilterplatte *f* filtration leaf
Handflaschenzug *m* block and tackle
Handform *f* hand mold
Handformmaschine *f* hand molding machine
Handgebläse *n* hand blowpipe
Handgebrauch *m* daily use
handgefertigt hand-made
handgesteuert manually controlled
Handgießlöffel *m* hand ladle
Handgriff *m* handle, grip, (Kunstgriff) knack, manipulation

Handhabe *f* manipulation
handhaben to handle, to manage, to manipulate, to operate
Handhabung *f* manipulation, handling, operation, use, **leichte** ~ convenient handling
Handianolsäure *f* handianolic acid
Handkarren *m* hand-barrow, hand cart
Handkelle *f* hand ladle
Handkippvorrichtung *f* hand tipping mechanism
Handkratzbürste *f* small wire brush
Handkurbel *f* cranked handle
Handlanger *m* hand (Am. E.), helper
Handlauf *m* hand rail
handlich handy
Handluftpumpe *f* hand air pump
Handmeißel *m* hand chisel
Handmuster *n* small sample
Handpresse *f* hand press
Handpreßluftstampfer *m* (Techn) hand pneumatic rammer
handpuddeln (Metall) to puddle by hand
Handrad *n* hand wheel
Handrädchen *n* micrometer scale knob, small handwheel
Handregelung *f* (Regeltechn) manual control
Handreibahle *f* hand reamer
Handreiniger *m* hand cleanser
Handschalter *m* hand switch
Handscheidung *f* separation by hand
Handschere *f* scissors, hand shears
Handschmierung *f* hand lubrication
Handschmiervorrichtung *f* lubricator actuated by hand
Handschuhschutzkammer *f* (Atom) glove box
Handschutz *m* handguard
Handschutzvorrichtung *f* hand safety guard
Handschweißen *n* hand welding
Handschweißung *f* manual welding
Handspindelpresse *f* hand screw press
Handstampfer *m* hand rammer
Handsteller *m* hand adjuster
Handsteuerung *f* hand regulation, hand gear
Handtorf *m* hand-cut peat
Handtuch *n* towel
Handumschalter *m* hand-operated change-over switch
Handvorgaberegler *m* manual loading station
Handwaage *f* hand scales
handwarm lukewarm
Handwerk *n* handicraft, trade
Handwerker *m* craftsman
handwerksmäßig workmanlike
Handwerkzeug *n* tool[s], implements, instruments, set of tools
Handwinde *f* hand winch
Hanf *m* (Bot) hemp
Hanfbürste *f* hemp brush
Hanfdichtung *f* hemp packing
Hanfeinlage *f* hemp layer, hemp core

Hanffaser f hemp fiber
Hanfkörner n pl hempseed
Hanflitze f hemp strand
Hanfnesselkraut n (Bot) hemp nettle, galeopsis
Hanföl n hemp oil, hempseed oil
Hanfriemen m hemp belt
Hanfsäure f (obs) linoleic acid
Hanfsamen m (Bot) hempseed
Hanfseil n hemp rope, hemp cord
Hanftau n hemp rope
Hanfumwickelung f hemp packing
Hanksit m (Min) hanksite
Hannayit m (Min) hannayite
Hansagelb n hansa yellow
Hanssen-Säure f Hanssen acid
Hantelmodell n dumbbell model
hantieren to handle, to manage, to manipulate, to operate, to work
Haplobakterium n haplo-bacterium, non-filamentous bacterium
haploid (Biol) haploid
Haploidie f (Biol) haploidy
Haptoglobin n haptoglobin
Hardystonit m (Min) hardystonite
Harfensieb n harp screen
Harmalan n harmalan
Harmalin n harmaline, 3,4-dihydroharmine, harmine dihydride
Harmalinrot n harmala red
Harman n harman, aribine, loturine
Harmin n harmine, yajeine
Harminsäure f harminic acid
Harmol n harmol
Harmolsäure f harmolic acid
harmonisch harmonic, harmonious
Harmotom m (Min) harmotome, cross stone
Harmyrin n harmyrine
Harn m urine
Harnabsatz m urinary sediment
harnabtreibend (Pharm) diuretic[al]
Harnanalyse f urine analysis
harnartig urinose, urinous
Harnbestandteil m urinary constituent
Harnblase f bladder
Harngärung f fermentation of urine
Harnindican n urine indicane
Harnkraut n (Bot) dyer's rocket, herniaria, rupture wort
Harnmittel n (Pharm) diuretic
Harnporphyrin n uroporphyrine
Harnsäure f uric acid
Harnsalz n urinary salt
Harnsand m (Med) urinary calculus, urinary gravel
harnsauer urate
Harnsediment n urinary sediment
Harnstein m urinary calculus
Harnstoff m urea, carbamide
Harnstoffchlorid n carbamyl chloride, carbamide chloride
Harnstoffharz n aminoplast resin
Harnstoffformaldehyd m urea formaldehyde
Harnstoffpreßmasse f aminoplast molding compound
Harnstoffsynthese f urea synthesis
Harnstoffzyklus m urea cycle
harntreibend diuretic
Harnuntersuchung f urine analysis, testing of urine, urinalysis, urine test
Harnwaage f urinometer
Harnzucker m sugar in urine
Harringtonit m (Min) harringtonite
Harrisit m (Min) harrisite
hart hard, firm, rigid, solid, ~ **gießen** to case-harden, to chill-cast, ~ **werden** to harden
Hartblei n hard or antimonial lead
Hartbrandstein m clinker, hard burned brick
Harteisen n hard iron
Harterz n (Min) quartzy copper ore
Hartfaser f hard fiber
Hartfaserplatte f hard board, molded fiber board
Hartfaserplattenlack m wall board coating
Hartfilter n hardened filter, hard filter
Hartfloß n specular iron
Hartfolie f rigid sheet
hartgebrannt hard-baked, hard-burned, hardened by burning
hartgefroren hard frozen
hartgelötet brazed, hard soldered
hartgewalzt hard-rolled
Hartgewebe n laminated fabric
Hartglas n hardened glass, hard glass
Hartglasbecher m hard glass beaker
Hartgummi m hard rubber, ebonite, vulcanite, vulcanized rubber
Hartgummiartikel m pl ebonite goods
Hartgummiisolator m ebonite insulator
Hartgummimehl n ebonite powder
Hartgummiplatte f ebonite plate
Hartgummirohr n hard rubber tube
Hartgummiwaren pl hard rubber goods
Hartguß m (Metall) chill casting, chilled cast iron
Hartgußform f chill casting mold
Hartgußroheisen n chill foundry pig iron
Hartgußteil m chill casting
Hartgußtiegel m casehardening crucible
Hartgußwalze f chilled iron roll
Hartharz n hard resin, solid resin
Hartholz n hardwood
Hartit n hartite
Hartkautschuk m ebonite, hardened caoutchouc, hard rubber
Hartkobalterz n (Min) skutterudite
Hartkobaltkies m (Min) skutterudite
Hartkorn n hard grain

Hartkupfer *n* hard [drawn] copper
Hartkupferdraht *m* hard-drawn copper wire
Hartlötbarkeit *f* brazability
Hartlötdraht *m* hard solder wire
hartlöten to braze, to hardsolder
Hartlöten *n* brazing, hard-soldering
Hartlötung *f* brazing, hard-soldering
Hartlötwasser *n* hard-soldering fluid
Hartlot *n* (Metall) brazing solder, hard solder, **löten mit** ~ to braze
Hartmachen *n* hardening
Hartmangan *n* (Min) braunite, manganese spar, rhodonite
Hartmanganerz *n* (Min) black manganese ore
Hartmanganstahlguß *m* braunite cast
Hartmeißel *m* cold chisel
Hartmessing *n* hard-drawn brass
Hartmetall *n* cutting alloy, hard metal
Hartmetallbestückung *f* carbide tipping
Hartmetallschneide *f,* **Werkzeug mit** ~ carbide-tipped tool
Hartmetallwerkzeug *n* carbide[-tipped] tool
Hartpapier *n* hard paper, kraft paper, laminated paper
Hartpapierisolation *f* hard paper insulation
Hartpappe *f* boxboard, cardboard
Hartparaffin *n* hard paraffin
Hartperlit *m* (Min) hard pearlite, troostite
Hartporzellan *n* hard[-fired] porcelain
Hartsalz *n* (Min) hard salt, impure sylvite
Hartschellack *m* hard lac resin
hartschlagen to hammer-harden, to cold-hammer
Hartschneidemetall *n* hard-cutting alloy
Hartspat *m* (Min) andalusite
Hartstahl *m* hard steel
Hartsteingut *n* feldspathic ware, hard whiteware
Harttantalerz *n* (Min) tantalite, columbite
Harttrockenglanzöl *n* hard-drying brilliant oil, hard-drying gloss oil
Harttrockenöl *n* hard-drying oil
hartverchromen to chromium-plate
Hartverchromung *f* (Elektrochem) hard chromium plating
Hartvernickelung *f* hard nickel plating
Hartversilberung *f* hard silver plating
Hartwachs *n* hard wax
Hartwalze *f* chilled roll
Hartwerden *n* hardening, solidifying
Hartzerkleinerung *f* crushing hard materials
Hartzink *n* hard zinc
Hartzinn *n* hard metal, hard pewter
Harz *n* resin, rosin, (Geigenharz) colophony, (Gummiharz) gum, **gemeines** ~ pine resin, **hitzehärtbares** ~ thermosetting resin, **ionenaustauschendes** ~ ion-exchange resin, **komplexierendes** ~ chelating resin, **künstliches** ~ artificial resin, synthetic resin
harzähnlich resin-like, resinoid, resinous, rosiny
Harzalkohol *m* resin alcohol

harzartig resin-like, resinoid, resinous
Harzbildung *f* formation of resin, resinification
Harzcerat *n* rosin cerate
Harzdestillationsanlage *f* resin distilling plant
harzelektrisch resinous electric, negative electric
Harzelektrizität *f* negative electricity, resinous electricity
harzen to be sticky, to extract resin from, to gather resin, (Holz) to exude resinous matter, (Techn) to resin, to resinate
Harzessenz *f* resin essence, rosin spirit
Harzester *f* ester gum
Harzfackel *f* resin torch
Harzfarbe *f* resin color
Harzfirnis *m* resin varnish
Harzfluß *m* flow of resin
harzförmig resiniform, resinous
harzfrei free from resin, non-resinous
Harzgalle *f* resin deposit (in wood), resin gall
Harzgang *m* resin channel
Harzgas *n* resin gas
harzgebunden resin-bonded
Harzgehalt *m* resin content
Harzgerbstoff *m* resinous tanning agent
Harzgewinnung *f* tapping of resin
Harzglanz *m* resinous luster
Harzgummi *n* resin rubber
harzhaltig containing resin, resiniferous, resinous
harzig resinous, resiny, ~ **machen** to resinify, ~ **werden** to become resinified
Harzkanal *m* resin channel
Harzkarbollösung *f* carbolic acid solution of resin
Harzkessel *m* resin melting pan
Harzkiefer *f* (Bot) pitch pine
Harzkitt *m* resin cement, resinous cement
Harzkleber *m* resin adhesive
Harzklebstoff *m* [synthetic] resin adhesive
Harzkörper *m* resinous substance, resinic body
Harzkohle *f* resinous coal
Harzkuchen *m* disc of resin
Harzlack *m* resin varnish
Harzleim *m* resin glue, resin size
Harzlösung *f* resin milk
Harzlösungsmittel *n* resin solvent
Harzmasse *f* resinous compound
Harzmühle *f* varnish crusher, varnish mill
Harznaphtha *n* resin oil
Harznest *n* resin pocket
Harzöl *n* resin oil, rosin oil
Harzpaste *f* paste resin
Harzpech *n* common rosin, liquid pitch
Harzpol *m* negative pole
Harzprodukt *n* resin product
harzreich [highly] resinous, rich in resin
Harzsäure *f* resinic acid, colophonic acid, resin acid
Harzsaft *m* resinous juice

Harzsalbe *f* rosin cerate
harzsauer resinate, colophonic
Harzschmelzanlage *f* resin melting plant
Harzschmelzen *n* resin melting
Harzseife *f* resin soap, rosin soap
Harzsirup *m* syrup resin
Harzspiritus *m* rosin spirit
Harzstange *f* resin rod
Harzstein *m* resinalite
Harzstoff *m* resinous substance
Harztasche *f* resin pocket
Harzteer *m* resinous tar
Harzträger *m* resin binder
Harzverarbeitung *f* resin processing plant
harzverleimt resin-bonded
Harzwasser *n* water resin
Harzzement *m* resinous cement
Haschisch *n* hashish, Indian hemp
Haselnußfarbe *f* nut brown
haselnußgroß hazelnut size
Haselnußöl *n* hazelnut oil
Haselstaude *f* (Bot) hazelnut
Haselwurz *f* (Bot) hazelwort, asarabacca, asarum
Haspe *f* hasp, clincher, cramp iron, hinge
Haspel *f* hank dyeing holder, sheaf, (Spule) spool, (Techn) winch, windlass
Haspelgrube *f* paddle pit
Haspelkufe *f* winch beck, winch
haspeln to reel; to wind
Haspelrad *n* chain pulley
Haspelseil *n* winch rope, windlass rope
Haspelwasser *n* reeling water
Hastanecin *n* hastanecine
Hastingsit *m* (Min) hastingsite
Hasubanol *n* hasubanol
Hatchettenin *n* hatchettenine, mineral tallow, rock tallow
Hatchettin *n* hatchettin
Hatchettolith *m* (Min) hatchettolite
Hatchetts Braun *n* Hatchett's brown
Haube *f* hood, bonnet, cap, (Dest) bubble cap
Haubenausgang *m* socket of vent device
haubengeglüht batch annealed
Haubenglühofen *m* hood type annealing furnace
Hauch *m* condensed moisture, bloom on varnish, breath, exhalation, haze on surfaces, tinge
hauchdünn extremely thin
Hauchecornit *m* (Min) hauchecornite
hauen to chop; to cut; to hew, to strike
Hauer *m* cutter; hewer
Hauerit *m* (Min) hauerite
Haufen *m* heap, accumulation, batch, cluster, pile
Haufenlaugen *n* heap leaching
Haufenröstung *f* heap roasting, roasting in heaps
Haufenverkohlung *f* charcoal burning in long piles
Haughtonit *m* (Min) haughtonite

Hauhechelwurzel *f* (Bot) restharrow root
Haumeißel *m* chisel, gouge
Hauptabsperrarmatur *f* main gate valve
Hauptachse *f* main axis, principal axis, (Math) major axis
Hauptagens *n* principal agent
Hauptanteil *m* main constituent, chief portion, principal constituent
Hauptantrieb *m* main gear
Hauptazimut *m* principal azimuth
Hauptbestandteil *m* chief constituent, main part
Hauptbindemittel *n* principal binding medium
Hauptbrechungszahl *f* (Opt) principal refractive index
Hauptbrennpunkt *m* (Opt) principal focal point, principal focus
Hauptbrennweite *f* (Opt) principal focal distance
Hauptdüse *f* main jet, main nozzle
Hauptebene *f* (Opt) principal plane
Hauptebenenabstand *m* distance between principal planes
Haupteinbruchstelle *f* main point of penetration
Haupteinfallswinkel *m* (Opt) angle of principal incidence
Hauptfach *n* main subject
Hauptfrontverspannung *f* bracing of the main furnace front
Hauptgleichung *f* (Math) fundamental equation
Hauptgruppe *f* (Element) main group
Haupthubgeschwindigkeit *f* main hoist speed
Hauptkolbenhub *m* down stroke
Hauptkondensator *m* main condenser
Hauptlaboratorium *n* chief laboratory
Hauptlager *n* headquarters, (Techn) main bearing
Hauptlegierungsbestandteil *m* main alloying component, main alloy ingredient
Hauptleitung *f* main line, mains, supply mains, (Rohrleitung) main piping, (Telef) trunk line
Hauptleitungsschiene *f* main busbar
Hauptlichtleitung *f* lighting mains
Hauptlichtmaschine *f* main dynamo
Hauptlinie *f* main line
Hauptnahrung *f* staple food
Hauptnenner *m* (Math) common denominator
Hauptniederlassung *f* main depot
Hauptnormale *f* einer Kurve principal normal
Hauptpatent *n* main patent, parent patent
Hauptphase *f* (Elektr) main phase
Hauptprodukt *n* main product
Hauptpunkt *m* principal point, chief point
Hauptpunktbrechwert *m* principal point refraction
Hauptpunkttriangulation *f* principal point triangulation
Hauptquantenzahl *f* principal quantum number
Hauptquelle *f* main source
Hauptreaktion *f* main reaction

Hauptregister *n* general index, main index, main register
Hauptreihe *f* (Chem) main series
Hauptrichtung *f* main orientation
Hauptrohr *n* main tube, main[s]
Hauptsäule *f* main column
Hauptsatz *m* fundamental law, axiom, (Thermodyn) law of thermodynamics
Hauptschacht *m* (Bergb) main shaft
Hauptschalter *m* main switch, master switch
Hauptschaltpult *n* central switch desk, main switch desk
Hauptschlüssel *m* master key
Hauptschlußbogenlampe *f* series arc lamp
Hauptschlußerregung *f* series excitation
Hauptschlußmotor *m* series[-wound] motor
Hauptschwingungsrichtung *f* principal vibration direction
Hauptserie *f* chief series
Hauptsicherung *f* main fuse
Hauptspannung *f* principal stress, (Elektr) main voltage
Hauptspeisepunkt *m* main feeding point
Hauptspektrum *n* principal spectrum
Hauptspirale *f* primary coil
Hauptstamm *m* main trunk
Hauptstempel *m* main ram
Hauptstrang *m* **Förderband** main trunk
Hauptstrecke *f* main line
Hauptströmung *f* main flow
Hauptstrom *m* main flow, main stream, (Elektr) main current
Hauptstromdynamo *m* (Elektr) series dynamo
Hauptstromerregung *f* series excitation
Hauptstromkreis *m* main circuit, principal circuit
Hauptstrommaschine *f* series dynamo
Hauptstrommotor *m* series motor
Hauptstromwiderstand *m* (Elektr) main circuit resistance
Hauptsymmetrieachse *f* principal axis of symmetry
Hauptträger *m* main girder
Haupträgheitsmoment *n* principal moment of inertia
Haupttragfläche *f* main plane
Hauptvalenz *f* (Chem) principal valence, principal valency
Hauptverdachtsmoment *n* leading suspect
Hauptverdampfer *m* main vaporizer
Hauptverwendungsgebiet *n* chief use
Hauptwelle *f* main shaft, driving shaft
Hauptwerk *n* (Unternehmen) main plant
Hauptwindleitung *f* main air duct, air main, blast main
Hauptwirkung *f* main action
Hauptzahl *f* (Math) cardinal number
Hauptzuführungsleitung *f* supply mains

Hauptzug *m* main pass, principal direction, principal feature
Hausenblase *f* fish glue, isinglass
Hausenblasefolie *f* sheet isinglass
Haushaltkonservierung *f* domestic preservation
Hausmannit *m* (Min) hausmannite
Hausschwamm *m* dry rot
Hausungeziefer *n* (Zool) house vermin
Haut *f* skin, (dünne Haut) film, (Kruste) crust, (Membran) membrane, (Tierhaut) hide
Hautabfälle *m pl* skin parings
Hautatmung *f* cutaneous respiration
Hautbildung *f* film formation, (Metall) skinning
Hautcreme *f* skin cream
Hauteffekt *m* (Elektr) skin effect
Hautentgiftungsmittel *n* skin decontamination agent
Hautentgiftungssalbe *f* skin decontamination ointment
Hautgewebe *n* (Med) cutaneous tissue
Hautgift *n* skin poison
Hautkrankheit *f* skin disease
Hautkrebs *m* (Med) skin carcinoma
Hautleim *m* glue from hides, hide glue
Hautnährcreme *f* (Kosm) skin nutrient
Hautöl *n* skin oil
Hautparasit *m* skin parasite, dermatozoon, ectozoon
Hautpflegemittel *n* cosmetic
Hautpilz *m* skin fungus, dermatophyte, ectophyte
Hautschutzmittel *n* skin-protecting agent
Hautschutzpräparat *n* skin-protecting preparation
Hautthermometer *n* skin thermometer
Hauttransplantation *f* (Med) skin grafting
Hautverhinderungsmittel *n* anti-skinning agent
Hautverhütungsmittel *n* anti-skinning agent
Hautwirkung *f* skin effect
Hauyn *m* (Min) hauyne
Hb Abk. für Hämoglobin
HbCO Abk. für Kohlenoxidhämoglobin
H-Bombe *f* H-bomb, hydrogen bomb
Heaviside'sche Regel *f* Heaviside's expansion rule
Heavisideschicht *f* Kennely-Heaviside layer, Heaviside region, ionosphere
Hebbewegung *f* vertical upward motion
Hebearm *m* lifting arm, lever
Hebeband *n* truss
Hebebaum *m* heaver, lever, lifter
Hebebock *m* jack
Hebedaumen *m* cam, lifter, tappet
Hebeeisen *n* crowbar
Hebegerät *n* lifting apparatus
Hebegerüst *n* lifting frame
Hebehaspel *f* winch
Hebekopf *m* tappet, cam, lifter

Hebekraft *f* leverage, lifting capacity, lifting force, lifting power
Hebekran *m* hoisting crane
Hebel *m* lever, **doppelarmiger** ~ two-armed lever, **einarmiger** ~ one-armed lever
Hebelade *f* lifting jack
Hebelanlasser *m* lever starter
Hebelanordnung *f* arrangement of levers, system of levers
Hebelarm *m* arm of a lever, lever arm
Hebelauslösung *f* lever release
Hebelausschalter *m* lever switch
Hebelbremse *f* lever brake
Hebelgesetz *n* (Phys) equation of moments, law of the lever
Hebelkraft *f* leverage
Hebelpresse *f* lever press
Hebelpreßwerk *n* lever press
Hebelschalter *m* lever or fork shifter, lever switch
Hebelschere *f* lever shears
Hebelsteuerung *f* lever control, spring catch
Hebelübersetzung *f* lever transmission
Hebelumschalter *m* lever switch, double throw, lever commutator
Hebelunterlage *f* fulcrum prop of a lever
Hebelverhältnis *n* leverage
Hebelvorrichtung *f* lever mechanism
Hebelwaage *f* lever balance, beam balance, beam scale, lever scale
Hebelwerk *n* leverage, lever arrangement, system of levers
Hebelwirkung *f* lever action, leverage
Hebemagnet *m* hoisting magnet, lifting magnet, vertical magnet
Hebemaschine *f* hoisting engine, hoisting machinery, winding engine
heben to lift, to elevate, to hoist, to raise
Hebepunkt *m* fulcrum
Heber *m* lever, lifter, (Chem) pipette, siphon, siphon tube
Heberbarometer *n* siphon barometer
Hebergefäß *n* siphon vessel
Heberleitung *f* siphon piping
hebern to pipette, to siphon
Heberpumpe *f* syphon pump
Heberrohr *n* siphon
Hebeschraube *f* lifting screw
Hebestange *f* crowbar, handspike
Hebetisch *m* lifting table
Hebevorrichtung *f* lifting device, hoist, lever, lifting mechanism
Hebewäsche *f* siphon separator
Hebewerk *n* drawworks (drilling), hoisting mechanism, lift, lifting arrangement
Hebewinde *f* jack, lifting winch, windlass
Hebezeug *n* lifting apparatus, lifting appliance
Hebezwinge *f* lifting bar
Hebronit *m* (Min) hebronite

Heckausreißer *m* rear-mounted ripper
Hecogenin *n* hecogenin
Hecogensäure *f* hecogenoic acid
Hedaquinium *n* hedaquinium
Hedenbergit *m* (Min) hedenbergite
Hederacosid *n* hederacoside
Hederagenin *n* hederagenin
Hederagonsäure *f* hederagonic acid
Hederichöl *n* hedge mustard oil
Hederin *n* hederine
Hederinsäure *f* hederic acid
Hederose *f* hederose
Hedyphan *m* (Min) hedyphane
Hefe *f* yeast, barm, leaven, lees, **obergärige** ~ top yeast, **untergärige** ~ bottom yeast
Hefeadenylsäure *f* yeast adenylic acid
hefeähnlich yeast-like, yeasty
Hefeanlage *f* yeast plant
Hefeapparat *m* yeast propagating apparatus, yeast propagator
Hefeart *f* type of yeast
Hefeaufziehvorrichtung *f* yeast cultivating apparatus
Hefeernte *f* crop of yeast
Hefeextrakt *m* yeast extract
Hefefett *n* yeast fat
Hefegabe *f* amount of yeast
Hefekultur *f* yeast culture
Hefenbitter *n* yeast bitter
Hefenextrakt *m* yeast extract
Hefenkeim *m* yeast germ
Hefensirup *m* yeast syrup
Hefenvermehrung *f* yeast growth, yeast increase
Hefepfropfen *m* yeast stopper
Hefepilz *m* yeast fungus, saccharomycete, yeast fungus
Hefepreßsaft *m* fermentation yeast juice
Hefepropagierungsapparat *m* yeast propagator
Heferaum *m* yeast room
Hefereinzucht *f* pure yeast culture
Hefereinzuchtapparat *m* pure yeasting machine
Hefeschaum *m* yeast scum
Hefezelle *f* yeast cell
Hefezucht *f* yeast culture
hefig barmy, yeast-like, yeasty
Hefnerkerze *f* (Opt) Hefner candle
Heft *n* booklet, bulletin, (Feile) haft, (Griff) handle, (Schwert etc.) hilt
heften to sew, to stitch
Heften *n* **mit Klammern** staple fastening
heftig vigorous, fierce; forcible, impetuous, intense, vehement, violent
Heftigkeit *f* intensity, vehemence, violence
Heftklammer *f* staple
Heftmaschine *f* stapler
Heftniet *m* binding rivet, dummy rivet, tacking rivet
Heftpflaster *n* (Med) adhesive plaster, adhesive tape, sticking plaster

Heftraum *m* stapling room
Heftschweißen *n* stich welding, tack welding
Heftverschluß *m* stapled closure
Heftzwecke *f* thumbtack, drawing pin, peg for drawing boards
Hegonon *n* hegonon
Heidenstein *m* (Min) ericite, heath stone
heilbar curable
heilen to cure, to heal
heilkräftig curative, therapeutic[al]
Heilkraft *f* curative power, therapeutic value
Heilkunde *f* medical science, medicine
Heilmittel *n* (Med) drug, medicament, medicine, remedy, therapeutic agent
Heilmittelchemie *f* pharmaceutical chemistry
Heilmittellehre *f* pharmacology
Heilpflanze *f* medical plant
Heilpflaster *n* (Med) healing plaster
Heilquelle *f* healing spring, medicinal well, mineral spring
Heilsalbe *f* (Pharm) healing ointment
Heilserum *n* antitoxic serum, healing serum
Heilstoff *m* curative, medicine, remedy
Heilwasser *n* curative water
Heilwirkung *f* therapeutic effect
Heilzweck *m* therapeutic purpose
Heintzit *m* (Min) heintzite
H-Eisen *n* (Techn) double-T-iron
Heisenbergsche Bewegungsgleichung *f* (Quant) Heisenberg equation of motion
Heisenbergsches Austauschintegral *n* (Magn) antiferromagnetic exchange integral
Heisenbergsche Unschärfe-Beziehung *f* Heisenberg uncertainty principle
Heisinglack *m* heising varnish
heiß ablöschen to quench hot, ~ **entfetten** to degrease hot
Heißabscheider *m* hot catchpot
Heißanstrich *m* hot plastic paint
Heißbearbeitung *f* hot working
Heißbiegefestigkeit *f* hot transverse strength
Heißblasen *n* hot-blast, heating to white heat
heißbrüchig hot-brittle, brittle in the hot state, hot-short
Heißchromatographie *f* hot chromatography
Heißdampf *m* superheated steam
Heißdampfarmaturen *pl* superheated steam fittings
Heißdampfentnahme *f* hot steam outlet
Heißdampfkühler *m* superheated steam cooler
Heißdampfventil *n* hot steam valve
Heißfixieren *n* heat-setting
Heißgas *n* hot gas
Heißgasschweißen *n* hot gas welding
Heißgassinterverfahren *n* hot gas sintering
Heißhärter *m* hot hardener
Heißkanalverteiler heated manifold
Heißklebelack *m* heat sealing lacquer, heat sealing coating

Heißkleber *m* hot-setting adhesive, heat-sealing adhesive
Heißlaboratorium *n* hot laboratory
Heißlaufmeldungsfarbe *f* heat indicating paint, thermorecording paint
heißlöten to hot-solder
Heißluft *f* hot air
Heißluftbad *n* hot-air bath
Heißluftmaschine *f* hot-air engine
Heißluftmotor *m* hot-air motor
Heißluftstromtrockner *m* convection drier
Heißlufttrichter *m* hot-air funnel
Heißlufttrocknung *f* hot-air drying
Heißluftvulkanisation *f* hot-air curing
Heißprägefarbe *f* hot stamping foil
Heißprägelack *m* hot stamping lacquer
Heißprägen *n* hot stamping
Heißpresse *f* hot-press
Heißpreßverfahren *n* (Polyester) matched metal die molding
Heißschmelzmasse *f* hotmelt coating
Heißsiegelapparat *m* hot sealing apparatus
heißsiegelbar heat-sealable, hot-sealable
Heißsiegelfähigkeit *f* heat sealing property
Heißsiegelfolie *f* heat-sealing film
Heißsiegelkleber *m* hot sealing adhesive
Heißsiegellack *m* heat-sealing lacquer
Heißsiegeln *n* [compression] heat sealing, heat sealing
Heißsiegelpresse *f* heat sealing press
Heißsiegelverfahren *n* hotsealing
Heißsiegelwachs *n* hotsealing wax
Heißspritzen *n* hot spraying
Heißspritzlack *m* hot coat, hot [spray] lacquer
Heißspritzverfahren hot lacquering process
Heißsprühverfahren *n* hot spraying, spray-drying with hot air
Heißspüler *m* hot rinser
Heißverfestigung *f* hot work hardening
heißverformbar hot-forming
Heißverkleben *n* heat-sealing
Heißverschweißen *n* heat-sealing
Heißverstrecken *n* hot-stretching
Heißwasser *n* hot water
Heißwasserbehälter *m* hot water tank
Heißwasserbereiter *m* hot water supplier
Heißwasserheizkörper *m* hot water radiator
Heißwasserheizung *f* hot water heating
Heißwasserpumpe *f* hot water pump
Heißwasserrücklauf *m* hot water return to boiler
Heißwassertrichter *m* hot water funnel
Heißwasservorlauf *m* hot water discharge from boiler
Heißwindkupolofen *m* hot-blast cupola
Heißwindleitung *f* hot air main
Heißwindschieber *m* hot-blast valve
Heizapparat *m* heating appliance
Heizbadflüssigkeit *f* heating bath liquid, heat transfer medium

Heizband *n* heater band, curing band, heating tape, strip heater
heizbar capable of being heated, heatable
Heizbatterie *f* (Elektr) filament battery
Heizdampf *m* heating steam, steam for heating purposes
Heizdraht *m* heater wire, resistance wire
Heizeffekt *m* heating effect, fuel efficiency
Heizelementschweißen *n* heating tool welding
heizen to heat, to fire, to stoke
Heizer *m* stoker, fireman, heater
Heizerstand *m* firing floor, stoking platform
Heizfaden *m* heating filament, ~ **mit Oxidschicht** oxide-coated heating filament
Heizfadenspannung *f* heater filament voltage
Heizfadenstrom *m* heater current
Heizfähigkeit *f* heating capacity
Heizfelge *f* curing rim
Heizfläche *f* heating surface
Heizflächenbelastung *f* heat transfer per unit surface area
Heizflächenzerstörung *f* burnout
Heizflamme *f* heating flame
Heizflüssigkeit *f* heating liquid, heat exchange medium
Heizgas *n* gas for heating, combustion gas, fuel gas, gaseous fuel
Heizgasvorwärmer *m* economizer
Heizgitter *n* heating grid
Heizgranate *f* heating tube
Heizhaube *f* heating jacket
Heizkammer *f* heating chamber, fire box
Heizkanal *m* heating channel, heater tunnel, heating flue, heating passage, steam channel
Heizkeil *m* welding tool
Heizkeilschweißen *n* heated tool welding, heated wedge welding
Heizkessel *m* boiler, autoclave
Heizkörper *m* radiator, heater, heating body, heating element
Heizkörperfarbe *f* radiator paint
Heizkörperverkleidung *f* radiator lining
Heizkopf *m* heater head
Heizkraft *f* calorific value, calorific power, heating power
Heizkranz *m* ring burner
Heizkreis *m* load circuit
Heizlampe *f* heating lamp
Heizleistung *f* watt density (extruder heaters)
Heizleiter *m* heating conductor
Heizloch *n* stoke hole, fire door
Heizmantel *m* heating jacket
Heizmaterial *n* fuel, heating material
Heizmedium *n* heating medium
Heizmittel *n* fuel, heating material
Heiznische *f* recess for radiators
Heizoberfläche *f* heating surface
Heizöl *n* fuel oil, heating oil
Heizpatrone *f* cartridge heater

Heizpilz *m* (Labor) heating mantle
Heizplatte *f* heating plate, hotplate, steam plate
Heizplattenschweißen *n* hot-plate welding
Heizraum *m* combustion chamber, fire chamber, fire place, heating space
Heizregister *n* heating elements
Heizröhrenkessel *m* fire tube boiler
Heizrohr *n* fire tube, heating flue
Heizrohrbündel *n* heating tube bundle
Heizschlange *f* heating coil, heating worm, steam coil
Heizschlangensystem *n* system of heating coils
Heizschlauch *m* (Gummi) curing bag
Heizspirale *f* heater spiral, heating coil
Heizstab *m* heating rod
Heizstoff *m* fuel
Heizstrom *m* heating current, filament current
Heiztrommel *f* heater drum
Heiztür *f* fire door, stoke hole
Heizung *f* heating, **induktive** ~ induction heating
Heizunruhe *f* fluctuation in heating current
Heizvorrichtung *f* heater, heating apparatus, heating plant
Heizwanne *f* heater trough
Heizwendel *f* coiled filament, heating spiral
Heizwerk *n* heating station
Heizwert *m* calorific value, calorific effect, calorific power, heat value, **oberer** ~ (Therm) gross calorific value, upper calorific value, **unterer** ~ lower calorific value, net calorific value
Heizwertbestimmung *f* calorific value determination
Heizwertmesser *m* calorific value meter, calorimeter
Heizwertregler *m* calorific value regulator
Heizwertuntersuchung *f* calorimetric test
Heizwicklung *f* filament winding
Heizwiderstand *m* (Elektr) filament rheostat
Heizwirkung *f* heating effect
Heizzone *f* heating zone
Heizzug *m* heating flue
Heizzylinder *m* heating cylinder
Hektar *n* (Maß) hectare
Hektin *n* hectine
Hektograph *m* hectograph
Hektographenblatt *n* hectographing paper
Hektographie *f* hectography
Hektokopierpapier *n* hectograph carbon paper
Hektowattstunde *f* hectowatt hour
Helcosol *n* helcosol, bismuth pyrogallate
Helenalin *n* helenalin
Helenien *n* helenien
Helenin *n* helenin
Heleurin *n* heleurine
Helianthin *n* helianthin, methyl orange
Helianthren *n* helianthrene
Helianthron *n* helianthrone

Helianthsäure f helianthic acid
Helicin n helicin
Helicoidin n helicoidin
Heliochromie f heliochromy, color photography, photochromatism
Heliochromoskop n heliochromoscope
Heliodor m (Min) heliodor
Helioechtgelb n helio fast yellow
Helioechtrot n helio fast red
Heliofarbstoff m helio dye
Heliograph m heliograph
Heliographie f heliography
Heliographiepapier n black line paper
Heliomesser m heliometer
Heliometer n heliometer
Heliophyllit m (Min) heliophyllite
Helioskop n helioscope
Heliotherapie f (Med) heliotherapy, sun cure
Heliotridan n heliotridane
Heliotridin n heliotridine
Heliotrin n heliotrine
Heliotrop m (Astr, Min) heliotrope
Heliotropin n heliotropin, piperonal
Heliotypie f (Phot) heliotyping
Helium n helium (Symb. He)
Heliumalter n helium age
Heliumatom n helium atom
Heliumblasenkammer f helium bubble chamber
Heliumgas n helium gas
Heliumgewinnung f recovery of helium
Heliumionenstoß m helium ion collision
Heliumkern m helium nucleus
Heliumröhre f helium tube
Heliumschicht f helium film
Heliumwellenfunktion f helium wave function
Helix f helix
Helixstruktur f (Biochem) helix structure, **doppelstrangige** ~ twin-threaded helix structure
hell bright, clear, light
Hellandit m (Min) hellandite
Hellanpassung f (Opt) bright adaptation
Hellbezugswert m brightness reference value, luminosity
hellblau light blue
Helleborein n helleboreine
Helleboretin n helleboretin
Helleborin n helleborin
Hellebrin n hellebrin
hellfarbig light-colored
hellgelb light yellow
hellgrau light gray
Helligkeit f brightness, light value, (Klarheit) clearness, (Leuchtkraft) luminous intensity, (Phys) brilliance, light intensity, ~ **einer Linse** (Opt) speed of a lens
Helligkeitsänderung f brightness change
Helligkeitsangleichung f brightness brightness compensation

Helligkeitsempfindung f brightness sensation
Helligkeitskurve f, **absolute** ~ absolute luminosity curve
Helligkeitsmesser m light-meter, luxmeter
Helligkeitsverteilung f brightness distribution
Helligkeitswert m brightness value, density or shading value
hellrot bright red
Hellrotglut f bright red heat
Hellstrahler m Briehl radiator
Helltran m clear light-yellow fish oil
Helm m head [of a still], (Hammer) shaft helve, handle, (Schutzhelm) helmet
Helmholtz-Schicht f Helmholtz double layer
Helminthosporin n helminthosporin
Helmitol n helmitol
Helvetan m helvetane
Helveticosid n helveticoside
Helvin m (Min) helvine
Hemellitenol n hemellitenol
Hemellithsäure f hemellitic acid, 2,3-xylic acid
Hemellitol n 1,2,3-trimethyl benzene, hemellitol
Hemellitylsäure f hemellitylic acid, 2,3-dimethylbenzoic acid
Hemiacetal n hemiacetal
Hemialbumose f hemialbumose, propeptone, semialbumose
Hemicellulose f hemicellulose
Hemieder n (Krist) hemihedron
Hemiedrie f (Krist) hemihedrism
hemiedrisch (Krist) hemihedral
Hemikolloid n hemicolloid
Hemimellitol n hemimellitol, hemimellitene
Hemimellitsäure f hemimellitic acid, benzene-1,2,3-tricarboxylic acid
hemimorph (Krist) hemimorphic, hemimorphous
Hemimorphie f (Krist) hemimorphism
Hemimorphismus m hemimorphism
Hemimorphit m (Min) hemimorphite, electric calamine, hydrous silicate of zinc, siliceous calamine
Hemipinsäure f hemipinic acid, 3,4-dimethoxyphthalic acid, hemipic acid
Hemipyocyanin n hemipyocyanin
Hemitoxiferin n hemitoxiferine
hemitrop (Krist) hemitrope, twinned
Hemitropie f (Krist) hemitropy
hemitropisch hemitrope, hemitropal
hemizyklisch hemicyclic
Hemlockharz n hemlock resin
hemmen to block, (Chem) to inhibit, (hindern) to hinder, (verlangsamen) to retard
Hemmen n blocking, inhibiting
Hemmkeil m recoil buffer, wedge
Hemmkette f drag chain
Hemmmittel n inhibiting agent, inhibiting substance, inhibitor
Hemmstoff m inhibitor, (Bot) plant growth retarder

Hemmung *f* hindrance, restraint, stopping, (Chem) inhibition, (Verlangsamung) retardation, **allosterische** ~ allosteric inhibition, **irreversible** ~ irreversible inhibition, **kompetitive** ~ competitive inhibition, **nichtkompetitive** ~ noncompetitive inhibition
Hemmungsstoff *m* inhibitor
Hemmungszentrum *n* inhibitory center
Hemmwirkung *f* inhibitory effect
Hempel-Bürette *f* Hempel [gas] burette
Heneikosan *n* heneicosane
Heneikosansäure *f* heneicosanoic acid
Heneikosylen *n* heneicosylene
Henkel *m* handle
Henna *f* (Färbemittel) henna
Hentriakontan *n* hentriacontane
Henwoodit *m* (Min) henwoodite
Heparen *n* heparene
Heparin *n* heparin
Heparprobe *f* (Anal) hepar test
Hepatektomie *f* (Med) hepatectomy
Hepatinerz *n* (Min) chrysocolla, cuprite
Hepatit *m* hepatite
Hepatitis *f* (Med) hepatitis
Hepatolyse *f* (Leberzellenzerfall) hepatolysis
Hepatozyt *m* hepatocyte
Heptacen *n* heptacene
Heptacontan *n* heptacontane
Heptacyclen *n* heptacyclene
Heptadecadien *n* heptadecadiene
Heptadecan *n* heptadecane
Heptadecansäure *f* heptadecanoic acid, margaric acid
Heptadecylamin *n* heptadecylamine
Heptadecylen *n* heptadecene, heptadecylene
Heptadien *n* heptadiene
Heptafulvalen *n* heptafulvalene
Heptahydrat *n* heptahydrate
Heptakosan *n* heptacosane
Heptaldehyd *m* heptaldehyde, heptanal, heptyl aldehyde
Heptaldoxim *n* heptaldoxime
Heptalen *n* heptalene
Heptamethylen *n* heptamethylene, cycloheptane, suberane
Heptaminol *n* heptaminol
Heptan *n* heptane, dipropylmethane, hexylmethane, methylhexane
Heptanaphthen *n* heptanaphthene
Heptanol *n* heptanol, heptyl alcohol
Heptanoylphenol *n* heptanoylphenol
Heptarsäure *f* heptaric acid
Heptatrien *n* heptatriene
Hepten *n* heptene, heptylene
Heptenylenglykol *n* heptenylene glycol
Heptin *n* heptine, heptyne
Heptindol *n* heptindole
Heptinsäure *f* heptinic acid

Heptode *f* (Elektr) heptode
Heptose *f* heptose
Heptoxim *n* heptoxime
Heptyl- heptyl
Heptylaldehyd *m* heptaldehyde, heptanal, heptoic aldehyde, heptyl aldehyde, heptylic aldehyde
Heptylamin *n* heptylamine
Heptylbromid *n* heptyl bromide
Heptylen *n* heptylene, heptene
Heptylsäure *f* heptylic acid, heptanoic acid, heptoic acid
herabrieseln to trickle down
herabsetzen to decrease, to lower, to reduce
Herabsetzung *f* lowering, reduction, ~ **des Werts** depreciation
herabsteigen to descend
herabtransformieren (Elektr) to step down
Heraclin *n* heraclin, bergamot camphor, bergaptene
Heraklitplatte *f* (HN) coarse fiber board
Herapathit *n* herapathite, quinine sulfate periodide
heraufsetzen to increase
herausbrechen to burst forth
herausdiffundieren to diffuse out
herausfließen to ooze out
Herausgeber *m* editor
herausgreifen to pick out
herauskommen to emerge, to issue
herauslösen to extract, to dissolve out, to eliminate
herausnehmbar extractable, removable
herausnehmen to take out, to eliminate; to remove
herausschlagen to kick out
herausschleudern to eject; to expel, to fling out, to shoot out, to spatter, to throw out
heraussickern to ooze out, to trickle out
herausspülen to rinse out, to wash out
herausstellen, sich ~ to prove
Herausstoßen *n* knocking-out
herausströmen to ooze out, to pour out
heraussuchen to pick out, to single out
heraustreten (Keilriemen) to ride out
herb bitter, harsh, raw, rough, sharp, tart
Herbacetin *n* herbacetin
Herbheit *f* acerbity, bitterness, tartness
Herbipolin *n* herbipoline
Herbizid *n* herbicide, weed killer, **Vorauflauf-** ~ pre-emergence herbicide
Herbstzeitlose *f* (Bot) meadow saffron
Herbstzeitlosensamen *m* colchicum seeds
Herclavin *n* herclavin
Herculin *n* herculin
Hercynit *m* (Min) hercynite, iron spinel
Herd *m* hearth, cooker, oven, stove, (Mittelpunkt) center, focus
Herdansätze *pl* hearth accretions

Herdasche *f* hearth ashes
Herdbelastung *f* furnace load
Herdblei *n* furnace lead
Herdbrücke *f* fire bridge
Herdeinsatz *m* hearth casing
Herdeisen *n* hearth plate, poker
Herderit *m* (Min) herderite
Herdfläche *f* hearth surface
Herdflächenleistung *f* hearth area efficiency
Herdform *f* hearth mold, open sand mold
Herdformerei *f* hearth molding, open sand molding
Herdfrischarbeit *f* hearth process, refinery process
Herdfrischeisen *n* bloomery iron, hearth refined iron, knobbled iron
Herdfrischen *n* open-hearth refining, litharge reduction, open-hearth process
Herdfrischprozeß *m* refinery process, Siemens Martin process
Herdfrischroheisen *n* charcoal hearth cast iron, refinery pig iron
Herdfrischschlacke *f* refinery cinders or slag
Herdfrischstahl *m* open-hearth steel, refined steel, charcoal hearth steel
Herdfutter *n* hearth lining
Herdgehalt *m* silver contained in hearth
Herdgewölbe *n* arch of the furnace, furnace crown, hearth arch, roof of the furnace
Herdguß *m* open sand casting
Herdlöffel *m* hearth ladle
Herdmauer *f* cutoff wall
Herdprobe *f* hearth assay
Herdraum *m* heating chamber
Herdringmaschine *f* hearth ring machine
Herdschlacke *f* hearth cinder
Herdsohle *f* hearth level
Herdspannung *f* furnace voltage
Herdstahl *m* charcoal hearth steel, fine steel
Herdtiefe *f* depth of the hearth
Herdtiefofen *m* soaking pit furnace, vertical heating furnace
hergestellt made, manufactured, prepared, **galvanisch** ~ electroplated
Heringsöl *n* herring oil
Herkulesschiene *f* hercules rail
Herkunft *f* origin
Hermannit *m* (Min) hermannite
Hermelin *m* ermine
Hermesfinger *m* (Bot) hermodactyl
hermetisch hermetical, ~ **verschlossen** hermetically sealed
Herniarin *n* herniarin, 7-methoxy coumarin
Heroin *n* heroin[e], diacetylmorphine
Héroult-Lichtbogenofen *m* Héroult electric-arc furnace
Héroult-Lichtbogen-Widerstandsofen *m* Héroult direct-heating arc furnace, Héroult electrode-hearth arc furnace
Héroult-Schachtofen *m* Héroult ore-smelting furnace
Héroult-Widerstandsofen *m* Héroult resistance furnace
Herrengrundit *m* (Min) herrengrundite, urvolgyite
Herrerit *m* (Min) herrerite
herrichten to prepare, to arrange
herrühren (von) to issue [from], to originate [from]
Herschelit *m* (Min) herschelite
herstammen to be derived from, to come from
herstellen to construct, to establish, to manufacture, to prepare, to process, to produce, (Chem) to prepare or synthesize (a compound), **eine Verbindung** ~ (allgemein) to establish a connection, **künstlich** ~ to synthesize;, **sachgemäß** ~ to make with appropriateness;, **serienmäßig** ~ to produce in quantity
Hersteller *m* producer, maker, manufacturer
Herstellerfirma *f* manufacturing plant, originating firm
Herstellerverschluß *m* (z. B. Karton) manufacturer's joint, body joint
Herstellung *f* manufacture, preparation, production, **fabrikationsmäßige** ~ processing, **partieweise** ~ batch production, **serienmäßige** ~ large-scale production, mass production, quantity production, **wirtschaftliche** ~ economical manufacture
Herstellungsdauer *f* operating time
Herstellungsgang *m* course of manufacture; manufacturing method, process of manufacture
Herstellungsgenauigkeit *f* accuracy of manufacture
Herstellungskosten *pl* cost of production
Herstellungsmaschinen *fpl* manufacturing machinery
Herstellungsmethode *f* method of preparation
Herstellungssumme *f* cost of production
Herstellungsverfahren *n* fabricating technique, manufacturing process, method of preparation
Herstellungsvorschrift *f* formula
Herstellungsweise *f* manner of manufacture, manufacturing method, method of production
Hertz *n* (Elektr) cycle per second, hertz
Hertz-Frequenzeinheit *f* Hertz frequency unit
Hertzsche Welle *f* Hertzian wave
Herumspritzen *n* splashing
herunterdrücken to depress, to lower
herunterkühlen to cool down
heruntersetzen to reduce
hervorbringen to bring forth, to create, to develop, to produce, to yield
hervorgehen [aus] to issue [from], to originate
hervorkommen to arise

hervorragen to project, to be salient, to overtop, to protrude, to stand out
hervorragend outstanding
hervorrufen to develop, to give rise to, to produce
hervorstehen to protrude
Herzanregungsmittel n (Pharm) cardiac stimulans
Herzbeutel m pericardium
Herzgift n cardiac poison
Herzglykosid n cardiac glycoside
Herzklappe f (Med) cardiac valve
Herztransplantation f (Med) heart transplant
Herzverpflanzung f (Med) heart transplant
Herzynin n herzynine
Hesperetin n hesperetin
Hesperetinsäure f (Hesperitinsäure) hesperetic acid, 3-hydroxy-4-methoxycinnamic acid, hesperitinic acid, isoferulic acid
Hesperiden n hesperidene, carvene, citrene, d-limonene
Hesperidin n hesperidin, citrin, vitamin P
Hesperitin n hesperitin
Hesperitinsäure f (Hesperetinsäure) hesperetic acid, hesperitinic acid, isoferulic acid, m-hydroxy-p-methoxycinnamic acid
Hesperonalcalcium n hesperonal calcium, calcium saccharose phosphate
Hessischbordeaux n Hessian bordeaux
Hessischpurpur n Hessian purple
Hessit m (Min) hessite, native silver telluride
Hessonit m (Min) hessonite, cinnamon stone
Hetairit m (Min) hetairite
Heteroatom n hetero-atom
heteroatomig heteroatomic, heteratomic
Heteroauxin n heteroauxin
Heterochinin n heteroquinine
heterochromatisch heterochromatic
Heterochromosom n (Allosom) heterochromosome, allosome
Heterocyclen pl heterocyclic compounds
heterodispers heterodisperse
Heterogamet m (Biol) anisogamete, heterogamete, heterozygote
heterogen heterogeneous
Heterogenit m heterogenite
Heterogenität f heterogeneity
Heterogenreaktor m heterogeneous reactor
Heterokodein n heterocodeine
Heteroladung f heterocharge
heterolog heterologous
Heterolupan n heterolupane
Heterolyse f heterolysis
heteromorph heteromorphic, polymorphic
Heteromorphie f heteromorphy
Heteromorphit m (Min) heteromorphite
Heterophyllin n heterophylline
heteropolar heteropolar

Heteropolykondensation f heteropolycondensation
Heteropolymerisation f heteropolymerization, copolymerization
Heteropolysaccharid n heteropolysaccharide
Heteropolysäure f heteropoly acid
Heterosis f (Biol) heterosis
Heterosit m (Min) heterosite
heterostatisch heterostatic
heterotroph heterotrophic
Heteroxanthin n heteroxanthine
heterozygot (Genpaar) heterzygous
Heterozygot m (Biol) heterozygote
heterozyklisch heterocyclic
Hetisin n hetisine
Hetokresol n hetocresol, metacresol cinnamate
Hetol n hetol, sodium cinnamate
Het-Säure f chlorendic acid, HET acid, hexachlorendomethylene tetrahydrophthalic acid
Heubachit m (Min) heubachite
Heulandit m (Min) heulandite
Heveabaum m (Bot) hevea tree, rubber tree
Heveen n heveene
Hewettit m (Min) hewettite
Hexaäthylbenzol n hexaethylbenzene
Hexaäthyldisiloxan n hexaethyldisiloxane
Hexabenzylcyclotrisiloxan n hexabenzylcyclotrisiloxane
Hexabromidzahl f hexabromide number, hexabromide value
Hexacen n hexacene
Hexachloräthan n hexachloroethane
Hexachlorbenzol n hexachlorobenzene
Hexachlordisilan n hexachlorodisilane
Hexachlorodisiloxan n hexachlorodisiloxane
Hexachlorophen n hexachlorophene
Hexachloroplatinsäure f hexachloro platinic acid
Hexachronsäure f hexachronic acid
Hexacontan n hexacontane
Hexacosanol n hexacosanol
hexacyclisch hexacyclic
Hexadecadien n hexadecadiene
Hexadecadiin n hexadecadiine
Hexadecan n hexadecane, cetane, dioctyl
Hexadecanol n cetyl alcohol, hexadecanol
Hexadecin n hexadecine, hexadecyne
Hexadecylalkohol m cetyl alcohol, hexadecyl alcohol
Hexadecylen n hexadecylene, cetene, hexadecene
Hexadien n hexadiene
Hexadiin n hexadiine, bipropargyl, hexadiyne
Hexaeder n (Krist) hexahedron
hexaedrisch hexahedral
Hexafluoracetylaceton n hexafluoroacetylacetone
Hexafluorokieselsäure f hexafluorosilicic acid

hexagonal (Krist) hexagonal
hexagonalholoedrisch (Krist) hexagonal holohedral
hexagonalpyramidal (Krist) hexagonal pyramidal
hexagonaltrapezoedrisch (Krist) hexagonal trapezohedral
Hexahelicen n hexahelicene
Hexahydrat n hexahydrate
Hexahydrobenzol n hexahydrobenzene, cyclohexane, hexamethylene
Hexahydrofarnesol n hexahydrofarnesol
Hexahydrohumulen n hexahydrohumulene
Hexahydropentaoxybenzol n hexahydropentahydroxybenzene, pentahydroxy cyclohexane
Hexahydropyridin n hexahydropyridine, piperidine
Hexahydrosalicylsäure f hexahydrosalicylic acid
Hexajoddisilan n hexaiodisilane
Hexakisoktaeder n (Krist) hexakisoctahedron
hexakisoktaedrisch (Krist) hexakisoctahedral
Hexakistetraeder n (Krist) hexakistetrahedron
Hexakontan n hexacontane, hexakontane
Hexakosan n hexacosane
Hexakosansäure f hexacosanic acid, hexacosanoic acid
Hexal n hexal, hexamethylen[e]amine sulfosalicylate
Hexalin n hexalin, cyclohexanol, hexahydrophenol
Hexalupin n hexalupine
Hexamekol n hexamecol
Hexamethoniumbromid n hexamethonium bromide
Hexamethoxydisiloxan n hexamethoxydisiloxane
Hexamethylbenzol n hexamethylbenzene
Hexamethylcyclotrisiloxan n hexamethylcyclotrisiloxane
Hexamethyldisilan n hexamethyldisilane
Hexamethyldisiloxan n hexamethyldisiloxane
Hexamethylen n hexamethylene, cyclohexane, hexahydrobenzene
Hexamethylenamin n hexamethyleneamine, formine
Hexamethylenbromid n hexamethylene bromide
Hexamethylendiamin n hexamethylene diamine
Hexamethylendiisocyanat n hexamethylene diisocyanate
Hexamethylenglykol n 1,6-dihydroxyhexane, hexamethylene glycol, hexane-1,6-diol
Hexamethylenimin n hexamethylene imine
Hexamethylentetramin n hexamethylenetetramine, hexamine, hexine, urotropine
Hexamethylolmelamin n hexamethylolmelamine
hexametrisch hexametric
Hexan n hexane, hexyl hydride

Hexanaphthen n hexanaphthene
Hexanitroäthan n hexanitroethane
Hexanitrobiphenyl n hexanitrobiphenyl
Hexanitrodiphenyl n hexanitrobiphenyl
Hexanol n hexanol
Hexanon n hexanone
Hexaoxyanthrachinon n hexahydroxyanthraquinone
Hexaoxybenzol n hexahydroxybenzene
Hexaphen n hexaphene
Hexaphenyläthan n hexaphenylethane
Hexaphenyldisiloxan n hexaphenyldisiloxane
Hexapyrin n hexapyrine
hexasymmetrisch hexasymmetric
Hexatrien n hexatriene
Hexavanadinsäure f hexavanadic acid
Hexen n hexene, hexylene
Hexenmehl n lycopodium
Hexeton n hexetone
Hexin n hexine, hexyne
Hexinsäure f hexinic acid
Hexitol n hexitol
Hexobarbital n hexobarbital
Hexode f (Elektr) hexode
Hexodenkappe f top cap of hexode tube
Hexogen n hexogen, trimethylene trinitramine
Hexokinase f (Biochem) hexokinase
Hexon n hexone
Hexonsäure f hexonic acid
Hexophan n hexophan
Hexoran n hexoran, carbon trichloride, hexachloroethane
Hexose f hexose
Hexosediphosphat n hexose diphosphate
Hexosediphosphorsäure f hexosediphosphoric acid
Hexosemonophosphatdehydrase f (Biochem) hexosemonophosphate dehydrogenase
Hexosephosphat n hexose phosphate
Hexosephosphorsäure f hexosephosphoric acid
Hexosidase f (Biochem) hexosidase
Hexuronsäure f hexuronic acid
Hexyl- hexyl
Hexylalkohol m hexyl alcohol
Hexylamin n hexyl amine
Hexylbromid n hexyl bromide
Hexylcain n hexylcaine
Hexylen n hexylene, hexene
Hexylenglykol n hexylene glycol
Hexylresorcin n hexylresorcinol
Heyderiochinon n heyderioquinone
Heyderiol n heyderiol
HFS Abk. für Hyperfeinstruktur
HF-Titration Abk. für Hochfrequenztitration
Hg-Bogen m mercury arc, mercury bridge
Hg-Druck m mercurial pressure
Hiascinsäure f hiascinic acid
Hibbenit m (Min) hibbenite
Hibiscussäure f hibiscus acid

Hibschit *m* (Min) hibschite
Hiddenit *m* (Min) hiddenite
Hieratit *m* (Min) hieratite
Hilfsanode *f* (Elektr) auxiliary anode
Hilfsantrieb *m* servodrive
Hilfsarbeiter *m* laborer, unskilled worker
Hilfsbeize *f* auxiliary mordant, by-mordant
Hilfsbetrachtung *f* auxiliary consideration
Hilfseinrichtung *f* auxiliary equipment, device
Hilfselektrode *f* auxiliary electrode, guard ring, secondary electrode (radio)
Hilfsenzym *n* auxiliary enzyme
Hilfsgerät *n* device
Hilfsgröße *f* auxiliary quantity
Hilfskessel *m* auxiliary boiler
Hilfskolben *m* auxiliary piston
Hilfskondensator *m* auxiliary condenser
Hilfskraft *f* auxiliary force
Hilfsleitung *f* auxiliary main
Hilfslenker *m* idler arm
Hilfslösungsmittel *n* solubilizer
Hilfsmaschine *f* auxiliary engine, spare engine, stand-by engine
Hilfsmittel *n* auxiliary product, aid, (Werkzeug) device
Hilfsmotor *m* booster motor
Hilfsphase *f* artificial phase, auxiliary phase
Hilfspipette *f* auxiliary pipette
Hilfspol *m* auxiliary pole
Hilfsprodukt *n* auxiliary product
Hilfspumpe *f* booster pump
Hilfsquelle *f* resource
Hilfsrelais *n* auxiliary relay
Hilfssatz *m* (Math) lemma
Hilfsschalter *m* auxiliary switch
Hilfsschieber *m* auxiliary valve
Hilfsskala *f* segmental scale
Hilfsspannung *f* auxiliary voltage
Hilfsstellgröße *f* (Regeltechn) auxilliary controlled variable
Hilfsstempel *m* assisting plug
Hilfsstoff *m* auxiliary product, adjuvant
Hilfsstrom *m* auxiliary current
Hilfstransformator *m* auxiliary transformer
Hilfsverstärker *m* servoamplifier
Hilfsvorrichtung *f* auxiliary attachment
Hillängsit *m* (Min) hillaengsite
Himbeeressenz *f* raspberry essence
Himbeerlikör *m* raspberry brandy
Himbeersirup *m* raspberry syrup
Himbeerspat *m* (Min) rhodochrosite
Himbelin *n* himbeline
Himmelblau *n* skyblue, azure, cerulean blue
Himmelslichtpolarisation *f* skylight polarization
Himmelsmechanik *f* (Astr) celestial mechanics
Himmelsmehl *n* earthly gypsum
Himmelsschreiber *m* (Werbeflugzeug) skywriter

hindern to block, to hinder, to inhibit, to prevent
Hindernis *n* hindrance, obstacle
Hinderung *f* hindrance, hindering, impediment, inhibition, **sterische** ~ steric hindrance
hindurchtreten to pass through
hineindiffundieren to diffuse into
hineindringen to penetrate
hineinpressen to press into, to push
Hineinstreuung *f* inscattering
hineintauchen to dip into, to submerge partially or entirely
hinlänglich sufficient
Hinokiflavon *n* hinokiflavone
Hinokinin *n* hinokinin
Hinokiöl *n* hinoki oil
Hinokisäure *f* hinokic acid
Hinokitol *n* hinokitol
Hinopurpurin *n* hinopurpurin
Hinreaktion *f* forward reaction
hinreichend sufficient
Hinsdalit *m* hinsdalite
Hinterdrehbank *f* relieving and cam profile lathe
hintereinander consecutive, in series, in tandem, serially
hintereinanderschalten to connect in series
Hintereinanderschaltung *f* (Elektr) series connection, (Motor) tandem connection
Hintergießmetall *n* backing metal
Hintergrund *m* background
Hintergrundspeicher *m* background memory
Hinterkante *f* rear edge
hinterlassen to leave behind, to leave as residue
hinterlegen to deposit
Hintersäule *f* rear stand
Hinterschneidung *f* back taper, undercut
Hinterzacken *m* back plate
Hintzeit *m* (Min) hintzeite
hinüberreißen (Dest) to carry over
hin- und herbewegen to reciprocate, to move to and fro, to rock, to shuttle, ~ **und zurück** reversible (reaction), there and back
Hin- und Herbewegung *f* reciprocating motion
Hin- und Herbiegeversuch *m* reverse bending test
hinuntertransformieren (Spannung) to step down [the voltage]
Hinweis *m* hint, reference
hinweisen to refer, to direct, to indicate
hinzufügen to add [to]
Hinzufügen *n* adding
Hinzufügung *f* addition
hinzurechnen to add
Hiochisäure *f* hiochic acid
Hiortdahlit *m* (Min) hiortdahlite
Hippeastrin *n* hippeastrine
Hippulin *n* hippulin
Hippurat *n* hippurate, salt or ester of hippuric acid
Hippuricase *f* (Biochem) hippuricase
Hippurit *m* (Min) hippurite, radiolite

Hippuritenkalk *m* (Min) hippuritic limestone
Hippursäure *f* hippuric acid, benzaminoacetic acid, benzoylglycine, urobenzoic acid
hippursauer hippurate
Hippuryl- hippuryl
Hippuryllysinmethylester *m* hippuryllysine methyl ester
Hiptagensäure *f* hiptagenic acid
Hiptagin *n* hiptagin
Hircin *n* hircine
Hirnanhangdrüse *f* pituitary gland
Hirndurchblutung *f* (Med) cerebral blood supply
Hirnhaut *f* (Med) meninges
Hirnhautentzündung *f* (Med) meningitis
Hirnlähmung *f* (Med) apoplexy
Hirnschädel *m* cranium
Hirsch-Filtertrichter *m* Hirsch funnel
Hirschhornsalz *n* commercial ammonium carbonate
Hirschtalg *m* suet of deer
Hirseerz *n* (Min) granular argillaceous iron ore, oolitic hematite
Hisingerit *m* (Min) hisingerite
Hispidin *n* hispidin
Histamin *n* histamine, β-imidazolyl ethylamine
Histaminase *f* (Biochem) histaminase
Histamindihydrochlorid *n* histamine dihydrochloride
Histaminoxidase *f* (Biochem) histamine oxidase
Histidase *f* (Biochem) histidase
Histidin *n* histidin[e]
Histidindihydrochlorid *n* histidine dihydrochloride
Histidinol *n* histidinol
Histochemie *f* histochemistry
Histogenese *f* (Biol) histogenesis
Histogenie *f* (Biol) histogenesis
Histohämatin *n* histohaematin
Histologie *f* histology
histologisch histological
Histolyse *f* (Gewebszerfall) histolysis, breaking down of tissue
Histon *n* histone
Histoplasma *n* (Bakt) histoplasma
Histosit *m* (Gewebsparasit) tissue parasite, histosite
Hitzdraht *m* (Elektr) hot or heated wire
Hitzdrahtgalvanometer *n* hot-wire galvanometer
Hitzeausgleich *m* heat balance
hitzebeständig heatproof, heat-resistant, heat-resisting, heat-stable, refractory, thermostable
Hitzebeständigkeit *f* heat resistance, heat-resisting quality, heat stability
Hitzedenaturierung *f* heat denaturation
Hitzedurchlässigkeit *f* thermal transmission
Hitzeeinwirkung *f* action of heat
hitzeempfindlich sensitive to heat
Hitzeempfindlichkeit *f* sensitivity to heat

Hitzeerzeugung *f* generation of heat
Hitzefestigkeit *f* heat resistance
Hitzegegenwert *m* equivalent of heat
Hitzegrad *m* degree of heat, intensity of heat
Hitzegradmesser *m* pyrometer
Hitzegradmessung *f* pyrometry
hitzehärtbar heat-setting, thermosetting
Hitzehärten *n* thermosetting
hitzehärtend heat-setting, thermohardening, thermosetting
Hitzemesser *m* pyrometer
Hitzemessung *f* heat measurement, pyrometry
hitzereaktiv heat-reactive
Hitzeschockprobe *f* thermal shock test
Hitzestrahlungsmesser *m* pyrometer
Hitzewirkungsgrad *m* thermal efficiency
Hjelmit *m* (Min) hielmite, hjelmite
Hobbock *m* [full aperture] drum
Hobel *m* plane
Hobelbank *f* planing bench
Hobeleisen *n* plane iron
Hobelmaschine *f* planing machine, planing fixture
hobeln to plane, to smooth
Hobeln *n* planing
Hobelspäne *n pl* [wood] shavings
Hobelvorrichtung *f* planing fixture
Hobelwelle *f* planer block
hoch high, tall, (Math) raised to the power, x **hoch 2** x [raised] to the second [power]
Hochätzung *f* acrography, relief engraving
hochaktiv (Atom) highly active, hot
hochbeansprucht heavy-duty, high-duty, highly stressed
Hochbehälter *m* elevated reservoir, high level tank, overhead reservoir, overhead tank
hochbrisant highly explosive
Hochdruck *m* high pressure, (Buchdr) relief printing, (Med) hypertension
Hochdruckanlage *f* high-pressure plant
Hochdruckarmaturen *pl* high-pressure fittings, high-pressure mountings
Hochdruckbogen *m* (Elektr) high-pressure arc
Hochdruckbrenner *m* high-pressure burner
Hochdruckdampf *m* high-pressure steam
Hochdruckdampfkessel *m* high-pressure [steam] boiler
Hochdruckdampfkocher *m* high-pressure steam cooker
Hochdruckdichtung *f* high-pressure seal, compressed asbestos fiber joint
Hochdruckfilterpresse *f* high pressure ram filters
Hochdruckflasche *f* high-pressure cylinder
Hochdruck-Flüssigkeitschromatographie *f* high-pressure liquid chromatography
Hochdruckförderung *f* high-pressure transport or conveyance
Hochdruckgebiet *n* (Meteor) high-pressure area
Hochdruckgebläse *n* high-pressure blower

Hochdruckglimmentladung – Hochleistungshomogenisiermaschine

Hochdruckglimmentladung *f* high-pressure glow discharge
Hochdruckheißdampfkessel *m* high-pressure steam boiler
Hochdruckhydrierung *f* high-pressure hydrogenation
Hochdruckkapselgebläse *n* high-pressure rotary blower
Hochdruckkessel *m* high-pressure boiler, high-pressure tank
Hochdruckkondensator *m* pressure type capacitor
Hochdruckkreis *m* high-pressure cycle
Hochdrucklampe *f* high-pressure lamp
Hochdruckleitung *f* high-pressure line, high-pressure piping
Hochdruckluftsystem *n* high-pressure air system
Hochdruckmantel *m* high-pressure casing
Hochdruckmeßgerät *n* high-pressure measuring equipment
Hochdruckpolyäthylen *n* high pressure polyethylene
Hochdruckpumpe *f* high-pressure pump
Hochdruckreaktor *m* high-pressure reactor
Hochdruckreifen *m* high-pressure tire
Hochdruckrohrleitung *f* power piping
Hochdrucksauerstoff *m* compressed oxygen
Hochdruckschaufel *f* high-pressure blade
Hochdruckschlauch *m* high-pressure hose
Hochdruckschraubengebläse *n* high-pressure spiral blower
Hochdruckstufe *f* high-pressure stage
Hochdrucksynthese *f* high-pressure synthesis
Hochdruckventil *n* high-pressure valve
Hochdruckverfahren *n* high-pressure process
Hochdruckzylinder *m* high-pressure cylinder
Hochemail *n* embossed enamel
hochempfindlich highly sensitive, supersensitive
Hochenergiephysik *f* high energy nuclear physics
hochentwickelt highly developed
hochexplosiv brisant, highly explosive
hochfahren to accelerate
hochfeuerfest highly fireproof, highly refractory
hochflüchtig (Chem) highly volatile
hochfördern to elevate
Hochförderung *f* elevation
hochfrequent of high frequency
Hochfrequenz *f* high frequency, radio frequency
Hochfrequenzbereich *m* high frequency range
Hochfrequenzdrossel *f* high-frequency choke
Hochfrequenzeisenkern *m* high-frequency iron core
Hochfrequenzgenerator *m* oscillator
Hochfrequenzhärtung *f* (Techn) induction hardening
Hochfrequenzinduktionsofen *m* high-frequency induction furnace
Hochfrequenzkabel *n* high-frequency cable
Hochfrequenznähen *n* stitch welding

Hochfrequenzplasmabrenner *m* high-frequency plasma torch
Hochfrequenzschweißen *n* electronic heat sealing, high-frequency welding
Hochfrequenzsiegeln *n* electronic heat sealing
Hochfrequenzspektroskopie *f* high frequency spectroscopy
Hochfrequenzstrom *m* high-frequency current
Hochfrequenztechnik *f* high-frequency [electrical] engineering
Hochfrequenztitration *f* high frequency titration
Hochfrequenztransformator *m* high-frequency transformer
Hochfrequenztrockner *m* dielectric drier
Hochfrequenzverleimung *f* high-frequency bonding
hochgesättigt (Chem) highly saturated
Hochgeschwindigkeitsfräsen *n* high-speed milling
hochgespannt highly stressed, highly superheated, high-pressure, high-tension
hochglänzend glossy
Hochglanz *m* bright luster, full gloss, [high] brilliancy, high finish, high gloss, high polish
Hochglanzbleikristallglas *n* high polish lead crystal glass
Hochglanzerzeugung *f* burnishing, fine polishing
Hochglanzfolie *f* glazing sheet
Hochglanzgebung *f* polishing
Hochglanzpaste *f* extra glossy paste
hochglanzpolieren to mirror finish, to burnish, to finish extra bright
Hochglanzpolitur *f* brilliant polish, high luster polish
hochgradig high-grade, highly concentrated, intense, of a high degree or grade
Hochhubsicherheitsventil *n* high lift safety valve
Hochhubwagen *m* high-lift platform truck
hochkant edgewise, on edge, upright
hochkantig edgewise, on edge, upright, ~ **stellen** to place edgeways, to up-edge
Hochkantstellung *f* upright position
hochkomprimiert highly compressed
Hochkonjunktur *f* boom
hochkonzentriert highly concentrated
hochlegiert highly alloyed, high-alloy
Hochleistung *f* high duty, high efficiency
Hochleistungsantrieb *m* heavy duty drive
Hochleistungsapparatur *f* high performance apparatus
Hochleistungsflächenfräsmaschine *f* heavy duty surface milling machine
Hochleistungsfüllkörper *pl* (Dest) heavy-duty packings
Hochleistungsgleichrichterröhre *f* power rectifying valve
Hochleistungshomogenisiermaschine *f* heavy duty homogenizer

Hochleistungsmaschine *f* highduty machine, highspeed machine
Hochleistungsmotor *m* high-power motor
Hochleistungsöl *n* heavy-duty oil
Hochleistungsreaktor *m* highpower pile, highpower reactor
Hochleistungsstahl *m* high-quality tool steel
Hochleistungstier *n* (Rind) heavy milker
Hochleistungsverformung *f* high energy working
Hochleistungsverstärker *m* high gain amplifier
Hochleitung *f* (Elektr) overhead wire
hochmolekular high-molecular, macromolecular, of high molecular weight
Hochofen *m* blast furnace, ~ **mit Kippgefäßbegichtung** blast-furnace with skip hoist, ~ **mit Kübelbegichtung** blast-furnace with bucket hoist, **den ~ abstechen** to tap the blast furnace, **den ~ anblasen** to blow into the furnace, to put the furnace in blast, **elektrischer ~** electric-smelting furnace
Hochofenabstich *m* blast-furnace casting
Hochofenanlage *f* blast furnace [equipment]
Hochofenarbeiter *m* furnace hand, furnace man
Hochofenarmaturen *f pl* blast-furnace fittings
Hochofenaufzug *m* blast-furnace hoist
Hochofenbegichtung *f* blast-furnace charging
Hochofenbeschickung *f* blast-furnace charge or charging
Hochofenfutter *n* blast-furnace lining
Hochofengang *m* blast-furnace operation
Hochofengas *n* blast-furnace gas, waste gas
Hochofengasreinigungsanlage *f* blast-furnace gas purifying plant
Hochofengasschieber *m* blast-furnace gas valve
Hochofengebläse *n* blast-furnace blowing engine
Hochofengestell *n* blast-furnace crucible, blast-furnace hearth or well
Hochofengicht *f* blast-furnace throat or mouth
Hochofenguß *m* iron cast out of a blast furnace
Hochofenkoks *m* blast-furnace coke
Hochofenpanzer *m* blast-furnace jacket
Hochofenprofil *n* blast-furnace lines
Hochofenprozeß *m* blast-furnace process
Hochofenrumpf *m* blast furnace without appendices
Hochofenschlacke *f* blast-furnace slag
Hochofenschlackensand *m* blast-furnace slag sand
Hochofenschmelze *f* blast-furnace smelting
Hochofensteine *pl* blast-furnace bricks
Hochofentragring *m* blast-furnace lintel plate, blast-furnace ring or mantle
Hochofenverfahren *n* blast-furnace process
Hochofenvorherd *m* blast-furnace settler
Hochofenwind *m* furnace blast
Hochofenzeichnung *f* blast-furnace drawing
Hochofenzement *m* blast-furnace cement, cement consisting of Portland cement and finely ground blast-furnace slag

hochohmig high-impedance, high-ohmic
Hochplatte *f* relief printing plate
hochpolymer highly polymerized, high-polymer
hochradioaktiv highly radioactive, hot
Hochraumlager *n* high-bay warehouse
Hochregal *n* high rack
Hochregallager *n* high-rack warehouse, high-bay warehouse
hochrot bright red, brick red, deep red
hochrund convex
hochschmelzend difficultly meltable, having a high melting point, high-melting
Hochschule *f*, **technische ~** institute of technology, technical university
Hochschultertrommel *f* crown drum
hochsiedend having a high boiling point, high-boiling
Hochsinterung *f* final sintering operation
Hochspannung *f* high voltage, high potential, high pressure, high tension
Hochspannungsanlage *f* high-voltage plant
Hochspannungselektrophorese *f* high-voltage electrophoresis
Hochspannungsentladungsröhre *f* high-voltage discharge tube
Hochspannungserzeuger *m* high-voltage dynamo, high-voltage generator
Hochspannungserzeugungsanlage *f* high-voltage unit
Hochspannungsflamme *f* high-voltage arc
Hochspannungsflammenreaktion *f* high-voltage arc reaction
Hochspannungsfreileitung *f* high-voltage overhead transmission line
Hochspannungsgefahr *f* high-voltage danger
Hochspannungsgehäuse *n* high-voltage shell
Hochspannungsisolator *m* high-voltage insulator
Hochspannungskabel *n* high-voltage cable
Hochspannungsklemme *f* high-voltage clamp, high-voltage terminal
Hochspannungsleitung *f* high-voltage conductor, high-voltage line
Hochspannungsmaschine *f* high-voltage machine
Hochspannungsnetzgerät *n* high-voltage power pack
Hochspannungspol *m* high-voltage terminal
Hochspannungsschalterrohr *n* circuit-breaker chamber for high-voltage switchgear
Hochspannungsstarkstrombogen *m* high-voltage power arc
Hochspannungsstrom *m* high-voltage current, high-voltage current
Hochspannungstechnik *f* high-voltage engineering
Hochspannungstransformator *m* high-voltage transformer
Hochspannungszündung *f* ignition
Hochspant *n* high frame
Hochspantsystem *n* high frame system

hochstellen to raise, to put upright
Hochstellen *n* raising
Hochstromkohlebogen *m* high-intensity arc
Hochstromtrafo *m* high current transformer
hochtemperaturbeständig resistant to high temperatures
Hochtemperaturchlorierung *f* high-temperature chlorination
Hochtemperaturfett *n* high-temperature grease
Hochtemperaturlegierung *f* high-temperature alloy
Hochtemperaturreaktor *m* high-temperature reactor
Hochtemperaturwerkstoff *m* high-temperature material
hochtourig high-speed, high-velocity
Hochvakuum *n* high vacuum
Hochvakuum-Diffusionspumpe *f* high-vacuum diffusion pump
Hochvakuumelektronenröhre *f* high-vacuum electron tube (Br. E. valve)
Hochvakuumfett *n* high vacuum grease
Hochvakuumölpumpe *f* high vacuum oil pump
Hochvakuumpumpe *f* high-vacuum pump
Hochvakuumröhre *f* (Elektr) high-vacuum tube
Hochvakuumtechnik *f* high-vacuum engineering, high-vacuum technology
Hochvakuumverstärker *m* high-vacuum amplifier
Hochvakuumzelle *f* high-vacuum phototube
hochverdichtend high-compressing
hochverdichtet highly compressed
hochveredelt permanently finished
Hochveredelung *f* high finish[ing]
Hochveredlung *f* permanent finishing
Hochveredlungsmittel *n* (Text) permanent finishing
hochvergärend highly fermenting
hochverschleißfest highly wear-resistant
Hochverstärker *m* high frequency amplifier
hochviskos highly viscous, very viscid
hochwertig high grade, of first class quality
hochwichtig highly important
hochwinden to heave, to hoist, to jack up, to lift up, to raise
hochwirksam highly effective
Hochzahl *f* (Math) exponent
hochziehen to lift up, to wind up
Hochziehen *n* lifting, wrinkling
Hodograf *m* hodograph
höchst maximal, maximum, peak
Höchstausschlag *m* (Schwingung) maximum amplitude
Höchstbeanspruchung *f* highest stress, maximum stress
Höchstbedarf *m* **an Kraft** maximum power demand
Höchstbelastung *f* maximum duty, maximum load, peak load, ultimate bearing stress, **zulässige** ~ rated load
Höchstdosis *f* (Pharm) maximum dose
Höchstdrehzahl *f* maximum speed
Höchstdruck *m* maximum pressure
Höchstdruckkessel *m* superhigh pressure boiler
Höchstdruckpumpe *f* pump for extreme pressures
höchstempfindlich extremely sensitive
Höchstentladung *f* maximum discharge
Höchstfrequenz *f* maximum frequency, ultra high frequency
Höchstgasdruck *m* maximum gas pressure
Höchstgehalt *m* maximum content
Höchstgeschwindigkeit *f* top or maximum speed
Höchstgewicht *n* maximum weight
Höchstladedruck *m* maximum boost
Höchstladegewicht *n* maximum weight of load
Höchstlast *f* maximum load or stress
Höchstlastreibung *f* friction at maximum load
Höchstleistung *f* maximum efficiency or power, crest power, extreme record performance, maximum output, peak load
Höchstmaß *n* maximum
Höchstreichweite *f* maximum range
höchstsiedend highest boiling
Höchstspannung *f* maximum voltage
Höchsttemperatur *f* maximum temperature
Höchsttemperaturbestrahlung *f* high-temperature irradiation
Höchsttemperaturdestillation *f* high-temperature distillation
Höchstvakuum *n* ultrahigh vacuum
Höchstverdampfungsfähigkeit *f* maximum evaporative capacity
Höchstwert *m* peak or maximum value
Höchstwertigkeit *f* (Element) maximum valence
Höchstzahl *f* maximum or highest number
höchstzulässig maximum permissible, limiting, maximum admissible
Höcker *m* hump
höckerig knobby, knotty, rough
Högbomit *m* (Min) hoegbomite
Höhe *f* altitude, height, ~ **des Potentialwalles** barrier height, ~ **eines Dreiecks** altitude of a triangle, ~ **eines Tones** pitch of a tone, **lichte** ~ height in the clear, inside height
Höhenbestimmung *f* height determination
Höhenförderer *m* elevator, lift conveyer
Höhengewinnfunktion *f* height gain function
Höhenisothermenkarte *f* high altitude isothermal chart
Höhenlage *f* altitude
Höhenmaßstab *m* vertical scale
Höhenmesser *m* altimeter, altitude meter, hypsometer
Höhenmessung *f* altimetry

Höhensonne *f* artificial sunlight, ultra-violet lamp
Höhenstrahlen *pl* cosmic rays, **primäre** ~ primary cosmic rays, **sekundäre** ~ secondary cosmic rays
Höhenstrahlmeson *n* (Atom) cosmic ray meson
Höhenstrahlneutron *n* cosmic ray neutron
Höhenstrahlteilchen *n* cosmic ray particle
Höhenstrahlung *f* cosmic radiation
Höhenstrahluntersuchung *f* cosmic ray investigation
Höhenthermometer *n* hypso-thermometer
Höhenverteilung *f* vertical distribution
Höhenwinkel *m* angle of elevation, angular height, azimuth angle, vertical visual angle
Höhepunkt *m* highest point, climax, crest, critical or culminating point, peak, summit, top
höherwertig of higher value, superior, (Chem) of higher valence, multivalent
Höhlengips *m* granular gypsum
Höhlenkalk *m* drop stone
Höhlung *f* hollow, cavity, (Aushöhlung) excavation, (Techn) groove, recess
Höllenöl *n* caster oil, curcas oil, lowest grade of olive oil
Höllenstein *m* silver nitrate, argentic nitrate, lapis infernalis, lunar caustic
Höllensteinbad *n* silver nitrate bath
Höllensteinhalter *m* silver nitrate holder
Höllensteinlösung *f* silver nitrate solution
hörbar audible
Hörbarkeitsgrenze *f* limit of audibility
Hörbereich *m* range of audibility, (Radio) broad-casting range
Hörer *m* receiver
Hörsaal *m* auditorium, lecture hall
Hörschwelle *f* threshold of hearing
Hörzeichen *n* acoustic signal
Hof *m* (Opt) corona, halo, ring
Hoffmannstropfen *pl* Hoffmann's tincture
Hofmann-Eliminierung *f* (Chem) Hofmann elimination
Hofmannsche Umlagerung *f* Hofmann rearrangement
Hofmannscher Abbau *m* Hofmann degradation
hohl empty, hollow, (Opt) concave
Hohlanode *f* (Elektr) hollow anode
Hohlbohrer *m* auger
Hohlbohrstahl *m* hollow boring tool
Hohldechsel *f* hollow adze
Hohldorn *m* hollow mandrel
Hohleisen *n* hollow chisel, gouge
hohlerhaben concavo-convex
Hohlfeile *f* round file
Hohlform *f* female die, shell form
hohlgegossen hollow-cast
hohlgeschliffen concave-ground, hollow-ground
Hohlgewinde *n* female thread
hohlgießen to cast upon a core, to hollow-cast
Hohlglas *n* concave glass, hollow glassware
Hohlglaswaren *f pl* hollow glassware
Hohlguß *m* hollow casting
Hohlheizung *f* (Prüftechnik) curing blow
Hohlhobel *m* hollow plane
Hohlkant *n* hollow chamfer
Hohlkanteisen *n* fluted bar iron
Hohlkathode *f* concave cathode, (Elektr) cylindrical cathode
Hohlkehle *f* (Masch) channel, flute, fluting, hollow groove, rabbet, recess
Hohlkehlhobel *m* rabbet plane
Hohlkehlschweißung *f* concave fillet weld
Hohlkeil *m* hollow key
Hohlkörper *m* hollow body
Hohlkugel *f* hollow sphere
Hohlleiter *m* (Elektr) waveguide
Hohllinse *f* concave lens, diverging lens
Hohlmaß *n* measure of capacity, (z. B. für Holz) dry measure
Hohlmeißel *m* gouge, hollow chisel
Hohlprisma *n* hollow prism
Hohlprofil *n* hollow section
Hohl[pump]gestänge *n* (Ölgewinnung) hollow rod
Hohlraum *m* cavity, blow hole, cell, empty chamber, hollow space, interstice, opening, pocket, pore space, space cavity
Hohlraumbildung *f* (Hydrodynamik) cavitation
Hohlraumfeld *n* cavity field
Hohlraumkammer *f* empty chamber
Hohlraumresonator *m* cavity resonator
Hohlraumresonatorröhre *f* resonant cavity tube, magnetron
Hohlraumstrahlung *f* cavity radiation, black [body] radiation
Hohlraumvolumen *n* (Füllkörperkolonne) void fraction
Hohlraumwachstum *n* cavity growth
Hohlreibahle *f* shell reamer
Hohlröhre *f* hollow tube
Hohlrost *m* hollow grate
Hohlroststab *m* hollow gratebar
Hohlrührer *m* ejector mixer
hohlrund concave
Hohlrundung *f* concavity
Hohlschaber *m* grooved scraper
Hohlschnitt *m* hollow section
Hohlschraube *f* female screw
Hohlseil *n* hollow cable, hollow rope
Hohlspat *m* (Min) hollow spar, chiastolite
Hohlspiegel *m* concave mirror, **parabolischer** ~ parabolic mirror
Hohlspiegelkathode *f* concave-mirror cathode
Hohlstein *m* hollow stone, gutter stone
Hohlstopfen *m* hollow stopper
Hohlwalze *f* hollow cylinder, hollow roll
Hohlwelle *f* hollow shaft

Hohlzirkel m inside calipers
Hohlzylinder m hollow cylinder
Hohlzylinderkathode f (Elektr) hollow cylindrical cathode
Hohmannit m (Min) hohmannite
Hokutolith m (Min) hokutolite
Holamin n holamine
Holaphyllamin n holaphyllamine
Holländer m (Pap) beater, beating engine, hollander, pulp engine, rag engine
Holländerblau n Dutch blue, Holland blue
Holländerkasten m vat
holländern to hollander, to pulp in a hollander, to pulp rags in a hollander
Holländerwaage f Dutch balance
Holländerweiß n Dutch white
Hollandit m (Min) hollandite
Hollerithsystem n perforated card system
Holme pl side bars or stanchions
Holmfräsmaschine f spar milling machine
Holmit m (Min) holmite, chrysophane
Holmium n holmium (Symb. Ho)
Holmquistit m (Min) holmquistite
Holocain n holocaine, phenacaine
Holoeder n (Krist) holohedron, holohedral crystal
Holoedrie f (Krist) holohedrism, holohedry
holoedrisch (Krist) holohedral
Holoenzym n (Biochem) holo-enzyme
Holographie f holography
holokristallin holocrystalline
holomorph holomorphic
Holomycin n holomycin
Holothin n holothin
Holozymase f (Biochem) holozymase
Holunder m (Bot) elder
Holunderbeere f (Bot) elder berry
Holunderbeerenöl n elderberry oil
Holunderblütentee m elder tea
Holunderwein m elder wine
Holz n wood, (Bauholz) lumber, timber, **anstehendes** ~ standing timber, **astfreies** ~ branchless wood, **astreiches** ~ branchy wood, **faules** ~ decayed wood, **feinfaseriges** ~ fine-grained wood, **frisches** ~ green wood, live wood, **gebeiztes** ~ stained wood, **getränktes** ~ impregnated wood, **grünes** ~ green wood, live wood, **morsches** ~ rotten wood, **nachbehandeltes** ~ improved wood, **rotfaules** ~ decayed wood, **versteinertes** ~ lithoxyl[e], petrified wood
Holzabfälle pl chips of wood, wood waste
Holzachat m (Min) wood agate
Holzäther m (obs) dimethyl ether, methyl ether
Holzalkohol m (obs) methanol, carbinol, methyl alcohol, wood alcohol, wood spirit
Holzamiant m (Min) ligneous asbestos, ligniform amianthus
holzartig ligneous, woody, xyloid

Holzasbest m ligneous asbestos
Holzasche f wood ashes
Holzaschenlauge f wood ash lye
Holzaschenwasser n pearl ash solution, wood ash liquor
Holzaufbereitung f conservation of wood, impregnation of wood, preparation of wood
Holzbacke f wooden jaw
Holzbau m wood construction
Holzbauweise f wood construction
Holzbearbeitung f wood-working
Holzbearbeitungsmaschine f wood-working machine
Holzbeize f wood mordant, wood stain
Holzbeizen n wood staining
Holzbindemittel n wood fiber bonding medium
Holzbohrer m wood auger, wood-boring tool
Holzbrandmalerei f pyrogravure
Holzbranntwein m wood spirit
Holzbruch m wood failure
Holzcellulose f wood cellulose
Holzchemie f chemistry of the wood
Holzdestillation f wood distillation
Holzdestillieranlage f wood distilling plant
Holzdruck m xylography
Holzdübel m wooden peg
Holzeinsatzkasten m removable wooden box
Holzessig m wood vinegar
Holzfäule f dry rot
Holzfaser f ligneous fiber, wood fiber, wood pulp
Holzfasergarn n wood fiber yarn
Holzfaserpapier n wood pulp paper
Holzfaserplatte f composition board, wood-fiber board
Holzfaserstoff m cellulose, lignin, lignocellulose, xylogen
Holzfeuer n wood fire
Holzfeuerung f wood-fired furnace
Holzfirnis m wood varnish
Holzform f wooden form
Holzfräser m (Techn) wood-milling cutter
holzfrei woodfree, (Pap) without cellulose
Holzfurnier n wood veneer
Holzgas n wood gas
Holzgasgenerator m wood gas generator
Holzgeist m methanol, carbinol, methyl alcohol, wood alcohol, wood spirit
Holzgerinne n wood trough
Holzgewebe n (Bot) xylem
Holzgrundierung f wood primer
Holzgummi m wood gum, xylan
Holzhammer m mallet
Holzharz n wood resin
Holzhorde f wood hurdle
holzig ligneous, woody
Holzigkeit f lignosity, woodenness
Holzimprägnieranlage f wood impregnation plant

Holzimprägnierapparat *m* wood preserving apparatus
Holzimprägnieren *n* wood preserving
Holzimprägnierung *f* lumber preservation, wood impregnation
Holzindustrie *f* wood [working] industry
Holzisolator *m* wooden insulator
Holzkalk *m* (obs) calcium acetate
Holzkasten *m* wooden box, wooden chest
Holzkiste *f* wooden box
Holzkitt *m* joiner's putty, wood cement
Holzkohle *f* charcoal, ~ **brennen** to carbonize wood
holzkohlenartig charcoal-like
Holzkohlenasche *f* charcoal ashes, pearl ash
Holzkohleneisen *n* charcoal iron, fined iron, refined iron
Holzkohlenfeuer *n* charcoal fire
Holzkohlenfeuerung *f* heating with charcoal
Holzkohlenfrischstahl *m* charcoal-hearth steel
Holzkohlenklein *n* charcoal dust
Holzkohlenlösche *f* charcoal breeze, charcoal dust
Holzkohlenmehl *n* charcoal powder, finely ground charcoal
Holzkohlenmeiler *m* charcoal heap, charring mound
Holzkohlenpulver *n* charcoal powder
Holzkohlenroheisen *n* charcoal hearth cast iron, charcoal pig iron, pig iron for refining
Holzkohlenschwärze *f* charcoal black
Holzkohlenstaub *m* charcoal dust, charcoal powder
Holzkohlenteer *m* wood tar
Holzkonservierung *f* wood preservation
Holzkunde *f* dendrology
Holz-Kunststoff-Kombination *f* wood plastic composite
Holzkupfer *n* fibrous olivenite, (Min) native copper arsenate, wood copper
Holzlack *m* wood lacquer
Holzleim *m* wood glue, xylocolla
Holzleiste *f* cleat
Holzlutte *f* wooden air conduit, wooden air duct
Holzmarmor *m* (Min) ligneous marble
Holzmaserung *f* veining
Holzmasse *f* wood pulp, wood paste
Holzmaßstab *m* wooden rule
Holzmehl *n* saw dust, wood dust, wood meal, wood powder
Holzmehlmühle *f* wood meal mill
Holznagel *m* peg
Holzöl *n* tung oil, wood [tar] oil
Holzölfirnis *m* wood oil varnish, gurjun balsam
Holzölsatz *m* wood oil sediment
Holzopal *m* (Min) ligneous opal, semiopal, wood opal
Holzpapier *n* wood paper
Holzpappe *f* wood board

Holzplattenband *n* wood-slat conveyor
Holzpolierlack *m* wood finish
Holzpulver *n* wood powder
Holzrahmenwerk *n* timber framework
Holzruß *m* woodsoot
Holzsäure *f* pyroligneous acid
Holzschalung *f* (Dach) roof boarding, wooden deck
Holzscheibe *f* wooden disc
Holzscheibenrauh *n* wood float finish
Holzscheit *n* fire wood
Holzschliff *m* mechanical pulp, wood pulp
holzschlifffrei free from pulp
Holzschliffpapier *n* mechanical wood pulp paper
Holzschraube *f* wood screw
Holzschutzmittel *n* wood preservative
Holzschwarz *n* wood black
Holzspäne *pl*, **kleine weiche** ~ excelsior
Holzspan *m* wood chip, wood shaving
Holzspanplatte *f* chip board
Holzspiritus *m* methanol, carbinol, methyl alcohol, wood alcohol
Holzsplitter *m* sliver
Holzstandöl *n* tung standoil
Holzstein *m* petrified wood
Holzsteinkohle *f* lignite
Holzstiel *m* wooden handle
Holzstift *m* [wooden] peg
Holzstoff *m* lignin, lignocellulose, mechanical wood pulp
Holzstoffaserungsmaschine *f* wood pulp shredding machine
holzstofffrei free from lignin, free from wood pulp
holzstoffhaltig containing lignin
Holzteer *m* wood tar
Holzteeröl *n* wood tar oil
Holzteilchen *n* ligneous matter
Holztorf *m* wood peat
Holztränkung *f* impregnation of wood
Holztrockenanlage *f* timber drying plant
holzverarbeitend wood-working
Holzverarbeitung *f* wood-working
Holzveredlung *f* wood processing
Holzvergaser *m* wood distilling apparatus
Holzvergasung *f* distillation of wood
Holzverkohlung *f* carbonization of wood, charcoal burning, charring of wood
Holzverzuckerung *f* hydrolysis of wood, wood saccharification
Holzwelle *f* wooden shaft
Holzwerk *n* wood construction
Holzwerkstoffe *pl* particle board
Holzwolle *f* wood fiber, excelsior, wood wool
Holzwollfilter *n* wood wool filter
Holzwollseil *n* wood wool rope
Holzzellstoff *m* chemical wood pulp, lignocellulose, wood cellulose

Holzzement *m* wood cement
holzzerstörend lignicidal
Holzzeug *n* wood cellulose, wood pulp
Holzzimt *m* cassia bark
Holzzinnerz *n* (Min) fibrous cassiterite, fibrous tin ore, wood tin
Holzzucker *m* wood sugar, xylose
Holzzunder *m* tinder, touch wood
Homarin *n* homarine
Homatropin *n* homatropine
Homboldtit *m* (Min) homboldtite, datolite
Homilit *m* (Min) homilite
Homoallantoin *n* homoallantoin
Homoallantoinsäure *f* homoallantoic acid
Homoanissäure *f* homoanisic acid
Homoanthranilsäure *f* homoanthranilic acid
Homoanthroxansäure *f* homoanthroxanic acid
Homoapocamphersäure *f* homoapocamphoric acid
Homoarginin *n* homoarginine
Homoasparagin *n* homoasparagine
Homoberberinbase *f* homoberberine base
Homobetain *n* homobetaine
Homobrenzcatechin *n* homopyrocatechol
Homocamphenilon *n* homocamphenilone
Homocampher *m* homocamphor
Homocamphersäure *f* homocamphoric acid
Homocaronsäure *f* homocaronic acid
Homochelidonin *n* homochelidonine
Homochinolin *n* homoquinoline
Homochinolinsäure *f* homoquinolinic acid
Homocholin *n* homocholine
homochrom homochromatic, homochromous, of uniform color
Homochrysanthemsäure *f* homochrysanthemic acid
Homoconiin *n* homoconiine, homoconine
Homocuminsäure *f* homocumic acid, homocuminic acid
Homocystamin *n* homocystamine
Homocystein *n* homocysteine
Homocystin *n* homocystine
homodispers homodisperse
Homoeder *n* (Krist) homohedron
Homoedrie *f* (Krist) homohedrism
homoedrisch (Krist) homohedral
homöomorph (Krist) homeomorphous, isomorphous
Homöomorphie *f* (Krist) homeomorphism
Homöopathie *f* homoeopathy
homöopathisch homeopathic
homöopolar homeopolar
homöotekt homeotect
Homoeriodictyol *n* homoeriodictyol
Homofenchol *n* homofenchol
Homogallusaldehyd *m* homogallaldehyde
homogen homogeneous, uniform as to grain size
Homogeneisen *n* homogeneous iron
Homogenisator *m* homogenizer

Homogenisierapparat *m* homogenizer
homogenisieren to homogenize, to make homogeneous
Homogenisieren *n* homogenizing
Homogenisiermaschine *f* homogenizer
Homogenisiersilo *m* homogenizing silo
Homogenisierung *f* homogenization
Homogenisierzone *f* melt zone, metering section, metering zone
Homogenität *f* homogeneity, homogeneousness
Homogenkohle *f* solid carbon
Homogenreaktor *m* homogeneous reactor
Homogenstahl *m* homogeneous steel
Homogentisinsäure *f* homogentisic acid
Homogeraniol *n* homogeraniol
Homogeraniumsäure *f* homogeranic acid
Homoguajacol *n* homoguaiacol, 2-methoxy-1-hydroxy-4-methylbenzene, creosol
Homoheliotropin *n* homoheliotropin
Homohordenin *n* homohordenine
Homokaffeesäure *f* homocaffeic acid
Homolävulinsäure *f* homolevulinic acid
Homoleucin *n* homoleucine
Homolinalool *n* homolinalool
Homolkasche Base *f* Homolka's base
homolog homologous
Homologe *n* homologue
Homologie *f* homology
Homolyse *f* homolysis
Homolysin *n* homolysine
Homomenthen *n* homomenthene
Homomenthon *n* homomenthone
Homomerochinen *n* homomeroquinene
Homomesiton *n* homomesitone
Homomesityloxid *n* homomesityl oxide
Homomorphismus *m* (Math) homomorphism
Homomorpholin *n* homomorpholine
Homomuscarin *n* homomuscarine
Homonataloin *n* homonataloin
Homoneurin *n* homoneurine
Homonicotinsäure *f* homonicotinic acid
Homonorcamphersäure *f* homonorcamphoric acid
homonym homonymic, homonymous
Homophoron *n* homophorone
Homophthalimid *n* homophthalimide
Homophthalsäure *f* homophthalic acid
Homopinen *n* homopinene
Homopinenol *n* homopinenol
Homopinol *n* homopinol
Homopiperidin *n* homopiperidine
Homopiperonal *n* homopiperonal
Homopiperonylalkohol *m* homopiperonyl alcohol
Homopiperonylamin *n* homopiperonyl amine
Homopiperonylsäure *f* homopiperonylic acid
homopolymer homopolymer
Homopolymerisation *f* homopolymerization

Homopolysaccharid *n* homopolysaccharide
Homopterocarpin *n* homopterocarpine
Homosalicylaldehyd *m* homosalicylaldehyde
Homosalicylsäure *f* homosalicylic acid
Homosaligenin *n* homosaligenin
Homosekikasäure *f* homosekikaic acid
Homoserin *n* homoserine
Homospezifität *f* homospecificity
Homoterephthalsäure *f* homoterephthalic acid
Homoterpenylsäure *f* homoterpenylic acid
Homoterpineol *n* homoterpineol
Homotetrophan *n* homotetrophan
Homotropin *n* homotropine, homatropine, tropine mandelate
Homovanillin *n* homovanillin
Homovanillinsäure *f* homovanillic acid
Homoveratrol *n* homoveratrole
Homoveratrylalkohol *m* homoveratryl alcohol
Homoverbanen *n* homoverbanene
homozygot (Genpaar) homozygous (gene pair)
Homozygot *m* (Biol) homozygote
homozyklisch homocyclic
Honig *m* honey
honigähnlich honey-like, melleous
Honiggeruch *m* odor of honey
Honiggeschmack *m* taste of honey
Honigscheibe *f* honey comb
Honigseim *m* clarified honey, liquid honey, virgin honey
Honigstein *m* (Min) honeystone, mellite
Honigtopas *m* (Min) honeystone, mellite
Honigwasser *n* honey water, hydromel
Honigwein *m* honeywine, mead, mulse
Hookesches Gesetz *n* Hooke's law
Hopeit *m* (Min) hopeite
Hopen *n* hopene
Hopenol *n* hopenol
Hopenon *n* hopenone
Hopfen *m* hops
Hopfenabkochung *f* (Brau) decoction of hops
hopfenähnlich like hops, lupuline
Hopfenaroma *n* aroma of hops
Hopfenaufguß *m* infusion of hops
Hopfenbau *m* hop growing
Hopfenbier *n* (Brau) hopped beer
Hopfenbitter *n* hop bitters, lupulin
Hopfenbittersäure *f* hop bitter acid, lupulinic acid
Hopfenbitterstoff *m* hop bitter
Hopfendarre *f* hopkiln
Hopfenhefe *f* superficial yeast, surface yeast
Hopfenkaltraum *m* hop cooling room
Hopfenkühlanlage *f* hop cooling plant
Hopfenkühlraum *m* hop cooling room
Hopfenlager *n* hop storage
Hopfenlagerkühlung *f* hop store refrigeration
Hopfenmehl *n* hop dust, lupulin
Hopfenöl *n* hop oil
Hopfensäure *f* hop bitter acid, lupulinic acid

Horbachit *m* (Min) horbachite
Horchgerät *n* sound locator
Horde *f* shalf pass
Hordein *n* hordeine, hordenine
Hordenin *n* hordeine, hordenine
Hordenkontakt *m* grid-type catalyst
Hordenkontaktkessel *m* tray converter
Hordenofen *m* quench tube reactor, shelf-type reactor [or cenvertor]
Hordensublimator *m* tray sublimer
Hordentrockner *m* shelf dryer, truck dryer
Horizont *m* horizon
horizontal horizontal
Horizontalachse *f* horizontal axis
Horizontalauflösung *f* horizontal resolution
Horizontale *f* horizontal line
Horizontalebene *f* horizontal plane
Horizontalförderband *n* horizontal belt conveyer
Horizontalreihe *f* (einer Matrix) horizontal row (of a matrix)
Horizontalrohrverdampfer *m* horizontal-tube evaporator
Horizontalschlämmer *m* hindered-settling classifier
Horizontalschnitt *m* horizontal section
Horizontalverschiebung *f* horizontal displacement, horizontal shift
Hormon *n* hormone, **adrenocorticotropes** ~ adrenocorticotropic hormone, **antidiuretisches** ~ antidiuretic hormone, **diabetogenes** ~ diabetogenic hormone, **gelbkörpererzeugendes** ~ corpus luteum hormone, progesterone, **gonadotropes** ~ gonadotropic hormone, **parathyroides** ~ parathyroid hormone
Hormonabsonderung *f* hormone secretion
hormonal hormonal, hormonic
Hormonal *n* peristaltic hormone
hormonarm hypohormonal
Hormonbehandlung *f* (Med) hormonal treatment
Hormongabe *f* (Med) administration of a hormone
Hormongleichgewicht *n* hormone balance
Hormonhaushalt *m* hormone balance
Hormonpräparat *n* hormone preparation
Hormonwirkung *f* hormonal action
Horn *n* horn
Hornabfall *m* horn shavings
Hornachat *m* (Min) horny agate
hornähnlich horn-like
hornartig horny
Hornbeize *f* maceration of horn
Hornblatt *n* horn sheet
Hornblei *n* (Min) hornlead, cromfordite, native lead chloride carbonate, phosgenite
Hornbleierz *n* (Min) phosgenite
Hornblendefels *m* (Min) amphibolite
Hornblendeschiefer *m* (Min) hornblende schist, hornblende slate, schistous amphibolite

Hornchlorsilber *n* (Min) cerargyrite, corneous silver, horn silver, native silver chloride
Hornerz *n* (Min) cerargyrite, corneous silver, horn silver, native silver chloride
Hornfeile *f* horn file, horn rasp
Hornfels *m* (Min) hornblende rock, hornstone
Hornflint *m* (Min) silicious schist
Horngewebe *n* horny tissue
Hornhaut *f* horny skin
hornig horny, corneous, hornlike
hornisieren to hornify
Hornisieren *n* hornifying
Hornisierung *f* hornification
Hornkobalt *m* (Min) asbolite, earthy cobalt
Hornkohle *f* charred horn, horn charcoal, horn coal
Hornmangan *n* (Min) photicite
Hornmehl *n* horn meal
Hornpergament *n* horn parchment
Hornporphyr *m* (Min) horn stone porphyry
Hornquecksilber *n* (Min) mercurial horn-ore, calomel, horn quicksilver, native mercurous chloride
Hornring *m* (der Felge) side ring
Hornschabsel *pl* horn shavings
Hornschiefer *m* (Min) horn slate, schistous amphibolite
Hornsilber *n* (Min) cerargyrite, corneous silver, horn silver, native silver chloride
Hornspäne *m pl* horn shavings
Hornstein *m* (Min) hornstone, chalcedony, chert, rockflint
hornsteinartig cherty
Hornsteinkies *m* chert gravel
Hornsteinporphyr *m* (Min) hornstone porphyry
Hornsteinwacke *f* petrosilicious wacke
Hornstoff *m* horny substance, keratin
Horsfordit *m* (Min) horsfordite
Hortensienblau *n* ferric ferrocyanide, Prussian blue
Hortiamin *n* hortiamine
Hortonolith *m* (Min) hortonolite
Hosemiazid *n* hosemiazide
Hosenmischer *m* twin-shell or Vee mixer
Houghit *m* houghite
Howlith *m* (Min) howlite
HOZ Abk. für Hochofenzement
HPC-Ruß *m* hard processing channel black
HR Abk. für Rockwell-Härte
H-Säure *f* H acid
Hub *m* (Kolben) stroke, (Maschinentisch) travel, (Pumpe etc.) lift, **doppelter** ~ up and down stroke
Hubbegrenzer *m* collar, stop
Hubbegrenzung *f* length of stroke
Hubbewegung *f* lifting motion, reciprocating motion
Hubbodenzentrifuge *f* underdriven centrifuge

Hubhöhe *f* height of lift, height of stroke, length of stroke
Hubkolbenverdichter *m* reciprocating compressor
Hubkraft *f* lifting force, (Kran) hoisting capacity
Hubkurve *f* cam
Hublänge *f* length of stroke
Hubleistung *f* lifting capacity, (Kran) hoisting capacity
Hubofen *m* oscillating grate oven
Hubpumpe *f* lifting pump
Hubraum *m* piston displacement, stroke volume
Hubschrauber *m* helicopter
Hubstange *f* lifter rod
Hubventil *n* lift valve
Hubverhältnis *n* stroke ratio
Hubvolumen *n* (Mech) [piston] displacement
Hubwagen *m* elevating platform truck
Hubwechsel *m* reversal of stroke
Hubwerk *n* hoisting unit, hoist drive, hoisting gear train
Hubzähler *m* stroke counter
Hubzahl *f* number of strokes (per minute)
Hudsonit *m* (Min) hudsonite
Hübnerit *m* (Min) huebnerite
Hügelit *m* (Min) hugelite
Hülle *f* (Einhüllung) envelope, jacket, (Schale) shell
Hüllenelektron *n* orbital electron
Hüllenspektren *n pl* shell spectra
Hüllprotein *n* coat protein
Hüllrohr *n* casing tube, jacket tube
Hülse *f* shell, casing, jacket, sheath, (Bot) hull, (Einsatzhülse) adaptor, (Etui) case, (Techn) bushing, sleeve
hülsen to shell
Hülsenfrüchte *pl* leguminous plants
Hülsenkupplung *f* sleeve coupling, friction clip
Hülsenschliff *m* socket joint
Hülsenwickelmaschine *f* core winding machine
Hütte *f* metallurgical or smelting plant, blast furnace plant, ironworks, mill works
Hüttenafter *n* dross, residuum, slag from foundry
Hüttenaluminium *n* commercially pure aluminium, primary aluminum pig
Hüttenanlage *f* smelting plant
Hüttenarbeit *f* founding or foundry work, smelting
Hüttenarbeiter *m* foundry worker
Hüttenbedarf *m* smelting works requirements
Hüttengare *f* refinery
Hüttengas *n* smelter gas
Hüttenglas *n* pot metal
Hüttenindustrie *f* metallurgical industry
Hüttenkalk *m* blast furnace lime
Hüttenkoks *m* metallurgical coke, smelting coke
Hüttenkunde *f* metallurgy

Hüttenmehl *n* flowers of arsenic, white arsenic
Hüttennichts *n* white tutty, impure zinc oxide, nihilum album
Hüttenrauch *m* metallurgical smoke, arsenic trioxide, flue dust, smelter smoke
Hüttenspeise *f* ores to be smelted
Hüttenstein *m* slag brick
Hüttentechnik *f* metallurgical engineering
Hüttentrichter *m* conical funnel
Hüttenwerk *n* smelting works, iron works, metallurgical plant
Hüttenwerkanlagen *f pl* metallurgical works or plants
Hüttenwerksabwasser *n* waste water from steel mills
Hüttenwesen *n* metallurgy, smelting
Hüttenwolle *f* slag wool
Hüttenzement *m* slag cement
Hüttenzinn *n* grain tin
Hufeisen *n* horseshoe
Hufeisenmagnet *m* horseshoe magnet
Huffett *n* neat's foot oil
Hufstabeisen *n* horseshoe iron
Hullit *m* (Min) hullite
Hulsit *m* (Min) hulsite
Hulupon *n* hulupone
Humat *n* humate, salt or ester of humic acid
Humbertiol *n* humbertiol
Humboldtilith *m* (Min) humboldtilite
Hume-Rothery-Regel *f* Hume-Rothery rule
Humin *n* humin
Huminsäure *f* humic acid
Humit *m* (Min) humite
Humulan *n* humulane
Humulen *n* humulene
Humulin *n* humulin, lupulin
Humulinsäure *f* humulic acid
Humulochinon *n* humuloquinone
Humulohydrochinon *n* humulohydroquinone
Humulon *n* humulon
Humus *m* humus, vegetable soil
humusartig humous
Humuserde *f* humus earth, vegetable soil
Humuskohle *f* humic carbon, humus coal
Humuskolloid *n* humus colloid
Humuslösung *f* humus solution
humusreich rich in humus
Humusschicht *f* humus layer, layer of mold
hundertgradig centigrade
hundertjährig centennial, secular
Hundertsatz *m* percentage
Hundertstel *n* one hundredth part, per cent
Hundertstelle *f* hundred's place
Hunderttausendstel *n* hundred thousandth [part]
hundertteilig centesimal, centigrade
Hungerflotation *f* starvation flotation
Hunnemannin *n* hunnemanine
Hupe *f* horn
Hureaulit *m* (Min) hureaulite

Huronit *m* huronite
Hussakit *m* (Min) hussakite
Husten *m* cough
Hustenbonbon *m* (Pharm) cough drop
Hustenmittel *n* (Pharm) cough remedy
Hutchinsonit *m* (Min) hutchinsonite
Hutmutter *f* cap nut, cover nut
huttenmännisch metallurgical
Hutzucker *m* loaf sugar
HV Abk. für Vickershärte
Hyacinth *m* (Min) hyacinth, jacinth, transparent red zircon
Hyacinthenöl *n* hyacinth oil, jacinth oil
Hyacinthgranat *m* (Min) cinnamon stone, essonite
Hyacinthkristall *m* hyacinthine crystal
Hyänasäure *f* hyenasic acid, hyenic acid
hyalin hyaline, resembling glass
Hyalit *m* (Min) hyalite
Hyalitglas *n* hyalite
Hyalobiuronsäure *f* hyalobiouronic acid
Hyalographie *f* hyalography
hyalographisch hyalographic
Hyaloidin *n* hyaloidine
Hyalomelan *m* hyalomelane
Hyalophan *m* (Min) hyalophane, barium feldspar
Hyalophotographie *f* hyalophotography
Hyalosiderit *m* (Min) hyalosiderite
Hyalotechnik *f* hyalotechnics
hyalotechnisch hyalotechnical
Hyalotekit *m* hyalotekite
Hyalurgie *f* hyalurgy
Hyaluronidase *f* (Biochem) hyaluronidase
Hyaluronsäure *f* hyaluronic acid
Hybrid *n* hybrid
Hybridation *f* (Biol) hybridization
Hybridbindung *f* hybrid binding
hybridisieren (Biol) hybridize
Hybridisierung *f* hybridization
Hydantoin *n* hydantoin, glycolylurea, imidazoledione
Hydantoinsäure *f* hydantoic acid, carbamidoacetic acid, carbamylglycine, glycoluric acid
Hydnocarpusöl *n* hydnocarpus oil
Hydnocarpussäure *f* hydnocarpic acid
Hydnoresinotannol *n* hydnoresinotannol
Hydracetin *n* hydracetin, pyrodin
Hydracetylaceton *n* hydracetylacetone
Hydracrylsäure *f* hydracrylic acid, 3-hydroxypropanoic acid
Hydracrylsäurealdehyd *m* hydracrylic aldehyde, hydracrolein
Hydralazin *n* hydralazine
Hydramid *n* hydramide
Hydramin *n* hydramine
Hydrangenol *n* hydrangenol
Hydrangin *n* hydrangine

Hydrant *m* hydrant
Hydrargillit *m* (Min) hydrargillite, gibbsite, native aluminum hydroxide
Hydrargyrol *n* hydrargyrol, mercury p-phenyl thionate
Hydrargyrum *n* hydrargyrum, mercury, quicksilver
Hydrastin *n* hydrastine
Hydrastinin *n* hydrastinine
Hydrastininhydrochlorid *n* hydrastinine hydrochloride
Hydrastinsulfat *n* hydrastine sulfate
Hydrastisrhizom *n* hydrastis rhizome
Hydrastsäure *f* hydrastic acid
Hydrat *n* hydrate
Hydratase *f* (Biochem) hydratase
Hydratation *f* hydration
Hydratationswärme *f* heat of hydration, hyration heat
Hydratbildung *f* hydration
Hydratcellulose *f* cellulose hydrate, hydrated cellulose
hydrathaltig hydrate-containing, hydrated
hydratisch hydrated
hydratisieren to hydrate
Hydratisierung *f* hydration
Hydratisierungsgrad *m* degree of hydration
Hydratisomerie *f* hydrate isomerism
Hydratropaalkohol *m* hydratropic alcohol
Hydratropasäure *f* hydratropic acid
Hydratsäure *f* acid hydrate, hydrated acid
Hydratwasser *n* hydration water, water of hydration
Hydratzellulose *f* (Chem) hydrated cellulose
Hydraulik *f* hydraulics
Hydraulikbagger *m* hydraulic excavator
Hydraulikflüssigkeit *f* hydraulic fluid
hydraulisch hydraulic
Hydrazicarbonyl *n* hydrazicarbonyl
Hydrazid *n* hydrazide
Hydraziessigsäure *f* hydraziacetic acid
Hydrazimethylen *n* hydrazimethylene
Hydrazin *n* hydrazine, diamidogen, diamine
Hydrazinbifluorid *n* acid hydrazine fluoride, hydrazine bifluoride, hydrazine hydrogen fluoride
Hydrazinchlorid *n* hydrazine chloride, hydrazine dihydrochloride
Hydrazindihydrochlorid *n* hydrazine dihydrochloride
Hydrazindimalonsäure *f* hydrazine dimalonic acid
Hydrazingelb *n* hydrazine yellow
Hydrazinhydrat *n* hydrazine hydrate
Hydrazinsulfat *n* hydrazine sulfate
Hydrazoameisensäure *f* hydrazoformic acid
Hydrazobenzol *n* hydrazobenzene, diphenylhydrazine
Hydrazodicarbonamid *n* hydrazodicarbonamide, biurea, hydrazoformamide
Hydrazoformamid *n* hydrazoformamide, biurea, hydrazodicarbonamide
Hydrazoformhydrazid *n* hydrazoformhydrazide, hydrazodicarbonhydrazide
Hydrazokörper *m* hydrazo compound
Hydrazomethan *n* hydrazomethane
Hydrazon *n* hydrazone
Hydrazotoluol *n* hydrazotoluene
Hydrazoverbindung *f* hydrazo compound
Hydrid *n* hydride
Hydrid-Übertragung *f* hydride transfer
Hydrieranlage *f* hydrogenation plant
Hydrierapparat *m* hydrogenation apparatus
Hydrierbenzin *n* hydrogenation process gasoline
hydrieren to hydrogenate
Hydrierraffinat *n* hydro-refined product
Hydrierung *f* hydrogenation, **katalytische** ~ catalytic hydrogenation, **partielle** ~ partial hydrogenation, **stereoelektronisch gesteuerte** ~ (Stereochem) stereoelectronically controlled hydrogenation
Hydrierungsverfahren *n* hydrogenation process
Hydrierwerk *n* hydrogenation plant
Hydrin *n* hydrine
Hydrindacen *n* hydrindacene
Hydrindan *n* hydrindane
Hydrinden *n* hydrindene, indan
Hydrindinsäure *f* hydrindic acid, o-aminomandelic acid
Hydrindol *n* hydrindole
Hydrindon *n* hydrindone, indanone, indone
Hydroanisoin *n* hydroanisoin
Hydroanthracen *n* hydroanthracene
Hydroapatit *m* hydroapatite
hydroaromatisch hydroaromatic
Hydrobagger *m* hydraulic excavator
Hydrobarometer *n* hydrobarometer
Hydrobenzamid *n* hydrobenzamide
Hydrobenzoin *n* hydrobenzoin
Hydroberberin *n* hydroberberine
Hydrobilirubin *n* hydrobilirubin, urobilin
Hydrobiologie *f* hydrobiology
Hydrobixin *n* hydrobixin
Hydroboracit *m* (Min) hydroboracite
Hydrobornylen *n* hydrobornylene
Hydroborocalcit *m* (Min) hydroborocalcite, hayesenite, hydrous borate of lime
Hydrobromid *n* hydrobromide, bromhydrate
Hydrobromsäure *f* hydrobromic acid, hydrogen bromide
hydrobromsauer hydrobromide
Hydrocamphen *n* hydrocamphene
Hydrocampnospermonol *n* hydrocampnospermonol
Hydrocarbostyril *n* hydrocarbostyril
Hydrocardanol *n* hydrocardanol
Hydrocellobial *n* hydrocellobial

Hydrocellulose *f* hydrocellulose
Hydrocelluloseacetat *n* hydrocellulose acetate
Hydrocephaelin *n* hydrocephalin
Hydrocerit *m* hydrocerite
Hydrocerussit *m* (Min) hydrocerussite
Hydrochalkon *n* hydrochalcone
Hydrochelidonsäure *f* hydrochelidonic acid, 4-oxoheptanedioic acid
Hydrochinan *n* hydroquinane
Hydrochinen *n* hydroquinene
Hydrochinin *n* hydroquinine, methylhydrocupreine
Hydrochinon *n* hydroquinone, hydroquinol, para-dihydroxybenzene
Hydrochinondimethyläther *m* hydroquinol dimethyl ether
Hydrochinonentwickler *m* hydroquinone developer
Hydrochinonmonomethyläther *m* hydroquinol monomethyl ether
Hydrochinotoxin *n* hydroquinotoxine
Hydrochinoxalin *n* hydroquinoxaline
Hydrochlorid *n* hydrochloride
Hydrochlorsäure *f* hydrochloric acid, hydrogen chloride
Hydrochromon *n* hydrochromone
Hydrocinchonicin *n* hydrocinchonicine, hydrocinchotoxine
Hydrocinchonidin *n* hydrocinchonidine, hydrochinidine
Hydrocinchonin *n* hydrocinchonine
Hydrocinchotoxin *n* hydrocinchotoxine, hydrocinchonicine
Hydrocinnamid *n* hydrocinnamide
Hydrocinnamoin *n* hydrocinnamoin
Hydrocollidin *n* hydrocollidine
Hydroconchinin *n* hydroconchinine, hydroquinidine
Hydrocortison *n* hydrocortisone
Hydrocotarnin *n* hydrocotarnine
Hydrocotoin *n* hydrocotoin
Hydrocrackreaktion *f* hydrocracking, hydrogenation cracking reaction
Hydrocresol *n* hydrocresol
Hydrocumarilsäure *f* hydrocoumarilic acid
Hydrocumarin *n* hydrocoumarin
Hydrocumaron *n* hydrocoumarone
Hydrocumarsäure *f* hydrocoumaric acid
Hydrocuprean *n* hydrocupreane
Hydrocupreen *n* hydrocypreene
Hydrocyanid *n* hydrocyanide
Hydrocyansäure *f* hydrogen cyanide, formonitrile, hydrocyanic acid, Prussic acid
Hydrodiffusion *f* hydrodiffusion
Hydrodolomit *m* (Min) hydrodolomite
Hydrodynamik *f* hydrodynamics
hydrodynamisch hydrodynamic, hydrodynamical
Hydrodynamometer *n* hydrodynamometer
hydroelektrisch hydroelectric

Hydroelement *n* hydroelement, hydrocell
Hydroextraktor *m* hydroextractor
Hydrofluocerit *m* (Min) hydrofluocerite
Hydrofluorid *n* hydrofluoride
Hydrofluorsäure *f* hydrofluoric acid, hydrogen fluoride
Hydroformieren *n* (Chem) hydroforming
Hydroformylierung *f* hydroformylation
hydrogalvanisch hydrogalvanic
Hydrogel *n* hydrogel
Hydrogen *n* hydrogen
Hydrogenase *f* (Biochem) hydrogenase
Hydrogencarbonat *n* acid carbonate, bicarbonate, hydrogen carbonate
Hydrogeniumborat *n* boric acid
Hydrogeniumchlorat *n* chloric acid
Hydrogeniumoxid *n* water
Hydrogenlicht *n* oxyhydrogen light, calcium light, drummond's light, lime light
Hydrogenolyse *f* hydrogenolysis
Hydrogenperoxid *n* hydrogen dioxide, hydrogen peroxide
Hydrogenpol *m* negative pole
Hydrogenpolysulfid *n* hydrogen polysulfide
Hydrogensalz *n* hydrogen salt
Hydrogenschwefel *m* hydrogen sulfide
Hydrogensulfat *n* hydrogen sulfate, acid sulfate, bisulfate
Hydrogensulfid *n* hydrosulfide
Hydroginkgol *n* hydroginkgol
Hydrographie *f* hydrography
Hydrohämatit *m* (Min) hydrohematite, crystalline hydrated ferric oxide
Hydrohydrastinin *n* hydrohydrastinine
Hydroit *m* hydroite
Hydrojodsäure *f* hydrogen iodide
Hydrojuglon *n* hydrojuglone
Hydrokaffeesäure *f* hydrocaffeic acid, hydrocoffeic acid
Hydrokautschuk *m* hydrocaoutchouc, hydrorubber
Hydrokette *f* hydrocell, hydroelement
Hydrol *n* hydrol
Hydrolactal *n* hydrolactal
Hydrolapachol *n* hydrolapachol
Hydrolase *f* (Biochem) hydrolase
Hydrolith *m* hydrolith, calcium hydride
Hydrologie *f* hydrology
hydrologisch hydrologic
Hydrolomatiol *n* hydrolomatiol
Hydrolysat *n* hydrolysate
Hydrolyse *f* hydrolysis
Hydrolysegeschwindigkeit *f* hydrolysis rate
Hydrolysenfällung *f* hydrolysis precipitation
hydrolysierbar hydrolyzable
hydrolysieren to hydrolyze, to undergo hydrolysis
hydrolysierend hydrolyzing
hydrolytisch hydrolytic

Hydromagnesit *m* (Min) hydromagnesite
Hydromagnocalcit *m* (Min) hydromagnocalcite
Hydromechanik *f* hydromechanics
hydromechanisch hydromechanical
Hydromelanothallit *m* (Min) hydromelanothallite
Hydrometallurgie *f* hydrometallurgy
Hydrometer *n* hydrometer, densimeter, hydraulic gauge
Hydrometrie *f* hydrometry
hydrometrisch hydrometric
Hydrometrograph *m* hydrometrograph
Hydromotor *m* hydromotor
Hydronalium *n* hydronalium
Hydronblau *n* hydron blue
Hydroniumion *n* hydronium ion
Hydrooctinum *n* hydrooctinum
Hydroorotsäure *f* hydroorotic acid
hydropathisch hydropathic
Hydroperoxid *n* hydroperoxide
Hydropersulfid *n* hydropersulfide
Hydrophan *m* (Min) hydrophane, transparent opal
hydrophil hydrophilic
Hydrophilierung *f* (Text) hydrophilizing
hydrophob hydrophobe, hydrophobic, moisture-repellent, non-absorbent, non-hygroscopic
Hydrophobierung *f* (Text) hydrophobizing, waterproofing
Hydrophobierungsmittel *n* water-repellent agent
Hydrophthalsäure *f* hydrophthalic acid
Hydrophyt *m* (Wasserpflanze) hydrophyte
Hydropiperinsäure *f* hydropiperic acid, hydropiperinic acid
Hydropiperoin *n* hydropiperoin
Hydropit *m* hydropite
hydropneumatisch hydropneumatic
Hydroprotopin *n* hydroprotopine
Hydropyrin *n* hydropyrine, lithium acetylsalicylate
Hydroresorcin *n* hydroresorcinol
Hydrorhombinin *n* hydrorhombinine
Hydroselenid *n* hydroselenide, selenide
Hydroselensäure *f* hydroselenic acid
Hydrosilikat *n* hydrosilicate
Hydrosol *n* hydrosol
Hydrosorbinsäure *f* hydrosorbic acid, hexenic acid, hexenoic acid
Hydrospaltung *f* hydrocracking, hydrogenolysis
Hydrosphäre *f* hydrosphere
Hydrostatik *f* hydrostatics
hydrostatisch hydrostatic
Hydrosulfid *n* hydrosulfide, sulfhydrate
Hydrosulfit *n* hydrosulfite, dithionite
Hydrotalkit *m* (Min) hydrotalcite
Hydrotechnik *f* hydrotechnics
hydrotechnisch hydrotechnical
Hydrotellurid *n* hydrotelluride, tellurhydrate

Hydrotellursäure *f* hydrotelluric acid, hydrogentelluride
hydrothermal hydrothermal
Hydrotimeter *n* (Techn) hydrotimeter
Hydrotoluoin *n* hydrotoluoin
Hydrotroilit *m* hydrotroilite
Hydrotropie *f* hydrotropy
Hydrouracil *n* hydrouracil
Hydrourushiol *n* hydrourushiol
Hydroverbindung *f* hydro compound
Hydroxamsäure *f* hydroxamic acid
Hydroxanthommatin *n* hydroxanthommatin
Hydroxid *n* hydroxide
Hydroxokomplex *m* hydroxide complex
Hydroxonsäure *f* hydroxonic acid
Hydroxyamin *n* hydroxy-amine
Hydroxyaminosäure *f* hydroxy-amino acid
Hydroxyamphetamin *n* hydroxyamphetamine
Hydroxyanthrachinon *n* hydroxyanthraquinone
Hydroxyanthranilsäure *f* hydroxyanthranilic acid
Hydroxyaphyllin *n* hydroxyaphylline
Hydroxyazobenzol *n* hydroxyazobenzene
Hydroxyazoverbindung *f* hydroxyazocompound
Hydroxybenzaldehyd *m* hydroxybenzaldehyde
Hydroxybenzanthron *n* hydroxybenzanthrone
Hydroxybenzoesäure *f* hydroxybenzoic acid
Hydroxybenzylalkohol *m* hydroxybenzyl alcohol
Hydroxybenzylcyanid *n* hydroxybenzyl cyanide
Hydroxybuttersäure *f* hydroxybutyric acid
Hydroxychinolin *n* hydroxyquinoline
Hydroxychlorochin *n* hydroxychloroquine
Hydroxydecansäure *f* hydroxydecanoic acid
Hydroxydesoxycorticosteron *n* hydroxydesoxycorticosterone
Hydroxydiphenyl *n* hydroxydiphenyl
Hydroxydiphenylamin *n* hydroxydiphenylamine
Hydroxydroseron *n* hydroxydroserone
Hydroxydul *n* (obs) hydroxide
Hydroxyhumulinsäure *f* hydroxyhumulinic acid
Hydroxyhydrochinon *n* hydroxyhydroquinone
Hydroxyketon *n* hydroxyketone
Hydroxykobalamin *n* hydroxycobalamine
Hydroxyl- *n* hydroxyl
Hydroxylalanin *n* hydroxylalanine
Hydroxylamin *n* hydroxylamine, oxyammonia
Hydroxylaminchlorhydrat *n* hydroxylamine hydrochloride
Hydroxylaminhydrochlorid *n* hydroxylamine hydrochloride
Hydroxylammoniumverbindung *f* hydroxylammonium compound
Hydroxylapatit *m* hydroxylapatite
Hydroxylase *f* (Biochem) hydroxylase
Hydroxylgruppe *f* hydroxyl group, hydroxyl radical, **alkoholische** ~ alcoholic hydroxyl group
hydroxylhaltig containing hydroxyl
hydroxylieren to hydroxylate

Hydroxylierung *f* hydroxylation
Hydroxylion *n* hydroxyl ion
Hydroxylionenkonzentration *f* hydroxyl ion concentration
Hydroxylpalmiton *n* hydroxylpalmitone
Hydroxylradikal *n* hydroxyl radical, hydroxyl group
Hydroxylysin *n* hydroxylysine
Hydroxylzahl *f* hydroxyl number
Hydroxymethoxybenzaldehyd *m* hydroxymethoxy benzaldehyde
Hydroxynervonsäure *f* hydroxynervonic acid
Hydroxynitrobenzylchlorid *n* hydroxynitrobenzyl chloride
Hydroxypentadecylsäure *f* hydroxypentadecylic acid
Hydroxyperezon *n* hydroxyperezone
Hydroxyphenylessigsäure *f* hydroxyphenylacetic acid
Hydroxyphenylnaphthylamin *n* hydroxyphenyl naphthylamine
Hydroxypinsäure *f* hydroxypinic acid
Hydroxyprolin *n* hydroxyproline
Hydroxypropionsäure *f* hydroxypropionic acid
Hydroxypropiophenon *n* hydroxypropiophenone
Hydroxypyren *n* hydroxypyrene
Hydroxypyridin *n* hydroxypyridine
Hydroxysäure *f* hydroxy acid
Hydroxytryptophan *n* hydroxytryptophane
Hydroxyvalin *n* hydroxyvaline
Hydrozelle *f* hydrocell, hydroelement
Hydrozimtaldehyd *m* hydrocinnamic aldehyde, hydrocinnamaldehyde
Hydrozimtalkohol *m* hydrocinnamic alcohol, hydrocinnamyl alcohol
Hydrozimtsäure *f* hydrocinnamic acid
Hydrozinkit *m* (Min) hydrozincite, zinc bloom
Hydrozyklon *m* hydrocyclone
Hydrür *n* (obs) hydride
Hydurilsäure *f* hydurilic acid
Hyenanchin *n* hyenanchin
Hygiene *f* hygiene, (Gesundheitslehre) hygienics, sanitary science
hygienisch hygienic, sanitary
Hygrin *n* hygrine
Hygrinsäure *f* hygrinic acid
Hygrofix *n* hygrofix
Hygrograph *m* hygrograph; recording hygrometer
Hygrometer *n* hygrometer
Hygrometrie *f* hygrometry
hygrometrisch hygrometric
Hygromycin *n* hygromycin
Hygrophyllit *m* (Min) hygrophyllite
Hygroskop *n* hygroscope
Hygroskopie *f* hygroscopy, hygrometry
hygroskopisch hygroscopic
Hygroskopizität *f* hygroscopic capacity, hygroscopicity

Hygroskopizitätszahl *f* hygroscopic coefficient
hylotrop hylotropic
Hyocholsäure *f* hyocholic acid
Hyoscin *n* hyoscine, scopolamine
Hyoscinbromhydrat *n* hyoscine hydrobromide
Hyoscinhydrobromid *n* hyoscine hydrobromide
Hyoscinsulfat *n* hyoscine sulfate
Hyoscyamin *n* hyoscyamine, daturine
Hyoscyaminbromhydrat *n* hyoscyamine hydrobromide
Hyoscyaminchlorhydrat *n* hyoscyamine hydrochloride
Hyoscyaminhydrobromid *n* hyoscyamine hydrobromide
Hyoscyaminsulfat *n* hyoscyamine sulfate
Hyp Abk. für Hydroxyprolin
Hypaphorin *n* hypaphorine
Hypargyrit *m* (Min) hypargyrite
Hyperacidität *f* hyperacidity, superacidity
Hyperämie *f* (Med) hyperemia
Hyperbel *f* (Math) hyperbola
Hyperbelbahn *f* hyperbolic orbit
Hyperbelfunktion *f* (Math) hyperbolic function
Hyperbelinverse *f* hyperbolic inverse
Hyperbelrad *n* hyperbolical gear
hyperbolisch hyperbolic
Hyperboloid *n* hyperboloid
Hypercalcämie *f* (Med) hypercalcemia
Hyperchlorat *n* perchlorate, salt or ester of perchloric acid
Hyperchlorid *n* perchloride
hyperchrom hyperchrome
Hyperchromie *f* hyperchromism, (Med) hyperchromicity
hypereutektisch hypereutectic
hyperfein hyperfine
Hyperfeinaufspaltung *f* hyperfine structure
Hyperfeinspektrum *n* hyperfine spectrum
Hyperfeinstruktur *f* hyperfine structure
Hyperfeinstrukturkopplung *f* hyperfine structure coupling
Hyperfeinstrukturmultiplett *n* hyperfine structure multiplet
Hyperfeinstrukturoperator *m* hyperfine structure operator
Hyperfiltration *f* hyperfiltration, reverse osmosis
Hyperfläche *f* hypersurface
hypergeometrisch hypergeometric
Hyperglykämie *f* (Med) hyperglycemia
Hypergol *n* hypergol
Hypericin *n* hypericin
Hyperin *n* hyperin, hyperoside
Hyperinsulinismus *m* (Med) hyperinsulinism
Hyperjodat *n* periodate, salt or ester of periodic acid
Hyperkern *m* (Atom) hypernucleus
hyperkomplex hypercomplex
Hyperkonjugation *f* hyperconjugation
Hyperol *n* hyperol, ortizon

Hyperon *n* (Atom) hyperon, superproton
Hyperoxyd *n* hyperoxide, peroxide
Hypersensibilisierung *f* (Phot) hypersensitization
Hypersthen *m* (Min) hypersthene, paulit
Hypersthenfels *m* (Min) hypersthène rock
hypersthenhaltig hypersthenic
Hypertension *f* (Med) hypertension
Hyperthyreose *f* (Med) hyperthyreosis
Hypertonie *f* hypertension
hypertonisch (Pharm) hypertonic
Hypervitaminose *f* (Med) hypervitaminosis
Hyphomyzeten *pl* (Fadenpilze) hyphomycetes
Hypnal *n* hypnal, chloralantipyrine
Hypnon *n* hypnone, acetophenone
Hypnoticum *n* (Med) hypnotic, soporific
hypnotisch hypnotic
Hypoblast *m* entoderm
Hypobromit *n* hypobromite, salt or ester of hypobromous acid
Hypochlorit *n* hypochlorite, salt or ester of hypochlorous acid
Hypochloritlösung *f* hypochlorite solution, hypochlorite liquor
Hypochlorsäure *f* hypochlorous acid
Hypochromie *f* (Med) hypochromicity
Hypodiphosphat *n* hypodiphosphate
Hypogäasäure *f* hypogaeic acid, 7-hexadecenoic acid
Hypogallussäure *f* hypogallic acid
hypogen hypogene
Hypoglykämie *f* (Med) hypoglycemia
Hypojodit *n* hypoiodite, salt or ester of hypoiodous acid
Hyponitrit *n* hyponitrite, salt or ester of hyponitrous acid
Hypophosphit *n* hypophosphite, salt or ester of hypophosphorous acid
Hypophosphorsäure *f* hypophosphoric acid
Hypophyse *f* hypophysis, pituitary gland
Hypophysenhinterlappen *m* (Anat) posterior pituitary lobe, posthypophysis
Hypophysenhormon *n* pituitary hormone
Hypophysenüberfunktion *f* hyperhypophysism, hyperpituitarism
Hypophysenvorderlappen *m* anterior lobe of the pituitary gland
Hyposiderämie *f* (Med) hyposideremia
Hyposterol *n* hyposterol
Hyposulfit *n* hyposulfite, dithionite
Hypotaurin *n* hypotaurine
Hypotension *f* (Med) hypotension
Hypotenuse *f* (Math) hypotenuse
Hypothalamus *m* (Med) hypothalamus
Hypothermie *f* (Biol) hypothermia
Hypothese *f* hypothesis
hypothetisch hypothetic[ally]
Hypothyroidismus *m* hypothyroidism
Hypotrochoide *f* hypotrochoid

Hypovitaminose *f* hypovitaminosis, vitamin deficiency
Hypoxanthin *n* hypoxanthine, 6-hydroxy purine, sarkine
hypsochrom hypsochrome, hypsochromic
Hyptolid *n* hyptolide
Hyraldit *n* hyraldite
Hyrgol *n* hyrgol
Hystatit *m* hystatite
Hystazarin *n* hystazarin
Hysterese *f* hysteresis, **dielektrische** ~ dielectric hysteresis, **magnetische** ~ magnetic hysteresis
Hysteresearbeit *f* (Magn) hysteresis energy
Hystereseeffekt *m*, **durch den** ~ **verursachte Verzerrung** (Magn) hysteresis distortion
Hysteresekoeffizient *m* coefficient of hysteresis, hysteresis coefficient
Hystereseschleife *f* hysteresis cycle, hysteresis loop
Hystereseverlust *m* hysteresis loss
Hysteresis *f* (Hysterese) hysteresis
Hythizin *n* hythizine

I

Iatrochemie *f* iatrochemistry
iatrochemisch iatrochemical
Iatrol *n* iatrol
Iba-Säure *f* iba-acid
Iberin *n* iberin
Iberit *m* iberite
Ibervirin *n* ibervirin
Ibit *n* ibit, bismuth hydroxyiodotannate
Ibochin *n* iboquine
Ichibasäure *f* ichiba acid
Ichnographie *f* ichnography
Ichthalbin *n* ichthalbin, ichthyol albuminate
Ichthargan *n* ichthargan, ichthyol silver
Ichthulin *n* ichthulin
Ichthyocoll *n* ichthyocolla, fish glue, isinglass
Ichthyoform *n* ichthyoform, ichthyol formaldehyde
Ichthyol *n* ichthyol, ammonium ichthyolate, ammonium ichthyolsulfonate, anysin
Ichthyoleiweiß *n* ichthyol albuminate, albumin ichthyolate, ichthalbin
Ichthyolersatz *m* ichthyol substitute
Ichthyolit *n* ichthyolite
Ichthyolpräparat *n* ichthyol preparation
Ichthyolpuder *m* ichthyol powder
Ichthyolrohöl *n* crude oil for ichthyol
Ichthyolseife *f* ichthyol soap
Ichthyophthalmit *m* ichthyophthalmite, apophyllite
Ichthyopterin *n* ichthyopterin
Icicabalsam *m* icicabalm
Icicaharz *n* icica resin
Icican *n* icicane
Icocaöl *n* icoca oil
Icthiamin *n* icthiamine
Icylfarbstoff *m* icyl dyestuff
Idaein *n* idaein
Iddingsit *m* (Min) iddingsite
ideal ideal, ~ **leitend** of ideal conductance, of perfect conductance
Idealkristall *m* ideal crystal, perfect crystal
Idealzustand *m* ideal state, perfect condition
identifizieren to identify, recognize
Identifizierung *f* identification
identisch identical
Identität *f* identity
Identitätsabstand *m* (Krist) identity period
Identitätsoperator *m* identity operator
Idiochromasie *f* idiochromasy
idiochromatisch idiochromatic
idioelektrisch idioelectric
Idiokinese *f* change of the germ plasma
idiokinetisch changing the germ plasma, idiokinetic
Idiomer *n* idiomere, chromomere
idiomorph (Krist) idiomorphic, automorphic

idiophan (Krist) idiophanous
Idiophanismus *m* (Krist) idiophanism
Idioplasma *n* idioplasm, germ plasm
Idit *m* idite, iditol
Idokras *m* (Min) idocrase, vesuvian, vesuvianite
Idonsäure *f* idonic acid
Idosaminsäure *f* idosaminic acid
Idose *f* idose
Idozuckersäure *f* idaric acid, idosaccharic acid
Idrialin *n* idrialene
Idrialit *m* (Min) idrialite
Iduronsäure *f* iduronic acid
IE Abk. für Internationale Einheit
Igasursäure *f* igasuric acid
Igelstein *m* (Min) echinite
Igelstromit *m* (Min) igelstromite
Iglesiasit *m* iglesiasite
Ignitron *n* ignitron
Ignotin *n* ignotine, carnosine
Ihleit *m* (Min) ihleite
Ikonometer *n* (Phot) iconometer, view meter
Ikonoskopie *f* iconoscopy
Ikosaeder *n* (Krist) icosahedron
ikosaedrisch (Krist) icosahedral
Ikositetraeder *m* (Krist) icositetrahedron
Ilexanthin *n* ilexanthin
Ilicin *n* ilicin
Ilicylalkohol *m* ilicyl alcohol
Ilizin *n* ilicin
Illipebutter *f* illipé butter, Borneo tallow
Illipefett *n* illipé fat
Illipeöl *n* illipé oil
Illium *n* (Leg) illium
Ilmenit *m* (Min) ilmenite, geikielite, menacantite
Ilmenium *n* ilmenium
Ilmenorutil *m* (Min) ilmenorutil
Ilsemannit *m* (Min) ilsemannite
Iltisfett *n* fitchet fat, polecat fat
Ilvait *m* (Min) ilvaite, yenite
Imabenzil *n* imabenzil
imaginär imaginary
Imaginärteil *m* (einer komplexen Zahl) imaginary part (of a complex number)
Imasatin *n* imasatin, isamic acid lactam
Imasatinsäure *f* imasatinic acid
Imbricarsäure *f* imbricaric acid
Imerinit *m* (Min) imerinite
Imesatin *n* imesatin, 3-iminooxindole
Imid *n* imid[e]
Imidazol *n* 1,3-diazole, imidazole, glyoxaline
Imidazolbase *f* imidazole base
Imidazolgruppe *f* imidazole group
Imidazolidin *n* imidazolidine
Imidazolin *n* imidazoline
Imidazolon *n* imidazolone
Imidazolring *m* imidazole ring

Imidbase *f* imide base
Imidoäther *m* imido ether
Imidocarbamid *n* imidocarbamide, guanidine
Imidocarbonverbindung *f* imidocarbon compound
Imidogruppe *f* imido group, imidogen
Imidoharnstoff *m* guanidine
Imidokohlensäure *f* imidocarbonic acid
Imidosulfonsäure *f* imidosulfonic acid, imidodisulfuric acid
Imidoverbindung *f* imido compound
Imidoxanthin *n* imidoxanthin, guanine
Imidsäureester *m* imido ester
Imin *n* imine
Iminbildung *f* imine formation
Iminodiameisensäure *f* iminodicarboxylic acid, iminodiformic acid
Iminodibenzyl *n* iminodibenzyl
Iminodiessigsäure *f* iminodiacetic acid
Iminodipropionsäure *f* iminodipropionic acid
Iminoindigo *m* iminoindigo
Iminopyrin *n* iminopyrine
Imipramin *n* imipramine
Imitation *f* copy, imitation
imitieren to imitate
Imker *m* bee-keeper
Immatrikulation *f* [university] enrol[l]ment
immatrikulieren to enrol[l]
Immedialgelb *n* immedial yellow
Immedialreinblau *n* immedial pure blue
Immergrün *n* (Bot) periwinkle
Immersionsflüssigkeit *f* immersion fluid
Immersionskondensor *m* (Opt) immersion condenser
Immersionsmethode *f* immersion method
Immersionsobjektiv *n* (Opt) oil-immersion objective
Immersionsöl *n* immersion oil
Immissionen *pl* nuisance level
Immunantikörper *m* immune antibody
immunbiochemisch immunobiochemical
Immunbiologie *f* immunobiology
immunbiologisch immunobiological
Immunchemie *f* immunochemistry
Immuneinheit *f* immunizing unit, I. U.
Immuneiweiß *n* immune protein
Immunelektrophorese *f* immuno-electrophoresis
immunisieren to immunize
Immunisierung *f* immunization, **frühzeitige** ~ pre-immunization, **passive** ~ passive immunization
Immunisierungseffekt *m* immunological effect
Immunisierungseinheit *f* immunizing unit, I.U.
Immunisierungsverfahren *n* immunization procedure
Immunität *f* immunity, ~ **gegen Bakterien** antibacterial immunity, antimicrobic immunity, **angeborene** ~ active immunity, congenital immunity, inherent immunity, innate immunity, **bleibende** ~ permanent immunity, **ererbte** ~ inherited immunity, **erworbene** ~ acquired immunity
Immunitätseinheit *f* antitoxin unit
Immunitätsforschung *f* immunology
immunitätshemmend anti-immune
Immunitätslehre *f* immunology
Immunitätsschwäche *f* hypo-immunity
Immunkörper *m* antibody
Immunkörperspiegel *m* antibody level of the blood
Immunochemie *f* immunochemistry
Immunoelektrophorese *f* immuno-electrophoresis
immuno-elektrophoretisch immuno-electrophoretic
Immunoglobulin *n* immune globuline, immunoglobulin
Immunologe *m* immunologist
Immunologie *f* immunology
immunologisch immunologic
Immunoreaktion *f* immune response
Immunsuppressiva *pl* immunosuppressives
Impedanz *f* impedance, apparent resistance, inductive resistance, **kapazitive** ~ capacitance
Impedanzwandler *m* impedance converter
Impellermischer *m* impeller mixer
Imperatorin *n* imperatorin, peucedanin
Imperialblau *n* imperial blue
impermeabel impermeable
Impermeabilität *f* impermeability
impfen to inject (with crystals), (Med) to inoculate, to vaccinate
Impfen *n* (Krist) inoculation
Impfkristall *m* seed crystal
Impfmethode *f* **zur Racemat-Spaltung** (Stereochem) inoculation method for the separation of enantiomers
Impfserum *n* vaccine
Impfstoff *m* vaccine
Impfstoffgewinnung *f* vaccine production
Impfstoffherstellung *f* vaccine production
Impfung *f* (Krist) seeding, (Med) inoculation, vaccination
implodieren to implode
Implosion *f* implosion
Imprägnationsöl *n* impregnating oil
Imprägnierapparate *m pl* impregnating equipment
imprägnieren to impregnate, to varnish, (wasserfest machen) to waterproof
Imprägnieren *n* impregnation, (Text) waterproofing
Imprägnierflüssigkeit *f* impregnating fluid
Imprägnierharz *n* lac
Imprägnierkessel *m* impregnating tank
Imprägnierlack *m* coating varnish
Imprägniermaschine *f* impregnating machine, lac smearing machine

Imprägniermittel *n* impregnating agent, impregnating preparation, (Text) waterproofing agent
Imprägnier- oder Sättigungsmittel *n* saturant
Imprägnieröl *n* impregnating oil
Imprägnierpfanne *f* impregnating pan
Imprägniertrog *m* impregnating vat
Imprägnierung *f* impregnation, wet-out (of glass fiber mats by polyester), **flammenfeste** ~ flame-proof impregnation
Imprägnierungsmittel *n* impregnating agent, (Text) waterproofing agent
Imprägnierverfahren *n* method of impregnation
improvisieren to improvise
Impsonit *m* (Min) impsonite
Impuls *m* (Anstoß) impulse, (Bewegungsgröße) momentum, (Elektr) pulse
Impulsabtrenner *m* pulse clipper, pulse stripper
Impulsachse *f* axis of rotation
Impulsänderung *f* momentum change
Impulsamplitude *f* (Telev) peak of pulse voltage
Impulsantwort *f* (Math) impulse response
Impulsauslösung *f* pulse triggering
Impulsbegrenzer *m* pulse clipper, pulse stripper
Impulsbegrenzerstufe *f* clipper stage
Impulsbestimmung *f* momentum determination
Impulsbetrieb *m* pulsed operation
Impulsbildungsstufe *f* pulse-forming stage
Impulsbreite *f* impulse width or length, pulse duration
Impulsdauer *f* duration of impulse
Impulseinsatz *m* setting in of impulse
Impulserregung *f* impulse excitation
Impulsfolge *f* series of pulses, impulse sequence, pulse operation
Impulsformer *m* impulse generator
Impulsformgebung *f* pulse shaping
Impulsfortpflanzung *f* pulse traveling
Impulsfrequenzuntersetzer *m* frequency divider
Impulsgabe *f* impulse transmission or emission, impulsing, pulsing
Impulsgeber *m* impulse generator
Impulsgrößenspektrum *n* pulse height spectrum
Impulsmessung *f* momentum measurement
Impulsquantelung *f* quantization of momentum
Impulsraumdarstellung *f* momentum space representation
Impulsraumintegral *n* momentum space integral
Impulsrechner *m* computer
Impulsregler *m* (Elektr) energy regulator
Impulssatz *m* momentum principle, theorem of momentum, (Math) theorem of impulse
Impulsübertragung *f* momentum transfer
Impulsübertragungsreaktion *f* impulsive reaction
Impuls- und Energieaustausch *m* momentum and energy exchange
Impulsverhalten *n* pulse response
impulsverlängernd pulse lengthening
Impulsverstärker *m* pulse amplifier

Impulsverteilung *f* momentum distribution
Impulsverzögerung *f* impulse delay
imstande sein to be able, to be in a position (to)
inaktiv inactive, inert, neutral, **optisch** ~ optically inactive
inaktivieren to inactivate
Inaktivierung *f* inactivation, deactivation, (Opt) racemization
Inaktivität *f* inactivity
Inangriffnahme *f* attack, initiation, start[ing]
Inaugenscheinnahme *f* inspection
inbegriffen included, inclusive
Inbetriebnahme *f* time of beginning of operation, putting into operation, start-up
Inbetriebsetzung *f* starting, initiating, rendering operative, setting in motion, start up
Incanin *n* incanine
Incaninsäure *f* incaninic acid
Incarnatrin *n* incarnatrin
Incarnatylalkohol *m* incarnatyl alcohol
Inchromverfahren *n* chromizing process
Indacen *n* indacene
Indamin *n* indamine, phenylene blue
Indan *n* indan, hydrindene
Indanol *n* indanol, hydroxyhydrindene
Indanon *n* indanone, hydrindone, indone, oxohydrindene
Indanthrazin *n* indanthrazine
Indanthren *n* (HN) indanthrene
Indanthron *n* indanthrone
Indazin *n* indazine
Indazol *n* indazole, benzopyrazole
Indazolchinon *n* indazole quinone
Inden *n* indene
Indenharz *n* indene resin
Index *m* (pl Indices) index, (pl Indices; Math) subscript, ~ **bei Maßstäben** index mark, sighting mark, **hochgestellter** ~ superscript, **mit einem** ~ **versehen** to index, **tiefgestellter** ~ subscript
Indexfehler *m* index error, sighting mark error
Indexkorrektion *f* index correction
Indexstrich *m* index mark or line, gauge mark
Indexzeiger *m* dial pointer, indicator needle
Indexziffer *f* index figure
Indiafaser *f* indian fiber
Indianit *m* indianite
Indican *n* indican
indifferent indifferent, (Gas) inert, (Gleichgewicht) neutral
Indifferenzlinie *f* dead line
Indifferenzpunkt *m* neutral point
Indifferenzzone *f* (Elektr) neutral zone, (Magn) magnetic equator
Indigblau *n* indigo blue
Indigkraut *n* (Bot) common dyer's woad, indigo plant
Indigo *m* indigo, indigotin

Indigoauflösung f indigo solution
Indigoauszug m indigo extract
indigoblau indigo blue
Indigoblau n indigo, indigo blue, indigotin
Indigoblauschwefelsäure f indigosulfuric acid, indigo disulfonic acid, indigosulfonic acid
Indigocarmin n indigocarmine, coerulinsulfuric acid, soluble indigo
Indigoderivat n indigo derivative
Indigodisulfonsäure f indigo disulfonic acid
Indigodruck m indigo print, indigo printing
Indigoersatz m indigo substitute
Indigoextrakt m indigo extract
Indigofarbe f indigo color
Indigofarbstoff m indigo blue, indigotin
Indigogütemesser m indigometer
Indigoid n indigoid
Indigokarmin n indigo carmine
Indigokomposition f indigo composition
Indigoküpe f indigo bath
Indigoleim m indigo gluten
Indigolith m (Min) blue tourmaline, indicolite
Indigolösung f indigo solution
Indigomessung f indigometry
Indigopurpur m indigo purple
Indigorot n indigo red, indirubin
Indigosäure f indigotic acid
Indigosalz n indigo salt, indigotate
Indigoschwefelsäure f indigodisulfonic acid, indigosulfonic acid, indigosulfuric acid
indigoschwefelsauer indigosulfate, indigodisulfonate, indigosulfonate
Indigostein m (Min) blue tourmaline, indicolite
Indigostoff m indigo blue, indigotin
Indigosulfosäure f indigosulfonic acid, indigosulfuric acid
Indigosuspension f indigo suspension
Indigosynthese f indigo synthesis
Indigotin n indigotin, indigo, indigo blue
Indigotinktur f indigo tincture
Indigoweiß n (Indigweiß) indigo white, leuco indigo, reduced indigo
Indigweiß n indigo white, reduced indigo
Indikan n indican
Indikator m indicator, indicating instrument or agent, **inaktiver** ~ inactive tracer, **radioaktiver** ~ radioactive tracer
Indikatoranhaltevorrichtung f detent gear of an indicator
Indikatorantrieb m indicator drive
Indikatoratom n tracer atom
Indikatorbereich m indicator range
Indikatordiagramm n indicator diagram
Indikatorelement n tracer element
Indikatorenmethode f tracer method
Indikatorfeder f indicator spring
Indikatorinstrument n indicating instrument, indicator
Indikatorkolben m indicator piston

Indikatorpapier n indicator paper
Indikatorrohr n indicator pipe
Indikatortechnik f tracer technique
Indikatortheorie f, **Ostwaldsche** ~ Ostwald's theory of indicators
Indikatorverfahren n tracer method
Indikatorzylinder m indicator cylinder
Indikatrix f (Math) indicatrix
Indikolit m (Min) indicolite, blue tourmaline
Indin n indin
indirekt indirect, secondary, ~ **beheizt** indirectly fired, ~ **wirkend** indirectly acting
Indirektdruck m offset print
Indirubin n indirubin, indigo red
Indischblau n Indian blue
Indischgelb n Indian yellow, cobalt yellow
Indium n indium (Symb. In)
Indiumantimonid n indium antimonide
Indiumchelat n indium chelate
Indiumchlorid n indium chloride
Indiumgehalt m indium content
Indiumjodid n indium iodide
Indiummonoxid n (Indium(II)-oxid) indium(II) oxide, indium monoxide
Indiumoxid n indium oxide
Indiumselenid n indium selenide
Indiumsuboxid n (Indium(I)-oxid) indium(I) oxide, indium suboxide
Indiumsulfat n indium sulfate
Indiumsulfid n indium sulfide, (Indium(III)-sulfid) indium(III) sulfide, indium sesquisulfide
Indiumtellurid n indium telluride
Indiumverbindung f indium compound
Individualität f individuality
individuell individual[ly]
Indizes m pl indices, (hochgestellt) superscripts, (tiefgestellt) subscripts
Indizienbeweis m circumstantial evidence
indizieren to indicate
Indizierhahn m indicator cock
indiziert indicated
Indizierung f indexing, indication
Indoanilin n indoaniline
Indochinolin n indoquinoline
Indochromogen n indochromogen
Indoform n indoform
Indol n indole, benzopyrrole, ketole
Indolbuttersäure f indolebutyric acid
Indolenin n indolenine
Indolessigsäure f indole acetic acid
Indolin n indoline, 2,3-dihydroindole
Indolinon n indolinone
Indolizidin n indolizidine
Indolizin n indolizine
Indolon n indolone
Indolpropionsäure f indolepropionic acid
Indolyl- indolyl
Indolylessigsäure f indoleacetic acid

Indon *n* indone
Indophenazin *n* indophenazine
Indophenin *n* indophenine
Indophenol *n* indophenol
Indophenolblau *n* indophenol blue
Indophenoloxidase *f* (Biochem) indophenol oxidase
Indoxyl *n* indoxyl, 3-hydroxyindole
Indoxylsäure *f* indoxylic acid
Indoxylschwefelsäure *f* indoxylsulfuric acid
Induktanz *f* (Magn) inductance
Induktion *f* induction, **elektrische** ~ electrostatic induction, **gegenseitige** ~ mutual induction, **hormonale** ~ hormonal induction, **koordinative** ~ coordinate induction, **magnetische** ~ magnetic flux, magnetic induction, **photochemische** ~ Draper effect, photochemical induction, **überflüssige** ~ (Biochem) gratuitious induction, **verteilte** ~ continuous loading, distributed inductance
Induktionsapparat *m* induction apparatus, induction coil, inductor
Induktionselektrizität *f* induction electricity
Induktionserscheinung *f* induction phenomenon
Induktionsfluß *m* induction flux, electrostatic flux
induktionsfrei induction-free, non-inductive
Induktionsgesetz *n* Faraday's law of electromagnetic induction, law of induction
Induktionsheizung *f* induction heater
Induktionskapazität *f* inductive capacity
Induktionskoeffizient *m* coefficient of induction
Induktionskoerzitivkraft *f* (Magn) intrinsic coercive force
Induktionskreis *m* inductive circuit
Induktionskurve *f* induction curve
Induktionslinie *f* line of induction
Induktionsmaschine *f* induction machine
Induktionsmotor *m* induction motor, asynchronous motor
Induktionsofen *m* induction furnace, induced oven
Induktionsperiode *f* induction period
Induktionsrolle *f* induction coil
Induktionsspannung *f* induced electromotive force
Induktionsspule *f* induction coil, conducting coil, inductor
Induktionsstrom *m* induction current, induced current
Induktionsvermögen *n* inductive capacity
Induktionswärme *f* induction heat
Induktionswiderstand *m* inductive resistance
induktiv (Elektr) inductive, inductional, ~ **gekoppelt** inductively coupled
Induktivität *f* inductance, inductivity
Induktometer *n* inductometer
Induktophon *n* inductophone
Induktor *m* inductor

Induktorium *n* induction coil
Induktorrad *n* inductor wheel
Induktorspule *f* induction coil
Induktoskop *n* inductoscope
Indulin *n* induline
Indulinscharlach *m* induline scarlet
indurieren to harden, to indurate
Industrie *f* industry, **feinmechanische** ~ manufacture of precision instruments, **holzverarbeitende** ~ woodworking industry, **kosmetische** ~ cosmetic industry
Industrieabfallstoff *m* industrial waste product
Industrieabgas *n* industrial waste gas
Industrieabwasser *n* industrial waste water, waste water from factories
Industrieanlage *f* industrial plant
Industriebetrieb *m* industrial operation, (Fabrik) industrial plant
Industriechemikalien *pl* industrial chemicals
Industrieerzeugnis *n* industrial product
Industrieforschung *f* industrial research
Industriegas *n* manufactured gas
Industriegummi *m* industrial rubber
Industriekarren *m* industrial truck
industriell industrial
Industrienorm *f* industrial standard [specification]
Industrieofen *m* industrial furnace
Industriereaktor *m* industrial reactor
Industriezweig *m* branch of industry
induzieren to induce
ineinanderfließen (verschmelzen) to coalesce
ineinandergeschachtelt telescoped
ineinandergreifend interlocking, telescoping, (Zahnräder) intermeshing
ineinanderpassen to fit into each other
ineinanderschieben to telescope
inert inert, inactive
Inertanz *f* inertance
Inertgas *n* inert gas
Inertie *f* inertia, inertness
Inesit *m* (Min) inesite, rhodotilite
Infektion *f* (Med) infection
Infektionserreger *m* infective agent
Infektionsgefahr *f* danger of infection, risk of infection
Infektionsherd *m* focus of infection
Infektionskrankheit *f* (Med) infectious disease
Infektionsquelle *f* source of infection
Infektionsresistenz *f* resistance to infection
Infektreaktion *f* reaction of infection
Infilco-Verfahren *n* infilco process
Infiltration *f* infiltration
infiltrieren to infiltrate
Infinitesimalrechnung *f* (Math) infinitesimal calculus
infizieren to infect
Inflexion *f* inflexion

Influenz – Innenansicht

Influenz f (Elektr) electrostatic induction, influence
Influenzelektrizität f influence electricity, static inductional electricity
Influenzmaschine f induction machine
Informatik f computer science
Information f information, data
infrarot infrared
Infrarot n, **entferntes** ~ far infrared
Infrarotanalysator m infrared analyzer
Infrarotbeseitigung f infrared removal
Infrarotdetektor m infrared detector
Infrarotheizung f infrared heating
Infrarotspektroskopie f infrared spectroscopy
Infrarotstrahler m infrared radiator
Infrarotstrahlung f infrared radiation
Infrarottrockner m infrared drier
Infrarottrocknung f infrared drying
Infraschall m infra sonics
Infraschwerewelle f infra gravity wave
infundieren (Med) to infuse
Infusion f infusion
Infusionsflasche f infusion bottle
Infusorienerde f infusorial earth, diatomaceous earth, kieselguhr
ingangsetzen to set in motion, to start
Ingangsetzen n, **stoßfreies** ~ starting without shock, **unvorhergesehenes** ~ accidental resumption of work
Ingangsetzung f setting in motion, starting, start-up
Ingenieur m, **beratender** ~ consulting engineer
Ingot m (Techn) ingot
Ingotstahl m ingot steel
Ingredienz n ingredient, component, constituent
Inguß m ingot mold
Ingwer m (Bot) ginger
ingwerartig gingerous
Ingwergewürz n ginger
Ingwerlikör m ginger brandy
Ingwerpulver n ginger powder, pulverized ginger
Ingwerstein m hardened calcariferous clay
Ingwertinktur f tincture of ginger
Inhaber m proprietor
inhalieren to inhale
Inhalt m content[s], (eines Körpers) capacity, volume, (Füllkörperkolonne) hold-up
Inhaltsangabe f table of contents
Inhaltsmarke f graduation mark
inhibieren to inhibit
Inhibierung f inhibition
Inhibitionsphase f inhibitory phase
Inhibitor m inhibitor, **kompetitiver** ~ competitive inhibitor
inhomogen inhomogeneous
Inhomogenität f inhomogeneity, heterogeneousness
Inhomogenitätskorrektion f inhomogeneity correction

Initialexplosivstoff m priming explosive, primer
Initialkraft f initial force, initiating power
Initialsprengstoff m initial detonating agent, priming explosive
Initialzünder m initiator
Initialzündmittel n initial explosive [substance], initial igniting agent, primer, priming composition
Initiator m initiator
initiieren to initiate
Initiierung f initiation
Injektion f **i. d. Blutbahn** intravenous injection
Injektionsflasche f injection vial
Injektionslösung f injectable solution
Injektionsmittel n injection preparation
Injektionsnadel f hypodermic needle, injection needle
Injektionsspritze f syringe
Injektor m injector
Injektorbrenner m premix burner, (Schweiß) low-pressure torch
Injektordüse f injector nozzle or cone
Injektorventil n injector valve
injizieren to inject
inkaufnehmen to make allowance for
Inklination f (Magn) inclination
Inklinationsnadel f inclinometer, (Schiff) dipping needle
Inklinationswinkel m angle of inclination
Inklusion f (Einschließung) inclusion
inkohärent incoherent
Inkohärenz f incoherence
Inkohlen n aging of coal
Inkohlung f mineralization, (Bergb) coalification, (in der Kokerei) carbonization
Inkohlungsprofil n section through rank
Inkompatibilität f incompatibility
Inkompatibilitätstensor m incompatibility tensor
inkompressibel incompressible
Inkompressibilität f incompressibility
inkongruent incongruent
Inkongruenz f (Stereochem) incongruence
inkonsequent inconsistent
Inkonsequenz f inconsistency
inkonstant inconstant, unstable
Inkonstanz f inconstancy, instability
Inkrafttreten n coming into force
Inkrustation f incrustation
inkrustieren to incrust
inkrustierend encrusting
Inkrustierung f incrustation
Inkubation f incubation
Inkubationszeit f incubation time
Inkunabel f wood stereotype
innehalten to pause, to stop
Innenabmessung f inside dimension
Innenanlage f internal plant
Innenanode f inner anode
Innenansicht f interior view

Innenanstrich m interior coating
Innenanstrichfarbe f indoor paint
Innenauskleidung f internal lining
Innenbackenbremse f internal expanding brake
Innendorn m inner mandrel
Innendruckversuch m internal pressure test
Innenentstäubung f internal dust extraction
Innenfeuerung f internal furnace
Innenfläche f internal surface
Innengewinde n internal thread, female thread, inside thread
Innenheizung f internal heating
Innenkorrosion f internal corrosion
Innenlack m interior varnish
Innenleitung f internal wiring
Innenlunker m blowhole
Innenmaß n inner dimension, internal measure
Innenmetallisierung f inner metallization
Innenmischer m internal mixer
Innenorbitalkomplex m inner orbital complex
Innenpolanker m radial coil armature
Innenpoldynamo m internal pole dynamo
Innenpolgestell n inner pole frame
Innenpolmantel m inner pole casing
Innenpunkt m inner point
Innensack m liner, lining
Innenschaltgerät n integrating accelerometer
Innenschicht f core sheet
Innenschleifmaschine f internal grinding machine
Innenschleifvorrichtung f internal grinding fixture
Innenschliff m internal grinding
Innenschmarotzer m (Biol) endoparasite
Innenseele f (Reifen) liner
Innenstrehler m inside chaser, inside chasing tool
Innentaster m inside calipers
Innentemperatur f internal temperature
Innenumwegschaltung f integral by-pass
Innenverzahnung f internal gears
Innenvierkant n recessed square
Innenwand f inner wall
Innenweite f inside diameter
Innenwiderstand m (Elektr) internal resistance
innenzentriert body-centered
inneratomar intraatomic
Innere n interior, core
innermolekular intramolecular
innewohnend inherent, innate
innig gemischt homogeneously mixed
inokulieren to inoculate
Inokulum n inoculum
Inophyllsäure f inophyllic acid
Inosin n inosine, hypoxanthin riboside
Inosindiphosphat n inosine diphosphate, IDP
Inosinmonophosphat n inosine monophosphate
Inosinphosphorylase f (Biochem) inosine phosphorylase

Inosinsäure f inosinic acid, inosinephosphoric acid
Inosintriphosphat n inosine triphosphate, ITP
Inosit m inositol, inosite
Inosose f inosose
Inoyit m inoyite
Insekt n insect, pest
Insektenausrottung f insect control, disinsectisation
Insektenbekämpfung f insect control
Insektenkunde f entomology, insectology
Insektenlarve f insect larva (pl larvae)
Insektenpulver n insecticide powder, insect powder
Insektenpuppe f chrysalis (pl chrysalises o. chrysalides)
Insektenstich m insect bite
Insektenvertilgungsmittel n insecticide
Insektenvertreibungsmittel n insectifuge
Insektenwachs n insect wax
insektizid insecticide, insecticidal
Insektizid n insect exterminator, insecticide
Inselmodell n island model
Insipin n insipin
Inspizierung f inspection
instabil unstable, instable, labile
Instabilität f instability
Installateur m plumber, (Elektriker) electrician
Installation f installation, (Gas, Wasser) plumbing
installieren to install, to equip
instandgesetzt overhauled
instandhalten to maintain, to keep up
Instandhaltung f maintenance, servicing, upkeep
Instandhaltungsarbeiten f pl maintenance routine work
Instandhaltungskosten pl maintenance costs
instandsetzen to repair, to recondition, to refit
Instandsetzung f repair [work], reconditioning, restoration
Instandsetzungsarbeiten f pl repairs, repair work
instandsetzungsfähig reparable, **nicht mehr** ~ beyond repair
Instanz f authority
instationär transient, unsteady
Instrument n, **registrierendes** ~ recorder, recording instrument
Instrumentalanalyse f instrumental analysis
Instrumentalfehler m instrumental error
Instrumentenbrett n control panel, instrument board
Instrumentengehäuse n instrument housing
Instrumentenhaube f instrument cover
Instrumentierung f instrumentation
Insulanolin n insulanoline
Insularin n insularine
Insulin n insulin, **gebundenes** ~ bound insulin, **pflanzliches** ~ vegetable insulin
Insulinausschüttung f insulin secretion

Insulinbildung *f* insulinogenesis
Insulineinheit *f* insulin unit
Insulinmangel *m* insulin deficiency, hypoinsulinism
Insulinmangeldiabetes *m* insulin deficiency diabetes
Insulinmangelkrankheit *f* hypoinsulinism
Insulinnachweis *m* test for insulin
Insulinschock *m* (Med) insulin shock [therapy]
Insulinzufuhr *f* administration of insulin
Intaglio *n* intaglio
Intarvin *n* intarvin, glycerol trimargarate, glyceryl margarate, margarin
Integerrinecinsäure *f* integerrinecic acid
integrabel integrable
Integrabilitätsbedingung *f* integrability condition
Integral *n* integral, **bestimmtes** ~ definite integral, **Doppel-** ~ double integral, **elliptisches** ~ elliptical integral, **Flächen-** ~ integral over a plane area, **Linien-** ~ line integral, **mehrfaches** ~ multiple integral, **Oberflächen-** ~ surface integral, **partikuläres** ~ particular integral, **singuläres** ~ singular integral, **unbestimmtes** ~ indefinite integral, **uneigentliches** ~ improper intergral, **Volumen** ~ volume integral
Integralbegriff *m* (Math) concept of the integral
Integralbeziehung *f* integral relation
Integraldarstellung *f* integral representation
Integralformel *f* integral formula
Integralgleichung *f* integral equation
Integralinvariante *f* integral invariant
Integralkern *m* integral kernel
Integralkurve *f* integral curve
Integralrechnung *f* integral calculus
Integralregelung *f* integral control, floating control
Integralsatz *m* integral theorem
Integralspannung *f* summated pressure
Integralstromstärke *f* summated current, vector sum of currents
Integralzahl *f* integer
Integralzeichen *n* integration sign
Integrand *m* integrand
Integration *f* integration, ~ **durch Substitution** integration by substitution, ~ **v. Vektorfunktionen** integration of vector functions, **graphische** ~ graphical integration, **kombinierte** ~ conglomerate integration, **numerische** ~ numerical integration, **partielle** ~ integration by parts
Integrationsgerät *n* integrator, integration device
Integrationsgrenzen *pl* limits of integration
Integrationskonstante *f* integration constant
Integrator *m* integrator
integrierbar integrable
integrieren to integrate
integrierend integrating, integral, integrant

Integriergerät *n* integrator
Integrierschaltung *f* integrating circuit
Integrierwerk *n* integrating gear
Integritätsbereich *m* domain of integrity, integral domain
Integrodifferentialgleichung *f* integro differential equation
Intensimeter *n* intensimeter
Intensität *f* intensiveness, (Phys, Elektr) intensity
Intensitätsanomalie *f* intensity anomaly
Intensitätsbereich *m* intensity region
Intensitätskonstanz *f* (Elektr) power output stability
Intensitätsmaximum *n* maximum intensity
Intensitätsniveau *n* level of intensity
Intensitätsregel *f* **für ein Multiplett** (Spektr) multiplet intensity rule
Intensitätsschwächung *f* intensity decrease
Intensitätssummensatz *m* intensity sum rule
Intensitometer *n* intensitometer
Intensivkühler *m* multiple coil condenser
Intensivmischer *m* intensifier-type mixer, intensive mixer
Intercellularsubstanz *f* intercellular substance
Interferenz *f* interference
Interferenzapparatur *f* diffraction camera
Interferenzaufnahme *f* interference photograph
Interferenzbild *n* (Opt) interference pattern
Interferenzdoppelbrechung *f* interference in double refraction
Interferenzgoniometer *n* x-ray diffractometer
Interferenzmeßverfahren *n* interferometry
Interferenzschwund *m* interference fading
Interferenzspektrogramm *n* interference spectrogram
Interferenzspektrum *n* interference spectrum
Interferenzstreifen *m* interference fringe
Interferenzwinkel *m* interference angle
interferieren to interfere
Interferometer *n* interferometer
Interferometrie *f* interferometry
Interferon *n* interferon
Interhalogenverbindung *f* interhalide compound
interkristallin intercrystalline
interlaminar interlaminar
intermediär intermediate, (Stoffwechsel) intermediary
Intermediärstoffwechsel *m* intermediary metabolism
intermetallisch intermetallic
Intermittenzeffekt *m* (Opt) intermittency effect
intermittierend intermittent, interrupted, periodical, pulsating
intermolekular intermolecular
Internationale Einheit *f* (I.E.) international unit, I.U.
Interpack-Füllkörper *m* interpack packing
interplanar interplanar

Interpolation *f* interpolation
interpolieren to interpolate
Interpunktionszeichen *n* punctuation mark
Interrenin *n* (Nebennierenrindenhormon) interrenin
Intervall *n* interval
intervenieren to intervene, to interfere, to interpose
interzellular intercellular
Intoxikation *f* intoxication
intraatomar intra-atomic
intrakutan intracutaneous
Intramin *n* intramine, contramine
intramolekular intramolecular
intramuskulär intramuscular
intravenös intravenous
Intrinsikfaktor *m* intrinsic factor
Intumeszenz *f* (Med, Krist) intumescence
Inula *f* inula, alant root, elecampane, elfwort, helenium, horseheal
Inulase *f* (Biochem) inulinase
Inulin *n* inulin, alantin, alant starch
Inulinase *f* (Biochm) inulinase
Invar *n* (HN) invar
invariabel invariable
invariant invariant
Invariante *f* invariant
Invarianzprinzip *n* (Math) principle of invariance
Invarstahl *m* Invar steel
Inventur *f* inventory, stock-taking
Inversion *f* (Math, Chem) inversion
Inversionsachse *f* (Krist) inversion axis
Inversionspunkt *m* inversion point, transition point
Inversionsschicht *f* inversion layer
Inversionsspektrum *n* (Spektr) reversal spectrum
Inversionstemperatur *f* inversion temperature
Inversionszentrum *n* center of inversion
Invertase *f* (Biochem) invertase, invertin, raffinase, saccharase, sucrase
invertierbar invertible
invertieren to invert
Invertierung *f* inversion
Invertin *n* invertase, invertin, saccharase, sucrase
Invertseife *f* cationic detergent, invert soap
Invertzucker *m* invert sugar
in vitro (Lat) in glass, ~ **vivo** (Lat) in life
Involution *f* involution
Inzidenzwinkel *m* angle of incidence, (Opt) angle of incidence
Inzucht *f* inbreeding
Iolith *m* (Min) iolite, cordierite, dichroite, water sapphire
Ion *n* ion, **hydratisiertes** ~ hydrated ion, aquo ion
Ionenabgabe *f* ion loss
Ionenabstand *m* interionic distance
Ionenadsorption *f* ion adsorption
Ionenaktivität *f* ion activity

Ionenaktivitätskoeffizient *m* ion activity coefficient
Ionenaustausch *m* ion exchange
Ionenaustauschchromatographie *f* ion exchange chromatography
Ionenaustauscher *m* ion exchanger
Ionenaustauschermembran *f* ion exchange membrane
Ionenaustauschgleichgewicht *n* ion exchange equilibrium
Ionenbahn *f* ion path
Ionenbeweglichkeit *f* ionic mobility
Ionenbewegung *f* ionic migration, ion[ic] movement, migration of ions
ionenbildend ionogenic
Ionenbildung *f* formation of ions, production of ions
Ionenbildungsgeschwindigkeit *f* velocity of formation of ions, velocity of ionization
Ionenbindung *f* ionic bond, heteropolar bond
Ionenbremsfeld *n* ion retarding field
Ionencharakter *m* ionic character
Ionendichte *f* ionic density
Ionendipolkomplex *m* ion-dipole complex
Ionendosimeter *n* ionic quantimeter
Ionendosismesser *m* ionic quantimeter
Ionenenergie *f* ionic energy
Ionenentladung *f* ionic discharge
ionenerzeugend ionogenic
Ionenfalle *f* (Rdr) ion trap
Ionenfarbe *f* ion color
Ionenflotation *f* ion flotation
Ionenform *f* ionic form
Ionengehalt *m* ionic concentration, ionic strength
Ionengeschwindigkeit *f* ionic velocity
Ionengitter *n* ion lattice
Ionengitterstrom *m*, **negativer** ~ reverse or negative grid current
Ionengleichgewicht *n* ionic equilibrium
Ionengleichung *f* equation of ions, ionic equation
Ionengrenzwert *m* ion limit
Ionenhydrat *n* hydrated ion
Ioneninkrement *n* ionic increment
Ionenkette *f* ion chain
Ionenkettenpolymerisation *f* ionic chain polymerization
Ionenkonzentration *f* ionic concentration, ionic strength
Ionenkristall *m* (Krist) ionic crystal
Ionenladung *f* ionic charge
Ionenladungszahl *f* ion charge number
Ionenleitfähigkeit *f* electrical conductivity of ions, ion conductivity
Ionenleitung *f* ionic conductivity
Ionenmizellen *pl* ionic micelles
Ionenpaar *n* pair of ions
Ionenprodukt *n* ionic product

Ionenpumpe *f* ion pump
Ionenradius *m* ionic radius
Ionenrauschen *n* (Elektr) gas noise
Ionenreaktion *f* ionic reaction
Ionenreibung *f* ionic friction
Ionenröhre *f* gas tube
Ionenrückkupp[e]lung *f* ion feed-back
Ionenrundlauf *m* circulation of ions
Ionensättigungsstrom *m* ion saturation current
Ionenschalter *m* gas-filled relay
Ionenspaltung *f* ionization, ionic cleavage
Ionenstärke *f* ionic strength, ionic concentration
Ionenstrahl *m* ionic ray
Ionenstrom *m* ion current
Ionenteilchendichte *f* ion particle density
Ionentheorie *f* ionic theory
Ionentransport *m* ion transport
Ionenüberführungszahl *f* (Elektrochem) transport number of ions
Ionenübergang *m* ion transition
Ionenverarmung *f* depletion of ions
Ionenverdichtung *f* concentration of ions
Ionenverzögerungsverfahren *n* ion retardation process
Ionenwanderung *f* ion[ic] migration, traveling of ions
Ionenwanderungsgeschwindigkeit *f* ion drift velocity
Ionenwertigkeit *f* ionic valency
Ionenwolke *f* ion cluster, ionic atmosphere
Ionenzerstäuberpumpe *f* ionic atomization pump
Ionenzustand *m* ionic state
Ion-Ion-Rekombination *f* ion-ion recombination
Ionisation *f* ionization, ionisation, ~ **vierter Ordnung** quarternary ionization
Ionisationsenergie *f* ionization energy
Ionisationsereignis *n* ionization event, ionizing event
Ionisationsgleichgewicht *n* ionization balance
Ionisationsgrad *m* degree of ionization
Ionisationskammer *f* ionization chamber
Ionisationskonstante *f* ionization constant
Ionisationsmanometer *n* ionization gauge, ion gauge
Ionisationsmesser *m* ion counter
Ionisationspotential *n* ionization potential
Ionisationsschicht *f* ionized layer
Ionisationsspannung *f* ionization voltage
Ionisationsstrom *m* ionization current
Ionisationsvakuummeter *n* ionization vacuum gauge
Ionisationswärme *f* heat of ionization
Ionisator *m* ionizer
ionisch ionic
ionisierbar ionizable
Ionisierbarkeit *f* ionisability
ionisieren to ionize
Ionisierung *f* ionization, electrolytic dissociation, ionic cleavage, ionizing

Ionisierungsarbeit *f* work of ionization
Ionisierungsausbeute *f* ionization efficiency
Ionisierungsbahn *f* ionization track
Ionisierungsbruchteil *m* ionization fraction
Ionisierungselektrode *f* ionizing electrode
Ionisierungsenergie *f* ionization energy
Ionisierungsgrenze *f* (Phys) ionisation limit
Ionisierungsmittel *n* ionizer
Ionisierungspotential *n* ionization potential
Ionisierungsquerschnitt *m* ionization cross section
Ionisierungsspannung *f* ionization voltage
Ionisierungsstärke *f* total ionization
Ionisierungsvermögen *n* ionizing power
Ionisierungswärme *f* heat of ionization
Ionisierungswelle *f* ionizing wave
Ionium *n* ionium (Symb. Io)
ionogen ionogenic
Ionophorese *f* ionophoresis
Ionosphäre *f* ionosphere
Ionosphärengeräusche *n pl* ionospheric soundings
Ionosphärenleitfähigkeit *f* ionospheric conductivity
Ionosphärenschall *m* ionosphere sound
Iontophorese *f* iontophoresis
Ipecacuanhasäure *f* ipecacuanhic acid
Ipecacuanhawurzel *f* ipecacuanha root
Ipomeansäure *f* ipomeanic acid
Ipomsäure *f* ipomic acid, sebacic acid
Iproniazid *n* iproniazid
Ipuranol *n* ipuranol
irden earthen, clay
Irdenware *f* earthenware, crockery
irdisch terrestrial
Iren *n* irene
Iretol *n* iretol
Irideszenz *f* (Opt) iridescence
Iridgold *n* gold-iridium alloy
Iridin *n* iridin
Iridium *n* iridium (Symb. Ir)
Iridiumäthylendiaminchelat *n* iridium ethylenediamine chelate
Iridiumchelat *n* iridium chelate
Iridiumchlorid *n* iridium chloride
Iridiumoxid *n* iridium oxide
Iridiumschwamm *m* iridium sponge, spongy iridium
Iridodicarbonsäure *f* iridodicarboxylic acid
Iridoskyrin *n* iridoskyrin
Iriphan *n* iriphan, strontium 2-phenylquinoline-4-carboxylate
Irisation *f* iridescence
Irisblau *n* iris blue
Irisblende *f* (Opt) iris diaphragm, iris shutter
irisieren to be iridescent, to iridesce
Irisieren *n* iridescence
irisierend iridescent
Irisiersalz *n* irisating salt

Irisierung f iridescence
Irisin n irisin
Irisöl n iris oil, orris oil
Irisviolett n iris violet
Iron n irone
irreal unreal
irregulär (Krist) irregular
irreparabel beyond repair, irreparable
irreversibel irreversible
Irreversibilität f irreversibility
irritieren to irritate
irrtümlich erroneous[ly], wrong[ly]
Irrtum m error, mistake
Irrtumswahrscheinlichkeit f error probability
Isabellfarbe f buff, cream color
Isaconitsäure f isaconitic acid, isoaconitic acid
Isäthionsäure f isethionic acid, ethylenehydrinsulfonic acid, oxyethylsulfonic acid
Isamid n isamide, amasatine
Isamsäure f isamic acid
Isanöl n isano oil, ongokea oil
Isanolsäure f isanolic acid
Isansäure f isanic acid
Isaphensäure f isaphenic acid
Isatan n isatan, hydroxybioxindol
Isatanthron n isatanthrone
Isathophan n isathophan
Isatin n isatin, isatinic acid lactam
Isatinecinsäure f isatinecic acid
Isatinsäure f isatinic acid, aminophenylglyoxylic acid, isatic acid
Isatinverbindung f isatin derivative
Isatogen n isatogen
Isatogensäure f isatogenic acid
Isatoid n isatoid
Isatol n isatol
Isatosäure f isatoic acid, N-carboxyanthranilic acid
Isatoxim n isatoxime, nitrosoindoxyl
Isatronsäure f isatronic acid
Isatropasäure f isatropic acid
Isatyd n isatyde
Isentrope f isentrope
Isentropenanalyse f isentropic analysis
isentropisch isentropic
Ishikawait m (Min) ishikawaite
Isoadenin n isoadenine
Isoäpfelsäure f isomalic acid
Isoagglutinin n iso-agglutinin, isoh[a]emagglutinin
Isoalloxazin n isoalloxazine
Isoamidon n isoamidone
Isoamyl- n isoamyl
Isoamylacetat n isoamyl acetate
Isoamylalkohol m isoamyl alcohol
Isoamylbromid n isoamyl bromide
Isoamylchlorid n isoamyl chloride
Isoamylen n isoamylene

Isoamylessigsäure f isoamylacetic acid
Isoandrosteron n isoandrosterone
Isoanethol n isoanethol
Isoantigen n iso-antigen
Isoantikörper m iso-antibody
Isoantipyrin n isoantipyrine
Isoascorbinsäure f isoascorbic acid
Isoasparagin n isoasparagine
Isobar n (Phys, Chem) isobar
Isobarbitursäure f isobarbituric acid
Isobare f (Meteor) line of constant pressure, (Metereologie) isobar
Isobarenschar f group of isobars
isobar[isch] isobaric
Isobeberin n isobeberine
Isobenzil n isobenzil
Isobernsteinsäure f isosuccinic acid
Isobianthron n isobianthrone
Isoborneol n isoborneol
Isobornylacetat n isobornyl acetate
Isobrenzschleimsäure f isopyromucic acid
Isobutan n isobutane
Isobuttersäureanhydrid n isobutyric anhydride
Isobuttersäurenitril n isobutyric nitrile, isobutyronitrile, isopropyl cyanide
isobuttersauer isobutyrate
Isobutyl- isobutyl
Isobutylacetat n isobutyl acetate
Isobutylalkohol m isobutyl alcohol, 2-methyl-1-propanol, isobutanol
Isobutylamin n isobutyl amine
Isobutylbenzoat n isobutyl benzoate
Isobutylbromid n isobutyl bromide
Isobutylchlorid n isobutyl chloride
Isobutylen n isobutene, isobutylene
Isobutylenbromid n isobutylene bromide, isobutylene dibromide
Isobutyliden n isobutylidene
Isobutyljodid n 1-iodo-2-methylpropane, isobutyl iodide
Isobutylmercaptan n isobutyl mercaptan
Isobutylpräparat n isobutyl preparation
Isobutylrhodanid n isobutyl thiocyanate
Isobutylsalicylat n isobutyl salicylate
Isobutyraldehyd m isobutyraldehyde, isobutyric aldehyde
Isobutyryl- isobutyryl
Isobutyrylchlorid n isobutyryl chloride
Isocadinen n isocadinene
Isocain n isocaine
Isocamphan n isocamphane
Isocamphersäure f isocamphoric acid
Isocamphoronsäure f isocamphoronic acid
Isocantharidin n isocantharidine
Isocapronsäure f isocaproic acid
Isocetinsäure f isocetinic acid
Isochamsäure f isochamic acid
Isochavicinsäure f isochavicinic acid
Isochinaldinsäure f isoquinaldic acid

Isochinamin – Isoleutherin

Isochinamin *n* isoquinamine
Isochinolin *n* isoquinoline
Isochlorin *n* isochlorine
Isochlorogensäure *f* isochlorogenic acid
Isocholesterin *n* isocholesterol
isochor isochoric
Isochore *f* isochore, Van't Hoff'sche ~ Van't Hoff's isochore
isochrom isochromatic
Isochroman *n* isochroman
isochromatisch isochromatic
isochron isochronous
Isochrone *f* isochronous curve
isochronisch isochronous
Isocinchomeronsäure *f* isocinchomeronic acid
Isocineol *n* isocineole
Isocitronensäure *f* isocitric acid
Isocoffein *n* isocoffeine
Isocolchicin *n* isocolchicine
Isocorybulbin *n* isocorybulbine
Isocorydin *n* isocorydine
Isocrotonsäure *f* isocrotonic acid
Isocumaran *n* isocoumaran
Isocumaranon *n* isocoumaranone
Isocumarin *n* isocoumarin, isocumarin
Isocumol *n* isocumene
Isocyanat *n* isocyanate, carbimide
Isocyanid *n* isocyanide, carbylamine, isonitrile
Isocyanin *n* isocyanin
Isocyansäure *f* isocyanic acid
Isocyanursäure *f* isocyanuric acid
isocyclisch isocyclic
Isocytidin *n* isocytidine
Isocytosin *n* isocytosine
Isodehydracetsäure *f* isodehydroacetic acid
Isodiapher *n* isodiaphere
Isodimorphie *f* (Krist) isodimorphism, double isomorphism
Isodimorphismus *m* isodimorphism
isodispers isodisperse
Isodosenschreiber *m* isodose recorder
Isodosis *f* isodose
Isodrin *n* isodrin
Isodulcit *m* isodulcite, rhamnose
Isodural *n* isodural
Isoduridin *n* isoduridine
Isodurol *n* isodurene
Isoduryl- isoduryl, trimethylbenzyl
Isodyname *f* isodynamic line
isoelektrisch isoelectric
Isoemetin *n* isoemetine
Isoenzym *n* (Biochem) isozyme
Isoephedrin *n* isoephedrine
Isoerucasäure *f* isoerucic acid, brassidic acid
Isoeugenit[ol] *n* isoeugenitol
Isoeugenol *n* isoeugenol
Isoeugenylacetat *n* isoeugenyl acetate
Isofebrifugin *n* isofebrifugin
Isofenchol *n* isofenchol
Isofenchosäure *f* isofenchoic acid
Isoferulasäure *f* isoferulic acid, hesperitinic acid
Isoflavon *n* isoflavone
Isoform *n* isoform
Isofucosterin *n* isofucosterol
Isofulminat *n* isofulminate
Isogalbulin *n* isogalbulin
Isogalloflavin *n* isogalloflavin
Isogeronsäure *f* isogeronic acid
Isogladiolsäure *f* isogladiolic acid
Isoglucal *n* isoglucal
Isoglutamin *n* isoglutamine
Isoglutathion *n* isoglutathione
isogonal isogonal, equiangular
Isogonie *f* (Krist) isogony
Isogramm *n* isogram
Isographie *f* isography
Isoguanin *n* isoguanine
Isohämanthamin *n* isohemanthamine
Isohemipinsäure *f* isohemipic acid
Isohexylnaphthazarin *n* isohexylnaphthazarin
Isohistidin *n* isohistidine
Isohomobrenzcatechin *n* isohomopyrocatechol
Isohumulinsäure *f* isohumulinic acid
Isohydrie *f* isohydry
isohydrisch isohydric
Isohydrobenzoin *n* isohydrobenzoin
Isoindigo *n* isoindigo
Isoindol *n* isoindole
Isoindolin *n* isoindoline
Isokainsäure *f* isokainic acid
Isokline *f* isoclinal line
Isokoproporphyrin *n* isocoproporphyrin
Isokryptoxanthin *n* isocryptoxanthin
Isolation *f* (Chem) isolation, (Elektr) insulation
Isolationsanstrich *m* insulating paint
Isolationsband *n* insulating tape
Isolationsfaktor *m* insulation coefficient
Isolationsfehler *m* defect in insulation
Isolationshülle *f* insulating covering
Isolationskammer *f* insulating chamber, insulating jacket
Isolationslack *m* insulation varnish
Isolationsmittel *n* (Elektr) insulator
Isolationsprüfer *m* insulation detector, insulation tester
Isolationsschicht *f* insulating layer
Isolationsstoff *m* insulating material, insulator, non-conducting composition
Isolationsvermögen *n* insulating power
Isolationswiderstand *m* insulation resistance
Isolationszustand *m* state of insulation
Isolator *m* insulator, non-conductor
Isolatoroberfläche *f* insulator surface
Isolaurelin *n* isolaureline
Isoleucin *n* isoleucine, 2-amino-3-methylpentanoic acid
Isoleucinsäure *f* isoleucinic acid
Isoleutherin *n* isoleutherin

Isolierband *n* insulating tape
isolieren (abtrennen) to isolate, (Elektr) insulate
isolierend isolating, (Elektr) insulating
Isolierfähigkeit *f* insulating property, insulating quality
Isolierfirnis *m* insulating varnish
Isoliergefäß *n* vacuum-insulated vessel
Isoliergewebe *n* insulating fabric
Isolierglocke *f* bell[-shaped] insulator, insulating bell
Isoliergriff *m* insulating handle
Isolierhülle *f* insulating covering
Isolierlack *m* insulating varnish
Isoliermasse *f* insulating material, non-conducting composition
Isoliermaterial *n* insulating material
Isoliermittel *n* insulating material, non-conducting material
Isolierpapier *n* insulating paper
Isolierpappe *f* insulating cardboard
Isolierpech *n* insulating pitch
Isolierplatte *f* insulating plate
Isolierpreßstoff *m* molded insulating material
Isolierrohr *n* insulating tube
Isolierrohrwickelmaschine *f* insulating tube coiling machine
Isolierrost *m* insulating grid, die grid
Isolierschicht *f* insulating layer, insulator
Isolierschlauch *m* insulating hose, insulating tubing, protective tubing
Isolierschnur *f* insulating cord
Isolierschutz *m* insulation
Isolierstoff *m* [molded] insulating material, insulant
isoliert (abgetrennt) isolated, (Elektr) insulated
Isolierung *f* (Chem) isolation, (Elektr) insulation
Isolierungsmaterial *n* insulation material
Isolierungsmethode *f* (Chem) isolation procedure
Isolimonen *n* isolimonene
isolog isologous
Isolomatiol *n* isolomatiol
Isolysergin *n* isolysergine
Isolysergol *n* isolysergol
Isolysergsäure *f* isolysergic acid
Isolysin *n* isolysine
Isomaltit *m* isomaltitol
Isomaltose *f* isomaltose
Isomenthon *n* isomenthone
isomer isomeric, isomerous
Isomer *n* isomer
Isomerase *f* (Biochem) isomerase
Isomere *pl*, **optische** ~ optical isomers
Isomerengemisch *n* mixture of isomers
Isomerie *f* isomerism, **geometrische** ~ (Stereochem) geometric[al] isomerism, **optische** ~ (Krist) enantiomerism, optical isomerism
Isomeriemöglichkeit *f* possibility of isomerism
Isomerisation *f* isomerization

isomerisieren to isomerize
Isomerisierung *f* isomerization, isomeric change, isomerizing
Isomerismus *m* isomerism
isometamer isometameric
isometrisch isometric, isometrical, (Math) regular
isomorph isomorphic, isomorphous
Isomorphie *f* (Krist) isomorphism
Isomorphiebeziehung *f* isomorphism
Isomorphin *n* isomorphine
Isomorphismus *m* isomorphism
isomorphotrop isomorphotrope
Isomycomycin *n* isomycomycin
Isoniazid *n* isoniazid
Isonicotinsäure *f* isonicotinic acid
Isonitril *n* isonitrile, carbylamine, isocyanide
Isonitrosoverbindung *f* isonitroso compound
Isooctan *n* isooctane
Isoocten *n* isooctene
Isoodyssinsäure *f* isoodyssic acid
Isoölsäure *f* isooleic acid
Isoöstron *n* isoestrone
Isooxindigo *n* isooxindigo
Isopentan *n* isopentane
Isopentylalkohol *m* isopentyl alcohol
isoperimetrisch isoperimetric
isoperistaltisch isoperistaltic
Isophorol *n* isophorol
Isophthalimid *n* isophthalimide
Isophthalsäure *f* isophthalic acid
Isopikrinsäure *f* isopicric acid
Isopiperinsäure *f* isopiperic acid
isopleth isopleth
Isoplethe *f* (Math) isopleth
isopolymorph (Krist) isopolymorphic
Isopolymorphie *f* isopolymorphism
Isopolysäure *f* isopoly acid
Isoporenfokus *m* isoporic focus
Isopral *n* isopral, trichloroisopropyl alcohol
Isopren *n* isoprene, 2-methyl-1,3-butadiene
Isoprenalin *n* isoprenaline
Isoprenkautschuk *m* isoprene caoutchouc
Isoprenoid *n* isoprenoid
Isoprenregel *f* isoprene rule
Isoprimverose *f* isoprimeverose
Isopropenyl- isopropenyl
Isopropyl- isopropyl
Isopropylacetylaceton *n* isopropylacetylacetone
Isopropyläthergemisch *n* isopropyl ether blend
Isopropylalkohol *m* isopropanol, isopropyl alcohol
Isopropylbenzoesäure *f* isopropylbenzoic acid
Isopropylbenzol *n* isopropyl benzene
Isopropylbromid *n* isopropyl bromide, 2-bromopropane
Isopropylessigsäure *f* isopropylacetic acid, isovaleric acid
Isopropyljodid *n* isopropyl iodide, 2-iodopropane

Isopropylmalonsäure f isopropylmalonic acid
Isopseudocholesterin n isopseudocholesterol
Isopulegol n isopulegol
Isopurpurogallon n isopurpurogallone
Isopurpursäure f isopuric acid
Isopyronen n isopyronene
Isoreserp[in]säure f isoreserpic acid
Isoreserpon n isoreserpone
Isoretinen n isoretinene
Isorhamnose f isorhamnose
Isoriboflavin n isoriboflavine
Isorotation f (Stereochem) isorotation
Isosaccharin n isosaccharin
Isosaccharinsäure f isosaccharinic acid
Isosafrol n isosafrole
Isoserin n isoserine
isosmotisch isosmotic
isostatisch isostatic
Isostilben n isostilbene
Isostrophanthsäure f isostrophanthic acid
isostrukturell isostructural
Isostrychnin n isostrychnine
Isostrychninsäure f isostrychnic acid
Isosulfocyansäure f isosulfocyanic acid, isothiocyanic acid, sulfocarbimide
isotaktisch isotactic
Isotestosteron n isotestosterone
Isotetralin n isotetralin
isotherm isothermal
Isotherme f isotherm, isothermal curve, line of constant temperature
isothermisch isothermal
Isothiazol n isothiazole
Isothiocyanat n isothiocyanate, salt or ester of isothiocyanic acid, sulfocirbamide
Isothiocyanester m ester of isothiocyanic acid
Isothiocyansäure f isothiocyanic acid, isosulfocyanic acid, sulfocarbimide
Isoton n isotone
Isotonie f isotonicity
isotonisch isotonic
isotop isotopic
Isotop n isotope, **stabiles** ~ stable isotope
Isotopenabtrennungsanlage f isotope separation plant
Isotopenanreicherung f isotope enrichment
Isotopenaustauschreaktion f isotopic exchange reaction
Isotopenchemie f isotope chemistry
Isotopeneffekt m isotope effect
Isotopenersetzung f isotope replacement
Isotopenforschung f research on isotopes
Isotopengemisch n isotopic mixture
Isotopengewicht n isotope weight, isotopic mass
Isotopengleichgewicht n isotopic equilibrium
Isotopenhäufigkeit f isotope abundance
Isotopenhäufigkeitsmessung f isotopic abundance measurement
Isotopenhäufigkeitsverhältnis n isotopic ratio

Isotopenherstellung f isotope production
Isotopenindikator m tracer isotope, tracer
Isotopenmarkierung f isotopic labeling
Isotopenspin m isotopic spin
Isotopenspinquantenzahl f (Quant) isotopic spin quantum number
Isotopentrennung f isotope fractionation, isotope separation, **elektromagnetische** ~ electromagnetic isotope separation
Isotopentrennungsanlage f (Calutron) calutron
Isotopentrennvorgang m isotope separation process
Isotopenverbindung f isotopic compound
Isotopenverdünnungsanalyse f (Atom) isotope dilution analysis
Isotopen-Verdünnungsmethode f isotope dilution method
Isotopenverhältnis n isotopic ratio
Isotopenzusammensetzung f isotopic composition
Isotopieeffekt m isotope effect
Isotopiespin m isotopic spin
Isotopieverschiebung f isotope effect or shift, optical isotope shift
Isotrehalose f isotrehalose
Isotron n isotron
isotrop isotropic
Isotropie f isotropy
Isotropiefaktor m isotropy factor
isotropisch isotropic
Isotryptamin n isotryptamine
Isotubasäure f isotubaic acid
Isovaleraldehyd m isovaleraldehyde, 3-methylbutanal
Isovaleriansäure f isovaleric acid
Isovaleriansäureester m isovalerate
Isovalerylchlorid n isovaleryl chloride
Isovalin n isovaline
Isovanillinsäure f isovanillic acid
Isoviolanthron n isoviolanthrone
Isoxazol n isoxazole
Isoxazolin n isoxazoline
Isozentrum n isocenter
Isozimtsäure f isocinnamic acid
Isozitronensäure f isocitric acid
Ist-Maß n actual measurement, actual size or dimension
Istmenge f actual quantity
Istwert m actual value, ideal value, instantaneous value (Am. E.), true value, (Regelgröße) control point
Itabirit m itabirite
Itacolumit m itacolumite
Itaconsäure f itaconic acid, methylenebutanedioic acid, methylenesuccinic acid
Itaconsäurediäthylester m diethyl itaconate
Italienischrot n Italian red
Itamalsäure f itamalic acid

Iterationsprozeß *m* iteration process
Iterationsverfahren *n* iteration method
iterieren to iterate
It-Platten *pl* asbestos-rubber high-pressure seals
Itrol *n* itrol, silver citrate
Ittnerit *m* (Min) ittnerite
Ivaöl *n* iva oil, musk milfoil oil
Ixen *n* ixene
Ixolon *n* ixolone
Ixon *n* ixone

J

Jaborandiöl *n* (Pharm) jaborandi oil
Jacobin *n* jacobine
Jacobsit *m* (Min) jacobsite
Jacodin *n* jacodine
Jaconec[in]säure *f* jaconecic acid
Jade *f* (Min) jade, nephrite
Jadeit *m* (Min) jadeite, greenstone
Jaffé-Base *f* Jaffé's base
Jahreserzeugung *f* yearly output, yearly capacity
Jahreshauptversammlung *f* annual general meeting
Jahrestag *m* anniversary
Jalape *f* (Bot) jalap
Jalapenharz *n* jalap resin
Jalapin *n* jalapin, orizabin
Jalapinsäure *f* jalapic acid
Jalapknolle *f* (Bot) jalap root
Jalapseife *f* jalap soap
Jalousiesichter *m* louvre classifier
Jalpait *m* (Min) jalpaite
Jalpinolsäure *f* jalpinolic acid
Jamaicabitterholz *n* bitter wood, quassia wood
Jamaicapfeffer *m* (Bot) Jamaica pepper
Jamaicin *n* jamaicine
Jambaöl *n* jamba oil
Jambulfrucht *f* (Bot) jambul, Java plum
Jambulol *n* jambulol
Jamesonit *m* (Min) jamesonite, feather ore, native lead antimony sulfide
Jangolacton *n* yangolactone
Jangonin *n* yangonin
Januselektrode *f* Janus electrode
Janusgrün *n* (Färb) Janus green
Japaconin *n* japaconine
Japaconitin *n* japaconitine, acetylbenzoyl japaconine
Japancampher *m* Japan camphor, laurel camphor
Japanholz *n* Japan wood
Japanlack *m* Japan lacquer, Japan varnish, mit ~ überziehen to japan
Japansäure *f* japanic acid
Japantalg *m* Japan tallow, Japan wax
Japanwachs *n* Japan wax, Japan tallow
Japonsäure *f* japonic acid
Jargon *m* (Min) jargon, jargoon, zircon
Jarosit *m* (Min) jarosite
Jarrowit *m* jarrowite
Jasminaldehyd *m* jasminaldehyde
Jasminöl *n* jasmine oil
Jasmon *n* jasmone
Jaspachat *m* (Min) jasper agate
Jaspis *m* (Min) jasper, bloodstone, Indian stone, touchstone
Jaspisachat *m* (Min) jasper agate
jaspisartig jaspery, like jasper

Jaspisfarbe *f* color of jasper
jaspisfarben jasper-colored
Jaspisgut *n* jasper ware
Jaspisonyx *m* (Min) jasponyx
Jaspisopal *m* (Min) jasper opal
Jaspisporzellan *n* jasperated china
Jaspissteingut *n* jasper pottery, jasper ware
Jasponyx *m* (Min) jasponyx
Jaspopal *m* (Min) jasper opal
Jatamansäure *f* jatamansic acid
Jatrorrhizin *n* jatrorrhizine
Jauche *f* liquid manure
Jaulingit *m* jaulingite
Javanicin *n* javanicin
Javazimt *m* (Bot) Java cinnamon
Javellesche Lauge *f* eau de Javelle, Javelle's disinfecting liquor, Javelle water
Jecolein *n* jecolein
Jecorin *n* jecorin
Jefferisit *m* jefferisite
Jeffersonit *m* jeffersonite
Jenaer Glas *n* (HN) Jena glass, Jena ware
Jenit *m* jenite
Jenkinsit *m* (Min) jenkinsite
Jeremejewit *m* (Min) jeremejewite
Jervan *n* jervane
Jervasäure *f* jervaic acid
Jervin *n* jervine
Jesaconitin *n* jesaconitine
Jesuitenpulver *n* Jesuits' powder
Jesuitenrinde *f* Jesuits' bark, cinchona bark
Jet-Färberei *f* jet dyeing
Jett *m n* jet
jettähnlich jetty
Jettglas *n* jet glass
jettschwarz jet black
Jigger *m* (Färb) jig, jigger
Jiggerfärber *m* jig dyer
Jobsches Verfahren *n* Job's method
Jod *n* iodine (Symb. I)
Jodacetat *n* iodoacetate
Jodaceton *n* acetone iodide
Jodadditionsprodukt *n* iodine addition product
Jodäther *m* iodoether
Jodäthyl *n* ethyl iodide, iodoethane
Jodalkyl *n* alkyl iodide, iodoalkane
Jodallyl *n* allyl iodide
Jodammonium *n* ammonium iodide
Jodampulle *f* iodine ampoule
Jodamylum *n* starch iodide
Jodanil *n* iodanil
Jodanilin *n* iodoaniline, iodaniline, iodoaminobenzene
Jodanlage *f* iodine plant
Jodantimon *n* antimony iodide
Jodargyrit *m* (Min) iodargyrite, iodyrite

Jodarsen *n* arsenic iodide
Jodarsenik *n* arsenic iodide
jodartig iodine-like
Jodaseptol *n* (Med) iodoaseptol
Jodat *n* iodate, salt or ester of iodic acid
Jodatometrie *f* (Anal) iodatometry
Jodaufnahme *f* iodine intake
Jodazid *n* iodine azide
Jodbalsam *m* iodine balsam
Jodbarium *n* barium iodide
Jodbenzoesäure *f* iodobenzoic acid
Jodbenzol *n* iodobenzene
Jodblei *n* lead iodide
Jodbromid *n* iodine bromide, iodine(III) bromide, iodine tribromide
Jodbromür *n* iodine(I) bromide, iodine monobromide
Jodcadmium *n* cadmium iodide
Jodcalcium *n* calcium iodide
Jodchinolin *n* iodoquinoline
Jodchlor *n* iodine chloride
Jodchlorid *n* iodine chloride, iodine(III) chloride, iodine trichloride
Jodchlorür *n* iodine(I) chloride, iodine monochloride
Jodcollodium *n* iodized collodion
Jodcyan *n* cyanogen iodide, iodine cyanide
Joddampf *m* iodine vapor
Joddioxid *n* iodine dioxide
Jodeisen *n* iron iodide
Jodeosin *n* iodeosin, erythrosin, tetraiodofluorescein
Jodgehalt *m* iodine content
Jodgold *n* gold iodide
Jodgorgosäure *f* iodogorgoic acid, 3,5-diiodotyrosine
Jodgrün *n* iodine green
jodhaltig iodiferous
Jodhydrin *n* iodohydrin
Jodid *n* iodide
jodieren to iodate, to iodize, (Chem) to iodinate
Jodieren *n* iodination, iodinizing
Jodierung *f* iodination
jodig iodous
Jodinin *n* iodinin
Jodinrot *n* iodine scarlet
Jodipin *n* iodipin
Jodismus *m* (Med) iodine poisoning, iodism
Jodisotop *n* iodine isotope
Jodit *n* iodite
Jodival *n* jodival
Jodkali *n* potassium iodide
Jodkalistärkepapier *n* potassium iodide starchpaper
Jodkalium *n* potassium iodide
Jodkaliumpapier *n* potassium iodide paper
Jodkalzium *n* calcium iodide
Jodkohlenstoff *m* carbon tetraiodide
Jodkupfer *n* copper iodide

Jodlebertran *m* iodated cod-liver oil
Jodlithium *n* lithium iodide
Jodlösung *f* iodine solution
Jodmagnesium *n* magnesium iodide
Jodmangan *n* manganese iodide
Jodmethyl *n* iodomethane, methyl iodide
Jodmittel *n* remedy containing iodine, (Pharm) iodiferous remedy
Jodmonobromid *n* iodine(I) bromide, iodine monobromide
Jodmonochlorid *n* iodine (I) chloride, iodine monochloride
Jodnatrium *n* sodium iodide
Jodnatron *n* sodium iodide
Jodoanisol *n* iodanisole, iodoanisole
Jodobenzol *n* iodobenzene
Jodoform *n* iodoform, triiodomethane
Jodoformgaze *f* iodoform gauze
Jodoformin *n* iodoformin
Jodoformprobe *f* iodoform test
Jodol *n* iodol, tetraiodopyrrole
Jodolseife *f* iodol soap
Jodometrie *f* iodometry
jodometrisch iodometric
Jodonium *n* iodonium
Jodonium-Ion *n* iodonium ion
Jodoniumverbindung *f* iodonium compound
Jodophen *n* iodophen, nosophen, tetraiodophenolphthalein
Jodophthalein *n* iodophthalein
Jodopsin *n* iodopsin
Jodopyrin *n* iodopyrine
Jodosobenzol *n* iodosobenzene
Jodosol *n* iodosol, thymol iodide
Jodosterin *n* jodosterol
Jodothyrin *n* iodothyrin
Jodoxynaphthochinon *n* iodohydroxynaphthoquinone
Jodpapier *n* iodized paper
Jodpentafluorid *n* iodine pentafluoride
Jodpentoxid *n* iodine pentoxide, iodic anhydride
Jodphenol *n* iodophenol
Jodphenyldimethylpyrazolon *n* iodophenyldimethylpyrazolone
Jodphosgen *n* carbonyl iodide
Jodphosphonium *n* phosphonium iodide
Jodphosphor *m* phosphorus iodide
Jodphthalsäure *f* iodophthalic acid
Jodpräparat *n* iodine preparation, iodine product
Jodprobe *f* iodine test
Jodquecksilber *n* mercury iodide
Jodquelle *f* iodine spring
Jodradium *n* radium iodide
Jodsäure *f* iodic acid
Jodsäureanhydrid *n* iodic anhydride, iodine pentoxide
Jodsalbe *f* (Pharm) iodine ointment
Jodsalicylsäure *f* iodosalicylic acid

Jodsalz n iodide
Jodschale f iodine cup
Jodschwefel m sulfur iodide
Jodseife f iodine soap
Jodsilicium n silicon iodide
Jodsoda f sodium iodide
Jodstärke f starch iodide, iodized starch
Jodstärkepapier n potassium iodide starch paper, iodo-starch paper, starch iodide paper
Jodstärkeprobe f starch iodide test
Jodstickstoff m nitrogen iodide
Jodthiouracil n iodothiouracil
Jodthymol n thymol iodide, iodosol
Jodtinktur f tincture of iodine, tribromiodine
Jodtoluol n iodotoluene
Jodtrichlorid n iodine(III) chloride, iodine trichloride
Jodür n (obs) iodide
Jodverbindung f iodine compound
Jodvergiftung f iodine poisoning, iodism
Jodwasser n iodine water
Jodwasserstoff m hydrogen iodide, hydriodic acid
Jodwasserstoffäther m ethyl iodide
Jodwasserstoffsäure f hydriodic acid
Jodwasserstoffsalz n iodide
jodwasserstoffsauer iodide
Jodwismut n bismuth iodide
Jodyrit m (Min) iodyrite, native silver iodide
Jodzahl f iodine number, Hübl number, iodine [absorption] value, Wijs number
Jodzimtsäure f iodocinnamic acid
Jodzink n zinc iodide
Jodzinkpapier n zinc iodide [starch] paper
Jodzinkstärkelösung f zinc iodide starch solution
Jodzinn n tin iodide
Jodzinnober m red mercuric iodide
Joghurt n yoghurt
Johannisbeersaft m currant juice
Johannisbeerwein m currant wine
Johannisblüte f (Bot) hypericon flower
Johannisbrot n (Bot) carob bean, locust bean, St. John's bread
Johanniskraut n (Bot) hypericon, St. John's wort
Johannit m (Min) johannite, uranvitriol
Johnstonit m johnstonite
Johnstrupit m (Min) johnstrupite
Jonan n ionane
Jonidin n ionidine
Jonol n ionol
Jonon n ionone
Jononsäure f iononic acid
Jopansäure f iopanoic acid
Jophensäure f iophenoic acid
Jordanit m (Min) jordanite, native arsenic and lead sulfide

Jordisit m (Min) jordisite
Joseit m (Min) joseite, bismuth tellurite
Josen n iosene
Josephinit m (Min) josephinite
Jothion n iothion, diiodohydroxypropane, diiodopropyl alcohol
Joule n (Maß) joule
Joulesche Wärme f Joulean heat
Joulescher Wärmeverlust m Joule's heat loss, ohmic loss, resistance loss
Joule-Thomson-Effekt m (Therm) Joule-Thomson effect
Juchtenleder n Russian leather
Juddit m juddite
Judenharz n jew's pitch, asphalt
Judenpech n jew's pitch, asphalt
Juglon n juglone, 5-hydroxy-1,4-naphthoquinone, nucin
Juglonsäure f juglonic acid
Julep m julep
Julin n juline
Julocrotinsäure f julocrotic acid
Julolidin n julolidine
Julolin n juloline
Jungbier n new beer
Jungfernblei n virgin lead
Jungfernglas n mica, selenite
Jungferngold n native gold
Jungfernmarmor m virgin marble
Jungfernmetall n native metal
Jungfernquecksilber n native mercury, virgin mercury
Jungfernschwefel m native sulfur, virgin sulfur
Jungfernwachs n virgin wax
Junipen n junipene
Juniperinsäure f 16-hydroxyhexadecanoic acid, 16-hydroxypalmitic acid, juniperic acid
Juniperol n juniperol
justieren (einstellen) to adjust
Justierring m adjusting ring
Justierschraube f adjusting screw
Justierstift m adjusting stud
Justierung f adjustment, standardization
Justierwaage f adjusting balance
Jute f (Bot) jute
Juteasche f jute ash
Jutefaser f (Bot) jute fiber
Jutehechelgarn n heckled jute yarn
Jutekohlschwarz n jute carbon black
Juteschwarz n jute black
Juwelierborax m jeweller's borax, octahedral borax
Juxtapositionszwilling m (Krist) contact twin
JZ – Abk. für Jodzahl

K

Kabel *n* cable, (Tau) tow, (Techn) rope [cable], **besponnenes** ~ wrapped cable, **gestopftes** ~ vaseline-filled cable, **unterirdisches** ~ underground cable, **unterseeisches** ~ submarine cable
Kabelanschluß *m* cable connection
Kabelband *n* wire insulating ribbon, wire insulating tape
Kabelbrett *n* cable board
Kabelcord *m* cable cord
Kabelcordkeilriemen *m* cable cord belt
Kabelendverschluß *m* cable terminal, end box, terminal box, terminal head
Kabelformstück *n* cable section
Kabelimprägnieranlage *f* cable impregnation plant
Kabeljau *m* (Zool) cod
Kabeljaulebertran *m* cod liver oil
Kabelklemme *f* cable clamp
Kabelkran *m* cable crane
Kabellegung *f* laying of a cable
Kabelleitung *f* cable
Kabelmantel *m* cable covering, cable sheathing, wire coating
Kabelmasse *f* cable compound, sheathing compound, wire covering compound
kabeln to cable, to telex
Kabelöl *n* cable oil
Kabelprüfgerät *n* cable testing equipment
Kabelschacht *m* manhole
Kabelschelle *f* cable sleeve
Kabelschuh *m* cable shoe
Kabelschutz *m* cable protection, protective covering of cables
Kabeltrockenanlage *f* cable drying plant
Kabelüberzug *m* wire coating
Kabelummantelung *f* cable sheathing, sheathing of cables
Kabelumspinnung *f* electric wire covering
Kabelverzweigung *f* cable branching
Kabelwachs *n* cable wax
Kabelwinde *f* cable winch, rope winch
Kachel *f* glazed tile, ceramic tile, Dutch tile
Kachelpresse *f* pot press
Kadamsamenfett *n* kadam seed oil
Kadamsamenöl *n* kadam seed oil
Kadaver *m* corpse, carcass
Kadaverin *n* cadaverin, pentamethylene diamine
Kadeöl *n* (Kadinöl) cade oil
kadmieren to cadmium-plate
Kadmieren *n* cadmium plating
Kadmium *n* cadmium (s. auch Cadmium)
Käfer *m* (Zool) beetle, bug
Käfig *m* cage
Käfiganker *m* squirrel cage rotor, short circuit rotor

Käfigeffekt *m* cage effect
Käfigverbindung *f* (Chem) cage compound
Käfig-Wirkung *f* (Stereochem) cage effect
Kälte *f* cold, chilliness, coldness
Kälteaggregat *n* refrigerator, refrigerating machine
Kälteanlage *f* refrigeration plant
kältebeständig cold-resistant, antifreezing
Kältebeständigkeit *f* antifreezing property, cold resistance, low temperature property
Kältebiegeschlagwert *m* brittleness temperature
Kältebruchfestigkeit *f* break resistance at low temperature, tensile strength at low temperature
Kältechemie *f* cryochemistry
kälteempfindlich susceptible to cold
kälteerzeugend frigorific, refrigerative
Kälteerzeugung *f* refrigeration
Kälteerzeugungsmaschine *f* refrigerating machine
Kältefärbeverfahren *n* cold dyeing process
Kältefernversorgung *f* refrigeration distribution network
kältefest frost-resisting
Kältefestigkeit *f* cold resistance, strength at low temperatures
Kältegemisch *n* freezing mixture
Kältegrad *m* degree below zero
Kälteisoliermittel *n* cold insulator
Kälteisolierung *f* cold insulation
Kälteleistung *f* refrigerating capacity
Kältemaschine *f* refrigerating machine, refrigeration equipment, refrigerator
Kältemesser *m* frigorimeter
Kältemischung *f* freezing mixture, frigorific mixture
Kältemittel *n* refrigerant, refrigerating agent
Kälteregler *m* cryostat
Kälteschutzmittel *n* cold insulator, (Gefrierschutzmittel) antifreeze, antifreezing agent
Kältespeicherung *f* accumulation of cold
Kältetechnik *f* refrigeration [engeneering]
Kälteträger *m* cooling medium
Kältetrübung *f* (Bier) chill haze
Kältevergoldung *f* cold-gilding
Kälteverhalten *n* low temperature behavior
Kämmererit *m* (Min) kaemmererite
Kännelkohle *f* (Bergb) cannel coal
Käse *m* cheese, (Metall) crucible stand
käseartig caseous, cheesy
Käsebildung *f* caseation
Käsebutter *f* cheese curds
Käselab *n* rennet
Käseleim *m* casein glue, cheese glue
käsen to coagulate, to curd, to curdle

Käsen *n* coagulation, curdling
Käserei *f* cheese factory
Käsestein *m* rough diamond
Käsestoff *m* casein
Käsewasser *n* whey
käsig caseous, cheesy, curdy
käuflich commercial, commercially available
Kaffalsäure *f* caffalic acid, coffalic acid
Kaffee *m* coffee
Kaffeearoma *n* coffee flavor
Kaffeebitter *n* caffeine, coffeine
Kaffeebohne *f* coffee bean
kaffeebraun coffee-colored
Kaffee-Ersatz *m* coffee substitute, coffee surrogate
Kaffeegerbsäure *f* caffetannic acid
Kaffeesäure *f* caffeic acid
Kaffeesurrogat *n* cereal substitute for coffee
Kaffeevergiftung *f* (Med) caffeinism
Kaffeezusatz *m* ingredient added to coffee
Kaffeidin *n* caffeidine
Kaffein *n* 1,3,7-trimetylxanthine, caffeine, coffeine, theine
Kaffeinbenzoat *n* caffeine benzoate
Kaffeinbromhydrat *n* caffeine hydrobromide
Kaffeinchlorhydrat *n* caffeine hydrochloride
Kaffeincitrat *n* caffeine citrate
Kaffolid *n* caffolide
Kaffolin *n* 1,3,6-trimethylallantoin, caffoline
Kaffursäure *f* caffuric acid
Kahinkawurzel *f* (Bot) cahinca root
kahl gehen to need no flux
Kahm *m* film, mold, scum
kahmen to mold, to turn moldy
kahmig moldy, ropy, stale
Kain[in]säure *f* kainic acid
Kainit *m* (Min) kainite
Kainosit *m* kainosite
Kairin *n* kairine, methyloxytetrahydroquinoline
Kairolin *n* kairoline
Kaiserblau *n* smalt
Kaisergelb *n* mineral yellow, Paris yellow
Kaisergrün *n* imperial green, Paris green
Kaiseröl *n* fine kerosene, Kaiser oil
Kaiserrot *n* imperial red, colcothar, English red, red iron oxide
Kajaputöl *n* (Kajeputöl) cajuput oil
Kajeput *n* cajeput, cajuput
Kakao *m* cocoa
Kakaoaroma *n* cocoa flavor
Kakaobohne *f* (Bot) cocoa bean
Kakaobutter *f* cocoa butter
Kakaoöl *n* cocoa oil, cocoa butter
Kakaorin *n* cocaorine
Kakodyl *n* cacodyl
Kakodylat *n* cacodylate
Kakodylbromid *n* cacodyl bromide, bromodimethylarsine, dimethylarsenic monobromide

Kakodylcarbid *n* cacodyl carbide
Kakodylchlorid *n* cacodyl chloride, chlorodimethylarsine, dimethylarsenic monochloride
Kakodyloxid *n* cacodyl oxide, alkarsine, dicacodyl oxide
Kakodylpräparat *n* cacodyl preparation
Kakodylsäure *f* cacodylic acid, alkargen, dimethylarsinic acid
kakodylsauer cacodylate
Kakodylwasserstoff *m* cacodyl hydride, dimethylarsine
Kakothelin *n* cacotheline
Kakotheliniumhydroxid *n* cacothelinium hydroxide
Kakoxen *m* (Min) cacoxene, cacoxenite, kakoxene
Kaktus *m* (Bot) cactus
Kalait *m* (Min) cal[l]aite, kal[l]aite
Kalamin *m* (Min) calamine
Kalamus *m* (Bot) calamus
Kalander *m* calender, calendering machine
Kalanderauftragsverfahren *n* calender coating
Kalanderfärbung *f* calender dyeing, calender staining
kalandern to calender
Kalanderpapier *n* calender paper
Kalanderwalze *f* calender roll
Kalandrierbarkeit *f* calendering
kalandrieren to calender
Kalandrieren *n* calendering
kalandriert friction-glazed
Kalandrierwalze *f* calender
Kalbfleisch *n* veal
Kaledonit *m* (Min) caledonite
Kaleszenz *f* calescence
Kaleszenzpunkt *m* point of calescence
kalfatern to calk, to caulk
Kalgoorlit *m* kalgoorlite
Kali *n* potassium hydroxide, (Ätzkali) caustic potash, **blausaures** ~ potassium cyanide, **chlorsaures** ~ potassium chloride, **kohlensaures** ~ potassium carbonate, **oxalsaures** ~ potassium oxalate
Kalialaun *m* potassium aluminum sulfate, common alum, potash alum, potassium alum, (Min) kalinite
Kaliammoniaksuperphosphat *n* potassium ammonium superphosphate
Kaliammonsalpeter *m* potassium ammonium nitrate
Kaliaturholz *n* caliatour wood
Kaliber *n* caliber, diameter of bore, gauge, (Metall) groove, roll opening, roll pass
Kalibergwerk *n* potash mine
Kaliberlehre *f* caliber gauge
kalibermäßig true to scale
Kaliberring *m* ring gauge
Kaliberwalze *f* grooved roll

Kaliberzirkel *m* calipers
Kalibicarbonat *n* acid potassium carbonate, potassium bicarbonate, potassium hydrogen carbonate
Kalibioxalat *n* acid potassium oxalate, potassium binoxalate, potassium hydrogen oxalate, salt of sorrel
Kaliblau *n* Prussian blue
Kalibleiglas *n* potash lead glass
kalibrieren to calibrate, to gauge, to graduate, to size, to standardize
Kalibrieren *n* calibrating, roll designing, ~ **der Walzen** design of rolls, roll drafting
Kalibrierpipette *f* calibrating pipette
Kalibrierring *m* calibrating ring
Kalibrierung *f* calibration, graduation scale
Kalibrierungszeichnung *f* pass template drawing
Kalibrierwalze *f* sizing roll
Kalidüngemittel *n* potash fertilizer
Kalidünger *m* potash fertilizer
Kalidüngesalz *n* potash fertilizer, potassium salt for fertilizing
Kalieisenalaun *m* potash iron alum, potassium ferric sulfate
Kalieisencyanür *m* (Kalium-eisen(II)-cyanid) potassium cyanoferrate(II), potassium ferrocyanide, yellow prussiate of potash
Kalifeldspat *m* (Min) native potassium aluminum silicate, orthoclase, potash feldspar, sunstone
Kaliform *f* potash mold
Kalifornium *n* californium (Symb. Cf)
kalifrei potash-free
Kaliglas *n* potash glass
Kaliglimmer *m* (Min) potash mica, muscovite
Kalignost *n* kalignost
kalihaltig containing potash, potassic, potassiferous
Kalihydrat *n* potassium hydroxide, caustic potash, hydrate of potash, potassium hydrate
Kalikalk *m* potash lime
Kaliko *m* (Text) calico, cotton
Kalikodruck *m* (Text) calico printing
Kalikugel *f* potash bulb
Kalilauge *f* potassium hydroxide [solution], caustic potash solution
Kalimagnesia *f* potash magnesia
Kalimetall *n* metallic potassium, potassium metal
Kalinatronfeldspat *m* (Min) potassium sodium feldspar
Kalinit *m* (Min) kalinite, native potassium aluminum sulfate
Kalipflanze *f* (Bot) glasswort, marsh samphire
Kalipräparat *n* potassium preparation
Kalisalpeter *m* potassium nitrate, nitrate of potash, [potash] niter, saltpeter
Kalisalpetersuperphosphat *n* potassium nitrate superphosphate

Kalisalz *n* potash salt, potassium salt
Kalisalzlager *n* deposit of potassium salt, bed of potash salt
Kalischmelze *f* potash fusion, potash melt
Kalischmierseife *f* potash soft soap
Kalischwefelleber *f* (obs) potassium sulfide, hepar sulfuris, liver of sulfur
Kaliseife *f* potash soap, soft soap
Kalisulfat *n* potassium sulfate
Kalitonerde *f* potassium aluminate, aluminate of potash
Kalium *n* potassium, (Symb. K)
Kaliumacetat *n* potassium acetate
Kaliumalaun *m* potassium alum, potash alum
Kaliumaluminat *n* potassium aluminate
Kaliumaluminiumsulfat *n* potassium aluminum sulfate, potash alum
Kaliumammoniumtartrat *n* potassium ammonium tartrate
Kaliumantimonat *n* potassium antimonate
Kaliumantimonyltartrat *n* potassium antimonyltartrate, tartar emetic
Kaliumarsenat *n* potassium arsenate
Kaliumarsenit *n* potassium arsenite
Kaliumaurocyanid *n* potassium aurocyanide
Kaliumbicarbonat *n* potassium bicarbonate, acid potassium carbonate, bicarbonate of potash, potassium hydrogen carbonate
Kaliumbichromat *n* potassium bichromate, bichromate of potash, potassium dichromate
Kaliumbifluorid *n* potassium bifluoride
Kaliumbioxalat *n* acid potassium oxalate, potassium bi[n]oxalate, potassium hydrogen oxalate, sorrel salt
Kaliumbisulfat *n* acid potassium sulfate, potassium bisulfate
Kaliumbisulfit *n* acid potassium sulfite, potassium bisulfite, potassium hydrogen sulfite
Kaliumbitartrat *n* acid potassium tartrate, cream of tartar, potassium bitartrate, potassium hydrogen tartrate
Kaliumboratglas *n* potassium borate glass
Kaliumborfluorid *n* potassium borofluoride, potassium fluoborate
Kaliumbrechweinstein *m* potassium antimonyl tartrate, tartar emetic
Kaliumbromat *n* potassium bromate
Kaliumbromid *n* potassium bromide
Kaliumcarbonat *n* potassium carbonate, potash
Kaliumchlorat *n* potassium chlorate
Kaliumchlorid *n* potassium chloride
Kaliumchlorit *n* potassium chlorite
Kaliumchloroplatinat *n* potassium chloroplatinate
Kaliumchromalaun *m* chromic potassium alum, potassium chromium sulfate
Kaliumchromat *n* potassium chromate
Kaliumcitrat *n* potassium citrate

Kaliumcyanat *n* potassium cyanate
Kaliumcyanid *n* potassium cyanide
Kaliumdampf *m* potassium vapor
Kaliumdichromat *n* potassium dichromate, red potassium chromate
Kaliumdigermanat *n* potassium digermanate
Kaliumdihydroarsenat *n* potassium dihydrogen arsenate
Kaliumdihydrophosphat *n* monobasic potassium phosphate, potassium dihydrogen phosphate
Kaliumdoppelsalz *n* potassium double salt
Kaliumeisencyanür *n* (Kalium-eisen(II)-cyanid) ferrous potassium cyanide, potassium ferrocyanide, yellow potassium prussiate
Kaliumeisen(III)-cyanid *n* ferric potassium cyanide, potassium ferricyanide, red potassium prussiate, red prussiate of potash
Kaliumferrat *n* potassium ferrate
Kaliumferricyanid *n* potassium ferricyanide, red potassium prussiate
Kaliumferrioxalat *n* ferripotassium oxalate, potassium ferric oxalate
Kaliumferrisulfat *n* ferripotassium sulfate, potassium ferric sulfate
Kaliumferrocyanid *n* ferrous potassium cyanide, potassium ferrocyanide, yellow potassium prussiate
Kaliumfluorid *n* potassium fluoride
Kaliumglycerophosphat *n* potassium glycerophosphate
Kaliumgoldbromür *n* potassium aurobromide
Kaliumgoldchlorid *n* potassium aurichloride, potassium chloroaurate(III)
Kaliumgoldcyanür *n* potassium cyanoaurate(I)
Kaliumhalogenid *n* potassium halide
kaliumhaltig potassic, potassiferous
Kaliumhydrat *n* potassium hydroxide
Kaliumhydrogenkarbonat *n* potassium bicarbonate
Kaliumhydroxid *n* potassium hydroxide, caustic potash, potassium hydrate
Kaliumhypochlorit *n* potassium hypochlorite, eau de Javelle
Kaliumhypochloritlösung *f* eau de Javelle, Javelle's disinfecting liquor
Kalium iodatum *n* (Lat) potassium iodide
Kaliumiridichlorid *n* potassium iridichloride, iridium potassium chloride, potassium chloroiridate
Kaliumjodid *n* potassium iodide
Kaliumjodidsalbe *f* (Pharm) potassium iodide ointment
Kaliumjodidstärke *f* potassium iodide starch
Kaliumkarbonat *n* potassium carbonate
Kaliumkobaltnitrit *n* cobaltic potassium nitrite, potassium cobaltinitrite
Kaliumkohlenoxid *n* potassium hexacarbonyl, potassium carboxide

Kaliumkupfercyanür *n* (Kalium-kupfer(I)-cyanid) potassium cuprocyanide, potassium cyanocuprate(I)
Kaliummagnesiumsulfat *n* potassium magnesium sulfate
Kaliummanganat *n* potassium manganate
Kaliummetall *n* metallic potassium, potassium metal
Kaliummonosulfid *n* potassium monosulfide, potassium sulfide
Kaliummonoxid *n* potassium monoxide
Kaliumnatriumcarbonat *n* potassium sodium carbonate
Kaliumnatriumtartrat *n* potassium sodium tartrate, Rochelle salt, Seignette's salt
Kaliumniobat *n* potassium niobate
Kaliumnitrat *n* potassium nitrate, niter, saltpeter
Kalium nitricum *n* (Lat) potassium nitrate
Kaliumoxalat *n* potassium oxalate
Kaliumoxid *n* potassium oxide
Kaliumoxidhydrat *n* (obs) potassium hydroxide, caustic potash, hydrate of potash, potassium hydrate
Kaliumpalladiumchlorür *n* potassium palladochloride, potassium chloropalladite
Kaliumperchlorat *n* potassium perchlorate
Kaliumperjodat *n* potassium periodate
Kaliumpermanganat *n* potassium permanganate
Kaliumpersulfat *n* potassium persulfate
Kaliumphenolsulfonat *n* potassium phenolsulfonate
Kaliumphosphat *n* potassium phosphate
Kaliumplatinchlorid *n* (Kalium-platin(IV)-chlorid) potassium platinichloride, potassium chloroplatinate
Kaliumplatinchlorür *n* (Kalium-platin(II)-chlorid) potassium platinochloride, potassium chloroplatinite
Kaliumplatincyanür *n* (Kalium-platin(II)-cyanid) potassium platinocyanide, potassium cyanoplatinite
Kaliumplatinochlorid *n* (Kalium-platin(II)-chlorid) potassium platinochloride, potassium chloroplatinite
Kaliumpolysulfid *n* potassium polysulfide
Kaliumporphinchelat *n* potassium porphin chelate
Kaliumpyrosulfit *n* potassium pyrosulfite, potassium metabisulfite
Kaliumquecksilbercyanid *n* (Kalium-quecksilber(II)-cyanid) mercuric potassium cyanide, potassium mercuricyanide
Kaliumquecksilberjodid *n* potassium mercury iodide, mercuric potassium iodide
Kaliumregulus *m* potassium regulus
Kaliumrhodanat *n* (obs) potassium thiocyanate
Kaliumrhodanid *n* potassium thiocyanate
Kaliumrhodanür *n* (obs) potassium thiocyanate
Kaliumsalz *n* potassium salt

Kaliumseife *f* potassium soap, soft soap
Kaliumselenat *n* potassium selenate
Kaliumsilbercyanür *n* (Kalium-silber-cyanid) potassium argentocyanide
Kaliumsilikat *n* potassium silicate
Kaliumsilikofluorid *n* potassium silicofluoride, potassium fluosilicate
Kaliumsulfat *n* potassium sulfate
Kaliumsulfhydrat *n* potassium hydrogen sulfide, potassium sulfhydrate
Kaliumsulfid *n* potassium sulfide, potassium monosulfide
Kaliumsulfit *n* potassium sulfite
Kaliumsulfocyanat *n* potassium thiocyanate
Kaliumsulfoguajacol *n* potassium guaiacol sulfonate
Kaliumtantalat *n* potassium tantalate
Kaliumtartrat *n* potassium tartrate
Kaliumtetraoxalat *n* potassium tetroxalate
Kaliumthiosulfat *n* potassium thiosulfate
Kaliumtrisulfid *n* potassium trisulfide
Kaliumverbindung *f* potassium compound
Kaliumwasserglas *n* potash waterglass, potassium silicate
Kaliumwasserstoff *m* potassium hydride
Kaliumwolframat *n* potassium tungstate
Kaliumxanthogenat *n* potassium xanthogenate
Kaliverbindung *f* potassium compound
Kaliwasserglas *n* potash-waterglass, potassium silicate
Kaliwerk *n* potash works
Kalk *m* lime, chalk, limestone, ~ **brennen** to burn lime, ~ **löschen** to slake lime, **den ~ totbrennen** to overburn lime, **fetter ~** fat lime, **gebrannter ~** caustic lime, quicklime, **gelöschter ~** slaked lime, **kohlensaurer ~** carbonate of lime, **mit ~ bewerfen** to rough-cast, **ungelöschter ~** quick-lime, unslaked lime, **verwitterter (abgestorbener) ~** dead lime, lime slaked in the air
Kalk-12-Hydroxyfett *n* calcium-12-hydroxy grease
Kalkablagerung *f* calcareous sediment, (Physiol) calcification
Kalkader *f* limestone vein
Kalkäscher *m* tanner's pit, lime pit, lixiviated ashes, soap waste
Kalkalabaster *m* (Min) calcareous alabaster
Kalkammonsalpeter *m* lime ammonium nitrate, calnitro
Kalkanstrich *m* lime paint, white-wash
Kalkanwurf *m* parget; plaster
Kalkaphanit *m* calcareous aphanite
kalkarm deficient in lime
kalkartig calcareous, lime-like
Kalkartigkeit *f* calcareousness, chalkiness
Kalkbad *n* lime bath
Kalkbaryt *m* (Min) calcareous barite

Kalkbeschickbühne *f* lime charging floor, lime charging gallery
Kalkbeständigkeit *f* lime resistance
Kalkbeton *m* lime concrete
Kalkbeuche *f* lime bucking lye, lime boil
Kalkbildung *f* (Med) calcification
Kalkbindevermögen *n* (Waschmittel) calcium chelating capacity
Kalkblau *n* blue verditer, blue ashes
Kalkboden *m* calcareous soil, lime soil
Kalkborat *n* calcium borate
Kalkbrei *m* lime cream, lime paste
Kalkbrennen *n* calcination of limestone
Kalkbrenner *m* lime burner
Kalkbrennerei *f* lime-kiln
Kalkbrennofen *m* lime-burning kiln, lime kiln
Kalkbruch *m* limestone quarry
Kalkbrühe *f* lime liquor, lime water, milk of lime, whitewash
Kalkeisenaugit *m* (Min) hedenbergite
Kalkeisenstein *m* (Min) calcareous iron stone, ferrugineous limestone
kalken to lime, to mix with lime, to soak in lime, (mit Kalk bewerfen) to roughcast, (tünchen) to whitewash
Kalken *n* **des Drahtes** chalking the wire
Kalkerde *f* calcareous earth, calcium oxide, lime
kalkerdig calcareous
Kalkfarbe *f* whitewash paint, distemper color, lime wash, lime[wash] paint
Kalkfaß *n* lime pit
Kalkfeldspat *m* (Min) lime feldspar, anorthite
Kalkfett *n* calcium grease, lime grease
kalkförmig calciform
kalkfrei free from lime
Kalkgebirge *n* calcareous mountain-chain
Kalkgestein *n* (Min) limestone rocks
Kalkglas *n* common glass, lime glass
Kalkglimmer *m* (Min) lime mica, margarite
Kalkglimmerschiefer *m* (Min) green marble, cipolin
Kalkgranat *m* (Min) green garnet, andradite, grossularite, lime garnet
Kalkgrube *f* lime pit
Kalkguß *m* watery lime mortar
Kalkhärte *f* calcium hardness
kalkhaltig calcareous, calciferous, containing lime
Kalkharmotom *m* christianite, lime harmotome, phillipsite
Kalkharnstoff *m* urea limestone
Kalkhaushalt *m* (Physiol) calcium balance
Kalkhütte *f* lime kiln
Kalkhydrat *n* calcium hydroxide, hydrate of lime, slaked lime
kalkig calcareous, chalky, limy
Kalkkalisulfat *n* (Min) syngenite
Kalkkasten *m* lime chest
Kalkkelle *f* lime trowel

Kalkkies – Kalorienwert

Kalkkies *m* calcareous gravel
Kalkkiesel *m* calcareous silex
Kalkkitt *m* calcareous cement, lime cement
Kalkkohlegemisch *n* mixture of lime and coal
Kalkkomplexfett *n* calcium complex grease
Kalkkonglomerat *n* calcareous breccia
Kalklager *n* calcareous deposit
Kalklauge *f* calcium hydroxide solution
Kalklicht *n* calcium light, Drummond light, limelight
Kalkloch *n* lime basin, lime pit
Kalklöschapparat *m* lime slaking apparatus
Kalklöschen *n* lime slaking
Kalklösung *f* calcium hydroxide solution, solution of lime
Kalkmalachit *m* (Min) calcareous malachite
Kalkmangel *m* (Med) calcium deficiency, hypocalcia
Kalkmergel *m* calcareous marl, lime marl
Kalkmesser *m* calcimeter
Kalkmilch *f* cream of lime, lime milk, lime wash, lime water, whitewash
Kalkmörtel *m* lime mortar
Kalkmühle *f* lime mill, lime crusher
Kalknatronzeolith *m* (Min) lime and soda mesotype, mesolite
Kalkniederschlag *m* lime precipitate, precipitated calcium carbonate
Kalkocker *m* (Min) calcareous ocher
Kalkofen *m* lime kiln
Kalkofengas *n* lime-kiln gas
Kalkoligoklas *m* (Min) labradorite
Kalkoolith *m* (Min) calcareous oolite
Kalkphyllit *m* (Min) calcareous phyllite
Kalkpulver *n* lime powder
Kalkrahm *m* cream of lime
kalkreich calcareous, rich in lime
Kalksalpeter *m* calcium nitrate
Kalksalz *n* calcium salt, lime salt
Kalksand *m* calcareous sand, lime sand
Kalksandstein *m* (Bauw) sand-lime brick, (Min) calcareous sandstone, lime sandstone
Kalksauerbad *n* lime sour bath
Kalkschaum *m* lime scum
Kalkschiefer *m* (Min) calcareous slate
Kalkschieferton *m* (Min) calcareous slate clay
Kalkschlacke *f* lime slag
Kalkschlamm *m* lime mud, lime sludge
Kalkschleier *m* lime coating
Kalkschwefelleber *f* (obs) calcium sulfide, hepar calcis, lime hepar
Kalkseife *f* calcium soap, lime soap
Kalksilikat *n* calcium silicate
Kalksilo *m* lime silo, lime bin
Kalksinter *m* calcareous sinter, calcareous tufa, calc sinter
Kalkspat *m* (Min) calcite, calcium carbonate, calcspar, iceland spar
Kalkspatinterferometer *n* calcite interferometer

Kalkstein *m* (Min) limestone, chalk, **sandiger** ~ marl-stone
Kalksteinbruch *m* limestone quarry
Kalksteingut *n* calcareous whiteware, limestone whiteware
Kalksteinzuschlag *m* calcareous flux, limestone flux
Kalkstickstoff *m* calcium cyanamide, lime-nitrogen, nitrolim[e]
Kalkstickstoffdünger *m* calcium cyanamide fertilizer
Kalksulfat *n* (obs) calcium sulfate
Kalksuperphosphat *n* calcium superphosphate
Kalktalk *m* (Min) calcareous talc, dolomite
Kalktiegel *m* lime crucible
Kalktongranat *m* (Min) lime garnet, grossularite
Kalktropfstein *m* calcareous sinter, stalactite
Kalktünche *f* whitewash
Kalktuff *m* (Min) calcareous tufa, calc tuff, tufaceous limestone
Kalkül *m, n* (Math) calculus
Kalkulation *f* calculation, computation, estimate
kalkulieren to calculate, to compute
Kalkuranglimmer *m* (Min) calcareous uranite, autunite
Kalkuranit *m* (Min) autunite
Kalkverbindung *f* calcium compound
Kalkverseifung *f* saponification by lime
Kalkwand *f* plaster wall
Kalkwasser *n* lime water, milk of lime
Kalkweinstein *m* calcareous tartar
Kalkweiße *f* lime wash, lime water
Kalkwerk *n* lime works
Kalkzementmörtel *m* lime-cement mortar
Kalkzuschlag *m* calcareous flux, limestone flux
Kallait *m* (Min) callaite, kal[l]aite
Kallikrein *n* (Biochem) callicrein
Kallilith *m* (Min) callilite
Kallochrom *m* callochrome
Kalmeghin *n* calmeghine
Kalmopyrin *n* calmopyrine
Kalmus *m* (Bot) calamus
Kalmuscampher *m* calamus camphor
Kalmusextrakt *m* extract of acorus
Kalmusöl *n* calamus oil
Kalmustinktur *f* tincture of acorus
Kalmuswurzel *f* (Bot) calamus root
Kalomel *n* calomel, mercurous chloride, mercury(I) chloride, mercury monochloride
Kalomelelektrode *f* calomel electrode
Kalomelfaserelektrode *f* fiber-type calomel electrode
Kalorie *f* calorie, caloric unit, thermal unit, **große** ~ kilogram calorie, **kleine** ~ gram calorie
Kalorienbedarf *m* calorie requirement
Kaloriengehalt *m* caloric content
kalorienreich rich in calories
Kalorienwert *m* calorific value

Kalorimeter *n* calorimeter
Kalorimeterrohr *n* calorimeter tube
Kalorimeterzubehör *n* calorimeter accessories, calorimeter equipment
Kalorimetrie *f* calorimetry
kalorimetrisch calorimetric
kalorisch caloric, calorific
kalorisieren to calorize
Kalorisieren *n* calorizing
Kalorisierung *f* calorizing
Kalottenmodell *n* space filling atom model
Kalotypie *f* calotypy
kalt cold, chilly, cool, frigid, ~ **blasen** to coldblast, ~ **entfetten** to degrease cold, ~ **gesättigt** saturated in the cold state, ~ **geschmiedet** cold forged, ~ **gewalzt** cold rolled, ~ **gezogen** cold drawn, ~ **verformbar** cold forming, workable in cold state
kaltabbindend cold-setting, cold-pressing
Kaltansatzlack *m* cold-cut varnish
Kaltanstrich *m* cold [plastic] paint
Kaltasphalt *m* cold mix, bituminous emulsion
Kaltaushärtung *f* cold tempering, cold age-hardening, cold aging
Kaltbad *n* cold bath
Kaltbeanspruchung *f* cold straining
Kaltbearbeitung *f* cold working
kaltbiegen to cold-bend
Kaltbiegen *n* cold bending
Kaltbiegeprobe *f* cold bending test
Kaltbruch *m* cold-brittleness, cold-shortness
kaltbrüchig brittle from cold, cold-short
Kaltbrüchigkeit *f* cold-brittleness, cold-shortness
Kaltdach *n* ventilated roof
Kaltdampf *m* cold vapor
Kaltdampfmaschine *f* vapor-compression refrigeration machine
Kalteinsatz *m* cold charging
Kaltextraktor *m* cold extractor
Kaltfärben *n* cold-dyeing
Kaltformen *n* (Met) cold working
Kaltformgebung *f* cold working
Kaltfrischen *n* fining in two operations
Kaltgasmaschine *f* gas refrigeration machine
kaltgetrocknet freeze-dried, lyophilized
kaltgewalzt cold-rolled
kaltgezogen cold-drawn
Kaltglasur *f* cold glaze
Kaltguß *m* cold casting, spoiled casting
kaltgussig cast with interruptions
kalthämmern to cold-hammer, to cold-work
kalthärten to cold-harden
kalthärten *n* cold-hardening, cold quenching
kalthärtend cold-setting
Kalthärter *m* cold hardener
Kalthärtung *f* cold setting
Kaltkreissäge *f* cold saw
Kaltlack *m* cold-cut varnish
Kaltlagerung *f* cold storage

Kaltleim *m* cold glue
kaltlöslich cold-soluble
Kaltlöt[e]stelle *f* cold junction, cold end
kaltpressen (Kunstharz) to cold-mold, (Text) to cold-dress, to cold-press
Kaltpressen *n* cold pressing
Kaltpreßmasse *f* cold molding compound
Kaltpreßschweißung *f* cold-press welding
Kaltpreßverfahren *n* (Polyester) low-temperature matched die molding, cold press molding
Kaltprobe *f* cold test
Kaltprofil *n* cold rolled section
kaltprofilieren to cold-form
Kaltreckung *f* cold-straining
Kaltrissigkeit *f* cold-shortness
Kaltsägemaschine *f* cold saw
Kaltschmieden *n* cold-hammering, hammer-hardening
Kaltschrumpfen *n* cold shrinking
Kaltschweiße *f* crack, flaw
Kaltschweißstelle *f* weld mark
kaltsiegelfähig cold sealable
Kaltsiegelung *f* cold sealing, selfsealing
kaltspritzen to cold-spray
Kaltsprödigkeit *f* cold brittleness
Kaltsterilisierung *f* cold sterilization
kaltstrecken to cold-stretch
Kaltstrecken *n* cold stretching
Kaltverarbeitung *f* cold working
Kaltverfestigung *f* cold-work hardening, cold-working, hardening by cold working, strain-hardening
Kaltverfestigungsgrenze *f* limit of strain-hardening, limit of work-hardening
Kaltverformbarkeit *f* cold working property
kaltverformen to cold-form, to cold-work
Kaltverformen *n* cold forming, cold working
Kaltverformung *f* cold deformation, cold forming, cold shaping, cold working, [plastic] cold working
kaltvergießen to cast or pour cold
Kaltvergoldung *f* cold gilding
Kaltverstreckbarkeit *f* cold drawability
Kaltvulkanisat *n* cold vulcanized product
Kaltvulkanisation *f* cold cure, cold curing, cold vulcanizing
kaltvulkanisieren to cold-vulcanize
Kaltwäsche *f* low-temperature scrubbing
kaltwalzen to cold-roll
Kaltwalzerei *f* cold rolling mill, cold rolling practice
Kaltwalzwerk *n* cold rolling mill, cold strip mill
Kaltwasserbad *n* cold-water bath
Kaltwasserfarbe *f* cold-water paint
Kaltwerden *n* cooling, getting cold
Kaltwiderstandsfähigkeit *f* cold resistance
Kaltwindbetrieb *m* cold blast practice
Kaltwindkupolofen *m* cold blast cupola
Kaltwindschieber *m* cold blast slide

kaltziehen to cold-draw
Kaltziehen *n* cold drawing
Kaluszit *m* (Min) caluszite, syngenite
Kalzimeter *n* calcimeter
Kalzinieranlage *f* calcination plant
kalzinieren to calcine
Kalzinierofen *m* calcining furnace, calcining kiln
Kalzinierung *f* calcination
Kalzit *m* (Min) calcite
Kamacit *m* (Min) kamacite
Kamala *f* (Pharm) kamala, kamela, kamila
Kamarezit *m* (Min) kamarezite, native basic copper sulfate
Kamazit *m* kamazite
Kambalholz *n* (Bot) camwood
Kamelhaar *n* camel hair
Kamera *f* (Phot) camera
Kamholz *n* (Bot) camwood
Kamille *f* (Bot) camomile, **echte ~** (Bot) german camomile, **falsche ~** (Bot) scentless camomile
Kamillenöl *n* camomile oil
Kamin *m* chimney, (Kolonne) downcomer, vapor outlet
Kaminbühne *f* chimney platform
Kaminkühler *m* chimney cooler
Kaminruß *m* chimney soot
Kaminschalter *m* stack switch
Kaminverband *m* (Säcke) stacked in chimney formation
Kamlolensäure *f* kamlolenic acid
Kamm *m* comb
Kammer *f* chamber
Kammerfilterpresse *f* chamber filter press, recessed plate press
Kammerkristall *m* (Chem) chamber crystal
Kammerofen *m* chamber furnace, box furnace
Kammerreaktor *m* chamber reactor
Kammersäure *f* chamber acid
Kammertrockner *m* chamber drier, drying chamber
Kammerverfahren *n* (Chem) lead chamber process
Kammerzentrifuge *f* multi-chamber centrifuge
Kammgarn *n* (Text) worsted yarn
Kammgarnspinnerei *f* worsted spinning
Kammkies *m* (Min) marcasite
Kammlänge *f* crest length
Kammlager *n* collar thrust bearing
Kammlinienanalyse *f* **nach Hoerl** (Statist) ridge line analysis
Kammogenin *n* kammogenin
Kammstahl *m* rack tool
Kammwalze *f* pinion, broad-faced steel gear, cogged cylinder
Kammwalzengetriebe *n* grooved roller gear
Kammwalzenständer *m* spindle box, spindle housing
Kammwalzenzapfen *m* pinion neck

Kammwalzgerüst *n* pinion housing
Kammzahn *m* chaser tooth
Kammzapfen *m* collar journal
Kampfer *m* camphor
Kampfersalbe *f* (Pharm) camphor ointment
Kampferspiritus *m* spirits of camphor
Kampfgas *n* poison gas, war gas
Kamphen *n* camphene
Kampher *m* (s. auch Campher u. Kampfer) camphor
kamphern to camphorate
Kampheröl *n* camphor oil
Kamphersäure *f* camphoric acid
Kampylit *m* (Min) campylite
Kanadabalsam *m* Canada balsam, balsam of fir
Kanadol *n* canadol, impure hexane
Kanal *m* (Abfluß) drain, sewer, (Abflußrinne) groove, (künstlicher) canal, (natürlicher) channel, (TV) channel
Kanalbildung *f* channeling
Kanaldämpfer *m* flash ager
Kanalform *f* channel mold
Kanalisationsabwasser *n* drain water, sewage water
kanalisieren to canalize, to drain through sewers
Kanalisieren *n* canalizing, draining by means of sewers
Kanalisierung *f* canalization, sewerage system
Kanalradsichter *m* centripetal classifier
Kanalruß *m* channel black
Kanalsichtrad *n* centripetal classifier
Kanalstrahlenanalyse *f* analysis of canal rays, positive-ray analysis
Kanalstrahlentladung *f* canal ray discharge
Kanalstrahlionenquelle *f* canal ray discharge ion source
Kanalstrahlröhre *f* canal ray tube
Kanalströmung *f* channel flow, fluid flow in open channels
Kanaltrockner *m* tunnel dryer
Kanalwasser *n* sewage [water]
Kanalwasserpumpe *f* sewage water pump
Kanamycin *n* kanamycin
kanariengelb canary yellow
Kanarin *n* canarin
kandieren (Gastr) to candy
Kandieren *n* candying
Kandiertrommel *f* coating drum
Kandiolin *n* candiolin
Kandiszucker *m* sugar candy
Kandit *m* (Min) candite, black spinel
Kaneel *m* (Bot) cinnamon bark
Kaneelbruch *m* broken cinnamon
Kaneelgranat *m* (Min) cinnamon stone, essonite
Kaneelrinde *f* (Bot) canella bark
Kaneelwachs *n* cinnamon wax
Kanirin *n* kanirin, trimethylamine oxide
Kanister *m* can, canister, container

kannelieren to chamfer, to channel, to flute, to groove
Kannelierung *f* chamfering, channeling, fluting
Kanonenmetall *n* gun metal
Kanonenofen *m* sealed tube furnace
Kanonenrohr *n* sealed tube
Kanosamin *n* kanosamine
Kante *f* border, edge, ridge, (Web) selvage
kanten to set on edge, to tilt
Kanteneffekt *m* edge effect
Kantenemission *f* edge emission
Kantenfestigkeit *f* (Tabletten) resistance to chipping
Kantenfräsmaschine *f* edge-milling machine
Kantenlänge *f* edge length
Kantenriß *m* edge crack, cross crack
Kantenspannung *f* (Web) selvage tension
Kantenstein *m* border stone
Kantenverschiebung *f* edge shift
Kantenwinkel *m* (Krist) edge angle
Kantvorrichtung *f* tilter
Kanüle *f* (Med) cannula
Kanyabutter *f* kanya butter
Kanzel *f* working platform
kanzerogen (Med) cancerogenic, carcinogenic
Kanzuiol *n* kanzuiol
Kaolin *n* (Min) kaolin, bolus alba, china clay, porcelain clay
kaolinisieren to kaolinize
Kaolinisierung *f* kaolinization
Kaolinit *m* (Min) kaolinite
Kaolinsandstein *m* (Min) kaoliniferous sandstone
Kapazität *f* capacity, (Kondensator) capacitance, **magnetische** ~ magnetic capacity, magnetic susceptibility
Kapazitätsabnahme *f* decrease of capacity, diminution of capacity
Kapazitätsausgleich *m* capacity balance
Kapazitätsbereich *m* (Elektr) capacitance range
Kapazitätseinbuße *f* loss of capacity
Kapazitätseinheit *f* unit of capacity
kapazitätsfrei noncapacitive
Kapazitätsgröße *f* capacity value
Kapazitätsindex *m* capacity index
Kapazitätskoeffizient *m* coefficient of capacity
Kapazitätsprobe *f* capacity test
Kapazitätsreaktanz *f* capacitance, capacity reactance, reactance due to capacity
Kapazitätsreserve *f* reserve capacity
Kapazitätsschwund *m* diminution of capacity, loss of capacity
Kapazitätsspannung *f* (Elektr) capacity voltage
Kapazitanz *f* (Elektr) capacitance, capacitive reactance
kapazitiv capacitive
Kapazitron *n* capacitron
Kapelle *f* (Metall) cupel, cupel furnace, refining hearth

Kapellenasche *f* bone ashes, test ashes
Kapellenform *f* cupel mold
Kapellengold *n* fine gold
Kapellenofen *m* cupelling furnace, assay furnace, sublimation furnace
Kapellenprobe *f* cupel test, cupellation
Kapellensilber *n* fine silver
Kapellenstativ *n* cupel stand
Kapellenzug *m* loss in cupellation
kapillar capillary
Kapillaraffinität *f* capillary affinity
kapillaraktiv capillary active, surface active
Kapillaraktivität *f* capillary activity
Kapillaranalyse *f* capillary analysis
Kapillaranziehung *f* capillary attraction
Kapillarattraktion *f* capillary attraction
Kapillardepression *f* capillary depression
Kapillardruck *m* capillary pressure
Kapillare *f* capillary, capillary tube
Kapillarelektrode *f* capillary electrode
Kapillarelektrometer *n* capillary electrometer
Kapillarfläschchen *n* [small] capillary bottle
Kapillarflüssigkeit *f* capillary fluid
kapillarförmig capillary, capillaceous
Kapillarimeter *n* capillarimeter
Kapillarität *f* capillarity, capillary attraction
Kapillaritätsansteigung *f* capillary rise
Kapillaritätsanziehung *f* capillary attraction
Kapillaritätskoeffizient *m* coefficient of capillarity
Kapillaritätskonstante *f* capillary constant
Kapillarkondensation *f* capillary condensation
Kapillarkraft *f* capillary force
Kapillarpotential *n* capillary potential
Kapillarröhrchen *n* small capillary tube
Kapillarröhre *f* capillary tube
Kapillarrohr *n* capillary tube
Kapillarsäule *f* capillary column
Kapillarsenkung *f* capillary fall
Kapillarspaltzelle *f* (Elektrolyse) capillary gap cell
Kapillarspannung *f* capillary pressure, capillary stress, capillary tension
Kapillartheorie *f* capillary theory
Kapillarviskosimeter *n* capillary viscosimeter
Kapillarwirkung *f* capillary action, capillary effect
Kapitalrendite *f* earnings yield
Kaplanschaufel *f* Kaplan bucket
Kapnoskop *n* capnoscope, smoke gauge
Kapok *m* Indian wadding, vegetable down
Kapoköl *n* kapok oil
Kappe *f* cap, bonnet, removable hood, removable top
Kappenanschlag *m* stop collar of cap
Kappenflasche *f* bottle with cap
kappenförmig hood-shaped
Kappenständer *m* open topped housing
Kappit *m* cappite

Kappnaht *f* sealing run
Kapsel *f* capsule, (Gieß) chill, (Hülse) case, ~ zum Stückkugelguß shot mold
kapselartig capsular, caselike
kapselförmig capsular, case-shaped
Kapselguß *m* case-casting; casting in chills, chill casting, chilled work
Kapselhochdruckgebläse *n* Martin blower
Kapsellack *m* capsule lacquer
Kapselton *m* (Keram) sagger clay
Kapselverschluß *m* clip-on cap
Kapselwerk *n* cased-in blower
kaputt broken, lost, spoilt
Kapuzinerkresse *f* (Bot) Indian cress, nasturtium
Kapuzinerpulver *n* lice-powder
Karabinerhaken *m* snap hook, spring safety hook
karätig (karatig) carat
Karaffe *f* decanter
Karamel *m* caramel
karamelisieren to caramelize, to form caramel
Karanjisäure *f* karanjic acid
Karanjol *n* karanjol
Karat *n* carat
Karatgewicht *n* carat weight, troy-weight
Karatgold *n* alloyed gold
Karathan *n* karathane
karatieren to alloy gold or silver
Karatieren *n* alloying of gold or silver
Karatierung *f* alloying of gold or silver
Karbamid *n* carbamide
Karbid *n* (s. Carbid) carbide
Karbidkohle *f* carbide carbon
Karbidlampe *f* carbide lamp
Karbidofen *m* carbide furnace
Karbidschlacke *f* carbide slag
Karbodynamit *n* carbodynamite
Karbol *n* (s. Carbol) carbol
Karbolineum *n* carbolineum
Karbolsäure *f* carbolic acid, phenol
Karbolseife *f* carbolic soap
Karbonat *n* carbonate, salt of carbonic acid
Karbonisation *f* carbonization, (Versetzen mit Kohlensäure) carbonation, (Zucker) saturation
Karbonisationsofen *m* carbonizing stove
Karbonisierapparat *m* carbonizing apparatus
karbonisieren to carbonize, to carburize
Karbonisierofen *m* carbonizing stove
Karbonit *n* carbonite
Karborund *n* carborundum, silicide of carbon
Karborundmühle *f* carborundum mill
Karborundum *n* carborundum
Karborundumdarstellung *f* carborundum manufacture
Karborundumofen *m* carborundum furnace
Karborundumstein *m* carborundum brick
Karbosiliciumoxid *n* siloxicon
Karbozink *n* zinc carbonate

karbozyklisch carbocyclic
Karburator *m* carburetor, carburettor
Kardamom *m* cardamom, cardamon, cardamum
Kardamomöl *n* cardamom oil
Kardangelenk *n* Cardan joint, universal joint
Kardanwelle *f* Cardan shaft
Kardierabfall *m* (Wolle) card waste
kardieren (Wolle) to card
Kardieren *n* (Wolle) carding
Kardinalpunkt *m* (Opt) cardinal point
Kardinalrot *n* cardinal red, deep scarlet, red violet
Kardinalzahl *f* (Grundzahl, Math) cardinal number
Kardiogramm *n* (Med) cardiogram
Kardioide *f* cardioid
Kardobenediktenbitter *m* blessed thistle bitter
Kardobenediktenextrakt *m* extract of blessed thistle
Kardobenediktenkraut *n* (Bot) blessed thistle
Karduspapier *n* (Techn) cartridge paper
Karelinit *m* (Min) karelinite
Karfunkel *m* (Med) carbuncle, (Min) almandine
Karfunkelstein *m* (Min) almandine, almandite, garnet carbuncle, pyrope
karieren to check, to checker, to chequer (Br. E.)
kariert checked, chequered (Br. E.)
Karies *f* (Zahnmed) caries
Karkasse *f* (Kfz) carcass, body of the tire, casing
Karkassenbruch *m* (Kfz) carcass break
Karkassfestigkeit *f* casing strength, durability of carcass
Karlsbader Salz *n* Carlsbad salt
Karlsbadzwilling *m* (Krist) Carlsbad twin
Karmalaunlösung *f* carmalum solution
Karmelitergeist *m* spirit of balm
Karmeliterwasser *n* carmelite water
Karmesin *n* crimson
karmesinfarbig crimson-colored
Karmesinlack *m* crimson lake
karmesinrot crimson, ~ färben to crimson
Karmin *n* carmine
Karminlack *m* carmine lake
Karminrot *n* carmine red
Karminsäure *f* carminic acid, cochinilin
Karnallit *m* (Min) carnallite
Karnaubasäure *f* carnaubic acid
Karnaubawachs *n* carnauba wax, Brasil wax
Karneol *m* (Min) cornelian, bright red chalcedony, carneol (obs)
Karnotit *m* (Min) carnotite
Karo *n* square, check, chequer (Br. E.)
Karobe *f* (Bot) carob bean
Karosserie *f* (Kfz) autobody, [automobile] body, car body
Karosserieblech *n* (Kfz) autobody sheet
Karpholith *m* (Min) carpholite
Karphosiderit *m* (Min) carphosiderite

Karrag[h]eenmoos *n* (Bot) carrag[h]een, Irish moss
Karrenbegichtung *f* barrow charging
Karstenit *m* karstenite
Karte *f* card, (Landkarte) map, (Seekarte) chart
Kartei *f* card file, card index or catalog[ue], filing cabinet
Karteikarte *f* index card, filing card, record card
Karteikasten *m* filing cabinet
Karteisystem *n* card record system
Kartoffel *f* potato
Kartoffelbranntwein *m* potato brandy, potato spirit
Kartoffelfuselöl *n* potato fusel-oil
Kartoffelkäfer *m* (Zool) Colorado beetle, potato-bug
Kartoffelmehl *n* potato flour
Kartoffelschale *f* potato peel
Kartoffelstärke *f* potato starch
Kartoffelwalzmehl *n* dextrinized potato flour
Kartoffelzucker *m* glucose from potato starch, potato sugar
Karton *m* cardboard, boxboard, pasteboard box, (mehrschichtig) pasteboard, (Schachtel) carton, **beschichteter** ~ coated board, **einlagiger** ~ single-ply board, **gegautschter** ~ lined board, patent-coated board, **geklebter** ~ pasteboard, pasted board, **gelackter** ~ lacquered board; varnished board, **gemusterter** ~ grained board, **gepreßter** ~ pressed board, **kaschierter** ~ laminated board, combined board, lined board, **luftgetrockneter** ~ air-dried board, **maschinengeglätteter** ~ machine-glazed board, **schimmelfester** ~ mildew-proof board, mold-resistant board, **verstärkter** ~ reinforced board
Kartonagen *pl* cardboard or pasteboard articles
Kartonagenindustrie *f* box manufacturing and paperworking industry
Kartonheftmaschine *f* card wire stitcher
Kartonrückseite *f* bottom liner of board
Kartothek *f* card file, card index
Karussellbad *n* (Galv) revolving bath
Karussellextrakteur *m* rotary extractor
Karussellnutschenfilter *n* tilting pan filter
Karyinit *m* (Min) karyinite
Karyopilit *m* karyopilite
karzinogen (Med) carcinogenic, cancerogenic
Karzinogenese *f* (Med) carcinogenesis
Karzinolyse *f* carcinolysis
karzinolytisch carcinolytic
Karzinom *n* (Med) carcinoma, cancer
kaschieren to back, to glue, to laminate, (Text) to bond
Kaschieren *n* (Pap) lamination coating, ~ **auf der Walze** roll laminating
kaschiert backed
Kaschierung *f* lamination, doubling

Kaschierwachs *n* laminating wax, lining wax
Kaschmir *m* (Text) cashmere
Kaschmirhaar *n* (Text) cashmere hair, cashmere wool
Kaschmirwolle *f* (Text) cashmere wool
Kascholong *m* (Min) cacholong
Kascholongopal *m* (Min) cacholong opal
Kascholongquarz *m* (Min) cacholong quartz
Kaseinammoniak *n* ammonium caseinate, eucasin
Kaseinborax *m* casein borax
Kaseinfarbe *f* casein paint, washable distemper
Kaseinfaser *f* casein fiber, casein wool
Kasein-Kunsthorn *n* casein formaldehyde resin, galalith
Kaseinleim *m* casein glue
Kaseinnatrium *n* sodium caseinate
kaseinsauer caseinate
Kaseinseide *f* (Text) casein silk
Kaskade *f* cascade, ~ **von Reaktoren** cascade of reactors
Kaskadenentmagnetisierung *f* cascade demagnetization
Kaskadengleichrichter *m* cascade rectifier
Kaskadenmethode *f* **der Isotopentrennung** cascade method of isotope separation
Kaskadenmühle *f* cascade mill
Kaskadennachweis *m* cascade detection
Kaskadenregelung *f* cascade control
Kaskadenröhre *f* cascade tube
Kaskadenschaltung *f* cascade connection or circuit
Kaskadenschauer *m* cascade shower
Kaskadenübergang *m* cascade transition
Kaskadenverfahren *n* cascade process
Kaskadenvorwärmer *m* cascade preheater
Kaskadenwäscher *m* cascade scrubber
Kaskadenzerfall *m* cascade decay
Kasolit *n* kasolite
Kassaunzucker *m* cassonade, muscovado sugar, raw sugar
Kassawamehl *n* cassava, manioc
Kasseler Blau *n* Cassel blue
Kasseler Braun *n* Cassel brown
Kasseler Gelb *n* Cassel yellow, lead oxychloride
Kasseler Grün *n* Cassel green, barium manganate
Kasserolle *f* casserole, skillet (Am. E.)
Kassette *f* (Phot) plate holder, cassette, dark slide
Kassettendecke *f* coffered ceiling
Kassiakolben *m* cassia flask
Kassiopeium *n* cassiopeium (Symb. Lu), lutecium
Kassiterit *m* (Min) cassiterite, tinstone
Kastanienauszug *m* chestnut extract
kastanienbraun chestnut-brown, auburn
Kastanienbraun *n* auburn, chestnut brown

Kastanienfluidextrakt *m* liquid extract of chestnut
Kasten *m* box, case, ~ **des Hochofens** crucible hearth of the blast furnace
Kastenausleger *m* box jib [crane]
Kasten-Einsatzhärteverfahren *n* box case hardening process, pack hardening process
Kastenformerei *f* molding in boxes
Kastengebläse *n* box blower
Kastenglühung *f* box annealing
Kastenguß *m* [box] casting, flask casting
Kastenhärtung *f* case hardening
Kastenkeimapparat *m* germinating box
Kastenpotential *n* box potential
Kastenstockthermometer *n* straight box thermometer
Kastenwinkelthermometer *n* elbow box thermometer
katabatisch katabatic
Katabolismus *m* (Biol) catabolism
Katabolit *m* (Physiol) catabolite
katadioptrisch catadioptric
Kataklase *f* (Geol) cataclase
Katalase *f* (Biochem) catalase
Katalaseaktivität *f* (Biochem) catalase activity
Katalasewirkung *f* catalase action
Katalog *m* catalog (Am. E.), catalogue (Br. E.)
katalogisieren to tabulate
Katalysator *m* catalyst, catalytic agent, catalyzer, **fest-angeordneter** ~ fixed bed catalyst
Katalysatorbelastung *f* space velocity
Katalysatorgift *n* catalyst poison
Katalysatoroberfläche *f* catalyst surface area
Katalysatorträger *m* catalyst support
Katalysatorvergiftung *f* poisoning of the catalyst
Katalyse *f* catalysis, **heterogene** ~ heterogeneous catalysis, **homogene** ~ homogeneous catalysis, **negative** ~ negative catalysis, **photochemische** ~ photochemical catalysis
katalysieren to catalyze
Katalytgift *n* anti-catalyst, catalytic poison
katalytisch catalytic, ~ **beschleunigen** to accelerate catalytically
Kataphorese *f* cataphoresis
Kataplasma *n* cataplasma
Katechin *n* (Färb) catechu, Japan earth, terra japonica
Katechingerbstoff *m* catechol tan
Katechinsäure *f* catechin, catechuic acid
Katechol *n* catechol, pyrocatechin
Katechugerbsäure *f* catechutannic acid, tannin of catechu
Katechusäure *f* catechuic acid
Kategorie *f* category
Kathartin *n* cathartin
Kathartinsäure *f* cathartic acid
Kathepsin *n* (Biochem) cathepsin

Kathete *f* **eines rechtwinkligen Dreiecks** leg of a rectangular triangle
Kathode *f* cathode, negative electrode, **direkt geheizte** ~ directly heated cathode, ~ **einer Photozelle** photocathode, **glühende** ~ incandescent cathode, **indirekt geheizte** ~ indirectly heated cathode, isopotential cathode, **kalte** ~ cold cathode
Kathodenabnutzung *f* cathode disintegration
Kathodenanheizzeit *f* cathode heating time, thermal time constant
Kathodenbelag *m* cathode coating
Kathodenberechnung *f* cathode evaluation
Kathodendichte *f* cathode density
Kathodendunkelraum *m* cathode dark space
Kathodenfaden *m* cathode filament
Kathodenfall *m* cathode drop
Kathodenfläche *f* cathode surface
Kathodenflüssigkeit *f* catholyte
Kathodenfluoreszenz *f* cathode fluorescence
Kathodenglimmlicht *n* cathode glow
Kathodenkupfer *n* cathode copper
Kathodenleuchten *n* cathode glow, cathode luminescence
Kathodenlicht *n* cathode glow
Kathodenmechanismus *m* cathode mechanism
Kathodenniederschlag *m* cathode deposit
Kathodenoszillograph *m* cathode-ray oscillograph
Kathodenraum *m* cathode region, cathode space
Kathodenröhre *f* cathode tube, thermionic valve
Kathodenrost *m* cathode grid
Kathodenspannung *f* cathode potential
Kathodenstange *f* cathode bar
Kathodenstrahl *m* cathode ray, negative ray
Kathodenstrahlbündel *n* cathode-ray beam
Kathodenstrahlerzeuger *m* electron gun
Kathodenstrahloszillograph *m* cathode-ray oscillograph, Braun tube
Kathodenstrahlröhre *f* cathode ray tube
Kathodenstrom *m* cathode current
Kathodenverstärker *m* cathode amplifier
Kathodenvorspannung *f* cathode bias
Kathodenwiderstand *m* cathode resistance
Kathodenzerstäubung *f* cathode evaporation, cathode sputtering, cathodic sputtering
kathodisch cathodic
Kathodolumineszenz *f* cathodoluminescence
Katholyt *m* catholyte
Kation *n* cation
kationaktiv cation-active, cationic
Kationbase *f* cation base
Kationenaustausch *m* cation exchange
Kationenaustauscher *m* cationic exchanger
Kationenaustauschtrennung *f* cation exchange separation
Kationenlücke *f* cation hole
Kationenwanderung *f* cation migration
Kationsäure *f* cation acid

Katophorit *m* (Min) catophorite
Katovit *n* catovit
Kattun *m* (Text) calico, cotton [cloth]
Kattundruckerei *f* calico printing, cotton-printing
Kattunfärberei *f* cotton dyeing
Kattunpapier *n* chintz paper
Katuranin *n* katuranin
Katze *f* (Transport) cat, trolley carriage, ~ **mit Vorgelege** geared trolley
Katzenauge *n* (Min) cat's eye, (Rückstrahler) reflector
Katzenaugenharz *n* dammar resin
Katzenausleger *m* (Kran) trolley jib
Katzenbaldrian *m* common valerian
Katzenblei *n* (Min) mica
Katzenglas *n* (Min) cat gold, gypsum spar, sparry gypsum, yellow mica
Katzenglimmer *m* (Min) cat gold, yellow mica
Katzengold *n* (Min) cat gold, gold glimmer, pyrite, yellow glimmer, yellow mica
Katzenkraut *n* (Bot) cat thyme
Katzensilber *n* (Min) cat silver, colorless mica
Katzenstein *m* (Min) light and porous chalkstone
Kaugummi *m* chewing gum
Kaupfeffer *m* betel [pepper]
Kaupren *n* cauprene
Kausalität *f* causality, causation
Kausalitätsforderung *f* causality requirement
Kausalitätsprinzip *n* causality principle
Kaustifizieranlage *f* causticizer
kaustifizieren to causticize, to caustify
Kaustik *f* caustic[s]
Kaustikfläche *f* caustic surface
Kaustikspitze *f* caustic tip
Kaustikum *n* (Chem) caustic, corrosive [substance], mordant
kaustisch caustic, corrosive, mordant
kaustizieren to causticize, to bite, to eat
Kaustizierung *f* caustification, cauterization
Kaustizität *f* causticity
Kaustobiolith *m* caustobiolith
Kautabak *m* chewing tobacco
Kauter *m* (Med) cautery
Kauterisation *f* (Med) thermocautery
Kautschin *n* caoutchene
Kautschucin *n* caoutchoucin
Kautschuk *m* caoutchouc, rubber, ~ **ohne Nerv** weak rubber, ~ **schwefeln** to vulcanize rubber
Kautschukabfall *m* caoutchouc waste, rubber waste
kautschukähnlich rubberlike
Kautschukband *n* rubber band
Kautschukbaum *m* caoutchouc tree, rubber tree
Kautschukersatzstoff *m* caoutchouc substitute
Kautschukfarbe *f* rubber paint
Kautschukfirnis *m* rubber varnish
Kautschukgewebe *n* rubber fabric
Kautschukgewinnung *f* rubber extraction
Kautschukgift *n* rubber poison
Kautschukkitt *m* caoutchouc cement, rubber cement
Kautschukkuchen *m* rubber cake, rubber sheet
Kautschuklack *m* rubber varnish
Kautschuklatex *m* rubber latex
Kautschuklösung *f* caoutchouc solution, rubber solution
Kautschukmasse *f* caoutchouc paste
Kautschukmilch *f* caoutchouc milk
Kautschukmilchsaft *m* rubber latex, rubber milk
Kautschuköl *n* caoutchoucin, rubber oil
Kautschuksaft *m* rubber latex
Kautschukschlauch *m* caoutchouc hose, caoutchouc tube, rubber hose
Kautschukstopfen *m* rubber cork
Kautschukteig *m* rubber mass
Kautschuktetrabromid *n* caoutchouc tetrabromide
Kautschukware *f* caoutchouc ware
Kaviar *m* caviar
Kavitation *f* cavitation
kavitationsbeständig resistant to cavitation
Kavolinit *m* cavolinite
Kawa-Kawa *f* kava kava
Kawasäure *f* kawaic acid
Kawawurzel *f* (Bot) kava root
Kefir *m* kefir
Kegel *m* cone, (abgeschrägter Teil) bevel, (verjüngter Teil) taper, **abgestumpfter** ~ truncated cone
Kegelachse *f* axis of cone
kegelähnlich conelike, conic[al], conoid, conoidal
Kegelblende *f* collimating cone
Kegelbrecher *m* cone crusher, conical grinder
Kegelbremse *f* cone brake
Kegelfallpunkt-Prüfung *f* (Email) cone-fusion test
Kegelfließpunkt *m* **nach Höppler** Höppler hardness
kegelförmig cone-shaped, conical, coniform
Kegelförmigkeit *f* conicity
Kegelgestalt *f* conic form
Kegelhöhe *f* (Buchdr) type size
kegelig conic[al], cone-shaped, coniform, (abgeschrägt) beveled, (sich verjüngend) tapered
Kegelkern *m* **mit Verlängerung** skirted cone
Kegelkonduktor *m* conical conductor
Kegelkopf *m* span head
Kegelkreiselmischer *m* cone impeller
Kegellinie *f* parabola
Kegelmühle *f* conical grinder, jordan mill
Kegel-Platte-Viskosimeter *n* cone and plate viscometer
Kegelpunkt *m* (einer Fläche) conical point

Kegelrad *n* bevel [spur] gear, bevel wheel, cone wheel
Kegelradmaschine *f* bevel[l]ing machine
Kegelreibrad *n* bevel friction wheel
Kegelrollenlager *n* taper roller bearing
Kegelscheibe *f* cone pulley
Kegelschnellrührer *m* cone impeller
Kegelschnitt *m* conic section
Kegelstift *m* taper pin
Kegeltrommel *f* conical drum
Kegelventil *n* conical or miter valve
kehlen to chamfer, to channel, to flute
Kehlhobel *m* fluting plane
Kehlkopf *m* larynx
Kehlnaht *f* fillet weld
Kehlrad *n* gorge wheel, groove wheel
Kehlschweißung *f* corner weld, filled weld
Kehlung *f* chamfering, fluting, grooving
Kehrbild *n* (Opt) inverted image
Kehrgetriebe *n* reversing gear
Kehricht *m* sweepings, waste
Kehrsalpeter *m* saltpeter sweepings
Kehrsalz *n* salt sweepings
Kehrseite *f* reverse side
Kehrwalzwerk *n* reversing rolling mill
Kehrwelle *f* reverse shaft
Kehrwerk *n* reversing gear
Kehrwert *m* reciprocal [value]
Kehrwertintegration *f* (Math) reciprocal integration
Keil *m* wedge, (Hemmvorrichtung) chock, (Längskeil) key, (Querkeil) cotter
keilähnlich wedge-shaped, cuneiform, wedgelike, (Bot) cuneate
Keilanordnung *f* key arrangement
Keileinguß *m* conical gate, conical runner, wedge inlet
keilen to wedge, to key, to split
keilförmig wedge-shaped, cuneiform, wedgelike
Keilhauit *m* (Min) keilhauite, yttrotitanite
Keilnut *f* keyseat, keyway
Keilnutenstoßmaschine *f* keywaying machine
Keilnutenziehmaschine *f* keyseating machine
Keilphotometer *n* wedge photometer
Keilriemen *m* V-belt, cone belt, **endlicher** ~ open-end V-belt, **endloser** ~ endless V-belt
Keilriemenantrieb *m* V-belt drive, cone belt drive
Keilriemenscheibe *f* V-belt pulley
Keilschlitz *m* keyseat
Keilstahl *m* key steel
Keilstrich-Formel *f* (Stereochem) flying wedge formula
Keilstück *n* wedge, wedgepiece
Keilverbindung *f* keying
Keilverzahnung *f* splines
Keilwelle *f* splined shaft
Keilwellenschleifmaschine *f* spline grinder
Keilwinkel *m* wedge angle

Keim *m* germ, (Ausgang) origin, (Bakt) bacillus, (Bot) bud, shoot, sprout, (Embryo) embryo, (Krist) crystal nucleus
Keimapparat *m* germinating apparatus
Keimbildner *m* nucleating agent
Keimbildung *f* formation of nuclei, formation of the crystal nucleus, nucleation, (Biol) germination
Keimblatt *n* (Bot) seedleaf
keimdicht germ-proof
Keimdrüse *f* germ gland, gonad
keimen to germinate, (Bot) to bud, to sprout
Keimen *n* germinating
keimend germinant
Keimentwicklung *f* development of the germ
keimfähig germinative
Keimfähigkeit *f* power of germination
keimförmig germiform, embryonal
keimfrei sterile, aseptic, free of germs, germ-free, germless
Keimgehalt *m* bacterial count
Keimhemmungsmittel *n* germination inhibitor
Keimkraft *f* germinating power, vitality
Keimkristall *m* crystal nucleus, seed crystal
Keimling *m* germ, seedling
Keimspat *m* (Min) crystallized zeolite
keimtötend germicidal, germicide, sterilizing
Keimtöter *m* germ killer, sterilizer
Keimtötung *f* sterilization
Keimträger *m* [germ] carrier
keimunfähig incapable of germinating, sterile
Keimung *f* germination
Keimzählung *f* bacterial counting
Keimzahl *f* bacterial count
Keimzelle *f* (Bot) gamete, germ cell
Kekulésche Benzolformel *f* Kekulé formula
Kelch *m* cup, (Bot) calyx, flower-cup
kelchförmig cup-shaped
Kelen *n* (HN) kelene, ethyl chloride
Kellerboden *m* basement floor
Kellerfußboden *m* cellar floor
Kellerlaboratorium *n* cellar laboratory
Kellerschwamm *m* wet rot
Keltapapier *n* gelatino chloride paper
Kelter *f* wine-press
Kelterbütte *f* presstub, pressvat
keltern to press grapes
Keltern *n* pressing of grapes
Kelterpresse *f* curb press, wine press
Kelterwein *m* wine from extracted juice
Kelvin-Effekt *m* (Elektr) Kelvin effect, skin effect
Kelvingrad *m* degree Kelvin
Kelvinskala *f* Kelvin scale
Kelvintemperatur *f* Kelvin temperature
Kelvintemperaturskala *f* Kelvin temperature scale
Kelyphit *m* (Min) kelyphite
Kenndaten *pl* characteristic data or values

Kennfarbe *f* color marking, marker color, tracer color
Kennfrequenz *f* assigned frequency
Kenngerät *n* identification apparatus
Kenngottit *m* kenngottite
Kenngröße *f* characteristic number, characteristic value or data
Kenngrößengleichung *f* characteristic equation
Kennkarte *f* identification card
Kennlinie *f* characteristic curve, (Regeltechn) steady-state characteristic (Am. E.), characteristic (Br. E.)
Kennlinienfeld *n* family of characteristics
Kennliniengleichung *f* characteristic equation
Kennnummer *f* identification number
kenntlich distinguishable, clear, distinct, (erkennbar) recognizable
Kenntlichmachung *f* characterization, marking
Kenntnis *f* knowledge, **~ nehmen** to take notice [of], **in ~ setzen** to inform, to advise
Kennwert *m* characteristic value
Kennwiderstand *m* (Elektr) characteristic impedance
Kennzahl *f* characteristic number, constant
Kennzeichen *n* characteristic, criterion, feature, index, mark, sign, (Med) symptom
kennzeichnen to mark
Kennzeichnung *f* characterization, identification, label, mark, marking
Kennziffer *f* characteristic number, code, index
Kennziffergleichung *f* coefficient equation
Kentrolith *m* (Min) kentrolite
Kephalin *n* cephalin, kephalin
Keracyanin *n* ceracyanin
keramchemisch chemicoceramic
Keramik *f* ceramics, pottery
Keramikmetallgemisch *n* ceramet, cermet
Keramikwaren *pl* ceramic goods, pottery
keramisch ceramic
Kerargyrit *m* (Min) cerargyrite
Kerasin *n* kerasin
Kerasit *m* (Min) cerasite
Keratin *n* keratin
Keratingewebe *n* keratin tissue, horny tissue
keratolisieren to keratolize
Kerbbiegeprobe *f* nick bend test, notched bar bend test, shock test on notched bar
Kerbbildung *f* (auch Ritzen) scoring
Kerbe *f* notch, dent, indentation, kerf, (Aussparung) recess, (Nut) slot
Kerbeinflußzahl *f* notch factor
Kerbelöl *n* chervil oil
Kerbempfindlichkeit *f* notch sensitivity
kerben to indent, to notch, to score
Kerbenfügung *f* slit and tongue joint
Kerbfaltversuch *m* notched-bar bend test
Kerbfestigkeit *f* impact resistance
Kerbnute *f* keyway
Kerbproblem *n* indentation problem

Kerbschlag *m* impact
Kerbschlagbiegeversuch *m* notched-bar impact bending test
Kerbschlagfestigkeit *f* impact strength, indentation value, notched-bar strength
Kerbschlagprobe *f* indentation test, notch bending test, notched-bar impact test, notch-impact test, shock test on notched bar
Kerbschlagversuch *m* indentation test, notch bending test, notched-bar impact test
Kerbschlagzähigkeit *f* notched-bar impact strength, notch impact strength
Kerbstift *m* slotted pin
Kerbzähigkeit *f* impact value, Izod impact strength, notched-bar toughness, notch impact strength
Kerbzähigkeitsprobe *f* notched-bar impact test
Kerbzähigkeitsprüfung *f* notched-bar impact test
Kerbzugprobe *f* notched bar tensile test
Kerbzugversuch *m* notched-bar tensile test
Kermes *m* (Färb) kermes
Kermesbeere *f* (Bot) kermes berry, fox glove, pigeon berry
Kermesin *n* crimson
Kermesit *m* (Min) kermesite, lyrostibnite, native red antimony sulfide
Kermeskörner *n pl* kermes grains
kermesrot scarlet
Kermesrot *n* kermes red, cochenille, scarlet
Kermessäure *f* kermisic acid
Kermesscharlach *m* kermes scarlet, French scarlet
Kern *m* (Atom) nucleus, (Induktionsspule, Gieß) core, (Mittelpunkt) center, heart, (Obst) kernel, pip, pit, stone, **~ mit halber Chromosomenzahl** (Biol) haploid nucleus, **~ mit Protonen und Neutronen von ungerader Zahl** (Atom) odd-odd nucleus, **(g-g-Kern)** even-even nucleus, **~ mit Protonen von ungerader und Neutronen von gerader Zahl** (Atom) odd-even nucleus, **angeregter ~** excited nucleus, **doppelt gerader ~** (Atom) even-even nucleus, **künstlich radioaktiver ~** (Atom) artificial radioactive nucleus, **positiv geladener ~** (Atom) positively charged nucleus
Kernabstand *m* (Atom) internuclear distance
Kernachse *f* (Atom) nuclear axis
Kernanlegierung *m* bead setter, spider
Kernanregung *f* (Atom) nuclear excitation
Kernanziehung *f* (Atom) nuclear attraction
Kernaufbau *m* (Atom) nuclear structure
Kernausrichtung *f* (Atom) nuclear alignment
Kernbeschießung *f* (Atom) nuclear bombardment
Kernbeschuß *m* (Atom) nuclear bombardment
Kernbestandteil *m* (Atom) nuclear component
Kernbildung *f* nucleation, nucleus formation
Kernbindemittel *n* core binder

Kernbindungsenergie f (Atom) nuclear binding energy
Kernbogen m core sheet
Kernbrennstoff m (Atom) nuclear fuel, reactor fuel, **denaturierter** ~ (Atom) denatured nuclear fuel
Kernbruchstück n (Atom) nuclear fragment
Kernchemie f (Atom) nuclear chemistry
kernchloriert (Chem) chlorinated in the nucleus
Kerndichte f nuclear density
Kerndrall m (Atom) nuclear spin
Kerndrehbank f core lathe
Kerndrehung f nuclear spin
Kerndrucker m core mark
Kerndurchmesser m (Atom) diameter of nucleus, nuclear diameter, (Gewinde) core diameter, minor diameter
Kerne pl, **kondensierte** ~ (Chem) condensed nuclei
Kernechtrot n nuclear fast red
Kerneigenschaft f nuclear property
Kerneinfang m (Atom) nuclear capture
Kerneinschnürung f (Techn) core groove, core recess
Kerneisen n arbor, core iron, mottled white pig iron
Kernemulsion f (Atom) nuclear emulsion
Kernenergie f nuclear energy
Kernenergieniveau n nuclear energy level
Kernenergieniveaudiagramm n nuclear energy-level diagram
Kernentmagnetisierung f (Atom) nuclear demagnetization
Kernexplosion f nuclear explosion
Kernfäule f heart-rot
Kernfahne f flipper
Kernfahnenanrollvorrichtung f flipper stitcher
Kernfarbstoff m (Hist) nuclear stain
Kernfeld n (Atom) nuclear field, **abstoßendes** ~ repulsing nuclear field
kernfern peripheral
Kernferromagnetismus m nuclear ferromagnetism
Kernform f (Gieß) primitive form
Kernformerei f (Gieß) core molding
Kernforschung f nuclear research
Kernfragmente pl (Atom) nuclear fragments
Kernfusion f (Atom) nuclear fusion
Kernfusionsreaktor m (Atom) nuclear fusion reactor
Kerngabel f core fork
Kerngerippe n core frame
Kerngröße f size of nucleus
Kernguß m cored casting, cored work
Kernhalbmesser m core radius, (Atom) nuclear radius
Kernhaus n (Frucht) core
Kernholz n heartwood

Kerninduktionsspektrograph m nuclear induction spectrograph
Kernindustrie f (Atom) nuclear industry
Kernisobar n nuclear isobar
Kernisomer n nuclear isomer
Kernisomerie f (Atom) nuclear isomerism
Kernit m (Min) kernite
Kernkasten m core box
Kernkettenreaktion f [nuclear] chain reaction
Kernkopplungskonstante f nuclear coupling constant
Kernkraft f (Atom) nuclear force
Kernkraftwerk n nuclear power plant, nuclear power station, power reactor
Kernkreislauf m core circuit
Kernladung f nuclear charge
Kernladungsverteilung f nuclear charge distribution
Kernladungszahl f (Atom) atomic number, nuclear charge number
Kernlinse f (Opt) nuclear lens
Kernlochstift m core pin
Kernluft f air in the core
Kernmacher m (Gieß) core maker, pattern maker
Kernmacherei f core molding
kernmagnetisch nuclear magnetic
Kernmagnetmoment n nuclear magneton
Kernmagneton n (Atom) nuclear magneton
Kernmarke f (Formerei) core mark, top print
Kernmaß n bore, caliber
Kernmasse f core compound
Kernmaterie f nuclear matter
Kernmembran f nuclear membrane
Kernmetallurgie f nuclear metallurgy
Kernmodell n (Atom) model of nucleus, nucleus model
Kernmoment n nuclear moment, **magnetisches** ~ nuclear magnetic moment
Kernnähe f nuclear region
Kernniveau n nuclear level
Kernphotoeffekt m (Atom) nuclear photoelectric effect, photonuclear effect
Kernphotoreaktion f photonuclear reaction
Kernphotozerfall m (Atom) nuclear photodisintegration, nuclear photodissociation
Kernphysik f nuclear physics, nucleonics
Kernphysiker m nuclear physicist
Kernplasma n (Biol) nucleoplasm
Kernpolymerie f nuclear polymerism
Kernpotential n nuclear potential
Kernprotein n nucleoprotein
Kernproton n (Atom) nuclear proton
Kernquadrupolkopplung f nuclear quadrupole coupling
Kernrahmen m (Gieß) core frame
Kernrakete f nuclear energy rocket
Kernraketenantrieb m nuclear rocket propulsion
Kernreaktion f nuclear reaction

Kernreaktor *m* nuclear pile or reactor, atomic pile, nuclear chain reactor
Kernregel *f* nuclear formula
Kernreiter *m* (Reifen) bead reinforcement, bead filler
Kernresonanz *f* (Spektr) nuclear [magnetic] resonance, **magnetische** ~ nuclear magnetic resonance
Kernresonanzenergie *f* nuclear resonance energy
Kernresonanzfluoreszenz *f* nuclear resonance fluorescence
Kernresonanzspektrograph *m* nuclear magnetic resonance spectrograph
Kernresonanzspektrometrie *f* nuclear magnetic resonance
Kernresonanzspektrum *n* nuclear magnetic resonance spectrum, NMR-spectrum
Kernröstung *f* core roasting
Kernsand *m* core sand
Kernsandbeschaffenheit *f* condition of core sand
Kernschablone *f* core strickle, core templet, strickle board
Kernschacht *m* lining of shaft
Kernschale *f* (Atom) nuclear shell
Kernschalenmodell *n* (Atom) nuclear shell model
Kernschliff *m* (Glas) male ground joint
Kernseife *f* hard soap, laundry soap, ordinary soap
Kernspaltreaktion *f* (Atom) fission chain reaction
Kernspaltung *f* (Atom) nuclear fission
Kernspaltungsenergie *f* (Atom) nuclear fission energy
Kernspaltungsfragment *n* fission fragment
Kernspaltungskettenreaktion *f* fission chain reaction
Kernspaltungsreaktor *m* fission reactor
Kernspeicher *m* core memory
Kernspektroskopie *f* (Atom) nuclear spectroscopy
Kernspektrum *n* nuclear spectrum
Kernspin *m* (Atom) nuclear spin
Kernspindel *f* core spindle
Kernspinkopplung *f* nuclear spin coupling
Kernspinquantenzahl *f* nuclear-spin quantum number
Kernsplitter *m* nuclear fragment
Kernstoß *m* nuclear collision
Kernstrahlung *f* nuclear radiation
Kernstreuung *f* nuclear scattering
Kernstruktur *f* nuclear structure
Kernstück *n* principal item, (Gieß) core part
Kernsuszeptibilität *f* nuclear susceptibility
Kernsymmetrie *f* nuclear symmetry
Kernsynthese *f* nuclear synthesis
Kerntechnik *f* nuclear engineering, nuclear technology
Kerntechnologie *f* nucleonics
Kernteilchen *n* nuclear particle, nucleon

Kernteilung *f* nuclear division
Kerntheorie *f* nuclear theory
Kerntransformator *m* core transformer
Kerntrockenofen *m* core drying stove
Kernübergang *m* nuclear transition
kernumgebend circumnuclear
Kernumgruppierung *f* nuclear rearrangement
Kernummantelung *f* (Reifen) bead wrap
Kernumwandlung *f* nuclear transformation, nuclear transmutation, **erzwungene** ~ enforced nuclear transformation, **künstliche** ~ artificial nuclear transformation
Kernumwicklung *f* (Reifen) bead cover fabric
Kernverdampfung *f* (Atom) nuclear evaporation
Kernvereinigung *f* (Atom) nuclear fusion
Kernverformung *f* nuclear distortion
Kernverknüpfung *f* linkage between nuclei
Kernverschmelzung *f* nuclear fusion
Kernwaffe *f* atomic weapon, nuclear weapon
Kernwechselwirkung *f* nuclear interaction
Kernzerfall *m* (Atom) nuclear decay
Kernzertrümmerung *f* (Atom) nuclear disintegration, nuclear fission
Kernzone *f* core
Kernzustand *m* nuclear state
Kerolith *m* (Min) cerolite
Kerosin *n* kerosene, kerosine
Kerrzelle *f* (Phys) Kerr cell
Kerstenit *m* (Min) kerstenite
Kertschenit *m* kertschenite
Kerze *f* candle, (Zündkerze) spark plug, ~ **hohen Wärmewertes** cold [spark] plug, ~ **niedrigen Wärmewertes** hot [spark] plug
Kerzenfilter *n* cartridge filter
Kerzenhalter *m* candle stick
Kerzenschlüssel *m* spark plug spanner, spark plug wrench
Kerzenstärke *f* (Phys) candle power, luminous intensity
Kerzenstecker *m* (Auto) spark plug socket
Kerzenstein *m* electrode of the spark plug
Kerzenzündung *f* spark plug ignition
Kessel *m* basin, boiler, kettle, reservoir, tank, **Flammrohr-Rauchrohr-** ~ radiant-type boiler, **Großwasserraum-** ~ water-tube boiler, **Naturumlauf-** ~ natural circulation boiler, **selbstregelnder** ~ automatically controlled boiler, **stehender** ~ vertical boiler, **ummantelter** ~ jacketed boiler, **Zwangsdurchlauf-** ~ forced-circulation boiler, **Zwangsumlauf-** ~ forced-flow once-through boiler
Kesselablaßhahn *m* boiler blow-off cock, mud cock
Kesselanlage *f* boiler plant
Kesselasche *f* boiler slag
Kesselbekleidung *f* boiler casing, boiler jacket
Kesselblech *n* boiler plate
Kesselboden *m* boiler end

Kesseldampf – Kettenreaktionsanlage

Kesseldampf *m* boiler steam
Kesseldruck *m* boiler pressure
Kesselexplosion *f* boiler explosion
Kesselfeuerung *f* boiler furnace
Kesselgas *n* boiler flue gas
Kesselhaus *n* boiler house
Kesselisolierung *f* boiler insulation
Kesselleistung *f* boiler output
Kesselniederschlag *m* boiler scale, deposit on boiler surface
Kesselspeisewasser *n* boiler feed water
Kesselstein *m* boiler scale, fur in boilers, incrustation
Kesselsteinabklopfer *m* boiler scaling tool or hammer
Kesselsteinablagerung *f* deposit of scale, incrustation
Kesselsteinbeseitigungsmittel *n* disincrustant, scale remover, scale solvent
Kesselsteinbildner *m* scale-producing salt
Kesselsteingegenmittel *n* antiincrustant, antiscale-forming agent, boiler compound, disincrustant, scale solvent
Kesselsteinkruste *f* layer of scale
Kesselsteinlösemittel *n* disincrustant, scale solvent
Kesselsteinmittel *n* antiincrustant, boiler compound, scale solvent
Kesselsteinpulver *n* antiincrustant powder
Kesselsteinschicht *f* layer of scale
Kesselsteinverhütung *f* prevention of scale formation
Kesselsteinverhütungsmittel *n* boiler anti-scaling composition
Kesselwagen *m* rail tank car, tank wagon
Kesselwasser *n* boiler [feed] water
Kesselzug *m* boiler draft
Ketazin *n* ketazine
Keten *n* ketene
Ketimin *n* ketimine
Ketin *n* ketine
Ketoamin *n* ketoamine
Ketoazidose *f* (Med) keto-acidosis
Keto-Enol-Gleichgewicht *n* keto-enol equilibrium
Keto-Enol-Tautomerie *f* keto-enol tautomerism
Ketoester *m* keto ester
Ketoform *f* keto form
Ketogenese *f* ketogenesis
Ketoglutaratdehydrogenase *f* (Biochem) ketoglutarate dehydrogenase
Ketoglutarsäure *f* ketoglutaric acid
Ketogruppe *f* keto group
Ketogulonsäure *f* ketogulonic acid
Ketohexokinase *f* (Biochem) ketohexokinase
Ketohexose *f* ketohexose
Ketol *n* ketol, ketone alcohol
Keton *n*, **aromatisches** ~ aromatic ketone
Ketonalkohol *m* ketone alcohol, ketol

ketonartig ketone-like, ketonic
Ketonbildung *f* ketogenesis
Ketonharz *n* ketone-resin
Ketonimid *n* ketone imide
Ketonkörper *m* ketone body
Ketonkörperausscheidung *f* (Urin) ketonuria
Ketonöl *n* ketone oil
Ketonspaltung *f* ketone splitting, ketonic cleavage
Ketonurie *f* (Med) ketonuria
Ketopantosäure *f* ketopantoic acid
Ketopentamethylen *n* cyclopentanone, ketopentamethylene
Ketopentose *f* ketopentose
Ketopinsäure *f* ketopinic acid
Ketosäure *f* keto acid, ketonic acid
Ketose *f* ketosugar, ketose, (Med) ketosis
Ketostearinsäure *f* ketostearic acid
Ketoverbindung *f* keto compound
Ketoxim *n* ketoxime
Ketozucker *m* ketose, ketosugar
Kettbaum *m* warp beam
Kette *f* chain, range, series, (Web) warp, ~ **mit gedrehten Gliedern** twist link chain, **galvanische** ~ galvanic cell, **gerade** ~ straight chain, **verzweigte** ~ branched chain
Kettelmaschine *f* (Text) linking machine
ketteln (Text) to link
Kettelschiene *f* linking bar
ketten to connect, to join, to link
Kettenabbruch *m* (Chem) chain breaking, chain termination
Kettenantrieb *m* chain drive
Kettenbecherwerk *n* chain bucket elevator
kettenbildend (Bakt) chain-forming
Kettenbruch *m* fracture of chain, (Math) continued fraction
Ketteneffekt *m* (Web) warp effect
Kettenfärbeapparat *m* warp dyeing machine
Kettenflaschenzug *m* chain pulley block
Kettenförderer *m* chain conveyor
kettenförmig chain-like
Kettenfräse *f* chain cutter molding machine
Kettenfräsmaschine *f* chain cutter molding machine
Kettenglied *n* link [of a chain]
Kettenhebevorrichtung *f* chain hoist
Kettenhebewerk *n* chain hoist
Kettenisomerie *f* chain isomerism
Kettenkasten *m* gear case
Kettenlänge *f* chain length, length of chain
Kettenmolekül *n* chain molecule
Kettenpumpe *f* chain pump
Kettenrad *n* chain wheel, sprocket gear, sprocket wheel
Kettenradantrieb *m* chain drive
Kettenreaktion *f* chain reaction, **gelenkte** ~ controlled chain reaction
Kettenreaktionsanlage *f* chain-reacting plant

Kettenregel *f* chain rule
Kettenschutz *m* (Fahrrad) [bicycle] gear case
Kettensilikat *n* silicate with chain structure
Kettenspanner *m* chain drive spanner
Kettenstruktur *f* chain structure
Kettenübertragung *f* chain transfer
Kettenübertragungsreagenz *n* chain transfer agent
Kettenverhakung *f* chain entanglement
Kettenverschleiß *m* wear of chain
Kettenverzweigung *f* chain branching
Kettenwachstum *n* chain growth
Kettenwiegeapparat *m* chain weighing column
Kettenwinde *f* chain winch
Kettfach *n* warp shed
kettscheren to warp
Kettstreifen *m* (Web) warp stripe
Kettstreifigkeit *f* (Web) warp stripiness
Kettstuhlwirkerei *f* warp knitting
Keuchhusten *m* (Med) whooping cough
Keule *f* club, (des Mörsers) pestle
Kevatron *n* kevatron
Keweenawit *m* keweenawite
KH Abk. für Karbonathärte
Khaki *n* (Erdbraun) khaki
Khakifarbe *f* khaki color
khakifarben khaki-colored
Khellacton *n* khellactone
Khellin *n* khellin
Khellinchinon *n* khellinquinone
Khellinin *n* khellinin
Khellinon *n* khellinone
Khellol *n* khellol
K-Hülle *f* K-shell
Kibdelophan *m* kibdelophane
Kichererbsensäure *f* (obs) ciceric acid, mixture of oxalic and malic acid
Kiefernharz *n* pine resin, rosin
Kiefernspinner *m* pine spinner
Kielbolzen *m* keel bolt
Kielwasser *n* wake
Kieme *f* (Zool) gill
Kiemenatmung *f* (Zool) gill breathing
Kien *m* resinous pine
Kienharz *n* pine resin, rosin
Kienholzöl *n* pinewood oil
Kienöl *n* oil of turpentine, pine oil
Kienruß *m* pine soot, lampblack, soot black
Kienteer *m* pine tar
Kienteerpech *n* pine pitch
Kies *m* gravel, **grober** ~ pebble, rubble
Kiesabbrand *m* calcined pyrites, purple ore, pyrite cinder, roasted pyrites
Kiesablagerung *f* deposition of gravel
Kiesapfel *m* (Min) kidney-shaped pyrites, pyrites in balls
kiesartig gravelly, gritty
Kiesboden *m* gravel soil
Kiesbrenner *m* pyrite kiln, pyrites burner

Kiesel *m* pebble, flint, flintstone, silex, silica
Kieselalgen *pl* diatomacae
kieselartig flinty, siliceous
Kieselblende *f* (Min) siliceous blende
Kieselcerit *m* (Min) cerite
Kieseleisenstein *m* (Min) siliceous iron ore
Kieselerde *f* (Min) siliceous earth, flinty earth, silica
kieselerdehaltig siliceous, siliciferous
Kieselfels *m* flintrock
Kieselfluoraluminium *n* aluminum silicofluoride, aluminum fluosilicate
Kieselfluorbarium *n* barium silicofluoride, barium fluosilicate
Kieselfluorblei *n* lead silicofluoride, lead fluosilicate
Kieselfluorcalcium *n* calcium silicofluoride, calcium fluosilicate
Kieselfluorid *n* silicon fluoride, silicon tetrafluoride
Kieselfluorkalium *n* potassium silicofluoride, potassium fluosilicate
Kieselfluormangan *n* manganese silicofluoride, manganese fluosilicate
Kieselfluornatrium *n* sodium silicofluoride, sodium fluosilicate
Kieselfluorsäure *f* silicofluoric acid, fluosilicic acid
Kieselfluorstrontium *n* strontium silicofluoride, strontium fluosilicate
Kieselfluorverbindung *f* silicofluoride, compound of silicofluoric acid, fluosilicate, silicon-fluorine compound
Kieselfluorwasserstoffsäure *f* silicofluoric acid, fluosilicic acid
kieselfluorwasserstoffsauer silicofluoride, fluosilicate
Kieselfluorzink *n* zinc fluosilicate
Kieselfluß *m* siliceous flux
Kieselflußsäure *f* silicofluoric acid, fluosilicic acid
Kieselgalmei *m* (Min) hydrous zinc silicate, siliceous calamine, siliceous smithsonite
Kieselgel *n* silica gel
Kieselgestein *n* siliceous rock, quartz rock
Kieselgewinnungsanlage *f* gravel extraction plant
Kieselgips *m* (Min) vulpinite
Kieselglas *n* flint glass
Kieselgrund *m* pebbly ground
Kieselgur *f* kieselguhr, diatomaceous earth, infusorial earth
Kieselhärte *f* flint hardness
kieselhaltig containing silica, siliceous, siliciferous
kieselhart flinthard, hard as flint
Kieselkalk *m* siliceous limestone
Kieselkalkeisen *n* (Min) ilvaite
kieselkalkhaltig silicocalcareous
Kieselkalkschiefer *m* (Min) siliceous shale

Kieselkalkspat *m* (Min) wollastonite
Kieselkalkstein *m* (Min) siliceous limestone
Kieselkreide *f* siliceous chalk
Kieselkristall *m* rock crystal, silica crystal
Kieselkupfer *n* (Min) chrysocolla
Kieselmagnesit *m* (Min) siliceous magnesite
Kieselmalachit *m* (Min) chrysocolla
Kieselmangan *n* siliceous manganese, rhodonite
Kieselmassewärmer *m* (Regenerator) pebble heater
Kieselmehl *n* pounded pebbles
Kieselmetall *n* metallic silicon, silicon metal
Kieselpulver *n* pebble powder
Kieselsäure *f* silicic acid
Kieselsäureanhydrid *n* silicic anhydride, silicon dioxide
Kieselsäuregehalt *m* **des Kalksteins** silica content of limestone
Kieselsäuregel *n* silica gel
kieselsäurehaltig containing silicic acid, siliceous
kieselsäurereich rich in silicic acid
Kieselsand *m* gravel
Kieselsandstein *m* siliceous sandstone
kieselsauer silicate
Kieselschiefer *m* (Min) flinty slate, siliceous schist
Kieselsinter *m* quartz sinter, siliceous sinter
Kieselspat *m* (Min) siliceous feldspar, albite
Kieselstein *m* pebble, flint stone, gravel stone
Kieselton *m* (Min) argillite, clay slate
Kieseltuff *m* siliceous sinter
Kieselweiß *n* silex white, siliceous whitening
Kieselwismut *n* bismuth blende, bismuth silicate, eulytite
Kieselwolframsäure *f* silicotungstic acid
Kieselzink *n* (Min) calamine
Kieselzinkerz *n* (Min) calamine, siliceous zinc ore, willemite
Kieserit *m* (Min) native magnesium sulfate
Kiesfilter *n* gravel filter, pebble filter
Kiesgrube *f* gravel pit
kieshaltig gravelly, pyritiferous
kiesig gritty, flinty, gravelly
Kieskupfererz *n* (Min) white copper ore
Kiesofen *m* pyrite burner, pyrite kiln
Kiespreßdeckung *f* board with layer of gravel embedded in bitumen
Kiessand *m* gravelly sand
Kiesschörl *m* (Min) zinciferous spinel
Kiessieb *n* gravel screen
Kieszink *n* (Min) calamine, native zinc silicate
Kiffkuchen *m* tan ball, tan cake
Kilbrickenit *m* kilbrickenite
Killinit *m* killinite
Kiln *m* kiln
Kiloampere *n* (Elektr) kiloampere
Kilodyn *n* kilodyne
Kilogauß *n* (Magn) kilogauss

Kilogramm *n* kilogram (Am. E.), kilogramme (Br. E.)
Kilogrammkalorie *f* large calorie, kilo-calorie, kilogram-calorie
Kilohertz *n* (Elektr) kilohertz
Kilokalorie *f* kilocalorie
Kilometerleistung *f* mileage
Kilometerstein *m* milestone
Kiloohm *n* (Elektr) kiloohm
Kilopond *n* kilogram weight
Kilovolt *n* kilovolt
Kilowatt *n* (Elektr) kilowatt
Kilowattstunde *f* (Elektr) kilowatt-hour
Kimme *f* border, edge, (Kerbe) notch
kimmen to fit with a brim, (einkerben) to notch
Kinderkrankheiten *pl* (Med) teething troubles
Kinderlähmung *f* (Med) infantile paralysis, polio [myelitis]
Kinematik *f* kinematics
Kinematographie *f* (obs) cinematography
Kinetik *f* kinetics;, ~ **im stationären Zustand** steady-state kinetics
Kinetin *n* kinetin
kinetisch kinetic
Kinetit *n* kinetite
Kinogerbsäure *f* kinotannic acid
Kinogummi *n* kino, kino gum
Kinoharz *n* kino, kino gum
Kinoin *n* kinoin
kippbar tiltable
Kippbatterie *f* reversible battery
Kippbecher *m* tipping bucket
Kippbestreben *n* tendency to tip
Kippbewegung *f* tilting or tipping motion
Kippbühne *f* tipping stage
kippen to tilt, to tip up
Kippen *n* **der Formen** tipping the molds.
Kipper *m* tipping device
Kippfrequenz *f* relaxation frequency
Kippgenerator *m* relaxation oscillator, scanning generator, sweep generator
Kippgerät *n* relaxation oscillator
Kipphebel *m* rocker arm, valve rocker
Kipphebelverschleiß *m* rocker arm wear, valve train wear
Kipphebelwelle *f* rocker arm shaft
Kippkessel *m* tilting pan
Kippkopfpresse *f* tilting head press
Kippkraft *f* tilting force (bulldozer)
Kippmischer *m* tilting mixer
Kippofen *m* tilting open hearth furnace, tipping furnace
Kipppfanne *f* seesaw pan, swingpan
Kippresse *f* tilting head press
Kippreiter *m* pivoting slider
Kippschalter *m* (Elektr) tumbler switch
Kippschaltung *f* flip-flop circuit, trigger circuit
Kippscher Apparat *m* (Chem) Kipp's apparatus, Kipp's generator

Kippschwingung *f* relaxation oscillation
Kippschwingungsdauer *f* relaxation period
Kippsicherheit *f* tilting safety, ~ **der Ruhelage** stability in the position of rest
Kippvorrichtung *f* tipping device, tilting apparatus
Kippwäschesieb *n* swing sieve
Kippwagen *m* dump wagon
Kippwannenfilter *n* tilting pan filter
Kirnvorrichtung *f* churning machine
Kirschbranntwein *m* cherry brandy, cherry spirit
Kirsche *f* cherry
Kirschensirup *m* cherry syrup
kirschfarben cherry-colored
Kirschgummi *n* cherry gum
Kirschlikör *m* cherry brandy
Kirschlorbeer *m* cherry laurel
kirschrot cherry red, cerise
Kirschrot *n* bright red, cherry red
Kirschrum *m* cherry rum
Kirschsaft *m* cherry juice
Kirschwasser *n* kirsch
Kissen *n* cushion, bolster, pad, pillow
Kissenverzeichnung *f* pincushion distortion
Kiste *f* box, chest
Kitol *n* kitol
Kitt *m* cement, lute, (Dichtmasse) sealing cement, (Email) filler, (Glaserkitt) putty
kittartig cementlike, puttylike
kittbar cementable
kitten to cement, to glue, to lute, (Glas) to putty
Kitterde *f* cement clay, luting clay
Kittindustrie *f* putty industry
Kittwachs *n* bee glue, propolis
Kjeldahlkolben *m* Kjeldahl flask
Kjeldahl-Methode *f* Kjeldahl method
Kläranlage *f* clarification plant, clarifier, clarifying basin, sewage plant, water treatment plant
Klärapparat *m* clarifying apparatus, settling apparatus
Klärbad *n* clearing bath
Klärbecken *n* settling tank, clearer
Klärbehälter *m* settling tank
Klärbottich *m* clearing vat, settling tank, settling vat
Kläre *f* clarifier, clarified liquid, clearing liquor, clear liquid, thin paste
klären to allow to settle, to clarify, to clean, to clear [up]; to purify, to settle
Klären *n* clarifying, clearing, settling
klärend clarifying, clearing
Klärfaß *n* (Brau) settler, settling cask
Klärfilter *n* filter
Klärflasche *f* decanting bottle, decanting flask
Klärgas *n* sewer gas
Klärgasanlage *f* sewer gas plant
Klärgefäß *n* clarifier, clarifying apparatus, settling tank

Klärgrube *f* clarification tank, settling chamber
Klärkessel *m* clarifier
Klärkübel *m* bleacher
Klärmittel *n* clarifier, clarifying agent, clearing agent
Klärpfanne *f* clearing pan, dissolving pan, melting pan
Klärschlamm *m* sewage sludge
Klärsel *n* clarified sugar, clear or fine liquor sugar
Klärstaub *m* test ashes, bone ash
Klärsumpf *m* clarifying sump, settling sump
Klärteich *m* settling tank
Klärtopf *m* clarifier
Klärung *f* clarification, clearing, fining, purification, settling, (Bier) chillproofing of beer
Klärungsfaß *n* clarifier, clearing cask
Klärungsmittel *n* clarifier, clearing agent, fining agent
Klärungsprozeß *m* process of clarification
Klärzentrifuge *f* centrifugal classifier, clarifying centrifuge, decanting centrifuge
klaffen to gape
Klafter *f* (altes Längenmaß) fathom, (Raummaß f. Holz) cord
Klagepatent *n* (Jur) patent in suit
klamm close, tight
Klammer *f* brace, clasp, cramp, (Büro-Klammer) clip, (Chem) clamp, (Heftklammer) staple
Klammerausdruck *m* (Math) expression in parentheses (brackets, braces)
Klammergabel *f* clamping fork
klammern to clamp, to cramp, to rivet
Klammern *pl*, **die ~ auflösen** to remove the parentheses (brackets, braces), **eckige ~** [square] brackets, **geschweifte ~** braces, **in ~ setzen** to put in[to] parentheses (brackets, braces), **runde ~** parentheses
Klammerrelation *f* bracket relation
Klammerring *m* retaining ring
Klammerzahn *m* clasped tooth
Klang *m* sound
Klangfarbemittel *n* tone shading means
Klangplatte *f* sonic plate
Klangzinn *n* fine tin
Klappdeckel *m* hinged cover
Klappe *f* lid, cover, flap, flap valve, trap door, (Ofen) drop door
Klappenventil *n* flap valve
Klappenverschluß *m* flap valve
klappern to clatter, to rattle
Klapperschlange *f* rattle snake
Klapperstein *m* clapperstone, eaglestone
Klappschraube *f* hinged bolt, swing bolt
Klappventil *n* flap valve
Klaprotholit *m* (Min) klaprotholite, native copper bismuth sulfide

klar clear, distinct, (durchsichtig) transparent, (rein) pure, (sauber) clean
Klarheit f clearness, distinctness, (Durchsichtigkeit) transparency
klarkochen to boil till clear
Klarlack m clear lacquer, varnish
Klarschriftbeleg m (Informatik) document for optical character reader
Klarsichtbehälter m high clarity container
Klarsichtfolie f transparent sheet
Klarsichtpackung f transparent pack
Klarspülmittel n clear rinsing agent
Klarstellung f elucidation, explanation
Klasse f class, grade, quality
Klassenbreite f class range
Klasseneinteilung f classification
Klassengrenze f class limit
Klassenhäufigkeit f class frequency
Klassieranlage f grading or classifying plant
Klassierapparat m classifier
Klassierdekanter m centrifuge decanter
klassieren to classify, to grade, to screen, to size, to sort
Klassieren n grading by sifting, sizing by sifting
Klassierer m classifier, ore sorter
Klassiersieb n (Metall) classifying screen; grading screen
Klassierung f classification
Klassifikation f classification
Klassifikator m classifier
klassifizieren to classify, to grade
klastisch (Geol) clastic
Klathrat n (Chem) clathrate compound, inclusion complex
Klaubarbeit f (Metall) hand-picking
Klaubband n (Metall) picking belt
klauben (Bergb) to pick, to sort
Klauberz n picking ore
Klaubetisch m picking table, sorting table
Klaue f claw
Klauenmagnet m claw magnet
Klauenmehl n (Metall) hoof flour, hoof meal, neat's foot meal
Klauenöl n neat's foot oil
Klausthalit m (Min) clausthalite, native lead selenide
Klebdispersion f adhesive dispersion
Klebeband n adhesive tape, tape adhesive
Klebefähigkeit f adhesiveness
Klebefestigkeit f adhesive strength, bonding strength
Klebefilm m film glue, glue film
Klebefolie f film glue, glue film
Klebekraft f adhesive or binding power
Klebelack m dope, sizing glue
Klebemasse f adhesive substance, cement
Klebemittel n adhesive, agglutinant, cement, gum

kleben to stick, to adhere, to cement, to paste, to pitch, (kleistern) to glue
Kleben n blocking
klebend sticking, adherent, adhering; adhesive
Klebepresse f pasting press
Kleber m glue, adhesive, (in Weizenmehl) gluten, **Haft-** ~ pressure-sensitive adhesive, **kaltabbindender** ~ cold setting adhesive, **Schmelz-** ~ hotmelt adhesive
Kleberbrot n gluten bread
Kleberleim m gluten glue, vegetable gelatin
Klebestreifen m adhesive tape, sealing tape, self-adhesive tape
Klebestreifenverschluß m taped closure
Klebewachs n sticking wax, wax cement
Klebezettel m gummed label, sticker
Klebfähigkeit f adhesion, binding property, cohesiveness
Klebfalz m adhesive tape
Klebfestigkeitsprüfung f bonding test
Klebfolie f adhesive film
Klebharz n resin adhesive
Klebkraft f adhesive power
Kleblack m adhesive lacquer
Klebmischung f mixed glue
Klebmittel n adhesive
Klebpflaster n adhesive plaster, sticking plaster
klebrig sticky, adhesive, glutinous, pasty, ropy, tacky
Klebrigkeit f stickiness, adhesiveness, tackiness, **Mittel zur Erhöhung der** ~ tackifier
Klebrigmacher m tackifier
Klebsand m molding sand, plastic refractory clay
Klebschicht f adhesive layer
Klebschiefer m (Min) adhesive slate, clay slate, slate clay
Klebschmutz m adhering dirt
Klebstoff m (f. Leder) cement, (Kitt) putty, (Leim) glue, (synthetisch) adhesive, adhesive substance, ~ **mit hohem Feststoffanteil** high solids adhesive, ~ **mit wasserabweisender Eigenschaft** moisture resistant adhesive, **schnell abbindender** ~ fast setting adhesive
Klebstoffindustrie f adhesives industry
Klebstoffkitt m cement
Klebstreifen m adhesive tape
Klebwachs n bee glue, propolis, sealing wax
Klecksdichte f blob density
Kleesäure f (obs) oxalic acid
Kleesalz n acid potassium oxalate, potassium bi[n]oxalate, potassium hydrogen oxalate, sorrel salt
Kleesamenöl n clover oil
kleesauer oxalate
Kleie f bran
Kleiebrot n bread made of bran
Kleiemelasse f bran molasses
kleienartig branny, farinaceous

Kleienbad *n* bran dye bath
Kleienbeize *f* bran liquid
Kleienessig *m* bran vinegar
Kleienwasser *n* bran water
kleiig branny
klein little, small, **unendlich** ~ infinitesimal
Kleinanzeige *f* (Werbung) classified ad
Kleinbessemerbirne *f* small bessemer convertor
Kleinbildkamera *f* (Phot) miniature camera
Klein-Computer *m* mini-computer
Kleineisen *n* light section iron
Kleineisenzeug *n* small iron ware
Kleinerwerden *n* decreasing, diminution, shrinkage
Kleinfilm *m* miniature film
Kleinfilter *m* fine filter
Kleinflansch *m* small flange
Kleinförderanlage *f* small conveyer
Kleingefüge *n* fine structure, fine texture; microstructure
Kleinguß *m* small casting
Kleinklärwerk *n* septic tank
kleinkörnig small-grained
Kleinkornmischung *f* (Beton) fine grain mixture
kleinkristallin finely crystalline
Kleinkupolofen *m* small cupola [furnace]
Kleinkuppelofen *m* small cupola [furnace]
Kleinlader *m* trickle charger
Kleinlastaufzug *m* low-capacity hoist
Kleinlebewesen *n* microorganism, microzoon
Kleinlichtbildkunst *f* miniature photography
kleinluckig close-grained, fine-meshed, of fine porosity
Kleinod *n* gem, jewel, treasure
Kleinröhre *f* miniature tube
Kleinstmaß *n* minimum limit, **auf das** ~ **zurückführen** to minimize
Kleinstnadelventil *n* miniature needle valve
kleinstückig small-sized, light-sized
Kleinstwert *m* minimum, minimum value
Kleinversuch *m* micro-scale test, small scale test
Kleinwinkelkorngrenze *f* small-angle grain boundary
Kleinwinkelstreuung *f* small angle scattering
kleinzackig small-notched, small-toothed
Kleister *m* glue, gum, paste, size, starch paste
kleisterig pasty, sticky
kleistern to paste
Klementit *m* klementite
Klemmbrett *n* terminal board
Klemme *f* clamp, clip, terminal
Klemmisolator *m* cleat insulator
Klemmkegel *m* centering cone, cone clamp
Klemmkegelverbindung *f* compressed cone connection
Klemmkonus *m* tapered cone
Klemmkupplung *f* clamp coupling, compression coupling
Klemmlänge *f* length of grip

Klemmplatte *f* clamp[ing] plate
Klemmschraube *f* set screw, adjusting screw, binding screw, screw clamp, (Elektr) clamping screw, terminal screw
Klemmspannung *f* terminal potential difference, terminal voltage
Klemmuffe *f* clamp, clamping sleeve
Klemmverbindung *f* compression joint
Klemmvorrichtung *f* clamping device, clutch
Klemmwalzenbock *m* nip roll stand
Klempner *m* plumber
Klempnerarbeiten *f pl* plumbing
Klette *f* (Bot) bur, burdock
Kletterfilmverdampfer *m* climbing-film evaporator
Kletterkran *m* climbing crane
Kletterverdampfer *m* rising film evaporator
Klima *n* climate
Klimaanlage *f* air-conditioning plant
Klimabeständigkeit *f* resistance to climatic conditions
Klimageräte *n pl* air-conditioning equipment
Klimalehre *f* climatology
Klimaprüfschrank *m* climatic chamber
Klimaraum *m* conditioning room
Klimaschrank *m* conditioning cabinet
klimatisch climatic
klimatisieren to air-condition
klimatisiert conditioned
Klimatisierung *f* air conditioning
Klimatologie *f* climatology
Klinge *f* blade, cutting blade
Klingelleitung *f* bell wire
klingfrei (Rad) hiss-free, nonmicrophonic
Klingstein *m* (Min) clink basalt, clink stone, phonolite
Klinik *f* (Med) clinics
klinisch clinical
Klinke *f* pawl, ~ **mit Rolle** pawl with roller, roller pawl
Klinkenauslösung *f* pawl release
Klinkenkupplung *f* pawl coupling
Klinker *m* clinker
Klinkerbeton *m* clinker concrete
Klinkerverwertungsmaschine *f* clinker utilization machine
Klinoachse *f* clino axis
Klinochlor *m* (Min) clinochlore, chlinochlorite, ripidolite
Klinodiagonale *f* clinodiagonal
Klinoedrit *m* clinohedrite, (Min) clinoedrite
Klinoenstatit *m* (Min) clinoenstatite
Klinohumit *m* (Min) clinohumite
Klinoklas *m* (Min) clinoclase, clinoclasite
Klinopyramide *f* (Krist) clinopyramid
Klinozoisit *m* (Min) clinozoisite
Klipsteinit *m* klipsteinite
Klirrfaktor *m* (Radio) coefficient of harmonic distortion, distortion factor

Klischee n cliché, block, printer's sterio
klischieren (Galvano) to electrotype
Kloake f common sewer, drain, sink
Kloakengas n sewer gas
Kloakenwasser n sewage water
Klobensäge f frame-saw
Klöpfel m mallet
Klöppel m beater, clapper, knocker
Klöppelhebel m strike lever
Klöppelmaschine f braiding machine
Klopfbremse f antiknock agent
Klopfdetektor m knockmeter
Klopfen n (Motor) knocking, ~ **des Motors** knocking of the motor, **Beschleunigungs-** ~ acceleration, front end knock, low speed knock, **Hochgeschwindigkeits-** ~ high speed knock
klopffest antiknock[ing], knock-resistant, knock-stable
klopffrei antiknock
Klopfmittel n (Antiklopfmittel) antiknock agent
Klopfwert m (s. Oktanzahl) octane number
Klostridium n clostridium
Klotz m block, log, stump, ~ **des Reifenprofils** cleat
Klotzbad n (Text) pad bath
Klotzdämpfverfahren n pad steam technique
Klotzen n (Text) impregnating, padding
Klotzverfahren n (Färberei) padding process, (Text) padding
Klovosen n klovosene
Klümpchen n little lump, (Med) nodule
klümperig clotted, clotty, curdy, lumpy
Klumpen m coarse lump, (Erde) clod
Klumpenbildung f formation of lumps
Klumpengold n ingot gold
klumpig clotted, clotty, lumpy
Klunker m clod, clot
Kluppe f threading die, die-stock, tongs
Kluppenbacken pl screw dies
Kluppenspannrahmen m clip stenter
Kluppzange f forceps
Knabbelkoks m crushed coke
Knäuel m n ball, coil, (Wolle) skein
knäueln to ball
Knagge f (Knaggen, m) cam, catch, lug
Knaggensteuerung f cam gear
Knall m detonation, explosion, report
Knallblei n fulminating lead
Knallbonbon m cracker
knallen to detonate, to explode, to go off with a report
Knallerbse f detonating ball
Knallgas n detonating gas, oxyhydrogen gas
Knallgasbildung f formation of oxyhydrogen gas
Knallgasflamme f oxyhydrogen flame
Knallgasgebläse n oxyhydrogen blowpipe
Knallgaskette f oxyhydrogen cell
Knallgaslicht n oxyhydrogen light

Knallgasschweißung f oxyhydrogen welding
Knallgold n fulminating gold, gold fulminate
Knallicht n oxyhydrogen light
Knallkörper m detonator
Knallkraft f explosive force
Knallpulver n detonating powder, fulminating powder, percussion powder, (Fulminat) fulminate
Knallpyrometer n explosion pyrometer
Knallquecksilber n fulminating mercury, mercuric fulminate
knallrot fiery red, scarlet
Knallsäure f fulminic acid
Knallsalpeter m ammonium nitrate, German saltpeter
knallsauer fulminate
Knallsilber n silver fulminate
Knallzucker m nitrosaccharose
Knallzünder m detonator
Knallzündmittel n detonating priming
Knallzündschnur f detonating cord, instantaneous fuse
knapp concise, brief, close-fitting, narrow, scanty, scarce, short, tight
Knappheit f scarcity, shortage
knarren to crack; to crackle; to jar, to rattle
Knauf m knob, head, top
Knautschlack[leder n] m crinkle leather
Knebel m clog, crossbar, cudgel, gag, stick, toggle
Knebelit m (Min) knebelite
Knebelkopf m bar head, tommy head
Knebelpresse f toggle press
Knebelschalter m (Elektr) jack switch
Knebelschraube f thumbscrew, tommy screw
Kneifzange f end cutting pliers, nippers, pincers
Knetapparat m kneading mill
knetbar kneadable, easily moldable, plastic
Knetbarkeit f kneadability
kneten to knead, to mold, to squeeze, (Gummi) to masticate, **mischen und** ~ to pug
Kneter m kneader, dough mixing machine, kneading machine
Knetfaß n pugmill
Knetgummi m n plasticine
Knetlegierung f malleable or forgeable alloy, plastic or wrought alloy
Knetmaschine f kneader, kneading machine
Knetmasse f model[l]ing material or clay
Knetpumpe f kneader pump
Knetwelle f rotor shaft
Knick m sharp bend, break, crack, flaw
Knickausleger m articulated jib
Knickband n kink band
Knickbanddichte f kink band density
Knickbeanspruchung f buckling stress, axial compression, breaking load
knicken to break, (falten) to fold
Knickerscheinung f buckling

Knickfestigkeit *f* breaking resistance, buckling strength, (Papier) folding strength, resistance to folding
Knicklast *f* buckling load, collapse load, maximum load
Knicklenkung *f* articulated steering
Knickpunkt *m* breaking point, point of inversion
Knickspannung *f* buckling stress
Knickung *f* breaking, buckling, cracking
Knickversuch *m* buckling or crippling test
Kniehebel *m* bellcrank lever, bent lever, toggle joint
Kniehebelbrecher *m* toggle crusher
Kniehebelpresse *f* toggle [lever] press
Kniehebelschließsystem *n* knee-toggle [clamping]
Knierohr *n* bent pipe, elbow bend
Kniestück *n* angle, elbow, elbow pipe
Knieverbindung *f* (Techn) elbow joint
Kniff *m* pinch, (Falte) crease, (Trick) trick
Kniffmaschine *f* crimping machine, folding machine
knirschen to crackle, to crunch
knistern (Chem) to crepitate, (z. B. Papier) to crackle, (z. B. Seide) to rustle
Knistern *n* crackling, crepitating, crepitation, rustling
knisternd crepitant
Knistersalz *n* decrepitating salt
Knitter *m* crease, fold, wrinkle
knitterfest crease-proof, crease-resistant, noncreasing, wrinkle-resistant
Knitterfestausrüstung *f* creaseproofing finish
Knitterfestigkeit *f* crease resistance, wrinkle resistance
Knitterfestmachen *n* crease proofing
knitterfrei crease-proof, crush-proof
Knitterfreiausrüstung *f* crease-resist finish
knitterig creased, crumpled, wrinkled
knittern to crease, to crumple
Knoblauch *m* (Bot) garlic
knoblauchartig garlic-like
Knoblaucherz *n* (Min) scorodite
Knoblauchgas *n* lost, mustard gas
Knoblauchöl *n* garlic oil
Knochen *m* bone
Knochenasche *f* bone ash[es], impure calcium phosphate
Knochenatrophie *f* (Med) bone atrophy
knochenbildend osteogenous
Knochenbildung *f* bone formation, osteogenesis
Knochendämpfapparat *m* bone steamer, bone steaming apparatus
Knochendüngemehl *n* bone [manure] meal
Knochendünger *m* bone manure
Knochendüngung *f* bone meal manuring
Knochenerde *f* bone earth, bone ash
Knochenfett *n* bone fat
Knochengallerte *f* bone gelatin, ossein
Knochengerüst *n* skeleton

Knochenkarzinom *n* (Med) osteocarcinoma
Knochenkohle *f* animal charcoal, bone black, bone charcoal
Knochenkohlefilter *n* char [coal] filter
Knochenkohleglühofen *m* bone-black furnace
Knochenkrebs *m* (Med) bone cancer, osteocarcinoma
Knochenlehre *f* (Med) osteology
Knochenleim *m* bone glue, gelatin of bones, osteocolla
Knochenmark *n* (Med) [bone] marrow
Knochenmarkleukozyt *m* myeloblast
Knochenmarklymphozyt *m* lymphoblast, myelolymphocyte
Knochenmarkmonozyt *m* myelomonocyte
Knochenmarktumor *m* (Med) medullary tumor
Knochenmarkzelle *f* marrow cell, myeloid cell
Knochenmehl *n* bone dust, bone meal
Knochenmehlammonsalpeter *m* bone meal ammonium nitrate
Knochenöl *n* bone oil, Dippel's oil, neat's foot oil
Knochenschaden *m* (Med) bone injury
Knochenschwarz *n* bone black, ivory black
Knochenstein *m* (Min) osteolite
Knochensubstanz *f* callus, ostein
Knochentalg *m* bone tallow
Knochenverkohlungsofen *m* bone carbonizing oven
Knochenzelle *f* bone cell, osteocyte
knochig bony
Knöterich *m* knot grass
Knolle *f* (Bot) bulb, nodule
Knollenopal *m* (Min) menilite
Knopf *m* button, stud, (Griff) handle, (Knauf) knob
Knopfsteuerung *f* botton control, botton steering
Knorpel *m* cartilage
knorpelartig cartilaginous
Knorpelband *n* cartilaginous tissue, cartilage
Knorpelbildung *f* (Biol) chondrification
Knorpelleim *m* chondrin
Knorpeltumor *m* chondroma
Knorpelzelle *f* cartilage cell, chondroblast, chondrocyte
Knospe *f* bud
Knospenbildung *f* formation of buds, (Biol) gemmation
Knoten *m* knot, nodule, (Glas) cord [in glass], vitreous inhomogeneity
Knotenabstand *m* distance between points of support
Knotenbrecher *m* (Zuck) disintegrator, sugar breaker
Knotenfänger *m* (Pap) screener, strainer
Knotenfestigkeit *f* knot tenacity, loop strength
Knotenlack *m* knotting varnish
Knotenlinie *f* (Akust) nodal line

Knotenpunkt *m* center, intersection, joint, junction, junction point, (Opt) nodal point
Knotenpunktbewegung *f* motion of point of intersection
Knotenpunktgeschwindigkeit *f* velocity of point of intersection
Knotensatz *m* zero rule
Knotenverbindung *f* joint connection
knotig knotty, nodular
knüpfen to tie, to attach, to knot
Knüppelerwärmung *f* bar heating
knusprig crisp
Koagulans *n* coagulant
Koagulat *n* coagulate, clot, coagulum
Koagulation *f* coagulation, **mechanische** ~ mechanical coagulation
Koagulationsmittel *n* coagulant
Koagulationswärme *f* heat of coagulation
koagulierbar coagulable
koagulieren to coagulate
Koagulieren *n* coagulating
Koagulierschutzmittel *n* anti-coagulation agent
koaguliert coagulated
Koagulierung *f* coagulation
koagulierungsfähig coagulable
Koagulierungsfähigkeit *f* coagulating property
Koagulierungsflüssigkeit *f* coagulating liquid
Koagulierungsmittel *n* coagulant, coagulating agent
Koaleszenz *f* coalescence
koalisieren to coalesce, to combine, to unite
koaxial coaxial
Koazervat *n* coacervate
Koazervation *f* coacervation
Kobalt *n* cobalt (Symb. Co)
Kobaltacetat *n* cobalt acetate
Kobaltalanin *n* cobalt alanine
Kobaltaluminat *n* cobalt aluminate
Kobaltammin *n* cobaltammine
Kobaltammoniumsulfat *n* cobalt ammonium sulfate
Kobaltantimonid *n* cobalt antimonide
Kobaltarsenat *n* cobalt arsenate
Kobaltarsenkies *m* (Min) cobalt pyrites, danaite, glaucodot, mispickel
kobaltartig cobalt-like
Kobaltbad *n* cobalt bath
Kobaltbeize *f* cobalt mordant
Kobaltblau *n* cobalt blue, smalt, Thenard's blue
Kobaltblüte *f* (Min) erythrite, phycsite, red cobalt
Kobaltblume *f* (Min) cobalt bloom, erythrite, phycsite, red cobalt
Kobaltbombe *f* cobalt bomb
Kobaltborid *n* cobalt boride, cobalt monoboride
Kobaltbromür *n* (Kobalt(II)-bromid) cobalt dibromide, cobalt(II) bromide, cobaltous bromide
Kobaltbronze *f* cobalt bronze

Kobaltcarbonat *n* cobalt carbonate
Kobaltcarbonylwasserstoff *m* cobalt hydrogen carbonyl
Kobaltchlorür *n* (Kobalt(II)-chlorid) cobalt dichloride, cobalt(II) chloride, cobaltous chloride
Kobaltchromstahl *m* cobalt-chromium steel
Kobalterz *n* (Min) cobalt ore
Kobaltfarbe *f* cobalt pigment, powder blue
Kobaltfarbstoff *m* cobalt coloring matter, cobalt pigment
Kobaltflasche *f* cobalt glass bottle
Kobaltfluorid *n* cobalt fluoride, (Kobalt(III)-fluorid) cobaltic fluoride, cobalt(III) fluoride, cobalt trifluoride
Kobaltgelb *n* cobalt yellow
Kobaltgitter *n* cobalt lattice
Kobaltglanz *m* (Min) cobalt glance, cobaltite
Kobaltglas *n* cobalt glass
Kobaltgraupen *pl* amorphous gray cobalt
Kobaltgrün *n* cobalt green, cobalt zincate, Rinmann's green, smalt green
kobalthaltig cobaltiferous
Kobalthydroxid *n* cobalt hydroxide
Kobaltiaksalz *n* cobaltammine salt, cobaltiac salt
Kobaltichlorid *n* (Kobalt(III)-chlorid) cobaltic chloride, cobaltichloride, cobalt(III) chloride
Kobalticyanid *n* cobaltic cyanide, cobalticyanide, cobalt(III) cyanide
Kobalticyankalium *n* potassium cobalticyanide, potassium cyanocobaltate(III)
Kobalticyanwasserstoff[säure] cobalticyanic acid, cyanocobaltic(III) acid
Kobalt(III)-chlorid *n* cobaltic chloride, cobalt(III) chloride, cobalt trichloride
Kobalt(III)-Salz *n* cobaltic salt, cobalt(III) salt
Kobalt(III)-Verbindung *f* cobaltic compound, cobalt(III) compound
Kobalt(II)-Salz *n* cobalt(II) salt, cobaltous salt
Kobalt(II)-Verbindung *f* cobalt(II) compound, cobaltous compound
Kobaltikaliumnitrit *n* cobaltic potassium nitrite, potassium cobaltinitrite, potassium nitrocobaltate(III)
Kobaltin *m* (Min) cobalt glance, cobaltine, cobaltite
Kobaltioxid *n* cobaltic oxide, cobalt(III) oxide, cobalt sesquioxide
Kobaltit *m* (Min) cobaltite, cobalt glance, cobaltine
Kobaltiverbindung *f* (Kobalt(III)-Verbindung) cobaltic compound, cobalt(III) compound
Kobaltjodür *n* (Kobalt(II)-jodid) cobalt diiodide, cobalt(II) iodide, cobaltous iodide
Kobaltkaliumnitrit *n* cobalt potassium nitrite, potassium nitrocobaltate
Kobaltkies *m* (Min) cobalt pyrites
Kobaltlegierung *f* cobalt alloy

Kobaltmanganerz *n* (Min) asbolane, asbolite, earth cobalt
Kobaltmulm *m* (Min) black cobalt ocher
Kobaltnickelkies *m* (Min) linnaeite
Kobaltnickelpyrit *m* (Min) cobalt nickelpyrites
Kobaltnitrat *n* cobalt nitrate
Kobaltnitrit *n* cobalt nitrite
Kobaltocalcit *m* cobaltocalcite
Kobaltochlorid *n* (Kobalt(II)-chlorid) cobalt(II) chloride, cobaltous chloride
Kobaltocyanwasserstoffsäure *f* cobaltocyanic acid, cyanocobaltic(II) acid
Kobaltomenit *m* (Min) cobaltomenite
Kobaltonitrat *n* cobalt(II) nitrate, cobaltous nitrate
Kobaltooxid *n* (Kobalt(II)-oxid) cobalt(II) oxide, cobaltous oxide, cobalt protoxide
Kobaltorhodanwasserstoffsäure *f* cobaltothiocyanic acid, thiocyanatocobaltic(II) acid
Kobaltosalz *n* (Kobalt(II)-salz) cobalt(II) salt, cobaltous salt
Kobaltosulfat *n* cobalt(II) sulfate, cobaltous sulfate
Kobaltoverbindung *f* (Kobalt(II)-Verbindung) cobalt(II) compound, cobaltous compound
Kobaltoxalat *n* cobalt oxalate
Kobaltoxid *n* cobalt oxide, (Kobalt(III)-oxid) cobaltic oxide, cobalt(III) oxide
Kobaltoxidhydrat *n* (obs) cobalt hydroxide
Kobaltoxydul *n* (Kobalt(II)-oxid) cobalt(II) oxide, cobalt monoxide, cobaltous oxide, cobalt protoxide
Kobaltoxyduloxid *n* cobalt(II)(III) oxide, cobaltocobaltic oxide
Kobaltoxydulsalz *n* (obs) cobalt(II) salt, cobaltous salt
Kobaltoxydulverbindung *f* (obs) cobalt(II) compound, cobaltous compound
Kobaltoxydulzinkoxid *n* (obs) cobalt(II) zincate, cobalt green, cobaltous zincate, Rinmann's green
Kobaltoxydulzinnoxid *n* (obs) cobalt(II) stannate, cobaltous stannate
Kobaltphosphid *n* cobalt phosphide
Kobaltplattierung *f* cobalt plating
Kobaltschwärze *f* (Min) asbolane, asbolite, black cobalt ore, cobalt black
Kobaltsesquioxid *n* cobaltic oxide, cobalt(III) oxide, cobalt sesquioxide, dicobalt trioxide
Kobaltsikkativ *n* cobalt siccative
Kobaltsilikat *n* cobalt silicate
Kobaltsilizid *n* cobalt silicide
Kobaltspat *m* (Min) spherocobaltite
Kobaltspeise *f* cobalt speiss
Kobaltspiegel *m* (Min) specular cobalt, transparent cobalt ore
Kobaltsulfat *n* cobalt sulfate
Kobaltsulfid *n* cobalt sulfide

Kobalttetracarbonyl *n* cobalt tetracarbonyl
Kobaltvitriol *m* cobalt vitriol, rose vitriol
Kobaltwismutfahlerz *n* (Min) tennantite
Kobaltzinkat *n* cobalt zincate
Kobellit *m* (Min) kobellite
Kochapparat *m* cooking apparatus
Kochbecher *m* beaker
kochbeständig fast to boiling
Kochbeutel *m* boilable pouch
Kochelit *m* kochelite
kochen to cook, (leicht sieden) to simmer, (sieden) to boil
Kochen *n* boiling, (Aufwallen) ebullition
kochend at the boil, boiling
Kocher *m* cooker, boiler, boiling apparatus, digester, reboiler (rectification column), **rotierender** ~ revolving boiler, revolving digester
Kocherlauge *f* cooking lye, boiling agent
Kocheruptionsperiode *f* boil eruption period
kochfest boil-proof
Kochfestigkeit *f* boil strength
Kochflasche *f* boiling flask
Kochgerät *n* cooking utensils
Kochgeschirr *n* cooking utensils
Kochkessel *m* boiling vessel, boiler, cooker
Kochkontrollapparat *m* pan boiling control apparatus
Kochofen *m* stove
Kochpresse *f* block press
Kochprobe *f* boiling test
Kochpuddeln *n* slag puddling
Kochpunkt *m* boiling point
Kochsalz *n* common salt, sodium chloride
Kochsalzelektrolyse *f* electrolysis of sodium chloride
Kochsalzgehalt *m* sodium chloride content
kochsalzhaltig containing common salt, containing sodium chloride
Kochsalzinfusion *f* (Med) saline infusion
Kochsalzlösung *f* sodium chloride solution, brine, common salt solution, **physiologische** ~ physiological salt solution
Kochtopf *m* boiler, pot
Kochverlust *m* loss by boiling
Kochzeit *f* boiling period
Kodäthylin *n* codethyline, morphine ethylate
Kodamin *n* codamine
Kodein *n* codeine, morphine methyl ether
Kodeinbrommethylat *n* codeine bromomethylate
Kodeinhydrochlorid *n* codeine hydrochloride
Kodeinon *n* codeinone
Kodeinphosphat *n* codeine phosphate
Koechlinit *m* (Min) koechlinite
Köder *m* bait
Ködergift *n* poison bait
Koeffizient *m* coefficient
Kölbchenform *f* bulb shape
Kölnischbraun *n* Cologne brown

Kölnisches Wasser *n* Cologne water, Eau de Cologne
Könenit *m* (Min) koenenite
Königsblau *n* cobalt blue, cobalt ultramarine, king's blue, royal blue
Königsbolzen *m* king pin
Königsgelb *n* chinese yellow, chrome yellow, king's yellow (arsenic trisulfide), massicot (lead monoxide)
Königsgrün *n* Paris green
Königssalbe *f* basilicon ointment, resin ointment, rosin cerate
Königswasser *n* aqua regia, nitro-hydrochloric acid
Koenzym *n* (Biochem) coenzyme
Köper *m* (Text) twill
Köperbindung *f* (Text) twill
Körnchen *n* granule
Körnelung *f* granulation
körnen to granulate, to corn, to grain, to grind
Körner *m* center mark, gauge point, prick punch (Am. E.), punch mark
Körnerquarz *m* (Min) quartzite
Körnerscharlach *m* Venetian scarlet
Körnerspitze *f* lathe center
Körnerzinn *n* grained tin, grain tin, granular tin
körnig grained, granular, granulated, gritty, in grains
Körnigkeit *f* granularity
Körnigkeitsmesser *m* granulometer
Körnung *f* grain, granulation
Körper *m* body, **fester** ~ solid, **flüssiger** ~ liquid, **gasförmiger** ~ gaseous body, **im** ~ **gebildet** formed in vivo, **schwarzer** ~ black body, **starrer** ~ (Mech) rigid body
Körperdesodorant *n* (Kosm) body deodorant
Körperflüssigkeit *f* body fluid
Körperfunktion *f* body function
körperlich bodily, physical
Körpermaß *n* cubic measure
Körperoberfläche *f* body surface, surface of a body
Körperpflege *f* personal hygiene
Körperpflegemittel *n* cosmetic
Körperpuder *m* talcum powder
Körperschalldämmung *f* absorption or attenuation of sound in solids
Körperstöße *m pl* body collisions
Körperstrom *m* current in solid conductor
Körperteil *m* part of the body
Körpertemperatur *f* body temperature
Körpertoxin *n* (Eigentoxin) esotoxin
Körperverschiebung *f* body displacement
Körperwärme *f* body heat
Koerzitivfeld *n* coercive field
Koerzitivkraft *f* coercive force, coercivity, retentivity
Koerzitivkraftmesser *m* coercimeter
Koerzivität *f* coercive force, coercivity

Koexistenz *f* coexistence
Koexistenzgleichung *f* (f. zweiphas. Zweistoffsystem) coexistence equation
koexistieren to coexist
Koffein *n* caffeine, coffeine
Koffeinvergiftung *f* caffeinism, coffeinism
Kofferpresse *f* block press
Kogasin *n* kogasin
Kognak *m* cognac, brandy (from wine)
kohärent coherent
Kohärenz *f* coherence, coherency
Kohärenzbegriff *m* concept of coherence
Kohärenzstrahlung *f* coherent radiation
Kohärer *m* (Elektr) coherer
Kohärerröhre *f* coherer tube
kohärieren to cohere
Kohäsion *f* coherence, cohesion, cohesiveness
Kohäsionsdruck *m* (Phys) cohesion pressure
Kohäsionseigenschaft *f* cohesive property
Kohäsionsfestigkeit *f* cohesion strength
Kohäsionskraft *f* (Phys) cohesional force, cohesion energy, cohesiveness
Kohäsionsvermögen *n* cohesiveness
kohäsiv cohesive
Kohle *f* (Holz-Kohle) charcoal, (Kohlenstoff) carbon, (Min) coal, ~ **zermahlen** to pulverize coal, **A-** ~ coal for gas and vapor adsorption, **amorphe** ~ amorphous carbon, **aschereiche** ~ high-ash coal, **backende** ~ caking coal, bituminous coal, **E-** ~ coal for decoloring, **feingemahlene** ~ pulverized coal, **fette** ~ fat coal, strongly caking coal, **Form-** ~ molded cylinders of coal, **G-** ~ coal for gasmasks, **geringwertige** ~ poor coal, **grobstückige** ~ lump coal, **halbfette** ~ semibituminous coal, **kurzflammige** ~ short flaming coal, steam coal, furnace coal, **magere** ~ dry burning coal, free-ash coal, non caking coal, **Pulver-** ~ coal in powder form, **schlackenreine** ~ non clinkering coal, **weiche** ~ cherry coal, **WR-** ~ coal for water purification
Kohleanode *f* carbon anode
Kohleblock *m* carbon block
Kohlebogen *m* carbon arc
Kohlebürste *f* (Elektr) carbon brush
Kohleelektrode *f* carbon electrode
Kohleelement *n* carbon cell
Kohleentgasung *f* (Metall) coal carbonization
Kohlefaden *m* carbon filament
Kohlefadenlampe *f* carbon filament lamp
Kohlefeuerung *f* coal burning [furnace], coal firing
Kohlefilter *n* charcoal filter
kohlefrei carbon-free, noncarbonaceous
Kohlegrus *m* slack, small coal
Kohlehüllen *pl* (Elektronenmikroskopie) carbon replica
Kohlehydrieranlage *f* coal hydrogenizing plant
Kohlehydrierung *f* coal hydrogenation

Kohlekontakt *m* carbon contact
Kohlelichtbogen *m* (Techn) carbon arc
Kohlelichtbogenschweißung *f* carbon-arc welding
Kohlemahlanlage *f* coal pulverizer
kohlen to carbonize, to carburize, to char
Kohlenabbau *m* (Bergb) working of coal
Kohlenader *f* (Bergb) coal vein
Kohlenanalyse *f* coal analysis
Kohlenanzünder *m* coal igniter, fire lighter
kohlenartig coal-like
Kohlenasche *f* coal ashes
Kohlenaufbereitung *f* coal dressing
Kohlenbergbau *m* coal mining
Kohlenblende *f* anthracite, glance coal
Kohlenblume *f* bituminous clay
Kohlenbogenlampe *f* carbon arc lamp
Kohlenbohrer *m* carbon borer
Kohlenbohrmaschine *f* coal auger
Kohlenbrandschiefer *m* bituminous shale
Kohlenbrecher *m* coal breaker, coal crusher
Kohlenbrennen *n* charcoal burning
Kohlenbürste *f* carbon brush
Kohlendioxid *n* carbon dioxide, carbonic anhydride
Kohlendioxidassimilation *f* carbon dioxide assimilation
Kohlendioxidschnee *m* carbon dioxide snow, dry ice
Kohlendisulfid *n* carbon disulfide
Kohlendithiolsäure *f* dithiolcarbonic acid
Kohlendruck *m* pigment printing, polychromatic process
Kohlendunst *m* smoke of coal, vapor of burning coal
Kohleneisenstein *m* carbonaceous iron stone, blackband
Kohlenelektrode *f* carbon electrode
Kohlenfaden *m* carbon filament
Kohlenfadenlampe *f* carbon filament lamp
Kohlenfeuerung *f* coal firing
Kohlenfilter *n* charcoal filter, boneblack filter
Kohlenförderband *n* belt conveyor for coal
Kohlenfutter *n* carbonaceous lining
Kohlengalmei *m* (Min) calamine, smithsonite
Kohlengas *n* carbon monoxide, coal gas
Kohlengestübe *n* coal breeze, slack, small coal
Kohlengicht *f* coal charge
Kohlengröße *f* size of coal
Kohlengrube *f* coal mine
Kohlengrus *m* slack, slack coal, small coal
Kohlenhalde *f* coal heap, coal pile
kohlenhaltig carboniferous, carbonaceous, containing coal
Kohlenhornblende *f* anthracite, black hornblende
Kohlenhydrat *n* carbohydrate
Kohlenhydratabbau *m* carbohydrate catabolism
Kohlenhydratdepot *n* carbohydrate depot
kohlenhydratreich rich in carbohydrates

Kohlenhydratstoffwechsel *m* carbohydrate metabolism
Kohlenhydratstoffwechselsteuerung *f* glycoregulation
Kohlenkalk *m* carboniferous limestone
Kohlenkalkspat *m* (Min) anthraconite
Kohlenkalkstein *m* carboniferous limestone
Kohlenkipper *m* coal tipper
Kohlenklein *n* slack, small coal
Kohlenmeiler *m* charcoal pile
Kohlenmonoxid *n* carbon monoxide
Kohlenmulm *m* coal breeze, slack, small coal
Kohlenoxid *n* carbon oxide, (Kohlenmonoxid) carbon monoxide
Kohlenoxidanzeiger *m* carbon oxide detector, carbon oxide indicator
Kohlenoxidapparat *m* carbon monoxide apparatus
Kohlenoxideisen *n* iron carbonyl
Kohlenoxidentwicklung *f* formation of carbon monoxide
Kohlenoxidgas *n* carbon oxide gas
Kohlenoxidhämoglobin *n* carbonyl hemoglobin
Kohlenoxidnickel *n* nickel carbonyl, nickel carboxide, nickel tetracarbonyl
kohlenoxidreich rich in carbon monoxide, with high carbon monoxide content
Kohlenoxidschreiber *m* carbon monoxide recorder or monitor, CO-recorder
Kohlenoxidvergiftung *f* carbon oxide poisoning
Kohlenoxychlorid *n* carbon oxychloride, carbonyl chloride, phosgene
Kohlenoxysulfid *n* carbon oxysulfide, carbonyl sulfide
Kohlenpulver *n* charcoal powder, ground charcoal
Kohlensäure *f* carbonic acid
Kohlensäureabscheider *m* carbon dioxide separator
Kohlensäureanhydrid *n* carbon dioxide, carbonic anhydride
Kohlensäureantrieb *m* carbonic acid gas drive
Kohlensäureanzeiger *m* carbon dioxide indicator
Kohlensäurearmaturen *fpl* carbon dioxide fittings
Kohlensäureassimilation *f* assimilation of carbon dioxide
Kohlensäureaufnahme *f* absorption of carbon dioxide
Kohlensäurebrot *n* aerated bread
Kohlensäurechlorid *n* carbonyl chloride, phosgene
Kohlensäuredampf *m* carbonic acid gas
Kohlensäurediäthylester *m* diethyl carbonate
Kohlensäureentwicklung *f* formation of carbonic acid
Kohlensäureester *m* carbonic ester
Kohlensäuregas *n* carbon dioxide gas, carbonic acid gas

Kohlensäuregehalt *m* carbon dioxide content
Kohlensäurehärtung *f* carbonic acid hardening
kohlensäurehaltig carbonated, containing carbon dioxide, containing carbonic acid
Kohlensäurekompressor *m* carbonic acid compressor
Kohlensäurelöscher *m* carbon dioxide fire extinguisher
Kohlensäuremesser *m* (Blut) carbonometer, (Phys) anthracometer
Kohlensäureschnee *m* carbon dioxide snow, dry ice
Kohlensäureschreiber *m* carbon dioxide recorder or monitor, CO_2-recorder
Kohlensäurestrom *m* carbonic acid current
Kohlensäuretonerde *f* (kohlensaure Tonerde) aluminum carbonate
Kohlensäurewasser *n* carbonated water, soda water
Kohlensandstein *m* (Bergb) carboniferous sandstone
kohlensauer aerated, (Chem) carbonate, salt or ester of carbonic acid
Kohlenscheider *m* coal separating plant
Kohlenschiefer *m* bituminous shale, coal slate
Kohlenschlacke *f* cinder
Kohlenschlamm *m* coal mud, coal slimes, coal washings
Kohlenschwarz *n* charcoal black, coal black
Kohlenspat *m* (Min) anthraconite
Kohlenstampfmaschine *f* coal stamping machine
Kohlenstampfmasse *f* compressed carbon mass
Kohlenstaub *m* coal dust, culm
Kohlenstaubanlage *f* coal pulverizing plant
Kohlenstaubexplosion *f* coal dust explosion
Kohlenstaubfeuerung *f* coal dust furnace
Kohlenstickstoff *m* cyanogen, dicyanogen, ethane dinitrile, oxalonitrile
Kohlenstoff *m* carbon (Symb. C), ~ **Ein-Kohlenstoff-Fragment** one-carbon fragment, **amphoterer** ~ amphoteric carbon, **gebundener** ~ fixed carbon, **radioaktiver** ~ radioactive carbon, radiocarbon
kohlenstoffarm low-carbon
Kohlenstoffart *f* variety of carbon
Kohlenstoffatom *n* carbon atom, **asymmetrisches** ~ asymmetric carbon atom, **endständiges** ~ terminal carbon atom
Kohlenstoffausscheidung *f* separation of carbon
Kohlenstoffbestimmung *f* determination of carbon
Kohlenstoffbindung *f* carbon bond, carbon linkage
Kohlenstoffcalcium *n* calcium carbide
Kohlenstoffchlorid *n* carbon tetrachloride, tetrachloromethane
Kohlenstoffdisulfid *n* carbon disulfide
Kohlenstoffdoppelbindung *f* carbon double bond
Kohlenstoffeisen *n* iron carbide

Kohlenstoffentziehung *f* carbon removal, decarbonization, decarburization
Kohlenstoffgehalt *m* carbon content
kohlenstoffhaltig carboniferous, carbonaceous, containing carbon
Kohlenstoffkern *m* carbon nucleus
Kohlenstoffkette *f* carbon chain, chain of carbon atoms
Kohlenstoffkreislauf *m* carbon metabolism
Kohlenstoffmetall *n* [metal] carbide, metallic carbide
Kohlenstoffnitrid *n* carbon nitride
Kohlenstoffoxybromid *n* carbon oxybromide, carbonyl bromide
Kohlenstoffoxychlorid *n* carbon oxychloride, carbonyl chloride, phosgene
Kohlenstoffquelle *f* carbon source
Kohlenstoffsilizium *n* carbon silicide, silicon carbide
Kohlenstoffskelett *n* carbon skeleton, skeleton of carbon atoms
Kohlenstoffstahl *m* carbon steel
Kohlenstoffstein *m* carbon brick
Kohlenstoffstickstofftitan *n* titanium carbonitride
Kohlenstoffsubnitrid *n* carbon subnitride, acetylene dinitrile
Kohlenstoffsuboxid *n* carbon suboxide
Kohlenstoffsulfid *n* carbon disulfide
Kohlenstofftetrachlorid *n* carbon tetrachloride, phenoxin, tetrachloromethane
Kohlenstoffuhr *f* carbon clock
Kohlenstoffverbindung *f* carbon compound, ((Karbid)) carbide
Kohlenstoffverbrauch *m* consumption of carbon
Kohlenstoffwerkzeugstahl *m* carbon tool steel
Kohlenstoffziegel *m* carbon brick
Kohlensturzbahn *f* coal chute
Kohlensuboxid *n* carbon suboxide
Kohlenteer *m* coal tar, gas tar
Kohlenteeröl *n* coal tar oil
Kohlentiegel *m* carbon crucible
Kohlentrichter *m* coal hopper
Kohlenverbrauch *m* coal consumption
Kohlenvergasung *f* coal distillation, coking, gasification of coal
Kohlenwaschmaschine *f* coal washer
kohlenwasserstoffartig hydrocarbonaceous
Kohlenwasserstoffbestimmung *f* carbohydrate determination
Kohlenwasserstoffe *pl* hydrocarbons, **aliphatische** ~ aliphatic hydrocarbons, **aromatische** ~ aromatic hydrocarbons, **gesättigte** ~ saturated hydrocarbons, **ungesättigte** ~ unsaturated hydrocarbons, **zyklische** ~ cyclic hydrocarbons
Kohlenwasserstoffgas *n* hydrocarbon gas
kohlenwasserstoffhaltig hydrocarbonaceous
Kohlenwasserstofflampe *f* hydrocarbon lamp

Kohlenwasserstoffverbindung *f* hydrocarbon
Kohlenzeche *f* (Bergb) coal mine
Kohlenzerkleinerung *f* coal breaking, coal crushing
Kohlepapier *n* carbon paper
Kohlesandgemisch *n* carbon sand mixture
Kohleschiffchen *n* carbon boat
Kohleschleifbügel *m* carbon slip bow
Kohleschleifstück *n* carbon sliding piece
Kohlestab *m* (Elektr) carbon rod
Kohlestift *m* carbon, carbon pencil, charcoal pencil
Kohletabletten *pl* (Pharm) charcoal tablets
Kohlezinkelement *n* carbon-zinc cell
Kohlung *f* carbonization, carburization, (Stahl) cementation
Kohlungsgrad *m* degree of carbonization
Kohlungsgranulat *n* carburizing granulate
Kohlungssalz *n* carburizing salt
Kohlungsstoff *m* carburization material
Kohlungszone *f* carbonizing zone, carburization zone
Kohlweißling *m* cabbage butterfly
Koinzidenzanalysator *m* coincidence analyzer
Koinzidenzschaltung *f* coincidence circuit
Kojisäure *f* kojic acid
Kokerei *f* coking plant, carbonization plant, coal carbonizing plant, coke-oven plant, (Betrieb) coking practice
Kokereiabwasser *n* coke-oven plant effluent
Kokereianlage *f* carbonization plant
Kokereigas *n* coke oven gas, town gas
Kokereitechnik *f* coal carbonizing practice
Kokereiwesen *n* coking practice, coke oven plant operation
Kokille *f* (Gieß) cast iron mold, chill mold, (Stahlwerk) ingot mold
Kokillenform *f* chill mold
Kokillengespann *n* group of chill molds
Kokillenguß *m* case-hardened casting, chill casting, (Stahlwerk) ingot casting
Kokillenlack *m* ingot mold varnish
Kokillentisch *m* ingot support
Kokkolith *m* (Min) coccolith
Ko-Kneter *m* Ko-kneader
Kokon *m* cocoon
Kokosfaser *f* coir
Kokosfett *n* coconut fat
Kokosnuß *f* coco[nut]
Kokosschrot *n* coconut meal
Koks *m* coke, ~ **ablöschen** to wet coke with water, ~ **ohne Erz** charge of coke without ore, **stückiger** ~ hard lumpy coke
Koksabfall *m* coke waste, scrap coke
Koksausbeute *f* yield of coke
Koksausdruckmaschine *f* coke pusher, coke pushing machine
Koksbereitung *f* coking
Koksbrennofen *m* coke-furnace, coking-kiln

Kokserzeugung *f* production of coke
Koksfilter *m* coke filter
Koksgabel *f* coke fork
Koksgas *n* coke gas, town gas
Koksgicht *f* coke charge
Koksgrus *m* coke breeze, coking duff
Koksheizung *f* heating with coke
Kokshochofen *m* coke [blast] furnace
Koksklassieranlage *f* coke sizing plant
Koksklein *n* coke dross
Kokskohle *f* coking coal
Kokskorb *m* coke basket
Kokskuchen *m* coke cake
Kokslagerplatz *m* coke storing place
Koksofen *m* coke oven
Koksofenbatterie *f* battery of coke oven
Koksofengas *n* coke oven gas
Koksofentür *f* door of coke oven
Kokspulver *n* coke powder
Koksroheisen *n* coke pig iron
Koksschicht *f* layer of coke
Koksstaub *m* coke dust
Kokswäscher *m* coke scrubber, coke washer
Kokumbutter *f* kokum-butter
Kokusagin *n* kokusagine
Kokusagininsäure *f* kokusagininic acid
Kolanuß *f* cola nut
Kolatur *f* filtrate, strained liquid
Kolben *m* bulb, (Chem) flask, glass flask, (Lötkolben) soldering iron, (Mech) piston, plunger, ram, (Retorte) alembic, retort, still, ~ **mit Ansatzrohr** tubulated flask, **eingeschliffener** ~ ground-in piston
Kolbenansatz *m* piston extension
Kolbenaufgang *m* piston ascent, upstroke
Kolbenbewegung *f* piston stroke
Kolbenblasenströmung *f* plug flow
Kolbenbolzen *m* gudgeon pin, piston pin
Kolbenbolzenlager *n* piston pin bushing
Kolbendampfmaschine *f* piston steam engine, reciprocating engine
Kolbendichtungsring *m* compression ring
Kolbendruck *m* piston pressure
Kolbenfläche *f* piston area
Kolbenflügel *m* piston blade, piston vane
Kolbenflüssigkeitszähler *m* displacement-type motor
Kolbenhals *m* (Chem) neck of a flask
Kolbenhingang *m* forward stroke
Kolbenhub *m* piston stroke, stroke of piston
Kolbenhubraum *m* piston stroke volume
Kolbenkompressor *m* reciprocating compressor
Kolbenkraftdiagramm *n* piston pressure diagram
Kolbenmanometer *n* piston gauge
Kolbenmembran *f* piston diaphragm
Kolbenniedergang *m* downstroke of a piston
Kolbenpressen *n* hydraulic extrusion

Kolbenpumpe f piston pump, plunger pump, reciprocating pump, **schnellaufende** ~ high-speed piston pump
Kolbenring m (Mech) piston ring
Kolbenschlüssel m piston wrench
Kolbenspiel n clearance of piston
Kolbenstange f piston rod
Kolbensteuerung f valve gear
Kolbenstrangpresse f ram extruder
Kolbenströmung f plug-flow
Kolbenstromreaktor m plug-flow reactor
Kolbenträger m (Chem) flask stand
Kolbenverdichter m reciprocating compressor
Kolbenweg m piston stroke, piston travel, stroke of piston
Kolbenweglinie f line indicating piston path
Kolbenzylinder m plug cylinder, plunger cylinder
Kolbingit m kolbingite
Kolchizin n colchicine
Kolibakterie f coli [bacillus]
Kolierapparat m filter, strainer, straining apparatus
kolieren (filtern) to filter, to percolate, to strain
Kolieren n filtering, straining
Kolierrahmen m filter frame
Koliertuch n filter cloth
Kollagen n collagen
Kollagenase f (Biochem) collagenase
Kollargol n collargol, colloidal silver
Kollektivmodell n collective model
Kollektor m (Elektr) commutator, collector, (Sammelscheibe) collector
Kollektorabdrehvorrichtung f commutator turning device
Kollektorabmessung f commutator dimension
Kollektordurchmesser m diameter of commutator
Kollektorendring m commutator endring
Kollektorgröße f size of commutator
Kollektorisolation f commutator insulation
Kollektorkonstruktion f commutator design
Kollektorkühlung f cooling of the commutator
Kollektorlänge f length of commutator
Kollektorlamellenzahl f number of commutator bars
Kollektormotor m commutator motor
Kollektormutter f commutator nut
Kollektorring m commutator ring
Kollektorschleifpapier n commutator polishing paper
Kollektorschleifvorrichtung f commutator grinding device
Kollektorschmiere f commutator lubricant
Kollektorspannung f commutator voltage
Kollergang m edge mill, edge runner mill, edge-runner mixer, muller mixer, pan grinder, pan mill, pug mill
Kollermühle f edge mill, pan grinder

kollern to grind and mix, to grind on the edge runner
Kollerung f grinding on the edge runner, mixture ground on the edge runner
Kollerzeit f mulling time
Kolleschale f Kolle culture flask
kollidieren to collide, to clash, (Auto) to crash [against]
Kollidin n collidine
Kollimation f collimation
Kollimator m collimator
Kollineation f collineation
Kollision f collision, **unelastische** ~ inelastic collision
Kollisionsdichte f (Atom) collision density
Kollisionszahl f (Atom) collision frequency per unit
Kollodion n collodion
Kollodium n collodion, collodium
kollodiumähnlich collodion-like
Kollodiumbaumwolle f collodion cotton
Kollodiumbild n collodiotype
Kollodiumersatz m collodion substitute
Kollodiumfasermasse f collodion wool
Kollodiumleitung f collodion passage
Kollodiumlösung f collodion solution
Kollodiumplatte f collodion plate
Kollodiumschicht f collodion film
Kollodiumseide f collodion silk, nitrate rayon
Kollodiumüberzug m collodion film
Kollodiumverfahren n collodion process
Kollodiumwolle f collodion cotton, collodion wool, collodium wool, nitrocellulose, pyroxylin, soluble gun cotton
Kolloid n colloid, colloidal substance, **hydrophiles** ~ hydrophilic colloid, **hydrophobes** ~ hydrophobic colloid, **lyophiles** ~ lyophilic colloid, **lyophobes** ~ lyophobic colloid, **natürliches** ~ natural colloid
kolloidal colloid, colloidal
Kolloidbeständigkeit f colloid stability
Kolloidcharakter m colloidality
Kolloidchemie f colloid chemistry
kolloidchemisch colloidochemical
Kolloiddiffusion f colloid diffusion
kolloiddispers colloidally disperse
Kolloidfarbe f colloidal color
Kolloidgel n colloidal gel
Kolloidgleichrichter m colloid rectifier
Kolloidkohle f colloidal coal
Kolloidmühle f colloid mill
Kolloidschwefel m colloidal sulfur
Kolloidstoff m colloid, colloidal substance
Kolloidteilchen n colloidal particle
Kolloidzusatz m addition of a colloid
Kolloidzustand m colloidal state
Kollokation f collocation
Kollotypie f (Phot) collotypy

Kollyrit *m* (Min) collyrite, allophane
Kolombinrot *n* columbine red
Kolonne *f* column, **atmosphärische** ~ atmospheric tower or still, **chromatographische** ~ chromatographic column, **Vakuum-** ~ vacuum tower
Kolonnenapparat *m* column apparatus
Kolonnenboden *m* plate, tray
Kolonnenende *n* column head
Kolonnenfüllmaterial *n* packing for fractionating columns
Kolonnenkopf *m* head of column
Kolonnenmantel *m* column shell
Kolonnenpackung *f* (Dest) column packing
Kolonnensumpf *m* bottom of column, sump of column
Kolonnenvolumen *n* column volume
Kolophen *n* colophene
Kolopholsäure *f* colophonic acid, colopholic acid
Kolophoneisenerz *n* (Min) pitchy iron ore, pitticite
Kolophonit *m* (Min) colophonite
Kolophonium *n* colophonium, colophony, pine resin, rosin
Kolophonon *n* colophonone
Kolophonsäure *f* colopholic acid, colophonic acid
Koloquintin *n* colocynthin
kolorieren to color
Kolorieren *n* coloration
Kolorimeter *n* colorimeter
Kolorimetrie *f* colorimetry
kolorimetrisch colorimetric
Kolorit *n* color, coloring, shade
Kolostrum *n* (Med) colostrum
Kolumbiagrün *n* Columbia green
Kolumbiaschwarz *n* Columbia black
Kolumbit *m* (Min) columbite
Kolumnenmaß *n* rule, scale
Komansäure *f* comanic acid
Komarit *m* (Min) komarite
Kombibeizmittel *n* combination seed disinfecting agent
Kombidose *f* composite container
Kombination *f* combination, ~ **k-ter Klasse** combination of n dissimilar numbers taken k at a time
Kombinationsfolie *f* laminated film
Kombinationslackfarbe *f* nitrosynthetic lacquer
Kombinationsleitwert *m* combined conductivity
Kombinationsmatrize *f* (Math) compound matrix
Kombinationswiderstand *m* combined resistance
Kombinatorik *f* combinatorial analysis, theory of combinations
kombinieren to combine
Komenaminsäure *f* comenamic acid
Komensäure *f* comenic acid
Kometenbahn *f* cometary orbit

Komma *n* comma, (Math) decimal point
Kommissionierbehälter *m* order-pick container
Kommissionierbereich *m* order-picking section
Kommissionierfördersystem *n* rack-running order-picking system
Kommissioniergerät *n* order picker
Kommissionierlager *n* order-picking warehouse
Kommissionierstapler *m* order stacker
Kommissionierstollen *pl* order-picking bays
kommunizierend communicating
kommutativ commutative
Kommutativität *f* (Math) commutative law
Kommutator *m* (Math, Elektr) commutator
Kommutatorgleichrichter *m* commutator rectifier, permutator
kommutieren to commute
kommutierend commutating
Kommutierungsversuche *pl* redistribution experiments
Komparator *m* (Techn) comparator
Komparatorschaltung *f* (Elektr) comparator, comparator circuit
Kompaß *m* compass
Kompaßabweichung *f* compass deviation
Kompatibilität *f* compatibility
Kompatibilitätsbedingung *f* compatibility condition
Kompatibilitätsrelation *f* compatibility relation
Kompensation *f* compensation
Kompensationsmethode *f* compensation method
Kompensationspol *m* compensating pole
Kompensationsregler *m* potentiometric controller
Kompensationsspannung *f* (Elektr) balancing voltage
Kompensationstheorem *n* (Elektr) compensation theorem
Kompensationsverfahren *n* (Elektr) compensation method
Kompensationswiderstand *m* (Elektr) compensating resistance
Kompensator *m* compensator, (Elektr) potentiometer
kompensieren to compensate
komplementär complementary
Komplementärfarbe *f* complementary color
Komplementärfarbigkeit *f* (Opt) complementary chromaticity
Komplementärmenge *f* complementary set
Komplementärwinkel *m* (Math) complementary angle
Komplementarität *f* complementarity
komplett complete, entire
komplex complex, ~ **binden** to complex, ~ **konjugiert** complex conjugated
Komplex *m* complex, (Chelat) chelate, **aktivierter** ~ activated complex, **beständiger** ~ stable complex, **inerter** ~ inert

complex, **ionogener** ~ ionogenic complex, **unbeständiger** ~ unstable complex
Komplexbildner m chelating agent, complexing agent
Komplexbildung f complex formation, complexing
Komplexchemie f complex chemistry
Komplexion n complex ion
komplexometrisch complexometric
Komplexon n complexon
Komplexreaktion f complex chemical reaction
Komplexsalz n complex salt
Komplexverbindung f complex compound
Komplexzahl f (Math) complex number
Komplikation f complication
komplizieren to complicate
kompliziert complicated, intricate
Komponente f component, (einer Legierung) constituent, **flüchtige** ~ volatile component
Komponentenwiderstand m component resistance
Kompositionsmetall n composition metal
Kompositionssatz m detonating composition
Kompost m (Agr) compost
Kompoundmotor m (Elektr) compound [-wound] motor
kompressibel compressible
Kompressibilität f compressibility, **isentropische** ~ isentropic compressibility, **isotherme** ~ isothermal compressibility
Kompressibilitätseinfluß m compressibility effect, effect of compressibility
Kompressibilitätsfaktor m (Mech) compressibility factor
Kompression f compression, **adiabatische** ~ (Therm) adiabatic compression
Kompressionsdruck m compression pressure
Kompressionsentspannung f decompression
Kompressionsgrad m degree of compression
Kompressionskältemaschine f compression refrigerating machine
Kompressionskühlschrank m compression refrigerator
Kompressionsmodul m bulk modulus
Kompressionsmotor m supercharged engine
Kompressionsraum m compression space
Kompressionsvakuummeter n compression vacuum gauge
Kompressionsverhältnis n compression ratio
Kompressionsverlust m loss of compression
Kompressionswärme f compression heat, heat of compression
Kompressionswiderstand m compression resistance
Kompressionszündung f ignition by compression
Kompressor m compressor
Kompressorenöl n compressor oil
Kompressormotor m supercharged motor
komprimierbar compressible

Komprimierbarkeit f compressibility
komprimieren to compress, (Holz) to compregnate
Komprimieren n compressing
komprimierend compressing
Komprimierung f compression
Konarit m konarite
Konche f conche
konchieren (Schokolade) to conche
Kondensabscheider m condensate separator
Kondensanz f capacitance, reactance
Kondensat n condensate, (Kondensationsprodukt) condensation product
Kondensatabscheider m condensate separator
Kondensatbehandlung f condensate treatment
Kondensation f condensation, **zwischenmolekulare** ~ intermolecular condensation
Kondensationsanlage f condenser unit, condensing plant
Kondensationsapparat m condenser, condensing apparatus
Kondensationsblase f condensing kettle
Kondensationsdampfmaschine f condensing engine
Kondensationsgefäß n condensing vessel, receiver
Kondensationsharz n condensation resin
Kondensationskammer f condensation chamber, condensing chamber
Kondensationskern m condensation nucleus
Kondensations-Polymerisation f condensation polymerization
Kondensationsprodukt n condensation product, (Kunststoff) condensation polymer
Kondensationspumpe f condensation pump
Kondensationspunkt m condensation point
Kondensationsraum m condensing chamber
Kondensationsrohr n condensation tube, condensing tube
Kondensationsturm m condensing tower, cooling tower
Kondensationsverlust m condensation loss, loss due to condensation
Kondensationswärme f heat of condensation
Kondensationswasser n condensed water, condensing water
Kondensationswasserabscheider m steam trap
Kondensator m condenser, (Elektr) capacitor, **leistungsstarker** ~ heavy-duty condenser
Kondensatorelektroskop n condenser electroscope
Kondensatorelement n condenser battery
Kondensatorentladung f discharge of a capacitor
Kondensatorkasten m condenser receiver
Kondensatorleistung f capacitor rating
Kondensatorpapier n condenser paper
Kondensatorplatte f condenser plate

Kondensatorröhre *f* condenser pipe, condenser tube
Kondensatorrohr *n* condenser pipe, condenser tube
Kondensatortemperatur *f* condenser temperature
Kondensatorwirkung *f* condenser effect
Kondensatpumpe *f* condensate pump, condensed steam pump
Kondensatsammelbehälter *m* condensate collector
Kondensatsammelgefäß *n* condensation trap, condensation accumulator vessel
Kondensatteiler *m* condensate divider
kondensierbar condensable
Kondensierbarkeit *f* condensability
kondensieren to condense, to liquefy, (eindicken) to concentrate
Kondensieren *n* condensing
kondensierend condensing
kondensiert condensed
Kondensierung *f* condensation
Kondensmilch *f* condensed milk, evaporated milk
Kondensor *m* (Opt) condenser
Kondensorlinse *f* (Opt) condensing lens, focus lens
Kondensstreifen *m* (Flugzeug) condensation trail
Kondenstopf *m* [steam] trap
Kondenswasser *n* condensation water, condensed water, condensing water
Kondenswasserentöler *m* condensate deoiling plant
Kondenswasserfilter *n* filter for condensation water
Konditionieranlage *f* (Text) conditioning plant
Konditionieranstalt *f* conditioning house, testing chamber
Konditionierapparat *m* conditioner
konditionieren to condition
Konditionieren *n* conditioning
Konditioniervorrichtung *f* conditioner
Konditionierwaage *f* conditioning balance
Konduktanz *f* (Elektr) conductance
Konduktionsmotor *m* conductive motor
Konduktometer *n* conductimeter, conductometer
Konduktometrie *f* conductometric analysis
konduktometrisch conductimetric, conductometric
Konduktor *m* conductor
Kondurangofluidextrakt *m* liquid extract of condurango
Kondurit *m* conduritol
Konfekt *n* confectionery, sweets
Konfektionierung *f* ancillary processing
Konfektionierungsabteilung *f* packaging department
Konfidenzintervall *n* (Statist) confidence limits
Konfidenzzahl *f* (Statist) confidence level

Konfiguration *f* configuration, **absolute** ~ absolute configuration, **ebene** ~ configuration in the plane, **geknickte** ~ puckered configuration, **räumliche** ~ configuration in space, **relative** ~ relative configuration
Konfigurationsbaustein *m* configurational base unit
Konfigurationsbestimmung *f* determination of configuration
Konfigurationsentropie *f* configurational entropy
Konfigurationsisomer *n* geometric isomer
Konfigurationsraum *m* configuration space
Konfigurationsumkehrung *f* inversion of configuration
Konfigurationsverwandtschaft *f* configurational relationship
Konfigurationswechselwirkung *f* configuration interaction
Konfitüre *f* jam
Konformation *f* conformation, **doppelt windschiefe** ~ doubly skewed conformation, **ekliptische** ~ eclipsed conformation, **gestaffelte** ~ staggered conformation
Konformationsänderung *f* (Biochem) conformational change
Konformationsanalyse *f* conformational analysis
Konformationswechsel *m* conformational interconversion
Konformer *n* conformational isomer, conformer
kongenital (Med) congenital, inborn
Konglomerat *n* (Geol) conglomerate
Konglomeratgefüge *n* conglomerate structure
konglomerieren to conglomerate
Konglutinin *n* conglutinin
Kongoblau *n* Congo blue, trypan blue
Kongoechtblau *n* Congo fast blue
Kongofarbe *f* Congo color
Kongogummi *n* Congo rubber
Kongopapier *n* Congo paper
Kongorot *n* Congo red
Kongorotlösung *f* Congo red solution
kongruent congruent
Kongruenz *f* (Math) congruence
Kongruenzannahme *f* assumption of congruence
Kongruenzschmelzpunkt *m* (Phys) congruent melting point
Kongruenztransformation *f* (Math) congruent transformation
kongruieren to agree, to be congruous, to coincide
Konichalcit *m* (Min) conichalcite
Koniin *n* coniine, 2-propylpiperidine, cicutine, conine
Konimeter *n* conimeter, konimeter
konimetrisch konimetric
Koninckit *m* (Min) koninckite
konisch conic, conical, coniform, tapered
Konit *n* conite, konite
Konizität *f* conicity, conical taper, tapering

Konjugation *f* conjugation
Konjugationseffekt *m* conjugative effect
konjugiert (Chem) conjugated, ~ **komplex** conjugate complex
konkav concave
Konkavgitter *n* concave grating
Konkavgitterspektrograph *m* concave gitter spectrograph
konkav-konvex (Opt) concavo-convex
Konkavlinse *f* (Opt) concave lens, diverging lens, negative lens
Konkavspiegel *m* concave mirror
Konkurrent *m* competitor
Konkurrenz *f* competition, rivalry
konkurrenzfähig competitive, able to compete, marketable
konkurrenzlos non-competitive, exclusive
konkurrieren to compete
konkurrierend competing, competitive
Konnodalkurve *f* tie lines
Konormale *f* conormal
Konsequenz *f* consequence
Konserve *f* tinned or canned food, (Eingemachtes) preserve[s]
Konservendose *f* open top can, tin [can]
Konservenfabrik *f* cannery, canning factory
Konservenhersteller *m* canner
Konservenindustrie *f* canning industry
Konservenvergiftung *f* poisoning caused by canned food
konservieren to conserve, to preserve, (in Dosen) to can
Konservieren *n* preserving
Konservierung *f* conservation, preservation
Konservierungsmethode *f* preservative method
Konservierungsmittel *n* microbicide, preservative, preserving agent
Konservierungsverfahren *n* preserving process
konsistent consistent, compact, firm, solid
Konsistenz *f* consistency
Konsistenzmaß *n* measure of consistency
Konsistenzmesser *m* consistometer
Konsistenzparameter *m* consistency variable
Konsole *f* console, bracket, truss
konsolidieren to consolidate
Konsolkran *m* bracket crane
Konsonanz *f* (Akust) consonance
konstant constant, invariable
Konstantan *n* constantan
Konstante *f* constant, ~ **der Oberflächenspannung** (Phys) coefficient of surface tension, **Avogadrosche** ~ Avogadro constant, **Faradaysche** ~ Faraday constant, **Loschmidtsche** ~ Loschmidt constant, Avogadro constant
Konstanterregung *f* constant excitation
konstant halten to keep constant, to maintain constant
Konstanthaltung *f* stabilization, stabilizing

Konstanz *f* constancy
Konstanzelement *n* element of constancy
konstatieren to ascertain, to state, to verify
Konstitution *f* constitution
Konstitutionseinfluß *m* constitutional influence
Konstitutionsformel *f* constitutional formula
Konstitutionsisomerie *f* constitutional isomerism
Konstitutionswasser *n* water of constitution
konstringieren (Med) to constringe
konstringierend constringent
konstruierbar constructable
konstruieren to construct, to design
Konstrukteur *m* designer, designing engineer
Konstruktion *f* construction
Konstruktionsfehler *m* error of construction
Konstruktionsglied *n* unit or element of construction
Konstruktionsmerkblatt *n* engineering table
Konstruktionsschiene *f* structural rail
Konstruktionsteil *n* structural member
Konstruktionsverleimung *f* general joinery
Konstruktionswerkstoff *m* constructional material
Konsum *m* consumption
Konsumdichte *f* density of consumption
Konsument *m* consumer
Konsumgebiet *n* area of consumption, area of supply
Konsumschwerpunkt *m* center of consumption [area]
Konsumstelle *f* place of supply
Konsumzentrum *n* center of consumption
Kontakt *m* catalyst, contact, **den** ~ **schließen** to close the contact, **federnder** ~ flexible contact, spring contact
Kontaktabdruckmethode *f* contact print method
Kontaktabstand *m* contact spacing
Kontaktanlage *f* contact plant
Kontaktansiedevernicklung *f* hot contact nickelling
Kontaktbelastung *f* catalyst severity or throughput
Kontaktbestrahlung *f* (Med) x-ray contact therapy
Kontaktboden *m* contact plate
Kontaktbügel *m* (Elektr) bow collector
Kontaktdraht *m* contact wire
Kontaktelektrode *f* contact electrode
Kontaktfeder *f* contact spring
Kontaktfinger *m* contact finger, contact blade
Kontaktfläche *f* area of contact, contact interface, contact surface
Kontaktgalvanisierung *f* contact plating, galvanizing by contact
Kontaktgeber *m* contactor
Kontaktgift *n* catalyst poison, (Schädlingsbekämpfung) contact poison
Kontaktgleichrichter *m* crystal detector or rectifier, contact detector

Kontaktgrundverfahren *n* reactive ground coat technique
Kontakthemmung *f* contact inhibition
Kontaktkatalyse *f* contact or surface catalysis
Kontaktkessel *m* converter
Kontaktklebstoff *m* contact adhesive
Kontaktknopf *m* contact button
Kontaktkorrosion *f* contact corrosion (corrosion at a contact with a second metal)
Kontaktleistung *f* space time yield
Kontaktmittel *n* catalyst, contact substance
Kontaktofen *m* catalytic furnace, reactor
Kontaktplatte *f* contact plate
Kontaktpotential *n* contact potential
Kontaktpressen *n* impression molding
Kontaktpulver *n* contact powder
Kontaktraum *m* contact or catalyst space
Kontaktrauschen *n* (Elektr) contact murmur
Kontaktregler *m* contact-type regulator
Kontaktrolle *f* contact roller, trolley, trolley wheel
Kontaktsäure *f* contact acid
Kontaktschalter *m* (Elektr) microswitch
Kontaktschlitten *m* contact carriage, brush rod
Kontaktspannung *f* contact voltage
Kontaktstelle *f* contact point
Kontaktstöpsel *m* contact plug
Kontaktstrom *m* contact current
Kontakttheorie *f* contact theory
Kontaktthermometer *n* contact thermometer
Kontakttrockner *m* contact dryer
Kontaktunstetigkeit *f* contact discontinuity
Kontaktverfahren *n* catalytic process, contact process
Kontaktvergiftung *f* catalyst poisoning
Kontaktvergoldung *f* gilding by contact
Kontaktverkupferung *f* contact copper-plating
Kontaktversilberung *f* silvering by contact
Kontaktweite *f* contact clearance
Kontaktwiderstand *m* contact resistance
kontaktwirksam catalytic
Kontaktwirkung *f* contact action, contact effect
Kontaktzeiger *m* contact pointer
Kontamination *f* contamination, pollution
kontaminieren to contaminate
Kontinentalrand *m* continental margin
Kontinentalscholle *f* (Geol) continental block
Kontinentalverschiebung *f* continental drift
kontinuierlich continuous
Kontinuität *f* continuity
Kontinuitätsgleichung *f* continuity equation, principle of continuity
Kontinuum *n* continuous spectrum, continuum
kontrahieren to contract
Kontrahieren *n* contracting
Kontraindikation *f* (Med) contra-indication
Kontraktion *f* contraction
Kontraktionstheorie *f* contraction theory
Kontrast *m* contrast

kontrastarm (Phot) low-contrast
Kontrastdarstellung *f* (Med) contrast radiograph
Kontrastempfindlichkeit *f* contrast sensibility
Kontrastfärbung *f* contrast staining, counter staining
Kontrastfilter *n* contrast filter
kontrastieren (Biol) to stain
Kontrastlösung *f* (Röntgenologie) contrast solution
Kontrastmethode *f* (Röntgenologie) contrast medium method
Kontrastmittel *n* contrast medium
Kontrastphotometer *n* contrast photometer
kontrastreich high-contrast
Kontrastverfahren *n* (Röntgenologie) contrast roentgenography
Kontrastverhältnis *n* contrast ratio
Kontrastwiedergabe *f* contrast rendition
Kontravalenz *f* contravalence
kontravariant contravariant
Kontrollampe *f* control lamp, signal lamp, telltale light
Kontrollanalyse *f* check analysis, control analysis
Kontrollanlage *f* checking equipment
Kontrollauslöseknopf *m* control reset button
Kontrollbestimmung *f* check or control determination, duplicate determination
Kontrolle *f* checking, control, inspection, monitoring, supervision, (Steuerung) control
Kontrolleur *m* controller
Kontrollgerät *n* monitor
Kontrollgewicht *n* check weight
kontrollierbar controllable
kontrollieren to check, to control, to verify, (beaufsichtigen) to supervise
Kontrollkolben *m* check flask
Kontrollmanometer *n* controlling pressure gauge
Kontrollpyrometer *n* control pyrometer
Kontrollreihe *f* control series
Kontrollrost *m* control grid
Kontrollschalter *m* safety switch
Kontrollstab *m* standard bar
Kontrolltier *n* control animal
Kontrollversuch *m* check test
Kontrollvorrichtung *f* control[ling] device, control gear
Kontur *f* contour, outline, ~ **des Gesenks** impression of a die
Konturen- und Hautpackungen *pl* bubble and skin packaging
Konturpackung *f* skin pack
Konus *m* cone
Konuskreiselmischer *m* cone impeller
Konusmembran *f* cone membrane
Konusmühle *f* conical mill
Konusrollenlager *n* taper roller bearing
Konus[sink]scheider *m* (Schwertrübetrennung) conical dense-medium vessel

Konvektion – koordinativ

Konvektion *f* convection, **erzwungene** ~ forced convection, **freie** ~ free convection
Konvektionsheizung *f* convection heating
Konvektionsstrom *m* convection current
Konvektionswärme *f* convection heat
konvektiv convective
konventionell conventional[ly]
konvergent convergent
Konvergenz *f* convergence, convergency, ~ **einer Reihe** (Math) convergence of a series
Konvergenzdruck *m* convergence pressure
Konvergenzgebiet *n* region of convergence
Konvergenzintervall *f* (v. Potenzreihen) range of convergence (of power series)
Konvergenzkreis *m* circle of convergence
Konvergenzwinkel *m* convergence angle
konvergieren to converge [to]
Konversionsfaktor *m* (Atom) conversion factor
Konversionskoeffizient *m* conversion ratio
Konversionssalpeter *m* conversion saltpeter
Konverter *m* converter, **feststehender** ~ fixed or stationary converter, **kippbarer** ~ tilting or tipping converter, **liegender** ~ barrel or horizontal converter
Konverteranlage *f* converter plant
Konverterauskleidung *f* converter lining
Konverterauswürfe *m pl* converter waste
Konverterbetrieb *m* (Metall) converter practice
Konverterbirne *f* converter
Konverterboden *m* converter bottom
Konverterbühne *f* converter charging platform or gallery
Konvertereinsatz *m* converter charge
Konverterfilm *m* converter film
Konverterfutter *n* converter lining
Konverterhalle *f* converter shed, converter shop
Konverterkupfer *n* converter copper
Konverterprozeß *m* converter process
Konverterreaktor *m* converter reactor
Konverterring *m* converter trunnion ring
Konvertgas *n* converted gas
konvertierbar convertible
konvertieren to convert
Konvertierung *f* conversion
konvex convex
konvex-konkav (Opt) convexo-concave
Konvexlinse *f* (Opt) converging lens, convex lens, positive lens
Konvexspiegel *m* (Opt) convex mirror, convex reflector
Konzentrat *n* concentrate
Konzentration *f* concentration, ~ **der Lösung** concentration of solution, strength of solution, **molare** ~ molar concentration
Konzentrationsabhängigkeit *f* dependence of concentration
Konzentrationsänderung *f* alteration of concentration, concentration change
Konzentrationsanlage *f* concentrating plant
Konzentrationsapparat *m* concentrator
Konzentrationsbereich *m* range of concentration
Konzentrationseinheit *f* concentration unit
Konzentrationselement *n* concentration cell
Konzentrationsgefälle *n* concentration gradient
Konzentrationskette *f* (Elektrochem) concentration cell
Konzentrationspolarisation *f* concentration polarization
Konzentrationsschmelzen *n* concentration melting
Konzentrationsstrom *m* concentration current
Konzentrationsüberschuß *m* concentration excess
Konzentrationsverhältnis *n* ratio of concentration
Konzentrationsverteilung *f* concentration distribution
konzentrieren to concentrate
Konzentrieren *n* concentrating
konzentriert concentrated
Konzentrierung *f* concentration, enrichment, (Sättigung) saturation
Konzession *f.* concession, licence (Br. E.), license (Am. E.), permission
konzessionieren to grant a concession or license, to license
Konzessionsbeschränkung *f* restriction of license or concession
konzis concise
kooperieren to cooperate
Koordinate *f* coordinate
Koordinatenachse *f* coordinate axis
Koordinatenachsen *f pl* system of coordinates
Koordinatenanfangspunkt *m* origin of coordinates
Koordinatenebene *f* coordinate plane
Koordinatenmanipulator *m* rectilinear manipulator
Koordinatennullpunkt *m* origin of coordinates
Koordinatenschreiber *m* coordinate recorder
Koordinatensystem *n* coordinate system, system of coordinates
Koordinatentransformation *f* transformation of coordinates
Koordinationsart *f* state of coordination
Koordinationsformel *f* coordination formula
Koordinationsgitter *n* coordination lattice
Koordinationsisomerie *f* coordination isomerism
Koordinationslehre *f* coordination theory
Koordinationspolyeder *n* (Krist) coordination polyhedron
Koordinationsrichtung *f* coordinate direction
Koordinationstheorie *f* coordination theory
Koordinationsverbindung *f* (Chem) coordination compound
Koordinationszahl *f* (Krist) coordination number, index of coordination
koordinativ coordinative

koordinieren to coordinate
Koordinierung *f* coordination
Kopal *m* (trop. Harz) copal, kopal
Kopalersatz *m* copal substitute
Kopalfirnis *m* copal varnish
Kopalin *n* (Min) copalin, copalite, mineral resin
Kopalinbalsam *m* copalin balsam
Kopallack *m* copal lacquer, copal varnish
Kopalöl *n* copal oil
Kopalsäure *f* copalic acid
Kopalschmelzanlage *f* copal resin melting plant
Kopalveresterungsanlage *f* copal esterification plant
Kopellidin *n* copellidine, 2-ethyl-6-methylpiperidine
Kopf *m* head, top, (Kolonne) head of column, ~ **einer Tabelle** heading of a table
Kopfbrenner *m* end burner
Kopfbügel *m* head-band
Kopfdünger *m* top fertilizer
Kopfkontakt *m* upper connection
Kopfkreis *m* addendum circle (gears), top circle
kopflastig topheavy, noseheavy
Kopflastigkeit *f* topheaviness, noseheaviness
Kopfplatte *f* head plate, cross head, lips (for sheeting or film dies), top plate
Kopfprodukt *n* fore-runnings, (Dest) head product, top product
Kopfraum *m* (bei gefüllten Packungen) headspace
Kopfschließdruck *m* closed-in pressure (at well head)
Kopfschmerzen *pl* headache
Kopfschraube *f* screw with head, bolt with head, cap screw, ~ **mit Schlitz** grub screw, ~ **mit Zylinderkopf** square head cap screw
Kopfschutz *m* helmet
Kopfspiel *n* [tip] clearance (of gear)
Kopfsteinpflaster *n* [cobble] stone pavement, rubble pavement
Kopftemperatur *f* (Destill) still temperature
Kopftrommel *f* head drum, head pulley
Kopie *f* copy, carbon copy, duplicate, (Phot) print, reproduction
Kopierapparat *m* copying apparatus
kopieren to copy, to make a tracing, (Phot) to print
Kopierfräsmaschine *f* copying mill, copying milling machine; duplicating mill, duplicating milling machine
Kopiergerät *n* copying apparatus, **photographisches** ~ photocopying apparatus, photostatic copying machine
Kopierpapier *n* copying paper
Kopiertinte *f* copying ink
Kopierverfahren *n* copying process
Kopiervorrichtung *f* coppying attachment
Koppel *f* (Riemen) coupler
Koppelausleger *m* coupled jib

Koppelgetriebe *n* linkage gear
koppeln to couple, to connect, to join, to tie, to unite
Koppeln *n* coupling, connecting, joining
Koppit *m* (Min) coppite
Kopplung *f* coupling, connection, linkage, **wilde** ~ (Elektr) cross coupling
Kopplungsfaktor *m* coupling factor
Kopplungskonstante *f* coupling constant
Kopplungsmatrix *f* dynamic or coupling matrix
Kopra *f* copra
Kopraöl *n* coconut oil
Koprolith *m* (Geol) coprolite
koprophil (Bakt) coprophilic, coprophilous
Koproporphyrin *n* coproporphyrin[e]
Koprostan *n* coprostane
Koprostanol *n* coprostanol, coprosterol, dihydrocholesterol, stercorol
Koprostansäure *f* coprostanic acid
Kopsan *n* kopsane
Kopsin *n* kopsine
Kopsinan *n* kopsinane
Kopsinilam *n* kopsinilam
Kopsinin *n* kopsinin
Koralle *f* coral
Korallenachat *m* (Min) coral agate
Korallenerz *n* (Min) coral ore
Korallenfarbe *f* coral color
korallenfarben coral-colored
Korallenlack *m* coral lac
Korallenpulver *n* coral powder
Korallenrot *n* coral red
Korallenversteinerung *f* corallite
Korallin *n* corallin
Korb *m* basket
Korbbandextrakteur *m* basket extractor
Korbflasche *f* carboy, bottle in a wicker case, demijohn
Korbkreiselmischer *m* cyclone impeller
Korbpresse *f* curb press, wine press
Korbsichter *m* classifier with circumferential screen
Korbzentrifuge *f* basket centrifuge
Kordel *f* cord, twine
Kordelriemen *m* rope belt
Kordofangummi *n* cordofan gum, kordofan gum
Kordonettseide *f* (Text) purse silk
Korinthe *f* dried currant
Kork *m* cork, **gepreßter** ~ agglomerated cork
korkähnlich cork-like, corky, suberose, suberous
korkartig cork-like, corky
Korkasbest *m* mountain cork, rock cork
Korkbaum *m* cork tree
Korkbohrer *m* cork borer
Korkbohrerschärfer *m* cork borer sharpener
Korkbohrmaschine *f* cork boring machine
Korken *m* cork [stopper]
Korkenzieher *m* corkscrew
korkig cork-like, corky, suberose

Korkisolation – Korrosion

Korkisolation *f* cork insulation
Korkkohle *f* burnt cork, cork charcoal
Korkmehl *n* cork powder
Korkmesser *n* cork knife
Korkplatte *f* (Bauw) cork plate, cork slab
Korkpresse *f* cork press
Korksäure *f* suberic acid, octanedioic acid
korksauer suberate
Korkscheibe *f* cork disc, cork pulley, sheet cork
Korkschleifscheibe *f* cork polishing wheel
Korkschwarz *n* cork black
Korkstein *m* cork board, cork brick
Korkstoff *m* suberin
Korkstopfen *m* cork [stopper]
Korkverarbeitung *f* cork processing
Korkzange *f* cork pliers
Korn *n* (Getreide) cereals, corn, grain, (Kristall) crystal, grain, ~ **einer photographischen Schicht** grain of a photographic layer, **feines** ~ fine grain, **grobes** ~ coarse grain
Kornalkohol *m* grain alcohol, grain spirit
Kornanhäufung *f* grain aggregation
Kornbeschaffenheit *f* character of grain
Kornbildung *f* granulation
Kornblei *n* grain lead, assay lead
Kornblumenblau *n* cornflower blue
Kornbrand *m* blight, smut
Kornbranntwein *m* corn brandy, grain spirits
Korndichte *f* grain density
Korndurchmesser *m* grain diameter
Korneisen *n* granular fracture iron
Kornfäule *f* blight, smut
Kornfeinheit *f* fineness of grain
Kornfreiheit *f* absence of coarse particles, freedom from bits
Kornfuselöl *n* fusel oil from grain
Korngefüge *n* grain structure
Korngrenzdiffusion *f* grain-boundary diffusion
Korngrenze *f* (Krist, Metall) crystal boundary, grain boundary
Korngrenzendiffusion *f* grain boundary diffusion
Korngrenzenergie *f* energy of grain boundary
Korngrenzenkorrosion *f* intergranular corrosion, intercrystalline corrosion
Korngrenzenrelaxation *f* (Krist) grain-boundary relaxation
Korngrenzenwanderung *f* grain boundary migration
Korngröße *f* grain size, granular size, particle size, size of grain, ~ **des Sandes** size of sand grain
Korngrößenbestimmung *f* particle size determination
Korngrößentrennung *f* grading, screening, sizing
Korngrößenverteilung *f* particle size distribution, grading curve
Kornit *n* cornite
Kornklassierung *f* screening, screen sizing
Kornkupfer *n* granulated copper

Kornpuddeln *n* puddling for crystalline iron
Kornscheide *f* cut size (screening)
Kornschnaps *m* corn brandy, corn spirits, grain spirits, whisky
Kornstahl *m* granulated steel
Kornstruktur *f* grain structure, granulated structure
Kornutin *n* cornutine
Kornverfeinerung *f* grain refinement, crystalline refinement
Kornvergrößerung *f* coarsening of grain, grain coarsening
Kornwaage *f* grain scales
Kornwachstum *n* crystalline growth, grain growth, growth of crystals
Kornzahl *f* grain number
Kornzinn *n* grain tin
Kornzucker *m* granulated sugar
Korona *f* corona, ~ **der Zündung** preonset corona
Koronabogen *m* corona arc
Koronadurchbruch *m* pulse corona
Koronaeffekt *m* corona effect
Koronaentladung *f* corona, corona discharge
Koronalinie *f* coronal line
Korpuskel *n* corpuscle, particle
korpuskular corpuscular
Korpuskularstrahlung *f* corpuscular radiation
Korpuskulartheorie *f* corpuscular theory, ~ **des Lichtes** (Opt) corpuscular theory of light
korrekt correct
Korrektionsfaktor *m* correction factor
Korrektivmittel *n* (Med) corrective
Korrektur *f* **der Badlösung** adjustment of the bath
Korrelation *f* correlation
Korrelationseinfluß *m* correlation effect
Korrelationsenergie *f* correlation energy
Korrelationskoeffizient *m* correlation factor
korrelieren to correlate
Korrespondenz *f* correspondence
Korrespondenzprinzip *n* (Quant) correspondence principle
korrespondieren to correspond
korrespondierend corresponding, (Lösung) isohydric
korrodierbar corrodible, corroding
Korrodierbarkeit *f* corrodibility
korrodieren to corrode
Korrodieren *n* corroding
korrodierend corrosive, corroding
korrodiert corroded
Korrosion *f* corrosion, rusting, ~ **unter gleichzeitiger mechanischer Beanspruchung** corrosion under mechanical stress, **atmosphärische** ~ atmospheric corrosion, **elektrochemische** ~ electrochemical corrosion, **Hochtemperatur-** ~ high-temperature corrosion, **interkristalline** ~

intercrystalline corrosion, intergranular corrosion, **Lochfraß-** ~ fretting corrosion, pitting corrosion, **örtliche (lokale)** ~ localized or selective corrosion, **selektive** ~ selective corrosion, **Tieftemperatur-** ~ low-temperature corrosion, **transkristalline** ~ transcrystalline corrosion, intracrystalline corrosion
Korrosionsangriff m attack by corrosion
korrosionsbeständig corrosion-proof, corrosion-resistant, non-corrodible, resistant to corrosion, (Stahl) stainless
Korrosionsbeständigkeit f corrosion-resisting quality, rustproof property, rust-resisting property, stainless property
Korrosionseinfluß m corrosive effect
Korrosionselement n corrosion cell
korrosionsempfindlich corrosion-prone, susceptible to corrosion
Korrosionsempfindlichkeit f corrodibility, sensitiveness to corrosion
Korrosionsermüdung f corrosion fatigue
Korrosionserscheinung f corrosion phenomenon, form of corrosion
korrosionsfest corrosion resistant
Korrosionsfestigkeit f corrosion resistance
korrosionsfördernd stimulating corrosion
korrosionsfrei noncorroding, noncorrosive, rustless, stainless
korrosionshemmend inhibiting corrosion
Korrosionsinhibitor m corrosion inhibitor
Korrosionsmittel n corrosive, corrosive agent
Korrosionsnarbe f corrosion pit
Korrosionsneigung f susceptibility to corrosion, tendency to corrosion, tendency to rust
Korrosionsproblem n corrosion problem
Korrosionsprüfung f corrosion testing
Korrosionsschutz m anticorrosive, corrosion protection, rust protection
Korrosionsschutzfarbe f anticorrosive paint
Korrosionsschutzmittel n anticorrosion additive, anticorrosive [agent], corrosion inhibitor, corrosion preventive
Korrosionsschutzöl n corrosion preventive oil
Korrosionsschutzpapier n vapor-phase inhibitor paper
Korrosionsschutzstab m stick of corrosion inhibitor
korrosionsverhindernd anticorrosive
Korrosionsverhinderung f corrosion inhibition
Korrosionsvorgang m corrosive action, corrosive effect
korrosiv corrosive
Korticin n corticine
Korund m corundum, **kleinkörniger** ~ emery
Korundellit m corundellite
Korundophilit m (Min) corundophilite
Korundscheibe f corundum disc
koruszieren to coruscate, to gleam, to sparkle
koruszierend coruscating

Korynit m korynite
Koschenillerot n cochineal red
Kosekante f cosecant
Kosin n kosin
Kosinus m (Math) cosine, **hyperbolischer** ~ hyperbolical cosine
Kosinusgesetz n (Opt) cosine emission law
Kosinussatz m (Math) cosine law
Kosmetik f cosmetics
Kosmetika n pl cosmetics
kosmetisch cosmetic
kosmisch cosmic
Kosmochemie f cosmic chemistry
Kosmonautik f astronautics, cosmonautics
Kosmos m cosmos
Kosmotron n cosmotron
Kosoextrakt m cusso extract
Kossel-Sommerfeldsches Verschiebungsgesetz n (Spektr) Kossel-Sommerfeld displacement law, law of spectroscopic displacement
Kost f, **kalorienreiche** ~ high-calorie diet, **vegetarische** ~ vegetarian diet
Kosten pl costs, expenses
Kostenaufwand m expenditure
Kostenschätzung f costs computation, estimate
Kostenvergleich m comparison of costs
Kostenvoranschlag m cost estimate
kostspielig costly, expensive
Kot m dung, excrement, faeces, feces, manure, mud, sludge
Kotangens m (Math) cotangent
Kotangente f (Math) cotangent
Kotangentensatz m (Math) cotangent law
Kotarnin n cotarnine
Kotarninhydrochlorid n cotarnine hydrochloride, stypticine
Kotarnon n cotarnone
Kotarnsäure f cotarnic acid
Kotflügel m mud guard
Kotonisieren n (Text) cottonizing, degumming
kovalent covalent
Kovalenz f covalence, covalency
kovariant covariant
Kovariante f covariant
Kovarianz f covariance (of random variables)
Kovolumen n covolume
KPG-Feinbürette f precision bore burette
KPG-Rührwelle f precision glass stirrer shaft
Krabbe f (Zool) shrimp
Krablit m krablite
krachen to crack, to crash, to rustle
Krackbenzin n cracked gasoline
kracken to crack
Kracken n cracking, **katalytisches** ~ catalytic cracking, **thermisches** ~ thermal cracking
Krackgas n cracked gas
Krackung f cracking [process]
Krackverfahren n cracking process

Kräfte *pl* **die in einem Punkt angreifen** (Mech) concurrent forces, ~ **geringer Reichweite** short-range forces, **chemische** ~ chemical forces, **Coulombsche** ~ Coulomb forces, **elektrostatische** ~ electrostatic forces, **entgegengesetzte** ~ forces having opposite directions, **gleichgerichtete** ~ forces having the same direction
Kräfteausgleich *m* force equilibrium
Kräftediagramm *n* diagram of forces
Kräftedreieck *n* triangle of forces
Kräftegleichgewicht *n* equilibrium of forces
Kräfteparallelogramm *n* parallelogram of forces
Kräftepolygon *n* force polygon
Kräftezerlegung *f* resolution of forces
kräftig strong, energetic, potent, powerful, robust, vigorous, (Speise) nourishing
kräftigen to fortify, to reinforce, to strengthen
Kräftigungsmittel *n* strengthening mean
Krähenauge *n* (Bot) nux vomica
Krählarm *m* (Eindicker) raking arm
Kränklichkeit *f* sickliness
Krätzblei *n* slag lead
Krätze *f* waste metal
Krätzfrischen *n* refining of waste metal
Krätzkupfer *n* copper from waste
Krätzmessing *n* brass clippings
Krätzsalbe *f* (Pharm) itch ointment
Krätzschlacke *f* slag of liquation
Krätzschlich *m* (Gieß) slick of waste metal
Krätzwurzel *f* (Bot) white hellebore root
Kräusellack *m* ripple varnish, shrivel varnish, wrinkle finish, wrinkle varnish
kräuseln to crimp, to curl, to goffer, to pucker
Kräuseln *n* crimping, (Farbe) rivelling
Kräuselung *f* crimp, wrinkle forming
Kräuselverschluß *m* crimp seal
Kräuterabsud *m* decoction of herbs
kräuterartig herbal
Kräuteraufguß *m* herb infusion
Kräuterauszug *m* herb extract
Kräuteressig *m* aromatic vinegar, medicated vinegar
Kräuterlikör *m* herb liqueur
Kräutermittel *n* (Pharm) herbal remedy
Kräuterpflaster *n* cataplasm of herbs
Kräutersaft *m* juice of herbs
Kräutersalbe *f* (Pharm) herbal ointment, herb salve
Kräutersalz *n* vegetable salt
Kräuterschnaps *m* herb spirits, medicated spirit
Kräutertee *m* herbal tea
Kräuterwein *m* medicated wine
Kraft *f* force, energy, power, strength, (Anstrengung) effort, ~ **der Zündung** igniting power, **abstoßende** ~ repellent force, **äußere** ~ external force, **angreifende** ~ acting force, **beliebig gerichtete** ~ force acting in any direction, **bewegende** ~ motive power, **eine** ~ ausüben to exert a force, **elektromotorische** ~ electromotive force, **innere** ~ internal force, **magnetomotorische** ~ magnetomotive force, **photoelektromotorische** ~ photoelectromotive force, **resultierende** ~ resultant force, **zwischenmolekulare** ~ intermolecular force
Kraftanlage *f* power plant, power station
Kraftantrieb *m* power drive
Kraftarm *m* **eines Hebels** force arm of a lever
Kraftaufwand *m* expenditure of force
Kraftausbeute *f* power output
Kraftbedarf *m* power consumption
Krafteck *n* force parallelogram
Krafteinheit *f* unit of force
Krafteinschätzung *f* power rating
Kraftersparnis *f* power saving
Krafterzeugung *f* power generation
Kraftfahrzeug *n* automobile, motor vehicle
Kraftfeld *n* magnetic field, field of force
Kraftflußdichte *f* flux density
Kraftflußvariation *f* variation of magnetic flux
Kraftflußverteilung *f* distribution of magnetic flux
Kraftgas *n* power gas, producer gas
Kraftgasanlage *f* power gas plant, gas producer
Kraftgaserzeuger *m* gas producer
Kraftgaserzeugungsanlage *f* power gas generating plant
Kraftkonstante *f* bending force constant
Kraftkonsument *m* power consumer
Kraftleitung *f* power line, power mains
Kraftlinie *f* line of force, **elektrische** ~ electric line of force, **magnetische** ~ line of magnetic force, magnetic line of force
Kraftlinienablenkung *f* deflection of lines of force
Kraftlinienbild *n* diagram of magnetic field
Kraftlinienbündel *n* bundle of lines of force
Kraftlinienfluß *m* flux of lines of force, magnetic flux
Kraftlinienrichtung *f* direction of lines of force
Kraftlinienstrom *m* magnetic flux
Kraftlinienübergang *m* passage of lines of force
Kraftlinienweg *m* path of the lines of force
Kraftlinienzusammenziehung *f* contraction of lines of force
Kraftmesser *m* dynamometer
Kraftmessung *f* measurement of power
Kraftmoment *n* moment of force
Kraft-Papier *n* kraft paper
Kraftpunkt *m* force point
Kraftradius *m* force radius
Kraftrichtung *f* direction of force
Kraftsammler *m* hydraulic accumulator
kraftsparend power-saving
Kraftspeicherwerk *n* hydroelectric power station
Kraftspiritus *m* motor spirit
Kraftsprit *m* alcohol for engine operation

Kraftstoff *m* fuel, motor gasoline, (Diesel) diesel fuel, (Vergaser) fuel, **klopffester** ~ high octane number fuel, knock resistant motor spirit, **Normal-** ~ regular grade motor spirit, **Super-** ~ premium grade motor spirit, **Vergaser-** ~ motor gasoline, motor spirit
Kraftstoffabscheider *m* fuel trap
Kraftstoffanforderung *f* octane number requirement
Kraftstoffanzeiger *m* gas gauge, gasoline gauge
Kraftstoffberuhigungstopf *m* anti-swirl device (for petrol tank)
Kraftstoffdüse *f* fuel jet
Kraftstoffeinspritzung *f* fuel injection
Kraftstoffersparnis *f* fuel saving, fuel economy
Kraftstoffgemisch *n* fuel mixture
Kraftstoffilter *n* fuel filter
Kraftstoff-Luft-Gemisch *n* fuel-air mixture
Kraftstoffmesser *m* fuel pressure gauge, gas gauge
Kraftstoffreiniger *m* fuel filter, fuel strainer
Kraftstoffverbrauch *m* fuel consumption
Kraftstoffzufuhr *f* fuel supply
Kraftstoffzusätze *pl* gasoline additives
Kraftstrom *m* electric power, power current
Krafttransformator *m* power transformer
Kraftübertragung *f* transmission of power, power transmission, transmission of energy
Kraftverbrauch *m* power consumption
Kraftverteilung *f* distribution of power, distribution of energy, power distribution
Kraftvervielfachung *f* multiplication of power
Kraftwagen *m* automobile, motor vehicle, (Lastwagen) truck
Kraftwerk *n* power plant, power station, **Atom-** ~ nuclear power station
Kraftzellstoff *m* kraft paper
Kraftzerlegung *f* decomposition of a force
Kragensteckvorrichtung *f* heavy-duty plug-and-socket connection
Krakelierlack *m* crackle varnish
Krampe *f* clasp, cramp, cramp iron, staple, ~ **mit Stiftspitze** pointed staple
krampen to clasp, to cramp, to fasten
Krampf *m* (Med) convulsion, cramp, spasm
Krampfader *f* (Med) varicose vein
krampfartig spasmodic
krampflösend spasmolytic
Krampfmittel *n* (Pharm) antispasmodic
krampfstillend antispasmodic
Krampstock *m* skimmer
Kran *m* crane, **Ausleger-** ~ jib crane, **Auslegerdreh-** ~ derricking jib crane, **Bock-** ~ gantry crane, **Dreh-** ~ rotary crane, **Drehwipp-** ~ rotary luffing crane, **Einträger-** ~ single-girder crane, **Greifer-Wipp-** ~ grabbing luffing crane, **Kletter-** ~ climbing crane, **Konsol-** ~ bracket crane, **Raupen-** ~ crawler crane, **Schwenk-** ~ slewing crane, **Schwimmer-** ~ floating crane, **Turmdreh-** ~ rotary tower crane, **Wipp-** ~ luffing crane, **Zweiträger-** ~ two-girder crane
Kranbalken *m* crane beam
Kranfahren *n* crane travel
Krangehänge *n* crane link
Kranhaken *m* crane hook
krank ill, sick
Krankabel *n* crane cable
Krankette *f* crane chain
krankhaft morbid, pathological
Krankheit *f* disease, illness, sickness, **ansteckende** ~ contagious disease, infectious disease, **chronische** ~ chronic disease, **übertragbare** ~ communicable disease
krankheitserregend causing disease, pathogenic
krankheitserzeugend pathogenic
Krankheitsherd *m* (Med) focus of a disease
Krankheitskeim *m* (Med) germ [of a disease]
Krankheitslehre *f* pathology
Krankheitssymptom *n* symptom
Krankheitsträger *m* [disease] carrier
krankheitsübertragend infectious
Krankheitsursache *f* cause of a disease
Krankheitsverlauf *m* (Med) course of a disease
Kranlaufwinde *f* hoist
Kranmagnet *m* crane magnet
Kranpfanne *f* crane ladle
Kranpreßluftstampfer *m* crane pneumatic rammer
Kranseil *n* crane cable, crane rope
Krantraverse *f* crane spreader
Krantzit *m* (Min) krantzite
Kranz *m* (der Riemenscheibe) rim
Kranzmodell *n* rim pattern
Krapp *m* (Bot) madder [root], **färben mit** ~ to madder
krappartig like madder
Krappbleiche *f* madder bleach
Krappbraun *n* madder-brown
Krappcarmin *n* madder carmine, red madder lake
Krappfärben *n* madder dyeing
Krappfarbe *f* madder color, madder dye
Krappfarbstoff *m* alizarin, madder dye
Krappgelb *n* madder yellow, xanthin
Krappkarmin *n* madder carmine
Krapplack *m* madder lake
Krapp-Purpur *m* madder purple, Field's purple
krapprot madder-colored, alizarin-red
Krapprot *n* alizarin, madder red
Krappwurzel *f* (Bot) madder root
Krater *m* crater, (Farbe) pinhole
Kraterbildung *f* (Korrosion) pinholing, pitting
Kratzbürste *f* scraper, scratch brush, stiff brush, wire brush
Kratze *f* iron rake, rabble, (Wolle) card
Kratzeisen *n* (Techn) scraping iron, doctor, scraper

kratzen to scrape, to scratch
Kratzen *n* scraping, scratch brushing, scratching
Kratzentuch *n* carding cloth
Kratzentür *f* puddling door
Kratzer *m* rake, scraper, (Schramme) scratch
Kratzerförderer *m* scraper conveyor
Kratzerführung *f* scraper guide
Kratzerscheibe *f* scraper disc, scraper plate
kratzfest marproof, nonmarring, scratch-resistant
Kratzfestigkeit *f* marresistance, scratch resistance
Kratzhärte *f* scratch resistance
Kratzkühler *m* scraped shell cooler
Kratzlader *m* scraper loader
Kratzmaschine *f* scratch brushing machine
Kratzwärmeaustauscher *m* scraped-surface heat exchanger
Kratzwasser *n* brushing liquid
Kraurit *m* (Min) kraurite
Kreatin *n* creatine, methylguanidoacetic acid
Kreatinin *n* creatinine
Kreatinphosphat *n* creatine phosphate
Kreatinphosphokinase *f* (Biochem) creatine phosphate kinase
Kreatinsynthese *f* creatine synthesis
Kreaton *n* creatone
Krebs *m* (Med) cancer, carcinoma, ~ **der Luftwege** cancer of the respiratory system, **rußbedingter** ~ soot cancer, **teerbedingter** ~ pitch-worker's cancer
krebsähnlich (Zool) cancriform, cancroid
krebsartig (Med) cancerous
krebsauslösend (Med) carcinogenic
Krebsbekämpfung *f* (Med) cancer control, combat against cancer
krebsbildend cancerogenic, carcinogenic
Krebsbildung *f* (Med) formation of a carcinoma
Krebsentstehung *f* carcinogenesis
Krebserkennung *f* (Med) cancer diagnosis
Krebserkrankung *f* (Med) carcinosis
krebserregend cancerogenic, carcinogenic
Krebserreger *m* carcinogen, carcinogenic agent
krebserzeugend carcinogenic
Krebserzeuger *m* carcinogen
Krebsforscher *m* cancerologist
Krebsforschung *f* cancer research
Krebsforschungsinstitut *n* cancer research institute
Krebsgeschwür *n* cancerous ulcer
Krebsgeschwulst *f* (Med) carcinoma
Krebsgewebe *n* (Med) cancerous tissue, cancer tissue
Krebsgewebebruchstück *n* cancerous tissue fragment
krebshemmend carcinostatic
Krebsherd *m* (Med) initial carcinoma
Krebsmetastase *f* (Med) metastasis of cancer
Krebsprophylaxe *f* (Med) cancer prophylaxis

Krebstiere *pl* crustaceans
Krebsursache *f* (Med) cause of cancer
Krebsverdacht *m* (Med) suspicion of cancer
krebsverdächtig indicative of cancer
Krebsverhütung *f* (Med) cancer prevention
Krebsvorbeugung *f* (Med) cancer prophylaxis
Krebswachstum *n* cancer growth, growth of a cancer
Krebswurzel *f* (Bot) bistort, squawroot
Krebszelle *f* cancer cell, cancerous cell
krebszellenzerstörend carcinolytic
Kreide *f* chalk, calcium carbonate, carbonate of lime, **geschlämmte** ~ whitewash
kreideartig chalky, cretaceous, like chalk
Kreideboden *m* (Geol) chalky soil
Kreideformation *f* (Geol) cretaceous system, **untere** ~ lower cretaceous system or stage
Kreidegebilde *n* (Geol) chalk formation, cretaceous structure
Kreideglas *n* white glass
Kreidegrund *m* chalk ground, whiting
Kreidegur *f* earthy calcium carbonate, mineral agaric
kreidehaltig chalky, containing chalk, cretaceous
Kreidemehl *n* powdered chalk
Kreidemergel *m* chalk marl
Kreidemühle *f* chalk mill
Kreiden *n* chalking
Kreidepapier *n* art paper, baryta paper, chalk overlay paper, enameled paper
Kreidepulver *n* chalk powder, pipe clay, powdered chalk
Kreideschiefer *m* chalk slate
Kreidestein *m* chalkstone
Kreidestift *m* chalk, crayon
kreideweiß chalky white, deathly pale
Kreis *m* circle, (Atom) orbit, (Elektr) circuit, **integrierender** ~ (Elektr) integrating circuit
Kreisabschnitt *m* segment
Kreisausschnitt *m* sector
Kreisbahn *f* circular path, (Atom) orbit
Kreisbewegung *f* circular motion or movement, gyration, revolution, rotation
Kreisblatt *n* circular chart
Kreisblattschreiber *m* circular chart recorder
Kreisbogen *m* arc, curvature
Kreisdurchmesser *m* diameter of a circle
Kreiseinteilung *f* circular graduation
Kreisel *m* gyroscope, top, (Luftf) gyrostabilizer
Kreiselbewegung *f* gyration, gyroscopic motion
Kreiselbrecher *m* gyratory crusher
Kreiselelektron *n* (Atom) spinning electron
Kreiselfrequenz *f* gyrofrequency
Kreiselgebläse *n* turboblower, centrifugal supercharger
Kreiselkompaß *m* gyrocompass
Kreiselkompressor *m* centrifugal compressor
Kreiselkurvenzeiger *m* gyroscopic turn indicator
Kreisellochblende *f* circular aperture

Kreiselmischer *m* gyro-mixer, impeller, impeller mixer
Kreiselpumpe *f* centrifugal pump
Kreiselrad *n* impeller, turbine
Kreiselrührer *m* impeller
Kreiselschwingung *f* gyro-oscillation
Kreiselsichter *m* centrifugal classifier
Kreiselverdichter *m* turbocompressor
Kreiselvorverdichter *m* centrifugal supercharger
Kreiselwechselwirkung *f* gyro-interaction
Kreiselwirkung *f* gyroscopic effect
kreisen to circle, to circulate, to move in a circle, to revolve, to rotate
kreisend rotating, circulating, gyratory, revolving, rotary, turning
Kreisfläche *f* circular area, circular surface
Kreisförderer *m* closed-circuit conveyor
kreisförmig circular, round, ~ **gebogen** bent in a circle
Kreisform *f* circular shape
Kreisfrequenz *f* angular frequency, angular velocity
Kreisfunktion *f* (Math) circular function
Kreisgang *m* circular motion, revolution, rotation
Kreisheizschlauch *m* full circle curing tube
Kreisinhalt *m* area of a circle, circular area
Kreiskolbenmotor *m* (Wankel) rotary piston engine
Kreiskolbenpumpe *f* rotary pump, rotating piston pump
Kreislauf *m* cycle, circuit, circulation, cyclic process, revolution, rotation, **geschlossener** ~ closed circuit, **großer** ~ (Med) systemic circulation, **kleiner** ~ (Med) pulmonary circulation
Kreislaufbehälter *m* circulating tank
Kreislauffunktion *f* (Med) circulatory function
Kreislaufkollaps *m* (Med) circulatory collapse
Kreislaufkrankheit *f* (Med) circulatory disease
Kreislaufmahlung *f* closed-circuit grinding
Kreislaufmittel *n* (Pharm) circulatory preparation, vasopressor
Kreislaufreaktor *m* circulating fuel reactor, recycle reactor
Kreislaufschwäche *f* (Med) circulatory insufficiency
Kreislaufstörung *f* (Med) circulatory disturbance
Kreislaufversagen *n* (Med) circulatory collapse
Kreislinie *f* circular line
Kreislochplatte *f* circular disc with a hole
Kreismesser *n* circular knife
Kreismessung *f* cyclometry
Kreismittelpunkt *m* center of a circle
Kreisprozeß *m* continuous process, cycle, cyclic process, **Carnotscher** ~ Carnot's cycle
Kreisquerschnitt *m* circular cross section
Kreisrinne *f* annular cavity, circular trough
kreisrund circular

Kreissäge *f* circular saw
Kreisscheibe *f* circular disk
Kreisschicht *f* circular layer
Kreisschreiber *m* circular chart recorder
Kreissegment *n* segment of a circle
Kreissektor *m* sector of a circle
Kreisströmung *f* circulation
Kreisstrom *m* circular current
Kreisteilung *f* circular scale
Kreisthermometer *n* circular scale thermometer
Kreisumfang *m* circumference [of a circle], perimeter of a circle, periphery
Kreisumlauf *m* circulation, cycle
Kreisvorgang *m* cyclic process
Kreittonit *m* kreittonite
Kremersit *m* (Min) kremersite
Kremnitzerweiß *n* Kremnitz white
Kremortartari *m* cream of tartar, potassium tartrate
Krempblech *n* flanged plate
Krempeleinstellung *f* card setting
krempeln (Text) to card
Kremserweiß *n* Kremnitz white
krenelieren (obs Zinnen aufsetzen) to crenelate
Krennerit *m* (Min) krennerite
Krensäure *f* crenic acid, apocrenic acid
Kreolin *n* creolin, analgin
Kreosol *n* creosol, 2-methoxy-4-methylphenol
Kreosot *n* creosote
Kreosotal *n* creosotal, creosote carbonate
Kreosotcarbonat *n* creosotal, creosote carbonate
kreosothaltig containing creosote, creosotic
Kreosotöl *n* creosote oil
Kreosottränkung *f* creosoting
Kreosotwasser *n* creosote water
Krepp *m* (Text) crepe, crape
Kreppapier *n* crepe paper
Kreppkalander *m* (Text) creping calander
Kreppseide *f* (Text) crepe de Chine
Kresalol *n* cresalol
Kresatin *n* cresatin, m-tolyl acetate
Kresaurin *n* cresaurin, trimethylaurin
Kresidin *n* cresidine, aminocresol
Kresol *n* cresol, cresylic acid, methylphenol
Kresolharz *n* cresol resin, cresylic resin
Kresolindophenol *n* cresol indophenol
Kresolnatron *n* sodium cresylate
Kresolpuder *m* cresol powder
Kresolpurpur *f* cresol purple, m-cresol-sulfone-phthalein
Kresolrot *n* cresol red
Kresolsäure *f* cresylic acid
Kresolseife *f* cresol soap
Kresolseifenlösung *f* cresol-soap solution, liquid cresol soap
Kresorcin *n* cresorcinol, 2,4-dihydroxytoluene, 4-methylresorcinol
Kresosteril *n* cresosteril
Kresotingelb *n* cresotine yellow

Kresotinsäure *f* cresotic acid, cresotinic acid, methylsalicylic acid
kresotinsauer cresotate
Kresse *f* (Bot) cress
Kressenöl *n* cress oil
Kressensamenöl *n* cress seed oil
Kresyl- cresyl
Kresylacetat *n* cresyl acetate, tolyl acetate
Kresyläthyläther *m* cresyl ethyl ether, ethyl cresylate
Kresyläthylester *m* ethyl cresylate, cresyl ethyl ether
Kresylalkohol *m* cresylic alcohol
Kresylcarbonat *n* cresyl carbonate, tolyl carbonate
Kresylmethyläther *m* cresyl methyl ether, methyl cresylate
Kresylmethylester *m* methyl cresylate, cresyl methyl ether
Kresylphosphat *n* cresyl phosphate, tolyl phosphate, tricresyl phosphate, tritolyl phosphate
Kresylsäure *f* cresylic acid, cresol
Kresylschwefelsäure *f* cresylsulfuric acid, tolylsulfuric acid
Kresyltoluolsulfonat *n* cresyl toluenesulfonate, tolyl toluenesulfonate
Kresylviolett *n* cresyl violet
Kreuzbalkenrührer *m* cross-blade mixer
Kreuzbandscheider *m* cross-belt separator
Kreuzbeerensirup *m* syrup of buckthorn, syrup of Persian berries
Kreuzdornbeere *f* (Bot) buckthorn berry, Persian berry
Kreuzdornbeerensirup *m* syrup of buckthorn or Persian berries
Kreuzdornöl *n* buckthorn oil
Kreuzeisen *n* cross iron
Kreuzgelenk *n* universal joint
Kreuzgelenkkupplung *f* cross link universal coupling, universal joint
Kreuzköper *m* (Text) cross twill, crow foot, crowfoot weave
Kreuzkonjugation *f* cross conjugation
Kreuzkopf *m* crosshead
Kreuzlochkopf *m* doubleholed capstan head
Kreuzmaß *n* T-square
Kreuzmuffe *f* crossing sleeve, double boss
Kreuzpolarisation *f* (Elektr) cross polarization
Kreuzpunkt *m* point of intersection
Kreuzschalter *m* commutator switch
Kreuzschlitten *m* cross slide, compound slide
kreuzschraffieren to crosshatch
Kreuzspule *f* cross-wound spool, cheese, crossed coil, ~ **mit wilder Wicklung** random cheese
Kreuzstein *m* chiastolite, cross stone, harmotome, staurolite
Kreuzstrom *m* cross current, cross flow (hydrodynamics)

Kreuzstromboden *m* (Dest) cross-flow tray
Kreuzstromwärmeaustauscher *m* cross-flow heat exchanger
Kreuzstück *n* cross, cross piece, flanged cross piece, four-way cross, pipe cross
Kreuztisch *m* cross-table, stage for scanning
Kreuztischmikrometer *n* stage micrometer
Kreuzung *f* crossing, (Bastardisierung) hybridization
Kreuzungspunkt *m* cross-over point, crosspoint
Kreuzungsversuch *m* crossover experiment
Kreuzventil *n* crossing valve
Kreuzverbindung *f* cross connection, cross joint
kreuzweise geschichtet cross laminated
Kriechdehnung *f* creep
kriechen to crawl, to creep
Kriechen *n* crawling, creep, creeping, surface leakage, ~ **dünner Flüssigkeitsschichten** pl film creep
Kriechfestigkeit *f* creep resistance, creep strength, resistance to tracking
Kriechgeschwindigkeit *f* creep rate
Kriechgrenze *f* creep limit, creep strength
Kriechgrenzenspannung *f* limiting creep stress
Kriechmodul *n* apparent modulus
Kriechprobe *f* (Anal) creep test
Kriechspannung *f* leakage voltage
Kriechstrecke *f* creep distance
Kriechstrom *m* creeping current, leakage current, tracking current
kriechstromfest tracking-resistant
Kriechstromfestigkeit *f* tracking resistance
Kriechverhalten *n* creep behavior, creeping property
Kriechversuch *m* creep test
Kriechweg *m* leakage path for electric current
Kriechwegbildung *f* tracking, ~ **durch Gleitfunken** spark tracking, ~ **durch leitende Ablagerung** deposit tracking, ~ **durch Lichtbogen** arc tracking
Kriechwelle *f* creeping wave
Kriegsindustrie *f* war industry
krimpen to wet, to moisten, to shrink
Kristall *m* crystal, **einachsiger** ~ uniaxial crystal, **einfach isodesmischer** ~ simple isodesmic crystal, **flüssiger** ~ anisotropic liquid, liquid crystal, **hemiedrischer** ~ hemihedral crystal, **idiochromatischer** ~ idiochromatic crystal;, **piezoelektrischer** ~ piezoelectric crystal, **wasserfreier** ~ anhydrous crystal, **zweiachsiger** ~ bi[n]axial crystal
Kristallachat *m* (Min) translucent agate
Kristallachse *f* crystal axis, crystallographic axis
Kristallachsenrichtung *f* crystallographic-axis orientation
kristallähnlich crystalline, crystalloid
Kristallanalyse *f* crystal analysis
Kristallansatz *m* crystalline deposit

Kristallanschuß *m* crop of crystals, crystallization
kristallartig crystal-like, crystalline, crystalloid
Kristallausbildung *f* crystalline growth, formation of crystals
Kristallbau *m* crystalline structure or form, crystal structure
Kristallbaufehler *m* crystal defect
Kristallbereich *m* crystal domain
Kristallbiegungsschwingung *f* crystal flexural vibration
Kristallbild *n* crystallogram, crystal pattern
Kristallbildung *f* crystalline deposit, crystalline growth, crystallization, formation of crystals, (Zuck) granulation
Kristallbildungslehre *f* crystallogeny
Kristallblock *m* crystal slab
Kristallblüte *f* flower of crystals
Kristallchemie *f* crystal chemistry
Kristalldetektor *m* (Atom) crystal detector
Kristalldruse *f* cluster of crystals, crystal druse
Kristallebene *f* crystallographic plane, crystal plane or face
Kristallehre *f* crystallography
kristallelektrisch piezoelectric
Kristallelektrizität *f* piezoelectricity
kristallen crystalline
Kristallfehler *m* imperfection of a crystal
Kristallfeld *n* crystalline field
Kristallfeldaufspaltung *f* crystalline field splitting
Kristallfeuchtigkeit *f* crystalline humidity
Kristallfilter *n* crystal filter
Kristallfläche *f* face of a crystal, facet, ~ **mit Miller-Indices** (Krist) standard plane, unit plane
Kristallflüssigkeit *f* liquid of crystallization
Kristallform *f* crystal form, crystalline state
Kristallforscher *m* crystallographer
Kristallgitter *n* crystal lattice, ((Spektr)) crystal grating, **gestörtes** ~ disordered crystal lattice
Kristallgitterabstand *m* crystal lattice spacing
Kristallgittermodell *n* crystal lattice model
Kristallgitterskala *f* crystal lattice scale
Kristallglas *n* crystal glass, flint glass, lead glass
Kristallglasur *f* crystalline glaze, frosted crystal glaze
Kristallgleichrichter *m* (Elektr) crystal rectifier
Kristallgummi *n* crystal gum
Kristallhäutchen *n* crystalline film
Kristallhaut *f* crystalline crust
Kristallhyperfeinstruktur *f* crystal hyperfine structure
kristallin crystalline
Kristallin *n* crystallin
Kristalline *f* celluloid varnish
kristallinisch crystalline, transparent
Kristallinse *f* crystal lens, crystalline lens

Kristallisation *f* crystallization, **fraktionierte** ~ fractional crystallization, **primäre** ~ primary crystallization, **sekundäre** ~ secondary crystallization
Kristallisationsanlage *f* crystallizing plant
Kristallisationsapparat *m* crystallizer
Kristallisationsbeginn *m* crystallization point, chilling point, cloud point
Kristallisationsdifferentiation *f* fractional crystallization
kristallisationsfähig crystallizable
Kristallisationsfähigkeit *f* crystallizability
Kristallisationsgefäß *n* crystallization vessel, crystallizer
Kristallisationsgeschwindigkeit *f* rate of crystalline growth
Kristallisationskern *m* center of crystallization, nucleus of crystallization
Kristallisationskolonne *f* crystallization column
Kristallisationsneigung *f* tendency to crystallize
Kristallisationspfanne *f* crystallizing pan
Kristallisationsschale *f* crystallizing dish
Kristallisationsvermögen *n* crystallizability
Kristallisationsverzögerung *f* crystal growth inhibition
Kristallisationsvorgang *m* crystallizing process
Kristallisationswärme *f* heat of crystallization
Kristallisationswasser *n* water of crystallization
Kristallisationszentrum *n* center of crystallization, nucleus of crystallization
Kristallisator *m*, **Sprüh-** ~ jet crystallizer, **Wirbel-** ~ vortex crystallizer, **Zerstäubungs-** ~ atomizing crystallizer
Kristallisieranlage *f* crystallization plant
Kristallisierband *n* belt crystallizer
kristallisierbar crystallizable
Kristallisierbarkeit *f* crystallizability
kristallisieren to crystallize
Kristallisieren *n* crystallizing
kristallisierend crystallizing
Kristallisierlack *m* crystalline varnish
Kristallisierschale *f* crystallizing dish, crystallizing pan
Kristallisierschnecke *f* crystallization with spiral
kristallisiert crystallized
Kristallisierung *f* crystallization
Kristallisierwalze *f* crystallizing rolls
Kristallisierwiege *f* Wulff-Bock crystallizer
Kristallit *m* crystallite, imperfectly formed crystal
Kristallkante *f* edge of a crystal
Kristallkeim *m* crystal germ, crystal nucleus, seed crystal
Kristallkeimsuspension *f* seeding suspension
Kristallkern *m* crystal center, crystal core, nucleus of a crystal
Kristallkernbildung *f* crystallization, nucleation
Kristallkiesel *m* crystal pebble
Kristallklasse *f* crystallographic class

Kristallkollimator *m* crystal collimator
Kristallkorn *n* crystal grain, crystalline grain
Kristallkornbildung *f* formation of a crystalline grain
Kristallkunde *f* crystallography
Kristallmehl *n* powdered crystal
Kristallmessung *f* crystallometry
Kristallmonochromator *m* crystal monochromator
Kristallmorphologie *f* crystal morphology
Kristalloberfläche *f* crystal surface
Kristallöl *n* (obs) gasoline
Kristallogramm *n* crystallogram
Kristallograph *m* crystallographer
Kristallographie *f* crystallography
kristallographisch crystallographic
kristalloid crystalloid, crystalloidal
Kristalloid *n* crystalloid
Kristallolumineszenz *f* crystalloluminescence
Kristallometrie *f* crystallometry
kristallometrisch crystallometric
kristallophysikalisch crystallophysical
Kristalloptik *f* crystal optics
Kristallorientierung *f* crystal orientation
Kristallose *f* crystallose, sodium salt of saccharin
Kristallotechnik *f* crystallotechnics
Kristallperiodizität *f* crystal periodicity
Kristallphotoeffekt *m* crystal photoelectric effect
Kristallphysik *f* crystal physics
Kristallprisma *n* pressed glass prism
Kristallpulver *n* powdered crystal
Kristallquarz *m* quartz crystal
Kristallrichtung *f* crystal orientation
Kristallsäure *f* crystallizing fuming sulfuric acid
Kristallschicht *f* crystalline layer
Kristallschnitt *m* crystal cut
Kristallschwingung *f* crystal oscillation, crystal vibration
Kristallsoda *f* crystallized soda, soda crystals
Kristallspektrometer *n* crystal spectrometer, Bragg spectrometer
Kristallspektrum *n* crystal spectrum
Kristallstruktur *f* crystal structure
Kristallstrukturanalyse *f* crystal structure determination
Kristallstrukturuntersuchung *f* crystal [structure] analysis
Kristallsymmetrie *f* crystal symmetry
Kristallsystem *n* crystallographic system, crystal system
Kristallszintillator *m* crystal scintillator
Kristallterm *m* crystal term
Kristalltriode *f* crystal triode
Kristallviolett *n* crystal violet
Kristallwachstum *n* crystalline growth
Kristallwasser *n* water of crystallization
kristallwasserfrei anhydrous, free from water of crystallization, ~ **machen** to dehydrate

Kristallwechselwirkung *f* crystalline field interaction
Kristallwinkel *m* crystal angle
Kristallwinkelmessung *f* [crystal] goniometry
Kristallzähler *m* crystal conduction counter
Kristallziehen *n* crystal pulling, ~ **aus der Schmelze** crystal pulling from melt
Kristallziehofen *m* crystal pulling machine
Kristallziehverfahren *n* crystal pulling method
Kristallzinn *n* grain tin
Kristallzucker *m* crystallized sugar
Kristallzüchtung *f* crystal growth
Kristallzwilling *m* twin crystal, compound twin
Kriterium *n* criterion
Krith *n* (Maß) crith, krith
Kritik *f* criticism
kritisch critical, delicate
kritisieren to criticize
Kröhnkit *m* (Min) kröhnkite
Kröse *f* (eines Fasses) croze
Krötengift *n* toad venom
Krokalith *m* crocalite
Krokidolith *m* (Min) crocidolite
Krokoit *m* (Min) crocoite, crocoisite, crocolite
Krokonsäure *f* crocic acid, croconic acid
krokonsauer crocoate, croconate
Krokydolith *m* (Min) crocidolite
Kromycin *n* kromycin
Kromycol *n* kromycol
Kronenbrenner *m* crown burner, ring burner
Kronenregellast *f* crown block rating (drilling)
Kronenstift *m* crown pin
Kronglas *n* (Opt) crown glass
Kronglasprisma *n* crown glass prism
Kronkümmel *m* (Bot) cummin
Kronzinn *n* standard tin
Kropfhalsmeßkolben *m* funnel neck volumetric flask
Kropfscheibe *f* eccentric disk or sheave
Krotonaldehyd *m* crotonaldehyde
Krotonöl *n* croton oil
Krotonsäure *f* crotonic acid
Krudosilber *n* unrefined silver
Krücke *f* fire rake, (Rohrstütze) bracket
Krügit *m* (Min) krügite
krümelig crumbly, friable
Krümelmaschine *f* pelletizer
krümeln to crumble
Krümelspritzmaschine *f* extruder pelletizer
Krümelzucker *m* grained sugar, granular sugar
Krümmelmasse *f* dry blend
krümmen to bend, to buckle, to crook, to curve, to warp, **sich** ~ to bend, to wind
Krümmer *m* quarter bend [pipe], ~ **mit Flanschen pl** flanged bend
Krümmerkern *m* curved core
Krümmerverlust *m* loss in bends
Krümmling *m* bend

Krümmung *f* amount of curving, bend, curve, dogleg, (Verzerrung) contortion, (z. B. einer Schiene) camber, ~ **einer Kurve** curvature
Krümmungsachse *f* axis of curvature
Krümmungshalbmesser *m* radius of bend
Krümmungskreis *m* circle of curvature
Krümmungslinie *f* line of curvature
Krümmungsmittelpunkt *m* center of curvature
Krümmungsradius *m* curvature radius
Krümmungswinkel *m* angle of curvature
Krug *m* jar, jug, mug, pitcher
krumm crooked, bent, curved, (verdreht) twisted
krummlinig curvilinear
Krummzirkel *m* bow compasses
krumpfecht (Text) shrink-resistant
krumpfen (Text) to shrink
Krumpfen *n* (Text) shrinking
Krumpffestigkeit *f* shrink resistance
Krumpffestmachen *n* (Text) shrink proofing
krumpffrei (Text) non-shrinking
Kruste *f* crust, incrustation, scale
Krustenbildung *f* incrustation, scale formation
Krustengestein *n* crustal rocks
Krustentier *n* crustacean
krustieren to incrust
krustig covered with a crust, crustaceous, crusted, crusty
Kryobiologie *f* cryobiology
Kryochemie *f* cryochemistry, low-temperature chemistry
Kryochirurgie *f* cryosurgery
Kryogenerator *m* low-temperature refrigerator
Kryogenin *n* cryogenin, 3-semicarbazidobenzamide, kryogenin
Kryohydrat *n* cryohydrate, cryosel
kryohydratisch cryohydric
Kryokonit *m* cryoconite
Kryolith *m* (Min) cryolite, native sodium aluminum fluoride
Kryolithporzellan *n* fusible porcelain
Kryolithsoda *f* (Min) cryolite soda
kryomagnetisch cryomagnetic
Kryomedizin *f* low-temperature application in medicine
Kryophor *m* cryophorus
Kryophyllit *m* (Min) cryophyllite
Kryopumpe *f* cryopump
Kryopumpen *n* cryopumping, cryotrapping
Kryoskopie *f* cryoscopy
kryoskopisch cryoscopic
Kryostat *m* cryostat
Kryptobrucinolon *n* cryptobrucinolone
Kryptocyanin *n* kryptocyanine
Kryptogenin *n* cryptogenin
kryptokristallin cryptocrystalline, microcrystalline
Kryptolepin *n* cryptolepine
Kryptolith *m* cryptolite
Kryptomeriol *n* cryptomeriol

Kryptomorphit *m* cryptomorphite
Krypton *n* krypton (Symb. Kr)
Kryptopiden *n* cryptopidene
Kryptopidinsäure *f* cryptopidic acid
Kryptopidiol *n* cryptopidiol
Kryptopimarsäure *f* cryptopimaric acid
Kryptopin *n* cryptopine
Kryptopleurin *n* cryptopleurine
Kryptopyrrol *n* cryptopyrrole
Kryptoskop *n* cryptoscope
Kryptostrobin *n* cryptostrobin
Kryptoxanthin *n* cryptoxanthin
Krysolgan *n* crysolgan
K-Säure *f* K-acid
K-Schale *f* (Atom) K-shell
Kubaholz *n* cuba wood, fustic wood
Kubatabak *m* havana tobacco
kubieren (Math) to cube, to raise to the third power
Kubikberechnung *f* cubature
Kubikinhalt *m* cubic capacity
Kubikmeter *n* cubic meter
Kubikmillimeter *n* cubic millimeter
Kubikwurzel *f* (Math) cubic root
Kubikwurzelziehen *n* extraction of cuberoot
Kubikzentimeter *n* cubic centimeter
kubisch cubic, ~ **dichtest gepackt** cubic closest packed, ~ **flächenzentriert** (Krist) cubic face-centered, ~ **raumzentriert** (Krist) cubic body-centered
Kubizierapparat *m* gauging apparatus
Kubizierung *f* (Math) raising to third power, volume calculation
Kubooktaeder *n* cubo octahedron
Kubus *m* (Potenz) cube, (Würfel) cube, hexahedron
Kuchenabwurf *m* cake discharge
Kuchenbrecher *m* [oil-]cake breaker
Kuchenformmaschine *f* [oil-]cake shaping machine
Kübel *m* bucket, pail, tank, tub, vat
Kübelaufzug *m* bucket elevator
Küchengerät *n* kitchen utensil
Kügelchen *n* bead, pellet
Kühlaggregat *n* refrigerating aggregate; refrigeration plant, refrigerator
Kühlanlage *f* refrigeration plant, cooling plant, quenching tower
Kühlapparat *m* condenser, cooling apparatus, refrigerator
Kühlbad *n* cooling bath, quenching bath
Kühlband *n* cooling conveyor
Kühlbehälter *m* cooling tank
Kühlbett *n* (Metall) cooling bed
Kühlelement *n* cooling disc
kühlen to cool; to chill, to reduce the temperature, to refrigerate, (Metall) to quench
Kühlen *n* refrigerating, chilling, cooling, (Metall) quenching

Kühler *m* condenser, cooler, (Metall) quencher, **gerader ~** straight cooler
Kühlerfläche *f* cooler area, cooler surface
Kühlergrill *m* radiator grill
Kühlerhaube *f* radiator cover
Kühlerjalousie *f* radiator shutter
Kühlermantel *m* condenser jacket
Kühlerretorte *f* condenser retort
Kühlerrost *m* radiator grill
Kühlerschlauch *m* (Kfz) radiator hose
Kühlfalle *f* freezing trap, cold trap, cooling trap
Kühlfaß *n* cooling vat, cooler
Kühlfläche *f* cooling surface
Kühlflüssigkeit *f* coolant, cooling fluid, cooling liquid
Kühlform *f* cooled tuyère
Kühlgefäß *n* cooling tank
Kühlhaus *n* cold storage house
Kühlkammer *f* cooling chamber
Kühlkanal *m* cooling channel
Kühlkasten *m* cooler, cooling box
Kühlkreislauf *m* coolant circuit, cooling cycle
Kühllagerung *f* cold storage
Kühlleistung *f* cooling performance or power
Kühlluft *f* cooling air
Kühlmantel *m* coolant jacket, cooling jacket
Kühlmantelriß *m* crack in the cooling jacket
Kühlmaschine *f* refrigerator, refrigerating machinery
Kühlmethode *f* method of cooling
Kühlmittel *n* coolant, cooling agent, cooling medium, cryogen, freezing mixture, refrigerant
Kühlmitteldurchlauf *m* coolant passage
Kühlöl *n* cooling oil, coolant, (Maschinenbau) cutting oil
Kühlofen *m* (Glas) annealing furnace, cooling furnace
Kühlpfanne *f* cooling pan
Kühlplatte *f* cooling disc
Kühlraum *m* cold room
Kühlrohr *n* cooling tube, condenser, refrigerating pipe
Kühlsalz *n* refrigerating salt
Kühlscheibe *f* cooling disc
Kühlschiff *n* cooling copper, cooling pan, cooling tray
Kühlschlange *f* coil condenser, condenser coil, cooling coil
Kühlschnecke *f* condenser coil, cooling coil
Kühlschrank *m* refrigerator
Kühlschrankinnenbehälter *pl* food compartments of refrigerators
Kühlspirale *f* cooling coil
Kühlstoff *m* coolant, cooling agent, freezing mixture
Kühlsubstanz *f* cooling agent
Kühltank *m* cooling tank
Kühltasche *f* cooling chamber

Kühlteich *m* cooling pond
Kühltrog *m* cooling trough
Kühltrommel *f* cooling drum
Kühltrübung *f* (Brauw) chill haze
Kühlturm *m* cooling tower
Kühlung *f* refrigeration, chilling, cooling, **~ durch adiabatische Entmagnetisierung** (Therm) cooling by adiabatic demagnetization, **~ durch Verdampfung** evaporation cooling
Kühlungskoeffizient *m* cooling coefficient
Kühlverfahren *n* cooling method
Kühlvorrichtung *f* cooling arrangement, cooling device; cooling fixture
Kühlwagen *m* refrigerator truck
Kühlwalze *f* cooling roll[er]
Kühlwasser *n* cooling water, coolant
Kühlwasserbehälter *m* cooling water tank
Kühlwassermenge *f* quantity of cooling water
Kühlwassertemperatur *f* temperature of cooling water
Kühlwasserverbrauch *m* consumption of cooling water
Kühlwerk *n* cooling plant
Kühlwirkung *f* cooling effect
Kühlzentrifuge *f* cooling centrifuge
Kühlzylinder *m* cooling cylinder, condenser
Küken *n* (Chem) plug of a cock, stopcock
Kümmel *m* (Bot) caraway seeds
Kümmelbranntwein *m* kümmel, spirit flavored with caraway seeds
Kümmelöl *n* caraway oil, cumin oil
kündigen to fire, to give notice, to give warning
künstlich artificial, imitated, synthetic[al]
Küpe *f* boiler, vat, **ammoniakalische ~** ammonia vat
küpen to vat
Küpenblau *n* indigo blue, vat blue
Küpendruck *m* vat print
Küpenfärben *n* vat dyeing
Küpenfärberei *f* vat dyeing
Küpenfarbstoff *m* vat dye, vat dyestuff
Küpenpräparat *n* vat dye preparation
Küpensäurelösung *f* vat acid leuco solution
Küpenverzögerer *m* vat retardant
Kürbiskernöl *n* pumpkin seed oil
kürzen (Math) to cancel
Kürzung *f* (Math) cancellation
Küvette *f* cuvette, (Glasschale) bulb
Küvettenverschluß *m* cuvette cover
Kufe *f* beck, reservoir, vat
Kugel *f* ball, globe, sphere, (Gewehrkugel) bullet, (Thermometer-Kugel) bulb
Kugelabschnitt *m* segment of a sphere
kugelähnlich spherical, spheroidal
kugelartig globular, spherical
Kugelbakterien *pl* spherical bacteria
Kugelblitz *m* ball lightning
Kugelbolzen *m* ball pin
Kugeldiorit *m* (Min) globular diorite

Kugeldreieck *n* spherical triangle
Kugeldruckhärte *f* ball-pressure hardness, indentation hardness
Kugeldruckhärteuntersuchung *f* static [ball] indentation test
Kugeldrucklager *n* thrust ball bearing
Kugeldruckprobe *f* ball-pressure or ball-indentation test, ball-hardness test, ball-thrust test, static hardness test
Kugeldurchmesser *m* diameter of a sphere
Kugeldurchschlaggerät *n* ball piercing apparatus
Kugeleindruck *m* spherical indentation
Kugelelektrode *f* sparking ball
Kugelfall *m* (in Kugelmühle) cataracting
Kugelfallhärte *f* scleroscope hardness
Kugelfallprobe *f* falling ball test, scleroscope hardness test
Kugelfallviskosimeter *n* dropping ball viscometer
Kugelfilter *m* spherical filter
Kugelfläche *f* spherical surface
Kugelflasche *f* balloon flask
kugelförmig ball-shaped, globe-shaped, globose, globular, spherical
Kugelform *f* ball shape, spherical mold
Kugelfunktion *f* spherical function, spherical harmonic
Kugelfunktionsentwicklung *f* spherical harmonic analysis
Kugelgelenk *n* ball [and socket] joint
Kugelgestalt *f* spherical shape, sphericity
Kugelgewölbe *n* cupola
Kugelhärteprobe *f* ball hardness test
Kugelhärteprüfer *m* ball hardness testing machine
Kugelhahn *m* ball stop-cock
Kugelhahnpipette *f* pipette with globe stopcock, weighing pipette with stopcock
kugelig globular, spherical, spheroidal
Kugeljaspis *m* (Min) Egyptian pebble
Kugelkalotte *f* spherical cap, spherical segment
Kugelkardan *n* universal joint
Kugelkondensator *m* spherical condenser
Kugelkonduktor *m* spherical conductor
Kugelkoordinate *f* spherical polar coordinate
Kugelkreisel *m* spherical top
Kugelkühler *m* ball condenser, globe condenser, spherical condenser
Kugelkupplung *f* spherical coupling, spherical joint
Kugellack *m* fine lake in balls, round lake
Kugellager *n* ball bearing, **einreihiges** ~ single row ball bearing, **konisches** ~ cup and cone bearing, **selbsteinstellendes** ~ self-aligning ball bearing
Kugellagerfett *n* ball bearing grease
Kugellagergehäuse *n* ball bearing housing
Kugellagerring *m* ball race
Kugellehre *f* ball-gauge

Kugelmagnet *m* spherical magnet
Kugelmolekül *n* globular molecule, ball molecule, spherical molecule
Kugelmühle *f* ball mill, pebble mill, pot mill
kugeln to roll, to treat in the ball mill
Kugeln *n* mill mixing
Kugelneigungsmesser *m* ball inclinometer
Kugelniveau *n* bulb level
Kugeloberfläche *f* surface of a sphere
Kugelpackung *f* (Krist) packing of the spheres, sphere packing, **dichteste** ~ closest sphere packing, **hexagonal dichteste** ~ closest packed hexagonal lattice
Kugelpanzergalvanometer *n* galvanometer with ballshaped shield
Kugelprotein *n* spherical protein
Kugelpyranometer *n* spherical pyranometer
Kugelröhre *f* bulb tube
Kugelrohr *n* bulb tube
kugelrund round, globe-shaped
Kugelschale *f* spherical shell
Kugelscharnier *n* ball and socket joint
Kugelschliffverbindung *f* spherical joint
Kugelschnitt *m* spherical section
Kugelschreiber *m* ball point pen[cil]
Kugelspiegel *m* spherical mirror
Kugelspurlager *n* ball thrust bearing
Kugel-Stab-Modell *n* (Stereochem) ball and stick model
Kugelstopfen *m* bulb stopper
Kugelstrahlen *n* shot peening
Kugelstrahler *m* spherical radiator
Kugelsymmetrie *f* (Atom) spherical symmetry
Kugel-T-Stück *n* flanged ball tee, flanged globe tee
Kugelventil *n* ball valve, globe valve
Kugelverschlußdüse *f* ball check nozzle
Kugelvorlage *f* spherical receiver
Kugelzelle *f* round cell, spherocyte
Kuhpocken *pl* (Med) cowpox
Kulmination *f* culmination
kulminieren to culminate, to reach the highest point
Kultur *f* (Bakt) culture
Kulturboden *m* (Bakt) culture medium
Kulturextrakt *m* (Bakt) culture extract
Kulturfiltrat *n* (Bakt) culture filtrate
Kulturflasche *f* (Bakt) culture bottle, culture flask
Kulturgefäß *n* (Bakt) culture flask
Kulturhefe *f* culture yeast
Kulturkolben *m* (Bakt) culture flask
Kulturpflanze *f* (Bot) cultivated plant
Kulturröhrchen *n* culture tube
Kulturschale *f* (Bakt) Petri dish, culture disk
Kulturversuch *m* cultivation experiment
Kumalinsäure *f* cumalinic acid
Kumarin *n* coumarin, cumarin, o-hydroxycinnamic lactone

Kumaronharz *n* coumarone resin
Kumarsäure *f* coumaric acid, coumarinic acid, cumaric acid, hydroxycinnamic acid
Kumidin *n* 1-amino-4-isopropylbenzene, cumidine, aminocumene
Kuminsäure *f* cumic acid, cuminic acid, p-isopropylbenzoic acid
Kumol *n* cumene, isopropyl benzene
Kumulation *f* accumulation, cumulation
kumulativ accumulative, cumulative
Kumulene *pl* cumulenes
kumulierend cumulating, cumulative
Kundendienst *m* service
Kunstbronze *f* art bronze
Kunstdarm *m* artificial gut, synthetic sausage casing
Kunstdruckpapier *n* art paper
Kunstdünger *m* fertilizer, artificial manure
Kunsteis *n* artificial ice, manufactured ice
Kunstelfenbein *n* abolit
Kunstfaden *m* artificial fiber, synthetic filament
kunstfärben to art-dye
Kunstfärben *n* art-dyeing
Kunstfaser *f* artificial fiber, synthetic fiber
Kunstglas *n* ornamental glass
Kunstgummi *n* synthetic rubber
Kunstharz *n* artificial resin, synthetic resin
Kunstharzausrüstung *f* resin finish
Kunstharzklebstoff *m* synthetic resin adhesive
Kunstharzlack *m* enamel
Kunstharzleim *m* synthetic resin cement
Kunstharzüberzugslack *m* synthetic resin finish
Kunstholz *n* compregnated wood, compressed wood, densified wood
Kunsthonig *m* artificial honey
Kunstkautschuk *m* synthetic rubber, elastomer
Kunstkohle *f* artificial carbon
Kunstleder *n* synthetic leather
Kunstpergament *n* parchment paper
Kunstschwamm *m* aerated plastic, foamed plastic, synthetic sponge material
Kunstseide *f* (Text) artificial silk, rayon
Kunstseidenfaden *m* rayon filament
Kunstseidengewebe *n* (Text) artificial silk fabric
Kunststoff *m* plastic material, plastic[s], plastomer, **aus ~ hergestellt** made of plastics
Kunststoffauskleidung *f* plastic lining
Kunststoffbehälter *m* plastic container, plastic case
kunststoffbeschichtet plastic coated
Kunststoffbeschichtung *f* plastic coating
Kunststoffdispersion *f* plastic dispersion, polymer dispersion
Kunststoffe *pl* **aus Äthylenharzen** ethylene plastics, **~ aus Furanharzen** furan plastics
Kunststoffgalvanisierung *f* electroplating of plastics
kunststoffimprägniert plastic-proofed
Kunststoffkabel *n* plastic-insulated cable
Kunststoffmetallisierung *f* metal-coating of plastics
Kunststoffolie *f* plastic sheet
Kunststoffplatte *f* plastic sheet
Kunststoffpresse *f* molding press
Kunststoffpreßmasse *f* plastic molding compound
Kunststoffrohr *n* plastic pipe
Kunststoffschweißen *n* welding of plastics
Kunststoffstopfen *m* plastic stopper
Kunststoffszintillator *m* scintillation plastic
Kunststofftank *m* plastic tank, **glasfaserverstärkter ~** plastic tank with glass fiber reinforced
Kunstwolle *f* artificial wool, manufactured wool
Kunzit *m* (Min) kunzite
Kuoxam *n* cuprammonium
Kupellation *f* cupellation
kupellieren to cupel
Kupfer *n* copper (Symb. Cu), **~ polen** to toughen, **galvanisch gefälltes ~** electrolytic copper, **hammergares ~** refined or toughpitch copper, **übergares ~** dry or underpoled copper
Kupferabfälle *m pl* copper scrap, copper waste, waste copper
Kupferacetat *n* copper(II) acetate, cupric acetate, neutral verdigris
Kupferacetatarsenit *n* copper acetoarsenite, cupric acetate arsenite, Paris green, Schweinfurt green
Kupferacetylacetonat *n* cupric acetylacetonate
Kupferacetylen *n* copper acetylide, cuprous acetylide
Kupferakkumulator *m* copper accumulator, copper storage battery
Kupferalanin *n* copper alanine, copper(II) aminopropionate
Kupferalaun *m* copper alum, copper aluminate, (Min) lapis divinus
Kupferalbuminat *n* copper albuminate
Kupferamalgam *n* copper amalgam
Kupferammoniaklösung *f* tetrammine copper(II) hydroxide solution, ammoniacal copper oxide solution, cuprammonia solution, cuprammonium hydroxide solution
Kupferammonium *n* cuprammonium
Kupferammoniumkunstseide *f* (Text) cuprammonium rayon
Kupferammoniumlösung *f* tetrammine copper(II) hydroxide solution, ammoniacal copper oxide solution, cuprammonium hydroxide solution
Kupferammoniumsulfat *n* copper ammonium sulfate, cuprammonium sulfate, cupric ammonium sulfate, tetrammine copper(II) sulfate
Kupferammoniumverbindung *f* cuprammonium compound

Kupferantimonglanz *m* (Min) chalcostibite, native copper sulfantimonide
Kupferarsenacetat *n* copper acetoarsenite, cupric acetate arsenite, Paris green, Schweinfurt green
Kupferarsenid *n* copper arsenide
Kupferarsenit *n* copper(II) arsenite, cupric arsenite
Kupferarsenitacetat *n* (s. Kupferarsenacetat)
kupferartig copper-like, coppery, cupreous
Kupferasbestdichtung *f* copper-asbestos gasket
Kupferasche *f* copper ash, copper scale
Kupferaushärtung *f* precipitation of excess of copper
Kupferautotypie *f* copper autotype process
Kupferbad *n* copper bath
Kupferband *n* copper band
Kupferbarren *m* copper bar
Kupferbeize *f* copper mordant
Kupferbenzoylacetonchelat *n* copperbenzoylacetone chelate
Kupferbeschlag *m* copper sheeting
Kupferbichlorid *n* copper dichloride, copper(II) chloride, cupric chloride
Kupferblatt *n* copper foil, sheet copper
Kupferblau *n* blue verdigris, chrysocolla, copper blue
Kupferblech *n* copper foil, copper plate, sheet copper
Kupferblei *n* alloy of copper and lead
Kupferbleiglanz *m* (Min) cuproplumbite
Kupferbleivitriol *n* (Min) linarite
Kupferblende *f* (Min) tennantite
Kupferblick *m* shine of copper
Kupferblüte *f* (Min) copper bloom, capillary cuprite, capillary red copper ore
Kupferbolzen *m* copper bolt
Kupferbraun *n* copper brown, earthy ferruginous cuprite, tile ore
Kupferbromat *n* copper(II) bromate, cupric bromate
Kupferbromid *n* copper bromide, (Kupfer(II)-bromid) copper(II) bromide, cupric bromide
Kupferbromür *n* (Kupfer(I)-bromid) copper(I) bromide, cuprous bromide
Kupferbügel *m* copper arch
Kupfercarbid *n* copper(I) acetylide, copper(I) carbide, cuprous acetylide, cuprous carbide
Kupfercarbonat *n* carbonate of copper, (Farbe) Bremen blue, (Min) azurite
Kupfercellulose *f* copper cellulose
Kupferchelat *n* copper chelate, cupric chelate
Kupferchlorat *n* copper(II) chlorate, cupric chlorate
Kupferchloridammoniak *n* ammonio-cupric chloride, copper ammonium chloride, cuprammonium chloride, cupric ammonium chloride, tetrammine copper(II) chloride

Kupferchlorür *n* (Kupfer(I)-chlorid) copper(I) chloride, cuprous chloride
Kupferchromat *n* copper(II) chromate, cupric chromate
Kupfercitrat *n* copper citrate, copper(II) citrate, cupric citrate
Kupfercyanür *n* (Kupfer(I)-cyanid) copper(I) cyanide, cuprous cyanide
Kupferdämpfer *m* copper damper
Kupferdämpfung *f* copper damping
Kupferdichtung *f* copper packing
Kupferdraht *m* copper wire
Kupferdrahtnetz *n* copper [wire] gauze
Kupferdrahtnetzrolle *f* roll of copper wire gauze
Kupferdrahtspule *f* copper wire coil
Kupferdrehspäne *pl* copper turnings
Kupferdruck *m* copper plate printing
Kupferdruckerschwärze *f* copper plate black, Frankfort black
Kupferdruckschwarz *n* copper plate black, Frankfort black
Kupferelektrode *f* copper electrode
Kupferenzym *n* (Biochem) copper enzyme
Kupfererz *n* (Min) copper ore
Kupferfahlerz *n* (Min) gray copper ore, tetrahedrite
Kupferfarbe *f* copper color, copper pigment
kupferfarben copper-colored
Kupferfedererz *n* (Min) capillary cuprite, copper bloom
Kupferfeilicht *n* copper filings
Kupferfeilspäne *pl* copper filings
Kupferfeinblei *n* chemical lead
Kupferferrocyanid *n* copper cyanoferrate(II), copper ferrocyanide
Kupferfolie *f* copper foil
Kupferfrischofen *m* copper finery
kupferführend copper-bearing, cupriferous
Kupfergalmei *m* (Min) calamine of copper
Kupfergalvanoplastikverfahren *n* copper galvanoplastic process
Kupfergare *f* copper refining, separation of copper
Kupfergarherd *m* copper furnace
Kupfergarmachen *n* copper refining
Kupfergarschlacke *f* copper slag
Kupfergehalt *m* copper content
Kupfergießerei *f* copper foundry
Kupferglanz *m* coppery luster, (Min) chalcocite, copper glance
Kupferglas *n* chalcocite
Kupferglimmer *m* (Min) chalcophyllite, micaceous copper
Kupferglycerin *n* copper glycerol
Kupferglycinat *n* cupric glycinate
Kupferglykokoll *n* copper glycine, copper aminoacetate, copper glycocoll
Kupfergold *n* Mannheim gold, similor

Kupfergranalien *pl* copper granules, granulated copper
Kupfergrün *n* copper green
kupferhaltig cupriferous, containing copper, cupreous
Kupferhammerschlag *m* copper scale
Kupferhaut *f* copper casing
Kupferhornerz *n* (Min) copper horn ore
Kupferhütte *f* copper smelting plant
Kupferhydroxid *n* copper hydroxide, (Kupfer(II)-hydroxid) copper(II) hydroxide, cupric hydroxide
Kupferhydroxydul *n* (Kupfer(I)-hydroxid) copper(I) hydroxide, cuprous hydroxide
kupferig copper-like, coppery, cupreous
Kupfer(II)-Salz *n* copper(II) salt, cupric salt
Kupfer(II)-Verbindung *f* copper(II) compound, cupric compound
Kupferindigo *m* indigo copper
Kupferion *n* copper ion
Kupfer(I)-Salz *n* copper(I) salt, cuprous salt
Kupfer(I)-Verbindung *f* copper(I) compound, cuprous compound
Kupferjodid *n* copper iodide, (Kupfer(II)-jodid) copper(II) iodide, cupric iodide
Kupferjodür *n* (Kupfer(I)-jodid) copper(I) iodide, cuprous iodide
Kupferkalk *m* copper calx
Kupferkies *m* (Min) chalcopyrite, copper pyrites
Kupferkitt *m* coppersmith's cement
Kupferknallsäure *f* copper fulminate, fulminating copper
Kupferkörner *n pl* granulated copper, beanshot
Kupferkohle *f* copper-plated carbon
Kupferkonstantanelement *n* copper-constantan couple
Kupferkunstseide *f* (Text) cuprammonium rayon
Kupferlasur *f* azurite, basic cupric carbonate, chessylite, copper blue, hydrated basic copper carbonate, mineral blue, mountain blue
Kupferlebererz *n* (Min) hepatic copper ore
Kupferlegierung *f* copper alloy
Kupferlichtdruck *m* helioengraving, heliogravure
Kupferlitze *f* copper strand
Kupferlösung *f* copper solution
Kupferlot *n* copper solder
Kupfermangan *n* cupromanganese
Kupfermanganerz *n* (Min) cupreous manganese ore
Kupfermantel *m* copper clad, copper jacket
Kupfermünze *f* copper coin
kupfern copper, copper-colored, coppery, cupreous
Kupfernatroncellulose *f* copper sodium cellulose
Kupfernickel *n* copper nickel alloy, nickeline, (Min) niccolite
Kupferniederschlag *m* copper deposit, copper precipitate

Kupfernitrat *n* copper(II) nitrate, cupric nitrate
Kupfernormalien *f pl* copper standards
Kupferocker *m* (Min) copper brown, earthy ferruginous cuprite, ocher of copper, tile ore
Kupferoleat *n* copper oleate
Kupferoxalat *n* copper oxalate, (Kupfer(II)-oxalat) cupric oxalate
Kupferoxid *n* copper oxide, (Kupfer(II)-oxid) copper(II) oxide, cupric oxide
Kupferoxidammoniakkunstseide *f* (Text) cuprammonium rayon
Kupferoxidammoniaklösung *f* ammoniacal copper oxide solution, cuprammonium hydroxide solution, tetrammine copper(II) hydroxide solution
Kupferoxidammoniakverbindung *f* ammoniacal copper oxide compound, copper oxide ammonia compound, cuprammonia compound, cuprammonium compound
Kupferoxidbrücke *f* (Elektr) copper oxide bridge
Kupferoxidhydrat *n* (obs) copper(II) hydroxide, cupric hydroxide
Kupferoxidplatte *f* cupric oxide plate
Kupferoxidsalz *n* (obs) copper(II) salt, cupric salt
Kupferoxidverbindung *f* copper(II) compound, cupric compound
Kupferoxychlorid *n* copper(II) oxychloride, copper oxychloride, cupric oxychloride
Kupferoxydul *n* (Kupfer(I)-oxid) copper(I) oxide, copper protoxide, cuprous oxide, red copper oxide
Kupferoxydulhydrat *n* (Kupfer(I)-hydroxid) copper(I) hydroxide, cuprous hydroxide
Kupferoxydulsulfit *n* (obs) copper(I) sulfite, cuprous sulfite
Kupferoxydulverbindung *f* (Kupfer(I)-Verbindung) copper(I) compound, cuprous compound
Kupferphenolsulfonat *n* copper phenolsulfonate
Kupferphosphid *n* copper(I) phosphide, cuprous phosphide
Kupferpol *m* copper pole
Kupferpresse *f* copper press
Kupferprobe *f* copper assay, copper sample, test for copper
Kupferradierung *f* copper etching
Kupferraffination *f* copper refining
Kupferraffinerie *f* copper refinery
Kupferrauch *m* copper fumes
Kupferregen *m* copper rain
Kupferreinigung *f* copper refining
Kupferresinat *n* copper resinate
Kupferreyon *n* (Text) cuprammonium rayon
Kupferrhodanat *n* copper thiocyanate
Kupferrhodanid *n* (Kupfer(II)-rhodanid) copper(II) thiocyanate, cupric thiocyanate
Kupferrhodanür *n* (Kupfer(I)-rhodanid) copper(I) thiocyanate, cuprous thiocyanate

Kupferröte *f* virgin copper
Kupferrohr *n* copper pipe, copper tube
Kupferrost *m* copper rust, verdigris
Kupferrostofen *m* copper calciner
kupferrot copper-colored, copper-red, coppery
Kupferrot *n* (Min) cuprite, red copper ore
Kupfersaccharat *n* copper saccharate
Kupfersalz *n* copper salt
Kupfersalzlösung *f* copper salt solution, solution of a copper salt
Kupfersamterz *n* (Min) velvet copper ore
Kupfersand *m* copper sand
Kupferschachtofen *m* copper blast furnace
Kupferschalentest *m* copper dish test
Kupferschaum *m* copper scum, tyrolite
Kupferscheibe *f* copper disk
Kupferschiefer *m* (Min) copper schist, cupriferous slate
Kupferschlacke *f* copper slag
Kupferschlag *m* copper ashes, copper scale
Kupferschlange *f* copper coil
Kupferschlich *m* copper slick
Kupferschmiedearbeit *f* copper forged work
Kupferschmiedestück *n* copper forging
Kupferschrot *m n* granulated copper
Kupferschwärze *f* black copper, black copper oxide, melaconite
Kupferschwamm *m* (Min) copper sponge, tyrolite
Kupferschwarz *n* copper black
Kupferschwefel *m* copper sulfide
Kupferschwefelkalk *m* copper containing limesulfur, cupreous calcium polysulfide
Kupferschweißung *f* copper welding
Kupferseide *f* (Text) cuprammonium rayon
Kupfersilber *n* copper silver alloy
Kupfersilberglanz *m* (Min) stromeyerite
Kupfersilikofluorid *n* copper silicofluoride, copper fluosilicate
Kupfersilizid *n* copper silicide
Kupfersinter *m* copper pyrites, copper scale
Kupfersmaragd *m* (Min) emerald copper, dioptase
Kupferspat *m* (Min) malachite
Kupferspinnfaser *f* cuprammonium staple fiber
Kupferspritzmittel *n* copper spray
Kupferstahldraht *m* copper-plated steel wire
Kupferstannat *n* copper stannate
Kupferstechen *n* copper engraving
Kupferstein *m* copper matte, crude copper
Kupfersteinröstofen *m* copper furnace, metal calciner
Kupferstich *m* copper engraving, copper plate print
Kupferstichplatte *f* plate for copper engraving
Kupferstreifenprobe *f* copper strip test
Kupfersulfat *n* copper sulfate, (Kupfer(II)-sulfat) blue vitriol, copper(II) sulfate, cupric sulfate

Kupfersulfatanlage *f* copper sulfate plant
Kupfersulfid *n* copper sulfide, (Kupfer(II)-sulfid) copper(II) sulfide, cupric sulfide
Kupfersulfophenylat *n* copper sulfophenolate, copper sulfocarbolate, copper sulfophenate
Kupfersulfür *n* (Kupfer(I)-sulfid) copper(I) sulfide, cuprous sulfide
Kupfertetramminsulfat *n* tetrammine copper(II) sulfate
Kupfertiefdruck *m* gravure printing
Kupferüberzug *m* copper coating, copper deposit
Kupferuranglimmer *m* (Min) chalcolite
Kupferuranit *m* (Min) chalcolite
Kupfervergiftung *f* copper poisoning
Kupferverhüttung *f* copper-ore smelting
Kupferverkleidung *f* copper lining
Kupfervitriol *n* blue vitriol, copper(II) sulfate, cupric sulfate
Kupfervitriolkristall *m* copper sulfate crystal
Kupfervoltameter *n* copper voltameter
Kupferwalze *f* copper cylinder
Kupferwasserstoff *m* copper hydride
Kupferwismuterz *n* (Min) klaprotholite, wittichenite
Kupferzahl *f* copper number, copper value
Kupferzement *m* cement copper
Kupferziegelerz *n* (Min) earthy ferruginous cuprite, tile ore
Kupferzinkakkumulator *m* copper zinc accumulator, copper zinc storage battery
Kupferzinkelement *n* copper-zinc cell
Kupferzuschlag *m* (Gieß) copper flux
Kupholith *m* kupholite
Kupolguß *m* cupola casting
Kupolhochofen *m* cupola blast furnace
Kupolofen *m* cupola, cupola furnace, cupola kiln, ~ **mit erweitertem Herd** compound cupola, reservoir cupola, tank cupola, ~ **mit Handbeschickung** hand-charged cupola, **basischer** ~ basic-lined cupola
Kupolofenabbrand *m* melting loss of a cupola furnace
Kupolofenauskleidung *f* cupola lining
Kupolofengattierung *f* cupola mixture
Kupolofengichtgas *n* waste top gas of a cupola furnace
Kupolofenguß *m* casting from the cupola furnace
Kupolofenschmelzen *n* melting in a cupola furnace
Kuppe *f* head, (Geol) summit, top, (Meniskus) meniscus
Kuppel *f* bonnet, cover, cupola, dome, hood
kuppeln to couple
Kuppelofen *m* cupola furnace
Kuppelofenschmelzen *n* melting in a cupola furnace

Kuppelrad *n* connecting gear
Kuppelstange *f* coupling rod, drag bar, draw bar
Kuppelzapfen *m* coupling pin
Kupplung *f* clutch, coupling, **biegsame (elastische)** ~ elastic coupling, flexible coupling, **Flüssigkeits-** ~ hydraulic clutch, **hydraulische** ~ hydraulic clutch, **lösbare** ~ clutch
Kupplungsbelag *m* clutch lining or facing
Kupplungsfußhebel *m* clutch pedal
Kupplungsgehäuse *n* clutch box, clutch housing
Kupplungskette *f* coupling chain
Kupplungskomponente *f* coupling component
Kupplungsmuffe *f* coupling box
Kupplungspedal *n* clutch pedal, clutch lever
Kupplungsspindel *f* coupling spindle
Kupplungsverkleidung *f* clutch facing
Kupplungswert *m* coupling value
Kupplungszapfen *m* wobbler
Kuprit *m* (Min) cuprite, red oxide of copper
Kupronelement *n* copper oxide cell
Kurbel *f* crank, crooked handle, winch, **beiderseitig gelagerte** ~ inside crank, **einseitig gelagerte** ~ outside crank, overhang crank
Kurbelabstand *m* crank distance
Kurbelachse *f* crank axle
Kurbelantrieb *m* crank drive
Kurbelarm *m* crank, lever arm
Kurbelbilge *f* crank bilge
Kurbelblatt *n* crank web
Kurbeldruck *m* crank pressure, pressure on lever or handle
Kurbelgriff *m* crank handle, crank pin
Kurbelhammer *m* hammer operated by a crank
Kurbelmechanismus *m* crank mechanism
kurbeln to crank, to turn a crank
Kurbelpresse *f* crank press
Kurbelpumpe *f* reciprocating pump
Kurbelschleife *f* crank guide
Kurbelstange *f* connecting rod
Kurbelwanne *f* crankcase
Kurbelwannenentlüftung *f* breather, crankcase ventilation, **positive (geschlossene)** ~ positive crankcase ventilation (PCV-System)
Kurbelwelle *f* crankshaft, crank axle
Kurbelwellendrehbank *f* crankshaft lathe
Kurbelwellenlager *n* crankshaft bearing, big end bearing
Kurbelwellenschaft *m* body of crankshaft
Kurbelwellenschleifmaschine *f* crankshaft grinder
Kurbelwerk *n* crank gear
Kurbelzapfen *m* crank pin
Kurchamin *n* kurchamine
Kurit *m* curite
Kurkuma *n* tumeric
Kurkumapapier *n* tumeric paper
Kurkumaprobe *f* tumeric test
Kuromatsuen *n* kuromatsuene

Kuromatsuol *n* kuromatsuol
Kurs *m* course
Kursus *m* course
Kurswinkel *m* azimuth of course
Kurve *f* curve, bend, (Zeichnung) graph, **ausgezogene** ~ solid curve, **logarithmische** ~ logarithmic curve, **punktierte** ~ dotted curve
Kurvenanpassung *f* (Math) curve fitting
Kurvenast *m* (Math) branch of a curve
Kurvenbandförderer *m* troughed-belt conveyor
Kurvenintegral *n* contour integral, curvilinear integral
Kurvenkipper *m* tipping device
Kurvenlineal *n* curved ruler, French curve
Kurvenschar *f* (Math) family of curves
Kurvenscheibe *f* rocker
Kurvenschiebung *f* (Kfz) cornering thrust
kurz short, brief, concise, ~ **gefaßt** concise
kurzarmig short-armed
Kurzbeize *f* short disinfection treatment
Kurzbewitterung *f* accelerated weathering
kurzbrüchig brittle
kurzfaserig short-fibered
kurzfristig for a short time
kurzgeschlossen short-circuited
Kurzgewinde *n* short thread
kurzhalsig short-necked
Kurzhalsrundkolben *m* shortnecked round-bottom flask
kurzkammig short-crested
Kurzkompressionsschnecke *f* screw with half-turn compression (i. e. screw that gives a sudden increase in compression at the beginning of the metering section by a reduction in the depth of flight)
kurzlebig short-lived
Kurzprüfung *f* accelerated test, short time test, snap test
Kurzreferat *n* abstract
Kurzreichweitengrenze *f* short-range limit
kurzreichweitig shortrange
Kurzschichtreaktor *m* short-bed reactor
kurzschließen to short-circuit
Kurzschließen *n* **der Zelle** shortcircuiting a cell
Kurzschließer *m* shortcircuiter
Kurzschluß *m* short-circuit, **vollständiger** ~ dead short-circuit
Kurzschlußabstich *m* short-circuit tapping
Kurzschlußanker *m* shortcircuit rotor, squirrel cage rotor
Kurzschlußbeseitigung *f* junction clean-up
Kurzschlußbremse *f* short-circuit brake
Kurzschlußbürste *f* short-circuit brush
Kurzschlußerregung *f* short-circuit excitation
Kurzschlußfunke *m* short-circuit spark
Kurzschlußkohle *f* short-circuit carbon
Kurzschlußkontakt *m* short-circuit contact
Kurzschlußleitung *f* bypass

Kurzschlußmotor *m* motor with short-circuit rotor
Kurzschlußofen *m* short-circuit furnace
Kurzschlußperiode *f* duration of short-circuit
Kurzschlußpotential *n* short-circuit potential
Kurzschlußring *m* short-circuit ring
Kurzschlußscheibe *f* short-circuit disk
Kurzschlußschleife *f* short-circuit loop
Kurzschlußspannung *f* short-circuit voltage
Kurzschlußstrom *m* short-circuit current
Kurzschlußstromkreis *m* short-circuit
Kurzschlußverlust *m* short-circuit loss
Kurzschlußwiderstand *m* short-circuit resistance
Kurzschlußwindung *f* short-circuit loop
Kurzschlußzeit *f* duration of short-circuit
Kurzstrahlung *f* short-wave radiation
Kurztrommelofen *m* short rotary furnace
Kurzversuch *m* accelerated test, short time test
Kurzwegdestillation *f* flash distillation, molecular distillation
Kurzwegdestillator *m* shortway distiller
Kurzwegfraktionierung *f* simple fractionation
Kurzwegverdampfer *m* molecular evaporator
Kurzwelle *f* shortwave
Kurzwellenapparat *m* shortwave set
kurzwellig shortwave
Kurzzeit-Crackverfahren *n* high-severity cracking
kurzzeitig for a short time, momentary, short-term, temporary
Kussin *n* koussine, kosin
Kussoblüten *f pl* (Bot) cusso [flowers], brayera, kousso flowers
Kussopulver *n* brayera powder, cusso powder, kousso powder
Kutkin *n* kutkin
Kwangosid *n* kwangoside
Kyanäthin *n* cyanethine, kyanethine
Kyanidin *n* 1,3,5-triazine, kyanidine, cyanidine
Kyanisation *f* kyanization
kyanisieren to kyanize
Kyanisieren *n* kyanizing
Kyanisierung *f* kyanization
Kyanit *m* (Min) cyanite, kyanite
Kyanmethin *n* kyanmethin
Kyaphenin *n* kyaphenine
Kybernetik *f* cybernetics
Kymograph *m* kymograph
Kynurenin *n* kynurenine
Kynureninase *f* (Biochem) kynureninase
Kynureningelb *n* kynurenine yellow
Kynurensäure *f* kynurenic acid, 4-hydroxyquinaldic acid
Kynurin *n* kynurin, 4-hydroxyquinoline
Kyrosit *m* kyrosite

L

Lab *n* lab, lab ferment, rennase
Labdan *n* labdane
Labdangummi *n* labdanum, ladanum
Labdanharz *n* labdanum, ladanum
Labdanolsäure *f* labdanolic acid
Labdanum *n* labdanum, ladanum
Labferment *n* (Biochem) rennase, chymase, lab ferment
labil labile, instable, unstable
Labilität *f* lability, instability, unstability
Labilitätszustand *m* state of instability, state of lability
Labkasein *n* rennet casein
Labmagen *m* rennet bag
Labor *n* laboratory, lab
Laborabguß *m* laboratory sink
Laborabzug *m* fume hood, fume cupboard
Laborant[in] *m (f)* laboratory assistant, laboratory technician
Laborantrieb *m* laboratory drive
Laboratorieneinrichtung *f* laboratory equipment
Laboratorium *n* laboratory, lab, **heißes** ~ hot laboratory
Laboratoriumseinrichtung *f* laboratory installation or equipment
Laboratoriumsexperiment *n* laboratory experiment
Laboratoriumsgeräte *pl* laboratory utensils
Laboratoriumsmaßstab *m* laboratory size
Laboratoriumsofen *m* laboratory furnace
Laboratoriumstechniker *m* laboratory technician
Laboratoriumstisch *m* laboratory bench
Laboratoriumsüberhitzer *m* laboratory superheater
Laboratoriumsunfall *m* laboratory accident
Laboratoriumsversuch *m* laboratory test, laboratory trial
Laboratoriumswaage *f* analytical balance, laboratory balance
Laborausbeute *f* laboratory yield
Laborbefund *m* laboratory findings
Laborgeräte *n pl* **aus Glas** scientific glassware
Laborhelfer *m* laboratory worker
laborieren to practice chemistry
Laborklemme *f* laboratory clamp
Labor[kraft]wagen *m* laboratory lorry (Br. E.), laboratory truck (Am. E.), laboratory van
Labormischer *m* laboratory mixer
Labormöbel *pl* laboratory furniture
Labormühle *f* laboratory [ball] mill, laboratory crusher
Laborprüfung *f* laboratory test, analysis, laboratory investigation
Laborpumpe *f* laboratory pump
Laborrührer *m* laboratory stirrer
Labortisch *m* [laboratory] bench
Labortrockenschrank *m* laboratory drying oven, drying cupboard
Laboruntersuchung *f* laboratory examination, laboratory inspection, laboratory investigation, laboratory study, laboratory test[ing]
Laborversuch *m* laboratory test, bench test, laboratory trial
Laborwalze *f* laboratory roller mill
Labrador *m* (Min) labradorite
Labradorblende *f* (Min) labrador hornblende, hypersthene, schillerspar
Labradorfeldspat *m* (Min) Labrador feldspar, labradorite, Labrador stone, saussurite
Labradorit *m* (Min) labradorite, Labrador feldspar, Labrador stone, saussurite
Labradorstein *m* (Min) Labrador feldspar, labradorite, Labrador stone, saussurite
Laburnin *n* laburnine
Labyrinthdichtung *f* labyrinth seal
Labyrinthspalt *m* labyrinth gap
Laccain *n* laccaine
Laccainsäure *f* laccaic acid, laccainic acid
Laccase *f* (Biochem) laccase
Laccersäure *f* lacceroic acid, dotriacontanic acid
Laccol *n* laccol
Lachgas *n* (Chem, Med) laughing gas, dental gas, nitrogen monoxide, nitrous oxide
Lachnophyllumester *m* lachnophyllum ester
Lachnophyllumsäure *f* lachnophyllic acid
Lachs *m* (Zool) salmon
Lachsersatz *m* smoked salmon substitute
lachsfarben salmon-colored
Lachsöl *n* salmon oil
Lack *m* lacquer, coating composition, varnish, (Asphaltgrundlage) japan, (Emaillelack) enamel, (Farblack) lake, **abziehbarer** ~ strip coating, **einfettiger** ~ varnish with oil length of 1:1, **glänzend auftrocknender** ~ (auf Polyesterbasis) non-air-inhibited [polyester] finish, **hochpigmentierter** ~ heavily loaded paint, paint with high pigment loading, **säurefester** ~ acid-proof varnish
Lackabbeizmittel *n* varnish remover
Lackader *f* varnished conductor
Lackaderdraht *m* varnished wire
Lackanstrich *m* coat of varnish
lackartig varnish-like
Lackauftragswalze *f* fountain roll
Lackbasis *f* lacquer base
Lackbeize *f* varnish stain
Lackbenzin *n* solvent naphtha, turpentine substitute, white spirit
Lackbildner *m* lacquer former, lake former
Lackdraht *m* enamelled wire
Lackemulsion *f* emulsion varnish or paint

Lackentferner *m* paint or varnish remover
Lackester *m* lac ester, varnish ester
Lackextrakt *m* lac extract
Lackfabrik *f* varnish works, paint factory, paint plant
Lackfarbe *f* varnish color, varnish paint
Lackfarbstoff *m* dyestuff for coloring lacquers, lake dyestuff, [pigment] toner
Lackfilmunterbrechungen *pl* imperfections of the lacquer layer
Lackfirnis *m* shellac varnish
Lackgewebe *n* varnished fabric
Lackhantel *f* film applicator
Lackharz *n* coating resin, copal, gum lac, paint resin, varnishing resin
Lackhaut *f* lacquer film
Lackhilfsmittel *n* lacquer auxiliary
lackieren to lacquer, to enamel, to paint, to varnish
Lackierer *m* lacquerer, varnisher
Lackiererei *f* paint shop
Lackiermaschine *f* coating machine, varnishing machine
Lackierpinsel *m* varnish brush
Lackierspritzverfahren *n* paint-spraying system
Lackierung *f* lacquer coating, varnish coating, **ausgesparte** ~ recessed coating, **elektrostatische** ~ lacquering by electrodeposition
Lackierungsmaschine *f* coater
Lackindustrie *f* paint, varnish and lacquer industry
Lackkörper *m* binder, bodied oil
Lacklack *m* lac lake
Lacklasurfarbe *f* transparent varnish color, transparent paint
Lackleder *n* patent leather
Lackleinöl *n* linseed oil for varnish, refined linseed oil
Lackmoid *n* lacmoid
Lackmus *m* litmus, lacmus, turnsole
Lackmusblau *n* litmus blue, turnsole blue
Lackmuspapier *n* litmus paper
Lackmustinktur *f* litmus liquor, litmus solution
Lackpapier *n* coated paper, varnished paper
Lackpigment *n* lake pigment
Lackrohstoffe *pl* paint and lacquer raw materials, raw materials in lacquer formulation
Lacksäure *f* laccaic acid, laccainic acid
Lacksatz *m* varnish sediment
Lackschicht *f* lacquer coat, layer of varnish
Lackschichtschleudergerät *n* centrifugal applicator
Lackschwarz *n* black lake
Lackteilchen *pl* bits
Lacküberzug *m* coat of lacquer
Lackverdünner *m* lacquer solvent, varnish thinner

Lackvergoldung *f* japanner's gilding
Lackverlaufsstörung *f*, **als Porenbildung sichtbare** ~ eye-holing
lac-Operon *n* (Biochem) lac operone
lac-Repressor *m* (Biochem) lac repressor
Lacroixit *m* (Min) lacroixite
Lactacidase *f* (Biochem) lactacidase
Lactacidogen *n* lactacidogen
Lactal *n* lactal
Lactalbumin *n* lactalbumin, milk albumin
Lactam *n* lactam, lactan
Lactamid *n* lactamide, 2-hydroxypropanamide, lactic [acid] amide
Lactaminsäure *f* lactaminic acid
Lactarazulen *n* lactarazulene
Lactarinsäure *f* lactarinic acid, 6-ketostearic acid, 6-oxooctadecanoic acid
Lactarsäure *f* lactaric acid, stearic acid
Lactase *f* (Biochem) lactase
Lactat *n* lactate, salt or ester of lactic acid
Lactatdehydrogenase *f* (Biochem) lactate dehydrogenase
Lacthydroxamsäure *f* lacthydroxamic acid
Lactid *n* lactide
Lactim *n* lactim, lactin
Lactit *m* lactitol
Lactobionsäure *f* lactobionic acid
Lactobiose *f* lactobiose, glucose-galactoside, lactose, milk-sugar
Lactobutyrometer *n* lactobutyrometer
Lactodensimeter *n* lactodensimeter, lactometer
Lactoflavin *n* lactoflavin
Lactoglobulin *n* lactoglobulin
Lactolid *n* lactolide
Lacton *n* lactone
Lactonase *f* (Biochem) lactonase
Lactonbindung *f* lactonic linkage
Lactonitril *n* lactonitrile, 2-hydroxypropane nitrile
Lactonsäure *f* lactonic acid, galactonic acid
Lactophenin *n* lactophenine, 4-lactylamino-phenyl ethyl ether
Lactosamin *n* lactosamine
Lactosazon *n* lactosazone
Lactose *f* lactose, glucose-galactoside, lactobiose, milk-sugar
Lactoskop *n* lactoscope
Lactoson *n* lactosone
Lactucarium *n* lactucarium, lettuce opium
Lactulose *f* lactulose
Lactyl- lactyl
Lactylierung *f* lactylation
Lactylmilchsäure *f* lactylolactic acid
Lacunit *m* lacunite
Ladan *n* labdanum, ladanum
Ladangummi *n* labdanum, ladanum
Ladanum *n* labdanum, ladanum
Ladeaggregat *n* charging set
Ladeanschluß *m* charging connection

Ladebeanspruchung *f* charging load
Ladebühne *f* charging or loading platform or ramp
Ladedauer *f* (Batterie) charging period
Ladedichte *f* charge density, charging density
Ladedruck *m* induction pressure, supercharge pressure
Ladedruckregler *m* boost control, manifold pressure control
Ladedruckschreiber *m* manifold pressure monitor or recorder
Ladefähigkeit *f* loading capacity
Ladegerät *n* battery charger, feeder, loader, ~ **mit Eigenfahrantrieb** self-propelled loader
Ladegewicht *n* charge, gross weight, load, load capacity
Ladegleichrichter *m* (Elektr) charging rectifier
Ladekapazität *f* charge capacity
Ladekoeffizient *m* coefficient of charge
Ladekondensator *m* charging capacitor
Ladekran *m* hoisting crane, loading crane
Ladeleitung *f* intake manifold
Lademarke *f* load line
Lademulde *f* charging box
laden to load, (Elektr) to charge
Laden *m* shop, store
Laden *n* **der Batterie** battery charging, ~ **des Kondensators** charging of a condenser, ~ **eines Akkumulators** battery charging
Ladenguß *m* box casting, casting in flasks
Ladenpreis *m* retail price
Ladentisch *m* counter
Ladeportal *n* loading portal
Ladepotential *n* charging potential
Lader *m* supercharger, blower, charging blower
Laderampe *f* loading platform, loading ramp, platform
Laderaupe *f* crawler
Ladergang *m* supercharger gear
Ladergehäuse *n* blower casing
Laderutsche *f* loading chute
Ladeschaufel *f* bucket
Ladespannung *f* charging voltage
Ladestärke *f* strength of charging current
Ladestation *f* (Batterie) charging station
Ladestrom *m* charging current
Ladestromstärke *f* strength of charging current
Ladeübersetzung *f* blower ratio
Ladeverlust *m* loss of charge
Ladevorrichtung *f* loading device, breech mechanism, (Elektr) charging equipment
Ladewiderstand *m* charging resistance, resistance during charge
Ladung *f* (Elektr) charge, (Last) freight-charging, load, loading, ~ **eines Elektrons** charge of an electron, **die** ~ **verdichten** to intensify the charge, **elektrische** ~ electric charge, electric loading, **entgegengesetzte** ~ opposite charge, **fiktive** ~ fictitious charge, **formale** ~ formal charge, **freie** ~ free charge, **induzierte** ~ induced charge, **negative** ~ negative charge, **positive** ~ positive charge, **spezifische** ~ specific charge, **statische** ~ static charge
Ladung-Radius-Verhältnis *n* charge radius ratio
Ladungsänderung *f* change of charge
Ladungsart *f* kind of charge
Ladungsausgleich *m* charge equalization
Ladungsaustausch *m* charge transfer
Ladungsaustauschspektrum *n* (Spektr) charge-transfer spectrum
Ladungsbereitstellung *f* charge supply
Ladungsdichte *f* density of charge
Ladungsdurchtrittsfaktor *m* charge transfer factor
Ladungseinheit *f* [unit of] charge
Ladungsflächendichte *f* surface density of charge
ladungsfrei zero charge
Ladungsinvarianz *f* charge independence
Ladungskoeffizient *m* coefficient of charge
Ladungsmasseverhältnis *n* charge to mass ratio
Ladungsmultiplett *n* charge multiplet
Ladungsraum *m* charge space
Ladungssäule *f* secondary battery
Ladungssinn *m* nature of a charge
Ladungsspeicherung *f* charge storage (solid-state phys.)
Ladungsspin *m* charge spin
Ladungsstrom *m* charging current
Ladungssymmetrie *f* charge symmetry
Ladungsträger *m* charge carrier
Ladungsträgerdichte *f* [charge] carrier density
Ladungsträgereinbau *m* charge carrier injection
Ladungsträgerproduktion *f* charge carrier generation
Ladungstransport *m* charge transport
Ladungstrennung *f* (Phys) charge separation
Ladungsübergang *m* (Phys) charge transfer
Ladungsüberschuß *m* excess of charge
Ladungsübertragung *f* charge transfer
ladungsunabhängig charge independent
Ladungsunabhängigkeit *f* charge independence
Ladungsverformung *f* charge distortion
Ladungsvermögen *n* [charge] capacity
Ladungsverteilung *f* charge distribution
Ladungsvervielfachung *f* multiplication of charge
Ladungsvorzeichenbestimmung *f* sign of charge determination
Ladungswolke *f* charge cloud
Ladungszahl *f* charge number
lähmen to paralyze, to cripple
Lähmung *f* paralysis, paralyzation
Länge *f* length, (Zeit) duration, **wirksame** ~ (des Keilriemens) pitch length
Längenabweichung *f* longitudinal deviation
Längenänderung *f* longitudinal deformation, deformation in extension

Längenausdehnung *f* linear dimension, linear extension
Längenausdehnungskoeffizient *m* coefficient of linear expansion
Längenbestimmung *f* longitude determination
Längeneinheit *f* unit of length
Längengrad *m* (Geogr) degree of longitude
Längenmaß *n* linear measure, length dimension, (Meßgerät) length gauge
Längenmessung *f* measurement of length, linear measurement, longimetry
Längenmetazentrum *n* longitudinal metacenter
Längentoleranz *f* length allowance
längentreu isometric, length-preserving
Längenübertragung *f* transfer of the length
Längenunterschied *m* difference in length
länglich oblong, ~ **rund** oval [-shaped]
längs lengthwise
Längsachse *f* longitudinal axis
Längsansicht *f* longitudinal view
Längsausdehnung *f* longitudinal expansion
Längsbedeckungsmaschine *f* longitudinal covering machine
Längsbeschleunigung *f* longitudinal acceleration
Längsblasverhältnis *n* draw ratio, longitudinal blow-up ratio
Längsdehnung *f* longitudinal extension
Längsdruck *m* longitudinal pressure
Längsfurche *f* longitudinal groove
Längsinduktion *f* longitudinal induction
Längslager *n* side bearing
Längsmagnetisierung *f* longitudinal magnetization
Längsnaht *f*, **gefalzte** ~ (Dose) interlocked sideseam
Längsnut *f* longitudinal groove, longitudinal slot
Längsrändel *pl* longitudinal knurls
Längsreckvorrichtung *f* "along" stretching device
Längsrichtung *f* longitudinal direction, lengthwise direction, m/c direction
Längsrippe *f* longitudinal rib
Längsriß *m* longitudinal crack
Längsschlitz *m* elongated slot
Längsschneider *m* slitter
Längsschnitt *m* longitudinal cut
Längsschwelle *f* longitudinal tie
Längsschwingung *f* longitudinal oscillation
Längsspant *n* longitudinal frame
Längsspritzkopf *m* axial head
Längsteilung *f* longitudinal spacing, longitudinal pitch
Längsträger *m* longitudinal girder, stringer
Längssummantelung *f* (Drahtkern) straight wrap
Längsvergrößerung *f* (Opt) longitudinal magnification
Längsvorschub *m* longitudinal feed
läppen (Techn) to lap
Läppen *n* (Techn) lapping
Läppmaschine *f* lapping machine

Läppmittel *n* lapping abrasive
Läppscheibe *f* lapping wheel
Lärchenbaumharz *n* larch resin, venetian turpentine
Lärchenholzöl *n* larch wood oil
Lärchenpech *n* larch pitch
Lärchenschwamm *m* agaric of the larch, purging agaric
Lärchenstoff *m* coniferin
Lärm *m* noise, roar
Lärmbekämpfung *f* combating of noise
Lärmbelästigung *f* noise disturbance
lärmempfindlich sensitive to noise
Lärmempfindlichkeit *f* sensitivity to noise, (Med) dysacousia
Lärmintensität *f* noise intensity [or level]
Lärmquelle *f* source of noise
Lärmschädigung *f* injury caused by noise
Lärmschutz *m* noise prevention
Lärmschutzabdeckung *f* (für Flugtriebwerk) acoustic panel
Lärmverbot *n* noise ban
Läufer *m* runner
Läuferbildung *f* running
Läufergewicht *n* sliding weight
Läuferphase *f* rotor phase
Läuferspannung *f* rotor voltage
Läuferstrom *m* rotor current
Läusepulver *n* lice-powder
Läuterbeize *f* white liquor
Läuterboden *m* false bottom, strainer
Läuterbottich *m* clarifying tub, straining vat
Läuterfeuer *n* refining fire
Läuterflasche *f* wash[ing] bottle
Läuterkessel *m* defecator, refining pan
Läutermethode *f* refining method
läutern to purify, to clarify, to percolate, to strain, (Met) to cupel, to refine, (rektifizieren) to rectify
Läutern *n* percolating, straining
läuternd percolating, straining
Läuterofen *m* (Metall) refining kiln
Läuterpfanne *f* clarifier, clearing pan, defecator, dissolving pan
Läutersud *m* first boiling of salt
Läutertuch *n* filter cloth
Läuterung *f* (Chem) clarification, rectification, refining, washing, (Metall) refinement, (Techn) purification, straining
Läuterungsgefäß *n* purifying vessel, melting pot, refining boiler
Läuterungskessel *m* purifying vessel, melting pot, refining boiler
Läuterungslauge *f* refining lye
Läuterungsmittel *n* purifying agent
Läuterungsprozeß *m* refining process
Läuterverfahren *n* (Gold, Silber) scorification
Lävan *n* levan

lävogyr levorotatory, levogyratory, levogyric, rotating to the left
Lävopimarsäure *f* levopimaric acid
Lävulin *n* levulin, synanthrose
Lävulinaldehyd *m* levulinic aldehyde, 4-oxopentanal, levulinaldehyde
Lävulinat *n* levulinate
Lävulinsäure *f* levulinic acid, 4-oxopentanoic acid, levulic acid
lävulinsauer levulinate
Lävulochloralose *f* levulochloralose
Lävulosan *n* levulosan, fructosan
Lävulose *f* levulose, fructose, fruit sugar
Lävulosin *n* levulosin, levulin, syanthrose
Lage *f* (Bergb) bed, layer, stratum, (Platz) location, position, (Schicht) layer, ply, **dünne** ~ film, lamination, sheet, **energetische** ~ energy position, **energetische** ~ **des Defektelektronen-leitenden Zustandes** energy of conductive hole state, **geneigte** ~ inclined position
Lageänderung *f* change of position, disalignment
Lageenergie *f* (Mech) potential energy
Lagegruppe *f* position group
Lagenhaftung *f* ply adhesion
Lagenholz *n* laminated wood, plywood
Lagenleim *m* veneer glue
Lagentrennung *f* ply separation
Lageplan *m* plan of layout, site plan
Lager *n* storehouse, (Geol) bed, deposit, seam, (Schicht) layer, stratum, (Techn) bearing, ~ **aus Schichtstoff** laminated bearing, **das** ~ **ausbuchsen** to bush the bearing, **das** ~ **ausfüttern** to line the bearing, **das** ~ **einschaben** to scrape the bearing, **das** ~ **läuft warm** the bearing runs hot, **das** ~ **nachstellen** to adjust the bearing, **das** ~ **schmieren** to grease the bearing, to lubricate the bearing, to oil the bearing, **einmalig beim Einbau mit Fett geschmiertes** ~ grease-packed bearing, **federndes** ~ spring bearing, **selbstschmierendes** ~ self-lubricating bearing
Lagerabnutzung *f* wear of bearing
Lagerarm *m* bearing bracket
Lageraufnahme *f* inventory
Lagerbalken *m* bearing beam, bearing carrier
Lagerbefestigung *f* bearing retainer
lagerbeständig stable on storage, storable
Lagerbeständigkeit *f* keeping properties, pot life, shelf life, storage life, (Konserve) can stability
Lagerbestand *m* inventory, stock
Lagerblock *m* pillow block
Lagerbuchse *f* bearing bushing
Lagerfachkartei *f* stock card index
Lagerfähigkeit *f* storing property, shelf life, storage life, storage stability
Lagerfläche *f* bearing surface
Lagerfutter *n* bearing bush
Lagergefäß *n* storage vessel

Lagergehäuse *n* bearing cage or casing
Lagergetreide *n* lodged corn
Lagerhaus *n* store, storehouse, warehouse
Lagerkäfig *m* bearing cage
Lagerkeller *m* storage basement, store cellar
Lagerkran *m* stock crane
Lagerkühlung *f* cooling of bearing
Lagerkugel *f* bearing ball
Lagerlegierung *f* antifriction alloy, bearing alloy
Lagermetall *n* antifriction metal, babbitt [metal], bearing metal, white metal, **zinkreiches** ~ zinc-base bearing metal
lagern to shelve, to store, to warehouse, (Techn) to support, to be rested, to rest, **Holz** ~ to season timber
Lagern *n* (Getreide) lodging
Lagernadel *f* bearing needle
Lageroberfläche *f* bearing surface
Lagerplatz *m* depot, stockpile, yard
Lagerprüfmaschine *f* bearing testing machine
Lagerrahmen *m* bearing frame
Lagerreibung *f* bearing friction
Lagerrolle *f* bearing roller
Lagerschale *f* bearing bush, bearing shell
Lagerschmierfett *n* bearing grease
Lagerschmierung *f* lubrication of the bearing
Lagerspiel *n* bearing clearance, slack
Lagerstätte *f* (Bergb) bed, deposit, seam [of ores], **angeschwemmte** ~ alluvial deposit
Lagerstättendruck *m* reservoir pressure
Lagerstättenkunde *f* science of mineral deposits, theory of deposit of ores
Lagerstättenphysik *f* reservoir engineering
Lagertank *m* storage tank
Lagerung *f* storage, storing, (Anordnung) arrangement, (Mech) resting
Lagerungsdichte *f* degree of compaction
Lagerverschleiß *m* wear of the bearing
Lagerversuch *m* storage test, **beschleunigter** ~ accelerated storage test
Lagervorrat *m* stock
Lagerzugriff *m* accessibility
Lagonit *n* lagonite
Lagrangesche Funktion *f* (Math) Lagrangian function
lahmlegen to render ineffective, to neutralize, to paralyze, to tie up
Laib *m* loaf
Laich *m* (Zool) spawn
Laichfisch *m* (Zool) seed fish
Laichzeit *f* spawning time
Laifan *n* laifan
Lake *f* brine, pickle
Lakeprober *m* salinometer
Lakkolith *m* laccolite
Lakmoid *n* lacmoid
Lakritze *f* licorice, liquorice
Lakritzenholz *n* licorice root
Lakritzenwasser *n* licorice water

Laktaldehydreduktase f (Biochem) lactaldehyde reductase
Laktam n lactam
Laktat n lactate
Laktation f (Med) lactation
Laktationshormon n lactogenic hormone, prolactin
Laktatracemase f (Biochem) lactate racemase
Laktobiose f lactobiose, milk-sugar
Laktoglobulin n lactoglobulin
Laktoprotein n lactoprotein
Laktose f lactose, milk sugar
Lallementiaöl n lallementia oil
Lambert n (Opt) lambert (unit of brightness)
Lambert-Beersches Gesetz n Lambert-Beer law
Lambertin n lambertine
Lambertit m (Min) lambertite
Lambertsches Gesetz n Lambert's law
lamellar foliated, lamellar, lamellated, lamellose, laminated, plate-like
Lamelle f lamina, layer
lamellenartig lamellar, laminated
Lamellenbremse f multiple-disc brake, multiple-plate brake
Lamellenkühler m (Kfz) cellular-type radiator
Lamellenmagnet m compound magnet, laminated magnet
Lamellenröhre f lamellar tube
Lamellenspan m discontinuous curled chip
Lamellenstruktur f lamellar structure, lamination
Lamellentechnik f vane technique
Lamellenverdampfer m lamella evaporator
Lamellenwärmeaustauscher m lamellar exchanger, segment heat exchanger
lamellieren (Techn) to laminate
Lametta f n silver tinsel
laminar (Strömung) laminar, steady, streamline
Laminaribiit m laminaribiitol
Laminaribiose f laminaribiose
Laminarströmung f (Mech) laminar flow
Lammkies m (Min) marcasite
Lampe f lamp, (Elektron) tube, ~ **mit Reflektor** reflecting lamp, **gasgefüllte** ~ gas-filled lamp
Lampenbrennstunde f lamp hour
Lampenfassung f lamp holder, lamp socket
Lampengehäuse n lamp house
Lampenofen m lamp furnace
Lampenruß m lamp black, carbon black, channel black
Lampenschwarz n lamp black, carbon black
Lampenteller m lamp plate
Lanacylviolett n lanacyl violet
Lanan n lanane
Lanarkit m (Min) lanarkite
Lancasterit m lancasterite
Lanceol n lanceol
Lanceolatin n lanceolatin

Land n land, (Erdboden) ground, soil, **neugewonnenes** ~ reclaimed ground
Landebahn f (Luftf) landing strip, runway
Landegeschwindigkeit f landing speed
Landeklappe f landing flap
landen to land
Landgewinnung f reclamation
Landröte f (Bot) madder
Landschaftsstein m (Min) florence stone, florentine marble
Landung f landing
Landvermessung f geodesy
Landwirtschaft f agriculture
landwirtschaftlich agricultural[ly]
Landwirtschaftskunde f husbandry
Landwirtschaftslehre f agricultural science
Landwirtschaftsministerium n Department of Agriculture (Am. E.), Fisheries and Food (Br. E.), Ministry of Agriculture
Landwirtschaftsplanung f agricultural planning
Landwirtschaftsschule f agricultural school
lang long
langarmig long-armed
Langbanit m (Min) langbanite
Langbeinit m (Min) langbeinite
Langerhanssche Inseln pl (Biol) islets of Langerhans
langfaserig long-fibered
langflammig long-flamed, long-flaming
Langflammrohr n long-flame tube
Langformat n oblong size
langfristig long-term
langhalsig long-necked
langkettig long-chain
langlebig long-lived
Langlebigkeit f longevity
Langlochbrenner m long-slot burner
Langlochfräser m end mill, **Zweischneider-** ~ two lipped end mill
Langlochfräsmaschine f slot milling machine
Langmuir Mulde f Langmuir trough
Langmuirsche Adsorptionsisotherme f Langmuir's adsorption isotherm
langölig long-oil
Langrohrverdampfer m long-tube evaporator
Langrost m long grate
langsam slow, ~ **auslaufen** to ooze out, ~ **kochen** to simmer, ~ **ziehend** (Färb) slow dyeing
Langsambinder m slow-setting cement
Langsamlauf m low-speed run, slow running
Langschleifen n long grinding
Langsieb n (Pap) endless wire
Langsiebmaschine f fourdrinier
Langstreckenflug m long-distance flight
Langwelle f long wave
Langwellenbereich m long-wave range
Langwellenempfänger m long-wave receiver
Langwellensender m long-wave transmitter

langwellig long-wave, of long wave length
langwierig long-term, tedious, wearisome
Langzeitbeanspruchung f sustained loading
Lanital n lanital
Lanocerinsäure f lanoceric acid
Lanolin n lanolin, hydrous wool fat, lanum
Lanolincreme f lanolin cream
Lanolinerzeugung f lanolin manufacture
Lanolinsäure f lanolin acid, wool fat acid
Lanopalminsäure f lanopalmic acid
Lanostan n lanostane
Lanosterin n lanosterol
Lanosterol n lanosterol
Lansfordit m (Min) lansfordite
Lantanaöl n lantana oil
Lantanursäure f lantanuric acid
Lanthan n lanthanum (Symb. La)
Lanthanammoniumnitrat n lanthanum ammonium nitrate
Lanthancarbonat n lanthanum carbonate
Lanthanchlorid n lanthanum chloride
Lanthanid n lanthanide
Lanthaniden f pl lanthanide elements, lanthanides
Lanthanidenreihe f lanthanide group
Lanthankaliumsulfat n lanthanum potassium sulfate
Lanthannitrat n lanthanum nitrate
Lanthanoxid n lanthanum oxide
Lanthansulfat n lanthanum sulfate
Lanthansulfid n lanthanum sulfide
Lanthionin n lanthionine
Lanthopin n lanthopine, lantol
Lapachenol n lapachenole
Lapachol n lapachoic acid, lapachol, targusic acid
Lapachon n lapachone
Lapachosäure f lapachoic acid, lapachol, targusic acid
Lapinon n lapinone
Lapis m (Lat) stone
Lapislazuli m (Min) lapis lazuli, azurestone, lazurite
Laplacesche Differentialgleichung f Laplace's differential equation
Laplace-Transformation f Laplacian transformation
Lappaconitin n lappaconitine
Lappengriffstopfen m pennyhead [or coin] stopper
Lappenkopf m lug head, thumb head
Larderellit m (Min) larderellite
Lardöl n lard oil
Laricinolsäure f laricinolic acid
Laricinsäure f laricinic acid, maltol
Lariciresinol n lariciresinol
Larixinsäure f larixic acid
Larmorpräzession f (Atom) larmor precession
Larve f (Biol) larva, chrysalis, grub

Larvicid n larvicide
Lasche f flap, (an Schienen) fish-plate, (Stoßplatte) butt-strap
laschen (Schiff) to lash, (Techn) to butt-strap
Laschenbolzen m fish-plate bolt
Laschenkaliber n fish-plate pass
Laschennietung f butt-joint rivetting
Laschenverbindung f strap connection
Laser m laser [light beam], **Festkörper-** ~ solid laser, **Gas-** ~ gas laser, **Rubin-** ~ ruby laser
Laseranlage f laser equipment
Laseranregung f laser source
Laserdiode f laser diode
Lasergerät n laser apparatus
Laserimpuls m laser impulse
Laserkristall m laser crystal
Laserol n laserol
Laseron n laserone
Laserpitin n laserpitin
Laserstrahl m laser beam
Lasertechnik f laser method
Laserverstärker m laser multiplier
lasieren to glaze
lasierend transparent
Lasierfarbe f glazing color
Lasierung f glazing
Lasiocarpin n lasiocarpine
Lasiocarp[in]säure f lasiocarpic acid
Lasionit m (Min) lasionite
Last f load, burden, weight, **bewegliche** ~ live load, **ruhende** ~ dead load
Laständerung f load change
Lastangriffspunkt m point of application of load
Lastarm m weight arm
Lastbügel m triangular lifting eye
Lastdehnungskurve f load-extension curve
Lastenaufzug m freight elevator (Am. E.), goods lift (Br. E.)
Lastex n lastex
Lastexfaden m lastex yarn
Lastfläche f loaded area
Lasthebemagnet m lifting magnet
Lastkraftwagen m (Lkw) truck (Am. E.), lorry (Br. E.)
Lastschaltgetriebe n load-shift gearing
Lastunabhängigkeit f load independence
Lastverteilung f load distribution
Lastwagenfelge f truck rim
Lastwert m load value
Lasur f glaze
Lasurblau n ultramarine
Lasurfähigkeit f opacity
Lasurfarbe f glazing composition
Lasurgrün n green bice
Lasurit m (Min) azurite, lapis lazuli
Lasurlack m clear varnish, transparent lacquer, [transparent] varnish
Lasurquarz m (Min) siderite
Lasurspat m (Min) azure spar, lazulite

Lasurstein *m* (Min) azure stone, lapis lazuli, lazurite, native ultramarine
latent latent
Latenz *f* latency
Laterit *m* (Min) laterite
Latex *m* (Bot) latex
Latexanstrich *m* latex paint
Latexmühle *f* latex mill
Latexschaum *m* foam rubber
Lathosterin *n* lathosterol
Lathyrusfaktor *m* lathyrus factor
Latrobit *m* latrobite
Lattengradierwerk *n* lattice cooling stack
Lattenkiste *f* crate
Lattichopium *n* lettuce opium
Latwerge *f* (Bot, Pharm) electuary
Laub *n* foliage, leaves
Laubanit *m* laubanite
Laubapfel *m* gallnut, oak-apple
Laubbaum *m* deciduous tree
Lauberde *f* leaf-mold
Laubwald *m* deciduous wood
Lauch *m* (Bot) leek
lauchgrün leek-green
Lauchhederichöl *n* garlic mustard oil
Laudanidin *n* laudanidine, tritopine
Laudanin *n* laudanine
Laudanosin *n* laudanosine
Laudanosolin *n* laudanosoline
Laudanum *n* laudanum, tincture of opium
Laue-Diagramm *n* Laue diagram
Laue-Flecke *pl* Laue spots
Laue-Verfahren *n* Laue method
Lauf *m* course, run, (Fließen) current
Laufbahn *f* path
Laufbandtrockner *m* screen-belt dryer
Laufbuchse *f* liner
Laufbühne *f* platform
Laufdecke *f* (Reifen) tread
Laufdeckenwulst *m* bead [of tire]
Laufdurchmesser *m* (Keilriemen) belt diameter
Laufeigenschaften *pl* running properties, (eines Lagers) antifrictional qualities (of a bearing)
Lauffeldröhre *f* travelling wave tube
Laufffläche *f* (Reifen) tire casing, tread
Laufflächenkrümmung *f* tread radius
Laufflächenplatte *f* tread gum
Laufflächenprofil *n* tread design, pattern design
Laufflächenreparatur *f* spot repair
Laufflächenzone *f* crown region
Laufgeschwindigkeit *f* (Elektr) sweep speed
Laufgewicht *n* sliding weight
Laufgewichtswaage *f* sliding weight balance, balance with sliding weight, steel yard
Laufglasur *f* flow glaze, running glaze
Laufkatze *f* [crane] trolley, [crane] carriage
Laufkatzenausleger *m* trolley jib
Laufkatzenfahrbahn *f* trolley track
Laufkatz-Knickausleger *m* articulated trolley jib

Laufkran *m* [overhead-]travelling crane
Laufkranz *m* girth ring
Laufmasche *f* (Text) ladder, runner
laufmaschenfest ladder-proof, run-proof
Laufrad *n* impeller, track wheel, vane wheel, (Dampfturbine) rotor, (hydraul. Turbine) runner
Laufradflügel *m* impeller blade
Laufrichtung *f*, **quer zur ~** cross-grain
Laufring *m* bearing race, girth ring
Laufrolle *f* caster, travelling roller, (Stromabnehmerrolle) trolley
Laufschiene *f* [guide] rail, sliding rail
Laufseite *f* **des Riemens** contact side of the belt
Laufstreifen *m* (Reifen) tread
Laufwelle *f* line shaft
Laufwerk *n* driving gear
Laufwinkel *m* (Laufzeitwinkel) transit phase angle
Laufzeit *f* operating time, test time, (einer Maschine) cycling time, (Lebensdauer) service life, (Motor) running time
Laufzeitmassenspektrograph *m* time-of-flight mass spectrograph
Laufzeitmessung *f* time-of-flight measurement
Laufzeitmethode *f* time-of-flight technique
Laufzeitspektrograph *m* velocitron
Laufzeitspektrometer *n* time-of-flight spectrometer
Lauge *f* (Chem) lye, alkali, base, lixivium, (Salzlauge) brine, (Seifenlauge) suds, **die ~ schärfen** to strengthen the lye
Laugeanlage *f* leaching plant, lixiviation plant
Laugebehälter *m* leaching vat or tank
laugebeständig alkali-proof, resistant to alkali[e]s
Laugebottich *m* leaching vat, lixiviation vat
Laugekessel *m* lye vessel
Laugekühler *m* sludge cooler
laugen to leach, to buck, to lixiviate, to steep in lye
Laugenabfüllapparat *m* alkaline liquor emptying apparatus
Laugenanlage *f* leaching plant, lixiviation plant
laugenartig alkaline, lixivious, resembling lye
Laugenasche *f* buck ashes, potash, soapboiler's waste
Laugenaschensieber *m* potash-sifter
Laugenbad *n* alkaline bath, liquor, lye bath, steep
Laugenbehälter *m* lye container, lye tank
laugenbeständig alkaliproof
laugenecht alkali-proof, fast to lye
Laugenfaß *n* leaching vat, lye vat
laugenfest alkali-resistant, alkali-resisting
Laugenfestigkeit *f* alkali resistance
Laugenflüssigkeit *f* alkaline solution, liquor, lye [bath]
Laugenkammer *f* alkali compartment
Laugenkessel *m* lye vat, caustic pot

Laugenkreislauf *m* circulation of the lye
Laugenkreislaufgeschwindigkeit *f* velocity of circulation of the lye
Laugenleitung *f* alkali conduit
Laugenmesser *m* alkalimeter
Laugenpumpe *f* alkali pump
Laugenregenerierung *f* lye recovery
Laugenreinigung *f* purification of the liquor
Laugenrückstand *m* leach residue
Laugensalz *n* alkali, alkaline salt, lixivial salt
Laugenschälen *n* chemical peeling
Laugenstein *m* caustic soda
Laugentemperatur *f* brine temperature, lye temperature
Laugentuch *n* bucking cloth
Laugenumlauf *m* alkali or lye circulation
Laugenumwälzanlage *f* liquor circulation plant
laugenunlöslich insoluble in alkali
Laugenverarbeitung *f* lye processing
Laugenverdampfer *m* lye evaporator
Laugenverteiler *m* liquor distributor
Laugenwaage *f* brine gauge, lye hydrometer
Laugenwäsche *f* caustic washing
Laugenwasser *n* weak lye, buckwater
laugenwiderstandsfähig alkali-resistant, lye-resisting
Laugenzelle *f* alkali compartment
Laugenzuführung *f* feed of lye, introduction of lye
Laugenzulauf *m* alkali feed
Laugenzusammensetzung *f* lye composition
Laugerückstand *m* leach residue, lixiviation residue, slimes
Laugung *f* leaching, lixiviation
laugungsfähig leachable
Lauralkonium *n* lauralkonium
Lauramid *n* lauramide
Lauran *n* laurane
Laurat *n* laurate
Laurelin *n* laureline
Lauren *n* laurene, pinene
Laurenon *n* laurenone
Laurentsäure *f* laurent acid
Laurin *n* laurin, glyceryl laurate, trilaurin
Laurinaldehyd *m* lauric aldehyde, dodecanal, lauraldehyde
Laurinsäure *f* lauric acid, dodecanoic acid
Laurinsäureester *m* laurate
Laurionit *m* (Min) laurionite
Laurit *m* (Min) laurite, native ruthenium sulfide
Laurodimyristin *n* laurodimyristine, glyceryl laurate dimyristate
Laurodistearin *n* laurodistearin, glyceryl laurate distearate
Lauroleinsäure *f* lauroleic acid
Laurolen *n* 1,2,3-trimethyl-1-cyclopentene, laurolene
Laurolitsin *n* laurolitsine
Lauron *n* 12-tricosanone, laurone

Lauronitril *n* lauronitrile
Lauronolsäure *f* lauronolic acid, laurolene-3-carboxylic acid
Lauronstearin *n* laurostearin, glyceryl laurate stearate
Laurotetanin *n* laurotetanine
Lauroylalanin *n* lauroylalanine
Laurylalanylglycin *n* laurylalanylglycine
Laurylalkohol *m* 1-dodecanol, lauryl alcohol, n-dodecyl alcohol
Laurylglycin *n* laurylglycine
Laus *f* louse (pl. lice)
laut loud, (lärmend) noisy
Laut *m* (Akust) sound, tone
Lautarit *m* (Min) lautarite
Lautempfindung *f* sound perception
lauter pure, (Met) unalloyed
lautgetreu high-fidelity, orthophonic
Lautinduktor *m* ringing inductor
Lautit *m* lautite
Lautsprecher *m* loudspeaker, megaphone
Lautsprecheranlage *f* loud-speaking unit, public address system
Lautsprechermembrane *f* speaker cone
Lautstärke *f* sound intensity, sound volume
Lautstärkemesser *m* sound intensity meter, volume indicator
Lautstärkemessung *f* sound measurement
Lautstärkepegel *m* level of sound, loudness level
Lautstärkeregler *m* volume regulator
Lautverstärker *m* (Akust) sound amplifier
lauwarm lukewarm, tepid
Lava *f* lava, **erstarrte** ~ solidified lava
lavaähnlich lava-like
Lavagestein *n* lava rocks
Lavaglas *n* vitreous lava, volcanic glass
Lavaguß *m* volcanic ashes
Lavandulol *n* lavandulol
Lavandulylsäure *f* lavandulic acid
Lavastein *m* lava rock
Lavastrom *m* lava flow, stream of lava
Lavendel *m* (Bot) lavender
Lavendelblau *n* lavender blue
Lavendelblüten *pl* (Bot) lavender flowers
Lavendelfarbe *f* lavender
lavendelfarben lavender-colored
Lavendelgeist *m* extract of lavender
Lavendelöl *n* lavender oil
Lavendelwasser *n* lavender water
Lavenit *m* (Min) lavenite
Lavezstein *m* potstone, steatite
Lavierfeuer *n* slow fire
Lawine *f* avalanche
Lawinendurchschlag *m* avalanche breakdown
Lawinenschwankung *f* avalanche fluctuation
Lawrencit *m* (Min) lawrencite
Lawrencium *n* lawrencium (Symb. Lr)
Lawson *n* lawsone
Lawsonit *m* (Min) lawsonite

Laxans *n* (Abführmittel, Pharm) laxative
Laxativum *n* (Pharm) laxative
Laxiermittel *n* laxative, cathartic, purgative
Lazulinblau *n* lazuline blue
Lazulit[h] *m* (Min) lazulite, blue spar
Lazurstein *m* (Min) lapis lazuli
LBV-Verfahren *n* (Schmelzverf) vacuum electric arc
Leadhillit *m* (Min) leadhillite
Leatheroid *n* leatheroid
Lebendvirusvakzine *f* live virus vaccine
Lebensdauer *f* durability, life, life-time, service life, **lange** ~ longevity, **mittlere** ~ average life, mean life, **vergleichbare** ~ comparative lifetime
lebenserhaltend life-sustaining
Lebenserwartung *f* life expectancy
lebensfördernd vital
lebensgefährlich highly dangerous
Lebensmittel *pl* food, foodstuffs
Lebensmittelchemie *f* food chemistry
Lebensmittelchemiker *m* food[stuff] chemist
Lebensmittelfälschung *f* food adulteration
Lebensmittelfarbstoff *m* food color
Lebensmittelgesetz *n* food law
Lebensmittelindustrie *f* food industry
Lebensmittelkonservierung *f* conservation of food, preservation of food
Lebensmitteltechnologie *f* food technology
Lebensmitteluntersuchung *f* food analysis
Lebensmittelvergiftung *f* food poisoning
Lebensmittelverpackung *f* food wrapping
Lebensmittelzusatz *m* food additive
lebensnotwendig vital
lebensvernichtend life-destroying
Lebensvorgang *m* vital process
lebenswichtig vital
Lebenszyklus *m* (Biol) life cycle
Leber *f* liver
Leberblende *f* (Min) zincblende
Lebereisenerz *n* (Min) pyrrhotite
Leberenzym *n* (Biochem) liver enzyme
Lebererz *n* (Min) hepatic ore, hepatic cinnabar, liver-brown cinnabar
Leberkies *m* (Min) hepatic pyrites, magnetic pyrites, pyrrhotite
Leberkobalt *m* (Min) hepatic cobalt
Leberkraut *n* (Bot) hepatica
Lebermittel *n* (Pharm) hepatic remedy, liver remedy
Leberopal *m* (Min) brown opaque opal, menilite
Leberschwefel *m* hepar sulfuris, liver of sulfur
Leberstärke *f* glycogen, animal starch, liver starch
Leberstein *m* hepatite, liverstone
Leberstoffwechsel *m* liver metabolism
Lebertran *m* cod-liver oil, fish-liver oil
Leberzelle *f* hepatic cell, liver cell
Lebewesen *n* living being, living organism

lebhaft werden to become vigorous
Lebhonig *m* clarified honey
Lebkuchen *m* Nuremberg gingerbread
Leblanc-Verfahren *n* Leblanc process
Lecanorolsäure *f* lecanorolic acid
Lecanorsäure *f* lecanoric acid, diorsellinic acid
Le Chateliersches Prinzip *n* (des kleinsten Zwanges) Le Chatelier's principle (of least restraint)
Lecithalbumin *n* lecithalbumin
Lecithin *n* lecithin
Lecithinase *f* (Biochem) lecithinase
leck leaking, leaky, ~ **werden** to leak
Leck *n* leak, leakage
Leckbehälter *m* collector tank
Leckdampf *m* leakage steam
Lecken *n* leakage
Leckfluß *m* leakage flow
Leckprüfung *f* leak test
lecksicher leak-proof
Leckstelle *f* leakage
Leckströmung *f* leakage flow
Leckstrom *m* leakage flow, (Elektr) leakage current
Lecksuche *f* detection of leakages, leak detection
Lecksucher *m* leakage tester, leak detector
Lecksuchgerät *n* leak detector
Lecksuchröhre *f* leak detector tube
Lecksuchsonde *f* leak detector head
Leckverlust *m* leakage rate
Leckwasserpumpe *f* bilge pump
Leclanché-Element *n* Leclanché cell
Lecontit *m* (Min) lecontite
Ledeburit *m* ledeburite
Leden *n* ledene
Leder *n* leather, **fettgares** ~ oildressed leather, **genarbtes (körniges)** ~ chagrin, shagreen, **Wasch-** ~ chamois leather, shammy leather
Lederabschabsel *n* leather waste, scrap leather
lederähnlich leather-like, coriaceous, leathery
lederartig leather-like, coriaceous, leathery
Lederbereitung *f* leather dressing
Lederdeckfarbe *f* leather-coating color
Lederdichtung *f* leather gasket, leather packing, leather sealing
Lederersatz *m* leather substitute
Lederfarbe *f* leather color
lederfarben leather-colored, buff, buff-colored
Lederfaserwerkstoff *m* leather board, leather substitute
Ledergelb *n* buff, (Chem) chrysaniline
Ledergrube *f* mastering-pit
Lederhohlkehle *f* leather hollow
Lederimitation *f* imitation leather
Lederindustrie *f* leather industry
Lederkissen *n* leather pad
Lederkobalt *m* (Min) yellow earthy cobalt
Lederkohle *f* leather charcoal
Lederlack *m* leather finish, leather varnish

Lederleim – Lehmschicht 466

Lederleim *m* leather glue, skin glue
Ledermanschette *f* cup leather
Ledernarbung *f* grain
Lederöl *n* leather [dressing] oil, dubbing
Lederpflegemittel *n* leather dressing
Lederriemen *m* leather strap, thong
Lederring *m* leather ring
Lederscheibe *f* leather washer
Lederschmiere *f* degras, dubbing
Lederstulp *m* leather cup
Ledersubstanz *f* coriaceous substance
Ledertreibriemen *m* leather driving belt, (Auto) vee belt
Lederwichse *f* leather polish
Lederzurichtung *f* leather dressing, leather finishing
Ledienosid *n* ledienoside
Ledol *n* ledol, ledum camphor
Ledsäure *f* ledic acid
leer empty, void, (unbesetzt) vacant
Leerblasen *n* (Dest) blowing
Leere *f* emptiness, vacancy, vacuum, void
leeren to empty, to clear, to void, (Phys) to evacuate
Leergang *m* idle motion, idle running
Leergangreibung *f* no-load friction
Leergewicht *n* dead weight, tare
Leerlast *f* dead load
Leerlauf *m* idle motion, idling, no-load, running without load
Leerlaufadmittanz *f* idle admittance
Leerlaufbuchse *f* idle bush
Leerlaufdrehzahl *f* idling or idle speed
Leerlaufdüse *f* idle nozzle
leerlaufen to run idle, to run without load
Leerlaufreibung *f* idle friction
Leerlaufrolle *f* idle wheel, idler
Leerlaufspannung *f* open-circuit voltage
Leerlaufspritzzeit *f* dry cycle time
Leerlaufstrom *m* no-load current
Leerlaufventil *n* idle valve
Leerlaufverbrauch *m* consumption under idling condition
Leerlaufversuch *m* no-load test
Leerpackung *f* display package, dummy
Leerraum *m* vacuum
Leer-Rohrnetz *n* standby channels
Leerscheibe *f* loose pulley
Leerstelle *f* vacancy, hole, interstice
Leerstellenbildung *f* vacancy formation
Leerstellendiffusion *f* vacancy diffusion
Leerstellenkonzentration *f* vacancy concentration
Leerstellenpaar *n* vacancy pair
Leerstellenwanderung *f* vacancy migration
Leertakt *m* (Motor) idle stroke
Leerung *f* emptying, clearing, evacuation
Leerversuch *m* blank experiment, blank test, no-load test

Leervolumen *n* (Hydrodynamik; Füllkörperkolonne) void fraction
Leerzeile *f* (Buchdr) white line [between passes]
Legeeisen *n* hearth plate
legierbar alloyable
Legierbarkeit *f* alloyability, alloying property
legieren to alloy, (mit Quecksilber) to amalgamate
Legieren *n* alloyage, alloying, ~ **mit Quecksilber** amalgamation
Legierstahl *m* alloy steel
Legierung *f* alloy, composition metal, (mit Quecksilber) amalgam, ~ **auf Aluminiumgrundlage** aluminum-based alloy, **antimagnetische** ~ non-magnetic alloy, **antimonhaltige** ~ antimony alloy, **binäre** ~ binary alloy, **hochschmelzende** ~ high-melting-point alloy, **hoch warmefeste** ~ superalloy, **niedrigschmelzende** ~ low-melting-point alloy, **paramagnetische** ~ paramagnetic alloy, **schmelzbare** ~ fusible alloy
Legierungsanteil *m* alloying contribution
Legierungsbestandteil *m* alloying component, alloying constituent
Legierungsbildung *f* formation of alloy
Legierungsblech *n* alloy sheet
Legierungselement *n* alloying component
Legierungsflecke *pl* freckles
Legierungsgehalt *m* alloying content
Legierungskörper *m* alloy constituent
Legierungspyrometer *n* cupel pyrometer
Legierungsstahl *m* alloy steel
Legierungsverfahren *n* alloying method
Legierungszusatz *m* alloying addition
Legierungszustand *m* alloyed state
Legumin *n* legumin, avenin
Lehm *m* loam, clay
lehmartig loamy
Lehmbaustein *m* unburnt sun-dried brick
Lehmform *f* loam mold
Lehmformerei *f* loam molding
Lehmformguß *m* loam molded casting
Lehmgrube *f* clay pit, loam pit
Lehmguß *m* loam casting, iron cast in a loam mold
lehmhaltig loamy, argillaceous, clayey
lehmig argillaceous, clayey, loamy, muddy
Lehmkern *m* loam core
Lehmkitt *m* loam lute
Lehmknetmaschine *f* pugmill
Lehmkuchen *m*, **den** ~ **herrichten** to prepare the loam cake
Lehmmergel *m* clay marl, loamy marl
Lehmmesser *n* loam beater
Lehmmörtel *m* clay mortar, loam mortar
Lehmplatte *f* clay plate
Lehmschicht *f* loam coat

Lehmstein *m* loam brick, sun-dried brick, unburnt brick
Lehmtünche *f* loaming
lehm- und sandhaltig argilloarenaceous
Lehmziegel *m* clay brick
Lehre *f* (Meßgerät) ga[u]ge, pattern, templet, (Theorie) theory, doctrine, science, ~ **vom Galvanismus** galvanology
Lehrenablesung *f* gauge reading
Lehrenkörper *m* blanks
Lehrentoleranz *f* gauge tolerance
Lehrgang *m* course, curriculum
Lehrling *m* apprentice
Lehrmeinung *f* hypothesis
Lehrmodell *n* demonstration model, instruction model
Lehrringe *pl* ring gauges
Lehrsatz *m* proposition, theorem
Lehrstoff *m* subjects taught
Lehuntit *m* lehuntite
Leiche *f* cadaver, corpse
Leichenalkaloid *n* ptomaine
Leichengift *n* ptomaine
Leichenöffnung *f* autopsy
Leichenwachs *n* adipocere
leicht easy, light, slight, ~ **brennbar** highly inflammable, ~ **entfernbar** easily removable, ~ **entzündlich** easily inflammable, ~ **oxydierbar** easily oxidizable, ~ **reduzierbar** easily reducible, ~ **schmelzbar** easily fusible, low-melting, ~ **verderblich** perishable, ~ **zerbröckelnd** friable, ~ **zersetzlich** easily decomposable
Leichtbau *m* lightweight construction
Leichtbauweise *f* lightweight construction method
Leichtbenzin *n* light gasoline, light distillate fuel (LDF), light petrol, white spirit
Leichtbeton *m* light-weight concrete
Leichtbrand *m* underburnt material (cement)
leichtentzündlich highly inflammable
leichtflüchtig readily volatile
Leichtflüssigkeit *f* mobility
Leichtflußmetallegierung *f* easily fusible alloy
Leichtlegierung *f* light alloy
leichtlöslich (Chem) easily soluble, readily soluble
Leichtmetall *n* light [weight] metal
Leichtmetallbearbeitung *f* light-metal working
Leichtmetallguß *m* light-metal casting
Leichtmetall-Legierung *f* light [metal] alloy
Leichtöl *n* light oil, spindle oil
Leichtradsatz *m* light wheel set
Leichtschmelzbarkeit *f* easy fusibility
leichtsiedend low-boiling, of low boiling point, volatile
Leichtstoffbauweise *f* light density construction
Leichtvernicklung *f* light nickelling
Leichtwasserreaktor *m* light water reactor

Leichtzuschlagbeton *m* lightweight aggregate concrete
Leidener Flasche *f* (Elektr) Leyden jar, electrical jar
Leidenfrostsches Phänomen *n* Leidenfrost phenomenon
Leidyit *m* leidyite
Leierziehbank *f* drawing machine with spool
Leifit *m* (Min) leifite
Leim *m* glue, mucilage, (Pap) size, **flüssiger** ~ liquid glue, mucilage, **kalthärtender** ~ cold-setting adhesive
Leimanlage *f* glue plant
Leimanstrich *m* coat of glue
leimartig gluey, gelatinous, glutinous
Leimauftrag *m* glue spread, **beidseitiger** ~ double layer glue spread, **einseitiger** ~ single layer glue spread
Leimauftragsmaschine *f* glue spreader
Leimbrühe *f* size
leimen to glue, to cover with glue, to gum, to size
Leimen *n* gluing, sizing
Leimfarbe *f* limewash, non-washable distemper, non-washable water paint
Leimfestigkeit *f* resistance due to sizing, bonding strength
Leimflüssigkeit *f* size
Leimform *f* gelatine mold
Leimfuge *f* glue joint, glue line
Leimgallerte *f* glue jelly, gelatin jelly
Leimgewebe *n* (Bindegewebssubstanz) collagen
Leimgradprüfung *f* sizing test
Leimgrund *m* glue-priming
Leimgut *n* glue stock
leimig gluey, gluish, glutinous, sticky
Leimkitt *m* joiner's cement, putty
Leimleder *n* leather cuttings, parings of skin
Leimlösung *f* glue solution
Leimmaschine *f* glue machine, gluing machine
Leimpresse *f* glue press
Leimspachtel *f* glue spreader
Leimstelle *f* glued joint
Leimstoff *m* gluten, gummy material, sizing material
Leimsüß *n* (obs) glycine, gelatin sugar, glycocoll
Leimung *f* (Pap) gluing, sizing
Leimvergoldung *f* gilding on water-size
Leimwasser *n* glue water, glue solution, size
Leimwerk *n* sizing press
Leimzucker *m* (obs) gelatin sugar, glycine, glycocoll
Lein *m* (Bot) flax
Leindotteröl *n* cameline oil
Leinen *n* (Text) linen
Leinenbindung *f* (Text) plain weave
Leinenpapier *n* linen [rag] paper
Leinenschuh *m* canvas shoe
Leinkraut *n* (Bot) toad flax
Leinkuchen *m* linseed oil cake

Leinöl *n* linseed oil
Leinölfarbe *f* linseed oil paint
Leinölfettsäure *f* linseed oil fatty acid
Leinölfirnis *m* boiled linseed oil, linseed oil varnish, lithographic varnish
Leinöllack *m* boiled linseed oil, linseed oil varnish
Leinölsäure *f* linoleic acid, 9,12-octadecadienoic acid, linolic acid
leinölsauer linoleate
Leinöl-Standöl *n* linseed standoil
Leinsalbe *f* flaxweed salve
Leinsamen *m* (Bot) linseed
Leinsamenabkochung *f* linseed decoction
Leinwand *f* (Text) canvas, cloth, linen
Leipzigergelb *n* Leipzig yellow, chrome yellow, lead chromate
Leiste *f* (Blendleiste) strip, (dünne) slat, (Querleiste) cleat
leisten to perform
Leistenzelle *f* strip cell
Leistung *f* output, performance, [working] capacity, (Ausbeute) production, yield, (Bagger) stripping capacity, (Masch) rating, (Nutzeffekt) effective power, efficiency, **abgegebene** ~ power output, **aufgenommene** ~ power input, **scheinbare** ~ apparent power
Leistungsabfall *m* power drop, loss of power
Leistungsaufnahme *f* [power] input, power consumption
Leistungsaufwand *m* power input
Leistungsbedarf *m* power requirements, (Elektr) power demand
Leistungsbereich *m* performance level, range of capacity
Leistungsbremse *f* power brake
Leistungsdreieck *n* power diagram
Leistungsempfindlichkeit *f* power sensitivity
leistungsfähig efficient
Leistungsfähigkeit *f* efficiency, capability, performance, productiveness, (Betrieb) capacity, productivity
Leistungsfähigkeitsprobe *f* performance testing
Leistungsfaktor *m* performance coefficient, power factor
Leistungsgabe *f* power output
Leistungsgleichrichter *m* power rectifier
Leistungsgrenze *f* limit of capacity, performance limit
Leistungskapazität *f* capacity
Leistungskennlinie *f* power or performance curve
Leistungskonstante *f* power constant
Leistungskurve *f* power curve
Leistungsmaximum *n* output maximum
Leistungsmesser *m* output meter, (Elektr) wattmeter

Leistungsmessung *f* power or performance measurement
Leistungsreaktor *m* power pile, power reactor
Leistungsregelung *f* control of output
Leistungsspannung *f* active voltage
Leistungsspielraum *m* performance margin
Leistungssteigerung *f* increase in output or efficiency or performance, output boost
Leistungsstrom *m* effective current
Leistungsverlust *m* power loss
Leistungsverminderung *f* power cutback
Leistungsvermögen *n* functional capacity
Leistungsverstärker *m* power amplifier
Leistungsverstärkung *f* power amplification
Leistungsvorwahlschalter *m* demand setter
Leistungszahl *f* performance value, power coefficient
Leistungszulage *f* incentive wage
Leitblech *n* (Techn) baffle, metal guide, templet
Leitdüse *f* guide nozzle
Leitelektron *n* conduction electron
leiten to conduct, to guide, to lead, to manage
leitend conducting, conductive, ~ **machen** to render conductive
Leiter *f* ladder
Leiter *m* (Elektr) conductor, **schlechter** ~ poor conductor
Leiterbewegung *f* movement of a conductor
Leiterelement *n* conductor element
Leiterkreis *m* [conductor] circuit
Leiterplatte *f* printed circuit board
Leiterwiderstand *m* conductor resistance
Leitfaden *m* manual
leitfähig conducting, conductive
Leitfähigkeit *f* conductivity, conductance, conductibility, conducting capacity, **dielektrische** ~ dielectric conductance, **lichtelektrische** ~ photoconductivity, **magnetische** ~ magnetic conductance, [magnetic] permeability, permeance, **molare** ~ molar conductivity, **richtungsabhängige** ~ asymmetrical conductivity, **spezifische** ~ [specific] conductivity, **unipolare** ~ unilateral or unidirectional conductivity
Leitfähigkeitsabfall *m* conductivity decay
Leitfähigkeitsband *n* conductivity band
Leitfähigkeitsbestimmung *f* conductivity measurement, determination of conductivity
Leitfähigkeitselektron *n* conductivity electron
Leitfähigkeitsgefäß *n* conductivity cell, vessel for measuring conductivity
Leitfähigkeitsmeßbrücke *f* conductivity indicator
Leitfähigkeitsmesser *m* conductivity meter, conductometer
Leitfähigkeitswasser *n* conductivity water
Leitfähigkeitszähler *m* conductivity counter
Leitfläche *f* conducting surface
Leitflosse *f* fin

Leitisotop *n* isotopic indicator, [isotopic] tracer
Leitisotopenmethode *f* radioisotope technique, tracer technique
Leitkurve *f* (Geom) directrix
Leitlack *m* conducting paint
Leitplanken *pl* crash barriers
Leitplatte *f* control plate
Leitrad *n* guide wheel
Leitradbuchse *f* diaphragm bush, diaphragm liner
Leitring *m* guide wheel
Leitrolle *f* guide pulley, guide roller
Leitsalz *n* (Elektrolyse) conducting salt
Leitschaufel *f* guide vane
Leitschaufelrad *n* vaned disk
Leitscheibe *f* diaphragm
Leitschiene *f* guide rail, (Elektr) conductor rail
Leitspindel *f* guide spindle, lathe spindle, leading screw
Leitspindeldrehbank *f* engine lathe
Leitstange *f* radius bar, radius rod
Leitstein *m* guide stone
Leitstrahl *m* pilot beam
Leitstrick *m* guy, stay
Leitung *f* (Fortpflanzung) transmission, (Führung) direction, guidance, (Leitungsmittel) conduit, main, (Röhrenleitung) piping, tubing, (Wärme, Elektr) conduction, (Zuleitung) feeder, ~ **durch Röhren** pipe line, **eine ~ mit ihrem Wellenwiderstand abschließen** to match leads, **metallische ~** metallic conduction, **unipolare ~** unidirectional conductance, **verlustlose ~** (Elektr) dissipationless line
Leitungsabzweigung *f* (Elektr) branching of a conductor
Leitungsband *n* (Elektr) conduction band
Leitungsdraht *m* conducting wire
Leitungsdruck *m* line pressure
Leitungseigenschaft *f* conduction property
Leitungselektron *n* conduction electron
leitungsfähig conductive
Leitungskabel *n* [line] cable
Leitungskoeffizient *m* coefficient of conductivity, conductivity ratio
Leitungsmonteur *m* pipe fitter
Leitungsnetz *n* network of mains, pipe line system, supply network, (Elektr) power mains
Leitungsphase *f* line phase
Leitungsprozeß *m* conduction process
Leitungsquerschnitt *m* cross section of main or pipe
Leitungsrohr *n* delivery tube, line, pipe
Leitungssalz *n* conducting salt
Leitungsspannung *f* line voltage, voltage of the line
Leitungsstange *f* conducting bar
Leitungsstangenklemme *f* rod connecting clamp
Leitungsstrom *m* conduction current
Leitungsstromdichte *f* conduction current density
Leitungssystem *n* piping
Leitungsverlust *m* line loss
Leitungsvermögen *n* conductibility
Leitungswärme *f* heat of conduction
Leitungswasser *n* tap water
Leitungswiderstand *m* line resistance
Leitungszweig *m* branch conductor
Leitvermögen *n* conductance, conductivity, **akustisches ~** acoustic conductivity, **spezifisches ~** [specific] conductivity
Leitwalze *f* guide roll[er], backing roll, control roll
Leitwert *m* conductivity
Leitwertsparameter *m* conductance parameter
Leitwertsverhältnis *n* (Elektr) conductance ratio
Leitzunge *f* baffle, flap
Lektor *m* lecturer
Lemongrasöl *n* lemongrass oil
Lemonol *n* lemonol, geranyl alcohol
Lenardröhre *f* Lenard tube
Lengenbachit *m* (Min) lengenbachite
Lenkarm *m* control arm
lenkbar (Chem) controllable, maneuverable
Lenkblech *n* baffle
lenken to guide, to direct, (regeln) to control, to govern
Lenkgestänge *n* steering linkage
Lenkrad *n* steering wheel, adjusting wheel
Lenkrolle *f* guide roller
Lenksäule *f* steering column
Lenkspurstangenkopf *m* tie rod head assembly
Lenkstange *f* handle bar
Lenkvorrichtung *f* steering device, steering gear
Lentin *n* lentine
Lenzin *m* lenzine, annaline, ground gypsum
Lenzpumpe *f* bilge pump, drain pump
Lenzschieber *m* bilge valve, drain valve
Leonecopalinsäure *f* leonecopalinic acid
Leonecopalsäure *f* leonecopalic acid
Leonhardit *m* leonhardite
Leonit *m* (Min) leonite
Leontin *n* leontin, caulosaponin
Leopoldit *m* leopoldite
Lepargylsäure *f* lepargylic acid
Lepiden *n* lepidene, tetraphenylfurfurane
Lepidin *n* lepidine, 4-methylquinoline
Lepidinsäure *f* lepidinic acid
Lepidokrokit *m* (Min) lepidokrokite
Lepidolith *m* (Min) lepidolite, lithium mica
Lepidomelan *m* lepidomelane
Lepidon *n* lepidone, 2-hydroxy-4-methylquinoline, 2-hydroxylepidine
Lepra *f* (Med) leprosy
Lepranthin *n* lepranthine
Leprapinsäure *f* leprapinic acid
Leptaflorin *n* leptaflorine
Leptochlorit *m* (Min) leptochlorite

Leptocladin – Leukacen

Leptocladin *n* leptocladine
Leptodactylin *n* leptodactyline
Leptogenin *n* leptogenin
Lepton *n* lepton
Leptosid *n* leptoside
Leptospermol *n* leptospermol
Leptospermon *n* leptospermone
Leptynit *m* leptynite
Lerbachit *m* lerbachite
lesbar legible, readable
Leseband *n* (Bergb) picking belt, sorting band
lesen to read, (Bergb) to pick, to sort
Lesen *n* reading
Lesetisch *m* (Bergb) picking or sorting table
Lespedin *n* lespedin
letal deadly, fatal, lethal
Letaldosis *f* lethal amount or dose
Lethargie *f* (Med) lethargy
Letternmetall *n* type metal
Lettsomit *m* (Min) lettsomite
Leucaenin *n* leucenine
Leucaenol *n* leucenol
Leuchtanode *f* luminous anode
Leuchtanstrich *m* luminescent coating
Leuchtbakterien *pl* luminous bacteria, photobacteria
Leuchtbild *n* luminescent image
Leuchtdauer *f* illuminated period, light period, phosphorescence period
Leuchtdecke *f* luminous ceiling
Leuchteinsatz *m* light panel
Leuchtelektron *n* emitting electron, photoelectron
leuchten to emit light, to glow, to luminesce, to shine, to sparkle
Leuchten *n* glowing, luminescence, luminosity, shining, **chemisches** ~ chemiluminescence
Leuchtenbergit *m* (Min) leuchtenbergite
leuchtend bright, glowing, illuminating, luminescent, luminous, shining
Leuchtenergie *f* luminous energy
Leuchterscheinung *f* luminous phenomenon
Leuchtfaden *m* luminous filament
Leuchtfähigkeit *f* luminosity
Leuchtfarbe *f* luminous color, fluorescent ink, luminous paint
Leuchtgas *n* city gas, coal gas, lighting gas
Leuchtgassauerstofflampe *f* gas oxygen lamp
Leuchtgasvergiftung *f* town-gas poisoning
Leuchtkörper *m* illuminator
Leuchtkraft *f* luminous intensity, brilliance, illuminating power, light value, luminosity
Leuchtmanometer *n* illuminated manometer
Leuchtmasse *f* glowing substance
Leuchtmaterial *n* luminescent material, phosphorescent substance
Leuchtöl *n* kerosene, kerosine
Leuchtpatrone *f* flare cartridge
Leuchtpetroleum *n* [lamp] kerosine, paraffin oil

Leuchtpigment *n* luminous pigment
Leuchtprobe *f* (für Zinn; Anal) flame test (for tin)
Leuchtpunkt *m* flash point
Leuchtquarz *m* (Min) luminous quartz
Leuchtreklame *f* electric sign advertising
Leuchtröhre *f* fluorescent tube, neon tube
Leuchtsäule *f* luminous column
Leuchtschirm *m* fluorescent screen, luminous or luminescent screen
Leuchtschrift *f* luminous lettered text
Leuchtspat *m* (Min) luminous spar
Leuchtspiritus *m* illuminating alcohol, mixture of alcohol and oil of turpentine
Leuchtspur *f* tracer streak
Leuchtstab *m* luminous bar
Leuchtstärke *f* brightness, luminosity, luminous intensity
Leuchtstein *m* lithophosphor, phosphorescent stone
Leuchtstoff *m* fluorescent or phosphorescent material, luminous substance
Leuchtstoffhandlampe *f* fluorescent hand lamp
Leuchtstofflampe *f* fluorescent lamp
Leuchtstoffröhre *f* fluorescent tube
Leuchtwert *m* illuminating value, luminosity
Leuchtwirkung *f* illuminating effect, luminous effect
Leuchtzeiger *m* luminous hand
Leuchtzifferblatt *n* luminous dial
Leucin *n* leucine, 2-amino-4-methylpentanoic acid, 2-aminoisocaproic acid
Leucinaminopeptidase *f* (Biochem) leucine aminopeptidase
Leucinaminotransferase *f* (Biochem) leucine aminotransferase
Leucinol *n* leucinol, 4-amino-2-methyl-5-pentanol
Leucinsäure *f* leucinic acid, 2-hydroxy-4-methylpentanoic acid
Leucit *m* (Min) leucite, white garnet
Leucitoid *n* leucitoid
Leucomelon *n* leucomelone
Leucrose *f* leucrose
Leucylalanin *n* leucylalanine
Leucylalanylalanin *n* leucylalanylalanine
Leucylasparagin *n* leucylasparagine
Leucylasparaginsäure *f* leucylaspartic acid
Leucylcystin *n* leucylcystine
Leucyldiglycylglycin *n* leucyl diglycyl glycine
Leucylglycylalanin *n* leucylglycyl alanine
Leucylglycylasparaginsäure *f* leucylglycylaspartic acid
Leucylglycylleucin *n* leucylglycyl leucine
Leucylleucin *n* leucylleucine
Leucylprolin *n* leucylproline
Leucyltriglycylglycin *n* leucyltriglycyl glycine
Leucyltryptophan *n* leucyltryptophan[e]
Leukacen *n* leucacene

Leukämie *f* leuc[a]emia, leuk[a]emia
leukämisch leuk[a]emic
Leukanilin *n* leucaniline, triaminotriphenylmethane
Leukaurin *n* leucaurine, trihydroxytriphenylmethane
Leukoagglutinin *n* leuco-agglutinin
Leukoalizarin *n* leucoalizarin
Leukoatromentin *n* leucoatromentin
Leukoauramin *n* leucoauramine
Leukobase *f* leuco base, leuko base
Leukoblast *m* leucoblast
Leukochalcit *m* (Min) leucochalcite
Leukochinizarin *n* leucoquinizarin
Leukocyklit *m* leucocyclite
Leukoellagsäure *f* leucoellagic acid
Leukoform *f* leuco form
Leukogallocyanin *n* leucogallocyanin
Leukoindigo *n* leucoindigo, indigo white
Leukoindophenol *n* leucoindophenol
Leukoküpenfarbstoff *m* leuco vat dyestuff
Leukolgelb *n* leucol yellow
Leukolin *n* leucoline, quinoline
Leukolyse *f* (Leukozytenauflösung) leucocytolysis, leucolysis
Leukomain *n* leucomaine
Leukomalachitgrün *n* leucomalachite green
Leukometer *n* leucometer
Leukonin *n* leuconin, sodium antimonate
Leukonsäure *f* leuconic acid, penta-oxocyclopentane
Leukopelargonidin *n* leucopelargonidin
Leukophan *m* (Min) leucophane
Leukoplast *n* (Med) adhesive tape
Leukopterin *n* leucopterin
Leukorosanilin *n* leucorosaniline
Leukorosolsäure *f* leucorosolic acid
Leukosphenit *m* (Min) leucosphenite
Leukothioindigo *n* leucothioindigo
Leukothionin *n* leucothionine, diaminothiodiphenylamine
Leukotil *m* (Min) leucotil
Leukotrop *n* leucotrope, phenyldimethyl benzylammonium chloride
Leukoverbindung *f* leuco-compound
Leukoxen *m* (Min) leucoxene
Leukozyt *m* (weißes Blutkörperchen) leucocyte, white blood corpusule
Leukozytenanstieg *m* increase of the leucocyte count, leucocytosis
Leukozytenarmut *f* hypoleucocytosis, leucopenia
Leukozytenbildung *f* formation of leucocytes
Leukozytenvermehrung *f* leucocytosis
Leukozytenverminderung *f* leucocytopenia, leucopenia
Leukozytenzerfall *m* breaking down of leucocytes, leucocytolysis
Leukozytogenese *f* leucocytogenesis
Leukozytolyse *f* (Med) leucocytolysis

Leukozytose *f* (Med) leucocytosis
Leuzit *m* (Min) leucite
Levarterenol *n* levarterenol
Leverierit *m* (Min) leverierite
levigieren to levigate, to pulverize, to rub to dust
Levomepromazin *n* levomepromazine
Levomethorphan *n* levomethorphan
Levorphan[ol] *n* levorphanol
Levothyroxin *n* levothyroxine
Levyn *m* (Min) levyne, levynite
Lewisbase *f* Lewis base
Lewisit *m* (Min) lewisite
Lewissäure *f* Lewis acid
Leydenerblau *n* Leyden blue, cobalt blue, King's blue
Leydener Flasche *f* (Elektr) Leyden jar
Lezithin *n* lecithin
Lezithinspiegel *m* (Med) lecithin level
LH Abk. für „luteinisierendes Hormon"
Lherzolyth *m* lherzolyte
L-Hülle *f* (Atom) L-shell
Libanoncedernöl *n* Lebanon cedar oil
Libelle *f* (Wasserwaage) [water] level, clinometer, spirit level
Libethenit *m* (Min) libethenite
Libidibi *n* libidibi, lividivi
Libocedren *n* libocedrene
Licansäure *f* licanic acid
Lichenin *n* lichenin, lichen starch, moss starch, reserve cellulose
Lichesterinsäure *f* lichesterinic acid
licht bright, light
Licht *n* light, **austretendes** ~ emergent light, **direktes** ~ direct light, **durchfallendes** ~ transmitted light, **durchgelassenes** ~ (Opt) transmitted light, **einfallendes** ~ incident light, **einfarbiges** ~ monochromatic light, **eingestrahltes** ~ incident light, **elliptisch polarisiertes** ~ elliptically polarized light, **kaltes** ~ cold light, **kontinuierliches** ~ continuous light, **künstliches** ~ artificial light, **langwelliges** ~ light of long wavelength, **linear-polarisiertes** ~ (Opt) plane-polarized light, **monochromatisches** ~ monochromatic light, **polarisiertes** ~ polarized light, **ultraviolettes** ~ ultra violet light, **zirkular polarisiertes** ~ circularly polarized light
Lichtabsorption *f* absorption of light, light absorption
Lichtadaption *f* (Auge) light adaptation
Lichtaggregat *n* (Elektr) lighting set
Lichtatmung *f* (Biochem) light respiration
Lichtausbeute *f* light efficiency, efficiency of a luminous source, light yield
Lichtausschluß *m* absence of light, exclusion of light, **unter** ~ in the absence of light
Lichtausstrahlung *f* radiation of light
Lichtbahn *f* light path

lichtbeständig lightproof, photostable, stable to light
Lichtbeständigkeit f light resistance
Lichtbeugung f diffraction of light
Lichtbild n photo, photograph, photographic picture
Lichtbildkunst f photography
Lichtbildmeßkunst f photogrammetry
Lichtbildschirm m projection screen
Lichtblende f (Phot) diaphragm
Lichtblitzentladung f flash discharge
Lichtbogen m [electric] arc, voltaic arc, **flackernder** ~ unsteady arc
Lichtbogenabstich m arc tapping
Lichtbogenbeheizung f arc heating
Lichtbogenbereich m arc zone
Lichtbogenelektroofen m electric arc furnace
Lichtbogenentfaltung f arc development
Lichtbogenentladung f arc discharge
Lichtbogenerhitzung f arc heating
Lichtbogenerzeuger m arc generator
Lichtbogenfestigkeit f arc resistance
Lichtbogengenerator m arc generator, arc source
Lichtbogengleichrichter m arc rectifier
Lichtbogengut n materials exposed to action of arc
Lichtbogenheizung f arc heating, heating by the arc
Lichtbogenkohlebruchstücke n pl fragments of arc carbons
Lichtbogenlampe f arc lamp
Lichtbogenofen m [electric] arc furnace
Lichtbogenplasmabrenner m plasma arc torch
Lichtbogen-Reduktionsofen m arc reduction furnace
Lichtbogenrohr n arc discharge tube
Lichtbogenscheibe f disc arc
Lichtbogen-Schmelzofen m arc furnace
Lichtbogenschweißautomat m automatic arc welding machine
Lichtbogenschweißelektrode f arc welding electrode
Lichtbogenschweißung f arc welding
Lichtbogenschweißverfahren n arc-welding method
Lichtbogenspannung f arc voltage
Lichtbogenstahlofen m electric steel melting furnace
Lichtbogenstrecke f arc gap
Lichtbogenwiderstand m arc resistance
Lichtbogenwiderstandserhitzung f heating by arc resitance
Lichtbogen-Widerstandsofen m direct-arc furnace
Lichtbogenzündung f arc ignition
lichtbrechend refracting, refractive
Lichtbrechung f refraction of ligth, [optical] refraction
Lichtbrechungsindex m refractive index

Lichtbrechungsvermögen n optical refractive power, refractivity
Lichtbündel n light beam, light bundle
lichtchemisch photochemical
Lichtchlorierung f photochlorination
Lichtdämpfung f subduing of light
Lichtdecke f light ceiling
lichtdicht light-proof, light-tight
lichtdielektrisch photodielectric
Lichtdruck m photographic printing, photogravure, photomechanical printing, (Heliotypie) heliographic printing
lichtdurchlässig light-transmitting, light-permeable, light-transmissive, permeable to light rays, translucent, transparent
Lichtdurchlässigkeit f light transmission, luminous transmittance, transparency
Lichtdurchlässigkeitsmesser m diaphanometer
lichtecht fast to light, fade-proof, light-fast, light-proof, light-resisting
Lichtechtheit f light fastness
Lichtechtheitsprüfung f lightfastness testing
Lichteffekt m luminous effect
Lichteinfall m incidence of light
Lichteinheit f unit of light, candle power
Lichteintrittsschacht m light admission port
Lichteinwirkung f action of light, photoelectrical action
lichtelektrisch photoelectric
Lichtelektrizität f photoelectricity
lichtempfindlich optically sensitive, sensitive to light, (Phot) sensitized, ~ **machen** to photosensitize, to sensitize
Lichtempfindlichkeit f light sensitivity, photosensitivity
Lichtenergie f light energy, luminous energy
lichtentwickelnd photogenic
Lichterscheinung f luminous phenomenon, optical phenomenon, phenomenon caused by light
lichterzeugend producing light, luminiferous, (Biol) photogenic
lichtfarben light-colored
Lichtfarbendruck m color phototypy, heliotype color printing, photomechanical color printing
Lichtfilter n light filter
Lichtfleck m light spot, (Phot) halation, **abtastender** ~ scanning light spot
Lichtfortpflanzung f transmission of light
lichtgebend illuminating, luminiferous, luminous
Lichtgeschwindigkeit f light velocity, speed of light
Lichtgeschwindigkeitsmessung f velocity or speed of light measurement
Lichtglanz m brightness, brilliant luster, radiance
Lichtgrün n light green, (Chem) methyl green
Lichthof m (Astr) halo, (Phot) halation

Lichtimpuls *m* light pulse
Lichtintensität *f* intensity of light, light intensity
Lichtintensitätsschwankung *f* light intensity variation
Lichtinterferenz *f* light interference
Lichtisomerisation *f* light isomerization
Lichtjahr *n* light year
lichtkatalytisch light-catalytic
Lichtkegel *m* cone of light, luminous cone
Lichtkörper *m* luminous body
Lichtkonsument *m* light consumer
Lichtkreis *m* circle of light, halo
Lichtkupferdruck *m* photogravure
Lichtleimdruck *m* collotypy
Lichtleitung *f* lighting circuit, lighting main
lichtliebend (Bakt) photophilous
Lichtmagnetismus *m* photomagnetism
Lichtmaterie *f* luminous matter
Lichtmenge *f* quantity of light
Lichtmesser *m* photometer
Lichtmessung *f* photometry
Lichtmikroskop *n* optical or light microscope
Lichtmikroskopie *f* optical microscopy
Lichtnetz *n* lighting network
Lichtpausapparat *m* [photo]copying apparatus, (f. Blaupause) blueprint apparatus
Lichtpause *f* photocopy, photoprint, (Blaupause) blueprint
Lichtpausgerät *n* [photo] copying apparatus
Lichtpauspapier *n* photocopy[ing] paper, (f. Blaupause) blueprinting paper
Lichtpausverfahren *n* blueprinting process, heliographic printing
Lichtpunkt *m* luminous spot, point of light
Lichtpunktlinienschreiber *m* light point line recorder
Lichtquant *n* light quant, photon
Lichtquelle *f* light source, illuminant, luminous source, source of light, **monochromatische** ~ (Opt) monochromatic illuminator, monochromatic light source, monochromator
Lichtraster *n* light diffuser
Lichtreaktion *f* (Biochem) light reaction
Lichtreflex *m* light reflection
Lichtreiz *m* luminous stimulus
Lichtrisse *m pl* (Prüftechnik) atmospheric cracking
Lichtrot *n* light red
Lichtschalter *m* light switch
Lichtschein *m* gleam, glow, radiance, shine
Lichtschimmer *m* gleam, glimmer
lichtschluckend light-absorbing, optically absorptive
Lichtschnitt *m* light intersection
Lichtschranke *f* electric eye, light barrier
Lichtschutz *m* protection against light, shade
Lichtschutzmittel *n* screening agent, sunscreen chemical
lichtschwach of low light intensity

Lichtseite *f* bright side
Lichtsignal *n* light signal, signal flare
Lichtsondentechnik *f* light probe technique
Lichtspalt *m* light slit, light gap
Lichtspektrum *n* light spectrum
Lichtspiegel *m* reflector
Lichtspiel *n* motion picture
Lichtstärke *f* light intensity, luminosity, (einer Glühbirne) candle-power
Lichtstärkemesser *m* lightmeter, photometer
lichtstark bright, of high light intensity
Lichtstrahl *m* light beam, luminous ray, ray of light
Lichtstrahlenbündel *n* beam of light
Lichtstrahlung *f* light radiation, radiation of light
Lichtstrecke *f* light path
Lichtstreuung *f* light scattering
Lichtstreuungsmethode *f* light scattering method
Lichttechnik *f* lighting technique
Lichttherapie *f* (Med) phototherapy
Lichttransformator *m* lighting transformer
lichtundurchlässig impervious to light, non-diaphanous, opaque
Lichtundurchlässigkeit *f* imperviousness to light, opacity, opaqueness
lichtunecht not fast to light, not light-proof
lichtunempfindlich light-insensitive, photostable
Lichtventil *n* light valve
Lichtverteilung *f* distribution of light
Lichtweg *m* light path, optical path
Lichtweite *f* clearance, inner width, inside diameter, width in the clear
Lichtwelle *f* light wave
Lichtwirkung *f* action of light, luminous effect
Lichtzeichen *n* illuminated sign
Lichtzeigergalvanometer *n* light spot galvanometer
Lichtzentrum *n* photocenter
Lichtzerlegung *f* diffraction of light
Lichtzerstreuung *f* dispersion of light, light diffusion
Lichtzerstreuungskraft *f* light diffusive power, light dispersive power
Lichtzerstreuungsvermögen *n* light diffusive capacity, light dispersive capacity
Lichtzutritt *m* admission of light, admittance of light
Licker *m* fat-liquor
Lidschatten *m* eye shadow
Lidstrichemulsion *f* emulsified cream eye shadow
Liebenerit *m* (Min) liebenerite
Liebigit *m* (Min) liebigite
Liebigkühler *m* Liebig condenser
Liebigscher Kühler *m* Liebig condenser
Liebstöckelöl *n* lovage oil
Liebstöckelwurzel *f* (Bot) lovage root
Lieferangebot *n* supply tender

Lieferant *m* supplier, contractor, dealer
Lieferart *f* condition of supply, kind of supply (means of transport)
Lieferbedingungen *f pl* specifications, terms of delivery or supply
Lieferer *m* supplier
Lieferfirma *f* supplier
Lieferfreigabeschein *m* release note
Lieferfrist *f* date of supply, time of delivery
Liefergrad *m* (eines Verdichters) volume efficiency
Lieferlage *f*, **ungünstige** ~ short supply
Liefermenge *f* quantity delivered, supplied quantity
liefern to supply, to deliver, to furnish
Lieferplan *m* supply program, supply schedule
Lieferschein *m* delivery bill, invoice, shipping slip
Lieferumfang *m* scope of delivery
Lieferung *f* delivery, shipment, supply, (Schiff) cargo, ~ **frei Hamburg** C.I.F. Hamburg, ~ **frei Schiff** free on board (F.O.B.)
Lieferungsbedingungen *pl* conditions of supply, terms of supply
Liegen *n* **an der Luft** contact with air, ~ **in Wasser** immersion in water
Lievrit *m* (Min) lievrite
Lift *m* elevator (Am. E.); lift (Br. E.)
Ligand *m* ligand
Ligandenaustausch *m* ligand replacement
Ligandenfeldtheorie *f* ligand-field theory
Ligase *f* (Biochem) ligase
Ligatur *f* ligature
Lignin *n* lignin
Ligninkörper *m* ligneous matter
Ligninkunststoff *m* lignin plastic
Lignit *m* lignite, peat coal
Lignocellulose *f* lignocellulose
Lignocerinsäure *f* lignoceric acid, tetracosanoic acid
Lignocerylalkohol *m* lignoceryl alcohol
Lignon *n* lignone
Lignonsulfonat *n* lignone sulfonate
Lignose *f* lignose
Lignosulfit *n* lignosulfite, lignosulfin
Lignosulfonsäure *f* lignosulfonic acid
Lignozellulose *f* ligno-cellulose
Ligroin *n* ligroin[e]
Ligroingaslampe *f* ligroin gas lamp
Ligurit *m* ligurite
Liguster *m* (Bot) common privet, ligustrum
Ligusticumlacton *n* ligusticum lactone
Likör *m* liqueur
likörartig liqueur-like
lila lilac-colored
Lila *n* lilac color, pale violet
Lilacin *n* lilacin, ligustrin, syringin
Lilafarbe *f* lilac color
Lilagenin *n* lilagenin

Lilie *f* (Bot) lily
Liliengewächs *n* (Bot) iridaceous plant
Lillianit *m* (Min) lillianite
Lillit *m* lillite
Lilolidin *n* lilolidine, tetrahydrolilole
Limaholz *n* (Bot) lima wood
Limbus *m* (Gradkreis) graduated circle, limb
Limen *n* limene
Limes *m* (Math) limit, limes
Limeswerte *pl* limiting values
Limette *f* (Bot) lime fruit
Limettenessenz *f* lime juice
Limettin *n* limettin, 5,7-dimethoxycumarin, citroptene
Limettöl *n* lime oil, limette oil
Limnit *m* limnite
Limocitrin *n* limocitrin
Limocitrol *n* limocitrol
Limonade *f* lemonade
Limonadenpulver *n* lemonade powder
Limonen *n* limonene
Limonetrit *n* limonetrite, p-menthane-1,2,8,9-tetrol
Limonit *m* (Min) limonite, bog ore, brown hematite
Limurit *m* limurite
Linaloeöl *n* linaloa oil, linaloe oil
Linalool *n* linalool, coriandrol
Linaloolen *n* linaloolene
Linalylacetat *n* linalyl acetate, 3,7-dimethyl-1,6-octadien-3-yl acetate
Linalylchlorid *n* linalyl chloride, 6-chloro-2,6-dimethyl-2,7-octadiene
Linamarin *n* linamarin
Linarigenin *n* linarigenin
Linarin *n* linarine
Linarit *m* (Min) linarite
Lindackerit *m* lindackerite
Lindelofidin *n* lindelofidine
Lindenblüten *f pl* (Bot) linden flowers
Lindenkohle *f* linden charcoal
Lindenöl *n* linden oil, basswood oil
Linderazulen *n* linderazulene
Linderin *n* linderene
lindern to relieve, to alleviate, to moderate
Lindersäure *f* linderic acid
Linderungsbalsam *m* (Pharm) balm, lenitive, palliative balsam
Linderungsmittel *n* (Pharm) palliative, soothing remedy
Linderungssalbe *f* (Pharm) lenitive ointment, liniment
Lindesit *m* lindesite
Lindsayit *m* lindsayite
Lineal *n* rule, ruler, straightedge
Linealführung *f* guiding system
linear linear, ~ **abhängig** linearly dependent, ~ **polarisiert** plane-polarized, ~ **unabhängig** linearly independent

Linearbeschleuniger *m* linear accelerator
Linearbeschleunigung *f* linear acceleration
linearisiert linearized
Linearisierung *f* linearization
Linearität *f* linearity
Linelaidinsäure *f* linelaidic acid
Lineweaver-Burk-Diagramm *n* (Kin) Lineweaver-Burk plot
Linie *f* line, ~ **gleichen Drucks** constant pressure line, isobar, ~ **gleichen Potentials** equipotential line, **ausgezogene** ~ full line, solid line, **gestrichelte** ~ broken line, dashed line, **krumme** ~ curved line, **punktierte** ~ dotted line, **strichpunktierte** ~ chain-dotted line, dash-dotted line, **verbotene** ~ (Spektr) forbidden line, **verbreiterte** ~ (Spektr) enhanced line, **zusammengesetzte** ~ (Spektr) multiplet
Linien *pl* **gleicher Kraftstärke** isodynamic lines, ~ **gleicher Neigung** isoclinal lines, ~ **gleicher Temperatur** isothermal lines, isotherms
Linienabstand *m* (Spektr) line interval
Linienaufspaltung *f* (Spektr) line splitting
Linienblitz *m* streak lightning
Linienbreite *f* (Spektr) line width
Liniendichte *f* (Spektr) line density
Liniendipol *m* line dipole
Linienform *f* line shape
Liniengruppe *f* group of lines
Linienintegral *n* line integral
Linienpaar *n* (Spektr) line pair
Linienprofil *n* profile of a line
linienreich (Spektr) rich in lines, with a large number of lines
Linienreihe *f* (Spektr) series of lines
Linienschreiber *m* continuous line recorder
Linienspannung *f* line voltage
Linienspektrum *n* line spectrum
Linienstrom *m* interlinked current, line current
Linienumkehr *f* line reversal, reversal dip
Linienverbreiterung *f* (Spektr) line broadening
Linienzug *m* curve
Liniment *n* liniment
Linin *n* linin
Linkrusta *f* (HN) lincrusta
Linkscampher *m* levocamphor
linksdrehend anticlockwise, counterclockwise, l[a]evorotatory, levogyric
Linksdrehung *f* anticlockwise rotation, counterclockwise motion, l[a]evorotation, levogyration, (Kfz) left turn, (Polarisation) left-handed polarization
linksfärben to dye inside out, to dye on the wrong side
linksgängig counterclockwise
Linksgewinde *n* left-hand[ed] thread
Linksmilchsäure *f* levolactic acid
Linkspolarisation *f* left-handed polarization, levopolarization

Linksquarz *m* (Min) left-handed quartz, levogyrate quartz
Linkssäure *f* levo-acid
Linksschraube *f* left-handed propeller
Linkssystem *n* left-handed system
Linksweinsäure *f* levotartaric acid
Linneit *m* (Min) linnaeite, cobalt nickel pyrite
Linnen *n* (Text) linen, canvas
Linnésches System *n* Linn[a]ean system
Linocaffein *n* linocaffein
Linocinnamarin *n* linocinnamarin
Linoleat *n* linoleate, linolenate
Linolein *n* linoleine
Linolensäure *f* linolenic acid, 9,12,15-octadecatrienoic acid
Linolenylalkohol *m* linolenyl alcohol
Linoleodistearin linoleodistearin, glyceryl linoleate distearate
Linoleon *n* linoleone
Linoleum *n* linoleum
Linoleumersatz *m* linoleum substitute
Linolin *n* linolin, glyceryl trilinolate, trilinolin
Linolsäure *f* 9,12-octadecadienoic acid, linoleic acid, linolic acid
Linosit *m* linosite
Linoxyn *n* linoxyn, solid oxidized linseed oil
Linse *f* (Bergb) small lump of ore, (Bot) lentil, (Opt) lens, ~ **mit veränderlichem Fokus** variable focus lens, **achromatische** ~ achromatic lens, **beschichtete** ~ coated lens, **bifokale** ~ bifocal lens, **bikonkave** ~ biconcave lens, **bikonvexe** ~ biconvex lens, **dezentrierte** ~ decentered lens, **doppelkonkave** ~ biconcave lens, concavoconcave lens, **doppelkonvexe** ~ biconvex lens, **elektrische** ~ electrostatic lens, **geblaute** ~ coated lens, **gefaßte** ~ mounted lens, **konkave** ~ concave lens, diverging lens, **konvexe** ~ condensing lens, convex lens, **konvexkonkave** ~ convexoconcave lens, **kurzbrennweitige** ~ short-focus lens, **langbrennweitige** ~ long-focus lens, **lichtstarke** ~ high-speed lens, **plankonkave** ~ plano-concave lens, **plankonvexe** ~ plano-convex lens, **zusammengesetzte** ~ (Opt) compound lens
Linsenabweichung *f* aberration of a lens
linsenartig lenticular, lentiform
Linsenastigmatismus *m* lenticular astigmatism
Linsenerz *n* (Min) liroconite, oolitic limonite, oolitic ore
Linsenfehler *m* (Opt) lens error
linsenförmig lenticular, lentiform, lentoid
Linsenglas *n* lens glass, optical glass
Linsengleichung *f* lens equation
linsengroß lentil-sized
Linsenkrümmung *f* lens curvature
Linsenkühler *m* lens cooler
Linsenkupfer *n* (Min) liroconite

Linsenpaar *n* pair of lenses
Linsenrasterverfahren *n* (Opt) lenticular process
Linsenschleifen *n* grinding of lenses
Linsensystem *n* lens combination, system of lenses
Linsentrübung *f* clouding of a lens, opacity of a lens
Linusinsäure *f* linusinic acid
Liothyronin *n* liothyronine
Liovil *n* liovil
Lipämie *f* (Med) lipaemia
Liparit *m* liparite
Lipase *f* (Biochem) lipase
Lipid *n* lipid, **nicht verseifbares** ~ nonsaponifiable lipid, **verseifbares** ~ saponifiable lipid
Lipidkatabolismus *m* lipid catabolism
Lipidmembran *f* lipid membrane
Lipidstoffwechsel *m* (Physiol) lipid metabolism
Lipoamid *n* lipoamide
Lipoblast *m* (Fettzelle) lipoblast
Lipochrom *n* lipochrome
Lipogenese *f* (Physiol) lipogenesis
lipoid lipide, lipoid
Lipoid *n* lipoid
Lipol *n* lipol
Lipolyse *f* lipolysis
lipolytisch lipolytic
Liponsäure *f* lipoic acid
lipophil (fettaffin) lipophilic
lipophob lipophobic
Lipoproteid *n* lipoprotein
Lipoprotein *n* lipoprotein
Liposom *n* (Fetttröpfchen in Zellen) liposome
Lipoxydase *f* (Biochem) lipoxydase
Lipozyt *m* lipocyte, fat cell
Lippenblütler *m* (Bot) labiate
Lippendichtung *f* cylindrical gasket
Lippenpfeife *f* pipe organ
liquid [financially] payable, liquid
liquidmagmatisch liquid magmatic
Liquidometer *n* liquidometer
Liquiduskurve *f* liquidus curve
Lirokonit *m* (Min) liroconite
Liskeardit *m* (Min) liskeardite
Lissajous-Figuren *pl* Lissajous figures
Lissaminfarbstoff *m* lissamine dyestuff
Lissolamin *n* lissolamine
Liste *f* catalog[ue], schedule, (Tabelle) table, ~ **technischer Angaben** data sheet
Listoformseife *f* listoform soap
Liter *n* liter (Am. E.), litre (Br. E.)
Literaturangabe *f* reference, bibliography
Literaturdokumentation *m* literature documentation
Literaturerfassung *f,* **chemische** ~ chemical documentation
Literaturstudium *n* literature search
Literatursuche *f* literature search

Literaturzusammenstellung *f* bibliography
Litergewicht *n* weight per liter
Literinhalt *m* capacity in liters
Literkolben *m* liter flask
Lithion *n* lithia, lithium oxide
Lithionglimmer *m* (Min) lepidolite, lithia mica
Lithionit *m* lithionite
Lithophilit *m* (Min) lithiophilite
Lithophorit *m* (Min) lithiophorite
Lithium *n* lithium (Symb. Li)
Lithium-12-hydroxyfett *n* lithium-12-hydroxy grease
Lithiumacetat *n* lithium acetate
Lithiumäthylanilid *n* lithium ethylanilide
Lithiumalkyl *n* lithium alkyl
Lithiumaluminiumhydrid *n* lithium aluminum hydride
Lithiumamid *n* lithium amide
Lithiumarsenat *n* lithium arsenate
Lithiumbenzoat *n* lithium benzoate
Lithiumbicarbonat *n* acid lithium carbonate, lithium bicarbonate, lithium hydrogen carbonate
Lithiumboratglas *n* lithium borate glass
Lithiumborhydrid *n* lithium borohydride
Lithiumchelat *n* lithium chelate
Lithiumchlorid *n* lithium chloride
Lithiumerz *n* (Min) lithium ore
Lithiumfeldspat *m* (Min) lithia feldspar
Lithiumfett *n* lithium grease
Lithiumfluorid *n* lithium fluoride
Lithiumgehalt *m* lithium content
Lithiumglimmer *m* (Min) lithia mica, lepidolite
Lithiumhydrid *n* lithium hydride
Lithiumjodid *n* lithium iodide
Lithiumnitrid *n* lithium nitride
Lithiumoxid *n* lithium oxide, lithia
Lithiumsalicylat *n* lithium salicylate
Lithiumsilikatglas *n* lithium silicate glass
Lithiumsulfat *n* lithium sulfate
Lithiumtantalat *n* lithium tantalate
Lithiumturmalin *m* (Min) apyrite
Lithiumurat *n* lithium urate
Lithiumverbindung *f* lithium compound
Lithocholsäure *f* lithocholic acid
Lithodur *n* (Pigment) lithodur
Lithographie *f* lithography
Lithographiefarbe *f* lithographic color
lithographieren to lithograph
lithographisch lithographic
Litholechtgelb *n* lithol fast yellow
Litholein *n* litholeine
Litholscharlach *n* lithol scarlet
Lithosiderit *m* (Min) lithosiderite
Lithosphäre *f* lithosphere
Litze *f* braid, strand, stranded wire, wire
Litzendraht *m* (Elektr) braided wire
Litzenlack *m* braid lacquer
Liveingit *m* (Min) liveingite

Livingstonit *m* (Min) livingstonite
Lizenz *f* license (Am. E.), licence (Br. E.), royalty
Lizenzabkommen *n* license contact
Lizenzbau *m* production or construction under license
Lizenzentzug *m* withdrawal of license
Lizenzerteilung *f* licensing [Am. E.]
lizenzfrei free of royalty
Lizenzgeber *m* licenser, licensor (Am. E.)
Lizenzgebühr *f* (Jur) license fee (Am. E.), licence fee (Br. E.), royalty
lizenzieren to licence, to permit to make or sell under license
lizenziert licensed (Am. E.), licenced (Br. E.)
Lizenznehmer *m* licensee
lizenzweise on a royalty basis
Loangocopalinsäure *f* loangocopalinic acid
Loangocopaloresen *n* loangocopaloresene
Loangocopalsäure *f* loangocopalic acid
Lobarsäure *f* lobaric acid
Lobelan *n* lobelane
Lobelanidin *n* lobelanidine
Lobelanin *n* lobelanine
Lobelidin *n* lobelidine, dl-lobelin
Lobelidiol *n* lobelidiol
Lobelie *f* (Bot) lobelia
Lobelienkraut *n* (Bot) lobelia
Lobelientinktur *f* lobelia tincture
Lobelin *n* lobeline
Lobelinsäure *f* lobelinic acid
Lobelionol *n* lobelionol
Lobelol *n* lobelol
Lobelon *n* lobelone
Loch *n* cavity, hole, perforation, (Grube) pit
Lochabstand *m* spacing of holes
Lochband *n* perforated belt, (Comp) printer carriage tape
Lochbandkanal *m* (Comp) tape channel
Lochbandvorschub *m* (Comp) tape-controlled carriage
Lochblende *f* perforated screen, pin hole diaphragm
Lochbohrer *m* auger bit
Lochdorn *m* piercer, piercing mandrel
Locheisen *n* punch
Lochelektrode *f* eines Elektronenmikroskops apertured electrode disk of an electron microscope
lochen to make a hole, to perforate, to pierce, to punch
Locher *m* (Comp) card punch
Lochfraß *m* pinholes, pinpoint corrosion, (Met) pitting
Lochfraßkorrosion *f* pitting corrosion, fretting corrosion, pinholing
Lochkamera *f* pinhole camera
Lochkarte *f* punch card
Lochkartenspalte *f* (Comp) punch card column

Lochkartenvorderseite *f* (Comp) card face
Lochkreis *m* bolt circle, pitch circle, pitch diameter
Lochlehre *f* hole gauge
Lochmaschine *f* perforating machine, stamping machine
Lochplatte *f* perforated plate
Lochprüfer *m* (Comp) numerical verifier
Lochring *m* matrix, [bed] die
Lochscheibe *f* breaker plate
Lochscheibenanode *f* apertured disk anode
Lochschriftübersetzer *m* (Comp) alphabetic interpreter
Lochschweißung *f* plug welding
Lochsirene *f* perforated siren [disk]
Lochstanze *f* perforating machine, punching die, punching machine, punch press
Lochstanzmaschine *f* perforating machine, punching machine
Lochstempel *m* piercing die, punch
Lochstempelstahl *m* steel for punches
Lochstift *m* core pin, hole-forming pin, plain core pin
Lochstiftplatte *f* core pin plate
Lochstreifen *m* punched tape, (Comp) paper tape
Lochstreifeneinheit *f* (Comp) paper tape unit
Lochstreifenkarte *f* (Comp) edge-punched card
Lochstreifenleser *m* (Comp) paper tape reader comp
Lochstreifen-Magnetbandumwandler *m* (Comp) paper tape to magnetic tape converter
Lochsuchgerät *n* pinhole detector
Lochtaster *m* inside calipers
Lochtiefe *f* depth of cavity
Loch- und Schlitznaht *f* slot weld
Lochung *f* perforation
Lochweite *f* inside diameter, width of hole
Lochzahlprüfer *f* (Comp) hole count check
Lochziegel *m* perforated brick
locker loose, (porös) porous, (schwammig) spongy, (weich) mellow, soft
lockern to loosen, to make loose, to relax, to slacken
Lockerung *f* loosening
Lockerungsgeschwindigkeit *f* incipient fluidizing velocity (fluidized beds)
Lockerungspunkt *m* fluidizing point
Lockstoff *m* attractant
Locuturin *n* locuturine
Lodal *n* lodal
löcherig perforated, full of holes, honeycombed, porous
Löcherleitung *f* gap conductivity
löchern to perforate, to punch holes
Löcherschwamm *m* larch agaric, white agaric
Löffel *m* spoon, (Schöpflöffel) ladle
Löffelbagger *m* shovel excavator
Löffelrührer *m* spoon agitator or stirrer

Löllingit *m* (Min) lollingite
lösbar detachable, (Chem) soluble
Lösbarkeit *f* (Chem) solubility
Löscharbeit *f* (Metall) charcoal fining process, tempering
Lösche *f* charcoal dust, cinder, culm
Löscheigenschaften *f pl* quenching properties
löschen to quench, to extinguish, (ausradieren) to erase, (Comp) to blank, (Kalk) to slake, (Math) to cancel, (Metall) to temper, ~ **auf Null** (Comp) to zero
Löschen *n* quenching, (Kalk) slaking, ~ **des Kokses** damping down of the coke, quenching of the coke
Löschfähigkeit *f* ability to extinguish
Löschfeld *n* (Comp) erase field
Löschfrist *f* period of unloading
Löschfunken *m* quenched spark
Löschfunkenstrecke *f* extinguishable spark gap
Löschgerät *n* extinguisher
Löschimpuls *m* (Comp) quench pulse
Löschkalk *m* quicklime, slaked lime
Löschkohle *f* quenched charcoal
Löschkopf *m* (Comp) erase head
Löschkreis *m* quench[ing] circuit
Löschmittel *n* extinguishing substance
Löschpapier *n* blotting paper
Löschsand *m* sand for extinguishing fires
Löschschalter *m* (Comp) acknowledge switch
Löschschaltung *f* quench[ing] circuit
Löschschaum *m* fire-extinguishing foam
Löschturm *m* damping-down tower, quenching tower
Löschung *f* quenching, erasing, extinguishing, obliterating, (Kalk) slaking
Löschwasser *n* (Met) tempering water
Löschwirkung *f* extinguishing effect
Löseapparat *m* dissolving apparatus
Lösegeschwindigkeit *f* rate of dissolving
Lösekessel *m* dissolving drum, dissolving tank; quencher
Lösemittel *n* solvent
Lösemittelwiedergewinnung *f* solvent recovery
lösen (ablösen) to detach, to disconnect, to disengage, to undo, (Bremse) to release, (Chem) to dissolve, (lockern) to loosen, to release, to slacken, to untie, to untwist, (losschrauben) to unscrew, (Math) to solve
Lösen *n* dissolving
Lösevermögen *n* dissolving power, solvency, solvent power
löslich soluble, ~ **machen** to solubilize
Löslichkeit *f* solubility, **beschränkte** ~ limited solubility, partial solubility, **molare** ~ molar solubility
Löslichkeitsbeeinflussung *f* influence on solubility
Löslichkeitsbestimmung *f* determination of solubility
Löslichkeitsdruck *m* solubility pressure
Löslichkeitsgrenze *f* limit of solubility
Löslichkeitskurve *f* curve of solubility
Löslichkeitsprodukt *n* solubility product, ionic product
Löslichkeitsunterschied *m* difference in solubility
Löslichkeitsverhältnis *n* relation of solubility
Löslichkeitsverminderung *f* decrease of solubility
Löß *m* (Geol) loess
Lösung *f* loosening, release, separation, (Chem) dissolution, solution, **äquimolekulare** ~ equimolecular solution, **ätherische** ~ ethereal solution, **alkoholische** ~ alcoholic solution, **feste** ~ (Krist) solid solution, **gesättigte** ~ saturated solution, **ideale** ~ ideal solution, **in** ~ **gehen** to dissolve, **kolloidale** ~ colloidal solution, sol, **konzentrierte** ~ concentrated solution, **molare** ~ molar solution, **nichtideale** ~ non-ideal solution, 1/100 **normale** ~ centinormal solution, decinormal solution, hundredth normal solution, normal solution, one tenth normal solution, **übersättigte** ~ supersaturated solution, **ungesättigte** ~ unsaturated solution, **verdünnte** ~ dilute solution, **wäßrige** ~ aqueous solution
Lösungen *f pl* **mit gleichem Dampfdruck** (Phys) isobaric solutions, isopiestic solutions
Lösungsanode *f* soluble anode
Lösungsbenzol *n* [light] solvent naphtha
Lösungsdichte *f* concentration
Lösungsdruck *m* solution pressure
Lösungseffekt *m* solvent effect
Lösungselektrode *f* solvent electrode
Lösungsenthalpie *f* solution enthalpy
Lösungsentropie *f* entropy of solution
Lösungserscheinung *f* phenomenon of solution
Lösungsfähigkeit *f* dissolving capacity, dissolving power, solvent power
Lösungsgleichgewicht *n* solution equilibrium
Lösungsglühen *n* (Metall) solution annealing
Lösungskraft *f* solvent power
Lösungskristall *m* solution-grown crystal
Lösungsmittel *n* solvent, **differenzierendes** ~ selective solvent, **dissoziierendes** ~ dissociating solvent, **flüchtiges** ~ volatile solvent, **nichtwäßriges** ~ non-aqueous solvent, **polares** ~ polar solvent, **selektives** ~ extractive agent, selective solvent, **unpolares** ~ nonpolar solvent
lösungsmittelarm with low solvent content
lösungsmittelbeständig fast to solvent[s]
Lösungsmitteldampf *m* solvent vapor
Lösungsmittelechtheit *f* fastness to solvents, resistance against solvent[s]
Lösungsmittelextraktion *f* solvent extraction
lösungsmittelfrei solvent-free
Lösungsmittelgemisch *n* solvent mixture

Lösungsmittelkleber *m* solution adhesive
Lösungsmittelpolymerisation *f* solution polymerization
Lösungsmittelrückgewinnung *f* recovery of solvent[s], solvent recovery
Lösungsmittelrückstand *m* residual solvent
Lösungsmittelüberschuß *m* solvent surplus
Lösungspolymerisation *f* polymerization in solution, solution polymerization
Lösungspotential *n* solution potential
Lösungsprozeß *m* solubilization process
Lösungsstärke *f* strength of solution
Lösungstheorie *f* solution theory
Lösungsvermittler *m* dissolving intermediary, solubilizer
Lösungsvermögen *n* dissolving power
Lösungsversuch *m* dissolving test
Lösungswärme *f* heat of solution
Lösungswasser *n* solvent water
Lötanalyse *f* blowpipe analysis
Lötapparat *m* lead soldering apparatus
Lötasche *f* sal ammoniac for soldering
lötbar capable of being soldered, solderable
Lötbesteck *n* complete blowpipe
Lötblock *m* soldering block
Lötbrenner *m* gas blowpipe
Lötbrett *n* soldering board
Lötdeckeldose *f* snap-on end can
Löteisen *n* soldering iron
löten (hartlöten) to braze, (weichlöten) to solder
Löten *n* (Hartlöten) brazing, hardsoldering, (Weichlöten) soldering
Lötflamme *f* blowpipe flame
Lötfuge *f* soldered joint, soldering seam
Lötgestell *n* soldering frame
Lötinseln *pl* cold soldering, cold spots
Lötkolben *m* soldering iron
Lötkupfer *n* soldering copper
Lötlampe *f* [blow] torch, soldering lamp
Lötmaterial *n* soldering material
Lötmetall *n* soldering metal
Lötmittel *n* solder
Lötnaht *f* soldering seam
Lötofen *m* soldering furnace
Lötpfanne *f* soldering pan
Lötprobe *f* blowpipe test
Lötpulver *n* soldering compound
Lötrohr *n* blowpipe
Lötrohranalyse *f* blowpipe analysis
Lötrohrapparat *m* blowpipe apparatus
Lötrohrflamme *f* blowpipe flame
Lötrohrfluß *m* blowpipe flux
Lötrohrgebläse *n* blowpipe with bellows
Lötrohrlampe *f* blowpipe lamp
Lötrohrprobe *f* blowpipe analysis, blowpipe test
Lötrohrprüfgerätschaft *f* blowpipe testing outfit
Lötrohrreagens *n* blowpipe reagent
Lötrohrspitze *f* nozzle of a blowpipe, blowpipe nipple

Lötrohruntersuchung *f* blowpipe analysis
Lötrohrversuch *m* blowpipe experiment
Lötsäure *f* soldering acid
Lötsalz *n* soldering salt, flux
Lötstelle *f* [soldered] joint or junction, brazing, soldered seam, soldering seam, **heiße** ~ hot end, hot junction, **kalte** ~ cold end, cold junction, dry junction, **schlechte** ~ dry joint
Lötung *f* soldering, sealing, (Hartlötung) brazing, **autogene** ~ autogenous soldering
Lötverbindung *f* [soldered] joint, [soldered] junction
Lötversuch *m* blowpipe test
Lötwasser *n* soldering fluid
Lötzange *f* soldering tweezers
Lötzink *n* soldering zinc
Lötzinn *n* tin solder; pewter for soldering, plumber's solder
Löweit *m* (Min) löweite
Löwenzahnbitter *n* taraxacin
Löwenzahnextrakt *m* taraxacum extract
Löwenzahnwurzel *f* (Bot) dandelion root
Löwigit *m* löwigite
Loganin *n* loganin
Loganit *m* loganite
Logarithmenpapier *n* log[arithmic] paper
Logarithmentafel *f* (Math) logarithmic table, table of logarithms
logarithmieren to take the logarithm [of]
logarithmisch logarithmic[al]
Logarithmus *m* logarithm, log, **Briggscher** ~ Brigg's logarithm, decadic logarithm, **dekadischer** ~ Brigg's logarithm, decadic logarithm, **fünfstelliger** ~ five-figure logarithm, **natürlicher** ~ natural logarithm
Logarithmusfunktion *f* logarithmic function
Lohbeize *f* tan liquor
Lohbrühe *f* bark liquor, tan[ning], tan pickle
Lohbrühleder *n* ooze leather
Lohe *f* tan bark
lohen to prepare with tan, to steep in tan
Loherde *f* tan earth
Lohfarbe *f* tan color
Lohfaß *n* tan vat
lohgar bark-tanned
Lohgare *f* currying, tanning
Lohgerberei *f* [bark] tannery, tan yard
Lohgrube *f* tan vat
Lohn *m* wages
Lohneinheit *f* unit for wages
Lohnempfänger *m* wage earner, hourly paid employee, workman
Lohnfärber *m* commission dyer
Lohnkosten *pl* labor cost, wage cost
Lohn- und Gehaltsabrechnung *f* payroll accounting
Lohpulver *n* tan powder
Lohrinde *f* tan bark, oakbark for tanning

Loiponsäure f loiponic acid, piperidine-3,4-dicarboxylic acid
Lokalanaesthesie f (Med) local an[a]esthesia
Lokalanaesthetikum n (Med) local anaesthetic
Lokalelement n local cell
Lokalisation f localization
lokalisieren to localize
Lokalisierung f localization
Lokalisierungssatz m localization theorem
Lokalkorrosion f localized corrosion, selective corrosion
Lokalwirkung f local action
Lokaose f locaose
Lokomotive f locomotive, engine
Lomatiol n lomatiol
Lonchidit m lonchidite
Longibornan n longibornane
Longiborneol n longiborneol, juniperol
Longidion n longidione
Longifdion n longifdione
Longifolsäure f longifolic acid
Longitudinalbedeckung f longitudinal covering
Longitudinalschwingung f longitudinal oscillation, longitudinal vibration
Longitudinalwelle f longitudinal wave
Lophan n lophane
Lophenol n lophenol
Lophin n lophine, 2,4,5-triphenylimidazole
Lophocerin n lophocerine
Lophoit m lophoite
Lophophorin n lophophorine
Lorandit m (Min) lorandite
Loranskit m (Min) loranskite
Lorbeer m (Bot) laurel, bay
lorbeerartig lauric
Lorbeerblatt n (Bot) bay leaf, laurel leaf
Lorbeercampher m laurin
Lorbeere f (Bot) bayberry, laurel berry
Lorbeeröl n bay oil, laurel oil
Lorbeerspiritus m bay rum
Lorentztransformation f Lorentz transformation
Lorenzenit m lorenzenite
Loretin n loretin, yatren
losbrechen to break out
Losbrechkraft f (Bagger) breakaway torque
Loschmidtsche Konstante f Loschmidt number, Avogadro number
Loschmidtsche Zahl f Loschmidt number, Avogadro number
lose loose, detached, (beweglich) movable, ~ **werden** to loosen
Loseblattsammelmappe f looseleaf binder
Losflansch m loose flange
loskitten to unlute
loskuppeln to disconnect, to disengage, to uncouple
loslösen to detach, to separate
Loslösung f detachment, separation
loslöten to unsolder

Losophan n losophan, eso-triiodo-m-cresol
losreißen (Elektronen) to tear loose
Losrolle f dancer roll, free roller, loose roller
Losscheibe f (bei Riemenantrieb) loose pulley
losschrauben to unscrew
lossprengen to blast off
lostrennen to detach, to separate
Lot n solder, soldering material, (mit niedrigem Zinngehalt) low tin solder, (Senkblei) plumb, plumb bob, plumb line, plummet
Lotabweichung f deflection from the vertical, plumb line deflection
Lotblei n lead solder, sounding lead
loten to fathom, to plumb, to sound, to take soundings
Lotleine f lead line
Lotoflavin n lotoflavin
lotrecht perpendicular, upright, vertical
Lotrechte f vertical
Lotschnur f plumb line
Lotschwankung f deflection of the vertical
Lotspritzer pl solder splashes
Lotusin n lotusin
Loxachina f loxa bark, cinchona bark
Loxodrome f (Math) loxodromic line
LPH Abk. für „lipotropes Hormon"
L-Reihe f (Phys) L-series
L-Schale f (Atom) L-shell
LTH Abk. für luteotropes Hormon
Luargol n luargol
Lubanol n lubanol, coniferol
Lublinit m lublinite
Lucanthon n lucanthone
Luchsöl n lynx fat
Luchsstein m (Min) thunderstone
Lucidol n lucidol, benzoyl peroxide, dibenzoyl peroxide
Luciferase f (Biochem) luciferase
Luciferin n luciferin
Lucigenin n lucigenin
Lucinit m (Min) lucinite, native aluminum phosphate
Lucit n lucite
Luckit m luckite
Lucoperdin n lycoperdine
Ludlamit m (Min) ludlamite
Ludwigit m (Min) ludwigite
Lücke f vacancy, deficiency, gap, interstice, void
lückenhaft defective, having gaps, incomplete
Lückenlänge f gap length
Lückenlängenmessung f gap length measurement
Lückenlängenverteilung f gap length distribution
lückenlos uninterrupted
Lückenvolumen n (Füllkörperkolonne) void fraction
lüften to ventilate, to aerate, to air
Lüften n (des Werkzeugs) breathing
Lüfter m fan, ventilator
Lüfterantrieb m fan drive

Lüfterflügel *m* fan blade
Lüfterkupplung *f* fan coupling
Lüfterschaufel *f* fan blade
Lüftung *f* aeration, airing, vent, ventilation
Lüftungsanlage *f* ventilating system
Lüftungsapparat *m* ventilator
Lüftungsart *f* system of ventilation
Lüftungsrasterdecke *f* false ceiling
Lüftungsschacht *m* air shaft, ventilating shaft
Lüftungsvorrichtung *f* ventilation appliance
Lüneburgit *m* (Min) luneburgite
Lüsterfarbe *f* luster color
Lüsterglasur *f* luster glaze, lustrous glaze
Lüsterklemme *f* porcelain insulator, terminal strip
Luft *f* air, atmosphere, (Spiel) play, **flüssige** ~ liquid air, **härten an der** ~ to airharden, **verdünnte** ~ rarefied air
Luftablaßventil *n* air escape valve, air vent
Luftabsaugvorrichtung *f* air suction ventilator
Luftabschluß *m* hermetic seal, air-tight seal, exclusion of air
Luftabschneider *m* air separator
Luftabzug *m* air vent
Luftaktivität *f* (Atom) airborne radioactivity
Luftanalyse *f* analysis of the atmosphere
Luftanfeuchter *m* air humidifier, air moistener
luftartig aeriform, air-like, gaseous
Luftaufnahme *f* aerial photograph
Luftaufnehmer *m* air receiver
Luftausdehnungsmaschine *f* hot-air engine
Luftausschluß *m* exclusion of air
Luftaustausch *m* (Lunge) respiratory exchange
Luftaustritt *m* air outlet, (Gaschromat) air diffuser
Luftbad *n* air bath
Luftbefeuchter *m* air moistener
Luftbeschaffenheit *f* condition of the air
luftbeständig air-resistant, not affected by air, stable in air
Luftbeständigkeit *f* resistance to air, stability in air
Luftbild *n* (Luftf) aerial photograph
Luftblase *f* air bubble, air cavity, air pocket, **eingeschlossene** ~ air cavity, entrapped air bubble
Luftblasen-Viskosimeter *n* bubble visco[si]meter
Luftbleiche *f* open-air bleaching
Luftbrechungsvermögen *n* air refractive power
Luftbremse *f* air brake
Luftbrücke *f* air bridge
Luftbürste *f* air knife
Luftbürstenauftrag *m* air-brush coating, air-knife coating
Luftdämpfung *f* air damping
Luftdampfgemisch *n* steam and air mixture
Luftdarre *f* [hot-] air dryer
luftdicht airtight, air-proof, ~ **machen** to air-proof, ~ **verschließen** to seal hermetically

Luftdichte *f* air density
Luftdichtheit *f* airtightness, impermeability to air
Luftdichtigkeitsmesser *m* air densimeter
Luftdruck *m* atmospheric pressure, barometric pressure, pneumatic pressure, (Techn) air pressure
Luftdruckaufzug *m* pneumatic lift
Luftdruckhammer *m* compressed air hammer
Luftdruckmaschine *f* air pressure engine
Luftdruckmeissel *m* pneumatic chisel
Luftdruckmesser *m* barometer
Luftdruckprüfer *m* air gauge
Luftdruckpumpe *f* air compressor pump
Luftdruckschreiber *m* barograph
Luftdruckschwankung *f* barometric variation
Luftdruckstand *m* barometric reading
Luftdrucküberwachung *f* air pressure maintenance
Luftdruckunterschied *m* difference of atmospheric pressure
Luftdüse *f* air jet, air nozzle, choke tube, tuyère
Luftdurchfeuchter *m* airdamping machine, air moistener
Luftdurchflußmenge *f* rate of air flow
Luftdurchflußmesser *m* air flow meter
luftdurchlässig permeable to air
Luftdurchlässigkeit *f* air permeability, venting property
Luftdurchschlagspannung *f* breakdown voltage in air
Lufteinblasung *f* air injection
Lufteinlaß *m* air inlet, air intake
Lufteinlaßschraube *f* air inlet screw
Lufteinlaßventil *n* air inlet valve, air inlet manifold
Lufteinschluß *m* air pocket, entrapped air, inclusion of air, trapping of air
Lufteinströmklappe *f* air inlet valve
Lufteinströmöffnung *f* air inlet hole
Lufteintritt *m* air inlet
Lufteinwirkung *f* atmospheric action, influence of air
Luftelektrizität *f* atmospheric electricity
Luftelektrometer *n* air electrometer
luftempfindlich affected by air, sensitive to air
Luftentkeimungsanlage *f* air sterilization plant
Luftentladung *f* air discharge
luftentzündlich (Chem) pyrophoric
Lufterhärtung *f* air hardening
Lufterhitzer *m* air heater
Lufterscheinung *f* atmospheric phenomenon
Luftexpansionsmaschine *f* expansion engine for air, hot air engine
Luftfahrt *f* aeronautics, aviation
Luftfederbalg *m* **mit drei Ausbauchungen** three-lobe bellow
Luftfeuchte *f* humidity

Luftfeuchteregelanlage f air humidity regulating installation
Luftfeuchtigkeit f atmospheric moisture, dampness of the atmosphere, humidity of air, hygrometric condition, **relative** ~ relative humidity
Luftfeuchtigkeitsgrad m relative humidity [of the air]
Luftfeuchtigkeitsmesser m hygrometer
Luftfilter n air filter, (Drehfilter) automatic air filter, (Reihenfilter) bag filter, (Rundfilter) radial bag filter, **Drallschicht-** ~ granular bed dust separator, **Flächen-** ~ screen cloth filter, **Rollband-** ~ automatic air filter, **Schwebstoff-** ~ air filter
Luftfilteröl n air filter oil
Luftförderrinne f air conveying passage
luftförmig aeriform, gaseous
Luftfracht f air freight
luftfrei free from air
Luftgebläse n air fan, blower
luftgefedert air-cushioned, air-suspended
luftgefüllt inflated, pneumatic
Luftgehalt m air content
luftgekühlt air-cooled
Luftgemisch n air mixture
luftgetrocknet air-dried
Luftgüte f quality of air for respiration
Luftgütemesser m (Gasprüfgerät) eudiometer
lufthärten to air-harden
Lufthärtestahl m air-hardening steel
Lufthärtung f air hardening
lufthaltig containing air, aerated, aeriferous
Lufthammer m air hammer, pneumatic hammer
Lufthaut f oxide film
Luftheizung f [hot-]air heating
Luftheizungsapparat m hot-air heating apparatus
Luftheizungsofen m hot-air heater
Luftheizvorrichtung f hot-air heating apparatus
Lufthitzemesser m air pyrometer
Lufthülle f atmosphere
Luftinduktion f air induction, induction in air
Luftinfektion f air-borne infection
Luftinjektor m pressure injector
Luftionisation f ionization of air
Luftisolation f air insulation
Luftkalk m hydrated lime, plaster of paris
Luftkammer f air chamber
Luftkanal m air duct, ventilating duct
Luftkessel m air chamber, air receiver
Luftkissen n air buffer, air cushion
Luftkissenboot n air cushion boat, hover craft
Luftklappe f air flap, (des Vergasers) choke, throttle plate, throttle valve
Luftkompressionsmaschine f air compressing engine
Luftkompressor m air compressor
Luftkondensator m air condenser

Luft-Kraftstoffgemisch n fuel-air mixture
Luftkühlanlage f air cooling plant
Luftkühler m air condenser, air cooler
Luftkühlung f air cooling
Luftlager n air[-cushioned] bearing
luftleer evacuated, deflated, void or exhausted of air, ~ **machen** to evacuate, to empty, to exhaust
Luftleere f vacuum, air exhaustion
Luftleeremesser m vacuum gauge
luftleitend air-conducting
Luftleitung f (Luftkanal) air duct
Luftloch n air pocket, vent hole
Luftmagnetfeld n magnetic field in air
Luftmalz n air-dried malt
Luftmangel m deficiency of air, lack of air
Luftmanometer n air manometer
Luftmaschine f atmospheric engine
Luftmenge f air volume, quantity of air
Luftmengenmesser m air meter
Luftmesser m (Phys) aerometer, air gauge
Luftmischung f mixture of air
Luftmörtel m air mortar, lime mortar, non-hydraulic mortar
Luftoxydation f air oxidation
Luftozonisierung f ozonization of air
Luftozonisierungsanlage f plant for ozonization of air
Luftpore f air void
Luftporenanteil m void content
Luftporenbeton m air entrained concrete
Luftpuffer m air buffer, air cushion
Luftpumpe f air pump, pneumatic pump
Luftpumpenglocke f air pump bell, air pump receiver
Luftpumpenkolben m air pump piston
Luftpumpenteller m air pump disc, air pump plate
Luftpumpenzylinder m air pump cylinder
Luftpumpenzylinderwand f air pump liner
Luftpyrometer n air pyrometer
Luftradioaktivität f radioactivity of the air
Luftrakel f floating knife
Luftrakelstreichmaschine f floating knife coater
Luftrakelstreichverfahren n floating knife coating
Luftreibung f air friction
luftreinigend [air] disinfecting, air-purifying
Luftreiniger m air purifier
Luftreinigung f air disinfection, air purification, ventilation
Luftreinigungsanlage f air cleaning plant, air conditioning plant, air filter unit
Luftreinigungsmaschine f [air] disinfecting apparatus
Luftrektifikationsanlage f air separation plant
Luftröhre f (Med) trachea, windpipe
Luftrührer m air agitator, airlift agitator
Luftsättigung f saturation of air

Luftsäule *f* air column
Luftsauerstoff *m* atmospheric oxygen, aerial oxygen
Luftsaugepumpe *f* aspirator
Luftsauger *m* aspirator
Luftsaugeraggregat *n* aspirating unit
Luftsaugeröhre *f* air suction pipe
Luftschicht *f* air film
Luftschlauch *m* (Reifen) inner tube
Luftschraube *f* air screw, propeller
Luftschwefel *m* aerial sulfur, atmospheric sulfur
Luftspalt *m* air gap
Luftspiegelung *f* (Opt) atmospheric reflection, fata morgana, mirage
Luftstählung *f* (Met) air hardening, air quenching
Luftstelle *f* dry spot
Luftstickstoff *m* atmospheric nitrogen, aerial nitrogen
Luftstickstoffverarbeitung *f* (Luftstickstoff-Fixierung) fixation of atmospheric nitrogen
Luftstoß *m* air blast
Luftstrahl *m* air blast, air jet
Luftstrecke *f* distance in air
Luftstrom *m* air current, air stream, **aufsteigender** ~ upcurrent
Luftstrommühle *f* ball mill with air drier
Lufttemperatur *f* air temperature
Luftthermometer *n* air thermometer
Lufttrennungsanlage *f* air separation plant
lufttrocken air-dried, air-dry, (Holz) air-seasoned
lufttrocknen to air-dry, (Holz) to air-season
Lufttrocknen *n* air conditioning, air drying
Lufttrockner *m* air dryer, air-drying apparatus, air heater, desiccator
Lufttrocknung *f* air-drying
Lufttrocknungsanlage *f* air-drying plant
Luftüberdruck *m* positive air pressure
Luftüberhitzer *m* blast superheater
Luftüberschuß *m* excess of air
Luftüberschußzahl *f* excess air coefficient
Luftübertragung *f* aerial transmission, air-borne transmission
Luftüberwachungsanlage *f* air monitor
Luftüberwachungsgerät *n* air monitor
Luftumlauf *m* circulation of air
Luftumwälzung *f* air circulation
luftundurchlässig impermeable to air
Luftundurchlässigkeit *f* air impermeability
Luftunterdruck *m* air depression, vacuum
Luftveränderung *f* change of air
Luftverbesserungsmittel *n* air improver
Luftverdichter *m* air compressor, air condenser
Luftverdichtung *f* air compression
Luftverdichtungsmaschine *f* air compressor
Luftverdichtungsmesser *m* atmospheric density gauge

Luftverdichtungspumpe *f* condensing air-pump
luftverdorben damaged by air
Luftverdrängung *f* air displacement
luftverdünnt evacuated, rarefied [of air]
Luftverdünnung *f* partial vacuum, rarefaction [of air]
Luftverdünnungsmesser *m* vacuum gauge
Luftverflüssigungsanlage *f* air liquefaction plant
Luftverflüssigungsmaschine *f* air liquefier
Luftverpestung *f* air pollution, (durch Radioaktivität) radioactive pollution of the air
Luftverseuchung *f* air-borne contamination, air contamination, air pollution
Luftverteilungsrohr *n* air distributing pipe
Luftverunreinigung *f* air contamination, air pollution
Luftvorwärmer *m* air preheater
Luftvulkanisation *f* air cure
Luftwaffe *f* air force
Luftwasser *n* atmospheric water
Luftwechsel *m* change of air, renewal of air
Luftwelle *f* air-wave
Luftwichte *f* (obs) density of the air, specific gravity of the air, specific weight of air
Luftwiderstand *m* atmospheric resistance, aerodynamic drag, air resistance
Luftzahl *f* excess air coefficient
Luftzerlegung *f* air separation, separation of air into constituents
Luftzerlegungsanlage *f* air separation plant
Luftzirkulation *f* circulation of air
Luftzünder *m* (Pyrophor) pyrophorus
Luftzuführung *f* air supply, air duct, air input, air vent
Luftzuführungskanal *m* air duct, air passage
Luftzuführungsöffnung *f* air inlet
Luftzufuhr *f* air inlet, air supply, (Feuer) draft (Am. E.), draught (Br. E.)
Luftzug *m* [air] draft, current of air, draught (Br. E.)
Luftzustand *m* atmospheric condition
Luftzutritt *m* admission of air, air inlet, air intake
Luftzwischenraum *m* air gap
Lugolsche Lösung *f* Lugol solution
Lukullit *m* lucullite
Lumazin *n* lumazine
Lumen *n* (Maß) lumen
Lumichrom *n* lumichrome
Lumicolchicin *n* lumicholchicine
Lumiflavin *n* lumiflavine
Lumiisolysergsäure *f* lumiisolysergic acid
Luminal *n* luminal, phenobarbital
Lumineszenz *f* luminescence
Lumineszenzanregung *f* luminescence excitation
Lumineszenzmikroskop *n* luminescence microscope
lumineszieren to luminesce
Luminol *n* luminol

Luminophor *n* luminophore
Lumiöstron *n* lumiestrone
Lumistanol *n* lumistanol
Lumisterin *n* lumisterol
Lumisterol *n* lumisterol
Lumpen *m pl* (Pap) rags
Lumpenauskohlung *f* carbonizing of rags
Lumpenbleiche *f* rag bleaching
Lumpenbrei *m* (Papier) pulp from rags, first stuff
Lumpenbütte *f* pulp vat
Lumpenpapier *n* rag paper
Lumpenstoff *m* (Pap) rag pulp
Lunacridin *n* lunacridine
Lunamarin *n* lunamarine
Lunasia[-Alkaloid] *n* lunasia
Lunasin *n* lunasine
Lungenentzündung *f* (Med) pneumonia
Lungenkraut *n* (Bot) lungwort
Lungenkrebs *m* (Med) lung cancer
Lunin *n* lunine
Lunker *m* (durch Schwindung d. Gußstücks) shrinkhole, (Gaseinschluß) blowhole, (Gieß) funnel, pipe, (Kunstharz) bubble, **primärer** ~ pipe (of ingots), **sekundärer** ~ axial sponginess, secondary pipe (in an ingot)
Lunkerbildung *f* [development of] piping, shrinking
lunkerfrei free of shrinkholes, pipeless
Lunkerhohlraum *m* shrinkage cavity, shrinkhole
lunkerig honeycombed, piped
Lunkerkopf *m* top end of a pipe
lunkerlos free of shrinkholes, pipeless
lunkern to pipe
Lunkerstelle *f* porous point, porous spot, shrinkhole
Lunkerung *f* blowhole formation, cavitation (in metals), liquid contraction, piping, porosity, shrinkage
Lunkerverhütung *f* pipe elimination
Lunkerverhütungsmittel *n* pipe eliminator, pipe eradicator
Lunnit *m* (Min) lunnite
Lunte *f* slowmatch
Lupan *n* lupane
Lupanin *n* lupanine
Lupanolin *n* lupanoline
Lupe *f* (Opt) magnifier, magnifying glass, magnifying lens
Lupen *n* lupene
Lupeol *n* lupeol
Lupeon *n* lupeone
Lupeose *f* lupeose
Lupetidin *n* lupetidine, 2,6-dimethylpiperidine
Lupinan *n* lupinane
Lupinidin *n* lupinidine
Lupinin *n* lupinine
Lupininsäure *f* lupininic acid
Luppe *f* (Metall) lump, ball of iron, bloom

Luppeneisen *n* ball iron, puddled iron, wrought iron
Luppenfeuer *n* smelting furnace
Luppenfrischfeuer *n* bloomery fire
Luppenhammer *m* ball press
Luppenmühle *f* rotary squeezer
Luppenstahl *m* bloom steel
Luppenwalzwerk *n* blooming mill, blooming rolls, roughing rolls, roughing train
Lupucarbonsäure *f* lupucarboxylic acid
Lupulin *n* lupulin, humulin
Lupulinsäure *f* lupulinic acid, lupulic acid
Lupulon *n* lupulon
Lupuloxinsäure *f* lupuloxinic acid
Lussatit *m* (Min) lussatite
Lustgas *n* (obs) laughing gas, nitrous oxide
Lutein *n* lutein
Luteohormon *n* corpus luteum hormone
Luteokobaltchlorid *n* luteocobaltic chloride, hexaammine cobalt(III) chloride
Luteol *n* luteol
Luteolin *n* luteolin, isofisetin, tetrahydroxyflavone
Luteolinidin *n* luteolinidine
Luteotropin *n* luteotropin
Luteoverbindung *f* luteo compound
Lutetium *n* lutetium, lucecium (Symb. Lu), (obs) cassiopeium
Lutezit *m* lutecite
Lutidin *n* lutidine, dimethylpyridine
Lutidinsäure *f* lutidinic acid, pyridine dicarboxylic acid
Lutidon *n* lutidone, 2,4-dimethyl-6-pyridone
Lutte *f* (Bergb) [air] duct
Lutter *m* low wine, singlings
Lutterblase *f* low-wine still
luttern to distil[l] singlings, to distil[l] weak brandy
Lux *n* (Maß) lux, meter-candle
Luxemburger Roheisen *n* Luxemburg pig iron
Luxmeter *n* light meter, lux meter, photometer
Luxsche Gaswaage *f* Lux gas balance
Luxullian *m* (Min) luxullian
Luziferase *f* (Biochem) luciferase
Luzonit *m* (Min) luzonite
Lyase *f* (Biochem) lyase
Lycetol *n* lycetol, 2,5-dimethylpiperazine tartrate, lupetazin tartrate, lysitol
Lycoctonin *n* lycoctonine
Lycopersen *n* lycopersene
Lycophyll *n* lycophyll
Lycopin *n* lycopene, lycopin
Lycopodium *n* (Bot) lycopodium, clubmoss, vegetable sulfur
Lycopodiumsamen *m* (Bot) lycopodium dust
Lycoren *n* lycorene
Lycorin *n* lycorine, narcissine
Lycotetraose *f* lycotetraose
Lycoxanthin *n* lycoxanthin

Lydit *m* (Min) Lydian stone, touchstone
Lygosin *n* lygosin, disalicylalacetone
Lygosinnatrium *n* sodium lygosinate
Lygosinpräparat *n* lygosin preparation
Lymanserie *f* (Spektr) Lyman series
Lymphdrüse *f* (Med) lymphatic gland, lymph node
Lymphe *f* (Med) lymph
Lymphgefäß *n* lymphatic vessel, lymph duct, lymph vessel
Lymphoblast *m* lymphoblast
Lymphozyt *m* (Med) lymphocyte
Lymphozytenarmut *f* (Med) hypolymp[a]emia, lymphocytopenia
Lymphozytenbildung *f* lymphogenesis
Lymphozytenmangel *m* lymphocytopenia
Lymphozytose *f* (Med) lymphatic leuk[a]emia, lymphocyth[a]emia
Lymphzelle *f* (Biol) lymphocyte
lyophil lyophilic
Lyophilisation *f* lyophilization
lyophob lyophobic
Lyosol *n* lyosol
lyotrop lyotropic, (Salz) lyotropic
Lysalbinsäure *f* lysalbinic acid
Lyse *f* lysis
Lysergen *n* lysergene
Lysergid *n* lysergide
Lysergin *n* lysergine
Lysergol *n* lysergol
Lysergsäure *f* lysergic acid
Lysidin *n* lysidine, 2-methyl-4,5-dihydroimidazole
Lysidintartrat *n* lysidine tartrate
Lysin *n* lysine, 2,6-diaminohexanoic acid
Lysindihydrochlorid *n* lysine dihydrochloride
Lysinogen *n* lysinogen
Lysinracemase *f* (Biochem) lysine racemase
Lysobakterium *n* lysobacterium
Lysoform *n* lysoform
Lysolecithin *n* lysolecithin
Lysosom-Körperchen *n* lysosome
Lysozithin *n* lysocithine
Lysozym *n* (Biochem) lysozyme, muramidase
Lyxit *m* lyxitol
Lyxoflavin *n* lyxoflavine
Lyxomethylit *m* lyxomethylitol
Lyxonsäure *f* lyxonic acid
Lyxosamin *n* lyxosamine
Lyxose *f* lyxose

M

Maalien *n* maaliene
Macarit *n* macarite
Macerale *pl* macerals
Machaerinsäure *f* machaerinic acid
Machaersäure *f* machaeric acid
Mache-Einheit *f* (Radioaktivität) mache unit
machen to make, to cause, to perform
Machilen *n* machilene, atractylene, eudesmene
Machilol *n* machilol, atractylol, eudesmol, uncineol
Machzahl *f* Mach number
Macilensäure *f* macilenic acid
Macilolsäure *f* macilolic acid
Macisöl *n* mace oil
Maclaurinsche Reihe *f* Maclaurin['s] series
MacLeod-Druckmesser *m* MacLeod gauge
Macleyin *n* macleyine, fumarine, protopine
Maclurin *n* maclurin
Macrogol *n* (Kurzbezeichnung für Polyäthylenglykol) macrogol
Macropon *n* macropone
Maculin *n* maculine
Maculosidin *n* maculosidine
Madaralban *n* madaralbane
Maddrellsches Salz *n* Maddrell's salt
Madenschraube *f* grub screw, headless screw, setscrew
Madenwurm *m* (Zool) pinworm, thread-worm
Madiaöl *n* madia oil
Madreporit *m* madreporite
Mälzerei *f* malt house, malting
männlich male, masculine
mäßig moderate, reasonable, temperate, (mittelmäßig) mediocre
mäßigen to moderate, to temper
Mafenid *n* mafenide
Mafuratalg *m* mafura butter, mafura tallow
Magazinseil *n* storage rope
Magdalarot *n* magdala red, naphthalene red
Magen *m* (Anat) stomach
Magenbitter *m* stomachic bitter
Magenbrennen *n* heart-burn
Magengeschwür *n* (Med) gastric ulcer
Magensäure *f* gastric acid, stomach acid, stomach acidity
Magensäuregehalt *m* acidity of the stomach
Magensäureüberproduktion *f* (Med) hyperacidity [of the stomach]
Magensaft *m* (Med) gastric juice
Magentarot *n* magenta red, magenta fuchsin
Magentasäure *f* magenta acid
Magentropfen *m pl* (Pharm) stomachic drops
mager lean, meager; (Bergb) low-grade, poor
Magerbeton *m* lean concrete
Magerkalk *m* poor lime
Magerkohle *f* semianthracite, dry burning coal, lean coal, nonbaking coal, nonbituminous coal, noncaking coal
Magerkoks *m* lean coke
Magermilch *f* low-fat milk, skimmed milk
Magerungsmittel *n* lean clay, lean material
magisch magic
Magma *n* (Geol) magma
magmatisch magmatic
Magnalium *n* (HN) magnalium
Magnesia *f* bitter earth, [calcined] magnesia, magnesium oxide, **gebrannte** ~ burnt magnesia, **saure** ~ bicarbonate of magnesia
Magnesiaabscheidung *f* separation of magnesia
Magnesiabad *n* magnesia bath
Magnesiableichflüssigkeit *f* magnesium hypochlorite solution
Magnesiaeisenglimmer *m* (Min) magnesium iron mica
Magnesiaglimmer *m* magnesium mica
Magnesiahärte *f* hardness due to magnesia, magnesia hardness
magnesiahaltig containing magnesia, magnesian
Magnesiakalk *m* (Min) magnesian limestone, dolomite
Magnesiamilch *f* (Pharm) milk of magnesia
Magnesiamischung *f* magnesia mixture
Magnesiamixtur *f* magnesia mixture
Magnesiarot *f* (Farbstoff) magnesia red
Magnesiasalpeter *m* magnesium nitrate, nitromagnesite
Magnesiaseife *f* magnesia soap
Magnesiazement *m* magnesia cement
Magnesiochromit *m* (Min) magnesiochromite
Magnesioferrit *m* (Min) magn[esi]oferrite
Magnesioludwigit *m* (Min) magnesioludwigite
Magnesit *m* (Min) magnesite, native magnesium carbonate
Magnesitspat *m* (Min) magnesite, native magnesium carbonate
Magnesitstein *m* magnesite, magnesite brick
Magnesitziegel *m* magnesite brick
Magnesium *n* magnesium (Symb. Mg)
Magnesiumacetat *n* magnesium acetate
Magnesiumalkyl *n* magnesium alkyl
Magnesiumaluminat *n* magnesium aluminate
Magnesiumamalgam *n* magnesium amalgam
Magnesiumantimonid *n* magnesium antimonide
Magnesiumband *n* magnesium ribbon, magnesium tape
Magnesiumblitzlicht *n* magnesium flashlight
Magnesiumborat *n* magnesium borate
Magnesiumbranntkalk *m* magnesium quicklime
Magnesiumbromid *n* magnesium bromide
Magnesiumbronze *f* magnesium bronze
Magnesiumcarbonat *n* magnesium carbonate

Magnesiumchlorid *n* magnesium chloride
Magnesiumcitrat *n* magnesium citrate
Magnesiumdraht *m* magnesium wire
Magnesiumgehalt *m* magnesium content
Magnesiumglycerophosphat *n* magnesium glycerophosphate, magnesium glycerinophosphate
Magnesiumhärte *f* magnesium hardness
magnesiumhaltig containing magnesium, magnesian
Magnesiumhydrat *n* magnesium hydrate, magnesium hydroxide
Magnesiumhydrophosphat *n* acid magnesium phosphate, magnesium hydrogen phosphate
Magnesiumhydroxid *n* magnesium hydroxide
Magnesiumjodid *n* magnesium iodide
Magnesiumlactat *n* magnesium lactate
Magnesiumlampe *f* magnesium lamp
Magnesiumlegierung *f* magnesium alloy
Magnesiumlicht *n* magnesium flashlight, magnesium light
Magnesiumlöschkalk *m* hydrated magnesium lime
Magnesiummergel *m* (Agr) magnesium agricultural lime
Magnesium-Mischkalk *m* mixed magnesium lime
Magnesium-Monel *n* (Leg) magnesium monel
Magnesiumnitrat *n* magnesium nitrate
magnesiumorganisch magnesium-organic, organomagnesium
Magnesiumoxid *n* magnesium oxide, calcined magnesia, magnesia, (Min) bitter earth
magnesiumoxidhaltig containing magnesia
Magnesiumpalmitat *n* magnesium palmitate
Magnesiumpektolith *m* magnesium pectolite
Magnesiumphosphat *n* magnesium phosphate
Magnesiumricinat *n* magnesium ricinate
Magnesiumsalz *n* magnesium salt
Magnesiumsilikat *n* magnesium silicate
Magnesiumsilikatniederschlag *m* magnesium silicate precipitate
Magnesiumsilikofluorid *n* magnesium silicofluoride
Magnesiumstearat *n* magnesium stearate
Magnesiumsulfat *n* magnesium sulfate
Magnesiumsulfatkalzinieranlage *f* magnesium sulfate calcining plant
Magnesiumsuperoxid *n* magnesium dioxide, magnesium peroxide, magnesium superoxide
Magnesiumtartrat *n* magnesium tartrate
Magnesiumverbindung *f* magnesium compound
Magneson *n* magneson
Magnet *m* magnet, ~ **einer Bogenlampe** magnet core of an arc lamp, **fremderregter** ~ electromagnet, nonpermanent magnet, **induzierender** ~ inducing magnet, **induzierter** ~ induced magnet, **künstlicher** ~ artificial magnet, **natürlicher** ~ loadstone, natural magnet, **permanenter** ~ permanent magnet, **temporärer** ~ temporary magnet
Magnetabscheider *m* magnetic separator
Magnetachse *f* magnetic axis
Magnetanker *m* magnet armature
Magnetband *n* magnetic tape
Magnetbremse *f* magnetic brake, selenoid brake
Magnetbündel *n* compound magnet
Magneteisen *n* (Min) magnetic iron ore, magnet iron, magnetite
Magneteisenerz *n* (Min) magnetic iron ore, magnet iron, magnetite
Magneteisensand *m* (Min) magnetic iron sand
Magneteisenstein *m* (Min) magnetic iron ore, magnetite
magnetelektrisch magneto-electric
Magnetelektrisiermaschine *f* magnetoelectric machine
Magnetelektrizität *f* magnetoelectricity
Magnetfeld *n* magnetic field, **transversales** ~ transverse magnetic field, **ungleichförmiges** ~ non-uniform magnetic field
Magnetfeldröhre *f* magnetron
Magnetfluß *m* magnetic flux
magnetgesteuert magnetically controlled
magnethaltig magnetic
Magnetinduktion *f* magnetic induction
Magnetinduktor *m* magneto-electric generator
magnetisch magnetic, ~ **erregbar** magnetizable, ~ **machen** to magnetize, ~ **werden** to become magnetic
magnetisierbar magnetizable
Magnetisierbarkeit *f* magnetizability
magnetisieren to magnetize
Magnetisieren *n* magnetizing
magnetisierfähig magnetizable
Magnetisierfähigkeit *f* magnetizability
Magnetisiergerät *n* magnetizing apparatus
Magnetisierung *f* intensity of magnetization, magnetic intensity, magnetization, magnetizing, ~ **bis zur Sättigung** saturated magnetization, **remanente** ~ residual magnetization, **überlagerte** ~ superposed magnetization
Magnetisierungsart *f* method of magnetization
Magnetisierungsintensität *f* intrinsic induction
Magnetisierungskoeffizient *m* coefficient of magnetization
Magnetisierungskurve *f* magnetization curve
Magnetisierungslinie *f* line of magnetism, magnetization line
Magnetisierungsrichtung *f* direction of magnetization
Magnetisierungsschleife *f* hysteresis loop
Magnetisierungsspirale *f* magnetizing coil
Magnetisierungsspule *f* magnetizing coil
Magnetisierungsstärke *f* intensity of magnetization

Magnetisierungszahl *f* coefficient of magnetization, permeability, permeance
Magnetisierungszyklus *m* cycle of magnetization, magnetic cycle
Magnetismus *m* magnetism, ~ **erzeugend** magnetiferous, **freier** ~ free magnetism, surface magnetization, **remanenter** ~ remanent magnetism, residual magnetism
Magnetit *m* (Min) magnetite, magnetic iron ore
Magnetjoch *n* magnet yoke
Magnetkern *m* magnet core
Magnetkernspeicher *m* magnetic core memory
Magnetkernzündung *f* magnetic spark plug ignition
Magnetkies *m* (Min) magnetic pyrite, pyrrhotite
Magnetkraftlinie *f* magnetic line of force, **achsiale** ~ axial line of magnetic force
Magnetkran *m* magnet crane
Magnetmesser *m* magnetometer
Magnetmoment *n* magnetic moment
Magnetmotor *m* magneto-electric motor
Magnetnadel *f* magnetic needle, compass needle
Magnetochemie *f* magneto-chemistry
magnetoelektrisch magnetoelectric
Magnetograph *m* magnetograph
Magnetohydrodynamik *f* (Abk. MHD) magnetohydrodynamics
magnetokalorisch magnetocaloric
Magnetometer *n* magnetometer
magnetometrisch magnetometric
Magnetomotor *m* magnetomotor
Magneton *n* (Phys) magneton
magnetooptisch magnetooptical
Magnetophon *n* magnetophone
Magnetophonband *n* magnetic tape, tape of magnetic recorder
Magnetopyrit *m* (Min) magnetic pyrite, magnetopyrite, pyrrhotite
Magnetorotationsspektrum *n* magnetic rotation spectrum
Magnetoskop *n* magnetoscope
Magnetostatik *f* magnetostatics
Magnetostriktion *f* magnetostriction
magnetostriktiv magnetostrictive
Magnetpol *m* magnetic pole, magnet pole
Magnetringscheider *m* circular magnet separator
Magnetrolle *f* magnetic idling roll
Magnetron *n* magnetron
Magnetrührer *m* magnetic stirrer, magnetic stirring unit
Magnetschalter *m* magnetic switch, selenoid switch
Magnetscheider *m* magnetic separator
Magnetscheidung *f* magnetic separation
Magnetspule *f* magnetic coil
Magnetstab *m* bar magnet, magnetic bar, straight magnet
Magnetstahl *m* magnet steel

Magnetstein *m* (Min) lode-stone, magnetic iron ore, magnetite
Magnetsystem *n* magnetic system
Magnetton *m* magnetic sound
Magnettonband *n* magnetic [recording] tape, tape for magnetic sound recording
Magnetträger *m* magnet carrier
Magnetunterbrecher *m* magnet interrupter
Magnetventil *n* magnetic valve
Magnetwicklung *f* magnet winding
Magnetwicklungswiderstand *m* field resistance
Magnetzündung *f* magneto-ignition
Magnochromit *m* (Min) magnochromite
Magnocurarin *n* magnocurarine
Magnoferrit *m* (Min) magnoferrite
Magnolamin *n* magnolamine
Magnolaminsäure *f* magnolaminic acid
Magnolith *m* (Min) magnolite
Magnolol *n* magnolol
Magon *n* magon
Mahagongummi *n* acajou gum
Mahagonharz *n* acajou resin
Mahagonholzöl *n* mahogany wood oil
Mahagoni *n* mahogany
mahagonibraun mahogany brown
Mahlanlage *f* milling plant, crushing plant, grinding mill, grinding plant, pulverizing equipment, (Pap) beating plant
Mahlarbeit *f* milling, grinding, (Pap) beating
Mahlbahn *f* grinding path
Mahleinheit *f* grinding unit
mahlen to grind, to mill, to triturate, (fein) to pulverize, (grob) to crush
Mahlfeinheit *f* degree of grinding, fineness of grinding
Mahlfeinheitsprüfgerät *n* testing equipment for grinding fineness
Mahlgang *m* milling cycle
Mahlgrad *m* degree of grinding, (Pap) degree of freeness
Mahlgut *n* grinding stock, grist, material to be ground, milling paste, (Farben) mill base
Mahlholländer *m* beater, breaker
Mahlraum *m* milling section
Mahlstein *m* millstone
Mahlteile *pl* crushing parts
Mahlung *f* grinding, milling, (Pap) beating
Mahlverfahren *n* milling process
Mahlwerk *n* breaker
Maillard-Reaktion *f* Maillard reaction
Mais *m* corn (Am. E.), Indian corn, maize
Maisch *m* mash
Maische *f* mash, grape must
Maischekolben *m* brewers mash flask
maischen to mash
Maischhefe *f* mash yeast
Maischthermometer *n* mash tub thermometer
Maischtrog *m* mingler
Maischwasser *n* mash liquor

Maisdextrose f corn sugar
Maisin n maisin, maize protein
Maiskeimöl corn oil
Maiskleber m maize gluten, zein
Maiskolben m corn cob
Maismehl n corn flour, Indian meal
Maisöl n corn oil, maize oil
[Mais]schlempe f dried distiller's solubles
Maisstärke f corn starch, maize starch
Maiszucker m corn sugar
Majestowurzel f (Färb) madder
Majolika f majolica
Majolikafarbe f majolica color
Majoran m [sweet] marjoram
Majoranöl n marjoram oil
MAK Abk. für „maximale Arbeitsplatz-Konzentration"
Makassaröl n Macassar oil
Mako m (Text) Egyptian cotton
Makroachse f (Krist) macroaxis
Makroanalyse f macroanalysis
Makroapparat m macroapparatus
Makrobakterium n macrobacterium
Makrobiose f macrobiosis
Makroblast m macroblast, megaloblast
Makrochemie f macrochemistry
Makroelementaranalyse f elementary macro-analysis
Makrogamet m (Biol) macrogamete, megagamete
Makroglobulin n macroglobulin
Makrokinetik f macrokinetics
Makrokosmos m macrocosm
makrokristallin macrocrystalline
Makrometeorologie f macrometeorology
Makromethode f macro-method
Makromolekül n macromolecule,
  kugelförmiges ~ spherical macromolecule,
  lineares ~ linear macromolecule,
  stäbchenförmiges ~ rod-like macromolecule,
  vernetztes ~ cross-linked macromolecule
makromolekular macromolecular
Makrophage m (Med) macrophage
Makrophotographie f macrophotography
Makrorheologie f macrorheology
makroskopisch macroscopic
Makrostruktur f macrostructure
Makrotomin n macrotomine
Makrotominsäure f macrotomic acid
Makroturbulenz f macroturbulence
Makrozyt m (Biol) macrocyte, gigantocyte, megalocyte
Makulatur f waste paper, scrap, waste sheets
Makulaturbogen m waste sheet
Malabargummi n mochras
Malabartalg m malabar tallow
Malachit m (Min) malachite
Malachitgrün n malachite green, benzal green, powdered malachite, Victoria green

Malachitkiesel m chrysocolla
Malakin n malakin, N-salicylal-p-phenetidine
Malakkanuß f Malacca nut, marking nut
Malakolith m (Min) malacolite, augite
Malakon m (Min) malakon
Malamid n malamide, 2-hydroxybutane diamide, malic amide
Malamidsäure f malamic acid, malamidic acid, malic acid monoamide
Malanilsäure f malanilic acid
Malaria f (Med) malaria
Malariabekämpfung f (Med) fight against malaria, malaria control
Malariaerreger m (Zool) malaria[l] parasite
Malariafieber n (Med) malarial fever
Malarin n malarin, acetophenone phenetidine
Malat n malate, salt or ester of malic acid
Malatdehydrogenase f (Biochem) malate dehydrogenase
Malatsynthetase f (Biochem) malate synthetase
Malcolmiin n malcolmiin
Maldonit n (Min) maldonite
Maleat n maleate, salt or ester of maleic acid
Maleatisomerase f (Biochem) maleate isomerase
Maleinaldehyd m maleic aldehyde, malealdehyde
Maleinanilsäure f maleanilic acid
Maleinatharz n maleic resin
Maleinharz n maleic resin
Maleinimid n maleinimide, maleimid
Maleinsäure f maleic acid, cis-butanedioic acid, ethylenedicarboxylic acid
Maleinsäureanhydrid n maleic anhydride, cis-butanedioic anhydride
Maleinsäurediäthylester m diethyl maleate
Maleinsäureharz n maleic resin
Maleinsäurehydrazid n maleic hydrazide
maleinsauer maleate
Maleinursäure f maleinuric acid
malen to paint
Maler m painter, decorator
Malerarbeit f paint work
Maleremail n painter's enamel
Malerfarbe f painter's color, artist's color
Malerfirnis m painter's varnish
Malerglasur f painter's glazing
Malergold n painter's gold, shell gold
Malergoldfarbe f ormolu, painter's gold
Malergrundierung f painter's priming
Malerlauge f painter's lye
Malersilber n painter's silver
Malerwalze f paint roller
Malettorinde f maletto bark
Maleylsulfathiazol n maleylsulfathiazole
Malignität f (Med) malignancy
Mallein n (Impfstoff) mallein
mallen to mold
Mallkante f molding edge
Malol n malol, prunol, ursolic acid, urson

Malomalsäure – Manganammoniumphosphat

Malomalsäure *f* malomalic acid
Malonaldehyd *m* malonaldehyde, malonic aldehyde
Malonaldehydsäure *f* malonaldehydic acid, formylacetic acid
Malonamid *n* malonamide, malonic amide, propanediamide
Malonamidsäure *f* malonamic acid, malonamidic acid
Malonanilid *n* malonanilide
Malonat *n* malonate, salt or ester of malonic acid
Malonester *m* malonate, malonic ester
Malonsäure *f* malonic acid, methane dicarboxylic acid, propanedioic acid
Malonsäurediäthylester *m* ethyl malonate, malonic acid diethyl ester
malonsauer malonate
Malonursäure *f* malonuric acid
Malonylchlorid *n* malonyl chloride
Malonyl-Coenzym A *n* (Biochem) malonyl-coenzyme A
Malonylharnstoff *m* malonylurea, barbituric acid, malonurea
Malonyltransacetylase *f* (Biochem) malonyl transacetylase
Malschlicker *m* paint slips
Maltase *f* (Biochem) maltase, genease
Malthacit *m* malthacite
Maltin *n* (Biochem) maltin, malt diastase, vegetable amylase
Maltit *m* maltitol
Maltobionsäure *f* maltobionic acid
Maltobiose *f* maltobiose, maltose
Maltodextrin *n* maltodextrin, amyloin
Maltogenamylase *f* (Biochem) maltogenic amylase
Maltol *n* maltol
Maltonsäure *f* maltonic acid, dextronic acid, d-gluconic acid
Maltosazon *n* maltosazone
Maltose *f* maltose, maltobiose, malt sugar
Maltoson *n* maltosone
Maltulose *f* maltulose
Malval[in]säure *f* malvalic acid
Malvenblätter *pl* (Bot) mallow leaves
Malvenfarbe *f* mauve
Malvenkraut *n* (Bot) mallow
Malvidin *n* malvidin
Malvin *n* malvin
Malz *n* malt, ~ **darren** to cure malt
Malzaufguß *m* infusion of malt
Malzauszug *m* malt extract
Malzbereitung *f* malting
Malzbier *n* malt beer, ale
Malzbonbon *m* cough lozenge, malt candy
Malzbottich *m* malt vat
Malzdarre *f* malt kiln
Malzdarrhorde *f* malt drying tray
Malzdiastase *f* (Biochem) malt diastase, maltin, vegetable amylase
Malzeichen *n* (Math) sign of multiplication
malzen (mälzen) to malt
Malzentkeimungsmaschine *f* malting machine
Malzextrakt *m* malt extract
Malzfabrik *f* maltery
Malzgerste *f* malt barley
Malzhaus *n* malt house
Malzkaffee *m* malt coffee
Malzkeime *pl* green malt, malt sprouts
Malzpulveranlage *f* malt-powder plant
Malzstärke *f* malt starch
Malzsurrogat *n* substitute for malt
Malztenne *f* malt floor
Malztennenkühlung *f* malt floor cooling
Malztreber *pl* spent malt, brewer's grains, malt husks, malt residuum
Malzzucker *m* maltobiose, maltose, malt sugar
Mammein *n* mammein
Mammutpumpe *f* air lift pump
Mandarindruck *m* mandarining
Mandarine *f* (Bot) mandarin
Mandel *f* (Bot) almond, (Med) tonsil
Mandelat *n* mandelate (salt or ester of mandelic acid)
Mandelbutter *f* almond butter
Mandelentzündung *f* (Med) tonsilitis
Mandelkern *m* almond [kernel]
Mandelkleie *f* almond meal, bran of almonds
Mandelmilch *f* almond milk, emulsion of almonds
Mandelnitril *n* mandelonitrile
Mandelnitrilase *f* (Biochem) mandelonitrilase
Mandelöl *n* almond oil
Mandelpaste *f* almond paste
Mandelsäure *f* mandelic acid, amygdalic acid, hydroxy-phenylacetic acid, phenylglycollic acid
Mandelsäureamid *n* mandelic amide, mandelamide
Mandelsäurenitril *n* mandelonitrile, benzalcyanohydrin, benzaldehyde cyanohydrin, mandelic nitrile
Mandelsalbe *f* almond ointment
Mandelseife *f* almond [oil] soap
Mandelsirup *m* syrup of almonds
Mandelstein *m* amygdaloid
Mandelstoff *m* amygdalin
Mandelstorax *m* amygdaloid storax
Mandioka *f* cassava
Mandragora *f* (Bot) mandragora
Mangabeiragummi *n* mangabeira rubber
Mangan *n* manganese (Symb. Mn)
Manganacetat *n* manganese acetate
Manganalaun *m* manganese alum, potassium manganic sulfate
Manganammoniumphosphat *n* manganese ammonium phosphate

Manganat *n* manganate, salt or ester of manganic acid
Manganbadmethode *f* manganese bath method
Manganblau *n* (Zementblau) manganese blue
Manganblende *f* (Min) native manganous sulfide, alabandite
Manganbraun *n* manganese bister, manganese brown
Manganbronze *f* manganese bronze
Mangancarbid *n* manganese carbide
Manganchlorür *n* (Mangan(II)-chlorid) manganese dichloride, manganese(II) chloride, manganous chloride
Mangandioxid *n* black manganese oxide, manganese binoxide, manganese(IV) oxide, (Min) pyrolusite
Manganeisen *n* (Min) ferromanganese
Manganeisenstein *m* (Min) triplite
Manganepidot *m* (Min) piedmontite
Manganerz *n* (Min) manganese ore
Mangangehalt *m* manganese content
Manganglanz *m* (Min) alabandite
Manganglycerophosphat *n* manganous glycerophosphate, manganous glycerinophosphate
Mangangranat *m* (Min) manganese garnet, spessartite
Mangangrün *n* manganese green, Cassel's green
manganhaltig containing manganese, manganiferous
Manganhartstahl *m* austenitic manganese steel
Manganheptoxid *n* manganese heptoxide
Manganhydroxid *n* manganese hydroxide
Manganhydroxydul *n* manganese(II) hydroxide, manganous hydroxide
Manganhyperoxid *n* manganese dioxide, manganese binoxide, manganese(IV) oxide, manganese peroxide
Mangani- manganese(III), manganic
Manganicyanwasserstoffsäure *f* cyanomanganic(III) acid, manganicyanic acid
Manganige Säure *f* manganous acid
Manganihydroxid *n* manganese(III) hydroxide, manganic hydroxide
Mangan(III)-oxid *n* manganese sesquioxide, manganic oxide
Mangan(III)-Salz *n* manganese(III) salt, manganic salt
Mangan(III)-Verbindung *f* manganese(III) compound, manganic compound
Mangan(II)-oxid *n* manganese monoxide, manganese protoxide, manganous oxide, (Min) manganosite
Mangan(II)-Salz *n* manganese(II) salt, manganous salt
Mangan(II)-Verbindung *f* manganese(II) compound, manganous compound
Manganin *n* manganin

Manganiphosphat *n* manganese(III) phosphate, manganic phosphate
Manganisalz *n* manganese(III) salt, manganic salt
Manganisulfat *n* manganese(III) sulfate, manganic sulfate
Manganit *m* (Min) grey manganese ore, manganite, native manganic oxide hydrate
Manganiverbindung *f* manganese(III) compound, manganic compound
Mangan(IV)-oxid *n* manganese dioxide, manganese binoxide
Manganjodür *n* (Mangan(II)-jodid) manganese(II) iodide, manganous iodide
Mangankies *m* (Min) hauerite, native manganic sulfide
Mangankiesel *m* (Min) native manganous metasilicate, rhodonite
Manganknolle *f* (Min) manganese nodule
Mangankupfer *n* cupromanganese
Mangankupfererz *n* (Min) crednerite
Manganlactat *n* manganese lactate, manganous lactate
Manganlegierung *f* manganese alloy
Mangannitrat *n* manganese nitrate
Mangano- manganese(II), manganous
Manganoacetat *n* manganese(II) acetate, manganous acetate
Manganoaxinit *m* (Min) manganoaxinite
Manganoborat *n* manganese borate, manganese(II) tetraborate, manganese siccative, manganous borate
Manganocalcit *m* (Min) manganocalcite
Manganocarbonat *n* manganese(II) carbonate, manganous carbonate
Manganocolumbit *m* (Min) manganocolumbite
Manganocyanwasserstoffsäure *f* cyanomanganic(II) acid, mangancyanic acid
Manganoferrum *n* ferromanganese
Manganohydroxid *n* manganese(II) hydroxide, manganous hydroxide
Manganoion *n* manganese(II) ion, manganous ion
Manganomanganit *n* manganomanganite
Manganometrie *f* manganometry
manganometrisch manganometric
Manganooleat *n* manganese(II) oleate, manganous oleate
Manganooxalat *n* manganese(II) oxalate, manganous oxalate
Manganophosphat *n* manganese(II) phosphate, manganous phosphate
Manganosalz *n* manganese(II) salt, manganous salt
Manganosit *m* (Min) manganosite, native manganous oxide
Manganosphärit *m* (Min) manganospherite
Manganostibiit *m* (Min) manganostibiite

Manganosulfat *n* manganese(II) sulfate, manganous sulfate
Manganotantalit *m* (Min) manganotantalite
Manganoverbindung *f* manganese(II) compound, manganous compound
Manganoxalat *n* manganese oxalate, manganese(II) oxalate, manganous oxalate
Manganoxidhydrat *n* (obs) manganese(III) hydroxide, manganic hydroxide
Manganoxidoxydul *n* (obs) manganese(II)(III) oxide, manganic manganous oxide, mangano-manganic oxide
Manganoxydul *n* (obs) manganese(II) oxide, manganese monoxide, manganese protoxide, manganous oxide
Manganoxydulhydrat *n* (obs) manganese(II) hydroxide, manganous hydroxide
Manganoxyduloxid *n* (obs) manganese(II)(III) oxide, manganic manganous oxide, mangano-manganic oxide
Manganoxydulsalz *n* (obs) manganese(II) salt, manganous salt
Manganoxydulverbindung *f* (obs) manganese(II) compound, manganous compound
Manganpecherz *n* (Min) triplite
Manganpektolith *m* (Min) manganpectolite
Manganperoxid *n* manganese dioxide, manganese binoxide, manganese(IV) oxide
Manganphosphat *n* manganese phosphate, (Mangan(II)-phosphat) manganese(II) phosphate, manganous phosphate
Manganpräparat *n* manganese preparation
Manganpyrophosphat *n* manganese pyrophosphate
Mangansäure *f* manganic acid, manganic(VI) acid
Mangansäureanhydrid *n* manganese trioxide, manganese(VI) oxide, manganic anhydride
mangansauer manganate(VI)
Manganschaum *m* bog manganese, [manganese] wad
Manganschwarz *n* manganese black
Manganselenid *n* manganese selenide, manganese(II) selenide, manganous selenide
Mangansesquioxid *n* manganese sesquioxide, manganese(III) oxide, manganic oxide
Mangansilikofluorid *n* manganese fluosilicate, manganese(II) silicofluoride
Mangansiliziumstahl *m* silico-manganese steel
Manganspat *m* (Min) manganese spar, native manganous carbonate, rhodochrosite
Manganspektrum *n* manganese spectrum
Manganstahl *m* manganese steel
Mangansulfid *n* manganese sulfide
Mangansulfür *n* (Mangan(II)-sulfid) manganese(II) sulfide, manganous sulfide
Mangansuperoxid *n* manganese dioxide, manganese(IV) oxide, manganese peroxide

Mangantetrachlorid *n* manganese(IV) chloride, manganese tetrachloride
Mangantongranat *m* (Min) spessartite
Mangantrifluorid *n* manganese(III) fluoride, manganese trifluoride, manganic fluoride
Mangantrioxid *n* manganese trioxide, manganese(VI) oxide, manganic anhydride
Manganverbindung *f* manganese compound
Mangan(VII)-oxid *n* manganese heptoxide
Manganviolett *n* manganese violet, Nuremberg violet
Mangan(VI)-oxid *n* manganese trioxide, manganic anhydride
Manganvitriol *n* manganese vitriol
Manganweiß *n* manganese white, manganous carbonate
Manganzinkspat *m* (Min) manganese zinc spar
Mangel *m* deficiency, lack, shortage, (Fehler) defect
mangelhaft deficient, poor, (fehlerhaft) defective, faulty, imperfect
Mangelkrankheit *f* (Med) deficiency disease
Mangelleiter *m* defect conductor
Mangelleitung *f* (Elektron) electron-defect conductivity
Mangeln *n* (Text) calendering
Mangiferin *n* mangiferin
Mangin *n* mangin
Mangoldsche Säure *f* Mangold's acid
Mangostin *n* mangostin
Mangrovenrinde *f* mangrove bark
Manihotöl *n* manihot oil
Manilahanf *m* (Text) Manila hemp, abaca
Manilapapier *n* Manila paper
Maniok *m* (Bot) manioc, cassava
Maniokstärke *f* cassava starch
Manna *n f* (Bot, Med) manna
Mannan *n* mannan
Mannasirup *m* syrup of manna
Mannazucker *m* manna sugar, mannite, mannitol
Manneotetrose *f* manneotetrose
Mannich-Reaktion *f* Mannich reaction
Mannid *n* mannide
mannigfach manifold
Mannigfaltigkeit diversity, variety
Manninotriose *f* manninotriose
Mannit *m* mannitol, mannite
Mannitan *n* mannitan
Mannitborsäure *f* mannitol boric acid
Mannitester *m* mannite ester, mannitol ester
Mannitol *n* mannitol, mannite
Mannloch *n* (Techn) manhole
Mannochloralose *f* mannochloralose
Mannoheptonsäure *f* mannoheptonic acid
Mannoheptose *f* mannoheptose
Mannomustin *n* mannomustine
Mannonsäure *f* mannonic acid
Mannosamin *n* mannosamine

Mannose *f* mannose
Mannosidase *f* (Biochem) mannosidase
Mannozuckersäure *f* mannosaccharic acid
Mannuron *n* mannurone
Mannuronsäure *f* mannuronic acid
Manometer *n* manometer, pressure gauge
Manometerbeobachtung *f* manometric observation
Manometerdruck *m* gauge pressure
Manometerfeder *f* pressure gauge spring
Manometergehäuse *n* pressure gauge case
Manometerrohr *n* pressure gauge pipe
Manometerskala *f* manometer scale
Manometerstand *m* pressure gauge position
manometrisch manometric
Manondonit *m* manondonite
Mansarde *f* (Techn) drying chamber, hotflue
Manschette *f* sleeve, (Bund) collar, (Dichtung) packing, (Kolben) cup
Manschettendichtung *f* concentric gasket, cup leather packing
Manschettendipol *m* sleeve dipole
Mansfelder Kupferschiefer *m* (Min) Mansfeld copper ore
Mantel *m* (Atom) blanket, (Gehäuse) case, casing, (Schutzhülle) sheath, shielding, (Umhüllung) covering, envelope, jacket
Mantelblech *n* jacket sheet iron
Mantelbrecher *m* rolling ring crusher
Manteleisen *n* mantle sheet steel
Mantelelektrode *f* covered electrode, sheathed electrode, **kegelige** ~ hollow-cone electrode
Mantelkühlung *f* jacket cooling
Mantelkupfer *n* jacket copper
Mantelmagnet *m* iron-clad magnet, shell-type magnet
Manteltransformator *m* shell transformer
Manthin *n* manthine
Mantisse *f* (Math) mantissa
Manuken *n* manukene
MAO Abk. für „Monoaminoxidase"
Mappin *n* mappine
Marantastärke *f* maranta
Mararaharz *n* marara, caranna, gum carane
Marcasit *m* (Min) marcasite, binarite, coal brass, coxcomb, native iron disulfide, radiated pyrite, white iron
Maretin *n* maretin, m-tolylsemicarbazide
Marfanil *n* marfanil
Margarine *f* margarine
Margarinesäure *f* margaric acid, daturic acid, heptadecanoic acid
Margarinfett *n* oleomargarine
Margarit *m* (Min) margarite, native calcium aluminum silicate
Margarodit *m* margarodite
Margaron *n* margaron, dihexadecyl ether
Margarosanit *m* (Min) margarosanite

Margosöl *n* azedarach oil, margosa oil, neem oil, veepa oil, veppam fat
Marienglas *n* (Min) mica
Marignacit *m* (Min) marignacite
Marihuana *n* (Pharm) marijuana, marihuana
Marineblau *n* marine blue, navy blue
Marineleim *m* marine glue
Marineöl *n* marine oil
marinieren to marinate, to pickle
Maripafett *n* maripa fat
Mark *n* marrow, core, medulla, pulp, (Knochen-Mark) bone marrow, (z. B. eines Haares) pith
markartig medullary, myeloid, pith-like
Markasit *m* (Min) marcasite, binarite, coal brass, coxcomb, native iron disulfide, radiated pyrite, white iron
Markasitglanz *m* (Min) native bismuth telluride, tellur bismuth, tetradymite
Marke *f* brand, mark, stamp, (Handelsmarke) trade mark
Markenname *m* brand
Markenschutz *m* protection of registered trade marks
Markfett *n* marrow fat
Markflüssigkeit *f* (Med) cerebrospinal fluid
Markgewebe *n* (Med) myeloid tissue
markieren to mark, to label; to stamp
Markierung *f* label, mark, ~ **der Preßplatte** platen mark, ~ **des Ausdrückstifts** ejector pin mark, ~ **des Werkzeuges** mold mark
Markierungsexperiment *n* tagging experiment
Markierungsfarbe *f* signal paint
Markierungsstreifen *m* memory or spider line
Markierungsstrich *m* graduation mark
Markierungstechnik *f* (Isotopen) labelling technique
Markierungsverbindung *f* tagging compound
Markinsel *f* group of pith cells
Markisenstoff *m* (Text) awning, canvas
Markkrebs *m* (Med) medullary cancer
marklos marrowless
Markogensäure *f* markogenic acid
Markownikoffsche Regel *f* Markovnikov rule
Markröhrchen *n* medullary tube
Markstoff *m* medullary substance, medullin
Marksubstanz *f* medullary substance, medullin
Markt *m* market
Marktforschung *f* market research
Marktwirtschaft *f*, **soziale** ~ free-enterprise system
Markush-Formel *f* Markush structure
Marlit *m* (Min) marlite
marlitartig marlitic
Marmatit *m* (Min) marmatite, native zinc iron manganese sulfide
Marmelade *f* jam
Marmelosin *n* marmelosin
Marmesilsäure *f* marmesilic acid

Marmin *n* marmin
Marmolith *m* (Min) marmolite
Marmor *m* marble, **Carrara** ~ Carrara marble
marmorähnlich marble-like, marmoraceous, marmoreal
marmorartig marble-like, marmoraceous, marmoreal
Marmorbruch *m* marble quarry
Marmoreffekt *m* marbling effect
marmorfarbig marble-colored
Marmorgips *m* (Marmorzement) marble gypsum
Marmorglas *n* marble glass
marmoriert marbled, marbleized, mottled, veined
Marmorierung *f* marbelizing, marbleizing, marmoration, mottles, mottling, (Dosen) sulfur staining
Marmorierwasser *n* (Pap) gum water for marbling paper
Marmorkalk *m* marble lime
Marmorplatte *f* marble plate
Marmorweiß *n* chalk whiting, marble white
Marmorzement *m* artificial marble; marble cement
Marokkoleder *n* Morocco leather
Marrianolsäure *f* marrianolic acid
Marrubanol *n* marrubanol
Marrubenol *n* marrubenol
Marrubinsäure *f* marrubic acid
Mars *m* (Astr) Mars
Marsbewohner *m* inhabitant of Mars
Marseiller Seife *f* castile soap, Marseilles soap
Marshit *m* (Min) marshite
Marshsche Arsenprobe *f* Marsh['s] test
Marsjahr *n* (Astr) Martian year
Martensit *m* martensite
Martensitbildung *f* martensite formation
Martensitpunkt *m* martensite point
Martineisen *n* open-hearth iron
Martinflußeisen *n* open-hearth iron
Martinflußstahl *m* open-hearth steel
Martingebläse *n* Martin blower
Martinit *m* (Min) martinite
Martinofen *m* Siemens-Martin furnace
Martinroheisen *n* open-hearth pig iron
Martinstahl *m* Martin steel, open-hearth steel
Martinstahlhütte *f* open-hearth steel works, Siemens-Martin steel works
Martit *m* (Min) martite, native ferric oxide
Martiusgelb *n* Martius yellow
Marzipan *n* marzipan
Mascagnin *m* (Min) mascagnine, mascagnite, mineral ammonium sulfate
Masche *f* (z. B. eines Netzes) mesh, (z. B. Strickmasche) stitch
Maschendraht *m* wire mesh, wire netting
maschenfest (Text) ladderproof, locknit, runproof
Maschenfestigkeit *f* loop tenacity

Maschenfestmittel *n* ladder-proofing agent
Maschensieb *n* mesh screen, mesh sieve
Maschenweite *f* mesh size
Maschenzahl *f* number of meshes
Maschine *f* engine, machine, (Motor) motor, ~ **für Riemenbetrieb** belt-driven machine, ~ **mit Handbetrieb** hand-power machine, ~ **zum Umlegen der Ummantelung** flipping machine, **die ~ abstellen** to stop the engine, **die ~ anlassen** to start the engine, **hydraulische** ~ hydraulic machine or engine, **landwirtschaftliche** ~ farming or agricultural machine
maschinell by machine, mechanical, ~ **angetrieben** driven by machine, power-driven, ~ **hergestellt** machine-made
Maschinenanlage *f* machine unit
Maschinenantrieb *m* machine drive
Maschinenarbeit *f* machine work
Maschinenbau *m* machine construction, mechanical engineering
Maschinenbauer *m* mechanical engineer
Maschinendefekt *m* failure of machinery
Maschinendruck *m* roller printing
Maschinenfabrik *f* machine factory
maschinenfertig completely machined, machine-finished
Maschinenfett *n* lubricating grease
Maschinengas *n* engine gas, power gas
maschinengezogen machine-drawn
Maschinenguß *m* machine casting
Maschinenhaus *n* turbine room
Maschineningenieur *m* mechanical engineer
Maschinenkurbel *f* crank
maschinenmäßig mechanical
Maschinenöl *n* lubricating oil, machine oil
Maschinenpapier *n* machine paper
Maschinenpark *m* machinery, train of machines
Maschinenpuddeln *n* machine puddling, mechanical puddling
Maschinensatz *m* group of machines
Maschinenstrom *m* generator current
Maschinentorf *m* machine-cut peat
Maschinerie *f* machine, mechanism
Maschinist *m* engine operator, mechanic engineer
Maser (Abk. für „Microwave Amplification by Stimulated Emission of Radiation) maser
Maser *f* (Holz) streak, vein
Maserdruckverfahren *n* printing grain effects on wood
Maserholz *n* speckled wood, streaky or curled wood, veined wood
maseriert mottled
masern to grain, to vein
Masern *pl* measles
Masernimpfstoff *m* (Med) measles vaccine
Masernserum *n* measles vaccine

Maserung *f* graining, mottling, (Holz) [wood] grain
Maske *f* mask, **Gesichts-** ~ face guard
Maskelynit *m* (Min) maskelynite
maskieren to mask
Maskierung *f* masking, sequestering
Maskierungsmittel *n* complexing agent
Masonin *n* masonine
Masonite-Verfahren *n* Masonite process
Masopin *n* masopin
Maß *n* dimension, measure, size, (Einheit) unit, (Meßgerät) gauge, (Verhältnis) ratio, (Volumen) volume, **genaues** ~ exact measurement
Maßabweichung *f* dimensional deviation
Maßanalyse *f* analysis by titration, titrating analysis, titrimetry, volumetric analysis
maßanalytisch titrimetric, volumetric
Maßbeständigkeit *f* dimensional stability
Masse *f* mass, (Keram) paste, (Menge) bulk, quantity, (Metall) dry sand, (Pap) pulp, (Stoff) material, substance, **breiige** ~ slurry substance, **kritische** ~ critical mass, **reduzierte** ~ (Mech) reduced mass, **schwammige** ~ sponge-like mass, **träge** ~ inert mass
Masseakkumulator *m* accumulator with pasted plates
Massebestimmung *f* determination of mass
Masse-Energie-Äquivalentgleichung *f* equation of mass-energy equivalence
massefrei massless
Masseguß *m* dry sand casting
Massekabel *n* power cable
Massekammer *f* plenum chamber
Massel *f* (Metall) pig [of iron], bloom
Masselbett *n* (Metall) pig bed, casting bed
Masselbrecher *m* pig breaker
Masseleisen *n* pig iron
Massenabsorption *f* mass absorption
Massenabsorptionskoeffizient *m* mass coefficient of absorption
Massenäquivalent *n* mass equivalent
Massenanwendung *f* mass application
Massenanziehung *f* gravitational attraction, mass attraction
Massenanziehungsgesetz *n* law of mass attraction
Massenartikel *pl* mass-produced goods
Massenbeförderung *f* transport of masses
Massenbeschleunigung *f* mass acceleration
Massenbremsvermögen *n* mass stopping power
Massendefekt *m* mass defect
Massendichte *f* mass density
Masseneinheit *f* mass unit, unit mass
Massenerhaltung *f* conservation of mass
Massenerkrankung *f* (Med) epidemic
Massenerzeugung *f* mass production, large output

Massenfabrikation *f* mass production
Massenfertigung *f* mass production, quantity production
Massenflußdichte *f* mass flow density
Massengleichung *f* mass equation
Massengüter *pl* bulk articles
Massenherstellung *f* mass production, wholesale manufacturing
Massenimpfung *f* mass vaccination
Massenkern *m* mass nucleus
Massenkonzentration *f* concentration of mass
Massenkorrektur *f* mass correction, ~ **für Austausch** exchange mass correction
Massenkraft *f* force due to gravity, body force
Massenmittelpunkt *m* mass center
Massenproduktion *f* mass production, quantity production
Massenschätzung *f* mass estimation
Massenschwächungskoeffizient *m* mass attenuation coefficient, mass absorption coefficient, mass extinction coefficient
Massenschwund *m* mass disappearance, mass loss
Massenspektrogramm *n* mass spectrogramm
Massenspektrograph *m* mass spectrograph, ~ **mit Geschwindigkeitsfokussierung** (Spektr) velocity focusing mass spectrograph
Massenspektrographie *f* mass spectrography
Massenspektrometer *n* mass spectrometer, **trochoidales** ~ (Spektr) trochoidal mass analyzer
massenspektrometrisch mass-spectrometric
Massenspektroskopie *f* mass spectroscopy
Massenspektrum *n* (Spektr) mass spectrum
Massenstrahler *m* mass radiator
Massenstreukoeffizient *m* mass scattering coefficient
Massenstromdichte *f* mass flow density
Massensynchrometer *n* mass synchrometer
Massenträgheit *f* inertia
Massentransport *m* mass transportation
Massenüberführungsfaktor *m* mass transfer coefficient
Massenübergang *m* mass transfer
Massenunterschied *m* mass difference
Massenvergiftung *f* (Med) mass poisoning
Massenverhältnis *n* mass ratio
Massenverlust *m* mass loss
Massenverteilung *f* distribution of mass
Massenwirkung *f* mass action
Massenwirkungsgesetz *n* law of mass action
Massenzahl *f* mass number, isotopic number
Masseschluß *m* [accidental] ground
Masseteilchen *n* mass particle
Masseteilchengeschwindigkeit *f* particle velocity
Massetransport *m* mass transfer
Massicot *m* (Min) massicot, native lead monoxide
massiv solid, stout, ~ **gießen** to cast solid

Massivholz *n* solid timber
maßeingefärbt self-colored
Maßeinheit *f* unit of measurement
Maßeinteilung *f* graduation
Maßflasche *f* measuring flask
Maßflüssigkeit *f* standard solution
Maßformel *f* standard formula
Maßgenauigkeit *f* accuracy requirements, conformity, gauge uniformity
maßgleich isometric
Maßgröße *f* dimensional factor
Maßkolben *m* volumetric flask
Massoi-Lacton *n* Massoia lactone
Maßpfropfen *m* fire clay plug
Maßregel *f* measure, step
Maßröhre *f* buret[te], graduated tube, measuring tube
Maßstab *m* scale, (Lineal) measuring rule, measuring stick, yardstick, **großtechnischer** ~ industrial scale, **in vergrößertem** ~ at an enlarged scale, **in verkleinertem** ~ at a reduced scale
maßstabgerecht in scale, true to scale
Maßstabsvergrößerung *f* (chem. Anlagenbau) scale-up
Maßstock *m* measuring stick
Maßsystem *n* measuring system, system of measurement, **absolutes** ~ absolute system of units, **metrisches** ~ international metric measures
Maßtoleranz *f* dimensional tolerance
Maßzeichnung *f* dimensioned drawing
Masticadienonsäure *f* masticadienonic acid
Masticin *n* masticin
Masticinsäure *f* masticinic acid, masticolic acid, masticonic acid
Mastikation *f* (Gummi) mastication
Mastix *m* [gum] mastic, mastiche
Mastixbranntwein *m* mastic liquor
Mastixfirnis *m* mastic varnish
Mastixharz *n* mastic resin
Mastixkitt *m* mastic cement
Mastizieren *n* (Gummi) masticating, mastication
Mastizierwalzwerk *n* breakdown mill
Mastleuchte *f* column lamp
Mastzelle *f* mast cell, labrocyte, mastocyte
Masurium *n* (obs) masurium, technetium
Masut *n* masut oil, mazut
Matairesinol *n* matairesinol
Matairesinolsäure *f* matairesinolic acid
Matatabilacton *n* matatabilactone
Matéblätter *n pl* Brazil tea, Jesuit tea, Paraguay tea, maté leaves
Mater *f* flong, matrix, mold
Material *n* **kurzer Halbwertzeit** short halflife material, **erschöpftes** ~ (Atom) depleted material, **nichtspaltbares** ~ (Atom) non-fissionable material
Materialausnutzung *f* material economy
Materialauswahlproblem *n* (Atom) compatibility problem
Materialbeanspruchung *f* material stress
Materialbedarf *m* material requirements
Materialbewegung *f* (Färb) agitation of the material
Materialbilanz *f* material balance
Materialeigenschaft *f* property of material
Materialeinsatz *m* material inventory
Materialermüdung *f* fatigue of material
Materialersparnis *f* saving in material
Materialfehler *m* fault in material, flaw in material
Materialfestigkeit *f* strength of material
Materialprüfgerät *n* material testing apparatus
Materialprüfung *f* testing of material
Materialprüfungsreaktor *m* materials testing reactor
Materialrolle *f* roll of fabric
Materialspannung *f* strain or tension of material
Materialüberhitzung *f* overheating of material
Materialverbrauch *m* material consumption
Materialvorrat *m* stock
Materialzuführung *f* feed inlet, ~ **durch Gefälle** gravity feed
Materie *f* matter, substance, **Aufbau der** ~ structure of matter
Materienwelle *f* de Broglie wave, matter wave
Materiestrahlen *m pl* corpuscular rays
Matezit *m* matezitol
Mathematik *f* mathematics, **angewandte** ~ applied mathematics, **reine** ~ pure mathematics
Mathematiker *m* mathematician
mathematisch mathematical
Maticin *n* maticin
Matikoblätter *n pl* matico leaves
Matikoöl *n* matico oil
Matildit *m* (Min) matildite, native silver bismuth sulfide
Matlockit *m* (Min) matlockite, native lead chloride and oxide
Matricariacampher *m* matricaria camphor, l-camphor
Matricariaester *m* matricaria ester
Matricarianol *n* matricarianol
Matricariasäure *f* matricaric acid
Matricarin *n* matricarin
Matridin *n* matridine
Matrin *n* matrine
Matrinidin *n* matrinidine
Matrinsäure *f* matrinic acid
Matrix *f* matrix
Matrize *f* matrix, chase, [female] die, female mold, master, mold [cavity], (aus einem Stück) cavity block, (Biochem) template, **Null-** ~

null or zero matrix, **orthogonale** ~ orthogonal matrix, **quadratische** ~ square matrix, **Rang einer** ~ order of a matrix, **reguläre** ~ regular matrix, **schiefsymmetrische** ~ skew symmetrical matrix, **singuläre** ~ singular matrix, **symmetrische** ~ symmetrical matrix, **vertauschbare** ~ commutable matrix
Matrizeneinsatz *m* mold insert, [bottom] plug
Matrizenhalter *m* matrix holder
Matrizenplatte *f* cavity plate, retainer plate
Matrizenrahmen *m* frame bolster
Matrizenreaktion *f* template reaction
Matrizenteil *n* split cavity block, plug
Matrizenwalze *f* bottom roll
Matsch- und Schneereifen *m pl* mud and snow tire
matt dull, dead, delustered, mat, (Glas) frosted, ground, (Licht) dim, faint, (Met) tarnished, (Pap) non-calendered, ~ **brennen** to dip dead, to dull, ~ **machen** to dull, to tarnish
Mattätzmittel *n* mat etching agent
Mattanstrich *m* dull finish
Mattanstrichfarbe *f* flat paint
Mattbeize *f* pickle for rendering a dull surface
mattblau dull blue, pale blue
Mattblech *n* pickled sheet
Mattbrenne *f* dead dip, dull pickle, dull pickling
Mattbrennen *n* tarnishing
Matte *f* mat
Matteisen *n* white pig iron
Mattenpreßverfahren *n* mat molding
Matteucinol *n* matteucinol
Mattfarbe *f* mat[t] color
mattgelb cream-colored, pale yellow
mattgeschliffen frosted, ground
Mattgewebe *n* dull fabric, mat fabric
Mattglanz *m* dull finish, dull luster
Mattglas *n* frosted glass, ground glass
Mattglasur *f* mat glaze
Mattgold *n* dead gold
mattgrau neutral gray
mattgrün dead green
Mattheit *f* dullness, dimness, faintness
mattierbar tarnishable
Mattierbürste *f* brush for producing a mat[te] surface
mattieren to deaden, to deluster, to dull, to tarnish, (Glas) to frost
Mattieren *n* delustering, ~ **des Messings** dead-dipping of brass
Mattiersand *m* matting sand
mattiert delustered, matted, (Glas) frosted
Mattierung *f* delustering, dulling, dull polish, matting, tarnishing
Mattierungsmittel *n* delusterant, delustering agent, dulling agent, flatting agent, gloss-reducing agent
Mattklarlack *m* dull clear varnish
Mattkohle *f* dull coal, durain
Mattlack *m* dull clear varnish, flat or dull lacquer, flat varnish or finish
Mattpolitur *f* flat polish
mattrot dim red
Mattscheibe *f* (Opt) ground glass disk
Mattschlagen *n* production of a mat[te] surface
mattschleifen to dull-grind
Mattstreifen *m* sandblasted rim
mattvergoldet dead-gilt
Mattvergoldung *f* dead gilding, mat[te] gilding
Mattversilberung *f* dull silvering
mattweiß dead white, dull white
matt werden to become tarnished
Maucherit *m* (Min) maucherite
Mauer *f* wall
Mauereckleiste *f* wall corner fillet
Mauerfraß *m* rot of walls
Mauerkranz *m* brick rim, wall rim
Mauerkraut *n* (Bot) wall pellitory
mauern to build, to mason
Mauersalpeter *m* wall saltpeter, calcareous niter
Mauersattel *m* wall saddle
Mauersockel *m* brick wall, stone wall
Mauerstein *m* [wall] brick
Mauerziegel *m* building brick, wall brick
Mauguinit *n* mauguinite
Maulbeerbaum *m* (Bot) mulberry tree
Mauleonit *m* mauleonite
Maulwurfpumpe *f* canned pump, multiplestage pump
Maurer *m* bricklayer, mason
Maurerschutzlack *m* plaster coat
Mauvanilin *n* mauvaniline
Mauve *n* aniline purple, mauvein, [Perkin's] mauve
Mauvein *n* aniline purple, mauvein, Perkin's mauve
maximal maximal, maximum
Maximalabweichung *f* maximum deviation
Maximalamplitude *f* maximum amplitude
Maximalausbeute *f* maximum output
Maximaldosis *f* (Med) maximum dose
Maximaldruck *m* maximal pressure
Maximalgeschwindigkeit *f* maximum speed
Maximalgewicht *n* maximum weight
Maximalleistung *f* maximum capacity, maximum output
Maximalspannung *f* (Elektr) maximum voltage
Maximalstrom *m* (Elektr) maximum current
Maximalthermometer *n* maximum thermometer
Maximalvalenz *f* maximum valence, maxivalence
Maximalwert *m* maximum value
Maximalwertigkeit *f* maximum valence, maxivalence
Maximum *n* maximum, peak
Maxit *m* (Min) maxite, leadhillite
Maxivalenz *f* maximum valence, maxivalence
Maxwellbrücke *f* Maxwell bridge

Maxwellsche Beziehung f (Elektr, Magn) Maxwell relation
Maxwellsche Korkzieherregel f Maxwell's corkscrew rule
Maxwellsche Verteilung f Maxwellian distribution
Mayonnaise f mayonnaise
Mazapilit m mazapilite
Mazarinblau n mazarine blue
Mazeration f maceration
mazerieren to macerate
Mazerieren n maceration, retting
Mazeriergefäß n macerator
MBS Abk. für „Methylmethacrylat-Butadien-Styrol"
MBT Abk. für „Mercaptobenzothiazol"
MCD Abk. für „Magnetischer Circulardichroismus"
mCi Abk. für Millicurie
Mecamylamin n mecamylamine
Mechanik f mechanics, (Mechanismus) mechanism, ~ **fester Körper** mechanics of rigid bodies, ~ **flüssiger Körper** hydraulics, ~ **gasförmiger Körper** aerodynamics
Mechaniker m mechanic, mechanician, (Maschinenschlosser) mechanist
mechanisch mechanical, ~ **betätigt** mechanically controlled, power-driven
mechanisieren to mechanize
Mechanisierung f mechanization
Mechanismus m mechanism
Mechanochemie f mechanochemistry
Meclozin n meclozine
Meconsäure f meconic acid
Medikament n (Pharm) drug, medicament, medicine, pharmaceutical product, remedy
Medinal n (HN) medinal, sodium 5,5-diethylbarbiturate
Medium n medium, **brechendes** ~ refractive medium
Medizin f medicine, (Mittel) drug, medicament, remedy, (Wissenschaft) medical science, **forensische** ~ forensic medicine, **gerichtliche** ~ forensic medicine, **physikalische** ~ physical medicine
Medizinbernstein m mineral amber
Medizinglas n medicine glass, phial
medizinisch medical, medicinal
medizinisch-botanisch medicobotanical
medizinisch-chemisch medicochemical
Medrylamin n medrylamine
Meereskunde f oceanography
Meer[es]leuchten n marine phosphorescence
Meeresoptik f marine optics
Meeresspiegel m sea level
Meerkies m shingle
Meerkrapp m (Bot) madder
Meerrettich m (Bot) horseradish
Meerrettichöl n horseradish oil
Meersand m seasand
Meerschaum m meerschaum, sea foam
Meerschlamm m sea ooze
Meerschweinchen n (Zool) guinea-pig
Meertorf m sea peat
Meerwasser n sea water, ocean water
Meerwasserbeständigkeit f seawater resistance
Meerwasserentsalzung f desalination of seawater
Meerzwiebel f (Bot) squill
Megabar n megabar
Megabasit m megabasite
Megabromit m megabromite
Megacarpin n megacarpine
Megaerg n 1.000.000 ergs, megaerg
Megafarad n (Elektr) 1.000.000 farads, megafarad, macrofarad
Megahertz n (Elektr) megacycle
Megaphen n megaphen
Megavolt n (Elektr) 1.000.000 volts, megavolt
Megohm n (Elektr) 1.000.000 ohms, megohm
Megohmit n megohmite
Megohmmeter n megohmmeter
Mehl n flour, meal, (pulverförmiger Stoff) dust, powder
mehlartig farinaceous, mealy
Mehlartigkeit f mealiness
Mehlgips m earthy gypsum
mehlhaltig farinaceous, farinose
mehlig farinaceous; floury, mealy
Mehlkreide f earthy calcite, white stone marl
Mehlprüfer m aleurometer, flour tester
Mehlstaub m flour dust, mill dust, stive
Mehltau m mildew
Mehrarbeit f overtime work
mehratomig polyatomic
Mehraufwand m additional costs or effort, extra expenses
Mehrbasigkeit f polybasicity
mehrbasisch polybasic
mehrdimensional multidimensional
Mehrelektrodenröhre f multi-electrode tube
Mehrelektrodensystem n multi-electrode system
Mehrelektronenspektrum n many-electron spectrum
Mehretagenofen m multiple heater
mehrfach manifold, multiple, multiplex, ~ **geladen** multiple charged
Mehrfachbalkenrührer m multiple paddle agitator
Mehrfachbindung f multiple bond
Mehrfachbrenner m multi-flame burner
Mehrfache[s] n (Math) multiple
Mehrfachfasermaterial n multiple fiber material
Mehrfachform f multiple cavity mold, composite mold, multi-impression mold
Mehrfachgleitung f multiple slip
Mehrfachkapillare f capillary tubes with several bores
Mehrfachkreissäge f multiple circular saw

Mehrfachpendel *n* multiple pendulum
Mehrfachschaltung *f* multi-coupling
Mehrfachschicht *f* multilayer
Mehrfachstreuung *f* plural scattering
Mehrfachstromerzeuger *m* multi-current dynamo
Mehrfachwerkzeug *n* **mit getrennten Füllräumen** separate pot mold
Mehrfachzertrümmerung *f* (Phys) spallation
Mehrfarbendruck *m* multi-color print, process printing, (Vorgang) multi-color printing
Mehrfarbenlack *m* multi-color finish
mehrfarbig multi-colored, polychromatic, (Krist) pleochroic
Mehrfarbigkeit *f* polychromatism, polychromy, (Krist) pleochroism
Mehrgehalt *m* excess content
Mehrgewicht *n* excess weight
Mehrgitterröhre *f* multiple grid tube
Mehrheit *f* majority
Mehrkammermühle *f* compartment mill
mehrkernig polynuclear
Mehrkolbenanordnung *f* multiple piston arrangement
Mehrkomponentendestillation *f* multicomponent distillation
Mehrlagenfolie *f* structured film
Mehrleiterkabel *n* multicore cable
Mehrleitersystem *n* multiple wire system
Mehrnährstoff *m* multinutrient
Mehrnährstoffdünger *m* multinutrient fertilizer
Mehrphasenanker *m* polyphase (or multiphase) armature
Mehrphasengenerator *m* polyphase (or multiphase) generator
Mehrphasenmotor *m* multicycle engine, polyphase motor
Mehrphasenofen *m* polyphase furnace
Mehrphasenschaltung *f* connection of polyphase circuits
Mehrphasenstrom *m* multiphase current
Mehrphasenstromanlage *f* multiphase current plant, polyphase plant
Mehrphasenstromerzeuger *m* multiphase current generator
Mehrphasensystem *n* multiphase system, **angeglichenes** ~ balanced multiphase system
Mehrphasenumformer *m* multiphase converter
Mehrphasenwechselstrom *m* multiphase current, polyphase current
mehrphasig multiphase, polyphase
mehrpolig multipolar
Mehrpreis *m* additional charge
mehrsäurig polyacid
Mehrschicht-Breitschlitzdüse *f* multi-manifold die
Mehrschichtenglas *n* laminated glass
Mehrschichtfolie *f* structured film

mehrschichtig multilayered, many-layered, multiply
Mehrschneckenpresse *f* multiscrew extruder
Mehrstoffgemisch *n* multicomponent mixture, multicomponent system
Mehrstoffgleichgewicht *n* multicomponent equilibrium
Mehrstufenentmagnetisierung *f* multiple-stage demagnetization
mehrstufig manystage, multistage
mehrteilig complex, consisting of several parts, multipart, multiple-part
Mehrtiegelverfahren *n* multiple crucible method
Mehrwalzenstuhl *m* cluster-roll mill
Mehrwegbehälter *m* returnable container
mehrwellig polyphase
mehrwertig (Alkohol) polyhydric, (Chem) multivalent, polyvalent
Mehrwertigkeit *f* (Chem) multivalence, polyvalence
Mehrzahl *f* plural, (Mehrheit) majority, plurality
mehrzellig multicellular
Mehrzentrenbindung *f* multicentered bonding
Mehrzentrenreaktion *f* multicentered reaction
Mehrzweck *m* multipurpose, dual purpose, multiple purpose
Mehrzweckofen *m* multipurpose furnace
Mehrzylindertrockner *m* multicylinder drier
mehrzylindrig multicylinder, polycylindrical
Meiler *m* kiln, pile
Meilerholz *n* charcoal wood
Meilerkohle *f* charcoal
Meilerkoks *m* heap coke
Meilerverkohlung *f* heap charring
Meilerverkokung *f* pile coking
Meiose *f* (Biol) meiosis
Meißelschablone *f* bit gauge
Meißelstahl *m* chisel steel, chisel temper
Meister *m* master, foreman
Meisterwurzelöl *n* masterwort oil
Mekkabalsam *m* Mecca balsam, balm of Gilead, balm of Mecca
Mekocyanin *n* mecocyanine
Mekonidin *n* meconidin
Mekonin *n* meconin, dimethoxyphthalide, opianyl
Mekoninsäure *f* meconinic acid, 5,6-dimethoxy-2-hydroxymethylbenzoic acid
Mekonsäure *f* meconic acid, 3-hydroxy-4-pyrone-2,6-dicarboxylic acid
mekonsauer meconate
Melacacidin *n* melacacidin
Melakonit *m* (Min) melaconite, native copper oxide
Melaleucaöl *n* melaleuca oil
Melam *n* melam
Melamazin *n* melamazine
Melamin *n* melamine, 2,4,6-triamino-1,3,5-triazine, cyanurotriamide

Melaminformaldehyd *m* melamine formaldehyde
Melaminformaldehydharz *n* melamine formaldehyde resin
Melaminharz *n* melamine [formaldehyde] resin
Melampyrin *n* melampyrine, dulcite, dulcitol
Melangallussäure *f* melanogallic acid
Melange *f* melange, mixture
Melangeeffekt *m* melange effect
Melanglanz *m* (Min) melane glance, black silverglance, brittle silver ore, stephanite
Melanglimmer *m* (Min) chalcodite, stilpnomelane
Melanilin *n* melaniline, N,N'-diphenylguanidine
Melanine *pl* (Pigmente) melanins
Melanit *m* melanite
Melanocerit *m* (Min) melanocerite
Melanochalcit *m* (Min) melanochalcite
Melanochroit *m* (Min) melanochroite
Melanocyten *pl* (Pigmentzellen) melanocytes
Melanoidin *n* melanoidin
Melanokarzinom *n* (Med) melanocarcinoma
Melanolith *m* (Min) melanolite
Melanophlogit *m* (Min) melanophlogite
Melanothallit *m* (Min) melanothallite
Melanotropin *n* (Hormon) melanotropin
Melanterit *m* (Min) melanterite, copperas
Melanurensäure *f* melanureic acid, 2,4-dihydroxy-6-amino-1,3,5-triazine, ammelide, cyanuramide
Melaphyr *m* (Min) melaphyre, augite porphyry, porphyry
Melasse *f* molasses
Melasseentzuckerung *f* extraction of sugar from molasses
Melassemaische *f* molasses mash or wash
Melassenkufe *f* vat for molasses
Melassensäure *f* mellassic acid
Melassepumpe *f* molasses pump
Melasseschnitzel *n pl* molasses pulp
Melatonin *n* melatonin
Meldolablau *n* Meldola blue
Melem *n* melem
Melen *n* melene
Melezitose *f* melezitose, melicitose, melizitose
Melibiase *f* (Biochem) melibiase
Melibiit *m* melibiitol
Melibionsäure *f* melibionic acid
Melibiosazon *n* melibiosazone
Melibiose *f* melibiose
Melibioson *n* melibiosone
Melibiulose *f* melibiulose
Melicitose *f* melicitose
Melicopidin *n* melicopidine
Melicopin *n* melicopine
Melidoessigsäure *f* melidoacetic acid
melieren to blend, to mingle, to mix
Melilith *m* (Min) melilite, melilith

Melilotin *n* melilotin
Melilotol *n* melilotol, 2-oxochroman, 3,4-dihydrocoumarin
Melilotsäure *f* melilotic acid, o-hydroxyhydrocinnamic acid
Melinit *m* (Min) melinite
Melinonin *n* melinonine
Melisimplexin *n* melisimplexin
Melisimplin *n* melisimplin
Melisse *f* (Bot) balm [mint]
Melissenblätter *pl* (Bot) balm leaves, melissa leaves
Melissengeist *m* carmelite water
Melissenöl *n* balm mint oil, melissa oil, verbena oil
Melissentee *m* infusion of balm
Melissinsäure *f* melissic acid
Melisson *n* melissone
Melissylalkohol *m* melissyl alcohol
Meliternin *n* meliternin
Melitose *f* melitose
Melitriose *f* melitriose
Melittin *n* (Peptid) melittin
Melizitose *f* melicitose, melezitose, melizitose
Melkfett *n* milking grease
Mellein *n* mellein
Mellibiase *f* mellibiase
Mellimid *n* mellimide
Mellit *m* mellite, (Min) honeystone, (Pharm) medicated honey
Melliten *n* mellitene, hexamethylbenzene
Mellitol *n* mellitol
Mellitsäure *f* mellitic acid, benzenehexacarboxylic acid
Mellon *n* mellon[e]
Mellophansäure *f* mellophanic acid
Melon *n* melon, mellon[e]
Melonenöl *n* melon oil, melon seed oil
Melonit *m* (Min) melonite
Melotte-Metall *n* Melotte fusible alloy
Melphalan *n* melphalan
Meltau *m* mildew
Membran *f* membrane, diaphragm, **elastische** ~ elastic membrane, **halbdurchlässige** ~ semi-permeable diaphragm, semi-permeable membrane, **interzelluläre** ~ intercellular membrane, **mehrschichtige** ~ compound membrane, **permeable** ~ permeable membrane, **semipermeable** ~ semi-permeable membrane
Membrandichtung *f* diaphragm seal
Membrandurchlässigkeit *f* membrane permeability
Membranelektrode *f* membrane electrode
Membranfilter *n* diaphragm, membranous filter
Membrangleichgewicht *n* membrane or Donnan equilibrium
Membranlipid *n* membrane lipid

Membranmanometer *n* für Vakuumtechnik diaphragm vaccum gauge
Membranpotential *n* membrane potential
Membranprotein *n* membrane protein
Membranpumpe *f* diaphragm pump
Membranschwingung *f* diaphragm oscillation
Membransetzmaschine *f* diaphragm-actuated jig
Membranventil *n* diaphragm valve
Membranverdichter *m* diaphragm compressor
Membranzusammensetzung *f* membrane composition
Menaccanit *m* (Min) menac[c]anite, menacconite, menachanite, menakanite, titanic iron ore, titaniferous magnetic iron oxide
Menadiol *n* menadiol
Menadion (Vitamin K$_3$) menadione, 2-methyl naphthoquinone
Menage *f* (Gewürzständer) cruet stand
Menaphthylamin *n* menaphthylamine, naphthylmethylamine
Menaphthylbromid *n* menaphthylbromide, naphthylmethyl bromide
Mendelevium *n* mendelevium (Symb. Md o. Mv)
Mendelsches Gesetz *n* Mendel's law
Mendipit *m* (Min) mendipite, chlorspar
Mendozit *m* (Min) mendozite
Meneghinit *m* (Min) meneghinite, native lead antimony sulfide
Menge *f* quantity, amount, mass
mengen to blend, to mix
Mengen *n* blending, mixing
Mengenbestimmung *f* quantitative determination, (Chem) quantitative analysis
Mengenbilanz *f* mass balance
Mengeneinheit *f* unit of quantity
Mengenfaktor *m* multiplicity factor
Mengenleistung *f* product capacity
mengenmäßig quantitative
Mengenmesser *m* volumeter, flowmeter, volumenometer
Mengenregelung *f* basis rate system
Mengenregler *m* volume control, volume regulator
Mengenschreiber *m* flow recorder
Mengenuntersuchung *f* quantitative examination
Mengenverhältnis *n* quantitative ratio, mass ratio, proportion, quantitative composition
Menger *m* blender, mixer
Mengfutter *n* mixed feed
Menggestein *n* conglomerate
Mengit *m* mengite
Mengkapsel *f* mixing capsule
Mengsel *n* medley, mixture
Mengspat *m* (Min) foliated spar
Mengspatel *m* mixing spatula
Mengteil *m* constituent, ingredient
Mengungsverhältnis *n* proportion of ingredients
Menhadenöl *n* menhaden oil

Menilit *m* menilite
Meniskus *m* meniscus
Menisperin *n* menisperine
Menispermin *n* menispermine
Mennige *f* minium, lead(II) (IV) oxide, plumbous plumbate, plumbous plumbic oxide, red lead [oxide], trilead tetroxide
Mennigeanstrich *m* coat of red lead
Mennigefarbe *f* red lead paint
mennigefarben minium-colored
Mennigekitt *m* red lead cement, red lead putty
Mennigerot *n* minium, red lead
Mensurglas *n* graduated glass, measuring glass
Menthacampher *m* menthol, mint camphor
Menthan *n* menthane, 1-methyl-4-isopropylcyclohexane
Menthanol *n* menthanol, hydroxymenthane
Menthanon *n* menthanone, oxomenthane
Menthen *n* menthene
Menthenol *n* menthenol, hydroxymenthene
Menthenon *n* menthenone, oxomenthene
Menthofuran *n* menthofuran
Menthol *n* menthol, 3-hydroxymenthane, menthacamphor, mint camphor
Mentholid *n* mentholide
Menthon *n* menthone, 3-oxomenthane
Menthonol *n* menthonol
Menthospirin *n* menthospirine, l-menthyl acetylsalicylate
Menthyl- menthyl
Menthylacetat *n* menthyl acetate
Menthylamin *n* menthylamine
Menthylaminhydrochlorid *n* menthylamine hydrochloride
Menthylester *m* menthyl ester
Menthyllactat *n* menthyl lactate
Menthylnitrobenzoat *n* menthyl nitrobenzoate
Menthylvalerat *n* menthyl valerate
Menyanthin *n* menyanthin, celastin
Menyanthol *n* menyanthol
Meperidin *n* meperidine
Mephenesin *n* mephenesin
Mephentermin *n* mephentermine
Mephenytoin *n* mephenytoin
Meprobamat *n* meprobamate
Meprylcain *n* meprylcaine
Mepyramin *n* mepyramine
Merallurid *n* meralluride
Meranzin *n* meranzin
Meratin *n* meratin
Mercaptamin *n* mercaptamine
Mercaptan *n* [ethyl] mercaptan, thiol
Mercaptid *n* mercaptide, metal mercaptan
Mercaptoäthanolamin *n* mercaptoethanolamine
Mercaptoäthylamin *n* mercaptoethylamine
Mercaptobenzimidazol *n* mercaptobenzimidazole
Mercaptobenzothiazol *n* mercaptobenzothiazole
Mercaptochinolin *n* mercaptoquinoline

Mercaptol – Merotropie

Mercaptol *n* mercaptol
Mercaptomerin *n* mercaptomerin
Mercaptopurin *n* mercaptopurine
Mercaptursäure *f* mercapturic acid
Mercerisation *f* (Baumwollveredelungsverf.) mercerization
Mercerisierhilfsmittel *n* mercerization auxiliary
Mercuderamid *n* mercuderamide
Mercur *m* (Astr) Mercury (s. auch Merkur)
Mercur *n* (obs) mercury, quicksilver (s. auch Merkur)
Mercurialmittel *n* (Pharm) mercurial
Mercurialpille *f* blue pill
Mercurialsalbe *f* blue ointment, mercurial ointment
Mercuribenzoat *n* mercuric benzoate, mercury(II) benzoate
Mercurichlorid *n* mercuric chloride, corrosive sublimate, mercury bichloride, mercury dichloride, mercury(II) chloride
Mercuricyanid *n* mercuric cyanide, mercury(II) cyanide
Mercuricyanwasserstoffsäure *f* cyanomercuric(II) acid, mercuricyanic acid
Mercurijodid *n* mercuric iodide, mercury biiodide, mercury diiodide, mercury(II) iodide
Mercurilactat *n* mercuric lactate
Mercurimetrie *f* mercurimetry
Mercurinitrat *n* mercuric nitrate, mercury(II) nitrate
Mercurioxid *n* mercuric oxide, mercury(II) oxide
Mercurioxycyanid *n* mercuric oxycyanide, mercury(II) oxycyanide
Mercuriphenolat *n* mercuric phenolate, mercuric carbolate, mercuric phenate, mercuric phenoxide, mercury(II) phenolate
Mercuriphosphat *n* mercuric phosphate, mercury(II) phosphate
Mercurirhodanid *n* mercuric rhodanide, mercuric thiocyanate, mercury(II) thiocyanate
Mercurisalicylat *n* mercuric salicylate
Mercurisaligenin *n* mercurisaligenin
Mercurisulfat *n* mercuric sulfate, mercury(II) sulfate
Mercurisulfid *n* mercuric sulfide, mercury(II) sulfide
Mercuriverbindung *f* (Quecksilber(II)-Verbindung) mercuric compound, mercury(II) compound
Mercuroacetat *n* mercurous acetate, mercury(I) acetate
Mercurobutol *n* mercurobutol
Mercurochlorid *n* mercurous chloride, calomel, mercury(I) chloride, mercury monochloride, mercury protochloride
Mercurochrom *n* mercurochrome, disodium salt of 2,7-dibromo-4-hydroxy mercury fluorescein, merbromin
Mercurojodid *n* mercurous iodide, mercury(I) iodide, mercury monoiodide
Mercuronitrat *n* mercurous nitrate, mercury(I) nitrate
Mercurophen *n* mercurophen, sodium hydroxy-mercury-o-nitrophenolate
Mercurophosphat *n* mercurous phosphate, mercury(I) phosphate
Mercurosulfat *n* mercurous sulfate, mercury(I) sulfate
Mercuroverbindung *f* mercurous compound, mercury(I) compound
Mergal *n* mergal
Mergel *m* marl, **künstlicher** ~ malm
mergelartig marly
Mergelerde *f* earthy marl
Mergelgrube *f* marl pit
mergelhaltig containing marl, marly
Mergelkalk *m* marly limestone
Mergeln *n* marling
Mergelsand *m* marly sand
Mergelsandstein *m* (Min) marly sandstone
Mergelschiefer *m* (Min) marlslate, slaty marl
Mergelstein *m* marlstone
Mergelton *m* argillaceous marl, marly clay
Merichinon *n* meriquinone
Meridianspannung *f* vertical stress on the center line
Meridionaltransport *m* meridional flux
Merimin *n* merimine
Merinorot *n* merino red, Turkey red
Merkblatt *n* data sheet, special leaflet
Merkmal *n* feature, indication, mark, sign
Merkur *m* (Astr) Mercury (s. auch Mercur)
Merkur *n* (obs) mercury, quicksilver (s. auch Mercur)
Merkurblende *f* (Min) cinnabar, native red mercury sulfide, vermil[l]ion
Merkurhornerz *n* (Min) horn quicksilver
Merkuri s. Mercuri
merkurial mercurial
Merkurialien *pl* mercurials
merkurieren to combine with mercury, to mercurize
Merkuriverbindung *f* mercuric compound, mercury(II) compound
Merkuro s. Mercuro
Merkuroverbindung *f* mercurous compound, mercury(I) compound
Merkurverbindung *f* mercury compound
Merochinen *n* meroquinene, 3-vinyl-piperidyl-(4)-acetic acid
Merocyanin *n* merocyanine
Merolignin *n* merolignin
meromorph (Math) meromorphic
Merosinigrin *n* merosinigrin
Merotropie *f* merotropism, merotropy

Merrifield-Technik *f* (Festphasen-Peptidsynthese) Merrifield technique
Mersalyl *n* mersalyl
Mersol *n* mersol
Mersolat *n* mersolate
Merthiolat *n* merthiolate
Merzerisation *f* (Text) mercerization [process]
Merzerisieranlage *f* (Text) mercerizer
merzerisieren (Text) to mercerize
Merzerisieren *n* (Text) mercerizing
Merzerisiernetzmittel *n* wetting agent for mercerizing
Merzerisierverfahren *n* (Text) mercerizing process
Mesaconsäure *f* mesaconic acid, methylfumaric acid
Mesantensäure *f* mesantenic acid
Mescalin *n* mescaline
Mescalinsulfat *n* mescaline sulfate
Mesembran *n* mesembran
Mesembren *n* mesembrene
Mesembrin *n* mesembrine
Mesembrol *n* mesembrol
Mesidin *n* mesidine, 2,4,6-trimethylaniline
Mesitenlacton *n* mesitene lactone, 4,6-dimethyl-coumalin
Mesitin[spat] *m* (Min) mesitine spar, breunnerite, mesitite, native magnesium iron carbonate
Mesitoesäure *f* mesitoic acid
Mesitol *n* mesitol, 2,4,6-trimethylphenol, mesitylene alcohol
Mesitonsäure *f* mesitonic acid, 2,2-dimethyl-4-oxopentanoic acid, 2,2-dimethyllevulinic acid
Mesityl- mesityl
Mesit[yl]aldehyd *m* mesitylaldehyde
Mesitylen *n* 1,3,5,-trimethylbenzene, mesitylene, mesitylol
Mesitylensäure *f* mesitylenic acid, 3,5-dimethylbenzoic acid, mesitylinic acid
Mesityloxid *n* mesityl oxide, isopropylidene-acetone
Mesitylsäure *f* mesitylic acid
Mesoatom *n* meso atom
Mesobilirubin *n* mesobilirubin
Mesochlorin *n* mesochlorine
Mesocorydalin *n* mesocorydaline
Mesoerythrit *m* mesoerythritol
Mesoform *f* meso form
Mesohäm *n* mesohem
Mesoinosit *m* mesoinositol
mesoionisch mesoionic
Mesokolloid *n* (Phys) mesocolloid
Mesolith *m* (Min) mesolite
mesomer mesomer
Mesomerie *f* mesomerism
Mesomerieeffekt *m* mesomeric effect

Mesomerieenergie *f* delocalization energy, resonance energy
mesomorph mesomorphous
Meson *n* meson, (obs) mesotron
Mesonenatom *n* mesonic atom
Mesonenausbeute *f* meson yield
Mesonenbahn *f* meson track
Mesoneneinfang *m* meson capture
Mesonenerzeugung *f* meson production
Mesonenfeld *n* meson field
Mesonennachweis *m* meson detection
Mesonenphysik *f* meson physics
Mesonentheorie *f* (Atom) meson theory
Mesonenzerfall *m* meson decay
Mesonniveau *n* mesonic level
Mesonpaardämpfung *f* meson pair damping
Mesonprotonstreuung *f* meson-proton scattering
Mesonterm *m* mesonic level, mesonic term
Mesophase *f* mesophase
mesophil mesophilic
Mesophyllin *n* mesophylline
Mesoporphyrin *n* mesoporphyrine
Mesoporphyrinogen *n* mesoporphyrinogene
Mesopyrophäophorbid *n* mesopyropheophorbide
Mesopyrrochlorin *n* mesopyrrochlorine
Mesorcin *n* mesorcin, mesorcinol
Mesotan *n* mesotan, ericin, methoxymethyl salicylate
Mesotartrat *n* mesotartrate
Mesothorium *n* mesothorium
Mesotron *n* (obs) meson, mesotron
Mesotronenschauer *m* mesotron shower
Mesotropie *f* mesotropism
Mesotyp *m* (Min) mesotype, featherzeolite
Mesoweinsäure *f* mesotartaric acid
Mesoxalaldehydsäure *f* mesoxalaldehydic acid
Mesoxalsäure *f* mesoxalic acid, oxomalonic acid, oxopropanedioic acid
mesoxalsauer mesoxalate
Mesoxalyl- mesoxalyl
Mesoxalylharnstoff *m* mesoxalylurea, alloxan
Mesoxophenin *n* mesoxophenin
Mesoyohimbin *n* mesoyohimbine
Mesozoikum *n* (Geol) Mesozoic
Messelith *m* (Min) messelite
messen to measure, (loten) to measure the depth of water, to sound, (mit einer Festlehre) to gauge
Messen *n* measuring, metering
Messenger-Ribonucleinsäure *f* messenger ribonucleic acid
Messer *n* knife, (Klinge) blade, (Techn) cutter, **eingesetztes** ~ inserted knife
Messerbrecher *m* knife crusher
Messergriff *m* knife handle
Messerhalter *m* knife holder
Messerschale *f* knife scale
Messerschlitten *m* knife sledge
Messerschneide *f* knife edge

Messerstütze *f* knife support
Messerträger *m* knife carrier
Messertrommel *f* cross-cutting knife drum
Messerwalze *f* knife roll, beater roll, knife drum
Messerwelle *f* knife block
Messing *n* [yellow] brass, ~ **abbrennen** to pickle brass
Messingabstrich *m* yellow brass scum
Messinganode *f* brass anode
messingartig brassy
Messingartigkeit *f* brassiness
Messingbad *n* brass bath
Messingblatt *n* brass foil
Messingblech *n* brass sheet
Messingblüte *f* (Min) aurichalcite
Messingdraht *m* brass wire
Messingdrahtnetzkathode *f* brass gauze cathode
messingfarben brass-colored
Messingkratzbürste *f* brass wire brush
Messinglot *n* hardsolder
Messingsäule *f* brass column
Messingschlaglot *n* brass solder
Messung *f* measurement, gauging, **eine ~ machen** to take a measurement
Messungsbegriff *m* measurement concept
Meßzylinder *m* graduated cylinder, measuring cylinder
Mesulfen *n* mesulfen
Mesylchlorid *n* mesylchloride
Meßanordnung *f* measuring arrangement
Meßapparat *m* measuring apparatus
Meßausrüstung *f* measuring equipment
Meßband *n* measuring tape, tape measure
meßbar measurable
Meßbecher *m* measuring beaker, measuring cup
Meßbereich *m* measuring range, range of measurement, scale range, span
Meßbildverfahren *n* photogrammetry
Meßbrücke *f* measuring bridge, resistance bridge
Meßbürette *f* graduated buret[te]
Meßdose *f* measuring cylinder
Meßdüse *f* calibrated nozzle
Meßergebnis *n* experimental result, measuring result, result of measurement
Meßfehler *m* measuring error, measuring fault
Meßfehlergrenze *f* error limit
Meßflasche *f* graduated flask, measuring flask
Meßflüssigkeit *f* measuring fluid
Meßfühler *m* (Regeltechn) primary or detecting element
Meßgas *n* test gas
Meßgefäß *n* graduated jar, measuring vessel
Meßgenauigkeit *f* accuracy of measurement
Meßgerät *n* gauge, measuring apparatus
Meßglied *n* measuring device, (Regeltechn) measuring means or unit
Meßgröße *f* quantity to be measured
Meßheber *m* graduated pipet[te], measuring pipette

Meßinstrument *n* measuring instrument
Meßkette *f* gauging chain, measuring chain
Meßklotz *m* gauge block
Meßkolben *m* graduated flask, measuring flask, volumetric flask
Meßkollimator *m* (Opt) focal collimator
Meßkopf *m* probe, sensing element, test head
Meßkreis *m* measuring circuit
Meßkreiseichung *f* standardization of the circuit
Meßlänge *f* gauge length
Meßlehre *f* measuring gauge
Meßleitungsverfahren *n* transmission line method
Meßmaschine *f* measuring machine
Meßmethode *f* measuring method, measuring system
Meßmikroskop *n* measuring microscope, reading microscope, scanning microscope
Meßokular *n* (Opt) micrometric eyepiece
Meßort *m* measuring location, measuring point
Meßpipette *f* graduated pipet[te], measuring pipette
Meßpult *n* measuring desk
Meßpunkt *m* experimental point, measuring point, (Ausscheidung einzelner Meßpunkte) exclusion of single observations
Meßreihe *f* series of measurement
Meßröhre *f* buret[te], measuring tube, (Phys) target tube
Meßrohr *n* measuring tube
Meßschraube *f* micrometer screw
Meßschritteinstellung *f* span adjustment
Meßstab *m* (Öl) dipstick
Meßstelle *f* measuring point, control point, observation point
Meßstellenumschalter *m* measuring point selector
Meßstrecke *f* measured length, measuring section
Meßstreifen *m* measuring tape
Meßstromverstärker *m* signal current amplifier
Meßtechnik *f* (Atom) counting technique
Meßtransformator *m* measuring transformer
Meßuhr *f* meter, metering clockwork
Meßumformer *m* (Regeltechn) measuring transmitter or transducer
Meßverfahren *n* method of measurement, test procedure
Meßverstärker *m* amplifier
Meßvorrichtung *f* measuring device
Meßwarte *f* control room
Meßwerk *n* measuring element
Meßwerkzeug *n* measuring instrument, measuring tool
Meßwert *m* experimental value, measured value, reading, test value
Meßwertdrucker *m* digital printer
Meßwertsammler *m* data logger
Meßwertspeicher *m* data logger

Meßwertwandler *m* transducer, transmitter
Meßwiderstand *m* measuring resistance
Meßzelle *f* measuring cell, analyzer
Meßzentrale *f* control room
Met *m* mead, hydromel
Metaantimonsäure *f* meta[a]ntimonic acid
Metaarsensäure *f* meta[a]rsenic acid
Metabiose *f* metabiosis
Metabisulfit *n* metabisulfite
metabolisch metabolic
Metabolismus *m* metabolism
Metabolit *m* metabolite
Metaborat *n* metaborate
Metaborsäure *f* metaboric acid
Metabrenztraubensäure *f* metapyroracemic acid
Metabrushit *m* (Min) metabrushite
Metaceton *n* metacetone, 3-pentanone, diethylketone
Metacetonsäure *f* (obs) propionic acid
Metachloral *n* metachloral
Metachloranilin *n* m-chloroaniline, 3-chloro-aminobenzene, 3-chlorophenylamine
Metachlorbenzoesäure *f* metachlorobenzoic acid
Metachromasie *f* metachromatism
Metachrombeize *f* metachrome mordant
Metachromfarbe *f* metachrome dye
Metachromgelb *n* metachrome yellow
Metacinnabarit *m* (Min) metacinnabarite, native black mercuric sulfide
Metacrolein *n* metacrolein
Metacyclin *n* metacycline
Metacyclophan *n* metacyclophane
Metacymol *n* m-cymene, 1-methyl-3-isopropylbenzene, isocymene
Metadichlorbenzol *n* m-dichlorobenzene
Metadioxybenzol *n* 1,3-dihydroxybenzene, 1,3-benzenediol, resorcinol
metadirigierend (Chem) meta-directing
Metaeisenoxid *n* metaferric oxide
Metaferrihydrat *n* metaferric hydroxide
Metaformaldehyd *m* metaformaldehyde
Metagallussäure *f* metagallic acid
Metagenese *f* (Biol) metagenesis
Metagenin *n* metagenin
Metahemipinsäure *f* metahemipic acid
Metakieselsäure *f* metasilicic acid
metakieselsauer metasilicate
Metakohlensäure *f* metacarbonic acid
Metakresol *n* m-cresol, m-hydroxytoluene, m-methylphenol
Metakresotinsäure *f* m-cresotic acid, 2,4-cresotic acid, 2-hydroxy-4-methylbenzoic acid
metakristallin metacrystalline
Metaldehyd *m* metaldehyde, meta acetaldehyde
Metall *n* metal, ~ **abbeizen** to scour metal, ~ **von kleiner Überspannung** metal of small overvoltage, **edles** ~ noble metal, precious metal, **legierbares** ~ alloyable metal, **unedles** ~ base metal

Metallabfall *m* metal chips, scrap metal, waste metal
Metallabscheidung *f* deposition of metal
Metallacetylid *n* metal acetylide
Metallack *m* metal coating, metal lacquer
metallähnlich metallic, like metal, metalliform, metalline
Metallalkyl *n* metal alkyl
Metallamid *n* metal amide
Metallammoniakverbindung *f* metal ammonia compound, ammoniated metal, metal ammoniate, metallammine
Metallarbeiter *m* metal worker
metallarm poor in metal
Metallarmierung *f* metal reinforcement
metallartig metallic, like metal, metalline, metalloid
Metallartigkeit *f* metallicity
Metallasche *f* metallic ashes
Metallatom *n* metal atom
Metallauflage *f* metal-plating
Metallaufnahme *f* (v. Lebensmitteln) metal pick-up
Metallausbeute *f* yield of metal
metallausscheidend metal-precipitating
Metallaustauschgeschwindigkeit *f* metal exchange rate
Metallazid *n* metal azide, metallic azide
Metallbad *n* metallic bath, metallic solution
Metallbearbeitung *f* metal working, metal cutting, metal processing, **spanabhebende** ~ metal cutting, **spanlose** ~ metal working
Metallbedampfung *f* metallizing
Metallbelag *m* metal coating, metallization
Metallbeschickung *f* alloyage
Metallbeschwerung *f* loading with metallic salts, weighting with metallic salts
Metallbindung *f* metallic bond
Metallblatt *n* metal foil, sheet of metal
Metallbleiche *f* albification
Metallbronze *f* metallic pigment
Metallcarbid *n* metal carbide, metal acetylide
Metallcarbonyl *n* metal carbonyl, metal carboxide
Metallchelat *n* metal chelate
Metallchemie *f* metallochemistry
Metallchlorid *n* metal[lic] chloride
Metallchromie *f* metalcoloring, metallochromy
Metallcyanid *n* metal[lic] cyanide
Metalldampf *m* metal vapor
Metalldampflampe *f* metal vapor lamp
Metalldarstellung *f* extraction of metals, metallurgy
Metalldichtung *f* metallic packing
Metalldraht *m* metal wire
Metalldrahtbürste *f* wire brush
Metalldrahtentladung *f* exploded wire
Metalldrahtentladungskontinuum *n* exploded wire continuum

Metalldrehbank — Metallsiliziumverbindung

Metalldrehbank *f* metal finishing lathe
Metalleffektfäden *pl* metal effect threads
Metallegierung *f* metal alloy, metallic alloy
Metalleigenschaft *f* metallic property, metallicity
Metalleinkristall *m* metal single crystal
Metallelektrode *f* metal electrode
metallen metallic, metalline
Metallentfettung *f* metal degreasing
Metallentgasung *f* metal degassing
Metallerz *n* metallic ore, metal ore
Metallfaden *m* metal filament
Metallfadenlampe *f* metal filament lamp
Metallfärbung *f* metallic coloring, metallochromy
Metallfarbe *f* metal[lic] color
metallfarbig metal-colored
Metallfarbkörper *m* metallic pigment
Metallfaser *f* metallic thread
Metallflansch *m* metal flange
metallförmig metalliform
Metallfolie *f* metal foil, metallic foil
Metallfolienüberzug *m* coating of metal foil
metallführend metalliferous
Metallgehalt *m* metal content[s]
Metallgekrätz *n* dross, waste metal
Metallgemisch *n* metallic mixture, mixture of metals
Metallgewinnung *f* metal production, extraction of metals, metallurgy
Metallgießerei *f* metal foundry
metallglänzend having a metallic luster, lustrous
Metallglanz *m* metallic luster, unvollkommener ~ pseudometallic luster
Metallglanzfarbe *f* lusterwash
Metall-Glas-Kitt *m* metal-to-glass seal
Metallgold *n* Dutch foil, Dutch gold
Metallgrundierung *f* metal primer
Metallguß *m* cast metal
metallhaltig metalliferous
Metallhitzemesser *m* metal pyrometer
Metallhüttenstaub *m* metallurgical dust
Metallhydrid *n* metallic hydride
Metallhydroxid *n* metal hydroxide, metallic hydroxide
Metallierung *f* (Chem) metalation
Metallin *n* metallin
Metallion *n* metal ion
metallisch metallic, ~ glänzend having a metallic luster
Metallisieranlage *f* (für Kunststoffe) vacuum coating machine
metallisieren to metallize, to metal-coat, to metal-plate
Metallisieren *n* metal[l]izing, metal-coating, metal-plating
Metallisierung *f* metallization
Metallisierungsverfahren *n* metal-coating process

Metallkalk *m* metallic calx
Metallkarbid *n* metal carbide
Metallkleben *n* metal bonding
Metallklemme *f* metal clamp
Metallkönig *m* metallic regulus, metal button
Metallkomplexsalz *n* complex metal salt
Metallkomposition *f* metallic composition
Metallkorn *n* granulated metal, metal grain
Metallkrätze *f* metal filings, metal waste
Metallkunde *f* metallography, (Erzscheidekunde) metallurgy
Metallmatrize *f* metal matrix
Metallmischung *f* metallic composition, metal mixture
Metallmischungswaage *f* alloy-balance
Metallmohr *m* metallic moire
Metallnebel *m* atomized metal, metal mist
Metallnetzeinlage *f* reinforcing mesh
Metallniederschlag *m* metal deposit, metallic precipitate
Metallochemie *f* metallochemistry
Metallochromie *f* metal coloring, metallochromy
Metallöffel *m*, durchlöcherter ~ perforated metal ladle
Metallographie *f* metallography
metallographisch metallographic
metalloid metalloid, non-metallic
Metalloid *n* metalloid, non-metallic element
Metalloidoxid *n* metalloid oxide
Metalloporphin *n* metalloporphine
metallorganisch metal[lo]organic, organometallic
Metalloxid *n* metallic oxide, metal oxide
Metalloxidkathode *f* metallic oxide cathode
Metallpackung *f* metal packing
Metallpapier *n* metallic paper
Metallpfanne *f* metal ladle
Metallpinsel *m* metallic brush
Metallporphin *n* metalloporphine
Metallprotein *n* metalloprotein
Metallpuffer *m* metal buffer
Metallpyrometer *n* metal pyrometer
Metallring *m* metal ring, grommet
Metallrohr *n* metal pipe
Metallrückstand *m* metal residue
Metallsäge *f* metal-cutting saw
Metallsalzkontakt *m* metal salt contact
Metallsand *m* metallic sand, grit
Metallschere *f* metal-cutting pliers
Metallschicht *f* layer of metal, metal film
Metallschirm *m* metallic screen
Metallschlacke *f* metal slag
Metallschlauch *m* flexible metal tube
Metallschmelze *f* fusion of metals, molten metal
Metallschwamm *m* (Atom) biscuit
Metallseife *f* metallic soap, silver soap
Metallsilber *n* Britannia metal, false leaf silver, imitation silver foil, white Dutch metal
Metallsiliziumverbindung *f* metal silicon compound

Metallspäne *pl* metal chips, metal filings, turnings
Metallspiegel *m* metallic mirror, metal mirror
Metallspitze *f* metallic point
Metallsplitter *m* metal fragment, metal splinter
Metallspritzmethode *f* metal spray method
Metallspritzverfahren *n* metal spraying, metal[l]ization
Metallstaub *m* metal dust
Metallstaublunge *f* (Med) siderosis
Metallstearat *n* metal stearate
Metallstück *n* metal fragment
Metall-Substrat-Komplex *m* (Biochem) metal-substrate complex
Metallsud *m* hot galvanizing bath
Metallsulfid *n* metallic sulfide, metal sulfide
Metallsulfidniederschlag *m* precipitate of metallic sulfide
Metalltemperofen *m* annealing furnace
Metallthermometer *n* metal[lic] thermometer
Metallüberzug *m* metal coating, metal plating
metallumsponnen metal-spun, metal-braided, metal-clad
Metalluntersuchung *f* metallographic examination
Metallurg[e] *m* metallurgist
Metallurgie *f* metallurgy
metallurgisch metallurgic
metallverarbeitend metal-working
Metallverarbeitung *f* metal working
Metallverbindung *f* (Chem) metal[lic] compound, (Mischung) combination of metals
Metallverlust *m* loss of metal
Metallverschmelzung *f* alloy, alloying
metamer metameric
Metamerie *f* metamerism
Metamerismus *m* metamerism
metamorph metamorphic, metamorphous
metamorphisch metamorphic
Metamorphismus *m* (Geol) metamorphism
Metamorphose *f* (Biol) metamorphosis
metamorphosieren to metamorphose
Metamyelozyt *m* metamyelocyte
Metanethol *n* metanethol, solid dianethol
Metanicotin *n* metanicotine
Metanilgelb *n* metaniline yellow, metanil yellow
Metanilingelb *n* metaniline yellow, metanil yellow
Metanilsäure *f* metanilic acid, m-aminobenzenesulfonic acid, m-anilinesulfonic acid
Metanitranilin *n* m-nitr[o]aniline, 3-nitro-1-aminobenzene, 3-nitrophenylamine
Metanitrotoluol *n* m-nitrotoluene, 3-nitro-1-methylbenzene
Metantimonsäure *f* met[a]antimonic acid
Metapepton *n* metapeptone
Metaphase *f* metaphase
Metaphen *n* metaphen

Metaphenylenblau *n* metaphenylene blue
Metaphenylendiamin *n* 1,3-benzene diamine, 1,3-diaminobenzene, m-phenylene diamine
Metaphosphat *n* metaphosphate
Metaphosphinsäure *f* metaphosphinic acid
metaphosphorig metaphosphorous
Metaphosphorsäure *f* metaphosphoric acid
metaphosphorsauer metaphosphate
Metaphthalsäure *f* 1,3-benzenedicarboxylic acid, m-phthalic acid, isophthalic acid
Metapilocarpin *n* metapilocarpine
Metapipton *n* metapiptone
Metaplasma *n* metaplasm
Metaraminol *n* metaraminol
Metarheologie *f* metarheology
Metasaccharin *n* metasaccharin
Metasaccharinsäure *f* metasaccharic acid
Metasaccharonsäure *f* metasaccharonic acid
Metasaccharopentose *f* metasaccharopentose
Metasäure *f* meta acid
Metasantonin *n* metasantonin
Metasilikat *n* metasilicate
Metasolvan *n* metasolvan
metasomatisch metasomatic
metastabil metastable
Metastabilität *f* metastability
metaständig in meta position
Metastase *f* (Med) metastasis
Metastellung *f* meta position
Metastyrol *n* metastyrene, metastyrolene
Metatitansäure *f* metatitanic acid
Metatoluidin *n* m-toluidine, 3-amino-1-methylbenzene, m-methylaniline
Metatorbernit *m* (Min) metatorbernite
Metavanadat *n* metavanadate, salt or ester of metavanadic acid
Metavanadinsäure *f* metavanadic acid
Metaverbindung *f* meta compound
Metaweinsäure *f* metatartaric acid
Metawolframsäure *f* metatungstic acid
Metaxit *m* (Min) metaxite, fibrous serpentine
Metaxylidin *n* metaxylidine
Metaxylol *n* metaxylene, m-dimethylbenzene
metazentrisch metacentric
Metazentrum *n* metacenter
Metazeunerit *m* (Min) metazeunerite
Metazinnober *m* (Min) metacinnabarite
Metazinnsäure *f* metastannic acid
metazinnsauer metastannate
Metazirkonsäure *f* metazirconic acid
Metazuckersäure *f* metasaccharic acid
Meteloidin *n* meteloidine
Meteor *m* meteor
meteorähnlich meteor-like
Meteoreisen *n* meteoric iron
meteorisch meteoric
Meteorit *m* meteorite
Meteorlicht *n* meteoric light

Meteorograph *m* meteorograph
Meteorographie *f* meteorography
meteorographisch meteorographic
Meteorologie *f* meteorology
meteorologisch meteorologic
Meteorometer *n* meteorometer
Meteorschweif *m* meteor trail
Meteorstahl *m* meteoric steel
Meteorstein *m* meteoric stone, aerolite, atmospheric stone
Meter *n* meter (Am. E.), metre (Br. E.)
Meterkilogramm *n* kilogrammeter, meter kilogram[me]
Metermaß *n* tape-measure, (Maßstab) meter scale, (Maßsystem) metric measure
Methacetin *n* methacetin, p-acetanisidine, p-methoxyacetanilide
Methacholin *n* methacholine
Methacrolein *n* methacrolein
Methacrylat *n* methacrylate
Methacrylsäure *f* methacrylic acid
Methacrylsäureester *m* methacrylate, methacrylic ester
Methacrylsäuremethylester *m* methacrylic methylester, methyl methacrylate
Methadon *n* methadone
Methämoglobin *n* methemoglobin
Methamphetamin *n* methamphetamine
Methamphetaminhydrochlorid *n* methamphetamine hydrochloride
Methan *n* methane, methyl hydride
Methanisierung *f* methanation
Methannitronsäure *f* methane nitronic acid
Methanol *n* methanol, methyl alcohol, (obs) carbinol, wood alcohol
Methanolanlage *f* methanol plant
Methansiliconsäure *f* methane siliconic acid
Methansulfinsäure *f* methane sulfinic acid
Methansulfonsäure *f* methane sulfonic acid
Methazonsäure *f* met[h]azonic acid
Methen *n* methene, methylene
Methin *n* methine, methenyl
Methingruppe *f* methine group
Methiodal *n* methiodal
Methionin *n* methionine
Methionol *n* methionol
Methionsäure *f* methionic acid, methene disulfonic acid, methylene disulfonic acid
Methode *f* method, manner, way, ~ **der kleinsten Fehlerquadrate** method of least squares, ~ **des steilsten Abstiegs** method of steepest descent, ~ **des steilsten Anstiegs** method of steepest ascent
Methodik *f* methodology
methodisch methodical[ly]
Methon *n* methone
Methopropylbenzol *n* methopropylbenzene, isobutylbenzene
Methoxy- methoxy[l]

Methoxyderivat *n* methoxy derivative
Methoxyessigsäure *f* methoxyacetic acid
Methoxyl- methoxy, methoxyl
Methoxylbestimmung *f* methoxy[l] determination
Methoxy[l]gruppe *f* methoxy[l] group
Methoxylierung *f* methoxylation
Methoxyphedrin *n* methoxyphedrine
Methoxyphenamin *n* methoxyphenamine
Methoxypropionsäure *f* methoxypropionic acid
Methronol *n* methronol
Methronsäure *f* methronic acid
Methyl- methyl
Methylacetanilid *n* methylacetanilide, exalgin
Methylacetat *n* methyl acetate
Methylacrylat *n* methyl acrylate
Methyladnamin *n* methyladnamine
Methyläther *m* methyl ether, dimethyl ether, methoxymethane
Methyläthyläther *m* methyl ethyl ether
Methyläthylketon *n* methyl ethyl ketone
Methyläthylpyrrol *n* methyl ethyl pyrrole
Methylal *n* methylal, dimethoxymethane, formal
Methylalkohol *m* methanol, methyl alcohol, (obs) carbinol, wood alcohol, **denaturierter** ~ methylated spirit
methylalkoholisch methanolic, methyl alcoholic
Methylamin *n* methylamine, aminomethane
Methylaminhydrochlorid *n* methylamine hydrochloride
Methylanilin *n* methylaniline
Methylanthrachinon *n* methylanthraquinone
Methylanthranilat *n* methyl anthranilate
Methylarsonsäure *f* methylarsonic acid, methanearsonic acid
Methylat *n* methylate, methoxide
Methylbenzanthracen *n* methyl benzanthracene
Methylbenzoat *n* methyl benzoate
Methylbenzoylaceton *n* methyl benzoylacetone
Methylbenzphenanthren *n* methyl benzophenanthrene
Methylbenzylchlorid *n* methyl benzyl chloride
Methylbenzylglyoxim *n* methyl benzylglyoxime
Methylbenzylketon *n* methyl benzyl ketone
Methylbutan *n* methyl butane
Methylbutanal *n* methyl butanal
Methylbutanol *n* methyl butanol
Methylbuten *n* methyl butene
Methylbutenolid *n* methyl butenolide
Methylbutyläther *m* methyl butyl ether
Methylbutylcarbinol *n* methyl butyl carbinol
Methylbutylessigsäure *f* methyl butylacetic acid
Methylbutylketon *n* methyl butyl ketone
Methylcarbazol *n* methyl carbazole
Methylcarbitol *n* methyl carbitol
Methylcarbylamin *n* methyl carbylamine, methyl isocyanide
Methylcellosolve *f* methyl cellosolve
Methylcellulose *f* methyl cellulose, tylose

Methylchinizarin *n* methyl quinizarin
Methylchinolin *n* methyl quinoline
Methylchloracetat *n* methyl chloroacetate
Methylchlorid *n* methyl chloride, chloromethane
Methylchlorsilan *n* methyl chlorosilane
Methylcholanthren *n* methyl cholanthrene
Methylcinnamat *n* methyl cinnamate
Methylconiin *n* methyl coniine
Methylcrotonsäure *f* methyl crotonic acid
Methylcyanacetat *n* methyl cyanoacetate
Methylcyanid *n* methyl cyanide, acetonitrile, cyanomethane, ethanenitrile
Methylcyclohexan *n* methyl cyclohexane
Methylcyclohexanol *n* methyl cyclohexanol
Methylcytosin *n* methyl cytosine
Methyldecalon *n* methyl decalone
Methyldiäthylmethan *n* methyl diethylmethane
Methyldiarylketon *n* methyl diaryl ketone
Methylen *n* methylene, methene
Methylenaminoacetonitril *n* methylene aminoacetonitrile
Methylenblau *n* methylene blue
Methylenchlorid *n* methylene chloride
Methylendigallussäure *f* methylene digallic acid
Methylenditannin *n* methylene ditannin, tannoform
Methylenglykol *n* methylene glycol
Methylengruppe *f* methylene group
Methylenguajakol *n* methylene guaiacol
Methylenimin *n* methylenimine, azomethine, methylene imine
Methylenjodid *n* methylene iodide, diiodomethane, methylene diiodide
Methylensulfat *n* methylene sulfate
Methylenviolett *n* methylene violet
Methylenzuckersäure *f* methylene saccharic acid
Methyleosin *n* methyl eosin
Methylester *m* methylester
Methylformanilid *n* methyl formanilide
Methylfructopyranosid *n* methyl fructopyranoside
Methylgentiobiose *f* methyl gentiobiose
Methylglucamin *n* methyl glucamine
Methylglucosid *n* methyl glucoside
Methylglykol *n* methyl glycol
Methylglykosid *n* methyl glycoside
Methylglyoxal *n* methyl glyoxal
Methylgruppe *f* methyl group
Methylgruppenakzeptor *m* methyl group acceptor
Methylgruppendonor *m* methyl group donor
Methylgruppenübertragung *f* methyl group transfer
Methylguanidin *n* methyl guanidine
Methylhämin *n* methyl hemin
Methylhalogenid *n* methyl halide
Methylheptenol *n* methyl heptenol
Methylheptylketon *n* methyl heptyl ketone

Methylhexalin *n* methyl cyclohexanol, methyl hexalin
Methylhexan *n* methyl hexane
Methylhydrid *n* methane, methyl hydride
Methylhydrochinon *n* methyl hydroquinone
Methylhydroxyacetophenon *n* methyl hydroxyacetophenone
methylierbar capable of being methylated
methylieren to methylate
methyliert methylated
Methylierung *f* methylation, **erschöpfende** ~ exhaustive methylation
Methylisobutyläther *m* methyl isobutyl ether
Methylisobutylketon *n* methyl isobutyl ketone
Methylisochinolin *n* methyl isoquinoline
Methylisopropylbenzol *n* methyl isopropylbenzene
Methylisopropylketon *n* methyl isopropyl ketone
Methylisopropylnaphthalin *n* methyl isopropylnaphthalene
Methylisopropylphenanthren *n* methyl isopropylphenanthrene
Methylisopropylphenol *n* methyl isopropylphenol
Methyljodid *n* methyl iodide, iodomethane
Methylkautschuk *m* methyl caoutchouc, methyl rubber
Methylketol *n* methyl ketol
Methylketolgelb *n* methyl ketol yellow
Methylmalonat *n* methyl malonate, dimethyl malonate
Methylmalonsäure *f* methyl malonic acid
Methylmannopyranosid *n* methyl mannopyranoside
Methylmethacrylat *n* methyl methacrylate
Methylmethoxybenzochinon *n* methyl methoxy[l]-benzoquinone
Methylmorphimethin *n* methyl morphimethine
Methylnaphthalin *n* methyl naphthalene
Methylnaphthalinsulfonsäure *f* methyl naphthalene sulfonic acid
Methylnaphthochinon *n* methyl naphthoquinone
Methylnaphthoylpropionsäure *f* methyl naphthoylpropionic acid
Methylnaphthylketon *n* methyl naphthyl ketone
Methylnitrobenzoat *n* methyl nitrobenzoate
Methylnonylketon *n* methyl nonyl ketone
Methyloctansäure *f* methyl octanoic acid
Methylolmelamin *n* methylol melamine
Methylolmethylenharnstoff *m* methylol methylene urea
Methylorange *n* methyl orange
Methylostärke *f* methyl starch
Methylotannin *n* methyl tannin
Methyloxalat *n* methyl oxalate, dimethyl oxalate
Methyloxidhydrat *n* (obs) methyl alcohol
Methyloxin *n* methyl oxine
Methylpentan *n* methyl pentane
Methylpentanal *n* methyl pentanal

Methylpentanol *n* methyl pentanol
Methylpenten *n* methyl pentene
Methylpentynol *n* methylpentynol
Methylphenäthylaminsulfat *n* methyl phenethylamine sulfate
Methylphenanthren *n* methyl phenanthrene
Methylphenanthrenchinon *n* methyl phenanthrene quinone
Methylphenidat *n* methylphenidate
Methylphenobarbital *n* methylphenobarbital
Methylphenylhydrazin *n* methyl phenyl hydrazine
Methylphenylketon *n* methyl phenyl ketone
Methylphenylnitrosamin *n* methyl phenyl nitrosamine
Methylphenylsilikon *n* methyl phenyl silicone
Methylphosphin *n* methyl phosphine
Methylphosphonsäure *f* methyl phosphonic acid, methane phosphonic acid
Methylphytylnaphthochinon *n* methyl phytylnaphthoquinone
Methylpikramid *n* methyl picramide
Methylpropylcarbinol *n* methyl propylcarbinol
Methylpropylketon *n* methyl propyl ketone
Methylpsychotrin *n* methylpsychotrine
Methylpyrazolcarbonsäure *f* methyl pyrazolecarboxylic acid
Methylpyridin *n* methyl pyridine
Methylpyridiniumhydroxid *n* methyl pyridinium hydroxide
Methylpyridon *n* methyl pyridone
Methylpyrrol *n* methyl pyrrole
Methylradikal *n* methyl radical
Methylreductinsäure *f* methyl reductinic acid
Methylredukton *n* methylreductone
Methylrhodanid *n* methyl thiocyanate
Methylrhodanür *n* (obs) methyl thiocyanate
Methylrosanilin *n* methylrosaniline
Methylrot *n* methyl red
Methylsalicylat *n* methyl salicylate
Methylsalicylsäure *f* methyl salicylic acid
Methylsenföl *n* methyl mustard oil, methyl isothiocyanate
Methylsilan *n* methyl silane
Methylsilikat *n* methyl silicate
Methylsilikon *n* methyl silicone
Methylsilikonharz *n* methyl silicone resin
Methylsilikonkautschuk *m* methyl silicone rubber
Methylsilikonöl *n* methyl silicone oil
Methylstearinsäure *f* methyl stearic acid
Methylsulfat *n* dimethyl sulfate, methyl sulfate
Methylsulfhydrat *n* methyl mercaptan, mercaptomethane, methane thiol, methyl hydrogen sulfide
Methylsulfid *n* dimethyl sulfide, methyl sulfide
Methylsulfinsäure *f* methanesulfinic acid
Methylsulfon *n* dimethyl sulfone, methyl sulfone
Methylsulfonsäure *f* methanesulfonic acid

Methylsulfoxid *n* dimethyl sulfoxide, methyl sulfoxide
Methyltartrat *n* methyl tartrate
Methyltetralon *n* methyl tetralone
Methylthiazolcarbonsäure *f* methyl thiazole carboxylic acid
Methylthienylketon *n* methyl thienyl ketone
Methylthiouracil *n* methylthiouracil
Methylthymolblau *n* methylthymol blue
Methyltoluolsulfonat *n* methyl toluene sulfonate
Methyltrichlorsilan *n* methyl trichlorosilane
Methyltrihydroxyanthrachinon *n* methyl trihydroxyanthraquinone
Methyltrinitrophenylnitramin *n* methyl trinitrophenylnitramine
Methylumbelliferon *n* methyl umbelliferon
Methylundecylketon *n* methyl undecyl ketone
Methyluracil *n* methyl uracil
Methylurethan *n* methyl urethane, methyl carbamate, urethylan
Methylverbindung *f* methyl compound
Methylvinylketon *n* methyl vinyl ketone
Methylviolett *n* methyl violet
Methylviologen *n* methylviologen
Methylwasserstoff *m* methane, methyl hydride
Methylxanthogensäure *f* methyl xanthogenic acid
Methylzahl *f* methyl number
Methylzellulose *f* methyl cellulose, tylose
Methymycin *n* methymycin
Methysticin *n* methysticine, kavain
Methysticol *n* methysticol
Meticillin *n* (Antibiotikum) meticillin
Metiram *n* (Fungizid) metiram
Metisazon *n* metisazone
Metixen *n* (Pharm) metixene
Metobromuron *n* (Herbizid) metobromurone
Metochinon *n* metoquinone
Metofenazat *n* (Med) metofenazate
Metol *n* metol, pictol, p-methylaminophenol, rhodol
Metoleinsäure *f* metoleic acid
Metonal *n* metonal
metonisch (Astr) metonic
Metoxazin *n* metoxazine
Metozin *n* metozine, antipyrine
Metrik *f* metrics, theory of meter
Metriol *n* metriol
metrisch metric
Metronom *n* metronom
Mevaldsäure *f* mevaldic acid
Mevalonolacton *n* mevalonic lactone
Mevalonsäure *f* mevalonic acid
Meyerhofferit *m* (Min) meyerhofferite, native calcium borate
Meymacit *m* (Min) meymacite
Mezcalin *n* mescaline, mezcaline
MF Abk. für Melamin-Formaldehyd-Kunstharz
MFK Abk. für metallfaser-verstärkter Kunststoff

MG *Abk. für Molekulargewicht*
Miargyrit *m* (Min) miargyrite, silver antimony glance
Miascit *m* miascite
Miasma *n* miasma, aerial poison, effluvium
Miazin *n* miazine
Mica *m* (Min) mica, glimmer
Micelle *f* micell[e], (pl micellae)
Micellenbildung *f* micelle formation
Michaelit *m* michaelite
Michaelreaktion *f* Michael's reaction
Michaelsonit *m* (Min) michaelsonite
Michlers Base *f* Michler's base
Michlers Keton *n* Michler ketone, p-diaminobenzophenone
Micranthin *n* micranthine
Microbromit *m* microbromite
Microphyllinsäure *f* microphyllic acid
Miedziankit *n* miedziankite
Miersit *m* (Min) miersite
Miesit *m* miesite
Mietenthermometer *n* silo thermometer
Mignonfassung *f* (Elektr) intermediate socket
Migränestift *m* (Pharm) headache pencil
MIG-Rührer *m* multi-stage impulse ribbon blender
MIG-Schweißen *n* (Abk. für Metall-Inertgas-Schweißen) MIG welding
Mikadogelb *n* mikado yellow
Mikanit *n* micanite
Mikanitleinwand *f* micanite linen
Mikanitpapier *n* micanite paper
Mikroamperemeter *n* micro-amperemeter
Mikroanalyse *f* (Anal) microanalysis
Mikroanalysenwaage *f* microanalysis balance
Mikroapparat *m* microapparatus
Mikroaufnahme *f* photomicrograph
Mikroautoradiographie *f* microautoradiography
Mikrobar *n* microbar
Mikrobarograph *m* microbarograph
Mikrobe *f* (Biol) microbe, microorganism
Mikrobenzüchtung *f* microbe culture
Mikrobild *n* (Phot) microphotograph
Mikrobin *n* microbin
Mikrobiologe *m* microbiologist
Mikrobiologie *f* microbiology
mikrobiologisch microbiological
mikrobisch microbial, (durch Mikroben verursacht) microbic
Mikrobrenner *m* microburner
Mikrobromit *m* microbromite
Mikrobürette *f* microburet[te]
Mikrochemie *f* microchemistry
mikrochemisch microchemical
Mikrochronometer *n* microchronometer
Mikrocidin *n* microcidin
Mikrocurie *n* microcurie
Mikrodehnung *f* microstrain
Mikrodensitometer *n* microdensitometer

Mikroeinstellung *f* microset
Mikroelementaranalyse *f* elementary microanalysis
Mikroextraktor *m* microextractor
Mikrofarad *n* (Elektr) microfarad
Mikrofarbstoff *m* microscopical stain
Mikrofilm *m* (Phot) microfilm
Mikrofilter *n* microfilter
Mikrogamet *m* (Biol) microgamete
Mikrogametozyt *m* microgametocyte
Mikrogefüge *n* microstructure
Mikrogramm *n* microgram (Am. E.), microgramme (Br. E.)
Mikrographie *f* micrography
Mikrohm *n* (Elektr) microhm
Mikrohochdruckgerät *n* micro high-pressure equipment
Mikrokern *m* micronucleus
Mikrokinetik *f* microkinetics
Mikroklin *m* (Min) microcline, amazonite
Mikroklinperthit *m* (Min) microclinperthite
mikrokosmisch microcosmic
Mikrokosmos *m* microcosm
Mikrokristall *m* microcrystal
mikrokristallin microcrystalline
Mikroliter *m* microliter, lambda
Mikrolith *m* (Min) microlite
Mikromerograph *m* micromerograph
Mikrometer *n* micrometer
Mikrometereinstellung *f* micrometer adjustment
Mikrometerschraube *f* micrometer screw
Mikromethode *f* micro-method
Mikrometrie *f* micrometry
mikrometrisch micrometric
Mikromikron *n* micromicron
Mikromol *n* micromole
Mikromolekül *n* micromolecule
Mikron *n* micron
Mikronährstoff *m* micronutrient
Mikronisieren *n* micronizing
Mikronukleus *m* micronucleus
Mikroorganismus *m* micro-organism
Mikroparasit *m* microparasite
Mikroperthit *m* (Min) microperthite
Mikrophage *m* microphage
Mikrophon *n* microphone
Mikrophoto *n* microphotograph, photomicrograph
Mikrophotographie *f* microphotograph, microphotography
mikrophotographisch microphotographic
Mikrophotometer *n* microphotometer
Mikropipette *f* micropipet[te]
Mikropolieren *n* micropolishing
Mikropore *f* micropore
Mikroprüfgeräte *n pl* microtesting equipment
Mikrorheologie *f* microrheology
Mikroriß *m* microcrack
Mikroröhre *f* microtube

Mikroröntgenbild *n* microradiograph
Mikroröntgenstrahl *m* microbeam of X-rays
Mikroschalter *m* micro-switch
Mikrosekunde *f* microsecond
Mikrosiebung *f* microsieving
Mikroskala *f* microscale
Mikroskop *n* microscope, ~ **zur Untersuchung lebenden Materials** biomicroscope, **binokulares** ~ binocular microscope, **mehrlinsiges** ~ compound microscope
Mikroskopbild *n* microscope image
Mikroskopgestell *n* microscope stand
Mikroskopie *f* microscopy
mikroskopieren to microscope, to examine microscopically
Mikroskopieren *n* microscopical testing
mikroskopisch microscopic
Mikrosom *n* microsome
Mikrosommit *m* (Min) microsommite
Mikrospektroskop *n* microspectroscope
Mikrostativ *n* microstand
Mikrostruktur *f* microstructure
Mikrotin *m* microtine
Mikrotom *n* microtome
Mikrotomie *f* microtomy
Mikrotommesser *n* microtome blade
Mikrotron *n* (Atom) microtron
Mikroturbulenz *f* microturbulence
Mikroverkapselung *f* microencapsulation
Mikrovolt *n* microvolt
mikrovolumetrisch microvolumetric
Mikrowaage *f* microbalance
Mikrowachs *n* microcrystalline wax, microwax
Mikrowelle *f* microwave, ultra-short wave
Mikrowellenausbreitung *f* microwave propagation
Mikrowellenbereich *m* microwave region
Mikrowellendielektrometer *n* microwave dielectrometer
Mikrowellendurchschlag *m* microwave breakdown
Mikrowellenentladung *f* microwave discharge
Mikrowelleninterferometrie *f* microwave interferometry
Mikrowellenmessung *f* microwave measurement
Mikrowellenresonanz *f* microwave resonance
Mikrowellenspektroskopie *f* microwave spectroscopy
Mikrowellenspektrum *n* microwave spectrum
Mikrowellentechnik *f* microwave technique
Milarit *m* (Min) milarite
Milbe *f* (Zool) mite, acarid
Milbenbekämpfungsmittel *n* (Pharm) acaricide
Milbeninfektion *f* mite infection
Milch *f* milk, (Emulsion) emulsion, **abgerahmte** ~ skimmed milk, **die** ~ **entrahmen** to skim the milk, **evaporierte** ~ evaporated milk, **pasteurisierte** ~ pasteurized milk, **pflanzliche** ~ vegetable milk, **sterilisierte** ~ sterilized milk
Milchachat *m* (Min) milk-white agate
milchartig milky, lacteal
Milchdrüse *f* mammary gland
Milcheiweiß *n* milk protein
milchfarbig milk-colored
Milchfettgehalt *m* fat content of milk
Milchgärung *f* milk fermentation
Milchglas *n* opal glass, frosted glass, glass porcelain, milk glass
milchhaltig lactiferous
milchig milky, lacteal
Milchindikator *m* milk indicator
Milchjaspis *m* (Min) galactite
Milchkondensieranlage *f* milk condensing plant
Milchkristall *m* milk crystal
Milchkühlanlage *f* milk cooling plant
Milchkühler *m* milk cooler
Milchlüfter *m* milk aerator
Milchmesser *m* lactometer, galactometer
Milchopal *m* (Min) milk-white opal
Milchpilz *m* milk agaric
Milchporzellan *n* glass porcelain
Milchpulver *n* milk powder
Milchquarz *m* (Min) milky quartz
Milchsäure *f* lactic acid, 2-hydroxypropanoic acid
Milchsäureäthylester *m* ethyl lactate
Milchsäurealdehyd *m* lactaldehyde, lactic aldehyde
Milchsäureanlage *f* lactic acid plant
Milchsäurebazillus *m* (Bakt) lactobacillus (pl lactobacilli)
Milchsäurebildung *f* formation of lactic acid
Milchsäuredehydrase *f* (Biochem) lactic dehydrogenase
Milchsäureester *m* ester of lactic acid, lactate
Milchsäureferment *n* (Biochem) lactic ferment
Milchsäuregärung *f* lactic fermentation
Milchsäurenitril *n* lactic nitrile, acetaldehyde cyanohydrin, hydroxypropanenitrile, lactonitrile
Milchsaft *m* (Bot) chyle, latex, milky juice
Milchsaphir *m* (Min) white cloudy sapphire
milchsauer lactate
Milchserum *n* milk serum
Milchstein *m* (Min) galactite
Milchstraße *f* (Astr) galaxy
Milchstraßensystem *n* (Astr) galactic system, galaxy
Milchwein *m* kumiss
Milchweiß *n* milk white
Milchzucker *m* milk sugar, lactobiose, lactose
Milchzuckeranlage *f* milk sugar plant
mild mild, mellow
mildern to alleviate, to mitigate, to moderate, to soften
mildernd (Med) lenitive

Milderung *f* damping, mitigation, moderation, tempering
Milderungsmittel *n* (Med) demulcent, lenitive, mitigant
Miliolitenkalk *m* (Min) miliolitic chalk
Millerit *m* (Min) millerite, nickel pyrites
Milliampere *n* (Elektr) milliampere
Milliarde *f* billion (Am. E.), milliard (Br. E.), thousand millions
Millibar *n* millibar
Millicurie *n* millicurie
Milligramm *n* milligram (Am. E.), milligramme (Br. E.)
Milliliter *m n* milliliter (Am. E.), millilitre (Br. E.)
Millimeter *m n* millimeter
Millimetergewinde *n* metric thread
Millimeterpapier *n* millimeter graph paper, plotting paper
Millimikron *n* millimicron, micromillimeter
Millimol *n* (Chem) millimole
Millionstel *n* millionth [part]
Milliröntgen *n* milliroentgen
Millisekunde *f* milli-second
Millivolt *n* (Elektr) millivolt
Millons-Reagens *n* Millon's reagent
Miloriblau *n* milori blue
Miloschin *m* miloschin
Milz *f* (Anat) milt, spleen
Mimetesit *m* (Min) mimetesite, mimetite
Mimetit *m* (Min) mimetesite, mimetite
Mimosaöl *n* mimosa oil
Mimosengummi *n* gum arabic
Mimosenrinde *f* mimosa bark
Mimosin *n* mimosine
Minalin *n* minaline
Minasit *m* minasite
Minasragrit *m* (Min) minasragrite
Minderdruck *m* diminished pressure, reduced pressure
Mindergewicht *n* deficiency in weight, underweight
Minderheit *f* minority
mindern to decrease
Minderung *f* decrease, diminution
minderwertig inferior, low-grade
Minderwertigkeit *f* inferiority, inferior quality
Mindestanforderung *f* minimum requirement, specification
Mindestarbeit *f* minimum work
Mindestdruck *m* minimum pressure
Mindestdurchbiegung *f* minimum deflection
Mindestgeschwindigkeit *f* minimum speed
Mindestleistung *f* minimum efficiency, (Ertrag) minimum output
Mindestüberspannung *f* (Elektr) minimum overvoltage
Mine *f* (Bergb) mine, (Bleistift) lead, (Kugelschreiber) refill

Mineral *n* mineral
Mineralalkali *n* mineral alkali
Mineralanalyse *f* mineral analysis
Mineralbad *n* mineral bath
Mineralbeize *f* mineral mordant
Mineralblau *n* mineral blue
Mineralchemie *f* mineral chemistry
Mineralfarbe *f* mineral color, mineral pigment
Mineralfarbstoff *m* mineral coloring matter
Mineralfaser *f* mineral fiber
Mineralfettwachs *n* mineral tallow, rock fat
Mineralfeuerrot molybdate fiery red
Mineralgang *m* (Bergb) mineral vein
Mineralgerbung *f* mineral tanning
Mineralharz *n* gedanite
Mineraliensammlung *f* collection of minerals
Mineralisation *f* mineralization
mineralisch mineral
Mineralkermes *m* (Min) kermesite, kermes mineral
Mineralkohle *f* mineral carbon
Mineralmasse *f* (Straßenbau) mineral matter
Mineralmohr *m* (Min) black mercury sulfide, ethiops mineral
Mineralöl *n* mineral oil
Mineralölraffinat *n* refined mineral oil
Mineralogie *f* mineralogy
mineralogisch mineralogical
Mineralpech *n* mineral pitch, asphalt
Mineralquelle *f* mineral spring
Mineralrot *n* (Min) cinnabar
Mineralsäure *f* mineral acid
Mineralschmieröl *n* mineral lubricating oil
Mineralschwarz *n* mineral black, carboniferous clay slate, ground graphite
Mineralstoffgehalt *m* mineral content[s]
Mineralstoffwechsel *m* metabolism of minerals
Mineraltalg *m* mineral tallow, hatchettite, ozocerite
Mineralturpeth *m* (Min) turpeth mineral
Mineralwachs *n* mineral wax, ozocerite
Mineralwasser *n* mineral water
Mineralweiß *n* mineral white, permanent white
Mineralwolle *f* mineral or slag wool, rock wool
Minette *f* (Min) minette, minet ore, oolitic iron ore
Minguetit *m* minguetite
Minimaldruck *m* lowest pressure
Minimalgehalt *m* minimum content
Minimalkurve *f* minimal curve
Minimalpolynom *n* minimal polynomial
Minimalspannung *f* minimum voltage
Minimalstrom *m* minimum current
Minimalthermometer *n* minimum thermometer
Minimalwert *m* minimum value
Minimaxkonzept *n* **b. Versuchsplanung** minimax concept (in factorial design)
Minimum *n* minimum

Minimumstrahldefinition – Mischpumpe

Minimumstrahldefinition *f* definition of minimum beam
Minimumstrahlkennzeichnung *f* marking of minimum beam
Minimum-Thermometer *n* minimum thermometer
Minioluteinsäure *f* minioluteic acid
Minium *n* (Min) minium, red lead
Minivalenz *f* minimum valence, minivalence
Minorante *f* minorant
Minoritätsträger *m* minority carrier
Minuselektrizität *f* negative electricity
Minuselektrode *f* negative electrode, negative plate
Minusplatte *f* (Akku) negative plate
Minuspol *m* negative pole
Minustemperatur *f* sub-freezing temperature
Minuszeichen *n* (Math) minus sign, negative sign
Minute *f* minute
minutenlang for several minutes
Minutenzeiger *m* minute hand
Minze *f* (Bot) mint
Miotikum *n* (Pharm) miotic
Mirabilit *m* (Min) mirabilite, native sodium sulfate
Miracil *n* miracil
Mirbanessenz *f* mirbane essence, mirbane oil, nitrobenzene
Mirbanöl *n* mirbane essence, mirbane oil, nitrobenzene
Miren *n* mirene
Miscella *f* miscella
Mischanlage *f* mixing installation
Mischapparat *m* mixer, mixing apparatus
mischbar miscible, mixable
Mischbarkeit *f* miscibility
Mischbatterie *f* single control shower mixer, [water] mixer, ~ **für Bad** mixing unit for bathwater
Mischbauweise *f* composite construction
Mischbehälter *m* mixing vessel
Mischbettaustauscher *m* mixed-bed exchanger
Mischbleichen *n* mixed bleaching
Mischbottich *m* mixing trough
Mischbrenner *m* circular burner
Mischdünger *m* compost, compound fertilizer, mixed fertilizer
Mischdüse *f* combining nozzle, mixing jet
Mischeisen *n* mixed iron
Mischelemente *n pl* mixed elements
mischen to mix, to blend, to compound, to mingle
Mischen *n* blending, mixing
Mischer *m* blender, (Metall) pig iron mixer, **Druck-** ~ pressurized mixer, **einfacher** ~ simple mixer, **Fließbett-** ~ fluidized bed mixer, **Gegenstrom-Zwangs-** ~ forced circulation mixer, **Hosen-** ~ twin-shell mixer, Vee mixer, **Innen-** ~ intensive mixer, **Kreisel-** ~ impeller or gyro mixer, **Pfannen-** ~ pan mixer, **Planeten-** ~ planetary mixer, **Scherscheiben-** ~ shearing disc mixer, **Schleuder-** ~ centrifugal mixer, **Schnecken-** ~ vertical screw mixer (for silos), **schrägstehender** ~ tilted mixer, **Schwerkraft-** ~ gravity mixer, **statischer** ~ static mixer, **Strahl-** ~ jet mixer, **Umlaufschnecken-** ~ vertical screw mixer, **Wankscheiben-** ~ swash plate mixer, **Zwangs-** ~ ribbon blender mixer
Mischer-Abscheider *m* mixer-settler
Mischerbühne *f* mixer platform
Mischergebäude *n* mixing shed
Mischerhaus *n* mixing shed
Mischerz *n* mixed ore
Mischfarbe *f* combination color, mixed color
Mischfilm *m* mobile film
Mischfilmkolonne *f* mobile film column
Mischfilmtrommel *f* mobile film drum
Mischflußschnellrührer *m* mixed flow impeller
Mischfutter *n* composite feed
Mischgarn *n* (Text) blended yarn
Mischgas *n* mixed gas, semiwater gas
Mischgefäß *n* mixing vessel
Mischgespinst *n* blended yarns
Mischgewebe *n* (Text) blended fabrics, union fabric
Mischhalde *f* mixing dump
Mischholländer *m* mixing beater
Mischkammer *f* mixing chamber (Gas) surge chamber
Mischkatalysator *m* mixed catalyst
Mischkessel *m* mixing vessel
Mischkneter *m* kneader mixer
Mischkoeffizient *m* eddy diffusion coefficient
Mischkohle *f* run of mine coal
Mischkondensation *f* co-condensation, direct contact condensation
Mischkristall *m* (Krist) mixed crystal
Mischkristallbildung *f* mixed crystal formation
Mischkristallegierung *f* mixed-crystal alloy, solid-solution alloy
Mischkristallreihe *f* series of mixed crystals
Mischkugel *f* mixing bell
Mischkultur *f* mixed culture
Mischlack *m* mixing varnish
Mischluftrührer *m* air-lift agitator
Mischmaschine *f* mixing machine
Mischmetall *n* mixed metal, alloy, metal composition, (Cer-Mischmetall) misch metal
Mischmühle *f* mixing mill
Mischplatte *f* mixing plate
Mischpolymer *n* copolymer, mixed polymer
Mischpolymerisat *n* copolymer
Mischpolymerisation *f* copolymerization, heteropolymerization
Mischpumpe *f* proportioning pump

Mischraum *m* compounding room, mixing chamber
Mischregler *m* mixing control
Mischreibung *f* mixed lubrication
Mischröhre *f* (Elektr) mixing valve, converter tube
Mischrohr *n* mixing tube
Mischsäure *f* (obs) mixed acid, nitrosulfuric acid
Mischsalz *n* mixed salt, double salt
Mischschmelzpunkt *m* mixed melting point
Mischschnecke *f* mixing screw
Mischspiritus *m* mixed spirit
Mischtrichter *m* mixing cone
Mischtrommel *f* rotary mixer, tumbling mixer
Misch- und Granuliermaschine *f* mixing and granulating machine
Mischung *f* mixture, blend, compound, (Metall) alloy, (Pap) mixtion, ~ **für Laufstreifenunterplatte** underbase stock, ~ **mit konstantem Siedepunkt** (Therm) constant boiling mixture, **äquivalente** ~ equivalent mixture, **angereicherte** ~ master batch, **eutektische** ~ eutectic mixture
Mischungsbestandteil *m* ingredient of the mixture
Mischungsenthalpie *f* enthalpy of mixing
Mischungsentropie *f* entropy of mixing, (Phys) entropy of mixing
Mischungsentwickler *m* compounder
Mischungserscheinung *f* phenomenon of mixture
mischungsfähig miscible, mixable
Mischungsherstellung *f* compounding procedure
Mischungslücke *f* miscibility gap, (in flüssig-flüssig-Systemen) miscibility gap
Mischungsprobe *f* sample of mixture
Mischungsregel *f* law of mixtures, rule of mixtures
Mischungsverhältnis *n* composition of a mixture, mixing proportion, ratio of the components of a mixture
Mischungsvolumen *n* volumetric proportions in mixtures
Mischungsvorgang *m* mixing process
Mischungsvorwärmer *m* heater in contact with exhaust
Mischungsvorwärmung *f* heating in contact with exhaust
Mischungswärme *f* heat of mixing
Mischvakzine *f* (Med) mixed vaccine
Mischventil *n* mixing valve
Mischvermögen *n* mixing power
Mischwald *m* (Bot) mixed forest
Mischwalze *f* mixing rolls, mixing mill
Mischzement *m* mixed cement
Mißbrauch *m* abuse, misuse
Mißdeutung *f* misinterpretation
Mißerfolg *m* failure
Mißfärbung *f* discoloration
mißfarbig discolored

Mißverhältnis *n* disproportion
Mist *m* dung
Mistbeet *n* hotbed
Mistpulver *n* poudrette, powdered manure
MIT Abk. f. Massachusetts Institute of Technology
Mitabtrennung *f* coseparation
Mitarbeiter *m* co-worker, collaborator, colleague, contributor, **freier** ~ free lance
Mitbestimmung *f* codetermination
mitbewegen to convect
Mitbewegung *f* convection, dragging
Mitchellit *m* mitchellite
mitfällen to coprecipitate
Mitfällung *f* coprecipitation
mitführen to carry along
Mitführung *f* entrainment
Mitglied *n* member
Mitgliedschaft *f* membership
Mithilfe *f* assistance, cooperation
Mithörer *m* fellow listener, monitor
Mithridat *n* mithridate
Mitisgrün *n* mitis green, Paris green
Mitizid *n* miticide
Mitläufer *m* backing, backing material, liner
Mitnehmerbolzen *m* carrier bolt
Mitnehmerglied *n* catch member
Mitnehmerscheibe *f* entrainer disk
Mitnehmerstift *m* coupling through-pin, driving pin, follower pin
Mitochondrien *pl* mitochondria
Mitochondriendurchlässigkeit *f* mitochondria permeability
Mitochondrienmembran *f* mitochondria membrane
Mitochondrienverteilung *f* mitochondria distribution
Mitomycin *n* mitomycin
Mitose *f* mitosis
Mitosegift *n* mitotic poison
Mitragynin *n* mitragyne, mitragynine
Mitraphyllol *n* mitraphyllol
Mitraversin *n* mitraversine
mitreißen to carry along, (Dest) to carry over, (Niederschlag) to carry down
Mitreißen *n* (Rektifikation) entrainment
Mitreißgrenze *f* entrainment point
Mitschwingungsgezeit *f* co-oscillational tide
Mitte *f* middle, (Mittelpunkt) center, (Mittelweg) medium
Mittel *n* (Arznei) remedy, (Medium) agent, medium, (Mittelwert) average, mean [value], ~ **gegen Schimmelbildung** fungicide, **äußerlich anzuwendendes** ~ (Pharm) remedy for external application, **antiseptisches** ~ (Med) antiseptic [agent], **arithmetisches** ~ (Math) arithmetical mean, **beruhigendes** ~ (Pharm) sedative, **blutdrucksenkendes** ~ (Pharm) hypotensor, **blutstillendes** ~ (Pharm)

h[a]emostatic, styptic, **das ~ nehmen** (Math) to [take the] average, **gefäßerweiterndes ~** (Pharm) vasodilator, **geometrisches ~** (Math) geometrical mean, **harntreibendes ~** (Pharm) diuretic, **keimtötendes ~** antiseptic, germicide, preservative, **oberflächenaktives ~** surface-active agent, surfactant
Mittelabgriff m center tap
Mittelachse f center axis, center line
Mittelbenzin n medium benzine, medium heavy gasoline (Am. E.), medium heavy petrol (Br. E.)
Mittelbinder m medium setting cement
Mittelblech n medium plate
Mittelelektrode f intermediate electrode
Mittelfarbe f intermediate color, middle-color, middle-tint, secondary color
mittelfeinkörnig medium grained
Mittelfrequenz f medium frequency
Mittelhals m center neck
mittelhart medium hard
Mittelkasten m middle box
Mittellage f central position
Mittellauf m (Dest) middle fraction, middle runnings
Mittelleiter m (Elektr) middle conductor, neutral wire, third wire
Mittellenker m transverse link
Mittellinie f axis
mittelmäßig mediocre
Mittelmäßigkeit f mediocrity
Mittelöl n medium-heavy oil, middle oil
Mittelpunkt m center, central point, middle, midpoint
Mittelrille f center groove
Mittelrippe f center rib
Mittelsenkrechte f median perpendicular, mid vertical
Mittelspannung f average tension, (Elektr) average voltage
Mittelspannungsanteil m stress ratio
mittelständig central
Mittelstellung f center position, intermediate position
Mittelstück n central piece, intermediate, middle piece
Mittelung f averaging
Mittelwalze f middle roll
Mittelwasser n mean water level
mittelweich medium soft
Mittelwellen f pl medium waves
Mittelwellenbereich m medium wave range
Mittelwert m average, mean value, middle value, **arithmetischer ~** arithmetic mean, **geometrischer ~** geometric mean, **harmonischer ~** harmonic mean, **quadratischer ~** root mean square
Mittelwertsatz m law of averages, mean value theorem

Mittelzug m medium drawing
mittenrichtig centered
mittensymmetrisch centrosymmetrical
mittig centric
Mixit m (Min) mixite
Mixtur f (Chem) mixture
Mizelle f micell[e]
Mizzonit m (Min) mizzonite
MKS-System n (Abk. für Meter-Kilogramm-Sekunde) MKS system
MMA Abk. f. Methylmethacrylat
Mobilisierung f mobilization
Mochastein m (Min) mocha stone, moss agate
Mochylalkohol m mochyl alcohol
Modacrylfasern f pl modacrylic fibers
Modalfasern f pl modal fibers
Modeartikel m fancy article
Modell n model, master model, matrix, mold, pattern
Modellack m pattern varnish, pattern blacking
Modellanlage f pilot plant
Modellbrett n stamping board
Modellgips m mo[u]lding plaster
Modellhälfte f pattern half
Modellhammer m molder's hammer
Modellheber m lifting handle
modellieren to model; to shape
Modellplatte f match plate, molding board, pattern plate, plate with patterns
Modellsand m molding sand
Modelltheorie f model theory
Modellträger m tooth block holder
Modellübertragung f scaling up
Modellumriß m outline of pattern
Moder m decay, mildew, mold, putridity, rottenness
Moderduft m moldy smell, mustiness, smell of decay
Modererde f humus, mold, rotten earth
Modererz n (Min) bog iron ore
Moderfäulepilz m soft rot
moderieren (Atom) to moderate
Moderierofen m base burner, stove
Moderierung f (Atom) deceleration, moderation, slowing down
moderig decayed, moldy, musty, putrid
modern to decay, to mildew, to molder (Am. E.), to moulder (Br. E.), to rot
modernisieren to modernize
Moderstein m rotten stone
Modertorf m moldy peat, drag turf
Modeton m (Farbe) season shade
Modifikation f modification, **allotrope ~** allotropic change
modifizierbar modifiable
modifizieren to modify
Modifiziermittel n modifier
Modifizierung f modification
modisch fashionable, fancy

Modul *m* modulus, module
Modulargleichung *f* modular equation
Modulation *f* (Elektron) modulation
Modulationsfrequenz *f* modulation frequency
Modulationsregler *m* modulation control
Modulationsstufe *f* modulator stage
Modulatorkristall *m* modulator crystal
Modulfunktion *f* modular function
Möbel *n* furniture
Möbelfolien *f pl* veneer strips
Möbellack *m* furniture varnish, furniture finish
Möbelpolitur *f* furniture polish
Möglichkeit *f* possibility, potentiality
Möhrenfarbstoff *m* car[r]otene, car[r]otin
Möhrenöl *n* carrot oil
Möhrensaft *m* juice of carrots
Möller *m* [blast furnace] burden, charge, mixture of ores and fluxes
Mönch *m* (Metall) die, punch, stamp
Mönchskolben *m* hydraulic plunger, hydraulic ram
Mörser *m* mortar
Mörserkeule *f* (Pharm) pestle, pounder
Mörtel *m* mortar, **hydraulischer** ~ hydraulic mortar, water mortar, **schnellbindender** ~ quickly hardening mortar
Mörtelmischmaschine *f* (Bauw) pan mixer
Mörtelsand *m* mortar sand
Mößbauer Effekt *m* Mössbauer effect
Mogelpackung *f* deceptive package, „bluff" package
Mohair *n* mohair
Mohawkit *m* mohawkite
Mohn *m* (Bot) poppy
Mohnöl *n* poppy oil
Mohnsäure *f* meconic acid, hydroxypyrocomane dicarboxylic acid
Mohnsamen *m* (Bot) poppy seed
mohnsauer meconate
Mohnsirup *m* syrup of poppy capsules
Mohr *m* (Min) black mercury sulfide, ethiops
Mohrrübenfarbstoff *m* carotene, caritol, carotin
Mohrrübenspiritus *m* cárrot spirit
Mohrsches Salz *n* Mohr's salt, ferrous ammonium sulfate
Mohrsche Waage *f* (Mohr-Westphalsche Waage) Mohr's balance
Mohssche Härte *f* Mohs' hardness
Mohssche Härteskala *f* Mohs' scale of hardness
Moissanit *m* (Min) moissanite
Mokayaöl *n* mocaya oil
Mol *n* mole, gram molecule
Molalität *f* molality
molar molar
Molarität *f* molarity
Moldawit *m* (Min) moldawite, water chrysolite
Mole *f* mole
Molekel *f* molecule
Molekeldissymmetrie *f* molecular dissymmetry

Molekelmodell *n* molecular model
Molekül *n* molecule, **aktiviertes** ~ activated molecule, **polares** ~ polar molecule, **zweiatomiges** ~ biatomic molecule
Molekülabbildung *f* (Phys) molecular diagram
Moleküladhäsion *f* molecular adhesion
Molekülaggregat *n* molecular aggregate, molecular cluster
Molekülanordnung *f* molecular arrangement, molecular structure
Molekülasymmetrie *f* molecular asymmetry
Molekülaufbau *m* molecular structure
Molekülbahn *f* molecular orbit
Molekülbau *m* molecular structure
Molekülbildung *f* molecule formation
Moleküldrehung *f* molecular rotation
Molekülemissionskontinuum *n* (Spektr) continuous molecular emission spectrum
Molekülgasion *n* molecular gas ion
Molekülgitter *n* molecular lattice
Molekülkomplex *m* aggregate of molecules, molecular cluster
Molekülmodell *n* molecular model
Molekülorbital *n* molecular orbital
Molekülrotation *f* molecular rotation
Molekülrotationsspektrum *n* molecular rotation spectrum
Molekülrückstoß *m* molecular recoil
Molekülrumpf *m* core of a molecule
Molekülschicht *f* layer of molecules, molecular layer
Molekülschwingung *f* vibration of a molecule
Molekülspektrum *n* molecular spectrum
Molekülumlagerung *f* molecular transformation
Molekülverband *m* molecular structure
Molekülverbindung *f* molecular complex
Molekülzustand *m* molecular state
molekular molecular
Molekularabstoßung *f* molecular repulsion
Molekularanziehung *f* molecular attraction
Molekularbewegung *f* Brownian motion, molecular movement
Molekularbiologie *f* molecular biology
Molekularbrechungsvermögen *n* molecular refraction
Molekulardestillation *f* molecular distillation
Molekulardispersion *f* molecular dispersion
Molekulardrehungsvermögen *n* molecular rotatory power
Molekulardruck *m* molecular pressure
Molekulardurchmesser *m* molecular diameter
Molekularenergie *f* molecular energy
Molekularformel *f* molecular formula
Molekulargewicht *n* molecular weight
Molekulargewichtsbestimmung *f* molecular weight determination
Molekulargewichtsverteilung *f* molecular weight distribution
Molekulargröße *f* molecular magnitude

Molekularität *f* molecularity
Molekularkraft *f* molecular force
Molekularleitfähigkeit *f* molecular conductivity
Molekularmagnetismus *m* molecular magnetism
Molekularrefraktion *f* molecular refraction
Molekularreibung *f* molecular friction
Molekularsieb *n* molecular sieve
Molekularsphäre *f* molecular sphere
Molekularstörung *f* molecular disturbance
Molekularstrahl *m* molecular beam, molecular ray
Molekularstrahlexperiment *n* molecular beam experiment
Molekularstrahlmethode *f* molecular beam method
Molekularstreuung *f* molecular scattering
Molekularströmung *f*, **thermische** ~ free molecule diffusion, Knudsen flow
Molekularstrom *m* molecular electric current
Molekularverzögerung *f* molecular retardation
Molekularvolumen *n* molecular volume
Molekularwärme *f* molecular heat
Molekularwirkung *f* molecular effect
Molekularzertrümmerung *f* molecular fission, fission of molecules
Molekularzustand *m* molecular state
Molenbruch *m* molar fraction
Molgewicht *n* gram-molecular weight
Molgröße *f* molar magnitude
Molke *f* whey
Molkensäure *f* (obs) lactic acid, 2-hydroxypropanoic acid
Molkerei *f* dairy
Molkonzentration *f* molar concentration
Mollisin *n* mollisin
Mollösung *f* molar solution
Molluscid *n* molluscide
Molluskizide *pl* molluscicides, molluscacides
Molotowcocktail *m* Molotov cocktail
Molpolarisation *f* molecular polarization
Molprozent *n* mole percent
Molquant *n* molar quantum
Molrefraktion *f* molar refractivity
Molrotation *f* molecular rotation
Molverhältnis *n* molar ratio
Molvolumen *n* molar volume, molecular volume
Molwärme *f* molar heat
Molybdän *n* molybdenum (Symb. Mo)
Molybdänblau *n* molybdenum blue
Molybdänbleierz *n* (Min) wulfenite
Molybdänbleispat *m* (Min) wulfenite
Molybdänborid *n* molybdenum boride
Molybdänchlorid *n* molybdenum chloride
Molybdändijodid *n* molybdenous iodide, molybdenum diiodide, molybdenum(II) iodide, molybdous iodide
Molybdändioxid *n* molybdenum dioxide
Molybdäneisen *n* ferromolybdenum
Molybdänfaden *m* molybdenum filament

Molybdängehalt *m* molybdenum content
Molybdänglanz *m* (Min) molybdenum glance, molybdenite
Molybdänglas *n* molybdenum glass
Molybdängruppe *f* molybdenum group
molybdänhaltig molybdeniferous
Molybdän(III)-Salz *n* molybdenum(III) salt
Molybdän(III)-Verbindung *f* molybdenum(III) compound, molybdic compound
Molybdän(II)-Salz *n* molybdenous salt, molybdenum(II) salt, molybdous salt
Molybdän(II)-Verbindung *f* molybdenous compound, molybdenum(II) compound, molybdous compound
Molybdänkies *m* (Min) molybdenite
Molybdänlegierung *f* molybdenum alloy
Molybdänmetall *n* metallic molybdenum, molybdenum metal
Molybdänocker *m* (Min) molybdic ocher, molybdine, molybdite
Molybdänoxid *n* molybdenum oxide
Molybdänsäure *f* molybdic acid
Molybdänsäureanhydrid *n* molybdenum trioxide, molybdic anhydride, molybdic oxide
molybdänsauer molybdate
Molybdänsilber *n* molybdenum silver
Molybdänstahl *m* molybdenum steel
Molybdänsulfid *n* molybdenum sulfide
molybdänsulfidfrei free from molybdenum sulfide
Molybdäntrioxid *n* molybdenum trioxide, molybdic anhydride, molybdic oxide
Molybdat *n* molybdate, salt or ester of molybdic acid
Molybdit *m* (Min) molybdic ocher, molybdite, native molybdenum trioxide
Molybdomenit *m* (Min) molybdomenite
Molybdophyllit *m* (Min) molybdophyllite
Molybdosodalith *m* (Min) molybdosodalite
Molysit *m* (Min) molysite
Moment *m* (Augenblick) instant, moment
Moment *n* (Phys) moment, momentum, **angulares** ~ angular momentum, **kinetisches** ~ kinetic momentum, **lineares** ~ linear momentum, impulse, **magnetisches** ~ magnetic momentum, **statisches** ~ static momentum
momentan instantaneous, momentary
Momentaufnahme *f* (Phot) instantaneous photograph, snapshot
Momentauslöser *m* (Phot) instantaneous release
Monacetin *n* monacetin, glycerol monoacetate
Monade *f* (Zool) monad
Monamid *n* monamide, monoamide
Monamin *n* monamine, monoamine
Monardaöl *n* monarda oil, horsemint oil
Monardin *n* monardin
Monarson *n* monarsone
Monastralpigment *n* monastral pigment

Monatsschrift *f* monthly publication
Monazit *m* (Min) monazite
Monazitsand *m* monazite sand
Mondgas *n* Mond gas
Mondgasgenerator *m* Mond gas producer
Mondgestein *n* lunar rocks
Mondstein *m* (Min) moonstone
Monelmetall *n* Monel [metal]
Monesiaextrakt *m* monesia extract
Monesiarinde *f* monesia bark
Monesin *n* monesin
Monetit *m* monetite
Monheimit *m* monheimite
Moniereisen *n* reinforcement iron
Monimolit[h] *m* (Min) monimolite
Monit *m* monite
Monitor *m* monitor
Monitron *n* monitron
Monninin *n* monninine
Monoacetat *n* monoacetate
Monoacetin *n* monoacetin, glycerol monoacetate, monacetin
Monoäthylin *n* monoethylin, glycerol monoethyl ether
Monoamid *n* monoamide, monamide
Monoamin *n* monoamine, monamine
Monoaminoxidase *f* (Biochem) monoamine oxidase
monoatomar monatomic
Monobariumsilikat *n* monobarium silicate
monobasisch monobasic
Monobenzon *n* monobenzone
Monoblast *m* (junger Monozyt) monoblast
Monobrandfarbe *f* one-fire color
Monobromcampher *m* bromocamphor, monobromated camphor
Monobromisovalerylharnstoff *m* monobromoisovaleryl urea
Monobrompropionsäure *f* monobromopropionic acid
Monocalciumphosphat *n* monocalcium phosphate
Monocaproin *n* monocaproin
Monocarbonsäure *f* monocarboxylic acid
Monochloramin *n* monochloroamine, chlor[o]amine
Monochloressigsäure *f* monochlor[o]acetic acid
Monochlorhydrat *n* monochlorhydrate, monohydrochloride
Monochlorhydrin *n* monochlor[o]hydrin
Monochlorsilan *n* monochlorosilane
Monochord *n* monochord
monochrom monochrome, monochromic
Monochromasie *f* (Opt) monochromatism, spectral purity
Monochromatfilter *m* monochromatic filter
monochromatisch monochromatic, monochroic
Monochromatisierung *f* monochromatization

Monochromator *m* monochromatic filter, monochromator
Monochrometer *n* monochrometer
Monochromgelb *n* monochrome yellow
Monocrotalin *n* monocrotaline
Monocrotal[in]säure *f* monocrotalic acid
Monocrot[in]säure *f* monocrotic acid
monocyclisch monocyclic
Monodenteriobenzol *n* monodenteriobenzene
monoenergetisch monoenergetic
monofil monofil
Monoformin *n* monoformin, glycerol monoformate
monogen monogenic
Monoglycerid *n* monoglyceride
Monographie *f* monograph
monoheteroatomig monoheteroatomic
Monohydrat *n* monohydrate
Monohydroxylbenzol *n* monohydroxybenzene, hydroxybenzene, phenol
monoisotop monoisotopic
Monokaliumphosphat *n* monopotassium phosphate
monoklin monoclinic
Monokristall *m* monocrystal
Monolaurin *n* monolaurin
Monolitpigment *n* monolite pigment
Monolupin *n* monolupine
monomer monomer[ic]
Monomer *n* monomer
Monomerie *f* monomerism
Monomethylamin *n* monomethylamine, methylamine
Monomethylolharnstoff *m* monomethylol urea
monomolekular monomolecular, uni-molecular
monomorph monomorphic
Monomyristin *n* monomyristin
Mononatriumcarbonat *n* acid sodium carbonate, monosodium carbonate, sodium bicarbonate, sodium hydrogen carbonate
Mononatriumcitrat *n* monosodium citrate
Mononatriumphosphat *n* monosodium phosphate
Mononatriumsalz *n* monosodium salt
Mononatriumsulfat *n* acid sodium sulfate, monosodium sulfate, sodium bisulfate, sodium hydrogen sulfate
Mononatriumsulfit *n* acid sodium sulfite, monosodium sulfite, sodium bisulfite, sodium hydrogen sulfite
Mononitrat *n* mononitrate
Mononitrophenol *n* mononitrophenol
Mononucleotid *n* mononucleotide
Monoolein *n* monoolein
Monopalmitin *n* monopalmitin
Monoperphthalsäure *f* monoperphthalic acid
monophasisch monophase, single-phase
Monophenetidincitrat *n* monophenetidine citrate
Monophosphothiamin *n* monophosphothiamine

Monopol *n* monopoly
Monorhein *n* monorhein
Monorheinanthron *n* monorheinanthrone
Monosaccharid *n* monosaccharide
Monosilan *n* monosilane
Monosilikat *n* monosilicate
Monostearin *n* monostearin
Monosubstitutionsprodukt *n* mono-substituted product
Monosulfonsäure *f* monosulfonic acid
Monoterpen *n* monoterpene
monoton monotonous
Monotonie *f* monotony
monotrop monotropic
Monotropie *f* monotropy
Monotropitosid *n* montropitoside
monovalent monovalent, univalent
Monowolframcarbid *n* monotungsten carbide
Monoxid *n* monoxide
Monozygot *m* monozygote
Monozyt *m* (pl Monozyten) monocyte (pl monocytes), (pl monocytes)
Monozyten-Lymphozytenverhältnis *n* monocyte-lymphocyte rate
Monradit *m* monradite
Monrolith *m* monrolite
Montage *f* assemblage, assembly, erection, machine mounting, setting up
Montageband *n* assembly line
Montagebau *m* prefabrication
Montagehalle *f* assembling workshop
Montagekleber *m* assembly adhesive
Montageklebstoff *m* assembly adhesive
Montageleim *m* joint glue, structural adhesive
Montageleimung *f* construction gluing, secondary gluing
Montanalkohol *m* montanyl alcohol, nonacosanol
Montanharz *n* montan resin
Montanin *n* montanine
Montanindustrie *f* mining industry
Montanit *m* (Min) montanite
Montanol *n* montanol
Montanon *n* montanone
Montanpech *n* montan pitch
Montansäure *f* montanic acid
Montanwachs *n* montan wax, lignite wax
Montanylalkohol *m* montanyl alcohol, nonacosanol
Montebrasit *m* montebrasite
Montejus *m* (Druckgefäß) monte-jus
Monteur *m* (am Fließband) assembler, (Elektr) mechanician, (für Maschinen) erector, (Mechaniker) mechanic
montieren to assemble, to erect, to fit, to set up
Montierung *f* assemblage, erection, fitting, mounting, setting up
Montmorillonit *m* (Min) montmorillonite
Montroydit *m* (Min) montroydite

Monuron *n* monuron
Moorbraunkohle *f* moor coal
Moorerde *f* bog earth, peaty soil
Moorkohle *f* moor coal
Moortorf *m* moor peat
Moorwasser *n* peat water
Moosachat *m* (Min) moss agate, mocha stone
Moosbitter *n* lichenin
Moosgallerte *f* cetraria, iceland moss
Moosgrün *n* moss green
Moosgummi *n* cellular rubber, expanded rubber
Moosstärke *f* lichenin
Moräne *f* (Geol) moraine
Morasterz *n* bog ore
Mordenit *m* (Min) mordenite
Morenosit *m* (Min) morenosite, nickel vitriol
Morganit *m* (Min) morganite
Morin *n* morin
Morinde *f* morinda
Morindon *n* morindone
Moringagerbsäure *f* moringatannic acid, maclurin
Moringaöl *n* behen oil
Moringin *n* moringine
Morinit *m* (Min) morinite
Morion *m* (Min) morion
Morolsäure *f* morolic acid
Moronal *n* moronal
Moroxit *m* (Min) moroxite
Morphan *n* morphan
Morphanthridin *n* morphanthridine
Morphenol *n* morphenol
Morphigenin *n* morphigenine
Morphimethin *n* morphimethine
Morphin *n* morphine, morphia, morphina, morphinum
Morphinacetat *n* morphine acetate
Morphinhydrochlorid *n* morphine hydrochloride
Morphinhydrojodid *n* morphine hydroiodide
Morphinismus *m* (Med) morphine poisoning, morphinism
Morphinmekonat *n* morphine meconate
Morphinon *n* morphinone
Morphinsäure *f* morphic acid
Morphinsulfat *n* morphine sulfate
Morphinvalerianat *n* morphine valerate
Morphium *n* morphium, morphia, morphina, morphine
Morphiumvergiftung *f* (Med) morphine poisoning, morphinism
Morphogenese *f* morphogenesis
Morphol *n* morphol
Morpholchinon *n* morpholquinone
Morpholin *n* morpholine
Morphologie *f* (Biol) morphology
morphologisch (Biol) morphologic
Morpholon *n* morpholone
Morphoran *n* morphoran

Morphosan n morphosane, morphine methylbromide
Morphothebain n morphothebaine
Morphotropie f morphotropy
Morrhualsäure f morrhuic acid
morsch decayed, decaying, rotten
Morschheit f rottenness
Morsealphabet n Morse code
Morseapparat m Morse apparatus
Mortalität f mortality
Morvenit m (Min) morvenite
Mosaikplatte f mosaic tile
Mosaikstruktur f (Krist) mosaic structure
Mosandrit m (Min) mosandrite
Moschus m musk, moschus
Moschusketon n musk ketone
Moschuskörneröl n musk seed oil
Moschuskorn n musk seed
Moschuskraut n (Bot) cat thyme
Moschuswurzel f (Bot, Pharm) sumbul root
Moschuswurzelöl n sumbul oil
Moseleysches Gesetz n Moseley's law
Mosesit m (Min) mosesite
Moskito m (Zool) mosquito
Moslen n moslene
Mossit m (Min) mossite
Most m must, (Apfelmost) cider
Mostrich m mustard
Motor m motor, engine, **der ~ läuft leer** the motor runs idle
Motoranker m motor armature
Motoranlasser m motor starter
Motorantrieb m motor drive
Motorbenzin n gas, gasoline (Am. E.), petrol (Br. E.)
Motordrehzahl f motor speed, (pro Minute) revolutions per minute
Motorenfabrik f motor works
Motorengeräusch n engine noise
Motorenöl n motor oil, lubricating oil
Motorenwicklung f motor winding
Motorhaube f cowling
Motorleistung f motor power
Motorpanne f engine failure
Motorthermometer n cylinder block thermometer
Motorumformer m motor convertor
Motorwelle f motor shaft
Mottenpulver n moth powder
Mottenschutzmittel n moth preventive, moth proofing agent
mottensicher mothproof
Mottramit m (Min) mottramite
moussieren to effervesce, to fizz, to froth, to sparkle
Moussieren n effervescence, sparkling
moussierend effervescent, frothing, sparkling
Movrin n movrine
MOZ Abk. für Motoroktanzahl

MPI Abk. für Max-Planck-Institut
Mucamid n mucamide
Mucedin n mucedine
Mucigen n mucigen
Mucin n mucin
Mucinogen n mucinogen
Mucobromsäure f mucobromic acid, dibromomalealdehydic acid
Mucochlorsäure f mucochloric acid, dichloromalealdehydic acid
Mucoid n mucoid
Mucoidin n mucoidin
Mucoinosit m mucoinositol
Mucoitinschwefelsäure f mucoitin sulfuric acid
Mucoitinsulfat n mucoitin sulfate
Muconsäure f muconic acid, 2,4-hexadienedioic acid
Mucopeptid n mucopeptide
Mucopolysaccharid n mucopolysaccharide
Mucopolysaccharidase f (Biochem) mucopolysaccharidase
Mucoproteid n (Mucoprotein) mucoprotein, glycoprotein
Mucosin n mucosin
Mucunaöl n mucuna oil
Mucusan n mucusane
Mudarin n mudarin
Mudarsäure f mudaric acid
Mückenvertreibungsmittel n mosquito repellent
Mühle f (fein) mill, pulverizer, (grob) crusher, **Aerofall- ~** aerofall mill, **Autogen- ~** autogenous tumbling mill, **Dreiwalzenring- ~** ring-roller mill, **Einrollen- ~** single-roller ring mill, **Federkraft-Kugel- ~** contrarotation ball-race-type pulverizing mill, **Federrollen- ~** ring-roller mill, **Fliehkraft-Kugel- ~** centrifugal ball mill, **Glocken- ~** cone mill, **Hammer- ~** hammer mill, **Kaskaden- ~** cascade mill, **Kolloid- ~** colloid mill, **Konus- ~** conical mill, **Kugel- ~** ball mill, **Luftstrom- ~** ball mill with air dryer, **Mehrkammer- ~** compartment mill, **Pendel- ~** centrifugal roll mill, **Planeten- ~** planetary mill, **Prall- ~** impact mill, **Ring- ~** ring-roller mill, **Rollen- ~** centrifugal roll mill, **Rührwerks- ~** stirred ball mill, **Scheiben- ~** disc attrition mill, **Schlag- ~** hammer mill, **Schneid- ~** rotary cutters, **Schüssel- ~** bowl mill, **Schwing- ~** vibrating ball mill, **Siebkugel- ~** ball mill with closed circuit mechanical classification, **Strahl- ~** fluid energy or jet mill, **Trommel- ~** tube mill, **Walz- ~** tumbling mill, **Walzring- ~** centrifugal roll mill, **Zweikammer- ~** two-compartment mill, **Zweirollen-Scheiben- ~** two-roll attrition mill
Mühleisen n stone spindle
Mühlenfeuerung f stoker firing
Mühlenkalkstein m millstone rock

Mühlennachprodukte – Muschelschieber

Mühlennachprodukte *pl* flour mill by-products
Mühlsteinquarz *m* silex for millstones
Müll *m* garbage, refuse, waste
Müllabfuhr *f* garbage removal, garbage disposal
Müllabladeplatz *m* garbage dump or pit
Müllfeuerung *f* refuse [destructor] furnace
Müllkrapp *m* mull madder
Müllverbrennungsofen *m* incinerator
Müllverdichter *m* compacter
münden to open into, to discharge
Mündung *f* orifice, (Röhre) mouth, opening, ~ der Birne throat of the converter
Münzbeschickung *f* alloyage
Münzgehalt *m* standard of alloy
Münzgewicht *n* standard weight
Münzmetall *n* coinage metal
Münzstahl *m* brescian steel
mürbe brittle, friable, short, (reif) mellow, ripe
Muffe *f* bosshead
Muffel *f* (Metall) muffle, ~ **aus Eisen** cast iron muffle, iron muffle, ~ **aus Schamotte** fire clay muffle
Muffelfarbe *f* burnt-in color
Muffelofen *m* assay furnace, muffle furnace, muffle roaster
Muffelverfahren *n* distillation process, retort process
Muffendruckrohr *n* delivery socket-pipe
Muffenhub *m* lift of sleeve
Muffenhülse *f* coupling [box]
Muffenkupplung *f* buttmuff coupling, cased butt coupling
Muffenrohr *n* socket[ed] pipe
Muffenrohrverbindung *f* spigot joint
Muffenverbindung *f* pipe union, sleeve joint
Muhuhu-Öl *n* muhuhu oil
mulchen to mulch
Mulde *f* trough, basin, (Bergb) cavity, hole
Muldenblei *n* pig lead
Muldenchargierkran *m* trough charging crane
Muldengurtförderer *m* troughed conveyor belt
Muldenkipper *m* dumper
Muldenkopf *m* head of ingot mold
Muldenrolle *f* trough roller
Muldensetzkran *m* trough charging crane
Muldentrockner *m* trough dryer
Mullanit *m* mullanite
Mullerit *m* (Min) mullerite, sylvanite
Mullicit *m* (Min) mullicite, vivianite
Mulm *m* ore dust, (Fäule) moldiness
mulmen to pulverize
mulmig moldy
Multidekameter *n* multi decameter
Multienzymkomplex *m* multienzyme complex
Multienzymsystem *n* (Biochem) multi-enzyme system
Multigruppentheorie *f* multi-group theory
Multiplett *n* multiplet
Multiplettanteil *m* multiplet component

Multiplettaufspaltung *f* (Atom) multiplet splitting
Multiplettstruktur *f* multiplet structure
Multiplex-Vorteil *m* Fellgett advantage
Multiplikand *m* (Math) multiplicand
Multiplikation *f* multiplication
Multiplikationsfaktor *m* (Atom) reproductive factor, (Math) multiplication [or multiplying] factor
Multiplikationstheorem *n* multiplication theorem
Multiplikationszeichen *n* sign of multiplication
Multiplikator *m* multiplier
Multiplikatorenbereich *m* domain of multipliers
Multipolentwicklung *f* multipole expansion
Multipolfeld *n* multipole field
Multipolordnung *f* multipolarity
Multipolpotential *n* multipole potential
Multipolstrahlung *f* multipole radiation
Multipolübergang *m* multipole transition
Multirotation *f* multirotation
Multivektor *m* multivector
Multivitamin *n* multivitamin
Multivitaminpräparat *n* multivitamin preparation
Multizellularvoltmeter *n* multicellular voltmeter
Multizyklon *m* multicyclone
Mundstück *n* mouthpiece, [die] orifice, nozzle
Munduloxsäure *f* munduloxic acid
Mundwasser *n* gargle, mouth wash
Munition *f* ammunition
Muntzmetall *n* Muntz metal
Murami[ni]dase *f* (Biochem) murami[ni]dase
Muraminsäure *f* muramic acid
Murchisonit *m* (Min) murchisonite
Murexan *n* murexane, aminobarbituric acid, uramil
Murexid *n* murexide, ammonium purpurate
Murexidprobe *f* (Anal) murexide assay
Murexin *n* murexine
Muriat *n* (obs) chloride
Muriazit *m* (Min) muriacite, anhydrite
Muromontit *m* muromontite
Muropeptid *n* muropeptide
Murrayin *n* murrayin
Muscarin *n* muscarine
Muscaron *n* muscarone
Muscarufin *n* muscarufin
Muschelerde *f* shell marl
Muschelerz *n* conchoidal iron ore
Muschelgold *n* mosaic gold, ormolu
Muschelgrus *m* alluvial shell deposit
muschelig conchoidal, shelly
Muschelkalk *m* (Min) shell limestone
Muschelkalkstein *m* (Min) shell limestone
Muschelmarmor *m* (Min) shell marble, fire marble
Muschelmergel *m* shell marl
Muschelschieber *m* mushroom valve

Muschelseidenstein *m* (Min) byssolite
Muschelversteinerung *f* ostracite
Muschketowit *m* muschketowite
Musivgold *n* crystallized stannic sulfide, mosaic gold
Musivsilber *n* mosaic silver
Muskatbalsam *m* mace oil, myristica oil, nutmeg butter
Muskatblüte *f* (Bot) mace
Muskatblütenöl *n* mace oil
Muskatbutter *f* mace butter, nutmeg butter
Muskatfett *n* mace butter, nutmeg butter
Muskatnuß *f* nutmeg
Muskatnußöl *n* nutmeg oil
Muskatöl *n* mace oil, nutmeg oil
Muskel *m* muscle, **ruhender** ~ resting muscle, **synchroner** ~ synchronous muscle
Muskeladenylsäure *f* muscle adenylic acid, AMP
Muskelanregung *f* muscle excitation
Muskeleiweiß *n* myosin
Muskelentspannung *f* muscle relaxation
Muskelextrakt *m* muscle extract
Muskelfarbstoff *m* muscle pigment, myoglobin
Muskelfaser *f* muscle fiber, muscular fiber
Muskelgewebe *n* muscle tissue
Muskelkontraktion *f* muscle contraction
Muskelleistung *f* muscular performance
Muskelphosphorylase *f* (Biochem) muscle phosphorylase
Muskelstoff *m* (obs) sarcosine, methylaminoacetic acid, methylglycine, methylglycocoll
Muskelzelle *f* muscle cell, myocyte
Muskelzucker *m* (obs) muscle sugar, inositol
Muskon *n* muscone, 3-methyl-cyclopentadecanone, muskine
Muskovit *m* (Min) muscovite, phengite
Muskulatur *f* muscular system
Muskulin *n* musculine
Musselin *m* (Text) muslin
Musselinglas *n* muslin glass
Mussit *m* (Min) mussite, diopside
Muster *n* sample, pattern, specimen
Musterausnehmer *m* designer
Musterbetrieb *m* pilot plant
Musterbeutel *m* sample bag
Musterexemplar *n* type specimen
Mustermaterial *n* pattern
Musternahme *f* batch sampling
Musterrad *n* patterning wheel
Musterschutz *m* protection of designs
Musterseite *f* (Buchdr) specimen page
Musterstopperhebel *m* pattern selector lever
Mustertrommel *f* pattern drum
Mutagen *n* mutagen
Mutante *f* (Biol) mutant
Mutarotation *f* mutarotation
Mutase *f* (Biochem) mutase
Mutation *f* (Biol) mutation

Mutationsgeschwindigkeit *f* mutation rate
Mutatochrom *n* mutatochrome
Mutatoxanthin *n* mutatoxanthin
Muthmannit *m* (Min) muthmannite
Mutter *f* (Techn) nut
Mutterblech *n* cathode core sheet
Mutterboden *m* native soil
Mutteressig *m* mother of vinegar
Mutterfaß *n* mother vat
Muttergestein *n* (Erdöl) [oil] source rock
Muttergewinde *n* female thread
Mutterharz *n* galbanum
Mutterhefe *f* mother yeast, parent yeast
Mutterisotop *n* parent isotope
Mutterkaliber *n* matrix
Mutterkorn *n* (Bot) ergot
Mutterkornalkaloid *n* ergot alkaloid
Mutterkornbrand *m* ergotism
Mutterkornextrakt *m* ergot extract, extract of ergot
Mutterkornfluidextrakt *m* liquid ergot extract
Mutterkornsäure *f* ergotic acid
Mutterkornvergiftung *f* (Med) ergotism
Mutterkraut *n* (Bot) feverfew, mother wort
Mutterkrautöl *n* feverfew oil
Mutterlauge *f* mother liquor
Mutterlaugensalz *n* bath salt
Muttersubstanz *f* mother substance, parent substance
Mutterzelle *f* mother cell, parent cell
Myanesin *n* myanesin
Mycaminit *m* mycaminitol
Mycaminose *f* mycaminose
Mycarose *f* mycarose
Mycobakterium *n* mycobacterium
Mycocerosinsäure *f* mycocerosic acid
Mycoctonin *n* mycoctonine
Mycoin *n* mycoine
Mycolipensäure *f* mycolipenic acid
Mycomycin *n* mycomycin
Mycophenolsäure *f* mycophenolic acid
Mycoproteid *n* (Mykoprotein) mycoprotein
Mycosamin *n* mycosamine
Mydriasin *n* mydriasine
Mydriatin *n* mydriatine
Myelin *n* myelin
Myeloperoxidase *f* myeloperoxidase
Mykodextran *n* mycodextrane
Mykogalaktan *n* mycogalactan
Mykol *n* mycol
Mykologie *f* mycology
Mykophenolsäure *f* mycophenolic acid
Mykose *f* (Med) mycose
Mykosterin *n* mycosterol
My-Meson *n* mu meson, muon, my-meson
Myofibrille *f* myofibril
Myoglobin *n* myoglobin
Myohämatin *n* myoh[a]ematin
Myoinosamin *n* myoinosamine

Myoinosit *m* myoinosit
Myokinase *f* (Biochem) myokinase
Myon *n* muon, mu meson
Myoneneinfang *m* muon capture
Myonenpaarerzeugung *f* muon pair production
Myosin *n* myosin
Myosmin *n* myosmine
Myrcen *n* myrcene, 2-methyl-6-methylene-2,7-octadiene
Myrcenol *n* myrcenol
Myrental *n* myrental
Myricawachs *n* myrcia wax
Myricetin *n* myricetin, oxyquercetin
Myricetrin *n* myricetrin
Myricin *n* myricin, myricyl palmitate
Myrickit *m* myrickite
Myricyl- myricyl
Myricylalkohol *m* myricyl alcohol, melissic alcohol
Myristicin *n* myristicin
Myristicinaldehyd *m* myristicin aldehyde
Myristicinglykol *n* myristicin glycol
Myristicinsäure *f* myristicinic acid
Myristicol *n* myristicol
Myristin *n* myristin, glycerol myristate, laurel wax, myrtle wax
Myristinsäure *f* myristic acid, tetradecanoic acid
Myristodistearin *n* myristo distearine, glycerol distearate myristate
Myristoleinsäure *f* myristoleic acid
Myristolsäure *f* myristolic acid
Myriston *n* myristone, myristic ketone, tridecyl ketone
Myristylalanin *n* myristoylalanine
Myristylaldehyd *m* myristoylic aldehyde
Myristylalkohol *m* myristyl alcohol, 1-tetradecanol
Myronsäure *f* myronic acid
myronsauer myronate
Myrosin *n* (Biochem) myrosine, myrosase, myrosinase
Myroxin *n* myroxin
Myrrhe *f* (aromat. Harz) myrrh
Myrrhenbalsam *m* myrrh balm
Myrrhenöl *n* myrrh oil
Myrrhentinktur *f* myrrh tincture
Myrtanol *n* myrtanol
Myrtenal *n* myrtenal, myrtenic aldehyde
Myrtenessenz *f* essence of myrtle berries
Myrtengrün *n* myrtle green
Myrtenöl *n* myrtle oil
Myrtenol *n* myrtenol
Myrtensäure *f* myrtenic acid
Myrtenwachs *n* myrtle wax, bayberry wax
Myrticolorin *n* myrticolorin
Myrtillidin *n* myrtillidin
Myrtillin *n* myrtillin
Myrtillogensäure *f* myrtillogenic acid
Myrtol *n* myrtol

Mytilit *m* mytilite, methyl inositol, mytilitol
Mytilotoxin *n* mytilotoxin
Myxobakterien *pl* (Biol) myxobacteria
Myxoxanthin *n* myxoxanthin
Myzel[ium] *n* (Biol) mycelium
Myzeten *pl* mycetes

N

nachahmen to copy, to duplicate, to imitate, to simulate
Nachahmung *f* imitation, simulation
Nacharbeit *f* fashioning
nacharbeiten to copy
Nachbaratom *n* neighbo[u]ring atom
Nachbarelement *n* neighbo[u]r element
Nachbarfrequenz *f* side frequency
Nachbargruppeneffekt *m* neighboring group participation
Nachbarschaft *f* neighborhood
Nachbarstellung *f* (Atom) neighboring position
nachbehandeln to aftertreat, to post-treat, to treat subsequently
Nachbehandlung *f* additional treatment, aftertreatment, secondary treatment
nachbeizen to re-dye
Nachbelichtung *f* (Phot) postexposure
Nachbeschickung *f* after-charging, subsequent charging
Nachbeschleunigung *f* postacceleration
nachbilden to copy, to imitate, to reproduce
Nachbleiche *f* subsequent bleaching
Nachbrand *m* second baking
nachbrennen to burn again, to smolder
Nachbrennen *n* after-burning
nachchargieren to recharge
nachchlorieren to post-chlorinate
nachchloriert postchlorinated
Nachchromieren *n* afterchroming
Nachchromierfarbstoff *m* afterchrome dyestuff
Nachdampf *m* after-damp, choke damp
Nachdruck *m* (Buchdr) reprint
Nachdruckzeit *f* plunger [or screw] forward time
nachdunkeln to become darker, to darken, to sadden
nacheichen to recalibrate, to check
Nacheichung *f* recalibration
nacheilen to follow, to lag
Nacheilung *f* lag, **magnetische** ~ (Magn) magnetic lag, **zeitliche** ~ time lag
Nacheilungswinkel *m* angle of lag
Nacheilwinkel *m* (Elektr) angle of lag
nacherhitzen (Met) to temper
nachfärben to dye again, to re-dye
Nachfilter *m* second filter
Nachfiltern *n* final filtering
nachfixieren (Phot) to afterfix
nachformen to copy, to reproduce
Nachformen *n* postforming
Nachformfräsmaschine *f* die sinking mill
nachforschen to investigate, to inquire [into]
Nachforschung *f* investigation, research
nachfüllen to replenish, to add to, to charge, to fill up
Nachfüllpackung *f* refill

Nachfüllung *f* refilling
Nachfüllventil *n* refillable valve
nachgären to ferment again
Nachgärung *f* after-fermentation, post-fermentation, secondary fermentation
Nachgärungsbottich *m* cleansing vat
Nachgärungsfaß *n* cleansing vat
Nachgärungskasten *m* settling vessel
Nachgärungskufe *f* cleansing vat
nachgeben to yield
nachgerben to tan a second time, to curry again
Nachgeruch *m* after-odor, after-smell
Nachgeschmack *m* aftertaste
nachgießen to add more [liquid]
Nachglimmen *n* afterglow
nachglühen (Met) to reanneal
Nachglühen *n* (Met) reannealing
Nachglühtemperatur *f* tempering temperature
Nachguß *m* after-mash, second wort
Nachhärtung *f* (Kunstharz) afterbake
Nachhärtungsfrist *f* conditioning period, aging time
Nachhinken *n* (zeitlich) time lag
Nachklärbecken *n* secondary (or final) settling tank
Nachklang *m* (Akust) reverberation
Nachkristallisation *f* after-crystallization
nachkupfern to after-copper
Nachladung *f* (Akku) second charge
Nachlässigkeit *f* negligence
nachlassen (aufhören) to cease, (lockern) to relax, to release, to slacken, (Met) to anneal, to reheat, to soften, to temper [down]
Nachlassen *n* (Farbe) fading, (Met) annealing, softening, tempering
Nachlauf *m* after-run, last runnings, second runnings
Nachlaufwerk *n* follower
Nachleuchtbild *n* residual image
Nachleuchtdauer *f* afterglow time
nachleuchten to afterglow, to phosphoresce
Nachleuchten *n* after-glow
Nachluft *f* additional air, supplementary air
Nachmehl *n* middlings
nachmessen to control a measurement, to measure again
Nachprodukt *n* afterproduct, low product, second product
nachprüfen to check, to control, to reexamine
Nachprüfung *f* checking [over], reexamination
nachrechnen to calculate again, to check
nachregeln to readjust
Nachreifen *n* subsequent ripening
Nachreiniger *m* final purifier, repurifier
Nachreinigung *f* after-purification, final cleaning

Nachricht f communication, information, message
Nachrichtensystem n system of communication
nachrösten to roast again
nachsalzen to add more salt
Nachschlagetabelle f reference table
Nachschlagewerk n encyclopedia, reference work
nachschmieren to lubricate again
Nachschubmagnet m feed magnet
Nachschwaden m after damp, choke damp
nachschwelen to continue smoldering
Nachschwindung f reheat change
nachseigern to refine again
nachsenden to forward
Nachsetzlöffel m ladle for adding something
Nachsetzung f after-charging, subsequent charging
Nachspannung f (Keilriemen) take-up
nachspülen to rinse [out]
nachstellbar adjustable
nachstellen to adjust, to match, to reset
Nachstellung f matching, readjustment, reset
Nachstellvorrichtung f readjusting device
Nachstellzeit f (Regeltechn) integral action time (Br. E.), reset time (Am. E.)
Nachstrom m (Mech) after-flow
Nachsud m second boiling, second evaporation
Nachtblau n night blue
nachtblind night-blind
Nachtblindheit f nightblindness
Nachteil m disadvantage
nachteilig disadvantageous, inconvenient
Nachtempern n after-annealing
Nachtgrün n night green, iodine green
Nachtrag m addendum, addition, supplement
Nachtrocknen n final or second drying process
Nachtschicht f night shift
Nachtstrom m night current
Nachverstrecken n after-stretching
Nachvulkanisation f continuance of cure, overcure
Nachwärme f (Reaktor) after-heat
nachwaschen to rewash
Nachweis m proof, detection, test
nachweisbar detectable
Nachweisempfindlichkeit f assay sensitivity, detection sensitivity
nachweisen to prove, to detect, to discover, to establish
Nachweisgrenze f detection limit
Nachweisinstrument n detector
Nachweismethode f detection method
nachwirken to show an aftereffect
Nachwirkung f aftereffect, (Elektr) proximity effect, **dielektrische** ~ dielectric fatigue, dielectric hysteresis, **elastische** ~ elastic afterworking, **magnetische** ~ magnetic aftereffect, magnetic fatigue, **plastische** ~ afterflow, elastic aftereffect, plastic flow persistence
Nachwürze f after-mash; second wort
nachzählen to count again
Nachzündung f retarded ignition, late ignition, re-ignition
NAD Abk. für Nikotinamid-adenin-dinucleotid
Nadel f needle, (Steck- und Haarnadel) pin, (z. B. Ätznadel) stylus
Nadelabweichung f deflection of the needle
Nadelausleger m luffing jib
Nadelausreißfestigkeit f stitch holding property
Nadelausschlag m deflection of the needle, needle throw
Nadelbiopsie f (Med) needle biopsy
Nadelboden m needle bottom, perforated bottom
Nadelbrenner m pin-point burner
Nadeleisenerz n (Min) needle iron ore
Nadelerz n (Min) needle ore
Nadelfeinregulierventil n precision needle valve
nadelförmig needle-shaped, (Bot Biol) acicular
Nadelgalvanometer n (Elektr) needle galvanometer, moving magnet galvanometer
Nadelholzkohle f pine charcoal, soft-wood charcoal
Nadelkohle f acicular lignite
Nadelkristall m (Krist) needle-shaped crystal
Nadellager n needle bearing
Nadelöhr n eye of a needle
Nadelpunktanguß m (Gieß) pin-point gate, pin-point gating
Nadelschmierer m needle lubricator
Nadelschmiergefäß n needle lubricator
Nadelsonde f needle probe
Nadelspannrahmen m pin stenter
Nadelspitze f point of a needle
Nadelstein m (Min) needle stone, needle zeolite
Nadelstrahlung f needle radiation
Nadelventil n needle valve, (Gaschromat) restrictor
Nadelwald m coniferous forest
Nadelwehr n pin weir
Nadelzeolith m (Min) needle zeolite, natrolite
Nadelzinnerz n (Min) acicular cassiterite, needle tin
NADH Abk. für reduzierte Form des Nikotinamid-adenin-dinucleotid
Nadireagens n Nadi reagent
Nadorit m (Min) nadorite
Nägeleinpfeffer m (Bot) allspice, pimento
Nägeleinwein m wine spiced with cloves
Naegit m (Min) naegite
Nähen n sewing, stitching
Näherung f (Math) approximation, ~ **mit fester Kopplung** strong-coupling approximation, **adiabatische** ~ adiabatic approximation, **schrittweise** ~ stepwise approximation
Näherungsformel f approximation formula

Näherungsgleichung f approximate equation
Näherungsgröße f approximate quantity
Näherungslösung f (Math) approximation
Näherungsmethode f approximation method
Näherungsrechnung f approximate calculation
Näherungsschalter m proximity switch
Näherungsverfahren n approximate method, trial and error method
Näherungswert m approximate value
Nährboden m (Kultur) culture medium, nutrient medium
Nährbodenflasche f nutrient bottle
nähren to feed, to nourish
Nährflüssigkeit f nutrient liquid, nutritive liquid
Nährgehalt m nutritional contents
Nährgewebe n (Bot) endosperm
Nährhefe f nutrient yeast, nutritive yeast
Nährkraft f nutritive power
Nährlösung f nutritive solution
Nährmittel n nutrient, food, nourishment, nutriment
Nährpräparat n nutritive preparation
Nährsalz n nutritive salt
Nährstoff m nutrient, nutritive substance
nährstoffarm oligotrophic
Nährstoffgehalt m nutrient content[s]
nährstoffreich eutrophic
Nährwert m food value, nutritive value
Näpfchen n cup, little bowl
Nässe f wetness, dampness, humidity, percentage of moisture
nässen to wet, to damp, to moisten
Nässen n damping, moistening, wetting
Nagel m nail
Nagelbarkeit f nail-holding property
Nagelbürste f nail brush
Nageleisen n nail iron, heading tool
Nagelerz n (Min) columnar argillaceous red iron ore
Nagelhaut f cuticle
Nagelkaliber n nail pass
Nagellack m (Kosm) nail polish
Nagellackentferner m (Kosm) nail polish remover
nagelneu brand-new
Nagelstein m (Min) onyx
Nagetier n (Zool) rodent
Nagyagit m (Min) nagyagite, tellurium glance
nah close, near
Nahaufnahme f (Phot) close-up
Naheinstellung f (Opt) short-range focusing
naheliegend obvious, (räumlich) adjacent
Nahordnung f short-range order
Nahordnungsparameter m short-range order parameter
Nahordnungstheorie f short-range order theory
Nahpunkt m (Opt) near point
nahrhaft nutritious, nutritive
Nahrhaftigkeit f nutritiousness

Nahrung f nourishment, nutriment
Nahrungsaufnahme f ingestion [of food]
Nahrungsbedarf m nutritional requirements
Nahrungsmangel m alimentary deficiency
Nahrungsmittel n food, foodstuff, nutriment
Nahrungsmittelchemie f food chemistry
Nahrungsmittelchemiker m food chemist
Nahrungsmittelfälschung f adulteration of food, food adulteration
Nahrungsmittelhygiene f food hygiene
Nahrungsmittelindustrie f food [processing] industry, foodstuffs industry
Nahrungsmittelkontrolle f food control, food inspection
Nahrungsmittelkunde f science of nutrition
Nahrungsstoff m food, foodstuff, nutrient, nutriment, nutritive substance
Nahrungszufuhr f food intake
Naht f seam, (Schweiß) weld
nahtlos seamless, (Schweiß) weldless
Nahtschweißen n seam welding
Nahtschweißmaschine f seam welder
Nahunordnung f short-range disorder
Nahwirkung f close-range action
Nakarat n (Färb) nacarat, smalt
Nakaratfarbe f nacarat
Nakrit m (Min) nakrite, keolinite
Nalorphin n nalorphine
Namakochrom n namakochrome
Nandinin n nandinine
Nantokit m (Min) nantokite
Napellin n napelline, benzaconine, benzoylconine, picraconitine
Naphtha f naphtha
Naphthacen n naphthacene
Naphthacenchinon n naphthacene quinone
Naphthacridin n naphthacridine, dibenzacridine
Naphthacylschwarz n naphthacyl black
Naphthalan n naphthalane
Naphthaldazin n naphthaldazine
Naphthaldehyd m naphthaldehyde
Naphthaldehydsäure f naphthaldehydic acid
Naphthalen n naphthalene, naphthalin, naphthene
Naphthalid n naphthalide
Naphthalimid n naphthalimide
Naphthalin n naphthalene, naphthalin, naphthene, tar camphor, ~ enthaltend containing naphthalene
Naphthalinabscheider m naphthalene separator
Naphthalinblau n naphthalene blue
Naphthalincampher m naphthalene camphor
Naphthalinderivat n naphthalene derivative
Naphthalinfarbe f (Färb) naphthalene dye
Naphthalinfarbstoff m naphthalene dyestuff
Naphthalinhydrid n naphthalene hydride
Naphthalinindigo m naphthalene indigo
Naphthalinrot n naphthalene red, magdala red
Naphthalinsäure f naphthalenic acid

Naphthalinsulfonsäure *f* naphthalenesulfonic acid
naphthalinsulfonsauer naphthalenesulfonate
Naphthalol *n* naphthalol, betol, naphthyl salicylate, salinaphthol
Naphthalsäure *f* naphthalic acid
Naphthamin *n* naphthamine, hexamethylene amine
Naphthan *n* naphthane, decahydronaphthalene, decalin, dekalin
Naphthanen *n* naphthanene
Naphthanisol *n* naphthanisol
Naphthanol *n* naphthanol
Naphthanthracen *n* naphthanthracene, benzanthrene
Naphthanthrachinon *n* naphthanthraquinone
Naphthanthracridin *n* naphthanthracridine
Naphthanthroxansäure *f* naphthanthroxanic acid
Naphtharson *n* naphtharson
Naphtharückstand *m* masut, mazut
Naphthazarin *n* naphthazarine, alizarin black
Naphthazin *n* naphthazine, anthrapyridine
Naphthen *n* naphthene
Naphthenat *n* naphthenate
Naphthensäure *f* naphthenic acid
Naphthhydrinden *n* naphthhydrindene
Naphthidin *n* naphthidine
Naphthimidazol *n* naphthimidazole
Naphthindazol *n* naphthindazole
Naphthinden *n* naphthindene
Naphthindigo *n* naphthindigo
Naphthindol *n* naphthindole
Naphthindolin *n* naphthindoline
Naphthindon *n* naphthindone
Naphthindoxyl- naphthindoxyl
Naphthionat *n* naphthionate
Naphthionsäure *f* naphthionic acid, 1-naphthylamine-4-sulfonic acid
Naphthionsalz *n* naphthionate
naphthionsauer naphthionate
Naphthisatin *n* naphthisatin
Naphthisatinchlorid *n* naphthisatin chloride
Naphthoacridin *n* naphthoacridine
Naphthobrenzcatechin *n* 1,2-dihydroxynaphthalene, 1,2-naphthalenediol, naphthopyrocatechol
Naphthocarbostyril *n* naphthocarbostyril
Naphthochinaldin *n* naphthoquinaldine
Naphthochinhydron *n* naphthoquinhydrone
Naphthochinolin *n* naphthoquinoline
Naphthochinon *n* naphthoquinone, dihydrodiketonaphthalene
Naphthochromanon *n* naphthochromanone
Naphthoealdehyd *m* naphthaldehyde, naphthalene carbonal, naphthoic aldehyde
Naphthoesäure *f* naphthalene carboxylic acid, naphthoic acid
Naphthofluoren *n* naphthofluorene
Naphthofuchson *n* naphthofuchsone
Naphthofuran *n* naphthofuran, benzocoumarane
Naphthohydrochinon *n* naphthohydroquinone
Naphthol *n* naphthol, hydroxynaphthalene
Naphthollallyläther *m* naphthol allyl ether
Naphtholbenzoat *n* naphthol benzoate, benzoylnaphthol, naphthyl benzoate
Naphtholblau *n* naphthol blue
Naphtholcarbonsäure *f* naphtholcarboxylic acid, hydroxynaphthoic acid
Naphtholfarbstoff *m* naphthol dye[stuff]
Naphtholgelb *n* naphthol yellow, citronin
Naphthollösung *f* naphthol solution
Naphtholnatrium *n* sodium naphtholate
Naphtholorange *n* naphthol orange
Naphtholphthalein *n* naphtholphthalein
Naphtholquecksilber *n* mercury naphtholate
Naphtholsäure *f* naphthol acid, hydroxynaphthoic acid
Naphtholsalicylat *n* naphthol salicylate
Naphtholsalol *n* naphtholsalol
Naphtholschwarz *n* naphthol black
Naphtholsulfonat *n* naphthol sulfonate
Naphtholsulfonsäure *f* naphtholsulfonic acid
Naphtholwismut *n* bismuth naphtholate
Naphthonitril *n* naphthonitrile, naphthalene carbonitrile, naphthyl cyanide
Naphthophenazine *n* naphthophenazine, phenonaphthazine
Naphthophenofluorindin *n* naphthophenofluorindine
Naphthopikrinsäure *f* naphthopicric acid
Naphthopurpurin *n* naphthopurpurin
Naphthopyran *n* naphthopyrane
Naphthopyrazin *n* naphthopyrazine, benzoquinoxaline
Naphthopyron *n* naphthopyrone
Naphthoresorcin *n* 1,3-dihydroxynaphthalene, 1,3-naphthalenediol, naphthoresorcinol
Naphthostyril *n* naphthostyril
Naphthotetrazol *n* naphthotetrazole
Naphthothianthren *n* naphthothianthrene
Naphthothiazin *n* naphthothiazine
Naphthothiazol *n* naphthothiazole
Naphthothioflavon *n* naphthothioflavone
Naphthothioindigo *n* naphthothioindigo
Naphthothioxol *n* naphthothioxole
Naphthotriazin *n* naphthotriazine
Naphthotriazol *n* naphthotriazole
Naphthoxazin *n* naphthoxazine, phenoxazine
Naphthoxazol *n* naphthoxazole
Naphthoylbenzoesäure *f* naphthoylbenzoic acid
Naphthoylchlorid *n* naphthoyl chloride
Naphthoylpropionsäure *f* naphthoylpropionic acid
Naphthoyltrifluoraceton *n* naphthoyltrifluoroacetone
Naphthpiperazin *n* naphthopiperazine
Naphthursäure *f* naphthuric acid

Naphthyl- naphthyl
Naphthylamin *n* naphthylamine, naphthalidine
Naphthylaminchlorhydrat *n* naphthylamine hydrochloride
Naphthylamingelb *n* naphthylamine yellow, Manchester yellow
Naphthylaminrot *n* naphthylamine red
Naphthylaminsulfonsäure *f* naphthylamine sulfonic acid
Naphthylblau *n* naphthyl blue
Naphthylendiamin *n* naphthylene diamine, diaminonaphthalene
Naphthylensulfonylid *n* naphthylene sulfonylide
Naphthylphenylamin *n* naphthyl phenylamine, phenyl naphthylamine
Naphthylsalicylat *n* naphthyl salicylate
Naphthylthioharnstoff *m* naphthylthiourea
Naphthyltrichlorsilan *n* naphthyltrichlorosilane
Naphthyridin *n* 1,8-benzodiazine, naphthyridine
Napoleonit *m* (Min) napoleonite, corsite
Nappaleder *n* nappa [leather]
Naramycin *n* naramycin
Narbe *f* scar, seam, (Leder) grain, (z. B. Med) mark
narben to mark, to scar, (Leder) to grain
Narbenbildung *f* scarring, (Metall) pitting
Narbenleder *n* grain leather
narbenlos unmarked, unscarred
Narbenseite *f* (Leder) grain side
narbig rough, scarred, (Leder) grained
Narbkalander *m* graining calender
Narbung *f* (Leder) grain
Narcein *n* narceine, narcine
Narceinhydrochlorid *n* narceine hydrochloride
Narceonsäure *f* narceonic acid
Narcissamin *n* narcissamine
Narcissidin *n* narcissidine
Narcissin *n* narcissine, lycorine
Narcotin *n* narcotine, opianine
Narcotinhydrochlorid *n* narcotine hydrochloride
Narcotolin *n* narcotoline
Nargol *n* nargol, silver nucleinate
Naringenin *n* naringenin, 4,5,7-trihydroxyflavanone
Naringeninsäure *f* naringeninic acid
Naringetol *n* naringetol
Naringin *n* naringin, aurantium
Narkophin *n* narcophine
Narkose *f* (Med) an[a]esthesia
Narkoseäther *m* ether for an[a]esthesia
Narkotikum *n* (Med) narcotic
Narkotin *n* narcotine, opianine
narkotisch an[a]esthetic, narcotic
narkotisieren (Labor) to narcotize, (Med) to an[a]esthetize, (mit Äther) to etherize
Narkotisieren *n* narcotizing
narkotisierend narcotizing
Narkotisierung *f* narcotization
narrensicher foolproof

Nartazin *n* nartazine
Narwedin *n* narwedine
Nase *f* nose, molded lug
Nasengasse *f* conduit pipe of the tuyère
Nasenkeil *m* gibhead key
Nasenschlacke *f* fusible dross
Nasenschmelzen *n* melting with a nose
Nasonit *m* (Min) nasonite
naß wet, (feucht) damp, humid, moist
Naßabriebfestigkeit *f* wet-scuffing resistance
Naßanalyse *f* wet analysis
Naßappretur *f* wet finishing
Naßaufbereitung *f* flotation, washing, wet cleaning
Naßauftrag *m* wet process
Naßauftragung *f* wet application
Naßbatterie *f* (Elektr) wet battery
Naßbehandlung *f* wet cleaning, wet processing
Naßbehandlungsechtheit *f* fastness to wet treatment
Naßbeize *f* wet dressing
Naßbleiche *f* wet bleach, sour bleaching, souring
Naßdampf *m* wet steam
naßecht fast to wet treatment, water-resistant
Naßechtheit *f* fastness to wetting, wet fastness property
Naßfestigkeit *f* wet strength
Naßfilter *n* wet filter
Naßgehalt *m* humidity, moisture contents
Naßgutzuführung *f* wet feed inlet
Naßhitzeausschrumpfung *f* wet heat shrinking
Naßkalk *m* milk of lime
naßkalt wet and cold
Naßkatalyseverfahren *n* wet catalysis process, wet contact process
Naßkompressionsmaschine *f* wet compression machine
Naßläufer *m* wet-dial water meter
Naßluftpumpe *f* wet air pump
Naßmahlen *n* wet grinding
Naßmahlung *f* wet grinding
Naßmetallurgie *f* hydrometallurgy
Naßmühle *f* mill for wet grinding
Naßölförderkapazität *f* (Ölgewinnung) total fluid withdrawal capacity
Naßoxydation *f* wet oxidation
Naßpochen *n* wet crushing
Naßpochwerk *n* wet stamp mill
Naßpressen *n* wet pressing
Naßprobe *f* moisture test
Naßpuddeln *n* wet puddling
Naßreinigung *f* washing, wet purification
Naßreißfestigkeit *f* wet tensile strength
Naßscheidung *f* wet separation
Naßscheuerfestigkeit *f* wet abrasion resistance
Naßsortierung *f* wet picking
Naßspinnen *n* wet-spinning
Naßverfahren *n* wet processing
Naßverklebung *f* wet bonding

Naßwandkolonne *f* wetted-wall column
Naßwischfestigkeit *f* wet rub fastness
Naßzerkleinerung *f* wet crushing
Nasturan *m* (Min) nasturan, pitch blende
naszieren to begin to exist
naszierend beginning to exist, (Chem) nascent
Natalensin *n* natalensine
Nataloe-Emodin *n* nataloe emodin
Nataloin *n* nataloin
Natanson-Regel *f* Natanson's rule
nativ native, natural
Natramid *n* sodamide, sodium amide
Natrium *n* sodium (Symb. Na), **essigsaures ~** sodium acetate, **kieselsaures ~** sodium silicate, **kohlensaures ~** sodium carbonate, **metallisches ~** sodium metal, **phosphorsaures ~** sodium phosphate, **salpetersaures ~** sodium nitrate
Natriumableitungsrohr *n* sodium leading-off tube
Natriumacetat *n* sodium acetate
Natrium aceticum *n* (Lat) sodium acetate
Natriumacetylid *n* sodium acetylide
Natriumäthylat *n* sodium ethylate, sodium ethoxide
Natriumäthylsulfat *n* sodium ethyl sulfate
Natriumalaun *m* sodium alum, aluminum sodium sulfate, soda alum
Natriumalkoholat *n* sodium alcoholate
Natriumalkylat *n* sodium alkylate, sodium alkoxide
Natriumaluminat *n* sodium aluminate
Natriumamalgam *n* sodium amalgam
Natriumamid *n* sodium amide, sodamide
Natriumammoniumphosphat *n* sodium ammonium phosphate, acid sodium ammonium phosphate, microcosmic salt
Natriumammoniumsulfat *n* sodium ammonium sulfate
Natriumamytal *n* sodium amytal
Natriumanhydropersulfat *n* anhydrous sodium persulfate
Natriumarsenat *n* sodium arsenate
Natriumarsenit *n* sodium arsenite
Natriumascorbat *n* sodium ascorbate
Natriumazid *n* sodium azide
Natriumbenzoat *n* sodium benzoate
Natrium benzoicum *n* (Lat) sodium benzoate
Natriumbenzolsulfonat *n* sodium benzene sulfonate
Natriumbicarbonat *n* sodium bicarbonate, acid sodium carbonate, baking soda, sodium hydrogen carbonate
Natriumbichromat *n* sodium bichromate, sodium dichromate
Natriumbisulfat *n* acid sodium sulfate, sodium bisulfate, sodium hydrogen sulfate
Natriumbisulfit *n* acid sodium sulfite, sodium bisulfite, sodium hydrogen sulfite

Natriumbitartrat *n* acid sodium tartrate, sodium bitartrate, sodium hydrogen tartrate
Natriumborat *n* sodium borate
Natriumborhydrid *n* sodium borohydride
Natriumbrechweinstein *m* antimony sodium tartrate, sodium antimonyl tartrate
Natriumbromid *n* sodium bromide
Natriumcarbeniat *n* sodium carbeniate
Natriumcarbonat *n* sodium carbonate
Natrium carbonicum *n* (Lat) soda, sodium carbonate
Natriumcaseinat *n* sodium caseinate
Natriumchlorat *n* sodium chlorate
Natriumchlorid *n* sodium chloride, common salt
Natriumcholat *n* sodium cholate
Natriumcholeinat *n* sodium choleinate
Natriumcinnamat *n* sodium cinnamate
Natriumcyanamid *n* sodium cyanamide
Natriumcyanid *n* sodium cyanide
Natriumcyclamat *n* sodium cyclamate
Natriumdampf *m* sodium vapor
Natriumdampf-Hochdrucklampe *f* sodium vapor high-pressure lamp
Natriumdampflampe *f* sodium-vapor lamp
Natriumdehydrocholat *n* sodium dehydrocholate
Natriumdiäthylbarbiturat *n* sodium diethylbarbiturate
Natriumdichromat *n* sodium bichromate, sodium dichromate
Natriumdijodsalicylat *n* sodium diiodosalicylate
Natriumdiuranat *n* sodium diuranate
Natriumdraht *m* sodium wire
Natriumelektrolyse *f* sodium electrolysis
Natriumferrisaccharat *n* ferric sodium saccharate, sodium ferric saccharate
Natriumflamme *f* sodium flame
Natriumfluorid *n* sodium fluoride
Natriumfluorsilikat *n* sodium fluorosilicate
Natriumformiat *n* sodium formate
Natriumformiatanlage *f* sodium formate plant
Natriumglycerophosphat *n* sodium glycerophosphate
Natriumgoldchlorid *n* gold sodium chloride, sodium aurochloride, sodium chloroaurate(III)
Natrium-Graphit-Reaktor *m* sodium-graphite reactor
natriumhaltig containing sodium
Natriumhexametaphosphat *n* sodium hexametaphosphate
Natriumhippurat *n* sodium hippurate
Natriumhydrazid *n* sodium hydrazide
Natriumhydrid *n* sodium hydride
Natriumhydrogencarbonat *n* sodium bicarbonate, acid sodium carbonate, baking soda, sodium hydrogen carbonate
Natriumhydrogensulfat *n* sodium hydrogensulfate

Natriumhydrogensulfid *n* sodium hydrogensulfide
Natriumhydro[gen]sulfit *n* sodium hydrogensulfite
Natriumhydrosulfit *n* sodium hydrosulfite
Natriumhydroxid *n* sodium hydroxide
Natriumhyperoxid *n* sodium dioxide, sodium peroxide
Natriumhypochlorit *n* sodium hypochlorite
Natriumhypophosphit *n* sodium hypophosphite
Natriumhyposulfit *n* sodium hyposulfite
Natriumion *n* sodium ion
Natriumjodat *n* sodium iodate
Natriumjodid *n* sodium iodide
Natriumkaliumtartrat *n* potassium sodium tartrate, sodium potassium tartrate
Natriumkarbonat *n* sodium carbonate
Natriumlampe *f* sodium vapor lamp
Natriumlinie *f* (Spektr) sodium line
Natriummalonat *n* sodium malonate, disodium malonate
Natriummetaarsenit *n* sodium metaarsenite
Natriummetagermanat *n* sodium metagermanate
Natriummetall *n* metallic sodium, sodium metal
Natriummetanilat *n* sodium metanilate
Natriummetaphosphat *n* sodium metaphosphate
Natriummethylsulfat *n* sodium methyl sulfate
Natriummonochromat *n* sodium [mono]chromate
Natriumniobat *n* sodium niobate
Natriumnitrat *n* sodium nitrate, chile saltpeter
Natriumnitrit *n* sodium nitrite
Natriumnitrobenzoat *n* sodium nitrobenzoate
Natriumnitrobenzolsulfonat *n* sodium nitrobenzene sulfonate
Natriumnitroprussiat *n* sodium nitroprussiate, sodium nitroferricyanide, sodium nitroprusside
Natriumnukleinat *n* sodium nucleinate
Natriumoleat *n* sodium oleate
Natriumoxidhydrat *n* (obs) sodium hydroxide
Natriumpantothenat *n* sodium pantothenate
Natriumperoxid *n* sodium peroxide
Natriumphenolat *n* sodium phenolate, sodium carbolate, sodium phenate
Natriumphenolsulfonat *n* sodium phenolsulfonate
Natriumphenyläthylbarbiturat *n* sodium phenylethyl barbiturate
Natriumphosphid *n* sodium phosphide
Natriumphosphit *n* sodium phosphite
Natriumplatinchlorid *n* sodium chloroplatinate(IV), sodium platinum chloride
Natriumplumbat *n* sodium plumbate
Natriumpolyphosphat *n* sodium polyphosphate
Natriumpolysulfid *n* sodium polysulfide
Natriumporphinchelat *n* sodium porphine chelate
Natriumpresse *f* sodium press
Natriumpumpe *f* (Biochem) sodium pump
Natriumpyrophosphat *n* sodium pyrophosphate
Natriumpyruvat *n* sodium pyruvate
Natriumsaccharinat *n* sodium saccharinate, saccharin sodium
Natriumsalicylat *n* sodium salicylate
Natriumsalz *n* sodium salt
Natriumschwefelverbindung *f* sodium sulfur compound
Natriumsesquicarbonat *n* sodium sesquicarbonate
Natriumsilikat *n* sodium silicate
Natriumstärkepräparat *n* sodium starch preparation
Natriumstannat *n* sodium stannate
Natriumstoffwechsel *m* (Physiol) sodium metabolism
Natriumsuccinat *n* sodium succinate
Natriumsulfat *n* sodium sulfate
Natriumsulfid *n* sodium sulfide, disodium sulfide
Natriumsulfit *n* sodium sulfite
Natriumsuperoxid *n* sodium peroxide, sodium dioxide
Natriumtetrametaphosphat *n* sodium tetrametaphosphate
Natriumtetraphenylporphin *n* sodium tetraphenylporphine
Natriumthiosulfat *n* sodium thiosulfate
Natriumtoluolsulfonat *n* sodium toluenesulfonate
Natriumtrimetaphosphat *n* sodium trimetaphosphate
Natriumtripolyphosphat *n* sodium tripolyphosphate
Natriumuranat *n* sodium uranate
Natriumvalerianat *n* sodium valerate
Natriumwolframat *n* sodium tungstate, sodium wolframate
Natriumzange *f* sodium pliers, sodium tongs
Natroborocalcit *m* (Min) natroborocalcite
Natrocalcit *m* (Min) natrocalcite, gaylussite
Natrolith *m* (Min) natrolite
Natron *n* sodium hydroxide, [caustic] soda, sodium carbonate, **doppelkohlensaures** ~ sodium bicarbonate, **kaustisches** ~ caustic soda, sodium hydroxide
Natronätzlauge *f* caustic soda solution, sodium hydroxide solution
Natronalaun *m* aluminum sodium sulfate, soda alum, sodium alum
Natronaluminat *n* sodium aluminate
Natronamalgam *n* sodium amalgam
Natronammonsalpeter *m* sodium ammonium nitrate
Natronanorthit *m* (Min) sodaic anorthite
Natronbleichlauge *f* soda bleaching lye
Natroncellulose *f* soda cellulose, soda pulp, sodium cellulose

Natroneisenalaun *m* sodium iron alum, sodium iron sulfate
Natronfeldspat *m* (Min) soda feldspar, albite
Natronglas *n* soda glass
Natronglimmer *m* (Min) paragonite
natronhaltig containing soda
Natronhydrat *n* sodium hydroxide, caustic soda
Natronhydratlösung *f* sodium hydroxide solution, caustic soda solution
Natronhyperoxid *n* sodium dioxide, sodium peroxide
Natronkalk *m* soda lime
Natronkalkfeldspat *m* (Min) andesine
Natronkalkglas *n* soda lime glass
Natronkraftpapier *n* sulfate kraft paper
Natronkupferlösung *f* soda and copper solution
Natronlauge *f* sodium hydroxide solution, caustic soda solution, **alkoholische** ~ alcoholic sodium hydroxide solution
Natronmesotyp *m* (Min) natrolite, needle stone
Natronmikroklin *m* (Min) sodium microcline
Natronorthoklas *m* (Min) sodium orthoclase
Natronpräparat *n* soda preparation
Natronsalpeter *m* sodium nitrate, chile saltpeter, soda niter
Natronsalz *n* sodium salt, soda salt
Natronschwefelleber *f* (obs) soda liver of sulfur, sodium sulfide
Natronsee *m* soda lake
Natronseife *f* hard soap, soda soap
Natronstaub *m* caustic soda dust
Natronwasserglas *n* soda waterglass, sodium silicate
Natronweinstein *m* sodium tartrate
Natronzellstoff *m* soda cellulose, soda pulp, sodium cellulose
Natrophilit *m* (Min) natrophilite
natürlich natural, native
Natur *f* nature, **in der** ~ **vorkommend** naturally occurring
Naturalgewicht *n* natural weight
Naturbenzin *n* natural gasoline
Naturerscheinung *f* natural phenomenon, natural process
Naturfarbe *f* natural color
naturfarben natural colored, self-colored
Naturfarbstoff *m* natural dyestuff
Naturfaser *f* natural fiber
Naturfett *m* natural fat, natural grease
Naturfett *n* (Wollfett) suint, wool grease
Naturforscher *m* scientist, naturalist
Naturforschung *f* natural science, scientific research
Naturgas *n* natural gas
Naturgerbstoff *m* natural tannin
Naturgesetz *n* law of nature, natural law
naturgetreu true to nature
Naturhärte *f* natural hardness

Naturharz *n* natural resin, gum rosin, vegetable resin
Naturkautschuk *m* natural rubber, crude rubber
Naturkraft *f* natural power
Naturkunde *f* natural science
Naturphilosophie *f* philosophy of nature
Naturprodukt *n* natural product
Naturseide *f* natural silk
Naturseidenabfall *m* waste from natural silk
Naturstein *m* natural stone
Naturstoff *m* natural substance, native substance
Naturumlaufkessel *m* natural circulation boiler
Natururan *n* natural uranium
Natururan-Graphit-Reaktor *m* natural uranium graphite reactor
Natururan-Reaktor *m* natural-uranium reactor
Naturwissenschaft *f* natural science
naturwissenschaftlich scientific, from the view point of natural science[s], pertaining to natural science
Naturzustand *m* natural state
Naumannit *m* (Min) naumannite
Navigation *f* navigation
Neapelgelb *n* Naples yellow, antimony yellow
Neapelgrün *n* chrome green, emerald green
Neapelrot *n* Naples red
Nebel *m* fog, (Astron) nebula, (leichter) haze, mist, (mit Rauch) smog
Nebelabscheidung *f* liquid mist removal
nebelartig foggy, mist-like
Nebelbildung *f* formation of mist
Nebelbildungsgrenzwert *m* cloud limit
Nebelblendscheibe *f* fog disc
Nebelhülle *f* nebular shell
Nebelkammer *f* [Wilson's] cloud chamber
Nebelkammeraufnahme *f* cloud chamber photography
Nebelkammerexpansion *f* cloud chamber expansion
Nebellinie *f* nebular line
Nebelsäure *f* smoke-screen acid
Nebelschleierverfahren *n* fog veil process
Nebelspur *f* cloud track, ~ **kosmischer Strahlen** cosmic ray track
Nebelspurbahn *f* cloud track
Nebelströmung *f* mist flow
Neben- accessory, additional, side
Nebenachse *f* (Krist) secondary axis
Nebenanlagen *pl* off-sites
Nebenapparate *pl* auxiliary appliances
Nebenauslaß *m* by-pass
Nebenausstrahlung *f* spurious radiation
Nebenbestandteil *m* minor constituent, secondary component, secondary constituent
Nebenbetrieb *m* secondary process, subsidiary business
Nebenbindung *f* secondary bond
Nebeneffekt *m* secondary effect

Nebeneigenschaft *f* secondary quality
nebeneinander side by side, adjacent, ~ **angeordnet** arranged side by side
nebeneinandergeschaltet parallel-connected
Nebeneinanderschaltung *f* connection in parallel, parallel connection
Nebenentladung *f* secondary discharge
Nebenerzeugnis *n* by-product
Nebenfarbe *f* secondary color
Nebenfrequenz *f* side frequency
Nebenfunktion *f* secondary function, tributary function
Nebengeräusch *n* background noise
Nebengestein *n* gang, gangue
Nebengruppe *f* subgroup
Nebenimpuls *m* satellite pulse
Nebenkanal *m* side drain, feeder
Nebenkasten *m* auxiliary container
Nebenleitung *f* by-pass
Nebenniere *f* adrenal gland
Nebennierenhormon *n* adrenal hormone
Nebennierenrinde *f* (Anat) adrenal cortex, suprarenal cortex
Nebennierenrindenextrakt *m* adrenocortical extract
Nebennierenrindenhormon *n* adrenal cortex hormone
Nebenprodukt *n* by-product
Nebenquantenzahl *f* secondary quantum number, subordinate quantum number
Nebenreaktion *f* side reaction, secondary reaction
Nebenreihe *f* auxiliary series
Nebenrohr *n* branch pipe, by-pass
Nebenschaltung *f* (Elektr) parallel [or shunt] connection
Nebenschilddrüsenhormon *n* parathyroid hormone
Nebenschluß *m* (Elektr) by-pass, shunt, **einen ~ bilden** to shunt, **im ~ zu** in shunt with, shunted across, **in den ~ legen** to [put in] shunt, **induktiver ~** inductive shunt, **magnetischer ~** [electro]magnetic leak, inductive shunt, magnetic shunt, **ohne ~** unshunted
Nebenschlußbogenlampe *f* shunt-wound arc lamp
Nebenschlußerregung *f* shunt excitation
Nebenschlußleitung *f* by-pass
Nebenschlußmagnet *m* shunt magnet
Nebenschlußmaschine *f* shunt dynamo
Nebenschlußmotor *m* shunt-wound motor
Nebenschlußregelung *f* shunt regulation
Nebenschlußschaltung *f* shunt connection
Nebenschlußstrom *m* shunt current
Nebenschlußstromkreis *m* shunt circuit
Nebenschlußverhältnis *n* shunt ratio
Nebenspeisehahn *m* supplementary feed cock

Nebenströmung *f* secondary flow, subsidiary flow
Nebenstrom *m* induction current
Nebenvalenz *f* secondary valence
Nebenweg *m* by-pass
Nebenwiderstand *m* shunt resistance
Nebenwirkung *f* secondary effect, secondary action, side effect
Nebenzweig *m* lateral branch
Nebularin *n* nebularine
Necrosamin *n* necrosamine
negativ negative, minus, ~ **elektrisch** electronegative, cathodal, negatively electric, resino-electric, ~ **geladen** negatively charged
Negativ *n* (Phot) negative
Negativität *f* negativeness, (Elektr) negativity
Negativpapier *n* (Phot) negative paper
Negativschaukasten *m* film illuminator, viewing screen
Negativverfahren *n* (Techn) female mold method, forming into female mold, plug assist forming
Negatron *n* negatron
Negatronenemission *f* (Atom) negatron emission
neigen to incline, to bend, (kippen) to tilt
Neigevorrichtung *f* tilting device
Neigung *f* slope, gradient, inclination, slant, tendency, trend
Neigungsebene *f* incline, slope
Neigungsmesser *m* inclinometer, tiltmeter
Neigungsschießen *n* dip shooting
Neigungswaage *f* inclination balance
Neigungswinkel *m* angle of inclination, angle of incline; draft angle
NE-Legierung *f* nonferrous alloy
Nelke *f* (Bot) clove
Nelkenöl *n* oil of cloves
Nelkenpfeffer *m* allspice, pimento
Nelkenstein *m* (Min) iolite
Nelkenwurzel *f* (Bot) avens root
Nemalith *m* (Min) nemalite, brucite
Nemaphyllit *m* (Min) nemaphyllite
Nemaqualith *m* nemaqualite
nematisch (Krist) nematic
Nematocid *n* nematocide
NE-Metall *n* nonferrous metal
Nemotin *n* nemotin
Nemotinsäure *f* nemotinic acid
Nennbelastung *f* rated load
Nenndrehzahl *f* rated speed
Nenndruck *m* static pressure
Nenner *m* (Math) denominator, **kleinster gemeinsamer ~** lowest common denominator
Nennfrequenz *f* (Elektr) rated frequency
Nennkapazität *f* rated capacity
Nennlast *f* nominal load
Nennleistung *f* nominal or rated capacity, normal output
Nennmaß *n* nominal size

Nennspannung *f* nominal voltage, rated voltage
Nenntemperatur *f* rated temperature
Nennweite *f* nominal width, nominal size
Nennwert *m* nominal value, face value, par value
Nennwertfehler *m* error of calibration
Neoabietinsäure *f* neoabietic acid
Neoagarobiose *f* neoagarobiose
Neoagarobit *m* neoagarobitol
Neoamylalkohol *m* neoamyl alcohol
Neoarsphenamin *n* neoarsphenamine
Neoasparaginsäure *f* neoaspartic acid
Neobotogenin *n* neobotogenin
Neocarthamin *n* neocarthamin
Neocerotinsäure *f* neocerotic acid
Neochanin *n* neochanin
Neochlorogensäure *f* neochlorogenic acid
Neocinchophen *n* neocinchophen
Neocolemannit *m* neocolemannite
Neocuproin *n* neocuproine
Neocyanin *n* neocyanine
Neodiarsenol *n* neodiarsenol
Neodym *n* neodymium (Symb. Nd)
Neodymcarbid *n* neodymium carbide
Neodymchlorid *n* neodymium chloride
Neodymcitrat *n* neodymium citrate
Neodymgehalt *m* neodymium content
Neodymoxid *n* neodymium oxide, neodymia
Neodymsulfat *n* neodymium sulfate
Neodymsulfid *n* neodymium sulfide
Neoergosterin *n* neoergosterol
Neogen *n* neogen
Neoglycerin *n* neoglycerol
Neohecogenin *n* neohecogenin
Neoherculin *n* neoherculin
Neohexan *n* neohexane, 2,2-dimethylbutane
Neoinosamin *n* neoinosamine
Neoinosit *m* neoinositol
Neoisoverbanol *n* neoisoverbanol
Neolactose *f* neolactose
Neolanblau *n* neolan blue
Neolin *n* neoline
Neolith *m* neolite
Neolithikum *n* (Geol) neolithic period
neolithisch neolithic
Neomenthol *n* neomenthol
Neomycin *n* neomycin
Neon *n* neon (Symb. Ne)
Neonbeleuchtung *f* neon light[ing]
Neongehalt *m* neon content
Neonindikator *m* neon indicator
Neonlampe *f* neon lamp
Neonlicht *n* neon light
Neonröhre *f* neon tube
Neonverflüssigungsmaschine *f* neon liquefier
Neopellin *n* neopelline
Neopentan *n* neopentane, 2,2-dimethylpropane
Neopentylalkohol *m* neopentyl alcohol
Neopentylbromid *n* neopentyl bromide
Neopentylglykol *n* neopentyl glycol

Neopentyl-Umlagerung *f* (Chem) neopentyl rearrangement
Neophylchlorid *n* neophyl chloride
Neophytadien *n* neophytadiene
Neopin *n* neopine
Neoplasma *n* (Med) neoplasm
Neopren *n* neoprene, duprene
Neoreserpsäure *f* neoreserpic acid
Neoretinen *n* neoretinene
Neosaman *n* neosaman
Neosin *n* neosine
Neostigmin *n* neostigmine
Neostrychnin *n* neostrychnine
Neotannyl- neotannyl
Neotantalit *m* (Min) neotantalite
Neoteben *n* neoteben
Neotrehalose *f* neotrehalose
Neotyp *m* (Min) neotype
Neotyrosin *n* neotyrosine
Neovitamin *n* neovitamin
Neozoikum *n* (Geol) Neozoic period
neozoisch (Geol) neozoic
Nepetalinsäure *f* nepetalinic acid
Nepetalsäure *f* nepetalic acid
Nepetolsäure *f* nepetolic acid
Nepetonsäure *f* nepetonic acid
Nepetsäure *f* nepetic acid
Nephelin *m* (Min) nepheline, nephelite
nephelinartig nephelinic
Nephelinbasalt *m* (Min) nephelinite
Nephelinit *m* (Min) nephelinite
Nephelometer *n* nephelometer
Nephelometrie *f* nephelometric analysis, turbidimetric analysis
nephelometrisch nephelometric
Nephrin *n* nephrine
Nephrit *m* (Min) nephrite, axstone, jade
Nephritoid *m* nephritoid
Nephrosteransäure *f* nephrosteranic acid
Nephrosterinsäure *f* nephrosterinic acid
Nepodin *n* nepodin
Nepouit *m* (Min) nepouite
Neptun *m* (Astr) Neptune
Neptunblau *n* neptune blue
Neptunit *m* (Min) neptunite
Neptunium *n* neptunium (Symb. Np)
Neral *n* neral, β-citral
Neriantin *n* neriantin
Nernstbrenner *m* Nernst burner
Nernsteffekt *m* Nernst effect
Nernstlampe *f* Nernst lamp
Nernstsches Verteilungsgesetz *n* distribution law of Nernst, Nernst theory
Nernstsches Wärmetheorem *n* Nernst heat theorem
Nernststift *m* Nernst glower, Nernst rod
Nerol *n* nerol
Nerolicampher *m* neroli camphor
Nerolidol *n* nerolidol, peruviol

Nerolin *n* nerolin
Neroliöl *n* neroli oil, orange flower oil
Nerv *m* nerve, ~ **des autonomen Systems** autonomic nerve, **sympathischer** ~ sympathetic nerve, gangliated nerve
Nervenberuhigungsmittel *n* (Pharm) nervous depressant, sedative
Nervenfaser *f* nerve fiber
Nervenfibrille *f* neurofibril
Nervengas *n* nerve gas
Nervengewebe *n* nerve tissue
Nervengift *n* neuroaralysant, neurotoxin
Nervenknoten *m* ganglion
Nervenlähmung *f* (Med) neuroparalysis
Nervenlehre *f* (Med) neurology
Nervenleiden *n* (Med) nervous disease, neuropathy
Nervensystem *n*, **autonomes** ~ autonomic nervous system, **parasympathisches** ~ parasympathetic nervous system, **peripheres** ~ peripheral nervous system, **sympathisches** ~ sympathetic nervous system, **vegetatives** ~ autonomic nervous system, **zentrales** ~ central nervous system
Nervenzelle *f* nerve cell, neurocyte, neuron
Nervenzentrum *n* nerve center
Nervon *n* nervon
Nervonsäure *f* nervonic acid, selacholeic acid, tetracosenoic acid
Nervosität *f* nervousness, nervous irritation
Nerylalkohol *m* neryl alcohol
Nerz *m* mink
Nerzöl *n* mink fat
Nesquehonit *m* (Min) nesquehonite
Nessel *f* (Bot) nettle
Nesselfieber *n* nettle-rash
Nesselgift *n* nettle poison
Nesselpolierscheibe *f* muslin polishing wheel
Nesselseide *f* (Text) nettle silk
Nesslers Reagens *n* Nessler's reagent, Nessler's solution
Nettogewicht *n* net weight
Nettolast *f* net load
Nettotransport *m* net transport
Netz *n* network, gauze, net, (Elektr) grid, **Leitungs-** ~ (Elektr) main
Netzanode *f* grid anode
Netzanschluß *m* (Elektr) power connection, electric main, mains connection
Netzanschlußgerät *n* power pack
Netzapparat *m* damping machine
Netzbarkeit *f* wettability
Netzbeize *f* (Färb) oil mordant
Netzboden *m* (Dest) wire-mesh tray
Netzebene *f* (Krist) lattice plane, atomic plane
Netzebenenabstand *m* interplanar spacing
Netzelektrode *f* net-shaped electrode, wire gauze electrode

netzen to moisten, to damp, to wet
netzförmig net-shaped, reticular, reticulate[d]
Netzgerät *n* (Elektr) power pack, power supply unit
Netzgleichrichter *m* (Elektr) full-wave rectifier
Netzhaut *f* (Med) retina
Netzkatalysator *m* catalyst gauze
Netzkessel *m* steeping vat
Netzkonstanthalter *m* (Elektr) voltage stabilizer
Netzmittel *n* wetting agent
Netzplantechnik *f* critical path analysis
Netzpolymer *n* network polymer
Netzschwefel *m* wettable sulfur
Netzspannung *f* (Elektr) line voltage, mains voltage
Netzspule *f* mains coil
Netzstörung *f* (Elektr) line disturbance
Netzstrom *m* line current
Netzstruktur *f* reticular structure
Netzteil *n* (Elektr) power supply unit
Netzwerk *n* network
neuartig novel
Neubau *m* building under construction, new building
Neubelegen *n* (Walze) recoating, relining
Neubergblau *n* Neuberg blue
Neubildung *f* new formation
Neubildungsgeschwindigkeit *f* regeneration velocity, velocity of regeneration
Neublau *n* new blue, Saxon blue
Neuburger Kieselkreide *f* (Min) Neuburg silicious chalk
Neuburger Kieselweiß *n* Neuburg silicious whiting
Neucoccin *n* neococcin, new coccin
Neudefinition *f* new definition, redefinition
Neueichung *f* recalibration
Neueinstellung *f* readjustment
Neuerung *f* innovation
Neufuchsin *n* new fuchsin
Neugelb *n* new yellow, massicot
Neugestaltung *f* reformation, reorganization
Neugewürz *n* (Bot) allspice, pimento
Neugold *n* Mannheim gold
Neugrün *n* new green
Neuheit *f* novelty
Neukonstruktion *f* new construction, revised design
Neumessing *n* malleable brass, yellow brass
Neuneck *n* enneagon, nonagon
neunflächig enneahedral
Neunflächner *m* (Krist) enneahedron
neunhubig nine-stroke
Neunzigflächner *m* (Krist) enneacontahedron
Neuralgie *f* (Med) neuralgia
Neuralthein *n* neuraltheine
Neuraminidase *f* (Biochem) neuraminidase
Neuraminsäure *f* neuraminic acid
Neuridin *n* neuridin

Neurin *n* neurine
Neurobiologie *f* neurobiology
Neurochemie *f* neurochemistry
Neurodin *n* neurodine
Neurofebrin *n* neurofebrin
Neurohormon *n* neurohormone
Neuroleptikum *n* (Pharm) neuroleptic
Neuron *n* neuron, neurocyte
Neuronal *n* neuronal, bromodiethyl acetamide, diethyl bromacetamide
Neurophysiologie *f* neurophysiology
Neuroplasma *n* neuroplasm
Neurosporin *n* neurosporene
Neurot *n* new red
Neurotropin *n* neurotropine
Neurovirus *m* neurovirus
Neusilber *n* argentan, German silver, nickel silver
Neusolidgrün *n* new solid green
neutral neutral, indifferent, inert, nonpolarized, (unparteiisch) impartial, ~ **reagieren** (Chem) to react neutral, to show a neutral reaction
Neutralbase *f* neutral base
Neutralblau *n* neutral blue
Neutralfarbe *f* neutral tint
Neutralfilter *n* (Opt) neutral density filter
Neutralisation *f* neutralization
Neutralisationsenthalpie *f* neutralization enthalpy
Neutralisationstitration *f* neutralizing titration
Neutralisationswärme *f* (Therm) heat of neutralization
Neutralisationszahl *f* neutralization number
Neutralisieranlage *f* neutralization plant
neutralisieren to neutralize
neutralisierend neutralizing
Neutralisierung *f* neutralization
Neutralisierungswärme *f* heat of neutralization
Neutralität *f* neutrality
Neutralon *n* neutralon
Neutralpunkt *m* neutral point, zero point
Neutralrot *n* neutral red
Neutralsäure *f* neutral acid
Neutralsalz *n* neutral salt
Neutralviolett *n* neutral violet
Neutretto *n* neutral meson
Neutrino *n* (Atom) neutrino
Neutron *n* neutron (s. auch Neutronen), ~ **der kosmischen Strahlung** cosmic ray neutron, ~ **hoher Energie** high energy neutron, **epithermisches** ~ epithermal neutron, **mittelschnelles** ~ intermediate neutron, **verzögertes** ~ (Atom) delayed [fission] neutron
Neutronen *pl*, **langsame** ~ slow neutrons, thermal neutrons, **schnelle** ~ high-speed neutrons, **thermische** ~ slow neutrons, thermal neutrons, **vagabundierende** ~ stray neutrons

Neutronenablagerung *f* neutron deposit
Neutronenabschirmung *f* neutron shielding
Neutronenabschwächung *f* neutron attenuation
Neutronenabsorber *m* neutron absorber
neutronenabsorbierend neutron-absorbing
Neutronenabsorption *f* neutron absorption
Neutronenaktivierungsanalyse *f* neutron activation analysis
Neutronen-Alpha-Reaktion *f* neutron-alpha reaction
Neutronenalter *n* (Atom) neutron age
Neutronenanlagerung *f* neutron deposit
Neutronenausbeute *f* neutron yield, neutron efficiency
Neutronenbehandlung *f* neutron treatment
Neutronenbeschießung *f* neutron bombardment
Neutronenbeschuß *m* neutron bombardment
Neutronenbestrahlung *f* neutron irradiation
Neutronenbeugung *f* neutron diffraction
Neutronenbeugungsaufnahme *f* neutron diffraction pattern
Neutronenbindungsenergie *f* (Atom) neutron binding energy
Neutronenbombardement *n* neutron bombardment
Neutronenbündel *n* neutron beam
Neutronendichte *f* (Atom) neutron density
Neutronendichteverteilung *f* neutron density distribution
Neutronendiffusion *f* neutron diffusion
Neutronendosismessung *f* neutron dosage measurement
neutronendurchlässig neutron-transparent
Neutroneneinfang *m* neutron capture, **parasitärer** ~ parasitic neutron capture
Neutronenemission *f*, **verzögerte** ~ (Atom) delayed neutron emission
Neutronenenergie *f* neutron energy
Neutronenfänger *m* neutron absorber, neutron robber
Neutronenfluß *m* neutron flux, ~ **im Reaktorzentrum** central pile flux of neutrons
Neutronenflußmessung *f* neutron flux measurement
Neutronen-Gamma-Reaktion *f* neutron-gamma reaction
Neutronengenerator *m* neutron generator
Neutronengeschoß *n* neutron bullet
Neutronengeschwindigkeit *f* (Atom) neutron velocity
Neutronennachweis *m* neutron detection
Neutronennachweistechnik *f* neutron detection technique
Neutronen-Neutronen-Reaktion *f* neutron-neutron reaction
Neutronenoptik *f* neutron optics
Neutronenphysik *f* neutron physics
Neutronenpolarisation *f* neutron polarization
Neutronenquelle *f* source of neutrons

Neutronenquellstärke *f* neutron source strength
Neutronenquerschnitt *m* neutron cross section
Neutronenradiographie *f* neutron radiography
Neutronenreflektor *m* neutron reflector
Neutronenresonanz *f* neutron resonance
Neutronenresonanzlinie *f* neutron resonance line
Neutronenschutz *m* neutron shield
Neutronenschwächung *f* (Atom) neutron attenuation
neutronenspeiend shooting out neutrons
Neutronenspektrometer *n* neutron spectrometer
Neutronenspektroskopie *f* neutron spectroscopy
Neutronenspektrum *n* neutron spectrum
Neutronenstrahl *m* neutron beam
Neutronenstrahlung *f* neutron radiation
Neutronenstreuung *f* neutron scattering
Neutronenstrom *m* (Atom) neutron current
Neutronentransmission *f* neutron transmission
Neutronenüberwachungsgerät *n* neutron monitor
Neutronenverlust *m* loss of neutrons
Neutronenverschluß *m* neutron shutter
Neutronenvervielfachung *f* multiplicity of neutrons
Neutronenwellenlänge *f* neutron wave-length
Neutronenzähler *m* neutron counter
Neutronenzerfall *m* neutron decay
Neutron-Proton-Reaktion *f* neutron-proton reaction
Neuurotropin *n* new urotropine
Neuviktoriablau *n* new Victoria blue
Neuweiß *n* permanent white
Neuwiederblau *n* blue verditer
Neuwiedergrün *n* Paris green
Nevillesäure *f* 1-naphthol-4-sulfonic acid, Neville acid
Newberryit *m* (Min) newberryite
Newporthit *m* newporthite
Newton *n* (Maß) newton
Newtonit *n* newtonite
Newton-Legierung *f* Newton's alloy
Newtonsche Ringe (Opt) Newton['s] rings
Newtonsche Flüssigkeit *f* Newtonian fluid
Newtonsche Strömung *f* Newtonian flow
NF-Messung *f* (Elektr) audiometry
NF-Spannung *f* (Elektr) low-frequency voltage
NF-Verstärker *m* (Elektr) low-frequency amplifier
Ngai-Campher *m* Ngai-camphor
Niacin *n* niacin
Niauliöl *n* niauli oil
Niccolit *m* (Min) niccolite
Nichin *n* nichine
Nichtadditivität *f* non-additivity
nicht aktiviert non-activated
nicht anzeigend non-indicating
nichtausfasernd non-fraying
nichtbrennbar non-flammable, non-inflammable
Nichtcarbonathärte *f* (Wasser) non-carbonate hardness

Nichteisenmetall *n* non-ferrous metal
Nichtelektrolyt *m* non-electrolyte
nichtentflammbar flame-resistant, non-inflammable
Nichtentwicklung *f* non-development
nichterkennbar kristallinisch indistinctly crystalline
Nichterscheinen *n* default, non-appearance
nichteuklidisch non-euclidian
nichtexistenzfähig incapable of existence
nichtexplosiv nonexplosive
nichtfaserig nonfibrous
nichtfilzend non-felting
nichtflüchtig non-volatile
nichtgasförmig non-gaseous
nichthomogen inhomogeneous, nonhomogeneous
nichtideal nonideal
nichtig futile, invalid, null, void
Nichtigerklärung *f* annulment
Nichtigkeit *f* futility, invalidity, nullity
nichtkompetitiv (Kinetik) non-competitive
nichtkreidend non-chalking
nichtkristallin noncrystalline
nichtkugelförmig aspherical
nichtleitend non-conducting, (isolierend) insulating
Nichtleiter *m* non-conductor, insulator
nichtleuchtend non-luminous
nichtlinear non-linear
Nichtlinearität *f* non-linearity
nichtlöslich insoluble, nonsoluble
Nichtmetall *n* non-metal, metalloid, non-metallic element
nichtmetallisch non-metallic
nichtmischbar immiscible, incapable of mixing
Nichtmischbarkeit *f* immiscibility, nonmiscibility
Nicht-Newtonsches Medium *n* non-Newtonian medium
nichtoxydierbar unoxidizable
nichtperiodisch aperiodic, nonperiodic
nichtreduzierbar irreducible, not reducible
nichtreversibel irreversible
nichtrostend non-corrosive, non-rusting, rustproof, stainless
nichtrußend sootless
nichts nothing
nichtschäumend nonfoaming
nichtschmelzbar infusible, non-fusible, non-meltable
nichtschrumpfend non-shrinkable
nichtseparierbar non-separable
nichtsplitternd splinter-proof
nichtstationär nonsteady, dynamic
nichtsymmetrisch asymmetric, non-symmetric, unilateral
nichttrocknend non-drying
nichtumkehrbar irreversible, nonreversible

nichtunterscheidbar indistinguishable
nichtverfärbend non-staining
nichtverfilzend non-felting
nichtvergasbar nongasifiable, nonvolatile
Nichtwärmeleiter *m* non-conductor of heat
nichtwäss[e]rig non-aqueous
Nickel *n* nickel (Symb. Ni), **feinverteiltes** ~ finely divided nickel, **kolloidales** ~ colloidal nickel
Nickelammoniumsulfat *n* ammonium nickel(II) sulfate, nickel[ous] ammonium sulfate
Nickelammoniumverbindung *f* nickel ammonium compound
Nickelantimonglanz *m* (Min) native nickel sulfantimonide, ullmannite
Nickelantimonkies *m* (Min) native nickel sulfantimonide, ullmannite
Nickelarsenglanz *m* (Min) gersdorffite, plessite
Nickelarsenik *n* nickel arsenide
Nickelarsenikkies *m* (Min) gersdorffite
nickelartig nickel-like
Nickelauskleidung *f* nickel lining
Nickelbad *n* nickel bath
Nickelbeize *f* nickel mordant
Nickelbeschlag *m* nickel fitting, nickel mounting
Nickelblech *n* nickel sheet
Nickelblüte *f* (Min) annabergite, nickel bloom
Nickelbronze *f* nickel bronze
Nickelcarbonat *n* nickel(II) carbonate, nickelous carbonate
Nickelcarbonyl *n* nickel [tetra]carbonyl
Nickelchelat *n* nickel chelate
Nickelchinaldinat *n* nickel quinaldinate
Nickelchlorür *n* (Nickel(II)-chlorid) nickel chloride, nickel(II) chloride, nickelous chloride
Nickelchrom *n* nickel chromium
Nickelcyanidkomplex *m* nickel cyanide complex
Nickelcyanür *n* (Nickel(II)-cyanid) nickel(II) cyanide, nickelous cyanide
Nickeleisen *n* ferronickel, nickel iron
Nickelerz *n* nickel ore
Nickelfeilspäne *m pl* nickel filings
Nickelflußeisen *n* nickel steel
Nickelformiat *n* nickel formate
Nickelgalvanoplastik *f* nickel galvanoplastic process
Nickelgefäß *n* nickel vessel
Nickelgehalt *m* nickel content
Nickelgewinnung *f* nickel production
Nickelglanz *m* (Min) nickel glance, gersdorffite
Nickelgrün *n* nickel green
Nickelgymnit *m* (Min) nickel gymnite, genthite
Nickelhärtungsverfahren *n* nickel hardening method
nickelhaltig containing nickel, nickeliferous
Nickelhydroxydul *n* (Nickel(II)-hydroxid) nickel(II) hydroxide, nickelous hydroxide

Nickelicyanwasserstoffsäure *f* cyanonickelic acid, nickelicyanic acid
Nickel(III)-oxid *n* nickelic oxide, nickel(III) oxide
Nickel(III)-oxidhydrat *n* nickelic hydroxide, nickel(III) hydroxide
Nickel(III)-Salz *n* nickelic salt, nickel(III) salt
Nickel(III)-Verbindung *f* nickelic compound, nickel(III) compound
Nickel(II)-oxid *n* nickel(II) oxide, nickelous oxide
Nickel(II)-Salz *n* nickel(II) salt, nickelous salt
Nickel(II)-Verbindung *f* nickel(II) compound, nickelous compound
Nickelin *m* neckeline
Nickeliverbindung *f* (Nickel(III)-Verbindung) nickelic compound, nickel(III) compound
Nickeljodat *n* nickel iodate
Nickelkies *m* (Min) nickel pyrites, millerite
Nickelkobaltkies *m* (Min) cobalt nickel pyrite, linnaeite
Nickelkobaltlegierung *f* nickel cobalt alloy
Nickelkohlenoxid *n* nickel carbonyl, nickel tetracarbonyl
Nickelkupfer *n* nickel copper
Nickelmangan *n* nickel manganese
Nickelmonoxid *n* nickel(II) oxide, nickel monoxide, nickelous oxide
Nickelnitrat *n* nickel nitrate
Nickelocyanwasserstoffsäure *f* cyanonickelous acid, nickelocyanic acid
Nickelonickelioxid *n* nickel(II)(III) oxide, nickelous nickelic oxide
Nickeloxalat *n* nickel oxalate
Nickeloxidammoniaklösung *f* ammoniacal nickelic oxide solution, nickel(III) oxide ammonia solution
Nickeloxydul *n* (Nickel(II)-oxid) nickel(II) oxide, nickel monoxide, nickelous oxide
Nickeloxydulhydrat *n* (obs) nickel(II) hydroxide, nickelous hydroxide
Nickeloxydulsalz *n* (obs) nickel(II) salt, nickelous salt
Nickeloxydulsulfat *n* (obs) nickel(II) sulfate, nickelous sulfate
Nickelschwamm *m* spongy nickel
Nickelsmaragd *m* (Min) emerald nickel, zaratite
Nickelspeise *f* (Min) nickel speiss
Nickelspießglanz *m* (Min) ullmannite
Nickelstahl *m* nickeliferous steel, nickel steel
Nickelstahlguß *m* nickel steel casting
Nickelstein *m* (Min) nickel mat, nickelmatte
Nickelsud *m* boiling nickel bath
Nickelsulfat *n* nickel sulfate, (Nickel(II)-sulfat) nickel(II) sulfate, nickelous sulfate, nickel vitriol
Nickelsulfid *n* nickel sulfide
Nickelsulfür *n* (Nickel(II)-sulfid) nickel(II) sulfide, nickelous sulfide

Nickeltetracarbonyl *n* nickel tetracarbonyl
Nickelthioglykolat *n* nickelous thioglycolate
Nickeltiegel *m* nickel crucible
Nickelüberzug *m* nickel plating
Nickelvitriol *n* nickel vitriol, nickel(II) sulfate, nickelous sulfate
Nickelzusatz *m* addition of nickel
Nicolsches Prisma *n* (Opt) Nicol's prism
Nicomorphin *n* nicomorphine
Nicopholin *n* nicopholine
Nicotein *n* nicotein[e]
Nicotellin *n* nicotelline
Nicotianin *n* nicotianine
Nicotimin *n* nicotimine
Nicotin *n* nicotin[e]
Nicotinamid *n* nicotinamide
nicotinfrei free from nicotine
Nicotingehalt *m* nicotine content
Nicotinsäure *f* nicotinic acid, 2-pyridine carboxylic acid, niacin
Nicotinsalz *n* nicotine salt
Nicotinursäure *f* nicotinuric acid
Nicotinvergiftung *f* (Med) nicotine poisoning, nicotinism
Nicotinylalkohol *m* nicotinylalcohol
Nicotyrin *n* nicotyrine
Niederblasen *n* blowing-out
Niederdruck *m* low pressure
Niederdruckdampf *m* low-pressure steam
Niederdruckdampfmaschine *f* low-pressure steam engine
Niederdruckentladung *f* low-pressure discharge
Niederdruckkessel *m* low-pressure boiler, low-pressure tank, low-pressure vessel
Niederdruckkolben *m* low-pressure plunger
Niederdrucklüfter *m* low-pressure fan
Niederdruckpolyäthylen *n* high density polyethylene, low-pressure polyethylene
Niederdruckpressen *n* low-pressure molding
Niederdruckpreßverfahren *n* low-pressure molding
Niederdruckreifen *m* low-pressure tire
Niederdrucksäule *f* low-pressure column
Niederdruckschichtstoff *m* low-pressure plastic
Niederdruckverfahren *n* (Polyester) pressure bag molding
niederdrücken (Hebel) to press or force down (a lever), (Knopf) to push [down] (a button)
Niederflußreaktor *m* lowflux reactor
Niederflußreaktorkern *m* lowflux reactor core
niederfrequent low-frequency
Niederfrequenz *f* low frequency, audiofrequency
Niederfrequenzstrom *m* low-frequency current
Niederfrequenztransformator *m* audio-transformer
Niedergang *m* (Kolben) down[ward] stroke
Niedergehen *n* **der Gicht** dropping of the charge
Niederhalter *m* holddown plate
Niederhubwagen *m* low-lift truck

Niederlegung *f* lowering, (Abdankung) abdication
Niederleistungsreaktor *m* low-power reactor
niedermolekular low-molecular
niederohmig low-resistance
niederreißen to demolish, to tear down, (z. B. Niederschlag) to carry down
Niederschlag *m* precipitate, deposit, sediment, (durch Kondensation) condensate, (Kesselstein) incrustation, (Regen) precipitation, rainfall, **aktiver** ~ active deposit, active precipitate, **atmosphärischer** ~ [atmospheric] precipitation, **feinkörniger** ~ fine-granular precipitate, **flockiger** ~ flaky precipitate, flocculent precipitate, **galvanischer** ~ electrodeposit, **käsiger** ~ curds, curdy precipitate, **periodischer** ~ periodic precipitation, **schlammiger** ~ muddy deposit
niederschlagbar precipitable
Niederschlagbarkeit *f* precipitability
niederschlagen (Chem) to precipitate, to sediment, (fest) to deposit, to settle, (flüssig) to condense
Niederschlagen *n* precipitation, sedimentation
Niederschlagsapparat *m* precipitator
Niederschlagsarbeit *f* precipitation process, reduction process
niederschlagsarm poor in precipitate
Niederschlagselektrode *f* precipitation electrode
Niederschlagsgefäß *n* precipitation vessel
Niederschlagsgewicht *n* weight of the deposit
Niederschlagsmittel *n* precipitant, precipitating agent
Niederschlagsraum *m* condensating space
Niederschlagstrog *m* deposit tank, precipitation tank
Niederschlagsverfahren *n* precipitating process
Niederschlagswärme *f* heat of condensation
Niederschlagswasser *n* condensation water, aqueous condensate, condensed water, water of condensation
Niederschlagswasserabscheider *m* steam trap
niederschmelzen to melt down, to smelt
Niederspannung *f* (Elektr) low tension, low voltage
Niederspannungsanlage *f* low-tension plant
Niederspannungserzeuger *m* low-voltage generator, low-tension dynamo
Niederspannungsgehäuse *n* low-tension shell
Niederspannungsisolator *m* low-tension insulator
Niederspannungskabel *n* low-tension cable
Niederspannungskraftquelle *f* low-potential source of supply
Niederspannungsleitung *f* low-tension line
Niederspannungsstrom *m* low-voltage current, low-tension current
Niederspannungszündung *f* low-tension ignition

Niedertemperaturfraktionierung f low-temperature rectification
Niedertemperaturreaktor m low-temperature reactor
niedertourig low-speed, slow-speed, with a low number of revolutions
niedertransformieren (Elektr) to step down
Niedervoltbogen m low-voltage arc
niedervoltig low-voltage
niedrig low
niedrigschmelzend low-melting
niedrigsiedend low-boiling
Niedrigsieder m low-boiling solvent
niedrigviskos of low viscosity
Niere f (Geol) kidney ore, nodule, (Med) kidney[s]
Nierenerz n (Min) kidney ore, nodular ore
nierenförmig kidney-shaped, reniform
Nierenmittel n (Pharm) kidney remedy, nephritic
Nierenstein m (Med) nephritic stone, (Min) kidney stone, nephrite, spherulite
Nieseln n drizzle
Nieskraut n (Bot) sneeze wort
Nieswurz[el] f (Bot) hellebore
Nieswurztinktur f tincture of green hellebore
Niet m rivet
Niete f rivet
Nieteisen n rivet iron
nieten to rivet
Nieten n riveting
Nietkluppe f riveting tongs
Nietkopf m rivet head
Nietmutter f stop nut
Nietschaft m rivet shank
Nietsenkkopf m countersunk rivet head
Nietteilung f spacing of rivets
Nietung f rivet joint, riveting
Nietverbindung f rivet[ed] joint
Nietwerkzeug n riveting tool
Nietzange f riveting tongs
Nigellaöl n nigella oil, small fennel oil
Nigerin n nigerine
Nigerose f nigerose
Nigranilin n nigraniline, aniline black, nigrosine
Nigrescit m nigrescite
Nigrin n nigrine
Nigrometer n nigrometer
Nigrosin n nigrosine, aniline black, nigraniline
Nikethamid n nikethamide
Nikotin n nicotine
Nikotinamid n nicotinamide
Nikotinamidadenindinukleotid n nicotinamide adenine dinucleotide
Nikotinamidmononukleotid n nicotinamide mononucleotide
nikotinarm with a low nicotine content
nikotinempfindlich sensitive to nicotine
nikotinfrei free from nicotine

Nikotingehalt m nicotine content
Nikotinsäure f nicotinic acid
Nikotinvergiftung f nicotine poisoning, nicotinism
Nilblau n nile blue
Ninhydrin n ninhydrin
Ninhydrinreaktion f ninhydrine reaction
Niob n niobium (Symb. Nb), (obs) columbium
Niobat n niobate, (obs) columbate
Niob(III)-Oxid n niobium(III) oxide, niobous oxide
Niob(III)-Verbindung f niobous compound, niobium(III) compound, (obs) columbous compound
Niobit m (Min) niobite, columbite
Niobium n niobium (Symb. Nb), (obs) columbium
Nioboxid n niobium oxide, (obs) columbium oxide
Niobpentafluorid n niobic fluoride, niobium pentafluoride, niobium(V) fluoride, (obs) columbic fluoride, columbium pentafluoride
Niobpentoxid n niobium pentoxide, niobic anhydride, niobic oxide, niobium(V) oxide, (obs) columbium pentoxide
Niobsäure f niobic acid, (obs) columbic acid
niobsauer niobate, (obs) columbate
Niob(V)-Oxid n niobic oxide, niobium(V) oxide
Niob(V)-Verbindung f niobic compound, niobium(V) compound, (obs) columbic compound
Niobwasserstoff m niobium hydride
Nioxim n nioxime
Nipecotinsäure f nipecotic acid
Niphos-Verfahren n (Korrosionsschutz) Niphos process
Nippel m nipple
Nirosta n (HN) rustless steel
Nirvanin n nirvanine
Nirvanol n nirvanol
Niton n (obs) niton, radium emanation, radon
Nitragin n nitragin
Nitramid n nitramide
Nitramin n nitramine
Nitranilin n nitraniline
Nitranilinchlorhydrat n nitraniline hydrochloride
Nitranilinorange n nitraniline orange
Nitranilinrot n nitraniline red
Nitranilinsulfonsäure f nitranilinesulfonic acid
Nitranilsäure f nitranilic acid, dinitrodihydroxy-benzoquinone
Nitranisol n nitranisol
Nitrat n nitrate, salt or ester of nitric acid
Nitratanlage f nitrate plant
Nitratbeize f nitrate mordant
Nitratfilm m (Phot) nitrate film
Nitratin n nitratine
Nitratmaschine f nitrate machine

Nitratocholin *n* choline nitrate
Nitratreduktase *f* (Biochem) nitrate reductase
Nitrazingelb *n* nitrazine yellow
Nitren *n* nitrene
Nitrenium-Ion *n* nitrenium ion
Nitrid *n* nitride
Nitridieren *n* nitriding
Nitrieranlage *f* nitrating plant, (Metall) nitriding equipment
Nitrierapparat *m* nitrating apparatus
nitrierbar (Bakt) nitrifiable
Nitrierdauer *f* duration of the nitrating process
nitrieren to nitrate, (Bakt) to nitrify, (Met) to nitride
Nitrieren *n* (Bakt) nitrifying, (Chem) nitration, (Met) nitriding, nitrogen-hardening
nitrierend nitrating, (Bakt) nitrifying
Nitriergemisch *n* nitrating mixture
nitrierhärten to nitride
Nitrierhärten *n* nitrogen-hardening, nitriding
Nitrierhärteverfahren *n* nitriding process, nitrogen-hardening process
Nitrierhärtung *f* nitrogen-hardening, nitriding process
Nitrierkasten *m* nitriding box
Nitrierofen *m* nitration furnace, nitriding furnace
Nitrierpapier *n* nitration paper
Nitriersäure *f* nitrating acid
Nitrierstahl *m* nitrosteel, nitersteel
Nitriertemperatur *f* nitrating temperature
Nitriertiefe *f* depth of nitration
Nitriertopf *m* nitrating vessel
Nitrierung *f* (Bakt) nitrification, (Chem) nitration, (Met) nitridation, nitriding
Nitrierungsgrad *m* degree of nitration, nitridation
Nitrierungshärte *f* nitridation hardness
Nitrierungsprodukt *n* nitration product
Nitrierungsstufe *f* stage of nitration
Nitrierungsvorgang *m* nitrating process
Nitrierverfahren *n* (Met) nitriding process
Nitrierwirkung *f* (Met) nitriding action
Nitrierzentrifuge *f* nitrating centrifuge
Nitrifikation *f* nitrification
nitrifizieren (Bakt) to nitrify
Nitril *n* nitrile
Nitrilbase *f* nitrile base
Nitrilgruppe *f* nitrile group
Nitrilkautschuk *m* nitrile rubber
Nitrilophosphorsäure *f* phosphoronitridic acid
Nitrilotriessigsäure *f* nitrilotriacetic acid
Nitrin *n* nitrine
Nitrit *n* nitrite
nitritfrei free from nitrite[s], nitrite-free
Nitritglätte *f* lead(II) nitrite
Nitritmutterlauge *f* nitrite mother liquor
Nitritogruppe *f* nitrito group, oxynitroso radical
Nitritpökelsalz *n* nitrite pickling salt

Nitritreduktase *f* (Biochem) nitrite reductase
Nitroacetophenon *n* nitro acetophenone
Nitroäthan *n* nitroethane
Nitroalkan *n* nitroalkane
Nitroanilin *n* nitroaniline
Nitroanisol *n* nitroanisole
Nitroanthrachinon *n* nitroanthraquinone
Nitrobakterien *pl* nitrobacteria
Nitrobenzaldehyd *m* nitrobenzaldehyde
Nitrobenzoesäure *f* nitrobenzoic acid
Nitrobenzoesäureanhydrid *n* benzo-nitro-benzoic anhydride, benzoate of nitrobenzoyl
Nitrobenzoesäurepropylester *m* propyl nitrobenzoate
Nitrobenzol *n* nitrobenzene, mirbane oil
Nitrobenzoldiazoniumchlorid *n* nitrobenzene diazonium chloride
Nitrobenzolsulfochlorid *n* nitrobenzenesulfonyl chloride
Nitrobenzylchlorid *n* nitrobenzyl chloride
Nitrobutan *n* nitrobutane
Nitrocellulose *f* nitrocellulose, cellulose nitrate, nitrated cellulose
Nitrocellulosedeckfarbe *f* nitrocellulose coating color
Nitrochinolin *n* nitroquinoline
Nitrochinon *n* nitroquinone
Nitrochlorbenzol *n* nitrochlorobenzene
Nitrochlortoluol *n* nitrochlorotoluene
Nitrococussinsäure *f* nitrococussic acid
Nitrocymol *n* nitrocymene
Nitroderivat *n* nitro derivative
Nitrodimethylanilin *n* nitrodimethyl aniline
Nitrodiphenyl *n* nitrodiphenyl
Nitroessigsäure *f* nitroacetic acid
Nitrofettsäure *f* nitrated fatty acid, nitro fatty acid
Nitroform *n* nitroform, trinitromethane
Nitrofural *n* nitrofural
Nitrogelatine *f* nitrogelatin, blasting gelatin
Nitrogenium *n* (Lat) nitrogen
Nitroglauberit *m* (Min) nitroglauberite
Nitroglycerin *n* nitroglycerol, glyceryl nitrate, nitroglycerin, trinitrin, trinitroglycerine
Nitrogruppe *f* nitro group
Nitroguanidin *n* nitroguanidine
Nitrohalogenbenzol *n* halonitrobenzene
Nitroharnstoff *m* nitrourea
Nitrokörper *m* nitro compound, nitro substance
Nitrokörperausscheidung *f* (Med) azoturia
Nitrokohlenstoff *m* tetranitromethane
Nitrokupfer *n* nitro copper
Nitrolack *m* nitro[cellulose] lacquer
Nitrolsäure *f* nitrolic acid
Nitromagnesit *m* nitromagnesite
Nitromannit *m* mannitol nitrate, nitromannite
Nitrometall *n* nitrometal
Nitrometer *n* nitrometer
Nitromethan *n* nitromethane, nitrocarbol

Nitromethylanilin *n* nitromethyl aniline
Nitromethylanthrachinon *n* nitromethyl anthraquinone
Nitromethylnaphthalin *n* nitromethyl naphthalene
Nitromethylpropan *n* nitromethyl propane
Nitron *n* nitron
Nitronaphthalin *n* nitronaphthalene
Nitronaphthalinsulfonsäure *f* nitronaphthalene sulfonic acid
Nitronaphthalintrisulfonsäure *f* nitronaphthalene trisulfonic acid
Nitronaphthochinon *n* nitronaphthoquinone
Nitronitrosobenzol *n* nitronitrosobenzene
Nitronitrosofarbstoff *m* nitro-nitroso dye
Nitroniumion *n* nitronium ion
Nitronnitrat *n* nitron nitrate
Nitronsäure *f* nitronic acid
Nitroparaffin *n* nitroparaffin
Nitrophenol *n* nitrophenol
Nitrophenylacetylen *n* nitrophenylacetylene
Nitrophenylcarbamylchlorid *n* nitrophenylcarbamyl chloride
Nitrophenylessigsäure *f* nitrophenylacetic acid
Nitrophenylhydrazin *n* nitrophenylhydrazine
Nitrophenylpropiolsäure *f* nitrophenylpropiolic acid
nitrophil nitrophilous
Nitrophthalsäure *f* nitrophthalic acid
Nitropropan *n* nitropropane
Nitropropen *n* nitropropene
Nitroprussidnatrium *n* sodium nitroferricyanide, sodium nitroprussiate, sodium nitroprusside
Nitroprussidverbindung *f* nitroprussiate, nitroprusside
Nitroprussidwasserstoffsäure *f* nitroprussic acid
Nitropulver *n* nitro powder, smokeless powder
Nitropyren *n* nitropyrene
Nitropyridin *n* nitropyridine
Nitrosalicylsäure *f* nitrosalicylic acid
Nitrosamin *n* nitrosamine
Nitrosaminrot *n* nitrosamine red
Nitroschwefelsäure *f* nitrosulfuric acid
Nitrosebakterium *n* nitrifying bacterium
Nitroseide *f* cellulose nitrate rayon, nitrocellulose rayon, nitrosilk
Nitrosesäure *f* nitrose acid
nitrosieren to treat with nitrous acid
Nitrosierung *f* nitrosation
Nitrosisulfosäure *f* nitrosisulfonic acid
Nitrosit *n* nitrosite
Nitrosobase *f* nitroso base
Nitrosobenzol *n* nitrosobenzene
Nitrosoblau *n* nitroso blue
Nitrosobutan *n* nitrosobutane
Nitrosodimethylamin *n* nitrosodimethylamine
Nitrosofarbstoff *m* nitroso dyestuff
Nitrosogruppe *f* nitroso group, nitroso radical

Nitrosohydroxylaminoxid *n* nitrosohydroxylamine oxide
Nitrosokautschuk *m* nitroso rubber
Nitrosomethylanilin *n* nitrosomethylaniline
Nitrosomethyltoluidin *n* nitrosomethyl toluidine
Nitrosomethylurethan *n* nitrosomethyl urethane
Nitrosonaphthol *n* nitrosonaphthol
Nitrosophenol *n* nitrosophenol
Nitrosophenylhydrazin *n* nitrosophenylhydrazine
Nitrosophenylhydroxylamin *n* nitrosophenyl hydroxylamine
Nitrosoresorcin *n* nitrosoresorcinol
Nitrososulfonsäure *f* nitrososulfonic acid
Nitrosourethan *n* nitrosourethane
Nitrosoverbindung *f* nitroso compound
Nitrosprengstoff *m* nitro explosive
Nitrostärke *f* nitro starch, xyloidine
Nitrosulfamid *n* nitrosulfamide
Nitrosulfathiazol *n* nitrosulfathiazole
Nitrosulfonsäure *f* nitrosulfonic acid
Nitrosylchlorid *n* nitrosyl chloride
Nitrosylsäure *f* nitrosylic acid, hyponitrous acid
Nitrosylschwefelsäure *f* nitrosylsulfuric acid
Nitrotoluidin *n* nitrotoluidine
Nitrotoluol *n* nitrotoluene
Nitrotoluolsulfonsäure *f* nitrotoluene sulfonic acid
Nitrourethan *n* nitrourethane
Nitroverbindung *f* nitro compound
Nitroweinsäure *f* nitrotartaric acid
Nitroxid *n* nitrogen oxide
Nitroxylgruppe *f* nitroxyl radical
Nitroxylol *n* nitroxylene
Nitrozellulose *f* nitrocellulose, cellulose nitrate, explosive cotton
Nitrozimtsäure *f* nitrocinnamic acid
Nitrozimtsäureäthylester *m* ethyl nitrocinnamate
Nitrozimtsäureester *m* ester of nitrocinnamic acid
Nitrür *n* (obs) nitride
Nitrylchlorid *n* nitr[ox]yl chloride
Nitrylsäure *f* (obs) nitrous acid
Nivalsäure *f* nivalic acid
Niveau *n* level, niveau, ~ **der Energie** energy level
Niveauabstand *m* level spacing
Niveauanzeiger *m* (Füllstand) level indicator
Niveaufläche *f* equipotential surface
Niveauflasche *f* level[l]ing bottle
Niveauhöhe *f* height of level
Niveaukugel *f* levelling bulb
Niveauschema *n* level scheme
Niveaustufe *f* (Atom) energy level, term
Niveauunterschied *m* difference of level
Niveauverbreiterung *f* level broadening
nivellieren to bring to a level, to level
Nivelliergefäß *n* leveling bulb
Nivellierschraube *f* levelling screw

Nivellierung f level[l]ing
Nivellierwaage f spirit level, water level
Nivenit m (Min) nivenite
Nivitin n nivitin
Nixierohr n nixie tube
n-Leiter m electron excess conductor, excess semiconductor, n-semiconductor
n-Leitung f n-conductivity
NMR Abk. für „Nuclear Magnetic Resonance"
Nobelium n nobelium (Symb. No)
Nocardamin n nocardamin
nocken to cleat, to core
Nocken m cam
Nockenantrieb m cam drive
Nockenpresse f cam press
Nockenrad n cam wheel
Nockensteuerung f cam control
Nockenwelle f cam shaft
Noctal n noctal
Nohlit m nohlite
Nomenklatur f nomenclature, ~ **organischer Verbindungen** nomenclature of organic compounds, **chemische** ~ chemical nomenclature, **Genfer** ~ Geneva nomenclature, **Stocksche** ~ Stock nomenclature
Nominalkerze f nominal candle power
Nominalwert m nominal value
nominell nominal[ly]
Nomogramm n nomogram, nomograph, self-computing chart
Nomographie f nomography
Nonadekan n nonadecane
Nonadien n nonadiene
Nonadilacton n nonadilactone
Nonakosan n nonacosane
Nonamethylenglykol n nonamethylene glycol
Nonan n nonane
Nonatriakontan n nonatriacontane
Nondecylsäure f nondecylic acid
Nonin n nonine, nonyne
Nonius m vernier
Noniusablesung f vernier reading
Noniuseinteilung f vernier scale
Noniusmodell n model of vernier
Noniusskala f vernier scale
Nonne f (Metall) matrix
Nontronit m (Min) nontronite
nonvariant invariant
Nonyl- nonyl
Nonylaldehyd m nonyl aldehyde, pelargonaldehyde
Nonylalkohol m nonyl alcohol
Nonylen n nonylene, nonene
Nonylphenol n nonylphenol
Nonylsäure f nonylic acid, pelargonic acid
Nootkaton n nootkatone
Nopadien n nopadiene, nopene
Nopalin n nopaline
Nopen n nopene, nopadiene
Nopinan n nopinane
Nopinen n nopinene
Nopinol n nopinol
Nopinon n nopinone
Nopinsäure f nopinic acid
Nopol n nopol
Noppe f (an der Profilrippe) suppressor button
Noradrenalin n noradrenaline
Norapomorphin n norapomorphine
Noraporphin n noraporphine
Noratropin n noratropine
Norbixin n norbixin
Norbornan n norbornane
Norbornen n norbornene
Norborneol n norborneol
Norcamphan n norcamphane
Norcamphen n norcamphene
Norcampher m norcamphor
Norcamphersäure f norcamphoric acid
Norcamphidin n norcamphidine
Norcaran n norcarane
Norcaryophyllensäure f norcaryophyllenic acid
Norcorydalin n norcorydaline
Nordenskiöldin m (Min) nordenskioldine
Nordesoxyephedrin n nordesoxyephedrine
Nordhauser Schwefelsäure f fuming sulfuric acid, Nordhausen acid, oleum
Nordlicht n (Astr) aurora borealis, northern lights, northlight
Nordlichtlinie f (Astr) green auroral line
Nordmarkit m nordmarkite
Nordstern m (Astr) north polar star, North Star
Norecgonin n norecgonine
Norgeraniumsäure f norgeranic acid
Norgesalpeter m Norge saltpeter, calcium nitrate, niter
Norhydrastinin n norhydrastinine
Norhyoscyamin n norhyoscyamine
Norit m (Geol) norite
Norkodein n norcodeine
Norleucin n norleucine, 2-aminohexanoic acid
Norlupinan n norlupinane
Norlupinon n norlupinone
Norm f norm, rule, standard, standard specification, **Deutsche Industrie-** ~ (DIN) German industrial standard
normal normal, regular, routine, standard, usual
Normalangabe f standard specification
Normalatmosphäre f standard atmosphere
Normalausführung f standard design, standard type, unit construction
Normalausrüstung f standard equipment
Normalbeanspruchung f normal stress
Normalbedingungen f pl standard conditions, (b. Gasen) standard temperature and pressure
Normalbelastung f normal or standard load
Normalbeschleunigung f normal acceleration
Normalbutan n n-butane, normal butane

Normaldichte *f* normal density
Normaldodecan *n* n-dodecane, normal dodecane
Normaldrehmoment *n* normal torque
Normaldruck *m* standard pressure, normal pressure
Normaldruckrektifikation *f* rectification at normal pressures
Normale *f* normal (to a curve), (Geom) normal [line]
Normalebene *f* normal plane (to a curve)
Normaleikosan *n* n-eicosane, normal eicosane
Normalelektrode *f* standard electrode
Normalelement *n* (Elektr) normal element, standard cell
Normalentropie *f* (Phys) standard entropy
normalerweise normally
Normalfestigkeit *f* (Text) conditioned tenacity
normalfeucht (Text) conditioned
Normalfeuchtigkeit *f* standard moisture
Normalfilter *n* standard filter
Normalformat *n* standard dimension, standard size
Normalfrequenz *f* standard frequency
Normalgas *n* standard gas
Normalgeschwindigkeit *f* allowed speed
Normalgewicht *n* standard weight
Normalgewinde *n* standard thread
Normalgewindelehre *f* standard thread gauge
Normalglas *n* ordinary glass, soda-lime glass, standard glass
normalglühen to normalize
Normalglühen *n* (Met) normalizing
Normalgröße *f* standard size
Normalheptan *n* n-heptane, normal heptane
Normalheptylsäure *f* normal heptylic acid, enanthic acid, n-heptanoic acid
normalisieren to standardize, to normalize, to regulate
Normalisierung *f* normalization, standardization
Normalität *f* normality
Normal-[Kadmium-] Element *n* standard Weston cadmium cell
Normalkalilauge *f* normal potassium hydroxide solution, N-potassium hydroxide solution
Normalkalomelelektrode *f* normal calomel electrode
Normalkerze *f* normal candle, standard candle power
Normalkonzentration *f* normal concentration
Normalkorn *n* correct product (screening)
Normalleistung *f* standard capacity
Normallösung *f* (Chem) normal solution, standard solution
Normalmaß *n* standard measure
Normalnonadekan *n* n-nonadecane, normal nonadecane
Normalpentan *n* normal pentane, n-pentane
Normalpotential *n* standard potential, ~ eines Metalls normal electrode potential of a metal

Normalprofil *n* standard cross section
Normalrohr *n* straight pipe
Normalsäure *f* N-acid, normal acid
Normalschwingung *f* normal mode or vibration
Normalspannung *f* normal stress, normal tension, (Elektr) normal voltage
Normalspektrum *n* (Spektr) normal spectrum, diffraction spectrum
Normalstärke *f* normal strength, standard strength
Normalstrahler *m* standard radiator
Normaltabelle *f* standard table
Normaltemperatur *f* normal temperature, standard temperature
Normaltetradekan *n* normal tetradecane
Normalumsatz *m* (Metabolismus) normal metabolic rate
Normalundekan *n* normal undecane
Normalverteilung *f* normal distribution
Normalwassergehalt *m* normal moisture content
Normalwasserstoffelektrode *f* normal [or standard] hydrogen electrode
Normalwiderstand *m* standard resistance
Normalzelle *f* standard cell
Normalzustand *m* normal state, standard condition
Normblatt *n* data sheet, industrial specification sheet
normen to standardize, to calibrate, to normalize
Normen *f pl* standards, rules, specifications
Normenaufstellung *f* normalization, standardization
Normenausschuß *m*, Deutscher ~ German standards commission
Normenentwurf *m* tentative standard
Normfarbwert *m* tristimulus value
Normfarbwertanteil *m* chromaticity coordinate
normgerecht according to standard specification, conforming to standards, normal, standard
normieren to standardize, to calibrate, to gauge, to regulate
Normlichtart *f* (Opt) standard illuminant
Normmaß *n* standard size
Normoblast *m* (Med) normoblast
Normorphin *n* normorphine
Normozyt *m* (Med) normocyte
Normschliff *m* standard ground joint
Normstimmton *m* (Akust) standard pitch
Normung *f* standardization, calibration, normalization
Normuscarin *n* normuscarine
Normuscaron *n* normuscarone
Normuscon *n* normuscone
Normvorschrift *f* standard specification
Normwasserstoffelektrode *f* standard hydrogen electrode
normwidrig abnormal, adverse
Nornarcein *n* nornarceine
Nornicotin *n* nornicotine

Noropiansäure f noropianic acid
Norpenaldsäure f norpenaldic acid
Norphedrin n norphedrin
Norpinsäure f norpinic acid
Norprogesteron n norprogesterone
Norsolanellsäure f norsolanellic acid, biloidanic acid
Northebain n northebaine
Northupit m (Min) northupite
Nortricyclen n nortricyclene
Nortropan n nortropane
Nortropin n nortropine
Nortropinon n nortropinone
Norvalin n norvaline, α-amino valeric acid
Norvalinamid n norvaline amide
Nosean m (Min) nosean, noselite
Nosophen n nosophen, iodophen, tetraiodophenolphthalein
Nosophennatrium n sodium nosophen, antinosin
Notabschaltung f emergency shut-down
Notaggregat n stand-by system
Notausgang m emergency exit
Notbehelf m expedient, makeshift
Notbeleuchtung f emergency lighting
Note f grade, mark
Notiz f memo[randum], note
Notizbuch n note book, memorandum book
Notsignal n danger signal, emergency signal
notwendig necessary
Notwendigkeit f necessity
Nova f (Astr) nova
Novaausbruch m nova outburst
Novacin n novacine
Novain n novaine
Noviose f noviose
Novobiocin n novobiocin
Novocain n novocaine
Novojodin n novo-iodine
Novolak n novolak
Novovanillin n novovanillin
NPK-Dünger m (Abk. für Stickstoff, Phosphor, Kalium) NPK fertilizer
NQR-Spektroskopie f (Kernquadrupolresonanz-Spektroskopie) NQR spectroscopy
Nuance f shade, hue, nuance, tint, tone
nuancieren to vary, to modulate, to shade, to tint, to tone
Nucin n nucin, juglone
Nuclease f (Biochem) nuclease
Nucleinsäure s. **Nukleinsäure**
Nudinsäure f nudic acid
Nürnberger Schere f lazy tongs
Nürnberger Violett n Nuremberg violet
nützlich useful
nuklear nuclear
Nuklearforschung f nuclear research
Nuklearmedizin f nuclear medicine
Nuklearreinheit f nuclear purity

Nuklearteilchen f subatomic particle
Nuklein n nuclein
Nukleinsäure f nucleic acid, (obs) nucleinic acid
nukleinsauer nucleinate
Nukleogenese f nucleogenesis
Nukleolus m nucleolus
Nukleon n (Atom) nucleon
Nukleonenbreite f nucleon width
Nukleonenzahl f (Atom) nucleon number, mass number
nukleophil nucleophilic
Nukleoproteid n nucleoprotein
Nukleoprotein n nucleoprotein
Nukleosid n nucleoside
Nukleosidase f (Biochem) nucleosidase
Nukleosidkinase f (Biochem) nucleoside kinase
Nukleosidmonophosphat n nucleoside monophosphate
Nukleosidtransferase f (Biochem) nucleoside transferase
Nukleosin n nucleosin
Nukleotid n nucleotide
Nukleotidase f (Biochem) nucleotidase
Nukleotid-Einheit f nucleotide unit
Nukleus m nucleus
Nuklid n nuclide
null und nichtig null and void
Null f zero, naught, nought, **auf ~ einstellen** to adjust to zero
Nullablesung f zero reading
Nullage f zero position
Nulleffekt m (Radioaktivität) background radiation
Nulleffektimpuls m background count
Nulleinstellung f zero adjustment
Nulleiter m (Elektr) neutral wire, third wire, zero wire, **geerdeter ~** earthed neutral wire
Nulleiterkabel n earth conductor cable
Nulleitung f earth conductor
Nullelement n null element
Nullenzirkel m [spring] bow compasses, spring bows
Nullfeder f zero spring
Nullfolge f (Math) zero series
Nullindikator m null indicator
Nullinie f reference line, neutral axis, zero axis, zero line, (Spektr) band center
Nullinstrument n zero instrument
Nullisokline f aclinic line
Nullkontrolle f zeropoint control
Nullmethode f zero method
Nullmoment n zero moment
Nulloperator m null operator
Nullpotential n zero potential, potential of zero charge
Nullpunkt m (Diagramm) initial point, origin, (Elektr) neutral point, (Instrument) zero range setting, (Temperatur) zero [point], **absoluter ~** absolute zero

Nullpunktabweichung f zero deviation, zero error
Nullpunktanomalie f origin distortion
Nullpunkteinstellung f zero adjustment
Nullpunktempfindlichkeit f zero-level sensitivity
Nullpunktenergie f zero point energy
Nullpunktkompensation f zero adjustment control
Nullpunktkontrolle f zero adjustment control
Nullpunktmethode f zero method, dead stop method
Nullpunktschieber m zero setting
Nullpunktsenthalpie f enthalpy at absolute zero
Nullpunktsentropie f entropy at absolute zero
Nullpunktsvolumen n zero point volume
Nullpunktunruhe f zero point motion
Nullpunktverschiebung f zero drift
Nullreichweite f zero range
Nullspannung f (Elektr) zero voltage, zero potential
Nullstellung f zero position, neutral position, (Abschaltstellung) off position
Nullteiler m zero divisor
Nullwert m zero value, ~ **der Stromstärke** zero current, zero value of the current
nullwertig zerovalent
Numeait m (Numeit, Min) numeaite, garnierite
numerieren to number
numerisch numerical
Numerometrie f (Titrationsmethode) numerometry
Numerus m antilog, antilogarithm, inverse logarithm
Nummer f number, figure, size, (Ausgabe) issue
Nupharin n nupharine
Nußbaumholz n walnut
Nußierit m nussierite
Nußkernmehl n nutmeal
Nußkohle f chestnut coal, nut coal
Nußmühle f nut mill
Nußöl n nut oil
Nut f groove, (Keilnut) keyway, (Kerbnut) notch, (Langnut) slot, (Schwalbenschwanzverbindung) mortise
Nute f (s. Nut)
nuten to groove, to channel, to slot
Nuten pl **fräsen** to mill grooves, to mill keyways, ~ **stoßen** to cut grooves, to cut keyways
Nutenfräser m slot cutter, slotting machine
Nutenstoßmaschine f slotting machine
Nutenteil m sunk key
Nutrose f nutrose
Nutschapparat m suction filter apparatus
Nutsche f filtering flask, suction filter, suction strainer
Nutschenfilter n pan filter, suction filter
Nuttalit m nuttalite
Nutverschlußkeil m slot wedge

Nutzanwendung f appliance, application, use
Nutzarbeit f effective work, useful work
nutzbar effective
Nutzbarmachung f utilization
Nutzbeanspruchung f working stress
nutzbringend profitable, useful
Nutzdurchmesser m useful diameter
Nutzeffekt m efficiency [effect], net effect, net efficiency, performance, useful effect, useful effort (of a machine)
Nutzen m advantage, benefit, gain, profit, ~ **ziehen** to derive advantage [from]
Nutzenergie f net energy
Nutzholz n lumber, timber
Nutzkapazität f useful capacity
Nutzlast f payload
Nutzleistung f net efficiency, output, yield
nutzlos useless
Nutzquerschnitt m useful cross section
Nutzspannung f useful voltage
Nutzstrom m useful current
Nutzung f utilization, use
Nutzungsdauer f serviceable life
Nutzungswert m efficiency
Nutzwärme f effective heat
Nutzwasser n non-drinkable water, water for industrial purposes
Nutzwert m efficiency
Nutzwertanalyse f efficiency study
Nutzwiderstand m useful resistance
Nux vomica f (Pharm) nux vomica
Nyctal n nyctal
Nydrazid n nydrazid
Nylanders Reaktion f Almen-Nylander test
Nylon n nylon
Nylongewebe n (Text) nylon fabric
Nylonkunststoff m nylon plastic
Nylonrohr n nylon cane
Nylonschnitzel m nylon flake
nylonverstärkt nylon-reinforced
Ny-Meson n neutral meson

O

Obduktion *f* post-mortem
oben above, at the top, overhead
Oben-Aufgabe-Filter *n* top-feed filter
Oberbekleidung *f* (Text) overwear
Oberboden *m* flooring
Oberdruck *m* pressure of top roll
Oberdruckpresse *f* downstroke press, top ram press
Oberfeuer *n* updraft fire, upper heat
Oberfläche *f* surface [area], **glatte** ~ even surface, **löcherige** ~ pitted surface, **poröse** ~ porous surface, **rauhe** ~ rough surface, (Papier) dry finish, **wellige** ~ corrugated surface, **wirksame** ~ effective area
Oberflächenabdruck *m* surface replica
Oberflächenableitung *f* (Elektr) surface leakage
Oberflächenadsorption *f* adsorptive power of the surface
Oberflächenaffinität *f* surface affinity
oberflächenaktiv surface-active, capillary active, surface-reactive
Oberflächenaktivität *f* surface activity
Oberflächenanästhetikum *n* surface an[a]esthetic
Oberflächenausführung *f* surface finish
Oberflächenaussehen *n* superficial appearance, appearance of surface
Oberflächenbau *m* surface structure
Oberflächenbearbeitung *f* surface treatment
Oberflächenbehandlung *f* surface treatment, (Straßenbau) surface coating
Oberflächenbeladung *f* charging of the surface, sensitization of the surface, surface activation
Oberflächenberieselung *f* surface cooling
Oberflächenbeschaffenheit *f* character of the surface, nature of surface, surface condition, surface finish, surface properties
Oberflächenbestimmung *f* surface-area determination
Oberflächenbogen *m* surface sheet
Oberflächenchemie *f* surface chemistry
Oberflächendichte *f* surface density
Oberflächendruck *m* surface pressure
Oberflächeneffekt *m* surface effect
Oberflächeneinfluß *m* surface effect, surface influence
Oberflächeneinheit *f* unit of surface, unit of area
Oberflächenenergie *f* surface energy
Oberflächenenergieparameter *m* surface energy parameter
Oberflächenentkohlung *f* surface decarburization
Oberflächenentladung *f* surface discharge
Oberflächenentwicklung *f* surface development
Oberflächenerneuerung *f* surface renewal
Oberflächenerscheinung *f* surface effect

Oberflächenfehler *m* surface defect, surface imperfection, surface irregularity
Oberflächenfestigkeit *f* surface stability, surface strength
Oberflächenfilm *m* surface film
oberflächengehärtet surface-hardened
Oberflächengestalt *f* shape of surface
Oberflächengestaltung *f* surface texturing
Oberflächengitter *n* surface lattice
Oberflächenglanz *m* surface luster
Oberflächengüte *f* surface quality, [surface] finish
Oberflächenhärte *f* surface hardness
Oberflächenhärtemaschine *f* surface hardening machine
Oberflächenhärtung *f* surface hardening, case hardening, flame hardening, surface converting, torch hardening
Oberflächenintegral *n* surface integral
Oberflächenionisierung *f* surface ionization
Oberflächenkohlung *f* surface carburizing, cementation
Oberflächenkomplex *m* surface complex
Oberflächenkondensation *f* surface condensation, condensation by contact
Oberflächenkondensator *m* surface condenser
Oberflächenkonservierung *f* surface preservation
Oberflächenkonzentration *f* surface concentration
Oberflächenkorrosion *f* flat etching type of corrosion
Oberflächenkraft *f* surface force
Oberflächenkratzer *m* surface scratcher
Oberflächenkruste *f* surface crust
Oberflächenkühler *m* surface cooler
Oberflächenkühlung *f* surface cooling
Oberflächenkultur *f* surface culture
Oberflächenlack *m* finish varnish
Oberflächenladung *f* surface charge
Oberflächenladungsdichte *f* (Phys) surface-charge density
Oberflächenleim *m* (Pap) surface size
Oberflächenleimung *f* (Pap) paper coating, surface sizing
Oberflächenleitfähigkeit *f* surface conductivity
Oberflächenleitung *f* surface conduction
Oberflächenleitwert *m* surface conductance
Oberflächenmagnetisierung *f* surface magnetization
Oberflächenmethode *f* surface test
Oberflächenorientierung *f* surface orientation
Oberflächenoxydation *f* superficial oxydation
Oberflächenpotential *n* [einer Lösung] surface potential [of a solution]
Oberflächenrauhheit *f* surface roughness
Oberflächenrauhigkeit *f* surface roughness

Oberflächenreibung f surface friction
Oberflächenreinheit f cleanliness of surface
Oberflächenriß m surface crack
Oberflächenschliff m surface grinding
Oberflächenspannung f surface tension, **Erniedrigung der ~** lowering of surface tension
Oberflächenspannungsmesser m surface tension meter
Oberflächenstrom m superficial current, surface current
Oberflächenstruktur f superficial structure, surface pattern
Oberflächentemperatur f surface temperature
Oberflächenverbindung f surface compound
Oberflächenverbrennung f surface combustion
Oberflächenveredelung f surface finishing, surface refinement
Oberflächenvergaser m surface carburettor
Oberflächenverseuchung f surface contamination
Oberflächenverteilung f surface distribution
Oberflächenverunreinigung f surface contamination
Oberflächenvlies n (glass fiber) overlay mat, surfacing mat
Oberflächenvorbehandlung f surface preparation
Oberflächenwanderung f surface migration
Oberflächenwasser n (Geol) surface water
Oberflächenwechselwirkung f surface interaction
Oberflächenwiderstand m surface resistance
oberflächenwirksam surface-active
Oberflächenwirkung f surface action
Oberflächenzementierung f case hardening
Oberflächenzusatzpotential n (Phys) additional surface potential
Oberflächenzustand m state of surface, surface finish, surface state
oberflächlich superficial
Oberflächlichkeit f superficiality
obergärig top-fermenting, (Brau) fermented from top
Obergärung f (Brau) top fermentation
Obergestell n (Ofen) top of crucible, top of hearth
Obergewicht n top weight
Oberhärte f surface hardness
Oberhaut f (Med) epidermis
Oberhautgewebe n (Med) epidermic tissue
Oberhefe f (Brau) head yeast, surface yeast, top fermentation yeast
oberirdisch above ground, overhead
Oberkante f top edge, upper edge, upper rim
Oberkasten m (Gieß) cope
Oberkessel m steam drum
Oberkolbenpresse f downstroke press, top ram press
Oberlast f top weight
oberlastig top-heavy
Oberlauf m headwaters

Oberleitung f (Aufsicht) supervision, (Elektr) overhead wiring, aerial line, overhead contact system
Oberlicht n toplight, fanlight, skylight
Oberschenkelknochen m (Anat) femur
Oberschicht f finish, top coat, top layer, upper layer, (Geol) upper stratum
Oberschwingung f harmonic oscillation
Oberseite f top, upper side
Oberspannung f (Elektr) higher harmonic voltage
Oberstrom m (Elektr) higher harmonic current
Oberteil n top part, upper part
Obertrumförderer m upperside run conveyor
Oberwalze f top roll
Oberwasser n headwater, upstream water
Oberwinddüse f top blowing tuyère
Oberwindfrischen n (Metall) refining steel by top-blowing
Objektbereich m specimen region
Objektiv n (Opt) objective, lens, object glass, **kurzbrennweitiges ~** objective of short focal length, **lichtstarkes ~** objective of great light transmitting capacity, **zweiteiliges ~** two-lens object-glass
Objektivbrennpunkt m focus of the objective
Objektivdeckel m lens cap
Objektivfassung f lens holder, lens tube
Objektklammer f object clamp
Objektsucher m object finder
Objekttisch m (Opt) microscope stage, stage of the slide
Objektträger m microscopic slide, slide holder, specimen holder
Objektträgerkasten m microscope slide box
Objektträgertrommel f object carrier drum
objekttreu object-preserving
Oblaten pl wafers
Oblitin n oblitine, carnitine ethyl ester
Obscurin n obscurine
Observatorium n observatory
Obsidian m (Min) obsidian, volcanic glass rock
Obst n fruit
Obstbau m fruit growing, fruit cultivation
Obstbranntwein m fruit brandy, fruit spirit
Obstkonserven f pl canned fruit, preserved fruit
Obstsaft m fruit juice
Obststeige f lugs (Am. E.), punnets (Br. E.)
Obstwein m fruit wine
Obtusatsäure f obtusatic acid
Obtusifolin n obtusifolin
Obtusilsäure f obtusilic acid
Ochracin n ochracin
Ochroit m ochroite
Ochsenzungenkraut n (Bot) bugloss
Ocimen n ocimene, dimethyl-octatriene
Ocker m (Min) ocher (Am. E.), ochre (Br. E.), paint rock
ockerfarben ocherous

ockergelb ocher-yellow
Ockergelb *n* (Min) ocher yellow, limonite
ockerhaltig containing ocher, ocherous
Ocrein *n* ocreine
Octaacetylsaccharose *f* octaacetyl sucrose
Octaamylose *f* octaamylose
Octacosanol *n* octacosanol
Octadecan *n* octadecane
Octadecansäure *f* octadecanoic acid, octadecylic acid, stearic acid
Octadecatriensäure *f* octadecatrienoic acid
Octadecin *n* octadecine, octadecyne
Octadecylalkohol *m* octadecanol, octadecyl alcohol
Octadecylbromid *n* octadecyl bromide
Octalin *n* octalin
Octalupin *n* octalupine
Octamethyllactose *f* octamethyl lactose
Octamethylmaltose *f* octamethyl maltose
Octamethylsaccharose *f* octamethyl sucrose
Octamylamin *n* octamylamine
Octan *n* octane, dibutyl
Octanol *n* octanol
Octantenregel *f* octant rule
Octanzahl *f* octane number
Octastearylsaccharose *f* octastearyl sucrose
Octatrienol *n* octatrienol
Octen *n* octene
Octencarbonsäure *f* octenic acid
Octopamin *n* octopamine
Octopin *n* octopine
Octose *f* octose
Octyl- octyl
Octylaldehyd *m* octylaldehyde, caprylaldehyde, caprylic aldehyde, octanal
Octylalkohol *m* octanol, octyl alcohol
Octylamin *n* octylamine
Octylen *n* octylene, octene
Octylnitrit *n* octyl nitrite
Octylphenol *n* octylphenol
Octylsäure *f* octylic acid, caprylic acid, hexylacetic acid, octanoic acid
Odyssin *n* odyssin
Odyssinsäure *f* odyssic acid
Oechsle-Grad *m* degree Oechsle
Ödem *n* oedema
Ödlandbegrünung *f* waste land recovery project
öffnen to open
Öffnung *f* opening, gap, hole, mouth, orifice, port, (Opt) aperture, (Schlitz, Spalt) slit, **kreisförmige** ~ annular opening
Öffnungsfunke *m* spark at break, spark on opening
Öffnungsinduktionsstrom *m* induction current at breaking
Öffnungsverhältnis *n* (Dest) open area, (Opt) aperture ratio, relative aperture
Öffnungswinkel *m* bevel angle
Öhr *n* ear, eye

Ökologie *f* bionomics, ecology
Ökospecies *f* (Biol) ecospecies
Öl *n* oil, **ätherisches** ~ essential oil, volatile oil, **dickes** ~ heavy oil, **dünnflüssiges** ~ light oil, **fluoreszierendes** ~ fluorescent oil, **gehärtetes** ~ hydrogenated oil, fixed oil, **geschwefeltes** ~ sulfurized oil, **kältebeständiges** ~ cold-test oil, frost-resisting oil, **leicht flüchtiges** ~ volatile oil, **mineralisches** ~ mineral oil, **nichttrocknendes** ~ non-drying oil, **pflanzliches** ~ vegetable oil, **sulfoniertes** ~ sulphated oil, **tierisches** ~ animal oil, **trocknendes** ~ drying oil, **viskoses** ~ viscous oil
Ölabdichtung *f* oil seal
Ölablaß *m* oil drain[er], oil outlet
Ölabscheider *m* oil separator, oil trap
Ölabscheidung *f* oil separation
Ölabsorption *f* oil absorption
Ölabstreicher *m* oil wiper
Ölacidimeter *n* oil acidimeter
Ölanstrich *m* coat of oil paint
Ölanstrichfarbe *f* oil base paint
ölarm containing little oil
ölartig oily, oil-like, oleaginous
Ölauffang[schale *f*]*m* bowl for overflowing oil
Ölauffangwanne *f* oil catchpan
Ölaufnahmefähigkeit *f* oil absorption [capacity]
Ölausscheider *m* oil extractor, oil separator
Ölbad *n* oil bath
Ölbaum *m* (Bot) olive tree
Ölbaumharz *n* elemi
Ölbehälter *m* oil tank, oil basin, oil reservoir
Ölbeize *f* oil mordant
ölbeständig oil-resistant
ölbildend oil-forming
Ölbindevermögen *n* oil-binding property
Ölblau *n* oil blue, indigo copper, Saxon blue, smalt
Ölbleiche *f* oil bleaching
Ölbleichen *n* oil bleaching
Ölbodensatz *m* oilfoot
Ölbohrer *m* oil drill
Ölbrenner *m* oil burner
Öldämpfung *f* oil damping
Öldampf *m* oil smoke, oil vapor
Öldampfstrahlpumpe *f* oil ejector
Öldesodorisierung *f* oil deodorization
Öldestillat *n* distillate of mineral oil
Öldestillationsverfahren *n* oil distilling process
öldicht oil-tight
Öldichtungsring *m* oil retainer ring
Öldruck *m* oil pressure, (Bild) oleograph, (Bilddruckverfahren) oleography
Öldruckanzeiger *m* oil-pressure gauge
Öldruckbild *n* oleograph
Öldruck-Dampfdruckzerstäuber *m* steam atomizer

Öldruckleitung *f* oil pressure pipe
Öldunst *m* oil fume, oil vapor
Öleinspritzung *f* oil injection
ölen to oil, to grease, to lubricate, (imprägnieren) to oil-impregnate
Ölentziehanlage *f* oil extracting plant
Ölerglas *n* glass lubricator, glass oil cup
Ölersatz *m* oil substitute
Ölextraktion *f* oil extraction
Ölextraktionsanlage *f* oil extraction plant, deoiling plant
Ölextraktor *m* oil extractor
Ölfänger *m* oil catcher, drip pan, oil dish
Ölfangbehälter *m* oil container
Ölfarbe *f* oil color, oil paint, **glänzende** ~ full-gloss oil paint, **hitzebeständige** ~ heat-resisting oil paint, **matte** ~ flat oil paint
Ölfarbenanstrich *m* coat of oil paint
Ölfarbendruck *m* oleograph, (Verfahren) oleography
Ölfarbstift *m* oil color pencil
Ölfeuerung *f* oil furnace, oil burning, oil firing, oil heating
Ölfilm *m* oil film
Ölfilter *n* oil filter
Ölfirnis *m* oil varnish, boiled oil
Ölflammofen *m* oil-burning furnace, oil-fired [air] furnace
Ölfleck *m* oil spot; oil stain
Ölförderpumpe *f* oil transfer pump
Ölfüllung *f* oil filling
Ölgas *n* oil gas
Ölgaserzeugungsanlage *f* oil-gas generating plant
Ölgasretorte *f* oil-gas retort
Ölgasteer *m* oil-gas tar
ölgefeuert oil-burning, oil-fired
Ölgeschmack *m* oily taste
Ölgewinnung *f* oil production, (aus Samen etc.) oil extraction
ölglänzend shiny with oil
Ölglas *n* lubricator glass
Ölgrund *m* oil primer, first coat of oil, oil ground, oil priming
Ölgrundierung *f* priming oil paint
Ölhärtung *f* oil hardening, (Chem) hydrogenation or hardening of oil
Ölhärtungsstahl *m* oil-hardening steel
ölhaltig containing oil, oleiferous
Ölharz *n* oil resin, oleoresin
Ölhaut *f* oil film, oil skin
Ölhefe *f* oil dregs, oil lees
Ölheizanlage *f* oil heating plant
Ölheizung *f* oil heating
ölig oily, oleaginous
ölimprägniert oil-impregnated
Ölindustrie *f* oil industry
Öl-in-Wasser-Emulsion *f* oil-in-water emulsion
Ölisolation *f* oil insulation

Ölisolator *m* oil insulator
Ölisolierung *f* oil insulation
Ölkammer *f* oil chamber, oil reservoir
Ölkanal *m* oil drain channel
Ölkanister *m* oil container
Ölkanne *f* oil can
Ölkautschuk *m* vulcanized linseed oil, factice (HN)
Ölkitt *m* putty
Ölkoks *m* petroleum coke
Ölkreide *f* oil chalk
Ölkreidestift *m* oil crayon
Ölkreislaufbehälter *m* oil circulating tank
Ölkruste *f* oil crust
Ölkuchen *m* oil cake
Oellacherit *m* (Min) oellacherite
Öllack *m* oil enamel, oil varnish, oleoresinous varnish
Ölleder *n* chamois leather, oil leather
Ölleitung *f* oil pipe, oil tube, pipe line
Öllieferländer *pl* oil-producing countries
öllöslich oil-soluble
Ölmanometer *n* oil pressure gauge
Ölmesser *m* oleometer
Ölmeßstab *m* oil dipstick, oil level gauge or indicator
Ölmotor *m* oil engine
Ölmühle *f* oil mill
Ölnute *f* oil groove
Ölocker *m* yellow ocher
Ölpapier *n* oiled paper, oil paper, transparent paper
Ölpaste *f* oil paste, oil color, paste paint
Ölpest *f* oil pollution
Ölpresse *f* oil press
Ölprüfer *m* oil tester
Ölpumpe *f* oil pump
Ölraffination *f* oil refining
Ölraffinieren *n* oil refining, purification of oil
Ölrauch *m* oil vapor
ölreich long-oil
Ölreiniger *m* oil filter, oil purifier
Ölreinigung *f* oil filtration, oil purification
Ölreinigungsvorrichtung *f* oil filter
Ölring *m* oil ring
Ölrückgewinnungsanlage *f* oil recovery plant, oil rectifier
Ölrücklaufrohr *n* oil return pipe
Ölrückstand *m* oil dregs, oil residue
Ölruß *m* oil black, lamp black
Ölsaat *f* (Bot) oil seed
Ölsäure *f* oleic acid, 9-octadecenoic acid, oleinic acid
Ölsäureglycerid *n* glycerol trioleate, glyceryl [tri]oleate, olein, triolein
ölsäurehaltig containing oleic acid
Ölsäureseife *f* oleic acid soap, olein soap
Ölsammler *m* save-oil, oil trough
Ölsatz *m* oil sediment

ölsauer oleate
Ölschaum *m* oily scum
Ölschieber *m* oil sluice valve
Ölschiefer *m* (Min) oil shale
Ölschlagen *n* oil pressing
Ölschlamm *m* oil mud, oil residue, oil sediment, oil sludge
Ölschleuder *f* oil extractor, oil whizzer
Ölschleuderring *m* oil thrower
Ölschmierung *f* oil lubrication
Ölschwarz *n* oil black, lamp black
Ölschwemmverfahren *n* oil flotation process
Ölseide *f* oiled silk
Ölseife *f* oil soap, oleic acid soap, olein soap
Ölsodaseife *f* Castile soap
Ölspaltgas *n* gas from oil gasification
Ölsparapparat *m* economical oiling apparatus
Ölspritze *f* lubricating syringe
Ölstandsanzeiger *m* oil level gauge
Ölstandskontrolle *f* oil level check
Ölstandsmesser *m* oil gauge
Ölstaub *m* oil spray, oil vapor
Ölstein *m* (Min) oilstone
Ölstoßdämpfer *m* oil shock absorber
Ölstrahlgebläse *n* oil jet blower
Öltanker *m* oil tankship, oiler (Am. E.), oil tanker
Öltinte *f* oily ink
Öltransformator *m* oil[-cooled] transformer
Öltrockner *m* oil siccative
Öltrocknung *f* dehydration of oil
Öltröpfchen *n* oil drop
Öltropfgefäß *n* sight feed lubricator
Ölüberdruckventil *n* oil relief valve
Ölüberlaufrohr *n* oil overflow pipe
Ölumlauf *m* circulation of oil, oil circulation
Ölumlaufschmierung *f* oil circulation lubrication
Ölung *f* lubrication, oiling
Ölvase *f* lubricator, oil cup
Ölverarbeitung *f* oil processing
Ölverbrauch *m* oil consumption
Ölverdickung *f* inspissation of oil, thickening of oil
Ölvergoldung *f* oil gilding, varnish gilding
Ölverträglichkeit *f* oil compatibility
Ölvorstreichfarbe *f* undercoat oil paint
Ölwaage *f* oleometer, oil hydrometer
Ölwäsche *f* (Gas- und Flüssigkeitsreinigung) oil absorption
Ölwasser *n* oily water
Ölwechsel *m* oil change
Ölweiß *n* oil white
Ölwerk *n* oil factory, oil mill
Ölzahl *f* oil absorption
Ölzerstäuber *m* oil atomizer
Ölzirkulation *f* circulation of oil
Ölzucker *m* (obs) glycerol
Ölzufluß *m* oil feed, oil supply
Ölzusatz *m* addition of oil

Önanthäther *m* enanthic ether
Önanthaldehyd *m* enanthal, enanthaldehyde, enanthic aldehyde, heptanal, heptyl aldehyde, n-heptaldehyde
Önanthalkohol *m* 1-heptanol, enanthol, n-heptyl alcohol
Önanthetol *n* enanthetol
Önanthin *n* enanthin, heptine, oenanthine, trienanthin
Önanthoin *n* enanthoin, n-hexyl-enanthylcarbinol, tetradecan-8-al-7-one
Önanthol *n* enanthole, heptanol
Önanthon *n* enanthone, 7-tridecanone, dihexylketone
Önanthotoxin *n* enanthotoxin
Önanthsäure *f* enanthic acid, enanthylic acid, heptanoic acid, n-heptoic acid
önanthsauer enanthate
Önanthyl- enanthyl
Önanthylsäure *f* enanthic acid, enanthylic acid, heptanoic acid, heptoic acid
Önidin *n* enidin
Önin *n* enin
Önometer *n* oenometer, wine areometer
Oerstedit *m* (Min) oerstedite
örtlich local, (lagemäßig) positional
Öse *f* catch, eye (of a needle), eyelet, loop, ring
Östradiol *n* estradiol, oestradiol
Östradiolbenzoat *n* estradiol benzoate
Östran *n* estrane
Östriol *n* estriol, oestriol
Östrodienol *n* estrodienol
Östrogen *n* estrogen
Östron *n* estrone (Am. E.), oestrone
OFC-Verfahren *n* one-flow cascade cycle
Ofen *m* (Chem, Metall) furnace, oven, (Glasherstellung) calcar, (zum Backen, Braten) stove, (zum Brennen) kiln, **~ mit niedergehender Flamme** downdraft-type furnace, **den ~ abstellen** to shut down the furnace, **den ~ anbohren** to tap the furnace, **den ~ beschicken** to charge the furnace, **elektrischer ~** electric furnace, **gasgefeuerter ~** gas-fired furnace, gas furnace, **herdbeheizter ~** conducting-hearth furnace, **hüttenmännischer ~** metallurgical furnace
Ofenabdeckung *f* furnace roof
Ofenabgas *n* furnace exhaust gas
Ofenauskleidung *f* furnace lining
Ofenausmauerung *f* brick lining of a furnace
Ofenbeschickung *f* charging the furnace, furnace charge
Ofenbetrieb *m* furnace operation, furnace practice
Ofenblei *n* pig lead
Ofenbrand *m* zinc deposit in furnace
Ofendruck *m* furnace pressure

Ofendurchmesser m diameter of furnace
Ofeneindeckung f furnace covering
Ofenemaillelack m baking enamel, stove enamel; stoving enamel
Ofenfarbe f black lead, graphite
Ofenfassung f furnace capacity
Ofenfrischerei f puddling
Ofenfüllung f oven charge
Ofenfundament n furnace foundation
Ofenfutter n furnace lining
Ofengalmei m furnace cadmia, tutty
Ofengang m working of a furnace
Ofengangstörung f disturbance of the furnace process
Ofengas n furnace gas
Ofengekrätz n furnace ends
Ofenglanz m stove polish
Ofengut n furnace product
Ofenhalle f furnace house
Ofenhaltbarkeit f furnace life
Ofeninnere[s] n interior of furnace
Ofenkopf m furnace top
Ofenkrücke f glazier's scraper, stoker
Ofenleistung f furnace efficiency, furnace output
Ofenmantel m furnace shell
Ofenöffnung f mouth of a furnace, opening of a kiln
Ofenquerschnitt m cross section of a furnace
Ofenrand m rim of furnace
Ofenraum m furnace chamber
Ofenreise f life of furnace
Ofenrost m furnace grate
Ofenruß m furnace black, furnace soot, oven soot
Ofensau f furnace pig, furnace sow
Ofenschacht m shaft of furnace
Ofenschlacke f furnace slag
Ofenschwärze f powdered black lead
Ofenschwamm m furnace cadmia, tutty
Ofensohle f furnace bottom, hearth level, oven floor
Ofentrocknung f oven drying
Ofenumformer m furnace transformer
Ofenverkokung f coking in ovens
Ofenverspannung f bracing of the furnace
Ofenwirkungsgrad m furnace efficiency
offen open
Offenbandhalbleiter m open band semiconductor
offenbaren (Patent) to disclose
Offenbarung f (Patent) disclosure
Offenlegungsschrift f (Pat) laid-open specification
offensichtlich apparent, evident
Offretit m offretite
Offsetdruck m offset printing
Offsetfarbe f offset ink
Offsetkopf m offset diehead

Offsetverfahren n offset printing process
Ohm n (Maß) ohm
Ohmmeter n ohmmeter
Ohmscher Spannungsabfall m ohmic drop
Ohmscher Verlust m ohmic loss, resistance loss
Ohmscher Widerstand m ohmic resistance
Ohmsches Gesetz n (Elektr) Ohm's law
Ohr n ear
OHZ Abk. für Hydroxylzahl
Oiazin n oiazine
Oisanit m oisanit
Oiticiaöl n oiticia oil
Okanin n okanin
okkludieren to occlude
Okklusion f occlusion
Okklusionsvermögen n occlusion capacity
Oktaeder n (Krist) octahedron
Oktaederstruktur f octahedron structure
Oktaedersymmetrie f (Krist) octahedral symmetry
oktaedrisch octahedral
Oktaedrit m (Min) native titanium dioxide, octahedrite
Oktakosan n octacosane
oktakovalent octacovalent
Oktalin n octaline, octahydronaphthalene
Oktanten-Diagramm n octant diagram
Oktanten-Regel f octant rule
Oktanthren n octanthrene, octahydrophenanthrene
Oktanthrenol n octanthrenol, hydroxyoctanthrene
Oktanthrenon n octanthrenone, oxooctanthrene
Oktanwert m octane number, antiknock value, octane rating
Oktanzahl f octane number
Oktett n (Chem) octet[te]
Oktettregel f octet rule
Okthracen n octhracene, octahydroanthracene
Okthracenol n octhracenol, hydroxyocthracene
Okthracenon n octhracenone, oxoocthracene
Oktopolanregung f octopole excitation, octupole excitation
Oktopolübergang m octopole transition, octupole transition
Oktose f octose
Oktupol m octopole, octupole
Okular n eyepiece, eyeglass, ocular
Okularblende f eye-piece diaphragm
Okulareinstellung f focussing of the eye-piece
Okularfotozelle f eye piece cell
Okularglas n microscopic eyepiece
Oldhamit m (Min) oldhamite
Oleanderblätter n pl (Bot) oleander leaves
Oleandrin n oleandrin
Oleandronsäure f oleandronic acid
Oleandrose f oleandrose
Oleanol n oleanol
Oleanolsäure f oleanolic acid, caryophyllin

Oleanon *n* oleanone
Oleat *n* oleate
Olefin *n* olefin[e]
Olefin-Additionsreaktion *f* addition reaction of an olefin
Olefinalkohol *m* olefine alcohol, olefinic alcohol
Olefinhalogenid *n* olefine halide
olefinisch olefinic
Olefinketon *n* olefin[e] ketone, olefinic ketone
Olefinmetallkomplex *m* olefin[e] metal complex
Olein *n* olein, glycerol trioleate, glyceryl [tri]oleate, triolein
Oleinalkohol *m* oleyl alcohol, 9-octadecen-1-ol
Oleinsäure *f* oleic acid, 9-octadecenoic acid, oleinic acid, red oil
oleinsauer oleate
Oleinseife *f* olein soap, red-oil soap
Oleodibutyrin *n* oleodibutyrin, dibutyroolein, glycerol dibutyrate oleate, glyceryl dibutyrate oleate
Oleodipalmitin *n* oleodipalmitin, dipalmitoolein
Oleodistearin *n* oleodistearin
Oleokreosot *n* oleo-creosote, creosote oleate
Oleomargarine *n* oleomargarine, margarine
Oleon *n* oleone
Oleopalmitostearin *n* oleopalmitostearin, glycerol palmitate stearate oleate, palmitostearoolein
oleophob oleophobic
Oleophobierung *f* oleophobizing
oleophosphorsauer oleophosphoric
Oleoresin *n* oleoresin
Oleosolfarbe *f* oil-soluble dye
Oleostearinsäure *f* oleostearic acid
Oleum *n* fuming sulfuric acid, oleum
Oleyläthansulfonsäure *f* oleyl ethanesulfonic acid
Oleylalkohol *m* oleyl alcohol
Oleylnatriumsulfat *n* oleyl sodium sulfate
Olibanöl *n* olibanum oil
Oligoklas *m* (Min) oligoclase
oligomer oligomeric
Oligomerisation *f* oligomerization
Oligomycin *n* oligomycin
Oligonspat *m* (Min) oligonspar
Oligonukleotid *n* oligonucleotide
Oligopeptid *n* oligopeptide
Oligosaccharid *n* oligosaccharide
Olivacin *n* olivacine
Olive *f* (Bot) olive
Olivenbaumgummi *n* olive tree rubber
Olivenblende *f* (Min) olivine
Olivenerz *n* (Min) olive ore, native copper arsenate, olivenite, wood copper
olivenfarben olive-colored
olivengrün olive green
Olivenit *m* (Min) olivenite, native copper arsenate, olive ore, wood copper
Olivenkernöl *n* olive-kernel oil
Olivenöl *n* olive oil
Olivetol *n* olivetol
Olivetolcarbonsäure *f* olivetolcarboxylic acid
Olivetorsäure *f* olivetoric acid
Olivil *n* olivil
Olivin *m* (Min) olivine, chrysolite, hyalosiderite
Ombuin *n* ombuin
Omega-Meson *n* omega-meson
Omegatron *n* omegatron
Ommochrom *n* ommochrome
Omphazit *m* (Min) omphacite, augite
Onegit *m* onegite
Onium-Salz *n* onium salt
Oniumstruktur *f* onium structure
Onkoit *m* oncoite
onkotisch (kolloid-osmotisch) oncotic
Onocerin *n* onocerin, onocol
Onocol *n* onocerin, onocol
Onofrit *m* (Min) onofrite, native mercury sulfide selenide
Onokol *n* onocerin, onocol
Ononetin *n* ononetin
Ononin *n* ononin
Ononit *m* ononitol
Onychit *m* onychite
Onyx *m* (Min) onyx, chalcedony quartz
Onyxachat *m* (Min) onyx-agate
Onyxmarmor *m* (Min) onyx-marble
Ooflavin *n* ooflavin
Oolith *m* (Min) oolite, peastone
Oolithenkalk *m* (Min) oolite
Oolithformation *f* (Geol) oolite formation
oolithhaltig oolitiferous
Ooporphyrin *n* ooporphyrine
Oosit *m* oosite
Oosporein *n* oosporein
opak opaque, translucent
Opal *m* (Min) opal
Opalachat *m* (Min) opal agate
Opalallophan *m* (Min) opal allophane, schroetterite
opalartig opaline, opal-like
Opalblau *n* opal blue
Opaleszenz *f* opalescence
opaleszieren to opalesce
Opalfarbe *f* opal color
Opalfirnis *m* opal varnish
Opalglanz *m* opalescence, opaline luster
Opalglas *n* opal glass, frosted glass
Opalglasplatte *f* opal glass plate
opalhaltig opaliferous
Opalin *n* opaline
opalisieren to opalesce, to opalize, to render opalescent
Opalisieren *n* opalescence
opalisierend opalescent
Opaljaspis *m* (Min) opal jasper
Opalmutter *f* opal matrix

Opazität f opacity
Operation f operation, method, **chirurgische ~** (Med) surgical operation, **mathematische ~** mathematical operation
Operatorenbereich m operator domain
Operatorenkalkül n operational calculus
Operatorisomorphie f operator isomorphism
Operment n (Min) orpiment, auripigment, King's yellow, native arsenic trisulfide
Operon n (Biochem) operon
Opferanode f sacrificial anode
Opheliasäure f ophelic acid
Ophiocalcit m (Min) ophiocalcite
Ophiotoxin n ophiotoxin
Ophioxylin n ophioxylin
Ophit m ophite
Opian n opiane
Opiansäure f opianic acid
Opiat n opiate
Opiazon n opiazone, 7,8-dimethoxy-1-phthalazone
Opium n opium
Opiumalkaloid n opium alkaloid
Opiumextrakt m extract of opium
Opiumgeruch m opium smell
Opiumpflaster n opium plaster
Opiumsäure f meconic acid, 3-hydroxy-4-pyrone-2,6-dicarboxylic acid
Opiumsirup m opiated syrup
Opiumsucht f addiction to opium, opiumism
Opiumsüchtiger m opium addict, opiomaniac
Opiumtinktur f tincture of opium, laudanum
Opiumvergiftung f (Med) opium poisoning
Opiumwasser n opium water
Opopanax m (Pharm) opopanax
Opsopyrrol n opsopyrrole
Optik f (Lehre v. Licht) optics, (Linsensystem) optical system
Optiker m optician
optimal optimal, optimum
Optimalitätsprinzip n optimality principle
Optimeter n optical indicator, optimeter
Optimierung f optimization, **~ nach der kleinsten Mischarbeit** minimum power consumption for a given mixing time
Optimum n optimum
Optionsrecht n option
optisch optic[al], visual, **~ aktiv** optically active, **~ inaktiv** optically inactive, **~ negativ** optically negative, **~ positiv** optically positive
Optochinsäure f optoquinic acid
Optoelektronik f (Elektrooptik) optoelectronics
oral oral
Orange f (Bot) orange
Orangefarbe f orange color
Orangegelb n orange yellow
Orangemennige f orange minium, orange lead
Orangenöl n orange [peel] oil
Orangensamenöl n orange seed oil
Orangenschalenöl n orange peel oil
Orangeocker m (Min) orange ocher
Orangerot n orange red
Orangit m (Min) orangite, native hydrated thorium silicate
Orbital n (Atom) orbital, **antibindendes ~** anti-bonding orbital
Orbitale pl, **überlappende ~** overlapping orbitals
Orbitaltheorie f orbital theory
Orcein n orcein
Orcin n orcin, 3,5-dihydroxytoluene, 5-methylresorcinol, orcinol
Orcirufamin n orcirufamin, 7-amino-4-methyl-2-phenoxazone
Orcylaldehyd m orcyl aldehyde, 4,6-dihydroxy-2-methylbenzaldehyde
ordentlich orderly, normal, regular
Ordinalzahl f ordinal number
Ordinate f ordinate, y-coordinate
Ordinatenachse f axis of ordinate, ordinate axis, y-axis
ordnen to order, to arrange, **der Größe nach ~** to classify, to grade
Ordnung f order, arrangement, classification, (Folge) series, succession, **~ der Reaktion** order of reaction
Ordnungsparameter m order parameter
Ordnungsumwandlung f ordering transition
Ordnungsunordnungsumwandlung f order-disorder transformation
Ordnungszahl f atomic number, ordinal number, periodic number
Orexin n orexin; 3-phenyl-3,4-dihydroquinazoline
Orexinchlorhydrat n orexin hydrochloride
Orexintannat n orexin tannate
Organelle f (Biol) organelle
Organiker m organic chemist
Organisation f organisation
organisch organic
organisch-chemisch organic-chemical
organisieren to organise, to organize
Organismen pl, **lebende ~** living organisms
Organismus m organism
Organogel n organogel
Organographie f (Med) organography
organographisch (Med) organographic
Organohalogensilan n organohalosilane
organoleptisch organoleptic[ally]
Organologie f (Med) organology
organologisch (Med) organologic
Organometall n organometal, organometallic compound
organometallisch organometallic
organoplastisch organoplastic
Organosilandiol n organosilanediol
Organosiliziumverbindung f organosilicon compound

Organosiloxan *n* organosiloxane
Organosol *n* organosol
Orgelmetall *n* organ-pipe metal
orientalisch oriental
orientierbar orientable
orientieren to orientate, to align, to direct
orientiert oriented, aligned, directed
Orientierung *f* orientation, (Richtungssinn) direction, **bevorzugte** ~ preferred orientation
orientierungsabhängig orientation-responsive, orientation-sensitive
Orientierungsabhängigkeit *f* orientation dependence
Orientierungsbeziehung *f* orientation relation
Orientierungsdreieck *n* orientation triangle
Orientierungswert *m* guide value
original original
Originalpackung *f* original packing, standard packing
Orixin *n* orixine
Orlean *n* orlean, annatto, annotta
Orleanfärberei *f* dying with annatto
Orleanorange *n* (Färb) annatto orange
Orleanrot *n* bixin
Orlon *n* orlon
Ormosin *n* ormosine
Ormosinin *n* ormosinine
Ornithin *n* ornithine, 2,5-diaminopentanoic acid
Ornithintransaminase *f* (Biochem) ornithine transaminase
Ornithintranscarbamylase *f* (Biochem) ornithine transcarbamylase
Ornithinzyklus *m* ornithine cycle
Ornithursäure *f* ornithuric acid
Orobol *n* orobol
Orogenese *f* (Geol) orogenesis
Orotidinphosphat *n* orotidine phosphate
Orotidylsäure *f* orotidylic acid
Orotsäure *f* orotic acid
Oroxylin *n* oroxylin
Orphenadrin *n* orphenadrine
Orsat-Blase *f* rubber expansion bag
Orsatscher Apparat *m* Orsat apparatus
Orseille *f* orseille, archil, orchil, orselle
Orseilleextrakt *m* archil extract, archil paste, orchil extract
Orseillekarmin *n* archil carmine
Orseillin *n* orseillin, orsellin, rocellin
Orseillinsäure *f* (Orsellinsäure) orsellinic acid, 4,6-dihydroxy-2-methylbenzoic acid
Orsellinat *n* orsellinate, salt or ester of orsellinic acid
Orsudan *n* orsudan, sodium 4-acetamido-2-methylbenzene arsenate
Ort *m* position, location, place, (Fleck) spot, (Math) locus, **geometrischer** ~ [geometrical] locus
Orthanilsäure *f* orthanilic acid
Orthit *m* (Min) orthite, allanite

Orthoachse *f* (Krist) orthoaxis, orthodiagonal axis
Orthoameisensäure *f* orthoformic acid
Orthoameisensäureester *m* orthoformate, orthoformic ester
ortho-Aminobenzoesäure *f* o-aminobenzoic acid, anthranilic acid
Orthoantimonsäure *f* orthoantimonic acid
Orthoarsensäure *f* orthoarsenic acid
orthobasisch orthobasic
Orthoborsäure *f* orthoboric acid
ortho-Chinon *n* o-quinone
ortho-Chlorbenzaldehyd *m* o-chlorobenzaldehyde, 2-chlorobenzenecarbonal
ortho-Chlorbenzoesäure *f* o-chlorobenzoic acid
ortho-Chlortoluol *n* o-chlorotoluene, 2-chloro-1-methylbenzene
orthochromatisch orthochromatic
Orthochromplatte *f* orthochromatic plate
ortho-Cymol *n* 1-methyl-2-isopropylbenzene, o-cymene
orthodiagonal (Krist) orthodiagonal
Orthodiagonale *f* (Krist) orthoaxis, orthodiagonal axis
orthodirigierend ortho-directing
Orthoessigsäure *f* orthoacetic acid
Orthoform *n* orthoform, methyl-4-hydroxy-3-aminobenzoate
orthogonal orthogonal, rectangular
Orthogonalisierungsprozeß *m* orthogonalization process
Orthogonalität *f* orthogonality
Orthogonalitätseigenschaft *f* orthogonality property
Orthogonalitätsrelation *f* orthogonality relation
Orthogonalsystem *n* orthogonal system
Orthokieselsäure *f* orthosilicic acid
Orthoklas *m* (Min) orthoclase, potash feldspar, sunstone
Orthoklasporphyr *m* (Min) orthoclase porphyry
Orthokohlensäure *f* orthocarbonic acid
Orthokohlensäureester *m* orthocarbonic ester
ortho-Kresol *n* o-cresol, o-hydroxytoluene, o-methylphenol
ortho-Nitranilin *n* o-nitraniline, 2-nitro-1-aminobenzene
ortho-Nitranisol *n* o-nitranisol, 2-nitro-1-methoxybenzene
ortho-Nitrobenzaldehyd *m* o-nitrobenzaldehyde, 2-nitrobenzenecarbonal
ortho-Nitrobenzoesäure *f* o-nitrobenzoic acid
ortho-Nitrotoluol *n* o-nitrotoluene, 2-nitro-1-methylbenzene
orthopädisch orthop[a]edic
Ortho-Para-Umwandlung *f* (Wasserstoff) ortho-para conversion
ortho-Phenylendiamin *n* 1,2-benzenediamine, 1,2-diaminobenzene, o-phenylene diamine

Orthophosphorsäure f orthophosphoric acid
ortho-Phthalsäure f 1,2-benzenedicarboxylic acid, o-phthalic acid
Orthoporphyr m (Min) orthoporphyry
Orthopositronium n orthopositronium
Orthoprisma n (Krist) orthoprism
orthorhombisch (Krist) orthorhombic
Orthosäure f ortho acid
Orthosalpetersäure f orthonitric acid
Orthoschwarz n ortho black
Orthosilikat n orthosilicate, salt or ester of orthosilicic acid
Orthosiliziumameisensäure f ortho silicoformic acid
Orthoskop n (Med) orthoscope
Orthostellung f (Chem) o-position, ortho position
ortho-Toluidin n o-toluidine, 2-amino-1-methylbenzene, 2-aminotoluene, o-tolylamine
ortho-Toluolsulfamid n o-toluenesulfonamide
ortho-Toluolsulfochlorid n o-toluenesulfonylchloride
Orthotropie f orthotropy
Orthoverbindung f o-compound, ortho compound
Orthowasserstoff m ortho hydrogen
Orthozimtsäure f orthocinnamic acid
Ortizon n ortizon, urea hydrogen peroxide
Ortol n ortol, 2-hydroxy-1-methylaminobenzene, 2-methylaminophenol
Ortsisomerie f position isomerism
Ortskoordinate f space coordinate
Ortsmessung f position measurement
Ortsvektor m position vector
Ortswechsel m change of position, migration
Ortung f location, orientation, position fixing, (Peilung) direction finding
Orvillit m orvillite
Oryzit m oryzite
Orzin n orcin, 3,5-dihydroxy-1-methylbenzene, 3,5-dihydroxytoluene, 5-methyl-1,3-benzenediol, orcinol
Osajin n osajin
Osazon n osazone
Oscin n oscine, 3,7-oxido-6-tropanol, scopoline
Oskulation f (Math) osculation
Osman n osmane
Osmat n osmate, salt or ester of osmic acid
Osmelit m osmelite
Osmiridium n iridosmine, osmiridium
Osmium n osmium (Symb. Os)
Osmiumfaden m osmium filament
Osmiumgehalt m osmium content
osmiumhaltig containing osmium
Osmium(II)-Verbindung f osmium(II) compound, osmous compound

Osmium(IV)-Verbindung f osmic compound, osmium(IV) compound
Osmiumlampe f osmium lamp
Osmiumlegierung f osmium alloy, osmide
Osmiumoxid n osmium oxide, (Osmium(IV)-oxid) osmium dioxide, osmium(IV) oxide
Osmiumoxydul n (Osmium(II)-oxid) osmium(II) oxide, osmium monoxide, osmium protoxide, osmous oxide
Osmiumperoxid n osmium peroxide, osmium tetroxide, osmium(VIII) oxide
Osmiumsäure f osmic acid
Osmiumsuperoxid n osmium peroxide, osmium tetroxide, osmium(VIII) oxide
Osmiumtetr[a]oxid n osmium tetroxide, osmium peroxide, osmium(VIII) oxide
Osmium(VIII)-oxid n osmium tetroxide
Osmolarität f osmolarity
Osmometer n osmometer
Osmometrie f osmometry
Osmondit m osmondite
Osmose f osmosis, **umgekehrte** ~ reverse osmosis
Osmoseapparat m osmosis apparatus
Osmosemesser m osmometer
osmotisch osmotic
Osmundeisen n Osmond iron, Osmund iron, Swedish bar iron
Osmundofen m Osmund furnace
Oson n osone
Osotriazol n 1,2,3-triazole, osotriazole
Ossein n ossein, collagen
Ossifikation f (Med) ossification
Osteochemie f osteochemistry
Osteogenese f (Knochenbildung) osteogenesis, osteogeny
Osteokarzinom n (Med) osteocarcinoma
Osteolith m (Min) osteolite
Osteolyse f (Knochenauflösung) osteolysis
Osthenol n osthenol
Osthol n osthol
Ostholsäure f ostholic acid
Ostranit m (Min) ostranite, zircon
Ostreasterin n ostreasterol
Ostruthin n ostruthin
Ostruthol n ostruthol
Ostwaldsche Indikatortheorie f Ostwald's theory of indicators
Ostwaldsches Verdünnungsgesetz n Ostwald's dilution law
Osyritrin n osyritrin
Oszillation f oscillation
Oszillationsquantenzahl f vibrational quantum number
Oszillationsumformer m oscillatory transformer
Oszillator m oscillator, **piezoelektrischer** ~ piezoelectric oscillator
Oszillatorfrequenz f oscillator frequency

Oszillatorröhre f oscillator tube
oszillieren to oscillate, to vibrate
oszillierend oscillating, oscillatory, vibrating, vibratory
Oszillierung f oscillation, vibration
Oszillogramm n oscillogram, oscillograph curve
Oszillograph m oscillograph, ~ **mit Braunscher Röhre** cathode ray oscillograph, **elektrostatischer** ~ electrostatic oscillograph
Oszillographentechnik f oscillographic technique
Oszillographenverstärker m oscillograph amplifier
Oszilloskop n oscilloscope
Otavit m (Min) otavite, native cadmium carbonate
Otobit n otobite
Otolith m (Med) otolith, ear-stone
Ottomotor m Otto engine
Ottrelith m (Min) ottrelite
Ouabain n ouabain, crystallized strophantin, uabain
oval oval
Ovalbumin n ovalbumin, egg albumin
Ovalen n ovalene
Ovalkaliber n oval pass
Ovalradzähler m gear-type meter, oval gear meter
Ovanen n ovanene
Ovizid n ovicide
Ovolarvizid n ovolarvicide
Ovoverdin n ovoverdin
Ovulation f (Biol) ovulation
Ovulationshemmer m ovulation inhibitor
Ovulit m ovulite
Oxadiazol n oxadiazole
Oxäthylamin n hydroxyethylamine
Oxalat n oxalate, salt or ester of oxalic acid
Oxalataustausch m oxalate exchange
Oxalatentwickler m oxalate developer
Oxalatochelat n oxalato chelate
Oxalbernsteinsäure f oxal[o]succinic acid
Oxalcitraconsäure f oxalcitraconic acid
Oxalen n oxalene
Oxalendiuramidoxim n oxalene diuramidoxime
Oxalessigsäure f oxaloacetic acid, oxobutanedioic acid
Oxalessigsäureester m oxalacetic ester, diethyl oxalacetate
Oxalessigsäuretransaminase f (Biochem) oxalo-acetic transaminase
Oxalester m oxalic ester
Oxalit m (Min) oxalite, humboldite
Oxalnitril n cyanogen, oxalonitrile
Oxalsäure f oxalic acid, ethanedioic acid
Oxalsäureanlage f oxalic acid plant
Oxalsäurediäthylester m diethyl oxalate
Oxalsäureester m oxalate
Oxalsäurefabrik f oxalic acid factory

Oxalsäuresalz n oxalate, salt of oxalic acid
Oxalurie f (Med) oxaluria
Oxalursäure f oxaluric acid, carbamyloxamic acid, oxalic monoureide
Oxalyl- oxalyl
Oxalylchlorid n oxalyl chloride
Oxalylharnstoff m oxalylurea, parabanic acid
Oxamethan n oxamethane, ethyl oxamate
Oxamid n oxamide, ethanediamide, oxalamide
Oxamidsäure f oxalic acid monoamide, oxamic acid, oxaminic acid
Oxaminblau n oxamine blue
Oxaminfarbstoff m oxamine dye
Oxamycin n oxamycin
Oxanamid n oxanamide
Oxanilid n oxanilide
Oxanilsäure f oxanilic acid, oxalic acid monoanilide, phenyloxamic acid
Oxanthranol n 10-hydroxyanthrone, oxanthranol, 9,10-anthradiol, anthrahydroquinone
Oxanthron n oxanthrone
Oxaphor n oxaphor, 50% alcoholic solution of 3-hydroxycamphor
Oxapropanium n oxapropanium
Oxazin n oxazine
Oxazinfarbstoff m oxazine dye
Oxazol n oxazole
Oxazolidin n oxazolidine
Oxazolidon n oxazolidone
Oxazolin n oxazoline
Oxazolon n oxazolone
Oxdiazin n oxdiazine
Oxdiazol n oxdiazole
Oxepan n oxepane
Oxepin n oxepin
Oxetan n oxetane
Oxeton n oxetone
Oxhaverit m oxhaverite
Oxid n oxide, **amphoteres** ~ amphoteric oxide, **basisches** ~ basic oxide, **geglühtes** ~ fused oxide, **höheres** ~ higher oxide, **in ein** ~ **verwandeln** to oxidize, **niedrigeres** ~ lower oxide, **saures** ~ acidic oxide
Oxidans n (pl Oxidantien) oxidant, oxidizer, oxidizing agent, oxidizing substance
oxidartig like an oxide, oxide-like
Oxidase f (Biochem) oxidase
Oxidation f oxidation, **anodische** ~ anodic oxidation, **beta-** ~ **der Fettsäuren** beta-oxydation of fatty acids, **biologische** ~ biological oxidation, **katalytische** ~ catalytic oxidation, **Oberflächen-** ~ surface oxidation, **selektive** ~ selective oxidation
Oxidationsbeständigkeit f oxidation resistance, oxidation stability
Oxidationsbleiche f peroxide bleaching
oxidationsfähig oxidizable
Oxidationsfähigkeit f oxidizability

Oxidationsflamme — Oxychlorkupfer

Oxidationsflamme *f* oxidizing flame
Oxidationsgrad *m* degree of oxidation
Oxidationsinhibitor *m* oxidation inhibitor
Oxidationskatalysator *m* oxidizing catalyst
Oxidationsmittel *n* oxidizing agent, oxidant, oxidizer
Oxidationsofen *m* oxidizing furnace, oxidizing oven
Oxidationspotential *n* oxidation potential
Oxidationsprodukt *n* oxidation product, product of oxidation
Oxidationsprozeß *m* oxidation process
Oxidationsraum *m* oxidation chamber
Oxidations-Reduktions-Elektrode *f* oxidation-reduction electrode
Oxidations-Reduktions-Kette *f* oxidation-reduction cell
Oxidations-Reduktions-Paar *n* oxidation-reduction pair
Oxidations-Reduktions-Potential *n* oxidation-reduction potential, redox potential
Oxidations-Reduktions-Reaktion *f* oxidation-reduction reaction, redox reaction
Oxidationsschlacke *f* oxidizing slag
Oxidationsstufe *f* oxidation step, degree of oxidation, stage of oxidation
Oxidationsvorgang *m* oxidation process
Oxidationswärme *f* heat of oxidation
Oxidationswirkung *f* oxidizing action, oxidizing effect
Oxidationszahl *f* oxidation number
Oxidationszustand *m* oxydation state
oxidierbar oxidizable, capable of oxidation
Oxidierbarkeit *f* oxidizability
oxidieren to oxidize
Oxidieren *n* oxidizing, oxidation process
oxidierend oxidizing
oxidisch containing oxide, oxidic, oxygenic
Oxidoreduktion *f* oxidation-reduction
Oxidbelag *m* oxide film
Oxidbeschlag *m* oxide coating
oxidbildend oxide-forming
Oxidbraun *n* oxide brown
Oxideinschluß *m* oxide inclusion
Oxidelektrode *f* oxide electrode
Oxidfarbe *f* oxide color
oxidfrei free from oxide[s]
Oxidgelb *n* oxide yellow
Oxidhäutchen *n* oxide film, pellicle of oxygen
oxidhaltig containing oxides, oxidic, oxidiferous
Oxidhaut *f* film of oxide, oxidized surface
Oxidhydrat *n* hydroxide, hydrated oxide
Oxidkathode *f* oxide cathode, oxide-coated filament, Wehnelt cathode
Oxidkathodenröhre *f* cathode tube with oxide-coated filament
Oxidmarialith *m* (Min) oxide marialite
Oxidmeionit *m* (Min) oxide meionite
Oxidocker *m* oxide ocher

Oxidoreduktase *f* (Biochem) oxido-reductase
Oxidrot *n* purple iron oxide, red iron oxide, Turkey red
Oxidschicht *f* layer of oxide, oxide film
oxidüberzogen oxide-coated
Oxidüberzug *m* oxide coating
Oxidwachs *n* oxide wax
Oxim *n* oxime
Oximid *n* oximide, oxalimide
Oxin *n* oxine, 8-hydroxyquinoline
Oxinat *n* oxinate
Oxindigo *n* oxindigo
Oxindol *n* oxindole, 2-ketoindoline
Oxiran *n* 1,2-epoxyethane, oxirane, ethylene oxide
Oxoadrenochrom *n* oxoadrenochrome
Oxoaristsäure *f* oxoaristic acid
Oxoctenol *n* oxoctenol
Oxoformel *f* (Zucker) planar sugar formula
Oxohämanthidin *n* oxohemanthidine
Oxon *n* oxone, sodium peroxide
Oxonit *n* oxonite
Oxoniumsalz *n* oxonium salt
Oxonsäure *f* oxonic acid
Oxophenarsin *n* oxophenarsine
Oxosäure *f* ketonic acid
Oxosynthese *f* oxosynthesis
Oxozon *n* oxozone
Oxthian *n* oxthiane
Oxthiazol *n* oxthiazole
Oxthin *n* oxthine
Oxthiol *n* oxthiole
Oxyacanthin *n* hydroxyacanthine, vinetine
Oxyacridin *n* hydroxyacridine
Oxyaldehyd *m* hydroxy aldehyde, oxyaldehyde
Oxyammoniak *n* (obs) hydroxylamine, oxyammonia
Oxyammoniumverbindung *f* hydroxylammonium compound, oxyammonium compound
Oxyanthrachinon *n* hydroxyanthraquinone
Oxyazoverbindung *f* hydroxyazo compound
Oxybenzaldehyd *m* hydroxybenzaldehyde
Oxybenzoesäure *f* hydroxybenzoic acid
Oxybenzol *n* hydroxybenzene, phenol
Oxybenzophenon *n* hydroxybenzophenone
Oxybernsteinsäure *f* hydroxysuccinic acid, malic acid
Oxybiotin *n* oxybiotin
Oxybuttersäure *f* hydroxybutyric acid
Oxybuttersäurealdehyd *m* (Oxybutyraldehyd) hydroxybutanal, hydroxy butyraldehyde
Oxycampher *m* hydroxycamphor, oxycamphor
Oxycarbonsäure *f* hydroxycarboxylic acid
Oxycellulose *f* oxycellulose
Oxychinolin *n* hydroxyquinoline, quinolinol
Oxychinon *n* hydroxyquinone
Oxychlorid *n* oxychloride
Oxychlorierung *f* oxychlorination
Oxychlorkupfer *n* copper oxychloride

Oxycinchophen *n* oxycinchophen
Oxycodon *n* (Analgetikum) oxycodone
Oxycyan *n* oxycyanogen
Oxycyanquecksilber *n* mercuric oxycyanide, mercury(II) oxycyanide
Oxyd *n* oxide (s. auch Oxid)
Oxydation *f* oxidation (s. auch Oxidation)
Oxydiazol *n* oxadiazole, oxdiazole
Oxydiphenylamin *n* hydroxydiphenylamine
Oxydiphenylmethan *n* hydroxydiphenylmethane
Oxydul *n* lower oxide
Oxyessigsäure *f* hydroxyacetic acid
Oxyfettsäure *f* hydroxyfatty acid
Oxyfluorid *n* oxyfluoride
Oxygen *n* oxygen (Symb. O)
Oxygenase *f* (Biochem) oxygenase
Oxygengas *n* oxygen gas
Oxygenierung *f* oxygenation
Oxygenium *n* (Lat) oxygenium (Symb. O)
Oxyhämoglobin *n* oxyhemoglobin
Oxyketon *n* hydroxyketone, ketol
Oxyketoncarbonsäure *f* hydroxyketo[ne]carboxylic acid
Oxymel *n* oxymel
Oxymenthylsäure *f* hydroxymenthylic acid
Oxymethylbenzoesäure *f* hydroxymethylbenzoic acid
Oxymorphon *n* oxymorphone
Oxynaphthalin *n* hydroxynaphthalene, naphthol
Oxynaphthoesäure *f* hydroxynaphthoic acid, naphtholcarboxylic acid
Oxynarcotin *n* oxynarcotine
Oxynicotin *n* oxynicotine
Oxypolygelatine *f* oxypolygelatin
Oxyprolin *n* hydroxyproline, 3-hydroxy-2-pyrrolidine carboxylic acid
Oxypropionsäure *f* hydroxypropionic acid, lactic acid
Oxysäure *f* hydroxy acid, oxacid, oxyacid
Oxysalz *n* oxy salt, salt of hydroxy acid
Oxyschwefelsäure *f* oxysulfuric acid
Oxyspartein *n* oxysparteine
Oxystearinsäure *f* hydroxystearic acid
Oxythiamin *n* oxythiamine
Oxytocin *n* ocytocin
Oxytocinase *f* (Biochem) oxytocinase
Oxytoluol *n* hydroxytoluene, cresol
Oxyverbindung *f* hydroxy compound, oxy compound
Oxyzellulose *f* oxycellulose
Oxyzimtsäure *f* hydroxycinnamic acid
Ozarkit *m* ozarkite
ozeanisch oceanic
Ozobenzol *n* ozobenzene
Ozokerit *m* ozocerite, fossil wax, mineral wax
Ozon *n* ozone, **befreien von** ~ to free from ozone, **behandeln mit** ~ to ozonize
Ozonanlage *f* ozone plant
Ozonapparat *m* ozone apparatus

Ozonbildung *f* ozone formation
Ozonbleiche *f* ozone bleach
Ozondarstellungsmethode *f* method of producing ozone
Ozonentwickler *m* ozone generator
Ozonentwicklung *f* ozonification, ozone generation, ozone production
ozonerzeugend ozoniferous, generating ozone, producing ozone
Ozonerzeuger *m* ozonizer, ozonator, ozone generator
Ozonerzeugung *f* ozone generation, ozone production, ozonification
Ozongehalt *m* ozone content, ~ **der Luft** amount of ozone [in the air]
Ozonglimmerröhre *f* mica ozone tube
ozonhaltig containing ozone, ozoniferous
Ozonid *n* ozonide
Ozonisation *f* ozonization
Ozonisator *m* ozonizer
ozonisieren to ozonize
Ozonisieren *n* ozonization
ozonisiert ozonized
Ozonisierung *f* ozonization
Ozonkontinuum *n* continuous ozone spectrum
Ozonlüftung *f* treatment of air with ozone
Ozonmesser *m* ozonometer
Ozonmessung *f* ozonometry
Ozonolyse *f* ozonolysis
ozonometrisch ozonometric
Ozonoskop *n* ozonoscope
Ozonpapier *n* ozone paper
Ozonreagenspapier *n* ozone [test] paper
Ozonröhre *f* ozone tube
Ozonröhrenelement *n* tubular ozonizer section
Ozonsauerstoff *m* oxygen in the form of ozone, ozonized oxygen
Ozonspaltung *f* ozonolysis
Ozonventilator *m* ozone fan
Ozonwasser *n* ozone[-containing] water
Ozonzerstörung *f* ozone annihilation

P

p- s. para-
p. a. for analysis
Paar *n* pair, couple
Paarbildung *f* pair formation, pair creation
paaren to pair, to couple, (Phys) to conjugate
Paarerzeugung *f* pair formation, pair production
Paarumwandlung *f* pair conversion
Paarung *f* pairing, coupling, (Phys) conjugation
Paarungsenergie *f* pairing energy
Paarvernichtung *f* (Atom) pair annihilation
paarweise in pairs
P-Abweichung *f* (Regeltechn) proportional offset
Pachnolith *m* (Min) pachnolite
Pachymsäure *f* pachymic acid
Packbleiche *f* kier bleaching
packen to pack, (einwickeln) to wrap [upp], **in Pakete ~** to parcel
Packfong *n* German silver, packfong, paktong
Packkorb *m* hamper
Packmaterial *n* packing material
Packpapier *n* wrapping paper, packing paper, (als Papiersorte) brown paper, (festes, braunes) kraft paper
Packpresse *f* box press, packing press
Packung *f* package, (Med, Techn) packing
Packungsanteil *m* packing fraction, (Phys) packing component
Packungsart *f* (Krist) arrangement of particles, packing of particles
Packungsdichte *f* packing density
Packungseffekt *m* concentrative effect, packing effect
Packungserscheinung *f* packing phenomenon
Paddel *n* paddle
Paddelrührer *m* paddle agitator, paddle mixer
Pädiatrie *f* (Med) pediatrics
Päonie *f* (Bot) peony [flower]
Päoniensamen *m* (Bot) peony seed
Päonin *n* peonine
Päonol *n* peonol
Pagodenstein *m* pagoda stone
Paket *n* package, parcel
paketieren to pack, (Metall) to fag[g]ot
Paketieren *n* piling, (Metall) faggoting (Br. E.), fagoting (Am. E.)
Palacheit *m* palacheite
Paläolithikum *n* (Geol) paleolithic period
Paläomagnetismus *m* paleomagnetism
Paläozoikum *n* (Geol) Paleozoic [age]
Palagonit *m* palagonite
Palatin *n* (Phthalsäureweichmacher) palatinol
Palatinit *m* palatinite
Palatinose *f* palatinose
Palatinrot *n* palatine red
Palette *f* palette
Palettensparer *m* pallet saver

Palettenwendeklammer *f* pallet turning clamp
Palingenese *f* (Entwicklung nach dem biogenetischen Grundgesetz) palingenesis
Palisanderholzöl *n* palisander wood oil
Palit *n* palite
Palladgold *n* (Min) palladium gold, porpezite
Palladium *n* palladium (Symb. Pd), **kolloides ~** colloidal palladium, electropalladiol (Am. E.)
Palladiumamalgam *n* palladium amalgam
Palladiumasbest *m* palladium asbestos, palladized asbestos
Palladiumbromür *n* (Palladium(II)-bromid) palladium(II) bromide, palladous bromide
Palladiumerz *n* (Min) palladium ore
Palladiumgehalt *m* palladium content
Palladiumgold *n* palladium gold, (Min) porpezite
Palladium(II)-oxid *n* pallad[i]ous oxide
Palladium(II)-Verbindung *f* pallad[i]ous compound, palladium(II) compound
Palladium(IV)-chlorid *n* palladic chloride, palladium(IV) chloride
Palladium(IV)-chlorwasserstoff *m* chloropalladic acid
Palladium(IV)-oxid *n* palladic oxide
Palladium(IV)-Verbindung *f* palladic compound, palladium(IV) compound
Palladiumjodür *n* (Palladium(II)-jodid) palladium(II) iodide, palladous iodide
Palladiumlegierung *f* palladium alloy
Palladiummohr *m* palladium black
Palladiumnitrat *n* palladium nitrate
Palladiumoxid *n* (Palladium(IV)-oxid) palladic oxide, palladium dioxide, palladium(IV) oxide
Palladiumoxydul *n* (Palladium(II)-oxid) palladium(II) oxide, palladium monoxide, palladous oxide
Palladiumoxydulnitrat *n* (obs) palladium(II) nitrate, palladous nitrate
Palladiumoxydulsalz *n* (obs) palladium(II) salt, palladous salt
palladiumsauer palladic
Palladiumschwamm *m* palladium sponge
Palladiumschwarz *n* palladium black
Palladiumsilicid *n* palladium silicide
Palladiumwasserstoff *m* palladium hydride
Palladochlorid *n* palladium(II) chloride, palladous chloride
Palladochlorwasserstoffsäure *f* chloropalladous acid
Palladohydroxid *n* palladium(II) hydroxide, palladous hydroxide
Pallasit *m* (Min) pallasite
Pallringe *pl* Pall rings
Palmarosaöl *n* palmarosa oil

Palmatin *n* palmatine
Palmatrubin *n* palmate rubine
Palmbranntwein *m* palm toddy
Palme *f* (Bot) palm
Palmellin *n* palmellin
Palmenbutter *f* palm oil
Palmenöl *n* palm oil
Palmerit *m* palmerite
Palmfett *n* palm oil
Palmhonig *m* palm honey
Palmitat *n* palmitate, cetylate, hexadecanate
Palmitenon *n* palmitenone
Palmitin *n* palmitin, glycerol palmitate
Palmitinkerze *f* palmitin candle
Palmitinsäure *f* palmitic acid, hexadecanoic acid, palmic acid
Palmitinsäurecetylester *m* cetyl palmitate
Palmitinsäureester *m* palmitate
Palmitinsäuresalz *n* palmitate, cetylate, hexadecanate
Palmitinseife *f* palmitin soap
Palmitodichlorhydrin *n* palmitodichlorohydrin, dichlorohydrin palmitate
Palmitodilaurin *n* palmitodilaurin, glycerol dilaurate palmitate
Palmitodiolein *n* palmitodiolein, glycerol dioleate palmitate
Palmitodistearin *n* palmitodistearin, glycerol distearate palmitate
Palmitoleinsäure *f* palmitoleic acid
Palmitolsäure *f* palmitolic acid, 7-hexadecynoic acid
Palmiton *n* 16-hentriacontanone, palmitone
Palmitoylcoenzym A *n* palmitoyl coenzyme A
Palmityl- palmityl
Palmitylalanin *n* palmitylalanine
Palmitylaldehyd *m* palmitic aldehyde, hexadecanal, palmitaldehyde
Palmitylglycin *n* palmitylglycine
Palmityltyrosin *n* palmityltyrosine
Palmkernmehl *n* palm nut meal
Palmkernöl *n* palm [kernel] oil, palm nut oil
Palmmehl *n* sago
Palmnußöl *n* coconut oil
Palmöl *n* palm oil
Palmölseife *f* palm oil soap
Paltreubin *n* paltreubine
Pamosäure *f* pamoic acid
Panacon *n* panacon
Panamabindung *f* basket weave
Panamagewebe *n* (Text) basket weave
Panamarinde *f* Panama bark, quillai bark
Panaxgummi *n* opopanax
Panchromasie *f* (Phot) panchromatism
panchromatisch (Phot) panchromatic
Pancratin *n* pancratine
Pandanstroh *n* pandan straw
Pandermit *m* (Min) pandermite
PAN-Faser *f* polyacrylonitrile fiber

Panflavin *n* panflavine
Paniculatin *n* paniculatine
Panit *m* panitol
Panklastit *n* panclastite
Pankreas *n* (Anat) pancreas
Pankreasamylase *f* (Biochem) pancreatic amylase, amylopsin
Pankreasdiastase *f* (Biochem) pancreatic diastase
Pankreasdrüse *f* (Anat) pancreas gland
Pankreasenzym *n* (Biochem) pancreatic enzyme
Pankreasfermentgemisch *n* pancreatic enzyme mixture
Pankreasfunktion *f* (Med) pancreatic function
Pankreaslipase *f* (Biochem) pancreatic lipase, pancreatolipase
Pankreasnucleinsäure *f* pancreatic nucleic acid, pancreas nucleic acid
Pankreassaft *m* pancreatic juice
Pankreatin *n* pancreatin
Pankreozymin *n* (Gewebshormon) pancreozymin
Pannarsäure *f* pannaric acid
Panne *f* accident, breakdown, failure
pannensicher puncture-proof
Panose *f* panose
Pansen *m* rumen
Pantethein *n* pantetheine
Pantethin *n* pantethine
Panthenol *n* panthenol
Panthothensäure *f* pantothenic acid
Pantoat *n* pantoate
Pantoffeltierchen *n* (Zool) paramecium, param[o]ecium, slipper animalcule
Pantograph *m* pantograph
Pantokain *n* (Pharm) pantocain
Pantonin *n* pantonine
Pantopon *n* (Pharm) pantopon
pantoponsüchtig addicted to pantopon
Pantosäure *f* pantoic acid
Pantoskop *n* pantoscope
pantoskopisch pantoscopic
Pantothein *n* pantotheine
Pantothin *n* pantothine
Panzeraktinometer *n* shielded actinometer
Panzergalvanometer *n* iron clad galvanometer
Panzerglas *n* armoured glass, bullet-proof glass
Panzerkarton *m* plated box
Panzerrohr *n* armed pipe
Panzerung *f* armor plating
Papageigrün *n* parrot green
Papain *n* papain, papayotin
Papaverin *n* papaverine
Papaverolin *n* papaveroline
Papaya[frucht] *f* (Bot) papaya
Papier *n* paper, **Alt-** ~ waste paper, **aluminium-kaschiertes** ~ aluminum-foiled paper, **Bütten-** ~ handmade paper, **einseitig gestrichenes** ~ one-side-coated paper,

Papier — Parabuxin

fettdichtes ~ grease-proof paper,
feuerfestes ~ fire-resisting paper,
flammfestes ~ flame resisting paper, **Fließ-** ~ absorbent paper, blotting paper, **gedecktes** ~ coated paper, **geleimtes** ~ sized paper, **geöltes** ~ oiled paper, **gestrichenes** ~ coated paper, glazed paper, **glattes** ~ satiny paper, **gummiertes** ~ gummed paper, **handgeschöpftes** ~ handmade paper, **holzfreies** ~ paper without wood-pulp, wood-free paper, **imprägniertes** ~ impregnated paper, **lichtdichtes** ~ black-out paper, **lichtempfindliches** ~ photographic paper, sensitized paper, **luftdichtes** ~ airproof paper, **metallisiertes** ~ metallized paper, **naßfestes** ~ wet-strength paper, **photographisches** ~ photographic paper, **rostschützendes** ~ antitarnish paper, **Schmirgel-** ~ abrasive paper, carborundum paper, smoothing paper, **schwer entflammbares** ~ flame retardant paper, **ungeleimtes** ~ unsized paper, blotting paper, **veredeltes** ~ processed paper, **wasserfestes** ~ waterproof paper, **Well-** ~ corrugated paper, **Zigaretten-** ~ cigarette paper
Papierabfall *m* waste paper
Papieraufbereitungsprozeß *m* pulping process
Papierauskleidung *f* paper lining
Papierausschuß *m* waste paper
Papierbahn *f* endless paper, paper web
Papierband *n* paper tape
Papierbecher *m* paper cup
Papierbeschwerer *m* paperweight
Papierbeutel *m* paper bag
Papierbrei *m* [paper] pulp, **chemischer** ~ chemical pulp
Papierchromatogramm *n* paper chromatogram
Papierchromatographie *f* paper [strip] chromatography, **absteigende** ~ descending paper chromatography, **aufsteigende** ~ ascending paper chromatography
Papierelektrophorese *f* paper electrophoresis
Papierfabrik *f* paper mill
Papierfabrikation *f* paper making, paper manufacture
Papierfaden *m* paper thread
Papierfilter *n* paper filter
Papierfüllstoff *m* paper filler
Papiergarn *n* paper yarn
Papiergewebe *n* paper cloth
Papiergewicht *n* paper weight
Papierhilfsmittel *n* papermaking auxiliary
Papierindustrie *f* paper industry
Papierkalander *m* paper calender, supercalender
Papierkohle *f* paper coal
Papierlack *m* paper lacquer, paper coating
Papierleimung *f* paper sizing
Papiermaché *n* paper maché
Papiermaschine *f* paper[-making]machine

Papiermasse *f* [paper] pulp, **halbchemische** ~ semi-chemical paper pulp, **mechanische** ~ mechanical paper pulp
Papiermassefänger *m* pulp catcher
Papiermesser *n* paper knife
Papierprüfgerät *n* paper testing equipment
Papierprüfung *f* paper testing
Papierrohr *n* paper tube
Papierrolle *f* paper roll
Papiersack *m* paper bag, paper sack
Papierscheibe *f* paper disk
Papierschnitzel *n pl* bits of paper
Papierstoffgarn *n* pulp yarn
Papierstreichvorrichtung *f* coating apparatus for paper
Papierstreifen *m* paper strip
Papierüberzug *m* paper coating
Papierunterlage *f* paper carrier
papierverarbeitend paper-working
Papierverarbeitungsanlage *f* paper processing plant
Papierveredelung *f* paper finishing, paper processing
Papiervorschub *m* chart speed, paper chart feed
Papierwatte *f* cellucotton
Papierwolf *m* paper shredding machine
Papierzunge *f* paper tab
Papinscher Topf *m* Papin's digestor
Pappe *f* cardboard, hard paper, paperboard, paste board, **bekieste** ~ gravel-coated board, **dichtere** ~ millboard, **paraffinierte** ~ paraffined cardboard
Pappel *f* (Bot) poplar
Pappelbalsam *m* tacamahac
Pappelöl *n* poplar oil
Pappelsalbe *f* poplar ointment
Papphülse *f* cardbord case
pappig pasty, sticky
Pappschachtel *f* cardboard box, pasteboard box
Pappschere *f* card cutter
Paprika *m* (Bot) bell pepper, paprica, paprika
Paprikaöl *n* paprica oil
Papyrin *n* papyrine
para-Amidoacetanilid *n* p-aminoacetanilide
para-Amidobenzoesäureäthylester *m* ethyl p-aminobenzoate
para-Amidosalicylsäure *f* p-aminosalicylic acid
para-Aminobenzoesäure *f* para-aminobenzoic acid
para-Aminophenol *n* p-aminophenol
Parabansäure *f* parabanic acid, imidazoletrione, oxalylurea
Parabel *f* parabola
Parablau *n* para blue
Parabolspiegel *m* paraboloidal reflector
Parabrenztraubensäure *f* parapyruvic acid
para-Bromacetanilid *n* p-bromoacetanilide
Parabutter *f* para butter
Parabuxin *n* parabuxin

Paracasein *n* paracasein
Paracetamol *n* paracetamol
para-Chinon *n* p-quinone
para-Chloranilin *n* p-chloraniline
para-Chlorbenzoesäure *f* p-chlor[o]benzoic acid
para-Chlornitrobenzol *n* p-chloronitrobenzene
para-Chlortoluol *n* p-chlorotoluene
Paraconsäure *f* paraconic acid
Paracotoin *n* paracotoin
Paracyan *n* paracyanogen
Paracyclophan *n* paracyclophane
para-Cymol *n* p-cymene
Paradeuterium *n* paradeuterium
para-Dichlorbenzol *n* p-dichlorobenzene
Paradieskorn *n* (Bot) paradise grain
para-Dihydroxybenzol *n* hydroquinol, p-dihydroxybenzene
paradirigierend (Chem) para-directing
Paraffin *n* (Chem) paraffin, paraffin wax, festes ~ paraffin wax, solid paraffin
Paraffinbad *n* paraffin bath
paraffinieren to coat with paraffin, to paraffin, to wax with paraffin
Paraffinkaschierung *f* wax backing
Paraffinkerze *f* paraffin candle
Paraffinkohlenwasserstoff *m* alkane, paraffin [hydrocarbon]
Paraffinlack *m* paraffin varnish
Paraffinöl *n* liquid paraffin, paraffin oil, white oil
Paraffinpapier *n* paraffin paper
Paraffinpfropfen *m* plug of paraffin wax
Paraffinreihe *f* paraffin series
Paraffinrest *m* (Chem) paraffin residue
Paraffinsäure *f* saturated aliphatic acid, paraffinic acid
Paraffinsalbe *f* paraffin ointment
Paraffinschmieröl *n* paraffin lubricating oil
Paraffintränkung *f* paraffin impregnation
Paraffinwachs *n* paraffin [wax]
Paraform *n* paraform
Paraformaldehyd *m* paraformaldehyde
Parafuchsin *n* parafuchsin
Paragenese *f* (Geol) paragenesis
paragenetisch paragenetic
Paraglobulin *n* paraglobulin
Paragonit *m* (Min) paragonite
Paragummi *n* para rubber
Paragummiöl *n* para rubber oil
Parahopeit *m* (Min) parahopeite
para-Hydroxybenzoesäure *f* p-hydroxybenzoic acid
Parakautschuk *m* para rubber
para-Kresol *n* p-cresol
Parakresse *f* para cress
Parakristall *m* paracrystal
Paralaurionit *m* (Min) paralaurionite
Paralbumin *n* paralbumin
Paraldehyd *m* paraacetaldehyde, paraldehyde

Paraldol *n* paraldol
parallaktisch parallactic
Parallaxe *f* parallax
Parallaxenausgleich *m* parallex compensation
parallaxenfrei free from parallax
parallel laufen to run parallel, ~ **verschieben** to translate
Parallelabsperrschieber *m* parallel-faced sluice valve
Parallele *f* parallel [line]
Parallelführung *f* parallel guide, parallel motion
parallelgeschaltet (Elektr) connected in parallel, parallel-connected, shunted
Parallelhammer *m* drop hammer
Parallelität *n* parallelism
Parallelklinke *f* bridging jack
Parallelkreis *m* (Elektr) parallel circuit
Parallellauf *m* concurrent flow
Parallellineal *n* parallel ruler
Parallelogramm *n* parallelogram, ~ **der Kräfte** parallelogram of forces
Parallelreaktion *f* parallel reaction
Parallelreihe *f* parallel row
Parallelreißer *m* marking gauge, surface gauge
Parallelreißstock *m* surface gauge
Parallelresonanz *f* parallel resonance
parallelschalten (Elektr) to connect in parallel
Parallelschaltung *f* (Elektr) parallel connection, shunt connection
Parallelstelle *f* parallel passage
Parallelstrom *m* parallel flow
Parallelstrom-Gasbrenner *m* parallel-flow gas burner
Parallelverschiebung *f* parallel displacement, parallel shift, [parallel] translation, translatory motion
Parallelversuch *m* parallel experiment
Parallelverwachsung *f* (Krist) twins with parallel axis
Parallelwiderstand *m* parallel resistance, (Scheinwiderstand) shunt impedance
Paralogit *m* paralogite
Paraluminit *m* (Min) paraluminite
Paralyse *f* (Med) paralysis
paramagnetisch paramagnetic
Paramagnetismus *m* paramagnetism
Paramandelsäure *f* paramandelic acid
Parameter *m* parameter
Parameterdarstellung *f* parametric representation
Parametergleichung *f* parametric equation
Parameterintegral *n* parameter integral
Parameterlinie *f* parameter curve
Parameteroptimierung *f* optimization of parameters
Parametrierungstechnik *f* parametrization technique
Paramid *n* paramide, mellimide

Paramilchsäure *f* paralactic acid, sarcolactic acid
Paraminbraun *n* paramine brown
paramorph (Krist) paramorphic, paramorphous
Paramorphin *n* paramorphine, morphine dimethyl ether, thebaine
Paramorphismus *m* paramorphism
Paramorphose *f* (Krist) paramorphosis
Paramucosin *n* paramucosin
Paranil *n* paranil
para-Nitracetanilid *n* p-nitroacetanilide
para-Nitranilin *n* 1,4-nitraniline, p-nitraniline
para-Nitrotoluol *n* p-nitrotoluene
Parankerit *m* parankerite
Paranußöl *n* para palm oil, Brazil-nut oil, cabbage palm oil
Parapektinsäure *f* parapectic acid
Parapepton *n* parapeptone
para-Phenetidin *n* p-phenetidine
para-Phenolsulfonsäure *f* p-phenolsulfonic acid
para-Phenylendiamin *n* p-phenylenediamine
Pararosanilin *n* pararosaniline
Pararosolsäure *f* pararosolic acid, aurin
Parasaccharinsäure *f* para-saccharic acid
para-Säure *f* p-acid, para acid
Parasit *m* parasite, **an einen spezifischen Wirt gebundener** ~ specific parasite, **im Innern des Wirtes lebender** ~ endoparasite, entozoon
parasitenartig parasitic
Parasitenbefall *m* parasitic disease
parasitenhaft parasitic
Parasitenmittel *n* (Pharm) antiparasitic
parasitieren to parasitize
parasitisch parasitic
Parasitologie *f* parasitology
Parasorbinsäure *f* parasorbic acid
Parastellung *f* para-position, p-position
Parastilbit *m* parastilbite
Paratakamit *m* (Min) paratacamite
Parathiazin *n* parathiazine
Parathion *n* parathion
Parathormon *n* parathyroid hormone
Parathyphus *m* (Med) paratyphoid [fever]
Paratit *m* paratitol
para-Toluidin *n* p-toluidine
para-Toluolsulfamid *n* p-toluenesulfonamide
para-Toluolsulfochlorid *n* p-toluenesulfochloride
Paratose *f* paratose
Paratyp *m* (Biol) paratype
Paratyphusvakzine *f* paratyphoid vaccine
para-Verbindung *f* para compound, p-compound
Paravivianit *m* (Min) paravivianite
Parawasserstoff *m* para hydrogen
Paraweinsäure *f* paratartaric acid
Paraxanthin *n* 1,7-dimethylxanthine, paraxanthine
Paraxialbahn *f* paraxial orbit

para-Xylidin *n* p-xylidine
para-Xylol *n* p-xylene
parazentrisch paracentric
Parazustand *m* para state
Paredrit *m* paredrite
Pareirawurzel *f* (Bot) white pareira
Parfüm *n* perfume, scent
Parfümerie *f* manufacture of perfumes
parfümieren to perfume, to scent
Pargasit *m* (Min) pargasite
Paricin *n* paricine
Paridin *n* paridin
Parietin *n* parietin
Parillin *n* parillin, parillic acid
Parinarsäure *f* parinaric acid
Pariser Blau *n* Paris blue, ferric ferrocyanide
Pariser Gelb *n* Paris yellow, lead chromate
Pariser Grün *n* Paris green, copper acetoarsenite
Pariser Lack *m* Paris lake, carmine lake
Pariser Rot *n* Paris red, minium
Pariser Schwarz *n* Paris black, lamp black
Parisit *m* (Min) parisite
Parität *f* parity
Paritätsänderung *f* parity change
Paritätsauswahlregel *f* parity selection rule
Paritätserhaltung *f* parity conservation
parke[ri]sieren to parkerize
Parkesieren *n* (Metall) Parkes process
Parkettbodenversiegelung *f* parquetry sealing
Parkiaöl *n* parkia oil
Parmetol *n* parmetol
Parodontose *f* (früher Paradentose) parodontosis
Parodyn *n* parodyne
Paromamin *n* paromamine
Paromose *f* paromose
Paroxazin *n* paroxazine
Paroxypropion *n* paroxypropione
Parquin *n* parquine
Partialbruch *m* (Math) partial fraction
Partialbruchregel *f* (Math) partial fraction rule
Partialbruchzerlegung *f* (Math) decomposition into partial fractions
Partialdruck *m* partial pressure
Partialdruckgesetz *n* Dalton's law of partial pressures
Partialquerschnitt *m* partial cross section
partialsynthetisch semi-synthetic
Partialvalenz *f* partial valence
Partie *f* (Reaktionsführung) batch
partiell partial
Partikel *f* particle
Partikelgröße *f* particle size
partikulär particular
partikularisieren to particularize
Parvolin *n* parvoline
parzellieren to parcel out
Paschenserie *f* (Spektr) Paschen series
Pascoit *m* (Min) pascoite
Passauit *m* passauite

passen to match, **der Größe nach** ~ to fit
passend fitting, (geeignet) appropriate, suitable
passieren to happen, to occur
passiv inert, passive
passivieren to inactivate, to passivate, to render passive
Passivierung f passivation
Passivierungsmittel n passivating agent
Passivierungspotential n passivation potential
Passivität f passivity, inactivity, inertness, passiveness
Paßring m adapter ring, shim
Paßrohr n adapting piece, adaptor, template pipe
Paßschraube f fitted bolt
Paßstift m adjusting pin, dowel, fitting pin, set pin
Paßstück n adjusting piece, fitting piece, **eingesetztes** ~ insert adapter
Paßtoleranz f fitting tolerance
Passung f fit
Passungsrost m fretting corrosion
Passungssystem n standard gauge
Paste f paste
Pastellfarbe f pencil color
Pastellstift m crayon [pencil]
Pastellton m pastel shade
Pasten n pasting process
Pastenanreibemaschine f paste grinding machine
pastenartig paste-like
Pastenfarbe f paste paint
Pastenharz n paste resin, grinding type resin
Pasteureffekt m (Biochem) Pasteur effect
Pasteurisation f pasteurization
Pasteurisieranlage f pasteurizing plant
Pasteurisierapparat m pasteurizer
pasteurisieren to pasteurize
Pasteurisieren n pasteurizing
Pasteurisierung f pasteurization
Pasteurkolben m culture flask
Pastille f (Pharm) medicated lozenge, pastille, tablet, troche
Pastillenpresse f tablet press
Pasting-Verfahren n pasting process
Patavaöl n pataba oil
Patellit m patellite
Patent n patent, ~ **anmelden oder beantragen** to apply for a patent, **abgelaufenes** ~ expired patent, **angemeldetes** ~ filed patent, patent applied for, pending patent, **ein** ~ **anmelden** to apply for a patent, **ein** ~ **erwirken** to secure a patent [for], **ein** ~ **ist erteilt worden** a patent has been granted, **ein** ~ **verletzen** to infringe on a patent
Patentamt n patent office
Patentanmelder m patent applicant
Patentanmeldung f patent application
Patentanspruch m patent claim
Patentanwalt m patent attorney (Am. E.)

Patentbeschreibung f patent specification
Patentblau n patent blue, triphenylmethane dye
Patenteinspruch m opposition to the granting of a patent
Patenterneuerung f patent renewal
Patenterteilung f patent grant[ing]
patentfähig patentable
Patentfähigkeit f patentability
Patentgeber m patentor
Patentgebühr f patent fee, (Jahresgebühr) patent annuity
Patentgegenstand m subject matter of a patent
Patentgelb n patent yellow, mineral yellow
Patentgesetz n (Jur) patent act, patent law
Patentgesetzgebung f patent legislation
Patentgesuch n patent application
patentierbar patentable
patentieren to grant a patent for s. th., to patent
patentiert patented, commissioned, **im In- und Ausland** ~ patented at home and abroad
Patentinhaber m patentee, patent holder
Patentkernnagel m patent core nail
Patentklage f (Jur) patent suit
Patentkohle f briquet[te], coal cake
Patentnickel n patent nickel alloy
Patentrecht n (Jur) patent right
Patentregister n patent register, patent rolls
Patentschrift f patent specification
Patentschutz m protection of or by a patent
Patentsucher m patent applicant
Patentträger m patentee
Patentverfahren n patent procedure
patentverletzend patent-infringing
Patentverletzung f patent infringement
Patentverwertung f patent exploitation
Patentzeichnung f patent drawing
Patentzement m Roman cement
Patentzündschnur f safety fuse
Paternoster m bucket elevator, paternoster elevator
pathogen (Med) pathogenic
Pathologe m (Med) pathologist
Pathologie f (Med) pathology, **allgemeine** ~ general pathology, **physiologische** ~ physiological pathology, **vergleichende** ~ comparative pathology
Pathometabolismus m pathometabolism
Patina f patina, green verditer
patinieren to cover with patina
Patinierung f coating of copper with green verditer
Patrinit m (Min) patrinite
Patrize f patrix, bottom mold, male die, male mold, plug
Patrone f shell
Patronenheizkörper m cartridge heater
Patronit m (Min) patronite
Patschulen n patchoulene
Patschuli n (Bot) patchouli

Patschulialkohol *m* patchouli alcohol
Patschuliduft *m* smell of patchouli
Patschuliessenz *f* essence of patchouli
Patschuliöl *n* patchouli oil
Pattersonit *m* (Min) pattersonite
Pattinsonieren *n* Pattinson's process
Patulin *n* patulin
Paucin *n* paucine
Pauli-Prinzip *n* Pauli principle
Paulit *m* paulite
Pauliverbot *n* Pauli exclusion principle
Pauschalakkord *m* lump sum contract
Pauschale *f* lump sum
Pauschalgebühr *f* flat rate, lump sum
Pauschalsumme *f* global amount, lump sum
Pause *f* (Unterbrechung) intermission, interval, rest, stop, (Zeichnung) print, tracing
pausen to print, to trace
Pausfilm *m* tracing film
Pauslampe *f* blue-print lamp
Pausleinen *n* tracing cloth
Pausleinwand *f* tracing cloth
Pauspapier *n* tracing paper, blue-print paper
Pauszeichnung *f* tracing
Pavin *n* pavine
Paytin *n* paytine
Pearceit *m* (Min) pearceite
Pech *n* pitch
pechartig pitch-like, bituminous, pitchy
Pechblende *f* (Min) pitch blende, pitchy iron ore
Pechdraht *m* pitched thread, shoemakers thread
pecheln to extract pitch
pechen to coat with pitch, to pitch
Pecherde *f* bituminous earth
Pecherz *n* (Min) pitch blende
Pechfaserrohr *n* bituminized fiber pipe
Pechgang *m* asphalt rock
Pechgranat *m* (Min) pitch garnet, colophonite
pechig pitchy
Pechkohle *f* bituminous coal, pitch coal
Pechkuchen *m* pitch cake
Pechkugel *f* pitch ball
Pechkupfer *n* chrysocolla
Pechöl *n* pitch oil, tar oil
Pechofen *m* pitch kiln, pitch oven
Pechpflaster *n* (Straßenbau) asphalt paving, tar paving
pechschwarz pitch black
Pechstein *m* (Min) pitch stone
Pechtorf *m* pitch peat
Pechuran *n* (Min) pitch blende
Peckhamit *m* (Min) peckhamite
Pedalgummi *n* pedal pad
Pegan *n* pegan
Peganit *m* (Min) peganite
Pegel *m* depth gauge, depth indicator, water gauge
Pegelstand *m* water (oil etc.) level
Pegen *n* pegene

Pegenon *n* pegenone
Pegmatit *m* (Min) pegmatite, giant granite
Pegmatitanhydrit *m* (Min) pegmatite anhydrite
pegmatitartig pegmatitic, pegmatoid
Pegmatolith *m* pegmatolite
Pegnin *n* pegnin
Pektase *f* (Biochem) pectase
Pektin *n* pectin
Pektinanlage *f* pectin plant
Pektinase *f* (Biochem) pectinase
Pektindepolymerase *f* (Biochem) pectin depolymerase
Pektinit *m* pectinite
Pektinose *f* pectinose, arabinose, gum sugar, pectin sugar
Pektinsäure *f* pectic acid
pektinsauer pectate, pectic
pektinspaltend pectolytic
Pektolith *m* (Min) pectolite
Pektose *f* pectose
Pektosinsäure *f* pectosinic acid
Pelagosit *m* pelagosite
Pelargon *n* pelargone, 9-heptodecanone
Pelargonäther *m* pelargonic ether
Pelargonaldehyd *m* pelargonaldehyde, n-nonylic aldehyde, nonanal
Pelargonat *n* pelargonate, salt or ester of pelargonic acid
Pelargonin *n* pelargonin, pelargonidin glucoside
Pelargonsäure *f* pelargonic acid, nonanoic acid
pelargonsauer pelargonate, nonylate, pelargonic
Pelargonyl- pelargonyl, nonanoyl
Pelentansäure *f* pelentanoic acid
Peligotröhre *f* peligot tube
Pellagra *n* (Med) pellagra
Pelle *f* husk, peel, skin
Pelletierband *n* conveyor pelletizer
Pelletieren *n* pelleting
Pelletierin *n* pelletierine, punicine
Pelletierkonus *m* cone pelletizer
Pelletierteller *m* balling pan, disc pelletizer
Pelletiertrommel *f* drum pelletizer
Pellitorin *n* pellitorine
Pellotin *n* pellotine
Pelokonit *m* peloconite
Pelton-Turbine *f* (Mech) Pelton wheel, impulse turbine, impulse wheel
Penald[in]säure *f* penaldic acid
Penam *n* penam
Pencatit *m* pencatite
Pendel *n*, **ballistisches** ~ ballistic pendulum, **kegelförmiges** ~ (Mech) conical pendulum, **mathematisches oder einfaches** ~ mathematic or simple pendulum, **physisches oder physikalisches oder zusammengesetztes** ~ physical or compound pendulum
Pendelaufhängung *f* pendulum suspension
Pendelausschlag *m* amplitude of pendulum swing, pendulum deflection

Pendelbelüfter m swing diffuser
Pendelbewegung f pendulum motion
Pendelhalter m pivoting clamp
Pendelhammer m pendulum hammer
Pendelkugellager n self-aligning ball bearing
Pendelmotor m (Transmission) dynamometer
Pendelmühle f bowl mill, centrifugal roll mill
pendeln to pendulate, to oscillate, to vibrate
pendelnd swinging
Pendelregler m pendulum governor
Pendelschlagwerk n impact pendulum
Penduluhr f pendulum clock
Penetration f penetration
Penetrationsmodell n penetration model
Penetrationszwilling m (Krist) penetration twin
Penetrometer n penetrometer
Penetrometerzahl f penetration index
Penfieldit m (Min) penfieldite
Penicillamin n penicillamine
Penicillansäure f penicillanic acid
Penicillensäure f penicillenic acid
Penicillin n penicillin
Penicillinase f (Biochem) penicillinase
penicillinempfindlich sensitive to penicillin
Penicillinkolben m penicillin flask
penicillinresistent penicillin-resistant, resistant to penicillin
Penicillinspiegel m penicillin level
penicillinunempfindlich insensitive to penicillin
Penicilliopsin n penicilliopsin
Penicillosäure f penicilloic acid
Penillasäure f penilloic acid
Penilloaldehyd m penilloaldehyde
Penillonsäure f penillonic acid
Penillsäure f penillic acid
Pennin m (Min) pennine, penninite
Pennit m pennite
Pennol n pennol
Pennon n pennone
Pennyroyalöl n pennyroyal oil
Pensum n lesson, task
Pentabromphosphor m phosphorus pentabromide, phosphorus(V) bromide
pentacarbocyclisch pentacarbocyclic
Pentacen n pentacene
Pentacetylglucose f pentacetylglucose
Pentachloräthan n pentachloroethane
Pentachlorthiophenol n pentachlorothiophenol
Pentacontan n pentacontane
Pentacosan n pentacosane
Pentacyansäure f pentacyanic acid
pentacyclisch pentacyclic
Pentadecan n pentadecane
Pentadecandicarbonsäure f pentadecane dicarboxylic acid
Pentadecansäure f pentadecanoic acid, isocetic acid, n-pentadecoic acid, n-pentadecylic acid
Pentadien n pentadiene
Pentaeder n pentahedron

pentaedrisch pentahedral
Pentaerythrit m pentaerythritol
Pentaerythrittetranitrat n pentaerythritol tetranitrate (PETN)
Pentaglycerin n pentaglycerol
Pentagon n (Fünfeck) pentagon
pentagonalhemiedrisch pentagonal hemihedral
Pentagondodekaeder n pentagon dodecahedron
Pentagonikositetraeder n pentagon icositetrahedron
Pentagridröhre f pentagrid valve or tube
Pentahomoserin n pentahomoserine
pentakarbozyklisch pentacarbocyclic
Pentakosan n pentacosane
Pental n pental, trimethylethylene
Pentalen n pentalene
Pentalin n pentaline, pentachloroethane
Pentamethonium n pentamethonium
Pentamethylbenzol n pentamethylbenzene
Pentamethylen n pentamethylene
Pentamethylenbromid n pentamethylene bromide
Pentamethylendiamin n pentamethylene diamine
Pentamethylensulfid n pentamethylene sulfide
Pentamethylglucose f pentamethylglucose
Pentan n pentane, normales ~ normal pentane, n-pentane
Pentanal n pentanal, amyl aldehyde
Pentandiol n pentanediol
Pentannormalthermometer n pentane normal thermometer
Pentanol n pentanol, amyl alcohol
Pentanon n pentanone
Pentanthiol n pentanethiol, amylmercaptan
Pentanthrimid n pentanthrimide
Pentapeptid n penta peptide
Pentaphen n pentaphene
Pentaphenylphosphor m pentaphenylphosphorus
Pentasulfid n pentasulfide
Pentathionat n pentathionate, salt or ester of pentathionic acid
Pentathionsäure f pentathionic acid
Pentatriakontan n pentatriacontane
Pentazin n pentazine
Penten n pentene
Pentetrazol n pentetrazole
Penthiazolidin n penthiazolidine
Penthiazolin n penthiazoline
Penthiophen n penthiophene, penthiofurane
Pentin n pentine, pentyne
Pentit m pentitol, pentite
Pentlandit m (Min) pentlandite
Pentode f (Elektr) pentode, five-electrode tube
Pentonsäure f pentonic acid
Pentosan n pentosan
Pentose f pentose
Pentoseisomerase f (Biochem) pentose isomerase
Pentosephosphat n pentose phosphate
Pentosezyklus m (Biochem) pentose cycle

Pentoxazolin *n* pentoxazoline
Pentoxid *n* pentoxide
Pentulose *f* pentulose
Pentylalkohol *m* pentyl alcohol
Penwithit *m* penwithite
Peperit *m* peperite
Peplolit *m* peplolite
Pepsin *n* (Biochem) pepsin
pepsinhaltig containing pepsin
Pepsinogen *n* (Biochem) pepsinogen
Peptid *n* peptide, **synthetisches** ~ synthetic peptide
Peptidanalyse *f* peptide analysis
Peptidase *f* (Biochem) peptidase
Peptidbildung *f* peptide formation
Peptidbindung *f* peptide bond
Peptidhydrolyse *f* peptide hydrolysis
Peptidoglykan *n* peptidoglycan
Peptidsynthese *f* peptide synthesis
Peptidtrennung *f* peptide separation
Peptidylpuromycin *n* peptidyl puromycin
Peptidyltransferase *f* (Biochem) peptidyl transferase
Peptisation *f* peptization
peptisieren to peptize, to peptonize
Pepton *n* peptone
peptonartig peptonoid
peptonerzeugend peptogenic
peptonhaltig peptonic
peptonisieren to peptonize
Peptonisierung *f* peptonization
Peptotoxin *n* peptotoxine
Peracidität *f* superacidity
Perameisensäure *f* performic acid, peroxyformic acid
Perameisensäureoxidation *f* performic acid oxidation
Perautan *n* perautan
Perbenzoesäure *f* perbenzoic acid, peroxybenzoic acid
Perborat *n* perborate, peroxyborate, salt or ester of perboric acid
Perborax *m* sodium perborate, sodium peroxyborate
Perborsäure *f* perboric acid, peroxyboric acid
Perbromsäure *f* perbromic acid
Perbuttersäure *f* perbutyric acid, peroxybutyric acid
Percain *n* percaine
Perchagummi *n* gutta percha
Perchloräthan *n* perchloroethane, hexachloroethane
Perchloräthylen *n* perchloroethylene, tetrachloroethene, tetrachloroethylene
Perchlorat *n* perchlorate, salt or ester of perchloric acid
Perchlorbenzol *n* perchlorobenzene, hexachlorobenzene
Perchlorbenzophenon *n* perchlorobenzophenone, decachlorobenzophenone
Perchlormethylmercaptan *n* perchloromethyl mercaptan
Perchlorsäure *f* perchloric acid
Perchlorsäureanhydrid *n* perchloric anhydride, chlorine heptoxide
Perchlorsäuremonohydrat *n* monohydrated perchloric acid
Perchromat *n* perchromate, salt or ester of perchromic acid
Perchromsäure *f* perchromic acid, peroxychromic acid
Percylit *m* (Min) percylite
Pereirin *n* pereirine
Peressigsäure *f* peracetic acid, peroxyacetic acid
Perezon *n* perezon
perforieren to perforate
Perforiermaschine *f* perforating machine
Perfusion *f* perfusion
Perfusionssystem *n* perfusion system
Pergament *n* parchment
pergamentähnlich parchment-like
pergamentartig parchment-like
Pergamentbereitung *f* parchment making
Pergamentfarbe *f* parchment color
pergamentieren to parchmentize
Pergamentiermaschine *f* parchmentizing machine
Pergamentleim *m* parchment glue
Pergamentpapier *n* parchment paper
Pergamentschlauch *m* parchment pipe
Pergamin *n* glassine, imitation parchment, parchmyn
Pergaminpapier *n* glassine paper
Pergenol *n* pergenol
perhydrieren to perhydrogenate, to perhydrogenize
Perhydrocarotin *n* perhydrocarotene
Perhydroindol *n* perhydroindole
Perhydrol *n* perhydrol, hydrogen peroxide solution
Perhydrolycopin *n* perhydrolycopene
Perhydronaphthalin *n* (Dekalin) perhydronaphthalene, decahydronaphthalene
Periacenaphthindan *n* periacenaphthindane
Periastrondrehung *f* (Astr) advance of periastron
Peribenzanthren *n* [peri]benzanthrene
Peribenzanthron *n* [peri]benzanthrone
Pericyclisierung *f* pericyclization
Pericyclocamphan *n* pericyclocamphane
Peridot *m* (Min) peridot, chrysolite
Perihelium *n* (Astron) perihelion
Periklas *m* (Min) periclase, periclasite
perikondensiert pericondensed
Perillaaldehyd *m* perillaldehyde
Perillaalkohol *m* perillic alcohol
Perillaöl *n* perilla oil
Perillasäure *f* perillic acid

Perillen *n* perillene
Perimidin *n* perimidine
Perimidon *n* perimidone
Perimorphose *f* (Krist) perimorphosis
Perinaphthan *n* perinaphthane
Perinaphthanon *n* perinaphthanone
Perinaphthoxazin *n* perinaphthoxazine
Periode *f* period, (Elektr) cycle, phase, **halbe** ~ half a period, **kleine** ~ short period, **lange** ~ long period, **mittlere** ~ medium period
Periodenanzeiger *m* period indicator
Periodendauer *f* duration of a period, time of oscillation
Periodenparallelogramm *n* periodic parallelogram
Periodensystem *n* (Chem) periodic system, periodic arrangement, periodic table
Periodenuhr *f* harmonic dial
Periodenzahl *f* frequency, number of cycles, number of periods
periodisch periodic[al], intermittent, recurrent
Periodizität *f* periodicity
Periodizitätsbedingung *f* periodicity condition
Periodizitätsgrad *m* degree of periodicity
Periodizitätsvolumen *n* periodic volume
Peripherie *f* periphery
peripher[isch] peripheral
Periplocin *n* periplocin
Periplogenin *n* periplogenin
Perisäure *f* peric acid
Peristaltik *f* (Med) peristalsis
Peristaltin *n* peristaltin
Peristerit *m* peristerite
Peritektikum *n* peritectic period
peritektisch peritectic
Perjodat *n* periodate, salt or ester of periodic acid
Perjodatabbau *m* periodate degradation
Perjodsäure *f* periodic acid
perjodsauer periodate
Perkarbonat *n* percarbonate
Perkinsche Reaktion *f* Perkin's reaction
Perkohlensäure *f* percarbonic acid, peroxycarbonic acid
perkohlensauer percarbonate, peroxycarbonate
Perkolation *f* percolation
Perkolator *m* percolator
Perkussion *f* percussion
Perkussionsfigur *f* percussion figure
Perkussionskraft *f* percussive force, ballistic power
Perkussionszentrum *n* center of percussion
Perkussionszünder *m* percussion primer
Perkussionszündhütchen *n* percussion cap
Perkussionszündung *f* percussion priming
perkutan percutaneous
perlartig pearl-like
Perlasche *f* pearl ash, pure potash
Perle *f* pearl, gem, jewel, (Glas, Chem) bead

Perlen *n* pearling
Perlenglanz *m* pearly luster, luster of pearls, nacreous luster
Perlenprobe *f* (Anal) bead test
perlfarben pearl-colored
Perlform *f* pearl form
Perlglimmer *m* (Min) pearl mica, margarite
Perlit *m* pearly constituent, (Min) pearlite, perlite
Perlkohle *f* pea coal
Perlkupfer *n* copper in grains
Perlmoos *n* (Bot) pearl moss, carragheen moss
Perlmutteffekt *m* nacreous effect
Perlmutter *f* mother of pearl, nacre, pearl-shell
perlmutterartig nacreous
Perlmutterblech *n* crystallized tin plate
Perlmutterglanz *m* nacreous luster, pearly luster
Perlmutter-Pigmentteilchen *pl* nacreous particles
Perlpolymerisate *pl* bead polymers
Perlpolymerisation *f* bead polymerization, grain polymerization
Perlröhre *f* [glass] bead tube
Perlrohr *n* bead tube
perlschnurartig like a string of beads
Perlseide *f* ardassine
Perlsinter *m* pearl sinter
Perlspat *m* (Min) pearl spar
Perlstein *m* pearl stone
Perlweiß *n* pearl white, pearl powder
permanent permanent
Permanentgas *n* permanent gas
Permanentgrün *n* permanent green
Permanentrot *n* permanent red
Permanentweiß *n* permanent white
Permanenz *f* permanence
Permanganat *n* permanganate
Permangansäure *f* permanganic acid
permeabel permeable
Permeabilität *f* permeability, magnetic inductive capacity, magnetic inductivity, ~ **bei kleinen Feldstärken** permeability at low magnetizing forces, ~ **der Kapillarwände** permeability of the capillary walls, **reversible** ~ reversible permeability, **umkehrbare** ~ reversible permeability
Permeabilitätsänderung *f* change of permeability
Permease *f* (Biochem) permease
Permselektivität *f* permselectivity
Permutation *f* permutation
permutieren (Math) to permute, to exchange
Permutit *m* (Min) permutite
Permutit-Verfahren *n* permutite process
Pernambukholz *n* Pernambuco wood
Pernambukogummi *n* Pernambuco rubber
Pernambukokautschuk *m* Pernambuco rubber, mangabeira
Pernigranilin *n* pernigraniline
Pernitrosecampher *m* pernitrosocamphor

Peronin *n* peronine, benzylmorphine hydrochloride
Peropyren *n* peropyrene
Peroxid *n* peroxide, superoxide
Peroxidase *f* (Biochem) peroxidase
Peroxidbleichechtheit *f* fastness to peroxide bleaching
Peroxidierung *f* peroxidation
Peroxidkatalyse *f* peroxide catalysis
Peroxid-Umlagerung *f* peroxide rearrangement
Peroxodischwefelsäure *f* peroxodisulfuric acid
Peroxogruppe *f* peroxo-group
Peroxokobaltkomplex *m* peroxocobaltic complex
Peroxo[mono]phosphorsäure *f* peroxo[mono]phosphoric acid
Peroxo[mono]schwefelsäure *f* peroxo[mono]sulfuric acid
Peroxophosphat *n* peroxophosphate
Peroxosulfat *n* peroxosulfate
Peroxydase *f* (Biochem) peroxidase
Peroxyderivat *n* peroxy derivative
Peroxydiphosphorsäure *f* peroxydiphosphoric acid
Peroxydol *n* peroxydol, sodium perborate
Peroxygenierung *f* peroxygenation
Peroxymonoschwefelsäure *f* peroxymonosulfuric acid
Peroxysäure *f* peroxy acid
Perpendikel *n* (Lot) plumb line, plummet, (Pendel) pendulum, (Senkrechte) vertical line
perpendikulär perpendicular
Perpetuum mobile *n* perpetuum mobile
Perphosphat *n* per[oxydi]phosphate
Perphthalsäure *f* perphthalic acid
Perpropionsäure *f* peroxypropionic acid, perpropionic acid
Perrhenat *n* perrhenate
Perrutheniat *n* perruthenate
Persäure *f* peracid, peroxy acid
Persalz *n* peroxy salt, persalt
Perschwefelsäure *f* peroxysulfuric acid, persulfuric acid
Perseaöl *n* Persea oil
Perseit *n* perseite, perseitol
Perseulose *f* perseulose
Persiko *m* (Likör) persico[t]
Persischbraun *n* Persian brown
Perspektive *f* perspective
perspektivisch perspective
Persulfat *n* peroxysulfate, persulfate, salt or ester of per[oxy]sulfuric acid
Persulfocyansäure *f* persulfocyanic acid, perthiocyanic acid
Persulfomolybdänsäure *f* persulfomolybdic acid, perthiomolybdic acid
Perthiokohlensäure *f* perthiocarbonic acid
Perthit *m* (Min) perthite
perthitähnlich perthitic

Pertinax *n* (HN) pertinax
Pertonal *n* pertonal
Perubalsam *m* Peru balsam
Peruol *n* peruol
Peruresinotannol *n* Peru resino tannol
Perurinde *f* Peruvian bark, cinchona
Perusilber *n* German silver
Peruvin *n* peruvin, cinnamic alcohol
Peruviol *n* peruviol, nerolidol
Pervitin *n* pervitin
Perylen *n* perylene
Perylenchinon *n* perylene quinone
Pest *f* plague
Pestflecken *m* plague spot
Pestgeruch *m* pestilential smell
Pestilenz *f* pestilence, plague
pestilenzialisch pestilential
Pestilenzkraut *n* (Bot) goat's rue
Petalit *m* (Min) petalite
Petersilienöl *n* parsley oil
Petersilienwurzel *f* (Bot) parsley root
Peterskraut *n* (Bot) goat's rue, wall pellitory
Petitgrainöl *n* petitgrain oil
Petrefakt *n* fossil, petrifaction
Petrichloral *n* petrichloral
Petrischale *f* Petri dish
Petrochemie *f* petrochemistry, petrochemical industry
Petroläther *m* petroleum ether, ligroin
Petrolatum *n* petrolatum, mineral fat
Petrolchemie *f* petroleum chemistry
Petroleum *n* petroleum, crude oil, mineral oil, rock oil; (raffiniert) kerosene, kerosine
Petroleumäther *m* petroleum ether
Petroleumasphalt *m* asphalt, earth pitch, petroleum pitch
Petroleumbenzin *n* petroleum spirit
Petroleumbrenner *m* kerosene burner
Petroleumdampf *m* petroleum vapor
Petroleumdestillierapparat *m* petroleum distilling apparatus
Petroleumeinspritzer *m* petroleum injector
Petroleumfeuerung *f* oil-fired furnace
Petroleumgas *n* petroleum gas
petroleumhaltig containing petroleum
Petroleumheizung *f* heating with petroleum
Petroleumkochapparat *m* petroleum stove
Petroleumkocher *m* petroleum stove
Petroleumkrackverfahren *n* petroleum cracking process
Petroleumlampe *f* kerosene lamp
Petroleummotor *m* petroleum engine
Petroleumnebel *m* petroleum spray
Petroleumprüfer *m* petroleum tester
Petroleumquelle *f* oil well, petroleum spring
Petroleumraffinerie *f* petroleum refining plant
Petroleumvergaser *m* petrol carburettor
Petrolin *n* petroline
petrolisieren to petrolize

Petrologie *f* petrology
Petrolsäure *f* petrolic acid
Petroselaidinsäure *f* petroselaidic acid
Petroselinsäure *f* petroselinic acid
Petzit *m* (Min) petzite
Peucedanin *n* peucedanin, imperatorin
Pfadeisen *n* pivot, socket
Pfanne *f* pan, (Brau) copper, (Gieß) ladle, (Lager) seat
Pfannenausguß *m* lip of ladle
Pfannenbügel *m* ladle handle, ladle support
Pfannenführung *f* ladle guide
Pfannengehänge *n* ladle bail
Pfanneninhalt *m* ladle capacity
Pfannenmischer *m* pan mixer
Pfannenrand *m* rim of ladle
Pfannensäure *f* evaporated acid, pan acid
Pfannenstein *m* boiler scale, pan scale
Pfannenwerk *n* salt works
Pfeffer *m* pepper
Pfefferalkaloid *n* pepper alkaloid
pfefferartig peppery
Pfefferminze *f* peppermint
Pfefferminzgeruch *m* peppermint scent, peppermint smell
Pfefferminzschnaps *m* peppermint brandy
Pfefferöl *n* pepper oil
Pfefferwasser *n* pepper water
Pfeife *f* pipe, (Akust) whistle, (Glashütte) blowing iron, **pneumatische** ~ pneumatic pipe
pfeifen to whistle
Pfeifenfirnis *m* pipe glaze, pipe varnish
Pfeifenstein *m* (Min) pipestone, catlinite
Pfeifenton *m* pipe clay
Pfeifventil *n* whistle valve
Pfeilgift *n* arrow poison
Pfeilrad *n* double helical gear, herringbone gear
Pfeilradgetriebe *n* herringbone gear
Pfeilverzahnung *f* double helical gear
Pfennigstein *m* (Min) lenticular iron ore
Pferdebohne *f* (Bot) horsebean
Pferdefett *n* horse fat, horse grease
Pferdemarkfett *n* horse marrow fat
Pferdeserum *n* horse serum
Pferdestärke *f* horsepower, HP
Pfirsich *m* (Bot) peach
Pfirsichblütenfarbe *f* peach blossom color
pfirsichfarben peach-colored
Pfirsichkern *m* peach kernel
Pfirsichkernöl *n* peach kernel oil
Pflanzenalbumin *n* vegetable albumin
Pflanzenasche *f* vegetable ashes
Pflanzenauszug *m* vegetable extract
Pflanzenbutter *f* vegetable butter
Pflanzencasein *n* vegetable casein, legumin
Pflanzenchemie *f* phytochemistry, plant chemistry
Pflanzeneiweiß *n* phytoprotein, vegetable protein

Pflanzenerde *f* garden mold, vegetable mold
Pflanzenextrakt *m* plant extract
Pflanzenfarbe *f* vegetable dye
Pflanzenfarbstoff *m* vegetable coloring matter
Pflanzenfaser *f* vegetable fiber
Pflanzenfett *n* vegetable fat, vegetable oil
Pflanzenfettseife *f* vegetable fat soap, vegetable oil soap
Pflanzenfibrin *n* vegetable fibrin
pflanzenfressend herbivorous
Pflanzengallerte *f* plant gelatin, pectin, vegetable jelly
Pflanzengift *n* phytotoxin, vegetable poison
Pflanzenharz *n* vegetable resin
Pflanzenhormon *n* phytohormone, plant hormone
Pflanzenkrankheit *f* plant disease
Pflanzenkunde *f* botany, phytology
Pflanzenleim *m* vegetable glue, vegetable adhesive, vegetable gum
Pflanzenöl *n* plant oil, vegetable fat
Pflanzenparasit *m* phytoparasite
Pflanzenphysiologie *f* plant physiology
Pflanzensäure *f* vegetable acid
Pflanzensaft *m* plant juice
Pflanzensalz *n* vegetable salt
Pflanzenschädling *m* plant pest
Pflanzenschleim *m* mucilage
Pflanzenschutz *m* crop protection, plant protection
Pflanzenschutzmittel *n* crop protection product, pesticide, plant preservation chemical, plant protective [product]
Pflanzenschwarz *n* vine black
Pflanzenseide *f* vegetable silk
Pflanzentalg *m* vegetable tallow
Pflanzenwachs *n* vegetable wax
Pflanzenwachstum *n* (Vegetation) vegetation
Pflanzenwuchsstoff *m* plant growth substance
Pflanzenzüchtung *f* cultivation of plants
pflanzlich vegetable
Pflaster *n* (Med) plaster, (Reifenreparatur) patch, (Straßenbau) pavement
Pflasterstein *m* curb stone
Pflatschen *n* lick-roll process
Pflaume *f* (Bot) plum
Pflege *f* care, maintenance, (Biol) cultivation
Pflegeleicht-Ausrüstung *f* easy care finishing
Pflegemittel *n* polish
pflügen to plough, to plow (Am. E.)
Pflug *m* plough (Br. E.), plow (Am. E.)
Pflugmesser *n* colter
Pfropfcopolymer *n* graft copolymer
Pfropfen (Flasche etc.) cork, stopper, (Med) plug, thrombus, wad
Pfropfenströmung *f* plug flow
Pfropfpolymer *n* graft polymer
Pfropfpolymerisat *n* graft polymer
Pfropfpolymerisation *f* graft polymerization

Pfropfung *f* grafting
Pfropfwachs *n* grafting wax
Pfund *n* (Maß) pound (lb.)
Pfundleder *n* sole leather
pH *m* (Wasserstoffionen-Exponent) pH [value] (hydrogen ion exponent)
Phäanthin *n* pheanthine
Phänomen *n* phenomenon
phänomenologisch phenomenological[ly]
Phänotypus *m* phenotype
Phäochlorophyll *n* pheochlorophyll
Phäophorbid *n* pheophorbide
Phäophorbin *n* pheophorbine
Phäophytin *n* pheophytine
Phäoporphyrin *n* pheoporphyrin
Phagen *pl* phages
Phagenbehandlung *f* (Med) bacteriophage treatment
Phagozyte *f* (Biol, Med) phagocyte
Phagozytenzerfall *m* phagocytolysis
Phagozytose *f* (Med) phagocytosis
Phakolit *m* (Min) phacolite
p-Halbleiter *m* p-type semiconductor
Phanodorm *n* phanodorm, cyclobarbital
phantasievoll imaginative
phantastisch fantastic
Pharmakochemie *f* pharmaceutical chemistry, pharmacochemistry
Pharmakokinetik *f* pharmacokinetics
Pharmakolith *m* (Min) pharmacolite, arsenate of lime
Pharmakologe *m* pharmacologist
Pharmakologie *f* pharmacology
pharmakologisch pharmacological
Pharmakopöe *f* (amtl. Arzneibuch) pharmacopoeia
Pharmakosiderit *m* (Min) pharmacosiderite
Pharmazeut *m* pharmaceutical chemist, pharmacist
Pharmazeutik *f* pharmaceutics
pharmazeutisch pharmaceutical
Pharmazie *f* pharmacy
Phase *f* phase, **außer** ~ dephased, out of phase, **disperse** ~ disperse phase, **entgegengesetzte** ~ opposite phase, **in** ~ cophasal, in phase, in step, **stationäre** ~ stationary phase
phasenabhängig phase-dependent
Phasenänderung *f* phase change, phase transition
Phasenanschnittsteuerung *f* phase shift control
Phasenanzeiger *m* (Elektr) phase indicator
Phasenausgleich *m* phase compensation, phase correction
Phasenbeziehung *f* phase relation
Phasendiagramm *n* phase diagram, phase pattern
Phasendifferenz *f* phase difference
phasenempfindlich phase sensitive
Phasenentmischung *f* phase separation

Phasengeschwindigkeit *f* phase velocity
Phasengesetz *n* phase rule
Phasengitter *n* phase grating
phasengleich in phase
Phasengleichgewicht *n* (Dest) phase equilibrium
Phasengleichheit *f* phase balance, phase coincidence
Phasengrenze *f* phase boundary, phase interface
Phasengrenzpotential *n* phase boundary potential
Phasengrenzschicht *f* interface
Phasenkontrastbild *n* phase contrast image
Phasenkontrastmikroskop *n* phase-contrast microscope
Phasenkontrastverfahren *n* (Mikroskopie) phase contrast method
Phasenmaß *n* phase constant
Phasenmesser *m* power-factor indicator, (Elektr) phasemeter
Phasennacheilung *f* phase lag
Phasenraum *m* phase space
Phasenregel *f* phase rule
Phasenregler *m* phase regulator
Phasenregulierung *f* phase regulation
Phasenresonanz *f* phase resonance
Phasenschieber *m* phase changer, phase shifter
Phasenschwund *m* phase fading
Phasenspannung *f* phase voltage
Phasensprung *m* phase jump
Phasenstrom *m* phase current
Phasentitration *f* phase titration
Phasentransformation *f* phase transformation
Phasentrennung *f* phase separation
Phasenübereinstimmung *f* phase coincidence
Phasenübergang *m* phase transition
Phasenumkehr *f* phase inversion
Phasenumkehrer *m* phase inverter
Phasenumkehrröhre *f* (Elektron) phase inverter [tube]
Phasenumschaltung *f* phase adjustment
Phasenumwandlung *f* phase change, ~ **erster Ordnung** first order phase transition
phasenungleich out of phase
Phasenungleichheit *f* phase unbalance
Phasenunterschied *m* difference of phases, phase difference
Phasenverhalten *n* phase behavio[u]r
Phasenverkettung *f* interconnection of phases, interlinking of phases
Phasenverschiebung *f* phase shift[ing], dephased condition, phase displacement, phase lag, phase shift
Phasenverschiebungswinkel *m* angle of phase difference, angle of phase displacement
phasenverschoben out-of-phase
Phasenverteilung *f* phase distribution
Phasenverzerrung *f* phase distortion, phase shift
Phasenverzögerung *f* phase lag[ging], phase retardation

Phasenverzögerungswinkel *m* angle of phase lag
Phasenvoreilung *f* phase advance, phase lead
Phasenvoreilungswinkel *m* angle of phase lead
Phasenwechselpunkt *m* phase transition point
Phasenwiderstand *m* phase resistance
Phasenwinkel *m* phase angle
Phasenwinkelanalyse *f* phase shift analysis
Phasenzahl *f* number of phases
Phaseolunatin *n* phaseolunatin
Phasotron *n* phasotron
PHB Abk. für para-Hydroxybenzoesäure
pH-Elektrode *f* pH electrode
Phellandral *n* phellandral
Phellandren *n* phellandrene
Phellandrinsäure *f* phellandric acid
Phellogensäure *f* phellogenic acid
Phellonsäure *f* phellonic acid
Phenacemid *n* phenacemide
Phenacetin *n* phenacetin, acetophenetide
Phenacetolin *n* phenacetolin, phenacetein
Phenacetornithursäure *f* phenacetornithuric acid
Phenacetursäure *f* phenaceturic acid
Phenacyl- phenacyl
Phenacylamin *n* phenacylamine
Phenacylbromid *n* phenacyl bromide
Phenäthyl- phenethyl
Phenakit *m* (Min) phenacite
Phenalan *n* phenalan
Phenalen *n* phenalene
Phenamin *n* phenamine, phenocoll
Phenaminblau *n* phenamine blue
Phenanthren *n* phenanthrene
Phenanthrenchinon *n* phenanthrenequinone
Phenanthrendibromid *n* phenanthrenedibromide
Phenanthrensulfonsäure *f* phenanthrenesulfonic acid
Phenanthridin *n* phenanthridine
Phenanthridon *n* phenanthridone
Phenanthrol *n* phenanthrol, hydroxyphenanthrene
Phenanthrolin *n* phenanthroline
Phenanthron *n* phenanthrone
Phenanthroylpropionsäure *f* phenanthroylpropionic acid
Phenanthrylamin *n* phenanthrylamine
Phenarsazin *n* phenarsazine
Phenazin *n* phenazine
Phenazinmethosulfat *n* phenazine methosulfate
Phenazon *n* phenazone
Phenazoniumhydroxid *n* phenazonium hydroxide
Phendioxin *n* phendioxin, dibenzodioxin, diphenylene dioxide
Pheneserin *n* pheneserine
Phenetidin *n* phenetidine, aminophenetole
Phenetol *n* phenetole, ethoxybenzene, phenyl ethyl ether
Phenetyl- phenetyl
Phengit *m* (Min) phengite, muscovite

Phenhomazin *n* phenhomazine
Phenicarbazid *n* phenicarbazide
Pheniodol *n* pheniodol
Phenmorpholin *n* phenmorpholine
Phenobarbital *n* phenobarbital
Phenobutiodil *n* phenobutiodil
Phenochinon *n* phenoquinone
Phenocyanin *n* phenocyanin
Phenokoll *n* phenocoll, aminoaceto phenetidine, phenamine
Phenokollchlorhydrat *n* phenocoll hydrochloride
Phenokollchlorid *n* phenocoll chloride
Phenokollsalicylat *n* phenocoll salicylate, salocoll
Phenol *n* phenol, carbolic acid, hydroxybenzene
Phenolacetein *n* phenol acetine
Phenoläther *m* phenol ether, phenyl ether
Phenolallyläther *m* phenolallyl ether
Phenolaluminium *n* aluminum phenolate, aluminum phenoxide
Phenolanilinharz *n* phenolic aniline resin
Phenolapparat *m* phenol apparatus
Phenolarsonsäure *f* phenolarsonic acid
Phenolat *n* phenolate, carbolate, phenate, phenoxide, phenylate
Phenolausscheidung *f* (Med) phenoluria
Phenolcalcium *n* calcium phenolate, calcium phenoxide
Phenolcarbonsäure *f* phenolcarboxylic acid, hydroxybenzoic acid
Phenoldisulfonsäure *f* phenoldisulfonic acid
Phenolester *m* phenyl ester
Phenolfarbstoff *m* phenol dye
Phenolformaldehydharz *n* phenol formaldehyde resin
Phenolgehalt *m* phenol content
Phenolglykosid *n* phenol glycoside
Phenolharz *n* phenolic resin
Phenolharzkleber *m* phenolic cement
Phenolharzklebstoff *m* phenolic [resin] adhesive
Phenolharzkunststoff *m* phenolic plastic
Phenolharzlack *m* phenolic varnish
Phenolharzleim *m* phenolic adhesive
Phenolharzpreßmischung *f* phenolic molding compound or composition
Phenolharzschaum *m* phenolic foam
Phenolharzschichtstoff *m* phenolic laminated sheet
Phenolhomologe[s] *n* phenol homolog[ue]
Phenolhydroxyl *n* phenol hydroxyl, phenolic hydroxyl
phenolisch phenolic
Phenolkalium *n* potassium phenolate, potassium phenoxide
Phenollösung *f* carbolic acid solution
Phenolnatrium *n* sodium phenolate, sodium phenoxide
Phenolöl *n* carbolic oil

Phenoloxidase f (Biochem) phenol oxidase
Phenolphthalein n phenolphthalein
Phenolphthaleinpapier n phenolphthalein paper
Phenolquecksilber n mercury phenolate, mercury phenoxide
Phenolrot n phenol red
Phenolschwefelsäure f phenylsulfuric acid, phenolsulfuric acid, phenyl hydrogen sulfate
Phenolsulfonat n phenolsulfonate
Phenolsulfonphthalein n phenolsulfonephthalein
Phenolsulfonsäure f phenolsulfonic acid, sulfocarbolic acid, sulfophenic acid
phenolsulfonsauer phenolsulfonate
Phenolwismut n bismuth phenolate, bismuth phenoxide
Phenonaphthacridin n phenonaphthacridine, benzacridine
Phenonaphthazin n phenonaphthazine, benzophenazine, naphthophenazine
Phenonium-Ion n phenonium ion
Phenophenanthrazin n phenophenanthrazine, dibenzophenazine, dinaphthazine, phenanthrophenazine
Phenophosphazinsäure f phenophosphazinic acid
Phenoplast m phenolic resin, phenoplast
Phenopyrin n phenopyrine
Phenosafranin n phenosafranine, 3,6-diamino-phenylphenazine chloride
Phenose f phenose, hexahydroxy hexahydrobenzene
Phenostal n phenostal, diphenyl oxalate
Phenosuccin n phenosuccin, pyrantin
Phenothiazin n phenothiazine
Phenoval n phenoval
Phenoxarsin n phenoxarsine
Phenox[a]thiin n phenoxathiin
Phenoxazin n phenoxazine, naphthoxazine
Phenoxthin n phenoxthine, phenothioxin
Phenoxyessigsäure f phenoxyacetic acid
Phenoxyphenylsilikon n phenoxyphenyl silicone
Phenoxypropylbromid n phenoxypropyl bromide
Phenselenazin n phenselenazine
Phenselenazon n phenselenazone
Phenthiazin n phenthiazine, phenothiazine
Phenuron n phenurone
Phenyl- phenyl
Phenylacetaldehyd m phenylacetaldehyde
Phenylacetamid n phenylacetamide
Phenylaceton n phenylacetone
Phenylacetursäure f phenylaceturic acid
Phenylacetylcarbinol n phenylacetylcarbinol
Phenylacridin n phenylacridine
Phenylacrolein n phenyl acrolein
Phenylacrylsäure f (Zimtsäure) phenylacrylic acid, cinnamic acid
Phenyläther m phenyl ether
Phenyläthyläther m phenyl ethyl ether, ethoxybenzene, phenetole

Phenyläthylalkohol m phenyl ethanol, phenylethyl alcohol
Phenyläthylamin n phenylethyl amine
Phenyläthylformamid n phenylethyl formamide
Phenylalanin n phenylalanine
Phenylalkohol m phenol, phenyl alcohol
Phenylalphanaphthylamin n phenyl-alpha-naphthylamine
Phenylamidoazobenzol n phenylaminoazobenzene
Phenylamin n (Anilin) aniline, phenylamine
Phenylaminopropan n phenyl aminopropane
Phenylaminopropanmonophosphat n phenylaminopropane monophosphate
Phenylaminopropansulfat n phenylamino propane sulfate
Phenylaminschwarz n phenylamine black
Phenylantimonsäure f benzenestibonic acid, phenylstibonic acid
Phenylarsenchlorür n (obs) dichlorophenylarsine, phenylarsenous chloride
Phenylarsin n phenylarsine
Phenylarsinoxid n phenylarsine oxide
Phenylarsinsäure f phenylarsonic acid
Phenylat n phenolate, phenate, phenylate
Phenylazid n phenyl azide
Phenylbenzoat n phenyl benzoate
Phenylbenzochinon n phenyl benzoquinone
Phenylbetanaphthylamin n phenyl betanaphthyl amine
Phenylborchlorid n phenylboron chloride
Phenylbraun n phenyl brown
Phenylbutadien n phenylbutadiene
Phenylbuten n phenylbutene
Phenylbuttersäure f phenylbutyric acid
Phenylcapronsäure f phenylcaproic acid
Phenylchlorid n chlorobenzene, phenyl chloride
Phenylchloroform n phenylchloroform, benzotrichloride
Phenylchlorsilan n phenylchlorosilane
Phenylcyanat n phenyl cyanate, cyanatobenzene
Phenyldiamin n phenyl diamine
Phenyldithiocarbaminat n phenyldithiocarbamate
Phenyldithiocarbaminsäure f phenyldithiocarbamic acid
Phenylen- phenylene
Phenylenblau n phenylene blue
Phenylenbraun n phenylene brown
Phenylendiamin n phenylene diamine, benzene diamine, diaminobenzene
Phenylendiazosulfid n phenylene diazosulfide
Phenylendisulfid n phenylene disulfide
Phenylenharnstoff m phenylene urea
Phenylenrest m phenylene residue
Phenylensulfonylid f phenylene sulfonylide
Phenylessigsäure f phenylacetic acid
Phenylessigsäurechlorid n phenylacetyl chloride

Phenylessigsäurenitril *n* phenylacetonitrile
phenylessigsauer phenylacetate
Phenylfettsäure *f* phenylated fatty acid, phenyl fatty acid
Phenylfluoron *n* phenylfluorone
Phenylglycin *n* phenylglycine, phenylglycocoll
Phenylglykolsäure *f* (Mandelsäure) phenylglycolic acid
Phenylharnstoff *m* phenylurea, phenyl carbamide
Phenylheptatrienal *n* phenylheptatrienal
Phenylhydrazin *n* phenylhydrazine
Phenylhydrazinsulfosäure *f* phenylhydrazinesulfonic acid
Phenylhydrazon *n* phenylhydrazone
Phenylhydroxylamin *n* phenylhydroxylamine
phenylieren to introduce a phenyl group, to phenylate
Phenylisocyanat *n* phenyl isocyanate
Phenylisopropylamin *n* phenylisopropylamine
Phenylisopropylaminsulfat *n* phenylisopropylamine sulfate
Phenylisothiocyanat *n* phenylisothiocyanate
Phenyljodidchlorid *n* phenyl iodochloride
Phenylkakodyl *n* phenylcacodyl
Phenylketonurie *f* (Med) phenylketonuria
Phenyllithium *n* phenyllithium
Phenylmagnesiumbromid *n* phenylmagnesium bromide
Phenylmalonat *n* phenylmalonate
Phenylmethylketon *n* (Acetophenon) acetophenone, phenyl methyl ketone
Phenylmethylpyrazolon *n* phenylmethylpyrazolone
Phenylmilchsäure *f* phenyllactic acid
Phenylnaphthylamin *n* phenylnaphthylamine
Phenylnitroäthan *n* phenylnitroethane
Phenylnitromethan *n* phenylnitromethane
Phenylon *n* phenylone
Phenylorange *n* phenyl orange
Phenylosazon *n* phenylosazone
Phenylparaconsäure *f* phenylparaconic acid
Phenylpentadienal *n* phenylpentadienal
Phenylphenazoniumfarbstoff *m* phenylphenazonium dye
Phenylpiperidin *n* phenylpiperidine
Phenylpolyenal *n* phenylpolyenal
Phenylpropanolamin *n* phenylpropanolamine
Phenylpropiolsäure *f* phenylpropiolic acid
Phenylpropionat *n* phenylpropionate
Phenylpropionsäure *f* phenylpropionic acid
Phenylpropylmalonsäure *f* phenylpropylmalonic acid
Phenylpyridin *n* phenylpyridine
Phenylpyruvat *n* phenyl pyruvate
Phenylsalicylat *n* phenyl salicylate
phenylsauer phenate
Phenylschwefelsäure *f* phenylsulfuric acid, phenylhydrogen sulfate

Phenylsenföl *n* phenyl isothiocyanate, phenyl mustard oil
Phenylserin *n* phenylserine
Phenylsilikat *n* phenyl silicate
Phenylsilikon *n* phenyl silicone
Phenylsiliziumchlorid *n* phenylsilicon chloride, trichlorophenylsilane
Phenylsulfaminsäure *f* phenylsulfamic acid
Phenylthiohydantoin *n* phenylthiohydantoin
Phenylthiohydantoinsäure *f* phenylthiohydantoic acid
Phenyltoluolsulfonat *n* phenyltoluenesulfonate
Phenyltrichlorsilan *n* phenyltrichlorosilane
Phenyltrimethylammoniumnitrat *n* phenyltrimethylammonium nitrate
Phenylundecapentaenal *n* phenylundecapentaenal
Phenylurethan *n* phenylurethane, ethyl phenylcarbamate
Phenylvaleriansäure *f* phenylvaleric acid
Pheophytin *n* pheophytin
Pheromon *n* pheromone
pH-Gradient *m* pH gradient
Philadelphit *m* (Min) philadelphite
Phillipsit *m* (Min) phillipsite
Phillygenin *n* phillygenine
Philosophenwolle *f* (obs) philosopher's wool, sublimed zinc oxide
Phiole *f* [flat-bottomed] vial
phlegmatisieren to phlegmatize (to make e.g. organic peroxides less liable to decomposition by means of inert compounds)
Phlobaphen *n* phlobaphene
Phlogistontheorie *f* phlogiston theory
Phlogopit *m* (Min) phlogopite
Phloionolsäure *f* phloionolic acid
Phloionsäure *f* phloionic acid
Phloracetophenon *n* phloracetophenone
Phlorchinyl- phloroquinyl
Phloretin *n* phloretin
Phloretinsäure *f* phloretic acid
Phlorizin *n* phlorizin
Phloroacetophenon *n* phloroacetophenone
Phloroglucid *n* phloroglucide, phloroglucine
Phlorol *n* phlorol, o-ethylphenol
Phlorrhizin *n* phlorrhizin
Phloxin *n* phloxin, tetrabromodichlorofluorescein
pH-Meßgerät *n* pH-meter
pH-Messung *f* pH-measurement, determination of the pH-value
pH-Meter *n* pH-meter
Phobiermittel *n* repellent
Phocaenin *n* phocenin, glycerol trivalerate, trivalerin
Phönizin *n* phenicin
Phönizit *m* phenicite
Pholedrin *n* pholedrine
Pholerit *m* (Min) pholerite

Phon *n* (Maß) phon
Phonolith *m* (Min) phonolite
Phonon *n* phonon
pH-Optimum *n* pH optimum
Phormium *n* phormium, New Zealand flax
Phoron *n* phorone, diisopropylidene acetone
Phoronsäure *f* phoronic acid
Phosgen *n* phosgene, carbon oxychloride, carbonyl chloride
Phosgenbildung *f* phosgene formation
Phosgenit *m* (Min) phosgenite, hornlead
Phosgenzerfall *m* phosgene decomposition
Phosokresol *n* phosocresol
Phosphagen *n* phosphagen
Phosphanthren *n* phosphanthrene
Phosphat *n* phosphate, salt or ester of phosphoric acid, **primäres** ~ (Dihydrogenphosphat) dihydrogen phosphate, primary phosphate, **sekundäres** ~ (Monohydrogenphosphat) monohydrogen phosphate, secondary phosphate, **tertiäres** ~ tertiary phosphate
Phosphatacidität *f* phosphate acidity
Phosphatanlage *f* phosphate plant
Phosphatase *f* (Biochem) phosphatase, **alkalische** ~ alkaline phosphatase
Phosphatbad *n* phosphate bath
Phosphatbindung *f*, **energiereiche** ~ energy-rich phosphate bond
Phosphatdüngemittel *n* phosphate fertilizer
Phosphatdünger *m* phosphate fertilizer
phosphatführend phosphate-bearing, phosphatiferous
Phosphatid *n* phosphatide
Phosphatidyläthanolamin *n* phosphatidyl ethanolamine
Phosphatidylcholin *n* phosphatidyl choline
Phosphatidylserin *n* phosphatidyl serine
phosphatieren to phosphate, (Techn) to phosphatize
Phosphatierung *f* phosphating, phosphatization
phosphatisch phosphatic
Phosphatkreide *f* phosphate chalk
Phosphatmehl *n* ground phosphate, phosphate powder
Phosphatmühle *f* phosphate mill
Phosphatpuffer *m* phosphate buffer
Phosphatübertragung *f* phosphate transfer
Phosphatüberzug *m* phosphate coating, phosphate treatment
Phosphazobenzol *n* phosphazobenzene
Phosphensäure *f* phosphenic acid
Phosphenyl- phosphenyl
Phosphenylchlorid *n* phosphenyl chloride
Phosphenylsäure *f* phosphenylic acid, benzenephosphonic acid, phenylphosphorous acid
Phosphid *n* phosphide

Phosphin *n* (Ledergelb) chrysaniline, leather yellow, (Phosphorwasserstoff) hydrogen phosphide, phosphine
phosphinige Säure *f* phosphinic acid
Phosphit *n* phosphite
Phosphoarginin *n* phosphoarginine
Phosphocerit *m* (Min) phosphocerite
Phosphocholin *n* phosphocholine
Phosphodiesterase *f* (Biochem) phosphodiesterase
Phosphoenolbrenztraubensäure *f* phosphoenolpyruvic acid
Phosphofructokinase *f* (Biochem) phosphofructokinase
Phosphoglucoisomerase *f* (Biochem) phosphoglucoisomerase
Phosphoglucomutase *f* (Biochem) phosphoglucomutase
Phosphogluconsäure *f* phosphogluconic acid
Phosphoglyceratkinase *f* (Biochem) phosphoglycerate kinase
Phosphoglyceratmutase *f* (Biochem) phosphoglyceromutase
Phosphoglycerinaldehyd *m* phosphoglyceraldehyde
Phosphoglycerinsäure *f* phosphoglyceric acid
Phosphohomoserin *n* phosphohomoserine
Phosphokinase *f* (Biochem) phosphokinase
Phosphokreatin *n* phosphocreatine
Phospholipase *f* (Biochem) phospholipase
Phospholipid *n* phospholipid
Phospholipoid *n* phospholipid[e]
Phosphomonoesterase *f* (Biochem) phosphomonoesterase
Phosphonium- phosphonium
Phosphoniumbase *f* phosphonium base
Phosphoniumchlorid *n* phosphonium chloride
Phosphoniumjodid *n* phosphonium iodide
Phosphonsäure *f* phosphonic acid
Phosphoprotein *n* phosphoprotein
Phosphor *m* phosphorus (Symb. P), **farbloser** ~ white or yellow phosphorus, **mit** ~ **verbinden** to phosphorate, to phosphorize, **roter** ~ red phosphorus, **schwarzer** ~ black phosphorus, **violetter** ~ violet phosphorus, **von** ~ **befreien** to dephosphorize, **weißer** ~ white or yellow phosphorus
Phosphorabscheidung *f* separation of phosphorus, elimination of phosphorus, removal of phosphorus
Phosphoran *n*, X ~ phosphorane
phosphorartig phosphorous
Phosphoraufnahme *f* phosphorus intake
Phosphorbazillus *m* (Leuchtbazillus) phosphobacterium
Phosphorbestimmung *f* determination of phosphorus
Phosphorbleierz *n* (Min) pyromorphite
Phosphorbleispat *m* (Min)) pyromorphite

Phosphorbronze phosphor bronze
Phosphorchalcit m (Min) phosphorchalcite, pseudomalachite
Phosphordampf m phosphorus vapor
Phosphordünger m phosphate fertilizer
Phosphoreisen n ferrophosphorus, iron phosphide
Phosphoreisennickel n ferrophosphorus nickel
Phosphoreisensinter m diadochite
Phosphorentzündung f phosphorus inflammation
Phosphoreszenz f phosphorescence
phosphoreszenzerzeugend phosphorogenic
Phosphoreszenzspektrum n phosphorescence spectrum
phosphoreszieren to phosphoresce
Phosphoreszieren n phosphorescence
phosphoreszierend phosphorescent
phosphorfrei free from phosphorus
Phosphorgehalt m phosphorus content
Phosphorgeruch m phosphorous odor
phosphorhaltig phosphorous, containing phosphorus, phosphorated, phosphoric, (phosphathaltig) phosphatic
Phosphoribit m phosphoribitol
Phosphoribomutase f (Biochem) phosphoribomutase
Phosphoribose f phosphoribose
phosphorig phosphorous
phosphorige Säure f phosphorous acid, orthophosphorous acid
Phosphorigsäureanhydrid n phosphorous anhydride, phosphorus trioxide
phosphorigsauer phosphorous
Phosphor(III)-bromid n (Phosphorbromür) phosphorous bromide, phosphorus(III) bromide, phosphorus tribromide
Phosphor(III)-chlorid n (Phosphorchlorür) phosphorous chloride, phosphorus(III) chloride, phosphorus trichloride
Phosphor(III)-jodid n (Phosphorjodür) phosphorous iodide, phosphorus(III) iodide, phosphorus triiodide
Phosphor(III)-oxid n (Phosphoroxydul) phosphorous(III) oxide, phosphorus trioxide
phosphorisieren to phosphorate, to phosphorize
Phosphorit m (Min) phosphorite, rock phosphate
Phosphorkalk m calcium phosphide
Phosphorkristall m phosphorus crystal
Phosphorkupfer n (Chem) copper phosphide, (Min) libethenite
Phosphorkupfererz n (Min) libethenite
Phosphorlöffel m deflagrating spoon, phosphorus spoon
Phosphorluft f phosphorized air
Phosphormasse f phosphorus composition, phosphorus paste
Phosphormetall n metallic phosphide, metal phosphide

Phosphormolybdänsäure f phosphomolybdic acid
Phosphornatrium n sodium phosphide
Phosphornickeleisen n (Min) nickel ferrophosphorus, schreibersite
Phosphornitrid n phosphorus nitride
Phosphoröl n (Pharm) phosphorated oil
Phosphorographie f phosphorography
Phosphorolyse f phosphorolysis
Phosphoroxychlorid n phosphorus oxychloride, phosphoryl chloride
Phosphorpaste f phosphorus paste
Phosphorpentabromid n phosphoric bromide, phosphorus pentabromide, phosphorus(V) bromide
Phosphorpentachlorid n phosphoric chloride, phosphorus pentachloride, phosphorus(V) chloride
Phosphorpentahalogen n phosphorus pentahalide
Phosphorpentasulfid n phosphoric sulfide, phosphorus pentasulfide, phosphorus(V) sulfide
Phosphorpentoxid n phosphorus pentoxide, phosphoric anhydride, phosphoric oxide, phosphorus(V) oxide
Phosphorproteid n phosphoprotein
Phosphorpulver n phosphorus powder
Phosphorroheisen n phosphoric pig iron
Phosphorsäure f phosphoric acid, **sirupöse** ~ syrupy phosphoric acid
Phosphorsäureanhydrid n phosphoric anhydride, phosphoric oxide, phosphorus pentoxide, phosphorus(V) oxide
Phosphorsäureanlage f phosphoric acid plant
Phosphorsäureester m phosphoric ester
Phosphorsalz n acid sodium ammonium phosphate, microcosmic salt
Phosphorsalzperle f microcosmic salt bead
phosphorsauer phosphate
Phosphorsesquisulfid n phosphorous sulfide, phosphorus(III) sulfide, phosphorus sesquisulfide, phosphorus trisulfide
Phosphorspion m phosphorus spy
Phosphorstahl m phosphorus steel
Phosphorstickstoff m phosphoretted nitrogen
Phosphorstoffwechsel m (Physiol) phosphorus metabolism
Phosphorsulfobromid n phosphorus sulfobromide
Phosphorsulfochlorid n phosphorus sulfochloride
Phosphortribromid n phosphorus tribromide, phosphorous bromide, phosphorus(III) bromide
Phosphortrichlorid n phosphorous chloride, phosphorus(III) chloride, phosphorus trichloride
Phosphortrihalogen n phosphorous trihalide

Phosphortrioxid *n* phosphorous oxide, phosphorus(III) oxide, phosphorus trioxide
Phosphortrisulfid *n* phosphorus(III) sulfide, phosphorus trisulfide
Phosphoruranylit *n* phosphoruranylite
Phosphor(V)-bromid *n* phosphoric bromide, phosphorus pentabromide, phosphorus(V) bromide
Phosphor(V)-chlorid *n* phosphoric chloride, phosphorus pentachloride, phosphorus(V) chloride
Phosphorverbindung *f* phosphorus compound
Phosphorvergiftung *f* (Med) phosphorus poisoning
Phosphor(V)-jodid *n* phosphoric iodide, phosphorus pentaiodide, phosphorus(V) iodide
Phosphor(V)-oxid *n* phosphoric oxide, phosphorus pentoxide, phosphorus(V) oxide
Phosphor(V)-sulfid *n* phosphoric sulfide, phosphorus pentasulfide, phosphorus(V) sulfide
Phosphorwasserstoff *m* phosphorus hydride, hydrogen phosphide, phosphine, phosphoretted hydrogen
Phosphorweinsäure *f* phosphotartaric acid
Phosphorwolframat *n* phospho-tungstate
Phosphorwolframsäure *f* phosphotungstic acid
Phosphoryl- phosphoryl
Phosphorylase *f* (Biochem) phosphorylase
Phosphorylasekinase *f* (Biochem) phosphorylase kinase
Phosphorylasephosphatase *f* (Biochem) phosphorylase phosphatase
Phosphorylchlorid *n* phosphoryl chloride
Phosphorylierung *f* phosphorylation, **oxydative** ~ oxidative phosphorylation
Phosphorzink *n* phosphor-zinc, zinc phosphide
Phosphorzinn *n* phosphor-tin, tin phosphide
Phosphorzündholz *n* phosphorus match
Phosphoserin *n* phosphoserine
Phosphosiderit *m* (Min) phosphosiderite
Phosphosphingosid *n* phosphosphingoside
Phosphotransacetylase *f* (Biochem) phosphotransacetylase
Phosphotransferase *f* (Biochem) phosphotransferase
Phostamsäure *f* phostamic acid
Phostonsäure *f* phostonic acid
Phot *n* (Opt) phot
Photen *n* photene
Photoablösung *f* photo detachment
Photoabsorptionsband *n* photoabsorption band
Photoabzug *m* photo, print
Photoanregungsquerschnitt *m* photoexcitation cross section
Photoatmung *f* photorespiration
Photobakterium *n* (Leuchtbakterium) photobacterium

Photobiologie *f* photobiolgy
Photochemie *f* photochemistry
Photochemigraphie *f* zinc etching, zincography
photochemisch photochemical
Photochemisches Äquivalentgesetz *n* [Einstein's] law of photochemical equivalents
Photochlorierung *f* photochlorination
Photochrom *n* photochrome
photochromatisch photochromatic
Photochromie *f* photochromatism, photography in natural colors
Photodeuteron *n* photodeuteron
photodielektrisch photodielectric
photodynamisch photodynamic
Photoeffekt *m* photo [electric] effect, **äußerer** ~ external photoelectric effect, Hallwachs effect, photoelectric emission, photoemissive effect, **innerer** ~ inner photoelectric effect, photoconductive effect, photoelectric[al] effect, **normaler** ~ normal photoelectric emission, **selektiver** ~ selective photoelectric emission
photoelektrisch photoelectric[al]
Photoelektrizität *f* photoelectricity
photoelektromagnetisch photoelectromagnetic
Photoelektron *n* photoelectron
Photoelektronenröhre *f* phototube
Photoelektronenvervielfacher *m* photoelectron multiplier
Photoelement *n* photocell
Photoemission *f* photo-emission
Photoemissionsstrom *m* photoemission current
Photofluorographie *f* photofluorography
Photogalvanographie *f* photogalvanography
photogalvanomagnetisch photogalvanomagnetic
photogen photogenic
Photogen *n* photogen
Photogenlampe *f* photogen lamp
Photogrammetrie *f* photogrammetry
Photograph *m* photographer
Photographie *f* photography, (Bild) photograph
photographieren to photograph, to take pictures
photographisch photographic
Photohalleffekt *m* photomagnetoelectric effect
Photohalogen *n* photohalide
Photoinitiator *m* photoinitiator
Photoionisierung *f* photoionization
Photoionisierungsausbeute *f* photoionization efficiency
Photo-Isomerisierung *f* photochemical isomerization
Photokatalyse *f* photocatalysis
Photokathode *f* photo-cathode
Photokernspaltung *f* photo-disintegration
Photokopie *f* copy, photo[-graphic] copy, photoprint
Photoleitempfindlichkeit *f* photoconductive sensitivity
Photoleiter *m* photo-conductor

Photoleitfähigkeit *f* photoconductivity
Photoleitfähigkeitszelle *f* photo-conductivity cell
Photoleitung *f* photoconduction
Photolithograph *m* photolithographer
Photolithographie *f* photolithography
photolithographisch photolithographic
Photolumineszenz *f* photoluminescence
Photolyse *f* photolysis
photomagnetisch photomagnetic
Photomagnetismus *m* photo-magnetism
Photomaximumausbeute *f* photopeak efficiency
Photomeson *n* photomeson
Photometer *n* photometer
Photometerlampe *f* photometer lamp
Photometrie *f* photometry
photometrieren to measure photometrically
photometrisch photometric
Photomikrographie *f* photomicrography
Photomultiplier *m* photomultiplier
Photon *n* photon, light quantum
Photonenabsorption *f* photoelectric absorption, photon absorption
Photonenstrahl *m* beam of photons
Photoneutron *n* photoneutron
Photonitrosierung *f* photonitrosation
Photonukleon *n* photonucleon
Photooximierung *f* photooximation
Photooxydation *f* photooxidation
Photooxygenierung *f* photooxygenation
Photopapier *n* photographic paper
photophil (lichtliebend) photophilous
Photophorese *f* photophoresis
Photophosphorylierung *f* photophosphorylation
Photoplatte *f* photographic plate
Photopolymerisation *f* photopolymerization
Photoproton *n* photoproton
Photorahmen *m* photo mount
Photoreaktion *f* photoreaction
Photoreaktor *m* photoreactor
Photoreduktion *f* photoreduction
Photosantonin *n* photosantonin
Photosantonsäure *f* photosantonic acid
Photosensibilisator *m* photosensitizer
photosensitiv photo-active, photo-sensitive
Photoskop *n* photoscope
Photospaltung *f* photofission
Photosphäre *f* photosphere
Photostrom *m* photocell current, photo[electric] current
Photostromabfall *m* photocurrent decay
Photostromanregung *f* photocurrent stimulation
Photosynthese *f* photosynthesis
photosynthetisch photosynthetic
Phototaxis *f* phototaxis
Phototropie *f* phototropism, phototropy
phototropisch phototropic
Phototypie *f* phototypy
Phototypographie *f* phototypography
Photoverstärker *m* photomultiplier

Photozelle *f* photo[electric] cell, electric eye, photo[electric] tube, ~ mit äußerem lichtelektrischen Effekt emission cell, photoemissive cell, gasgefüllte ~ gas-filled photocell, gas phototube
Photozerfall *m* photo disintegration
Photozersetzung *f* photo dissociation
Photozinkographie *f* photozincography
photozinkographisch photozincographic
Phrenosin *n* phrenosin
Phrenosinsäure *f* phrenosinic acid
pH-Skala *f* pH scale
Phthalacen *n* phthalacene
Phthalaldehydsäure *f* phthalaldehydic acid
Phthalamid *n* phthalamide
Phthalamidsäure *f* phthalamic acid
Phthalan *n* phthalan
Phthalanil *n* phthalanil, N-phenylphthalimide
Phthalanilid *n* phthalanilide
Phthalanilsäure *f* phthalanilic acid
Phthalat *n* phthalate
Phthalazin *n* phthalazine
Phthalazon *n* phthalazone
Phthaldehyd *m* phthalaldehyde, phthalic aldehyde
Phthalein *n* phthalein
Phthalid *n* phthalide
Phthalidenessigsäure *f* phthalidene acetic acid
Phthalimid *n* phthalimide
Phthalimidin *n* phthalimidine
Phthalimidoessigsäure *f* phthalimidoacetic acid
Phthalin *n* phthaline
Phthalmonopersäure *f* monoperoxyphthalic acid, monoperphthalic acid
Phthalocyanin *n* phthalocyanine
Phthalocyaninblau *n* phthalocyanine blue
Phthalocyaninfarbstoff *m* phthalocynine dye
Phthalocyaningrün *n* phthalocyanine green
Phthalonimid *n* phthalonimide
Phthalonitril *n* phthalonitrile
Phthalonsäure *f* phthalonic acid
Phthalophenon *n* phthalophenone
Phthaloxim *n* phthaloxime
Phthalsäure *f* 1,2-benzene dicarboxylic acid, phthalic acid, alizarinic acid, o-benzene dicarboxylic acid
Phthalsäureanhydrid *n* phthalic anhydride, phthalandione
Phthalsäuredimethylester *m* dimethyl phthalate
Phthalsäureester *m* phthalate, phthalic ester
Phthalsäureharz *n* phthalic resin
phthalsauer phthalate, phthalic
Phthalyl- phthalyl
Phthalylchlorid *n* 1,2-benzenedicarbonyl chloride, phthalyl chloride
Phthalylglycin *n* phthalylglycine
Phthalylglycylchlorid *n* phthalylglycyl chloride
Phthalylhydrazid *n* phthalylhydrazide
Phthalylsulfathiazol *n* phthalylsulfathiazole

Phthalylsynthese — Pigment

Phthalylsynthese f phthalyl synthesis
Phthiokol n phthiocol
Phthionsäure f phthioic acid
Phulaxit n phulaxite
Phulwarabutter f phulwara butter
pH-Verschiebung f pH-shift
pH-Wert m pH-value
pH-Wert-Regler m pH regulation instrument
Phycit m phycitol
Phycomyceten pl (Algenpilze) phycomycetes
Phygon n phygon
Phykobilin n phycobilin
Phykochrom n phycochrome
Phykocyan n phycocyanogen
Phykocyanin n phycocyanin
Phykoerythrin n phycoerythrin
Phykophäin n phycophein
Phyllit m (Min) phyllite, chloritoid, mica slate
Phyllochinon n (Vitamin K) phylloquinone, vitamin K
Phyllochlorin n phyllochlorine
Phylloerythrin n phylloerythrin
Phyllohämin n phyllohemin
Phylloporphyrin n phylloporphyrin
Phyllopyrrol n phyllopyrrole
Phyllorin n phyllorin
Phyostigmin n phyostigmine
Physalin n physalin
Physciasäure f physcic acid, chrysophyscin, parietin, physcion
Physcion n physcion, chrysophyscin, parietin, physcic acid
Physeter|in|säure f physeteric acid
Physetolsäure f physetoleic acid, hypogaeic acid
Physik f physics
physikalisch physical
physikalisch-chemisch physical chemical, physicochemical
Physiker m physicist
Physikochemiker m physical chemist
Physiographie f physiography
Physiologie f physiology, **angewandte** ~ applied physiology, **experimentelle** ~ experimental physiology, **pathologische** ~ pathological physiology, pathophysiology, **vergleichende** ~ comparative physiology
physiologisch physiological
Physodalin n physodalin
Physodalsäure f physodalic acid
Physol n physol
Physostigmin n physostigmine, eserine
Physostigmol n physostigmol
Physoxanthin n physoxanthin
Phytadien n phytadiene
Phytan n phytane
Phytanol n phytanol
Phytase f (pflanzl Esterase) phytase
Phyten n phytene
Phytensäure f phytenic acid

Phytin n phytin
Phytinsäure f phytic acid
Phytochemie f phytochemistry
phytochemisch phytochemical
Phytoen n phytoene
Phytohormon n phytohormone
Phytol n phytol
Phytolaccatoxin n phytolacca toxin
Phyton n phytone
Phytoparasit m (Pflanzenparasit) phytoparasite
phytophag (pflanzenfressend) phytophagous, herbivorous
Phytoplasma n phytoplasm
Phytosterin n phytosterin, phytosterol
Phytosterolin n phytosterolin
phytotoxisch phytotoxic
Piaselenol n piaselenole
Piauzit m piauzite
Piazin n piazine, p-diazine
Piazothiol n piazothiole
PIB Abk für Polyisobutylen
Picamar n picamar
Picein n picein
Picen n picene
Picenchinon n picene quinone
pichen to pitch
Pichpech n common pitch
Pichurimbohne f pichurim bean
Pichurimtalgsäure f pichurim oil
Pichwachs n bee glue, propolis
Picit m picite
Pickelbrühe f pickel bath
Pickeln n (Häute) pickling
Pickelsäure f pickling acid
Pickeringit m (Min) pickeringite
Picofarad n pico-farad, micro-micro farad
Picolin n picoline, methylpyridine
Picolineisenprotoporphyrin n picoline ferroprotoporphyrin
Picolinsäure f picolinic acid, pyridinecarboxylic acid
Picotit m (Min) picotite, chrome spinel
Picrocrocin n picrocrocin
Picylen n picylene, picenefluorene
PID-Regler m proportional and integral (or floating) and derivative action controller, proportional plus reset plus rate action controller
Piezoeffekt m piezoelectric effect
piezoelektrisch piezoelectric
Piezoelektrizität f piezoelectricity
Piezokristall m (Krist) piezoelectric crystal, **bimorpher** ~ (Elektr) bimorph cell
Piezomessung f piezometry
Piezometer n (Druckmesser) piezometer
Piezotropie f piezotropy
Pigment n coloring matter, pigment, (Beize) lake, **halbmineralisches** ~ metallo-organic pigment

Pigmentbakterie *f* chromogenic bacterium, pigment bacterium
Pigmentbindemittel *n* pigment binder
Pigmentbraun *n* pigment brown
Pigmentchromgelb *n* pigment chrome yellow
Pigmentdruck *m* pigment printing
Pigmentechtrot *n* pigment fast red
Pigmentfarbstoff *m* organic pigment, pigment toner
Pigmentierbad *n* pigmenting bath
pigmentieren to pigment
Pigmentierung *f* pigmentation
Pigmentscharlach *n* pigment scarlet
Pigmentteig *m* pigment paste
Pigmentverteiler *m* pigment disperser
Pigmentverträglichkeit *f* colorant acceptance
Pigmentzelle *f* chromatophore, pigment[ed] cell
Pikamar *n* picamar, propylpyrogallol dimethyl ether
Pikolin *n* picoline, methylpyridine
Pikramid *n* picramide, picrylamine
Pikramin *n* picramine
Pikraminsäure *f* picramic acid
pikraminsauer picramate
Pikranilid *n* picranilide
Pikrat *n* picrate, carbazotate
Pikrin *n* picrin
Pikrinsäure *f* picric acid, 2,4,6-trinitrophenol, chrysolepic acid, picranisic acid, picronitric acid
Pikrinsäurekomplex *m* picric acid complex
pikrinsauer picrate
Pikrit *m* picrite
Pikrocin *n* picrocine
Pikrocininsäure *f* picrocininic acid
Pikrocinsäure *f* picrocinic acid
Pikroilmenit *m* picroilmenite
Pikrol *n* picrol, diiodoresorcin potassium monosulfonate
Pikrolith *m* (Min) picrolite
Pikrolonsäure *f* picrolonic acid
Pikromerit *m* (Min) picromerite
Pikromycin *n* picromycin
Pikropharmakolith *m* (Min) picropharmacolite
Pikropodophyllin *n* picropodophyllin
Pikroroccellin *n* picroroccellin
Pikrosäure *f* picroic acid
Pikrosminsteatit *m* (Min) picrosmine steatite
Pikrotin *n* picrotin
Pikrotinsäure *f* picrotinic acid
Pikrotitanit *m* (Min) picrotitanite
Pikrotoxin *n* picrotoxin
Pikryl- picryl
Pikrylacetat *n* picryl acetate
Pikrylchlorid *n* picryl chloride
Pilarit *m* pilarite
Pilébrechwerk *n* pilé sugar crusher
Pilékläre *f* pilé liquor
Pilézucker *m* pilé sugar

Piliermaschine *f* milling machine, soap mill
Pilinit *m* pilinite
Pilit *m* pilite
Pille *f* pill, pellet
Pillenbrett *n* pill board
pillenförmig pill-shaped
Pillenform *f* mold for making pills
Pillenmasse *f* material for making pills
Pillenschachtel *f* pill box
Pillingsprüfmaschine *f* pilling tester
Pilocarpidin *n* pilocarpidine
Pilocarpin *n* pilocarpine
Pilocarpinhydrochlorid *n* pilocarpine hydrochloride
Pilocarpinnitrat *n* pilocarpine nitrate
Pilocarpinsäure *f* pilocarpic acid
Pilopsäure *f* pilopic acid
Pilosin *n* pilosine
Pilosinin *n* pilosinine
Pilotenventil *n* pilot-operated valve
Pilotlicht *n* (Mikroskop) pilot lamp
Pilz *m* fungus, mushroom, **eßbarer** ~ mushroom, **pathogener** ~ disease fungus
Pilzbildung *f* fungoid growth
Pilzcellulose *f* fungus cellulose
pilzförmig fungiform
Pilzgeflecht *n* mycelium
Pilzisolator *m* mushroom insulator
Pilzkrankheit *f* (Med) mycosis
Pilzkunde *f* mycology
Pilzmischer *m* mushroom mixer
pilztötend fungicidal
Pilzvergiftung *f* (Med) mushroom poisoning, mycetism
Pilzwucherung *f* fungoid growth
Pimanthren *n* pimanthrene
Pimarabietinsäure *f* pimarabietic acid
Pimarsäure *f* pimaric acid
Pi[π]-Bindung *f* π-bond
Pi[π]-Elektron *n* π-electron
Pi[π]-Elektronensystem *n* π-electron system
Pimelinaldehyd *m* pimelic aldehyde
Pimelinketon *n* pimelinketone, cyclohexanone
Pimelinsäure *f* heptanedioic acid, pimelic acid
pimelinsauer pimelate
Pimelit[h] *m* (Min) pimelite
Piment *m n* piment
Pimentpfeffer *m* allspice
Pimentrum *m* bay rum
Pi-Meson *n* pi meson, pion
Pi[π]-Komplex *m* π-complex
Pimpinellasaponin *n* pimpinella saponin
Pimpinellenöl *n* pimpinella oil
Pimpinellin *n* pimpinellin
Pinabietin *n* pinabietin
Pinabietinsäure *f* pinabietic acid
Pinacyanol *n* pinacyanol
Pinakiolith *m* (Min) pinaciolite
Pinakoid *n* (Krist) pinacoid

Pinakol *n* pinacol, 2,3-dimethyl-2,3-butanediol, tetramethyl ethyleneglycol
Pinakolin *n* pinacolin
Pinakolinalkohol *m* pinacolyl alcohol, 3,3-dimethyl-2-butanol
Pinakolinchlorid *n* pinacolyl chloride
Pinakolin-Umlagerung *f* pinacoline rearrangement, pinacolone rearrangement
Pinan *n* pinane
Pinanol *n* pinanol
Pinastrinsäure *f* pinastric acid
Pinaverdol *n* pinaverdol
Pinch-Effekt *m* (Atom) pinch effect
Pinen *n* pinene
Pinenchlorhydrat *n* pinene hydrochloride
Pineytalg *m* piney tallow
Pinguit *m* (Min) pinguite
Pinicolsäure *f* pinicolic acid
Pinidin *n* pinidine
Pinienkern *m* pine kernel
Pinientalg *m* pine tallow
Pinifolsäure *f* pinifolic acid
Pininsäure *f* pininic acid
Pinit *m* pinitol, pentahydroxycyclohexane, pinite
Pinksalz *n* pinksalt, ammonium hexachlorostannate, ammonium stannic chloride, tin-ammonium chloride
Pinnoit *m* (Min) pinnoite
Pinocamphan *n* pinocamphane
Pinocamphersäure *f* pinocamphoric acid
Pinocampholensäure *f* pinocampholenic acid
Pinocamphon *n* pinocamphone
Pinocarveol *n* pinocarveol
Pinocarvon *n* pinocarvone
Pinol *n* pinol, pine camphor
Pinolen *n* pinolene
Pinolit *m* pinolite
Pinolol *n* pinolol
Pinolon *n* pinolone
Pinonaldehyd *m* pinonaldehyde
Pinononsäure *f* pinononic acid
Pinonsäure *f* pinonic acid
Pinophansäure *f* pinophanic acid
Pinotöl *n* pinot oil
Pinozytose *f* pinocytosis
Pinsäure *f* pinic acid
Pinsel *m* brush
Pinselhaar *n* bristle, brush hair
Pinselin *n* pinselin
Pinselinsäure *f* pinselic acid
Pinselvergoldung *f* brush gilding
Pinselverkupferung *f* brush coppering
Pintadoit *m* (Min) pintadoite
Pinzette *f* forceps, pincers, tweezers
Pion *n* pi meson, pion
Pipecolein *n* pipecoleine
Pipecolin *n* pipecoline, methylpiperidine

Pipecolinsäure *f* pipecolinic acid, piperidine-N-carboxylic acid
Pipecolyl- pipecolyl
Piperazin *n* piperazine, diethylenediamine, ethylenimine, hexahydropyrazine, piperazidine
Piperettin *n* piperettine
Piperettinsäure *f* piperettic acid
Piperidazin *n* piperidazine
Piperidin *n* piperidine, hexahydropyridine
Piperidinblau *n* piperidine blue
Piperidiniumhydroxid *n* piperidinium hydroxide
Piperidon *n* piperidone
Piperidylhydrazin *n* piperidyl hydrazine
Piperidylurethan *n* piperidyl urethane
Piperil *n* piperil
Piperilsäure *f* piperilic acid
Piperimidin *n* piperimidine
Piperin *n* piperine, piperyl-piperidine
Piperinsäure *f* piperic acid, piperinic acid
Piperiton *n* piperitone
Piperolidin *n* piperolidine
Piperonal *n* piperonal, heliotropin, piperonyl aldehyde
Piperonylalkohol *m* piperonyl alcohol
Piperonylchlorid *n* piperonyl chloride
Piperonylsäure *f* piperonylic acid, heliotropic acid, methylene protocatechuic acid
Piperoxan *n* piperoxan
Piperyl- piperyl
Piperylen *n* piperylene, pentadiene
Piperylurethan *n* piperylurethane
Pipette *f* pipet[te]
Pipettenflasche *f* pipette bottle
Pipettengestell *n* pipette holder, pipette stand
Pipettenständer *m* pipette holder, pipette stand
pipettieren to pipet[te], to measure with a pipette, to transfer with a pipette
Pipettiergerät *n* pipetting apparatus
Pipitzahoinsäure *f* pipitzahoic acid, perezon
Pipitzol *n* pipitzol
Pipradrol *n* pipradrol
Piprinhydrinat *n* piprinhydrinate
PI-Regler *m* proportional and integral (or reset) action controller, proportional plus floating action controller
Pirssonit *m* (Min) pirssonite
Pisangwachs *n* pisang wax, plantain wax
Pisanit *m* (Min) pisanite
Piscidin *n* piscidin
Piscid[in]säure *f* piscidic acid
Pisolith *m* (Min) pisolite
pisolithartig pisolitic
pisolithhaltig pisolitiferous
Pissophan *m* pissophane
Pistazien *pl* (Frucht) pistachio-nuts, pistachios
Pistaziengrün *n* pistachio green
Pistazienöl *n* pistachio oil
Pistazit *m* (Min) pistacite, epidote

Pistill *n* pestle, pounder
Pistolenröhre *f* pistol pipe
Pistomesit *m* (Min) pistomesite
Pitafaser *f* pita fiber
Pitahanf *m* pita hemp, Manila hemp
Pittacol *n* pittacol, eupittonic acid
Pittizit *m* (Min) pitticite, scorodite
Pivalinaldehyd *m* pivalaldehyde, 2,2-dimethylpropanal, trimethyl acetaldehyde
Pivalinsäure *f* pivalic acid, 2,2-dimethylpropanoic acid, trimethylacetic acid
Pivaloin *n* pivaloin
Pivalon *n* pivalone
Pivalophenon *n* pivalophenone
Pivotelement *n* (Math) pivot element
Pizein *n* pizein
Placebo *n* (Pharm) placebo
Plättchen *n* flake, pellet, (Anat) lamina (pl. laminae), (Blut) platelet, (Knochen) scale
plätten to flatten
Plagiocitrit *m* plagiocitrite
Plagioklas *m* (Min) plagioclase
plagioklastisch plagioclastic
Plagionit *m* (Min) plagionite
Plakat *n* poster
Plakatfarbe *f* color for printing placards, lithographic color
Plakatsäule *f* poster pillar
Plan *m* plan, design, layout, ~ **Skizze** draft
Plancheit *m* (Min) plancheite
Planchers Base *f* Plancher's base
Plancksches Strahlungsgesetz *n* Planck's law of radiation
Plancksches Wirkungsquantum *n* Planck's constant, Planck's [elementary] quantum of action
Plandrehbank *f* face lathe
Planerit *m* (Min) planerite
Planet *m* (Astr) planet
planetarisch (Astr) planetary
Planetarium *n* planetarium
Planetenbahn *f* (Astr) planetary orbit
Planetengetriebe *n* planetary gear
Planetenmischer *m* planetary-type mixer, vertical paint mixer
Planetenmühle *f* planetary mill
Planetenrührer *m* planetary agitator
Planetenrührwerk *n* planetary [paddle] mixer, planetary stirrer
Planetensystem *n* planetary system
Planfilter *n* table filter
Planflansch *m* face flange, plane flange
Planfräsen *n* face milling
Planfräser *m* face mill
Planfräsmaschine *f* face milling machine
Plangitter *n* plane grating
Planglas *n* (Opt) plano-spherical lens
planieren to plane, to planish, to smooth

Planierlöffel *m* skimmer
Planierraupe *f* bulldozer
Planierwasser *n* gluewater, size
Planimeter *n* planimeter, integrator
Planimetrie *f* plane geometry
planimetrieren to planimeter
Planke *f* plank
Plankollektor *m* horizontal commutator
plankonkav plano-concave
plankonvex plano-convex
Plankton *n* (Biol) plankton
planmäßig methodical, systematic[al]
Planoferrit *m* (Min) planoferrite
planparallel plane-parallel
Planrad *n* face wheel
Planroller *m* face stitcher
Planrost *m* flat grate, horizontal grate
Planrostfeuerung *f* furnace for steam boiler
Planscheibe *f* face plate
Planschleifen *n* surface grinding
Planschleifmaschine *f* surface grinder
Planschliff *m* plane ground joint
Planschliffverbindung *f* flat flange joint
Planschneider *m* cutting machine
Planschnitt *m* planing cut
Plansichter *m* plansifter
Plansieb *n* flat sieve, gyratory screen
Planspiegel *m* plane mirror
Plansymmetrie *f* plane symmetry
plansymmetrisch planisymmetric
Planteobiose *f* planteobiose
Planteose *f* planteose
Planung *f* planning, project, schedule
Planungskosten *pl* engineering costs
planvoll carefully planned
Planzug *m* cross feed
Plaque *f* (Med) plaque
Plasma *n* plasm, plasma
Plasmaantrieb *m* plasma propulsion
Plasmabrenner *m* plasma burner; plasma torch
Plasmachemie *f* plasma chemistry
Plasmaersatzstoff *m* blood substitute
Plasmakaskadenbrenner *m* plasma cascade torch
Plasmakugel *f* plasma sphere
Plasmamembran *n* plasma membrane
Plasmaphysik *f* plasma physics
Plasmaprotein *n* plasmaprotein
Plasmaschicht *f* plasma layer, äußere ~ exoplasm, **innere** ~ endoplasm, entoplasm
Plasmaschneiden *n* plasma arc cutting
Plasmaschweißen *n* plasma welding
Plasmaschwingung *f* plasma oscillation
Plasmaspritzverfahren *n* plasma jet spraying
Plasmastrahl *m* plasma beam
Plasmaströmung *f* (Bot) cyclosis
Plasmawechselwirkung *f* plasma interaction
Plasmazelle *f* plasma cell, plasmacyte
Plasmochin *n* plasmochin, aminoquin, pamaquin, plasmoquine

Plasmodium *n* (Biol) plasmodium
Plasmolyse plasmolysis
plasmolytisch plasmolytic
Plaste *pl* plastics, **druckhärtbare** ~ pressure setting plastics, **härtbare** ~ duroplastics, plastics capable of being hardened, **hitzehärtbare** ~ thermosetting plastics
plastifizieren to plasticize, to plastify
Plastifizierung *f* plasticizing
Plastifizierzone *f* melting zone
Plastigel *n* plastigel
Plastikator *m* plastifier
Plastikfolie *f* plastic foil
Plastikkleber *m* synthetic resin adhesive
Plastikmetall *n* plastic metal
Plastilin *n* plastilina, plastiline
plastisch plastic
Plastisol *n* plastisol
Plastit *n* plastite
plastizieren to plasticise, to plasticize (Am. E.)
Plastizier[ungs]mittel *n* plasticizer
Plastizität *f* plasticity, (Dreidimensionalität) three-dimensional quality
Plastizitätsgrenze *f* plastic limit
Plastizitätsmesser *m* plastometer
Plastochinon *n* plastoquinone
Plastocyanin *n* plastocyanin
Plastograph *m* plastograph
Plastometer *n* plastometer
Plateauneigung *f* plateau slope
Platin *n* platinum (Symb. Pt)
Platinammoniumchlorid *n* ammonium chloroplatinate, platinammonium chloride
Platinapparat *m* platinum apparatus
platinartig platinoid
Platinasbest *m* platinized asbestos
Platinat *n* platinate, salt or ester of platinic acid
Platinbarren *m* platinum ingot
Platinblase *f* platinum still
Platinblech *n* platinum foil, platinum sheet
Platinblende *f* platinum screen
Platinbronze *f* platinum bronze
Platinchlorid *n* platinum chloride
Platinchlorwasserstoffsäure *f* chloroplatinic acid
Platincyanbarium *n* barium platinocyanide, barium cyanoplatinate(II)
Platincyanür *n* (Platin(II)-cyanid) platinous cyanide, platinum dicyanide, platinum(II) cyanide
Platincyanwasserstoff *m* cyanoplatinic acid, platinocyanic acid
Platindraht *m* platinum wire
Platindrahtöse *f* platinum-wire loop
Platindruck *m* platinotype process
Platine *f* plate bar, sheet bar
Platinelektrode *f* platinum electrode
platinenthaltend platiniferous
Platinersatz *m* platinum substitute
Platinerz *n* platinum ore

Platinerzkorn *n* grain of platinum ore
Platinfeuerzeug *n* Döbereiner's lamp, Döbereiner's matchbox
Platingefäß *n* platinum vessel
Platingehalt *m* platinum content
Platingerät *n* platinum ware
platinhaltig containing platinum, platiniferous
Platinhydrosol *n* platinum hydrosol
Platinichlorwasserstoff *m* chloroplatinic(IV) acid
Platinicyanwasserstoffsäure *f* cyanoplatinic(IV) acid, platinicyanic acid
platinieren to platinize, to coat with platinum, to combine with platinum, to platinum-plate
Platinieren *n* platinizing
platiniert platinized, platinum-plated
Platinierung *f* platinization
Platin(II)-chlorid *n* (Platinchlorür) platinous chloride, platinum dichloride, platinum(II) chloride
Platin(II)-hydroxid *n* platinous hydroxide
Platin(II)-oxid *n* (Platinoxydul) platinous oxide, platinum(II) oxide, platinum monoxide
Platinirhodanwasserstoffsäure *f* thiocyanatoplatinic(IV) acid
Platiniridium *n* platinum iridium
Platinisalz *n* (Platin(IV)-salz) platinic salt, platinum(IV) salt
Platin(IV)-chlorid *n* platinic chloride, platinum(IV) chloride, platinum tetrachloride
Platiniverbindung *f* (Platin(IV)-Verbindung) platinic compound, platinum(IV) compound
Platin(IV)-hydroxid *n* platinic hydroxide
Platin(IV)-oxid *n* platinic oxide, platinum(IV) oxide
Platinkaliumsalz *n* platinum potassium salt
Platinkegel *m* platinum cone
Platinkessel *m* platinum boiler
Platinkohle *f* platinized charcoal
Platinkontakt *m* platinum contact
Platinkonus *m* platinum cone
Platinlegierung *f* platinum alloy
Platinlöffel *m* platinum spoon
Platinlösung *f* platinum solution
Platinmesser *n* platinum knife
Platinmohr *m* platinum black, spongy platinum
Platinmuffel *f* platinum muffle
Platinnadel *f* platinum pin
Platinnickelelement *n* platinum nickel cell
Platinochlorid *n* platinous chloride, platinum dichloride, platinum(II) chloride
Platinochlorwasserstoffsäure *f* chloroplatinic(II) acid, chloroplatinous acid
Platinocyanwasserstoffsäure *f* cyanoplatinic(II) acid, platinocyanic acid
Platinoid *n* platinoid
Platinorhodanwasserstoffsäure *f* thiocyanatoplatinic(II) acid, thiocyanatoplatinous acid

Platinoverbindung f (Platin(II)-Verbindung) platinous compound, platinum(II) compound
Platinoxydulverbindung f (obs) platinous compound, platinum(II) compound
Platinpapier n platinotype paper
Platinplattierung f platinum plating
Platinreihe f platinum group
Platinretorte f platinum retort, platinum still
Platinrhodiumdraht m platinum rhodium wire
Platinrhodiumelement n platinum rhodium couple
Platinrückstand m platinum residue
Platinsäure f platinic acid
platinsauer platinate
Platinschale f platinum basin, platinum dish
Platinscheidung f platinum refining
Platinschiffchen n platinum boat
Platinschwamm m platinum sponge, spongy platinum
Platinschwarz n platinum black
Platinsilicid n platinum silicide
Platinspatel m platinum spatula
Platinspirale f platinum coil, platinum spiral
Platinspitze f platinum point
Platinsulfid n platinum sulfide
Platinsulfür n (obs) platinous sulfide, platinum(II) sulfide
Platintetrachlorid n platinum tetrachloride
Platintiegel m platinum crucible
Platinüberzug m platinum coating
Platinveraschungsschale f platinum incineration dish
Platinverbindung f platinum compound
Platinzinkelement n platinum zinc element
Platte f plate, sheet, (Holz) board, (Stein-Platte) slab
Platten n laminating
Plattenaustauscher m plate exchanger
Plattenband n plate conveyer, apron conveyer
Plattenbandförderer m apron conveyor
Plattenelektrisiermaschine f plate-type electrostatic machine
Plattenelektrode f plate electrode
Plattenfeder f plate spring
Plattenfedermanometer n spring pressure gauge
Plattenfilter n filter press, platen filter
Plattenförderer m platform conveyer
Plattenformerei f plate molding
Plattengefrierapparat m plate freezer
Plattengleichung f plate equation
Plattenglimmer m (Min) sheet mica
Plattengröße f plate size
Plattengummi n sheet rubber
Plattenhalter m (Phot) plate holder
Plattenkassette f (Phot) plate holder
Plattenkautschuk m India rubber sheet, sheet rubber
Plattenkondensator m plate condenser
Plattenkorn n (Phot) grain of emulsion

Plattenkultur f plate culture
Plattenkupp[e]lung f plate clutch
Plattenpreßfilter n hydraulic filter press
Plattensengen n singeing on hot plates
Plattenthermometer n unsheathed thermometer on scale plate
Plattenventil n plate valve
Plattenverdampfer m plate evaporator
Plattenwärmeaustauscher m plate-type heat exchanger, brazed-plate heat exchanger
Plattform f platform
Plattformkipper m tipping platform
Plattformwagen m platform car
plattieren to plate, (Metall) to clad
Plattieren n plating, smoothing, (Metall) cladding, (Schweiß) weld laminating
plattiert plated
Plattierung f plating, (Holz) veneering, (Met) cladding
Plattierungsanlage f plating plant
Plattierungsbad n plating bath
Plattnerit m (Min) plattnerite
Plattseide f slack silk
Plattsilber n white copper
Plattwalze f smoothing roll
Platynecin n platynecine
Platynecinsäure f platynecic acid
Platynit m (Min) platynite
Platz m place, ground, location, site
Platzbedarf m space requirement
platzen to burst, to explode, (Schweißnaht) to crack
Platzen n (Reifen) blowout, bursting
Platzmangel m lack of space; restricted space
Platzpatrone f blank cartridge
Platzquecksilber n (obs) fulminating mercury, mercuric fulminate
Platzscheibe f bursting disk, safety diaphragm
Platzwechsel m interchange of sites, migration
Pleiadien n pleiadiene
Pleistozän n Pleistocene [period]
p-Leiter m p-conductor
p-Leitung f p-type semi-conductivity
Plenargyrit m (Min) plenargyrite
Pleochroismus m (Krist) pleochroism, pleochromatism
pleochroitisch pleochroic, pleochromatic
Pleonast m (Min) pleonast
Pleonektit m pleonectite
Plessit m (Min) plessite
Pleuellager n connecting rod bearing
Pleuelstange f connecting rod
Plexigelpaste f plexigel paste
Plexiglas n (HN) plexiglass
Plinian m plinian
Plinthit m plinthite
Plioform n plioform wax
plissieren to pleat
Plissieren n pleating

plötzlich abrupt, sudden[ly]
Plombe *f* lead seal, (Zahnmed) filling
plombieren to seal, to lead, to plug
Plombieren *n* lead sealing
Plombierit *m* plombierite
Plüsch *m* (Text) plush
Plüschleder *n* suede, velvet leather
Plumbagin *n* plumbagin, methyljuglone
Plumbat *n* plumbate, salt or ester of plumbic acid
Plumbioxid *n* lead dioxide, lead(IV) oxide, plumbic oxide
Plumbisalz *n* lead(IV) salt, plumbic salt
Plumbit *n* plumbite
Plumbiverbindung *f* (Blei(IV)-Verbindung) lead(IV) compound, plumbic compound
Plumbocalcit *m* (Min) plumbocalcite
Plumboferrit *m* (Min) plumboferrite
Plumbogummit *m* (Min) plumbogummite
Plumboniobit *m* (Min) plumboniobite
Plumbosalz *n* lead(II) salt, plumbous salt
Plumboverbindung *f* (Blei(II)-Verbindung) lead(II) compound, plumbous compound
Plumbum *n* (Lat) lead
Plumierarinde *f* plumieria bark
Plumieräsure *f* plumieric acid
Plumierid *n* plumieride
Plumosit *m* (Min) plumosite
Plungerpumpe *f* plunger pump
Plus-Elektrizität *f* positive electricity
Plus-Minus-Skala *f* plus minus scale
Plusminustoleranz *f* plus minus tolerance
Pluspol *m* positive pole
Pluszeichen *n* plus sign
Pluto *m* (Astr) Pluto
Plutonium *n* plutonium (Symb. Pu)
Plutoniumbrutreaktor *m* plutonium breeder
Plutoniumreaktor *m* plutonium pile
Plutonylion *n* plutonyl ion
Pneumatik *f* pneumatics, pneumatic system
Pneumatik *m* (Luftreifen) pneumatic tire
Pneumatikreifen *m* pneumatic tire
pneumatisch pneumatic
Pneumatolyse *f* (Geol) pneumatolysis
Pneumobakterium *n* Friedländer's bacillus
Pneumokokkeninfektion *f* pneumococcosis
p-n-Übergang *m* p-n-transition
Pocheisen *n* beating iron
pochen to crush, to pound, to stamp
Pocherz *n* milling ore, stamping ore
Pochgänge *m pl* inferior ore
Pochgestein *n* stamp rock
Pochmehl *n* pulverized ore
Pochmühle *f* stamp mill
Pochschlamm *m* ore slime
Pochschlich *m* ore slime
Pochschuh *m* beating iron
Pochstempel *m* stamp, stamper, stamping iron
Pochtrog *m* stamp mortar

Pochtrübe *f* slime, [stamp] pulp
Pochwerk *n* stamp mill
Pocken *pl* (Med) [small] pox, variola
Pockenerreger *m* smallpox virus
Pockenimpfung *f* smallpox vaccination
Pockenstein *m* (Min) variolite
Pockenvirus *m* smallpox virus
Pockholz *n* pock wood
Podocarpinsäure *f* podocarpic acid
Podolit *m* (Min) podolite
Podophyllin *n* podophyllin
Podophyllinharz *n* podophyllin resin
Podophyllinsäure *f* podophyllic acid
Podophyllotoxin *n* podophyllotoxin
Podophyllsäure *f* podophyllic acid
Podophyllum *n* podophyllum, mandrake, may-apple
Pökel *m* pickle, pickling brine
Pökeldauer *f* pickling time, curing time
Pökelfaß *n* pickling tub
Pökelfleisch *n* pickled meat, salted meat
Pökelgeruch *m* pickling odor
Pökelkeller *m* pickling cellar
Pökelkufe *f* pickling vat
pökeln to pickle
Pökeln *n* pickling
Poise *n* (Viskositätseinheit) poise (unit of viscosity)
Poiseuillesches Gesetz *n* Poiseuille's formula
Poissonsche Gleichung *f* Poisson's equation
Poissonsche Konstante *f* Poisson's constant
Pol *m* pole, **induzierender** ~ inducing pole, **negativer** ~ cathode, negative pole, **positiver** ~ anode, positive pole
Polabstand *m* distance between poles
Polanker *m* pole armature
Polanziehung *f* polar attraction
polar polar
Polarachse *f* (Geom) polar axis
Polarbahn *f* polar orbit
Polardiagramm *n* polar diagram
Polarimeter *n* polarimeter
Polarimetrie *f* polarimetry
polarimetrisch polarimetric
Polarisation *f* polarization, **elliptische** ~ elliptic polarization, **lineare** ~ plane or linear polarization, **magnetische** ~ magnetic polarization, **zirkulare** ~ circular polarization
Polarisationsapparat *m* polarization apparatus
Polarisationsebene *f* plane of polarization
Polarisationselektrode *f* polarization electrode
Polarisationsenergie *f* polarization energy
Polarisationserscheinung *f* phenomenon of polarization
Polarisationsfarbe *f* polarization color
Polarisationsgitter *n* polarizing grating
Polarisationsinstrument *n* polarizing instrument
Polarisationskapazität *f* polarization capacity
Polarisationskonstante *f* polarization constant

Polarisationskurve *f* polarization curve
Polarisationsmikroskop *n* polarizing microscope
Polarisationsnachweis *m* polarization detection
Polarisationsprisma *n* polarizer, polarizing prism
Polarisationsschwund *m* polarization fading
Polarisationsspannung *f* polarization voltage
Polarisationsspiegel *m* polarizer
Polarisationsstrom *m* polarization current
Polarisationswiderstand *m* polarization resistance
Polarisationswinkel *m* angle of polarization, Brewster angle
Polarisationszustand *m* polarization state, state of polarization
Polarisator *m* polarizer
polarisierbar polarizable
Polarisierbarkeit *f* polarizability, **elektrochemische** ~ (Elektr) ionic polarizability
Polarisierbarkeitstensor *m* polarizability tensor
polarisieren to polarize
Polarisieren *n* polarizing
polarisierend polarizing
polarisiert polarized, **elliptisch** ~ elliptically polarized, **geradlinig** ~ plane-polarized
Polarisierung *f* polarization, **magnetische** ~ magnetic polarization
Polariskop *n* polariscope
Polarität *f* polarity
Polarkoordinate *f* (Math) polar coordinate
Polarkoordinaten *pl,* **sphärische** ~ geographic coordinates
Polarkreis *m* arctic zone, polar circle, polar region
Polarogramm *n* (Elektr) polarogram
Polarograph *m* polarograph
Polarographie *f* polarography
polarographisch polarographic
Polarplanimeter *n* polar planimeter
Polbildung *f* pole formation, polarization
Polbogenlänge *f* length of pole arc
Polbüchse *f* pole ring
Pole *pl,* **gleichnamige** ~ like poles, similar poles, **ungleichnamige** ~ opposite poles
Poleiöl *n* pennyroyal oil
polen to pole, to render polar, (Kupfer) to toughen copper
Polen *n* **des Bleis** poling of lead
Polende *n* pole end
Polentfernung *f* distance between poles
Polerzeugung *f* pole excitation, production of poles
Poleyöl *n* oil of pennyroyal
Polform *f* pole shape
Polianit *m* (Min) polianite, pyrolusite
polierbar polishable
Polierblech *n* polishing plate
Poliereisen *n* burnisher

polieren to polish, to burnish, (glätten) to smooth, (Walzgut) to planish, **elektrolytisch** ~ to anode-brighten, to electrobrighten
Polieren *n* polishing
Polierer *m* burnisher, polisher
polierfähig capable of taking a polish
Polierfilz *m* polishing cloth
Polierfilzscheibe *f* felt polishing wheel
Polierflüssigkeit *f* liquid polish, polishing liquid
Poliergold *n* burnishing gold
Polierkalk *m* polishing chalk
Polierkomposition *f* composition of polishing material, polishing paste
Poliermaschine *f* polishing machine
Poliermasse *f* polishing composition, polishing powder
Poliermittel *n* polish, polishing material, polishing medium, polishing powder
Polierplatin *n* burnishing platinum
Polierpulver *n* polishing powder
Polierrot *n* ferric oxide, jewellers red
Polierscheibe *f* polishing disc, polishing wheel
Polierschiefer *m* polishing slate, tripoli
Poliersilber *n* burnishing silver
Polierstahl *m* burnisher, polisher
Polierstaub *m* polishing dust
Polierstein *m* burnishing stone, polishing stone
poliert polished
Poliertrommel *f* polishing drum
Poliertuch *n* polishing cloth
Polierwachs *n* polishing wax
Polierwalze *f* burnishing roll
Poligalin *n* polygalin, polygalic acid
Poliment *n* gilding size, gold size
Polinduktion *f* pole induction
Polio *f* (Med) polio [myelitis]
Polioimpfstoff *m* poliomyelitis vaccine
Polioimpfung *f* polio inoculation
Poliomyelitis *f* (Med) polio [myelitis]
Polioserum *n* polio serum
Poliovakzine *f* polio vaccine
Politur *f* gloss, polish, sheen
politurfähig capable of taking polish
Politurmasse *f* polishing paste
Polituröl *n* polishing oil
Politurspiritus *m* polishing spirit
Polkante *f* pole edge
Polkern *m* pole core
Polklemme *f* (Elektr) pole terminal
Pollänge *f* length of pole
Pollen *m* (Blütenstaub) pollen
Pollucit *m* (Min) pollucite, pollux
Pollücke *f* interpolar distance
Pollux *m* (Min) pollucite, pollux
Polmitte *f* pole center
Polonium *n* polonium (Symb. Po)
Polpaar *n* pair of poles
Polpapier *n* pole paper
Polrad *n* magnet wheel

Polreagenspapier n polarity paper, pole paper
Polring m pole ring
Polsättigung f pole saturation
Polschraube f pole shoe
Polschuhbohrung f pole shoe bore
Polschuhecke f pole shoe tip
Polschuhspitze f pole shoe tip
Polspannung f polar tension
Polstärke f pole strength
Polster n bolster, cushion, pad
Polsterung f padding, upholstery
Polsucher m pole finder
Polsuchpapier n pole finding paper
Poltern n (Metall) removing the pickling acid
Polumkehr f pole reversal, polarity reversal
Polumkehrbarkeit f reversibility of poles
Polumschaltung f pole changing [control]
Polung f polarity, polarization
Polverdunklung f polar blackout
Polversetzung f pole dislocation
Polwechsel m change of poles
Polwechsler m pole-changing switch
Polyacenaphthylen n polyacenaphthylene
Polyacrylamid n polyacrylamide
Polyacrylamidelektrophorese f polyacrylamide electrophoresis
Polyacrylat n polyacrylate
Polyacrylfaser f polyacryl fiber
Polyacrylharz n polyacrylic resin
Polyacrylnitril n polyacrylonitrile
Polyacrylnitrilfaser f polyacrylonitrile fiber
Polyacrylsäure f polyacrylic acid
Polyacrylsäureester m polyacrylate
Polyaddition f polyaddition
Polyadelphit m (Min) polyadelphite
Polyäther m polyether
Polyäthylen n polyethylene
Polyäthylenfilm m polyethylene film
Polyäthylenfolie f polyethylene sheeting
Polyäthylenglykol n polyethylene glycol
Polyäthylenpolyamin n polyethylene polyamine
Polyaffinitätstheorie f polyaffinity theory
Polyalanin n polyalanine
Polyalkohol m polyalcohol, polyhydric alcohol, polyhydroxy alcohol
Polyalthsäure f polyalthic acid
Polyamid n polyamide
Polyamidfaser f polyamide fiber
Polyamidkunststoff m nylon plastic
Polyamidseide f polyamide continuous filament
Polyamin n polyamine
Polyargyrit m (Min) polyargyrite
Polyarsenit m (Min) polyarsenite
Polyazofarbstoff m polyazo dyestuff
Polybasit m (Min) polybasite
Polybuten n polybutene, polybutylene
Polybutylen n polybutene, polybutylene
Polychloropren n polychloroprene

Polychlortrifluoräthylen n polychlorotrifluoroethylene
Polychroismus m (Opt) polychroism
Polychroit n polychroite
polychrom[atisch] polychromatic, polychrome
Polychromie f polychromy
Polychromsäure f polychromic acid
polycyclisch polycyclic
polydispers polydisperse
Polydispersität f polydispersity
Polydymit m (Min) polydymite
Polyeder n polyhedron
polyedrisch polyhedral
Polyelektrolyt m polyelectrolyte
polyenergetisch polyenergetic
Polyester m polyester
Polyesterfaser f polyester fiber
Polyestergewebe n polyester fabric
Polyesterharz n polyester resin
Polyesterurethan n polyester urethane
polyfunktionell polyfunctional
Polygala f polygala, milkwort
Polygalasäure f polygalic acid, polygalin
Polygalin n polygalic acid, polygalin
Polyglucuronsäure f polyglucuronic acid
Polyglycerin n polyglycerol
Polyglykol n polyglycol
Polyglykolid n polyglycolide
Polygon n polygon
Polygonabbildung f polygon mapping
Polygonisierung f polygonization
Polygonmessung f polygoniometry
Polygonzugverfahren n point slope method
Polyhalit m (Min) polyhalite
Polyhalogenid n polyhalide
polyheteroatomig polyheteroatomic
Polyhydroxyketon n polyhydroxy ketone
Polyhydroxy[l]aldehyd m polyhydroxy aldehyde
Polyisobutylen n polyisobutylene
Polykieselsäure f polysilicic acid
Polykondensat n condensation polymer, polycondensation product
Polykondensation f polycondensation
Polykondensationsprodukt n polycondensation product
Polykras m (Min) polycrase
Polykrasit m (Min) polycrasite
polykristallin polycrystalline
Polylithionit m (Min) polylithionite
polymer polymer, polymeric
Polymerase f (Biochem) polymerase
Polymerbenzin n polymer gasoline
Polymer[e] n polymer, polymeride (Br. E.)
Polymere pl, **lineare** ~ linear polymers
Polymerie f polymerism
Polymerisat n polymer, polymeride, polymerizate
Polymerisation f polymerization, ~ **in Masse** block polymerization, bulk polymerization,

mass polymerization, **lichtinduzierte** ~ photopolymerization, **lineare** ~ linear polymerization, **stereoregulierte** ~ stereoregulated polymerization
Polymerisationsanlage f polymerization reactor
Polymerisationsemulsion f polymerization emulsion
Polymerisationsgrad m degree of polymerization
Polymerisationsinhibitor m polymerization inhibitor
Polymerisationskatalysator m polymerization catalyst
Polymerisationsprodukt n polymer, polymerizate
polymerisierbar polymerizable
polymerisieren to polymerize
Polymethacrylsäurealkylester m polyalkyl methacrylate
Polymethakrylsäureester m polymethacrylate
Polymethylen n polymethylene
Polymethylendiamin n polymethylene diamine
Polymethylenhalogenid n polymethylene halide
Polymignit m (Min) polymignite
Polymnit m polymnite
Polymolybdänsäure f polymolybdic acid
polymorph polymorphic, polymorphous
Polymorphie f polymorphism, polymorphy
Polymorphismus m polymorphism, polymorphy
Polymyxin n polymyxin
Polyneuridinsäure f polyneuridinic acid
Polynitrokomplex m polynitro complex
Polynom n (Math) polynomial, multinomial
Polynucleotid n polynucleotide
Polyol n polyol, polyhydric alcohol
Polyomavirus n polyoma virus
Polyose f (Polysaccharose) poly[satchar]ose
Polyoxyäthylen n polyoxyethylene
Polyoxymethylen n polyoxymethylene
Polypapier n polypaper
Polypeptid n polypeptide
Polypeptidase f (Biochem) polypeptidase
Polypeptidkette f polypeptide chain
Polyphalit m polyphalit
Polyphenoloxydase f (Biochem) polyphenoloxidase
Polyphosphat n polyphosphate
Polyphosphatase f (Biochem) polyphosphatase
Polyphosphorsäure f polyphosphoric acid
Polyporensäure f polyporenic acid
Polyporsäure f polyporic acid
Polypropylen n polypropylene
Polypropylenglykol n polypropylene glycol
Polyribonukleotid n polyribose nucleotide
Polyribosom n polyribosom, polysome
Polysaccharase f (Biochem) polysacharase
Polysaccharid n polysaccharide, **sulfatisiertes** ~ sulfated polysaccharide
Polysaccharidsynthese f polysaccharide synthesis
Polysäure f poly acid
Polysalicylid n polysalicylide

Polyschwefel m, **catena-** ~ catena-polysulfur
Polyschwefelwasserstoff m hydrogen polysulfide
Polysiloxan n polysiloxan
Polysom n polysome
Polystyrol n polystyrene
Polystyrolharz n polystyrene resin
Polysulfid n polysulfide
Polysulfidkautschuk m polysulfide rubber
Polysymmetrie f polysymmetry
Polytechnik f polytechnics
Polytechnikum n polytechnical school, technical highschool
Polytelit m polytelite
Polyterpen n polyterpene
Polytetrafluoräthylen n polytetrafluoroethylene
Polythionat n polythionate
Polythionsäure f polythionic acid
Polytrifluorchloräthylen n polytrifluorochloroethylene
polytrop polytropic
Polyvinylacetal n polyvinyl acetal
Polyvinylacetat n polyvinyl acetate
Polyvinylalkohol m polyvinyl alcohol
Polyvinylalkyläther m polyvinyl alkyl ether
Polyvinylbutyral n polyvinyl butyral
Polyvinylcarbazol n polyvinyl carbazole
Polyvinylchlorid n polyvinyl chloride, PVC, **nachchloriertes** ~ postchlorinated PVC, **nicht nachchloriertes** ~ not postchlorinated PVC
Polyvinylcyanid n polyvinyl cyanide
Polyvinylmethyläther m polyvinyl methyl ether
Polyvinylpyrrolidon n polyvinyl pyrrolidone
Polyzimtsäure f polycinnamic acid
POM Abk. für Polyoxymethylen
Pomade f pomade
Pomeranze f (Bot) orange
pomeranzenartig orange-like
Pomeranzenbitter n hesperidin
Pomeranzenblütenöl n orange flower oil, neroli oil
Pomeranzenbranntwein m orange brandy
Pomeranzenlikör m orange bitter
Pomeranzenöl n orange peel oil
Pomeranzenschale f orange peel
Pompejanischrot n Pompeian red, Pompey red
Ponceaurot n Ponceau red
Pongamöl n pongam oil
Ponsolblau n ponsol blue
Popelin m (Text) poplin
Populen n populene
Populin n populin, **mit** ~ **behandeln** to populinate
Pore f pore, (Hohlraum) void, (Holz) grain, pore
Porenanteil m proportion of voids
Porenbeton m cellular concrete, light-weight concrete
Porenbildung f pinholing, pitting
Porendichtigkeit f hold-out
Porenflüssigkeit f pore fluid

Porenflüssigkeitsinelastizität *f* pore fluid inelasticity
Porenfüller *m* sealer
Porengips *m* light-weight plaster
Porengröße *f* pore size, size of pore
Porenkörper *m* (Kunststoff) aerated plastic, foamed plastic
Porenleitfähigkeit *f* pore conductivity
Porenprüfgerät *n* porosity determination equipment
Porenprüfmittel *n* pore testing agent
Porenraum *m* pore space
Porenschließer *m* sealer
Porenstoff *m* aerated plastic, foamed plastic
Porenverhütungsmittel *n* anti-pit agent
Porenweite *f* pore size, porosity
Porenzahl *f* void ratio
porig porous, cellular
Porigkeit *f* porosity, porousness
porös porous, permeable, poriferous
Poroidin *n* poroidine
Poroplast *n* foamed plastic
Porosimetrie *f* porosimetry
Porosität *f* porosity, porousness
Porositätsgrad *m* degree of porosity
Porpezit *m* (Min) porpezite
Porphin *n* porphin[e]
Porphingerüst *n* porphin ring, porphin structure
Porphinring *m* porphin ring
Porphobilinogen *n* porphobilinogen
Porphyr *m* (Min) porphyry, **schwarzer** ~ black porphyry, nero antico
porphyrähnlich porphyritic
porphyrartig porphyritic
Porphyrazin *n* porphyrazine
Porphyrexid *n* porphyrexide
Porphyrilin *n* porphyrilin
Porphyrilsäure *f* porphyrilic acid
Porphyrin *n* porphyrin
Porphyrinbiosynthese *f* porphyrin biosynthesis
Porphyrinring *m* porphyrine ring
Porphyrinstoffwechsel *m* porphyrin metabolism
Porphyrit *m* (Min) porphyrite
Porphyroid *n* porphyroid
Porphyropsin *n* porphyropsin
Porphyroxin *n* porphyroxine
Porphyrschiefer *m* (Min) porphyritic schist
Portalkran *m* portal jib crane
Portalstapler *m* straddle carrier
Portion *f* portion
portionsweise by small amounts, in portions
Portlandhochofenzement *m* portland blast furnace cement
Portlandkalk *m* portland limestone
Portlandklinker *m* portland clinker
Portlandzement *m* portland cement
Porto *n* postage
Porzellan *n* porcelain, china, **chemisch-technisches** ~ porcelain for chemical-technical purposes, **hartes** ~ hard porcelain
Porzellanabdampfschale *f* china evaporating basin, porcelain evaporating basin
porzellanartig porcelaneous
Porzellanbecher *m* porcelain beaker
Porzellanbehälter *m* porcelain container, porcelain tank
Porzellanbrennofen *m* porcelain kiln
Porzellanerde *f* China clay, kaolin, porcelain clay
Porzellanfabrikation *f* porcelain manufacture
Porzellangerät *n* porcelain ware
Porzellanglocke *f* porcelain cup
Porzellanisolator *m* porcelain insulator
Porzellanit *m* (Min) porcelanite, porcelain jasper
Porzellanjaspis *m* (Min) porcelain jasper, porcellanite
Porzellankasserolle *f* porcelain pan or casserole
Porzellankitt *m* porcelain cement
Porzellanmalerei *f* China painting
Porzellanmanufaktur *f* manufacture of porcelain
Porzellanmasse *f* porcelain paste
Porzellanmörtel *m* pozzuolana mortar
Porzellanring *m* porcelain ring
Porzellanröhre *f* porcelain tube
Porzellanrohr *n* porcelain tube
Porzellanschale *f* porcelain cup
Porzellanschiffchen *n* porcelain boat
Porzellan-Schmelzfarbe *f* porcelain color
Porzellanspat *m* (Min) porcelain spar, scapolite
Porzellantiegel *m* porcelain crucible
Porzellanton *m* China clay, kaolin, porcelain clay
Porzellantrichter *m* porcelain funnel
Porzellanwanne *f* porcelain trough
Porzellanwaren *pl* chinaware
Positionsgenauigkeit *f* positional accuracy
positiv positive, ~ **geladen** positively charged
Positiv *n* (Phot) positive
positiv-elektrisch positive electric, positively electrical
Positivform *f* positive mold
Positivverfahren *n* male mold method
Positron *n* positron, antielectron, positive electron
Positronenbahn *f* positron track
Positronenemission *f* (Atom) positron emission
Positronenstrahlung *f* positron emission
Positronenzerfall *m* (Atom) positron decay, positron disintegration
Positronium *n* positronium
Positronstrahler *m* positron emitter
Poskin *n* poskine
Posten *m* item
Postulat *n* postulate
postulieren to postulate

Potential *n* potential, ~ **der Erde** earth potential, **chemisches** ~ chemical potential, **elektrisches** ~ electrical potential, **elektrophoretisches** ~ electrophoretic potential, **thermodynamisches** ~ thermodynamic potential
Potentialabfall *m* drop of potential, fall of potential, potential gradient
Potentialanstieg *m* increase in potential
Potentialausgleich *m* compensation of potential
Potentialberg *m* potential barrier
potentialbildend potential-forming
Potentialdifferenz *f* potential difference
Potentialfeld *n* potential field
Potentialfläche *f* potential surface, **konstante** ~ equipotential surface, isopotential surface
Potentialfunktion *f* **der Elektrizität** electric potential
Potentialgefälle *n* potential gradient
Potentialgleichung *f* potential equation
Potentialgradient *m* potential gradient
Potentialhügel *m* potential barrier
Potentiallage *f* potential position
Potentialmulde *f* (Elektron) potential trough, potential well
Potentialrand *m* (Elektron) potential barrier
Potentialregler *m* (Elektr) potential [or voltage] regulator
Potentialschwelle *f* potential barrier, potential threshold
Potentialsprung *m* potential jump, change in potential
Potentialtheorie *f* theory of potential
Potentialtopf *m* potential well
Potentialtopftiefe *f* potential well depth
Potentialunterschied *m* potential difference
Potentialverschiebung *f* displacement in potential
Potentialverteilung *f* potential distribution
Potentialwall *m* potential barrier
Potentialwert *m* potential value
potentiell potential
Potentiometer *n* potentiometer, pH-meter, ~ **für Titration** titrimeter
Potentiometrie *f* potentiometry
potentiometrisch potentiometric
Potenz *f* potency, (Math) exponent, power, **dritte** ~ cube, third power, **in die -nte** ~ **erheben** to raise to the -th power, **zur dritten** ~ **erheben** to cube, to raise to the third power, **zur zweiten** ~ **erheben** to raise to the second power, to square, **zweite** ~ second power, square
Potenzansatz *m* exponential equation
potenzieren to raise to a higher power
Potenzprodukt *n* power product
Potenzreihe *f* power series
Potenzreihenentwicklung *f* power series expansion

Potenzverteilung *f* exponential distribution
Pottasche *f* potash, potassium carbonate
Pottaschebrennen *n* potash calcining
Pottaschefluß *m* crude potash
Pottascheküpe *f* potash vat
Pottaschelauge *f* potash lye
Pottaschelösung *f* potash solution
Pottaschesiederei *f* potash factory
Pottaschewasser *n* potash water
Pottlot *n* graphite, black lead, plumbago
Pottmetall *n* pot metal
Pouzacit *m* pouzacite
Powellit *m* (Min) powellite
pozzuolanartig pozzuolanic
Pozz[u]olanerde *f* pozzuolana
PP Abk. für Polypropylen
ppb Abk. für „parts per billion" = 0,001 ppm
Prädissoziation *f* predissociation
Prädissoziationsspektrum *n* predissociation spectrum
präformiert preformed
Prägedruck *m* embossed printing
Prägefolie *f* embossed sheet
Prägekalander *m* embossing calender, stamping calender
Prägelack *m* stamping lacquer
prägen to stamp, (hohlprägen) to emboss, (massivprägen) to coin, (z. B. Blech) to stamp
Prägen *n* embossing, coining, stamping, (der Formteilkonturen im Werkzeug mit dem Prägestempel) hobbing
Prägeplatte *f* embossing plate
Prägepresse *f* hobbing press, stamping press
Prägerahmen *m* chase ring
Prägering *m* chase ring
Prägestanze *f* stamping die
Prägestempel *m* hob; punch
Prägewalze *f* embossing roll
Prägewerk *n* stamping press
Prägung *f* embossing
Prähämataminsäure *f* prehemataminic acid
Präkambrium *n* Precambrian
Präkondensation *f* precondensation
Prämie *f* bonus, premium, prize
prämieren to accord an award, to give a prize
pränumerando beforehand, in advance
Präparat *n* preparation
Präparateraum *m* specimen room
Präparateröhrchen *n* preparation tube
präparieren to prepare, (konservieren) to preserve, (Med) to dissect
Präparierlupe *f* dissecting lens
Präpariermikroskop *n* dissecting microscope
Präpariersalz *n* preparing salt, sodium stannate
Präservativ *n* preservative
Präserve *f* preserves
präservieren to preserve
Präservieren *n* preserving
Präservierung *f* preservation

Präservierungsmittel *n* preservative
Präzession *f* precession, ~ **des Kerns** nuclear precession
Präzipitat *n* precipitate, **rotes** ~ (Chem) mercuric oxide, red precipitate, **weißes schmelzbares** ~ (Chem) diammine mercuric chloride, **weißes unschmelzbares** ~ (Chem) amido-mercuric chloride, white fusible precipitate, white infusible precipitate
Präzipitation *f* precipitation
Präzipitationswärme *f* heat of precipitation
Präzipitierbottich *m* precipitating vat
präzipitieren to precipitate
Präzipitieren *n* precipitating
Präzipitierfaß *n* precipitation cask
Präzipitiergefäß *n* precipitating vessel
Präzipitin *n* precipitin
präzise precise
präzisieren to define, to formulate more precisely, to make precise
Präzision *f* precision
Präzisionsdrehbank *f* precision lathe
Präzisionsfertigguß *m* precision finished casting
Präzisionsgebläse *n* precision blower
Präzisionsglasrohr *n* precision glass tube
Präzisionsguß *m* precision casting
Präzisionsinstrument *n* precision instrument, precision tool
Präzisionsmechanik *f* precision mechanics
Präzisionsmessung *f* accurate measuring, precision measuring
Präzisionswert *m* precision value
Praktikant *m* trainee
Praktikum *n* practical course, laboratory course
praktisch practical, expedient, handy
praktizieren to exercise, (Med) to practice, to practise (Br. E.)
Prall *m* collision, impact, shock
Prallbeanspruchung *f* impact loading
Prallblech *n* baffle
Prallbrecher *m* impact crusher
Pralldüse *f* impactor nozzle
Prallelastizität *f* rebound resilience
Prallelektronen *pl* rebound electrons
prallen to dash, to impinge ([on, upon, against]), to strike
Prallfläche *f* surface of impingement
Prallhaube *f* baffle
Prallmühle *f* impact mill
Prallringzentrifuge *f* baffle-ring centrifuge
Prallschirm *m* baffle plate, deflector
Pralltellermühle *f* impeller breaker [mill]
Prallzerkleinerung *f* impact crushing
Prallzerkleinerungsmaschine *f* impact pulverizer
Pramocain *n* pramocaine
Pranke *f* claw, clutch
Praseodym *n* praseodymium (Symb. Pr)
Praseodymcarbid *n* praseodymium carbide
Praseodymchlorid *n* praseodymium chloride
Praseodymdioxid *n* praseodymium dioxide
Praseodymgehalt *m* praseodymium content
Praseodymoxid *n* praseodymium oxide, praseodymia, praseodymium(III) oxide, praseodymium sesquioxide
Praseodymselenat *n* praseodymium selenate
Praseodymsulfat *n* praseodymium sulfate
Praseodymsulfid *n* praseodymium sulfide, praseodymium(III) sulfide
Praseolith *m* (Min) praseolite
Praseosalz *n* praseo salt
Prasopal *m* (Min) prase opal
Praxis *f* (Med, pl Praxen) practice
praxisbezogen relating to practice
Precipitron *n* precipitron
Predazzit *m* predazzite
Prednisolon *n* prednisolone
Prednison *n* prednisone
Pregnan *n* pregnane
Pregnandiol *n* pregnanediol
Pregnandion *n* pregnanedione
Pregnenolon *n* pregnenolone
Pregrattit *m* (Min) pregrattite
Prehnit *m* (Min) prehnite
Prehnitenol *n* prehnitenol
Prehnitol *n* prehnitol
Prehnitolsulfonsäure *f* prehnitenesulfonic acid
Prehnitsäure *f* 1,2,3,4-benzene-tetracarboxylic acid, prehnitic acid
Preis *m* price, charge
Preisaufschlag *m* additional charge
Preiselbeere *f* cranberry
Preisliste *f* price list
Preisschild *n* price tag
Preiszettel *m* tag
Prellbolzen *m* recoil spindle
Prellklotz *m* recoil, spring beam
Prellung *f* recoil
Prellvorrichtung *f* bumping mechanism, cushioning
Prenol *n* prenol
Prenylchlorid *n* prenyl chloride
Prephensäure *f* prephenic acid
Preßbahn *f* lamination
Preßbleche *pl* cauls
Preßdruck *m* mold[ing] pressure
Presse *f* press, ~ **mit Einzelantrieb** self-contained press, ~ **mit Gummisack** bag molding, ~ **zum Kaschieren** laminating press
Preßeffekt *m* pressure effect
pressen to press, to compress, to jam, to mold, to squeeze, to stamp, (Kohle) to briquet
Pressen *n* compression, pressing, ~ **mit Hochfrequenzvorwärmung** high-frequency molding, ~ **unter hohem Druck** high-pressure molding
Pressenholm *m* strain rod; column
Pressenunterteil *n* bolster

Preßfehler *m* molding fault; (Blase) blister, bubble, (Loch) pit; (Streifen) segregation
Preßfilter *n* press[ure] filter
Preßflüssigkeit *f* pressure fluid
Preßform *f* [compression] mold, matrix, press mold, **mehrteilige** ~ split mold
Preßformmaschine *f* compression mold, molding press, press molding machine
Preßgas *n* compressed gas
Preßglas *n* molded glass, pressed glass
Preßglimmer *m* pressed mica
Preßgrat *m* flash
Preßgratlinie *f* flashline
Preßgrundzeit *f* basic press time
Preßguß *m* press[ure] casting
Preßhartglas *n* pressed hard glass
Preßharz *n* molding resin
Preßhefe *f* pressed yeast
Preßholm *m* press cross beam
Preßholz *n* compregnated wood, densified wood, high density wood, high duty wood, laminated wood
Preßholzkleber *m* veneer glue
Preßkissen *n* pressure pad, forming pad, rubber pad
Preßkohle *f* coal briquette, pressed charcoal
Preßkolben *m* press plunger, main ram, press ram
Preßkorken *m* press cork
Preßkuchen *m* press cake, expeller cake
Preßling *m* molded piece, pressed article
Preßluft *f* compressed air
Preßluftanschluß *m* air pressure manifold
Preßluftarmatur *f* compressed air mounting
preßluftbetätigt air-operated
Preßluftbohrer *m* pneumatic drill
Preßlufteintritt *m* air pressure inlet
Preßluftflasche *f* compressed-air cylinder
Preßlufthammer *m* pneumatic hammer
Preßluftmotor *m* compressed air motor
Preßluftpistole *f* blow gun
Preßluftrührwerk *n* compressed air stirrer
Preßluftstampfmaschine *f* pneumatic rammer
Preßmassage *f* **mit Gewebe-Füllstoff** fabric-filled molding compound
Preßmasse *f* molding composition, molding compound, molding material
Preßmischung *f* molding composition, molding compound
Preßpartie *f* press section
Preßplatte *f* press plate
Preßpulver *n* molding powder
Preßpumpe *f* press[ure] pump
Preßrückstand *m* pressed pulp
Preßsäule *f* column of a press
Preßsaft *m* pressed juice
Preßschichtholz *n* densified laminated wood
Preßschmierpumpe *f* grease gun
Preßschmierung *f* pressure lubrication
Preßschweißen *n* press[ure] welding
Preßschwitzverfahren *n* press sweating process
Preßspan *m* press board
Preßspanplatten *pl* chipboard
Preßspritzen *n* plunger molding, flow molding, pot type molding
Preßspritzform *f* plunger mold, transfer mold
Preßspritzkolben *m* pot plunger, transfer plunger
Preßstempel *m* force plug, molding plug, [press] die
Preßtechnik *f* molding technique
Preßteil *n* molded piece, molded article, pressed part
Preßtisch *m* table press
Preßtorf *m* pressed peat
Pressung *f* impression, pressing
Preßverfahren *n* molding technique
Preßvorgang *m* molding cycle
Preßwalze *f* pressure roll
Preßwasser *n* high-pressure water, (Zuck) pulp press water
Preßwerkzeug *n* molding tool, press tool
Preßzeit *f* clamping time (cold bonding), press time (hot bonding)
Preßzylinder *m* compression cylinder, press cylinder
Preußischblau *n* Prussian blue, Berlin blue, ferric ferrocyanide
Preußischbraun *n* Prussian brown
Primachin *n* primaquine
primär primary
Primärätzung *f* primary etching
Primäranregung *f* primary excitation
Primärelektron *n* primary electron
Primärelement *n* primary cell
Primärfarbe *f* elementary color, primary color
Primärformation *f* (Geol) primary period
Primärgefüge *n* primary structure
Primärimpuls *m* primary pulse
Primärkrebs *m* (Med) primary cancer
Primärkreis *m* (Elektr) primary circuit
Primärkristallisation *f* primary crystallization
Primärluft *f* primary air
Primärneutron *n* primary neutron
Primärreaktion *f* primary reaction
Primärspannung *f* (Elektr) primary voltage
Primärspektrum *n* (Spektr) primary spectrum
Primärspule *f* primary coil
Primärstation *f* generating station, supply station
Primärstrahlung *f* primary radiation
Primärstrom *m* (Elektr) inducing current, primary current
Primärstromkreis *m* primary circuit
Primärteilchen *n* primary particle
Primel *f* (Bot) primrose, cowslip flower
primitiv primitive
Primulin *n* primulin

Primulinbase f primulin base
Primulindisulfonsäure f primulindisulfonic acid
Primulinfarbstoff m primulin dye
Primulinrot n primulin red
Primverin n primverin
Primzahl f (Math) prime number
Prinzip n principle, ~ **der Gleichverteilung der Energie** principle of equipartition of energy, ~ **der kleinsten Wirkung** (Mech) least-energy principle, ~ **des beweglichen Gleichgewichtes** (von Van't Hoff) Van't Hoff's principle of mobile equilibrium, ~ **des geringsten Zwanges** (von Le Chatelier) principle of least action
prinzipiell on principle
Prinzipschaltung f skeleton diagram
Prinzipschnitt m diagrammatic section
Prinzmetall n prince metal
Prinz-Säure f Prinz acid
Prioritätsbeanspruchung f priority claim
Prioritätsrecht n (Jur) priority right
Prisma n (Opt) prism, **bildumkehrendes** ~ image-inverting prism, **geradsichtiges** ~ direct-vision prism
Prismatin m (Min) prismatine
prismatisch prismatic
Prismatoid n (Geom) prism[at]oid
Prismenfläche f prism surface
Prismenspektrum n prismatic spectrum
Privatmitteilung f private communication
pro analysi for analysis
Probarbital n probarbital
Probe f (Kontrolle) check, (Muster) pattern, sample, specimen, (Prüfung) examination, (Versuch) assay, experiment, test, testing, trial, ~ **durch Abtreiben** cupellation, ~ **ziehen** to sample, **eine** ~ **entnehmen** to take a sample
Probeabzug m (Buchdr) specimen print
Probeauswahl f selection of test pieces
Probebelastung f loading test, test load
Probebeschweißung f test weld
Probebetrieb m test run
Probedichte f density of specimen
Probedruck m test pressure
Probeentnahme f sampling
Probeflasche f specimen bottle
Probegewicht n standard weight, test weight
Probeglas n sample bottle
Probegut n sample [material], sampling material
Probelauf m test run, trial run
Probenahme f sampling, taking samples, testing, test portion
Probenehmer m sampler, sampling equipment
Probeneingabe f (Gaschromat) sampling
Probeneinlaßheizung f (Gaschromat) injection port heater
Probengeber m **für Gase** (Gaschromat) gas sampling valve
Probengebung f (Gaschromat) injection block
Probenteilung f sample splitting

Probenvorbereitung f sample preparation
Probepapier n test paper
Proberechnung f trial and error method, trial calculation
Probeschachtel f sample box
Probeseite f specimen page
Probesilber n standard silver
Probestab m test bar, test rod
Probestoff m sampling material, sample
Probestreifen m test strip
Probestück n sample, specimen, test piece
Probevorrichtung f testing arrangement, testing device
Probewaage f assay balance
probeweise on approval, tentative
Probezeit f testing time
Probeziehen n sampling
Probezinn n common tin, standard tin
Probierblei n assay lead
Probierbleisieb n test lead screen
probieren to try, to test, (Metall) to assay
Probiergewicht n (Metall) assay weight
Probiergold n standard gold
Probierhahn m gauge cock, test cock
Probiermetall n test metal
Problematik f problematic situation
Problemstellung f formulation of the problem
Procain n procaine
Procarboxypeptidase f (Biochem) procarboxypeptidase
Procellose f procellose
Prochlorit m (Min) prochlorite
Procionfarbstoff m procion dyestuff
Prodeconium n prodeconium
Produkt n product, **Abbau-** ~ degradation product, **biochemisches** ~ biochemical product, **calciniertes** ~ calcined product, **pflanzliches** ~ plant product, **Reaktions-** ~ reaction product, **Spalt-** ~ split product
Produktdarstellung f product representation
Produktion f production, generation, manufacture, output, yield, **Tages-** ~ daily output
Produktionsanlage f production unit
Produktionsdrehbank f production type lathe
Produktionserlaubnis f production permit
Produktionsfabrik f large-scale factory
Produktionsmöglichkeit f production potential
Produktionsprogramm n manufacturing activities
Produktionsreaktor m production reactor
produktiv productive
Produktivität f productivity
Produktkennzeichnung f marking on products
Produzent m manufacturer, producer
produzieren to produce
Proenzym n (Biochem) proenzyme

Profil n profile, contour, side projection, ~ **der Lauffläche** pattern, **gespritztes** ~ extruded shape
Profilalbum n profile album
Profilfaser f profile fiber
profilieren to profile
Profilkalander m profiling calender
Profilrisse pl channel cracking, groove cracking
Profilrohr n profile[d] tube
Profilrolle f disc roller
Profilskala f edgewise scale
Profilstab m extruded bar
Profilstahl m sectional [or structural] steel, sectional steel
Profilstreifen m bead filler, bead reinforcement
Profiltiefe f non-skid depth, non-skid tread thickness
Profilverjüngung f tapering-in section
Profilverschiebung f profile shift
Profilverschiebungsfaktor m correction for profile shift
Profilwalze f profile mill cutter
Profilwalzen n section rolling
Proflavin n (HN) proflavine, diaminoacridine
Progesteron n progesterone, progestin
Progestin n progesterone, progestin
Programm n program, plan, scheme
programmgesteuert program-controlled
Programmiereinrichtung f program[m]ing device
programmieren to program
Programmieren n programming
Programmierer m programmer
Programmiersprache f program[m]ing language
Programmiersystem n program[m]ing system
programmiert program[m]ed
Programmierung f (Math) programming, **lineare** ~ linear programming
Programmregelung f program control
Progression f (Math) progression
Proinsulin n proinsulin
Projekt n project, scheme
Projektabwicklung f project execution
projektieren to project, to plan, to scheme
Projektierung f projecting, planning, preliminary engineering
Projektion f projection
Projektionsablesung f projection reading scale
Projektionsapparat m projector, projecting apparatus
Projektionsebene f projection plane
Projektionsschirm m projection screen
projektiv projective
Projektivität f projectivity
projizieren to project
Prolaktin n prolactin
Prolamin n prolamine
Prolektit m prolectite
Proliferation f proliferation

Proliferationsphase f (Physiol) proliferation phase
Prolin n prolin[e], pyrrolidine carboxylic acid
Prolinase f (Biochem) prolinase
Prolindehydrogenase f (Biochem) proline dehydrogenase
Prolinol n prolinol
Promazin n promazine
Promethazin n promethazine
Promethium n promethium (Symb. Pm)
Promille n [part] per thousand
promovieren to attain a doctorate
Prontosil n prontosil
Propäsin n propaesin, propyl amino benzoate
Propan n propane
Propanal n propanal
Propandiol n dihydroxypropane
Propangas n propane gas
Propanol n propanol, propyl alcohol
Propargyl- propargyl
Propargylaldehyd m propargyl aldehyde
Propargylalkohol m propargyl alcohol, propinol
Propargylsäure f propargylic acid, propiolic acid, propynoic acid
propargylsauer propargylate, propargylic
Propellerachse f axis of a propeller
Propellermischer m propeller mixer
Propellerpumpe f propeller pump
Propellerrührer m propeller mixer
Propen n propene
Propenal n propenal, acrolein, acrylic aldehyde, allyl aldehyde, ethylene aldehyde
Propenol n propenol
Propenyl- propenyl
Propenylalkohol m propenyl alcohol
Propenyliden n propenylidene
Prophetin n prophetin
Prophylaxe f (Med) prophylaxis, prevention
Propin n propine
Propiolacton n propiolactone
Propiolaldehyd m propiolaldehyde, propargylaldehyde, propynal
Propiolsäure f propiolic acid, acetylenecarboxylic acid, carboxyacetylene, propargylic acid, propynoic acid
propiolsauer propargylate, propiolate
Propion n propion
Propionaldazin n propionaldazine
Propionaldehyd m propanal, propionaldehyde
Propionamid n propionamide, propanamide
Propionamidin n propionamidine
Propionat n propionate
Propionitril n propionitrile, ethylcyanide, propane nitrile
Propionpersäure f peroxypropionic acid, perpropionic acid
Propionsäure f propionic acid, propanoic acid, pseudacetic acid
Propionsäureanhydrid n propionic anhydride

Propionsäureester *m* propionate
propionsauer propionate, propionic
Propionyl- propionyl
Propionylchlorid *n* propionyl chloride, propanoyl chloride
Propionyl-CoA *n* propionyl-CoA
Propionylperoxid *n* propionyl peroxide
Propiophenon *n* propiophenone
Propolisharz *n* propolis resin
Proponal *n* proponal, di-iso-propylbarbituric acid
Proportion *f* proportion
proportional proportional, proportionate, **direkt** ~ directly proportional, **umgekehrt** ~ inversely proportional
Proportionalbereich *m* proportionality range, throttling range, (Regeltechn) proportional band
Proportionalität *f* proportionality
Proportionalitätsannahme *f* assumption of proportionality
Proportionalitätsfaktor *m* proportionality factor or coefficient
Proportionalzählrohr *n* proportional counter
Proportionieren *n* proportioning
Propyl- propyl
Propylacetat *n* propyl acetate
Propylacetylen *n* 1-pentyne, propylacetylene
Propyläther *m* propyl ether
Propylal *n* propylal
Propylalkohol *m* propyl alcohol
Propylamin *n* propylamine, aminopropane, propane amine
Propylbenzol *n* propylbenzene
Propylbromid *n* propyl bromide
Propylchlorid *n* propyl chloride
Propylen *n* propylene, propene
Propylenchlorhydrin *n* propylenechlorohydrin
Propylendiamin *n* propylenediamine
Propylenglykol *n* propylene glycol, propanediol
Propylenimin *n* propylenimine
Propylenoxid *n* propylene oxide, propene oxide
Propylensulfid *n* propylene sulfide, propene sulfide
Propylformiat *n* propyl formate
Propyliden *n* propylidene
Propyljodid *n* propyl iodide
Propylmercaptan *n* propyl mercaptan
Propyloxin *n* propyloxine
Propylsilicon *n* propylsilicone
Propylthiouracil *n* propylthiouracil
Propylwasserstoff *m* propane, propyl hydride
Prorennin *n* prorennin
Prosapogenin *n* prosapogenin
Prosiloxan *n* prosiloxane
Prospekt *m* prospectus, advertising folder, leaflet
Prospektionszähler *m* radioactive ore detector
Prostagladine *pl* prostagladins

Prostatadrüse *f* prostata gland
prosthetisch prosthetic
Protactinium *n* protactinium (Symb. Pa)
Protagon *n* protagon
Protamin *n* protamine
Protaminase *f* (Biochem) protaminase
Protargol *n* protargol
Proteacin *n* proteacin, leucodrin
Protease *f* (Biochem) protease
Proteid *n* proteid, protein
Protein *n* protein, **denaturiertes** ~ denatured protein, **fibrilläres** ~ fibrous protein, **globuläres** ~ globular protein, **konjugiertes** ~ conjugated protein, **konstitutives** ~ constitutive protein, **kristallines** ~ crystalline protein, **natives** ~ native protein, **oligomeres** ~ oligomeric protein, **polymeres** ~ polymeric protein
Proteinabbau *m* protein degradation
Proteinanalyse *f* protein analysis
Proteinase *f* (Biochem) proteinase
Proteindenaturierung *f* protein denaturation
Proteinfällung *f* protein precipitation
Proteinfraktionierung *f* protein fractionation
Proteingerüst *n* protein back-bone
proteinhaltig containing protein
Proteinhydrolyse *f* protein hydrolysis
Proteinkörper *m* protein, protein substance
Proteinkomplex *m* protein complex
Proteinkonformation *f* protein conformation
Proteinochromogen *n* proteinochromogen
Proteinreinigung *f* protein purification
Proteinrenaturierung *f* protein renaturation
Proteinspiegel *m* (Med) protein level
Proteinstruktur *f* protein structure
Proteinsynthese *f* protein synthesis
Proteintrennung *f* protein separation
Proteinumsatz *m* protein turnover
Proteinuntereinheit *f* protein subunit
Proteinverbindung *f* protein compound
Proteinzusammensetzung *f* protein composition
Proteohormon *n* proteohormone
Proteolyse *f* proteolysis
Proteolyt *m* proteolyte
proteolytisch proteolytic
Prothrombin *n* prothrombin
Prothrombokinase *f* (Biochem) prothrombokinase
Protium *n* protium
Protoanemonin *n* protoanemonin
Protobastit *m* (Min) protobastite
Protoberberin *n* protoberberine
Protocatechualdehyd *m* protocatechualdehyde, 3,4-dihydroxybenzaldehyde
Protocatechusäure *f* protocatechuic acid, 3,4-dihydroxybenzoic acid
Protocetrarsäure *f* protocetraric acid
Protocrocin *n* protocrocin
Protoechinulinsäure *f* protoechinulinic acid

Protoglucal *n* protoglucal
Protohäm *n* protohem
Protoklas *m* protoclase
Protokoll *n* record[s]
protokollieren to register
Proton *n* proton, **schnelles** ~ high-speed proton, **thermisches** ~ thermal proton
Protonelektronmassenverhältnis *n* proton-electron mass ratio
Protonenabstoß *m* proton repulsion
Protonenaffinität *f* proton affinity
Protonenakzeptor *m* proton acceptor
Protonenaustausch *m* proton exchange
Protonenbahn *f* proton path, (Spur) proton track
Protonenbeschießung *f* proton bombardment
Protonenbeschleuniger *m* proton accelerator
Protonenbeschleunigung *f* **nach Cockcraft und Walton** Cockcraft-Walton proton acceleration
Protonendonator *m* proton donor
Protonengeschoß *n* proton bullet
Protonenmasse *f* proton mass
Protonenpräzessionsfrequenz *f* proton precession frequency
Protonenreichweite *f* proton range
Protonenspektrum *n* proton spectrum
Protonenspin *m* proton spin
Protonensprung *m* proton jump
Protonenspur *f* proton track
Protonensynchrotron *n* proton synchrotron
Protonenüberschuß *m* proton excess
Protonenübertragung *f* proton transfer
Protonenzahl *f* proton number, (Atom) number of protons
Protonierung *f* protonation
Protopapaverin *n* protopapaverine
protophil (Lösungsmittel) protophilic
Protopin *n* protopine, fumarine, macleyine
Protoplasma *n* protoplasm, protoplasma
Protoplast *m* protoplast
Protoporphyrin *n* protoporphyrin
Protoquercit *m* protoquercitol
Prototropie *f* prototropy
Prototyp *m* prototype
Protoveratrin *n* protoveratrine
Protovermiculit *m* (Min) protovermiculite
Protoxid *n* protoxide
Protozoon *n* (pl Protozoen) protozoon (pl protozoa)
Proustit *m* (Min) proustite
provisorisch make-shift, emergency, temporary
Provitamin *n* provitamin, vitamin precursor
Provitamin A *n* provitamin A, β-carotine
Proximitätsregel *f* proximity rule
Prozedur *f* procedure
Prozent *n* per cent
Prozentgehalt *m* percentage, per cent content
prozentig per cent
Prozentsatz *m* percentage

Prozeß *m* operation, process, (Jur) law suit, **adiabatischer** ~ adiabatic process, **biologischer** ~ biological process, **chemischer** ~ chemical process, **gesteuerter** ~ controlled process, **kontinuierlicher** ~ continuous process
Prozeßanalyse *f* process analysis
Prozeßanalysentechnik *f* on-stream analysis
Prozeßentwicklung *f* process engineering
Prozeßoptimierung *f* process optimization
Prozeßrechner *m* process control computer
Prüfabzug *m* (Buchdr) proof
Prüfanstalt *f* testing laboratory
Prüfapparat *m* tester
prüfbar testable
Prüfbericht *m* test report
Prüfbestimmung *f* test specification
Prüfdaten *n pl* test data
Prüfdraht *m* test wire
Prüfdruck *m* testing pressure
prüfen (ausprobieren) to test, (bestätigen) to confirm, to verify, (Chem) to analyze, to assay, (erforschen) to investigate, (nachprüfen) to check, to inspect, (untersuchen) to examine
Prüfer *m* tester, assayer, examiner
Prüfergebnis *n* test result
Prüfgerät *n* analyzer, tester, testing equipment
Prüflaboratorium *n* testing laboratory
Prüfling *m* examinee, (Objekt) sample, test piece
Prüfmaß *n* check gauge, standard gauge
Prüfmethode *f* method of determination
Prüfpendel *n* testing pendulum
Prüfschein *m* test certificate
Prüfspannung *f* (Elektr) experimental voltage, testing voltage, (Mech) proof stress
Prüfstand *m* exposure rack, rig, test[ing] bench, test[ing] stand
Prüfstellung *f* testing position
Prüfstrecke *f* inspection table
Prüfstück *n* object submitted for testing
Prüftransformator *m* testing transformer
Prüfung *f* test[ing], checking, examination
Prüfungsanstalt *f* testing laboratory
Prüfungsart *f* testing method
Prüfungsbericht *m* test report
Prüfungsergebnis *n* result of experiment, test result
Prüfungsmittel *n* testing agent
Prüfungsverfahren *n* examination procedure
Prüfungsvorschriften *pl* directions for testing, testing rules, test specification[s]
Prüfungszeugnis *n* test certificate
Prüfventil *n* check valve
Prüfverfahren *n* test[ing] method, testing process, (Jur) prosecution
Prüfvorschriften *pl* test specifications
Prüfwesen *n* testing
Prulaurasin *n* prulaurasin

Prunasin *n* prunasin
Prunellensalz *n* molten potassium nitrate, prunella salt, sal prunella
Prunetol *n* prunetol, genistein
Prunin *n* prunin
Prussiat *n* prussiate, **gelbes** ~ potassium cyanoferrate(II), potassium ferrocyanide, **rotes** ~ potassium cyanoferrate(III), potassium ferricyanide
PS (Pferdestärke) horsepower, HP
Psammit *m* (Geol) psammite
Pseudoaconitin *n* pseudaconitine
Pseudoadenosin *n* pseudoadenosine
Pseudoalkali *n* pseudo alkali
Pseudoamethyst *m* (Min) pseudoamethyst
Pseudoapatit *m* (Min) pseudoapatite
Pseudoaspidin *n* pseudoaspidin
pseudo-asymmetrisch pseudo-asymmetric
Pseudobase *f* pseudo-base
Pseudobrookit *m* (Min) pseudobrookite
Pseudobutylalkohol *m* pseudobutyl alcohol, tertiary butyl alcohol
Pseudobutylchlorid *n* pseudobutyl chloride, tertiary butyl chloride
Pseudocholesterin *n* pseudocholesterol
Pseudochrysolith *m* (Min) pseudochrysolite
Pseudoconhydrin *n* pseudoconhydrine
Pseudocumenol *n* pseudocumenol
Pseudocumidin *n* pseudocumidine, 2,4,5-trimethylaniline
Pseudocumol *n* 1,2,4-trimethylbenzene, pseudocumene
Pseudocumolsulfonsäure *f* pseudocumenesulfonic acid
Pseudocumyl *n* pseudocumyl
Pseudocyanin *n* pseudocyanine
Pseudodiazoessigsäure *f* pseudodiazoacetic acid
Pseudodiosgenin *n* pseudodiosgenin
pseudoeuklidisch pseudoeuclidian
Pseudogaylussit *m* (Min) pseudogaylussite
Pseudogerbstoff *m* pseudo-tannin
Pseudoglucal *n* pseudoglucal
Pseudogranat *m* (Min) pseudogarnet
Pseudohalogen *n* pseudo halide, pseudohalogen
Pseudoharnsäure *f* pseudouric acid, 5-carbamidobarbituric acid
Pseudoharnstoff *m* isourea, pseudourea
pseudohexagonal (Krist) pseudohexagonal
Pseudojonon *n* pseudoionone
Pseudokatalysator *m* pseudocatalyzer
Pseudokatalyse *f* pseudocatalysis
pseudokatalytisch pseudocatalytic
Pseudokristall *m* (Krist) pseudo-crystal, pseudo-crystallite
pseudokristallin pseudocrystalline
Pseudoleucin *n* pseudoleucine
Pseudolibethenit *m* (Min) pseudolibethenite
Pseudolimonen *n* pseudolimonene
Pseudomalachit *m* (Min) pseudomalachite
Pseudomerie *f* pseudomery
Pseudomonasbakterium *n* pseudomonas
pseudomonotrop pseudomonotropic
Pseudomonotropie *f* pseudomonotropy
pseudomorph pseudomorphous
Pseudomorphie *f* pseudomorphy
Pseudomorphose *f* pseudomorphosis
Pseudonortropin *n* norpseudotropine
Pseudopelletierin *n* pseudopelletierine
Pseudophit *m* pseudophite
Pseudopinen *n* pseudopinene
pseudoplastisch pseudoplastic
Pseudopurpurin *n* pseudopurpurin
pseudoracemisch pseudoracemic
Pseudosaccharin *n* pseudosaccharin
Pseudosäure *f* pseudo acid
Pseudosmaragd *m* (Min) pseudoemerald
Pseudosphäre *f* pseudosphere
Pseudostrychnin *n* pseudostrychnine
Pseudosulfocyanwasserstoff *m* pseudothiocyanic acid
Pseudosymmetrie *f* pseudosymmetry
Pseudotensor *m* pseudotensor
pseudoternär pseudoternary
pseudotetragonal (Krist) pseudotetragonal
Pseudotropin *n* pseudotropine
Pseudoxanthin *n* pseudoxanthine
Psicain *n* psicaine
Psicose *f* psicose
Psilomelan *m* (Min) psilomelane, black iron ore
Psittazinit *m* (Min) psittacinite
Psoralen *n* psoralen
Psoralsäure *f* psoralic acid
Psoromsäure *f* psoromic acid
Psychopharmaka *pl* psychopharmacologic agents
Psychosedativum *n* (Pharm) tranquillizer
Psychotrin *n* psychotrine
Psychrometer *n* psychrometer
psychrometrisch psychrometric
Psyllaalkohol *m* psylla alcohol
Psyllasäure *f* psylla acid
Pteridin *n* pteridine
Pterin *n* pterin
Pterocarpin *n* pterocarpine
Pteroinsäure *f* pteroic acid
Pterolith *m* pterolite
Pteropterin *n* pteropterin
Pterostilben *n* pterostilbene
Pteroylglutaminsäure *f* pteroylglutamic acid
Ptilolith *m* (Min) ptilolite
Ptomain *n* ptomain[e]
Ptyalin *n* (Biochem) ptyaline, ptyalase, salivin
Ptychotisöl *n* ptychotis oil
Puberulonsäure *f* puberulonic acid
Puberulsäure *f* puberulic acid
Publikation *f* publication
Pucherit *m* (Min) pucherite

Puddeldrehofen *m* rotary puddling furnace or machine
Puddeleisen *n* puddled iron, puddled steel
Puddelherd *m* puddling hearth, puddling basin
Puddelhütte *f* puddling works
Puddelluppe *f* puddle ball
puddeln (Met) to puddle
Puddeln *n* puddling, ~ **auf Korn** puddling for crystalline iron
Puddelofen *m* puddling furnace
Puddelprozeß *m* puddling process
Puddelroheisen *n* pig iron for puddling, forge pig iron, mill iron
Puddelschlacke *f* puddling slag, bulldog metal, puddle cinder, tap cinder, tappings
Puddelspiegel *m* specular forge pig iron
Puddelstahl *m* puddled steel, puddled iron
Puddelverfahren *n* puddling process
Puddelwerk *n* puddling works
Puddingstein *m* pudding stone, conglomerate
Puddler *m* puddler
Puder *m* powder
Pudergold *n* dusting gold
puderig powdery
pudern to powder
Puderplatin *n* dusting platinum
Pudersilber *n* dusting silver
Puderzucker *m* confectioner's sugar, icing sugar, powdered sugar
Pülpe *f* pulp
Pünktchenbildung *f* (Farben) bittiness
Puffer *m* shock absorber, (Chem) buffer, (Stoßdämpfer) bumper
Pufferbatterie *f* buffer battery
Pufferfeder *f* buffer spring
Puffergemisch *n* buffer reagent
Pufferkapazität *f* buffer capacity
Pufferkontakt *m* buffer contact
Pufferlösung *f* buffer solution
puffern to buffer
Puffersäure *f* buffer acid
Pufferspannung *f* buffer voltage
Puffersystem *n* buffer system
Pufferung *f* buffering
Pufferwirkung *f* buffer effect, buffer action, buffering action
Pufierit *m* pufierite
Pukatein *n* pukateine
Pulegan *n* pulegane
Pulegon *n* pulegone
Pulpe *f* (Pülpe) pulp, wood pulp
Pulsation *f* pulsation
Pulsationsbett *n* pulsed bed
Pulsator *m* pulsator
pulsieren to pulsate
Pulsieren *n* pulsation
pulsierend pulsating
Pulsierkolonne *f* pulsating column
Pulsometer *n* pulsometer

Pult *n* desk
Pultdach *n* hipped roof
Pulver *n* powder, ~ **zum Räuchern** fumigating powder, **brisantes** ~ explosive powder, **fein gemahlenes** ~ finely ground powder, **feinkörniges** ~ fine-grained powder, small-grained powder, **grob gemahlenes** ~ coarsely ground powder, **großkörniges** ~ coarse-grain powder, **rauchloses** ~ smokeless powder
pulverartig powdery
Pulverbeschichtung *f* powder coating
Pulverdampf *m* gunpowder smoke
Pulverdiagramm *n* (Krist) powder pattern
Pulvererz *n* gunpowder ore
Pulverexplosion *f* powder explosion
Pulverfabrikation *f* gunpowder manufacture
Pulverfaß *n* powder barrel, powder keg
Pulverflasche *f* reagent bottle
pulverförmig powdery, pulverulent
Pulverglättfaß *n* polishing cask
Pulverglas *n* wide-mouthed bottle
Pulverglühanode *f* powder anode
Pulverholz *n* wood for gun powder
pulverig powdery, dust-like, dusty, pulverulent
Pulverisieranlage *f* pulverizing equipment
pulverisierbar pulverable, pulverizable
pulverisieren to pulverize, to grind, to powder
Pulverisieren *n* pulverizing, powdering
Pulverisiermaschine *f* pulverizing machine
pulverisiert powdered, pulverized
Pulverkohle *f* activated charcoal in powder form, powdered coal
Pulverkorn *n* grain of powder
Pulverkuchen *m* gun powder press cake
Pulverladung *f* powder charge
Pulverlöscher *m* (Feuerlöscher) powder fire extinguisher
Pulvermetallurgie *f* powder metallurgy
Pulvermischer *m* powder blender
Pulvermörser *m* mortar for powdering, chemist's mortar, powder mortar
pulvern to powder, to pulverize
Pulverpreßkuchen *m* (Gieß) [powder] press cake
Pulverpreßling *m* compressed-powder charge
Pulverprobe *f* powder test
Pulverprobiervorrichtung *f* powder testing device
Pulverschwamm *m* powdered tinder
Pulversintern *n* powder sintering
Pulvertonne *f* powder barrel, powder cask, powder keg
pulvertrocken dry as powder, powder dry
Pulververfahren *n* (Krist) powder method of analysis
Pulvinsäure *f* pulvic acid
Pumpanlage *f* pumping set, pumping station
Pumpe *f* pump, **Adsorptions-** ~ adsorption pump, **Axial-** ~ axial-flow pump, **Dampfstrahl-** ~ ejector pump, **Diffusions-** ~

diffusion pump, **Dosierkolben-** ~ reciprocation proportioning pump, **Exzenterschnecken-** ~ eccentric single-rotor screw pump, **Flüssigkeitsring-** ~ liquid ring [or seal] pump, **Freistrom-** ~ non clogging pump, **Gasballast-** ~ gas ballast pump, **Höchstdruck-** ~ pump for extreme pressures, **Ionenzerstäuber-** ~ ionic atomization pump, **Kolben-** ~ reciprocating pump, **Kondensations-** ~ condensation pump, **Kreisel-** ~ centrifugal pump, **Kreiskolben-** ~ rotating piston pump, rotary pump, **Kryo-** ~ cryopump, **Mammut-** ~ air-lift pump, **Membran-** ~ diaphragm pump, **Öldampfstrahl-** ~ oil ejector pump, **Propeller-** ~ propeller pump, **Quecksilberdampfstrahl-** ~ mercury ejector pump, **Scheibenkolben-** ~ solid-piston pump, **Schlauch-** ~ peristaltic pump, **Schnecken-** ~ screw pump, **Schraubenspindel-** ~ screw displacement pump, **Seitenkanal-** ~ side-channel pump, **Sperrschieber-** ~ single-lobe pump, **Spindel-** ~ single-rotor screw pump, **Strahl-** ~ jet pump, **Tauchkolben-** ~ submerged-piston pump, **Treibmittel-** ~ vapor pump, **Trennflügel-** ~ single-lobe pump, **Trochoiden-** ~ trochoidal pump, **Turbomolekular-** ~ turbomolecular pump, **Vakuum-** ~ vacuum pump, **Verdränger-** ~ positive displacement pump, **Wasserdampfstrahl-** ~ steam-jet ejector, **Wasserring-** ~ liquid seal pump, **Wasserstrahl-** ~ aspirator
pumpen to pump, (evakuieren) to evacuate
Pumpenanlage f pumping plant
Pumpenbauart f type of pump
Pumpenflügel m pump impeller
Pumpengelenk n pump coupling
Pumpengestänge n pump rods
Pumpenhaus n pump house
Pumpenkammer f pump room
Pumpenkörper m pump body
Pumpenkolben m pump piston, pump plunger
Pumpenstange f sucker rod
Pumpensteiger m venting channel
Pumpenstock m pricker, stirring rod
Pumpenventil n pump valve, clack valve
Pumprad n pump wheel
Pumpstand m pump stand
Punicin n punicine, pelletierine
Punkt m point, (Einzelheit) item, (Fleck) dot, (Stelle) spot, **durch denselben** ~ **gehend** concurrent, conpunctual, **isoelektrischer** ~ isoelectric point, **kritischer** ~ critical point, **toter** ~ dead center, dead point
Punktanguß m pin-point gate
Punktanordnung f arrangement of points
Punktberührung f point contact
Punktdipol n point dipole

Punktdrucker m multi-point recorder
Punktentladung f point discharge
Punktfehler m point defect
Punktfehlordnung f point defect
Punktflächentransformation f point surface transformation
punktförmig point-shaped, punctiform, punctual
punktgeschweißt spot-welded
Punktgitter n point lattice, **ebenes** ~ point lattice in the plane, **räumliches** ~ point lattice in space
Punktgruppe f point group
punktiert dotted
Punktkontakt m point contact
Punktkurve f dotted curve
Punktladung f point charge
Punktlage f point position
Punktleimverfahren n spot glueing
Punktlichtquelle f point source of light
punktschweißen to spot-weld
Punktschweißen n spot welding
Punktschweißmaschine f spot welder, spot welding machine
Punktschweißung f spot welding
Punktschweißverfahren n spot welding
Punkttransformation f point transformation
Punktwirkungsgrad m (Rektifikation) point efficiency
Puppe f (Extrusion) billet for extruding, (Gummi) roll of uncured rubber, (Insekt) chrysalis
PUR Abk. für Polyurethan
Purgierkraut n hedge hyssop, purgative herb
Purgiersalz n purgative salt
Purifikation f purification
Purin n purin[e]
Purinbase f purine base
Purinbiosynthese f purine biosynthesis
Purinderivat n purine derivative
Purinnukleotid n purin nucleotide
Purinoesäure f purinoic acid
Purinphosphoribosyltransferase f (Biochem) purine phosphoribosyl transferase
purissimum (Lat) purest
Puromycin n puromycin
Purpur m purple
Purpurat n purpurate
Purpuratindikator m purpurate indicator
Purpurcarmin n purple carmine, ammonium purpurate, murexide
Purpureaglykosid n purpurea glycoside
Purpurerz n purple ore
Purpurfärbung f purple coloring
Purpurfarbe f purple
purpurfarben purple-colored
Purpurholz n purple wood
Purpurin n purpurine
Purpurit m (Min) purpurite
Purpurlack m madder purple, purple lake

purpurn purple
Purpurogallin *n* purpurogallin
Purpurogenon *n* purpurogenone
Purpuroxanthen *n* purpuroxanthene, xanthopurpurin
purpurrot purple-colored
Purpurrot *n* purple red
Purpursäure *f* purpuric acid
Purpurschiefer *m* (Min) purplish schist
Purpurschwefelsäure *f* sulfopurpuric acid
Purpurviolett *n* purple violet
Purpurviolettfarbe *f* purple violet color
Puschkinit *m* puschkinite
Putrescin *n* 1,4-butanediamine, putrescine, butylenediamine, tetramethylenediamine
Putz *m* (Bauw) plaster, rough-casting
Putzbaumwolle *f* cotton waste, waste cotton
Putzbrettchen *n* clearer board
putzen to clean, to cleanse, to dress, (durch Sandstrahlen) to sandblast, (Häute) to short-hair, (polieren) to polish
Putzen *n* **des Blockes** cleaning the block
Putzer *m* cleaner
Putzkalk *m* stucco
Putzlappen *m* cleaning rag, scouring cloth
Putzleder *n* chamois leather
Putzmaschine *f* cleansing machine
Putzmittel *n* cleaning or polishing powder, cleansing material, detergent, polish, scouring material
Putzöffnung *f* cleaning hole
Putzöl *n* dry cleaning oil, polishing oil
Putzpulver *n* polishing powder
Putzwerg *n* cleaning waste
Putzwerkzeug *n* cleansing tool
Putzwolle *f* cleaning wool, polishing wool, waste wool
Puzzolanerde *f* Puzzolan earth
PVAC Abk. für Polyvinylacetat
PVAL Abk. für Polyvinylalkohol
PVC Abk. für Polyvinylchlorid
PVP Abk. für Polyvinylpyrrolidon
Pyknit *m* (Min) pycnite
Pyknochlorit *m* (Min) pycnochlorite
Pyknometer *n* pycnometer, specific gravity bottle
pyknometrisch pycnometric
Pyknotrop *m* (Min) pycnotrop
Pyocyanin *n* pyocyanin
Pyoktanin *n* pyoctanine, pyoktanin
Pyolipinsäure *f* pyolipic acid
Pyracen *n* pyracene
Pyraconitin *n* pyraconitine
Pyracridon *n* pyracridone
Pyracyclen *n* pyracyclene
Pyrallolith *m* pyrallolite
Pyramide *f* pyramid, **abgestumpfte** ~ truncated pyramid
pyramidenförmig pyramidal

Pyramidenoktaeder *n* pyramidal octahedron
Pyramidenspat *m* (Min) pyramidal carbonate of lime
Pyramidentetraeder *n* trigonal dodecahedron
Pyramidon *n* pyramidon, amidopyrine, dimethylaminophenyl-dimethyl pyrazolone
Pyran *n* pyran
Pyranen *n* pyranene
Pyranose *f* pyranose
Pyranosid *n* pyranoside
Pyranthren *n* pyranthrene
Pyranthridin *n* pyranthridine
Pyranthridon *n* pyranthridone
Pyranthron *n* pyranthrone
Pyrantin *n* pyrantin, phenosuccin
Pyranyl- pyranyl
Pyrargillit *m* pyrargillite
Pyrargyrit *m* (Min) pyrargyrite
Pyrazin *n* 1,4-diazine, pyrazine
Pyrazinsäure *f* pyrazinoic acid
Pyrazol *n* pyrazole
Pyrazolblau *n* pyrazole blue
Pyrazolidin *n* pyrazolidine
Pyrazolidon *n* pyrazolidone
Pyrazolin *n* pyrazoline, dihydropyrazole
Pyrazolon *n* pyrazolone
Pyrazolonfarbstoff *m* pyrazolone dye
Pyrazolsäure *f* pyrazolic acid
Pyrelen *n* pyrelene
Pyren *n* pyrene
Pyrenchinon *n* pyrenequinone
Pyrethrin *n* pyrethrin
Pyrethrinsäure *f* pyrethric acid
Pyrethrol *n* pyrethrol
Pyrexglas *n* (HN) pyrex [glass]
Pyrgom *m* pyrgom
Pyrheliometer *n* pyrheliometer
Pyridazin *n* 1,2-diazine, pyridazine
Pyridazon *n* pyridazone
Pyridil *n* pyridil
Pyridilsäure *f* pyridilic acid
Pyridin *n* pyridinp[ep]
Pyridincarbonsäureamid *n* pyridine carboxylic acid amide
Pyridinderivat *n* pyridine derivative
Pyridiniumhydroxid *n* pyridinium hydroxide
Pyridinkern *m* pyridine nucleus, pyridine ring
Pyridinmesoporphyrin *n* pyridine mesoporphyrin
Pyridinoxid *n* pyridine oxide
Pyridinrot *n* pyridine red
Pyridinsulfonsäure *f* pyridine sulfonic acid
Pyridofluoren *n* pyridofluorene
Pyridon *n* pyridone
Pyridophthalan *n* pyridophthalane
Pyridophthalid *n* pyridophthalide
Pyridostilben *n* pyridostilbene
Pyridoxal *n* pyridoxal
Pyridoxalphosphat *n* pyridoxal phosphate

Pyridoxamin *n* pyridoxamin
Pyridoxin *n* pyridoxin[e], vitamin B₆
Pyridyl- pyridyl
Pyridylazonaphthol *n* pyridylazonaphthol
Pyriliumsalz *n* pyrilium salt
Pyrimidin *n* pyrimidine
Pyrimidinbase *f* pyrimidine base
Pyrimidinring *m* pyrimidine ring
Pyrimidon *n* pyrimidone
Pyrimidyl- pyrimidyl
Pyrindan *n* pyrindan
Pyrindol *n* pyrindol
Pyrit *m* (Min) iron pyrites
Pyritabbrand *m* pyrites cinder
pyritartig pyritic
pyrithaltig pyritiferous
Pyrithiamin *n* pyrithiamine
Pyritofen *m* pyrites burner, pyrite[s] furnace
Pyritoid *n* pyritoid
Pyro *n* pyro, collodion cotton, pyroxylin, soluble gun cotton
Pyroabietinsäure *f* pyroabietic acid
Pyroalizarinsäure *f* pyroalizaric acid, phthalic anhydride
Pyroantimonat *n* pyroantimonate
Pyroantimonsäure *f* pyroantimonic acid
Pyroarsensäure *f* pyroarsenic acid
Pyroaurit *m* (Min) pyroaurite
Pyrobelonit *m* (Min) pyrobelonite
Pyrocatechin *n* 1,2-benzenediol, 1,2-dihydroxybenzene, pyrocatechol, pyrocatechin
Pyrochinin *n* pyroquinine
Pyrochlor *m* (Min) pyrochlore
Pyrochlormikrolit *m* pyrochloremicrolite
Pyrochroit *m* (Min) pyrochroite
Pyrochromat *n* pyrochromate, dichromate
Pyrocinchonsäure *f* pyrocinchonic acid
Pyrocoll *n* pyrocoll
Pyrodin *n* pyrodin, acetylphenylhydrazine
pyroelektrisch pyroelectric
Pyroelektrizität *f* pyroelectricity
Pyroentwickler *m* pyro-developer
Pyrogallol *n* 1,2,3-benzenetriol, 1,2,3-trihydroxybenzene, pyrogallol, pyrogallic acid
Pyrogallolphthalein *n* pyrogallol phthalein, gallein
Pyrogalloltriacetat *n* pyrogalloltriacetate
Pyrogallolwismut *n* bismuth pyrogallate
Pyrogallussäure *f* 1,2,3-benzenetriol, 1,2,3-trihydroxybenzene, pyrogallic acid, pyrogallol
pyrogallussauer pyrogallate, pyrogallic
Pyrogen *n* pyrogen
Pyrogenfarbstoff *m* pyrogen dye
pyrogenfrei pyrogen-free
Pyroglutaminsäure *f* pyroglutamic acid

Pyrokatechin *n* 1,2-benzenediol, 1,2-dihydroxybenzene, pyrocatechol, pyrocatechin
Pyroklasit *m* pyroclasite
Pyrokohlensäure *f* pyrocarbonic acid
Pyrokoll *n* pyrocoll
Pyrokoman *n* 1,4-pyrone, pyrocomane
Pyrolignit *m* pyrolignite
Pyrolusit *m* (Min) pyrolusite, polianite
Pyrolyse *f* pyrolysis
pyrolysieren to pyrolyze
pyromagnetisch pyromagnetic
Pyromekonsäure *f* pyromeconic acid, pyrocomenic acid
Pyromellitsäure *f* pyromellitic acid
Pyrometallurgie *f* pyrometallurgy
Pyrometer *n* pyrometer
Pyrometerausgleichsleitung *f* pyrometer compensating circuit
pyromorph pyromorphous
Pyromorphit *m* (Min) pyromorphite
Pyromucat *n* pyromucate, salt or ester of pyromucic acid
Pyromuconsäure *f* pyromucic acid, furan-2-carboxylic acid, furoic acid
Pyromucyl *n* pyromucyl, 2-furoyl
Pyron *n* pyrone
Pyronaphtha *n* pyronaphtha
Pyronin *n* pyronine
Pyroninfarbstoff *m* pyronine dye
Pyronon *n* pyronone
Pyrop *m* (Min) pyrope
Pyrophäophorbid *n* pyropheophorbide
Pyrophanit *m* (Min) pyrophanite
pyrophor[isch] pyrophoric
Pyrophosphat *n* diphosphate, pyrophosphate
Pyrophosphatase *f* (Biochem) pyrophosphatase
pyrophosphorige Säure *f* pyrophosphorous acid
Pyrophosphorsäure *f* pyrophosphoric acid
pyrophosphorsauer pyrophosphate
Pyrophotographie *f* pyrophotography
Pyrophyllit *m* (Min) pyrophyllite
Pyrophysalit *m* (Min) pyrophysalite
Pyropissit *m* pyropissite
Pyrorthit *m* pyrorthite
Pyrosäure *f* pyro acid
Pyroschleimsäure *f* (Brenzschleimsäure) pyromucic acid, furan-2-carboxylic acid
pyroschleimsauer pyromucate, pyromucic
pyroschwefelig pyrosulfurous, pyrosulfite
Pyroschwefelsäure *f* pyrosulfuric acid, disulfuric acid
Pyrosklerit *m* pyrosclerite
Pyroskop *n* pyroscope
Pyrosmalit *m* (Min) pyrosmalite
Pyrosmaragd *m* (Min) chlorophane
Pyrostilbit *m* (Min) pyrostilbite, kermesite
Pyrostilpnit *m* (Min) pyrostilpnite
Pyrosulfat *n* pyrosulfate

Pyrosulfit *n* pyrosulfite
Pyrosulfurylchlorid *n* pyrosulfuryl chloride
Pyrotartrat *n* pyrotartrate
Pyrotechnik *f* pyrotechnics
pyrotechnisch pyrotechnical
Pyrotraubensäure *f* (Brenztraubensäure) pyruvic acid, pyroracemic acid
Pyrotritarsäure *f* pyrotritaric acid, uvic acid
Pyroweinsäure *f* pyrotartaric acid
Pyroxen *m* (Min) pyroxene
pyroxenartig pyroxenic
pyroxenhaltig pyroxeniferous
Pyroxoniumsalz *n* pyroxonium salt
Pyroxylin *n* (Schießbaumwolle) pyroxylin, collodion cotton, soluble gun cotton
Pyroxylinfaden *m* pyroxylin filament, cellulose nitrate filament
Pyroxylinlack *m* pyroxylin varnish
Pyrrhoarsenit *m* (Min) pyrrhoarsenite
Pyrrhosiderit *m* (Min) pyrrhosiderite
Pyrril *n* pyrril
Pyrrocolin *n* pyrrocoline
Pyrrodiazol *n* pyrrodiazole
Pyrrol *n* pyrrol[e], azole
Pyrrolblau *n* pyrrol[e] blue
Pyrrolidin *n* pyrrolidine, tetrahydropyrrol[e]
Pyrrolidon *n* pyrrolidone
Pyrrolin *n* pyrroline, dihydropyrrol[e]
Pyrrolincarbonsäure *f* pyrroline carboxylic acid
Pyrrolizidin *n* pyrrolizidine
Pyrrolon *n* pyrrolone
Pyrromethen *n* pyrromethene
Pyrroporphyrin *n* pyrroporphyrin
Pyrroyl- pyrroyl
Pyrryl- pyrryl
Pyrthiophanthron *n* pyrthiophanthrone
Pyruvat *n* pyruvate
Pyruvatdecarboxylase *f* (Biochem) pyruvate decarboxylase
Pyruvatdehydrogenase *f* (Biochem) pyruvate dehydrogenase
Pyruvatdehydrogenasekomplex *m* (Biochem) pyruvate dehydrogenase complex
Pyruvatkinase *f* (Biochem) pyruvate kinase
Pyruvatoxydase *f* (Biochem) pyruvate oxidase
Pyruvyl- pyruvyl
Pyruvylalanin *n* pyruvylalanine
Pyruvylglycin *n* pyruvylglycine
Pyrylen *n* pyrylene
pythagoreisch Pythagorean
pythagoreischer Lehrsatz *m* (Math) Pythagorean theorem
Pyvuril *n* pyvuril
p-Zustand *m* p-state

Q

Q-Faktor *m* magnification factor
Quader *m* block of stone, freestone
Quaderformation *f* chalk formation
Quadersandstein *m* sandstone [in blocks]
Quadrangel *n* (Viereck) quadrangle, square
quadrangulär quadrangular, square
Quadrant *m* quadrant, quarter of a circle
Quadranteisen *n* quadrant iron
Quadrantelektrometer *n* quadrant electrometer
Quadrat *n* square
Quadratausgleich *m* square adjustment
Quadratdezimeter *n* square decimeter
Quadrate *pl*, **Methode der kleinsten** ~ method of least squares
Quadrateisen *n* square [bar] iron
quadratförmig quadratic, square
quadratisch quadratic, square, ~ **abnehmen** to decrease quadratically
Quadratkeil *m* square key
Quadratkilometer *n* square kilometer
Quadratmeter *n* square meter
Quadratmetergewicht *n* (Papier) grams per square meter, substance
Quadratmillimeter *n* square millimeter
Quadratsäure *f* squaric acid
Quadratsumme *f* sum of squares
Quadratur *f* quadrature
Quadratwurzel *f* (Math) square root, ~ **ziehen** to extract the square root
Quadratzahl *f* square [number]
Quadratzentimeter *n* square centimeter
Quadricyclen *n* quadricyclene
quadrieren to raise to the second power, to square
quadrilateral four-sided, quadrilateral
Quadrinom *n* expression containing four terms
Quadroxid *n* (obs) quadroxide, tetroxide
Quadrupelintensität *f* quadrupole intensity
Quadrupelkopplungsterm *m* quadrupole coupling term
Quadrupelmoment *n* (Elektr, Mag) quadrupole moment
Quadrupelpunkt *m* quadruple point
Quadrupelschwingung *f* quadrupole vibration
Quadrupelstrahlung *f* quadrupole radiation
Quadrupelübergang *m* quadrupole transition
Quadrupelverzerrung *f* quadrupole distortion
Quadrupelwechselwirkung *f* quadrupole interaction
Quadrupolkraft *f* quadrupole force
Qualimeter *n* qualimeter
Qualität *f* quality, grade
Qualitätsabweichung *f* deviation in quality
Qualitätseisen *n* refined iron, special iron
Qualitätserhaltung *f* quality preservation, quality retention

Qualitätsgrenze *f*, **annehmbare** ~ acceptable quality level
Qualitätskoeffizient *m* coefficient of quality
Qualitätskontrolle *f* quality control
Qualitätsleder *n* high-class leather
Qualitätsmaß *n* quality constant
Qualitätsprüfung *f* quality test
Qualitätsstahl *m* high-grade steel, special steel
Qualitätsüberwachung *f* quality control
Qualitätsunterschied *m* difference in quality
Qualitätsverbesserung *f* improvement of quality
Qualitätsveredelung *f* improving the quality
qualitativ qualitative
Qualle *f* (Zool) jelly-fish
Qualm *m* thick smoke
Qualmabzugsrohr *n* chimney, smoke disperser
qualmen to smolder, to emit thick smoke
Qualmfeuer *n* smoldering fire
Quant *n* (Phys) quantum
quanteln to quantize
Quantelung *f* quantization
Quanten *pl* (Phys) quanta
Quantenausbeute *f* quantum yield
Quantenbahn *f* quantum orbit, quantum path
Quantenbedingung *f* quantum condition
Quantenbiochemie *f* quantum biochemistry
Quantenbiologie *f* quantum biology
Quantenchemie *f* quantum chemistry
Quanteneffekt *m* quantum effect
Quantenelektrodynamik *f* quantum electrodynamics
Quantenfehler *m* quantum defect
Quantengewicht *n* quantum weight
quantenhaft pertaining to the quantum theroy, quantic
Quantenhypothese *f* quantum hypothesis
Quantenmechanik *f* quantum mechanics
quantenmechanisch quantum mechanical
Quantenphysik *f* quantum physics
Quantenresonanz *f* quantum resonance
Quantensprung *m* quantum jump, quantum transition
Quantenstatistik *f* (Quant) quantum statistics
Quantenstreuung *f* quantum scattering
Quantentheorie *f* quantum theory
Quantenübergang *m* quantum transition
Quantenverlust *m* quantum leakage
Quantenzahl, innere ~ inner quantum number
Quantenzahl *f* quantum number, **azimutale** ~ (Quant) azimuthal quantum number
Quantenzahlenvergrößerung *f* **bei Elektronen** electron promotion
Quantenzustand *m* quantum state
Quantimeter *n* (Rönt) quantimeter
quantisiert quantized
Quantisierung *f* quantization

Quantität *f* quantity, amount
Quantitätsbestimmung *f* quantification, quantitative determination
Quantitätsfaktor *m* quantitative factor, factor of capacity
Quantitätsgalvanometer *n* quantity galvanometer
quantitativ quantitative
Quantivalenz *f* quantivalence
Quantosom *n* (Biol) quantosome
Quantum *n* (Quantität) portion, quantity, quantum
Quark *m* cottage cheese, curds
quarkartig curdy
quartär quaternary, consisting of four, fourfold
Quartärstruktur *f* quaternary structure
Quartation *f* quartation, separation of silver from gold
Quartett *n* quartet[te]
Quartieren *n* **des Probeguts** quartering the sample
Quarz *m* quartz, rock crystal, ~ **abschrecken** to cool the quartz, **geschmolzener** ~ fused quartz, melted quartz, **optisch inaktiver** ~ racemic quartz, **piezoelektrischer** ~ piezoelectric quartz
Quarzachat *m* (Min) quartzy agate
quarzähnlich quartzose, quartzous, quartzy
quarzartig quartzose, quartzous, quarz-like
Quarzchromatlinse *f* chromatic quartz lens
Quarzdreieck *n* silica triangle
quarzelektrisch piezoelectric
Quarzfaden *m* quartz filament, quartz thread
Quarzfadendruckmesser *m* quartz-fiber manometer
Quarzfadenmanometer *n* quartz-fiber manometer
Quarzfarbe *f* quartz color
Quarzgang *m* quartz vein
Quarzgefäß *n* quartz utensil
Quarzgerät *n* quartz ware
Quarzglas *n* quartz glass, fused silica
Quarzgleichrichter *m* quartz or crystal rectifier
Quarzgut *n* quartz ware, fused silica
quarzhaltig quartziferous
quarzig quartziferous, quartzose, quartzous, quartzy
Quarzin *n* quartzine
Quarzit *m* (Min) quartzite
Quarzkeil *m* quartz wedge
Quarzkiesel *m* quartz gravel
Quarzkörner *n pl* quartz grains
Quarzkohlengemisch *n* quartz carbon mixture
Quarzkolben *m* (Chem) silica flask
Quarzkristall *m* (Min) quartz crystal, rock crystal
Quarzküvette *f* quartz cuvette
Quarzlampe *f* quartz lamp
Quarzlinse *f* quartz lens

Quarzmehl *n* quartz powder
Quarzoszillator *m* quartz oscillator
Quarzpisolith *m* (Min) quartz pisolite
Quarzporphyr *m* (Min) quartz porphyry
Quarzrohr *n* silica tubing
Quarzsand *m* (Min) quartz sand, arenaceous quartz, quartzose sand
Quarzsandstein *m* quartzy sandstone
Quarzschamottestein *m* silica refractory
Quarzschiefer *m* (Min) quartzose schist, quartz slate
Quarzsinter *m* quartz sinter, siliceous sinter
Quarztrachyt *m* quartz trachyte
Quarzuhr *f* quartz-clock, quartz crystal clock
Quarzwatte *f* quartz wool
Quarzwolle *f* quartz wool
Quasiatom *n* (Phys) quasi-atom
quasi-axial pseudoaxial
quasichemisch quasi chemical
quasielastisch quasielastic
Quasigleichgewicht *n* quasi equilibrium
Quasiimpuls *m* quasi momentum
Quasileiter *m* (Elektr) quasi-conductor
quasilinear quasilinear
Quasilinearisieren *n* quasilinearization
Quasimolekül *n* quasi-molecule
Quasineutralität *f* quasi neutrality
quasiplastisch pseudoplastic
quasistabil quasistable
quasistationär quasistationary
quasistatisch quasistatic
quasiviskos quasiviscous
Quassiabitter *n* quassin
Quassiaextrakt *m* extract of quassia wood
Quassiaholz *n* quassia wood, bitter wood
Quassin *n* quassin
Quassit *n* quassite
quaternär composed of four parts, quaternary
Quaternärstahl *m* quaternary steel
Quaterrylen *n* quaterrylene
Quebrachamin *n* quebrachamine
Quebrachin *n* quebrachine
Quebrachit *m* 1-inositol monomethyl ether, quebrachitol, quebrachite
Quebracho *n* quebracho
Quebrachoextrakt *m* quebracho extract
Quebrachogerbsäure *f* quebracho-tannic acid
Quebrachoholz *n* quebracho wood
Quebrachorinde *f* quebracho bark
Quecke *f* (Bot) couch-grass
Queckenwurzel *f* (Bot) carex root, couchgrass root
Quecksilber *n* mercury, hydrargyrum (Symb. Hg), quicksilver, **blausaures** ~ mercury cyanide, **knallsaures** ~ fulminating mercury, mercury fulminate, **kolloidales** ~ colloidal mercury, electromercurol (Am. E.), **legieren mit** ~ to amalgamate
Quecksilberalkyl *n* mercury alkyl

Quecksilberantimonsulfid *n* mercury antimony sulfide, mercury thioantimonate
quecksilberartig mercurial
Quecksilberauflage *f* coating of mercury
Quecksilberbad *n* mercury bath
Quecksilberbalsam *m* mercurial balsam
Quecksilberbarometer *n* mercury barometer
Quecksilberbenzoat *n* mercuric benzoate, mercury(II) benzoate
Quecksilberbranderz *n* inflammable cinnabar
Quecksilberbromür *n* (obs) mercurous bromide, mercury(I) bromide
Quecksilberbüchse *f* mercury box
Quecksilberchloridamid *n* mercury amide chloride, mercuriammonium chloride, mercuric ammonium chloride, white precipitate
Quecksilberchlorojodid *n* mercuric chloroiodide
Quecksilberchlorür *n* (Quecksilber(I)-chlorid) mercurous chloride, calomel, mercury(I) chloride, mercury monochloride
Quecksilbercyanür *n* (Quecksilber(I)-cyanid) mercurous cyanide, mercury(I) cyanide, mercury monocyanide
Quecksilbercyanwasserstoffsäure *f* cyanomercuric acid
Quecksilberdampf *m* mercury vapor
Quecksilberdampfgleichrichter *m* mercury [vapor] rectifier
Quecksilberdampfglühgleichrichter *m* hot-cathode mercury-vapor rectifier
Quecksilberdampflampe *f* mercury vapor lamp
Quecksilberdampfröhre *f* (mit Glühkathode) mercury-vapor tube (with thyratron)
Quecksilber-Dampfstrahlpumpe *f* mercury ejector
Quecksilberdampfstrom *m* mercury vapor stream
Quecksilberdiäthyl *n* mercury diethyl, mercury ethide
Quecksilberdimethyl *n* mercury dimethyl
Quecksilberdruck *m* mercury pressure
Quecksilbererz *n* mercury ore, quicksilver ore
Quecksilberfaden *m* mercury column, mercury thread
Quecksilberfahlerz *n* tetrahedrite containing mercury
Quecksilberfalle *f* mercury trap
quecksilberführend mercuriferous, containing mercury, mercurial
Quecksilberfüllung *f* mercury filling
Quecksilberfulminat *n* mercury fulminate
Quecksilbergefäß *n* mercury vessel
Quecksilbergehalt *m* mercury content
Quecksilberglanz *m* (Min) onofrite
Quecksilbergleichrichter *m* (Elektr) mercury-vapor rectifier
Quecksilberglidin *n* mercury glidin
Quecksilbergur *f* native mercury

Quecksilberhalogen *n* mercury halide
quecksilberhaltig containing mercury, mercurial, mercuriferous
Quecksilberhochdruckentladungsröhre *f* high-pressure mercury discharge tube
Quecksilberhochdrucklampe *f* high pressure mercury vapor lamp
Quecksilberhochvakuumpumpe *f* mercury vapor pump
Quecksilberhornerz *n* horn quicksilver, native calomel
quecksilberig mercurial
Quecksilber(II)-chlorid *n* mercuric chloride, corrosive sublimate, mercury dichloride, mercury(II) chloride
Quecksilber(II)-oxid *n* mercuric oxide, red precipitate
Quecksilber(II)-Verbindung *f* mercuric compound, mercury(II) compound
Quecksilber(I)-oxid *n* mercurous oxide, black precipitate
Quecksilber(I)-Verbindung *f* mercurous compound, mercury(I) compound
Quecksilberjodat *n* mercury iodate
Quecksilberjodid *n* mercury iodide, (Quecksilber(II)-jodid) mercuric iodide, mercury diiodide, mercury(II) iodide
Quecksilberjodür *n* (Quecksilber(I)-jodid) mercurous iodide, mercury(I) iodide, mercury monoiodide
Quecksilberkapsel *f* mercury bulb
Quecksilberkippschalter *m* mercury circuit breaker, mercury switch
Quecksilberkontakt *m* mercury contact
Quecksilberkontinuum *n* continuous mercury spectrum
Quecksilberkuppe *f* mercury meniscus
Quecksilberlampe *f* mercury lamp
Quecksilberlebererz *n* hepatic cinnabar, hepatic mercury ore
Quecksilberlegierung *f* amalgam, mercury alloy
Quecksilberlichtbogen *m* mercury arc
Quecksilberlösung *f* mercury solution
Quecksilbermanometer *n* mercury manometer, mercury pressure gauge
Quecksilbermeniskus *m* meniscus of mercury
Quecksilbermittel *n* (Pharm) mercurial, mercurial preparation, mercurial remedey
Quecksilbermohr *m* (Min) black mercuric sulfide
Quecksilberniederschlag *m* mercury precipitate
Quecksilbernitrat *n* mercury nitrate
Quecksilberofen *m* mercury furnace
Quecksilberoxycyanid *n* mercuric oxycyanide, mercury(II) oxycyanide
Quecksilberoxydul *n* (obs) mercurous oxide, mercury(I) oxide
Quecksilberoxydulsalz *n* (obs) mercurous salt, mercury(I) salt

Quecksilberpeptonat *n* mercury peptonate
Quecksilberpflaster *n* mercurial plaster
Quecksilberphenylat *n* mercury phenylate, mercuric carbolate, mercuric phenolate, mercury phenate, mercury phenoxide, phenol mercury
Quecksilberphosphat *n* mercury phosphate
Quecksilberpille *f* (Pharm) mercurial pill, blue pill
Quecksilberporphinchelat *n* mercuric porphin chelate
Quecksilberpräparat *n* (Pharm) mercurial preparation
Quecksilberpräzipitatsalbe *f* ammoniated mercury ointment, white precipitate ointment
Quecksilberpumpe *f* mercury pump
Quecksilberpunktlampe *f* mercury point lamp
Quecksilber-Quarzlampe *f* quartz mercury vapor lamp
Quecksilberrhodanid *n* mercury thiocyanate, (Quecksilber(II)-rhodanid) mercury(II) thiocyanate
Quecksilberrhodanür *n* (Quecksilber(I)-rhodanid) mercurous thiocyanate, mercury(I) thiocyanate
Quecksilber-Rückschlagventil *n* mercury safety valve
Quecksilberruß *m* mercurial soot
Quecksilbersättigung *f* mercurialization
Quecksilbersäule *f* mercury column, barometric column, mercury pressure
Quecksilbersäulendruck *m* **in Millimetern** pressure in terms of millimeters of mercury
Quecksilbersalbe *f* (Pharm) mercurial ointment, mercury ointment
Quecksilbersalicylat *n* mercury salicylate, mercuric salicylate, mercury(II) salicylate
Quecksilbersalpeter *m* mercury nitrate
Quecksilbersalz *n* mercury salt
Quecksilberschalter *m* mercury circuit breaker; mercury switch
Quecksilberschwarz *n* black mercury
Quecksilberseife *f* mercurial soap
Quecksilberspat *m* (Min) horn quicksilver, native mercurous chloride
Quecksilberstand *m* mercurial level
Quecksilberstein *m* (Min) native quicksilver
Quecksilbersublimat *n* corrosive sublimate, mercuric chloride, mercury dichloride, mercury(II) chloride
Quecksilbersuccinimid *n* mercury succinimide, mercuric succinimide, mercury(II) succinimide
Quecksilbersulfat *n* mercury sulfate
Quecksilbersulfid *n* mercury sulfide, (Quecksilber(II)-sulfid) mercuric sulfide, mercury(II) sulfide
Quecksilbersulfür *n* (Quecksilber(I)-sulfid) mercurous sulfide, mercury(I) sulfide

Quecksilbertannat *n* mercury tannate
Quecksilberthermometer *n* mercurial thermometer
Quecksilbertropfelektrode *f* dropping mercury electrode
Quecksilberverbindung *f* mercury compound
Quecksilberverfahren *n* mercury process
Quecksilbervergiftung *f* (Med) mercurial poisoning, hydrargyrism
Quecksilbervergoldung *f* mercurial gilding
Quecksilbervernickelung *f* mercurial nickel plating
Quecksilberverschluß *m* mercury seal
Quecksilberversilberung *f* silvering by amalgamation
Quecksilberverstärker *m* mercury intensifier
Quecksilbervitriol *n* mercury vitriol, sulfate of mercury
Quecksilberwanne *f* mercury trough
Quecksilberwasser *n* mercurial solution
Quecksilberzange *f* mercury tongs
Quellbeständigkeit *f* resistance to swelling
Quellbottich *m* steeping tub, steeping vat
Quelle *f* source, (Gewässer) spring, (Öl) well, **punktförmige** ~ point source
quellen (aufquellen) to swell, (einweichen) to soak
Quellen *n* swelling, moisture expansion, ~ **der Dichtung** gasket swelling
Quellenangabe *f* reference
Quellerz *n* limonite
Quellfähigkeit *f* swelling capacity
Quellfassung *f* well shaft
quellfest non-swelling, swell-resistant
Quellhilfsmittel *n* swelling auxiliary
Quellmehl *n* swell-starch flour
Quellsäure *f* apocrenic acid, crenic acid
Quellsalz *n* spring salt, well salt
Quellsole *f* spring brine, well brine
Quellung *f* swelling, expansion, soaking
Quellungskolloid *n* swelling colloid
Quellungswärme *f* swelling heat
Quellverhalten *n* swelling behavior
Quellvermögen *n* swelling capacity
Quellwasser *n* spring water
Quendel *m* (Bot) wild thyme
Quendelöl *n* thyme oil
Quenstedtit *m* (Min) quenstedtite
quer cross[wise]; transversal, transverse, (schräg) diagonal, (seitlich) lateral, (zur Faserrichtung) cross-grain
Querachse *f* transversal axis, transverse axis
Querbeanspruchung *f* transverse strain
Querberippung *f* transverse finning
Querbewehrung *f* lateral reinforcement
Querbiegeversuch *m* transverse bend test, cross-breaking test
Querbinde *f* cross tie, traverse
Querblasverhältnis *n* transverse blow-up ratio

Querbohrung *f* cross boring, transverse borehole
Quercetagetin *n* quercetagetin
Quercetin *n* quercetin, meletin
Quercetinsäure *f* quercetinic acid
Querceton *n* quercetone
Quercimeritrin *n* quercimeritrine
Quercin *n* quercin, quercinitol
Quercit *m* quercitol, acorn sugar, cyclohexanepentol, pentahydroxycyclohexane, quercite
Quercitrin *n* quercitrin
Quercitron *n* quercitron
Quercitronlack *m* quercitron lake
Quercitronrinde *f* quercitron bark, yellow-oak bark
Querdehnung *f* elongation in cross direction, lateral expansion, transverse extension
Querdehnungsziffer *f* (Poissonsche Konstante) Poisson's ratio
Querdrift *f* cross drift
Quereffekt *m* (Krist) transversal effect
Querfalte *f* transverse fold
Querfeld *n* (Elektr) cross field
Querfestigkeit *f* lateral strength, transversal strength
Querförderband *n* cross belt
Querformat *n* (Buchdr) broadside
Quergabelstapler *m* side-loading forklift
Quergleiten *n* cross slip
Quergleitlinie *f* cross slip line
Quergleitung *f* cross slip
Querhahn *m* clip, pinch cock, spring clamp
Querhaupt *n* cross head, tie
Querholz *n* wood cut against the grain
Querinduktion *f* cross induction
Querkopfziehform *f* cross head die
Querkreissäge *f* cross-cut circular saw
querlaufend transverse
Querlinie *f* cross line
Quermagnetfeld *n* transverse magnetic field
quermagnetisieren to cross-magnetize
Quermagnetisierung *f* cross magnetization
Quermagnetisierungseffekt *m* cross magnetizing effect
Querprofil *n* cross section
Querrichtung *f* transverse direction, cross[wise] direction
Querriegel *m* cross bar, anchor, cotter
Querriß *m* cross crack
Querruder *n* (Quersteuer) aileron
Querschneider *m* cross cutter, guillotine
Querschnitt *m* cross section, profile, transverse section
Querschnittsabnahme *f* reduction of cross-sectional area
Querschnittsfläche *f* cross sectional area
Querschnittsveränderung *f* change of cross-sectional area, variation of cross section
Querschnittsverminderung *f* reduction of cross-sectional area
Querschnittsverringerung *f* reduction of cross section
Querschnittszeichnung *f* sectional drawing
Querschwelle *f* cross tie
Querschwingung *f* transverse vibration, lateral oscillation
Querspießglanz *m* (Min) jamesonite
Querspritzkopf *m* cross [extruder] head
Querstab *m* cross bar
Quersteg *m* cross piece
Querströmung *f* cross current
Querstrom *m* (Hydrodynamik) cross flow
Querstromsichter *m* cross-current classifier
Querstück *n* cross member
Quersumme *f* sum of the digits
Querteilung *f* transverse spacing
Querträger *m* cross bar, cross beam, cross girder
Querverankerung *f* cross supports
Querverbindung *f* cross link, cross linkage
Querverweis *m* cross-reference
Querwiderstand *m* (Elektr) cross resistance, reactance
Querzug *m* cross flue, transverse flue
Querzusammenziehung *f* transverse contraction
Quetenit *m* (Min) quetenite
Quetsche *f* crushing tool, wringer
quetschen to squeeze, (zerdrücken) to crush
Quetscher *m* squeezer, squeegee (Am. E.)
Quetschgrenze *f* compression yield point, yield point in compression
Quetschhahn *m* pinchcock, clip, pinch clamp
Quetschmaschine *f* crushing machine, squeezer, squeezing machine
Quetschmühle *f* crushing mill, crusher
Quetschtube *f* collapsible tube
Quetschventil *n* pinch valve
Quetschwalze *f* squeeze roller
Quetschwirkung *f* squeezing effect
Quickarbeit *f* amalgamation
Quickbeize *f* amalgamating bath, amalgamating fluid
Quickbrei *m* amalgam
quicken to amalgamate, to amalgamize
Quickerz *n* mercurial ore
Quickfaß *n* amalgamating barrel
Quickgold *n* gold amalgam
Quickmetall *n* amalgamated metal
Quickwasser *n* quick water
Quietol *n* quietol
quietschen to squeak
Quillajarinde *f* quillai[a] bark
Quillajasäure *f* quillaic acid
Quinquiphenyl *n* quinquiphenyl
Quintessenz *f* concentrated extract, quintessence
Quintupelpunkt *m* quintuple point
quirlen to twirl, to agitate, to stir, to whirl
Quirlkies *m* (Min) safflorite

Quisqueit *m* (Min) quisqueite
Quitte *f* (Bot) quince
Quittenöl *n* quince oil
Quittensaft *m* quince juice
Quittenschleim *m* quince mucilage
Quittung *f* receipt
Quotient *m* (Math) quotient
Quotientenmesser *m* quotient meter
quotisieren to quote

R

Rabatt *m* deduction, discount, rebate
Rabelaisin *n* rabelaisin
Rabenglimmer *m* (Min) cryphyllite
Rabies *f* (Tiermed) hydrophobia, rabies
Racemat *n* (Stereochem) racemate, racemic compound, salt or ester of dl-tartaric acid
Racemethorphan *n* racemethorphan
Racemform *f* racemic modification
Racemie *f* racemism
racemisch racemic
racemisieren to racemize
Racemisierung *f* racemization, ~ **durch Waldensche Umkehr** racemization by Walden inversion, **thermische** ~ thermal racemization
Racemkörper *m* racemic substance
Racemoramid *n* racemoramide
Racemorphan *n* racemorphan
Rachen *m* throat, (Med) pharynx
Rachenabstrich *m* (Med) throat swab
Rachenblütler *m pl* (Bot) scrophulariac[e]ae
Rachenhöhle *f* pharynx
Rachenlehre *f* caliper gauge, calipers, external gauge
Rachitis *f* (Englische Krankheit) rachitis, rickets
rachitisch rickety, rachitic
Rad *n* wheel
Radabstand *m* wheelbase
Radantrieb *m* wheel drive
Radar *m n* (Elektr) radar
Radaranlage *f* radar installation, radar unit
Radarantenne *f* radar antenna
Radarbildschirm *m* radar screen
Radargerät *n* radar [apparatus]
radargesteuert radar-controlled
Radarkontrolle *f* radar control
Radaroberflächenreichweite *f* radar surface range
Radarschirm *m* radar-screen
Radarsicherung *f* insulation against radar detection
Radarsuchgerät *n* air search radar
Radartechnik *f* radar engineering
Radarwarnsystem *n* radar warning system
Radbohrmaschine *f* wheel bore
Raddeanin *n* raddeanine
Raddozer *m* wheeldozer
Raddrehbank *f* wheel turning lathe
Radiacmeter *n* (Radioaktivität) radiacmeter
radial radial
Radialausdehnung *f* radial expansion
Radialbelastung *f* radial load
Radialbeschleunigung *f* radial acceleration
Radialbewegung *f* radial motion
Radialbohrer *m* radial drill
Radialbohrmaschine *f* radial drilling machine
Radialeigenfunktion *f* radial eigenfunction

Radialgeschwindigkeit *f* radial velocity
Radialintegral *n* radial integral
Radialkammer *f* radial motion chamber
Radialkraft *f* radial force
Radiallager *n* radial bearing
Radialmethode *f* radial method
Radialpropeller *m* radial propeller
Radialriß *m* radial crack
Radialschlag *m* (Kfz) radial runout
Radialschwingung *f* radial mode
Radialstrom *m* radial flow
Radialstromwäscher *m* radial-flow scrubber
Radialverdichter *m* radial-flow compressor
Radiator *m* radiator
Radiatorventil *n* radiator valve
Radicinin *n* radicinin
Radienquotienten *pl* radius ratio
radieren (Kunst) to etch, (mit Gummi) to erase, to rub out, (Reifen) to abrade
Radierfirnis *m* etching varnish
Radiergummi *m* india rubber, eraser
Radierschablone *f* erasing pattern
Radierung *f* etching
Radikal *n* radical, **aliphatisches** ~ (aliphat. Rest) aliphatic radical, aliphatic residue;, **aromatisches** ~ (aromat. Rest) aromatic radical, aryl radical, **freies** ~ free radical, **kurzlebiges** ~ short-lived radical, **langlebiges** ~ long-lived radical
Radikalfänger *m* radical inhibitor
Radikalkettenreaktion *f* radical chain reaction
Radikalpolymerisation *f* radical polymerization
Radikalwanderung *f* migration of radicals
Radio *m* (Apparat) radio [set], (Funk) broadcasting, radio, wireless (Br. E.)
Radioaktinium *n* radioactinium (Symb. RdAc)
radioaktiv radioactive, ~ **machen** to render radioactive
Radioaktivität *f* radioactivity, **induzierte** ~ induced radioactivity, **künstliche** ~ artificial radioactivity
Radioaktivitätsnachweis *m* radioactivity detection
Radioapparat *m* radio [set], wireless (Br. E.)
Radiobiologie *f* radiobiology
radiobiologisch radiobiological
Radioblei *n* radiolead
Radiochemie *f* radiochemistry
Radiochemiker *m* radiochemist
Radiochrometer *n* radiochrometer
Radioelement *n* radioactive element, radioelement
Radioemanation *f* radium emanation, niton, radon
Radiofrequenz *f* radiofrequency

Radiofrequenzspektrometer *n* radiofrequency spectrometer
Radiogehäuse *n* radio cabinet
radiogen radiogenic
Radiogramm *n* (Phys) radiogram, radiograph
Radiographie *f* radiography
Radioindikator *m* radioactive tracer, radio-tracer
Radiointerferometer *n* radio-interferometer
Radioisotop *n* radioactive isotope, radioisotope
Radiojod *n* radioactive iodine
Radiokohlenstoff *m* radiocarbon
Radiokohlenstoff-Datierung *f* radiocarbon-dating
Radiolith *m* (Min) radiolite
Radiologe *m* radiologist, roentgenologist
Radiologie *f* radiology
radiologisch radiological
Radiolumineszenz *f* radioluminescence
Radiolyse *f* radiolysis
radiomarkiert isotope-labelled
Radiometer *n* radiometer
radiometrisch radiometric
Radionuklid *n* radioactive nuclide, radionuclide
Radiophosphor *m* radioactive phosphorus
Radiophotographie *f* radiophotograph[y]
Radioschwefel *m* radiosulfur
Radiosensibilität *f* radiosensitivity
Radiostrontium *n* radiostrontium
Radiotellur *n* radio tellurium, polonium
Radiotherapie *f* radiotherapy
Radiothorium *n* radiothorium (Symb. RdTh)
Radiothoriumgehalt *m* radiothorium content
Radiowiderstand *m* radio resistor
Radium *n* radium (Symb. Ra)
Radiumatom *n* radium atom
Radiumbehandlung *f* (Med) radiotherapy
Radiumbromid *n* radium bromide
Radiumeinwirkung *f* effect of radium
Radiumemanation *f* radium emanation, niton, radon
Radiumgewinnungsanlage *f* radium refinery
radiumhaltig containing radium
Radiumjodid *n* radium iodide
Radiumpräparat *n* radium specimen
Radiumprobe *f* sample of radium
Radiumquelle *f* radium source, supply of radium
radiumresistent radium-resistant
Radiumstrahlen *m pl* radium rays
Radiumstrahlung *f* radium emanation, radium radiation, radon
Radiumtherapie *f* (Med) radium therapy
radiumverseucht radium-contaminated
Radiumzerfall *m* [spontaneous] disintegration of radium
Radius *m* radius
Radiusvektor *m* radius vector
Radix *f* (Math) root
radizieren (Math) to extract a root

Radizieren *n* (Math) extraction of the root
Radkranz *m* wheel rim
Radlader *m* wheel loader, payloader
Radlauf *m* rotation, run, travel
Radlenker *m* guard rail
Radlinie *f* cycloid
Radon *n* radon (Symb. Rn), niton, radium emanation
Radonhohlnadel *f* radon seed
Radsatz *m* wheel set
Radsatzdrehbank *f* wheel lathe
Radscheibe *f* wheel disc
Radstand *m* (Kfz) wheel base
Radstern *m* wheel spider
Radteilung *f* pitch of the wheel
Radwelle *f* gear shaft
Rädelerz *n* (Min) antimonial lead ore, bournonite, wheel ore
Räderformmaschine *f* gear wheel molding machine
Räderwalzwerk *n* disc mill, wheel rolling mill
Räderwerk *n* gear[ing], wheel-work
Rändelkopf *m* milled head
Rändelmutter *f* knurl nut
Rätter *m* (Bergb) riddle, shaking table
rättern (Bergb) to riddle, to screen
Räucheressenz *f* aromatic essence, fragrant essence, fumigating essence
Räucheressig *m* aromatic vinegar
Räucherkerzchen *n* fumigating candle
Räuchermittel *n* fumigant, gaseous insecticide
räuchern (desinfizieren) to fumigate, (rauchtrocknen) to smoke
Räuchern *n* fumigating, fumigation
Räucherpapier *n* fumigating paper
Räucherpulver *n* fumigating powder
räumen to clear, to empty, to remove
räumlich in space, spatial, three-dimensional, (Chem) steric
Räummaschine *f* broaching machine, (Min) coal loader
Räumvorrichtung *f* broaching device
Rafaelit *m* rafaelite
Raffinade *f* (Zucker) refined sugar, **flüssige** ~ liquid sugar
Raffinadekupfer *n* refined copper
Raffinadezucker *m* refined sugar
Raffinat *n* refined product, (Öl) raffinate
Raffination *f* refinement, refining, ~ **im Schmelzfluß** furnace refining, **elektrolytische** ~ electrolytic refinery
Raffinationsabfall *m* [refinery] waste
Raffinationsanlage *f* refinery, refining plant
Raffinationsbehandlung *f* refinement, refining treatment
Raffinationsprozeß *m* refining process
Raffinationsschmelzen *n* refining smelting
Raffinationsverfahren *n* refining process
Raffinationsvorgang *m* refining process

Raffinatkupfer *n* refined copper
Raffinatsilber *n* refined silver
Raffinatzink *n* refined spelter
Raffinerie *f* finery, refinery
Raffineur *m* refiner
Raffinieranlage *f* refining plant
raffinieren to purify, to refine
Raffinieren *n* refining
Raffinierofen *m* (Metall) refining furnace
Raffinierprozeß *m* refining process
Raffinierstahl *m* refined steel, merchant bar
raffiniert refined
Raffinierung *f* refining
Raffinose *f* raffinose, melitose, melitriose
Rahm *m* cream
Rahmbildung *f* formation of cream
Rahmen *m* frame, (Schuh) welt, (Skelett) skeleton, (Spannrahmen) stretcher, (Umfang) scope
Rahmenantenne *f* loop antenna
Rahmenfilter *m* frame filter
Rahmenfilterpresse *f* frame filter press, plate-and-frame press
Rahmenhammer *m* drop hammer
Rahmenpresse *f* frame filter press
Rahmenquerspant *m* traverse frame
Rahmenrührer *m* gate paddle agitator
Rahmentrockner *m* tenter drier
Rahmenverstärkung *f* yoke block
Rahmgefriermaschine *f* cream freezer
Rahmgewinnung *f* extraction of cream
Rahmkühler *m* cream cooler
Raimondit *m* (Min) raimondite
Rainfarnblüte *f* (Bot) tansy flower
Rainfarnkraut *n* (Bot) tansy
Rainfarnöl *n* tansy-oil
Rakel *f* doctor, doctor blade, [doctor] knife
Rakelappretur *f* (Text) doctor finish
Rakelstreichverfahren *n* knife coating
Rakete *f* rocket, ~ **mit festem Brennstoff** dry-fuelled rocket, solid-propellant rocket, ~ **mit flüssigem Brennstoff** liquid-fuelled rocket, liquid-propellant rocket, **Boden-Boden-** ~ surface-to-surface missile, SSM, **Boden-Luft-** ~ surface-to-air missile, SAM, **Luft-Unterwasser-** ~ air-to-underwater missile
Raketenabschußbasis *f* rocket launching site
Raketenantrieb *m* rocket drive, rocket propulsion
Raketenforschung *f* rocketry, rocket research
Raketenkopf *m* rocket head
Raketenraumschiff *n* rocket-propelled space-ship
Raketenreaktor *m* rocket pile
Raketentreibstoff *m* rocket propellant
Ralstonit *m* (Min) ralstonite
Ramalinolsäure *f* ramalinolic acid
Ramalsäure *f* ramalic acid
Ramaneffekt *m* Raman effect

Ramanlinie *f* (Spektr) Raman line
Ramanspektrum *n* (Spektr) Raman spectrum
Ramanverschiebung *f* (Spektr) Raman shift
Rambutantalg *m* rambutan tallow
Ramie *f* (Bot) China grass, ramie
Ramiefaser *f* China grass, ramie fiber
Ramiegarn *n* ramie yarn
Ramirit *m* ramirite
Rammelsbergit *m* (Min) rammelsbergite
Rampe *f* ramp
Ramsayfett *n* (Vakuumfett) vacuum grease
Rand *m* border, brim, edge, (Buchdr) margin, (Gefäß) rim, ~ **der Walze** collar of the roll
Randanit *m* randanite
Randanmerkung *f* marginal note
Randbedingung *f* boundary condition, limiting condition, ~ **der Periodizität** periodic boundary condition
Randbehandlung *f* boundary treatment
Randbemerkung *f* marginal note
Randdichte *f* boundary density
Randeffekt *m* boundary effect
Randeinzug *m* (Extrusion) neck-in
Randentkohlung *f* rim decarbonization
Randerscheinung *f* side issue
Randgängigkeit *f* maldistribution; wall effect on flow
Randgärung *f* rim fermentation
Randgummierung *f* edge gumming
Randit *m* randite
Randknospe *f* nodule at the edges
Randkoeffizient *m* marginal coefficient
Randkollokation *f* boundary collocation
Randleiste *f* strip
Randlochkarten *pl* marginal punched cards
Randproblem *n* boundary problem
Randpunkt *m* frontier point
Randschärfe *f* (Opt) marginal sharpness
Randstrom *m* current near the edge of the field
randverschmiert having diffuse edges
Randwert *m* boundary value
Randwertaufgabe *f* boundary value problem
Randwertproblem *n* (Math) boundary value problem
Randwinkel *m* **der Oberflächenspannung** boundary angle (of surface tension), rim angle (of surface tension)
Rang *m* order, class, degree, grade, quality, rank, row, (Theater) tier
Ranit *m* ranite
Ranque-Effekt *m* Ranque effect
Ranunculin *n* ranunculin
ranzig rancid; rank
Ranzigkeit *f* rancidity; rancidness; rankness
Ranzigwerden *n* rancidity, tendency to become rancid
Raolin *n* raolin
Rapanon *n* rapanone
Raphanin *n* raphanin

Raphilit *m* raphilite
Rapidechtfarbe *f* rapid fast dyestuff
Rapid-Eis *n* rapid ice
Rapidstahl *m* high-speed steel
Rapinsäure *f* rapinic acid
Raps *m* rape, colza
Rapsöl *n* colza oil, rape oil, rape seed oil (refined grade)
Rasanz *f* rapidity
rasch fast, quick, rapid
raschbindend quick setting
Raschig-Ringe *pl* (HN) Raschig rings
Raschig-Synthese *f* Raschig synthesis
Rasen[eisen]erz *n* bog [iron] ore, hematite, limonite, marsh ore, swamp ore
Rasenerzreiniger *m* bog iron ore purifier
Rasierklinge *f* razor blade
Rasierschaum *m* (Kosmetik) shaving lather
Rasierwasser *n* (Kosmetik) pre-shave lotion, shaving lotion
Raspel *f* rasp
raspeln to rasp
Raspeln *n* grating, levigation, pulverizing
Raspelspäne *m pl* raspings
Raspit *m* (Min) raspite
Rasse *f* (Mensch) race, (Tier) breed, stock, **gekreuzte** ~ crossbreed
Rast *f* (Hochofen) bosh
Rastdruck *m* pressure on the bosh
Rastenscheibe *f* slotted disc, star wheel
Raster *m* grating, light diffuser, screen, (f. Spektren) line screen, (i. d. Braunschen Röhre) rast (in the cathode ray tube), (Liniennetz) grid of lines, (Telev.) mosaic
Rasterätzung *f* autotypy
Rasterelektronenmikroskop *n* scanning electron microscope
Rasterkathode *f* (Telev) photocathode
Rastermikroskop *n* raster [scan] microscope, screen microscope
Rasterschirm *m* scanning screen
Rastersystem *n* screen system
Rasterton *m* (Buchdr) tint
Rasterverhältnis *n* grid ratio
Rastformen *f pl* (Hochofen) sections of boshes
Rasthebel *m* stop lever
Rasthöhe *f* height of bosh
Rastmantel *m* shell of bosh
Rastmauerung *f* bosh brickwork
Rastolyt *m* rastolyte
Raststeine *m pl* (Hochofen) bosh stones
Rastwinkel *m* angle of bosh
Ratanhia *f* r[h]atany, krameria
Ratanhiatinktur *f* rhatany tincture
Ratanhiawurzel *f* rhatany root, krameria
Ratanhin *n* r[h]atanhine, N-methyl-l-tyrosine
Rathit *m* (Min) rathite
rational rational

Rationalisierung *f* rationalization, simplification, systematization
Rationalität *f* rationality
Rationalitätsgesetz *n* (der Kristallparameter) rationality law (of crystal parameters)
rationell economical, rational, reasonable
Rattengift *n* rat poison, ratsbane
Rattermarken *pl* chatter marks
Raubasin *n* raubasine
Rauch *m* fume, smoke, ~ **und Nebel** smog
Rauchabzug *m* fume exhaust, ventilation
Rauchabzugskanal *m* flue
Rauchabzugsöffnung *f* chimney hole
Rauchachat *m* (Min) smoky agate
rauchartig smoky
Rauchbekämpfung *f* smoke abatement
Rauchbelästigung *f* smoke nuisance
Rauchbildung *f* development of smoke, formation of smoke, generation of smoke
rauchdicht smokeproof, smoketight
Rauchdichte *f* smoke density, thickness of smoke
Rauchdichtemesser *m* smoke density meter
rauchen to fume, to smoke
Rauchen *n* fuming, smoking
rauchend fuming, smoking
Rauchentwicklung *f* generation or development of smoke
Raucherhusten *m* (Med) smoker's cough
Raucherkrebs *m* (Med) smoker's cancer
Raucherzeuger *m* smoke generator
Rauchexplosion *f* flue gas explosion
Rauchfärbung *f* coloration of smoke
Rauchfang *m* chimney hood, flue
rauchfarben smoke-colored
Rauchfleisch *n* smoked meat
rauchfrei smokeless
rauchgar smoke-cured, smoke-dried
Rauchgas *n* burnt gas, exhaust gas, flue gas, furnace gas
Rauchgasanalysator *m* flue gas analyzer
Rauchgasanalyse *f* flue gas analysis
rauchgasdicht effective as smoke barrier
Rauchgasentstäuber *m* dust separator [for flue gas]
Rauchgasklappe *f* flue gas flap
Rauchgasprüfer *m* flue gas tester
Rauchgasreinigung *f* cleaning of flue gas
Rauchgasthermometer *n* exhaust gas thermometer
Rauchgasverwertung *f* flue gas utilization
Rauchgasvorwärmer *m* flue-gas preheater, economizer
Rauchgasvorwärmung *f* flue gas heating
Rauchgaszusammensetzung *f* composition of flue gases
Rauchglas *n* gray glass, tinted glass
Rauchhelm *m* smoke helmet
rauchig smoky, blackened with smoke, smelling of smoke

Rauchkammer *f* smoke box, smoke chamber
Rauchkammergas *n* smoke box gas
Rauchkanal *m* chimney flue
Rauchkohle *f* incompletely carbonized charcoal
Rauchlosigkeit *f* absence of smoke
Rauchmeldegerät *n* smoke monitor
Rauchmelder *m* smoke alarm
Rauchpulver *n* fumigating powder
Rauchpunkt *m* smoke point
Rauchpunktbestimmung *f* determination of smoke point
Rauchquarz *m* (Min) smoky quartz
Rauchrohrkessel *m* convection-type boiler, fire-tube boiler, smoke tube boiler
rauchschwach giving little smoke
Rauchschwaden *pl* vapors of smoke
rauchschwarz smoky black, sooty
Rauchskala *f* smoke scale
Rauchspurmunition *f* tracer bullet
Rauchstärke *f* smoke density, thickness of smoke
Rauchtopas *m* (Min) smoky topaz, smoke quartz, smokestone
Rauchuntersuchungsapparat *m* flue gas testing apparatus
Rauchverbrennung *f* smoke combustion
Rauchverbrennungseinrichtung *f* smoke consuming device
Rauchverdünnung *f* diminution of smoke
Rauchverhütung *f* smoke prevention
Rauchverminderung *f* diminution of smoke
rauchverzehrend smoke consuming, fumivorous
Rauchverzehrer *m* smoke consumer
Rauchverzehrung *f* smoke consumption, smoke combustion
Rauchwaren *pl* furs, peltry
Raugustin *n* raugustine
rauh rough, hard, rude, unpolished, (unbearbeitet) raw
rauhen to roughen, (Tuch) to tease, (Wolle) to card
Rauhen *n* buffing, graining, granulating, roughening, roughing
Rauhfaser *f* (Text) raised fabric
Rauhigkeit *f* coarseness, roughness
Rauhigkeitsspektrum *n* roughness spectrum
Rauhimbin *n* rauhimbine
Rauhkratze *f* rasp
Rauhmaschine *f* napping mill, rasp; tire buffing machine
Rauhreif *m* hoar frost, white frost
Rauhtiefe *f* depth of score, maximum roughness number, roughness, roughness-height
Rauhwaschtrommel *f* preliminary washing drum
Raum *m* room, chamber; space, (Rauminhalt) volume, (Spielraum) clearance, (Zwischenraum) interspace, **aktivitätsfreier** ~ (Atom) clean area, cold area, **luftverdünnter** ~ space filled with rarefied air, **schalltoter** ~ anechoic room, **toter** ~ dead space
Raumausdehnung *f* spatial expansion; volume expansion
Raumbedarf *m* space required, spatial requirement
Raumberechnung *f* volume calculation
Raumbeständigkeit *f* constancy of volume, incompressibility
Raumbewetterung *f* air conditioning
Raumbild *n* space diagram, stereoscopic picture
Raumdiagonale *f* space diagonal
Raumdichte *f* density by volume, ~ **des Magnetismus** magnetic density in space
Raumdimension *f* space dimension
Raumdrehung *f* space rotation
Raumeinheit *f* unit of volume, unit of space
Raumerfüllung *f* (Atom) space filling
Raumersparnis *f* saving in space, economy in space
Raumfähre *f*, **wiederverwendbare** ~ space shuttle
Raumfahrt *f* astronautics, cosmonautics, interplanetary aviation
Raumfahrzeug *n* spacecraft, spaceship
Raumfokussierung *f* space focus[s]ing
Raumformel *f* space formula, spatial formula
Raumforschung *f* space research
Raumgebilde *n* three-dimensional structure, space diagram
Raumgehalt *m* volumetric content, content by volume
Raumgeometrie *f* stereometry, solid geometry
Raumgewicht *n* weight per unit of volume, [bulk] density, unit weight, volumetric weight
Raumgitter *n* (Krist) atomic structure, crystal lattice, space lattice, three-dimensional lattice, **flächenzentriertes** ~ face-centered lattice, **kubisches** ~ (Krist) cubic space lattice
Raumgitterebene *f* space lattice plane
Raumgitterstruktur *f* (Krist) space-lattice structure
Raumgleichung *f* continuity equation
Raumgröße *f* volume, capacity, content
Raumgruppe *f* (Krist) space group
Raumgruppensymbolik *f* space group symbols
Raumgruppentabelle *f* table of space groups
Rauminhalt *m* volume, capacity, cubical contents
Rauminhaltmesser *m* volume meter
raumisomer stereoisomeric
Raumisomer *n* stereoisomer
Raumisomerie *f* stereoisomerism, spatial isomerism
Raumitorin *n* raumitorine
Raumkoordinate *f* space coordinate
Raumkrümmung *f* space curvature
Raumkühlung *f* space cooling
Raumkurve *f* space curve, three-dimensional curve

Raumladeverzerrung *f* space-charge distortion
Raumladung *f* space charge, spatial charge
Raumladungsanhäufung *f* space-charge accumulation
raumladungsbegrenzt space-charge limited
Raumladungsdetektor *m* space-charge detector
Raumladungsdichte *f* space-charge density
Raumladungseffekt *m* space-charge effect
Raumladungsgitter *n* space-charge grid
Raumladungskapazität *f* space-charge capacity
Raumladungsverzerrung *f* space-charge distortion
Raumladungswolke *f* concentration or accumulation of space charges, space-charge cloud
Raumladungszone *f* space-charge zone (solid-state phys)
Raummaß *n* cubic measure, measure of capacity, solid measure, volume
Raummeßbild *n* stereogram
Raummesser *m* volume meter
Raummeter *n* cubic meter
Raumrichtung *f* direction in space
Raumschaufelzentrifuge *f* raker blade centrifuge
Raumschiff *n* space ship or shuttle
raumsparend space-saving
Raumspiegelung *f* space reflection
Raumteil *m* [part by] volume
Raumtemperatur *f* ambient temperature, room temperature
Raumtemperaturregler *m* room thermo-regulator
Raumtransformation *f* space transformation
Raumverminderung *f* decrease in volume, volume contraction
Raum-Volumen-Geschwindigkeit *f* volume-space velocity
Raumwelle *f* bodily wave, sky wave, space wave
Raumwinkel *m* solid angle
Raum-Zeit-Ausbeute *f* space-time yield, yield within a certain time and space
Raumzeitstruktur *f* space-time structure
raumzentriert (Krist) body-centered, space-centered
Raunescin *n* raunescine
Raunescinsäure *f* raunescic acid
Raunormin *n* raunormine
Raunormsäure *f* raunormic acid
Raupe *f* (Zool) caterpillar
Raupenbildung *f* (Kunstharz) foldback
Raupendozer *m* caterpillar dozer
Raupenfahrzeug *n* tracked vehicle
Raupenkette *f* caterpillar
Raupenkran *m* crawler crane
Raupenmuster *n* diamond design
Raupin *n* raupine
Rauschen *n* noise, random noise, (Störung) interference
Rauschfaktor *m* noise factor

Rauschgelb *n* auripigment, King's yellow, orpiment, yellow arsenic blende
Rauschgift *n* dope, drug, narcotic
Rauschgifthandel *m* drug traffic
Rauschgiftsucht *f* drug addiction
Rauschgiftsüchtiger *m* drug addict, dope-fiend (slang)
Rauschmessung *f* noise measurement
Rauschmittel *n* intoxicant
Rauschrot *n* (Min) arsenic disulfide, realgar, red arsenic glass, red arsenic [sulfide], red orpiment, ruby arsenic
Rauschsilber *n* imitation silver foil, silver tinsel
Rauschspannung *f* (Elektron) noise voltage
Rauschzahl *f* noise figure
Raute *f* diamond, rhombus, (Bot) rue
Rautenflächner *m* (Krist) rhombohedron
rautenförmig diamond-shaped, rhombic, rhomboidal
Rautenöl *n* rue oil
Rautenspat *m* (Min) rhomb spar
Rauwolscan *n* rauwolscane
Rauwolscin *n* rauwolscine
Rauwolscon *n* rauwolscone
Rauwolsinsäure *f* rauwolsinic acid
Ravenilin *n* ravenilin
Ravisonöl *n* ravison oil
Rayleighsche Streustrahlung *f* Rayleigh scattering
Reabsorption *f* reabsorption
Reagens *n* (Reagenz) reagent, **mikroanalytisches** ~ microanalytical reagent
Reagensfarbe *f* test color
Reagensglas *n* test tube, test glass
Reagensglasgestell *n* test tube stand
Reagensglashalter *m* test tube holder
Reagensglasversuch *m* test tube research
Reagenskelch *m* test cup, test glass
Reagenslösung *f* reagent solution, test solution
Reagenspapier *n* reagent paper, test paper
Reagensröhrchen *n* test tube
Reagentien *pl* reagents
Reagentienflasche *f* reagent bottle, reagent flask
Reagentienraum *m* reagent room
Reagenz *n* reagent
Reagenzdosierung *f* reagent feeding
Reagenzglas *n* test-tube
Reagenzglasbürste *f* test-tube brush
Reagenzglasgestell *n* test-tube rack
Reagenzglashalter *m* test-tube holder
Reagenzien *pl* reagents
Reagenzpapier *n* test paper
reagieren to react
Reagieren *n* reacting
reagierend reacting, reactive
Reaktanz *f* (Elektr) reactance, **induktive** ~ inductance [reactance], positive reactance, **kapazitive** ~ capacitance, capacity reactance
Reaktanzfaktor *m* reactance coefficient

Reaktion f reaction, ~ **erster Ordnung** first-order reaction, **allergische** ~ (Med) allergic reaction, **anaplerotische** ~ anaplerotic reaction, **basische** ~ basic reaction, **bimolekulare** ~ bimolecular reaction, **elektrochemische** ~ electrochemical reaction, **endergonische** ~ endergonic reaction, **exergonische** ~ exergonic reaction, **exotherme** ~ exothermic reaction, **gekoppelte** ~ coupled reaction, **geschwindigkeitsbestimmende** ~ rate-limiting reaction, **heterogene** ~ heterogeneous reaction, **homogene** ~ homogeneous reaction, **irreversible** ~ irreversible reaction, **monomolekulare** ~ monomolecular reaction, **neutrale** ~ neutral reaction, **photochemische** ~ photochemical reaction, **reversible** ~ reversible reaction, **saure** ~ acid reaction, **schnellverlaufende** ~ rapid reaction, **sekundäre** ~ secondary reaction, **stereokinetisch gesteuerte** ~ stereokinetically controlled reaction, **stürmische** ~ vigorous reaction, **trimolekulare** ~ trimolecular reaction, **umkehrbare** ~ reversible reaction, **unvollständige** ~ incomplete reaction, balanced reaction, **verzögerte** ~ delayed reaction, **zusammengesetzte** ~ consecutive reaction, coupled reaction
Reaktionsablauf m course of reaction
Reaktionsarbeit f reaction process
Reaktionsbereich m reaction region
Reaktionsdauer f reaction time
Reaktionsdruck m reaction pressure
Reaktionsdurchschreibepapier n color-react paper
Reaktionsenergie f reaction energy
Reaktionsenthalpie f enthalpy of reaction
Reaktionsentropie f reaction entropy
Reaktionserzeugnis n product of reaction
reaktionsfähig reactive, capable of reacting, capable of reaction
Reaktionsfähigkeit f reactivity, capability of reaction, reacting capacity
Reaktionsfarbstoff m reactive dyestuff
Reaktionsflüssigkeit f reaction liquid
Reaktionsgas n reaction gas
Reaktionsgefäß n reaction vessel
Reaktionsgemisch n reaction mixture
Reaktionsgeschwindigkeit f reaction rate, velocity of reaction
Reaktionsgeschwindigkeitskonstante f rate constant
Reaktionsgleichgewicht n equilibrium of reaction
Reaktionsgleichung f reaction equation
Reaktionsgrundverfahren n (Polyester) reaction ground coat process
Reaktionshemmung f reaction inhibition
Reaktionsisochore f isochor[e] of reaction

Reaktionsisotherme f reaction isotherm, isotherm of reaction
Reaktionskammer f reaction chamber
Reaktionskette f reaction chain
Reaktionskinetik f reaction kinetics, chemical kinetics
Reaktionsknäuel n reaction cluster
Reaktionslack m two-component finish
Reaktionslawine f reaction avalanche
Reaktionslenkung f reaction control
reaktionslos non-reactive
Reaktionsmasse f reaction mass
Reaktionsmechanismus m reaction mechanism
Reaktionsmittel n reagent, agent, reactant
Reaktionsmotor m reaction motor
Reaktionsordnung f order of reaction
Reaktionspartner m coreactant, reactant
Reaktionsprimer m wash primer, etch primer, metal conditioner
Reaktionsprodukt n reaction product
Reaktionsraum m reaction space
Reaktionsreihe f pl **der Metalle** electrochemical series of metals
Reaktionsschema n reaction scheme
Reaktionstechnik f reaction technology
Reaktionsteilnehmer m reactant
Reaktionstemperatur f temperature of reaction
reaktionsträge inactive, inert, sluggish
Reaktionsträger m carrier of reaction
Reaktionsträgheit f inactivity, inertness
Reaktionsturm m reaction tower
Reaktionstyp m reaction type
Reaktionsverlauf m course of reaction
Reaktionsvermögen n reactivity
Reaktionsvolumen n reaction volume
Reaktionswärme f heat of reaction
Reaktionswiderstand m resistance to reaction
Reaktionswirkung f reaction effect
Reaktionszeit f reaction time, response time
reaktiv reactive
Reaktivfarbstoff m reactive dye
reaktivieren to reactivate
Reaktivierung f reactivation
Reaktivität f reactivity
Reaktor m reactor, atomic pile, ~ **für Forschungszwecke** research reactor, ~ **mit angereichertem Uran** enriched uranium reactor, ~ **mit flüssigem Brennstoff** liquid-fuel reactor, ~ **ohne Reflektor** bare pile, **adiabatischer** ~ adiabatic reactor, **angereicherter** ~ enriched fuel reactor, **berylliummoderierter** ~ beryllium-moderated reactor, **beweglicher** ~ mobile reactor, CO_2-**gekühlter** ~ carbon-dioxide cooled reactor, **deuteriummoderierter** ~ deuterium-moderated reactor, **Diffusionsflammen-** ~ diffusion flame reactor, **Dünnschicht-** ~ thin-film (wetted wall column) reactor, **epithermischer** ~ epithermal

reactor, **gasgekühlter** ~ gas-cooled reactor, **graphitmoderierter** ~ graphite-moderated reactor, **homogener** ~ homogeneous reactor, **ideal gemischter** ~ ideal mixed reactor, **kalter** ~ cold reactor, **Kammer-** ~ chamber reactor, **Kolbenstrom-** ~ plug-flow reactor, **Kreislauf-** ~ recycle reactor, **mittelschneller** ~ intermediate reactor, **primärer** ~ primary reactor, **regenerativer** ~ breeder [reactor], **Rohr-** ~ tubular reactor, tube reactor, flow tube reactor, **Schlaufen-** ~ recycle reactor, **Schnecken-** ~ screw reactor, **schwerwassermoderierter** ~ heavy-water moderated reactor, **sekundärer** ~ secondary reactor, **Strahldüsen-** ~ jet tube reactor
Reaktorabfälle pl reactor waste
Reaktorabschirmung f reactor shielding
Reaktorbeschickung f reactor loading
Reaktorbrutmantel m reactor blanket
Reaktorgift n reactor poison
Reaktorgitter n active lattice
Reaktorkaskade f cascade of reactors
Reaktorkern m reactor core
Reaktorleistung f reactor power
Reaktormantel m reactor shell
Reaktorphysik f reactor physics
Reaktorschaltwarte f reactor control room
Reaktorsteuerung f reactor control
Reaktortechnik f reactor technology
Reaktortechnologie f reactor technology, reactor engineering
real real, actual
Realgar m (Min) realgar, arsenic disulfide, red arsenic glass, red arsenic [sulfide], red orpiment, ruby arsenic
Realgasfaktor m compressibility factor
realisierbar realizable
Realität f reality
Realkristall m imperfect crystal
Réaumurskala f Réaumur scale
Réaumurthermometer n Réaumur thermometer
Rebenmeltau m grape mildew
Rebenschwarz n vine black, Frankfort black, vegetable black
Recanescin n recanescine
Recanescinalkohol m recanescic alcohol
Recanescinsäure f recanescic acid
Rechen m rake
Rechenanlage f computor
Rechenautomat m automatic calculating machine
Rechenfehler m error in calculation, miscalculation
Rechenmaschine f calculating machine, calculator, computing machine, counting machine, **elektronische** ~ [electronic] computer
Rechenoperation f arithmetic operation

Rechenschaltung f set-up for a calculating machine
Rechenscheibe f circular slide rule
Rechenschieber m (Math) slide rule, calculating rule
Rechenstab m slide rule, calculating rule
rechnen to calculate, to compute, to count
Rechnen n calculation, computation
Rechner m computer, calculator
Rechnung f calculation, computation
rechnungsmäßig according to calculation
Rechteck n rectangle
Rechteckdose f rectangular can
rechteckig rectangular, oblong
Rechteckimpuls m square signal, square-topped pulse
Rechteckkurve f rectangular curve
Rechteckspannung f square wave voltage
Rechtehandregel f (Phys) right-hand rule
rechtmäßig lawful, legitimate
Rechtsablenkung f deflection to the right
Rechtsborneol n d-borneol
Rechtscampher m dextrocamphor
rechtsdrehend dextrorotatory, dextrogyric
Rechtsdrehung f clockwise rotation, right-handed rotation
rechtsgängig right-handed, (Gewinde) righthand threaded
Rechtsgewinde n righthand thread
Rechtsgültigkeit f validity
Rechtspolarisation f dextrorotation, right-handed polarization
Rechtssäure f dextro-acid
Rechtssystem n right-handed system
Rechtsweinsäure f dextrotartaric acid, d-tartaric acid
Rechtwinkelnomogramm n right angle nomogram
rechtwinklig right-angled
rechtwinklig gebogen bent at a right angle
Reckalterung f strain aging
recken to extend, to lengthen, to rack, to stretch
Recken n stretching
Reckfähigkeit f stretchability
Reckmaschine f stretching machine
Reckmodul m modulus of stretch
Reckprobe f pull test, stretch test, tensile test
Reckspannung f [tensile] strain, tensile stress
Reckung f stretching, racking
Reddingit m (Min) reddingite
Redestillation f redistillation, rerunning
redestillieren to redistil[l]
Redlerfilter n drag chain filter
Redoxelektrode f oxidation-reduction electrode, redox electrode
Redoxgleichgewicht n redox equilibrium
Redoxindikator m redox indicator
Redoxmessung f redox measurement
Redoxpolymerisation f redox polymerization

Redoxpotential *n* oxidation-reduction potential, redox potential
Redoxreaktion *f* redox reaction
Redox-System *n* redox system
Redruthit *m* (Min) redruthite
Reduktase *f* (Biochem) reductase
Reduktinsäure *f* reductic acid
Reduktion *f* reduction, **elektrolytische** ~ electrolytic reduction
Reduktionsäquivalent *n* reduction equivalent
Reduktionsarbeit *f* reduction process
reduktionsfähig capable of reduction
Reduktionsflamme *f* reducing flame, reduction flame
Reduktionsgas *n* reducing gas, reduction gas
Reduktionsgeschwindigkeit *f* velocity of reduction
Reduktionsgetriebe *n* reducing gears, speed reducer
Reduktionshütte *f* reduction works
Reduktionskoeffizient *m* coefficient of reduction
Reduktionskohle *f* reducing carbon
Reduktionskraft *f* reducing power
Reduktionsmittel *n* reducing agent, deoxidizing agent, reducer, reducing agent, reductive agent, **energisches** ~ strong reducing agent
Reduktionsofen *m* (Metall) reduction furnace
Reduktions-Oxydationskette *f* reduction-oxidation cell
Reduktionspotential *n* reduction potential
Reduktionsschema *n* reduction scheme
Reduktionsschlacke *f* reducing slag, deoxidizing slag, reduction slag, white slag
Reduktionsschmelzen *n* reduction smelting
Reduktionsstufe *f* reduction stage
Reduktionsteilung *f* (Biol) reduction division, meiosis, (Phys) reduction graduation
Reduktionstemperatur *f* reduction temperature
Reduktionsventil *n* reducing valve, regulating valve
Reduktionsvermögen *n* reducing power
Reduktionsvorgang *m* reducing process
Reduktionswärme *f* heat of reduction
Reduktionszone *f* reduction zone
reduktiv reductive, by reduction
Reduktodehydrocholsäure *f* reductodehydrocholic acid, 3-hydroxy-7,12-diketocholanic acid
Redukton *n* reductone
reduzierbar reducible
Reduzierbarkeit *f* reducibility
reduzieren to reduce
Reduzieren *n* reducing, reduction
reduzierend reducing, reductive
Reduzierfähigkeit *f* reducing power
Reduzierflansch *m* reducing flange
Reduziergetriebe *n* reducing gear, reduction gear
Reduzierkupplung *f* reducing coupling
Reduziermittel *n* reducing agent

Reduzierofen *m* reduction furnace
Reduzierstärke *f* reducing power
Reduzierstück *n* reducer, reducing fitting, reducing piece
reduziert reduced
Reduziertransformator *m* reducing transformer, step-down transformer
Reduzier-T-Stück *n* reducing tee[-piece]
Reduzierung *f* reduction
Reduzierventil *n* reducing valve, reduction valve
Referat *n* lecture, report
Referenz *f* recommendation, reference
Referenzelektrode *f* reference electrode
Referenzellipsoid *n* reference ellipsoid
Referenzlinie *f* reference line
Refiner *m* refiner
reflektieren (Opt) to reflect
Reflektor *m* reflector, (Scheinwerfer) projector, ~ **von Neutronen** neutron reflector
Reflex *m* reflex, reflection; reflexion
Reflexbewegung *f* reflex [motion]
Reflexfolie *f* reflecting film
Reflexgalvanometer *n* mirror galvanometer, reflex galvanometer
Reflexion *f* reflection, reflex, reflexion, ~ **des Lichtes** reflection of light
Reflexionsablesung *f* reading by mirror
Reflexionsdämpfung *f* reflectivity
Reflexionsebene *f* reflecting surface
Reflexionsgalvanometer *n* reflex galvanometer
Reflexionsgitter *n*, **konkaves** ~ (Opt) concave reflection grating
Reflexionsgoniometer *n* reflecting goniometer
Reflexionskoeffizient *m* reflection coefficient, reflectivity
Reflexionskugel *f* sphere of reflection
Reflexionsmesser *m* reflection meter, reflectometer
Reflexionsmikroskop *n* reflecting microscope
Reflexionspolarisator *m* reflection polarizer
Reflexionsprisma *n* reflecting prism
Reflexionsvermögen *n* reflecting power, reflection factor
Reflexionswinkel *m* angle of reflection or reflexion
Reflexklystron *n* (Elektr) reflex klystron
Reflexreichtum *m* reflection abundance
Reformierbenzin *n* reformed gasoline
Reformierung *f* reforming
Reformwaren *f pl* reform goods
Refraktion *f* (Opt) refraction, **spezifische** ~ specific refraction
Refraktionsäquivalent *n* equivalent of refraction
Refraktionsindex *m* index of refraction
Refraktionskoeffizient *m* coefficient of refraction
Refraktionswinkel *m* angle of refraction
Refraktometer *n* (Opt) refractometer
refraktometrisch refractometric

Refraktor *m* (Opt) refractor, prismatic glass
Regal *n* shelf, bookcase, rack, **einhängbares** ~ hook-in rack
Regalfächer *pl* shelves in a rack
Regalförderzeug *n* rack stacker
Regalgang *m* [rack] aisle
Regalgasse *f* [rack] aisle
Regel *f* rule, principle, standard;, ~ **der gegenseitigen Ausschließung** (Spektr) rule of mutual exclusion;, **allgemeine** ~ general rule, **empirische** ~ empirical rule
Regelabweichung, bleibend ~ (Regeltechn) offset, steady-state error, or sustained deviation
Regelabweichung *f* exception, **bleibende** ~ offset
Regelarmaturen *f pl* regulating flow fittings
Regelation *f* (von Eis) refreezing, regelation
Regelbandbreite *f* deviation from set value
regelbar adjustable, controllable, variable
Regelbarkeit *f* adjustability, controllability
Regelbereich *m* range of adjustment, range of control, scale range
Regelgenauigkeit *f* control accuracy
Regelgetriebe *n* PIV drive, regulable gear unit, variable speed gear[ing], **stufenloses** ~ infinitely variable speed transmission
Regelglied *n* control element
Regelgröße *f* controlled variable (Am. E.), controlled condition, **bleibende** ~ controlled variable (Am. E.), controlled condition (Brit. E.)
Regelknopf *m* control knob
Regelkreis *m* closed loop, control loop, control system (Brit. E.), feedback control system (Am. E.)
Regelkreisglieder *pl* control loop elements
regelmäßig regular
Regelmäßigkeit *f* regularity
Regelmotor *m* variable speed motor
regeln to regulate, to adjust, to control, to govern
Regelpunkt *m* control point
Regelraum *m* control room
regelrecht regular, normal
Regelschalter *m* (Elektr) regulating switch
Regelschaltung *f* control circuit
Regelschema *n* control diagram
Regelschleife *f* (Tauchmaschine) control festoon accumulator
Regelsignalgeber *m* control impulse transmitter
Regelspannung *f* control voltage
Regelstrecke *f* open loop, control system of a process (Am. E.)
Regelstufe *f* regulating step
Regeltechnik *f* measurement and control
Regeltransformator *m* (Elektr) variable transformer
Regelunempfindlichkeit *f* (Regeltechnik) dead band (Am. E.), dead zone (Br. E.)

Regelung *f* regulation, adjustment, (Regeltechnik) automatic control, feedback control, **automatische** ~ automatic control, automatic regulation, **schwebende** ~ floating control, **stetige** ~ modulating control
Regelungstechnik *f* automatic control technology, control engineering
Regelungsvorgang *m* regulation process
Regelventil *n* control valve, regulating valve
Regelverstärker *m* control amplifier, variable [gain] amplifier
Regelvorgang *m* control process
Regelvorrichtung *f* adjusting device, control element, control equipment, controls
Regelweise *f* control form
Regelwiderstand *m* adjustable resistance, regulating resistance
regelwidrig contrary to rule, abnormal, irregular
Regelzeit *f* (Regeltechn) recovery time
Regenboden *m* (Dest) shower plate
Regenbogenachat *m* (Min) rainbow agate
Regenbogenerz *n* (Min) iridescent lead sulfate
Regenbogenhaut *f* (Anat) iris
Regenbogenquarz *m* (Min) rainbow quartz
Regenerat *n* regenerated material, (Gummi) reclaim, reclaimed rubber
Regeneratcellulose *f* regenerated cellulose
Regeneratgummi *n* reclaimed rubber, regenerated rubber
Regeneration *f* regeneration, (Gummi) devulcanization, reclaiming, (Katalysator) revivification
Regenerationsfähigkeit *f* regenerative power
Regenerationsprodukt *n* product of regeneration
Regenerationsprozeß *m* process of regeneration
Regenerationsstoffwechsel *m* regenerative metabolism
Regenerationsvermögen *n* recovery power
Regenerationsvorgang *m* regenerative process
Regenerativelement *n* regenerative cell
Regenerativfeuerung *f* regenerative firing, regenerative furnace, regenerator
Regenerativkammer *f* regenerative chamber
Regenerativofen *m* regenerative [gas] furnace
Regenerativprinzip *n* regenerative principle
Regeneratkautschuk *m* devulcanized rubber, reclaimed rubber
Regeneratkessel *m* digester
Regenerator *m* regenerator
Regenerieranlage *f* regeneration plant
regenerieren to regenerate, to reactivate, (Gummi) to devulcanize, to reclaim
Regeneriermittel *n* (Gummi) reclaiming agent
Regenerierung *f* regeneration
Regenmesser *m* rain gauge
Regenschirmglocke *f* (Dest) umbrella-type bubble cap
Regenschirmglockenboden *m* (Dest) umbrella-type bubble-cap tray

Regentropfen *m* rain drop[let]
Regenwasser *n* rain water
Register *n* register, index, record
Registerkupplung *f* coupler switch unit
Registratursystem *n* filing system
Registrieranlage *f* recording arrangement
registrieren to record, to register
Registriergalvanometer *n* recording galvanometer
Registriergerät *n* recording instrument
Registrierinstrument *n* recording instrument, registering instrument
Registriermanometer *n* recording manometer, steam-pressure register
Registrierpapier *n* recording paper, chart paper
Registrierpyrometer *n* recording pyrometer
Registriertafel *f* recorder panel
Registrierthermometer *n* recording thermometer, self-registering thermometer, temperature recorder
Registrierung *f* registration, recording, scanning
Registriervorrichtung *f* recording mechanism
Registrierwaage *f* recording scales
Regler *m* control equipment, controlling means, control system, (Drehzahl) governor, speed-setting device, (Elektr) controller, regulator, (f. Wärme) thermostat, (Kunststoffe) modifier, (Reglerventil) regulating valve, valve governor, **integral wirkender** ~ integral (or floating) action controller
Reglermuffe *f* governor sleeve
Reglersubstanz *f* regulating substance; regulator
Reglerventil *n* governor valve, regulating valve
Reglerwiderstand *m* rheostat
regnen (Bodenhydraulik) to weep
Regnen *n* (Bodenhydraulik) weeping
Regressionslinie *f* regression line
Regressionsmodell *n* regression model
regulär regular
Regularisierung *f* (Quant) regularization
Regulation *f* (Biol) regulation
Regulator *m* regulator, governor
Regulatorenzym *n* regulatory enzyme
Regulatorfeder *f* governor spring
Regulatorgen *n* regulatory gene
Regulatorklappe *f* governor slide, governor valve
Regulatorstreifen *m* control strip
Regulatorwelle *f* governor shaft
regulierbar adjustable, controllable, regulable
Regulierbarkeit *f* adjustability, controllability
regulieren to regulate, to control, to govern, (justieren) to adjust
Regulierhahn *m* regulating cock
Regulierschieber *m* regulating slide, regulating damper, regulating valve
Regulierschraube *f* regulating screw
Regulierspannung *f* regulating voltage

Regulierstange *f* regulating rod
Reguliertransformator *m* regulating transformer
Regulierung *f* regulation, control, governing
Regulierventil *n* control valve, regulating valve
Reguliervorrichtung *f* regulating device, regulating apparatus
Regulierwiderstand *m* rheostat, adjustable resistance, compensating resistance, regulating resistance
regulinisch reguline, consisting of pure metal
Regulus *m* regulus
regungslos motionless, still
rehfarben fawn-colored
Rehleder *n* doeskin
Rehmannsäure *f* rehmannic acid
Reibahle *f* broach, reamer, reamering tool
Reibapparat *m* grinding apparatus, triturating apparatus
Reibe *f* grater, rasp
reibecht fast to rubbing
Reibechtheit *f* rubbing fastness
Reibeigenschaft *f* frictional property
Reibeisen *n* grater, rasp, reamering tool
reiben to grate, to grind, to rasp, to rub, (polieren) to polish, (scheuern) to scour, (zermalmen) to triturate, (zerreiben) to crush
Reibeputz *m* textured plaster
Reibfestigkeit *f* resistance to abrasion
Reibfläche *f* frictional area, frictional surface, rubbing surface
Reibkissen *n* rubber pad
Reibkorrosion *f* fretting corrosion (destruction of metal surfaces by the combined action of corrosion and very slight frictional movement of the surfaces in contact)
Reiblöten *n* tinning
Reibrad *n* friction wheel or pulley or disk
Reibradantrieb *m* friction drive
Reibring *m* friction ring
Reibrolle *f* friction disk
Reibschale *f* (Mörser) mortar
Reibscheibe *f* friction disk, friction wheel
Reibstauförderer *m* friction accumulating conveyor
Reibstein *m* grindstone
Reibteller *m* friction disk
Reibtrommel *f* friction drum, friction roller
Reibung *f* friction, abrasion, grinding, rubbing, **gleitende** ~ sliding friction, **innere** ~ internal friction, (Flüssigkeit) viscosity, **rollende** ~ rolling friction
Reibungsabnutzung *f* frictional wear
Reibungsabzug *m* friction let-off
Reibungsantrieb *m* friction drive
Reibungsarbeit *f* frictional work
Reibungsbremse *f* friction brake
Reibungsdruckabfall *m* friction pressure drop
Reibungsdurchmesser *m* frictional diameter

Reibungselektrisiermaschine *f* electrical friction machine, friction electrostatic machine
Reibungselektrizität *f* frictional electricity
Reibungserwärmung *f* friction heating
Reibungsfaktor *m* coefficient of friction, friction factor
Reibungsfläche *f* friction surface, rubbing surface
reibungsfrei frictionless, without friction
Reibungshammer *m* friction hammer
Reibungskegel *m* friction cone
Reibungskoeffizient *m* coefficient of friction, friction[al] coefficient
Reibungskraft *f* frictional force
Reibungskupplung *f* friction clutch, friction coupling
reibungslos frictionless
Reibungsmaschine *f* friction machine
Reibungsprobe *f* friction test
Reibungsschicht *f* friction layer
Reibungsschweißen *n* friction welding, spin welding
Reibungsspannung *f* friction stress
Reibungsverlust *m* friction[al] loss
Reibungswärme *f* frictional heat, heat due to friction
Reibungswiderstand *m* frictional resistance
Reibungswinkel *m* angle of friction
Reibungswirkung *f* rubbing action
Reibungszahl *f* coefficient of friction, friction[al] coefficient
Reibwalze *f* friction roller
Reibwalzwerk *n* friction roll
Reibzeug *n* rubber pad
reich rich, (reichlich) abundant, (wohlhabend) wealthy
Reichardtit *m* (Min) reichardtite
Reichblei *n* argentiferous lead, rich lead
Reichfrischen *n* (Kupfer) enriching of copper by refining
reichhaltig comprehensive
reichlich abundant, ample
Reichschaum *m* zinc crust
Reichschlacke *f* rich slag
Reichschmelzen *n* precious metal smelting
Reichweite *f* range, radius of operation, reach, scope, (Atom) energy range, energy region, **maximale** ~ (Atom) maximum range
Reichweiteneffekt *m* range effect
Reichweitenenergie *f* range energy
Reichweitenmessung *f* range measurement
Reichweitenverkürzung *f* range reduction
reif ripe; mature
Reif *m* white frost, (Reifen) collar, hoop, ring
Reife *f* maturity, ripeness
Reifegrad *m* degree of ripeness
Reifeisen *n* hoop iron, strip iron
reifen to become mature, to become ripe, to mature, to ripen
Reifen *m* (Kfz) tire, tyre (Br. E.), (Luftreifen) pneumatic, (Mech) collar, hoop, ring, ~ **in Übergröße** oversize tire, **profilierter** ~ non-skid tire, **schlauchloser** ~ tubeless tire
Reifenabschleifmaschine *f* buffing lathe
Reifenbreite *f* (Kfz) section diameter
Reifendecke *f* (Kfz) tire casing
Reifenhöhe *f* (Kfz) section height
Reifenprofil *n* tread
Reifenrohling *m* green tire
Reifenrunderneuerungsmaterialien *n pl* recapping material, tire capping materials
Reifenteilreparatur *f* (Kfz) sectional repair
Reifenwalzwerk *n* tire mill, strip mill
Reifenwickeltrommel *f* building-up form
Reifezeit *f* maturing time
Reifungsdauer *f* time for ripening
Reihe *f* row, line, order, (Folge) sequence, succession, (Math) progression, series, (Spalte) column, **arithmetische** ~ arithmetic progression, arithmetic series, **harmonische** ~ harmonic series, **homologe** ~ homologous series;, **in** ~ **geschaltet** serially connected, **unendliche** ~ (Math) infinite series
reihen to rank, to arrange
Reihenanordnung *f* tandem arrangement
Reihenbad *n* series bath
Reihenblockofen *m* series block furnace
Reihendynamo *m* series dynamo
Reihenentwicklung *f* (Math) progression
Reihenfertigung *f* series production
Reihenfilter *n* bag filter
Reihenfolge *f* order, sequence, succession, turn, **zeitliche** ~ chronological sequence
Reihenparallelschaltung *f* series parallel connection
Reihenschaltung *f* connection in series, cascade connection, series connection
Reihenschlußmotor *m* series-wound motor
Reihenuntersuchung *f* routine testing
rein (gereinigt) purified, (klar) clear, (sauber) clean, tidy, (unvermischt) pure, ~ **darstellen** to isolate
Reinasche *f* potash
Reinbenzol *n* rectified benzene
Reinblau *n* pure blue
Reindarstellung *f* purification, ~ **des Gases** production of pure gas
Reinelement *n* (ohne Isotope) pure element (without isotopes)
Reinfektion *f* (Med) reinfection, superinfection
Reingas *n* pure gas, treated gas
Reingewicht *n* net weight
Reingewinn *m* net profit, net gain, net yield
Reinheit *f* clearness, (Chem) purity, **spektrale** ~ spectral purity
Reinheitsgrad *m* [degree of] purity, degree of fineness, percentage purity
Reinheitsprobe *f* test for purity

reinigen to clean, to cleanse, to purge, to scour, to scrub, to wash, (Chem) to purify, (desinfizieren) to disinfect, (Dest) to rectify, (entfetten) to ungrease, (Met) to refine
Reinigen *n* cleaning, cleansing, washing, (Chem) purifying, (Met) refining, ~ **der Wolle** scouring of wool
Reiniger *m* purifier, defecator
Reinigermasse *f* cleanser, detergent, purifying material
Reinigung *f* cleaning, cleansing, washing, (Chem) purification, ~ **des Bades** clearing the bath, ~ **organischer Verbindungen** purification of organic compounds;, ~ **von Enzymen** purification of enzymes, **chemische** ~ dry cleaning, **elektrolytische** ~ electrolytical refining
Reinigungsanlage *f* purification plant, purifying plant, refiner
Reinigungsapparat *m* purifier, purifying apparatus
Reinigungsbad *n* cleaning bath
Reinigungsbassin *n* cleaning basin, depositing tank
Reinigungsbecken *n* cleaning basin
Reinigungsbehälter *m* filtering basin, filter tank
Reinigungsbottich *m* purifying tank
Reinigungsbrettchen *n* clearer board
Reinigungscreme *f* (Kosm) cleansing cream
Reinigungselektrode *f* sweeping electrode
Reinigungsfeld *n* cleaning field, clearing field
Reinigungsfeuer *n* refining fire
Reinigungsflasche *f* wash bottle
Reinigungsgeräte *n pl* cleaning equipment
Reinigungsgraben *m* cleaning trench
Reinigungskasten *m* mud box, settling box
Reinigungsmasse *f* purifying mass
Reinigungsmethode *f* method of purification
Reinigungsmittel *n* cleanser, detergent, purifier, purifying agent
Reinigungsöffnung *f* cleaning hole, clean-out, man hole
Reinigungsprozeß *m* process of purification, purification process
Reinigungssalbe *f* purifying salve
Reinigungstank *m* leaching tank
Reinigungstür *f* cleaning door
Reinit *m* (Min) reinite
Reinkohle *f* pure carbon
Reinkolonne *f* pure product column
Reinkultur *f* pure culture
Reinlecithin *n* pure lecithin
Reinnickel *n* pure nickel
reinrassig (Tier) thoroughbred
Reinseide *f* all-silk, pure silk
reinst purest, high-grade
Reinsubstanz *f* pure substance
Reinxylol *n* rectified xylene
Reis *m* rice

Reisbranntwein *m* arrack, rice spirit
Reishülle *f* rice hull
Reisöl *n* rice oil
Reissit *m* reissite
Reisstärke *f* rice starch
Reißbanddose *f* tear-strip can
Reißbaumwolle *f* reused cotton [from rags]
Reißblei *n* blacklead, graphite, plumbago
Reißbleitiegel *m* black lead crucible, graphite crucible
Reißboden *m* marking-off board
Reißbrett *n* drawing board
Reißdehnung *f* elongation at break
reißen to crack, to split, (zerreißen) to break, to burst, to tear
Reißen *n* **der Fäden** (Reifen) breaking of the cords
Reißfeder *f* drawing pen
Reißfestigkeit *f* tear resistance, rupture strength, tensile strength, (Fasern) tenacity
Reißfestigkeitsprobe *f* tearing-strength test
Reißgelb *n* auripigment, King's yellow, orpiment, yellow arsenic blende
Reißkohle *f* charcoal [crayon]
Reißlack *m* crackle finish
Reißnadel *f* etching tool, scribing tool
Reißnagel *m* drawing pin, thumbtack (Am. E.)
Reißöl *n* rag pulling oil
Reißprobe *f* tearing test, breaking test
Reißscheibe *f* bursting disc, rupture disc, safety diaphragm
Reißschiene *f* tee square, T-square
Reißverschluß *m* zipper
Reißversuch *m* tearing test
Reißweg *m* path of tear
Reißwolle *f* reclaimed wool, shoddy
Reißzeug *n* drawing instruments, drawing set
Reiter *m* rider
Reiterband *n* bias band
Reiterlineal *n* rider scale
Reitersalbe *f* blue ointment, mercury ointment
Reiterverschiebung *f* movement of the rider
Reiterversetzung *f* rider adjustment
Reiz *m* irritation, (Anregung) stimulus
Reizbarkeit *f* irritability
Reizdünger *m* growth stimulating fertilizer
reizempfindlich susceptible to stimuli
reizen to irritate, (anregen) to stimulate
reizend irritant, (anregend) stimulating
reizerregend (Med) irritative
Reizfortpflanzung *f* stimulus conduction
Reizgas *n* sneeze gas, tear gas
Reizhemmung *f* retardation of stimulation
Reizmittel *n* irritant, (Anregungsmittel) stimulant
Reizstoff *m* stimulating substance
Reizung *f* irritation, excitation, provocation, (Anregung) stimulation

Reizwirkung *f* irritant action, (Anregung) stimulating effect, **anregende** ~ stimulating effect, **störende** ~ irritating effect
Reklamation *f* complaint, objection
Reklame *f* advertisement, advertising, ~ **machen** to advertise
reklamieren to complain, to object
Rekombination *f* recombination, **bevorzugte** ~ preferential recombination
Rekombinationsgeschwindigkeit *f* (Elektr) recombination velocity
Rekombinationskoeffizient *m* recombination coefficient
Rekombinationskontinuum *n* recombination continuum or band
Rekombinationszentrum *n* recombination center
rekombinieren to recombine
Rekonstruktion *f* reconstruction
Rekonzentration *f* reconcentration
Rekristallisation *f* recrystallization
Rekristallisationszwillinge *pl* (Krist) recrystallization twins
rekristallisieren to recrystallize
Rektifikation *f* rectification, **azeotrope** ~ azeotropic rectification, **extraktive** ~ extractive distillation, **stetige** ~ continuous rectification
Rektifikationsapparat *m* rectifier, rectifying apparatus
Rektifikationskolonne *f* rectifying column, fractionating column
Rektifikationssäule *f* rectifying column
Rektifikationszone *f* rectification section
Rektifizieranlage *f* rectification plant
Rektifizierapparat *m* rectifier, rectifying apparatus
rektifizierbar rectifiable
Rektifizierboden *m* plate, tray
rektifizieren to rectify, to fractionate
Rektifizieren *n* rectifying
Rektifiziersäule *f* rectifying column
rektifiziert rectified
Rektifizierung *f* rectification
Rekuperator *m* (Metall) recuperator, reversing exchanger
Rekursionsformel *f* recursion formula
Relais *n* relay, ~ **mit Schnellauslösung** instantaneous-release relay, quick-operating relay, ~ **mit verzögerter Auslösung** slow-release relay, time-delay relay
Relation *f* relation
relativ relative, comparative
Relativbewegung *f* relative motion
Relativgenauigkeit *f* relative accuracy
relativistisch relativistic
Relativität *f* relativity
Relativitätsgesetz *n* law of relativity
Relativitätsprinzip *n* principle of relativity, relativity principle

Relativitätstheorie *f* theory of relativity, **spezielle** ~ restricted theory of relativity
Relativverschiebung *f* relative displacement
Relaxans *n* (Pharm) relaxant
Relaxation *f* relaxation
Relaxationsverfahren *n* (Iterationsverf) method of relaxation
Relaxationsverhalten *n* relaxation behavior
Relaxationszeit *f* relaxation time
Relief *n* relief, embossment
Reliefdruck *m* relief printing
Relugit *n* relugite
remanent remaining, residual, (Phys) remanent
Remanenz *f* (Phys) remanence, residual magnetism
Remanenzspannung *f* residual voltage
Remingtonit *m* remingtonite
Renardit *m* (Min) renardite
Renaturierung *f* renaturation
Rengel *m* tapping iron
Renghol *n* renghol
Renieratin *n* renieratene
Renneisen *n* direct-process malleable iron
Rennfeuer *n* bloomery fire, bloomery forge
Rennfeuereisen *n* soft iron
Rennfeuerschlacke *f* [catalan] furnace slag, tap cinder
Rennin *n* rennin, rennase
Rennreifen *m* (Kfz) racing tire
Renormierung *f* **der Masse** (Quant) renormalization of mass
Renovasculin *n* renovasculine, sterilized 10% lactose solution
renovieren to renovate, to redecorate, to renew
Renovierung *f* renovation
Renoxydin *n* renoxydine
Renoxyd[in]säure *f* renoxydic acid
Rensselärit *m* (Min) rensselaerite
Repandin *n* repandine
Repandulinsäure *f* repandulinic acid
Reparatur *f* repair [work]
reparaturbedürftig in need of repair
Reparaturmasse *f* patching compound
Reparaturplättchen *n* **zum Kaltauflegen** cold patch
Reparaturwerkstatt *f* repair workshop
reparieren to repair
repassieren (Seide) to boil off a second time
Replikation *f* replication, **disperse** ~ disperse replication, **konservative** ~ conservative replication, **semikonservative** ~ semi-conservative replication
Replikationsgeschwindigkeit *f* replication rate
Reppe-Synthese *f* (Chem) Reppe synthesis
Repräsentant *m* representative
Repräsentativwerbung *f* institutional advertising
Repression *f* (Biochem) repression, **genetische** ~ genetic repression, **katabolische** ~ catabolic

repression, catabolite repression, **koordinative** ~ coordinative repression
Repressor m repressor
Repressor-Induktor-Komplex m repressor-inducer complex
Reproduktion f reproduction
Reproduktionsverfahren n process of reproduction, reproducing method
reproduzierbar reproducible
Reproduzierbarkeit f reproducibility
reproduzieren to reproduce
Repsöl n (Rapsöl) colza oil, rape[seed] oil
Resacetophenon n resacetophenone, 2,4-dihydroxyacetophenone
Resaldol n resaldol
Resaurin n resaurin
Resazurin n resazurin, 7-hydroxy-2-phenoxazone-10-oxide
Rescinnamin n rescinnamine
Reseda f (Bot) reseda, mignonette
Resedagrün n mignonette green
Resedaöl n mignonette oil, reseda oil
Reserpan n reserpan
Reserpilin n reserpiline
Reserpin n reserpine
Reserpinalkohol m reserpic alcohol
Reserpindiol n reserpinediol
Reserp[in]säure f reserpic acid
Reserpon n reserpone
Reserpoxidin n reserpoxydine
Reservage f (Schutzbeize) resist paste
Reservebatterie f spare battery, standby battery
Reservedruck m resist printing
Reservekohlenhydrat n reserve carbohydrate
Reservemaschine f stand-by engine
Reservespeicher m (Comp) standby store
Reservetank m reserve tank
Reserveteil n spare part
Reservoir n reservoir, storage basin, tank
Resinat n abietate, resinate
Resineon n resineon
Resinit m (Min) resinite, pitch-stone
Resinol n resinol
Resinotannol n resinotannol
Resistenz f resistance
Resistsalz n resist salt
Resocyclopharol n resocyclopharol
Resodiacetophenon n resodiacetophenone, 4,6-diacetylresorcinol
Resoflavin n resoflavin
Resogalangin n resogalangin
Resokämpferid n resokaempferide
Resol n resol
Resolharz n resol resin
Resolingelb n resolin yellow
Resonanz f resonance, (Eigenschwingung) natural frequency, **in** ~ in tune, resonant, **kernmagnetische** ~ nuclear magnetic resonance

Resonanzband n resonance band
Resonanzbedingung f resonance condition
Resonanzbereich m resonance range
Resonanzeinfang m resonance capture
Resonanzenergie f resonance energy
Resonanzerschütterung f resonant vibration
Resonanzfluoreszenz f resonance fluorescence
Resonanzfrequenz f resonance frequency
Resonanzfrequenzänderung f resonant frequency variation
Resonanzglied n resonance term
Resonanzhinderung f, **sterische** ~ steric inhibition of resonance
Resonanzhybrid n resonance hybrid
Resonanzkreis m resonance or resonating circuit
Resonanzlinie f (Spektrum) resonance line
Resonanzneutron n (Atom) resonance neutron
Resonanzniveau n resonance level
Resonanzphoton n resonance photon
Resonanzschwingung f resonance vibration, sympathetic vibration
Resonanzsieb n resonance screen
Resonanzspannung f resonance potential
Resonanzspektrum n (Spektr) resonance spectrum
resonanzstabilisiert resonance-stabilized
Resonanzstrahlung f resonance radiation
Resonanzstreuer m resonance scatterer
Resonanzstreuung f resonance scattering
Resonanzstromkreis m resonance circuit
Resonanzstruktur f resonating structure
Resonanzüberlagerung f resonance overlap
Resonanzübertragung f resonance transfer
Resonanzverstärker m resonance amplifier
Resonanzwand f baffle
Resonanzwiderstand m resonance resistance
Resonanzwirkung f resonance effect
Resonanzzustand m resonance state
Resonator m resonator
Resoorobol n resoorobol
Resorber m re-absorber
resorbierbar reabsorbable, absorptive
resorbieren to resorb
Resorbieren n resorption
Resorbin n resorbin
Resorcin n 1,3-benzenediol, 1,3-dihydroxybenzene, resorcinol, resorcin
Resorcinblau n resorcin blue
Resorcinol-Phenol-Kleber m resorcinol-phenolic adhesive
Resorcinphthalein n resorcinol phthalein, 3,6-dihydroxy fluoran, fluorescein
Resorcinvergiftung f (Med) resorcinism
Resorcit m 1,3-cyclohexanediol, resorcitol, hexahydroresorcinol
Resorcylaldehyd m resorcyl aldehyde
Resorcylsäure f resorcylic acid
resorcylsauer resorcylate
Resorption f resorption

Resorufin *n* resorufin, 7-hydroxy-2-phenoxazone
Resosantal *n* resosantal
respiratorisch respiratory
Rest *m* remainder, remnant, rest, (Chem) radical, residue
Restablenkung *f* residual deviation
Restabweichung *f* (Opt) residual aberration
Reständerung *f* residual drift
Restaktivität *f* residual activity
restaurieren to restore
Restbelastung *f* residual charge
Restblech *n* residual [anode] sheet
Restdruck *m* residual pressure
Restfeld *n* remanent field
Restfeuchtigkeit *f* residual moisture [content]
Restgas *n* residual gas
Restglied *n* **einer unendlichen Reihe** (Math) remainder of an infinite series
Resthärte *f* (Wasser) permanent hardness, residual hardness
Restintensität *f* residual intensity
Restkondensationskern *m* (Atom) reevaporation nucleus
Restladung *f* residual charge
Restlauge *f* residual liquor or lye
restlich remaining, residual
Restlösung *f* residual liquid
restlos absolute, complete, total, without residue
Restmagnetismus *m* residual magnetism
Restneutron *n* residual neutron
Restöl *n* residual oil, flux oil
Restschmelze *f* residual melt
Restspannung *f* residual stress
Restspannungszustand *m* residual state of stress
Reststabkräfte *f pl* residual bar forces
Reststrahlung *f* (Atom) residual [nuclear] radiation
Reststrom *m* residual current
Restvalenz *f* residual valence
Restwärme *f* (Reaktor) afterheat
Restwiderstand *m* residual resistance
resublimieren to resublimate
Resultante *f* resultant
Resultat *n* result
resultieren to result
Resultierende *f* resultant
Resveratrol *n* resveratrol
Retarder *m* retarder
Retardierungsfaktor *m* retardation factor
Retardierungskorrektur *f* retardation correction
Retardierungszeit *f* retardation time
Reten *n* 1-methyl-7-isopropylphenanthrene, retene
Retenchinon *n* retene quinone
Retenperhydrid *n* perhydroretene, retene perhydride
Retention *f* retention, ~ **der Konfiguration** (Stereochem) retention of configuration

Retentionsindex *m* (Gaschromat) retention index
Rethrin *n* rethrin
Reticulin *n* reticulin
Reticulozyten *pl* reticulocytes
Retinal *n* retinal
Retinalith *m* (Retinasphalt) retinasphalt
Retinen *n* retinene
Retinol *n* retinol, codol, resin oil
Retorte *f* retort, alembic
Retortenbank *f* retort bench
Retortenbatterie *f* set of retorts
Retortenbauch *m* bulb of a retort
Retortengestell *n* retort stand
Retortengraphit *m* retort graphite
Retortenhalter *m* retort stand
Retortenklemme *f* retort clamp
Retortenkohle *f* retort coal, gas carbon, retort carbon
Retortenkontaktverfahren *n* retort contact process
Retortenmündung *f* mouth of a retort
Retortenofen *m* retort furnace
Retortenrückstand *m* residue of retorts
Retortenvergasung *f* gasification in retorts
Retortenverkokung *f* retort coking
Retortenvorstoß *m* adapter, condenser
Retronecanol *n* retronecanol
Retronecanon *n* retronecanone
Retronecin *n* retronecine
Retronecinsäure *f* retronecic acid
retten to save, (erhalten) to preserve
Retusche *f* retouch[ing]
retuschieren to retouch
Retuschierlack *m* retouching varnish
reverberieren to reverberate
Reverberierofen *m* reverberatory furnace
reversibel reversible
Reversibilität *f* reversibility
Reversibilitätsprinzip *n* (Opt) principle of reversibility
Reversierblechstraße *f* reversing plate train
Reversierblechwalzwerk *n* reversing plate mill
Reversierwalzwerk *n* reversing rolling mill
Reversionspendel *n* reversible pendulum
Revision *f* checking, testing
Revolver *m* revolver, (Mikroskop) nose piece
Revolverdrehbank *f* capstan lathe, revolver lathe, turret lathe
Revolverstanze *f* rotating punch
Reynoldsche Zahl *f* Reynold's number
Reyon *m* artificial silk, rayon
Reyoncellulosefaser *f* (Text) viscose rayon
Rezbanyit *m* rezbanyite
Rezensent *m* reviewer
rezensieren to criticize, to review
Rezension *f* criticism, review
Rezept *n* formula, recipe, (Pharm) prescription
Rezeptivität *f* receptivity
Rezeptor *m* receiver, (Biochem) receptor

rezeptpflichtig by prescription only, Rx only
rezessiv recessive
Rezipient *m* bell-jar, receiver, recipient
Rezipiententeller *m* bell-jar plate
reziprok reciprocal
Reziprokwert *m* reciprocal, ~ **der Dispersionskraft** constringence
Reziprozität *f* reciprocity
Reziprozitätstheorem *n* reciprocity theorem
RGT-Regel *f* (Reaktionsgeschwindigkeit-Temperatur-Regel) RVT rule (reaction velocity-time rule)
Rhabarber *m* (Bot) rhubarb
Rhabarberon *n* rhabarberone, 3-hydroxymethyl-chrysazin, alocemodin, isoemodin
Rhabarbersirup *m* rhubarb syrup
Rhabarbertinktur *f* rhubarb tincture
Rhabdionit *m* (Min) rhabdionite
Rhabdit *m* (Min) rhabdite
Rhabdolith *m* (Min) rhabdolith
Rhabdophan *m* (Min) rhabdophane, phosphocerite
Rhätizit *m* (Min) rhaeticite
Rhagit *m* (Min) rhagite
Rhamnal *n* 1,5-oxido-1-hexene-3,4-diol, rhamnal
Rhamnazin *n* rhamnazin
Rhamnegin *n* rhamnegin, xanthoramnin
Rhamnetin *n* rhamnetin
Rhamnicogenin *n* rhamnicogenol, pentahydroxy-methyl-9-anthrone
Rhamnicosid *n* rhamnicoside
Rhamnin *n* rhamnin
Rhamnit *m* rhamnitol, rhamnite
Rhamnocitrin *n* rhamnocitrin
Rhamnofluorin *n* rhamnofluorin
Rhamnoheptose *f* rhamnoheptose
Rhamnohexose *f* rhamnohexose
Rhamnol *n* rhamnol
Rhamnonsäure *f* rhamnonic acid, 2,3,4,5-tetrahydroxycaproic acid, 2,3,4,5-tetrahydroxyhexanoic acid
Rhamnose *f* rhamnose
Rhamnosid *n* rhamnoside
Rhamnosterin *n* rhamnosterin
Rhamnoxanthin *n* rhamnoxanthin
Rhamnulose *f* rhamnulose
Rhaponticin *n* rhaponticin
Rhapontigenin *n* rhapontigenin
Rhapontin *n* rhapontin
Rhein *n* rhein
Rheinamid *n* rhein amide
Rheinchlorid *n* rhein chloride
Rhenium *n* rhenium (Symb. Re)
Rheniumheptoxid *n* rhenium heptoxide
Rhenium(VI)-oxid *n* rhenic [acid] anhydride, rhenium trioxide
Rheochrysin *n* rheochrysin

Rheologie *f* rheology
rheologisch rheological
Rheometer *n* rheometer
rheopektisch rheopectic
Rheopexie *f* (Kolloidchem) rheopexy
Rheostat *m* rheostat, variable resistance
rheostatisch rheostatic[ally]
Rhesusaffe *m* rhesus monkey
Rhesusfaktor *m* rhesus-factor, Rh-factor
Rhesusfaktorbestimmung *f* Rh-testing
rhesus[faktor]-positiv Rh-positive
Rhetsin *n* rhetsine
Rhetsinin *n* rhetsinine
rheumatisch (Med) rheumatic
Rheumatismus *m* (Med) rheumatism
Rheumemodin *n* rheumemodin
Rh-Faktor *m* (Rhesusfaktor) rhesus factor, Rh factor
Rhigolen *n* rhigolene
Rhinanthin *n* rhinanthin
Rhinologie *f* (Med) rhinology
Rhizocarpsäure *f* rhizocarpic acid
Rhizoid *n* (Bot) rhizoid
Rhizom *n* (Bot) rhizome, rootstock
Rhizoninsäure *f* rhizoninic acid
Rhizonsäure *f* rhizonic acid, barbatic acid
Rhizopterin *n* rhizopterine
Rhodacen *n* rhodacene
Rhodaform *n* rhodaform, hexamethylenetetramine methyl thiocyanate
Rhodamin *n* rhodamine, 3,6-diaminofluoran
Rhodaminplatte *f* rhodamine plate
Rhodan *n* thiocyanogen
Rhodanaluminium *n* aluminium thiocyanate
Rhodanammonium *n* ammonium thiocyanate
Rhodanat *n* thiocyanate, rhodanide, salt or ester of thiocyanic acid, sulfocyanate
Rhodanbarium *n* barium thiocyanate
Rhodancalcium *n* calcium thiocyanate
Rhodaneisen *n* iron thiocyanate, ferric thiocyanate, iron(III) thiocyanate
Rhodanid *n* rhodanide, salt or ester of thiocyanic acid, sulfocyanate, thiocyanate
Rhodanin *n* rhodanine, rhodanic acid
Rhodankali[um] *n* potassium thiocyanate
Rhodankupfer *n* copper thiocyanate; copper(II) thiocyanate, cupric thiocyanate
Rhodanlösung *f* thiocyanate solution
Rhodanmethyl *n* methyl thiocyanate
Rhodannatrium *n* sodium thiocyanate
Rhodanquecksilber *n* mercury thiocyanate
Rhodansalz *n* thiocyanate, rhodanide, salt of thiocyanic acid, sulfocyanate
Rhodantonerde *f* (obs) aluminum thiocyanate
Rhodanür *n* (obs) thiocyanate, rhodanide, sulfocyanate
Rhodanverbindung *f* thiocyanogen compound

Rhodanwasserstoff *m* hydrogen thiocyanate, hydrogen sulfocyanate, rhodanic acid, sulfocyanic acid, thiocyanic acid
Rhodanzinn *n* tin thiocyanate, stannic thiocyanate, tin(IV) thiocyanate
Rhodeasapogenin *n* rhodeasapogenin
Rhodeit *m* rhodeite, rhodeol
Rhodeohexonsäure *f* rhodeohexonic acid, 2,3,4,5,6-pentahydroxyenanthic acid, 2,3,4,5,6-pentahydroxyheptanoic acid
Rhodeonsäure *f* rhodeonic acid, 2,3,4,5-tetrahydroxycaproic acid; 2,3,4,5-tetrahydroxyhexanoic acid
Rhodeose *f* rhodeose
Rhodeotetrose *f* rhodeose-tetrose, d-lyxomethylose
Rhodexin *n* rhodexin
Rhodinal *n* rhodinal
rhodiniert rhodium-plated
Rhodinol *n* rhodinol
Rhodinsäure *f* rhodinic acid
Rhodium *n* rhodium (Symb. Rh), **kolloidales** ~ colloidal rhodium, electrorhodial (Am. E.)
Rhodiumbad *n* rhodium bath
Rhodiumchlorid *n* rhodium chloride, rhodium trichloride
Rhodiumgehalt *m* rhodium content
Rhodiumgold *n* (Min) rhodite, rhodium-gold
Rhodiummetall *n* metallic rhodium, rhodium [metal]
Rhodiumoxid *n* rhodium oxide
Rhodiumplattierung *f* rhodanizing
Rhodiumsalz *n* rhodium salt
Rhodiumverbindung *f* rhodium compound
Rhodizit *m* (Min) rhodizite, native calcium borate
Rhodizonsäure *f* rhodizonic acid, dihydroxydiquinoyl
Rhodochlorin *n* rhodochlorine
Rhodochrosit *m* (Min) rhodochrosite, native manganese carbonate, red manganese
Rhodocladonsäure *f* rhodocladonic acid, hexahydroxymethylanthraquinone
Rhododendrol *n* rhododendrol
Rhodol *n* rhodol
Rhodolith *m* (Min) rhodolite
Rhodonit *m* (Min) rhodonite, manganese spar, native manganese silicate
Rhodophit *m* rhodophite
Rhodophyllit *m* rhodophyllite
Rhodopin *n* rhodopin
Rhodoporphyrin *n* rhodoporphyrine
Rhodopsin *n* rhodopsin
Rhodosamin *n* rhodosamine
Rhodotilith *m* rhodotilite
Rhodoxanthin *n* rhodoxanthin, thujorhodin
Rhodulin *n* rhoduline
Rhodusit *m* rhodusite
Rhöadin *n* rhoeadine

Rhöagenin *n* rhoeagenine
Rhönit *m* (Min) rhoenite
Rhoifolin *n* rhoifolin
Rhombarsenit *m* (Min) rhombarsenite, claudetite
Rhombendodekaeder *n* (Krist) rhombic dodecahedron
rhombenförmig rhombic, rhomb-shaped
Rhombenglimmer *m* (Min) rhombic mica, biotite, phlogopite
Rhombifolin *n* rhombifoline
Rhombinin *n* rhombinine
rhombisch rhombic, rhomb-sphaped
rhombischhemiedrisch rhombohemihedral
rhombischhemimorph rhombohemimorphous
rhombischholoedrisch rhomboholohedral
rhombischpyramidal rhombopyramidal
Rhomboeder *n* (Krist) rhombohedron
Rhomboederfüllsteine *m pl* rhomboid filling stones
rhomboedrisch (Krist) rhombohedral
Rhomboid *n* rhomboid
rhomboidisch rhomboid, rhomboidal
Rhomboklas *m* (Min) rhomboclase
Rhombus *m* diamond, rhombus
Rhusester *m* rhus ester
Rhuslack *m* rhus varnish
Rhyakolith *m* rhyacolite
Rhyolith *m* rhyolite
rhythmisch rhythmic
Rhythmus *m* rhythm
Ribamin *n* ribamine
Ribazol *n* ribazole
Ribit *m* ribitol
Ribodesose *f* ribodesose
Riboflavin *n* riboflavin, lactoflavin, vitamin B$_2$
Riboflavinphosphat *n* riboflavin phosphate
Riboketose *f* riboketose
Ribonsäure *f* ribonic acid
Ribonuclease *f* (Biochem) ribonuclease
Ribonucleinsäure *f* ribonucleic acid, **virale** ~ viral ribonucleic acid
Ribosamin *n* ribosamine
Ribose *f* ribose
Ribosom *n* ribosome
Ribosomenausgangskomplex *m* ribosome initiation complex
Ribosomendissoziation *f* ribosome dissociation
Ribosomenfunktion *f* ribosome function
Ribosomenrekonstitution *f* ribosome reconstitution
Ribosomenverteilung *f* ribosome distribution
Ribulose *f* ribulose
Ribulosediphosphat *n* ribulose diphosphate
Richellit *m* (Min) richellite, native iron calcium phosphate
Richtapparat *m* trueing apparatus
Richtblei *n* plumb bob, plumb line
Richtcharakteristik *f* directional diagram

Richtdorn *m* mandrel
Richtehalle *f* straightening shop
richten to adjust, to repair, (ausrichten) to straighten
Richterit *m* (Min) richterite
Richtfähigkeit *f* directional property
Richtform *f* **beim Akkukasten** shrink form
Richtgröße *f* directional quantity
richtig correct, right, true
Richtigkeit *f* accuracy, correctness; properness
Richtigstellung *f* correction, rectification
Richtleistung *f* apparent output or power
Richtleitwert *m* admittance
Richtlinie *f* direction, guide-line, *(pl)* code of practice
Richtlot *n* plumb bob
Richtmagnet *m* controlling magnet, directing magnet
Richtmaschine *f* straightening machine
Richtmaß *n* standard of measure
Richtplatte *f* straightening plate
Richtpresse *f* straightening press
Richtrezeptur *f* recommended formulation, suggested formulation
Richtscheit *n* rule
Richtschiene *f* levelling rule
Richtschnur *f* guide, plumb line
Richtschraube *f* adjusting screw
Richttyptiefe *f* reference type strength
Richtung *f* direction, (Ausrichtung) adjustment, alignment, (Entwicklung) trend, (Orientierung) orientation, **achsiale** ~ axial direction, **bevorzugte** ~ (Opt) privileged direction, **optische** ~ optical direction, **senkrechte** ~ vertical direction
Richtungsabhängigkeit *f* directional dependence
Richtungsänderung *f* change in direction
Richtungsangabe *f* indication of direction
Richtungsanzeige *f* indication of direction
Richtungsanzeiger *m* direction indicator
Richtungsdurchschlag *m* directional breakdown
richtungsfokussierend direction focus[s]ing
Richtungsfokussierung *f* directional focus[s]ing
Richtungskoinzidenz *f* directional coincidence
Richtungsquantelung *f* directional quantization
Richtungsumkehr *f* reversal [of direction]
richtungsunabhängig independent of direction
Richtungsunschärfe *f* direction uncertainty
Richtungsverteilung *f* directional distribution
Richtungswechsel *m* change of direction
Richtungswinkel *m* angle of direction
Richtvorrichtung *f* shrinkage block or jig
Richtwaage *f* level
Richtwalze *f* straightening roll
Richtwerkzeug *n* dressing tool
Richtzähler *m* (Atom) directional counter
Ricin *n* ricin
Ricinelaidinsäure *f* 12-hydroxy-9-octadecenoic acid, ricinelaidic acid
Ricinenöl *n* dehydrated castor oil
Ricinensäure *f* ricinenic acid
Ricinin *n* 1-methyl-4-methoxy-3-cyano-2-pyridone, ricinine
Ricinolein *n* ricinolein, glyceryl or glycerol [tri]ricinoleate, triricinolein
Ricinoleinsäure *f* ricinoleic acid
Ricinoleylalkohol *m* ricinoleyl alcohol
Ricinolsäure *f* 12-hydroxy-9-octadecenoic acid, ricinol[e]ic acid, ricinic acid
Ricinsäure *f* 12-hydroxy-9-octadecenoic acid, ricinic acid, ricinol[e]ic acid
Ricinstearolsäure *f* ricinostearolic acid
Ricinuselaidinsäure *f* 12-hydroxy-9-octadecenoic acid, ricinelaidic acid
Ricinusöl *n* castor oil
Ricinus (s. auch unter „Rizinus")
Rickardit *m* (Min) rickardite
Riddellinsäure *f* riddellic acid
Riebeckit *m* (Min) riebeckite
riechen to smell
Riechstoff *m* odoriferous substance, perfume
riefeln to flute, to groove
Riefenbildung *f* scoring
Riegel *m* (Schloß) bolt
Riemannsche Integrierbarkeit *f* Riemann's criterion
Riemen *m* belt, strap
Riemenantrieb *m* belt drive
Riemenbreite *f* width of belt
Riemendicke *f* belt thickness
Riemenführer *m* belt guide
Riemengabel *f* belt guide
Riemenhaken *m* hook belt fastener
Riemenklammer *f* hook belt fastener
Riemenkralle *f* claw belt fastener
Riemenlänge *f* belt length
Riemenleder *n* belting
Riemenleiter *m* belt guide
Riemenpumpe *f* belt-driven pump
Riemenrollenformmaschine *f* pulley molding machine
Riemenscheibe *f* band pulley, belt pulley, **feste** ~ tight pulley, **geteilte** ~ split pulley, **lose** ~ loose pulley
Riemenscheibenformmaschine *f* pulley molding machine
Riemenscheibenmodellsatz *m* set of pulley patterns
Riemenschleifmaschine *f* belt-driven grinding machine
Riemenübertragung *f* belt transmission
Riemenverschiebung *f* belt shifting
Ries *n* (Papiermaß) ream
Rieselapparatur *f* trickling apparatus
Rieselelektrode *f* dropping electrode
rieselfähig (Pulver) free-flowing
Rieselfilm *m* falling film

Rieselfilm-Destillationsturm *m* wetted-wall tower
Rieselfilmelektrode *f* dropping electrode
Rieselfilmkolonne *f* falling film column, wetted-wall column
Rieselfläche *f* irrigation surface
Rieselkondensator *m* spray condenser
Rieselkühler *m* open-surface cooler
rieseln to trickle
Rieseln *n* trickling
Rieselreaktor *m* trickle-bed reactor
Rieselrost *m* grate with external water cooling
Rieselströmung *f* trickle flow
Rieseltrockner *m* cascade dryer, flighted dryer, spray drier
Rieselwärmeaustauscher *m* trickled-surface heat exchanger
Rieselwasser *n* irrigation water, trickling water
Rieselwerk *n* cooling tower
Riesenblutkörperchen *n* megalocyte
Riesenchromosom *n* giant chromosome
Riesenerythrozyt *m* giant erythrocyte
Riesenmolekül *n* giant molecule, high polymer
Riesenwuchs *m* giant growth
Riesenzelle *f* giant cell
Riet *n* (Web) comb
Rietblatt *n* (Web) comb
Riffelblech *n* checkered sheet (Am. E.), chequered plate
Riffelhorde *f* corrugated heat-transfer strip for regenerators
riffeln to corrugate
Riffeltrichter *m* ribbed funnel
Riffelung *f* corrugation, groove, grooving
Riffelwalze *f* corrugated roll
Rikolit *m* ricolite
Rillbarkeit *f* creasability
Rille *f* groove
rillen to crease
Rillenbreite *f*, **obere** ~ top width of groove
Rillengrund *m* grooved part
Rillenisolator *m* grooved insulator
Rillenkugellager *n* deep-groove ball bearing
Rillenreibrad *n* grooved friction wheel
Rillenschiene *f* grooved rail
Rillentrommel *f* grooved drum
Rillenwalzentrockner *m* grooved drum drier
Rillenwinkel *m* (Keilriemen) groove angle
Rilltiefe *f* depth of crease
Rillung *f* crease
Rinde *f* (Baum) bark, cortex, (Erde etc.) crust
Rindenbildung *f* (Bot) cortication
Rindenborax *m* jeweller's borax
Rindenhormon *n* cortical hormone
Rindenstein *m* (Min) tufaceous limestone
Rindenstoff *m* corticin
Rindenzerfaserungsmaschine *f* bark shredding machine
Rinderfett *n* beef suet

Rindermarkfett *n* beef marrow fat
Rindertalg *m* beef tallow
Ring *m* ring, (Kettenglied) link, (Öse) hoop, loop, (Unterlegscheibe) washer, ~ **der Walze** collar of the roll, **aromatischer** ~ (Chem) aromatic ring, **benzoider** ~ (Chem) benzoid ring, **ebener** ~ (Chem) uniplanar ring, **fünfgliedriger** ~ (Chem) five-membered ring, **spannungsfreier** ~ (Chem) strainless ring
Ringachat *m* (Min) ring agate
Ringacylierung *f* ring acylation
Ringaufdornversuch *m* ring expanding or drift test
Ringaufspaltung *f* ring cleavage, ring rupture
Ringbandscheider *m* circular-magnet belt separator
Ringbeschlag *m* ferrule
Ringbildung *f* (Chem) cyclization, ring closure, ring formation
Ringbrenner *m* ring burner
Ringbrennpunkt *m* ring focus
Ringeinguß *m* circular gate runner
Ringelektrode *f* annular electrode
Ringelektromagnet *m* ring-shaped electromagnet
Ringersche Lösung *f* Ringer's solution
Ringerweiterung *f* ring extension
Ringetagentrockner *m* turbo-drier
Ringfaltversuch *m* ring folding test
ringförmig annular, circular, ring-shaped, (Chem) cyclic
Ringform *f* circular form, ring shape
Ringfüllkörper *m pl* rings
Ringgebläse *n* side-channel compressor
Ringgenerator *m* annular producer
Ringgitter *n* ring grating
Ringhomologe[s] *n* cyclic homolog[ue]
Ringkanal *m* ring channel
Ringkessel *m* circular tank
Ringketon *n* cyclic ketone
Ring-Ketten-Struktur *f* ring-cluster structure
Ring-Ketten-Tautomerie *f* (Chem) ring-chain tautomerism
Ringkolbenzähler *m* rotary piston meter
Ringleitung *f* ring conduit
Ringmagnet *m* annular magnet, ring magnet
Ringmechanismus *m* ring mechanism
Ringmutter *f* eye nut
Ringnut *f* annular groove
Ringöffnung *f* (Chem) ring opening
Ringofen *m* annular furnace
Ringpresse *f* pot press
Ringsäule *f* (Gaschromat) coiled column
Ringschaltung *f* loop system
Ringschluß *m* (Chem) cyclization, ring closure
Ringschraube *f* eyebolt
Ringsolenoid *n* annular solenoid
Ringspalt *m* annular clearance, annulus
Ringspaltung *f* (Chem) ring cleavage
Ringspaltweite *f* annulus width

Ringströmung *f* annular flow, film flow
Ringstruktur *f* (Chem) ring structure
Ringsystem *n* (Chem) cyclic system
Ringummantelungsmaschine *f* ring covering machine
Ringverbindung *f* (Chem) cyclic compound, ring compound
Ringvorlesung *f* interdisciplinary course of lectures
Ringwaage *f* ring balance
Ringwalzenmühle *f* ring roll mill
Ringwalzenpresse *f* ring roll press
Ringzugfestigkeit *f* annular tensile strength
Ringzugversuch *m* ring tensile test
Rinkit *m* (Min) rinkite
Rinne *f* (Abflußrinne) channel, drain, drainage, outlet, pipe, (Dachrinne) gutter, (Riefe) furrow, groove
Rinneit *m* (Min) rinneite
rinnen (fließen) to flow, to run, (leck sein) to leak, (rieseln) to trickle
Rinnen *n* (Strömen) flowing
rinnenförmig channeled, fluted, grooved
Rinnenofen *m* submerged-channel induction furnace
Rinnentrog *m* shaking trough
Rionit *m* rionite
Ripidolith *m* (Min) ripidolite
Riponit *m* riponite
Rippe *f* rib, (Gewölberippe) groin, (Kühlrippe) cooling fin, ~ **im Seitendekor** (Reifen) decorative rib
Rippelblech *n* checkered plate
Rippenkühler *m* ribbed cooler
Rippenoberfläche *f* extended surface
Rippenplattenwärmeaustauscher *m* brazed-plate-fin heat exchanger
Rippenrohr *n* finned pipe, finned tube
Rippenrohrwärmeaustauscher *m* extended-surface heat exchanger, finned-tube heat exchanger
Rippentrichter *m* ribbed funnel
Risiko *n* risk, venture
Risinsäure *f* risic acid. rissic acid
rissig cracked, crevassed, creviced, fissured, torn, (Holz) shaken
Ristin *n* ristin, alcoholic solution of hydroxyethyl benzoate
Riß *m* (Bruch) fracture, rupture, (durch Reißen) rip, tear, (Holz) chink, shake, (Leck) leakage, (Lücke) gap, (Spalte) crevasse, (Spaltung) fissure, (Sprung) crack, flaw, (Steingut) craze, (Zeichnung) draft, plan, **interkristalliner** ~ intercrystalline crack
Rißauffangtemperatur *f* crack arrest temperature (CAT)
Rißausbreitung *f* crack propagation
Rißausweitungskraft *f* crack extension force
Rißbildung *f* crack formation, cracking, (Lack) checking, ~ **im Profil** channel cracking, groove cracking
Rißebene *f* plane of projection
Rißfestigkeit *f* crack strength
rißfrei free from cracks, free from flaws
Rißfreiheit *f* freedom from cracking
Rißtafel *f* plane of projection
Rißverfahren *n* method of projection
Rißwachstum *n* (Prüftechnik) crack growth
Rittingerit *m* (Min) rittingerite
Ritz *m* (Kratzer) scratch
ritzbar susceptible to scratching
Ritzbarkeit *f* susceptibility to scratching
Ritzel *n* pinion, bevel gear, gear wheel
Ritzelwelle *f* pinion shaft
ritzen to scratch, to slit
Ritzen *n* (Kerbbildung) scoring
Ritzhärte *f* sclerometric hardness, scratch hardness, scratch resistance
Ritzhärteprobe *f* surface-scratching test
Ritzhärteprüfer *m* sclerometer, scratch tester
Ritzhärteprüfung *f* sclerometer test
Ritzhärtezahl *f* scratch-hardness number
Ritzmaschine *f* scoring machine, scratching machine
Ritzverfahren *n* scratch method
Ritzversuch *m* scratch test
rivalisieren to rival, to oppose
Rivanol *n* rivanol
Rivotit *m* (Min) rivotite, native basic copper antimony carbonate
Rizinoleat *n* ricinoleate, ricinate
Rizinus *m* (Bot) castor-oil plant (s. auch unter „Ricinus")
Rizinusöl *n* castor oil, **geschwefeltes** ~ sulfonated castor oil
Rizinusölseife *f* castor oil soap
Rizinussamen *m* castor beans
RNS-Biosynthese *f* RNA biosynthesis
RNS-Polymerase *f* (Biochem) RNA polymerase
RNS-Synthese *f* RNA synthesis
Robbe *f* (Zool) seal
Robbentran *m* seal oil
Robiniaöl *n* robinia oil
Robinin *n* robinin
Robinobiose *f* robinobiose
Robinucleinsäure *f*, **ribosomale** ~ ribosomal ribonucleic acid
Roblingit *m* (Min) roblingite
Roboter *m* robot
Roburit *m* (Sprengstoff) roburite
Roccellin *n* roccelline, orse[i]llin
Roccellsäure *f* roccellic acid
Rochellesalz *n* Rochelle salt, potassium sodium tartrate
Rockwellhärte *f* Rockwell hardness
Rockwellhärteprüfung *f* Rockwell-hardness test
Rockwellhärtezahl *f* Rockwell-hardness number
Rodalith *m* rodalite

roden to clear, to root out
Rodinal *n* rodinal
Rodung *f* cleared woodland
Röhre *f* pipe, conduit, tube, (Med) duct, **abgeschmolzene** ~ sealed-off tube, **Braunsche** ~ Braun tube, cathode ray tube, **gasgefüllte** ~ gas-content tube, gas-filled tube, **luftgekühlte** ~ air-cooled tube, **nahtlose** ~ seamless pipe, **wassergekühlte** ~ water-cooled tube
röhrenartig tubular, fistular, tube-like
Röhrenbestückung *f* tube complement
Röhrencassie *f* (Bot) cassia pulp
Röhrendestillationsofen *m* pipe still
Röhrenfilter *m* pipe filter
röhrenförmig tubular
Röhrengang *m* pipe line, conduit, main
Röhrengenerator *m* tube generator
Röhrengleichrichter *m* tube rectifier
Röhrenguß *m* tube casting
Röhrenkessel *m* tubular boiler
Röhrenklemme *f* tube clip
Röhrenkondensator *m* tubular condenser, pipe condenser, surface condenser
Röhrenkontaktkessel *m* tube converter
Röhrenkühler *m* tubular cooler
Röhrenlampe *f* tubular lamp
Röhrenlibelle *f* tube level
Röhrenlot *n* pipe solder
Röhrenmagnet *m* tubular magnet
Röhrenmuffe *f* pipe socket
Röhrenofen *m* tube furnace, tube reactor, tubular furnace
Röhrenozonisator *m* tube ozonizer
Röhrenradius *m* tube radius
Röhrenseparator *m* tubular bowl separator
Röhrensicherung *f* tubular fuse
Röhrenspaltofen *m* steam cracker
Röhrenspaltverfahren *n* steam cracking process
Röhrenstrang *m* coil of piping
Röhrenträger *m* tube support
Röhrentrommeltrockner *m* drum drier with heating tubes
Röhrenventil *n* tube valve
Röhrenverdampfer *m* tube evaporator, tube still, tubular vaporizer
Röhrenverstärker *m* tube amplifier
Röhrenvoltmeter *n* valve voltmeter
Röhrenvorsatz *m* valve preamplifier
Röhrenwalzen *n* tube rolling
Röhrenwerk *n* piping, tubing
Röhrenwischer *m* tube brush
Röhrenzentrifuge *f* tubular bowl centrifuge
Roemerin *n* remerine
Römerit *m* (Min) romerite
Römerkerze *f* Roman candle
Röntgen *n* (Röntgeneinheit) roentgen [unit]
Röntgenabsorptionskontinuum *n* X-ray absorption band

Röntgenabsorptionsspektrum *n* X-ray absorption spectrum
Röntgenanalyse *f* X-ray analysis
Röntgenanlage *f* X-ray equipment
Röntgenapparat *m* röntgen apparatus, X-ray apparatus
Röntgenaufnahme *f* X-ray [picture], radiograph, X-ray film
Röntgenbefund *m* radiological result[s]
Röntgenbehandlung *f* (Med) X-ray treatment
Röntgenbestrahlung *f* roentgenotherapy, radiotherapy
Röntgenbeugungsaufnahme *f* X-ray diffraction pattern
Röntgenbeugungsmethode *f* X-ray diffraction method
Röntgenbild *n* X-ray picture, radiograph
Röntgenblitz *m* X-ray flash
Röntgenblitzinterferenz *f* X-ray flash interference
Röntgendiagnostik *f* X-ray diagnostics
Röntgendiagram *n* X-ray diagram, X-ray diffraction pattern
Röntgeneinheit *f* roentgen [unit]
Röntgeneinrichtung *f* X-ray equipment
Röntgen-Emissionsspektrum *n* (Spektr) X-ray emission spectrum
Röntgenfeinstruktur *f* (Krist) X-ray structure
Röntgenfilm *m* roentgen film, X-ray film
Röntgenlaboratorium *n* X-ray laboratory
Röntgenlinie *f* X-ray line
Röntgenmesser *m* roentgenometer
Röntgenmeßgerät *n* roentgenometer
Röntgennegativ *n* X-ray negative
Röntgenogramm *n* roentgenogram, X-ray [picture]
Röntgenographie *f* roentgenography
röntgenographisch radiographic
Röntgenologe *m* roentgenologist
Röntgenologie *f* roentgenology
Röntgenoskop *n* roentgenoscope
Röntgenoskopie *f* roentgenoscopy
Röntgenpositiv *n* X-ray positive
Röntgenröhre *f* X-ray tube
Röntgenschirm *m* fluorescent screen
Röntgenspektrograph *m* X-ray spectrograph
Röntgenspektrum *n* Roentgen spectrum
Röntgenstrahl *m* Roentgen ray, X-ray
Röntgenstrahlaufnahme *f* exograph
Röntgenstrahlen *pl*, **harte** ~ hard X-rays, **weiche** ~ soft X-rays
Röntgenstrahlenbündel *n* X-ray beam
röntgenstrahlendurchlässig radiolucent, radioparent
Röntgenstrahlendurchlässigkeit *f* radiolucency, radioparency
Röntgenstrahlenmesser *m* roentgenometer, X-ray radiometer
röntgenstrahlenundurchlässig radiopaque

Röntgenstrahlphoton – Roheisencharge

Röntgenstrahlphoton *n* X-ray photon
Röntgenstrahlung *f* X-radiation
Röntgenstreuung *f* **von Kristallen** *pl* X-ray diffraction of crystals
Röntgenstrukturanalyse *f* X-ray structural analysis, direct X-ray analysis
Röntgenterm *m* X-ray level
Röntgentherapie *f* radiotherapy
Röntgenuntersuchung *f* X-ray examination
Röpperit *m* (Min) roepperite
rösch brittle, crisp
Röstanlage *f* roasting plant, calcining plant, roasting installation
Röstapparat *m* calcining apparatus, roasting apparatus
Röstarbeit *f* (Metall) roasting process
Röstbetrieb *m* roasting practice
Röstbett *n* roasting bed
rösten to sinter, (Brot) to toast, (dörren) to torrefy, (Erze) to calcine, to roast
Rösten *n* roasting, calcining, grilling, sintering, (Dörren) torrefaction, (Flachs etc.) retting, ~ **der Eisenerze** *m pl* calcination of iron ores, roasting of iron ores, **chlorierendes** ~ chlorinating roasting, **oxydierendes** ~ oxidation roasting
Röster *m* calciner, roaster
Rösterzeugnis *n* calcined product; roasting product
Röstgas *n* calcination gas, flue gas; gas from roasting, roaster gas
Röstgasreinigungsanlage *f* calciner gas cleaning plant
Röstgut *n* ore to be calcined, ore to be roasted, roasting charge
Röstherd *m* hearth roaster, roasting hearth
Röstkammer *f* roasting chamber
Röstmalz *n* roasted malt
Röstofen *m* calciner, calcining furnace, roaster, roasting furnace, roasting kiln
Röstprobe *f* calcination assay, calcination test
Röstprodukt *n* calcined product, product of roasting
Röstprozeß *m* calcining process, roasting process
Röstreaktionsarbeit *f* roast-reaction process
Röstreaktionsverfahren *n* roasting-reaction method
Röstreduktionsarbeit *f* roast-reduction process
Röstreduktionsverfahren *n* roast-reduction process
Röstrückstand *m* residue from roasting, roasting residue
Röstscherben *m* roasting dish, scorifier
Röstschlacke *f* slag from roasting
Röstschmelzen *n* roasting smelting
Röststaub *m* dust of roasted ore

Rösttemperatur *f* calcining temperature, roasting temperature
Röstung *f* roasting, (Dörren) torrefaction, (Metall) calcination, calcining, (Text) retting
Röstverfahren *n* method of calcination, roasting process
Röstverlust *m* loss due to roasting
Röstwasser *n* steeping water
Röstzuschlag *m* flux for roasting
Rößlerit *m* (Min) rosslerite
Röte *f* redness, (Färberröte) madder
Rötel *m* (Min) red chalk, ruddle
Rötelerde *f* (Min) red ocher, ruddle
Rötelfarbe *f* color made of red chalk
Röteljaspis *m* (Min) blood-colored jasper
röten to redden
rötlich reddish, ~ **braun** reddish brown, russet
Röttisit *m* (Min) rottisite, native hydrated nickel silicate
Rogen *m* (Zool) roe, spawn
Rogenstein *m* (Min) roestone, oolite, oolitic lime
Roggen *m* (Bot) rye
Roggenbrot *n* rye bread
Roggenkleber *m* gluten contained in rye
Roggenkornbrand *m* rye blight, rye smut
Roggenmehl *n* rye flour
Roggenöl *n* rye oil
Roggenstärke *f* rye starch
roh (Öl, Metalle, Leder) crude, raw, unrefined, (Stein, Diamant) rough, unwrought, (Wolle) natural, raw
Rohabwasser *n* crude waste water
Rohaluminium *n* crude aluminum
Rohantimon *n* crude antimony, unrefined antimony
Roharbeit *f* (Metall) raw smelting
Rohasbest *m* raw asbestos
Rohaufbereitung *f* preliminary preparation
Rohaufbrechen *n* first breaking-up
Rohbarren *m* muck bar, puddle[d] bar
Rohbaumwolle *f* raw cotton
Rohbenzol *n* crude benzene
Rohblech *n* raw sheet iron
Rohblei *n* crude lead
Rohbraunkohle *f* crude lignite
Rohbrom *n* raw bromine
Rohdiamant *m* rough diamond
Rohdolomit *m* (Min) crude dolomite, raw dolomite
Roheisen *n* pig iron, ~ **braten** to broil pig iron, ~ **für den Temperguß** malleable pig iron, **feinkörniges** ~ close-grained pig iron, close pig iron, **geläutertes** ~ pure pig iron, **graues** ~ graphitic pig iron, gray pig iron, **phosphorsaures** ~ phosphorous pig iron, **schaumiges** ~ highly carbonated iron, pig-iron saturated with carbon, **schwarzes** ~ kishy pig-iron, **weißes** ~ white pig iron
Roheisenbad *n* pig iron bath
Roheisencharge *f* pig iron charge

Roheisenerzeugung f pig-iron production
Roheisen-Erz-Verfahren n pig and ore process, pig iron-ore process
Roheisengans f iron pig
Roheisengichtsatz m pig iron charge
Roheisenguß m cast iron casting
Roheisenlager n pig iron storage
Roheisenmassel f pig [iron]
Roheisenmischer m pig iron mixer, blast-furnace mixer, hot-metal mixer
Roheisenschlacke f dross of pig iron
Roheisen-Schrott-Verfahren n pig iron-scrap process
Rohentwurf m sketch, rough drawing
Roherdöl n crude oil
Roherz n crude ore, raw ore
Roherzeugnis n raw product
Rohfaser f crude fiber
Rohfettsäure f crude fatty acid
Rohfilzbahnen pl dry fell-sheeting
Rohfrischen n (Metall) first process of finery, first refining, gray iron fining
Rohfrischperiode f (Eisen) boiling stage [of iron]
Rohfrischschlacke f fining slag
Rohgang m cold working, irregular working
rohgar partially refined
Rohgas n crude gas, raw gas
Rohgeschmack m crude taste
Rohgewebe n (Text) raw fabric
Rohgewicht n gross weight
Rohglas n raw glass, roughcast glass
Rohgoldanode f crude gold anode
Rohgummi m crude rubber, raw rubber, unvulcanized rubber
Rohgummimilch f rubber latex
Rohguß m rough casting, undressed casting
Rohhaut f raw hide
Rohhautritzel n rawhide pinion
Rohhomogenat n (Biochem) crude homogenate
Rohkarbidmischung f crude carbide mixture
Rohkautschuk m crude rubber, raw rubber, unvulcanized rubber
Rohkohle f raw coal, rough coal
Rohkresol n crude cresol
Rohkupfer n crude copper, blister copper
Rohlanolin n crude lanoline
Rohlauge f crude lye, crude liquor, crude lixivium
Rohleder n untanned leather
Rohleinöl n raw linseed oil
Rohling m blank, (Holz) rough wood, (zum Verpressen) slug, **geschmiedeter** ~ rough forging
Rohluppe f puddle ball
Rohmasse f (Zement) raw mixture
Rohmaterial n raw material
Rohmessing n crude brass
Rohmetall n crude metal

Rohnaphtha n crude naphtha, crude petroleum, native naphtha
Rohnaphthalin n crude naphthalene
Rohöl n crude oil, **gemischtbasisches** ~ mixed base crude oil, **naphthenbasisches** ~ naphthenic base crude oil, **paraffinbasisches** ~ paraffin base crude oil
Rohölemulsion f crude oil emulsion
Rohofen m furnace for melting native ore
Rohpapier n base paper
Rohparaffin n crude paraffin
Rohpetroleum n crude petroleum, crude naphtha, native naphtha
Rohpikrinsäure f crude picric acid
Rohplatin n crude platinum
Rohprodukt n crude product, raw product
Rohr n pipe, tube, (Bot) cane, reed, (Leitung) conduit, ~ **mit Füllkörpern** packed tube, ~ **mit Klebnaht** cemented tube, ~ **ziehen** to draw pipes, to draw tubes, **bearbeitetes** ~ **aus Schichtpreßstoff** machine laminated tube, **druckloses** ~ non-pressure pipe, **formgepreßtes** ~ **aus Schichtpreßstoff** molded laminated tube, **gewickeltes** ~ rolled tube, **knieförmig gebogenes** ~ elbow tube
Rohrabschneider m tube cutter
Rohrabzweigstück n y-branch
Rohranliegethermometer n pipe attachment thermometer
Rohranordnung f pipe arrangement
Rohransatz m pipe connection, pipe junction
Rohranschluß m pipe connection
Rohrarmaturen pl pipe fittings, tube fittings
Rohrausdehnungsstück n pipe expansion joint
Rohrausgleicher m pipe compensator
Rohrauskleidung f pipe lining
Rohrbefestigung f pipe fastening, pipe support
Rohrboden m tube support plate
Rohrbogen m right angle bend
Rohrbruch m crack in pipes, pipe burst
Rohrbrücke f gantry, overhead pipelines
Rohrbündel n tube bundle
Rohrbündelkondensator m shell-and-tube condenser
Rohrbündelkühler m multiple-tube cooler
Rohrbündelverdampfer m tube bundle evaporator
Rohrbündelwärmeaustauscher m shell-and-tube heat exchanger, tube bundle heat exchanger
Rohrdraht m conduit wire
Rohrdruckfilter n tubular pressure filter
Rohrdurchmesser m pipe diameter
Rohrentladung f pipe discharge
Rohrflansch m pipe flange
rohrförmig tubular
Rohrgewinde n pipe thread
Rohrhaken m wall hook, pipe support
Rohrhalter m pipe hanger, pipe support
Rohrisolierung f tubular insulation

Rohrkerbzugversuch *m* notched-tube tensile test
Rohrkern *m* tubular core
Rohrknie *n* elbow
[Rohr]kontistraße *f* continuous finishing mill for pipe
Rohrkratzer *m* tube scrape
Rohrkrümmer *m* pipe elbow
Rohrkupplung *f* pipe coupling
Rohrlänge *f* length of pipe
Rohrleitung *f* pipe line, conduit, main, piping
Rohrleitungsanlage *f* piping layout
Rohrleitungsarmaturen *f pl* pipe line fittings
Rohrmaterial *n* piping, tubing
Rohrmühle *f* pebble mill, rod mill, tube mill, (Zucker) cane mill
Rohrnetz *n* piping, distributing pipes, pipe-system
Rohrofen *m* tube furnace
Rohrpost *f* pneumatic tube system
Rohrpostanlage *f* pneumatic tube conveyor
Rohrpostbehälter *m* shuttle for pneumatic tube conveyor
Rohrpresse *f* extrusion press for tubes, tube extruding press, tube extrusion machine
Rohrreaktor *m* tubular reactor, tube reactor
Rohrreibung *f* pipe friction
Rohrring *m* ring conduit
Rohrrührer *m* pipe ejector mixer
Rohrsaft *m* (Zucker) cane juice
Rohrschelle *f* conduit clip
Rohrschlange *f* coiled pipe, [pipe] coil, spiral [coil of pipe], worm
Rohrschlüssel *m* tubular socket wrench
Rohrschraubstock *m* tube vice
Rohrschuh *m* casing shoe
Rohrsteckschlüssel *m* tubular box spanner
Rohrstock *m* cane
Rohrstöpsel *m* tube plug
Rohrstrang *m* coil of piping
Rohrströmung *f* pipe flow
Rohrströmungsziffer *f* skin friction coefficient (in pipe flow)
Rohrtrockner *m* flash dryer
Rohrverbindung *f* pipe joint, pipe connector, tube joint
Rohrverdampfer *m* tube still
Rohrverschraubung *f* screwed connection, screw pipe coupling, screw pipe joint
Rohrverstopfung *f* pipe choking
Rohrverzweigung *f* manifold
Rohrweiche *f* pipe junction
Rohrweite *f* inside diameter of a pipe or tube
Rohrwerk *n* piping, tubing
Rohrwickelmaschine *f* tube winding machine
Rohrzange *f* pipe tongs, pipe wrench
Rohrzerstäuber *m* tubular atomizer, tubular sprayer
Rohrziehstraße *f* tube drawing mill
Rohrzucker *m* cane sugar, saccharose, sucrose

Rohrzuckerfabrik *f* cane sugar mill
Rohsalpeter *m* crude saltpeter
Rohschiene *f* muck bar
Rohschlacke *f* fining slag, poor slag, raw slag, tap cinder
Rohschmelze *f* crude melt
Rohschmelzen *n* raw smelting, smelting of crude metals
Rohschwefel *m* crude sulfur
Rohseide *f* (Text) raw silk
Rohseidenfaden *m* (Text) raw silk thread
Rohsieb *n* coarse sieve
Rohsilber *n* crude silver, unrefined silver
Rohsoda *f* crude soda, crude sodium carbonate, soda ash
Rohsole *f* raw brine
Rohspiritus *m* crude spirit, raw spirit
Rohstahl *m* natural steel, raw steel
Rohstahleisen *n* pig iron
Rohstein *m* crude matte, coarse metal
Rohsteinschlacke *f* coarse metal slag
Rohsteinschmelzen *n* coarse metal smelting
Rohstoff *m* raw material, starting material
Rohstyrol *n* crude styrene
Rohviscose *f* raw viscose
Rohwasser *n* natural water, untreated water
Rohwolle *f* raw wool
Rohziegel *m* unburnt brick, unburnt tile
Rohzink *n* rough spelter, rough zinc
Rohzucker *m* raw sugar, unrefined sugar
Rohzuckermelasse *f* factory molasses
Rohzuckernachprodukt *n* second raw sugar
Rohzustand *m* crude state, raw state
Rolladen *m* roller shutter
Rollapparat *m* rewinder
Rollband-Luftfilter *n* automatic air filter
Rollbandmaß *n* spring flex rule
Rolle *f* roller, (Garn etc.) coil, (Mangel) calender, mangle, (Papier etc.) roll, (Scheibe) pulley, pulley wheel, reel, sheave, (Walze) roll[er] cylinder, **feste** ~ fixed pulley, simple pulley, **lose (oder bewegliche)** ~ loose [or movable] pulley
rollen to roll, (Papier) to curl
Rollenbahn *f* gravity roller conveyor
Rollenblech *n* roll sheet iron
Rollenblei *n* rolled lead
rollenförmig cylindrical, roll-shaped
Rollenkasten *m* roller vat
Rollenkette *f* roller chain
Rollenkettenförderer *m* roller-flight conveyor
Rollenklinke *f* roller pawl
Rollenkufe *f* roller vat
Rollenlager *n* roller bearing, **konisches** ~ taper roller bearing, **zylindrisches** ~ parallel roller bearing
Rollenmühle *f* centrifugal roll mill
Rollenpapier *n* continuous paper
Rollenpresse *f* rolling press

Rollenprüfstand *m* chassis dynamometer, test bench
Rollentisch *m* lifting table
Rollenzug *m* block and tackle, pulley block
Rollfaß *n* churn, cleansing drum, rotary mixer, tumbling barrel
Rollfaßverzinnung *f* tumbler tin-plating
Rollflasche *f* [narrow-necked] cylindrical bottle
Rollgang *m* roller gear bed
Rollholz *n* rolling pin
Rollkalander *m* roller calender
Rollkolbenpumpe *f* peristaltic pump
Rollmühle *f* rolling mill
Rollofen *m* continuous furnace heating
Rollquetscher *m* muller mixer
Rollrandflasche *f* beaded rim bottle, bottle with rolled flange
Rollreibung *f* (Mech) rolling friction, rolling resistance
Rollschlitten *m* roller guide
Rollstanze *f* rolling press
Rollwagen *m* truck, lorry (Br. E.), trolley
Rollwiderstand *m* rolling resistance
Rollwulst *m* (Walzwerk) bank
Romanzement *m* natural cement, Roman cement
Romanzowit *m* romanzovite
Romeit *m* (Min) romeite, native calcium antimonate
Rongalit *n* rongalite
Rootspumpe *f* Roots pump
rosafarben pink, rose, rosy
Rosamin *n* rosamine
Rosanilin *n* rosaniline, bis-p-aminophenyl-4-amino-m-tolylcarbinol
Rosasit *m* (Min) rosasite
Rosavergoldung *f* pink gilding
Roscherit *m* (Min) roscherite
Roscoelit *m* (Min) roscoelite, vanadium-mica ore
Rosein *n* (Anilinrot) roseine
Roselith *m* (Min) roselite
Rosellan *m* (Min) rosellane, decomposed anorthite, rosite
Rosenbalsam *m* balm of roses
Rosenblätter *n pl* rose petals
Rosenduft *m* fragrance of roses
Rosenessig *m* rose vinegar
Rosenfarbe *f* rose color, rosy tint
Rosengeruch *m* odor of roses
Rosenholz *n* palisander, rosewood
Rosenholzöl *n* rosewood oil
Rosenhonig *m* rose honey
Rosenkupfer *n* rose copper, refined copper
Rosenöl *n* rose oil
Rosenolsäure *f* rosenolic acid
Rosenquarz *m* (Min) rose quartz
Rosenschwamm *m* (Bot) bedeguar
Rosenspat *m* (Min) rose spar, rhodochrosite
Rosenstahl *m* rose steel, superfine steel

Rosenzinn *n* finest English tin
Roseokobaltsalz *n* roseocobaltic salt
Rosesches Metall *n* Rose's metal
Rosetiegel *m* Rose crucible
Rosette *f* rosette
Rosettenherd *m* copper refining hearth
Rosettenkupfer *n* copper in disks, rosette copper
rosettieren to produce rosette copper
Rosiersalz *n* pinking salt
Rosindol *n* rosindole
Rosindon *n* rosindone, rosindulone
Rosindulin *n* rosinduline
Rosine *f* raisin, (Korinthe) currant
Rosinidin *n* rosinidin
Rosinol *n* rosinol, rosin oil
Rosit *m* (Min) rosite, decomposed anorthite, rosellane
Rosmarin *m* (Bot) rosemary
Rosmarinblätter *n pl* rosemary leaves
Rosmarinöl *n* oil of rosemary
Rosmarinsäure *f* rosmarinic acid
Rosmarinsalbe *f* rosemary ointment
Rosoesäure *f* rosoic acid
Rosolsäure *f* rosolic acid, corallin
Rost *m* (Feuerrost) grate, grid, screen, (Korrosion) rust, **beweglicher** ~ shaking grate, **den** ~ **mechanisch beschicken** to stoke mechanically, **drehbarer** ~ revolving grate, rotary grate, **fester** ~ fixed grate, stationary grate, **geteilter** ~ sectional grate, **liegender** ~ spread footing, **schräger** ~ sloping grate
Rostanfressung *f* corrosion, ~ **bis zu tiefen Grübchen** *n pl* honeycomb corrosion, **grübchenartige** ~ pitting, tubercular corrosion, **langadrige** ~ channeling corrosion
Rostangriff *m* corrosive action, corrosive attack
Rostansatz *m* deposit of rust, formation of rust
Rostbelastung *f* grate load
Rostbeschickung *f* charging, stoking
rostbeständig corrosion-proof, non-corrodible, non-rusting, rustproof, rust-resistant, rust resisting, stainless
Rostbeständigkeit *f* corrosion resistance, non-corrodibility, rust resisting quality, stainless property
Rostbildner *m* corroding agent
Rostbildung *f* corrosion, rust formation
Rostboden *m* (Dest) wire mesh tray or plate
rostbraun rusty brown
Rostbraun *n* rust-brown
rostempfindlich corrodible, sensitive to corrosion
Rostempfindlichkeit *f* corrodibility, sensitiveness to corrosion
rosten to rust, to corrode, to get rusty, to oxidize
Rosten *n* corroding, rusting
Rostentferner *m* rust remover
Rostentfernung *f* rust removal
Rostentfernungsmittel *n* rust-removing agent
Rostfarbe *f* rust color

rostfarben rust-colored
Rostfestigkeit f corrosion resistance, noncorrodibility
Rostfeuer n grate fire
Rostfeuerung f grate firing, stoking, **selbsttätige** ~ stoker firing
Rostfläche f grate area, grate surface
Rostfleck m rust stain
Rostfraß m corrosion
rostfrei stainless, antirust, free from rust, rust-free
Rostfuge f grate opening, airspace between the grate
Rostgefahr f danger of corrosion
Rostglied n fire link
Rostgrübchen n pit
Rosthornit m rosthornite
rostig rusty
Rostkitt m rust cement, iron[-rust] cement
Rostkohle f red hydrogenous charcoal
Rostkühlung f grate cooling
Rostluftgebläse n roaster air fan
Rostmittel n antirust agent, rust inhibitor, rust preventive
Rostnarbe f corrosion pit
Rostschicht f layer of rust
Rostschlagen n removal of clinker
rostschützend rust-preventing, rust-proofing
Rostschutz m rust-proofing, rust protection
Rostschutzanstrich m anti-rust paint, rust preventive paint
Rostschutzfarbe f anti-corrosion paint, anti-rust paint, rust preventive paint
Rostschutzfett n rustproofing grease
Rostschutzmittel n anticorrosive agent, anti-rust agent, rust inhibitor, rust preventive
Rostschutzverfahren n rustproofing process
Rostschweiß m reddish yolk
Rostschwelle f fire bar bearer, grid bar bearer
rostsicher corrosion resistant, rust-proof, stainless
Rostsicherheit f non-corrodibility, rust-proofing, stainless property
Rostspalt m grate opening, air space between the grate
Rostspat m (Min) red sparry iron ore
Rostspinnverfahren n grid spinning
Roststab m fire bar, grate bar
Rostträger m fire bar bearer, grid bar bearer
rostverhütend anti-corrosive, rust-preventing
Rostverhütung f rust inhibition, rust prevention
Roßfenchel m water fennel, (Bot) horsebane
Roßkastanienöl n horse chestnut oil
rot red
Rot n red [color]
Rotadrosselmesser m throttle rotameter
Rota-Filmverdampfer m rotary film evaporator
Rotalge f (Bot) red alga
Rotalixdrehanodenröhre f rotalix tube

Rotamesser m flow meter, rotameter
Rotarsennickel n (Min) arsenical nickel
Rotation f revolution, rotation, ~ **der Schwingungsfläche** (Krist) rotation of the vibration plane, **freie** ~ free rotation
Rotationsabplattung f rotational flattening
Rotationsabsorptionsfrequenz f rotational absorption frequency
Rotationsachse f axis of rotation, rotational axis
Rotationsbandenspektrum n rotational band spectrum
Rotationsbewegung f rotation, rotatory motion
Rotationsdispersion f (Opt) dispersion of rotation, rotary dispersion
Rotationsdispersionskurve f rotatory dispersion curve
Rotationsdruck m rotogravure
Rotationsdünnschichtverdampfer m rotating wetted wall evaporator
Rotationsellipsoid n ellipsoid of revolution, ellipsoid of rotation
Rotationsenergie f energy of rotation
Rotationsfläche f surface of revolution
Rotationsfreiheit f rotational freedom
Rotationsgeschwindigkeit f speed of rotation, velocity of rotation
Rotations-Inversionsachse f (Krist) rotation-inversion axis
Rotationsisomere pl rotational isomers
Rotationskolonne f rotary column
Rotationskonstante f (Spektr) rotational constant
Rotationsniveau n rotational level
Rotationsparaboloid n paraboloid of revolution
Rotationspolarisation f rotation polarization
Rotationspressen n rotation molding
Rotationspumpe f (Kreiselpumpe) rota[to]ry pump
Rotationsquantenzahl f rotational quantum number
Rotationsradius m radius of rotation
Rotationsrakete f rotatory rocket
Rotationsrichtung f direction of rotation, sense of rotation
Rotationsschweißen n spin welding
Rotationsschwingungsspektrum n rotation vibrational spectrum, vibration rotation spectrum
Rotationsspektrum n rotational spectrum
Rotationssterilisation f rotational sterilisation
Rotationsstreuung f rotary dispersion
Rotationssymmetrie f symmetry of rotation
Rotationsverbreiterung f rotational broadening
Rotationsverdampfer m rotation evaporator
Rotationsverzerrung f rotational distortion
Rotationsviskosimeter n rotational viscosimeter
Rotationszerstäuber m centrifugal disc atomizer
Rotbeize f (Färb) red liquor, red mordant
rotblau reddish blue

Rotbleierz *n* crocoisite, crocoite, native lead chromate, red lead ore, red lead spar
rotbraun reddish brown
Rotbruch *m* hot-shortness, red shortness
Rotbruchprobe *f* redheat test
rotbrüchig hot-short, redshort
Rotbrüchigkeit *f* red-shortness
Roteisen *n* (Min) hematite pig
Roteisenerz *n* hematite, red iron ore
Roteisenocker *m* (Min) earthy hematite, red iron ocher
Roteisenstein *m* (Min) red hematite, red iron stone
Rotenolol *n* rotenolol
Rotenolon *n* rotenolone
Rotenon *n* rotenone, derrin, nicoulin, tubain, tubatoxin
Rotenonon *n* rotenonone
Rotensäure *f* rotenic acid, isotubaic acid
Roteol *n* roteol
Roterde *f* red earth
rotfärben to color red
Rotfärbung *f* red coloration
Rotfäule *f* heart rot, red rot
rotfarbig red-colored
rotfaul thoroughly decayed
Rotfilter *n* red filter
rotfleckig red-spotted, red-stained
Rotfreilicht *n* red-free light
rotgar (Gerb) tanned
rotgelb orange-colored, reddish yellow
Rotgerberei *f* bark tannery
Rotgießerei *f* red copper foundry
Rotgiltigerz *n* (Rotgilderz) red silver ore
rotglühen to heat red hot, to heat to red heat
rotglühend glowing red, red-hot
Rotglühhärte *f* red hardness
Rotglühhitze *f* red heat
Rotglut *f* red heat, **auf ~ erhitzen** to heat to red heat, **dunkle ~** dull red, low red
Rotgluthärte *f* red hardness
Rotgluthitze *f* red heat
Rotgold *n* red gold
Rotgültigerz *n* (Rotgiltigerz) red silver ore, **dunkles ~** aerosite, pyrargyrite, ruby silver, **lichtes ~** proustite
Rotguß *m* red cast, gun metal, red brass, red bronze
Rotheizen *n* redheat process
Rothitze *f* red heat
Rothoffit *m* (Min) rothoffite
rotieren to revolve, to rotate, (kreiseln) to spin
rotierend revolving; rotating, rotational, rota[to]ry
Rotierofen *m* rotating furnace
Rotkohle *f* red charcoal
Rotkupfer *n* (Min) red copper, chalcotrichite, red oxide of copper
Rotlack *m* red enamel

Rotlauge *f* red lye
Rotlicht *n* red light
Rotnickelkies *m* (Min) niccolite
Rotor *m* rotor, head
Rotorstrom *m* rotor current
Rotpause *f* red print
Rotschlamm *m* red sludge
Rotsilbererz *n* red silver ore
Rotspat *m* (Min) rhodonite
Rotspießglanz *m* (Min) kermesite, native red antimony sulfide
Rotstein *m* (Min) native manganese silicate, red manganese, rhodonite
Rotte *f* calciner
Rottegrube *f* retting pit
rotten to ret, to steep
Rotten *n* retting, steeping
Rottlerin *n* rottlerin
Rottmethode *f* method of retting, steeping method
Rottöner *m* red toner
Rotundin *n* rotundine
Rotvergoldung *f* red gilding
Rotverschiebung *f* red shift, shift of spectral lines toward red [or longer] waves
Rotviolett *n* red purple
Rotwarmbiegeprobe *f* hot-bending test
Rotwarmhärte *f* red hardness
Rotweinfarbe *f* claret, color of red wine
Rotzinkerz *n* red zinc ore, zincite
Rouleauxdruckmaschine *f* cylinder printing machine
R-Säure *f* (Färb) R-acid
Ruban *n* ruban
Rubatoxanon *n* rubatoxanone
Rubazonsäure *f* rubazonic acid
Rubeansäure *f* rubeanic acid
Rubeanwasserstoff[säure] rubeane, rubeane hydride, rubeanic acid
Rubellit *m* (Min) rubellite, tourmaline
Ruberythrinsäure *f* ruberythric acid, alizarin-2-primveroside, ruberythrinic acid, rubianic acid
Rubiadin *n* rubiadin, 2,4-dihydroxy-1-methylanthraquinone, 4-methylpurpuroxanthin
Rubiadinglucosid *n* rubiadine glucoside
Rubicell *m* (Min) rubicelle
Rubicen *n* rubicene
Rubichlorsäure *f* rubichloric acid
Rubichrom *n* rubichrome
Rubidin *n* rubidine, lycopin
Rubidium *n* rubidium (Symb. Rb)
Rubidiumalaun *m* rubidium alum, rubidium aluminum sulfate
Rubidiumbromid *n* rubidium bromide
rubidiumhaltig containing rubidium
Rubidiumhydrid *n* rubidium hydride
Rubidiumjodat *n* rubidium iodate

Rubidiumjodid – Rückprallhärte

Rubidiumjodid *n* rubidium iodide
Rubidiumplatinchlorid *n* rubidium chloroplatinate
Rubidiumstrontiumalter *n* rubidium strontium age
Rubidiumsulfat *n* rubidium sulfate
Rubidiumsuperoxid *n* rubidium dioxide, rubidium peroxide
Rubidiumverbindung *f* rubidium compound
Rubiginol *n* rubiginol
Rubiginsäure *f* rubiginic acid
Rubin *m* (Min) ruby
Rubinblende *f* (Min) ruby blende, pyrargyrite
Rubinfarbe *f* ruby color
rubinfarben ruby[-colored]
Rubinglas *n* ruby glass
Rubinglimmer *m* (Min) ruby-mica, goethite
Rubingranat *m* (Min) rock ruby
rubinrot ruby[-colored]
Rubinschwefel *m* (Min) red arsenic blende, ruby sulfur, realgar, red arsenic glass, red arsenic [sulfide], red orpiment, ruby arsenic
Rubinspinell *m* (Min) ruby spinel
Rubixanthin *n* rubixanthin
Rubremetin *n* rubremetine, dihydroemetine
Rubren *n* rubrene
Rubreserin *n* rubreserine
Rubrit *m* rubrite
Rubroskyrin *n* rubroskyrin
Ruck *m* jerk, jolt, sudden turn
ruckartig jerky
ruckweise jerking, jerky, (pulsierend) pulsating, (unterbrochen) discontinuous, intermittent
Ruder *n* rudder
Ruderquadrant *m* rudder quadrant
Ruderrahmen *m* rudder frame
Rübe *f* (Zucker) [sugar] beet
Rübenmelasse *f* beetroot molasses
Rübensaft *m* beet juice
Rübensirup *m* beet syrup
Rübenspiritus *m* spirit from beetroot molasses
Rübenzucker *m* beet[root] sugar
Rüböl *n* [refined] colza oil
Rübölkuchen *m* rapeseed cake
Rübsenöl *n* colza oil, rape[seed] oil
Rückansicht *f* rear view
Rückbewegung *f* return motion, (Kolben) return stroke
rückbilden to form back
Rückbildung *f* reformation
Rückdampf *m* (Zucker) exhaust steam
rückdiffundieren to diffuse back
Rückdiffusion *f* back-diffusion
Rückdiffusionsverlust *m* back diffusion loss
Rückdrehmoment *n* self-aligning torque
Rückdruckfeder *f* return spring
Rückdruckkolben *m* draw-back ram, opening ram, pull-back ram
Rückdruckstift *m* [ejector plate] return pin

Rückenmark *n* (Anat) spinal cord
Rückenmarkentzündung *f* (Med) myelitis
Rückentladung *f* back discharge
Rückfall *m* relapse
Rückfederung *f* resilience, recovered energy
Rückfluß *m* backflow, return flow, (Rektifikation) reflux
Rückflußkühler *m* reflux condenser, backflow condenser
Rückformungstemperatur *f* recovery temperature
rückführen to recycle
Rückführgeschwindigkeit *f* reset rate
Rückführgröße *f* reset amount
Rückführleitung *f* return line
Rückführung *f* nachgebende ~ (Regeltechn) elastic feed back
Rückführung *f* feedback, ~ **mit Vorhalt** rate reset
Rückführwirkung *f* reset action
Rückgang *m* retraction, return motion, (Abnahme) decrease, (Kolben) backstroke, ~ **der Leitfähigkeit** decrease of the conductivity
rückgewinnbar recoverable
rückgewinnen to reclaim, to recover, to salvage, (Gummi) to reclaim
Rückgewinnung *f* reclamation, recovery, regaining, regeneration, (Elektr, Energie) recuperation, (Gummi) reclaiming, ~ **des Lösungsmittels** solvent recovery
Rückgewinnungsanlage *f* recovery plant, save-all
Rückgutbunker *m* bin for return fines
Rückhaltespannung *f* hold back tension
Rückhalteträger *m* hold back carrier
Rückkoppelungssignal *n* feedback signal
Rückkopplung *f* reaction coupling, regenerative coupling, (Elektr) feedback
Rückkopplung *f* reaction condenser
Rückkopplungskreis *m* feedback circuit, regenerative circuit
Rückkopplungsleitung *f* feedback
Rückkopplungsschaltung *f* feedback circuit
rückläufig backward
Rücklage *f* reserve
Rücklauf *m* reverse or backward movement, (Kolben) return stroke, (Rektifikation) reflux
Rücklaufanschlag *m* backward stop dog
Rücklaufdoppelrad *n* reverse double gear wheel
Rücklaufkondensator *m* reflux condenser
Rücklaufkühler *m* reflux condenser
Rücklaufschlamm *m* return sludge
Rücklaufverhältnis *n* (Rektifikation) reflux ratio
Rücklaufzerstäubung *f* return-flow atomization
Rückleitung *f* feedback, recirculation, (Techn) return tube
Rückprall *m* repercussion, resilience
Rückprallatom *n* recoil atom
Rückprallelastizität *f* rebound elasticity, resilience
Rückprallhärte *f* scleroscope hardness

Rückprallprüfgerät *n* resilience tester
Rückreaktion *f* back[ward] reaction, reverse reaction
Rückschlag *m* rebound, recoil, setback, (Flamme) backfiring, flashback
Rückschlagen *n* backfiring
Rückschlaghemmung *f* prevention of backfiring
Rückschlagsicherung *f* flame trap
Rückschlagventil *n* back pressure valve, check valve, non-return valve, safety valve; vacuum breaker
Rückschluß *m* conclusion, inference
Rückschmelzverfahren *n* melt-back method
Rückseite *f* (Metallurgie) pouring side, ~ **des Gewebes** back of the fabric, reverse side
Rückseitenappretur *f* (Text) back finish
Rücksicht *f* consideration, respect
Rücksog *m* backlash
Rücksprung *m* rebound, resilience
Rücksprungverfahren *n* rebound process
Rückspülsystem *n* (Gaschrom) back-purge system
Rückstände *m pl* tailings
rückständig residual, residuary
Rückstand *m* dregs, refuse, remainder, residue, residuum, (Sieb) oversize product, **hochsiedender** ~ high-boiling residue, **unlöslicher** ~ insoluble residue
Rückstandbildung *f* formation of residuum
Rückstandsanalyse *f* residue analysis
Rückstandsbrenner *m* burner for oil residues
rückstandsfrei free from residue; without sediment
Rückstandsöl *n* residual oil
Rückstein *m* back stone, crucible bottom
Rückstellfeder *f* return spring
Rückstellmoment *n* self-aligning torque
Rückstellung *f* **von Hand** manual reset
Rückstellzeit *f* restoring time
Rückstoß *m* repulsion, (Mot) recoil
Rückstoßatom *n* recoil atom
Rückstoßbruchstück *n* recoil fragment
Rückstoßdüse *f* reaction-propulsion jet
Rückstoßelektron *n* recoil electron, Compton electron
Rückstoßenergie *f* recoil energy
Rückstoßfeder *f* recoil spring
Rückstoßionisierung *f* recoil ionization
Rückstoßkern *m* (Atom) recoil nucleus
Rückstoßkraft *f* power of recoil, repulsion power
Rückstoßmethode *f* recoil method
Rückstoßmotor *m* jet propulsion engine
Rückstoßnachweis *m* recoil detection
Rückstoßproton *n* recoil proton
Rückstoßspektrum *n* (Atom) recoil spectrum
Rückstoßstift *m* return pin
Rückstoßteilchen *n* **bei Spaltung** fission recoil particle

Rückstoßzähler *m* recoil counter
rückstrahlen (Opt) to reflect
Rückstrahler *m* (Opt) reflector
Rückstrahlmikroskopie *f* reflection electron microscopy
Rückstrahlung *f* reflection, reflexion
Rückstrahlungsgoniometer *n* reflective goniometer
Rückstreusättigung *f* backscattering saturation
Rückstreuung *f* back scattering
Rückströmsperre *f* nonreturn valve
Rückstrom *m* backflow, backwash, (Elektr) return current, reverse current
Rücktitration *f* back-titration
Rücktitrieren *n* (Chem) back titration
Rücktrumm *m* (Keilr) return reach, return run
Rückverformung *f* reformation
Rückvermischen *n* backmixing
Rückvermischung *f* back mixing
Rückverwandlung *f* retransformation
rückwärts backward, ~ **gestreut** back-scattered
Rückwärtsbewegung *f* retrograde motion
Rückwand *f* rear wall
Rückwasser *n* back water
rückwirkend reactive, retroactive
Rückwirkung *f* reaction, reactive effect, retroaction
Rückzündung *f* backfire
Rückzugfeder *f* return spring
Rückzugkolben *m* pull-back ram
Rühranode *f* moving anode
Rührapparat *m* stirrer, agitator, stirring apparatus
Rührarm *m* agitating arm, stirring arm
Rührautoklav *m* agitator autoclave, autoclave with stirrer
Rührbütte *f* stirring tub, pulp chest
Rührdruckgefäß *n* stirred pressure vessel
Rühreisen *n* stirrer, poker, rake
rühren to stir, to agitate
Rühren *n* stirring, stirring agitation
Rührer *m* stirrer, agitator, mixer, **Anker-** ~ anchor mixer, **Blatt-** ~ blade mixer, **Dreikant-** ~ triangular mixer, **elektromagnetischer** ~ electromagnetic stirrer, **Gitter-** ~ gate paddle mixer, **Hohl-** ~ ejector mixer, **Impeller-** ~ impellor mixer, **Kreuzbalken-** ~ cross-blade mixer, **mechanischer** ~ mechanical stirrer, **MIG-** ~ multi-stage impulse ribbon blender, **Propeller-** ~ propellor mixer, **Rohr-** ~ pipe ejector mixer, **Schaufel-** ~ paddle mixer, **Scheiben-** ~ disc mixer, **Turbinen-** ~ turbine mixer, **Wendel-** ~ spiral of helical blade mixer, **Zweiwellen-** ~ twin-rotor mixer
Rührflügel *m* agitating vane, agitator arm
Rührgebläse *n* blower agitator
Rührgeschwindigkeit *f* stirring rate
Rührhaken *m* iron rake, rabble

Rührholz *n* wooden paddle, wooden stirrer
Rührkessel *m* agitator vessel
Rührkesselkaskaden *pl* stirred vessel cascades
Rührkrücke *f* scrape
Rührmaschine *f* agitating machine, stirring machine
Rührmotor *m* stirrer motor, stirring motor
Rührschnecke *f* spiral stirrer
Rührschweißen *n* twist welding
Rührstab *m* stirring rod or pole, glass rod, stirrer
Rührstativ *n* stirrer stand
Rührtrockner *m* rotary dryer
Rührverfahren *n* (beim Schlämmen) tossing
Rührvorrichtung *f* stirring device, agitator, stirrer
Rührwelle *f* stirring gear
Rührwerk *n* stirring device, agitator, stirrer, stirring apparatus, stirring unit
Rührwerkmotor *m* stirring motor
Rührwerksantrieb *m* stirrer drive
Rührwerksapparat *m* agitation-type flotation machine
Rührwerksflügel *m* blade of agitator
Rührwerkskessel *m* mixing kettle
Rührwerksmühle *f* stirred ball mill
Rührwerkstrockner *m* dryer with agitator
Rührwerkswelle *f* agitator shaft
Rührzylinder *m* cylinder with stirrer
Rüssel *m* **der Form** tuyère nozzle
Rüstbaum *m* scaffold pole
Rüster *f* (Holz) elm
Rüsterrinde *f* elm bark
Rüstkosten *pl* set-up costs
Rüststange *f* scaffold pole
Rüstungsindustrie *f* war material industry
Rüttelapparat *m* vibrator
Rüttelbeton *m* vibrated concrete
Rüttelbewegung *f* shaking motion
Rüttelformmaschine *f* jarring molding machine
Rüttelgewicht *n* apparent density, loose weight
Rüttelkolben *m* jarring piston
Rüttelmaschine *f* jolting machine
rütteln to shake, to jolt, to vibrate
Rüttelprobe *f* shatter test
Rüttelsieb *n* shaking sieve, vibrating sieve
Rütteltisch *m* shaking table, vibrating table
Rüttelzeug *n* shaking apparatus
Rüttler *m* shaker, vibrator
Ruficoccin *n* ruficoccin
Rufigallussäure *f* rufigallic acid, rufigallol
Rufin *n* rufin
Rufiopin *n* 1,2,5,6-tetrahydroxyanthraquinone, rufiopin
Rufol *n* 1,5-anthracenediol, 1,5-anthradiol, 1,5-dihydroxyanthracene, rufol
Ruhe *f* rest
Ruhebelastung *f* dead load, static load
Ruhedruck *m* static pressure
Ruhedruckverhältnis *n* static pressure ratio
Ruheenergie *f* rest energy, static energy
Ruhekontakt *m* spacing contact
Ruhelage *f* rest position, equilibrium position, steady position
Ruhemasse *f* rest mass, ~ **des Elektrons** rest mass of the electron
ruhen to rest
ruhend at rest, stationary, (Gewässer) stagnant, (statisch) static
Ruhepotential *n* equilibrium potential, rest potential
Ruhespannung *f* voltage in open circuit
Ruhestellung *f* static position, neutral position
ruhig noiseless, quiet, silent
Ruhr *f* (Med) dysentery
Ruhramöbe *f* Entamoeba hystolytica
Ruhrbazillus *m* dysentery bacillus
Rum *m* rum
Rumbrennerei *f* rum distillery
Rumicin *n* rumicin
Rumpf *m* trunk, body (vessel), (Atom) residue of the atomic nucleus, (Luftf) fuselage
Rumpfbeplankung *f* hull plating
Rumpfelektron *n* inner electron
Rumpfholm *m* longeron
Rumpfisomerie *f* core isomerism
Rumpfit *m* (Min) rumpfite
Rumpfmodell *n* core model
Rumpfrohrholm *m* longeron tube
rund round, (kreisförmig) circular, (kugelförmig) spherical
Rundbeschicker *m* circular feeder
Rundbiegen *n* pipe forming
Rundbrecher *m* giratory crusher
Rundbrenner *m* ring burner
Runddämpfer *m* cottage steamer
Rundeisen *n* round iron
runderhaben convex, (auf beiden Seiten) biconvex
runderneuern (Reifen) to recap, to retread
Runderneuerung *f* (Reifen) recapping, full-capping, retreading
Rundfilter *m* radial bag filter, round filter
Rundfunk *m* broadcasting, radio, wireless (Br. E.)
Rundfunkapparat *m* radio [set], wireless (Br. E.)
Rundfunkempfänger *m* radio receiver, wireless receiver (Br. E.)
Rundgewinde *n* round thread
Rundholz *n* scaffold pole, spar
Rundkanal *m* circular passage
Rundkeil *m* round key
Rundkolben *m* (Chem) round[-bottom] flask
Rundläufer *m* rotary pelleting machine
Rundlauffehler *m* degree of unbalance eccentricity
Rundlochbrenner *m* circular slot burner
Rundlochsieb *n* round hole sieve or screen
Rundofen *m* circular furnace
Rundriemen *m* round belt

Rundschieber *m* circular slide valve
Rundschlag *m* circular twist
Rundschleifen *n* plain grinding
Rundschleifmaschine *f* cylindrical grinder
Rundseil *n* round cable, round rope
Rundstrahlbrenner *m* fan-tailed burner
Rundthermostat *m* round thermostat
Rundung *f* rounding
Rundungshalbmesser *m* radius of curvature
Rundwirkerei *f* (Web) circular knitting
Runkelrübe *f* beetroot
Runkelrübenmelasse *f* molasses of beetroot
Runkelrübenzucker *m* beetroot sugar, sucrose
Runzelbildung *f* wrinkling, crawling, formation of wrinkles
Runzellack *m* brocade finish, ripple finish, wrinkle finish
runzeln to wrinkle, to crinkle
Rupfen *m* bagging
Ruß *m* carbon black, lampblack, soot, sootblack, (Gummichemie) carbon black
Rußansatz *m* deposit of soot
rußartig fuliginous, like soot
Rußbildung *f* formation of soot
Rußbrennerei *f* manufacture of lampblack
Rußdampf *m* fuliginous fumes, sooty vapor
rußen to soot, to produce soot
rußend producing soot, smoking while burning
rußfarben of sooty color
Rußflocke *f* soot flake
Rußhütte *f* lampblack factory
rußig fuliginous, sooty
Rußkammer *f* smoke condenser
Rußkobalt *m* (Min) asbolane, asbolite, black earthy cobalt ore
Rußkohle *f* earthy pit coal, sooty coal
Rußkreide *f* black chalk
Rußschwarz *n* carbon black, lampblack, soot, sootblack
Rußsilber *n* (Min) black silver ore
Rußvorlage *f* soot collector, soot receiver
Rutaecarpin *n* rutecarpine
Ruthenat *n* ruthenate, salt or ester of ruthenic acid
Ruthenium *n* ruthenium (Symb. Ru)
Rutheniumgehalt *m* ruthenium content
Ruthenium(III)-oxid *n* ruthenious oxide
Ruthenium(IV)-oxid *n* ruthenic oxide
Rutheniumoxydul *n* (obs) ruthenium(II) oxide, ruthenium monoxide
Rutheniumsäure *f* ruthenic acid
Rutheniumverbindung *f* ruthenium compound
Rutherford *n* (Radioaktivitätseinheit) rutherford (unit of radioactivity)
Rutherfordin *m* (Min) rutherfordine, uranyl carbonate
Rutherfordit *m* (Min) rutherfordite
Rutil *m* (Min) rutile, native titanium dioxide
Rutilit *m* rutilite

Rutin *n* rutin, myrticolorin, osyrithrin, sophorin
Rutinose *f* rutinose
Rutinsäure *f* rutinic acid
Rutosid *n* rutoside
Rutsche *f* chute, shoot
rutschen to slide, to glide, to shift
Rutschen *n*, **seitliches** ~ side slip
rutschfest anti-skid, non-skid
Rutschfestigkeit *f* anti-skid property
rutschig sliding
rutschsicher non-skid
Rutschsicherheit *f* slip resistance
Rutylen *n* rutylene
Rydbergkonstante *f* Rydberg constant
Rysglas *n* (Min) mica, specular stone

S

Saat *f* seed
Saatbeizmittel *n* seed dressing
Saatbestellung *f* (Landw) sowing
Saatgut *n* seed[-corn]
Saatgutbeize *f* seed dressing
Saatzucht *f* seed growing
Sabadillsäure *f* cevadic acid, tiglic acid
Sabadillsamen *m* cevadilla, sabadilla seed
Sabadin *n* sabadine
Sabadinin *n* sabadinine
Sabinaketon *n* sabinaketone
Sabinan *n* sabinane
Sabinen *n* sabinene
Sabinensäure *f* sabinenic acid
Sabininsäure *f* sabinic acid
Sabinol *n* sabinol, 6-hydroxysabinene
Sabromin *n* sabromin, calcium dibromobehenate
Saccharase *f* (Biochem) saccharase, invertase, sucrase
Saccharat *n* saccharate, sucrate
Saccharetin *n* saccharetin
Saccharid *n* saccharide
Saccharimeter *n* saccharimeter
Saccharimetrie *f* saccharimetry
Saccharin *n* benzosulfimide, benzoylsulfonic imide, gluside, saccharin[e], saccharinol, sulfabenzoic acid imide, sykose
saccharinhaltig containing saccharin
Saccharinsäure *f* saccharinic acid
Saccharit *m* saccharite
Saccharobiose *f* saccharobiose, saccharose, sucrose
Saccharometer *n* saccharometer
Saccharomyzet *m* saccharomycete (pl saccharomyces)
Saccharonsäure *f* saccharonic acid
Saccharose *f* saccharobiose, saccharose, sucrose
Saccharosephosphorylase *f* (Biochem) saccharose phosphorylase
Sacharat *n* (Saccharat) sucrate
Sachbeweis *m* concrete evidence, material proof
Sachgebiet *n* domain, field
sachgemäß pertinent, appropriate
Sachkenntnis *f* experience, expert knowledge
sachkundig experienced, expert
Sachkundiger *m* expert, specialist
Sachlage *f* state of affairs
sachlich objective
Sachregister *n* subject index, table of contents
Sachverhalt *m* circumstances, state of affairs
sachverständig experienced, expert
Sachverständiger *m* authority, expert, specialist
Sack *m* sack, bag, ~ **in Kastenform** gusseted sack, **freitragender** ~ heavy-duty sack
Sackausklopfmaschine *f* sack beating machine

Sackfilter *n* bag filter
Sackförderer *m* bag conveyor, sack conveyor
Sackfüllmaschine *f* sack filling machine
Sackgasse *f* blind alley
Sackkalk *m* slaked lime
Sackleinwand *f* bagging, coarse canvas, sacking
Sackloch *n* blind hole, ~ **mit Gewinde** tapped blind hole
Sadebaum *m* (Bot) savin
Sadebaumextrakt *m* extract of savin
Sadebaumöl *n* savin oil
Säbelkolben *m* (Chem) saber flask
Sächsischblau *n* Saxon blue, cobalt blue, indigo
Säcke *pl*, **freitragende** ~ heavy-duty sacks
säen (Landw) to sow
Säge *f* saw
sägeartig ausgezackt serrated
Sägeblatt *n* saw blade, saw web
Sägebock-Formel *f* (Stereochem) sawhorse representation
Sägebogen *m* saw bow, saw frame
Sägefurnier *n* saw veneer
Sägemaschine *f* saw[ing machine]
Sägemehl *n* sawdust
Sägengewinde *n* buttress thread
Sägeprofilgewinde *n* buttress thread
Sägespäne *pl* sawdust
Sägezahngewinde *n* buttress thread
Säkulardeterminante *f* secular determinant
Säkulargleichung *f* secular equation
Säkularverzögerung *f* secular retardation
sämisch chamois[-dressed], oil-tanned
sämischgerben to chamois
Sämischgerben *n* chamois-dressing, chamoising
Sämischleder *n* chamois [leather], oil-tanned leather, shammy leather, wash leather
sättigen to saturate, to impregnate
Sättigen *n* saturating
sättigend saturating
Sättiger *m* saturator
Sättigung *f* saturation, impregnation
Sättigungsaktivität *f* (Atom) saturated activity
Sättigungsapparat *m* saturating apparatus, saturator
Sättigungsdruck *m* saturation pressure
Sättigungserfordernis *n* saturation requirement
sättigungsfähig capable of saturation, impregnable
Sättigungsfähigkeit *f* saturation capacity
Sättigungsfeldstärke *f* strength of magnetic field at saturation
Sättigungsgrad *m* degree of saturation, degree of concentration
Sättigungsgrenze *f* limit of saturation, saturation limit, saturation point
Sättigungsisomerie *f* saturation isomerism

Sättigungskapazität f saturation capacity
Sättigungskoeffizient m saturation factor
Sättigungskonzentration f concentration of a saturated solution
Sättigungskräfte pl saturating forces
Sättigungskurve f saturation curve
Sättigungsmagnetisierung f saturation magnetization
Sättigungsmittel n saturating agent, saturant
Sättigungspunkt m saturation point
Sättigungsstromstärke f saturation current
Sättigungstemperatur f saturation temperature
Sättigungsverhältnis n (Luftfeuchtigkeit) relative humidity of air
Sättigungsvermögen n saturability
Sättigungswert m saturation value
Sättigungszustand m state of saturation
säubern to clean, to cleanse, to purify
säuerlich acidulous, sourish, ~ **machen** to acidify, to acidulate
Säuerlichkeit f slight acidity, subacidity
Säuerling m sparkling mineral water, acidulous mineral water, sour wine
säuern to acidify, to acidulate, to sour, (Teig) to leaven
Säuern n acidifying, acidulating
Säuerung f acidification, acidulation, (Teig) leavening
säuerungsfähig acidifiable
Säuerungsgefäß n souring vessel
Säuerungsmittel n acidifier, acidifying agent
Säugetier n mammal
Säugling m infant, baby, (Plast) pulp product
Säuglingsnahrungsmittel n baby food
Säule f stanchion, tower, (Chem) column, (Elektr) pile, (Pfeiler) pillar, **chromatographische** ~ chromatographic column
Säulenarm m pillar bracket
Säulenchromatographie f column chromatography, liquid-solid chromatography
Säuleneisen n quadrant iron
säulenförmig columnar
Säulenfuß m base plate
Säulenhängelager n wall bracket, console
Säulenpresse f column press
Säulenrekombination f columnar recombination
Säulenschaft m trunk
Säulenwaage f column balance
Säulenwiderstand m columnar resistance
Säure f acid, acidity, acidness, sourness, (z. B. Früchte) tartness, **aromatische** ~ aromatic acid, **arsenige** ~ arsenious acid, **Carosche** ~ Caro's acid, persulfuric acid, **dreibasische** ~ tribasic acid, **einwertige** ~ univalent acid, **hypochlorige** ~ hypochlorous acid, **hypophosphorige** ~ hypophosphorous acid, **konzentrierte** ~ concentrated acid, **mehrbasische** ~ polybasic acid, **organische** ~ organic acid, **phosphorige** ~ phosphorous acid, **salpetrige** ~ nitrous acid, **sauerstofffreie** ~ hydracid, oxygen-free acid, **schwache** ~ weak acid, **schwefelige** ~ sulfurous acid, **starke** ~ strong acid, **übersättigt mit** ~ superacidulated, **ungesättigte** ~ unsaturated acid, **verdünnte** ~ diluted acid, **wässrige** ~ aqueous acid, **wasserfreie** ~ anhydrous acid
Säureabbau m degradation with acid
Säureabfall m acid sludge
Säureabfluß m acid outlet
Säureabluftanlage f acid fume venting plant
Säureabsatzbehälter m acid settling drum
Säureabscheider m acid separator
Säureäquivalent n acid equivalent
Säureakzeptor m acid acceptor
Säurealizarinblau n acid alizarin blue
Säureamid n acid amide
Säureangriff m acid corrosion, attack by acid
Säureanhydrid n acid anhydride
Säureanlage f acid plant
Säurearmaturen pl acid-proof fittings
säureartig acid-like
Säureauffrischung f renewal of acid
Säureaufschluß m decomposition by acid
Säureaustauscher m acid exchanger
Säureazid n acid azide
Säurebad n acid bath, pickle
Säureballon m acid carboy
Säure-Base-Gleichgewicht n acid-base equilibrium or balance
Säure-Base-Indikator m acid-base indicator
Säure-Base-Katalyse f acid-base catalysis
Säure-Base-Regulation f acid-base regulation
Säure-Base-Stoffwechsel m acid-base metabolism
Säurebehälter m acid container
Säurebehandlung f acid treatment
säurebeständig acid-proof, acid-resisting
Säurebeständigkeit f resistance to acids
Säurebestimmung f acid determination
säurebildend acid forming
Säurebildung f acid formation, acidification
Säurechlorid n acid chloride
Säuredampf m acid fume, acid vapor
Säurederivat n acid derivative
Säuredichte f acid density, acid concentration, specific gravity of acid
Säuredichtemesser m hydrometer
Säureechtheit f fastness to acid[s], resistance to acid[s]
Säureeindampfanlage f acid concentration plant
säureempfindlich sensitive to acids
Säureempfindlichkeit f sensitiveness to acids
Säureerhitzer m acid heater
Säurefällbad n acid precipitating bath
Säurefarbstoff m acid dyestuff

säurefest acid-proof, acid-resistant, fast to acid, impervious to acids
Säurefestigkeit f acid resistance
Säurefluorid n acid fluoride
säurefrei acid-free, free from acid
Säurefuchsin n acid fuchsine
Säuregärung f acid fermentation
Säuregehalt m acid content, acidity
Säuregelb n acid yellow
Säuregemisch n acid mixture
Säuregrad m degree of acidity
säurehärtend acid-curing
Säurehärter m acid catalyst
Säurehalogenid n acid halide
säurehaltig acidiferous, containing acid
Säureheber m acid siphon
Säurehydrat n acid hydrate
Säurehydrazid n acid hydrazide
Säurehydrolyse f acid hydrolysis, acidolysis
Säureimid n acid imide
Säurekamin m acid chimney
Säurekatalyse f acid catalysis
Säurekessel m acid tank
Säurekesselwagen m acid tank wagon
Säurekitt m acid-proof cement
säurekochecht fast to boiling acid
Säurekreiselpumpe f centrifugal acid pump
Säurekühler m acid cooler
Säurekufe f acid vat
säurelöslich acid soluble
Säuremesser m acidimeter, acid meter
Säuremessung f acidimetry
Säuremörtel m acid-proof mortar
Säurenebelwascher m acid fume scrubber
Säureproton n acid proton
Säureprüfer m acid tester, hydrometer
Säureraffination f (von Ölen) acid refining (of oils)
Säureregenerat n (Gummi) acid reclaim rubber
Säurerest m acid residue
Säurerinne f acid gutter
Säure-Rückgewinnungsanlage f acid recovery plant
Säurerückstand m acid residue, residual acid
Säurerührer m acid agitator
Säureschieber m acid gate valve
Säureschlammpumpe f acid sludge pump
Säureschlauch m acid-proof hose, acid-proof tube, acid-proof tubing
Säureschleuder f acid hydroextractor
Säureschutz m acid protective equipment
Säureschutzanstrich m acid-proof paint coating
Säureschutzbau m acid-proof structures
Säureschutzfett n (Akku) acidproof grease
Säurespaltung f acid cleavage
Säurespeicher m acid storage vessel
Säurespindel f hydrometer
Säurespinnverfahren n acid spinning process
Säurespur f trace of acid

Säureständer m acid cistern
Säuretrog m acid vat
Säureturm m acid tower
Säureüberschuß m excess of acid
Säureumrechnungstabelle f acid dilution table
Säureverarmung f impoverishment of acid
Säurevergiftung f acid poisoning
Säurewiderstand m resistance to acid
säurewidrig antacid, antiacid
Säurewirkung f action of acids
Säurezahl f acid number (Am. E.), acid value (Br. E.)
Säurezentrifuge f acid centrifuge
Säurezusatz m addition of acid
Saffian n morocco (leather), saffian leather
Saffianimitation f imitation morocco
Saflor m (Bot) safflower
Saflorgelb n safflower yellow
Safloröl n safflower oil
Saflorrot n safflower red, carthamin
Safran m saffron
safranähnlich saffron-like, saffrony
Safranal n safranal
Safranbronze f tungsten bronze, wolfram bronze
Safranfarbe f saffron color
safranfarben saffron-colored
safrangelb saffron [yellow]
safranhaltig containing saffron
Safranin n safranine, phenosafranine
Safraninol n safraninol
Safranlack m carthamus paint
Safranöl n saffron oil
Safranol n safranol
Safransurrogat n saffron substitute
Safrol n safrole, shikimol
Safrosin n safrosin
Saft m (Holz) sap, (Obst, Fleisch) juice, **Magen-** ~ gastric juice, **Pankreas-** ~ pancreatic juice
saftfrisch (Holz) green
Saftgehalt m sap content
saftig juicy, succulent, (Holz) sappy, (Metall) wet
Saftigkeit f juiciness, sappiness
saftreich rich in juice, rich in sap, succulent
Sagapengummi n sagapenum
Sagenit m (Min) sagenite
Sago m sago
Sagradaextrakt m extract of cascara sagrada
Sagradarinde f cascara sagrada, sacred bark
Saisonfarbe f fashion shade, season shade
Saite f chord, string, (Darm) catgut string
Saitenaufspanner m string fitter
Sajodin n saiodin, calcium iodobehenate
Sakebiose f sakebiose
Salacetamid n salacetamide
Salacetol n salacetol, acetonyl salicylate, salantol
Salatöl n salad oil

Salazinsäure *f* salazinic acid
Salazosulfamid *n* salazosulfamide
Salbe *f* liniment, ointment, salve
Salbei *m* (Bot) sage, salvia
Salbeiöl *n* sage oil
salbenartig salve-like, salvy, unctuous
Salbengrundlage *f* ointment base
Salbenspatel *m* salve spatula
Salcomin *n* salcomine
Salep *m* salep
Salicil *n* salicil
Salicin *n* salicin, saligenin
Salicyl- salicyl
Salicyläthylester *m* ethyl salicylate
Salicylaldehyd *m* salicylic aldehyde, salicylal, salicylaldehyde
Salicylalkohol *m* salicyl alcohol
Salicylamid *n* salicylamide, 2-hydroxybenzamide, salamide
Salicylamylester *m* amyl salicylate
Salicylanilid *n* salicylanilide, salifebrin
Salicylat *n* salicylate, salt or ester of salicylic acid
Salicylessigsäure *f* salicylacetic acid
Salicylgelb *n* salicyl yellow
Salicylid *n* salicylide, tetrasalicylide
salicylieren to salicylate
Salicylnatron *n* sodium salicylate
Salicylosalicylsäure *f* salicylosalicylic acid, diplosal
Salicyloyl *n* salicyloyl
Salicylpräparat *n* salicylic preparation, salicylate
Salicylsäure *f* salicylic acid, o-hydroxybenzoic acid
Salicylsäureäthylester *m* ethyl salicylate
Salicylsäureseife *f* salicylic soap
Salicylsäurezahnpulver *n* salicylic tooth powder
salicylsauer salicylate
Salicylstreupulver *n* salicylated talc
Salicyltalg *m* salicylated tallow
Salicylursäure *f* salicyluric acid
Salicylwatte *f* salicylic cotton
Saliformin *n* saliformine, hexamethylenetetramine salicylate, urotropine salicylate
Saligenin *n* saligenin, salicin, salicyl alcohol
Salimenthol *n* salimenthol, mentholsalicylate, samol
Salinaphthol *n* salinaphthol, betol
Salinazid *n* salinazid
Saline *f* saline, saltern, salt works
Salinenwasser *n* saline water
Salinigrin *n* salinigrin
salinisch briny, saline
Salinometer *n* salinometer
Salipurpol *n* salipurpol
Salipyrin *n* salipyrine, pyrosal, salpyrin, tyrosal
Saliretin *n* saliretin

Salit *n* salit, bornyl salicylate
Salmiak *m* ammonium chloride, sal ammoniac
Salmiakblumen *f pl* flowers of sal ammoniac
Salmiakelement *n* Leclanché cell, sal ammoniac cell
Salmiakgeist *m* ammonia water, ammonium hydroxide, **alkalischer** ~ aqueous ammonia
Salmiaklösung *f* solution of salammoniac, **ameisensaure** ~ ammonium chloride solution
Salmin *n* salmine
Salmit *m* salmite
Salmonella *f* (Bakt) Salmonella (pl Salmonellae)
Salmonellabefall *m* (Med) salmonellosis
Salmonellen *pl* (Bakt) salmonella bacilli, salmonellae, salmonellas
Salmonellenerkrankung *f* (Med) salmonellosis
Salmonellose *f* (Med) salmonellosis
Salochinin *n* saloquinine, quinine salicylate, salicylquinine
Salol *n* (HN) salol, phenyl salicylate
Salolrot *n* salol red
Salophen *n* salophen, p-acetamidophenyl salicylate
Salosalicylid *n* salosalicylide, disalicylide
Salpeter *m* sodium nitrate, niter, (Chilesalpeter) saltpeter (Am. E.), saltpetre (Br. E.), (Kalisalpeter) potassium nitrate, ~ **brechen** to pulverize saltpeter, **natürlicher** ~ wall saltpeter, **roher** ~ crude saltpeter
salpeterartig like saltpeter or niter, nitrous
Salpeterauflösung *f* solution of niter
Salpeterblumen *pl* saltpeter flowers, niter efflorescence
Salpeterdruse *f* (Min) crystallized saltpeter
Salpeterdunst *m* nitrous fumes
Salpetererde *f* nitrous earth, saltpeter earth
Salpetererzeugung *f* niter production
Salpeterfraß *m* corrosion by niter
Salpetergrube *f* saltpeter mine
Salpetergütemesser *m* nitrometer
Salpeterhafen *m* niter pot, saltpeter pot
salpeterhaltig containing saltpeter
Salpeterhütte *f* niter works
Salpeterkessel *m* saltpeter boiler
Salpeterlauge *f* saltpeter lye
Salpeterlösung *f* saltpeter solution
Salpeterpapier *n* niter paper, nitrous paper
Salpeterprobe *f* nitrate sample, nitrate test, saltpeter test
Salpetersäure *f* nitric acid, **konzentrierte** ~ concentrated nitric acid, aqua fortis, **rauchende** ~ fuming nitric acid
Salpetersäureanhydrid *n* nitric anhydride, nitrogen pentoxide, nitrogen(V) oxide
Salpetersäureanlage *f* nitric acid plant
Salpetersäurebad *n* nitric acid bath
Salpetersäuredampf *m* nitric acid vapor
Salpetersäureester *m* ester of nitric acid
salpetersäurehaltig containing nitric acid

Salpetersäuresalz *n* nitrate, salt of nitric acid
salpetersauer nitric, (Salz) nitrate
Salpeterschaum *m* calcium nitrate efflorescence, wall-saltpeter
Salpeterschwefelsäure *f* nitrosulfuric acid, nitrosylsulfuric acid
salpeterschwefelsauer nitrosulfate
Salpetersieder *m* kelp burner
Salpetersiederei *f* saltpeter works
Salpetersud *m* boiling of saltpeter
Salpetersuperphosphat *n* saltpeter superphosphate mixture
Salpeterwaage *f* nitrometer
Salpeterwasser *n* nitrous water
salpetrig containing saltpeter, nitrous
Salpetrigsäureanhydrid *n* nitrous anhydride, dinitrogen trioxide, nitrogen sesquioxide
salpetrigsauer nitrous, (Salz) nitrite
Salsolidin *n* salsolidine
Salsolin *n* salsoline
Salumin *n* salumin, alumin[i]um salicylate
Salvarsan *n* (Arsphenamin, HN) salvarsan, arsphenamine
Salz *n* salt, (Kochsalz) table salt, (Steinsalz) rock salt, **alkalisches** ~ alkaline salt, **ameisensaures** ~ formate, **äpfelsaures** ~ malate, **basisches** ~ basic salt, subsalt, **borsaures** ~ borate, **citronensaures** ~ citrate, **essigsaures** ~ acetate, **flußsaures** ~ fluoride, **gerbsaures** ~ tannate, **inneres** ~, **kohlensaures** ~ carbonate, **kristallisiertes** ~ crystalline salt, **molybdänsaures** ~ molybdate, **ölsaures** ~ oleate, **oxalsaures** ~ oxalate, **palmitinsaures** ~ palmitate, **phosphorigsaures** ~ phosphite, **phosphorsaures** ~ phosphate, **propionsaures** ~ propionate, **saures** ~ acid salt, hydrogen salt, **schwefeligsaures** ~ sulfite, **schwefelsaures** ~ sulfate, **überschüssiges** ~ excess salt, **undissoziiertes** ~ undissociated salt, **unterchloriges** ~ hypochlorite, **wasserfreies** ~ anhydrous salt
Salzablagerung *f* salt deposit
Salzader *f* salt vein
salzähnlich salt-like
Salzansammlung *f* salt deposit
salzartig salt-like, saline
Salzasche *f* salt-ash
Salzbad *n* salt bath, brine bath
Salzbadeinsatzhärtung *f* salt-bath casehardening
Salzbadtiegel *m* salt bath pot
Salzbedarf *m* (Physiol) salt requirement
Salzbehälter *m* salt tank
Salzbeize *f* salt dressing
Salzbereitung *f* salt making
Salzbergwerk *n* [rock-]salt mine
salzbildend salifiable, salt-forming
Salzbildner *m* halogen, salt former, salt-forming substance

Salzbildung *f* salification, salt formation, **innere** ~ internal salt formation
salzbildungsunfähig insalifiable, nonsalifiable
Salzblumen *f pl* flowers of salt
Salzboden *m* saline soil
Salzbrücke *f* salt bridge
Salzbrühe *f* brine
Salzbrunnen *m* saline spring, salt spring
Salzchemie *f* chemistry of salts
Salzdampf *m* salt steam, salt vapor
Salzdecke *f* cover impregnated with salts
Salzdiffusionshof *m* salt aureola of diffusion
Salzeffekt *m* salt effect, **linearer** ~ linear salt effect
salzen to salt
Salzerde *f* salt earth
Salzermittlung *f* halometric test
Salzfabrikation *f* salt production, salification
Salzfluß *m* saline flux, (Gieß) salt-flux
salzförmig saliniform
salzfrei salt-free
Salzgehalt *m* salinity, salt contents
Salzgehaltdiagramm *n* salinity diagram
Salzgehaltmesser *m* halometer, salimeter, salinometer
Salzgehaltmessung *f* halometry
Salzgeschmack *m* salty taste
salzgetränkt soaked in salt solution
Salzgewinnung *f* salt production, salt recovery
Salzglasur *f* salt glaze
Salzgrube *f* salt mine, salt pit
salzhaltig saliferous, salt-bearing
Salzhaltigkeit *f* salineness, salinity, saltiness
Salzhaushalt *m* (Physiol) salt equilibrium
Salzhaut *f* salt crust, salt-film
salzig salty, briny, saline
Salzigkeit *f* saltiness, salineness, salinity
Salzkruste *f* salt crust
Salzkübel *m* salt bucket
Salzkupfererz *n* (Min) atacamite
Salzlager *n* salt bed, salt deposit, salt layer
Salzlake *f* pickle, salt brine
Salzlauge *f* brine, pickle, salt brine
Salzlaugung *f* brine leaching
Salzlösung *f* salt solution, brine, saline solution
Salzmasse *f* saline concretion
Salzmesser *m* salimeter, salinometer
Salzmessung *f* halometry, salimetry, salinometry
Salzmutterlauge *f* bittern
Salzniederschlag *m* salt deposit, saline deposit
Salzpaar *n* salt pair
Salzpaare *pl*, **reziproke** ~ reciprocal salt pairs
Salzpfannenstein *m* salt pan scale
Salzpflanze *f* (Bot) halophyte
Salzprobe *f* salt sample, salt test
Salzquelle *f* saline spring, salt-spring, salt well
salzreich rich in salt
Salzrinde *f* salt crust
Salzrückstand *m* salt residue

Salzsäure *f* hydrochloric acid, muriatic acid (commercial grades)
Salzsäureanlage *f* hydrochloric acid plant
Salzsäurebehälter *m* hydrochloric acid container
Salzsäurebeize *f* hydrochloric acid pickle
Salzsäurehydrat *n* hydrated hydrochloric acid
Salzsäurenebel *m* hydrochloric acid mist
salzsauer hydrochloric, (Salz) chloride
Salzschicht *f* layer of salt
Salzschlag *m* (Min) common granulated quartz
Salzschmelze *f* fused salt, salt melt
Salzsieder *m* salt boiler, salt maker
Salzsole *f* [salt] brine, salt spring, salt water
Salz-Sprüh-Test *m* (Korrosion) salt-spray test
Salzstein *m* (Kesselstein) boiler scale
Salzteich *m* brine pond
Salzteilchen *n* saline particle
Salzton *m* saliferous clay, salt clay
Salztrockenofen *m* salt drying oven
Salzwaage *f* brine gauge, salinometer
Salzwasser *n* salt water, brine
Salzwasserbad *n* salt water bath
Salzwasserkühler *m* brine cooler
Salzwasserkühlung *f* brine cooling, salt water refrigeration
Salzzusatz *m* salt addition
Salzzuschlag *m* addition of salt, salting
Samandarin *n* samandarine
Samarium *n* samarium (Symb. Sm)
Samariumcarbid *n* samarium carbide
Samariumchlorid *n* samarium chloride, (Samarium(III)-chlorid) samaric chloride, samarium(III) chloride, samarium trichloride
Samarium(III)-Verbindung *f* samaric compound
Samarium(II)-Verbindung *f* samarous compound
Samariumoxid *n* samarium oxide, (Samarium(III)-oxid) samaric oxide, samarium(III) oxide, samarium sesquioxide
Samariumsulfat *n* samarium sulfate
Samarskit *m* (Min) samarskite
Samatose *f* samatose
Sambesischwarz *n* Sambesi black
Sambunigrin *n* sambunigrin
Samen *m* (Bot) seed, (Zool) sperm
Samenbeizung *f* seed disinfection
Samenhülse *f* hull, husk, pod
Samenkern *m* (Bot) seed-kernel, (Zool) spermatic nucleus
Samenschale *f* seed husk
samentötend spermatocidal, spermicidal
Samenzelle *f* spermatic cell, spermatozoon (pl spermatozoa), sperm cell
Samin *n* samin
Samiresit *m* samiresite
Sammelanode *f* collector anode, gathering anode
Sammelbatterie *f* storage battery
Sammelbecken *n* collecting vessel, reservoir
Sammelbegriff *m* broader term, collective term

Sammelbehälter *m* collecting tank or trough, collector, reservoir, storage tank
Sammelbrunnen *m* reservoir
Sammelelektrode *f* collecting electrode, gathering electrode
Sammelfuchs *m* (Metall) collecting flue
Sammelgefäß *n* collecting vessel, receiver, receptacle, reservoir
Sammelherd *m* (Metall) iron receiver
Sammelleitung *f* common main, collecting pipe, (Masch) manifold
Sammellinse *f* (Opt) convergent lens, convex lens, positive lens
sammeln to collect, to gather, (fokusieren) to focus, (konzentrieren) to concentrate
Sammelpackung *f* multipack
Sammelpunkt *m* assembly point, (Brennpunkt) focus
Sammelraum *m* collecting chamber
Sammelring *m* collecting ring
Sammelrinne *f*, **ringförmige** ~ collecting ring
Sammelrohr *n* collecting pipe
Sammelschiene *f* bus bar, collecting bar
Sammelsirup *m* syrup from split sugar
Sammeltrichter *m* loading hopper
Sammelvorlage *f* collector vessel
Sammelzylinder *m* collecting cylinder
Sammetblende *f* (Min) goethite
Sammeterz *n* (Min) cyanotrichite
Sammler *m* collector, (Batterie) accumulator, storage battery
Sammlerbatterie *f* storage battery
Sammlersäure *f* accumulator acid
Sammlerschäumer *m* frothing agent
Sammlung *f* collection
Samoit *m* samoite
Samol *n* samol, salimenthol
Samsonit *m* (Min) samsonite, silver manganese antimony sulfide
Samt *m* (Text) velvet
Samtappretur *f* (Text) velvet finish
samtartig velvety
samtglänzend velvet-like, velvety
samtschwarz velvet black
Sand *m* sand, **fein gesiebter** ~ facing sand, **fetter** ~ loamy sand, **feuerfester** ~ refractory sand, **magerer** ~ sand containing little clay
Sandarakharz *n* sandarac[h] resin
sandartig sandlike, sandy
Sandauslaufrohr *n* sand discharge pipe
Sandbad *n* sand bath
Sandbadschale *f* sand bath dish
Sandbergerit *m* sandbergerite
Sandboden *m* sandy soil
Sanddüse *f* sand nozzle
Sandelholz *n* sandalwood, **afrikanisches** ~ barwood, camwood
Sandelholzöl *n* sandalwood oil, santal oil

Sandelrot – Sarkokollin

Sandelrot *n* santalenic acid, santalic acid, santalin
sanden to sand
sandfarben sand-colored
Sandfilter *n* sand filter
Sandfiltration *f* sand filtration
Sandform *f* sand mold
Sandformerei *f* sand molding
sandführend sand-bearing
Sandgebläse *n* sand blast [apparatus]
Sandglimmer *m* (Min) mica
Sandguß *m* sand casting
Sandharz *n* sandaracine
sandig sandy, arenaceous, gritty
Sandkalkstein *m* (Min) quartziferous carbonate of lime
Sandkern *m* [foundry] sand core
Sandkohle *f* free burning coal, noncaking coal
Sandkorn *n* grain of sand
Sandkruste *f* sand skin
Sandleiste *f* sand edge
Sandmergel *m* sandy marl, lime gravel
Sandmeyer-Reaktion *f* (Chem) Sandmeyer['s diazo-] reaction
Sandpapier *n* sand paper
sandreich sandy, gravelly, rich in sand
Sandsackmodell *n* **des Atomkerns** hypothesis of compound nucleus
Sandschicht *f* layer of sand, sand stratum
Sandschiefer *m* (Min) sandy shale, schistous sandstone
Sandschleifstein *m* sand grinding stone
Sandsegge *f* (Bot) carex root, sea sedge
Sandstein *m* sandstone, **bunter** ~ variegated sandstone
sandsteinartig resembling sandstone
Sandsteinerde *f* grit
Sandsteinlager *n* (Geol) sandstone stratum
Sandstrahl *m* sand blast
Sandstrahlanlage *f* sandblast plant
Sandstrahldüse *f* sandblast nozzle
sandstrahlen to sandblast, to shotblast
Sandstrahlen *n* sandblasting, sanding
Sandstrahlgebläse *n* sandblast apparatus, sandblast machine, sandblast unit
Sandstrahlschutzgerät *n* sandblast protective equipment
Sanduhr *f* sand glass
Sandverdichtung *f* sand packing
Sandzement *m* sand cement
Sandzucker *m* brown sugar
Sanforausrüstung *f* (Text) sanfor finish
Sanforisieren *n* (Text) sanforizing [process]
sanft mild, soft
Sanguinarin *n* sanguinarine
Sanidin *m* (Min) sanidine
Sanikel *m* (Bot) sanicle [leaves]
sanitär sanitary
Sanoform *n* sanoform, methyl diiodosalicylate

Sanshoamid *n* sanshoamide
Sanshool *n* sanshool
Santalal *n* santalal
Santalbinsäure *f* santalbic acid
Santalen *n* santalene
Santalin *n* (Santal) santal, santalenic acid, santalic acid, santalin
Santalinfarbstoff *m* red saunders
Santalol *n* santalol
Santalsäure *f* santalenic acid, santal, santalic acid, santalin
Santen *n* santene
Santenhydrat *n* santene hydrate
Santenol *n* santenol
Santenonalkohol *m* santenonalcohol
Santensäure *f* santenic acid
Santonin *n* santonin
Santoninnatrium *n* sodium santoninate
Santoninsäure *f* santoninic acid
santoninsauer (Salz) santoninate
Santonsäure *f* santonic acid
Santorinerde *f* santorin earth
Santowax *n* santowax
Saphir *m* (Min) sapphire, **grüner** ~ green corundum, **roter** ~ oriental ruby, **weißer** ~ leucosapphire
saphirähnlich sapphirine
Saphirblau *n* sapphirine blue
Saphirfluß *m* (Min) blue Derbyshire spar
Saphirin *m* sapphirine
Saphirquarz *m* (Min) blue quartz
Saphirrubin *m* (Min) spinel ruby
Sapimin *n* sapimine
Sapogenin *n* sapogenine
Saponarin *n* saponarin
saponifizieren to saponify
Saponifizieren *n* saponification
Saponin *n* saponin
Saponit *m* (Min) saponite
Saponittalk *m* steatite
Saporval *n* saporval
Saposalicylsalbe *f* saposalicylic ointment
Sapotalin *n* sapotalene
Sapovaseline *f* sapovaseline
Sap[p]anholz *n* sapan wood
Saprobien *pl* saprobiotic organisms
Sapropel *m* sapropel
Sapropelkohle *f* sapropelic coal
Saprophyten *pl* saprophytes
Saprozoen *pl* saprozoic organisms
Saran *n* saran
Sarcosin *n* sarcosine
Sardonyx *m* (Min) sardonyx
Sargasterin *n* sargasterol
Sarin *n* sarin
Sarkin *n* sarcine, hypoxanthine, sarkine
Sarkinit *m* (Min) sarkinite
Sarkokol *n* sarcocol
Sarkokollin *n* sarcocollin

Sarkolit *m* (Min) sarcolite
Sarkolysin *n* sarcolysine
Sarkom *n* (Med) sarcoma (pl sarcomas or sarcomata)
Sarkomycin *n* sarkomycin
Sarkosid *n* sarcoside
Sarkosin *n* sarcosine, methylaminoacetic acid, methylglycine
Sarkosinanhydrid *n* sarcosine anhydride
Sarkosinnitril *n* sarcosine nitrile
Sarkosinsäure *f* sarcosine acid
Sarkosporidien *pl* sarcosporidia
Sarmentogenin *n* sarmentogenin
Sarmentose *f* sarmentose
Sarracinsäure *f* sarracinic acid
Sarsaparille *f* (Bot) sarsaparilla
Sarsaparillwurzel *f* (Pharm) sarsaparilla
Sarsasaponin *n* sarsasaponin
Sartorit *m* (Min) sartorite
Sassafrasholz *n* sassafras root
Sassolin *m* (Min) sassoline, native boric acid, sassolite
Satellit *m* (Astr) satellite
Satin *m* (Text) satin
Satinage *f* glazing
Satingewebe *n* (Text) satin weave
satinieren (Pap) to glaze
Satinierkalander *m* (Leder) glazing calender, (Pap) [super]calender
satiniert glazed
Satinierwalze *f* calender roll, glazing roller
Satinpapier *n* glazed paper
Sativinsäure *f* sativic acid
satt (Farbe) deep, rich
sattblau deep blue
Sattdampf *m* saturated steam, saturated vapor
Sattel *m* saddle, collar beam, ~ **der Falltür** discharge door top
Sattelfederdraht *m* saddle spring wire
Sattelpunkt *m* saddle point
Sattelpunktskonfiguration *f* configuration of saddle point
Sattelschlange *f* saddle-shaped coil
Sattelstück *n* (Innenmischer) sliding door top
Sattelwaage *f* saddle scale
Saturateur *m* (Zucker) saturation tank, saturator
Saturation *f* (Zucker) saturation
Saturationsgefäß *n* saturation vessel, saturator
Saturationsöl *n* saturation oil
Saturationsscheidung *f* (Zucker) purification by carbonation
Saturationsschlamm *m* lime sediment, sediment from carbonation
saturieren to saturate, to impregnate, (Zucker) to carbonate sugar
Saturnzinnober *m* (Min) Saturn cinnabar, minium, Saturn red
Satz *m* (Bodensatz) dregs, grounds, (Charge) batch, charge, (Garnitur) set, (Grundsatz) axiom, (Lehrsatz) principle, theorem, (Niederschlag) deposit, precipitate, sediment, settlings, (Papier) pack, (Preis) price, rate, (Sprung) leap, (Zusammensetzung) composition, mixture, ~ **der spektroskopischen Verschiebung** (Spektr) law of spectroscopic displacement, Kossel-Sommerfeld displacement law, ~ **von den konstanten Wärmesummen von Hess** law of constant summation of heat of Hess, ~ **von Pythagoras** Pythagorean theorem
Satzbetrieb *m* batch operation, stagewise operation
Satzkoks *m* charge coke
Satzmehl *n* farina, fecula, starch flour
satzweise batchwise, intermittently
Sau *f* (Metall) sowmetal
sauber clean, (Chem) pure
Sauberkeit *f* cleanliness, neatness
sauer acid, sour, tart, (Chem) acidic, ~ **werden** (Milch) to curdle
Sauerbad *n* sour bath
Sauerblei *n* (obs) lead chromate
Sauerbleiche *f* sour bleaching
Sauerbrühe *f* acid liquor
Sauerbrunnen *m* acidulous spring water
Sauerdorn *m* (Bot) barberry, berberry
Sauerdornbitter *m* berberine
Sauerfutter *n* (Silofutter) silage [fodder]
Sauerhonig *m* oxymel
Sauerklee *m* clover sorrel, wood sorrel
Sauerkleesalz *n* (obs) acid potassium oxalate, potassium bioxalate, potassium hydrogen oxalate, salt of sorrel
Sauermilch *f* curdled milk
Sauerstoff *m* oxygen (Symb. O), ~ **entziehen** to deoxidize, to deoxygenate, ~ **entziehend** removing oxygen
Sauerstoffabgabe *f* release of oxygen, generation of oxygen
Sauerstoffableitungsröhre *f* oxygen outlet tube
Sauerstoffänger *m* oxygen absorbent
sauerstoffangereichert oxygen-enriched
Sauerstoffanlage *f* oxygen plant
Sauerstoffanreicherung *f* enriching with oxygen, oxygenation
sauerstoffarm lacking in oxygen
Sauerstoff[atmungs]gerät *n* oxygen inhaling apparatus
Sauerstoffaufnahme *f* oxygen absorption
Sauerstoffausschluß *m* absence of oxygen
Sauerstoffbedarf *m*, **biochemischer** ~ (B.S.B.) biochemical oxygen demand (B.O.D)
Sauerstoffbedürfnis *n* oxygen requirement
sauerstoffbeladen oxygenated
Sauerstoffbeladung *f* **der Anode** charging the anode with oxygen
Sauerstoffbildung *f* oxygen formation
Sauerstoffbombe *f* oxygen cylinder

Sauerstoffbrücke – Saugvorrichtung

Sauerstoffbrücke *f* oxygen bridge
Sauerstoffdruck *m* oxygen pressure
Sauerstoffelektrode *f* oxygen electrode
Sauerstoffentwicklung *f* generation of oxygen
Sauerstoffentziehung *f* deoxidation
Sauerstoffentzug *m* deoxygenation, oxygen removal
Sauerstofferzeuger *m* oxygen generator, oxygen producer
Sauerstoffflasche *f* oxygen cylinder
Sauerstoffgas *n* gaseous oxygen, oxygen gas
Sauerstoffgehalt *m* oxygen content
Sauerstoffgewinnung *f* production of oxygen
Sauerstoffgewinnungsanlage *f* oxygen plant
Sauerstoffgewinnungsapparat *m* oxygen generator
sauerstoffhaltig containing oxygen
Sauerstoffkompression *f* oxygen compression
Sauerstoff-Konverterstahl *m* oxygen converter steel
Sauerstoff-Konverterverfahren *n* oxygen steelmaking process
Sauerstoffkonzentration *f* oxygen concentration
Sauerstoffkreislauf *m* oxygen cycle
Sauerstoffmangel *m* lack of oxygen, oxygen deficiency
Sauerstoffmesser *m* oxygen analyzer, oxygen meter
Sauerstoffpol *m* oxygen pole, anode
Sauerstoffprüfer *m* oxygen analyzer
sauerstofffrei oxygen-free
Sauerstoffsäure *f* oxyacid, oxygen acid
Sauerstoffschreiber *m* oxygen recorder
Sauerstoffschweißung *f* oxyhydrogen welding
Sauerstoffstrahl *m* oxygen jet
Sauerstoffträger *m* oxygen carrier
Sauerstofftransport *m* migration of oxygen
Sauerstoffüberträger *m* oxygen carrier
Sauerstoffverbindung *f* oxygen compound, (Oxid) oxide
Sauerstoffverbrauch *m* oxygen demand
Sauerstoffverschiebung *f* oxygen displacement
Sauerstoffzuleitungsröhre *f* oxygen inlet tube
Sauerteig *m* leaven, sour dough
Sauerwasser *n* acidulated water, sour water
Sauerwein *m* sour wine
Sauerwerden *n* souring, (essigsauer) acetification
Sauganschluß *m* vacuum connection
Saugapparat *m* suction apparatus, aspirator
Saugbürste *f* suction brush
Saugdruck *m* suction pressure
Saugdüse *f* suction nozzle, venturi tube
saugen to suck
Saugen *n* sucking, suction
Sauger *m* aspirator, exhauster, sucker, suction apparatus
saugfähig absorbent, absorptive
Saugfähigkeit *f* absorbency, absorptive capacity, absorptivity

Saugfestigkeit *f* suction strength
Saugfilter *m* suction filter, porcelain funnel, suction strainer
Saugfilterpresse *f* suction filter press
Saugflasche *f* filtering flask, aspirator [bottle], suction bottle
Sauggas *n* suction gas
Sauggasanlage *f* suction gas plant
Sauggaserzeuger *m* suction gas producer
Sauggasmotor *m* suction gas motor
Sauggeschwindigkeit *f* pumping speed
Saugglas *n* suction bottle
Saugheber *m* siphon
Saughöhe *f* suction height, suction head
Saughub *m* suction stroke
Saugkamm *m* collector comb
Saugkanal *m* intake duct, suction passage, suction port
Saugkasten *m* suction box
Saugkolben *m* valve piston, (Chem) suction flask
Saugkorb *m* strainer, suction basket
Saugkraft *f* suction force
Saugkrümmer *m* suction bend
Saugkupolofen *m* fan cupola, suction cupola
Saugkuppelofen *m* fan cupola, suction cupola
Saugleistung *f* suction power
Saugleitung *f* suction conduit, suction line, suction pipes, suction piping
Saugluft *f* suction air
Saugluftanlage *f* suction air plant, vacuum plant
Saugluftförderanlage *f* suction air conveyor
Saugluftförderer *m* vacuum conveyor tube
Saugmagnet *m* plunger magnet
Saugmassel *f* feed-head
Saugöffnung *f* intake channel
Saugpapier *n* absorbent paper
Saugpipette *f* suction pipet[te]
Saugpumpe *f* lift pump
Saugraum *m* suction chamber
Saugring *m* collecting ring
Saugröhrchen *n* capillary
Saugrohr *n* suction pipe
Saugrohranschluß *m* suction pipe connection
Saugrüssel *m* suction tube
Saugseite *f* suction side, collecting side, intake side
Saugspitze *f* collecting point
Saugstäbchen *n* filterstick
Saugstrahlpumpe *f* ejector, jet suction pump
Saugstutzen *m* intake port, suction pipe socket
Saugtiegel *m* suction crucible
Saug- und Druckpumpe *f* lift and force pump, sucking and forcing pump
Saugventil *n* suction valve
Saugverfahren *n* emulsion process, forming into female mold, vacuum forming
Saugvorgang *m* suction process
Saugvorrichtung *f* suction device

Saugwäsche *f* vacuum wash
Saugwalze *f* coucher
Saugwiderstand *m* resistance to suction
Saugwind *m* suction air
Saugwindkessel *m* suction air vessel, suction air chamber
Saugwirkung *f* sucking action, suction effect
Saugzellenfilter *n* suction cell filter
Saugzug *m* induced draught or draft
Saugzuganlage *f* induced draft installation
Saum *m* seam, edge
Sauschwanz *m* (Zwirnmaschine) pigtail guide
Saussurit *m* (Min) saussurite
Savinin *n* savinin
Sayboltsekunde *f* Saybolt second
Saynit *m* (Min) saynite, bismuth nickel
Scacchit *m* (Min) scacchite
Scandium *n* scandium (Symb. Sc)
Scandiumoxid *n* scandium oxide
Scandiumsulfat *n* scandium sulfate
Sceleranecinsäure *f* sceleranecic acid
Sceleratin *n* sceleratine
Schabemesser *n* scraper knife, scraping knife
schaben to scrape
Schabenpulver *n* white arsenic
Schaber *m* scraper
Schabevorrichtung *f* scraper, scraping device
Schabinpapier *n* Dutch metal paper
Schablone *f* pattern, model, template, templet, [templet] stencil
Schablonendruck *m* stencilling
Schablonenformerei *f* templet molding
Schablonenhalter *m* arm support of template, tooth block holder
Schablonenlack *m* screen varnish
Schablonenmaschine *f* designing machine
Schablonenstreifen *m* controlling strip
Schabmesser *n* scraper
Schabsel *n pl* parings, scrapings, shavings
Schacht *m* (Bergb) pit, shaft, well, ~ **abteufen** to sink the shaft
Schachtanlage *f* shaft installation, shaft plant
Schachtaufsatz *m* shaft top
Schachtauskleidung *f* shaft lining, furnace lining
Schachtausmauerung *f* shaft lining
Schachtdurchmesser *m* shaft diameter
Schachtel *f* box
Schachtelpresse *f* pot press
Schachtglühofen *m* shaft furnace for reheating
Schachthärteofen *m* shaft furnace for hardening
Schachtofen *m* blast furnace, cupola furnace, pit furnace; shaft furnace, [shaft] kiln
Schachtofenarbeit *f* blast-furnace smelting
Schachtofenschmelzen *n* blast-furnace smelting
Schachtpumpe *f* shaft pump
Schachtstein *m* shaft brick
Schachttrockner *m* cylinder dryer, tunnel dryer

Schaden *m* damage, injury, ~ **durch Teilchenbeschuß** (Atom) bombardment damage
Schadenersatz *m* damage compensation
Schadenersatzanspruch *m* compensation claim
schadhaft faulty
schädigen to damage, to hurt, to injure
Schädigung *f* damage
schädlich destructive, harmful, injurious, (Dämpfe) noxious, (nachteilig) detrimental
Schädlichkeit *f* injuriousness, malignancy
Schädling *m* insect, pest, vermin
Schädlingsbekämpfung *f* pest control
Schädlingsbekämpfungsmittel *n* pesticide, insecticide, pest control agent
schäften to splice
Schäkel *m* shackle
Schälchen *n* [little] cup
schälen to peel, to husk, to pare, to shell
Schälen *n* peeling, (Met) spalling
Schälfolie *f* sliced film
Schälmaschine *f* detreader, detreading machine, rotary cutter
Schärfe *f* sharpness, (Chem) acridity, causticity, (Phot) clearness
schärfen to sharpen
Schärfer *m* sharpener
schätzen to appraise, to estimate, to evaluate, to value
Schätzung *f* estimate, valuation
Schätzwert *m* estimated value
schäumen to bubble, to foam, to froth, to scum, to sparkle, (aufschäumen) to effervesce, to fizz, (Plast) to expand
Schäumen *n* foaming, scumming, (Gärung) effervescence, ~ **der Schmelze** frothing of the melt
schäumend effervescent, foaming
Schäumer *m* frother, frothing agent
Schäumigkeit *f* foaminess
Schafgarbe *f* (Bot) milfoil
Schaft *m* shaft, shank
Schaftdurchmesser *m* shank diameter
Schaftfräser *m* shank cutter
Schaftung *f* scarf joint, tapered overlap
Schaftverzahnung *f* pin gear
Schake *f* chain link, link of chain
Schakenstein *m* link stone
schal flat, insipid, stale
Schale *f* (Elektronen) shell, (Früchte etc.) husk, peel, shell, (Gefäß) bowl, container, cup, dish, (Kruste) crust, (Muschel) shell, (Waage) scale pan, (Wärmeaustauscher) shell, **abgeschlossene** ~ (Atom) complete electron shell, **äußere** ~ outer shell, **innere** ~ inner shell, **vollbesetzte** ~ (Atom) closed shell
Schalenabschluß *m* (Chem) completion of a shell
Schalenanemometer *n* cup anemometer

Schalenarretierung f (Waage) pan arrest (of a balance)
Schalenbauteil n monocoque structure
Schalenblende f (Min) fibrous sphalerite
Schalenbügel m pan holder
Schalenbügelhöhe f height of pan holder
Schaleneisenstein m botryoidal iron ore, kidney iron ore
Schalenelektrode f dished electrode
Schalenflügel m cup vane
schalenförmig bowl-shaped, cup-shaped
Schalenform f cup shape, pan style, (Gieß) chill mold
Schalenguß m casehardened casting, chill casting, die casting
Schalengußstahl m casehardened steel, die cast steel
Schalenkalk m (Min) laminar and globular aragonite
Schalenkreuzanemometer n cup anemometer
Schalenkreuzwindmesser m cup anemometer
Schalenkupplung f split coupling
Schalenmodell n (Atom) shell model
Schalenuntergruppe f subshell
Schalenverteilung f shell distribution
Schall m sound, (Hall) reverberation, (Resonanz) resonance
Schallabsorption f sound absorption
Schallabsorptionsgrad m sound absorption coefficient
Schallanalyse f sound analysis
Schallaufnahme f sound recording
Schallaufzeichnung f sound recording
Schallausbreitung f sound propagation
Schallbeugung f (Akust) sound diffraction
Schallbrechung f (Akust) refraction of sound
Schalldämmaß n sound reduction index, sound transmission loss
schalldämpfend sound-absorbing, deadening, sound-damping, sound-proofing
Schalldämpfer m sound absorber, silencer, (Kfz) muffler
Schalldämpfung f silencing, sound attenuation
schalldicht soundproof
Schalldruck m acoustic pressure, sound pressure
Schalldurchlässigkeit f sound permeability, sound transmission
Schallfortpflanzung f propagation of sound, sound propagation
Schallfrequenz f acoustic frequency, audio-frequency
Schallgeschwindigkeit f velocity of sound
Schallintensität f sound intensity
Schallisolation f sound insulation
schallisolierend sound-insulating
Schallisoliermittel n acoustic insulator
Schallisolierplatte f acoustic panel
Schallisolierung f sound insulation
Schallmeßgerät n phonometer

Schallmessung f acoustic measurement, phonometry
Schallpegel m sound level
Schallplatte f record
Schallquantenentropie f phonon entropy
Schallquantum n phonon, quantum of sound
Schallquelle f source of sound
Schallreiz m sound stimulus
Schallschluckdecke f sound deadening ceiling
schallschluckend sound-absorbing
Schallschluckgrad m sound absorption coefficient
Schallschlucklack m acoustic paint, antinoise paint, sound absorbing paint, sound deadening paint
Schallschluckplatte f baffle
Schallschutz m sound insultation
Schallschutzplatte f acoustic panel
Schallsieb n sonic screen
Schallspektrum n (Akust) sound spectrum
Schallstärke f intensity of sound
Schallwahrnehmung f sound perception
Schallwelle f acoustic wave, sound wave
Schallzerstäuber m sonic atomizer
Schaltanordnung f circuit diagram
Schaltbild n circuit diagram, switch-board diagram, wiring diagram
Schaltblech n shift plate
Schaltbogen m switch arc
Schaltbrett n switchboard, [electrical control] panel
Schaltdose f switch box
Schaltdraht m jumper wire
Schaltdrahtlack m control cable lacquer
schalten to switch, to control, (ausschalten) to disconnect, to switch off, (Getriebe) to shift
Schalten n switching, (Kfz) gear changing
Schalter m switch, circuit-breaker, controller, ~ **mit Umkehrwirkung** reversed switch, **dreipoliger** ~ triple-pole switch, **einpoliger** ~ single-pole switch
Schalterdose f switch box
Schalterkappe f switch covering
Schalterkasten m switch box
Schalterrohr n heavy-duty circuit breaker
Schaltersockel m switch base
Schaltfinger m control finger
Schalthebel m clutch lever, shift lever
Schaltjahr n leap year
Schaltkasten m switch box
Schaltknopf m control knob
Schaltkuppelung f clutch [coupling], shifting coupling
Schaltrad n ratchet wheel
Schaltröhre f relay tube
Schaltscheibe f indexing disk
Schaltschema n switch board diagram, wiring diagram
Schaltschieber m sliding switch

Schaltschrank *m* control cabinet, switch cabinet
Schaltskizze *f* circuit diagram
Schaltstange *f* gearshift bar
Schaltsteuerung *f* feed control
Schalttätigkeit *f* switching action
Schalttafel *f* switch board, control panel, instrument board, operating panel, switch panel, switch plate
Schaltuhr *f* phototimer, switch clock, time switch, ~ **mit Zeitdrucker** readout timer
Schaltung *f* gear shift, network, shifting, switching
Schaltungsart *f* **der Bäder pl** method of connecting the baths
Schaltungstechnik *f* circuit technique
Schaltventil *n* control valve, pilot valve
Schaltvorrichtung *f* switch device, switch mechanism
Schaltwerk *n* control[ling] mechanism
Schamotte *f* chamotte; [burned] fireclay, fire brick, fire-proof clay
Schamotteretorte *f* fireclay retort
Schamottestein *m* fire[clay] brick, refractory brick
Schamottetiegel *m* chamotte crucible
Schapbachit *m* (Min) schapbachite
Schappeseide *f* schappe silk
Schappespinnerei *f* schappe spinning
Schar *f* (Math) family
scharf sharp, (beißend) acrid, biting, burning, (beizend) mordant, (Geruch) pungent, (rauh) harsh, (schrill) piercing, (spitz) acute, pointed, ~ **eingestellt** focused, in focus, ~ **schmeckend** acrid, sharp, tart
scharfbegrenzt sharp edged, sharply defined
Scharfeinstellung *f* (Opt) exact focussing
Scharffeuerfarbe *f* fire-proof color, sharp fire paint
scharfkantig sharp-edged, angular
scharf riechend pungent
Scharlach *m* scarlet, (Med) scarlet fever
Scharlachfärben *n* scarlet dyeing
Scharlachfarbe *f* scarlet color
scharlachfarben scarlet
Scharlachrot *n* scarlet, cochineal
Scharnier *n* hinge [joint]
Scharnierband *n* articulated conveyor belt
Scharnierpresse *f* tilting head press
Scharnierunterlage *f* hinge pad
Scharre *f* rake, scraper
Scharte *f* fissure, notch
Schatten *m* shade, shadow
Schattenbild *n* shadow image, shadow picture
Schattenbildverfahren *n* silhouette procedure
Schattenglas *n* shade glass
Schattenkreuzröhre *f* maltese cross tube
Schattenmikroskopie *f* shadow projection microscopy

Schattenphotometer *n* shadow photometer, Rumford's photometer
Schattenstelle *f* blind spot
Schattenstift *m* shadow pencil
Schattiergewebe *n* shadow cloth
Schattierung *f* hue, shade, tinge, tint, tone
Schaubild *n* graph, diagram, flow chart
Schauer *pl,* **kosmische** ~ cosmic showers
Schaufel *f* (Radblatt) paddle, (Schippe) scoop, shovel, (Turbine) blade, bucket
Schaufelmulde *f* bucket trough
Schaufelrad *n* turbine [blade], bladed wheel; impeller, vane wheel
Schaufelradbagger *m* bucket wheel excavator
Schaufelradmischer *m* turbine mixer
Schaufelrührer *m* paddle mixer, paddle stirrer
Schaufeltrockner *m* blade dryer, paddle dryer
Schaufelung *f* blading, buckets
Schaufelwelle *f* rotor shaft
Schaufelwinkel *m* blade angle
Schauglas *n* sight glass
Schaukammer *f* inspection room
Schaukelofen *m* tilting open hearth furnace
Schauloch *n* observation hole, inspection hole, inspection window, peep hole
Schaum schlagen to churn
Schaum *m* foam, froth, scum, (Gischt) spray, (Seife) lather, suds, **fester** ~ rigid foam
Schaumabstreifer *m* [foam] skimmer
schaumähnlich foam-like, foamy, frothy
schaumbekämpfend antifoam, antifroth
schaumbeständig foam holding
Schaumbeton *m* aeroconcrete, foam concrete, porous concrete
Schaumbildung *f* foam formation, foaming, frothing
Schaumbildungsvermögen *n* foaming quality, frothing quality
Schaumbrecher *m* foam breaker
Schaumentwicklung *f* foaming, frothing
Schaumerz *n* bog manganese
Schaumerzeuger *m* (Metall) frothing reagent
Schaumfähigkeit *f* foaming property
Schaumfängeraufsatz *m* anti-foaming splash head
Schaumfang *m* skimmer gate [in casting]
Schaumfeuerlöscher *m* foam extinguisher
Schaumgemisch *n* emulsion foam
Schaumgips *m* foliated gypsum
Schaumgold *n* Dutch metal, imitation gold
Schaumgummi *m* foam rubber, sponge rubber
Schaumhahn *m* scum cock
Schaumhaken *m* skimmer
Schaumhöhe *f* height of foam
schaumig (Gieß) porous
Schaumkalk *m* (Min) aragonite, slate spar, [sparry] aphrite
Schaumkelle *f* skimming ladle, skimmer
Schaumkleber *m* foam glue

Schaumkopf *m* scum riser
Schaumkorb *m* scum basket
Schaum[kunst]stoff *m* aerated plastic, foam[y] plastic
Schaumlöffel *m* skimmer, skimming spoon
schaumlos foamless, frothless
Schaummischkammer *f* foam mixing chamber
Schaummittel *n* expanding agent, foaming agent
Schaumöl *n* antifroth oil
Schaumröhre *f* foam tube, scum pipe
Schaumschlagmaschine *f* foamer, frother
Schaumschwärze *f* animal charcoal dust
Schaumschwimmaufbereitung *f* flotation
Schaumseife *f* lathering soap
Schaumspat *m* (Min) analcite, laumontite
Schaumstabilisator *m* froth stabilizing agent
Schaumstoff *m* aerated plastics, foam plastics
Schaumstoffkörper *m* plastic foam
Schaumströmung *f* foam flow, froth flow
Schaumstück *n* porous piece
Schaumton *m* fuller's earth
Schaumtrichter *m* scum riser
Schaumverhütung *f* prevention of foam
Schaumverhütungsmittel *n* antifoam, antifoaming agent, antifrothing agent
Schaumverstärker *m* (Siebbodenkolonne) bubbling promoter
Schaumwein *m* champagne, sparkling wine
Schaumzerstörung *f* defoaming, foam inhibition
Schauöffnung *f* inspection hole, inspection panel, peep hole, sight hole
Schaupackung *f* display packing, dummy
Schauzeichen *n* indicator signal
Scheck *m* check, cheque (Br. E.)
scheckig checkered, dappled, speckled, spotted
Scheelbleierz *n* lead tungstate
Scheelerz *n* scheelite
Scheelesches Grün *n* Scheele's green, acid copper arsenite
Scheelit *m* (Min) scheelite
Scheelsäure *f* (obs) tungstic acid
Scheelspat *m* (Min) scheelite
Schefferit *m* (Min) schefferite
Scheibe *f* disc, disk, plate, (Flaschenzug) sheave, (Glas) pane, (Schleifscheibe) wheel, (Schnitte) slice, (Unterlegscheibe) washer, (Zahlenscheibe) dial
Scheibenausgleicher *m* disk compensator
Scheibenblei *n* glazier lead
Scheibeneisen *n* pig disk
Scheibenfilter *n* disk filter
scheibenförmig disc-shaped
Scheibenhalter *m* disk holder
Scheibenhaspel *f* disk winch
Scheibenholländer *m* (Pap) centrifugal rag engine
Scheibenkolben *m* disc piston, solid piston
Scheibenkolbenpumpe *f* solid-piston pump
Scheibenkonduktor *m* disc-shaped conductor

Scheibenkreiselmischer *m* disk impeller [mixer]
Scheibenkühler *m* disc cooler
Scheibenkupfer *n* rose copper, rosette copper
Scheibenmaschine *f* disc machine
Scheibenmühle *f* disc attrition mill
Scheibenpilze *pl* discomycetes
Scheibenregenerator *m* plate regenerator
Scheibenreißen *n* conversion into rosettes
Scheibenrost *m* disc separator
Scheibenrührer *m* disc mixer
Scheibenschützer *m* disk cover
Scheibenspule *f* double-flanged bobbin
Scheibenstein *m* discolith
Scheibenventil *n* disc valve
Scheibenwaschanlage *f* windscreen wash assembly
Scheibenwischer *m* windscreen wiper
Scheibenzelle *f* disk cell
Scheibenzerstäuber *m* disc atomizer
Scheibenziehbank *f* drawing machine with spool
scheidbar separable
Scheidbarkeit *f* separability
Scheide *f* sheath
Scheideanlage *f* screening device, separating device
Scheideanstalt *f* refinery
Scheidebürette *f* separating buret[te]
Scheideerz *n* picked ore, screened ore
Scheideflüssigkeit *f* separating liquid
Scheidegefäß *n* separating glass, separating vessel, separator trap
Scheidegold *n* gold purified by parting
Scheidegut *n* material to be separated
Scheidekapelle *f* cupel
Scheidekolben *m* separating flask
Scheidekuchen *m* liquation disk
Scheidelinie *f* boundary line, line of demarcation
Scheidemagnet *m* screening magnet, sorting magnet
Scheidemehl *n* (Met) dust of picked ore
Scheidemittel *n* parting agent, separating agent
scheiden to separate, to clear, to screen, (Chem) to analyze, to isolate, (mechanisch) to pick, to sort, (Met) to refine, (trennen) to part
Scheiden *n* separating, clarifying, clearing
Scheideofen *m* almond furnace, parting furnace
Scheidepfanne *f* clarifier, defecating pan, defecation tank, defecator
Scheidepresse *f* press separator
Scheidepunkt *m* separation point
Scheider *m* separator, grader
Scheiderz *n* screened ore
Scheideschlamm *m* (Zucker) defecation scum, defecation slime, molasses
Scheidesieb *n* separating sieve
Scheidesilber *n* parting silver
Scheidetrichter *m* separating funnel, separatory funnel

Scheide- und Schüttelvorrichtung *f* separating and shaking apparatus
Scheideverfahren *n* (Met) refining process, parting process, separating process
Scheidevorgang *m* separating process
Scheidevorrichtung *f* separator, parting device, screening device, separating apparatus, sorting device
Scheidewand *f* separating wall, baffle plate, partition [wall], (Bot) septum, (elektrolytisch) diaphragm, **halbdurchlässige** ~ semipermeable partition
Scheidewasser *n* aqua fortis, nitric acid
Scheidmünzsilber *n* billon silver
Scheidung *f* separation, clarification, clearing, parting, (Zucker) defecation, **nasse** ~ wet separation
Scheidungsfilter *m* defecator
Scheidungsmittel *n* separating agent, parting agent
Schein *m* brightness, (Aussehen) appearance, look, (Lichtschein) shine
scheinbar apparent, fictitious, obvious
Scheinbombage *f* (Dose) springer
Scheindiffusionskoeffizient *m* eddy diffusivity
Scheingold *n* imitation gold
Scheinleistung *f* apparent output or power
Scheinleitfähigkeit *f* eddy conductivity
Scheinleitwert *m* admittance
Scheinreibung *f* eddy viscosity
Scheinwerfer *m* head lamp, search light, spot light
Scheinwerferlicht *n* flood light
Scheinwiderstand *m* apparent resistance
Scheitel *m* (Winkel, Kegel) vertex, ~ **des Bodens** top of the base
Scheitelabstand *m* zenith distance
Scheitelbogen *m* azimuth
Scheitelfaktor *m* amplitude factor, peak factor
Scheitelkante *f* culminating edge
Scheitelpunkt *m* vertex, zenith
scheitelrecht (obs) perpendicular, vertical
Scheitelwert *m* maximum value, peak [value]
Scheitelwinkel *m* azimuth, crest angle, vertex angle, vertical angle
scheitern to fail
Schellack *m* shellac[k], garnet lac (Am. E.), **gebleichter** ~ bleached shellac, white shellac
schellacken to shellac
Schellackfirnis *m* shellac varnish
Schellackpolitur *f* French polish
Schellan *n* shellane, compressed natural gas
Schelleisen *n* riveting set
Schellhammer *m* cup-shaped hammer, set hammer, snap hammer
Schellharz *n* white rosin
Schellolsäure *f* shellolic acid
Schema *n* scheme, diagram, (Fließbild) flow sheet, (Muster) pattern, (System) plan, system

Schemabild *n* diagrammatic section
schematisch schematic, diagrammatic[ally]
Schemazeichnung *f* flow sheet
Schemel *m* stool
Schenkel *m* branch, side leg, side piece, (Winkel) arm, leg, (Zirkel) leg
Schenkellänge *f* shank length
Schenkelquerschnitt *m* cross-section of arms
Scherbeanspruchung *f* shear, shearing force, shearing stress, tangential stress
Scherbeneis *n* flake ice
Scherblatt *n* shear blade, comb
Scherbolzen *m* shear pin
Schere *f* [pair of] scissors, shears
scheren to shear, to crop, to warp
Scheren *n* shearing
Scherfestigkeit *f* resistance to shearing, shear[ing] strength, shear[ing] stress
Schergrenze *f* shearing limit
Scherkraft *f* shearing force, shearing stress
Scherleiste *f* scuff rib
Schermodul *m* modulus of shearing
Schernikit *m* schernikite
Scherscheibenmischer *m* shearing disc mixer
Scherschwingung *f* shear mode
Scherspannung *f* shearing stress, shearing force, shearing resilience, tangential force
Scherung *f* shear
Scherungsbruch *m* shear fracture
Scherungsdruck *m* shearing pressure, shearing stress
Scherungsfestigkeit *f* shearing strength
Scherungsinstabilität *f* shearing instability
Scherungskoeffizient *m* rate of shear
Scherungswelle *f* shear wave
Scherwirkung *f* shear action
Scherzugversuch *m* tensile shear stress
scheuerfest scrubbable
Scheuerfestigkeit *f* abrasion property, fastness to scraping, scrub resistance, scuff resistance, wear resistance
Scheuergerät *n* scourer
Scheuerleiste *f* base board
Scheuermittel *n* abrasive
Scheuermühle *f* tumbling barrel
scheuern to scour, (mit Sand) to scour [with sand], (polieren) to polish, (reiben) to chafe, to rub
Scheuerpulver *n* scouring powder
Scheuertrommel *f* tumbling barrel
Schicht *f* layer, film, (Arbeitsschicht) shift, (Geol) stratum, (Holz) lamina, ply, (Met) coating, **absorbierende** ~ absorbing layer, **atmosphärische** ~ atmospheric layer, **durchlässige** ~ permeable layer, **einatomige** ~ monoatomic layer, **lichtelektrische** ~ photoelectric layer or film, **lichtempfindliche** ~ light-sensitive surface or layer, **monoatomare** ~ monoatomic layer,

monomolekulare ~ monolayer, monomolecular film, unimolecular film
Schichtaufnahme f (Gewebe) laminography
Schichtdicke f thickness of the layer
Schichtdickenmessung f film thickness measurement
schichten to arrange in layers, to laminate, to layer
Schichtenabstand m interlayer distance
Schichtenbildung f lamination
Schichtenfilter n sheet filter
Schichtenfiltration f (Zucker) layer filtration
Schichtenfolge f (Geol) stratigraphy
Schichtengefüge n lamellar structure
Schichtengitter n layer lattice, layer lattices
Schichtentrennung f delamination
schichtenweise in layers, stratified
Schichtgefüge n laminated structure
Schichtgitter n (Krist) layer lattice
Schichtgitterstruktur f layer-lattice structure
Schichtglätte f melting litharge
Schichthöhe f height of layer, height of bed, height of packing
Schichtholz n laminated wood, (Sperrholz) plywood
schichtig lamellar, laminated, plate-like
Schichtkorrosion f layer-corrosion
Schichtleistung f output per shift
Schichtlinien pl (Krist) layer lines
Schichtliniendiagramm n (Röntgen) rotational X-ray diagram
Schichtplatte f laminated sheet
Schichtpreßstoff m laminate, laminated plastic
Schichtsieden n film boiling
Schichtspaltung f delamination
Schichtstoff m laminate, laminated material, laminated plastic
Schichtstoffaußenhaut f laminate skin
Schichtstoffplatte f laminated plastic panel
Schichtstofftechnik f lamination
Schichtströmung f stratified flow
Schichtträger m substrate
Schichtung f layering, stratification
Schichtverband m interlaminar bonding
Schichtwachstum n film growth
Schichtwolke f stratus cloud
Schiebebühne f traveling platform, traverser
Schiebedach n sliding roof, sunshine roof
Schiebedeckel m slip cap
Schiebefenster n sash window, sliding window
schiebefest nonslip, ~ **machen** to impart a nonslip finish
Schiebegitter n sliding grating
Schiebekarren m wheel barrow
Schiebekassette f sliding box
Schiebekontakt m slide contact, slider
Schiebemaststapler m reach truck
schieben to slide, to push, to shove

Schieber m slider, safety bolt, slide bar, slide carriage, slide valve, sluice valve, (Kassettenschieber) slide
Schieberahmen m sash
Schieberdiagramm n slide valve diagram
Schieberegister n (Comp) shift register
Schiebergehäuse n gate valve housing
Schieberlager n pushrod pin bearing
Schieberost m sliding grate
Schieberstange f slide rod, slide valve spindle
Schiebersteuerung f slide valve gear
Schieberwerkzeug n bar mold
Schiebeschachtel f slide box
Schiebestempel m sliding punch
Schiebetür f sliding door
Schiebewiderstand m rheostat, slide resistance
Schiebgewicht n sliding weight
Schieblehre f caliper slide, slide gauge, vernier callipers
Schiebung f shear strain
schief (schräg) bevel, oblique, out of line, slant[ing]
Schiefer m (Min) slate, schist, shale
schieferähnlich schistous, slate-like
Schieferalaun m feather alum
schieferartig schistous, slate-like, slaty
Schieferblei n (Min) slaty lead
Schieferbottich m slate vat
Schiefergips m (Min) foliated gypsum
Schiefergrün n green carbonate of copper
schieferhaltig schistous, slaty
Schieferkalkstein m (Min) sparry limestone
Schieferkohle f slaty coal, splint coal
Schieferkreide f (Min) graphitic clay
Schiefermarmor m (Min) hard calcareous slate
Schiefermehl n powdered slate
Schiefermergel m marl slate
schiefern to exfoliate, to scale off
Schieferöl n schist oil, shale oil
Schieferplatte f slate slab
Schieferschwelöl n shale oil
Schieferspat m (Min) slate spar
Schieferteer m shale tar
Schieferton m (Min) slate clay
schiefflächig (Krist) plagihedral
schieflaufend tilted
schiefrig foliated, schistous, slate-like, slaty
schiefsymmetrisch skew-symmetric[al]
schiefwinklig oblique angular
Schiene f rail, bar, beam, track, (Med) splint, ~ **des Bades** busbar of the bath
Schienenkaliber n rail pass
Schienenlasche f fish plate, track joint
Schienennagel m [rail] spike
Schienenrost m fire bar, grate bar
Schienenstoßzwischenlagen pl rail joint baseplates
Schierling m (Bot) hemlock
Schierlingsgift n conium, hemlock poison

Schierlingssäure *f* conic[ic] acid, coniic acid
Schießbaumwolle *f* explosive cotton, cellulose trinitrate, gun cotton, nitrocellulose, pyroxyline
schießen to fire, to shoot, (Bergb) to blast
Schießkabel *n* fuse wire
Schießofen *m* bomb oven, Carius oven, sealed tube furnace
Schießpulver *n* gun powder
Schiffahrt *f* navigation
Schiffbau *m* shipbuilding
Schiffbaumaterial *n* shipbuilding material
Schiffbruch *m* shipwreck
Schiffchen *n* little ship, (Labor) boat, (Web) shuttle
Schiffsbelader *m* shiploader
Schiffsche Base *f* Schiff's base
Schiffsfirnis *m* marine varnish
Schiffskitt *m* cement
Schiffsleim *m* marine glue
Schiffspech *n* common black pitch
Schikane *f* baffle
Schild *n* name plate, sign board, (Preis-Schild) tag
Schilddrüse *f* (Med) thyroid gland
Schilddrüsenfunktion *f* thyroid activity, thyroid function
Schilddrüsenhormon *n* thyroxine
Schilddrüsenkrebs *m* (Med) thyroid cancer
Schilddrüsenüberfunktion *f* (Med) hyperthyroidism
Schilddrüsenunterfunktion *f* (Med) hypothyrosis, thyropenia
Schildkrötenöl *n* turtle oil
Schildpatt *n* tortoise shell
Schildzapfen *m* gudgeon, trunnion
Schillerfarbe *f* changeable color, schiller color
schillerfarbig iridescent
Schillerfels *m* (Min) gabbro, limesoda feldspar
Schillerglanz *m* iridescent luster, changeable luster
schillern to iridesce, to change colors
Schillern *n* iridescence, opalescence
schillernd iridescent, opalescent, scintillating, (Web) shot-colored
Schillerquarz *m* (Min) cat's eye, changeable feldspar
Schillerspat *m* (Min) schiller spar
Schillerstoff *m* iridescent substance
Schimm *m* [electromagnetic] shim
Schimmel *m* mildew, moldiness, mo[u]ld
schimmelartig mold-like, moldy, musty
Schimmelbildung *f* mold formation, mold growth, molding
schimmelfest mildew-resistant
schimmelfleckig mildewed
Schimmelgeruch *m* moldy smell
schimmelig moldy, musty
schimmeln to become moldy, to mold

Schimmelpilz *m* mold [fungus]
Schimmer *m* glimmer, luster
schimmern to glimmer, to glisten
schimmernd lustrous, resplendent
Schinusöl *n* shinus oil
Schippe *f* shovel
Schirm *m* screen, shield[ing] diaphragm
Schirmanguß *m* fan gate
Schirmfaktor *m* shielding factor
Schirmgitter *n* screen grid
Schirmwirkung *f* screening action, screening effect
Schistosomiasis *f* (Bilharziose) bilharziasis, schistosomiasis
Schizomyzeten *pl* (Spaltpilze) Schizomycetes
Schlacke *f* slag, clinker, dross, [rich finery] cinder, scoria, (Puddelabstich) tap cinder, ~ **abziehen** to draw off slag, to rake out slag, to remove slag, to skim off slag, to slag [off], **dünnflüssige** ~ fluid slag, wet slag, **eisenarme** ~ slag poor in iron, **saure** ~ acid slag, **schmelzbare** ~ fusible slag, **Thomas-** ~ basic slag, phosphate slag, Thomas slag, **zähflüssige** ~ dry slag, semipasty slag, sticky slag, viscous slag
schlacken to slag, to clinker, to form slag, to scorify
Schlackenabfluß *m* slag discharge
Schlackenabscheider *m* slag skimmer
Schlackenabsonderung *f* slag separation
Schlackenabstich *m* tapping the slag
Schlackenabstichrinne *f* slag spout
Schlackenansatz *m* adhesion of the slag
Schlackenarbeit *f* slag smelting, working of slag
schlackenartig scoriaceous, drossy, slaggy, slag-like
Schlackenauge *n* cinder tap, slag hole
Schlackenbeton *m* slag concrete
Schlackenbett *n* cinder bed
schlackenbildend slag-forming
Schlackenbildner *m* flux for making slag, slag-forming constituent
Schlackenbildung *f* formation of slag, formation of clinker, scorification
Schlackenbildungsperiode *f* slag forming period
schlackenblau scoriaceous blue
Schlackenblech *n* slag plate
Schlackenblei *n* slag lead
Schlackenblock *m* slag block
schlackend clinkering
Schlackendecke *f* slag cover
Schlackendepot *n* (Atom) active storage area
Schlackeneinschluß *m* slag or cinder inclusion, trapped slag
Schlackeneisen *n* cinder iron, slag iron
Schlackenerz *n* vitreous silver ore
Schlackenfaden *m* slag thread
Schlackenfänger *m* slag trap
Schlackenfang *m* slag chamber, slag pocket

schlackenförmig scoriform
Schlackenform *f* slag tuyère
schlackenfrei free from slag, (Kohle) non-clinkering
Schlackenfrischen *n* pig boiling
Schlackengehalt *m* clinker content[s]
Schlackengranulation *f* slag granulation
Schlackengrube *f* slag pit
Schlackenhalde *f* slag dump
schlackenhaltig containing slag, slag bearing
Schlackenherd *m* slag furnace, slag hearth
Schlackenkammer *f* slag chamber, slag pocket
Schlackenkobalt *m* (Min) safflorite
Schlackenkruste *f* slag crust
Schlackenkuchen *m* clinker cake, slag cake
Schlackenkübel *m* slag bucket
Schlackenkugel *f* slag ball
Schlackenlava *f* scoriaceous lava
Schlackenloch *n* slag hole
Schlackenmehl *n* ground slag, Thomas meal
Schlackenpfanne *f* slag ladle
Schlackenpresse *f* slag press
Schlackenpuddeln *n* slag puddling, pig boiling
schlackenreich rich in slag, scoriaceous, slaggy
Schlackenrost *m* clinker grate, dumping grate
Schlackensand *m* slag sand
Schlackenschicht *f* layer of slag
Schlackenschweißverfahren *n* electroslag welding
Schlackenstein *m* slag stone, cinder stone, slag brick
Schlackenstichloch *n* cinder notch, cinder tap, flushing hole, slag hole or notch
Schlackenstrahl *m* slag flow
Schlackenstreifen *m* slag streak
Schlackentrift *f* flow of slag
Schlackenüberlauf *m* skimmer
Schlackenverschmelzung *f* fusion of slags, melting of slags
Schlackenwagen *m* slag wagon
Schlackenwechsel *m* change of slag
Schlackenwolle *f* slag wool, mineral wool, rock wool
Schlackenzacken *m* fore plate
Schlackenzahl *f* (CaO/SiO$_2$) slag ratio
Schlackenzement *m* cinder cement, slag cement
Schlackenziegel *m* slag brick
Schlackenzusatz *m* cinder charging
Schlackenzuschlag *m* slag addition
schlackig cindery, clinkery, drossy, scoriaceous, slaggy
Schlägerstange *f* beater bar
Schlägerwalze *f* beater bar
Schlämmanalyse *f* elutriation, sedimentation analysis
Schlämmapparat *m* elutriating apparatus, cleansing apparatus, elutriator
Schlämmarbeit *f* elutriating
Schlämmaschine *f* decanting machine
Schlämmbottich *m* washing tub

schlämmen to clear of mud, to elutriate, to levigate, to wash, (Bergb) to buddle, (Öl) to bail
Schlämmen *n* decanting, washing
Schlämmer *m* elutriator, **Falten-** ~ fluted classifier, **Horizontal-** ~ hindered-settling classifier
Schlämmfaß *n* washing tub
Schlämmgerät *n* elutriator
Schlämmkreide *f* precipitated chalk, whiting
Schlämmung *f* elutriation, decantation, levigation, sedimentation, washing, (Bergb) buddling
Schlämmverfahren *n* elutriating process, sedimentation, washing process
Schlämmvorrichtung *f* cleansing device, washing apparatus
schlaff loose, flabby, slack, (Seil) sagging
Schlafkrankheit *f* (Med) sleeping sickness
Schlaflosigkeit *f* insomnia
Schlafmittel *n* (Pharm) hypnotic, sleeping drug, soporific
Schlag *m* stroke, beat, bump, impact, knock, percussion, rap, **elektrischer** ~ electric shock
Schlagader *f* (Anat) artery
Schlagarbeit *f* impact energy, striking energy, **spezifische** ~ specific energy of blow
schlagartig sudden[ly]
Schlagbeanspruchung *f* blow stress, impact stress, shock stress
Schlagbiegefestigkeit *f* impact bending strength
Schlagbiegeprobe *f* impact bending test, blow bending test, shock bending test
Schlagbiegeversuch *m* impact bending test
Schlagbohrer *m* percussion borer
Schlagbohrmaschine *f* electric hammer drill
Schlagbrecher *m* impact crusher
Schlagdruckbeanspruchung *f* compressive impact stress, dynamic compression stress
Schlagdruckversuch *m* dynamic compression test, impact compression test
Schlagelastizität *f* impact elasticity
Schlagempfindlichkeit *f* sensitivity to percussion
schlagen to strike, to beat, to hit, (Öl) to press
schlagfest impact-resistant
Schlagfestigkeit *f* impact resistance, impact strength, resistance to shock, shatterproofness
Schlagfigur *f* percussion figure
Schlagfinne *f* striker, striking edge, striking pendulum
Schlagfläche *f* striking surface
Schlaghärte *f* impact hardness
Schlaghärteprüfer *m* impact hardness tester, dynamic hardness testing machine
Schlaghärteprüfung *f* impact hardness test
Schlagholländer *m* breaker beater
Schlagkreuzmühle *f* cross beater mill
Schlagküpe *f* beating vat
Schlaglot *n* hard solder

Schlagmessing *n* beaten brass
Schlagmühle *f* beater mill, centrifugal mill, hammer mill
Schlagpendel *n* striking pendulum, shock pendulum, striker
Schlagpressen *n* impact molding
Schlagpreßverfahren *n* impact mo[u]lding
Schlagprobe *f* impact test, shock test
Schlagprüfgerät *n* impact tester
Schlagradhandpresse *f* hand wheel stamping press
Schlagrolle *f* tappet roller
Schlagschlüssel *m* hammering spanner, wrench hammer
Schlagschraube *f* drive screw
Schlagsieb *n* vibrating screen
Schlagspaltung *f* impact cleaving
Schlagstärke *f* strength of blow
Schlagversuch *m* impact test, shock test
Schlagwerkzeug *n* straightening hammer
Schlagwetter *n* (Bergb) fire damp
Schlagwetteranzeiger *m* fire damp indicator, safety lamp
Schlagwiderstand *m* resistance to impact
Schlagwirkung *f* effect of stroke, strength of blow
Schlagwort *n* slogan
Schlagzähigkeit *f* impact resistance, impact strength
Schlagzahl *f* number of blows
Schlagzeile *f* headline
Schlagzerreißversuch *m* impact tearing test, dynamic tensile test, impact pulling test, impact tension test, tensile shock test
Schlagzugbeanspruchung *f* impact tensile stress
Schlagzugversuch *m* impact tensile test
Schlamm *m* mire, mud, slime, sludge, slurry, slush, (Kesselschlamm) deposit, incrustation, silt, (Schlick) concentrates, schlich, (Zucker) scum
Schlammablagerung *f* deposit of mud
Schlammablaßhahn *m* mud cock
Schlammabscheider *m* slime separator
Schlammabscheidung *f* desludging
Schlammaufbereitung *f* preparation of the sludge
Schlammbeutel *m* sludge bag
Schlammboden *m* muddy ground
Schlammeißel *m* mud bit
Schlammentwässerung *f* sludge draining
Schlammfilter *n* mud filter
Schlammgrube *f* slime pit
Schlammhahn *m* mud cock
schlammhaltig muddy, oozy
Schlammherd *m* slime pit, slime table
schlammig muddy, miry, oozy, silty, slimy, sludgy
Schlammkasten *m* mud box
Schlammkohle *f* coal mud

Schlammkontaktanlage *f* reactor-clarifier for water treatment
Schlammkonzentrationsmesser *m* sludge concentration meter
Schlammpfännchen *n* scum pan
Schlammpumpe *f* sludge pump
Schlammreaktor *m* slurry reactor
Schlammscheider *m* slime separator
Schlammschicht *f* slime layer
Schlammseparator *m* desludger
Schlammteller *m* slime plate
Schlammventil *n* sludge valve
Schlammwassereindicker *m* slurry thickener
Schlange *f* worm, coil, serpentine [pipe], spiral condenser, (Zool) snake
Schlangenabsorber *m* coil-type absorber
Schlangengift *n* snake poison, venom
Schlangenkühler *m* coil condenser, coil-cooler, spiral condenser
Schlangenrohr *n* coil, spiral tube, worm
Schlangenstein *m* (Min) ophite; serpentine, snake stone
Schlangenventil *n* valve with two bends
Schlangenwurzel *f* (Bot) snake root
Schlauch *m* hose, flexible tubing, [rubber] tubing, **pannensicherer** ~ puncture-sealing safety tube
Schlauchansatz *m* rubber tube connection
Schlauchanschluß *m* hose coupling
Schlaucharmierungsmaschine *f* hose armoring machine
Schlauchbeutel *m* tubular bag
Schlauchbiegeermüdungsprüfung *f* fatigue tube test
Schlauchklemme *f* tube clamp, tubing clip
Schlauchkupplung *f* coupling for rubber tubing, hose coupling, rubber coupling
Schlauchmaschine *f* tube extruding press, tube extrusion machine, tubing machine
Schlauchmaterial *n* flexible tubing
Schlauchpilze *pl* (Askomyzeten) ascomycetes
Schlauchpumpe *f* elastic tube pump
Schlauchreparaturgummi *m* vulcanizing tube repair gum
Schlauchschelle *f* hose clip
Schlauchspritzform *f* extrusion die for tubing
Schlauchtülle *f* nozzle
Schlauchverbindung *f* rubber tube connection
Schlauchwelle *f* hose nipple
Schlauchwickelmaschine *f* hose wrapping machine, tube winding machine
Schlauchzusammensetzmaschine *f* tube splicer
Schlaufe *f* loop
Schlaufenreaktor *m* recycle reactor, loop reactor
schlecht bad, ill, poor, (verdorben) rotten
Schlechtwetterperiode *f* period of poor weather
Schlehdorn *m* (Bot) blackthorn, sloe tree
Schlehe *f* (Bot) sloe
Schleier *m* (Phot) fog, haze

Schleierbildung *f* cloudiness, hazing, turbidity
Schleifapparat *m* grinder
Schleifarbeit *f* grindling operation
Schleifautomat *m* automatic grinding machine
Schleifbock *m* grinding trestle
Schleifbrett *n* grinding board
Schleifbürste *f* buffing brush
Schleifdrahtkompensator *m* slide wire potentiometer
Schleife *f* bow, link, loop
schleifen to grind; to sand, (Glas mattschleifen) to frost, (schärfen) to sharpen, (schmirgeln) to polish, to smooth
Schleifen *n* grinding
Schleifenbildung *f* loop formation
Schleifengalvanometer *n* string galvanometer, (Elektr) loop galvanometer
Schleifenlinie *f* (Math) lemniscate
Schleifenoszillograph *m* loop oscillator
Schleifentrockner *m* loop drier
Schleiffläche *f* grinding surface, wearing [sur]face
Schleifgrund *m* rubbing primer, sanding sealer
Schleifhärte *f* resistance to polish
Schleifkante *f* buffing rib
Schleifkomposition *f* grinding composition
Schleifkontakt *m* (Elektr) sliding contact
Schleifkorn *n* abrasive grain
Schleiflack *m* dull finish lacquer, flatting varnish
Schleifleinen *n* abrasive cloth
Schleifmaschine *f* grinding machine, grinder
Schleifmaß *n* grinding width
Schleifmittel *n* abrasive, grinding material, polishing material
Schleifmittelindustrie *f* abrasives industry
Schleifmotor *m* grinding motor
Schleifpapier *n* abrasive paper, emery paper
Schleifpaste *f* abrasive paste
Schleifpulver *n* grinding powder
Schleifriemen *m* belt for grinding machine
Schleifring *m* collecting ring, slip ring
Schleifringanker *m* slip ring armature
Schleifringläufer *m* slip ring rotor
Schleifsand *m* abrasive sand
Schleifscheibe *f* grinding disk, abrasive disk, [abrasive] grinding wheel, abrasive wheel
Schleifscheibenharz *n* grinding wheel resin
Schleifschmirgel *m* grinding emery
Schleifstaub *m* dust from grinding, abrasive dust
Schleifstein *m* grindstone, whetstone
Schleifstuhl *m* grinding bench
Schleif- und Poliermaschine *f* grinding and polishing machine
Schleifvorgang *m* grinding operation
Schleifwerkzeug *n* grinding tool
Schleifwirkung *f* abrasive action, abrasive power, attrition

Schleim *m* slime, (Pflanzen-Schleim) mucilage, (Zool) mucus
Schleimabsonderung *f* mucous secretion
schleimartig slimy
Schleimbeutel *m* (Med) [mucous] bursa
schleimbildend muciferous, mucigenous, (Bakt) slime-forming
Schleimbildung *f* formation of mucus
Schleimgewebe *n* mucous tissue
Schleimharz *n* gum resin
Schleimhaut *f* mucosa (pl mucosae), mucous membrane
schleimig mucilaginous, mucous, slimy, (zähflüssig) viscid
schleimlösend (Med) expectorant, mucolytic
Schleimsäure *f* mucic acid
schleimsauer mucate
Schlempe *f* dried distiller' solubles, (Brauw) vinasse, (Dest) bottom product
Schlempeverdampfer *m* residual liquor evaporator, reboiler
schleppen to drag, to tug
Schlepper *m* (Schleppboot) tugboat, (Traktor) tractor
Schleppkettenförderer *m* drag-chain conveyor
Schleppkreisförderer *m* power-and-free conveyor
Schleppmittel *n* (Rektifikation) entrainer, separating agent
Schleppwalze *f* friction roll
Schleppzangenziehbank *f* drawing machine with pliers
Schleuder *f* centrifuge, centrifugal apparatus, hydro-extractor; sling
Schleuderbandförderer *m* centrifugal belt conveyor
Schleuderenergie *f* toss energy (in impact testing)
Schleudergebläse *n* centrifugal blower, fan blower
Schleuderguß *m* centrifugal casting, rotary casting
Schleuderhonig *m* strained honey
Schleudermaschine *f* centrifugal machine, centrifuge, hydro-extractor
Schleudermischer *m* centrifugal mixer
Schleudermühle *f* centrifugal mill, disintegrator
schleudern to centrifuge, to hydroextract, to strain, (werfen) to fling, to hurl
Schleudern *n* centrifugal casting
Schleuderpsychrometer *n* whirling hygrometer
Schleudertrockner *m* centrifugal hydroextractor, whizzer
Schleuderverfahren *n* (Kunststoff) rotational molding of plastics
Schlich *m* (Bergb) pulp
Schlichtebad *n* sizing bath
schlichtecht (Web) fast to sizing
Schlichteisen *n* scraper

schlichten to dress, to finish, to plane, to planish, to smooth, (Gieß) to blackwash, (Web) to size
Schlichtfeile f smooth cut file
Schlichtleim m (Text) size
Schlichtmaschine f (Techn) finishing machine, (Text) sizing machine
Schlichtmittel n (Web) sizing agent, sizing material
Schlichtstahl m smoothing tool
Schlick m mud, ooze
Schliere f streak, stria
Schlieren pl schlieren, (auf der Oberfläche) surface waviness
Schlierenaufnahme f (Phot) schlieren photograph
Schlierenbildung f (Opt) striation
Schlierenblende f schlieren diaphragm
Schlierenoptik f (Opt) schlieren optics
Schlierenphotographie f schlieren photography
Schlierenverfahren n schlieren method
Schließbewegung f closing motion
Schließdruck m clamping pressure, close-off rating, locking pressure
schließen to close, (zuschließen) to lock
Schließen n **der Form** mold closure
Schließkopf m closing head
Schließkraft f locking pressure
Schließleistung f clamping capacity
Schließpresse f transfer press
Schließring m closing ring
Schließung f closing, shutting, ~ **durch Eigengewicht** gravity closing
Schließungsfunke m spark before contact, spark on closing
Schließvorrichtung f locking device
Schliff m ground joint, (Glätte) polish
Schliffbild n polished section
Schliffdeckel m ground cover
Schliffgerät n (Chem) ground joint apparatus
Schliffriefen pl wire drawing
Schliffstopfen m ground glass stopper
Schlinge f loop, noose, sling
Schlingenfestigkeit f loop strength
schlingern to roll
Schlippesches Salz n Schlippe's salt; sodium sulfantimonate
Schlitten m carrier, sled, sledge, (Masch) carriage
Schlittenbahn f (Techn) sledge railing
Schlittenmikrotom n sliding microtome
Schlitz m slit, slash, (Einwurf) slot
Schlitzantenne f slotted antenna
Schlitzboden m perforated tray, Turbogrid tray
Schlitzbreite f slot width
Schlitzbrenner m batwing burner
Schlitzdüse f air jet, slit orifice
schlitzen to slit, to cleave, to split, (kerben) to notch
Schlitzfestigkeit f resistance to tear

Schlitzform f slot die
Schlitzfräser m mill grinder
Schlitzführung f slot guide
Schlitzlochkarten pl slotted cards
Schlitzmundstück n slit orifice
Schlitzsieb n slit sieve
Schlitztiefe f slot depth
Schlitzverschluß m slotted shutter
Schlitzwandgreifer m slotted [wall] grab
Schlitzwerkzeug n slot die
Schlitzzerstäuber m slot atomizer, slot sprayer
Schlosser m fitter, mechanic
Schlosserei f fitter's shop
Schloß n lock
Schloßfuß m bolt base
Schlot m chimney, flue, smoke stack
Schluckvermögen n absorption capacity, downcomer performance
schlüpfen to slip
Schlüpfen n slipping, ~ **der Räder** skidding of the wheels, ~ **des Riemens** slipping of the belt
Schlüpfgeschwindigkeit f slip speed
schlüpfrig slippery
Schlüpfrigkeit f lubricity
Schlüssel m key, (Code) cipher, code, **Schrauben-** ~ spanner (Br. E.), wrench (Am. E.)
Schlüsselalaun m feathery alum
Schlüsselfilter n horizontal rotary filter
Schlüsselfläche f **der Mutter** flat face of nut, key face of nut
Schlüsselkomponente f key component
Schlüsselmühle f bowl mill
Schlüsselrosette f escutcheon
Schlüsselschraube f spanner bolt, wrench bolt
Schlüsselstellung f key position
Schluß m (Ende) end, (Erdschluß) earth connection, (Math) inference, (Nebenschluß) shunt
Schlußfolgerung f conclusion
Schlußkaliber n final pass, last pass
Schlußreiniger m final purifier
Schlußstrich m topcoat
Schlußwort n final remark
schmackgar dressed with sumach
schmackhaft palatable, tasty
Schmälze f oil
Schmälzen n lubricating, lubrication, ~ **der Wolle** wool oiling
Schmälzmasse f softener, softening material
schmal narrow
Schmalfilm m (Phot) narrow film
Schmalfilmkamera f narrow film camera
Schmalfilmlampe f miniature film lamp
Schmalkeilriemen m narrow V-belt, wedge type V-belt
Schmalt m (Email) enamel
Schmaltblau n cobalt blue, smalt
Schmalte f (Kobaltschelze) smalt

Schmalz n lard, melted fat
schmalzig lardaceous, lardy
Schmalzöl n lard oil, wool oil
schmarotzend parasitic[al]
Schmarotzer m (Zool) parasite
schmecken to taste
Schmelz m (Email) enamel
Schmelzanlage f (Met) melting plant, smelting plant
Schmelzarbeit f smelting process
Schmelzbad n melting bath
schmelzbar fusible, smeltabe, **leicht** ~ easily fusible, with a low melting point, **schwer** ~ refractory, with a high melting point
Schmelzbarkeit f fusibility, meltability
Schmelzbasalt m cast basalt
Schmelzbereich m melting range
Schmelzblau n smalt
Schmelzbuch n charge book
Schmelzdiagramm n melting point diagram
Schmelzdraht m fuse wire
Schmelzdruck m melting pressure
Schmelzdruckkurve f melting point pressure curve
Schmelze f (Met) melt, molten mass, smelt, (Schmelzen) fusion, melting, smelting, (Schmelzhütte) smelting works
Schmelzeinsatz m fuse
Schmelzelektrolyse f fused salt electrolysis
schmelzen to melt, to fuse, (Metalle) to become liquid, to smelt
Schmelzen n melting, fusing, fusion, liquation, (Met) smelting, ~ **ohne Oxydation** melting under a white slag, melting without oxidation, white-slag melt-down
Schmelzenthalpie f melting enthalpy
Schmelzentropie f (Phys) entropy of fusion
Schmelzer m furnaceman
Schmelzerde f fusible clay
Schmelzerei f smelting house
Schmelzfarbe f enamel color, vitrifiable color
Schmelzfeuer n fusing fire
Schmelzfeuerung f slag tap furnace
Schmelzflammofen m reverberatory furnace
schmelzflüssig molten
Schmelzfluß m fused mass, melt, smelting flux
Schmelzflußelektrolyse f fusion electrolysis, smelting flux electrolysis
Schmelzführung f conduct of the melting operation, smelting practice or schedule
Schmelzgang m fusion or working process, process of melting, progress of melting
Schmelzgeschwindigkeit f melting rate
Schmelzgewicht n weight of charge
Schmelzglas n fusible glass, enamel
Schmelzglasur f enamel
Schmelzgleichgewicht n fusion equilibrium
Schmelzgrad m melting point, fusing point
Schmelzgut n furnace charge

Schmelzharz n casting resin
Schmelzherd m smelting or furnace hearth, heating hearth, melting stove
Schmelzhitze f fusion temperature, melting temperature, (Schweiß) welding heat
Schmelzindex m melt index
Schmelzintervall n melting range
Schmelzkegel m fusible cone, melting cone
Schmelzkessel m fusion kettle, melting pan, melting pot
Schmelzkleber m holt-melt adhesive, hot-sealable glue, thermoplastic adhesive
Schmelzklebstoff m thermoplastic adhesive, thermosetting adhesive
Schmelzkörper m melting substance
Schmelzkoks m foundry coke, smelter coke, smelting coke
Schmelzkontakt m fused catalyst
Schmelzkupfer n copper dross
Schmelzkurve f fusion curve
Schmelzlegierung f fusible alloy
Schmelzlinge pl melt-grown crystals
Schmelzmeister m head melter
Schmelzmittel n flux, fusing assistant, liquefier
Schmelzofen m melting furnace, (Met) smelting furnace
Schmelzpfanne f fusion pot
Schmelzpunkt m melting point, fuse point, fusing point, fusion point, softening point, **inkongruenter** ~ incongruent melting point
Schmelzpunktapparat m (mit Kapillarröhrchen) [capillary tube] melting-point apparatus
Schmelzpunktbestimmungsapparat m melting point apparatus
Schmelzpunkterniedrigung f depression of melting point, lowering of the melting point
Schmelzpunktröhrchen n melting point capillary or tube
Schmelzreise f life of furnace
Schmelzrinne f annular groove, annular hearth
Schmelzröhrchen n melting point tube or capillary
Schmelzschleuder f melting centrifuge
Schmelzschweißung f fusion welding
Schmelzsicherung f fusible cutout, safety cutout
Schmelzspannung f fusing voltage
Schmelzspinnen n melt extrusion, melt spinning
Schmelzstahl m German steel, natural steel, rough steel
Schmelzstein m dipyre, mizzonite
Schmelzstöpsel m fuse plug
Schmelzstoff m charge
Schmelzstreifen m fuse strip
Schmelztauchverfahren n hot galvanizing
Schmelztemperatur f melting or fusion temperature
Schmelztiegel m melting crucible
Schmelztiegelzange f crucible tongs

Schmelzung *f* melting, fusion, liquation, (Met) smelting
Schmelzungsgang *m* melting operation
Schmelzverbindung *f* fused joint
Schmelzverfahren *n* melt film coating, (Met) melting process, method of smelting
Schmelzverlauf *m* process of smelting, state of smelting
Schmelzviskosimeter *n* fusion visco[si]meter
Schmelzwärme *f* heat of fusion, melting heat, **latente** ~ (Therm) latent heat of fusion
Schmelzwasser *n* melted snow or ice, melting water
Schmelzzange *f* crucible tongs
Schmelzzeit *f* charge-to-tap time; time of melting or smelting
Schmelzzement *m* aluminous cement
Schmelzzone *f* fusing zone, melting zone
Schmelzzuschlag *m* furnace addition
Schmelzzyklon *m* cyclone smelting furnace
Schmer *n* fat, grease, suet
Schmerstein *m* soapstone
Schmerz *m* ache, pain
Schmerzbetäubung *f* analgesia
Schmerzbetäubungsmittel *n* (Pharm) analgesic, painkiller
schmerzhaft painful
schmerzlindernd lenitive
Schmerzlinderung *f* relief of pain
schmerzlos painless
Schmerzmittel *n* (Pharm) analgetic; painkiller
schmerzstillend analgesic, analgetic
Schmetterling *m* (Zool) butterfly
Schmetterlingsblütler *m* (Bot) papilionaceous flower
Schmetterlingsbrenner *m* [bat] wingburner
Schmied *m* [black]smith, forger, hammersmith
schmiedbar forgeable, malleable
Schmiedbarkeit *f* forgeability, forging property, forging quality, malleability
Schmiede *f* forge
Schmiedeeisen *n* maleable iron, forging steel, low carbon steel, wrought iron
Schmiedeeisenabfälle *m pl* wrought iron scrap
Schmiedeeisenring *m* wrought iron ring
schmiedeeisern wrought-iron
Schmiedeesse *f* forge
Schmiedefeuer *n* forge fire
Schmiedegesenk *n* drop forge die
Schmiedeherd *m* smith's hearth
Schmiedekohle *f* gas coal, smith's coal
Schmiedelegierung *f* wrought alloy
schmieden to forge, to hammer
Schmieden *n* forging
Schmiedeofen *m* forging furnace
Schmiedepresse *f* forging press
Schmiedeprobe *f* forging test
Schmiederoheisen *n* forge pigs
Schmiedestahl *m* forged steel

Schmiedestück *n* forging
Schmiedetemperatur *f* forging temperature
Schmiege *f* bevel, protractor
schmiegsam flexible, pliant
Schmiegsamkeit *f* flexibility, pliancy
Schmiegungsebene *f* osculating plane
Schmierapparat *m* lubricator
Schmierbuchse *f* grease cup
Schmierbüchse *f* oil cup
Schmiere *f* lubricant, grease, oil, (Schmutz) slush
Schmiereigenschaft *f* lubricating property
schmieren to grease, to lubricate, to oil
Schmieren *n* greasing, lubricating
schmierfähig capable of lubricating
Schmierfähigkeit *f* lubricating power, lubricity, oiliness
Schmierfett *n* [lubricating] grease
Schmierfettdepot *n* grease reservoir
Schmierfilm *m* oil film
Schmierflüssigkeit *f* lubricating liquid, lubricating oil
Schmierhahn *m* grease cock, lubrication cock
schmierig greasy, oily
Schmierigkeit *f* (Papier) low freeness
Schmierkanne *f* oil can
Schmierkopf *m* smear head
Schmierloch *n* oil hole
Schmiermaterial *n* lubricant, lubricating material
Schmiermittel *n* lubricant
Schmiermittelprüfgerät *n* lubricant testing equipment
Schmiernippel *m* lubrication nipple
Schmiernut *f* oil groove, oil way
Schmieröl *n* lubricating oil, **mineralisches** ~ mineral lubricating oil
Schmierölpumpe *f* lubricating pump
Schmierölring *m* oil ring
Schmierölschicht *f* film of lubricating oil
Schmierpistole *f* grease gun
Schmierpumpe *f* lubricating pump
Schmierseife *f* potash soap, soft soap
Schmierstelle *f* lubricating point
Schmierstutzen *m* lubrication connection
Schmierung *f* lubrication, oiling, smearing
Schmiervase *f* grease cup, lubricator, oil cup
Schmiervorrichtung *f* lubricating device, lubricator
Schmierwert *m* lubricating value
Schmierwirkung *f* lubricating effect
Schminke *f* make-up
Schmirgel *m* emery, abrasive [powder]
Schmirgelleinen *n* abrasive cloth, emery cloth
Schmirgelleinwand *f* emery cloth
schmirgeln to abrade, to grind with emery, to rub with emery, (Holz) to sand
Schmirgeln *n* polishing, glazing
Schmirgelpapier *n* emery paper, sand paper
Schmirgelpulver *n* emery powder

Schmirgelscheibe *f* abrasive disc, emery disc, emery wheel
Schmirgelschlammapparat *m* apparatus for decanting emery
Schmirgelstaub *m* emery dust
Schmirgeltuch *n* emery cloth
schmoren to stew
Schmucküberzug *m* decorative coating
Schmutz *m* dirt, filth, mud, (Ablagerung) sediment, sludge
Schmutzfänger *m* dirt trap, tail flap
schmutzig dirty, filthy
schmutzlösend dirt dissolving
Schmutzwasser *n* dirty water, waste water
Schmutzwasserbehälter *m* waste water container
Schmutzwasserkläranlage *f* waste water clarifying plant
Schnabelklemme *f* crocodile clip
Schnappschalter *m* quick action switch
Schnappverschluß *m* spring catch
Schnapsbrennerei *f* distillery, stillhouse
Schnarrventil *n* blow valve, safety valve
Schnauze *f* nozzle, snout, (Gefäß) spout
Schnecke *f* screw, spiral gears, worm [gear], (der Strangpresse) extruding screw, (Zool) snail, **doppelgängige** ~ double-flighted screw, **eingängige** ~ single-flighted screw, **kernprogressive** ~ variable root screw
Schneckenachse *f* worm shaft
Schneckenantrieb *m* worm gear drive
Schneckendrehzahl *f* screw speed
Schneckendurchmesser *m* screw diameter
Schneckenextrakteur *m* screw-conveyor extractor
Schneckenförderer *m* worm [or screw] conveyor
schneckenförmig helical
Schneckengang *m* screw flight
Schneckengehäuse *n* extrusion chamber
Schneckengetriebe *n* worm gear[ing]
Schneckengewinde *n* screw thread
Schneckenkolbenspritzgußmaschine *f* reciprocating screw injection mo[u]lding machine
Schneckenmaschine *f* screw machine
Schneckenmischer *m* screw mixer
Schneckenpresse *f* screw extruder, worm extruder
Schneckenpressen *n* screw extrusion
Schneckenpumpe *f* screw pump
Schneckenrad *n* spiral wheel, worm gear[wheel], worm wheel
Schneckenreaktor *m* screw reactor
Schneckenrohrförderer *m* fixed-screw conveyor
Schneckenschleuse *f* screw feeder
Schnecken-Siebzentrifuge *f* scroll conveyor centrifuge
Schneckenspiel *n* radial clearance
Schneckenspritzgußmaschine *f* screw type injection molding machine

Schneckenstrangpresse *f* screw extruder, screw type extrusion machine
Schneckentempo *n* snail's pace
Schneckentrockner *m* screw drier, worm drier
Schneckentrommeltrockner *m* screw-conveyor drum drier
Schneckenverdampfer *m* screw-flights evaporator
Schneckenwärmeaustauscher *m* screw-flights heat exchanger
Schneckenwelle *f* worm shaft
Schnee *m* snow
schneeartig snow-like
Schneebergit *m* (Min) schneebergite
Schneewasser *n* snow water, melted snow
schneeweiß snow white
Schneidabfall *m* clippings, trimmings
Schneidapparat *m*, **autogener** ~ oxyacetylene cutter
Schneidbalken *m* throat bar
Schneidbrenner *m* blowpipe, cutting torch
Schneide *f* cutting edge, bit, knife edge, **scharfe** ~ sharp edge, **stumpfe** ~ blunt edge
Schneidemaschine *f* cutting machine
schneiden to cut, (in Scheiben) to slice, (sich schneiden) to cross, to intersect, ~ **und stürzen** to cut and blend
Schneiden *n* cutting, ~ **von Kraftlinien** cutting of lines of force, **autogenes** ~ autogenous cutting, oxyacetylene cutting
schneidend cutting, crossing, intersecting, sharp
Schneidenlager *n* knife edge support
Schneidewinkel *m* angle of cut bias
Schneidezange *f* clippers, cutting pliers
Schneidflamme *f* cutting flame
Schneidhärte *f* (Werkzeug) abrasive hardness, cutting hardness
Schneidkante *f* cutting edge
Schneidlineal *n* cutting rule
Schneidmaschine *f* cutting machine, cutter; mechanical slicer
Schneidmaß *n* trim size
Schneidmesser *n* cutter blade, knife edge
Schneidmetall *n* cutting alloy, metal-cutting material
Schneidmühle *f* cutting mill, impeller breaker [mill], rotary knife cutter
Schneidschraube *f* tapping screw
Schneidvorgang *m* cutting process
Schneidwerkzeug *n* cutting tool
Schneidwinkel *m* cutting angle
Schneidzahn *m* cutter tooth
Schneidzerkleinerung *f* size reduction by cutters
Schneidzeug *n* edge tools
schnell fast, high-speed, quick, rapid
schnell[ab]bindend quick-setting
Schnell[ab]kühlung *f* (Kältetechnik) quick freezing

Schnellablaß *m* emergency release, jettisoning gear
schnellabtastend fast-scanning
Schnellalterungsversuch *m* accelerated aging test
Schnellanalyse *f* rapid analysis
Schnellanlaufen *n* highspeed start
Schnellarbeitsstahl *m* high-speed steel
schnellaufend high-speed
Schnellaufladung *f* rapid charge
Schnellautomatenstahl *m*, **kohlenstoffarmer** ~ low carbon free-cutting steel
Schnellautomatenweichstahl *m* soft free-machining steel
Schnellbeize *f* rapid dipping liquid
schnellbewegt high-speed
Schnellbewetterung *f* accelerated weathering
Schnellbinder *m* quick-setting cement
Schnellbleiche *f* quick chemical bleaching
Schnellbrutreaktor *m* fast breeding reactor
Schnelldrehbank *f* high-speed lathe
schnelldrehend high-rotative
Schnelldrehstahl *m* high-speed tool steel
Schnellentlader *m* rapid discharger
Schnellentladung *f* rapid discharge
schnellerbewegt faster moving
Schnellessig *m* quickvinegar
Schnellfilter *n* quick-run filter
Schnellfiltrieren *n* rapid filtration
schnellfliegend high-energy
schnellfließend fast flowing
Schnellfluß *m* quick flux
Schnellgärung *f* accelerated fermentation, quick fermentation
Schnellgalvanoplastik *f* rapid galvanoplastic process
Schnellgang *m* overdrive
Schnellgefrieren *n* quick freezing
Schnellgefrierraum *m* rapid freezing room, sharp freezer
Schnellgerbmethode *f* quick tanning method
Schnellgerbung *f* rapid tanning
Schnellgrießmethode *f* grain test
schnellhärtend (Kunstharz) quick-curing
Schnelligkeit *f* velocity, rapidity, rate, speed
Schnellklemme *f* quick clamp
Schnellkochtopf *m* autoclave, pressure cooker
Schnellkorrosionsversuch *m* accelerated corrosion test
Schnellkraft *f* elasticity, springiness
Schnellkühlung *f* rapid cooling system
Schnellkupfergalvanoplastikbad *n* rapid electroplating copper bath
Schnellmischer *m* high-speed mixer
Schnellneutronen-Brutreaktor *m* (Atom) fast neutron breeder reactor
Schnellneutronenfluß *m* fast neutron flux
schnellöffnend quick-opening
Schnellot *n* fusible metal, quick solder, soft solder

Schnellpresse *f* rapid acting press
Schnellpreßmasse *f* fast-curing molding compound
Schnellrührer *m* high-speed stirrer, high-speed mixer, impeller [mixer]
Schnellschlußkupplung *f* rapid coupling
Schnellschlußventil *n* quick action stop valve
Schnellschneidestahl *m* high-speed steel
Schnellspaltung *f* (Atom) fast fission
Schnellspaltungseffekt *m* (Atom) fast fission effect
Schnellspaltungsfaktor *m* fast fission factor
Schnellstahl *m* high-speed [tool] steel
schnelltrocknend fast drying, quick drying
Schnelltrockner *m* flash drier
Schnelltrocknung *f* rapid drying
Schnellumlaufverdampfer *m* rapid action evaporator, rapid circulation evaporator
Schnellverfahren *n* quick procedure, rapid method
Schnellvernicklung *f* rapid nickelling
Schnellverstellung *f* quick action adjustment
Schnellviskosimeter *n* rapid visco[si]meter
Schnellwaage *f* quick balance, rapid balance, steel-yard
Schnellwägen *n* rapid weighing
Schnellwalzwerk *n* high-speed rolling mill
Schnellwasserbestimmer *m* rapid moisture tester
schnellwirkend quick-acting
Schnitt *m* (Dest) cut, (Einschnitt) incision, (Schneiden) cut, ~ **in der Lauffläche** tread cut
Schnittbandführung *f* ribbon conveyor
Schnittbrenner *m* slit burner, batwing-burner, slot burner
Schnittebene *f* sectional plane
Schnittfestigkeit *f* cut resistance
Schnittfläche *f* intersecting plane
Schnittgeschwindigkeit *f* cutting velocity
Schnittkraft *f* cutting force (lathe cutting)
Schnittleistung *f* cutting efficiency
Schnittmodell *n* sectional model
Schnittmuster *n* pattern
Schnittpunkt *m* intersecting point, intersection
Schnittpunktdurchmesser *m* (Reifen) wheel intersection diameter
Schnittrand *m* cut edge
Schnittwachstum *n* (Reifen) cut growth
Schnittweite *f* width of cut
Schnittwinkel *m* (Reifen) bias angle, ~ **des Gewebes** green ply angle
Schnitzel *m n* chip, clipping, paring, scrap, shaving, turning, (Zucker) beet slice
schnitzen to carve
Schnitzmesser *n* carving knife
schnüffeln to sniff
Schnürrad *n* round-belt pulley
Schnur *f* cord, string
schnurgerade straight
Schnurgewicht *n* cord weight

Schnurrolle – Schraube

Schnurrolle *f* cord pulley
Schnurverstärker *m* cord circuit repeater
Schnurwirtel *m* cord pulley
Schock *m* shock
Schockfixierung *f* shock setting
Schock[wellen]reaktor *m* (Gastechnologie) shock wave reactor, shock tube
Schöllkopfsche Säure *f* Schoellkopf's acid
Schöllkraut *n* (Bot) celandine
Schöllkrautöl *n* celandine oil, chelidonium oil
schönen to shade, to tone
Schönfärberei *f* dyeing in high colors
Schöngrün *n* copper aceto-arsenite, Paris green
Schönheitsmittel *n* cosmetic
Schönheitspflege *f* cosmetics
Schönheitswasser *n* (Kosm) cosmetic wash
Schönit *m* (Min) schoenite
Schönungsfarbstoff *m* brightening dyestuff
Schönungsflüssigkeit *f* brightening fluid
Schöpfbecher *m* feed scoop
schöpferisch creative, productive
Schöpfgefäß *n* scoop
Schöpflöffel *m* ladle, scoop
Schöpfpapier *n* handmade paper
Schöpfprobe *f* ladle sample, test sample
Schöpfrad *n* bucket wheel, scoop wheel
Schöpfraum *m* pump chamber
Schöpfthermometer *n* scooping thermometer
Schöpfzelle *f* bucket
Schörl *m* (Geol) schorl
schörlähnlich schorlaceous
Schörlblende *f* (Min) horn-blende
Schörlit *m* (Min) schorlite, pycnite
Schörlkristall *m* schorl crystal
Schörlschiefer *m* (Min) schistous horn-blende
Schokolade *f* chocolate
Scholle *f* (Landw) clod
Schollenbrecher *m* (Landw) clod breaker
schonend careful, cautious, considerate, gentle, mild
Schonung *f* care, ~ **des Ofenfutters** care of the furnace lining
Schoopsches Metallspritzverfahren *n* Schoop's metal spraying process
Schopper *m* chopper
Schore *f* inner bar, tie, traverse
Schorlomit *m* (Min) schorlomite
Schornstein *m* chimney, chimney stack, flue, smoke stack
Schornsteinlüfter *m* chimney ventilator
Schornsteinventilator *m* chimney ventilator
Schornsteinzug *m* chimney draft or draught
Schote *f* husk, pod, shell
Schotenpfeffer *m* capsicum, red pepper
Schotten-Baumannsche Reaktion *f* (Chem) Schotten-Baumann reaction
Schotter *m* crushed rock, gravel
Schottky-Effekt *m* Schottky effect, autoelectric effect

schräg oblique, skew, slanting, sloping, (diagonal) diagonal, transverse, (geneigt) inclined
Schräganalogie *f* diagonal relationship
Schrägaufzug *m* inclined elevator, inclined hoist, inclined lift, ~ **mit Kippgefäß** skip hoist with tipping bucket
Schrägbahn *f* inclined track
Schräge *f* obliquity, slant, slope
Schrägfilm *m* inclined film, oblique film
Schrägförderband *n* inclined belt conveyor
Schrägförderer *m* inclined conveyer
schräggestellt inclined, tilted
Schrägkugellager *n* oblique ball bearing
Schrägkugelmühle *f* inclined ball mill
Schräglage *f* inclined arrangement, slanting position, sloping position
Schräglaufwinkel *m* (Reifen) slip angle
Schrägmaß *n* bevel, bevel protractor
Schrägrohr *n* oblique tube
Schrägrohrverdampfer *m* inclined-tube evaporator
Schrägrohrzugmesser *m* oblique draft gauge
Schrägrost *m* sloping grate
Schrägschneider *m* bias cutter, bias cutting machine
Schrägschnittwinkel *m* (Reifen) bias angle
Schrägschulterring *m* bead seat band
Schrägsitzventil *n* valve with inclined seat, bevel-fit valve, oblique seat valve
Schrägspalt *m* inclined slit
Schrägspritzkopf *m* angle extrusion head, oblique extruder head
schrägstellbar inclinable, tiltable
Schrägstrahlen *pl* (Opt) skew rays
Schrägtrommelmischer *m* tilted [cylinder] mixer, tilted drum mixer
Schrägverstellung cross-axing, crossing (of calender rolls)
schrägverzahnt helical gear
Schrägwalzwerke *pl* skew rolling mill
schrägwinklig bevel, oblique
Schrägzahnstirnrad *n* helical spur gear, twisted tooth spur gear
Schrämlader *m* (Bergb) cutter-loader
schraffieren to hatch, to hachure, to shade
schraffiert cross-hatched, section-lined, shaded
Schraffierung *f* hatching
Schramme *f* scar, scratch
Schrank *m* cabinet, closet, cupboard, locker
Schranktrockner *m* drying cabinet
Schrapperseil *n* scraper rope
Schraubdeckel *m* screw cap
Schraube *f* (mit Mutter) bolt, (ohne Mutter) screw, (Propeller) propeller, ~ **ohne Ende** endless screw, **doppelgängige** ~ two-thread screw, **eingepaßte** ~ fitted bolt, reamed bolt, **flachgängige** ~ square-thread screw, **linksgängige** ~ left-hand screw

schrauben to screw
Schraubenachse f (Krist) screw axis
Schraubenbakterium n Spirillum (pl Spirilla)
Schraubenfeder f helical spring, spiral spring
Schraubenfläche f helicoid, helicoidal surface
Schraubengebläse n helical blower
Schraubengewinde n screw thread, worm
Schraubenklemme f screw clamp
Schraubenkolbenpumpe f helical piston pump
Schraubenkopf m screw head, **halbrunder** ~ [half-]round screw head, **versenkter** ~ flat [screw] head, sunk [screw] head
Schraubenlochkreis m bolt hole circle, bolt [pitch] circle
Schraubenlüfter m propeller blower
Schraubenpumpe f screw pump, axial pump
Schraubenquetschhahn m screw pinchcock
Schraubenradgetriebe n norm gear
Schraubenrohr n spiral tube
Schraubenrührer m propeller mixer
Schraubenschlüssel m monkey wrench, screw spanner, spanner (Br. E.), wrench (Am. E.)
Schraubenschub m propeller thrust
Schraubensicherungslack m thread locking compound
Schraubensichter m helical classifier
Schraubenspindel f lead screw spindle
Schraubenspindelpumpe f screw displacement pump
Schraubenverbindung f bolted joint, screw connection
Schraubenverdichter m screw compressor
Schraubenverschluß m screw[ed] plug
Schraubenversetzung f screw dislocation
Schraubenzieher m screw-driver
Schraubheber m jack-screw (Am. E.), screw-jack (Br. E.)
Schraubklemme f screw clamp
Schraublehre f micrometer [calipers]
Schraublehrenhalter m micrometer stand
Schraubmuffe f sleeve nut, thread[ed] sleeve
Schraubmutter f nut
Schraubstock m vise
Schraubungssinn m direction of screwing
Schraubverbindung f screwed pipe joint
Schraubverschluß m screw closure, ~ **mit Originalitätssicherung** pilferproof cap
Schraubzwinge f clamp [iron], screw clamp
schreiben to write, (Maschine) to typewrite, (registrieren) to record, to register
Schreiber m (Registriervorrichtung) recorder, recording appratus, ~ **mit Integrator** integrating recorder, ~ **mit Regelsatz** recorder controller
Schreiberschlitten m writing carrier
Schreibfestigkeit f (Lacke) pressure-marking resistance, (Scheuerfestigkeit b. Druckfarben) scuff resistance
Schreibgerät n recorder

Schreibkreide f common chalk, writing chalk
Schreibmaschine f typewriter
Schreibpapier n writing paper
Schreibspur f impression
Schreibstein m grapholite
Schreibthermometer n recording thermometer
Schreibtinte f writing ink
Schreibunterlage f blotting pad
Schreibweise f printing style
Schrenzpapier n chip paper
Schrift f, **gesperrte** ~ (Buchdr) spaced type
Schrifterz n graphic tellurium, sylvanite
Schriftgold n (Min) graphic gold, sylvanite
Schriftgranit m graphic granite
Schriftmetall n type metal
Schriftprobe f type specimen
Schriftsteller m author, writer
Schrifttellur n (Min) sylvanite
Schrittgröße f step size
schrittweise gradual[ly], progressively, step by step, successive[ly]
Schrittweite f step size
Schröckingerit m (Min) schrockingerite
Schrödingerfunktion f Schrödinger equation
Schrödingersche Schwingungsgleichung f Schrödinger oscillation equation
Schroteffekt m (Elektr) Schottky effect, shot noise
schroten to chip, to crush, to grind roughly
Schroterei f malt crushing room
Schrotkühlanlage f meal cooling unit
Schrotkühlung f meal cooling
Schrotmeißel m cold chisel
Schrotmühle f rough grinding mill
Schrotmulde f scrap iron box
Schrotpressen n shot molding
Schrott m metal scrap, scrap iron, scrap metal, waste
Schrottaufbereitungsanlage f scrap preparation plant
Schrott-Roheisenverfahren n pig-and-scrap process
Schrottverhüttung f scrap melting
Schrottzugabe f additon of scrap, scrapping
schrüen (Porzellan) to give the first baking
Schrumpfanfälligkeit f collapse
schrumpfen to shrink, to contract, to shrivel
Schrumpffolie f shrink film
Schrumpfgewinde n shrinking thread
Schrumpflack m shrinking lacquer
Schrumpfrahmen m overfeed tenter
Schrumpfstück n shrinkage block
Schrumpfübermaß n contraction allowance
Schrumpfung f shrinkage, contraction, shortening, shrinking, shriveling, (Biol) involution
Schrumpfverpackung f shrink packaging
Schrumpfvorrichtung f shrinkage block
Schruppstahl m roughing tool

Schub *m* shear, thrust
Schubbeanspruchung *f* shearing stress, tangential stress
Schubbelastung *f* thrust loading
Schubdehnung *f* shear strain
Schubelastizität *f* elasticity in shear
Schubelastizitätsmodul *m* modulus of shearing, [shearing] modulus of elasticity
Schubfestigkeit *f* critical shearstress, shear[ing] strength
Schubförderrinne *f* pushing trough
Schubgabelstapler *m* retractable-fork reach truck
Schubkraft *f* shearing force, thrust force
Schublade *f* drawer
Schublehre *f* caliper square
Schubleistung *f* thrust performance
Schublinie *f* shearing stress line
Schubmodul *m* shear modulus, rigidity modulus
Schubrahmenstapler *m* retractable-most reach truck
Schubrinne *f* pushing trough
Schubrost *m* chain-grate stoker
Schubschneckenspritzgußmaschine *f* reciprocating-screw injection molding machine
Schubspannung *f* shearing [force], shear stress, tangential stress
schubweise in batches
Schubwendetrockner *m* reciprocating drier
Schubwiderstand *m* resistance to shear, shearing strength
Schubzahl *f* shear coefficient, strain coefficient
Schubzentrifuge *f* automatic batch centrifugal pusher, reciprocal pusher centrifuge
schüren to poke, to rake the fire, to stir the fire, to stoke
schürfen (Min) to prospect
Schürffläche *f* thrust plane
Schüröffnung *f* stirring hole
Schürrost *m* stoking grate
Schürstange *f* poker, crow bar, rabble
Schüssel *f* dish
Schüttbeton *m* cast concrete, poured concrete
Schüttdichte *f* bulk density
Schüttelapparat *m* shaking apparatus, rocking appliance
Schüttelautoklav *m* agitated autoclave, shaking autoclave
Schüttelbewegung *f* shaking motion, shaking movement
Schüttelfrequenz *f* shaking speed
Schüttelherd *m* vibrating table
Schüttelmaschine *f* shaking machine, mechanical shaker
schütteln to shake, to agitate
Schütteln *n* shaking, agitating
Schüttelrätter *m* jig table
Schüttelrinne *f* shaking trough

Schüttelrost *m* shaking grate, oscillating grate
Schüttelrostfeuerung *f* furnace with shaking grate, furnace with movable fire bars
Schüttelrutsche *f* shaker conveyor, shaking shoot
Schüttelsieb *n* shaking sieve, riddle, shaker screen, shaker sifter, swinging sieve, vibrating screen
Schütteltisch *m* vibrating table
Schüttelwirkung *f* shaking effect
Schütterz *n* live ore
Schüttfeuerung *f* self-feeding furnace
Schüttgewicht *n* apparent density, bulk density, bulk weight, loose weight
Schütthöhe *f* (Füllkörperkolonne) height of packing
Schüttlerwellenlager *n* bearing for reciprocating crankshaft
Schüttrohr *n* discharge pipe
Schüttrost *m* chute grate, continuous charging grate
Schüttsintern *n* pour sintering
Schüttungshöhe *f* height of bed
Schüttvolumen *n* bulk volume
Schüttvorrichtung *f* stock distributing gear
Schütz *m* (Relais) relay
schützen to guard [against], to protect [from], to shelter [from]
schützend protective
Schützit *m* (Min) celestine
Schuhindustrie *f* boot and shoe industry
Schuhkreme *f* shoe polish
Schuhpech *n* cobbler's wax
Schuhschmiere *f* dubbing
Schuhschwärze *f* shoe blacking
Schuhwachs *n* shoe polish
Schuhwichse *f* shoe polish, blacking
Schukosteckdose *f* (Elektr, HN) protective contact socket
Schulterpartie *f* shoulder area
Schulterschräge *f* (Reifen) bead seat taper
Schumanngebiet *n* (Spektr) Schumann region
Schungit *m* (Min) schungite
Schuppe *f* flake, scale, (Kopfschuppe) dandruff
schuppen to flake, to scale
Schuppenbildung *f* scaling
Schuppengraphit *m* (Min) flaky graphite
Schuppenstein *m* (Min) lepidolite
schuppig scaly, laminated, squamous
Schurwolle *f* fleece
Schuß *m* shot, (Bergb) blast, charge, (Knall) report, (Web) pick, weft
Schußfaden *m* (Web) cross weave
schußfrei (Web) weftless
Schußgewicht *n* shot weight
Schußleistung *f* shot capacity
Schußspule *f* pirn, (Web) pirn, weft bobbin
schußspulen to pirn
Schußspulmaschine *f* pirn winder

Schußstreifen *m* (Web) weft bar
Schußwaffe *f* fire arm
Schusterpech *n* cobbler's wax
Schutt *m* rubbish, refuse
Schutthaufen *m* rubbish dump
Schutz *m* protection, safeguard, shelter, shielding, **kathodischer** ~ cathodic protection
Schutzanstrich *m* anticorrosive coat of paint, protective coating, seal
Schutzbedeckung *f* shielding
Schutzbelag *m* protective coat
Schutzblech *n* guard plate, screen, (Auto) mudguard
Schutzbrille *f* protecting glasses, protecting goggles, protecting spectacles
Schutzdecke *f* (Papier) paper coating
Schutzdosis *f* (Med) prophylactic dose
Schutzelektrode *f* guard electrode
Schutzfirnis *m* protecting varnish
Schutzfolie *f* backing
Schutzgas *n* inert gas, buffer gas, protective gas, safety gas
Schutzgasanlage *f* inert gas plant
Schutzgasofen *m* artificial atmosphere furnace
schutzgeimpft vaccinated
Schutzgewebe *n* protective tissue
Schutzgitter *n* protective grating, [screen] grid
Schutzglas *n* protecting glass, glass-guard
Schutzhandschuh *m* hand saver
Schutzhaube *f* protecting cap, protecting cover
Schutzhülle *f* protective covering
Schutzhülse *f* [protective] casing
Schutzimpfung *f* vaccination, immunization, prophylactic inoculation
Schutzisolierung *f* protective isolation
Schutzkammer *f* shielded enclosure
Schutzkappe *f* protecting cap
Schutzkleidung *f* protective clothing
Schutzkolloid *n* protective colloid
Schutzkontakt *m* (Elektr) protective contact
Schutzlack *m* protecting lacquer
Schutzmantel *m* protective jacket
Schutzmarke *f* trade mark, **eingetragene** ~ registered trademark
Schutzmaske *f* face guard, protecting mask
Schutzmaßnahme *f* safety measure
Schutzmittel *n* preservative
Schutznetz *n* safety net
Schutzpapier *n* release paper (with pressure-sensitive adhesives)
Schutzplatte *f* protecting plate, baffle plate
Schutzpolster *n* wadding
Schutzring *m* guard ring, (Elektr) guard ring
Schutzrohr *n* protective tube
Schutzschalter *m* safety switch, protective switch
Schutzschicht *f* protective layer, insulating film or layer
Schutzstrich *m* top coat
Schutztrichter *m* protective funnel, safety funnel

Schutzüberzug *m* protective coating, exterior coating
Schutzverschluß *m* shielding plug
Schutzvorrichtung *f* protective device, safety device
Schutzvorschriften *f pl* safety regulations
Schutzwall *m* protective rampart
Schutzwand *f* protective wall, barrier
Schutzwirkung *f* protective effect
Schutzzeichen *n* trade mark
Schwabbel *f* buff, lap
Schwabbelmaschine *f* mechanical mop
Schwabbelmittel *n* buffing agent
Schwabbeln *n* buffing
Schwabbelscheibe *f* buffing wheel
schwach weak, light, slight, (gelinde) gentle, mild, (matt) faint, ~ **alkalisch** slightly alkaline, ~ **bestrahlt** lightly irradiated, ~ **bitter** slightly bitter, ~ **erhitzen** to heat gently
Schwachgas *n* blast furnace gas, lean gas, poor gas, weak gas
Schwachstrom *m* low-voltage current, weak current
Schwachstromglocke *f* low-voltage bell
Schwachstromkabel *n* communication cable, weak-current cable
Schwachstromleitung *f* weak current line
Schwachstromtechnik *f* weak-current engineering
Schwaden *m* choke damp, firedamp
schwächen to weaken, (vermindern) decrease, diminish, to lessen
Schwächung *f* weakening, (Dämpfung) attenuation, damp[en]ing, (Verringerung) diminution
Schwächungskoeffizient *m* attenuation coefficient
Schwärze *f* blackness, blacking, lamp black, smoke black
Schwärzegrad *m* degree of blackness
schwärzen to blacken, to darken, (Gieß) to black wash
Schwärzung *f* blackening, (Phot) density
Schwärzungsabstufung *f* (Opt) density graduation
Schwärzungsbereich *m* density range
Schwärzungsgrenze *f* (Phot) blackening limit
Schwärzungskurve *f* (Phot) characteristic [film] curve
Schwärzungsmesser *m* densitometer
Schwärzungsphotometer *n* photographic plate photometer
Schwal *m* rich finery cinder, sinter slag
Schwalarbeit *f* slag washing, slag washing process
schwalbenschwanzförmig dove-tailed
Schwalbenschwanzzwilling *m* (Krist) butterfly twin, fishtail twin

Schwalbenstein (Min) swallow stone, lenticular agate stone
Schwalboden *m* slag bed, slag bottom
Schwamm *m* sponge, (Pilz) fungus
Schwammfilter *m* sponge filter
schwammförmig spongy
Schwammgummi *m* expanded rubber, sponge rubber
Schwammholz *n* decayed wood, spongy wood
schwammig spongy, porous, (Holz) decayed, fungous, rotten, (Papier) bibulous
Schwammigkeit *f* sponginess, ~ **des Gußeisens** sponginess of cast iron
Schwammkohle *f* sponge charcoal
Schwammkunststoff *m* foamed plastic
Schwammkupfer *n* copper sponge, spongy copper
Schwammseife *f* porous soap
Schwammstein *m* mushroom stone, spongite
Schwammstoff *m* fungin
Schwammtorf *m* spongy peat
Schwangerschaftstest *m* pregnancy test
schwanken to fluctuate, to oscillate, to vary
Schwanken *n* fluctuation, oscillation
schwankend variable, (fluktuierend) fluctuating, (unsicher) labile, unsteady
Schwankung *f* oscillation, fluctuation, (Amplitude) amplitude variation
Schwankungsbereich *m* range of variations
Schwankungsfrequenz *f* frequency of flutter
Schwankungsquadrat *n* square fluctuation
Schwanz *m* tail
Schwarm *m* (Elektronen) shower [of electrons]
Schwarmbildung *f* (Moleküle) agglomeration, clustering
schwarz black, ~ **streichen** to japan
Schwarzbeere *f* (Bot) bilberry, blueberry
schwarzblau blue black
Schwarzblech *n* blackplate, black sheet [iron]
Schwarzbottich *m* mercury bucket
schwarzbraun black brown, dark brown
schwarzbrüchig black-short
Schwarzdruck *m* (Buchdr) printing in black
Schwarzeisen *n* high silicon pig iron, pig iron rich in silicon
Schwarzerz *n* (Min) black silver ore, stephanite, tetrahedrite
Schwarzfärben *n* black dyeing
Schwarzfärberei *f* black dyeing
Schwarzfäule *f* black rot
schwarzfarbig black-colored
Schwarzglasfilter *n* black glass filter
Schwarzgüldigerz *n* (Min) black silver glance ore, stephanite
Schwarzkernguß *m* black heart malleable iron
Schwarzkohle *f* black [char]coal
Schwarzkümmel *m* nigella seeds
Schwarzkupfer *n* black copper, smelt copper

Schwarzkupfererz *n* black copper ore, melaconite
Schwarzkupferschlacke *f* coarse copper slag
Schwarzlauge *f* black liquor
Schwarzmanganerz *n* black manganese ore, hausmannite, psilomelane
Schwarzmehl *n* (Roggenmehl) rye flour
Schwarznickelbad *n* black nickelling solution
Schwarznickelüberzug *m* black nickel coating
Schwarzöl *n* black oil, dark boiled linseed oil
Schwarzpech *n* black pitch
Schwarzschmelz *m* black enamel, niello
Schwarzsilberglanz *m* (Min) stephanite
Schwarzspießglanzerz *n* bournonite
Schwarzstein *m* (Min) black iron mica
Schwarzstrahler *m* black body radiator
Schwarzweißfilm *m* black-and-white film
Schwarzweißphoto *n* black-and-white photograph
Schwarzwurzel *f* (Bot) baneberry root
Schwazit *m* (Min) schwazite, spaniolite
Schwebebühne *f* suspended platform
Schwebedichte *f* buoyant density
Schwebefähigkeit *f* suspension power
Schwebefahrzeug *n* hovercraft
Schwebegastrockner *m* supporting gas dryer
Schwebegeschwindigkeit *f* free-falling velocity, terminal velocity
Schwebekörper *m* suspended body
Schwebekrackanlage *f* (Öl) cat cracker
Schwebekühler *m* fluid-bed cooler
Schwebelage *f* suspended position, hanging position
Schwebemagnet *m* floating magnet
Schwebemethode *f* (zur Dichtebestimmung) suspension method
Schwebemittel *n* antisettling agent, flotation agent, suspending agent
schweben to be suspended, to float, to hover
schwebend suspended, in suspension, pending
Schwebeschmelzofen *m* suspension furnace
Schwebestoff *m* suspended matter, suspended substance
Schwebeteilchen *n* suspended particle
Schwebevergasung *f* fluidized gasification
Schwebstoff-Filter *n* air filter
Schwedisches Grün *n* Swedish green, Scheele's green
Schwefel *m* Symb. S sulfur (Am. E.), brimstone, sulphur (Br. E.), **monokliner** ~ monoclinic sulfur, **plastischer** ~ plastic sulfur, **rhombischer** ~ rhombic sulfur
Schwefelabdruck *m* sulfur impression, brimstone impression, sulfur print
Schwefelabscheidung *f* separation of sulfur
Schwefeläther *m* sulfuric ether
Schwefelalkali *n* (obs) alkali sulfide
Schwefelalkohol *m* thioalcohol, thiol, (obs) mercaptan, hydrosulfide

Schwefelallyl *n* allyl sulfide, diallyl sulfide
Schwefelammon *n*, **farbloses** ~ colorless ammonium sulfide, **gelbes** ~ yellow ammonium sulfide
Schwefelantimon *n* antimony sulfide
Schwefelantimonblei *n* (Min) antimony lead sulfide, boulangerite
Schwefelantimonsäure *f* sulfoantimonic acid, thioantimonic acid
schwefelarm poor in sulfur
Schwefelarsen *n* arsenic sulfide
Schwefelarsensäure *f* thioarsenic acid
Schwefelauflösung *f* sulfur solution
Schwefelausscheidung *f* separation of sulfur
Schwefelbad *n* (Phot) sulfide toning
Schwefelbakterien *pl* sulfur bacteria
Schwefelbarium *n* barium sulfide
Schwefelbariumofen *m* barium sulfide furnace
Schwefelberg *m* sulfur pit
Schwefelbestimmung *f* (Anal) sulfur determination
Schwefelblei *n* lead sulfide
Schwefelblüte *f* flowers of sulfur, sublimated sulfur
Schwefelbrennofen *m* sulfur kiln
Schwefelbromid *n* sulfur bromide, sulfur dibromide
Schwefelbromür *n* (obs) sulfur monobromide
Schwefelbrücke *f* sulfur bridge
Schwefelcadmium *n* cadmium sulfide
Schwefelcalcium *n* calcium sulfide
Schwefelchlorid *n* sulfur chloride, sulfur dichloride
Schwefelchlorür *n* (obs) sulfur monochloride
Schwefelchromsäure *f* chromosulfuric acid
Schwefelcyan *n* cyanogen sulfide, (obs) thiocyanogen
Schwefelcyanammonium *n* (obs) ammonium thiocyanate
Schwefelcyankalium *n* (obs) potassium thiocyanate
Schwefeldampf *m* sulfur vapor
Schwefeldestillationsofen *m* sulfur distilling furnace
Schwefeldichlorid *n* sulfur bichloride, sulfur dichloride
Schwefeldioxid *n* sulfur dioxide, sulfurous acid anhydride
Schwefeldioxidgewinnungsanlage *f* sulfur dioxide production plant
Schwefeldunst *m* sulfur[ous] vapor
Schwefeleisen *n* iron sulfide
Schwefelentfernung *f* desulfurization, desulfurizing, removal of sulfur
Schwefelerde *f* sulfur earth
Schwefelerz *n* sulfur ore
schwefelessigsauer (Salz) thioacetate
Schwefelextraktionsanlage *f* sulfur extraction plant

Schwefelfaden *m* sulfured wick
Schwefelfarbe *f* sulfur color, sulfur dye
schwefelfarben sulfur-colored
Schwefelfarbstoff *m* sulfur dye
schwefelfrei free from sulfur
Schwefelgehalt *m* sulfur content
Schwefelgelb *n* sulfur yellow
Schwefelgermanium *n* (obs) germanium sulfide
Schwefelgeruch *m* sulfur odor
schwefelgesäuert sulfated, treated with sulfuric acid
Schwefelgestein *n* sulfur-bearing rock
Schwefelgewinnungsanlage *f* sulfur production plant
Schwefelgold *n* (obs) gold sulfide
Schwefelgrube *f* sulfur mine, brimstone pit, sulfur pit
Schwefelhalogen *n* sulfur halide
schwefelhaltig containing sulfur
Schwefelhaltigkeit *f* sulfur[e]ousness
Schwefelharnstoff *m* thiourea
Schwefelheptoxid *n* sulfur heptoxide
Schwefelhütte *f* sulfur refinery
schwef[e]lig sulfur[e]ous (Am. E.), sulfurous, sulphur[e]ous (Br. E.)
schwefelige Säure *f* sulfurous acid
Schwefelindium *n* (obs) indium sulfide
Schwefeliridium *n* (obs) iridium sulfide
Schwefelkalium *n* (obs) potassium sulfide
Schwefelkalk *m* lime sulfide, lime sulfur
Schwefelkammer *f* sulfur chamber, sulfuring room, sulfur stove
Schwefelkarbolsäure *f* phenolsulfonic acid, sulfocarbolic acid, sulfophenol
Schwefelkies *m* (Min) ferrous sulfide, iron pyrites
schwefelkieshaltig pyritiferous
Schwefelkiesröstung *f* pyrites roasting
Schwefelkobalt *m* (Min) cobalt sulfide, linnaeite
Schwefelkohle *f* high-sulfur coal, sulfurous coal
Schwefelkohlensäure *f* sulfocarbonic acid, trithiocarbonic acid
Schwefelkohlenstoff *m* carbon disulfide
Schwefelkohlenstoffanlage *f* carbon disulfide plant
Schwefelkohlenstoffgewinnung *f* production of carbon disulfide
Schwefelkolben *m* retort for distilling sulfur, sulfur distilling retort
Schwefelkupfer *n* copper sulfide
Schwefelläuterofen *m* sulfur refining furnace
Schwefelleber *f* liver of sulfur
Schwefelleinöl *n* balsam of sulfur
Schwefelluft *f* sulfur[e]ous gas
Schwefelmagnesium *n* (obs) magnesium sulfide
Schwefelmangan *n* (obs) manganese sulfide
Schwefelmolybdän *n* (obs) molybdenum sulfide
Schwefelmonochlorid *n* sulfur monochloride
schwefeln to sulfur, to sulfurate, to sulfurize

Schwefeln *n* sulfurating, sulfuring, sulfurizing
Schwefelnatrium *n* (obs) sodium sulfide
Schwefelnickel *n* (obs) nickel sulfide
Schwefelniederschlag *m* sulfur precipitate
Schwefelofen *m* sulfur kiln
Schwefeloxid *n* sulfur oxide
Schwefelpfanne *f* sulfur melting pan
Schwefelphosphor *m* phosphorus sulfide
Schwefelplatin *n* platinum sulfide
Schwefelpocken *pl* sulfur pockmarks
Schwefelprobe *f* sulfur test
Schwefelpulver *n* powdered sulfur
Schwefelquecksilber *n* (obs) mercury sulfide
Schwefelquelle *f* sulfur spring
Schwefelraffinerie *f* sulfur refinery
Schwefelrauch *m* sulfur fume
Schwefelreinigung *f* sulfur purification
Schwefelrubin *m* (Min) realgar, ruby sulfur
Schwefelsäure *f* sulfuric acid, **rauchende** ~ fuming sulfuric acid, oleum
Schwefelsäureanhydrid *n* anhydrous sulfuric acid, sulfuric anhydride, sulfur trioxide
Schwefelsäureanhydridanlage *f* sulfuric acid anhydride plant
Schwefelsäureanlage *f* sulfuric acid plant
Schwefelsäureballon *m* sulfuric acid carboy, sulfuric acid demijohn
Schwefelsäuredämpfe *pl* fumes of sulfuric acid
Schwefelsäuredimethylester *m* dimethyl sulfate
Schwefelsäureester *m* sulfate ester
Schwefelsäurefabrikation *f* manufacture of sulfuric acid
schwefelsäurefrei free from sulfuric acid
Schwefelsäurehydrat *n* hydrated sulfuric acid, sulfuric acid hydrate
Schwefelsäuremonohydrat *n* sulfuric acid monohydrate
Schwefelsäuresalz *n* sulfate, salt of sulfuric acid
Schwefelsalbe *f* (Pharm) sulfur ointment
Schwefelsalz *n* sulfur salt, (Sulfat) sulfate, (Thiosalz) thio salt
schwefelsauer sulfuric, (Salz) sulfate
Schwefelschlacke *f* sulfur dross
Schwefelschwarz *n* sulfur black
Schwefelselen *n* selenium sulfide
Schwefelsesquioxid *n* sulfur sesquioxide
Schwefelsilber *n* (Min) argentite, silver glance, (obs) silver sulfide
Schwefelspießglanz *m* (Min) stibnite
Schwefelstange *f* roll of sulfur
Schwefelstickstoff *m* nitrogen sulfide
Schwefeltetrachlorid *n* sulfur tetrachloride
Schwefeltetroxid *n* sulfur tetroxide
Schwefelthallium *n* (obs) thallium sulfide
Schwefeltönung *f* sulfide toning
Schwefeltonerde *f* (obs) aluminum sulfide
Schwefeltrioxid *n* sulfur trioxide
Schwefelung *f* sulfuration, sulfuring, sulfurization, sulfur treatment

Schwefelverbindung *f* sulfur compound, (Sulfid) sulfide
Schwefelverbrennungsofen *m* sulfur combustion furnace
Schwefelverfärbung *f* (Dosen) sulfur staining
Schwefelverlust *m* sulfur loss
Schwefelwasser *n* sulfur[eous] water
Schwefelwasserstoff *m* hydrogen sulfide, sulfuretted hydrogen
Schwefelwasserstoffgruppe *f* (Anal) hydrogen sulfide group
Schwefelwasserstoffmesser *m* hydrogen sulfide meter
Schwefelwasserstoffniederschlag *m* hydrogen sulfide precipitate
Schwefelwasserstoffsäure *f* hydrogen sulfide, hydrosulfuric acid
Schwefelwasserstoffsalz *n* sulfide
schwefelwasserstoffsauer (Salz) sulfide
Schwefelwasserstoffverbindung *f* compound of hydrogen sulfide
Schwefelwerk *n* sulfur refinery
Schwefelwismut *n* (obs) bismuth sulfide
Schwefelwurzel *f* (Bot) brimstonewort root, peucedanum root
Schwefelzink *n* (obs) zinc sulfide
Schwefelzinkweiß *n* zinc sulfide white, lithopone, zincolith
Schwefelzinn *n* tin sulfide
Schwefligsäure *f* sulfurous acid
Schwefligsäureanhydrid *n* sulfur dioxide, sulfurous anhydride
Schwefligsäuregas *n* sulfur dioxide gas
schwefligsauer (Salz) sulfite (Am. E.), sulphite (Br. E.)
Schweinefett *n* lard
Schweinefleisch *n* pork
Schweinfurter Grün *n* Paris green, Schweinfurt green
Schweinsgummi *n* hog-gum
Schweiß *m* sweat, perspiration, (Wolle) yolk [of wool], (Wollschweiß) suint
Schweißaggregat *n* welding unit
Schweißanlage *f* welding plant
Schweißarbeit *f* welding
Schweißasche *f* raw potash, suint ash
schweißbar weldable
Schweißbarkeit *f* weldability
Schweißbildung *f* perspiration
Schweißbrenner *m* welding torch, welding gun
Schweißdraht *m* welding wire, filler rod, welding rod
Schweißdüse *f* welding tip
schweißecht perspiration-resistant
Schweißechtheit *f* perspiration fastness
Schweißeisen *n* weld iron, wrought iron
Schweißeisenblech *n* wrought iron plate
Schweißelektrode *f* welding electrode

schweißen to weld, (Elektr) to electroweld, (Kunststoff) to heat-seal
Schweißen *n* welding, (in einer Vorrichtung) jig welding, (Kunststoff) heat-sealing, **dielektrisches** ~ high-frequency welding, **geradliniges** ~ linear welding
Schweißer *m* welder
Schweißfehler *m* defect in welding
Schweißfeuer *n* welding fire
Schweißgehalt *m* (Wolle) amount of yolk, suint content
Schweißgerät *n* welding torch, welding gun
schweißhärten to weld-harden
Schweißhitze *f* welding heat
Schweißkolben *m* welding handle
Schweißlampe *f* welding torch
Schweißlinie *f* weld line, weld mark
Schweißmarkierung *f* weld mark
Schweißmaschine *f* welding machine
Schweißmetall *n* welding metal
Schweißmittel *n* welding compound, welding flux
Schweißnaht *f* flow line, seam weld, soldered joint, welded joint, welding seam
Schweißnahtprüfgerät *n* weld-seam tester
Schweißofen *m* welding furnace
Schweißofenschlacke *f* reheating furnace slag, tap cinder
Schweißpaket *n* fagot, pile
Schweißplattieren *n* cladding materials by welding
Schweißplattierung *f* weld deposited coating
Schweißprobe *f* welding test
Schweißpulver *n* (Pharm) diaphoretic powder
Schweißrissigkeit *f* weld cracking
Schweißsand *m* welding sand
Schweißsatz *m* welding set
Schweißschmiedeeisen *n* weld iron
Schweißstab *m* filler rod, welding rod
Schweißstahl *m* weldable steel, weld[ing] steel
Schweißstelle *f* location of weld, welded joint, welding position, weld[mark], **fehlerhafte** ~ weld mark
Schweißtemperatur *f* welding heat
schweißtreibend (Pharm) diaphoretic, sudorific
Schweißung *f* welding, **aluminothermische** ~ thermit welding, **autogene** ~ autogenous welding, **elektrische** ~ arc welding, electric welding, electrowelding, **hydrooxygene** ~ oxyhydrogen welding
Schweißverfahren *n* welding process
Schweißwachs *n* (Wolle) yolk wax
Schweißwolle *f* greasy wool
Schweizerit *n* schweizerite
Schwelanlage *f* [low-temperature] carbonization plant
Schwelbenzin *n* gasoline from coal
Schwelbrand *m* (Bergb) smoldering fire

schwelen to smolder, to burn slowly, to carbonize at low temperature, to distill at low temperature
Schwelen *n* low-temperature carbonization, smoldering, smothering
Schwelerei *f* low-temperature carbonization
Schwelgas *n* carbonization gas, gas from low-temperature distillation, incompletely burnt gas
Schwelkohle *f* brown coal for low-temperature retort process
Schwelkoks *m* low-temperature carbonization coke, semi-coke
Schwellbeize *f* swelling liquor
Schwelle *f* threshold
schwellen to swell
Schwellendetektor *m* threshold detector
Schwellenempfindlichkeit *f* threshold sensitivity
Schwellenenergie *f* (Atom) threshold energy
Schwellengesetz *n* threshold law
Schwellenkaliber *n* sleeper pass
Schwellenpotential *n* threshold potential
Schwellenspannung *f* threshold potential
Schwellenstrom *m* (Elektr) threshold current
Schwellenwert *m* threshold value, **lichtelektrischer** ~ photoelectric threshold
Schwellenwertkurve *f* threshold field curve
Schwellenwertmessung *f* threshold measurement
Schwellfarbe *f* tanner's liquor
Schwellfestigkeit *f* inflation strength
Schwellhöhe *f* barrier height
Schwellkraft *f* (Harz) swelling capacity
Schwellung *f* swelling
Schwelperiode *f* period of slow combustion
Schwelraum *m* carbonizing chamber, carbonizing space, coking chamber
Schwelretorte *f* retort for dry distillation
Schwelteer *m* low-temperature tar, tar extracted by low-temperature process
Schwelung *f* low-temperature carbonization, slow burning, smoldering
Schwelvorgang *m* carbonization process
Schwelzone *f* carbonization zone, smoldering zone
Schwelzylinder *m* coking cylinder, partial carbonizer
Schwemmaufbereitung *f* flotation
schwemmen to wash
Schwemmfilterung *f* sewage purification by filtration
Schwemmland *n* reclaimed ground
Schwemmverfahren *n* flotation
Schwemmwasser *n* wash water
Schwengel *m* handle of peel
Schwengellager *n* saddle bearing
Schwenkarm *m* swinging arm
Schwenkaufnahme *f* oscillating exposure
schwenkbar pivoted, swivel-mounted
Schwenkbohrmaschine *f* swinging driller

schwenken to swing, to swivel
Schwenkgetriebe *n* turning gear
Schwenkrohr *n* swing pipe
Schwenktaster *m* swinging key
schwer (Mech) heavy, ponderous, weighty, (schwierig) difficult, hard, ~ **entzündlich** not easily inflammable, ~ **schmelzbar** difficult to fuse, difficult to melt, refractory
Schwerachse *f* axis through center of gravity
Schwerbenzin *n* heavy gasoline, heavy petrol
Schwerbeton *m* dense concrete
Schwerbleierz *n* (Min) heavy lead ore, plattnerite
Schwerchemikalie *f* heavy chemical
Schwere *f* gravitation, gravity, heaviness, weight
Schwereanomalie *f* gravity anomaly
Schwereanziehung *f* gravitational attraction
Schwerebeobachtung *f* gravity observation
Schwerebeschleunigung *f* acceleration of gravity, gravitational acceleration
Schwerefeld *n* gravitational field, gravity field
Schwereflüssigkeit *f* heavy liquid
Schwereformel *f* gravity formula
Schwerelinie *f* neutral axis
schwerelos weightless
Schweremesser *m* gravimeter
Schwerenetz *n* gravity network
Schwererde *f* heavy earth, barium oxide, baryta
Schwerevermessung *f* gravity survey
schwerflüchtig difficultly volatile, difficult to volatilize, not volatile
Schwerflüchtigkeit *f* low volatility
schwerflüssig viscous, heavy, viscid
Schwerflüssigkeit *f* high viscosity
Schwergut *n* heavy goods
Schwerindustrie *f* heavy industry
Schwerkraft *f* [force of] gravity
Schwerkraftbahn *f* gravity line
Schwerkraftfeld *n* gravity field
Schwerkraftmesser *m* gravimeter
Schwerkraftmischer *m* gravity mixer
Schwerkraftzentrum *n* center of gravity
schwerlöslich difficultly soluble, difficult to dissolve
Schwerlöslichkeit *f* low solubility
Schwermetall *n* heavy metal
Schwermetallsalz *n* salt of a heavy metal
Schweröl *n* heavy oil
Schwerpunkt *m* center of gravity, center of mass
Schwerpunktbestimmung *f* determination of the center of gravity
Schwerpunktsbewegung *f* center of mass motion
Schwerpunktshalbmesser *m* radius of center of gravity
Schwerpunktskoordinate *f* center of mass coordinate
Schwerpunktslage *f* location of the center of gravity
Schwerpunktsradius *m* radius of center of gravity

Schwerpunktssystem *n* center of gravity system
Schwerpunktsverlagerung *f* eccentricity of center of gravity
schwerrostend rustproof, stainless
Schwerspat *m* (Min) heavy spar, barite[s], barium sulfate, barytes
schwerspathaltig barytic
Schwerstein *m* (Min) scheelite, tungstate of lime
Schwertantalerz *n* tantalite
Schwertrübetrennung *f* dense-media separation, heavy-medium separation
Schweruranerz *n* uranite
Schwerwasserreaktor *m* heavy-water pile
schwerwiegend serious, (gewichtig) weighty
schwierig difficult, complex, intricate
Schwimmachse *f* axis of flotation, buoyancy axis
Schwimmaufbereitung *f* flotation
Schwimmebene *f* plane of buoyancy
schwimmen to float, to swim
Schwimmen *n* floating, swimming
schwimmend floating, supernatant
Schwimmer *m* [ball] float
Schwimmernadelventil *n* float needle valve
Schwimmerventil *n* float valve
schwimmfähig buoyant
Schwimmfähigkeit *f* buoyancy
Schwimmhäute *f* overflow (injection molding)
Schwimmkran *m* floating crane
Schwimmkugel *f* buoyancy bulb, float
Schwimmreibung *f* fluid lubrication
Schwimmsand *m* quicksand
Schwimmsandschicht *f* layer of quicksand
Schwimm-Sink-Aufbereitung *f* sink-and-float separation
Schwimmstein *m* floatstone
Schwimmstoffzerkleinerung *f* disintegration of floating material
Schwimmthermometer *n* floating thermometer
Schwimmvermögen *n* buoyancy
schwinden (abnehmen) to decrease, (schrumpfen) to contract, to shrink
Schwinden *n* decrease, contraction, loss, shrinkage
Schwindlunker *m* shrinkage cavity
Schwindmaß *n* amount of contraction, amount of shrinkage, degree of shrinkage, (Lehre) shrinkage gauge
Schwindung *f* shrinkage
Schwindungsgrad *m* degree of shrinkage
Schwingachse *f* oscillating axle, independent front suspension
Schwinge *f* rocker arm, rocking arm, rotating shaft
schwingen to swing, to oscillate, to undulate, to vibrate
Schwingen *n* cycling, vibrating
schwingend swinging
Schwingförderer *m* vibrating conveyor
Schwinghebel *m* rocking lever

Schwingherd *m* vibrating table
Schwingkreis *m* oscillatory circuit
Schwingkristall *m* crystal oscillator
Schwingkristallmethode *f* (Krist) oscillating crystal method
Schwingmetallagerung *f* rubber bonded mounting
Schwingmühle *f* vibration mill, swing [sledge] mill
Schwingquarz *m* (Elektr) piezoelectric crystal, quartz resonator
Schwingrätter *m* jig table, vibrating table
Schwingrinne *f* rocking trough, vibrating trough
Schwingsieb *n* shaking sieve, swinging sieve, vibrating screen
Schwingsiebschleuder *f* centrifugal sieve
Schwingsiebzentrifuge *f* oscillating screen centrifuge
Schwingung *f* oscillation, vibration, (Wellenbewegung) undulation, **abklingende** ~ dying oscillation or vibration, **aperiodische** ~ aperiodic oscillation, **erzwungene** ~ forced oscillation, forced vibration, **freie** ~ free oscillation, free vibration, **gedämpfte** ~ damped oscillation, **gedämpfte, harmonische** ~ damped harmonic oscillation, **gleichförmige** ~ continuous oscillation, **harmonische** ~ harmonic oscillation, **ungedämpfte** ~ continuous oscillation, sustained oscillation, undamped oscillation
Schwingungsachse *f* axis of vibration or oscillation
Schwingungsamplitude *f* oscillation amplitude
Schwingungsanalysator *m* harmonic analyzer
Schwingungsbauch *m* antinode, maximum amplitude
Schwingungsdämpfung *f* vibration damping
Schwingungsdauer *f* duration of oscillation, period of oscillation[s]
Schwingungsebene *f* plane of oscillation
Schwingungsenergie *f* oscillation energy
Schwingungserreger *m* vibration exciter
Schwingungserregung *f* vibration excitation
schwingungsfähig oscillatory, vibratory
schwingungsfrei non-oscillating, free from vibrations
Schwingungsfrequenz *f* oscillation frequency
Schwingungsgleichung *f* wave equation
Schwingungsinstabilität *f* oscillatory instability
Schwingungsisolierung *f* vibration insulation
Schwingungsknoten *m* vibration or oscillation node
Schwingungskreis *m* oscillating circuit
Schwingungsmechanismus *m* oscillation mechanism
Schwingungsmesser *m* vibration indicator, vibration measuring apparatus
Schwingungsmöglichkeit *f* oscillation possibility

Schwingungsperiode *f* oscillation time, period of oscillation
Schwingungsphase *f* oscillation phase, vibration phase
Schwingungsquant *n* vibration quantum
Schwingungsquantenzahl *f* vibrational quantum number
Schwingungsrichtung *f* vibration direction
Schwingungsrißkorrosion *f* corrosion fatigue
Schwingungsspektrum *n* vibrational spectrum
Schwingungssystem *n*, **entartetes** ~ degenerate oscillating system
Schwingungsüberlagerung *f* superposition of vibrations, vibration superposition
Schwingungsweite *f* amplitude
Schwingungszahl *f* frequency number, oscillation frequency, vibration frequency
Schwingungszentrum *n* center of oscillation
Schwingungszustand *m* vibrational state
Schwingviskosimeter *n* oscillating disk viscosimeter
Schwingwand *f* membrane, diaphragm
schwitzen to perspire, to sweat
Schwitzen *n* perspiration, sweating
Schwitzlacke *pl* reflow or flow-back finishes
Schwitzwasserkorrosion *f* corrosion by condensed moisture
Schwöde *f* lime paste, place where hides are limed
Schwödefaß *n* lime vat
Schwödegrube *f* lime pit
Schwödemasse *f* lime cream
schwöden to lime
Schwöden *n* liming
Schwund *m* (Abklingen) fading, (durch Zusammenziehen) contraction, (Verlust) leakage
Schwundausgleich *m* (Radio) antifading control
Schwundriß *m* contraction crack
Schwundstelle *f* shrinkage spot
Schwungbewegung *f* vibratory movement
Schwungkraft *f* centrifugal force
Schwungmaschine *f* centrifugal machine
Schwungmoment *n* moment of gyration
Schwungrad *n* (Mech) flywheel
Schwungradantrieb *m* flywheel drive
Schwungraddampfpumpe *f* flywheel steam pump
Schwungradkranz *m* flywheel rim
Schwungring *m* rotating ring
Scillabiose *f* scillabiose
Scillain *n* scillain
Scillaren *n* scillaren
Scillaridin *n* scillaridin
Scillaridinsäure *f* scillaridinic acid
Scilliglaucosid *n* scilliglaucoside
Scillin *n* scillin
Scillipikrin *n* scillipicrin
Scillirosid *n* scilliroside
Scillitin *n* scillitin

Scillitoxin n scillitoxin
Sclaren n sclaren
Sclerotinol n sclerotinol
Scolecit m (Min) scolecite
Scombrin n scombrin
Scoparin n scoparin
Scoparon n scoparone
Scopin n scopine
Scopolamin n scopolamine
Scopolaminhydrochlorid n scopolamine hydrochloride
Scopolaminmethylbromid n scopolamine methylbromide
Scopolaminmethylnitrat n scopolamine methylnitrate
Scopolein n scopoleine
Scopoletin n scopoletin
Scopolin n scopoline
Scutamol n scutamol
Scutellarin n scutellarin
Scyllit m scyllitol
Scylloquercit m scylloquercitol
Sebacat n sebacate
Sebacil n sebacil
Sebacinsäure f sebacic acid, ipomic acid
sebacinsauer (Salz) sebacate
Sebacoin n sebacoin
Sebaconitril n sebaconitrile
Sebazinsäureamid n sebacamide
Secaclavin n secaclavine
Secalin n secaline, trimethylamine
sechsatomig hexatomic
Sechseck n hexagon
sechseckig hexagonal, sexangular
Sechsfachkabel n six-way cable
sechsflächig hexahedral
Sechsflächner m (Krist) hexahedron
sechsgliedrig six-membered
Sechskant m hexagon, ~ **mit Ansatz** hexagon head with collar
Sechskanteisen n hexagon iron
sechskantig hexagonal
Sechskantmutter f hexagonal nut
Sechsphasensternschaltung f six-phase star connection
Sechsphasensystem n six-phase system
Sechsring m (Chem) six-membered ring
sechsseitig hexagonal, six-sided
sechswertig hexavalent
sechswinklig hexangular
Secobarbital n secobarbital
Secretin n secretin
Sedamin n sedamine
Sedanolid n sedanolide
Sedanolsäure f sedanolic acid
Sedanschwarz n Sedan black
sedativ (Med) sedative
Sedativsalz n sedative salt
Sedativum n (Pharm) sedative

Sediment n sediment, deposit, layer, precipitation
sedimentär sedimentary
Sedimentation f sedimentation, precipitation, settling
Sedimentationsgeschwindigkeit f (Zentrifugation) sedimentation velocity, velocity of settling
Sedimentationsgleichgewicht n sedimentation equilibrium
Sedimentationskoeffizient m sedimentation coefficient
Sedimentationssäule f sedimentation column
Sedimentationsversuch m sedimentation test
Sedimentbildung f sedimentation
Sedimentgestein n sedimentary rocks
sedimentieren to sediment, to settle
Sedimentierhilfsmittel n sedimentation aid
Sedinin n sedinine
Sedoheptit m sedoheptitol
Sedoheptose f sedoheptose
Sedoheptulosan n sedoheptulosan
Sedoheptulose f sedoheptulose
Sedoheptuloson n sedoheptulosone
Sedridin n sedridine
Seebachit m (Min) seebachite
Seebeckeffekt m Seebeck effect, thermoelectric effect
Seesalz n sea salt
Seesalzraffinieranlage f sea salt refinery plant
Seesand m sea sand
Seetang m sea tang, seaweed
Seewasser n salt or sea water
seewasserbeständig salt-water-proof
Seewasserdestillierapparat m sea water distillation apparatus
Seewasserentsalzung f sea water desalination
Segerkegel m Seger cone, fusible cone, melting cone
Segment n (Kreisabschnitt) segment
Segregationsgrad m segregation factor
Sehfehler m defect of vision
Sehfeld n field of vision
Sehkraft f visual faculty, sight
Sehne f (Anat) tendon, sinew, (Kreis) chord
Sehnerv m (Anat) optic nerve
sehnig fibrous
Sehpurpur m rhodopsin
Sehreiz m optical stimulus
Sehschärfe f power of vision, seeing capacity
Sehschlitz m viewing slot
Sehschwelle f visual threshold
Sehweite f visual distance
seicht shallow
Seichtwasserwelle f shallow water wave
Seide f (Text) silk, **beschwerte** ~ weighted silk, **chinesische** ~ China silk, Chinese silk, **echte** ~ natural silk, **künstliche** ~ artificial silk, rayon
Seidelbastrinde f mezereon bark

Seidenglanz *m* silky lustre
Seidenglanzkalander *m* (Pap) glazing calender
Seidenpapier *n* tissue paper
Seidenraupe *f* bombyx, silkworm
Seife *f* soap, **Kern-** ~ hard soap, laundry soap (Am. E.), sodium soap, **Schmier-** ~ potash soap, soft soap
Seifenblase *f* soap bubble
Seifenemulsion *f* soap emulsion
Seifenerde *f* fuller's earth, smectite
Seifenersatz *m* soap substitute
Seifenerz *n* alluvial ore
Seifenfabrik *f* soap works
Seifenflocke *f* soap flake
Seifengehalt *m* soap content
seifenhaltig saponaceous
Seifenkocher *m* soap boiler
Seifenkraut *n* (Bot) soap weed
Seifenlauge *f* soap solution, soap suds
Seifenleim *m* soap glue, soap paste
Seifenlösung *f* soap solution
Seifenpflaster *n* soap plaster
Seifenpulver *n* soap powder
Seifenrinde *f* soap bark
Seifenschaum *m* soap lather
Seifensiederei *f* soap works
Seifensiederlauge *f* soap boiler's lye, caustic lye
Seifenstein *m* caustic soda, soapstone
Seifenstoff *m* saponin
Seifensud *m* soap charge
Seifenton *m* fuller's earth, saponaceous clay, smectite
Seifenverbrauch *m* soap consumption
Seifenveredelungsmittel *n* soap improver
Seifenwasser *n* soap suds, soap water
Seifenzinn *n* stream tin
seifig saponaceous, soapy
Seigerblei *n* liquation lead
Seigerherd *m* liquation hearth
Seigerkrätze *f* scrapings of liquation
seigern (Metall) to liquate, to refine, to segregate
Seigern *n* liquation, refining, segregating
Seigerprozeß *m* liquation process
Seigerrückstand *m* liquation residue
Seigerschlacke *f* liquation slag
Seigerung *f* liquation, segregation
Seigerungshärtung *f* precipitation hardening
Seigerungszone *f* line of segregation (in an ingot)
Seignettesalz *n* Seignette salt, sodium potassium tartrate
Seihebeutel *m* filtering bag
Seihefaß *n* filtering cask
Seihefilz *m* filtering felt
Seihegefäß *n* filtering vessel, straining vessel
seihen to filter, to percolate, to strain
Seihen *n* filtering, percolating, straining
Seiher *m* strainer, filter, percolator
Seiherpresse *f* cage press, strainer press
Seihetuch *n* strainer

Seil *n* rope, (Kabel) cable, **ein ~ abwickeln** to unwind a rope, **ein ~ aufwickeln** to wind up a rope
Seilantrieb *m* rope drive
Seilfallhammer *m* rope drop hammer
Seilrolle *f* rope pulley, rope sheave
Seilscheibe *f* winding drum, winding rope pulley
Seilscheibengerüst *n* rope pulley frame
Seilschlinge *f* rope loop
Seilschloß *n* cable joint
Seilschmierer *m* rope lubricator
Seilspleißung *f* rope splicing
Seiltrommel *f* cable drum, rope drum, rope pulley, winding drum
Seilwinde *f* rope winch
Seilwindung *f* coil of rope, turn of rope
Seim *m* glutinous liquid, mucilage, viscous liquid
seimig glutinous, mucilaginous
Seismograph *m* seismograph
Seismologie *f* seismology
Seismometer *n* seismometer
Seismometrie *f* seismometry
seismometrisch seismometric[al]
Seismoskop *n* seismoscope
Seite *f* side, (Buch) page, (Krist) face
Seiten *pl*, **gegenüberliegende ~** opposite sides
Seitenansicht *f* side view
Seitenausleger *m* side jib
Seitendrehmeißel *m* side cutting tool
Seitendruck *m* lateral pressure
Seitenfläche *f* lateral face, side surface
Seitengabelstapler *m* side loader, side-loading forklift
Seitenhals *m* side neck
Seiteninduktion *f* lateral induction
Seitenkammer *f* side chamber
Seitenkanalpumpe *f* side-channel pump
Seitenkette *f* side chain
Seitenmischluftrührer *m* side airlift agitator
Seitenriß *m* side elevation
Seitenstahl *m* side tool
Seitenstrom *m* (Destillation) sidestream
seitenverkehrt side-inverted
Seitenverschiebung *f* lateral deflection
Seitenverspannung *f* (Ofen) lateral bracing of furnace
Seitenwand *f* side wall
Seitenzug *m* side flue
seitlich lateral
seitwärts aside, laterally, sideways
Sekans *m* (Math) secant
Sekante *f* (Math) secant
Sekikasäure *f* sekikaic acid
Sekret *n* secretion
Sekretabsonderung *f* secretion
Sekretin *n* secretin
Sekretion *f* secretion
Sekt *m* champagne, dry wine

Sektor *m* sector
Sektorfeld *n* sector field
sekundär secondary
Sekundärausstrahlung *f* secondary emission
Sekundärdurchbruch *m* secondary burst
Sekundärelektron *n* secondary electron
Sekundärelektronenemission *f* secondary electron emission
Sekundärelektronenresonanz *f* secondary electron resonance
Sekundärelektronenverstärker *m* secondary electron multiplier
Sekundärelement *n* secondary cell, storage cell
Sekundäremission *f* reemission, secondary flow
Sekundäremissionsfaktor *m* secondary emission factor
Sekundäremissionskathode *f* secondary emission cathode
Sekundäremissionsstrom *m* secondary emission current
Sekundärfluoreszenz *f* secondary fluorescence, sensitized fluorescence
Sekundärinfektion *f* secondary infection
Sekundärkreis *m* (Elektr) secondary circuit
Sekundärluft *f* secondary air
Sekundärneutron *n* secondary neutron
Sekundärreaktion *f* secondary reaction
Sekundärreaktor *m* (Atom) secondary reactor, enriched pile
Sekundärspannung *f* secondary stress, (Elektr) induced voltage, secondary voltage
Sekundärspektrum *n* secondary spectrum
Sekundärstrahler *m* (Opt) selective radiator
Sekundärstrahlung *f* secondary radiation
Sekundärströmung *f* secondary flow
Sekundärstrom *m* induced current
Sekundärstruktur *f* (Chem) secondary structure
Sekundärteilchen *n* secondary particle
Sekundärwicklung *f* secondary winding
Sekunde *f* second
Sekundenpendel *n* seconds pendulum
Selacholeinsäure *f* selacholeic acid
Selachylalkohol *m* selachyl alcohol, glycerol monooctadecenyl ether
Seladonit *m* (Min) celadonite
Selagin *n* selagine
Selaginol *n* selaginol
Selbit *m* selbite
selbstabdichtend self-sealing
selbstabgleichend self-balancing
Selbstabholung *f* customer pick-up
selbstabschirmend self-shielding
Selbstabstoßung *f* self-repulsion
selbständig independent, ~ **erhärten** to harden by itself
Selbstalterung *f* aging at room temperature
Selbstauslöser *m* (Phot) automatic release
Selbstausschalter *m* automatic circuit breaker
Selbstaustausch *m* self-exchange

Selbstbindemittel *n* self-binding medium
selbstdichtend self-sealing
Selbstdiffusion *f* autodiffusion, self-diffusion
Selbst[ein]färbung *f* in-plant coloring
Selbstenergiediagramm *n* self-energy diagram
Selbstentladegefäß *n* self-discharger
Selbstentladen *n* self-discharging
Selbstentlader *m* self-discharger
Selbstentladestrom *m* self-discharge current, leakage current
Selbstentladung *f* self-discharge, (Akku) spontaneous discharge
Selbstentleerer *m* self-discharger
Selbstentleerung *f* automatic discharge
selbstentzündlich spontaneously inflammable
Selbstentzündung *f* self-ignition, spontaneous ignition, spontaneous inflammation
Selbstentzündungstemperatur *f* spontaneous ignition temperature, S.I.T.
selbsterhaltend self-sustaining
Selbsterhitzung *f* self-heating
selbsterregend self-exciting
Selbsterregung *f* self-excitation
Selbsterwärmung *f* spontaneous heating, spontaneous evolution of heat
selbstgängig self-fluxing
Selbstglanzemulsion *f* dry-bright emulsion
Selbstgreiferbetrieb *m* automatic gripping action
selbsthärtend self-curing
selbsthaftend (Klebband) self-adhesive
Selbsthemmung *f* **des Gewindes** non-reversibility of the thread
Selbstinduktion *f* self-induction
selbstinduktionsfrei non-inductive
Selbstinduktionsgesetz *n* law of self-induction
Selbstinduktionskoeffizient *m* coefficient of self-induction
Selbstinduktionsspannung *f* self-induction voltage, voltage of self-induction
Selbstinduktionsspule *f* self-inductance coil
selbstinduktiv self-inductive
Selbstinduktivität *f* self-inductance
selbstinduziert self-induced
Selbstionisation *f* autoionization
selbstklebend self-adherent, self-adhering, self-sealing
Selbstkleber *m* pressure sensitive adhesive
Selbstklebung *f* cold sealing, selfsealing
Selbstkosten *pl* cost-price
selbstleuchtend self-illuminating
selbstlöschend self-quenching
Selbstlötung *f* autogenic soldering
Selbstöler *m* self lubricator
Selbstoxydation *f* autoxidation, self-oxidation
selbstregelnd self-regulating, automatically regulating
Selbstregelung *f* (Reaktor) self-regulation
selbstregistrierend self-recording, self-reading

Selbstregulierung f automatic regulation, feedback regulation
selbstreinigend self-cleaning
selbstschmierend self-lubricating
Selbstschmierung f self-lubrication
selbstschreibend self-recording, self-registering
selbstsperrend self-locking
Selbststeuerung f automatic regulation, (Biol) self-regulation
selbsttätig automatic
selbsttragend unsupported
Selbstüberhitzung f self-superheating
Selbstumkehr f self-inversion, self-reversal
Selbstumkehrung f (von Spektrallinien) self-reversal (of spectral lines)
selbstverständlich obvious, self-evident
selbstverzehrend self-consuming
selbstvulkanisierend self-curing
Selbstzersetzung f spontaneous decomposition
Selbstzeugung f spontaneous generation
selbstzündend pyrophoric, self-igniting
Selbstzünder m automatic lighter, self-igniter
Selbstzündung f self-ignition
Selektionstheorie f theory of natural selection
selektiv selective
Selektivabsorption f selective absorption
Selektivität f selectivity
Selektivstrahler m selective emitter
Selen n selenium (Symb. Se)
Selenäthyl n ethyl selenide, diethyl selenide
Selenat n selenate, salt or ester of selenic acid
Selenazin n selenazine
Selenazol n selenazole
Selenblei n (Min) lead selenide, clausthalite
Selenbleikupfer n (Min) lead copper selenide
Selenbromür n (obs) selenium monobromide
Selenchlorid n selenium chloride, selenium tetrachloride
Selenchlorür n (obs) selenium monochloride
Selencyanid n selenium cyanide
Selencyansäure f selenocyanic acid
Selendehydrierung f selenium dehydrogenation
Selendioxid n selenium dioxide, selenious anhydride
Seleneisen n iron selenide, ferrous selenide, iron(II) selenide
Selenempfänger m selenium receiver
Selenerz n selenium ore
Selenfilter m (Phot) selenium glass
Selengehalt m selenium content
Selengitter n selenium lattice
Selenglas n selenium glass
Selengleichrichter m selenium rectifier
Selenhalogen n selenium halide
selenhaltig seleniferous
Selenharnstoff m selenurea
Selenhexafluorid n selenium hexafluoride
Selenhydrat n hydroselenide
Selenid n selenide

selenig selenious
Selenigsäureanhydrid n selenious anhydride, selenium dioxide
selenigsauer (Salz) selenite
Selenindigo n selenindigo
Seleninsäure f selenin acid
Selenit m (Min) selenite
Selenit n salt or ester of selenious acid
selenithaltig selenitiferous
Selenkompaß m selenic compass
Selenkupfer n (Min) copper selenide, berzelianite
Selenkupferblei n (Min) copper lead selenide
Selenmetall n metallic selenide, metal selenide
Selenobenzoesäure f selenobenzoic acid
Selenocyan n selenocyanogen
Selenofuran n selenofuran, selenophene
Selenonaphthen n selenonaphthene
Selenophen n selenofuran, selenophene
Selenophenol n selenophenol, phenyl hydrogen selenide, phenyl hydroselenide
Selenopyronin n selenopyronine, selenoxanthene
Selenosalicylsäure f selenosalicylic acid
Selenoxanthen n selenoxanthene, selenopyronine
Selenoxid n selenium oxide
Selenquecksilber n mercury selenide, (Min) tiemannite
Selenquecksilberblei n (Min) lehrbachite
Selensäure f selenic acid
Selensäurehydrat n selenic acid hydrate, hydrated selenic acid
selensauer (Salz) selenate
Selenschicht f selenium film, selenium layer
Selenschlamm m selenium slime or mud or sludge
Selenschwefel m (Min) selensulfur, volcanite
Selenschwefelquecksilber n (Min) onofrite
Selensilber n (Min) silver selenide, naumannite
Selensol n selenium sol
Selensperrschichtzelle f selenium barrier cell
Selentetrachlorid n selenium tetrachloride
Selentrioxid n selenium trioxide, selenic anhydride
Selenüberzug m selenium coating
Selenür n (obs) selenide
Selenverbindung f selenium compound, (Selenid) selenide
Selenwasserstoff m hydrogen selenide
Selenwasserstoffsäure f hydroselenic acid
Selenwismutglanz m (Min) selenium bismuth glance
Selenzelle f selenium cell
Selinan n selinane
Selinen n selinene
Sellait m (Min) sellaite
selten rare, seldom
Seltenerdmetall n rare earth metal
Selterswasser n Seltzer [water]

Semibenzol *n* semibenzene
Semicarbazid *n* semicarbazide, aminourea
Semicarbazidhydrochlorid *n* (Semicarbazidchlorhydrat) semicarbazide hydrochloride
Semicarbazon *n* semicarbazone
Semichromgerbung *f* semi-chrome tanning
semicyclisch semicyclic
Semidin *n* semidine
Semidin-Umlagerung *f* semidine rearrangement
Semihydrat *n* semihydrate
Semikolloid *n* semicolloid, hemicolloid
Seminose *f* seminose
Semioxamazid *n* semioxamazide
semipermeabel semipermeable
semipolar semi-polar
Sempervirin *n* sempervirine
Semseyit *m* (Min) semseyite
Senait *m* (Min) senaite
Senarmontit *m* (Min) senarmontite
Sendeantenne *f* transmitting aerial
Sendeapparat *m* transmitter
Sendebereich *m* transmission range
Sendeempfangsapparat *m* transceiver, two-way radio
senden to send, to transmit
Sender *m* transmitter, (Station) transmitting station
Senderöhre *f* transmitter valve or tube
Sendestation *f* transmitting station
Sende- und Empfangsgerät *n* transceiver, transmitting and receiving set, two-way radio
Senecifolidin *n* senecifolidine
Senecifolin *n* senecifoline
Senecin *n* senecine
Senecinsäure *f* senecic acid
Senecioalkaloid *n* senecio alkaloid
Senecionin *n* senecionine
Seneciosäure *n* senecioic acid
Seneciphyllin *n* seneciphylline
Seneciphyll[in]säure *f* seneciphyllic acid
Senegaextrakt *m* senega extract
Senegafluidextrakt *m* liquid extract of senega
Senegalgummi *n* Senegal gum
Senegasirup *m* senega syrup, gum arabic
Senegawurzel *f* (Bot) senega root
Senegin *n* senegin
Senf *m* mustard
Senfgas *n* mustard gas, yperite
Senfgassulfon *n* mustard gas sulfone
Senföl *n* mustard oil, allyl isothiocyanate
Senfpulver *n* ground mustard
sengen to parch, to scorch, (Gewebe) to singe
Senkblei *n* plumb [bob], plummet, sounding lead
Senkbühne *f* lowering stage
Senkel *m* plumb bob
senken to lower, to depress, to sink
Senkkörper *m* diving body
Senklot *n* plumb bob

senkrecht perpendicular, upright, vertical, ~ **aufeinander** rectangular, ~ **auf einer Fläche** perpendicular to a surface
Senkrechtantrieb *m* vertical drive
Senkrechtauflösung *f* (Telev) vertical resolution
Senkrechte *f* perpendicular [line]
Senkrechtförderer *m* vertical conveyor
Senkrechthobel *m* vertical shaper
Senkrechtintensität *f* vertical intensity
Senkrechtstärke *f* vertical intensity
Senkrechtstellung *f* vertical position
Senkschraube *f* countersunk screw, flat-head screw (Am. E.)
Senkspindel *f* specific gravity spindle
Senkvorrichtung *f* lowering device
Senkwaage *f* hydrometer, areometer, plumb rule
Sennesblätter *pl* (Bot) senna leaves
Sennidin *n* sennidine
Sennit *m* sennitol
Sennosid *n* sennoside
Sensibilisator *m* photosensitizer, sensitizer
sensibilisieren (Phot) to sensitize
Sensibilisierung *f* sensitization
Sensibilisierungsfarbstoff *m* sensitizing dye
Sensibilisierungsmaximum *n* sensitizing maximum
Sensibilität *f* sensitivity, sensitiveness
sensitiv sensitive
Sensitivgrün *n* sensitive green
Sensitokolorimeter *n* sensitocolorimeter
Sensitometrie *f* (Phot) sensitometry
sensorisch sensory
Separatorzentrifuge *f* centrifugal separator
separierbar separable
separieren to separate, (aufbereiten) to dress
Sepia *f* sepia
sepiabraun sepia brown
Sepiolith *m* (Min) sepiolite, meerschaum
Sepsin *n* sepsine
Sepsis *f* (Med) sepsis, blood poisoning, septic disease
septisch septic
Sequenzanalyse *f* sequence analysis
Sequenzhomologie *f* sequence homology
Sequoyit *m* sequoyitol
Seredin *n* seredine
Serendibit *m* serendibite
Sericin *n* sericin, silk gelatin, silk glue
Sericinschicht *f* sericin coating, sericin layer
Sericit *m* (Min) sericite
sericitartig sericitic
Sericitschiefer *m* (Min) sericite schist
Serie *f* series
Seriencharakteristik *f* series characteristic
Serienerregung *f* series excitation
Serienerzeugung *f* mass production
Serienfertigung *f* series production
Seriengrenze *f* (Spektrum) series limit
serienmäßig in series, regular, standard

Serienmaschine *f* serial machine
Serienmotor *m* series motor
Serienofen *m* series furnace
Serienschalter *m* series switch, multicircuit switch
Serienschaltung *f* series connection
Serienschnitt *m* serial sectioning
Serienschnittmikrotom *n* serial sectioning microtome
Serienspektrum *n* series spectrum
Serienwiderstand *m* series rheostat
Serin *n* serine
Serindehydratase *f* (Biochem) serine dehydratase
Serindesaminase *f* serine deaminase
Serinol *n* serinol
Serizin *n* sericin[e]
Serodiagnose *f* (Med) serodiagnosis; serum diagnosis
Serolactaminsäure *f* serolactaminic acid
Serologie *f* serology
serologisch serologic, immunologic
Seromycin *n* seromycin
seronegativ seronegative
seropositiv seropositive
Serotonin *n* serotonin
Serpentin *m* (Min) serpentine
Serpentinasbest *m* (Min) serpentine asbestos, amianthus, fibrous serpentine
serpentinhaltig containing serpentine
Serpentinit *m* (Min) serpentinite
Serpentinmarmor *m* (Min) lizard stone
Serpentinsäure *f* serpentinic acid
Serpentinstein *m* (Min) serpentine stone
Serratam[in]säure *f* serratamic acid
Serum *n* serum, **Immun-** ~ immune serum, **Normal-** ~ normal serum
Serumalbumin *n* serum albumin
Serumamylase *f* (Biochem) amylase in the blood serum
Serumeiweiß *n* serum protein
Serumglobulin *n* serum globulin
Serumlipoproteid *n* serum lipoprotein
Serviertablett *n* serving tray
Serviettenpapier *n* napkin paper
Servoantrieb *m* servodrive
Servokomponente *f* servocomponent
Servomechanismus *m* servomechanism
Servopotentiometer *n* servopotentiometer
Servoregelsystem *n* servosystem
Servosteuerung *f* servocontrol
Servoverstärker *m* servoamplifier
Sesamin *n* sesamin
Sesamöl *n* sesame oil
Sesamölseife *f* sesame oil soap
Sesamol *n* sesamol
Sesamolin *n* sesamolin
Seselin *n* seselin
Sesquibenihen *n* sesquibenihene
Sesquicarbonat *n* sesquicarbonate

Sesquichlorid *n* sesquichloride
Sesquioxid *n* sesquioxide
Sesquisalz *n* sesquisalt
Sesquiterpen *n* sesquiterpene
Sesselform *f* (Stereochem) chair-form
Setacylblau *n* setacyl blue
Setzboden *m* charging platform, charge level
Setzbottich *m* settling vat
setzen to place, to put, to set, (sich setzen, abscheiden) to precipitate, to settle, **außer Betrieb** ~ to put out of action, to stop, **in Betrieb** ~ to set to work
Setzen *n* (Buchdr) typesetting, ~ **der Beschickung** subsidence of the charge
Setzkasten *m* settling tank
Setzkopf *m* set head, swaged head
Seuche *f* (Med) epidemic, plague
Sexangulit *m* sexangulite
Sexiphenyl *n* (obs) hexaphenyl, sexiphenyl
Sextol *n* sextol
Sextolphthalat *n* sextol phthalate
Sextolstearat *n* sextol stearate
Sextupelpunkt *m* sextuple point
Sexualhormon *n* sex hormone
Seybertit *m* (Min) seybertite
sezernieren to secrete, (Med) to excrete
sezieren to dissect
Shantung *m* (Text) shantung
Shattuckit *m* (Min) shattuckite
Sherardisier-Verfahren *n* sherardizing process
Sheridanit *m* (Min) sheridanite
Shibuol *n* shibuol
Shikimisäure *f* shikimic acid
Shikonin *n* shikonin
Shirlan *n* shirlan
Shisonin *n* shisonin
Shoddy *n* shoddy, waste wool
Shogaol *n* shogaol
Shonansäure *f* shonanic acid
Shore-Fallprobe *f* Shore's dynamic indentation test
Shore-Härte *f* shore hardness
Shore-Härteprüfer *m* Shore durometer
Shunt *m* (Elektr) shunt
shunten (Elektr) to put in shunt, to shunt
Sialinsäure *f* sialic acid
Sialyltransferase *f* (Biochem) sialyl transferase
siamesisch Siamese
Siaresinolsäure *f* siaresinolic acid
Siberit *m* (Min) siberite
Sichelform *f* sickle form
Sichelzelle *f* sickle cell, drepanocyte
Sichelzellenhämoglobin *n* sickle-cell h[a]emoglobin
Sicherheit *f* safety, security, (Gewißheit) certainty
Sicherheitsabschaltung *f* safety shut-down, emergency shut-down, rapid shut-down
Sicherheitsabstand *m* safety zone

Sicherheitsblockierung – Siebkoks

Sicherheitsblockierung *f* safety interlock
Sicherheitseinrichtung *f* safety device
Sicherheitsfaktor *m* safety factor
Sicherheitsflasche *f* safety bottle
Sicherheitsglas *n* safety glass, triplex glass, (Schichtglas) compound or laminated glass
Sicherheitshaken *m* safety hook
Sicherheitskappe *f* safety cap
Sicherheitsklappe *f* safety valve, safety device
Sicherheitskreis *m* safety circuit
Sicherheitslampe *f* Davy lamp, flame-proof lamp
Sicherheitsleiter *f* safety ladder
Sicherheitsmaßnahme *f* factor of safety, measure of security, precaution, safety measure
Sicherheitsmembran *f* safety diaphragm
Sicherheitsregler *m* safety regulator
Sicherheitsröhre *f* safety tube
Sicherheitsrohr *n* safety pipe, safety tube
Sicherheitsschalter *m* safety switch
Sicherheitsschloß *n* latch
Sicherheitssprengstoff *m* safety explosive
Sicherheitsstange *f* bell safety rod
Sicherheitsstillsetzung *f* safety shut-down
Sicherheitsventil *n* safety valve, [pressure-]relief valve
Sicherheitsventilbelastung *f* safety valve weight
Sicherheitsverschluß *m* safety lock
Sicherheitsvorrichtung *f* safety device, safety valve
Sicherheitsvorschrift *f* safety rule
Sicherheitszone *f* safety zone
Sicherheitszünder *m* safety fuse
sichern to secure, to make sure, to safeguard
Sicherung *f* safeguard, (Elektr) [safety] fuse, (Sicherheitsvorrichtung) safety device, security precaution
Sicherungsdose *f* fuse box
Sicherungseinsatz *m* fuse insert
Sicherungskasten *m* fuse box
Sicherungsrücklage *f* reserve fund
Sicherungsscheibe *f* locking washer
Sicherungsschraube *f* safety screw, locking screw, securing screw
Sicherungsstöpsel *m* fuse plug
sichtbar visible, ~ **machen** to visualize, to make visually perceptible, to render visible
Sichtbarkeit *f* visibility
Sichtbeton *m* exposed concrete
Sichter *m* classifier, (Bergb) sifting device, **Jalousie-** ~ louvre classifier, **Kanalrad-** ~ screen classifier, **Korb-** ~ classifier with circumferential screen, **Plan-** ~ plansifter, **Querstrom-** ~ cross-current classifier, **Schrauben-** ~ helical classifier, **Spiral-** ~ spiral classifier, **Steigrohr-** ~ gravity classifier, **Streu-** ~ classifier with closed air circuit, **Turbo-** ~ whizzer
Sichtfeinheit *f* degree of fineness by sieving

Sichtleistung *f* performance of a classifier
Sichtlochkarten *f pl* optical coincidence cards
Sichtmesser *m* visibility meter
Sichtpackung *f* skin pack
Sichtröhre *f* sighting tube
Sichtung *f* sieving
Sichtungsanlage *f* separating plant
Sichtweite *f* range of vision
Sicke *f* (Dose) bead, crimp
Sickenmaschine *f* (f. Dosen) beading machine
Sickeranlage *f* sump hole
Sickergrube *f* dry well, sewage pit, soakaway
sickern to leak, to ooze, to percolate, to trickle
Sickerungsstrecke *f* percolation range
Sickerwasser *n* percolating water, sewage
Sicklerit *m* (Min) sicklerite
Sideringelb *n* siderin yellow
Siderit *m* (Min) siderite, spathic iron
Siderokonit *m* (Min) sideroconite
Sideroskop *n* sideroscope
Siderosphäre *f* siderosphere
Sidioquarzglasgerät *n* sidio quartz glass utensil
Sidotsche Blende *f* Sidot's blende
Sieb *n* sieve, screen, screener, sifter, strainer, (Bergb) ratter, riddle, (Elektr) eliminator, **grobes** ~ wide meshed sieve, **Ober-** ~ top wire, **Rund-** ~ rotary screen
Siebanalyse *f* screening analysis, sieve analysis
Siebanlage *f* screening plant
Siebbandtrockner *m* screen belt dryer
Siebbelag *m* screen bottom, screen lining
Siebboden *m* sieve plate, sieve tray
Siebbodenkolonne *f* **für flüssig-flüssig Extraktion** reciprocating plate extraction column, **pulsierende** ~ pulse-type sieve plate column
Siebdekanter *m* screen centrifuge decanter
Siebdraht *m* screening wire
Siebdruck *m* (Buchdr) screen printing
Siebdruckfarbe *f* screen printing color
sieben to sift, to screen, to sieve, to size
siebenatomig heptatomic
Siebeneck *n* heptagon
siebeneckig heptagonal
siebenfach seven-fold
siebenflächig heptahedral
Siebenflächner *m* heptahedron
siebengliedrig seven-membered
Siebenring *m* (Chem) seven-membered ring
siebenwertig heptavalent, septivalent
Siebenwertigkeit *f* heptavalence
Siebfilter *m* sieving filter, filtering screen, strainer
Siebfraktion *f* screening fraction
Siebgewebe *n* screen cloth
Siebgütegrad *m* screen efficiency
Siebgut *n* screening material
Siebkammer *f* straining chamber
Siebkette *f* band-pass filter
Siebkoks *m* sifted coke

Siebkugelmühle f screening [ball] mill
Siebmaschine f sieving machine
Siebpackung f screen pack
Siebpartie f wire section
Siebplansichter m screen classifier
Siebplatte f sieve plate, filter plate
Siebplattenkolonne f sieve plate column
Siebrost m sieve grate
Siebrückstand m screening[s], sieve residue
Siebsatz m set of sieves
Siebschale f straining dish, strainer
Siebschlagmühle f screen beater mill
Siebschleuder f centrifugal filter
Siebtrockner m screen dryer, sieve dryer
Siebtrommel f rotary sieve
Siebtrommelzentrifuge f perforated basket centrifuge
Siebung f sifting
Siebvorrichtung f sifting device, screener, strainer
Siedeabfall m boiling sediment
Siedeanalyse f analysis by boiling, analysis by fractional distillation
Siedebereich m boiling range
Siedeblech n boiling plate, boiling sheet
Siedegefäß n distillation flask, boiler, boiling vessel
Siedegrad m boiling point
Siedegrenze f boiling range
Siedehitze f boiling heat
Siedekapillare f boiling capillary
Siedekessel m boiler
Siedekolben m distillation flask
Siedekühlen n (v. Reaktoren) evaporative cooling
Siedekurve f boiling [point] curve
Siedelauge f boiling lye, evaporating liquor
sieden to boil [gently], to seethe, (gelinde) to simmer
Sieden n boiling, ebullition, (gelinde Sieden) simmering, ~ **bei freier Konvektion** free convection boiling, pool boiling, ~ **bei Zwangskonvektion** forced convection boiling, ~ **mit Dampferzeugung** boiling with steam generation, **örtliches** ~ local boiling, surface boiling, **stoßweises** ~ bumping
Siedepfanne f evaporating boiler
Siedeprodukt n boiling product
Siedepunkt m boiling point
Siedepunktbestimmungsapparat m boiling point apparatus
Siedepunktserhöhung f raising of the boiling point
Siederohr n boiler pipe, heating tube
Siederohrkessel m boiler with fire
Siederückstand m evaporation residue
Siedesalz n (obs) sodium chloride
Siedesole f brine
Siedestein m boiling stone, boiling chip

Siedetemperatur f boiling temperature
Siedethermometer n thermobarometer
Siedeverhalten n distillation characteristics
Siedeverzug m bumping, delay in boiling, retardation of ebullition
Siedeverzugsspirale f anti-bumping spiral, ebullition regulating coil
Siedewasserreaktor m boiling water reactor
Siegel n seal
Siegellack m sealing wax
Siegellackstange f stick of sealing wax
Siegelstein m (Min) agate
Siegelwachs n [soft] sealing wax
Siegenit m (Min) siegenite
Siemens-Einheit f Siemens mercury unit
Siemens-Martin-Betrieb m open-hearth practice
Siemens-Martin-Ofen m open-hearth furnace
Siemens-Martin-Prozeß m open-hearth process, pig-and-scrap process, Siemens-Martin process
Siemens-Martin-Roheisen n open-hearth pig iron
Siemens-Martin-Stahl n open-hearth steel
Siemens-Martin-Stahlprozeß m open-hearth process
Siemens-Prozeß m pig and ore process
Siemens-Regenerativfeuerung f Siemens regenerative open-hearth furnace
Siemenssche Wechselklappe f Siemens butterfly valve
Siemensstahl m open-hearth steel
Sigma-Bindung f sigma bond, σ-bond
Sigma-Komplex m sigma complex, σ-complex
Sigmareaktor m sigma pile
Signal n signal, **akustisches** ~ acoustic signal
Signalanlage f signal installation, signalling plant
Signaleinrichtung f signal alarm
Signalgabe f signalling, signal transmission
Signalisiereinrichtung f alarm device
Signalkontaktgeber m signalling instrument
Signallampe f signal lamp
Signalstreifen m signal strip
Signalthermometer n alarm thermometer
Signiereinrichtung f marker
signieren to mark, to sign, to stamp
Signierfarbstoff m sighting dyestuff
Signiergelb n sighting yellow
Signierstift m marking pin
Signiertinte f marking ink
Signifikanzzahl f (Statistik) level of significance
Signumfunktion f (Math) signum function
Sikkativ n siccative
Sikkativfirnis m siccative varnish
Silan n silane
Silantriol n silanetriol
Silber n silver (Symb. Ag), **gediegenes** ~ native silver, **gemünztes** ~ coined silver, silver coin, **knallsaures** ~ silver fulminate, **kolloidales** ~

colloidal silver, electrocollargol (Am. E.), **massives** ~ solid silver
Silberacetat *n* silver acetate
Silberacetylid *n* silver acetylide
Silberätzstein *m* silver nitrate
Silberalbuminat *n* silver albuminate
Silberamalgam *n* silver amalgam
Silberantimonglanz *m* (Min) silver antimony glance, miargyrite
Silberantimonid *n* silver antimonide
Silberantimonsulfid *n* silver antimony sulfide, silver thioantimonate
Silberarbeit *f* silver work
Silberarsenat *n* silver arsenate
silberartig silvery, argentine, silver
Silberasbest *m* (Min) silver asbestos
Silberazid *n* silver azide
Silberbichromat *n* silver bichromate, silver dichromate
Silberblatt *n* silver foil
Silberblende *f* (Min) proustite
Silberblick *m* silver brightening
Silberbrennen *n* silver refining
Silberbrennherd *m* silver refining hearth
Silberbromid *n* silver bromide
Silbercaproat *n* silver caproate
Silbercaprylat *n* silver caprylate
Silbercarbid *n* silver carbide
Silberchlorat *n* silver chlorate
Silberchlorbromid *n* silver chloride bromide
Silbercyanid *n* silver cyanide
Silbercyanwasserstoffsäure *f* cyanoargentic acid
Silberdithionat *n* silver dithionate
Silberdraht *m* silver wire
Silbereinkristall *m* silver single crystal
Silbererz *n* silver ore
Silberfällungsmittel *n* silver precipitant
Silberfahlerz *n* (Min) argentiferous tetrahedrite
Silberfarbe *f* silver [color]
silberfarben silver-colored
Silberfeilspäne *pl* silver filings
Silberfluorid *n* silver fluoride
Silberfolie *f* silver foil
Silberformiat *n* silver formate
silberführend argentiferous
Silbergalvanoplastik *f* silver galvanoplastic process
Silbergare *f* silver refining
Silbergehalt *m* silver content
silberglänzend silvery
Silberglanz *m* (Min) silver glance, argentite
Silberglimmer *m* (Min) common mica, muscovite
Silbergold *n* argentiferous gold
silbergrau silver grey
Silberhalogenid *n* silver halide
silberhaltig argentiferous, containing silver
Silberhornerz *n* (Min) horn silver, cerargyrite
Silberhütte *f* silver foundry, silver works

Silberhydroxid *n* silver hydroxide
silberig argentine, silvery
Silberjodat *n* silver iodate
Silberjodid *n* silver iodide
Silberkies *m* (Min) argentopyrite, sternbergite, white arsenical pyrite
Silberkorn *n* grain of silver
Silberkupferglanz *m* (Min) silver copper glance, stromeyerite
Silberlack *m* silver [stoving] varnish, clear face varnish, silver enamel, tin lacquer
Silberlactat *n* silver lactate
Silberlegierung *f* silver alloy
Silberlot *n* silver solder
Silbermalat *n* silver malate
Silbermalonat *n* silver malonate
Silbermesser *m* argentometer
Silbermine *f* (Bergb) silver mine
silbern silver[y]
Silberniederschlag *m* deposit of silver, silver precipitate
Silbernitrat *n* silver nitrate
Silberönanthat *n* silver enanthate
Silberorthophosphat *n* silver orthophosphate
Silberoxalat *n* silver oxalate
Silberoxid *n* silver oxide
Silberoxidsalz *n* silver oxysalt
Silberoxidul *n* (obs) silver suboxide
Silberoxidverbindung *f* compound of silver oxide
Silberpalmitat *n* silver palmitate
Silberpapier *n* silver paper, (Stanniolpapier) tin foil
Silberpappelöl *n* white poplar tree oil
Silberperoxid *n* silver peroxide
Silberphosphat *n* silver phosphate
Silberplattierung *f* silver cladding, silver plating
Silberprobe *f* silver assay, test for silver
Silberpropionat *n* silver propionate
Silberprotein *n* silver protein
Silberpulver *n* powdered silver, silver powder
Silberpyrophosphat *n* silver pyrophosphate
Silberraffination *f* silver refining
Silberraffinationsvorrichtung *f* apparatus for silver refining
Silbersalbe *f* colloid silver ointment
Silbersalpeter *m* (obs) silver nitrate
Silbersand *m* argentiferous sand, silver sand
Silbersau *f* silver brick, silver ingot
Silberscheideanstalt *f* silver refinery
Silberscheidung *f* silver refining, separation of silver, silver separation
Silberschein *m* silvery luster
Silberschlaglot *n* silver solder
Silberschlamm *m* silver tailings
Silberschwärze *f* (Min) earthy argentite, earthy silver glance
Silberselenid *n* silver selenide
Silberselenit *n* silver selenite
Silberspiegel *m* silver-coated mirror

Silberspießglanz *m* (Min) discrasite, dyscrasite
Silberstahl *m* silver steel
Silbersud *m* hot silvering bath
Silbersulfid *n* silver sulfide
Silbersuperoxid *n* silver peroxide
Silbertannenöl *n* pine oil
Silbertiegel *m* silver crucible
Silbervoltameter *n* silver voltameter
Silberwaage *f* silver balance
Silberwasser *n* (obs) aqua fortis, nitric acid
Silberweiß *n* silver white
Silberweißbad *n* hot light silvering bath
Silberweißsud *m* hot light silvering bath
Silberwismutglanz *m* (Min) argentobismuthite
Silfbergit *m* (Min) silfbergite
Silican *n* silicane, hydrogen silicide, silane, silicomethane, silicon hydride
Silicat *n* silicate, salt or ester of silicic acid
Silicatschlacke *f* silicate slag
Silicicolin *n* silicicolin
Silicid *n* silicide
Siliciolith *m* siliciolite
Siliciophit *m* siliciophite
Silicium *n* (Silizium) silicon; silicium (Symb. Si)
Siliciumäthan *n* disilane, silicoethane
Siliciumborid *n* silicon boride
Siliciumbronze *f* silicon bronze
Siliciumbronzedraht *m* silicon bronze wire
Siliciumcarbid *n* silicon carbide, carborundum
Siliciumchlorbromid *n* silicon chloride bromide
Siliciumchlorid *n* silicon chloride
Siliciumchloroform *n* silicochloroform, trichlorosilane
Siliciumdioxid *n* silicon dioxide, silica
Siliciumeisen *n* ferrosilicon, iron silicide
Siliciumfluorid *n* silicon fluoride, fluosilicate
Siliciumfluorverbindung *f* fluosilicate, silicon fluorine compound
Siliciumfluorwasserstoffsäure *f* silicofluoric acid, fluosilicic acid, silicoflSiliciumfluorwasserstoff
Siliciumgehalt *m* silicon content
siliciumhaltig siliceous
Siliciumjodid *n* silicon iodide
Siliciumjodoform *n* silicon iodoform
Siliciumkarbid *n* silicon carbide, carborundum
Siliciumkreide *f* siliceous chalk
Siliciumkupfer *n* silicon copper
Siliciummagnesium *n* magnesium silicide
Siliciummanganstahl *m* silicon manganese steel
Siliciummethan *n* silane, silicomethane, silicon tetrahydride
Siliciummonoxid *n* silicon monoxide
Siliciumoxalsäure *f* silicooxalic acid
Siliciumoxid *n* silicon dioxide, silica, silicic anhydride
Siliciumoxidhydrat *n* (obs) hydrated silicon dioxide, silicic acid

Siliciumphenyltrichlorid *n* phenyl silicon trichloride, phenyl trichlorosilane
Siliciumspiegel *m* ferrosilico manganese
Siliciumstahl *m* silicon steel
Siliciumtetrabromid *n* silicon tetrabromide
Siliciumtetrafluorid *n* silicon tetrafluoride
Siliciumtetrajodid *n* silicon tetraiodide
Siliciumverbindung *f* silicon compound, (Silicid) silicide
Siliciumwasserstoff *m* hydrogen silicide, silane, silicon hydride
Silicoäthan *n* disilane, silicoethane
Silicoameisensäure *f* silicoformic acid
Silicoborcalcit *m* silicoborocalcite
Silicoessigsäure *f* silicoacetic acid
Silicofluorid *n* silicofluoride, fluosilicate, salt or ester of fluosilicic acid, salt or ester of silicofluoric acid
Silicol *n* silicol
Silicomangan *n* manganese silicide, silicomanganese
Silicon *n* silicone
Siliconharz *n* silicone resin
Siliconkautschuk *m* silicone rubber
Siliconöl *n* silicone oil
Siliconsäure *f* siliconic acid
Siliconschmierfett *n* silicone grease
Silicooxalsäure *f* silicooxalic acid
Silicopropan *n* silicopropane
Silicospiegel *m* ferrosilicomanganese, silicious ferromanganese
Silika *f* silica, silicic anhydride, silicon dioxide
Silikaaufbereitungsanlage *f* silicon dressing plant
Silikagel *n* silica gel
Silikat *n* silicate, salt or ester of silicic acid
Silikin *n* cellulose nitrate rayon
Silikol *n* silicol
Silikon *n* s. Silicon
Silikose *f* (Med) silicosis
Silit *n* (HN) silite, silicon carbide
silizieren to siliconize
Silizierungsgrad *m* degree of silicification, degree of silication
Silizium *n* silicon, (s. auch Silicium) silicium (Symb. Si)
Sillimanit *m* (Min) sillimanite
Silo *m* silo
Siloxan *n* siloxane
Siloxen *n* siloxen
Siloxicon *n* (HN) siloxicon
Silundum *n* silundum, silicon carbide
Silvan *n* sylvan, 2-methylfuran
Silvestren *n* silvestrene, sylvestrene
Silylamin *n* silylamine
Simonellit *n* simonellite
Simulation *f* simulation
Simulator *m* simulator
simultan simultaneous

Sinalbin *n* sinalbin
Sinamin *n* sinamine, allylcyanamide
Sinapin *n* sinapine
Sinapinalkohol *m* sinapyl alcohol
Sinapinsäure *f* sinapic acid
sinapinsauer sinapic, (Salz) sinapate
Sinapisin *n* sinapisine
Sinapolin *n* sinapolin, diallylurea
Singulett *n* singlet
Singulettsystem *n* singlet system
Singulettzustand *m* singlet state
Sinigrin *n* sinigrin, potassium myronate
Sinistrin *n* sinistrin
sinken to sink, (abnehmen) to decrease, (absinken) to drop, to fall
Sinken *n* drop, sinking, (Abnahme) decrease
Sinkgeschwindigkeit *f* sedimentation velocity, settling velocity
Sinkmittel *n* (Flotation) depressant
Sinkscheideverfahren *n* sink and float process
Sinkscheidung *f* heavy media separation
Sinkschwimmtrennung *f* flotation, heavy media separation
Sinkstoff *m* sediment, deposit, precipitate
Sinn *m* sense, (Bedeutung) meaning, significance, (Richtung) direction
Sinnbild *n* symbol
Sinnesprüfung *f* organoleptic test, sensory testing
Sinomenin *n* sinomenine
Sinosid *n* sinoside
Sinter *m* sinter, scale
Sinteranlage *f* sintering plant
Sinterbunker *m* bin for product
Sintereisen *n* sintered powder metal
Sinterhartmetall *n* cemented carbide
Sinterkathode *f* powder cathode
Sinterkohle *f* sintering coal, cherry coal, hard coal
Sinterkuchen *m* sintered cake
Sinterkühler *m* sinter cooler
Sintermetall *n* powder metal, sintered metal
Sintermetallurgie *f* powder metallurgy
sintern to sinter, to bake, to cake, to clinker, to frit, to fuse, to slag, (Keramik) to vitrify
Sintern *n* sintering, baking, clinkering, fritting, fusing, vitrification
Sinterofen *m* sintering furnace
Sinterprozeß *m* slag process
Sinterquarz *m* siliceous sinter
Sinterröstung *f* flash roasting, sinter roasting
Sinterröstverfahren *n* sintering process
Sinterschlacke *f* clinker
Sinterung *f* sintering, baking, clinkering, fritting, fusing, vitrification
Sinterungshitze *f* sintering heat, incrustation heat
Sinterverfahren *n* (Polyäthylen) Engel process
Sintervorgang *m* sintering process, slag process

Sinterwerkstoff *m* sintered material
Sinus *m* (Math) sine, ~ **einer Zahl** sine of a number, **hyperbolischer** ~ hyperbolic sine, sinh
Sinusbedingung *f* sine condition
Sinusbussole *f* sine compass, sine galvanometer
Sinusentzerrung *f* sinus correction
sinusförmig sine-shaped, sinusoidal
Sinusfunktion *f* (Math) sine function
Sinus hyperbolicus *m* hyperbolic sine
Sinuskurve *f* sine curve, sinusoid
Sinuslinie *f* sine curve
Sinusreihe *f* sine series
Sinussatz *m* (Math) theorem of the sine
Sinusspannung *f* sine [curve] voltage
Sinusstrom *m* sine current
Sinustransformation *f* sine transformation
Siomin *n* siomine, hexamethylenetetramine tetraiodide
Siphon *m* siphon, syphon
siphonieren to siphon
Siphonrohr *n* syphon pipe, U-pipe
Siphonüberlauf *m* siphon spillway
Sipylit *m* (Min) sipylite
Sirene *f* siren
Sirenenton *m* siren sound
Sirenenventil *n* siren valve
Siriusmetallampe *f* sirius metal lamp
Sirup *m* syrup, sirup, treacle
sirupartig sirupy, treacly, viscous
Sirupdichte *f* sirupy consistency
Sirupdicke *f* sirupy consistency
siruphaltig containing sirup, containing treacle
Sirupkonsistenz *f* consistency of sirup
sirupös sirupy
Sisalagenin *n* sisalagenin
Sisalhanf *m* sisal hemp
Sismondin *m* (Min) sismondine, sismondite
sistieren (Med) to stop
Sitostan *n* sitostane
Sitostanol *n* sitostanol
Sitostanon *n* sitostanone
Sitosten *n* sitostene
Sitosterin *n* sitosterol
Sitosterol *n* sitosterol
Sitz *m* seat, (Passung) fit
sitzen (passen) to fit
Skala *f* scale, graduation, **doppelseitig kalibrierte** ~ double graduated scale
Skalarfeld *n* scalar field
Skalarprodukt *n* (zweier Vektoren) scalar product
Skalenablesung *f* scale reading
Skalenaräometer *n* graduated hydrometer
Skaleneinteilung *f* graduation of scale
Skalenintervall *n* scale interval, scale division
Skalenmikroskop *n* scale microscope
Skalenoeder *n* (Krist) scalenohedron
skalenoedrisch (Krist) scalenohedral

Skalenscheibe *f* dial
Skalenteilung *f* scale division
Skalpell *n* scalpel
Skalping *n* scalping (screening to remove foreign matter)
Skammonienharz *n* scammony
Skammonin *n* scammonin
Skandium *n* scandium (Symb. Sc)
Skapolith *m* (Min) scapolite, wernerite
Skarbroit *m* scarbroite
Skatol *n* scatol, methylindole, skatole
Skatoxyl *n* skatoxyl
Skelett *n* skeleton
Skelettmuskel *m* sceletal muscle
skimmen to skim
Skimmen *n* skimming
Skimmetin *n* skimmetine
Skimmianin *n* skimmianine
Skimmianinsäure *f* skimmianinic acid
Skimmin *n* skimmin
Skimmschicht *f* skim coat
Skizze *f* outline, sketch, (Entwurf) design, draft
skizzieren to outline, to sketch
Skleretinit *m* scleretinite
Skleroklas *m* (Min) scleroclas
Sklerometer *n* sclerometer
Skleroprotein *n* scleroprotein
Sklerose *f* (Med) sclerosis
Skleroskop *n* scleroscope, hardness drop tester
Skleroskophärte *f* scleroscope hardness, rebound hardness
Sklerotinsäure *f* sclerotinic acid, ergot acid, sclerotic acid
Skolezit *m* (Min) scolecite, needlestone
Skolopsit *m* (Min) scolopsite
Skopolamin *n* scopolamine
Skorbut *m* (Med) scurvy, scorbutus
skorbutisch (Med) scorbutic
Skorie *f* cinder, dross, scoria, slag
Skorifikation *f* scorification
Skorodit *m* (Min) scorodite, pitticite
Skrubber *m* gas washer, scrubber
Skutterudit *m* (Min) skutterudite
Skyrin *n* skyrin
Slotterfarbe *f* ink for corrugated board
Smalogenin *n* smalogenin
Smalte *f* smalt
Smaltin *m* (Min) smaltine, smaltite
Smaltit *m* (Min) smaltine, smaltite
Smaragd *m* (Min) emerald
smaragden emerald
smaragdfarben emerald, smaragdine
Smaragdgrün *n* emerald green
Smaragdit *m* (Min) smaragdite
Smaragdmalachit *m* (Min) emerald malachite, euchroite
Smaragdspat *m* (Min) amazonite, amazon stone, green feldspar
Smegma *n* smegma, greasy secretion

Smilacin *n* smilacin, pariglin, salseparin
Smilagenin *n* smilagenin
Smirnovin *n* smirnovine
Smithit *m* (Min) smithite
Smithsonit *m* (Min) smithsonite
Sobralit *m* sobralite
Sobrerol *n* sobrerol, pinol hydrate
Sockel *m* base
Soda *f* soda, sodium carbonate, **kalzinierte** ~ calcined soda, sodium carbonate, **kaustische** ~ caustic soda, sodium hydroxide
sodaalkalisch alkaline with soda
Sodaasche *f* soda ash
Sodaauszug *m* soda extract[ion]
Sodabad *n* soda bath
Sodablau *n* ultramarine
Sodadarstellung *f* preparation of soda
Sodafabrik *f* soda factory, soda works
Sodagefäß *n* soda tank
Sodagehalt *m* soda content
sodahaltig containing soda
Sodaküpe *f* soda vat
Sodalauge *f* soda lye
Sodalith *m* (Min) sodalite
Sodalösung *f* soda solution
Sodamehl *n* powdered sodium carbonate monohydrate
Sodapulver *n* soda powder, effervescent powder, powdered sodium carbonate, Seidlitz powder
Sodarückgewinnung *f* (Papier) soda recovery
Sodarückstand *m* soda residue
Sodastein *m* caustic soda, sodium hydroxide
Sodawasser *n* soda water
Soddy-Fajansscher Verschiebungssatz *m* displacement law of Soddy, Fajans and Russel
Soddyit *m* (Min) soddyite
Sörensen-Phosphatlösung *f* Sörensen standard phosphate solution
Sofortwirkung *f* immediate effect
Sog *m* suction
Sogdianose *f* sogdianose
soggen to crystallize out, to salt down
Soggenpfanne *f* crystallizing pan
Sohldruck *m* base pressure, bottom pressure
Sohlendichtung *f* bottom sealing
Sohlenkanal *m* bottom flue
Sohlplatte *f* base plate, bedplate, bottom plate, sole plate
Sojabohne *f* soy bean, soja bean
Sojaextraktionsschrot *n* soybean oil meal
Sojaöl *n* (Sojabohnenöl) soy [bean] oil, soja [bean] oil
Sojasapogenol *n* soyasapogenol
Sol *n* sol, **ausgeflocktes** ~ flocculated sol, **thixotropes** ~ thixotropic sol
Solabiose *f* solabiose
Soladulcidin *n* soladulcidine
Solandrin *n* solandrine
Solanellsäure *f* solanellic acid

Solanesol *n* solanesol
Solanidin *n* solanidine
Solanin *n* solanin
Solanocapsin *n* solanocapsine
Solanorubin *n* solanorubin, licopene
Solarisation *f* (Phot) solarization
Solarstearin *n* solarstearin
Solasulfon *n* solasulfone
Solatriose *f* solatriose
Solbehälter *m* brine cistern, brine pit
Sole *f* brine, salt spring, salt water
Soleableitung *f* brine outlet
Soleaustritt *m* brine outlet
Solebereiter *m* brine mixer
Soleeindämpfer *m* brine concentrator
Solefilter *m* brine strainer
Solekühler *m* brine cooler
Solekühlung *f* brine cooling
Solenoid *n* solenoid
Solezulauf *m* brine inlet
Solezuleitung *f* brine inlet
Solfaß *n* brine tub
Solfatarit *m* (Min) solfatarite
Solheber *m* saltwater elevator
Soliduskurve *f* solidus curve
Solidvernicklung *f* solid nickelplating
Solidversilberung *f* solid silverplating
Solketal *n* solketal
Solldurchmesser *m* nominal diameter
Solleistung *f* nominal output, rated output
Sollfrequenz *f* rated frequency
Sollgewicht *n* standard weight
Sollmaß *n* nominal size, rated size
Sollmenge *f* theoretical quantity
Sollwert *m* theoretical value, desired value, face value, reference input, set point, **effektiver** ~ effective set point
Sollwerteinstellung *f* set point adjustment
Solochromfarbstoff *m* solochrome dyestuff
Solorsäure *f* soloric acid
Solquelle *f* brine spring
Solsalz *n* brine salt, spring salt
Solurol *n* solurol, nucleotin phosphoric acid, thyminic acid
Solvatation *f* solvation
Solvatationsenergie *f* energy of solvation
Solvathülle *f* solvating envelope
solvatisieren to convert into a sol, to solvate
solvatisiert solvated
Solvatisierung *f* solvation
Solvaysoda *f* Solvay soda, ammonia soda
Solvayverfahren *n* Solvay process, [Solvay's] ammonia soda process
Solvens *n* solvent
Solventnaphtha *n* solvent naphtha
solvieren to dissolve
solvierend dissolving, solvent
Solvolyse *f* solvolysis
Solwaage *f* salimeter, brine gauge, salinometer

Solwasser *n* brine, saltwater
Somalin *n* somalin
Somatotrophin *n* somatotrophin
Sombrerit *m* (Min) sombrerite
Sommervillit *m* sommervillite
Somnirol *n* somnirol
Somnitol *n* somnitol
Somnoform *n* somnoform
Sonde *f* probe
Sondenausdehnung *f* probe area
Sondencharakteristik *f* probe characteristic
Sondentechnik *f* probe technique
Sonderanfertigung *f* special production, special construction
Sonderausführung *f* special design
Sonderausrüstung *f* special equipment, extras
Sonderdruck *m* reprint
Sondererregung *f* separate excitation
Sondergewinde *n* special thread
Sondergröße *f* special size
Sondergußeisen *n* special cast iron
Sonderkonstruktion *f* special construction
sondern to separate
Sonderprodukt *n* special product
Sonderprospekt *m* special brochure
Sonderrechenstab *m* special slide rule
Sondersatz *m* special charge
Sonderstahl *m* alloy steel, special steel
Sondervorschriften *pl* special rules
Sonderzweckvorrichtung *f* special purpose fixture
sondieren to probe, to sound
Sonnenbestrahlung *f* solar radiation
Sonnenblumenöl *n* sunflower oil
Sonnenbräune *f* suntan
Sonnenchromosphäre *f* solar chromosphere
Sonnenenergie *f* solar energy
Sonnenfinsternis *f* (Astr) solar eclipse
Sonnenfleck *m* (Astr) sun spot
Sonnenkontinuum *n* sun continuum
Sonnenlicht *n* sunlight
Sonnenmikroskop *n* solar microscope
Sonnenphotosphäre *f* solar photosphere
Sonnenschutzfarbe *f* shading paint
Sonnenspektrum *n* solar spectrum
Sonnenstein *m* (Min) sunstone
Sonnenstrahlung *f* solar radiation
Sonnensystem *n* solar system
Sonnenuhr *f* sundial
Sonnenwärme *f* solar heat
Sonnenwende *f* solstice
Sophocarpidin *n* sophocarpidine
Sophoranol *n* sophoranol
Sophoricol *n* sophoricol
Sophorin *n* sophorine
Sophorit *m* sophoritol
Sophorose *f* sophorose
Soranjidiol *n* soranjidiol
Sorbaldehyd *m* sorbic aldehyde, sorbaldehyde

Sorbat *n* sorbate, salt or ester of sorbic acid
Sorbicillin *n* sorbicillin
Sorbierit *m* sorbieritol
Sorbin *n* 1,3,4,5,6-pentahydroxy-2-hexanone, sorbin, sorbinose, sorbose
Sorbinaldehyd *m* sorbaldehyde, sorbic aldehyde
Sorbinose *f* 1,3,4,5,6-pentahydroxy-2-hexanone, sorbinose, sorbin, sorbose
Sorbinsäure *f* sorbic acid, 2,4-hexadienoic acid, hexadienic acid
sorbinsauer (Salz) sorbate
Sorbit *m* (Chem) sorbitol, sorbite, (Metall) sorbite
Sorbitdehydrogenase *f* (Biochem) sorbit dehydrogenase
Sorbosazon *n* sorbosazone
Sorbose *f* sorbose
Sorburonsäure *f* sorburonic acid
Sorbylalkohol *m* sorbyl alcohol
Sorelzement *m* Sorel cement, magnesia cement
Sorghumöl *n* sorghum oil
Sorigenin *n* sorigenin
Sorinin *n* sorinin
Sorption *f* sorption
Sorptionsgleichgewicht *n* sorption equilibrium
Sorptionsisotherme *f* isotherme of sorption
Sorte *f* quality, grade, kind, sort, specimen, type
Sortieranlage *f* separating plant, picking plant
sortieren to sort, to size
Sortieren *n* separating, **rohes ~ der Erze** rough separating [of ores]
Sortiermaschine *f* grader, sorting machine
Sortiertisch *m* sorting table
Sortierung *f* **nach der Größe** sizing;, **elektrische ~** electrostatic separation;, **mechanische ~** mechanical sorting
Sortiment *n* assortment, range
Soupleseide *f* souple silk, half boiled silk
souplieren (Seide) to souple [silk], to half boil
Soupieren *n* (Seide) soupling, partial boiling
Soxhletapparat *m* Soxhlet apparatus
Soxhlet-Extraktor *m* Soxhlet extractor
Soyanal *n* soyanal
Sozojodol *n* soz[o]iodol, diiodo-p-phenolsulfonic acid, sozoiodolic acid
Sozojodolsäure *f* diiodo-p-phenolsulfonic acid, soz[o]iodol
Sozojodolsalz *n* sozoiodol salt
Sozolsäure *f* sozolic acid, aseptol, o-phenolsulfonic acid
spacheln to prime
Spachtel *m* spatula
Spachtelfarbe *f* filling-up color
Spachtelfußboden *m* jointless floor
Spachtelkitt *m* putty
Spachtelmasse *f* filler, mastic
Spachtelwalze *f* reverse roll coater
Späne *m pl* chips, shavings

Späneentölung *f* chip-oil extraction
Spätgärung *f* delayed fermentation
Spätzündung *f* late ignition, retarded ignition
Spallation *f* (Atom) spallation
Spalt *m* (Bergb) cleavage, (Lücke) gap, (Öffnung) gap, opening, (Opt) slit, (Riß) crack, fissure, rent, split, (Schlitz) slot
Spaltalgen *pl* myxophyceae, schizophyceae
Spaltanlage *f* cracking plant
Spaltausbeutekurve *f* fission yield curve
Spaltausleuchtung *f* slit illumination
spaltbar cleavable, fissile, fissionable, separable
Spaltbarkeit *f* cleavability, fissility, fissionability, ((Abblätterung)) delamination
Spaltbild *n* (Opt) slit image
Spaltblende *f* (Opt) collimating slit
Spaltbreite *f* (Opt) slit width, width of slit
Spaltbruchstück *n* fission fragment
Spaltbrüchigkeit *f* cleavage brittleness
Spaltbüchse *f* split bushing
Spaltdestillation *f* (von Rohöl) cracking (of crude oil)
Spaltdichtung *f* controlled gap seal
Spaltdorn *m* split mandrel
Spalte *f* gap, crevice, narrow opening, (Buchdr) column
Spaltebene *f* cleavage plane
spalten to break down, to cleave, to crack, to fissure, to separate, to slit, to split, (Chem) to dissociate, **eine chem. Bindg. ~** to break a chemical bond
Spalten *n* splitting [up], (Erdöl) cracking
Spaltenergie *f* (Atom) fission energy
Spaltenindex *m* (Buchdr) column index
Spaltfestigkeit *f* interlaminar strength
Spaltfläche *f* cleavage plane, cleavage face, cleavage surface
Spaltfragment *n* (Atom) fission fragment
Spaltgas *n* cracked gas, product gas (cracking process)
Spaltgift *n* (Atom) fission poison
Spalthöhe *f* slit height
Spaltkammer *f* (Atom) fission chamber
Spaltkollimation *f* slit collimation
Spaltkorrosion *f* crevice corrosion
Spaltmaterial *n* (Atom) core material, fissile material, fission material
Spaltneutron *n* (Atom) fission neutron
Spaltpilz *m* fission fungus
Spaltpilzgärung *f* schizomycetic fermentation
Spaltplatte *f* split wall tile
Spaltpolmotor *m* split phase motor, split pole motor
Spaltprodukt *n* fission product, fragment
Spaltproduktvergiftung *f* (Atom) fission-product poisoning
Spaltprozeß *m* fission process
Spaltreaktor *m* (Atom) fission reactor
Spaltrohrmotor *m* canned motor

Spaltrohrwärmeaustauscher *m* split-tube heat exchanger
Spaltrohstoff *m* (Atom) fertile material
Spaltschwelle *f* fission threshold
Spaltstoff *m* reactor fuel, reactor feed
Spaltstoffabfälle *m pl* reactor waste
Spaltstoffausnutzung *f* burn-up
Spaltstoffblock *m* fuel slug
Spaltstoffhülse *f* cartridge
Spaltstoffkanal *m* fuel channel
Spaltstoffstab *m* cartridge, fuel rod
Spaltstück *n* cleavage product, fission product, split product
Spalttrümmer *pl* (Atom) fission fragments
Spaltüberwachungsgerät *n* fission monitor
Spaltung *f* cleaving, breaking up, scission, splitting, (Atom) fission, (Biol) division, (Chem) dissociation, ~ **des Atomkerns** nuclear fission, ~ **mit langsamen Neutronen** slow-neutron fission, ~ **von Erdöl** cracking of petroleum, ~ **von Fettsäuren** splitting of fatty acids
Spaltungsebene *f* cleavage plane
Spaltungsenergie *f* (Atom) fission energy
Spaltungskristall *m* cleavage crystal
Spaltungsnachweis *m* fission detection
Spaltungsprodukt *n* cleavage product, (Atom) fission product
Spaltungsquerschnitt *m* fission cross section
Spaltungsuntersuchung *f* fission investigation
Spaltvelour *m* split suede
Spaltzählung *f* fission counting
Spaltzerfallskette *f* fission decay chain
Spaltzone *f* (Atom) active zone
Span *m* chip, splint, (Bohrspan) bore chip, boring, (Drehspan) cutting, turning, (Feilspan) filing, (Hobelspan) shaving, (Splitter) splinter
Spanabfluß *m* chip flow
spanabhebend metal-cutting, cutting, machining
Spanbildung *f* chip formation
Spanbrechernute *f* chip breaker
Spangolith *m* (Min) spangolite
Spanholz *n* chip wood
Spanholzplatte *f* chip board
Spann *m* instep
Spannbacke *f* clamping jaw, contact jaw, wedge grip
Spannbandsystem *n* suspension band system, taut suspension system
spannbar extensible, stretchable, tensile
Spannbarkeit *f* stretchability, tensibility
Spannbeton *m* prestressed concrete
Spannbüchse *f* retainer ring
Spannbügel *m* tension screw
Spanndruck *m* clamping pressure
Spanne *f* (Spielraum) margin, tolerance, (Zeitspanne) interval, span
Spanneinrichtung *f* gripping arrangement

spannen to strain, to stress, (Appretur) to tenter, (strecken) to stretch
Spannen *n* stretching, tentering
Spannfutter *n* chuck [jaws], lathe chuck
Spanngetriebe *n* chucking mechanism
Spanngewicht *n* tension weight
Spannhebel *m* clamping lever
Spannkraft *f* elasticity, clamping power, extensibility, (Dampf) expansibility
Spannleiste *f* mount
Spannmutter *f* retainer ring
Spannrahmen *m* clamping frame, stenter, tenter, tentering frame
Spannrahmentrockner *m* tenter drier
Spannriemen *m* clamping strap
Spannring *m* clamping ring; retainer ring
Spannrolle *f* (Treibriemen) idler
Spannschloß *n* stretching bolt, stretching screw, turnbuckle
Spannschraube *f* tightening screw, clamp bolt, clamp screw
Spannsprödigkeit *f* tension brittleness
Spannstock *m* vise
Spannung *f* tension, (Elektr) potential, voltage, (Spannen) stretching, (Zugbeanspruchung) tensile stress, ~ **en beheben** to alleviate stresses, to relieve stresses, **elektrische** ~ voltage, **induzierte** ~ induced voltage, **innere** ~ internal stress, molded-in stress, **mit hoher** ~ high-tension, (Elektr) high-voltage, **sterische** ~ steric strain
Spannungsabfall *m* voltage drop, potential difference, **kritischer** ~ disintegration voltage, **Ohmscher** ~ ohmic drop of voltage
spannungsabhängig depending on the voltage
Spannungsänderung *f* (Elektr) change of voltage, voltage fluctuation, voltage variation, (Metall) change of strain
Spannungsanhäufung *f* aggregate tension
Spannungsanstieg *m* (Elektr) potential rise
Spannungsanzeiger *m* strain indicator
Spannungsausgleich *m* (Elektr) voltage compensation
Spannungscharakteristik *f* (Elektr) voltage characteristic
Spannungsdehnungsbeziehung *f* stress-strain relation
Spannungsdehnungsdiagramm *n* stress-strain diagram
Spannungsdeviation *f* stress deviation
Spannungsdiagramm *n* stress graph
Spannungsdifferenz *f* potential difference, (Elektr) voltage difference
Spannungsdoppelbrechung *f* strain double refraction, stress birefringence
Spannungsdruckdiagramm *n* compression-stress[-deformation] diagram
Spannungseinheit *f* tension unit, (Elektr) voltage unit

Spannungsenergie f potential energy
spannungserhöhend (Elektr) booster, step-up
Spannungserhöher m battery booster, voltage step-up means
Spannungserhöhung f (Elektr) voltage increase
spannungserniedrigend negative booster, step-down
Spannungserniedrigung f voltage decrease
Spannungsfaktor m strain factor
Spannungsfeder f tension spring
Spannungsfeld n stress field
Spannungsfläche f stress surface
spannungsfrei free of tension, strainless, stress-free, ~ **glühen** to anneal stress-free, ~ **machen** to temper, (durch Ausglühen) to stress-anneal
Spannungsfunktion f (Elektr) stress function
Spannungsgefälle n voltage drop, (Elektr) voltage gradient, electric potential gradient, potential drop
Spannungsgleichhalter m voltage stabilizer
Spannungsgleichung f (Elektr) voltage equation
Spannungsgrad m degree of tension
Spannungsgradient m voltage or potential gradient
Spannungsgrenze f limit of voltage
Spannungshöchstwert m (Elektr) peak voltage
Spannungsintensitätsfaktor m stress intensity factor
Spannungsknotenpunkt m voltage or potential nodal point
Spannungskoeffizient m [thermal] expansion coefficient
Spannungskomponente f voltage component
Spannungskonstanthalter m stabilized power supply, voltage stabilizer
Spannungskonzentration f stress concentration
Spannungskorrosion f stress corrosion, (Polyäthylen) stress cracking
Spannungskurve f voltage curve
Spannungsladung f voltage charge
Spannungslinie f pressure curve
spannungslos without tension, free from tension, (Elektr) dead
Spannungsmesser m tension indicator, (Elektr) voltmeter
Spannungsoptik f photoelasticity
spannungsoptisch photoelastic
Spannungsphase f voltage phase
Spannungsphoto n strain photograph
Spannungsprüfeinrichtung f tensiometer
Spannungsprüfer m strain gauge, (Elektr) polarity indicator, voltage indicator
Spannungsregelröhre f voltage regulator tube
Spannungsregler m voltage regulator
Spannungsregulierung f voltage regulation
Spannungsreihe f electrochemical series, electromotive series, **elektrochemische** ~ series of elements (in accordance with electrochemical potentials)
Spannungsrelaxation f stress relaxation
Spannungsresonanz f tension resonance
Spannungsriß m stress crack, stress crazing
Spannungsrißbildung f stress cracking
Spannungsrißkorrosion f stress corrosion cracking, (Stahl) sulfide stress cracking
Spannungsrückgang m (Elektr) fall of voltage
Spannungsschwankung f (Elektr) fluctuation of voltage
Spannungssequenzkomponente f component of voltage sequence
Spannungssicherheit f (Elektr) spark-over strength
Spannungssprung m (Elektr) sudden voltage difference
Spannungsstabilisator m (Elektr) voltage stabilizer
Spannungsteiler m potential divider
Spannungsteilertransformator m balancing transformer
Spannungstensor m stress tensor
Spannungstheorie f theory of strain, strain theory
Spannungsüberschuß m (Elektr) excess of voltage
Spannungsunterschied m potential difference, tension difference
Spannungsvektor m voltage vector
Spannungsverfestigung f (Met) strain hardening
Spannungsverlust m pressure loss, (Elektr) voltage drop
Spannungsverstärker m booster
Spannungsverstärkung f voltage amplification
Spannungszusammenbruch m voltage collapse
Spannute f groove
Spannvorrichtung f clamping arrangement, chucking device, clamping device
Spannwalze f clamp roll, pull rod
Spannweite f span
Spannwerkzeug n clamping tool, work holding device
Spant n former, frame
Spantenaufsetzmaschine f frame mounting machine
Spantiefe f depth of cut
Spantring m former, frame
Spanwinkel m rake angle
Sparassol n sparassol
Sparbeize f pickling bath containing an inhibitor, restrainer
Sparbeizzusatz m pickling inhibitor
Sparbrenner m pilot burner
Sparflämmchen n pilot flame
Sparflamme f pilot flame
Spargelstein m asparagus stone
Sparren m rafter, scantling
sparsam economic[al]

Sparsamkeit f economy
Spartalit m (Min) spartalite, native zinc oxide
Spartein n sparteine, lupinidine
Spartransformator m economizing transformer
Spartyrin n spartyrine
Sparventil n throttling valve
Spasmus m spasm
Spat m (Min) spar
Spateisenstein m (Min) siderite
Spatel m spatula, surfacer, trowel
spatelförmig spatula-shaped
Spatenstich m (Landw) cut with spade
Spaterz n (Min) spathic ore
Spathiopyrit m (Min) spathiopyrite
Spathulatin n spathulatine
spatig sparry, spathic
Spatsäure f (obs) hydrofluoric acid
Spearminzöl n spearmint oil
speckartig steatitic
Speckkalk m fat lime
Specköl n lard oil
Speckschmierung f lard lubrication
Speckstein m steatite, soapstone, talc
Specktorf m bituminous peat, pitch peat
Specktran m train oil
Specularit m (Min) specularite, specular hematite, specular iron
speerförmig spear-shaped
Speerform f spear shape
Speerkies m (Min) spear pyrites
Speichel m saliva, spittle
Speicheldrüse f (Anat) salivary gland
Speichendraht m spoke wire
Speicher m granary, reservoir, storage tank, (Elektr) accumulator
Speicheranlage f storage plant
Speichereffekt m memory effect
Speicherelektrode f energy-storage electrode
speichern to store, to accumulate
Speichern n storage, storing
Speicherofen m regenerative furnace
Speicherröhre f storage tube
Speichertankmischer m retention mixer
Speichertrommel f (Elektron) memory drum
Speicherung f storage, accumulation
Speichervermögen n storage capacity
Speicherwechselrichter m series inverter
Speise f (Speis m, Bauw) mortar
Speisedrossel f feed coil
Speiseeis n icecream
Speisegas n feed gas
Speisekabel n feed cable
Speisekreis m supply circuit
Speiseleitung f feeder, feed line, feed piping
speisen to feed, to supply, (laden) to charge, to load
Speiseöl n edible oil
Speisepumpe f feed [water] pump
Speisepunkt m feeding point

Speiser m feeder
Speiseröhre f (Anat) [o]esophagus
Speisetemperatur f feed temperature
Speisewasser n feed lye or liquor, feed water
Speisewasseranschluß m feed water connection
Speiskobalt m (Min) arsenical cobalt ore, grey cobalt ore, smaltine
spektral spectral
Spektralanalyse f spectral analysis, spectroscopical analysis, spectrum analysis
spektralanalytisch spectroanalytic, spectrometric, spectroscopic
Spektralapparat m spectrometer
Spektralbeobachtung f spectroscopic observation
Spektralbereich m spectral range, spectral region
Spektralcharakteristik f (Spektr) spectral characteristic
Spektraldurchlaß m (Spektr) spectral transmission
Spektralempfindlichkeit f (Phot) spectral sensitivity
Spektralfarbe f spectral color
Spektrallinie f spectral line
Spektrallinienfaltung f folding of spectral lines
Spektrallinienverschiebung f (Spektr) shift of spectral lines
Spektralphotometer n spectrophotometer
Spektralreihe f spectral series
Spektralrohr n spectral tube
Spektralserie f (Spektr) series of spectrum lines
Spektralverschiebung f spectral shift
Spektralverteilung f spectral distribution
Spektrenprojektor m spectra projector
Spektrogramm n spectrogram
Spektrograph m spectrograph
Spektroheliogramm n spectroheliogram
Spektrohelioskop n spectrohelioscope
Spektrometer n spectrometer
Spektrometermethode f spectrometer method
Spektrometrie f spectrometry, **hochauflösende** ~ high-resolution spectrometry
spektrometrisch spectrometric
Spektroskop n spectroscope
Spektroskopie f spectroscopy
spektroskopisch spectroscopic
Spektrum n spectrum, ~ **der magnetischen Resonanz** (Spektr) magnetic resonance spectrum, ~ **mit dunklen Linien** dark-line spectrum, **Absorptions-** ~ absorption spectrum, **Banden-** ~ band spectrum, **Beugungs-** ~ diffraction spectrum, **diskontinuierliches** ~ discontinuous spectrum, line spectrum, **Emissions-** ~ emission spectrum, **Flammen-** ~ flame spectrum, **IR-** ~ infra-red spectrum, **kontinuierliches** ~ continuous spectrum, continuum, **Linien-** ~ line spectrum, **Massen-** ~ mass spectrum,

Raman- ~ Raman spectrum, **Röntgen-** ~ roentgen spectrum, X-ray spectrum, **UV-** ~ ultra-violet spectrum
Spencemetall n spence metal
Spencerit m (Min) spencerite
Spender m donor
Spenderblut n donor's blood
Spengler m plumber
Sperma n sperm[a], sperminal fluid
Spermacetöl n sperm oil
Spermatogenese f spermatogenesis
Spermatophyten pl (Bot) spermatophyta
Spermatozoen pl spermatozoa
Spermazet n spermaceti
Spermidin n spermidine
Spermin n spermine
Spermizid (Pharm) spermicidal
Spermöl n sperm oil
Spermostrychnin n spermostrychnine
Sperrad n ratchet wheel
Sperrbolzen m strain bolt
Sperrelais n guard relay, locking relay
sperren (absperren) to cut off, (blockieren) to block, (zusperren) to lock, to shut
Sperrfilter n (Opt) suppression filter
Sperrflüssigkeit f liquid seal, sealing liquid, **Dichtung durch** ~ fluid packing
Sperrhaken m ratchet
Sperrholz n plywood
Sperrholzkleber m ply glue, veneer glue
Sperrholzleim m ply glue
Sperrholzspannring m plywood clamping ring
sperrig bulky, wide spreading
Sperrklinke f pawl, ratchet
Sperrschicht f blocking layer, interlining, (Anode) insulating anodic coating
Sperrschichtmaterial n barrier material
Sperrschichtphotoeffekt m barrier layer photoelectric effect
Sperrschichttheorie f barrier layer theory
Sperrschichttransistor m barrier transistor
Sperrschichtzelle f barrier-type cell, photovoltaic cell, rectifier cell
Sperrschieberpumpe f single-lobe pump
Sperrschiene f lock bar, treadle bar
Sperrschwinger m blocking oscillator
Sperrspannung f blocking voltage, inverse voltage
Sperrstrom m (Elektr) backward or inverse current
Sperrstromkreis m rejector circuit
Sperrventil n check valve, stop valve
Sperrwasser n sealing water
Sperrwirkung f valve effect, surface insulation effect
Sperrylith m (Min) sperrylite
Sperrzahn m ratchet tooth
Sperrzeit f dead time
Sperrzelle f barrier-type cell

Spessartin m (Min) spessartite, aluminum garnet
Spezialausführung f special design, special type
Spezialausrüstung f special equipment
Spezialgips m special gypsum
Spezialgußeisen n special cast iron
spezialisieren to specialize
Spezialist m specialist, expert
Spezialkabel n special lead
Speziallegierung f special alloy
Spezialreagentien pl special reagents
Spezialstahl m special steel
Spezies f species
Spezifikation f specification
spezifisch specific
Spezifität f specificity
Sphäre f sphere
sphärisch spheric[al]
Sphärit m spherite
Sphärizität f sphericity
Sphaerocarpin n spherocarpine
Sphäroid n spheroid
Sphärokolloid n sphero-colloid
sphärokonisch spheroconical
Sphärolith m (Geol) spherolite
Sphärometer n spherometer
Sphaerophorin n spherophorin
Sphärophorol n spherophorol
Sphärophysin n spherophysine
Sphärosiderit m (Min) spherosiderite, clay ironstone
Sphalerit m (Min) sphalerite, zinc blende
Sphen m (Min) sphene, titanite
Sphenoid n (Krist) sphenoid
sphenoidisch sphenoidal
Sphingin n sphingine
Sphingoin n sphingoin
Sphingolipid n sphingolipid
Sphingomyelin n sphingomyelin
Sphingosin n sphingosine
Sphondin n sphondin
Sphragid m (Min) sphragide
Spicken n **der Wolle** wool oiling
Spicköl n wool oil, spinning oil
Spiculisporsäure f spiculisporic acid
Spiegel m (einer Flüssigkeit) level, (Min) specular iron, spiegel, (reflekt. Fläche) mirror, reflector, ~ **der Badlösung** level of the bath solution
Spiegelablesung f reading by reflection
Spiegelbild n mirror image, reflected image
Spiegelbildfunktion f mirror image function
spiegelbildisomer enantiomorphic
Spiegelbildisomerie f enanthiomorphism
spiegelblank high luster finished
Spiegelblende f (Min) specular galena
Spiegelbronze f mirror bronze
Spiegelebene f mirror plane, reflection plane, (Krist) plane of symmetry

Spiegeleisen n specular [cast] iron, (Min) specular iron hematite
Spiegelerz n specular iron ore
Spiegelgalvanometer n mirror galvanometer, reflecting galvanometer
Spiegelglanz m reflecting luster, (Min) wehrlite
Spiegelglas n mirror glass, plate glass
spiegelglatt dead-smooth
Spiegelgleichheit f mirror symmetry
Spiegelkies m (Min) specular pyrite
Spiegelkobalt m (Min) specular cobalt ore
Spiegelmetall n speculum metal
Spiegelmikroskop n reflecting microscope
Spiegelöffnung f reflector aperture
Spiegelschweißen n hot-shoe welding
Spiegelskala f mirror scale
Spiegelstein m (Min) specular stone
Spiegelsymmetrie f axial symmetry
Spiegelung f (Opt) reflection, mirage
Spiegelungsinvarianz f reflection invariance
Spiegelungssymmetrie f symmetry of reflection
spiegelverkehrt mirror-inverted
spiegelverkleidet mirror-panelled
Spiel n (Spielraum) clearance, (toter Gang) backlash, ~ **des Eisens** surface play of iron
Spielart f variant, modification, variation, variety
spielend leicht very easy
spielfrei free from backlash
Spielraum m (Bewegungsfreiheit) play, (Fehler) tolerance
Spießbleierz n (Min) bournonite
Spießblende f (Min) antimony blende, kermesite
Spießblumen f pl (Min) flowers of antimony
Spießbutter f (obs) antimony trichloride, butter of antimony
Spießerz n antimony ore
Spießglanz m (Min) antimony glance, stibnite
spießglanzartig antimonial
Spießglanzasche f antimony ash
Spießglanzkermes m (Min) kermesite, kermes mineral
Spießglanzkönig m regulus of antimony
Spießglanzmetall n metallic antimony
Spießglanzsafran m crocus of antimony
Spießglanzsilber n (Min) dyscrasite
Spießglaserz n (Min) stibnite
Spießkobalt m (Min) smaltine, smaltite
Spießocker m (Min) antimony ocher
Spießoxid n (obs) antimony trioxide
Spießschwefel m (obs) antimony sulfide
Spießweinstein m (obs) tartrated antimony
Spießweiß n (obs) antimony trioxide, antimony white
Spießzinnober m (Min) kermesite
Spigeliawurzel f (Bot) spigelia root
Spiköl n spike oil
Spilanthol n spilanthol
Spin m (Phys) spin

Spinabhängigkeit f dependence on spin
Spinacin n spinacine
Spinasterin n spinasterol
Spinausrichtung f spin alignment
Spinaustauschwechselwirkung f spin exchange interaction
Spinazarin n spinazarin
Spinbahn f [spin] orbit
Spinbahnaufspaltung f spin orbit splitting
Spinbahnkopplung f spin [orbit] coupling
Spinbahnwechselwirkung f spin orbit interaction
Spindel f spindle, (Aräometer) areometer, (Hydrometer) hydrometer, (Welle) arbor
Spindelbaum m (Bot) euonymus, spindle tree
Spindelflansch m spindle flange
Spindelführung f stem guide
Spindelfuß m spindle socket
Spindellagerung f spindle bearing
Spindelpresse f spindle press, hand wheel stamping press, screw press
Spindelpumpe f single-rotor screw pump
Spindelring m spacing collar
Spindelventil n spindle valve
Spindelzelle f spindle cell
Spindichte f spin density
Spindrehimpuls m (Atom) spin angular moment
Spindublett n (Atom) spin doublet
Spin-Echo-Verfahren n (Atom) spin-echo-method, method of free nuclear induction
Spineinfluß m spin effect
Spinell m (Krist) spinel
Spinellzwilling m (Krist) spinel twin
Spinenergie f spin energy
Spinerhaltung f spin conservation
Spinfunktion f spin function
Spingitterrelaxation f spin lattice relaxation
Spingröße f spin magnitude
Spinimpulskoppelung f spin-orbit coupling
Spinkopplung f spin coupling
Spinmagnetismus m spin magnetism
Spinmatrize f spin matrix
Spinmoment n spin momentum
Spinmomentdichte f spin momentum density
Spinmultiplett n (Atom) spin multiplet
Spin-Multiplizität f spin multiplicity
Spinnbad n spin bath
Spinndüse f spinning nozzle, (Kunstfaserherstellung) spinneret
Spinne f (Zool) spider
spinnen to spin
Spinnen n spinning
Spinnerei f spinning mill
Spinnfaden m strand
Spinnfarbenkarte f shade card
Spinnfaser f staple fiber
Spinnflüssigkeit f spinning dope
spinngefärbt mass-dyed, dope-dyed
Spinngeschwindigkeit f rate of extrusion

Spinnkanne *f* spinning can
Spinnkopf *m* spinning head
Spinnlösung *f* spinning solution
Spinnmaschine *f* spinning machine
Spinnmischung *f* spinning blend
Spinnpaste *f* spinning paste
Spinnring *m* spinning ring
Spinnspule *f* spinning bobbin
Spinntemperatur *f* (Kunststoff) extrusion temperature
Spinnverfahren *n* spinning system
Spinnzylinder *m* spinning roller
Spinochrom *n* spinochrome
Spinoperator *m* spin operator
Spinorfeld *n* spinor field
Spinorkalkül *n* spinor calculus
Spinpolarisation *f* spin polarization
Spinpopulation *f* spin population
Spinquantenzahl *f* (Atom) spin quantum number
Spinrichtung *f* (Atom) spin orientation
Spin-Spin-Wechselwirkung *f* (Atom) spin-spin interaction
Spinstreuung *f* spin flip scattering
Spinthariskop *n* spinthariscope
Spinulosin *n* spinulosin
Spinumkehr *f* interspin crossing
Spinumklappung *f* spin flip
Spinverdopplung *f* spin doubling
Spinverteilung *f* spin distribution
Spinwellenmethode *f* spin wave method
Spinzahl *f* (Atom) spin number
Spion *m* feeler gauge
Spionenatom *n* spy atom, tracer
Spionenprobe *f* spy sample
Spiräin *n* spiraein
Spiraeosid *n* spiraeoside
Spiralauftragwalze *f* spiral applicator
Spiralbahnspektrometer *n* spiral orbit spectrometer
Spiralbandtrockner *m* screw-conveyor drier
Spiralbohrer *m* twist drill
Spiralbohrerschleifmaschine *f* [twist] drill grinder
Spirale *f* spiral, coil, coiled filament
Spiralenzirkel *m* volute compasses
Spiralfeder *f* spiral spring, buffer spring, coil spring
spiralförmig spiral, coiled, helical
Spiralkühler *m* coil-condenser
Spirallänge *f* (Bohrer) flute length
Spiralnebel *m* spiral nebula
Spiralrakel *f* spiral applicator
Spiralrohr *n* spiral tube
Spiralrohrwärmeaustauscher *m* spiral-tube heat exchanger
Spiralsichter *m* spiral classifier
Spiralstopfen *m* spiral stopper
spiralverzahnt spirally fluted
Spiralwärmeaustauscher *m* spiral heat exchanger
Spiralwaschflasche *f* spiral washing bottle

Spiralwelle *f* spirally wound flexibel shaft
Spiralwindsichter *m* spiral classifier
Spiralzahnkegelrad *n* spiral bevel gear
Spirituosen *pl* alcoholic liquors
Spiritus *m* spirit, alcohol, ethanol, **vergällter (denaturierter)** ~ denatured or methylated spirits
spiritusartig spirituous, alcoholic
Spirituslack *m* spirit varnish
Spirituslampe *f* alcohol lamp, spirit lamp
Spiritusrektifikation *f* spirits rectification
Spiritusverdampfer *m* alcohol evaporator, alcohol vaporizer
Spirituswaage *f* alcoholometer; spirit level
Spirocid *n* spirocide
Spirocyclan *n* spirocyclan, spiran, spiro compound
spirocyclisch spirocyclic
Spirographisporphyrin *n* spirographisporphyrin
Spiroheptan *n* spiroheptane
Spirohexan *n* spirohexane
Spiropentan *n* spiropentane
Spirosal *n* spirosal, glycol monosalicylate, glysal
Spiro-Verbindung *f* spiro-compound
spitz pointed, acute, sharp, ~ **zulaufen** to taper
Spitzbogenkaliber *n* splayed circular pass
Spitzbohrer *m* pointed drill, bit
Spitze *f* peak, crest, summit, top, (Ende) tip, (für Stäbe) apex, (Geom) vertex, (Nadel) point, **verstählte** ~ steel point, hardened point
Spitzenabstand *m* length between centers
Spitzenausleger *m* flying jib
Spitzenbelastung *f* peak load
Spitzendiode *f* point contact diode
Spitzendrehbank *f* center lathe
Spitzenelektrode *f* needle electrode, point electrode
Spitzenentladung *f* point discharge
Spitzengeschwindigkeit *f* peak velocity, top speed
Spitzenkontakt *m* point contact
Spitzenlagerung *f* point support, point suspension
Spitzenlast *f* peak load
Spitzenlehre *f* crest template, point gauge, thread template
Spitzenleistung *f* maximum capacity, maximum output, top performance
Spitzenmaterial *n* point material
Spitzenradius *m* nose radius
Spitzenspannung *f* (Elektr) peak voltage, crest voltage
Spitzenspiel *n* crest clearance
Spitzenstrom *m* peak current
Spitzentransistor *m* point contact transistor
Spitzenwert *m* maximum [value]
Spitzenwirkung *f* point effect
Spitzenzähler *m* alpha particle counter, geiger counter

Spitzgewinde n sharp V-thread
spitzig pointed, sharp
Spitzkolben m tapered flask
Spitzstampfer m pegging rammer
spitzwinklig acute-angled
Spitzzapfen m conical pivot, pointed journal
Spitzzirkel m dividers
Spitzzyklon m (Bergb) conical hydrocyclone
spleißen to splice, to cleave, to split, (Kupfer) to refine copper
Splint m cotter [pin], split pin
Splintholz n sap wood
splissen to splice
Splitter m chip, splinter, (Bruchstück) fragment
splitterfrei non-splintering, (Glas) non-shattering, shatter-proof
splittern to shatter, to splinter
splittersicher splinter-proof, (Glas) shatter-proof
Splitterung f splitting [up]
Spodiophyllit m spodiophyllite
Spodiosit m spodiosite
Spodium n bone black, bone charcoal
Spodumen m (Min) spodumene, triphane
Spongesterin n spongesterol
Spongin n spongin
Spongit m spongite
Spongosin n spongosine
Spongothymidin n spongothymidine
Spongouridin n spongouridine
spontan spontaneous
Spontanspaltung f spontaneous fission
Spontanverdampfung f flash distillation
Spontanzerfall m spontaneous decay
Spore f spore
Sporenbildung f spore formation
sporentötend sporicidal
spratzen (spratzeln) to scatter, to spatter, to spit, to splash, to spurt
Spratzen n spitting, spurting
Spratzkupfer n copper rain
Spreitbarkeit f spreadability
Spreitungskoeffizient m (Therm) spreading coefficient
Spreizdorn m expanding mandrel
spreizen to spread, to expand, to force apart
Spreizfeder f expanding spring
Spreizflügel m folding blade
Spreizgewinde n expanding screw thread
Spreizkeil m expanding wedge
Spreizring m expanding ring
Spreizwirkung f expanding action
sprengen to blast, (aufsprengen) to burst open, to force open, (besprengen) to sprinkle, to water, (spalten) to fissure
sprengfähig explosive
Sprengflüssigkeit f explosive liquid
Sprengfüllung f explosive charge
Sprenggelatine f explosive gelatin, gelatin dynamite
Sprengkapsel f detonating cap, detonator
Sprengkapselzünder m detonator
Sprengkörper m explosive charge
Sprengkohle f cracking coal
Sprengkraft f explosive power
Sprengladung f bursting charge, explosive charge
Sprengloch n blast hole
Sprengmittel n explosive, blasting agent
Sprengpatrone f explosive cartridge
Sprengpulver n explosive powder, blasting powder
Sprengstoff m explosive, dynamite
Sprengung f explosion, blasting, blowing up, bursting, (Bruch) rupture
Sprengvorgang m blasting process
Sprengwirkung f explosive action, explosive effect
Sprengzünder m detonator
Spreu f chaff
springen (bersten) to break, to burst, (einen Sprung bekommen) to crack, (schnellen) to bounce, to jump, to leap
Springen n bursting, breaking, cracking
Springkraft f elasticity, resilience
Springstift m bouncing pin (antiknock test)
Sprinkleranlage f sprinkler plant, sprinkler system
Sprit m alcohol, ethyl alcohol, spirit[s]
sprithaltig containing spirit, spirituous
Spritlack m spirit lacquer, spirit varnish
spritlöslich spirit-soluble
Spritzapparat m sprayer, spraying device, sprinkler
Spritzbarkeit f extrudability
Spritzbarmacher m extrusion aid
Spritzbeton m gunite
Spritzdruck m injection [molding] pressure, spraying pressure
Spritzdüse f injection [molding] nozzle, jet molding nozzle, spray nozzle
Spritze f [injection] syringe, (Injektion) injection
spritzen to spatter, to splash, to spray, to spurt, (injizieren) to inject, (strangpressen) to extrude
Spritzen n spattering, squirting, (Injektion) injection, (Kunststoff) injection molding, **angußloses** ~ direct-injection molding
Spritzfähigkeit f injection property
Spritzfehler m injection defect
Spritzflasche f wash[ing] bottle, squeeze bottle
Spritzform f extrusion die, injection die, jet mold, (Kunststoff) injection mold
spritzgießen to injection mold
Spritzgießen n injection molding
Spritzguß m injection [die] casting, injection molding
Spritzgußform f hot runner mold, injection mold, **angußlose** ~ runnerless injection mold

Spritzgußlegierung f die-casting alloy
Spritzgußmaschine f injection molding machine
Spritzgußmasse f injection compound
Spritzgußtechnik f injection molding practice, injection molding technique
Spritzgußteil n casting, injection molded part
Spritzgußverfahren n injection molding process
Spritzgußwerkzeug n injection mold
Spritzgußzylinder m injection molding cylinder
Spritzhärtung f hardening by sprinkling
Spritzkabine f spray booth
Spritzkolben m injection plunger, pot plunger
Spritzkopf m die head, extruder head, extruding head
Spritzlack m spray[ing] lacquer, spray-on finish
spritzlackieren to spray-lacquer
Spritzlackierung f spray lacquering
Spritzling m casting, injected article
Spritzlingsfläche f injected area
Spritzlötverfahren n jet solder method
Spritzmaschine f extruder
Spritzmasse f spraying mixture
Spritzmetallisieren n metal spraying
Spritzmittel n spray
Spritzmulde f sprue bushing
Spritzpistole f spray gun, wire gun
Spritzpresse f extrusion press, injection [molding] press
Spritzpressen n transfer molding
Spritzpreßling m molded piece
Spritzröhre f spray pipe, syringe
Spritzschmierung f splash lubrication
Spritzteller m distributing pan (cooling tower)
Spritzturm m spray tower (for crystallization)
Spritzung f [injection] shot
Spritzverfahren n jet molding, spraying technique, spray painting, (Kunststoff) injection molding
Spritzverschluß m spray closure
Spritzvolumen n injection capacity, shot capacity
Spritzvorgang m injection
Spritzwasser n spray water
Spritzwerkzeug n extrusion die
Spritzwinkel m cast angle (of manifold die)
Spritzzyklus m injection cycle, (Verarbeitungsgeschwindigkeit) injection cycle
Sprödbruch m brittle fracture
spröde brittle, friable, short
Sprödglaserz n (Min) brittle silver glance, brittle silver ore, stephanite
Sprödigkeit f brittleness, friability, shortness
Sprödigkeitspunkt m brittle point
Sproß m (Bot) shoot, sprout
Sprosse f support
Sprottenöl n sprat oil
Sprudel m soda water
Sprudelbett n aggregate fluidization

sprudeln (schäumen) to bubble
Sprudeln n bubbling
Sprudelsalz n Carlsbad salt
Sprudelstein m deposit from hot springs
Sprühapparat m atomizer, sprayer, (für Absorption) scrubber
Sprühdose f spray can, aerosol can
Sprühelektrode f corona discharge electrode
sprühen to spray, to drizzle, to sprinkle
Sprühentladung f brush discharge, corona [discharge], spray discharge
Sprühkappe f (Aerosoldose) actuator cap
Sprühkautschuk m spray dried rubber
Sprühkolonne f spray column, spray tower
Sprühkopf m (Aerosoldose) actuator
Sprühkopfzuführung f (Aerosoldose) actuator feeder
Sprühkristallisator m jet crystallizer
Sprühnebel m drizzling fog, wet fog
Sprühpistole f atomizer, spray gun
Sprühregen m drizzle
Sprühstelle f spray point
Sprühtrockner m spray dryer
Sprühtrocknung f spray drying
Sprühturm m (Verfahrenstech) spray tower
Sprühverfahren n (Färberei) spraying process
Sprühwäscher m spray washer
Sprung m (Riß, Spalt) crack, fissure, rent, (Schnellen) bounce, jump[ing], (Unstetigkeit) discontinuity
Sprungbildung f jog formation
Sprungeingang m step input
Sprungelastizität f resilience
Sprungerscheinung f skip phenomenon
Sprungfeder f elastic spring
Sprungfrequenz f transition frequency
Sprunghärte f rebound hardness
sprunghaft discontinuous, nonsequential
Sprunghaftigkeit f inconsistency, violent fluctuation
Sprungmechanismus m jump mechanism
Sprungwahrscheinlichkeit f transition probability
Sprungweite f range
Spülanlage f washing plant
Spülbohrschlauch m rotary hose
spülen to rinse, to clean, to flush, to scavenge, to wash
Spülen n rinsing, flushing
Spülflüssigkeit f rinsing liquid
Spülgas n cleansing gas
Spülgasschwelofen m (Lurgi) gas-recycle coking oven
Spülhalle f rinsing room
Spülkasten m flush box
Spülluft f purging air
Spülmaschine f dish-washer, rinsing machine
Spülmittel n rinsing agent
Spülraum m washing room
Spültisch m scullery, sink

Spülung *f* rinsing, cleaning, cleansing, flushing, scavenging
Spülungseinfluß *m* washout
Spülventil *n* flushing gate, scavenging valve
Spülwasser *n* flushing water, rinsing water, wash water
Spülzeit *f* purge period, scavenger time
Spürgerät *n* detector
Spule *f* spool, reel, (Kern zum Aufwickeln) bobbin, (Wicklung) coil, **eisenlose** ~ air-core coil
spulen to reel, to spool, to wind
Spulenachse *f* axis of coil
Spulenfeld *n* coil field
Spulengalvanometer *n* coil galvanometer
Spulengatter *n* creel
Spulenhalter *m* coil carrier
Spulenkern *m* bobbin, core of a coil, ~ **für Magnetband** hub for magnetic tape
Spulenlänge *f* bobbin height
Spulenspinnverfahren *n* bobbin spinning
Spulenstromversorgung *f* coil current supply
Spulenweite *f*, **mittlere** ~ bobbin overall diameter
Spulöl *n* spooling oil
Spund *m* bung, (Gieß) shutter
Spur *f* (Bahn) trace, track, trail, (kl. Menge) trace
Spurarbeit *f* trace work
Spureigenschaft *f* track property
Spurenchemie *f* trace[r] chemistry
Spurenelement *n* trace[r] element
Spurenfinder *m* tracer
Spurennachweis *m* trace detection
Spurensucher *m* tracer
Spurenuntersuchung *f* tracer study
spurenweise in traces
Spurerweiterung *f* amplification of gauge
Spurlager *n* step bearing, thrust bearing
Spurprüfgerät *n* wheel alignment tester
Spurprüfung *f* alignment test
Spurrit *m* (Min) spurrite
Spurrolle *f* track wheel, troughing idler
Spurschlacke *f* concentration slag
Spurstange *f* [steering axle] tie rod, track rod
Spurstein *m* concentrated matte
Spurverzerrung *f* tracking error
Spurzapfen *m* king pin, pivot [journal], swivel pin
Squalan *n* squalane
Squalen *n* squalene, spinacene
Squamatsäure *f* squamatic acid
S-Säure *f* (Färb) S-acid
Stab *m* bar, rod, (Mitarbeiter) staff, **unbearbeiteter** ~ unwrought bar
Stabantenne *f* flag pole antenna
Stabdehnung *f* bar expansion
Stabeisen *n* bar iron, round iron
Stabelektrode *f* rod electrode

Stabfederindikator *m* beam spring indicator
Stabform *f* rod shape
stabil steady, (Chem) stable, (fest) firm, rigid, sturdy, (haltbar) durable
Stabilisation *f* stabilization
Stabilisationsröhre *f* stabilizer tube
Stabilisator *m* stabilizer
stabilisieren to stabilize, to regularize
Stabilisiermittel *n* stabilizer
Stabilisierung *f* stabilizing
Stabilisierungsbad *n* stabilizing bath
Stabilisierungswiderstand *m* stabilization resistance
Stabilit *n* stabilite
Stabilität *f* stability, solidity
Stabilitätsbedingung *f* stability condition
Stabilitätsgrad *m* degree of stability
Stabilitätsgrenze *f* limit of stability
Stabilitätskonstante *f* stability constant
Stabilitätslinie *f* stability curve
Stabilitätsprüfung *f* stability check
Stabilitätsregel *f* stability rule
Stabisolator *m* rod insulator
Stabkern *m* center of bar
Stabmagnet *m* bar magnet
Stabmaterial *n* rod stock
Stabmühle *f* rod mill
Stabquerschnitt *m* cross section of a bar
Stabthermometer *n* solid-stem thermometer, thermometer graduated on the stem
Stabthermostat *m* rod thermostat
Stabwalzwerk *n* bar rolling mill
Stabzelle *f* (Histol) rod cell, staff cell
Stabziehen *n* drawing of rod
Stacheldraht *m* barbed wire
Stachellänge *f* spike length
Stachelwalze *f* toothed roll, rough roll
Stachydrin *n* stachydrine
Stachyose *f* stachyose
Stadelröstung *f* stall roasting
Stadium *n* stage, state
Stadtgas *n* town gas
Stäbchen *n* rod
Stäbchenbakterie *f* (Bakt) rod-shaped bacterium
stäbchenförmig rod-shaped
Stäbchenzelle *f* (Histol) rod cell, staff cell
stählen to steel, to harden
Stählung *f* conversion into steel, steeling hardening
Ständer *m* stand, holder, pedestal
Ständerklemme *f* (Chem) clamp for stand
Ständersäule *f* pillar, post
Ständerstrom *m* stator current
ständig constant, continuous; permanent, steady
Stärke *f* (Dicke) size, thickness, (Intensität) intensity, (Kohlenhydrat) starch, (Kraft) force, power, strength, ~ **einer Linse** (Opt) lens power
Stärkeabbau *m* starch degradation

Stärkeanlage *f* starch plant
Stärkeart *f* variety of starch
stärkeartig amylaceous, amyloid
Stärkebildner *m* amyloplast
Stärkeblau *n* starch blue
Stärkeeinheit *f* starch equivalent
Stärkefermentation *f* amylo-fermentation, amylolytic fermentation
stärkeführend amylaceous, containing starch, starchy
Stärkegehalt *m* starch content
Stärkegel *n* starch gel
Stärkegel-Elektrophorese *f* starch gel electrophoresis
Stärkeglanz *m* starch gloss
Stärkegrad *m* degree of strength, intensity
Stärkegummi *n* starch gum, dextrin
stärkehaltig amylaceous, containing starch, starchy
Stärkekleister *m* starch paste
Stärkekorn *n* starch granule
Stärkelösung *f* starch solution
Stärkemehl *n* starch flour, amylum, powdered starch, starch powder
stärkemehlhaltig amylaceous, containing starch, starchy
Stärkemesser *m* amylometer
stärken to strengthen, (bestärken) to confirm, (mit Stärke) to starch
Stärken *n* strengthening, (mit Stärke) starching
Stärkepapier *n* starch paper
Stärkepulver *n* powdered starch, starch powder
Stärkereaktion *f* starch test
Stärkesirup *m* starch syrup
Stärkespaltung *f* amylolysis
Stärkeüberschuß *m* excess of starch
Stärkewasser *n* starch water
Stärkezucker *m* starch sugar, starch syrup
Staffelit *m* staffelite
Staffelkessel *m* multiple stage boiler
staffeln to arrange in lines, to grade, (Arbeitszeit etc.) to stagger
Staffelrost *m* multistage grate
Staffelwalze *f* stepped roll
stagnieren to stagnate
stagnierend stagnant
Stahl *m* steel, ~ **brennen** to temper steel, ~ **dekarbonisieren** to soften steel, ~ **garmachen** to refine steel, ~ **gerben** to refine steel, ~ **mit hohem oder niedrigem Kohlenstoffgehalt** high- or low-carbon steel, ~ **plätten** to draw out the steel, **abgeschreckter und angelassener** ~ quenched and drawn steel, **austenitischer** ~ austenitic steel, **beruhigter** ~ killed steel, **beunruhigter** ~ wild steel, **blasiger** ~ blister steel, **gehärteter** ~ hardened steel, heat-treated steel, **halbberuhigter** ~ semi-killed steel, **hitzbeständiger** ~ heat-resistant steel, **korrosionsbeständiger** ~ stainless steel, **legierter** ~ alloy steel, **nichtrostender** ~ rustproof steel, stainless steel, **niedriggekohlter** ~ dead soft steel, low-carbon steel, **niedriglegierter** ~ alloy-treated steel, low-alloy steel, **rostbeständiger** ~ noncorroding steel, rustless steel, stainless steel, **rostfreier** ~ stainless steel, **schweißbarer** ~ weldable steel, **übereutektoider** ~ hypereutectoid steel, **übergarer** ~ burnt steel, overblown steel, **überhitzter** ~ overheated steel, **unlegierter** ~ carbon steel, plain steel, simple steel, **unmittelbar vergießbarer** ~ steel ready for casting, **unruhiger** ~ effervescent steel, running steel, **vergüteter** ~ annealed steel, heat-treated steel, **warmfester** ~ steel for high temperature, steel with a high creep limit, steel with good high-temperature characteristics, **weicher** ~ mild steel, soft steel
stahlähnlich steel-like
Stahlanode *f* steel anode
Stahlbereitung *f* (Metall) steelmaking
Stahlbereitungsprozeß *m* steel process
Stahlbeton *m* steel concrete
Stahlblau *n* steel blue
Stahlblech *n* sheet steel, steel plate, steel sheet
Stahlblock *m* steel ingot, bloom
Stahlbrenner *m* multiple-intertube burner
Stahlbronze *f* steel bronze
Stahlcordreifen *m* wire cord tire
Stahldeul *m* puddled steel ball, lump of puddled steel
Stahldraht *m* steel wire
Stahldrehspäne *m pl* steel turnings
Stahldübel *m* steel dowel
Stahleisen *n* steel pig, open hearth pig iron
Stahlerz *n* steel ore, (Min) siderite
Stahlerzeugung *f* steel manufacture, steel production
Stahlflasche *f* steel bottle, steel cylinder
Stahlformguß *m* steel casting
Stahlgewinnung *f* steel production
Stahlgießerei *f* steel foundry
stahlgrau steel gray
Stahlguß *m* steel cast[ing], cast steel
Stahlgußarmaturen *f pl* cast steel mountings
Stahlgußgehäuse *n* cast steel casing
Stahlhärte *f* hardness of steel
Stahlhärtung *f* hardening of steel, tempering of steel
stahlhart hard as steel
Stahlhochbauten *m pl* steel superstructures
Stahlkratzbürste *f* steel wire brush
Stahlkugel *f* steel ball
Stahllegierung *f* steel alloy
Stahlmagnet *m* steel magnet
Stahlmesserchen *n* steel scalpel
Stahlmörser *m* steel mortar
Stahlofen *m* steel furnace

Stahlprobe *f* steel sample
Stahlpuddeln *n* puddling of steel, steel puddling
Stahlpulver *n* iron powder
Stahlrahmen *m* steel frame
Stahlreif[en] *m* steel hoop
Stahlroheisen *n* open hearth pig iron
Stahlrohrgestell *n* tubular steel frame
Stahlschablone *f* steel rule
Stahlscheibe *f* steel disk
Stahlschiene *f* steel rail
Stahlschneide *f* steel edge
Stahlschrott *m* steel scrap
Stahlseilgurtband *n* wire-reinforced belt
Stahlstein *m* (Min) chalybeate tartar, sparry iron
Stahlstößel *m* steel nose of breaker
Stahlstreifen *m* steel strip
Stahlträger *m* steel beam
Stahlwasser *n* chalybeate water
Stahlwein *m* iron wine
Stahlwerkskokille *f* ingot mold
Stahlwolle *f* steel wool
Stahlzylinder *m* steel cylinder
Stalagmit *m* stalagmite
stalagmitartig stalagmitical
Stalagmometer *n* stalagmometer
Stalagmometrie *f* stalagmometry
Stalaktit *m* stalactite
stalaktitisch stalactitic
Stamm *m* stem, (Bakt) strain, (Baum) trunk
Stammflotte *f* stock liquor
Stammgesenk *n* fixed plate
Stammkörper *m* parent substance
Stammküpe *f* stock vat
Stammlösung *f* stock solution
Stammplatte *f* fixed plate
Stammsäure *f* parent acid
Stammsatz *m* (Informatik) master record
Stammschmelze *f* concentrated compound
Stammsubstanz *f* parent compound, parent substance
Stammverbindung *f* parent compound
Stammwürze *f* (Brau) original wort
Stammzelle *f* mother cell, parent cell
Stampfasphalt *m* tamped asphalt
Stampfbeton *m* stamped concrete, tamped concrete
Stampfer *m* rammer, stamper, tamper
Stampfform *f* stamping form
Stampfgewicht *n* tamped weight
Stampfmaschine *f* stamping machine
Stampfmasse *f* tamping compound
Stampfvolumen *n* bulking volume, ramming volume
Stampfvolumenmesser *m* tamped volume measuring appliance
Stand *m* (Ablesestellung) reading, (Niveau) level, (Standort) position, stand, (Stellung) position, ~ **des Thermometers** thermometer reading

Standanzeiger *m* [liquid] level indicator
Standard *m* standard
Standardabweichung *f* (Statistik) standard deviation
Standardausführung *f* standard design, conventional design
Standardeigenschaft *f* standard property
Standardeinstellung *f* standard adjustment, standard setting
Standardelektrode *f* standard electrode
Standardelement *n* (Elektr) standard cell
Standardenthalpie *f* standard enthalpy
Standardentropie *f* standard entropy
Standardfehler *m* standard error
Standardfehlerberechnung *f* calculation of standard errors
Standardisierung *f* standardization
Standardkalomelelektrode *f* normal calomel electrode
Standardkapazität *f* standard capacity
Standardlichtquelle *f* (Opt) standard light source
Standardlösung *f* standard solution
Standardmodell *n* standard type
Standardnormalverteilung *f* (Statistik) standard normal distribution
Standardpotential *n* (Elektrochem) standard potential
Standardpräparat *n* (Spektr) internal standard
Standardreaktionsarbeit *f* standard work of reaction
Standardreaktionsenthalpie *f* standard enthalpy of reaction
Standardwasserstoffelektrode *f* standard hydrogen electrode
standfest stable
Standfestigkeit *f* resistance to continuous stress
Standfestigkeitsapparat *m* stability apparatus
Standflasche *f* flask with flat bottom, flat bottomed flask
Standgefäß *n* stock vessel, storage vessel, (Chem) jar
Standhöhe *f* liquid level
Standmesser *m* [liquid] level indicator
Standöl *n* bodied oil, stand oil
Standölkochanlage *f* stand oil boiling plant
Standortfrage *f* plant location study
standortsgebunden stationary
Standpfanne *f* fixed ladle
Standpunkt *m* point of view, standpoint
Standrohr *n* down comer, stand pipe, vertical pipe
Standzeit *f* (Chem) shelf life, (Konserven) can stability
Stange *f* bar, rod bar, (Kosmetik) stick, (Met) ingot, (Schwefel) roll
Stangenabstich *m* tapping with a crowbar
Stangeneisen *n* bar iron, round iron
Stangenkali *n* stick potash
Stangenkitt *m* stick cement

Stangenkohle *f* columnar coal
Stangenschwefel *m* roll sulfur, stick sulfur
Stangenspat *m* (Min) columnar barite
Stangensteuerung *f* rod control
Stangentriebwerk *n* rack gear, rack and pinion
Stangenverbindungsklemme *f* rod connecting clamp
Stangenvorschub *m* bar feed
Stangenziehbank *f* rod drawing bench
Stangenzinn *n* bar tin
Stangenzug *m* drawing bars
Stannat *n* stannate, salt or ester of stannic acid
Stannibromid *n* (Zinn(IV)-bromid) stannic bromide, tin(IV) bromide, tin tetrabromide
Stannichlorid *n* stannic chloride, tin(IV) chloride, tin tetrachloride
Stannichlorwasserstoffsäure *f* chlorostannic acid
Stannichromat *n* stannic chromate, tin(IV) chromate
Stannihydroxid *n* stannic hydroxide, tin(IV) hydroxide
Stannijodid *n* stannic iodide, tin tetraiodide
Stannin *m* (Min) stannine, stannite, tin pyrites
Stanniol *n* tin foil
Stanniollack *m* tinfoil lacquer
Stanniolpapier *n* tin foil
Stannioxid *n* (Zinn(IV)-oxid) stannic oxide, tin dioxide, tin(IV) oxide
Stannisalz *n* (Zinn(IV)-salz) stannic salt, tin(IV)salt
Stannisulfid *n* stannic sulfide, tin disulfide, tin(IV) sulfide
Stannit *m* stannine, stannite, tin pyrites
Stannoacetat *n* (Zinn(II)-acetat) stannous acetate, tin(II) acetate
Stannochlorwasserstoffsäure *f* chlorostannous acid
Stannochromat *n* stannous chromate, tin(II) chromate
Stannohydroxid *n* stannous hydroxide, tin(II) hydroxide
Stannojodid *n* stannous iodide, tin diiodide, tin(II) iodide
Stannojodwasserstoffsäure *f* iodostannous acid
Stannooxalat *n* stannous oxalate, tin(II) oxalate
Stannooxid *n* stannous oxide, tin(II) oxide, tin monoxide
Stannosalz *n* (Zinn(II)-Salz) stannous salt, tin(II) salt
Stannosulfat *n* stannous sulfate, tin(II) sulfate
Stannosulfid *n* stannous sulfide, tin(II) sulfide, tin monosulfide
Stannotartrat *n* stannous tartrate, tin(II) tartrate
Stannum *n* (Lat) tin
Stanzabfälle *pl* punchings, stampings, waste from stamping
Stanzait *m* stanzaite
Stanzautomat *m*, **vollautomatischer** ~ automatic die cutter

Stanze *f* punch [press], die-cutter, punching machine, stamp
stanzen (lochstanzen) to punch, (löchern) to perforate, (prägen) to stamp
Stanzen *n* punching, blanking, die cutting, pressing, stamping
Stanzerlager *n* punch bar bearing
Stanzfähigkeit *f* punching quality
Stanzform *f* blanking die, chop-out die, drop hammer die, knock-out die, matrix mold, steel rule die
Stanzlack *m* stamping enamel
Stanzlinie *f* punch line
Stanzmaschine *f* punching machine, stamping-out machine
Stanzmesser *n* cutting die
Stanzplatte *f* punching support
Stanzporzellan *n* porcelain for punching
Stanzpresse *f* clicking press, cutting-out press
Stanzteil *n* stamped part
Stanzvorrichtung *f* stamping device
Stanzwerkzeug *n* blanking tool, stamping tool
Stapel *m* pile, stack
Stapelfaser *f* staple fiber
Stapelfehler *m* (Krist) stacking disorder, stacking fault
Stapelfehlerbildung *f* stacking fault formation
Stapelfehlerenergie *f* stacking fault energy
Stapelförderer *m* staple conveyor
Stapelgestell *n* stacking rack
Stapellänge *f* staple length
Stapellift *m* stacking lift
stapeln to stack, to accumulate, to store
Stapeloperator *m* stacking operator
Stapelpresse *f* multi-daylight press
Stapelstrecke *f* marshalling section
Stapelung *f* piling, stacking
Staphisagrin *n* staphisagrine, staphisaine
Staphisagroin *n* staphisagroine
Staphylokokkeninfektion *f* (Med) staphylococcal infection
Staphylokokkenstamm *m* staphylococcal strain
stark (intensiv) intensive, (kräftig) powerful, strong, ~ **durchsetzt** having a high content [of], ~ **oxydierend** strongly oxidizing
Starkbier *n* (dunkles) stout beer
Starkfeld-Trommel-[Rillen-]Scheider *m* high-intensity magnetic groove drum separator
Starkisolator *m* (Elektr) insulator for power current
Starkstrom *m* high voltage current, heavy current, power current, strong current
Starkstromglocke *f* highvoltage bell
Starkstromkabel *n* power cable
Starkstromleitung *f* power circuit, power line
Starkstromnetz *n* (Elektr) power system
Starkstromtechnik *f* power engineering, heavy current engineering

Starkvernickelung – Staudüse 702

Starkvernick[e]lung *f* heavy nickelling
Starkversilberung *f* heavy silverplating
starkwandig thick-walled, stout-walled
starkwirkend drastic, efficacious, highly active, powerful
starr rigid, inflexible, (Körper) solid
Starre *f* inflexibility, rigidity, stiffness
Starrheit *f* rigidity, stiffness
Starrkörperverschiebung *f* rigid body displacement
Starrkoppeln *n* rigid coupling
Starrkrampf *m* (Med) tetanus
Starrschmiere *f* solid lubricant
Startbahn *f* (Luftf) runway
Startreagens *n* (Kettenreaktion) trigger
Startreaktion *f* initiating reaction
Startstellung *f* starting position
Startventil *n* starting valve
Staßfurtit *m* (Min) stassfurtite
Statcoulomb *n* statcoulomb
Statik *f* statics
stationär fixed, static, stationary, steady
Stationsspannung *f* station voltage
statisch static, ~ bestimmbar statically determinable
Statistik *f* statistics
Statistiker *m* statistician
statistisch statistical, ~ verteilt randomly distributed
Stativ *n* stand, support
Stativfuß *m* stand base
Stativreiter *m* sliding support
Stativstange *f* stand rod
Stator *m* (Elektr) stator
Statorstrom *m* stator current
Status nascendi (Lat) nascent state, status nascendi
Statvolt *n* statvolt
Stau *m* dynamic air pressure
Staub *m* dust, (Blütenstaub) pollen, (Pulver) powder
Staubabdichtung *f* dust stop
Staubablagerung *f* dust deposit
Staubabscheider *m* dust catcher, dust collector, dust separator
Staubabscheideverfahren *n* dust-separating method
Staubabscheidung *f* dust collection, dust separation
Staubalken *m* choker bar (for regulating flow of material in die of extruder for sheets)
staubartig dust-like, dusty, (pulverig) powdery, pulverulent
Staubauswurf *m* dust emission
Staubbelästigung *f* dust nuisance
Staubbild *n*, elektrisches ~ electrical dust figure, Lichtenberg figure
Staubbildung *f* dust formation, dusting
Staubbrenner *m* pulverized coal burner

staubdicht dustproof, dust-tight
Staubdichtung *f* dustproof packing
Staubentferner *m* dust remover
Staubentwicklung *f* formation of dust
Staubexplosion *f* dust explosion
Staubfänger *m* dust catcher, dust collector
Staubfang *m* dust collector, dust box, dust catcher, dust chamber
Staubfeuerung *f* pulverized coal firing
Staubfilter *n* dust filter
Staubfließverfahren *n* fluidization process
Staubförderanlage *f* dust conveyor
staubförmig dusty, (pulverig) pulverized
staubfrei dust-free
Staubgefäß *n* (Bot) stamen
Staubgehalt *m* (der Luft) dust content of the air
Staubgemisch *n* dust mixture
Staubkammer *f* dust box, dust chamber, dust settling chamber, gravity settling chamber
Staubkern *m* dust core
Staubkohle *f* coal dust, culm, dust coal, powdered coal
Staubkorn *n* coarse dust particle
Staublech *n* baffle [plate], baffle member
Staubmaske *f* respirator
Staubmeßgerät *n* dust measuring instrument, dust meter
Staubröstofen *m* dust roaster
Staubsack *m* dust bin, dust pocket
Staubsammler *m* dust collector
Staubsauger *m* vacuum cleaner
Staubschwelung *f* low temperature carbonization of fines
staubsicher dustproof
Staubteilchen *n* dust particle
staubtrocken dry as dust, bone dry, dust dry
Staubverhütung *f* dust prevention
Staubzucker *m* (Puderzucker) confectioner's sugar, sugar powder
stauchen to compress, to shorten by forging, to upset
Stauchen *n* compression, ~ in Längsrichtung end-to-end compression, ~ in Querrichtung side-to-side compression, vertikales ~ top-to-bottom compression
Stauchkaliber *n* up-set pass
Stauchmaschine *f* upsetting machine
Stauchprobe *f* bulging test, shock crushing test, upsetting test
Stauchrichtung *f* compression direction
Stauchtextur *f* upsetting texture
Stauchung *f* compression; shortening, upsetting
Stauchungsübergang *m* upset
Stauchversuch *m* bulging test, compression test, crushing test, upsetting test
Stauchwert *m* crusher index
Stauchwirkung *f* crushing effect, upsetting effect
Staudruck *m* back pressure, dynamic pressure
Staudüse *f* Pitot nozzle

stauen to stow, (Wasser) to dam
Stauen *n* stowage, stowing
Staufferbüchse *f* grease cup
Staufferfett *n* cup grease
Stauflansch *m* reducing flange, throttle
Stauförderer *m* accumulation conveyor, power-and-free overhead conveyor, powerized storage line
Staugrenze *f* (Dest) loading point
Stauhöhe *f* height of damming
Staukopf *m* accumulator, constricting head, **Blasanlage mit ~** blow molding unit with plunger-type extruder
Staumauer *f* dam
Staupunkt *m* stagnation point, flooding point, (Dest) upper load limit, **~ der Flüssigkeitsströmung** fluid flow stagnation point
Staurand *m* [calibrated] orifice, throttle diaphragm
Stauröhre *f* Pitot tube
Staurolith *m* (Min) staurolite
Stauscheibe *f* baffle plate, breaker plate, die restriction
Stausee *m* artificial lake, storage lake
Stauung *f* accumulation; damming, stowage
Stauventil *n* stop valve
Stearaldehyd *m* stearaldehyde
Stearamid *n* stearamide
Stearat *n* stearate, salt or ester of stearic acid
Stearin *n* stearin, glycerol tristearate, tristearin
Stearinform *f* stearin mold
Stearingoudron *m* stearin pitch
Stearinkerze *f* stearin candle
Stearinkuchen *m* stearin cake
Stearinpech *n* stearin pitch
Stearinsäure *f* stearic acid, octodecanoic acid
stearinsauer (Salz) stearate
Stearinseife *f* stearin soap, common soap
Stearinteer *m* stearin pitch
Stearodibutyrin *n* stearodibutyrin, glycerol dibutyrate stearate
Stearodichlorhydrin *n* stearodichlorohydrin, glycerol dichlorohydrin stearate
Stearodipalmitin *n* stearodipalmitin, glycerol dipalmitate stearate
Stearolsäure *f* stearolic acid, 9-octadecynoic acid
Stearon *n* 18-pentatriacontanone, stearone
Stearopten *n* stearoptene, oleoptene
Stearoxylsäure *f* stearoxylic acid, 9,10-dioxooctadecanoic acid
Stearyl- stearyl, octadecanoyl
Stearylalanin *n* stearylalanine
Stearylglycin *n* stearylglycine, stearylaminoethanoic acid
Steatit *m* (Min) steatite
Steatitmasse *f* steatite compound
Stechapfel *m* (Bot) thornapple

Stechapfelextrakt *m* extract of stramonium
Stechapfelsamen *m* stramonium seed
stechen to pierce, to hole, to stab, (Insekt) to sting
stechend piercing, (Geruch) penetrating, pungent
Stechheber *m* pipette, [plunging] siphon, siphon tube
Stechmücke *f* mosquito
Stechpalme *f* (Bot) holly
Stechtorf *m* cut peat
Steckdose *f* (Elektr) [plug] socket, plug box, wall plug
Stecker *m* (Elektr) plug
Steckeranschluß *m* plug connection
Steckerbuchse *f* plug socket
Steckfassung *f* plug-in socket
Steckhülse *f* socket
Steckkontakt *m* plug box, (in der Wand) wall socket
Stecknadel *f* pin
Stecknadelprogrammierung *f* (Comp) pinboard programming
Steckschlüssel *m* socket wrench
Steg *m* (Spritzguß) legs of spider (in blow-molding head), runner, (Techn) inner bar, support
Stegkette *f* studlink chain
Stegverpackung *f* pack subdivided into compartments
Stehbock *m* bearing block, floor frame, floor stand
Stehbolzen *m* dowel pin, stay bolt
Stehbütte *f* stock tub
stehen to stand, **~ lassen** to allow to stand, to let stand, **senkrecht ~** to be perpendicular to
stehend standing, (aufrecht) upright, (stagnierend) stagnant, (stationär) stationary
Stehknecht *m* standing vise
Stehkolben *m* (Chem) flat bottom[ed] flask
Stehlager *n* bracket, pedestal bearing, pillow block (Am. E.), vertical bearing
steif stiff, inflexible, rigid
steifen to stiffen, to toughen
Steifheit *f* stiffness, rigidity
Steifheitsgrad *m* degree of stiffness
Steifigkeit *f* stiffness, rigidity
Steifigkeitsmodul *m* stiffness modulus
Steifleinen *n* backing
Steifmittel *n* stiffening agent
Steigband *n* inclined belt
steigen to ascend, to mount, to rise, (zunehmen) to increase
steigend gießen (Gieß) to bottom-cast, to bottom-pour
Steiger *m* (Gieß) riser
Steigeröhre *f* rising main
Steigerung *f* increase, raising, rise
Steiggeschwindigkeit *f* (v. Gasblasen) rising velocity

Steighöhe *f* height [of rise], elevation, **kapillare** ~ capillary rise
Steighöhenmaßstab *m* vertical range scale
Steighöhenmethode *f* (zur Teilchengrößenbestimmung) sedimentation-equilibrium method (for the determination of particle size)
Steigkanal *m* uptake
Steigleiter *f* ladder
Steigleitung *f* ascending pipe
Steigrohr *n* ascending tube, delivery pipe, elevating tube, feedpipe, riser, rising pipe, standpipe, (Aerosoldose) dip tube
Steigrohrleitung *f* ascending pipe line
Steigrohrsichter *m* gravity classifier
Steigung *f* ascent, rise, (Gefälle) grade, gradient, (Neigung) incline, slope, ~ **der Schraube** pitch of a screw, **gleichbleibende** ~ constant pitch
Steigungsfehler *m* pitch error
Steigungswinkel *m* angle of inclination, helix angle
Steigwind *m* ascending current
Steigwinkel *m* angle of elevation
Steilfördergurt *m* high-angle conveyor belt
Steilheit *f* steepness
Steilrohrkessel *m* vertical tube boiler
Steilrohrverdampfer *m* short evaporator coil
Stein *m* stone, brick, (Fels) rock, **feuerfester** ~ refractory brick
Steinalaun *m* rock alum, aluminite
steinartig stone-like, stony
Steinasche *f* stone ashes
Steinbrecher *m* stone breaker
Steinbrennofen *m* kiln, oven
Steinbrucherzeugnis *n* quarry product
Steinbühlergelb *n* barium chromate, barium yellow
Steinbutter *f* (Min) rock butter, native alum
Steindruck *m* lithograph, lithographic print, (Verfahren) lithography
Steindruckkalkstein *m* lithographic limestone
Steinfarbe *f* stone color
Steingrau *n* gray pigment from clay slate, stone gray
Steingut *n* crockery, earthenware, pottery, stoneware, **braunes** ~ brown ware, **feines** ~ opaque porcelain, stone china, wedgewood
Steingutgeschirr *n* pottery
Steingutwanne *f* earthenware tank
Steinhärteanlage *f* brick hardening plant
steinhart hard as stone
Steinheilit *m* (Min) steinheilite
Steinholz *n* woodstone
Steinkitt *m* cement for stone, lithocolla, mastic cement
Steinklee *m* (Bot) melilot, white clover
Steinkohle *f* anthracite, hard coal, mineral coal, pit coal

Steinkohlenbenzin *n* coal gasoline
Steinkohlenbergwerk *n* coal mine
Steinkohlenbrikett *n* coal briquet[te]
Steinkohlendestillation *f* distillation of bituminous coal, carbonization of bituminous coal
Steinkohlengas *n* coal gas
Steinkohlengaserzeugung *f* coal-gas manufacture
Steinkohlenleuchtgas *n* coal gas, illuminating gas from coal
Steinkohlenmehl *n* fine coal dust
Steinkohlenöl *n* coal tar oil
Steinkohlenschiefer *m* coal bearing shale
Steinkohlenschlacke *f* cinders
Steinkohlenschwelteer *m* low-temperature coal tar
Steinkohlenschwelung *f* coal carbonization
Steinkohlenteer *m* bituminous coal tar
Steinkohlenteeröl *n* coal tar oil, coal tar naphtha
Steinkohlenteerpech *n* coal tar pitch
Steinkohlenverkokung *f* coking [of coal]
Steinkohlenvorkommen *n* bituminous coal deposit
Steinleim *m* cement for stone, lithocolla
Steinmannit *m* (Min) steinmannite
Steinmark *n* (Min) lithomarge, marrow stone
Steinmehl *n* stone powder
Steinmergel *m* stone marl
Steinmörtel *m* cement, hard mortar
Steinöl *n* petroleum, rock oil
Steinpech *n* stone pitch, compact native bitumen, hard asphalt, hard pitch
Steinpfeiler *m* stone pillar
Steinporzellan *n* hard porcelain
Steinring *m* stone ring
Steinsalz *n* rocksalt
Steinschraube *f* stone bolt
Steinsockel *m* stone wall, brick wall
Steinwolle *f* rock wool
Steinzement *m* concrete
Steinzeug *n* pottery, stoneware
Steinzeuggefäß *n* stoneware vessel
Steinzeugrohr *n* earthenware pipe, vitrified clay pipe
Stellantrieb *m* adjusting drive (for control pump)
Stellbereich *m* (Regeltechn) operating range of the final control element
Stellbuchse *f* adjusting sleeve
Stelle *f* spot, place, point, (bei Zahlen) digit, **matte** ~ dead spot, dull spot
Stellenverschiebung *f* (Comp) arithmetic shift
Stellenwert *m* (Math) local value, place value
Stellerit *m* (Min) stellerite
Stellgeschwindigkeit *f* floating speed
Stellglied *n* final control device
Stellgröße *f* controller output, manipulated variable
Stellhebel *m* adjusting lever

Stellit *m* (Legierung) stellite
Stellkeil *m* adjusting key, adjusting wedge, wedge bolt
Stellmittel *n* (Email) suspension agent
Stellmotor *m* positioner, power unit motor operator (Am. E.), servomotor
Stellmutter *f* adjusting nut, lock nut
Stellöl *n* flux oil
Stellorgan *n* (Regeltechnik) final control element, positioner, regulating unit
Stellring *m* adjusting collar, tension ring
Stellschlitten *m* adjusting slide block
Stellschlüssel *m* monkey wrench
Stellschraube *f* adjustable screw, adjusting screw, set screw, (Begrenzungsschraube) check screw, (Nivellierschraube) levelling screw
Stellstift *m* (Mech) adjusting pin, capstan spike, tommy
Stellstück *n* adjusting piece
Stellung *f* position, ~ **auf Deckung** (Stereochem) eclipsed conformation, ~ **auf Lücke** (Stereochem) staggered conformation, **liegende** ~ horizontal position
Stellungnahme *f* comment[s]
stellungsisomer place isomer, position isomer
Stellungsisomer *n* (Chem) positional isomer
Stellungsisomerie *f* place isomerism, position isomerism
Stellvorrichtung *f* adjusting device, positioner, switch lever
Stellwerk *n* positioner
Stellwerksanlage *f* switch post installation
Stellwinkel *m* bevel protractor
Stellzeit *f* time required for adjustment
Stelznerit *m* (Min) stelznerite
Stemmaschine *f* mortising machine
Stemmeisen *n* crowbar, mortise chisel
stemmen to ca[u]lk
Stemmen *n* fullering
Stemmfuge *f* calked joint
Stemmhammer *m* ca[u]lking hammer
Stemmkante *f* peening ridge
Stemmplatte *f* force plate, plunger retainer
Stemmsetze *f* calking tool
Stempel *m* stamp, die, force plug, plunger, punch, ram, ~ **eines Preßwerkzeuges** hob, **beweglicher** ~ floating punch, **loser** ~ loose punch, **pendelnder** ~ floating punch
Stempeldruck *m* ram pressure
Stempelfarbe *f* stamping ink
Stempelkissen *n* ink pad
stempeln to stamp, to mark
Stempelplatte *f* die plate
stengelig stalked
Stephanin *n* stephanine
Stephanit *m* (Min) stephanite
Steran *n* sterane
Steranthren *n* steranthrene
Sterblichkeit *f* lethality, mortality

Stercobilin *n* stercobilin
Stercorit *m* stercorite
Sterculiaöl *n* sterculia oil
Sterculiasäure *f* sterculic acid
Sterculsäure *f* sterculic acid
Stereochemie *f* stereochemistry
stereochemisch stereochemical
Stereogrammbeschreibung *f* stereogram description
stereographisch stereographic
Stereoid *n* stereoid
stereoisomer stereo-isomeric
Stereoisomer *n* stereoisomer
Stereoisomerase *f* (Biochem) stereoisomerase
Stereoisomerie *f* stereoisomerism, geometrical isomerism
Stereometer *n* stereometer
Stereometrie *f* stereometry, solid geometry
stereometrisch stereometric
Stereoselektivität *f* stereoselectivity
Stereoskop *n* stereoscope
stereoskopisch stereoscopic
Stereospektrogramm *n* (Spektr) stereospectrogram
Stereospezifität *f* stereospecificity
steril sterile, sterilized, (aseptisch) aseptic
Sterilfiltration *f* sterile filtration
Sterilisation *f* sterilization
Sterilisationsmittel *n* sterilizing agent
Sterilisator *m* sterilizer
sterilisieren to sterilize
Sterilität *f* sterility
Sterin *n* sterol
sterisch steric
Sternatmosphäre *f* stellar atmosphere
Sternbergit *m* (Min) sternbergite, iron silver glance
Sternbildung *f* star formation
Sterndämpfer *m* star steamer
Sterndreieckschalter *m* (Elektr) star-delta switch
Sternflächner *m* star polyhedron
sterngeschaltet (Elektr) star-connected
Sternhaufen *m* (Astr) star cluster, stellar cluster
Sternit *m* sternite
Sternjahr *n* (Astr) sidereal year
Sternkunde *f* astronomy
Sternpunkt *m* (Elektr) star point
Sternsaphir *m* (Min) star sapphire, star stone
Sternschaltung *f* (Elektr) star circuit, star connection, Y-connection
Sternscher Vervielfacher *m* Stern multiplier
Sternschnuppe *f* meteor
Sternspannung *f* (Elektr) star voltage
Sternspektroskop *n* astrospectroscope
Sternspektrum *n* stellar spectrum
Sternstrom *m* star current, Y-current
Sternsystem *n* (Astr) galaxy
Sternzelle *f* (Histol) star cell, astrocyte, Kupffer's cell, spider cell

Steroid – stickstofffrei

Steroid *n* steroid, sterid
Steroidgerüst *n* steroid skeleton
Steroidhormon *n* steroid hormone
Sterosan *n* sterosan
Sterrometall *n* sterro metal
Stethoskop *n* (Med) stethoscope
stetig constant, continuous, stable
Stetigförderer *m* continuous-flow conveyor
Stetigkeit *f* steadiness, continuity, stability, uniformity, ~ **einer Funktion** continuity of a function
Stetigkeitsbedingung *f* condition for continuity
Stetigkeitsgrenze *f* continuity limit
Steuer *n* (Steuerrad) steering wheel, (Steuerruder) rudder, (Steuerwerk) controls
Steuerapparat *m* controlling mechanism
Steuerarm *m* master arm
Steuerbefehlspeicher *m* (Comp) control register, program register
Steuerbühne *f* control platform; operating pulpit
Steuerelektrode *f* trigger electrode
Steuerfrequenz *f* pilot frequency
Steuergerät *n* control mechanism
Steuergitter *n* control grid
Steuergitterröhre *f* control grid tube
Steuerhebel *m* steering arm
Steuerimpuls *m* drive pulse
Steuerkante *f* leading edge
steuern to control, to regulate
Steuern *n* **der Presse** controlling the press
Steuernocke *f* cam
Steuerorgan *n* control
Steuerpult *n* control board, regulation desk
Steuerrad *n* steering wheel
Steuerradachse *f* steering wheel axle
Steuerrelais *n* (Elektr) pilot relay
Steuerschieber *m* operating valve, regulating slide valve
Steuerspannung *f* command voltage
Steuerstange *f* (Atomtechnik) control rod
Steuerstromkreis *m* (Elektr) control circuit
Steuerung *f* control [gear], regulation, (Regeltechn) feed forward control, open-loop control, **einfache offene** ~ (Comp) open-loop control, **örtliche** ~ (Comp) local control, **photoelektrische** ~ photoelectric control, **rückführungslose** ~ (Comp) open-loop control
Steuerungsantrieb *m* valve gear
Steuerungsdaumen *m* sequence switch cam
Steuerungsdiagramm *n* distribution diagram
Steuerungseinrichtung *f* steering arrangement
Steuerungsgestänge *n* control rods, valve gear
Steuerungshahn *m* distributing cock
Steuerungsvorgang *m* control operation
Steuerventil *n* check valve
Steuerwalze *f* timing shaft
Steuerwelle *f* cam shaft
Steuerwerk *n* control unit

Steuerwinkel *m* angle of contact
Steuerzylinder *m* control cylinder
Stevinsäure *f* stevic acid
Stewartit *m* (Min) stewartite
Stibaminglucosid *n* stibamine glucoside
Stibiat *n* antimonate, stib[i]ate
Stibiconit *m* (Min) stibiconite, antimony ocher
Stibin *n* stibine, antimonous hydride
Stibinoanilin *n* stibinoaniline
Stibinobenzol *n* stibinobenzene
Stibiotantalit *m* (Min) stibiotantalite
Stibith *m* (Min) stibiconite, stibite
Stibium *n* (Lat) antimony
Stiblith *m* stiblite
Stibnit *m* (Min) stibnite, antimony glance
Stibonsäure *f* stibonic acid
Stibosamin *n* stibosamine
Stich *m* (Kupferstich) engraving, (Nadelstich) stitch
Stichauge *n* tap hole
Stichausreißfestigkeit *f* stitch tear resistance
Stichbahn *f* branch conveyor or line
Sticheinreißfestigkeit *f* puncture resistance
Sticheisen *n* tapping bar, stocker's rod
Stichel *m* burin, graver, graving tool, style, stylus
Stichflamme *f* blowpipe flame, explosive flame, fine pointed flame, flareback, jet of flame, (Zündlampe) pilot light
Stichloch *n* (Metall) tap hole
Stichlochstopfmaschine *f* taphole gun, blast-furnace gun, clay gun, mud gun
Stichlochversetzung *f* taphole displacement
Stichöffnung *f* (Metall) tapping hole
Stichpfropf *m* tap hole plug
Stichprobe *f* random test, sample at random, spot check, (Metall) assay of tapped metal
Stichrohrbrücke *f* overhead pipe junction
Stichtit *m* (Min) stichtite
Stichtorf *m* dug peat
Stichwort *n* catchword, clue, keyword
Stickoxid *n* nitric oxide
Stickoxidgleichgewicht *n* nitric oxide equilibrium
Stickoxydul *n* (obs) nitrous oxide
Stickstoff *m* nitrogen (Symb. N)
Stickstoffaufnahme *f* nitrogen absorption
Stickstoffausscheidung *f* nitrogen excretion
Stickstoffbakterien *pl* nitrifying bacteria
Stickstoffbestimmung *f* nitrogen determination, ~ **nach Kjeldahl** Kjeldahl nitrogen method
Stickstoffbor *n* boron nitride
Stickstoffbrücke *f* nitrogen bridge
Stickstoffcalcium *n* calcium nitride
Stickstoffchlorid *n* nitrogen [tri]chloride
Stickstoffdioxid *n* nitrogen dioxide
Stickstoffdünger *m* nitrogen[ous] fertilizer
stickstofffrei free from nitrogen, non-nitrogenous

Stickstoffgabe f (Agr) nitrogen supply
Stickstoffgehalt m nitrogen content
Stickstoffgewinnung f (Biol) fixation of nitrogen
Stickstoffhalogen n nitrogen halide
stickstoffhaltig containing nitrogen, nitrogen-containing, nitrogenous
Stickstoffixierung f (Biol) nitrogen fixation, **enzymatische** ~ enzymatic nitrogen fixation, **symbiontische** ~ symbiontic nitrogen fixation
Stickstoff-Kohlenstoff-Kreislauf m carbon-nitrogen cycle
Stickstoffkompressor m nitrogen compressor
Stickstoffkreislauf m nitrogen cycle
Stickstofflithium n lithium nitride
Stickstoffmagnesium n magnesium nitride
Stickstoffnatrium n sodium nitride
Stickstoffoxid n nitrogen oxide
Stickstoffoxydation f oxidation of nitrogen
Stickstoffoxydul n (obs) dinitrogen monoxide, nitrous oxide
Stickstoffpentoxid n dinitrogen pentoxide, nitric anhydride
Stickstoffperoxid n nitrogen peroxide
Stickstoffquecksilber n mercury nitride
Stickstoffretention f nitrogen retention
Stickstoffsättigung f saturation with nitrogen
Stickstoffsäure f nitrogen-containing acid
Stickstoffsammler m (Bot) nitrogen-fixing plant
Stickstoffsesquioxid n dinitrogen trioxide, nitrogen sesquioxide, nitrous anhydride
Stickstoffsilber n silver nitride
Stickstoffsilicid n nitrogen silicide, silicon nitride
Stickstoffstoffwechsel m nitrogen metabolism
Stickstoffstrom m nitrogen flow
Stickstofftetroxid n nitrogen tetroxide, dinitrogen tetroxide
Stickstofftrioxid n nitrogen trioxide, dinitrogen trioxide, nitrogen sesquioxide, nitrous anhydride
Stickstoffwasserstoff m hydrazoic acid, hydrogen trinitride, nitrogen hydride
Stickstoffwasserstoffsäure f hydrazoic acid, hydronitric acid
Stiefmütterchen n (Bot) pansy, viola tricolor
Stiel m (Bot) shank, stem, (Griff) handle, helve, (Nagel) shaft
Stielhammer m shaft hammer
Stift m (einer Welle) journal (of a crankshaft), (Nagel) peg, pin, (Schreibstift) pencil
Stiftelektrode f pin electrode
Stiftfräser m hollow mill
Stiftkette f bushing chain, riveted chain
Stiftmühle f pinned disk mill, disk attrition mill, [iron] disk mill, peg impactor mill
Stiftnietung f pin riveting
Stiftschraube f set-screw, stud [bolt]
Stiftung f foundation
Stigma n stigma

Stigmastan n stigmastane
Stigmasterin n stigmasterol
stigmatisch (Opt) stigmatic
Stigmator m (Elektronenmikroskop) stigmator
Stil m style
Stilart f style
Stilbazol n stilbazole
Stilbazolin n stilbazoline
Stilben n stilbene, bibenzal, diphenylethylene
Stilbendulcin n stilbenedulcin
Stilbengelb n stilbene yellow
Stilbenleuchtstoff m stilbene phosphor
Stilbit m (Min) stilbite
Stilböstrol n stilbestrol
still quiet, silent, (ruhig) calm, (unbeweglich) motionless
Stillingiaöl n stillingia oil
Stillingsäure f stillingic acid
Stillopsidin n stillopsidin
Stillsetzen n shut-down, shut-off
Stillstand m standstill
Stillstandszeit f shut-off time, down-time
stillstehen to stand still, to stop
stillstehend stagnant, (stationär) stationary
Stilpnomelan m (Min) stilpnomelane
Stilpnosiderit m (Min) stilpnosiderite
Stimmband n (Anat) vocal cord
stimmen (Akust) to tune, (richtig sein) to be correct, to be right
Stimmgabel f tuning fork
Stimulans n (Pharm) stimulant, stimulus
Stinkasant m (Pharm) asa fetida
Stinkflußspat m (Min) bituminous fluorite, bituminous fluor spar, fetid fluor spar
Stinkkalk m (Min) anthraconite, bituminous limestone, stinkstone
Stinkmergel m bituminous marl, fetid marl
Stinkquarz m (Min) bituminous quartz, fetid quartz
Stinkstein m (Min) anthraconite, fetid stone, stinkstone
Stinkzinnober m (Min) fetid cinnabar
Stipitatsäure f stipitatic acid
Stirnansicht f frontal view
Stirnfläche f front face, front surface, top surface
Stirnfräser m front milling tool, end mill, end-milling cutter
Stirnplatte f head plate
Stirnrad n (Masch) spur gear, spur wheel
Stirnradantrieb m (Masch) spur wheel drive
Stirnradgetriebe n spur gear[ing]
Stirnradmotor m gear head motor
Stirnseite f front [side], facade
Stirnwand f (Bauw) face wall
Stirnzapfen m journal on end of shaft
Stirrholz n wooden spatula, stirrer
Stizolobsäure f stizolobic acid
stockdunkel pitch-dark

stocken – Stoffwechsel

stocken (gerinnen) to clot, to congeal, to curdle, (stillstehen) to stand still, to stop
Stockpunkt *m* gel point, pour point, setting point, solidifying point
Stockpunktbestimmungsapparat *m* solidification point apparatus
Stockthermometer *n* straight thermometer
Stockung *f* interruption, stagnation
Stockwerk *n* floor, storey
Stöchiometrie *f* (Chem) stoichiometry
stöchiometrisch (Chem) stoichiometric[al]
Stöpsel *m* stopper, (aus Kork) cork, (Elektr) plug
Stöpselflasche *f* stoppered bottle, stoppered flask
Stöpselhahn *m* cock stop
Stöpselkasten *m* (Elektr) resistance box
Stöpselkopf *m* plug head
stöpseln (Chem) to cork, to stopper, (Elektr) to plug [in]
Stöpselsicherung *f* plug fuse
Störabsperrer *m* interference suppressor
Störbandleitung *f* impurity band conduction
Störbeseitigung *f* (Comp) debugging
stören to disturb, to interrupt, to trouble
Störfeld *n* interference field
Störgeräusch *n* noise
Störgröße *f* disturbance value, interference factor
Störgrößenaufschaltung *f* (Regeltechn) disturbance feed-forward control
Störhintergrund *m* background noise
Störmeßgerät *n* interference measuring apparatus
Störpegel *m* interference level
Störphase *f* distortion phase
Störschutz *m* interference elimination, (Radio) noise eliminator
Störstelle *f* defect, point of disturbance
Störstellenbeweglichkeit *f* defect mobility
Störstrahlung *f* interference radiation
Störstrom *m* (Elektr) interference current, parasite current, stray current
Störsuchaufgabe *f* trouble-location problem
Störung *f* disturbance, breakdown, disarrangement, disorder, failure, interference, trouble, (Unterbrechung) interruption, **magnetische** ~ magnetic disturbance, magnetic perturbation, **stationäre** ~ local disturbance
Störungsbeseitigung *f* trouble shooting
störungsempfindlich perturbation sensitive
Störungsenergie *f* perturbation energy
störungsfrei trouble free
Störungskontrollgerät *n* detector box
Störungsmethode *f* perturbation method
Störungsreaktion *f* interference reaction
Störungsrechnung *f* disturbance calculation, perturbation calculation
Störungssuche *f* trouble shooting
Störungstheorie *f* perturbation theory
störungsunempfindlich perturbation insensitive
Störungszentrum *n* center of disturbance, trouble spot
Störwellenmethode *f* distorted wave method
Stössel *m* pounder, rammer, stamper, (Chem) pestle
Stösselstange *f* push rod
Stoff *m* (Material) material, (Materie) matter, (Pap) pulp, stuff, (Substanz) substance, (Text) cloth, fabric, material, **amphoterer** ~ amphoteric compound, **bildsamer** ~ plastic material, **feuerfester** ~ refractory material, **gelöster** ~ solute, **radioaktiver** ~ radioactive material, radioactive substance
Stoffabflußnute *f* flash groove, spew groove, spew way
Stoffänger *m* (Pap) pulp catcher, save-all, settling cone, stuff catcher
Stoffaufbau *m* structure of matter
Stoffauskleidung *f* fabric lining
Stoffaustausch *m* mass transfer
Stoffaustauschprozeß *m* mass transfer
Stoffaustauschwiderstand *m* mass transfer resistance
Stoffbespannung *f* fabric covering
Stoffbilanz *f* material balance
Stoffdurchgangswiderstand *m* resistance to mass transfer
Stoffdurchgangszahl *f* overall mass transfer coefficient
Stoffgemisch *n* mixture [of substances]
Stoffgleichheit *f* isogeny
Stoffilter *m* cloth filter, filter cloth
Stoffkreiselpumpe *f* pulp-handling centrifugal pump
Stofflack *m* (Text) textile coating
stofflich material
stofflos immaterial
Stoffmühle *f* (Pap) beater, refiner
Stofförderanlage *f* (Pap) pulp conveyor
Stoffprüfung *f* examination of material
Stoffrückseite *f* (Text) back of fabric
Stoffteilchen *n* particle
Stofftransport *m* mass transfer
Stoffübergang *m* mass transfer, ~ **am umströmten Tropfen** mass transfer from continuous phase to droplets, ~ **mit gleichzeitiger Reaktion** mass transfer accompanied by reaction
Stoffübergangsgeschwindigkeit *f* rate of mass transfer
Stoffübergangskoeffizient *m* mass transfer coefficient
Stoffübergangszahl *f* mass transfer coefficient
Stoffwechsel *m* (Physiol) metabolism, **abbauender** ~ catabolism, **aufbauender** ~ anabolism, **erhöhter** ~ (Med)

hypermetabolism, **gestörter** ~ disordered metabolism, **Gewebe-** ~ tissue metabolism, **intermediärer** ~ intermediary metabolism, **umkehrbarer** ~ reversible metabolism
Stoffwechselablauf m metabolic process
Stoffwechselanomalie f metabolic abnormality
Stoffwechselblockierung f metabolic inhibition
Stoffwechselerkrankung f metabolic disease
Stoffwechselgröße f metabolic rate
Stoffwechselregulation f metabolic regulation
Stoffwechselweg m metabolic pathway
Stoffwechselzwischenprodukt n metabolic intermediate
Stoffwert m physical characteristic
Stoffzusammensetzung f composition of material, (Pap) furnish of paper
Stokesit m (Min) stokesite
Stokesscher Satz m Stokes' theorem
Stollen m (Bergb) gallery tunnel, tunnel, (eines Reifenprofils) lug (of a tire profile)
Stollenbau m (Bergb) mining by galleries
Stolpenit m stolpenite
Stolzit m (Min) stolzite, lead tungstate
Stopfbüchse f (Masch) stuffing box, stuffing bush
Stopfbüchsenausgleicher m stuffing box compensator
Stopfbüchsenschraube f (Masch) gland stud, stuffing box stud
stopfen to stop a leak
Stopfen m stopper, screwed plug, stopper plug, threaded plug
Stopfenausguß m stopper nozzle
Stopfenbett n stopper socket
Stopfenhebevorrichtung f stopper lifting device
Stopfmaschine f stuffing machine
Stopp m, **willkürlicher** ~ (Comp) request stop
Stoppbefehl m, **bedingter** ~ (Comp) conditional breakpoint instruction
Stopplicht n stop light
Stoppmutter f stop nut
Stoppuhr f stop watch
Storax m storax [resin], styrax
Storaxöl n storax oil
Storaxsalbe f storax ointment
Storaxsaponin n storax saponin
Storaxzimtsäure f storax cinnamic acid
Storchschnabel m (Zeichengerät) pantograph
Storchschnabelarm m pantograph arm
Stoß m blow, push, shock, stroke, thrust, (Anhäufung) heap, pile, (Aufprall) impact, (Impuls) impulse, (Rückschlag) recoil, (Zusammenstoß) bump, collision, percussion, ~ **erster Art** (Atom) collision of the first kind, ~ **mit Impulsübertragung** momentum transfer collision, ~ **zwischen Elektronen** electron-electron collision, **unelastischer** ~ inelastic collision

Stoßanregung f collision excitation, collision impulse, shock excitation
Stoßapproximation f impulse approximation
stoßartig jerky
Stoßausgleicher m stroke compensator
Stoßbelastung f shock load
Stoßblech n buttstrap, coverplate
stoßbohren (Bergb) to bore by percussion
Stoßbohrer m (Bergb) percussion drill
Stoßbruch m concussion burst, concussion fracture, impact break
stoßdämpfend shock-absorbing
Stoßdämpfer m (Auto) shock absorber
Stoßdämpfung f shock-absorption
Stoßdauer f collision period, duration of collision, shock period
Stoßdefekt m impact failure
Stoßdichte f (Atom) collision density
Stoßdurchschlag m impulse breakdown
Stoßeinfluß m effect of collisions
Stoßelastizität f impact resilience, shock elasticity
Stoßempfindlichkeit f susceptibility to shock
stoßen to push, to knock, to thrust, (zerkleinern) to pound, ~ **auf** to come upon
Stoßen n pushing
Stoßenergie f collision energy
Stoßentaktivierung f collision deactivation
Stoßentladung f impulse discharge, pulsed discharge
Stoßerregung f impulse excitation, shock excitation
Stoßfaktor m collision factor, (Therm) frequency factor
stoßfest shock-proof
Stoßfestigkeit f shock proofness, impact strength, resistance to shock
Stoßfluoreszenz f (Opt) impact fluorescence
Stoßfortpflanzung f pulse propagation
Stoßgalvanometer n ballistic galvanometer
stoßgeschützt shock-proof
Stoßgröße f pulse size
Stoßgrößenverteilung f pulse size distribution
Stoßhäufigkeit f collision frequency
Stoßionisation f ionization by collision, impact ionization
Stoßkalander m (Pap) beating mill
Stoßkomplex m collision complex
Stoßkraft f impact force, impulse, percussive force
Stoßleiste f protective strip
Stoßmoment n collision momentum
Stoßmühle f impact pulverizer
Stoßparameter m impact parameter
Stoßpartikel n collision particle
Stoßpartner m collision partner
Stoßpolarisation f impact polarization
Stoßproblem n collision problem
Stoßquerschnitt m collision cross section

Stoßreaktor *m* (Gastechnologie) shock wave reactor, shock tube
Stoßrohr *n* shock tube
Stoßsäge *f* slotting saw
Stoßschutz *m* wadding
Stoßspannung *f* surge voltage
Stoßstärke *f* collision strength, shock strength
Stoßübergang *m* collision transition
stoßunempfindlich insensitive of shock
Stoßventil *n* relief valve
Stoßverbreiterung *f* impact broadening
Stoßverdampfung *f* flash evaporation
Stoßverletzung *f* bruise damage, impact bruise
Stoßversuch *m* impact experiment
Stoßvervielfachung *f* collision multiplication
Stoßvorgang *m* collision mechanism
Stoßwahrscheinlichkeit *f* probability of collision
stoßweise in batches, intermittent, pulsating
Stoßwelle *f* shock wave, (Explosion) percussion wave
Stoßzahl *f* (Atom) frequency number, number of collisions
Stoß-Zug-Mechanismus *m* push-pull mechanism
Stovain *n* stovaine
Stovarsol *n* stovarsol, acetarsone
Strähne *f* strand
Strähnenbildung *f* (Dest) channeling
straff taut, tight
straffen to tighten
Straffheit *f* tightness
Strahl *m* (Blitz) flash, (Lichtstrahl) ray, beam, (Wasserstrahl) jet, stream,
  außerordentlicher ~ extraordinary ray,
  durchfallender ~ transcident ray, transmitted ray, **einfallender** ~ incident ray,
  gebündelter ~ focused beam, **ordentlicher** ~ ordinary ray, **reflektierter** ~ reflected ray,
  zusammengesetzter ~ (Phys) complex ray
Strahlablenker *m* ray deflector
Strahlablenkung *f* beam deflection
Strahlabschneiderfahne *f* flapper
Strahlantrieb *m* jet propulsion
Strahlbaryt *m* (Min) radiated barite, Bologna stone
Strahlbrenner *m* jet burner
Strahldicke *f* thickness of jet
Strahldüsenreaktor *m* jet tube reactor
strahlen to radiate, to emit rays of light
Strahlen *pl*, **infrarote** ~ infrared rays,
  kosmische ~ cosmic rays, **ultraviolette** ~ ultraviolet rays
Strahlenabschirmung *f* radiation shielding
Strahlenaktivität *f* radiation activity
Strahlenanalysator *m* radiation analyser
Strahlenauslöschung *f* radiation elimination
Strahlenaustrittsfenster *n* (Röntgen) transparent window, X-ray window
Strahlenbahn *f* light path

Strahlenbehandlung *f* radiation treatment, (Med) ray therapy
Strahlenbereich *m* range of radiation
Strahlenblende *f* (Min) wurtzite
Strahlenbrechung *f* diffraction of rays, refraction [of rays]
Strahlenbrechungsmesser *m* refractometer
Strahlenbremsung *f* ray retardation
Strahlenbündel *n* (Strahlenbüschel) beam of rays, bundle of rays, **heterogenes** ~ heterogeneous beam
Strahlenchemie *f* radiation chemistry
strahlend radiant, radiating
Strahlendosis *f* radiation dosage, radiation dose
Strahleneinfall *m* incidence of rays
Strahlenemission *f* radiation
strahlenempfindlich radiosensitive
Strahlenenergie *f* radiation energy
Strahlenfilter *m* ray filter
strahlenförmig radiate[d], actinoid
Strahlengang *m* (Opt) path of rays, course of rays
strahlengefährdet exposed to radiation hazards
Strahlengefahr *f* radiation hazard
strahlengeschädigt damaged by radiation
Strahlenglimmer *m* (Min) striated mica
Strahlenhärte *f* ray hardness, hardness of radiation
Strahlenhärtemesser *m* penetrometer
Strahleninterferometer *n* ray interferometer
Strahlenkonservierung *f* radiation preservation
Strahlenkrebs *m* (Med) X-ray cancer
Strahlenkupfer *n* (Min) clinoclasite
Strahlenlehre *f* radiology
Strahlenmesser *m* radiometer, (aktinische Strahlen) actinometer, (Sonnenstrahlung) solarimeter
Strahlenmischer *m* jet mixer
Strahlennachweisgerät *n* radiation detection instrument
Strahlenöffnung *f* beam hole
Strahlenpilz *m* (Biol) actinomyces
Strahlenquant *n* ray quantum
Strahlenquarz *m* (Min) fibrous quartz
strahlenresistent radiation resistant
Strahlenresistenz *f* radioresistance
Strahlenring *m* halo
Strahlenschaden *m* radiation damage
Strahlenschutz *m* radiation protection
Strahlenschutzröhre *f* anti-radiation tube
Strahlenschutztherapie *f* antiradiation therapy
Strahlenstärkemesser *m* radiation intensitometer
Strahlenstein *m* (Min) radiolite, actinolite, amianthus
Strahlensterilisierung *f* radiation sterilization
Strahlenteiler *m* beam splitter
Strahlentherapie *f* radiation treatment, (Med) radiation therapy
Strahlenüberwachungsgerät *n* radiation monitor

strahlenundurchlässig radiopaque
Strahlenundurchlässigkeit *f* radiopacity, opacity for radiation
Strahlenwarngerät *n* radiation monitor
Strahlenwirkungslehre *f* actinology
Strahlenzerfall *m* ray disintegration
Strahler *m* radiator; emitter, irradiator
Strahlerz *n* (Min) aphanesite, clinoclasite
Strahlgips *m* (Min) fibrous gypsum
Strahlhärtung *f* stream hardening
strahlig radiated
Strahlkeil *m* belemnite
Strahlkies *m* (Min) marcasite
Strahlkörper *m* radiator
Strahlmischen *n* mixing by liquid jets
Strahlmischer *m* jet agitator, jet blender, jet mixer
Strahlmittel *n* blasting agent
Strahlmühle *f* fluid energy mill, jet mill
Strahlprallmühle *f* nozzle pulverizer
Strahlpumpe *f* jet pump
Strahlregler *m* jet regulator
Strahlrohr *n* (Sandstrahlgebläse) nozzle holder
Strahlrückstoßprinzip *n* jet propulsion principle
Strahlsauger *m* ejector
Strahlschörl *m* (Min) radiated tourmaline
Strahlstärke *f* radiant intensity
Strahlstein *m* (Min) actinolite, amianthus
Strahlsteinschiefer *m* (Min) actinolitic schist
Strahlstrom *m* jet stream
Strahltriebwerk *n* jet power plant, jet unit
Strahlturbine *f* Pelton turbine
Strahlung *f* radiation, **Fluoreszenz-** ~ fluorescence, fluorescent radiation, **heterogene** ~ heterogeneous radiation, **Höhen-** ~ cosmic radiation, **homogene** ~ homogeneous radiation, **ionisierende** ~ ionizing radiation;, **kohärente** ~ coherent radiation, **kosmische** ~ cosmic radiation, **magnetische** ~ magnetic [field] radiation, **Primär-** ~ primary radiation, **radioaktive** ~ radioactive radiation, **schwarze** ~ black[-body] radiation, **Sekundär-** ~ secondary radiation, **Ultraviolett-** ~ ultraviolet radiation
strahlungsabschirmend radiation shielding
Strahlungsabschirmung *f* radiation shield, radiation protection
Strahlungsabsorption *f* radiation absorption
Strahlungsanisotropie *f* ray anisotropy
Strahlungsausbeute *f* radiating efficiency
Strahlungsbehandlung *f* radiation treatment
Strahlungsbereich *m* range of radiation
Strahlungsbreite *f* radiation width
Strahlungscharakteristik *f* radiation characteristic
Strahlungschemie *f* radiation chemistry
Strahlungsdämpfung *f* radiation damping, radiation resistance

Strahlungsdiagramm *n* radiation diagram, radiation chart
Strahlungsdichte *f* radiation density
Strahlungsdosis *f* radiation dosage
Strahlungsdurchgang *m* radiation transmission
Strahlungseffekt *m* radiation effect
Strahlungseinwirkung *f* radiation bombardment
Strahlungsempfindlichkeit *f* radiosensitivity
Strahlungsenergie *f* radiation energy
strahlungsfähig radiative
Strahlungsfestigkeit *f* radioresistance, radiotolerance
Strahlungsfläche *f* emitting area, emitting surface
Strahlungsfluß *m* radiation flux
strahlungsfrei free of radiation, radiationless
Strahlungsgefährdung *f* danger of radiation
Strahlungsgesetz *n* radiation law
Strahlungsglied *n* radiation term
Strahlungsheizung *f* convection heating, radiation heating
strahlungsinduziert radiation induced, radiation initiated
Strahlungsintensität *f* intensity of radiation
Strahlungsintensitätsmesser *m* scintillation counter
Strahlungskatalyse *f* radiation catalysis
Strahlungskessel *m* radiation boiler
Strahlungskoeffizient *m* radiation coefficient
Strahlungskorrektur *f* radiative correction
Strahlungslebensdauer *f* radiative lifetime
Strahlungsleistung *f* radiation efficiency
strahlungslos non-radiative, radiationless
Strahlungsmeßgerät *n* radiation meter, radiation measuring equipment
Strahlungsmeßinstrument *n* radiation [measuring] instrument
Strahlungsnachweisgerät *n* radiation detector
Strahlungsnorm *f* standard of radiation
Strahlungsnormal *n* radiation standard
Strahlungsnormale *f* (Opt) standard light source
Strahlungsofen *m* radiation furnace, convection heater
Strahlungspotential *n* radiation potential
Strahlungspyrometer *n* radiation pyrometer
Strahlungsquelle *f* source of radiation, **punktförmige** ~ (Opt) point source
Strahlungsreichweite *f* (Atom) radiation length
Strahlungsrekombination *f* radiative recombination
Strahlungsschirm *m* radiation screen, radiation shield
Strahlungsschwächung *f* radiation attenuation
Strahlungsstrom *m* radiation flux
Strahlungssucher *m* radioscope
Strahlungstemperatur *f* (Therm) radiation temperature
Strahlungstheorie *f* radiation theory
Strahlungstrockner *m* radiation dryer

Strahlungsübergang – Strepsilin

Strahlungsübergang *m* radiation transition, radiative transition
Strahlungsüberschuß *m* radiation surplus
Strahlungsüberwachung *f* radiation survey
Strahlungsverlust *m* loss by radiation
Strahlungsvermögen *n* radiating power
Strahlungswärme *f* heat of radiation, radiant heat
Strahlungswarngerät *n* radiation monitoring device
Strahlungswiderstand *m* radiation resistance
Strahlungszählrohr *n* radiation counter tube
Strahlwäscher *m* jet scrubber
Strahlzeolith *m* (Min) foliated zeolite, stilbite
Stralit *m* stralite
Stramin *m* (Text) canvas
Strang *m* (Web) hank
Strangaufweitung *f* memory effect (extrusion)
Strangfärbeapparat *m* hank dyeing apparatus
stranggepreßt extruded
Strangguß *m* continuous casting
Strangpresse *f* plodder, (Met) extrusion press, extruder, extruding press
strangpressen (Met) to extrude
Strangpressen *n* extrusion, extrusion molding
Strangpreßform *f* extrusion die
Straß *m* strass
Straßenbau *m* road construction
Straßenbelag *m* road surface
Straßenerprobung *f* road test
Straßenkehricht *m* street sweepings
Straßenmarkierung *f* road marker
Straßenmarkierungsfarbe *f* road paint, traffic paint, zone marking paint
Straßenmarkierungsfarben *pl* traffic paints
Straßfurter Salze *n pl* Strassfurt salts
Stratopeit *m* stratopeite
Stratosphäre *f* stratosphere
Strebe *f* cross brace, prop, stay, strut
streben to aim [at], to strive, to struggle
streckbar extensile, stretchable, (Metall) ductile
Streckbarkeit *f* extensibility, stretchability, (Metall) ductility
Strecke *f* distance
strecken to stretch, to draw, to elongate, to extend, (verdünnen) to dilute
streckenweise by sections
Streckenzug *m* straight-edge
Streckfestigkeit *f* elongation resistance, resistance to stretching
Streckform *f* stretch die
Streckformen *n* stretch forming
Streckgrenze *f* elastic limit, tensile yield, yield point, yield strength
Streckgrenzeneffekt *m* yield phenomenon
Streckgrenzenerscheinung *f* yield phenomenon
Streckhärtung *f* strain hardening
Streckkraft *f* stretching force
Streckmaschine *f* (Blech) stretcher leveller

Streckmetall *n* expanded metal, extended metal
Streckmittel *n* extender, filler, (Verdünnung) diluent
Streckprobe *f* stretch test
Streckrahmen *m* clamping frame
Streckspannung *f* tensile stress at yield
Streckspinnen *n* stretch spinning
Streckung *f* stretching, drawing, elongation
Streckungsmittel *n* extender, adulterant, filler, (Verdünnung) diluent
Streckungs-Schrumpfungs-Kurve *f* stretch shrinkage curve
Streckziehform *f* stretch die
Streckziehverfahren *n* drawing process
Strehler *m* chaser, chasing tool
Streichanlage *f* coater, coating machine
Streichbarkeit *f* brushability
Streicheisen *n* skimmer
streichen (aufstreichen) to coat; to spread
Streichen *n* brushing, coating, painting, (Annullieren) cancellation
streichfähig brushable
Streichfähigkeit *f* brushability, brushing property
Streichfeuer *n* fire of reverberatory furnace
Streichgarn *n* (Text) carded yarn, woollen yarn
Streichgeschwindigkeit *f* spreading velocity
Streichgewicht *n* coat weight
Streichholz *n* match
Streichholzschachtel *f* matchbox
Streichkäse *m* cheese spread
Streichlack *m* brushing lacquer, brushing varnish, coating compound
Streichmaschine *f* coating machine, spread coater, spreading machine, (Gummi) rubber spreader
Streichmasse *f* coating mass, coating substance
Streichmesser *n* spreading knife
Streichmischung *f* coating mixture
Streichpresse *f* coating press
Streichputz *m* brushable textured finish
Streichtorf *m* molded peat
Streichverfahren *n* spread coating
Streichviskosität *f* brush viscosity
Streifen *m* streak, strip, stripe, (Zone) zone
Streifenbild *n* (Opt) fringe pattern
Streifenbildung *f* streak formation, streakiness
Streifenblattschreiber *m* strip chart recorder
streifenförmig streaky, striped
Streifenform *f* strip form
Streifenheizkörper *m* strip heater
Streifenleiter *m* strip line
Streifensicherung *f* strip fuse
Streifigkeit *f* stripiness
Streifkasten *m* mud box, settling box
Streitobjekt *n* issue
strenggenommen strictly speaking
Strengit *m* (Min) strengite
Strepsilin *n* strepsilin

712

Streptamin *n* streptamine
Streptidin *n* streptidine
Streptobiosamin *n* streptobiosamine
Streptokinase *f* (Biochem) streptokinase
Streptokokkus *m* (pl Streptokokken) streptococcus (pl streptococci)
Streptolysin *n* streptolysin
Streptomycin *n* streptomycin
Streptose *f* streptose
Streptovitacin *n* streptovitacin
Streuamplitude *f* scattering amplitude
Streubereich *m* dispersion range, scatter range, spreading range
Streubrechung *f* dispersed refraction
Streubüchse *f* shaker, box with perforated cap
Streudiffusion *f* eddy diffusion
Streudose *f* sprinkler can
Streudüse *f* spray nozzle
Streueffekt *m* effect of scattering
Streuelektron *n* scatter electron
streuen to disperse, (Elektr) to leak, to stray, (Licht) to scatter
Streufaktor *m* scattering factor, dispersion coefficient, leakage coefficient
Streufeld *n* stray field
Streufeldabschirmung *f* stray field screening
Streufluß *m* stray flux
Streuinduktion *f* stray induction
Streukoeffizient *m* scattering coefficient, leakage coefficient
Streukontinuum *n* scattering continuum
Streukopplung *f* stray coupling, spurious coupling
Streulicht *n* stray light
Streulinse *f* dispersion lens
Streumatrix *f* scattering matrix
Streumessung *f* scattering measurement
Streuphase *f* dispersion phase
Streuquerschnitt *m* scattering cross section
Streureaktanz *f* stray reactance
Streusand *m* dry sand
Streuschwund *m* scatter fading
Streusichter *m* classifier with closed air circuit
Streuspannung *f* stray voltage
Streustrahlen *pl* scattered rays, stray rays
Streustrahlenschutz *m* protection against scattered radiation
Streustrahlung *f* scattered radiation, stray radiation
Streustrom *m* stray current, eddy current, vagabond current
Streuteilchen *n* scattered particle
Streuung *f* scattering, deviation, (Math) mean error, (Strahlen) dispersion, ~ **an Grenzflächen** boundary scattering, ~ **der Kraftlinien** magnetic leakage, ~ **einer Lichtquelle** dispersion or scattering of a light source, spread of a light source, ~ **im weiten Winkel** large-angle scattering, **akustische** ~ (Akust) acoustic scattering, **elektromagnetische** ~ electromagnetic leakage, **lineare** ~ (Spektr) linear dispersion, **magnetische** ~ magnetic cross flux, magnetic dispersion, magnetic leakage, **unelastische** ~ inelastic scattering
Streuungsbereich *m* zone of dispersion
Streuungsdiagramm *n* (Opt) diffusion indicatrix
streuungslos without leakage
Streuungsquerschnitt *m* scattering cross section
Streuungswinkel *m* dispersion angle, scattering angle
Streuungszahl *f* leakage coefficient
Streuvermögen *n* scattering power
Streuwiderstand *m* leakage resistance
Streuwindsichter *m* whizzer
Streuwinkel *m* scattering angle
Streuzentrum *n* scattering center
Strich *m* dash, line, (Teilstrich) division line
Strichbreite *f* width of scratch
Strichelektrode *f* filament electrode
stricheln to mark with a dotted line
Strichfokus *m* (Opt) linear focus
Strichgitter *n* linear grating, (Opt) graticule
Strichklischee *n* line block
strichliert dash-lined
Strichlinie *f* dotted line
strichpunktiert indicated by a dot-dash line
Strich-Punkt-Verfahren *n* (Elektron) dash-dot method
Strick *m* rope
stricken to knit
Strickwaren *f pl* (Text) knitted fabrics
Strigovit *m* (Min) strigovite
Strip-Förderer *m* strip winder
strippen to strip
Strippen *n* stripping
Strobosid *n* stroboside
strömen to stream, to flow, to pour
Strömen *n* flowing
Strömung *f* current, flow, flux, stream, **Binghamsche** ~ Bingham flow, **gleichförmige** ~ (Mech) uniform flow, **kompressible** ~ compressible flow, **laminare** ~ laminar flow, **plastische** ~ plastic flow, **stationäre** ~ stationary flow, steady flow;, **turbulente** ~ turbulent flow, **wirbelfreie** ~ irrotational fluid motion
Strömungsbild *n* flow pattern
Strömungsdiagramm *n* flow diagram
Strömungsdivergenz *f* (Mech) divergence of fluid
Strömungsgeschwindigkeit *f* velocity of flow
Strömungsgleichung *f* flow equation
Strömungslehre *f* rheology, (Flüssigkeiten) hydrodynamics, (Gase) aerodynamics
Strömungsmesser *m* flow meter
Strömungsmittel *n* flow medium
Strömungsmodell *n* flow model
Strömungsrichtung *f* direction of flow

Strömungsrohr *n* flow tube, tubular reactor, **ideales** ~ plug-flow reactor
Strömungsspirale *f* spiral of air stream
Strömungstyp *m* flow type
Strömungsvorgang *m* mechanism of flow, process of flow
Strömungswächter *m* flow guard
Strogonowit *m* strogonowite
Strohcellulose *f* straw cellulose
strohfarbig straw-colored
Strohfaser *f* straw fiber
Strohgeflecht *n* straw plait
Strohhalm *m* straw
Strohpappe *f* strawboard
Strohseilspinnmaschine *f* straw rope spinning machine
Strohstein *m* (Min) strawstone, carpholite
Strohzellstoff *m* (Pap) straw pulp
Strom *m* (Elektr) current, (Strömung) current, flow, stream, ~ **geringer Frequenz** low-frequency current, ~ **gleicher Richtung** unidirectional current, **abklingender** ~ decaying current, **den** ~ **sperren** to cut off the current, **einwelliger** ~ single-phase current, **Gleich-** ~ direct current, **gleichgerichteter** ~ direct current, rectified current, **Hochspannungs-** ~ high-voltage current, **induzierender** ~ inducing current, **induzierter** ~ induced current, **lichtelektrischer** ~ photoelectric current, **Niederspannungs-** ~ low-voltage current, **phasenverschobener** ~ out-of-phase current, dephased current, phase-displaced current, **pulsierender** ~ pulsating current, ripple current, **thermoelektrischer** ~ thermocurrent, **unregelmäßiger** ~ fluctuating current, **unterbrochener** ~ interrupted current, **vagabundierender** ~ current from external source[s], **Wechsel-** ~ alternating current
Stromabgabefähigkeit *f* current discharge capacity
Stromabnahme *f* decrease of current
Stromabnehmer *m* (Elektr) current collector, brush, (Verbraucher) consumer
Stromänderung *f* current change
Stromanlage *f* electric power plant
Stromanstieg *m* current rise
Stromanzeiger *m* current indicator, (Ampèremeter) amperemeter, (Galvanometer) galvanometer
Stromart *f* type of current
Stromaufnahmefähigkeit *f* charge capacity
Stromausbeute *f* current efficiency, current yield, electrodeposition equivalent
Stromausfall *m* black-out, absence of current
Stromausgleich *m* compensation of current
Strombahn *f* current path
Strombelastung *f* power load
Strombrecher *m* baffle

Stromdämpfer *m* current damper
Stromdämpfung *f* current attenuation, damping of current
Stromdiagramm *n* current diagram
Stromdichte *f* current density, **kathodische** ~ cathodic current density
Stromdichteverteilung *f* distribution of current density
Stromdreieck *n* current triangle
stromdurchflossen current carrying, traversed by a current
Stromdurchführung *f* current lead-in
Stromdurchgang *m* current passage
Stromeinführungselektrode *f* (Elektr) electrode carrying current
Stromeinheit *f* current unit
Stromeintrittszone *f* negative or cathodic area
Stromempfindlichkeit *f* current sensitivity
Stromentnahme *f* consumption of current
Stromerzeuger *m* [current] generator
Stromerzeugung *f* generation of current
Stromerzeugungsaggregat *n* current generating set
Stromerzeugungsanlage *f* current generating plant, electricity generating plant
Stromeyerit *m* (Min) stromeyerite
Stromfeld *n* (Elektr) current field, magnetic field due to a current
Stromförderer *m* continuous conveyor
stromführend charged, current carrying, live
Stromgefälle *n* (Elektr) potential difference along a current
Stromgeschwindigkeit *f* current velocity
Stromgleichrichter *m* current rectifier
Stromimpuls *m* (Elektr) pulsed voltage
Stromintensität *f* current intensity, amperage, current strength
Stromkapazität *f* current capacity
Stromklassierer *m* hydraulic classifier
Stromkomponente *f* current component
Stromkonsum *m* consumption of current
Stromkonzentration *f* current concentration
Stromkreis *m* (Elektr) circuit, **äußerer** ~ external circuit, **den** ~ **öffnen** to break the circuit, **den** ~ **schließen** to close the circuit, **geschlossener** ~ closed circuit, **in den** ~ **einschalten** to put into the circuit, **innerer** ~ internal circuit, **offener** ~ open circuit
Stromkreiselement *n* element of the circuit
Stromlauf *m* flow of current
Stromlaufschema *n* (Elektr) diagram of connections
Stromleitung *f* (Elektr) power line, **unterirdische** ~ underground line
Stromlieferung *f* power supply, supply of current
Stromlinie *f* streamline
stromlinienförmig streamlined

Stromlinienform f streamline form, streamlined contour, streamline shape
Stromlinienstreuung f dispersion of the current paths
stromlinig streamlined
stromlos currentless, dead, out of circuit
Stromlosigkeit f absence of current
Strommesser m amperemeter, ammeter, fluxmeter
Stromnetz n current network, electric supply lines, network
Stromphase f (Elektr) current phase
Stromquelle f power source, source of current
Stromregler m current regulator
Stromrichtung f (Elektr) direction of current
Stromrichtungsanzeiger m polarity indicator
Stromsammler m accumulator, storage battery
Stromschleife f current loop
Stromschluß m closing of the circuit
Stromschwächung f weakening of the current
Stromschwankung f fluctuation of current
Stromsequenzkomponente f component of current sequence
Stromspannung f voltage
Stromspannungskennlinie f volt-ampère characteristic, current-voltage characteristic
Stromstabilisierung f current stabilization
Stromstärke f amperage, current intensity, strength of current, **gleichbleibende** ~ constant current
Stromstoß m current impulse, current pulsation, rush of current, **kurzer** ~ pulse
Stromteilung f subdivision of the current
Stromtransformator m current transformer
Stromtransport m transport of the current
Stromtrockner m pneumatic-conveyor drier
Stromübergang m current conduction
Stromübertragung f current transmission
Stromuhr f (Stromzähler) current meter
Stromumkehrung f reversal of current, commutation of current
Strom- und Spannungsmesser m volt- and ammeter
Stromunterbrecher m circuit breaker; circuit cut-out switch
Stromunterbrechung f interruption of the current
Stromvektor m current vector
Stromverbrauch m consumption of current
Stromverbraucher m power consumer
Stromverdrängung f displacement of current
Stromverlust m leakage of current, loss of current
Stromversorgung f power supply, **proportional geregelte** ~ proportional power
Stromverstärkung f current amplification
Stromverteilung f current distribution
Stromvervielfachung f current multiplication
Stromverzögerung f retardation of current
Stromverzweigung f branching of current

Stromwärmeverlust m ohmic loss, resistance loss
Stromwandler m current transformer
Stromwechsel m (Elektr) alternation
Stromweg m path of current
Stromwender m commutator, current reverser
Stromwenderisolation f commutator insulation
Stromwendung f commutation, reversal of current
Stromzähler m current meter
Stromzuführungskabel n power feed cable
Stromzuführungsschelle f current conducting clamp
Stromzunahme f increase of current
Stromzweig m branch circuit
Strontian m (Min) strontia, strontium oxide
Strontianerde f (Min) strontia, strontium oxide
Strontianit m (Min) strontianite, native strontium carbonate
Strontiansalpeter m strontium nitrate
Strontianzucker m strontium sucrate
Strontium n strontium (Symb. Sr)
Strontiumacetat n strontium acetate
Strontiumalter n strontium age
Strontiumcarbid n strontium carbide, strontium acetylide
Strontiumcarbonat n strontium carbonate
Strontiumchlorid n strontium chloride
Strontiumgehalt m strontium content
Strontiumhydroxid n strontium hydroxide
Strontiumisotop n strontium isotope
Strontiumjodid n strontium iodide
Strontiumnitrat n strontium nitrate
Strontiumoxid n strontium oxide, strontia
Strontiumoxidhydrat n (obs) strontium hydroxide, hydrated strontium oxide
Strontiumperborat n strontium perborate
Strontiumsalpeter m strontium nitrate
Strontiumsulfat n strontium sulfate
Strontiumwasserstoff m strontium hydride
Strontiumweiß n strontium white
Strontiumwolframat n strontium tungstate
Strophanthidin n strophanthidin
Strophanthidinsäure f strophanthidinic acid
Strophanthin n strophanthin, strophanthum
Strophanthinsäure f strophanthinic acid
Strophanthobiose f strophanthobiose
Strophanthosid n strophanthoside
Strophanthotriose f strophanthotriose
Strophanthsäure f strophanthic acid
Strophanthum n strophanthin, strophanthum
Strophanthusöl n strophanthus seed oil
Strüverit m (Min) struverite
Struktur f grain, texture, (Chem) structure, **adrige** ~ nerve structure, **aromatische** ~ aromatic structure, **benzoide** ~ benzoid structure, **blättrige** ~ foliated structure, **ebene** ~ planar structure, **faserige** ~ fibrous structure, **feinschuppige** ~ fine-scaled structure, **gewinkelte** ~ angular structure,

körnige ~ granular structure, **lamellare** ~ lamellar structure, **primäre** ~ primary structure, **schieferartige** ~ slaty structure, **sekundäre** ~ secondary structure, **tertiäre** ~ tertiary structure, **zellenartige** ~ cellular structure, honeycombed structure
strukturabhängig structure dependent
Strukturänderung *f* change in structure, change in texture, (Chem) structural change
Strukturanalyse *f* (Chem) structural analysis
Strukturaufbau *m* structure, structural design
Strukturbestimmung *f* determination of structure
Strukturbeweis *m* proof of structure
Strukturchemie *f* structural chemistry
Struktureinheit *f* structural unit
strukturell structural
strukturempfindlich structure sensitive
Strukturermüdung *f* structural fatigue
Strukturfaktor *m* structure factor
Strukturfestigkeit *f* tear resistance
Strukturformel *f* (Chem) structural formula
Strukturisomerie *f* (Chem) structural isomerism
strukturlos amorphous, structureless
Strukturprotein *n* structural protein
Strukturschema *n* configuration
Strukturtyp *m* (Krist) structural type
Strukturveränderung *f* change of texture, (Chem) change of structure
Strukturverschiedenheit *f* difference in structure
strukturviskos structurally viscose
Strukturviskosität *f* structural viscosity
Strukturwandel *m* change in structure
Strunk *m* stock, stump, trunk
struppig shaggy, (borstig) bristly
Struvit *m* (Min) struvite
Strychnidin *n* strychnidine
Strychnin *n* strychnine, vauqueline
Strychninacetat *n* strychnine acetate
Strychninhydrochlorid *n* strychnine hydrochloride
Strychninnitrat *n* strychnine nitrate
Strychninsäure *f* strychnic acid
Strychnospermin *n* strychnospermine
Stuck *m* stucco
Stuckgips *m* sculptor's plaster, stucco
Studerit *m* studerite
Studie *f* study
Stück *n* piece, (Bruchteil) fragment, (Metall) bloom, (Seife) cake, (Zucker) lump
Stückarbeiter *m* piece worker, job worker
Stückerz *n* lump ore
Stückfärberei *f* piece dyeing
stückgefärbt piece dyed
Stückgehalt *m* (Erz) proportion of lump ore
Stückgewicht *n* weight of a single piece
Stückigmachen *n* size enlargement
Stückinhalt *m* number of pieces
Stückkiesofen *m* gravel-filled kiln
Stückkohle *f* large coal, lump coal

Stücklohn *m* wage per piece
Stückquarz *m* lump quartz, quartz in lumps
Stückseife *f* hard soap
Stückware *f* piece goods
Stückzahl *f* piece number
Stülpenbeutelzentrifuge *f* horizontal filter-bag centrifuge
Stürzbühne *f* tipping stage
stürzen to tumble
Stürzflamme *f* reverberatory flame
Stürzprobe *f* shatter test (coke)
Stürzrinne *f* chute
Stützbalken *m* brace, stringer
Stütze *f* pillar, prop, stanchion, stand, support post
Stützgewebe *n* (Histol) stroma, pseudocartilage, supporting tissue
Stützhöhe *f* vertical span
Stützkern *m* retaining core
Stützplatte *f* bracket plate
Stützpunkt *m* point of support, base point, bearing surface, (Drehpunkt) fulcrum
Stützring *m* thrust ring
Stützrolle *f* idler
Stützwalze *f* backing roll, supporting roll
Stufe *f* (Grad) degree, grade, graduation, (Schritt) step, (Stadium) stage, (Treppe) stair, step
Stufenblende *f* (Phot) diaphragm with several grades
Stufeneluierung *f* stepwise elution
Stufenerz *n* (Min) graded ore, pure ore
Stufenfolge *f* graduation
Stufenfunktion *f* step function
Stufengenerator *m* cascade generator
Stufengetriebe *n* multi-step reduction gear
Stufengitter *n* echelon grating
Stufenheizung *f* (Gummi) vulcanization in stages, step-up cure
Stufenhöhe *f* step height
Stufenkaskade *f* square cascade
Stufenkeil *m* (Opt) stepped photometric absorption wegde
Stufenkolbenpumpe *f* differential pump
Stufenkompressor *m* stage compressor
Stufenlänge *f* step length
Stufenlinse *f* (Opt) echelon lens, step lens
stufenlos stepless, ~ **regelbar** infinitely variable
Stufenmechanismus *m* stepwise mechanism
Stufenmotor *m* step-up motor
Stufennut *f* step groove
Stufenprisma *n* (Opt) echelon prism
Stufenprozeß *m* step process
Stufenrakete *f* multi-stage rocket
Stufenreaktion *f* successive reaction
Stufenregelung *f* step control, zone control
Stufenregler *m* step controller
Stufenschalter *m* step switch
Stufenschaltung *f* cascade connection

Stufenscheibe *f* step pulley
Stufenschneidemaschine *f* progressive cutting machine
Stufentransformator *m* step-down transformer
Stufenverfahren *n* cascade process
Stufenversetzung *f* edge dislocation
Stufenwalze *f* stepped roll
stufenweise stepwise, by degrees, gradually, in stages, step by step
Stufenzahl *f* number of stages
Stufenzahnrad *n* cluster gear, stepped gear
stumpf (Farbe) dull, (Schneide) blunt, (Winkel) obtuse, ~ **angefugt** butt joined, ~ **löten** to butt-solder, ~ **schweißen** to butt-weld, ~ **zusammensetzen** to butt splice
Stumpf *m* trunk
stumpfkantig blunt-edged
Stumpfkegel *m* truncated cone
Stumpfnaht *f* butt joint, butt weld
Stumpfschweißen *n* butt welding
Stumpfstoß *m* butt joint, butt weld
Stumpfstoßschweißung *f* butt welding
stumpfwinklig obtuse angled
Stunde *f* hour
Stundendrehzahl *f* revolutions per hour
Stundengeschwindigkeit *f* hourly rate
Stundenleistung *f* output per hour
Stundenplan *m* time-table
Stuppeasäure *f* stuppeaic acid, stuppeic acid
Sturin *n* sturin
Sturmhut *m* (Bot) aconite
Sturz *m* crash, fall, sudden drop, (Felge) camber
Stutzen *m* adapter
Stylopin *n* stylopine
Stylotyp *m* (Min) stylotype
Styphninsäure *f* styphnic acid, 2,4-dihydroxy-1,3,5-trinitrobenzene
styphninsauer (Salz) styphnate
Stypticin *n* stypticine, cotarnine hydrochloride
Stypticit *m* stypticite
Styptikum *n* (Pharm, Med) styptic
Styptol *n* styptol, cotarnine phthalate
Styracin *n* styracine, cinnamyl cinnamate
Styracit *n* styracitol, styracite
Styracol *n* styracol
Styrax *m* (Balsam) storax, (Baum) styrax
Styren-Butadien *n* styrene-butadiene
Styrol *n* styrene, cinnamene, phenylethylene, styrol, styrolene, vinylbenzene
Styrolharz *n* styrene resin
styrolisiert styrenated
Styrolkunststoff *m* styrene plastic
Styron *n* styron, cinnamyl alcohol
Styryl- styryl
Styrylketon *n* styryl ketone
Subaphyllin *n* subaphylline
Subathizon *n* subathizone
subatomar subatomic
Subchlorid *n* subchloride

Subdeterminante *f* (Math) sub-determinant
Suberan *n* suberane, cycloheptane
Suberin *n* suberin, cork fat
Suberinsäure *f* suberic acid, octanedioic acid
Suberoin *n* suberoin
Suberol *n* suberol, cycloheptanol
Suberon *n* suberone, cycloheptanone
Suberonsäure *f* suberonic acid
Suberylalkohol *m* suberyl alcohol
Suberylarginin *n* suberylarginine
Subhalogenid *n* subhalide
Subhaloid *n* basic halide
subjektiv subjective[ly]
subkutan subcutaneous[ly]
Sublamin *n* sublamine, mercuric sulfate ethylenediamine
Sublimat *n* (Chem) mercuric chloride, (sublimierter Stoff) sublimate (condensed substance)
Sublimatbad *n* sublimate bath
Sublimation *f* sublimation
Sublimationsapparat *m* sublimation apparatus
Sublimationsdruck *m* sublimation pressure
Sublimationsenthalpie *f* sublimation enthalpy
Sublimationskern *m* sublimation nucleus
Sublimationskoeffizient *m* sublimation coefficient
Sublimationskühlung *f* sublimation cooling
Sublimationskurve *f* sublimation curve
Sublimationspunkt *m* sublimation point
Sublimationstemperatur *f* sublimation temperature
Sublimationswärme *f* heat of sublimation
Sublimatlösung *f* mercuric chloride solution
Sublimator *m* sublimer, **Dünnfilm-** ~ thin-film sublimer, **Horden-** ~ tray sublimer, **Turbinen-** ~ turbine sublimer
Sublimierapparate *pl* sublimation equipment
sublimierbar sublimable
Sublimierbarkeit *f* sublimability
Sublimierechtheit *f* fastness to subliming
sublimieren to sublimate, to sublime
Sublimieren *n* sublimating
Sublimiergefäß *n* sublimating vessel
Sublimierofen *m* sublimator, subliming furnace
Sublimierretorte *f* sublimation retort
Sublimiertopf *m* subliming pot
Sublimierung *f* sublimation
Submikron *n* submicron
submikroskopisch submicroscopic
Submikrostruktur *f* submicrostructure
Subnitrat *n* subnitrate
Subnormale *f* subnormal
suborbital suborbital
Suboxid *n* suboxide, lowest oxide
Substantivität *f* substantivity
Substanz *f* substance, material, matter, **anionenaktive** ~ anionic substance, **chemisch reine** ~ chemically pure substance,

farblose ~ colorless substance, **hochpolymere** ~ highly polymerized substance, **kationenaktive** ~ cationic substance
Substanzmenge *f* amount of substance
Substanzveränderung *f* change of substance
Substanzverlust *m* loss of material
Substituent *m* substituent
substituierbar replaceable
substituieren to substitute, to replace
substituierend substituting, ~ **wirken** to react by substitution
Substituierung *f* substitution
Substitution *f* substitution, ~ **von Radikalen** radical substitution, **elektrophile** ~ electrophilic substitution, **nucleophile** ~ nucleophilic substitution
Substitutionsprodukt *n* substitution product
Substitutionsreaktion *f* substitution reaction
Substrat *n* carrier, substrate, (Biochem) substrate, (Bot) substratum
Substrathemmung *f* (Biochem) substrate inhibition
Substratsättigung *f* (Biochem) substrate saturation
Subtangente *f* subtangent
Subtilin *n* subtilin
Subtilisin *n* (Biochem) subtilisin
Subtrahend *m* (Math) subtrahend
subtrahieren to subtract
Subtraktion *f* (Math) subtraction
Subtraktionsfarbe *f* subtraction color, subtractive color
Subtraktivfilter *m* (Phot) subtractor
Subtraktivverfahren *n* (Phot) subtractive process, three-color subtraction printing
Succinamid *n* succinamide
Succinamidsäure *f* succinamic acid
Succinanhydrid *n* succinic anhydride
Succinanilid *n* succinanilide
Succinat *n* succinate, salt or ester of succinic acid
Succinatdehydrogenase *f* (Biochem) succinate dehydrogenase
Succinatthiokinase *f* (Biochem) succinate thiokinase
Succindialdehyd *m* succindialdehyde, butanedial
Succinhydrazid *n* succinhydrazide
Succinimid *n* succinimide
Succinimidin *n* succinimidine
Succinit *m* (Min) succinite
Succinoabietinolsäure *f* succinoabietinolic acid
Succinoabietinsäure *f* succinoabietic acid
Succinoabietol *n* succinoabietol
Succinoarginin *n* (Biochem) succino-arginine
Succinodehydrogenase *f* (Biochem) succinodehydrogenase
Succinonitril *n* succinonitrile, ethylene cyanide
Succinoresen *n* succinoresene

Succinoresinol *n* succinoresinol
Succinoylsulfathiazol *n* succinoylsulfathiazole
Succinsäure *f* succinic acid, butanedioic acid
Succinyl- succinyl, butanedioyl
Succinylbernsteinsäure *f* succinylosuccinic acid
Succinylchlorid *n* succinyl chloride
Succinylcholin *n* succinylcholine
Succinyl-CoA *n* (Biochem) succinyl coenzyme A, succinyl CoA
Succinylfluorescein *n* succinylfluorescein
Suchbohrung *f* exploration well
suchen to look [for], to search, to seek
Sucher *m* (Opt) viewer, view finder
Sucherobjektiv *n* view finder lens
Sucherokular *n* (Opt) focusing lens
Suchmethode *f* search method
Suchradar *n* search radar
Sucht *f* addiction, (Med) disease, epidemic
Sucrase *f* (Biochem) sucrase, invertase
Sucrat *n* sucrate
Sucrol *n* sucrol, dulcin, p-ethoxyphenylurea
Sucrose *f* sucrose, saccharose
Sucrosedichtegradient *m* (Biochem) sucrose [density] gradient
Sucrosegradientenzentrifugation *f* sucrose gradient centrifugation
Sud *m* (Färb) mordant
Sudan *n* sudan
Sudhaus *n* (Brau) brewing house
Sudvergoldung *f* hot gilding
Sudvernick[e]lung *f* hot nickel plating
Sudwerk *n* boiling apparatus
Sudzeit *f* (Brau) period of brewing
süchtig addicted
Sülze *f* gelatine, jelly
süßen to sweeten
Süßerde *f* beryllia, glucina
Süßholz *n* licorice
Süßholzextrakt *m* licorice extract
süßlich sweetish
süßsäuerlich sourish sweet
Süßstoff *m* sweetening agent, (Saccharin) saccharin
Süßwasser *n* fresh water
Süßwasserbehälter *m* fresh water tank
Süßwasserfisch *m* fresh water fish
Sukzession *f* succession
Sulfacetamid *n* sulfacetamide
Sulfachrysoidin *n* sulfachrysoidine
Sulfadiazin *n* sulfadiazine
Sulfadicramid *n* sulfadicramide
Sulfaethidol *n* sulfaethidole
Sulfaguanidin *n* sulfaguanidine
Sulfamethoxypyridazin *n* sulfamethoxypyridazine
Sulfamid *n* sulfamide
Sulfamidsäure *f* sulfamic acid
Sulfanilamid *n* sulfanilamide
Sulfanilid *n* sulfanilide

Sulfanilsäure *f* sulfanilic acid
sulfanilsauer sulfanilate
Sulfanthrol *n* sulfanthrol
Sulfantimonat *n* sulfantimonate, thioantimonate
Sulfantimonsäure *f* thioantimonic acid
Sulfapyridin *n* sulfapyridine
Sulfapyrimidin *n* sulfapyrimidine
Sulfarsenat *n* sulfarsenate, thioarsenate
Sulfarsensäure *f* sulfarsenic acid, thioarsenic acid
Sulfat *n* sulfate, salt or ester of sulfuric acid, (Br. E.) sulphate
Sulfatase *f* (Biochem) sulfatase
Sulfatation *f* sulfatization
Sulfatcellulose *f* sulfate pulp
Sulfatgehalt *m* sulfate content
sulfathaltig containing sulfate
Sulfathiazol *n* sulfathiazole
sulfatieren to sulfate, (durch Rösten etc.) to sulfatize
Sulfatieren *n* sulfating, (durch Rösten etc.) sulfatizing
Sulfatierung *f* sulfation, (durch Rösten) sulfatization
Sulfation *n* sulfate ion
Sulfatisierungsmittel *n* sulfating agent
Sulfatlaugeeindampfanlage *f* sulfate liquor concentrator
Sulfatpapier *n* wood pulp paper
Sulfatstoff *m* (Pap) sulfate pulp
Sulfatwasser *n* sulfate water
Sulfatzellstoff *m* (Pap) sulfate cellulose, sulfate pulp
Sulfcarbaminsäure *f* thiocarbamic acid
Sulfcarbanil *n* thiocarbanil, phenyl isothiocyanate
Sulfensäure *f* sulfenic acid
Sulfhydrat *n* hydrogen sulfide, acid sulfide, hydrosulfide, sulfhydrate
Sulfhydryl- *n* sulfhydryl, mercapto
Sulfid *n* sulfide, salt of hydrogen sulfide
Sulfidal *n* sulfidal, colloidal sulfur
Sulfideinschluß *m* sulfide inclusion
Sulfidierung *f* sulfidization
sulfidisch sulfidic, containing sulfide
sulfieren to sulfurize, to sulfurate
Sulfierung *f* sulfuration, sulfurization
Sulfimid *n* sulfimide
Sulfin *n* sulfine, sulfonium compound
Sulfinsäure *f* sulfinic acid
Sulfinsalz *n* sulfonium salt
Sulfit *n* sulfite, salt of sulfurous acid, (Br. E.) sulphite
Sulfitablauge *f* sulfite waste liquor
Sulfitcellulose *f* (Pap) sulfite cellulose, sulfite pulp
sulfitieren to sulphite
Sulfitkocher *m* (Pap) sulfite digestor
Sulfitkochlauge *f* sulfite liquor

Sulfitlauge *f* sulfite liquor, sulfite lye, sulfite solution
Sulfitlaugeanlage *f* (Pap) sulfite liquor plant
Sulfitoxydase *f* (Biochem) sulfite oxidase
Sulfitprozeß *m* (Pap) sulfite process
Sulfitreduktase *f* (Biochem) sulfite reductase
Sulfitverfahren *n* (Pap) sulfite process
Sulfitzellstoff *m* (Pap) sulfite cellulose, sulfite pulp
Sulfobase *f* sulfur base
Sulfobenzid *n* sulfobenzide, diphenyl sulfone
Sulfobenzimid *n* sulfobenzimide, saccharin
Sulfobenzoesäure *f* sulfobenzoic acid
sulfobenzoesauer (Salz) sulfobenzoate
Sulfoborit *m* (Min) sulfoborite
Sulfocarbanilid *n* sulfocarbanilide, diphenylthiourea, thiocarbanilide
Sulfocarbolsäure *f* sulfocarbolic acid, phenolsulfonic acid, sulfophenol
sulfocarbolsauer sulfocarbolate, sulfophenolate
Sulfocarbonat *n* sulfocarbonate, thiocarbonate, trithiocarbonate
Sulfocarbonsäure *f* sulfocarboxylic acid
Sulfochlorid *n* sulfochloride, sulfonyl chloride
Sulfocyan *n* thiocyanogen, sulfocyanogen
Sulfocyanat *n* thiocyanate, sulfocyanate
Sulfocyaneisen *n* iron thiocyanate
Sulfocyanid *n* thiocyanate, sulfocyanate, sulfocyanide
Sulfocyankalium *n* (obs) potassium sulfocyanate
Sulfocyansäure *f* thiocyanic acid, sulfocyanic acid
sulfocyansauer (Salz) thiocyanate, sulfocyanate
Sulfocyanverbindung *f* thiocyanate, sulfocyanogen compound, thiocyanogen compound
Sulfoessigsäure *f* sulfoacetic acid
Sulfoform *n* sulfoform, triphenylstibine sulfide
Sulfogruppe *f* sulfo-group, sulfonic group
Sulfoharnstoff *m* thiourea, sulfourea
Sulfohydrat *n* hydrosulfide, (obs) sulfhydrate, hydrogen sulfide
Sulfoichtyolsäure *f* sulfoichtyolic acid
Sulfokohlensäure *f* sulfocarbonic acid, trithiocarbonic acid
Sulfolan *n* sulfolane
Sulfolyse *f* sulfolysis
Sulfon *n* sulfone
Sulfonal *n* sulfonal, sulfonmethane
Sulfonamid *n* sulfonamide
Sulfonat *n* sulfonate
Sulfoncarbonsäure *f* sulfone carboxylic acid
Sulfondichloramidobenzoesäure *f* dichloramidobenzosulfonic acid
Sulfonieranlage *f* sulfonation plant
sulfonierbar sulfonatable
sulfonieren to sulfonate
Sulfonieren *n* sulfonating
Sulfonierung *f* sulfonation

Sulfoniumradikal – Suspensionskolloid

Sulfoniumradikal *n* sulfine, sulfonium radical
Sulfoniumverbindung *f* sulfonium compound
Sulfonsäure *f* sulfonic acid
Sulfonsäureamid *n* sulfonamide
Sulfonsäurechlorid *n* sulfochloride
Sulfonyldiessigsäure *f* sulfonyldiacetic acid
Sulfopersäure *f* persulfuric acid, peroxysulfuric acid
sulfophenolsauer sulfophenolate, sulfocarbolate
Sulfophthalsäure *f* sulfophthalic acid
Sulforaphen *n* sulforaphen
Sulfosalicylaldehyd *m* sulfosalicylaldehyde
Sulfosalicylsäure *f* sulfosalicylic acid
sulfosalicylsauer sulfosalicylate
Sulfosol *n* sulfosol
Sulfoxidradikal *n* sulfoxide radical
Sulfoxylsäure *f* sulfoxylic acid
Sulfozimtsäure *f* sulfocinnamic acid
Sulfür *n* (obs) lower sulfide, protosulfide
Sulfuretin *n* sulfuretin
sulfurieren to sulfurate, to sulfurize
Sulfurieren *n* sulfurating, sulfurizing
Sulfurierung *f* sulfonation, sulfuration, sulfurization
Sulfurolschwarz *n* sulfurol black
Sulfuryl- sulfuryl
Sulfurylchlorid *n* sulfuryl chloride
Sulochrin *n* sulochrin
Sulvanit *m* (Min) sulvanite
Sumach *m* (Bot, Gerb) sumac[h]
Sumachgerbung *f* sumac[h] tanning
Sumachtrübe *f* sumac[h] liquor
Sumaresinol *n* sumaresinol, sumaresinolic acid
Sumatrol *n* sumatrol
Sumbulwurzel *f* (Bot) sumbul root, musk root
Summand *m* (Math) addend
Summationsformel *f* summation formula
Summationszeichen *n* (Math) summation sign
Summe *f* sum, **algebraische** ~ algebraic[al] sum
Summenformel *f* summation formula, empirical formula
Summengleichung *f* summation equation
Summensatz *m* sum rule
Summenübertragung *f* (Comp) total transfer
Summenzähler *m* (Comp) summation meter
Summer *m* buzzer
Summererregung *f* buzzer excitation
summieren to sum, to add up, to total
Summierung *f* summation, **absatzweise** ~ (Comp) intermittent integration, **selektive** ~ (Comp) selective summarizing
Summierungsband *n* (Spektr) summation band
Summierungsprinzip *n* principle of additivity
Sumpf *m* sump, marsh, swamp, (Dest) bottom
Sumpfablaßventil *n* waste valve
Sumpfeisenstein *m* (Min) bog iron ore
Sumpferz *n* (Min) bog [iron] ore, limonite, marsh ore
Sumpfgas *n* marsh gas

Sumpfprodukt *n* (Dest) bottom product, residue
Sumpfreaktor *m* packed bubble column
Sumpfverfahren *n* liquid phase process
Superbenzin *n* premium gasoline
Superbombe *f* superbomb
superelastisch superelastic
Superkalandrieren *n* (Pap) supercalendering
Superlegierung *f* (Hochtemperaturlegierung) high temperature alloy
Superleitfähigkeit *f* superconductivity
Supermultiplett *n* supermultiplet
Superoxid *n* superoxide, peroxide
Superoxidbildung *f* formation of peroxide
Superoxidschlamm *m* peroxide sediment
Superpalit *n* superpalite
Superphosphat *n* superphosphate
Superpolyamid *n* superpolyamide
Superpolyester *m* superpolyester
Superpolymer[e] *n* superpolymer
Superposition *f* superposition, superimposement
Superpositionsprinzip *n* superposition principle, ~ **der Optik** (Opt) principle of optical superposition
supersaturieren to supersaturate
Supersulfid *n* persulfide
Superwaffe *f* superweapon
Superzentrifuge *f* supercentrifuge, tubular bowl centrifuge
Supinidin *n* supinidine
Supinin *n* supinine
supplementär supplementary
Supplementärwinkel *m* supplementary angle
Suppressor *m* (Biochem) suppressor
Suppressormutation *f* suppressor mutation
supraflüssig superfluid
Suprafluid *n* superfluid
Suprafluidität *f* superfluidity
Supraleitelektron *n* superconduction electron
supraleitend superconductive
Supraleiter *m* superconductor, supraconductor, **nichtidealer** ~ (Elektr) hard superconductor, non-ideal superconductor
Supraleitfähigkeit *f* superconductivity, supraconduction, supraconductivity
Supraleitungsübergang *m* superconductive transition
supramolekular supramolecular
Suprasterin *n* suprasterol
Suramin *n* suramin
Surinamin *n* surinamine
Surpalit *n* surpalite, diphosgene
Surrogat *n* substitute
Susannit *m* susannite
suspendieren to suspend
Suspendiermaschine *f* suspension machine
Suspension *f* suspension, **grob disperse** ~ coarsely dispersed suspension, **wäßrige** ~ aqueous suspension, water slurry
Suspensionskolloid *n* suspensoid

Suspensionspolymerisation *f* suspension polymerization
Suspensionsreaktor *m* slurry reactor
Suspensoid *n* suspensoid
Suszeptanz *f* susceptance
Suszeptibilität *f* susceptibility, coefficient of induced magnetization
Suszeptibilitätsmaximum *n* susceptibility maximum
Suxamethonium *n* suxamethonium
Suxethonium *n* suxethonium
Svanbergit *m* (Min) svanbergite
Svedberg-Einheit *f* (Ultrazentrifugation) Svedberg unit
Swartziol *n* swartziol
Swertinin *n* swertinin
Swietenose *f* swietenose
Sychnodymit *m* sychnodymite
Sycocerylalkohol *m* sycoceryl alcohol
Sydnon *n* sydnone
Syenilit *m* syenilite
Syenit *m* (Min) syenite
syenithaltig syenitic
Sylvan *n* sylvan
Sylvanerz *n* (Min) sylvanite
Sylvanit *m* (Min) sylvanite
Sylvestren *n* sylvestrene
Sylvin *m* (Min) sylvine, isomorphous native mixture of K and Na chlorides
Sylvinit *m* (Min) sylvinite
Sylvinsäure *f* sylvic acid, abietic acid
Sylvit *m* (Min) sylvite, native potassium chloride
Symbiont *m* (Biol) symbiont
symbio[n]tisch symbiotic
Symbiose *f* symbiosis
Symbol *n* symbol
Symmetrie *f* symmetry, **äußere** ~ external symmetry, **höchste** ~ highest symmetry, **innere** ~ internal symmetry, **statistische** ~ statistic symmetry, **x-fache** ~ x-fold symmetry
Symmetrieachse *f* axis of symmetry
Symmetrieebene *f* plane of symmetry
Symmetrieeigenschaft *f* property of symmetry
Symmetrieelement *n* element of symmetry
Symmetriegrad *m* degree of symmetry
Symmetrieoperator *m* symmetry operator
Symmetriesymbolik *f* symmetry symbols
Symmetriezahl *f* symmetry factor
Symmetriezentrum *n* center of symmetry
symmetrisch symmetric[al]
Symons-Brecher *m* Symons standard cone crusher
Sympatholytin *n* sympatholytin
Sympatol *n* sympatol
Symptom *n* symptom
Synaldoxim *n* synaldoxime
Synanthrose *f* synanthrose
Syncain *n* procaine, syncaine

synchron synchronous, ~ **laufen** to run synchronous
Synchrongeschwindigkeit *f* synchronous speed
Synchronisation *f* synchronization
synchronisieren to synchronize
Synchronisierung *f* synchronization, timing
Synchronismus *m* synchronism
Synchronmotor *m* synchronous motor
Synchron-Phasenschieber *m* synchro phase shifter, phase-shifting transformer
Synchronrechner *m* synchronous computer
Synchronübertrager *m* synchro transmitter
Synchronuhr *f* synchronous clock
Synchroskopmethode *f* triggered oscilloscope method
Synchrotron *n* synchrotron
Syndiazotat *n* syndiazotate
Syndikat *n* syndicate
syndiotaktisch syndiotactic
Syndrom *n* (Med) syndrome
Synephrin *n* synephrine
Synergismus *m* synergism, mutual action
Syn-Form *f* syn-isomer
Syngenit *m* (Min) syngenite
syn-Konfiguration *f* (Stereochem) syn-configuration
Synthalin *n* synthalin
Synthease *f* (Biochem) synthetase
Synthese *f* synthesis, ~ **ausführen** to synthesize, **asymmetrische** ~ asymmetric synthesis, **stereoselektive** ~ stereoselective synthesis
Synthesefaser *f* synthetic fiber
Synthesegas *n* synthesis gas, water gas
Synthesegasanlage *f* synthesis gas plant
synthetisch synthetic, artificial, ~ **herstellen** to prepare synthetically
synthetisieren to synthesize
Synthetisierung *f* synthesizing
Syntonin *n* syntonin, muscle fibrin, parapeptone
Syoyu-Aldehyd *m* syoyu-aldehyde
Syringaalkohol *m* syringyl alcohol
Syringasäure *f* syringic acid
Syringenin *n* syringenin
Syringin *n* syringin, ligustrin, lilacin
System *n* system, method, scheme, **abgeschlossenes** ~ closed system, **antisymmetrisches** ~ antisymmetric system, **azeotropes** ~ azeotropic system, **binäres** ~ binary system, two-component system, **dioptrisches** ~ (Opt) dioptric system, **heterogenes** ~ heterogeneous system, **homogenes** ~ homogeneous system, **katoptrisches** ~ (Opt) catoptric system, **kondensiertes** ~ condensed system, **konjugiertes** ~ conjugated system, **kubisches** ~ cubic system, **metrisches** ~ metric system, **monoklines** ~ monoclinic system, **periodisches** ~ periodic system, periodic table, **rhombisches** ~ rhombic

system, **tetragonales** ~ tetragonal system, **triklines** ~ triclinic system
Systemanalyse f system analysis
systematisch systematic[al], methodical
Systemspeicher m (Comp) system residence
Szilard-Chalmers-Verfahren n Szilard-Chalmers method
Szillarenin n scillarenin
Szintigramm n scintigram
Szintillation f scintillation
Szintillationsansprechvermögen n scintillation response
Szintillationsflüssigkeit f scintillation liquid
Szintillationsmesser m scintillation counter
Szintillationsspektrometer n scintillation spectrometer
Szintillationsspektrometrie f scintillation spectrometry
Szintillationssubstanz f scintillant
Szintillationszähler m scintillation counter
Szintillator m scintillation counter, scintillator
szintillieren to scintillate
Szintilloskop n scintilloscope

T

Tabacoresen *n* tabacoresene
Tabacoresinol *n* tabacoresinol
Tabakmosaikvirus *m* tobacco mosaic virus
Tabaksamenöl *n* tobacco seed oil
Tabakteer *m* tobacco tar
Tabaschir *m* (Tabaxir) tabashir, bamboo secretion, tabasheer, tabashis
Tabbyit *m* tabbyite
tabellarisch tabular, ~ **zusammenstellen** to tabulate
tabellarisieren to tabularize, to tabulate
Tabellarisierung *f* tabulation
Tabelle *f* table, chart, index, list, register
Tabergit *m* tabergite
Tablett *n* tray
Tablette *f* pill, pellet, tablet
Tablettengröße *f* pellet size
Tablettenherstellung *f* tablet production
Tablettenimplantate *pl* hypodermic tablets
Tablettenkomprimiermaschine *f* automatic tableting machine
Tablettenpresse *f* pelleting press, tablet compressing machine
Tablettenzerfallbarkeitsprüfer *m* tablet disintegration tester
Tablettenzucker *m* tablet sugar
tablettieren to pellet
Tablettierfähigkeit *f* pelleting property
Tablettiermaschine *f* pelleting press, tablet compressing machine, tableting machine
Tablettierpresse *f* pelleting press, tablets press
Tablettierwerkzeug *n* preforming tool
Tabtoxinin *n* tabtoxinine
Tabun *n* tabun
Tacamahak *n* (Harz) tacamahac
Tachhydrit *m* (Min) tachhydrite
Tachiol *n* tachiol, silver fluoride
Tachograph *m* tachograph, velocity recorder
Tachometer *m* speedometer, tachometer
Tachyaphaltit *m* tachyaphaltite
Tachylyt *m* tachylite
Tachysterol *n* tachysterol
Tacrin *n* tacrine
Täfelung *f* panelling, wainscot, wall board
Tänzerrolle *f* dancer roll, floating roll
Tänzerwalze *f* dancer roll (for tension control)
Tätigkeit *f* occupation, work
Täuschung *f*, **optische** ~ optical delusion
Tafel *f* table, (Blech) plate, sheet, (Glas) pane, (Holz) board, (Stein) slab, tablet
Tafelbasalt *m* (Min) tabular basalt
Tafelbreite *f* sheet width
tafelförmig tabular
Tafelöl *n* salad oil
Tafelschiefer *m* (Min) roofing slate

Tafelspat *m* (Min) tabular spar, scale stone, wollastonite
Tagatose *f* tagatose
Tagebau *m* (Bergb) open cast mining, open digging
Tagebaugewinnung *f* open-cast working
Tagesleistung *f* daily output
Tageslicht *n* daylight
Tageslichtleuchtfarben *pl* daylight fluorescent colors
Tagesproduktion *f* daily output
Tagilit *m* (Min) tagilite
Tagschicht *f* (Bergb) day shift
Tag- und Nachtbetrieb *m* 24 hours service, day and night service
Tagundnachtgleiche *f* equinox
Tagung *f* conference, convention, meeting
Tagwasser *n* (Bergb) surface water
Taifun *m* typhoon
Tainiolith *m* (Min) tainiolite
Takatonin *n* takatonine
Taka[w]obase *f* Taka[w]o base
Takonit *m* (Min) taconite
Takt *m* beat, (Motor) stroke
Taktoid *n* (Kolloidchem) tactoid
Tal *n* valley, (einer Kurve) trough
Talbot-Bänder *pl* (Spektr) Talbot bands
Talg *m* tallow, suet, **chinesischer** ~ Chinese tallow, vegetable wax
talgartig sebaceous, tallow-like, tallowy
Talgdrüse *f* (Med) sebaceous gland
talghaltig containing tallow, sebaceous
Talgöl *n* tallow oil
Talgsäure *f* (obs) stearic acid
Talgseife *f* tallow soap
Talgstein *m* (Min) soapstone, steatite
Talgstoff *m* (obs) stearin
Talit *m* talitol, talite
Talk *m* (Min) talc, talcum
Talkapatit *m* (Min) talcapatite
talkartig talcose, talcous
Talkglimmer *m* (Min) lamellar talc
Talkgneis *m* (Min) talc gneiss
Talkgranat *m* (Min) talc garnet
Talkpulver *n* talcum powder
Talkquarz *m* (Min) talcose quartz
Talkschiefer *m* (Min) slaty talc, talc schist
Talkspat *m* (Min) magnesite
Talkstein *m* (Min) soapstone, steatite
Talkton *m* talcous clay
Talkum *n* talc[um], pulverized soapstone
Talkumiermaschine *f* soap stone machine
Talkumpuder *m* powdered soapstone, talc[um] powder
Tallianin *n* tallianine
Tallöl *n* tall oil

Tallöllack – Tartronylharnstoff

Tallöllack *m* tall oil varnish
Tallölsäure *f* talloleic acid
Talmi *n* talmi [gold], gold-like brass
Talolacton *n* talolactone
Talomethylose *f* talomethylose
Talonsäure *f* talonic acid
Talosamin *n* talosamine
Taloschleimsäure *f* talomucic acid
Talose *f* talose
Tamanit *m* (Min) tamanite
Tamarinde *f* (Pharm) tamarind
Tamarugit *m* (Min) tamarugite
Tampicin *n* tampicin
Tampicolsäure *f* tampicolic acid
Tampon *m* pad, plug, tampon
Tanaceten *n* tanacetene
Tanaceton *n* tanacetone
Tanacetophoron *n* tanacetophorone
Tandemmaschine *f* tandem engine
Tandempumpe *f* tandem pump
Tang *m* kelp, tang, (Bot) seaweed
Tangens *m* (Math) tangent
Tangente *f* tangent
Tangentensatz *m* (Math) tangent law, tangent theorem
Tangentenschnittpunkt *m* point of intersection of tangent[s]
tangential tangential
Tangentialbeanspruchung *f* tangential stress
Tangentialbeschleunigung *f* tangential acceleration
Tangentialebene *f* tangential plane
Tangentialgeschwindigkeit *f* tangential velocity
Tangentialguß *m* (Gieß) tangential casting
Tangentialkraft *f* (Mech) tangential force
Tangentialschnitt *m* tangential cut
Tangentialspannung *f* tangential stress
Tanghinin *n* tanghinine
Tank *m* tank, container, reservoir, storage basin, vessel
Tankanlage *f* storage tank installation
Tankanstrich *m* tank coating
tanken to fill up, to refuel
Tanken *n* refueling
Tanklager *n* (Öl) oil storage tank
Tankstelle *f* filling station, gas[oline] station, service station
Tankwagen *m* tank wagon
Tannal *n* tannal, aluminum tannate
Tannalbin *n* tannalbin, tannin albumate
Tannase *f* (Biochem) tannase
Tannat *n* tannate, salt or ester of tannic acid
Tanne *f* fir, pine, (Edeltanne) silver fir
Tannenzapfenöl *n* fir cone oil, pine cone oil
tannieren to tan, to mordant with tannic acid
Tannigen *n* tannigen, acetyltannin
Tannin *n* tannin, tannic acid
Tanninalbuminat *n* tannin albumate, tannalbin
Tanninbleisalbe *f* (Pharm) lead tannate ointment

Tannindigallussäure *f* tannin digallic acid
Tanninsalbe *f* (Pharm) tannin ointment
Tannoform *n* tannoform, methylene ditannin
Tannogen *n* tannogen, acetyltannin, tannigen
Tannon *n* tannon, tannopin, urotropine tannin
Tannopin *n* tannon, tannopin, urotropine tannin
Tannyl- tannyl
Tantal *n* tantalum (Symb. Ta)
Tantalat *n* tantalate, salt of tantalic acid
Tantaldioxid *n* tantalum dioxide
Tantalerz *n* (Min) tantalum ore
Tantalfluorid *n* tantalum fluoride, (Tantal(V)-flourid) tantalum pentafluoride, (Tantal(V)-fluorid) tantalum(V) fluoride
tantalig (Chem) tantalous
Tantal(III)-Verbindung *f* tantalous compound
Tantalit *m* (Min) tantalite
Tantalmetall *n* tantalum metal
Tantalpentachlorid *n* tantalic chloride, tantalum pentachloride
Tantalpentoxid *n* (Tantal(V)-oxid) tantalic anhydride, tantalic oxide, tantalum pentoxide, tantalum(V) oxide
Tantalsäure *f* tantalic acid
tantalsauer (Salz) tantalate
Tantaltetroxid *n* (Tantal(IV)-oxid) ditantalum tetroxide, tantalum tetroxide
Tantal(V)-Verbindung *f* tantalic compound
Tapete *f* wall paper
Tapetenunterlage *f* wallpaper underlay
Tapioka *f* tapioca
Tapiolith *m* (Min) tapiolite
tappen to tap, (herumtappen) to fumble, to grope
Tara *f* dead weight, tare
Tarapacait *m* (Min) tarapacaite
Taraxacerin *n* taraxacerin
Taraxanthin *n* taraxanthin
Tarbuttit *m* (Min) tarbuttite
Target *n* (Atomphys) target
Targetfläche *f* target area
Targetstrom *m* target current
Tarierbecher *m* tare cup
tarieren to tare
Tarierschrot *m* bird shot, tare shot
Tariervorrichtung *f* taring device
Tarierwaage *f* tare balance
Tarif *m* rate
Tarirsäure *f* tariric acid, 5-octadecynoic acid
Tarnanstrich *m* camouflage [paint], dazzle paint
tarnen to camouflage
Tarnfarbe *f* camouflage paint
tartarisieren to tartarize
Tartar[us] *m* (Weinstein) tartar
Tartrat *n* tartrate, salt of tartaric acid
Tartrazin *n* tartrazine
Tartronsäure *f* (Oxymalonsäure) tartronic acid, oxymalonic acid
Tartronursäure *f* tartronuric acid
Tartronylharnstoff *m* tartronylurea

Tasche *f* bag, pocket
Taschenamperemeter *n* pocket ammeter
Taschenausgabe *f* pocket edition
Taschenbuch *n* pocket book
Taschenformat *n* pocket size
Taschenlampe *f* flash light
Taschenvoltmeter *n* pocket voltmeter
Tasmanit *m* tasmanite
Tasmanon *n* tasmanone
Taspininsäure *f* taspininic acid
Taste *f* key
tasten to feel, to touch
Taster *m* calipers, tracer
Tasterlehre *f* caliper gauge
Tasterzirkel *m* calipers
Tastkopf *m* probing head
Taststrahl *m* scanning beam
tatsächlich actual, real, true
Tau *m* dew
Tau *n* rope
taub deaf, (Bergb) dead, sterile
Taubeschlag *m* dew
Taubkohle *f* (Bergb) blind coal
Tauchbandscheider *m* submerged belt separator
Tauchbatterie *f* immersion battery, plunge battery
Tauchbehälter *m* dip tank
Tauchelektrode *f* dipping electrode
Tauchelement *n* plunge cell
tauchen to dip, to immerse, to plunge, to steep
Taucherglocke *f* diving bell
Taucherkolbenpumpe *f* plunger pump
Tauchfarbe *f* dipping paint
Tauchflammverfahren *n* submerged-flame process
Tauchform *f* dipping mold, master steel pattern
Tauchformen *n* solvent molding
Tauchgalvanisierung *f* immersion galvanization, galvanizing by dipping
Tauchhärtung *f* dip-hardening
Tauchhartlöten *n* dip brazing
Tauchkolben *m* plunger piston
Tauchkolbenpumpe *f* plunger piston pump
Tauchkondensator *m* submerged-coil condenser
Tauchlack *m* dipping lacquer, dipping varnish
Tauchlackierapparat *m* dip and drain equipment
Tauchlöten *n* dip soldering
Tauchmaschine *f* dipping machine
Tauchpumpe *f* immersion pump
Tauchrahmen *m* dipping frame
Tauchrohr *n* dip pipe, immersion stem
Tauchrohrverdampfer *m* submerged-tube evaporator
Tauchschmierung *f* splash lubrication
Tauchsieder *m* immersion heater
Tauchtank *m* dip tank
Tauchverdampfer *m* submerged evaporator
Tauchverfahren *n* dipping process

Tauchverkupferung *f* copper-plating by immersion
Tauchversuch *m* immersion test
tauchverzinken to hot-galvanize
Tauchverzinnung *f* dip-tinning
Tauchwaage *f* aerometer, hydrometer
Tauchwachs *n* dip coating wax
Tauchwalze *f* fountain roll, immersion roll
Tauchzähler *m* dipping counter
Tauchzylinder *m* plunge cylinder, plunger
tauen to thaw
Tauen *n* thawing
Taufliege *f* (Zool) drosophila
taugen to be fit, to be of value, to be useful
tauglich able, fit, qualified
Tauglichkeit *f* fitness, usefulness
tauig dewy
Taukurve *f* dew point curve
Taulast *f* towing load
Taumelkorn *n* (Bot) darnel
Taumelmischer *m* dry-blend mixer, dry-coloring mixer
Taumelsieb *n* gyratory screen with supplementary whipping action
Taumeltrockner *m* tumbler dryer, tumbling dryer
Taumelzentrifuge *f* tumbler centrifuge
Taupunkt *m* dew point
Taurin *n* taurine, aminoethanesulfonic acid
Taurocarbamidsäure *f* taurocarbamic acid
Taurocholat *n* taurocholate, salt or ester of taurocholic acid
Taurocholeinsäure *f* taurocholeic acid
Taurocholsäure *f* taurocholic acid
Taurocholsalz *n* taurocholate, salt of taurocholic acid
Taurocyamin *n* taurocyamine
Taurolithocholsäure *f* taurolithocholic acid
Taurylsäure *f* taurylic acid
Tausch *m* exchange
Tausendstel *n* thousandth [part]
Tautolith *m* tautolite
tautomer tautomeric
Tautomer[e] *n* (Chem) tautomeric derivative
Tautomerie *f* (Chem) tautomerism
Taxicatigenin *n* taxicatigenin
Taxicatin *n* taxicatin
Taxin *n* taxine
Taxis *f* (Biol) taxis
Taylor-Reihenentwicklung *f* (Math) Taylor-series expansion
Taylorsche Reihe *f* (Math) Taylor's series
T-Düse *f* T die
Teakholz *n* teak [wood]
Teallit *m* (Min) teallite
Tebelon *n* tebelon, isobutyl oleate
Technetium *n* technetium, (Symb. Tc) masurium (obs)

Technicolorverfahren *n* (HN) technicolor process
Technik *f* technology, engineering, technical science, (Verfahren) technique, **chemische** ~ chemical [and process] engineering
Techniker *m* technician
Technikum *n* pilot plant, technical school
technisch technological, (fachgemäß) technical
technisch *f* industrial
technisch-chemisch technochemical
Technologie *f* technology, **chemische** ~ chemical technology
technologisch technological
Teclubrenner *m* (Chem) teclu burner
Tecomin *n* tecomin, lapachoic acid, lapachol, targusic acid
Tectochinon *n* tectoquinone
Tectochrysin *n* techtochrysin
Tee *m*, **harntreibender** ~ (Pharm) diuretic tea
Teeaufguß *m* infusion of tea
Teer *m* tar
Teerabscheider *m* tar extractor, tar separator
Teeranstrich *m* coat of tar, tar coating
teerartig tarry
Teerasphalt *m* tar asphalt, coal tar pitch
Teerband *n* tarred tape
Teerbestandteil *m* tar constituent
Teerbildung *f* tar formation
Teerbitter *n* picamar, propylpyrogallol dimethyl ether
Teerdampf *m* tar fume, tar vapor
Teerderivat *n* derivate of tar
Teerdestillation *f* [coal-]tar distillation
Teerdunst *m* tar fume[s]
teeren to tar
Teeren *n* tarring
Teerfarbstoff *m* coal-tar dye
Teerfeuerung *f* tar furnace
Teergas *n* tar gas
Teergehalt *m* tar content
Teergrube *f* tar pit
Teerhefe *f* tar dregs
Teerkocherei *f* tar boiling plant
Teerkrebs *m* (Med) tar cancer
Teerkruste *f* tar incrustation
Teernebel *m* tar mist, tar vapor
Teeröl *n* [coal] tar oil
Teerpapier *n* tar paper
Teerpappe *f* tarred board
Teerpech *n* tar pitch, artificial asphalt, coal tar asphalt
Teerrückstand *m* tarry residue
Teersammler *m* tar collector
Teersatz *m* tar dregs, tar sediment
Teerscheider *m* tar separator
Teerschutzüberzug *m* tar enamel coating
Teerschwelapparat *m* tar distilling apparatus
Teerüberzug *m* tar coating, tar incrustation
Teerung *f* tarring, tar spraying

Teerwasser *n* tar water
Teerwerg *n* tarred oakum
Teflon *n* (HN) teflon, polytetrafluor ethylene
Teichmannsche Häminkristalle *pl* Teichmann's crystals
Teichonsäure *f* teichoic acid
Teig *m* dough, paste, (Pap) pulp
teigartig pasty
Teigfarbe *f* (Pap) pulp color
teigig doughy, pasty
Teigknetmaschine *f* dough kneading machine
Teigpressen *n* dough molding
Teigpreßverfahren *n* dough molding process
Teigschaber *m* dough scraper
Teil *m* portion, (Anteil) share, (Bruchstück) fragment
Teil *m n* part
Teilansicht *f* partial view
Teilbarkeit *f* divisibility
Teilchen *n* particle, **beschleunigtes** ~ accelerated particle, **ionisierendes** ~ ionizing particle, **subatomares** ~ subatomic particle
Teilchenbeschleuniger *m* particle accelerator
Teilchenbeschuß *m* particle bombardment
Teilchenbewegung *f* particle motion
Teilchenbild *n* particle image
Teilchendurchmesser *m*, **mittlerer** ~ mean particle diameter
Teilchenenergie *f* particle energy
Teilchengeschwindigkeit *f* particle velocity
Teilchengestalt *f* particle shape
Teilchengröße *f* particle size
Teilchenimpuls *m* particle momentum
Teilchenladung *f* particle charge
Teilchennachweis *m* particle detection
Teilchenstoß *m* particle impact
Teilchenstruktur *f* particle structure
Teilchenverschiebung *f* displacement of particles, particle shift
Teilchenzahl *f* number of particles
Teilchenzahldichte *f* particle density
Teilchenzusammenpressung *f* compression of particles
Teilchenzustand *m* particle state
Teildruck *m* partial pressure
teilen to divide, to split, (abtrennen) to separate
Teiler *m* (Math) divisor, **gemeinsamer** ~ common divisor
Teilerfolg *m* partial success
Teilfolge *f* subsequence
Teilgitter *n* sublattice
Teilhärtung *f* selective hardening
Teilkondensation *f* partial condensation
Teilkreis *m* graduated circle, (Zahnrad) pitch circle
Teilkreisdurchmesser *m* pitch diameter
Teilladung *f* partial cargo, partial load
teillöslich partly soluble
Teilmenge *f* (Chem) aliquot

Teilnahme *f* participation
teilnehmen to participate, to take part
Teilreaktion *f* (Chem) partial reaction
Teilröstung *f* partial roasting
Teilschablone *f* dividing template
Teilschale *f* partial shell
Teilscheibe *f* graduated plate
Teilschwingung *f* partial oscillation, partial vibration
Teilstrich *m* (Math) division mark
Teilstrom *m* branch current
Teilung *f* (Abteilung) partitioning, (Einteilung) graduation, (Math) division, (Spaltung) splitting
Teilungsdichte *f* cut size
Teilungsfehler *m* (Math) error of division
Teilungslinie *f* parting line
Teilungszahl *f* grade efficiency
Teilungszeichen *n* (Math) division sign
Teilversetzung *f* partial dislocation
teilverträglich partly compatible
teilweise partial
T-Eisen *n* T-bar, tee iron, T-iron
Tekticit *m* tecticite
Tektit *m* (Min) tectite, anstralite, bottlestone, chrysolite
Tektonik *f* (Geol) tectonics
tektonisch (Geol) tectonic
Telefon *n* phone, telephone
Telefonleitung *f* telephone circuit
Telegramm *n* telegram, **drahtloses** ~ radiogram
Telegraphie *f* telegraphy, **drahtlose** ~ radiotelegraphy, wireless telegraphy
telegraphieren to cable, to telegraph
Telemeter *n* telemeter, range finder
Telemotor *m* telemotor
Teleobjektiv *n* (Phot) telephotographic lens
Telepathin *n* telepathine
Telephotographie *f* telephotography
Telephotometrie *f* telephotometry
Teleskop *n* telescope
Teleskopbagger *m* telescopic excavator
teleskopisch telescopic[al]
Teleskopwagen *m* telescope carriage
Teletherapie *f* teletherapy
Telfairiasäure *f* telfairic acid
Teller *m* plate, tray, (Ventil) seat
Teller-Düsenzentrifuge *f* nozzle-discharge centrifuge
Tellermesser *n* circular slitting-knife
Tellermischer *m* pan mixer
Tellerofen *m* rotary hearth furnace
Tellerreiber *m* muller
Tellerseparator *m* disc separator
Tellerskala *f* circular scale
Tellertrockner *m* disk dryer, plate dryer
Tellertrommeltrockner *m* rotary shelf drier
Tellerventil *n* disk valve
Tellerwäscher *m* disc separator

Tellerzelle *f* plate cell
Tellerzentrifuge *f* disc centrifuge
Tellinit *m* tellinite
Tellur *n* tellurium (Symb. Te)
Telluralkyl *n* alkyl telluride, dialkyl telluride
Tellurat *n* tellurate, salt of telluric acid
Tellurblei *n* lead telluride
Tellurcyansäure *f* tellurocyanic acid
Tellurdiäthyl *n* [di]ethyl telluride
Tellurdioxid *n* tellurium dioxide, tellurous acid anhydride
Tellurerz *n* (Min) tellurium ore
tellurführend telluriferous
Tellurgehalt *m* tellurium content
Tellurglanz *m* (Min) tellurium glance, nagyagite
Tellurgold *n* gold telluride
Tellurgoldsilber *n* gold silver telluride, (Min) sylvanite
Tellurhalogen *n* tellurium halide
tellurhaltig telluriferous
Tellurid *n* telluride
tellurig (Chem) tellurous
Tellurigsäureanhydrid *n* tellurium dioxide, tellurous anhydride
Tellur(II)-Verbindung *f* telluride compound
Tellurit *m* (Min) tellurite
Tellurit *n* tellurite, salt or ester of tellurous acid
Tellurium *n* tellurium
Tellur(IV)-Verbindung *f* tellurous compound
Tellurkohlenstoff *m* carbon telluride
Tellurmetall *n* metal[lic] telluride
Tellurmonoxid *n* tellurium monoxide
Tellurnickel *n* (Min) nickel telluride, melonite
Tellurophenol *n* tellurophenol
Tellursäure *f* telluric acid
Tellursäurehydrat *n* telluric acid hydrate
Tellursalz *n* tellurium salt
tellursauer (Salz) tellurate
Tellursilber *n* silver telluride
Tellursilberblei *n* (Min) lead silver telluride
Tellursilberblende *f* (Min) stutzite
Tellurtrioxid *n* tellurium trioxide, telluric anhydride
Tellurür *n* (obs) lower telluride
Tellurverbindung *f* telluride, tellurium compound
Tellur(VI)-Verbindung *f* telluric compound
Tellurwasserstoff *m* hydrogen telluride
Tellurwasserstoffsäure *f* hydrogen telluride, hydrotelluric acid
tellurwasserstoffsauer (Salz) telluride
Tellurwismut *n* (Min) bismuth telluride, tetradymite
Teloidin *n* teloidine
Teloidinon *n* teloidinone
Telomer[e] *n* telomer
Telomerisation *f* telomerization
telomerisieren to telomerize
Teloschistin *n* teloschistin

Temiskamit *m* (Min) temiscamite
Temperafarbe *f* distemper
Temperatur *f*, **absolute** ~ absolute temperature, **erhöhte** ~ elevated temperature, **kritische** ~ critical temperature, **maximale** ~ maximum temperature, **mittlere** ~ average temperature, mean temperature, **niedrigste** ~ minimum temperature
Temperaturabfall *m* drop in temperature
temperaturabhängig temperature dependent
Temperaturabhängigkeit *f* temperature dependence
Temperaturabnahme *f* temperature decrease, temperature drop, **sprunghafte** ~ decalescence
Temperaturänderung *f* change of temperature, variation of temperature
Temperaturalarmvorrichtung *f* temperature alarm device
Temperaturanstieg *m* rise of temperature
Temperaturausgleich *m* temperature compensation, equalization of temperature
Temperaturbeeinflussung *f* temperature effect, temperature influence
Temperaturbehandlung *f* heat treatment, (Met) tempering
Temperaturbeobachtung *f* temperature observation
Temperaturbereich *m* temperature range
Temperaturbeständigkeit *f* heat resistance
Temperatureffekt *m* temperature effect
Temperatureinfluß *m* influence of temperature, temperature effect
Temperatureinstellung *f* temperature adjustment, temperature setting
temperaturempfindlich heat sensitive
temperaturerhöhend thermoexcitory
Temperaturerhöhung *f* rise in temperature, temperature increase
Temperaturerniedrigung *f* temperature drop
Temperaturfaktor *m* temperature coefficient
Temperaturformstabilität *f* heat distortion point
Temperaturfühler *m* thermoelement, temperature tracer, thermocouple
Temperaturgefälle *f* temperature drop
Temperaturgradient *m* (Therm) temperature gradient
Temperaturgrenze *f* temperature limit
Temperaturintervall *n* temperature interval
Temperaturjahresmittel *n* mean annual temperature
Temperaturkoeffizient *m* temperature coefficient
Temperaturkonstanz *f* constancy of temperature
Temperaturleitfähigkeit *f* diffusivity
Temperaturmaximum *n* maximum temperature
Temperaturmessung *f* temperature measurement, thermometry
Temperaturminimum *n* minimum temperature
Temperaturregelung *f* thermoregulation

Temperaturregistrierapparat *m* temperature registration device
Temperaturregler *m* temperature control, temperature regulator
Temperaturschichtung *f* thermal stratification
Temperaturschreiber *m* temperature recorder
Temperaturschwankung *f* fluctuation in temperature, temperature variation
Temperatursenkung *f* lowering of temperature, temperature reduction
Temperaturskala *f* temperature scale, ~ **nach Fahrenheit** Fahrenheit temperature scale
Temperatursprung *m* temperature jump
Temperatursteigerung *f* rise in temperature
Temperaturstrahlung *f* thermal radiation
temperaturunabhängig temperature-independent
Temperaturverhalten *n* thermal property
Temperaturverlauf *m* (Therm) heating pattern
Temperaturwächter *m* temperature control
Temperaturwechselbeständigkeit *f* thermal spalling resistance
Temperaturwechseltest *m* thermal cycling test, thermocycling test
Temperaturzunahme *f* temperature increase
Temperbedingung *f* (Metall) annealing condition
Tempereisen *n* malleable cast iron
Temperguß *m* malleable casting
Tempergußeisen *n* malleable cast iron, malleable pig iron
Tempergußstück *n* malleable iron casting
Temperiermantel *m* temper jacket
Temperkohle *f* temper carbon, graphitic carbon
Temperkohleabscheidung *f* graphitization
Temperkohlebildung *f* formation of temper carbon
Tempermittel *n* tempering agent, tempering material
tempern to temper[harden], to age artificially, to anneal, to reheat, (Plast) stoving
Tempern *n* tempering, annealing, slow cooling
Temperofen *m* annealing furnace
Temperroheisen *n* malleable pig iron
Temperschrott *m* malleable scrap
Temperstahlguß *m* malleable cast iron
Temperung *f* annealing
Temperverfahren *n* annealing process
Tempo *n* speed, pace, tempo
temporär temporary
Tenakel *n* filtering frame
Tenazität *f* tenacity
Tendenz *f* **fallende** ~ downward tendency
Tennantit *m* (Min) tennantite
Tenorit *m* (Min) tenorite, native black copper oxide
Tenosin *n* tenosine
Tensid *n* surfactant
Tensor *m* tensor
Tensoralgebra *f* tensor algebra

Tensorbedingung *f* tensor condition
Tensordichte *f* tensor density
Tensorkraft *f* tensor force
Tensoroperator *m* tensor operator
Tensorwechselwirkung *f* tensor force interaction
Tenuazonsäure *f* tenuazonic acid
Tenuiorin *n* tenuiorin
Tephroit *m* (Min) tephroite
Tephrosin *n* tephrosin
Tephrosinsäure *f* tephrosic acid
Teppichgrundgewebe *n* carpet backing
Teppichunterlage *f* rug underlay
Teraconsäure *f* teraconic acid
Terapinsäure *f* terapinic acid
Teratolith *m* (Min) teratolite
Terbinerde *f* terbia, terbium oxide
Terbium *n* terbium (Symb. Tb)
Terbutolharz *n* terbutol resin
Tereben *n* terebene
Terebinsäure *f* terebic acid, 2,2-dimethylparaconic acid, terebinic acid
Terecamphen *n* terecamphene
Terephthalaldehyd *m* terephthal aldehyde
Terephthalalkohol *m* terephthalyl alcohol
Terephthalgrün *n* terephthal green
Terephthalonsäure *f* terephthalonic acid
Terephthalophenon *n* terephthalophenone
Terephthalopinakon *n* terephthalopinacone
Terephthalsäure *f* terephthalic acid
Teresantalol *n* teresantalol
Teresantalsäure *f* teresantalic acid
Terlinguait *m* (Min) terlinguaite
Term *m* term, **anomaler** ~ (Spektr) anomalous term, displaced term, **metastabiler** ~ metastable term
Termanzahl *f* (Atom) level numbers
Termaufspaltung *f* (Atom) term splitting
Termbeeinflussung *f* (Atom) term influence, level displacement
Termdichte *f* (Atom) level density
Termenergie *f* (Atom) term energy
Termierit *m* (Min) termierite
Terminkontrolle *f* expediting
Terminologie *f* terminology
Terminolsäure *f* terminolic acid
Terminplan *m* deadline, time schedule
Terminus *m* **terminus technicus** (Lat) technical term
Termlage *f* term value
Termschema *n* term diagram, [energy] level diagram
Termstruktur *f* level structure
Termsystematik *f* level systematics
Termwert *m* term value
ternär ternary
Terpacid *n* terpacid
Terpan *n* terpane
Terpen *n* terpene

terpenähnlich terpene-like
Terpenchemie *f* terpene chemistry
terpenfrei terpene-free, terpeneless
Terpengruppe *f* terpene group
Terpenolsäure *f* terpenolic acid
Terpensäure *f* terpenic acid
Terpentin *n* turpentine
Terpentinalkohol *m* oil of turpentine, rosin oil, spirits of turpentine
Terpentinbeize *f* mordant based on turpentine
Terpentincampher *m* turpentine camphor
Terpentinersatz *m* turpentine substitute
Terpentinfarbe *f* turpentine varnish
Terpentinfirnis *m* turpentine varnish
Terpentinharz *n* turpentine resin, colophony
Terpentinlack *m* turpentine varnish
Terpentinöl *n* oil of turpentine, rosin oil, spirits of turpentine
Terpentinpech *n* turpentine pitch
Terpentinsäure *f* terebic acid, 2,2-dimethylparaconic acid, terebinic acid
Terpentinsalbe *f* (Pharm) turpentine ointment
Terpentinsurrogat *n* turpentine substitute
Terpenylsäure *f* terpenylic acid
Terphenyl *n* diphenylbenzene, terphenyl
Terpilen *n* terpilene, terpinylene
Terpilenhydrid *n* terpilene hydride
Terpin *n* terpin[e], dihydroxymenthane
Terpinen *n* terpinene
Terpineol *n* terpilenol, terpineol
Terpinhydrat *n* terpine hydrate
Terpinol *n* terpinol
Terpinolen *n* terpinolene
Terrakotta *f* (Keram) terra cotta
terrassenartig step-shaped
Terrazzo *m* (Zementmosaik) terrazzo, Venetian mosaic
Terrein *n* terrein
Terreinsäure *f* terreic acid
Terrestrinsäure *f* terrestric acid
Terrylen *n* (HN) terrylene, dacron
tertiär tertiary
Tertiärformation *f* (Geol) tertiary system
Tertiärstruktur *f* tertiary structure
Tesafilm *m* Scotch tape, cellotape
Tesselit *m* tesselite
tesseral (Krist) tesseral, cubic, isometric
Tesseralkies *m* (Min) skutterudite, smaltite
Tesseralsystem *n* isometric system
Test *m* examination, test
Testan *n* testane
Testbenzin *n* mineral varnish, white spirit
Testgütekurve *f* (Statist) power distribution curve
Testkörper *m* test body, test piece
Testosteron *n* testosterone
Teststreifenapparatur *f* test strip apparatus
Testteilchen *n* probe particle
Testverteilung *f* (Statist) test distribution curve

Tetanie *f* (Med) tetany
Tetanin *n* tetanine
Tetanthren *n* tetanthrene, tetrahydrophenanthrene
Tetanthrenon *n* tetanthrenone
Tetanus *m* (Med) tetanus
Tetanusantitoxin *n* tetanus antitoxin, antitetanic serum
Tetanusbazillus *m* tetanus bacillus, clostridium tetani
Tetanusschutzimpfung *f* (Med) tetanus inoculation
Tetanusspritze *f* (Med) antitetanic injection
Tetanustoxin *n* tetanus toxin
Tetartoeder *n* (Krist) tetartohedron
Tetartoedrie *f* (Krist) tetartohedry
tetartoedrisch (Krist) tetartohedral
Tetraacetyldextrose *f* tetraacetyldextrose
Tetraäthylblei *n* tetraethyl lead
Tetraäthylenpentamin *n* tetraethylene pentamine
Tetrabarbital *n* tetrabarbital
Tetrabase *f* tetrabase, Michler's base
Tetraborsäure *f* tetraboric acid
Tetrabromäthan *n* tetrabromo ethane
Tetrabromstearinsäure *f* tetrabromostearic acid
Tetracain *n* tetracaine
Tetracarbinol *n* tetracarbinol
Tetracen *n* tetracene
Tetrachloräthan *n* tetrachlor[o]ethane, (HN) cellon
Tetrachlorbenzol *n* tetrachlorobenzene
Tetrachlorchinon *n* tetrachloroquinone, chloranil, tetrachloro-benzoquinone
Tetrachlorkohlenstoff *m* carbon tetrachloride
Tetrachlorsilicium *n* silicon tetrachloride, tetrachlorosilane
Tetrachlorzinn *n* tin tetrachloride
Tetracontan *n* tetracontane
Tetracosan *n* tetracosane
Tetracyclon *n* tetracyclone
Tetracylin *n* tetracylin
Tetradecan *n* tetradecane
Tetraeder *n* (Krist) tetrahedron
Tetraedersymmetrie *f* tetrahedral symmetry
tetraedrisch tetrahedral
Tetraedrit *m* (Min) tetrahedrite
Tetrafluorkohlenstoff *m* carbon tetrafluoride
tetrafunktionell tetrafunctional
tetragonal tetragonal
Tetrahexakontan *n* tetrahexacontane
Tetrahydrat *n* tetrahydrate
Tetrahydroabietinsäure *f* tetrahydroabietic acid
Tetrahydrobiopterin *n* tetrahydrobiopterin
Tetrahydrochinolin *n* tetrahydroquinoline
Tetrahydrofolsäure *f* tetrahydrofolic acid
Tetrahydrofuran *n* tetrahydrofuran
Tetrahydrofurfurylalkohol *m* tetrahydrofurfuryl alcohol
Tetrahydropyrrol *n* tetrahydropyrrole

Tetrahydroserpentin *n* tetrahydroserpentine
Tetrajodkohlenstoff *m* carbon tetraiodide
Tetrakishexaeder *n* (Krist) tetrakishexahedron
Tetrakontan *n* tetracontane
Tetrakosan *n* tetracosane
tetrakovalent tetracovalent
Tetralin *n* (HN) tetralin[e], tetrahydronaphthalene
Tetralol *n* tetralol
Tetralon *n* tetralone
Tetralupin *n* tetralupine
tetramer tetrameric
Tetramethyläthylen *n* tetramethylethylene
Tetramethylammoniumhydroxid *n* tetramethylammonium hydroxide
Tetramethylammoniumjodid *n* tetramethylammonium iodide
Tetramethylbase *f* tetramethylated base
Tetramethylen *n* tetramethylene, cyclobutane
Tetramethylenbromid *n* tetramethylene bromide
Tetramethylendiamin *n* tetramethylene diamine
Tetramethylenglykol *n* tetramethylene glycol
Tetramethylglucose *f* tetramethylglucose
Tetramethylmethan *n* tetramethylmethane
Tetramin *n* tetramine
Tetramsäure *f* tetramic acid
Tetranatriumpyrophosphat *n* tetrasodium pyrophosphate
Tetranitrochrysazin *n* tetranitrochrysazin, chrysammic acid
Tetranitromethan *n* tetranitromethane
Tetranitromethylanilin *n* tetranitromethylaniline
Tetranitrophenol *n* tetranitrophenol
Tetranthrimid *n* tetranthrimide
Tetraphen *n* tetraphene
Tetraphenoxysilan *n* tetraphenoxysilane
Tetraphenyläthylen *n* tetraphenylethylene
Tetraphenylen *n* tetraphenylene
Tetraphenylhydrazin *n* tetraphenylhydrazine
Tetraphenylmethan *n* tetraphenylmethane
Tetraphenylsilicium *n* silicon tetraphenyl
tetrasymmetrisch tetrasymmetric
Tetrathionat *n* tetrathionate, salt or ester of tetrathionic acid
Tetrathionsäure *f* tetrathionic acid
Tetratriakontan *n* tetratriacontane
Tetrazen *n* tetrazene
Tetrazin *n* tetrazine
Tetrazol *n* tetrazole
Tetrazon *n* tetrazone
Tetrazotsäure *f* tetrazotic acid
Tetrinsäure *f* tetrinic acid
Tetrode *f* tetrode
Tetrolsäure *f* tetrolic acid, butynoic acid
Tetronal *n* tetronal
Tetronsäure *f* tetronic acid
Tetrophan *n* (HN) tetrophan, dihydrobenzacridine carboxylic acid
Tetrose *f* tetrose

Tetroxan *n* tetroxane
Tetroxid *n* tetroxide
Tetroylbenzoesäure *f* tetroylbenzoic acid
Tetroylpropionsäure *f* tetroylpropionic acid
Tetryl *n* tetryl, 2,4,6-trinitrophenylmethylnitramine, tetralite
Tetrylammonium *n* tetrylammonium
Tetryzolin *n* tetryzoline
Tetuin *n* tetuin
Teucrin *n* teucrin
Teufelsdreck *m* (Pharm) asafetida
Teufelsstein *m* (Min) belemnite
teufen (Bergb) to bore, to deepen, to sink
Texasit *m* (Min) texasite
Textilfaser *f* textile fiber, textile thread
Textilgewebe *n* textile fabric
Textilhilfsmittel *n* textile auxiliary
Textilien *pl* textiles
Textilindustrie *f* textile industry
Textilit *n* textilite yarn
Textilose *f* testilose yarn
Textilvered[e]lung *f* textile dressing, textile finishing
Textilvered[e]lungsmittel *n* textile finishing agent
Textur *f* texture, structure
T-förmig T-shaped
Thalenit *m* (Min) thalenite
Thalidomid *n* thalidomide
Thalleiochinolin *n* thalleioquinoline
Thalliion *n* (Thallium(III)-ion) thallic ion, thallium(III) ion
Thallin *n* thalline, tetrahydro-4-methoxyquinoline
Thallinsulfat *n* thalline sulfate, tetrahydro-4-methoxyquinoline sulfate
Thallioxid *n* (Thallium(III)-oxid) thallic oxide, thallium(III) oxide
Thallisalz *n* thallic salt, thallium(III) salt
Thallium *n* thallium (Symb. Tl)
Thallium[aluminium]alaun *m* thallium alum, aluminum thallium(I) sulfate, thallous aluminum sulfate
Thalliumbromür *n* (obs für Thallium(I)-bromid) thallium(I) bromide, thallium monobromide, thallous bromide
Thalliumchlorid *n* thallium chloride, (Thallium(I)-chlorid) thallium(I) chloride, thallium monochloride, thallous chloride, (Thallium(III)-chlorid) thallic chloride, thallium(III) chloride, thallium trichloride
Thalliumformiat *n* thallium formate
Thalliumgehalt *m* thallium content
Thalliumglas *n* thallium glass
Thalliumhydroxydul *n* (obs für Thallium(I)-hydroxid) thallium(I) hydroxide, thallous hydroxide
Thalliumjodid *n* thallium iodide, (Thallium(II)-jodid) thallic iodide, thallium(III) iodide, thallium triiodide

Thalliumjodür *n* (obs für Thallium(I)-jodid) thallium(I) iodide, thallium monoiodide, thallous iodide
Thalliummalonat *n* thallium malonate
Thalliumoxid *n* thallium oxide, (Thallium(III)-oxid) thallic oxide, thallium(III) oxide, thallium sesquioxide
Thalliumoxydul *n* (obs f. Thallium(I)-oxid) thallium(I) oxide, thallous oxide
Thalliumoxydulcarbonat *n* (obs f. Thallium(I)-carbonat) thallium(I) carbonate, thallous carbonate
Thalliumoxydulchlorat *n* (obs) thallium(I) chlorate, thallous chlorate
Thalliumoxydulnitrat *n* (obs) thallium(I) nitrate, thallous nitrate
Thalliumselenat *n* thallium selenate
Thalliumsulfid *n* thallium sulfide
Thalliumsulfür *n* (obs f. Thallium(I)-sulfid) thallium(I) sulfide, thallous sulfide
Thalliverbindung *f* (Thallium(III)-Verbindung) thallic compound, thallium(III) compound
Thallochlorat *n* (Thallium(I)-chlorat) thallium(I) chlorate, thallous chlorate
Thallochlorid *n* (Thallium(I)-chlorid) thallium(I) chloride, thallium monochloride, thallous chloride
Thalloion *n* (Thallium(I)-ion) thallium(I) ion, thallous ion
Thallojodat *n* thallium(I) iodate, thallous iodate
Thallooxid *n* thallium(I) oxide, thallous oxide
Thalloverbindung *f* (Thallium(I)-Verbindung) thallium(I) compound, thallous compound
Thalmin *n* thalmine
Thamnolsäure *f* thamnolic acid
Thapsiasäure *f* thapsic acid
Thaumasit *m* (Min) thaumasite
Theanin *n* theanine
Thebain *n* thebaine, dimethyl morphine
Thebainhydrochlorid *n* thebaine hydrochloride
Thebenin *n* thebenine
Thein *n* 1,3,7-trimethyl-2,6-dihydroxypurine, 1,3,7-trimethylxanthine, theine, caffeine
Theke *f* shop counter
Thelephorsäure *f* thelephoric acid
Thenaldehyd *m* thenaldehyde
Thenalidin *n* thenalidine
Thenardit *m* (Min) thenardite, native sodium sulfate
Thenardsblau *n* Thenard's blue
Thenoesäure *f* thenoic acid
Thenoylaceton *n* thenoylacetone
Thenylamin *n* thenylamine
Theobromin *n* theobromine, 3,7-dimethyl-2,6-dihydroxypurine, 3,7-dimethylxanthine
Theocin *n* theocine, theophylline
Theodolit *m* theodolite
Theolactin *n* theolactine

Theophorin *n* theobromine sodium formate, theophorine
Theophyllin *n* 1,3-dimethyl-2,6-dihydroxypurine, 1,3-dimethylxanthine, theophylline
theoretisch theoretic[al]
Theorie *f* theory
Theralith *m* (Geol) theralite
Therapeut *m* therapist
Therapeutik *f* therapeutics
Therapeutikum *n* therapeutic
therapeutisch therapeutic
Therapie *f* (Med) therapy, treatment
Thermalhärtung *f* high-temperature quenching, hot-quenching method
Thermalquelle *f* thermal spring, hot spring
Thermalschwefelbad *n* thermal sulfur bath
Thermalwasser *n* thermal water
Therme *f* hot spring, thermal spring
Thermion *n* thermion
Thermionenaussendung *f* thermionic emission
Thermionenröhre *f* thermionic valve
Thermionenstrom *m* thermionic current
thermisch thermal, thermic
Thermistor *m* thermistor
Thermit *n* thermit[e], mixture of aluminum and ferric oxide
Thermitschweißung *f* thermite welding
Thermitverfahren *n* aluminothermic process, thermite process
Thermoamperemeter *n* thermo-ammeter
Thermoanalyse *f* thermal analysis
Thermobarometer *n* thermobarometer
Thermobatterie *f* thermobattery, thermopile
Thermochemie *f* thermochemistry
thermochemisch thermochemical
Thermochromfarbe *f* thermochrome paint
Thermochromstift *m* thermochrome crayon
Thermodiffusion *f* thermodiffusion, thermal diffusion
Thermodiffusionsverhältnis *n* thermal diffusion ratio
Thermodin *n* thermodin
Thermodynamik *f* thermodynamics, **chemische** ~ chemical thermodynamics
thermodynamisch thermodynamic
Thermoeffekt *m* thermoelectric effect
thermoelastisch thermoelastic
Thermoelastizität *f* thermoelasticity
Thermoelastizitätskoeffizient *m* (Therm) thermoelastic coefficient
thermoelektrisch thermoelectric
Thermoelektrizität *f* thermo-electricity
Thermoelement *n* thermocouple, thermoelectric couple, thermoelement
Thermofarbe *f* thermocolor
Thermofarbenskala *f* color scale of temperature
Thermofixieren *n* heat setting, thermofixation, thermosetting
Thermofixierung *f* heat setting, thermosetting
Thermogalvanometer *n* thermoelectric galvanometer
Thermogenese *f* thermogenesis, production of heat
Thermograph *m* thermograph
Thermokaustik *f* thermocautery
Thermokette *f* thermocouple, thermoelement
Thermokleber *m* thermoplastic adhesive
Thermokompensator *m* thermocompensator
Thermokompression *f* thermocompression
Thermokraft *f* thermal force, thermoelectric force, thermoelectric power
thermolabil thermolabile
thermolumineszent thermoluminescent
Thermolumineszenz *f* thermoluminescence
Thermolyse *f* thermolysis
thermolytisch thermolytic
thermomagnetisch thermomagnetic
Thermomanometer *n* heat-pressure gauge, thermogauge
Thermometer *n* thermometer, **aufzeichnendes** ~ recording thermometer
Thermometerfaden *m* mercury thread, thermometric column
Thermometerfassung *f* thermometer mounting
Thermometerkugel *f* thermometer bulb
Thermometerrohr *n* thermometer tube
Thermometersäule *f* thermometer column
Thermometerschutzrohr *n* thermometer sheath
Thermometerskala *f* thermometer scale
Thermometerstand *m* thermometric reading
Thermometertasche *f* thermometer pocket
Thermometrie *f* thermometry
thermometrisch thermometric
Thermomotor *m* thermoelectric motor, thermomagnetic motor
Thermonatrit *m* (Min) thermonatrite, native monohydrated sodium carbonate
thermonuklear thermonuclear
Thermopaar *n* thermocouple, thermoelectric couple, thermoelement
thermophil (Bakt) thermophil[ic]
Thermophon *n* thermophone
Thermophyllit *m* (Min) thermophyllite
Thermophysik *f* thermophysics
Thermoplast *m* thermoplastic
thermoplastisch thermoplastic, heatdeformable
Thermopren *n* thermoprene
Thermopsin *n* thermopsine
Thermoregler *m* thermoregulator, thermostatic control
Thermoregulation *f* heat regulation, thermoregulation heat regulation
Thermoregulator *m* thermoregulator, thermostat
Thermorelais *n* thermal relay, thermorelay
thermoresistent thermostable, heat-resistant
Thermorezeptor *m* thermoreceptor
Thermoschalter *m* thermal circuit breaker
Thermoschreiber *m* thermograph

Thermosflasche *f* thermos [bottle]
Thermosiphonheizung *f* (Kolonne) thermosiphon reboiler
Thermosiphonprinzip *n* thermosiphon principle
Thermoskop *n* thermoscope
Thermospannung *f* thermoelectric power, Seebeck voltage
thermostabil thermostable, heat resistant, thermoresistant
Thermostat *m* thermostat, automatic heat regulator, (für tiefe Temperaturen) cryostat
thermostatisch thermostatic
Thermostrom *m* thermoelectric current
Thermotron *m* thermotron
Thermoverformung *f* thermo-forming
Thesin *n* thesine
Thesinsäure *f* thesinic acid
Thetafunktion *f* theta function
Theta-Meson (Theton) theta meson
Theta-Polarisation *f* (Elektr) theta polarization
Theton *n* theta meson
Theveresin *n* theveresin
Thevetin *n* thevetin
Thevetose *f* thevetose
Thiacetarsamid *n* thiacetarsamide
Thiacetsäure *f* thioacetic acid, thiacetic acid
Thiachromon *n* thiachromone
Thialdin *n* thialdine
Thiamid *n* thiamide, thioamide
Thiamin *n* thiamin[e], aneurin, vitamin B₁
Thiaminase *f* (Biochem) thiaminase
Thiaminhydrochlorid *n* thiamine hydrochloride
Thiaminmononitrat *n* thiamine mononitrate
Thiaminpyrophosphat *n* thiamine pyrophosphate, carboxylase, cozymase II, TPP
Thiamorpholin *n* thiamorpholine
Thianaphthen *n* thianaphthene
Thianthren *n* thianthrene, dibenzo-1,4-dithiin, diphenylene disulfide
Thiapyran *n* thiapyran
Thiaxanthen *n* thiaxanthene
Thiazan *n* thiazane
Thiazin *n* thiazine
Thiazol *n* thiazole
Thiazolidin *n* thiazolidine
Thiazolin *n* thiazoline
Thiazolyl- thiazolyl
Thiazosulfon *n* thiazosulfone
Thienyl- thienyl
Thienylchlorid *n* thienyl chloride
Thienylmethylamin *n* thienylmethylamine
Thiet *n* thiete
Thietan *n* thietane
Thiiran *n* thiirane
Thinolith *m* (Min) thinolite
Thioaceton *n* thioacetone
Thioäpfelsäure *f* thiomalic acid
Thioäther *m* thioether

Thioalkohol *m* mercaptan, thioalcohol
Thioameisensäure *f* thioformic acid
Thioanhydrid *n* thioanhydride, anhydride of a thioacid
Thioanilin *n* thioaniline
Thioanisol *n* thioanisol
Thioantimonsäure *f* thioantimonic acid
Thioarsenat *n* thioarsenate, salt or ester of thioarsenic acid
Thioarsenigsäure *f* thioarsenous acid
thioarsenigsauer (Salz) thioarsenite
Thioarsenit *n* thioarsenite, salt or ester of thioarsenous acid
Thioarsensäure *f* thioarsenic acid
thioarsensauer (Salz) thioarsenate
Thiobakterien *pl* thiobacteria, sulfur bacteria
Thiobarbitursäure *f* thiobarbituric acid
Thiobenzanilid *n* thiobenzanilide
Thiobenzoesäure *f* thiobenzoic acid
Thiocarbamid *n* thiocarbamide, thiourea
Thiocarbamidsäure *f* thiocarbamic acid
Thiocarbanilid *n* thiocarbanilide
Thiocarbazid *n* thiocarbazide
Thiocarbin *n* thiocarbin
Thiocarbinol *n* thiocarbinol
Thiocarbonyl- thiocarbonyl
Thiochinanthren *n* thioquinanthrene
Thiochroman *n* thiochroman, dihydrobenzothiopyran
Thiochromen *n* thiochromene, benzothiopyran
Thiochromon *n* thiochromone, benzothiopyrone
Thiocol *n* (HN) potassium guaiacol sulfonate, thiocol
Thiocresol *n* thiocresol
Thioctinsäure *f* thioctic acid
Thiocyan *n* thiocyanogen
Thiocyanin *n* thiocyanin
Thiocyankalium *n* (obs) potassium thiocyanate
Thiocyansäure *f* thiocyanic acid
Thiocyanverbindung *f* thiocyanate, thiocyanogen compound
Thiodiazin *n* thiodiazine, diazthine
Thiodiazol *n* thiodiazole
Thiodiessigsäure *f* thiodiacetic acid
Thiodiglykol *n* thiodiglycol, dihydroxyethylene sulfide
Thiodiglykolsäure *f* thiodiglycolic acid
Thiodiphenylamin *n* thiodiphenylamine, phenothiazine
Thiodipropionsäure *f* thiodipropionic acid
Thioessigsäure *f* thioacetic acid
Thioflavon *n* thioflavone
Thioform *n* thioform, basic bismuth dithiosalicylate
Thioformanilid *n* thioformanilide
Thiogenfarbstoff *m* thiogen dye, sulfur dyestuff
Thiogenschwarz *n* thiogen black
Thiogermaniumsäure *f* thiogermanic acid
Thioglucose *f* thioglucose

Thioglykol *n* thioglycol
Thioglykolsäure *f* thioglycolic acid
Thioharnstoff *m* thiourea, sulfourea, thiocarbamide
Thioharnstoffharz *n* thio-urea resin
Thiohydantoin *n* thiohydantoin
Thioindigo *n* thioindigo
Thioindigorot *n* thioindigo red
Thioindol *n* thioindole
Thioindoxyl *n* thioindoxyl
Thiokakodylsäure *f* thiocacodylic acid
Thioketon *n* thioketone
Thiokinase *f* (Biochem) thiokinase
Thiokohlensäure *f* thiocarbonic acid
Thiokol *n* (HN) thiokol, polyethylene tetrasulfide
Thiokresol *n* thiocresol
Thiol *n* thiol, mercaptan, thioalcohol
Thiolan *n* thiolane
Thiolase *f* (Biochem) thiolase
Thiolsäure *f* thiolic acid
Thiolvalin *n* thiolvaline
Thiolyse *f* thiolysis
thiolytisch thiolytic
Thiomalonsäure *f* thiomalonic acid
Thiomilchsäure *f* thiolactic acid
Thionalid *n* thionalid
Thionaphthen *n* thionaphthene
Thionaphthol *n* thionaphthol, naphthyl mercaptan
Thioncarbaminsäure *f* thionocarbamic acid
Thionessal *n* thionessal, tetraphenylthiophen
Thionin *n* thionine, Lauth's violet
Thioninfarbstoff *m* thionine dye, thiazine dye
Thionsäure *f* thionic acid
Thionursäure *f* thionuric acid, 5-sulfo-5-aminobarbituric acid
Thionyl- thionyl
Thionylbromid *n* thionyl bromide
Thionylchlorid *n* thionyl chloride
Thiooxin *n* thiooxine
Thiooxindol *n* thiooxindole
Thiopegan *n* thiopegan
Thiophan *n* thiophane
Thiophanthren *n* thiophanthrene
Thiophanthron *n* thiophanthrone
Thiophen *n* thiophene
Thiophenin *n* thiophenine, aminothiophene
Thiophenol *n* thiophenol
Thiophosgen *n* thiophosgene, thiocarbonyl chloride
Thiophosphat *n* thiophosphate
Thiophosphorsäure *f* thiophosphoric acid
Thiophthalid *n* thiophthalide
Thiophthen *n* thiophthene, bithiophene
Thiopyran *n* thiopyran
Thiosäure *f* thioacid
Thiosalicylsäure *f* thiosalicylic acid
Thiosalz *n* thio-salt

Thioschwefelsäure *f* thiosulfuric acid
thioschwefelsauer thiosulfate
Thiosemicarbazid *n* thiosemicarbazide
Thioserin *n* thioserine
Thiosinamin *n* thiosinamine, allylsulfocarbamide, allylthiourea
Thiostannat *n* thiostannate, salt or ester of thiostannic acid
Thiosulfat *n* thiosulfate, salt or ester of thiosulfuric acid
Thiosulfosäure *f* thiosulfonic acid
Thiotenol *n* thiotenol
Thiotetrabarbital *n* thiotetrabarbital
Thiothiamin *n* thiothiamine
Thiouracil *n* thiouracil
Thiourazol *n* thiourazole
Thiourethan *n* thiourethane
Thiouridin *n* thiouridine
Thioxan *n* thioxane
Thioxanthen *n* thioxanthene
Thioxanthon *n* thioxanthone, benzophenone sulfide
Thioxen *n* thioxene, dimethylthiophene
Thioxin *n* thioxine
Thioxinschwarz *n* thioxine black
Thioxol *n* thioxole
Thiozinnsäure *f* thiostannic acid
thiozinnsauer (Salz) thiostannate
Thiuram *n* thiuram
Thiuramdisulfid *n* thiuram disulfide
thixotrop thixotropic
Thixotrop *n* (Kolloid) thixotrope
Thixotropie *f* thixotropy
Thollosid *n* tholloside
Thollosidsäure *f* thollosidic acid
Thomasbirne *f* (Metall) Thomas converter
Thomaseisen *n* (Metall) Thomas steel, basic iron, basic steel, mild steel
Thomasflußeisen *n* (Metall) Thomas low-carbon steel
Thomasflußstahl *m* basic converter steel, Thomas steel
Thomaskonverter *m* (Metall) Thomas converter
Thomasmehl *n* [ground] basic slag, Thomas meal
Thomasroheisen *n* basic Bessemer pig iron, basic pig [iron], Thomas pig iron
Thomasschlacke *f* basic slag, Thomas slag
Thomasstahl *m* basic converter steel, basic steel, mild steel, Thomas steel
Thomasstahlwerk *n* basic steel works
Thomas-Verfahren *n* basic Bessemer process, Thomas process
Thomsenolith *m* (Min) thomsenolite
Thomsoneffekt *m* Thomson effect
Thomsonit *m* (Min) thomsonite
Thorcarbid *n* thorium carbide
Thorerde *f* thoria, thorium dioxide, thorium oxide
Thorerz *n* (Min) thorium ore

Thorianit *m* (Min) thorianite
Thorit *m* (Min) thorite
Thorium *n* thorium (Symb. Th)
Thorium-Brutreaktor *m* thorium breeder
Thoriumemanation *f* thorium emanation, thoron (Symb. Tn)
Thoriumfaden *m* thoriated filament
Thoriumgehalt *m* thorium content
Thoriumniederschlag *m* thorium precipitate
Thoriumnitrat *n* thorium nitrate
Thoriumoxid *n* thorium oxide, thorium dioxide, (Min) thoria
Thoriumreihe *f* thorium series
Thoriumröhre *f* thoriated filament valve
Thoriumsulfat *n* thorium sulfate
Thoriumsulfid *n* thorium sulfide, thorium disulfide
Thoriumverbindung *f* thorium compound
Thorium-Zerfallsreihe *f* thorium radioactive series
Thorolit *m* thorolite
Thoron *n* (Thoriumemanation) thoron
Thortveitit *m* (Min) thortveitite
Thraulith *m* thraulite
Threit *m* threitol
threo-Form *f* (Stereochem) threo form, threo modification
Threonin *n* threonine
Threoninaldolase *f* (Biochem) threonine aldolase
Threonindesaminase *f* (Biochem) threonine deaminase
Threoninol *n* threoninol
Threonsäure *f* threonic acid
Threopentulose *f* threopentulose
Threosamin *n* threosamine
Threose *f* threose
Thrombin *n* thrombin, thrombase, zymoplasm
Thrombokinase *f* (Biochem) thrombokinase
Thrombolith *m* thrombolite
Thromboplastin *n* (Biochem) thromboplastin, thrombokinase
Thrombose *f* (Med) thrombosis
Thrombozym *n* thrombozym, prothrombin, thrombokinase
Thrombozyt *m* thrombocyte, blood platelet
Thrombozytenanstieg *m* (Med) thrombocytosis
Thrombozytenwert *m* (Med) thrombocyte value
Thujaketon *n* thujaketone
Thujaketonsäure *f* thujaketonic acid
Thujamenthol *n* thujamenthol
Thujan *n* thujane
Thujaöl *n* thuja oil
Thujen *n* thujene, tanacetene
Thujetin *n* thujetin
Thujetinsäure *f* thujetic acid
Thujigenin *n* thujigenin
Thujin *n* thujin
Thujol *n* thujol, absinthol
Thujon *n* thujone, tanacetone

Thujyl- thujyl
Thujylalkohol *m* thujyl alcohol
Thujylamin *n* thujylamine
Thulit *m* (Min) thulite
Thulium *n* thulium (Symb. Tm o. Tu)
Thuringit *m* (Min) thuringite
Thylakoid *n* thylakoid
Thylakoidmembran *f* thylakoid membrane
Thymen *n* thymene
thymianartig thyme-like
Thymiancampher *m* thyme camphor, 3-methyl-6-isopropylphenol, thymol
Thymianfluidextrakt *m* liquid extract of thyme
Thymiansäure *f* thymianic acid
Thymidin *n* thymidine
Thymidindiphosphat *n* thymidine diphosphate, TDP
Thymidinmonophosphat *n* thymidine monophosphate
Thymidintriphosphat *n* thymidine triphosphate
Thymidylsäure *f* thymidylic acid
Thymin *n* thymine, 5-methyluracil
Thyminose *f* thyminose
Thymochinon *n* thymoquinone
Thymoform *n* thymoform
Thymol *n* thymol, 3-methyl-6-isopropylphenol, thyme camphor
Thymolblau *n* thymol blue, thymol sulfonphthalein
Thymolindophenol *n* thymolindophenol
Thymolphthalein *n* thymolphthalein
Thymolsulfonphthalein *n* thymol sulfonphthalein, thymol blue
Thymomenthol *n* thymomenthol
Thymonucleinsäure *f* thymonucleic acid
Thymotal *n* thymotal, thymyl carbamate
Thymotinaldehyd *m* thymotic aldehyde
Thymotinsäure *f* thimotic acid
Thymotol *n* thymotol, thymyl iodide
Thymus *m* (Med) thymus
Thymusdrüse *f* (Med) thymus gland
Thynnin *n* thynnin
Thyratron *n* thyratron
Thyresol *n* thyresol
Thyroameisensäure *f* thyroformic acid
Thyrobuttersäure *f* thyrobutyric acid
Thyroessigsäure *f* thyroacetic acid
Thyroglobulin *n* thyroglobulin
Thyroidin *n* thyroidine
Thyrojodin *n* (obs) thyroiodine, thyroxine
Thyronamin *n* thyronamine
Thyronin *n* thyronine
Thyronucleoproteid *n* thyronucleoprotein
Thyropropionsäure *f* thyropropionic acid
Thyroxin *n* thyroxine, thyroiodine
tief deep, low, (Farbe) dark
tiefätzen to intaglio, to overetch
Tiefätzprobe *f* deep-etch test
Tiefätzung *f* deep etching

Tiefbau *m* (Bergb) deep mine workings
Tiefbettfelge *f* drop center rim, well base rim
tiefbraun dark brown
Tiefdruck *m* (Buchdr) gravure [printing], (Meteor) depression
Tiefdruckmaschine *f* gravure printing machine, photogravure printing machine
Tiefdruckplatte *f* intagliated printing plate
Tiefe *f* depth, ~ **der Färbung** depth of the color
Tiefendimension *f* dimension in depth
Tiefeneinstellung *f* (Opt) depth adjustment
Tiefenfilter *n* deep bed filter
Tiefenkarte *f* bathymetric chart
Tiefenlehre *f* depth gauge
Tiefenmesser *m* depth indicator, (Nav) sea gauge
Tiefenschärfe *f* (Phot) depth of focus
Tiefenstufe *f*, **geothermische** ~ geothermal gradient
Tiefenunschärfe *f* (Opt) lack of depth of focus
Tiefenwirkung *f* penetrative effect, (eines Abbeizmittels) penetration
tiefgefärbt deeply colored
Tiefgefrierschrank *m* deep freezer
Tiefkälteverfahren *n* low temperature process
Tiefkühlanlage *f* deep freezer, deep freezing apparatus
Tiefkühlapparat *m* deep freezer, deep freezing apparatus
tiefkühlen to freeze, to refrigerate
Tiefkühlen *n* deep freezing
Tiefkühlkost *f* deep frozen food
Tiefkühltruhe *f* deep freezer
Tiefkühlung *f* deep freezing, deep cooling
Tiefkühlverfahren *n* deep-freezing method
Tiefofen *m* reheating furnace, soaking pit
Tiefofenkran *m* crane for the reheating furnace, soaking pit crane
Tiefrelief *n* (Buchdr) intaglio
tiefschmelzend low-melting
tiefsiedend low-boiling
Tiefsttemperatur *f* lowest temperature
Tiefstwert *m* minimum value
Tiefteer *m* low-temperature tar
Tieftemperatur *f* low temperature
Tieftemperatur-Entgraten *n* cryotrimming
Tieftemperaturforschung *f* cryogenics
Tieftemperaturkautschuk *m* cold rubber
Tieftemperaturkonvertierung *f* low-temperature shift conversion
Tieftemperaturmahlen *n* cryogrinding
Tieftemperaturoxidation *f* low temperature oxidation
Tieftemperaturpolymerisat *n* low temperature polymer
Tieftemperatur-Schleudern *n* cryotumbling
Tieftemperaturtechnik *f* cryogenics
Tieftemperaturteer *m* low-temperature tar
Tieftemperaturvergasung *f* low temperature carbonization
Tieftemperaturverkokung *f* low temperature carbonization
Tiefungsversuch *m* cupping test, deep-drawing test
Tiefziehbarkeit *f* drawability
Tiefziehblech *n* deep-drawn sheet metal
tiefziehen to deep-draw
Tiefziehen *n* deep drawing, vacuum forming, wall-ironing, ~ **mit Gleitmittel** grease forming
Tiefziehfähigkeit *f* deep drawing property
Tiefziehmaschine *f* thermoforming machine
Tiefziehpresse *f* deep-draw press
Tiefziehprobe *f* cupping test
Tiefziehverfahren *n* deep drawing
Tiegel *m* (Metall) crucible, melting pot
Tiegelbauch *m* crucible belly
Tiegelbrennofen *m* crucible oven
Tiegeldeckel *m* crucible lid
Tiegeldreieck *n* crucible triangle
Tiegelfabrik *f* crucible works
Tiegelflußstahl *m* crucible cast steel
Tiegelfutter *n* crucible lining
Tiegelgießerei *f* (Tiegelguß) casting in crucibles
Tiegelgußstahl *m* crucible cast steel
Tiegelherstellung *f* manufacture of crucibles
Tiegelhohlform *f* crucible mold
Tiegelinhalt *m* crucible charge, crucible contents
Tiegelmasse *f* crucible material, refractory material for crucibles
Tiegelofen *m* crucible furnace, pot furnace
Tiegelpresse *f* crucible press
Tiegelprobe *f* crucible test
Tiegelring *m* crucible ring
Tiegelschachtofen *m* [shaft] crucible furnace
Tiegelscherben *f pl* old crucible material
Tiegelschmelzerei *f* crucible melting plant
Tiegelschmelzofen *m* crucible melting furnace, pot furnace
Tiegelschmelzverfahren *n* crucible melting process
Tiegelstahl *m* crucible [cast] steel
Tiegelstahldarstellung *f* crucible steel process, pot steel process
Tiegelstahlhütte *f* crucible steel works
Tiegeltrockner *m* crucible drying apparatus
Tiegelverkokung *f* crucible coking test
Tiegelzange *f* crucible tongs
Tiemannit *m* (Min) tiemannite
Tiemann-Reimersche Reaktion *f* Tiemann-Reimer reaction
Tierart *f* species
Tierarzneimittel *n* veterinary medicine
Tierarzt *m* veterinarian
Tierchemie *f* animal chemistry
Tierfaser *f* animal fiber
Tierfett *n* animal fat
tierisch animal
Tierklinik *f* veterinary hospital
Tierkörperasche *f* animal ashes

Tierkohle *f* animal charcoal
Tierkreis *m* (Astr) zodiac
Tierparasit *m* zooparasite
Tierversuch *m* animal experiment
Tierzucht *f* breeding of animals
Tigererz *n* (Min) stephanite
Tiglinaldehyd *m* tiglic aldehyde, 2-methyl-2-butenal, tiglaldehyde
Tiglinalkohol *m* tiglic alcohol
Tiglinat *n* tiglate, salt or ester of tiglic acid
Tiglinsäure *f* tiglic acid
Tigloidin *n* tigloidine
Tigogenin *n* tigogenin
Tikonium *n* ticonium
Tilasit *m* (Min) tilasite
Tilgung *f* (Amortisation) amortization, (Ausrottung) extermination, (Schulden) repayment, settlement, (Streichung) cancelling
Tiliacorin *n* tiliacorine
Tillmans Reagens *n* Tillman's reagent
Tinkal *m* (Min) tincal, native borax
Tinktur *f* tincture
Tinte *f* ink, **magnetische** ~ magnetic ink
Tintenaufnahmefähigkeit *f* ink receptivity
Tintenblau *n* ink blue
Tintenfleck *m* ink spot, ink stain
Tintenprobe *f* ink test
Tintenpulver *n* ink powder
Tintenstein *m* inkstone
Tipper *m* **des Vergasers** carburetor primer
Tisch *m* table, (Opt) stage, **drehbarer** ~ turntable
Tischbelag *m* table covering
Tischdrehbank *f* bench lathe
Tischfläche *f* table area
Tischgerät *n* table model
Tischhub *m* table travel
Tischklemme *f* bench clamp
Tischlerleim *m* joiner's glue
Tischlerplatte *f* block board, core board
Tischleuchte *f* table lamp
Tischplatte *f* table top
Tischpresse *f* rotary tabletting press
Titan *n* titanium (Symb. Ti)
Titanammoniumformiat *n* titanium ammonium formate
Titanammon[ium]oxalat *n* titanium ammonium oxalate
Titanat *n* titanate
Titanaugit *m* (Min) titanaugite
Titanchlorid *n* titanium chloride, titanic chloride
Titandioxid *n* (Min) titania, (Titan-IV-oxid)) titanium dioxide, titanic anhydride, titanic oxide, titanium(IV) oxide
Titandioxidfüllstoff *m* titanium dioxide filling material
Titaneisen *n* titaniferous iron
Titaneisenerz *n* (Min) ilmenite, titanic iron ore
Titaneisensand *m* titaniferous iron sand

Titaneisenstein *m* (Min) ilmenite, titanic iron ore
Titanerz *n* (Min) titaniferous ore, titanium ore
Titanfluorwasserstoffsäure *f* fluotitanic acid
Titanflußsäure *f* fluotitanic acid
Titanformiat *n* titanium formate
titanführend titaniferous
Titangehalt *m* titanium content
Titangelb *n* titan yellow
Titangranat *m* (Min) rutile
Titanhalogen *n* titanium halide
titanhaltig containing titanium, titaniferous
Titanisalz *n* (Titan(IV)-salz) titanic salt, titanium(IV) salt
Titanit *m* (Min) titanite
Titanit *n* (Hartmetall) titanite
Titanium *n* titanium (Symb. Ti)
Titaniumsalz *n* titanium salt
Titankaliumoxalat *n* titanium potassium oxalate
Titankiesel *m* (Min) brookite
Titanlösung *f* titanium solution
Titanmetall *n* metallic titanium, titanium metal
Titanolith *m* (Min) titanolite
Titanolivin *m* (Min) titanolivine
Titanomorphit *m* (Min) titanomorphite
Titanoxid *n* titanium oxide, (Min) titania, (Titan(IV)oxid) titanic anhydride, titanic oxide, titanium dioxide, titanium(IV) oxide, **kristallines** ~ (Min) anatase, **schwarzes** ~ (Min) ilmenorutile
Titanoxidhydrat *n* hydrous titanic oxide, hydrous titanium dioxide, titanic acid
Titanpigment *n* titanium pigment
Titansäure *f* titanic acid, hydrous titanic oxide, hydrous titanium dioxide, titanic hydroxide
Titansäureanhydrid *n* titanic anhydride, titanium dioxide
Titansalz *n* titanium salt, (Titan(III)-salz) titanium(III) salt, titanous salt, (Titan(IV)-salz) titanic salt
titansauer (Salz) titanate
Titanstahl *m* titanium steel
Titanthermit *n* titanium thermit
Titanverdampferpumpe *f* titanium vapor pump
Titanweiß *n* titanium white
Titel *m* title
Titelblatt *n* front page, title page
Titer *m* titer, (Seide) silk titer, ~ **einer Lösung** standard strength of a solution, titration standard
Titerabweichung *f* (Text) denier variation
Titerflüssigkeit *f* standard solution, titration solution
Titration *f* titration, **bromometrische** ~ bromometry, **elektrometrische** ~ electrometric titration, potentiometric titration, **jodometrische** ~ iodometric titration, iodometry, **potentiometrische** ~ potentiometric titration
Titrationskurve *f* titration curve

Titrationsverfahren *n* titration method
Titrieranalyse *f* analysis by titration, volumetric analysis
Titrierapparat *m* titrating apparatus, titration apparatus, volumetric apparatus
titrieren to titrate
Titrieren *n* titrating, titration
Titrierflüssigkeit *f* standard solution, titrating solution
Titrierlösung *f* standard solution
Titriermethode *f* titration method
Titriersäure *f* standard acid, titrating acid
Titrierung *f* titration
Titriervorrichtung *f* titrating apparatus
Titrierzahl *f* titer, titration value
Titrierzelle *f* titration cell
Titrimetrie *f* titrimetry
titrimetrisch titrimetric, volumetric
Titriplex *n* (HN) titriplex (commercial name for ethylene diamine tetraacetic acid)
Tiza *m* tiza, ulexite
toasten to toast
Tobiassäure *f* Tobias acid
Tocamphyl *n* tocamphyl
Tochteraktivität *f* daughter activity
Tochtergesellschaft *f* subsidiary company
Tochterkern *m* daughter nucleus
Tochterkolonie *f* (Bakt) daughter colony
Tochterkultur *f* (Bakt) subculture
Tochterzelle *f* (Biol) daughter cell
Tocol *n* tocol
Tocopherol *n* tocopherol, vitamin E
Tocopheroxid *n* tocopheroxide
Tocopherylchinon *n* tocopherylquinone
Tocopurpur *n* tocopurple
tödlich deadly, lethal, mortal
tönen (Akust) to sound, (Färb) to shade, to tint
Töner *m* toner
tönern argillaceous, clayey, of clay
Tönung *f* (Färb) shading, toning, (Farbton) hue, shade, tint, (Malerei) tonality
Tönungskraft *f* (Färb) tinting strength
Töpferei *f* pottery, ceramic art
Töpfererde *f* potter's clay
Töpfererz *n* potter's ore
Töpferscheibe *f* potter's wheel
Töpferton *m* potter's clay
Töpferware *f* pottery
Toilettenartikel *m* toilet article
Tokorogenin *n* tokorogenin
Tokorogensäure *f* tokorogenic acid
Tolan *n* tolane, diphenylacetylene
Tolanrot *n* tolane red
Tolazolin *n* tolazoline
Tolbutamid *n* tolbutamide
Toledoblau *n* toledo blue
Toleranz *f* tolerance, allowable variation, allowance, permissible limits
Toleranzbereich *m* allowable limits

Toleranzdosis *f* tolerance dose, threshold dose
Toleranzfeld *n* zone of tolerance, extent of tolerance
Toleranzgrenze *f* control limit
Toleranzwerte *pl* allowable limits
Tolidin *n* tolidine, dimethylbenzidine
Tolil *n* tolil, dimethylbenzil
Tolit *n* tolite, trinitrotoluene
Tollkirsche *f* (Bot) belladonna, deadly night-shade
Tollkirschenextrakt *m* extract of belladonna
Tollkrautwurzel *f* belladonna root
Tollwut *f* rabies
Tollwutimpfung *f* antirabic vaccination
Tollwutvakzine *f* rabies vaccine
Tolonium *n* tolonium
Tolpronin *n* tolpronine
Tolualdehyd *m* tolualdehyde
Toluamid *n* toluamide, carbamyltoluene
Tolubalsam *m* [balsam of] tolu
Toluchinaldin *n* toluquinaldine
Toluchinhydron *n* toluquinhydrone
Toluchinol *n* toluquinol
Toluchinolin *n* toluquinoline
Toluchinon *n* toluquinone
Toluidin *n* toluidine
Toluidinblau *n* toluidine blue
Toluidinorange *n* toluidine orange
Toluidinsulfonsäure *f* toluidine sulfonic acid
Tolunitril *n* tolunitrile, cyanotoluene
Toluol *n* toluene, methylbenzene
Toluoldiamin *n* toluenediamine
Toluoldiazoniumhydroxid *n* toluenediazonium hydroxide
Toluolmoschus *m* toluene musk
Toluolsulfamid *n* toluenesulfonamide
Toluolsulfinsäure *f* toluenesulfinic acid
Toluolsulfochlorid *n* toluenesulfochloride, toluenesulfonyl chloride
Toluolsulfonamid *n* toluenesulfonamide
Toluolsulfonanilid *n* toluenesulfonanilide
Toluolsulfonsäure *f* toluenesulfonic acid
Toluolsulfonylchlorid *n* toluenesulfonyl chloride
Toluphenon *n* toluphenone
Toluresitannol *n* toluresitannol
Tolusafranin *n* tolusafranine
Tolutinktur *f* tolu tincture
Toluylen *n* toluylene
Toluylenblau *n* toluylene blue
Toluylendiamin *n* toluylene diamine
Toluylenrot *n* toluylene red
Toluylenrotbase *f* toluylene red base
Toluylsäure *f* toluic acid, methylbenzoic acid
Tolyl- tolyl
Tolylcarbinol *n* tolylcarbinol
Tolylhydrazin *n* tolylhydrazine
Tolylmethylketon *n* tolylmethyl ketone
Tolylphenylketon *n* tolyl phenyl ketone
Tolylschwarz *n* tolyl black

Tolypyrin *n* tolypyrine
Tomatidin *n* tomatidine
Tomatin *n* tomatine
Tombak *m* (Legierung) tombac
Tombakmetall *n* tombac metal
Tombakrohr *n* tombac pipe
Ton *m* clay, kaolin, (Akust) tone, (Färb) tint, (Schall) sound, ~ **schlämmen** to wash the clay, **bildsamer** ~ pipe clay, plastic clay, **eisenschüssiger** ~ clay containing iron, **fetter** ~ fuller's earth, soapy clay, **feuerfester** ~ fire clay, refractory clay, **gebrannter** ~ terracotta, **kalkhaltiger** ~ calcareous clay, lime marl, **magerer** ~ meager clay
Tonabnehmer *m* pick-up (of a phonograph), phonograph cartridge
tonartig argillaceous, clayey
Tonband *n* recording tape
Tonbeize *f* red liquor
Tonbindemittel *n* clay cement
Tonboden *m* clay soil
Tonbrei *m* (Keram) clay slip
Tondämpfer *m* sound damper
Tondreieck *n* (Chem) pipeclay triangle
Toneisenstein *m* (Min) clay ironstone
tonen (Phot) to tone
Tonen *n* (Phot) toning
Toner *m* (Phot) toner
Tonerde *f* argillaceous earth, alum earth, alumina, aluminium oxide, **essigsaure** ~ aluminum acetate, **essigweinsaure** ~ aluminum acetotartrate, **geschlämmte** ~ emulsified alumina, **kieselsaure** ~ aluminum silicate, **schwefelsaure** ~ aluminum sulfate
Tonerdeanlage *f* alumina plant
Tonerdebeize *f* aluminum mordant
Tonerdegehalt *m* alumina content
tonerdehaltig aluminiferous, containing alumina
Tonerdehydrat *n* hydrated alumina, (obs) aluminum hydroxide
Tonerdesalz *n* (obs) aluminum salt
Tonerreger *m* sound producer
Tonfarbe *f* (Akust) timbre
tonfarbig clay-colored
Tonfilter *n* clay filter
Tonfixierbad *n* (Phot) toning and fixing bath
Tonfixiersalz *n* (Phot) toning and fixing salt
Tonflanschenrohr *n* earthenware flanged tube
Tonfrequenz *f* sound frequency
Tonfrequenzspektrometer *n* audio-frequency spectrometer
Tongefäß *n* clay vessel, earthen vessel
Tongips *m* argillaceous gypsum
tongipshaltig argillogypseous
Tongut *n* pottery, porous earthenware
Tonhöhe *f* (Akust) pitch
tonig argillaceous, clayey
Tonindustrie *f* clay industry

Tonkabohne *f* tonka bean
Tonkalk *m* (Min) argillaceous limestone, argillocalcite
Tonkalkstein *m* marly limestone
tonkieselhaltig argillosiliceous
Tonkiste *f* fire-clay box
Tonkühlschlange *f* ceramic cooling coil, earthenware cooling coil
Tonmergel *m* clay marl
Tonmischer *m* clay mixer
Tonmörtel *m* fireclay mortar
Tonmuffel *f* clay retort
Tonne *f* barrel, cask, (Maß) ton = 1000 kg
Tonnengehalt *m* tonnage
Tonofen *m* clay furnace
Tonplatte *f* clay disk
Tonporphyr *m* (Min) argillaceous porphyry
Tonregler *m* (Akust) tone controller
Tonretorte *f* clay retort
Tonrille *f* (Akust) sound track
Tonsand *m* argillaceous sand
Tonsandstein *m* argillaceous sandstone
Tonscheibe *f* clay disk
Tonschicht *f* clay layer
Tonschiefer *m* (Min) argillaceous schist, argillaceous slate, argillite, clay slate
tonschieferhaltig schistoargillitic
Tonschlämme *f* clay wash
Tonschneider *m* clay cutter, clay mixer
Tonseife *f* aluminous soap
Tonspeise *f* clay mortar, clay slip, tempered clay
Tonspektrum *n* (Akust) sound spectrum
Tonstein *m* claystone
Tonteller *m* clay plate, clay dish
Tontiegel *m* [fire] clay crucible
ton- und eisenhaltig argilloferruginous
Tonverarbeitung *f* clay processing
Tonwaren *f pl* earthenware, pottery
Tonwiedergabe *f* (Akust) sound reproduction
Tonwiedergabetechnik *f* (Akust) sound-reproduction technique
Tonzelle *f* clay-cell
Tonzuschlag *m* aluminous flux
Tonzylinder *m* porous pot
Topas *m* (Min) topaz
Topasfluß *m* artificial topaz
Topazolith *m* (Min) topazolite
Topfglühofen *m* pot annealing furnace
Topfmagnet *m* potshaped magnet
Topfzeit *f* can life, potlife, pot time
Topochemie *f* topochemistry
topologisch (Geom) topological
Toramin *n* toramin
Torbanit *m* torbanite
Torbernit *m* (Min) torbernite
Torf *m* peat
torfartig peat-like, peaty
Torfasche *f* peat ashes

Torfbrikett – Trägereinfang

Torfbrikett *n* peat briquet[te]
Torfeisenerz *n* (Min) bog iron ore
Torferde *f* peat mold
Torfgas *n* peat gas
Torfgasgenerator *m* peat gas producer
Torfgeruch *m* peaty odor
Torfgewinnung *f* peat digging
torfhaltig containing peat
Torfmehl *n* powdered peat
Torfmull *m* peat dust
Torfplatte *f* peat slab
Torfstaub *m* peat dust
Torfstein *m* peat brick
Torgummi *n* bassora rubber
Tormentillextrakt *m* tormentil extract
Tormentillgerbsäure *f* tormentil tannic acid
Toroidkondensator *m* toroidal condenser
Torreol *n* torreol
Torsion *f* torsion, torsional stress, twist
Torsionsaufhängung *f* torsion suspension
Torsionsbeanspruchung *f* torsional stress
Torsionschwingzentrifuge *f* torsional vibrator
Torsionsdrehwaage *f* torsion balance
Torsionselastizität *f* torque elasticity
Torsionselektrometer *n* torsion electrometer
Torsionsfeder *f* torsion spring
Torsionsfestigkeit *f* torsional strength
torsionsfrei torsionfree
Torsionsgrenze *f* yield point of torsional shear
Torsionskonstante *f* torsion constant
Torsionskraft *f* torsional force, twisting force
Torsionsmodul *m* shear modulus, torsional rigidity
Torsionsmoment *n* torsional or twisting moment
Torsionspendel *n* torsion pendulum
Torsionsprüfmaschine *f* torsional tester
Torsionsschwinggerät *n* oscillating twisting machine
Torsionsschwingung *f* torsional wave
Torsionsschwingungsapparatur *f* torsional oscillation apparatus
Torsionsschwingungsfrequenz *f* torsional frequency
Torsionsschwingungsversuch *m* torsional pendulum test
Torsionsspannung *f* torsional stress
Torsionsversuch *m* torsion test, twisting test
Torsionswaage *f* torsion balance
Torsionswiderstand *m* resistance to torsion, twisting resistance
Torsionswinkel *m* torsion angle, angle of torque, angle of twist
Torstahl *m* tor steel
Tosylchlorid *n* tosyl chloride, p-toluene sulfonyl chloride
Tosylschutzgruppe *f* tosyl blocking group
tot dead, (Bergb) sterile
total total, complete
Totalixröhre *f* totalix tube

Totalnutzeffekt *m* over-all efficiency
Totalreflektometer *n* total reflectometer
Totalreflexion *f* total reflection
Totalreflexprisma *n* totally reflexing prism
Totalrücklauf *m* (Rektifikation) total reflux
Totalrückstrahlwinkel *m* (Opt) angle of total reflection
Totalspiegelung *f* (Opt) total reflection
Totalsynthese *f* total synthesis
totbrennen to dead-burn, to overburn (lime)
Totbrennen *n* dead roasting, overburning
Totenkopf *m* (Min) colcothar
Totgang *m* backlash
totgegerbt overtanned
totgemahlen overground
Totgewicht *n* (Eigengewicht) dead weight
Totliegende[s] *n* (Bergb) deads
totmahlen to overgrind
Totpunktlage *f* dead center position
Totrösten *n* overroasting
totsicher foolproof
Totvolumen *n* stagnant volume
Totzeit *f* (Regeltechnik) dead time (Am. E.), distance-velocity log (Br. E.)
Tourenzähler *m* revolution counter, speed counter, speed indicator, speedometer
Tourenzahl *f* number of revolutions
Toxämie *f* (Med) blood poisoning, tox[a]emia
Toxalbumin *n* toxalbumen, toxalbumin
Toxiferin *n* toxiferine
Toxigen *n* toxigene
Toxikologie *f* toxicology
toxikologisch toxicological
Toxin *n* toxin
toxisch toxic, poisonous
Toxoflavin *n* toxoflavin
Toxoid *n* toxoid
Toxopyrimidin *n* toxopyrimidine
Trabant *m* (Astr) satellite
Tracerchemie *f* tracer chemistry
Tracertechnik *f* tracer technique
Tracerverfahren *n* tracer methodology
Tracheide *f* (Bot) tracheid
Tracht *f* (von Kristallen) [crystal] habit
Trachyt *m* (Min) trachyte
Trachytporphyr *m* (Min) trachyte porphyry
träge lazy, sluggish, (Chem) inactive, inert
Träger *m* arbor, beam, carrier [material], girder, support, supporting base, (Chem) carrier, substrate, ~ **des Trichters** hopper bracket
Trägerauffüllung *f* carrier replenishment
Trägerbeweglichkeit *f* carrier mobility
Trägerbrücke *f* girder bridge
Trägerbündel *n* carrier beam
Trägerdampfdestillation *f* carrier distillation
Trägerdichte *f* carrier density
Trägerdiffusion *f* carrier diffusion
Trägereinfang *m* carrier trapping

Trägerelektrode *f* carrier electrode, supporting electrode
Trägerelektrolyt *m* supporting electrolyte
Trägerfällung *f* carrier precipitation
Trägerfolie *f* carrier foil
Trägerfrequenz *f* carrier frequency
Trägergas *n* carrier gas
Trägergassublimation *f* entrainer or carrier sublimation
Trägerinjektion *f* carrier injection
Trägerkaliber *n* girder pass
Trägerlebensdauer *f* carrier lifetime
Trägerlösung *f* carrier solution
trägerlos unsupported
Trägermaterial *n* (Gaschromat) solid support
Trägernetz *n* supporting girders
Trägeröl *n* flotation oil
Trägerprotein *n* (Biochem) carrier protein
Trägerpunkt *m* base point
Trägerspannung *f* carrier voltage
Trägersubstanz *f* carrier [substance], emulsion matrix
Trägheit *f* (Beharrung) inertia, (Chem) inactivity, inertness, **chemische** ~ chemical inertness
Trägheitsachse *f* axis of inertia
trägheitsfrei inertialess
Trägheitsgesetz *n* law of inertia
Trägheitskraft *f* inertial force
Trägheitsmittelpunkt *m* center of inertia
Trägheitsmoment *n* moment[um] of inertia
Trägheitsvermögen *n* inertia
Trägheitswelle *f* inertial wave
Trägheitswiderstand *m* inertial resistance
Tränendrüse *f* lachrymal gland
Tränengas *n* tear gas
Tränkbad *n* impregnating bath
Tränkbehälter *m* impregnating trough, impregnating vat
tränken to soak, to impregnate, to infiltrate, to varnish
Tränkharz *n* resin varnish
Tränkkessel *m* impregnating vessel
Tränklack *m* impregnating varnish, impregnation varnish
Tränkmasse *f* impregnating liquid, impregnating material
Tränkung *f* impregnation, soaking, steeping
Tränkungsdauer *f* impregnation time
Tränkverfahren *n* impregnating (process)
träufeln to drip, to drop, to trickle
Tragant *m* (Bot) tragacanth
Tragantgummi *n* [gum] tragacanth, gum dragon
Tragantin *n* tragantine
Tragantschleim *m* tragacanth mucilage
Tragantstoff *m* bassorin
Tragarm *m* supporting arm, cantilever, carrying arm
Tragbalken *m* supporting girder

tragbar portable
Trageigenschaft *f* (Text) wearing property
tragen to carry, (Kleidung) to wear, (stützen) to support
Tragfähigkeit *f* bearing capacity, bearing strength, lifting power, [load] carrying capacity
Tragfeder *f* bearing spring
Tragfläche *f* aerofoil, plane
Tragfuß *m* foot, support
Traggerüst *n* (Histol) supporting tissue
Tragkraft *f* carrying power, charging capacity, portative force, supporting power
Tragkranz *m* bracket rim
Tragkreuz *n* spider
Traglager *n* supporting bearing
Tragpratze *f* supporting bracket
Tragrahmen *m* supporting frame
Tragriemen *m* carrying strap
Tragring *m* bearing ring
Tragrolle *f* carrier roller, idler
Tragscheibe *f* (Treibriemen) idler
Tragseil *n* carrying cable
Tragseite *f* (Transportbänder) carrying side
Tragspindel *f* supporting coil
Tragstange *f* bearing rod, supporting rod
Tragtrommel *f* (Treibriemen) idler
Tragzapfen *m* pivot, bearing journal, pin, supporting journal
Traktionsakkumulator *m* traction accumulator
Traktrix *f* (Math) tractrix
Tran *m* fishoil, train oil
Tranfettsäure *f* fish oil fatty acid
Trangeruch *m* smell of train oil
tranhaltig containing train oil
tranig containing train oil
Tranquil[l]izer *m* (Pharm) tranquil[l]izer, sedative
Transacetylase *f* (Biochem) transacetylase
Transacetylierung *f* (Biochem) transacetylation
Transaldolase *f* (Biochem) transaldolase
Transaminase *f* (Biochem) transaminase
transaminieren to transaminate
Transaminierung *f* (Biochem) transamination
Transanordnung *f* trans arrangement
trans-Brückenisomer *n* (Stereochem) trans-bridged isomer
Transdichloräthan *n* transdichlorethane
Transdichloräthylen *n* transdichlorethylene
Transeife *f* fish-oil soap, train-oil soap
Transferase *f* (Biochem) transferase
Transfer-Druckkammer *f* transfer pot
Transferkolben *m* transfer plunger
Transferpressen *n* transfer molding
Transfer-Ribonucleinsäure *f* transfer ribonucleic acid, transfer RNA
Transfer-RNS s. Transfer-Ribonucleinsäure
trans-Form *f* (Stereochem) trans-form, trans-isomer

Transformation f transformation, **lineare** ~ (Math) linear transformation, **reguläre** ~ (Math) regular transformation
Transformationsmatrix f transformation matrix
Transformator m (Elektr) transformer
Transformatorenstation f transformer station
Transformatorenstrom m transformer current
Transformatorregelung f regulating of transformer
Transformatorspule f transformer coil
transformieren to transform
Transformierungsverhältnis n ratio of transformation
Transistor m (Elektr) transistor
Transistorgerät n transistor apparatus
Transketolase f (Biochem) transketolase
Transkription f (Biochem) transcription
Transkriptionshemmung f (Biochem) inhibition of transcription
Translation f translation
Translationsbewegung f translational motion
Translationsebene f (Krist) translation plane
Translationsenergie f translational energy
Translationsfläche f (Krist) translation plane
Translationsgeschwindigkeit f translatory velocity
Translationsgruppe f (Krist) translation group
Translationsregulation f (Biochem) regulation of translation
translatorisch translational
transluzent translucent
Transmethylierung f transmethylation
Transmission f transmission
Transmissionsanlage f power transmission plant
Transmissionsantrieb m transmission drive
Transmissionshammer m belt-driven hammer
Transmissionsquerschnitt m transmission cross section
transparent transparent, diaphanous
Transparentlack m clear varnish
Transparentleder n transparent leather
Transparenz f transparency, transmissivity (for light), transmittancy
Transparenzmesser m diaphanometer
Transpeptidase f (Biochem) transpeptidase
Transpeptidierung f transpeptidation
Transphosphatase f transphosphatase
Transphosphorylase f transphosphorylase
Transphosphorylierung f transphosphorylation
Transport m transport[ation], shipment, transfer, transmission, **aktiver** ~ active transport, **erleichterter** ~ faciliated transport, **gekoppelter** ~ coupled transport
Transportband n conveyer [belt], carrier roller
Transporteur m shipper, carrier, (Geom) protractor
Transportfilz m (Papier) conveyor felt
Transportgefäß n transport vessel
transportieren to transport

Transportkarren m transfer car
Transportkasten m transport case, portable box, transport box
Transportketten pl conveyor chains
Transportkontrolle f transfer check
Transportmittel n means of transportation
Transportprotein n transport protein
Transposition f (Math) transformation
Trans-Stellung f (Stereochem) trans position
Transsulfurase f (Biochem) trans-sulfurase
Transthiolierung f transthiolation
Transuran n transuranic element, transuranium element
Transuranelemente pl transuranium elements
transversal transversal, transverse
Transversale f transversal
Transversalitätsbedingung f transversality condition
Transversalschwingung f transverse oscillation
Transversalwelle f transverse wave
transzendent transcendental
Trapez n trapezoid
trapezförmig trapeziform, trapezoidal
Trapezgewinde n acme thread, tapered thread
Trapezoeder n (Krist) trapezohedron
Trapezregel f trapezoidal rule
Trapezring m trapezoidal washer
Trapporphyr m (Min) porphyritic trap
Trappsandstein m (Min) trap sandstone
Trapptuff m trap tuff
Traß m trass, volcanic tuff
Traßbeton m trass concrete
Traßpapier n grey wrapping paper
Traubenachat m (Min) botryoid agate
traubenartig grapelike
Traubenblei n (Min) mimetite
Traubenessig m grape vinegar
Traubengärung f fermentation of grapes
Traubenpresse f wine press
Traubensäure f (obs. für Weinsäure) tartaric acid
Traubensaft m grape juice
Traubenstein m (Min) botryolite
Traubenzucker m d-glucose, grape sugar
Traufe f down spout
Traumaticin n traumaticin[e]
Traumatinsäure f traumatic acid
Traversellit m (Min) traversellite
Travertin m (Min) travertine
Treberbranntwein m marc brandy
treffen to hit, to strike, (begegnen) to meet
Treffer m hit
Treffertheorie f hit theory, target theory
Trefferwahrscheinlichkeit f probability of hits
Treffplatte f target
Treffpunkt m point of impact, point of incidence
Trehalase f (Biochem) trehalase
Trehalose f trehalose
Treibachse f driving axle

Treibarbeit f (Metall) cupellation
Treibdampfpumpe f booster type diffusion pump
Treibdüse f booster nozzle
Treibeisen n white pig iron
treiben to drive, to push, (gären) to ferment, (hämmern) to emboss
Treibgas n power gas, propellant
Treibhaus n green-house, hot-house
Treibmittel n propellant, (Pumpe) pump fluid, (Schaumstoff) aerating agent, blowing agent, expanding agent, foaming agent
Treibmittelfüllung f (Pumpe) pump fluid filling
Treibmittelpumpe f vapor pump, jet pump
Treibmittelrücklauf m (Pumpe) pump fluid return pipe
Treiböl n (Motor) motor fuel oil
Treibofen m cupel[l]ing furnace, refining furnace
Treibprozeß m (Metall) cupellation, refining
Treibriemen m drive belt, driving belt, transmission belting
Treibschwefel m (obs) native sulfur
Treibstoff m fuel, power fuel, propellant
Treibstoffpumpe f fuel pump
Treibstoffraum m fuel cell
Treibstoffsynthese f motor fuel synthesis
Treibstoffzusatz m additive for fuels
Treibstrahl m driving jet, power jet
Treibverfahren n (Metall) cupellation, refining process
Treibzapfen m crank pin
Tremolit m (Min) tremolite
Tremuloidin n tremuloidin
Trennanlage f, **chemische** ~ chemical separation plant
trennbar separable
Trennblech n baffle
Trenndüsenverfahren n separation nozzle process
Trenneinsatz m partition
trennen to separate, to sever, (auflösen) to resolve, (Elektr) to disconnect
Trennen n separating, dividing, (Elektr) disconnecting
Trenner m separator
Trennfaktor m separation factor, **theoretischer** ~ (Atom) ideal separation factor
Trennfestigkeit f resistance to separation
Trennflügelpumpe f single-lobe pump
Trennflüssigkeit f (Gaschromat) liquid phase
Trennfüllung f (Gaschromat) stationary phase
Trennfuge f partition line, parting compound line, ~ **der Blasform** blow mold parting line, ~ **der Form** mold parting line
Trenngrad m grade efficiency
Trenngrenze f cut size (screening)
Trennisolator m disconnecting insulator
Trennkammer f chromatography tank
Trennkoeffizient m (Dest) distribution ratio
Trennkurve f grade efficiency curve

Trennleistung f column efficiency
Trennlinie f separating line
Trennmittel n mold release, parting compound, release agent
Trennmuffe f disconnecting box
Trennpapier n (Klebebänder) release paper
Trennrohr n (Atom) separator tube
Trennsäule f separating column
Trennschärfe f selectivity, separating capacity, separation sharpness
Trennschleuder f centrifuge separator
Trennschnitt m (Schweiß) separating cut
Trennstellung f spacing position
Trennstufenzahl f number of theoretical tray
Trennung f disintegration, parting, severing, (Chem.) separation, (durch Filtrieren) filtering, (Teilung) division, **absorptionschromatographische** ~ separation by absorption chromatography, **chromatographische** ~ chromatographic separation, **elektroanalytische** ~ electro-analytical separation, **elektrolytische** ~ electrolytic separation
Trennungsapparat m separator
Trennungsebene f division plane
Trennungserzeugnis n separation product, product of separation
Trennungsfläche f interface
Trennungsflüssigkeit f separation liquid
Trennungsgang m separation process, **analytischer** ~ analytical procedure
Trennungsmethode f method of separation
Trennungsmittel n means of separation
Trennungspotential n separation potential
Trennungsschicht f separating layer
Trennungsverfahren n separation method, separation process
Trennungsvermögen n selectivity
Trennungsvorgang m process of separation
Trennungswärme f heat of separation
Trennventil n separating valve
Trennverfahren n (für Dispersionen) settling
Trennvermögen n (Gaschromat) resolving power
Trennvorrichtung f separator
Trennwand f partition [wall]
Trennwanddiffusion f gaseous diffusion
Trennwiderstand m cohesive resistance
Trennwirkung f (Dest) efficiency
Treooxazolin n threooxazoline
Treppenbelag m stair covering
Treppenrost m step[ped] grate
Treppenrostgenerator m step[ped] grate producer
Treppenschutzschiene f stair protection rail
Trespe f (Bot) brome grass
Trester pl grounds, husks, residue
Tresterbranntwein m grape marc brandy
Tresterwein m grape marc wine
Tretamin n tretamine
treten to step, to tread

Tretgebläse *n* foot bellows
Tretkontakt *m* floor contact
Triacanthin *n* triacanthine
Triacetat *n* triacetate
Triacetin *n* triacetin, glycerol triacetate, glyceryl acetate
Triacetondialkohol *m* triacetone dialcohol
Triacetondiamin *n* triacetone diamine
Triacontanol *n* triacontanol
Triacylglycerid *n* tri[acyl]glyceride
Triäthanolamin *n* triethanolamine
Triäthylamin *n* triethylamine
Triäthylbenzol *n* triethylbenzene
Triäthylendiamin *n* triethylenediamine
Triäthylenglykol *n* triethylene glycol
Triäthylenrhodamin *n* triethylene rhodamine
Triäthylphosphin *n* triethylphosphine
Triäthylsilan *n* triethylsilane
Triakisoktaeder *n* (Krist) triakisoctahedron
Triakontan *n* triacontane
Triakontylen *n* triacontylene
Trialkylsilan *n* trialkylsilane
Triaminobenzoesäure *f* triaminobenzoic acid
Trianthrimid *n* trianthrimide
Triarylmethyl-Carbanion *n* triarylmethyl carbanion
Triarylmethyl-Carboniumion *n* triarylmethyl carbonium ion
Triarylmethyl-Radikal *n* triarylmethyl radical
Triazin *n* triazine
Triazo- triazo, azide
Triazobenzol *n* triazobenzene, phenyl azide
Triazol *n* triazole
Triazolidin *n* triazolidine
Triazolin *n* triazoline
Triazolon *n* triazolone
Triazolyl- triazolyl
Tribenzylamin *n* tribenzyl amine
Tribromanilin *n* tribromoaniline
Tribromessigsäure *f* tribromoacetic acid
Tribromhydrin *n* tribromohydrin
Tribromphenol *n* tribromophenol
Tribromphenolat *n* tribromophenolate
Tribromphosphor *m* phosphorus tribromide
Tribüne *f* platform, gallery, stand
Tributyrin *n* tributyrin, glycerol tributyrate, glyceryl butyrate
Tricalciumphosphat *n* neutral calcium phosphate, tricalcium phosphate
Tricaprin *n* tricaprin, glycerol tricaprate, glyceryl caprate
Tricaproin *n* tricaproin, glycerol tricaproate, glyceryl caproate
Tricaprylin *n* tricaprylin, glycerol tricaprylate, glyceryl caprylate
Tricarballylsäure *f* tricarballylic acid, propane tricarboxylic acid
Tricarbonsäurecyclus *m* tricarboxylic acid cycle, citric acid cycle, Krebs cycle

Tricatin *n* tricatin
Trichalcit *m* (Min) trichalcite
Trichine *f* trichina
trichinös (Med) trichinous
Trichinose *f* (Med) trichinosis
Trichinoskoplampe *f* trichinoscope lamp
Trichit *m* trichite
Trichloräthan *n* trichloroethane
Trichloräthylen *n* trichloroethene, trichloroethylene
Trichloräthylphosphat *n* trichlorethyl phosphate
Trichlorbutyliden *n* trichlorobutylidene
Trichlorcyan *n* cyanuric chloride, trichlorocyanidine
Trichloressigsäure *f* trichloroacetic acid
trichloressigsauer (Salz) trichloroacetate
Trichlorhydrin *n* trichlorohydrin
Trichlorsilan *n* trichlorosilane
Trichobakterium *n* trichobacterium
Trichocer[e]in *n* trichocereine
Trichodesminsäure *f* trichodesmic acid
Trichosansäure *f* trichosanic acid
Trichroismus *m* (Opt) trichroism
Trichromasie *f* trichromatism
trichromatisch trichromatic
Trichter *m* funnel
Trichterbildung *f* formation of pipes, funnel formation
Trichtereinlage *f* filter cone
Trichtereinsatz *m* filter cone
trichterförmig funnel shaped, infundibular
Trichtergestell *n* stand for funnel
Trichterhals *m* neck of a funnel
Trichterhalter *m* (Chem) funnel holder, funnel stand
Trichterkolben *m* funnel flask
Trichtermühle *f* cone mill, hopper mill
trichtern to pour through a funnel
Trichterrohr *n* funnel pipe, funnel tube
Trichterstativ *n* (Chem) funnel stand
Trichtertrockenofen *m* funnel kiln
Trichtertrockner *m* hopper drier
Tricin *n* tricin
Tricyanchlorid *n* tricyanogen chloride, cyanuric chloride, trichlorocyanidine
Tricyansäure *f* tricyanic acid, cyanuric acid, trihydroxycyanidine
Tricyansäuretriamid *n* 2,4,6-triamino-s-triazine, cyanurotriamide, melamine
Tricyansäureverbindung *f* compound of cyanuric acid, cyanurate
Tricyclamol *n* tricyclamol
Tricyclazin *n* tricyclazine
Tricyclen *n* tricyclene
Tricycl[en]al *n* tricycl[en]al
Tricyclensäure *f* tricyclenic acid
Tricymylamin *n* tricymylamine
Tridecan *n* tridecane
Tridecylaldehyd *m* tridecylaldehyde, tridecanal

Tridecylen *n* tridecylene
Tridecylsäure *f* tridecylic acid, tridecoic acid
Tridymit *m* (Min) tridymite
Triebel *n* [drive] pinion
Triebelwelle *f* pinion shaft
Triebfeder *f* driving spring, main spring
Triebkraft *f* driving power, impulsive force, motive power, propelling force
Triebsand *m* drift sand, quick sand, shifting sand
Triebscheibe *f* [driving] pulley
Triebstahl *m* pinion steel
Triebwerk *n* drive, [gear] transmission
triefen to drip, to drop
Trielaidin *n* trielaidin
Triferrin *n* triferrin, iron paranucleinate
Trifluoracetylaceton *n* trifluoroacetyl acetone
Trifluormonochloräthylen *n* trifluoromonochloroethylene
Triformol *n* triformol, paraformaldehyde
Trift *f* drift, (Abtrift) leeway
Triftstrom *m* drift current
Triggerschaltung *f* (Elektr) trigger circuit
Triglycerid *n* triglyceride
Triglykol *n* triglycol
trigonal (Krist) trigonal
trigonalbipyramidal trigonalbipyramidal
Trigondodekaeder *n* (Krist) trigondodecahedron
Trigonellin *n* trigonelline, nicotinic methylbetaine
Trigonometer *m* trigonometer
Trigonometrie *f* (Math) trigonometry
trigonometrisch trigonometric
Trihexosan *n* trihexosan
Trihydroxybenzol *n* trihydroxybenzene
Trihydroxymethylanthrachinon *n* trihydroxy methyl anthraquinone
Trihydroxypurin *n* trihydroxypurine
Triisopropylphenol *n* triisopropylphenol
triklin (Krist) triclinic
triklinholoedrisch (Krist) tricline holohedral
Trikosan *n* tricosane
Trikosylalkohol *m* tricosyl alcohol, tricosanol
Trikot *m* (Text) jersey cloth, tricot
Trikotagen *pl* (Text) knitted goods
Trikresol *n* tricresol
Trikresylphosphat *n* tricresyl phosphate
Trilactin *n* trilactin
Trilaurin *n* trilaurin, glycerol trilaurate, glyceryl laurate
Trilexfelge *f* trilex rim
Trilinolein *n* trilinolein, glycerol trilinolate, glyceryl linolate
Trilinolenin *n* trilinolenin
Trilinolin *n* trilinolein, glycerol trilinolate, glyceryl linolate, trilinolin
Trilit *n* trilite, trinitrotoluene
Trillion *f* one billion billions (Am. E.), quintillion (Am. E.), trillion (Br. E.)

Trilon *n* trilon
Trimargarin *n* trimargarin, glycerol trimargarate, glyceryl margarate
Trimellitsäure *f* trimellitic acid
trimer trimeric
Trimer *n* trimer
Trimerit *m* (Min) trimerite
Trimesinsäure *f* trimesic acid, trimesitinic acid
Trimesitinsäure *f* trimesitic acid
Trimetaphan *n* trimetaphan
Trimetaphosphorsäure *f* trimetaphosphoric acid
Trimethadion *n* trimethadione
Trimethindinium *n* trimethindinium
Trimethoxybenzoesäure *f* trimethoxybenzoic acid
Trimethylacetophenon *n* trimethylacetophenone
Trimethylamin *n* trimethylamine
Trimethylaminoxydase *f* (Biochem) trimethylamine oxidase
Trimethylammoniumbromid *n* trimethylammonium bromide
Trimethylbenzol *n* trimethylbenzene
Trimethylbernsteinsäure *f* trimethylsuccinic acid
Trimethylbor *n* trimethylboron
Trimethylbutan *n* trimethylbutane
Trimethylcellulose *f* trimethylcellulose
Trimethylchlorsilan *n* trimethylchlorosilane
Trimethylcyclohexan *n* trimethylcyclohexane
Trimethylen *n* trimethylene, cyclopropane
Trimethylenbromhydrin *n* trimethylene bromohydrin
Trimethylencyanid *n* trimethylene cyanide
Trimethylendiamin *n* trimethylene diamine
Trimethylenglykol *n* trimethylene glycol
Trimethylenimin *n* 1,3-propylene imine, trimethylene imine, tetrahydroform
Trimethylenoxid *n* trimethylene oxide
Trimethylester *m* trimethyl ester
Trimethylglucose *f* trimethylglucose
Trimethylhydrochinon *n* trimethylhydroquinone
Trimethylmethan *n* trimethylmethane
Trimethylnaphthalin *n* trimethylnaphthalene
Trimethylpentan *n* trimethylpentane
Trimethylsilan *n* trimethylsilane
Trimethylsilanol *n* trimethylsilanol
Trimethylsilylgruppe *f* trimethylsilyl group
Trimethylstibin *n* trimethylstibine, antimony trimethyl
Trimethylxanthin *n* trimethylxanthine, caffeine
Trimmkante *f* trim line
Trimmstab *m* shim rod
trimolekular trimolecular
trimorph trimorphic, trimorphous
Trimorphie *f* (Krist) trimorphism
Trimorphin *n* trimorphine
Trimyristin *n* trimyristine, glycerol trimyristate, glyceryl myristate
Trinatriumcitrat *n* trisodium citrate
Trinatriumphosphat *n* trisodium phosphate

Trinitrid – Trithian

Trinitrid *n* triazo, trinitride
Trinitrin *n* trinitrin, glycerol trinitrate, glyceryl nitrate, nitroglycerin
Trinitroanthrachinon *n* trinitroanthraquinone
Trinitrobenzoesäure *f* trinitrobenzoic acid
Trinitrobenzol *n* trinitrobenzene
Trinitrobutyltoluol *n* trinitrobutyltoluene
Trinitrochlorbenzol *n* trinitrochlorobenzene
Trinitrofluorenon *n* trinitrofluorenone
Trinitrophenol *n* trinitrophenol, picric acid
Trinitrophenylacridin *n* trinitrophenylacridine
Trinitroresorcin *n* trinitroresorcinol
Trinitrostilben *n* trinitrostilbene
Trinitrotoluol *n* trinitrotoluene, trinol, trotyl
Trinitroxylol *n* trinitroxylene
trinkbar drinkable, potable
Trinkerit *m* trinkerite
Trinkwasser *n* drinking water, fresh water, potable water
Trinkwasserversorgung *f* drinking water supply
Trinol *n* trinitrotoluene, trinol, trotyl
Trinucleotid *n* trinucleotide
Trioblockwalzwerk *n* three high blooming mill
Trioctylphosphat *n* trioctyl phosphate
Triode *f* triode, three-electrode tube
Triolein *n* triolein, glycerol trioleate, glyceryl oleate, olein
Trional *n* trional
Triose *f* triose
Trioseisomerase *f* (Biochem) triose isomerase
Triosemutase *f* (Biochem) triose mutase
Triosephosphat *n* triose phosphate
Triosephosphatdehydrogenase *f* (Biochem) triosephosphate dehydrogenase
Triosephosphatisomerase *f* (Biochem) triosephosphate isomerase
Trioseredukton *n* triose-reductone
Triowalzwerk *n* three high mill
Trioxan *n* trioxan[e], paraformaldehyde
Trioxid *n* trioxide
Trioxybenzol *n* benzenetriol, trihydroxybenzene
Trioxymethylen *n* trioxymethylene
Tripalmitin *n* tripalmitin, glycerol tripalmitate, glyceryl palmitate
Tripel *m* tripoli, rottenstone
Tripelargonin *n* tripelargonin
Tripelennamin *n* tripelennamine
Tripelerde *f* tripoli
Tripelhelix *f* triple helix
Tripelkalk *m* silicious limestone
Tripelpunkt *m* triple point
Tripelpunktsdruck *m* triple-point pressure
Tripelsalz *n* triple salt
Tripelschiefer *m* (Min) tripoli slate
Triphan *m* (Min) triphane, spodumene
Triphenin *n* triphenin, N-propionylphenetidine
Triphenylamin *n* triphenylamine
Triphenylbenzol *n* triphenylbenzene
Triphenylbromsilan *n* triphenylbromosilane
Triphenylcarbinol *n* triphenylcarbinol
Triphenylen *n* triphenylene
Triphenylguanidin *n* triphenyl guanidine
Triphenylmethan *n* triphenylmethane
Triphenylmethancarbonsäure *f* triphenylmethane-carboxylic acid
Triphenylmethanfarbstoff *m* triphenylmethane dye
Triphenylmethyl- triphenylmethyl, trityl
Triphenylmethylradikal *n* triphenylmethyl radical
Triphenylphosphat *n* triphenylphosphate
Triphenylphosphin *n* triphenylphosphine
Triphenylphosphit *n* triphenylphosphite
Triphenylphosphoniumbromid *n* triphenylphosphonium bromide
Triphenylrosanilin *n* triphenylrosaniline
Triphenylsilan *n* triphenylsilane
Triphenylsilanol *n* triphenylsilanol
Triphenylsiliciumchlorid *n* triphenylsilicon chloride, chlorotriphenylsilane
Triphenylstibin *n* antimony triphenyl, triphenylstibine
Triphosphopyridinnucleotid *n* (NADP, TPN) nicotineamide adenine dinucleotide phosphate, triphospho-pyridine nucleotide
Triphylit *m* (Min) triphylite
Triplett *n* triplet
Triplettserie *f* (Spektr) triplet series
Triplettspektrum *n* triplet spectrum
Triplettsystem *n* triplet system
Triplettzustand *m* triplet state
Triplit *m* (Min) triplite
Triploidit *m* (Min) triploidite
Tripolyphosphorsäure *f* tripolyphosphoric acid
Trippkeit *m* (Min) trippkeite
Triprolidin *n* triprolidine
Tripropionin *n* tripropionin, glycerol tripropionate, glyceryl propionate
Tripropylamin *n* tripropylamine
Tripropylsilan *n* tripropylsilane
Triptan *n* triptane
Triptycen *n* triptycene
Tripyrrol *n* tripyrrole
Triricinolein *n* triricinolein, glycerol triricinoleate, glyceryl ricinoleate
TRIS (Abk) trishydroxymethylamino methane
Trisaccharid *n* trisaccharide
Trisazofarbstoff *m* trisazo dye
Trisilan *n* trisilane
Trisilylamin *n* trisilylamine
Tristearin *n* tristearin, glycerol tristearate, glyceryl stearate
Trisulfan *n* trisulfane
trisymmetrisch trisymmetric
Tritan *n* tritane
Triterpen *n* triterpene
Triterpenoid *n* triterpenoid
Trithian *n* trithiane

Trithioacetophenon *n* trithioacetophenone
Trithiobenzaldehyd *m* trithiobenzaldehyde
Trithiocyanursäure *f* [tri]thiocyanuric acid
Trithioformaldehyd *m* trithioformaldehyde
Trithiokohlensäure *f* trithiocarbonic acid
Trithiol *n* trithiol
Trithiolan *n* trithiolane
Trithion *n* trithione
Trithionsäure *f* trithionic acid
Tritisporin *n* tritisporin
Tritium *n* tritium (Symb. T), **mit ~ behandelt** tritiated
Tritiumbetastrahlen *m pl* tritium beta rays
tritiummarkiert tritiated
Tritochorit *m* tritochorite
Tritol *n* tritol, trinitrotoluene
Triton *n* triton
Tritopin *n* tritopine
Trittbrett *n* running board
trittfest non-skid, non-slip
Trittschalldämmung *f* impact noise insulation
Trityl- trityl, triphenylmethyl
Tritylalkohol *m* trityl alcohol
Tritylchlorid *n* trityl chloride, triphenylchloromethane
Triundecylin *n* triundecylin
Trivalerin *n* trivalerin, glycerol trivalerate, glyceryl valerate
trivariant trivariant
trivial trivial
Trivialname *m* trivial name
Trochoidalanalysator *m* trochoidal analyzer
Trochoide *f* (Math) trochoid
Trochoidenpumpe *f* trochoidal pump
Trochotron *n* trochotron
trocken dry, (dürr) arid, (Holz) well-seasoned, (öde) barren, **~ verzinken** to sherardize, **absolut ~** absolutely dry, bone dry
Trockenanlage *f* drying plant
Trockenapparat *m* dryer, drying apparatus
Trockenappretur *f* dry finishing
Trockenaufbereitung *f* dry or mechanical dressing process
Trockenaufschluß *m* dry-sinter process
Trockenauftragsmenge *f* (Lack) coating weight, dry weight
Trockenauszug *m* dry extract
Trockenbatterie *f* drycell battery
Trockenbeize *f* dry dressing
Trockenblech *n* dryer tray
Trockenbleiche *f* dry bleaching
Trockenbrikett *n* dry pressed briquet[te]
Trockendampf *m* dry steam, superheated steam
Trockendarre *f* drying house
Trockendestillation *f* destructive distillation
Trockenei *n* dry egg
Trockeneinfärbung *f* dry coloring
Trockeneis *n* dry ice, carbon dioxide snow, solid carbon dioxide
Trockeneistrog *m* dry ice pan
Trockenelement *n* dry cell [battery]
Trockenentgasung *f* dry distillation
Trockenfarbe *f* dry color, pastel color
Trockenfestigkeit *f* dry strength
Trockenfilter *n* dry filter
Trockenfilz *m* drying felt
Trockenfirnis *m* drying oil, siccative varnish
Trockenflasche *f* drying bottle, drying flask
Trockenfüllung *f* dry filling
Trockengas *n* dry gas
Trockengehalt *m* dry content
Trockengemüse *n* dried vegetables
Trockengestell *n* drying rack, drain board, drying frame, drying stand
Trockengewebe *n* (Text) water-repellent fabric
Trockengewicht *n* dry weight
Trockengut *n* dry substance, material to be dried
Trockenhefe *f* dry yeast
Trockenheit *f* dryness
Trockenheitsgrad *m* degree of dryness
Trockenhilfsmittel *n* drying auxiliary
Trockenkammer *f* drying chamber, drying room, hot-air chamber
Trockenkanal *m* drying tunnel
Trockenkasten *m* dryer, drying chamber, drying closet
Trockenkautschukgehalt *m* dry rubber content
trockenklebrig tacky dry
Trockenklebrigkeit *f* dry tack
Trockenkompressionsmaschine *f* dry compression machine
Trockenkondensor *m* dry condenser
Trockenlackmasse *f* dry lacquer compound
Trockenläufer *m* dry-dial water meter
Trockenlauf *m* dry operation
Trockenmahlen *n* dry grinding
Trockenmahlung *f* dry grinding
Trockenmilch *f* dry milk, milk powder
Trockenmittel *n* dehumidifier, desiccant, drying agent, (Chem) dehydrating agent, (Farbe) siccative
Trockennährboden *m* desiccated medium
Trockenöl *n* drying oil, siccative oil
Trockenofen *m* drying oven, baking oven, drying kiln, drying stove, (Holz) seasoning kiln
Trockenpartie *f* dryer section, drying section
Trockenpatrone *f* dehydration unit, drying cell
Trockenpistole *f* [Abderhalden] drying pistol, drying apparatus
Trockenplatte *f* dry[ing] plate
Trockenpolymerisation *f* bulk polymerization
Trockenpräparat *n* dry preparation
Trockenprobe *f* dry assay
Trockenpuddeln *n* dry puddling
Trockenpulver *n* drying powder, siccative
Trockenrahmen *m* drying frame
Trockenraum *m* drying chamber, drying room

Trockenreiniger – Trommelrad

Trockenreiniger *m* dry cleanser
Trockenreinigung *f* dry cleaning, dry purification
Trockenröhre *f* drying tube
Trockenrückstand *m* dry residue
Trockenschale *f* drying dish
Trockenschleifen *n* dry grinding
Trockenschleuder *f* drying centrifuge
Trockenschnecke *f* worm type dryer
Trockenschrank *m* [cabinet] dryer, desiccator cabinet, drying cabinet, drying closet, hot-air cabinet
Trockensiebung *f* dry sifting
Trockenspinnverfahren *n* dry spinning
Trockensterilisator *m* drying sterilizer
Trockenstoff *m* desiccant, drying agent, drying substance, (getrockneter Stoff) dry substance
Trockenstoffgehalt *m* solids content
Trockentrommel *f* rotary dryer, drying cylinder, drying drum
Trockentunnel *m* (Keram) tunnel dryer
Trockenturm *m* drying tower, turret dryer (leather), (Calciumchlorid) calcium chloride cylinder
Trocken- und Röstapparat *m* drying and roasting apparatus
Trockenverfahren *n* dry[ing] process
Trockenvermahlung *f* (Bergb) dry crushing, dry grinding
Trockenwalze *f* drying drum, drying roller
Trockenzeit *f* drying time
Trockenzelle *f* dry cell
Trockenzentrifuge *f* centrifugal dryer
Trockenzone *f* drying zone
Trockenzylinder *m* cylinder dryer, drying cylinder, drying drum
Trockne *f* dryness, **bis zur ~ eindampfen** to evaporate to dryness
trocknen to dry, to desiccate
Trocknen *n* drying, dehumidification, desiccating, (Holz) wood seasoning, **~ an der Luft** air drying, drying in the air, **~ der Elektrode** drying of the electrode, **~ der Formen** drying of the molds, **~ durch Zerstäuben** spray-drying
trocknend drying
Trockner *m* dryer, desiccator, drainer, drier, **Band- ~** belt drier, **Drallrohr- ~** spiral tube pneumatic drier, **Düsen- ~** jet drier, **Horden- ~** shelf drier, **Kammer- ~** chamber drier, **Riesel- ~** spray drier, **Rillenwalzen- ~** grooved drum drier, **Ringetagen- ~** turbodrier, **Röhrentrommel- ~** drum with heating tubes, **Roto-Louvre- ~** Roto-Louvre drier, **Schleuder- ~** centrifuge drier, **Schneckentrommel- ~** screw-conveyor drum drier, **Schubwende- ~** reciprocating drier, **Spiralband- ~** screw-conveyor drier, **Strom- ~** pneumatic-conveyor drier, **Teller- ~** disc drier, **Tellertrommel- ~** rotary shelf drier, **Trommel- ~** drum drier, **Tunnel- ~** tunnel drier, **Walzen- ~** roller drier, **Wirbelschicht- ~** fluidized-bed drier, **Wirbelstoß- ~** pulsating-bed drier
Trocknung *f* drying, desiccation, (Holz) wood seasoning
Trocknungsanlage *f* dryer, drying apparatus, drying equipment
Trocknungsapparatur *f* dryer, drying apparatus
Trocknungsbeschleuniger *m* drying accelerator
Trocknungsmittel *n* siccative, dehydrating agent, desiccant, drying agent
Trögerit *m* (Min) troegerite, trogerite
Trögersche Base *f* Troeger's base
T-Röhre *f* T-pipe
Tröpfchen *n* droplet, small drop
Tröpfchenbildung *f* formation of droplets
Tröpfchenmodell *n* liquid-drop model
Tröpfchenverhalten *n* liquid-drop behavior
tröpfeln to drip, to drop, to trickle
Trog *m* trough, vat, (Metall) box, **elektrolytischer ~** electrolytic tank
Trogförderband *n* conveying trough, trough conveyor
trogförmig trough shaped
Trogfutter *n* trough lining
Trogkettenförderer *m* drag conveyor
Trogkneter *m* trough kneader
Trogmischer *m* trough mixer
Trogpresse *f* pot press
T-Rohr *n* T-pipe, T-tube
Troilit *m* (Min) troilite
Trolit *m* (Kunststoff) trolite
Trolleit *m* (Min) trolleite
Trollixanthin *n* trollixanthin
Trommel *f* drum, (Walze) cylinder, roller
Trommelapparat *m* (Galv) galvanizing drum, drum apparatus
Trommeldrehfilter *n* rotary drum filter
Trommeleinstellung *f* drum set
Trommelfilter *n* drum filter, revolving filter, rotary filter
Trommelgalvanisierung *f* (Galv) barrel plating
Trommelhaspel *f* drum winch
Trommelkamera *f* drum camera
Trommelkippanlage *f* drum dumper
Trommelklebgummi *m* (Reifen) core cement
Trommelkonverter *m* barrel converter
Trommelkreiselrührer *m* rotor cage impeller
Trommelmischer *m* rotary mixer, barrel mixer, cylinder mixer, drum mixer, tumbling mixer, **aufrechtstehender ~** end-over-end type drum mixer
Trommelmühle *f* barrel mill, tumbling barrel, tumbling mill
Trommeln *n* tumbling
Trommelofen *m* rotary furnace, rotary roaster
Trommelrad *n* drum wheel

Trommelregenerator *m* drum regenerator
Trommelrotor *m* drum rotor
Trommelsaugfilter *n* single-compartment drum filter
Trommelscheider *m* drum separator
Trommelschichtenfilter *n* drum type sheet filter
Trommelsextant *m* drum sextant
Trommelsieb *n* cylindrical sieve
Trommel[sink]scheider *m* revolving-drum dense-medium vessel (dense-medium separation)
Trommelstern *m* drum spider
Trommeltrockner *m* drum dryer, rotary dryer
Trommelwascher *m* drum washer
Trommelwaschmaschine *f* drum washer
Trommelzelle *f* drum cell
Trommelzellenfilter *n* multi-compartment drum filter
Tronasalz *n* trona salt
Troostit *m* (Min) troostite
Tropaaldehyd *m* tropaic aldehyde, tropic aldehyde
Tropacocain *n* tropacocaine, benzoylpseudotropeine
Tropäolin *n* tropeolin
Tropan *n* tropan
Tropanol *n* tropanol
Tropanon *n* tropanone
Tropasäure *f* tropic acid, tropaic acid
Tropein *n* tropein[e]
Tropen *pl* tropics
Tropenfarbe *f* tropical paint
tropenfest resistant to tropical conditions
Tropenfestigkeit *f* stability to tropical conditions
tropengeschützt tropicalized
Tropenisolierung *f* insulation for the tropics
Tropenklima *n* tropical climate
Tropenpflanze *f* (Bot) tropic plant
tropfbar capable of forming drops
Tropfbecher *m* drip cup
Tropfbehälter *m* drip[ping] cup, drip[ping] pan
Tropfbrett *n* draining board, dripping board, plate drainer
Tropfdüse *f* drip nozzle
Tropfeinsatz *m* drop dispenser
Tropfelektrode *f* dropping electrode
tropfen to drop
Tropfen *m* drop
Tropfenabscheider *m* mist eliminator
Tropfenbildung *f* formation of drops
Tropfenfänger *m* drop collector, mist eliminator
Tropfenfängeraufsatz *m* splash head adapter
tropfenförmig drop-shaped, tear-shaped
Tropfenform *f* form of a drop, **in** ~ in form of drops
Tropfenkondensation *f* drop condensation
Tropfenwachstum *n* droplet growth
tropfenweise drop by drop, dropwise
Tropfenzähler *m* drop counter

Tropfflasche *f* drop[ping] bottle
Tropfhahn *m* drip cock
Tropfkapillare *f* drop capillary
Tropfnasenbildung *f* (Lack) fat-edge formation
Tropföler *m* dropping lubricator, sight-feed lubricator
Tropfpunkt *m* drop[ping] point, liquefying point
Tropfring *m* drip ring
Tropfschale *f* collecting tray, drip pan
Tropfschmiergefäß *n* dropping lubricator
Tropfschmierung *f* dropfeed oiling
Tropfschwefel *m* drop sulfur
Tropfstein *m* dripstone, (herabhängender) stalactite, (nach oben wachsender) stalagmite
tropfsteinartig stalactitic
Tropfsteinbildung *f* stalactitic formation
tropfsteinförmig stalactiform
Tropftrichter *m* separating funnel, drip funnel, dropping funnel
Tropfverschluß *m* dropper closure
Tropfwasser *n* drip water
Tropfzink *n* drop zinc
Tropfzinn *n* drop tin, granulated tin
Tropidin *n* tropidine
Tropigenin *n* tropigenine
Tropiglin *n* tropigline
Tropiliden *n* 1,3,5-cycloheptatriene, tropilidene
Tropilidin *n* tropilidin
Tropin *n* tropine, 3-hydroxytropan, 3-tropanol
Tropinon *n* tropinone
Tropinsäure *f* tropinic acid
Tropokollagen *n* tropocollagen
Tropolon *n* tropolone
Tropomyosin *n* tropomyosin
Tropon *n* tropone
Troponin *n* troponin
Troposphäre *f* troposphere
Tropylium-Ion *n* tropylium ion
Trotyl *n* trotyl, trinitrotoluene
trüb[e] cloudy, turbid, (dunkel, düster) murky, (matt) dull, (verwischt) blurred
Trübe *f* turbidity, cloudiness, dullness, (Bergb) dross, (Pap) pulp
trüben to become turbid, to tarnish
Trübglas *n* frosted glass
Trübung *f* turbidity, cloudiness, milkiness, (Lackfilm) blushing
Trübungsfaktor *m* turbidity factor
Trübungsmesser *m* turbidimeter
Trübungsmessung *f* (Phys) nephelometry
Trübungsmittel *n* (Email) dulling agent, opacifier
Trübungspunkt *m* cloud point
Trübungstemperatur *f* turbidity point
Trüffel *f* (Bot) truffle
Trümmerachat *m* (Min) brecciated agate
Trunksucht *f* dipsomania
Truxellin *n* truxelline
Truxen *n* truxene

Truxillin *n* truxilline
Truxillsäure *f* truxillic acid
Truxinsäure *f* truxinic acid
Trypaflavin *n* trypaflavin[e], acriflavine
Trypanblau *n* trypan blue, diamine blue
Trypanrot *n* trypan red
Trypanviolett *n* trypan violet
Tryparsamid *n* tryparsamide
Trypsin *n* (Biochem) trypsin, (obs) trypsase
Trypsinogen *n* (Biochem) trypsinogen
Tryptamin *n* tryptamine
Tryptathionin *n* tryptathionine
Tryptazan *n* tryptazan
Trypton *n* tryptone
Tryptophan *n* tryptophan[e], indolylalanine
Tryptophanase *f* (Biochem) tryptophanase
Tryptophandecarboxylase *f* (Biochem) tryptophan decarboxylase
Tryptophanol *n* tryptophanol
Tryptophanoxydase *f* (Biochem) tryptophan oxidase
Tryptophansynthetase *f* (Biochem) tryptophan synthetase
Tryptophol *n* tryptophol
Tschermigit *m* (Min) tschermigite
Tschewkinit *m* tschewkinite
T-Schiene *f* T-rail
Tschugajeffs Reagens *n* Tschugajew's reagent, dimethylglyoxime
Tsetsefliege *f* tsetse fly
T-Stück *n* T-piece, flanged tee
Tsuzusäure *f* tsuzuic acid
T-Träger *m* T-bar, T-iron
Tuaminoheptan *n* tuaminoheptane
Tubanol *n* tubanol
Tubasäure *f* tubaic acid
Tubazid *n* tubazid
Tubenfüllmaschine *f* tube filling machine
Tubenklammer *f* tube clip
Tuberkel *m* (Med) tubercle, tuberculum
Tuberkelbazillus *m* tubercle bacillus, Koch's bacillus
Tuberkulin *n* tuberculin
tuberkulös (Med) tubercular, tuberculous
Tuberkulose *f* (Med) tuberculosis
Tuberkulostearinsäure *f* tuberculostearic acid
Tuch *n* (Text) cloth, fabric
Tuchdrucker *m* (Text) cloth printer, textile printer
Tuchfilter *n* cloth filter, fabric collector
Tuchlutte *f* canvas air conduit, cloth air conduit
Tucholit *n* tucholite
Tuchscharlach *n* cloth scarlet
Tuchschwabbel *f* cloth lap
Tucumöl *n* tucum oil
Tümpelblech *n* tymp sheet steel
Tümpeleisen *n* tymp plate
Tümpelstein *m* tymp stone
Tünche *f* lime-wash, whitewash

tünchen to lime wash, to whitewash
Tüncher *m* white-washer
Tünchfarbe *f* limewash paint, water color
Tünchkalk *m* lime for whitewashing
Tünchmaschine *f* white washing machine
Tüpfelanalyse *f* spot analysis, analysis by spot tests
Tüpfelmethode *f* spot method
tüpfeln to spot, to test by spotting
Tüpfelplatte *f* spot plate
Tüpfelprobe *f* spot test
Tüpfelreaktion *f* spot test
Tüpfelstein *m* (Min) stignite
Türfüllung *f* door panel
Türkis *m* (Min) turquoise
Türkisblau *n* turquoise blue
Türkischbraun *n* Turkey brown
Türkischrot *n* Turkey red
Türverschluß *m* door latch
Tuesit *m* tuesite
Tüte *f* paper bag
Tuff *m* (Min) tufa, tuff
tuffartig tufaceous
Tufferde *f* tufaceous earth
Tuffkalk *m* tufaceous limestone
Tuffstein *m* tuffstone
tuffsteinartig tufaceous
Tuffwacke *f* trap tuff
Tulametall *n* tula metal
Tumor *m* (Med) tumo[u]r, **bösartiger** ~ (Med) malignant tumor, **experimenteller** ~ experimental tumor
tumorbildend (Med) oncogenous
Tumorbildung *f* (Med) oncogenesis
Tumorwachstum *n* growth of the tumor
Tumorzelle *f* tumor cell
Tumorzellenverschleppung *f* (Med) dislocation of tumor cells
tumorzellenzerstörend oncolytic
Tumulosinsäure *f* tumulosic acid
Tungöl *n* tung oil, Chinese wood oil
Tunnel *m* tunnel
Tunnelboden *m* tunnel-cap tray
Tunneleffekt *m* (Atom) tunneling effect
Tunnelofen *m* tunnel kiln
Tunneltrockner *m* tunnel dryer
Tupffarbe *f* stippling paint
Turanit *m* (Min) turanite
Turanose *f* turanose
Turbidimeter *n* turbidimeter
Turbidimetrie *f* turbidimetry
Turbine *f* turbine
Turbinenanlage *f* turbine plant
Turbinenbohrer *m* turbodrill
Turbinendüse *f* turbine nozzle
Turbinen-Durchflußmesser *m* turbometer or turbine flowmeter
Turbinendynamo *m* turbodynamo, turbo generator

Turbinenläufer *m* turbine rotor
Turbinenlager *n* turbine mounting
Turbinenleistung *f* turbine output
Turbinenmischer *m* turbine impeller [mixer], radial flow impeller, turbine mixer
Turbinenrührer *m* turbine agitator, turbine mixer
Turbinenschaufel *f* turbine blade, vane
Turbinensublimator *m* turbine sublimer
Turbinentrockner *m* turbo-dryer
Turbinentrommel *f* turbine rotor
Turbinenwelle *f* turbine shaft
Turbinenwirkungsgrad *m* turbine efficiency
Turboalternator *m* turbo-alternator
Turboantrieb *m* turbojet drive
Turbobrenner *m* turbo burner, turbulent burner
Turbodynamo *m* turbodynamo, turbogenerator
Turbogebläse *n* turbo-blower
Turbogenerator *m* turbogenerator
Turbomaschine *f* turbo-engine or motor
Turbomischer *m* turbo-mixer, impact [wheel] mixer, turbine mixer
Turbomolekularpumpe *f* turbomolecular pump
Turbosichter *m* whizzer
Turbosieber *m* turbo-sifter
Turbotrockner *m* turbo-dryer
Turboverdichter *m* turbocompressor
Turboviskosimeter *n* turboviscosimeter
turbulent turbulent
Turbulenz *f* turbulence
Turbulenzeinfluß *m* turbulence effect
Turbulenzenergie *f* eddy energy
Turbulenzfluß *m* eddy flux
Turbulenzgrad *m* degree of turbulence
Turbulenzkoeffizient *m* (Viskosität) eddy viscosity coefficient
Turbulenztransport *m* eddy flux
Turicin *n* turicine
Turjit *m* turjite
Turkischrotöl *n* Turkey red oil
Turkisgrün *n* turquoise green
Turm *m* tower, column, turret
Turmalin *m* (Min) tourmalin[e], **farbloser** ~ achroite
Turmalingranit *m* (Min) tourmaline granite
Turmalinzange *f* tourmaline tongs
Turmdrehkran *m* rotary tower crane
Turmeron *n* turmerone
Turmextraktor *m* mixer-settler tower
Turmlauge *f* (Pap) tower liquor
Turmverflüssiger *m* vertical shell-and-tube condenser
Turmwäscher *m* scrubber
Turnbulls Blau *n* ferrous cyanoferrate(III), ferrous ferricyanide, Turnbull's blue
Turnerit *m* (Min) turnerite
Turnus *m* cycle, turn
Tusche *f*, **chinesische** ~ Indian or China ink
Tuschfarbe *f* watercolor
Tussalvin *n* tussalvine

Tutocain *n* tutocaine, butamin
T-Ventil *n* spindle or flap-valve
Twitchells Reagens *n* Twitchell's reagent
Tychit *m* (Min) tychite
Tylophorin *n* tylophorine
Tyndalleffekt *m* Tyndall effect
Tyndallkegel *m* Tyndall cone
Type *f* (Buchstabe, Letter) type
Typenbeschränkung *f* variety reduction
typenfremd (Blut) incompatible
typengleich isotypical
Typenmetall *n* type or letter metal
Typenschild *n* type plate
Typhotoxin *n* typhotoxin
Typhus *m* (Med) typhoid fever
Typhusbakterien *pl* typhoid bacilli, Salmonella typhi
Typhusimpfstoff *m* antityphoid vaccine
Typhusschutzimpfung *f* (Med) antityphoid vaccination
Typhusvakzine *f* typhoid vaccine
typisiert standardized
Typisierung *f* standardization
Tyramin *n* tyramine, hydroxyphenyl-ethylamine
Tyrit *m* (Min) tyrite
Tyrocidin *n* tyrocidine
Tyrolit *m* (Min) tyrolite, copper froth
Tyrosin *n* tyrosin[e], p-hydroxyphenylalanine
Tyrosinase *f* (Biochem) tyrosinase
Tyrosindecarboxylase *f* (Biochem) tyrosine decarboxylase
Tyrosinol *n* tyrosinol
Tyrosinosis *f* tyrosinosis
Tyrosintransaminase *f* (Biochem) tyrosine transaminase
Tyrosol *n* tyrosol, p-hydroxyphenetyl alcohol
Tyrotoxicon *n* tyrotoxicon, tyrotoxin
Tyrotoxin *n* tyrotoxin, tyrotoxicon
Tysonit *m* (Min) tysonite
Tyvelose *f* tyvelose

U

u. a. (Abk) among others
Ubbelohde-Viscosimeter n **mit hängendem Kugelniveau** suspended-level Ubbelohde viscometer
Ubichinon n ubiquinone
U-Bogen m U-tube
Übelkeit f nausea, sickness
übelriechend ill-smelling, fetid, malodorous
überätzen to overetch
überarbeiten, etwas ~ to revise, to touch up
überbeanspruchen to overstrain, to overstress
Überbeanspruchung f overstrain[ing], overstress[ing]
überbelasten to overcharge, to overload
Überbelastung f overcharge, overload
überbelichten (Phot) to overexpose
Überbelichtung f (Phot) overexposure
überblauen to give a bluish tint
Überbleibsel n residue
Überblick m survey
überbrücken (Elektr) to bridge-over
überbürden to overburden, to overload
überdecken to cover, (überlappen) to overlap
Überdecken n overlapping, shadowing
Überdeckung f overlap
Überdeckungshöhe f (Gaschromat) height of fill
überdestillieren to distil[l] over
Überdestillieren n distilling over
Überdosierung f overfeeding
Überdruck m excess pressure, overpressure, pressure above the atmospheric, **hydrostatischer** ~ excess hydrostatic pressure
Überdruckbehälter m excess pressure container
überdrucken to overprint, to print over
Überdruckgebiet n superpressure zone
Überdrucklack m overprint varnish
Überdruckturbine f reaction turbine
Überdruckventil n pressure-relief valve, blow-off valve, expansion valve
übereinanderlagern to superimpose, to superpose
Übereinanderlagerung f superposition
übereinanderliegen to be super[im]posed
übereinanderliegend superimposed
Übereinanderpressen n layer molding, stack molding
Übereinanderschweißung f lap welding
übereinstimmen to agree
Übereinstimmung f agreement, conformity
überempfindlich hypersensitive, supersensitive, ultrasensitive
Überempfindlichkeit f hypersensitiveness, supersensitivity, ultrasensitivity
übereutektisch (übereutektoid) hypereutectic
Übereutektoid n hypereutectoid
überfärben to overcolor, to overdye, to top
Überfärben n cross dyeing

überfarbecht (Färb) fast to cross-dyeing
überflüssig superfluous
Überflutung f flooding
Überform f (Gieß) coat, mantle
überführbar transformable
Überführbarkeit f transformability
überführen to transfer, to convert, to transform
Überführung f transport, (Brücke) bridge, viaduct, (Chem) conversion, **elektrochemische** ~ electrochemical transport
Überführungswärme f heat of transition
Überführungszahl f transport number
Überfunktion f excess function, hyperfunction, overactivity
Übergabekopf m transfer gear
übergären to overferment
Übergang m crossing, junction, passage, (Chem, Phys) transition, (Übertragung) transfer, ~ **innerhalb einer Schale** (Atom) intrashell transition, **erlaubter** ~ permitted transition, **erzwungener** ~ forced or nonspontaneous transition, **strahlenloser** ~ transition without radiation, **verbotener** ~ forbidden transition
Übergangsbereich m transition region
Übergangsbestandteil m transition substance
Übergangsdose f contact box
Übergangseinheit f (Füllkörperkolonne) height equivalent to theorical plate (HETP), transfer unit, **gasseitige** ~ (Dest) overall gas-phase transfer unit
Übergangselement n transition element
Übergangserscheinung f transition phenomenon
Übergangsfarbe f transition color
Übergangsflansch m adapter flange
Übergangsfunktion f (Regeltechn) step function response (Brit. E.), time response, transient response (Am. E.)
Übergangshalbwertzeit f transition half life
Übergangskolloid n hemicolloid
Übergangskonus m adapter shaft, transmission shaft
Übergangsleitfähigkeit f conductivity of transfer
Übergangsleitwert m transconductance, transfer characteristic
Übergangsmetall n transition metal
Übergangsmetallverbindung f transition metal compound
Übergangspunkt m transition point, transformation point
Übergangsrohr n adapter pipe, reducing pipe
Übergangsstadium n transitional state
Übergangsstück n [joint] adapter, reducing pipe
Übergangstemperatur f transition temperature
Übergangsverbot n forbidden transition
Übergangsverhalten n transient response
Übergangswärme f heat of transition

Übergangswahrscheinlichkeit f transition probability
Übergangswiderstand m transition resistance, contact resistance
Übergangszeit f transitional period, transit time
Übergangszelle f (Biol) transitional cell
Übergangszone f transition zone
Übergangszustand m transition state, intermediate state, transient condition, transient state
übergar (Metall) dry, overrefined
übergehen (Dest) to distil[l], to pass over
Übergeschwindigkeit f excess velocity, superspeed
Übergewicht n excess weight, overweight
übergießen to pour over
Überheizen n superheating
Überhitze f excess heat, waste heat
überhitzen to superheat, (Dampf) to overheat
Überhitzer m (Dampfmasch) superheater
Überhitzeranlage f superheating plant
Überhitzerheizfläche f superheater surface
Überhitzerrohr n superheater tube or pipe
Überhitzerschlange f superheater coil
Überhitzung f superheating
Überhitzungsgrad m degree of superheating
Überhitzungstemperatur f temperature of superheating
Überhitzungswärme f superheating heat
überholen to overhaul
Überholen n overhauling
Überholförderer m by-passing conveyor
Überholung f overhauling
überholungsbedürftig in need of overhauling
Überkieselung f silicification
überkochen to boil over
Überkompressionsverhältnis n overcompression ratio
überkritisch supercritical
überkrusten to incrust
Überkrustung f incrustation
überladen to overcharge, to supercharge
Überladung f overcharging
überlagern to superimpose, to superpose
überlagert superimposed
Überlagerung f superposition, overlapping
Überlagerungsempfänger m heterodyne radio receiver
Überlagerungsmethode f heterodyne method
überlappen to overlap
Überlappen n overlapping
überlappend overlapping
Überlappstoß m lap weld
überlappt geschweißt lap welded
überlappt gelötet lap-seamed
Überlapptnaht f (Dose) lap seam
Überlappung f overlap[ping], ~ **der Ladungsdichte** overlap charge density, ~ **von Orbitalen** overlapping of orbitals

Überlappungsbereich m transition range, transition region
Überlappungsmodell n overlap model
Überlast f overcharge, overload, surcharge
Überlastauslöser m overload release
Überlastbarkeit f overload capacity
überlastet sein to be overloaded
Überlaststrom m overload current
Überlastung f overloading
Überlastungsschutz m overload protection, (Elektr) overcurrent relay
Überlauf m overflow, spillway, weir (column)
Überlaufabfall m spill waste
Überlaufbehälter m expansion reservoir
überlaufen to overflow
Überlaufen n overflowing
Überlaufgefäß n overflow vessel
Überlaufkontrolle f unloading control
Überlaufleitung f overflow line
Überlaufnähte pl (Spritzguß) flashing
Überlaufrohr n overflow pipe, overflow tube
Überlaufventil n overflow trap
Überlaufwasserstandsmelder m overflow alarm
Überlaufwehr n [overflow] weir
Überlaufzentrifuge f overflow centrifuge
Überlebende[r m] f survivor
überleiten to lead over, to pass over
Überleiten n leading over, passage
Überleitung f transfer
übermäßig excessive
Übermangansäure f (Permangansäure) permanganic acid
übermangansauer (permangansauer) permanganate
Übermaß n excess, amount of oversize, overmeasure, surplus
übermitteln to transmit
übernormal supernormal
Überoxydierung f overoxidation, peroxidation
Überpotential n excess potential, overpotential, overvoltage
Überproduktion f overproduction, (Drüse) oversecretion
überprüfen to check, to examine
Überprüfung f check, examination, revision
überquellen to overflow
überreißen to entrain
Überreißen n entrainment
Überrelaxationsmethode f method of overrelaxation
Überrest m remainder, residue, rest
überrollbar crushproof
Übersäen n overseeding
übersättigen to oversaturate, to supersaturate
Übersättigen n supersaturating
übersättigt supersaturated
Übersättigung f oversaturation, supersaturation
übersäuern to overacidify, to superacidulate

Übersäuerung – übertreffen

Übersäuerung f overacidification, superacidulation
überschäumen to foam over, to froth over
Überschäumen n (b. Öffnen v. Flaschen etc.) gushing
Überschallgeschwindigkeit f supersonic speed, supersonic velocity
Überschallwelle f supersonic wave, ultrasonic wave
überschichten to cover with a layer, to layer, to stratify
überschichtet covered with a layer, stratified
Überschieber m bolted joint
Überschießen n overshooting
Überschlag m (Elektr) flashover, sparkover
Überschlagsspannung f flashover voltage, sparkover voltage
Überschlagsweg m sparkover path
Überschmelzung f superfusion, enameling
Überschraubmuffe f screw-on socket
überschreiten to exceed, to override; to transgress
Überschreitung f **des Innendrucks** (Reifen) overinflation
Überschrift f heading, caption
überschüssig excessive, excess [of], in excess, overabundant, surplus, (verbleibend) remaining
Überschuß m excess, surplus
Überschußenthalpie f excess enthalpy
Überschußentropie f excess entropy
Überschußladung f excess charge
Überschußlast f excess load
Überschußneutron n excess neutron
Überschußvolumen n excess volume
überschwefelsauer (obs) persulfate, (Salz) peroxysulfate
Überschwemmung f flood
Überschwingung f overshoot
Überseekabel n submarine cable, transmarine cable
übersetzen (Getriebe) to gear, (Metall) to overcharge the furnace
Übersetzung f (Masch) transmission, (Opt) optical transmission
Übersetzungsgetriebe n gear transmission
Übersetzungsverhältnis n ratio of transmission, transformation ratio
Übersicht f summary, survey, synopsis
übersichtlich well-arranged, well-ordered
Übersichtstabelle f chart, synoptical table
Überspannung f overpressure, overstraining, (Elektr) excessive voltage
Überspannungsschutz m (Elektr) overvoltage protection
Überspannungssicherung f overvoltage protection
überspringen to jump over, (Elektr) to flash over, to spark

Überspringen n flashing over, sparking over
überspritzecht non-bleeding
Überspritzen n spraying
Überspritzmaschine f (techn. Schläuche) covering machine
übersprudeln to bubble over, to overflow
Überstand m excess length
übersteigen to surmount, to exceed
Übersteigerung f increased rate
Übersteuerung f oversteering
überstreichen to brush, to smear, to spread
überströmen to flow over
Überströmkanal m by-pass, return passage
Überströmrohr n overflow pipe
Überströmventil n overflow valve
Überstrom m excess current, overcurrent
Überstrukturgitter n superlattice
Überstrukturreflex m superlattice reflection
Überstülphaube f slip cover
Überstunden pl overtime, overwork
übersynchron hypersynchronous
Übertemperatur f excess temperature
Überträger m carrier, transmitter
Überträgerbürste f collecting brush
Überträgertheorem n transformer theorem
Übertrag m (Comp) carry, **autonomer** ~ (Comp) self-instructed carry, **gesteuerter** ~ (Comp) separately instructed carry, **vollständiger** ~ (Comp) complete carry
übertragbar communicable, (Med) contagious, infectious, transmissible
übertragen (Rad) to broadcast, (weitergeben) to transmit
Übertragung f carrying over, transfer [ence], (Weitergabe) transmission, **direkte** ~ direct reading, direct transmission
Übertragungsbefehl m (Comp) cue
Übertragungsfaktor m (Regeltechnik) proportional control factor (Am. E.), steady state gain (Br. E.)
Übertragungsfunktion f (Regeltechn) transfer function (Am. E.), transmission characteristic (Brit. E.)
Übertragungsgerät n transcriber
Übertragungsglied n impedance-matching circuit link
Übertragungskatalyse f transport catalysis
Übertragungskonstante f chain transfer constant (in polymerization)
Übertragungsleitung f transmission line
Übertragungsmittel n transmission agent
Übertragungspotential n transfer potential
Übertragungsprinzip n principle of transfer
Übertragungsrate f (Comput) transfer rate
Übertragungsverhältnis n transfer rate
Übertragungsverhalten n transient response
übertreffen to exceed, to beat, to excel, to outbalance, to outmatch

übertreiben to exaggerate, (Dest) to distil over, to drive over
Übertreibung f exaggeration
übertreten to infringe, to trespass
übertrocknen to overdry
übertünchen to whitewash
Überwachsung f (Krist) overgrowth
Überwachung f control, checking, monitoring, supervision
Überwachungsanlage f control plant, control system
Überwachungsgerät n controller, monitor, survey instrument
Überwachungsgeräte n pl monitoring equipment
Überwachungspersonal n supervisory personnel
Überwachungsprogramm n (Comp) tracing routine
Überwälzen n overmilling
überweit over-dimensioned
überwiegend predominant, preponderant
überwintern (Zool) to hibernate
Überwinterung f hibernation
Überwurfmutter f cap nut, screw cap
Überwurfschraube f cap screw
überzählig over-abundant
überziehen to coat, to cover, (sich überziehen) to become coated
Überzug m coating, cover[ing], dip coat; film, layer, (Galv) plating, (Kruste) crust, incrustation, **abstreifbarer** ~ strippable coating, **galvanischer** ~ electrodeposit, electroplating, **gelartiger** ~ gel coat finish, **glänzender** ~ gloss finish, **glasartiger** ~ enamel, **halbglänzender** ~ semigloss finish, **matter** ~ flat finish
Überzugaktivierung f coating activation
Überzugharz n coating resin
Überzugsemaille f finishing enamel
Überzugsfluß m coating flux
Überzugslack m coating varnish, overprint varnish, top coat
Überzugsleitfähigkeit f coating conductivity
Überzugsmasse f coating compound
Überzugsmischung f coating compound
Überzugsverfahren n coating process
Übung f exercise, training
U-Eisen n U-iron, channel-iron
U-förmig U-shaped
Uhligit m (Min) uhligite
Uhr f watch, (Meßwerk) meter, (Zeitmesser) timer
Uhrglas n watch glass
Uhrzeigergegensinn m anti-clockwise direction
Uhrzeigersinn m, gegen den ~ counter-clockwise, im ~ clockwise [direction]
Ukita-Säure f Ukita's acid
Ulein n uleine
Ulexin n ulexine
Ulexit m (Min) ulexite

Ullmannit m (Min) ullmannite
Ulmin n ulmin
Ulminbraun n ulmin brown
Ulminsäure f ulmic acid, geic acid
Ulrichit m (Min) ulrichite
Ultrachinin n ultraquinine
ultraecht ultra fast
Ultrafeinrelais n ultrafine relay
Ultrafilter m ultrafilter
Ultrafiltration f ultrafiltration
Ultra-Gaszentrifuge f gas ultracentrifuge
Ultraionisierung f ultraionization
Ultrakondensor m (Opt) ultracondenser
Ultrakurzwelle f microwave, ultrashort wave
Ultrakurzwellensender m ultrashort wave transmitter
Ultramarin n ultramarine, (Min) azure stone
ultramarin[blau] ultramarine
Ultramarin[e]blau n ultramarine blue
Ultramarinfarbe f ultramarine color
ultramikrochemisch ultramicrochemical
Ultramikroskop n ultramicroscope
Ultramikroskopie f ultramicroscopy
ultramikroskopisch ultramicroscopic
Ultrapore f ultrapore
ultrarot infrared
Ultrarotabsorption f infrared absorption
Ultrarotabsorptionsschreiber m infrared absorption recorder
Ultrarotabsorptionsspektrum n infrared absorption spectrum
Ultrarotdurchlässigkeit f infrared permeability, infrared transmittance
Ultrarotemission f infrared emission
Ultrarotfilter m infrared filter
Ultrarotproblem n infrared problem
Ultrarotspektroskopie f infrared spectroscopy
Ultrarotspektrum n infrared spectrum
Ultrarotsperre f infrared block
ultrarotundurchlässig opaque to infrared
Ultraschall m supersonics, ultrasonics
Ultraschallabschwächung f ultrasonic attenuation
Ultraschallapparat m ultrasonic apparatus
Ultraschallbiometrie f ultrasonic biometry
Ultraschallecholotung f supersonic sounding
Ultraschallehre f ultrasonics
Ultraschallerzeuger m ultrasonic generator, ultrasound generator
Ultraschallfrequenz f supersonic frequency
Ultraschallgerät n ultrasonic apparatus
Ultraschallgeräte n pl supersonic equipment
Ultraschallgerbung f supersonic tannage
Ultraschallgeschwindigkeit f supersonic speed, ultrasonic velocity
Ultraschallprüfgerät n supersonic flaw detector
Ultraschallwelle f supersonic wave
Ultraschwerewelle f ultra gravity wave
Ultrastrahl m cosmic ray

Ultrastrahlung *f* cosmic radiation
ultraviolett ultraviolet
Ultraviolett *n* ultraviolet
Ultraviolettbestrahlung *f* ultraviolet radiation
ultraviolettdurchlässig transparent to ultraviolet light
Ultraviolettinhibitor *m* ultraviolet inhibitor
Ultraviolettlampe *f* ultraviolet lamp, artificial sunlight lamp
Ultraviolettlicht *n* ultraviolet light
Ultraviolettstrahler *m* ultraviolet ray emitter
ultraweich ultra-soft
Ultrazentrifuge *f* ultracentrifuge, **analytische** ~ analytical ultracentrifuge, **präparative** ~ preparative ultracentrifuge
umändern to alter, to change, to convert, (korrigieren) to correct
Umänderung *f* alteration, change, transformation
Umaminierung *f* (Transaminierung) transamination
umarbeiten to remodel, to rework
Umbau *m* rebuilding, reconstruction
umbauen to rebuild
Umbauten *pl* modifications
Umbellatin *n* umbellatine
Umbelliferon *n* umbelliferone, 4-hydroxycoumarin
Umbelliferose *f* umbelliferose
Umbelliprenin *n* umbelliprenin
Umbellsäure *f* umbellic acid, p-hydroxycoumaric acid
Umbellularsäure *f* umbellularic acid
Umbellulon *n* umbellulone
Umber *m* (Min) umber
Umbererde *f* umber
umbiegen to bend over
umbilden to transform
Umbilicarsäure *f* umbilicaric acid
Umbilicin *n* umbilicin
Umbördelung *f* flanging
Umbugelektrode *f* beading electrode
Umbugschieber *m* doubler (for plastics sealing)
umdestillieren to redistil
umdisponieren to reorganize
Umdotierung *f* **der Schmelze** redoping of melt
umdrehen to turn (over), (rotieren) to revolve, to rotate
Umdrehung *f* revolution, rotation
Umdrehungen *pl* **in der Minute** revolutions per minute
Umdrehungsachse *f* axis of rotation
Umdrehungsanzeiger *m* revolution indicator, revolution telltale
Umdrehungsellipsoid *n* rotation ellipsoid, **abgeplattetes** ~ oblate ellipsoid, **verlängertes** ~ prolate ellipsoid
Umdrehungsgeschwindigkeit *f* speed of rotation
Umdrehungshyperboloid *n* rotation hyperboloid

Umdrehungsmesser *m* tachometer
Umdrehungsparaboloid *n* rotation paraboloid
Umdrehungszähler *m* revolution counter
Umdrehungszahl *f* number of revolutions
Umdruckfarbe *f* reprinting ink
Umesterung *f* ester interchange, transesterification
umfällen (Chem) to [dissolve and] reprecipitate
Umfällung *f* reprecipitation
umfärben (Färb) to redye
umfallen to fall over, to tip over
Umfang *m* circumference, perimeter, periphery
umfangreich voluminous, ample, extensive
Umfangsgeschwindigkeit *f* **im Wälzkreis** pitch line velocity
Umfangsriß *m* circumferential crack
Umfangsspannung *f* circumferential stress, hoop stress
umfassend comprehensive, ample
umflechten to braid
umformen to transform
Umformer *m* (Elektr) transformer
Umformerröhre *f* converter tube
Umformung *f* change, transformation, (Strom) conversion
Umfrage *f* inquiry
Umführungsleitung *f* by-pass (pipe)
Umgangsleitung *f* by-pass
umgeben to surround
Umgebung *f* neighborhood
Umgebungstemperatur *f* ambient temperature
Umgehungsleitung *f* by-pass [duct], by-pass line
Umgehungsventil *n* by-pass valve
umgekehrt converse, inverse, reciprocal, (Bild) inverted (image), (Ordnung) reverse, ~ **proportional** inversely proportional
Umgekehrte[s] *n* reverse, (Ordnung) opposite
umgesetzt (Chem) converted
umgießen to decant, to transfer
Umgrenzung *f* boundary, limitation, restriction
umgruppieren to rearrange, to [re]group
Umgruppierung *f* rearrangement, regrouping, ~ **im Kern** nuclear rearrangement, **atomare** ~ atomic rearrangement
umhüllen to cover, to encase, to envelop
Umhüllung *f* covering, envelope, jacket, sheath, sheathing, shell
Umkarton *m* wrapping
Umkehr *f* reversal
umkehrbar reversible, capable of being turned, invertibel, **nicht** ~ irreversible, nonreturn
Umkehrbarkeit *f* reversibility
Umkehrbild *n* inverted image, reversed image
umkehren to invert, to reverse, (umdrehen) to turn [around]
Umkehrerscheinung *f* (Phot) reverse phenomenon, solarization
Umkehrfunktion *f* inverse function
Umkehrintegral *n* inversion integral

Umkehrkontaktgrundverfahren *n* reverse reactive ground coat technique
Umkehrprisma *n* inverting prism, right-angled prism
Umkehrpunkt *m* turning point, point of regression
Umkehrsatz *m* inversion theorem
Umkehrschalter *m* reversing switch
Umkehrspiegel *m* inverting mirror, reversing mirror
Umkehrspülung *f* (Mot) reverse scavenging
Umkehrstation *f* return station
Umkehrung *f* reversion, inversion, switching
Umkehrverfahren *n* reversion procedure
Umkehrwalzenbeschichtung *f* reverse roll coating
Umkehrwalzwerk *n* reversing mill; reversing [rolling] mill
umkippen to flip, to overturn, to tilt over, to tip over
Umklappen *n* **des Spins** spin flip
Umklappprozeß *m* (Elektr) flip-over process
umkleiden to coat
Umkleidungsplatte *f* casing
Umklöppelung *f* braiding
umkonstruieren to reconstruct, to rebuild, to redesign
Umkreis *m* circuit, circumference
umkreisen to circle around, to rotate [around]
Umkreisgeschwindigkeit *f* peripheral velocity
Umkristallisation *f* recrystallization
umkristallisieren to recrystallize
Umkristallisieren *n* recrystallizing
Umkristallisierung *f* recrystallization
umladen *n* to unload
Umladen *n* unloading
Umladung *f* charge exchange, unloading
Umladungsquerschnitt *m* charge exchange cross section
umlagern to rearrange, to regroup, to transpose, (umgeben) to surround
Umlagerung *f* (Chem) rearrangement, transposition, ~ **bei Friedel-Crafts-Reaktionen** Friedel-Crafts rearrangement, ~ **über ein Carbonium-Ion** rearrangement via a carbonium ion, ~ **unter Nachbargruppenbeteiligung** rearrangement with participation of neighboring groups, ~ **von Alkylgruppen** shift of alkyl groups, **Beckmannsche** ~ Beckmann rearrangement, **Benzidin-** ~ benzidine rearrangement, **Claisen-** ~ Claisen rearrangement, **Curtius-** ~ Curtius rearrangement, **Friessche** ~ Fries rearrangement, **intramolekulare** ~ intramolecular change, intramolecular rearrangement, **radikalische** ~ radical rearrangement, **Wagner-Meerwein-** ~ Wagner-Meerwein rearrangement
Umlauf *m* circulation, circuit, cycle, rotation

Umlaufbahn *f* (des Elektrons) orbit (of the electron)
Umlaufbewegung *f* revolution, rotary motion, rotation, (Phys) orbital motion
Umlaufbiegeprüfung *f* fatigue test with rotating beam
umlaufen to rotate
Umlauffrequenz *f* rotational frequency
Umlaufgebläse *n* rotary blower
Umlaufgeschwindigkeit *f* rotational speed, speed of rotation
Umlaufkanal *m* by-pass
Umlaufkessel *m* circulation boiler, rotating steam generator
Umlaufkratzbürste *f* rotating scratch brush
Umlaufkühlung *f* rotation cooling
Umlauflager *n* storage circuit
Umlaufleitung *f* (Atom) circulation loop
Umlaufmenge *f* circulating amount
Umlaufpumpe *f* circulating pump, rotary pump
Umlaufregler *m* by-pass regulator, circulation regulator
Umlaufschmierung *f* circulating system lubrication, circulatory lubrication
Umlaufschneckenmischer *m* vertical screw mixer
Umlaufspeicher *m* (Comput) circulating storage
Umlaufthermostat *m* rotary thermostat
Umlaufventil *n* by-pass valve, rotating valve
Umlaufverdampfer *m* circulating evaporator, circulatory cyclone evaportor, forced circulation evaporator
Umlaufzeit *f* period of revolution
Umlenkblech *n* baffle
umlenken to divert
Umlenkkopf *m* cable extruder head
Umlenkprisma *n* deviating prism
Umlenkrad *n* (Stahlketten) guide sprocket
Umlenkrelais *n* diverting relay
Umlenkstation *f* relay station
Umlenktrommel *f* idler, tail drum, tail pulley
Umlenkungsdüse *f* reversing nozzle
Umluft *f* circulating air
Umluftförderanlage *f* circulating air conveyor
Umluftheizung *f* heating by circulating air
Umluftsichter *m* whizzer
Umlufttrockner *m* circulating drier
Umluftverfahren *n* air-circulation system
ummagnetisieren to reverse the magnetism, to reverse the magnetization
Ummagnetisierung *f* magnetic reversal, reversal of magnetism, reversal of magnetization
ummanteln to sheathe
ummantelt sheathed
Ummantelung *f* coating, cover, jacket, sheathing, shell
Umordnung *f* (Chem) rearrangement
umorganisieren to reorganize, to rearrange
Umorientierung *f* reorientation
Umpolarisation *f* reversal of polarization

umpolarisieren – Umwandlungstemperatur

umpolarisieren to reverse the polarity
Umpolarisierung *f* reversal of polarity
umpolen to reverse the poles
Umpolung *f* reversion of the poles
Umpump *m* circulation by pumping
Umpumpbetrieb *m* pump circulation
umrechnen to convert
Umrechnung *f* conversion, recalculation, recomputation
Umrechnungsfaktor *m* conversion factor
Umrechnungskonstante *f* conversion constant
Umrechnungstabelle *f* conversion table, reduction table
Umrechnungstafel *f* conversion table
Umrechnungsverhältnis *n* conversion ratio
Umrechnungswert *m* conversion factor
Umrichter *m* transverter
Umriß *m* contour, outline, sketch
Umrißdiagramm *n* contour diagram
Umrollmaschine *f* rewinding machine
umrühren to stir [up], to agitate
Umsatz *m* turnover, exchange, (Allg) reaction, (Chem) conversion
Umsatzgeschwindigkeit *f* rate of reaction
Umsatzvariable *f* conversion variable
umschaltbar reversible
umschalten to switch, to reverse, to shift
Umschalter *m* reversing switch, throw-over switch, (Motor) commutator
Umschaltschütz *m* (Elektr) reversing contactor
Umschaltventil *n* reversing valve
Umschaltwärmeaustauscher *m* reversing heat exchanger
umschaufeln to stir with a shovel
umschichten to rearrange
Umschlag *m* (Brief) envelope, (Chem) change, (Med) compress, (Papier) folder, paper cover
umschlagen to turn [over]
Umschlag[s]punkt *m* change point, transition point, (Titration) final point of a titration
Umschlagzeit *f* shelf-life
umschmelzen to remelt, to recast
Umschmelzmetall *n* remelt metal, scrap metal
Umschmelzofen *m* remelting furnace
Umschmelzverfahren *n* remelting process
umschreiben to circumscribe
Umschreibung *f* circumscription
umschütten to pour into another container, to transfer by pouring, (verschütten) to spill
umsetzen (Chem) to cause to react
Umsetzen *n* converting
Umsetzung *f* (Chem) conversion, reaction, (Umstellen) transposition, (Umwandlung) transformation, **chemische** ~ chemical reaction
Umsetzungsgeschwindigkeit *f* velocity of conversion
Umsetzungszahl *f* (Atom) conversion ratio
umspannen (transformieren) to transform

Umspanner *m* (Transformator) transformer
Umspannleistung *f* transformer capacity
umspeichern (Comp) to exchange
Umspeicherung *f* (Comp) exchanging
umspulen to re-reel, to rewind
umständlich circumstantial, (beschwerlich) cumbersome
Umsteckverfahren *n* rejigging (electroplating)
Umstellhebel *m* reversing handle, reversing lever, shift lever
Umstellklappe *f* reversing valve
Umsteuerhebel *m* reversing handle
umsteuern to reverse
Umsteuerungsvorrichtung *f* reversing gear
Umsteuerventil *n* checker valve
Umsteuerwelle *f* reversing shaft
Umströmkanal *m* circulation channel, air way, circulation passage
umstürzend revolutionary
Umtausch *m* exchange, (Ersatz) replacement
Umverpackung *f* wrapping
umwälzend revolutionary
Umwälzgas *n* recycle gas
Umwälzpumpe *f* circulating pump, recirculation pump, transfer pump
Umwälzung *f* circulation
umwandelbar convertible
umwandeln to change, to convert, (Atom) to transmute, (Elektr) to transform
Umwandler *m* converter, (Elektr) transformer, ~ **in Digitalschreibweise** (Comp) quantizer
Umwandlung *f* change, conversion, transformation, (Atom) transmutation, ~ **der Atome** transmutation of atoms, **chemische** ~ chemical change, **radioaktive** ~ radioactive transformation, radioactive transmutation
Umwandlungsbereich *m* transformation range, transition range
Umwandlungsdiagramm *n* transformation diagram
Umwandlungsfähigkeit *f* capacity for change
Umwandlungsfaktor *m* conversion factor
Umwandlungsgeschwindigkeit *f* rate of transformation
Umwandlungskoeffizient *m* conversion ratio
Umwandlungskonstante *f* (Atom) disintegration constant
Umwandlungsprodukt *n* transformation product
Umwandlungsprozeß *m* process of conversion, transformation process, (Biol) metamorphosis
Umwandlungspunkt *m* transformation point, critical point, transition point, **magnetischer** ~ point of magnetic transformation
Umwandlungsspannung *f* transformation potential
Umwandlungsspektrum *n* conversion spectrum
Umwandlungstemperatur *f* transition temperature, equilibrium temperature

Umwandlungsverhältnis *n* conversion ratio, degree of transformation
Umwandlungsvorgang *m* conversion process
Umwandlungswärme *f* heat of transformation, critical heat
Umwandlungszone *f* transition zone
Umweganregung *f* indirect excitation
Umwegleitung *f* by-pass
Umwelt *f* environment
umweltbedingt environmental
umwerfen to knock over
umwickeln to rewind, to wind around, to wrap
unabhängig independent
Unabhängigkeit *f* independence
unangegriffen unacted upon, unaffected, unattacked
unangreifbar (Korrosion) noncorrosive
unaufgelöst (Chem) undissolved, (Opt) unresolved
unaufhaltsam continuous, irrepressible
unauflösbar insoluble
Unauflösbarkeit *f* (Chem) insolubility
unauflöslich insoluble
Unauflöslichkeit *f* insolubility
unausdehnbar inexpansible
unausführbar impracticable
unausgeglichen non-balanced, non-compensated, unbalanced
unausgerichtet unaligned
unauslöschbar inextinguishable
unauslöschlich inextinguishable
Unbalanz *f* unbalance
unbearbeitbar unmachinable
unbearbeitet crude, in blank condition, in blank form, raw, unwrought
unbedingt absolute, unconditional
unbegrenzt boundless, unlimited, (unbestimmt) indefinite, (unendlich) infinite
Unbegrenztheit *f* boundlessness
unbekannt unknown
Unbekannte *f n* unknown
unbelebt dead, inanimate, lifeless
unbelichtet unexposed
unbemannt (Satellit) unmanned
unberechenbar incalculable
unberücksichtigt disregarded, neglected
unbeschaded irrespective of
unbeschädigt undamaged
unbeschnitten untrimmed
unbeschränkt unlimited, indefinite, unbound
unbeschwert unpigmented
unbesetzt unoccupied, vacant
unbeständig inconstant, labile, unstable, unsteady
Unbeständigkeit *f* inconstancy, instability
unbestimmbar undeterminable, (undefinierbar) undefinable
Unbestimmbarkeitsrelation *f* indeterminacy principle

unbestimmt indefinite, indeterminate
Unbestimmte *f* (Math) variable
Unbestimmtheit *f* indefiniteness, indeterminacy
Unbestimmtheitseffekt *m* uncertainty effect
Unbestimmtheitsprinzip *n* (Quant) unvertainty principle, indeterminacy principle
unbewaffnet (Auge) naked
unbeweglich motionless, immobile, immovable, (fest) fixed, stationary
unbeweisbar unprovable
unbiegsam inflexible, rigid
unbrauchbar useless, unsuitable
Unbrauchbarmachen *n* making useless
unbrennbar incombustible, noncombustible, (feuerfest) fireproof, (nicht entzündbar) non-inflammable
Unbrennbarkeit *f* non-inflammability
uncalciniert uncalcined
Undecadien *n* undecadiene, hendecadiene
Undecadiin *n* undecadiyne, hendecadiyne
Undecan *n* undecane, hendecane
Undecanol *n* undecanol, hendecanol, hendecyl alcohol, undecyl alcohol
Undecansäure *f* undecanoic acid
Undecin *n* undecine, hendecyne
Undecyl- undecyl
Undecylaldehyd *m* undecylic aldehyde, hendecanal
Undecylbromid *n* undecyl bromide
Undecylen *n* undecylene, hendecene, undecene
Undecylensäure *f* undecylenic acid
Undecylsäure *f* undecylic acid, hendecanoic acid
undefiniert undefined
undehnbar inextensible, (Met) non-ductile, unmalleable
undenkbar unbelievable, inconceivable, incredible, unimaginable
undeutlich indistinct, (schwach) faint, vague
undicht leaking, leaky, untight, (durchlässig) pervious
Undichte *f* leakage
Undichtigkeit *f* leakage, leakiness, porosity
Undichtigkeitsgrad *m* degree of porosity
undissoziiert undissociated
Undulation *f* undulation
Undulationsstrom *m* undulatory current
Undulationstheorie *f* undulation theory
undurchdringlich impenetrable, impermeable, impervious
Undurchdringlichkeit *f* impenetrability, impermeability
undurchführbar impracticable, irrealizable
undurchlässig impenetrable, impermeable, impervious, (Licht) opaque, ~ **für Röntgenstrahlen** radiopaque
Undurchlässigkeit *f* impermeability, imperviousness
undurchsichtig non-transparent, opaque

Undurchsichtigkeit – Ungleichförmigkeit

Undurchsichtigkeit *f* non-transparency, opacity, opaqueness
uneben uneven, (rauh) rough
Unebenheit *f* unevenness, roughness
unecht not proof, spurious, (künstlich) artificial, false, imitation, (Math) improper
unedel (Met) base, electronegative
uneinheitlich inhomogeneous, irregular
unelastisch inelastic, nonelastic
unempfindlich insensitive, nonreactive, (Med) immune
Unempfindlichkeit *f* insensibility, insensitiveness, nonreactivity, nonresponsiveness, unsusceptibility
unendlich endless, infinite, interminable, nonterminating, unlimited, ~ **groß werden** to approach infinity, ~ **klein** infinitesimal, **auf** ~ **eingestellt** focused for infinity
Unendlicheinstellung *f* (Opt) infinity adjustment
Unendlichkeit *f* infinity
Unentbehrlichkeit *f* indispensability
unentflammbar flameproof, non-inflammable
Unentflammbarkeit *f* flameproofness, non-ignitability, non-inflammability
unentwegt persistent
unerbittlich inexorable
unerfahren inexperienced
unerklärbar (unerklärlich) inexplicable, unexplainable
unermeßlich immeasurable, immense
unerregbar unexcitable
unerschöpflich inexhaustible
Unerschöpflichkeit *f* inexhaustibility
unerträglich unbearable
unerwartet unexpected
Unfähigkeit *f* inability, incapacity
Unfall *m* accident, **leichter** ~ mishap
Unfallmeldeanlage *f* accident signalling system
Unfallschutz *m* accident prevention
Unfallstation *f* ambulance station, emergency ward, first-aid post
Unfallverhütung *f* accident prevention
Unfallverhütungsvorschriften *pl* regulations for prevention of accidents
unfermentiert unfermented
unfiltrierbar non-filterable
Unfruchtbarkeit *f* barrenness, (Biol) sterility
unfühlbar impalpable
ungebeugt (Opt) undiffracted
ungebleicht unbleached
ungebrannt (Ziegel) unburnt
ungebündelt (Opt) uncollimated
ungebunden free, uncombined
ungedämpft undamped
ungedreht untwisted
ungeeignet improper, inexpedient, unsuitable
ungefähr approximately
ungefährlich harmless

ungefärbt colorless, uncolored, unstained, (Tuch) raw, undyed
ungefiltert unfiltered
ungegerbt untanned
ungehärtet soft, unhardened
ungeladen neutral, not charged
ungeleimt (Papier) unsized
ungelöscht (Kalk) unslaked
ungelöst (Chem) undissolved, (z. B. Problem) unsolved
Ungelöste[s] *n* insoluble part
ungelötet unsoldered
ungemein extremely
ungemischt unmixed
ungenau inaccurate, inexact
Ungenauigkeit *f* inaccuracy, (Abweichung) deviation, (Ungewißheit) uncertainty
ungenügend insufficient, poor
ungeordnet disarranged, disordered, irregular, (zufällig) random
ungepaart unpaired
ungequantelt unquantized
ungerade not straight, (verbogen) crooked
ungerade-gerade (Atom) odd-even
ungerade-ungerade (Atom) odd-odd
ungerinnbar uncoagulable, uncongealable
ungeröstet unroasted, uncalcined
ungesättigt unsaturated
Ungesättigtheit *f* unsaturated state
ungesalzen unsalted
ungeschliffen rough
ungeschmolzen unfused
ungeteert untarred
ungetempert unannealed
ungewiß uncertain, doubtful, problematic
Ungewißheitsprinzip *n* indetermination principle, uncertainty principle
ungewöhnlich extraordinary, unusual
ungewürzt unseasoned
Ungeziefer *n* bug[s], pest[s], vermin
Ungezieferbekämpfung *f* vermin destruction, vermin o. pest extirpation
ungezügelt uncontrolled
ungiftig nonpoisonous, non-toxic
unglasiert unglazed, unvarnished
ungleich dissimilar, unequal, unlike
ungleichartig heterogeneous, not uniform, of opposite kind
Ungleichartigkeit *f* dissimilarity, heterogeneity
ungleichfarbig heterochromous, of different colors
ungleichförmig irregular, asymmetrical, inhomogeneous, non-uniform, unequal, unsymmetrical
Ungleichförmigkeit *f* dissimilarity, heterogeneity, inconstancy, inhomogeneity, nonuniformity

Ungleichförmigkeitsgrad *m* degree of irregularity
Ungleichgewicht *n* unbalance, disequilibrium, imbalance
Ungleichgewichtszustand *m* non-equilibrium state
Ungleichheit *f* dissimilarity, inequality, unevenness, ~ **nach Clausius** (Therm) inequality of Clausius
ungleichmäßig asymmetrical, discontinuous, disproportionate, heterogeneous, non-uniform, unsymmetrical
Ungleichmäßigkeit *f* inconstancy, nonuniformity
ungleichnamig unlike, (Elektr) opposite
Ungleichung *f* inequation
Ungleichverteilung *f* (Füllkörperkolonne) mal-distribution
ungleichwinkelig having unequal angles
ungültig invalid, void
Ungültigkeit *f* invalidity, nullity
Ungulinsäure *f* ungulinic acid
unhandlich clumsy, impractical, unhandy
unharmonisch disharmonious, inharmonious, unharmonious
unheilbar incurable
unhygroskopisch non-hygroscopic
unifarben of uniform color
unimolekular unimolecular
Unionverschraubung *f* union [screw]
Unipolardynamo *m* unipolar dynamo
Unipolarinduktion *f* unipolar induction
Unipolarität *f* unipolarity
Unistückware *f* plain shade piece goods
Universalarznei *f* (Pharm) cure-all, universal remedy
Universaldrehtisch *m* universal three-axis stage
Universaleisen *n pl* flitchplates
Universalgalvanometer *n* universal galvanometer
Universalgelenk *n* universal joint
Universalindikatorpapier *n* universal indicator paper
Universallack *m* jobbing varnish, universal varnish
Universallehre *f* universal gauge
Universalmanipulator *m* general purpose manipulator
Universalmanometer *n* universal manometer
Universalmaschine *f* multi-purpose machine
Universalmikroskop *n* universal microscope
Universalmittel *n* (Med) cure-all, universal remedy
Universalprüfmaschine *f* universal testing machine
Universal-Rechenmaschine *f* general purpose computer
Universalrechner *m* multi-purpose computer
Universalrezept *n* (Med) cureall, universal remedy

Universalschleifmaschine *f* universal grinding machine
Universalschraubenschlüssel *m* monkey wrench, universal screw spanner
Universalwalzwerk *n* universal rolling mill
Universalwerkzeug *n* all-purpose tool
universell multipurpose
Universum *n* universe
unizellulär (einzellig) unicellular
unkomprimierbar incompressible
unkondensierbar uncondensable
unkorrigiert uncorrected
unkorrodierbar incorrodible
Unkosten *pl* costs, expenses
Unkraut *n* weed[s]
Unkrautvertilgungsmittel *n* herbicide, weed killer
unkristallisierbar uncrystallizable
unkristallisiert not crystallized, uncrystallized
unlegiert unalloyed, plain
unlösbar unsolvable, (Chem) insoluble
unlöschbar (Kalk) unquenchable, unslakable
unlöslich insoluble
Unlösliche[s] *n* insoluble matter
Unlöslichkeit *f* insolubility
Unlöslichmachen *n* insolubilization, rendering insoluble
Unlöslichmacher *m* insolubilizer
Unlöslichwerden *n* becoming insoluble
unmagnetisch nonmagnetic
unmeßbar immeasurable
unmischbar immiscible, nonmiscible
Unmischbarkeit *f* immiscibility
unmittelbar direct, straight
unmittig off center
Unmittigkeit *f* off-center position
unmodifiziert unmodified
unordentlich disorderly, irregular
Unordnung *f* disorder
Unordnungsordnungsumwandlung *f* disorder order transition
unorientiert unoriented
unoxydierbar inoxidizable, nonoxidizable
unpaarig (ungepaart) unpaired, (ungerade) odd
unperiodisch aperiodic, nonperiodic
unplastifiziert unplasticized
unpolar non-polar, homopolar
unpolarisierbar nonpolarizable
unpolarisiert nonpolarized, unpolarized
unreduzierbar irreducible
unregelmäßig irregular, fluctuating
Unregelmäßigkeit *f* irregularity, anomaly, **magnetische** ~ magnetic anomaly
unrein (Chem) contaminated, impure, (unsauber) unclean, (unscharf) fuzzy
Unreinheit *f* contamination, fuzziness, impurity, uncleanness, (Kristall) cloudiness
unrelativistisch non-relativistic
Unruhe *f* restlessness, turbulence, unsteadiness

unruhig flackern to flare
unrund non-circular, eccentric, out of true, untrue, ~ **werden** to round off, to splay
Unrunddrehbank *f* eccentric lathe
Unrundschleifmaschine *f* eccentric grinder
unsatiniert unglazed
unschädlich harmless, innoxious
Unschärfe *f* diffusiveness
Unschärfebedingung *f* uncertainty condition
Unschärfe-Beziehung *f* **von Heisenberg** Heisenberg uncertainty principle
Unschärferelation *f* (Phys) uncertainty relation
Unschärfering *m* (Opt) apertural effect
unscharf diffuse, blurred, hazy, (Rand) fuzzy, (stumpf) unsharp, (undeutlich) indistinct, (ungenau) inexact
Unschlitt *n* tallow, suet
Unschlittseife *f* tallow soap
unschmelzbar infusible, refractory, unmeltable
Unschmelzbarkeit *f* infusibility
unschweißbar unweldable
Unselektivität *f* lack of selectivity
unsicher insecure, (gefährlich) unsafe, (instabil) unstable, (ungewiß) uncertain
Unsicherheit *f* instability, insecurity, uncertainty, unsafety
Unsicherheitsfaktor *m* uncertainty factor
Unsicherheitsrelation *f* uncertainty relation
unsichtbar invisible
Unsichtbarkeit *f* invisibility
unsorgfältig careless
unspaltbar uncleavable
unstabil instable, unstable
Unstabilität *f* instability, **thermische** ~ thermal instability
unstarr non-rigid
unstetig (Gang der Maschine) unsteady, (Math) discontinuous
Unstetigkeit *f* unsteadiness, (Math) discontinuity
Unstetigkeitsbedingung *f* discontinuity condition
Unstetigkeitsfläche *f* discontinuity surface, surface of discontinuity
Unstetigkeitslinie *f* discontinuity line
Unstetigkeitsstelle *f* (Math) point of discontinuity
Unstetigkeitswechselwirkung *f* discontinuity interaction
Unsymmetrie *f* asymmetrie, dissymmetry, unbalance
Unsymmetriegrad *m* degree of asymmetry
unsymmetrisch asymmetrical, unsymmetrical
untätig inactive, indifferent, inert
unteilbar indivisible
Unteilbarkeit *f* indivisibility
Unterausschuß *m* sub-commitee
Unterbau *m* base, foundation, substructure
unterbelichten (Phot) to underexpose
unterbelichtet (Phot) underexposed

Unterbelichtung *f* (Phot) underexposure
unterbrechen to interrupt, to disconnect, to discontinue
Unterbrecher *m* interrupter, breaker, **elektrolytischer** ~ electrolytic interrupter, **selbsttätiger** ~ automatic contact breaker
Unterbrecherkontakt *m* contact breaker
Unterbrechung *f* interruption, break, stop
Unterbrechungsbad *n* (Phot) stop bath
Unterbrechungsfunke *m* (Elektr) break spark
Unterbrechungsimpuls *m* break impulse
Unterbrechungsstelle *f* point of interruption
unterbrochen interrupted, intermittent, periodic
unterbromige Säure *f* hypobromous acid
unterbromigsauer (Salz) hypobromite
unterchlorig hypochlorous
unterchlorige Säure *f* hypochlorous acid
Unterchlorigsäureanhydrid *n* chlorine monoxide, hypochlorous anhydride
unterdimensioniert of too low capacity
Unterdosierung *f* underfeeding
Unterdruck *m* underpressure, depression, low pressure, partial vacuum, subpressure
Unterdruckgärung *f* (Brau) vacuum fermentation
Unterdruckgaserzeuger *m* suction gas producer
Unterdruckgebiet *n* low-pressure area, subpressure zone
Unterdruckkammer *f* vacuum chamber, depression box, depression chamber
Unterdruckmesser *m* vacuum gauge
Unterdruckpresse *f* bottom ram press
Unterdruckraum *m* vacuum space
Unterdruckverdampfer *m* vacuum evaporator
Unterdrückung *f* suppression
Unterdrückungsspannung *f* (Elektr) cut-off voltage
untereinander abhängig interdependent
Untereinheit *f* subunit
Unterernährung *f* malnutrition, underfeeding
untererregt underexcited
Untererregung *f* underexcitation
Untereutektikum *n* hypoeutectic
untereutektisch hypoeutectic
Untereutektoid *n* hypoeutectoid
Unterfeuerung *f* undergrate firing
Unterführung *f* underpass
Unterfunktion *f* (Med) hypoactivity, hypofunction, impaired function, weak function
untergärig (Brau) bottom fermented
Untergärung *f* (Brau) bottom fermentation
untergeordnet subordinate
Untergestell *n* (Metall) bottom of crucible, bottom of hearth
Untergewicht *n* underweight
Untergitter *n* sublattice
Unterglasur *f* underglaze
Unterglasurfarbe *f* underglaze color

Untergrund *m* background, base, (Erde) subsoil
Untergrundstrahlung *f* background radiation
Untergrundstreuung *f* background scattering
Untergrundüberwachung *f* background monitoring
Untergruppe *f* (Blut) subgroup, (Zool, Bot) subspecies
Unterhalt *m* maintenance, support
unterhalten (Reaktion, Verbrennung) to maintain, to keep going, to sustain
Unterhaltskosten *f pl* cost of maintenance
Unterhaltung *f* maintenance, upkeep
Unterhaltungsarbeiten *pl* maintenance work
Unterhaltungskosten *pl* cost of maintenance
Unterhautgewebe *n* (Med) hypodermis, subcutaneous tissue
Unterhefe *f* (Brau) bottom yeast, low fermentation yeast
unterhöhlen to undermine
unterirdisch subterranean
unterjodige Säure *f* hypoiodous acid
Unterkasten *m* bottom box, lower box, (Gieß) drag
Unterkolbenpresse *f* bottom ram press, upstroke press
unterkopieren (Phot) to underprint
Unterkorrektion *f* undercorrection
unterkritisch below the critical, subcritical
unterkühlen to undercool
unterkühlt supercooled
Unterkühlung *f* undercooling, subcooling, supercooling
Unterlage *f* base [plate], backing, foundation, support, (Histol) substratum, (Verstärkung) backer
Unterlagen *pl* data
Unterlagscheibe *f* washer
Unterlagsplatte *f* base plate, bed plate, stay plate
Unterlagsring *m* supporting ring
Unterlegscheibe *f* washer
Unternehmen *n* enterprise, undertaking
Unternehmer *m* contractor, (Arbeitgeber) employer
Unterniveau *n* sublevel
unternormal subnormal
unteroxydiert underoxidized
unterphosphorige Säure *f* hypophosphorous acid
Unterphosphorsäure *f* hypophosphoric acid
Unterprogramm *n* (Comp) subroutine
Unterpulverschweißverfahren *n* submerged arc welding process
Unterputz *m* scratch coat
Unterraum *m* subspace
unterrichten to instruct, to teach
Unterrostung *f* rust creep, under-rusting
untersättigt undersaturated
Untersättigung *f* undersaturation
Untersatz *m* **aus Hartgummi** ebonite support
Unterschale *f* (Atom) subshell

unterscheiden to differentiate, to discriminate, to distinguish
Unterscheidungsfaktor *m* (Opt) discrimination index
Unterscheidungsmerkmal *n* characteristic feature, distinctive feature
Unterschicht *f* bottom layer, first coat, ground coat, lower layer, priming coat
Unterschied *m* difference
unterschiedlich different
Unterschreitung *f* **des Innendrucks** (Reifen) underinflation
Unterschrift *f* signature
Unterschwefelsäure *f* hyposulfuric acid
Unterseekabel *n* submarine cable
Unterseite *f* underside
Untersetzungsgetriebe *n* reduction gear
Unterspannung *f* low tension, (Elektr) undervoltage
Unterstempel *m* bottom plug, bottom stamp
Untersteuern *n* understeering
Unterstruktur *f* substructure
Unterstück *n* bottom part, lower part
unterstützen to support, to bolster
Unterstützung *f* support
Unterstützungsplatte *f* supporting bed
Unterstützungspunkt *m* point of support
Unterstufe *f* sublevel
untersuchen to examine, to test, (nachforschen) to investigate
Untersuchung *f* (Ausforschung) inquiry, (Chem) analysis, (Forschung) investigation, research, study, (Med) examination, medical check-up, (Prüfung) test, **naßchemische** ~ wet test, **trockenchemische** ~ dry test
Untersuchungsapparat *m* analyzer
Untersuchungsausschuß *m* committee of inquiry, investigation board
Untersuchungsbereich *m* range of investigation
Untersuchungsgerät *n* testing equipment
Untersuchungslabor *n* research laboratory
Untersuchungslaboratorium *n* research laboratory
Untersuchungsmaterial *n* material for investigation
Untersuchungsmethode *f* method of examination
Untersuchungsobjekt *n* object to be examined
Untersuchungsreihe *f* series of tests
Untersuchungsverfahren *n* method of investigation
Untertag[e]bau *m* (Berg) underground mining, underground working
Untertagebetrieb *m* (Bergb) underground working
untertauchen to submerge
Unterteil *n* bottom part
unterteilt subdivided
Unterteilung *f* subdivision
Untertemperatur *f* insufficient temperature

Untertitel *m* subtitle
Unterton *m* (Farbe) undertone
Unterwalze *f* bottom roll
Unterwasseranstrich *m* underwater coating
Unterwasserfarbe *f* underwater paint, ship's bottom paint
Unterwasserhorchgerät *n* hydrophone
Unterwasserlack *m* marine finish
Unterwasserlager *n* submerged bearing
Unterwasserpumpe *f* submerged pump
Unterwasserspiegel *m* tail water level
unterwerfen to subject [to]
Unterwindfeuerung *f* forced draft furnace
unterworfen sein to be subjected [to]
Unterzellgewebe *n* (Med) subcutaneous cell tissue
Untiefe *f* shallowness
untrennbar inseparable
unübertrefflich unsurpassed
unumgänglich indispensable
ununterbrochen continuous[ly], uninterrupted
unveränderlich invariable, unalterable, unchangeable
unverändert unaltered, unchanged, unmodified
unverantwortlich irresponsible
unverbrennbar fireproof, incombustible, non-combustible
Unverbrennbarkeit *f* incombustibility
unverdaulich indigestible
unverderblich incorruptible
unverdichtbar incompressible, incondensable
unverdünnt undiluted
unverestert unesterified
unverfälscht unadulterated
unverformbar non-workable
unvergärbar unfermentable
unvergasbar ungasifiable
unvergleichbar incomparable
unverhältnismäßig disproportionate
unverhüttbar (Metall) unsmeltable, unworkable
unvermischbar immiscible
unveröffentlicht unpublished
unverschlackt unscorified
unversehrt undamaged
unverseifbar (Chem) unsaponifiable
unversiegbar inexhaustible
unverständlich inexplicable
unverträglich incompatible
Unverträglichkeit *f* incompatibility
unverzerrt undistorted
unverzüglich without delay
unverzweigt branchless, unbranched
unvollkommen uncompleted, defective
unvollständig imperfect, incomplete, partial
unvoreingenommen unprejudiced
unvorstellbar inconceivable, unimaginable
unwägbar imponderable, unweighable
Unwägbarkeit *f* imponderability, unweighability
unwahrscheinlich improbable, implausible

Unwahrscheinlichkeit *f* improbability
unwesentlich unimportant; unessential, (nicht zugehörig) extraneous
unwiderruflich irrevocable
unwillkürlich involuntary
unwirksam inactive, ineffective, inefficient
Unwirksamkeit *f* inactivity, ineffectiveness
unwirtschaftlich unecconomic
unwissend ignorant
unwissenschaftlich unscientific
Unwucht *f* unbalance, balance error
Unwuchtpaste *f* balance dough, balance stock
unzählbar countless, innumerable
Unzahl *f* immense number
Unze *f* (Maß) ounce (= 28,34 g)
unzerbrechlich unbreakable, nonbreakable, nonshattering, shatterproof
Unzerbrechlichkeit *f* shatterproofness
unzerlegbar indecomposable, indivisible
unzerreißbar untearable, indestructible
unzersetzbar indecomposable
unzerstörbar indestructible
Unzerstörbarkeit *f* indestructibility
unzerstört undecomposed, undestroyed
unzertrennbar (unzertrennlich) inseparable
Unzulänglichkeit *f* deficiency, insufficiency, shortcoming
unzusammendrückbar incompressible
Unzusammendrückbarkeit *f* incompressibility
unzusammenhängend incoherent
unzuverlässig unreliable
unzweckmäßig inappropriate, inexpedient, unsuitable
unzweideutig unambigous
Upas *n* upas
U-Profil *n* channel section
Uracil *n* uracil, 2,6-dihydroxypyrimidine
Uracylsäure *f* uracylic acid
Uralit *m* (Min) uralite
Uralorthit *m* (Min) uralorthite
uralt ancient
Uramil *n* uramil, aminobarbituric acid, murexan
Uramildiessigsäure *f* uramildiacetic acid
Uran *n* uranium (Symb. U), ~ **in Blöcken** lumped uranium, **angereichertes** ~ enriched uranium
Uranacetat *n* uranium acetate
Uranat *n* uranate, salt of uranic acid
Uranblei *n* (Min) uranium lead
Uranbrenner *m* uranium pile or reactor
Uranbrennstoff *m* uranium fuel
Urancarbid *n* uranium carbide, uranium acetylide
Urandiol *n* uranediol
Urandioxid *n* uranium dioxide, uranium(IV) oxide, uranous oxide
Uranerz *n* (Min) uranium ore
Urangehalt *m* uranium content
Urangelb *n* uranium yellow

Uranglas *n* uranium glass
Uranglasplatte *f* uranium glass plate
Uranglimmer *m* (Min) uranium glimmer, torbernite, uranite, uranium mica
Urangraphitreaktor *m* carbon uranium pile, uranium graphite reactor
Urangrün *n* uranochalcite
uranhaltig containing uranium, uraniferous
Uranhexafluorid *n* uranium hexafluoride, uranic fluoride, uranium(VI) fluoride
Uraniablau *n* urania blue
Uranid *n* uranide
uranig uranous
Uranin *n* uranin, sodium fluorescein
Uraninit *m* (Min) uraninite, pitchblende
Uranisotop *n* (Atom) uranium isotope
Uranit *m* (Min) uranite
uranithaltig uranitiferous
Uranium *n* uranium (Symb. U)
Uraniumchlorid *n* uranium chloride
Uraniverbindung *f* (Uran-(VI)-Verbindung) uranic compound, uranium(VI) compound
Uran(IV)-Verbindung *f* uranous compound
Urankern *m* uranium nucleus
Urankopieverfahren *n* uranium copying process
Uranmeiler *m* uranium pile or reactor
Uranmetall *n* metallic uranium, uranium metal
Urannatriumacetat *n* uranyl sodium acetate
Urannatriumchlorid *n* uranyl sodium chloride
Uranniobit *m* (Min) uranoniobite
Urannitrat *n* uranium nitrate
Uranochalcit *m* (Min) uranochalcite
Uranocircit *m* (Min) uranocircite
Uranocker *m* (Min) uranic ocher, uraconite
Uranohydroxid *n* (Uran-(IV)-hydroxid) uranium(IV) hydroxide, uranous hydroxide
Uranolepidit *m* (Min) uranolepidite
Uranooxid *n* (Uran-(IV)-oxid) uranium dioxide, uranium(IV) oxide, uranous oxide
Uranophan *m* (Min) uranophane
Uranopilit *m* (Min) uranopilite
Uranosalz *n* (Uran-(IV)-Salz) uranium(IV) salt, uranous salt
Uranosphärit *m* (Min) uranospherite
Uranospinit *m* (Min) uranospinite
Uranotantal *n* (Min) samarskite
Uranothallit *m* (Min) uranothallite
Uranothorit *m* (Min) uranothorite
Uranotil *m* (Min) uranotil
Uranouranioxid *n* uranium(IV)(VI) oxide, uranoso-uranic oxide
Uranoverbindung *f* (Uran-(IV)-Verbindung) uranium(IV) compound, uranous compound
Uranoxidnitrat *n* uranyl nitrate, uranium(VI) dioxydinitrate
Uranoxidoxydul *n* (obs) uranium(IV)(VI) oxide, uranoso-uranic oxide
Uranoxidsalz *n* (obs) uranyl

Uranoxydul *n* (obs) uranium dioxide, uranium(IV) oxide, uranous oxide
Uranoxyduloxid *n* (obs) uranium(IV)(VI) oxide, uranoso-uranic oxide
Uranoxydulsalz *n* (obs) uranium(IV) salt, uranous salt
Uranoxydulsulfat *n* (obs) uranium(IV) sulfate, uranous sulfate
Uranpechblende *f* (Min) pitchblende, uraninite
Uranpecherz *n* (Min) pitchblende, uraninite
Uranphosphat *n* uranium phosphate
Uranreaktor *m* uranium pile
Uranreihe *f* uranium series
Uransäure *f* uranic acid
Uransäureanhydrid *n* (Uran-(VI)-oxid) uranic anhydride, uranic oxide, uranium trioxide, uranium(VI) oxide
Uransalz *n* uranium salt
uransauer (Salz) uranate
Uranspaltprodukt *n* fission product of uranium
Uranspaltung *f* (Atom) uranium fission, uranium splitting
Uranstab *m* uranium slug
Uranstabhülse *f* uranium cartridge
Uranstrahlen *m pl* uranium rays
Uransuperoxid *n* uranium peroxide
Urantetrafluorid *n* uranium tetrafluoride
Uranverbindung *f* uranium compound
Uranverstärker *m* uranium intensifier
Uranvitriol *n* uranium vitriol, (Min) johannite, uranochalcite
Uran(VI)-Verbindung *f* uranic compound
Uranvorkommen *n* uranium deposit
Uranyl- uranyl
Uranylacetat *n* uranyl acetate
Uranylhydrogenphosphat *n* acid uranyl phosphate, uranyl hydrogen phosphate
Uranylnitrat *n* uranyl nitrate, uranium(VI) dioxydinitrate
Uranylradikal *n* uranyl radical
Uranylspektrum *n* uranyl spectrum
Uranylsulfat *n* uranyl sulfate
Uranzerfall *m* uranium decay
Uranzerfallsreihe *f* uranium decay series
Uranzirkit *m* (Min) uranocircite
Urari *n* urari, curare
Urat *n* urate, lithate, salt or ester of uric acid
Urazin *n* urazine, aminourazole
Urazol *n* urazole, diketotriazolidine
Urbanit *m* (Min) urbanite
Urbestandteil *m* elementary constituent
Ureabromin *n* ureabromine
Urease *f* (Biochem) urease
Urechochrom *n* urechochrome
Ureid *n* ureide
Ureotelismus *m* (Phys) ureotelism
Urethan *n* urethane, ethyl carbamate, ethylurethane
Urethanchloral *n* chloral urethane, uralin

Urethanharz *n* urethane resin
Urethylan *n* urethylan, methyl carbamate, methylurethane
Uretidin *n* uretidine, tetrahydrourete
Uretin *n* 1,2-dihydrourete, uretine
Urezin *n* urezin
Urform *f* primitive form, prototype
Urgneis *m* (Min) primitive gneiss
Urgranit *m* (Min) primitive granite
Urheber *m* author, creator
Urheberrecht *n* copyright
Uricase *f* (Biochem) uricase
Uridin *n* uridine, uracil-riboside
Uridindiphosphat *n* uridine diphosphate, UDP
Uridindiphosphatgalactose *f* (Biochem) uridine diphosphate galactose
Uridindiphosphatglucose *f* uridine diphosphate glucose
Uridinmonophosphat *n* uridine monophosphate, UMP
Uridinphosphorsäure *f* uridine phosphoric acid
Uridintriphosphat *n* uridine triphosphate
Uridylsäure *f* uridylic acid
Uridyltransferase *f* (Biochem) uridyl transferase
Urin *m* urine
urinartig urinous
Urinbodensatz *m* urinary sediment
Urinologie *f* urinology
Urinometer *n* urinometer, hydrometer for determining the specific gravity of urine
Urkraft *f* original force, original power
Urkunde *f* document
Urlösung *f* original solution
Urmaß *n* standard measure
Urmaterie *f* prime matter, ylem
Urmensch *m* primitive man
Urmeter *n* standard [or prototype] meter
Urobilin *n* urobilin
Urobilinogen *n* urobilinogen
Urocanase *f* (Biochem) urocanase
Urocanin *n* urocanin
Urocaninsäure *f* urocaninic acid
Urochrom *n* urochrome
Uroerythrin *n* uroerythrin
U-Rohr *n* U-tube
U-Rohr-Manometer *n* U-tube manometer
Urol *n* urol, urea quinate
Uron *n* uron
Uronsäure *f* uronic acid
Uroporphyrin *n* uroporphyrin
Urorosein *n* urorosein
Urosin *n* urosine, lithium quinate
Urothion *n* urothion
Urotropin *n* urotropine, hexamethylenetetramine
Uroxansäure *f* uroxanic acid
Urplasma *n* archiplasm, ylem
Ursache *f* cause
Ursan *n* ursane

Urschiefer *m* (Min) primitive slate, argillite
Ursol *n* ursol, p-phenylenediamine
Ursolsäure *f* ursolic acid
Urson *n* urson, malol, prunol, ursolic acid
ursprünglich initial, original, primeval
Ursprung *m* origin, ~ **des Koordinatensystems** origin of the coordinate system
Urstoff *m* primary matter, element, initial material, ylem
Ursubstanz *f* original material, original substance, primary matter
Urteer *m* crude tar, low temperature tar, original tar
Urtinktur *f* mother tincture
Urtiter *m* (Anal) original titer, titrimetric standard
Urtitersubstanz *f* (Anal) standard titrimetric substance
Urtonschiefer *m* (Min) primitive argillite
Urtyp *m* prototype
Urusen *n* urusene
Urushenol *n* urushenol
Urushinsäure *f* urushic acid, laccol
Urushiol *n* urushiol
Urusit *m* urusite
Urzelle *f* elementary cell
Urzellentheorie *f* monogenesis
Urzeugung *f* abiogenesis, spontaneous generation
Urzustand *m* initial state, original state
Usnetinsäure *f* usnetinic acid
Usnetsäure *f* usnetic acid
Usninsäure *f* usnic acid
Usnolsäure *f* usnolic acid
U-Stahl *m* channel steel
Ustilsäure *f* ustilic acid
Uvanit *m* (Min) uvanite
Uvaol *n* uvaol
Uvinsäure *f* uvic acid, dimethylfuranecarboxylic acid, pyrotritartaric acid, uvinic acid
Uviol *n* uviol, ultraviolet transmitting glass
Uviolglas *n* (Opt) uviol glass
Uvitinsäure *f* uvitic acid
Uvitonsäure *f* uvitonic acid
Uwarowit *m* (Min) uvarovite, uwarowite
Uzarin *n* uzarin

V

Vaccensäure *f* vaccenic acid
Vacciniin *n* vacciniin
vagabundierend eddy, stray
Vagabundierneutron *n* (Atom) stray neutron
vage vague
Vakerin *n* vakerin
Vakublitz *m* vacuum flash
Vakumeter *n* vacuum gauge, vacuum manometer
Vakuole *f* (Biol) vacuole
Vakuum *n* vacuum, **hohes** ~ high vacuum
Vakuumanlage *f* vacuum plant
Vakuumanschluß *m* vacuum attachment
Vakuumbedämpfung *f* vacuum deposition
Vakuumbeton *m* vacuum concrete
Vakuumbox *f* (Reifen) vacuum expander
Vakuumbremse *f* vacuum brake
Vakuumdestillation *f* vacuum distillation
Vakuumdestillierblase *f* vacuum still
vakuumdicht vacuum-tight, hermetically sealed
Vakuumelektrodynamik *f* electrodynamics in vacuum
Vakuumentnahme *f* vacuum take-off
Vakuumexsikkator *m* vacuum desiccator
Vakuumfett *n* vacuum grease
Vakuumfilter *n* vacuum filter
Vakuumfiltration *f* vacuum filtration
Vakuumformung *f* vacuum forming
Vakuumgärung *f* vacuum fermentation
Vakuumgefriertrockner *m* vacuum freeze drier
vakuumgeglüht vacuum annealed
Vakuumgitterspektrograph *m* vacuum grating spectrograph
Vakuumgleichrichter *m* (Elektr) vacuum rectifier
Vakuumglocke *f* vacuum bell jar
Vakuumgummisackverfahren *n* vacuum bag molding
Vakuumhahn *m* vacuum tap
Vakuumkammer *f* vacuum chamber
Vakuumkessel *m* vacuum tank
Vakuumkocher *m* cold boiler
Vakuumkolben *m* vacuum flask
Vakuumkorrektur *f* vacuum correction
Vakuumkühler *m* vacuum cooler
Vakuum-Lichtbogenofen *m* vacuum arc furnace
Vakuummantel *m* vacuum jacket
Vakuummesser *m* vacuometer, vacuum gauge
Vakuummeßgerät *n* vacuum gauge
Vakuummeßinstrument *n* vacuum gauge, vacuum measuring instrument
Vakuummeter *n* vacuum gauge, vacuum manometer
Vakuumöl *n* vacuum oil
Vakuumofen *m* vacuum furnace
Vakuum[photo]zelle *f* vacuum photoelectric cell
Vakuumphysik *f* vacuum physics

Vakuumpolarisationsdiagramm *n* vacuum polarization diagram
Vakuumprüfer *m* vacuum tester
Vakuumpumpe *f* vacuum pump
Vakuumregler *m* vacuum regulator
Vakuumreiniger *m* vacuum cleaner
Vakuumrektifikation *f* vacuum rectification or distillation
Vakuumröhre *f* vacuum tube
Vakuumschalter *m* vacuum switch
Vakuumschaufeltrockner *m* vacuum paddle drier
Vakuumschlauch *m* pressure tubing
Vakuumschmelzanlage *f* vacuum melting plant
Vakuumschrank *m* vacuum cabinet
Vakuumtank *m* vacuum tank
Vakuumtechnik *f* high-vacuum technique
Vakuumtiefkühltrockner *m* vacuum freeze dryer
Vakuumtiefziehverfahren *n* straight vacuum forming
Vakuumtrockenofen *m* vacuum drying oven
Vakuumtrockenschrank *m* vacuum shelf dryer
Vakuumtrockner *m* vacuum dryer
Vakuumtrommelfilter *n* vacuum drum filter
Vakuumtrommeltrockner *m* vacuum rotary dryer
Vakuumverdampfer *m* vacuum evaporator
Vakuumverdampfung *f* vacuum evaporation
Vakuumverdampfungsanlage *f* vacuum evaporation plant
Vakuumverdampfungsapparat *m* vacuum evaporator
Vakuumverfahren *n* vacuum process, (Polyester) vacuum bag molding
Vakuumverformung *f* vacuum forming
Vakuumverpackung *f* vacuum package
Vakuumverschluß *m* vacuum seal
Vakuumvorlage *f* vacuum receiver
Vakuumvorstoß *m* receiver adapter with vacuum connection
Vakuumwalzentrockner *m* vacuum drum dryer
Vakuumzelle *f* vacuum cell
Vakzination *f* vaccination
Vakzine *f* vaccine
Val *n* (Grammäquivalent) gram-equivalent
Valencianit *m* (Min) valencianite
Valentinit *m* (Min) valentinite, white antimony
Valenz *f* (Chem) valence, valency, **gesättigte** ~ saturated valency, **koordinative** ~ coordinative valency, **lokalisierte** ~ localized valency, **nicht abgesättigte** ~ unsaturated valency, **nichtlokalisierte** ~ delocalized valency
Valenzband *n* valancy band, valence band
Valenzbegriff *m* concept of valence
Valenzbindung *f* valence bond
Valenzbindungsmethode *f* valence bond method
Valenzeinheit *f* valence [unit]

Valenzelektron *n* valency electron
valenzgesteuert valency controlled
Valenzkraftfeld *n* valence force field
Valenzorbital *n* valence orbital
Valenzrichtung *f* directed valence
Valenzschale *f* (Chem) valence shell
Valenzschwingung *f* stretching frequency, stretching vibration
Valenzstrich *m* (Chem) valence dash
Valenzstrichformel *f* (Chem) valence-dash formula
Valenzstufe *f* valence stage
Valenzwinkel *m* valence angle
Valeraldazin *n* valeraldazine
Valeraldehyd *m* valeraldehyde, pentanal, valeral
Valeraldoxim *n* valeraldoxime
Valeramid *n* valeramide
Valeramidin *n* valeramidine
Valeranilid *n* valeranilide
Valeren *n* amylene, valerene
Valerianat *n* valerate, salt or ester of valeric acid, valerianate
Valerianöl *n* valerian oil
Valeriansäure *f* valeric acid, pentanoic acid, valerianic acid
Valeriansäureäthylester *m* ethyl valerate
Valeriansäureanhydrid *n* valeric anhydride
Valeriansäurementholester *m* menthol valerate, validol
Valeriansäuresalz *n* valerate, salt of valeric acid, valerianate
Valeridin *n* valeridin, sedatin, valerylphenetidine
Valeroidin *n* valeroidine
Valerol *n* valerol
Valerolakton *n* valerolactone
Valeron *n* valerone, diisobutyl ketone
Valeronitril *n* valeronitrile, butyl cyanide
Valerophenon *n* valerophenone
Valerydin *n* valeridin, sedatin, valerydin, valerylphenetidine
Valeryl- valeryl, pentanoyl
Valerylbromid *n* valeryl bromide, pentanoyl bromide
Valerylchlorid *n* valeryl chloride, pentanoyl chloride
Valerylen *n* valerylene, 2-pentyne, methylethylacetylene
Valerylnitril *n* valeryl nitrile
Validol *n* validol, menthol valerate
Valin *n* valine, 2-amino-3-methylbutanoic acid, aminoisovaleric acid
Valinol *n* valinol
Valinomycin *n* valinomycin
Valleit *m* valleite
Vallericit *m* (Min) vallericite
Vallesin *n* vallesine
Valoneasäure *f* valone[a]ic acid
Valtropin *n* valtropine

Valyl *n* valyl, diethylvaleramide
Vanadat *n* vanadate, salt or ester of vanadic acid
Vanadin *n* vanadium (Symb. V), ~ **enthaltend** containing vanadium, vanadiferous
Vanadinat *n* vanadate, salt or ester of vanadic acid
Vanadinbleierz *n* (Min) vanadinite
Vanadinbleispat *m* (Min) vanadinite
Vanadingehalt *m* vanadium content
vanadinhaltig containing vanadium, vanadiferous
vanadinig vanadous
Vanadinit *m* (Min) vanadinite
Vanadinsäure *f* vanadic acid
Vanadinsäureanhydrid *n* vanadic anhydride, vanadium pentoxide
vanadinsauer vanadic, (Salz) vanadate
Vanadinsesquioxid *n* vanadium sesquioxide, vanadic oxide
Vanadinstahl *m* vanadium steel
Vanadinstickstoff *m* vanadium nitride
Vanadisalz *n* (Vanadium-(III)-Salz) vanadium(III) salt
Vanadium *n* vanadium (Symb. V)
Vanadiumchlorid *n* vanadium chloride, (Vanadium-(III)-chlorid) vanadic chloride, vanadium(III) chloride, vanadium trichloride
Vanadiumchlorür *n* (Vanadium-(II)-chlorid) vanadium dichloride, vanadium(II) chloride, vanadous chloride
Vanadiumdichlorid *n* vanadium dichloride, vanadium(II) chloride, vanadous chloride
Vanadiumeisen *n* ferrovanadium
vanadiumhaltig vanadiferous
Vanadium(III)-Verbindung *f* vanadium(III) compound, vanadous compound
Vanadium(II)-Verbindung *f* vanadium(II) compound, vanadous compound
Vanadiumsäure *f* vanadic acid
Vanadiumstahl *m* vanadium steel
Vanadium(V)-Verbindung *f* vanadic compound, vanadium(V) compound
vanadometrisch vanadometric
Vanadosalz *n* (Vanadium-(II)-Salz) vanadium(II) salt, vanadous salt
Vanadosulfat *n* (Vanadium-(II)-sulfat) vanadium(II) sulfate, vanadous sulfate
Vanadylsalz *n* vanadyl salt
Vanadylsulfat *n* vanadyl sulfate
Vanderosid *n* vanderoside
Van der Waals-Radius *m* van der Waals radius
Van der Waalssche Kräfte *pl* van der Waals forces
Van der Waalssche Wechselwirkungen *pl* Van der Waals interactions
Vanille *f* (Bot) vanilla
Vanillearoma *n* vanilla flavor
Vanilleersatz *m* vanilla substitute
Vanillenschote *f* vanilla pod

Vanillin *n* vanillin, 3-methoxy-4-hydroxy benzaldehyde
Vanillinsäure *f* vanillic acid, 3-methoxy-4-hydroxybenzoic acid
Vanillinzucker *m* vanillin sugar
Vanillosid *n* vanilloside, glucovanilline
Vanillylalkohol *m* vanillyl alcohol
Vanthoffit *m* (Min) vanthoffite
Vanyldisulfamid *n* vanyldisulfamide
Vaporimeter *n* vaporimeter
Varek *n* varec, kelp
Vareksoda *f* kelp soda
variabel variable
Variabilität *f* variability
Variable *f* variable
Variante *f* variety, variant
Varianzanalyse *f* analysis of variance
Variation *f* (Biol) variation
Variationsableitung *f* variation derivative
Variationsprinzip *n* variation[al] principle
Variationsrechnung *f* calculus of variations
Variationsverfahren *n* variation principle
Varietät *f* variety
variieren to alter, to vary
Variolarsäure *f* variolaric acid
Variolit *m* (Min) variolite
Variscit *m* (Min) variscite, peganite
Varvicit *m* varvicite
Vasculose *f* impure lignin, vasculose
Vaselin *n* vaseline [oil], paraffin oil, petrolatum, petroleum jelly, vaseline
Vaseline *f* petrolatum, petroleum jelly
Vaselin[e]öl *n* vaseline oil
Vasenol *n* vasenol
Vashegyit *m* (Min) vashegyite
Vasicin *n* vasicine
Vasicinon *n* vasicinone
Vaskulose *f* vasculose
Vasoliment *n* vasoliment
Vasopressin *n* vasopressin
Vasotonin *n* vasotonine
Vauquelinit *m* (Min) vauquelinite
Veatchin *n* veatchine
vegetabilisch vegetable
Vegetalin *n* vegetaline
Vegetation *f* vegetation
Vegetationsgürtel *m* zone of vegetation
Vegetationsperiode *f* vegetation period
Vegetationszeit *f* vegetative period
vegetativ vegetative
Veilchen *n* (Bot) violet
veilchenblau violet blue
Vektor *m* vector
Vektoraddition *f* vector addition
Vektoranalyse *f* vector analysis
Vektordiagramm *n* vector diagram
Vektorensatz *m* vector set
Vektorfeld *n* vector[ial] field
Vektorfluß *m* flux of vector

Vektorgleichung *f* vector equation
Vektorgröße *f* vector quantity
vektoriell vectorial
Vektormodell *n* (des Atoms) vector model (of the atom)
Vektorraum *m* vector space
Vellein *n* vellein
Venasquit *m* venasquite
Vene *f* (Med) vein, vena
Venetianerseife *f* Venetian soap
Ventil *n* valve, air valve, vent, **doppelseitiges** ~ double-seat valve, **mechanisch gesteuertes** ~ mechanically operated valve
Ventilantrieb *m* valve actuator
Ventilation *f* ventilation
Ventilator *m* ventilator, [suction] fan
Ventilatorkühlturm *m* forced-circulation cooling tower
Ventilauslaß *m* valve outlet
Ventilblock *m* valve block
Ventilboden *m* (Rektifikation) valve tray
Ventildeckel *m* valve box cover
Ventileinsatz *m* valve core
Ventileinschleifen *n* valve grinding
Ventileinstellung *f* valve adjustment
Ventilentlastung *f* valve relief
Ventilfeder *f* valve spring
Ventilfederberechnung *f* valve spring computation
Ventilfett *n* valve seal
Ventilführung *f* valve guide
Ventilführungsrohr *n* valve liner
Ventilgehäuse *n* valve box, valve chamber, valve housing
Ventilgehäusedeckel *m* valve box cover
Ventilhaube *f* valve bonnet
Ventilhebel *m* valve lever
Ventilhub *m* valve lift
ventilieren to ventilate
Ventilierschraubung *f* valve joint
Ventilkäfig *m* valve cage
Ventilkegel *m* valve cone, valve disk
Ventilklappe *f* valve flap
Ventilklemmstück *n* valve washer lock
Ventilkörper *m* valve housing
Ventilkolben *m* piston valve
Ventilleder *n* valve leather
Ventiloberteil *n* valve bonnet
Ventilölkanne *f* valve oil can
Ventilsitz *m* valve seat
Ventilspiel *n* valve clearance, valve travel
Ventilspindel *f* valve stem
Ventilstange *f* sleeve rod, stopper rod, valve stem
Ventilsteuerung *f* valve regulation, valve gear
Ventilstoß *m* valve lift
Ventilteller *m* valve disc
Ventilverkleidung *f* valve gear casing
Ventilverschluß *m* valve seal

Ventilverschraubung — Verbindung

Ventilverschraubung *f* valve cap
Ventilzelle *f* valve cell
Venturibrenner *m* venturi burner
Venturidüse *f* venturi tube
Venturirohr *n* venturi tube
Venturiwäscher *m* venturi scrubber
Venturi-Wirbelschicht *f* venturi fluidized bed
Veracevin *n* veracevine
veränderlich variable
Veränderliche *f* (Math) variable, **abhängige** ~ (Math) dependent variable
verändern to alter, to change, to modify, to vary
Veränderung *f* change, alteration, variation, **bleibende** ~ permanent set, **eine** ~ **durchmachen** to undergo a change, **permanente** ~ permanent set
Veränderungsmöglichkeit *f* changeability
veräthern to etherify, to convert into an ether
verallgemeinern to generalize
Verallgemeinerung *f* generalization
verallgemeinerungsfähig generalizable
Verankerung *f* anchorage, ~ **des Anstrichs** adhesion of the coating
Veranschaulichung *f* illustration
Veranthridin *n* veranthridine
Verarbeitbarkeit *f* processibility, workability
Verarbeiter *m* fabricator
Verarbeitung *f* processing, fabricating, machining, treatment, ~ **von Plastisol** plastisol molding, **mechanische** ~ mechanical manipulation
Verarbeitungsanlage *f* processing plant
Verarbeitungshinweis *m* processing direction
Verarbeitungsmethode *f* method of processing
Verarbeitungsrichtlinie *f* processing instruction
Verarbeitungsstufe *f* stage of manufacture
Verarbeitungstemperatur *f* processing temperature
Verarbeitungstoleranz *f* machining allowance
verarmen to become poor, to exhaust, to impoverish
Verarmung *f* exhaustion, impoverishment, reduction of strength
veraschen to ash, to incinerate
Veraschen *n* incinerating
Verascher *m* incinerator
Veraschung *f* incineration
Veraschungsschale *f* (Chem) incineration dish
Veratral *n* veratral
Veratraldehyd *m* veratraldehyde, 3,4-dimethoxy benzaldehyde, 3,4-dimethoxybenzenecarbonal
Veratramin *n* veratramine
Veratril *n* veratril
Veratrin *n* veratrine, cevadine
Veratrinharz *n* veratrine resin
Veratrinsäure *f* veratric acid, 3,4-dimethoxybenzoic acid
Veratrinsalz *n* veratrate, salt of veratric acid
Veratrinsulfat *n* veratrine sulfate

Veratroidin *n* veratroidine
Veratrol *n* 1,2-dimethoxybenzene, veratrole
Veratrosin *n* veratrosine
Veratroylchlorid *n* veratroyl chloride
Veratrumalkaloid *n* veratrum alkaloid
Veratrylalkohol *m* veratryl alcohol
Veratrylchlorid *n* veratryl chloride
Verband *m* (Med) bandage, (Verbindung) union
Verbandkasten *m* first-aid box
Verbandsnorm *f* industry standard
Verbandstoff *m* (Med) bandaging material
Verbandstoffsterilisator *m* bandage sterilizer
Verbandwatte *f* (Pharm) absorbent cotton
Verbandzeug *n* (Med) bandaging material
Verbanol *n* verbanol
Verbanon *n* verbanone
Verbascose *f* verbascose
Verbazid *n* verbazide
Verben *n* verbene, verbena, vervain
Verbenalin *n* verbenalin, verbenaloside
Verbenalinsäure *f* verbenalinic acid
Verbenalosid *n* verbenalin, verbenaloside
Verbenaöl *n* verbena oil, vervain oil
Verbenol *n* verbenol
verbessern to correct, to improve
verbessert corrected, improved
Verbesserung *f* correction, improvement, upgrading
verbiegbar deformable
Verbiegung *f* deformation, buckling, warping
Verbindbarkeit *f* joinability
verbinden to attach, to bind [together], to connect; to couple, to join; to link, (vereinigen) to combine; to unite
Verbinden *n* combining, binding, connecting, coupling, joining, linking, uniting
Verbinderpresse *f* vulcanizing press
Verbindung *f* combination, joint, link, union, (Chem) compound, (chem. Vereinigung) binding, combining, (Verbinden) coupling, joining, linking, (Zusammenhang) connection, **alicyclische** ~ alicyclic compound, **aliphatische** ~ aliphatic compound, **amphotere** ~ amphoteric [compound], **asymmetrische** ~ asymmetric compound, **binäre** ~ binary compound, **cyclische** ~ cyclic compound, **exotherme** ~ exothermic compound, **heterocyclische** ~ heterocyclic compound, **in der Natur vorkommende** ~ naturally occurring compound, **intermetallische** ~ intermetallic compound, **isomere** ~ isomeric compound, **konjugierte** ~ conjugated compound, **mehrkernige** ~ polycyclic compound, **mesomere** ~ resonance compound, **metastabile** ~ metastable compound, **organische** ~ organic compound, **physikalische** ~ physical compound, **polare** ~ polar compound, **rechtsdrehende** ~ dextro-compound, **ungesättigte** ~ unsaturated

Verbindung — Verbrennungskapillare

compound, **unlösbare** ~ permanent joint,
unlösliche ~ (Chem) insoluble compound,
wasserdichte ~ watertight joint
Verbindungsblech *n* connecting plate, connecting strip
Verbindungsbolzen *m* tie bolt
Verbindungsbrücke *f* connecting bridge
Verbindungselement *n* joining element
Verbindungsfähigkeit *f* (Chem) affinity, combining ability
Verbindungsgefäß *n* (Med) intercommunicating vessel
Verbindungsgestell *n* connecting rack
Verbindungsgewicht *n* combining weight
Verbindungsgleichung *f* equation of combination
Verbindungsglied *n* link
Verbindungshahn *m* connecting cock, connecting tap
Verbindungskabel *n* connecting cable
Verbindungskanal *m* communicating gallery
Verbindungskitt *m* cement, sealing compound
Verbindungsklemme *f* connecting terminal, connector
Verbindungsleitung *f* connecting main, lead wire, transfer circuit, ~ **zum Vakuum** vacuum connection
Verbindungslinie *f* connecting line, tie line (in phase diagram)
Verbindungsmuffe *f* (Elektr) connecting sleeve, joint box, jointing sleeve, junction box, socket
Verbindungsnaht *f* joint
Verbindungsprozeß *m* linking process
Verbindungsröhre *f* connecting tube
Verbindungsschiene *f* connecting busbar
Verbindungsschnur *f* (Elektr) connecting cord
Verbindungsschweißen *n* weld joining
Verbindungsstange *f* connecting rod
Verbindungsstelle *f* joint, junction (point)
Verbindungsstück *n* connecting piece, connection, joint
Verbindungsstufe *f* combination stage
Verbindungsverhältnis *n* (Chem) combining proportion
Verbindungsvolumen *n* combining volume
Verbindungswärme *f* heat of combination
Verblasen *n* blowing, (Schmelze) air blasting, ~ **in flüssigem Zustand** bessemerizing, converting
Verblaserösten *n* blast roasting
verblassen to fade, to lose color
verbleiben to remain, to continue
verbleichen to fade
verbleien to cover with lead
Verbleien *n* lead coating
verbleit lead-coated
Verbleiung *f* lead coating, lead lining, lead plating
Verbleiungsanlage *f* lead coating plant
Verblockung *f* interlock

verborgen hidden, cryptic
verbrannt burned, burnt
Verbrauch *m* consumption
verbrauchen to consume, to spend, to use up, to work up
Verbraucher *m* consumer
Verbrauchsdichte *f* consumption density
Verbrauchsgebiet *n* area of consumption, region of consumption
Verbrauchsleitung *f* (Elektr) service main
Verbrauchsphase *f* load phase
Verbrauchsstelle *f* place of consumption, place of use
verbreiten to distribute, to diffuse, to spread, (ausstreuen) to disseminate, (fortpflanzen) to propagate
Verbreiterung *f* widening, broadening, enlarging
Verbreitung *f* distribution, spread, (Fortpflanzung) propagation
verbrennbar combustible, burnable
Verbrennbarkeit *f* combustibility
verbrennen to burn, (explosionsartig) to deflagrate, (verbrühen) to scald, (versengen) to scorch
Verbrennen *n* burning
Verbrennung *f* (Chem) combustion, ~ **ersten Grades** first degree burn, **beschleunigte** ~ accelerated combustion, **lebhafte** ~ rapid combustion, **plötzliche** ~ burning off, **rauchschwache** ~ combustion producing little smoke, **schleichende** ~ slow combustion, **träge** ~ slow combustion, **unvollkommene** ~ imperfect or incomplete combustion, **unvollständige** ~ imperfect combustion, incomplete combustion, **verzögerte** ~ retarded combustion, **vollständige** ~ complete combustion
Verbrennungsablauf *m* course of combustion
Verbrennungsanalyse *f* analysis by combustion, combustion analysis
Verbrennungsanlage *f* combustion plant
Verbrennungsdruck *m* combustion pressure
Verbrennungsdüse *f* combustion nozzle, combustion tuyère
Verbrennungsenergie *f* combustion energy
Verbrennungsenthalpie *f* combustion enthalpy
Verbrennungsergebnis *n* result of combustion
Verbrennungserscheinung *f* phenomenon of combustion
Verbrennungserzeugnis *n* product of combustion
Verbrennungsgas *n* combustible gas, flue gas, gas of combustion
Verbrennungsgeschwindigkeit *f* combustion rate, combustion velocity, rate of combustion
Verbrennungsintensität *f* intensity of combustion
Verbrennungskammer *f* combustion chamber
Verbrennungskapillare *f* capillary combustion tube

Verbrennungskraftmaschine *f* internal combustion engine
Verbrennungsluft *f* air for combustion
Verbrennungsmarkierung *f* burnt spot
Verbrennungsmaschine *f* internal combustion engine
Verbrennungsmethode *f* combustion method
Verbrennungsmotor *m* internal combustion engine
Verbrennungsofen *m* combustion furnace
Verbrennungspipette *f* combustion pipet[te]
Verbrennungsprodukt *n* combustion product
Verbrennungsprozeß *m* combustion process
Verbrennungsraum *m* combustion chamber, combustion space, fire chamber
Verbrennungsresultat *n* result of combustion
Verbrennungsröhre *f* combustion tube
Verbrennungsrohr *n* combustion tube
Verbrennungsrückstand *m* combustion residue
Verbrennungsschacht *m* combustion shaft
Verbrennungsschälchen *n* combustion boat
Verbrennungsschiffchen *n* combustion boat
Verbrennungsspannung *f* combustion pressure
Verbrennungsturbine *f* internal combustion turbine
Verbrennungswärme *f* heat of combustion
Verbrennungszone *f* combustion zone, oxidation zone
Verbrennungszustand *m* state of combustion
verbrühen to scald
Verbundbauweise *f* sandwich construction
Verbunddampfmaschine *f* compound steam engine
Verbunddampfpumpe *f* compound steam pump
Verbunderregung *f* compound excitation
Verbundfolie *f* composite film
Verbundfolienbeutel *m* laminated bag
Verbundglas *n* compound glass, laminated glass, safety glass
Verbundguß *m* (Gieß) compound casting
Verbundgußschiene *f* compound cast rail
Verbundkessel *m* combination boiler
Verbundkonstruktion *f* composite structure
Verbundlinse *f* (Opt) compound lens
Verbundmaschine *f* compound engine, compound dynamo
Verbundmühle *f* combination mill
Verbundplatte *f* sandwich panel
Verbundpreßstoff *m* combined plastic
Verbundrohr *n* compound pipe
Verbundstoff *m* combined material, composite material
Verbundwerkstoff *m* composite material, laminate
Verbutterung *f* **des Quecksilbers** sliming of mercury
verchromen to chrome-plate, to chromium-plate
verchromt chrome-plated, chromium-plated
Verchromung *f* chromium-plating

Verdachtsmoment *n* suspicious fact, suspicious clue
verdampfbar vaporizable, (flüchtig) volatile
Verdampfbarkeit *f* vaporizability, (Flüchtigkeit) volatility
verdampfen to vaporize, (verdunsten) to evaporate
Verdampfen *n* vaporizing, (Verdunsten) evaporating
Verdampfer *m* evaporator, vaporizer, **Dünnschicht-** ~ thin-film evaporator, **Durchlauf-** ~ continuous evaporator, **Fallfilm-** ~ failing-film evaporator, **Fallstrom-** ~ failing-film evaporator, **Horizontalrohr-** ~ horizontal-tube evaporator, **Kletterfilm-** ~ climbing-film evaporator, **Kurzweg-** ~ molecular evaporator, **Lammellen-** ~ lammella evaporator, **Langrohr-** ~ long-tube evaporator, **Platten-** ~ plate evaporator, **Röhren-** ~ tube evaporator, **Rota-Film-** ~ rotary film evaporator, **Rotations-** ~ rotary evaporator, **Schnecken-** ~ screw-flights evaporator, **Schrägrohr-** ~ inclined-tube evaporator, **Steilrohr-** ~ short-coil evaporator, **Tauchrohr-** ~ submerged-tube evaporator, **überfluteter** ~ flooded evaporator, **Umlauf-** ~ circulation evaporator, **Wendelrohr-** ~ coil evaporator
Verdampferrohr *n* evaporator tube
Verdampferrückstand *m* evaporator residue
Verdampferschlange *f* evaporator coil
Verdampferspeisepumpe *f* evaporator feed pump
Verdampfersystem *n* evaporator assembly
Verdampfpfanne *f* evaporation pan
Verdampfschale *f* evaporating dish, evaporating basin
Verdampfung *f* evaporation, vaporization, volatilization, ~ **durch Entspannung** flash evaporation, **stoßartige** ~ flash evaporation
Verdampfungsanlage *f* evaporating plant
Verdampfungsapparat *m* evaporator, vaporizer
Verdampfungsenthalpie *f* evaporation enthalpy
Verdampfungsentropie *f* (Phys) entropy of vaporization
Verdampfungsfähigkeit *f* evaporative capacity
Verdampfungsgeschwindigkeit *f* rate of evaporation, rate of vaporization
Verdampfungsgleichgewicht *n* equilibrium of evaporation
Verdampfungskoeffizient *m* coefficient of evaporation
Verdampfungskühlung *f* cooling by evaporation, evaporation cooling, evaporative cooling
Verdampfungskurve *f* evaporation curve
Verdampfungsleistung *f* vaporization capacity
Verdampfungsmaschine *f* evaporating machine
Verdampfungsmesser *m* evaporimeter
Verdampfungsoberfläche *f* evaporation surface

Verdampfungsofen *m* evaporator furnace
Verdampfungsprüfer *m* volatility test apparatus
Verdampfungspunkt *m* evaporation point, steam point
Verdampfungsreaktor *m* boiling water reactor
Verdampfungsregler *m* evaporation regulator
Verdampfungsröstung *f* volatilization roasting
Verdampfungsrückstand *m* evaporation residue
Verdampfungstemperatur *f* evaporation temperature, volatilization temperature
Verdampfungsverlust *m* loss by evaporation
Verdampfungsvermögen *n* evaporating capacity
Verdampfungswärme *f* heat of evaporation, heat of vaporization
Verdampfungswert *m* evaporative value
Verdampfungszahl *f* coefficient of evaporation
Verdampfungszeit *f* time of evaporation
Verdampfungsziffer *f* coefficient of evaporation
Verdampfungszylinder *m* expansion cylinder
verdauen to digest
verdaulich digestible
Verdaulichkeit *f* digestibility
Verdauung *f* digestion
Verdauungsenzym *n* digestive enzyme
Verdauungsorgan *n* digestive organ
Verdauungsprozeß *m* digestion, digestive process
Verdauungssystem *n* digestive system
Verdauungsvorgang *m* digestion process
verdecken to cover
Verdeckstoff *m* covering material, roofing material
verdeckt hidden, concealed
Verderb *m* decay, deterioration, ruin, spoilage
verderben to destroy, to get rotten, to spoil
verderblich perishable
verdichtbar (kondensierbar) condensable, (zusammendrückbar) compressible
Verdichtbarkeit *f* compressibility, (Kondensierbarkeit) condensability
verdichten to compact, (absiegeln) to seal, (kondensieren) to condense, (konzentrieren) to concentrate, (zusammendrücken) to compress
Verdichten *n* compressing, concentrating, condensing
Verdichter *m* compressor;, condenser, **Axial-** ~ axial-flow compressor, **Boxer-Kolben-** ~ double-piston compressor, **Drehkolben-** ~ rotary compressor, **Hubkolben-** ~ reciprocating compressor, **Kolben-** ~ reciprocating compressor, **Membran-** ~ diaphragm compressor, **Schrauben-** ~ screw compressor, **Vielzellen-** ~ sliding-vane rotary compressor
Verdichtergehäuse *n* compressor casing
verdichtet densified
Verdichtung *f* (Kondensation) condensation, (Konzentration) concentration, (Zusammendrückung) compression,
adiabatische ~ adiabatic compression,
isothermische ~ isothermal compression,
mehrstufige ~ multistage compression
Verdichtungsarbeit *f* compression work
Verdichtungsdruck *m* compression pressure
Verdichtungsfaktor *m* compressibility factor
Verdichtungsgrad *m* compression ratio, bulk factor
Verdichtungshub *m* compression stroke
Verdichtungspolarisation *f* concentration polarization
Verdichtungsstoßleuchten *n* shock wave luminescence
Verdichtungsverhältnis *n* compression ratio
Verdichtungswärme *f* heat of compression
Verdichtungswelle *f* compression wave
Verdichtungszone *f* compression zone
Verdichtungszündung *f* compression ignition
verdienen to earn
Verdienstspanne *f* margin of profit
verdoppeln, sich ~ to double, to duplicate
verdrängen to displace, to drive out, to expel, to supplant
Verdrängerpumpe *f* positive displacement pump
Verdrängung *f* displacement, (Entfernung) removal
Verdrängungschromatographie *f* displacement chromatography
Verdrängungskolben *m* displacer piston
Verdrängungsreaktion *f* substitution reaction
Verdrängungsstrom *m* displacement current
Verdrahtung *f* wiring
verdrehen to twist
Verdrehfestigkeit *f* torsional strength, torque resistance
Verdrehgrenze *f* yield point in torsion
Verdrehung *f* distortion, torque, torsion, torsional strain, twist
Verdrehungsbruch *m* torsion failure
Verdrehungselastizität *f* torsional elasticity
Verdrehungsfähigkeit *f* torsibility
Verdrehungsgrenze *f* yield point of torsional shear
Verdrehungskraft *f* torsional force
Verdrehungsmesser *m* torsiometer, torsion meter
Verdrehungsmodul *n* modulus of torsion, modulus of torsional shear
Verdrehungsmoment *n* torque, torsional moment, twisting moment
Verdrehungsprobe *f* torsion test
Verdrehungswinkel *m* angle of twist, torsional angle
verdünnbar dilutable, (Gas) rarefiable
Verdünnbarkeit *f* dilutability, (Gas) rarefiability
verdünnen to thin, (Gas) to rarefy, (Lösung) to dilute, (schwächen) to weaken
Verdünnen *n* (Gas) rarefying, (Lösung) diluting, thinning
verdünnt dilute, diluted, weak

Verdünnung – Verfeinerung

Verdünnung *f* dilution, thinning, (Gas) rarefaction
Verdünnungsanalyse *f* dilution analysis
Verdünnungsgesetz *n* dilution law
Verdünnungsgrad *m* degree of dilution
Verdünnungsmittel *n* diluent, diluting agent, extender, thinner
Verdünnungswärme *f* dilution heat, heat of dilution
verdunkeln to darken, to black out
Verdunkeln *n* darkening
Verdunk[e]lung *f* blackout, (Astr) eclipse
Verdunkelungsanstrich *m* blackout paint
Verdunkelungsvorhang *m* blackout curtain
verdunstbar evaporable, vaporizable
Verdunstbarkeit *f* evaporableness
verdunsten to evaporate, to vaporize
Verdunsten *n* evaporating
Verdunstung *f* evaporation, vaporization, volatilization
Verdunstungsapparat *m* evaporating apparatus
Verdunstungsdruck *m* vapor pressure
Verdunstungsfläche *f* evaporation surface
Verdunstungskälte *f* cold due to vaporization
Verdunstungskarburator *m* surface carburet[t]or
Verdunstungskondensator *m* evaporation condenser
Verdunstungskühlung *f* evaporative cooling
Verdunstungsmesser *m* atmometer
Verdunstungsverflüssiger *m* evaporative condenser
Verdunstungsverlust *m* loss by evaporation
Verdunstungswärme *f* heat of vaporization
Verdunstungszahl *f* evaporation number
Verdunstungszeit *f* evaporation time or rate
veredeln to improve, to refine, to upgrade
Veredeln *n* improving, age-hardening, enriching, refining, (Text) finishing
Vered[e]lung *f* improvement, enrichment, refinement, (Text) finishing, **Oberflächen-** ~ surface treatment
Veredelungsgerbstoff *m* auxiliary tannin for improving the tannage
Vered[e]lungsindustrie *f* (Text) finishing industry
Veredelungsstoff *m* additive
Veredelungsverfahren *n* refining process
Vered[e]lungsvorgang *m* refining process
Vereinbarkeit *f* compatibility
Vereinbarung *f* arrangement
vereinfachen to simplify
Vereinfachung *f* simplification
vereinheitlichen to standardize, to uniformalize
Vereinheitlichung *f* standardization, uniformalization
vereinigen to combine, to unite, (sammeln) to collect, to gather
Vereinigung *f* association, combination, union
vereinzelt isolated, solitary, sporadic
vereisen to cover with ice, to ice, to turn into ice

Vereisung *f* (Geol) glaciation, ice formation
vereisungsfrei ice-free, nonicing
Vereisungsschutzflüssigkeit *f* anti-icing fluid
verengen (sich zusammenziehen) to contract, **sich** ~ to narrow
Verengerungs-T-Stück *n* flanged reducing tee, unequal tee
Verengung *f* narrowing
vererben (Biol) to inherit
Vererbung *f* inheritance, (Biol) heredity
Vererbungsbiologie *f* genetics
Vererbungsgesetz *n* (Mendelsches Gesetz) Mendelian law, Mendel's law
Vererbungslehre *f* genetics
vererzbar mineralizable
vererzen to mineralize
Vererzung *f* mineralization
Vererzungsmittel *n* mineralizer
veresterbar esterifiable
verestern to esterify
Veresterung *f* esterification
Veresterungsanlage *f* esterification plant
verfälschen to adulterate
Verfälschung *f* adulteration
verfärben to change color, to decolorize
Verfärben *n* discoloration, fading
verfärbend discoloring
Verfärbung *f* discoloration
Verfahren *n* (Handlung) proceeding, (Methode) method, technique, (Techn, Chem) mode, operation, procedure, process, treatment, ~ **einstellen** to discontinue proceedings, **indirektes** ~ indirect method, **irreversibles** ~ irreversible process, **metallurgisches** ~ metallurgical process, **nasses** ~ wet process, **subtraktives** ~ (Opt) subtractive color process, **synthetisches** ~ synthetic process
Verfahrensanalyse *f* operational analysis
Verfahrensfehler *m* error of approximation
Verfahrensplanung *f* process design
Verfahrensprinzip *n* principle of process
Verfahrensregelung *f* process control
Verfahrensschema *n* [process] flow sheet
Verfahrenstechnik *f* chemical [and process] engineering, engineering chemistry, process technology
Verfahrenstechniker *m* chemical engineer
Verfahrensuntersuchung *f* operation research
Verfahrensweise *f* method, technique
Verfall *m* breakdown, deterioration, disintegration
Verfasser *m* author, writer
Verfassung *f* condition
verfaulbar putrescible
Verfaulbarkeit *f* putrescibility
verfaulen to decay, to putrefy, to rot
verfeinern to improve, to refine, to upgrade
Verfeinerung *f* improvement, refinement, upgrading

verfestigen to consolidate, to fasten, to make firm, to strengthen, (fest werden) to solidify, (versteifen) to stiffen
Verfestigung *f* consolidation, solidification, work hardening, ~ **durch Cottrell-Effekt** (Krist) Cottrell hardening, ~ **durch Recken** strain hardening
Verfestigungsgrad *m* degree of consolidation
Verfestigungskurve *f* work hardening curve
verfilzen to felt
Verfilzen *n* felting
verfilzt felted
verflachen to level off
verflanscht flanged
Verflanschung *f* well head
verflüchtigen, sich ~ to evaporate, to volatilize
Verflüchtigen *n* volatilization
Verflüchtigung *f* evaporation, volatilization, (fester Stoffe) sublimation
verflüssigen to liquefy, to fluidize
Verflüssigen *n* liquefying
Verflüssiger *m* condenser
Verflüssigung *f* liquefaction, liquefying, thinning, (Kondensation) condensation
Verflüssigungsanlage *f* liquefying plant
Verflüssigungsapparat *m* liquefier
Verflüssigungsleistung *f* plasticizing capacity
Verflüssigungsmittel *n* liquefying agent, thinning agent
Verflüssigungspunkt *m* melting point
verformbar deformable, moldable, plastic, workable, ~ **bleiben** to remain permanently plastic, **warm** ~ thermoplastic
Verformbarkeit *f* deformability, plasticity, (Met) malleability
verformen to deform, to shape
Verformung, bleibende ~ permanent set
Verformung *f* deformation, distortion, shaping, **plastische** ~ plastic deformation, plastic working, **spanabhebende** ~ cutting shaping, **spanlose** ~ noncutting shaping, **ungleichmäßige** ~ (Mech) non-uniform strain
Verformungsarbeit *f* work of deformation
Verformungsbruch *m* ductile fracture
Verformungsentfestigung *f* work softening
Verformungsgeschwindigkeit *f* strain rate
Verformungsmechanismus *m* mechanism of deformation
Verformungsvorgang *m* forming process
Verformungsweg *m* strain path
Verformungswerkzeug *n* forming tool
Verformungswiderstand *m* resistance to deformation, tensile strength
Verfrachtung *f* conveying, transport
verfrüht premature
verfügbar available
vergällen to denature, to embitter
Vergällen *n* denaturing
Vergällung *f* denaturation, denaturing

Vergällungsmittel *n* denaturant
vergärbar fermentable
Vergärbarkeit *f* (Brau) fermentability
vergären to ferment
Vergären *n* fermenting
Vergärung *f* fermentation
Vergärungsgrad *m* degree of fermentation
vergasbar gasifiable, vaporizable
Vergasbarkeit *f* gasifiability, vaporizability
vergasen to gasify, to vaporize
Vergasen *n* gasification, gasifying, gassing, vaporizing
Vergaser *m* gasifier, vaporizer, (Auto) carburet[t]or
Vergaserdüse *f* carburetor nozzle
Vergasergraphit *m* retort graphite
Vergaserkraftstoff *m* carburet[t]or fuel
Vergasermotor *m* carburetor engine
Vergasernadel *f* carburetor float spindle
Vergaserschwimmer *m* carburetor float
Vergaserventil *n* float valve
Vergasung *f* gasification, carburetion, destructive distillation, gassing, vaporization
Vergasungsanlage *f* gasification plant
Vergasungswärme *f* gasifying heat
vergießbar ready to cast
Vergießbarkeit *f* castability
vergießen to pour out, (verschütten) to shed, to spill
Vergießtemperatur *f* casting temperature, pouring temperature
vergiften to intoxicate, to poison, (verunreinigen) to contaminate
Vergiften *n* poisoning, intoxicating, (Verunreinigen) contaminating
Vergiftung *f* intoxication, poisoning, (Verunreinigung) contamination, ~ **des Katalysators** poisoning of the catalyst, ~ **durch Chloral** chloralism, **Fleisch-** ~ meat poisoning, **Gas-** ~ gas poisoning, **Nahrungsmittel-** ~ food poisoning, **Quecksilber-** ~ mercurialism, mercury poisoning
Vergiftungserscheinung *f* symptom of intoxication, symptom of poisoning
vergilben to turn yellow, to yellow
Vergilben *n* yellowing
Vergilbung *f* yellowing
vergipsen to plaster
verglasen to glaze, to vitrify
Verglasung *f* glazing
Vergleich *m* comparison, **logischer** ~ logical comparison, **zahlenmäßiger** ~ numerical comparison
vergleichbar comparable, commensurable
Vergleichbarkeit *f* comparability
vergleichen to compare
Vergleichselektrode *f* comparison electrode, reference electrode

Vergleichsfärbung f (Färb) comparative dyeing, comparison dyeing
Vergleichsflüssigkeit f reference fluid, standard liquid
Vergleichsfunktion f comparison function
Vergleichsgas n reference gas
Vergleichsgrundlage f basis of comparison
Vergleichskörper m reference body
Vergleichslehre f (Techn) master gauge, reference gauge
Vergleichslichtquelle f (Opt) comparison lamp
Vergleichslinie f reference line, comparison line
Vergleichslösung f standard solution, comparison solution
Vergleichsmaßstab m standard of comparison, standard of reference
Vergleichsmessung f comparison measuring
Vergleichsmethode f comparison method
Vergleichspegel m reference level
Vergleichsprobe f control sample or test, reference sample
Vergleichsschaltung f (Elektr) comparator [circuit]
Vergleichsspektrallinie f (Spektr) internal standard line
Vergleichsspektrum n (Spektr) comparison spectrum
Vergleichsstrahler m standard radiator
Vergleichsstrecke f comparison section
Vergleichsunterlage f basis of comparison
Vergleichsverfahren n comparative method
Vergleichsversuch m blank test, comparative experiment, comparison test
Vergleichswert m comparative value
Vergleichswiderstand m standard resistance
Vergleichszahl f comparative figure
vergolden to gild, to gold-plate
Vergolden n gilding, gold-plating
Vergoldestein m agate burnisher
vergoldet gilded, gold-plated
Vergoldung f gilding, gold-plating
Vergoldungsflüssigkeit f gilding solution
Vergoldungssalz n gilding salt
Vergoldungswachs n gilder's wax
vergoren (Brau) fermented
Vergrauung f graying, greying
vergrößern to enlarge, to magnify, (zunehmen) to increase
Vergrößerung f enlargement, magnification, scale-up
Vergrößerungsapparat m (Phot) enlarger, ~ **mit selbsttätiger Einstellung** autofocus enlarger
Vergrößerungsgerät n enlarger, magnifying apparatus
Vergrößerungsglas n magnifying glass
Vergrößerungslinse f magnifying lens
Vergrößerungsmaßstab m magnification scale, scale of enlargement

Vergrößerungsnutzgrad m (Opt) magnification effectiveness
Vergrößerungsspiegel m concave mirror, magnifying mirror
Vergünstigung f favor, privilege
Vergütbarkeit f tempering quality
vergüten (ausgleichen) to compensate, (Met) to anneal, to harden, to temper, (verbessern) to improve
Vergüten n compensation, (Met) age-hardening, aging, heat-treating, tempering
Vergüteofen m (Met) tempering furnace
Vergütung f (Stahl) hardening and tempering, heat treatment, quenching and tempering, thermal refinement
Vergütungsglühen n (Leichtmetall) solution heat treatment
Vergußharz n casting resin
Vergußkammer f sealing chamber
Vergußmasse f filler, sealing compound
Verhältnis n (Proportion) proportion, (zweier Größen) ratio, **im quadratischen** ~ in proportion to the square [of], **umgekehrtes** ~ inverse ratio
Verhältnisanteil m proportion, proportionate share
verhältnismäßig comparatively, (proportional) proportional, proportionate, (relativ) relative
Verhältniszahl f proportionality factor
verhärten to harden
Verhärten n hardening
Verhärtung f hardening, (Zellgewebe) sclerosis
verhalten, sich ~ to act, to behave
Verhalten n behavior, performance
verharren to remain
verharzen to become resinous, to resinify
Verharzen n resinifying
Verharzung f resinification
verhindern to prevent, to avoid, to hinder, to inhibit
Verhinderung f prevention
verholzen to lignify, to become wood
Verholzung f lignification
Verholzungsgrad m degree of lignification
verhüllen to cover, to veil
verhüten to prevent, to avert, (vermeiden) to avoid
verhüttbar (Bergb) ready for smelting, treatable, workable
verhütten (Metall) to smelt
Verhütten n (Metall) smelting
Verhüttung f (Metall) metallurgical operations, smelting
Verhüttungsvorgang m (Metall) smelting process
Verhütungsmittel n contraceptive, preventive
Verjährung f superannuation
verjagen to drive out, to expel
verjüngen (spitz zulaufen) to chamfer, to taper
Verjüngung f narrowing, taper

Verjüngungsnippel *m* reducing nipple
Verjüngungsrohrstutzen *m* reducer
verkäuflich salable
Verkäuflichkeit *f* salability
verkalken to calcify, to calcine, to lime
Verkalkung *f* (Biol) calcification, (Chem) calcination, liming
verkanten to tilt
verkapseln to encapsulate
Verkauf *m* sale
Verkaufsabteilung *f* sales department
Verkaufsnummer *f* catalog number, sales number
Verkehr *m* traffic, (Nachrichtenwesen) communications
Verkehrsmittel *n* means of transportation
Verkehrstechnik *f* transportation engineering
Verkehrsunfall *m* traffic accident
verkehrt inverse, reversed
verkeilen to key
verkeilt keyed
verketten to interlink, to link
verkettet interlinked, intermeshed
Verkettung *f* linkage, bonding, chain formation, interlinking
Verkettungspunkt *m* interlinking point, junction point
verkieseln to silicify
Verkieselung *f* silicification
verkitten to cement, to lute, to paste together, to putty, to seal
Verkitten *n* cementing, luting, sealing
Verkittung *f* cementation, cementing, luting, sealing
Verkittungsmaterial *n* plastering material
verklagen wegen to sue [for]
verklammern to brace
Verklammerung *f* staple fastening
Verklebbarkeit *f* bonding properties
verkleben to bond, to glue, to seal
verkleiden to case, to cover, to jacket, to line, (mit Blei) to lead
verkleidet panelled
Verkleidung *f* casing, covering, panelling, screen, sheathing
Verkleidungsblech *n* covering panel
Verkleidungsrohr *n* cover tube, protective tube
verkleinern to diminish, to reduce, (schrumpfen) to shrink
Verkleinerung *f* decrease, diminution, reduction
Verkleinerungsglas *n* diminishing glass
Verkleinerungsmaßstab *m* reduction scale, scale of reduction
verklemmen to jam
Verknappung *f* shortage
verknistern (Chem) to decrepitate
Verknisterungswasser *n* (Chem) decrepitation water

verknorpeln to become cartilaginous, to form gristles
verknorpelt cartilaginous
Verknorpelung *f* cartilaginification, chondrification
Verknüfung *f* linkage, combination, connection, union
verknüpfen to bind, to combine [with], to connect, to join, to tie [together]
verkochen (konzentrieren) to boil away, to boil down, to concentrate, (Zuck) to boil string-proof
Verkochen *n* (Zuck) pan boiling
verkörpern to embody, to personify
verkohlen to carbonize, to char, to convert into coal
Verkohlen *n* carbonizing, charring
verkohlt charred
Verkohlung *f* carbonization, charring, ~ **in Meilern** heap-charring
verkoken to coke, to carbonize
Verkokung *f* coking, carbonization [of coal], ~ **in Retorten** cylinder coking, ~ **von Steinkohle** coking of pit coal
Verkokungsanlage *f* coking plant
Verkokungsfähigkeit *f* coking capability, coking capacity
Verkokungsgeschwindigkeit *f* rate of coking
Verkokungskammer *f* coking chamber
Verkokungsprobe *f* coking test, crucible test
Verkokungsretorte *f* carbonization retort
Verkokungstemperatur *f* temperature of coking
Verkokungsvorgang *m* coking process
verkorken to cork
verkrusten to crust, to incrust
Verkrustung *f* incrustation
verkümmern to degenerate, to stunt
Verkümmerung *f* degeneration
verküpen (Färb) to vat
verkürzen to abbreviate, to abridge, to condense, to contract, to shorten
verkürzt abbreviated, contracted, diminished, shortened
Verkürzung *f* abbreviation, contraction, shortening
verkupfern to copper, to copper-plate, to plate with copper
Verkupfern *n* coppering, copper-plating
Verkupferung *f* coppering, copper-plating
Verladeanlage *f* loading plant
Verladebrücke *f* transporter bridge
verladen to load, to ship
Verlader *m* loader, transporter
Verladerampe *f* loading ramp, loading wharf
Verladung *f* loading, charging
verlängern to lengthen, to elongate, to extend, to prolong, (dehnen) to stretch
verlängert elongated, extended

Verlängerung – verpacken

Verlängerung *f* elongation, extension, lengthening, prolongation
Verlängerungshebel *m* lengthening lever
Verlängerungsreiter *m* extending slider
Verlängerungsschnur *f* (Elektr) extension cable
Verlängerungsstück *n* adapter, extension piece
verläßlich dependable
Verläßlichkeit *f* dependability
verlagern to dislocate, to displace, to move, to shift
Verlagerung *f* dislocation, displacement, shift[ing]
verlangsamen to decelerate, to moderate, to retard, to slow down
Verlangsamung *f* deceleration, moderation, slowing down
Verlauf *m* course, (Fluß) flux, (Lack) levelling, ~ **der Herstellung** course of manufacture, ~ **der Siedekurve** shape of the boiling curve, ~ **einer Rohrleitung** run of a pipe
verlaufen (Färb) to bleed color
Verlaufen *n* **der Farbe** (Färb) running of the dye
Verlauffähigkeit *f* (Lack) levelling property, spreadability
Verlaufmittel *n* (Lack) flow improver, levelling agent
Verlaufseigenschaften *pl* flow properties
verlegen to displace, to shift, **zeitlich** ~ to delay, to postpone
verleihen to confer, (Eigenschaft) to impart
verleimen to glue, to gum
Verleimen *n* gluing
Verleimung *f* bonding
Verleimzeit *f* assembly time
verletzen to damage, to hurt, to infringe
Verletzung *f* damage, infringement, ~ **durch Bestrahlung** radiation injury
verlieren to lose
verlöten (mit Blei) to plumb, to solder
Verlust *m* leakage, waste, ~ **bei der Destillation** distillation loss, **magnetischer** ~ magnetic loss
Verlustbeiwert *m* loss factor
Verlustdämpfung *f* loss damping
Verlustfaktor *m* loss factor, dissipation factor
verlustfrei free of loss
verlustlos without loss or waste
Verlustwärme *f* heat loss
Verlustwinkel *m* (Elektr) loss angle, power factor
vermahlen to crush, to grind, to mill, to pulverize
Vermahlung *f* grinding, pulverizing
Vermahlungsfähigkeit *f* grindability
vermehren to increase, to multiply
Vermehrung *f* growth, increase
Vermehrungsfaktor *m* multiplication factor
Vermehrungskonstante *f* multiplication constant
vermeiden to avoid
Vermeidung *f* avoidance
vermengen to blend, to mingle, to mix

Vermengung *f* blending, mixing, mixture
Vermerk *m* note, notice
vermessen to survey, to measure
Vermessung *f* survey[ing], measurement
Vermessungskunde *f* surveying
Vermiculit *m* (Min) vermiculite
vermindern, sich ~ to decrease, to diminish
vermindert diminished, reduced
Verminderung *f* decrease; diminution
Verminologie *f* verminology
vermischbar miscible
vermischen to blend, to compound, to intermingle, to mingle, to mix, (Metall) to alloy, (vergällen) to adulterate
Vermischen *n* adulterating, blending, compounding, mingling, mixing, (Metall) alloying
Vermischung *f* blend, mixture, (Biol) crossing, interbreeding, (Vergällung) adulteration
vermitteln to arrange
Vermittler *m* agent
Vermittlungsgebühr *f* commission, fee
Vermizid *n* vermicide
vermodern to decay, to mo[u]lder
vermodert mo[u]ldy, musty
Vermoderung *f* decay, decomposition, rotting
vernachlässigbar negligible
vernachlässigen to neglect, to disregard, to omit
Vernachlässigung *f* neglect, omission
verneinend negative
vernetzen to cross-link, to interlace, to intermesh
Vernetzer *m* cross-linker, cross-linking agent
vernetzt cross-linked
Vernetzung *f* network, cross linkage, formation of cross bonds, interlacing
Vernetzungsstelle *f* cross linkage position, cross-linking site
vernichten to destroy, to annihilate, to annul
Vernichtung *f* annihilation
Vernichtungsstrahlung *f* annihilation radiation
vernickeln to nickel, to nickel-plate
Vernickeln *n* nickeling, nickel-plating
vernickelt nickel-plated
Vernick[e]lung *f* nickeling, nickel-plating
Vernickelungsbad *n* nickel bath
Vernier *m* vernier
vernieten to rivet
Vernietung *f* riveting, (Nietverbindung) riveted joint
Vernin *n* vernine
Vernolsäure *f* vernolic acid
Veröffentlichung *f* publication
Veronal *n* (HN) veronal, barbital, diethylbarbituric acid
Veronesergrün *n* green earth, Verona green, viridian green
Verordnung *f* order, rule
Verosterin *n* verosterine
verpacken to pack, (einwickeln) to wrap

Verpacken *n* packaging, packing
Verpackung *f* packing, **handelsübliche** ~ commercial packaging, **lose** ~ bulk packing, **wiederverwendbare** ~ reusable container
Verpackungsband *n* strapping
Verpackungseinheit *f* packing unit
Verpackungsgewicht *n* tare weigth
Verpackungsmaterial *n* packaging material
Verpackungsmittel *n* packing material
verpflanzen to transplant
verpflichten to engage, to oblige
verpichen to pitch, to pitchcoat, to pitchline
Verpichung *f* pitch formation
verplatinieren to platinum-plate
Verplatinierung *f* platinum-plating, platinization, platinizing
verpochen to stamp
Verpochen *n* stamp milling
Verpreßbarkeit *f* moldability
verpuddeln to puddle
verpuffen to crackle, to deflagrate, to detonate, to explode, to puff off
Verpuffung *f* deflagration, detonation, explosion
Verpuffungsspannung *f* explosion pressure
Verpuffungsverfahren *n* explosion method
verpuppen (Biol) to change into a chrysalis
Verputz *m* plaster[ing]
verputzen to plaster, to dress
Verquecksilbern *n* (Chem) amalgamating
verquicken (obs) to amalgamate; to quicken
Verquicken *n* (obs) amalgamating, quickening
Verquickung *f* (obs) amalgamation, quickening
verreiben to grind [fine], to triturate
Verreibung *f* trituration, ~ **der Farbe** (Buchdr) ink distribution
Verreißen *n* **der Form** stripping of the mold
verriegeln to bolt, to lock
verringern to decrease, to diminish, to lessen, to lower, to minimize, to reduce
Verringerung *f* decrease, diminution, reduction
verrohren to case
Verrohrung *f* casing, tubing
verrosten to corrode, to rust
Verrosten *n* rusting, corroding, formation of rust
verrostet corroded, rusted, rusty
Verrostung *f* corrosion, rusting
Verrottung *f* rotting
Verrückung *f* displacement
verrußen to soot
Verrußen *n* sooting
Verrußung *f* (Katalysator) carbonization
Versäuerung *f* acidulation, souring
versagen to fail, to miss
Versagen *n* failure
Versammlung *f* assembly
Versandeimer *m* dispatch pail
versandfähig transportable
Versandfaß *n* shipping cask
Versandkiste *f* packing case

Versandschachtel *f* packing case
Versatznietung *f* staggered riveting
verschärfen (Bedingungen) to severe
verschaffen to procure
verschalen to board
Verschalung *f* boarding, planking, reveting
verschiebbar displaceable, sliding
Verschiebbarkeit *f* displaceability, movability
Verschiebebühne *f* shuttle platform
verschieben to displace, to move, to remove, to shift, (aufschieben) to defer, to delay, to postpone
Verschiebestempel *f* sliding punch
Verschiebewinde *f* shunting drum
Verschiebung *f* displacement, shift[ing], (Geol) dislocation, ~ **der Absorptionsbande** displacement of the absorption band, ~ **der Absorptionskanten** absorption edge shift, ~ **von Atomen** atomic displacement, **chemische** ~ chemical displacement
Verschiebungsfeld *n* displacement field
Verschiebungsgesetz *n* displacement law
Verschiebungssatz *m* **von Soddy, Fajans und Russel** displacement law of Soddy, Fajans and Russel
Verschiebungsstrom *m* displacement current
Verschiebungsvorgang *m* displacement process
verschieden different, dissimilar, distinct, diverse, unlike, varying
verschiedenartig different, diversified, heterogeneous, unlike
Verschiedene[s] *n* miscellaneous, sundries
verschiedenfarbig varicolored
Verschiedenheit *f* diversity, difference, variety
verschiedenphasig dephased
verschiedenzellig (Biol) heterocellular
Verschießen *n* (Färb, Text) fading
Verschießungsgrad *m* (Text) degree of discoloration, degree of fading
verschiffen to ship
Verschiffung *f* shipment
verschimmeln to get moldy
verschlacken to clinker, to flux, to scorify, to slag
Verschlackung *f* slagging
Verschlackungsfähigkeit *f* capacity for forming slag, fluxing power
Verschlackungsgefahr *f* danger of slagging
Verschlag *m* bin, crate
Verschlammung *f* accumulation of mud, silting
verschlechtern to deteriorate, to fall off in quality
Verschlechterung *f* deterioration
verschleiert camouflaged
Verschleiß *m* (Abrieb) abrasion, wear [and tear]
Verschleißband *n* wear strip
verschleißen to wear out
verschleißfest wear resistant
Verschleißfestigkeit *f* wear resistance
Verschleißhärte *f* resistance against wear [and tear]

Verschleißplatte f wearing plate
Verschleißprüfung f abrasion test, wear test
Verschleißscheibe f (Innenmischer) rotor end plate
Verschleißschutzbelag m anti-abrasive coating
Verschleißverhalten n wearing quality
verschleppen (Krankheit) to carry, (übertragen) to spread
verschließen to close, to lock [up], to seal, to shut (off), to stopper, (luftdicht) to seal off air-tight
Verschließen n sealing
Verschließmaschine f sealing machine, (Dosen) closing machine
verschlimmern to aggravate
Verschlingen n **der Seile** twisting of the ropes
verschlissen worn out
verschlucken to swallow, to absorb
verschlungen intertwining
Verschluß m closure; seal, (Phot) shutter, (Vorrichtung) closing device, lock, stopper, ~ **mit Originalitätssicherung** tamperproof closure, **hermetischer** ~ hermetic seal, **luftdichter** ~ air-tight seal, hermetic seal
Verschlußauslösung f (Phot) shutter release
Verschlußblende f (Phot) diaphragm, shutter
Verschlußdeckel m closing cover
Verschlußdüse f shut-off nozzle
Verschlußflansch m sealing flange
Verschlußhahn m stop cock
Verschlußkappe f cap [closure]
Verschlußkette f closure chain
Verschlußklappe f (Phot) shutter flap
Verschlußpfropfen m plug
Verschlußring m lock ring
Verschlußschraube f locking screw, screw plug, threaded nose cap
Verschlußstopfen m stopper
Verschlußstück n plug
verschmelzen to fuse together, to melt, to smelt, (Färb) to blend, (löten) to solder
Verschmelzung f fusion, melting, smelting, (Met) alloy
Verschmelzungsenergie f fusion energy
Verschmelzungskörper m (Met) alloy constituent
Verschmelzungsreaktion f fusion reaction
Verschmiedbarkeit f forgeability
verschmieden to forge
verschmieren to smear
Verschmierung f (Röntgenstrahlen) blurring
verschmoren (Elektr) to burn
verschmutzt contaminated
Verschmutzung f soiling, contamination, fouling, pollution
Verschmutzungsgefahr f danger of contamination
verschneiden (Spirituosen) to adulterate, to blend
Verschneiden n (Gummichemie) cross blending, (Spirituosen) adulterating, blending

Verschnitt m blended mixture, blending, cuttings, waste, (Aufhellung) tint tone (pigment plus extender), (Mahlen) fines, (Spirituosen) blend
Verschnittfarbe f color lake, lake pigment, reduced pigment
Verschnittmittel n adulterating agent, diluting agent, filler, (f. Lösungsmittel) diluent, ~ **für Pigmente** extender
Verschnittverhalten n dilution capacity
verschränkt crossed
Verschrammung f scratch
verschrauben to screw, (zusammenschrauben) to screw together
Verschraubung f screw fitting; screwing, (Gewinde) thread
Verschraubungsteil n screw fitting
verschrotten to scrap
Verschrottung f scrapping
verschweißen to weld [together], (Kunststoff) to heat-seal
Verschweißen n fusing, welding, (Kunststoff) heat-sealing, (Met) seizure (stoppage of a machine as a result of excessive friction)
Verschweißung f heat-seal, weld, **fugenlose** ~ seamless weld
verschwelen to carbonize or distill under vacuum at a low temperature
Verschwelung f low temperature carbonization
verschwenden to waste
verschwimmen (Opt) to become indistinct
verschwinden to disappear, to vanish, (langsam) to fade
Verschwinden n disappearance
Verschwindimpulsgeber m vanishing impulse transmitter
verseifbar saponifiable
Verseifbarkeit f saponifiability
verseifen (Chem) to saponify
Verseifen n (Chem) saponifying
Verseifung f (Chem) saponification
Verseifungskolben m (Chem) saponification flask
Verseifungszahl f (Chem) saponification number, saponification value, (Jodzahl) iodine value
versenden to dispatch, to send, to ship
Versendung f shipment
versengen to scorch
Versenkbohrer m (Techn) countersink
versenken to dip
Versenol n versenol
versetzbar removable
versetzen (Chem) to add, (neu anordnen) to move, to rearrange, to relocate, to shift, (vermischen) to compound, to mix, to treat
Versetzen n (Chem) addition
versetzt (in Zick-Zack) staggered, zig-zag

Versetzung *f* dislocation, rearrangement, relocation, shift
Versetzungsbewegung *f* dislocation motion
Versetzungsdichte *f* dislocation density
Versetzungsenergie *f* (Krist) energy of dislocation
Versetzungskern *m* dislocation kernel
Versetzungslinie *f* dislocation line
Versetzungsnetzwerk *n* dislocation network
Versetzungssprung *m* dislocation jog
Versetzungsverteilung *f* dislocation distribution
verseuchen to contaminate, to infect
Verseuchung *f* contamination, **radioaktive** ~ radioactive contamination
Verseuchungsstoff *m* contaminant
Versicherung *f* insurance
versiegeln to seal
Versiegelungsmittel *n* sealer
versiegen to dry up, to exhaust
versilbern to silver, to silver-plate
Versilbern *n* silvering, silver-plating
versilbert silver-plated
Versilberung *f* silvering, silver-plating
versinken to sink, to submerge
versorgen to supply
Versorgung *f* supply
verspannen to brace, to clamp, to stay
versperren to bar, to barricade, to block
verspritzen to spatter, to spill, to splash
verspröden to embrittle
Versprödung *f* development of brittleness, embrittlement, ~ **nach dem Schweißen** weld decay
Versprühen *n* spraying
verstählen to acierate, to convert into steel, to edge with steel, to point with steel
Verstählen *n* acierating, steel-plating
Verstählung *f* acieration, steel-plating, **galvanische** ~ electroplating with steel
verständlich comprehensible, intelligible, logical
Verständnis *n* understanding
verstärkbar (Elektr) amplifiable
Verstärkbarkeitsgrenze *f* maximum amplification
verstärken to reinforce, to strengthen, (Elektr) to amplify, (konzentrieren) to concentrate, (Opt) to intensify, (stimulieren) to stimulate, (versteifen) to stiffen
verstärkend multiplicative
Verstärker *m* (Elektr) amplifier, booster, (Phot) intensifier, **magnetischer** ~ transductor, **rückgekoppelter** ~ feedback amplifier
Verstärkereingang *m* (Elektr) amplifier input
Verstärkerelement *n* (Elektr) amplifying element
Verstärkerfüllstoff *m* reinforcing filler
Verstärkerkanal *m* (Elektr) booster channel
Verstärkerkette *f* (Elektr) amplifier cascade
Verstärkerkreis *m* (Elektr) amplifier circuit
Verstärkerpumpe *f* booster pump

Verstärkerregelung *f* (Elektr) gain control
Verstärkerrelais *n* pilot relay type
Verstärkerröhre *f* (Elektr) amplifying valve
Verstärkersäule *f* concentrating column (in distillation plant)
Verstärkerschirm *m* intensifying screen
Verstärkerstufe *f* amplification stage
Verstärkerteil *m* (Rektifikationskolonne) rectifying section
Verstärkerwirkung *f* (Elektr) amplifier action, amplifier effect
verstärkt concentrated, reinforced, strengthened, (Elektr) amplified
Verstärkung *f* reinforcement, stiffening, (Elektr) amplification, gain, (Kernresonanz) enhancement, (Konzentrierung) concentration, enrichment, ~ **des Photostroms** amplification [or gain] of the photocurrent, **lineare** ~ linear amplification
Verstärkungsbatterie *f* (Elektr) subsidiary battery
Verstärkungsfaktor *m* (Elektr) amplification factor; amplifier gain
Verstärkungsgrad *m* (Elektr) degree of amplification
Verstärkungsmaterial *n* reinforcing material
Verstärkungsrippe *f* strengthener
Verstärkungsteil *m* enriching zone
Verstärkungsverhältnis *n* enrichment ratio
verstäuben to atomize, to spray
Verstäuber *m* atomizer, sprayer
Verstäubungsapparat *m* atomizer, sprayer
Verstäubungsverlust *m* dust loss
versteifen to stiffen, to brace, to prop, to stay, to strut
Versteifung *f* reinforcement, stiffening, (Techn) splint
Versteifungsmittel *n* stiffener, antisoftener
Versteifungsstrebe *f* drag strut
Versteifungsstreifen *m* stiffener
versteinern to petrify
Versteinern *n* petrifying
versteinert petrified
Versteinerung *f* petrifaction
versteinerungsfähig petrifiable
verstellbar adjustable, movable
Verstellbarkeit *f* adjustability
Verstellbereich *m* regulating range, (Elektr) amplification range
Verstellen *n* adjustment
Verstellhebel *m* adjusting lever
Verstellmöglichkeit *f* provision for adjustment
Verstellnabe *f* controllable hub
Verstellung *f* displacement, rearrangement, relocation, shifting, (Justieren) adjustment
verstopfen to clog [up], to jam, (Bohrloch) to clay, to tamp, (ein Loch) to plug, to stop up, (Ritzen) to fill up
Verstopfen *n* choking

Verstopfung f choking, (Med) constipation, ~ **des Stichloches** freezing of the tap hole, stopping up of the tap
verstrahlen to radiate
verstrammen to strengthen
Verstrammung f strengthening
verstreben to strut
Verstrebung f strut[ting], brace
Verstreckbarkeit f stretchability
verstrecken to stretch, (Kunststoffäden) to draw, to neck down, to stretch
Verstrecken n stretching, drawing
verstreckt (kaltverstreckt) cold drawn
Verstreckung f necking [down], stretch
Verstreichbarkeit f brushability
verstreichen to spread, (auslaufen) to expire, (Fugen) to fill up, (Zeit) to pass or slip by
verstreuen to disperse, to scatter
verstricken to ensnare, to entangle
verstümmeln to deform, to disable, to garble, to maim, to mangle, to mutilate
verstummen to become silent
Versuch m experiment, assay, test
versuchen to try, to experiment, to test
Versuchplanung f, **faktorielle** ~ factorial design
Versuchsanlage f testing plant, experimental plant, pilot plant
Versuchsanordnung f experimental arrangement, experimental procedure
Versuchsanstalt f experimental station, laboratory
Versuchsaufbau m experimental setup
Versuchsauswertung f test evaluation
Versuchsbedingung f experimental condition, test condition
Versuchsbericht m test report
Versuchsbetrieb m pilot plant, experimental works
Versuchsdauer f duration of test, experimental time
Versuchsdurchführung f experimental procedure
Versuchseinrichtung f testing installation
Versuchsergebnis n result of experiment, test result
Versuchsfahrt f trial run
Versuchsfehler m experimental error
Versuchsfeld n experiment field, field for experiment, proving ground
Versuchskocher m experimental boiler
Versuchslaboratorium n test or experimental laboratory
Versuchslast f test load
versuchsmäßig experimentally
Versuchsmaßstab m experimental scale
Versuchsmaterial n experimental material, test material
Versuchsmaus f test mouse
Versuchsmethode f experimental method, tentative method

Versuchsmodell n pilot model, test or experimental model
Versuchspflanze f experimental plant
Versuchsphase f experimental stage
Versuchsplanung f design of experiments
Versuchsprodukt n development product
Versuchsprotokoll n lab journal, log sheet, test protocol
Versuchsraum m laboratory, room for experiments
Versuchsreaktor m experimental reactor, test reactor
Versuchsreihe f series of experiments, series of tests
Versuchsröhrchen n test tube
Versuchsspannung f (Elektr) experimental voltage, testing voltage
Versuchsstab m test bar, test rod
Versuchsstadium n experimental stage
Versuchsstück n test piece
Versuchstier n experimental animal
Versuchsvorgang m experimental procedure
Versuchsvorschrift f experimental direction
versuchsweise experimental[ly], tentative
versüßen to sweeten, to edulcorate
versüßt sweetened, edulcorated
Versüßung f sweetening, edulcoration
vertagen to delay
vertauschbar interchangeable, (kommutierbar) commutable
Vertauschbarkeit f interchangeability
vertauschen to exchange, to interchange, (permutieren) to permute
Vertauschung f exchange, commutation, interchange, permutation
Vertauschungsregel f commutation rule
Vertauschungsrelation f commutation relation
verteilen to distribute, to divide, (ausbreiten) to spread, (diffundieren) to diffuse, (verstreuen) to disperse, to scatter
Verteilen n distributing, diffusing, dispersing, dividing, spreading
Verteiler m dispenser, distributor, redistributor, spreader
Verteilerboden m (Dest) dristributor, (Kolonne) feed plate
Verteilerdüse f jet
Verteilerkanal m manifold
Verteilerkopf m separator head
Verteilerschacht m distributor shaft
Verteilertafel f distributing panel, distribution board
Verteilerwalze f frame roller
verteilt, fein ~ finely dispersed, suspended, **gleichmäßig** ~ evenly or continuously distributed, **punktförmig** ~ distributed in lumps, **stetig** ~ evenly or uniformly distributed

Verteilung *f* distribution, division, (Zerstreuung) dissipation, **hypergeometrische** ~ hypergeometric distribution, **räumliche** ~ spatial distribution
Verteilungschromatographie *f* distribution chromatography
Verteilungsfunktion *f* distribution function
Verteilungsgeschwindigkeit *f* distribution velocity
Verteilungsgesetz *n* law of distribution
Verteilungshahn *m* branch cock, distribution cock
Verteilungskoeffizient *m* distribution coefficient, (Chromatogr) partition coefficient
Verteilungskopf *m* manifold
Verteilungskurve *f* distribution curve
Verteilungsleitung *f* distributing main
Verteilungsmischer *m* (Zuck) distributing mixer, mixing trough
Verteilungsmittel *n* dispersing agent
Verteilungsquotient *m* distribution constant
Verteilungsröhre *f* distributing tube
Verteilungsrohr *n* distribution pipe
Verteilungsstelle *f* (Dest) liquid feed point
Verteilungsstück *n* distribution piece
Verteilungssummenkurve *f* cumulative frequency curve
Verteilungswerk *n* distributing station
Verteilungszentrale *f* distributing station
Vertiefung *f* deepening, cavity, depression, groove, hollow, recess
vertikal vertical
Vertikalablenkung *f* vertical deflection
Vertikalachse *f* vertical axis
Vertikalausbreitung *f* vertical diffusion
Vertikalbewegung *f* vertical movement
Vertikale *f* perpendicular [line], vertical [line]
Vertikalgalvanometer *n* (Elektr) balance galvanometer
Vertikalkolben *m* vertical plunger
Vertikalkomponente *f* vertical component
Vertikalpolarisation *f* vertical polarization
Vertikalschnitt *m* vertical cross section
Vertikalverschiebung *f* vertical shift
vertilgen to annihilate, to eradicate, to exterminate, to extirpate
Vertilgung *f* extermination
Vertilgungsmittel *n* eradicator, extirpator
Verträglichkeit *f* (Med) tolerancy, (Pharm) compatibility
Verträglichkeitsgrenze *f* tolerance limit
Verträglichkeitsprobe *f* tolerance test
Vertrag *m* contract
Vertragsforschung *f* contract research
vertreiben to drive away, to expel
vertretbar (ersetzbar) replaceable
vertreten to replace, to substitute
Vertreter *m* agent, representative
vertrocknen to desiccate, to dry up

Vertrocknung *f* desiccation
verunreinigen to contaminate, to pollute, to render impure, to soil, to vitiate
verunreinigt contaminated, polluted
Verunreinigung *f* contamination, impurity, pollution, ~ **der Luft** atmospheric pollution
Verunreinigungsherd *m* source of contamination, source of pollution
verursachen to cause, to produce
vervielfachen to manifold, to multiply
Vervielfacher *m* multiplier
Vervielfacherphotozelle *f* electron multiplier phototube, multiplier photoelectric cell
Vervielfachungsprozeß *m* multiplication process
vervielfältigen to multiply, to reproduce
Vervielfältiger *m* multiplier
Vervielfältigungsapparat *m* copying apparatus, duplicating apparatus
Vervielfältigungsmaschine *f* copying machine, duplicating machine, (mittels Hektograph) hectograph, (mittels Mimeograph) mimeograph
vervollkommnen to complete, to improve
Vervollkommnung *f* perfection
vervollständigen to complete
Vervollständigung *f* completion
Verwachsung *f* intercrescence, (Krist) intergrowth
Verwachsungsebene *f* (Krist) composition plane
Verwachsungsfläche *f* (Krist) composition surface
verwackeln (Phot) to blur
Verwässerungsschutzklausel *f* dilution clause
Verwaltungsgebäude *n* administration building
verwalzen to roll [into]
verwandeln to change, to convert, to transform, (Atom) to transmute, (Biol) to metamorphose, ~ **in Kohlenstoff** to carbonize
Verwandeln *n* changing, converting, transforming, (Atom) transmuting
Verwandlung *f* change, conversion, transformation, (Atom) transmutation, (Biol) metamorphosis
verwandt related [to]
Verwandtschaft *f* relationship, (Chem) affinity
verwaschen (Färb) to bleed color, (Phot) to blur
Verwaschung *f* washing, (Phot) blurring
verweigern to deny, to refuse
Verweigerung *f* denial, refusal
verweilen to remain, to stay
Verweilzeit *f* dwell time (injection molding), residence time, time of direct contact
Verweilzeitverteilung *f* residence time distribution
Verweis *m* reference
verweisen to refer to
verwendbar adaptable
Verwendbarkeit *f* usefulness, adaptability, usability, **vielseitige** ~ versatility of service

verwenden to use, to apply, to employ
Verwendung *f* use, application
Verwendungsbereich *m* adaptability range, field or range of application
Verwendungsfähigkeit *f* applicability, employability, usability, usefulness
Verwendungszweck *m* [intended] use, purpose
verwerfen to reject, (Niederschlag) to dispose [of], to throw away
Verwerfen *n* warping
Verwerfung *f* rejection
verwerten to utilize
Verwertung *f* utilization
verwesen to decay, to decompose slowly, to putrefy, to rot
Verwesen *n* decay, putrefying, rotting, slow decomposing
verweslich liable to decay
Verwesung *f* decay, putrefaction, rotting, slow decomposition
Verwick[e]lung *f* complication, entanglement
Verwindungsprobe *f* torsion test
verwirklichen to realize
Verwirklichung *f* realization
verwirren (Garn) to snarl
verwischt blurred
verwittern to decay, to decompose, to disintegrate, to weather
Verwittern *n* decaying, disintegrating, weathering
verwittert disintegrated, weathered
Verwitterung *f* weathering, disintegration, **~ der Oberfläche** surface weathering
Verwitterungsbestandteil *m* weathered constituent
Verwitterungsgebilde *n* product of disintegration
Verwitterungsprodukt *n* decomposition product
Verwitterungsschicht *f* weathered layer
Verwitterungsschutzmittel *n* anti-weathering agent
Verzahnmaschine *f* (Techn) gear cutting machine
verzapfen (Techn) to mortise
verzehren to consume, to use up
Verzeichnis *n* directory, register
verzerren to deform, to distort
verzerrt distorted, strained
Verzerrung *f* (Mech) deformation [strain], (Mech, Opt) distortion, **nichtlineare ~** amplitude distortion, nonlinear distortion, **stationäre ~** stationary distortion
Verzerrungsdeviation *f* strain deviation
Verzerrungsenergie *f* strain energy
verzerrungsfrei distortion-free, distortionless, nondistorting
Verzerrungsfreiheit *f* freedom from distortion
Verzerrungstensor *m* deformation tensor
Verzerrungstoleranz *f* tolerance of distortion
Verzerrungsverbreiterung *f* distortion broadening

verziehen to distort, to buckle, to deform, to warp
Verziehen *n* **des Ringes** buckling of the ring, **~ des Stahls** contraction of the steel
Verzierung *f* ornament, **~ über der Glasur** (Keram) overglaze decoration
verzinken to galvanize, to coat with zinc, **feuer- ~** to hot-galvanize
Verzinken *n* galvanizing, zinc-coating, (heißes Verzinken) hot galvanizing
Verzinkerei *f* galvanizing plant
Verzinkung *f* galvanization
Verzinkungswanne *f* galvanizing bath
verzinnen to tin-plate, to line with tin, to tin
Verzinnen *n* tin-coating, tinning, tin-plating
verzinnt tin-coated, tinned
Verzinnung *f* tin-plating, lining with tin, tinning, (galvanische Verzinnung) electro-tinning, **~ durch Ansieden** tin-plating in hot bath
Verzinnungsanstalt *f* tinplate works
Verzögerer *m* retarder, inhibitor
verzögern to retard, to delay, to inhibit, (verlangsamen) to decelerate
Verzögern *n* retarding
Verzögerung *f* retardation, lag, (Verlangsamung) deceleration, delayed action, **thermische ~** thermal lag
Verzögerungsdemodulator *m* delay demodulator
Verzögerungserscheinung *f* lag phenomenon
Verzögerungspotential *n* retardation potential
Verzögerungsrelais *n* time-lag relay
Verzögerungsschalter *m* delay switch
Verzögerungswinkel *m* angle of lag; retardation angle
Verzögerungszeit *f* period of retardation, time lag
verzuckern to saccharify, (überzuckern) to sugar-coat
Verzuckerung *f* saccharification
Verzunderung *f* high temperature scaling
verzweigen to branch, to ramify
verzweigt branched, arborescent, ramified
Verzweigung *f* branching [off], by-pass (tubing), ramification, **~ von Ketten** chain branching
Verzweigungsfaktor *m* branching factor
Verzweigungslinie *f* branch line
Verzweigungsmodell *n* branching model
Verzweigungspunkt *m* branching point, branch point
Verzweigungsstelle *f* branch point
Verzweigungsverhältnis *n* branching ratio
Verzweigungswahrscheinlichkeit *f* branching probability
Vestrylamin *n* vestrylamine
Vesuvian *m* (Min) vesuvian, vesuvianite
Vesuvin *n* (Farbstoff) vesuvin; aniline brown, Bismarck brown, Manchester brown, triamino azobenzene
Veterinär *m* veterinary, veterinary surgeon

Veterinärmedizin f veterinary science
Vetivalen n vetivalene
Vetivazulen n vetivazulene
Vetiveröl n vetiveria oil
Vetiveron n vetiverone
Vetivon n vetivone
V-förmig V-shaped
Viboquercit m viboquercitol
Vibration f vibration
Vibrationsdämpfung f vibration absorption
Vibrationselektrometer n vibration electrometer
Vibrationsenergie f vibrational energy
Vibrationsgalvanometer n (Elektr) vibration galvanometer
Vibrationsrührer m vibration mixer
Vibrationssieb n vibrating screen
Vibrator m vibrator
Vibratorelektrode f vibrator electrode
vibrieren to vibrate, to oscillate
Vibrieren n vibrating
Vibrograph m vibrograph
Viburnit[ol] n viburnitol
Viburnumrinde f viburnum bark, black haw bark
Vicianin n vicianine
Vicianose f vicianose
Vicilin n vicilin
Vicin n vicine
vicinal vicinal, neighboring [position]
Vicinaleffekt m vicinal effect
Vicinalfläche f (Krist) vicinal surface
Vicinalfunktion f vicinal function
Vickers-Härte f Vicker's or diamond pyramid hardness
Vidalschwarz n vidal black
Vieh n cattle, livestock
Viehfutter n fodder, forage
Viehzucht f cattle breeding, stock-farming
vieladrig multicore
vielatomig polyatomic
vielbasisch polybasic
vielblätterig polyfoil
Vieldrahtzählrohr n multiwire counter tube
Vieleck n polygon
vieleckig polygonal, multiangular
vielfach manifold, multiple
Vielfache[s] n multiple, ~ **einer Zahl** (Math) multiple of a number, **das kleinste gemeinsame** ~ least common multiple, **ganzes** ~ integral multiple, whole multiple, **gemeinsames** ~ (Math) common multiple, **gerades** ~ even multiple, **ungerades** ~ odd multiple
Vielfachformen pl multiple molds
Vielfachkabel n multicore cable
Vielfachkeilriemen m multiple V-belt
Vielfachkontakt m multiple contact
Vielfachleiter m multiconductor
Vielfachstoß m multiple collision
Vielfachstreuung f multiple scattering

Vielfachverzögerungsdiskriminator m multiple delay discriminator
Vielfachzähler m multiscaler
Vielfachzerlegung f (eines Atomkerns) multiple decay (of an atomic nucleus)
Vielfachzertrümmerung f spallation
vielfältig multiple
Vielfältigkeit f multiplicity
Vielfalt f multiplicity
vielfarbig polychromatic
Vielfarbigkeit f (Opt) polychromatism
Vielflächner m polyhedron
vielgestaltig multiform, polymorphous
Vielgestaltigkeit f polymorphism
vielgliedrig (Math) polynomial
vielkernig polynuclear
Vielkristall m polycrystal
Vielphasenstrom m (Elektr) multiphase current
Vielpolanker m multipolar armature
Vielpoldynamo m multipolar dynamo
vielpolig multipolar
vielschichtig multilayered
Vielschlitzmagnetron n multicavity magnetron
Vielschnittdrehbank f multiple cut lathe
vielseitig manysided, (Krist) polyhedral, (vielfach verwendbar) versatile
Vielseitigkeit f diversity, variety, versatility
Vielstellenschreiber m logger
Vielstoffgemisch n multicomponent mixture, multicomponent system
vielstufig multistage
vieltausendfach thousands of times
Vieltypenausbreitung f multimode propagation
vielversprechend very promising
Vielwegehahn m (Chem) multiple-way stopcock
vielwertig multivalent, polyvalent
Vielwertigkeit f multivalence, polyvalence, (Math) multiplicity
Vielzahl f multiplicity, plurality
Vielzellenverdichter m sliding-vane rotary compressor
vielzellig multicellular
vierachsig four-axle
vieratomig tetratomic
vierbasig quadribasic, tetrabasic
vierbasisch tetrabasic
vierbindig (Web) fourharness weave, fourshaft weave
vierdimensional four-dimensional
Vierdimensionalität f four-dimensionality
Viereck n quadrangle
viereckig quadrangular, four-cornered, (quadratisch) square
vierfach fourfold, quadruple
Vierfachbetrieb m quadruplex system
Vierfachchlorkohlenstoff m (obs) carbon tetrachloride
Vierfarbendruck m four-color printing
vierflächig tetrahedral, four-faced

Vierflächner *m* (Krist) tetrahedron
Vierganggetriebe *n* four-speed gear
viergliedrig four-membered
vierkant square
Vierkant *m* square
Vierkantdose *f* rectangular can
Vierkantdraht *m* square wire
Vierkanteisen *n* (Techn) square iron, square bar
Vierkantfeile *f* square file
Vierkantkaliber *n* diamond pass
Vierkantkeil *m* square key
Vierkantkopf *m* square head
Vierkantkopfschraube *f* (f. Holz) lag screw
Vierkantmutter *f* square nut
Vierkantscheibe *f* square washer
Vierkantschlüssel *m* (Techn) square wrench (Am. E.), square spanner (Br. E.)
Vierkantschraube *f* collar head screw, square-headed bolt
Vierkomponentenwaage *f* four-component balance
Vierleitersystem *n* (Elektr) four-wire system
viermolekular quadrimolecular
Vierphasenkreuzschaltung *f* (Elektr) four-phase star connection
Vierphasensternschaltung *f* (Elektr) four-phase star connection
Vierphasensystem *n* four-phase system
vierphasig (Elektr) four-phase
vierpolig fourpolar, fourpole
Vierring *m* four-membered ring
vierseitig four-sided, quadrangular
vierstellig four-digit, four-figure
Vierstoffgemisch *n* quaternary mixture
Vierstoffsystem *n* four-component system, quaternary system
Viertakt *m* four-cycle, four-stroke cycle
Viertaktmaschine *f* four-cycle engine
Viertaktmotor *m* four-cycle engine
Viertel *n* fourth, quarter
Viertelperiode *f* quarter-period
Viertelwellenlängenplättchen *n* quarter wave length plate
vierundzwanzigflächig icositetrahedral
Vierundzwanzigflächner *m* (Krist) icositetrahedron
Vierwalzenkalander *m* (Papier) four-roll calender
Vierwegehahn *m* (Kreuzhahn) four-way cock
Vierwege-Stapler *m* four-way fork lift truck
vierwertig tetravalent, quadrivalent
Vierwertigkeit *f* tetravalence, quadrivalence
vierzählig fourfold, quadruple, (Symmetrie) tetragonal
vierzahnig tetradentate
Vierzentrenaddition *f* (Stereochem) four-center addition
vierzigfach forty-fold
vignettieren (Opt) to vignette

Vignettierung *f* (Opt) vignetting [effect]
Vignettierungseffekt *m* (Opt) vignetting effect
Vignolschiene *f* flat bottom rail
Vigorit *m* vigorite
Vigoureuxdruck *m* vigoureux printing, melange printing
Viktoriablau *n* Victoria blue
Viktoriagelb *n* Victoria yellow, antinonnin
Villarsit *m* (Min) villarsite
Villiaumit *m* (Min) villiaumite
Vinamar *n* vinamar
Vinbarbital *n* vinbarbital
Vincain *n* vincaine
Vinhaticosäure *f* vinhaticoic acid
Vinoplaste *pl* vinyl plastics
Vinyl- vinyl, ethenyl
Vinylacetal *n* vinyl acetal
Vinylacetat *n* vinyl acetate
Vinylacetylen *n* vinyl acetylene
Vinylalkohol *m* vinyl alcohol, ethenol, vinol
Vinylbenzoat *n* vinyl benzoate
Vinylbenzol *n* vinyl benzene
Vinylbromid *n* vinyl bromide
Vinylbutyrat *n* vinyl butyrate
Vinylcarbazol *n* vinyl carbazol
Vinylchlorid *n* vinyl chloride, chloroethene, chloroethylene
Vinylcrotonat *n* vinyl crotonate
Vinylcyanür *n* (obs) vinyl cyanide, acrylonitrile
Vinylen- vinylene
Vinylessigsäure *f* vinylacetic acid
Vinylester *m* vinyl ester
Vinylformiat *n* vinyl formate
Vinylharz *n* vinyl resin
Vinylharzschaum *m* vinylic foam
Vinyliden- vinylidene
Vinylidenchlorid *n* vinylidene chloride
Vinylidenkunststoff *m* vinylidene plastic
Vinylidenplaste *pl* vinylidene plastics
Vinylierung *f* vinylation
Vinyljodid *n* vinyl iodide
Vinylkunststoff *m* vinyl plastic
Vinylpolymer[e] *n* vinyl polymer
Vinylpropionat *n* vinyl propionate
Vinylpyridin *n* vinyl pyridine
Vinylpyrrolidon *n* vinyl pyrrolidone
Vinylverbindung *f* vinyl compound
Vinyon *n* vinyon
Vioform *n* vioform, iodochlorohydroxyquinoline, nioform
Violanthren *n* violanthrene
Violanthron *n* violanthrone
Violaquercitrin *n* violaquercitrin
Violaxanthin *n* violaxanthin
violett violet
Violettblau *n* violet blue
Violettstein *m* (Min) iolite
Violin *n* violine

Violursäure *f* violuric acid, 5-isonitrosobarbituric acid
Violutin *n* violutin
Violxanthin *n* violxanthin
Virenkultur *f* virus culture
Virenssäure *f* virensic acid
Virial *n* virial
Virialgleichung *f* virial equation
Virialkoeffizient *m* virial coefficient
Virialsatz *m* virial theorem
Viridiflorin *n* viridiflorine
Viridiflorinsäure *f* viridifloric acid
Viridin *n* viridine
Virologe *m* virologist
Virologie *f* virology
Virose *f* (Med) virus disease
virotoxisch virotoxic
Virulenz *f* virulence
Virus *m* (pl Viren) virus (pl viruses)
Virusimpfstoff *m* virus vaccine
Viruskrankheit *f* (Med) virus disease, virosis
Viruskunde *f* virology
Virusproteid *n* (Virusprotein) virus protein
Virusstamm *m* viral strain, virus strain
Visammin *n* visammin
Visamminol *n* visamminol
Viscaoutschin *n* viscaoutschine
Viscin *n* viscin
Viscose *f* viscose
Visierapparat *m* sighting apparatus
Visierlinse *f* rear sighting lens
Visitiereisen *n* sounding rod
viskoelastisch visco-elastic
Viskoelastizität *f* visco-elasticity
viskoplastisch visco-plastic
viskos viscid, viscous
Viskose *f* (Viscose) viscose
Viskosefaden *m* viscose filament
Viskosefaser *f* viscose rayon
Viskoselösung *f* viscose [solution]
Viskosepumpe *f* viscose pump
Viskoserayon *n* viscose rayon
Viskoseschwamm *m* viscose sponge
Viskoseseide *f* viscose rayon
Viskosezellwolle *f* viscose staple fiber
Viskosimeter *n* viscosimeter, viscometer
Viskosimetrie *f* viscosimetry, viscometry
viskosimetrisch viscosimetric
Viskosität *f* viscosity, **absolute** ~ absolute viscosity, **anomale** ~ anomalous viscosity, **dynamische** ~ dynamic viscosity, **innere** ~ inherent viscosity, **plastische** ~ plastic viscosity, **scheinbare** ~ apparent viscosity
Viskositätsapparat *m* viscosimeter
Viskositätsbeziehung *f* viscosity relationship
Viskositätseffekt *m* viscosity effect
Viskositätsgrad *m* degree of viscosity
Viskositätsindex *m* viscosity index
Viskositätskonstante *f* viscosity coefficient

Viskositätskorrektur *f* correction for viscosity
Viskositätsmesser *m* viscosimeter, viscometer
Viskositätsverminderer *m* viscosity depressant
Viskositätszahl *f* Staudinger function
Visnagin *n* visnagin
visuell visual
Vitachrom *n* vitachrome
vital vital
Vitamin *n* vitamin[e], **antiskorbutisches** ~ antiscorbutic vitamin, **fettlösliches** ~ fat-soluble vitamin, **wasserlösliches** ~ water-soluble vitamin
Vitamineinheit *f* vitamin unit
Vitaminierung *f* vitamin enrichment
vitaminisieren to vitaminize
Vitaminkunde *f* vitaminology
Vitaminmangel *m* hypovitaminosis
vitaminreich rich in vitamins
Vitaminüberschuß *m* hypervitaminosis
Vitaminvorläufer *m* provitamin, previtamin
Vitellin *n* vitellin
Vitexin *n* vitexine
Vitiatin *n* vitiatine
Vitrifikation *f* vitrification
vitrifizieren to vitrify
Vitrinit *m* vitrinite
Vitriol *n* vitriol, **blaues** ~ blue vitriol, copper sulfate, vitriol of copper, **grünes** ~ copperas, ferrous sulfate, green vitriol, **weißes** ~ goslarite, white vitriol, zinc sulfate
vitriolartig vitriolic, vitriol-like
Vitriolbildung *f* formation of vitriol, vitriolation
Vitriolblei *n* lead vitriol
Vitriolbleierz *n* (Min) anglesite
Vitriolbleispat *m* (Min) anglesite
Vitriolerz *n* (Min) vitriolic ore
Vitriolgelb *n* jarosite
vitriolhaltig containing vitriol, vitriolic
Vitriolhütte *f* sulfuric acid works, vitriol works
vitriolisieren to vitriolate, to vitriolize
Vitriolisieren *n* vitriolating, vitriolizing
Vitriolkies *m* (Min) marcasite, white iron pyrites
Vitriolküpe *f* blue vat, copperas vat
Vitriollauge *f* vitriolic lye
Vitriollösung *f* solution of copper sulfate, solution of vitriol
Vitriolöl *n* (obs) oil of vitriol, sulfuric acid
Vitriolschiefer *m* (Min) pyritic schist, pyritic shale
Vitriolsiederei *f* sulfuric acid works, vitriol works
vitro, in ~ (Lat) in a test tube, in vitro
Vivarit *m* (Min) blue ochre, native Prussian blue
Vivianit *m* (Min) vivianite
vivo, in ~ (Lat) in the living organism, in vivo
vizinal vicinal, neighboring
Vizinalfläche *f* (Krist) vicinal face
V-Kerbe *f* vee-notch
Vlies *n* fleece

Voacanginsäure f voacangic acid
Voelckerit m (Min) voelckerite
Vogelkunde f ornithology
Vogelleim m bird glue, bird lime
Voglianit m (Min) voglianite
Voglit m (Min) voglite
Voigtit m (Min) voigtite
Volborthit m (Min) volborthite
Volemit m volemite, volemitol
voll full, entire, (randvoll) brimful
Vollast f full load
Vollasterregung f full load excitation
vollautomatisch fully automatic
Vollautomatisierung f [complete] automation
Vollbahnschiene f heavy rail, rail for trunk lines
Vollbauweise f (Massivbauweise) solid construction
Volldampf m full steam
volldimensioniert fully dimensioned
Volldüngemittel n complete fertilizer
Volldünger m compound fertilizer
vollelektrisch all-electric
vollenden to complete, to finish
vollendet completed, perfect
Vollendung f completion
Vollentsalzung f **von Wasser** complete softening of water
Vollfarbigkeit f color saturation
Vollfeuer n full heat
vollflächig (Krist) holohedral
Vollflächigkeit f (Krist) holohedry
Vollflächner m (Krist) holohedron
vollfüllen to fill [up]
Vollgerbstoff m synthetic replacement tannin
Vollgewebereifen m canvas tire
Vollgummireifen m solid rubber tire
Vollguß m solid casting
vollisoliert fully insulated
Vollkörper m solid body
vollkommen complete, perfect, ~ **mischbar** (Chem) completely miscible
Vollkonserve f fully preserved food
Vollkreis m complete circle, full circle
vollkristallin holocrystalline
Vollkugel f solid sphere
Vollkultur f (Biol) complete medium
Vollmantel m full jacket, solid jacket
Vollmilch f fat milk, unskimmed milk
Vollperiode f complete cycle
Vollpipette f (Anal) bulb pipet[te], plain pipet[te], transfer pipet[te], volumetric pipet[te]
vollplastisch fully plastic
Vollprofil n total cross section
Vollrad n solid wheel
Vollschwingung f full cycle
Vollseil n solid cable, solid rope
Vollsicht f full view
Vollsichtskala f full view scale
Vollspule f solid coil

vollständig complete, perfect, ~ **gar** (Metall) burned-off
Vollständigkeit f completeness
Vollständigkeitsrelation f completeness relation
Vollstopfen m (Vollstöpsel) solid stopper
vollsynthetisch fully synthetic
Vollton n (Pigment ohne Verschnittmittel) full strength, mass tone
Vollwandbauweise f plate girder construction
Vollwandzentrifuge f solid-bowl or imperforate-basket centrifuge
Vollweggleichrichter m (Elektr) full-wave rectifier
Vollziegel m solid brick
Volt n volt
Voltaelement n (Elektr) voltaic cell, voltaic element
Voltainduktor m volta induction coil
Voltait m (Min) voltaite
Voltakette f (Elektr) voltaic cell
Voltameter n (Elektr) voltameter
voltametrisch voltametric
Voltampere n (Elektr) voltampere, watt
Voltapotential n voltaic potential
Voltmesser m voltmeter
Voltmeter n voltmeter, ~ **mit Hilfsnebenwiderständen** voltmeter multiplier, **elektronisches** ~ **mit Spitzenablesung** electronic peak-reading voltmeter, **elektronisches** ~ thermionic voltmeter, valve voltmeter, **elektrostatisches** ~ electrostatic voltmeter, **vergleichendes** ~ slide-back voltmeter
Voltsekunde f (Maß) volt second
Voltspannung f (Elektr) voltage
Voltstärke f (Elektr) voltage
Volttransformator m voltage transformer
Voltverlust m (Elektr) voltage drop
Voltzahl f (Elektr) voltage
Voltzin m (Min) voltzine
Volumdilatation f cubic dilatation
Volumeinheit f unit of volume
Volumen n (pl Volumina) bulk, volume, **inkompressibles** ~ incompressible volume, **kritisches** ~ critical volume, **partielles spezifisches** ~ partial specific volume, **spezifisches** ~ specific volume, **unzusammendrückbares** ~ incompressible volume
Volumenabnahme f decrease in volume
Volumenänderung f change of volume, alteration in volume
Volumenanzeiger m volume indicator, volumescope
Volumenarbeit f constant pressure change
Volumenausdehnung f volume expansion
Volumendichte f volume density
Volumeneffekt m volume effect

Volumeneinheit *f* unit of volume
Volumenenergie *f* volume energy
Volumengewicht *n* volume weight, weight of unit volume
Volumenintegral *n* (Math) volume integral
Volumenionisierung *f* volume ionization
Volumenkontraktion *f* volume contraction
Volumenleitung *f* volume conduction
Volumenmesser *m* volumenometer, volumeter
Volumenometer *n* volume meter, volumenometer, volumeter
Volum[en]prozent *n* percent by volume
Volum[en]prozentgehalt *m* volume percent content, volumetric percentage
volum[en]prozentig percent by volume
Volum[en]prozentskala *f* volume percentage scale
Volum[en]teil *m* part by volume
Volum[en]verhältnis *n* volume proportion, volume ratio
Volumenverlust *m* loss in volume
Volumenvermehrung *f* increase in volume
Volumenverminderung *f* volume contraction
Volumenviskosität *f* volume or bulk viscosity
Volumenzähler *m* positive displacement meter
Volumenzunahme *f* increase in volume
Volumeter *n* volumenometer, volumeter
Volumetrie *f* (Anal) volumetric analysis
volumetrisch volumetric
Volumgewicht *n* specific gravity, volume weight, weight of unit volume
voluminös bulky, voluminous
Volumkontraktion *f* volume contraction
Volummesser *m* volumenometer, volumeter
Voluntal *n* voluntal, trichloroethylurethan
Vorabdruck *m* (Buchdr) preprint
Vorabscheider *m* (Öl) free water knock-out
Voralarm *m* preliminary alarm, warning signal
Voranreicherung *f* cobbing
Voranschlag *m* [provisional] estimate
Voranstrich *m* first coat
Vorarbeit *f* preliminary work, preparatory work
Vorauflaufherbizid *n* pre-emergence herbicide
vorausberechnen to calculate in advance, to precalculate, to predetermine
vorausberechnet precalculated
Vorausberechnung *f* precalculation, estimation
vorausbestimmen to predetermine, to determine in advance, to estimate
vorausgehen to antecede
Voraussage *f* forecast, prediction
voraussagen to forecast, to predict
Voraussetzung *f* prerequisite
Vorbedingung *f* precondition, prerequisite
Vorbehalt *m*, **unter ~ aller Rechte** all rights reserved
vorbehandeln to pretreat
Vorbehandlung *f* preliminary treatment, pretreatment, primary treatment
Vorbehandlungsmittel *n* preparative chemical

Vorbeize *f* preliminary mordant, weak mordant
Vorbeizen *n* first pickling
Vorbelastung *f* initial loading
vorbereiten to prepare
Vorbereitung *f* preparation, pretreatment
Vorbereitungsraum *m* preparation room
vorbeugen to prevent
vorbeugend (Med) prophylactic
Vorbeugungsmaßnahme *f* preventive measure
Vorbild *n* model, pattern
vorbilden to preform, to preshape
Vorbildung *f* qualification
Vorbrenne *f* pickling bath, preliminary pickle
Vorderachse *f* front axle
Vorderansicht *f* front view
Vorderantrieb *m* front-wheel drive
Vorderblende *f* (Opt) front diaphragm
Vordergrund *m* foreground
Vorderkante *f* leading edge
Vorderplatte *f* front plate
Vorderrad *n* front wheel
Vorderradausrichtung *f* front wheel alignment
Vorderradbremse *f* front-wheel brake
Vorderseite *f* face, front [side], (Metall) charging side
Vorderwalze *f* front roll
Vordrehen *n* roughing
Vordruck *m* preliminary pressure
voreilen to advance, to hasten, to lead
voreilig premature
Voreilwinkel *m* lead angle
vorenthalten to withhold
Vorentladung *f* predischarge
Vorentwässerung *f* preliminary desiccation
Vorentwurf *m* preliminary design
Vorerhitzung *f* preheating
Vorfabrikation *f* prefabrication
vorfertigen to prefabricate
Vorfertigung *f* prefabrication
Vorfeuerung *f* prefiring
Vorfilter *n* primary filter, roughing filter
Vorfiltern *n* preliminary filtering
Vorform *f* first shape
vorformen to preform
Vorformling *m* preform
Vorformring *m* shaping ring
Vorformverfahren *n* preform process
Vorfrischen *n* (Metall) primary purification
Vorfrischofen *m* primary furnace, refiner
vorführen to demonstrate
Vorführer *m* (Film, Dias) projectionist
Vorfülldruck *m* pilot pressure
Vorfüllventil *n* pilot valve
Vorgang *m* (Hergang) action, event, occurrence, (Prozeß) operation, procedure, process, reaction, **~ beim Verkoken** process of coking, **~ hoher Geschwindigkeit** high speed event, **aperiodischer ~** aperiodic phenomenon, **chemischer ~** chemical process, chemical

reaction, **einmaliger** ~ nonrecurrent action, unique action or phenomenon, **nicht umkehrbarer** ~ irreversible cycle, **periodischer** ~ cyclic operation, periodic operation, **umkehrbarer** ~ reversible process, **zeitlich veränderlicher** ~ action variable with time
Vorgarn *n* roving
Vorgelege *n* countershaft gear
Vorgelegewelle *f* countershaft, jack shaft
vorgemischt premixed
vorgerben to pre-tan
Vorgerbung *f* preliminary tannage, preliminary tanning
vorgereinigt pre-purified, purified in a preliminary way
vorgetäuscht simulated
vorgewärmt preheated
Vorglühen *n* annealing, preliminary heating
Vorglühherd *m* annealing furnace, forehearth
Vorglühofen *m* annealing furnace, annealing oven
Vorhalt *m* (Regeltechn) derivative action (Am. and Br. E.), rate action (Am. E.)
Vorhaltzeit *f* derivative action time (Brit. E.), rate time (Am. E.)
Vorhandensein *n* presence
Vorhang *m* curtain, sag (paint)
Vorhauserit *m* (Min) vorhauserite
vorheizen to preheat
vorher festlegen to preset
Vorherd *m* forehearth
vorhergehend preliminary
vorherrschen to prevail
vorherrschend prevailing, prevalent
Vorhersage *f* forecast, prediction
vorhersagen to forecast
vorhersehen to foresee
Vorholfeder *f* recoil spring
Vorkaliber *n* shaping groove
Vorkalkulation *f* precalculation, preliminary calculation
vorkalkulieren to precalculate
Vorkammer *f* (Motor) precombustion chamber
Vorkammerdurchspritzverfahren *n* runnerless molding
vorkeimen (Biol) to pregerminate
Vorknetung *f* preliminary mastication
vorkommen to occur, (erscheinen) to appear, (sich ereignen) to happen, **natürlich** ~ naturally occurring
Vorkommen *n* occurrence, (Min) deposit
Vorkondensat *n* precondensate
Vorkondensation *f* precondensation
Vorkondensator *m* preliminary condenser
vorkühlen to precool
Vorkühler *m* precooler, forecooler, primary cooler
Vorkühlung *f* precooling, forecooling

Vorlack *m* primer
Vorlackfarbe *f* undercoat
Vorläufer *m* forerunner, precursor
vorläufig preliminary, provisional
Vorlage *f* receiver, condensate tank, recipient vessel
Vorlauf *m* (Dest) first runnings, first light oil (coal tar), fore-runnings
Vorlaufbehälter *m* preliminary liquor tank
Vorlaugung *f* preliminary leaching
Vorleimen *n* (Papier) presizing
Vorlesung *f* lecture
Vorlesungsversuch *m* lecture demonstration
Vormagnetisierung *f* premagnetization
Vormaischbottich *m* steeping trough
Vormischbrenner *m* premix burner
vormischen to pre-mix
Vormischung *f* master batch
vorplastifizieren to plasticize on the surface
Vorplastifiziermethode *f* preplasticizing method
Vorpolarisierung *f* prepolarization
vorpolieren to give the first polish, to prepolish
Vorpreßverfahren *n*, **hydraulisches** ~ pipe-pushing
Vorprobe *f* preliminary test, quick test
Vorprodukt *n* starting product, (Dest) first runnings
Vorprüfung *f* preliminary test, previous examination
Vorpumpwerk *n* low-lift pumping station
vorraffinieren to improve (lead), to soften
Vorraffinierofen *m* softening furnace
vorragend projecting
Vorrang *m* priority
Vorrat *m* stock, supply
Vorratsbehälter *m* storage vessel
Vorratsflasche *f* stock bottle
Vorratsgefäß *n* reservoir, storage flask, storage vessel
Vorratslösung *f* (Chem) stock solution
Vorratsraum *m* store room
Vorreaktion *f* preliminary reaction, prereaction, ~ **im Brenngemisch** preflame ignition
Vorrecht *n* privilege
Vorrede *f* introduction
Vorreduktion *f* preliminary reduction, prereduction
vorreduzieren to prereduce
Vorregelung *f* precontrol
vorreinigen to pre-purify, to purify in a preliminary way
Vorreiniger *m* preliminary purifier
Vorreinigung *f* preliminary purification, preliminary cleaning
Vorrichtung *f* apparatus, appliance, device, facility, gadget
vorrösten to pre-roast
Vorrösten *n* (Metall) preroasting
Vorröstofen *m* preroasting furnace

vorrücken to advance, to move forward
Vorsatzbeton *m* concrete for facing, face mix
Vorsatzlinse *f* (Opt) front or supplementary lens
Vorschaltkasten *m* (Elektr) shunt box
Vorschaltturbine *f* superposed turbine
Vorschaltwiderstand *m* (Elektr) rheostat
vorschleifen to rough polish
Vorschneidfräser *m* roughing cutter, stocking cutter
Vorschrift *f* direction, instruction, regulation, rule, (Spezifikation) specification
Vorschriften *pl*, **gesundheitspolizeiliche** ~ department-of-health regulations,
   technische ~ instructions, specifications
vorschriftsmäßig according to regulations
Vorschub *m* feed
Vorschubänderung *f* change of feed
Vorschubbereich *m* range of feed
Vorschubdaumen *m* spacing cam
Vorschubgetriebe *n* feed-gear mechanism
Vorschubkopf *m* feeder head
Vorschubrolle *f* feed roll
Vorschubspannung *f* feed voltage
Vorschubwalze *f* feed roller
Vorschweißbund *m* welding-neck collar
Vorschweißflansch *m* welding-neck flange
Vorsicht *f* care, caution, precaution
vorsichtig careful, cautious
Vorsichtsmaßnahme *f* precaution
Vorsichtsmaßregel *f* precaution, precautionary measure, safety order
Vorsieb *n* forescreen
vorsintern to presinter
Vorsorge *f* provision
Vorsortieren *n* presorting, preclassifying
Vorspann *m* (recording tape) leader
Vorspannung *f* initial stress, initial tension, prestressing, (Elektr) bias, primary potential, (Mech) initial stress, **negative** ~ negative bias
vorspringen to project
Vorsprung *m* head start, lead, (Bosse) boss, (vorspringender Teil) projection
Vorspur *f* (Kfz) toe-in
Vorstand *m* board of directors
Vorsteckkeil *m* cotter
Vorsteckstift *m* stop pin
Vorsteuerdruck *m* pilot pressure
Vorstoff *m* raw material, semifinished goods
Vorstoß *m* adapter, outlet piece
Vorstraße *f* blooming mill, cogging mill, roughing mill
Vorstrecke *f* blooming mill, blooming strand, cogging mill, cogging strand, roughing mill
Vorstreckung *f* prestretch
Vorstreichen *n* prime-coating
Vorstreichfarbe *f* priming paint
Vorstufe *f* first step, primary step
vortäuschen to simulate
Vorteil *m* advantage

vorteilhaft advantageous
Vortexring *m* collar vortex, vortex ring
Vortrag lecture
Vortragende(r) *m* speaker
vortrocknen to predry
Vortrockner *m* predryer
Vortrocknung *f* predrying, preliminary drying
vorübergehend temporary, transient
Voruntersuchung *f* preliminary investigation
Vorvakuumbeständigkeit *f* stability of forevacuum
Vorverbrennung *f* precombustion
Vorverdampfer *m* pre-evaporator
Vorverdampfung *f* pre-evaporation
vorverdichten to precompress
Vorverstärker *m* preamplifier, input amplifier
Vorverstärkung *f* gain amplification, input amplification
Vorversuch *m* preliminary experiment, preliminary test
Vorvulkanisation *f* (Gummi) setting cure
vorwählen to pre-select
Vorwärmeherd *m* forehearth
Vorwärmemantel *m* heater shell
vorwärmen to preheat
Vorwärmen *n* preheating
Vorwärmeofen *m* preheating furnace, preheating oven
Vorwärmer *m* preheater, ~ **für Speisewasser** feed-water heater
Vorwärmewerk *n* warm-up mill
Vorwärmezone *f* preheating zone, heating-up zone, preliminary heating zone
Vorwärmkammer *f* preheating chamber
Vorwärmschrank *m* preheating cabinet
Vorwärmung *f* preheating, preliminary heating
vorwärtsstoßen to push forward
Vorwärtsstreuung *f* forward scattering
Vorwäsche *f* preliminary washing, (durch Absorption) first absorption
Vorwalze *f* billet roll, bloom roll
vorwalzen to cog [down], to roll
Vorwalzstrecke *f* roughing strand of rolls
Vorwalzwerk *n* blooming mill, cogging mill, roughing mill
Vorwaschen *n* preliminary washing
Vorwaschtrommel *f* preliminary washing drum
Vorwiderstand *m* (Elektr) preresistor, compensating resistance
Vorzeichen *n* indication, (Math) sign, **negatives** ~ negative sign, **positives** ~ positive sign
Vorzeichenbestimmung *f* (Math) determination of sign, sign determination
Vorzeichenfestsetzung *f* (Math) sign convention
Vorzeichenkonvention *f* (Math) sign convention
Vorzeichenregel *f* rule of signs
Vorzeichenumkehrung *f* reversal of signs
Vorzeichenwechsel *m* change of sign

vorzeitig premature
Vorzerkleinern *n* coarse to medium crushing; jaw crushing, preliminary breaking, primary crushing
vorziehen to draw forth, to draw forward
Vorziehstempel *m* assisting plug
vorzüglich excellent, choice, preferable, superior
Vorzüglichkeit *f* excellence, superiority
Vorzündung *f* advanced ignition, early ignition, pre-ignition
Vorzug *m* preference
Vorzugsmilch *f* certified milk
Vorzugsrichtung *f* preferred direction
V-Profil *n* angle section
V-Riemen *m* V-belt, vee belt
Vulkan *m* volcano
Vulkanasche *f* volcanic ash
Vulkanausbruch *m* volcanic eruption
Vulkanfiber *f* vulcanized fiber
Vulkanisat *n* cured good
Vulkanisation *f* vulcanization, curing
Vulkanisationsaktivator *m* activator for curing agents
Vulkanisationsbeschleuniger *m* (Gummi) vulcanizing agent
Vulkanisationsniveau *n* level of cure
Vulkanisator *m* vulcanizer
vulkanisch volcanic
Vulkanisieranstalt *f* vulcanization works
Vulkanisierapparat *m* vulcanizer
vulkanisierbar vulcanizable
vulkanisieren to vulcanize, to cure
Vulkanisieren *n* vulcanizing, curing
Vulkanisierkessel *m* vulcanizing kettle
Vulkanisiermittel *n* curing agent
Vulkanisierpresse *f* vulcanizing press
vulkanisiert vulcanized
Vulkanisierung *f* vulcanization
Vulkanit *n* (Hartgummi) hard rubber, vulcanite
Vulkanitfaser *f* vulcanized fiber
Vulkanologie *f* volcanology
Vulkanschörl *m* (Min) volcanic schorl
Vulkollan *n* vulcollan
Vulpinit *m* (Min) vulpinite
Vulpinsäure *f* vulpic acid, chrysopicrin, vulpinic acid
Vuzin *n* vuzine, isooctylhydrocupreine

W

Waage *f* balance, scales, weighing machine, **aerodynamische** ~ wind-tunnel [balance], **analytische** ~ analytical balance, precision balance, **automatische** ~ automatic balance, **hydrostatische** ~ hydrostatic balance, **magnetische** ~ magnetic balance, **metallometrische** ~ metallometric balance, **Mohrsche** ~ Mohr balance, Westphal balance, **Westphalsche** ~ Westphal balance, Mohr balance
Waagebalken *m* balance arm, balance beam
Waagekasten *m* balance case
Waagengehäuse *n* scale housing, balance case
Waag[e]rechte *f* horizontal [line]
Waag[e]rechtförderung *f* (Techn) level transportation
Waagerechtintensität *f* horizontal intensity
Waageschneide *f* balance blade
waagrecht horizontal
Waagschale *f* weighing dish, balance pan, pan for weights, scale pan
Wabe *f* honeycomb
Wabenstruktur *f* honeycomb structure
Wacholder *m* (Bot) juniper
Wacholderbranntwein *m* gin
Wacholdercampher *m* juniper camphor
Wacholdergeist *m* gin
Wacholderharz *n* juniper resin, gum juniper, juniper gum
Wacholderöl *n* juniper oil
Wacholderschnaps *m* gin
Wachs *n* wax, ~ **zur Verhinderung von Lichtrissen** sunproof wax, **gebleichtes** ~ bleached wax
wachsähnlich wax-like, waxy
wachsartig wax-like
Wachsbeize *f* wax stain
Wachsdraht *m* waxed wire, wax insulated wire
Wachsemulsion *f* wax emulsion
wachsen to grow, to expand, to increase, (einwachsen) to wax
Wachsen *n* growth, increase, (Einwachsen) waxing
Wachsfaden *m* waxed thread
Wachsfarbe *f* wax color
wachsfarben wax-colored
Wachsfirnis *m* wax varnish
Wachsform *f* wax mold
Wachsgehalt *m* wax contents
wachsgelb wax-yellow
Wachsgießtisch *m* wax pouring table
Wachskaschierung *f* wax backing
Wachskitt *m* luting wax, wax cement
Wachsleim *m* (Pap) wax size
Wachsleimung *f* (Pap) [paraffin-]wax sizing
Wachsmalerei *f* wax painting, encaustic painting

Wachsmatrize *f* wax matrix
Wachsmodell *n* wax model
Wachsöl *n* wax oil
Wachsopal *m* (Min) wax opal
Wachspapier *n* wax paper
Wachspoliermittel *n* wax polish
Wachspräparat *n* preparation in wax
Wachssäure *f* (obs) ceric acid
Wachssalbe *f* wax ointment, cerate
Wachsschmelze *f* (Wachsschmelzerei) wax melting house
Wachsschmelzkessel *m* wax melting pot
Wachsschmelztisch *m* wax pouring table
Wachsschweiß *m* waxy yolk
Wachstuch *n* oil cloth, wax cloth
Wachstuchlack *m* oil cloth finish, oil cloth varnish
Wachstum *n* growth, increase
Wachstumsfaktor *m* (Biol) growth factor
wachstumsfördernd (Biol) growth-promoting
Wachstumsfront *f* (Krist) growth front
Wachstumsgeschwindigkeit *f* rate of growth
wachstumshemmend growth-inhibiting
wachstumshindernd growth-inhibiting, (Bakt) bacteriostatic
Wachstumshormon *n* (Biochem) growth hormone
Wachstumskurve *f* growth curve
Wachstumsruhe *f* (Biol) dormancy
Wachstumsspirale *f* growth spiral
Wachstumszentrum *n* center of growth
Wacke *f* (Geol) wacke
Wackelkontakt *m* (Elektr) defective contact, loose connection
Wackenrodersche Flüssigkeit *f* Wackenroder's solution
Wad *n* (Min) wad, bog manganite
Waderz *n* (Min) wad ore
wägbar weighable, ponderable
Wägbarkeit *f* weighability, ponderability
Wägebürette *f* (Anal) weight buret[te]
Wägefläschchen *n* (Chem) weighing bottle
Wägegenauigkeit *f* weighing accuracy
Wägegläschen *n* weighing bottle
Wägeglas *n* weighing bottle
wägen to weigh, to balance
Wägepipette *f* (Anal) weighing pipet[te]
Wägeröhrchen *n* weighing tube
Wägesatz *m* set of weights
Wägeschiffchen *n* **aus Aluminium** aluminium weighing boat
Wägeserie *f* serial weighings
Wägetisch *m* weighing table
Wägevorrichtung *f* weighing appliance, weighing device
Wägezimmer *n* balance room, weighing room

Wägung f weighing
wählen (auswählen) to choose, to select, (Wählscheibe) to dial
Wählerscheibe f dial, selector disk
wälzen to roll, to revolve, to turn over
Wälzgas n circulated gas
Wälzkolbenpumpe f lobe pump
Wälzkreis m pitch circle
Wälzlager-Schnappkäfig m snap-ring needle bearing
Wärme f heat, warmth, ~ **aufspeichern** to store heat, ~ **ausstrahlen** to radiate heat, ~ **durchlassen** to transmit heat, **abgegebene** ~ heat conducted off, heat given off, **aufgespeicherte** ~ stored heat, **latente** ~ latent heat, **molare** ~ molar heat, **potentielle** ~ potential heat, **spezifische** ~ specific heat, heat capacity, **strahlende** ~ radiant heat, **zugeführte** ~ added heat, heat supplied, received heat
Wärmeabfall m heat drop
Wärmeabfuhr f (Ableitung) heat flow, heat removal
Wärmeabgabe f heat emission, (Verlust) loss of heat, **adiabatische** ~ adiabatic heat drop
Wärmeabgabefläche f heat emitting surface
wärmeabgebend exothermal
Wärmeableitung f dissipation of heat, heat abstraction
Wärmeabnahme f temperature decrease, temperature drop
Wärmeabsorption f heat absorption, thermal absorption
Wärmeäquivalent n equivalent of heat
Wärmeakkumulator m heat accumulator
Wärmearbeitswert m mechanical equivalent of heat
Wärmeaufnahme f absorption of heat
Wärmeaufnahmefähigkeit f heat absorption capacity, caloric receptivity
Wärmeaufspeicherung f accumulation of heat, heat storage
Wärmeausbeute f thermal efficiency, thermal [energy] yield
Wärmeausbreitung f thermal diffusion
Wärmeausbreitungsvermögen n thermal diffusivity
Wärmeausdehnung f heat expansion, thermal expansion
Wärmeausdehnungskoeffizient m coefficient of thermal expansion
Wärmeausdehnungsvermögen n thermal expansivity
Wärmeausdehnungszahl f coefficient of thermal expansion
Wärmeausgleich m heat compensation, heat equalization
Wärmeausgleichung f heat interchange
Wärmeausnutzung f utilization of heat

Wärmeausstrahlung f radiation of heat, interchange of heat
Wärmeaustausch m heat exchange, interchange of heat
Wärmeaustauscher m heat exchanger, **Doppelrohr-** ~ double-pipe heat exchanger, **Dünnschicht-** ~ thin film heat exchanger, **Film-** ~ thin film heat exchanger, **Gegenstrom-** ~ countercurrent heat exchanger, **Glattrohrbündel-** ~ shell and tube heat exchanger, **Gleichstrom-** ~ cocurrent heat exchanger, **Kratz-** ~ scraped surface heat exchanger, **Kreuzstrom-** ~ crossflow heat exchanger, **Lamellen-** ~ segment heat exchanger, **Platten-** ~ plate heat exchanger, **Riesel-** ~ trickled-surface heat exchanger, **Rippenrohr-** ~ extended surface heat exchanger, finned tube heat exchanger, **Rohrbündel-** ~ heat exchanger with tube bundles, **Schnecken-** ~ screw-flight heat exchanger, **Spiral-** ~ spiral heat exchanger
Wärmeaustauschkoeffizient m coefficient of heat transfer
Wärmebedarf m heat requirement
Wärmebehälter m heat reservoir
Wärmebehandlung f heat treatment, thermal treatment
wärmebeständig heatproof, heat-resistant
Wärmebeständigkeit f heatproof quality, heat resistance, heat stability, resistance to heat, thermal stability
Wärmebewegung f heat motion, thermal agitation, thermal motion
Wärmebilanz f heat balance, calorific balance, thermal balance
Wärmebildner m heat producer
Wärmebildung f production of heat
Wärmeblitz m heat flash
Wärmebrennpunkt m heat focus
Wärmedämmwert m thermal resistance
Wärmedehnung f thermal expansion, [heat] dilatation
Wärmedehnungsmesser m [heat] dilatometer
Wärmedehnzahl f thermal expansion coefficient
Wärmediffusion f thermodiffusion
Wärmedurchgang m heat transmission, heat transfer
Wärmedurchgangskoeffizient m overall coefficient of heat transfer, thermal transmittance
Wärmedurchgangszahl f overall coefficient of heat transfer, thermal transmittance
wärmedurchlässig diathermanous, diathermic
Wärmedurchlässigkeit f diathermancy, heat conductance
Wärmedurchlaßwiderstand m thermal resistance
Wärmedurchschlag m heat breakdown, thermal breakdown
Wärmeeinheit f thermal unit, unit of heat

Wärmeeinstrahlung *f* heat irradiation
wärmeempfindlich sensitive to heat
Wärmeenergie *f* heat energy, thermal energy
Wärmeentwicklung *f* development of heat, evolution of heat
Wärmeentziehung *f* abstraction of heat
Wärmeentzug *m* heat abstraction, cooling
Wärmeerhöhung *f* increase of heat
wärmeerzeugend exothermic, heat-producing
Wärmeerzeuger *m* heat generator, thermogenerator
Wärmeerzeugung *f* production of heat, generation of heat
Wärmefaktor *m* heat factor
Wärmefassungsvermögen *n* heat capacity
Wärmefluß *m* heat flux, thermal flow
Wärmeflußbild *n* heat flow layout
Wärmefortpflanzung *f* transference of heat
Wärmefreimachung *f* heat release
Wärmefühler *m* thermocouple, temperature bulb
Wärmegefälle *n* heat gradient, heat drop, temperature gradient
Wärmegleichgewicht *n* thermal equilibrium
Wärmegleichung *f* heat equation
Wärmegrad *m* degree of heat, temperature
Wärmegradient *m* temperature gradient
Wärmegrube *f* [heated] soaking pit
wärmehärtbar thermosetting
Wärmehaushalt *m* heat balance, heat economy
Wärmeimpulsschweißen *n* thermal impulse welding
Wärmeinhalt *m* heat capacity
Wärmeintensität *f* intensity of heat
Wärmeionisation *f* temperature ionization, thermal ionization
Wärmeisolation *f* heat insulation
Wärmeisolator *m* heat insulator
wärmeisolierend heat-insulating
wärmeisoliert heat-insulated, thermally insulated
Wärmeisolierung *f* heat insulation, thermal insulation
Wärmekapazität *f* heat capacity
Wärmekoeffizient *m* temperature coefficient
Wärmekonvektion *f* [heat] convection
Wärmekraft *f* thermal power
Wärmekraftanlage *f* thermal power station
Wärmekraftwerk *n* heat-generating station
Wärmekreislauf *m* thermal cycle
Wärmelagerung *f* high temperature aging
Wärmelampe *f* thermo-lamp
Wärmelehre *f* thermodynamics
Wärmeleistung *f* heat output, thermal efficiency, (Nutzeffekt) heat efficiency
wärmeleitend heat-conducting
Wärmeleiter *m* conductor of heat, transmitter of heat
Wärmeleitfähigkeit *f* thermal conductivity
Wärmeleitfähigkeitsdetektor *m* thermal-conductivity cell

Wärmeleitfähigkeitskoeffizient *m* thermal conductivity
Wärmeleitstift *m* heating pin
Wärmeleitung *f* conductance of heat, heat conduction, thermal conduction
Wärmeleitungskurve *f* heat-conductivity curve
Wärmeleitungsmanometer *n* thermal-conductivity gauge
Wärmeleitvermögen *n* (Wärmeleitungsvermögen) heat conductivity, thermal conductivity, **spezifisches** ~ (Therm) coefficient of thermal conductivity
Wärmeleitwiderstand *m* thermal resistance
Wärmeleitzahl *f* coefficient of thermal conductivity
wärmeliefernd exothermal
Wärmemagnetismus *m* thermo-magnetism
Wärmemaschine *f* heat engine
Wärmemenge *f* amount of heat, quantity of heat
Wärmemengenmesser *m* calorimeter
Wärmemengenmessung *f* calorimetry
Wärmemeßbombe *f* calorimetric bomb
Wärmemeßeinrichtung *f* heat measuring device
Wärmemesser *m* calorimeter
Wärmemeßgerät *n* (Hitzemesser) pyrometer
Wärmemessung *f* calorimetry, measurement of heat
wärmen to warm [up], to heat
Wärmenutzung *f* utilization of heat
Wärmeofen *m* [re]heating furnace
Wärmephänomen *n* thermal phenomenon
Wärmeplatte *f* hot plate
Wärmepol *m* heat pole
Wärmepumpe *f* heat pump
Wärmequelle *f* source of heat
Wärmerauschen *n* thermal agitation noise
Wärmeregler *m* thermostat, heat regulator, thermoregulator
Wärmeregulator *m* thermoregulator
Wärmereiz *m* thermal stimulus
Wärmeröhre *f* hot air pipe
Wärmerohr *n* heat pipe, heat tube
Wärmerückgewinnung *f* heat recovery
Wärmerückstrahlung *f* heat reflection
Wärmesammler *m* heat accumulator
Wärmeschalter *m* thermal switch
Wärmeschrank *m* drying oven, incubator
Wärmeschutz *m* heat insulator or insulation, thermic protection
Wärmeschutzhaube *f* heat insulating jacket
Wärmeschutzmantel *m* heat insulator casing
Wärmeschutzmasse *f* heat insulating material, heat insulator, non-conducting material
Wärmeschutzmittel *n* heat insulator, non-conducting material
Wärmeschutzplatte *f* thermal insulation board
Wärmeschutzvorrichtung *f* thermal shield
Wärmeschwankung *f* heat fluctuation, heat variation

Wärmeschwingung *f* heat vibration
Wärmesensibilisierungsmittel *n* heat sensitizing agent
Wärmeskala *f* heat scale
Wärmespannung *f* thermal stress
Wärmespeicher *m* heat reservoir, heat storage, regenerator
Wärmespeicherung *f* heat storage
Wärmespektrum *n* heat spectrum, thermal spectrum
Wärmestabilität *f* heat stability, thermal stability
Wärmestandfestigkeit *f* (Kerzen) bending point
Wärmestau *m* heat accumulation
Wärmestauung *f* accumulation of heat
Wärmestrahl *m* heat ray
Wärmestrahler *m* heat radiator
Wärmestrahlung *f* radiation of heat, radiant heat, thermal radiation
Wärmestrahlungsfühler *m* heat radition sensing device
Wärmestrahlungsheizung *f* radiant heating system
Wärmestrahlungskoeffizient *m* coefficient of radiation of heat
Wärmeströmung *f* convection of heat, heat flow, heat flux
Wärmestrom *m* flow of heat, heat flux
Wärmestromdichte *f* rate of heat flow per unit area
Wärmesummen *pl*, **Satz der konstanten ~ von Hess** law of constant summation of heat of Hess
Wärmetechnik *f* heat engineering
Wärmetheorem *n* **von Nernst** Nernst heat theorem
Wärmetönung *f* (Chem) evolution of heat, heat change, heat effect, (Reaktionswärme) heat of reaction
Wärmeträger *m* heat transfer fluid, heat transfer medium
Wärmetransport *m* heat transport
Wärmeübergang *m* heat transfer, passage of heat, transmission of heat
Wärmeübergangskoeffizient *m* heat transfer coefficient
Wärmeübergangszahl *f* heat transfer coefficient
Wärmeüberschuß *m* excess of heat
Wärmeübertragung *f* heat transfer, transmission of heat
Wärmeübertragungsmittel *n* heat transfer medium
Wärmeübertragungszahl *f* heat transfer coefficient
Wärmeumlauf *m* heat circulation
Wärmeumsatz *m* heat transformation
wärmeunbeständig thermolabile
Wärmeundichtigkeit *f* heat leak
wärmeundurchlässig athermanous, heat-impermeable, heat insulating

Wärmeunterschied *m* difference in temperature
Wärmeverbrauch *m* heat consumption
Wärmevergütung *f* heat-treatment
Wärmeverlauf *m* heat gradient
Wärmeverlust *m* loss of heat
Wärmeversprödung *f* thermal embrittlement
Wärmeverteilung *f* heat distribution
Wärmevervielfältiger *m* thermo-multiplier
Wärmeverzug *m* thermal lag
Wärmevorgang *m* thermal process
Wärmewert *m* calorific value, heat value
Wärmewiderstand *m* thermal resistance
Wärmewirkung *f* heat effect
Wärmewirkungsgrad *m* thermal efficiency, calorific efficiency
Wärmezerstreuung *f* heat dissipation
Wärmezirkulation *f* heat circulation
Wärmezufuhr *f* heat input, heat supply, supply of heat
Wäsche *f* (Text) linen, (Waschen) wash[ing]
Wäscher *m* washer, (für Gas) gas washing bottle, scrubber, **Düsen-** ~ jet scrubber, **Film-** ~ falling-film scrubber, **Venturi-** ~ venturi scrubber, **Zyklon-** ~ cyclone scrubber
wäss[e]rig aqueous, hydrous, watery, ~ **alkoholisch** aqueous alcoholic
Wässerigkeit *f* wateriness
wässern to water, (besprengen) to sprinkle, (eintauchen) to soak, (hydratisieren) to hydrate, (verdünnen) to dilute, **den Film ~** to wash the film
Wagenfeder *f* car spring
Wagengestell *n* carriage frame
Wagenheber *m* jack
Wagenladung *f* car load, truck load
Wagenpflegemittel *n* car polish
Wagenschmiere *f* axle grease, wagon grease, wheel grease
Wagnerit *m* (Min) wagnerite
Wagner-Meerwein-Umlagerung *f* (Chem) Wagner-Meerwein rearrangement
Wagonrücker *m* pinch bar
Wagonwaage *f* truck balance, wagon balance
Wahl *f* choice, alternative, (Auswahl) selection
Wahlmagnet *m* selecting magnet
wahlweise by choice, optional, selective
wahr true, correct, (Phys) intrinsic
Wahrheitsbekräftigung *f* affirmation
wahrnehmbar noticeable, perceptible, (sichtbar) visible, **mit dem Auge ~** visible;, **mit dem bloßen Auge ~** visible to the naked eye
Wahrnehmbarkeit *f* perceptibility
wahrnehmen to observe, to notice, to perceive
Wahrnehmung *f* observation, perception
Wahrnehmungsgrenze *f* limit of perception
Wahrnehmungsvermögen *n* perception
wahrscheinlich likely, probable, probably
Wahrscheinlichkeit *f* probability, likelihood, **bedingte ~** conditional probability

Wahrscheinlichkeitsdichte *f* probability density
Wahrscheinlichkeitsfaktor *m* probability factor
Wahrscheinlichkeitsgesetz *n* probability law
Wahrscheinlichkeitsintegral *n* probability integral
Wahrscheinlichkeitskurve *f* probability curve
Wahrscheinlichkeitsrechnung *f* calculus of probability
Wahrscheinlichkeitsverteilung *f* probability distribution
Wahrscheinlichkeitswelle *f* probability wave
Waid *n* (Färb) woad
Waidblau *n* woad blue
Walait *m* walaite
Walchowit *m* walchowite
Waldensche Umkehrung *f* (Chem) Walden inversion
Waldmeister *m* (Bot) woodruff
Waldmeisteröl *n* sweet woodruff oil
Waldwirtschaft *f* forestry, wood culture
Walfischspeck *m* blubber
Walfischtran *m* train oil, whale-oil
Walkbassin *n* trough
walkecht fast to milling
walken to mill, (Reifen) to flex
Walkererde *f* fuller's earth, soap rock
Walkerit *m* (Min) walkerite
Walkermüdung *f* fatigue under scrubbing
Walkplatte *f* rolling plate
Walkwiderstand *m* resistance to flexing, resistance to scrubbing
Walkzone *f* (Reifen) flexing area
wallen to boil, to bubble
Wallen *n* boiling, bubbling, ebullition
Wallplatte *f* baffle plate, dam plate
Wallstein *m* baffle stone, dam stone
Walnuß *f* (Bot) walnut
walnußgroß walnut sized
Walnußöl *n* walnut oil
Walnußschale *f* walnut shell
Walnußschalenmehl *n* walnut shell flour
Walone *f* (Gerb) valonia
Walpurgit *m* (Min) walpurgit
Walrat *m* spermaceti
Walratöl *n* sperm oil
Waltherit *m* (Min) waltherite
Waluewit *m* (Min) waluewite
Walzanode *f* rolled anode
Walzarbeit *f* rolling energy
Walzbahn *f* path of rolling, rolling curve
walzbar rollable
Walzbarkeit *f* rolling ability
Walzbarren *m* rolling ingot, rolling slab
Walzbetrieb *m* rolling mill operation, rolling mill practice
Walzblech *n* rolled sheet [metal]
Walzblei *n* rolled lead, sheet lead
Walzblock *m* bloom
Walzdorn *m* roll mandrel

Walzdruck *m* rolling pressure, **direkter** ~ direct rolling pressure, **indirekter** ~ side rolling pressure
Walze *f* roll, roller, (Kalander) calender, (Trommel) cylinder, drum, **direkt angetriebene** ~ direct coupled roll, **drehbare** ~ revolving roll, **vorgelagerte** ~ offset roll
Walzeisen *n* rolled iron
Walzeisenträger *m* rolled iron beam
walzen to roll [out], (Papier) to mill
Walzen *n* rolling, (Kunstst) calendering
Walzenabstand *m* distance between rolls
Walzenauftragmaschine *f* doctor roll, roll coater
Walzenauftragverfahren *n* roll coating
Walzenbart *m* burr, fin
Walzenbewegung *f* rotation of rolls
Walzenbezug *m* cover for roller, roller covering
Walzenbrecher *m* double-roll mill, gyratory crusher, rolling crusher
Walzenbreite *f*, **wirksame** ~ roll face
Walzenbruch *m* fracture of roll
Walzenbürste *f* roll brush
Walzendrehbank *f* (Techn) roll lathe
Walzendrehmoment *n* roll torque
Walzendruck *m* pressure of the rolls
Walzendruckpapier *n* roller printing paper
Walzendurchbiegung *f* roll bending
walzenförmig cylindrical
Walzenform *f* cylindrical shape
Walzenfräser *m* side milling cutter, (Techn) plain milling cutter
Walzenfurche *f* groove, pass
Walzenlack *m* roll coating varnish
Walzenlager *n* roller bearing
Walzenmasse *f* (Buchdr) roller composition
Walzenmesser *n* roll blade
Walzenmühle *f* roll[er] mill
Walzenmund *m* nip of the rolls
Walzenpresse *f* briquetting rolls, roller press
Walzenriefelung *f* grooving of rollers
Walzenrostfeuerung *f* furnace with cylinder grate
Walzenscheider *m* induced roll separator
Walzenschrägverstellung *f* roll crossing
Walzensinter *m* mill cinder, mill scale
Walzenspalt *m* nip of the rolls, roll clearance or gap
Walzenspalteinstellung *f* mill setting, roll-nip adjustment
Walzenständer *m* mill housing
Walzenstirnfräser *m* face cutter, plain milling cutter
Walzenstraße *f* rolling train
Walzenstrecke *f* strand of rolls
Walzenstreichverfahren *n* knife coating
Walzenstuhl *m* roller carriage arm, (Buchdr) cylinder support, roller frame
Walzentrockner *m* drum dryer, rolling dryer

Walzenverschleiß *m* wear of rolls
Walzenvorschub *m* roller feed
Walzenwehr *n* roller weir
Walzenwiderstand *m* drum rheostat
Walzenzapfen *m* roll neck, (Gummi) cylinder journal
Walzenzugmaschine *f* rolling mill engine
Walzfell *n* rolled sheet, rolling sheet
Walzfläche *f* rolling surface
Walzfolie *f* rolled sheet
Walzgerüst *n* rolling stand
Walzgut *n* rolling stock
Walzhaut *f* rolling sheet
Walzkessel *m* roll pot, roll vat
Walzkreis *m* rolling circle
Walzkupferplatte *f* rolled copper plate
Walzkurve *f* contact curve
Walzlackiermaschine *f* roller coater
walzlackiert roller-coated
Walzlager *n* antifriction bearing
Walzlegierung *f* rolled alloy
Walzmühle *f* tumbling mill
Walzplattieren *n* cladding
Walzprofil *n* rolled shape
Walzrichtung *f* grain direction
Walzringmühle *f* centrifugal roll mill
Walzsieb *n* revolving screen
Walzstraße *f* mill train, train of rolls
Walzstrecke *f* mill train, set of rolls, train of rolls
Walztafel *f* rolling sheet
Walztisch *m* roller table
Walzverfahren *n* roller coating, roller process, rotary furnace method
Walzvorgang *m* rolling operation
Walzwerk *n* calender, rollers, rolling mill, twin roll
Walzzinn *n* laminated tin
Walzzunder *m* (Met) roll scale
Wand *f* wall, (Zwischenwand) partition, **halbdurchlässige** ~ semipermeable wall, semipermeable diaphragm, semipermeable membrane, **lichtdurchlässige** ~ light admitting wall
Wandanschluß *m* (Elektr) wall socket
Wandarm *m* wall bracket
Wandauskleidung *f* (Metall) furnace lining
Wandbekleidung *f* wainscot, wall board, wall covering, wall panelling
Wanddicke *f* wall thickness
Wanddurchbruch *m* wall trimming
Wandeffekt *m* (Rheologie) wall effect
Wandeinfluß *m* (Rheologie) wall effect
wandelbar variable, changeable, inconstant
Wandenergie *f* (Krist) wall energy
Wanderanode *f* movable anode
Wanderbett *n* (technische Reaktionsführung) moving bed

wandern to migrate, to shift, to travel, (diffundieren) to diffuse
Wandern *n* migration, shifting
Wanderrost *m* traveling grate, chain grate, travelling-grate stoker, **runder** ~ circular-grate stoker
Wanderrostbeschickung *f* traveling stoker feed
Wandertisch *m* platform conveyer
Wanderung *f* migration, shifting
Wanderungsbeständigkeit *f* migration stability
Wanderungsgeschwindigkeit *f* migration velocity, drift velocity
Wanderungssinn *m* (Ionen) direction of migration, sense of migration
Wanderungsverlust *m* migration loss
Wanderwelle *f* progressive wave, running wave
Wanderwellengenerator *m* impulse generator, surge generator
Wanderzelle *f* migratory cell
Wandfarbe *f* wall paint
Wandgleitung *f* slippage along the wall
Wandgrundierung *f* wall primer
Wandhalter *m* wall bracket
Wandkachel *f* wall tile
Wandkatalyse *f* wall catalysis
Wandkonsole *f* wall console
Wandkran *m* wall crane
Wandladung *f* wall charge
Wandladungsdichte *f* wall charge density
Wandleiste *f* skirting board
Wandlüfter *m* wall fan
Wandlung *f* conversion, transformation
Wandplatte *f* [wall] panel, wall slab
Wandreibung *f* wall friction, skin friction
Wandrekombination *f* (Chem) wall recombination
Wandschalttafel *f* wall switch board
Wandschirm *m* screen
Wandschrank *m* closet (Am. E.), cupboard
Wandstärke *f* thickness of the wall, wall thickness
Wandstecker *m* (Elektr) wall plug
Wandstütze *f* wall bracket
Wandtäfelung *f* wainscot, wall panelling
Wandung *f* partition, wall partition
Wandverkleidung *f* **aus Blech** sheet metal lining
Wandverputz *m* plaster work
Wange *f* cheek, side wall
Wanne *f* tub, tank, trough, vat, **pneumatische** ~ pneumatic trough
Wannenform *f* (Stereochem) boat-form
Wannenofen *m* tank furnace
Wanze *f* (Zool) [bed] bug
Wapplerit *m* (Min) wapplerite
Wardit *m* (Min) wardite
Ware *f* merchandise, ware
Waren *f pl* goods, commodities
Wareneingang *m* stock receipt
Warenlager *n* storehouse

Warenzeichen *n* trade mark, brand
warm warm
warmabbindend hot setting
warmaushärten to age artificially
Warmbandstahl *m* hot-rolled steel strip
warmbearbeiten to hot-work, to work at red heat
Warmbearbeitung *f* hot-working
Warmbehandlung *f* heat treatment
Warmbiegeprobe *f* hot bending test
Warmbildsamkeit *f* (Met) forgeability, hot ductility
Warmblasegas *n* hot blast gas
Warmblaseperiode *f* hot-blast period
Warmbleiche *f* warm bleach
Warmbreitband *n* hot-rolled wide strip
warmbrüchig hot-brittle, hot-short, red-short
Warmbrüchigkeit *f* hot-shortness
warmdehnbar (Met) hot-ductile
Warmdehnbarkeit *f* (Met) hot ductility
warmfest heat-resistant
Warmfestigkeit *f* high temperature strength or stability, hot strength
Warmfließgrenze *f* high temperature yield strength
Warmformgebung *f* hot-forming, hot-shaping, hot-working
warmhärtbar thermosetting
Warmhalteplatte *f* hot plate
Warmklebstoff *m* hot-setting adhesive
warmlaufen (Mot) to run hot
Warmlaufen *n* **des Lagers** heating up of the bearing
Warmluft *f* warm air
warmpressen to hot-press
Warmpressen *n* hot-molding, hot-pressing
Warmprobe *f* heat test, hot test
Warmprofil *n* hot rolled section
warmrecken to hot strain
Warmrecken *n* hot straining, hot working
Warmsäge *f* hot saw
Warmschmieden *n* hot forging
Warmstreckgrenze *f* hot yield point
Warmverarbeitungsfähigkeit *f* hot workability
warmverformbar hot-workable, thermoplastic
Warmverformbarkeit *f* hot-forming property, thermoplasticity
Warmverformung *f* hot forming, hot working
Warmvergoldung *f* hot gilding
Warmvergütung *f* artificial aging
warmwalzen to hot-roll
Warmwalzen *n* hot rolling
Warmwasserbehälter *m* hot-water tank
Warmwasserhahn *m* hot-water tap
Warmwasserheizung *f* hot-water heating
Warmwasserpumpe *f* hot water pump
Warmwasserspeicher *m* hot-water tank, boiler
Warmwasserversorung *f* hot-water supply
Warmwindapparat *m* hot blast stove
Warmzerreißprobe *f* hot tensile test

warmziehen to hot draw
Warmziehen *n* hot drawing
Warnfilter *m* warning filter
Warngerät *n* monitor, warning device, warning indicator
Warnlicht *n* warning light
Warnungssignal *n* danger signal
Warnvorrichtung *f* alarm apparatus
Warrenit *m* (Min) warrenite
Warringtonit *m* (Min) warringtonite
warten to wait, (Techn) to attend, to service
Wartezeit *f* waiting time
Wartung *f* maintenance, attendance, service, **vorbeugende** ~ preventive maintenance
Wartungskosten *pl* maintenance cost
Wartungsprogramm *n* service routine
Wartungszeit *f* servicing time
Warwickit *m* (Min) warwickite
waschaktiv surface-active
Waschamber *m* common yellow amber
Waschbarkeit *f* washability
Waschbenzin *n* gasoline used for cleaning purposes
Waschblau *n* washing blue
Waschbottich *m* washing tub
waschecht fast to washing, (Text) dyed in the grain, fast, washable
waschen to wash, to clean, to scrub
Waschen *n* washing, scrubbing, ~ **der Erze** ore washing
Wascherde *f* fuller's earth, soap rock
Wascherz *n* wash ore, diluvial ore
waschfest washproof
Waschfestigkeit *f* fastness to washing, (Lacke) wet abrasion resistance
Waschflasche *f* wash[ing] bottle
Waschflotte *f* washing liquid
Waschflüssigkeit *f* washing liquid
Waschgold *n* placer gold
Waschholländer *m* (Pap) washing engine
Waschkasten *m* washing tank
Waschkristalle *pl* washing crystals, soda crystals
Waschlauge *f* washing liquor, scouring liquor
Waschlösung *f* wash solution
Waschmaschine *f* washing machine
Waschmittel *n* detergent
Waschöl *n* absorption oil
Waschpulver *n* detergent, washing powder
Waschputz *m* marble chip filled plaster, aggregate plaster
Waschrohstoff *m* detergent base material
Waschsäure *f* washing acid
Waschseife *f* washing soap
Waschsieb *n* calender
Waschturm *m* absorption column, scrubbing tower, washing column, washing tower
Waschung *f* washing, scrubbing
Waschverfahren *n* washing process

Waschvorgang *m* washing process, cleaning process, scrubbing process
Waschvorrichtung *f* washing device, scrubber, washer
Waschwasser *n* wash water, (Chem) washings
Waschwirkung *f* cleaning effect
Waschzinn *n* alluvial tin ore, steam tin
Waschzone *f* washing zone
Waschzyklon *m* dense-media cyclone
Washingtonit *m* (Min) washingtonite
Wasser *n* water, hydrogen oxide, ~ **entziehen** to dehydrate, to desiccate, to remove water, **chloriertes** ~ chlorinated water, **destilliertes** ~ distilled water, **eisenhaltiges** ~ chalybeate water, **extrazelluläres** ~ extracellular water, **fließendes** ~ running or flowing water, **gashaltiges** ~ aerated water, **gebundenes** ~ bound water, **gefiltertes** ~ filtered water, **hartes, kalkhaltiges** ~ hard calcareous water, **intrazelluläres** ~ intracellular water, **kohlensäurehaltiges** ~ carbonic water, **leichtes** ~ light water, normal water, ordinary water, **schweres** ~ deuterium oxide, heavy water, **stehendes** ~ stagnant water, **weiches** ~ soft water
Wasserabfluß *m* discharge of water, drainage
Wasserabflußrohr *n* water outlet pipe
Wasserabgabe *f* elimination of water
Wasserableitung *f* drainage
Wasserabsaugpumpe *f* water exhaust pump
Wasserabsaugung *f* water exhaust
Wasserabscheider *m* water separator, water trap
Wasserabscheidung *f* water separation
Wasserabsorptionsvermögen *n* water-absorbing capacity
Wasserabspaltung *f* splitting off water
wasserabstoßend hydrophobe, water repellent
wasserähnlich water-like, watery
Wasseranalyse *f* water analysis
Wasseranlagerung *f* hydration
wasseranziehend hygroscopic, moisture-attracting, water-attracting
Wasseranziehungsvermögen *n* hygroscopicity
Wasseraufbereitung *f* water treatment
Wasseraufnahme *f* water absorption, intake of water
Wasseraufnahmefähigkeit *f* absorption capacity for water, hygroscopicity
Wasseraufnahmevermögen *n* water absorbing capacity
wasseraufsaugend water-absorbing
Wasseraufsaugungsvermögen *n* hygroscopic quality, water absorptivity
Wasserauftrieb *m* buoyancy in water
Wasserauslaß *m* water outlet
Wasseraustritt *m* water elimination
Wasserbad *n* (Chem) water bath
Wasserbadring *m* water bath ring
Wasserballon *m* carboy

Wasserbecken *n* water tank
Wasserbedarf *m* water requirement, amount of water required, water demand
Wasserbehälter *m* reservoir, water container, water tank
Wasserbehandlung *f* water treatment
Wasserbeize *f* water stain
Wasserberieselung *f* water irrigation, water spraying
wasserbeständig water-proof, resistant to water, stable in water, water-resistant
Wasserbeständigkeit *f* waterproofness, water resistance, water resisting property
Wasserbestimmungsapparat *m* water determination apparatus
Wasserbidestillator *m* water bidistiller
wasserbindend hydrophilic, water-absorbent
Wasserbindevermögen *n* water-absorptive capacity
Wasserbindung *f* binding of water
Wasserbiologie *f* hydrobiology
Wasserblase *f* water bubble, vesicle, water blister
wasserblau water blue
Wasserbleiocker *m* (Min) molybdic ocher, molybdite
Wasserchrysolith *m* (Min) obsidian, volcanic glass
Wasserdampf *m* steam, water vapor, ~ **einblasen** to inject steam, ~ **schlägt sich nieder** water vapor condenses, **mit** ~ **abblasen** to distill with steam, **überhitzter** ~ superheated steam
Wasserdampfdestillation *f* steam distillation
wasserdampfdicht impermeable to water vapor
Wasserdampfdurchlässigkeit *f* moisture vapor transmission rate; water vapor permeability
Wasserdampfentwickler *m* steam boiler
Wasserdampferzeuger *m* steam boiler
Wasserdampfflüchtigkeit *f* volatility in steam
Wasserdampfstrahlsauger *m* steam ejector
Wasserdampfteildruck *m* partial pressure of steam
Wasserdampfzusatz *m* water-vapor addition
Wasserdestillierapparat *m* water distilling apparatus, water still
wasserdicht waterproof, impervious to water, leakproof, water-tight
Wasserdichte *f* density of water
Wasserdichtheit *f* waterproofness
Wasserdichtmachen *n* waterproofing, impermeabilization
Wasserdichtungsmittel *n* waterproofing compound
Wasserdruck *m* water pressure, hydraulic pressure
Wasserdruckaufzug *m* hydraulic lift
Wasserdruckbremse *f* hydraulic brake
Wasserdruckhöhe *f* hydraulic pressure head

Wasserdruckmanometer *n* hydraulic pressure gauge
Wasserdruckpresse *f* hydraulic press
Wasserdruckprobe *f* hydraulic test
Wasserdruckregelung *f* regulation of water pressure
Wasserdunst *m* water vapor
Wasserdurchbruch *m* water burst
wasserdurchlässig pervious to water
Wasserdurchlässigkeit *f* permeability for water
Wasserdynamo *m* hydroelectric generator
Wassereinfüllrohr *n* water inlet tube
Wassereinlaß *m* water inlet
Wassereinspritzung *f* injection of water
Wasserelektrolyse *f* electrolysis of water
Wasserelektrolyseur *m* water electrolyzer
wasserempfindlich sensitive to water
Wasserempfindlichkeit *f* sensitivity to water
Wasseremulsionsfarbe *f* water paint
Wasserenteisenung *f* extraction of iron from water
Wasserenteisenunganlage *f* water-softening plant
Wasserentgasung *f* water degassing
Wasserenthärter *m* water softener
Wasserenthärtung *f* water softening,
  chemische ~ chemical softening of water
Wasserenthärtungsanlage *f* water softening plant
Wasserentmanganung *f* demanganizing of water
Wasserentölung *f* separation of oil from water
Wasserentsalzung *f* demineralization of water, desalination of water, desalting of water
wasserentziehend dehydrating, desiccating
Wasserentziehung *f* dehydration, desiccation, removal of water
Wasserentziehungsmittel *n* dehydrating agent
Wassererosion *f* erosion by water
Wasserfarbe *f* water color, distemper, water paint
Wasserfenchel *m* (Bot) water fennel
wasserfest waterproof, resistant to water, watertight
Wasserfestigkeit *f* waterproofness
Wasserfestmachen *n* waterproofing
Wasserfilter *m* water filter
Wasserflasche *f* water bottle
Wasserfleck *m* (Pap) water spot
Wasserförderung *f* raising of water
wasserfrei anhydrous, dehydrated, desiccated, free from water
Wasserführung *f* water duct, water passage
Wassergas *n* water gas, **angereichertes** ~ enriched water gas
Wassergasabzug *m* water gas outlet
Wassergasanlage *f* water gas plant
Wassergaserzeuger *m* water gas producer
Wassergaserzeugung *f* production or generation of water gas
Wassergasteer *m* water-gas tar

Wassergefäß *n* water container, water tank, water vessel, (Med) lymphatic vessel
Wassergehalt *m* water content, humidity, moisture content, percentage of moisture,
  freier ~ free moisture [content]
Wassergehaltbestimmung *f* water content determination
wassergekühlt water-cooled, water-jacketed
wassergesättigt saturated with water
Wasserglas *n* (Chem) sodium silicate, water glass, (Gefäß) tumbler, water glass
Wasserglasanstrichfarbe *f* water glass paint
Wasserglasfarbe *f* silicate paint
Wasserglaskitt *m* water glass cement
Wasserglaslösung *f* water glass solution
Wasserglasseife *f* water glass soap
Wassergroßreinigungsanlage *f* water-screening plant
Wassergrün *n* watergreen
Wasserhärte *f* hardness of water
Wasserhärtung *f* water hardening
Wasserhärtungsstahl *m* water-hardening steel
Wasserhahn *m* faucet, tap
wasserhaltig hydrated, aqueous, containing water, hydrous
Wasserhaushalt *m* (Physiol) water balance, water equilibrium
Wasserhaut *f* water skin
wasserhell clear as water, transparent
Wasserinhalt *m* volume of water
Wasser-in-Öl-Emulsion *f* water-in-oil emulsion
Wasserkalander *m* water calender
Wasserkalk *m* water lime, hydraulic lime
Wasserkalorimeter *n* water calorimeter
Wasserkanal *m* water channel, water line
Wasserkasten *m* water tank, cooling trough, water trough
Wasserkessel *m* water tank, boiler
Wasserkläranlage *f* water clarifier
wasserklar clear as water, transparent, (durchsichtig) limpid
Wasserkraft *f* hydraulic power, water power
Wasserkraftanlage *f* water power plant
Wasserkraftgebläse *n* hydraulic blower
Wasserkraftmaschine *f* water power engine
Wasserkraftwerk *n* hydroelectric power plant
Wasserkühlanlage *f* water cooling plant
Wasserkühler *m* water cooler
Wasserkühlmantel *m* water cooling jacket
Wasserkühlung *f* water cooling
Wasserlack *m* water paint
Wasserleitung *f* water piping, water conduit, water main
wasserlöslich water-soluble, soluble in water
Wasserlöslichkeit *f* aqueous solubility, water solubility
Wassermangel *m* lack of water, (Physiol) water deficiency, hydropenia
Wassermantel *m* water jacket

Wassermantelofen *m* water-jacketed furnace
Wassermenge *f* amount of water
Wassermesser *m* water meter
Wassermessung *f* hydrometry
Wassermörtel *m* hydraulic mortar
Wassermolekül *n* water molecule
Wasserniederschlag *m* deposition of moisture
Wasserniveau *n* water level
Wasseroberfläche *f* water surface
Wasser-Öl-Farbe *f* oilbound water paint
Wasseropal *m* (Min) water opal, hyalite, hydrophane
Wasserpolster *n* water cushion
Wasserprüfung *f* water analysis, water testing
Wasserpumpe *f* water pump
Wasserrad *n* water wheel
Wasserreaktoranlage *f* aqueous reactor system
Wasserregulator *m* constant water level regulator
Wasserreinigung *f* water purification
Wasserreinigungsanlage *f* water purification plant
Wasserringpumpe *f* liquid seal pump
Wasserrinne *f* water trough
Wasserröhrenkühler *m* water-tube-type cooler
Wasserrohr *n* water pipe
Wasserrohrkessel *m* water tube boiler
Wasserrückkühlanlage *f* water recooling plant
Wasserrückkühlung *f* water recooling system
Wasserrückstand *m* water residue
Wassersäule *f* water column, column of water
Wassersäulendruck *m* water-column pressure, hydraulic pressure
Wassersäulendruckmesser *m* water-column pressure gauge
Wassersaphir *m* (Min) water sapphire
Wasserschicht *f* water layer
Wasserschieber *m* water sluice valve
Wasserschlauch *m* water hose, water tubing
Wasserschnecke *f* Archimedian screw, water screw
Wasserschwertlilie *f* (Bot) water lily
Wasserspeicher *m* water storage tank or basin
Wasserspeiser *m* feed water regulator
Wasserspiegel *m* (Wasserstand) water level, water surface
Wasserspiegellinie *f* line of water level
Wasserspritze *f* water syringe
Wassersprühregen *m* water spray
wasserstabilisiert water stabilized
Wasserstand *m* water level
Wasserstandsanzeiger *m* water-level indicator, water gauge
Wasserstandsmesser *m* water gauge, water level indicator
Wasserstandsmessung *f* water level measurement
Wasserstandsregler *m* water level regulator
Wasserstandsröhre *f* water gauge glass tube
Wasserstein *m* [boiler] scale

Wassersteinansatz *m* deposit of scale
Wasserstoff *m* hydrogen (Symb. H), ~ **in statu nascendi** nascent hydrogen, **aktivierter** ~ activated hydrogen, **atomarer** ~ atomic hydrogen, **befreien von** ~ to dehydrogenate, to dehydrogenize, **naszierender** ~ nascent hydrogen, **schwerer** ~ heavy hydrogen, (Deuterium) deuterium, (Tritium) tritium, **verbunden mit** ~ hydrogenated, hydrogenized
Wasserstoffableitungsröhre *f* hydrogen outlet tube
Wasserstoffabscheidung *f* hydrogen liberation
wasserstoffähnlich hydrogen-like, hydrogenic
Wasserstoffäquivalent *n* hydrogen equivalent
Wasserstoffakzeptor *m* hydrogen acceptor
Wasserstoffanlage *f* hydrogen plant
Wasserstoffanlagerung *f* hydrogenation
Wasserstoffatmosphäre *f* hydrogen atmosphere
Wasserstoffatom *n* hydrogen atom
Wasserstoffaufnahme *f* hydrogen absorption, hydrogen uptake
Wasserstoffaufnehmer *m* hydrogen acceptor
Wasserstoffaustauschreaktion *f* hydrogen exchange reaction
Wasserstoffbakterien *pl* hydrogen bacteria
Wasserstoffbehälter *m* hydrogen tank, hydrogen container
Wasserstoffbeladung *f* (Metall) charged with hydrogen
Wasserstoffbildung *f* formation of hydrogen, hydrogen formation
Wasserstoffbindung *f* hydrogen bond
Wasserstoffblasenkammer *f* hydrogen bubble chamber
Wasserstoffbombage *f* hydrogen swell
Wasserstoffbombe *f* hydrogen bomb, H-bomb
Wasserstoffbrückenbindung *f* hydrogen bond, **innermolekulare** ~ innermolecular hydrogen bond, **intramolekulare** ~ intramolecular hydrogen bond
Wasserstoffelektrode *f* hydrogen electrode
Wasserstoffentwickler *m* hydrogen generator
Wasserstoffentwicklung *f* generation of hydrogen
wasserstoffentziehend dehydrogenating
Wasserstoffentziehung *f* dehydrogenation
Wasserstofferzeuger *m* hydrogen generator
Wasserstofferzeugungsapparat *m* hydrogen generator
Wasserstoffflamme *f* hydrogen flame
Wasserstoffflasche *f* hydrogen cylinder, hydrogen bottle
Wasserstoffgas *n* hydrogen gas
Wasserstoffgasanlage *f* hydrogen gas plant
Wasserstoffgehalt *m* hydrogen content
Wasserstoffgemisch *n* hydrogen mixture
Wasserstoffgewinnungsanlage *f* hydrogen generating plant, hydrogen producing plant

wasserstoffhaltig containing hydrogen, hydrogenous
Wasserstoffion *n* hydrogen ion
Wasserstoffionenaktivität *f* hydrogen ion activity
Wasserstoffionenexponent *m* hydrogen ion exponent
Wasserstoffionenkonzentration *f* hydrogen ion concentration
Wasserstoffkern *m* hydrogen nucleus, proton
Wasserstoffknallgas *n* detonating gas
Wasserstoffkontinuum *n* hydrogen continuum
Wasserstoffkrankheit *f* (Met) hydrogen embrittlement
Wasserstofflampe *f* hydrogen lamp, Döbereiner's lamp
Wasserstofflichtbogen *m* hydrogen arc
Wasserstofflinie *f* (Spektr) hydrogen line
Wasserstofflötung *f* hydrogen soldering
Wasserstoffmolekül *n* hydrogen molecule
Wasserstoffperoxid *n* hydrogen peroxide
Wasserstoffreduktion *f* hydrogen reduction
wasserstoffreich rich in hydrogen
Wasserstoff-Sauerstoff-Schweißung *f* autogenous welding, oxyhydrogen welding
Wasserstoffserie *f* (Spektr) hydrogen series
Wasserstoffspektrum *n* hydrogen spectrum
Wasserstoffstrom *m* hydrogen flow, hydrogen stream
Wasserstoffsuperoxid *n* hydrogen peroxide, hydrogen superoxide
Wasserstoffüberspannung *f* hydrogen overvoltage
Wasserstoffüllung *f* hydrogen filling
Wasserstoffunkenstrecke *f* hydrogen spark gap
Wasserstoffverbindung *f* hydrogen compound, hydride
Wasserstoffversprödung *f* hydrogen embrittlement
Wasserstoffzahl *f* hydrogen-ion concentration
Wasserstoffzuleitungsröhre *f* hydrogen inlet tube
Wasserstrahl *m* water jet
Wasserstrahldüse *f* water jet nozzle
Wasserstrahlkondensator *m* water jet condenser
Wasserstrahlpumpe *f* water jet pump
Wassertechnik *f* hydraulics
Wasserthermometer *n* water thermometer
Wassertröpfchen *n* water droplet
Wassertrog *m* water trough
Wasserturbine *f* hydroturbine
Wasserturm *m* water tower
Wasseruhr *f* water gauge, water meter
Wasserumlaufkühlung *f* cooling by means of circulating water
wasserundurchlässig waterproof, impervious to water, watertight
Wasserundurchlässigkeit *f* imperviousness to water
wasserunlöslich insoluble in water

Wasserunlöslichkeit *f* water insolubility
Wasserverbrauch *m* consumption of water
Wasserverdrängung *f* displacement of water
Wasserverdunstung *f* evaporation of water
Wasserverschluß *m* water seal
Wasserverseuchung *f* water contamination, water pollution
Wasserversorgung *f* water supply
Wasservorlage *f* water receiver
Wasservorrat *m* water supply, water reserve
Wasservorwärmer *m* water preheater
Wasserwaage *f* water level, spirit level
Wasserwärmer *m* hot water apparatus
Wasserwanne *f* water trough
Wasserwerk *n* waterworks
Wasserwert *m* water equivalent
Wasserzähler *m* water meter
Wasserzeichen *n* (Pap) watermark, **geprägtes** ~ (Pap) impressed watermark
Wasserzeichenapparat *m* (Pap) watermark apparatus
Wasserzement *m* hydraulic cement
Wasserzersetzung *f* decomposition of water
Wasserzufluß *m* water inlet
Wasserzuflußrohr *n* water inlet pipe
Wasserzuleitungsrohr *n* water-supply pipe
Watt *n* (Elektr) watt, ampere-volt, volt-ampere
Watte *f* cotton [wool], wadding
Wattebausch *m* wad
Wattefilter *m* cotton filter
Wattepfropfen *m* cotton plug
Wattierung *f* wadding
Wattkomponente *f* watt component, energy component, power component
Wattleistung *f* wattage
wattlos (Elektr) wattless
Wattmesser *m* wattmeter
Wattmeter *n* wattmeter
Wattregler *m* energy regulator
Wattsekunde *f* (Elektr) watt-second
Wattstrom *m* watt current, active current
Wattstunde *f* watt-hour
Wattstundenzähler *m* (Elektr) watt-hour meter
Wattverbrauch *m* watt consumption, wattage
Wattverlust *m* energy loss in watts
Wattzahl *f* number of watts
Waugelb *n* luteolin
Wavellit *m* (Min) wavellite
Webart *f* (Web) [type of] weave
weben to weave
Weben *n* weaving
Webende *n* selvage, selvedge
Weber *n* (Magn) weber
Weberknoten *m* weaver's knot
Weberschiffchen *n* shuttle
Webkante *f* (Web) selvage, selvedge
Webnerit *m* (Min) webnerite
Webschütze *m* shuttle
Webskyit *m* webskyite

Websterit *m* (Min) websterite
Webstuhleinstellung *f* (Web) loom setting
Wechsel *m* change, shift, variation, (Austausch) exchange, (Strom) alternation
Wechselbeanspruchung *f* alternating stress, alternative stress, stress reversal
Wechselbewegung *f* intermittent motion, reciprocating motion
Wechselbeziehung *f* correlation, interrelation, mutual relation, relationship
Wechselbiegebeanspruchung *f* cyclic loading
Wechselbiegeprüfung *f* alternating bending test
Wechselbiegung *f* reversed bending
Wechselbruchspannung *f* endurance limit
Wechselfeld *n* (Elektr) alternating field,
  einphasiges ~ single-phase alternating field,
  magnetisches ~ alternate magnetic field,
  mehrphasiges ~ polyphase alternating field
Wechselfestigkeitsprüfung *f* alternate stress test, fatigue test
Wechselgetriebe *n* transmission gear, change gear
Wechselinduktion *f* mutual inductance, mutual induction
Wechselklappe *f* butterfly valve
Wechselkraftfluß *m* alternating flux
Wechsellicht *n* light intensity variations
Wechsellichtmessung *f* intermittent photometry
wechseln to change, to alter, to shift, to vary, (austauschen) to exchange, (Strom) to alternate, (untereinander austauschen) to interchange
wechselnd alternating, variable, varying
wechselpolar heteropolar
Wechselpolinduktion *f* heteropolar induction
Wechselpotential *n* alternating potential
Wechselschalter *m* two-way switch
Wechselschlagbiegeversuch *m* alternating-impact bending test
Wechselschlagversuch *m* alternating impact test, alternating-stress test
wechselseitig mutual, reciprocal, two-way
Wechselseitigkeit *f* reciprocity
Wechselspannung *f* (Elektr) alternating voltage, (Mech) alternating stress
Wechselspannungskomponente *f* alternating component of voltage
Wechselstrom *m* alternating current
Wechselstromakkumulator *m* alternating current accumulator
Wechselstromamperemeter *n* alternating current ammeter
Wechselstromanlage *f* alternating current plant
Wechselstrombogen *m* alternating current arc
Wechselstrombrücke *f* alternating current bridge
Wechselstromdynamo *m* alternating current generator, alternator
Wechselstromerregung *f* alternating current excitation

Wechselstromfeld *n* alternating field,
  mehrphasiges ~ polyphase alternating field
Wechselstromgenerator *m* alternating current generator, alternator
Wechselstromgleichrichter *m* alternating current rectifier
Wechselstromgröße *f* alternating current value
Wechselstromkollektormotor *m* alternating current commutator motor
Wechselstromkreis *m* alternating current circuit
Wechselstromleistung *f* alternating current output, alternating current power
Wechselstromleitung *f* alternating current line
Wechselstromlichtbogen *m* alternating current arc
Wechselstrommagnetisierung *f* magnetization by alternating current
Wechselstrommotor *m* alternating current motor
Wechselstromnebenschlußmotor *m* alternating current shunt motor
Wechselstromnetz *n* alternating current circuit
Wechselstromperiode *f* alternating current period
Wechselstromquelle *f* alternating current source
Wechselstromspannung *f* alternating current voltage
Wechselstromsystem *n* alternating current system
Wechselstromtechnik *f* alternating current engineering
Wechselstromtransformator *m* alternating current transformer
Wechselstromumformer *m* alternating current converter
Wechselstromverlust *m* alternating current loss
Wechselstromwiderstand *m* alternating current resistance, impedance
Wechseltauchversuch *m* alternate immersion test
Wechselumsetzung *f* reciprocal conversion
Wechselventil *n* change-over valve, two-way valve
Wechselverformung *f* alternating load deformation
wechselweise alternately, by turns, ~ **aufeinanderfolgen** to alternate
Wechselwirkung *f* alternating effect, correlation, interaction, reciprocal action
Wechselwirkungsaufspaltung *f* interaction splitting
Wechselwirkungsdarstellung *f* interaction representation
Wechselwirkungsenergie *f* mutual-potential energy
Wechselwirkungsgesetz *n* Newton's third law
Wechselwirkungskraft *f* interaction force
Wechselwirkungswiderstand *m* interference drag
Wechselzahl *f* frequency number, number of alternations, ~ **eines Ferments** turnover number of an enzyme

Wechselzersetzung *f* double decomposition
Weckamin *n* (Pharm) analeptic amine
Weckmittel *n* (Pharm) analeptic
Wedelolacton *n* wedelolactone
Wedelsäure *f* wedelic acid
Weg *m* way, passage, path, (Metabolismus) pathway, (Verfahren) method
wegbrennen to burn off
Wegedorn *m* (Bot) buckthorn
Wegeinheit *f* unit path
Wegerichkraut *n* (Bot) plantain
Wegfall *m* omission, suppression
wegkochen to boil away, to boil off
Weglänge *f* path [length], distance, **mittlere freie** ~ mean free path, average free path
Weglängenverteilung *f* path length distribution
Wegnahme *f* elimination, removal, seizure
wegnehmen to remove, to take away
wegräumen to clear away, to remove
Wegreißen *n* tearing away
wegschaffen to eliminate, to remove
wegschieben to shift
wegschneiden to cut away
wegspülen to wash away
Wegstrecke *f* path
Wegwartwurzel *f* (Bot) chicory root
wegwaschen to wash away, to wash off
wehen to blow
Wehneltblende *f* Wehnelt control grid
Wehr *n* weir
Wehranlage *f* weir installation
Wehrlit *m* (Min) wehrlite
Weibullit *m* (Min) weibullite
weich soft, mellow, non-rigid, smooth, tender, ~ **löten** to softsolder, to solder
Weichasphalt *m* soft asphalt
Weichblei *n* refined lead, soft lead
Weichbleiabdruck *m* soft lead impress
Weiche *f* bypass, switch
Weicheisen *n* soft iron
Weicheisenbündel *n* soft iron bar
Weicheisenkern *m* soft iron core
Weicheisenkies *m* (Min) marcasite
Weicherz *n* (Min) silver sulfide
weichfeuern (Metall) to melt down
Weichfolie *f* flexible sheet
Weichglühen *n* [soft] annealing, spheroidizing
Weichgummi *m* soft rubber
Weichguß *m* malleable [cast] iron, malleable pig iron
Weichharz *n* oleoresin, soft resin
Weichheit *f* softness
Weichheitsgrad *m* degree of softness
Weichholz *n* soft wood
Weichkernstahl *m* soft center[ed] steel
Weichkupfer *n* soft copper
Weichkupferdraht *m* soft copper wire
Weichlöten *n* brazing, soft soldering

Weichlot *n* soft solder, tin-lead solder, tin solder, **löten mit** ~ to softsolder, to solder
weichmachen to soften, (Kunststoff) to plasticize
Weichmachen *n* softening, fluxing, plasticizing (plastics)
Weichmacher *m* emollient, softener, softening agent, (Kunststoff) plasticizer, **gelatinierender** ~ plasticizer with solvent properties
weichmacherfrei unplasticized
Weichmachermischung *f* (Kunststoff) plasticizer blend
Weichmachmittel *n* (Kunststoff) plasticizing agent, (Wasser) water softener
Weichmachungsgrad *m* plasticizing rate
Weichmachungsmittel *n* softener
Weichmanganerz *n* (Min) pyrolusite
Weichmetall *n* soft metal
Weichpackung *f* flexible package
Weichparaffin *n* soft paraffin, soft wax
Weichporzellan *n* soft porcelain
Weich-PVC *n* plasticized PVC
Weichstahl *m* mild steel, soft steel
Weichwerden *n* softening
Weidenkorb *m* wicker basket
Weidenrinde *f* willow bark
Weihrauch *m* frankincense, incense, olibanum
Weihrauchharz *n* frankincense, incense resin
Weihrauchöl *n* frankincense oil
Wein *m* wine, ~ **mit Harz würzen** to resinate wine
Weinbau *m* cultivation of vine, viticulture
Weinbeere *f* grape
Weinberg *m* vineyard
Weinbrand *m* brandy made from wine
Weinernte *f* vintage
Weinessig *m* wine vinegar
Weinfälschung *f* adulteration of wine
Weinfaß *n* wine cask
Weinfuselöl *n* wine fusel oil
Weingärung *f* vinous fermentation
Weingehalt *m* wine content
Weingeist *m* [ethyl] alcohol, spirit of wine
weingeistartig alcoholic
Weingeistessig *m* brandy vinegar
weingelb wine yellow
weinhaltig containing wine
Weinhefe *f* wine yeast, wine lees
Weinhefenasche *f* calcined wine lees
Weinkeller *m* wine cellar
Weinkernöl *n* grape-seed oil
Weinklärung *f* clarifying of wine
Weinlagerung *f* wine storage
weinsäuerlich like sourish wine
Weinsäure *f* tartaric acid, **linksdrehende** ~ levorotatory tartaric acid
Weinsäurelösung *f* solution of tartaric acid
weinsauer tartrate
Weinschönung *f* treatment of wine

Weinstein *m* tartar
weinsteinartig tartareous, tartar-like
Weinsteinbildung *f* formation of tartar
Weinsteingehalt *m* tartar content
Weinsteinsalz *n* potash salt of tartar
Weinsteinwasser *n* tartar solution
Weintraube *f* grape, bunch of grapes
Weintreber *pl* marc of grapes
Weintrester *pl* marc of grapes
weisen to point
Weißakazienöl *n* white acacia oil
Weißanlaufen *n* blushing
Weißanstrich *m* whitewash, limewash
Weißblech *n* tinned sheet iron, tin plate
Weißblechabfall *m* tin plate scrap
Weißblechkessel *m* tinning vat
Weißbleiche *f* (Text) bleaching of cotton
Weißbleierz *n* (Min) white lead ore, cerussite
Weißbrennen *n* calcining at white heat
Weißbruch *m* (PVC) white break
weißbrüchig of pale fracture, of white fracture
Weißdornfluidextrakt *m* liquid extract of hawthorn
Weißeisen *n* white metal, white [pig] iron, **kleinspiegeliges** ~ white crystalline iron
Weißemail *n* porcelain enamel
weißen to whiten, (bleichen) to bleach, (Metall) to refine (iron), (tünchen) to whitewash
Weißerz *n* (Min) arsenopyrite, siderite
Weißfarbe *f* white pigment
weißfarbig white colored, (gebleicht) blanched
Weißgehalt *m* whiteness, (Email) reflectance
Weißgehaltmesser *m* white level indicator
weißgelb pale yellow
Weißgerberei *f* alum tanning, tawery
weißglühen to white heat
Weißglühen *n* incandescence
weißglühend incandescent, white hot
Weißglühhitze *f* incandescence, white heat
Weißglut *f* incandescence, white heat
Weißgold *n* platinum, white gold
Weißgolderz *n* (Min) sylvanite
Weißgrad *m* (Pap) whiteness
Weißgradmesser *m* leucometer
weißgrau light gray, pale gray
Weißgüldigerz *n* (Min) white silver ore, argentiferous tetrahedrite
Weißguß *m* white cast iron, white metal
Weißhitze *f* white heat
Weißhornblende *f* (Min) tremolite
Weißit *m* (Min) weissite
Weißkalk *m* white lime, fat lime
Weißkies *m* (Min) arsenopyrite
Weißkupfer *n* white copper, (Min) domeykite, (Plattsilber) plate silver
Weißkupfererz *n* (Min) white copper ore, cubanite
Weißlack *m* white finish, white enamel, white varnish
weißlich whitish
Weißlot *n* tin solder, soft solder
Weißmessing *n* white brass
Weißmetall *n* white metal, antifriction metal, babbit metal, **versilbertes** ~ argentine
Weißpause *f* white print
weißsieden (Silber) to blanch
Weißsieden *n* hot light silvering
Weißsiedlauge *f* blanching liquor
Weißsigit *m* weissigite
Weißspießglanzerz *n* (Min) white antimony, valentinite
Weißstein *m* white stone, granulite
Weißsud *m* blanching solution
Weißsudbad *n* hot light silvering bath
Weißsudverzinnung *f* tin-plating in hot bath
Weißtannenöl *n* pine oil, silver fir oil
Weißtellur *n* (Min) sylvanite
Weißtöner *m* (Text) bleaching agent
Weißtonerde *f* china clay, kaolin, white alumina, white argillaceous earth
Weißtonerhaltung *f* whiteness retention
Weißvernick[e]lung *f* white nickelling, nickel-plating
Weißvitriol *n* white vitriol, zinc sulfate
Weißwein *m* white wine
Weißzucker *m* white sugar
weit (breit) broad, wide, (entfernt) far, remote, (groß) large
Weite *f* width, (Amplitude) amplitude, **lichte** ~ clearance, inner width, inside diameter, width in the clear
weiten to broaden, to expand, to stretch
weiterbefördern to convey, to transport
Weiterbehandlung *f* further treatment, subsequent treatment
Weiterentwicklung *f* [further] development
weiterführen to carry on, to continue
weiterleiten to convey, to transfer
Weiterleitung *f* transmission
weiteroxydieren to oxidize further
weiterverarbeiten to process further, (Text) to convert, to improve, to refine
Weiterverarbeitung *f* further processing, subsequent treatment
Weiterverschmieden *n* final forging
Weithalsflasche *f* (Chem) wide-necked bottle, wide-mouth bottle
weithalsig wide-necked, wide-mouth
Weithalskolben *m* (Chem) wide-necked flask
weitmaschig coarse-meshed, wide-meshed
weitverbreitet widespread, widely circulated
Weitwinkelaufnahme *f* (Phot) wide-angle photograph
Weitwinkellinse *f* (Opt) wide-angle lens
Weitwinkelobjektiv *n* (Phot) wide aperture lens
Weizen *m* (Bot) wheat
Weizenkeimöl *n* wheat germ oil
Weizenkleie *f* wheat bran

Weizenmalz *n* malted wheat, wheat malt
Weizenmehl *n* wheat flour
Weizennachmehl *n* wheat bran
Weizenstärke *f* wheat starch
Weizenstroh *n* wheat straw
Weizenstrohzellstoff *m* wheat [straw] pulp
Weldon-Verfahren *n* (Chem) Weldon (chlorine process)
welken to fade, to wither
Wellblech *n* corrugated sheet, corrugated iron, corrugated plate
Wellblechpresse *f* press for corrugated sheets
Wellblechwalze *f* corrugating roll
Wellblechwalzwerk *n* corrugating rolling mill
Welle, genutete ~ shaft with key way
Welle *f* (Achse) shaft, arbor, spindle, (Opt, Elektr, Akust) wave, **einfallende** ~ incident wave, **elektromagnetische** ~ electromagnetic wave, **elliptisch polarisierte** ~ elliptically polarized wave, **fortschreitende** ~ traveling wave, **gekröpfte** ~ crank shaft, **glatte** ~ free shaft, **stehende** ~ stationary wave, **unendlich lange** ~ infinitely long wave, **wandernde** ~ progressive wave
wellen to corrugate
Wellenabsorption *f* absorption of waves
Wellenanregung *f* generation of waves
Wellenausbreitung *f* wave propagation
Wellenbahn *f* (Atom) wave path
Wellenbahnfunktion *f* (Atom) orbital wave function
Wellenband *n* wave band
Wellenbauch *m* wave loop
Wellenbereich *m* wave range
Wellenberg *m* peak of wave, wave crest
Wellenbewegung *f* undulation, undulatory or harmonic motion, wave motion
Wellenbild *n* oscillogram
Wellenbock *m* tailshaft bracket
Welleneigenschaft *f* wave character
Wellenerzeuger-Einrichtung *f* wavemaker
wellenförmig wave-like, undulated
Wellenfortpflanzung *f* propagation of waves, wave propagation
Wellenfrequenz *f* wave frequency
Wellenfront *f* wave front
Wellenfunktion *f* wave function
Wellenfunktionsformalismus *m* wave function formalism
Wellengeschwindigkeit *f* wave velocity
Wellengleichrichter *m* (Elektr) wave rectifier
Wellengleichung *f* wave equation
Wellenkamm *m* wave crest
Wellenknoten *m* (Phys) nodal point
Wellenkonstante *f* wave constant
Wellenkopf *m* wave front
Wellenkupp[e]lung *f* shaft coupling
Wellenlänge *f* wave length

Wellenlängenabhängigkeit *f* dependence on wave length
Wellenlängenbereich *m* region of wave length
Wellenlängengrenze *f* wave length limit
Wellenlager *n* main bearing, shaft bearing
Wellenlehre *f* wave theory
Wellenleitung *f* line shaft, shafting
Wellenleitwert *m* wave conductance
Wellenleitwert-Matrix *f* admittance matrix
Wellenlinie *f* wave line, sinus line, undulatory line
Wellenmechanik *f* wave mechanics
wellenmechanisch wave mechanical
Wellenmesser *m* wavemeter
Wellenmessung *f* wave measurement
Wellennatur *f* wave character[istic]
Wellenpaket *n* (Phys) wave packet
Wellenprofil *n* corrugation profile
Wellenschreiber *m* ondograph
Wellenschwingung *f* undulation
Wellensieb *n* wave filter
Wellensprektrum *n* frequency spectrum
Wellensteilheit *f* wave steepness
Wellenstörung *f* wave disturbance
Wellenstrahlengleichrichter *m* jet wave rectifier
Wellenstrom *m* (Elektr) pulsating current
Wellental *n* trough (of a wave)
Wellentheorie *f* **des Lichtes** (Opt) wave theory of light
Wellenvektor *m* wave vector
Wellenwiderstand *m* output impedance, wave drag
Wellenzahl *f* wave number
Wellenzapfen *m* shaft journal
wellig undulatory, wavy
Wellpapier *n* corrugated paper
Wellpappe *f* corrugated paper
Wellrohr *n* corrugated pipe
Wellsiebboden *m* (Kolonne) ripple tray
Wellsit *m* (Min) wellsite
Welt *f* world
Weltall *n* universe
Welterzeugung *f* world production
Weltkugel *f* cosmosphere
Weltraum *m* cosmic space
Weltraumrakete *f* space rocket
Weltraumschiff *n* space ship
Weltraumschiffahrt *f* astronautics, space navigation
Weltraumstation *f* space station
Weltraumstrahlung *f* cosmic rays
weltumspannend world-wide
Weltvorrat *m* world's supply
Wendebecken *n* turning basin
Wendegetriebe *n* reverse gear, reversing gear
Wendel *f* helix, spiral
Wendelantenne *f* corkscrew antenna
Wendelleiter *m* (Elektr) spiral conductor
Wendelrohrverdampfer *m* coil evaporator

Wendelrührer – Wert

Wendelrührer *m* helical ribbon agitator, spiral mixer
Wendelscheider *m* Humphrey separator
Wendelwuchtförderer *m* vibrating spiral elevator
wenden to reverse, to turn
Wendeofen *m* rotary puddling furnace
Wendeplatte *f* turnover pattern plate, turnover top table
Wendepol *m* reversing pole
Wendepresse *f* reverse[d] press
Wendeprisma *n* (Opt) reversing prism, Dove prism
Wendepunkt *m* turning point, point of inflection
Wender *m* manipulator, reverser, rotator
Wendeschalter *m* reversing switch
Wendeschaufel *f* turning blade
Wendetangente *f* inflectional tangent
Wendevorrichtung *f* rotating apparatus
Wendezylinder *m* rotating cylinder
Wendigkeit *f* maneuverability, manoeuvrability
Wendung *f* reversal, reversion, turn
Werbeabteilung *f* publicity department
Werbeartikel *m* advertising article
Werbedruckschrift *f* prospectus
Werbefirma *f* advertising agency
Werbeformel *f* slogan
Werbegeschenk *n* propaganda gift
werben to advertise
Werbung *f* advertising
Werdegang *m* process of manufacturing
werfen to throw, to cast, (sich verziehen) to buckle, to distort, to warp
Werfen *n* (Verziehen) warp[ing], **~ des Stahls** shrinkage of the steel
Werfung *f* warping
Werg *n* tow
Werk *n* (Erzeugnis) work, (Fabrik, Mechanismus) works
Werkbank *f* workbench
Werkblei *n* crude lead, raw lead, workable lead
Werksilber *n* silver from lead ore
Werksnummer *f* serial number
Werkstätte *f* workshop
Werkstatt *f* workshop
Werkstattprüfung *f* shop test
Werkstattzeichnung *f* workshop drawing, shop drawing
Werkstoff *m* raw material, **alkalibeständiger ~** alkali-resistant material, **einen ~ behandeln** to process a material, **einen ~ verarbeiten** to work with a material, **feuerfester ~** fire-proof material, **gerbsäurebeständiger ~** tannic acid-resisting material, **geschichteter ~** laminated material, **polymerer ~** polymeric material
Werkstoffanhäufung *f* material accumulation
Werkstoffbedarf *m* consumption of material
Werkstoffe *pl*, **thermoplastische ~** thermoplastics
Werkstoffermüdung *f* fatigue of material
Werkstoffestigkeit *f* strength of material
Werkstoffkunde *f* materials technology
Werkstoffprüfapparat *m* material testing apparatus
Werkstoffprüfung *f* material testing, **zerstörungsfreie ~** nondestructive testing of material
Werkstoffverfeinerung *f* processing of construction material
Werkstoffzuführung *f* feed of material
Werkstoleranz *f* work tolerance
Werkstück *n* piece of work, working part, work piece
Werkzeichnung *f* working drawing
Werkzeug *n* tool, implement, instrument, utensil, (Hilfsgerät) appliance, **~ aus Kunststoff** plastic tool, **~ für spanlose Formung** tool for noncutting shaping, **Abspannen eines ~ es** stripping, **scharf schneidendes ~** keen edge tool, **schnell laufendes ~** high-speed tool, **spanabhebendes ~** cutting tool, **zusammengesetzes ~** split mold
Werkzeughalter *m* tool holder, die block
Werkzeugkasten *m* tool kit
Werkzeugkiste *f* tool box, tool chest
Werkzeugmacher *m* mold maker
Werkzeugmacherdrehbank *f* tool lathe
Werkzeugmacherei *f* toolroom
Werkzeugmaschine *f* machine tool
Werkzeugmaschinenlack *m* machine tool finish
Werkzeugöffnungshub *m* opening stroke
Werkzeugprüfmaschine *f* material testing machine
Werkzeugsatz *m* tool set
Werkzeugschleifmaschine *f* tool [disc] grinder
Werkzeugschluß *m* closing joint
Werkzeugstahl *m* tool steel, **hochlegierter ~** special-alloy tool steel, **legierter ~** alloy tool steel, **unlegierter ~** carbon tool steel, unalloyed tool steel
Werkzeugtasche *f* kit bag
Werkzeugzuhaltekraft *f* mold locking force
Werkzink *n* raw zinc
Werkzinn *n* raw tin
Wermut *m* vermuth, absinthe
Wermutbitter *m* absinthin
Wermutkraut *n* (Bot) common wormwood
Wermutöl *n* wormwood oil
Wernerit *m* (Min) wernerite, scapolite
Wert *m* value, **angenäherter ~** approximate value, **beständiger ~** constant value, **experimenteller ~** experimental value, **höchstzulässiger ~** maximum safe value, **konstanter oder unveränderlicher ~** constant value, **kritischer ~** critical value, **numerischer ~** (Math) numerical value, **optimaler ~** optimum value, **reziproker ~**

reciprocal value, **wahrer** ~ true value, **wahrscheinlichster** ~ most probable value, **zulässiger** ~ admissible value, safe value
Wertbestimmung f determination of value
Werthmannit m (Min) werthmannite
wertig valent
Wertigkeit f valence, valency, **elektrochemische** ~ electrochemical valency, **negative** ~ negative valence, **stöchiometrische** ~ stoichiometric valence
Wertigkeitsbindung f valence bond
Wertigkeitselektron n valency electron
Wertigkeitsstufe f valence stage
Wertschätzung f estimation
Wesen n nature
wesentlich essential, substantial
Wesselys Anhydrid n Wessely's anhydride
Westonelement n Weston cell
Westrumit m westrumite
Wettbewerb m competition
Wettbewerber m competitor
wettbewerbsfähig competitive
Wetter n weather, (Schlagwetter) firedamp
Wetter pl (Bergb) firedamp
Wetterbeobachtung f meteorological observation, weather observation
wetterbeständig weatherproof, weather-resistant, weather resisting
Wetterbeständigkeit f weather resistance
Wetterfestigkeit f weather-proofness, weather resistance
Wetterkarte f meteorological map, weather map
Wetterkunde f meteorology
Wetterlampe f (Bergb) safety lamp
Wetterschleuse f (Bergb) air lock
Wetterschutzanstrich m weather protective coating, weather resistant coating
Wetterstein m (Min) belemnite
Wetterzünder m (Bergb) safety fuse
Wetzschiefer m novaculite
Wetzstein m whetstone, (Min) hone
Wharangin n wharangin
Wheatstonesche Brücke f (Elektr) Wheatstone['s] bridge
Wheelerit m wheelerite
Whewellit m (Min) whewellite
Whitneyit m (Min) whitneyite
Wichte f specific weight, density, specific gravity
wichtig important
Wickelfeder f spiral spring, volute spring
Wickelhammer m flexible belt drop stamp
wickeln to wind, to coil, to roll, to wrap
Wickelrolle f winding drum
Wickeltrommel f winding drum
Wickeltuch n fabric wrapper
Wick[e]lung f coil[ing], winding, wrapping
Wickelverfahren n filament winding

Wicklung f, **bifilare** ~ bifilar winding, dual-strand winding, **hochohmige** ~ high resistance winding, **kleine** ~ bobbin
Wicklungsquerschnitt m cross-sectional area of winding
Wicklungssinn m direction of winding
Widdren n widdrene
Widdrenal n widdrenal
Widdrensäure f widdrenic acid
Widerhall m echo, resonance, reverberation
Widerlager n abutment, counter flange
Widerschein m reflection
widersinnig absurd, in opposite direction
widerspiegeln to reflect
Widerspruch m contradiction
widerspruchsvoll contradictory, inconsistent
Widerstand m (Blindwiderstand) reactance, (Elektr) resistance, (Mech) drag, ~ **ausschalten** to cut out resistance, ~ **einschalten** to insert or switch in resistance, **äquivalenter** ~ equivalent resistance, **den** ~ **kurzschließen** to short-circuit the resistance, **dielektrischer** ~ dielectric resistance, **effektiver** ~ effective resistance, **elektrischer** ~ rheostat; **induktionsfreier** ~ non-inductive resistance, **induktiver** ~ inductive resistance, **induzierter** ~ induced resistance, induced drag, **innerer** ~ differential resistance, internal resistance, output resistance, plate resistance, **kapazitiver** ~ capacitance, **kritischer** ~ critical resistance, **lichtelektrischer** ~ photoresistance, **magnetischer** ~ magnetic resistance, **negativer** ~ negative resistance, **Ohmscher** ~ ohmic resistance, direct-current resistance, **punktförmig verteilter** ~ lumped resistance, **scheinbarer** ~ apparent resistance, **spannungsabhängiger** ~ exponential resistor, **spezifischer** ~ resistivity, **thermischer** ~ thermal resistance, **vorgeschalteter** ~ series resistance
Widerstandsanomalie f resistance anomaly
widerstandsbeheizt resistance-heated
Widerstandsbeiwert m resistance coefficient
Widerstandsbremsung f rheostatic braking
Widerstandsbrücke f (Elektr) resistance bridge, wheatstone bridge
Widerstandsdraht m (Elektr) resistance wire
Widerstandseinheit f (Elektr) unit of resistance
Widerstandselement n element of resistance, resistance grid
Widerstandserhitzung f resistance heating
Widerstandserhöhung f increase of resistance
widerstandsfähig resistant, resisting
Widerstandsfähigkeit f capacity for resistance, resistivity, ~ **gegen atmosphärische Korrosion** resistance to weathering, **dielektrische** ~ dielectric strength, **natürliche** ~ inherent resistance

Widerstandsgerät *n* resistor, (Rheostat) rheostat
Widerstandsgitter *n* resistance grid
Widerstandsheizung *f* resistance heating
Widerstandskapazität *f* resistance capacity
Widerstandskasten *m* resistance box
Widerstandskoeffizient *m* coefficient of resistivity, (eines Rührers) power number (of a mixer)
Widerstandskörper *m* resistance body, resistor
Widerstandskomponente *f* resistance component
Widerstandskraft *f* resisting force
Widerstandslegierung *f* resistance alloy
Widerstandslinie *f* resistance line
Widerstandsmesser *m* (Elektr) ohmmeter
Widerstandsmessung *f* resistance measurement, resistance measuring
Widerstandsmoment *n* moment of resistance
Widerstandsofen *m* resistance furnace
Widerstandsphotozelle *f* photo-resistive cell
Widerstandsplatte *f* resistance plate
Widerstands-Reduktionsofen *m* resistance reduction furnace
Widerstandsregler *m* rheostat
Widerstandsschaltung *f* resistive circuit
Widerstandsschweißung *f* incandescent welding, resistance welding
Widerstandsspule *f* resistance coil
Widerstandsthermometer *n* resistance thermometer
Widerstandsträger *m* resistance support
Widerstandsverstärker *m* resistance amplifier
Widerstandszahl *f* (Dest) resistance coefficient
Widerstandszelle *f* photoresistance cell
Widerstandszunahme *f* resistance increase
widerstehen to resist, to withstand
Widerwärtigkeit *f* adversity, unpleasantness
widmen to dedicate, to devote
Widmung *f* dedication, devotion
wieder auffüllen to refill, to replenish, ~ **aufladen** to recharge, ~ **auflösen** to redissolve, ~ **ausrichten** to realign, ~ **beschicken** to recharge, to refill, to reload, ~ **füllen** to replenish
Wiederanblasen *n* restarting
Wiederanreicherung *f* reconcentration
Wiederaufbau *m* reconstruction, rebuilding, regeneration
wiederaufbereiten to regenerate
Wiederaufbereitung *f* reprocessing
Wiederaufnahme *f* resumption
wiederaufnehmen to resume
Wiederausflockung *f* reprecipitation
Wiederausrichtung *f* realignment
Wiederbelastung *f* reloading
wiederbeleben to revive
Wiederbeleben *n* reviving, reactivating, regenerating
Wiederbelebung *f* revivification, reactivation, regeneration

Wiederbrauchbarmachen *n* regenerating
wiedereinschmelzen to recast, to remelt
Wiedereinschmelzen *n* remelting, recasting
wiedererhitzen to reheat
Wiedererwärmen *n* reheating
wiedererzeugen to regenerate
Wiedergabe *f* reproduction
Wiedergabetreue *f*, **hohe** ~ (Akust) high fidelity
wiedergewinnen to recover, to regenerate
Wiedergewinnen *n* recovering
Wiedergewinnung *f* recovery, reclaiming, regeneration, salvage
Wiederherstellung *f* restoration, (Wiederbelebung) revivification
Wiederherstellungsarbeit *f* repair work
Wiederhervorbringung *f* reproduction
wiederholen to repeat
Wiederholprogramm *n* (Comp) rerun routine, rollback routine
wiederholt repeated
Wiederholung *f* repetition, recurrence
Wiederholungsversuch *m* replicate
wiederinstandsetzen to repair
Wiederinstandsetzung *f* restoration
Wiederkäuer *m* (Zool) ruminant
wiederkauen to ruminate
wiederkehren to recur, to return
Wiederkristallisation *f* recrystallization
wiederkristallisieren to recrystallize
Wiedervereinigung *f* recombination, reunion
Wiedervereinigungsgeschwindigkeit *f* (Elektr) recombination velocity
Wiedervereinigungsgesetz *n* recombination law
Wiederverkauf *m* resale
Wiederverteiler *m* redistributor
wiederverwenden to re-use
Wiederverwendung *f* repeated use
wiederverwertbar re-usable
Wiederzusammenbauen *n* reassembly
Wiederzusammenfügen *n* reassembly
Wiegegläschen *n* (Chem) weighing bottle
wiegen to weigh
Wiegeröhrchen *n* weighing tube
Wiegeschale *f* scale pan
Wiegestechheber *m* weighing pipet[te]
Wiegevorrichtung *f* balance, scales
Wieland-Gumlich-Aldehyd *m* Wieland-Gumlich aldehyde
Wienergrün *n* Vienna green
Wiensche Konstante *f* Wien constant
Wiensches Strahlungsgesetz *n* Wien's radiation law
Wiensche Verschiebung *f* (Phys) Wien shift
Wiesenerz *n* (Min) bog [iron] ore, limonite, marsh ore
Wildbret *n* game
Wildleder *n* buckskin
Wildlederpolierscheibe *f* deerskin polishing disk
Willagenin *n* willagenin

Willardiin *n* willardiine
Willemit *m* (Min) willemite
willkürlich arbitrary, random
Willkürlichkeit *f* arbitrariness, randomness
Willyamit *m* (Min) willyamite
Wilsonkammer *f* expansion chamber, [Wilson] cloud chamber
Wilsonkammeraufnahme *f* cloud chamber photograph
Wilsonnebelspurmethode *f* Wilson cloud track method
Wiltshireit *m* (Min) wiltshireite
Wimperntusche *f* (Kosm) mascara
Wind *m* wind, air [blast], blast, breeze, ~ **mit geringer Pressung** soft blast, ~ **mit starker Pressung** cutting blast, **heißer** ~ hot blast, **kalter** ~ cold blast
Windanzeiger *m* wind indicator
Windberechnung *f* calculation of blast
winddicht windproof
Winddruckmesser *m* air pressure recorder
Winde *f* winch, windlass
Windeisen *n* tap wrench
winden to wind, (um eine Spule) to reel
Winderhitzer *m* air heater, hot blast stove
Winderhitzung *f* blast heating
Winderosion *f* wind erosion
Winderzeuger *m* wind generator
Windetau *n* winding rope
Windfrischapparat *m* (Metall) converter
Windfrischen *n* (Metall) converting
Windfrischverfahren *n* (Metall) Bessemer process, converter process, purifying process
Windgeschwindigkeit *f* wind velocity
Windgeschwindigkeitsmesser *m* anemometer, wind gauge, wind velocity indicator
windgesichtet wind-sifted
Windkammer *f* air chamber, air reservoir
Windkanal *m* air duct, wind tunnel, ~ **geschlossener Bauart** return-flow wind tunnel, ~ **offener Bauart** non-return-flow wind tunnel
Windkasten *m* blast box, wind box
Windkessel *m* blast tank, compression surge drum
Windklappe *f* air inlet valve
Windkranz *m* distribution pipe
Windlade *f* wind chest
Windleitung *f* air pipe, blast pipe
Windloch *n* blast inlet
Windmesser *m* anemometer, blast gauge, wind gauge, wind meter
Windmessung *f* anemometry
Windmühle *f* windmill
Windöffnung *f* (Metall) tuyere hole
Windofen *m* air furnace, wind furnace
Windpocken *pl* (Med) chicken-pox
windschief skew, warped
windschnittig streamlined, aerodynamic

Windschutzscheibe *f* wind screen, wind shield
Windsichten *n* air classification
Windsichter *m* air classifier, air separator
Windsichtung *f* air classification
Windspannung *f* blast pressure
Windstärke *f* wind velocity
Windstärkemesser *m* anemometer
Windstrom *m*, **gleichmäßiger** ~ continuous blast, uniform blast
windtrocken wind dried
Windtrocknung *f* blast drying, wind drying
Windung *f* turn, convolution, (Draht) coil, (Torsion) torsion
Windungswinkel *m* angle of torsion
Windungszahl *f* number of turns, number of windings
Windventil *n* blast valve
Windverbrauch *m* blast consumption
Windverteilung *f* air distribution, blast distribution
Windzacken *m* baffle plate
Windzuführung *f* blast inlet
Windzylinder *m* blast cylinder
Winkel *m* angle, (Dreieck) triangle, **rechter** ~ right angle, **spitzer** ~ acute angle, **stumpfer** ~ obtuse angle
Winkelabhängigkeit *f* angular dependence
Winkelabstand *m* angular distance
Winkelabweichung *f* angular deviation
Winkelarm *m* angle bracket, wall bracket
Winkelbeschleunigung *f* angular acceleration
Winkelbewegung *f* angular motion
Winkelblech *n* angle sheet iron
Winkeldeformation *f* angular deformation
Winkeldrehmoment *n* angular momentum
Winkeleinstellung *f* angular adjustment
Winkeleisen *n* (Techn) angle iron, **gleichschenkliges** ~ angle with equal sides
Winkeleisengestell *n* angle iron support
Winkelfräser *m* angle cutter
Winkelfrequenz *f* angular frequency
Winkelfunktion *f* (Math) trigonometric function
Winkelgeschwindigkeit *f* angular velocity
Winkelhahn *m* angle cock
Winkelhalbierende *f* bisector of an angle, bisecting line of an angle
Winkelheber *m* bent syphon tube
Winkelintervall *n* angular interval
Winkelkastenthermometer *n* elbow box thermometer
Winkelkatze *f* angle-type trolley
Winkelkoordinate *f* angular coordinate
Winkelmesser *m* (für Flächenwinkel) goniometer, (für Linienwinkel) protractor, (Theodolit) theodolite
Winkelpresse *f* angle molding press
Winkelrad *n* miter gear
Winkelring *m* angle ring
Winkelrohr *n* bent tube

Winkelschweißen *n* angular welding
Winkelspannung *f* angular strain
Winkelstück *n* connecting tube
Winkelsumme *f* angular sum
Winkelthermometer *n* bent thermometer, elbow thermometer
winkeltreu equiangular, isogonal
Winkelventil *n* corner valve
Winkelverdrehung *f* angular deflection
Winkelverschiebung *f* angular displacement
Winkelverteilung *f* angular distribution
Winkelwulsteisen *n* bulb angle
Winkelzahn *m* helical tooth, helical gearing
Winker *m* direction indicator
Winklerbürette *f* (Anal) Winkler buret[te]
Winkligkeit *f* angularity
Wintergrünöl *n* oil of wintergreen
Winterschlaf *m* (Biol) hibernation
Winzer *m* vine-grower, vintager
Wippbrücke *f* tipping bridge
Wippdrehkran *m* luffing-slewing crane, rotary luffing crane
Wippe *f* balancer, lever, rocker
Wippkran *m* luffing crane; luffing-jib crane
Wirbel *m* eddy, swivel, whirl, (Anat) vertebra (pl vertebrae)
Wirbelbettstromklassierer *m* fluidizing classifier
Wirbelbewegung *f* turbulence, swirling motion
Wirbelbrenner *m* turbulent burner, vortex burner
Wirbeldiffusion *f* eddy diffusion
Wirbeldüse *f* swirl nozzle
Wirbeleinfluß *m* turbulence effect
Wirbelfließverfahren *n* fluidized flow process
wirbelfrei nonvortical
Wirbelhaken *m* shackle hook, swivel hook
Wirbelkammer *f* turbulence chamber
Wirbelknochen *m* vertebra
Wirbelkristallisator *m* vortex crystallizer
Wirbelrohr *f* vortex tube
Wirbelsäule *f* (Med) spine, vertebral column
Wirbelschicht *f* fluid bed, fluidized bed (process engineering), turbulent layer (dynamics), (Klassierer) teeter bed
Wirbelschichtanlage *f* fluid[ized] bed apparatus, fluid[ized] bed plant
Wirbelschichtofen *m* fluid[ized] bed furnace
Wirbelschichtreaktor *m* fluidized bed reactor
Wirbelschichttrockner *m* fluidized bed dryer
Wirbelschichtverfahren *n* fluid[ized] bed process
Wirbelsichter *m* vortex type separator, whirl sorting plant
Wirbelsintern *n* pebble-bed sintering, whirl sintering
Wirbelsinterverfahren *n* fluidized bed process
Wirbelstabilisierung *f* whirl stabilization
Wirbelstoßtrockner *m* pulsating-bed drier, turbulent impact dryer
Wirbelströmung *f* eddy flow, vorticity

Wirbelstrom *m* (Elektr) eddy current
Wirbelstromkonstante *f* eddy current constant
Wirbelstromkreis *m* eddy current circuit
Wirbeltier *n* (Zool) vertebrate
Wirbelung *f* whirling
Wirbelwelle *f* rotational wave
Wirbelwert *m* vorticity, vorticity potential
Wirkdruck *m* differential pressure
Wirkdruckverfahren *n* active pressure method
Wirkdruckwandler *m* flow meter
wirken to act [on], to operate, to work, (Text) to knit
Wirken *n* activity
Wirkhärtung *f* (Krist) strain hardening, work hardening
Wirkkomponente *f* active component
Wirkleistung *f* (Elektr) effective power
Wirkleitwert *m* conductance
wirksam active, effective, (wirkungsvoll) efficacious, efficient
Wirksamkeit *f* (Leistungsfähigkeit) effectiveness, efficacy, efficiency, strength, (Tätigkeit) activity, **biologische** ~ biological aktivity, biological effectiveness
Wirkspannung *f* (Elektr) active voltage
Wirkstoff *m* active component, active ingredient
Wirkstrom *m* active current
Wirkung *f* action, effect, (Einfluß) influence, **antagonistische** ~ antagonistic action, **bakterizide** ~ bactericidal effect, **biologische** ~ biological effect, **deformierende** ~ deformation, **entmagnetisierende** ~ demagnetizing effect, **photochemische** ~ photochemical activity, **photoelektrische** ~ photoelectric effect, **physiologische** ~ physiological effect, **polarisierende** ~ polarizing power, **radioaktive** ~ radioactive effect, **radiobiologische** ~ radiobiological action, **reduzierende** ~ reducing action, **rostschützende** ~ rust-preventing quality, **sterische** ~ steric effect, **thermoelektrische** ~ thermoelectric effect, Seebeck effect
Wirkungsart *f* kind of action, mode of operation
Wirkungsbereich *m* range of effectiveness
Wirkungsbreite *f* range of action
Wirkungsdauer *f* duration of effect, persistency
Wirkungseinbuße *f* (Droge) loss of efficiency
wirkungsfähig effective, active, capable of acting, efficient
Wirkungsfähigkeit *f* effectiveness, activity, efficiency
Wirkungsfeld *n* field of action
Wirkungsfunktion *f* action function
Wirkungsgrad *m* efficiency, effectiveness, **adiabatischer** ~ adiabatic efficiency, **elastischer** ~ resilience, **gesamter** ~ gross efficiency, over-all efficiency, **manometrischer** ~ manometric efficiency,

mechanischer ~ mechanical efficiency,
optischer ~ optical efficiency
Wirkungsgradbestimmung *f* efficiency test
Wirkungsgröße *f* action magnitude, action quantity
Wirkungshalbmesser *m* effective radius
Wirkungskraft *f* effective force
Wirkungskreis *m* sphere of action
wirkungslos inactive, ineffective, inefficient, inert
Wirkungslosigkeit *f* inefficiency, inactivity, ineffectiveness
Wirkungsprinzip *n* action principle
Wirkungsquantum *n* (Phys) Planck's constant
Wirkungsquerschnitt *m* effective cross section, **differentieller** ~ differential cross section
Wirkungsrichtung *f* direction of action, effective direction
Wirkungssphäre *f* effective range, sphere of action
wirkungsvoll effective
Wirkungsweise *f* mode of operation, function, method of operation, mode of action, performance
Wirkungswert *m* effective value, efficiency
Wirkungszeit *f* reaction time
Wirkungszone *f* reaction zone
Wirkwaren *f pl* (Text) knitted goods
Wirkwiderstand *m* effective resistance, real resistance
Wirkzeit *f* reaction time
Wirt *m* (Biol) host, **als** ~ **dienen** (Biol) to act as a host
wirtschaftlich economical
Wirtschaftlichkeit *f* economy
Wirtschaftskrise *f* economic crisis
Wirtschaftslage *f* economic situation
Wirtschaftsverband *m* industrial federation
Wirtsmolekül *f* host molecule
Wirtsorganismus *m* (Biol) host organism
Wirtswechsel *m* (Biol) host change
Wirtszelle *f* (Biol) host cell
wischen to wipe, to rub
wischfest rub-fast
Wischwachsemulsion *f* wash polish
Wiserit *m* (Min) wiserite
Wismut *n* bismuth (Symb. Bi)
Wismutat *n* bismuthate
Wismutbenzoat *n* bismuth benzoate
Wismutbleierz *n* (Min) bismuth lead ore, schapbachite
Wismutblende *f* (Min) bismuth blende, eulytite
Wismutblüte *f* (Min) bismite, bismuth ocher
Wismutbronze *f* bismuth bronze
Wismutbutter *f* bismuthous chloride, bismuth trichloride, butter of bismuth
Wismutchlorid *n* bismuth chloride, (Wismut(III)-chlorid) bismuth(III) chloride, bismuthous chloride, bismuth trichloride
Wismutcinnamat *n* bismuth cinnamate
Wismutcitrat *n* bismuth citrate
wismuten to solder with bismuth
Wismuterz *n* bismuth ore
Wismutgallat *n* bismuth gallate
Wismutgehalt *m* bismuth content
wismutgetränkt bismuth loaded
Wismutglätte *f* bismuth litharge, bismuth oxide
Wismutglanz *m* (Min) bismuth glance, bismuthinite
Wismutgold *n* maldonite
wismuthaltig bismuthiferous
Wismuthydroxid *n* (Wismut(III)-hydroxid) bismuth hydroxide, bismuth(III) hydroxide, bismuthous hydroxide
Wismut(III)-Verbindung *f* bismuthous compound
Wismutit *m* (Min) bismuthite
Wismutkobaltkies *m* (Min) cheleutite
Wismutkupfererz *n* (Min) wittichenite
Wismutlot *n* bismuth solder
Wismutmetall *n* bismuth metal, metallic bismuth
Wismutnickelkobaltkies *m* (Min) grunauite
Wismutniederschlag *m* bismuth precipitate
Wismutnitrat *n* (Wismut(III)-nitrat) bismuth nitrate, bismuth(III) nitrate, bismuthous nitrate, bismuth trinitrate
Wismutocker *m* (Min) bismuth ocher, bismite
Wismutoxid *n* bismuth oxide, (Wismut(V)-oxid) bismuthic anhydride, bismuthic oxide, bismuth pentoxide, bismuth(V) oxide
Wismutoxychlorid *n* bismuth oxychloride, bismuthyl chloride
Wismutoxyjodid *n* bismuth oxyiodide, bismuthyl iodide
Wismutpentoxid *n* bismuth pentoxide, bismuthic anhydride, bismuthic oxide, bismuth(V) oxide
Wismutpyrogallat *n* bismuth pyrogallate, helcosol
Wismutsäure *f* bismuthic acid
Wismutsäureanhydrid *n* bismuthic anhydride, bismuthic oxide, bismuth pentoxide, bismuth(V) oxide
Wismutsalicylat *n* bismuth salicylate
wismutsauer (Salz) bismuthate
Wismutschwamm *m* spongy bismuth
Wismutselenid *n* bismuth selenide
Wismutsilber *n* (Min) schapbachite
Wismutspaltung *f* bismuth fission
Wismutspat *m* (Min) bismuthite
Wismutsubgallat *n* bismuth subgallate, dermatol
Wismuttellur *n* bismuth telluride, telluric bismuth, tetradymite
Wismuttetroxid *n* bismuth tetroxide
Wismutvalerianat *n* bismuth(III) valerate, bismuthous valerate, bismuth valerate
Wismut(V)-Verbindung *f* bismuthic compound
Wismutwasserstoff *m* bismuth hydride
Wismutweiß *n* bismuth white, bismuth subnitrate

Wissen *n* knowledge
Wissenschaft *f* science, **angewandte** ~ applied science
Wissenschaftler *m* scientist
wissenschaftlich scientific
Withamit *m* (Min) withamite
Witherit *m* (Min) witherite, native barium carbonate
witterungsbeständig weather-resisting
Witterungsbeständigkeit *f* resistance to atmospheric corrosion
Witterungseinfluß *m* weather factor
Witterungseinwirkung *f* atmospheric exposure
Witterungskunde *f* meteorology
Witterungsverhältnisse *pl* atmospheric conditions
Wittichenit *m* (Min) wittichenite
Wocheinit *m* wocheinite
Wöhlerit *m* (Min) woehlerite
wölben to vault
Wölben *n* doming
Wölbung *f* bulging, convexity, curvature
Wörthit *m* woerthite
Wogenwolke *f* (Meteor) billow cloud
Wogonin *n* wogonine
wohlausgebildet well-formed, well-shaped
Wohlgeruch *m* fragrance
wohlriechend aromatic, fragrant
wohlschmeckend palatable, savory
Wolchit *m* wolchite
Wolfachit *m* (Min) wolfachite
Wolfram *n* tungsten, wolfram (Symb. W)
Wolframat *n* tungstate, salt of tungstic acid
Wolframblau *n* wolfram blue, mineral blue
Wolframbleierz *n* (Min) lead tungstate, stolzite
Wolframbogenlampe *f* tungsten-arc lamp
Wolframbronze *f* tungsten bronze
Wolframcarbid *n* tungsten carbide
Wolframeinkristall *m* single crystal of tungsten
Wolframelektrode *f* tungsten electrode
Wolframerz *n* tungsten ore, wolfram ore
Wolframfaden *m* tungsten filament, tungsten wire, **thoriumhaltiger** ~ thoriated tungsten filament
Wolframgehalt *m* tungsten content
Wolframgelb *n* yellow tungsten bronze
Wolframgleichrichter *m* (Elektr) tungsten rectifier
wolframhaltig tungsteniferous
Wolframit *m* (Min) wolframite
Wolframlampe *f* tungsten lamp
Wolframmetall *n* metallic tungsten, metallic wolfram, tungsten metal, wolfram metal
Wolframocker *m* (Min) tungstic ocher, tungstite
Wolframsäure *f* tungstic acid
Wolframsäureanhydrid *n* tungstic anhydride, tungsten trioxide
wolframsauer (Salz) tungstate
Wolframspitze *f* tungsten tip

Wolframstahl *m* tungsten steel
Wolframtrioxid *n* tungsten trioxide, tungstic anhydride
Wolframweiß *n* tungsten white, barium tungstate
Wolfsbergit *m* (Min) wolfsbergite
Wolfsstahl *m* natural steel
Wolke *f* cloud, **radioaktive** ~ radioaktive cloud
Wolkenachat *m* (Min) clouded agate
Wolkenbildung *f* cloud formation
Wolkenentladung *f* cloud discharge
wolkig cloudy
Wolkonskoit *m* wolkonskoite
wollähnlich resembling wool, fleecy, wool-like, woolly
Wollasche *f* wool ashes
Wollastonit *m* (Min) wollastonite
Wolle *f* wool, ~ **einfetten** to grease the wool, to oil the wool, ~ **schmälzen** to grease the wool, to oil the wool, ~ **spicken** to oil the wool
Wollentfettung *f* degreasing of wool
Woll-Ersatzstoff *m* wool substitute
Wollfaser *f* wool fiber
Wollfett *n* wool fat, lanolin, wool grease
Wollfettemulsion *f* wool grease emulsion
Wollfettgewinnung *f* wool grease recovery
Wollfettpräparat *n* wool grease preparation
Wollfettschaum *m* wool grease scum
Wollfilz *m* wool felt
Wollgarn *n* (Web) spun wool, wool yarn
Wollgrün *n* wool green
Wollhornstoff *m* wool keratin
wollig woolly, fleecy, wool-like
Wollöl *n* wool oil
Wollschmiere *f* suint, wool yolk
Wollschweiß *m* suint, wool yolk
Wollschweißasche *f* potash from suint
Wollviolett *n* wool violet
Wollwachs *n* wool wax
Wollwäsche[rei] *f* wool scouring, wool washing
Wollwaschwasser *n* suds
Woodsches Metall *n* Wood's alloy
Woodwardit *m* (Min) woodwardite
Worobieffit *m* worobieffite
Woulffsche Flasche *f* (Chem) Woulff bottle
Wrackguß *m* spoiled casting
wuchern to grow exuberantly, to proliferate
Wucherung *f* growth, (Bot) exuberance, rank growth, (Med) tumo[u]r
Wuchs *m* (Gestalt, Form) figure, shape, (Wachstum) growth
Wuchsstoff *m* growth promoter, growth-promoting substance
Wucht *f* (Stoßkraft) impact, momentum
wünschenswert desirable
Würfel *m* cube, dice, (Krist) hexahedron
Würfelalaun *m* cubic alum
Würfelerz *n* (Min) cube ore, pharmacosiderite
Würfelfläche *f* cube face
würfelförmig cubic, cuboid, dice-shaped

Würfelform *f* cube shape
Würfelgips *m* anhydrite, cube spar
Würfelgitter *n* (Krist) cubical lattice
Würfelkante *f* cube edge
Würfelkohle *f* cobbles, cob coal, lump coal
Würfelkopf *m* cubic head
Würfelsalpeter *m* cubic niter, cubic saltpeter
Würfelspat *m* (Min) anhydrite, cube spar
Würfelzeolith *m* (Min) analcite, cubicite
Würfelzucker *m* cube sugar, lump sugar
Würze *f* seasoning, spice, (Brau) wort
Würz[e]-kühler *m* wort cooler
Würzekühlung *f* wort cooling
würzen to flavor, to season, to spice
Würzgeruch *m* odor of spices, scent of spices, spicy odor
Würzgeschmack *m* aromatic flavor, aromatic taste, spicy flavor, spicy taste
Würzkufe *f* (Brau) wort vat
Würzraum *m* malt chamber
Würzwein *m* medicated wine, spiced wine
Wulfenit *m* (Min) native lead molybdate, wulfenite
Wulst *m* reinforcement, torus, (Reifen) bead
Wulstanrollvorrichtung *f* bead stitcher
Wulstanschlußmaße *pl* bead base dimensions
Wulstaustrieb *m* lip
Wulstband *n* flap
Wulsteisen *n* bulb iron, bulb angles, bulb flats
Wulstferse *f* bead heel
Wulstkern *m* bead [core]
Wulstkernbelag *m* bead wrapper
Wulstpartie *f* bead area, bead region
Wulstrahmen *m* welt[ing]
Wulstreifen *m* clincher tire
Wulstspitze *f* bead toe
Wulstumlage *f* tie-in
Wulstzehe *f* bead toe
Wunde *f* wound, **offene** ~ open sore, **klaffende** ~ gash, **offene** ~ slight injury, **schwere** ~ severe injury
Wunder *n* miracle, marvel, wonder
Wundererde *f* (Min) lithomarge, stone marrow
Wundersalz *n* (obs) glauber's salt, salt mirabile, sodium sulfate decahydrate
Wunderwasser *n* (obs) aqua mirabilis
Wunderwerk *n* miracle
Wundfieber *n* (Med) wound fever
Wundpflaster *n* (Pharm) plaster
Wundsalbe *f* (Pharm) ointment for wounds
Wundstarrkrampf *m* (Med) tetanus
Wurf *m* throw, cast, casting, projection
Wurfbahn *f* trajectory, flight path
Wurfbecherwerk *n* rotating bucket elevator
Wurfgeschoß *n* missile, projectile
Wurflehre *f* ballistics
Wurflinie *f* trajectory
Wurfrad *n* scoop wheel
Wurfschaufel *f* hand scoop

Wurfsichter *m* winnower
Wurfsieb *n* shaking screen
Wurm *m* worm
Wurmgetriebe *n* worm gear
Wurmkraut *n* (Bot) anthelmintic herb, pinkroot, tansy
Wurmmittel *n* (Pharm) anthelmintic, vermicide, vermifuge
Wurmmoos *n* (Bot) Corsican moss, worm moss
Wurmrinde *f* worm bark
Wurmsamen *m* wormseed
Wurmstein *m* (Min) helmintholite
wurmstichig vermiculate, worm-eaten
Wurmtang *m* worm moss
wurmtötend vermicidal
wurmvertilgend anthelmintic, worm destroying
Wurstersches Salz *n* Wurster salt
Wurstgift *n* sausage poison
Wursthaut *f* sausage skin
Wurstvergiftung *f* sausage poisoning
Wurtzilit *m* wurtzilite
Wurtzit *m* (Min) wurtzite
Wurtzitgitter *n* (Krist) wurtzite lattice
Wurtzitstruktur *f* wurtzite structure
Wurtzittyp *m* wurtzite structure
Wurzel *f* root, radix, (Quadratwurzel) square root, **die** ~ **ziehen** to extract the root, **dritte** ~ cube root
Wurzel[aus]ziehen *n* (Math) extraction of a root
Wurzelexponent *m* (Math) index of a root, radical index
Wurzelfüllung *f* (Zahnmed) root filling
Wurzelgerbstoff *m* root tannin
Wurzelgröße *f* root element
Wurzelharz *n* wood resin
Wurzelkautschuk *m* root rubber
Wurzelknoten *m* node
Wurzeltorf *m* fibrous peat
Wurzelverzweigung *f* root branching
Wurzelzeichen *n* (Math) radical sign
wurzelziehen (Math) to extract a root

X

X-Achse f x-axis, abscissa
Xanthalin n xanthaline
Xanthamylsäure f xanthamylic acid, amyldisulfocarbonic acid
Xanthanwasserstoff m xanthan hydride, perthiocyanic acid
Xanthat n xanthate, xanthogenate
Xanthazol n xanthazol
Xanthein n xanthein
Xanthen n xanthene, dibenzo-1,4-pyran
Xanthenfarben pl xanthene dyestuffs
Xanthenol n xanthenol, 9-hydroxyxanthene, **offenes** ~ xanthydrol
Xanthenon n xanthenone, 9-ketoxanthene, xanthone
Xanthin n xanthin[e], 2,6-dihydroxypurine
Xanthindehydrase f (Biochem) xanthine dehydrogenase, Schardinger enzyme
Xanthinoxidase f (Biochem) xanthine oxidase
Xanthion n xanthione
Xanthit m xanthite
Xanthoapocyanin n xanthoapocyanine
Xanthoarsenit m (Min) xanthoarsenite
Xanthocillin n xanthocillin
Xanthogen n xanthogene
Xanthogenamid n xanthogen amide
Xanthogenat n xanthogenate, salt or ester of xanthic acid, xanthate
Xanthogensäure f xanthic acid, xanthogenic acid, xanthonic acid
xanthogensauer xanthate, xanthogenate
Xanthohumol n xanthohumol
Xantholith m xantholite
Xanthom n (Med) xanthoma
Xanthomatosis f (Med) xanthomatosis
Xanthon n xanthone, 9-ketoxanthene, xanthenone
Xanthoperol n xanthoperol
Xanthophyll n xanthophyll
Xanthophyllit m (Min) xanthophyllite
Xanthopikrin n xanthopicrin
Xanthopikrit n xanthopicrite
Xanthoprotein n (Biochem) xanthoprotein
Xanthoproteinreaktion f xanthoprotein reaction
xanthoproteinsauer xanthoproteate
Xanthopterin n xanthopterine
Xanthopuccin n xanthopuccine
Xanthopurpurin n 1,3-dihydroxy anthraquinone, xanthopurpurin
Xanthoraphin n xanthoraphine
Xanthorhamnin n xanthorhamnin
Xanthortit m xanthortite
Xanthosiderit m (Min) xanthosiderite
Xanthosin n xanthosin
Xanthotoxin n xanthotoxin
Xanthoxylen n xanthoxylene, xanthoxylin

Xanthoxylin n xanthoxylene, xanthoxylin
Xanthurensäure f xanthurenic acid
Xanthydrol n xanthydrol, 9-hydroxyxanthene
Xanthyl- xanthyl
Xanthylsäure f xanthylic acid
X-Einheit f x-unit
Xenolith m (Min) xenolite
xenomorph (Krist) xenomorphic
Xenon n xenon (Symb. X)
Xenonblasenkammer f xenon bubble chamber
Xenonbogen m xenon arc
Xenonhochdrucklampe f xenon high pressure lamp
Xenonlampe f xenon lamp
Xenonröhre f xenon tube
Xenotim m (Min) xenotime
Xenyl- xenyl, biphenyl-
Xenylamin n xenylamine
Xeroform n xeroform, bismuth tribromophenolate
Xerographie f xerography
xeromorph xeromorphic
xerophil xerophilic
Xerophyt m (Bot) xerophyte
Ximeninsäure f ximenynic acid
Ximensäure f ximenic acid
Xi-minus-Antiteilchen n xi-minus antiparticle
Xi-Teilchen n xi-particle
x-Koordinate f x-coordinate
Xonotlit m xonotlite
Xylal n xylal
Xylan n xylan
Xylarsäure f xylaric acid
Xylem n (Bot) xylem
Xylenol n xylenol, dimethylphenol
Xylenolharz n xylenol resin
Xylenolphthalein n xylenolphthalein
Xylenrot n xylene red
Xylidid n xylidide
Xylidin n xylidine
Xylidinrot n xylidine red
Xylidinsäure f xylidic acid, 4-methylisophthalic acid, xylidinic acid
Xylit m xylitol
Xylitan n xylitan
Xylobiose f xylobiose
Xylocain n xylocaine
Xylochinol n xyloquinol
Xylochinon n xyloquinone
Xylochloral n xylochloral
Xylochloralose f xylochloralose
Xyloflavin n xyloflavine
Xylohydrochinon n xylohydroquinol, dimethylhydroquinol
Xyloidin n xyloidine
Xyloketose f xyloketose

Xylol *n* xylene
Xylolith *m* (Min) xylolith, wood-stone
Xylolsulfonsäure *f* xylenesulfonic acid
Xylonsäure *f* xylonic acid
xylonsauer xylonate
Xylopin *n* xylopine
Xylopinin *n* xylopinine
Xylorcin *n* xylorcin, dimethylresorcinol, xylorcinol
Xylose *f* xylose, wood sugar
Xyloson *n* xylosone
Xylotil *m* (Min) xylotile, amianthus, ligniform asbestos, rockwood
Xylulose *f* xylulose
Xylulosephosphat *n* xylulose phosphate
Xylyl- xylyl
Xylylchlorid *n* xylyl chloride
Xylylen- xylylene
Xylylendiamin *n* xylylenediamine
Xylylendichlorid *n* xylylene dichloride
Xylylenharnstoff *m* xylylene urea
Xylylsäure *f* xylic acid, dimethylbenzoic acid
xylylsauer xylate

Y

Y-Achse *f* ordinate, Y-axis
Yajein *n* yajeine, harmine
Yanolith *m* yanolite
Yard *n* (Maß) yard
Yenit *m* yenite, ilvaite
Ylang-Ylangöl *n* ylang-ylang oil
Yohimbin *n* yohimbine, corynine, methyl yohimbate
Yohimbinsäure *f* yohimb[oa]ic acid
Yohimbon *n* yohimbone
Yonogen[in]säure *f* yonogenic acid
Yperit *n* yperite, 2,2-dichlorodiethyl sulfide, mustard gas
Y-Rohr *n* Y-tube
Y-Schaltung *f* (Elektr) Y-connection, star connection
Ysop *m* (Bot) hyssop
Ysopöl *n* hyssop oil
Y-Spannung *f* (Elektr) Y-voltage
Ytterbin *n* ytterbia, ytterbium oxide
Ytterbinerde *f* ytterbia, ytterbium oxide
Ytterbit *m* ytterbite
Ytterbium *n* ytterbium (Symb. Yb)
Ytterbiumgehalt *m* ytterbium content
Yttererde *f* yttrium earth, yttria, yttrium oxide
Ytterflußspat *m* (Min) yttrocerite
ytterhaltig containing yttrium, yttriferous
Ytteroxid *n* yttrium oxide, yttria, yttrium earth
Yttersalz *n* yttrium salt
Ytterspat *m* (Min) xenotime
Yttrialith *m* (Min) yttrialite
Yttrium *n* yttrium (Symb. Y)
Yttriumchlorid *n* yttrium chloride
Yttriumgehalt *m* yttrium content
Yttriumnitrat *n* yttrium nitrate
Yttrocerit *m* (Min) yttrocerite
Yttrocolumbit *m* (Min) yttrocolumbite
Yttrofluorit *m* (Min) yttrofluorite
Yttroilmenit *m* (Min) yttroilmenite
Yttrokrasit *m* (Min) yttrocrasite
Yttrotantalit *m* (Min) yttrotantalite
Yttrotitanit *m* (Min) yttrotitanite
Yuccafaser *f* yucca fiber
Yuccasaponin *n* yuccasaponin
Yukon *n* (obs) yukon, meson
Y-Zerstäuber steam atomizer, Y-atomizer

Z

Z-Achse *f* Z-axis
Zacken *m* tooth, jag, notch, serration, (Auszackung) indentation
Zackenrolle *f* toothed roller
Zackenwalze *f* rough roll
zackig indented, jagged, notched, serrated, toothed
zäh tough, cohesive, stringy, tenacious, viscous, (Met) ductile, ~ **machen** to toughen, ~ **werden** to toughen
Zähfestigkeit *f* tenacity, toughness
zähflüssig stringy, viscid, viscous
Zähflüssigkeit *f* viscosity, viscousness
Zähflüssigkeitsmesser *m* viscosimeter
Zähigkeit *f* tenacity, toughness, viscosity, viscousness, (Met) ductility, **spezifische** ~ specific viscosity
Zähigkeitskehrwert *m* fluidity, inverse of viscosity
Zähigkeitskoeffizient *m* coefficient of viscosity, viscosity factor
Zähigkeitsmesser *m* viscosimeter
Zähkupfer *n* tough pitch copper
Zählausbeute *f* (Atom) counting efficiency
Zähldiamant *m* counting diamond
zählen to count, to number
Zähler *m* counter; meter, (eines Bruches) numerator (of a fraction), (Elektron) scaler, ~ **mit Borfüllung** boron-filled counter
Zählerablesung *f* meter reading
Zählerausbeute *f* counter efficiency
Zählergehäuse *n* (Elektr) [electricity] meter box
Zählertafel *f* meter board
Zählerwerk *n* recording mechanism
Zählkammer *f* counting ionization chamber
Zählkapazität *f* count capacity
Zählrohr *n* counter [tube]
Zählrohrausbeute *f* counter efficiency
Zählrohrcharakteristik *f* counter characteristic, counting response
Zählrohreichung *f* counter calibration
Zählrohrlebensdauer *f* counter life
Zählrohrlöschkreis *m* counter quench circuit
Zählrohrsteuerung *f* counter control
Zählrohrtotzeit *f* counter dead time
Zählung *f* counting, computation, metering, registering
Zählverfahren *n* method of counting
Zählwerk *n* counter, counting device, counting mechanism, meter, register, **mechanisches** ~ mechanical register
zähnig dentate, indented, jagged, toothed
zähschlackig forming tough clinker or slag
Zängearbeit *f* shingling
zängen to hammer the bloom, to shingle
Zäpfchen *n* (Pharm) suppository

Zaffer *m* zaffer, impure cobalt oxide
Zahl *f* number, digit, figure, (Index) index, (Konstante) constant, **dreistellige** ~ three-figure number, three-place number, **eingeklammerte** ~ number in brackets, **ganze** ~ (Math) integer, **gebrochene** ~ fractional number, **gerade** ~ even number, **imaginäre** ~ imaginary number, **irrationale** ~ irrational number, **komplexe** ~ complex number, **konjugiert komplexe** ~ conjugated complex number, **natürliche** ~ natural number, **rationale** ~ rational number, **reelle** ~ real number, **ungerade** ~ odd number
Zahlenbeispiel *n* numerical example
Zahlendarstellung *f*, **binäre** ~ binary notation, **dezimale** ~ decimal notation, **duodezimale** ~ duodecimal notation, **hexadezimale** ~ hexadecimal notation, **oktale** ~ octal notation, octonary notation
Zahleneinheit *f* unit number
Zahlenfolge *f* sequence of numbers
Zahlengleichung *f* (Math) numerical equation
Zahlenindex *m* numerical index
zahlenmäßig numerical, quantitative
Zahlenmittel *n* number average
Zahlenreihe *f* series
Zahlensystem *n* number system, **binäres** ~ binary number system, **dekadisches** ~ decadic number system, **dezimales** ~ decimal number system
Zahlenverhältnis *n* numeral ratio
Zahlenwert *m* numerical value, **absoluter** ~ nondimensional quantity, nondimensional ratio
Zahn *m* tooth, (eines Rades) cog
zahnärztlich dental
Zahnarzt *m* dentist
Zahnbelag *m* dental plaque, bacterial plaque
Zahnbreite *f* breadth of tooth
Zahndicke *f* tooth thickness
Zahndruckdynamometer *n* gear dynamometer
Zahnfäule *f* (Zahnmed) caries
Zahnflanke *f* tooth flank, flank gear
Zahnflankenspiel *n* backlash
Zahnflankenfestigkeit *f* surface strength of gear teeth
Zahnfleisch *n* gum[s]
Zahnform *f* tooth profile
Zahnfüllung *f* tooth filling
Zahngesperre *n* ratchet and pawl
Zahngipsabdruck *m* dental mold
Zahnheilkunde *f* dentistry
Zahnkaries *f* dental caries
Zahnkette *f* gear chain
Zahnkettenrad *n* sprocket wheel

Zahnkitt *m* dental cement
Zahnkopf *m* tooth face
Zahnkranz *m* gear rim, toothed rim
Zahnkreisteilung *f* circular pitch
Zahnlücke *f* space of the tooth, tooth space
Zahnpasta *f* tooth-paste
Zahnplombe *f* [tooth] filling
Zahnprofil *n* tooth profile
Zahnpulver *n* tooth-powder
Zahnrad *n* cogwheel, gear, gear wheel, toothed wheel
Zahnradantrieb *m* gear drive
Zahnradfräsmaschine *f* gear cutting machine
Zahnradgetriebe *n* toothed gearing
Zahnradpumpe *f* gear pump
Zahnradübersetzung *f* transmission gear
Zahnräderformmaschine *f* gear wheel molding machine
Zahnreinigungsmittel *n* tooth-paste
Zahnriemen *m* (Keilriemen) cogged belt, toothed belt
Zahnriemenrad *n* tooth-belt gear
Zahnscheibe *f* ratchet wheel, crown gear, crown wheel, face gear, toothed disk, **federnde** ~ toothed spring disc
Zahnscheibenmühle *f* toothed attrition mill, toothed disk mill
Zahnschmelz *m* enamel
Zahnsegment *n* toothed segment
Zahnstange *f* rack, ~ **und Ritzel** rack and pinion
Zahnstein *m* dental calculus, odontolith, tartar
Zahnsubstanz *f* dentine, tooth substance
Zahnteilung *f* circular pitch
Zahntiefe *f* depth of tooth
Zahnung *f* indentation, toothing
Zahnweinstein *m* tartar on the teeth
Zahnwurzel *f* root of the tooth
Zange *f* pliers, tongs, (Kneifzange) pincers, (Med) forceps
Zapfen *m* pivot, trunnion, (Schraube) journal, (Zimmerei) tenon
Zapfenbohrer *m* pin drill, tap borer
Zapfendruck *m* pivot thrust
Zapfendüse *f* pivot nozzle
Zapfenkupplung *f* square jaw clutch
Zapfenlager *n* pivot bearing, socket, spindle bearing
Zapfenlagermetall *n* antifriction metal
Zapfenlagerung *f* pivoting
Zapfenreibung *f* journal friction
Zapfenschneidmaschine *f* tenoning machine
Zapfenverbindung *f* spigot joint
Zapfhahn *m* discharge nozzle, drain cock, tap
Zapfmesser *n* tapping knife
Zapfung *f* tapping
Zaponlack *m* zapon lacquer, zapon varnish
Zaratit *n* (Min) zaratite
Zeagonit *m* zeagonite
Zeaxanthin *n* zeaxanthin

Zebromal *n* zebromal, ethyl phenyldibromopropionate
Zeche *f* (Bergb) coal pit, mine
Zechenkohle *f* mine coal
Zeder *f* (Bot) cedar
Zedernharz *n* cedar resin
Zedernholzöl *n* cedar wood oil
Zeemaneffekt *m* Zeeman effect
Zehneck *n* decagon
Zehnerlogarithmus *m* logarithm to the base ten
Zehnerpotenz *f* decimal power, power of ten
Zehnersystem *n* decimal system
Zehnerwaage *f* decimal balance
Zehntel *n* tenth [part]
Zehntelgrad *m* tenth of a degree
Zehntelliter *n m* deciliter
Zehntellösung *f* tenth-normal solution
zehntelnormal decinormal
Zeichen *n* sign, mark, (Signal) signal, (Symbol) symbol, **chemisches** ~ chemical symbol
Zeichenautomat *m* plotter
Zeichenbrett *n* drawing board
Zeichendichte *f* (Computer) high-density storage
Zeichenerklärung *f* explanation of signs, explanation of symbols, key to symbols
Zeichengerät *n* drawing instrument, plotter
Zeichenkreide *f* drawing chalk
Zeichenpapier *n* design paper, drawing paper
Zeichenregel *f* (Math) rule of signs
Zeichenstift *m* drawing pencil
Zeichentinte *f* drawing ink, marking ink, (Kopiertinte) indelible ink
zeichnen to draw, to design, to draft, to sketch
Zeichnen *n*, **technisches** ~ engineering drawing
Zeichnung *f* drawing, (Diagramm) diagram, graph, (Entwurf) design, sketch
zeigen to demonstrate, to show
Zeiger *m* (Instrument) indicator, (Kompaß) needle, (Uhr) hand, **der** ~ **schlägt aus** the pointer moves
Zeigerablesung *f* indicator reading, pointer reading
Zeigerausschlag *m* needle deflection
Zeigerdiagramm *n* vector diagram
Zeigergalvanometer *n* (Elektr) needle galvanometer
Zeigermanometer *n* indicating pressure gauge
Zeigerstellung *f* pointer setting
Zeigerthermometer *n* dial thermometer
Zeigerwerk *n* clock-work
Zeilenindex *m* line index
Zeilenmatrix *f* row matrix
Zeilensummenprobe *f* (Math) check column method
Zein *n* zein
Zeiselsche Methode *f* Zeisel's method
Z-Eisen *n* Z-iron, Z-bar
Zeit *f* time, period, **die** ~ **abstoppen** to stop the time, to time, **offene** ~ open assembly time

(adhesives), wet edge time (paints), **produktive** ~ production time
zeitabhängig time-dependent
Zeitablenkung *f* time deflection
Zeitabstand *m* time interval
Zeitalter *n* era
Zeitangabe *f* date
Zeitauflösungsvermögen *n* time resolution
Zeitaufnahme *f* (Phot) time exposure
Zeitauslösung *f* time-release
Zeitbestimmung *f* determination of time
Zeitbruchdehnung *f* elongation at break in creep
Zeitbrucheinschnürung *f* reduction in area in creep
Zeitbruchlinie *f* time-for-fracture curve
Zeitdauer *f* duration, period of time
Zeitdehnlinie *f* creep curve
Zeitdehnspannung *f* time yield limit
Zeitdifferenz *f* time lag
Zeiteinheit *f* unit [of] time, **in der** ~ per unit of time
Zeitfestigkeit *f* fatigue strength for finite life
Zeitfolge *f* time sequence
Zeitgeist *m* spirit of the age
zeitgemäß up-to-date, modern, timely
zeitgenössisch contemporary
Zeitgenosse *m* contemporary
Zeitgesetz *n* (einer Reaktion) rate law
Zeitgleichung *f* equation of time
Zeitglied *n* (Regeltechn) time function element
Zeithärtung *f* age-hardening, time quenching
Zeitintervall *m* time interval
Zeitintervallmessung *f* events timing
Zeitkanalanalysator *m* time channel analyzer
Zeitkonstante *f* time constant
Zeitkriechgrenze *f* time yield limit
zeitlich aufeinanderfolgend chronological, ~ **veränderlich** variable with time
Zeitlinie *f* time line
Zeitlupenanordnung *f* slow-motion arrangement
Zeitlupenaufnahme *f* (Phot) slow-motion picture
Zeitmarkengeber *m* time stamping machine
Zeitmesser *m* timing device
Zeitmeßgerät *n* chronometer
Zeitmessung *f* chronometry
Zeitmittel *n* time average
Zeitmittelwert *m* mean time value
Zeitplanregelung (Regeltechn) program control (Brit. E.), time pattern control (Am. E.), time program control
zeitraubend time-consuming, tedious
Zeitraum *m* period, period of time
Zeitschalter *m* time switch
Zeitschreiber *m* time recorder
Zeitschrift *f* journal
Zeitskala *f* time scale
Zeitspanne *f* time interval
Zeitstandbruchdehnung *f* elongation at break in creep

Zeitstandbrucheinschnürung *f* reduction in area in creep
Zeitstandfestigkeit *f* creep strength
Zeitstandkriechgrenze *f* time yield limit
Zeitstandversuch *m* creep test
Zeitstand-Zugversuch *m* tensile creep test
Zeitumkehroperator *m* time reversal operator
Zeit-Umsatz-Kurve *f* (Kin) time-activity curve
Zeitvektor *m* time vector
Zeitverhalten *n* time-cycle operation
Zeitverschiebung *f* time shift
Zeitverzögerung *f* time lag
Zeitzünder *m* time fuse
Zellase *f* (Biochem) cellase
Zellatmung *f* (Biol) cellular respiration
Zellchromosomen *pl* cell chromosomes
Zelldifferenzierung *f* (Biol) cell differentiation
Zelldrahtglas *n* chickenwire glass
Zelle *f* cell, compartment, segment, (Elektr) battery, element, **aerobe** ~ aerobic cell, **anaerobe** ~ anaerobic cell, **basophile** ~ basophilic cell, **Blut-** ~ blood cell, **bösartige** ~ malignant cell, **diploide** ~ diploid cell, **Eiter-** ~ pus cell, **elektrolytische** ~ electrolytic cell, **Fett-** ~ fat cell, **gegengeschaltete** ~ countercell, **grampositive** ~ gram-positive cell, **haploide** ~ haploid cell, **Keim-** ~ germ cell, **kleine** ~ cellule, **Langerhanssche** ~ Langerhans' cell, **lichtelektrische** ~ photoelectric cell, photoelectric tube, **lichtempfindliche** ~ light-reactive cell, photo[electric] cell, **Mutter-** ~ mother cell, parent cell, **reife** ~ mature cell, **Sichel-** ~ sickle cell, **Stern-** ~ astrocyte, star cell, **Tochter-** ~ daughter cell, **Wand-** ~ parietal cell, **Zylinder-** ~ cylindrical cell
Zellabsterben *n* necrobiosis
Zellaktivität *f* cell activity
Zellapparat *m* single-cell apparatus
zellenartig cell-like, cellular
Zellenatmung *f* vesicular breathing
Zellenaufbau *m* cell structure
Zellenbau *m* cellular structure
Zellenbeton *m* cellular concrete
zellenbildend cell-forming, cytogenic
Zellenbildung *f* cytogenesis, cell formation
Zellenbiologie *f* cytobiology
Zelleneinschluß *m* cell inclusion
Zellenfilter *n* revolving filter
zellenförmig cellular
Zellenförmigkeit *f* cellularity
Zellengalmei *m* (Min) cellular calamine
Zellengleichrichter *m* honeycomb rectifier
Zellenisolator *m* cell insulator
Zellenkalk *m* cellular chalk
Zellenkühler *m* cellular cooler
Zellenmethode *f* cellular method
Zellenprüfung *f* cell test

Zellenrad *n* bucket wheel, impeller
Zellenradextrakteur *m* bucket-wheel extractor
Zellenradschleuse *f* rotary-vane feeder, star feeder
Zellenschalter *m* cell switch
zellentötend cytocidal
Zellferment *n* cell enzyme, cell ferment
zellförmig cellular
Zellforschung *f* cytology
Zellfraktionierung *f* cell fractionation
Zellglas *n* cellophane
Zellgummi *m* expanded rubber, foam rubber
zellig cellular
Zellkern *m* nucleus
Zellkern-Protein *n* nuclear protein
Zellkörper *m* cellular body
Zellkultur *f* cell culture, biocytoculture
Zellmasse *f* cellular substance
Zellmembran *f* cellular membrane
Zellmorphologie *f* cytomorphology
Zelloberfläche *f* cell surface
Zellobiose *f* cellobiose, cellose
Zelloidin *n* celloidin
Zelloidinpapier *n* (Phot) celloidin paper, printing-out paper
Zellon *n* cellone
Zellophan[papier] *n* cellophane [paper]
Zellorganelle *f* [cell] organelle
Zellparasit *m* cell parasite, cytozoon (pl cytozoa)
Zellphysiologie *f* cytophysiology
zellphysiologisch cytophysiological
Zellpigment *n* cell pigment
Zellplasma *n* cytoplasm
Zellproduktion *f* cytogenesis
Zellreaktion *f* cell reaction
Zellspaltung *f* (Biol) fission of cells, (Zellkernteilung) mitosis, (Zerstörung, Zerfall) lysis
Zellstoff *m* cellulose, **gebleichter** ~ bleached pulp, **halbchemischer** ~ semichemical pulp, **harter** ~ low boiled pulp, **hochwertiger** ~ high-grade pulp, **minderwertiger** ~ low-grade pulp, **veredelter** ~ processed pulp, refined pulp, **weicher** ~ high boiled pulp
Zellstoffaufschluß *m* pulping
Zellstoffballen *m* pulp bale
Zellstoffbleiche *f* pulp bleaching
Zellstoffgarn *n* (Text) cellulose yarn
Zellstoffharz *n* pitch
Zellstoffherstellung *f* pulp production
Zellstoffilter *m* wood pulp filter
Zellstoffilz *m* pulp felt
Zellstoffkarton *m* pulp board
Zellstoffkocher *m* pulp digester, cellulose digester, cellulose kier
Zellstofflösung *f* cellulose solution
Zellstoffmahlung *f* pulp beating
Zellstoffpapier *n* cellulose paper
Zellstoffpappe *f* pulp board
Zellstoffschleim *m* pulp slime
Zellstoffseide *f* (Text) cellulose silk
Zellstofftrockenmaschine *f* pulp drying machine
Zellstofftrocknung *f* cellulose drying, pulp drying
Zellstoffveredelung *f* pulp processing
Zellstoffwatte *f* artificial cotton, cellulose wadding, pulp wadding
Zellstruktur *f* cell structure
Zellteilung *f* (Biol) cell division, **direkte** ~ amitosis, direct cell division, **indirekte** ~ indirect cell division, mitosis
Zelltod *m* (Zerfall) cytolysis
zelltötend cytocidal
Zellulase *f* (Biochem) cellulase
Zelluloid *n* celluloid
Zelluloseabkömmling *m* cellulose derivative
Zellulosebrei *m* cellulose pulp
Zelluloselack *m* cellulose finish, cellulose lacquer
Zellveränderung *f* cytomorphosis
zellvernichtend cytocidal, cytoclastic
Zellverschmelzung *f* cell fusion, plasmogamy
Zellwand *f* cell wall
Zellwolle *f* spun rayon, staple fiber, staple rayon
Zellzerfall *m* cytolysis
Zeltleinwand *f* canvas
Zeltplane *f* awning
Zeltstoff *m* awning, canvas
Zement *m* cement, ~ **abbinden lassen** to allow cement to set, **dünnflüssiger** ~ cement grout, **langsam bindender** ~ slow-setting cement, **mit** ~ **verstreichen** to cement, **schnell bindender** ~ quick-setting cement, **totgebrannter** ~ dead cement
zementartig cement-like
Zementation *f* cementation
Zementationsgefäß *n* cementation vat
Zementbeton *m* cement concrete
Zementbrei *m* cement slurry
Zementbrennofen *m* cement kiln
Zementdrehofen *m* rotary cement kiln
Zementeinspritzung *f* cementation
Zementerzeugung *f* production of cement
Zementfabrik *f* cement factory, cement works
Zementfilz *m* cement felt
Zementgerbstahl *m* [double] shear steel
zementhärten to cement-temper
zementieren to cement, to caseharden
Zementieren *n* cementing, cementation [process], ~ **im Spritzverfahren** cementing by jet
Zementierkiste *f* cementing box
Zementiermittel *n* cementing agent
Zementierofen *m* cementation furnace, [steel] converting furnace
Zementierpulver *n* cementation powder, cementing powder

Zementierung *f* cementation, **stufenweise** ~ multiple-stage cementing
Zementindustrie *f* cement industry
Zementit *m* (Metall) cementite, carbide of iron, **körniger** ~ granular cementite, **kugeliger** ~ nodular cementite, spheroidal cementite
Zementitzerfall *m* cementite disintegration
Zementkalkmörtel *m* cement-lime mortar
Zementklinker *m* cement clinker
Zementkupfer *n* cement copper
Zementmetall *n* cement metal
Zementmilch *f* cement grout
Zementmörtel *m* cement mortar
Zementofen *m* cementation furnace, cement kiln
Zementsilber *n* cement silver
Zementsockel *m* cement foundation
Zementspritzapparat *m* cement gun
Zementstahl *m* cementation steel, blister steel, cement[ed] steel
Zementstahlbereitung *f* cementation of steel
Zementstahldarstellung *f* manufacture of cement steel
Zementstahlofen *m* steel cementing furnace
Zementstaub *m* cement dust
Zementstein *m* cement stone
Zementtrog *m* cement tank
Zementüberguß *m* covering of hydraulic mortar
Zementwasser *n* cementing water
Zenit *m* zenith
Zenitpartie *f* **des Reifens** crown of the tire
Zenitwinkel *m* (Reifen) cord angle
Zentigramm *n* centigram
Zentimeter *n m* centimeter
Zentimeter-Gramm-Sekunde-System *n* (CGS-System) centimeter-gram-second system, c.g.s. system
Zentimetermaß *n* inch-rule
Zentipoise *n* centipoise, c.p.
Zentner *m* centner
zentral central
Zentralachse *f* central axis
zentralbelastet center-loaded
Zentralbewegung *f* central force motion
Zentraleinheit *f* (Comput) central processing unit
Zentralfeld *n* central field
Zentralfeldnäherung *f* central field approximation
Zentralheizung *f* central heating, hot water heating, steam heating
Zentralintensität *f* central intensity
zentralisieren to centralize
Zentralisierung *f* centralization
Zentralit *n* centralite
Zentralität *f* centricity
Zentralkartei *f* master file
Zentralkörperchen *n* centriole, central body, centrosome, **doppeltes** ~ diplosome
Zentralkraft *f* central force

Zentrallinie *f* center line
Zentralnervensystem *n* central nervous system
Zentralpunkt *m* central point
Zentralregister *n* central register
Zentralrohr *n* central tube
Zentralschaltpult *n* main switch desk
Zentralschmierung *f* central lubrication, one-shot lubrication
Zentralsteuerung *f* centralized control
Zentralstück *n* center piece, core
Zentralsymmetrie *f* central symmetry
zentralsymmetrisch central-symmetric, centrosymmetric
Zentrierdrehbank *f* centering lathe
zentrieren to center
Zentriermaschine *f* (Gieß) centering molding machine
Zentrierrahmen *m* centering frame
Zentrierring *m* centering ring
Zentrierscheibe *f* centering disc
Zentrierung *f* centering, alignment, (Gieß) mold alignment
zentrifugal centrifugal
Zentrifugalabscheider *m* centrifugal separator
Zentrifugalbeschleunigung *f* centrifugal acceleration
Zentrifugalbewegung *f* centrifugal motion
Zentrifugalboden *m* (Dest) centrifuge tray
Zentrifugalextraktor *m* centrifugal extractor
Zentrifugalfeld *n* centrifugal field
Zentrifugalfilter *m* centrifugal filter
Zentrifugalgebläse *n* centrifugal blower
Zentrifugalguß *m* centrifugal casting
Zentrifugalkraft *f* centrifugal force
Zentrifugalpendel *n* centrifugal pendulum
Zentrifugalprozeß *m* centrifugal method
Zentrifugalpumpe *f* centrifugal pump, propeller type pump
Zentrifugalreinigungsfilter *n* pressure leaf filter
Zentrifugalschmierung *f* centrifugal lubrication
Zentrifugalsichter *m* centrifugal separator
Zentrifugalsortierer *m* centrifugal strainer, centrifiner
Zentrifugaltrockenmaschine *f* hydroextractor
Zentrifugalzerstäuber *m* centrifugal atomizer, centrifugal sprayer
Zentrifugation *f* centrifugation, **differentielle** ~ differential centrifugation, **zonale** ~ zonal centrifugation
Zentrifuge *f* centrifuge, hydroextractor, **Absaug-Filter-** ~ air-knife or suction-discharge centrifuge, **Absetz-** ~ sedimentation centrifuge, centrifugal settler, **Dekantier-** ~ decanting centrifuge, centrifugal decanter, **diskontinuierliche** ~ discontinuous centrifuge, **Doppelkegel-** ~ double-cone centrifuge, **Düsen-Teller-** ~ nozzle discharge centrifuge, **Entwässerungs-** ~ dewatering centrifuge, **Faltensieb-** ~ fluted

Zentrifuge — zerlegbar

screen centrifuge, **Filterband-** ~ filter belt centrifuge, **Freistrahl-** ~ impulse centrifuge, **Hänge[korb]-** ~ suspended centrifuge, **Klär-** ~ centrifugal classifier, **Korb-** ~ basket centrifuge, **Leitkanal-** ~ differential volute centrifuge, **Prallring-** ~ baffle-ring centrifuge, **Raumschaufel-** ~ raker blade centrifuge, **[Ring]kammer-** ~ multi-chamber centrifuge, **Röhren-** ~ tubular bowl centrifuge, supercentrifuge, **Schwingsieb-** ~ oscillating screen centrifuge, **Super-** ~ supercentrifuge, tubular bowl centrifuge, **Teller-** ~ disc centrifuge, **Teller-Düsen-** ~ nozzle discharge centrifuge, **Torsionschwing-** ~ torsional centrifuge, **Überlauf-** ~ overflow centrifuge, **Vollwand-** ~ solid bowl centrifuge
Zentrifugenglas n centrifuge tube
Zentrifugentrommel f drum of the centrifuge
zentrifugieren to centrifuge, to hydroextract
zentripetal centripetal
Zentripetalbeschleunigung f centripetal motion
Zentripetalkraft f centripetal force
zentrisch centric, central
Zentroassymmetrie f centro-dissymmetry
Zentrosom n (Biol) centrosome
Zentrum n center (Am. E.), centre (Br. E.), **aktives** ~ (Biochem) active center
Zeolith m (Min) zeolite, **blättriger** ~ foliated zeolite
zeolithaltig zeolitic
Zeolitherde f mealy zeolite
Zeolithgruppe f zeolite group
Zeophyllit m (Min) zeophyllite
Zephaelin n cephaeline
Zephiran n zephiran
zerätzen to destroy with caustics
zerbrechen to break [into pieces], to fracture
Zerbrechen n breakage, fracture
zerbrechlich breakable, fragile, (spröde) brittle
Zerbrechlichkeit f fragility, (Sprödigkeit) brittleness
zerbröckeln to crumble
zerdrücken to crush, to crumble, (zu Brei) to squash
Zerfall m disintegration, ruin, (Atom) decay, (Chem) decomposition, (in Ionen) dissociation, **radioaktiver** ~ radioactive decay
zerfallen to break down, to decompose, to disintegrate, to dissociate, to fall to pieces, (zerbröckeln) to crumble
Zerfallen n decomposing, disintegrating, dissociating, ~ **des Erzes** crumbling of the ore
Zerfallselektron n (Atom) decay electron resulting from disintegration
Zerfallsenergie f disintegration energy
Zerfallsfolge f (Atom) decay sequence
Zerfallsgeschwindigkeit f rate of decay, **radioaktive** ~ radioactive disintegration velocity

Zerfall[s]gesetz n decay law, disintegration law
Zerfallsgrad m degree of decomposition
Zerfallsgrenze f decomposition limit, dissociation limit
Zerfallskonstante f decay constant, decay coefficient, disintegration constant, **radioaktive** ~ radioactive disintegration constant
Zerfallskurve f decay curve, decomposition curve
Zerfallsmodul m decay modulus
Zerfallsprodukt n decomposition product, (Atom) decay product, fission product, (Ionen) dissociation product
Zerfallsprozeß m decomposition process
Zerfallsrate f disintegration rate
Zerfallsreaktion f decomposition reaction
Zerfallsreihe f decay series, **radioaktive** ~ radioactive decay series
Zerfallsschema n decay scheme
Zerfallsstufe f decomposition stage
Zerfallswärme f heat of decomposition
Zerfallswahrscheinlichkeit f decay probability, disintegration probability
Zerfallsweg m decay path, decay channel
Zerfallszeit f decay period, decay time, disintegration time
Zerfaserer m (Pap) shredders, unravelling machine
zerfasern to fray [out], (Pap) to shred, to unravel
Zerfasern n size reduction by cutters
Zerfließbarkeit f deliquescence, deliquescent property
zerfließen to deliquesce
Zerfließen n deliquescence
zerfließend deliquescent
zerfressen to cauterize, to corrode
Zerfressen n cauterizing, corroding
Zerfressung f cauterization, corrosion
Zerhacker m chopper
Zerkleinerer m disintegrator
zerkleinern to crush, to break up into small pieces, to disintegrate, to reduce to small pieces, to shred
Zerkleinerung f crushing, disintegration
Zerkleinerungsanlage f pulverizing plant, crushing plant, crushing unit, disintegration plant
Zerkleinerungsgrad m reduction ratio
Zerkleinerungsmaschine f crusher, crushing machine, disintegrating machine, grinding machine, pulverizer
Zerkleinerungsmühle f crushing mill
Zerkleinerungsplatte f crushing plate
Zerkleinerungswalzwerk n crushing mill
zerklüftet full of fissures
zerlegbar decomposable, dissectible, separable, (abtrennbar) detachable, (spaltbar) fissionable, (zusammenfaltbar) collapsible

Zerlegbarkeit *f* decomposability
zerlegen to decompose, to disassemble, to split up, to take apart, (Chem) to analyze, (Opt) to disperse, to resolve
Zerlegen *n* decomposing
Zerlegung *f* decomposition, disintegration, dissection, (Auflösung) resolution, (Trennung) separation, **spektrale** ~ spectral decomposition
zermahlen to grind, to pulverize
Zermahlen *n* grinding
zermalmen to crush, to grind
Zermürbung *f* fatigue test
Zerotinfarbe *f* cerotine color
Zerotinsäure *f* cerotic acid
zerplatzen to burst, to explode
zerpulvern to pulverize
Zerpulvern *n* pulverizing
zerquetschen to crush, (zu Brei) to squash
Zerquetschen *n* crushing
Zerrbild *n* distorted picture
zerreiben to grind, to mill, to powder, to pulverize, to triturate
Zerreiben *n* triturating
Zerreibung *f* attrition, pulverization, trituration
Zerreißbelastung *f* breaking load, breaking stress, ultimate tensile stress
Zerreißdiagramm *n* stress-strain diagram
zerreißen to rupture, to tear [up], (auseinanderreißen) to tear asunder or apart
zerreißfest tear-proof
Zerreißfestigkeit *f* tensile strength, resistance to tensile stress, tear resistance
Zerreißfrequenz *f* shatter oscillation
Zerreißgeschwindigkeit *f* tear speed
Zerreißgrenze *f* breaking limit, breaking point, destruction limit, tear[ing] limit
Zerreißmaschine *f* tension tester
Zerreißmodul *m* modulus of rupture
Zerreißplättchen *n* bursting disk
Zerreißprobe *f* breaking test, tensile test, tension test
Zerreißprüfung *f* tension testing
Zerreißring *m* tensile ring
Zerreißscheibe *f* rupture disk
Zerreißspannung *f* rupture stress
zerren to pull, (schleppen) to drag
zerschlagen to shatter, to break, to crush
zerschneiden to cut, to dissect, (in Scheiben) to slice
Zerschneidung *f* dissection
zerschnitzeln to shred
zersetzbar decomposable
Zersetzbarkeit *f* decomposability, destructibility
zersetzen to decompose, to disintegrate, (durch Elektrolyse) to electrolyze, (verfaulen) to decay
Zersetzen *n* decomposing, disintegrating
zersetzend disintegrative

zersetzlich decomposable, destructible, unstable
Zersetzung *f* decay, decomposition, disintegration, (durch Elektrolyse) electrolysis, (in Ionen) dissociation, **thermische** ~ thermal decomposition
Zersetzungsbottich *m* decomposition tank
Zersetzungsdestillation *f* destructive distillation, dry distillation
Zersetzungsgeschwindigkeit *f* rate of decomposition
Zersetzungskatalysator *m* decomposition catalyst
Zersetzungskolben *m* decomposition flask, reaction flask
Zersetzungskurve *f* curve of decomposition, decomposition curve
Zersetzungsmittel *n* decomposing agent
Zersetzungspotential *n* decomposition potential, potential of decomposition
Zersetzungsprodukt *n* decomposition product
Zersetzungspunkt *m* point of decomposition
Zersetzungsrückstand *m* decomposition residue
Zersetzungsspannung *f* (Galv) decomposition voltage
Zersetzungsvorgang *m* process of decomposition
Zersetzungswärme *f* heat of decomposition
Zersetzungswert *m* decomposition value
Zersetzungswiderstand *m* (Elektr) electrolytic resistance
zerspalten to cleave, to split
zerspanen to remove metal by cutting chips, to splinter off
Zerspanung *f* cutting operation
Zerspanungseigenschaft *f* cutting property
zersplittern to crack, (Glas) to shatter, (Holz) to splinter
zerspringen to burst, to crack, to explode
zerstäuben to spray, to atomize, to disperse, to pulverize
Zerstäuber *m* sprayer, atomizer, diffuser, pulverizer, **Dreh-** ~ rotary-cup atomizer, **Öldruck-Dampfdruck-** ~ steam atomizer, **Rotations-** ~ centrifugal disc atomizer, **Rücklauf-** ~ return-flow atomizer, **Schall-** ~ sonic atomizer
Zerstäuberkegel *m* atomizer cone
Zerstäuberpistole *f* spray gun
Zerstäuberventil *n* atomizing valve
Zerstäubung *f* atomization, dispersion, pulverization, **elektrische** ~ electrical disintegration
Zerstäubungsmittel *n* atomizing agent, spray
Zerstäubungsreiniger *m* atomizing purifier, spraying cleaner
Zerstäubungstrockner *m* spray dryer, suspended particle dryer
Zerstäubungstrocknung *f* spray drying
Zerstäubungsvergaser *m* spray carburetor, jet carburetor

zerstörbar destructible
zerstören to destroy, to demolish, to ruin
zerstörend destructive
Zerstörung *f* destruction, demolition, deterioration, ruin
Zerstörungseffekt *m* destructive effect
zerstörungsfrei non-destructive
zerstoßen to crush, to grind, to pound, to pulverize
zerstreuen to disperse, to dissipate, to scatter
Zerstreuung *f* dispersion, dissipation, scattering
Zerstreuungsbild *n* image formed by a divergent lens
Zerstreuungskoeffizient *m* coefficient of dispersion
Zerstreuungskreis *m* (Opt) apertural effect
Zerstreuungslinse *f* (Opt) dispersing lens, diverging lens
Zerstreuungsphotometer *n* dispersion photometer
Zerstreuungsvermögen *n* dispersive power, diffusibility, scatter effect, scattering power
Zerstückelung *f* breaking into small pieces, dismemberment
zerteilbar divisible
zerteilen to divide, to split
Zerteilungsgrad *m* degree of division, dispersion
zertrümmern to demolish, to destroy, to shatter, to smash, to wreck
Zertrümmerung *f* demolition, destruction, smashing, (Atom) nuclear fission, ~ **des Kernes** breaking of the core
Zertrümmerungswahrscheinlichkeit *f* smashing probability
Zetapotential *n* zeta potential, electrokinetic potential
Zethren *n* zethrene
Zettelmaschine *f* warping machine
zetteln (Web) to warp
Zetteln *n* warping
Zeug *n* matter, stuff, (Pap) pulp
Zeugdruck *m* textile printing
Zeugdruckerei *f* (Text) textile printing, textile-printing plant
Zeugfänger *m* (Pap) stuff catcher
Zeugfärberei *f* dye works
Zeugnis *n* certificate, testimony
Zeugregler *m* (Pap) pulp meter
Zeunerit *m* (Min) zeunerite
Zibet *m* civet
Zibetan *n* civetane
Zibeton *n* civetone
Zichorie *f* (Bot) chicory
Zichorienbraun *n* chicory brown
Zichorienkaffee *m* chicory coffee
Zichorienkraut *n* (Bot) chicory
Zichorienwurzel *f* chicory root
Zick-Zack-Konfiguration *f* (Stereochem) zigzag configuration

Ziegel *m* brick, **feuerfester** ~ fire brick, kiln brick;, **hartgebrannter** ~ clinker
Ziegelei *f* brickworks, brickyard
Ziegelerde *f* brick clay, brick earth
Ziegelerz *n* tile ore
Ziegelfarbe *f* brick color, brick red
ziegelfarbig brick colored, tile colored
Ziegelform *f* brick's shape
Ziegelmehl *n* brick dust
ziegelrot brick-red
Ziegelton *m* brick clay, tile clay, tile loam
Ziegelung *f* briquetting
Ziegelware *f* brick products
Ziegenstein *m* bezoar
Ziegentalg *m* goat tallow
Ziegler-Katalysator *m* Ziegler's catalyst
ziehbar ductile
Ziehbarkeit *f* ductility
Ziehbiegemaschine *f* draw former
Zieheisen *n* draw plate, (Draht) die
Ziehelektrode *f* sweeping electrode
ziehen to draw, (schleppen) to drag, to haul, (Wurzel) to extract, (zerren) to pull, **Blasen** ~ to raise blisters, **Fäden** ~ to be ropy, to be stringy, to string
Zieherscheinung *f* oscillation hysteresis phenomenon, coupling hysteresis effect
Ziehfähigkeit *f* drawing property
Ziehform *f* draw die, drawing die
Ziehgitter *n* draw grid
Ziehloch *n* draw hole
Ziehöffnung *f* discharging hole
Ziehpressen *n* swaging
Ziehring *m* clamping ring
Ziehschablone *f* strickling board
Ziehschleifmaschine *f* honing machine
Ziehspachtel *f* knifing surfacer
Ziehstempel *m* forming die
Ziehteil *n* drawn part
Ziehtiefe *f* extent of draw
Ziehverfahren *n* drawing [process]
Ziehvorgang *m* drawing process
Ziehwerkzeug *n* drawing die
Ziehzange *f* wire pliers
Ziel *n* aim, target, **unbewegliches** ~ fixed target
Zielatom *n* target atom
Zielfindung *f* (Radar) homing
Zielgeschwindigkeit *f* target speed
zielgesteuert route coded, routed
Zielgröße *f* response value
Zielkennzeichnung *f* destination positioning
Zielplatte *f* (Zyklotron) cyclotron target
Zielpunkt *m* target
Zielscheibe *f* target
Zierin *n* zierin
Zieron *n* zierone
Zierschild *n* garnish plate
Ziffer *f* figure, numeral, **arabische** ~ Arabic numeral, **römische** ~ Roman numeral

Zifferblatt *n* dial, dial face
ziffernmäßig numerical
Zigarettenpapier *n* cigarette paper
Zimmertemperatur *f* room temperature
Zimt *m* cinnamon
Zimtäther *m* cinnamic ether
Zimtaldehyd *m* cinnamaldehyde, cinnamic aldehyde
Zimtalkohol *m* cinnamyl alcohol
Zimtblüte *f* cinnamon flower, (Bot) cassia bud
Zimtblütenöl *n* cassia oil
Zimtbranntwein *m* spirit of cinnamon
Zimtbraun *n* cinnamon brown
zimtfarben cinnamon colored
Zimtöl *n* cinnamon oil, **chinesisches** ~ cassia oil, Chinese cinnamon oil
Zimtsäure *f* cinnamic acid
Zimtsäureäthylester *m* cinnamic ethyl ester, ethyl cinnamate
Zimtsäurebenzylester *m* benzyl cinnamate, cinnamic benzyl ester
zimtsauer cinnamic, (Salz) cinnamate
Zimtsirup *m* syrup of cinnamon
Zimtwasser *n* cinnamon water
Zinckenit *m* (Min) zinckenite
Zincon *n* zincon
Zingeron *n* zingerone
Zingiberen *n* 1-methyl-4-propenyl cyclohexane, zingiberene
Zingiberol *n* zingiberol
Zink *n* zinc (Symb. Zn), **gewachsenes** ~ native zinc
Zinkacetat *n* zinc acetate
Zinkäthyl *n* zinc ethyl, diethyl zinc, zinc diethyl, zinc ethide
Zinkäthylsulfat *n* zinc ethyl sulfate
Zinkätzung *f* zinc etching, zincography
Zinkakkumulator *m* zinc accumulator
Zinkalaun *m* zinc alum, aluminum zinc sulfate
Zinkalkyl *n* zinc alkyl
Zinkamalgam *n* zinc amalgam
Zinkamyl *n* zinc amyl, zinc diamyl
zinkartig zincky, zinc-like, zincoid
Zinkasche *f* zinc ash, zinc dross
Zinkat *n* zincate
Zinkbad *n* zinc bath
Zinkbeschlag *m* zinc coating
Zinkblech *n* sheet zinc, zinc plate
Zinkblechvernick[e]lung *f* nickel-plating of sheet zinc
Zinkblende *f* (Min) zinc blende, sphalerite
Zinkblendegitter *n* (Krist) zinc blende lattice
Zinkblendetyphalbleiter *m* zinc blende type semiconductor
Zinkblüte *f* zinc bloom, hydrozincite
Zinkblumen *f pl* flowers of zinc, zinc oxide
Zinkbutter *f* butter of zinc, zinc chloride
Zinkcarbamat *n* zinc carbamate
Zinkchlorid *n* zinc chloride

Zinkchloridlauge *f* zinc chloride lye
Zinkcitrat *n* zinc citrate
Zinkcyanid *n* zinc cyanide
Zinkcyanür *n* (obs) zincous cyanide
Zinkdestillierofen *m* zinc [distillation] furnace
Zinkdibraunit *m* (Min) zincdibraunite
Zinkeisenelement *n* zinc iron cell
Zinkeisenerz *n* (Min) franklinite
Zinkeisenspat *m* (Min) ferriferous smithsonite
Zinkeisenstein *m* (Min) franklinite
Zinkelektrode *f* zinc electrode
Zinkelektrolyse *f* zinc electrolysis
Zinkenit *m* (Min) zinkenite
Zinkentsilberung *f* zinc desilverization
Zinkenverstellgerät *n* prong adjuster (forklift)
Zinkerz *n* zinc ore
Zinkerzrösthütte *f* zinc ore roasting plant
Zinkfahlerz *n* (Min) tennantite
Zinkfarbe *f* zinc color, zinc paint
Zinkfeilspäne *m pl* zinc filings
Zinkferrocyanid *n* zinc cyanoferrate(II), zinc ferrocyanide, zinc ferrous cyanide
Zinkfluorid *n* zinc fluoride
Zinkfolie *f* zinc foil
Zinkformiat *n* zinc formate
Zinkgehalt *m* zinc content
Zinkgekrätz *n* zinc dross, zinc oxide
Zinkgelb *n* zinc yellow, zinc chromate
Zinkgewinnung *f* zinc production, extraction of zinc
Zinkglas *n* (Min) siliceous calamine
Zinkgranalien *f pl* granulated zinc
Zinkgrau *n* zinc gray
Zinkgrün *n* zinc green
zinkhaltig containing zinc, zinciferous
Zinkhochätzung *f* zincography
Zinkhütte *f* zinc smeltery, zinc works
Zinkhüttenanlage *f* zinc smelting installation
Zinkit *m* (Min) zinkite
Zinkjodat *n* zinc iodate
Zinkjodidstärkepapier *n* zinc iodide starch paper
Zinkkalk *m* zinc ash, zinc calx, zinc dross
Zinkkiesel *m* (Min) siliceous calamine
Zinkkitt *m* zinc cement
Zinkkohlenbatterie *f* zinc carbon battery
Zinkkontakt *m* zinc contact
Zinkkupfermelanterit *m* (Min) zinc copper melanterite
Zinklaktat *n* zinc lactate
Zinklegierung *f* zinc alloy
Zinkmehl *n* zinc powder
Zinkmetasilikat *n* zinc metasilicate
Zinkmethyl *n* zinc methyl, dimethyl zinc, zinc dimethyl
Zinknitrat *n* zinc nitrate
Zinkofen *m* zinc furnace
Zinkofenbruch *m* cadmia, tutty
Zinkographie *f* zincography
Zinkoleat *n* zinc oleate

zinkorganisch zinc-organic
Zinkorthosilikat n zinc orthosilicate
Zinkosit m (Min) zincosite
Zinkoxid n zinc oxide
Zinkoxidanlage f zinc oxide plant
Zinkpecherz n sphalerite
Zinkperhydrol n zinc perhydrol, zinc peroxide
Zinkphosphit n zinc phosphite
Zinkpol m zinc pole, cathode, negative pole
Zinkraffination f zinc refining
Zinkrauch m zinc fume
Zinksalbe f zinc ointment
Zinksalicylat n zinc salicylate
Zinksalz n zinc salt
zinksauer (Salz) zincate
Zinkschaum m zinc scum
Zinkscheibe f zinc disk
Zinkschnitzel n pl zinc filings
Zinkschwamm m spongy zinc, cadmia, tutty, zinc sponge
Zinkselenid n zinc selenide
Zinksilicofluorid n zinc silicofluoride, zinc fluosilicate
Zinksilikat n zinc silicate
Zinkspat m (Min) zincspar, calamine, smithsonite
Zinkspinell m zinc spinel, (Min) gahnite
Zinkspritzguß m zinc die-casting
Zinkstab m zinc rod
Zinkstaub m zinc dust
Zinkstaubdestillation f zinc dust distillation
Zinkstaubfarbe f zinc-rich paint
Zinkstearat n stearate of zinc
Zinkstreifen m strip of zinc
Zinksubgallat n zinc subgallate
Zinksubkarbonat n zinc subcarbonate
Zinksulfat n zinc sulfate, white vitriol
Zinksulfidpulver n zinc sulfide powder
Zinksulfidschirm m zinc-sulfide screen
Zinksuperoxid n zinc peroxide
Zinktetraphenylporphin n zinc tetraphenylporphine
Zinktripelsalz n zinc triple salt
Zinküberzug m zinc coating
Zinkverbindung f zinc compound
Zinkvernickelung f nickel-plating of zinc
Zinkvitriol n white vitriol, zinc sulfate
Zinkvitriolanlage f white vitriol plant
Zinkvoltameter n zinc voltameter
Zinkweiß n zinc white, zinc oxide
Zinkwolle f flowers of zinc
Zinkzinnamalgam n zinc tin amalgam
Zinn n tin (Symb. Sn), (für Hausgeräte) pewter
Zinnabfall m refuse of tin, tin waste
Zinnabstrich m tin scum
Zinnader f tin lode, tin vein
Zinnafter m tin ore refuse
Zinnamalgam n amalgam of tin

Zinnammoniumchlorid n tin ammonium chloride, ammonium chlorostannate(IV), pinksalt
zinnartig tin-like, tinny
Zinnasche f tin ashes, stannic oxide
Zinnbad n tinning bath
Zinnbaum m tin tree
Zinnbeize f tin mordant
Zinnbergwerk n stannary
Zinnblatt n tin foil
Zinnblech n sheet tin, tin plate
Zinnblumen f pl flowers of tin
Zinnbromür n (obs f. Zinn(II)-bromid) stannous bromide, tin dibromide, tin(II) bromide
Zinnbromwasserstoffsäure f bromostannic acid
Zinnbutter f butter of tin, stannic chloride
Zinnchlorammonium n ammonium chlorostannate(IV), pinksalt, tin ammonium chloride
Zinnchlorid n tin chloride, (Zinn(IV)-chlorid) stannic chloride, tin(IV) chloride, tin tetrachloride
Zinnchlorür n (obs f. Zinn(II)-chlorid) stannous chloride, tin dichloride, tin(II) chloride
Zinnchlorwasserstoffsäure f chlorostannic acid
zinnchlorwasserstoffsauer chlorostannate
Zinndibromid n stannous bromide, tin dibromide, tin(II) bromide
Zinndichlorid n stannous chloride, tin dichloride, tin(II) chloride
Zinndioxid n tin dioxide, stannic anhydride, stannic oxide, tin(IV) oxide
Zinndiphenylchlorid n diphenylstannic chloride
Zinndisulfid n stannic sulfide, mosaic gold, tin disulfide, tin(IV) sulfide
Zinndraht m tin wire
zinnen to tin
zinnern tin, pewter
Zinnerz n (Min) tin ore, cassiterite
Zinnerzformation f tin ore formation
Zinnerzsand m crop of tin
Zinnfeilicht n tin filings
Zinnfeilspäne m pl tin filings
Zinnfolie f tinfoil
zinnführend containing tin, stanniferous, tin bearing
Zinngehalt m tin content
Zinngekrätz n tin dross, tin refuse
Zinngerät n pewter, tin vessel
Zinngeschrei n (Chem) crackling of tin, tin cry
Zinngießer m tin founder
Zinngitter n tin grate
zinnglasiert tin-glazed
Zinnglasur f tin glaze, tin glazing
Zinngranalien f pl granulated tin
Zinngraupen f pl crystallized oxide of tin, tin crystals
Zinngrün n tin green
Zinnguß m tin cast

zinnhaltig containing tin, stanniferous, tin bearing
Zinnhydrid n stannane, tin hydride
Zinnhydroxid n tin hydroxide, (Zinn(IV)-hydroxid) stannic hydroxide, tin(IV) hydroxide
Zinnhydroxydul n (obs f. Zinn(II)-hydroxid) stannous hydroxide, tin(II) hydroxide
Zinn(II)-Verbindung f stannous compound
Zinnjodid n tin iodide, (Zinn(IV)-jodid) stannic iodide, tin(IV) iodide, tin tetraiodide
Zinnjodür n (obs f. Zinn(II)-jodid) stannous iodide, tin diiodide, tin(II) iodide
Zinnkalk m tin calx, stannic oxide, tin(IV) oxide
Zinnkies m (Min) tin pyrites, stannite
Zinnknirschen n crackling of tin, tin cry
Zinnkrätze f tin ashes, tin refuse
Zinnlegierung f tin alloy
Zinnlösung f tin solution
Zinnlot n tin solder
Zinnmonoxid n tin monoxide, stannous oxide, tin(II) oxide
Zinnober m (Min) cinnabar, red mercuric sulfide, red mercury sulfide, vermilion, **grüner** ~ cinnabar green
Zinnobererde f (Min) ore of cinnabar
Zinnobererz n cinnabar ore
Zinnoberfarbe f vermilion
Zinnoberrot n vermilion
Zinnoberscharlach n cinnabar scarlet
Zinnoberspat m (Min) crystallized cinnabar
Zinnoxid n tin oxide, (Zinn(IV)-oxid) stannic anhydride, stannic oxide, tin dioxide, tin(IV) oxide
zinnoxidhaltig containing tin oxide
Zinnoxidhydrat n stannic hydroxide, tin hydroxide, tin(IV) hydroxide
Zinnoxidnatrium n (obs) sodium stannate
Zinnoxidnatron n (obs) sodium stannate
Zinnoxidsalz n (obs) stannic salt, tin(IV) salt
Zinnoxidverbindung f stannic compound, tin(IV) compound
Zinnoxydul n (obs f. Zinn(II)-oxid) stannous oxide, tin(II) oxide, tin monoxide, tin protoxide
Zinnoxydulnatron n (obs) sodium stannite
Zinnoxydulsalz n (obs) stannous salt, tin(II) salt
Zinnoxydulverbindung f (obs) stannous compound, tin(II) compound
Zinnpest f tin pest, tin plague
Zinnphosphatbeschwerung f weighting with tin phosphate
Zinnprobe f tin sample, (Anal) tin test
Zinnprotoxid n tin protoxide, stannous oxide, tin(II) oxide, tin monoxide
zinnreich rich in tin
Zinnrhodanür n (obs) stannous thiocyanate, tin(II) thiocyanate
Zinnrohr n tin pipe

Zinnsäure f stannic acid
Zinnsäureanhydrid n stannic anhydride, stannic oxide, tin dioxide, tin(IV) oxide
Zinnsalmiak m ammonium chlorostannate(IV), pinksalt, tin ammonium chloride
Zinnsalz n tin salt
Zinnsand m grain tin
zinnsauer stannic, (Salz) stannate
Zinnsaum m tin selvedge
Zinnschlich m tin slimes, fine tin
Zinnschwamm m spongy tin
Zinnseife f stream tin
Zinnselenür n (obs) stannous selenide, tin(II) selenide, tin monoselenide
Zinnsoda f sodium stannate
Zinnstaub m tin dust
Zinnstein m (Min) cassiterite, tinstone
Zinnsulfid n tin sulfide, (Zinn(IV)-sulfid) stannic sulfide, tin disulfide, tin(IV) sulfide
Zinnsulfocyanür n (obs) stannous thiocyanate, tin(II) thiocyanate
Zinnsulfür n (obs f. Zinn(II)-sulfid) stannous sulfide, tin(II) sulfide, tin monosulfide, tin protosulfide
Zinntellurür n (obs) stannous telluride, tin(II) telluride, tin monotelluride
Zinntetrabromid n tin tetrabromide, stannic bromide, tin(IV) bromide
Zinntetrachlorid n tin tetrachloride, stannic chloride, tin(IV) chloride
Zinntetrafluorid n tin tetrafluoride, stannic fluoride, tin(IV) fluoride
Zinntetrajodid n tin tetraiodide, stannic iodide, tin(IV) iodide
Zinnverspiegelung f tinfoil coat[ing]
Zinnwaldit m (Min) zinnwaldite, lithionite
Zinnwasserstoff m stannane, tin hydride
Zinnwolle f mossy tin metal
Zippeit m (Min) zippeite
Zirbeldrüse f (Med) pineal gland
Zirkel m pair of compasses, (Stechzirkel) dividers
Zirkelit m (Min) zirkelite
Zirkelkasten m compass set
Zirkon m (Min) zircon, **gemeiner** ~ zirconite
Zirkonat n zirconate
Zirkonchlorid n zirconium chloride
Zirkonerde f (Min) zirconia, zirconium dioxide
Zirkonglas n zirconium glass
Zirkonit m (Min) zirconite
Zirkon[ium] n zirconium (Symb. Zr)
Zirkon[ium]dioxid n zirconium dioxide, (Min) zirconia
Zirkon[ium]gehalt m zirconium content
Zirkoniumhydrid n zirconium hydride
Zirkon[ium]hydroxid n zirconium hydroxide
Zirkon[ium]metall n metallic zirconium, zirconium metal
Zirkon[ium]nitrat n zirconium nitrate

Zirkoniumoxid – Zubringerband

Zirkon[ium]oxid *n* zirconic anhydride, zirconium dioxide, zirconium oxide, (Min) zirconia
Zirkon[ium]oxychlorid *n* zirconium oxychloride, zirconyl chloride
Zirkoniumphosphat *n* zirconium phosphate
Zirkoniumphosphid *n* zirconium phosphide
Zirkoniumsäure *f* zirconic acid, zirconium hydroxide
Zirkoniumsilicid *n* zirconium silicide
Zirkoniumsulfat *n* zirconium sulfate
Zirkoniumtetrafluorid *n* zirconium tetrafluoride, zirconium fluoride
Zirkon[ium]verbindung *f* zirconium compound
Zirkonlampe *f* zirconium lamp
Zirkonweiß *n* zirconium white
Zirkonyl- zirconyl
Zirkonylchlorid *n* zirconyl chloride
Zirkulardichroismus *m* circular dichroism
Zirkularpolarisation *f* circular polarization
Zirkularpolariskop *n* circular polariscope
Zirkulation *f* circulation
Zirkulationsapparat *m* circulation apparatus
Zirkulationsflüssigkeit *f* circulating liquid
Zirkulationsgeschwindigkeit *f* rate of circulation
Zirkulationspumpe *f* circulation pump
Zirkulationsreaktor *m* circulating fuel reactor
Zirkulationsströmung *f* circulation flow
Zirkulationssystem *n* circulating system, circulation system
zirkulieren to circulate
zirkumpolar circumpolar
zischen to fizz, to hiss, to sizzle, to whiz
Zischen *n* fizzing, hissing, sizzling
zischend fizzing, sizzling
Zischmechanismus *m* hissing mechanism
Zissoide *f* (Math) cissoid
Zisterne *f* cistern, well
Zitrakonsäure *f* citraconic acid, pyrocitric acid
Zitral *n* citral
Zitrazinsäure *f* citrazinic acid
Zitrone *f* (Bot) lemon
Zitronellaöl *n* citronella oil
Zitronenäther *m* citric ether
Zitronenbaum *m* lemon tree
Zitronenblüte *f* (Bot) lemon blossom
zitronenfarben citrine, lemon
zitronengelb citrine, lemon yellow
Zitronengelb *n* lemon yellow
Zitronenlimonade *f* lemonade
Zitronenmelisse *f* (Bot) lemon balm, balm mint
Zitronenöl *n* lemon oil
Zitronenpresse *f* lemon squeezers
Zitronensäure *f* citric acid
Zitronensäureester *m* citrate, citric ester
Zitronensäurezyklus *m* citric acid cycle, Krebs cycle, tricarboxylic acid cycle
Zitronensaft *m* lemon juice
zitronensauer (Salz, Ester) citrate
Zitronenschale *f* lemon peel

Zitrothymöl *n* lemon thyme oil
Zitrovorumfaktor *m* citrovorum factor
Zitrusfrucht *f* citrus fruit
Zitterbewegung *f* vibratory motion
Zitterelektrode *f* vibrating electrode
zittern to tremble, to quiver
Zitwer *m* (Bot) zedoary
Zitwersamen *m* santonica, wormseed, zedoary seed
Zitwersamenöl *n* wormseed oil
Zivilrecht *n* civil law
Zodialkallicht *n* (Astr) zodiacal light
Zölestin *m* (Min) celestine, strontium sulfate
Zoisit *m* (Min) zoisite, thulite
Zoll *m* duty, (Maß) inch
Zollamt *n* customs office
Zollgewinde *n* screw thread basing on the inch system
Zollmaß *n* measure in inches
zollpflichtig dutiable
zonal gemittelt zonally averaged
Zone *f* zone, boundary, ~ **konstanter Zusammensetzung** pinch zone
Zonenachse *f* zone axis
Zonenebene *f* (Krist) zone plane
Zonenelektrophorese *f* zone electrophoresis
Zonenfehler *m* zonal aberration
Zonenfolge *f* (Halbleiter) sequence of zones
Zonengefrierverfahren *n* zone refrigeration process
Zonenheizung *f* zone heating
Zonenindex *m* zone index
Zonenregel *f* rule of zones
Zonenreinigen *n* zone refining
Zonenschmelzen *n* zone melting, zone refining
Zonenschmelzverfahren *n* zone melting process, zone refining
Zonenverband *m* crystal zone
Zonenzentrifugation *f* zone [or zonal] centrifugation
Zonochlorit *m* (Min) zonochlorite
Zoochemie *f* animal chemistry, zoochemistry
Zoolith *m* (Min) zoolithe
zoolithisch zoolithic
Zoologie *f* zoology
zoologisch zoological
Zoomarinsäure *f* zoomaric acid, hexadecenoic acid
Zooparasit *m* zooparasite
Zorgit *m* zorgite
Zubehör *n* accessories, appliance
Zubehörteil *n* accessory [part]
Zuber *m* tub
zubereiten to prepare, to dress, to finish
zubereitet prepared, treated
Zubereitung *f* preparation, dressing, finishing
zubilligen to allow
Zubringer *m* feeder
Zubringerband *n* feeder, feeding conveyor

Zubringerleitung f feeder line
Zubringerpumpe f feed pump
Zucht f cultivation
Zucker m sugar, **brauner** ~ brown sugar, **linksdrehender** ~ levo-rotatory sugar, **rechtsdrehender** ~ dextro-rotatory sugar, **reduzierender** ~ reducing sugar, **verbrannter** ~ burnt sugar; caramel;, **verwandeln in** ~ to saccharize, **zu** ~ **werden** to saccharify
zuckerähnlich saccharoidal, sugar like
Zuckeralaun m alum sugar
Zuckeranalyse f sugar analysis
Zuckerarmut f (Blut) hypoglyc[a]emia, glycopenia
zuckerartig saccharine, saccharoidal, sugar-like, sugary
Zuckerartigkeit f saccharinity
Zuckerbildung f production of sugar, (Biol) glycogenesis, (Verzuckerung) saccharification
Zuckerbranntwein m sugared spirit
Zuckercouleur f caramel, tincture of burnt sugar
Zuckerdicksaft m molasses, treacle
Zuckererde f clay for refining sugar
Zuckerersatz m sugar substitute
zuckererzeugend (Biol) glycogenous
Zuckererzeugung f production of sugar
Zuckerfabrik f sugar factory
Zuckerfabrikation f sugar manufacture
Zuckergärung f fermentation of sugar, saccharine fermentation
Zuckergehalt m sugar content
Zuckergehaltsmesser m saccharimeter
Zuckergeschmack m sugary taste
Zuckerguß m icing
zuckerhaltig containing sugar, sacchariferous, saccharine
Zuckerhaltigkeit f saccharinity
Zuckerherstellung f sugar production
Zuckerhonig m sugar honey, crystallized honey, molasses, treacle
zuckerig saccharine, sugary
Zuckerin n zuckerin, saccharin
Zuckerindustrie f sugar industry
Zuckerinversion f inversion of sugar
Zuckerkandis m sugar candy
Zuckerkochthermometer n sugar refining thermometer
Zuckerkohle f charcoal for refining sugar, sugar coal
zuckerkrank diabetic
Zuckerkrankheit f (Med) diabetes
Zuckerlösung f sugar solution
Zuckermandel f sugared almond
Zuckermesser m saccharimeter
Zuckermessung f saccharimetry
Zuckermühle f sugar mill, cylinder press
Zuckerquetsche f sugar crusher
Zuckerraffinade f refined sugar

Zuckerraffinerie f sugar refinery
Zuckerrohr n sugar cane
Zuckerrohrrückstände m pl bagasse, cane trash
Zuckerrohrsaft m cane juice
Zuckerrübe f sugar beet, beet root (Am. E.)
Zuckerrübenschnitzel n sugar beet chip
Zuckersäure f saccharic acid, sugar acid
Zuckersaft m saccharine juice
Zuckersaftentfärbungsanlage f sugar syrup decolorizing plant
zuckersauer saccharate
Zuckerspiegel m (Blut) blood-sugar level
Zuckerstein m (Min) finely granular albite
Zuckerstoff m saccharine matter
Zuckerstoffwechsel m glycometabolism, saccharometabolism
Zuckertinktur f saccharine tincture
Zuckertrester m sugar residue
Zuckerverbindung f sugar compound
Zuckung f convulsion, twitching
zudrehen to close, (Hahn) to turn off
züchten to breed, to raise, (Bakterien) to grow, (Pflanzen) to cultivate
Züchter m breeder, grower
Züchtung f breeding
Zündakkumulator m (Elektr) ignition accumulator
Zündanode f ignition anode
Zündapparat m igniter, ignition apparatus, primer, priming apparatus
Zündband n fuse, quick match
zündbar flammable, ignitible, inflammable
Zündbarkeit f flammability, ignitibility, inflammability
Zündbatterie f ignition battery
Zündbedingung f condition for ignition
Zünddraht m (Bergb) cartridge wire
Zündeinrichtung f ignition device
Zündelektrode f ignition electrode
zünden to ignite, to catch fire, to kindle fire
Zünden n firing, igniting, lighting
Zünder m detonator, fuse, igniter, **elektrischer** ~ electric match or fuse, **empfindlicher** ~ instantaneous fuse, nondelay fuse
Zündergehäuse n fuse housing
Zündersatz m fuse composition
zündfähig ignitible
Zündfähigkeit f ignitibility, ignition quality
Zündflämmchen n pilot flame, igniting flame
Zündflamme f igniting flame, pilot flame
Zündflammenleitung f ignition flame passage
Zündfolge f ignition sequence, firing order
Zündfunken m ignition spark
Zündgeschwindigkeitsbestimmung f ignition velocity determination
Zündgruppe f ignition group
Zündholz n match
Zündholzindustrie f match industry

Zündhütchen *n* igniter cap, blasting cap, detonating cap, percussion cap, primer
Zündkabel *n* (Kfz) spark plug wire (Am. E.), ignition lead (Br. E.)
Zündkabellack *m* ignition cable lacquer
Zündkapsel *f* detonator
Zündkerze *f* (Kfz) spark plug
Zündkerzenkabel *n* ignition wire
Zündkraft *f* igniting power
Zündladung *f* detonating charge
Zündleitung *f* ignition pipe
Zündlücke *f* zone of non-ignition
Zündmasse *f* ignition compound, ignition mixture
Zündmoment *m* time of ignition, moment of sparking
Zündpatrone *f* primer, priming cartridge
Zündpille *f* fuse head
Zündprobe *f* ignition test
Zündpunkt *m* ignition point
Zündsatz *m* priming composition
Zündschalter *m* ignition switch
Zündschlüssel *m* (Kfz) ignition key
Zündschnur *f* firing tape, match cord
Zündschwamm *m* German tinder, punk
Zündspannung *f* breakdown voltage, ignition threshold
Zündsteuerung *f* ignition distribution
Zündstörung *f* ignition interference
Zündstoff *m* primer, igniting agent, inflammable material
Zündtemperatur *f* ignition temperature, ignition point
Zündung *f* (Auto) ignition, (Bergb) priming, **die ~ abstellen** to disconnect the ignition, **die ~ einstellen** to adjust the ignition, **verzögerte ~** late ignition, **vorzeitige ~** pre-ignition
Zündungsart *f* type of ignition
Zündungseinstellung *f* adjustment of ignition
Zündungsregulierung *f* adjustment of ignition
Zündungsstromkreis *m* (Elektr) ignition circuit
Zündungstemperatur *f* ignition temperature
Zündungszeit *f* ignition time
Zündverzug *m* ignition lag, ignition delay, ignition retardation
Zündvorgang *m* ignition
Zündwärme *f* ignition heat
Zündzeitpunkt *m* time of ignition
zufällig accidental, occasional
Zufall *m* accident, chance, random cause
Zufallsfolge *f* random series
Zufallsgesetz *n* law of chance
Zufallsgröße *f* random variable
Zufallsorientierung *f* random orientation
Zufallsstichprobe *f* random sample
Zufallswert *m* accidental value
Zufluß *m* feed, inflow, influent, supply
Zuflußeinsteller *m* flow setting

Zuflußkanal *m* (Gieß) runner
zufrieren to freeze up
Zufrieren *n* **des Abstichloches** freezing up of the tap hole
zufügen to add
zuführen to add, to feed, to supply
Zuführung *f* feed[ing], supply, (Elektr) lead-in, (Leitung) feed-line
Zuführungsdraht *m* (Elektr) lead wire
Zuführungskapillare *f* feed capillary
Zuführungskopf *m* delivery head
Zuführungsleitung *f* supply main
Zuführungsrohr *n* supply pipe
Zuführwalze *f* feed roller
Zufuhr *f* supply, addition, feed, inlet
Zug *m* drawing, pull, stress, tension, traction, (Luftzug) draft, draught, **~ und Druck** push and pull, **mechanischer ~** traction
Zugabe *f* allowance
Zugabegefäß *n* charge tank
Zugabevorrichtung *f* charging mechanism
zugänglich accessible, available
Zugänglichkeit *f* accessibility, availability
Zugang *m* access, (Eingang) entrance
Zugbeanspruchung *f* stretching strain, tensile stress
Zugdruckdauerprüfmaschine *f* tension-compression fatigue testing machine
Zugdruckverhältnis *n* ratio of tension to thrust
Zugdruckversuch *m* tension and compression test
zugeben to add
Zugelastizitätsmodul *m* modulus of elasticity in tension
Zugelektrode *f* pull electrode
zugemischt admixed
zugeordnet assigned
Zugermüdungsversuch *m* tensile fatigue test
zugeschmolzen sealed
Zugeständnis *n* concession
Zugfeder *f* tension spring
Zugfeld *n* (Festkörperphys) drain field
zugfest tension-proof
Zugfestigkeit *f* [resistance to] tensile stress, breaking strain, tensile strength [at maximum load], ultimate strength
Zugfestigkeitsprüfmaschine *f* tensile-strength testing machine
Zugfestigkeitsuntersuchung *f* tensile-strength test
Zugfestigkeitswert *m* tensile strength
Zughebel *m* traction lever
Zugießen *n* addition of liquid, pouring in
Zugkanal *m* (Feuerung) chimney flue, uptake
Zugkraft *f* drafting force, traction, tractive force
Zugleine *f* pull cord
Zugleistung *f* tractive power
Zugluft *f* draft (Am. E.), draught (Br. E.)
Zugmaschine *f* tractor
Zugmesser *m* draft gauge

Zugofen *m* air furnace
Zugprobe *f* pull test, tensile test
Zugprüfung *f* pull test, tensile test
Zugregelung *f* draft regulation
Zugregler *m* draft regulator
Zugriff[s]zeit *f* access time
zugrundeliegend fundamental
zugrunderichten to ruin
Zugschlagzähigkeit *f* tensile impact strength
Zugschwellfestigkeit *f* fatigue strength in tension
Zugseil *n* pulling rope, traction rope, (Seilbahn) haulage rope, traction rope
Zugspaltung *f* tension cleaving
Zugspannung *f* [intensity of] tensile stress, tensile strength, tension
Zugstab *m* tensile bar
Zugstärke *f* intensity of draft
Zugstange *f* tie rod
Zugtextur *f* tensile texture
Zug- und Druckkurve *f* strain-stress curve
Zug- und Druckspannung *f* tensile and compressive stress
Zugverformung *f* stretching strain
Zugversuch *m* tensile test, straining frame experiment, ~ **mit Dauerbeanspruchung** endurance tension test, fatigue tension test, ~ **mit ruhender Last** static tension test, ~ **mit Schlagbeanspruchung** impact tension test, ~ **mit statischer Beanspruchung** static tension test
Zugvorrichtung *f* draw gear, hitch (Am. E.)
Zugwalze *f* tensioning roll
Zuhilfenahme *f* assistance
Zuhörerschaft *f* audience
zukitten to cement [up]
zukorken to cork, to stopper
Zukunftsmöglichkeit *f* future possibility
zulässig permissible, safe
Zulässigkeit *f* admissibility, tolerance
Zulauf *m* run, (Bergb) feed trough, (Dest) feed, (Substanz) feed (substance)
Zulaufbelastung *f* feed load
Zulaufboden *m* (Kolonne) feed plate
Zulaufrohr *n* intake pipe
Zulaufstrom *m* inflow current
Zulaufstutzen *m* (Bodenkolonne) downcomer
Zulauftemperatur *f* admission temperature
zuleimen to glue up
Zuleitung *f* inlet, (Elektr) lead wire
Zuleitungsrohr *n* feed pipe, inlet tube, supply duct
Zuleitungsstromkreis *m* (Elektr) supply circuit
Zuleitungswiderstand *m* (Elektr) lead resistance
Zuluftkanal *m* air inlet conduit
zumachen to close, to shut
zumischen to admixture, to mix with
Zunahme *f* increase, gain, growth, **prozentuale** ~ percentage increase
Zunahmekurve *f* growth curve

Zunder *m* tinder, (Hammerschlag) hammer scale
Zunderasche *f* ashes of tinder
zunderbeständig non-scaling
Zunderbeständigkeit *f* non-scaling property, resistance to scaling
Zunderbildung *f* (Met) scale formation
zunderfest non-scaling
Zunderschwamm *m* German tinder
zunehmen to increase, to grow, to rise
zunehmend increasing, cumulative
Zunyit *m* (Min) zunyite
zuordnen to assign
Zuordnungsverfahren *n* method of assignment
zurechtschneiden to dress
Zurichte *f* dressing, trimming (leather)
zurichten to dress, to trim
Zurichtmasse *f* sizing material
Zurichtung *f* finishing
Zurichtungshalle *f* straightening shop
Zurlit *m* zurlite
Zurückbehalten *n* retention
Zurückdrängung *f* repression
zurückerhalten to recover, to reclaim
zurückfedern to spring back
zurückfließen to flow back
Zurückfließen *n* reflux
zurückführen to lead back, to trace back
zurückgehen to return, (abnehmen) to decline
zurückgestrahlt reflected
zurückgestreut back scattered
zurückgewinnbar recoverable, reclaimable
zurückgewinnen to recover, to reclaim, to recuperate
Zurückgewinnung *f* recovery, recuperation
zurückhalten to detain, to retain
zurücklassen to leave behind
zurückprallen to rebound
zurückschnellen to fling back
zurückstoßen to push back, to repel
zurückstrahlen to reflect
Zurückstrahlung *f* reflection
zurückströmen to flow back, to reflux
zurücktitrieren to titrate back
Zurücktitrieren *n* (Anal) back titrating
Zurücktitrierung *f* (Anal) back-titration
zurückwalzen to roll on return pass
zurückziehen to draw back, to pull back, to retract, to withdraw
zusätzlich additional, additive, extra, suplementary
Zusammenarbeit *f* cooperation
zusammenarbeiten to cooperate, to work together
zusammenbacken to bake, to cake, to frit
zusammenballen to agglomerate, to aggregate, to ball together, to conglomerate
Zusammenballen *n* agglomeration, balling, conglomeration
Zusammenballung *f* agglomeration, balling, coagulation, conglomeration

zusammenbrechen to break down, to collapse, to fall into pieces
zusammenbrennen to burn together, to frit
Zusammenbruch *m* breakdown, collapse
zusammendrückbar compressible
Zusammendrückbarkeit *f* compressibility
zusammendrücken to compress, to compact, to press together
zusammenfallen to coincide, (zusammenbrechen) to collapse
zusammenfalten to fold [up]
Zusammenfassung *f* summary
zusammenfließen to coalesce
Zusammenfließen *n* coalescence
zusammenfrieren to freeze together
zusammenfritten to frit
zusammenfügen to assemble, to connect, to couple, to fix together, to join
Zusammenfügen *n* combining, connecting, joining, uniting
zusammengedrängt concise
zusammengekittet cemented
zusammengesetzt combined, complex, composed, laminated (glass)
zusammengießen to pour together
zusammenhängen to be connected
zusammenhängend coherent, cohesive, connected
Zusammenhalt *m* cohesion, consistency
Zusammenhang *m* context
zusammenkitten to cement, to putty
Zusammenkitten *n* cementing, bonding
zusammenkleben to glue together, to agglutinate, to stick together
Zusammenklebung *f* agglutination, adhesion
zusammenklumpen to conglomerate, to agglutinate, to conglobate
zusammenkratzen to scrape
zusammenlaufen (Linien) to converge
zusammenlegbar collapsible
zusammenleimen to agglutinate, to glue [together]
zusammenlöten to solder [together]
Zusammenlöten *n* soldering
Zusammenprall *m* collision
zusammenprallen to collide
zusammenpreßbar compressable
zusammenpressen to compress
zusammenrechnen to add up
zusammenrühren to mix together by stirring, to stir together
Zusammenschluß *m* bond, grouping together
zusammenschmelzen to melt together, to fuse, to melt down
Zusammenschmelzen *n* fusing, melting
Zusammenschmelzverfahren *n* melting down process
zusammenschrauben to screw together
zusammenschrumpfen to shrink, to contract, to shrivel, to wrinkle
Zusammenschrumpfen *n* shrinkage
zusammenschütten to pour together, to mix
zusammenschweißen to weld together
Zusammenschweißen *n* welding together
zusammensetzen to put together; to assemble, to combine, to compose, to compound
Zusammensetzstelle *f* splice
Zusammensetzung *f* assembly, combination, (Chem) composition, **chemische** ~ chemical composition, **kritische** ~ critical composition, **prozentuale** ~ percentage composition, **variable** ~ variable composition
Zusammenspiel *n* coordination, interplay
zusammenstellen to join, to put together, to summarize, (z. B. Material) to compile
Zusammenstellung *f* assembly, combination, compilation, grouping, summary
Zusammenstoß *m* collision
zusammenstoßen to collide, to encounter
Zusammenstoßen *n* colliding
zusammentreffen to coincide, to meet, (konvergieren) to converge
Zusammentreffen *n* collision, encounter
Zusammentreffwahrscheinlichkeit *f* (Atom) probability of collision
zusammenwachsen to grow together
zusammenwirken to collaborate, to cooperate, to work together
Zusammenwirkung *f* combined action, combined effect, cooperation
zusammenzählen to add
zusammenziehbar contractible, contractile
zusammenziehen, sich ~ to contract, to shrink
Zusammenziehen *n* contracting, (Werfen) crawling, creeping
Zusammenziehung *f* contraction, shrinking
Zusatz *m* addition, (Anhang) appendix, supplement, (Chem) additive, admixture, (Nachtrag) addendum, **flüssiger** ~ liquid addition, **kohlender** ~ recarburizer
Zusatzaggregat *n* booster aggregate, addition set
Zusatzbatterie *f* (Elektr) booster battery
Zusatzbelastung *f* additional load
Zusatzbelüftung *f* compressed air conditioning
Zusatzbestimmung *f* supplementary regulation, supplementary rule
Zusatzdraht *m* stringer bead, welding bead
Zusatzdynamo *m* supplementary dynamo
Zusatzeisen *n* added iron
Zusatzelektrolyt *m* supplementary electrolyte
Zusatzelement *n* alloying element
zusatzfrei without additive
Zusatzgerät *n* additional apparatus, booster, supplementary apparatus
Zusatzlinse *f* (Opt) supplementary lens
Zusatzmenge *f* supplementary amount
Zusatzmetall *n* added metal, alloying metal
Zusatzmittel *n* added substance, addition agent, additive, admixed substance, blending agent, filler, supplementary agent

Zusatzmoment *n* additive moment
Zusatzpatent *n* patent of addition
Zusatzpotential *n* additional potential
Zusatzpumpe *f* booster pump
Zusatzspannung *f* additional voltage, boosting voltage
Zusatzstoff *m* added substance
Zusatzstromkreis *m* booster circuit
Zusatzsubstanz *f* added material
Zusatztransformator *m* (Elektr) booster transformer
Zusatztrockner *m* (Tauchanlage) booster dryer
Zusatzverstärker *m* (Elektr) booster amplifier
Zuschauerraum *m* auditorium
Zuschlag *m* addition, admixture, (Metall) flux
Zuschlagerz *n* (Metall) fluxing ore
Zuschlaghammer *m* sledge hammer
Zuschlagkalkstein *m* calcareous flux, limestone flux
zuschmelzen to close by melting, to seal
zuschneiden to cut to size, to trim
zuschrauben to screw on
Zuschußwasser *n* compensation water
zusetzen to add, (Met) to alloy, (vergällen) to adulterate
Zusetzen *n* **von Erzen** addition of ores
zuspitzen to sharpen, (sich verjüngen) to taper
Zuspitzung *f* tapering, pointing
Zustand *m* circumstance, condition, situation, status, (Chem) state, **abstoßender ~** (Atom) repulsing condition, **aktiver ~** active state, **amorpher ~** amorphous state, **angeregter ~** (Atom) excited state, **bindender ~** bonding condition, **entarteter ~** degenerate state, **fester ~** solid state, **flüssiger ~** liquid state, **freier ~** free state, uncombined state, **gasförmiger ~** gaseous state, **gediegener ~** native condition, **gleichbleibender ~** steady condition, steady state, **hochangeregter ~** highly excited state, **kolloider ~** colloidal condition, **kristalliner ~** crystalline state, crystallinity, **metastabiler ~** metastable state, **nichtbindender ~** (Atom) anti-bonding condition, **plastischer ~** plastic condition, **quasi-stationärer ~** quasi-stationary state, **stationärer ~** stationary state, steady state, **ungeordneter ~** disordered arrangement, disordered state, **vorübergehender ~** transient state
Zustandsänderung *f* change of state, **adiabatische ~** adiabatic change, adiabatic change of condition, **isentropische ~** isentropic chance of state, **isothermische ~** isothermal change of state, **isotrope ~** (Phys) isotropic change, **polytropische ~** polytropic change of state
Zustandsdiagramm *n* diagram of state, phase diagram
Zustandsdichte *f* density of state

Zustandsgleichung *f* equation of state, **kalorische ~** (Therm) Einstein equation for heat capacity
Zustandsvariable *f* variable of state
Zustandsveränderung *f* transition
Zustellung *f* **des Ofens** (Metall) relining the furnace, repairing the lining
Zustimmung *f* agreement, approval
Zuströmen *n* affluence
Zuströmgeschwindigkeit *f* inflow velocity
Zustrom *m* afflux, influx
Zutat *f* ingredient
Zuteilung *f* allotment, (Dosis) dosage
Zuteilvorrichtung *f* distributing equipment
zutreffend suitable, appropriate
Zutritt *m* access, admission, admittance
zutropfen lassen to add drop by drop
Zuwachs *m* growth, accretion, increase, increment
Zuwachsfaktor *m* build-up factor
Zwang *m* constraint, compulsion, **Prinzip des geringsten ~** principle of least constraint
Zwangsdurchlaufkessel *m* forced circulation boiler, once-through boiler
zwangsläufig constrained, forced
Zwangslizenzbestimmung *f* compulsory license regulation
Zwangsmischer *m* ribbon blender
Zwangsumlauf *m* forced circulation
Zwangsumlaufheizung *f* forced circulation heating
Zwangsumlaufkocher *m* thermal diffusion reboiler
Zwangsumlaufkühlung *f* forced circulation cooling
Zwangsumlaufverdampfer *m* forced circulation evaporator
zwanzigflächig (Krist) icosahedral
Zwanzigflächner *m* (Krist) icosahedron
Zweck *m* purpose
zweckdienlich suitable, conducive, purposeful
zweckentsprechend appropriate
zweckmäßig purposeful
zweiachsig biaxial
Zweiachsigkeit *f* biaxiality
zweiatomig diatomic
Zweibadverfahren *n* two-bath process
zweibasig dibasic
zweideutig ambiguous
Zweideutigkeit *f* ambiguity
zweidimensional two-dimensional
Zweidrahtaufhängung *f* bifilar suspension
Zweierschale *f* (Atom) doublet ring
Zweierstoß *m* double collision, two-body collison
zweifach double, twofold, **~ brechend** (Opt) birefractive
Zweifachbindung *f* double bond
Zweifachschwefelzinn *n* (obs) tin disulfide

Zweifachstreuexperiment – Zweiwegeventil

Zweifachstreuexperiment *n* double scattering experiment
Zweifachuntersetzer *m* binary scaler
Zweifachverbindung *f* binary compound
Zweifarbendruck *m* (Buchdr) two-color printing
zweifarbig dichromatic, dichroic, two-colored
Zweifarbigkeit *f* dichroism
Zweifelsfall *m* case of doubt
Zweiflammenrohrkessel *m* double flue boiler
Zweifluidtheorie *f* two-fluid theory
Zweig *m* branch
Zweiganggewinde *n* two-way thread
Zweigangschnecke *f* two-thread worm
zweigestaltig dimorphic, dimorphous
zweigeteilt bipartite
Zweigleitung *f* branch line
zweigliedrig (Math) binomial, (System) binary
Zweigniederlassung *f* branch
Zweigrohr *n* branch pipe
Zweigstrom *m* (Elektr) branch current
Zweigstromkreis *m* branch circuit
zweihalsig two-necked
Zweikammerkondenstopf *m* twin chamber steam trap
Zweikammermühle *f* two-compartment mill
zweikernig binuclear
Zweikernreifen *m* dual bead tire
Zweikörperformalismus *m* two-body formalism
Zweikörperkraft *f* two-body force
Zweikörperproblem *n* two-body problem
Zweikomponentenkleber *m* mixed adhesive
Zweikomponentenspritzvorrichtung *f* dual-spraying equipment, twin-feed spray equipment
Zweikomponentenwaage *f* two-component balance
Zweileiterkabel *n* twin cable
Zweileitersystem *n* two-wire system
zweimal twice
Zweimesonensystem *n* two-meson system
zweimetallisch bimetallic
Zweiphasenanker *m* two-phase armature
Zweiphasendreileitersystem *n* two-phase three wire system
Zweiphasendruckverfahren *n* two-phase printing process, blotch printing
Zweiphasendynamo *m* two-phase generator
Zweiphasengleichgewicht *n* two-phase equilibrium
Zweiphasenleitung *f* two-phase wiring
Zweiphasenmotor *m* two-phase motor
Zweiphasenströmung *f* (Hydrodynamik) two-phase flow
Zweiphasenstrom *m* two-phase [alternating] current
Zweiphasenstromerzeuger *m* two-phase generator
Zweiphasensystem *n* two-phase system

Zweiphasentransformator *m* two-phase transformer
Zweiphasenvierleitersystem *n* two-phase four wire system
zweiphasig two-phase, diphase
Zweipolgleichrichtung *f* diode rectification
zweipolig two-polar, bipolar, dipole, double pole
Zweipolröhre *f* diode tube
Zweipunktregelung *f* on-off control, two position control, two-step control (Br. E.)
Zweipunktregler *m* on-off controller (Am. E.), two-position controller (Am. E.), two-step controller
Zweiquantenvernichtungsstrahlung *f* two photon annihilation radiation
Zweirollen-Scheibenmühle *f* two-roll attrition mill
Zweisäulenapparat *m* two-column apparatus
Zweisäulen-Rektifikation *f* double-column rectification
zweisäurig diacid
Zweischichtenkondensator *m* two-plate capacitor
Zweischichtensinterverfahren *n* double-bed sintering
Zweischleifenoszillograph *m* double-wire loop oscillograph
zweiseitig bilateral, two-sided
zweistellig (Math) two-digit, two-figure
Zweistoffgemisch *n* binary mixture, binary system, two-component mixture, (Verfahrenstechn) binary mixture
Zweistofflegierung *f* binary alloy
Zweistoffsystem *n* (Verfahrenstechn) binary system
Zweistrahlbetatron *n* double-ray betatron
Zweistrahlgerät *n* double-beam instrument
Zweistrahlproblem *n* two-beam problem
Zweistufenprozeß *m* two-stage process
Zweistufenreduktion *f* two-step reduction
Zweistufenregler *m* two-stage governor
zweistufig two-stage
Zweitaktmaschine *f* two-cycle engine
Zweitbeschleuniger *m* secondary accelerator, additional accelerator
zweiteilig bipartite, consisting of two pieces
Zweitluft *f* secondary air
Zweitluftzufuhr *f* twin air supply
Zweiträgerkran *m* two-girder crane
Zweitverdampfer *m* second evaporator
Zweiwalzenmühle *f* twin roller mill, two-roller mill
Zweiwalzenständer *m* housing for two rolls
Zweiwalzenstuhl *m* two-roll mill
Zweiwalzentrockner *m* double-drum drier
Zweiwalzwerk *n* twin rolling mill
Zweiwegehahn *m* two-way cock
Zweiwegepumpe *f* two-way pump
Zweiwegeventil *n* two-way valve

Zweiwellenmikroskopie *f* two-wavelength microscopy
Zweiwellenrührgerät *n* twin-rotor mixer
zweiwertig (Chem) divalent, bivalent
Zweiwertigkeit *f* (Chem) bivalence
zweizählig two-fold, binary
Zweizentrenbindung *f* (Atom) two-center bond
Zweizentrenorbital *n* (Atom) bicentric orbital
Zweizentrenproblem *n* two-center problem
Zweizonenmodell *n* two-zone model
Zweizonenreaktor *m* two-zone reactor
Zweizylindermotor *m* two-cylinder engine
Zwerchfell *n* (Med) diaphragm
Zwergstufenschalter *m* midget step switch
Zwergwuchs *m* dwarfism
Zwickel *m* gusset
Zwiebel *f* onion, (Knolle) bulb
zwiebelartig bulb-like, bulbous
Zwiebelmarmor *m* (Min) cipolin
Zwiebelöl *n* onion oil
Zwielicht *n* twilight
Zwieselit *m* (Min) triplite
Zwilling *m* twin
Zwillingsachse *f* (Krist) twin axis
Zwillingsbereifung *f* dual tire
Zwillingsbildung *f* twin formation, twinning
Zwillingsblock *m* twin block
Zwillingsdampfmaschine *f* twin engine
Zwillingsebene *f* twinning plane
Zwillingsfläche *f* (Krist) twinning plane
Zwillingskristall *m* (Krist) twin crystal
Zwillingslinse *f* (Opt) binary lens
Zwillingsmaschine *f* twin engine
Zwillingspumpe *f* twin pump
Zwillingsreifen *pl* twin tires
Zwillingssalz *n* double salt
Zwillingsschreiber *m* duplex recorder
Zwillingstrommelmischer *m* twin cylinder mixer, twin-shell blender
Zwillingszylinder *m* twin cylinder
Zwinge *f* vise, ferrule
Zwingeisen *n* clamp iron
Zwirneinstellung *f* (Web) twist setting
Zwirnen *n* twisting
Zwischenabscheider *m* interstage trap
Zwischenanstrich *m* undercoat
Zwischenbaugewebe *n* (Reifen) breaker fabric
Zwischenbehälter *m* intermediate container, intermediate tank, surge tank
Zwischenbelüftung *f* intermediate air release
Zwischenbeschleuniger *m* interaccelerator
Zwischenboden *m* false bottom, partition
Zwischenelektrode *f* dynode
Zwischenelektrodenkapazität *f* interelectrode capacitance
Zwischenerzeugnis *n* intermediate product
Zwischenfarbe *f* intermediate color
Zwischenferment *n* (Biochem) zwischenferment (glucose 6-phosphate dehydrogenase)

Zwischenform *f* intermediate form
zwischengeschichtet interstratified
Zwischengetriebe *n* intermediate drive, intermediate gearing
Zwischengitteratom *n* (Krist) interstitial atom
Zwischengitterion *n* interstitial ion
Zwischengittermechanismus *m* interstitial mechanism
Zwischengitterpaar *n* interstitial pair
Zwischengitterplatz *m* (Krist) interstitial position, interstice
Zwischengitterwanderung *f* interstitial migration
Zwischenglied *n* connecting member, intermediary member, intermediate
Zwischenglühung *f* intermediate annealing, intermediate softening
Zwischengröße *f* intermediate size
Zwischenkern *m* intermediate nucleus
Zwischenkörper *m* intermediary body, intermediate body
Zwischenkühler *m* intermediate cooler
Zwischenkühlmittel *n* intermediate coolant
Zwischenkühlung *f* intermediate cooling
Zwischenlage *f* interleaf, intermediate layer, interposition
Zwischenlagegewebe *n* breaker fabric
Zwischenlandung *f* stop
zwischenmolekular intermolecular
Zwischenpause *f* interval
Zwischenphase *f* intermediate phase
Zwischenplatte *f* backing plate, floating plate, support plate
Zwischenpol *m* interpole
Zwischenprodukt *n* intermediate [product]
Zwischenraum *m* clearance, intermediate space, intermedium, interspace, interstice, (Lücke) gap, **atomarer** ~ atomic interspace
Zwischenraumanteil *m* (Hydrodynamik; Füllkörperkolonne) void fraction
Zwischenraumvolumen *n* (Hydrodynamik; Füllkörperkolonne) void fraction
Zwischenreaktion *f* (Chem) intermediate reaction
Zwischenring *m* spacer ring, middle ring
Zwischenschalter *m* intermediate switch
Zwischenschaltung *f* cut-in, interconnection
Zwischenschicht *f* interlayer, barrier sheet, interlining, intermediate layer
Zwischenschritt *m* intermediary step
zwischenständig intermediate
Zwischenstecker *m* adapter, attachment plug
Zwischenstoffwechsel *m* intermediary metabolism
Zwischenstück *n* adapter, connecting piece, distance piece
Zwischenstufe *f* intermediary stage, (chem. Verbindung) intermediate
Zwischentaste *f* (Schreibmaschine) spacer

Zwischenüberhitzung *f* intermediate superheating
Zwischenverbindung *f* intermediate [compound]
Zwischenwalzwerk *n* intermediate rolls
Zwischenwand *f* division wall, interstructure, partition [wall]
Zwischenwert *m* intermediate value
Zwischenwirt *m* intermediate host
Zwischenzeit *f* [time] interval
Zwischenzustand *m* intermediate state
Zwitterion *n* zwitter ion, amphoteric ion, dipolar ion
zwölfflächig (Krist) dodecahedral
Zwölfflächner *m* (Krist) dodecahedron
zyanig cyanous
zyanophil (Bakt) cyanophilous
Zygadenin *n* zygadenine
Zygadit *m* zygadite
Zyklisierung *f* cyclization
Zyklon *m* cyclone
Zyklonabscheider *m* cyclone separator
Zyklonfeuerung *f* cyclone firing
Zyklonrührer *m* cyclone impeller
Zyklonwärmeaustauscher *m* cyclone heat exchanger
Zyklonwäscher *m* cyclone scrubber
Zyklotron *n* cyclotron, **frequenzmoduliertes** ~ frequency-modulated cyclotron
Zyklotronresonanzexperiment *n* cyclotron resonance experiment
Zyklotronumlauffrequenz *f* cyclotron frequency
Zyklus *m* cycle
Zylinder *m* cylinder, (Walze) roller, **einfachwirkender** ~ single acting cylinder, **graduierter** ~ graduated cylinder, **hängender** ~ inverted cylinder, **luftgekühlter** ~ air-cooled cylinder, **stehender** ~ upright cylinder, **umlaufender** ~ revolving cylinder, **wassergekühlter** ~ water-cooled cylinder
Zylinderauskleidung *f* cylinder liner
Zylinderbohrmaschine *f* cylinder bore, cylinder boring machine
Zylinderbüchse *f* barrel sleeve
Zylinderdeckel *m* cylinder cover
zylinderförmig cylindrical
Zylinderfunktion *f* (Math) cylinder function
Zylindergebläse *n* cylindrical blower
Zylinderinhalt *m* cylinder volume
Zylinderkondensator *m* cylindrical condenser
Zylinderkonduktor *m* cylindrical conductor
Zylinderkopf *m* (Mot) cylinder head
Zylinderkopfdichtung *f* cylinder head joint
Zylinderlinse *f* (Opt) cylindrical lens
Zylindermagnet *m* cylindrical magnet, rod magnet
Zylindermischer *m* drum mixer
Zylinderöl *n* cylinder oil
Zylinderofen *m* cylinder furnace

Zylinderschleifmaschine *f* drum sander
Zylinderschmierung *f* cylinder lubrication
Zylinderstift *m* straight pin
Zylindertrockner *m* can dryer
Zylinderummantelung *f* cylinder clothing
Zylindervolumen *n* cubic capacity of the cylinder, cylinder volume, piston displacement
Zylinderzelle *f* cylindrical cell
Zymase *f* (Biochem) zymase
Zymogen *n* (Biochem) zymogen
Zymohexase *f* (Biochem) zymohexase
Zymologie *f* zymology
Zymonsäure *f* zymonic acid
Zymoskop *n* cymoscope
Zymosterin *n* zymosterol
Zypressenöl *n* cypress oil
Zyste *f* (Biol) cyst
Zytochemie *f* cytochemistry
Zytochromoxydase *f* (Cytochromoxydase .118526) cytochrome oxidase
Zytohormon *n* cytohormone
Zytokinase *f* (Biochem) cytokinase
Zytoklasis *f* cytoclasis
Zytologie *f* cytology
Zytolyse *f* cytolysis
Zytolysin *n* cytolysin
zytolytisch cytolytic
Zytometrie *f* cytometry
zytometrisch cytometric

Anhang
Appendix

Verzeichnis der in diesem Wörterbuch verwandten Abkürzungen
Abbreviations used in this Dictionary

Abk	Abkürzung	abbreviation
adj	Adjektiv	adjective
adv	Adverb	adverb
Aerodyn	Aerodynamik	aerodynamics
Agr	Landwirtschaft	agriculture
Akust	Akustik	acoustics
Alch	Alchemie	alchemy
allg	allgemein	generally
Am. E.	amerikanisches Englisch	American English
Anat	Anatomie	anatomy
Arch	Architektur	architecture
Astr	Astronomie	astronomy
Atom	Atomterminologie	atomic terminology
Bakt	Bakteriologie	bacteriology
Bauw	Bauwesen	civil engineering
Bergb	Bergbau	mining
Biochem	Biochemie	biochemistry
Biol	Biologie	biology
Bot	Botanik	botany
Brau	Brauereiwesen	brewing
Br. E.	britisches Englisch	British English
Buchdr	Buchdruck	typography
Chem	Chemie, chemisch	chemistry, chemical
Comp	Computertechnologie	computer, computer technology
Cyb	Kybernetik	cybernetics
Dest	Destillation	distillation
Elektr	Elektrizität	electricity
Elektrochem	Elektrochemie	electrochemistry
Elektron	Elektronik	electronics
f	weiblich	feminine
Färb	Färberei	dyeing
Galv	Galvanometrie	galvanometry
Gen	Genetik	genetics
Geogr	Geographie	geography
Geol	Geologie	geology
Geom	Geometrie	geometry
Gerb	Gerberei	tanning
Gummi	Gummiindustrie	rubber industry
Histol	Histologie	histology
HN	Handelsname (ohne Garantie)	tradename (without guarantee)
Hydrodyn	Hydrodynamik	hydrodynamics
Jur	Rechtswissenschaft	jurisprudence
Keram	Keramik	ceramics
Kfz	Kraftfahrzeug	motor vehicle
Kin	Kinetik	kinetics
Kosm	Kosmetik	cosmetics
Krist	Kristallographie	crystallography
Lat	Latein	Latin
Leg	Legierung	alloy
Luftf	Luftfahrt	aeronautics
m	maskulin, männlich	masculine
Mag	Magnetismus	magnetism
Masch	Maschinenwesen	mechanical engineering
Maß	Maßeinheit	measuring unit
Math	Mathematik	mathematics

Mech	Mechanik	mechanics
Met	Metall	metal
Metall	Metallurgie	metallurgy
Meteor	Meteorologie	meteorology
Mol. Biol	Molekularbiologie	molecular biology
Mot	Motor	motor
n	neutrum, sächlich	neuter
o	oder	or
obs	veraltet	obsolete
Opt	Optik	optics
Pap(ier)	Papierindustrie	paper industry
Pathol	Pathologie	pathology
Pharm	Pharmakologie	pharmacology
Phot	Photographie	photography
Phys	Physik	physics
Physiol	Physiologie	physiology
Plast	Kunststoffe	plastics
pl	plural	plural
Quant	Quantentheorie	quantum theory
Rad	Radio	radio
Rdr	Radar	radar
Rönt	Röntgenspektroskopie	X-ray spectroscopy
Schiff	Schiffahrt	maritime terminology
Schweiß	Schweißtechnik	welding
Spektr	Spektroskopie	spectroscopy
Statist	Statistik	statistics
Stereochem	Stereochemie	stereochemistry
Symb	Symbol	symbol
Techn	Technik	technology
Telef	Telefon	telephone
Telev	Fernsehen	television
Text	Textilindustrie	textile industry
Therm	Thermodynamik	thermodynamics
Vet	Tiermedizin	veterinary medicine
vgl	vergleiche	confer
Web	Weberei	weaving
Zahnmed	Zahnmedizin	dentistry
z. B.	zum Beispiel	for instance
Zool	Zoologie	zoology
Zuck	Zuckerindustrie	sugar industry

Chemische Elemente
Chemical Elements

Deutsch / German	Englisch / English	Symbol	Ordnungszahl / Atomic Number
Actinium	actinium	Ac	89
Aluminium	alumin[i]um	Al	13
Americium	americium	Am	95
Antimon	antimony	Sb	51
Argon	argon	Ar (A)	18
Arsen	arsenic	As	33
Astat	astatine	At	85
Barium	barium	Ba	56
Berkelium	berkelium	Bk	97
Beryllium	beryllium	Be	4
Bismut	bismuth	Bi	83
Blei	lead	Pb	82
Bor	boron	B	5
Brom	bromine	Br	35
Cadmium	cadmium	Cd	48
Caesium	cesium	Cs	55
Calcium	calcium	Ca	20
Californium	californium	Cf	98
Cer	cerium	Ce	58
Chlor	chlorine	Cl	17
Chrom	chromium	Cr	24
Cobalt	cobalt	Co	27
Curium	curium	Cm	96
Dysprosium	dysprosium	Dy	66
Einsteinium	einsteinium	Es (E)	99
Eisen	iron	Fe	26
Erbium	erbium	Er	68
Europium	europium	Eu	63
Fermium	fermium	Fm	100
Fluor	fluorine	F	9
Francium	francium	Fr	87
Gadolinium	gadolinium	Gd	64
Gallium	gallium	Ga	31
Germanium	germanium	Ge	32
Gold	gold	Au	79
Hafnium	hafnium	Hf	72
Helium	helium	He	2
Holmium	holmium	Ho	67
Indium	indium	In	49
Iod	iodine	I	53
Iridium	iridium	Ir	77
Kalium	potassium	K	19
Kohlenstoff	carbon	C	6
Kupfer	copper	Cu	29
Krypton	krypton	Kr	36
Lanthan	lanthanum	La	57
Lawrencium	lawrencium	Lr	103
Lithium	lithium	Li	3
Lutetium	lutetium	Lu	71

Deutsch / German	Englisch / English	Symbol	Ordnungszahl / Atomic Number
Magnesium	magnesium	Mg	12
Mangan	manganese	Mn	25
Mendelevium	mendelevium	Md (Mv)	101
Molybdän	molybdenum	Mo	42
Natrium	sodium	Na	11
Neodym	neodymium	Nd	60
Neon	neon	Ne	10
Neptunium	neptunium	Np	93
Nickel	nickel	Ni	28
Niob	niobium (columbium)	Nb	41
Nobelium	nobelium	No	102
Osmium	osmium	Os	76
Palladium	palladium	Pd	46
Phosphor	phosphorus	P	15
Platin	platinum	Pt	78
Plutonium	plutonium	Pu	94
Polonium	polonium	Po	84
Praseodym	praseodymium	Pr	59
Promethium	promethium	Pm	61
Protactinium	protactinium	Pa	91
Quecksilber	mercury	Hg	80
Radium	radium	Ra	88
Radon	radon	Rn	86
Rhenium	rhenium	Re	75
Rhodium	rhodium	Rh	45
Rubidium	rubidium	Rb	37
Ruthenium	ruthenium	Ru	44
Samarium	samarium	Sm	62
Sauerstoff	oxygen	O	8
Scandium	scandium	Sc	21
Schwefel	sulfur	S	16
Selen	selenium	Se	34
Silber	silver	Ag	47
Silicium	silicon	Si	14
Stickstoff	nitrogen	N	7
Strontium	strontium	Sr	38
Tantal	tantalum	Ta	73
Technetium	technetium	Tc	43
Tellur	tellurium	Te	52
Terbium	terbium	Tb	65
Thallium	thallium	Tl	81
Thorium	thorium	Th	90
Thulium	thulium	Tm	69
Titan	titanium	Ti	22
Uran	uranium	U	92
Vanadium	vanadium	V	23
Wasserstoff	hydrogen	H	1
Wolfram	tungsten	W	74
Xenon	xenon	Xe	54
Ytterbium	ytterbium	Yb	70
Yttrium	yttrium	Y	39
Zink	zinc	Zn	30
Zinn	tin	Sn	50
Zirconium	zirconium	Zr	40

Deutsche Maße und Gewichte
German Measures and Weights

1. Längenmaße – Linear Measure
1 Millimeter (mm)	=			0.039 inch
1 Zentimeter (cm)	=	10 mm	=	0.393 inch
1 Dezimeter (dm)	=	10 cm	=	3.937 inches
1 Meter (m)	=	100 cm	=	1.093 yards
1 Kilometer (km)	=	1000 m	=	0.621 mile

2. Flächenmaße – Square Measure
1 Quadratmillimeter (mm²)	=			0.0015	square inch
1 Quadratzentimeter (cm²)	=	100 mm²	=	0.1549	square inch
1 Quadratdezimeter (dm²)	=	100 cm²	=	15.499	square inches
1 Quadratmeter (m²)	=	100 dm²	=	1.195	square yards
1 Quadratkilometer (km²)	=			0.386	square mile
			=	247.11	acres
1 Ar (a)		= 100 m²	=	119.59	square yards
1 Hektar (ha)		= 100 a	=	2.47	acres

3. Raummaße – Cubic Measure
1 Kubikmillimeter (mm³)	=			0.61×10^{-4} cubic inch
1 Kubikzentimeter (cm³)	=	1000 mm³	=	0.61×10^{-1} cubic inch
1 Kubikdezimeter (dm³)	=	1000 cm³	=	61.02 cubic inches
1 Kubikmeter (m³)	=	1000 dm³	=	35.31 cubic feet
			=	1.307 cubic yards

4. Hohlmaße – Measure of Capacity
				Am. E.			Br. E.
1 Milliliter (ml)	=		16.2	minims	=	16.8	minims
1 Zentiliter (cl)	=	10 ml	= 0.338	fluid ounce	=	0.352	fluid ounce
1 Deziliter (dl)	=	100 ml	= 3.38	fluid ounces	=	0.352	fluid ounces
1 Liter (l)	=	1000 ml	= 2.1	pints	=	1.76	pints
			= 1.06	liquid quarts	=	0.88	quart
			= 0.26	gallon	=	0.22	gallon

5. Gewichte – Weight
1 Milligramm (mg)	=	0.015 grain		
1 Gramm (g)	=	15.432 grains		
1 Kilogramm (kg)	=	2.205 pounds (Avoirdupois)		
	=	2.679 pounds (Troy)		
1 Tonne (t)	=	1000 kg	=	1.102 short tons* (Am. E.)
			=	0.984 long ton* (Br. E.)
1 Pfund (Pfd.)	=	500 g	=	1.102 pounds*
1 Zentner (Ztr.)	=	100 Pfd.	=	1.102 hundredweights* (Am. E.)
			=	0.984 hundredweight* (Br. E.)
1 Doppelzentner (dz)	=	100 kg	=	2.204 hundredweights* (Am. E.)
			=	1.968 hundredweights* (Br. E.)

* Avoirdupois

Temperaturumrechnungstabelle (von −20 bis 100° Celsius)
Conversion Table of Temperatures (from −20 to 100° Celsius)

Celsius °C	Fahrenheit °F	Celsius °C	Fahrenheit °F	Celsius °C	Fahrenheit °F
100	212	58	136.4	18	64.4
98	208.4	56	132.8	16	60.8
96	204.8	54	129.2	14	57.2
94	201.2	52	125.6	12	53.6
92	197.6	50	122	10	50
90	194				
		48	118.4	8	46.4
88	190.4	46	114.8	6	42.8
86	186.8	44	111.2	4	39.2
84	183.2	42	107.6	2	35.6
82	179.6	40	104	0	32
80	176				
		38	100.4	−2	28.4
78	172,4	36	96.8	−4	24.8
76	168.8	34	93.2	−6	21.2
74	165.2	32	89.6	−8	17.6
72	161.6	30	86	−10	14
70	158				
		28	82.4	−12	10.4
68	154.4	26	78.8	−14	6.8
66	150.8	24	75.2	−16	3.2
64	147.2	22	71.6	−18	−0.4
62	143.6	20	68	−20	−4
60	140				

Temperaturumrechnungsregeln
Rules for Converting Temperatures

1. Celsius in Fahrenheit
   Celsius into Fahrenheit
   $x\,°C = (32 + \frac{9}{5}x)\,°F$

2. Fahrenheit in Celsius
   Fahrenheit into Celsius
   $x\,°F = (x - 32)\,\frac{5}{9}\,°C$

Umrechnungstabelle
nicht mehr zugelassener Einheiten in Einheiten des SI-Systems

Conversion Table
for Obsolete Units into Newly Adopted SI-Units

Überholte Einheiten Obsolete Units	SI-Einheiten SI-Units	
Ångström	1 Å	= 0,1 nm (Nanometer)
Atmosphäre, technische	1 at	= 0,980665 bar ≈ 1 bar = 98,0665 · 10³ Pa (Pascal)
Atmosphäre, physikalische	1 atm	= 1,01325 bar ≈ 1 bar = 101,325 · 10³ Pa (Pascal)
Curie	1 Ci	= 3,7 · 10¹⁰ · $\frac{1}{s}$
Dyn	1 dyn	= 10⁻⁵ N (Newton)
Erg	1 erg	= 10⁻⁷ J (Joule)
Gauß	1 G	= 10⁻⁴ T (Tesla) = 10⁻⁴ Wb/m² (Weber/m²)
Grad Kelvin	1° K	= 1 K (Kelvin)
Kalorie	1 cal	= 4,1868 J (Joule)
Kilopond	1 kp	= 9,80665 N ≈ 10 N (Newton)
Kilopond/cm²	1 kp/cm²	= 0,980665 bar ≈ 1 bar = 0,0980665 N/mm² ≈ 0,1 N/mm²
Maxwell	1 M	= 10⁻⁸ Wb (Weber)
Meter-Wassersäule	1 mWS	= 0,0980665 bar ≈ 0,1 bar = 9,80665 · 10³ Pa (Pascal)
Millimeter-Quecksilbersäule	1 mmHg	= 1,33322 mbar = 133,322 Pa (Pascal)
Micron, Mü	1 μ	= 1 μm (Mikrometer)
Millimü	1 mμ	= 1 nm (Nanometer)
Oersted	1 Oe	= 79,5775 A/m ≈ 80 A/m (Ampere/m)
Pferdestärke	1 PS	= 735,49875 W ≈ 736 W (Watt)
Poise	1 P	= 0,1 $\frac{N \cdot s}{m^2}$ = 0,1 Pa · s
Pond	1 p	= 9,80665 · 10⁻³ N ≈ 10⁻² N (Newton)
Röntgen	1 R	= 0,258 · 10⁻³ C/kg (Coulomb/kg)
Rutherford	1 rd	= 10⁻² · $\frac{J}{kg}$
Stokes	1 St	= 10⁻⁴ $\frac{m^2}{s}$
Torr	1 Torr	= 1,33322 mbar ≈ 133,322 Pa (Pascal)

Umrechnung von SI-Einheiten
Conversion of SI-Units

1 Newton	= 1 N = 1 kg m/s²
1 Pascal	= 1 Pa = 1 Newton/m² = 1 N/m² = 10 μbar
1 Bar	= 1 bar = 10⁵ N/m² = 0,1 N/mm² = 10⁵ Pa
1 Joule	= 1 J = 1 Wattsekunde = 1 Ws = 1 Newtonmeter = 1 Nm
1 Watt	= 1 W = 1 Newtonmeter/Sekunde = 1 Nm/s

Periodensystem der Elemente
Periodic Table of the Elements

1a	2a	3b	4b	5b	6b	7b	8			1b	2b	3a	4a	5a	6a	7a	0
1 **H** 1.0079																	2 **He** 4.00260
3 **Li** 6.94	4 **Be** 9.01218											5 **B** 10.81	6 **C** 12.011	7 **N** 14.0067	8 **O** 15.9994	9 **F** 18.998403	10 **Ne** 20.17
11 **Na** 22.98977	12 **Mg** 24.305											13 **Al** 26.98154	14 **Si** 28.0855	15 **P** 30.97376	16 **S** 32.06	17 **Cl** 35.453	18 **Ar** 39.948
19 **K** 39.0983	20 **Ca** 40.08	21 **Sc** 44.9559	22 **Ti** 47.90	23 **V** 50.941	24 **Cr** 51.996	25 **Mn** 54.9380	26 **Fe** 55.847	27 **Co** 58.9332	28 **Ni** 58.71	29 **Cu** 63.546	30 **Zn** 65.38	31 **Ga** 69.72	32 **Ge** 72.59	33 **As** 74.9216	34 **Se** 78.96	35 **Br** 79.904	36 **Kr** 83.80
37 **Rb** 85.467	38 **Sr** 87.62	39 **Y** 88.9059	40 **Zr** 91.22	41 **Nb** 92.9064	42 **Mo** 95.94	43 **Tc** 98.9062	44 **Ru** 101.07	45 **Rh** 102.9055	46 **Pd** 106.4	47 **Ag** 107.868	48 **Cd** 112.41	49 **In** 114.82	50 **Sn** 118.69	51 **Sb** 121.75	52 **Te** 127.60	53 **I** 126.9045	54 **Xe** 131.30
55 **Cs** 132.9054	56 **Ba** 137.33	57* **La** 138.9055	72 **Hf** 178.49	73 **Ta** 180.947	74 **W** 183.85	75 **Re** 186.2	76 **Os** 190.2	77 **Ir** 192.22	78 **Pt** 195.09	79 **Au** 196.9665	80 **Hg** 200.59	81 **Tl** 204.37	82 **Pb** 207.2	83 **Bi** 208.9808	84 **Po** (209)	85 **At** (210)	86 **Rn** (222)
87 **Fr** (223)	88 **Ra** 226.0254	89** **Ac** (227)	104 — (261)	105 — (262)													

*Lanthanoide *Lanthanoids	58 **Ce** 140.12	59 **Pr** 140.9077	60 **Nd** 144.24	61 **Pm** (145)	62 **Sm** 150.4	63 **Eu** 151.96	64 **Gd** 157.25	65 **Tb** 158.9254	66 **Dy** 162.50	67 **Ho** 164.9304	68 **Er** 167.26	69 **Tm** 168.9342	70 **Yb** 173.04	71 **Lu** 174.97
**Actinoide **Actinoids	90 **Th** 232.0381	91 **Pa** 231.0359	92 **U** 238.029	93 **Np** 237.0482	94 **Pu** (244)	95 **Am** (243)	96 **Cm** (247)	97 **Bk** (247)	98 **Cf** (251)	99 **Es** (254)	100 **Fm** (257)	101 **Md** (258)	102 **No** (259)	103 **Lr** (260)

Ziffern in Klammern sind die Massenzahlen des stabilsten Isotops des betreffenden Elements.
Numbers in parentheses are mass numbers of most stable isotope of that element.